Logic, Epistemology, and the Unity of Science

Volume 55

Logic, Epistemology, and the Unity of Science aims to reconsider the question of the unity of science in light of recent developments in logic. At present, no single logical, semantical or methodological framework dominates the philosophy of science. However, the editors of this series believe that formal frameworks, for example, constructive type theory, deontic logics, dialogical logics, epistemic logics, modal logics, and proof-theoretical semantics, have the potential to cast new light on basic issues in the discussion of the unity of science.

This series provides a venue where philosophers and logicians can apply specific systematic and historic insights to fundamental philosophical problems. While the series is open to a wide variety of perspectives, including the study and analysis of argumentation and the critical discussion of the relationship between logic and philosophy of science, the aim is to provide an integrated picture of the scientific enterprise in all its diversity.

This book series is indexed in SCOPUS.

For inquiries and submissions of proposals, authors can contact Christi Lue at
christi.lue@springer.com

Raffaele Pisano • Jean Dhombres •
Patricia Radelet de Grave •
Paolo Bussotti

Homage to Evangelista Torricelli's *Opera Geometrica* 1644–2024

Text, Transcription, Commentaries and Selected Essays as New Historical Insights

In Collaboration with Julie Robarts

Springer

Raffaele Pisano
HOPAST at IEMN, Department of Physics
Lille University–CNRS
Villeneuve d'Ascq, France

Jean Dhombres
CAK-EHEES
Paris, France

Patricia Radelet de Grave
Catholic Louvain La Neuve University
IRMP
Louvain-la-Neuve, Belgium

Paolo Bussotti
Udine University
DIUM
Udine, Italy

ISSN 2214-9775 ISSN 2214-9783 (electronic)
Logic, Epistemology, and the Unity of Science
ISBN 978-3-031-06965-9 ISBN 978-3-031-06963-5 (eBook)
https://doi.org/10.1007/978-3-031-06963-5

This Springer imprint is published by the registered company Springer Nature Switzerland AG
The registered company address is: Gewerbestrasse 11, 6330 Cham, Switzerland

Thought does not respect national frontiers. Yet scientific ideas are far from stateless citizens
(Thackray, A. 1970. Atoms and Powers. Cambridge MA: Harvard University Press, p. 4).

Audentes fortuna iuvat
(Vergilius, Æneis, X, 284)

Per aspera ad astra
(Cicero, De natura deorum III, 40; Vergilius, Æneis, IX, 641; Seneca, Hercules furens, II, 437–441)

Acknowledgments

We want to express our appreciation to all contributing authors for their efforts to produce essays of interest and of high quality worthy of this special volume. The result is excellent.

In 2009 the Torricelli Project was born thanks to Prof. Raffaele Pisano's personal initiative. Between 2009 and 2010 he established his early contacts, particularly in Florence where he met the outstanding Galilean and Torricellian scholar, Prof. Maurizio Torrini (1942–2019; Museo Galileo, Florence, Italy) who welcomed the Torricelli Project with profound interest. A correspondence (July 2010) exists, as well. Thus, we take this opportunity to offer our sincere acknowledgments to Prof. Maurizio Torrini and his assistant, Dr. Sara Bonechi (Museo Galileo, Florence, Italy). Then, Prof. Pisano developed the Torricelli Project with his friend and colleague, Prof. Paolo Bussotti (University of Udine, Italy), finally involving the current other co-authors, as well.

We take this opportunity to address a friendly warm appreciation to Prof. Martin David Davis, the world distinguished mathematician and the computer scientist in the fields of mathematical logic and computability theory. Unfortunately, he passed away at the age of 94 on January 1st, 2023, after delivering and proofread his manuscript. He was a known international interdisciplinary—intellectual scientist who developed advancements of *Post-Turing Model* and co-developed the *Davis–Putnam–Logemann–Loveland* (DPLL) algorithm within Boolean satisfiability solvers; including his interests in interdisciplinary topics like History of Mathematics.

Just as importantly, we address our warm acknowledgments to the *History of Science Collections* and *University of Oklahoma Libraries* at the *University of Oklahoma*, USA (Kerry V. Magruder, Stephen P. Weldon, and JoAnn Palmeri) for the *Opera Geometrica Courtesy Credit*. We also thank the *Library* at the *Max Planck Institute for the History of Science* in Berlin, Germany. Particularly, we are very grateful to Dr. Kerry V. Magruder and Prof. Stephen P. Weldon for their continuous collaboration in administrative management of *Opera geometrica in situ*.

We express our appreciation to Dr. Julie Robarts (University of Melbourne, Australia) for her observations on English language; eventual remaining English mistakes are our own. The English proofreading funding was kindly supported by Institute of Electronics, Microelectronics and Nanotechnology (IEMN), UMR 8520, Lille University–CNRS, France.

Thanks also to doctoral candidate Mr. Andrea Durlo (IEMN, University of Lille, France) for his editorial collaboration.

Thank you!

Finally, of great importance, we address our acknowledgments to Prof. Shahid Rahman, Springer Editor-in-Chief book Series, for his fine work and continuous assistance—and—collaboration during this extensive research-editorial project; and to the Springer reviewers for their excellent suggestions.

April 2023

Raffaele Pisano
Torricelli Research Project Leader

Remarks for the Reader

The research for this major entirely original Torricelli Project dates to the mid-2000s and was carried out by Prof. Raffaele Pisano. No accessible edited transcription (and English translation) of Torricelli's *Opera Geometrica* (1644) previously existed; *idem* for Torricelli's *Opere*. This research is devoted to the history and historical epistemology of science, integrating the history and epistemology of scientific methods and combining epistemological and historical approaches to clearly identify significant historical hypotheses. Therefore bibliographic references, the relationships between physics–mathematics and physics–geometry, and the role played by science in context are strongly stressed. In the future, the Torricelli Project also envisages a second and third part, respectively; the commented translation, side-by-side into English of the *Opera Geometrica*; and the commented translation, side-by-side of the whole *Opere di Evangelista Torricelli* (Torricelli's Works).

This volume resulting from our Torricelli Research Project consists of three main parts: *Introduction, Selected Essays* on the subject/field, and *Opera Geometrica Transcription*. For the latter, the aim has been to maintain the character of the original *Opera Geometrica* but also to make it easy to understand for the modern reader. We have corrected obvious printers' errors and indicated this in footnotes. In addition, on occasion, words have been added for continuity of phrasing, or to add a short in-text explanation, or to give an alternative modern term. The source used for our research and critical transcription of the *Opera Geometrica* is held at the *History of Science Collections—University of Oklahoma Libraries*, University of Oklahoma, USA. An official permission, including the use of the images, was requested and obtained. An English translation of the *Opera Geometrica* is planned for the near future appearing within Torricelli's *Opere*.

Therefore, this monograph and anthology consists of three main parts, resulting in a research—and—source book combined in one unique volume.

The first part includes a long chapter devoted to celebrating and introducing Torricelli, his *Opera geometrica*, and other main mathematical and geometrical works to the reader. Important subjects pertaining to the relationships of physics–mathematics and geometry applied to mechanics have been accurately presented.

The second part is composed of a selection of six essays on the subject. These are presented in a preliminary standalone chapter. The authors of four essay (hereafter *Authors*) are the same authors of this volume; additionally, two distinguished contributors (non–Authors' volume and hereafter *Invited Contributors*) were invited to write essays on the subject included in this volume. The papers appear in alphabetical order.

The third part presents the critical transcription of the *Opera geometrica* (1644). Each page and image is set side-by-side with the transcribed and edited text.

The book's essays attend to different individual ideas, scientific, historical, and epistemological accounts. Each of the eminent *Authors* and *Invited Contributors* are responsible for his/her/their own opinions, which should be regarded as reflecting their personal scientific background and expertise. They are also individually responsible for the images, reprints, quotations, acknowledgments, and all related permissions/approvals displayed/not displayed in their essays.

This volume has been independently blind peer-reviewed following *Springer*'s policy, including its Springer Basic Style used for references sections.

April 2023

Raffaele Pisano
Torricelli Research Project Leader

About the Authors and Contributors

Torricelli Research Project Leader

Raffaele Pisano
Full Professor, Head *History of Physics and Applied Science and Technology* Team (HOPAST), Institute of Electronics, Microelectronics and Nanotechnology (IEMN), UMR CNRS 8520—Lille University
France

Torricelli Project Scientific Committee

Raffaele Pisano
Full Professor, HOPAST at IEMN, Lille University—CNRS
France

Paolo Bussotti
Professor, DIUM, Udine University
Italy

Torricelli Project Scientific Editorial Staff

Raffaele Pisano
Full Professor, HOPAST at IEMN, Lille University—CNRS
France

Paolo Bussotti
Professor, DIUM, Udine University
Italy

Patricia Radelet de Grave
Emeritus Professor, Catholic University of Louvain–la–Neuve
Belgium

Julie Robarts
Dr. Researcher, University of Melbourne
Australia

Andrea Durlo
Ph.D. Student, IEMN, Lille University—CNRS
France

Authors' Book and Essays

Paolo Bussotti
Professor, DIUM, Udine University
Italy

Jean Dhombres
Emeritus Professor, DR, Centre Alexandre Koyré—CNRS
France

Raffaele Pisano
Full Professor, HOPAST at IEMN, Lille University—CNRS
France
and Associate to Archive Poincaré, Lorraine University
France

Patricia Radelet de Grave
Emeritus Professor, Catholic University of Louvain–la–Neuve
Belgium

Invited Contributor's Essays

Martin Davis (1928–2023)
Emeritus Professor, Courant Institute, New York University | Visiting Scholar, University of California, Berkeley
USA

J. B. Shank
Professor, University of Minnesota
USA

Financial Support Torricelli—History of Physics Project

IEMN
Institute of Electronics, Microelectronics and Nanotechnology, UMR CNRS 8520—Lille University
France

Contents

Part 1

Opera Geometrica, 1644

Celebrating Evangelista Torricelli's *Opera Geometrica* (1644–2024): Details in History and Historiography of Physics, Geometry and Mathematics

1

Raffaele Pisano and Paolo Bussotti

Abstract

In this essay we describe Torricelli's life and works in their scientific context, including the Archimedean heritage in Torricelli's works. We analyse the changes of Torricelli's works and explain the novelties of the edition we are offering. Then, we provide a picture of the most significant results obtained by Torricelli (1608–1647), particularly in mechanics and geometry. Furthermore, we also focus on the Torricelli's methodology, specifying how he proved two of his achievements, given their novelty and mathematical meaning: (a) the volume of the "solido acutissimo"; (b) the geometry of the logarithmic spiral. An ad hoc Content of the *Opera geometrica* is proposed as well. It is based on the original chronological sequences of the contents wrote by Torricelli. This is important to evaluate the birth of the scientific matters presented by Torricelli and the editorial process of the publication.

Keyword

Torricelli • Archimedes • Cavalieri • *Opera Geometrica* • Physics-mathematics relationship • Principle of virtual works • Logarithmic spiral • Infinitesimal calculus

1 Torricelli's Life and Works

Evangelista Torricelli (Roma, 1608—Firenze, 1647; hereafter also Evangelista) spanned a fruitful period for science. Thanks to the contributions of several scientists and mathematicians, in the second half of the 30s and in the 40s of the seventeenth century, mathematical technique was substantially improved. However, Archimedes (c. 287—c. 212 BC) remained a reference point for Torricelli. In this respect Torricelli is interesting because he represented a new, modern, and open-minded scientist, who tried to rediscover and extend the results of Greek mathematics. The influence of Archimedes's science (Archimedes 1897) and methodology (Archimedes 1912, 2018) on Torricelli is particularly significant and allows us to understand the inspiration of the great Syracusean, and Greek geometry and science for modern science (Pisano and Capecchi 2010a).

Torricelli was not a misunderstood genius. His value was fully recognised during his lifetime. From 1642, after Galileo's departure, until his death which occurred prematurely in 1647, probably due to typhus fever, he was the Court Mathematician of the (Gran Duke Ferdinand II) Granduca Ferdinando de' Medici (1549–1609). He was in contact with famous scientists and mathematicians, not only in Italy. Especially in the French mathematical milieu, his results in geometry were known and considered of primary importance. Torricelli had an epistolary exchange with some of the most skilled scholars

R. Pisano (✉)
HOPAST at IEMN, Lille University, Lille, France
e-mail: raffaele.pisano@univ-lille.fr

P. Bussotti
DIUM, Udine University, Udine, Italy
e-mail: paolo.bussotti@uniud.it

R. Pisano et al., *Homage to Evangelista Torricelli's Opera Geometrica 1644–2024*,
Logic, Epistemology, and the Unity of Science 55,
https://doi.org/10.1007/978-3-031-06963-5_1

of his epoch such as Pierre de Fermat (1601/7–1665; Bussotti and Pisano 2020) and Gilles Personne de Roberval (1602–1675). That is to say, he was an external member of that cultural-scientific milieu, whose point of convergence was Marin Mersenne (1588–1648). Torricelli had close relations with Bonaventura Cavalieri (1598–1647), the other great Galileian and geometer and the other main Italian mathematician whose *floruit* dates to the late first half of the seventeenth century. This notwithstanding, only a part of Torricelli's works were published in his lifetime. This occurred in 1644 when the Florence edition of Torricelli's *Opera geometrica* was published. As a matter of fact, that edition did not include the whole of Torricelli's geometrical production. When Torricelli died, he left several manuscript works and outlines of works. Some of them were published in the late seventeenth, eighteenth and nineteenth centuries, but it was not until the beginning of the twentieth century that a complete edition of Torricelli's works was published. However, this does not close this story because, despite being a fundamental point of reference, this edition has some weaknesses. Another edition was published in 1975. An epistolary exchange among the Galileo's disciples was published (Galluzzi and Torrini 1975) by Paolo Galluzzi and Maurizio Torrini (1942–2019), containing several letters by and to Torricelli (see also: Pisano and Bussotti 2012, 2015b, c, d, 2017a, b, c; Pisano, Agassi and Drozdova).

In what follows we outline the most significant facts to characterize the adventures of Torricelli's production because they stand as an example of the tortuous way in which the complete knowledge of the results and techniques of scholars can be achieved.[1]

1.1 Adventures and Misadventures

Torricelli was born in Rome on the 15th of October 1608. His parents came from Faenza. In particular, his mother's family was the owner of the farm *La Torricella* near Faenza (Emilia Romagna, Italy). His father, Gaspare Roberti came from Bertinoro—a village between Forlì and Cesena –. Evangelista adopted the name of his mother. His parents sent Evangelista to study in Faenza under the guide of his maternal uncle Gian Francesco (1567–?) who took the name Jacopo when he entered the monastic order of the Camaldolesi. Between 1623 and 1625 Torricelli was trained in mathematics at the Jesuit college of Faenza (though as an external member). He revealed a particular skill in mathematics. In the second half of 1625 Evangelista returned to Rome and met—probably thanks to the mediation of his uncle—the Benedictine Benedetto Castelli (1578–1643), a famous mathematician who was a pupil of Galileo Galilei (1564–1642). From 1626 Castelli was lecturer of Mathematics at the university of Rome at the *Palazzo della Sapienza* (also "Roma La Sapienza"). For six years Evangelista frequented the mathematical academy held by Castelli. Here he met other young scholars interested in mathematics such as Raffaello Magiotti (1597–1656; Berlingozzi 1902) and Antonio Nardi (1598–1648?). In his first letter to Galileo dated 11 September 1632 (Galilei 1890–1909, XIV, pp. 386–387) Torricelli informed the great scientist that, with Castelli, he had studied many important mathematical and scientific subjects: Euclid (Fl. 300 BC), Archimedes (c. 287—c. 212 BC), Apollonius of Perga (c. 240—c. 190 BC), Theodosius of Bithynia (c. 169—c. 100 BC) Ptolemy (c. 100—170 AD), Johannes de Sacrobosco (c. 1115—c. 1256), Nicolaus Copernicus, (1473–1543), Tycho Brahe (1546–1601), Christianus Severini Longomontanus (1562–1647) and Johannes Kepler (1571–1630). He wrote that he had also read and profoundly meditated on Galileo's *Dialogue Concerning the Two Chief World Systems* (*Dialogo sopra i due massimi sistemi del mondo*; hereafter *Dialogo*). It is likely that he was also instructed in algebra. Torricelli also frequently saw two Jesuit mathematicians of the Collegio Romano, Austrian Christoph Grienberger (1561–1636) and German Christopher Scheiner (1573–1650). He was particularly interested in their reactions to the reading of Galileo's *Dialogo* (Galilei 1632, 1954) which were positive, though Scheiner complained that Galileo was not respectful towards him. Obviously, he was referring to the controversy regarding the sunspots.

From 1632 to 1641 Torricelli was the secretary of monsignor Giovanni Battista Ciampoli (1589–1643), secretary to Pope Urban VIII (1568–1644). Ciampoli had fallen from grace of Urban VIII because he declared his support of Galileo's *Dialogo* and had favoured the publication of Galileo's work. Consequently, he was given several secondary tasks in different parts of the Papal States, and Torricelli travelled with him. In February 1641 Torricelli came back to Rome and began to frequent the

[1] Cfr. Loria and Belloni (Torricelli 1919, "Introduzione"; Torricelli 1975, "Introduzione" by Lanfranco Belloni). For the history of mathematics and Torricellean biography-*notizie*, see Loria (1897, 1903, 1909, 1915, 1917, 1919, 1922a, b, 1923, 1932, 1938, 1944, 1945, 1947), Podetti (1914), Agostini (1930a, b, c, 1931, 1949, 1950, 1951a, b, c, d, e), Bortolotti (1921, 1922, 1923, 1924, 1925a, b, 1928a, b, c, 1930, 1931, 1937, 1939a, b, 1947), Loria and Vassura (Torricelli 1919–1944), Vassura (Vassura 1945), Visani (1952), Procissi (Procissi 1952, 1953), Taton (1963), Belloni (Torricelli 1975), Fabbri (Fabbri 1947–1997). See also: Toscano 2008, 2014; Favino 2009, 2019; Rossini 1959; Bertoni 2003 and their suggested biographical references, as well.

mathematical circle around Castelli, with Magiotti and Nardi. Evangelista had written a treatise on the motion of projectiles. Castelli was so impressed by this work, which completed the theory expounded by Galilei in the third Day of *Discorsi e dimostrazioni matematiche* (1638; Galilei 1890–1909, VIII, pp. 39–362) that he sent it to Galileo. The old scientist invited Torricelli to his residence in Arcetri. Here, besides Galileo, Torricelli also encountered Vincenzo Viviani (1622–1703), who was also a guest of Galileo (Viviani 2001). In that period, Torricelli was offered the possibility of becoming the mathematician of the Grand Duchy of Tuscany (Granduca di Toscana) with a salary of 200 scudi. Thus, he began to frequenting high society. He could dedicate his time to research and became a member of several academies, the most prestigious among them being the famous *Accademia della Crusca*, where Evangelista gave several lectures and talks on diverse arguments, such as the problem of the impact, the origin of gravity and the nature of the winds. From 1641—at the suggestion of Galileo —Torricelli also began to construct lenses for telescopes.[2] In this field he became a true expert and this work was a source of further earnings. In 1644 his *Opera geometrica* was published in Firenze. With this work, Torricelli became a well-known scientist throughout Europe.

2 Presenting Torricelli's *Opera Geometrica*

Let us now give an overview of Torricelli's mathematical works, focusing, in particular, but not exclusively, on the *Opera geometrica* (Cfr. Loria's and Belloni's works):

- The Content of *Opera geometrica*;
- A Historiography of Torricelli's *Opera geometrica* and *Works*;
- *In quibus Archimedis Doctrina* in *Opera geometrica*;
- Details from a selection of Torricelli's *Opera geometrica* proofs on main reasoning as most original contributions to geometry applied to mechanics and infinitesimals.

2.1 The Content of *Opera Geometrica*

The *Opera Geometrica* written both in Latin and, a small part, in the Italian vernacular, has two main topics: Motion and Geometry. Taking into account our previous research, we divided it ad hoc in four main parts. For this reason, below we propose the Content of the *Opera geometrica* (Table 1) including the main sections of each part, as they are presented by Torricelli. For example, if a section (e.g., the appendices) belongs to arguments dealt with in the initial chapters, but in fact appears at the end of the *Opera geometrica*, then we chose to respect this editorial choice. This choice also takes into account the original editorial pagination, that is the numbering of pages that restarts at any time; the term "Book (*Liber*)" meant chapter. Hereafter we will continue to use the term Book to avoid any confusion with respect to the modern significance.

[2] Cfr. Galluzzi 1976; Miniati et al. 2002; Molesini 2010; Paternoster et al. 1996; Ronchi 1924, 1968.

Table 1 Contents of *Opera Geometrica* (393 numbered pages)

Sections	Parts	Title	Main content
Preliminary matters	Part 0	*Serenissimo Magno Duci Etruria* (pp. 3–4) *Proemium* (pp. 5–10). [N.B.: This section also fits with spheral solids. Thus, it may be also considered as the initial part of the topic below]	Preface Prologue, Six Polygon species, Specifications on definitions and suppositions
Chapters on the first elements of geometry, motion and mechanics	Part 1	*De sphaera et solidis sphaearalibus, Liber primus* (pp. 11–46) *De solidis sphaeralibus, Liber secundus* (pp. 47–94)	29 Propositions 52 Propositions
	Part 2	*De motu graviun naturaliter descendetium, Liber primus* (pp. 97–153) *De motu proiectorum, Liber secundus* (pp. 154–243)[3]	58 Propositions 37 Propositions
Chapters on Geometryand Infinitesimal	Part 3	*De dimensione parabolae solidique hyperboloci, problemata duo* (pp. 1–84)[4] *Ad Serenissima Principem Lopoldum ab Etruria* (pp. 3–4) *Ad Lectorem Proemium* (pp. 5–16) *Quadratura parabolae, Problema Primum* (pp. 17–84)[5] *Appendix De dimensione cycloidis* (pp. 85–90) *Appendix Scholium De cycloidibus aliarum specierum* (pp. 91–92)	Preface Prologue, 6 Suppositions–definitions 21 Propositions 3 Theorems Specifications
	Part 4	*De solido acuto hyperbolico, Problema alterum* (pp. 93–112) *De solido acuto hyperbolico, Problema Secudnum* (pp. 113–135)[6] *Appendix De dimensione Cochlea* (pp. [136]145–150)[7]	14 Examples, 1 Scholium, 1 Definition 29 Corollaries 1 Definition, 5 Lemmas, 1 Theorem, 1 Scholium

Part 1

De sphaera et solidis sphaeralibus, 2 Books

Torricelli analyses several figures generated by a regular polygon inscribed in or circumscribed to a circle which rotates around one of its axes of symmetry. Some of the results expounded by Torricelli were already known to Archimedes (Capecchi and Pisano 2007a, b). The most important method used by Torricelli was exhaustion. Though Archimedes was his explicit source of inspiration, the reasonings proposed by Torricelli are original. The material presented by Torricelli in this work can be dated to the year 1641.

Part 2

De motu graviun naturaliter descendetium et proiectorum, 2 Books

Torricelli presents his famous principle according to which two heavy bodies mutually linked cannot move spontaneously unless their centre of gravity descends (Pisano 2017; Pisano and Capecchi 2007a, b, 2010a). Torricelli solved several problems whose main source of inspiration were Galileo's works, but he invented new and more rigorous methods. The two books are followed by an *Appendix* entitled *De motu aquarum* where Torricelli formulated the law that given a liquid within a vessel with a hole at the bottom, the liquid falls out from that hole with a velocity proportional to the square root of the distance between the free surface of the liquid and the bottom of the vessel.

[3] N.B.: The final part is in Italian vernacular language (pp. 216–243).

[4] N.B.: The numbering of pages restarts from 1.

[5] N.B.: The pages number 55 and 56 are repeated. The numbering of pages is broken: pages 79 and 80 are missing. Nevertheless the content is saved.

[6] N.B.: The numbering of pages restarts from 129. Thus it would be pp. 129–135.

[7] N.B.: The numbering of pages starts with p. 136 and just after with p. 145, so the numeration continues consequently. The last page is 150, but, in fact, the Appendix is only seven pages long.

Part 3

De dimensione parabolae solidique hyperbolici, 2 Books, 2 Problems, 2 Appendices

This is a very original section. Torricelli used the technique of Cavalieri's indivisibles extending it beyond the rectilinear indivisibles, so creating the curved indivisibles. In fact, Cavalieri in one circumstance introduced curved indivisibles. However, he did not develop a dedicated theory, and there are some doubts whether he considered them completely legitimate (Cfr. Vita 1972; Andersen 1985, p. 336). The most important result is the proof of the famous theorem of the "acute hyperbolic solid" which has a finite volume though being generated by the rotation of a hyperbola around its asymptote. This contribution is significant also from a historical-methodological point of view because Torricelli claimed that the indivisibles might have been the heuristic procedure through which Archimedes reached his results. In fact, for the quadrature of the parabolic segment, Torricelli proved (Torricelli 1644, III, 1, *Quadratura parabolae*, p. 17; see also Proposition III, p. 32 *et sugg.*) the Archimedean theorem (Archimedes 1897, Propositions 24, p. 251) by means of two different methods and several procedures (see details below).

Part 4

De solido acuto hyperbolico, 2 Problems, 1 Appendix

Torricelli proved that the area of the cycloid produced in the first rotation is equal to three times the area of the generating circle. This result had been communicated to Cavalieri in 1643 (Torricelli 1919–1944, III, Letter to Cavalieri, February 28th, p. 108).

2.2 A Historiography of Torricelli's *Opera Geometrica* and *Works*

Besides his theoretical contributions, the other discovery for which Torricelli is famous concerns the barometric experiment (Cfr. Waard; West). It also dates to Torricelli's *anni mirabiles* 1641–1644. The precursor of the experiment can be referenced in the discussion between Giovanni Battista Baliani (1582–1666) and Galileo on the weight of air. In 1614 Baliani asked Galileo whether he knew a method to measure the specific weight of air. Galileo answered affirmatively in a letter to Baliani on 12 March 1614 (Galileo 1890–1909, XII, p. 35; Torricelli 1975, p. 36). Many years later, in 1630, Baliani told Galileo of the difficulty of raising water with a pump to the height of a hill. Galileo answered that this was the case because the column of water breaks under its own weight. He established 18 *braccia* to be the greatest height to which a column of water could be raised by a siphon. According to Galileo, such a maximal height was an indication of the vacuum resistance. In his opinion, the water's cohesion was due to the attraction produced by the vacuoles interposed between the particles of water. When the weight of the water column overcomes such a cohesion-force, the column breaks. In contrast to this, Baliani was convinced that the impossibility of raising water higher than 18 *braccia*, equivalent to 10.8 m, depended on the weight of air. He claimed that air weighs on the bodies, which are on the Earth's surface, like water weighs on the bodies, which are on the seabed. Torricelli knew this discussion. He was convinced that Baliani's arguments were correct and tried to determine exactly the pressure exerted by the atmosphere on a point of the Earth's surface in the direction joining the point to the earth's centre. Given the difficulty to form and experiment with a water column of 10.8 m, Torricelli conceived and developed, jointly with Viviani, his famous experiment with a column of quicksilver (mercury) of the same weight as a column of water 10.8 m high. Thus, he considered a column of mercury of height 1 and ¼ *braccia*, namely 76 cm. The experiment did not fully succeed because the vacuum in the tube constructed by Torricelli was not perfect. It was, however, sufficient to prove that all the ideas connected to the *horror vacui* were wrong. Pascal who, on the air's pressure, wrote the two fundamental memoires *Traitez de l'Equilibre des Liqueurs et de la pesanteur de la masse de l'air*[8] and executed the celebrated experiment of the *Puis-de-Dôme* was influenced by Torricelli's work, though he did not mention Torricelli. It is known that two letters to Michelangelo Ricci, in which Torricelli described his ideas and his experiment, were sent by Ricci himself to Mersenne who almost surely informed Pascal.

Torricelli was a versatile scientist, and his opinion was requested with regard to the drainage of the Valdichiana to free that zone from stagnant water. There were two opposite ideas: (1) to drain the river Chiana; (2) to raise, by flood, the level of marshy land to achieve the necessary slope so that water might flow towards the Arno. Alessandro del Borro (1600–1656),

[8] The *Traitez* (*Traités*) was published in 1663, one year after Pascal death (Cfr. Pascal 1998–2000).

who was the chief of the Granduca Ferdinando's army, and Famiano Michelini (1604–1665), lecturer of mathematics at the university of Pisa (he became professor from 1648) were favourable to the first solution. Whereas, Torricelli and Andrea Arrighetti (1594–1672), a disciple of Galileo, who was the intermediary in the correspondence between Galileo himself and his daughter Maria Celeste, thought the second solution to be far preferable. The uncertainties and acrinomy of this dispute had the consequence that no decision was taken until 1789 when the solution proposed by Torricelli was adopted.

The relations of Torricelli with the French mathematicians had been very good, but from 1646 the situation changed: Roberval vindicated his priority for the quadrature of the cycloid and accused Torricelli of plagiarism. Roberval and René Descartes (1596–1650) began to doubt that the Galileian law on the free fall of bodies and the parabolic trajectory of projectiles was correct because the experiments did not fit with the theoretical predictions. There was a brief epistolary exchange in which the answers from Torricelli were not decisively in favour of Galileo's law. Then, Torricelli began to write a systematic work to answer all the doubts of the French mathematicians entitled *Racconto d'alcuni problemi proposti e passati tra gli matematici di Francia et il Torricelli ne i quattro anni prossimamente passati* (Torricelli 1919–1944, III, pp. 1–32). However, death, after a brief illness, took Torricelli in the night between 24 and 25 October 1647, so he could not conclude his work.

Torricelli had appointed Ludovico Serenai, who was a close friend, testamentary heir in around 1640. He left precise instructions on the best technique to construct lenses as well as on the publication of his scientific manuscripts. He wanted their publication to be edited by Bonaventura Cavalieri and Michelangelo Ricci (1619–1682). But Cavalieri died shortly after Torricelli, on 30 November 1647. This was when the adventures and misadventures of Torricelli's manuscripts began.

Serenai tried to convince Ricci to edit Torricelli's works, through Magiotti as intermediary, and later writing directly to Ricci. In a letter written the 11 April 1648 Ricci communicated to Serenai that he was no longer interested in mathematics and that several family commitments prevented him undertaking the difficult task of editing Torricelli's manuscripts. Serenai then proposed the work to Magiotti, but he also declined. Finally, Serenai tried to involve Viviani in the difficult enterprise. After many doubts, Viviani decided that all Torricelli's manuscripts should be copied and sent to him because he did not want to have sole responsibility for their management. Serenai, who was a jurist and did not have any mathematical knowledge, accepted the heavy duty of copying the manuscripts and sending them to Viviani. However, Viviani did not produce the edition. He only published three theorems by Torricelli drawn from the booklet *De proportionibus* within his work *Quinto libro degli Elementi di Euclide ovvero Scienza universale delle Proporzioni spiegata con la dottrina di Galileo* (Podetti 1914; Viviani 1674). Loria developed different hypotheses with regard to Viviani's behaviour (Cfr. Loria in "Introduzione" to Torricelli 1919–1944, pp. XIX–XXIV): perhaps Viviani was engaged in his own research or had to dedicate much time to other tasks, or his almost maniacal precision induced him to write and rewrite the same passages. It must also be taken into account that his health was not good. As matter of fact, he did not publish Torricelli's works. Serenai understood that the enterprise appeared increasingly difficult. In 1674 he decided that, after his death, the manuscripts were to be given to Agostino Nelli (1661–1725) and, in case of his departure, to Ridolfo Paganelli (1618–1693) and, eventually, to Carlo Roberto Dati (1619–1676). Serenai died on 28 February 1685 and Viviani on 22 September 1703. The subsequent circumstances of Torricelli's manuscripts have some inexplicable aspects: when Agostino Nelli died, his son Giovanni Battista, who was a disciple of Viviani, convinced Viviani himself to accept the fiduciary deposit of the Torricelli's manuscripts. This is rather strange, given that Viviani had always refused to keep those manuscripts. But, as Loria claimed, what is even more astonishing is that Viviani left precise testamentary indications as to the whole of his patrimony, but he never mentioned Torricelli's manuscripts. As a consequence of this serious negligence of Viviani, the manuscripts passed to the abbot Jacopo Panzanini (fl. first half of the eighteenth century) nephew of Viviani and lecturer of mathematics in Firenze (also Florence). In the meantime, in 1715, Tommaso Bonaventuri (1675–1731), a Florentine erudite, member of the *Accademia della Crusca*, who jointly with Benedetto Bresciani (1658–1740), Giuseppe Averani (1662–1738) and Guido Grandi (1671–1742) edited the first Florentine edition of Galileo's works, published Torricelli's *Lezioni accademiche* (Torricelli 1715, 1823; see also Torricelli 1864). When, in 1733 Panzanini died, he left the manuscripts to his nephews Carlo and Angelo. They did not guess the value of that material and sold a part of it, together with some of Galileo's manuscripts, to a grocer. He used a manuscript of Galileo to wrap a piece of mortadella, which was bought by Giovan Battista Clemente Nelli (1725–1793), a Florentine academic, who was a bibliophile and an antiquarian. His most important work was *Vita e commercio letterario di Galileo*, published in Florence in 1793, though, in fear of ecclesiastic censorship, he indicated Lausanne as place of edition. This work is important because the biography of Galileo, based on a deep analysis of his letters, is reconstructed. Nelli, in that piece of paper in which his mortadella was wrapped, recognized Galileo's handwriting. Through the grocer, he contacted Panzanini and acquired the whole material, which included Torricelli's manuscripts. In a letter to Giovanni Lami (1697–1770) dated 15 September 1750 (Nelli 1753, II, p. 763, see also footnote 3) Nelli mentioned all the manuscripts he had bought. Comparing this list with that written by Serenai and Viviani, it is evident that almost all the material was saved. Nelli tried to organize an edition of the Torricelli's manuscripts, but without success.

In 1768 the works of Torricelli on the drainage of the Valdichiana (*Scrittura sopra la Chiana* et sgg. his replies on the subject) were published in the fourth tome of the multi-author work *Raccolta d'autori che trattano del moto delle acque* (AA.VV 1768, IV, pp. 115–118; pp. 125–142; pp. 143–147; pp. 157–166; see also Torricelli 1919–1944, IV). In his testament, Nelli required that his heirs sell the manuscripts to some private citizen only if the Great Duke of Tuscany refused to buy them. In 1818 Ferdinand III Great duke of Habsburg-Lorraine (1769–1824) acquired the manuscripts. They were preserved in the *Biblioteca Palatina*, created by Ferdinando III himself. In 1861 the *Biblioteca Palatina* was merged with the *Biblioteca Magliabechiana*, because of a provision by Francesco De Sanctis (1817–1883), the Minister of Education of the new-born Italian state. In 1864 a monument to Torricelli was inaugurated in Faenza and Giovanni Ghinassi (1809–1870) published a biography of Torricelli with some scientific letters (Ghinassi 1864). In 1875 Baldassarre Boncompagni (1821–1894) published a further selection of the scientific letters by Torricelli.[9] Charles Henry (1859–1926) published another letter by Torricelli. In our context, the work by Raffaello Caverni (1837–1900), *Storia del metodo sperimentale in Italia* (Caverni 1891–1900) is important because the author reported several sections of Torricelli's unpublished works. In 1900 Gino Benedetto Loria (1862–1954) edited Torricelli's studies on the logarithmic spiral (Loria 1900). In 1903 Loria himself planned to institute the National Edition of Torricelli's works, and at the eighth Section of the *Congresso internazionale di scienze storiche* (*International conference of historical sciences*) held in Roma (Rome), 1903, proposed to the Italian government that the *Accademia dei Lincei* to examine Torricelli's manuscripts in order to decide the best way to prepare the National Edition. The Education Minister accepted to finance this explorative work which was entrusted to Giovanni Vailati (1863–1909) in 1904. However, in 1905 Vailati became a member of the *Commissione per il riordinamento degli studi secondari in Italia* (*Commission for the rearrangement of the secondary studies in Italy*) and the idea of the National Edition of Torricelli's works ran aground. The following year, 1906, as a consequence of a proposal by Giuseppe Vassura (1866–1949) the Conference of the *Italian Physical Society* proposed to the Italian government to prepare an edition of Torricelli's works. Such a proposal had no answer from the Italian government. At this point the Municipality of Faenza, on the occasion of the tercentenary of Torricelli's birth, decided to publish his works. This enterprise was entrusted to Giuseppe Vassura. The work-plan was to publish three tomes, the first one containing the geometrical works, the second the *Lezioni accademiche* (Torricelli 1919–1944, II), the writing on mechanics and other various writings. Finally, the third volume should have contained Torricelli's scientific correspondence. Vassura worked for four years on the edition and completed the second and the third tomes. However, in 1912 Vassura had to leave Faenza and other tasks prevented him from continuing his work. To avoid the enterprise failing Vassura suggested that the Faenza Municipality contact Gino Loria to edit the *Opera geometrica* and the geometrical manuscripts. It seems that after many doubts and much indecisions Loria accepted, and in 1919 the *Opere di Evangelista Torricelli* were published in four volumes, 5 tomes by Stabilimento Tipografico G. Montanari in Faenza, edited by Gino Loria and Giuseppe Vassura (Torricelli 1919–1944). In 1944 a fourth volume (see below) edited solely by Vassura was added.

1919\. Vol. I, Part I. This is subtitled as *Geometria*. It is edited by Gino Loria. *De Sphaera et solidis sphaeralibus libri duo* (*Ivi*, pp. 1–87); *De dimensione parabolae* (*Ivi*, pp. 91–162) also including the two appendices on the dimension of the cycloids (*Ivi*, pp. 163–172); *De solido hyperbolico acuto* (*Ivi*, pp. 173–221) with the Appendix *De dimensione cochleae* (*Ivi*, pp. 223–230). These works belong to the 1644 *Opera geometrica*. Whereas the following writings were published for the first time: Appendix to the *Lemma XX* in *De dimensione parabola* (*Ivi*, pp. 231–238); *De Tactionibus* (*Ivi*, pp. 239–290); *De proportionibus liber* (*Ivi*, pp. 295–327); *De planis varia* (*Ivi*, pp. 329–345); *De solidis varia* (*Ivi*, pp. 347–363); *De circulo et adscriptis* (*Ivi*, pp. 365–375); *De comparatione perimetrorum cylindri, coni ac sphaerae* (*Ivi*, pp. 377–386); *De aequalitate perimetrorum cylindri, coni ac sphaerae varia.* (*Ivi*, pp. 387–407);

1919\. Vol. I, Part II. This part is also subtitled as *Geometria*. It is edited by Gino Loria. It only includes previously unpublished works. *Campo di tartufi* (*Ivi*, pp. 1–43); *Contro gl'infiniti* (*Ivi*, pp. 45–58); *Sugli isoperimetri* (*Ivi*, pp. 49–55); *De centro gravitatis sectoris circuli* (*Ivi*, pp. 59–77); *De maximis et minimis* (*Ivi*, pp. 79–97); *Nova per armillas stereometria* (*Ivi*, pp. 107–173); *De centro gravitatis planorum ac solidorum* (*Ivi*, pp. 175–226); *De infinitis hyperbolis* (*Ivi*, pp. 227–274); *De infinitis parabolis* (*Ivi*, pp. 275–328); *De cycloide*; (*Ivi*, pp. 329–334); *De*

[9] In the 8th tome of his *Bullettino* Boncompagni published "Intorno ad alcune lettere di Evangelista Torricelli, del P. Marino Mersenne e di Francesco Du Verdus"; "Lettere di Evangelista Torricelli al P. Marino Mersenne"; "Lettere del P. Marino Mersenne ad Evangelista Torricelli"; "Lettere di Francesco Du Verdus ad Evangelista Torricelli." (Boncompagni 1875, pp. 353–456).

Hemyhyperbola logaritmica (*Ivi*, pp. 335–347); *De infinitis spiralibus* (*Ivi*, pp. 349–399); *Sezioni coniche* (*Ivi*, pp. 403–414); *De indivisibilibus* (*Ivi*, pp. 417–432); *Miscellanea* (*Ivi*, pp. 433–440); Miscellanea. Appendice: *Lettera di Filaleti a Timauro Antiate, Della vera storia della cicloide e della famosissima esperienza dell'argento vivo*[10] (*Ivi*, pp. 441–482).

1919. Vol. II. It is edited by Giuseppe Vassura. *Lezioni accademiche* (*Ivi*, pp. 1–100); *De motu gravium naturaliter descendentium et proiectorum* (*Ivi*, pp. 101–102); *De motu gravium naturaliter descendentium* (*Ivi*, pp. 103–152); *De motu proiectorum* (*Ivi*, pp. 153–232); *De motu ac momentis varia* (*Ivi*, pp. 233–250, which are also included in the *Opera geometrica,* 1644); *Soluzione di un problema intorno alle cose che stanno nell'umido* (*Ivi*, pp. 251–262); *Scritture sulla bonificazione della Val di Chiana* (*Ivi*, pp. 263–310) as writings on the Valdichiana's draining and a brief, but interesting treatise on practical perspective *Prospettiva pratica* (*Ivi*, pp. 311-endbook).

1919. Vol. III. It is edited by Giuseppe Vassura. *Racconto d'alcuni problemi proposti e passati scambievolemente tra gli matematici di Francia, et il Torricelli ne i quattro anni prossimamente passati* (*Ivi*, pp. 1–32); *Carteggio scientifico* (*Ivi*, pp. 33–510); *Indice cronologico del carteggio* (*Ivi*, pp. 511–516); *Indice alfabatico del carteggio* (*Ivi*, pp. 517–endbook)

1944. Vol. IV. It is edited by Giuseppe Vassura in a single volume and consists of three internal editorial parts. *Prefazione al Vol. IV* (*Ivi,* pp. 9–18); *Documenti alla Vita* (*Ivi*, Part, I, pp. 19–122); *Documenti alla Opere* (*Ivi*, Part II, pp. 123–198); *Appendice*, including scientific correspondence (*Ivi*, Part III, pp. 199–217; among which are *Ricordi dettati da Torricelli a L. Serenai* with the last will of testament of Torricelli and *Notizie suggerite da Vincenzo Viviani a Lodovico Serenai*, which concern Torricelli's life); *Scritture tolte da un codice mss del XVII secolo* (*Ivi*, pp. 218–230); *Pubblicazioni moderne* (*Ivi*, pp. 231–338; they are modern essays by Mario Gliozzi, Amedeo Agostini and Ettore Bortolotti); *Pubblicazioni da consultare* (*Ivi*, pp. 339–endbook).

The Faenza edition is the most complete available and, therefore, is a fundamental reference point for exploring Torricelli. Nonetheless, it has been subject to several criticisms. In particular, Part II of Vol. I, edited by Loria, which contains Torricelli's geometrical manuscripts was criticized. Before his death, Torricelli did not have time to put his manuscripts into rational order, and Loria transcribed them with a very limited number of interventions. Ettore Bortolotti (1866–1947) claimed that the very few interventions by Loria are misleading and that his annotations are wrong. He wrote explicitly:

> The manuscripts remained available to the scholar in the National Library of Florence for a long time. After more than two centuries, they have been published in the Faenza edition of *Opere Geometriche*. However, the typical disorder of the unpublished writings was made worse by the editor's work, who did not penetrate the Author's thought. This notwithstanding, he wanted to intervene and omitted necessary things, while repeating several times (without realizing it) things of low or no value. Furthermore, his annotations are a mere phantasy. In contrast to this, the pure reproduction of the manuscript would have been a meritorious work.[11]

As Lanfranco Belloni (1944–2017) points out (Torricelli 1975, pp. 20–21), there were two aspects to Bortolotti's judgement: first of all, Bortolotti and Loria had two different opinions of Torricelli's geometrical works. Bortolotti thought that many elements of integral calculus were already present in Torricelli; whereas Loria considered Torricelli an ingenious mathematician, but considered his methods closer to those of Archimedes than to those of the inventors of infinitesimal calculus. Furthermore, Bortolotti accused Loria of having carried out his work of editor without any enthusiasm for penetrating the real meaning of Torricelli's methods. The Faenza edition had a troubled history, and it is true that Loria could have presented Torricelli's manuscripts with more care, in order to provide a rational order to Torricelli's works. However, the edition, such as it is, for the moment, is an irreplaceable document. In 1955, Ettore Carruccio (1908–1980) offered an improved edition in comparison to Loria's *De infinitis spiralibus* (Torricelli 1955). More recently, the Italian translation of the *Opera Geometrica*, edited by Belloni was published in 1975. This precious edition also contains an Italian translation of other geometrical writings by Torricelli. The most important ones concern indivisibles and the logarithmic spiral. Furthermore, the academic lectures, the letters on vacuum and the writings on the draining of the Valdichiana are also added. The historical notes by Belloni as well as his Introduction are documents that enrich this edition and make it a significant reference point for Torricelli research.

---

[10] Cfr. Torricelli [1644]1897.

[11] "I manoscritti di Torricelli sono rimasti a lungo nella Biblioteca Nazionale di Firenze, a disposizione degli studiosi; dopo più di due secoli hanno visto la luce, nella edizione faentina delle *Opere Geometriche*. Ma il caratteristico disordine degli scritti inediti fu aggravato dall'opera dell'editore, che non seppe penetrare nel pensiero dell'Autore, e pur volle metterci mano, omettendo cose necessarie, ripetendo a più riprese (senza nemmeno accorgersene) cose di scarso o nessun valore, e postillando fantasticamente. Mentre sarebbe stata opera meritoria, anche la semplice e pura genuina riproduzione del manoscritto, tal quale si trovava." (Bortolotti 1939a, b, p. 459; Authors' translation).

The Loria–Vassura edition of Torricelli's *Opere* does not offer a translation into a modern language. Torricelli's works are presented in their original Latin and vernacular Italian language: a crucial reference for historians of mathematics and physics. Our present Torricelli Springer editorial project (started in 2010 by Pisano's personal initiatives and his early contacts with Maurizio Torrini and other scholars in the field),[12] at this stage, includes a selection of outstanding papers on the subject and a commented transcription of the *Opera geometrica* only; all in a unique international volume. For the future, in the long term and if possible, we are thinking—and—scheduling the English commented translation of both the *Opera geometrica* and the whole of Torricelli's *Opere*; this is possible because distinguished scholars such as Loria, Vassura, Bortolotti, Agostini et al., before us, analysed and archived mountains of documents. We hope to do that for our posteriors, as well.

2.3 Torricelli *In Quibus Archimedis Doctrina*

In this section we discuss the role played by Archimedes's mechanics in Torricelli's works (Pisano and Capecchi 2010a, Capecchi and Pisano 2007a, b; et al. Pisano's works below). Archimedes was truly a scientific inspiration for Torricelli as he himself declared from the beginning of the *Opera geometrica* (Fig. a).

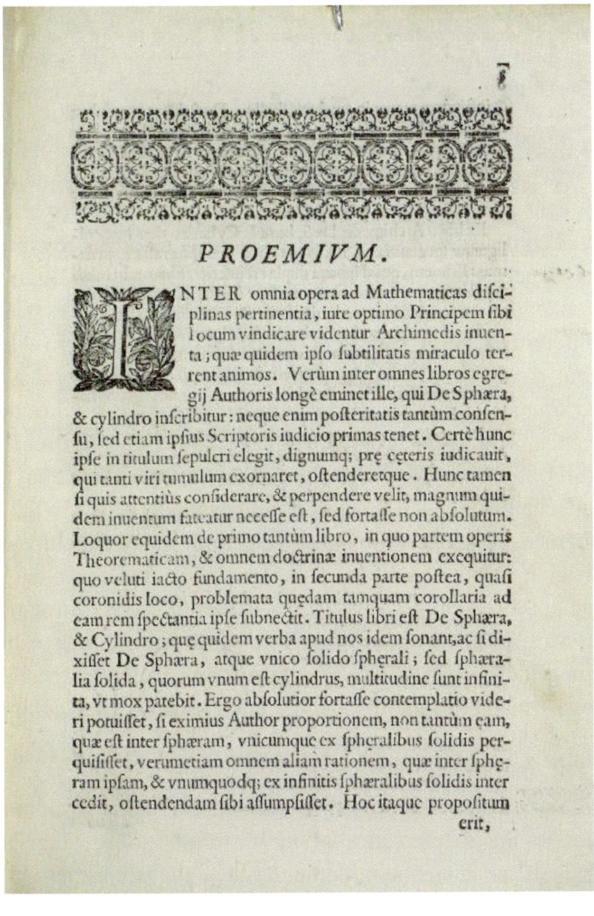

7

PROEMIVM.

INTER omnia opera ad Mathematicas diſciplinas pertinentia, iure optimo Principem ſibi locum vindicare videntur Archimedis inuenta; quæ quidem ipſo ſubtilitatis miraculo terrent animos. Verùm inter omnes libros egregij Authoris longè eminet ille, qui De Sphæra, & cylindro inſcribitur: neque enim poſteritatis tantùm conſenſu, ſed etiam ipſius Scriptoris iudicio primas tenet. Certè hunc ipſe in titulum ſepulcri elegit, dignumq; prę cęteris iudicauit, qui tanti viri tumulum exornaret, oſtenderetque. Hunc tamen ſi quis attentiùs conſiderare, & perpendere velit, magnum quidem inuentum fateatur neceſſe eſt, ſed fortaſſe non abſolutum. Loquor equidem de primo tantùm libro, in quo partem operis Theorematicam, & omnem doctrinæ inuentionem exequitur: quo veluti iacto fundamento, in ſecunda parte poſtea, quaſi coronidis loco, problemata quędam tamquam corollaria ad eam rem ſpectantia ipſe ſubnectit. Titulus libri eſt De Sphæra, & Cylindro; quę quidem verba apud nos idem ſonant, ac ſi dixiſſet De Sphæra, atque vnico ſolido ſphęrali; ſed ſphæralia ſolida, quorum vnum eſt cylindrus, multitudine ſunt infinita, vt mox patebit. Ergo abſolutior fortaſſe contemplatio videri potuiſſet, ſi eximius Author proportionem, non tantùm ęam, quæ eſt inter ſphæram, vnicumque ex ſphęralibus ſolidis perquiſiſſet, verumetiam omnem aliam rationem, quæ inter ſphęram ipſam, & vnumquodq; ex infinitis ſphæralibus ſolidis intercedit, oſtendendam ſibi aſſumpſiſſet. Hoc itaque propoſitum

erit,

"Among all the works pertaining to the Mathematical disciplines, those of Archimedes seem to claim the first place; indeed, with their miraculous subtlety they throw in dismay the minds."

Fig. a Torricelli Inspiring by Archimedes's Science (Torricelli 1644, *Proemium*, p. 7; Authors' translation). *Source* Courtesy University of Oklahoma, History of Science Collections

[12] In 2009 the Torricelli-Project was born thanks to Raffaele Pisano's personal initiative. In Florence, he met the outstanding Galilean and Torricellian scholar, Prof. Maurizio Torrini (1942–2019; Museo Galileo, Florence, Italy) who welcomed the Torricelli-Project with profound interest. A correspondence (July 2010) exists, as well. In the future, the Torricelli Project also envisages a second and third part respectively, the commented translation, side-by-side into English of the *Opera Geometrica* and the commented translation, side-by-side of the whole *Opere di Evangelista Torricelli* (Torricelli's Works).

Below the reader will find in the detail the influence exerted by Archimedes through the centuries and his heritage in Torricelli's works, including his approach to solving problems and conducting proofs.

From the history of physics—and—mathematics standpoint, the late 30s and the 40s of the seventeenth century represent a particularly rich period of research for the development of science. Mathematical techniques advanced rapidly, compared to the beginning of the century. Nevertheless, infinitesimal calculus (Bussotti 2019) was not yet available. A few years later this branch of mathematics will undergo a spectacular development which will allow scientists to solve specific problems (e.g., the calculations of areas, volumes, centre of gravity) and, very importantly, to deal with instantaneous physical–mathematical quantities (i.e., acceleration). Scholars understood that *finitary techniques* were not enough to solve certain problems addressed by the new science and the new mathematics. By *finitary technique*, we mean a technique through which it is possible, e.g., to prove the equivalence of two planes of solid figures through the *equidecomposition* in a finite number of parts. Beyond this, the scholars referred to the historically available technical model: the Archimedean. In this context, the *exhaustion method* was particularly significant for two main reasons:

(a) Many problems were solved by this method, as we will see below with regard to Torricelli;
(b) The exhaustion method and its potentially infinite procedures were a source of inspiration for techniques, which were slightly different from the exhaustion method itself.

For, Cavalieri, earliest credited inventor of the indivisibles method, expressed to Galilei in 1621, his interest in the field:

> […] I am proving some Archimedes's propositions, in a different manner with respect to him, and in particular those on the quadrature [of the segment of the] parabola, [and] also different from those of Y.L.; […].[13]

Archimedes had also been the inspiration of Kepler and Galileo themselves (Bussotti 2015; Pisano and Bussotti 2017a). For example the infinitary techniques used by Kepler, especially in the *Nova Stereometria doliorum* (1615), has the Syracusan scientist as a reference point (Pisano and Bussotti 2012, 2017c). When calculus became the standard paradigm of the infinitary procedures (which happened around the 80s of the seventeenth century), references to Archimedes became fewer and less direct. Torricelli's works (around 40s) represent one of the most accomplished results achivied without resorting to calculus.

From the point of view of historical epistemology, in some aspects Archimedes breaks the geometrical paradigm with respect to Euclid. If we interpret *normal science à la* Kuhn as Euclidean geometry, namely a discipline based upon a system of axioms, which rely on self-evidence, then Archimedes offers a different scientific paradigm. The birth of new modern science, whose necessity was not only to develop pure mathematics, but also to apply mathematics to the external world and to technique, adopted a paradigm *à la Archimedes* rather than *à la Euclid*. This means that the basic principles did not derive from self-evidence, only, but they were also drawn from experience. This is the case with Torricelli (see below), but it also happens within Tartaglia's *Science of weights*, i.e., Books VII–VIII of the *Quesiti et invenzione diverse* (Tartaglia [1554]1959; Pisano 2020, 2016; Pisano and Capecchi 2016) and in Galileo's mechanics, as well (Pisano and Bussotti 2012, 2015b, 2017a).

The concepts themselves of mathematical proof and of mathematical rigour were almost lost until the sixteenth century and they were reacquired, with difficulty, through a process which—with regard to mathematical disciplines other than calculus—stretched into the seventeenth century with the fundamental contributions of Fermat (Bussotti 2006; Bussotti and Pisano 2020) and Descartes (Bussotti and Pisano 2013; Schuster 2000, 2013a, b). In this phase, the Greek scholars represented a model of rigour and creativity, which was an important guide for the scientists living in the sixteenth and seventeenth century. As to calculus and more modern branches of mathematics, the rigourization–process was completed only in the nineteenth century (Pisano 2015, 2016, 2017; see also Pisano et al. 2021).

Archimedes was fundamental for Torricelli's works, and for this reason his influence on the Italian scientist is so significant from a historical–scientific standpoint, particularly with regard to the mathematics of indivisibles. Indeed, Torricelli's approach was to confront geometric matter both with the methods of the ancients, in particular the exhaustion method, and with indivisibles, so attempting to compare the two, as is clearly seen in his letters to Cavalieri (Torricelli 1919–1944, III). For example, we also note his pedagogical attitude to solving problems by different methods (Figs. b, c, d, e and f). He solved

[13] "[…] vado dimostrando alcune proposizioni d'Archimede diversamente da lui, et in particolare la quadratura della parabola, divers'ancora da quello di V.S; […]." (Letter from Cavalieri to Galilei, 1621, 15 December. Galilei 1890–1909, XIII, p. 81; Authors' translation).

in twenty-one different ways Archimedes's Proposition 24 squaring the parabolic segment[14]: 11 proofs by the method of exhaustion, 10 proofs by Cavalieri's method of the indivisibles (Cavalieri 1635, IV, Theorem I, Proposition I, p. 285):

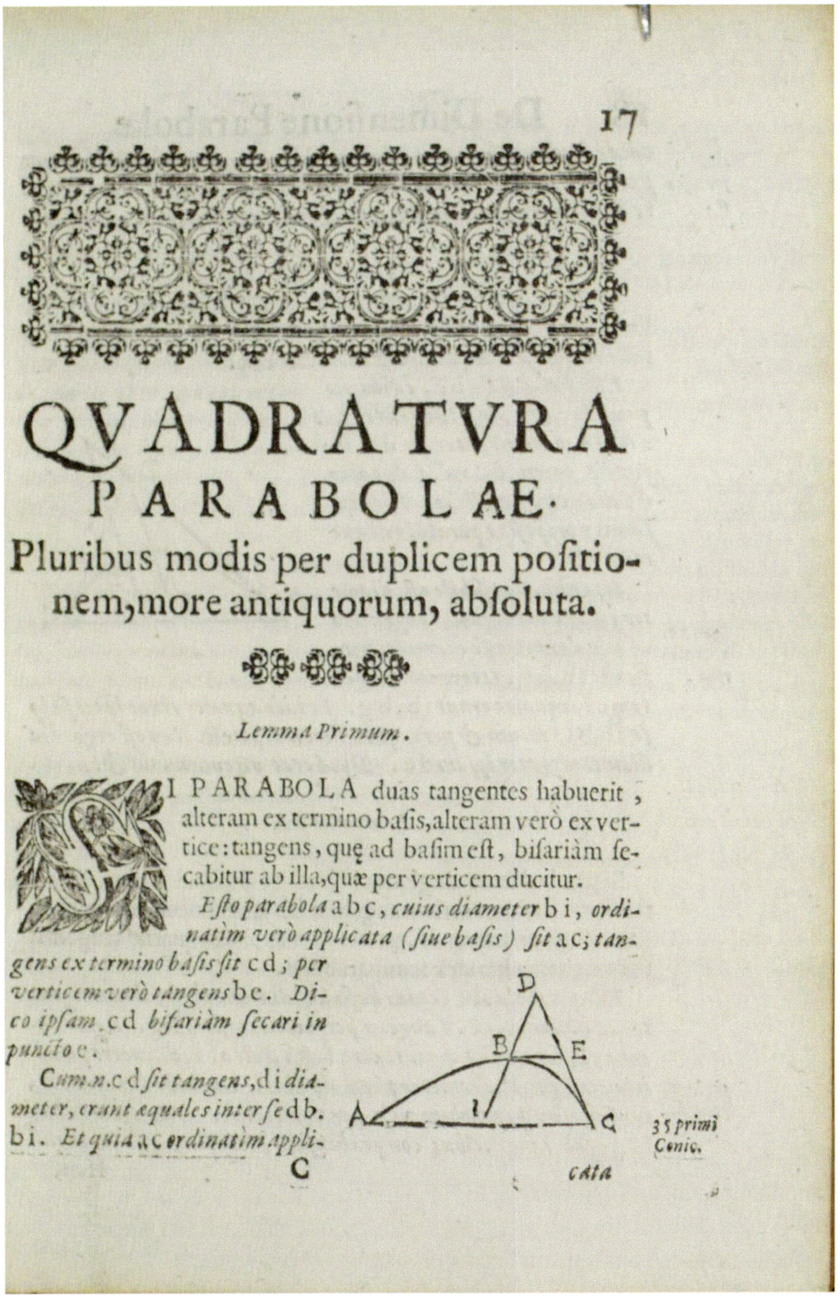

17

QVADRATVRA PARABOLAE.

Pluribus modis per duplicem positionem, more antiquorum, absoluta.

Lemma Primum.

SI PARABOLA duas tangentes habuerit, alteram ex termino basis, alteram verò ex vertice: tangens, quę ad basim est, bifariàm secabitur ab illa, quæ per verticem ducitur.

Esto parabola a b c, *cuius diameter* b i, *ordinatim verò applicata (siue basis) sit* a c; *tangens ex termino basis sit* c d; *per verticem verò tangens* b e. *Dico ipsam* c d *bifariàm secari in puncto* e.

Cum c d *sit tangens*, d i *diameter, erunt æquales inter se* d b, b i. *Et quia* a c *ordinatim appli-*

35 primi Conic.

C *cata*

Fig. b Torricelli's *Quadratura parabolae pluris modis*—Starting with Method of the Ancient Geometers (Torricelli 1644, III, p. 17). *Source* Courtesy University of Oklahoma, History of Science Collections

[14] Torricelli 1644, III, 1, *Quadratura parabolae pluris modis per duplicem positionem more antiquorum absoluta*, p. 17; *Lemma VII*, p. 28, *Proposition III*, p. 32 *et sugg*.; Archimedes 1897, Propositions (17, p. 246 and) 24, p. 251; see also Proposition 10, p. 162; *Id.* 1912, the *Method*, Propositions I–II, pp. 15–18). For Torricelli's quotation to Valerio see: Valerio 1606. On the subject, see also: Galilei 1890–1909, II, pp. 140–146, pp. 286–294; Vol. VIII, pp. 176–183; p. 306 et sgg. (Cfr. Bussotti 2019; Pisano and Capecchi 2010a).

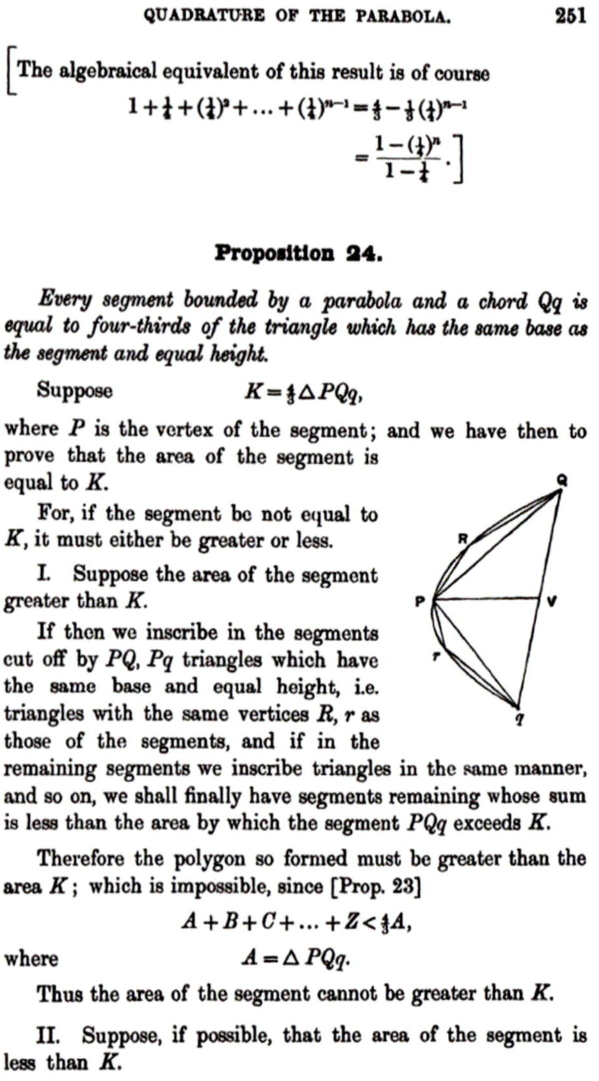

QUADRATURE OF THE PARABOLA. 251

[The algebraical equivalent of this result is of course

$$1+\tfrac{1}{4}+(\tfrac{1}{4})^2+\ldots+(\tfrac{1}{4})^{n-1}=\tfrac{4}{3}-\tfrac{1}{3}(\tfrac{1}{4})^{n-1}$$

$$=\frac{1-(\tfrac{1}{4})^n}{1-\tfrac{1}{4}}.]$$

Proposition 24.

Every segment bounded by a parabola and a chord Qq is equal to four-thirds of the triangle which has the same base as the segment and equal height.

Suppose $K=\tfrac{4}{3}\triangle PQq$,

where P is the vertex of the segment; and we have then to prove that the area of the segment is equal to K.

For, if the segment be not equal to K, it must either be greater or less.

I. Suppose the area of the segment greater than K.

If then we inscribe in the segments cut off by PQ, Pq triangles which have the same base and equal height, i.e. triangles with the same vertices R, r as those of the segments, and if in the remaining segments we inscribe triangles in the same manner, and so on, we shall finally have segments remaining whose sum is less than the area by which the segment PQq exceeds K.

Therefore the polygon so formed must be greater than the area K; which is impossible, since [Prop. 23]

$$A+B+C+\ldots+Z<\tfrac{4}{3}A,$$

where $A=\triangle PQq$.

Thus the area of the segment cannot be greater than K.

II. Suppose, if possible, that the area of the segment is less than K.

Fig. c Archimedes *Proposition 24* on Quadrature of the Parabola (Archimedes 1897, *Proposition 24*, p. 251). *Source* Google Books. Public domain

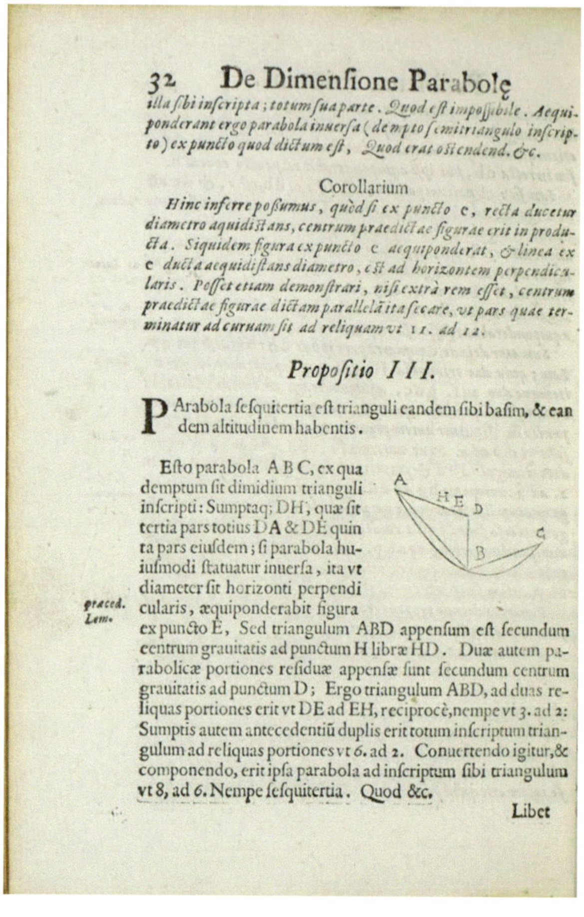

32 De Dimensione Parabolę

illa sibi inscripta; totum sua parte. Quod est impossibile. Aequiponderant ergo parabola inuersa (dempto semitriangulo inscripto) ex puncto quod dictum est, Quod erat ostendend. &c.

Corollarium

Hinc inferre possumus, quòd si ex puncto c, recta ducetur diametro aquidistans, centrum praedictae figurae erit in producta. Siquidem figura ex puncto c aequiponderat, & linea ex c ducta aequidistans diametro, est ad horizontem perpendicularis. Posset etiam demonstrari, nisi extrà rem esset, centrum praedictae figurae dictam parallelā ita secare, vt pars quae terminatur ad curuam sit ad reliquam vt 11. ad 12.

Propositio III.

PArabola sesquitertia est trianguli eandem sibi basim, & eandem altitudinem habentis.

Esto parabola A B C, ex qua demptum sit dimidium trianguli inscripti: Sumptaq; DH, quæ sit tertia pars totius DA & DE quinta pars eiusdem; si parabola huiusmodi statuatur inuersa, ita vt diameter sit horizonti perpendicularis, æquiponderabit figura ex puncto E, Sed triangulum ABD appensum est secundum centrum grauitatis ad punctum H libræ HD. Duæ autem parabolicæ portiones residuæ appensæ sunt secundum centrum grauitatis ad punctum D; Ergo triangulum ABD, ad duas reliquas portiones erit vt DE ad EH, reciprocè, nempe vt 3. ad 2: Sumptis autem antecedentiū duplis erit totum inscriptum triangulum ad reliquas portiones vt 6. ad 2. Conuertendo igitur, & componendo, erit ipsa parabola ad inscriptum sibi triangulum vt 8, ad 6. Nempe sesquitertia. Quod &c.

praced. Lem.

Libet

Fig. d Torricelli's *Proposition III* on Quadrature of the Parabola (Torricelli 1644, III, pp. 32–33). *Source* Courtesy University of Oklahoma, History of Science Collections

Archimedes's influence on Torricelli can also be detected in two general approaches of both scientists because Torricelli regarded axioms in a similar way to Archimedes's and differently from the Euclidean approach. Axioms have their origin not only in evidence, but also in experience. Therefore, they are different from, e.g., the epistemically rigid structure of the Euclidean axioms. Additionally, they could also be *local axioms*, namely unproved propositions valid in a given theory, whose validity could be non-universal, again in contrast to Euclid's axioms (Euclid 2007), which were, at that time, believed universally valid. Torricelli exploited Archimedes's techniques in an original manner. Therefore, knowledge of Archimedes's contribution is fundamental to the analysis of Torricelli's *Opera geometrica*.

Archimedes was (perhaps) the first scientist to set *rational criteria* for determining centres of gravity of bodies and his work contains physical concepts formalised on a mathematical basis (Capecchi and Pisano 2010b, 2011). In *Book I* of *On the Equilibrium of Planes* (Archimedes 1897, I, pp. 189–202) he, besides studying the rule governing the law of the lever, also finds the centres of gravity of various geometrical plane figures (Pisano and Capecchi 2010a; Pisano 2009a). Archimedes's typical method of arguing in mechanics was by the use of reduction *ad absurdum*. Torricelli in his study on the centres of gravity (Torricelli 1644, II, 1, pp. 97–153: see II, 2, pp. 154–243) also adopted this ancient Archimedean approach. Particularly, in *De motu graviun naturaliter descendetium et proiectorum, Liber primus* (*Ibidem*) Torricelli's famous premise is expounded (Torricelli 1644, II, 1, p. 99) and the early foundations of the *Principle of virtual laws* were modelled (Pisano 2017; Capecchi and Pisano 2007b, 2010a).

In *Archimedis Quadratura parabolae* (Archimedes 1897, pp. 233–252) Archimedes also first obtained his results using a mechanical approach and then reconsidered the arguments faced by means of a geometrical method in order to confirm the correctness of his results in a rigorous way. Similarly, Torricelli, with the compelling idea of duplicating the procedure (see above) devoted many pages to prove certain theorems on the *parabolic segment*, by following the geometry used in antiquity (Torricelli 1644, pp. 17–54) and then also proving the validity of his proofs with the *indivisibilium* method (Torricelli 1644, IV, pp. 93–135: Fig. e).

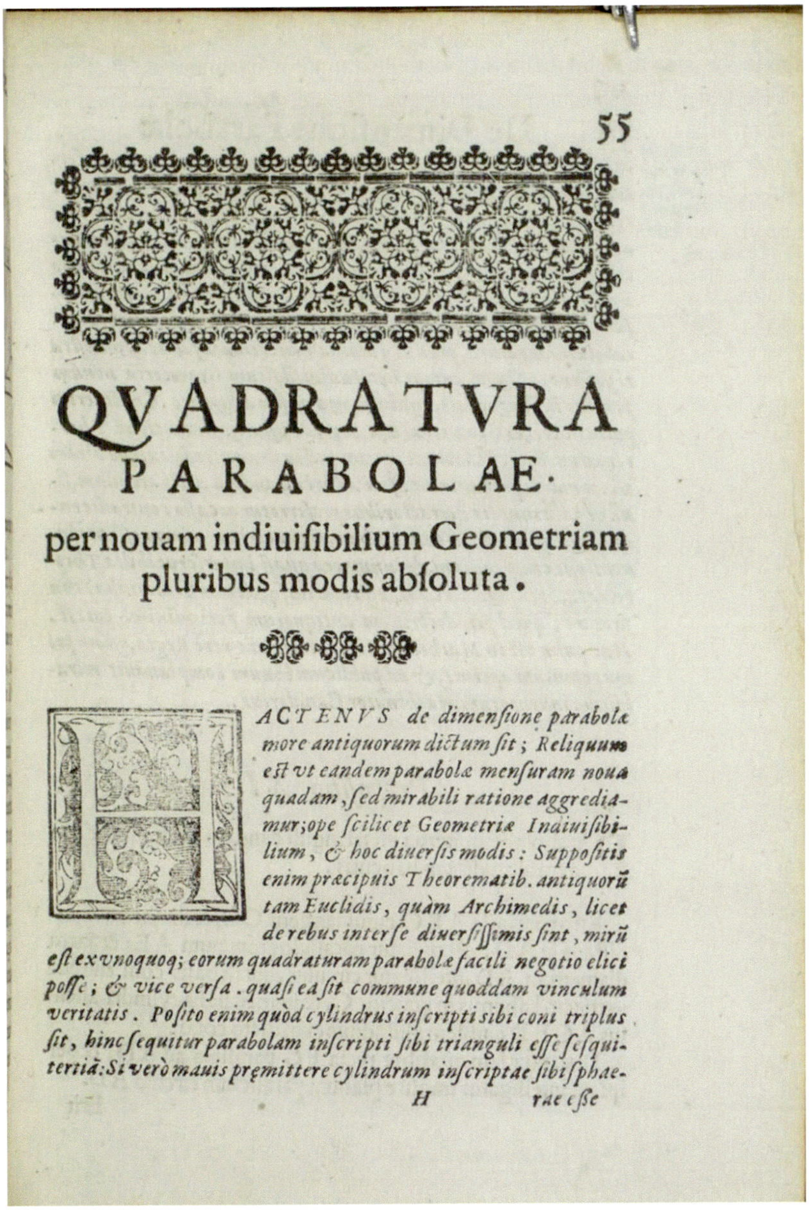
55

QVADRATVRA
PARABOLAE.
per nouam indiuisibilium Geometriam
pluribus modis absoluta.

HACTENVS de dimensione parabolæ more antiquorum dictum sit; Reliquum est vt eandem parabolæ mensuram noua quadam, sed mirabili ratione aggrediamur; ope scilicet Geometriæ Indiuisibilium, & hoc diuersis modis: Suppositis enim præcipuis Theorematib. antiquorū tam Euclidis, quàm Archimedis, licet de rebus inter se diuersissimis sint, mirū est ex vnoquoq; eorum quadraturam parabolæ facili negotio elici posse; & vice versa. quasi ea sit commune quoddam vinculum veritatis. Posito enim quòd cylindrus inscripti sibi coni triplus sit, hinc sequitur parabolam inscripti sibi trianguli esse sesquitertiā: Si verò mauis premittere cylindrum inscriptae sibi sphae-

H rae esse

Fig. e Torricelli's Quadratura parabolae by Method of the indivisibles (Torricelli 1644, III, p. 55 [N.B.: pages number 55 and 56 are repeated]; *Source* Courtesy University of Oklahoma, History of Science Collections

In this respect, it is also interesting to note that Torricelli explicitly and repeatedly underlined Archimedes's methods—and—reasonings (Cfr. Archimedes 1897, *Quadratura parabolae,* pp. 253–252; pp. 55–84; Torricelli 1644, III, *Quadratura parabolae per novam indivisibilium Geometriam pluribus modis absoluta*, p. 55). For example:

Concordance of the previous proof with the doctrine of Archimedes.[15]

Therefore, his main ideas were both to give the reader results and methods and also to claim that the indivisibles technique was not completely unknown to ancient Greek scholars. Heath, with Heiberg the main scholar and editor of Archimedes, wrote:

> [...] the historian of mathematics, in dealing with Archimedes obligations to his predecessors, has comparatively easy task before him. But it is necessary, first, to give some description of the use which Archimedes made of the general methods which had found acceptance with the earlier geometers, and secondly, to refer to some particular results which he mentions as having been previously discovered and as lying at the root of his own investigations or which he tacitly assumes as known. (Archimedes 1897, III, p. XI).

Torricelli seems to hold onto the idea that the method of demonstration of the ancients, such as Archimedes's method, was intentionally kept secret. He stated that the ancient geometers worked according to a method "[...] in invenzione [...]" suitable "[...] ad occultandum artis arcanum [...]" (Torricelli 1644, III, p. 55). However, Archimedean influence in Torricelli goes further. The books *De sphaera et solidis sphaeralibus* (Torricelli 1644, I, 1, *Liber primus*, pp. 11–46) present an enlargement of the Archimedean proofs in *On the sphere and cylinder* (Archimedes 1897, II, pp. 1–90). For, we read Torricelli's frontispiece on Archimedes's scientific heritage:

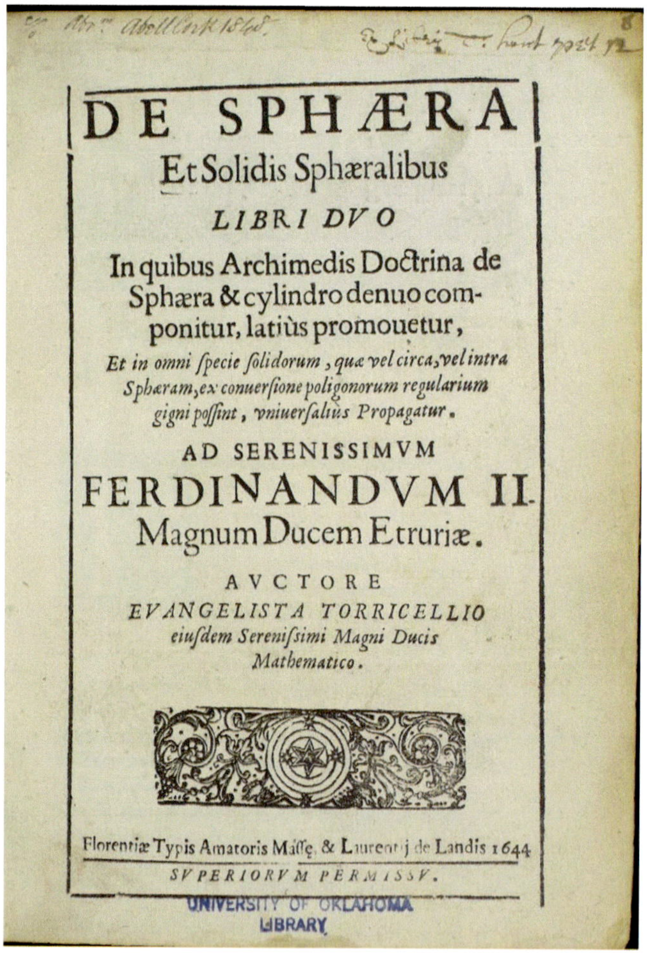
DE SPHÆRA
Et Solidis Sphæralibus
LIBRI DVO
In quibus Archimedis Doctrina de Sphæra & cylindro denuo componitur, latiùs promouetur,
Et in omni specie solidorum, quæ vel circa, vel intra Sphæram, ex conuersione poligonorum regularium gigni possint, vniuersaliùs propagatur.
AD SERENISSIMVM
FERDINANDVM II.
Magnum Ducem Etruriæ.
AVCTORE
EVANGELISTA TORRICELLIO
eiusdem Serenissimi Magni Ducis Mathematico.
Florentiæ Typis Amatoris Massæ & Laurentij de Landis 1644
SVPERIORVM PERMISSV.

Fig. f Torricelli claims his Inspiring by Archimedes's science: *In quibus Archimedis Doctrina* (Torricelli 1644, Frontispiece). *Source* Courtesy University of Oklahoma, History of Science Collections

[15] "Concordantia dantia praecedentis demostrationis cum doctrina Archimedis." (Torricelli 1644, IV, *De solido acuto hypebolico, Problema alterum*, p. 103; Authors' translation. See also pp. 95–102; pp. 108–112). In this Book, see also his claiming on Euclid's proof (*Ivi*, Example IV, p. 97, Example XI, p. 105).

In other domains, Torricelli also faced problems not yet solved by Archimedes, or by the other mathematicians of antiquity. But by adoptiong the same proofing style as Archimedes, he avoided to formulating axiomatic sentences of a theory (i.e., *á la* Euclid). In his words:

> *Scholium.* [...]. The truth of the preceding Theorem [*De solido hyperbolico acuto problema secundum*] is sufficiently clear in itself, and also confirmed by the examples proposed at the beginning of the book. But I think, in order that I may in this part satisfy the reader, who is not a friend of indivisibles [method], I will repeat this very demonstration at the end of the opera [*Opera geometrica*], by the usual way of demonstrating by ancient Geometers.[16]

We also remark that mechanical and geometrical argumentations presented by Archimedes in his *Method* (Archimedes 1912) were obviously unknown in Torricelli's time. Johan Ludvig Heiberg (1854–1928) only discovered it in 1906. Therefore, in Archimedes's writing there were lines of reasoning which—because of a lack of justification—were labelled as mysterious by most scholars. Thus, in such instances it was necessary to assure the reader of the validity of the thesis and also to offer a more heuristic approach, given the counterintuitive nature of exhaustion reasoning and *reductio ad absurdum*, by proving Archimedes's results with some other technique. It is well known that Archimedes in the *Method* (Archimedes 1912) faced a series of problems whose solution he anticipated by means of crucial propositions, i.e., based on the determination of centre of gravity and on the equilibrium of the lever. Then he proved them by *reductio ad absurdum*, using, in many cases the exhaustion method (Pisano and Capecchi 2010a). Indeed, Archimedes's himself did not attribute the same amount of certainty to his *Method* of proof as he attributed to other geometrical proofs; his reasoning on *Quadratura parabolae* (Archimedes 1897, Proposition 24, p. 251) is a typical example. At the beginning of the *Method*, addressing Eratosthenes (276–196 B.C.), Archimedes wrote:

> Seeing moreover in you, as I say, an earnest student, a man of considerable eminence in philosophy, and an admirer [of mathematical inquiry], I thought fit to write out for you and explain in detail in the same book the peculiarity of a certain method, by which it will be possible for you to get a start to enable you to investigate some of the problems in mathematics by means of mechanics. This procedure is, I am persuaded, no less useful even for the proof of the theorems themselves; for certain things first became clear to me by a mechanical method, although they had to be demonstrated by geometry afterwards because their investigation by the said method did not furnish an actual demonstration. (Archimedes 1912, p. 13).

When it comes to Torricelli as an Archimedean scholar, one of his methodologically characteristic proofs was the return to the demonstrative approach followed by the ancient Greeks, with the explicit description of the technique of reasoning actually used. But, what about his mathematical techniques? Besides the well-known *ad absurdum* proof he also used proportions with the classical techniques of the *permutando* and the *ex aequo*. In *De proportionibus liber*, Torricelli defined them explicitly (Torricelli 1919–1944, p. 314).

Generally speaking, Torricelli—like Galilei and Kepler (Pisano and Bussotti 2012, 2017a; Bussotti 2015)—was not interested in algebra and adhered to the language of proportions. In the *De Proportionibus liber* (Torricelli 1919–1944, pp. 295–327) adopted a mathematical language, using only the theory of proportions in geometry. In such a way, he naturally avoided the use of the *plus* or *minus*, whose logical status was still uncertain. Thus, he replaced the *plus* and *minus* by *dividing* (Torricelli 1919–1944, p. 313) and the *composing* (*Ivi*, p. 316). This choice allowed him to work always resorting to the ratio of segments. By following the ancients to sum up segments, he could have thought them as aligned and then translated and connected, making use of terms like "simul", "et" or "cum" (Torricelli 1919–1944, Proposition, XV, p. 318). We remark that the proofs by means of indivisibles are not *reductio ad absurdum*. Instead, in nearly all other proofs Torricelli used the technique typical of proportions, *dividendo, permutando* and *ex aequo*. The correct use of proportions was an important step towards the birth of modern science, in particular in its initial phases. Below we present Torricelli's proofs (Table 2) divided by methods and techniques (adapted from Pisano 2011).

---

[16] "*Scholium.* [...]. Veritatem præcedentis Theorematis [*De solido hyperbolico acuto problema secundum*] satis per se claram, & et per exempla ad initium libelli proposita confirmatam satis superq; puto Tamen ut in hac parte satisfaciam lectori etiam Indivisibilium parùm amico, iterabo hanc ipsam demonstrationem in calce operis, per solitam veterum Geometrarum viam demonstrandi longiorem quidem sed non ideo mihi certiorẽ." (Torricelli 1644, IV, 2, p. 116; Authors' translation).

Table 2 The Main Archimedean proofs by Torricelli in his *Opera Geometrica*

Items	Kind of proof	*Opera Geometrica* (Torricelli 1644)
		QUADRATURA PARABOLÆ PLURIBS MODIS [...] ANTIQUORUM absoluta
Lemma II,V,VI, X–XI, XII–XIII XVII, Propositio IV	*Ad absurdum* proof	*Ivi*, pp. 18–54
Lemma XIV	*Ex æquo*, *dividendo* and *permutando*	(*Ibidem*)
Lemma XVI, XVIII	*Ex aequo*	(*Ibidem*)
Lemma XIX	*Ex æquo* and *Ad absurdum*	(*Ibidem*)
Propositio III[17]	*Componendo*	(*Ibidem*)
Propositio V	*Ad absurdum* proof and *Componendo*	(*Ibidem*)
Propositio IX[18]	*Ex æquo* et *Ad absurdum*	(*Ibidem*)
		DE SPHAERA ET SOLIDIS SPHAEARALIBUS
Propositio XVIII	*Ad absurdum*	*Ivi*, I, pp. 28–29
Propositio XIX[19]	*Ad absurdum*	*Ivi*, I, pp. 30–33
Propositio XXII–XXIII[20]	*Ex æquo*	*Ivi*, I, pp. 35–36
Propositio IV	Proof done by Archimedes[21]	*Ivi*, II, pp. 51–52
		QUADRATURA PARABOLÆ PER NOVAM INDIVISIBILIUM GEOMETRIAM PLURIBUS MODIS ABSOLUTA[22]
Lemma XXII,XXIX[23]	*Ex æquo*	*Ivi*, pp. 61–72
Lemma XXX	Proof based on those done by	*Ivi*, pp. 74–77
Lemma XXXI	Archimedes, *Equiponderants*	DE SOLIDO ACUTO HYPERBOLICO PROBLEMA ALTERUM
		Ivi, I, pp. 93–112
Exemplum I–II, IV–V, X, XII–XIV	Proof *in concordantia præcedentis demonstrationis cum doctrina Archimedis*	*Ivi*, I, 95–108; I, pp. 113–135

(continued)

[17] In Proposition III, Torricelli quotes Luca Valerio's *De centris gravitatis solidorum libri tres* (Valerio 1604, II, p. 12). He wrote: "Libet hic demonstrare Lemma Lucae Valerij, nostro tamen modo, diversisque penitùs Mechanicae principijs. Ipse enim utitur propositione illa, quà ante demontraverat centrum gravitatis hemisphereij. Nos autem simili ratione, ac in praecedentibus, demontrabimus et Lemma, et ipsam Valerij conclusionem." (Torricelli 1644, *Quadratura parabolae*, p. 33).

[18] *Lemma* e *Propositio* are proved by means of *Componendo* and *Conversione*.

[19] "Quod est absurdum e contra principia Archimedis." (Torricelli 1644, *De sphaera et solidis sphaearalibus*, I, p. 31).

[20] "Tria haec Theoremata, quae sequuntur, ex Archimede desumpta sunt; quod quidem fecimus ne lector Archimedem adire cogeretur, sed universam hanc doctrinam in hoc libello haberet." (Torricelli 1644, *De sphaera et solidis sphaearalibus*, I, p. 35).

[21] Archimedes 1897, *On the sphere and cylinder*, II, Proposition 25, p. 32. A difference can be remarked: Archimedes proved the case-study, 4–multipolygons, while Torricelli proved the case-study 2–multipolygons.

[22] "Hactenus de dimensione parabolae more antiquorum dictum sit; Reliquum est eandem parabolae mensuram nova quedam, sed mirabili ratione aggrediamur; ope scilicet Geometriae Indivisibilium, et hoc diversis modis [...]" (Torricelli 1644, *Quadratura parabolae*, p. 55).

[23] For the other Lemmas/Propositions Torricelli adopted a geometrical construction by means of *componendo* and *dividendo*.

Table 2 (continued)

Other *Lemmas* and *corollaries*[24]	*Ex æquo* (Exemplum X)	
Exemplum III, XI	Dimostrazione *in concordantia cum theoremate Euclidis*	*Ivi*, I, p. 97; I, pp. 104–105 DE MOTU PROIECTORUM[25]
Lemma (From Prop. XXXIII)[26]	Proof of Archimedes's proposition on *Conoids and spheroids*[27]	*Ivi*, II, p. 187 DE PROPORTIONIBUS LIBER[28]
Propositio[29] III	*Ad absurdum*	Torricelli 1919–1944. I, p. 308; pp. 313–317)
Propositio VIII	*Dividendo*	
Propositio IX	*Permutando*	
Propositio X	*Permutando* and *Ex æquo*	
Propositio XI	*Ex æquo*	
Propositio XII	*Componendo* and *Conversion*	
Propositio XIII–XIV	*Permutando*	
Propositio XV	*Permutando* and *Componendo*	
Propositio XVI	*Permutando* and *Ad absurdum*	

The complex historical process called *rediscovery of the ancients* involves many aspects; one of them is the rediscovery of classical Greek geometry with its heritage of results and methods. Actually, the development of the methods and techniques of the ancient geometers were so important in the seventeenth century that they remained a crucial part of the mathematical apparatus used by the physicists until Newton (Pisano and Bussotti 2016, 2017b, Bussotti and Pisano 2014; Pisano 2023–2024).

3 A Selection from Torricelli's *Opera Geometrica* Proofs

Below we present a selection from *Opera geometrica* of the proofs and reasoning focusing on the original and significant contributions to geometry with respect to their application to mechanics and infinitesimals.

3.1 De Sphaera Et Solidis Sphaeralibus

In *De sphaera et solidis sphaeralibus*, the first work of the 1644 edition, Torricelli, extending the method of exhaustion (in *Archimedis Doctrina*, as above commented) proved a series of theorems on the solids obtained by the rotation of a regular polygon inscribed in, or circumscribed, to a circle. Thus, solids of rotation inscribed in or circumscribed to a sphere are studied. This programme is clearly expounded by Torricelli in the *Proemium* of his work.

Torricelli individuates six species of revolution solids: those of the first and second species are given by the rotation respectively along the apothem (*Catetum*, according to Torricelli) and the diagonal of an even-sided regular polygon inscribed in a circle. That of third species is given by the rotation of an odd-sided regular polygon inscribed in a circle and

[24] The others are not *ad absurdum* proofs. They are proved by geometrical construction.
[25] Torricelli 1644, *De motu proiectorum*, II, pp. 154–190.
[26] "Propositio Archimedis est sequens lemma in libro de Sphearoidibus et Conoidibus, quam tamen expeditius demontrabimus." (Torricelli 1644, *De motu proiectorum*, II, p. 187).
[27] Archimedes 1897, pp. 99–150.
[28] In the end of this book, Torricelli claimed for the readers the *concordantia* with Euclidean ones. (e.g., propositions XVII and XVIII, XXIV and XXV of *Elements, Book V*).
[29] The others are not *ad absurdum* proofs. They are proved by geometrical construction.

rotating around the apothem. Those of the fourth and fifth species are analogous to the ones of the first and second species, but they are circumscribed. That of the sixth species is analogous to the solid of the third species, but it is circumscribed. Torricelli seeks the ratio between the volumes of such solids and that of the sphere in which they are inscribed/circumscribed. In these works, there are several propositions whose proof is based on classical Euclidean methods without resorting to exhaustion, but many interesting statements, which concern the areas and the volumes of curvilinear or mixtilinear figures are proved by exhaustion. The first of this type of propositions is the third one, where Torricelli proves that given a right cylinder whose height:

$$AC = \frac{1}{4} AB$$

where AB is the diameter of the basis circle c (Fig. 1), the external surface of the cylinder is equal to that of c.

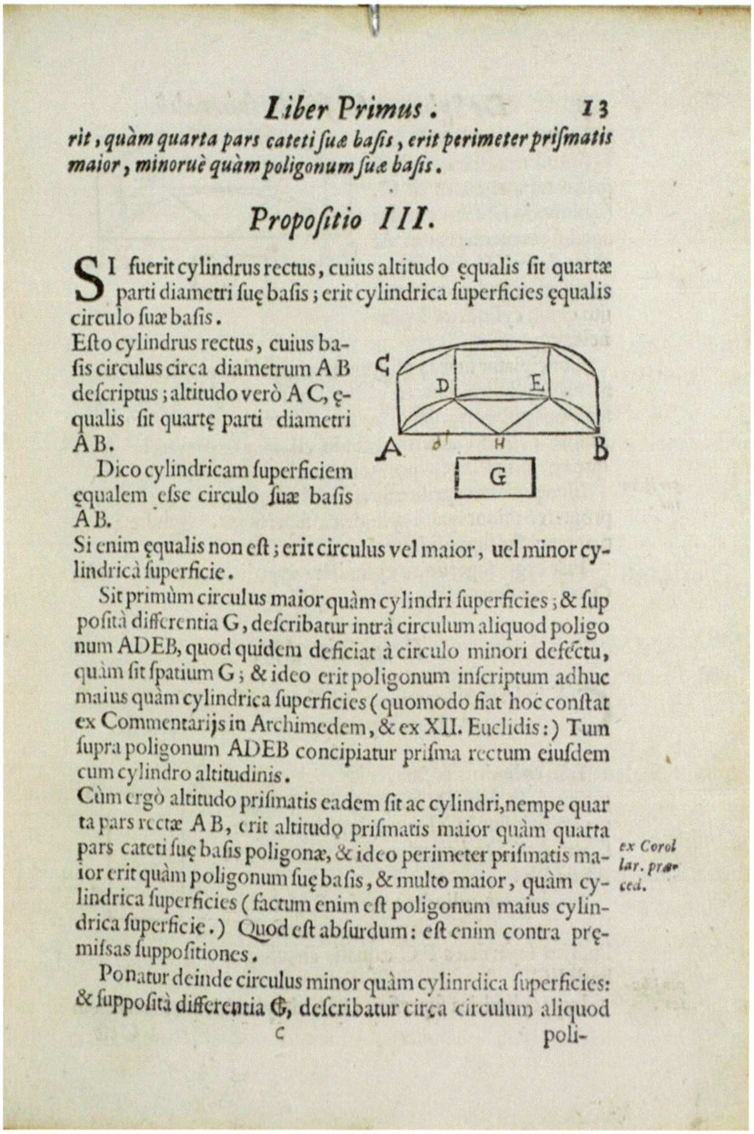
Liber Primus. 13

rit, quàm quarta pars cateti ſuæ baſis, erit perimeter priſmatis maior, minoruè quàm poligonum ſuæ baſis.

Propoſitio III.

SI fuerit cylindrus rectus, cuius altitudo ęqualis ſit quartæ parti diametri ſuę baſis; erit cylindrica ſuperficies ęqualis circulo ſuæ baſis.

Eſto cylindrus rectus, cuius baſis circulus circa diametrum A B deſcriptus; altitudo verò A C, ęqualis ſit quartę parti diametri A B.

Dico cylindricam ſuperficiem ęqualem eſſe circulo ſuæ baſis A B.

Si enim ęqualis non eſt; erit circulus vel maior, uel minor cylindricâ ſuperficie.

Sit primùm circulus maior quàm cylindri ſuperficies; & ſuppoſità differentia G, deſcribatur intrà circulum aliquod poligonum ADEB, quod quidem deficiat à circulo minori defectu, quàm ſit ſpatium G; & ideo erit poligonum inſcriptum adhuc maius quàm cylindrica ſuperficies (quomodo fiat hoc conſtat ex Commentarijs in Archimedem, & ex XII. Euclidis:) Tum ſupra poligonum ADEB concipiatur priſma rectum eiuſdem cum cylindro altitudinis.

Cùm ergò altitudo priſmatis eadem ſit ac cylindri, nempe quarta pars rectæ A B, erit altitudo priſmatis maior quàm quarta pars cateti ſuę baſis poligonæ, & ideo perimeter priſmatis maior erit quàm poligonum ſuę baſis, & multo maior, quàm cylindrica ſuperficies (factum enim eſt poligonum maius cylindrica ſuperficie.) Quod eſt abſurdum: eſt enim contra pręmiſſas ſuppoſitiones. *ex Corollar. præced.*

Ponatur deinde circulus minor quàm cylinrdica ſuperficies: & ſuppoſità differentia G, deſcribatur circa circulum aliquod

C poli-

Fig. 1 The diagram used by Torricelli to prove Proposition III (Torricelli 1644, I, 1, p. 13). *Source* Courtesy University of Oklahoma, History of Science Collections

In the corollaries to his propositions, Torricelli highlights which theorems of *De Sphaera et Cylindro* by Archimedes are deducible from the train of thought he was developing. This is a coherent confirmation of what Torricelli had claimed in his *Proemium*: his source of inspiration is indeed a work by Archimedes and his aim is to offer a more complete treatment than Archimedes's. The following proposition where exhaustion is used is the eighth one. It states that, given a right cone, the area of its basis is to the area of the external surface as the basis' radius is to the side of the cone (that is the segment of a generatrix between the vertex and the basis circle). In the following propositions Torricelli offers a series of relations between the external surfaces of cones and cylinders.

Starting from Proposition XI, through a Lemma stated before such proposition, Torricelli introduced the truncated cone, also creating a new vocabulary for its significant elements. Thus (Fig. 2), given a truncated cone and a cutting plane perpendicular to the axis, the straight line *EH* which joins the mean points of the oblique sides of the trapezium *ABCD* was called by Torricelli *Linea media aritmetica frusti conici* (arithmetical mean line of the truncated cone). The rectangle of dimension *EB* and *AB* was named *rectangulum proprium frusti conici* (proper rectangle of the truncated cone).

20 *De Sphæra, & solidis sphæralib.*

Lemma.

Si fuerit ABCD fruſtum coni recti, abſciſſum planis ad axem erectis (hoc enim modo ſemper intelligemus fruſta conica) ſecenturque latera AB, DC bifariam in punctis E, & H. iungaturq; EH. Dico rectam EH componi ex vtràque BL, AI, nempe ex ſemidiametris baſium oppoſitarum fruſti conici.

Iungantur BD, EI, LH; Et quoniam AI, ID æquales ſunt; item AE, EB, æquales: erunt parallelæ EI, BD. & ideo in parallelogrammo æqualia erunt latera ID, EM. Ob eandem cauſam æqualia ſunt BL, MH. Ergo tota EH æqualis erit ipſis ID, BL ſimul ſumptis. Quod erat, &c.

Definitiones.

Vocabimus impoſterum breuitatis cauſa lineam EH mediã Aritmeticam fruſti conici.

Rectangulum verò ſub EH & AB latere fruſti conici, dicemus rectangulum proprium fruſti conici.

Propoſitio XI.

CVrua ſuperficies fruſti conici, planis ad axem erectis abſciſſi, ad conicam quamlibet ſuperficiem, eſt vt rectangulum proprium fruſti, ad rectangulum ſub latere, & ſemidiametro baſis ipſius coni.

Eſto fruſtum conicũ ABCD abſciſſum planis ad axem erectis, ſitque conus quilibet EFG, cuius axis FH. Dico curuam fruſti AC ſuperficiem, ad curuã coni EFG ſuperficiem, eſſe, vt rectã gulum ſub AB, & ſub utraque AL, BI contentum, ad rectangulũ FEH,

Com-

Fig. 2 The figure used by Torricelli to introduce the basic elements of the truncated cone (Torricelli 1644, I, 1, p. 20). *Source* Courtesy University of Oklahoma, History of Science Collections

An important result, obtained without resorting to exhaustion, is given in Proposition XII. Here Torricelli proves that the external surface of a truncated cone is to that of a right cylinder as the proper rectangle of the truncated cone is to the rectangle having as dimensions the height of the cylinder and the diameter of the circle's basis. From this theorem, Proposition XVI of Archimedes's *De Sphaera et Cylindro* first book follows as a corollary: the external surface of a truncated cone is equal to the circle's whose radius is the mean proportional between the side of the truncated cone and its arithmetical mean. The first significant theorem, which concerns new rotation figures, is Proposition XIII (Fig. 3, particularly see p. 23). Here Torricelli considers two perpendicular diameters *AB* and *CD* of a circle and a tangent in a point *I* different from *A*,*B*,*C*,*D*. Produce the tangent so that *I* is the mean point of a segment which does not intersect *AB* produced. Then the external surface of the cylinder whose height is *GH* and whose basis circle has diameter equal to *AB* is the same as that of the truncated cone whose side is the tangent segment *ML* and whose basis' radius is *MN*.

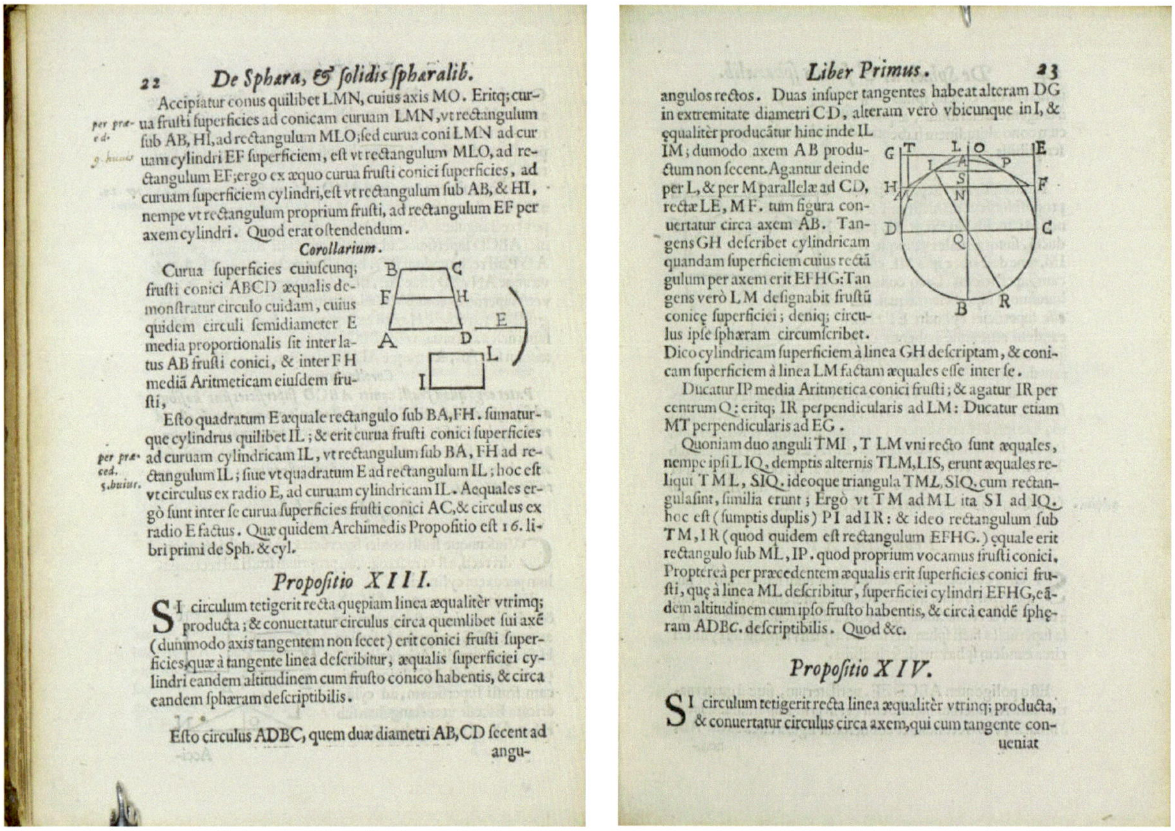

22 *De Sphæra, & ſolidis ſphæralib.*

Accipiatur conus quilibet LMN, cuius axis MO. Eritq; curua fruſti ſuperficies ad conicam curuam LMN, vt rectangulum ſub AB, HI, ad rectangulum MLO; ſed curua coni LMN ad curuam cylindri EF ſuperficiem, eſt vt rectangulum MLO, ad rectangulum EF; ergo ex æquo curua fruſti conici ſuperficies, ad curuam ſuperficiem cylindri, eſt vt rectangulum ſub AB, & HI, nempe vt rectangulum proprium fruſti, ad rectangulum EF per axem cylindri. Quod erat oſtendendum.

Corollarium.

Curua ſuperficies cuiuſcunq; fruſti conici ABCD æqualis demonſtratur circulo cuidam, cuius quidem circuli ſemidiameter E media proportionalis ſit inter latus AB fruſti conici, & inter FH mediã Aritmeticam eiuſdem fruſti,

Eſto quadratum E æquale rectangulo ſub BA, FH. ſumaturque cylindrus quilibet IL; & erit curua fruſti conici ſuperficies ad curuam cylindricam IL, vt rectangulum ſub BA, FH ad rectangulum IL; ſiue vt quadratum E ad rectangulum IL; hoc eſt vt circulus ex radio E, ad curuam cylindricam IL. Aequales ergò ſunt inter ſe curua ſuperficies fruſti conici AC, & circulus ex radio E factus. Quæ quidem Archimedis Propoſitio eſt 16. libri primi de Sph. & cyl.

per præced. 5. huius.

Propoſitio XIII.

SI circulum tetigerit recta quępiam linea æqualitèr vtrimq; producta; & conuertatur circulus circa quemlibet ſui axẽ (dummodo axis tangentem non ſecet) erit conici fruſti ſuperficies, quæ à tangente linea deſcribitur, æqualis ſuperficiei cylindri eandem altitudinem cum fruſto conico habentis, & circa eandem ſphæram deſcriptibilis.

Eſto circulus ADBC, quem duæ diametri AB, CD ſecent ad angu-

Liber Primus. 23

angulos rectos. Duas inſuper tangentes habeat alteram DG in extremitate diametri CD, alteram verò vbicunque in I, & ęqualitèr producãtur hinc inde IL IM; dumodo axem AB productum non ſecent. Agantur deinde per L, & per M parallelæ ad CD, rectæ LE, MF. tum figura conuertatur circa axem AB. Tangens GH deſcribet cylindricam quandam ſuperficiem cuius rectãgulum per axem erit EFHG: Tangens verò LM deſignabit fruſtũ conicę ſuperficiei; deniq; circulus ipſe ſphæram circumſcribet.

Dico cylindricam ſuperficiem à linea GH deſcriptam, & conicam ſuperficiem à linea LM factam æquales eſſe inter ſe.

Ducatur IP media Aritmetica conici fruſti; & agatur IR per centrum Q: eritq; IR perpendicularis ad LM: Ducatur etiam MT perpendicularis ad EG.

Quoniam duo anguli TMI, TLM vni recto ſunt æquales, nempe ipſi LIQ, demptis alternis TLM, LIS, erunt æquales reliqui TML, SIQ. ideoque triangula TML, SIQ. cum rectangula ſint, ſimilia erunt; Ergò vt TM ad ML ita SI ad IQ. hoc eſt (ſumptis duplis) PI ad IR: & ideo rectangulum ſub TM, IR (quod quidem eſt rectangulum EFHG.) ęquale erit rectangulo ſub ML, IP. quod proprium vocamus fruſti conici. Propterea per præcedentem æqualis erit ſuperficies conici fruſti, quę à linea ML deſcribitur, ſuperficiei cylindri EFHG, eãdem altitudinem cum ipſo fruſto habentis, & circà eandẽ ſphęram ADBC. deſcriptibilis. Quod &c.

Propoſitio XIV.

SI circulum tetigerit recta linea æqualitèr vtrinq; producta, & conuertatur circulus circa axem, qui cum tangente conueniat

Fig. 3 The figure used by Torricelli to prove Proposition XIII (Torricelli 1644, I, 1, pp. 22–23). *Source* Courtesy University of Oklahoma, History of Science Collections

The first proposition which concerns a "spheral solid" is the XVth one. Here Torricelli assumes that, given a circle *c* and a circumscribed even-sided regular polygon, the latter rotates around its diagonal. The surface of the spheral solid so obtained is equal to the external surface of the cylinder having the same height as the spheral solid and circumscribed to the sphere, given by the rotation of *c* around a diameter.

Proposition XVI states the analogous theorem if the polygon rotates around its apothem. While the Proposition XVII considers the situation of a circumscribed odd-sided regular polygon rotating around its apothem.

Now Torricelli can prove (Proposition XVIII) that the surface of a hemisphere is equal to the external surface of a cylinder having the same basis and height. The beautiful and long demonstration is given by exhaustion.

Proposition XIX represents another magnificent example concerning the application of the exhaustion method. Here Torricelli proves that the surface of any spherical segment is equal to the external surface of the cylinder circumscribed to the sphere and having the same height as the segment. Now Torricelli is ready to prove in two different manners—both different from Archimedes's—one of the most classical results proved in *De Sphaera et Cylindro*: the spherical surface is four times the surface of its great circle. The proofs are clear because they rely on the series of theorems previously proved by Torricelli. After that (Propositions XXIII and XXIV) Torricelli studies the external surface of a rhomboidal solid, that is a solid given by two cones having the same base circle and joined in that circle.

Among the immediately following propositions, the most interesting is Proposition XXVII where Torricelli proved: be given a truncated cone *ABCD* (Fig. 4) and a cone *BEC* having the same basis as the bigger basis of *ABCD* and its same height. The volume of the solid given by the difference of *ABCD* and *BEC* is equal to that of the cone whose basis has the same surface as the external surface of the truncated cone and whose height is equal to the perpendicular *EH* elevated from the vertex of the cone to the side of the truncated cone.

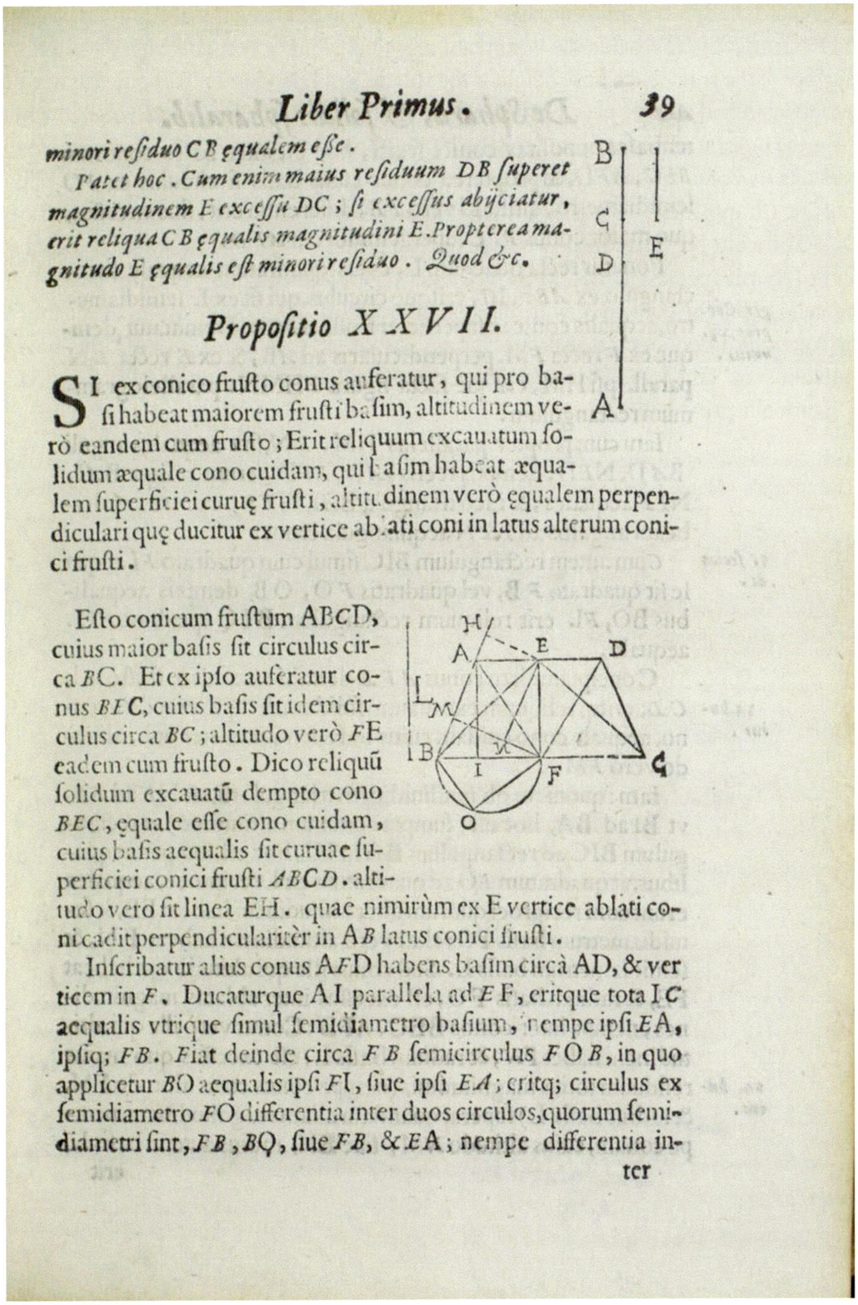
Liber Primus. 39

minori residuo CB æqualem esse.

Patet hoc. Cum enim maius residuum DB superet magnitudinem E excessu DC; si excessus abijciatur, erit reliqua CB æqualis magnitudini E. Propterea magnitudo E æqualis est minori residuo. Quod &c.

Propositio XXVII.

SI ex conico frusto conus auferatur, qui pro basi habeat maiorem frusti basim, altitudinem verò eandem cum frusto; Erit reliquum excauatum solidum æquale cono cuidam, qui basim habeat æqualem superficiei curuæ frusti, altitudinem verò æqualem perpendiculari quæ ducitur ex vertice ablati coni in latus alterum conici frusti.

Esto conicum frustum ABCD, cuius maior basis sit circulus circa BC. Et ex ipso auferatur conus BEC, cuius basis sit idem circulus circa BC; altitudo verò FE eadem cum frusto. Dico reliquũ solidum excauatũ dempto cono BEC, æquale esse cono cuidam, cuius basis aequalis sit curuae superficiei conici frusti ABCD. altitudo vero sit linea EH. quae nimirùm ex E vertice ablati coni cadit perpendiculariter in AB latus conici frusti.

Inscribatur alius conus AFD habens basim circà AD, & verticem in F. Ducaturque AI parallela ad EF, eritque tota IC aequalis vtrique simul semidiametro basium, nempe ipsi EA, ipsiq; FB. Fiat deinde circa FB semicirculus FOB, in quo applicetur BO aequalis ipsi FI, siue ipsi EA; eritq; circulus ex semidiametro FO differentia inter duos circulos, quorum semidiametri sint, FB, BQ, siue FB, & EA; nempe differentia inter

Fig. 4 The figure used by Torricelli to prove Proposition XXVII (Torricelli 1644, I, 1, p. 39). *Source* Courtesy University of Oklahoma, History of Science Collections

Torricelli offers two demonstrations of this theorem. In none of them is the exhaustion method used. These proofs, far from being trivial, are rather long and well argued.

Proposition XXIX is a beautiful example of infinitesimal geometry because Torricelli solves this problem: to inscribe in and to circumscribe to a revolution solid two figures composed of cylinders having the same height so that the difference between the volumes of the two figures is less than any given solid. Through this results, Torricelli proves as the thirtieth and last proposition of the first book, the classical Archimedes's result: the sphere has a quadruple volume in respect to the cone whose basis is a great circle of the sphere and whose height is equal to the sphere's radius. The proof is based on the exhaustion method.

The first three propositions of the second book concern the volume of cones circumscribed to a sphere. For example, the first one states that, given a cone circumscribed to a sphere, its volume is equal to that of the cone whose base's area is equal to the whole surface of the previous cone and whose height is equal to the radius of the sphere.

Propositions IV–XIV regard the spheral solids produced by the rotation of an even-sided regular polygon. Let us consider the first two. In Proposition IV Torricelli proves that the volume of a spheral solid of second species is equal to that of a cone whose surface is equal to the surface of the solid and whose height is the sphere's radius. Torricelli clarifies that, when the number of sides is a multiple of 4, the proposition had already been proved by Archimedes in *De Sphaera et Cylindro*, while when the number of sides is of the form $4n + 2$ the proof was missing. He furnished such a proof. Proposition V is dedicated to the spheral solid of fifth species. Torricelli proves that its volume is to the volume of the inscribed sphere as the surface of the solid is to the surface of the sphere. The same proportion (Proposition VI) holds between the diagonal of the solid and the diameter of the sphere.

From Proposition XV Torricelli begins to deal with odd-sided regular polygons. The first case analysed (Proposition XV) is that of a spheral solid of fourth species. Torricelli proved that its volume is equal to that of a cone whose basis is equal to the solid's surface and whose height coincides with the sphere's radius. The propositions until the XIXth one regard the solids generated by the rotation of odd-sided regular polygons. After that, a series of propositions begins where the volumes of the different spheral solids are compared. We will provide the reader with two examples (Fig. 5a).

In Proposition XX Torricelli proves that:

> In a sphere of volume C, be described an even sided spheral solid of volume S generated by the rotation around the diagonal of the generating polygon. The sphere's volume is to C-S in duplicate ratio of the sphere's diameter to the polygon's side.[30] (Fig. 5b).

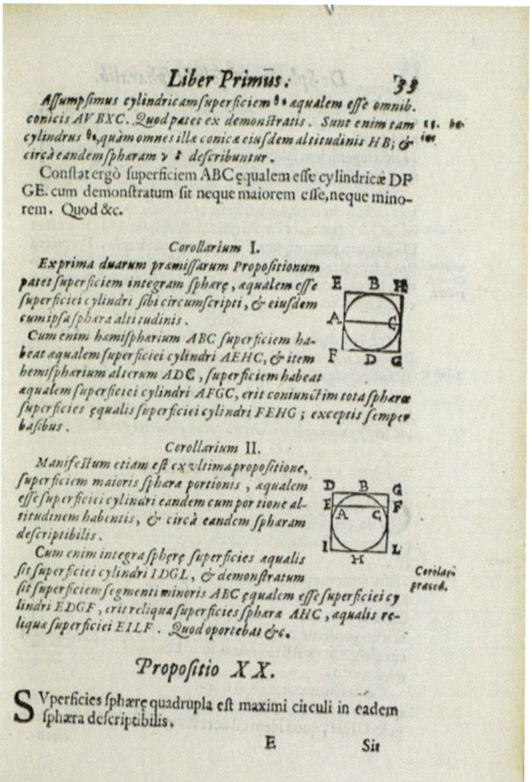

Liber Primus. 33

Aſſumpſimus cylindricam ſuperficiem æqualem eſſe omnib. conicis AVBXC. Quod patet ex demonſtratis. Sunt enim tam cylindrus, quàm omnes illæ conicæ eiuſdem altitudinis HB; & circà eandem ſphæram deſcribuntur.

Conſtat ergò ſuperficiem ABC æqualem eſſe cylindricæ DPGE. cum demonſtratum ſit neque maiorem eſſe, neque minorem. Quod &c.

Corollarium I.

Ex prima duarum præmiſſarum Propoſitionum patet ſuperficiem integram ſphæræ, æqualem eſſe ſuperficiei cylindri ſibi circumſcripti, & eiuſdem cum ipſa ſphæra altitudinis.

Cum enim hæmiſphærium ABC ſuperficiem habeat æqualem ſuperficiei cylindri AEHC, & item hemiſphærium alterum ADC, ſuperficiem habeat æqualem ſuperficiei cylindri AFGC, erit coniunctim tota ſphæræ ſuperficies æqualis ſuperficiei cylindri FEHG; exceptis ſemper baſibus.

Corollarium II.

Manifeſtum etiam eſt ex vltima propoſitione, ſuperficiem maioris ſphæræ portionis, æqualem eſſe ſuperficiei cylindri eandem cum portione altitudinem habentis, & circà eandem ſphæram deſcriptibilis.

Cum enim integræ ſphæræ ſuperficies æqualis ſit ſuperficiei cylindri IDGL, & demonſtratum ſit ſuperficiem ſegmenti minoris ABC æqualem eſſe ſuperficiei cylindri EDGF, erit reliqua ſuperficies ſphæræ AHC, æqualis reliquæ ſuperficiei EILF. Quod oportebat &c. Corolar. præced.

Propoſitio XX.

SVperficies ſphæræ quadrupla eſt maximi circuli in eadem ſphæra deſcriptibilis.

E Sit

Fig. 5a The Proposition XX (Torricelli 1644, I, 1, p. 33). *Source* Courtesy University of Oklahoma, History of Science Collections

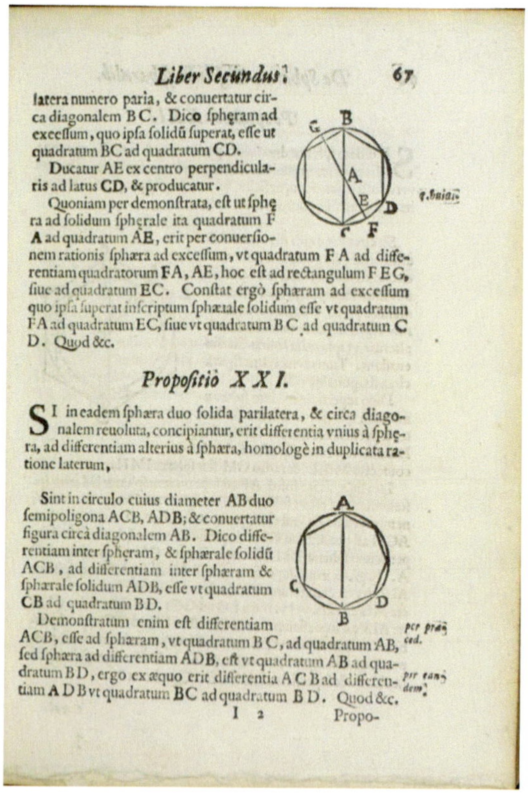

Liber Secundus. 67

latera numero paria, & conuertatur circa diagonalem BC. Dico ſphæram ad exceſſum, quo ipſa ſolidũ ſuperat, eſſe ut quadratum BC ad quadratum CD.

Ducatur AE ex centro perpendicularis ad latus CD, & producatur.

Quoniam per demonſtrata, eſt ut ſphæra ad ſolidum ſphærale ita quadratum FA ad quadratum AE, erit per conuerſionem rationis ſphæra ad exceſſum, vt quadratum FA ad differentiam quadratorum FA, AE, hoc eſt ad rectangulum FEG, ſiue ad quadratum EC. Conſtat ergò ſphæram ad exceſſum quo ipſa ſuperat inſcriptum ſphærale ſolidum eſſe vt quadratum FA ad quadratum EC, ſiue vt quadratum BC ad quadratum CD. Quod &c.

Propoſitiò XXI.

SI in eadem ſphæra duo ſolida parilatera, & circa diagonalem reuoluta, concipiantur, erit differentia vnius à ſphæra, ad differentiam alterius à ſphæra, homologè in duplicata ratione laterum,

Sint in circulo cuius diameter AB duo ſemipoligona ACB, ADB; & conuertatur figura circà diagonalem AB. Dico differentiam inter ſphæram, & ſphærale ſolidũ ACB, ad differentiam inter ſphæram & ſphærale ſolidum ADB, eſſe vt quadratum CB ad quadratum BD.

Demonſtratum enim eſt differentiam ACB, eſſe ad ſphæram, vt quadratum BC, ad quadratum AB, ſed ſphæra ad differentiam ADB, eſt vt quadratum AB ad quadratum BD, ergo ex æquo erit differentia ACB ad differentiam ADB vt quadratum BC ad quadratum BD. Quod &c. per præced. per eandem.

I 2 Propo-

Fig. 5b The diagram used by Torricelli to prove Proposition XX (Torricelli 1644, I, 1, p. 67). *Source* Courtesy University of Oklahoma, History of Science Collections

[30] "Si intra sphaeram descriptum sit sphaerale solidum parilaterum, circamque diagonalem revolutum, erit sphaera ad excessum, quo ipsa solidum superat, in duplicata ratione diametri sphaerae ad latus poligoni". (Torricelli 1644, I, 2, p. 66; Authors' translation).

Torricelli proves that the sphere's volume is to the volume of the difference between the sphere and the inscribed solid as BC^2 is to CD^2.

Proposition XXII is interesting because Torricelli compares the sphere's surface with those of an inscribed and a circumscribed even-sided spheral solid. This proposition states that given two similar even-sided solids, one inscribed in and the other circumscribed to the sphere, rotate these solids around their diagonal, the sphere's surface is mean proportional between the surfaces of the two solids.

The propositions that follow until the XXVIIth one are theorems on even-sided spheral solids. Their proofs are well argued.

Proposition XXVIII (Fig. 6) is noteworthy. It concerns even-sided solids, but it distinguishes between those whose number of sides is of the form $4n$ and those whose number of sides is of the form $4n + 2$. This theorem states that the volume of a spheral solid, generated by the rotation of a polygon, whose number of sides is of the form $4n + 2$, around its diagonal is twice the solid-rhombus inscribed in the solid.

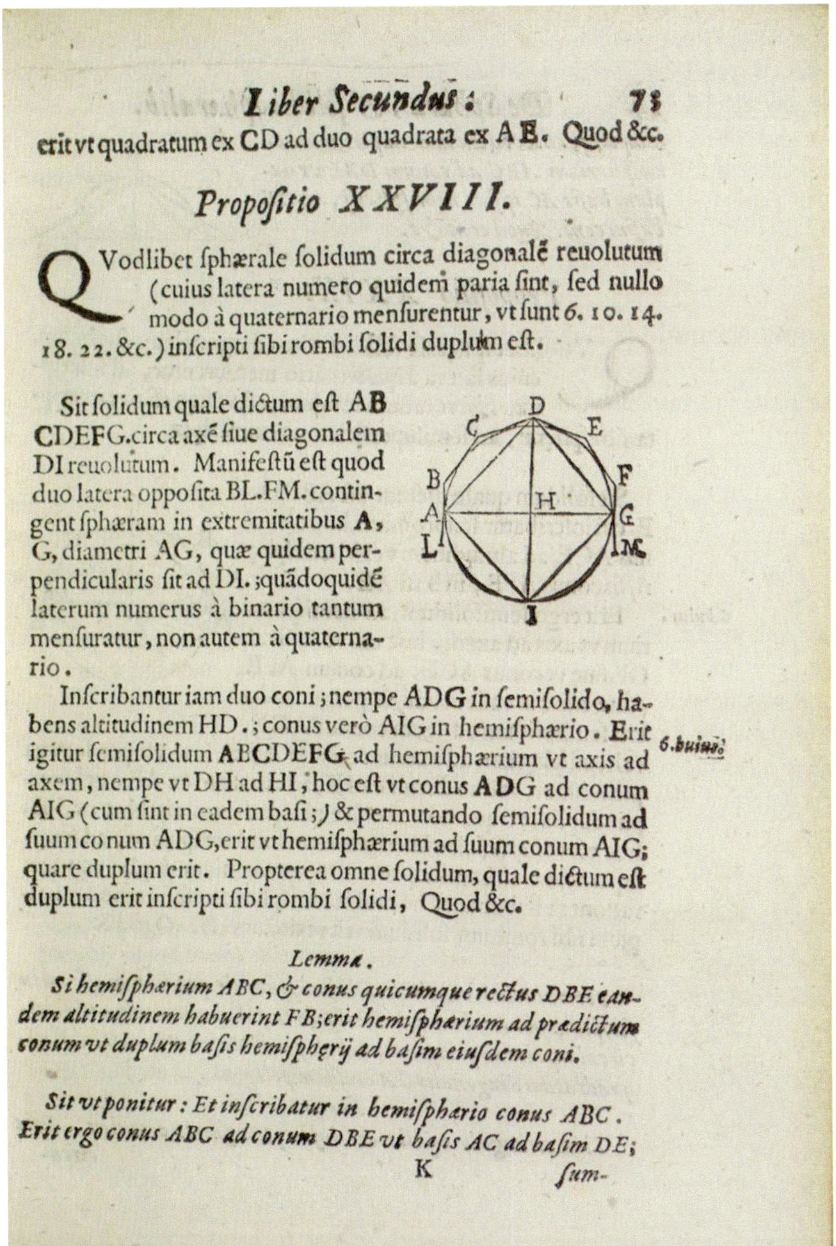

Liber Secundus. 73

erit vt quadratum ex CD ad duo quadrata ex AB. Quod &c.

Propositio XXVIII.

Qvodlibet sphærale solidum circa diagonalē reuolutum (cuius latera numero quidem paria sint, sed nullo modo à quaternario mensurentur, vt sunt 6. 10. 14. 18. 22. &c.) inscripti sibi rombi solidi duplum est.

Sit solidum quale dictum est AB CDEFG. circa axē siue diagonalem DI reuolutum. Manifestū est quod duo latera opposita BL. FM. contingent sphæram in extremitatibus A, G, diametri AG, quæ quidem perpendicularis sit ad DI.; quādoquidē laterum numerus à binario tantum mensuratur, non autem à quaternario.

Inscribantur iam duo coni; nempe ADG in semisolido, habens altitudinem HD.; conus verò AIG in hemisphærio. Erit igitur semisolidum ABCDEFG ad hemisphærium vt axis ad axem, nempe vt DH ad HI, hoc est vt conus ADG ad conum AIG (cum sint in eadem basi;) & permutando semisolidum ad suum conum ADG, erit vt hemisphærium ad suum conum AIG; quare duplum erit. Propterea omne solidum, quale dictum est duplum erit inscripti sibi rombi solidi, Quod &c. 6. huius.

Lemma.

Si hemisphærium ABC, & conus quicumque rectus DBE eandem altitudinem habuerint FB; erit hemisphærium ad prædictum conum vt duplum basis hemisphærij ad basim eiusdem coni.

Sit vt ponitur: Et inscribatur in hemisphærio conus ABC. Erit ergo conus ABC ad conum DBE vt basis AC ad basim DE;

K *sum-*

Fig. 6 The diagram used to Torricelli to prove Proposition 28. The even-sided solid is obtained by the rotation of the polygon *DEFGI* [...] around its diagonal *DI* and the solid-rhombus by the rotation of *ADGI* around the same diagonal (Torricelli 1644, I, 1, p. 73). *Source* Courtesy University of Oklahoma, History of Science Collections

Up to Proposition XXIXth Torricelli dealt with spheral solids. From Proposition XXX various theorems on the volumes of cones and cylinders inscribed in or circumscribed to a sphere are proved. With Proposition XXXV Torricelli begins to analyse the spheral solids generated by the rotation of a hexagon. For example, in Proposition XXXV he proves that the spheral solid's volume generated by the rotation of a hexagon around its apothem is seven times the cone's having the same basis and the same height. The reader can find specific and curious propositions, of clear demonstration, such as the following one (Proposition XLIII): given a regular hexagon, consider a second regular hexagon inscribed so that its vertices lie in the middle points of the circumscribed hexagon. Rotate the whole figure around the apothem of the circumscribed hexagon (and, hence around the diagonal of that inscribed). The volume of the circumscribed spheral solid is to that of the inscribed one as 14 to 9.

Among the remaining nine propositions XLIV–LII, the XLIVth one is particularly interesting. Torricelli compared the spheral solids generated by the rotation of a regular polygon around the diagonal and the apothem respectively. He proved that the volume of the first one is to that of the second one as twice the rectangle whose sides are the diagonal and the apothem is to the sum of the square of the diagonal and the square of the apothem. As a Corollary, it follows that the volume of the first solid is less than the other's. In the second book the method of exhaustion is never used. This work by Torricelli is written in the classical style of Greek geometry. In the *Proemium* Torricelli recognised that his source of inspiration was Archimedes. As a matter of fact, *De Sphaera et Solidis Sphaeralibus* might have been written by the Great Syracusan. Torricelli does not use any algebraic calculations. One could say that Archimedes in works such as the *Quadrature of parabolas* and *Conoids and spheroids* resorted to more algebraic calculations than did Torricelli in *De Sphaera et Solidis Sphaeralibus*! From a mathematical and historical perspective, now is the occasion to recall once again Archimedes's heritage in Torricelli. The latter used the method of exhaustion as rigorously as Archimedes and also extended Archimedes's results.

3.2 De Motu Gravium Naturaliter Descendentium

If Archimedes was the source of inspiration of *De Sphaera et Solidis Sphaeralibus*, Galileo inspired the second work inserted in *Opera geometrica*: *De Motu Gravium Naturaliter Descendentium*, although the Archimedean influence is conspicuous in this work, too.[31] Specifically, the third and the fourth Days of *Discorsi e dimostrazioni intorno a due nuove scienze* are the crucial reference points of this writing, which is divided into two books: in the first one the fundamental elements of parabolic motion are studied; in the second the motion of projectiles is examined. It must be pointed out that not all physicists were convinced that the trajectory of a projectile was a parabola. Among other purposes, Torricelli also aimed to show it was a parabola. In his introductory considerations, the author clarifies that his book was written for specialists who were familiar with Galileo's works. The fundamental concept was that of "*momento*" (*momentum*). In fact, it had been introduced by Benedetti, but studied systematically by Galileo. One might say that the momentum (Galluzzi 1979) of a weight is the measure of weight's effect in function of the position of the weight itself. The fundamental theorem is that stated first by Galileo as the sixth proposition of his *Le Mecaniche* which claims that the *momenta* of equal weights on differently inclined planes are as the perpendiculars drawn from the weights to the Earth's surface if the length of the planes where the weights lie is the same (Fig. 7).

[31] With regard to the influence exerted by Archimedes on Torricelli's *De Motu Gravium*, see Pisano 2009a; see also: 2009b, c, d; Pisano and Capecchi 2010a, b and Pisano and Bussotti 2014a, b; Pisano 2001.

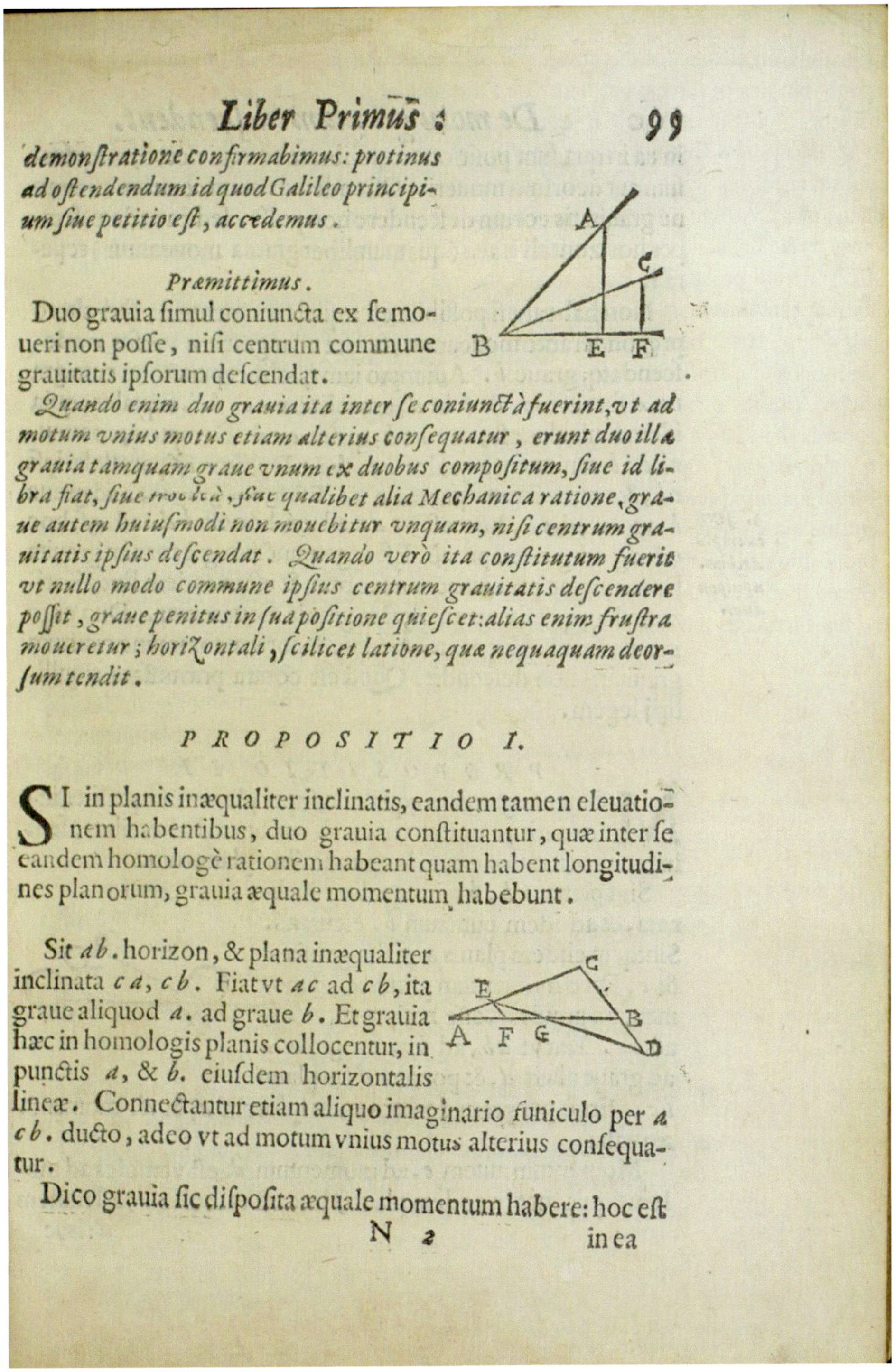
Liber Primus. 99

demonstratione confirmabimus: protinus ad ostendendum id quod Galileo principium siue petitio est, accedemus.

Præmittimus.

Duo grauia simul coniuncta ex se moueri non posse, nisi centrum commune grauitatis ipsorum descendat.

Quando enim duo grauia ita inter se coniuncta fuerint, vt ad motum vnius motus etiam alterius consequatur, erunt duo illa grauia tamquam graue vnum ex duobus compositum, siue id libra fiat, siue troclea, siue qualibet alia Mechanica ratione, graue autem huiusmodi non mouebitur vnquam, nisi centrum grauitatis ipsius descendat. Quando verò ita constitutum fuerit vt nullo modo commune ipsius centrum grauitatis descendere possit, graue penitus in sua positione quiescet: alias enim frustra moueretur; horizontali, scilicet latione, quæ nequaquam deorsum tendit.

PROPOSITIO I.

SI in planis inæqualiter inclinatis, eandem tamen eleuationem habentibus, duo grauia constituantur, quæ inter se eandem homologè rationem habeant quam habent longitudines planorum, grauia æquale momentum habebunt.

Sit *ab*. horizon, & plana inæqualiter inclinata *ca*, *cb*. Fiat vt *ac* ad *cb*, ita graue aliquod *a*. ad graue *b*. Et grauia hæc in homologis planis collocentur, in punctis *a*, & *b*. eiusdem horizontalis lineæ. Connectantur etiam aliquo imaginario funiculo per *acb*. ducto, adeo vt ad motum vnius motus alterius consequatur.

Dico grauia sic disposita æquale momentum habere: hoc est

N 2 in ea

Fig. 7 The fundamental theorem claims that, if $AB = BC$ and if in A and C two equal weights a and c lie, the momentum of a is to the momentum of c as AE is to CF (Torricelli 1644, II, 1, pp. 99; see figure top right side). *Source* Courtesy University of Oklahoma, History of Science Collections

Galileo presented this proposition as a given. Torricelli was going to prove it.

The *De Motu Gravium Naturaliter Descendentium* has to be regarded as an original contribution to the development of physics because Torricelli saw the seed of the whole science of movement in his famous principle:

Two heavy bodies joined together cannot move by themselves unless their common centre of gravity descends.[32]

Torricelli's idea is, thus, an element of the physics of principles which, at the end of the seventeenth century, was by far overtaken by Newton's physics of forces, but that, from the second half of the eighteenth century, became a fundamental way to face the problems of this discipline (Cfr. Pisano and Capecchi 2013). Torricelli's principle is directly tied to that of virtual velocities (Cfr. Pisano and Bussotti 2016, 2017b).

In the initial propositions of his work, Torricelli analyses the momenta of spheres or systems of spheres posed in different positions on an inclined plane. As a matter of fact, he distinguishes the heavy body's momentum in a total momentum and in its components determined by the sinus and the cosine of the angle of inclination of the inclined plane in which the heavy body is posed.

In Proposition III Torricelli proved the fundamental theorem by Galileo that the moments of two equally heavy bodies lying on two equally long, but unequally inclined planes, are as the perpendiculars from the highest vertex of the plane to the ground (Torricelli 1644, p. 104). In the proof, Torricelli's principle plays an indirect role because it is not exploited explicitly. However, Proposition II is used. The proof of this proposition relies upon Proposition I, whose demonstration is based on the principle. A series of propositions, which have a clear Galilean flavour, follows. For example, in Proposition IV, Torricelli proves that, given two inclined planes of the same altitude, the fall-times are as the planes' length. As Torricelli himself recognizes, this proposition is a consequence of Galileo's famous Theorem VI, Proposition VI of the third Day of *Discorsi e dimostrazioni*: given a vertical circle, if, from its highest point you draw any number of inclined planes so that their lengths cut the circumference, the fall-times along these planes are equal.

Starting from Proposition VI some clear properties of the parabola and of the representations of motions through parabolas are introduced. For example, in Proposition X, Torricelli proves that, given a body starting from rest and falling on inclined planes, the fall-time is as the straight lines applied to the parabolas in the considered point (Fig. 8).

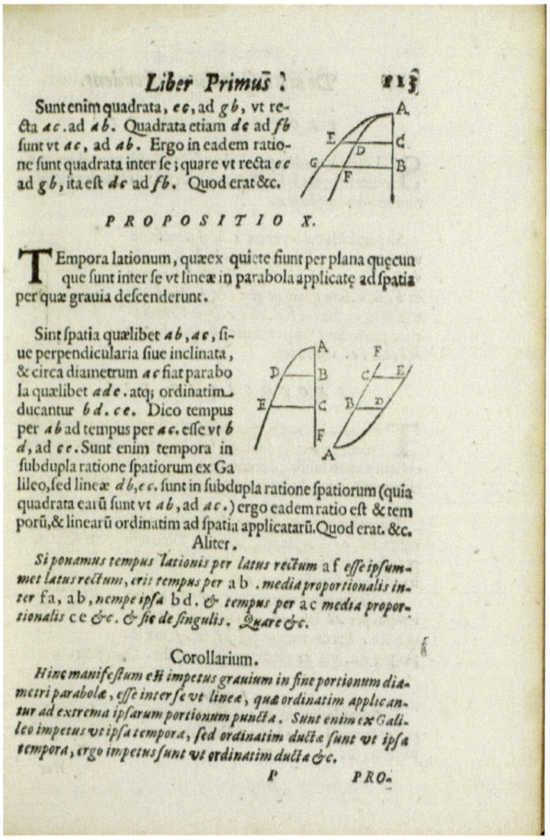
Liber Primus. 113

Sunt enim quadrata, *ec*, ad *gb*, vt recta *ac*. ad *ab*. Quadrata etiam *de* ad *fb* sunt vt *ac*, ad *ab*. Ergo in eadem ratione sunt quadrata inter se; quare vt recta *ec* ad *gb*, ita est *de* ad *fb*. Quod erat &c.

PROPOSITIO X.

TEmpora lationum, quæ ex quiete fiunt per plana quęcunque sunt inter se vt lineæ in parabola applicatę ad spatia per quæ grauia descenderunt.

Sint spatia quælibet *ab*, *ac*, siue perpendicularia siue inclinata, & circa diametrum *ac* fiat parabola quælibet *ade*. atq; ordinatim ducantur *bd*. *ce*. Dico tempus per *ab* ad tempus per *ac*. esse vt *bd*, ad *ce*. Sunt enim tempora in subdupla ratione spatiorum ex Galileo, sed lineæ *db*, *ec*. sunt in subdupla ratione spatiorum (quia quadrata earũ sunt vt *ab*, ad *ac*.) ergo eadem ratio est & temporũ, & linearũ ordinatim ad spatia applicatarũ. Quod erat. &c.

Aliter.

Si ponamus tempus lationis per latus rectum af *esse ipsummet latus rectum, erit tempus per* ab *. media proportionalis inter* fa, ab, *nempe ipsa* bd. *& tempus per* ac *media proportionalis* ce *&c. & sic de singulis. Quare &c.*

Corollarium.

Hinc manifestum est impetus grauium in fine portionum diametri parabolæ, esse inter se vt lineæ, quæ ordinatim applicantur ad extrema ipsarum portionum puncta. Sunt enim ex Galileo impetus vt ipsa tempora, sed ordinatim ductæ sunt vt ipsa tempora, ergo impetus sunt vt ordinatim ducta &c.

P PRO-

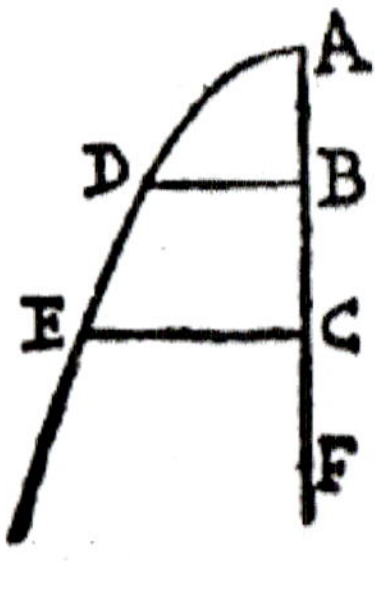

Consider a parabola *ADE*. Be *AF* the fall-height. The fall-time from *A* to *B* is to the fall-time from *A* to *C* as *BD* is to *EC*.

Fig. 8 The figure that Torricelli used to prove Proposition X (Torricelli 1644, II, 1, p. 113). *Source* Courtesy University of Oklahoma, History of Science Collections

[32] "Duo gravia simul coniuncta ex se moveri non posse, nisi centrum commune gravitatis ipsorum descendat". (Torricelli 1644, II, 1, p. 99; Authors' translation). On the state of the art before Torricelli's centre of gravity, see Pisano 2007, 2009a, b, c, d, 2011; Capecchi and Pisano 2007a, b, 2008, 2010a, b, Buscelli 2010. For an examination of this principle see below Pisano and Bussotti' article in this book.

Many of these propositions by Torricelli are based on Galileo's theorem that, in free fall, spaces are inversely as the time squared. It is remarkable that some scientists, for example Descartes and Roberval, were not convinced that Galileo's law was generally correct. Roberval on the 1st of January 1646 wrote to Torricelli that he did not believe in Galileo's laws and put forward some arguments. In his answer (1646, July 7th) Torricelli claimed, in substance, that Galileo's and his own wider considerations are correct at least from a mathematical point of view, that is: supposing Galileo's law to be true, then the consequences explored by Galileo and more broadly by Torricelli himself are true. If from a certain perspective, Torricelli's answer is undoubtedly disappointing because it is not an answer to Roberval's doubts, from another perspective, it is interesting. For, Torricelli seems interested in rational mechanics, almost independently of physical applications, whereas Roberval is wondering how nature behaves; namely in physics rather than in rational mechanics. With Proposition XVII the study of parabolic motion begins. This proposition is a problem (Fig. 9): given a parabola of vertex *B*, find the point *A* such that, if a body falls from *A* and, arrived in *B* with a certain *impetus*, such *impetus* is directed horizontally, the body's trajectory is the parabola.

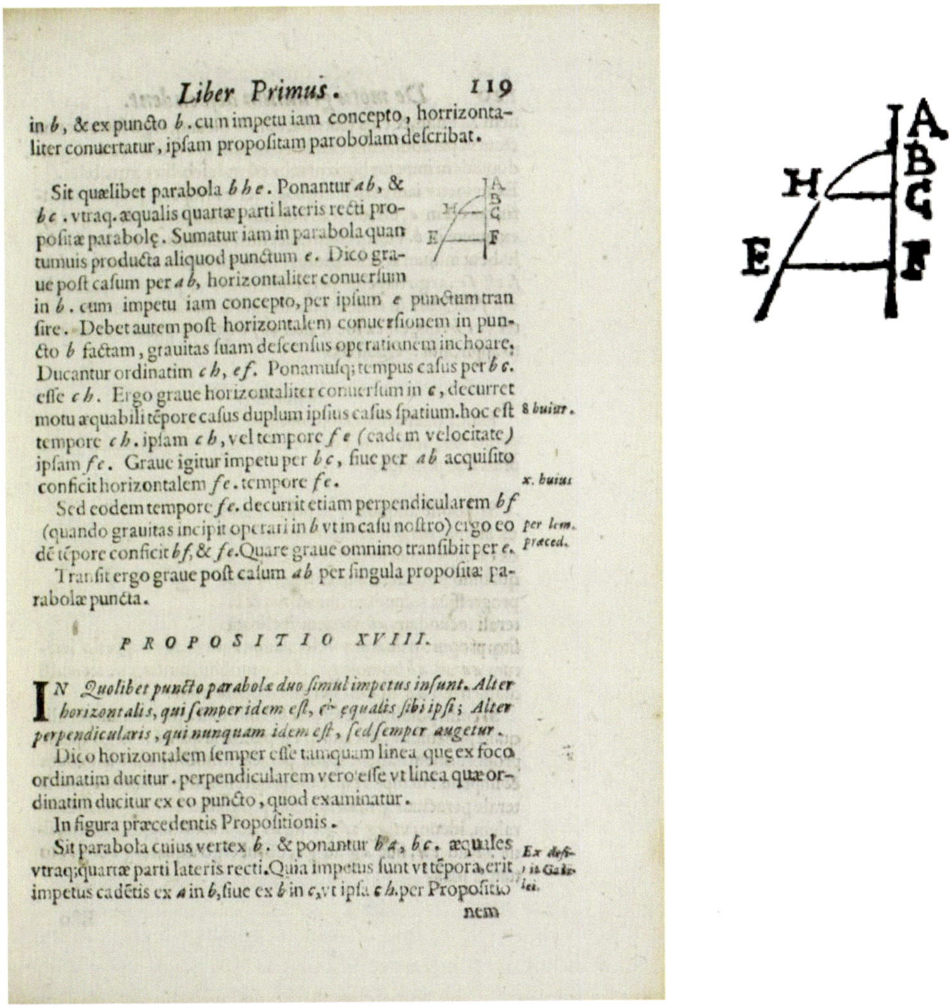

Liber Primus. 119

in *b*, & ex puncto *b*. cum impetu iam concepto, horrizontaliter conuertatur, ipſam propoſitam parobolam deſcribat.

Sit quælibet parabola *bhe*. Ponantur *ab*, & *bc*. vtraq. æqualis quartæ parti lateris recti propoſitæ parabolę. Sumatur iam in parabola quantumuis producta aliquod punctum *e*. Dico graue poſt caſum per *ab*, horizontaliter conuerſum in *b*. cum impetu iam concepto, per ipſum *e* punctum tranſire. Debet autem poſt horizontalem conuerſionem in puncto *b* factam, grauitas ſuam deſcenſus operationem inchoare; Ducantur ordinatim *ch*, *ef*. Ponamuſq; tempus caſus per *bc*. eſſe *ch*. Ergo graue horizontaliter conuerſum in *c*, decurret motu æquabili tẽpore caſus duplum ipſius caſus ſpatium. hoc eſt tempore *ch*. ipſam *cb*, vel tempore *fe* (eadem velocitate) ipſam *fe*. Graue igitur impetu per *bc*, ſiue per *ab* acquiſito conficit horizontalem *fe*. tempore *fe*. *8 huius.* *x. huius*

Sed eodem tempore *fe*. decurrit etiam perpendicularem *bf* (quando grauitas incipit operari in *b* vt in caſu noſtro) ergo eodẽ tẽpore conficit *bf*, & *fe*. Quare graue omnino tranſibit per *e*. *per lem. præced.*

Tranſit ergo graue poſt caſum *ab* per ſingula propoſitæ parabolæ puncta.

PROPOSITIO XVIII.

IN Quolibet puncto parabolæ duo ſimul impetus inſunt. Alter horizontalis, qui ſemper idem eſt, & equalis ſibi ipſi; Alter perpendicularis, qui nunquam idem eſt, ſed ſemper augetur.

Dico horizontalem ſemper eſſe tamquam linea quę ex foco ordinatim ducitur. perpendicularem vero eſſe vt linea quæ ordinatim ducitur ex eo puncto, quod examinatur.

In figura præcedentis Propoſitionis.

Sit parabola cuius vertex *b*. & ponantur *ba*, *bc*. æquales vtraq; quartæ parti lateris recti. Quia impetus ſunt vt tẽpora, erit impetus cadẽtis ex *a* in *b*, ſiue ex *b* in *c*, vt ipſa *ch*. per Propoſitionem *Ex defi. 1 in Galilei.*

Fig. 9 The figure used by Torricelli to solve the problem expounded in Proposition XVII (Torricelli 1644, II, 1, p. 119). *Source* Courtesy University of Oklahoma, History of Science Collections

It must be noted that the term *impetus* does not have a single precise meaning. In this case it means velocity.

In the following proposition, the classical Galilean theorem that in a parabolic motion the horizontal component of velocity is constant, while the vertical is uniformly accelerated is proved.

In the noteworthy scholium following Proposition XVIII, the parallelogram rule for velocities is proved by a remarkable *ad absurdum* reasoning. In the same scholium Torricelli teaches how to draw the tangent at a point of a parabola by means of kinematic considerations. Through a similar reasoning he also explains how to draw the tangent to an Archimedean spiral and also outlines the possibility to draw the tangent at a point of the cycloid. The solution of this last problem caused the priority dispute with Roberval. A long series of propositions follows in which several aspects of the motion along a parabola are analysed: velocities, times, etc. Torricelli also examines the case in which a body falls along the chords of a parabola and proves (Proposition XXVII, Fig. 10) that the fall-time along the chord of a parabola having an extreme in the vertex is directly as the length of the chord and inversely as the ordinate of the second extreme of the chord itself.

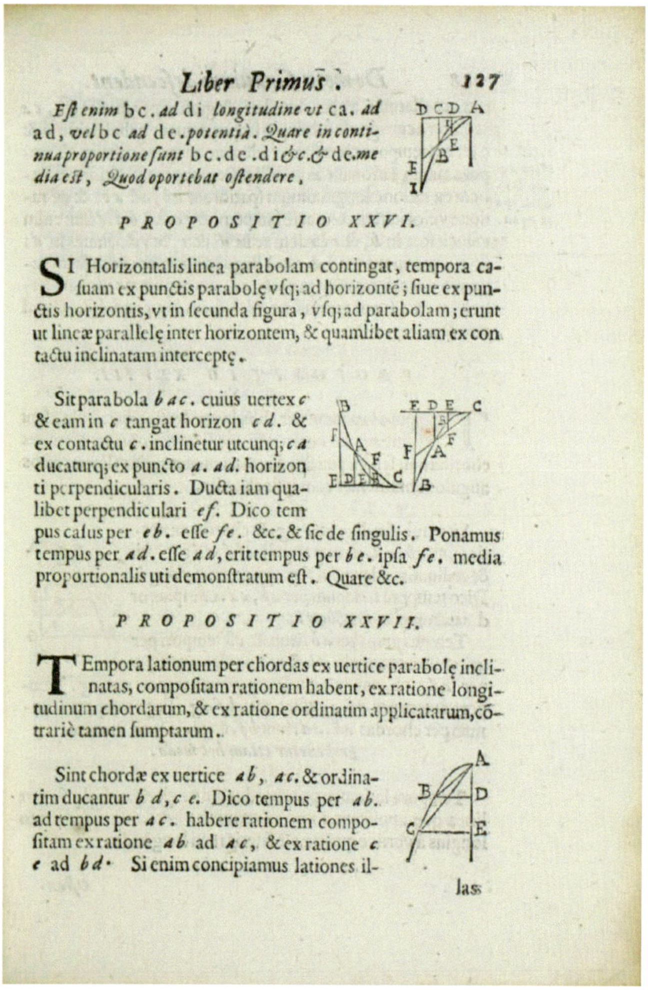
Liber Primus. 127

Eſt enim bc. *ad* di *longitudine vt* ca. *ad* ad, *vel* bc *ad* dc. *potentiâ. Quare in continua proportione ſunt* bc. dc. di *&c. &* dc. *media eſt, Quod oportebat oſtendere.*

PROPOSITIO XXVI.

SI Horizontalis linea parabolam contingat, tempora caſuum ex punctis parabolę vſq; ad horizontē; ſiue ex punctis horizontis, vt in ſecunda figura, vſq; ad parabolam; erunt ut lineę parallelę inter horizontem, & quamlibet aliam ex contactu inclinatam interceptę.

Sit parabola *bac.* cuius uertex *c* & eam in *c* tangat horizon *cd.* & ex contactu *c.* inclinetur utcunq; *ca* ducaturq; ex puncto *a. ad.* horizonti perpendicularis. Ducta iam quali bet perpendiculari *ef.* Dico tempus caſus per *eb.* eſſe *fe.* &c. & ſic de ſingulis. Ponamus tempus per *ad.* eſſe *ad*, erit tempus per *be.* ipſa *fe.* media proportionalis uti demonſtratum eſt. Quare &c.

PROPOSITIO XXVII.

TEmpora lationum per chordas ex uertice parabolę inclinatas, compoſitam rationem habent, ex ratione longitudinum chordarum, & ex ratione ordinatim applicatarum, cōtrariè tamen ſumptarum.

Sint chordæ ex uertice *ab*, *ac.* & ordinatim ducantur *bd*, *ce.* Dico tempus per *ab.* ad tempus per *ac.* habere rationem compoſitam ex ratione *ab* ad *ac*, & ex ratione *ce* ad *bd.* Si enim concipiamus lationes il-

las

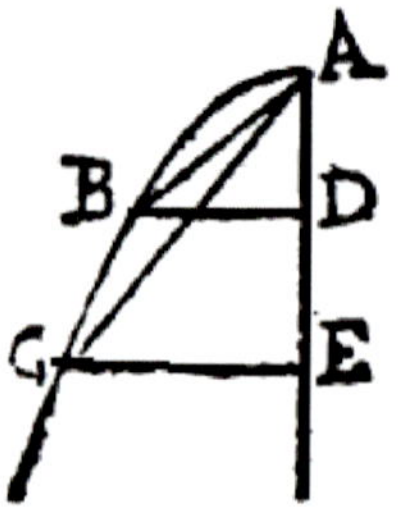

Consider a body falling along the chord AB and a body along the chord AC. Their fall-times are as $\frac{AB}{AG} \cdot \frac{CE}{BD}$.

Fig. 10 The diagram used by Torricelli to prove the Proposition XXVII (Torricelli 1644, II, 1, p. 127). *Source* Courtesy University of Oklahoma, History of Science Collections

Torricelli proves (Proposition XXVIII) that it is also possible to find a single segment which is as the fall-time along a chord. For, given the chord AB, it is enough to draw the straight line perpendicular to AB. Such a line will cut the parabola's axis in a point F. Draw the ordinate FH at the parabola in F. The fall-time along AB will be as FH (Fig. 11).

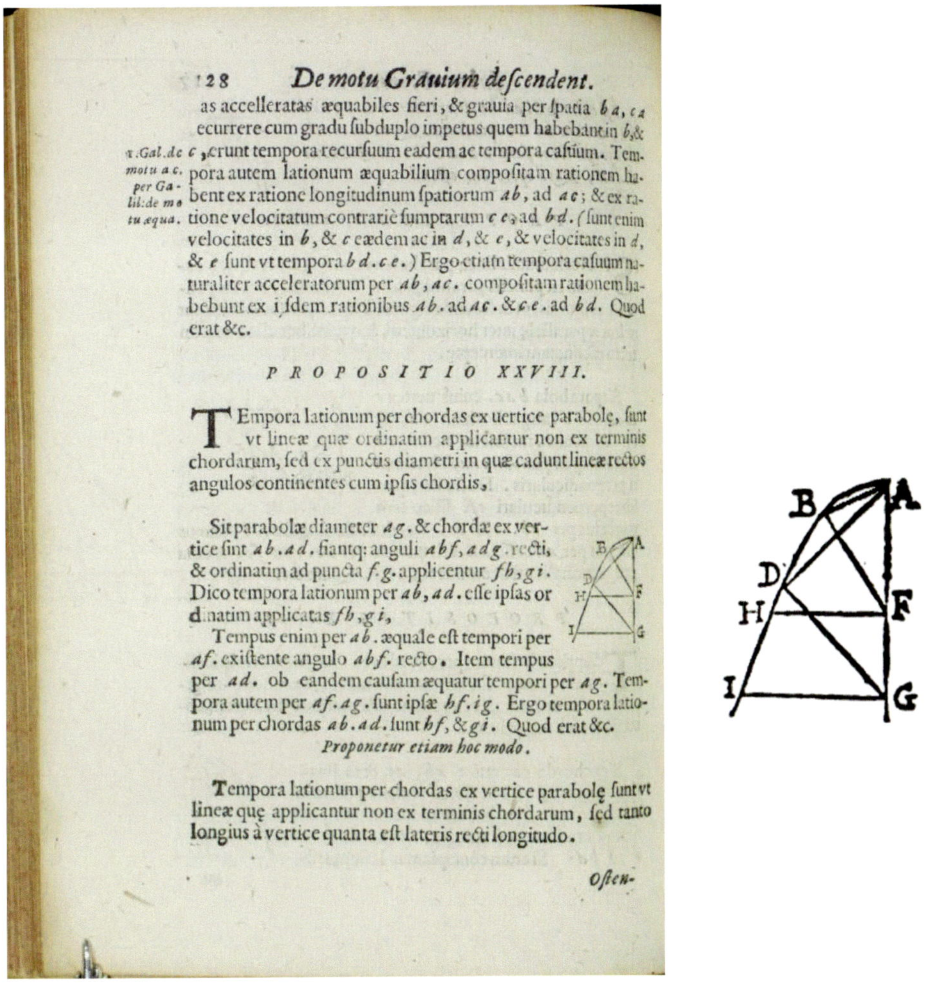

128 *De motu Grauium descendent.*

as accelleratas æquabiles fieri, & grauia per spatia *ba, ca* ecurrere cum gradu subduplo impetus quem habebant in *b,* & *c*, erunt tempora recursuum eadem ac tempora casuum. Tempora autem lationum æquabilium compositam rationem habent ex ratione longitudinum spatiorum *ab*, ad *ac*; & ex ratione velocitatum contrariè sumptarum *c e*, ad *bd*. (sunt enim velocitates in *b*, & *c* eædem ac in *d*, & *e*, & velocitates in *d*, & *e* sunt vt tempora *bd.ce.*) Ergo etiam tempora casuum naturaliter acceleratorum per *ab, ac*. compositam rationem habebunt ex iisdem rationibus *ab*. ad *ac*. & *ce*. ad *bd*. Quod erat &c.

x.Gal.de motu ac. per Galil:de motu æqua.

PROPOSITIO XXVIII.

Tempora lationum per chordas ex uertice parabolę, sunt vt lineæ quæ ordinatim applicantur non ex terminis chordarum, sed ex punctis diametri in quæ cadunt lineæ rectos angulos continentes cum ipsis chordis.

Sit parabolæ diameter *ag*. & chordæ ex vertice sint *ab.ad*. fiantq; anguli *abf, adg*. recti, & ordinatim ad puncta *f.g*. applicentur *fh, gi*. Dico tempora lationum per *ab, ad*. esse ipsas ordinatim applicatas *fh, gi*.

Tempus enim per *ab*. æquale est tempori per *af*. existente angulo *abf*. recto. Item tempus per *ad*. ob eandem causam æquatur tempori per *ag*. Tempora autem per *af.ag*. sunt ipsæ *hf.ig*. Ergo tempora lationum per chordas *ab.ad*. sunt *hf*, & *gi*. Quod erat &c.

Proponetur etiam hoc modo.

Tempora lationum per chordas ex vertice parabolę sunt vt lineæ quę applicantur non ex terminis chordarum, sed tanto longius à vertice quanta est lateris recti longitudo.

Osten-

Fig. 11 The diagram used by Torricelli to prove Proposition XXVIII (Torricelli 1644, II, 1, p. 128). *Source* Courtesy University of Oklahoma, History of Science Collections

From Proposition XXIX a series of theorems and problems are presented which concern the free fall along inclined planes. Most of these propositions are connected to the fundamental statement by Galileo on the fall-time along a chord of a circle. For example, in Proposition XXXV Torricelli considers a perpendicular of given altitude on the horizon and a horizontal segment a of which one extreme is the foot of the perpendicular. It is required to find an inclined plane, along which the fall time is the same as along the perpendicular. An obviously necessary and sufficient condition for the problem to have a solution is that a is not longer than half the perpendicular.

The three propositions XLI-XLIII are a very interesting digression (as Torricelli himself claims) on the motion along a hyperbola and an ellipse.

In Proposition XLI Torricelli considers a horizontal segment AB divided into four equal parts AC, CD, DE, EB. From C and E he draws a hyperbola whose foci are A and B respectively (Fig. 12). Consider a point F on a branch FE of the hyperbola. The fall time along FA is the same as the fall time along FB plus the time in which the segment AB is traversed with the speed acquired in B.

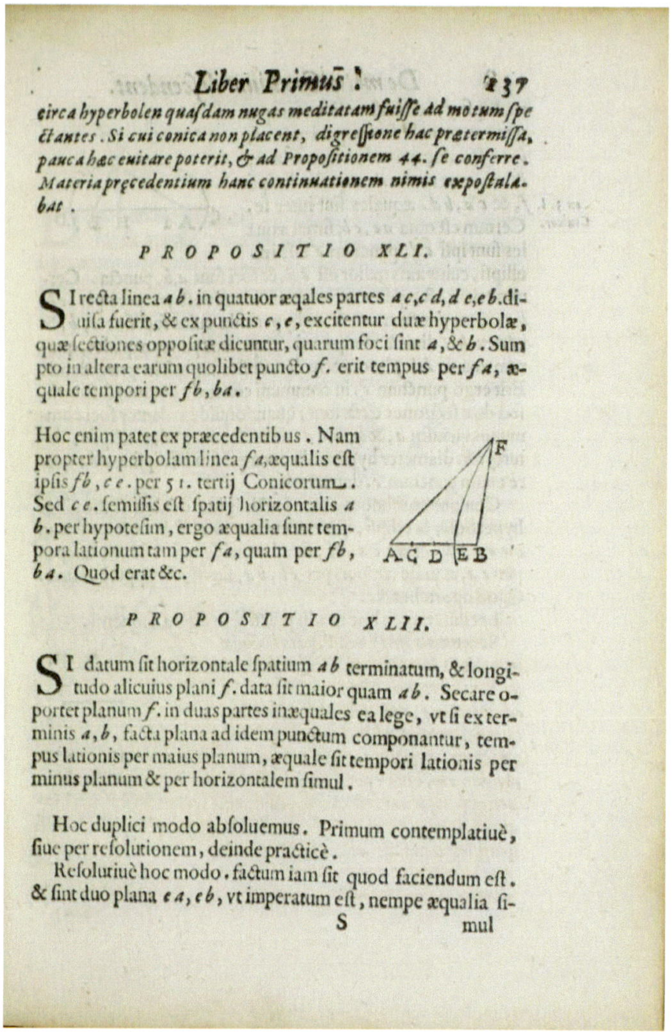

Liber Primus. 137

circa hyperbolen quasdam nugas meditatam fuisse ad motum spectantes. Si cui conica non placent, digressione hac prætermissa, pauca hæc euitare poterit, & ad Propositionem 44. se conferre. Materia precedentium hanc continuationem nimis expostulabat.

PROPOSITIO XLI.

SI recta linea *ab*. in quatuor æqales partes *ac*, *cd*, *de*, *eb*. diuisa fuerit, & ex punctis *c*, *e*, excitentur duæ hyperbolæ, quæ sectiones oppositæ dicuntur, quarum foci sint *a*, & *b*. Sumpto in altera earum quolibet puncto *f*. erit tempus per *fa*, æquale tempori per *fb*, *ba*.

Hoc enim patet ex præcedentibus. Nam propter hyperbolam linea *fa*, æqualis est ipsis *fb*, *ce*. per 51. tertij Conicorum. Sed *ce*. semissis est spatij horizontalis *ab*. per hypotesim, ergo æqualia sunt tempora lationum tam per *fa*, quam per *fb*, *ba*. Quod erat &c.

PROPOSITIO XLII.

SI datum sit horizontale spatium *ab* terminatum, & longitudo alicuius plani *f*. data sit maior quam *ab*. Secare oportet planum *f*. in duas partes inæquales ea lege, vt si ex terminis *a*, *b*, facta plana ad idem punctum componantur, tempus lationis per maius planum, æquale sit tempori lationis per minus planum & per horizontalem simul.

Hoc duplici modo absoluemus. Primum contemplatiuè, siue per resolutionem, deinde practicè.

Resolutiuè hoc modo. factum iam sit quod faciendum est. & sint duo plana *ea*, *eb*, vt imperatum est, nempe æqualia si-

S mul

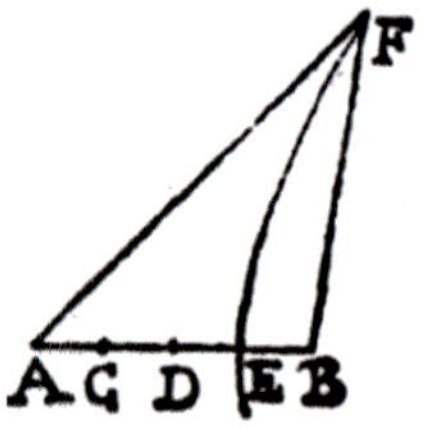

Fig. 12 The diagram used by Torricelli to prove Proposition XLI (Torricelli 1644, II, 1, p. 137). *Source* Courtesy University of Oklahoma, History of Science Collections

Proposition XLII is a fascinating problem (Fig. 13): be given the horizontal segment *AB* and the segment *F* longer than *AB*. It is required to cut *F* in two unequal sections so as to form an inclined plane whose basis is *AB* and whose vertex is *E* such that:

(1) *AE+EB=F*;
(2) The fall time along the longer side of the plane *AE* is equal to the fall time along *EB* plus the time necessary to traverse the segment *AB* with the velocity acquired in *B*.

Torricelli proves that *E* is the intersection point between the ellipse whose foci are *A* and *B* and whose major axis is *CD* = *F* and the hyperbola whose foci are *A* and *B* and whose vertex is in the point *H* obtained by dividing into four equal parts the segment *AB*. Thus, it is

$$BH = \frac{1}{4}AB.$$

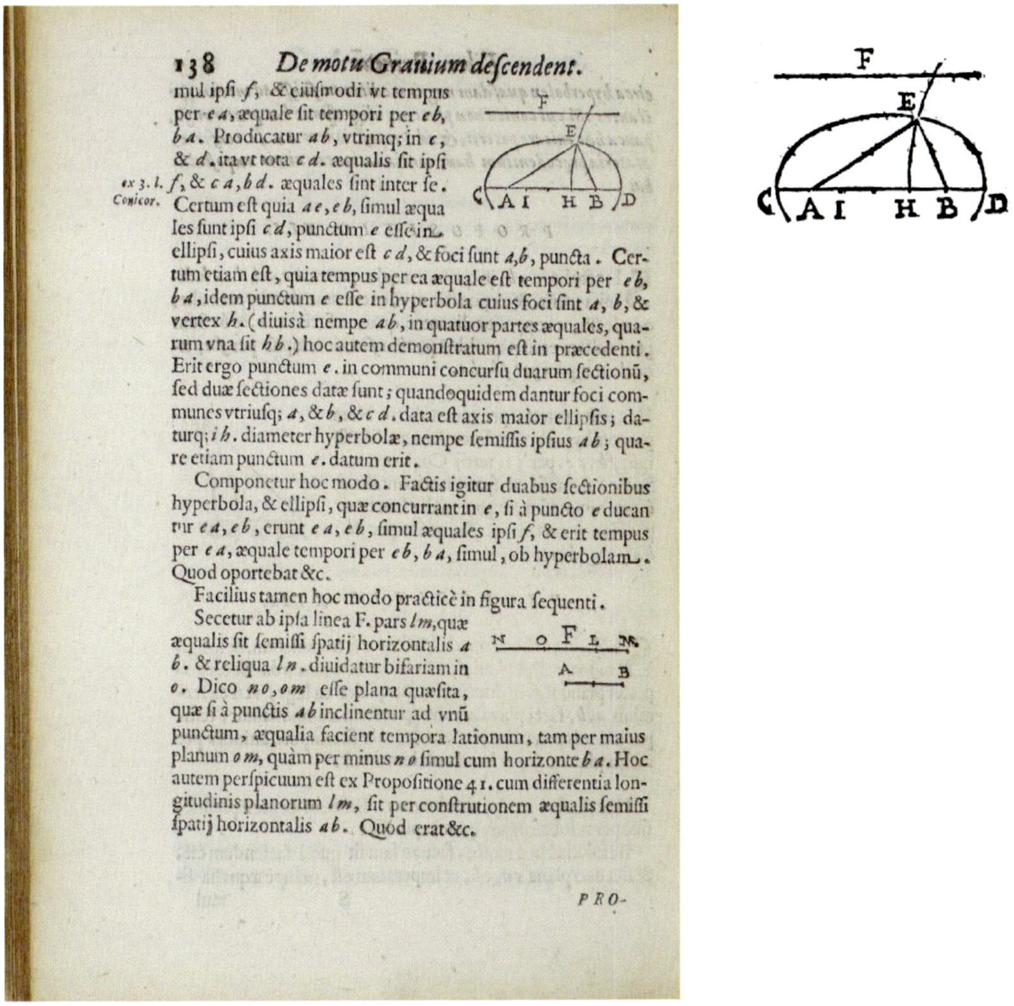
138 *De motu Grauium descendent.*

mul ipsi *f*, & eiusmodi vt tempus per *ea*, æquale sit tempori per *eb*, *ba*. Producatur *ab*, vtrimq; in *c*, & *d*. ita vt tota *cd*. æqualis sit ipsi *f*, & *ca*, *bd*. æquales sint inter se. Certum est quia *ae*, *eb*, simul æquales sunt ipsi *cd*, punctum *e* esse in ellipsi, cuius axis maior est *cd*, & foci sunt *a*,*b*, puncta. Certum etiam est, quia tempus per ea æquale est tempori per *eb*, *ba*, idem punctum *e* esse in hyperbola cuius foci sint *a*, *b*, & vertex *h*. (diuisa nempe *ab*, in quatuor partes æquales, quarum vna sit *hb*.) hoc autem demonstratum est in præcedenti. Erit ergo punctum *e*. in communi concursu duarum sectionũ, sed duæ sectiones datæ sunt; quandoquidem dantur foci communes vtriusq; *a*, & *b*, & *cd*. data est axis maior ellipsis; daturq; *ib*. diameter hyperbolæ, nempe semissis ipsius *ab*; quare etiam punctum *e*. datum erit.

ex 3. l. Conicor.

Componetur hoc modo. Factis igitur duabus sectionibus hyperbola, & ellipsi, quæ concurrant in *e*, si à puncto *e* ducantur *ea*, *eb*, erunt *ea*, *eb*, simul æquales ipsi *f*, & erit tempus per *ea*, æquale tempori per *eb*, *ba*, simul, ob hyperbolam. Quod oportebat &c.

Facilius tamen hoc modo practicè in figura sequenti. Secetur ab ipsa linea F. pars *lm*, quæ æqualis sit semissi spatij horizontalis *ab*. & reliqua *ln*. diuidatur bifariam in *o*. Dico *no*, *om* esse plana quæsita, quæ si à punctis *ab* inclinentur ad vnũ punctum, æqualia facient tempora lationum, tam per maius planum *om*, quàm per minus *no* simul cum horizonte *ba*. Hoc autem perspicuum est ex Propositione 41. cum differentia longitudinis planorum *lm*, sit per construtionem æqualis semissi spatij horizontalis *ab*. Quod erat &c.

PRO-

Fig. 13 The diagram used by Torricelli to prove Proposition XLII (Torricelli 1644, II, 1, p. 138). *Source* Courtesy University of Oklahoma, History of Science Collections

From Proposition XLIV Torricelli analyses the time fall along two sides of a triangle, considering them as inclined planes, while the third side is regarded as horizontal. For example, in Proposition XLIV he proves that there is only one right-angled triangle such that the fall-time along the hypotenuse is the same as that along a small side plus the time necessary to traverse the other (horizontal) small side with the speed acquired at the end of the fall along the first small side. This triangle is the Pythagorean triangle 3,4,5, where the small side of length 4 is that horizontal.

Among the other propositions of the first book, the 49th one is particularly remarkable. Here Torricelli considers (Fig. 14) a vertical circle and draws the tangent at its highest point *A*. Fall a body from *A* until the circle's centre *B*. In *B* be posed a plane, inclined downwards or upwards. Then, the body, with the speed acquired in *B*, will traverse the space *BD* equal to the radius plus the perpendicular drawn from the point where the plane cuts the circle to the tangent in *A*, in the same time in which it has fallen from *A* to *B*.

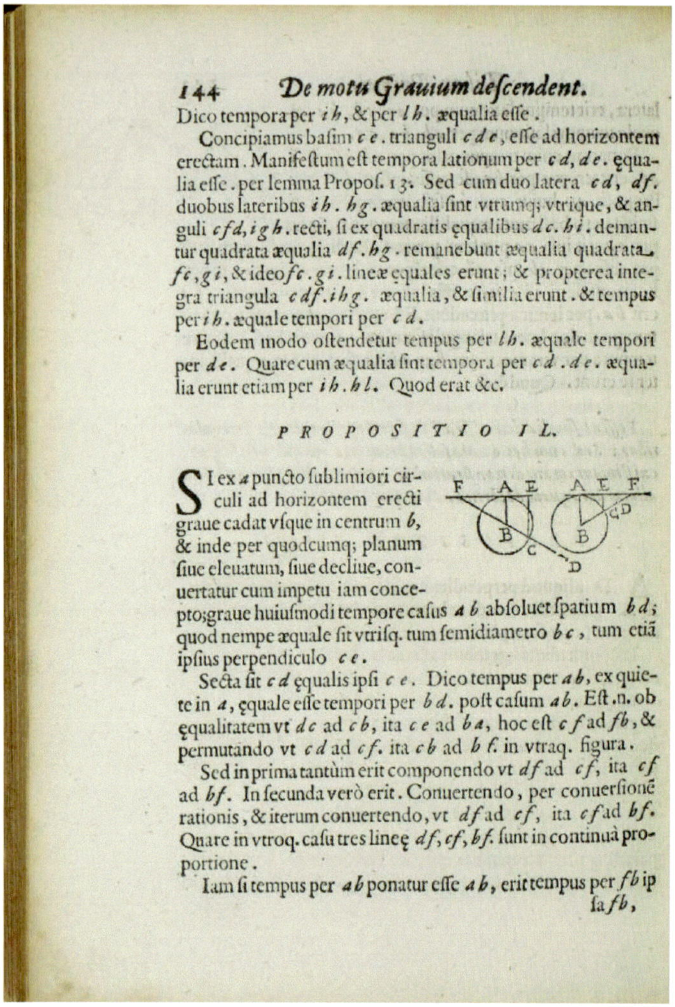

144 *De motu Grauium descendent.*

Dico tempora per *ih*, & per *lh*. æqualia esse.

Concipiamus basim *ce*. trianguli *cde*, esse ad horizontem erectam. Manifestum est tempora lationum per *cd*, *de*. ęqualia esse. per lemma Propos. 13. Sed cum duo latera *cd*, *df*. duobus lateribus *ih*. *hg*. æqualia sint vtrumq; vtrique, & anguli *cfd*, *igh*. recti, si ex quadratis ęqualibus *dc*. *hi*. demantur quadrata æqualia *df*. *hg*. remanebunt æqualia quadrata *fc*, *gi*, & ideo *fc*. *gi*. lineæ ęquales erunt; & propterea integra triangula *cdf*. *ihg*. æqualia, & similia erunt. & tempus per *ih*. æquale tempori per *cd*.

Eodem modo ostendetur tempus per *lh*. æquale tempori per *de*. Quare cum æqualia sint tempora per *cd*. *de*. æqualia erunt etiam per *ih*. *hl*. Quod erat &c.

PROPOSITIO IL.

SI ex *a* puncto sublimiori circuli ad horizontem erecti graue cadat vsque in centrum *b*, & inde per quodcumq; planum siue eleuatum, siue decliue, conuertatur cum impetu iam concepto; graue huiusmodi tempore casus *ab* absoluet spatium *bd*; quod nempe æquale sit vtrisq. tum semidiametro *bc*, tum etiã ipsius perpendiculo *ce*.

F A E A E F B C D B C D

Secta sit *cd* ęqualis ipsi *ce*. Dico tempus per *ab*, ex quiete in *a*, ęquale esse tempori per *bd*. post casum *ab*. Est .n. ob ęqualitatem vt *dc* ad *cb*, ita *ce* ad *ba*, hoc est *cf* ad *fb*, & permutando vt *cd* ad *cf*. ita *cb* ad *bf*. in vtraq. figura.

Sed in prima tantùm erit componendo vt *df* ad *cf*, ita *cf* ad *bf*. In secunda verò erit. Conuertendo, per conuersionẽ rationis, & iterum conuertendo, vt *df* ad *cf*, ita *cf* ad *bf*. Quare in vtroq. casu tres lineę *df*, *cf*, *bf*. sunt in continuà proportione.

Iam si tempus per *ab* ponatur esse *ab*, erit tempus per *fb* ipsa *fb*,

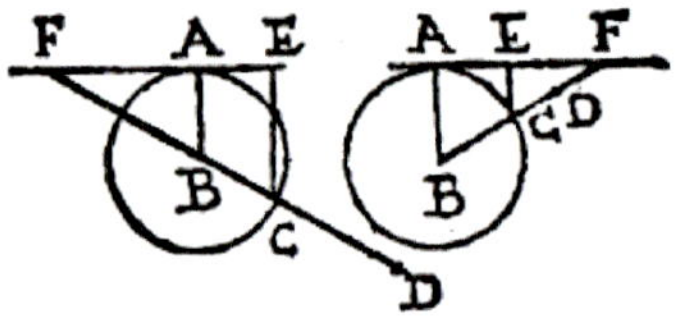

Fig. 14 The figure used by Torricelli to prove the Proposition IL. On the left, the plane is inclined downwards, on the right upwards (Torricelli 1644, II, 1, p. 144). *Source* Courtesy University of Oklahoma, History of Science Collections

Torricelli offered two different demonstrations of this theorem.

Problem LV concerns an interesting question of minimum. Torricelli considers as given a horizontal segment *AB*. Be also given the perpendicular in *A* (or in *B*) at *AB*. On such perpendicular, determine a point *C* such that the time to traverse *CA* + *AB* (or *CB* + *BA*) is the shortest. The answer is that *CA* (or *CB*) = 1/2*AB*. It must be noted that, in this case, too, the Galilean character of Torricelli's works is apparent, because Galileo in the third Day of *Discorsi e dimostrazioni* had also faced problems of minimal time, for example, in the theorems XIX, XX (Propositions XXX and XXXI respectively).

The LVIIIth and last proposition of the first book proves an interesting property of parabolic motion (Fig. 15): consider a segment *AB* of the parabola's axis and the time necessary for a body in free fall to traverse *AB* plus the ordinate *BC* with the speed acquired in *B*. This time is equal to that necessary for the body in free fall to traverse a segment *DB* as long as ¼ of the parabola's *latum rectum* plus the ordinate *BC* with the speed acquired in *B* falling from *D*.

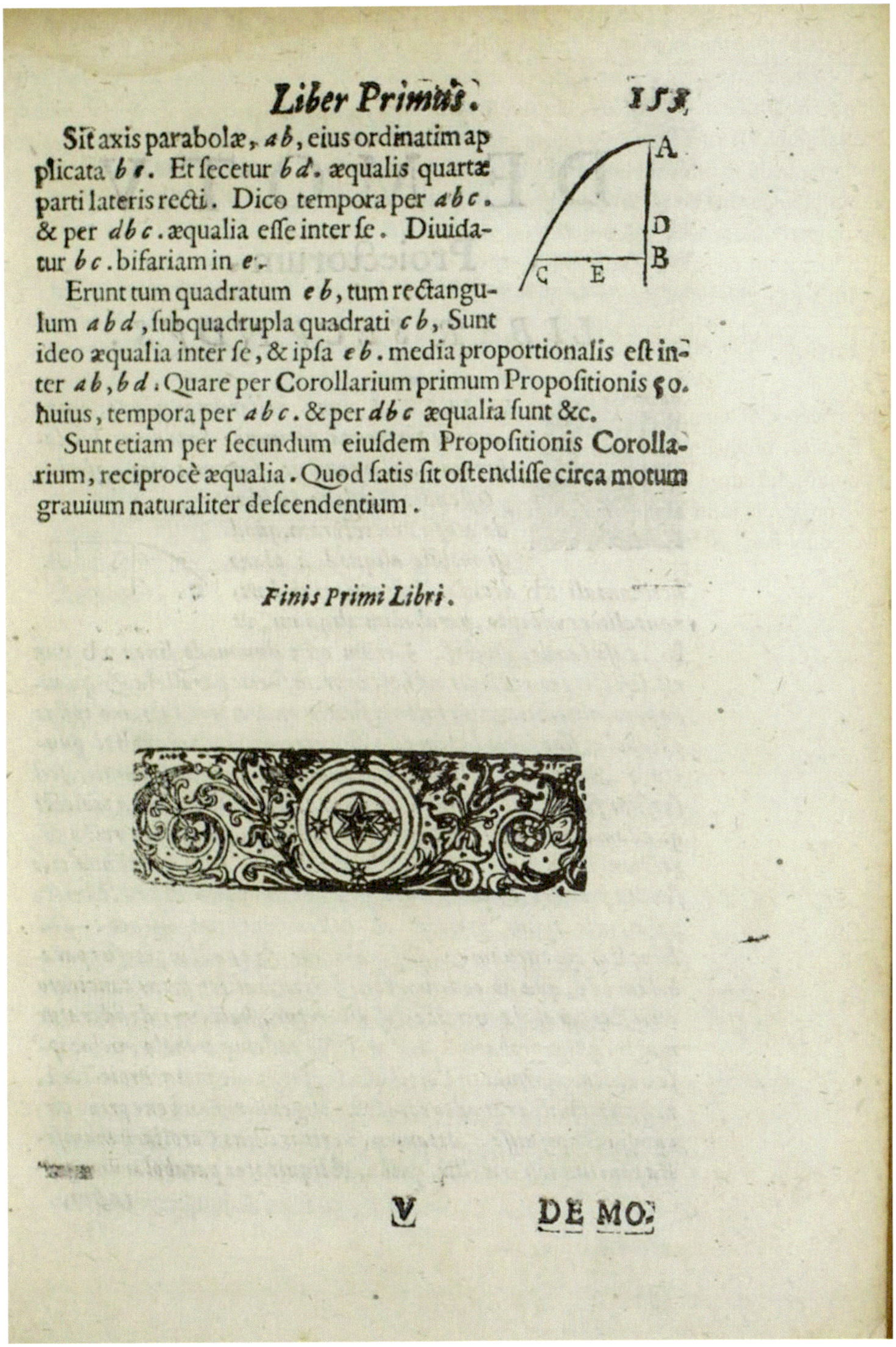

Liber Primus. 153

Sit axis parabolæ, *ab*, eius ordinatim ap plicata *bc*. Et ſecetur *bd*. æqualis quartæ parti lateris recti. Dico tempora per *abc*. & per *dbc*. æqualia eſſe inter ſe. Diuidatur *bc*. bifariam in *e*.

Erunt tum quadratum *eb*, tum rectangulum *abd*, ſubquadrupla quadrati *cb*, Sunt ideo æqualia inter ſe, & ipſa *eb*. media proportionalis eſt inter *ab*, *bd*. Quare per Corollarium primum Propoſitionis 50. huius, tempora per *abc*. & per *dbc* æqualia ſunt &c.

Sunt etiam per ſecundum eiuſdem Propoſitionis Corollarium, reciprocè æqualia. Quod ſatis ſit oſtendiſſe circa motum grauium naturaliter deſcendentium.

Finis Primi Libri.

V DE MO-

Fig. 15 The figure used by Torricelli to prove Proposition LVIII (Torricelli 1644, II, 1, p. 153). *Source* Courtesy University of Oklahoma, History of Science Collections

The first book of this work is conceived having the third Day of Galileo's *Discorsi e dimostrazioni* as its source of inspiration. This notwithstanding, there are many original ideas, foremost the Torricelli principle and its use. Some specific propositions, as those on the movement in conic sections different from a parabola, or the last ones we have referred to, are interesting and show that Torricelli had the intention to establish the science of mechanics more rigorously and broadly than Galileo, though remaining in the line of reasoning indicated by Galileo.

It seems to us necessary to mention at least one consideration developed by Belloni: in a letter to Mersenne written in January 1645 Torricelli expresses the conviction in the existence of a proportionality between the cause of motion's change (*momentum*) and velocity.[33] This means that Torricelli's—like Kepler's forces—produce velocities and not accelerations, which, in its turn, indicates that the inertia principle was not yet grasped in all its features.

The second book of *De motu gravium naturaliter descendentium* concerns the projectiles' motion and is inspired by the fourth Day of Galileo's *Discorsi e dimostrazioni*. Torricelli's aim is stated in his introductory considerations: in the Corollary of Theorem IV, Proposition VII of *Discorsi*'s fourth Day, Galileo stated that the trajectory of a projectile is a parabola, exactly as the motion of a body to which a horizontal (direction) velocity is impressed. Torricelli firmly believed in the truth of this proposition, but thought that so far it had been more supposed true than really demonstrated. Thus, his first aim is to offer a proof of this property. Afterwards, several characteristics of the motion of projectiles will be analysed relying upon such theorem, which Torricelli proves as Proposition III.

The Definition with which the book begins is interesting:

> The direction of projection is defined as that straight line which touches the curved line of the projectile in the first point of that curved line itself [...].[34]

Here there is the concept of tangent at a curve in its initial point, which is implicitly associated with the idea of an incipient motion. It is surely premature to speak of concepts close to those of infinitesimal analysis, but Torricelli's notion of tangent in the first point of the curve is interesting. The first remarkable proposition is the second one. Here Torricelli proves that the trajectory of a projectile in its highest point *B* cuts the perpendicular *BH* intercepted between the horizon and the line of direction into two equal parts *BC* and *CH* (Fig. 16).

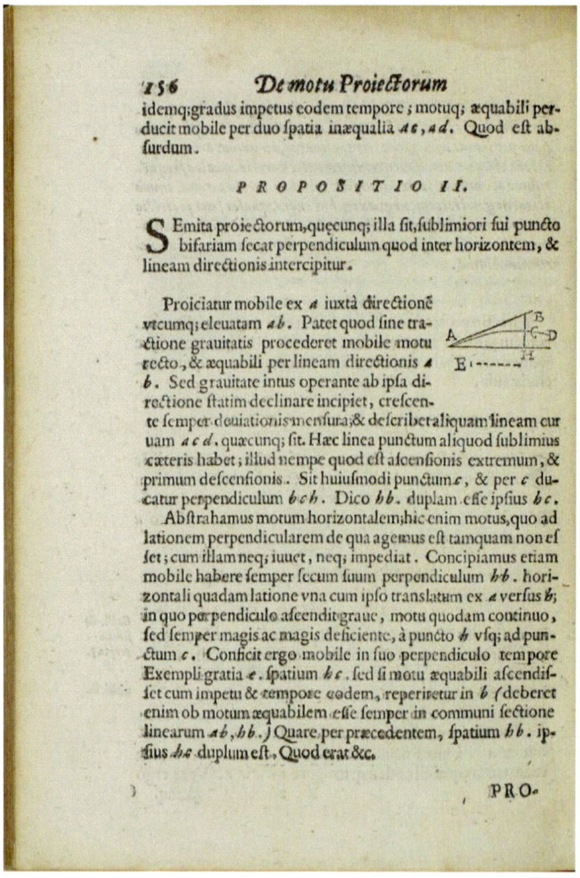

156 De motu Proiectorum

idemq;gradus impetus eodem tempore; motuq; æquabili perducit mobile per duo ſpatia inæqualia *ac*, *ad*. Quod eſt abſurdum.

PROPOSITIO II.

Semita proiectorum, quęcunq; illa ſit, ſublimiori ſui puncto bifariam ſecat perpendiculum quod inter horizontem, & lineam directionis intercipitur.

Proiciatur mobile ex *a* iuxtà directionē vtcumq; eleuatam *ab*. Patet quod ſine tractione grauitatis procederet mobile motu recto, & æquabili per lineam directionis *ab*. Sed grauitate intus operante ab ipſa directione ſtatim declinare incipiet, creſcente ſemper deuiationis menſura; & deſcribet aliquam lineam curuam *acd*, quæcunq; ſit. Hæc linea punctum aliquod ſublimius cæteris habet; illud nempe quod eſt aſcenſionis extremum, & primum deſcenſionis. Sit huiuſmodi punctum *c*, & per *c* ducatur perpendiculum *bch*. Dico *hb*. duplam eſſe ipſius *hc*.

Abſtrahamus motum horizontalem; hic enim motus, quo ad lationem perpendicularem de qua agemus eſt tamquam non eſſet; cum illam neq; iuuet, neq; impediat. Concipiamus etiam mobile habere ſemper ſecum ſuum perpendiculum *hb*. horizontali quadam latione vna cum ipſo translatum ex *a* verſus *h*; in quo perpendiculo aſcendit graue, motu quodam continuo, ſed ſemper magis ac magis deficiente, à puncto *h* vſq; ad punctum *c*. Conficit ergo mobile in ſuo perpendiculo tempore Exempli gratia *e*. ſpatium *hc*. ſed ſi motu æquabili aſcendiſſet cum impetu & tempore eodem, reperiretur in *b* (deberet enim ob motum æquabilem eſſe ſemper in communi ſectione linearum *ab*, *hb*.) Quare per præcedentem, ſpatium *hb*. ipſius *hc* duplum eſt. Quod erat &c.

PRO-

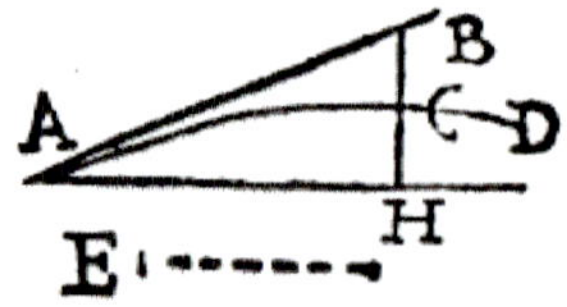

Fig. 16 The diagram used by Torricelli to prove proposition II (Torricelli 1644, II, 2, p. 156). *Source* Courtesy University of Oklahoma, History of Science Collections

[33] Belloni adds a note to his Italian translation of *De motu gravium naturaliter descendentium* (Torricelli 1975, pp. 164–166, note 7; Authors' translation) where the Italian translation of Torricelli's letter to Mersenne is referred to.

[34] "Directio projectionis dicitur linea recta quae tangit lineam curvam proiecti in primo punto eiusdem lineae curvae [...]." (Torricelli 1644, II, 2, p. 155; Authors' translation).

At the beginning of the demonstration, Torricelli claims that, without gravity, the projectile would continue its motion uniformly along the projection line. Belloni claims this to be a clear indication that Torricelli had a broader conception of inertia than Galileo because Galileo recognised only horizontal inertia (Torricelli 1975, p. 224, note 3). Without confronting now one of the most problematic questions of the whole history of science, namely what Galileo's concept of inertia was, it must be noted how elusive and difficult the notion of inertia is. It is true that Torricelli probably had a broader conception of rectilinear inertia than Galileo, but the fact that he believed a force to produce a velocity rather than a change of velocity, is a sign that a full comprehension of this concept had not yet been achieved. Proposition III is the fundamental statement which is at the basis of the further acquisitions explained in the book:

> The curved line described by a mobile thrown with any inclination is a parabola. It is that parabola which the mobile would describe if it were thrown horizontally from the vertex of the curve.[35]

Torricelli developed a long and detailed proof of this proposition, so that no doubt of its truth could remain. The propositions until the eight one can be interpreted as corollaries of the third one. The Proposition IX solves, instead, a significant problem.

Torricelli requires (Fig. 17):

> Be given the impetus *BA*, namely according to the definition, that acquired by a body which falls naturally from *B* to *A*. Be given the throw-direction *AI* and be the throw-speed that acquired in *A*. It is required to find the length, the height and the whole route of the parabola.[36]

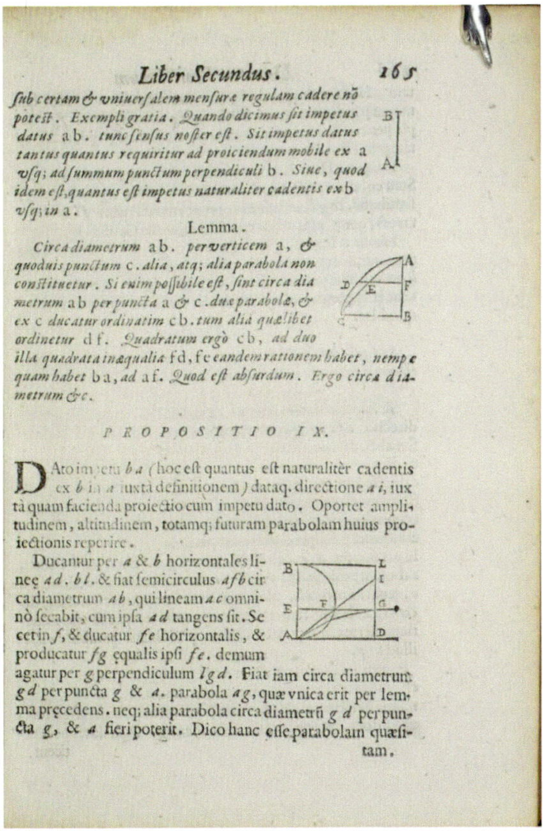

Liber Secundus. 165

sub certam & vniuersalem mensuræ regulam cadere nō potest. Exempli gratia. Quando dicimus sit impetus datus ab. *tunc sensus noster est. Sit impetus datus tantus quantus requiritur ad proiciendum mobile ex* a *vsq; ad summum punctum perpendiculi* b. *Sine, quod idem est, quantus est impetus naturaliter cadentis ex* b *vsq; in* a.

Lemma.

Circa diametrum ab. *per verticem* a, *& quoduis punctum* c. *alia, atq; alia parabola non constituetur. Si enim possibile est, sint circa diametrum* ab *per puncta* a *&* c. *duæ parabolæ, & ex* c *ducatur ordinatim* cb. *tum alia quælibet ordinetur* df. *Quadratum ergò* cb, *ad duo illa quadrata inæqualia* fd, fe *eandem rationem habet, nempe quam habet* ba, *ad* af. *Quod est absurdum. Ergo circa diametrum &c.*

PROPOSITIO IX.

DAto impetu *ba* (hoc est quantus est naturalitèr cadentis ex *b* in *a* iuxtà definitionem) dataq. directione *ai*, iuxtà quam facienda proiectio cum impetu dato. Oportet amplitudinem, altitudinem, totamq; futuram parabolam huius proiectionis reperire.

Ducantur per *a* & *b* horizontales lineę *ad*. *bl*. & fiat semicirculus *afb* circa diametrum *ab*, qui lineam *ac* omninò secabit, cum ipsa *ad* tangens sit. Secet in *f*, & ducatur *fe* horizontalis, & producatur *fg* ęqualis ipsi *fe*. demum agatur per *g* perpendiculum *lgd*. Fiat iam circa diametrum *gd* per puncta *g* & *a*. parabola *ag*, quæ vnica erit per lemma pręcedens. neq; alia parabola circa diametrū *g d* per puncta *g*, & *a* fieri poterit. Dico hanc esse parabolam quæsitam.

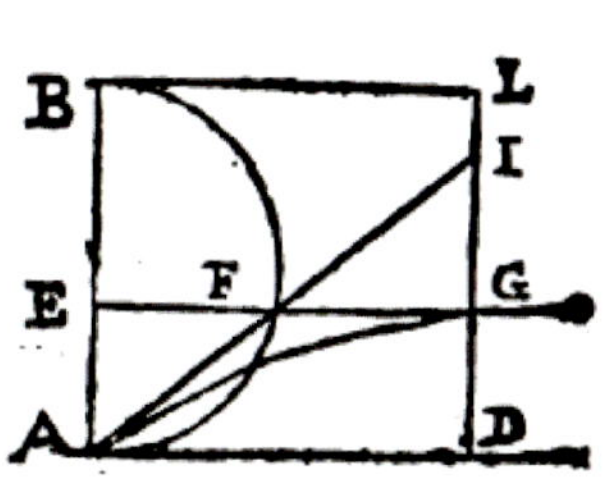

Fig. 17 The diagram used by Torricelli to prove Proposition IX (Torricelli 1644, II, 2, p. 165). *Source* Courtesy University of Oklahoma, History of Science Collections

[35] "Linea curva, quae describitur a mobile secundum quamlibet elevationem proiecto, parabola est, et prorsus, eadem, quam describeret mobile si cum horizontali impetu proiceretur a vertice eiusdem lineae curvae." (Torricelli 1644, p. 157; Authors' translation).

[36] "Dato impetu *ba* (hoc est quantus est naturaliter cadentis ex *b* in *a* iuxta definitionem) dataque directionem *ai*, iuxta quam facienda proiectio cum impetu dato. Oportet amplitudinem altitudinem totamque futuram parabolam huius proiectionis reperire." (Torricelli 1644, p. 165; Authors' translation).

Torricelli offers a beautiful geometrical solution to this problem. Furthermore, he adds a series of interesting commentaries on the different heights and horizontal distances from the point *A* reached by the projectile in function of the angle *DAI*. In this way Torricelli, as he explicitly states, completes a series of observations developed by Galileo in the fourth Day of *Discorsi e dimostrazioni*.

After this problem, a series of propositions follows in which, given two data (for example *impetus* and length, Proposition XI), it is required to find other data (i.e., height, Proposition XI). Torricelli also discusses the cases in which the proposed problem cannot be solved. His work continues analysing the properties of maximum and minimum of a throw whose direction is an angle of 45 degrees (Prop. 17). A series of propositions on the throw-time follows. All the propositions from the Xth to the XXth are of clear demonstration. Propositions XXI and XXII solve, instead, two more difficult problems: in the former Torricelli requires finding the constant horizontal velocity (*impetus*) as well as the variable vertical velocity in any point of the parabolic trajectory, given the initial velocity and the maximal height of the throw. In this proposition as well as in others, the throw-speed is not given directly but through the concept of *sublimitas*: suppose that (Fig. 17) the speed with which the projectile is thrown be that acquired by a body in free fall, which falls from *B* to the ground level *A*. Be *AE* the maximal height of the throw. The difference $BE = BA - EA$ is the *sublimitas* of the parabola.

In Proposition XXII it is required to find the entire *impetus* in a parabolic motion.

Until Proposition XXII Torricelli has sought properties which can be useful not only for mathematicians and natural philosophers, but also for engineers who intend to construct weapons. From Proposition XXIII a more difficult and theoretically far-reaching part of Torricelli's works begins because the author discovers several geometrical properties of parabolic motion which belong to pure rational mechanics. They do not hold a pragmatic value, but a highly theoretical one.

In the first of these propositions, it is proved that all the parabolas described with the same initial *impetus* have the same *lactus rectum*. In Proposition XXV, it is required to find the focus of a parabola, given:

1) Either the *impetus* and the throw-direction;
2) Or the *impetus* and the range of the throw;
3) Or the range and the direction.

From Proposition XXIX a series of theorems begins that concerns geometrical *loci* which are quadrics.

In the XXIXth Proposition it is proved that, if from a point *A*, a projectile is thrown with the same initial speed (*impetus*), but with different directions, the vertices of all the parabolas belong to a spheroid whose major axis is double the minor (Fig. 18).

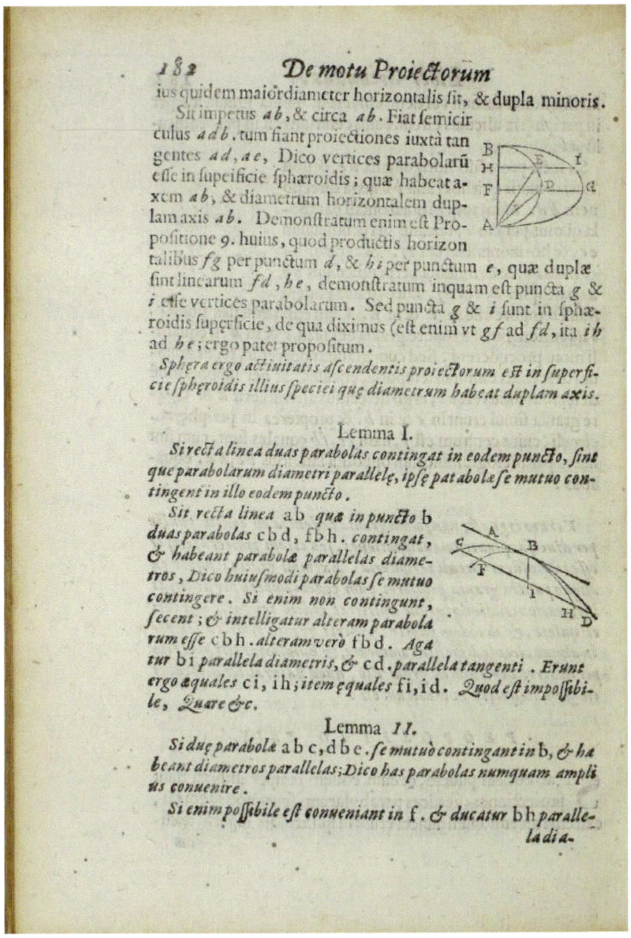

182 *De motu Proiectorum*

ius quidem maiordiameter horizontalis sit, & dupla minoris.

Sit impetus *ab*, & circa *ab*. Fiat semicirculus *adb*. tum fiant proiectiones iuxtà tangentes *ad*, *ae*, Dico vertices parabolarũ esse in supeificie sphæroidis; quæ habeat axem *ab*, & diametrum horizontalem duplam axis *ab*. Demonstratum enim est Propositione 9. huius, quod productis horizontalibus *fg* per punctum *d*, & *hi* per punctum *e*, quæ duplæ sint linearum *fd*, *he*, demonstratum inquam est puncta *g* & *i* esse vertices parabolarum. Sed puncta *g* & *i* sunt in sphæroidis superficie, de qua diximus (est enim vt *gf* ad *fd*, ita *ih* ad *he*; ergo patet propositum.

Sphera ergo actiuitatis ascendentis proiectorum est in superficie spheroidis illius speciei que diametrum habeat duplam axis.

Lemma I.

Si recta linea duas parabolas contingat in eodem puncto, sint que parabolarum diametri parallele, ipse patabolæ se mutuo contingent in illo eodem puncto.

Sit recta linea ab *quæ in puncto* b *duas parabolas* cbd, fbh. *contingat, & habeant parabolæ parallelas diametros, Dico huiusmodi parabolas se mutuo contingere. Si enim non contingunt, secent; & intelligatur alteram parabolarum esse* cbh. *alteram verò* fbd. *Agatur* bi *parallela diametris, &* cd. *parallela tangenti. Erunt ergo æquales* ci, ih; *item equales* fi, id. *Quod est impossibile, Quare &c.*

Lemma *II.*

Si due parabolæ abc, dbe. *se mutuò contingant in* b, *& habeant diametros parallelas; Dico has parabolas numquam amplius conuenire.*

Si enim possibile est conueniant in f. *& ducatur* bh *parallela dia-*

From the point *A*, the projectile is thrown. The straight lines *AD* and *AE* indicate the

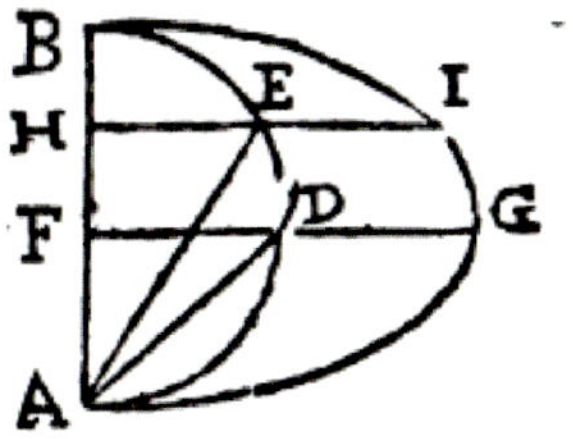

different directions of the throw. The segment *AB* represents the impetus and the minor axis of the spheroid on which the vertices of the parabolas lie.

Fig. 18 The diagram used by Torricelli to prove Proposition XXIX (Torricelli 1644, II, 2, p. 182). *Source* Courtesy University of Oklahoma, History of Science Collections

In Proposition XXX it is proved that, if you throw a projectile from a point *A*, with the same speed, but with different directions, all the parabolas will be tangent to a parabolic conoid (paraboloid of revolution). The *latus rectum* of the conoid is four times the throw-height if the projectile is thrown upwards. The proof offered by Torricelli is valid for the parabolas which are the trajectories of a projectile thrown upwards. Torricelli's analysis is completed as follows: the parabola of the horizontal throw does not touch the conoid, but approaches it indefinitely (Proposition XXXI). The parabolas which are trajectories of downward throws do not touch the paraboloid. However, if you produce the parabolic trajectory from the opposite side of the throw-point, they touch.

The book, which includes 37 propositions, is closed by various theorems and problems, but what we have described is sufficient as a guide to the reader.

The *De motu gravium* has a long Appendix on the motion of waters entitled *De motu aquarum*. The source of inspiration of this Appendix are the works of Benedetto Castelli, defined by Torricelli as "my master".[37] The main result of this important writing is the so-called "Torricelli law". It can be expressed like this: given a vessel full of water, drill a very little hole in the vessel's wall so that the vertical distance between the hole and the water's surface is h, the water will fall from the hole with a speed $v = \sqrt{2gh}$. Interestingly Torricelli assumes this law as a supposition which, at the beginning of the Appendix, he tries to prove experimentally. He writes:

> Let us suppose that the water, which exits violently from an orifice has, in that point the same *impetus* as a heavy body, or a drop of that water which falls from the liquid's surface until the orifice.[38]

[37] The most significant work by Castelli on water's movement is *Della misura dell'acque correnti* (Castelli 1628). This is the writing which inspired Torricelli.

[38] "Supponimus. Aquas violenter erumpentes in ipso eruptionis punto eunde impetum habere, quem haberet grave aliquod, sive ipsius acquae gutta una, si ex suprema eisudem aquae superficie usque ad orificium eruptionis naturaliter cecidisset." (Torricelli 1644, II, 2, p. 191; Authors' translation).

Torricelli claims that experience confirms this principle. For, connect the hole in the vessel A (suppose it is drilled at a vertical distance h from the water's surface) by means of a little horizontal tube with another vessel B, where a small hole is drilled at the same height of the hole in A. The water will ascend from the hole in B for the height h. However, Torricelli claims that an observation is paramount: the ascent of the water in B does not reach exactly the height h. He explains the reason why this happens. He continues claiming that from each hole the water falls with a parabolic trajectory. For, the initial drops which exit the tube are as projectiles. Therefore, they describe parabolas. The following drops follow the same path. Thence, the continuous water jet has a parabolic form (Torricelli 1644, II, 2, p. 193). Once again Torricelli states that the form of the jet does not seem exactly a parabola because no experiment can take place in the vacuum.

After that, he considers a vessel whose height is AB and where holes at E, D, C are drilled (Fig. 19). He solves the problem to find the amplitude (that is the horizontal distance from the hole in the vessel and the point in which the water lands) of any parabola described by the water which exits the holes.

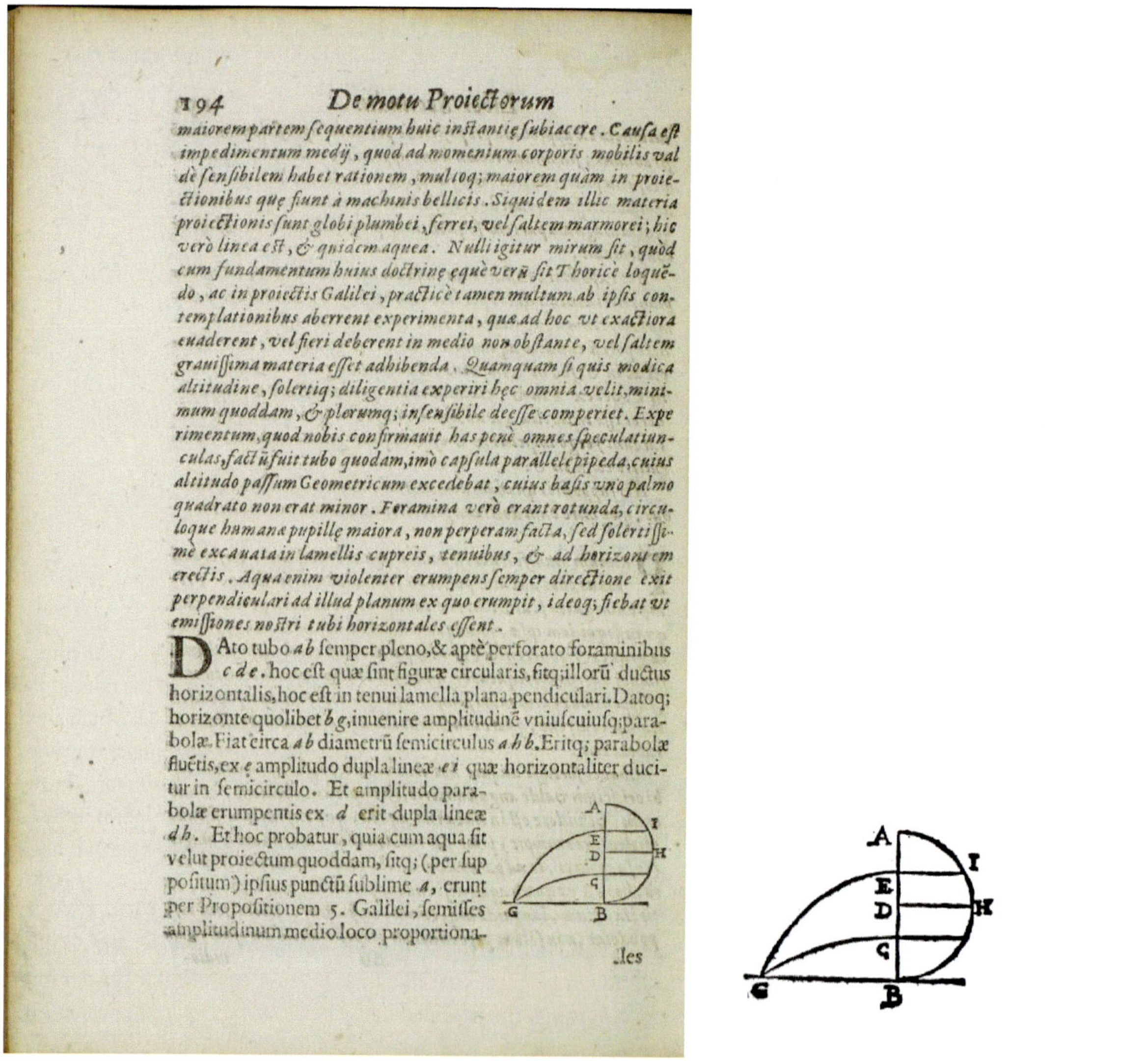

194 *De motu Proiectorum*

maiorem partem sequentium huic instantię subiacere. Causa est impedimentum medij, quod ad momentum corporis mobilis valdè sensibilem habet rationem, multoq; maiorem quàm in proiectionibus quę fiunt à machinis bellicis. Siquidem illic materia proiectionis sunt globi plumbei, ferrei, vel saltem marmorei; hic verò linea est, & quidem aquea. Nulli igitur mirum sit, quòd cum fundamentum huius doctrinę ęquè verū sit Thoricè loquēdo, ac in proiectis Galilei, practicè tamen multum ab ipsis contemplationibus aberrent experimenta, quæ ad hoc vt exactiora euaderent, vel fieri deberent in medio non obstante, vel saltem grauissima materia esset adhibenda. Quamquam si quis modica altitudine, solertiq; diligentia experiri hęc omnia velit, minimum quoddam, & plerumq; insensibile deesse comperiet. Experimentum, quod nobis confirmauit has penè omnes speculatiunculas, factū fuit tubo quodam, imò capsula parallelepipeda, cuius altitudo passum Geometricum excedebat, cuius basis vno palmo quadrato non erat minor. Foramina verò erant rotunda, circuloque humanę pupillę maiora, non perperam facta, sed solertissimè excauata in lamellis cupreis, tenuibus, & ad horizontem erectis. Aqua enim violenter erumpens semper directione exit perpendiculari ad illud planum ex quo erumpit, ideoq; fiebat vt emissiones nostri tubi horizontales essent.

DAto tubo *ab* semper pleno, & aptè perforato foraminibus *c d e*. hoc est quæ sint figuræ circularis, sitq; illorū ductus horizontalis, hoc est in tenui lamella plana pendiculari. Datoq; horizonte quolibet *b g*, inuenire amplitudinē vniuscuiusq; parabolæ. Fiat circa *ab* diametrū semicirculus *a h b*. Eritq; parabolæ fluētis, ex *e* amplitudo dupla lineæ *e i* quæ horizontaliter ducitur in semicirculo. Et amplitudo parabolæ erumpentis ex *d* erit dupla lineæ *d h*. Et hoc probatur, quia cum aqua sit velut proiectum quoddam, sitq; (per suppositum) ipsius punctū sublime *a*, erunt per Propositionem 5. Galilei, semisses amplitudinum medio loco proportiona-

-les

Fig. 19 The diagram used by Torricelli to prove the above-mentioned theorem. In the picture the amplitude of the parabola of the water exiting from E is GB (Torricelli 1644, II, 2, p. 194). *Source* Courtesy University of Oklahoma, History of Science Collections

Torricelli proves a series of beautiful theorems in which he connects his results on the motion of projectiles with the problems concerning the motion of water. For example, he is able to determine the speed with which the water exits the tube's holes, given the height of the tube.

He also poses other problems which are related to solid geometry. An interesting one, in part already considered by Castelli, concerns the forms of the solid assumed by the water exiting a hole drilled at the base of a big vessel where the water level is maintained constant (Torricelli 1644, II, 2, p. 197). Torricelli also considers other interesting situations. One, which also has a practical importance is the following: be given a tube inclined upwards in respect to the horizon with a little hole at the higher extremity. Be some water pushed in the tube from the lower extremity. It is given the point where the water jet exiting the tube lands. It is required to find the highest level reached by the jet (Torricelli 1644, II, 2, p. 198). As always in this text, Torricelli explains the best conditions to develop an experiment, which confirms the theoretical result. A further remarkable achievement concerns the quantity of water, which exits a vessel in the same time interval, but under different conditions: be given (Fig. 20) a cylindric or prismatic vessel with a little hole at the basis. Be the vessel first filled with water until the height *DA* (as the water comes out it must be replaced by more water so that the water level does not decrease) and afterwards until the height *FH*. Torricelli proves that, in the two cases, the quantity of water exiting the vessel in the time interval Δt is as the square root of *DA* to the square root of *FH*.

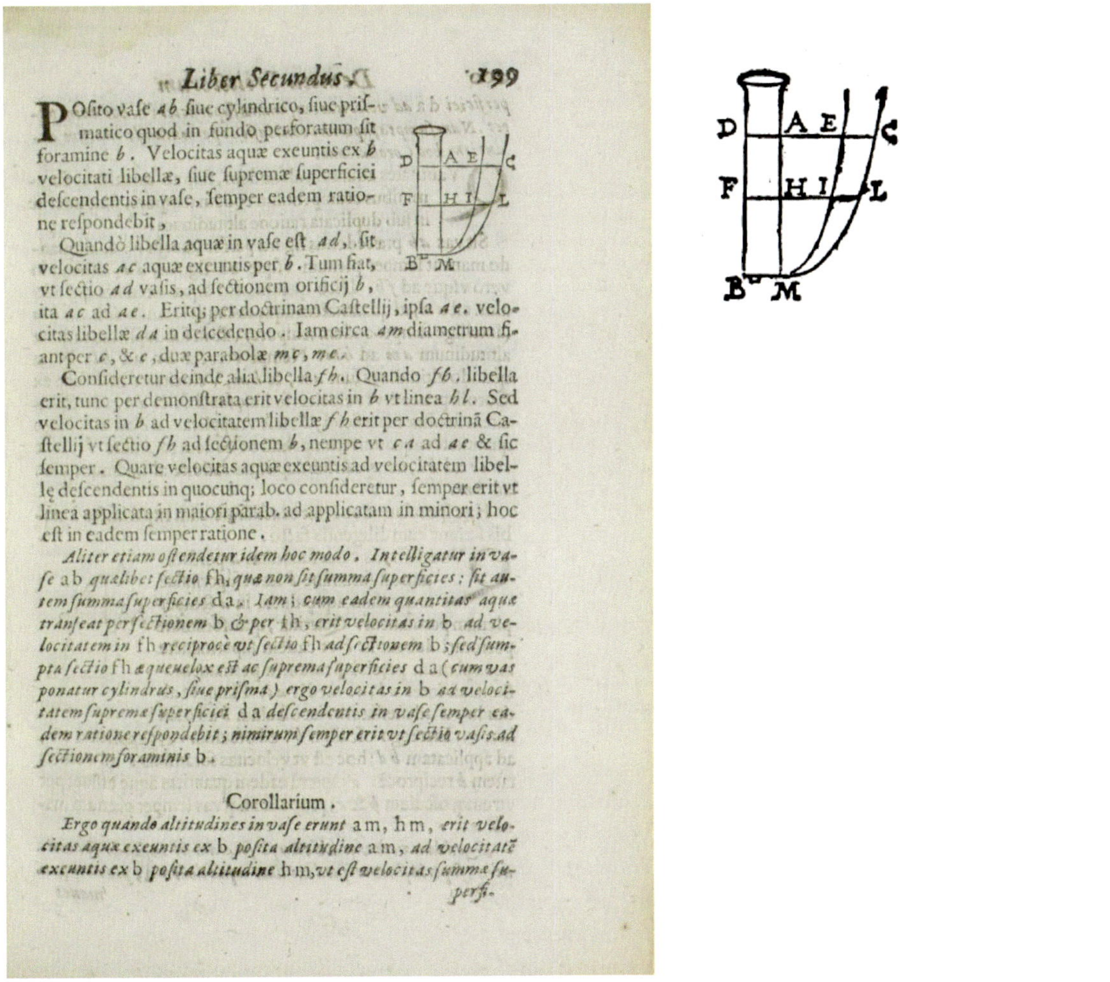

Liber Secundus. 199

POſito vaſe *ab* ſiue cylindrico, ſiue priſmatico quod in fundo perforatum ſit foramine *b*. Velocitas aquæ exeuntis ex *b* velocitati libellæ, ſiue ſupremæ ſuperficiei deſcendentis in vaſe, ſemper eadem ratione reſpondebit,

Quandò libella aquæ in vaſe eſt *ad*, ſit velocitas *ac* aquæ exeuntis per *b*. Tum fiat, vt ſectio *ad* vaſis, ad ſectionem orificij *b*, ita *ac* ad *ae*. Eritq; per doctrinam Caſtellij, ipſa *ae*. velocitas libellæ *da* in deſcendendo. Iam circa *am* diametrum fiant per *c*, & *e*, duæ parabolæ *mc*, *me*.

Conſideretur deinde alia libella *fh*. Quando *fh*. libella erit, tunc per demonſtrata erit velocitas in *b* vt linea *hl*. Sed velocitas in *b* ad velocitatem libellæ *fh* erit per doctrinā Caſtellij vt ſectio *fh* ad ſectionem *b*, nempe vt *ca* ad *ae* & ſic ſemper. Quare velocitas aquæ exeuntis ad velocitatem libellę deſcendentis in quocunq; loco conſideretur, ſemper erit vt linea applicata in maiori parab. ad applicatam in minori; hoc eſt in eadem ſemper ratione.

Aliter etiam oſtendetur idem hoc modo. Intelligatur in vaſe ab *qualibet ſectio* fh, *quæ non ſit ſumma ſuperficies; ſit autem ſumma ſuperficies* da. *Iam; cum eadem quantitas aquæ tranſeat per ſectionem* b *& per* fh, *erit velocitas in* b *ad velocitatem in* fh *reciprocè vt ſectio* fh *ad ſectionem* b; *ſed ſumpta ſectio* fh *æquevelox eſt ac ſuprema ſuperficies* da (*cum vas ponatur cylindrus, ſiue priſma*) *ergo velocitas in* b *ad velocitatem ſupremæ ſuperficiei* da *deſcendentis in vaſe ſemper eadem ratione reſpondebit; nimirum ſemper erit vt ſectio vaſis ad ſectionem foraminis* b.

Corollarium.

Ergo quando altitudines in vaſe erunt am, hm, *erit velocitas aquæ exeuntis ex* b *poſita altitudine* am, *ad velocitatē exeuntis ex* b *poſita altitudine* hm, *vt eſt velocitas ſummæ ſu-*

perfi-

Fig. 20 The diagram used by Torricelli to prove the theorem expounded in the running text (Torricelli 1644, II, 2, p. 199). *Source* Courtesy University of Oklahoma, History of Science Collections

The treatise on water is closed by two beautiful theorems. In the first one Torricelli considers a vessel whose form is a paraboloid of revolution, being the vertex the lowest point of the paraboloid. Be the vessel filled of water till a certain height. If a hole is drilled at the vertex and if, when the water exits, it is not replaced by other water exists, the motion with which the surface of the water descends is not uniform. The second proposition generalizes and offers a precise mathematical quantification of the speed with which the water's surface in a vessel of any form descends. Thus, be given a vessel of any form (Fig. 21). The speed with which the water's surface descends in *AE* is to the speed in *BD* as

$$\frac{\sqrt{FC}}{\sqrt{GC}} \cdot \frac{BD}{AE}.$$

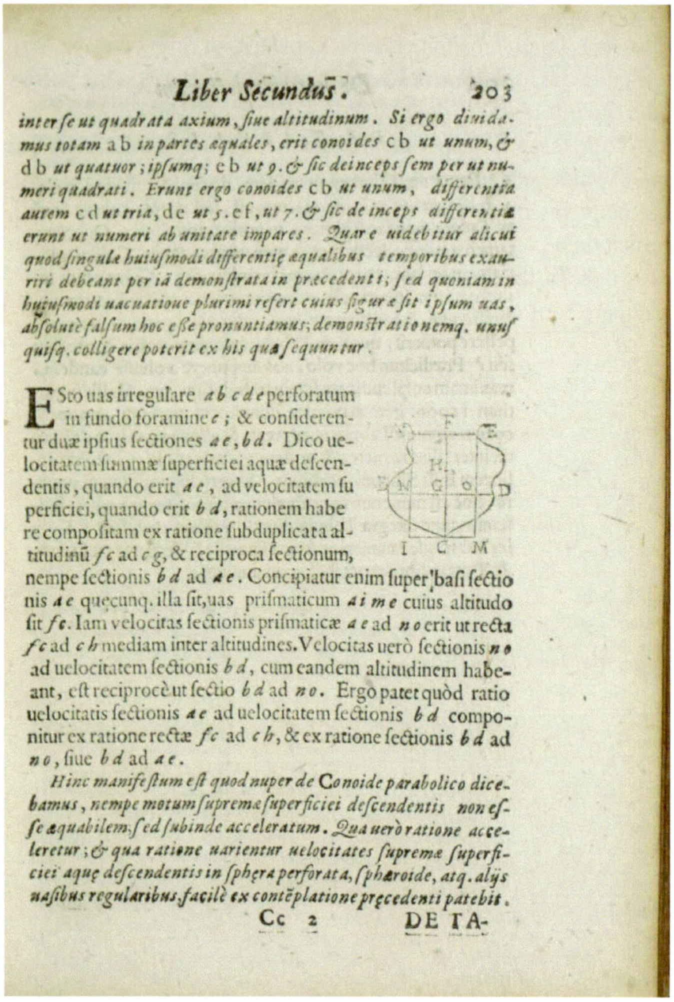

Liber Secundus. 203

inter ſe ut quadrata axium, ſiue altitudinum. Si ergo diuidamus totam ab *in partes æquales, erit conoides* cb *ut unum, &* db *ut quatuor; ipſumq;* eb *ut 9. & ſic deinceps ſemper ut numeri quadrati. Erunt ergo conoides* cb *ut unum, differentia autem* cd *ut tria,* de *ut 5.* ef*, ut 7. & ſic de inceps differentiæ erunt ut numeri ab unitate impares. Quare uidebitur alicui quod ſingulæ huiuſmodi differentię æqualibus temporibus exauriri debeant per iã demonſtrata in præcedenti; ſed quoniam in huiuſmodi uacuatione plurimi refert cuius figuræ ſit ipſum uas, abſolutè falſum hoc eſſe pronuntiamus; demonſtrationemq. unuſquiſq. colligere poterit ex his quæ ſequuntur.*

ESto uas irregulare *abcde* perforatum in fundo foramine *c*; & conſiderentur duæ ipſius ſectiones *ae*, *bd*. Dico uelocitatem ſummæ ſuperficiei aquæ deſcendentis, quando erit *ae*, ad velocitatem ſuperficiei, quando erit *bd*, rationem habere compoſitam ex ratione ſubduplicata altitudinũ *fc* ad *cg*, & reciproca ſectionum, nempe ſectionis *bd* ad *ae*. Concipiatur enim ſuper baſi ſectionis *ae* quęcunq. illa ſit, uas priſmaticum *aime* cuius altitudo ſit *fc*. Iam velocitas ſectionis priſmaticæ *ae* ad *no* erit ut recta *fc* ad *ch* mediam inter altitudines. Velocitas uerò ſectionis *no* ad uelocitatem ſectionis *bd*, cum eandem altitudinem habeant, eſt reciprocè ut ſectio *bd* ad *no*. Ergo patet quòd ratio uelocitatis ſectionis *ae* ad uelocitatem ſectionis *bd* componitur ex ratione rectæ *fc* ad *ch*, & ex ratione ſectionis *bd* ad *no*, ſiue *bd* ad *ae*.

Hinc manifeſtum eſt quod nuper de Conoide *parabolico dicebamus, nempe motum ſupremæ ſuperficiei deſcendentis non eſſe æquabilem, ſed ſubinde acceleratum. Qua uerò ratione acceleretur; & qua ratione uarientur uelocitates ſupremæ ſuperficiei aquę deſcendentis in ſphęra perforata, ſphæroide, atq. alijs uaſibus regularibus, facilè ex contẽplatione pręcedenti patebit.*

Cc 2 DE TA-

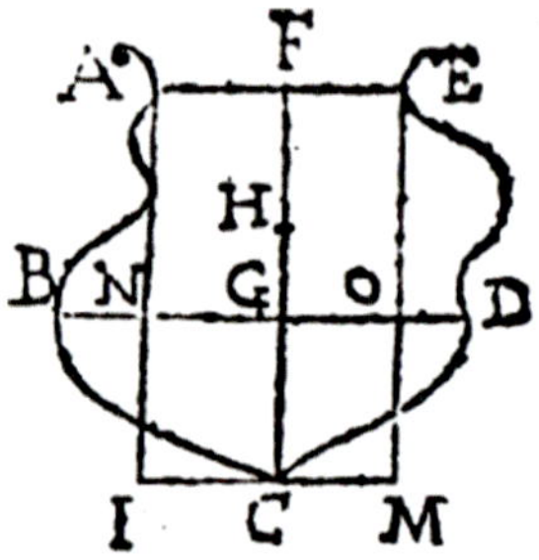

Fig. 21 The figure used by Torricelli to prove the theorem expounded in the running text (Torricelli 1644, II, 2, p. 203). *Source* Courtesy University of Oklahoma, History of Science Collections

The results obtained in *De Motu Aquarum* can be interpreted as an Appendix of Torricelli's studies on the fall of the heavy bodies because they depend on the achievements obtained in the two books *De motu gravium naturaliter descendentium.* No new principle is added. Torricelli's law, too, depends on the properties of falling bodies analysed in the two books. This notwithstanding, *De Motu Aquarum* represents a fundamental novelty in the study of the fluids' motion, where many important attainments, as we have seen, are explained. Properly speaking, Torricelli does not yet develop a theory of hydrodynamics, but his results are remarkable and represent a significant improvement in the knowledge of this discipline. From a methodological point of view, Torricelli is faithful to his geometrical approach. For, in the proofs developed in *De Motu Aquarum*, no algebraic means is used. Torricelli follows his classical style, which characterises the whole of his production. As to the epistemology of this work, the experimental aspect plays a far more important role than in the two books on the motions of the heavy bodies. Probably this depends one the fact that Torricelli's theoretical apparatus is not sufficient to prove all his results as theorems. Thus, he resorts more often than usual to the evidence of experiment. He demonstrates care in considering the conditions that might modify the experiment's results.

De Motu Gravium is closed by two further parts:

(1) a series of tables (Torricelli 1644, pp. 204–226) where Torricelli connects the elements of the heavy bodies' motion. For example (Fig. 22), he offers the relation between the initial inclination angle of a projectile's trajectory and the amplitude of the half-parabola which is the trajectory itself. The condition is that the projectile always starts with the same speed for every inclination angle and the maximal length of the half-parabola is posed equal to 10,000 (obtained when the inclination is 45 degrees).

Tabula continens Altitudines semiparabolarũ quarũ impetus idẽ sit cũ impetu præcedentis Tabulæ. Supponitur maxima altitudo part. 10000. Sunt autem numeri Tabulæ semisses sinuum versorum arcuum eleuationis duplorum.

GRAD Eleuat.	ALTIT. semipar.	GRAD Eleuat.	ALTIT. Semipar.	GRAD Eleuat.	ALTIT. Semipar.
1	3	31	2653	61	7650
2	12	32	2808	62	7796
3	27	33	2966	63	7939
4	49	34	3127	64	8078
5	76	35	3290	65	8214
6	109	36	3455	66	8346
7	149	37	3622	67	8473
8	194	38	3790	68	8597
9	245	39	3960	69	8716
10	302	40	4132	70	8830
11	364	41	4304	71	8940
12	432	42	4477	72	9045
13	506	43	4651	73	9145
14	585	44	4826	74	9240
15	670	45	5000	75	9330
16	760	46	5174	76	9415
17	855	47	5349	77	9494
18	955	48	5523	78	9568
19	1060	49	5696	79	9636
20	1170	50	5868	80	9698
21	1284	51	6040	81	9755
22	1403	52	6210	82	9806
23	1527	53	6378	83	9851
24	1654	54	6545	84	9891
25	1786	55	6710	85	9924
26	1922	56	6873	86	9951
27	2061	57	7034	87	9973
28	2204	58	7192	88	9988
29	2350	59	7347	89	9997
30	2500	60	7500	90	10000

Fig. 22 The first table constructed by Torricelli (1644, II, 2, p. 207). *Source* Courtesy University of Oklahoma, History of Science Collections

We will not deal any further with these tables and their explanations because they do not substantially alter the picture we are forming.

(2) a treatise in Italian entitled "On the set square [Della Squadra]" (Torricelli 1644, pp. 226–243). This might be considered a work of practical military geometry (sometimes also applied to land surveying) in which Torricelli shows how to construct a set square which can determine the correct elevation for a projectile to reach the target. This set square is conceived to be applied to every possible machine (Cfr. Pisano and Bussotti 2014a, b, 2015a). We also add that this chapter (*Liber*) is not marked as *Liber Secundum* as usually displayed in the previous pages; but it concludes all the previous chapters on motion:

> End of the Books on Motion.[39]

Therefore, as stated above, the reader should pay attention to the ordering of the page numbers of *Opera geometrica*. Torricelli clarifies that his treatise is addressed to the "Filosofi Geometri" (Philosopher Geometers, namely physicists and mathematicians) and not to the "Bombardieri Pratici" (practical military artillaryman, namely the military technicians, sometimes engineers, who were busy with war machines, who lacked a preliminary theoretical scientific background; Lorini 1596, 1609; Pisano and Robarts 2023; Pisano's works below; Capecchi and Pisano 2007a; Robinson 1994).

3.3 The Quadrature of the Parabola and the Cubature of the Acute Hyperbolic Solid

The final work of the *Opera geometrica* is entitled *De Dimensione Parabolae Solidique Hyperbolici Problemata duo cum Appendice De Dimensione Spatii Cycloidis et Cochleae*. With regard to the squaring of parabola, Torricelli explains that he is going to offer twenty means to solve the problem, some of them developed through the geometrical method (what we call exhaustion), some of them through the geometry of indivisibles. It is appropriate to remark that Torricelli knew (and was inspired by) Archimedes's *The quadrature of the parabola*. In this work two methods to square the parabolic segment are used: the former is mechanical because Archimedes resorted to the law of the lever and to the notion of centre of gravity. However, he did not picture the planar figures composed with an infinity of segments, which is typical of the *Method*. Obviously, Torricelli did not know the *Method*. The latter is purely geometrical.

The long *Praemium* has a remarkable methodological and epistemological value because Torricelli claims that the argument of squaring the parabola is old and the problem is completely solved. Nonetheless, a mathematical result can be important not only because it proves new theorems and solves new problems, but also because it shows the reach of a method or of a series of methods. This aspect is as important as the discovery of new theorems. Torricelli also adds a justification of his view, which is based on the utility of geometry in practical life: geometry is the most suitable discipline to sharpen ingenuity. One who is excellent in geometry will also excel in architecture, military art, arithmetical and measuring arts. These disciplines are fundamental for the health of a country both in peace and in war. Thus, he who is skilled in geometry will be also useful for the *res publica*. Since studies concerning the possibility of offering different proofs of the same theorem are among the most refined in geometry, they are among the most useful, also from a practical standpoint insofar as they help to develop geometrical skill (Torricelli 1644, III, 1, pp. 7–8). However, even if the utility of geometry for the *res publica* were not recognized, the theoretical efforts must not be despised. Otherwise, a miller should be considered more valuable than an artist. Therefore, it seems that Torricelli considered, in a sense, mathematics close to an art, but also endowed with a practical value, which the arts do not have.

The methodological importance of the *Praemium* also concerns more specific aspects of geometry: Torricelli defends mechanical methods in geometry.

[39] "Fine de' libri del moto" (Torricelli 1644, p. 243; Authors' translation).

At the beginning of his reasoning he refers to the ideas of those who deny the validity of such methods:

(a) The geometrical objects have no gravity, as the mechanical methods suppose;
(b) The threads through which the figures are attached at a lever are supposed to be equidistant, whereas they converge towards the Earth's centre.

Torricelli answers that, following this line of reasoning all the principles of geometry should be denied because, for example, it is false that circles have a centre, the sphere a surface and the cone a volume. It is clear that these properties are ascribed to geometrical figures by definition. Thus, it is also legitimate to assign a weight to such figures by definition. With regard to the second objection: it is true that, if considered on the Earth, the threads tend to converge towards the Earth's centre. However, if the lever is transported to the celestial regions, the threads will be parallel. Furthermore, for a geometer the fact that, in practice, the threads converge has no meaning because he can imagine (postulate) that they do not converge. From his point of view, this abstraction is perfectly legitimate. The picture, Torricelli claims, changes completely if we consider the real, physical figures. Then, it is necessary to take into account that the threads appended to a lever tend to converge, but if abstract, geometrical figures are considered, this is not a problem (Torricelli 1644, III, pp. 9–11).

Thus, from this *Praemium*, it is evident that Torricelli had an advanced image of geometry: it is an abstract discipline, based on axioms and definitions. If the axioms one introduces are not contradictory, it is legitimate to introduce and employ them. Torricelli surely has an abstract view of his discipline, not tied to anything empirical.

The Quadrature of the parabola opens with a chapter entitled *Suppositiones et definitines*. The first supposition is connected to Torricelli's principle. For, it is stated that, given any magnitude M hanging from a point which belongs to M itself, the gravity centre of M is not at rest, until it has arrived at the lowest point of its sphere (Torricelli 1644, III, pp. 11–12). This simply means that the suspension point, the gravity centre and the Earth's centre have to be collinear. Torricelli adds a brief argument which can, in fact, be interpreted as a proof. Some other very intuitive suppositions and definitions follow. There are six suppositions concerning the equilibrium of figures and two further which are not connected to the figures regarded as heavy bodies:

(1) given a tangent t to a parabola at a point P of the parabola, if you draw from the points $Q, R, S, \ldots$ of the tangent a series of straight lines parallel to the parabola's axis until they cut the parabola resp. in $Q', R', S', \ldots$ it holds $QQ' : RR' : SS', \ldots = \sqrt{PQ} : \sqrt{PR} : \sqrt{PS}, \ldots$. This property, as Torricelli claims, can be deduced from Proposition XX of the first book of Apollonius' *Conics*;
(2) be given a parabola, its axis and a chord AB perpendicular to the axis. From the points $C, D, E, \ldots$ of AB draw series of straight lines parallel to the axis, which cut the parabola in C', D', E', [....] It holds $CC' : AC \cdot CB = DD' : AD \cdot DB = EE' : AE \cdot EB = \ldots$.

After this preliminary section, the propositions begin. The first series of statements are five lemmas on the tangents to a parabola. The first three are rather clear, while it seems to us appropriate to report the fourth and the fifth ones because they connect the properties of the tangents to those of the figures considered as heavy bodies and because their proofs—especially in regard to the fifth Lemma are not trivial.

The fourth Lemma states that (Fig. 23) be given a parabola, a chord Ad and the tangents at A and at d. Envelop the parabola with any number of tangents drawn from points belonging to the parabolic arc Abd. The figure so generated is in equilibrium if it is suspended at the vertex B of the parabola.

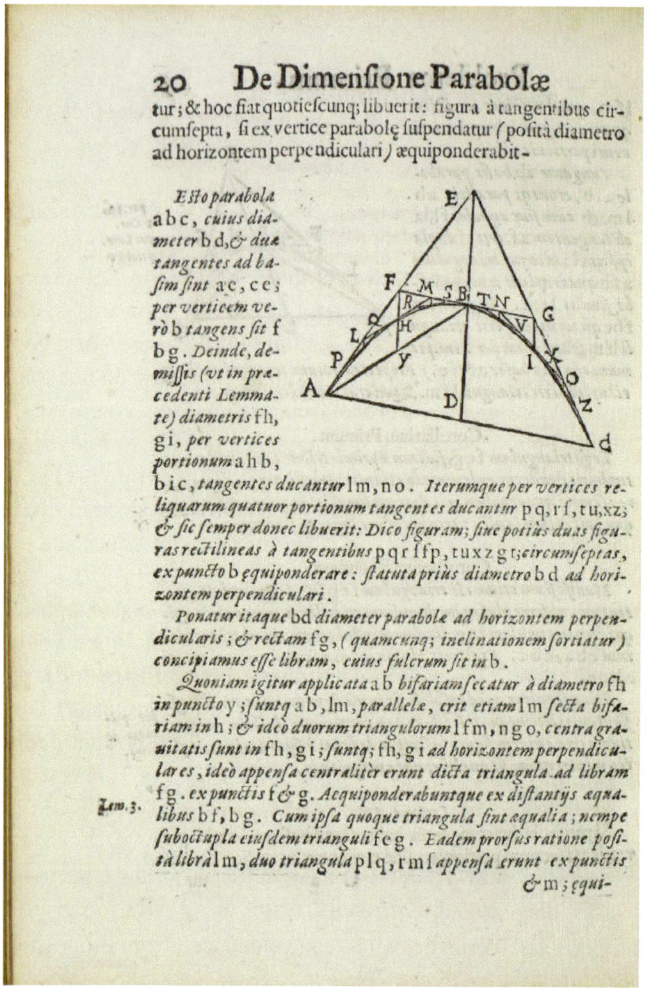

20 De Dimenſione Parabolæ

tur; & hoc fiat quotieſcunq; libuerit: figura à tangentibus circumſepta, ſi ex vertice parabolę ſuſpendatur (poſità diametro ad horizontem perpendiculari) æquiponderabit-

Eſto parabola a b c, *cuius diameter* b d, *& duæ tangentes ad baſim ſint* a c, c e; *per verticem verò* b *tangens ſit* f b g. *Deinde, demiſſis (vt in præcedenti Lemmate) diametris* f h, g i, *per vertices portionum* a h b, b i c, *tangentes ducantur* l m, n o. *Iterumque per vertices reliquarum quatuor portionum tangentes ducantur* p q, r ſ, t u, x z; *& ſic ſemper donec libuerit: Dico figuram; ſiue potiùs duas figuras rectilineas à tangentibus* p q r ſ f p, t u x z g t; *circumſeptas, ex puncto* b *ęquiponderare: ſtatuta priùs diametro* b d *ad horizontem perpendiculari.*

Ponatur itaque bd *diameter parabolæ ad horizontem perpendicularis; & rectam* f g, *(quamcunq; inclinationem ſortiatur) concipiamus eſſe libram, cuius fulcrum ſit in* b.

Quoniam igitur applicata a b *biſariam ſecatur à diametro* f h *in puncto* y; *ſuntq;* a b, lm, *parallelæ, crit etiam* l m *ſecta biſariam in* h; *& ideò duorum triangulorum* l f m, n g o, *centra grauitatis ſunt in* f h, g i; *ſuntq;* f h, g i *ad horizontem perpendiculares, ideò appenſa centraliter erunt dicta triangula ad libram* f g. *ex punctis* f *&* g. *Aequiponderabuntque ex diſtantijs æqualibus* b f, b g. *Cum ipſa quoque triangula ſint æqualia; nempe ſuboctupla eiuſdem trianguli* f e g. *Eadem prorſus ratione poſità librà* l m, *duo triangula* p l q, r m ſ *appenſa erunt ex punctis* Lem. 3.

& m; *ęqui-*

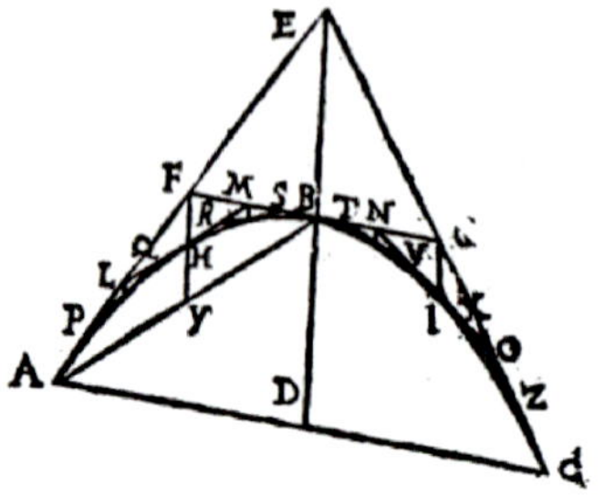

Draw the tangents in *A* and *d*. From the vertex draw the parallel to the basis *Ad* which cuts the two tangents in *F* and *G* resp. From these points draw the parallels to the axis which cut the parabola in *R* and in *I*. Draw the tangents *LM* and *NO* in these two points. From *L,M,N,O*, draw the parallels to the axis and the tangents at the points where such parallels cut the parabola and so on. The figure, or rather as Torricelli claims, the two rectilinear figures *PQRSTP* and *TUXZGT* are in equilibrium if they are suspended in *B*

Fig. 23 The figure used by Torricelli to prove Lemma IV (Torricelli 1644, III, 1, p. 20). *Source* Courtesy University of Oklahoma, History of Science Collections

The fifth Lemma considers a chord *AC* of a parabola whose vertex is *B*. Be given the tangent *c* at *C* and *b* at *B* and the parallel *a* to the axis at *A*. Be $F = c \cap b$ and $G = a \cap b$. Cut *FG* in *E*, so that $FE = 2EG$. The figure *ABCFG* is in equilibrium if suspended in *E* (Fig. 24). This Lemma is proved by a very ingenious reasoning based on the exhaustion method.

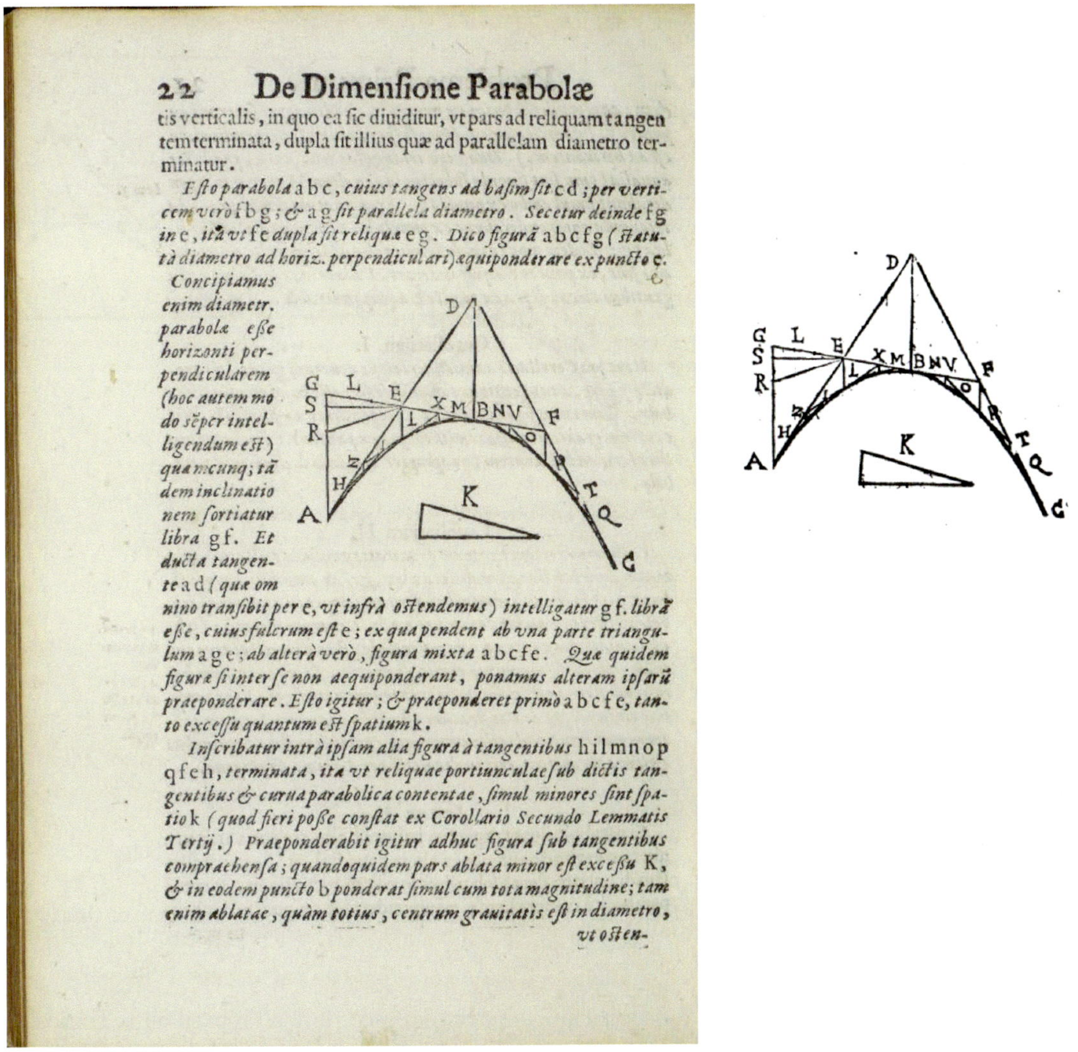

22 De Dimenſione Parabolæ

tis verticalis, in quo ea ſic diuiditur, vt pars ad reliquam tangentem terminata, dupla ſit illius quæ ad parallelam diametro terminatur.

Eſto parabola a b c, *cuius tangens ad baſim ſit* c d; *per verticem verò* f b g; *&* a g *ſit parallela diametro. Secetur deinde* f g *in* e, *ità vt* f e *dupla ſit reliquæ* e g. *Dico figurã* a b c f g (*ſtatutà diametro ad horiz. perpendiculari*) *æquiponderare ex puncto* e.

Concipiamus enim diametr. parabolæ eſſe horizonti perpendicularem (hoc autem modo sẽper intelligendum eſt) quamcunq; tãdem inclinationem ſortiatur libra g f. *Et ducta tangente* a d (*quæ omnino tranſibit per* e, *vt infrà oſtendemus*) *intelligatur* g f. *librã eſſe, cuius fulcrum eſt* e; *ex qua pendent ab vna parte triangulum* a g e; *ab alterà verò, figura mixta* a b c f e. *Quæ quidem figuræ ſi inter ſe non aequiponderant, ponamus alteram ipſarũ praeponderare. Eſto igitur; & praeponderet primò* a b c f e, *tanto exceſſu quantum eſt ſpatium* k.

Inſcribatur intrà ipſam alia figura à tangentibus h i l m n o p q f e h, *terminata, ita vt reliquae portiunculae ſub dictis tangentibus & curua parabolica contentae, ſimul minores ſint ſpatio* k (*quod fieri poſſe conſtat ex Corollario Secundo Lemmatis Tertij.*) *Praeponderabit igitur adhuc figura ſub tangentibus compraehenſa; quandoquidem pars ablata minor eſt exceſſu* K, *& in eodem puncto* b *ponderat ſimul cum tota magnitudine; tam enim ablatae, quàm totius, centrum grauitatis eſt in diametro,*

vt oſten-

Fig. 24 The diagram used by Torricelli to prove the Lemma V (Torricelli 1644, III, p. 22). *Source* Courtesy University of Oklahoma, History of Science Collections

The lemmas III and IV also have corollaries. The lemmas with their corollaries are used to offer the first proof that the parabolic segment is 4/3 the triangle having its same base and altitude (Torricelli 1644, III, Proposition I, pp. 24–25). In the proof of Proposition I, the exhaustion method is not directly used, but it is in the fundamental Lemma V.

Lemma VI is another crucial proposition proved by exhaustion. It states that, given two parabolic segments *ABC* and *DEF* as well as the tangents at the extremes of the bases, *AG*, *CG*; *DH*, *FH*, the surfaces of the trilinear figures *ABCG* and *DEFH* are as the triangles' *AGC* and *DHF* (Fig. 25).

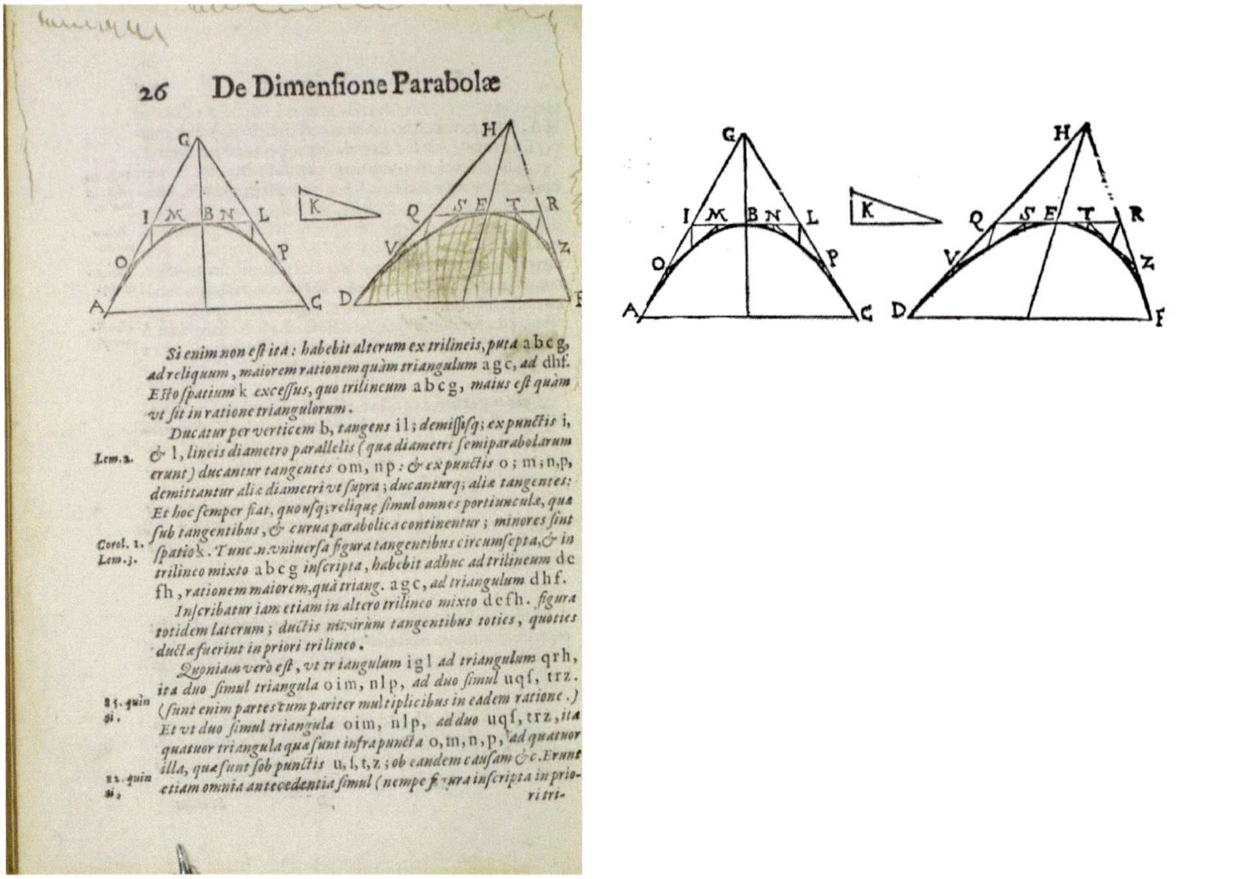

Fig. 25 The diagram used by Torricelli to prove Lemma VI (Torricelli 1644, III, p. 26). *Source* Courtesy University of Oklahoma, History of Science Collections

It must be noted that in this case the law of the lever equilibrium is not used. In the Proposition II Torricelli offers the second demonstration on the area of the parabolic segment. In this case the proof, based on Lemma VI, does not entail mechanical elements. Thus, it is in the style of the second proof given by Archimedes in *The quadrature of the parabola*. Torricelli's work continues with a series of four fascinating Lemmas, numbered VII, VIII, IX, X, the most interesting of which is the last one, proved by a long and admirable geometrical reasoning, where exhaustion plays a fundamental role. In Lemma X, Torricelli proves that:

> Lemma X. If from a parabola, half of the inscribed triangle is removed, the whole remaining mixtilinear figure is in equilibrium around the base's point of the remaining triangle which divides such a base so that its part whose extremity is on the curve is four times the part whose extremity is on the diameter.[40] (See also Fig. 26).

[40] "*Lemma X*. Si ex parabola auferatur dimidium trianguli inscripti, tota reliqua figura mixta aequiponderabit ex puncto basis reliqui trianguli, in quo sic ea dividitur, ut pars ad curvam terminata quadrupla sit illius, quae terminator a diametro." (Torricelli 1644, III, p. 30; Authors' translation; Torricelli's italic style).

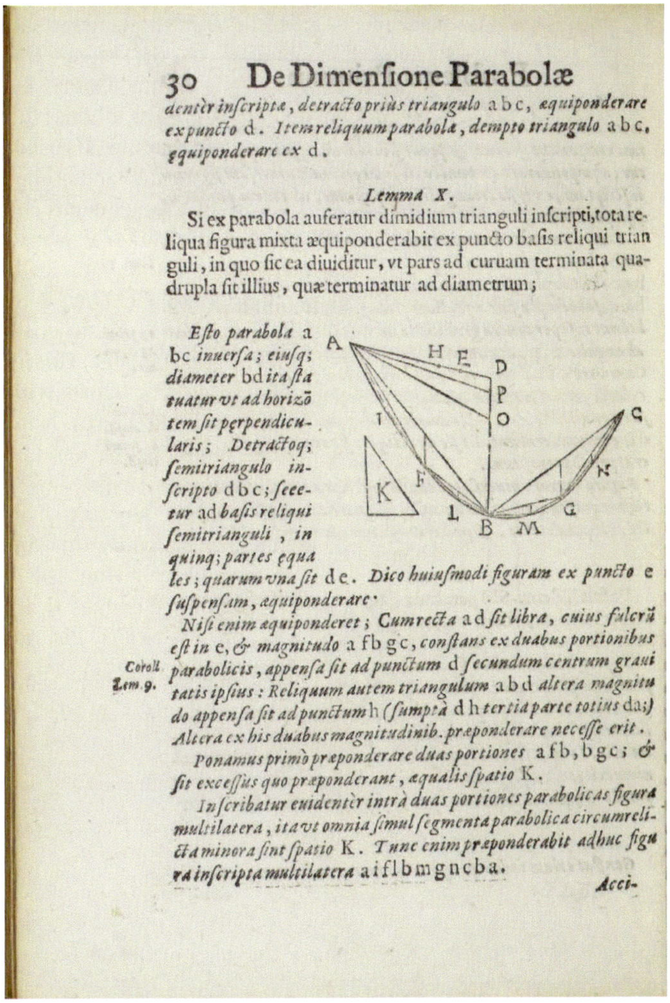

30 De Dimenſione Parabolæ

dentèr inſcriptæ, detracto priùs triangulo a b c, *æquiponderare ex puncto* d. *Item reliquum parabolæ, dempto triangulo* a b c, *æquiponderare ex* d.

Lemma X.

Si ex parabola auferatur dimidium trianguli inſcripti, tota reliqua figura mixta æquiponderabit ex puncto baſis reliqui trianguli, in quo ſic ea diuiditur, vt pars ad curuam terminata quadrupla ſit illius, quæ terminatur ad diametrum;

Eſto parabola a bc *inuerſa; eiuſq; diameter* bd *ita ſtatuatur vt ad horizõtem ſit perpendicularis; Detractoq; ſemitriangulo inſcripto* d b c; *ſeecetur* ad *baſis reliqui ſemitrianguli, in quinq; partes æquales; quarum vna ſit* d e. *Dico huiuſmodi figuram ex puncto* e *ſuſpenſam, æquiponderare.*

Niſi enim æquiponderet; Cum recta a d *ſit libra, cuius fulcrũ eſt in* e, *& magnitudo* a f b g c, *conſtans ex duabus portionibus*
Coroll. *parabolicis, appenſa ſit ad punctum* d *ſecundum centrum grauitatis ipſius: Reliquum autem triangulum* a b d *altera magnitu-*
Lem. 9. *do appenſa ſit ad punctum* h *(ſumptà* d h *tertia parte totius* da;) *Altera ex his duabus magnitudinib. præponderare neceſſe erit.*

Ponamus primò præponderare duas portiones a f b, b g c; *& ſit exceſſus quo præponderant, æqualis ſpatio* K.

Inſcribatur euidentèr intrà duas portiones parabolicas figura multilatera, ita vt omnia ſimul ſegmenta parabolica circumrelicta minora ſint ſpatio K. *Tunc enim præponderabit adhuc figura inſcripta multilatera* a i f l b m g n c b a.

Acci-

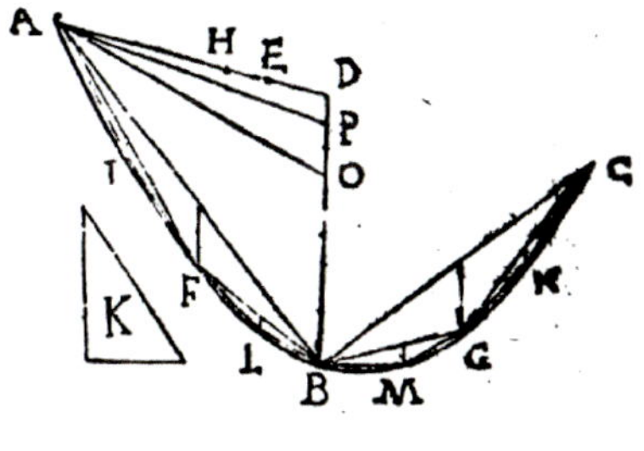

Explanation of Torricelli's statement: Be *ABC* the parabola, *BD* a diameter perpendicular to the horizon. Remove the triangle *DBC*. In the diagram this operation is already developed. Divide the remaining segment *AD* of the basis into five equal parts, so that *AE*=4*ED*.

Fig. 26 The diagram used by Torricelli to prove Lemma X (Torricelli 1644, III, p. 30). *Source* Courtesy University of Oklahoma, History of Science Collections

The further series of lemmas allowed Torricelli to reach, in Proposition III, another demonstration of the parabolic segment's area. Torricelli's work proceeds in the same manner we have explained. That is, a lemma or a series of lemmas on the equilibrium of particular figures connected to the parabolic segment is proved and, after any single series of lemmas, a new demonstration concerning the area of the parabolic segment is given. The exhaustion method is used to prove Lemma XI, which is exploited in Proposition IV, where two different manners to square the parabola on the basis of this lemma are offered. In Proposition V a further proof by exhaustion is given where there is no reference to mechanical means. Thus, this proof is in the style of Archimedes's second demonstration offered in *The quadrature of the parabola*. Lemma XII, which is used in Proposition VI, is proved by exhaustion, and it is not connected with mechanical properties. Thence, also the proof expounded in Proposition VI is not mechanical. Lemma XIII concerns the equilibrium of certain figures and is used in Proposition VII to offer a further quadrature. In this Lemma, the proof of which is very long, the exhaustion method is used. The proof explained in Proposition VIII is based on the series of lemmas XIV–XVII. They are very interesting. The first one concerns solid geometry: consider two cones *ABC* and *DEF*. Divide the sides of the triangles through the axes of the two cones in the same number of equal parts. Through the division points draw planes parallel to the bases of the cones. Each cone will be divided into a cone and a series of truncated cones. In each truncated cone inscribe a cylinder. All the cylinders in each cone will have the same height. The lemma claims that the volume of the cone *ABC* is to that of *DEF* as all the cylinders in *ABC* are to the cylinders in *DEF*. The proof is based on a direct method. Lemma XV, which is not proved by exhaustion, is an interesting proposition of infinitesimal character: be given a mixtilinear triangle included between a parabola, its tangent and a straight line parallel to the diameter. In the triangle, it is possible to inscribe a figure formed of parallelograms of equal height. This figure is less than the triangle by any given quantity (Fig. 27).

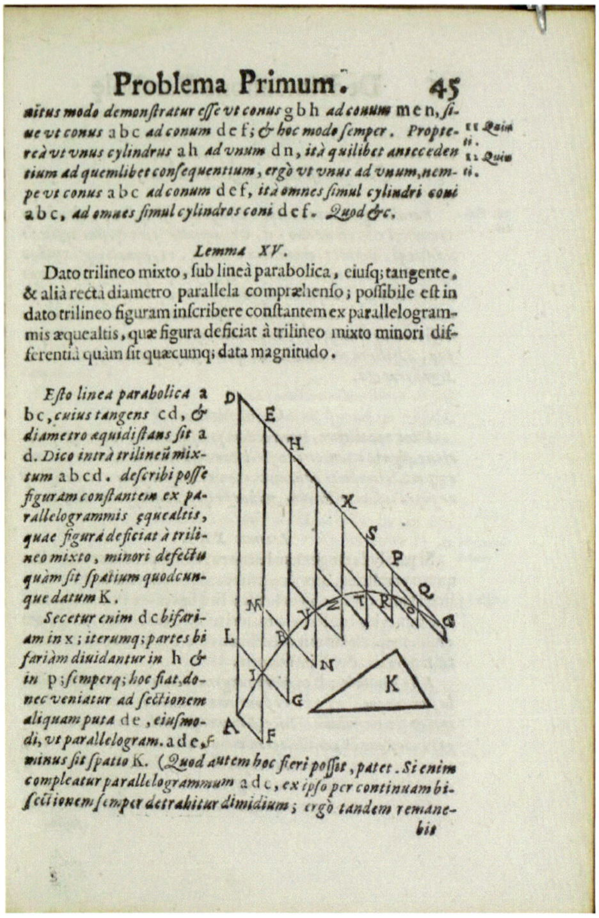

Problema Primum. 45

nitus modo demonstratur esse vt conus g b h *ad conum* m e n, *siue vt conus* a b c *ad conum* d e f; *& hoc modo semper. Propterea vt vnus cylindrus* a h *ad vnum* d n, *ità quilibet antecedentium ad quemlibet consequentium, ergò vt vnus ad vnum, nempe vt conus* a b c *ad conum* d e f, *ità omnes simul cylindri coni* a b c, *ad omnes simul cylindros coni* d e f. *Quod &c.* 11 Quin ti. 12 Quin ti.

Lemma XV.

Dato trilineo mixto, sub lineà parabolica, eiusq; tangente, & alià rectà diametro parallela compræhenso; possibile est in dato trilineo figuram inscribere constantem ex parallelogrammis æquealtis, quæ figura deficiat à trilineo mixto minori differentià quàm sit quæcumq; data magnitudo.

Esto linea parabolica a b c, *cuius tangens* c d, *& diametro æquidistans sit* a d. *Dico intrà trilineũ mixtum* a b c d. *describi posse figuram constantem ex parallelogrammis æquealtis, quae figura deficiat à trilineo mixto, minori defectu quàm sit spatium quodcunque datum* K.

Secetur enim d c *bifariam in* x; *iterumq; partes bifariàm diuidantur in* h *& in* p; *semperq; hoc fiat, donec veniatur ad sectionem aliquam puta* d e, *eiusmodi, vt parallelogram.* a d e, *f minus sit spatio* K. (*Quod autem hoc fieri possit, patet. Si enim compleatur parallelogrammum* a d c, *ex ipso per continuam bisectionem semper detrahitur dimidium; ergò tandem remanebit*

AO is the parabolic segment, *OD* is the tangent and *AD* the straight line parallel to the axis. The mixtilinear triangle is *AOD*. The figure formed by parallelograms is constructed like this: divide the tangent segment *OD* into two equal parts in *X*. Bisect each of the two parts resp. in *H* and in *P* and so on. Continue to bisect until reaching the segment *DE* which is the side of the parallelogram *DAFE* whose area is less than the given quantity *K*. Torricelli constructs a series of parallelograms *CO,OR,TR…*, whose sum, according to Torricelli's constructive criteria is equal to the parallelogram *DAFE*. This is

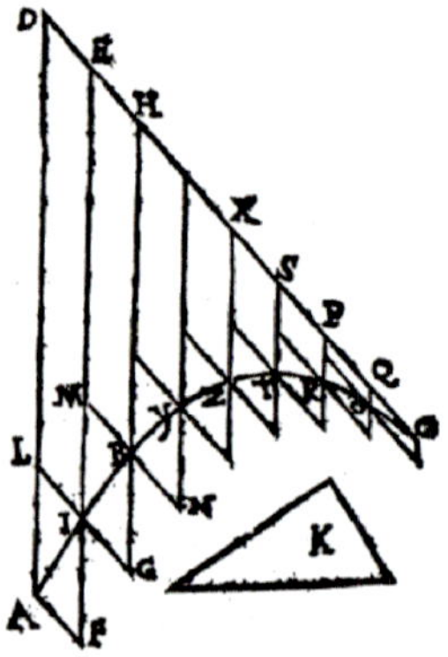

the inscribed figure with the required features. Note that this construction has several similarities with Newton's lemmas on integral calculus in the *Principia*. (Newton 1687, I, Section I, Lemmas II, II, IV, pp. 28-30 Cfr. Pisano and Bussotti 2023a).

Fig. 27 The diagram used by Torricelli to prove Lemma XV (Torricelli 1644, III, 45). *Source* Courtesy University of Oklahoma, History of Science Collections

A further Lemma in which the exhaustion method is used is the 17th one. Torricelli proves that, given the parabolic segment *ABC* of base *AC*, its tangent *CD* and two straight lines *BE* and *AD* parallel to the parabola's axis, the area of the mixtilinear triangle *ABCD* is to the area of the mixtilinear triangle *CBE* as the cube of *DC* is to the cube of *CE*. In this case, there is no reference to the mechanical method. Proposition VIII offers Torricelli's 9th proof by exhaustion of the quadrature of parabola. Proposition IX is introduced by two lemmas, numbered XVIII and XIX. The former might be called "arithmetical". It concerns the mutual relations among four series of magnitudes. The style of this Lemma is similar to that of Archimedes's arithmetical lemmas (see the first two propositions of *Conoids and spheroids*; the propositions 10 and 11 of the *Spirals*; Proposition 23 of *The quadrature of parabola*). Note that no algebraic notation is used (Cfr. Pisano and Bussotti 2015c, d). The demonstration is written in words, without any symbolization. This lemma is important because it is also used in the second part of *De mensura parabolae*, where the proofs through indivisibles are expounded. The Lemma states: be given four series of magnitudes. The first a and the third c be constant magnitudes. The second b_i and the fourth d_i variable, but under the condition that, for every pair b_j and d_j, it holds $a : b_j = c : d_j$, then it holds $\sum a : \sum b_j = \sum c : \sum d_j$.

The Lemma XIX states that, given a parabolic segment of basis *AB*, through the extreme *A* draw the tangent at the parabola and through *B* the parallel to the parabola's axis. Be *C* their intersection point. The area of the triangle *ABC* is three times the area of the parabolic segment. The proof is given through a reasoning by exhaustion without resorting to considerations on the lever. Thence, the purists too should judge this proof geometrical. The demonstration is long and needs many subtleties, of which Torricelli was capable. As a matter of fact, from this Lemma the proof concerning the area of the parabolic segment follows as a clear corollary, which is shown in Proposition IX. The first part of the book on the quadrature of the parabola is closed by Proposition X, which is a piece of geometrical virtuosity. The arithmetical Lemma XVIII is used as well as Lemma XV. Since no mechanical element is employed, this proof can, to all effects and purposes, be considered geometrical. In what follows, we present a table where we summarize which, among the eleven exhaustion quadratures proposed by Torricelli, are purely geometrical and which resort to mechanical instruments. In the first column the number of propositions is reported, in the second one the lemmas used to prove it, in the third one the mechanical or geometrical character of the proof. We also indicate where mechanical reasoning occurs. It is, nonetheless, appropriate to remember that, according to Torricelli, there is no substantial difference between a purely geometrical proof and a mechanical one. However, according to other mathematicians, the first of whom was Archimedes himself, such a difference exists. In the following, we present in Table 3 the Propositions upon Lemmas and kinds of Proofs.

Table 3 Propositions, Lemmas and Kinds of Proofs (Torricelli 1644, III)

Number of Proposition	Lemmas used	Kind of proof
Proposition I	Lemmas I–V	Mechanical Lever used in Lemma V
Proposition II	Lemma VI	Geometrical
Proposition III	Lemmas VII–X	Mechanical Lever used in Lemma X
Proposition IV Two proofs are completed	Lemma XI	Mechanical Lever is used in Lemma XI and in a further unnumbered lemma used in the second proof, which is mechanical, as well
Proposition V		Geometrical
Proposition VI	Lemma XII	Geometrical
Proposition VII	Lemma XIII	Mechanical Lever is used in Lemma XIII
Proposition VIII	Lemmas XIV–XVII	Geometrical
Proposition IX	Lemmas XVIII and XIX	Geometrical
Proposition X	Lemmas XV and XVIII	Geometrical

Before analysing the section in which Torricelli used the indivisibles method to square the parabola, an observation seems appropriate to us on the links between the exhaustion method and the infinitesimal, and, in particular, integral calculus. It is clear that exhaustion implies that the concept of limit is known. In some propositions this is particularly conspicuous. Torricelli's Lemma XV is an example, given its similarity, as far as the geometrical construction is concerned, to Newton's lemmas on the integral calculus expounded in the *Principia*. Indeed, Torricelli is looking for two quantities whose difference becomes arbitrarily small when another quantity, the side of a parallelogram, becomes arbitrarily small. The fact that the concept of function did not yet exist is not a problem in this respect. The most skilled geometers knew the notion of limit intuitively and knew how to use it without running into errors. However, there is a substantial difference between exhaustion method and integral calculus: the former is associated neither with a calculus nor with an algebraic symbolization of the geometrical quantities. This is a crucial aspect. There is no standard way to develop a reasoning by exhaustion. An invention is necessary in every single demonstration, though the general scheme of the proof is invariable. The need to use the *reductio ad absurdum* is the counterpart of the direct calculation. In integral calculus, you have a series of standard procedures, which allows a direct calculation of the required areas and volumes. Therefore, our position is in the middle between that of the scholars who see profound similarities between exhaustion method and integral calculus and those who consider them to be two completely different mathematical procedures.[41]

The brief introductory remarks to the second part of Torricelli's work on the square of the parabola are significant: first, Torricelli claims that the quadrature of the parabolic segment can be deduced from some of the most famous theorems proved by the ancients:

[41] The relation between exhaustion method and integral calculus has been debated for a long time in the literature on Archimedes. This Introduction is not the appropriate place to debate such a literature. Broadly speaking, it is possible to state that there are two tendencies. The first one underlines the similarities between the two methods, whereas the other highlights their irreducible differences. Without any claim to be exhaustive, we mention: (a) Bachmakova 1963 (The author offers a modernizing treatment of several of Archimedes's results), (b) Archimedes 1897; Heath 2002 (see Introduction), (c) Merzbach-Boyer 2011 (see Chap. 6), (d) Rufini 1926 (it is dedicated to Archimedes's mechanical method, but there are many considerations on exhaustion), (e) Zeuthen 1896, pp. 177–187, (f) Saito 2013. On our side, as clarified in the running text, we incline toward an historical intermediate position. More, see recently: Bussotti 2019; Pisano and Bussotti 2023a).

(1) The cylinder's volume is three times the inscribed cone's volume;
(2) The cylinder's volume is 3/2 the volume of the inscribed sphere;
(3) The cone's gravity centre G is on the axis so that the distance between G and the cone's vertex is three times the distance between G and the cone's base;
(4) After the first rotation the spiral's area is 1/3 of the first circle's area. Reciprocally, assuming the square of parabola, the four theorems follow. How are these deductions possible?

Torricelli's answer is that they are possible thanks to indivisibles. Torricelli's opinion on Cavalieri's geometry of indivisibles is clear: such geometry is admirable. It is not a new discovery: probably it is the heuristic through which the ancients arrived at their discoveries. Torricelli's words are so clear that no interpretation in necessary:

> I would not dare geometry of indivisibles to be a new invention. Rather, I incline to think that the ancient geometers resorted to this method to discover the most difficult theorems. After that, in the proof they preferred a different method both to hide the secrets of their art and to avoid offering to envious detractors any occasion for criticism. It is anyway certain that this geometry permits an admirable energy saving in the discovery of new truths and allows us to establish new and almost inscrutable theorems. This cannot be done through the ancients' methods.[42]

These are really prophetic words if we think of Heiberg's discovery of Archimedes's *Method*! But they are not only prophetic, they imply a legitimation of Cavalieri's method against the detractors, such as Guldin. Torricelli found no logical problem in indivisibles. They work, they allow us to solve difficult mathematical problems; thus, they are legitimate. We know that Cavalieri himself was, instead, very concerned about the logical validity of his indivisibles; he tried to justify their use, and also attempted to identify a correct and an incorrect conceptualization by distinguishing between *retto transito* and *obliquo transito* (Cfr. Andersen 1985; Giusti 1980). Torricelli's attitude was more pragmatic. He explained the principles in a series of lemmas (Torricelli 1644, III, p. 57) on Cavalieri's geometry, but, with the term *principles* he meant the *basic mathematical statements* rather than the *logical bases* (Pisano 2010a; Capecchi and Pisano 2007a, b).[43]

This section of Torricelli's work opens with the first demonstration of the parabola's quadrature given by the technique of indivisibles (Proposition XI). Let us see Torricelli's ways of reasoning (Fig. 28).

Be ABC the parabolic segment, CD the tangent at an extreme of its base and AD the parallel to the diameter drawn from the other extreme. Complete the parallelogram AE. Describe a circle around AD, regarded as the base both of the cone whose vertex is C and of the cylinder $ADEC$. Draw FG parallel to AD. It is an indivisible of the parallelogram $ADEC$ and IB is an indivisible of the mixtilinear triangle $ABDC$.

Through AG, draw a plane parallel to the circle AD. It holds:

$$FG : IB = DA : IB = DC^2 : IC^2 = DA^2 : IG^2$$

This proportion holds for any other straight line parallel to DA. For example if we draw $F'G'$ and indicate the points where it cuts resp. the tangent and the parabola by I' and B', it is:

$$F'G' : I'B' = DA : I'B' = DC^2 : I'C^2 = DA^2 : I'G'^2$$

Because of Lemma XVIII, all the first magnitudes, namely the parallelogram regarded as composed of the indivisibles DA, FG, $F'G'$,... are to all the second magnitudes, that is DA, IB, $I'B'$ regarded as composing the mixtilinear triangle $ABCD$, as all the third magnitudes, namely all the circles of diameter DA, which are the indivisibles of the cylinder, are to all the circles of variable diameter $I^{(i)}G^{(i)}$ which are the indivisibles of the cone. Therefore, the parallelogram is three times the mixtilinear triangle. Hence, the parabolic segment is 1/3 of the triangle ADC, which is 4 times the triangle whose basis is AC and whose height is that of the segment. Thus, the theorem is proved.

---

[42] "Quoad autem haec Indivisibilium Geometria noivum penitus inventum sit, equidem non ausim affirmare. Crediderim potius veteres Geometras hac methodo usos in inventione Theorematum difficillimorum, quamquam in demonstrationibus aliam viam magis probaverint, sive ad occultandum artis arcanum, sive ne ulla invidis detractoribus proferretur occasio contradicendi. Quicquid est, certum est hanc Geometriam mirum esse pro inventione compendium, et innumera quasi imperscrutabilia Theoremata, brevibus, directis, affermativisque; demonstrationibus confirmare; quod per doctrinam antiquorum fieri minime potest" (Torricelli 1644, III, p. 56; Authors' translation).

[43] Cfr. This does not mean that Torricelli was not interested in establishing the conditions under which the technique of indivisibles is correctly applicable. As we will see, he also offers a series of examples in which the indivisibles are applied incorrectly.

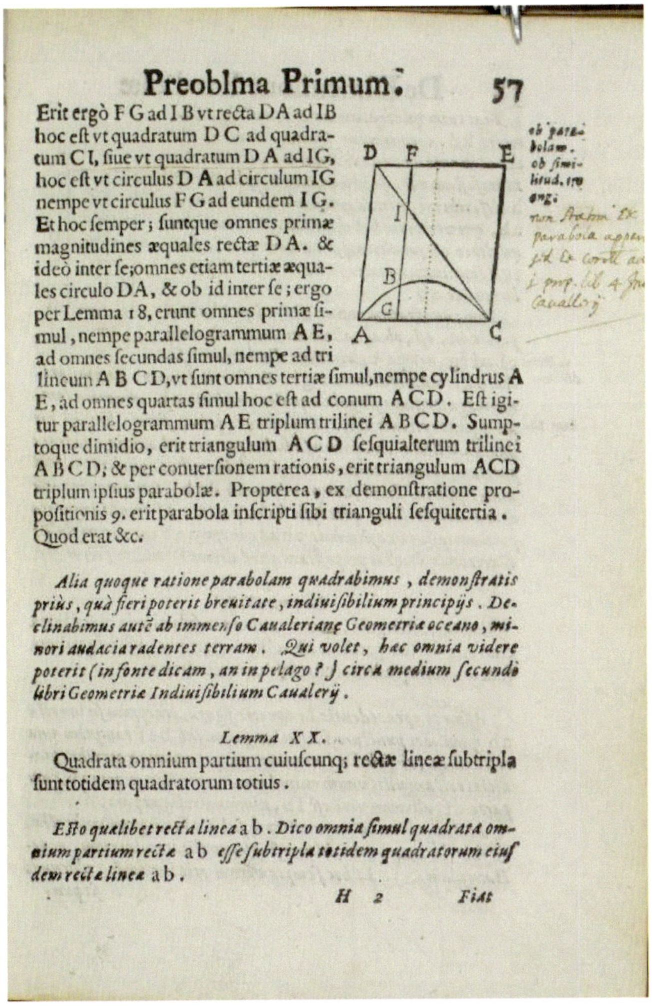

Preoblma Primum. 57

Erit ergò F G ad I B vt recta D A ad I B hoc eſt vt quadratum D C ad quadratum C I, ſiue vt quadratum D A ad I G, hoc eſt vt circulus D A ad circulum I G nempe vt circulus F G ad eundem I G. Et hoc ſemper; ſuntque omnes primæ magnitudines æquales rectæ D A. & ideò inter ſe; omnes etiam tertiæ æquales circulo D A, & ob id inter ſe; ergo per Lemma 18, erunt omnes primæ ſimul, nempe parallelogrammum A E, ad omnes ſecundas ſimul, nempe ad trilineum A B C D, vt ſunt omnes tertiæ ſimul, nempe cylindrus A E, ad omnes quartas ſimul hoc eſt ad conum A C D. Eſt igitur parallelogrammum A E triplum trilinei A B C D. Sumptoque dimidio, erit triangulum A C D ſeſquialterum trilinei A B C D; & per conuerſionem rationis, erit triangulum A C D triplum ipſius parabolæ. Propterea, ex demonſtratione propoſitionis 9. erit parabola inſcripti ſibi trianguli ſeſquitertia. Quod erat &c.

Alia quoque ratione parabolam quadrabimus, demonſtratis priùs, quà fieri poterit breuitate, indiuiſibilium principijs. Declinabimus autē ab immenſo Caualerianę Geometriæ oceano, minori audacia radentes terram. Qui volet, hæc omnia videre poterit (in fonte dicam, an in pelago?) circa medium ſecundi libri Geometriæ Indiuiſibilium Caualerij.

Lemma XX.

Quadrata omnium partium cuiuſcunq; rectæ lineæ ſubtripla ſunt totidem quadratorum totius.

Eſto qualibet recta linea a b. *Dico omnia ſimul quadrata omnium partium rectæ* a b *eſſe ſubtripla totidem quadratorum eiuſdem rectæ lineæ* a b.

H 2 Fiat

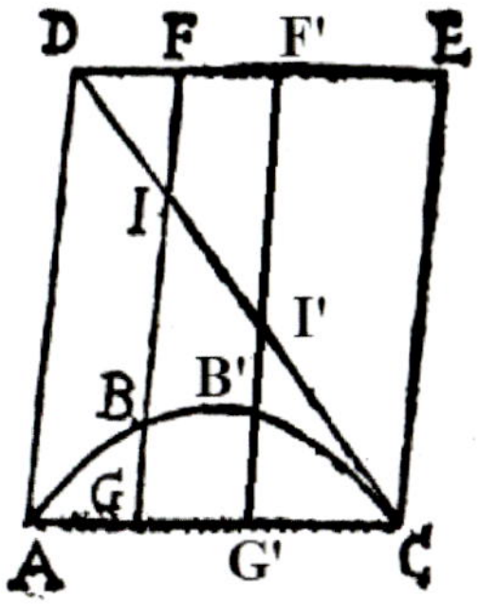

Fig. 28 Torricelli's diagram in which the segment *F'I'B'G'* has been added to make Torricelli's proof clearer (Torricelli 1644, III, p. 57). *Source* Courtesy University of Oklahoma, History of Science Collections

This is the first way in which Torricelli squared the parabola relying upon indivisibles. No lemma was added for this first demonstration. However, to obtain the other proofs, several further lemmas were conceived and proved by Torricelli.

The indivisibles method is applied both in the proofs of the lemmas and of the propositions. The proof of Proposition XII is based on two lemmas, the 20th and the 21st. The former is a very audacious statement because Torricelli considers a segment *AB* and the series of its subsegments, for example, those whose extreme is *A*. Let us denote them by *AX*. Torricelli believes that the concept "all the subsegments *AX*" is perfectly meaningful. Now to each subsegment associate the whole segment *AB*. Torricelli proved that the sum of all the squares of the subsegments *AX* is 1/3 the sum of the squares of all the equinumerous segments *AB*. He, thus, believed in the existence of the set of all the infinite subsegments *AX*, which is actually infinite. The idea that a continuum of geometrical dimension n is composed of an infinite set of continua of geometrical dimension n-1 is typical of the indivisibles method. However, here Torricelli is imagining a different situation because he is not regarding the segment *AB* as composed of its infinite points. Rather he is considering such a segment as the support of an actual infinity of subsegments whose variable length is *AX*. They do not compose the segment *AB*. This is a very interesting aspect of Torricelli's view.

Lemma XXI is another statement concerning indivisibles. Armed with these two lemmas and with Lemma XVIII, which is used in many proofs included in this second part of his quadrature of the parabola, in Proposition XII Torricelli offers the second demonstration based on indivisibles.

In Proposition XIII, only the XVIIIth Lemma is used. The proof is based on Archimedes's theorem that the cylinder's volume is 4/3 the volume of the inscribed sphere.

The following proposition in which Torricelli dealt with the quadrature of the parabola is interesting because it is based both on the indivisibles method and on the principle of the lever. Therefore, it is a mechanical quadrature through indivisibles. For, Lemma XXII proves that, given a lever *L*, suspend a series of magnitudes at *n* different points of the lever (the magnitudes are, in general, different). Suspend a second series of magnitudes at the same points. Be the magnitudes of the two series proportional. A single point of the lever will be the equilibrium centre of both the series of magnitudes. The second Lemma on which the proof of Proposition XIV relies is a geometrical statement based on the lever's equilibrium. This time the indivisibles are also used and the similarity between the proof of Torricelli's Lemma XXIII and those given by Archimedes in his *Mechanical Method* is impressive. Let us see: given a parabolic segment of base *AC*, draw the tangent in *C* and a parallel to the axis in *A*. Be *D* their intersection point. The mixtilinear triangle *ABCD* is in equilibrium if it is suspended in a point *H* of the tangent such that CH=3*HD* (Fig. 29).

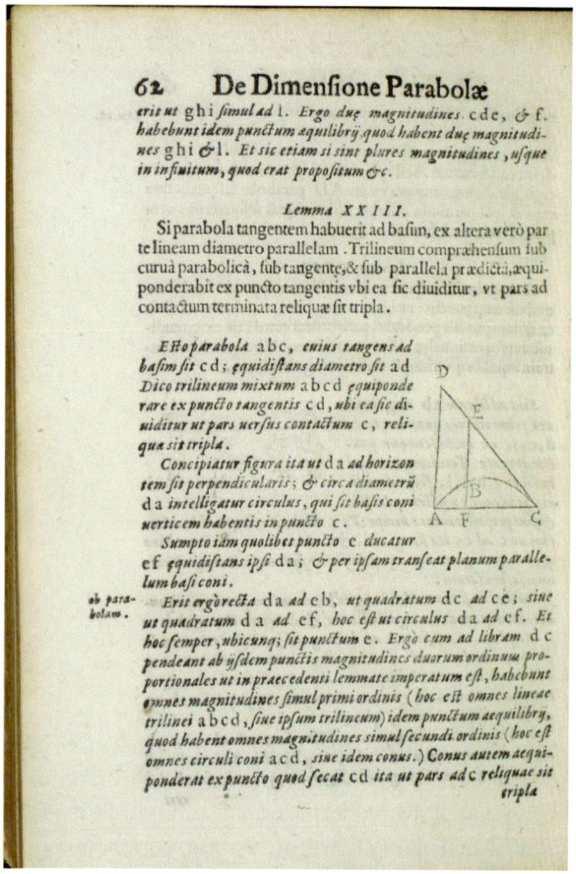
62 De Dimensione Parabolæ

erit ut g h i *simul ad* l. *Ergo duę magnitudines* c d e, *&* f. *habebunt idem punctum æquilibrij, quod habent duę magnitudines* g h i *&* l. *Et sic etiam si sint plures magnitudines, usque in infinitum, quod erat propositum &c.*

Lemma XXIII.

Si parabola tangentem habuerit ad basim, ex altera verò parte lineam diametro parallelam. Trilineum compræhensum sub curuà parabolicà, sub tangentę, & sub parallela prædicta, æquiponderabit ex puncto tangentis vbi ea sic diuiditur, vt pars ad contactum terminata reliquæ sit tripla.

Esto parabola a b c, *cuius tangens ad basim sit* c d; *ęquidistans diametro sit* a d *Dico trilineum mixtum* a b c d *ęquiponderare ex puncto tangentis* c d, *ubi ea sic diuiditur ut pars uersus contactum* c, *reliqua sit tripla.*

Concipiatur figura ita ut d a *ad horizontem sit perpendicularis; & circa diametrũ* d a *intelligatur circulus, qui sit basis coni uerticem habentis in puncto* c.

Sumpto iam quolibet puncto e *ducatur* e f *ęquidistans ipsi* d a; *& per ipsam transeat planum parallelum basi coni.*

ob parabolam. *Erit ergò recta* d a *ad* e b, *ut quadratum* d c *ad* c e; *siue ut quadratum* d a *ad* e f, *hoc est ut circulus* d a *ad* e f. *Et hoc semper, ubicunq; sit punctum* e. *Ergò cum ad libram* d c *pendeant ab ijsdem punctis magnitudines duorum ordinum proportionales ut in praecedenti lemmate imperatum est, habebunt omnes magnitudines simul primi ordinis (hoc est omnes lineae trilinei* a b c d, *siue ipsum trilineum) idem punctum aequilibrij, quod habent omnes magnitudines simul secundi ordinis (hoc est omnes circuli coni* a c d, *siue idem conus.) Conus autem aequiponderat ex puncto quod secat* c d *ita ut pars ad* c *reliquae sit tripla*

Proof of Lemma XXIII

Around *DA* as diameter draw a circle regarded as the base of a cone whose vertex is *C*. Take any point *E* on the tangent. Draw *EF* parallel to *DA* and through *EF* a plane parallel to that of the circle *DA*. It is

$$DA\!:\!EB = DC^2\!:\!CE^2.$$

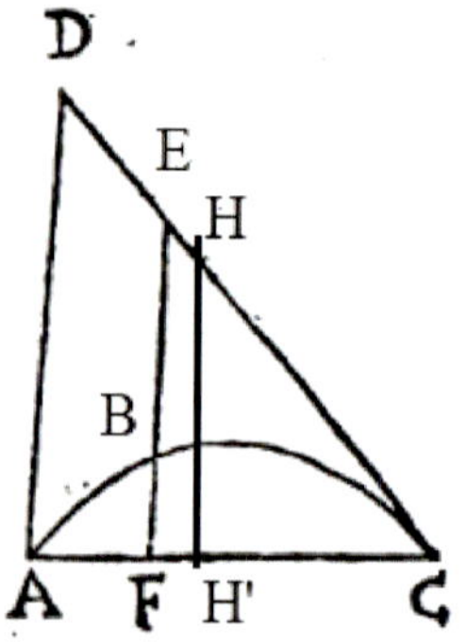

Thus, according to Lemma XXII, all the magnitudes of the first series, that is the segments composing the mixtilinear triangle *ABCD*, have the same equilibrium point as the magnitudes of the second series, which compose the cone. Since the cone is in equilibrium if it is suspended at the point *H* of *DC* such that *CH*=3*HD*, the same holds for the trilinear.

Fig. 29 Adapted Diagram for Proofing Lemma XXIII. For convenience of the reader, we have added the straight line *HH*' (see right side of the table) which does not exist in Torricelli's diagram (Torricelli 1644, III, p. 62). *Source* Courtesy University of Oklahoma, History of Science Collections

Through these two lemmas, in Proposition XIV Torricelli is able to square the parabola in two different manners, both based on the indivisibles method enriched by that of the lever's equilibrium.

The following Proposition XV reaches the quadrature of the parabola in two different ways. This time the technique associated by Torricelli with the indivisibles is no longer that of the lever's equilibrium, but that of the infinite series of proportional quantities. The lemmas on which the two proofs are based are numbered XXIV–XXVII. The most interesting among them is the XXVIth one because Torricelli asks to find a segment whose length is equal to that of a segment composed of an infinite number of decreasing segments. The Lemma states: given infinite straight lines (segments) in continued proportion, making up a decreasing progression, find the straight line equal to the sum of all the infinite given straight lines. In Torricelli's time, it was well known that a sum of infinite decreasing addends can produce a finite value.

For example, this was known for the geometrical series whose general term is less than one.[44] However, here it seems that Torricelli contemplates the possibility of developing a calculation with infinities, which is remarkable. Obviously, this calculation is subject to the usual limitation which connotes Torricelli's mathematics: no algebraic symbolism is used.

Lemma XXVIII is noteworthy from a conceptual point of view. It states that given two infinite series of segments $AB, CD, EF, \ldots$ and $BG, DH, FI, \ldots,$ each of them in continued geometrical proportion such that $AB : BG = CD : DH = EF : FI : \ldots,$ the sum of all the segments belonging to the first series is to the sum of all those belonging to the second series as $AB : BG$. Torricelli claims that this lemma is included in Euclid's *Elements* (Euclid 2007, V, 12).[45] However, he offered a different demonstration because almost all the mathematicians think that Euclid was referring to a finite number of magnitudes, while he devised a proof completely applicable to the case of an infinite number, too. This is a further indication that he considers it legitimate to develop a calculus with infinite quantities. This is not exactly equivalent to claiming that Torricelli is on the road which led to infinitesimal calculus because:

(1) He used no algebraic symbolization;
(2) The relation between actual and potential infinite does not seem completely clear. For, while claiming that he considers the sum of an infinite number of magnitudes, Torricelli seems to refer to something as the sum of all the magnitude of the series, whose number is actually infinite, rather than to the sum of an arbitrary number of such magnitudes (a potential conception of infinity). The infinitesimal calculus refers to such sums in the second meaning. Torricelli's ideas in this Lemma seems almost closer to the calculus of infinite series envisioned by Bolzano in his *Paradoxien des Unendliches* (Bolzano 1851, §§ 29–33, pp. 43–55).

In Proposition XVI Torricelli develops an interesting technique: he conceives an infinite series of triangles, included in the parabola, whose areas decrease in continued proportion and an infinite series of triangles out of the parabola whose areas also decrease in an analogous continued proportion (see Fig. 30). The sum of the triangles of the first series generates the parabola, the sum of the triangles of the second series generates the mixtilinear triangle whose two sides are the two concurrent tangents at the extremes of the parabolic segments' base and whose third side is the parabolic arc. Since the first triangle of the first series is double the first triangle of the second series, through a relatively clear reasoning, it is possible to obtain the parabola's quadrature.

[44] Cfr. *Tractatus de latitudinibus formarum* (1361; Oresme 1486) by Nicholas Oresme. In the eight question of the third book, entitled *De mensura et intensione in infinitum quarundam difformitatum* ("The measure and intension to infinity of some differences") he proves that the altitude of a finite surface can be arbitrarily increased, so that its form is changed, but not its surface. Oresme proved through a graphic method that the geometrical series of general term ½ converges.

[45] "If there are any number of magnitudes whatsoever (which are) proportional then as one of the leading (magnitudes is) to one of the following, so will all of the leading (magnitudes) be to all of the following." (Euclid 2007, p. 142).

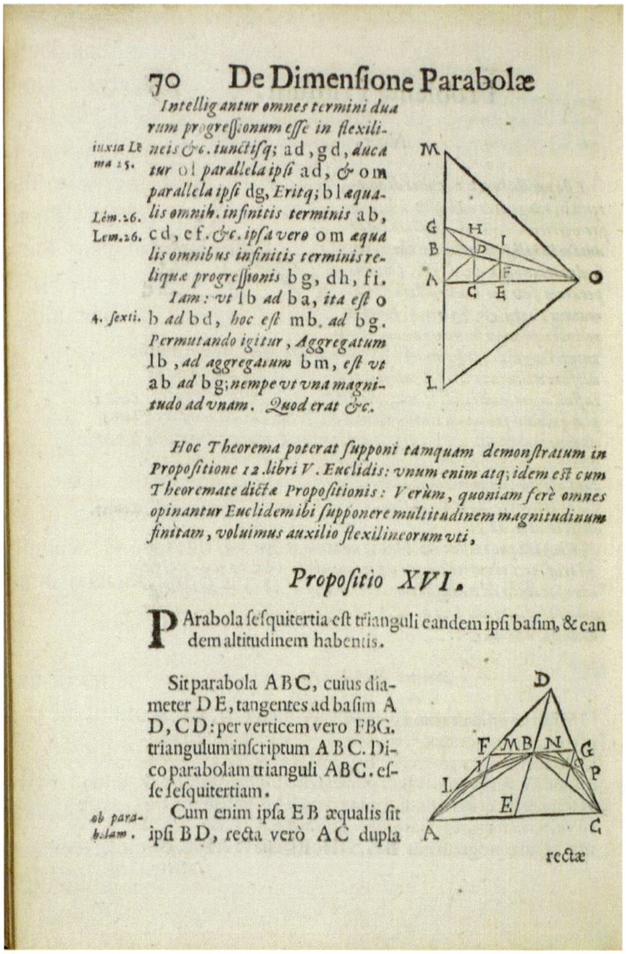

70 De Dimensione Parabolæ

Intelligantur omnes termini duarum progressionum esse in flexilineis &c. iunctisq; ad, gd, ducatur ol parallela ipsi ad, & om parallela ipsi dg, Eritq; bl æqualis omnib. infinitis terminis ab, cd, cf. &c. ipsa vero om æqualis omnibus infinitis terminis reliquæ progressionis bg, dh, fi.

iuxta Lemma 25.

Lem. 26.

Lem. 26.

Iam: vt lb ad ba, ita est ob ad bd, hoc est mb. ad bg. Permutando igitur, Aggregatum lb, ad aggregatum bm, est vt ab ad bg; nempe vt vna magnitudo ad vnam. Quod erat &c.

4. sexti.

Hoc Theorema poterat supponi tamquam demonstratum in Propositione 12. libri V. Euclidis: vnum enim atq; idem est cum Theoremate dictæ Propositionis: Verùm, quoniam ferè omnes opinantur Euclidem ibi supponere multitudinem magnitudinum finitam, voluimus auxilio flexilineorum vti,

Propositio XVI.

Parabola sesquitertia est trianguli eandem ipsi basim, & eandem altitudinem habentis.

Sit parabola ABC, cuius diameter DE, tangentes ad basim AD, CD: per verticem vero FBG. triangulum inscriptum ABC. Dico parabolam trianguli ABC. esse sesquitertiam.

Cum enim ipsa EB æqualis sit ipsi BD, recta verò AC dupla rectæ

ob parabolam.

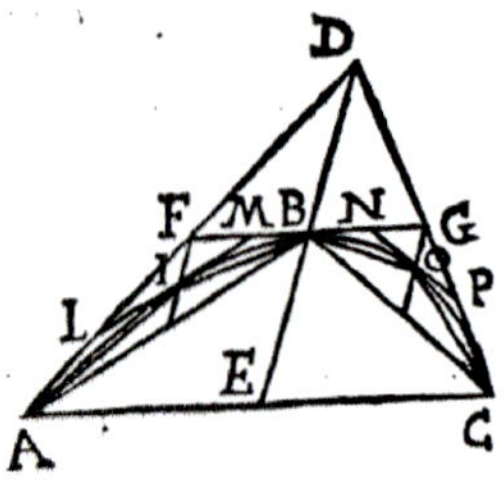

Fig. 30 The diagram used by Torricelli to prove Proposition XVI. The series of the triangles internal to the parabolic segments are *ABC*, *AIB*, *COB*, etc. The series of the external triangles is *DFG*, *LMF*, *PNG*, etc (Torricelli 1644, III, p. 70). *Source* Courtesy University of Oklahoma, History of Science Collections

In Proposition XVII the parabola is squared through the indivisibles technique by resorting to the property proved by Archimedes in his *Spirals* (Proposition XXIV) according to which the area of the first circle is three times the area of the spiral. Torricelli's proof also exploits a further statement on the series of proportional magnitudes (Lemma XXIX).

Propositions XVIII and XIX show two further methods to square the parabola which are based on mechanic considerations. Torricelli assumes as a premise Archimedes's seventh proposition proved in the second book of *On the equilibrium of planes*: the gravity centres of two similar parabolic segments divide the diameters in the same proportion. The Lemma which is premised, the XXXth one, shows that given a parabolic segment, its gravity centre divides its axis so that the part between the gravity centre and the vertex is 4/3 the other part. It is here that Archimedes's proposition is exploited.

The further method to square the parabola (Torricelli 1644, III, Proposition XIX, p. 78) through indivisibles is important to understand the role played by infinitesimal entities within Torricelli's geometry. This also includes the relationship between geometry–infinitesimals because it is based on a Lemma previously considered by Torricelli (Torricelli 1644, III, Proposition XVIII, Lemma XXXI, p. 76 [pp. 75–78]). At the end of the Proposition XVIII, and before Lemma XXXI, he claimed:

> Author, Most illustrious. Antonio Rocca eminent Geometer[46]

[46] "Authore Io. Antonio Roccha praestandi Geometra praestans Geometra." (Torricelli 1644, III, Proposition XVIII, p. 76; Authors' translation).

As a matter of fact, Giannantonio Rocca (1607–1656; Roda 2017) was the author of the *Lemma XXXI* which was consistent with the scientific approach of the "Schola Cavaleriana" (Torricelli 1644, Proposition XVIII, p. 76). *Lemma XXXI*, which can be regarded as a proposition included in the science of weights, stated (see also Fig. 31):

> If [we] suspend a flat figure at a lever, which cuts the figure itself, [then] the momenta of the figure's segments are as the revolution solids described by the rotation of such segments around the secant lever.[47]

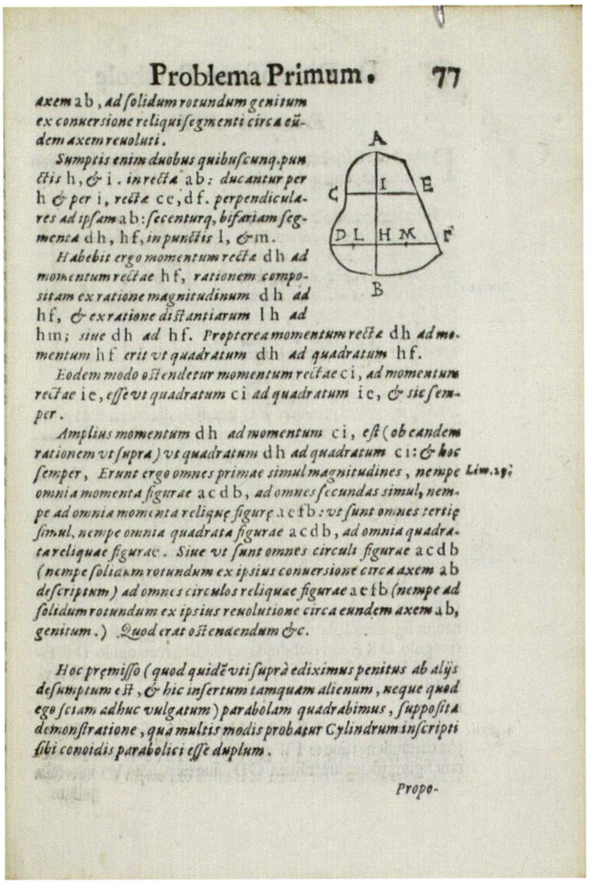

Problema Primum. 77

axem ab, *ad solidum rotundum genitum ex conuersione reliqui segmenti circa eũdem axem reuoluti.*

Sumptis enim duobus quibuscunq. punctis h, *&* i. *in recta* ab: *ducantur per* h *& per* i, *rectae* ce, df. *perpendiculares ad ipsam* ab: *secenturq, bifariam segmenta* dh, hf, *in punctis* l, *&* m.

Habebit ergo momentum recta dh *ad momentum rectae* hf, *rationem compositam ex ratione magnitudinum* dh *ad* hf, *& ex ratione distantiarum* lh *ad* hm; *siue* dh *ad* hf. *Propterea momentum recta* dh *ad momentum* hf *erit vt quadratum* dh *ad quadratum* hf.

Eodem modo ostendetur momentum rectae ci, *ad momentum rectae* ie, *esse vt quadratum* ci *ad quadratum* ie, *& sic semper.*

Amplius momentum dh *ad momentum* ci, *est (ob eandem rationem vt supra) vt quadratum* dh *ad quadratum* ci: *& hoc semper, Erunt ergo omnes primae simul magnitudines, nempe omnia momenta figurae* acdb, *ad omnes secundas simul, nempe ad omnia momenta reliquę figurę* aefb: *vt sunt omnes tertię simul, nempe omnia quadrata figurae* acdb, *ad omnia quadrata reliquae figurae. Siue vt sunt omnes circuli figurae* acdb *(nempe solidum rotundum ex ipsius conuersione circa axem* ab *descriptum) ad omnes circulos reliquae figurae* aefb *(nempe ad solidum rotundum ex ipsius reuolutione circa eundem axem* ab, *genitum.) Quod erat ostendendum &c.* Lem. 29.

Hoc pręmisso (quod quidẽ vti suprà ediximus penitus ab aliis desumptum est, & hic insertum tamquam alienum, neque quod ego sciam adhuc vulgatum) parabolam quadrabimus, supposita demonstratione, quã multis modis probatur Cylindrum inscripti sibi conoidis parabolici esse duplum.

Propo-

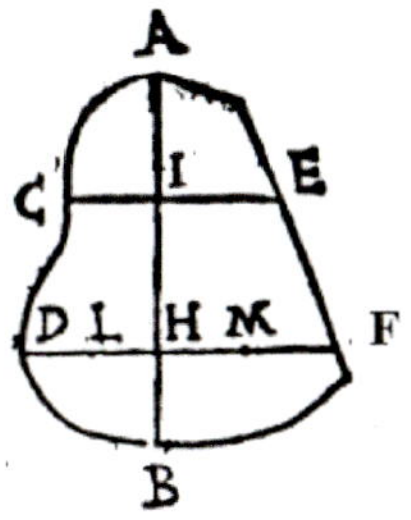

The Lemma states that the momentum of the segment *ACBD* is to the momentum of *AEFB* as the solid of revolution generated by the rotation of the segment *ACDB* around *AB* is to the solid of revolution generated by the rotation of *AEFB*.

Fig. 31 Be the flat figure *ACDBFE*. Be *AB* the lever. Rotate the figure around *AB* (Torricelli 1644, III, p. 77). *Source* Courtesy University of Oklahoma, History of Science Collections

The Propositions XX and XXI present the last two methods to square the parabola by indivisibles. Both the quadratures are mechanical and are based upon Lemmas XXXII and XXXIII (Torricelli 1644, III, pp. 81–82). The latter is particularly remarkable because two parabolic areas are compared. For, Torricelli proves that:

> Be given a parabolic segment *ABC* with axis *BH* and two further parabolic segments *AEC* and *AGC* having the same base as *ABC*. Be the axes equally inclined and equal to half *BH*, the parabolic area *ABC* is equal to the sum of the two parabolic areas *AECG* (*Ibidem*, Fig. 32).

---

[47] "Si figura plana super aliqua si recta linea figurami ipsam secante libretur, erunt momenta segmentorum figurae ut sunt solida rotunda ab ipsis segmentis, circa secantem lineam revolitis, descripta." (Torricelli 1644, III, Lemma XXXI, p. 76; Authors' translation). On the subject, Caverni also quoted Rocca (Caverni 1895, IV, p. 136).

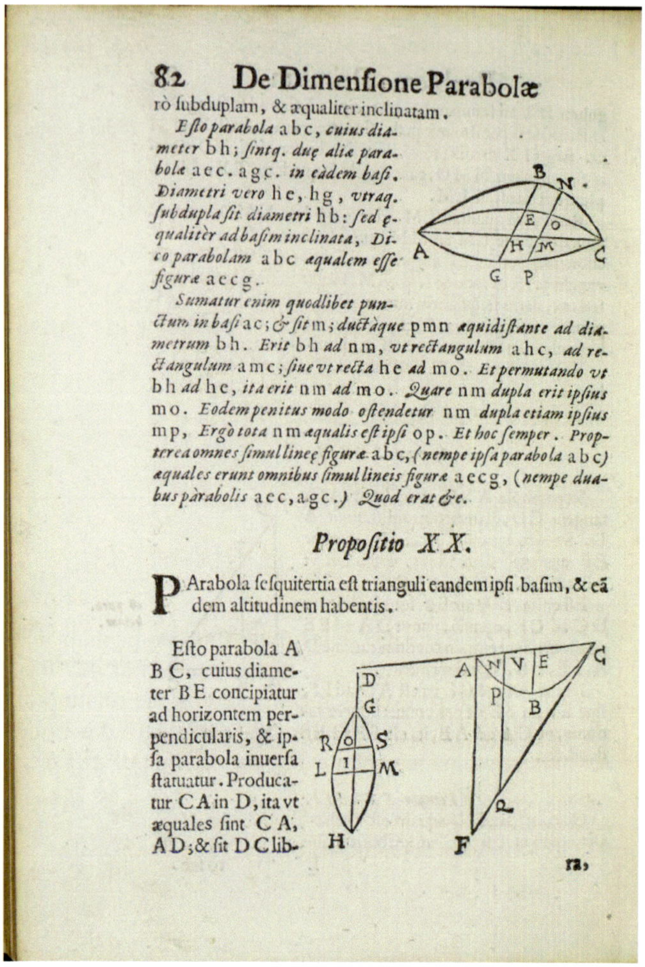

82 De Dimenſione Parabolæ

rò ſubduplam, & æqualiter inclinatam.

Eſto parabola a b c, *cuius diameter* b h; *ſintq. duę aliæ parabolæ* a e c. a g c. *in eàdem baſi. Diametri vero* h e, h g, *vtraq. ſubdupla ſit diametri* h b: *ſed ęqualitèr ad baſim inclinata, Dico parabolam* a b c *æqualem eſſe figuræ* a e c g.

Sumatur enim quodlibet punctum in baſi a c; *& ſit* m; *ductàque* p m n *æquidiſtante ad diametrum* b h. *Erit* b h *ad* n m, *vt rectangulum* a h c, *ad rectangulum* a m c; *ſiue vt recta* h e *ad* m o. *Et permutando vt* b h *ad* h e, *ita erit* n m *ad* m o. *Quare* n m *dupla erit ipſius* m o. *Eodem penitus modo oſtendetur* n m *dupla etiam ipſius* m p, *Ergò tota* n m *æqualis eſt ipſi* o p. *Et hoc ſemper. Propterea omnes ſimul lineę figuræ* a b c, *(nempe ipſa parabola* a b c*) æquales erunt omnibus ſimul lineis figuræ* a e c g, (*nempe duabus paràbolis* a e c, a g c.) *Quod erat &c.*

Propoſitio XX.

PArabola ſeſquitertia eſt trianguli eandem ipſi baſim, & eādem altitudinem habentis.

Eſto parabola A B C, cuius diameter B E concipiatur ad horizontem perpendicularis, & ipſa parabola inuerſa ſtatuatur. Producatur C A in D, ita vt æquales ſint C A, A D; & ſit D C lib-

ra,

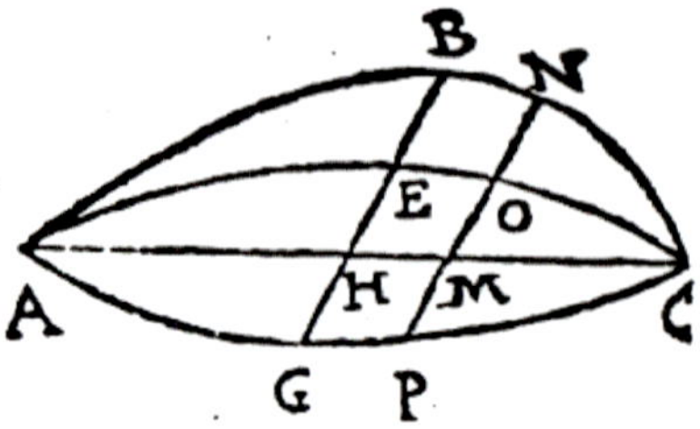

Fig. 32 The diagram used by Torricelli to prove the Lemma XXXIII (Torricelli 1644, III, p. 82). *Source* Courtesy University of Oklahoma, History of Science Collections

The quadrature of the parabola in Proposition XX is interesting because Torricelli produces the base *AC* of the parabola by a segment *AD* = *AC* and regards *AD* as a lever whose fulcrum is *A*. Suspend the parabola in its position and the figure *GLHM*, composed of two parabolas as in lemma XXXIII at the extreme *D* of the lever (Fig. 33). Starting from this construction, Torricelli is able to develop a beautiful reasoning *à la* Archimedes (we mean the mechanical method) which allows him to square the parabola.

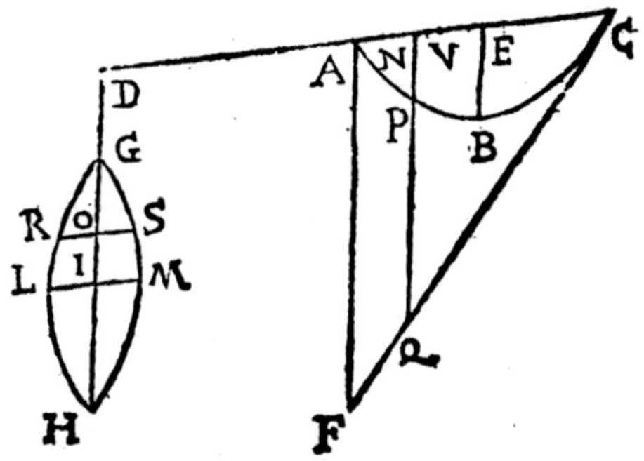

Fig. 33 The diagram used by Torricelli to prove the Proposition XX (Torricelli 1644, III, p. 82; see above also Fig. 32). *Source* Courtesy University of Oklahoma, History of Science Collections

In this second part of his book Torricelli offered eleven propositions dedicated to the quadrature of the parabola. Since in each of the XIVth and the XVth propositions, two different quadratures exist, one concludes that Torricelli squared the parabola in thirteen different ways by the indivisibles. As in the case of the exhaustion method, we refer to the following Table 4 to summarize Torricelli's work (see also above).

Table 4 Torricelli's methodological approach

Number of Proposition	Lemmas used	Kind of proof
Proposition XI	1) Lemma XVIII 2) Given a cone and a cylinder with the same base and height, the volume of the cylinder is three times that of the cone	Geometrical
Proposition XII	Lemmas XX and XXI	Geometrical
Proposition XIII	Archimedes's proposition that the volume of the cylinder circumscribing a sphere is 3/2 the sphere's volume	Geometrical
Proposition XIV The quadrature is given in two manners	Lemmas XXII and XXIII	Mechanical
Proposition XV The quadrature is given in two manners	Lemmas XXIV–XXVII	Geometrical
Proposition XVI	Lemma XXVIII	Geometrical
Proposition XVII	1) Lemma XIX 2) Archimedes's proposition that the area of the first circle is 3 times the area of the inscribed spiral	Geometrical
Proposition XVIII	1) Lemma XXX 2) Archimedes's theorem that the gravity centres of the parabolic segments divide the diameters in the same proportion	Mechanical
Proposition XIX	Lemma XXXI (Rocha's [Rocca's] Lemma)	Mechanical
Proposition XX	Lemmas XXXII and XXXIII	Mechanical
Proposition XXI	(1) Properties of the gravity centre of a cone (2) Properties of the gravity centre of a semi-parabola	Mechanical

There are several causes for interest underlying *De dimensione parabolae*:

(1) Mathematics is not only a discipline whose aim is to obtain new theorems and solve new problems. The methodological aspect is also important. Different demonstrations of the same proposition open different conceptual perspectives. For this reason, they are significant.
(2) The method of exhaustion and the method of indivisibles both involve infinity. The former with infinity in its potential form. The concept of limit exists, but there is no technique concerning calculation with limits, neither a general procedure to obtain quadrature and cubatures in a standard manner as will be the case with integral calculus. This is the

reason why it is necessary to resort to the *ab absurdum* reasoning. On the contrary, in the indivisibles technique, the concept of actual infinity seems involved. In a sense, this admirable technique seems a calculation with actual infinity. This aspect is particularly conspicuous in Torricelli, who extends such conception of infinity to the sum of series, as well.

(3) *De dimensione parabolae* is admirable from a conceptual point of view, but it is once again appropriate to point out that Torricelli lacked an arithmetical–algebraic technique, which is also typical of the other great Italian mathematicians, such as Cavalieri. Of course, this was an enormous disadvantage compared to the European mathematicians who used an algebraic symbolism. In Cavalieri's and Torricelli's epoch this gap was not yet completely tangible because of their ability and because algebraic technique applied—to summarize—to infinitesimal problems such as those of tangents, maxima, minima, quadratures and cubatures was still in the initial phase. However, in the following decades the gap became evident and Italian mathematics lost its importance within the evolution of European science.

This work by Torricelli concludes with two appendices dedicated to the cycloid. The former is very important because Torricelli proves that the area of the cycloid is three times the circle's area. He demonstrates this statement in three different manners: the first and the third through indivisibles, the second by exhaustion. From a historical point of view, it must be recalled that Torricelli had a priority dispute with Roberval as to the quadrature of cycloid. Torricelli, who had always been inclined to avoid any polemic, this time complained about Roberval's behaviour because, after having received a letter from Torricelli with the determination of the cycloid's gravity centre and the quadrature, Roberval claimed to have already reached the same results, though in inverse order.

Torricelli opens his Appendix writing that no mathematician had squared the cycloid. He attempts to offer a justification of their failure: if you try to determine, as evidently the mathematicians of the early seventeenth century did, the area of the cycloid experimentally, the experiment does not seem to offer precise indications. Thus, such mathematicians arrived at thinking that the ratio between the area of the generating circle and that of the cycloid could be irrational. These words by Torricelli probably means that they first tried to determine the cycloid's area experimentally. One method might be to suspend a tiny circle and a tiny cycloid generated by that circle and of its same material at the extremes of a lever and to determine the point of equilibrium. Obviously, these empirical methods can be useful to have an idea of the result, but are useless to gain an accurate result. The first proof works like this (Fig. 34):

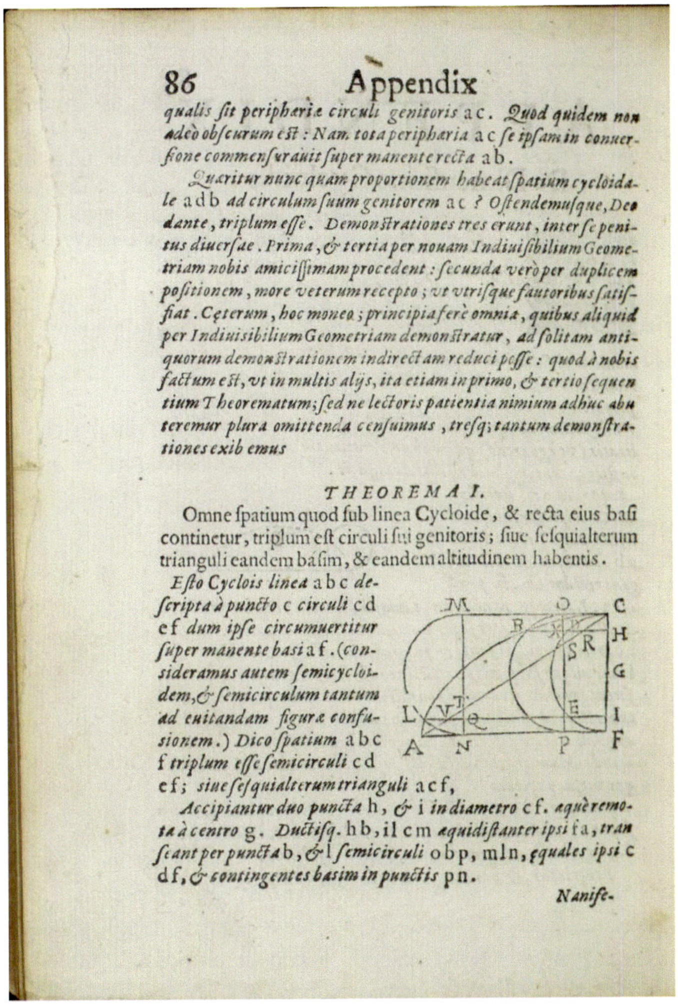

86 Appendix

qualis ſit peripheriæ circuli genitoris a c. *Quod quidem non adeò obſcurum eſt: Nam tota peripheria* a c *ſe ipſam in conuerſione commenſurauit ſuper manente rectā* a b.

Quæritur nunc quam proportionem habeat ſpatium cycloidale a d b *ad circulum ſuum genitorem* a c? *Oſtendemuſque, Deo dante, triplum eſſe. Demonſtrationes tres erunt, inter ſe penitus diuerſae. Prima, & tertia per nouam Indiuiſibilium Geometriam nobis amiciſſimam procedent: ſecunda verò per duplicem poſitionem, more veterum recepto; vt vtriſque fautoribus ſatiſfiat. Cęterum, hoc moneo; principia ferè omnia, quibus aliquid per Indiuiſibilium Geometriam demonſtratur, ad ſolitam antiquorum demonſtrationem indirectam reduci poſſe: quod à nobis factum eſt, vt in multis alijs, ita etiam in primo, & tertio ſequentium Theorematum; ſed ne lectoris patientia nimium adhuc abuteremur plura omittenda cenſuimus, treſq; tantum demonſtrationes exib emus*

THEOREMA I.

Omne ſpatium quod ſub linea Cycloide, & recta eius baſi continetur, triplum eſt circuli ſui genitoris; ſiue ſeſquialterum trianguli eandem baſim, & eandem altitudinem habentis.

Eſto Cyclois linea a b c *deſcripta à puncto* c *circuli* c d e f *dum ipſe circumuertitur ſuper manente baſi* a f. (*conſideramus autem ſemicycloidem, & ſemicirculum tantum ad euitandam figuræ confuſionem.*) *Dico ſpatium* a b c f *triplum eſſe ſemicirculi* c d e f; *ſiue ſeſquialterum trianguli* a c f,

Accipiantur duo puncta h, *&* i *in diametro* c f. *æquè remota à centro* g. *Ductiſq.* h b, i l c m *æquidiſtanter ipsi* f a, *tranſeant per puncta* b, *&* l *ſemicirculi* o b p, m l n, *æquales ipsi* c d f, *& contingentes baſim in punctis* p n.

Naniſe-

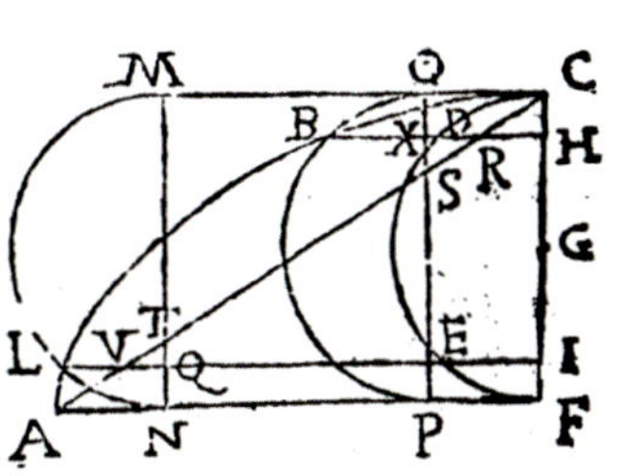

Fig. 34 The diagram used by Torricelli to square the cycloid (Torricelli 1644, III, p. 86). *Source* Courtesy University of Oklahoma, History of Science Collections

The demonstration is developed on the half cycloid *ABCF*, while *CDEF* is half the generator circle. On its diameter *CF*, take *H* and *I* so that $HG = GI$. Therefore, $CH = IF$. Draw CM, HB, IL parallel to *AF*. Through the points *B* and *L* where these parallels cut the cycloids, draw the half circles *OBP*, *MLN*, which touch the cycloid's base in *P* and *N* resp. It is $HD = IE = XB = QL$ and $\text{arc.}OB = \text{arc.}LN$. Furthermore, $CR = VA$. Given the definition of cycloid, the semi-circumference *MLN* is equal to *AF*, arc.*LN* = *AN*, arc.*LM* = *NF* and arc.*BO* = *PF*. Thus, $AN = PF$, consequently $AT = SC$ and, hence, $VT = SR$ and $VQ = XR$. Therefore, it is

$$LV + BR = LQ + BX = EI + DH$$

This is always true wherever you take the points *I* and *H* on *CF* provided that they are equidistant from the centre *G*. All the segments of the figure *ALBCA* are, thence, equal to all the segments of the semi-circle *CEDF*. The area of the triangle *ACF* is double the area of the semi-circle. Therefore, by composing, the area of the half-cycloid is 3/2 the area the triangle *ACF* and, hence, three times that of the semi-circle. This is the first published determination of the cycloid's area. We avoid the details of the two other proofs given by Torricelli, neither will we consider the scholium postponed by Torricelli to the *Appendix* (Torricelli 1644, III, *Appendix De dimensione cycloidis* pp. 85–90). Rather, we will analyse Torricelli's result which was probably the most surprising for the whole European mathematical milieu: the volume of the hyperbolic acute solid.

The essay *De solido acuto hyperbolico* is very rich from a methodological, conceptual and mathematical point of view. Let us summarize the main reasons of interest behind this work:

(1) Torricelli shows the existence of a solid whose extension is infinite, but whose volume is finite. He was perfectly aware of the novelty of his result. For, he writes:

> So far, in the Schools of Mathematics you find measures of figures which are limited on all sides. Among all the solids, of which the ancient and modern authors—with many difficulties—determined the measure, no one has an infinite extension, as far as I know. If you consider an infinitely extended solid or a flat figure, everyone thinks the size of this figure to be infinite. In fact, a solid of infinite length exists, which is so tiny that, however produced to infinity, does not exceed the size of a small cylinder. It is the solid generated by the hyperbola, to which this treatise is dedicated. So far, no one considered this solid, but it gives rise to a plurality of interesting theorems. If my judgement were not vailed by passion, I would say that in the whole history of geometry none of the figures considered so far is more interesting.[48]

Thus, from a mathematical standpoint the examination of an infinitely extended solid having a finite volume is a remarkable novelty.

(2) With regard to the methods used to prove his theorem, Torricelli offered two demonstrations: one based on exhaustion and the other one on indivisibles. However, here Torricelli clearly explains his predilection for indivisibles: the exhaustion method is perfect from a geometrical point of view, but it is not heuristic and, relying only on it, geometry had made little progress. The true scientific method is that of indivisibles. Torricelli is entirely explicit:

> As to the proof method, we will demonstrate a single remarkable theorem in two manners: with the indivisibles and following the ancients' method. Actually, my result was discovered through the geometry of indivisibles, which is the real scientific and, so to say, natural proof-method.[49]

Torricelli continues, claiming that the ancients' geometry was extremely limited because they did not know or did not use indivisibles, while the admirable Cavalieri opened the door to calculate areas and volumes of numerous and variegated classes of solids.

(3) To calculate the volume of the acute hyperbolic solid, Torricelli did not resort to the usual Cavalieri's indivisibles. One might call them "rectilinear". Rather, he invented and used the "curved indivisible". The idea is that a figure can be interpreted not only as an aggregate of rectilinear segments, but also of curvilinear ones. Torricelli pointed out that his idea was approved by Raffaello Magiotti. To explain his new method, Torricelli proposed fourteen examples of propositions, which were known, but were proved without using curved indivisibles. He specified when there was a concordance between his result and one of those obtained by Archimedes. We present two of his examples, the first and easiest one, relative to a flat figure and the 12th which, instead, concerns a solid figure.

The first example chosen by Torricelli regards the circle's area (Fig. 35). Archimedes in the first Proposition of *De mensura circuli* proved that the area of the circle *BD* is equal to that of the right triangle of which one small side *BC* is equal to the rectified circumference and the other small side *AB* equal to the radius.

[48] "Hactenus enim in Mathematicis Scholis repertae sunt dimensiones figurarum ab omni parte finem habentium; quandoquidem inter omnia solida, quae ab antiquis et modernis Auctoribus multiplici conatu ad mensuram redacta sunt, nullum adhuc, quod ego sciam, ullam dimensionem habuit extensione infinitam. Imo statim atque proponatur sive solidum aliquod, sive figura plana, cuius aliqua extensio in infinitam distantiam procedat, unusquisque cogitabit huiusmodi figuram infinite magnitudinis esse debere. Attamen solidum habet Geometria, longitudine quidem infinitum, sed tanta praeditum subtilitate, ut licet in infinitum producatur, exigui tamen cylindri molem non excedat. Tale erit solidum illud ab hyperbola genitum, quod huius libelli comtemplatione prosequemur; intactum hucusque ab aliis, et multiplici curiosaque Theorematum varietate faecundissimum; eo usque ut, nisi me fallat affectus, universa Geometria inter hactenus consideratas figuras nullam habeat curiositate abundationem." (Torricelli 1644, IV, p. 93; Authors' translation).

[49] "Quo ad methodum demonstrandi, unicum quidem, et praecipuum Theorema duplici conatu ostendemus, et per indivisibilia, et more Veterum. Quamquam (ut vera fateamur) primo inventum sit per Indivisibilium Geometriam; qui sane verus est demonstrandi modus scientificus, semper directus, et ipsi naturae germanus." (Torricelli 1644, IV, p. 94; Authors' translation).

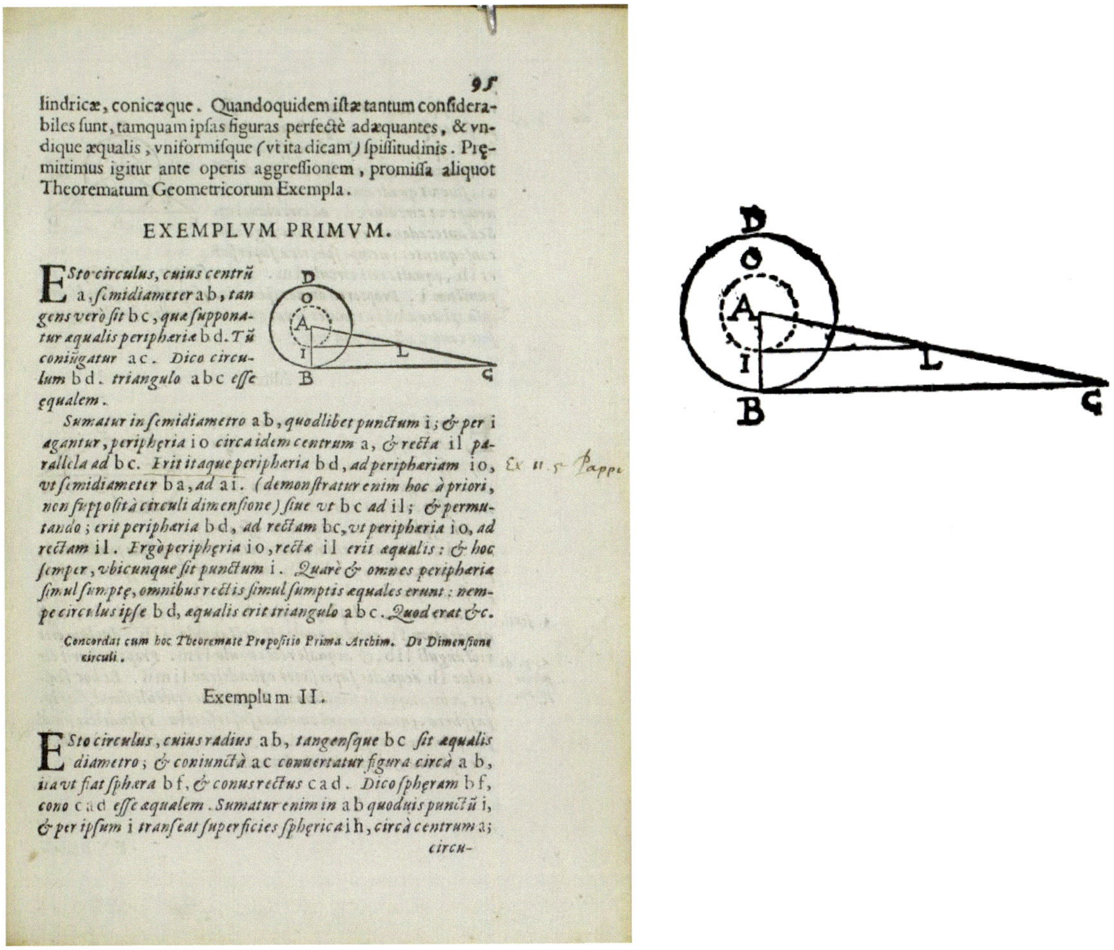

95

lindricæ, conicæque. Quandoquidem istæ tantum considerabiles sunt, tamquam ipsas figuras perfectè adæquantes, & vndique æqualis, vniformisque (vt ita dicam) spissitudinis. Pręmittimus igitur ante operis aggressionem, promissa aliquot Theorematum Geometricorum Exempla.

EXEMPLVM PRIMVM.

ESto circulus, cuius centrũ a, *semidiameter* ab, *tangens verò sit* bc, *quæ supponatur æqualis periphæriæ* bd. *Tũ coniũgatur* ac. *Dico circulum* bd. *triangulo* abc *esse ęqualem*.

Sumatur in semidiametro ab, *quodlibet punctum* i; *& per* i *agantur, periphęria* io *circa idem centrum* a, *& recta* il *parallela ad* bc. *Erit itaque periphæria* bd, *ad periphæriam* io, *vt semidiameter* ba, *ad* ai. *(demonstratur enim hoc à priori, non supposità circuli dimensione) siue vt* bc *ad* il; *& permutando; erit periphæria* bd, *ad rectam* bc, *vt periphæria* io, *ad rectam* il. *Ergò periphęria* io, *recta* il *erit æqualis: & hoc semper, vbicunque sit punctum* i. *Quarè & omnes periphæriæ simul sumptę, omnibus rectis simul sumptis æquales erunt: nempe circulus ipse* bd, *æqualis erit triangulo* abc. *Quod erat &c.* Ex 11. 5. Pappi

Concordat cum hoc Theoremate Propositio Prima Archim. De Dimensione circuli.

Exemplum II.

ESto circulus, cuius radius ab, *tangensque* bc *sit æqualis diametro; & coniunctà* ac *conuertatur figura circà* ab, *ita vt fiat sphæra* bf, *& conus rectus* cad. *Dico sphęram* bf, *cono* cad *esse æqualem. Sumatur enim in* ab *quoduis punctũ* i, *& per ipsum* i *transeat superficies sphęrica* ih, *circà centrum* a;

circu-

Fig. 35 The diagram used by Torricelli to square the circle through the curved indivisibles (Torricelli 1644, IV, p. 95). *Source* Courtesy University of Oklahoma, History of Science Collections

Torricelli reasons like this: on the radius *AB*, take a point *I*. Draw the circumference of radius *AI* and the segment *IL* parallel to *BC*, which is the tangent at *B*. Choose *BC* as long as the circumference *BD*. The circumference *BD* is to the circumference *IO* as *AB* is to *AI*,[50] that is as *BC* to *IL*. Therefore circ.BD : BC = circ.IO : IL. Thus, circ.IO = IL. Hence, all the circumferences constructed as *IO*, which compose the circle *BD* are respectively equal to all the segments as *IL* which compose the triangle *ABC*. Ergo, the area of the circle and that of the triangle are equal. The circumference *IO* is one of the curved indivisibles which compose the circle of diameter *BD*.

The XIIth example concerns the volume of the paraboloid of revolution (Fig. 36).

[50] Torricelli points out that this property does not depend on the quadrature of the circle (ibid., p. 95).

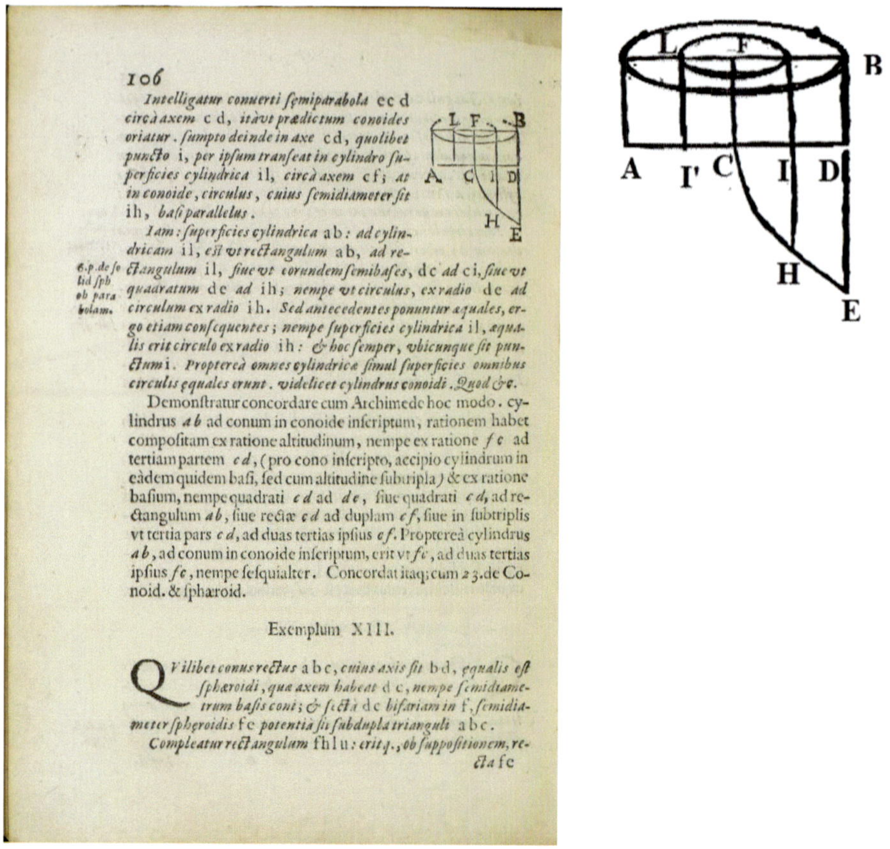

106

Intelligatur conuerti ſęmiparabola e c d *circà axem* c d, *itàvt prædictum conoides oriatur. ſumpto deinde in axe* c d, *quolibet puncto* i, *per ipſum tranſeat in cylindro ſuperficies cylindrica* i l, *circà axem* c f; *at in conoide, circulus, cuius ſemidiameter ſit* i h, *baſi parallelus.*

Iam: ſuperficies cylindrica a b*: ad cylindricam* i l, *eſt vt rectangulum* a b, *ad rectangulum* i l, *ſiue vt eorundem ſemibaſes*, d c *ad* c i, *ſiue vt quadratum* d e *ad* i h; *nempe vt circulus, ex radio* d e *ad circulum ex radio* i h. *Sed antecedentes ponuntur æquales, ergo etiam conſequentes; nempe ſuperficies cylindrica* i l, *æqualis erit circulo ex radio* i h*: & hoc ſemper, vbicunque ſit punctum* i. *Propterea omnes cylindricæ ſimul ſuperficies omnibus circulis ęquales erunt. videlicet cylindrus conoidi. Quod &c.*

6.p. de ſolid ſph ob parabolam.

Demonſtratur concordare cum Archimede hoc modo. cylindrus *a b* ad conum in conoide inſcriptum, rationem habet compoſitam ex ratione altitudinum, nempe ex ratione *f c* ad tertiam partem *c d*, (pro cono inſcripto, accipio cylindrum in eàdem quidem baſi, ſed cum altitudine ſubtripla) & ex ratione baſium, nempe quadrati *c d* ad *d e*, ſiue quadrati *c d*, ad rectangulum *a b*, ſiue rectæ *c d* ad duplam *c f*, ſiue in ſubtriplis vt tertia pars *c d*, ad duas tertias ipſius *c f*. Propterea cylindrus *a b*, ad conum in conoide inſcriptum, erit vt *f c*, ad duas tertias ipſius *f c*, nempe ſeſquialter. Concordat itaq; cum 23. de Conoid. & ſphæroid.

Exemplum XIII.

Q*Vilibet conus rectus* a b c, *cuius axis ſit* b d, *ęqualis eſt ſphæroidi, quæ axem habeat* d c, *nempe ſemidiametrum baſis coni; & ſectà* d c *biſariam in* f, *ſemidiameter ſphęroidis* f c *potentià ſit ſubdupla trianguli* a b c.

Compleatur rectangulum f h l u*: erit q., ob ſuppoſitionem, recta* f c

Fig. 36 Adapted. The diagram used by Torricelli to prove the example referred to in the running text. For convenience of the reader, we have completed the figure (right side of the table) (Torricelli 1644, IV, p. 106). *Source* Courtesy University of Oklahoma, History of Science Collections

Consider the half-parabola *EHC* of height *DC*. Take *AC* = *DC*. Construct *DB* so that $ED = \sqrt{AD \cdot BD}$. The theorem states that the right cylinder *ADB* of axis *CF* has the same volume as the paraboloid obtained by the rotation of the parabolic segment *CED* around the axis *CD*. Take a point *I* on *CD*, so that the cylindrical surface *IL* passes trough *I* as well as the semi-circle of radius *IH*. The cylindrical surface *AB* is to the cylindrical surface *IL* as $AD \times DB$ is to $I'I \times DB$, namely as *DC* is to *IC* and as ED^2 is to IH^2. Therefore, since

$$ED^2 = AD \times BD, \text{it is } IH^2 = I'I \times DB.$$

This means that the cylindric surfaces, which are the curved indivisibles that compose the cylinder, are equal to the circles which compose the paraboloid. Thus, the theorem is proved. Afterwards, Torricelli proves that this theorem agrees with Proposition XXIII of Archimedes's *Conoids and Spheroids*.

After the examples, Torricelli defines the "acute hyperbolic solid" as the figure generated by the rotation of a hyperbola around one of its asymptotes (Torricelli 1644, IV, p. 112). The fundamental theorem where Torricelli determines the volume of the acute hyperbolic solid is preceded by five lemmas. Their proofs are relatively clear and do not need any particular demonstrative technique as indivisibles or exhaustion.

Lemma I (Fig. 37) considers a hyperbola, whose asymptotes *AB* and *AC* are perpendicular. Rotate the figure around the asymptote *AB*, so generating the acute hyperbolic solid. Inscribe any rectangle *DEFG* passing through the asymptote in the solid. This rectangle is equal to the square of the hyperbola's semi–axis.

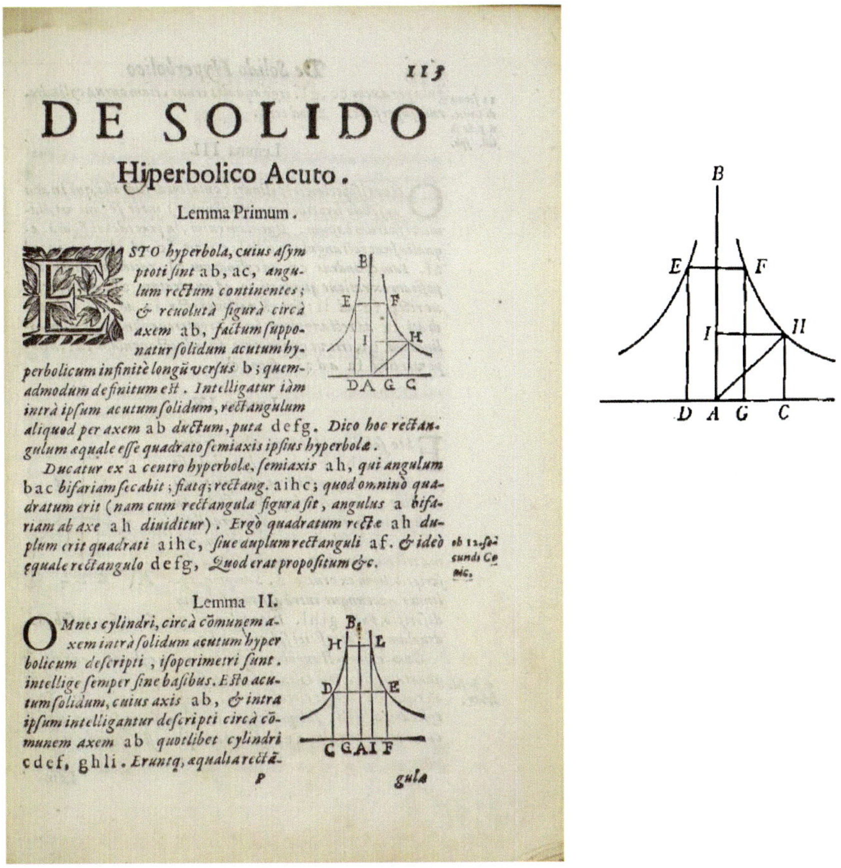
113

DE SOLIDO

Hiperbolico Acuto.

Lemma Primum.

ESTO hyperbola, cuius asymptoti sint ab, ac, angulum rectum continentes; & reuolutà figurà circà axem ab, factum supponatur solidum acutum hyperbolicum infinitè longũ versus b; quemadmodum definitum est. Intelligatur iàm intrà ipsum acutum solidum, rectangulum aliquod per axem ab ductum, puta defg. Dico hoc rectangulum æquale esse quadrato semiaxis ipsius hyperbolæ.

Ducatur ex a centro hyperbolæ, semiaxis ah, qui angulum bac bifariam secabit; fiatq; rectang. aihc; quod omninò quadratum erit (nam cum rectangula figura sit, angulus a bifariam ab axe ah diuiditur). Ergò quadratum rectæ ah duplum erit quadrati aihc, siue duplum rectanguli af. & ideò æquale rectangulo defg, Quod erat propositum &c. ob 12. secundi Conic.

Lemma II.

OMnes cylindri, circà cõmunem axem intrà solidum acutum hyperbolicum descripti, isoperimetri sunt. intellige semper sine basibus. Esto acutum solidum, cuius axis ab, & intra ipsum intelligantur descripti circà cõmunem axem ab quotlibet cylindri cdef, ghli. Eruntq; æqualia rectã-

P gula

Fig. 37 The diagram used by Torricelli to prove the Lemma I, but usable also in reference to the other Lemmas (Torricelli 1644, IV, p. 113). *Source* Courtesy University of Oklahoma, History of Science Collections

In Lemma II (Fig. 37) Torricelli proves that all the cylinders described in the acute hyperbolic solid around the rotation axis *AB* have the same external surface.

In Lemma III it is proved that the volumes of such cylinders are as the diameters of the bases.

In Lemma IV Torricelli demonstrates that the lateral surface of the cylinder *DEFG* is ¼ the surface of the sphere whose radius is equal to the hyperbola's semiaxis.

Finally, Lemma V (Torricelli 1644, IV, p. 115; see also Fig. 37), which derives directly from the previous one, claims that the lateral surface of the cylinder *DEFG* is equal to that of the circle whose radius is the hyperbola's semi–axis *AH*. Armed with the concept of curved indivisibles and with these five lemmas, Torricelli is able to determine the volume of the acute hyperbolic solid:

> Consider the infinitely long acute hyperbolic solid cut with a plane perpendicular to the axis. Its volume plus the volume of its base's cylinder is equal to the volume of the right cylinder whose basis is the verse side, namely the hyperbola's axis, and whose height is equal to the radius of the acute solid's base.[51]

[51] "Solidum acutum hyperbolicum infinite longum, sectum plano ad axem erecto, una cum cylindro suae basis, aequale est cylindro cuidam recto, cuius basis diameter sit latus versum, sive axis hyperbolae, altitudo vero sit aequalis semidiametro basis acuti solidi." (Torricelli 1644, IV, p. 115; Authors' translation).

Thus, Torricelli proved that (Fig. 38):

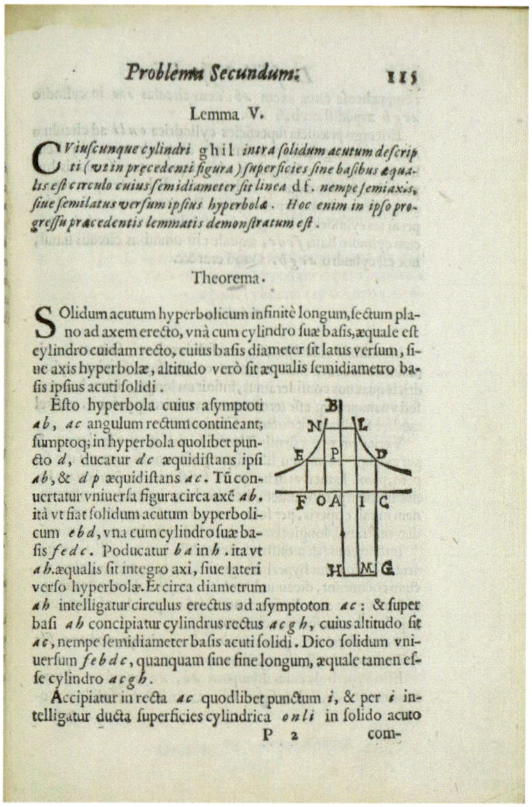

Problema Secundum. 115

Lemma V.

Cuiuscunque cylindri ghil *intra solidum acutum descripti (vt in precedenti figura) superficies sine basibus aequalis est circulo cuius semidiameter sit linea* df. *nempe semiaxis, siue semilatus versum ipsius hyperbolae. Hoc enim in ipso progressu praecedentis lemmatis demonstratum est.*

Theorema.

Solidum acutum hyperbolicum infinitè longum, sectum plano ad axem erecto, vnà cum cylindro suae basis, aequale est cylindro cuidam recto, cuius basis diameter sit latus versum, siue axis hyperbolae, altitudo verò sit aequalis semidiametro basis ipsius acuti solidi.

Esto hyperbola cuius asymptoti *ab*, *ac* angulum rectum contineant; sumptoq; in hyperbola quolibet puncto *d*, ducatur *dc* aequidistans ipsi *ab*, & *dp* aequidistans *ac*. Tũ conuertatur vniuersa figura circa axẽ *ab*. ità vt fiat solidum acutum hyperbolicum *ebd*, vna cum cylindro suae basis *fedc*. Poducatur *ba* in *h*. ita vt *ah* aequalis sit integro axi, siue lateri verso hyperbolae. Et circa diametrum *ah* intelligatur circulus erectus ad asymptoton *ac*: & super basi *ah* concipiatur cylindrus rectus *acgh*, cuius altitudo sit *ac*, nempe semidiameter basis acuti solidi. Dico solidum vniuersum *febdc*, quanquam sine fine longum, aequale tamen esse cylindro *acgh*.

Accipiatur in recta *ac* quodlibet punctum *i*, & per *i* intelligatur ducta superficies cylindrica *onli* in solido acuto

P 2 com-

Be given the hyperbola whose asymptotes *AB* and *AC* are perpendicular. Consider any point *D* on the hyperbola. Draw *ED* parallel to *AC* and *DP* parallel to *AC*. Rotate the figure around the asymptote *AB* as rotation axis. The acute hyperbolic solid *EDB* is generated. It has the circle of diameter *ED* as base and is infinitely extended along *PB* as axis. Now produce *AB* until *H* so that *AH* is equal to the hyperbola's axis. On the base *AH* construct the right cylinder *AHGC*, whose height is *AC*. The volume of the acute hyperbolic solid plus that of the cylinder *EDCF* is equal to the volume of the cylinder *AHGC*.

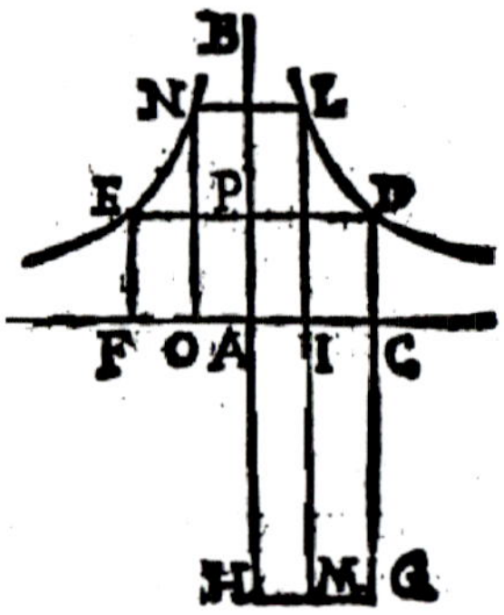

Fig. 38 The determination of the volume of the acute hyperbolic solid (Torricelli 1644, IV, p. 115). *Source* Courtesy University of Oklahoma, History of Science Collections

In order to prove the theorem Torricelli reasons like this: on *AC* he considers any point *I*; he draws the parallel to *AB* which cuts the hyperbola in *L* and constructs the rectangle of dimensions *AI* and *IL*. If this rectangle rotates around *AB*, the cylindrical surface *ONLI* is generated. In the cylinder *ACGH* consider the circle of diameter *IM*, where *M* is obtained producing *IL*. The cylindrical surface *ONLI* is to the surface of the circle whose diameter is *IM* as in the following proportion:

$$\mathrm{S}(ONLI) \text{ is to } \mathrm{S}(IM)\ OI \times IL \text{ is to } \left(\frac{IM}{2}\right)^2.$$

Because of Lemma V these surfaces are equal, for any position of the point *I*. Thus, considering the acute hyperbolic solid plus the cylinder at its base composed of indivisibles such as the cylinder *ONLI*, while the cylinder *ACGH* is composed of the circles whose diameter is *IM*, it follows that the volume of the two figures is equal. This fundamental theorem is followed by 29 corollaries where Torricelli proved a series of properties connected to the acute hyperbolic solid. The corollaries are preceded by a brief but interesting scholium where Torricelli expresses his astonishment for the result he had obtained. For, we read:

> It might appear unbelievable that, though this solid having an infinite length, no one of the considered cylindrical surfaces has an infinite length. All of them are finite, as each one having a mediocre familiarity with the doctrine of conic sections will clearly recognize.[52]

[52] "Incredibile videri potest, cum solidum hoc infinitam longitudinem habeat, nullam tamen ex illis superficiebus cylindricis quas nos consideramus, infinitam longitudinem habere; sed unamquamquamque esse terminatam; ut unicuique patebit, cui vel modice familiaris sit doctrina Conicorum." (Torricelli 1644, IV, p. 116; Authors' translation).

These words are similar to those written by Cantor to Dedekind in a letter on 29 June 1877 after having discovered the bijection between the bi-dimensional and the linear *continuum*: "I see it, but I do not believe it".[53]

Cavalieri, as the Dedekind of Torricelli, shared the astonishment for this discovery and expressed his admiration for Torricelli in a letter written on 17 December 1641 where we read:

> I received your letter when I was in bed with fever and gout. [...] Despite my sickness, I was able to enjoy the very tasty fruits of your ingenuity. I came to be in infinite admiration of that infinitely long hyperbolic solid, which is equal to a three dimensional finite body. When I communicated this result to some of my disciples, who are philosophers, they admitted this fact to appear them really marvellous and astonishing. [...]. I cannot understand how you were able to guess [literally to fish] so easily, in the infinite depth of that solid, its dimension. Indeed, the solid seems to me infinitely long, because the plane space which generates the solid appears to me infinitely long and each part of this plane space generates a part of the solid.[54]

Torricelli does not miss the occasion to confirm that the demonstration through indivisibles is perfectly legitimate from the geometrical standpoint. However, he also added a proof by exhaustion to satisfy the reader who raises doubts on the indivisibles (Torricelli 1644, III, 2, p. 116).

In the second and last part of his work, entitled, *De Dimensione Acuti solidi Hyperbolici iuxta methodum Antiquorum* (Torricelli 1644, IV. The numbering of pages restarts from 129. Thus it should be: p. 136 and pp. 129–135) Torricelli determines the volume of the hyperbolic solid through exhaustion. In the brief introductory consideration, he claims that the most difficult aspect of solving such a problem by exhaustion might appear the inscription and circumscription of a figure to a figure of infinite length. However, Torricelli, recalls that Roberval, too, solved such a problem with methods different from his own. He had received Roberval's letter on the cubature of the hyperbolic solid when the *Opera geometrica* was already in press, so that he could not publish Roberval's proof. Furthermore, from Roberval's letter, it was not clear whether he desired his proof to be published.

---

[53] In his letter to Dedekind, which obviously is written in German, Cantor wrote in French (to underline even more his astonishment for the result he had obtained): "[...] je la vois, mais je ne la crois pas [...]." (Cantor and Dedekind 1937, p. 34). After few days, on 2 June 1877, Dedekind answered to have found no lack or mistake in Cantor's proof. He wrote: "I developed once again the proof of your demonstration and I found no lack. I think your interesting proposition to be correct and I congratulate you" (Authors' translation; German text: "Ihren Beweis habe ich noch einmal geprüft, und ich habe keine Lücke darin entdeckt; ich glaube gewiss, dass Ihr interessanter Satz richtig ist, und beglückwünsche Sie zu demselben." (*Ivi*, p. 37).

[54] "Mi giunge la lettera di V.S.M. Rev.do in tempo che io stavo nel letto con la febbre, e gotta [...] ho però goduto al dispetto del male d' saporitissimi frutti del suo ingegno, essendomi riuscito infinitamente ammirabile quel solido iperbolico infinitamente lungo, ed uguale ad un copro, quanto a tutte e tre le dimensioni, finito, ed avendolo io comunicato ad alcuni miei scolari filosofi, hanno confessato parergli veramente meraviglioso, e stravagante, che ciò possa essere [...] non so come abbi pescato nell'infinita profondità di quel solido così facilmente la sua dimensione, poiché veramente a me pare infinitamente lungo, parendomi infinitamente lungo lo spazio piano, che lo genera, ad ogni parte di esso generando parte del solido". We point out that the term "filosofo" was then used also to indicate the mathematicians and the natural philosophers, namely the physicists. (Torricelli 1919–1944, III, pp. 65–66; Authors' translation. See also Belloni's Introduction in Torricelli 1975, p. 14; Gàbici and Toscano 2006, pp. 56–57).

Torricelli's proof is based upon seven lemmas. The most interesting of these lemmas is the sixth one, the only one proved by exhaustion (Fig. 39):

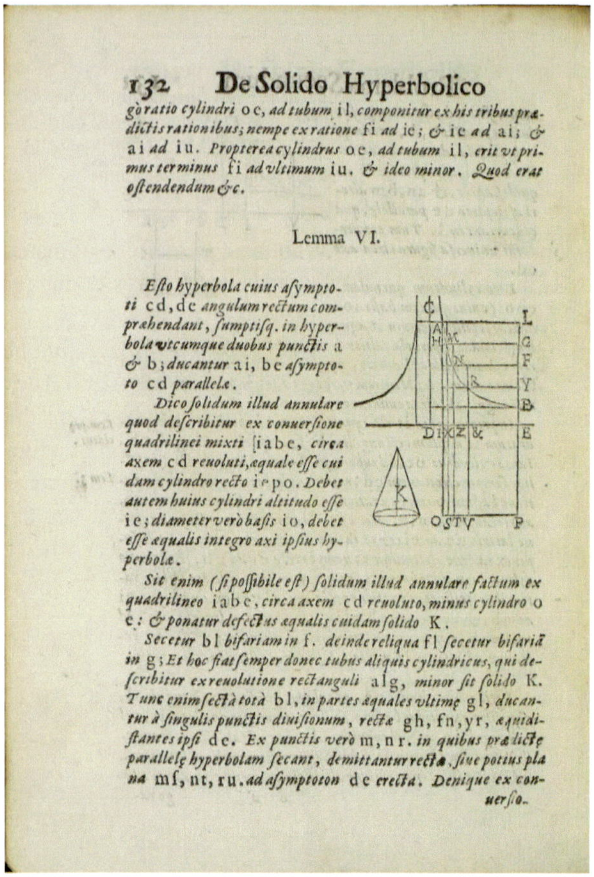
132 De Solido Hyperbolico

gò ratio cylindri oc, ad tubum il, componitur ex his tribus prædictis rationibus; nempe ex ratione fi ad ie; & ie ad ai; & ai ad iu. Propterea cylindrus oc, ad tubum il, erit vt primus terminus fi ad vltimum iu. & ideo minor. Quod erat ostendendum &c.

Lemma VI.

Esto hyperbola cuius asymptoti cd, de angulum rectum comprehendant, sumptisq. in hyperbola vtcumque duobus punctis a & b; ducantur ai, be asymptoto cd parallelæ.

Dico solidum illud annulare quod describitur ex conuersione quadrilinei mixti iabe, circa axem cd reuoluti, æquale esse cuidam cylindro recto iepo. Debet autem huius cylindri altitudo esse ie; diameter verò basis io, debet esse æqualis integro axi ipsius hyperbolæ.

Sit enim (si possibile est) solidum illud annulare factum ex quadrilineo iabe, circa axem cd reuoluto, minus cylindro oe: & ponatur defectus æqualis cuidam solido K.

Secetur bl bifariam in f. deinde reliqua fl secetur bifariã in g; Et hoc fiat semper donec tubus aliquis cylindricus, qui describitur ex reuolutione rectanguli alg, minor sit solido K. Tunc enim sectà totà bl, in partes æquales vltimæ gl, ducantur à singulis punctis diuisionum, rectæ gh, fn, yr, æquidistantes ipsi de. Ex punctis verò m, nr. in quibus prædictę parallelę hyperbolam secant, demittantur rectæ, siue potius plana mf, nt, ru. ad asymptoton de erecta. Denique ex conuersio-

Be given a hyperbola whose asymptotes *CD* and *EF* are perpendicular. Take any two points *A* and *B* on the hyperbola and draw the parallels to *CD*. Rotate the mixtiliner quadrilateral *IABE* around *CD*. The so obtained hollow solid has the same volume as the right cylinder *IEPO*.

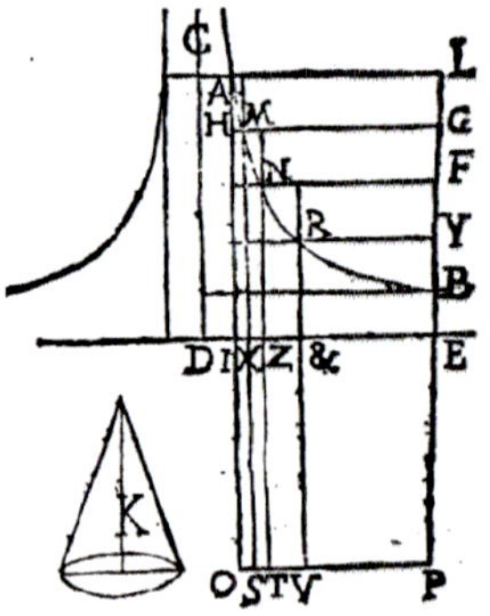

Fig. 39 Lemma VI and its diagram (Torricelli 1644, IV, p. 132). *Source* Courtesy University of Oklahoma, History of Science Collections

The proof of the theorem runs like this (Fig. 39): suppose the hyperbolic solid to be less than the cylinder *ACIH*, constructed as the cylinder *ACGH* of previous Fig. 38. Consider the cylinder *NCIL* equal to the hyperbolic solid. Produce *LN* until its intersection *M* with the hyperbola. The point *M* exists since *LN* is supposed parallel to the asymptote *AB*. Because of Lemma VI, the cylinder *NCIL* is equal to the hollow solid described by the rotation of the mixtilinear quadrilateral *NMDC* around *AB*. Therefore, it will be less than the hyperbolic solid, which is against the assumption. In a similar manner, Torricelli proves that the cylinder *NCIL* cannot be less that the hyperbolic solid. The demonstrative core of this theorem is Lemma VI proved by exhaustion.

Torricelli's *Opera geometrica* ends with an appendix concerning several geometrical properties of the cochlea.[55]

The main purpose of our Introduction is to explain Torricelli's results obtained in *Opera geometrica*. We have expounded them and commented on the most important methods he used. We have also seen what the similarities and the differences between exhaustion method and integral calculus are. Given their relevance for the history of geometry, we wish to offer the reader a view on two other important mathematical discoveries by Torricelli and add some considerations on indivisibles in relation to integral calculus. Thus, we will briefly analyse two works which do not belong to the *Opera Geometrica*: the memoire on the logarithmic spiral and that on the infinite hyperbolas.

[55] As above cited, *Opera Geometrica* often restarts the order of the numbering pages. When it comes to *Appendix De dimensione Cochlea*: pp. [136]145–150. The numbering of pages starts with p. 136 and just after with p. 145, so the numeration continues consequently. The last page is 150, but, in fact, the Appendix is only seven pages long.

4 Intermezzo 1645–1646: The Works on the Infinite Curves

This section is devoted to analysing and commenting upon relationship between Torricelli's geometrical methods and infinitesimal calculus. We will do it by means of two—at that time—unpublished works of Torricelli (both in Torricelli 1919–1944; *De Infinitis Spiralibus* in Torricelli 1975, pp. 507–548). In fact, these manuscripts were not included in *Opera geometrica* and neither were they published during Torricelli's lifetime. This adequately allows us to introduce Torricelli's calculus with infinite and infinitesimal magnitudes in the next section below.

A geometrical (also logarithmic) spiral is represented by a curve, which originates from a point, moving farther away as it revolves around the point. The argument is geometrically complex and fascinating. The traditions of works on spirals and their attractive representation in nature run from Antiquity to modern times (Pisano and Bussotti 2015c, d).

First, below we present *De infinitis spiralibus* (1645). It represents advanced research on spiral curves dealt with Torricelli. Two of the important goals achieved by Torricelli were: 1) the elegant dimensional passage to geometrical approach, to logarithmic spiral and 2) to have made clear its difference from two-dimensions Archimedean spiral (although Torricelli was inspired by it which consisted of equal arms only). We also remark the intellectual charm of these asymmetric curves, as presented by Torricelli in their logarithmic forms. It is important to note the infinite character of the logarithmic spiral, which, along with its generation around a fixed center–point, approaches or moves away from this center–point without ever reaching it. Thus this center–point can be considered as an asymptotic point. In modern terms, we could say that Torricelli pointed out that even if the geometry of the problem is typically central–asymptotic, that is it is necessary to rotate infinite times around the center before reaching it, the total distance travelled is instead finite. Therefore, the intellectual charm consisted in having shown that for the logarithmic spiral the *infinite* and the *finite* coexisted at least in mathematical reasoning. It is evident that *De infinitis spiralibus* belonged with Torricelli's *Opera geometrica* arguments as part of his concurrent research program, even thought, it was not published in 1644 with *Opera geometrica*.

Secondly, we will deal with another outstanding work: *De infinitis hyperbolis* (1646). We will analyse this interesting geometrical—mathematical essay as it is transcribed into modern language by Bortolotti from Torricelli's original work. Taking into account Bortolotti's arguments, below we propose our own standpoint on the subject. We will show how Torricelli developed a calculus with infinite and infinitesimal quantities. But, in our opinion, this Torricellean calculus cannot be considered as the precursor of infinitesimal calculus. For, we will point out several aspects which make Torricelli's work irreducible to the mentality and to the concepts which, a few years later, connoted infinitesimal calculus.

4.1 *De Infinitis Spiralibus* (1645)

Torricelli's *De infinitis spiralibus*is (1645) is an important geometrical essay where Torricelli solved very difficult problems. It was posthumously published both in *Opere di Evangelista Torricelli* by Gino Loria and Giuseppe Vassura (Torricelli 1919–1944, I, 2, pp. 349–399) and in *Opere scelte di Evangelista Torricelli* (Torricelli 1975, pp. 507–548) by Lanfranco Belloni. The latter is based on Carruccio's previous work (Torricelli 1955, 1975, p. 49, p. 488; see also Bortolotti 1925a, b, 1928b, 1939a, b). As the two editions are almost identicals we follow the former (Torricelli 1919–1944). However, when they diverge, we follow the reconstruction by Carruccio (1955) then edited by Belloni (Torricelli 1975, pp. 507–548).[56]

In the following we refer to Loria and Vassura's edition of Torricelli's *Opere* (Torricelli 1919–1944).

[56] It is to be noted out that Descartes already introduced the logarithmic spiral. In a letter to Mersenne written on 13 July 1638 Descartes deals with a problem of physics (Descartes 1897–1913, II, pp. 222–245). In this context Descartes introduces the logarithmic spiral (*Ivi*, pp. 232–233). But in this letter there is only a hint of the curve. By far more important is the letter Descartes sent to Mersenne on 12 September 1638 (*Ivi*, pp. 352–362). Descartes claimed that "[...] this spiral has many properties which allow us to recognize it [...]" ("Mais, pour cette spiral, elle a plusieurs proprietez qui la rendent assiez reconnaissable"; *Ivi*, p. 360). In particular: 1) if A is the origin of the spiral and B_i its points, there is the same proportion between the arc of spiral AB_i and the radius vector AB_i, which is a remarkable discovery; 2) the angle between the tangent and the radius vector is constant for every point of the spiral (*Ibidem*). There is, however, no demonstration or systematic study of this curve in Descartes, although he recognised some of its main features. It is well known that, after Torricelli, many mathematicians studied the equiangular spiral, first of all Jakob Bernoulli, who defined it *spira mirabilis*. However, in our context we do not think appropriate to face this question.

In the *De infinitis spiralibus* Torricelli describes (Fig. 40) the following curve, introducing its "First and true *definition*" (*Prima Definitio, et vera*; Torricelli 1919–1944, I, 2, p. 361):

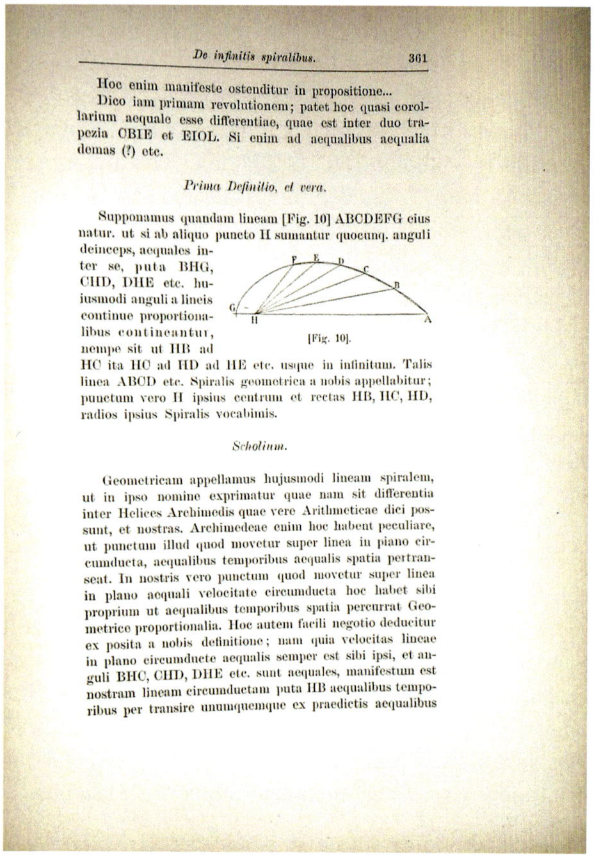

De infinitis spiralibus. 361

Hoc enim manifeste ostenditur in propositione...

Dico iam primam revolutionem; patet hoc quasi corollarium aequale esse differentiae, quae est inter duo trapezia CBIE et EIOL. Si enim ad aequalibus aequalia demas (?) etc.

Prima Definitio, et vera.

Supponamus quandam lineam [Fig. 10] ABCDEFG eius natur. ut si ab aliquo puncto H sumantur quocunq. anguli deinceps, aequales inter se, puta BHG, CHD, DHE etc. huiusmodi anguli a lineis continue proportionalibus contineantur, nempe sit ut HB ad HC ita HC ad HD ad HE etc. usque in infinitum. Talis linea ABCD etc. Spiralis geometrica a nobis appellabitur; punctum vero H ipsius centrum et rectas HB, HC, HD, radios ipsius Spiralis vocabimis.

[Fig. 10].

Scholium.

Geometricam appellamus hujusmodi lineam spiralem, ut in ipso nomine exprimatur quae nam sit differentia inter Helices Archimedis quae vere Arithmeticae dici possunt, et nostras. Archimedeae enim hoc habent peculiare, ut punctum illud quod movetur super linea in piano circumducta, aequalibus temporibus aequalis spatia pertranseat. In nostris vero punctum quod movetur super linea in plano aequali velocitate circumducta hoc habet sibi proprium ut aequalibus temporibus spatia percurrat Geometrice proportionalia. Hoc autem facili negotio deducitur ex posita a nobis definitione; nam quia velocitas lineae in plano circumducte aequalis semper est sibi ipsi, et anguli BHC, CHD, DHE etc. sunt aequales, manifestum est nostram lineam circumductam puta HB aequalibus temporibus per transire unumquemque ex praedictis aequalibus

Consider the line such that, if from a point H equal angles $BHC, CHD, DHE, \ldots$ are drawn, such angles have the sides in continued proportion. Namely it holds: $HB\!:\!HC = HC\!:\!HD = HD\!:\!HE, \ldots$. Such line is named by Torricelli a geometrical spiral (*spiralis geometrica*). The point H is its centre and the segments $HB, HC, HD, \ldots$ its radii.

Fig. 40 Torricelli's diagram on *Prima Definitio, et vera* (Torricelli 1919–1944, I, 2, p. 361; see also Torricelli 1975, p. 511). *Source* Public domain

The name chosen by Torricelli depends on a kinematic property of his curve, as he clarified in the scholium posed after the definition: the Archimedean spiral can be regarded as generated by a point moving uniformly on a straight line rotating uniformly around a centre. This point, hence, describes equal spaces in equal times, whereas in the geometrical spiral the spaces traversed in equal times are in geometrical proportion. This means that the point moves on the straight line of uniformly accelerated motion. This argument was clearly proved by Torricelli as follows: the straight line *GA* rotates uniformly and traverses equal angles in equal times. The sides of the angles are in continuous proportion, as well as their differences. But the differences of the radius vectors represent the spaces. Therefore, Torricelli gives a geometrical definition of his spiral from which its kinematic characterization also follows. The profound research by Torricelli continued showing a further fundamental difference between his spiral and Archimedes's one: if, in the latter, you consider a point *P*, the number of turns is finite towards the origin of the curve and infinite in the opposite direction. Whereas in the geometrical spiral, the number of turns is infinite in both directions. Torricelli gives this brilliant proof (Fig. 41):

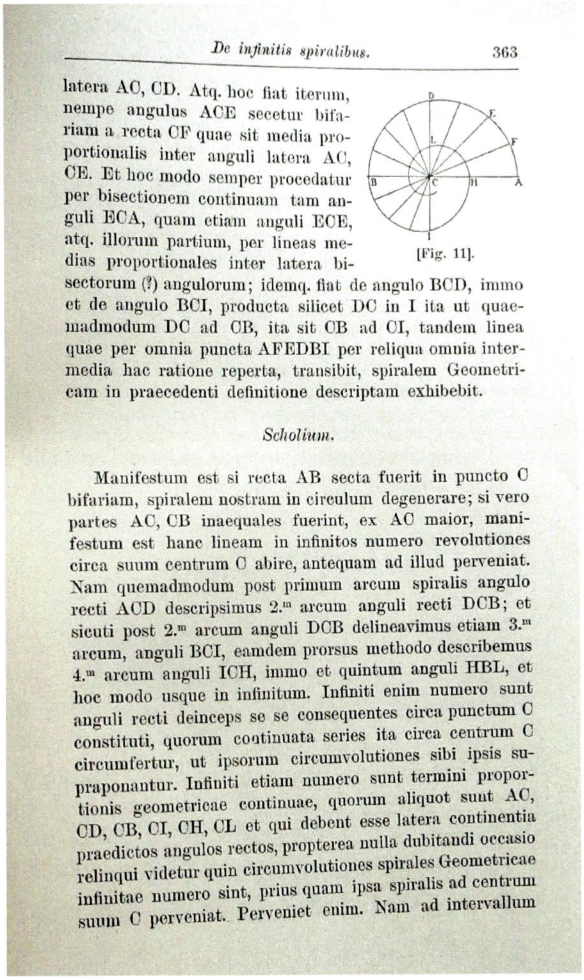

De infinitis spiralibus. 363

latera AC, CD. Atq. hoc fiat iterum, nempe angulus ACE secetur bifariam a recta CF quae sit media proportionalis inter anguli latera AC, CE. Et hoc modo semper procedatur per bisectionem continuam tam anguli ECA, quam etiam anguli ECE, atq. illorum partium, per lineas medias proportionales inter latera bisectorum (?) angulorum; idemq. fiat de angulo BCD, immo et de angulo BCI, producta silicet DC in I ita ut quaemadmodum DC ad CB, ita sit CB ad CI, tandem linea quae per omnia puncta AFEDBI per reliqua omnia intermedia hac ratione reperta, transibit, spiralem Geometricam in praecedenti definitione descriptam exhibebit.

Scholium.

Manifestum est si recta AB secta fuerit in puncto C bifariam, spiralem nostram in circulum degenerare; si vero partes AC, CB inaequales fuerint, ex AC maior, manifestum est hanc lineam in infinitos numero revolutiones circa suum centrum C abire, antequam ad illud perveniat. Nam quemadmodum post primum arcum spiralis angulo recti ACD descripsimus 2.^m arcum anguli recti DCB; et sicuti post 2.^m arcum anguli DCB delineavimus etiam 3.^m arcum, anguli BCI, eamdem prorsus methodo describemus 4.^m arcum anguli ICH, immo et quintum anguli HBL, et hoc modo usque in infinitum. Infiniti enim numero sunt anguli recti deinceps se se consequentes circa punctum C constituti, quorum continuata series ita circa centrum C circumfertur, ut ipsorum circumvolutiones sibi ipsis supraponantur. Infiniti etiam numero sunt termini proportionis geometricae continuae, quorum aliquot sunt AC, CD, CB, CI, CH, CL et qui debent esse latera continentia praedictos angulos rectos, propterea nulla dubitandi occasio relinqui videtur quin circumvolutiones spirales Geometricae infinitae numero sint, prius quam ipsa spiralis ad centrum suum C perveniat. Perveniet enim. Nam ad intervallum

Consider a segment *AB*. Divide it in *C*. Draw *CD* perpendicular to *AB* such that $AC{:}CD = CD{:}CB$. Bisect the angle *DCA* and construct *CE*, so that $CD{:}CE = CE{:}CA$. Bisect the angle *ECA* and construct *CF* such that $CE{:}CF = CF{:}CA$. Continue this operation indefinitely. Make the same operation for the angles *ECD* and *DCB*. It is, thus, clear that the geometrical spiral has an infinite number of turns from the part of its origin and from the other part. This is also a method to construct the spiral by points.

Fig. 41 Torricelli's Proof which show the difference between his spiral and Archimedes's one (Torricelli 1919–1944, I, 2, p. 363; see also Torricelli 1975, p. 514). *Source* Public domain

Now there is a noteworthy reasoning: Torricelli claims that the spiral, after an infinite number of rotations will reach the centre *C*. He argues like this: The reasoning previously shown proves that the spiral will arrive closer to the centre than any given finite distance. Thus, it will reach the centre. He argues:

> It reaches the centre: for, it arrives closer to the centre than any given interval. Therefore, it reaches the centre itself.[57]

Torricelli approached the concept of limit because, to use modern terms, he considers the limit of the spiral when the number of turns tends to infinity, and claims, correctly that such a limit is the centre of the spiral. On the other hand, he also concludes that the limit-value is reached at the end of the process, which is not necessarily true. Thus, we think one can legitimately conclude that Torricelli grasped many aspects of the concept of limit, but some important features escaped him.

Torricelli proves several fundamental geometrical properties of the logarithmic spiral: first of all, he demonstrates that, considering two points of the spiral, the straight line joining them is internal to the spiral. This implies that the curve has a tangent in every point. Afterwards, Torricelli proves two theorems—concerning the tangents to the spiral –, which, because of their importance, he names Castor I and Pollux 2. Castor (Torricelli 1919–1944, I, 2, pp. 366–367) claims (Fig. 42).

---

[57] "Perveniet enim. Nam ad intervallum perveniet quolibet dato intervallo minus; ergo et ad ipsum centrum perveniet." (Torricelli 1919–1944, I, 2, pp. 363–364; Authors' translation).

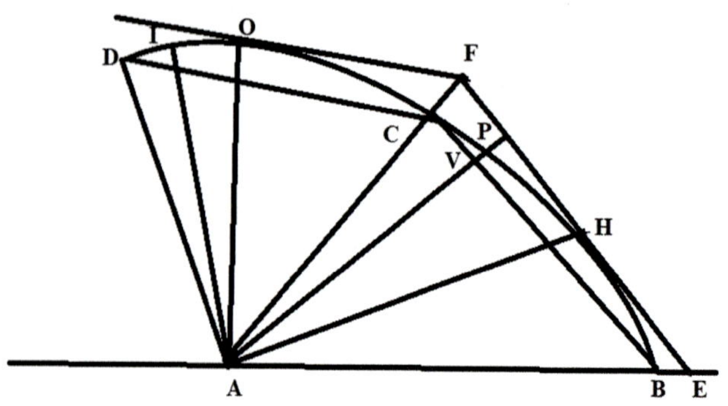

Be A the centre of the spiral, AB the greatest among the considered radii, be equal the angles BAC and CAD. Construct the polygonal BC, CD inscribed in the spiral. Be EF tangent to the spiral at H and parallel to BC. If, from the point F where the tangent cuts AC produced, you draw the parallel to CD, this straight line is tangent to the spiral in a point O such that the two angles CAO and BAH are equal.

Fig. 42 Our representation of Torricelli's diagram. (Torricelli 1919–1944, I, 2, p. 366; see also Torricelli 1975, p. 518)

Torricelli first proves that, constructed the two equal angles CAO and BAH, the straight line FO and CD are parallel. After that he demonstrates that FO is a tangent. The whole proof runs through classical Euclidean methods. The second part relies on an *ad absurdum* reasoning, which resorts neither to exhaustion nor to indivisibles.

This theorem (Castor) is the base to prove a fundamental property of the logarithmic spiral, namely the radius vector makes a constant angle with the tangent, for any point of the curve (Pollux), which is demonstrated by Torricelli in the following proposition. Torricelli was fully justified to call these theorems "Castor" and "Pollux"! After Castor and Pollux three other properties are proved. We will refer only to the first one because it is used in the spiral's rectification, which is one of the most important results obtained by Torricelli (Fig. 43):

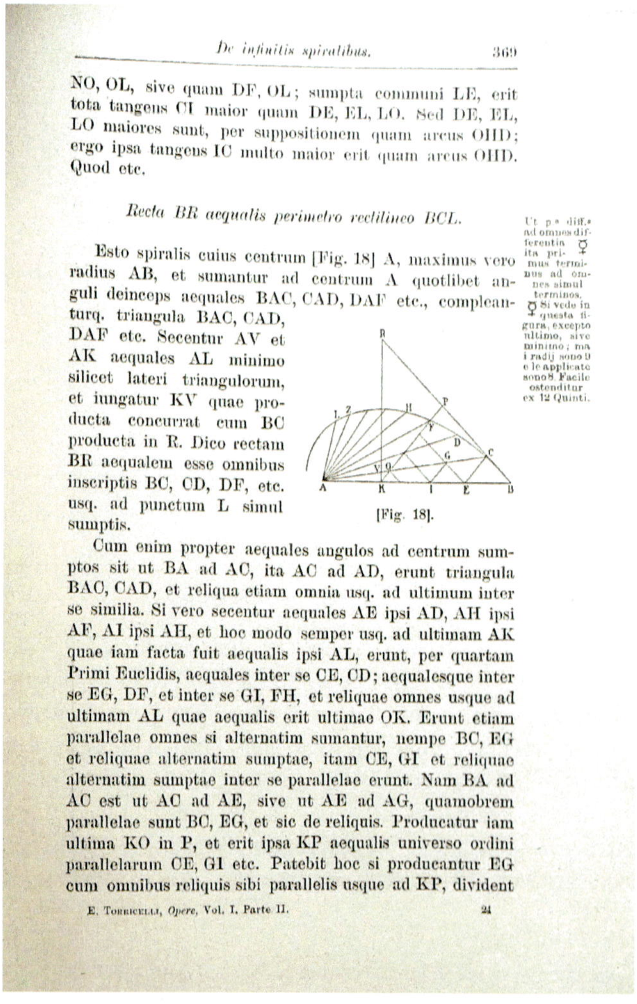

De infinitis spiralibus. 369

NO, OL, sive quam DF, OL; sumpta communi LE, erit tota tangens CI maior quam DE, EL, LO. Sed DE, EL, LO maiores sunt, per suppositionem quam arcus OHD; ergo ipsa tangens IC multo maior erit quam arcus OHD. Quod etc.

Recta BR aequalis perimetro rectilineo BCL.

Esto spiralis cuius centrum [Fig. 18] A, maximus vero radius AB, et sumantur ad centrum A quotlibet anguli deinceps aequales BAC, CAD, DAF etc., compleanturq. triangula BAC, CAD, DAF etc. Secentur AV et AK aequales AL minimo silicet lateri triangulorum, et iungatur KV quae producta concurrat cum BC producta in R. Dico rectam BR aequalem esse omnibus inscriptis BC, CD, DF, etc. usq. ad punctum L simul sumptis.

Ut p.ª diff.ª ad omnes differentia ☿ ita primus terminus ad omnes simul terminos.

☿ Si vede in questa figura, excepto ultimo, sive minimo; ma i radij sono 9 e le applicate sono 8. Facile ostenditur ex 12 Quinti.

[Fig. 18].

Cum enim propter aequales angulos ad centrum sumptos sit ut BA ad AC, ita AC ad AD, erunt triangula BAC, CAD, et reliqua etiam omnia usq. ad ultimum inter se similia. Si vero secentur aequales AE ipsi AD, AH ipsi AF, AI ipsi AH, et hoc modo semper usq. ad ultimam AK quae iam facta fuit aequalis ipsi AL, erunt, per quartam Primi Euclidis, aequales inter se CE, CD; aequalesque inter se EG, DF, et inter se GI, FH, et reliquae omnes usque ad ultimam AL quae aequalis erit ultimae OK. Erunt etiam parallelae omnes si alternatim sumantur, nempe BC, EG et reliquae alternatim sumptae, itam CE, GI et reliquae alternatim sumptae inter se parallelae erunt. Nam BA ad AC est ut AC ad AE, sive ut AE ad AG, quamobrem parallelae sunt BC, EG, et sic de reliquis. Producatur iam ultima KO in P, et erit ipsa KP aequalis universo ordini parallelarum CE, GI etc. Patebit hoc si producantur EG cum omnibus reliquis sibi parallelis usque ad KP, divident

E. Torricelli, *Opere*, Vol. I. Parte II. 24

Be AB the longest radius of the spiral sector BL. From the centre A draw any number of equal consecutive angles $BAC, CAD, DAF, \ldots$, so as to obtain the triangles $BAC, CAD, DAF, \ldots$. On AB and AC take the segments AK and AV equal to the smallest radius AL. The segment VK produced cuts the segment BC produced in R. The segment BR is equal to the sum of the inscribed polygonal $BC + CD + \cdots + ZL$.

Fig. 43 Torricelli's diagram on a fundamental property of the logarithmic spiral. (Torricelli 1919–1944, I, 2, p. 369; see also Torricelli 1975, p. 523). *Source* Public domain

The rectification of the geometrical spiral runs like this (Fig. 44):

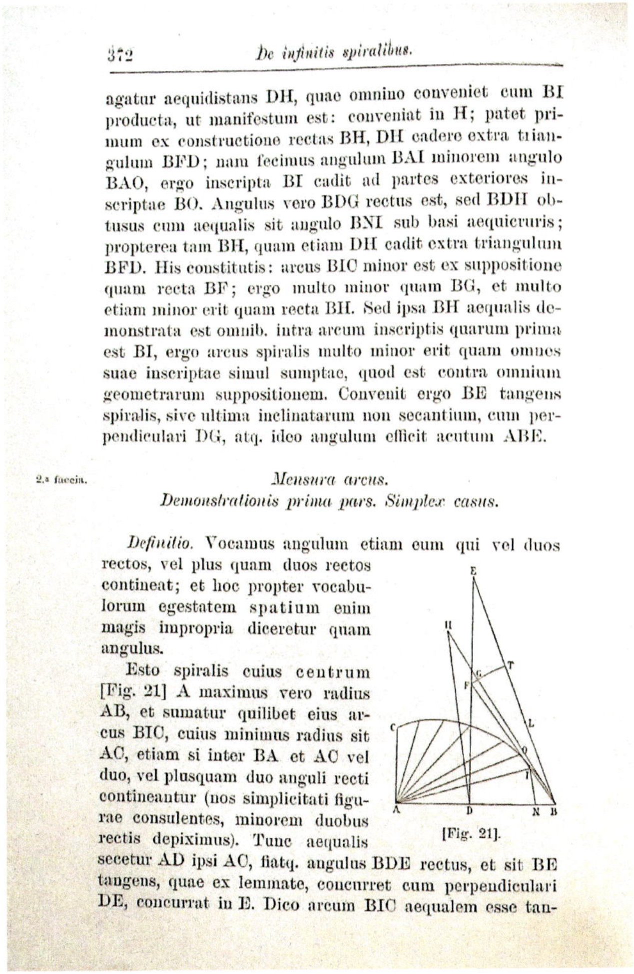

372 *De infinitis spiralibus.*

agatur aequidistans DH, quae omnino conveniet cum BI producta, ut manifestum est: conveniat in H; patet primum ex constructione rectas BH, DH cadere extra triangulum BFD; nam fecimus angulum BAI minorem angulo BAO, ergo inscripta BI cadit ad partes exteriores inscriptae BO. Angulus vero BDG rectus est, sed BDH obtusus cum aequalis sit angulo BNI sub basi aequicruris; propterea tam BH, quam etiam DH cadit extra triangulum BFD. His constitutis: arcus BIC minor est ex suppositione quam recta BF; ergo multo minor quam BG, et multo etiam minor erit quam recta BH. Sed ipsa BH aequalis demonstrata est omnib. intra arcum inscriptis quarum prima est BI, ergo arcus spiralis multo minor erit quam omnes suae inscriptae simul sumptae, quod est contra omnium geometrarum suppositionem. Convenit ergo BE tangens spiralis, sive ultima inclinatarum non secantium, cum perpendiculari DG, atq. ideo angulum efficit acutum ABE.

2.a faccia. *Mensura arcus.*

Demonstrationis prima pars. Simplex casus.

Definitio. Vocamus angulum etiam eum qui vel duos rectos, vel plus quam duos rectos contineat; et hoc propter vocabulorum egestatem spatium enim magis impropria diceretur quam angulus.

Esto spiralis cuius centrum [Fig. 21] A maximus vero radius AB, et sumatur quilibet eius arcus BIC, cuius minimus radius sit AC, etiam si inter BA et AC vel duo, vel plusquam duo anguli recti contineantur (nos simplicitati figurae consulentes, minorem duobus rectis depiximus). Tunc aequalis secetur AD ipsi AC, fiatq. angulus BDE rectus, et sit BE tangens, quae ex lemmate, concurret cum perpendiculari DE, concurrat in E. Dico arcum BIC aequalem esse tan-

Be *A* the centre of the spiral; *AB* the longest and *AC* the shortest radius of the considered arc. Be *AD*=*AC*. Construct the right angle *BDE*. Be *BE* the tangent, which, for a previous Lemma, cuts *DE*. The theorem claims that the arc *BC* is equal to *BE*. (With *BC* we will indicate the arc of spiral).

Fig. 44 Torricelli's diagram on the rectification of the geometrical spiral. (Torricelli 1919–1944, I, 2, p. 372; see also Torricelli 1975, p. 529). *Source* Public domain

Torricelli speaks of two proofs, one by exhaustion and the other by indivisibles. However, it was possible to find in Torricelli's manuscripts only the proof by exhaustion.

It is so conceived:

Suppose $BC < BE$.

From *BE* subtract *BL*=*BD*.

Bisect continuously *LE* until you obtain a segment $TE < BE - BC$.

Therefore $BT > BC$. From *B* draw *BF*, such that *F* belongs to *DE* and $BF = BT$.

Therefore, by construction $BC < BF < BE$.

Since *BE* is the tangent, *BF* cuts the spiral in *O*. Join *A* and *O*.

Bisect continuously the angle *BAC* until reaching the angle $BAI < BAO$.

Consider the polygonal inscribed in the spiral whose first side is *BI*, which, produced, belongs to the angle *EBF*. On *AB* take $AN = AI$. Join *N* and *I*.

Draw *DH* parallel to *IN*. The straight line *DH* is external to the angle *EDB*. For, *EDB* is right, while *HDB* is obtuse because it is equal to *INB* (*DH*//*IN*), which is adjacent to the base of the isosceles triangle *ANI*.

Thus, *DH* and *BI* cut in *H*. Since $BF > BC$, it is also $BG > BC$ and $BH > BC$.

However, the segment *BH*, because of the previous theorem, is equal to the sum of the inscribed polygonal, whose first side is *BI*.

Hence, the sum of the sides of the inscribed polygonal should be bigger than the arch of spiral, but this is absurd.

A consequence of Torricelli's theorem is that any segment of the geometrical spiral has a finite length, even when it includes the infinite turns of the spiral around its centre. With a similar reasoning Torricelli proved that neither is the arch bigger than the considered tangent segment.

In other parts of the *Infinite spirals*, Torricelli proves, through exhaustion, another astonishing result concerning the logarithmic spiral: given a spiral segment, whose perimeter has the centre as an extreme and a point on the spiral as the other extreme, the area of the segment is finite, despite the infinite turns of the spiral. The road that led Torricelli to the proof of his theorem is rather complex: first he proves some theorems on the areas of polygonal figures inscribed in the spiral and on the areas of particular triangles, one side of which is a segment of a tangent to the spiral.

Below we present Torricelli's interesting result on the quadrature of a surface. He proved a theorem with regard to the area of a spiral (Fig. 45).[58]

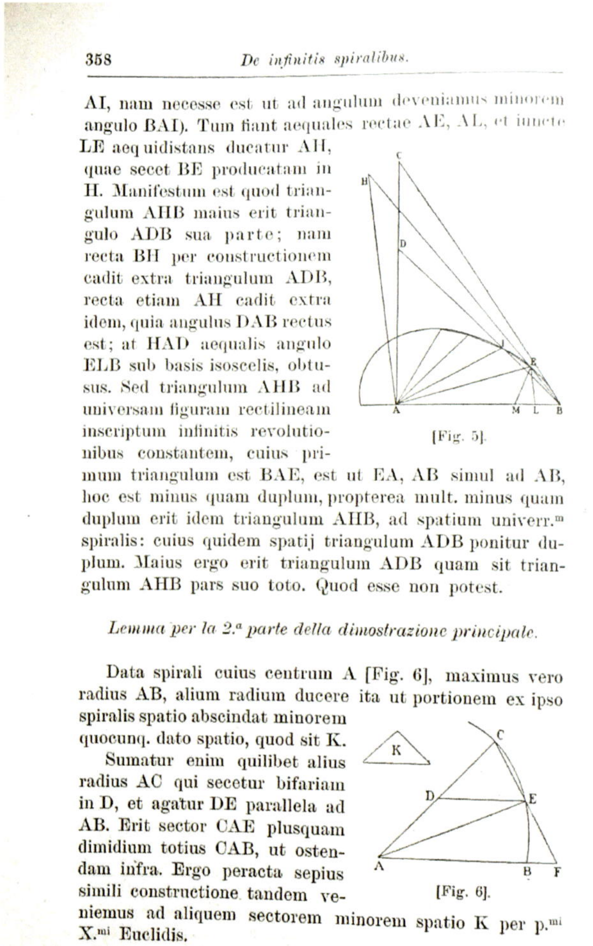
358 *De infinitis spiralibus.*

AI, nam necesse est ut ad angulum deveniamus minorem angulo BAI). Tum fiant aequales rectae AE, AL, et iuncte LE aequidistans ducatur AH, quae secet BE producatam in H. Manifestum est quod triangulum AHB maius erit triangulo ADB sua parte; nam recta BH per constructionem cadit extra triangulum ADB, recta etiam AH cadit extra idem, quia angulus DAB rectus est; at HAD aequalis angulo ELB sub basis isoscelis, obtusus. Sed triangulum AHB ad universam figuram rectilineam inscriptum infinitis revolutionibus constantem, cuius primum triangulum est BAE, est ut EA, AB simul ad AB, hoc est minus quam duplum, propterea mult. minus quam duplum erit idem triangulum AHB, ad spatium univerr.[m] spiralis: cuius quidem spatij triangulum ADB ponitur duplum. Maius ergo erit triangulum ADB quam sit triangulum AHB pars suo toto. Quod esse non potest.

[Fig. 5].

Lemma per la 2.ª parte della dimostrazione principale.

Data spirali cuius centrum A [Fig. 6], maximus vero radius AB, alium radium ducere ita ut portionem ex ipso spiralis spatio abscindat minorem quocunq. dato spatio, quod sit K.

Sumatur enim quilibet alius radius AC qui secetur bifariam in D, et agatur DE parallela ad AB. Erit sector CAE plusquam dimidium totius CAB, ut ostendam infra. Ergo peracta sepius simili constructione tandem veniemus ad aliquem sectorem minorem spatio K per p.[mi] X.[mi] Euclidis.

[Fig. 6].

Be *A* the centre of the spiral, *AB* the longest radius of the segment, and *BC* the tangent to the spiral. Draw *AC* perpendicular to *AB*. The whole surface including infinite revolutions of the spiral is equal to the triangle *ABC*.

Fig. 45 Torricelli's diagram (bottom page) on the theorem with regard to the area of the spiral. (Torricelli 1919–1944, I, 2, p. 358; see also Torricelli 1975, I, p. 542). *Source* Public domain

[58] We refer to Belloni, who based his editing of *De infinitis spiralibus* (Cfr. Torricelli 1975, p. 488) on Carruccio's work. Both were notable Torricellean scholars and editors (see also: Bortolotti 1925a, b, 1928b, 1939a, b). Historiographically interesting, for the editorial projects of Torricelli's *Opere* during the past century, is Belloni's remark to Loria about *De inifinitis spiralibus* (Torricelli 1975, p. 488; see also Belloni's tale on Bortolotti's attack on Loria, *Ivi*, p. 50 [pp. 49–51]). We continue to use Loria and Vassura's images (Torricelli 1919–1944). We note that Loria and Vassura presented the image and its content (Fig. 45) before (Torricelli 1919–1944, I, 2, p. 358) the theorems discussed above (Figs. 40, 41, 42, 43 and 44; *Ivi*, p. 361, p. 363, p. 366, p. 369, p. 372). Instead, Belloni–Carruccio provided it after (Torricelli 1975, p. 542) the above-cited theorems. We also point out a drawing–difference between Loria and Vassura's figure (Fig. 45) and Belloni's one (Torricelli 1975, p. 542). The reader is invited to pay attention to the point *H* in the two figures.

We do not reproduce the details of Torricelli's proof here, because we believe the details furnished so far are sufficient for the reader to gain an idea of Torricelli's methods and way of thinking. His two key methods are exhaustion and indivisibles, which he uses with great ability and extends to an extraordinary number of cases which prior to his work were not included in the range of these two procedures. The second aspect to underline is his skill in treating questions concerning infinity. He had no fear of facing figures which, in different manners, entail infinity and to find geometrical means to treat them. One might say that Torricelli was one of the first mathematicians to measure infinity and to discover many unexpected properties such as those relative to the volume of the acute hyperbolic solid or to the rectification and quadrature of the geometrical spiral. Finally, we offer some results obtained by Torricelli in his memoire *De infinitis hyperbolis* (Torricelli 1919–1944, I, 2, pp. 227–274) jointly with the commentaries by Bortolotti, because this will allow us to develop a brief discussion on Torricelli's methods.

4.2 *De Infinitis Hyperbolis* (Ca. 1646)

The idea behind this *De infinitis hyperbolis*, is to study the curves which, in modern terms, are written as $x^m y^n = k$, where m and n are integers. Without limitation, it is possible to suppose $|n| > |m|$. Torricelli, however, always refers to specific cases and never resorts to algebraic methods, but to geometrical ones. The memoire was far from being ready for publication. Therefore, the material is not well ordered and it is necessary to offer a line of interpretation. It is useless to present the propositions one after the other. In the following we take into account Belloni (and Carruccio) as Torricelli's *Opere* editors and Bortolotti (1925a, b, 1928b, 1939a, b; see also Torricelli 1919–1944, pp. 231–274).

Let us start to follow the excellent line of interpretation offered by Bortolotti. It is possible to identify a fundamental lemma, which, Bortolotti transcribes like this in modern terms (Bortolotti 1925b, pp. 946–948; see Fig. 46):

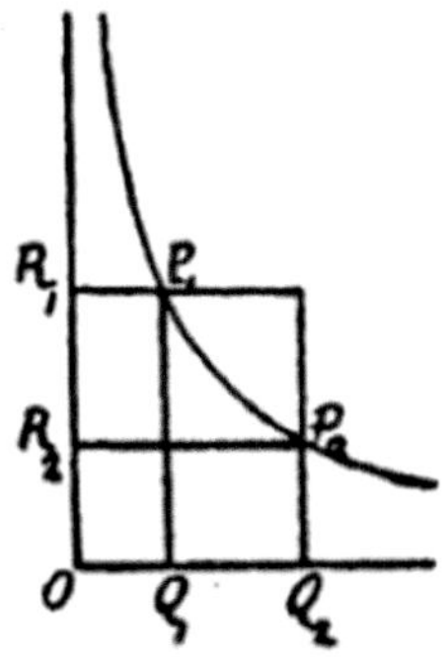

Fig. 46 The diagram used by Bortolotti to prove Torricelli's fundamental Lemma (Bortolotti 1925b, p. 946). *Source* Public domain

Be given the curve $x^m y^n = k$. Consider two points $P_1 \equiv (x_1, y_1), P_2 \equiv (x_2, y_2)$ belonging to it. Be $Q_1, Q_2; R_1, R_2$ their projections on the axes. Complete the rectangles $P_1Q_2, P_2Q_1, P_1R_2, P_2R_1$.

The Lemma can be divided in three main parts as in the following:

(1) The ratio of the inscribed rectangles is greater than that of the exponents, which is greater than that of the circumscribed rectangles. Namely:

$$\frac{P_2Q_1}{P_1R_2} > \frac{n}{m} > \frac{P_1Q_2}{P_2R_1}$$

(2) Rotate the figure around the y-axis. The ratio of the solids generated by the rotation of the inscribed rectangles, the exponents' ratio and that of the solids generated by the rotation of the circumscribed rectangles satisfy the following law:

$$\frac{\text{solid generated by } P_2Q_1}{\text{solid generated by } P_1R_2} > \frac{2n}{m} > \frac{\text{solid generated by } P_1Q_2}{\text{solid generated by } P_2R_1}$$

(3) Analogous relations hold if the rectangles rotate around the x-axis.

Bortolotti (Bortolotti 1925b) transcribes the geometrical reasonings into algebraic and general terms reaching the demonstrations in few passages. However, Torricelli analyses some specific cases and offers geometrical proofs which are rather complex. The most complete of these proofs is relative to the hyperbola.

$$xy^3 = k.$$

It is useful to report it for the reader to understand the profound difference between Torricelli's way of reasoning in respect to an algebraic approach (Fig. 47). In the following we come back to refer to Loria and Vassura as Torricelli's *Opere* editors (Torricelli 1919–1944).

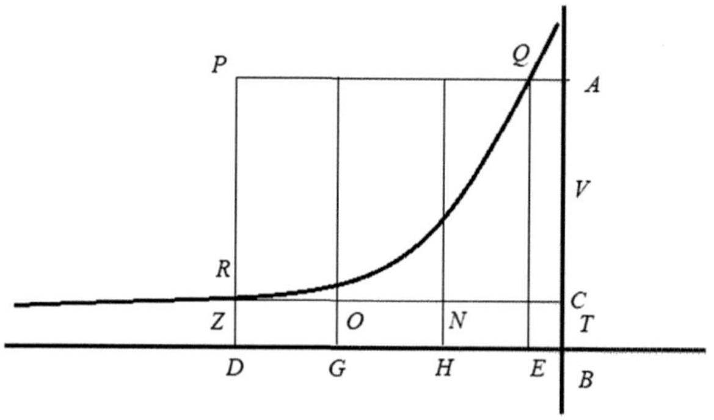

Fig. 47 Our representation of the figure used by Torricelli to explain the reasoning above. (Torricelli 1919–1944, I, 2, p. 246)

Consider two points $Q \equiv (E, A); R \equiv (D, C)$ belonging to the hyperbola. It is $AB^3 : BC^3 = DB : BE$. Consider two mean proportionals GB, HB between DB and EB. Since it holds.

$$DB : GB = GB : HB = HB : EB,$$

it is also

$$DG : GH = GH : HE = DB : GB = GB : HB = HB : EB$$

Because it is

$$DB : GB = GB : HB,$$

by subtracting you have

$$DB : (DB - GB) = GB : (GB - HB)$$

$$DB : GB = DG : GH$$

The proportion involving GH and HE is obtained in the same manner. With a further reasoning on the proportions, it is possible to prove that it is also:

$$AB : BC = DG : GH = GH : HE = DB : GB = GB : HB = HB : EB$$

Since $AB : BC = HB : EB$, the rectangles of diagonals AE and HC are equal. Thus, the rectangles AN and HN are also equal. Basing on this fundamental acquisition, with another argument, Torricelli demonstrates that the ratio between the rectangles DN to AN is greater than 3 (Torricelli wrote "4", but we are confident that he meant "3"). The proves concerning the items (2) and (3) of the fundamental Lemma (Cfr. Bortolotti 1925b, p. 946) are obtained by other arguments which also include the use of barycentres.

Torricelli analyses some other specific cases, for example $m = 3$, $n = 7$.

Bortolotti proved the lemma through algebra. His demonstrations cannot be ascribed to Torricelli because they belong to an order of ideas which is not that of Torricelli, who reasoned through geometry.

The next step of Torricelli's analysis of infinite hyperbolas concerns the quadratures (Fig. 48):

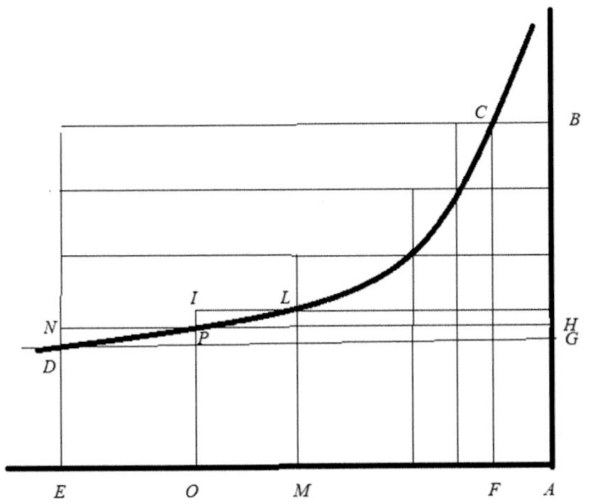

Consider the hyperbola *CD*. Given an orthogonal system of coordinates with origin *A* where the asymptotes *AE* and *AB* of the hyperbola mutually intersect, be (*F*,*B*) and (*E*,*G*) the coordinates of *C* and *D*[58]. The mixtilinear quadrilateral *EDCF* is to the mixtilinear quadrilateral *DCBG* as the exponent *n* of *BA* is to the exponent *m* of *EA*.

Fig. 48 Our representation of the figure used by Torricelli to explain the reasoning above. (Torricelli 1919–1944, I, 2, p. 256)[59]

Torricelli developed a reasoning by exhaustion in the following way:

Suppose that $\frac{EDCF}{DCBG} < \frac{n}{m}$ and that $\frac{EDCF}{D'C'B'G'} = \frac{n}{m}$.
Bisect continuously the rectangle *DLBG* till reaching a rectangle *DH* such that its area is less than $DCBG - D'C'B'G'$.
Inscribe in *DCBG* a figure composed of rectangles as high as *DH*. They are the rectangles *IG*, *LH*, etc.
Therefore, the inequality $\frac{EDCF}{\sum \text{rectagles}_{IG,LH,...}} < \frac{n}{m}$ also holds.
However, this is absurd. As a matter of fact, $\frac{DO}{IG} > \frac{n}{m}$, because of the fundamental Lemma (Bortolotti 1925b, p. 946).
Thus, the quadrilinear *EDIO* is such that $\frac{EDIO}{IG} > \frac{n}{m}$. The same reasoning holds for the rectangles *PM* and *LH* and for all the inscribed rectangles.
Therefore, the whole quadrilinear *EDCF*, which is bigger than the sum of the rectangles *DO*, *IM*, etc has a ratio greater than *n/m* in respect to the sum of the rectangles *IL*, *LH*, ..., which is absurd.

By means of an analogous reasoning, Torricelli also proved to be absurd the hypothesis

$$\frac{EDCF}{DCBG} > \frac{n}{m}$$

One of the elements in this ingenious proof—as it is typical of many reasonings by exhaustion—consists in regarding a mixtilinear figure as composed of rectilinear figures which approximate it better the more numerous they are. The idea behind these reasonings is that they are valid whatever the number of composing figures. One can say that they are valid even in the case that their number tends to infinity.

This is what Torricelli proved.

[59] Torricelli does not use this language typical of analytical geometry, but, in this case, a more modern language does not betray his thought. It must be pointed out that the diagram (Torricelli 1919–1944, I, 2, p. 256) has many defects. Instead of the letter *F* you find *P*. The letter *I* is not indicated. We have not seen the original diagram in the manuscript, but at least a note by the editors would have been necessary to explain the defects of such a figure. In our adapted diagram we edited it.

Bortolotti (1925a, b; see also 1928b, 1939a, b) started from Torricelli's result that

$$\frac{EDCF}{DCBG} = \frac{n}{m}$$

and developed a series of reasonings, which are not present in Torricelli's work, to reach the conclusion that

$$\int_{x_1}^{x_2} x^{\frac{m}{n}} dx = \frac{1}{\frac{m}{n}+1}\left(x_2^{\frac{m}{n}+1} - x_1^{\frac{m}{n}+1}\right)$$

Bortolotti does not claim explicitly that Torricelli achieved this result. But it is rather obvious that this is his idea. Through the fundamental Lemma (Bortolotti 1925b, p. 946) Torricelli is also able to determine the tangents at any point of a hyperbolic curve. His theorem states that (Fig. 49):

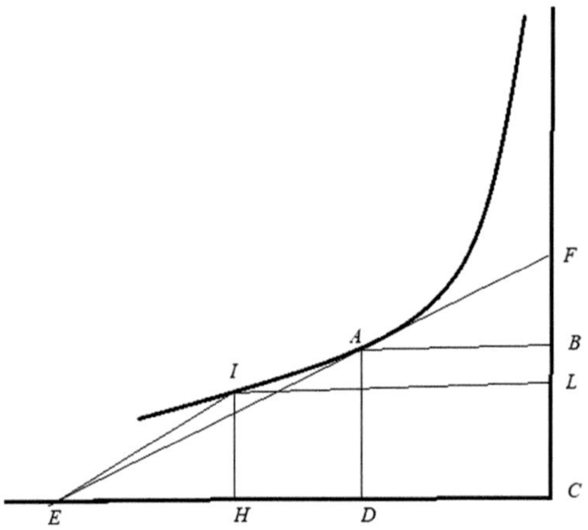

Given a hyperbola of asymptotes *EC* and *CF*, it is required to find the tangent at *A*. Draw *AB* and *AD* resp. parallel to the asymptotes. Choose the point *E* such that $ED\colon DE = CB\colon BF = n\colon m$. The straight line *EF* is the required tangent.

Fig. 49 Our representation of the figure used by Torricelli to explain the reasoning above. (Torricelli 1919–1944, I, 2, p. 257)

Bortolotti (Bortolotti 1925b) starting from the fundamental lemma by Torricelli develops the following interesting reasoning, which also includes some propositions proved in the memoire *De infinitis parabolis* (Torricelli 1919–1944, I, 2, pp. 275–328).

The context concerns the way to draw the tangent to a parabola of the form:

$$y^n = kx^m$$

Torricelli considered the parabola *ABC*. The quadrilinear *BCFE* is to the quadrilinear *BCGH* as *n* to *m*. Bisect *FE* in *I*. Draw the abscissa *ID* and from *D* the ordinate *DL*. The ratio *n*/*m* also holds between the two quadrilinears *DCEI* and *DCGL*. Now Torricelli claims:

> If we will do this division an infinite number of times or if we suppose that such a division is done, two lines *CE* and *CG* will remain instead of the two figures [the quadrilinears]. These lines will be in the same ratio as the exponents, not according to their length [longitudine] but according to their quantity.[60]

Bortolotti developed (Bortolotti 1925a, b) the following argumentation: if you consider *AE* and *AG* as the coordinates of a point namely (to use Torricelli's language, if you consider them in length) the equation $AE^n = kAG^m$ subsists. If you regard them as indivisibles, because of the conservation of the formal properties, the ratio *n*/*m* holds. Therefore, according to Torricelli, the indivisibles are composed of two factors: a finite factor, which is the ordinate or the abscissa of a point belonging to the curve and an infinitesimal factor. For the ordinate such a factor is the thickness of the abscissa and for the abscissa the thickness of the ordinate. Thus, these indivisibles would be written in modern symbolism as *ydx* and *xdy*. Hence,

---

[60] "Et se faremo, o supporremo fatta questa divisione infinite volte, resteranno in cambio di figure due linee CE, et CF [CG], le quali, non secondo la longitudine, ma secondo la quantità saranno nel medesimo modo come l'esponente all'esponente." (Torricelli 1919–1944, I, 2, p. 322; Authors' translation). We point out that there is a mistake: sometimes you find letter *F* instead of the letter *G* in the Torricelli's text. But it is obvious when *G* is necessary. The editors do not note this problem. In the translation, we have reported the correct letter. In the original Italian text both those correct and those incorrect.

these are elements of areas, operating with which it is possible to find the tangents to the hyperbolas or parabolas. Therefore, the relations between the elements of area can be written as

$$\frac{ydx}{xdy} = \frac{n}{m}$$

According to Bortolotti (Bortolotti 1925a, b, 1928b), Torricelli arrived very close to the concept of integral and of differential. Both Torricelli's use of exhaustion and of indivisibles can be interpreted as fundamental steps towards infinitesimal calculus[61]. Especially the previous relation is considered fundamental by Bortolotti because it reduces all the properties of a geometrical figure to a unique principle. This represented a novelty, which influenced Barrow and was at the basis of Leibniz's concept of differential (Bortolotti 1939a, b, pp. 466–467; Bussotti and Pisano 2017; Bussotti 2015). Thus, Bortolotti claims that:

(a) Though Torricelli does not speak of differential triangle, he supposes implicitly its existence (Bortolotti 1939a, b, p. 471);
(b) Torricelli's concept of definite integral was absolutely clear (Bortolotti 1939a, b, p. 471–472).

5 Infinitesimal Calculus and Indivisibles

In this section, we do not propose addressing all the nuances of the concept of indivisible. This would be too broad a perspective for our Introduction. Rather we propose these ideas focusing on Torricelli:

(1) From a conceptual point of view, indivisibles represent, so to speak, "a third road" to reach results concerning curvilinear or mixtilinear figures, as well as to solve physical problems. The two others are the exhaustion method and infinitesimal calculus.
(2) The concept of limit is not involved in indivisibles. This is what prevents us seeing in the indivisible technique a precursor of infinitesimal calculus. Obviously, this does not exclude that the inventors of infinitesimal calculus were inspired by Cavalieri and Torricelli, nor that the results expressed through indivisibles can be transformed into the language of differential and integral calculus.
(3) Torricelli's indivisibles are different from Cavalieri's. In the former, the concept of actual infinitesimal is involved, whereas in Cavalieri this is not the case.
(4) The method closer to indivisibles is that of Archimedes's barycentres.

An interesting account of Torricelli's ideas on indivisibles can be found in *De infinitis parabolis*. The following explanations concern the section entitled "Delle tangenti delle parabole infinite per lineas supplementares" ("On the tangents to the infinite parabolas through supplementary lines"; Torricelli 1919–1944, I, 2, pp. 320–323).

Torricelli claimed it to be false that all the indivisibles of the same kind are equal. Not all the points are equal. The lines have a width. The width of different lines is different. The surfaces have a thickness. The thickness of different surfaces is different. Torricelli's examples are clear: consider two concentric circles and, from the centre, draw a series of straight lines cutting both the external and the internal circumference. There is a bi-univocal correspondence between the points cut by such lines in the internal and in the external circumference. However, these points are not equal: be p_e, p_i resp. points belonging to the external and the internal circumference and d_e, d_i the respective diameters. It is:

$$p_e : p_i = d_e : d_i$$

[61] Bortolotti also addressed questions such as the fundamental theorem of calculus and indefinite integrals in Torricelli, but what we have expounded is sufficient to give a clear idea of his opinions. Bortolotti expressed the idea that Torricelli's conceptions are close to those of the inventors of infinitesimal calculus in several works. Although we do not share his opinion, his contribution is probably the most complete and informative on Torricelli (Bortolotti 1921, 1922, 1923, 1924, 1925a, b, 1928a, b, 1931, 1937, 1939a, b, 1947). An author who wrote important contributions on Torricelli is Amedeo Agostini. Basically, he followed Bortolotti's train of thought. (Agostini 1930a, b, c, 1931, 1949, 1950, 1951a, b, c, d, e).

Consider another significant example given by Torricelli as presented in the following figure (Fig. 50):

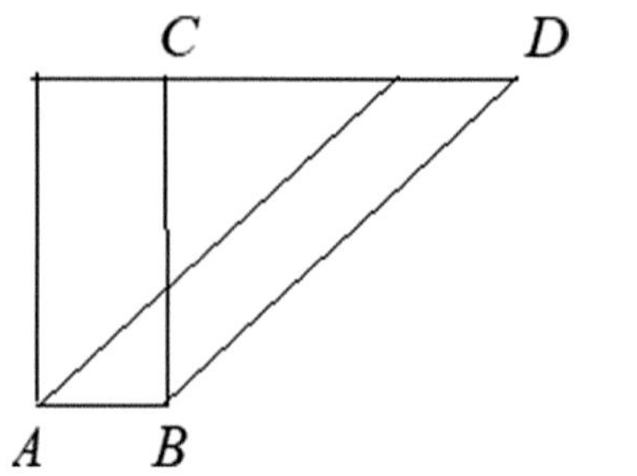

Be given two parallelograms on the same base *AB*. In both of them, draw the infinite lines parallel to the sides ("infinite linee parallele ai lati" (Torricelli 1919-1944, I, 2, p. 321). The aggregate of the straight lines parallel to *BC* will be equal to that of the straight lines parallel to *BD* because the parallelograms are equivalent. Their number is also equal. What is different is their respective length and width. Those in the parallelogram *AC* are wider and shorter than those in the parallelogram *AD*.

Fig. 50 On two parallelograms and infinite lines parallel to the sides (Torricelli 1919–1944, I, 2, p. 321). *Source* Public domain

Hence, it seems to us that no doubt can subsist: according to Torricelli, indivisibles are actually infinitesimal quantities, which compose the figures. The figures are the sum of their indivisibles. The indivisible points, lines and surfaces are not the Euclidean points, lines and surfaces. An infinite aggregate of Euclidean points does not compose a surface, whereas an infinite aggregate of indivisible points does. They are actual infinitesimals whose number is infinite in any finite segment. The same reasoning can be extended to indivisible lines (be they straight or curved) and surfaces.

Torricelli continues with a further interesting observation: Galileo was right while claiming that a point can be equal to a line. He is referring to the famous bowl of Galileo (Fig. 51):

74 DISCORSI E DIMOSTRAZIONI MATEMATICHE

SAGR. Ammirabil proposta veramente mi par cotesta: però sentiamone l'esplicazione e la dimostrazione.

SALV. È necessario farne la figura, perchè la prova è pura geometrica. Per tanto intendasi il mezzo cerchio AFB, il cui centro C, ed intorno ad esso il parallelogrammo rettangolo ADEB, e dal centro a i punti D, E siano tirate le rette linee CD, CE; figurandoci poi il semidiametro CF, perpendicolare a una delle due AB, DE, immobile, intendiamo intorno a quello girarsi tutta questa figura: è manifesto che dal rettangolo ADEB verrà descritto un cilindro, dal semicircolo AFB una mezza sfera, e dal triangolo CDE un cono. Inteso questo, voglio che ci immaginiamo esser levato via l'emisferio, lasciando però il cono e quello che rimarrà del cilindro, il quale, dalla figura che riterrà simile a una scodella, chiameremo pure scodella: della quale e del cono prima dimostreremo che sono eguali; e poi, un piano tirato parallelo al cerchio che è base della scodella, il cui diametro è la linea DE e centro F, dimostreremo, tal piano, che passasse, v. g., per la linea GN, segando la scodella ne i punti G, I, O, N, ed il cono ne' punti H, L, tagliare la parte del cono CHL eguale sempre alla parte della scodella, il cui profilo ci rappresentano i triangoli GAI, BON; e di più si proverà, la base ancora del medesimo cono, cioè il cerchio il cui diametro HL, esser eguale a quella circolar superficie che è base della parte della scodella, che è come se dicessimo un nastro di larghezza quanta è la linea GI (notate intanto che cosa sono le definizioni de i matematici, che sono una imposizion di nomi, o vogliam dire abbreviazioni di parlare, ordinate ed introdotte per levar lo stento tedioso che voi ed io sentiamo di presente per non aver convenuto insieme di chiamar, v. g., questa superficie, *nastro circolare*, e quel solido acutissimo della scodella *rasoio rotondo*): or comunque vi piaccia chiamargli, bastivi intendere che il piano prodotto per qualsivoglia distanza, pur che sia parallelo alla

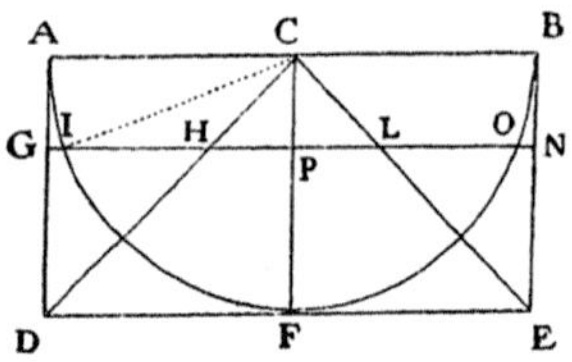

1. *cotesta: e però* — 5. *parellelogrammo rettangulo*, s — 7. *semidiametro CF immobile* — 9. *rettangulo*, s — 21. *per la linea GH, segando*, s — 27. *di larghezza quanto è* — 34. In G era stato scritto *paralello*, che GALILEO corresse di sua mano in *parallelo*. —

Be given the cylinder. Inscribe the cone of vertex *C*. Consider the hemisphere inscribed in the cylinder. Subtract the hemisphere from the cylinder. The remaining solid was called by Galileo "bowl" ("scodella"). He proved that the volume of the bowl is the same as that of the inscribed cone.

Fig. 51 A representation of Galileo's "scodella" (bowl) (Galilei 1890–1909, VIII, Day I, p. 74). *Source* French National Library—Gallica. Public domain

Galileo proved the proposition through indivisibles, demonstrating that each slide s_1 of the bowl has the same surface as the slide s of the cone. If the plane α cutting the figure moves from b_1 to b_2, in the bowl the circumference c remains, while in the cone the point O. Galileo claimed:

Now, since as these solids diminish equality is maintained between them up to the very last, we are justified in saying that, at the extreme and final end of this diminution, they are still equal and that one is not infinitely greater than the other. It appears therefore that we may equate the circumference of a large circle to a single point.[62]

Galileo seems to be astonished at this conclusion, possibly because he neither has an exact concept of limit nor an idea like Torricelli's that indivisibles entities are different from the corresponding Euclidean entities. Torricelli, instead, is not surprised, at all. We read:

What Galileo claims with regard to a point equal to a line is true. In our hyperbolic solid it is true that the infinitely long axis is equal to a cylindric surface, or to a given [quanto] circle.[63]

Torricelli is arguing that the indivisible given by the infinitely long axis of the solid is regarded as an infinitely long cylinder whose base circumference has an infinitesimal radius; infinitesimal, but not null. Multiplying the infinitesimal base circumference by an infinitely long height, it is possible to obtain a finite quantity equal to the other cylinders, which compose the hyperbolic solid. Thus, the length of the axis is an infinite *given* quantity and the length of the base circumference is an infinitesimal *given* quantity. There is no limit process. This is not only a theoretical assumption. As a matter of fact, the proofs by infinitesimals we have analysed do not involve considerations based on the concept of limit, which is, instead, the fundamental one in infinitesimal calculus. Bortolotti (1939a, b), as to the axis of the hyperbolic solid writes:

At limit, when x=0, the cylindric surface [involucro] degenerates in the axis of the conoid. Its length is infinite and its radius is null. However, because of the formal properties' conservation, we can consider, at the limit, too, the area of a cylindric surface [involucro] equal to that of the circle *OD* [one of the circles composing the cylinder at which the hyperbolic solid is equal]. As a matter of fact, the circles of the cylinder *OAED* [the circle at which the solid is equal] are to be regarded as discs of infinitesimal thickness. The cylindrical surfaces [involucri], as well, are not Euclidean surfaces, but have an infinitesimal thickness.[64]

By using the modern concept of limit, the identity between the volume of the hyperbolic solid V_{hs} and the cylinder's V_c means:

$$\forall \varepsilon > 0, \exists \delta > 0 / |x| < \delta \rightarrow |V_{hs} - V_c| < \varepsilon.$$

Therefore, the known volume of the cylinder is as good an approximation as you could desire of the hyperbolic solid's volume. There is no consideration regarding the axis of the solid. However, Torricelli does not reason in this manner. He considers the hyperbolic solid as composed of an infinity of hollow cylinders whose height is finite and whose thickness is infinitesimal. The finite cylinder of equal volume is instead composed of an infinity of cylinders whose base surfaces are finite and whose height is infinitesimal. Therefore, if the first paragraph of Bortolotti's quotation (see above; Bortolotti 1939a, b) is taken as referring to the modern concept of limit, it is, at best, obscure. If it is taken as interpreting as the scientific cognitive process which led Torricelli to his result, it is possibly correct. With regard to the second paragraph of the previous passage, if "infinitesimal" means "actual infinitesimal", the idea expressed by Bortolotti (*Ibidem*) is agreeable. But this seems in contradiction to the whole interpretation advocated by Bortolotti (*Ibidem*). Hence, by the term "infinitesimal" he means "potentially infinitesimal". In contrast to this, we have argued that Torricelli's conception has nothing to do with the notion of limit. An actually infinite number of objects whose magnitude is actually infinitesimal is at stake. Thus,

---

[62] "Or mentre che nella diminuzione de i due solidi si va, sino all'ultimo, mantenendo sempre tra essi la egualità, ben par conveniente il dire che gli altissimi ed ultimi termini di tali menomamenti restino tra di loro eguali, e non l'uno infinitamente maggior dell'altro: par dunque che la circonferenza di un cerchio immenso possa chiamarsi eguale a un sol punto." (Galilei 1890–1909, VIII, Day I, p. 75; Translation from Galilei 1954, p. 28).

[63] "Quello che dice il Galileo di un punto eguale ad una linea è vero e nel nostro solido hyperbolico è vero che l'asse infinitamente lungo è eguale ad una superficie cilindrica, overo ad un circolo quanto." (Torricelli 1919–1944, I, 2, p. 321: Authors' translation).

[64] "Al limite per x=0, l'involucro cilindrico degenera nell'asse del conoide, la sua lunghezza è infinita ed il raggio nullo. Ma per la conservazione delle proprietà formali, possiamo considerare, anche al limite, uguale l'area di un involucro cilindrico a quella del cerchio *OD*. Nel fatto poi, i cerchi del cilindro *OAED*, sono da considerarsi come dischi di spessore infinitesimo, ed anche gli involucri cilindrici, non sono superfici euclidee, ma hanno spessore infinitesimo." (Bortolotti 1939a, b, p. 470).

according to our view, Torricelli considers the figures as sums of infinitesimal magnitudes of the same kind. Torricelli's segments are sums of extended points, which, in fact, are infinitesimal segments. *Mutatis mutandis*, the same considerations hold for Torricelli's surfaces and solids. One can say that Torricelli developed, in part, a calculus with actually infinite and infinitesimal quantities: his indivisibles. It is a different road both from the exhaustion method and from infinitesimal calculus. While referring to the secondary literature, Mancosu and Vailati expressed an opinion similar to ours:

> [...] he [Torricelli] conceived of curved indivisibles as having a thickness; Cavalieri's indivisibles have a lower dimension than the figure they characterize, and therefore one cannot speak of a thickness of indivisibles in his account. Finally, Torricelli substituted for the complexity of Cavalieri's approach (in which the geometric figure and the class of indivisibles were considered as two separate magnitudes) a simple identification of the figure with its characterizing indivisibles (Mancosu and Vailati 1991, p. 53).

The two authors continue analysing Torricelli's statement and proof of the theorem on the acute hyperbolic solid. They express it like this:

> The theorem as it stands, is about the proportion between two figures; an infinitely long solid and a cylinder. Thus the notion of actually infinite length is present in the very statement of the theorem. Moreover, the proofs themselves, as they stand, make sense only if the hyperbolic solid is given as infinitely long *in actu*. For we are comparing the volume of a fixed determinate solid with that of a cylinder, and there is no indication that the volume of the hyperbolic solid is thought of as the limit of a succession of volumes converging to that of the cylinder (Mancosu and Vailati 1991, p. 57).

The same opinion is expressed by Alexander who writes:

> In fact, there were important differences between Cavalieri's and Torricelli's approaches to infinitesimals. Most critically, in Torricelli's method all the indivisible lines taken together really did make up the surface of a figure, and all the indivisible planes actually composed the volume of a solid. Cavalieri, as will be recalled, worked hard to avoid this identification, speaking of "all the lines" as if they were different from a plane and of "all the planes" as if they were different from a solid. But Torricelli had no such qualms. In his proof, he moves directly from "all the lines" to "the area itself" and from "all the planes" to "the volume itself", without bothering with the logical niceties that so concerned the elder. (Alexander 2014, p. 108).

Bascelli highlights that Torricelli's points are not indistinguishable entities, and neither are his linear indivisibles. Furthermore, Torricelli's indivisibles have the same dimension as the continuum they compose. They do not have a dimension less. After these agreeable considerations Bascelli claims:

> Hence, the geometry of indivisibles is the usual plane geometry applied to an infinitesimal scale where dimensions are minute and finite (Bascelli 2015, pp. 127–128).

It is not possible that the dimensions of Torricelli's indivisible are finite. They are actually infinitesimal. As Bascelli herself admits (Bascelli 2015 pp. 112–113) and as it is absolutely evident from Torricelli's work, a figure is composed of an infinite number of indivisibles. This would be impossible if their dimensions were finite. They have to be actually infinitesimal. Obviously admitting actually infinitesimal quantities implies several problems, but this notion is crucial for Torricelli's geometry (Cfr. De Gandt 1987, 1992).

Torricelli had a less problematic attitude than Cavalieri towards indivisibles. The latter addressed many logical nuances of indivisibles. He wrote with regard to their nature:

> I have used an artifice [the indivisibles] similar to that used by the algebraists to solve the given problems. Actually, though being the roots ineffable, absurd, and unknown, they add, subtract, multiply and divide them. They are convinced to have done their duty provided that they find the required information from the given problem. Thus, in the same manner, I have resorted to the multiplicity of the indivisibles (be they lines or planes) to investigate the measure of the continuous quantities. [...] This has been the case, though their number is unmentionable, absurd and unknown. It is, however, included in visible limits as to its size.[65]

[65] "*Artificio autem tali sum, quale ad propositas quaestiones ab solvendas Algebratici adhibere solent; qui quidem numerorum radices, quamvis ineffables surdas, ac ignotas, nihilominus simul aggregantes subtrahentes, multiplicantes, dividentes, dummodo proposite rei exeptatum sibi notitia enucleare valeant, sua satis obijsse numera sibi persuandent, Non aliter ergo indivisibilium sive linearum, sive planarum congerie (ijsdem ut in lib. 2. explicatur assumptis) licet quoad eorundem numerum innominabilis, surda, ac ignora, quoad magnitudinem tamen conspicuis limttibus clausa, ad continuorum investigandam mensuram usus sum, ut legenti in processu operis apparebit.*" (Cavalieri 1635, *Prefatio*, p. 3 (s. n.); Cavalieri italic style. Authors' translation).

Therefore, Cavalieri is claiming that his indivisibles are not necessarily interpretable as real mathematical entities, which compose the continuum. They can be regarded as artifices useful to reach a certain purpose. They are subject to precise rules. For example, it is necessary to distinguish "retto transito" and "obliquo transito".[66] However, they have the same status as the imaginary numbers used to solve the equations. They are a part of the symbolism. This means that the continuum is not composed of indivisibles, but it can be studied through them. Actually, it is not clear to grasp exactly what Cavalieri thought in regards to his indivisibles. It is certain that he assumed the quasi-operationalist position we have referred to, but what were the indivisibles from an ontological point of view, according to Cavalieri? The historians of mathematics have expressed different opinions: Guido Castelnuovo (1865–1952) suggested that, according to Cavalieri, a flat, finite area is composed of the infinite numbers of chords intercepted by a system of parallel straight lines. Each of these chords is regarded as an infinitesimal rectangle, an idea which we ascribe to Torricelli's indivisibles rather than to Cavalieri's. However, Castelnuovo continued, the Jesuat scientist denied resorting to actual infinitesimals and ascribes the interpretative difficulties involved in indivisibles to the ambiguities of Cavalieri's language. Castelnuovo claims that Cavalieri made no mistake, but that his indivisibles cannot be regarded as an anticipation of the concept of definite integral because there is no reference to the concept of limit, a step which was taken by Pietro Mengoli (1626–1686) thirty years before Newton (Cfr. Castelnuovo 1962, pp. 44–49). Lucio Lombardo Radice (1916–1982) in his interesting Introduction to Cavalieri 1966, claims two theses:

(a) The rationale behind the method of indivisibles is essentially different from that behind integral calculus (Lombardo Radice in Cavalieri 1966, p. 10);
(b) The indivisibles are the first example in which the mathematics of infinity is developed.

Cavalieri laid the initial bases of operations with infinite sets: he did not think that a continuum of degree n is the sum of an infinite aggregate of continua whose degree is n-1. However, the points of a line, the lines of a flat region and the planes of a solid are perfectly conceivable. Each of these three aggregates is composed of an infinite number of elements (Lombardo Radice in Cavalieri 1966, pp. 24–26 and pp. 652–653). Andersen, Giusti and Jullien avoided both the theses that Cavalieri was a precursor of integral calculus and of infinite sets (Andersen et al. 2015; Jullien 2015). There is no sum, as is the case with integral calculus, in Cavalieri's method. On the other hand, Cavalieri's method seems directly connected to the measure of a figure, which can hardly be obtained in a set-theoretical context without distorting some of Cavalieri's basic assumption. Thus, indivisibles enter the general—and admittedly not always transparent—notion of magnitude. After a refined analysis of the notion of the indivisible, the three authors attempt to draw a conclusion: although Cavalieri avoided speaking of the sum of indivisibles and even if most of his theorems do not require such an assumption, there are some basic propositions in which the idea that a figure is composed of the sum of its indivisibles seem necessary (Andersen et al. 2015, p. 55; Jullien 2015). The legacy of Cavalieri's indivisibles had already been given by Giusti and Andersen in their contributions published resp. in 1980 and 1985. Here the procedures of the Jesuat mathematician are transcribed into a clear calculus; the differences with integral calculus are pointed out; the collective and distributive method is clearly explained. Andersen also clarifies that, among the European mathematicians, Torricelli's conception of indivisibles rather than Cavalieri's was spread (Cfr. Andersen 1985, pp. 353–363).[67]

Torricelli's attitude is more realistic and pragmatic: the indivisibles really compose the continuum. This does not mean that he was not interested in the logic behind indivisibles' correct use. He wrote, e.g., a text in which a series of incorrect utilizations of indivisibles are presented (Torricelli 1919–1944, I, 2, pp. 415–432). For each example, he developed a set of logical and mathematical considerations aimed at revealing the mistakes hidden in the reasoning. But undoubtedly he had a realistic way of conceiving the indivisibles.

Finally, as we have shown, it must be noted that Torricelli's conception of indivisibles and of infinite and infinitesimal quantities is rather different from that of the inventors of infinitesimal calculus. In the same period in which Torricelli

[66] On the distinction between "retto transito" and "obliquo transito" see, e.g., Cavalieri's letter to Galileo on 2 October 1634 (Galilei 1890–1909, XVI, p. 138). This letter is also presented by Cavalieri (Cavalieri 1966, pp. 754–757: see also Andersen 1985; pp. 308–310; Mancosu 1996, p. 236).

[67] With regard to the spread of indivisibles the whole collective book Jullien 2015 is a precious document. From a mathematical perspective an interesting paper is Carroll, Dougherty and Perkins 2013, where the authors revisit in modern forms some techniques by Torricelli and Roberval, without betraying their original procedures. They also valuate if Guldin's criticisms were justified. Among the works dealing with Cavalieri's indivisibles we also mention De Gandt 1991, where a plurality of noteworthy questions are addressed. The author clarifies that the choice to admit the existence of indivisibles is, eventually, axiomatic; Festa 1992, where the author, among other themes, also faces the issue concerning the relation between a figure and its indivisibles; Palmieri 2009. This is a remarkable paper on the manner in which Cavalieri used the technique of superimposing figures.

conceived the indivisibles, the author who had ideas closer to those connoting calculus was surely Fermat (Bussotti and Pisano 2020). This does not exclude that, as already mentioned, the inventors of calculus were inspired by Cavalieri's and Torricelli's ideas. We do not deal with this subject because the literature is relatively abundant. We prefer to stress that a concept, which seems in filiation with the notion of indivisible exists in physics. It is that of the "minimal lines of given length" or "minimal surfaces of given thickness".

In the Proposition XXXIX of section VII, Book I of *Principia*'s first book Newton solved the following problem:

> Supposing a centripetal force of any kind, and granting the quadratures of the curvilinear figures; it is required to find the velocity of a body, ascending or descending in a right line, in the several places through which it passes; as also the time in which it will arrive at any place; And vice versa.[68]

In the course of the proof, Newton used this concept (Fig. 52):

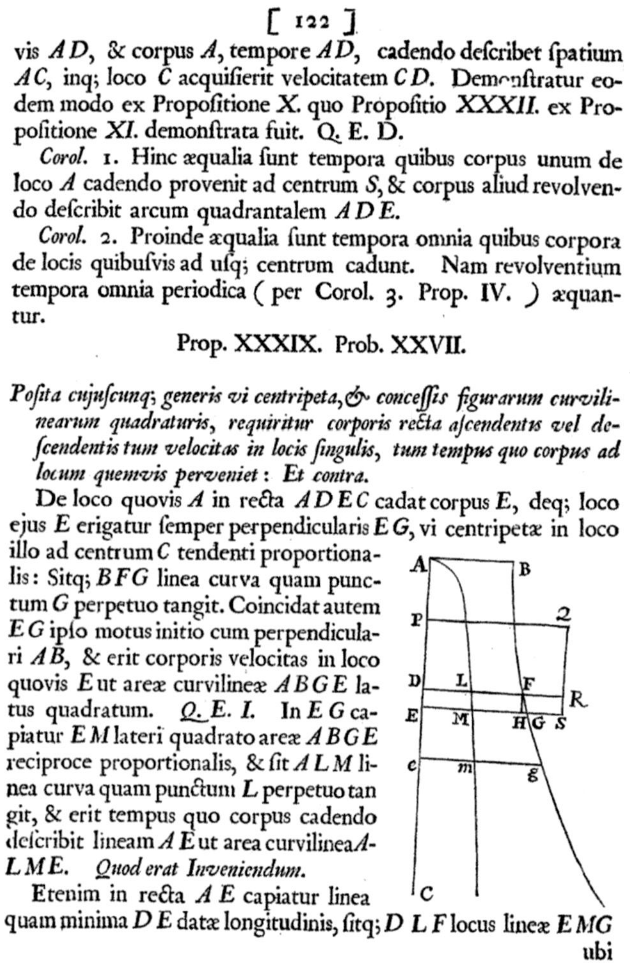

[122]

vis *AD*, & corpus *A*, tempore *AD*, cadendo deſcribet ſpatium *AC*, inq; loco *C* acquiſierit velocitatem *CD*. Demonſtratur eodem modo ex Propoſitione *X*. quo Propoſitio *XXXII*. ex Propoſitione *XI*. demonſtrata fuit. Q. E. D.

Corol. 1. Hinc æqualia ſunt tempora quibus corpus unum de loco *A* cadendo provenit ad centrum *S*, & corpus aliud revolvendo deſcribit arcum quadrantalem *ADE*.

Corol. 2. Proinde æqualia ſunt tempora omnia quibus corpora de locis quibuſvis ad uſq; centrum cadunt. Nam revolventium tempora omnia periodica (per Corol. 3. Prop. IV.) æquantur.

Prop. XXXIX. Prob. XXVII.

Poſita cujuſcunq; generis vi centripeta, & conceſſis figurarum curvilinearum quadraturis, requiritur corporis recta aſcendentis vel deſcendentis tum velocitas in locis ſingulis, tum tempus quo corpus ad locum quemvis perveniet: Et contra.

De loco quovis *A* in recta *ADEC* cadat corpus *E*, deq; loco ejus *E* erigatur ſemper perpendicularis *EG*, vi centripetæ in loco illo ad centrum *C* tendenti proportionalis: Sitq; *BFG* linea curva quam punctum *G* perpetuo tangit. Coincidat autem *EG* ipſo motus initio cum perpendiculari *AB*, & erit corporis velocitas in loco quovis *E* ut areæ curvilineæ *ABGE* latus quadratum. *Q. E. I.* In *EG* capiatur *EM* lateri quadrato areæ *ABGE* reciproce proportionalis, & ſit *ALM* linea curva quam punctum *L* perpetuo tangit, & erit tempus quo corpus cadendo deſcribit lineam *AE* ut area curvilinea *ALME*. *Quod erat Inveniendum.*

Etenim in recta *AE* capiatur linea quam minima *DE* datæ longitudinis, ſitq; *DLF* locus lineæ *EMG* ubi

In a phase of his reasoning, Newton wrote:

"In the straight line *AE*, be taken the as small as possible line *DE* of a given length and be *DLF* the place of the line *EMG* when the body was in *D*."[68]

Fig. 52 The diagram used by Newton in Proposition XXXIX, Problem XXVII. (Newton 1686, I, p. 122). *Source* French National Library—Gallica. Public domain[69]

[68] Newton [1726]1729, p. 163.

[69] "Etenim in recta *AE* capiatur linea quam minima *DE* datae longitudinis, sitque *DLF* locus lineae *EMG*, ubi corpus versabatur in *D*." (Newton 1686, I, pp. 120–123; Authors' translation).

A line "quam minima datae longitudinis" cannot be regarded as a potentially infinite quantity. It seems an indivisible which composes the straight line *AC* and, from a formal point of view, it is. This does not mean that Newton believed his lineola *DE* to be an actual infinitesimal. He did not believe in the existence of actually infinitesimal quantities. The lineola does not belong to the ontology of mathematics. It is an artifice—which is not in contradiction with the rest of Newton's concepts—to reach a certain purpose[70]. In this respect, Newton's "linea quam minima datae longitudinis" is closer to Cavalieri's indivisibles than to Torricelli's.

A comparison between the doctrine of indivisibles and Archimedes's mechanical method can be developed. As we have seen, Torricelli thought that, behind Archimedes's proofs by exhaustion, a heuristic method should exist. Heiberg's discovery in 1906 was a spectacular confirmation of Torricelli's ideas.[71] Furthermore, the mechanical method belongs to an order of thought close to that of indivisibles. As a matter of fact, Archimedes argued a flat figure to be composed of straight lines and a spatial figure composed of surfaces. It is enough to take into account one passage in the first proof given by Archimedes within the argument to square parabola (Fig. 53):

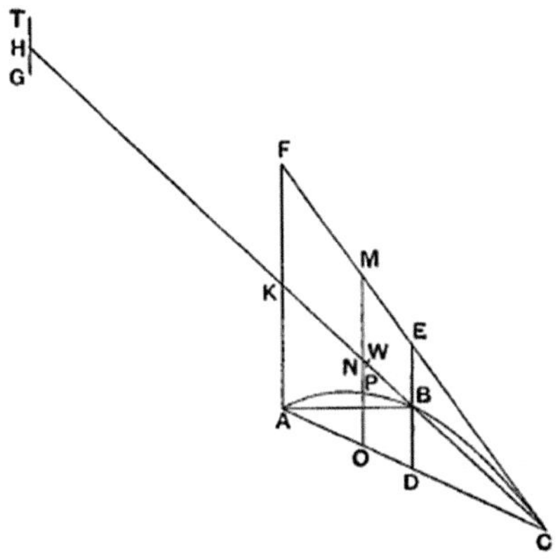

Fig. 53. "And, since the triangle *CFA* is made up of all the parallel lines like *MO*, and the segment *CBA* is made up of all the straight lines like *PO* within the curve [...]". (Archimedes 1912, p. 17).

Fig. 53 The diagram used by Archimedes (Archimedes 1912, p. 16). *Source* Public domain

Therefore, rectilinear figures are composed of an aggregate of straight lines. This idea is common to the doctrine of the indivisibles and to Archimedes's method. The concept to reach the solution of a problem resorting to the lever is a further fundamental feature shared by Archimedes's and some of Torricelli's procedures. Furthermore, the discoveries dating to the beginning of the twenty-first century of the palimpsest containing Archimedes's *Method* have shown us that Archimedes, at least in one circumstance—the cubature of the cylindrical hoof—used an authentic calculus with an actually infinite number of infinitesimal figures without resorting to the lever. This is close to what Torricelli did in many of the examples we have analysed. Therefore, there are several similarities between Torricelli's and Archimedes's methods. From an ontological point of view, it is very difficult to establish whether Archimedes thought that his indivisibles compose the figure or if they were an artifice to reach the required results. In other terms: was Archimedes's view on the decomposition and composition of figures closer to Cavalieri's or Torricelli's? (Cfr. Pisano and Capecchi 2010a; Capecchi and Pisano 2007a, b). The objects used by Archimedes are Euclidean entities, namely the lines have no width and the surfaces have no thickness. Since we exclude that Archimedes has fallen into such a serious mistake, we propend for the first hypothesis: Archimedes used his indivisibles as an artifice. He did not really believe that an infinite aggregate of continua of the order *n*-1 might compose a continuum of order *n*.

---

[70] We have offered a broad treatment of the way in which Newton used this concept and others similar in Pisano and Bussotti 2014a, b, 2017a, b, c; Bussotti 2015, pp. 50–53. It is to point out that in physics until the 19th century it was rather common to resort to infinitesimal quantities whose ontological status was not very clear, but whose use was.

[71] It is probable that the mechanical method was a guide in Archimedes's research. Therefore, under this respect, it has a heuristic value. However, as Acerbi stressed (Cfr. Archimedes 2018, p. 70; see also Archimedes 1897), Archimedes presented his procedure in a highly formalized manner. Thus, Archimedes's presentation was not that typical of a heuristic procedure.

6 Concluding Remarks

Torricelli is a significant example of the use the moderns made of the ancients' mathematical methods, rediscovering and employing their techniques, but adapting them to the needs of the new science.

In the first half of the seventeenth century some of the most skilled mathematicians tried to extend the results obtained by Archimedes to calculate the areas and the volumes of curvilinear and mixtilinear figures. They also aimed at transforming Archimedes's method of exhaustion into a more direct, malleable and general approach, which might allow them to overtake the results obtained by the great Syracusan. They were aware that to resort to infinitesimal methods in order to solve these problems was necessary. Therefore, the crucial question became how to develop a mathematics of infinite and infinitesimal quantities. At the beginning, we note Kepler's excellent works, but—for several aspects—his infinitesimal methods were unsatisfactory. Generally speaking, Bonaventura Cavalieri was one of the first mathematicians to offer a theory of infinitesimal quantities by means of his geometry of indivisibles. Seeting aside some logical obscurities in Cavalieri's concepts, his results were of primary importance.

The main features of Cavalieri's method were:

(1) The geometrical approach. His way of proceeding was still based on proportions rather than on equations. This aspect made his method certainly also applicable to physics (statics) at that time, just before Newtonian mechanics. (This does not mean that the proportions were more suitable to physics than equations).
(2) The scarce use of algebra;
(3) The lack of the concept of limit within indivisibles theory.

Evangelista Torricelli's contributions to geometry share the characteristics which connote Cavalieri's approach. Nonetheless, in Torricelli's work there is something more: he was both interested in solving the mathematical problems, and aimed at finding the tangible nature of the logical—mathematical structure behind geometry. According to his opinion, the exhaustion method, of which Torricelli was an expert and a *virtuoso*, was an excellent demonstrative procedure. However, beyond its lack of generality, it had a further problem: it did not grasp the nature of geometrical relations; it was extremely artificial. It proved the truth of a proposition, but did not go to the most profound root of geometry. In contrast to this, Torricelli was convinced that indivisibles reached the real grounds of the questions: they were a geometrical means through which the figures were decomposed into their atomic elements. For example, given two figures A and B. Let us suppose that all the metric properties (i.e., perimeter, area, volume, depending on it being a flat or a solid figure) of the figure A and of its indivisibles are known. Whereas, the metric properties of the figure B have to be found, while we know some features of its indivisibles and of their relations with those of the figure A. Then, under enough general conditions, it is possible to deduce the metric properties of B.

According to Torricelli, this way of proceeding is purely geometrical, rigorous, direct and grasps the foundations connecting the different figures. It is not a coincidence that Torricelli considered figures as composed of their indivisibles and that gave a fundamental impulse to the development of this doctrine (as Archimedes did with his figures where the masses suspended to a lever moved jointly with it, also called *rigid movement*). For, he reached astonishing results such as those concerning the acute hyperbolic solid. Therefore, Torricelli was a mathematician and a researcher who contributed to the development of several theories within mathematical sciences. He did not reduce geometry to a series of techniques: for him, geometry had proper methods and the importance of using a correct method was as crucial as the result one reaches. We think that this aspect is, perhaps, one of the reasons why Torricelli never cared about algebra. In fact, probably, for his aims, the use of algebra within geometry was and external an improper element. Without any doubt, this lack prevented him from fully appreciating the developments which were taking place abroad, especially in France.

The results and the methods used by Torricelli in geometry suffice to ensure him a primary position within the scientific panorama of the seventeenth century. His extraordinary talent was not restricted to the branches of geometry and mathematics. Torricelli was a brilliant physicist, as well. He showed his talent as a polyhydric scientist also giving fundamental contributions to the study of the fluids and, as a matter of fact, he first conceived atmospheric pressure in the accurate manner. Besides his known studies on the barometer, in *Opera geometrica*, he created (so–called nowadays) "Torricelli principle" in Mechanics (Pisano and Bussotti 2022) which is a crucial conceptual step in the theoretical physics of principles. It states that *two heavy bodies joined together cannot move by themselves unless their common centre of gravity descends* (Torricelli 1644, p. 99; *Id.*, 1919–1944, p. 105). The enunciation of Torricelli's principle actually arose from problems that were extraneous to the theory of the centre of gravity (Capecchi and Pisano 2007a). Specifically, it originated from the attempt to provide a more solid foundation for Galilean mechanics using this principle as a criterion of equilibrium in special cases only (Pisano and Bussotti 2023a, b). By working both as an Archimedean and a physicist, he dealt with the centres of gravity.

We briefly recall two fundamental steps developed by Torricelli in this theory:

(1) The passage from the case study of the centre of gravity of a single body to the case study of that of a system of bodies;
(2) The problem of convergence–parallelism with regard to the motions of heavy bodies falling toward the centre of the Earth.

Torricelli's idea was an analogical extension of the constraint imposed by the centrobaric theory, from the motion of the barycenter of a body to the motion of the barycenter of a system of rigid bodies. This allowed processing (before Newton's mechanics) a purely geometric point as if it were a point-like heavy body (in modern terms, *mass*). The Torricellean extension had both a qualitative nature and a quantitative one. It was supported by detailed geometrical–mathematical arguments. Particularly, his reasonings *ex suppositione* (Cfr. Torricelli 1644, pp. 9–15: Capecchi and Pisano 2010a) were logically consistent because they were confirmed by experience and results obtained by other *equilibrium criteria*, where the causes were investigated, too. He stated his principle as the foundation of dynamics and not of statics only. In fact, Torricelli enunciated his equilibrium principle for two bodies only, and then he extended it to situations of *unstable equilibrium* where the causes should have been discussed. To this end, more and less directly, he used infinitesimal displacements thereby also making clear the interplay between physics (mechanics) and mathematics (geometry). For this reason, his *equilibrium criterion* is substantially equivalent to the modern *principle of virtual works* (Pisano 2017) later on explicitly formulated in 1715 by Johann Bernoulli (1667–1748). Nevertheless and in any case, we remark that Torricelli enounced his principle before Christian Huygens (1629–1695) and the English School of physicists, i.e., Isaac Barrow (1630–1677) and Robert Hook (1635–1703).

Opera geometrica was on the borderline of the birth of modern science, so its scientific state was still in progress to find general principles in order to solve a set of problems with uniform methods; this granted, the Torricelli principle must be appreciated even more.

Finally, in order to inaugurate the celebration of the 380th *Opera geometrica* anniversary (1644–2024), we summarized a selection of Torricelli's most important studies and discoveries. Many others might be added, as, e.g., the invention of a working set square, his further mathematical results concerning the logarithmic spiral. Several of these achievements were obtained in his outstanding *Opera geometrica*. We also discussed the most remarkable results obtained by Torricelli after the publication of his *Opera geometrica*. These results remained unpublished during Torricelli's lifetime. They show that Torricelli was further refining his mathematical technique and making his views broader.

Last but not least, we also underline Torricelli's studies—and—collaboration with Galileo at the end of his life, particularly during the pre-print of *Discorsi and dimostrazioni matematiche intorno a due nuove scienze*.

In conclusion this extensive essay, dedicated to Torricelli's *Opera geometrica* anniversary (1644–2024)—within a celebrative huge volume on *Opera geometrica* –, introduces the magnificent main work in Mechanics, Mathematics and Geometry of a brillant Archimedean—and—Galilean scientist to the modern readers. Evangelista unfortunately died at the age of only 39, in the middle of his productive period as a physicist, geometer and mathematician.

Acknowledgments We thank the anonymous reviewers for their careful reading of our manuscript and their many insightful comments and suggestions. We also express gratitude to Dr. Julie Robarts (University of Melbourne, Australia) for her observations on English language; eventual remaining English mistakes are our own.

References

AA.VV (1768) Raccolta d'Áutori che Trattatano del Moto delle Acque. Seconda Edizione. Tomo Quarto. In Firenze MDCCLXVIII. Nella Stamperia di Sua Altezza Relae Con Licenza de' Superiori.
AA.VV (2009) Galileiana. Journal of Galilean Studies VI:3–70.
Agostini A (1930a) Dimostrazione di una proposizione di Torricelli sulla spirale logaritmica. Periodico di Matematiche 168(4–10):143–151.
Agostini A (1930b) Un brano inedito di Torricelli sulla rettificazione della spirale logaritmica. Il Bollettino di Matematica 170(2–9):XXV–XXXVIII.
Agostini A (1930c) Il problema inverso delle tangenti nell'opera di Torricelli. Archeion 12(1):33–37.
Agostini A (1931) Il problema inverso delle tangenti e l'integrazione indefinita nelle opere del Torricelli. Archeion 13:55–59.
Agostini A (1949) La memoria di Torricelli sopra la spirale logaritmica riordinata e completata. Accademia Navale, Livorno.
Agostini A (1950) Il metodo delle tangenti fondato sopra la dottrina dei moti nelle opere di Torricelli. Periodico di Matematiche 28(4):141–158.
Agostini A (1951a) I baricentri trovati da Torricelli. Bollettino Unione Matematici Italiani, pp. 149–159.
Agostini A (1951b) Massimi e minimi nella corrispondenza di Torricelli con Michelangelo Ricci. Atti del IV Congresso dell'U.M.I.

Agostini A (1951c) Il "De tactionibus" di Evangelista Torricelli. Bollettino Unione Matematici Italiani, pp. 319–321.
Agostini A (1951d) Problemi di massimo e minimo nella corrispondenza di E. Torricelli. Rivista Matematica dell'Università di Parma, pp. 265–275.
Agostini A (1951e) Gli studi di Torricelli sulla cicloide. Accademia Navale, Livorno.
Alexander A (2014) Infinitesimal: How a Dangerous Mathematical Theory Shaped the Modern World. Scientific American/Fakkar. Straus ad Giroux, New York.
Andersen K (1985) Cavalieri's method of indivisibles. Archive for History of Exact Sciences 31:291–367.
Andersen K, Giusti E, Jullien V (2015) Cavalieri's Indivisibles. In: Jullien V (ed). Seventeenth-Century Indivisibles Revisited. Springer, Dordrecht, pp. 31–55.
Archimedes (1897) The Works of Archimedes. Edited by T.L. Heath. Cambridge. The Cambridge University Press, Cambridge.
Archimedes (1912) The Method of Archimedes. Recently Discovered by Heiberg. A Supplement to The Works of Archimedes 1897. Edited by T. L. Heath. Cambridge. The Cambridge University Press, Cambridge.
Archimedes (2018) Metodo. Nel laboratorio del genio. Boringhieri, Torino.
Bachmakova IG (1963) Les Méthodes différentielles d'Archimède. Archive for History of Exact Sciences 2:87–107.
Bascelli T (2015) Torricelli's Indivisibles. In Jullien 2015, pp. 105–136.
Berlingozzi R (1902) Raffaello Maggiotti e la sua opera scientifica nel secolo XVII. Tipografia Varchi, Montevarchi.
Bertoni G (2003) La faentinit di Evangelista Torricelli e il suo vero luogo di nascita. Studi e ricerche del Liceo Torricelli. Ragazzini, Faenza, pp. 5–12.
Bolzano B (1851) Paradoxien des Unendliches. Reclam, Leipzig.
Boncompagni B (1875) Intorno ad alcune lettere di Evangelista Torricelli, del P. Marino Mersenne e di Francesco de Verdus. Bullettino di bibliografia e di storia delle scienze matematiche e fisiche VIII:353–456.
Bortolotti E (1921) Gli inviluppi di linee curve e i primordi del metodo inverso delle tangenti. Periodico di Matematiche 4/1:263–276.
Bortolotti E (1922) Le prime applicazioni del calcolo integrale alla determinazione del centro di gravità delle figure geometriche. Rendiconti della Reale Accademia di Bologna, Bologna.
Bortolotti E (1923) Notizia storica su la "Regola di Torricelli" per la misura dei volumi. Periodico di Matematiche 3/4.
Bortolotti E (1924) La scoperta e le successive generalizzazioni di un teorema fondamentale di calcolo integrale. Archeion 5:205–227.
Bortolotti E (1925a) La memoria "De infinitis hyperbolis" di Torricelli. Archeion 6/1:49–58; see also 6/2:139–152.
Bortolotti E (1925b) La memoria "De infinitis hyperbolis" di Torricelli. Leonardo da Vinci, Roma.
Bortolotti E (1928a) I progressi del metodo infinitesimale nell'Opera geometrica di Evangelista Torricelli. Periodico di Matematiche 8/1:18–59.
Bortolotti E (1928b) Le prime rettificazioni di un arco di curva nella memoria "De infinitis spiralibus" di Torricelli. Rendiconti della R. Accademia di Bologna.
Bortolotti E (1928c) Studi e ricerche sulla storia della matematica in Italia nei secoli XVI e XVII. Zanichelli, Bologna.
Bortolotti E (1930) Le coniche di Apollonia e il problema inverso delle tangenti di Torricelli. Archeion XII/3–4:267–271.
Bortolotti E (1931) L'integrazione indefinita e il teorema di inversione nell'opera di Torricelli. Archeion 1/13:60–64.
Bortolotti E (1937) Il metodo infinitesimale nell'Opera geometrica di Evangelista Torricelli. Atti del primo Congresso dell'Unione matematica italiana. Zanichelli, Bologna, pp. 459–477.
Bortolotti E (1939a) L'infinito e il limite nel Rinascimento italiano. Bollettino della Unione Matematica Italiana Serie II, Anno I.
Bortolotti E (1939b) L'Opera geometrica di Evangelista Torricelli. Monatsschrift für Mathematik und Physik 48:457–486.
Bortolotti E (1947) La storia della matematica nella università di Bologna. Zanichelli, Bologna.
Buscelli T (2010) I fondamenti della nuova scienza del moto: la cinematica di Galileo e la geometria di Torricelli. PhD Dissertation. Scuola di Dottorato di Ricerca in Filosofia. Indirizzo: Filosofia Teoretica. University of Padua, Padua.
Bussotti P (2015) The Complex Itinerary of Leibniz's Planetary Theory. Physical Convictions, Metaphysical Principles and Keplerian Inspiration. Springer–Birkhäuser, Basel.
Bussotti P (2019) The Calculations of Areas and Volumes Using the Method of Archimedes. Some Didactic Considerations. Journal of Baltic Science Education 18/6:812–815.
Bussotti P, Pisano R (2013) On the Conceptual Frames in René Descartes' Physical Works. AHS 2/3:106–125.
Bussotti P, Pisano R (2014) Newton's Philosophiae Naturalis Principia Mathematica « Jesuit » Edition: The Tenor of a Huge Work. Accademia Nazionale Lincei Rendiconti Lincei Matematica e Applicazioni – History of Mathematics 25:413–444.
Bussotti P, Pisano R (2017) Historical and Philosophical Details on Leibniz's Planetary Movements as Physical–Structural Model. In: Pisano, Fichant, Oliveira and Bussotti 2017, pp. 49–92.
Bussotti P, Pisano R (2020) Historical and Foundational Details on the Method of Infinite Descent: Every Prime Number of the Form 4n+1 is the Sum of Two Squares. Foundations of Science (Springer) 25:671–702.
Cantor G, Dedekind R (1937) Briefwechsel. Edited by E. Noether and J. Cavaillès. Hermann, Paris.
Capecchi D, Pisano R (2007a) La teoria dei baricentri di Torricelli come fondamento della statica. Physis XLIV:1–29.
Capecchi D, Pisano R (2007b) Il principio di Torricelli prima di Torricelli. In: Preziosi B (ed) Proceedings of XXIV SISFA Congress. Bibliopolis, Napoli, pp. 107–112.
Capecchi D, Pisano R (2008) La meccanica in Italia nei primi anni del Cinquecento. Il contributo di Niccolò Tartaglia. Proceedings of XXV SISFA Congress. University of Milano, Milano, C17.1–C17.6.
Capecchi D, Pisano R (2010a) Reflections on torricelli's principle in mechanics. Organon 42:81–98.
Capecchi D, Pisano R (2010b) Scienza e tecnica nell'architettura del Rinascimento. CISU, Roma.
Carroll T, Dougherty ST, Perkins D (2013) Infinitesimal and a Tale of Seventeenth-Century Mathematics. Mathematics Magazine 86/4:239–254.
Castelli B (1628) Della misura dell'acque correnti. Stamperia Camerale, Roma.
Castelnuovo G (1962) Le origini del calcolo infinitesimale nell'era moderna con scritti di Newton, Leibniz, Torricelli, Feltrinelli, Milano [First edition *Id.*, 1938, Zanichelli, Bologna].
Cavalieri B (1635) Geometria Indivisibilius continuo rum nuova quadam ratione promota. Ferrone, Bologna.

Cavalieri B (1966) Geometria degli indivisibili. UTET, Torino.
Caverni R (1891–1900) Storia del metodo sperimentale in Italia, 6 vols. Civelli, Firenze.
De Gandt F (1987) L'oeuvre de Torricelli: science galiléenne et nouvelle géométrie. Les Belles Lettres, Paris.
De Gandt F (1991) Cavalieri's Indivisible and Euclid's Canons. In: Barker P, Ariew R (eds) Revolution and continuity. Essays in the history and philosophy of early modern science. The Catholic University of America, Washington, pp. 157–182.
De Gandt F (1992) L'évolution de la théorie des indivisibles et l'apport de Torricelli. Biblioteca Nuncius Studi Testi X:103–118.
Descartes R (1897–1913) Oeuvres de Descartes. Vol 12. In: Adams C, Tannery P (eds) Paris; Discours de la méthode et Essais, Specimina philosophiae. vol VI Principia philosophiae, Latin version vol VIII, Principia philosophiae French translation vol IX; Physico-mathematica vol X, Le Monde ou Traité de la lumière, vol XI (Id, 1964–1974 par Rochot B, Costabel P, Beaude J et Gabbery A, Paris).
De Waard C (1936) L'expérience barométrique ses antécédents, et ses explications. Etude historique. Imprimerie nuovelle, Thouars.
Euclid (2007) Euclid's Elements of geometry. The Greek text of J.L. Heiberg (1883–1885) from Euclidis Elementa, edidit et Latine interpretatus est I. L. Heiberg, in aedibus B.G. Teubneri, 1883–1885. Edited and provided with a modern English translation by Richard Fitzpatrick. Via: http://farside.ph.utexas.edu/Books/Euclid/Elements.pdf.
Fabbri S (1947–1997) La Società Torricelliana di Scienze e Lettere di Faenza nel Cinquantenario della Fondazione. Bollettino della Società Torricelliana di Scienze e Lettere s.n.
Favino F (2009) In Urbe mathematicus: Torricelli a Roma. Galilaeana VI:1000–1032.
Favino F (2019) Entry: Torricelli. Dizionario biografico degli italiani. Vol. 96. Via: https://www.treccani.it/enciclopedia/evangelista-torricelli_%28Dizionario-Biografico%29/.
Festa E (1992) La notion d'«agrégat d'indivisibles» dans la constitution de la cinématique galiléenne : Cavalieri, Galilée, Torricelli. Revue d'histoire des sciences 45/2–3:307–336.
Gàbici S, Toscano F (2006) Scienziati di Romagna. Sironi, Milano.
Galilei G (1632) Dialogo sopra i due massimi sistemi del mondo. In: Galilei 1890–1909 VII:21–520.
Galilei G (1638) Discorsi e dimostrazioni matematiche intorno a due nuove scienze attenenti alla meccanica e i movimenti locali, del signor Galileo Galilei linceo. Elsevirii, Leida. In: Galilei 1890–1909, VIII, pp. 39–362. [1639 Viviani's additions: *Ivi,* VIII, pp. 215–219, pp. 442–445; see also: *Ivi, Frammenti attinenti ai Discorsi e dimostrazioni matematiche intorno a due nuove scienze* VIII, pp. 363–448].
Galilei G (1890–1909) Le opere di Galileo Galilei: Edizione nazionale sotto gli auspici di sua maestà il re d'Italia. Favaro A (ed), Barbèra, Firenze.
Galilei G (1954) Dialogues Concerning Two New Sciences, translated by Henry Crew & Alfonso de Salvio, with an introduction by Antonio Favaro, Dover Publications, Inc., New York [Originally published in 1904 by the MacMillan company].
Galluzzi P (1976) Evangelista Torricelli: Concezione della matematica e segreto degli occhiali. Annali dell'istituto e Museo di Storia della Scienza di Firenze I:71–95.
Galluzzi P (1979) Momento: Studi Galileiani. Ateneo e Bizzarri, Roma.
Galluzzi P, Torrini M (1975–1984) (eds) Le opere dei discepoli di Galileo Galilei. Carteggio. Giunti–Barbera, Firenze, 2 vols.
Ghinassi G (1864) Lettere fin qui inedite di Evangelista Torricelli precedute dalla vita di lui scritta da Giovanni Ghinassi. Pietro Conti, Faenza.
Giusti E (1980) Bonaventura Cavalieri and the Theory of Indivisibles. Cremonese, Bologna.
Gliozzi M (1931) Origini e sviluppi dell'esperienza Torricelliana. Giappichelli, Torino.
Jullien V (2015) (ed) Seventeenth-Century Indivisibles Revisited. Springer, Dordrecht.
Loria G (1897) Evangelista Torricelli e la prima rettificazione di una curva. Rendiconti della Regia Accademia dei Lincei: 5/6:318–323.
Loria G (1903) Un'impresa nazionale di universale interesse. Pubblicazione delle opere di Evangelista Torricelli. Atti II Congresso Internazioanle delle Scienze Storiche, pp. 23–27.
Loria G (1909) In memoria di Evangelista Torricelli. Discorso. Rivista Ligure di Scienze. Lettere ed arti, Genova.
Loria G (1915) L'infinito e l'infinitesimo secondo i matematici dell'antichità e secondo i matematici moderni anteriori al secolo XVIII. Scientia 12 [see also *Id*, 1916, Scientisa I].
Loria G (1917) Una generazione delle coniche a centro ideata da E. Torricelli. Periodico di Matematiche 32:189–192.
Loria G (1919) E. Torricelli nella storia della geometria. Rendiconti della Regia Accademia dei Lincei 5/28:409–415.
Loria G (1922a) Una lettera inedita di E. Torricelli al P. Mersenne. Archivi di storia della scienza 3:272–276.
Loria G (1922b) L'opera geometrica di Evangelista Torricelli. Bollettino di Matematica 2/1:I–VII.
Loria G (1923) Evangelista Torricelli. Gli scienziati italiani. Vol. I, Parte II, pp. 299–311.
Loria G (1932) Una lettera inedita di Evangelista Torricelli. Archeion XIV:12–14.
Loria G (1938) Le prétendu larcin de Torricelli. Archeion XIX.
Loria G (1944) Il III centenario del barometro. Bollettino della Società Torricelliana di Scienze e Letter s.n.
Loria G (1945) Prime linee di una bibliografia torricelliana. Bollettino della Società Torricelliana di Scienze e Letter s.n.
Loria G (1947) L'avverso destino di Evangelista Torricelli. Rendiconti dell'accademia Nazionale dei Lincei 8/3:665–669.
Lorini B (1596). Delle Fortificationi di Bvonaivto Lorini. Nobile Fiorentino. Libri Cinque. Ne' quali si mostra con le più facili regole la Scienza con la ratica, di fortificare le città, & altri luoghi sopra diversi siti [...] Nuovamente dati in luce. Con Priuilegio. In Vinegia, appresso Gio. Antonio Rampanzetto, MDXCVI. BNCF – Palatino 11. 3 .7. 37.
Lorini B (1609) Le Fortificationi di Bvonaivto Lorini. Nobile Fiorentino. Nnuovamente ristampate, Corrette & Ampliate [...] Con l'Aggiunta del Sesto Libro. Con Privilegi. In Venetia, presso Francesco Rampazetto. fortificare le città, & altri luoghi sopra diversi siti [...]. Nuovamente dati in luce. In Vinegia, MDCIX. Presso Francesco Rampanzetto.
Mancosu P (1996) Philosophy of Mathematics and Mathematical Practice in the Seventeenth Century. Oxford University Press, NY-Oxford.
Mancosu P, Vailati E (1991) Torricelli's Infinitely Long Solid and Its Philosophical Reception in the Seventeenth-Century. Isis 82/1:50–70.
Merzbach UC, Boyer CB (2011) A history of mathematics. 3rd edition. Wiley, Hoboken.
Middleton WEK (1963) The Place of Torricelli in the History of Barometer. Isis 54/1:11–28.
Middleton WEK (1964) The History of the Barometer. Johns Hopkins, Baltimore.
Miniati M, Van Helden A, Greco V, Molesini G (2002) Seventeenth-century telescope optics of Torricelli, Divini, and Campani. Applied Optics 41/4:644–647.

Molesini G (2010) Testing telescope optics of seventeenth-century Italy. In: Van Helde A, Dupré S, van Gent-R, Zuidervaart H (eds). The origins of the telescope. KNAW Press, Amsterdam, pp. 271–280.

Nelli GBC (1793) Vita e commercio letterario di Galileo Galilei Nobile e Patrizio Fiorentino Matematico e Filosofo Sopraordinario de' Grand Duchi di Toscana Cosimo e Ferdinando II, 2 vols. Gia' Ghetti Sinibaldi da Montecuccoli, Losanna.

Newton I (1687) Philosophiae naturalis principia mathematica. Imprimatur S. Pepys. Reg. Soc. Preses. Julii 5. 1686. Londini, Jussi Societatus Regiae ac Typis Josephi Streater. Prostat apud plures Bibliopolas. Anno MDCLXXXVII.

Newton I ([1726]1729) The mathematical principles of natural philosophy, by Sir Isaac Newton. Translated into English by Motte A, Symonds, London.

Nonnoi G (1997) Evangelista Torricelli, la pressione dell'aria e gli 'autori dei crepuscoli'. Bollettino della Società Torricelliana di Scienze e Lettere:17–45.

Oresme N (1486) Tractatus de latitudinibus formarum. Mathaeus Cerdonis, Padua.

Palmieri P (2009). Superimposition: on Cavalieri's practice of Mathematics. Archive for History of Exact Sciences 63:471–495.

Pascal B (1998–2000) Traités de l'équilibre des liqueurs et de la pesanteur de la mas de l'air, contenant l'explication des causes de divers effets de la Nature qui n'avaient point été bien connus jusques ici, et particulièrement ceux que l'on avait attribués à l'horreur du vide. In: Œuvres complètes de Blaise Pascal. Vol 1. Gallimard, Paris, pp. 459–543.

Paternoster G, Rinzivillo R, Schettino E (1996) Studio di una lente per cannocchiale di grandi dimensioni lavorata da Evangelista Torricelli. Nuncius XI:123–134.

Pisano R (2007) Brief history of centre of gravity theory. Epistemological notes. In Kokowski M (Ed). Proceedings of the 2nd ESHS Congress, Polish Academy of Science, Krakow, pp. 934–941.

Pisano R (2009a) Il ruolo della scienza archimedea nei lavori di meccanica di Galilei e di Torricelli. In: Giannetto E, Giannini G, Capecchi D, Pisano R, (eds). Da Archimede a Majorana: La fisica nel suo divenire. Guaraldi Editore, Rimini, pp. 65–74.

Pisano R (2009b) On method in Galileo Galilei' mechanics. In: Hunger H (ed). Proceedings of ESHS 3rd Conférence. Vienne: The Austrian Academy of Science, Wien, pp. 147–186.

Pisano R (2009c) Continuity and discontinuity. On method in Leonardo da Vinci' mechanics. Organon 41:165–182.

Pisano R (2009d) Galileo Galilei. Riflessioni epistemologiche sulla resistenza dei corpi. In Giannetto E, Giannini G, Toscano M (eds). Relatività, Quanti Chaos e altre Rivoluzioni della Fisica. Guaraldi Editore Rimini, pp. 61–72.

Pisano R (2011) Physics–Mathematics Relationship. Historical and Epistemological notes. In Barbin E, Kronfellner M, Tzanakis C, (eds), European Summer University History And Epistemology In Mathematics. Vienna: Verlag Holzhausen GmbH-Holzhausen Publishing Ltd., pp. 457–472.

Pisano R (2015) (ed) A Bridge between Conceptual Frameworks. History of Mechanism and Machine Science 27, Dordrecht, Springer.

Pisano R (2016) Details on the mathematical interplay between Leonardo da Vinci and Luca Pacioli? British Journal for the History of Mathematics 31/2:104–111.

Pisano R (2017) A Development of the Principle of Virtual Laws and its Framework in Lazare Carnot's Mechanics as Manifest Relationship between Physics and Mathematics. Transversal–International Journal for Historiography of science 2:166–203.

Pisano R (2020) A Tale of Tartaglia's Libro Sesto & Gionta in Quesiti et Inventioni diverse (1546–1554). Exploring the Historical and Cultural Foundations. Foundations of Science 25/2:477–50.

Pisano R (2023–2024) Galileo Galilei: The Two Manuscripts *On Mechanics*. Translations, Text and Commentaries. Dordrecht, Springer, *forthcoming*.

Pisano R, Bussotti P (2012) Galileo and Kepler. On Theoremata Circa Centrum Gravitatis Solidorum and Mysterium Cosmographicum. History Research 2/2:110–145.

Pisano R, Bussotti P (2014a) Historical and Epistemological Reflections on the Culture of Machines around the Renaissance: How Science and Technique Work? Acta Baltica Historiae et Philosophiae Scientiarum 2/2:20–42.

Pisano R, Bussotti P (2014b) Notes on mechanics and mathematics in Torricelli as physics mathematics relationships in the history of science. Problems of Education in the 21st Century International Journal. Special Issue 61/61:88–97.

Pisano R, Bussotti P (2015a) Historical and Epistemological Reflections on the Culture of Machines around the Renaissance: Machines, Machineries and Perpetual Motion. Acta Baltica Historiae et Philosophiae Scientiarum 3/1:69–87.

Pisano R, Bussotti P (2015b) Galileo in Padua: architecture, fortifications, mathematics and « practical » science, Lettera Matematica Pristem International:209–221.

Pisano R, Bussotti P (2015c) The Emergencies of Mechanics and Thermodynamics in the Western Society during 18th–19th Century. In: Pisano R (ed) 2015, pp. 399–436.

Pisano R, Bussotti P (2015d) Fibonacci and the Abacus Schools in Italy. Mathematical Conceptual Streams, Education and its Changing Relationship with Society. Almagest 6/2:126–165.

Pisano R, Bussotti P (2016) A Newtonian Tale Details on Notes and Proofs in Geneva Edition of Newton's Principia. Bulletin Journal of the British Society for the History of Mathematics 31/3:160–178.

Pisano R, Bussotti P (2017a) (eds) Homage to Galileo Galilei 1564–2014. Reading Iuvenilia Galilean Works within History and Historical Epistemology of Science. Philosophiae Scientiae 21/1.

Pisano R, Bussotti P (2017b) The Fiction of Infinitesimals in Newton's Works. On the Metaphoric Use of the Infinitesimals in Newton. In Marcacci F, Rossi MG (eds). Isonomia-Epistemologica. Reasoning, Metaphor and Science 9:141–160.

Pisano R, Bussotti P (2017c) On the Conceptualization of Force in Johannes Kepler's Corpus: an Interplay between Physics/Mathematics and Metaphysics. In: Pisano R, Agassi J and Drozdova D (eds). Hypothesis and Perspective in HPS. Homage to Alexandre Koyré 1892–1964. Springer, pp. 295–346.

Pisano R, Bussotti P (2022) Conceptual Frameworks on the Relationship between Physics–Mathematics in the Newton *Principia Geneva Edition* (1822). Foundations of Sciences (Springer) 27/3:1127–1182.

Pisano R, Bussotti P (2023a) Historical and Methodological Details on the *De motu gravium naturaliter descendentium* in Torricelli's Opera Gometrica (1644) In: Pisano R, Dhombres J, Radelet de Grave P, Bussotti P (eds). Homage to Evangelista Torricelli's *Opera Geometrica* 1644–2024. Text, Transcription, Commentaries and Selected Essays as New Historical Insights, pp. 189–224.

Pisano R, Bussotti P (2023b) The Physics–Mathematics Interplay in Newton *Principia* Geneva Edition (1822): A New Case Study of the Three-body Problem, Proposition LXVI–Theorem XXVI. Submitted to Philosophical Transactions A.

Pisano R, Capecchi, D (2008) La meccanica in Italia nei primi anni del Cinquecento. Il contributo di Niccolò Tartaglia. In: Tucci P (ed). Proceedings of XXV SISFA Congress. Milano, pp. C17.1–C17.6 via: http://www.brera.unimi.it/sisfa/atti/index.html.

Pisano R, Capecchi D (2009) Il ruolo della meccanica ne Le fortificationi di Buonaiuto Lorini In: D'Agostino, S. (ed). III Convegno di Storia dell'ingegneria, II. Napoli: Cuzzolin editore, pp. 797–808.

Pisano R, Capecchi D (2010a) On Archimedean Roots in Torricelli Mechanics. In: Ceccarelli M, Paipetis S (eds). The Genius of Archimedes. Springer, Dordrecht, pp. 17–28.

Pisano R, Capecchi D (2010b) Galileo Galilei: Notes on Trattato di Fortificazione. In: Altamore A., Antonini G. (eds), Galileo and the renaissance scientific discourse. Edizioni nuova cultura Roma, pp. 28–41.

Pisano R, Capecchi D (2012) Historical Reflections on Scale Ratio in Galilean Tratattto di Fortificazione. In: Ceccarelli M (ed) International Symposium on the History of Machines and Mechanisms, Springer, pp. 463–473.

Pisano R, Capecchi D (2013) Conceptual and mathematical structures of mechanical science between 18th and 19th centuries. Almagest 2/4:86–121.

Pisano R, Capecchi D (2016) Tartaglia's science weights. Mechanics in Sixteen-Century. Selections from Books VII–VIII. Dordrecht, Springer.

Pisano R, Agassi J, Drozdova D (eds) (2017a) Hypothesis and Perspective in History and Philosophy of Science. Homage to Alexandre Koyré 1892–1964. Dordrecht, Springer.

Pisano R, Fichant M, Bussotti P, Oliveira ARE (2017b) (eds) The Dialogue between Sciences, Philosophy and Engineering. New Historical and Epistemological Insights. Homage to Gottfried W. Leibniz 1646–1716. The College Publications, London.

Pisano R, Cooppersmith J, Peake M (2021) Essay on Machines in General (1786). Text, Translations and Commentaries. Lazare Carnot's Mechanics – Vol. I. Springer, Cham.

Pisano R, Robarts J (2023) Briefs on Buonaiuto Lorini's Fortificationi (1609): Geometry, Machines & Mechanics into Engineering during the Renaissance. In Ceccarelli M et al. 8th HMM Ankara, Springer, Cham, *forthcoming*.

Podetti F (1914) La teoria delle proporzioni in un manoscritto inedito di Evangelista Torricelli. Bollettino di Bibliografia e Storia delle Scienze Matematiche A/16–3:65–71.

Prager FD (1981) Berti's devices and Torricelli's barometer from 1641 to 1643. Annali dell'Istituto e Museo di storia della scienza di Firenze 2:35–53.

Procissi A (1952) Nuove ricerche bibliografiche torricelliane. Bollettino della Società Torricelliana di Scienze e Lettere 3:13–15.

Procissi A (1953) Su l'inviluppo delle parabole in Torricelli e sulla nozione di inviluppo di una famiglia di curve piane. Periodico di matematiche 4/31–1:113–116.

Robinson P (1994) Evangelista Torricelli. The Mathematical Gazette, 78, 481:37–47.

Roda M (2017) Entry: Rocca, Giovanni Antonio. Dizionario Biografico degli Italiani. Via https://www.treccani.it/enciclopedia/giovanni-antonio-rocca_(Dizionario-Biografico.

Ronchi V (ed) (1968) Scritti di ottica. Scritti di Tito Lucrezio Caro, Leonardo da Vinci, G. Rucellai, G. Fracastoro, G. Cardano, D. Barbaro, F. Maurolico, G. B. Della Porta, G. Galilei, F. Sizi, E. Torricelli, F. M. Grimaldi, G. B. Amici. Il Polifilo, Milano.

Ronchi V (1924) Sopra una lente di Evangelista Torricelli. Universo 5/2:1–3.

Rossini G (1959) La famiglia di Evangelista Torricelli. In: AA.VV. Convegno di studi torricelliani in occasione del 350 anniversario della nascita di Evangelista Torricelli, 19–20 ottobre 1958, Lega, Faenza. pp. 133–149.

Rougier L (2010) De Torricelli à Pascal. Philosophia Scientiae 14/2.

Rufini E (1926) Il "Metodo" di Archimede e le origini del calcolo infinitesimale nell'antichità. Zanichelli, Bologna.

Saito K (2013) Archimede e la doppia riduzione all'assurdo. Lettera Matematica Pristem 86:21–28.

Schuster JA (2000) Descartes opticien: the construction of the law of refraction and the manufacture of its physical rationales, 1618–1629. In: Gaukroger, Schuster and Sutton 2000, pp. 258–312.

Schuster JA (2013a) Descartes–Agonistes. Physico–mathematics, Method and Corpuscular–Mechanism 1618–1633. Springer, Dordrecht.

Schuster JA (2013b) Cartesian Physics. In: Fox R *et al.* (eds). The Oxford Handbook of the History of Physics, The Oxford University Press, Oxford, pp. 56–95.

Tartaglia N ([1554]1959) La nuova edizione dell'opera Quesiti et inventioni diverse de Nicolo Tartaglia brisciano Riproduzione in facsimile dell'edizione del 1554. Masotti A (ed.). Commentari dell'Ateneo di Brescia, Tipografia La Nuova cartografica, Brescia.

Taton R (1963) L'annonce de l'expérience barométrique en France. Revue d'histoire des sciences et leurs applications 16:77–83.

Torricelli E (1644) Opera geometrica. Florentiae. Masse & de Landis, Florentiae.

Torricelli E ([1644]1897) Esperienza dell'argento vivo, Accademia del Cimento. In Gustav Hellmann G (ed): Instrumenti per conoscer l'alterazioni dell'aria, Neudrucke von schriften und karten über meteorologieund erdmagnetismus, Asher & Co, Berlin.

Torricelli E (1715) Lezioni Accademiche d'Evangelista Torricelli Mattematico, e Filosofo del Sereniss. Ferdinando II Gran Duca di Toscana, Jacopo Guiducci, Santi Franchi (eds), S.A.R., Firenze.

Torricelli E (1823) Lezioni accademiche. In: Silvestri G (ed). Biblioteca scelta di opere italiane antiche e moderne vol. 139, Milano.

Torricelli E (1864) Lettere fin qui inedite di Evangelista Torricelli precedute dalla vita di lui. G. Ghinassi (ed.), Faenza.

Torricelli E (1919–1944) Opere di Evangelista Torricelli edite in occasione del III centenario della nascita col concorso del Comune di Faenza da Gino Loria e Giuseppe Vassura. Montanari, Faenza.

Torricelli E (1955) De infinitis spiralibus. Edited by Carruccio E. Domus Galilaeana, Pisa.

Torricelli E (1975) Opere scelte. Belloni L (ed). UTET, Torino.

Toscano F (2008) L'erede di Galileo. Vita breve e mirabile di Evangelista Torricelli. Sironi, Milano.

Toscano F (2014) Evangelista Torricelli: grandi scoperte e autoconsure. Nuova Civiltà delle Macchine 3/1:31–46.

Valerio L (1606) Quadratura parabolae per simplex falsum: et altera quam secunda Archimedis expiditior ad martium columnam. Apud Lepidum Facium, Romae.

Vassura G (1945) Umanità in Evangelista Torricelli. Pubblicato dalla commissione per le onoranze a Evangelista Torricelli III per l'anno 1944. Centenario della scoperta del barometro. Bollettino della Società Torricelliana di Scienze e Lettere:15–18.

Visani A (1952) Notizie sulla salute e sull'ultima malattia di Evangelista Torricelli. Bollettino della Società Torricelliana di Scienze e Lettere 3:5–12.

Vita V (1972) Gli indivisibili curvi di Bonaventura Cavalieri. Archimede XXIV/1–2:16–24.

Viviani V (1674) Quinto Libro degli Elementi d'Euclide ovvero Scienza Universale delle Proporzioni Spiegata colla Dottrina del Galileo. Alla Condotta, Firenze.

Viviani V (2001) Vita di Galileo. Salerno Editrice, Roma.

West JB (2013) Torricelli and the Ocean of Air: The First Measurement of Barometric Pressure. Physiology 28:66–73.

Zeuthen HG (1896) Geschichte der Mathematik im Altertum und Mittelalter. Höst, Kopenhagen.

Part 2
Selected Essays

Introducing Selected Essays on *Opera Geometrica* (1644)

2

Raffaele Pisano and Paolo Bussotti

Abstract

Two topics, Science and History, have been studied for centuries, if not millennia, often in disjointed ways. The realization that Science is nourished by its History is quite obvious nowadays and is a standpoint shared by all the contributors in this book. Given this fact, the following chapters were written by interdisciplinary scholars who are experts in this field. The admirably written contributions concern History of Science, Epistemology, Philosophy of Science, but are not restricted to these disciplines. They also discuss theoretical perspectives and ingenious results achieved by beautiful and clever minds working together or inspired by one another. Torricelli's history is a fascinating one, filled with struggles and insights that have been studied and examined in depth by internationally renowned specialists. The gripping stories reconstructed, as well as the epistemological and logical threads drawn from long and painstaking research, are presented in a clear and absorbing fashion, in each instance with a unique style and always with considerably talented analysis. The preceding celebratory historical and historiographical chapter on *Opera geometrica* (Torricelli in Opera geometrica. Florentiae. Masse & de Landis (1644); Torricelli in Opere di Evangelista Torricelli edite in occasione del III centenario della nascita col concorso del Comune di Faenza da Gino Loria e Giuseppe Vassura. Montanari (1919–1944)) is also a structured introductory essay to the contents of this extensive Springer research project. Below, the reader will find summaries of selected distinguished essays written in order to celebrate the magnificence of Torricelli's *Opera geometrica.* The whole volume has to be considered as an homage both to *Opera geometrica* and to Evangelista, a genial scientist.

Keywords

Torricelli • *Opera Geometrica*

1 Selected Essays

Martin Davis (1928–2023) (The United States of America) in "Encounters with Infinity: From Torricelli to Gödel" delves in the depths of the concept of infinity, some of its particularities, difficulties and paradoxes from Torricelli's trumpet to the infinite sets that contemporary set theorists study. This concept was, and still is, a complex matter for mathematicians and scientists. Exploring the relationship between finite and infinite in the historical context of the development of theories and newly found proofs, Martin Davis brilliantly manages to explain and describe the evolution that proceeded until Gödel and his discovery of undecidability. Ranging from Aristotle to Leibniz, Euler and Cantor to cite a few names, the related scientific theories and progress on the matter until the revolution wrought by Gödel's discovery in the twentieth century that

R. Pisano (✉)
HOPAST at IEMN, Lille University, Lille, France
e-mail: raffaele.pisano@univ-lille.fr

P. Bussotti
DIUM, Udine University, Udine, Italy
e-mail: paolo.bussotti@uniud.it

R. Pisano et al., *Homage to Evangelista Torricelli's Opera Geometrica 1644–2024*,
Logic, Epistemology, and the Unity of Science 55,
https://doi.org/10.1007/978-3-031-06963-5_2

shook the philosophical grounds is reminiscent of Torricelli's discovery in the seventeenth century and, to say the least, is mesmerizing and a delight to read.

Jean Dhombres (France) in "A new Cycloid Narrative Centered on Torricelli and Roberval to Understand the Diverse Ways of the Mathematical Revolution" raises general reflections about the practice of history and epistemology of mathematics. Starting with an argument regarding findings on the newly discovered cycloid curve in the seventeenth century that avoids the distractions of contemporary quarrels and biases that historians have perpetuated, Jean Dhombres provides an extraordinarily rich analysis of both the science and work accomplished and of the social bonds encompassing the scientific community and culture that reigned at that time before the "Mathematical Revolution". With this intricate original perspective, this analysis of the historical context, works, and relations between two scientists that never met and the results they obtained is highly remarkable, precise and greatly enlightening.

Raffaele Pisano (France) and Paolo Bussotti (Italy) in "Historical and Methodological Details on the *De motu gravium naturaliter descendentium* in Torricelli's *Opera geometrica* (1644)" discuss Torricelli's principle in mechanics and his attempt to construct a general view of a physics based on general principles. They compare Torricelli's principle and the virtual works principle before studying the role of the former within the history of virtual works principle. Finally, the two authors analyse the logic behind the demonstrations provided in physics relying on general principles such as that of Torricelli. Raffaele Pisano and Paolo Bussotti accomplish this through an extremely detailed mathematical and logical step by step analysis. Moreover, the data they gathered and compiled are especially important with respect to the research concerning early modern mathematics and make an important contribution to the History and Epistemology of Science.

The authors also composed an extensive introductory essay "Celebrating Evangelista Torricelli's *Opera geometrica* (1644–2024): Details in History and Historiography of Physics, Geometry and Mathematics". It is a meticulous and up-to-date historical account of Torricelli and his *Opera geometrica*, including an ad hoc addition related to the editorial process of the publication (1644). The authors describe Torricelli's life and works in their scientific context, including the Archimedean heritage in Torricelli's works, particularly in mechanics and geometry: several examples are analysed and discussed from historical and physics–mathematics standpoints.

Patricia Radelet de Grave (Belgium) in "Torricelli's Proof of Galileo's Assumption on the Velocity Acquired at the End of a Free Fall" discusses Torricelli's Proof. The author also stresses the profound influence exerted on Torricelli's *De motu* by the three months passed in the company of Galileo. The difficulties encountered by both scientists in the formalization of new concepts and ideas are also pointed out. Starting with a sound reminder of some biographical facts and with an emphasis on *Galilean momentum* and an "early notion of function", we are then presented with an excellent full, detailed, and precise historical analysis and account concerning the concepts of motion, statics, and dynamics and how they were studied and processed at that time. The comparison of Galileo's and Torricelli's proofs is very intriguing and thought-provoking, necessarily revisiting some of Galileo's previous texts in order to render a clear understanding of his vocabulary; thankfully, Patricia Radelet de Grave does an excellent job explaining everything we need in a coherent and straightforward manner.

J. B. Shank (The United States of America) in "Why Publish a Book of Geometry in Seventeenth-Century Europe? Science, Status, and Print Culture in Evangelista Torricelli's *Opera geometrica* (1644)" investigates the historical nature of Torricelli's rationale for publishing his mathematical work, including his only book *Opera geometrica*. Emphasizing the very different cultural norms and standards for publishing mathematics in the seventeenth century as compared with those that govern the same publishing today, the chapter reconstructs Torricelli's historical publication strategies and their influence in shaping his reputation and celebrity as a mathematician. Comparing Torricelli's rationale with the very different case of Galileo and his relationship to books and printed publications, Shank does not situate Torricelli as a mathematician following the "Galilean model" of his predecessor at the Medici court, but as a figure whose publishing practices were emblematic of wider trends, including those followed by contemporaries such as Roberval, Huygens, Newton, Leibniz, and Viviani. Framed this way, the publication history of Torricelli's *Opera geometrica* shows how Torricelli's publishing worked to establish his esteemed reputation with his sovereign patrons as a court mathematician by keeping him far from polemical fights.

Thanks to the cutting-edge research of prominent specialists, this collection represents a remarkable amount of work and a thorough analysis of Torricelli's background, evolution, and historical context as well as a testimony to his extensive influence on the scientific community and sciences. Regardless of the disciplinary field of the readers, be it Science, History or Epistemology, they are guaranteed an inspired and inspiring journey filled with discoveries, insights, revolutions and much more.

Acknowledgements We thank the anonymous reviewers for their careful reading of our manuscript and their many insightful comments and suggestions. We also express gratitude to Dr. Julie Robarts (University of Melbourne, Australia) for her observations on English language; eventual remaining English mistakes are our own.

References

Torricelli E (1644) Opera geometrica. Masse & de Landis, Florentiae.

Torricelli E (1919–1944) Opere di Evangelista Torricelli edite in occasione del III centenario della nascita col concorso del Comune di Faenza da Gino Loria e Giuseppe Vassura. Montanari, Faenza.

Encounters with Infinity: From Torricelli to Gödel

3

Martin Davis

I could be bounded in a nutshell, and count myself a king of infinite space
(Shakespeare [1600] 1980, Hamlet Act. 2, Scene 2, p. 1089)

Abstract

To the Pythagoreans, *to apeiron*, that which has no bounds, the infinite, "was something abhorrent" (Moore, The infinite, Routledge, p 19, 2001). Nevertheless the infinite thrust itself upon them once they became aware of the fact that there was no common length for which the sum of finitely many instances of it would be equal to both the length of the side and the length of the diagonal of a square. Again and again mathematicians have engaged warily in infinitary reasoning, and most often the methods were found to be useful and robust even when rigorous justification was only to be obtained at a later time. In this essay I will survey some of these encounters with the infinite, beginning with Torricelli's horn that was so shocking in the seventeenth century, and ending with the very enormous infinite sets that contemporary set theorists study without any hope of a proof that their existence is consistent with the axioms of ordinary mathematics.

Keywords

Torricelli • Gödel • Determinacy • Indivisibles • Infinitesimals • Infinity • Projective sets • Zermelo–Fraenkel axioms • Vienna Circle • Jesuits

1 Torricelli's Shocker

In 1641 the Italian mathematician Evangelista Torricelli (1608–1647) found that the volume of a certain solid of infinite extent is finite. The solid is obtained by rotating about an axis a plane figure with infinite area. Specifically, in modern terminology, it is the figure bounded by the hyperbola whose equation is $y = 1/x$, the line $x = 1$ and the horizontal asymptote of the hyperbola, namely the X-axis, whose area is expressed by the integral:

$$\int_1^\infty \frac{1}{x} dx$$

that diverges to infinity.

M. Davis (✉)
Courant Institute, New York University, New York, USA
e-mail: martin@eipye.com

University of California, Berkeley, USA

R. Pisano et al., *Homage to Evangelista Torricelli's Opera Geometrica 1644–2024*,
Logic, Epistemology, and the Unity of Science 55,
https://doi.org/10.1007/978-3-031-06963-5_3

Torricelli's solid is formed by rotating this figure about the X-axis. It is a routine exercise in elementary calculus to calculate the finite volume of this solid of revolution:

$$\pi \int_1^\infty \frac{1}{x^2} dx = \pi$$

However, at the time, it created a sensation because it contradicted prevalent views about the infinite. Torricelli himself remarked "[…] if one proposes to consider a solid, or a plane figure, infinitely extended, everybody immediately thinks that such a figure must be of infinite size.".[1] In 1649, Pierre Gassendi (1592–1655) wrote

> Mathematicians […] weave those famous demonstrations, some so extraordinary that they even exceed credibility, like what…Torricelli showed of a certain…solid infinitely long which nevertheless is equal to a finite cylinder.[2]

Writing in 1666, Isaac Barrow (1630–1677) found Torricelli's result contradicting what Aristotle had taught. He referred to Aristotle's dictum, "there is no proportion between the finite and the infinite":

> The truth of which statement, a very usual and well known axiom, has been in part broken by…modern geometricians [who] demonstrate […] equality of…solids protracted to infinity with other finite…solids which prodigy […] Torricelli exhibited first.[3]

Much can be learned from this example about the way in which mathematicians expand the applicability of existing methods to new problems and with how they deal with the philosophical problems that may arise. Torricelli used a technique called the method of *indivisibles*, a method pioneered by Bonaventura Cavalieri (1598–1647) that provided a short-cut for solving area and volume problems. Torricelli used this technique to prove that his infinite body had the same volume as a certain finite cylinder. The method conceived of each of the two bodies being compared as constituted of a continuum of plane figures. Although there was no rigorous foundation for this, Cavalieri and later Torricelli showed how effective it could be in easily obtaining interesting results. They were well aware of the Eudoxes-Archimedes method of exhaustion (which they called "the method of the ancients"), and used it to confirm their results and/or to convince skeptics.[4]

But, Torricelli insisted on the validity of the new method. He experimented with a novel mathematical technique, applied it to new examples, and discovered something new about the infinite.

Torricelli understood very well that these infinitesimal methods had to be used with care, that the "ancients" were not wrong to be wary of *to apeiron*. He knew that careless use could lead to irresoluble paradoxes and enjoyed presenting some of these (Alexander 2014, p. 112). But he also knew that used correctly these methods made it possible to quickly and easily solve problems that had seemed quite difficult. He demonstrated his virtuosity with the classical methods as well as the new ones by presenting 21 proofs of Archimedes's famous quadrature of the parabola (Alexander 2014, pp. 107–111).

The Church that had silenced Galileo Galilei (1564–1642) for daring to propose a replacement for Ptolemy's ancient cosmology which removed the earth from its position as the center of the universe was not silent either on efforts to use indivisibles instead of the method of exhaustion. The Jesuit order declared that indivisibles were not to be taught at its institutions. Torricelli was spared the difficulties facing mathematicians wanting to use these methods because he had the good fortune to be offered the professorship Galileo's death had made available at the University of Pisa (Alexander 2014, pp. 153–165).

2 Leibniz Exploits the Infinite

Not in the least daunted by the warnings from Aristotle and Thomas Aquinas (1225–1274) against completed infinities, Gottfried Leibniz (1646–1716) boldly proclaimed:

> I am so in favor of the actual infinite that instead of admitting that Nature abhors it, as is commonly said, I hold that Nature makes frequent use of it everywhere, in order to show more effectively the perfections of its Author.[5]

However he was not willing to countenance the possibility of a number that counted all of the natural numbers. As he said in a letter to Nicolas Malebranche (1638–1715):

---

[1] English translation in Mancosu (1996, p. 137), from Torricelli (1644, p. 173).

[2] English translation in Mancosu (1996, p. 137), from Gassendi ([1649] 1964, p. 264b).

[3] English translation in Mancosu (1996, p. 139), from Barrow ([1683] 1976, p. 255).

[4] The method of exhaustion worked by showing that the quantity concerned could be neither smaller nor larger than the suggested result. Therefore it was necessary to already have that result at hand, whereas with indivisibles the answer could be computed directly.

[5] Dauben (1979, p. 124). For the original French, see Cantor (1932, p. 179).

For any number there exists a corresponding even number which is its double. Hence the number of all numbers is not greater than the number of even numbers, that is, the whole is not greater than the part.[6]

Likewise, Galileo, observing the correspondence between each natural number and its square, had noted any attempt to speak of the number of all natural numbers would violate the Euclidean principle that the whole is greater than any of its parts. But this was two centuries before Cantor's realization that a coherent theory of the infinite would require abandoning that Euclidean axiom.

René Descartes (1596–1650) had shown how to treat geometric problems algebraically, and this enabled Newton and Leibniz to develop the calculus, transforming problems whose solution had required virtuosic ability with indivisibles or the like, to straightforward algebraic manipulation. Leibniz's free and easy use of infinitesimals in his development of the calculus confronted the infinite in a way that he found it difficult to explain coherently. On the one hand he treated infinitesimals as simply very small non-zero quantities subject to all the usual rules of algebra while on the other hand he allowed himself to delete them as though they were equal to zero. We can see this in his derivation of the rule for the differential of a product. Here *du* and *dv* are infinitesimal "differences". Leibniz wrote:

$$\begin{aligned} duv &= (u+du)(v+dv) - uv \\ &= udv + vdu + dudv \\ &= dv + vdu \end{aligned}$$

The product *dudv* was an infinitesimal of "higher order" than *du* or *dv*; it was sufficiently more infinitesimal than either, and so could just be dropped. It is not surprising that not everyone was satisfied. Thus, Michel Rolle (1652–1719) of the Paris Academy wrote:

In the first place one sees that since all these infinites of first order such a *dx* or *dy* have no real extension then all the infinites of the other sorts will also be nothing about absolute zeros in the calculus. All these infinite sequences of infinites, provided by the system, will only be nothings that are supposed to be infinitely contained within other nothings.[7]

Although this Leibnizean practice continued in general use, especially among physicists and engineers, among rigor-minded mathematicians, infinitesimals were abandoned as fundamentally unsound in favor of the Cauchy–Weierstrass theory of limits. It was only with Abraham Robinson's nonstandard analysis in the twentieth century that the use of infinitesimals was finally justified.

3 Euler Factors a Polynomial of Infinite Degree

In 1644 Pietro Mengoli (1626–1686), a student of Cavalieri, posed the problem of finding the sum of the infinite series $\sum_{n=1}^{\infty} 1/n^2$. Ninety years later Leonhard Euler (1707–1783) found the answer. His method depended on assuming that infinite power series would have certain of the crucial properties of an ordinary polynomial involving the relationships between its zeros and its coefficients. For example the polynomial $x^3 - 8x^2 + 19x - 12$ has the zeros 1, 3, and 4. Therefore

$$x^3 - 8x^2 + 19x - 12 = (x-1)(x-3)(x-4)$$

If we wish to verify this by multiplying the three factors we can see that the $-8x^2$ term is obtained by the addition $(-1x^2) + (-3x^2) + (-4x^2)$. So the coefficient of $-x^2$ is the *sum* of the zeros. Euler's idea was to somehow use this idea in summing $1/n^2$. His "polynomial" was

$$\frac{\sin x}{x} = 1 - \frac{x^2}{6} + \frac{x^4}{24} - \frac{x^6}{144} + \cdots$$

[6] Gerhardt (1978, I, p. 338). The translation from Latin is by Alexis Manaster Ramer.
[7] English translation in Mancosu (1996, p. 167) from Rolle (1703, p. 312).

whose zeros are $\{\pm\pi n | n = 1, 2, 3, \ldots\}$. But this can't work: the exponents in our example are decreasing from 3 to 0, whereas in the power series for $\sin x/x$ the exponents are increasing from 0 in infinity in jumps of 2. To see what to do, let's rewrite our polynomial as

$$x^3 - 8x^2 + 19x - 12 = -12\left(1 - \frac{19}{12}x + \frac{8}{12}x^2 - \frac{1}{12}x^3\right)$$

Of course the zeros are still 1, 3, and 4, but the factorization now looks like this:

$$1 - \frac{19}{12}x + \frac{8}{12}x^2 - \frac{1}{12}x^3 = \left(1 - \frac{x}{1}\right)\left(1 - \frac{x}{3}\right)\left(1 - \frac{x}{4}\right)$$

and the sum $1 + 1/3 + 1/4$ is the coefficient of $-x$. . So boldly (if hesitantly) following Euler, remembering not to neglect the negative zeros, we write[8]:

$$\begin{aligned} 1 - \frac{x^2}{6} + \frac{x^4}{24} - \frac{x^6}{144} + \cdots &= \left(1 - \frac{x}{\pi}\right)\left(1 + \frac{x}{\pi}\right)\left(1 - \frac{x}{2\pi}\right)\left(1 + \frac{x}{2\pi}\right)\left(1 - \frac{x}{3\pi}\right)\left(1 + \frac{x}{3\pi}\right)\ldots \\ &= \left(1 - \frac{x^2}{\pi^2}\right)\left(1 - \frac{x^2}{4\pi^2}\right)\left(1 - \frac{x^2}{9\pi^2}\right) \end{aligned}$$

So the coefficient of $-x^2$ which is $1/6$ should be

$$\frac{1}{\pi^2} + \frac{1}{4\pi^2} + \frac{1}{9\pi^2} + \cdots = \frac{1}{\pi^2}\left(1 + \frac{1}{4} + \frac{1}{9} + \cdots\right)$$

Thus we get Euler's result:

$$1 + \frac{1}{4} + \frac{1}{9} + \cdots = \frac{\pi^6}{6}$$

The story goes that Euler verified this result by what might well be called a "mathematical experiment". It was easy for Euler to compute a pretty good approximation to $\pi^2/6$. He could also calculate $\sum_{k=1}^{n} 1/k^2$ for some suitably large value of n and then compare the two numbers. He was known to be a prodigious calculator but that infinite series converges very slowly. From Wikipedia we have

$$\frac{\pi^2}{6} = 1.644934\ldots$$

I wrote and ran a program on my PC to compute $1 + 1/4 + 1/9 + \cdots + 1/n^2$ for various values of n; the results are shown in a table:

n	$1 + 1/4 + 1/9 + \cdots + 1/n^2$
100	1.63498390
1000	1.64439346
30,000	1.64490073

From this data it is clear that Euler would not have been able to obtain what he needed by direct calculation. Even adding 30,000 terms, far beyond what is humanly possible by direct calculation, would only have yielded four correct decimal digits. Instead he used a formula, found by Euler himself and, independently, by Colin MacLaurin (1698–1746), that approximates sums by integrals. He used this "Euler–MacLaurin formula" to calculate the sum of the series correct to 20

[8] Although Euler obtained this infinite product by using a doubtful analogy, Karl Weierstrass (1815–1897) eventually obtained it quite rigorously as a special case of a general theorem about complex analytic functions.

places. The agreement with approximations to $\pi^2/6$ gave Euler the confidence to announce the result to the mathematical public. However, using the Euler-MacLaurin formula required considerable care. It required truncating a divergent (so-called *asymptotic*) series at just the right point when the sum still provided a useful approximation before the series marches off to infinity. In Euler's time the rigorous treatment of such series had still not been developed, so in effect his "experimental" method for verify his heuristic method for summing $\sum_1^\infty 1/n^2$ was to use another heuristic method!

Though Euler was never able to make this argument rigorous (the necessary basis for such was yet to be developed), he did make sure to develop another fully convincing proof that the sum in question really is $\pi^2/6$.

Wasn't the heuristic argument fully convincing? Well no: the computational evidence was perfectly consistent with, say,

$$1+\frac{1}{4}+\frac{1}{9}+\cdots=\frac{\pi^6}{6}+\gamma$$

where γ is some real number with $0<\gamma<10^{-10^{100}}$.

4 Cantor Leaves Us a Conundrum

Cantor has provided us with not just one, but with two routes upwards to larger and larger infinities. First the ordinal route. In modern set theory, following John von Neumann (1903–1957). Each ordinal number is taken to be simply the set of its predecessors, and the cardinal number of a set is defined as the least ordinal number for which there is a one–one correspondence with that set. The infinite cardinal numbers are then Cantor's $\aleph$s[9]:

$$\aleph_0 \ \aleph_1 \ \aleph_2 \ \ldots \ \aleph_\omega \ \ldots$$

Cantor's cardinal route upwards also begins with $\aleph_0$:

$$\beth_0=\aleph_0;\ \beth_{\alpha+1}=2^{\beth_\alpha};\ \beth_\gamma=\cup_{\alpha<\gamma}\beth_\alpha \quad \text{for } \gamma \text{ a limit ordinal.}$$

What I am calling Cantor's conundrum is the problem of determining where the $\beth$s fit among the $\aleph$s. Cantor himself conjectured that $\beth_1=\aleph_1$, an assertion that became known as the *Continuum Hypothesis* (CH). It was the first of Hilbert's famous 1900 list of 23 problems. What makes this a conundrum not merely an unsolved problem is the work of Gödel and Paul Cohen (1934–2007) that established that the truth or falsity of this assertion can not be decided from the usual Zermelo–Fraenkel axioms for set theory.[10] Should one accept this as decisive and perhaps conclude (as some have) that the concepts of set theory are inherently so vague that CH simply cannot be said to possess a truth value? Or should one conclude that the Zermelo–Fraenkel axioms do an inadequate job and need to be strengthened?

The independence of CH from the Zermelo–Fraenkle axioms was proved by constructing models of the axioms in which CH is true as well as models in which it is false. These models can be viewed as consisting of countable infinite directed graphs.[11] Although viewed *externally* such models can be seen to be countably infinite structures with cardinality $\aleph_0$, using Cohen's method of*forcing*, they can be so constructed that *internally* $\beth_1$ can have almost any value one wishes. The natural generalization

$$\beth_\alpha=\aleph_\alpha \quad \text{for all } \alpha$$

of Cantor's CH was proposed by Philip Edward Bretrand Jourdain (1879–1919) in 1905 and is called the *Generalized Continuum Hypothesis* (GCH). In that connection, Cohen's methods enable the construction of such a countable model in which, say, $\beth_1=\aleph_{17}$ and $\beth_2=\aleph_{93}$. These models, though they faithfully satisfy the Zermelo–Fraenkel axioms, "bounded in a nutshell" indeed, surely do not do justice to Cantor's expansive vista. The large values achieved by the $\beth$s are obtained by insuring that the mappings that effect the one–one-correspondence with smaller $\aleph$s are omitted from the model.

[9] Details can be found in a number of excellent text books, e.g., see Enderton (1977).

[10] For full details of this work see Jech (1991).

[11] A *directed graph* or *digraph* is a set of objects called *vertices* together with a set of ordered pairs of these vertices called *edges*. When such a digraph is used to model the axioms of set theory, the vertices represent sets and the existence of an edge $\langle a,b\rangle$ is taken to represent the membership relation $a\in b$.

In his remarkable essay Kurt Friederich Gödel (1906–1978) (Gödel [1947] 1990, p. 64) strongly argued for the need for new axioms with whose aid CH could be decided. This essay appeared in 1947, and its revision in 1964, and no such axiom has yet come to light. But Gödel's approach has found vindication in a different direction, that of *descriptive set theory* which studies sets definable in the language of set theory and in particular the hierarchy of what are called *projective sets.* Investigations of projective sets leading up to the 1930s made little progress. When Paul Cohen's new techniques were brought to bear the source of the difficulties became evident: the Zermelo–Fraenkel axioms were inadequate for the purpose. And in this case, a new axiom did make all the difference.[12]

This new axiom makes use of the notion of *determinacy* of a given set A of real numbers, which is defined in terms of a sort of game. Players I and II alternately move by each specifying a binary digit 0 or 1. The game continues for $\aleph_0$ moves. The players thus specify the binary expansion of a real number x in the unit interval. If x is the fractional part of a member of A, then I wins; otherwise II wins. A is called *determined* if either there is a strategy for I to win or there is a strategy for II to win. The axiom of determinacy (PD) is the assertion: *Every projective set of real numbers is determined.* Adding PD to the Zermelo–Fraenkel axioms enables a rather attractive theory of the projective sets. For example PD implies that every projective set is Lebesgue measurable. This had previously only been proved for sets at the very bottom of the projective hierarchy. PD itself gained a certain legitimacy from the Martin-Steel proof that PD follows from a so called large cardinal axiom (more about that below).

5 Infinitesimals Achieve Legitimacy

Leibniz never provided his followers with clear cogent rules for working with infinitesimals. Bishop George Berkeley (1685–1753) had famously found that the new calculus was fatally tainted by the contradiction of a quantity that was treated as being zero or non-zero at the whim of the practitioner. Leibnizian infinitesimals embodied this contradiction in its most blatant form. After the important foundational work of Cantor, Richard Dedekind (1831–1916), and Weierstrass, infinitesimals were seen by mathematicians as both unnecessary and unworkable. This did not prevent physicists and engineers from freely using infinitesimals and enjoying their great convenience. Abraham Robinson's *nonstandard analysis* (Davis 2005; Robinson 1970) provided a perfectly rigorous way to work with infinitesimal and infinitely large quantities. He showed how to obtain, along side of the usual *standard* universe of sets, a nonstandard universe in which infinitesimal and infinite quantities flourished. A formal language was used as a bridge between these two domains via a *Transfer Principle* that asserts that a true sentence of either is also true of the other. Not only did this legitimize Leibnizian infinitesimal reasoning, it even made it possible to transform Euler's heuristic method for obtaining the sum $\sum 1/n^2$ into a perfectly rigorous proof.

6 Conclusion: Gödel's Legacy

This essay began with Torricelli's shocker. Gödel's discovery of arithmetic undecidability generated as profound a shock to the philosophical standpoints prevalent in the twentieth century as Torricelli's discovery did in the seventeenth. For the positivists of the Vienna Circle who wished to identify mathematical truth with provability, there was facing the existence of

---

[12] A primer of the terminology of descriptive set theory: Writing $\mathcal{R}$ for the set of real numbers and $\mathcal{R}^n$ for n-dimensional Euclidean space, the *Borel sets* of $\mathcal{R}^n$ are all the subsets of $\mathcal{R}^n$ that can be obtained from the open sets by applying a finite numbers of times the operations of countable union, countable intersection, and complement with respect to $\mathcal{R}^n$. For a set $C \subseteq \mathcal{R}^{n+1}$, Proj($C$) is defined as the set $A \subseteq \mathcal{R}^n$ defined by

$$\langle x_1, \ldots, x_n \rangle \in A \Leftrightarrow \exists y \in \mathcal{R}[\langle x_1, \ldots, x_n, y \rangle \in C]$$

The hierarchy of *projective sets* is defined simultaneously in all $\mathcal{R}^n$ as follows:

$\sum_0^1$	The set of Borel sets
$\prod_m^1$	$\left\{A \subseteq \mathcal{R}^n \mid \mathcal{R}^n - A \in \sum_m^1\right\}$
$\sum_{m+1}^1$	$\left\{A \subseteq \mathcal{R}^n \mid \exists \prod_m^1 [A = \mathrm{Proj}(B)]\right\}$

propositions that that were true but unprovable in a given system. For those working with Hilbert on proving the consistency of arithmetic (notably von Neumann) there was the realization that their effort was doomed. Gödel himself in a talk in 1933, just two years after the appearance of his shocker, expressed himself as follows (Gödel [1933] 1995):

> [...] we are confronted by a strange situation. We set out to find a formal system [of axioms] for mathematics and instead of that found an infinity of systems, and whichever system you choose [...] there is one [...] whose axioms are stronger. But [...] this character of our systems [...] is in perfect accord with certain facts which can be established quite independently [...]. For any formal system you can construct a proposition – in fact a proposition of the arithmetic of integers – which is certainly true if the given system is free from contradictions but cannot be proved in the given system. Now if the system under consideration (call it S) is based on the theory of types, it turns out that [...] this proposition becomes a provable theorem if you add to S the next higher type and the axioms concerning it. The construction of higher and higher types is necessary for proving theorems even of a relatively simple structure. (Gödel [1933] 1995), p. 47)

As Gödel recognized already in 1933, this hierarchy of higher types continues with infinite types. In his Gibbs lecture (Gödel [1951] 1995), continued the same theme, and even conjectured that the Riemann Hypothesis itself may require infinitary methods for its proof. Meanwhile the arithmetic character of undecidability has been clarified by further work. Formal systems will yield true but unprovable propositions that simply assert that a certain polynomial equation with integer coefficients has no solutions in natural numbers. If such a proposition is false, its falsity can be demonstrated using the elementary school algorithms for adding and multiplying.

The hierarchy of ever larger sets gives rise to a sequence of truths of this kind whose truth can only be established by making use of these "large cardinals". So far mathematicians can (and do) tranquilly ignore all of this because the specific examples of undecidability brought forth are not propositions about which they care very much. Gödel believed that the future will bring examples about which they do care.

References

Alexander A (2014) Infinitesimal: How a Dangerous Mathematical Theory Shaped the Modern World. Scientific American–Ferrar, Straus and Giroux, NY.

Barrow I ([1683] 1976) Lectiones Mathematicae. Whewell W (ed). The Mathematical Works. Verlag–Hildesheim, New York.

Cantor G (1932) Gesamelte Abhandlungen. Springer, Berlin.

Dauben J (1979) Georg Cantor: His mathematics and Philosophy of the Infinite. Princeton, NJ.

Davis M (2005) Applied Nonstandard Analysis. Dover, NY [Reprinted with corrections from *Id.*, 1977, Interscience-Wiley, NY].

Davis M (2012) The Universal Computer: The Road from Leibniz to Turing. CRC Press, Boca Raton.

Enderton H (1977) Elements of Set Theory, Academic Press, NY.

Gassendi P ([1649] 1964) Syntagma. Petrus Gassendi, Opera Omnia, I, Verlag, Stuttgart–Bas Cannstatt.

Gerhardt C (1978) (ed) Die Philosophischen Schriften von G. W. Leibniz. 7 Vols. Georg Olms Verlag, Hildesheim–NY.

Gödel K ([1933] 1995) The Present Situation in the Foundations of Mathematics. In: Feferman S *et al.* (eds). Kurt Gödel Collected Works. Vol. III. Oxford University Press, NY, pp. 45–53.

Gödel K ([1947] 1990) What is Cantor's Continuum Problem? In: Feferman S *et al.* (eds). Kurt Gödel Collected Works. Vol. II. Oxford University Press, NY, pp. 176–187 [See also revised and expanded version, *Id.*, pp. 254–270].

Gödel K ([1951] 1995) Some Basic Theorems on the Foundations of Mathematics and Their Implications. In: Feferman S *et al.* (eds). Kurt Gödel Collected Works. Vol. III. Oxford University Press, NY, pp. 304–323.

Jech T (1991) Set Theory. Academic Press, NY [Second edition with corrections from: *Id.*, 1978, Springer, Berlin–Heidelberg].

Mancosu P (1996) Philosophy of Mathematics & Mathematical Practice in the Seventeenth Century. Oxford University Press, NY–Oxford.

Moore AW (2001) The Infinite. Second Edition, Routledge, London.

Robinson A (1970) Non–Standard Analysis. North–Holland, Amsterdam.

Rolle M (1703) Du nouveau système de l'infini par M. Rolle. In: Histoire de l'Académie Royale des Sciences Avec les Memoires de Mathematique & de Physique pour la même Anne. A Paris, Chez Jean Boudot, Imprimerie Ordinaire du Roy, & de de l'Académie Royale des Sciences, pp. 312–336.

Shakespeare W ([1600] 1980) Hamlet. Bevington D (ed). The Complete Works of Shakespeare. Third Edition. Collins NY.

Torricelli E (1644) De Solido Hyperbolico Acuto Problema Secundum. In: *Id.*, (1919–1944) Opera geometrica [Vol. I, pp. 113–135]. Opere di Evangelista Torricelli. Loria G, Vassura G (eds). 4 Vols., Montanari, Faenza, pp. 191–221.

A New Cycloid Narrative Centered on Torricelli and Roberval to Understand the Diverse Ways of the Mathematical Revolution

4

Jean Dhombres

Abstract

In this paper, I intend to document the ongoing quarrel on the cycloid in order to analyse the diverse mathematical ways of the scientific revolution, and the judgments mathematicians gave about their reasons to improve some processes as well as to publish or not, and possibly to improve on their predecessors or rivals. What is particularly fascinating in the story as told is that Johann Bernoulli considered the case of the cycloid to be obsolete, due to the completely new *Calculus*. Therefore, the focus on Torricelli and Roberval, which was maintained by many authors, had far from obvious purposes that I wish to investigate by looking at various testimonies. In fact, this taking the opportunity of the republication of Torricelli's *Opera geometrica*, I too confront the history of science as done by mathematicians and by historians, both for methods, results, and posterity. As I consider so many reactions to this quarrel, I have mostly appended original translations in English of various documents in the form of annexes. Because of the important number of letters quoted, and of original sources, I have introduced some lists just before the annexes. The bibliography is at the end of the paper.

Keywords

Cycloid • Roulette • Mathematical revolution in the seventeenth century • Priority quarrels • Roberval • Torricelli • Mersenne • Wallis • Leibniz • Newton • Scientific ethos

1 Introduction. The Honesty of a Scientist and Its Relevance for Historians of Science

In his rector's address in Latin pronounced in 1701 at the University of Basel, the already aged Jacob Bernoulli gave a rather long explanation about a quarrel made with Torricelli by those whom he called "French mathematicians" (*Oratio de Historia Cycloidis*: Bernoulli Jacob [1701]1993, p. 261; Weil 1993), and particularly one named Roberval. There, Bernoulli attributed this hard judgement to Huygens, but it had in fact been written somewhat earlier by the British mathematician John Wallis who was enraged not to have won Pascal's challenge about the mathematical curve that will be our mathematical subject and around which we will construct a large scenario. We could multiply important names who are to intervene, even if roughly speaking, the question was rather a minor one: it was about Torricelli having discovered some unexpected properties of a new curve, called the cycloid, without mentioning Roberval's anteriority, which Torricelli did in fact know when he published in 1644. It does not look like something very new or very serious. Proud of a whole century that had just ended, and explicitly in the name of scientific research and of researchers, with a sense of the mathematical revolution that had been under way, Bernoulli dared to bluntly say the accusation towards Torricelli was a non-sense. He gave two different reasons that have little to do with the question raised: one had to take into consideration Torricelli's quality as an expert in the (new) sciences and not only mathematics, and the second was to recall his reputation as an "honest" man, devoid of prejudices and so unable to hide the truth.

J. Dhombres (✉)
Centre Alexandre Koyré/EHESS–CNRS, Paris, France
e-mail: jean.dhombres@ehess.fr

R. Pisano et al., *Homage to Evangelista Torricelli's Opera Geometrica 1644–2024*,
Logic, Epistemology, and the Unity of Science 55,
https://doi.org/10.1007/978-3-031-06963-5_4

The following judgment is in *Oratio de Historia Cycloidis*' Jacob Bernoulli (Bernoulli [1701]1993; Weil 1993) and said to belong to Huygens. It was in fact written by Wallis (Cfr. below Annexes 4 and 5 and correlated footnotes).

> In any case, we know from his works that Torricelli was a learned mathematician of value, and one who deserves the highest degree for his mathematics; I also believe him an honest man. And I do not see what he could be guilty of to the point of heating Roberval's bile. In 1644 Torricelli published his proofs on the area of the cycloid being the triple of the area of the generating circle. I do not see why it should have been forbidden! One does not deny that these proofs were his and it is not alleged that he claimed for himself anything taken from Roberval. (Bernoulli [1701]1993, p. 261).[1]

"*Ingenum*" is the Latin expression used to mean, and best translated by, "honest man". I thus allude to the broader context of the word in seventeenth century Europe, which perhaps is not the right one for this singular case. Bernoulli wished to confirm to the ideal of a mathematical career then possible, free from any bad behaviour, but perhaps also free from the prejudices of the old cultural world. We should not forget this kind of militancy, long before the Enlightenment when a different notion of progress would be invented and diffused in the new "public sphere". Motivations for a career and change towards something better in the sciences are historically and geographically dependent, and even constructed, as one so easily and lazily says. Lazily, because one forgets to mention who is responsible for the construction of the scientists' honesty. For this construction, could we think of the actions of scientists as forming groups, learned societies and creating lobbies, or of societies of the time traversed by ideological movements? Could it be the effect of the more or less public organization of ideas that may have been fashioned by scientist's results or scientist's propaganda, or just perhaps by historians, from all ages, through their own modes of classification? Certainly Bernoulli, in front of university students and of his colleagues as well, acted in conformity with what he thought, which is not enough to explain why he so thought. I have not translated the word *ingenum* by "candid", which describes a man's attitude pure of guile, but in a way ignorant, and so prompt to believe what is said by anyone as being true. The word candid will however be used in similar cases of priority disputes, for example by Newton in 1714 about Marquis de l'Hôpital, as the French academician ignored Newton's priority while writing in 1696 his *Analyse des infiniment petits*.[2] But we may also recall the splendid use of the word by Voltaire in moral philosophy. We may think that the qualification of honesty for a scientist, up to the idea of an ignorance for lack of information, should imply thoughts in general terms about the significance of new results, and how to present them to scientific peers, and also to a general public. As already said, Bernoulli's presentation of Torricelli was expressed in 1701, certainly according to some moral canons of his time; but it was to describe events about sixty years before. Is there any sign of discrepancy with his own time, which may mean some progress had been made, or was it the simple affirmation of some values about scientific activities that were always to be respected because of their disappearance? What were Bernoulli's purposes while writing about Torricelli and Roberval? Was it just another *fin de siècle* discourse in favour of mathematics, at a time when physics in the form of the Newtonians' view of the world looked far more important? What then was the actuality for Jacob[3] Bernoulli?

He considered the French as a gang, united in the defence of one of its members, and this definitively seemed to be bad behaviour; mainly because they were ignorant of results already obtained. That part of his discourse was a breach of the usually non-nationalistic pattern of the Republic of Letters that the well-known university professor in Basle was supposed to represent. Unless he meant that there should now be a specific part of this so-called Republic, composed of scientists only, and for whom national habits may just correspond to different scientific styles in the universal process of invention, and particularly when it came to the diffusion of results. Was it to imply there ought to be no difference in terms of the respect we owe to all scientists, as they are servants to truth? Even as a sort of ethnological verification of diverse scientists, such a national and esthetic distinction could not succeed in implying different judgments, particularly when it came to plagiarizing someone. Precisely when with Torricelli there was no theft or copying, contrary to what had been suggested by Blaise Pascal, a Frenchman who was then more famous for physics and for his religious views than for mathematics. One part of Bernoulli's speech was to quash such accounts of Torricelli. But then he had to accept the existence of "bad guys" within science. It did not mean they were doing bad science! Obviously, this qualification also meant there was a need for regulation of what was thought to be a career and a decent scientific life, that may go as far as attempting to delineate what a good life was for a scientist, gifted or not. Moreover, it also implied that scientists must be united to avoid such possible bad behaviours and in a heroic stance, putting away villains.

For Bernoulli to establish national distinctions might also have been his way of considering the new development of science in terms of social and cultural forces, and possibly an evaluation of the new differentiated role of the State on

[1] Jacob Bernoulli's text, *Oratio*, remained in manuscript form until Weyl's edition in 1993. The date "1701" is the date Jacob gave his read. Thus, hereafter, Bernoulli's *Oratio* will be quoted as Bernoulli [1701]1993; Weil 1993, if any.

[2] Newton in 1714 will qualify as candid the tone of *Analyse des infiniment petits, pour l'intelligence des lignes courbes* by de l'Hôpital (de l'Hôpital (1696). See Newton's judgment published as an appendix to *The Quarrel between Newton and Leibniz* (Hall 1980).

[3] Bernoulli always signed "Jacob', and I will maintain this orthography, and not "Jakob", which is the far more recent German use.

scientific issues, precisely once some national Academies had been created.[4] Should we deny Bernoulli some sociological reflections due to the fact that sociology only appeared as an organized knowledge in the late nineteenth century? Such a logical deduction has been dryly mocked by Bruno Latour in his paper about Ramses II (Latour 1988). The most recent Academy at this date existed in Berlin, founded by the sovereign of Prussia through the action of Leibniz[5] and this author was seen as a splendid illustration of the German world to which Jacob Bernoulli adhered including fidelity to Lutheran practice. So that, in spite of Latin remaining the *lingua franca* of learned persons implying a common use of *ingenum*, we perhaps have to think less in national terms than in cultural ones as fashioned by a language and a religion as well, and so perhaps creating a style less for scientific writings and certainly more for a scientific life in charge of establishing useful truths, unknown up to now. Following what Robert K. Merton did for experimental science concerning England and Puritans' ascetic views in the late seventeenth century, it is possible to think of a general ethos fashioned by Lutheranism. In addition, perhaps by using Leibniz, who expressed a very close connection to Catholicism, a religion to which both Torricelli and Roberval adhered, with no necessity to make it explicit. This Catholicism could then favour pure and transcendental mathematics (Merton 1973),[6] as a proof there was an abstract thought a priori and absolutely regulating the world. Far different from just the Cartesian warrant about human reason, ontologically free from being misled by a God as He cannot act as a liar.

If I do not pursue such explicative discourses, it is mainly because I have almost nothing more here to a priori bring on the specific subject of religion as related to my project to look on the cycloid through the eyes of those who wrote narratives on this curve to show something had happened to mathematics. I just mention on this religious question what was said about Pascal when he went to challenge[7] scientists with the cycloid. There are testimonies from historians that of those who surrounded Pascal in 1658 at least some thought his excellence in mathematics would support the true Catholic faith. The idea was never evoked in texts related to the challenge, and especially never under Pascal's control. But historians have frequently resorted to anecdote.

I do not forget that ethical values were in the process of being forged to establish scientific communities at a time when quarrels among scholars seem to develop once more as mathematics seemed to have to play a far stronger role in the sciences, and intellectual authorities were seriously demolished as such. These themes are present in Bernoulli's discourse as we will see, and will have to analyze. If such values had been important to the debates I have yet to present, quarrels may be revealing, and this is one of the reasons why historians used those quarrels. But then I have to add there was no public quarrel between Torricelli and Roberval at the very beginning. And if a part of my study is to show how and why public quarrels generate and develop, it is not to confirm all historical narratives about quarrels. Therefore, let me approach the subject by considering Jacob Bernoulli as a historical thinker within the transfer, or extension of scientific culture from Italy to France and England, but also towards the German speaking countries, including the Netherlands. He wished to propose a judgment on what was good scientific behaviour, using an old dispute, but we should not forget to say that the gang of Frenchmen Jacob Bernoulli was implying, was not composed of the ones around Mersenne in 1644, when the so-called quarrel is supposed to have begun. In fact, Jacob Bernoulli was alluding to another period, around 1660, when new considerations on the cycloid appeared, with Blaise Pascal as a leader of challenges to all European mathematicians, and perhaps under the hidden agenda to magnify the Catholic faith. Torricelli was no longer living, but Roberval (1602–1675) was still present, and not too far indeed from Pascal. So it is necessary to speak of another phase, if we maintain the

---

[4] The famous *Accademia dei Lincei*, to which Galileo belonged, cannot be considered at this date as having a national character, but far more as an intellectual institution to provide a place for science within the Catholic world in general, and so in a way opposed to universities becoming rather parochial, or too dependent on the censorship of the Counter Reform.

[5] The Prussian Academy of sciences was founded in Berlin in March 1700. The *Royal Society* in London first, and then immediately the *Académie royale des sciences* in Paris had been founded in the 1660s. If those academies certainly belong to the period of political absolutism, they cannot represent absolutism in terms of science, and by 1700 in Paris, elections were regularly organized on various issues, preserving anonymity. The notion of "official" science did not have a meaning before some public opinion was recognizable. Except one may argue there was something official about the rather unhistorical concept of "jesuit science". But at least it was not associated with any relevant academy.

[6] I know that the link to Catholicism and Lutheran practices, so strongly suggested by Leibniz, ruins the geo-strategical advantage of Merton's thesis about the Puritans in England and Pietism in Germany. However, I am faithful, if I may say so, to Merton's way of looking at "ethical motivations", by translating it to catholic scientists. I too avoid judgements on "Continental science" being compared in quality to "English science" as if such geographical qualifications had any signification, outside jingoism. I certainly will not use the expression "Jesuit science", that seems to negate the variety of behaviours of Jesuit scientists. In a book, I have previously devoted some attention to a "marginal" Jesuit like Gregorius à S^{to} Vincentio (Cfr. Dhombres and Radelet de Grave 2008).

[7] Challenges will be used again by the two Bernoulli brothers and Leibniz around the 1690s, also about the cycloid, but have never been related by historians to religious issues. It is more or less common acceptance that Leibniz was testing Calculus, with mainly scientific aims to help a better knowledge of it, and even to submit it to concrete issues.

qualification of a quarrel for the first instance. Bernoulli in his speech was mixing the two phases, and this is where we could think of him to be a bad historian, even if he raised significant epistemological issues. Could this imply that he had a sort of eternal vision for science? We will soon discover that this was not the case. Jacob Bernoulli inserted his actions—to write about the cycloid—within cultural issues about Moderns and Ancients. This was a way to historically discuss progress, and so to evoke a scientific revolution. In Bernoulli's point of view, such discussions implied quarrels. Not necessarily about results, but on how to use or to publicize such results and perhaps on the ways to get new results. This seems to us to be a rather good issue about the changes in mathematics that may be associated with what was so strongly called the scientific revolution by Alexandre Koyré and is to be questioned today, notably for the role mathematics played in it.

Jacob Bernoulli's rhetorical discourse to students was certainly not to settle all facts or to reach a final truth concerning Torricelli and Roberval. It has been, and perhaps still is, a source for many historical and epistemological questions. One may even say that the literary genre used by Jacob Bernoulli offered the opportunity to organize a reflection without too many mathematical technicalities on various issues on mathematics, its role, its progress, and its history, precisely at the turn of the century, before the real beginning of the Enlightenment and changes in the interpretation of progress, and at a time that one gifted historian called a *crise de conscience* (Hazard 1935).[8] This discourse led to considerations on discoveries, on inventions, on priorities, on methods, on the way Ancients had to be forgotten, on innovations, and also on the reasons for a mathematician to put his thoughts in written form for publication without any direct teaching aim. It was also about the reasons to write the history of mathematics as an adventure, so providing some dynamic tension to the narrative he was writing. However, it was focused on one theme only, the cycloid. Bernoulli's narrative established that this curve was able to fashion the genre of Torricelli's book (1644). Later Roberval's presentation owed much to Torricelli. We already enter in a kind of historical discourse.

At this point, we must state some facts. Torricelli had only added an Appendix with a very few pages on this curve, at the end of a rather long exposition about *De dimensione parabolæ*, where he expanded on Cavalieri's work on indivisibles. But there was another appendix on the "hyperbolic solid", and it is certain that this last addition established Torricelli's fame far more than the one on the cycloid, first in a specific French circle around Mersenne, then to a larger audience. In this second Appendix, a volume of revolution was shown to be of finite measure whereas the area of the plane surface between the meridian line and the axis was infinite.[9] Since Valerio and Kepler, and Guldin too, such results were looked for in an attempt to unite computations of areas, volumes, and centers of gravity,[10] and to-day we see this move as some first steps in preparation of the Calculus. If in his speech Bernoulli focused on the cycloid alone, this was perhaps to prevent this sort of archaeological view of Calculus. Mathematicians of the calibre of Jacob Bernoulli are very sensitive to wrong attributions for all sorts of discovery. Roberval too had claimed he worked not only with the cycloid, but with generalized cycloids, and also with volumes of revolution. This seemed irrelevant to Bernoulli, and so deserving of an explanation.

What Torricelli had written in his *Opera mathematica* about the speed of water running at the base of a tube proportional to the square root of the height of the water in the tube (this is now called Torricelli's law) was certainly seen as far more important. It was immediately popularized by Mersenne in a book published the same year.[11] Moreover, experiments with mercury for air pressure, discussed by Torricelli in letters in June 1644, were also performed with Mersenne when he went to Italy that year, leading back in France to the experiments by Pascal and others that began in October 1646. It gave some years later the basic narrative on experimental science, which is also an important stage in the quarrel on Ancients and Moderns (Cfr. Pascal 1963) . But there was no quarrel at this time between French and Italian scientists, even if there certainly was a serious quarrel in natural philosophy about the void, quite independent of nationalities of experimenters.[12] If

[8] In this book, Hazard is particularly clear in describing the accelerated rhythm in mental changes within French society. For less than a decade, minds turned away from Bossuet's kind of vision of regulated history to an explosion of scepticism. But Hazard was not much interested by the role played in this period by Newton's *Principia* or by the invention of the Calculus. James R. Jacob has been quite alone in manifesting the role of political changes in the English scientific revolution in his *Robert Boyle and the English Revolution* (Jacob 1977).

[9] It was directly linked to paradoxes on infinity, and even if computations may have seemed strange to non-specialists, the results appealed to philosophers in general (Cfr. Mancosu and Vailati 1991; contrary to earlier references, and as it is related to the Roberval/Torricelli case, this one is registered in the last section of the bibliography).

[10] For us, the results are simply explained via the behaviour for integration at infinity of inverse powers of the variable. This was very new for scientists at the time, and because of the kind of technicality, introducing directly mathematicians rather than philosophers to problems concerning infinity.

[11] See the list for primary references.

[12] The question of the void has been intensively treated by historians, and particularly in the recent years. It suffices to quote for the English-speaking world Shapin and Schaffer (1985) and Osler (2000). But in such narratives, mathematics did not seem to be an issue.

we noticed how Jacob Bernoulli emphasized the general scientific stature and ethical personality of Torricelli, this was not however to exonerate the scientist as having behaved as an honest man about the cycloid. Was it in order to restrict the arguments to one curve only? Indeed, around the last third part of the seventeenth century, a rather astonishing number of written histories appeared about this specific curve. There were two books, one by Carlo Dati, and another one by Johannes Groening, plus two rather long explanations on the history of the curve itself, one by Blaise Pascal and the other by John Wallis. The last one was obviously written to counterattack what Pascal affirmed, which was certainly based on a direct contact with Roberval. Therefore, the true Torricelli-Roberval quarrel only began around 1660, based both on accounts of earlier events and on new discoveries. Twenty years or so earlier there were many letters exchanged but no publicity, contrary to what happened in the 1660s when texts were published in order to show important novelties that were happening for the old science of Euclid. They were not the last arguments, and Calculus was still to come. Even if such interventions were centered on results, the idea guiding those historical accounts was to explain how such changes[13] might have occurred. It was therefore a cultural issue to know whether the changes came from Italy, with Torricelli as a hero, or from Roberval as another one from France. But with no help or recourse from ancient authors. Certainly, John Wallis was trying to show how England, and specifically himself, was the origin for the main changes. Bernoulli thought all such discussions were ridiculous.

> Although it does not seem to me worth arguing to find out who first found these Theorems, either Torricelli or Roberval, while what was discovered about the Cycloid at the time of his childhood does not seriously matter. (Bernoulli [1701]1993, p. 262)

Jacob Bernoulli stated that real findings on the cycloid only began in the 1690s, except perhaps what had been done earlier by Huygens in the 1670s as we will see. It is insufficient for us to-day to just focus on questions of priority. Therefore, if to investigate this kind of varied, and in a way idle, later historical thought is my aim, I certainly maintain the focus on what was being said about the Roberval-Torricelli relations from 1643 onwards, and therefore on the only book published by Torricelli, his *Opera geometrica* that is the subject of the present volume. To justify how I will proceed, as I do not wish to write a story about the various stories on this curve up to now, it is enough to say that I am trying to detect philosophical judgments about discoveries and their importance. But I need to hold together two different aspects. One is about the results obtained and possibly a selection of methods, and the second is the meaning for the manufacture of science, possibly the organization of the human mind, and finally the importance for mankind if some revolution has to be imagined. This general aim of Auguste Comte and of many others—gave the history of science its impetus (Comte 1830–1842).[14] Philosophy of science is always impoverished if cut off from history, as is so often said, but less often practiced. But I do not wish to form a sort of rational discourse on how epistemology and mathematics finally came together in the cycloid case, and brought about Calculus. My choice is precisely to read from actors and historians the difficulties they encountered in building a sort of united point of view on the curve, while they were precisely engaged in telling stories about "quarrels". A quarrel seems so contrary to the idea of unity. In October 1643, Torricelli himself explained in a letter to Roberval that he would avoid speaking of certain subjects, because they were too far from the main points he wished to stress in his forthcoming book.

> In truth, I placed this double theorem among those of secondary interest. It will in fact have no place in the trifle works that I must very soon publish, in which I put the greatest importance to the unity of the subject.[15] (Letter from Torricelli to Roberval, October 1st, 1643)

This unity on the subject raises many questions, and in a way arriving at more or less explicit answers depends the way historians organize their narratives. My aim is thus to also check some historians' honesty, and question why they remain silent about certain facts. They cannot all be just candid presentations. If one says that a revolution in mathematics is a change of one possible unity to another one, is it then difficult to determine what the unity of mathematics was when distributed in the *quadrivium* around 1640?

---

[13] From available texts, it is impossible to suggest there were religious issues, but it is quite perceptible that the question of a movement for progress was at stakes. Moreover, at least Jesuit teaching according to the *Ratio Studiorum* maintained nothing of Aristotle was to be changed or to the christening of his theories by Thomas Aquinas, the *Doctor communis* as Thomas de Aquino is sometimes called.

[14] See also my *Comments* in *Premiers cours de philosophie positive* (Comte 2007).

[15] Some references to the Latin text are given in the translation of this letter in Annex 1. There is a special list of correspondences in chronological order in the bibliographical section.

1.1 The Ignorant Despises What He Cannot Reach

My question could be felt too localized, in the sense that I wish to understand if Torricelli's Appendix on the cycloid helped to create, or not, this expected unity. It is certainly impossible to forget that Bernoulli's discourse—forty years after the second phase of the quarrel—was pronounced once it was clear something radically new had appeared with Calculus,[16] and precisely when the quarrel between Leibniz and Newton had just been launched via a book by Nicolas Fatio de Duillier.[17] This is certainly the reason why Bernoulli's text[18] was published in 1993, with no translation, in *Die Werke* of Jacob Bernoulli's *Oratio* 1701 (Bernoulli [1701]1993) under the scrutiny of André Weil (here translated as an annex n° 5). This last mathematician was acting in fact as a historian in the process of creating a commentary of the text. He then noticed with a keen interest Bernoulli's conception of the practice of mathematics being the reserve of a particularly small elite, and so once more, an ethical attitude,[19] like modesty, was questioned.

> For the more recent period, [Bernoulli] was satisfied (from modesty, he said) with a brief allusion to his own discoveries, before concluding with an exhortation to his listeners, if not to participate in mathematical research (given that very few people are fit to it, he says), at least not to deny her the just esteem she deserves. Then his speech ends with the sacramental word: *dixi*.[20]

Bernoulli's words were more violent, but at the same time he glorified some rare mathematicians in spite of their ridiculous ways of quarrelling. Does this sound rather new in the annals of the science of Euclid? For Bernoulli, mathematicians did represent an ideal of nobility, even those who were not born in the aristocracy, of which Evangelista Torricelli might be an example. Roberval only belonged to the gentry, but for him it was no social problem to earn one's living by being a professor. This is contrary to what will be the case for Marquis de l'Hôpital who could not decently receive a salary from the Academy of sciences, and so had to be put in an honorary position. There was however no problem for the *Imprimerie royale* to pay for the publication of his famous textbook, for which he received no money. So that what Bernoulli now wrote, was a revolution in terms of what society had to accept as important, i.e. an aristocracy based on mathematical excellence.

> You have thus heard, gentlemen listeners, the History of the Cycloid. No other curve at this time is more famous or more known among Mathematicians. You have seen it been taken care of by non vulgar men, but by remarkable masters, for the splendour of their birth, the renown of their knowledge, the repercussion of their virtues and their authority; and by far more unrivalled great masters. You have seen them bitterly fighting and not being dispossessed of anything without enduring it very badly. What, then, do you conclude, is hidden in such works but divine, capable of filling the minds of those who know it with incredible voluptuousness? Do not count on me, at least, either to appease your taste, or to go where these great men have already arrived at: it is only possible for an extremely small number. As for me, I think I would have not wasted my time if I obtain for some of you, who with strong laughs are ridiculing a science that they do not know, and such great men, that do not apply what is commonly said: "The ignorant despises what he cannot reach." (Bernoulli [1701]1993, pp. 265–266)

We must also think about the uses of the word ignorant. As we will see, some protagonists working on the cycloid were ignorant of results already obtained, like Pascal, but this may also be the case for Torricelli, at least according to Bernoulli. The first professional historian of mathematics, Etienne Montucla, did not quote Bernoulli's address in the first edition in 1758 of his *Histoire des mathématiques*, but he used the same sources to write at length on the cycloid. Plus, he added some more about Mersenne to whom I will have to come. But contrary to what Bernoulli explained in his speech, Montucla's book was written to popularize mathematics. Even when showing there were important changes in the very old science, if not

---

16 I will explain later how Calculus radically changed some questions related to the cycloid. But this cannot be done before having stated what were the problems.

17 An accusation of plagiarism was launched by Nicolas Fatio de Duillier in *Lineæ brevissimæ descensus investigatio geometrica duplex, cui addita est investigatio geometrica solidi rotundi in quo minima fiat resistentia* (De Duillier 1699, p. 18). The technical character of the book prevented the quarrel being live within the Republic of letters. This does not mean that its treatment has not contributed to fashion a "social behavior" of scientists, particularly at a time when it was necessary to adopt general rules of behaviour in scientific societies like the academies.

18 A translation of the speech is provided in Annex 5 to the present paper, so that the reference is abbreviated as Jacob Bernoulli's *Oratio* (Bernoulli [1701]1993). The bibliography has been divided into 5 different sections, by separating editions of complete works, and of correspondence. Then there is a list in chronological order of published letters relevant for the Torricelli/Roberval issue, and finally the usual separation of primary from secondary literature.

19 Modesty was a virtue supposedly respected within the Jesuit order, but certainly intervening as well about what a "honest" man was (Cfr. Meskens 2005). By providing an aristocratic status to mathematicians, Jacob Bernoulli exonerated them only to behave as "honest" men in what concerned their productions. Aristocrats are not required to be modest.

20 "Pour la période plus récente, il se contente (par modestie, dit-il) d'une brève allusion à ses propres découvertes, avant de conclure par une exhortation à ses auditeurs, sinon de participer à la recherche mathématique (ce qui n'est donné qu'à bien peu de gens dit-il), du moins à ne pas lui refuser la juste estime qu'elle mérite. Puis son discours se termine par le *dixi* sacramentel." (Bernoulli [1701]1993, p. 266).

revolutionary ones, he insisted they were not so difficult to grasp, at least by honest men. More than forty years later, the astronomer Lalande would revise in 1799 what Montucla wrote. So Lalande added his own usually sharp commentaries and explicitly focused on societal issues about the culture a good mathematician should have, and a kind of joy of life that Pascal did not seem to enjoy, due to his religious beliefs. Therefore, historians and commentators went on, until André Weil in 1993 (Weil 1993). Even if the precise mathematical subject of this quarrel was almost forgotten, and if the details of such events were not too often repeated,[21] until now the quarrel as such has taken a sort of paradigmatic importance. Various historians used it for very different objectives. There was an object, the curve, and an adventure about properties being explored. But there was also the idea of an enormous simplification due to a new mathematical way, namely Calculus. Historians largely perpetued what had been the issues of the four accounts (Cfr. Dati 1919; Gröning 1701) that appeared before 1700, with Bernoulli's text coming at this very moment and accounting for the new century at its very beginning. This in itself could justify considering this question once more. My interest is however more in the structure of narratives of this quarrel, and this must be related to the place Torricelli occupied in mathematics with his unique book, the *Opera geometrica* of 1644, and his appendix on the cycloid. It is not a book with one theme only, and perhaps it represents the first instance of a pure research publication, without a precise theme. Even if Torricelli insisted on unity of thought! On the contrary, all historians made it clear, Roberval's rare contributions were extremely focused on different but precise subjects. But they were printed a long time after the second phase of the quarrel, and certainly shaped over time to each have a formal unity. Therefore, the comparison between Torricelli and Roberval made by historians is a show of the essential tension between innovation and educative tradition that Thomas Kuhn explained in a sort of second step in his thoughts (Kuhn 1977, Chap. 9). If I recall that Auguste Comte had strongly affirmed that active scientists were not the best ones to let people understand progress in the sciences (Comte 1843), were the historians I have chosen, even acting as philosophers, more able to let us understand the ways of discoveries and of their relative importance?

1.2 Why then to Scrutinize Such Historical Texts?

Actuality obviously confronts every historian, as he cannot simply say that he tells some facts and may ignore that they have to be organized in the form of a fiction, linking things to the present, even if the link is tenuous, or worse, if today's activities do not seem to have anything to do with some past activities. At least to show how time runs and to provide some general interest to the mathematical sciences. My strategy is certainly not to systematically use all documents and views provided by actors like the mathematicians involved, as well as by persons acting as historians in order to encompass an encyclopaedic view of all the issues about the Torricelli-Roberval exchanges, and the transformation into a quarrel. But my investigation on the cycloid cannot be seen as just a mathematical survey of the first obtention of its properties (Cfr. Teixeira 1909, Chap. X): it has to be on why various narratives are structured according to certain schemes of discoveries. For example, and for the time of his discourse at least, the affirmation of a lack of heritage of the mathematical quarrel, lies at the heart of Jacob Bernoulli's rhetoric. Under the description of a historical action, I do not exclude the two direct actors themselves, Roberval and Torricelli. In fact both wrote letters to each other, and to their colleagues, where they justified their own conduct as scientists, more than just defending some technical and mathematical points, which fifty years later were largely obsolete due to a completely new approach with new methods as Bernoulli claimed. Obsolescence is a process, not something quite natural. If the names of Torricelli and of Roberval were absent from Newton's *Principia* in 1687, the cycloid was the object of many propositions in this book for which it is said that there is as little as possible mention of Calculus (Cfr. Newtons' first occurrence of cycloids: 1687, Prop. 49, Th. 17, I). Had objects transcended methods? This could be a way to maintain a historical unity in mathematics in spite of continuously new objects and new questions, some being forgotten rather than falsified.

A way of looking at such questions about writing history is to know how it is possible to forget in mathematics, a science where errors are particularly rare. The voluntary game of forgetfulness is contrary to the process of posterity, which is an important way for historians, at least for me (Dhombres 1988). Posterity does not mean direct filiation, but that some lines of

[21] Most general histories of mathematics give details about the Torricelli-Roberval issue, and I already quoted Montucla (1758) for his *History of mathematics*, and as well could quote Moritz Cantor who published more that a century later (Cantor, 1900).

thought are revivified and transformed in the same move. Thus, I am back, as in a loop, to a sort of self-analysis of this historical narrative. Torricelli's book has in fact for a long time been a victim of this game of forgetfulness. Very few authors explain what it was that he did on the cycloid, that may have been different from what Roberval did.[22] I intend to be here precise, to the point of translating and commenting Torricelli's appendix on the cycloid, in spite of my being far from a native English speaker.[23]

Even by restricting attention to the cycloid, Torricelli and Roberval appear to be two very different persons, in their behaviours as well as in their styles, and even in their background. In addition, this was always detected by our chosen "historians", and they made judgments on such differences. In a way, we may think of the two as different *Idea types* of the Weberian kind, to the point of having to distinguish among national styles.[24] So, one may understand why historians were so keen to write on these two adventurers, one having lived almost long enough to have seen the first ideas about the Calculus, and the other having disappeared so quickly.

As already stated, I take the testimony of Jacob Bernoulli as a piece of history of mathematics, rather than a mathematical text. I am comforted in this respect by the fact that this author stopped short of providing technical details for properties of the cycloid that were obtained by using Calculus, where he should be given a role from the years 1690 onwards. This was ironically questioned by André Weil, as we read it. What is the role of problem solving in history of mathematics, and particularly about the cycloid when the questions solved does not seem to be the first questions that were raised? Is the history of a quarrel just a debate about priorities and about properties that are true, without trying to see why they were looked for? To answer to such a question is perhaps close to what has been recently called the *practical turn in philosophy of science*.

Torricelli himself, in the already quoted letter to Roberval, claimed that he came to the cycloid "by chance". Could it be possible to understand this sentence, even to qualify it as purely rhetorical, without inquiring about what was found, which was not already known to him from various sources? I will later be thinking in terms of serendipity, which Robert K. Merton himself introduced in his sociological studies. Even if this last scholar worked this concept less for science than for the social sciences, to emphasize his own unexpected discovery about the link between experimental science in England and Puritanism in the second part of the seventeenth century (Merton 1968). To give Torricelli's "chance" a place, even as a possible narrative, I still need some precise analysis of proofs by Torricelli and by Roberval. In addition, I am confronted to the surprising fact that Torricelli added that this curve was studied by Galileo almost fifty years before, when Roberval claimed it came from Mersenne in 1615! Wallis will eventually claim the cycloid was investigated during the fifteenth century. Maybe such claims were just written to affirm the existence of the two groups of scientists, in Italy and in France? Was it not the existence of groups of researchers that renders stories about the quarrel so interesting? Can we then have a sociological view on such groups without simultaneously looking at the questions they were treating? Is the silence from historians about differences between Torricelli and Roberval on the cycloid a way to preserve a unity in the progress of mathematics? But then why is there an obsolescence of older results?

I cannot omit the fact that Torricelli and Roberval themselves had organized a sort of fiction about the curve that made it so fascinating. A "candid" answer by Torricelli to Roberval is clear about his general behaviour.

> You first ask me if I consider the center of gravity of the parabola a priori as a solved problem or if, ignorant, I seek it. I indeed would blush to place an unproved theorem in the midst of my other minor propositions, whose proof is provided. I have shown this theorem according to a single and concise proof. (Letter from Torricelli to Roberval, October 1st, 1643)

The late publication in 1693 of some of Roberval's papers, and of this old letter too, as well as the vehement letter already mentioned, belong to the same effort of clarification of what was a good scientific behaviour. In spite of a clear anachronism

---

[22] It appears that Roberval wrote a vehement letter to Torricelli in July 1646, let it be read by Mersenne, but he never sent it. It was left in his papers to be published by the Academy of sciences in 1693 (Roberval 1693a, b, c, d, e). I will still use it for some description of his discoveries, but refer to it in a specific way, according to *1644 Roberval's declaration*.

[23] See the translation of Torricelli's explanations on the cycloid in Annex 2 at the end of the present paper.

[24] The difference is in fact tenuous between national styles and priorities. It seems clear that French historians, beginning with Montucla, but Costabel and Itard as well during the last century, were quite happy to claim Roberval's priority without inquiring too much about the different methods used by Torricelli and Roberval. My purpose is to suggest how Roberval found an analytical way to work with the cycloid for certain problems, using the parametrization of curves and as well a translation change in the variable. Only this last aspect looked like Cavalieri's ways with indivisibles. But it was free from the notion of a flow of straight lines through figures. It may have been inspired by what more or less transpired about Cavalieri's methods before his publication in 1635, or could be interpreted as the "infinitesimalist" *Zeitgeist,* as we may notice similar ways by Kepler or St Vincent, and perhaps may as well explain what I will be calling the *Robervalian triangle*. A difficulty for historians is to distinguish between two different changes, indivisibles and equations, due to Cavalieri and Descartes if we wish to name two heroes. As the changes appeared almost simultaneously, our required analysis cannot preclude actors of the time to have been more or less influenced by both.

created by the long delay between 1644 and 1693. We may then listen once more to Jacob Bernoulli, claiming in 1701 that he did not understand why Roberval's papers were not amended, or at least clarified by the Académie des sciences under which label they appeared.

> This can be read in the pamphlets published on both sides, and the long letter of protest from Roberval to Torricelli can be found in the *Recueil de mathématiques et de physique* printed in Paris, eight years ago. Into which Roberval's discoveries are contained - some are talented enough, but in some others the author made unpardonable logical mistakes, so that I am surprised the editor had not noticed and suppressed them.[25] (Bernoulli [1701]1993)

At least, editors in 1693 made their point clear. They did not know when Roberval had written these texts, and they wished to leave testimony he worked on it till very late, even leaving some questions unanswered in the margin (Fig. 1). Such a testimony was refused by Jacob Bernoulli as unfit to good and positive science the way an academy should warrant it. Here arises the dialectical opposition between research and teaching, known since Archimedes as well as from Euclid's elements, which plays an important part in history of mathematics.

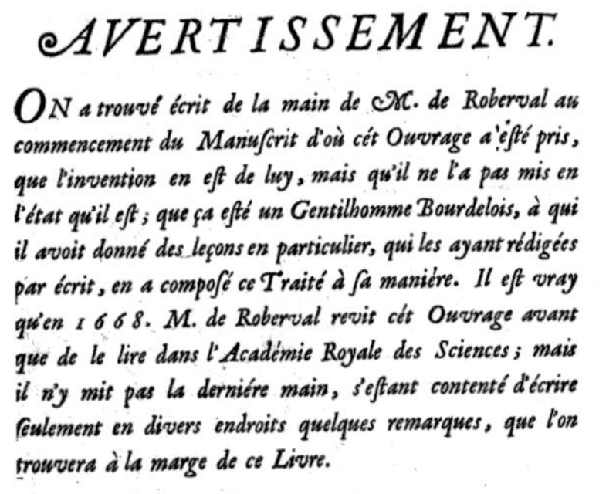
AVERTISSEMENT.

ON a trouvé écrit de la main de M. de Roberval au commencement du Manuſcrit d'où cét Ouvrage a eſté pris, que l'invention en eſt de luy, mais qu'il ne l'a pas mis en l'état qu'il eſt ; que ça eſté un Gentilhomme Bourdelois, à qui il avoit donné des leçons en particulier, qui les ayant rédigées par écrit, en a composé ce Traité à ſa maniére. Il eſt vray qu'en 1668. M. de Roberval revit cét Ouvrage avant que de le lire dans l'Académie Royale des Sciences ; mais il n'y mit pas la derniére main, s'eſtant contenté d'écrire ſeulement en divers endroits quelques remarques, que l'on trouvera à la marge de ce Livre.

Fig. 1 Note being put by the editor at the beginning of the edition of some of Roberval's papers (Roberval 1693e, III, p. 65). *Source* Bibliothèque Nationale de France—Gallica. Public domain

1.3 Oblivion in History of Mathematics

Making history is also a part of the process mathematicians are engaged while they publish. Once they have disappeared, was it not the duty of the newly instituted Academy having unpublished texts to try to accommodate those with the present situation? If the question of the existence of professional and supposedly unbiased historians of mathematics was thus implicitly raised, one cannot but notice that Jacob Bernoulli was almost completely uninterested by methods developed by either Torricelli or Roberval. He just wanted to take care of results. It is obviously not by ignorance. His attitude is precisely linked to his operation of oblivion of ways that have disappeared due to the coming of Calculus, as already noted. Therefore, the quarrel of the Ancients and the Moderns is a horizon for our inquest, and this was part of a strong cultural movement in Europe. But we cannot assert it was the same in the different countries, Italy, France, England and even Switzerland where Jacob Bernoulli lived, all concerned with the cycloid. For Jacob Bernoulli, both Torricelli and Roberval were certainly two Ancients! But they were not the giants on which shoulders Newton said he was mounted to see better.[26] Just look at the way Jacob Bernoulli despised humanist efforts of the sixteenth century:

> When two centuries ago and during the last one, Divine Providence had given mortals the free gift that is the discovery of Typographic Art, the crass barbarity and ignorance of Fine Arts, which infested the preceding centuries, began to retreat little by little and to dissipate, those who dealt with literature worked at first to save from death the writings and works of the Ancients. In which they believed that the treasury of all sciences was to be found. They reconstituted those that were lost, and rectified those who were corrupted. So that scholars employed all their time to perfect editions, versions, interpretations and commentaries, and they had no other occupation throughout the whole sixteenth century. So that as erudition during this period of time was restricted to the bounds of the study of languages and philology alone, positive sciences [*scientiæ reales*] were not increased in any way. This increase was reserved for the following century only, the seventeenth. Then, indeed, the secrets of the Ancients had already been sufficiently excavated and toiled so that the work which was to be consecrated to this task, might now be regarded as useless. And yet nothing was found there that marked a very great knowledge of Nature,

---

[25] Jacob Bernoulli will not precisely indicate those errors, and as well ignore *Robervalian triangle* about which more is to be said.
[26] In the cycloid case at least, we may see how Newton's use was closer to that of Roberval than that of Torricelli.

> or which could satisfy a mind eager for knowledge. Under these conditions, here and there more sagacious masters came, who set to themselves the task of knowing for themselves, and, leaving aside opinions of the Ancients, carefully scrutinized Nature by reading it in its very book.[27] It is, therefore, thanks to them that the sciences were not only recovered and sanitized in their first foundations, but still prolonged in an immense way beyond the antique limits already reached, and daily enriched with new increments. (Bernoulli [1701]1993)

A different process of oblivion is also to be seen in one of the famous papers by Blaise Pascal, his *Historia Trochoidis, sive Cycloidis, gallicè, la roulette* (Pascal 1992, *Histoire de la roulette*, III 1657–1662) as published in 1658, where he gave the three names under which the curve happened to be known in the seventeenth century. Pascal almost entirely avoided speaking about Torricelli's work, even if he recognized the Italian scientist used cycloid as a name for the curve under inspection. One may say that something analogous appeared against Roberval (who used the word trochoid) and Pascal (fond of the word roulette) in the other text by John Wallis, also a history of the cycloid. It accompanied a very long and technical text, *De cycloide*, that was published in 1659, accompanied by several letters. It is sure that I have to be far more cautious with those two last texts, by Pascal and Wallis, which came with the challenge launched in 1658 by Dettonville, alias Pascal. In spite of their technical issues, both texts are polemical. Precisely those two texts may explain the reason why Jacob Bernoulli wished to correct a history too obviously in favor on the one hand of Roberval or on the other of Torricelli. But was Wallis' choice a way to better exhibit his own inventions, or was Pascal's choice to mention Roberval a way to explain the origins of his own discoveries? Bernoulli was happy to reject them all. In other words, he wanted to adopt the posture of a historian tracing a road for the construction of which he had played a role, but that he did not want; but that he did not want to take into account because it appeared to be the only modern way. His criticism, often acerbic, was related to the preceding texts of Pascal and Wallis, but of Dati and Gröning as well, so that he construed a different historical narrative. Moderns of the early seventeenth century were to be forgotten in order to emphasize the fantastic changes in this era.

Coming from an actor who did not want to talk about his own action, why, then, did he write? I answer this question by attacking on several fronts, notably on the idea of a scientific revolution to which Torricelli is associated, less doubtless by his own works, than by the reputation of Galileo who covered him with his tutelary shadow. Jacob Bernoulli is almost implacable on this subject in terms of scientific notoriety that has the value of a radical change. The authority of Galileo is a warrant, and it extends to Torricelli and to his book.

> Galileo was without contest the first among all the mathematicians of his time, of a very penetrating and quite divine genius, and he was remarkable both to solve problems and to raise ones of a new genre in all disciplines, even if he excelled particularly in Optics and Astronomy. The army of the literati was in admiration of Galileo,[28] when with the new mechanism of his tubular machine, he brought the sky closer to their eyes and showed new satellites around Jupiter and new constellations, which he later called "Mediceans." They gave to this master, in every place, such fame and such glory that he was reputed the Archimedes of his age. (Bernoulli [1701]1993, p. 261)

We see two questions juxtaposed. The question of the change brought by the Moderns, of which Galileo is a standard-bearer, but also that of the authority of one only to cover a historical phenomenon, the scientific revolution extended to Bernoulli's time, but perhaps arriving then to an end.[29]

On the other hand, in his own story, Pascal manifested the role of Mersenne, not as a good mathematician, or a good physicist, but as a remarkable solicitor of questions, in the mood of a new modernity.

> He (Mersenne) had a particular talent for raising nice questions, in which, perhaps, he had no rival. But even if he had no such luck to solve those, while it is properly in this that all the honor consists, it is nevertheless true that we are obliged to him, in that he provided the opportunity of several beautiful discoveries, which perhaps would never have been made, had the scholars not been piqued by him.[30] (Pascal 1992, IV, p. 214)

Thus Pascal, using Mersenne's personality, and taking care of the modesty of Roberval who published so little, while providing a narrative about groups, is dependent on individuals. Pascal is not describing progress as due to the invention of some methods, and he is also reluctant to see the active intervention of so reputed masters. What was the situation around

---

[27] The expression, *naturæ libro*, the book of Nature, must be here a direct allusion to Galileo's book, *Il Saggiatore* (1623). Galileo had meant that this "book" was written in geometrical terms.

[28] Bernoulli's appreciation concerning Galileo's achievements in the sciences was oscillating between uncontested glory and an uncritical reception by scientists in general.

[29] The way Jacob Bernoulli is speaking of Galileo is quite similar to what Auguste Comte will later write. It is far from establishing Galileo as the symbol of all struggles against instituted religions, or as the herald of lay thought.

[30] "Il (Mersenne) avait un talent tout particulier pour former de belles questions; en quoi il n'avait peut-être pas de semblable: mais encore qu'il n'eût pas un pareil bonheur à les résoudre, et que ce soit proprement en ceci que consiste tout l'honneur, il est vrai néanmoins qu'on lui a obligation, et qu'il a donné l'occasion de plusieurs belles découvertes, qui peut-être n'auraient jamais été faites s'il n'y eût excité les savants." (Pascal 1992, IV, p. 214).

1640? Had Roberval any authority coming from his link with Mersenne? One sees at least that there was no equality of authority according to the two sides of the Alps, due to the status given to Galileo, and perhaps due to the strong impression a revolution had occurred.

I must add a warning here, as I mentioned Bernoulli's refusal of technical explanations, and still I use him as a guide through the threads of this plot linking Roberval and Torricelli, even years after the death of the second author. I will have to correct his silence on methods. Would it however be necessary to make a difference between narratives according to the way they use or not mathematical items, proofs, styles, and technical details? If I were to oppose two kinds of discourse on this single criterion of the use of mathematics, I would commit an error. By retaining only the texts making use of appropriate mathematical knowledge, I would eliminate in the name of a particular professionalism about the history of mathematics both mathematicians like Jacob Bernoulli in his discourse of 1701, and historians who organized a story to read for those who did not know mathematics. Part of the discussion on the cycloid may be read as an acculturation of mathematics in European society. It is not by chance that it took place during the long-term quarrel of Ancients and Moderns. Moreover, it is as well foolish to restrict what is at stake to either pure mathematics or to religious and political issues. Today I can think of Jean Mesnard, who in his monumental edition of Pascal's *Œuvres Complètes* (unfortunately unfinished) gave in 1992, just before André Weil's publication, an important place to the Torricelli/Roberval issue, with no other mathematical explanation beyond that derived from the text of Pascal himself (Pascal 1992, pp. 147–578).[31] He did not comment on it, except to translate Pascal's results in terms of integration. The aim of this literary historian on Pascal, was to show the intellectual effervescence that had developed from the time of Torricelli and Roberval. In a sense then, and this is welcomed, Jean Mesnard contradicted the work of Jacob Bernoulli about the lack of actuality of the Torricelli/Roberval issue.

Mesnard provided a large cultural echo to the quarrel, in order to organize an account of the changes occurring during the late seventeenth century. He inserted several authors in a movement in France, in Italy in England, and in the Netherlands also, from the 1630s to late in the 1660s. Therefore, authorities were multiplied. Moreover, he did not forget to mention the idea that the challenge by Pascal was a sort of good defense of the true Catholic religion. A family anecdote had indicated that Pascal might have solved the new problems with the roulette in an attempt to ignore some tooth ache, but did not want to publish anything at first: a Duke convinced him of the happy effect that a success of the scientific spirit would have for his other apologetic purposes (Pascal 1992, pp. 168–169).

What in fact could have been the issue of such a counsel given to Pascal? Was there a strong epistemological belief about the unity of truth and knowledge that could not be dissociated in different domains? Or could it be the proof that a tendency existed for a domination by the mathematical sciences of all kinds of knowledge, bearing the claims of the so-called Moderns? Such questions are relevant, but we should not forget that Pascal never in fact used his results on the cycloid to glorify Catholicism. Reputations of Galileo, Descartes, or Pascal were such, outside any social importance of the three so different scientists, that they were thought to open a new world, in spite of the fact that they gave different roles to the Catholic faith. This idea of a new world suited an acting historian like Bernoulli, who knew how to give his place to Leibniz in the formation of Calculus, and in any case was able to recall the presence of Newton as well for a precise new cosmology. But this was for another quarrel, not for the one that we are concerned with. Scientific authority, shared or not, is however at stake in the various texts we are using. It was a hidden issue in the first letter of Torricelli to Roberval, at the end of 1643, in a sort of a long tradition of problems to be respected in mathematics.

> I am not accustomed to dwell on figures that are not usual, nor to solids which, even if they are new, have not at least arisen from plane figures known since a long time. And this essentially because the result of my labors when all has been successful according to the desire of the mind, must also be able to obtain the unanimous approval of cultured men, and that no one must be there to be jealous of figures that I would myself have found. (Letter from Torricelli to Roberval, October 1st, 1643)

One does not gain authority by working on novelties, so Torricelli seems to write in a sort of mathematical conservatism. Was this due to the intellectual atmosphere reigning in Italy after Galileo's condemnation? The historical issue due to a decline of Italy because of religious censorship has been a leading factor in history of science in the nineteenth or even twentieth century (Cfr. Braudel [1989]1994). But was not the main question that of the validity of a new method, the method with indivisibles that Torricelli claimed to defend (Cfr. Bascelli 2015)? Just as Pascal did later. It is another form of the question of authority, this time about ways of doing mathematics that had not been used by the Ancients, or even were

---

[31] The choice to avoid technicalities is even justified by Jean Mesnard for his presentation of Pascal. The danger is to overemphasize the quality of Pascal's discoveries: on the cycloid, Pascal did not add a lot to what Torricelli and Roberval had already done. A way to ignore history of science in any biography is to uncritically take for granted the scientific status of one person.

refused by them. Therefore, Torricelli offered three proofs, as we will see, the well accepted one according to the Ancients, and two with the new method from Cavalieri.

> One of my ways of access proceeds from the doctrine of the indivisibles, which, if we were always addressing an enlightened reader, might be exposed in a few words. The other, which proceeds from the inscribed and circumscribed figures in the manner of the Ancients, is not really so quick, but it is simple and may be worthy of curiosity. (Letter from Torricelli to Roberval, October 1st, 1643)

This discourse is good rhetoric to say that the new method is both quick to understand and quick to execute, far quicker at least than the old method. It is also a process of oblivion. Without any doubt, Torricelli is launching this new posture of efficiency, that was not defended by Cavalieri who is immensely long in his famous 1635 book (Cavalieri 1635). But as we shall see it cannot be said that Torricelli is using the trick of thick indivisibles, that is more or less Riemann sums with constant steps that Pascal and Wallis will use, as well as Fermat, and that have paved the way to Calculus.[32] John Wallis will even pretend the same as Torricelli, but making use of a completely different method, based on an induction from integers to all kinds of numbers. He even thought it had to be generally installed in mathematics as the real new process. We then retroactively understand Bernoulli's reaction: both novelties from Torricelli and from Roberval, but from Wallis as well, do not matter any longer. Thus, in Bernoulli's opinion, it was not possible to present Roberval as an innovator. Even if he had the originality of an analytical point of view on the cycloid. In fact, Roberval felt unsatisfied not to be able to transform such inventions, like the ones we call *Robervalian triangles*, into something analogous to what the Ancients had done. Perhaps sometimes obsessed by Roberval's declaration, it however could be our temptation to emphasize his capacity for innovation, at least once having observed how Torricelli and some direct successors differed on the cycloid from what we may guess Roberval had done before 1644, the year Torricelli issued his book.

In providing a dossier like Pierre Costabel liked to present, but not in order to circumscribe the curve itself or to discuss later quarrels as such, I have to show various narrative potentialities of the Torricelli-Roberval relations up to an understanding of the ways to Calculus, which possibly implies possible hindrances to it.[33] This is so that I do not try to have Torricelli better judged than what was achieved by Bernoulli. I am not interested either by what Gaston Bachelard considered was our historical task, that is to judge what was then the best position to have had about Torricelli or Roberval. On the contrary, if I dare say so, I prefer to select different and sometimes contradictory issues raised by historians and mathematicians over the years once they had the story of Roberval and Torricelli in their mind. Even if they were sometimes dreaming of establishing the status for eternity of a mathematician like Torricelli, or if they looked for what it meant to be a "honest" mathematician precisely at a time when there was such a number of geniuses, as noticed by Jacob Bernoulli. By the way, the expression "honest" was directly taken by Jacob Bernoulli from Wallis' *de Cycloide*. Was honesty a different value for the Oxfordian Wallis who wrote before the Glorious Revolution of 1688 or for Bernoulli, who lived in the quiet comfort of Switzerland, far from the royal Court habits or even the morals of the European gentry as he was from a merchant family?

2 A Story Told in Terms of Problem Solving

Let us then investigate the epic qualification given to the curve by Montucla; it was perhaps done by the historian to lessen all discussions about the dramatis personæ. We cannot avoid this exercise here, and it may at least bring something new about the relations between Roberval and Torricelli, first organized by Mersenne.

2.1 The "Helena of Geometers"

It was Montucla in 1758 (Montucla 1758, 1799) who aptly gave Helena's name to the curve, the cycloid as we are used to say now, that had been studied by almost everyone from 1637 onwards, and was named roulette by Mersenne. The name Helena that Montucla used was certainly an allusion to the famous war she occasioned more than for the beauty of the curve itself! But contrary to what had been suggested, the public war was not at first between Roberval and Torricelli. There certainly was a difference between them for the origins of the curve, and thus the historian Montucla was essentially interested in knowing who had found this curve first. But this was not the main interest for Jacob Bernoulli, who wished to

[32] For last multi-voiced book to have appeared on indivisibles see Jullien (2015).

[33] There are several archives at the Koyré Center which are files on the Torricelli-Roberval so-called quarrel. One was organized by Jean Itard, but I could not find it, and another one by Pierre Costabel, whose papers have been well archived.

show that Galileo, who might have encountered this curve, was in fact not able to solve any problem about it. For some historians, failures of the greatest scientists are better ingredients for good scientific narrative.

> But precisely one hundred years ago today or one hundred and two years (as Torricelli testifies) , Galileo had first invented this curve in Pisa on the occasion of the construction of an arch bridge in this city. He gave it the name Cycloïs, following the example of Cissoid (and yet in conformity with the Greek language, it would be better to say Cycloid than Cyclois) He soon began to study the nature and the properties of this line, and more particularly to seek the measure of the area under the curve. It is quite strange in this inquiry that this penetrating Master was quite tortured, incessantly embarrassed [...] and he will never reach the truth by a geometrical reasoning. Finally, once he had sweated blood and water over this problem forty whole years without any result, and entirely despaired of any success, he transmitted the burden to Bonaventura Cavalieri by a letter which he wrote to him in the year thirty-nine of the last century, and he invited him to solve the knot, which he had not been able to untangle.[34] (Bernoulli [1701]1993)

Bernoulli certainly gave an explanation of what this curve was, although he seemed to doubt that something concrete had been done before Torricelli. He avoided mentioning Roberval there. But he did not explain at all why the search for the area of the arc was the first problem to be solved. He did not seem to know Mersenne had first seen a "half ellipse" in 1636 (see below Annex 3) and then had provided far more the next year, precisely the computation of the area of an arch. Therefore, the form of the curve, that apparently was the question up to 1637, did not seem to be the aim of any research according to Bernoulli. In the title of his 1644 book, Torricelli specifically mentioned the search of the space of the cycloid (*De dimensione spatio Cycloidalis*...), and nothing else

> The Cyclois (or Cycloid), which others call Trochoid, the French Roulette (or rather small wheel), is a Line Curve of the number of those called Mechanical, or Transcendant. It is that traced in the air by a nail fixed to the circumference of a wheel that evolves by turning on a plane. It is commonly believed that the one who first discovered it and was the father of this curve is Galileo. Although Wallis thinks that it was well before Galileo, and affirms that it was already traced in a work of cardinal de Cuse, in the collection of the Savilian Library which exists among his manuscripts proper, transferred about 1454 by donation of this cardinal. But it is described wrongly in the printed copies and subsequently, it is said to be mentioned again in about 1520 by Charles de Bovelles, a no less famous French geometer. (Bernoulli [1701]1993)

Bernoulli directly took such details from Wallis' d*e Cycloide* (Cfr. below extract, Annex 4). But it was the way for him as well as for the English mathematician and theologian to distinguish among Moderns of the seventeenth century. What was important for Bernoulli was the subject of another war, the use of infinitesimals in what was still called Geometry. In fact the subject was geometry of curves no longer submitted to the algebraic restriction Descartes had invented: in the text quoted above the word transcendant appears as an essential one. It had, therefore, been judged important for him to point out that any valuable work was indeed only possible by authors for whom the infinitely small was a serious and necessary issue. But also that the method could not succeed solely using the indivisibles as led by Cavalieri, considered as insufficient and necessarily outdated. Bernoulli hastened to show that Torricelli's success owed little to the indivisibles:

> It is thus that, happier in that than Cavalieri himself, the first of all who measured the area of the cycloid, or, if you will, unveiled its measure to the world, is Evangelista Torricelli. This mathematician of Florence used these few rudiments of Geometry of the Indivisibles, which Bonaventura Cavalieri had once exposed. Thanks to those, he found that the area of the Cycloid is exactly three times that of his circle. What Torricelli himself, after discovering it, took care to publish by printing it the following year in 1644.

We are too confronted with another historical question, about the tension created by the novelty of things that were not claimed as being infinitesimal, but clearly were not within the classical realm of mathematics. Could we say it was a war? Recently a sort of a bright novel affirmed there was a ban on infinite divisibility due to the Jesuit order, so linking in our mind the heroic stance about Galileo defending the Earth's motion and infinitesimalists (Alexander 2014).[35] The danger, so it is affirmed in this other narrative, was a theological one about atoms and indivisibility, so Christian transubstantiation could not be ultimately true. The main difficulty, once more, to sustain such a (good) plot is the fact that it is absent from any

---

[34] The mentioned letter from Galileo in Arcetri to Cavalieri in Bologna, is dated February, 24th of 1640, and preserved by Carlo Dati (1663) in his book (Dati 1919, pp. 3–4), then reproduced in Galileo's *Opere*, XVIII, pp. 153–154.

[35] Due to Leibniz, the term "infinitesimal", instead of terms like Calculus, Analysis, or more specialized Indivisibles in the title raises a problem. As the author stopped short around dates like the appearance of John Wallis's *Arithmetica Infinitorum* in 1656. It is true that this Wallis's book had been translated some years ago by the late Jacqueline Stedall, *Arithmetic of infinitesimals,* and she wished to show a path through the idea of computation included in the word arithmetic. Alexander's "thesis" developed a micro-history with Jesuits, which Torricelli and Roberval cannot be part of. Therefore, his narrative makes sense only if the revolutionary Calculus is shown as a sequence, and this fact is not seriously shown. It is contrary to the immense past that Lazare Carnot provided in 1797 with his *Réflexions sur la métaphysique du calcul infinitésimal.* Another recent book directed by Vincent Jullien too covered a rather long period of time, *Seventeenth century indivisibles revisited*; it gave me the possibility to speak about the historiography of indivisibles, and so to analyze those attempts pretending there was only one path towards the Calculus. This may justify why I am trying to show here that Roberval and Torricelli were not doing the same things, at least for the cycloid.

text with some mathematical content. In fact there was the main difficulty in any infinite composition of continuous magnitudes (*compositio continui*) that prevented using any unwarranted summation process. We clearly see that if this indeed was war, it sort of quietened, or was regularized within mathematics at the end of the seventeenth century. The invention of summation methods like infinite power series allowed the adaptation to transcendent curves of Descartes' algebraic inventions like the method of indeterminate coefficients. Let us now hear from Torricelli himself.

In October 1643, in a direct letter to Roberval, Torricelli who already had news from Mersenne, presented his version of the facts concerning the cycloid. For he wanted to explain that he quickly solved a problem that had been proposed to him by Galileo, through Cavalieri as if there was some age hierarchy to be respected. He successfully tried different methods to find the area of the arch ("Which he called the dimension of the curve"), and new methods with the indivisibles were secured by the recourse to the old method for areas and volumes due to Euclid, what would be soon called the method by exhaustion, avoiding or reducing infinitesimals of any kind to another structure by measuring things up to less than a given quantity, however small it might be. To link the research of an area to indivisibles seems quite normal for a young mathematician educated close to Cavalieri.

> The measure of the cycloid (for it is the name given forty-five years ago to the figure by Galileo, and perhaps only known today by you under the name of trochoid), came to me without hoping for it, and I would say almost without searching for it. I then proved it five times, and each time according to different principles. (Letter from Torricelli to Roberval, October 1st 1643)

No war indeed! Torricelli added that he failed to calculate the volume obtained when the arch is put into rotation about the axis of the curve. He just gave the explanation from Galileo for finding the area, and a negative answer for volumes, proving that in Italy results from Fermat and Descartes around 1638, with volumes of revolution and centers of gravity, had not been known. At the same time, Torricelli mentioned tangents to the curve, and in a very brief way claimed it had been shown to him by Viviani, the younger one as close to Galileo as himself. Thus, he was explaining that the problem no longer appears to be of real interest to him, and he will only devote some lines to it at the end of his appendix on the cycloid, before working on curves close to the cycloids. In his own account, which Jacob Bernoulli knew well, Blaise Pascal thought the origin of the curve was to be found in Mersenne around 1615,[36] with the question of its form as the first to be raised. This is an important piece of information. Pascal articulated his statement in two surprising stages. One was to reverse the question of origin: the curve is "[…] so common that after the straight line and the circle, there are no other so frequently found." (Pascal 1963, p. 117 et sgg., 1992, IV); so that one should be astonished that ancient geometers (*priscorum Seculorum geometras*) knew nothing at all about it! The argument of naturalness of the curve—a frequent one in mathematics but by necessity historically dependent—, is different from that of simplicity. Descartes had declared in 1638 that the curve could not be simple as it did not belong to the ones he accepted to consider for his *Géométrie*,[37] i.e. what we call algebraic curves, that is, defined via an equation of the form $P(x, y) = 0$ where P is a polynomial in two variables. In the same way he refused to give too many details on tangents. Could Torricelli be interested in discussing the difference between transcendental and algebraic curves? We are not aware to what extent Italian scientists knew about Descartes' position. Even if Mersenne went to Italy, and talked with Torricelli, we may think that the mercury experiment was more important to them both. In 1649 however, when Descartes'*Géométrie* will have been translated into Latin by Frans van Schooten, the translator will put some clear comments on the cycloid, precisely using the content of the Descartes' letter to Mersenne (Descartes 1637; 2nd ed. 1659, pp. 265–271). An indirect proof that this had not been understood earlier. Many colleagues worked on this project launched by van Schooten, just as may be seen from the title of the book, with another edition ten years later. Therefore, here was a new group of Cartesian mathematicians intervening, with Frenchmen in a minority, and so not a gang, distinct as well from Mersenne's group and obviously from the other group constituted by Galileo's disciples. From an epistemological and historical point of view, another period commenced. So that, ten years later, we may surmise that the interest put on the cycloid by Pascal was a way to surpass Descartes' limitation, and thus a way to declare war on the Cartesian presentation of mathematics as a well ordered and easy-going domain. But the war was not between Torricelli and Roberval. We cannot conclude for the moment the reason why Pascal entered the battlefield, but there is a strong feeling he did not know the situation for the curve at the very beginning of 1658 well (Cfr. Costabel 1964) and so had to be informed by Roberval. A simple explanation based on Pascal being informed through Roberval's animosity against Torricelli has the inconvenience of supposing that Roberval was not conscious that he had launched something that Torricelli had not understood, and that

[36] On that date, having left La Flèche college where it did not seem to have received a true scientific education even though it was the same college where Descartes was educated a few years later, Mersenne was working in the Nevers convent of the Minims friars, and learning theology.
[37] See the letter from Descartes to Mersenne, August 23rd 1638 (*Œuvres de Descartes*, II, pp. 313, I. 3–13).

could have been extended by van Schooten's group with no recognition of it by Roberval either. This is the tone of the *1644 Roberval Declaration.*

According to Pascal, after a first stage which was the recognition of the form of the curve, a second stage was to explain why Mersenne wrote to all Europe around 1637, to all those he believed capable, and thus to Galileo in his home prison, only to discuss the question of the area of an arch. The reduction of an inquiry and of a curve to the single problem on an area was part of a game that Descartes judged more or less infantile. Pascal did not mention the fact that Galileo never replied to any of the few letters Mersenne sent him (Dhombres 2011a) . There is a testimony of a letter by Galileo, where he claims that the entire Academy in Florence could not read the characters Mersenne had written[38]! At least this could be a proof of shared work in Italy on the curve in general, and not for the area only. In any case, Pascal concluded for the area only: None could succeed, and all despaired" (*at nemini res ex sententia cessit, omnesque de nodi illius dissolutione desperarunt*). Terms of war, but this time against the curve, contrary to what Homer described in his epic poem.

This is also what gave Pascal the possibility of introducing Roberval into the story at this very moment. For Mersenne, so Pascal said, seeing the young mathematician Roberval was able to solve several problems, asked him in the 1630s the question of finding things on the roulette. No more is specified. A phrase is often quoted when it is a matter of telling a story of science: the famous sentence of Pope, "Newton finally came". Here Pascal wrote: indeed, M. de Roberval succeeded (*Nec vero cum sua spes frustata est*). The date provided by Pascal for this success is the year 1634, which corresponds to the date when Roberval became a royal professor,[39] and perhaps may hint at a cultural and institutional confrontation with Italy, as the only source of scientific knowledge. However, we certainly cannot speak at that stage of an official knowledge, as later could be possible due to the creation of the *Academy of sciences.* In Pascal's presentation this had to be the date of Roberval's success: he proved that the space of roulette is "triple of the circle that generates it" (*Trochoïdis spatium spatii Rotae a qua describitur triple esse demonstravit* (Pascal 1992).

From a history so told, we cannot understand why the most urgent thing required by Mersenne was to find such an area, except if there was some idea of a challenge. Bernoulli mentioned the challenge, but only in order to obtain some glory, which seems dubious in Mersenne's case, even if Catholicism may receive its share. Was it not far more urgent to find a way to construct the curve, a task for which the value of the area seems less important, or too indirect? The restriction to one problem only is not evoked by Jacob Bernoulli. As a counterpoint, this raises the question of finding tangents, as one may think that tangents then were important for inventing the form of the curve. Could it be an exact construction of the curve? What does this mean for a curve like the roulette, that was linked at least by Mersenne to the still unsolved question of the quadrature of the circle, where exactness was the main problem? This last troublesome problem was avoided by Torricelli, as it was now by Jacob Bernoulli. The silence about the quadrature of the circle, that we consider as a positive move because it avoided a still unsolved problem, is somewhat strange for somebody like Bernoulli when delivering a speech to unspecialized students for whom the quadrature might have been something known. His silence may be a game to show that new questions were arising from Calculus, and so sending old problems to oblivion, or better said, the quadrature of the circle was to be studied along completely new lines that could be too difficult for students to grasp. Even Pascal's new ways had not been successful, and Bernoulli is then consistent with his posture not to mention the Calculus. Pascal's vocabulary about the roulette does not belong to the Euclidean vocabulary, but to that of mechanics. It was as if Pascal wanted this other vocabulary to show or even to prove the attribution of the discovery of the curve to Roberval himself. The latter had in fact published a work in 1636 on the movements on inclined planes (Roberval 1636). In addition, Torricelli in October 1643 recognized this as Roberval's particular skill, in order to suggest that he had nothing else but mechanics to explain the origin of an interest in the curve.

[38] It seems difficult to believe that the puzzle was less on the properties of the curve than a question of deciphering Mersenne's writing. Had moreover Mersenne linked the question of the measure of the arch to that of the Aristotle's wheel, and so to the far larger question of infinity? The answer seems to be yes, but then there is no mention at all in Torricelli's works. This shows a "modern" aspect of the mathematics being worked on; its problems became autonomous from any philosophical issue. This may be valid for Roberval as well, uniting the two authors in this respect. This conversely does not mean that mathematics had no role to play in philosophical issues, and Pascal's use of mathematical examples was frequent. Mersenne and Galileo in their scientific papers were involved in arguments about the Aristotle's wheel, as was Roberval in his letter to Mersenne dated January 1637. More on that will be described in Sect. 4.3, once we have seen what was really discussed.

[39] There is a tradition to affirming that Roberval was obsessed by having to regularly prove he was able to be above possible concurrents (never named) for the Ramus chair. This could explain his frequent delays in quickly answering precise mathematical questions. But was it not a way to avoid confronting a personal difficulty, that is to be able to conclude any work and let it go to the press?

I suppose that your so penetrating and fertile mind, has observed by itself, being in a sense the inventor of this figure: a line of this kind is indeed familiar in nature, and it arises from the composition of two movements, rectilinear and circular. (Letter from Torricelli to Roberval, Ist of October, 1643)

More specifically, we know that Roberval was working on the composition of movements. He even will write a book on the subject, that would be published later (Roberval 1693b). He was known to have used this composition to find tangents to the cycloid. But Roberval gave a categorically different testimony in 1646: to Torricelli, he claimed to have come to the composition of movements once having worked on the cycloid, and after this he had invented a theory of the infinite (*inventâ infiniti doctrine*).

And this theory of the infinite (let me use this vocabulary in this letter and afterwards forget about it) being quite well developed at that time, I applied myself to adopt its meaning for finding tangents to curves. And first of all by the strength of Analysis (vi Analyseos), I found a certain method which then just invented did not seem to me universal, although long after it was recognized as such. Certainly I sought this universality, and disdained (as every one) particular methods. It was our trochoids that gave me the opportunity to look closely at the composition of movements. Such was the occasion, and towards the year 1636, we publicly released a universal method of tangents that was deduced from it. This is what emerges from the theme of our lessons, and our most noble disciple M. du Verdus has transcribed them, and distributed them to many. That is why for a long time such a theory has been recognized as mine by public trust, and many testimonies are available. (*1646 Roberval's Declaration*, Epistola Aegidi Personerii de Roberval ad Evangelistam Torricellium, 1693, p. 297)

At least Torricelli's remarks, if taken seriously, show that people were aware of the mechanical turn of Roberval's thoughts about the cycloid. This aspect will disappear with Pascal. This leads to the question of purity of methods, for which geometry was at stake. We may think that it was also the case for Fermat using what I later will call the baroque frame of reference. Could we surmise, due to the silence as well from Jacob Bernoulli, that there was a debate among scientists regarding some lack of rigor that mechanics risked bringing to mathematics (Jullien 2015; Lenoble 1943)? That is, at least if mechanics was to be forgotten, as well as indivisibles (De Gandt 1984–1985). But everybody knew that Galileo himself had been interested by mechanics for its role in falling bodies. Moreover, most mathematicians, including Wallis, were interested in having an axiomatic presentation for mechanics, in the same way as for Euclid's Elements about which there were always discussions. Could differences be inferred in the appreciation of mechanics in the various scientific networks, and about what seemed natural? Having heard Pascal and Wallis, plus some other sentences from Torricelli, we have to turn to the earlier authors, Galileo and Mersenne, both interested in mechanical problems for their own purposes, and not necessarily because such problems raised mathematical questions. Both authors seemed to be the two chief nodes of the two networks we are examining, and they were later claimed to have both discovered the *Helena*, and both were said not to have succeeded in anything about it. The first war with no winner!

2.2 Mersenne Took Care of Letters, Playing the Role of Intermediary Between the Geometers of Each of the Two Nations

Intermediaries, networks (Bernoulli [1701]1993; Weil 1993) and direct or indirect relations are certainly some main subjects of new narratives for contemporary history of science, focusing on groups rather than on individuals, providing to history of science its gist of scientific activities and also to the term war a new meaning, different from a duel. Fortunately, former historians had published enough correspondence sources, so that one may have a rather good picture of the various scientific networks of the seventeenth century,[40] and I already mentioned three of them around Galileo, around Mersenne and around van Schooten. In this context, it is important to notice that from the very beginning, in his first letter to Roberval dated October 1st, 1643, Torricelli acknowledged his pleasure to be able to directly speak to Roberval. This was precisely to refuse any intermediary, or better he wrote *sin alio interprete*. He hated the very idea of a hierarchical network, as well as perhaps the idea of someone interpreting your own ideas. This reflection does not undermine the importance of networks, but may exemplify a sudden difficulty for their organization: they too are constructions. In 1643, Galileo being dead, and having been condemned for almost a decade, even if he was not sagacious enough as Pascal later suggested, Mersenne looked for an opportunity to establish new kinds of relations with those who were considered as Galileo's disciples. He never had such

[40] This is also the reason why I gave a separate list of such editions of Correspondences. I recall that all letters have not been kept, and for example around Mersenne there are gaps in exchanges with Italy, notably with Beaugrand, after Descartes' Geometry and during Mersenne's trip to Italy in 1644.

links with Galileo himself. A narrative in this sense is thus possible from the very tone of Torricelli's letter. In addition, this implies two kinds of initiatives. First a way to work on a new basis with these disciples, erasing the dangerous effects of Galileo's condemnation. Second to offer a possible transfer to France, or more generally to the Republic of Letters, for the continuation of what had been thought as revolutionary from Galileo. If one may easily erase today the idea of a scientific revolution, having been too much emphasized by now classical authors like Alexandre Koyré, it is however more difficult to erase the feeling in the 1640s that something essentially new had happened in terms of knowledge. Torricelli did write directly to Roberval, contrary to the polite usage of the Republic of Letters, which recommended following a certain order of conduct to make contacts. In fact, as royal professor in Paris Roberval had some months earlier written a letter to Mersenne (Letter from Roberval to Mersenne in July 1643) in which he was discussing results or problems listed by Torricelli, recently appointed professor at the University of Pisa. Mersenne had communicated to Roberval a text given by Torricelli to Jean-François Nicéron, then brought to Mersenne.[41] But now Torricelli argued, if one may say so, that like Roberval, and unlike Mersenne or Nicéron, he too was professor at a University. Therefore, they may discuss as colleagues of the *universitas*, with no specific groups or clans.

> I want to talk to you without any intermediary, renowned colleague (And who can deny you this title?). And although your letter is addressed to the famous Mersenne, I cannot restrain the impulse of my mind from running to you, I cannot prevent my mind from devoting itself entirely to you, as to the Apollo of Geometers. (Letter by Torricelli to Roberval, 1st of October 1643)

Roberval, then aged 32 years, had been chosen as a royal professor in 1634, but he had earlier been acting as a professor at the university college named after Maître Gervais. Torricelli, then aged 33 years, succeeded Galileo early in 1642 in the chair of mathematics of Pisa, where like Galileo, he had no regular teaching activities. But both men had a special status. Roberval was a royal professor, nominated by the king and free to give the science courses as he wished, without any interference of the University curriculum in Paris, or from his colleagues. Torricelli was more directly linked to a monarch, as he had been chosen by the Grand Duke Ferdinand II de Medici to be his "mathematician". Galileo, who died in 1642, held such a position since 1610, but officially had also been "the philosopher" at the Tuscan court. So far, so good for the oldest network, the university one, somewhat broken by religious identities that universities felt obliged to exhibit. But we noticed the exceptional place in the University Torricelli and Roberval occupied as scientists; it perhaps was their passport to communicate directly, as Pierre Bourdieu (Bourdieu 2001) might have described it. The situation was certainly not the same in that other network, if we may use the expression here, that was called the Republic of letters. Fame was in this case an important factor, making a sort of differentiation between personalities, or a dubbing and then creating a sense of an invisible aristocracy.[42] But, as can be read from Bernoulli's account in 1701, in the undefined but clearly recognized scientist part of this republic, perhaps due to localized academies, national issues were at stake. Even if academies immediately tried to establish international links, or better said inter-academic relations, appointing foreign correspondents or "membres étrangers". It was not only a question of prestige, but also the feeling scientific changes had to also imply technological changes. In terms of mood this may be described by different general names, Cartesianism or Newtonianism, and critically Jacob Bernoulli belonged to precisely both trends. In terms of reputation, and if one judges by publications, both Torricelli and Roberval had almost nothing to offer in 1643. A text by Roberval had been included in 1636 in the *Harmonie universelle*, a book published under the name of Mersenne, even though Roberval had also published the same text separately. This fact was acknowledged by Torricelli in 1643. Had Torricelli then seen the second volume of *Harmonie universelle*, where in a very short page, Mersenne spoke about the cycloid and its parametric equations, as we shall see, and Roberval's result about the area of one arch of the curve[43]? It does not seem so. Indeed contacts between networks were not working so well, and we may see some of the new practices by Mersenne as an attempt to better forge links. A possible defect may then be understood from thev Torricelli/Roberval issue. In fact, in terms of scientific patronage, the two young scholars were on and unequal level, and the title "Apollo" provided by Torricelli to Roberval is pure flattery. Torricelli was known to have been for some few months the collaborator of Galileo. His experience of the void in a tube with mercury, due perhaps to his link

---

[41] Letter from Jean-François Nicéron to Marin Mersenne about Torricelli's work, *Correspondance Mersenne*, XII, pp. 201–209; Letter from Roberval to Mersenne, July 1643, *Correspondance Mersenne*, XII, pp. 252–267.

[42] See the lively description of the choice of a "prince of letters" in this Republic by Marc Fumaroli, "Nicolas-Claude Fabri de Peiresc, prince de la République des Lettres, 4ème centenaire de la naissance de Gassendi", Conférence du mercredi 3 juin 1992 à la Maison Erasme à Bruxelles; printed in: 1996. *Rivista de litterature moderne e comparat*, 49/1.

[43] For English translation see below Annex 3.

with the *Accademia della Crusca* in Florence, was yet to happen. This certainly will have a tremendous importance in physics. Torricelli was also known to be close to another friend of Galileo, the now well-known Bonaventura Cavalieri who had published in 1635 a very specific "Geometry of what is the continuous by the indivisibles" (*Geometria indivisibilibus continuorum*). Cavalieri's book does not seem to have been read by many people,[44] and Torricelli's book, far shorter, will be the main vector for disseminating Cavalieri's ideas outside Italy at least. If Pascal is right in claiming Roberval successfully worked on the cycloid as early as 1634, then the French scholar could not habe been informed of Cavalieri's methods. But there is a difficulty only if one assumes that Roberval's ways in 1634 were with indivisibles in the manner Cavalieri wrote about them; this now has to be questioned. Montucla was eager to prove this date of 1634 by quoting Mersenne, a man whose honesty should not be questioned, like that of Torricelli for Jacob Bernoulli. The original word is in fact "candor",[45] and as we have already seen it also means a sort of ignorance. In his *1646 Roberval declaration*, we have read he fixed the date 1636 for the divulgation of the tangents to the cycloid. Roberval was essentially known through his link to Mersenne. The Minim monk was organizing regular meetings, where he received every scientist coming to Paris, and developed a large epistolary correspondence throughout Europe. A network was established there, outside the world of universities, with the specific trend of advancing by restricting the focus more and more to scientific problems. Mersenne had been writing a lot in the defense of the Catholic faith, urging for a far better link with scientists.[46] In the 1640s, he had changed his subjects of interest and was preparing books on purely scientific themes: his "physico mathematical reflections" would be published in 1644, including important remarks from Torricelli whom he met during his trip that year to Italy (Mersenne 1644) Mersenne. Without personal contact with Galileo, Mersenne had already published the first works of Galileo on mechanics in French (Mersenne 1634) Mersenne, as well as the last ones in 1639 (Mersenne 1639).[47] A rather strange way for establishing links! But Mersenne was not the only one to have direct contact with Italian scientists. In addition, at this point we must recount a strange story in Italy, for which we have very little documentation.

2.3 The Bad Guy is Named: The Other Story About the "Plagiarist" Jean de Beaugrand

Jacob Bernoulli ([1701]1993) summarized Pascal's booklet, in which an official like Jean de Beaugrand[48] was played the dirty role in these networks to which he did not belong.

> Mersenne first found this line around 1615 and, after a delay of about twenty years, in 1634 precisely, proposed it to Roberval. The latter having realized this wish invited the other geometers to participate in the research, but after having indicated to them the triple ratio of the cycloid to its circle. Very quickly, two of them found the proof, namely Fermat and Descartes, but Beaugrand, a plagiarist, sent these proofs to Galileo in 1638, along with other discoveries of third parties, removing the names of the authors and perhaps without attributing to himself such discoveries. However, he used terms such that, for not too attentive readers, it might seem that only modesty, in any case, prevented him from admitting to be the author: so, to somewhat disguise the whole thing, he changed the word Trochoid into Cycloid.[49] A little while after, Galileo and Beaugrand himself having died, Torricelli who had found Galileo's manuscripts, believed he could then take possession of all those with impunity, and annexed the discovery. He published a book in 1644, which was not received in France without

[44] For example, Wallis assured in 1656 in his preface to the *Arithmetica infinitorum* that he had never read Cavalieri, but only Torricelli. This scientist was far less interested than Cavalieri in discussing "all the lines" as being a magnitude susceptible of Book V in Euclid's Elements, which is an important point for the validation of the method. A part of his 1644 book was to convince his readers of the easiness to use indivisibles and also of a large variety of proofs they may provide. It was therefore a book to suggest inventions, with the cycloid as one example among many others. Even if Torricelli, as had done Cavalieri, mentioned difficulties due to an incorrect use of indivisibles.

[45] "Le P. Mersenne écrivant in 1647, donne la solution du problème de l'aire de la cycloïde, et la date de l'année 1634. On ne peut douter de la candeur de ce Père." (Montucla 1758, II, part. IV, livre I, p. 45).

[46] In 1624 for example, Mersenne wrote a book titled *L'impiété des déistes, athées et libertins de ce temps, combattue et renversée de point en point par des raisons tirées de la philosophie et de la théologie*. For him, science later became a far more important way to fight atheism and "libertine" spirit. Thirty-four years later, there could have been a sort of inverse epistemological situation if we were to interpret Pascal's qualification in the sciences as a warrant for his "honesty" in theology. This possible move, a scientific challenge by Pascal's friend in order to establish religious supremacy, was either despised or mocked by almost all historians.

[47] This book is about the book by Galileo which is at the root of his condemnation.

[48] Jean de Beaugrand held an official function as royal secretary, from the year 1635 onwards, under the supervision of Chancelier Seguier, and in particular was in charge of edition privileges.

[49] Pascal, who used the word "roulette" in the title of his history, seemed to infer that the word trochoid was Roberval's first choice (in 1634?) and that de Beaugrand was the one to have changed it into cycloid.

some laughter, since the book attributed to Torricelli what Roberval had been seen as the father among the livings for eight years already. And other affirmations of the same kind […].[50] (Bernoulli [1701]1993)

Pascal[51] specified that de Beaugrand in 1638 having "picked up" several solutions for the area of the arch of the roulette (it was called the "plan"), and as well the work of Fermat on the "excellent" method of *de maximis and de minimis*, sent both to Galileo, without explicitly specifying who authored the works. At least we have a confirmation of Mersenne's role in collecting papers of various authors on similar subjects, and also taking care of novelties on these years following Descartes' *Geometry* that came as bringing a serious change in mathematics. In this story however there is no mention of Descartes' Geometry! There was a serious difference as well in the network organized by Mersenne compared to previous groups, in the sense that there was now a center collecting papers and deciding what was important to be diffused, and certainly maintained author's reputation, but also expected some progress due to concurrence between authors, and between authorities as well. It was a typical social process. Pascal affirmed that there was at that moment an act of de Beaugrand, with the change of name for the curve, which became cycloid, and a possible omission of author's names. When Galileo died, his sort-of secretary Torricelli inherited all documents, and Pascal affirmed that Torricelli thought it possible that he could "take advantage (*ea omnia secure jam ad se transferri posse arbitratus est*; Pascal 1992, p. 227). Pascal implied Torricelli knew the papers he had were not from the Galileo school. There were serious breaches to the unwritten code in the Republic of Letters, but not necessarily within the new group Mersenne was creating, where ideas and concepts were prominent.

It is true that de Beaugrand went to Italy in 1635 and that he met Cavalieri in just the year when the method from indivisibles was published; he also met Galileo and usually it is believed that he essentially communicated about works by Fermat on *de maximis and de minimis*, as testified by a letter from Cavalieri.[52] We notice Fermat's name was not being left out, contrary to what perhaps happened for Roberval if one follows what Roberval claimed. Perhaps this strange story, told so long after the possible events, may later have played a role in the precise organisation of letters and priorities in the new academy of sciences, in Paris or in London (Dhombres 1996, 1999). It was especially against the accusation of appropriation by Torricelli that Jacob rebelled in 1701. Moreover, he appealed to Christiaan Huygens, then deceased, making him say words that are in fact due to Wallis:

> He[53] defends at some length Torricelli and restores to him what belongs to him. And certainly, I would prefer, he says, if only out of respect, that the author of the Histoire de la Roulette [Blaise Pascal] would have abstained from speaking of Torricelli rather than sully a man of great merit, already dead for several years. (Bernoulli [1701]1993)

[50] I leave in French what Pascal wrote, which is really close to what Bernoulli gave. "En 1638, feu M. de Beaugrand ayant ramassé les solutions du plan de la Roulette, dont il y avait plusieurs copies, avec une excellente méthode d*e maximis et de minimis* de M. de Fermat, il envoya l'une et l'autre à Galilée, sans en nommer les auteurs: il est vrai qu'il ne dit pas précisément que cela fût de lui; mais il écrivit de sorte qu'en n'y prenant pas garde de près, il semblait que ce n'était que par modestie qu'il n'y avait pas mis son nom; et, pour déguiser un peu les choses, il changea les premiers noms de *Roulette*, et *Trochoïde*, en celui de *Cycloïde.* Galilée mourut bientôt après, et M. de Beaugrand aussi. Torricelli succéda à Galilée, et, tous ses papiers lui étant venus entre les mains, il y trouva entre autres ces solutions de la Roulette sous le nom de cycloïde, écrites de la main de M. de Beaugrand, qui paraissait en être l'auteur; lequel étant mort, il crut qu'il y avait assez de temps passé pour faire que la mémoire en fût perdue, et ainsi il pensa en profiter." (Histoire de la roulette appelée autrement la trochoïde ou la cycloïde où l'on rapporte par quels degrés on est arrivé à la connaissance de la nature de cette ligne, daté par Pascal du 10 octobre 1658 (*Œuvres complètes de Pascal*, Jacques Chevalier (éd.), p. 191; Pascal 1992, p. 217).

[51] Pascal wrote in his title about the "nature" of the curve, rather than about some specific properties, while he will only be interested in areas, volumes, centers of gravity. Such a title, perhaps unconsciously, refers to a refusal of Descartes' limitation to algebraic curves to which the cycloid does not belong.

[52] This letter did not give mathematical details. Mathematical works by Fermat on tangents are usually opposed by historians to works by Descartes. And both authors at first quarreled on the subject, being reconciled though Desargues' intervention: this is a clear indication that Mersenne's network was able to sometimes solve deep epistemological problems. The cycloid raised a specific issue. Fermat had a construction using what he termed with a verb "*adaequare*", that is a computation using infinitely small quantities reduced to the first order as we may say. It was general, easy in the case of algebraic curves, but he had to work on particular properties of sines for the cycloid, and near 0 to make use of the equivalence of $\sin x$ and x. Descartes' work in his *Géométrie* in 1637 could not provide tangents to this curve as it was immediately accepted it was not an algebraic curve. In 1638, Descartes invented, specifically for the cycloid, a brilliant method making use of what we call an instantaneous center of rotation, which I will be discussing in Sect. 4, as it truly belongs to mechanics. Torricelli in his *Opera geometrica* did not mention or made use of such an opposition. We may think he was aware of Descartes' mechanical process, known perhaps to him through Viviani. But it can be thought that Viviani learned Roberval's diamond process to construct a tangent.

[53] If we follow what Bernoulli wrote the "he" referred to Huygens. But in fact he was using what Wallis had written. One may think it is just a defect of his memory. It is perhaps important for him to refer to Huygens, in the sense that this man, even if he was involved in the challenge, had acquired an international status in the scientific part of the Republic of letters, linked both to the French Academy of sciences in Paris and the Royal Society in London.

Nothing of this kind appeared in Torricelli's book in 1644. But as early as August 1640 there is a testimony by Desargues, a man indeed related to the Mersenne group, when he published his short *Brouillon project d'exemple d'une manière universelle du S.G.D.L touchant la pratique du trait à preuves.* It appeared as an appendix on the fourth page of this leaflet.[54] Desargues testified that some people forgot to make clear what they had obtained from Mersenne. Free of epistemological consequences, the story about Beaugrand confronts the lack of stability of networks. However, Desargues, and even Pascal, did not specify what precisely had been copied. Therefore, it led Bernoulli to ask a question not only of independent discoveries of the area of the arch of a cycloid, but also for the invention of different methods to find it. We thus come to this aspect, which is not the last one to be analyzed, and contrary to Jacob Bernoulli, my narrative has now to enter into the mathematics involved.

3 When the Story Involves Methods

Below, I have to compare the ways Torricelli and Roberval proved similar results.

3.1 It Is Possible That Torricelli and Roberval Have Found the Measure of an Arch, Each in His Way?

If it were possible to compare figures by Roberval and by Torricelli in the 1640s on the cycloid in view of the measurement of the area of its arch, it quite certainly might be verified that their manner of obtaining it necessarily differed. Unfortunately, we only have figures of Torricelli in 1644 to be opposed to those of Roberval, which were only public in 1693 (Fig. 2).

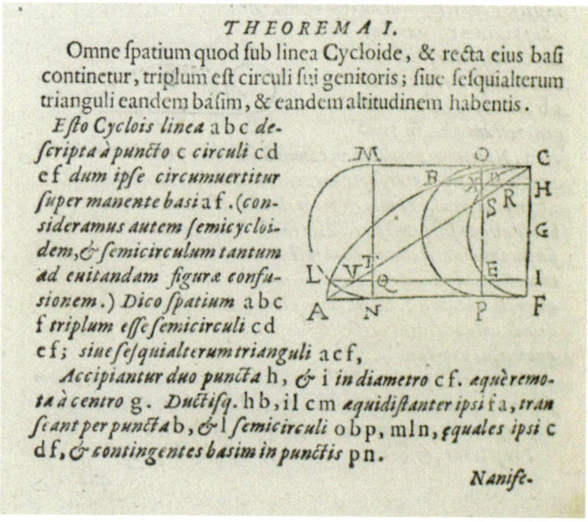
THEOREMA I.

Omne ſpatium quod ſub linea Cycloide, & recta eius baſi continetur, triplum eſt circuli ſui genitoris; ſiue ſeſquialterum trianguli eandem baſim, & eandem altitudinem habentis.

Eſto Cyclois linea a b c *deſcripta à puncto* c *circuli* c d e f *dum ipſe circumuertitur ſuper manente basi* a f. (*consideramus autem ſemicycloidem, & ſemicirculum tantum ad euitandam figuræ confuſionem.*) *Dico ſpatium* a b c f *triplum eſſe ſemicirculi* c d e f; *siue ſeſquialterum trianguli* a c f,

Accipiantur duo puncta h, *&* i *in diametro* c f. *æquè remota à centro* g. *Ductiſq.* h b, i l c m *æquidiſtanter ipsi* f a, *tranſeant per puncta* b, *&* l *ſemicirculi* o b p, m l n, *æquales ipsi* c d f, *& contingentes basim in punctis* p n.

Nanifeſ-

The figure provided by Torricelli for the first theorem in his appendix on the cycloid in *Opera geometrica* (1644). It is important to clearly set at which intersections of two straight lines are *X, S,* and *R* (as well as *V, T, Q*). It is also important to know that *G* is the middle point of *CF*, and the middle of *MN* is at the same horizontal level as *G*. It is not the intersection of the cycloid with the line *MN*. We in fact have to pay attention to the fact that for the semi circle *NLM* point *N* is not placed at the origin *A* of the curve limited to half an arch. *NLM* provides a generic position of the generating circle, and so *N* is what we may call a variable point, and *P* as well, then different from fixed points *A* or *F*.

Fig. 2 Torricelli (1644, p. 86). *Source* Courtesy University of Oklahoma, History of Science Collections

Different dates for sure, but quite different figures in terms of drawing. For Torricelli in his Theorem 1 of the appendix on the cycloid, the principle of demonstration is simple: it must be established that the area of what he calls a "biline" which is identified by *ALOCSTA* (Fig. 2), composed of the semi-arch of the cycloid and of the straight line *AC*, has for its area that of the semicircle, for example *NLM*, which is a generic position of the circle (or better said in this figuration a semi circle). Once this is established, and will see it later, the area of the right triangle *ACF* is the half-product of its height (twice the radius of the generating circle) by the base (half-perimeter of the generating circle according to Archimedes's result in "The Measure of the Circle"). Therefore, then it gives the area of the entire generating circle. Consequently, the triple of the area of the generating circle exactly furnishes the area of the whole arch.

Roberval's technique is different, even if it is based as well on half a cycloid (Fig. 3): it involves a curve, denoted *A1234567D* on our figure dated 1693, and a domain, a bicurviline[55] formed by the two curved lines from *A* to *D*, one is the

[54] This was noticed by Taton (1951) and for the text quotation Taton's *Desargues et le monde scientifique de son époque* see Dhombres and Sakarovitch (1994, p. 44).

[55] The word bicurviline does not appear by Roberval, or by Torricelli. This last author spoke of a biline (*bilinearis figura*). The neologism bicurviline is useful to make a clear distinction between the two figures by Roberval and by Torricelli. Last Torricelli spoke of a triline for another figure composed of circles and of straight lines named *arcuato*, what I translate by arc form.

cycloid and the other precisely is this intermediate curve, which is recognized to have been called by Roberval the "companion to the cycloid". We shall later be studying it as such (Sect. 3.3). Torricelli did not allude to it at all, which really marks their difference. The area of this bicurviline is however the same as that of the Torricelli's biline, that is half the area of the generating circle. Contrary to Torricelli, Roberval made no use of the generic position of the rolling circle for finding the area of the arch: he is using this generic position only to indicate how to construct a tangent to the curve. So that his figure in 1693 has multiple purposes, and perhaps may represent different stages between 1634 and 1675. Even in his *1646 Declaration*, Roberval was not explicating which constructions he then used for the tangent to the cycloid. The half circle of reference is in an immobile position to the left of his figure (Fig. 3). In order to be able to conclude for the value of the area of the arch, it is necessary for Roberval to prove the new curve *A1234567D* equally divides rectangle *ABDC*. In the same manner by which the diagonal *AC* obviously realizes the same equal division in Torricelli's figure (Fig. 2). The said diagonal did not appear in Roberval's figure (Fig. 3), while it was important for Torricelli in 1644. I am not sure that we can say that the Roberval's division in two equal parts is obvious. This symmetry is however known as that which made Roberval's figure appear to be simpler in the way Pascal will so pretend. The main difference for the finding of the area is that Roberval in 1693 maintains an immobile semi-circle, and no generic circle. It seems formally important, even if we have no proof that this was established by Roberval in the 1640s or earlier. I will later give a name for this kind of representation, but at least prefer to call it a frame of reference as will be explained later.

I am sure however that the symmetry, which the verse sine curve provokes, became obvious once functionally interpreted, that is once the graph of the verse sine function had been drawn.[56] It seems possible that this result was earlier obtained by Roberval, at least if one insists on the explanation by functions from curves like the verse sine, something that I will later discuss under the name of pre-functional representation. In fact, according to a simple computation to be soon explained, one has just to change θ into $\pi - \theta$ in the parametric equations of the verse sine curve to see that at abscissa $R\theta$ the ordinate is equal to the difference to $2R$ at abscissa $R(\pi - \theta)$. The computation is particularly evident:

$$2R - R(1 - \cos(\pi - \theta)) = R(1 - \cos\theta).$$

Once this symmetrical situation is established, but it requires a computation with trigonometric functions (even if one realizes it may be just a sinusoid), then the equal division of the area of the rectangle by the cycloid is obvious. Either through the indivisible point of view, or through the use of the graph point of view implying the functional idea. In the pre functional representation for looking at curves in rectangular axes, it is possible to change axes, and just replace the abscissa axis by a parallel one at a distance $2R$. Therefore, even the numbering of points at equal steps exhibits symmetry.

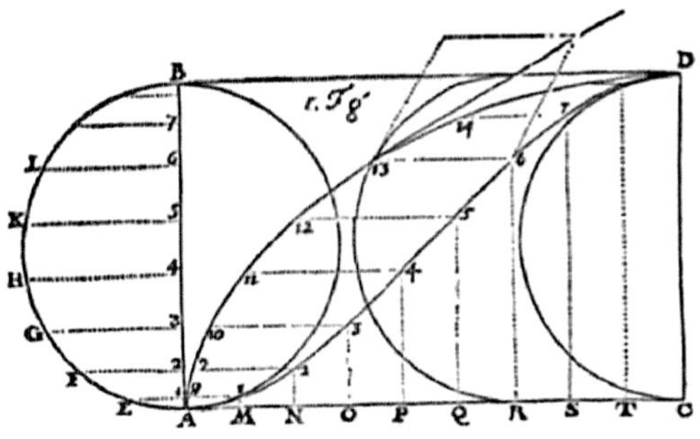

There are at least two purposes in this figure, the tangent and for this a generic circle is drawn, and the measure of the arch, for which the generic circle is not required, a fixed position being enough. It is not important for the proof to notice that the only uniform division of lines is on the horizontal line with points *A, M, N,* etc. to *D,* but it might have been an issue. We see for example the link 11-4 and H-4 on the same horizontal line. The tangent at any point of the cycloid is obtained as the diagonal of a diamond, the lengths of the equal sides being unimportant. Except in the case where the diamond is flat, which is the case in *A*, a cusp point with a vertical tangent to the cycloid. It is however not from this figure that we may understand how the intermediate line *A12...7D* is being constructed, in spite of the presence of this numbering of points.

Fig. 3 A figure provided in a text by Roberval published in 1693 by the Academy of Sciences (Roberval 1693c, p. 192; see also *De Trochoide*, same issue, p. 249). *Source* Bibliothèque Nationale de France—Gallica. Public domain

[56] My impression is that it is the first drawing in mathematics of a sinusoid is precisely the verse sinus curve (ill. 3 or 4). But they are dated 1693, and we have it drawn earlier by John Wallis in 1659, after he learned from Pascal in 1658 about what Roberval had done; also by Huygens the same year (ill. 33). For a sinusoid in astronomy see an occurrence in Peiresc's manuscripts dated 1611 (Dhombres 2006, pp. 14–74).

It will be readily admitted that Cavalieri's principle (Roberval 1693d, p. 108; see also ill. 10) which consists in proving the equalities of areas by displacing segments on the horizontal as possibly here on Roberval's figure (Fig. 3), easily provides the equality of the area of the semicircle and of the bicurviline. But it could have been possible to also apply this line of thought to the intermediate curve invented by Roberval, and so to obtain the area under the curve seen as a function, and thus to prove this area was half the area of the rectangle. Even in the context of indivisibles being doubly used by Roberval, as they were by Torricelli in 1644 (but not in such so simple way as we will see), at least we ascertain that the two authors used two different intermediate lines to obtain the same result. This clearly prevents the simple transmission from one to the other, even if there is the same value for two different areas as can be seen via Fig. 4, if one looks only at the circle and at the cycloid. One is so led to properly question the kind of indivisibles earlier used by both authors, and if something different might not have been thought by Roberval. But, first let us investigate Torricelli's proofs.

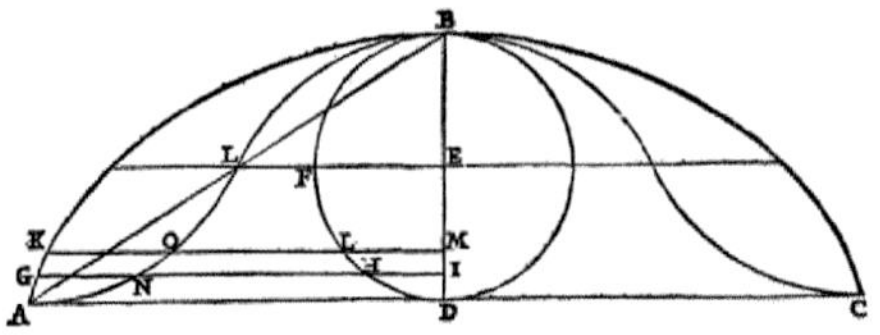

Fig. 4 Roberval's scheme published in 1693c (p. 108 in the academic text), exhibiting the possible fact that the diagonal, straight line *ALB* and the verse sine curve *ANOLB*, or the companion to the cycloid as named by Roberval, were making the same area once seen as function on the horizontal line *ADC*. Point *E* is the center of the generating circle, being put in an immobile and central position, and *L* on the middle line *LFE*, gives the point of inflexion of the verse sine curve, and as well the intersection with diagonal *AB*, something Roberval is proud to have proved. Was this fact known to Roberval before 1638, once at least Descartes investigated the form of various cycloids? He does not claim so in 1646. *Source* Bibliothèque Nationale de France—Gallica. Public domain

Torricelli's first proof of his theorem 1 requires a series of geometrical remarks on triangles, to be able to use Cavalieri's principle. And it requires too the intervention of a certain kind of algebra. Taken together, that is to say added, lengths *LV* and *BR* of the biline (Fig. 2) are equal to the added lengths *LQ* and *BX*, which are carried horizontally on the semicircle. Torricelli's geometrical proof started from two similar triangles *VTQ* and *RSX* that became equal (angles are equal two-by-two, due to the symmetry of the figure depending on the choice of points *H* and *I*, with middle point *G*, and there is an equality of two sides). The symmetry provided by point *G* is the trick used by Torricelli; the symmetry for Roberval's possible proof comes from a change of variable, and a change of axes as well according to the computation we have shown, coming from the parametric equations that will soon be explained. But there is too an algebraic calculation in Torricelli's work, with additions and hidden subtractions. In fact, if we start from $LV + BR$, and write $LV + VQ + BR - VQ$, then $LV + VQ = LQ$, and $BR - VQ = BR - XR$ (as $VQ = XR$ in the two equal triangles *VTQ* and *RSX*), and so as $BR - XR = BX$, we deduce $LV + BR = LQ + BX$. We do not therefore have a simple "division" of the two areas, but a double run, or flow[57] by segments of straight lines, up and down, towards a horizontal line, not represented by Torricelli, and starting from the point *G*. By using his principle, Cavalieri spoke in 1635 of a flow, but he had the sophistication of Torricelli's proof.

It seems impossible to invent the proof Torricelli presented in his theorem 1 without having a precise idea beforehand of an equality of areas made by two domains, the biline and the semi circle, which has been placed in an immobile position to the right of the figure (Fig. 2). In addition, one may suppose that this is the reason why Torricelli insisted on telling Galileo had thought about the value of the area of an arch of cycloid much earlier. There is no such sophistication in Roberval's use of Cavalieri's principle as seen in Fig. 3, a natural candidate for the true first one, the one that says what there is to find. At least if one forgets the additional semi circle in a generic position just used for explaining how to find the tangent. But in Roberval's case, we need to have beforehand some properties of the "companion to the cycloid". Are we helped by the use of numbers, which is another addition made in this too rich figure of Roberval (Fig. 3)? Was it a genuine way to explain the manner to draw a function, or had Roberval been influenced later by Wallis' ways of arithmetization in his *Arithmetica infinitorum* of 1656? On this figure that appeared in 1693, the division on the horizontal line is in equal parts, which is irrelevant for the functional point of view, but the division on the vertical line *AB* cannot be in equal parts. A careful inspection of Fig. 3 reveals that this non-uniform division in the vertical prevents, in view of the area of the semi circle being put to the left, the trick with thick indivisibles or Riemann sums with constant steps visible on the horizontal axis *AC* (Fig. 3), which Pascal among others so nicely explained in 1658. This visual fact in the figure suggests another interpretation for Roberval's first attempt.

[57] Cavalieri wrote explicitly in his "Geometry by indivisibles" in 1635 about "all the straight lines that are obtained by the transit according to a given direction" (*omnes lineas recti transitus figurae sumptas cum regula*).

Before coming to it, we notice that Torricelli had provided two other proofs in his 1644 book, but had never used the curve that was so important for Roberval, and that we will see mentioned as early as 1637 by Mersenne. The last of Torricelli's proofs is just an application of Cavalieri's principle using some symmetry of the figure. Torricelli's second proof is a rather simple proof by contradiction according to what was called the exhaustion method to show that the cycloid curve *ABC* divides into two equal parts the domain limited by a curve composed of parts of circles and parts of straight lines, called *arcuato* by Torricelli (it denotes a figure like *AGECD*, a sort of rectangle with two circular arcs, see Fig. 5a, b). There are no indivisibles being explicitly used, just a convenient division of this domain, and so an approximation according to the way Euclid used in book XII of the Elements for the area of a circle (look at the description of theorem II in Annex 2, with an area *K* to the left, used as a sort of epsilon in our modern technique of limits, and a length like $IE = HI = LH = AL$ that one has to take small enough to cope with the given *K*). The idea stems from a very simple geometrical fact (which may be as well a simple application of Cavalieri's principle): in Fig. 5a, b, the area of the curvilinear space *AGECFD* is that of the rectangle *AECD*.

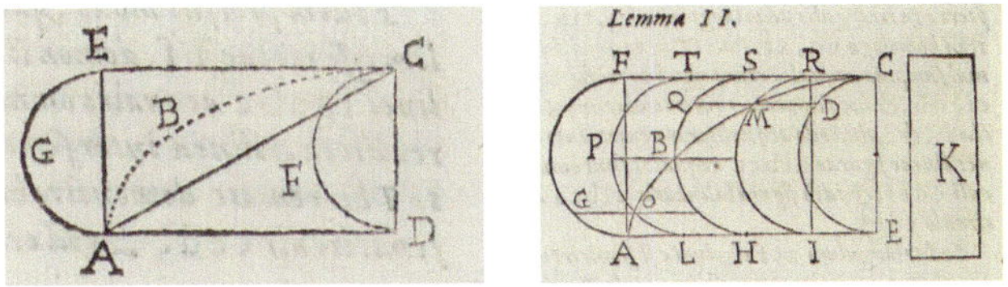

Figs. 5a and 5b To the left is the figure used for Lemma II in Torricelli's appendix on the cycloid in 1644, and to the right the one for theorem II. What is not at once perceptible in this figure is the fact that *EI* (= *PF*) is the small element by which *AF* (the perimeter of the generating circle) is being divided into equal parts. In the figure to the left, there is no generic rolling circle, and two immobile positions. With the figure to the right we have the impression of a movement in the cartoon way. And perhaps more if one notices that successions of semi circles are according to small lengths. We here may speak of curvilinear indivisibles, while in a classical way Torricelli spoke of inscribed and circumscribed figures (Torricelli 1644, pp. 88–89). *Source* Courtesy University of Oklahoma, History of Science Collections

Therefore, we have justified what Bernoulli just offered as a possibility, that there are different ways used by Torricelli and by Roberval. Let us then read another sentence by Jacob Bernoulli, as it may help us to discover an origin to Roberval's possible first method.

3.2 Today All Philosophers Agree That Nothing Is Known of Things Apart from the Various Relations Which They Have Between Themselves

By such a sentence (Bernoulli [1701]1993), Jacob Bernoulli who refused to refer to Calculus in his 1701 speech, was perhaps aiming at the way curves were now represented via equations due to the Cartesian revolution. He seems to hint at what we now call analytic geometry, for which a specific definition is not easy outside a simple one: analytic geometry is what is done on figures while using equations. This leaves the questions of using Calculus, tangents, etc. unclear. He solemnly introduced his point with a philosophical remark, claiming that we only know things through the relations they have one to another. Even if he arrived at the usual Aristotelian setting about the category of quantity, the more or the less being the realm governed by mathematics, one is astonished to read such a clear role devoted to analyticity as providing an ontology to objects, here to curves:

> Today all philosophers agree that nothing is known of things apart from the various relations which they have between themselves and of which (especially for material things) the principal is that by which they can be compared according to the more and to the less. It follows that, since there is nothing more capable of representing any curve than the so-called coordinates, we can easily guess the importance of the consideration of curves, and we shall not be surprised at seeing that their study mobilizes the whole of Geometry, and that the greatest Geometers have been opposed one another with always a great rivalry of ingenuity. (Bernoulli [1701]1993)

It is difficult not to read these sentences of 1701 without thinking to a strong fidelity to Descartes' thoughts, especially when explicit mention was made of coordinates. They alone enable the possibility to establish equations, which weave the network of relations that make a curve to exist. But Bernoulli made no allusion to the limitation that Descartes had put in 1637 in his *Géométrie*, that of considering only algebraic relations. On the contrary, since he is going to speak about the cycloid, Bernoulli seemed to indicate all the resources of geometry had to be made available in order to overcome the battle on this curve. Bernoulli is writing after Leibniz, with whom he was in contact; and Bernoulli had worked on curves and knew he had considerably extended Descartes' point of view. Nothing really like that is to be read later in Montucla, when he wrote in 1758; he began by recognizing that the cycloid is in itself sufficiently rich with properties to make up its real interest. He added that this wealth justifies in some way that disputes have arisen, but he avoided talking about the question of the analytical representation of the curve. For Montucla however, it was obvious that Descartes, to whom the question of the cycloid was submitted in 1638, had to consider that the problem was very much below him.

> Such was indeed his superiority over all geometers of his time, that the questions which occupied them most, cost him, for the most part, but a mediocre attention. It is easy to be convinced on this by reading his letters. (Montucla 1799–1802, partie IV, livre I, II, p. 55; see also *Id.*, 1758)

Why not! But nothing precise was said about the nature of Descartes' advantage using equations. On the contrary, if I may say, Montucla described another method to which I shall return, proper to Descartes and of the kinematic type. The intention of the instantaneous center of rotation precisely avoids the use of coordinates (Figs. 28 and 29). As neither Torricelli nor Roberval even in his posthumous work of 1693 seemed to make direct use of coordinates for the cycloid, one may think that Bernoulli's remark basically only concerned another stage of the work on the cycloid, once Calculus had been invented. On the contrary, and piqued by the remark written by Bernoulli, I think that it is necessary to make the effort to carefully read the very meticulous text due to Montucla. A few pages later, he in fact gave a reference to a text by Mersenne, printed in 1637,[58] which contains a contribution that Mersenne affirmed to be due to Roberval. It is particularly astonishing to read in this text what we may directly interpret as the parametric equations of the cycloid. I quote:

> For if verse sines are elevated on the line described on the horizontal plane by the circumference of the circle to which it is equal, and the versed sines are so much applied to the right sines, left and parallel to the plane, the curved line described by the circle will pass through all the ends of the right sines, as by so many points, which show the manner of describing it. (Mersenne 1637, XIe, II partie, p. 25; see below English Annex 3)

I insist on the word describe here being used, and I read it to mean "construct". We may transcribe, having just to remember that at that time, with a radius R and an angle θ, the verse sine denoted the following expression: $R(1 - \cos\theta)$, and naturally a right sine was just $R\sin\theta$. . So that when one puts above (*élève* as it is written) a verse sine according to an abscissa $R\theta$, and as it is said then puts to the left the right sine $R\sin\theta$, one in fact obtains the current point of the cycloid. It amounts to saying, with words, what we write in symbols when we say that the running point on the curve has for its two coordinates in two orthogonal axes, one being the AC in Fig. 4 on which the wheel is rolling and the other the middle axis, BD in Fig. 4:

$$R(\theta - \sin\theta)$$

$$R(1 - \cos\theta)$$

We do not have a direct function, from an abscissa to an ordinate, but precisely what will be called a parametrized curve.

I am comforted in the sense of this analytical description by the fact that Mersenne, who frequently abused figures, here drew nothing. His description of the curve leaves no doubt about the line that is described by the rotation of a circle on a straight line. Mersenne here does not ascribe this to Roberval. The construction of the curve relies on trigonometric functions, as was explicitly done in the Mersenne's text: this introduction inexorably brought a new trend in this kind of geometry. As the values of such functions, necessary to construct curves, are not exact in the sense that they are obtained from numerical tables where they are listed according to an explicit degree of approximation. This is different from the exactness as required by a geometrical construction with compasses and rulers, even if the quality of approximation can be far better from what is visually obtained by graphic methods for tracing lines. Henk Bos has brightly and with good reasons insisted on the change of exactness required in geometry from the use of polynomial algebra, and the extension of possible compasses for constructing points to algebraic curves (Bos 2001). The introduction of trigonometric functions also brought a serious change in mathematics, which was not at first theorized, and even was refused in geometry by Descartes himself, while he freely used trigonometric functions in physics, and particularly for optics (Dhombres 2011b). It is not enough to say it corresponded to the use of numerical tables, because tables provided not only numbers up to a certain approximation, but offered new objects, for which I deliberately used the word function, even if this name was not yet to be coined in mathematics.

I am sure that an objection has be made here, not in the sense that I gave formulae where Mersenne only explained in words, but to question signs. They were not regulated at this time for trigonometry, and so not used in general for

[58] Modern investigations have established that Mersenne's addition, perhaps written under Roberval's direction, was in fact printed in early January 1638. This changes nothing for the Torricelli/Roberval issue, as most of the letters sent to Mersenne on the subject were later this year.

trigonometric functions in the case of angles that do not belong to the first quarter of the circle. Trigonometric functions in fact were viewed as lengths, so we have to multiply our functions by the radius of the circle they are related to. If our sine remains positive as a function at least when one goes from 0 to π, and clearly a symmetry situation may be used after $\pi/2$, then our cosine changes its sign, a functional idea that was not then easily understood depending on the conceptions as lengths. This is true, but the objection is less important if it is a verse sine that is mentioned, the value of which then remains positive and even increasing in the domain of variations of the angle from[59] 0 to π. Thus, it is possible to accept the parametric equations as provided for half a turn of the cycloid having replaced the second one by the verse sine: the semi-circle is generally what is drawn, as we have seen in Figs. 2 and 3, and as can also be seen in John Wallis' booklet in 1659 (Fig. 7).

It seems strange that up to now we have only seen in figures what is for us only part of the cycloid, an arch, and not its extension, including the periodic aspect. It requires for angles an extension to all real values in the parametric equations. This aspect is however quite natural from a definition of the cycloid like the one Mersenne provided by mentioning the wheel of a wagon. But the same remark can be said about the verse sine curve that is not related to mechanics, but to trigonometry, and more specifically to numerical tables, and less to the geometry of the circle. So even if we have to be attentive to Descartes' drawing of the periodic aspect (Fig. 6, and as well in a figure in the group above see Fig. 31_2), we have to notice it is presented with the mechanical idea in mind. But we may argue as well that the periodicity aspect was clearly drawn by Descartes, and that it leads to the non algebraical aspect of the curve due to an infinite number of intersection points with some straight lines parallel to the axis. It is thus possible that the cycloid was the first instance making the importance of periodicity in mathematics clear.

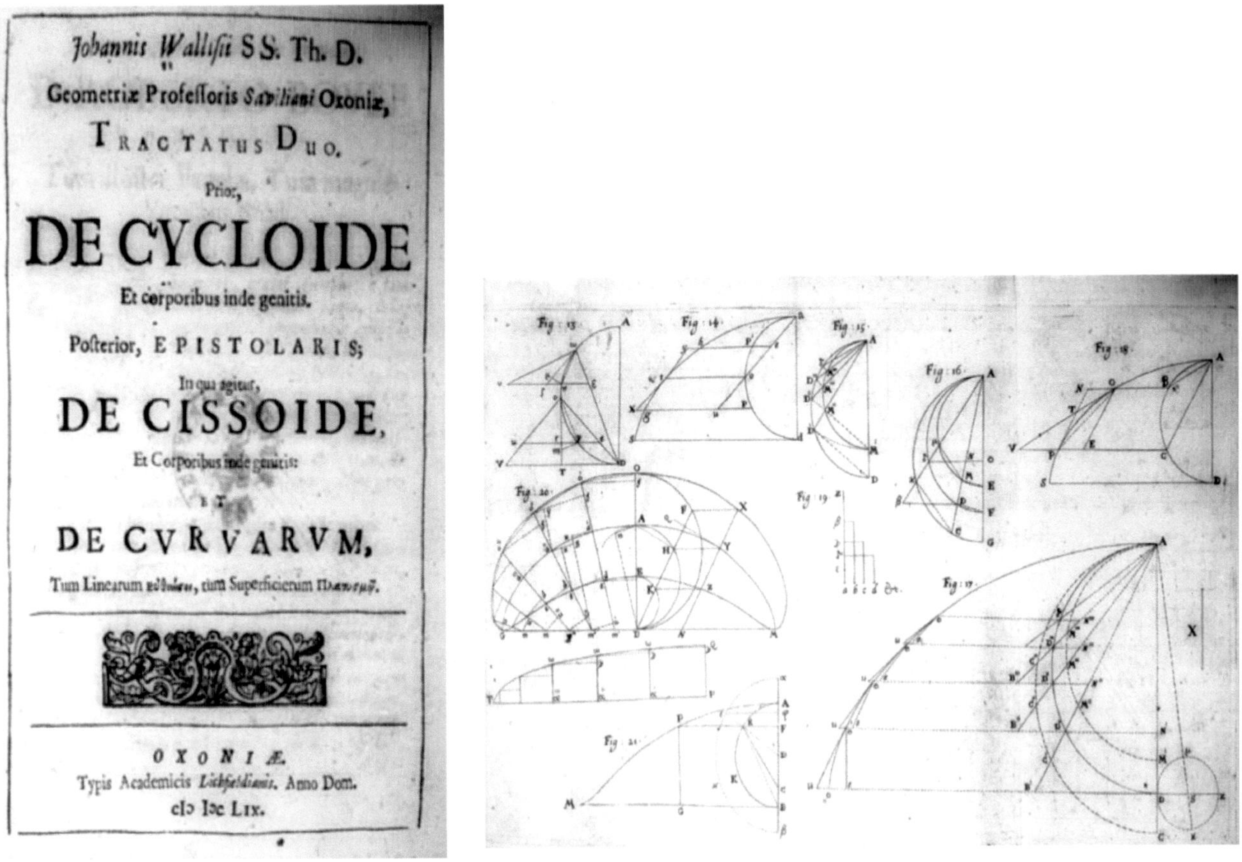

Fig. 6 The title page of *De Cycloide* by Wallis in 1659, and the second table for figures in this book. Only one figure shows a complete turn of the circle on the straight line (Fig. 20 on this table), but there is no representation of the cycloid with more than one arch. *Source* Bibliothèque Nationale de France—Gallica. Public domain

[59] We already noticed that by looking at half a cycloid, the angle q may vary only from 0 to p. But usually at this time it went from 0 to a right angle. We perhaps observe an occurrence of a larger variation, like the one which occurs in Descartes' Dioptrics, when he was working with the rainbow in 1637 and so had to deal with angles not necessarily restricted to being acute (Dhombres 2018).

From an epistemological point of view, it then seems necessary to provide some terminology. To make a distinction with the word function that will receive a considerable impulse due to Calculus, and as well a complete modification of the obtention of numerical tables, I will here name what Mersenne obtained a parametric representation of the cycloid. In addition, I will use pre-functional representation for what Roberval did with the verse sine function, as well as Descartes when speaking about the elongated cycloid. The terminology of parametric representation, instead of speaking indistinctly of analytic geometry (the last expression did not exist during the seventeenth century), will also be useful when used in contrast with two other historical representations of the cycloid, better named as will be explained, as the classical frame of reference and the baroque frame of reference. This terminology is not to establish impermeable distinctions, and one task is to understand how the functional idea intervened in the conceptual work about the parametric representation. This would be more difficult if I had chosen the possible terminology of parametrized curve, or even if I had used the word equations instead of "representation", and so put away some spatial effect. At least it cannot be regretted that one uses a precise terminology, even if it looks anachronistic, for explaining the Torricelli/Roberval issue and also trying to understand the main changes of the period.

Links also have to be found in the way the parametric representation was verbally obtained. It essentially depends on the variable $R\theta$ being put as an abscissa: this corresponds to the movement of the point of contact between the axis and the circle, which is rolling but without sliding. This last notion had just been invented by Mersenne in the second volume of *Harmonie universelle*, and I am not sure this really came from Roberval. It is not established that Mersenne, even helped by Roberval, really understood that the speed of the moving contact point on the axis was zero: there is an annihilation from the speed to the left of the contact point due to the rolling of the circle by the speed to the right due to the translation move of the circle.[60] Descartes, less than a year later, will understand this zero speed as creating an instantaneous center of rotation, and he will explain this in a letter to Mersenne.[61] Thus it yields an easy construction of the tangent at a point on the cycloid: it has to be perpendicular to the straight line linking the point on the cycloid to the corresponding contact point. This mechanical aspect, to which I will come later and which was perhaps known to Torricelli, is not the only thing that makes it possible to understand Roberval's research that I call a pre-functional representation. I wish to further discuss that functional part.

3.3 The Verse Sine Function, or the Curve in Orthogonal Axes

We have been so accustomed to the historical presentation of Cavalieri's indivisibles, that means to horizontally move geometric segments of straight lines, that we do not first recognize in Mersenne's writing the importance of the description by which he explained the verse sine function. Even if the first figure to appear with the graph of such a function will be in 1659 only (Fig. 33), or for Roberval in 1693e (Fig. 4), the vocabulary used by Mersenne is without any doubt what we call a graph construction, which we summarize with the two equations:

$$x = R\theta$$

$$y = R(1 - \cos\theta)$$

Even if it was not noticed at the time, this can be reduced to a pure functional version, with an equation where R is acting as a parameter, depending on the precision of the table by which trigonometric lines are computed:

$$y = R\left(1 - \cos\frac{x}{R}\right).$$

From this formula, we deduce as well the following one

$$y = 2R\sin^2\frac{x}{2R}$$

[60] It is not obvious for all readers, and Margaret E. Baron ([1969]1987) attributed the fixity to point N.

[61] Letter from Descartes to Mersenne, 23rd of August, 1638; *Œuvres de Descartes*, II, p. 307 et seq.

and obviously it provides positivity for y. In our way of observing translations, it is better to notice that the graph of the verse sine functions, i.e. the "companion curve", is just a sinusoid. It so easily explains for us inflexion points on Fig. 4. But then we have to historically consider that no graph for the sine function had previously been constructed. What could the advantage have been of Mersenne and of Roberval speaking of sinusoid, just adding a new name? The process by which graphs of elementary functions came to be known did depend on a very different process: Calculus. This coming of age is something that Bernoulli avoided in his narrative, and it is the reason why I introduced the notion of pre-functional representation.

If we come back to 1637, the problem was how to use parametric representation to draw curves, for instance the graph of the verse sine function and naturally the graph of the cycloid. Since the appearance of Calculus, derivatives give the possibility to discriminate zones of increase or of decrease for the functions involved: this is the basic trick learned in elementary Calculus courses. It amounts with the help of parametric representation, to determine points with horizontal or vertical tangents. Before Calculus, this seems to have been immediately understood by Descartes, precisely in his already mentioned letter of end of August 1638. Following Mersenne, Descartes studied generalized cycloids, curves whose parametrized equations could be written by adding a multiplicative coefficient t to radius R, that is:

$$x = R(1 - t\sin\theta)$$

$$y = R(1 - t\cos\theta)$$

When t is greater than 1, then the curve can no longer be reduced to the graph of a function as the curve involves loops. This is what Descartes is roughly drawing in his letter to Mersenne (Fig. 7).

A drawing of a curve sketched by Descartes in a letter to Mersenne in 1638, that cannot be reduced to a graph of a function, even if it has a parametric representation, precisely made of what we call functions.

Fig. 7 This drawing appears on page 310 in the volume 2 of Descartes' *Œuvres complètes*, August 23rd 1638 (Descartes 1996). *Source* Bibliothèque Nationale de France—Gallica. Public domain

There is another very important feature of this drawing in the fact that it did not represent a part of the generalized cycloid but shows periodicity. This generality for the usual cycloid is certainly said by Mersenne, but is not reproduced as coming from old age by Wallis in his letter to the Royal Society in 1697 (Fig. 21). Most, if not all the figures of the seventeenth century, are restricted to an arch only. Descartes is explicit enough to express that he worked with tangents to determine when a loop appears, that is when the abscissa goes back, or as he wrote, when the curve '*commence à se replier en dedans*'. It is a point where the tangent is vertical. It certainly could be worth looking for various devices used by mathematicians to construct such curves, and in this mechanics plays an important role, with an important step being made by Huygens in his *Horologium oscillatorium*. Certainly in 1673, even if he had known the parametric equations, Huygens could not use so easy computations with derivatives

$$dx = R(1 - \cos\theta)d\theta$$

$$dy = R\sin\theta d\theta$$

$$\frac{dy}{dx} = \frac{\sin\theta}{1 - \cos\theta} = \frac{1}{\tan\frac{\theta}{2}}$$

So that, a formula easily readable in geometric terms, proving the tangent at point B on the cycloid is parallel to the chord BE in the circle, as the angle of this chord with the vertical in B is precisely half the angle θ, and so providing the result with the inverse of the tangent (Fig. 11), or figure by Wallis numbered 4 in Fig. $6_{2)}$.

Nonetheless Huygens was able to construct tangents to the cycloid without knowing derivatives, and so to obtain a rather precise form of this curve (Fig. 11). There Huygens goes back to the old synthetic tradition in geometry drawing a convenient line, and then establishing that it has to be the required tangent. I don't think this was the way for Roberval to have known how to construct tangents to the graph of the verse sine function (Fig. 8). He seems to have a thought of a right triangle in T, similar to another right triangle SLB for the circle. Both triangles look like characteristic triangles to be found in Isaac Barrow's published texts not earlier than 1683, that is after Roberval's death. But I do not wish to use this terminology, and prefer to speak of *Robervalian triangles*. They are built along the horizontal and vertical axes, and provide by their hypotenuse the direction of the tangent to the curve to which they are linked. Today it is easy to follow what Roberval was doing by introducing an angle $\theta = LED$, and by writing $d\theta$ to explain the infinitely small. But I am not attributing any other role than that of a notation for the figure of the two right *Robervalian triangles*. Moreover, if obtained results may be interpretable in terms of derivation, this is not the effect of a conscious move by Roberval. It is easy to read values in the first right triangle associated to the circle, and considering that the angle with the horizontal basis of this triangle is θ:

$$RL = Rd\theta, LM = R\sin\theta, SL = R\cos\theta d\theta, RS = R\sin\theta d\theta$$

so that according to the equality between TO and RS, we immediately deduce $TO = R\sin\theta d\theta$. This result is convenient with our interpretation in infinitesimal terms with the verse sine, as we may read it in the vertical line. If this is not what Roberval was looking for, at least we see that he treats data on the two orthogonal axes as independent when variations are considered. Within this conception of an independence between coordinates, and due to the baroque frame of reference, or even to the rolling property without sliding, Roberval sets on the horizontal line TV (Tv in Fig. 8): he gets by composition the direction to the tangent to the curve of verse sine. In fact, the tangent to this direction, this time in its trigonometric meaning, and according to the orthogonal frame of reference is $\sin\theta$. This quantitative result is for us important, as it denotes the analytical game between derivations about sine and cosine functions. This is of no use to Roberval, even if he has then traced, for the first time, a trigonometric function: he only required a construction and so what he looked for was the direction of OV. Moreover, at this time a sine was a length and not a pure number, like that expressing the direction of a line. At least it is not the "sine function" which was described at this stage around 1638, even with a reducing factor t with the so-called *shortened cycloids*, but it is the verse sine function. Without the need of a representation, it could be induced from a translation on the abscissa of length $R\theta$ one gets the parametric representation of the circle.

$$x = R\sin\theta,$$

$$y = R(1 - \cos\theta)$$

The circle is the marking of the cycloid when one understands the graph of the verse sine function as a frame of reference, that is something immobile. Analogous in this case to the semicircle we just discussed for the obtention of a tangent by Huygens, with Figs. 11 and 12, where obviously the generating circle is immobile, and serves as a reference. Consequently, what is said to be stacking the sines in an "indivisible" vision, or better still, as a "sum" of these in an "integral" or thick indivisibles, and otherwise "infinitist" visions, may simply be an effect of the parametric representation. Seen in such way it has no relation at all with Torricelli's proof. This narrative of the parametric representation cannot completely destroy Pascal's story about Jean de Beaugrand, but it considerably reduces the idea of some similarity between the method of Roberval and Torricelli.

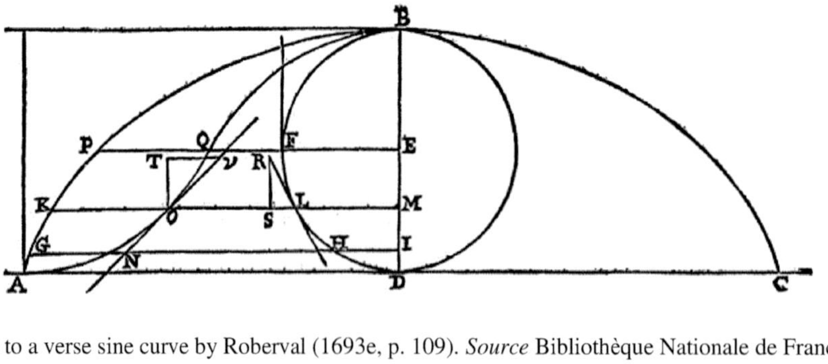

Fig. 8 Drawing of a tangent to a verse sine curve by Roberval (1693e, p. 109). *Source* Bibliothèque Nationale de France—Gallica. Public domain

Surprisingly Mersenne's description of the verse sine curve, or perhaps better said as Wallis' verse sine function, did not reappear quickly in the mathematical literature: we saw it by Huygens in 1659 (Fig. 30). It is then interesting to notice how Newton precisely used the verse sine when dealing with the cycloid in his *Principia* in 1687 (Fig. 9).

Proposition 48, theorem 16

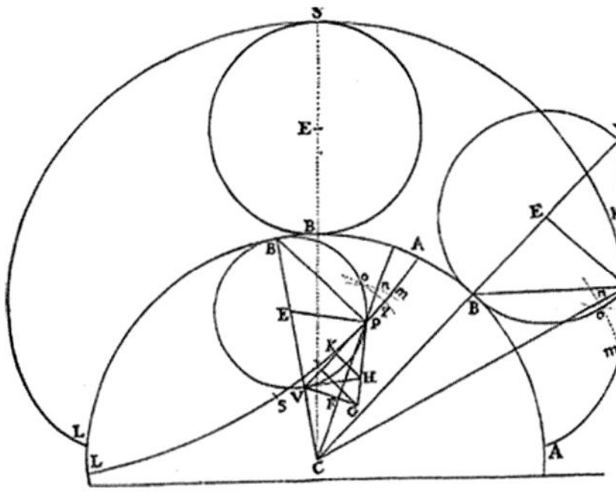

Fig. 9 Figures about generalized cycloids in Newton's *Principia* (Newton 1687, p. 148). *Source* Bibliothèque Nationale de France—Gallica. Public domain

> If a wheel stands upon the outside of a globe at right angles thereto, and revolving about its own axis goes forward in a great circle, the length of the curvilinear path which any point, given in the perimeter of the wheel, hath described, since, the time that it touched the globe (which curvilinear path we may call the cycloid, or epicycloid), will be to double the verse sine of half the arc which since that time has touched the globe in passing over it, as the sum of the diameters of the globe and the wheel to the semi-diameter of the globe.[62] (Newton 1687, I, Sect. X, p. 185)

One of our mentors in this inquiry, the historian Etienne Montucla, was particularly clear about what had been found first for the cycloid and that was precisely the parametric representation:

> Those things were the first that geometers were requiring from them to be solved.[63]

From this I may deduce one way Roberval had found the area of an arch, as he had just to show that translations to a fixed vertical axis, reduced the bicurviline parametric equation to that of the circle. In fact, I do not have to make a long discourse to explain how Cavalieri's principle worked. I just have to use one of the figure drawn by Cavalieri where he shows how an area delimited by two curves, so a bicurviline, is reduced to the area of a curve when limited by a straight (here vertical) line (Fig. 10).

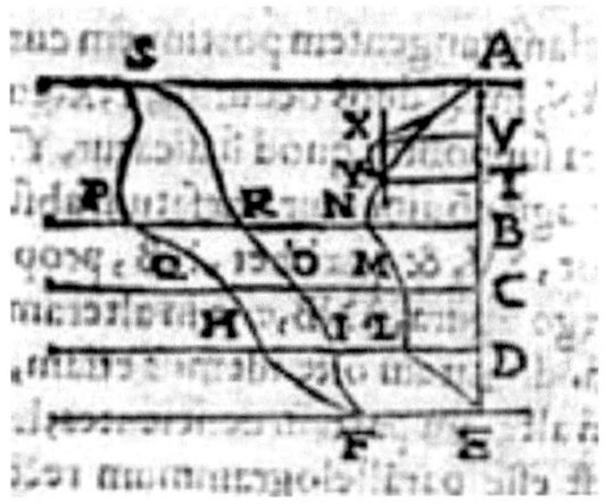

Fig. 10 Figure provided by Cavalieri in his Geometry of the continuous with indivisibles (Cavalieri 1635, VII, p. 17). Plate from the original edition 1635

[62] Newton's *Principia* as translated in English by Andrew Motte in 1729, with a preface by Roger Cotes. To which is added, *The Laws of the Moon's Motion, according to Gravity, by John Machin, Benjamin Motte, Londres* (Newton 1687, I, Sect. X, p. 185).

[63] "Ces choses là sont les premières que les géomètres se proposèrent" (Montucla 1799–1802, II, Partie IV, livre I, p. 53). This sentence was not written in the first edition in 1758 (Montucla 1758) and so later it was thought necessary to emphasize the analytical aspects coming from parametric equations.

Mersenne, who perhaps wrote by using some notes from Roberval, described the parametric representation above only as the "second way" to construct the curve. Mersenne first described a phenomenological way by registering on some paper the move made by a fixed point on a wheel that rolls without sliding over a straight line. So that he had in his view the periodic aspect. And immediately, without any break, he indicated a third way to construct the cycloid (Fig. 36), Mersenne speaks about the area of the arch of the curve, that is three times that of the circle which allows it to be constructed.

> Now, between the several properties of this line, that which the Sieur Roberval has encountered is very remarkable, namely that the space comprised between the straight line equal to the circumference of the circle, and the curved line, is three times the circle which describes the curved line. (Mersenne 1636)

This is the result that Torricelli wrote three times in 1644 in his appendix on the cycloid. But Torricelli never used the parametric representation, as we may ascertain in reading his appendix. Mersenne, perhaps because he was under the guidance of Roberval for mathematics at least, gave no proof in 1637 for the area result, and even if a construction is provided for the current point of the curve, we have been in a way taught to think that Roberval knew the trick of computation by indivisibles. But the result does not mean, in our language, that he measured the area by integrating the function

$$R \sin \theta \quad (\text{between } 0 \text{ and } \pi).$$

There is quite another game on variables and functions if we wish to see a path to Calculus. We now have to take some time to measure the distance between any integration result and both the parametric representation and the construction of the curve, and *Robervalian triangles* as well. It became the first question as Montucla said, or the true origin, if not of the curve, then of its entry to mathematics. It was said by Mersenne to be the second way, after the experimental process with a true rolling globe or even a circle, that we could also call the empirical one. But this last mention of empiricism is a mistake, in the sense that the movement has already to be theoretically understood as rolling without sliding.

3.4 What We Learn About Functions from the Step with Calculus?

As soon as the parametric representation of the curve is known, and by means of the essential rule for the change of variables, integral calculus makes it possible to simply find the area of the arch. Indeed, with

$$x = R(\theta - \sin \theta)$$

$$y = R(1 - \cos \theta)$$

we immediately get Roberval's result after the few usual transformations of Calculus, which are not reduced to the computation of primitives, and which are de facto impossible to be seen just as a continuation of the method by indivisibles[64]:

$$\int_0^{2\pi R} y dx = R^2 \int_0^{2\pi} (1 - \cos \theta)(1 - \cos \theta) d\theta =$$

$$R^2 \left[\int_0^{2\pi} d\theta + \int_0^{2\pi} \cos^2 \theta d\theta - \int_0^{2\pi} \cos \theta d\theta \right] = \frac{3}{2} 2\pi R^2 = 3\pi R^2$$

Two lines at most to get the said result. We may then read Mersenne somewhat further as he is investigating other kinds of cycloids, the so-called shortened cycloids,[65] that is to say those curves described by points inside the generating circle, if one maintains the mechanical origin of the curve. It suffices to write with a coefficient t ($0 < t < 1$) to mark the shortening of the radius of the circle, namely the distance Rt from the fixed point to the center of the circle known as the generator.

[64] It is here useful to remember that if one were to directly integrate a sine function over an interval, as if we were to apply indivisibles for integration, one could not get the area of the semi circle. Integration has to take care of the parametric representation of the circle, using abscissa as well as ordinate, while the method by indivisibles just has to use an equality of lengths, with only abscissa for example.

[65] Those shortened cycloids originate from the trajectory of a fixed point located inside the generating circle.

$$x = R(\theta - t\sin\theta)$$

$$y = R(1 - t\cos\theta)$$

Using Calculus of which Mersenne was ignorant, we find a simple expression directly linking the area of the arch to the area of the generating circle:

$$\int_0^{2\pi R} y dx = R^2 \int_0^{2\pi} (1 - t\cos\theta)(1 - t\cos\theta) d\theta$$

$$= R^2 \left[\int_0^{2\pi} d\theta + \frac{t^2}{2} \int_0^{2\pi} d\theta - 2t \int_0^{2\pi} \cos\theta d\theta \right] = \left(t^2 + 2\right)\pi R^2$$

The interpretation of this integration in order to provide the area under the curve is valid only if the curve is always above the rolling axis, namely precisely for $0 \leq t \leq 1$. It is probable that, for the sake of analogy, in the case different from the classical cycloid, through Mersenne, the royal professor wanted to measure not from the axis on which the circle is rolling, but from the parallel to this axis passing through the fixed point inside the generating circle in its immobile position like in Fig. 9. We thus have to subtract from the value just obtained that of the rectangle area $2\pi R(1 - t)R$. So that the new area is: $(t^2 + 2t)\pi R^2$. Mersenne wrote, no doubt now under the dictation of Roberval:

> But when the circumference is the sub-double of the straight line, the space between the straight line and the curve is fivefold the area of the circle that describes the said lines.[66]

This corresponds well to the value $t = ½$ (the sub-double expression) and of course the radius of the circle in play being half the radius of the circle which rolls without sliding. Thus

$$\frac{1}{2}\left(\frac{1}{2} + 2\right)\pi R^2 = 5\pi(R)^2.$$

More generally the value of the area is

$$\left(1 + \frac{2}{t}\right)\pi R^2$$

for the shortened cycloid. With $t = 2/5$, Mersenne computed the value $6\pi R^2$, and he stated it in a way that seems rather convoluted, because it alluded to the "straight line", which in fact is the perimeter of the generating circle (in our terms $2\pi R$), and to the circumference, this time the perimeter of the small circle (the previous value multiplied by t).

> If the straight line is two and a half times greater than the circumference, the space is sixfold that of the circle.[67]

Mersenne (or better Roberval?) took the opportunity to say that, although the distance from the fixed point to the center of the circle is reduced, the result will be greater than the area of the circle on which the fixed point is. It is a rather bizarre formulation to say that the area under the curve remains greater than the area of the circle in question, that of radius tR, of an always positive quantity, which we calculated precisely as equal to $2t\pi R^2$. It is impossible that Roberval did not also find it so. The complication of the exposition, apart from the possibility of bad will on Roberval's part explaining, aiming to damage Mersenne's network, nevertheless shows that the consideration of formulas became important, up to the consideration of limit, here making t to tend to 0. It belongs to this zone in-between parametric representation and pre-functional representation. We have to pay attention to the fact that such results were obtained no later than 1638, but without the use of integral calculus. Our way of using Calculus was to verify that the results were correct. May we know by which methods such results on generalized cycloids were obtained? One has usually thought that they were obtained from indivisibles. Moreover, this is not difficult to obtain. However, I have tried to show another issue, with the parametric representation,

---

[66] "Mais lorsque la circonférence est sous-double de la ligne droite, l'espace compris entre la ligne droite et la courbe est quintuple du cercle qui décrit les dites lignes." (Mersenne 1637, p. 25).

[67] "Si la ligne droite est deux fois & demie plus grande que la circonférence, l'espace est sextuple du cercle." (Mersenne 1637, p. 25).

indicating that Roberval might have worked in this analytical way. We cannot decide from the results only, except that we will notice that there is some difficulty in the explanation provided by Torricelli in 1644. But even if we do not conclude in one way or another, at least such results mean that the geometrical properties of the curves were less sought after than their analytical properties in Mersenne's circle: we may see this as a turning point in mathematical thinking, towards an emphasis put on computations, rather than on geometric representation. Or, better said, it seems to indicate that geometric representations now became dependent on computations. This took place for the cycloid before the publication of Descartes'*Geometry*, who with some force due to his algebraic choices exiled the curve from his considerations. If it seems difficult to attribute Descartes' refusal only to animosity towards Roberval, at least one cannot challenge Roberval, through Mersenne, since we do not have his own documents, and his original way. Moreover, it seems important to add that with the parametric representation, and the use of trigonometric tables, points of the roulette could easily be drawn. We have noticed that Mersenne gave no figure for his explanation about the roulette. We even could put trigonometric tables as relevant figures. I have preferred to mention a parametric representation and have shown it relied to a conception approaching functions, something I called pre-functional representation. Things would be far more complicated had Mersenne wished to evoke area measurements in the case where t is greater than 1, so with a larger circle than the generating one, or with what may be called an elongated cycloid. He did not proceed that way, even if he did not justify himself by the shape of the curve, which passes below the rolling axis, and especially has loops that prevent a precise meaning for the area under the curve (Fig. 31). Therefore, we discover that the question of the effective representation of the curve played a role, and so that computation of an area logically must come second to what we may call the functional representation. There is no way of knowing for sure how Mersenne, or Roberval, were aware of this difficulty for area, except one has to report that they avoided speaking nonsense. It is probable that Descartes knew their questions. In addition, in one single act in a letter in 1638, he found the shape of cycloids in general. But if he gave no other means of knowing how he proceeded, he duly indicated that he had the possibility of constructing a tangent at any point of the cycloid, or rather of all sorts of cycloids. He invented the instantaneous center of rotation, a notion that would remain confined for a long time to mechanics. This won't prevent Descartes from actually giving the elements that are necessary to determine the shape of cycloids, including the loops for any value of the parameter t greater than 1, and as we have seen to indicate that the "moment" of turning back, and therefore the beginning of the loop, happened when there was a vertical tangent. Evenf if I cannot directly enrol Descartes in 1638 in the search for the parametric representation of curves, and so by their equations only, that is to say by using analytical parameterization, when he only used the mechanical trick of the instantaneous center of rotation, but when he determined through vertical or horizontal tangents, he clearly worked within the pre-functional representation.

It is then intriguing to notice that Torricelli, in 1644, when he also came to other cycloids, did not seem to make a difference for the computation of an area between elongated or shortened cycloids. He tried to be very general in his treatment of the various curves, but failed to explain that there was a serious difference between the two kinds of cycloids. Loops make the difference, but Torricelli only spoke of a sinuous behaviour, so that we may have real doubts Torricelli was conscious of the different drawings, and even of the interest to look beyond problems of indivisibles. We may think that he just followed Mersenne's way without realizing its interest. On the contrary, Roberval perhaps, or Descartes for sure, took the occasion of these generalized cycloids to think about the meaning of the parametric representation and so organized thought about what can be called functions. This step does not seem to have been commented on by the historians we chose.

In his *1646 Declaration*, Roberval is unable to formulate the analytical turn, that most of the historians have noticed in general terms, in any way another than a real taste, and a formidable apprenticeship.

> And I must confess, then being 27 years old, even if by continuous exercise for ten years I had for task, not in a mediocre way, to learn, to teach, and to treat the mathematical things, and first really the Analytics in which to-day I still take great delights, I had not gathered nor had received the sufficient mind strengths that are required for this mode. Meanwhile, as I often pondered with myself by which safest way I could better penetrate the so suave mathematical knowledge, I decided to carefully examine the divine Archimedes, almost unique among ancient geometers. And through this meditation, I was able to capture this theory of infinity that can never be sufficiently praised. Then it was how I called what the very illustrious Cavalieri calls the theory of indivisibles. (*1646. Declaration*, Roberval 1693e)

3.5 "Candid" Silences in Narratives on Two Different Representations of the Cycloid

It may be surprising for a reader now to learn that our two natural and simple computations with differential calculus or with integral calculus for either finding tangents to the cycloid or the numerical value of the area for an arch, were not the ones used at the very beginning of Calculus. This deserves an explanation, at least if one wishes to have a narrative about cycloids without too many omissions, and if one does not act as Jacob Bernoulli. Moreover, there seems to be one way for an explanation, which amounts to saying that the parametric representation that the duo Mersenne/Roberval obtained through the use of the cartesian method with orthogonal coordinates had been forgotten for a while. I prefer first to think in terms of such an oblivion rather than to explore another trail, that of a geometrical reaction. Let us try to write along this new narrative, and to do so we have to go back to the question of finding tangents to the curve and to forget for a while the area question about the arch of a cycloid. From the beginning we have been surprised by the interest for this specific question when the cycloid was discovered.

Torricelli gently teased Roberval about the tangent to any point of the cycloid, for he knew that his French correspondent had developed a kinematic technique for tangents. It is very efficient in this case of a rolling circle without sliding on a straight line that makes the cycloid, and if it is a type of behaviour that we will study later (see below), it cannot at all come from just empiricism and observation, even with an organized experiment. Indeed, the infinitesimal movement of the rolling circle is reduced to a rotation around the point of contact of the circle on which the point of the cycloid is located, and this is not a fact, but a mental construction incapable of being effectively managed in an experiment. As already said, the tangent is in consequence orthogonal to the line which goes from this point of contact of the circle to the current point of the cycloid, and the orthogonality immediately gives the direction of the tangent as that of the chord, which in this circle runs from the current point to the summit of the circle, at the vertical to the point of contact. At least if one adopts Descartes' point of view, which is read in Fig. 29 provided later, without therefore having to draw the chord parallel to the tangent in the moving circle. However, this straight line is clearly visible in a drawing by Huygens, as can be seen with Fig. 11, that is to say, as early as 1659. But if we believe him, Roberval in 1693 disposed in 1640 of another way for the tangent, by the intervention of a diamond, as already seen in Fig. 3, and also in the figure that appears in Wallis in 1659 (Fig. 27) as well as in 1693. It is very likely that Torricelli referred to this mechanical way of Roberval in his letter, since he spoke of the composition of movements, while composition is not useful in Descartes' way, as the movement is reduced to just a rotation.

> The tangent of the line had already been shown to me by the then adolescent Florentine Vincenzo Viviani, a pupil of the famous Galileo. I suppose that your so penetrating and fertile mind has observed it by itself, being in a sense the inventor of this figure; a line of this kind is indeed familiar in Nature. And it arises from the composition of two movements, rectilinear and circular. (Letter from Torricelli to Roberval, 1st October 1643)

Composition of movements was essential to Galileo's theory about the fall of heavy bodies, with a composition of a uniform rectilinear motion and a vertical falling motion, and is undoubtedly the clearest sign of the scientific revolution. As one has to be truly a theoretician to imagine that the launching of bodies on inclined planes always depend on a rectilinear and uniform motion. Torricelli seemed to recall the necessity of bringing everything back to Galileo. If it is true that in 1644 Torricelli also alluded to a greater generality for the construction of the tangent, turning to generalized cycloids, perhaps suggesting that he may also have adopted Cartesian construction, we certainly have too quickly assimilated the construction of Roberval's diamond to a simple exploitation of the equality[68] of the length of an arc as $R\theta$ to the distance traveled on the axis by the generating circle, and thus to what enabled by means of the properties of the circle to obtain the parametric representation. If Torricelli's drawings in 1644 suggest this Cartesian thought, at least he deduced no equation. A manuscript figure of Huygens dating from 1659 (Fig. 11_1), later reproduced in the famous work of 1673 (Fig. 11_2), indicates a different representation of the cycloid, even if it is quite close. It is that which gives the equality of the length of the arc *AE* and of the length being put on the horizontal *EB*. Coordination is no longer made to a rolling circle, but to a stationary circle, fixed in the middle of the figure, and therefore with an origin at the point *A* at the top of this circle. The major difference with what is read in Mersenne is that it is no longer orthogonal co-ordinates that are used, but it is somehow the immovable circle, or better said semi circle, that makes a frame of reference. As with Mersenne, this other representation could lead to an equation in this frame of reference constituted by the immobile circle and the axis on which there is rolling, or rather a parallel to this axis passing through *A*. But it did not appear at first in this setting.

---

[68] The demonstration of this equality of lengths, one curvilinear and the other rectilinear, is almost obvious. Especially if one uses the *arcuato* so named by Torricelli. In Fig. 6, and with its notations that are not those of Fig. 11, by comparing the position of the generating circle with the point 0 and the immobile circle of diameter *CE* in this figure, it is easily shown that the rectilinear length *AE* is equal to the curvilinear length *AND*, therefore the length *LE* is equal to the curvilinear length *OT*. It is enough by the *arcuato* to bring everything back to the immobile circle. We can therefore imagine that the reference from the circle in a central position, which is not in Torricelli, nevertheless came from him. We see it at work with Huygens and with Wallis around the same date, but will see it as well by Fermat in 1638 (Fig. 24).

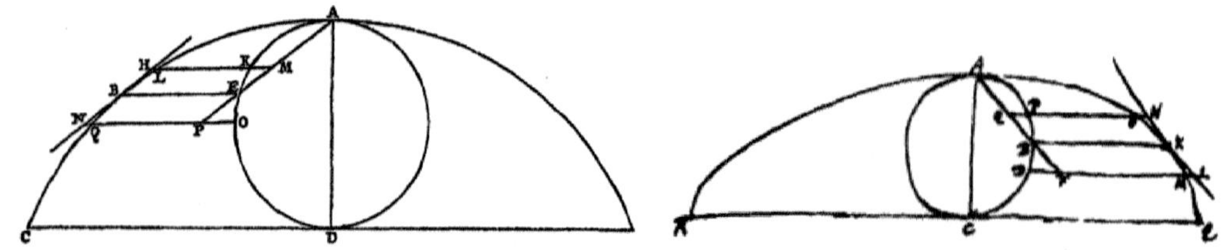

Fig. 11 A drawing from Huygens in *Horologium oscillatorium* in 1673 (Huygens 1934, XVIII, Prop. XIV, p. 152, left side) and an extract from a 1659–manuscript (Huygens 1920, XIV, p. 375, right side). *Source* Bibliothèque Nationale de France—Gallica. Public domain

Huygens did not use Roberval's diamond for the tangent, which he probably did not know. But to find the tangent to the running point, an obvious object of his drawing as early as 1659, he reasoned without indivisibles, and of course without differentials, in order to establish that the line *EA* gives the direction of the tangent at *B* to the cycloid. His direct demonstration is based on inequalities, similar to the way by which Viviani modified Apollonius' indirect way (by contradiction) to obtain the tangent to a parabola (Dhombres 2010). Huygens had not used the kinematics found by Descartes that he must have known, nor did he use the parametric representation. Huygens easily showed (Fig. 11) that the points *H*, or *M* on the parallel line passing through *B* are necessarily external to the cycloid, and thus proved the property of tangency.[69] Of course, the result is based on the lengths of arcs being placed on straight lines, with the equality of *BE* to the length of the arc *EA*: this equality is precisely what could serve as parameterization in this curvilinear reference frame. But it did not appear at this point in history. The obtention of this other parametric representation is as easy as that with the orthogonal frame of reference, the axis on which there is the rolling and the perpendicular *NM* to this straight line at the point of contact *N* with this axis (see Fig. 2 for the letters being used). The reference frame is here constituted by the axis of the rolling *AB* and the perpendicular to this axis, i.e. *DB*, but passing through the immobile position *D* (Fig. 4).

So that, with an angle phi corresponding to half theta of the previous parametrization

$$X = R\varphi + R\sin\varphi$$

$$Y = R(1 + \cos\varphi)$$

Huygens explained that he had been embarrassed to publish this, since the result on the tangent had already been established by Wren, and had even been inserted in the supplement to the text *De cycloide* due to Wallis (Wallis 1659, p. 71; see below Annex 4). In fact, it can be seen that the drawing with the generating circle set motionless at the center of the figure of the cycloid arch has been competing for some time with the design of the curve in orthogonal axes as explained by Mersenne, and what I called by using a singular: the parametric representation. I maintain the word "representation", but it should take a plural: parametric representations. I prefer to use another terminology to emphasize the difference of the two parametric representations. It is the tool that allows differences. There is a baroque frame of reference when the generating circle is placed in a stationary (and sometimes) central position, and a classical frame of reference in the case where the generating circle is grasped at a given position, allowing the practical construction by calculation and numerical tables, to obtain the running point of the cycloid from its coordinates relative to orthogonal axes. It should be noted that what I call classical, apparently coming first to Roberval, if it allows the construction of the curve, does not allow an obvious way to find tangents, at least in the state of minds around 1640. But neither did the baroque frame of reference for that matter. Except if one were to use Barrow's characteristic triangle, Fermat's adequality, or differential and even fluxional data. Then we may argue too simply that we are no longer within the Torricelli/Roberval debate. It is precisely here that the use of *Robervalian triangles* may open a possibility, about which almost certainly Torricelli was not aware, even if we still don't know how Torricelli was constructing tangents to the curve as explained to him by Viviani.

However, the baroque frame of reference intuitively provides the result on the area: there is a division of the arch of the cycloid by the central circle into three regions that appears equal. But it is possible as well that this particular parametric

[69] This came with Proposition XV in *Horologium oscillatorium* in 1673 (Huygens 1934, XVIII, p. 152; plate from the original, p. 39). It can be read in manuscripts from Huygens dated 1658, so that it was written not too long after Huygens heard about Pascal's challenge (Huygens 1920, XIV, p. 375). There is certainly more to look for in the relation between Huygens and Roberval.

representation is precisely an effect of this result antedated very early by Roberval,[70] and recognized as not being demonstrated before 1640 among the Italians. The two parametric representations can both suggest the diamond method of Roberval (Fig. 3), or as well by Wallis (Fig. 30) and both may carry the method of Fermat's adequality.[71] Wallis used the baroque frame of reference, which gave Huygens the means to confirm his construction of the tangent, and thus to start his study of the involute in the "Oscillating Clock", pushing toward a new way that will be called differential geometry. Manuscript drawings of Pascal in 1658 shows he used the baroque frame of reference (Pascal 1963, p. 105) and also the treatise published under the name Dettonville, and that dealt with roulette (Pascal 1963, p. 167).

The baroque frame of reference, and not the one I call classical, was still used by Johann Bernoulli in 1692 for his explanation of how to manipulate differential calculus. It was also taken up by the Marquis de l'Hôpital in his Analysis of the infinitely small,[72] published in 1696. But we must see that the latter author directly went to the consideration of functions, without the name, and we should rather speak of functions of functions (Fig. 34; Dhombres 2023a, b) . Is it really possible to say that we are still in the pre-functional representation, and that functions are not there? It is true that they are not mentioned. The baroque frame of reference is no more efficient than the classical one to use the new calculation, and its use in the 1690s can simply point to a habitus taken from Huygens, Wallis or Pascal about 1658. It is however surprising that the parametric equations were not used, the simplicity of which we already saw. It is therefore particularly revealing to mention here an extraordinary letter from Leibniz to Huygens of 1690 (Cfr. Leibniz 2012) in which Leibniz contacted his correspondent after many years of silence, and intended to make him understand that, in the meanwhile, and in the *Acta Eruditorum* of October 1684 in particular, he had explored a "new calculation".

Apparently used according to the figure, the baroque frame of reference with indication of the central semicircle, leads the tangent *TC* to be parallel to the chord *AE*. This is not plotted, but the text provided the explanation by calculating that the tangent is perpendicular to the traced *FE* chord (Fig. 12). And at this point, the parametric representation intervenes, according this time to the classical frame of reference, hence in rectangular coordinates. Leibniz arrived by the rules of his *Calculation* (here he said of his *Algebra*) to what he named, and we after him, a "differential equation". This completely new tool allows one to obtain the tangent to the cycloid. I give here only an English translation of the end of this letter written in French, and I follow the notations of Leibniz as found in the edition of Huygens's Works.[73]

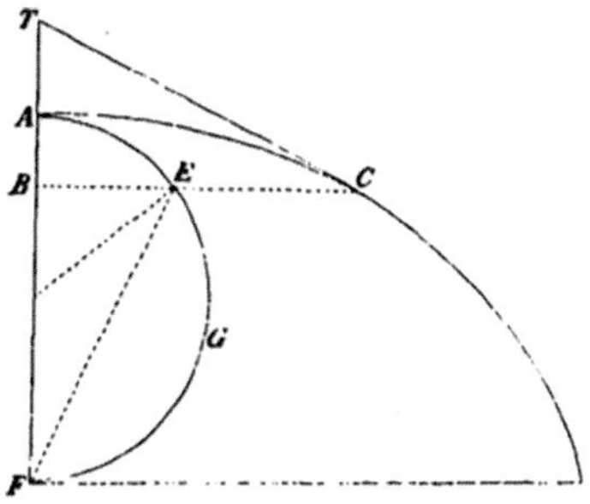

Fig. 12 Figure to be found in a letter from Leibniz to Huygens, July 25th, 1690 (Huygens 1901, IX, p. 448). *Source* Bibliothèque Nationale de France—Gallica. Public domain

From the same mean, I am reducing the analysis of curves that M. Des Cartes was calling Mechaniques, and for example Cycloids, by expressing via an equation the relation between x and y, the abscissa and the ordinate of the curve. For example the verse sine *AB* being x, then the circular arc *FGE* is by me so written $\int\left(adx : \sqrt{2ax - xx}\right)$, that is the arc is the sum of

[70] One should perhaps gain more information by carefully studying the letter sent in 1646 by Roberval to Torricelli, and published in 1693. But as Montucla said, it is a very "pedantic" letter, something like a dispute, but not declared as such, and not easy to confront with established facts. I noticed that its content was never quoted by the historians I selected.
[71] Cfr. letters from Fermat to Descartes: 1638, *Œuvres de Descartes*, II; *Œuvres de Fermat*, III.
[72] Cfr. paragraph numbered as 15 (Proposition II, p. 16 and sgg.: de l'Hôpital (1696)), which corresponds to the latin text from Bernoulli, and also Fig. 34 about which I will discuss below in relation to *Robervalian triangles*. A French synoptic version of these two texts is prepared in collaboration with Patricia Radelet de Grave.
[73] Letter from Leibniz to Huygens, July 25th, 1690 (*Œuvres complètes de Christiaan Huygens*, 9, Martinus Nijhoff, 1901, p. 448 *et seq*.; Huygens 1888–1950). The manuscript is in Leyden, and it was for the first time published in *Ch. Huge Exercitationes Mathematicae*, fasc I, p. 23, by P. J. Uylenbroek in 1833, then by C. I. Gerhardt, in *Leibnizens Schriften*, Bd II, p. 41, and in *Briefwecsel von G. W Leibniz*, p. 594. It is obviously to be found in the new Leibniz Edition.

elements of the circular curve, which are $adx : \sqrt{2ax - xx}$ (or $\frac{adx}{\sqrt{2ax-x^2}}$, as for me the double point indicate a division and so avoid the subscription of the divisor). In such a way that the elements of the circular curve are to dx, the elements corresponding to the abscissa, as the radius a to the right sines[74] $\sqrt{2ax - xx}$. This being settled, the ordinate of the Cycloid, being perpendicular to the axis, which we will name y, will be $\sqrt{2ax - xx} + \int (adx : \sqrt{2ax - xx}) = y$. By means of this equation, I find all properties of the cycloid without any use of a figure, as if it were an ordinary line. By looking for example for the differential equation of that equation, we may find tangents to the cycloid. Because of $d\sqrt{2ax - xx} = a - xdx : \sqrt{2ax - xx}$ according to the rule I provided in my Algebra, so that $dy = (2a - x)dx : \sqrt{2ax - xx}$ or $dy : dx = (2a - x) : \sqrt{2ax - xx}$. Thus, in the cycloid the ordinate is to the part of the axis between the ordinate and the tangent (that is dy is to dx), as $2a$-x, the verse-sine of the arc *FGE*, is to the right sine, that is *CB* is to *BT* as *FB* to *BE*.

It is still true that the word function is not to be found, even if Leibniz used it a lot in manuscripts in 1675, but it is replaced by the word equation, and by the analytical expression of one ordinate, that is y. At this stage, one has reached the functional representation. We too notice that Leibniz mixes differential and integral calculus that a teaching tradition has for so long separated.

Jacob Bernoulli is completely silent on these two types of identification, even though he has manifested the invention of parametric representation, without giving the name of a discoverer. Could we be certain that he wished to allude to no quarrels, for fear of having to enter into the quarrels of his own time? There was the bitter one he had with his brother Johann: they separated in 1694, and never met again. Remember that Johann used the cycloid as a sort of coat of arms (see ill. 13) when his *Opera mathematica* were published in 1742 (Bernoulli 1742b; see also 1742a). More important, there was the quarrel that had recently appeared, with the claim Leibniz had plagiarized Newton. We know that this went far in the academic world, with an official judgment from the *Royal Society* in 1713, that has continued to be an object of discussions to this day. Could Jacob, who died in 1704, have foreseen this turn of event? Then the notion of national groups would easily be raised in 1713, due to the fact that almost all British mathematicians backed behind Newton. It was a sort of separation, not really with German mathematicians, but a distinction from all the Continental ones.

In order to make an equilibrium between Torricelli and Roberval, even if the priority is without any doubt in favor of the French, a historian like Pierre Costabel who has studied another narrative on the cycloid due to Boscovich (Cfr. Costabel 1962), was persuaded that Torricelli had had his own way for constructing the tangent to the curve, as he had for the area of an arch. Boscovich was using the baroque frame of reference of the cycloid, and not the classical one, which is surprising at this date. Moreover, he used as well Huygens's way for tangents by comparing lengths and using inequalities, being free of infinitesimals. Both authors are not at all using *Robervalian triangles*, and they make use of the old technique to prove a tangent, with points of the curve outside the tangency being of the same side of the straight line.

For any social study of science, the interesting fact is that the two quarrels, between the Bernoullis and between Newton and Leibniz, occurred at a time when Academies of sciences existed. Precisely when such institutions were still fashioning their ways of conduct to reach a kind of judgment on good scientific procedures. We are faced with the same problem we encountered in the story about Jean de Beaugrand: good scientific communication between networks is not an obvious process. To claim that it is a social process is easy, but does not give a hint about the way it was achieved. Before the existence of academies, the only way for mathematicians to publicly quarrel was through books, and by presenting competitions. This is what Pascal did when he organized a prize about the roulette (the cycloid) in 1658, and refused to consider Wallis as a possible winner. In a similar way the two brothers of the Bernoulli family did this in the 90s, by launching challenges about curves like the Catenary or the Brachistochrone and the Tautochrone, which mathematicians answered, but only those who were aware of new methods whether they were in England or on the Continent. Challenges will disappear, but later replaced by the prize system. The Academy chose a subject, generally an applied one, and asked for solutions (Fig. 13).

[74] Leibniz, by mistake, wrote verse sine instead of right sine.

The four laudative verses by Voltaire were published before Bernoulli's death. The generating circle for the cycloid is in the middle position, what I called the baroque frame of reference.

Fig. 13 The cycloid as a coat of arms for Johann Bernoulli (Bernoulli Johann 1742b, I; Rüber designed what was engraved by G. F. Schmidt). *Source* Bibliothèque Nationale de France—Gallica. Public domain

It was mainly for some rhetorical effect that I felt surprised at not seeing the classical frame of reference for the cycloid at the very beginning of integral calculus. Indeed, although effective, this location would not be worth the new calculation. Since the area between the semi-cycloid and the semicircle would certainly be obtained by the integration

$$\int_0^{2R} YdX, \quad \text{with } X = R(1 - \cos\theta) \quad \text{and} \quad Y = R\sin\theta.$$

I have adopted for axes of coordinates those which present the generating circle as immovable in the middle, but do not need the baroque frame of reference, putting just X for the vertical abscissa and Y for the horizontal ordinate. In fact, it is useless to really compute this integral, since it cannot be but the area of the semicircle. This does not depend on a recourse to indivisibles, as so tells the parametric representation of the circle, that which I propose as being due to Roberval, and in other words, as the set of functions happens to be the same for the two curves. One verifies of course with the following integrals:

$$\int_0^{2R} YdX = \int_0^{\pi} R^2\sin^2\theta d\theta = \frac{R^2}{2}\int_0^{\pi}(1 - \cos 2\theta)d\theta = \frac{R^2}{2}\pi$$

One would expect that a mathematician who reveals something new, unlike a professor writing a textbook, must be able to provoke an effect on his reader, and so to strike his imagination, especially if, as with Bernoulli, he tries to provoke the idea of a revolution. This is an effect which is important to consider, if at least we take the idea of a social confrontation seriously. Perhaps it was due to the absence of such striking proofs, when he had new results, that Roberval was so reluctant to publish. Other reasons can be invoked. Quite often along the preceding explanations, I mentioned problems of disciplinary classification, with the question of algebra which allowed Descartes to reject the cycloid, that of trigonometry which manages the trigonometric representation of the cycloid but seemed to be refused by some geometers, or that of the use of the curves obtained from numerical calculation with tables. It is important to analyze this question in view of the "purity" of methods. Because it was discussed, it indicates cleavages within the scholarly community, and it could allow us to better understand the transition from the pre-functional representation to the function itself. It is possible to add here a Bachelardian remark in the sense that there could have been an epistemological obstacle, due to the intervention of the sine, and trigonometric functions as well as pure numbers.

4 Mathematics Alone?

It is then necessary to show precisely how scientists of the time tried to look at the cycloid from a pure mathematical point of view, avoiding both philosophical aspects and mechanical ones.

4.1 The Childhood of Geometry

Faithful to my method of tracking some historians according to questions I raise, I must switch to explanations about Galileo's "mechanical" initiatives. Jacob adopted a condescending attitude towards Galileo's way in this, and so spoke of the childhood of geometry in order to tell the genesis of calculus on the infinitely small, and some first naive ways.

> In this childhood of Geometry, we will rightly forgive a Master. By instituting a mechanical experiment, he explored the question through the areas of figures that were suspended to a lever. So he had suspected that this area was three times that of the generating circle, but he will never reach the truth by a geometrical reasoning. That if the error had been inconstant, if it had sometimes been less than triple, sometimes more, then the perspicacious mathematician affirmed that he would have pursued the search, rejecting the variations on the inequalities of matter and on the deckle edges of the cutting. (Bernoulli [1701]1993)

But Montucla was far more unfavorable in 1758.

> The following direction does not seem to me very honorable for Galileo. For if we believe Torricelli, he thought of weighing a cycloid on several thin and thick materials, to compare it to the circle [...]. (Montucla 1758)

It is easy to guess to what Montucla and Bernoulli are alluding, but each in his own way. A history of the cycloid had appeared in 1663, *Lettera to Filaleti*, by Carlo Dati (1619–1679), who had been a pupil of Torricelli. He was conducting a double inquiry, on the one hand on the cycloid, and on the other hand on the barometrical experiment (Dati 1663).[75] He was then linking two of the main glorious Torricelli's achievements, and registering these in the same unity of one discipline, named mathematical physics. For which the idea of using a weighting process could not be a priori respected, except as a heuristic form. Earlier in his letter dated October 1643, Torricelli explained to Roberval the mathematical ingenuity of Galileo, who had made many drawings of the cycloid, and envisioned to experiment in a physical way.

> Many years ago he proposed the problem to our admirable geometrician Cavalieri, and told him what he had confirmed to me and to many others, namely that he had once done an experiment. He suspended areas done with some matter on a small pair of scales to know the relation of the area of the cycloid to that of the circle which gave it birth. The result, by some inevitable fatality I cannot explain, was always less than the triple. Hence he had abandoned this enterprise, fearing that there would be some incommensurability.[76] That if the error had been inconstant, if it had sometimes been less than triple, sometimes more, then the perspicacious mathematician affirmed that he would have pursued the search, rejecting the variations on the inequalities of matter and on the deckle edges of the cutting. (Torricelli's letter to Roberval on the 1st of October 1643)

Torricelli, if he paid homage to Galileo's approach that may later pave the way to success, was not able to explain why Galileo's results were always inferior to three. But from then on, he could understand why the great scientist had given up trying any further. For Galileo could think that the dependence of the area of the arch of the cycloid to the area of the circle was not according to a simple relation, but to what he calls an incommensurable one. In a way this question of a computation, or the search for a formula instead of a simple geometric and rational result, is not that far from what we just saw with the parametric representation using trigonometry functions. Except that Torricelli did not seem to have been aware of the advantage of Mersenne's parametrization. This question has perhaps to be linked to Jacob Bernoulli's "philosophical" phrase at the beginning of his speech concerning relations between "things".

Galileo could also have been afraid that the result for which he was looking for the area of an arch of a cycloid could be brought in relation to the problem of what sort of a number the ratio of the area of a circle could be to the square of its radius. Mersenne was optimistically expecting that perhaps the value of the area would, with possible further work, bring something

[75] Dati was speaking both about the cycloid and the mercury experiment, while Bernoulli omitted the experiment on physics.

[76] As expressed by Torricelli, Galileo's idea is that one could have a value close to three, i.e. that is three times the area of the circle by default for the area of the arch, but not equal to three. It could have to be expressed by a formula giving a non-integer result. So that the area of the arch would be with no common measure to that of the generating circle of the cycloid.

new to the problem of the quadrature of the circle.[77] Torricelli never did raise such a long awaited solution to the problem, and he wished to have one point of view only, the horizon of computing areas. So he claimed their should be one theory only.

4.2 A Unity of Material as Torricelli Said, or Progress in All Domains as Mersenne Perceived?

It seems a priori impossible to recognize some unity of material (Cfr. Letter from Torricelli to Roberval, October 1st, 1643) in *Opera geometrica* by just inspecting the table of contents of Torricelli's book (Fig. 14).

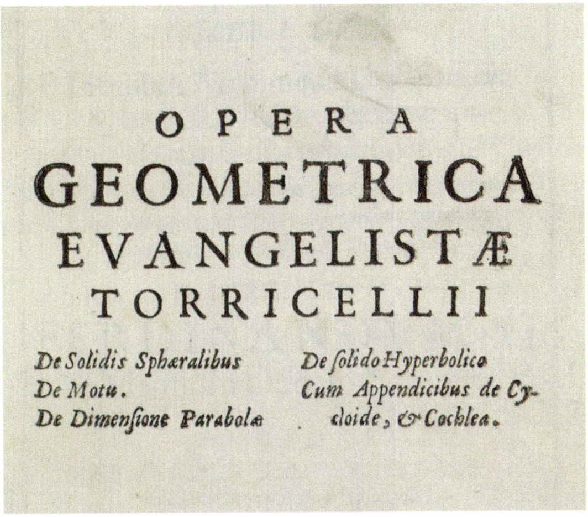
OPERA
GEOMETRICA
EVANGELISTÆ
TORRICELLII

De Solidis Sphæralibus
De Motu.
De Dimensione Parabolæ
De solido Hyperbolico
Cum Appendicibus de Cycloide, & Cochlea.

Fig. 14 The content of Torricelli's *Opera geometrica*, with three main parts, one being on mechanics, and two appendices, the last one being about the cycloid (Torricelli 1644). *Source* Courtesy University of Oklahoma, History of Science Collections

Clearly mechanics is being presented with new kinds of geometry, not necessarily qualified in one or another mathematical domain. At least it is certain that there is nothing related to the experiment on the void Torricelli was conducting at this time, and the author did not try to give any explanations about links between the three main parts of his book. The first part is about solid geometry, the second about movements according to the dynamics invented by Galileo and organized along the law for falling bodies (Souffrin 1993). In addition, the third part presents a great number of ways to obtain Archimedes's result for the area of a parabola, notably by using indivisibles, and also how to discover centers of gravity. In this last part, Torricelli was certainly in the mood of Descartes and Fermat who had previously been working (but in letters only, and not with published texts) on centers of gravity, and on what we interpret as integrals of powers of the unknown. Apart from the cycloid, which certainly attracted the attention of historians like Jacob Bernoulli and Etienne Montucla, most Torricelli's other results are interpretations of the same computation, that of the integral of a power function of x. Of course, the word "integral" did not appear (to be later defined by Leibniz), but it can be said that there is an estimate of the area under a curve expressible as said powers, and indeed a name had been found, generalized parabolas and hyperbolas, depending on whether one considers positive or negative powers of x and $1/x$. A text by Fermat, published in his *Opera* in 1679, but which can probably be dated from the years preceding 1640, was quite explicit on this subject. Even the appendix on the infinite hyperbolic figure by Torricelli entered in this general picture, with no physics involved. Torricelli's approach was far closer to Cavalieri's results using indivisibles, but he was arriving as well to this kind of computations on powers of the variable.

[77] Was it so ridiculous to think of the squaring of the circle if it were proved that the arch of a cycloid was directly connected to the circle? Descartes kept the cycloid out of the curves "admissible in my geometry", because he thought it was not algebraic. The reason was that to construct the curve it was required to have the length $2Rp$ carried on an axis, as the base of the arch of a cycloid. It was only later that the non-algebraic character was seen at a glance by the infinite number of intersections of the curve with a straight line. Obviously any parallel to the axis of the rolling, distant less than $2R$ to this axis, intersects the curve in an infinity of points, as happens in any periodic function, at least if one accepts not to restrict oneself to an arch, and exhibits the periodicity. The same is true for the curve of the verse sine, whose periodic extension is not made visible, unlike the cycloid when understood as a mechanical curve from a movement of a carriage.

This was a move towards analytical method, even in the indivisibles point of view. We have read Roberval explaining in 1646 that for him the analytical point of view came first.

The mention of Archimedes may provide a solution to the puzzle about Torricelli's composition. We may register Torricelli's book not really within the new genre settled by Mersenne under the qualification of mathematical physics, but far better as a revival or posterity of Archimedes's kind of mathematization. The unity in Torricelli's book lies in the use of "geometry", as is stated precisely in the title: it may certainly treat problems from physics, but in the axiomatized way like the Euclidean geometry as taught by Archimedes either for the lever or for the equilibrium of a boat or a floating body. In doing this, Torricelli was not acting the same way as Mersenne in his *Harmonie universelle*. The Minim friar introduced for example frequencies for sound, and so was by excellence establishing mathematical physics. It is just not the question of mathematization that is being debated, but the origin of knowledge; it did not stem from experiment but from an analytical exhibition of theory. Experiments have to come in a second phase. In 1646, Roberval claimed he preferred Archimedes to any of the other antique authors. Even if this was a topos of the time to use Archimedes, it does not mean that we may identify references to Archimedes by Mersenne and by Roberval. Discussions here about mathematics and physics according to Roberval is a subject requiring more than we are able to provide in the present analysis.

Mersenne's two volumes (1636 and 1637) have a complete title that is interesting in itself for the variety of things to be treated: *Harmonie universelle, contenant la théorie et la pratique de la musique, où il est traité de la Nature des Sons, & des Mouvemens, des Consonances, des Dissonances, des Genres, des Modes, de la Composition, de la voix, des Chants, & de toutes sortes d'Instruments Harmoniques.* But Mersenne tried to justify why he was going from one subject to another, providing explicit titles for some sections. Moreover, it is in the second book, in the first volume, a book devoted to a *De motu*, like Torricelli's second item, but adding *corporum* (*Du mouvement des corps*) that a distinction was made among possible movements of a circle, or better said of a globe. The physicist's mood always imposed realist surroundings. At the end of a proposition (Fig. 15), Mersenne questioned whether a rolling globe has more speed than a sliding one, as if the two notions were that clear.

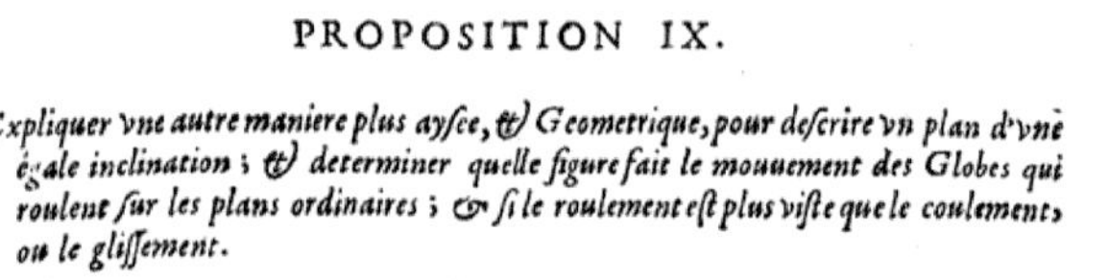
PROPOSITION IX.

Expliquer vne autre maniere plus aysee, & Geometrique, pour descrire vn plan d'vne égale inclination; & determiner quelle figure fait le mouuement des Globes qui roulent sur les plans ordinaires; & si le roulement est plus viste que le coulement, ou le glissement.

Proposition IX.
Explain another way more easy and more geometrical to describe a plane of equal inclination, and to determine what figure makes the motion of globes rolling on ordinary planes, and whether the rolling is faster than flowing or sliding.
(Author's translation)

Fig. 15 Mersenne's Proposition IX (Mersenne 1637). *Source* Bibliothèque Nationale de France—Gallica. Public domain

Clearly a new concept had appeared and Mersenne knew he was entering new territory. This question about the different ways for a rolling globe to behave on an axis came after Mersenne had examined Galileo's famous and revolutionary experiment on an inclined plane. The difference for the experiment was provoking different computation for the terrestrial g. Mersenne raised also the question of curves for which there is an experimental drawing. It is then transformed into a more geometric drawing, being explicitly constructed (Fig. 17). If the absolute geometrical definition of the inclination of a plane (represented by a line) on which a "globe" moves is at each point the angle made with respect to the line which joins this point to the center of the Earth, and if Mersenne sought strangely what should happen when one wanted to obtain the line of equal inclination to the center of the Earth, the last part of the proposition forgets this unreality, which was much in fashion and gave a long discussion on the occasion of the publication (Beaugrand 1636) of the *Geostatice* from Jean de Beaugrand in the same year 1636. These considerations are related to the loxodromic curves of naval cartography. It must therefore be accepted that, for Mersenne, mathematical research is not separated in very distinct subjects, and that it brings into play a sort of common front. This is a form of unity. Efficiency is not everywhere the same, and undeniably Mersenne is of a rather delirious optimism since he thought that the measurement of the area of the arch of the cycloid could lead, in a second time, to the quadrature of the circle. But Wallis also wanted the cycloid to have been encountered much earlier, around 1500, by authors interested in the quadrature, himself having invented a new form of arithmetic quadrature of the circle (Fig. 16).

In this new context, recognizing that it is difficult to make a difference between rolling and sliding, even if one considers an "effect" in the case of a ball, Mersenne immediately applied these mechanical differences to the theoretical case that allows roulette, which he does not yet name this way. He explained the difference between the center of the circle, which merely slides and the point, which touches the line while rolling. Because for him this curve can only be a half-ellipse.[78] He immediately thinks of deducing a new construction for this conic. That would solve a mathematical problem by physics alone! Mersenne noticed that there is an interesting consequence in this interpretation, since the ratio of the major axis to the minor axis, in fact the ratio of the perimeter of the circle to twice the radius that we note π is then deduced from the construction of the supposed ellipse. So Mersenne could say something about the squaring of the circle.

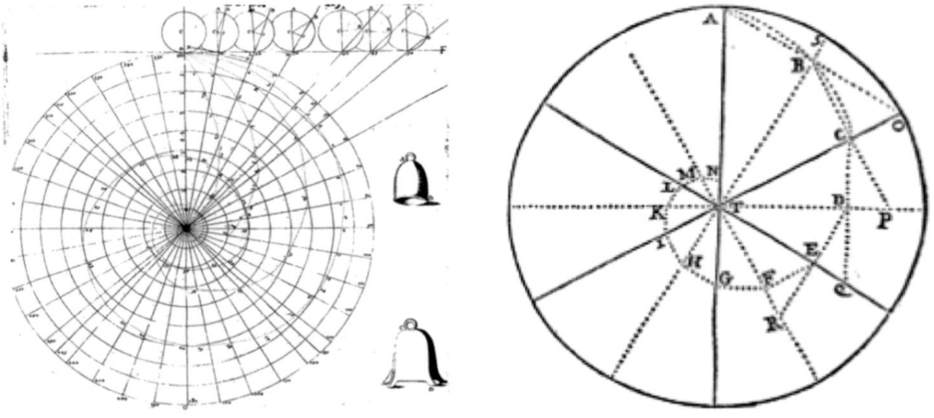

Figs. 16 and 17 Drawings by Mersenne on curves in spiral form in his *Harmonie universelle*, which shows the invention of a construction of an approximation for a curve (Mersenne 1637, p. 115 [right side], p. 119 [left side]). *Source* Bibliothèque Nationale de France—Gallica. Public domain

A year after, and for the second volume of *Harmonie universelle*, as a sort of 20th and last book, Mersenne corrected his affirmation on the ellipse. He now defined a "roulette". I have translated in an annex the two pages of *Observations physiques et mathématiques*, the eleventh one (see below Annex 3). There is no drawing either, but Roberval is mentioned as the Geometer who helped Mersenne, and we already saw how they invented the parametric representations of the curve. It is only at this point, that Marin Mersenne made a clear allusion to the paradox named Aristotle's wheel, which Galileo had just made fashionable.

4.3 Rolling and Sliding: Mersenne's Way of Eliminating Paradoxes on Infinity, on Divisibility and on the Continuous

One of the traces of what may be called a scientific revolution may paradoxically be an absence, and this ipso facto leads historians to deny it. Such is the way in which Marin Mersenne "solved" the old question of "Problems on mechanics", known as Aristotle's wheel, without even bothering to refer to it (Mersenne 1636, II, p. 120). He merely invented several modes of point movements on a straight line, with one very particular that he called rolling, and the others that are different kinds of sliding combined with rolling, to which he also gave the name of "coulement", or possibly flow. Mersenne knew that he had solved it on mechanical terms, that is by avoiding speaking of an infinity of points, using the difference between pure rolling and pure sliding. As we have seen, the parametrization directly uses the maintenance of a certain arc length which analytically explains why $R\theta$ appears in the equation. One should not confuse this with Cavalieri's way of dealing with indivisibles, the flow of a line in a figure along another one called *regula*, which appeared a year earlier, and which mathematically "united" Torricelli's presentation. We have read the proposition (Mersenne 1636, p. 119) on a body that falls, which Mersenne intended to "demonstrate" in 1636 in his immense encyclopedia on sounds and music that is the "Universal Harmony". Then he goes on:

[78] This is explained with no drawing by Mersenne (1636, II, p. 120).

> I leave many other conclusions which can be deduced from this line, in order to explain the second part of this proposition, which consists in knowing what is the figure which the ball which rolls on a plane describes. I say, rolling, especially as if it only slides, it describes as many straight lines as there are points in its half-circle, the longest and highest of which is perpendicular to the end of its axis.[79]

No drawing is being made, but it is easy to imagine from what is usually found with question 24 in editions of Problems on mechanics (Figs. 18 and 19).

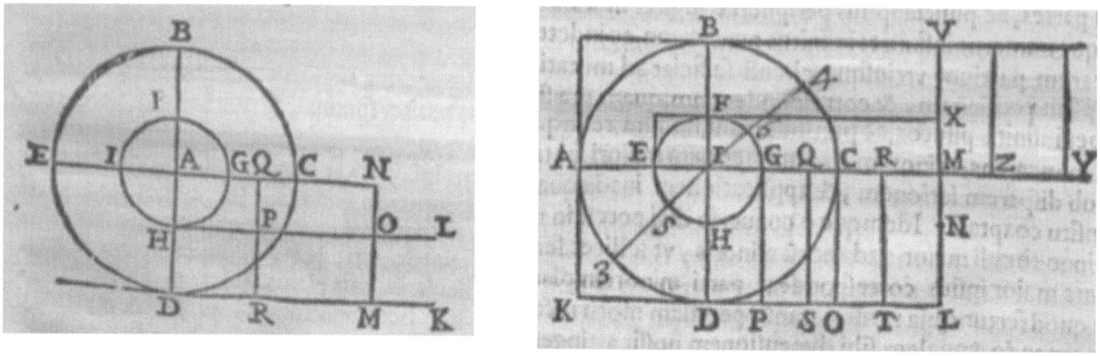

Fig. 18 The only two illustrations which accompany the commentary of the 24th Question in Joannis de Guevara's *In Aristoteles Mechanicae Commentariij* (De Guevara, 1627, p. 209 (left side] p. 216 [right side]) Rome, 1627, pp. 209 and 216). It is remarkable that nothing–visual makes it possible to distinguish what really rolls on an axis. The figure is purely geometrical. *Source* Max Planck Institute for the History of Science | CC-BY-SA. http://echo.mpiwg-berlin.mpg.de/MPIWG: NX6FH5F4 | Public Domain

We give another occurrence of question 24 in the book *Problems on mechanics* as it was translated in Latin in a Parisian edition (Fig. 19).

DVbitatur quam ob caufam maior circulus æqualem i minor cir- XXIIII
culo conuoluitur lineam, quando circa idem centrum fuerint po
fiti: feorfum autem reuoluti quemadmodū alterius magnitudo
ad magnitudinem fe habet alterius, fic & illorum ad fe inuicē fiunt lineæ.

"A difficulty is raised to know how it is possible that a greater circle when turning draws a path having the same length as a smaller one with the same center".

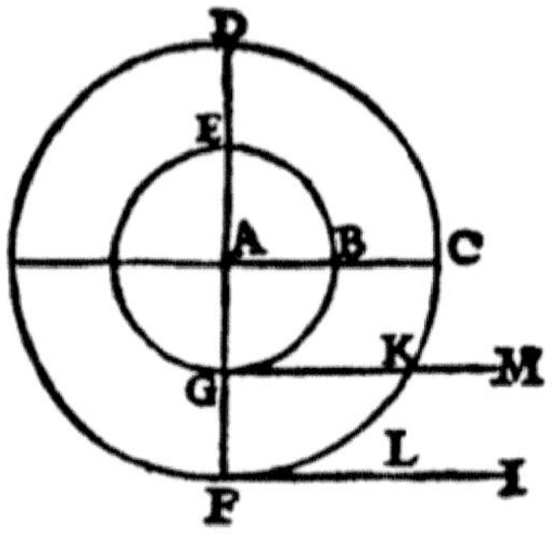

Fig. 19 Figure, Latin text, and English translation of the original Greek. Question 24, *Quæstiones mechanicæ, more explicita a Nicolao Leonico Thomaeo* (1530; Aristotle [1936]1955, p. 389). *Source*: Google Books, Public domain

[79] "Je laisse quantité d'autres conclusions que l'on peut déduire de cette ligne, afin d'expliquer la seconde partie de cette proposition, qui consiste à savoir quelle est la figure que décrit la boule qui roule sur un plan. Je dis, *qui roule*, d'autant que si elle glisse seulement, elle décrit autant de lignes droites qu'il y a de points dans sa demie circonférence, dont la plus longue et la plus haute est perpendiculaire à l'extrémité de son axe" (Mersenne 1636, p. 156).

From a figure found in a Latin edition of Aristotle's Problems on mechanics by Johannes de Guevara in 1626 we may explain Mersenne's solution (Fig. 18). If one supposes circle *DEBC* is rolling on a straight line *DK*, then circle *HIFG* will have a part sliding, and a part rolling, the difference amounting to the difference of the two perimeters of the two circles. And this can be said in the opposite way if it is the circle *HIFG* that is rolling without sliding. In his work published in 1638, the *Discorsi e dimostrazioni matematiche intorno to due nove scienze*, Galileo established that in spite of different perimeters there were as many points in the small circle as in the big one, as seen by the stretching bijection organized from the center, which is called in French "homothétie" since Michel Chasles (Fig. 20). Atomism, continuity, divisibility, all or almost all could intervene to explain the difficulty raised by Aristotle or his pupil, not without some irony. De Guevara in 1627 took 17 pages for his commentary on this issue. Mersenne in 1636 did not even comment, as he was sure of his fact. That is to say, his assurance came from a concept. One reads the elegant way by which Mersenne avoided the atomic conception of the point which had for so long been the paradox, using the vague notion of a "quasi point". Pure rolling, and therefore the different forms of sliding combined with rolling, was not reduced to a punctual contact; a measure for the difference is the length of the arc of the circle compared to the length on the rolling axis. It is a conservation result when there is no sliding. We may say that Mersenne (Fig. 19bis) had a way of experimenting in mathematics: this is also how he invented a sort of elliptical shape by considering a drawing for what we call the cycloid and which he later called as a roulette.

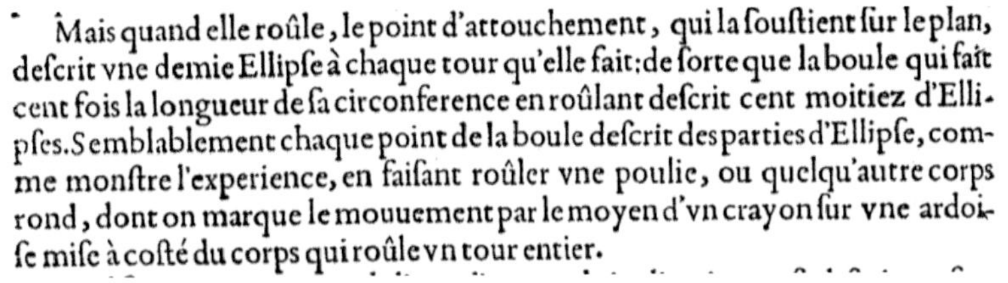
Mais quand elle roûle, le point d'attouchement, qui la ſouſtient ſur le plan,
deſcrit vne demie Ellipſe à chaque tour qu'elle fait: de ſorte que la boule qui fait
cent fois la longueur de ſa circonference en roûlant deſcrit cent moitiez d'Elli-
pſes. Semblablement chaque point de la boule deſcrit des parties d'Ellipſe, com-
me monſtre l'experience, en faiſant roûler vne poulie, ou quelqu'autre corps
rond, dont on marque le mouuement par le moyen d'vn crayon ſur vne ardoi-
ſe miſe à coſté du corps qui roûle vn tour entier.

Fig. 19bis Mersenne, *Harmonie universelle* (Mersenne 1636, I, II, p. 120). *Source* Bibliothèque Nationale de France—Gallica. Public domain

Mersenne at that time was able to discuss with Gilles Personne de Roberval, who just became a royal professor. Thus, the adventure of the curve was launched: the most astonishing thing is that the same year Descartes excluded it from curves that had the right to be treated in his Geometry. This did not prevent Mersenne from also launching the problem of the curves described by points inside the circle which rolls or outside this circle, providing elongated or shortened cycloids. The adjective for those curves refers only to the comparison of the different perimeters (Mersenne 1637, p. 25). That of the circle which rolls without sliding is precisely used as a standard of pure rolling. So that even if we may speak of a generalization for the cycloid, there is in fact a sort of necessity to measure the effects of sliding, or the lack of it for the elongated cycloid. Unlike his predecessors, Mersenne did not speak of points, but of lengths. Consequently, Mersenne realized that there is a part of the movement in the other direction when a point is fixed outside the circle that rolls. It is understandable why Mersenne insisted on attributing to his work both the qualifier of "physics" and that of "mathematics". In addition, it would be anomalous to consider that this work is distinct from his reflections on the cosmological problem, or from the considerations on Galileo's law of the fall of bodies (Mersenne 1636, p. 124).

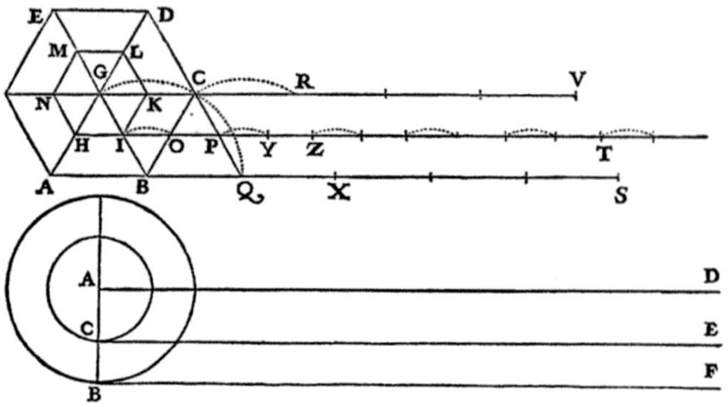

Fig. 20 Drawings in the Galileo's *Discorsi* (Galilei 1638, p. 69). *Source* Bibliothèque Nationale de France—Gallica. Public domain

As a counterpoint to the lack of drawing in Mersenne, Galileo's famous drawing in the *Discorsi* (Fig. 20) allows a hexagon to be seen instead of a circle, thus locating portions as *OP* not covered in the movement by the side of the small polygon.[80] This gave Galileo the idea of an "intermediate void" for the circle, conceived as a polygon with an infinite number of sides. Salviati in the dialogue evoked a "sliding", but certainly Mersenne did not care about the explanations with the "voids". Of course, Galileo's question will take on quite another value in the second part of the nineteenth century, with the definition by Dedekind as well as Cantor, of an infinite set as the one that has a bijection with a proper part. But it must be acknowledged that the kind "*in parti non quante*"[81] that Galileo used to make us think of "voids" is much less useful than the idea of pure sliding as opposed to pure rotation, a sliding that actually takes place for the center of the circle that rolls without sliding. Moreover, pure rolling as we have seen is at the basis of the parametric representation.

4.4 The Narrative from Drawings: Classical Versus Baroque Frame of Reference for the Cycloid

Letters between Roberval and Torricelli, before the creation of the great academies of science, were exchanged in an intermediate period between two long-term regimes for dissemination and practice and for the organization of intellectual networks. There is the regime of the printed scientific book, a fruit of the humanist world of the Renaissance, and there is the regime of letters destined to be read by several and involving several subjects. This second regime will be disciplined with the registration of the *Royal Society* in 1662, as at its beginning many letters will be published, possibly presented by a fellow. The subject was then unique, generally determined by a title. Along the eighteenth century, it will gradually lead to the standardized form of a scientific paper. The old style is still readable in 1697, with the letter Wallis wrote to confirm he had seen the cycloid in manuscripts far earlier, and which was published in the *Philosophical Transactions* (Fig. 21).

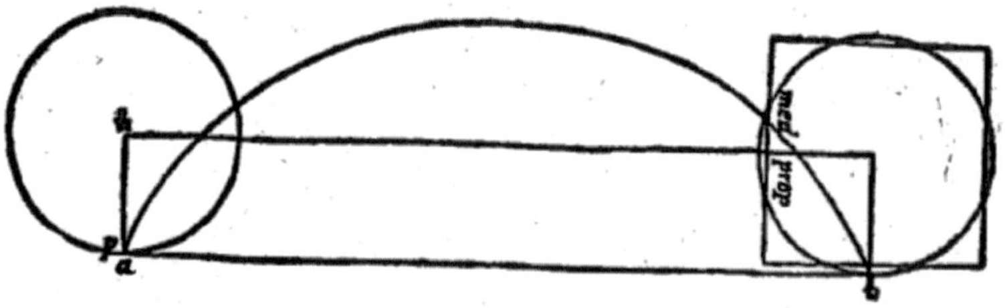

Fig. 21 The drawing attributed by John Wallis to the cardinal de Cusa, exhibiting one arch only for the cycloid, with a clear indication to the squaring of the circle. It is presented in a vertical way in the *Transactions* (Letter from Leibniz to Huygens, 25th, July 1690: Huygens 1920, XIV, p. 451; see also *Philosophical Transactions* 19/561:561–566). *Source* Bibliothèque Nationale de France—Gallica. Public domain

[80] The treatment by Galileo of the Aristotle's wheel, which Mersenne must have known by reason of the tribulations of the Galileo manuscript as far as Leyden, is to be found in the *Discorsi*, on the first day from page 69 (Galilei 1638). The use of a polygon seems to have helped Descartes as well for his finding tangent to the cycloid (Cfr. letter form Descartes to Mersenne, 23rd of August, 1638, in *Œuvres de Descartes*, vol. II, p. 308).

[81] Cfr. Costabel (1964) and the author's references to the classic texts on the Aristotle's wheel, e.g., Drabkin (1950). The literature on Aristotle's wheel is enormous.

It is then noteworthy that no book title can be found such as "mathematical course" before the publication in 1634 of the bilingual volumes of Pierre Hérigone. It will be modestly followed by some books, generally by Jesuits, like those of Kaspar Schott or Milliet de Chales. But this shows an absence of enthusiasm among the pedagogues, which would radically change in the following century. The dispute between Torricelli and Roberval has not been the object of a pedagogical commitment, at the time as well as today. This is not a minor historical omission. Jacob Bernoulli did not even play pedagogy in his 1701 speech. He was even ferociously ironic when he addressed students, merely repeating to them that they cannot judge quarrels between great scholars, even if they looked ridiculous: and he made use of the Latin adage, the "shoemaker should not rise higher than the sole". On the contrary, as we have read, Roberval in his 1646 Declaration emphasized his role as a professor giving lectures, doubting the role of Du Verdus writing notes precisely from Roberval's lessons. However, this new genre was not that fashionable, and we may see that Barrow's lectures proved not to be successful in terms of selling.

Writing a letter was obviously a usual practice for the "Republic of Letters". This expression, already used, denotes an informal, and almost mythical place, which in the present historiography of science is replacing, in good sociological consciousness, both the heroic narratives on modernity forged by a few geniuses and the weighty monographs focused on institutions. So much the better if contemporary historical research can thus better grasp intangible phenomena, which were repugnant both to the positivists who founded history of science and to the users of statistical data on the role of science in economics and industrial technology. Letters do not interest the analysts of history focusing on large structures like the development of capitalism, colonialist imperialism, or absolutism. The phenomena of which I speak are indeed the bonds of friendship (obviously the basis for the first Italian academies, but also for the meetings of Marin Mersenne, or meetings at *Gresham College* in London), common views on what is worth knowing, what its pursuit requires, views on the place of scientific knowledge in life, which also implies having a consciousness of what an era is constructing, as well as undoing. These phenomena are in fact quite traceable, and the letter is one of the important elements for such a study. As was manifested by Max Weber, who used such a letter by Franklin to speak of the mentality for research (Weber 1904–1905). Our predecessors as historians of science knew it, and since the nineteenth century they produced magnificent and erudite collections, from which, as we have seen in the present case, one can now easily find enough material for narratives. Whether with the edition in 18 volumes of Mersenne's *Correspondence* (completed in 1988), or with the *Carteggio* of the disciples of Galileo, and of course with the formidable national edition realized by Antonio Favaro, or with the Correspondence part which comes first in the collection of Descartes' works realized jointly by a literary man like Adolphe Adam and a scientist like Paul Tannery, then revised among others by Pierre Costabel and Bernard Rochot (Rochot 1965). Such a work remains to be done in the case of Roberval, whose manuscripts are rarely quoted.

In this intermediate position of a letter towards 1640, and beyond the psychological information and the usual forms of politeness, an essential point is the variety of the subjects being treated. In Descartes, for example, it is not uncommon to see a dozen subjects intervening in a single letter to Father Mersenne. This is also the case of Torricelli writing to Roberval: Jean Itard divided the letter of 1643 into five distinct points (Itard 1975). But what also appears is that there is usually no disorder, even if there is one bulk, as questions are generally treated in the way they were raised in another letter. For a letter is not isolated: it is an element of an exchange that may continue. This was the case between Torricelli and Roberval—and they were corresponding up to the early death of Evangelista Torricelli. Even if the *1646 Roberval Declaration* has to be taken as an extreme case. We see that as well with Fermat who, challenged on the cycloid by Mersenne in 1638, sent a first letter in which he expressed his distrust of the result on the area of the arch of the cycloid, and then modified his judgment in a second letter, and explained his solution according to the predicted result (Cfr. Letter from Fermat to Mersenne in 1638) . It is there that he employed what we have called the baroque frame of reference for the cycloid. But can we rely on estimated figures as they are drawn in some letters?

At least, one cannot a priori accuse the genre of a letter of not tolerating figures. As we see as well in Mersenne's writings to Descartes, and especially in the letter written on April 28, 1638, in which he announced, among other things, Roberval's result on the precise value of the area of the arch of a cycloid (Cfr. Descartes 1996, II, p. 116). If the tangent to each of the two extremities is still oblique, it became vertical in Descartes's argumentative answer on May 27, 1638 (Descartes 1996, II, p. 136). But a few months later, in another letter to Mersenne, Descartes no longer put vertical tangents at both extremities (Letter from Descartes to Mersenne, 27 July 1638, *Œuvres de Descartes*, II, p. 261). Does this a meaning? It seems difficult in any case that Galileo, if he had tried to weigh the area of the arch of the cycloid, as Torricelli maintained, could have committed such a mistake in drawing. Besides, when without any drawing, Mersenne spoke of the curve in question, he mistakenly assimilated it to half an ellipse, and at least this curve shows vertical tangents at both ends. But the inhibited use of such figures in correspondences is not true for all authors, and there are no figures in the exchanges between Torricelli and Roberval, except in the *1646 Declaration*. Descartes and Fermat encouraged by Mersenne may appear as exceptions, in their time. However, we have no difficulties from figures about the cycloid, even in letters, to distinguish the classical frame of reference from the baroque one.

A particularly interesting point is the insertion of figures in a letter when it could be regarded as a representation of reality, with mechanics in mind for example, therefore, the kind of experiments mathematicians may conduct. There is none in Torricelli's letter of October 1643, but a few months after, a drawing of a cycloid appears in a book, still quite awkward (Fig. 22). The curve is in dotted lines, as it should have to be for the representations of what was to be constructed in geometry, like the lines in a perspective drawing. It was rather new to do so, but it became quite popular. In this figure, however, tangents at points *A* and *B* do not appear vertical. Then the first stated theorem is a quadrature in the strict sense, since the relation is between a surface delimitated by curves and a rectilinear one, that of a triangle (Fig. 2). This turning back to rectilinear figures will soon be forgotten, and this certainly will make history, as an abandonment of the question of squaring to give place to questions relative to integration, that is to functions and their classification. The very presence of these cusps, as they are named since the end of the seventeenth century, characterizes the cycloid among the ones introduced by Mersenne, the elongated or the shortened cycloids, and certainly paved the way to a study of dynamical properties of families of curves, cusps in between sinuosity and loops. This will be one of the main features of Newton's classification of curves of the third degree. The intervention of other cycloids was not merely a show of mathematical generality. Indeed, these curves made it possible to highlight the notion of sliding, but they also validated the parametric representation, and as well preparing the way to a study of singular points in curves.

However, as we noticed with Torricelli, there was a difficulty in understanding any interest for these curves if it were just a question of measuring areas.

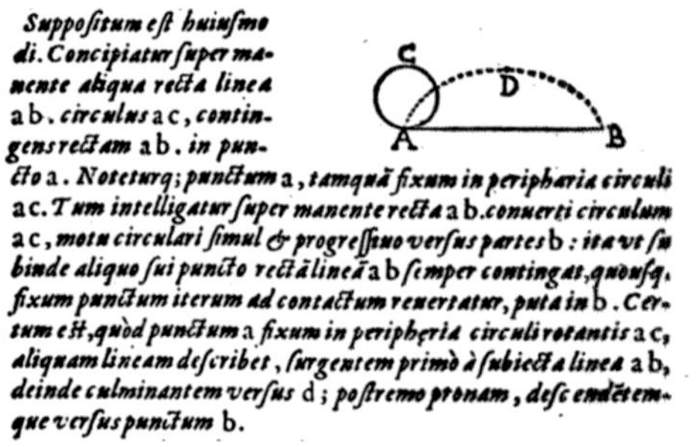
Suppositum eſt huiuſmodi. Concipiatur ſuper manente aliqua recta linea ab. *circulus* ac, *contingens rectam* ab. *in puncto* a. *Noteturq; punctum* a, *tamquã fixum in periphaeria circuli* ac. *Tum intelligatur ſuper manente recta* ab. *conuerti circulum* ac, *motu circulari ſimul & progreſſiuo verſus partes* b *: ita vt ſubinde aliquo ſui puncto rectã lineã* ab *ſemper contingat, quouſq. fixum punctum iterum ad contactum reuertatur, puta in* b. *Certum eſt, quòd punctum* a *fixum in peripheria circuli rotantis* ac, *aliquam lineam deſcribet, ſurgentem primò à ſubiecta linea* ab, *deinde culminantem verſus* d; *poſtremo pronam, deſcendẽtemque verſus punctum* b.

Fig. 22 The first diagram concerning the cycloid in Torricelli's *Opera Geometrica* (Torricelli, 1644, *Appendix, De Dimensions Cycloidis*, p. 85). *Source* Courtesy University of Oklahoma, History of Science Collections

We may do a sort of journey among figures. The cycloid in Torricelli's *Opera geometrica* only occupied a few pages (Torricelli, 1644, *Appendix, De Dimensions Cycloidis*, p. 85) no more than for a letter! It is noticeable that the rolling circle helped representing movement: just see Fig. 2. The indication for movement disappeared later (Figs. 23, 24 and 26), except in a drawing by Descartes in 1638 (Fig. 25). Huygens has both movement and a fixed semi-circle, his representation being insistent on tangents.

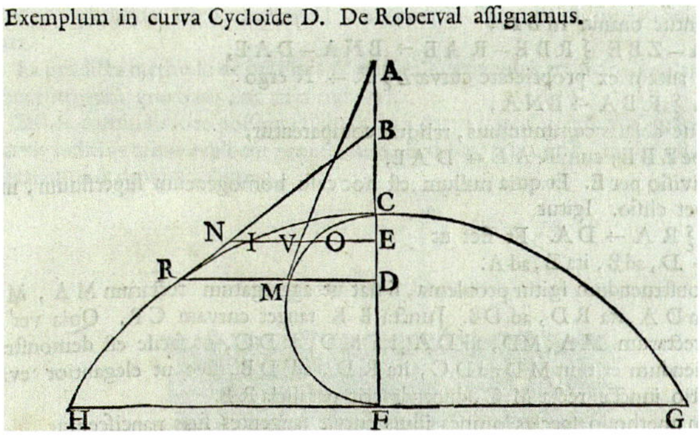

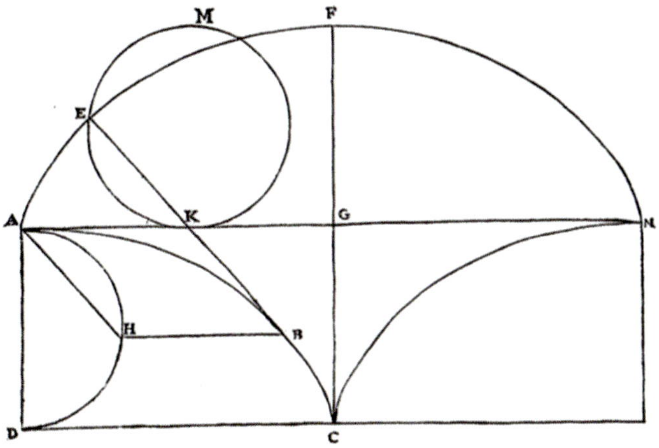

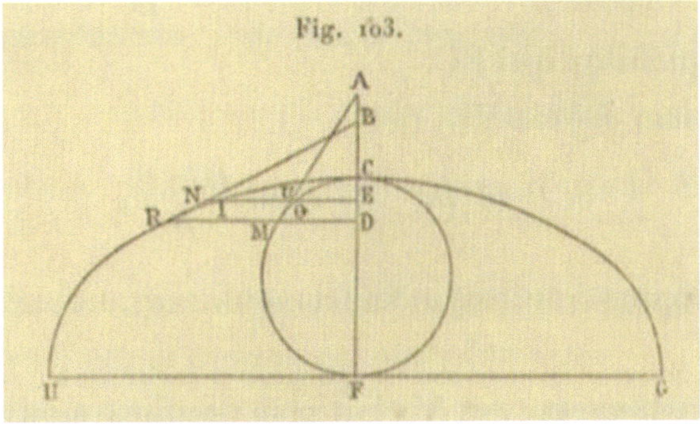

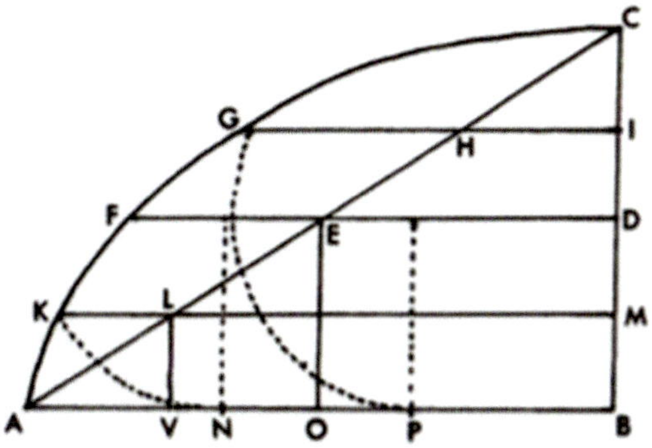

Figs. 23–26 (Left-Right-Right-Left) Four representations of the cycloid, which do not all show verticality at the extremities of the arch. Figures 23 and 25 are from Fermat. For the first, he said it was reproduced from a letter by Roberval, and the second is in his *Opera* posthumously issued in 1676 (*Œuvres*, vol. 1, p. 163; *Opera*, p. 69; a manuscript is available at the Bibliothèque Nationale de France: n° 3280, fol. 112–117). We see that Fermat offered an accentuated verticality for the behavior of the curve at the extremities of the arch. Figure 26 is taken from a letter of Descartes, July 27, 1638, in which he discussed the computation of the area (page 263 in the second volume of his *Œuvres*). The move is here suggested by two positions of the generating circle, both reduced to one arc, and in dotted lines. Figure 25 is in Huygens, in his *Horologium oscillatorium* in 1678, where he demonstrated from the parallelogram *AHNK* the extraordinary property of the tangent which comes to generate another cycloid (p. 66 of the original book). Huygens is using the classical frame of reference in the upper part of his figure, and the baroque one in the lower part. *Source* Bibliothèque Nationale de France—Gallica. Public domain

In 1693, Roberval's work became very careful, since it gave the means to construct the tangent at each point of the Cycloid (see above Fig. 3). But it is possible to read two figures here, one with the baroque frame of reference (immobile circle to the left,) and one with only a generic circle to find a tangent. It is the same with Fig. 27, due to Wallis (Figs. 28, 29 and 30).

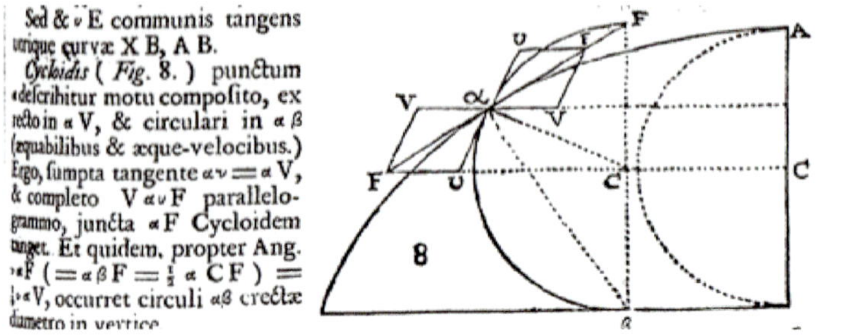

Sed & υ E communis tangens utrique curvæ X B, A B.
Cycloidis (Fig. 8.) punctum describitur motu composito, ex recto in α V, & circulari in α β (æquabilibus & æque-velocibus.) Ergo, sumpta tangente αυ = α V, & completo V α υ F parallelogrammo, juncta α F Cycloidem tanget. Et quidem, propter Ang. υαF (= α β F = ½ α C F) = ½ υ α V, occurret circuli α β rectæ diametro in vertice

Fig 27. It is not taken here from the first English edition in 1685 where there is mention of the cycloid but no figure, but from the Latin edition in 1693 in his *Algebra Tractatus. Historicus & Practicus.*

Fig. 27 The construction of the tangent to the cycloid with the diamond, in Wallis' *Algebra* (Wallis, 1693, p. 401). *Source* Bibliothèque Nationale de France—Gallica. Public domain

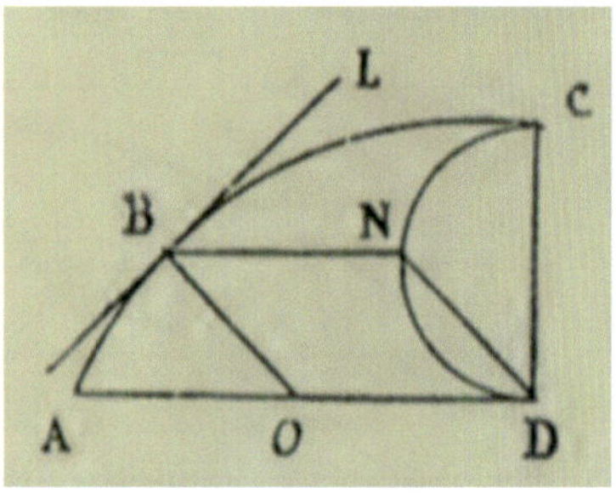

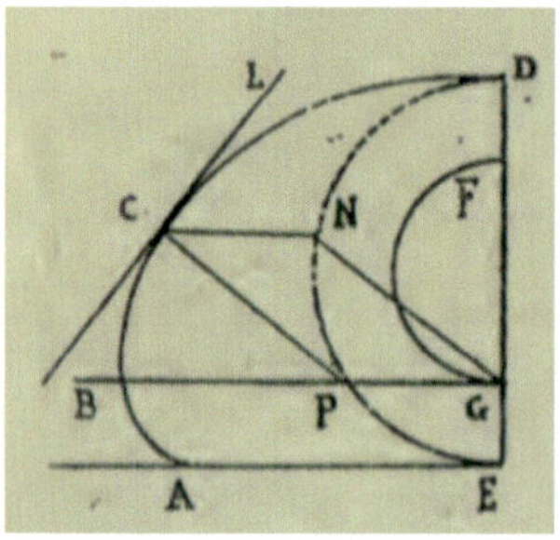

Figs. 28 and 29 Two representations of the cycloid, taken from the correspondence of Descartes, from a letter, August 23, 1638, where he indicated the construction of a tangent (the first is on p. 308 with a baroque representation and center *D*, and the second is p. 310 concerning an elongated cycloid and taking into account the instantaneous center of rotation in *G*). *Source* Bibliothèque Nationale de France—Gallica. Public domain

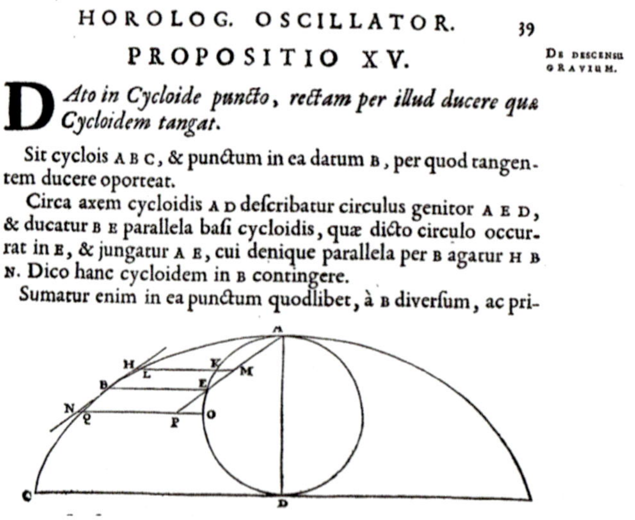

HOROLOG. OSCILLATOR. 39

PROPOSITIO XV. De descensu gravium.

Dato in Cycloide puncto, rectam per illud ducere quæ Cycloidem tangat.

Sit cyclois A B C, & punctum in ea datum B, per quod tangentem ducere oporteat.

Circa axem cycloidis A D describatur circulus genitor A E D, & ducatur B E parallela basi cycloidis, quæ dicto circulo occurrat in E, & jungatur A E, cui denique parallela per B agatur H B N. Dico hanc cycloidem in B contingere.

Sumatur enim in ea punctum quodlibet, à B diversum, ac pri-

Fig. 30 The construction of the tangent to the cycloid by parallelism in Huygens's text "Oscillating Clock (1678)". (Huygens 1678, p. 39; 1932, Vol. XVII) with the explanation using inequalities as it used to be in the old form of a method by exhaustion, and a baroque frame of reference. *Source*: Bibliothèque Nationale de France—Gallica. Public domain

It is then interesting to show how Pascal immediately adopted Wren's result for the rectification of a cycloid (Fig. 31_1), with both generic arcs of semi circles and an immobile one. He was also able to restrict the computation to an arc of a cycloid. In addition, a link was made with the rectification of an ellipse for the non classical cycloids, which is quite easy to do using a parametric representation. At least when one has the habit of classifying integrals. Pascal used another method.

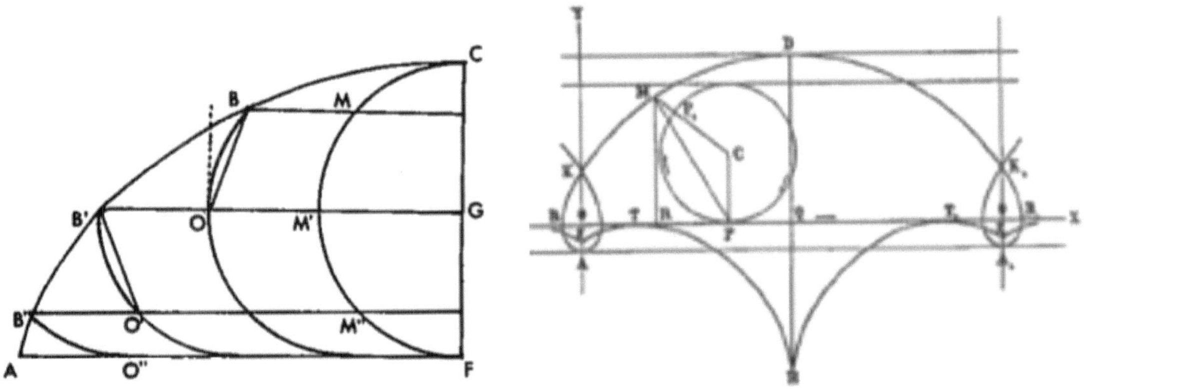

Fig. 31 To the left is a drawing from Pascal for the rectification of a cycloid (which came after Pascal had heard of Wren's rectification) and to the right is a drawing in the early twentieth century (Teixeira, 1609) of an elongated cycloid and its loops, with the generic circle to the usual cycloid that is rolling without sliding on an axis (Pascal [1693]1992; Teixeira, 1609). *Source* Bibliothèque Nationale de France—Gallica. Public domain

5 About Discoveries

We are now ready to say more about the conditions for various discoveries on the curve.

5.1 What About the Uses of Such a Curve?

Was the question about usefulness completely avoided by Jacob Bernoulli? He restricted himself almost entirely to mathematics, except at the end of his speech, where he raised the question of utility for ordinary life and referred to Huygens and his invention for the regulation of clocks.

> But above all, the inventions later made by the very ingenious Huygens, are worthy to pass to posterity, because of their far greater utility in ordinary life. He had first adapted the use of pendulums to the mechanisms of clock-making, but had remarked that, under all circumstances, the equality of oscillations could not be obtained, since it was observed that long elongations were slower than the short ones. He began to find out on what curve the pendulums should oscillate so that their vibrations could be made in exactly the same time, and he found that our cycloid enjoyed this property. But as there was still another difficulty, that the pendulum may oscillate along this curve (while the pendulum wire, when it is stretched straight, can give the pendulum no other movement than circular), he continued to seek what curvature of the wire, moreover, enabled this to be obtained. And he again found that this curvature was cycloidal. To this end he taught the artifice of suspending the pendulum between two plates of cycloidal form, in order to obtain, by their alternate winding and unwinding, a precisely equal movement of the mechanism. (Bernoulli [1701]1993)

There is no change in the tone of Bernoulli's speech while he gave enough technical details, straight from Huygens. Historians, and more often than not English speaking ones, consider that this link between theoretical science based on mathematics and technology is due to Newtonianism, and is the direct effect of the diffusion of a general mood from the English hero, at the Royal Society and beyond through coffee houses or public lectures (Ellis 2004). With Bernoulli, we see diffusion of science in a university setting, by somebody who cannot be seen as a Newton propagandist. The impulse is attributed to Huygens, who certainly tried to diffuse the same spirit while he was responsible at the Academy of sciences in Paris, and may also be considered as one of the vectors for what Paul Hazard described as a "crise de conscience". If Huygens had written how he had linked his mathematics to applications, there was nothing analogous to coffee-houses. It seems clear that the diffusion process was not the same, and perhaps one has to differentiate between the interest that may arise for cosmological interpretations like that of Newton, and clock regulation. In fact, Bernoulli is mixing two different texts with almost the same name by Huygens. One is the technical text in 1658 with the pendulum after the challenge organized by

Pascal and the prize refused to Wallis, and the other is the 1673 text that is theoretical from a mathematical point of view, leading to differential geometry with involutes and evolutes. A completely different utility of the cycloid was examined by Torricelli himself in 1644, when he described as trifles his own achievements to be soon published. This was somehow a typical posture of modesty, sometimes used in the republic of Letters, but certainly not able to attract large audience.

> It is true that I ought today to happily judge the fate of the trifles I wrote, and to no longer count them for nothing henceforth, since they have seemed to you worthy to bear your judgment and to be enriched by your observations. (Letter from Torricelli to Roberval, October 1st, 1643)

As already said, on mathematical grounds, the quarrel became serious in the sixties of the seventeenth century, due to Blaise Pascal and his challenge about the "roulette", involving this time the British John Wallis, and many others like Christiaan Huygens. This scientist was invited to Paris from 1666 to 1681 by the *Academy of sciences*. At this time, Roberval was still living, but he did not react in a public way, even once the foundations of Academies of science in Paris and in London gave him the possibility to do so in a "normalized" way. However, Roberval certainly was behind Pascal's *Histoire de la roulette*, even if not in an open way. Pascal's challenge was confined to a very restricted audience. It was a posthumous and academic publication of Roberval, fifty years after the first intervention, which once more revived the question. A letter sent in 1643 by Torricelli to Roberval was so published in 1693, certainly due to the influence of Fontenelle, soon to become permanent secretary of the institution created in 1666, when precise regulations, not exactly convivial, were made for publications, priorities, and the work of academicians. Such a regular means of publications like the *Histoire et Mémoires de l 'Académie des sciences* only began in 1699. Around 1700, the completely new context was that of academies of science. They were less rivals than well informed of each other. Priority issues were indeed very important, but it was mainly the time to specify how scientific issues had to be settled. In other words, should not we also reverse the sociological question: was the history of the Torricelli-Roberval quarrel a part of the discussion about having an academic organization for science? I do not mean that there were no other reasons, but I am precisely aiming at the role of actors as historians of a quarrel in the definative organization of academic life.

The manner in which Etienne Montucla discussed Torricelli, when he published his rather popular *Histoire des mathématiques* in 1758, is somewhat different from what Jacob Bernoulli did, as he lived in a very regulated academic sphere. Montucla thus wished to judge Roberval's as well as Torricelli's work, and so he was correcting Jacob Bernoulli's discourse on how things historically happened. In the same way as Bernoulli, he was interested in showing how new mathematical Galilean science was transmitted to France, before its installation in England and in Germany. So he was also interested in nationalistic biases and national styles, and one of his efforts was to correct bad judgements from Wallis on Roberval. At the same time, like Bernoulli, he was trying to understand the motivations for Wallis and Pascal to have written as they did.

Montucla was already rather hard on Roberval:

> M. de Roberval never had the art to express ideas with clarity and precision. (Montucla 1758, Part. IV, Liv. I)

But later, in 1799, and we may suppose that this was written by the astronomer Lalande, the judgment was altogether about an insensitivity of Roberval to anything outside mathematical proofs. He was supposed to have said after hearing some piece of theatre, a tragedy: What does it prove? Then the text goes to evoke a certain kind of asceticism of scientists in the seventeenth century, not far in fact from Merton's thesis.

> 3°. Finally, it may be said that at that time, when knowledge was more in depth than superficial, scholars were for the most part very concentrated in the circle of their studies, and extremely new on any other object.[82]

The reason might have been the lack of culture of Roberval! It is meant in the sense of the end of eighteenth century, that is for Lalande a lack of secularism, then we see the opposite to what Merton was trying to explain. Lalande was immediately attacking Pascal, well known as a strict Jansenist:

> But if we find Roberval ridiculous, because of his insensitivity to poetry, what shall we say of Pascal, of the celebrated Pascal, who has never seen anything in it more than in a bouquet to Iris. (Montucla 1758, II)

Therefore, Montucla was doing what I tried as well, that is to explain the context for his own secondary sources. Unlike Jacob Bernoulli, Montucla did not want to escape from the long view of history and the arrival of Calculus. He was even to

---

[82] "3°. Enfin, l'on peut dire que dans ce temps, où le savoir était plus en profondeur qu'en surface, les savants étaient pour la plupart fort concentrés dans le cercle de leurs études, et extrêmement neufs sur tout autre objet" (Montucla 1799–1802, II, Part. IV, Liv. I, 1799, p. 51).

provide an explanation of Roberval, which enabled him to understand the future lines of Newton's fluxions, while Torricelli, by subscribing to the indivisible genre, would presage the way of Leibniz. We thus find the great quarrel of the eighteenth century, that between the Leibnizians and Newtonians. Moritz Cantor, who published before 1900, devoted a whole chapter to the quarrel between Torricelli and Roberval (Cantor, pp. 876–891). Thus, we may as well speak of a quite different context when Ettore Bertolotti wrote in 1947 about Torricelli enlightening the University of Bologna with some kind of Calculus results, once Loria and Vassura had edited, beginning in 1919, Torricelli's *Opere*, with a fourth volume occurring in 1944. Just before most papers from Torricelli disappeared in the bombing of Faenza. I give some figures (Figs. 32, 33, 34 and 35), with just some comments in the legends, to show various forms of drawing. In addition, we end (Fig. 35) with the astonishing representation of the cycloid in 1624, in a thesis directed by Gregorius of S^{to} Vincentio, such that one may think there is an influence of Galileo in this drawing showing an experiment as well as the form of the cycloid. We are back to the questions of how organized experiments played a role in the fashioning of new mathematics, and this is a point where Mersenne and Roberval diverged.

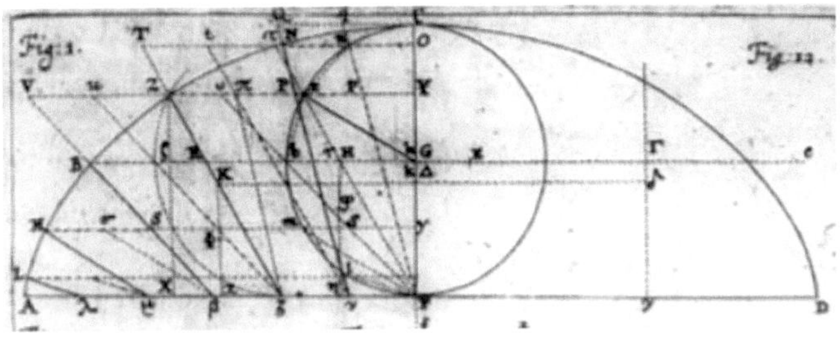

Fig. 32 A figure from John Wallis in his De Cycloide, 1659, using both a classical and baroque frame of reference to the curve, not restricted to half an arch, but with no extension outside the arch (Wallis 1659 [figures in the end of book from p. 123]). *Source* Bibliothèque Nationale de France—Gallica. Public domain

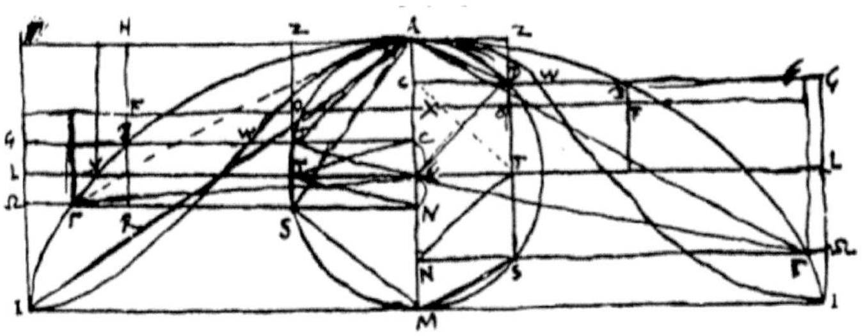

There is no generic position of the generating circle, but the verse sine function introduced by Roberval is drawn with the diagonal line of the rectangle.

Fig. 33 A manuscript figure drawn by Huygens around 1659 (Huygens 1920, XIV, p. 352). *Source* Bibliothèque Nationale de France—Gallica. Public domain

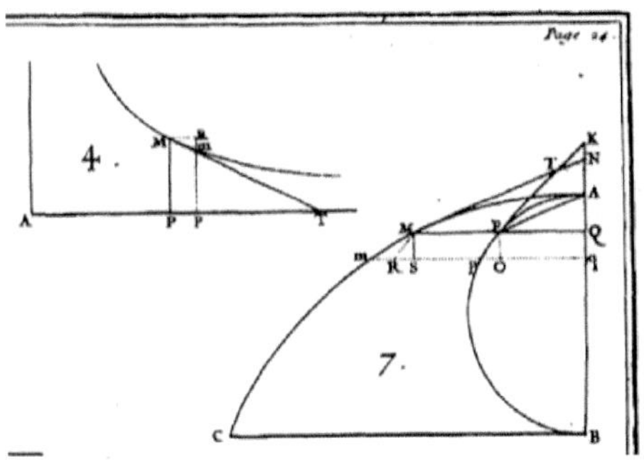

One sees the baroque frame of reference for the cycloid, and also the computation required to draw a tangent to the cycloid at point *M*, with the composition of two increments, *RS* due to the circle (first reference), and *mR* due to the other axis of reference. This is the way to geometrically represent a function of function. This finally gives tangent *MT*, afterwards found parallel to the chord *AP* in the circle. One has to notice that the last construction *mRM* may be related to Roberval's construction as seen in III. 3.

Fig. 34 A figure appearing in the first plate in *Analyse des infiniment petits* in 1696 (de l'Hôpital 1696 [figures in the end of book from p. 123]). *Source* Bibliothèque Nationale de France—Gallica. Public domain

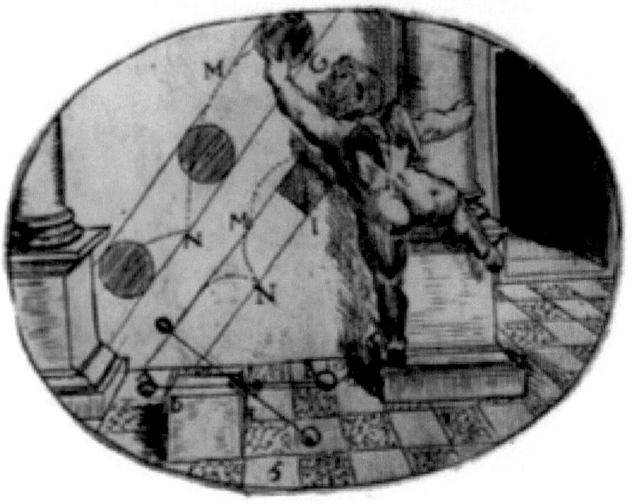

They are theses defended at the Jesuit college of Louvain in 1624. The putto brings down the ball, or rather indicates a movement; the text gives no name to the curve that is seen drawn and had a bad arrangement of the cusp tangents.

Fig. 35 An astonishing vignette, certainly the first which gave an experimental mean of drawing a cycloid (Gregorius of Sto Vincentio, Theorem 8, *Theoremata mathematica scientiæ static*). *Source* Courtesy of Brussel's Royal Library

5.2 Representations and Frames of Reference

It is now the place to discuss various ways of constructing the cycloid and its representations.

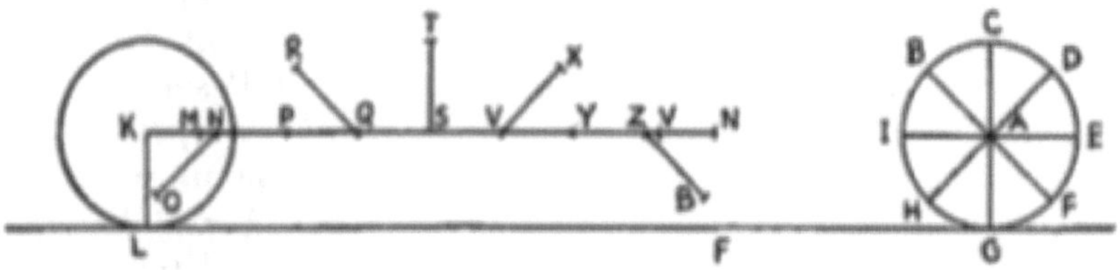

This figure will be almost identically repeated in the Roberval's papers, posthumously published in Paris in 1693.

Fig. 36 Figures that may be found in a letter by Roberval to Mersenne in January 1637, with the circle used in the right place (Letter by Roberval to Mersenne, January 1637). *Source* Bibliothèque Nationale de France—Gallica. Public domain

I have not up to now indicated the third method Mersenne used for the construction of the roulette, after the experimental one as a registration of a movement and the second one by way of the verse sine curve. This is provided with Fig. 36, and one clearly sees how one generates the curve in *O*, *M*, *R*, *T*, *K*, *V*, and *B*, once the circle (here to the right) has been divided in a certain number of angles. In this graphic method, there is theoretically no need of a trigonometric table, but just a good report of angles. So the circle here is used as a frame of reference for angles. This cannot lead very far.

If I have given precise definitions for the other two kinds of representation of the cycloid, a parametric representation, and a pre-functional one, but have distinguished two frames of reference, classical and baroque, it is certainly not to put forward a clear-cut opposition between analytics and geometry, or mechanics and geometry, or even trigonometry and geometry. These oppositions are however there, in germ if one thinks of the various ways that come after. I chose non-mathematical words to qualify the two frames of reference, because they are very much in tune with the times. If indeed the two representations intervened together and first publicly by Mersenne, I help to perceive by the distinction of frames of reference that the one which made posterity came first, and was slow to be appreciated. The baroque epithet marks in some way the opposition to the classical form, and we find the baroque frame of reference in Leibniz in 1690 as we have seen. But clearly Leibniz went to the classical frame of reference to present his "differential equations". This new path implied the decline of the baroque frame of reference.

I do not pretend that the two frames of reference do not originate from an analytic will: on the contrary, they make use of parametric representations. Any parametric representation expresses motion, rolling without sliding, which it translates into formulas of equality of curved and rectilinear lengths, but also by the figuration of the generating circle in generic position, and sometimes even at the rate of motion in the manner of a cartoon. This is the result of our journey through images (Figs. 23, 24, 25, 26, 27, 28, 29, 30, 31, 32, 33, 34 and 35). The two different frames of reference have, in one sense, the purpose of making the movement disappear to the reader's eyes, and too of diminishing the mechanical origin of the cycloid. The contrast between classical and baroque references does not, therefore, concern movement and immobility, as might be expected from the analogy with classification in history of art, but the way of locating a curve. There is, with the classical frame, the rectilinear orthogonal reference, obviously fixed, which carries within itself the concept of function, but only after some analysis is conducted. The condition to achieve this conception is to accept to look separately at an abscissa and an ordinate, which Leibniz differentials allow, as well as Newton's fluxions. The pre-functional representation was therefore not naturally a fruit of the parametric representation, but it could become so, after some work of which Roberval was so fond. There is, with baroque reference, a partly curvilinear marker (a circle of course), also fixed, which allows the play of a certain skill of calculation, very ancient and without indivisible or infinitesimal, as it can be read both by Huygens and by Boscovich. But also by Pascal, as he prepared the notion of curvilinear integral, whereas the classical reference adapted formidably to the integral of a function from Riemann sums with constant steps. In addition, Huygens directed towards differential geometry as well. One can then have the false impression, looking at the figures used, that the distinction between classical and baroque references faded in the years after 1658, insofar as the generating circle appeared in a generic position, sometimes repeated several times to simulate movement (see ill. 34). It is better to look analytically, that is, to grasp what allowed calculation and was reference. The decline was to come later.

The play with the circle was there from the origins, due to the parametric representation resulting from the text of Mersenne himself, which, after these considerations of posterity, makes it possible to return to the two protagonists of our narrative, Torricelli and Roberval. Getting rid of the idea of a uniformity of discoveries, which plagued the narratives of the historians we followed and criticized, Dati and Groening, but as well Bernoulli and Montucla. Independently even of his explicit recourse to indivisibles, Torricelli can trigger the point of view of the baroque frame of reference in that everything referred to the circle, and this is seen by the use of the *arcuato*. Roberval, and this is what I have tried to show in the absence of early seventeenth-century figures by this mathematician, went to the classical representation because it allowed the recognition of the circle from its parametric equations in a rectilinear reference frame and at the end of a work on the coordinates of the cycloid, and not by its direct use in the mechanical generation of the curve.

5.3 Composition of Movements and Robervalian Triangles

It is time now, and on the basis of the *1646 Roberval Declaration*, to discuss the kind of Roberval's inventions for finding tangents to the cycloid. Just the figure alone as provided in 1693 does not allow one to go directly to the *robervalian triangle* at the running point to the cycloid itself. But if, in the very spirit of what Roberval has explained, we try to proceed by searching horizontally at the running point of the cycloid for the base of a *robervalian triangle* by subtracting from the *TV* length of one of the *robervalian triangles* another length *SL*, which corresponds precisely to the variation of the length *LM*. Then in this way two components of the *robervalian triangle* of the cycloid, horizontally the length *TV-SL*, and always vertically *TO*, allow by composition to obtain the direction of the tangent to the cycloid. As these different lengths can be expressed from the angle by the trigonometric functions, the calculation can take place here, as in any analytical geometry, with the help of the classical frame of reference. So that the ratio of

$$TV{-}SL\ (= Rd\theta - R\cos\theta d\theta)\ \text{ to } TO\ (= R\sin\theta d\theta),$$

from which vanish the notation $d\theta$, that is

$$\frac{1-\cos\theta}{\sin\theta},$$

provides the tangent of the angle of the inclined direction to the vertical, one of the axes of the classical frame of reference. We should not forget the purpose of all such computations: to find the direction of the tangent to the curve. Two very simple transformations on trigonometric lines

$$\left(1-\cos\theta = 2\sin^2\frac{\theta}{2}, \sin\theta = 2\cos\frac{\theta}{2}\sin\frac{\theta}{2}\right)$$

effectively yield as the value of this ratio

$$\tan\frac{\theta}{2}, \quad \text{so that angle } MCD \text{ (ill.)}$$

is the half of angle θ. Which accounts for the tangent to the cycloid of its bisector property for the angle θ. Or of the parallelism of this tangent to the chord, or even the construction always associated with the name of Roberval, a rhombus on the horizontal and on a parallel to the tangent to the generating circle whose diagonal gives the tangent to the cycloid.

At least one may understand how the sinusoid was being thought of as a periodic function, via the passage of the verse sine curve. Roberval was eager to explain this far later, when he was preparing his documents to be published at the Academy of sciences. It was in fact the first analytic construction of a curve. Moreover, it led to the first public presentation of a sinusoid by John Wallis, precisely working on the cycloid (Fig. 37).

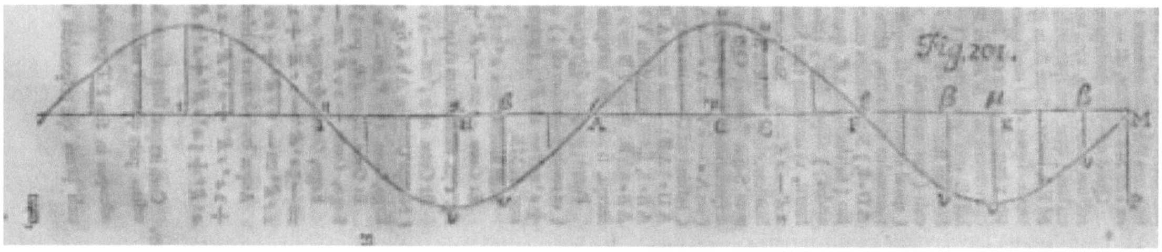

Fig. 37 The graph of the sine function as it appeared in John Wallis (1670, *De motu*) (Wallis 1695, I, p. 883). *Source* Bibliothèque Nationale de France—Gallica. Public domain

5.4 Another Narrative with Serendipity, or the Contrast Between Good Fortune, Recognition of an Invention and Long Search

Even by localizing our story around the 1640s, and taking care of the two representations, the parametric one and the pre-functional one, plus two different frames of reference, presented as a classical one and a baroque one, it is still possible to discuss different regimes for the discovery of unanticipated facts by chance, something which has recently good fortune among historians of science.[83] Without departing from the specific case of Roberval and Torricelli, I am using the concept of serendipity, arising from the literary world and the Middle East with the three princes of Serendip even if the word is British by essence,[84] with a good measure of criticism and also using a sort of opposite concept, that of long and hard study common to discovery.

Serendipity is obvious in 1636 with the unanticipated discovery by Mersenne of a pure mechanical difference between rolling and sliding of a circle on a straight line. It did not help the question of infinity when a family of points are considered, that Galileo had raised in his very famous *Discorsi* (published in 1638, but that Mersenne had perhaps been able to already read in a manuscript form in 1637). Such a question was maintained as a sign that there was always some danger in using reasoning about infinity. But Mersenne's mechanical mention of rolling without sliding solved an aspect of question 24 in the old Aristotelian kind of book entitled Problems on mechanics that have exerted such an influence on scientists and philosophers during the sixteenth century and the beginning of the next. His discovery induced Mersenne to investigate different curves, not only the cycloid as the curve of a point fixed on the rolling without sliding circumference, but also the two other cases for points within the circle, or outside the circle, providing an elongated or restricted cycloid. Nobody would have thought this might bring something interesting, had it not been for the Aristotelian setting that was completely transformed. By various means, and essentially by critically using old narratives, the present paper tried to explore many consequences and at least the two representations of the cycloid. And two frames of reference bearing the names of our two selected mathematicians, Roberval and Torricelli, but only one of them, the classical one finally survived.

[83] Robert K. Merton was certainly responsible for the use of the expression in studies in science, and he added the idea that from somewhat seen or perceived by chance, there is a correct induction to the truth. It had attracted various authors far before the multiplication of specific science studies (Cfr. Taton 1955).

[84] Horace Walpole was the one to coin the word in English in 1754.

Then the explanation by Mersenne, no later than very early 1638, of the parametric equations of the various cycloids, and the determination of a function by its graph in the case of the verse sine, may also be seen as a serendipity effect. In the sense it was not looked for. But the discovery happened at a date that we cannot determine for sure and 1634 is a possibility implying Roberval. But then the presented result was about an area, that of an arch of a cycloid, and did not imply the two posterities that we have just privileged. The verse sine curve, or better said the "compagne de la cycloïde", was not immediately seen as important, perhaps due to the use of trigonometric functions it requires, and the absence of a periodic representation, and this is important for our notion of pre-functional representation. We have good reason to doubt Roberval was aware that it was his main invention, and that there was some generality behind it. For instance, it does not seem possible that at this early date he might have thought that the area was accessible, not of the whole arch, but of any part of it, determined by a parallel line to the axis, as the task will be for Johann Bernoulli, and Leibniz, but then they had Calculus. Serendipity cannot characterize Roberval's true achievements for the area of the arch of the cycloid. In the sense he looked unsatisfied his whole life long and always sought for better proven results, without adapting himself around 1658 when really new results occurred with Pascal and Wallis and Wren as well. We have no hint of why he was interested in finding the area at the very beginning. If I have suggested that it came to him from an interpretation of the parametric representation and the pre-functional one, it is certainly not to affirm this came by chance. There was a lot of work behind it, and certainly Mersenne's explanation about rolling proved essential. Roberval explained in 1646 to Torricelli how he asked Fermat for help about the cycloid, and received this answer.

> I, he says, to invent, I worked; work on your own, because it is through this work that you will grasp that lies the main part about which to rejoice

And immediately Roberval expressed his own hard and conceptual work speaking of numbers, and of his ways to integration, as something essential.

> Encouraged by such a great man, what to do? I worked, and I resorted to the help of our infinites (because I had not yet learned that it was not only ours), and so for the first time I extended those to numbers. In fact, as soon as we know in numbers the relation to their greatest of the sum of powers of the same kind, taken indefinitely in natural order, I noticed that the plane surfaces of the paraboles can be usefully compared to their parallelograms, even the solids to their cylinders, and the spaces of the spirals to their circles. And this in all kinds of powers. That with certitude I iterated without difficulty. Immediately visible was the same sum of squares of all numbers taken indefinitely in the natural order, 1, 4, 9, 16, 25, etc., to the greatest of them taken as many times as there are squares in the sum. Which is to the cube of the very root of this greater square as 1 is to 3, or gives 1/3. The sum of the cubes taken in the same way to the greatest of them taken as many times as there are cubes, or at the quadrato-square, is seen as 1 to 4, that is, ¼; the sum of the quadrato-squares provides the same way 1/5. And so on to infinity.

But what can be said of the work that looks at infinitesimal properties, and about what we called *Robervalian triangles*? It seems difficult to claim it was done by chance. We have tried to describe the work as a true analysis of the parametric representation. And we have quoted his explanation to Fermat. The completely new idea of a length like *OT* free of any determination. I tend to think that the idea of a *Robervalian triangle* was helped by the classical frame of reference.

Descartes' discovery on the role of tangents came by chance, in the sense the problem was asked to him by Mersenne, about the way to construct a curve through its parametric equations. But our formulation is too abrupt, and Descartes' effect was to transform a general question about the form of the new curve, the roulette that Mersenne had explained in the context of the rolling without sliding and of the generalized cycloids. Nevertheless, he quickly solved the elongated cycloid case, showing that the curves have a loop, determining the place where the curve turns backwards (that is when there is a vertical tangent). What he did in 1638, and in a few lines, is precisely what any high school student is to-day doing when working on the graph of a function. Except the student benefits from Calculus and general rules about the functions that compose the parametric representation; all this was not available at the time. For tangents, Descartes went to the mechanical property of rolling without sliding, as it provides a point for which the speed is zero, so leading to a rotation around this point, what we call the instantaneous center of rotation. Moreover, Descartes immediately generalized what he had done with the cycloid to the rolling without sliding of any plane curve over another one. This is an invention, and to speak of pure chance would be to refuse to investigate the effective work of a mathematician. Was it possible for Descartes to have thought that the general rolling of curves one to another was the very general case for kinematics? There are however two natural questions about why Descartes' methods were not immediately diffused, the one about pre-functional representation using horizontal and vertical tangents, and as well the one about kinematic geometry. Could not the kind of aristocratic scorn he manifested against devoting time to prove properties about this curve have been to let people think there was a serendipity effect? Another narrative may intervene here, using a far older letter Descartes wrote to Mersenne at the end of 1630, and so certainly before Roberval had found anything about the roulette.

About the geometric lines that you were writing to me, I dare not exempting myself to speak about it in my Treatise. But consider it is but trifle, and I am really astonished that there may be someone to think others may ignore those.[85]

In the critical edition of Mersenne's *Correspondance*, this remark is immediately linked to the cycloid, which seems dubious in the sense that precisely in 1637 in his *Géométrie*, which must be the announced Treatise, this curve will be rejected for algebraic reasons. Moreover, for sure in 1630 Descartes could not speak about the cycloid as a well-known curve! But if we accept that his remark really applied to tangents to the cycloid, that we know he explained in 1638 along with the form of elongated cycloids by expressing the center of instantaneous rotation, then we would have a trivial explanation for serendipity in this case. In the sense that such problems are just not difficult, and the surprise only comes from the fact that nobody had ever taken the task before to work on it. This however contradicts the long delay of posterity for the two facts we mentioned. It also does raise the question to judging if Roberval was so wrong in specifying 1636 as the real beginning about *Robervalian triangles*. But our question does not take into account of the time it takes to digest new mathematics results when they are not anticipated. One may bring to mind that it took almost twenty years for European mathematicians to grasp what Descartes had done in his Geometry, and this was only due to a collaboration of scientists under the lead of Frans van Schooten. We cannot tell the same story about Roberval's results, as they were confined within his own rooms once Mersenne was no longer there to investigate. Even if some discussions with Blaise Pascal may have been useful in 1658, but perhaps more useful to Pascal than to Roberval. Moreover, when they were published in 1693, because something revolutionary had taken place with Calculus, it was too late for them to have any impact. The only thing we may try is to see some ideas having prepared for Calculus, and that was already done by Montucla when he explained some influence on Newton, or others about the influence on Leibniz and the transmutation process, related to the change of variables in integrals. But this is another story.

Torricelli in his letter of October 1643, where he must already have in mind his book of 1644, the only book published during his lifetime, did not write everything he had been thinking about. His direct answer to Roberval did not concern the "new geometry" of curves, with essentially the generalized parabolas. However he gave information on the "hyperbolic solid". Even though he had already composed a long work on the movement of a heavy body (this work will occupy many pages of his *Opera geometrica*, Fig. 14), which could directly interest Roberval, himself having composed in 1636 a Treatise on mechanics. This is one of the most notable challenges for the historian. How to recreate, even reconstruct, an intellectual atmosphere, if by its very narration he is limited to an aspect of things. As a historian Jacob Bernoulli dealt only with the cycloid. This curve, as everybody knows, results from a movement, and in this case belongs to the natural philosophy in the sense of Aristotle, even to the Mechanics in the sense that this expression had taken since at least Stevin and Maurolyco, and perhaps of a new science which could be called Mechanics in the singular. This did not appear directly in Torricelli's letter of 1643. Nevertheless, Torricelli made a sort of sign to Roberval, indicating that he knew that he had invented a method, different from that of the indivisibles, for the study of the tangent to curves, that it had to do with mechanics, a science obviously still under construction in the 1640s, with a distinction between a Galilean-like dynamic and a Stevin-like static. It could therefore be a subject of debate between Roberval and Torricelli. But if there is no possible effect of serendipity on Torricelli in connection with the cycloid, it is because he has sought the most diverse methods for sustaining Cavalieri's indivisibles. Then there is nothing surprising, or in the order of luck, that he then chose the area of the cycloid to work since Cavalieri established well the calculation of the areas as his objective. It is, on the contrary, his attention to all that can be useful to the indivisibles which becomes the very reason for success. Torricelli could not guess that Roberval had invented a way to construct tangents to a curve like the verse sinus by using his *triangles*.

In Pascal's story on roulette, written in 1658, it is mentioned that Mersenne, aware of Roberval's result about the area of an arch, proposed to different scholars to find it. Pascal wrote that three solutions arrived in 1635, which is impossible, and 1638 would be a better choice. Pascal went much further by assuring that Roberval's provided the best, not only because it was the most natural, but because it gave the possibility to go further.

In 1635, with the new aid[86] (*hoc novi adjuti subsidio*), there were two who gave proofs: their solutions were received almost at the same time, one by M. de Fermat, an adviser to the Parliament of Toulouse, the second of the late M. Descartes; and both of them different from each other, and still more so from that of M. de Roberval. In such a way, however, that in seeing them together, it is not difficult to recognize who the author was, for it is true that it has a peculiar character, and that it is taken by so beautiful and so simple a way that it is

[85] "Pour les lignes dont vous m'écrivez, je ne sçaurois m'exempter d'en parler suffisamment en mon Traitté; mais cela est si peu de choses, que je m'estonne qu'il y ait quelqu'un qui pense que les autres l'ignorent." Letter from Descartes to Mersenne, December 23rd, 1630.

[86] This so-called "new" resource provided by Mersenne to his correspondants is the precise value of the area of the arch. So Pascal in his narrative is attributing to Mersenne the orientation for research not on the form of the curve, but on its area, and with a simple result to be shown.

> known to be the most natural. And it is indeed in this way that he arrived at far more difficult dimensions on this subject, to which the other methods have not been able to serve. (Pascal 1963, 1992, p. 118)

Could we think *Robervalian triangles* are at stakes, or is it all about centers of gravity, or volumes related to the cycloid, about which I have not much spoken, when they are at the center of attention for Pascal. If what we just read from Roberval is a typical judgment from mathematicians who weigh history according to natural elegance, it is still necessary to find the texts that support his point of view. Jean Itard, who has devoted himself to a close study of the texts, can only verify the announcement of the results in letters of the year 1638, but not the details of the techniques involved, and the question of the form itself of the curve, by its tangents for example, had disappeared. It is in any case on April 28, 1638 that Mersenne announced the value of the area of the arch of the cycloid to Descartes (Letter from Mersenne to Descartes, 28th of April 1638), while Fermat spoke about it to Mersenne as soon as February (Letter from Fermat to Mersenne in February 1638). On July 27th, Fermat gave his proof, as well as Descartes,[87] and this is established in Mersenne's *Correspondance*. As for Roberval, we must wait for a letter from Du Verdus to Mersenne of June 3rd, 1644, so that his method may be explained (Cfr. Letter from Du Verdus to Mersenne) . It is therefore understandable that Torricelli in his letter of October 1643, finally made the following recommendation to Roberval:

> Farewell, renowned colleague, do accelerate the edition of your works in the interest of all those persons of knowledge. (Letter from Torricelli to Roberval, 1st October 1643)

It is interesting here to state what Roberval in his *1646 Declaration* was sure to have done, at least as it brings many things together in a synthetic way.

Pascal will explicitly say that Roberval was indifferent to the very idea of publication, "having taken so little care to make himself appear that he never had anything printed" (Pascal 1963, 1992, p. 119) Latin indicates: *tam parvum de fama*. Pascal could be said to be the victim of serendipity. He would have, by means of his previous work with "sums of line" or "sines", the tool that he adapted with genius and generality to the case of the cycloid since it goes beyond the sole consideration of the arch, and address to any portions of the cycloid, and questions of centers of gravity. However, on the history of this curve itself, Pascal did not seem to know much. He was in a sense dazzled by his own results to the point of thinking that he was the only one capable of reaching them. Hence the concurrence ... and the discovery that Roberval had already proved the area of an arch, and that Christopher Wren settled the question never raised before about the length of the cycloid along an arch. The latter obtained the surprising result that this length is eight times the radius of the generating circle, a commensurable result, thus not involving the number *pi*. One can then speak well of serendipity for this reading by Pascal of the work of Wren. To the extent that he immediately altered the terms of the contest he had earlier launched. But we cannot speak of unexpected luck, since Pascal understood that his sums of lines could easily lead to the result of Wren, and much more that we have difficulty evaluating. In any case, he did so independently of the parametric representation or the pre-functional representation.

6 Conclusion

It was in order to select a great number of questions of various scope to which I tried to provide answers, that I used remarks from Jacob Bernoulli' text written in 1701, from Etienne Montucla's comments as well in his *Histoire des mathématiques* in 1758, and from two authors who wrote on the history of the cycloid during the seventeenth century. For sure I used the documents from Roberval and Torricelli. I must emphasise the fact that I was not aiming to criticize the facts those historians gave but to compare these with what modern historiography brought due to the exploration of archives, which I worked at as well, trying to specialize on the original drawings. Bernoulli's document helped to raise general reflections about the practice of history and philosophy of mathematics, as there was a problem concerning two ways, and a possible plagiarism. Is it really possible to copy someone if ideas presented are not understood? We had no information about Beaugrand, but can we not also say that Pascal plagiarized Roberval while precisely writing his History about the cycloid. If I decided to exemplify the role of historians of science in organizing some discourses, which quite often deserve to be deconstructed, it is not to erase such productions, but to better analyse the two original authors from the various posterities that have been attached to them, as well as what has possibly been forgotten. I avoid an encyclopaedic mix of stories and narratives, or even mocking predecessors by just comparing their different views without analysing their motivations, their sources, and their heroes as

[87] Cfr. Letter from Fermat to Mersenne, 27th July 1638, after many other letters.

well. By quoting from historians, perhaps not according to their direct intentions, what I try to arrive at is a panel of motivations about writing a history of the cycloid during the seventeenth century.

Focused on a very particular moment beginning in 1644 about some mild priority issue between Roberval and Torricelli, my questions had been aimed at the possibility of narrative presentations in history of science that are able to cross disciplinary borders and encompass epistemological problems in a far more general scope, like cosmologies and imaginations for a whole century. Is it possible to write on mathematics as inserted in the process of time, without analyzing the content of the questions being raised and possibly solved? The place of theorists in societies like those of the seventeenth Century, with different models, from absolutist France, two times revolutionary England, and a dispersed Italy but united by Counter Reform, is certainly a challenge for a historian of mathematics who does not think it impossible to check the different mathematical problems being discussed, but is not much impressed by what are so quickly presented as solutions, forgetting motivations and even proofs. But I am helped by the huge number of papers, and books across so many centuries. To avoid distractions, I brought only few about Roberval and Torricelli. Even if historians are biased in other kinds of quarrels, at least they were all preoccupied by the possibility of finding reasons for what we may call the "revolutionary" advent of Calculus. Even if it took more than a century from Roberval and Cavalieri around 1644 to Euler's *Introductio in analysin infinitorum* of 1748. So that, in contrast to Bernoulli avoiding Calculus, I introduced many examples, such that the *Robervalian triangles*, or the way Leibniz explained his computations on the cycloid to Huygens who was a master on this curve. It helped to delineate various issues from various authors writing on the cycloid, and so to better appreciate the word "revolution".

I too have tried to disentangle different networks. For instance I discussed the meanings of Mersenne being called an intermediary, and perhaps the refusal by Torricelli to use such an intermediary. An author has the right not to conform to what present social studies taught he should have done. So that I questioned what relations mean between two scientists who never met, like Roberval and Torricelli. Both authors had difficulties with their publications, even if the two held high-ranking university positions and they left a huge number of unpublished manuscripts, including lessons as a new genre of teaching, except most of Torricelli were lost. For this reason I raised the question of the uses of letters and of books to circulate results, even if generally these are of interest for historians of the book only. I also had to take aim at the very possibility of national styles in mathematics, at in period before the revolution brought by the Calculus with Newton (born in 1642), Leibniz (born in 1646) and even Jacob Bernoulli (born in 1654). The three of them could not be participants to the quarrel in 1644.

From an epistemological point of view, the issue was certainly no longer to consider problems as important in themselves in mathematics, but it was to discuss the occurrence of methods, like the ones Leibniz and Newton both had created and about which they were saying almost nothing. It was even to discuss the advantage of a completely different language like algebra, or the better procedure to follow geometrical and traditional Euclidean ways of writing. This was already a point about the so-called Torricelli quarrel, to know if indivisibles could fashion a method, even when everybody wished to replace Cavalieri's indivisibles by "thick" indivisibles, leading as we see them today to Riemann sums. We will not know if Pascal really discussed these things with his Jansenist friends, but while in *Les Pensées* there are many references to mathematics, the cycloid and what Pascal did, never appear. It can be argued that after all Pascal had not been truly brilliant, since the Protestant and amateur Christopher Wren had obtained much more on the rectification of the cycloid, while the Anglican theologian John Wallis was not doing badly at all. The most convincing result of the deconstruction of the so-called Torricelli-Roberval narratives is to establish that the two authors did not have the same representation of the cycloid because of different frames of reference, one classical and the other baroque. They did not have the same posterity. The baroque representation, which played more with the figure than on the analytic, is present in Huygens, but also Leibniz and de l'Hôpital. Therefore, it places the history of the Calculus in new perspectives.

7 Bibliography

References are listed at the end of the present paper. Here the purpose is to provide three kinds of lists of documents; First comes a list of Correspondences, the a selection of references to letters used in the commentary, and last is a list of some references to Torricelli's *Opera geometrica*.

Editions of Correspondences

Le opere dei discepoli di Galileo Galilei, vol. 1, *Carteggio*, 1642–1648, Paolo Galluzzi, Maurizio Torini (éd.), Florence, Giunti Barbera, 1975.

Correspondance du P. Marin Mersenne, religieux minime: publiée et annotée par Cornelis de Waard; avec la collaboration de Bernard Rochot, puis Armand Beaulieu; édition entreprise sur l'initiative de Madame Paul Tannery et continuée par le C. N.R.S, Éditions du Centre national de la recherché scientifique (1932–1988), 18 vols.

Correspondance de Descartes, *Œuvres de Descartes*, Vols. I–II.

Selected References to Correspondences in a Chronological Order

Letter form Jean de Beaugrand to Mersenne, 28 June 1630, *Correspondance du P. Marin Mersenne*, n° 180, p. 504 *et seq.*

Letter form Jean de Beaugrand to Mersenne, 30 July 1630, *Correspondance du P. Marin Mersenne*, n° 180, p. 513 *et seq.*

Letter from Descartes to Mersenne dated December 23, 1630, n° 163, *Correspondance du P. Marin Mersenne*, n° 180, p. 594.

Letter from Roberval to Mersenne, 6 (or 16 January) 1637, *Correspondance du P. Marin Mersenne*, Cornelis de Waard (ed.), vol. VI, 1960, pp. 167–177. This letter has been copied by van Schooten (now at the Library of the University of Groningen, Mss 120, fol. 13 verso-15 verso), so it may explain why van Schooten used it in the first latin edition of Descartes' Geometry in 1649, and was too available by a copy that was preserved in Viviani's papers in Mss Galileinai, vol. III, 99 verso-103 verso (this may explain why Torricelli spoke of an information for tangents provided to him through Viviani).

Letter from Descartes to Mersenne, 29 January 1638, *Correspondance du P. Marin Mersenne*, vol. IX p. 88.

Letter 1 June 1638, *Correspondance du P. Marin Mersenne*, vol. VII, p. 240 *et seq.*

Letter from Fermat to Mersenne, 27 July 1638, *Correspondance du P. Marin Mersenne*, vol. IX p. 397–403.

Letter from Descartes to Mersenne, 23 August 1638, *Œuvres de Descartes*, tome 2, p. 307, etc.

Letter from Mersenne to Pell, 20 January 1640, *Correspondance du P. Marin Mersenne*, vol. IX, p. 91.

Letter form Descartes to Mersenne, 29 January 1640, n° 817, p. 88.

Letter from Galileo Galilei to Bonaventura Cavalieri, 24 February 1640, as quoted by Carlo Dati (pp. 3–4), and reproduced in *Le Opere di Galileo*, t. XVIII, pp. 153–154, also in *Correspondance du P. Marin Mersenne*, t. IX, p. 125.

Letter from Mersenne to John Pell, 7 March 1640, *Correspondance du P. Marin Mersenne*, vol. IX, pp. 181–185.

Letter from Roberval to Fermat, 1640, Letter from Roberval to Fermat, 6 August 1640, *Correspondance du P. Marin Mersenne*, p. 273 *et seq.*; *Oeuvres de Fermat*, t. II (1894), pp. 200–201.

Letter from Torricelli to Jean-François Nicéron (May or June 1643), *Correspondance du P. Marin Mersenne*, vol. XII.

Letter from Jean-François Nicéron to Marin Mersenne about a text by Torricelli, May–June 1643, *Correspondance du P. Marin Mersenne*, vol. XII, pp. 201–209.

Letter from Roberval to Marin Mersenne in July (?) 1643, *Correspondance du P. Marin Mersenne*, vol. XII, p. 252–267, and *Divers ouvrages de mathématiques et de physique par MMrs les savants de l 'Académie Royale des Sciences*, Imprimerie royale, Paris, 1693, pp. 278–282.

Letter from Torricelli to Roberval, 1st October 1643. It has been reproduced in *Correspondance du P. Marin Mersenne*, vol. XII, pp. 328–333 (see also t. XIV, Paris, 1988, p. 4). It is also to be found in *Divers ouvrages de mathématiques et de physique par MMrs les savants de l 'Académie Royale des Sciences*, Imprimerie royale, Paris, 1693, pp. 283–284. This letter in Latin has been translated into French by Jean Itard in his paper of 1975, and is here translated into English. It is also to be found in *Opere* di E. Torricelli, Faenza, 1919, t. III, pp. 148–149.

Letter from Torricelli to Roberval, Ist January 1646, *Opere di E. Torricelli*, Faenza, 1919, t. III, pp. 45–46.

Letter from Roberval to Torricelli, not dated but may be from June 1646, Epistola Aegidi Personerii de Roberval ad Evangelistam Torricellium, in *Divers ouvrages de mathématiques et de physique par MMrs les savants de l 'Académie Royale des Sciences*, Imprimerie royale, Paris, 1693, p. 284–302. This very long and strangely polemical letter had never been translated and studied in great details. It is referred here under the title *1646 Roberval's Declaration*.

Letter from Leibniz to Huygens, 25th, July1690. In *Œuvres de Huygens*, vol. 14, p. 451 *et seq.*

An Extract of a Letter from Dr. Wallis, of May 4, 1697, concerning the Cycloid known to Cardinal Cusanus, about the Year 1450 and to Carolus Bovillius about the year 1500, *Philosophical Transactions*, vol. 19, n° 561, pp. 561–566.

Selected Torricelli References

Torricelli E (1644) *Opera geometrica*, Masse & de Landis, Florentia.
Torricelli E (1919–1944) Opere di Evangelista Torricelli. Loria G, Vassura G (eds). 4 Vols., Montanari, Faenza.
Torricelli E (1975) *Opere scelte di Evangelista Torricelli.* Bellone L (ed). UTET, Padova.
Torricelli E (1715) *Lezioni academiche d'Evangelista Torricelli matematico*, e filosofo del sereniss. Ferdinando II Gran Duca di Toscana, Jacopo Giudici e Santi Franchi, S.A.R., Florence, 1715 [deuxième edition, Silvestri, Milan, 1823].

De cycloid. Torricelli 1644, pp. 85–92, pagination beginning with *De Dimensione parabolae.* In Torricelli 1919–1944, Opere, I, pp. 163–172 [see also Torricelli 1975].
De dimensione parabolae, solidique hyperbolici problemata duo: antiquam alterum [...] cum appendice De dimensione spatij cycloidis & cochleae. Torricelli 1644, separated pagination.
De motu gravium naturaliter descendentium et projectorum. In Torricelli 1644, pp. 95–243.
De solido hyperbolico. Torricelli 1644, pp. 113–135 [see also Torricelli 1975, I, p. 173; English translation by Struik DJ (1986) A Source Book in Mathematics, 1200–1800. Princeton University Press, NJ, pp. 227–231].
Racconto di alcuni problemi proposti et passati scambievolmente tra gli metematici di Francia et il Torricelli nei Quattro anni prossimamente passati. Torricelli 1975, III, pp. 7–32).

Annexes

For reader's convenience, below I have translated into English a selection of texts (mostly written in Latin or in French):

1. Translation of a letter from Torricelli to Roberval, October 1st, 1643.
2. Translation of an appendix on the cycloid in Torricelli's *Opera Geometrica.*
3. Translation of *Observation XI* in Mersenne's *Observations physiques et mathématiques, in Harmonie universelle,* second volume, dated 1637, but this part was perhaps published in 1638 (pp. 24–25).
4. Extracts translated from John Wallis' *De Cycloide*, 1659.
5. Translation of Jacob Bernoulli's Address in 1701 on the cycloid.

Annex 1. Translation of a Letter Sent by Torricelli to Roberval, October 1st, 1643

This letter was included in the various papers from Roberval published in 1693 by the French Academy of Sciences (vol. III, p. 283–285). This letter is also to be found in *Opere di Evangelista Torricelli* (Torricelli 1919–1944, III, pp. 148–149). The text of this Latin letter is also in the volume XII of *Correspondance du P. Marin Mersenne* (pp. 328–333). It has been also translated and commented into French (Itard 1975, pp. 258–260).

To the very renowned Roberval, with the distinguished greetings from Evangelista Torricelli[88]

I want to speak openly to you without any intermediary,[89] renowned colleague.[90] (And who can deny you this title?). And although your letter is addressed to the famous Mersenne, I cannot restrain the impulse of my mind to run to you. I cannot prevent my mind from devoting itself entirely to you as to the Apollo of Geometers. It is true that I ought today to happily judge the fate of the trifles[91] I wrote, and to no longer count those for nothing henceforth, since they have seemed to you worthy to attract your judgment and to be enriched by your observations.

[88] I have introduced some divisions for paragraphs, compared to the edition of this letter as it appeared in Latin in 1693. Because the present annex is not an edition of a series of letters, I freely use the possibility of questioning the meaning of the letter, in relation to my text.
[89] *Sin alio interprete.*
[90] According to Roberval's profession, I so interpret in academic terms the expression *Vir clarissime*, generally used in correspondance between members of the *Respublica litterarum.*
[91] *Nugas meas.*

You first ask me if I consider the center of gravity of the parabola a priori as a solved problem or if, ignorant, I seek it. I indeed would blush to place an unproved theorem in the midst of my other minor propositions, whose proofs were provided. I have proven this theorem according to a single and concise proof.[92]

But on this occasion I have been able to admire the fruitfulness of your mind about so many parabolas and their solids, understood not only from a geometrical point of view but also from a mechanical point of view[93] and so written to finally obtain the knowledge[94] of their measures.[95] On this point, I have, and will never have, anything to say. If, on the one hand, unless I am wrong, I believe that a theorem of this kind requires a very rigorous examination, on the other hand I am not accustomed to dwell on figures that are not usual, nor to solids which, even if they are new, have not at least arisen from plane figures known since a long time. And this essentially because the result of my labors when all has been successful according to the desire of the mind,[96] must also be able to obtain the unanimous approval of cultured men,[97] and that there must be no one to be jealous of figures that I would alone have found.[98]

The measure of the cycloid (for it is the name given to the figure by Galileo, forty-five years ago, perhaps only known today by you under the name of trochoid), came to me without expecting it, and I would say almost without searching it.[99] I then proved it five times, and each time according to different principles. I know nothing about the measure of solids of the cycloid.[100] The tangent of the line had already been shown to me by the then adolescent Florentine Vincenzo Viviani, a pupil of the famous Galileo[101] I suppose that your so penetrating and fertile mind has observed it by itself, being in a sense the inventor of this figure[102]: a line of this kind is indeed familiar in Nature. And it arises from the composition of two movements, rectilinear and circular.[103] In fact to this day witnesses are still to be found, to whom Galileo formerly communicated his unsuccessful labors on this figure. There are even a few pages of the famous mathematician, who already, in his youth, had drawn some figures[104] and initiated attempts to explain the subject. Many years ago he proposed the problem to our admirable geometrician Cavalieri, and told him what he had confirmed to me and to many others, namely that he had once done an experiment. He suspended areas made out of some matter on a small pair of scales to know the relation of the area of the cycloid to that of the circle which gave it birth. The result, by some inevitable fatality[105] I cannot explain, was always less than triple. Hence he had abandoned this enterprise,[106] fearing that there would be some incommensurability. That if the error had been inconstant, if it had sometimes been less than

---

[92] Torricelli's long work on parabolas is *De dimensione parabolae*, and to this text is added the appendix on the cycloid.

[93] Torricelli goes from results on curves to mechanical questions and he well knows Roberval has been working in such a direction. It is not clear whether he thinks Roberval had not been interested by curves as particular objects for the geometer. Moreover, there is no indication that Torricelli was aware Roberval had an analytic perception of the curve, through its parametrization as expressed by Mersenne (Annex 3). We may thus think that Torricelli was making a distinction between problems in mechanics and geometrical issues on curves, perhaps with an agenda regarding rigor. Later in this letter he will not favourably recall the so-called mechanical experiment of Galileo on the cycloid. Could we induce that what was earlier written by Roberval did not imply a direct use of the indivisibles?

[94] *Ad mensuram scientiamque redacta.*

[95] In spite of the mention of mechanics, there is no explanation why "measures" of the cycloid were looked for, and why the form of the curve was not first investigated. But perhaps because Torricelli already knew this form had been explained by Roberval and Mersenne.

[96] *Ex animi voto.*

[97] *Communem litterarum.*

[98] This part of the letter, typical of letters with the *Respublica litterarum*, is certainly written to refrain Roberval from being jealous. What follows affirms Galileo's anteriority on the cycloid. But there is something particular, if not strange, in Torricelli's writing about a conformity between the desire of a mathematician and the reception of his work.

[99] *Misi sese ultro obtulit non speranti, pene dixi non quaeriti.* Did Torricelli mean that his mathematics have noting to do with what Roberval did? My hypothesis is that Torricelli was placing great importance to the method by the indivisibles, and he perceived that Roberval's method, around 1636–1637, was based on something else, which could be a mechanical method. Could he know that this was the analytical character of the curve with function like sine and verse sine? Or did Torricelli mean that he had quite late been interested by the cycloid, perhaps due to Viviani's knowledge of the tangents? What was indeed Viviani's method for the tangent, and had it something to do with Descartes' center of instantaneous rotation as obtained by Descartes in 1638?

[100] Since the beginning of the seventeenth century, with Luca Valerio and Simon Stevin, there had been a strong movement in geometry towards finding centers of gravity of solids generated by curves. Torricelli clearly means here that he was not interested by this aspect in the case of the cycloid. Some pages later he will be looking about such centers for conic sections.

[101] Vincenzo Viviani (1622–1709) went to live with Galileo as soon as 1639, and was then under the scientific direction of Torricelli, to whom he would succeed in the position of mathematician and philosopher to the Grand Duke of Florence. But this happened a long time after Torricelli's death. Viviani would be one of the first members of the *Accademia del Cimento*, and he wrote in 1654 a rather surprising biography of Galileo, that was published in 1717 only. The fact that a letter from Roberval was found among the letters kept by Viviani may leave us to think that Viviani knew about the tangent result from Roberval, and that Torricelli is acknowledging this in the present letter. This is not then Descartes' method with the center of instantaneous rotation, which will be in a way public in 1649 only, when it will be included in commentaries by van Schooten in the Latin version of Descartes' Geometry. But it may as well be the announcement that there was a whole group of mathematicians working together in the tradition of Galileo, with mathematical physics as an objective, as Mersenne was the one to name. Whether it means a group opposite to the larger Mersenne's one in Paris does not seem to be a question. In 1644, Mersenne will be travelling to Italy, and will work in close relations to Torricelli, and others.

[102] *Quo ad auctorem hujus figurae* [...] I prefer to use in English the word "inventor" instead of "author".

[103] The mention of two movements, typical of Roberval's way of thinking, may indicate that Viviani was not aware of Descartes' method using the instantaneous center of rotation with only one movement.

[104] *picturas.*

[105] *Nescio quo fato.*

[106] *Ideo incoeptam contemplationem deservisse.*

the triple, sometimes more, then the perspicacious mathematician[107] affirmed that he would have pursued the research, rejecting the variations on the inequalities of matter and on the deckle edges of the cutting.

I establish the proposition relative to the solid described by any conic section rotating about its axis, and the center of gravity of this same solid in a single and brief proof, which supposes only a moderate knowledge of Apollonius.[108] In truth, I have placed this double theorem among those of secondary interest. It will in fact have no place for it in the little works that I must very soon publish,[109] in which I give the greatest importance to the unity of the subject.[110]

As for the hyperbolic solid[111] which no longer belongs to me - it is yours - I would die rather than conserve the hope of seeing a proof that is remarkable enough and scholarly enough to merit comparison with yours.[112]

I collect to-day the most precious fruit of my work simply because you, so illustrious and brilliant, have deigned to pick up some trifle of mine by such penetrating proofs and with such an abundance of knowledge. I begin by thanking you very warmly. Then, to satisfy to your request, I have to inform you that my method of demonstration for this solid is quite different from yours. One of my ways of access proceeds from the doctrine of the indivisibles, which, if we were always addressing an enlightened reader, might be exposed in few words. The other which proceeds from the inscribed and circumscribed figures in the manner of the Ancients, is not really so quick, but it is simple and may be worthy of curiosity.[113]

I have only found in your writing one thing that fits with my processes: it is the construction to be made in order to divide according to a given ratio a segment of an hyperbolic solid[114] Apart from this similar construction, the proofs are very different.

Besides, I recognize a much more evident hyperbola in the praises you give me than in the proofs in which you measure the hyperbolic solid itself. I would like to be worthy of it one day, so that I may, by reading your works that I avidly await, understand these, take the sweetest fruits of science and the inestimable riches of the mind, and enrich my own intelligence.

Farewell, renowned colleague, do accelerate the edition of your works in the interest of all those persons of knowledge.[115]

Florence, October 1st, 1643

Annex 2. Translation of the Appendix on the Measure of the Cycloid in Torricelli's *Opera Geometrica* (1644)

This text appeared at the end of *De dimensione parabolae*, which has its own pagination in *Opera geometrica* (Torricelli, 1644, pp. 85–92). The text deals with measure of the parabola. It begins after page 243 of the book. It has been also reproduced, first by Gino Loria (Torricelli, 1919–1944, I, pp. 163–169), and—then partially translated into Italian—by Lanfranco Belloni (Torricelli, 1975, pp. 410–415).

In this appendix, I am pleased to add the solution of a not-uninteresting problem which, at first glance, seems difficult if one considers the matter and the proposition. It has tormented the best mathematicians of our century and it escaped them for many years. In fact, the proof was denied to them due to a lack of experimentation in their hands. If one actually weighs on a scale the material spaces of the figure made by hand, I do not know by what fate the ratio,[116] which in reality is three, always

[107] *Linceaeus mathematicus*. There is a play on words as Galileo was a member of the *Accademia dei Lincei*, named from this animal, a lynx, being reputed for his perspicacity. But perhaps this was just a mention of his academic affiliation so as to insist on the reputation of Galileo.

[108] It is sort of usual enterprise during the seventeenth century to try simplifying Apollonius' work on conics, and so to modify the presentation of the great Greek geometer. Newton will give a personal presentation of what he needed from conics in his *Principia mathematica*.

[109] *Quae nunc propalare cogor.*

[110] Should one understand that at this time Torricelli was not sure he will add two appendixes on the cycloid and on the hyperbolic solid in his *Opera geometrica*?

[111] This allusion is to the then stupefying fact that in the rotation of part of a hyperbola around one of its asymptote, while the plane area to this asymptote is infinite, and the center of gravity at infinity, still the volume is finite.

[112] This very rhetorical language means that Torricelli presents his result on the hyperbolic solid as new, not depending on results by Descartes and Fermat in 1638 that were implying integration of powers of one variable. He offers it to Roberval, perhaps for this mathematician to find the relations with those earlier results, or to clarify the possible links.

[113] The proof provided by Torricelli in the appendix of the cycloid, is said to be according to the Ancients (his Theorem 2), because it organized a double proof by contradiction. But the approximation is quickly computed, as if it was already a routine, and that the main idea was to see "thick" indivisibles, or Riemann sums with constant steps, instead of Cavalieri's indivisibles using "all the lines". In fact in 1644, Torricelli will work with a more complicated process than Riemann sums with constant steps, using curvilinear trapezes. Thus here we have to notice this remark about an old method used as an object of curiosity, and not as a warrant of the validity of the method by indivisibles.

[114] Division of figures has been a subject of interest since Dee and Commandino published a text in 1570, commenting an old work by Al-Bagdedi. The question is, given a geometric figure, generally a rectilinear one or a circle or a sphere, to construct beginning from a given point a line or a plane cutting the figure in two parts according to a given ratio. Both Clavius and Stevin around 1605 devoted some long passages to this division, under the alternative of introducing or not some algebraic equations (of second degree).

[115] *Publicam litteratorum.*

[116] The Latin word used by Torricelli is "*proportio*", meaning the ratio of two areas, the area of the arch and the area of the circle. This ratio is found to be equal to three.

appeared less than three. It was then that, suspecting it was due to some incommensurable magnitude (which is what I believe[117]), the solution was despaired of being found, and the search of these mathematicians[118] was interrupted.[119]

The following assumptions are made. We suppose on a straight line *AB* a circle *AC* tangent to the straight line *AB* at a point *A*. And we then fix point *A* on the circumference of the circle. One imagines to rotate the circle *AC* on the fixed straight line *AB*, with a movement that is both circular and progressive[120] towards *B*, so that at each successive moment the straight line *AB* is always touched at a single point, and the fixed point *A* on the circle, after a climb and culmination at *D* descending towards *B*, is found again in contact with the line.[121]

Our predecessors have called this *ADB* line a cycloid, especially Galileo some 45 years ago. The line *AB* is called the base of the cycloid and the circle *AC* the generating circle of the cycloid.[122]

It results from the nature[123] of the cycloid the property that its base *AB* is equal to the perimeter of the generating circle *AC*. And really this is not so obscure. In fact, the entire circumference *AC*, in its rotation, is commensurable[124] to the fixed line *AB*. It is now necessary to find what ratio[125] the cycloidal space *ADB* has to [the space of] the generating circle *AC* (Fig. 38).

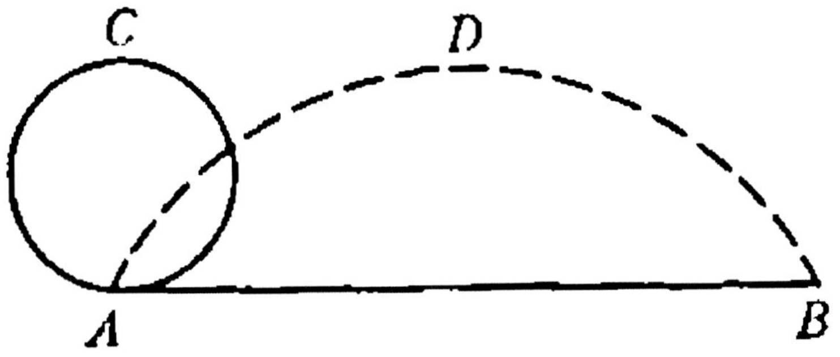

Fig. 38 Torricelli, 1644, *Appendix, De Dimensions Cycloidis*, p. 85. *Source* Courtesy University of Oklahoma, History of Science Collections

[117] Here Torricelli endorsed the explanation for Galileo's loss of interest in finding the area of the arch of the cycloid, but did not know for sure that this was the true reason. The idea was that the ratio to the area of the circle could be an irrational value, close to three, but under three.

[118] It is surprising that Torricelli does not mention here Galileo's name, and prefers to consider that there was an active group of mathematicians looking for results on the cycloid, and particularly for the value of the arch.

[119] *Instituta contemplativa ab illis dimissa sit*. The plural to mathematicians, when usually only Galileo is implied in the anecdote, could mean that Cavalieri was too involved, and even before Cavalieri some members of the *Accademia dei Lincei* were informed as well (according to the Torricelli's letter of October, 1st 1643). Apparently there is no document on the cycloid about this research done at the Lincean Academy. But it should be noted how, from 1643 onwards, the cycloid was regarded as an object to be studied by a group of scientists.

[120] It is clear that Torricelli is not aware of the mechanical description of the movement by Mersenne, rolling without sliding, but adheres to the idea of a composition of two movements, as will be later (?) explained by Roberval.

[121] The curve description is made here by using the mechanical vocabulary with two movements implied, and this is to be compared with the description by Mersenne in 1637 in *Harmonie universelle* far closer to how the graph of a function is described, without mentioning something like a climb or a descent, but an increase or a decrease.

[122] *Genitor cicloidis*. There is no attested text from Galileo that provides this name, but that of "*linea arcuate*" is given in Dati. It is possible that this earlier denomination explains why Torricelli is using "arcuato some lines later for a certain figure, and I translate it by "arcform". One has to notice that the cycloid is not represented as a periodical curve, but just an arch is drawn. Mersenne describes the periodicity, and Descartes in 1638, has drawn the periodic aspect of the curve.

[123] This "nature" comes from the rolling without sliding by Mersenne, and is read by Torricelli as an immediate application of the two movements into which he decomposed the move of the circle.

[124] The emphasis is on the commensurability of the ratio of the two lengths, resulting in fact in an equality. We must recall that the length of the base was not known to be commensurable with the radius of the circle, and this is precisely a more particular problem than that of squaring the circle. Quickly Torricelli has dismissed the question of how a curved length has to be put on a straight line: lengths appear as mere quantities.

[125] *Quam proportionem*.

We will prove, God willing, that the ratio is triple. There will be three proofs, all different.[126] The first and third will proceed from the new geometry of the indivisibles,[127] which I like very much.[128] The second will proceed from the double position,[129] according to the method of the Ancients, and in order to satisfy the partisans of the two methods. Moreover, I warn that all the principles by which any geometry of indivisibles is proved can be reduced to the mere indirect proof of the Ancients. This is what we have done, as in many other cases, for the first and third of the following theorems. However, in order not to exhaust the patience of the reader, we have abstained from many proofs and only exhibit three.

Theorem I

The entire space contained between the cycloid line and the line of its base is the triple of its generating circle. Or else is the sesquialter part (3/2) of the (area of the[130]) triangle, which has the same base and same height[131] [that of the cycloid or diameter of the circle] (Fig. 39).

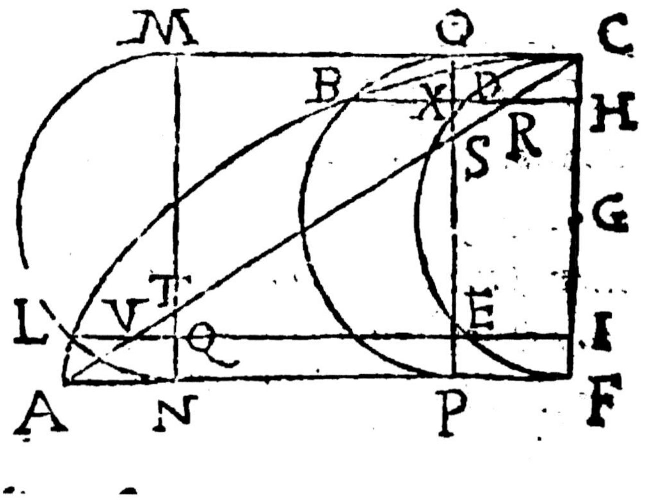

Fig. 39 Torricelli, 1644, *Appendix, De Dimensions Cycloidis*, p. 86. *Source* Courtesy University of Oklahoma, History of Science Collections

Given the cycloid *ABC* described up to point *C* of the circle *CDEF* when it rotates on its base *AF*.[132] (We consider only the half cycloid and the half circle to avoid confusion in the figure[133]). I say that space *ABCF* is the triple [of that] of the semi-circle *CDEF*. Or the sesquialterous part (3/2) of triangle *ACF*.

---

[126] In his letter to Roberval from October 1st 1643 (see the preceding text), Torricelli announced that he already had at his disposal five proofs. His habit is to multiply the number of proofs.

[127] Torricelli did not write "geometry from indivisibles" (*geometria Indivisibilibus*) as in the title of Cavalieri's book in 1635, but "geometry with indivisibles" (*indivisibilium geometriam*).

[128] *Nobis amicissimam.*

[129] The expression is for double false position. It denotes the double reasoning by contradiction that will be named the "deal of exhaustion" by Gregorius à S^{to} Vincentio in 1647, in his *Opus geometricum.*

[130] I will not repeat in parenthesis that question about a triangle here is just its area.

[131] The second manner of expressing the result is a true quadrature, as it reduces an area to that of a rectilinear figure, and so to a square. But in the same way as in Archimedes's Measure of the circle, which is not called Quadrature of the circle, this is not a quadrature by ruler and compass, as one has to know first *AF*, the perimeter of the circle in the second figure, or *AB* in the first one. This book from Archimedes shows in a rigorous way that the area of a circle is the area of a right triangle constructed on the line of a radius and on another line whose length is equal to the length of the circumference of the circle.

[132] In the figure, letters are uppercase letters, but in the original text lowercase. I prefer to write with italicized capital letters, and so V in the figure, written as a *u* in the original text, appears here as *V*. One has to see point *X* as being distinct from point *D*, and *X* is on the vertical line from 0, while *D* is on the horizontal line *BH*.

[133] The restriction to half the figure may be seen as a refusal to look at angles outside the range with positive sine.

Let the two points *H* and *I* be taken on the diameter *CF*, equidistant[134] from the center *G*. Let *HB*, *IL* and *CM* be parallel to *FA*, which pass through points *B* and *L* on the half circles *OBP* and *MLN* that are equal to *CDF* and are tangent to the base at points *P* and *N*.

It is evident from proposition 14 of Book III[135] that lines *HD*, *IE*, *XB* and *QL* are equal. Arcs *OB* and *LN* are still equal. Just as *CH* and *IF* are equal, *CR* and *VA* will be equal by parallelism.

The whole [half] circumference *MLN*, by the given definition of the cycloid, is equal to the line *AF*. Similarly, from the same cause, arc *LN* is equal to line *AN*, since arc *LN* commensurably extends on line *AN*. The remaining arc *LM* will therefore be equal to the remaining straight line *NF*. For the same reason, arc *BP* will be equal to line *AP*, and arc *BO* to the straight line *PF*.

Now the line *AN* is equal to the arc *LN*, or to the arc *BO*, or to the straight line *PF*. Then by parallelism, *AT* and *SC* will be equal. Since *CR* and *AV* will be equal, the remainders *VT* and *SR* will be equal. Thus in the equiangular triangles *VTQ* and *RSX* the homologous sides *VQ* and *XR* will be equal. It follows that, taken together,[136] the two straight lines *LV* and *BR* will be equal to the two straight lines *LQ* and *BX* [taken together]. And this will always be true, wherever we take the two points *H* and *I*, provided they are equally distant from the center.[137] All the lines[138] in figure *ALBCA* will therefore be equal to all the lines of the *CDEF* semicircle. Thus the bilinear[139] figure *ALBCA* will be equal to the semicircle *CDEF*.

But the triangle *ACF* is the double of the semicircle *CDEF*. (In fact, the triangle *ACF* is symmetrical to the triangle used in the first proposition of Archimedes's Measure of the circle,[140] the side *AF* being equal to the [length of the] semicircle, and the side *FC* equal to the diameter, whence it follows that the triangle *ACF* is equal to the area of the whole circle of diameter *CF*). Thus, by adding,[141] the entire cycloidal space will be equal to 3/2 of the inscribed triangle *ACF*, truly triple of the *CDEF* semicircle. What was to be proved…

Lemma 1

Let two semicircles *EIF*[142] and *GAD* be described on the opposite sides of a rectangle *AF*. The figure between the circumference and the remaining sides will be equal to the said rectangle.

A figure of this kind is called an *arcform*.[143] This terminology applies to the parts as well as to the entire figure, whatever might be the division by a line parallel to *FD* (Fig. 40).

[134] This kind of reasoning by symmetry uses the circle, but it is not explained by Torricelli. He seems to be the only one to play with it. Roberval was using another kind of symmetry as proved only by a paper published in 1693, the one which proves that the division by the sine verse function in two equal areas of the rectangle that is not completed here, double as an area of the triangle *AFC* in the figure for this theorem.

[135] This proposition refers to Euclid, book III of the Elements, at proposition 14 (I wrote "cord" instead of "line" in the Euclidean text): "In a circle, two equal cords are at equal distances from the center and those that are at equal distances are equal".

[136] It means that the two lengths are added. This language is not an algebraic one, but algebraic operations are required on lengths, which are not juxtaposed contrary to what is usual with Euclid. It seems useful to write down the various steps of the computation to avoid thinking there is a visual evidence: If one starts from $LV + BR$, and writes $LV + VQ + BR - VQ$, then $LV + VQ = LQ$, and $BR - VQ = BR - XR$ (as $VQ = XR$ in the two equal triangles *VTQ* and *RSX*), and so as $BR - XR = BX$, we finally get $LV + BR = LQ + BX$.

[137] Point *G*, the middle point of *CF* (and of *HI* as well) is not named in the text, and it has even been forgotten in Torricelli's *Opere* as well.

[138] *Omnes lineae.* For sure it is the manner of writing used by Cavalieri in 1635, except that Torricelli made use of two opposite transits through the two divided areas, using the fact that *H* and *I* are symmetric points with respect to *G*. Cavalieri's principle is thus quickly put to use, with no comment, as if it were something quite obvious.

[139] *Bilinearis figura.*

[140] The Latin adjective is "reciprocam", just to explain the symmetry via the diagonal *AC* of the rectangle, or the hypotenuse of the triangle *AFC*. For sure, the allusion is to the famous treatise of Archimedes on the measure of the circle.

[141] *componendo.* This vocabulary was reserved to proportion theory and had a meaning different from addition, but here, without doubt, it means an addition, as in elementary arithmetic.

[142] Letter *I*, on the left in the middle, is lacking on the figure, which unfortunately has been placed with Theorem I in Torricelli's *Opere*, but was correctly placed in the original *Opera*.

[143] *Vocetur autem talis figura Arcuatum.* This word seems to have been invented by Torricelli.

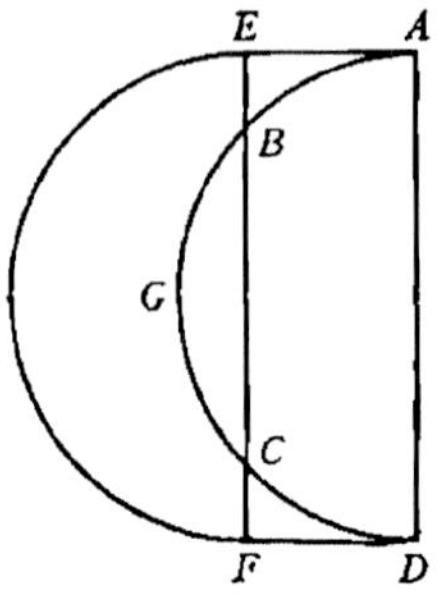

Fig. 40 Torricelli, 1644, *Appendix, De Dimensions Cycloidis*, p. 88.*Source* Courtesy University of Oklahoma, History of Science Collections

A way of proof. Since the semicircles are equal, by removing the common [circular] segment *BGC*, and adding the common trilines *EBA* and *CFD*, the result we wanted to prove is clear. In the case where the circular segment is zero, the proof is shorter and easier. It is also easy to show, with the same prostapheresis,[144] that the arcform, cut by a line parallel to *FD*, is equal to the rectangle, that has the same height and is constructed on the same basis.[145]

Lemma II

Let the cycloid *ABC* be described up to point *C* of the semicircle *CDE* when it turns on the straight line *AE*. We complete the rectangle *AFCE*, and that semicircle *AGF* be drawn around the diameter *AF*. I say that the cycloid *ABC* divides *arcform AGFCDE* into two equal parts (Fig. 41).

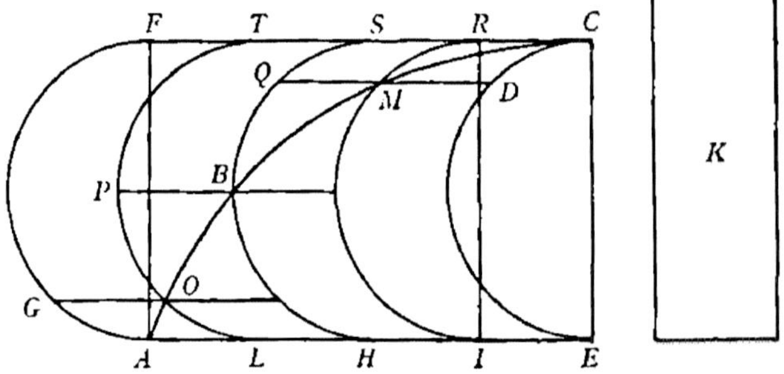

Fig. 41 Torricelli, 1644, *Appendix, De Dimensions Cycloidis*, p. 88. *Source* Courtesy University of Oklahoma, History of Science Collections

If this were not the case, one of the two trilines *FGABC* and *ABCDE* would be larger than half the *arcform*. Suppose that it is one of the two that is greater than half of the *arcform*, for example *ABCDE*. So that the excess of this triline on the halves of the *arcform* equals[146] to the space *K*.

[144] This vocabulary seems strange to us, as prostapheresis meant a trigonometrical formula replacing the product (of two cosines) by a subtraction. Here it just means an addition. Is there a sort of analogy with the operations conducted for the proof of the first theorem?
[145] This last result, a more general case, could easily be proved by Cavalieri's principle due to the transit of the constant length *EA*. But Torricelli is on the process of having a proof according to the Ancients, and perhaps he just wished to show how indivisibles could be replaced with no problem.
[146] As quite usual in this Euclidean method, space *K* is drawn in the form of a rectangular area.

Let *AE* be cut into two equal parts at point *H*, and again cut (in equal parts) *EH* at *I*, and do so[147] until we obtain a rectangle such[148] as *IEC*, which is less than space *K*. Let us divide then [the length] *AE* into parts equal to *IE*. Let the semicircles equal to *CDE* which are tangent to the base at these points *L*, *H*, *I*, and which intersect the cycloid at points *O*, *B*, M. From these points, we draw the lines *GO*, *PB*, and *QMD*, parallel to the base *AE*.

The *arcform OH* will[149] thus be equal to *GL*, while the *arcform BI* will be equal to the *arc form PH* and the *arcform ME* to the *arcform QI*. So that the whole figure inscribed[150] in the triline *ABCDE*, formed inside the *arcform*, will be equal to the figure circumscribed to the same triline, with an excess of the whole *arcform IMRCDE*. If to this circumscribed figure the *arcform IMRCDE* is added, it will exceed the inscribed by an excess equal to the said *arcform*, or to the rectangle *RE*. Thus, by an excess smaller than space *K*. Therefore, the figure inscribed in the triline will be even larger than the triline *FGABC*. But it is equal to the other figure composed of the *arcforms*, and inscribed in the triline *FGABC*. This inscribed figure, in fact, will be greater than its triline *FGABC*. The part will be greater than its whole, which is not possible.[151]

That the figures inscribed are equal is patent. In fact arc *OL* is equal to line *LA*, or to line *IE*, or to arc *RM* by the property of the cycloid. Therefore, the *arcform OH* will be equal to the *arcform MS*, and thus of all the others.

If, on the contrary, it is assumed that the triline *FGABC* is greater than half the *arcform AGFCDE*, both construction of the figure and proof are completely analogous. It is therefore concluded that the cycloid *ABC* is divided into two equal parts by the *arcform AGFCDE*. What was proposed.

Theorem II

The cycloidal space is three times that of the generating circle

Let be described the cycloid *ABC* up to the point *C* of the circle *CFD*. I say that the space *ABCD* is triple of the semicircle *CFD*.

Let the rectangle *ADCE* be constructed, let the semicircle *AGE* be constructed, let *AC* be drawn.

The triangle *ADC* is the double of the semicircle *CFD* (since its base *AD* is the perimeter of the semicircle *CFD* from a property of the cycloid itself). The height *DC* is the diameter of the circle. Therefore the rectangle *ED* will be the quadruple of the semicircle *CFD*.

The arcform *AGECFD* will be the quadruple of this semicircle. By the preceding lemma (II), the triline *ABCFD* will be twice the semicircle.[152] And, by adding, the space *ABCE* will be triple of the semicircle *CFD* (Fig. 42).

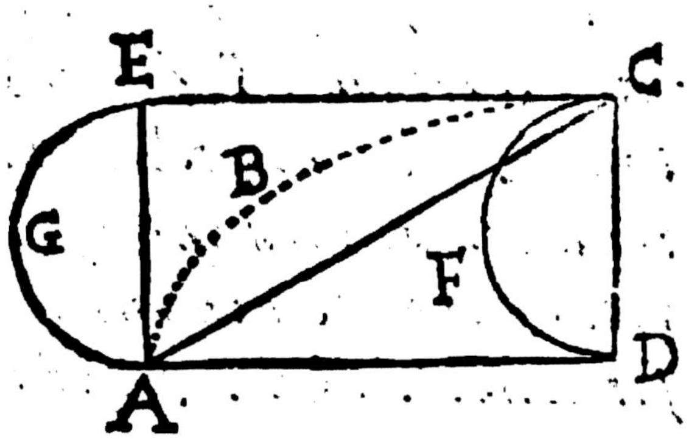

Fig. 42 Torricelli, 1644, *Appendix, De Dimensions Cycloidis*, p. 89. *Source* Courtesy University of Oklahoma, History of Science Collections

[147] *Et sic fiat.* This expression will be used a lot by Gregorius à S^to^ Vincentio, preparing an algorithmization of the method of exhaustion. He will add "semper" between "sic" and "fiat". Torricelli here is making two steps only to get point *I*, and length *IE* that organizes the division of line *AE* in different parts. He so avoids having a cumbersome number of steps.

[148] Contrary to the Euclidean habit, the rectangle is denoted by three points, and not only two.

[149] This time an arcform is denoted by only two points as for an ordinary rectangle.

[150] Torricelli is among the first authors to inscribe figures that are not rectilinear figures, or to say it differently his use of *IE* as a small part is more general to what is usually a Riemann sum. But he uses constant steps.

[151] This is clearly the point were a reasoning by contradiction is made. There seems to have been some strange arrangements in the proof so that it is not so easy to follow where the symmetry is playing its role (for inscribed or circumscribed figures). But is is clear that the play on inequalities is to lead to show that the constant *K* should be less than any quantity.

[152] Letter *G* seems to have been added on the figure just for the possibility to denote the area *EGABCE*, and its equality to the area *ABCFD*. The two areas added lead to the area of the rectangle *AECD*.

Theorem III

Every cycloidal space is three times that of its generating circle.

Let the given the cycloid *ABC* described up to point *C* of the semicircle *CED*. I say that space *ABCD* is the triple of the semicircle *CED*.

One completes the rectangle *AFCD*, then draws the semicircle *AGF*, and takes two points *H* and *I* on the diameter *CD* at equal distances from the center, *HL* and *IG* being drawn parallel to *AD*, cutting the cycloid at two points *B* and *O*. One finally draws, by *B* and by *0*, two semicircles *BPQ* and *MON*, as was done in the previous proposition. *GR* and *OV* being equal and *RO* being in common, then the line *GO* is equal to the line *RV*. Or equal to the line *AN*, or to the arc *ON* according to the property of the cycloid, or to the arc *PB*, or to the line *PC*, or *TH*, or *BS*.

In a manner analogous to that with which we have shown that line *GO* is equal to line BS, we show that all the lines of the triline *FGABC* are equal one by one[153] to all the lines of the triline *ABCED*. Hence the two trilines shall be equal.

Therefore, according to the previous theorem, it is shown that the cycloidal space is the triple of the semicircle *CED*. What was required (Fig. 43).

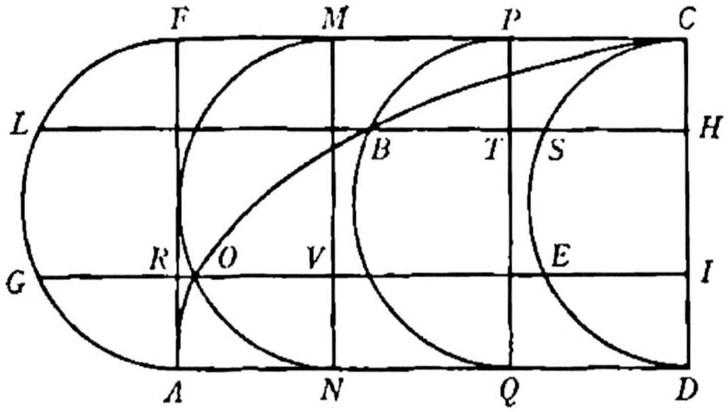

Fig. 43 Torricelli, 1644, *Appendix, De Dimensions Cycloidis*, p. 90. *Source* Courtesy University of Oklahoma, History of Science Collections

Scholia on cycloids of other species

So far we have spoken of the cycloid. It would be inopportune to prolong this discussion by proofs, for an annex would then give birth to a book.[154] Nevertheless, little could be said about this plane figure that it would not be improper to call the *primary cycloid*, as long as an infinity of other species of cycloid are generated from it.[155]

Let us suppose, in fact, that the figure (the first one) is not only the circumference of the circle which rotates uniformly,[156] but also the whole plane extended to infinity,[157] both inside and outside [of the circle]. It is clear that the center of the circle will describe a straight line parallel to *AB*. While a point inside the circumference will describe a cycloid shorter[158] than the

[153] The expression, *omnes singulae lineae*, is not a usual expression for indivisibles by Cavalieri. The principle of the proof, direct and simple, is then the same as the one a Roberval's text is showing in 1693. Except that the figures involved are not the same, Roberval using the graph of the verse sine function. Moreover Torricelli's trick (equality of *GO* and *BS* as soon as *GH* = *DI*) cannot easily work for a part of the arch. Whereas if one is able to work on the verse sine function, Roberval's trick is working for a part as well.

[154] This rhetoric stance may prove Torricelli added rather late the annex on the cycloid to his book. Or that he is imitating Descartes' refusal of multiplying proofs.

[155] In his *Observations physiques et mathématiques*, Mersenne had already discussed such generalized cycloids, and in my essay on the Roberval/Torricelli issue, I tried to explain that this was due to the origin of the curve in view of the treatment of Aristotle's wheel paradox. Torricelli had not discussed this paradox.

[156] *Aequabili conversione rotari.*

[157] The idea of one plane moving on another one seems to be genuinely due to Torricelli, and is not discussed in such terms by Roberval, or by Mersenne. It will be systematized as a geometric tool in the nineteenth century.

[158] *humiliores*. The vocabulary, short or long, is chosen by Torricelli to take care of the comparison of the two perimeters, but it would be simpler to evoke the radius, or distance from the fixed point concerned to the center of the rotating circle, or generating circle. Further on Torricelli will be conscious of this and will speak unfortunately of the "proper generating circle," whose radius is the distance from the fixed point to the center. At least their remains only one base.

primary cycloid[159] *ADB* (as incredible as it may appear), sinuous and not always turning its concavity on the same side.[160] Such curves will be described by points that lie close to the center of the rotating circle *AC*. Points which lie outside the circumference *AC* will in their turn describe cycloids longer than the first, growing indefinitely.[161]

One may call the *proper generating circle* of a cycloid the circle whose circumference is concentric with that of *AC* and passes through the point described by this cycloid. With this convention, all cycloids are instituted on an equal footing. The shortened cycloid has a base greater than that of the circumference of the [proper] generating circle, whilst the elongated has a smaller base.[162]

The relation which each of the cycloidal figures has with its triangle, or with its own proper generating circle, is always greater than unity, and varies at infinity. If we consider together and the triangle and the proper generating circle, the ratio will be one.

The space between a cycloid and its base is to the triangle constructed on the same base and having the same height as the perimeter of its own generating circle and twice the base of the cycloid to twice the base of the cycloid itself. While any cycloidal space is at its proper generating circle as double of the base of that same cycloid and the perimeter of the generating circle to the circumference of the circle.[163]

Here a [new] problem is raised: given any ratio greater than one, to find the cycloidal space, which is either a triangle or a circle, in the given ratio or on a given basis.

*The ratio of a cycloidal space to another cycloidal space (*even if not generated by the same primary cycloid*) is the composition of the ratio of heights and the ratio of the doubles of the bases and perimeters of the generating circles.*

It is certain that the tangent at any point of a cycloid can be found. First to the primary cycloid in particular, then generally, which is also useful for all the others. *The tangent at any point of a primary cycloid is the straight line which runs from the highest point of the generating circle to the said point.*[164]

The tangent at any point of a cycloid is drawn in the following manner. Pass its generating circle through the given point on the cycloid, which at the given point is tangent to a line that meets both the base of the cycloid and another line parallel to it. Like the radius of the circle proper to the radius of the primary circle, the tangent intercepted by the given point and the base, or the line parallel to another line, taken appropriately from the end of the tangent in the said base, or in the parallel line. Thus, from the extremity of this line, let us draw the tangent to the given point of the cycloid.[165]

From this figure we could have deduced not that few theorems, of some interest for those who deal with mechanics.[166] For the time being, we decided to spare the reader the trouble.

The End

---

[159] A reference like *ADB* is to the figure appearing before theorem 1.

[160] Descartes will provide a drawing for the "sinuous" curve of a shortened cycloid, in his letter do Mersenne (August 23rd, 1638).

[161] No figure of these other cycloids is provided by Torricelli, and this unlimited "growth" has just the meaning of a radius increasing. Figures had been provided by Descartes in August 1638 in a letter to Mersenne, dated August 23rd, 1638. They will appear in a printed text in the first Latin edition of the Geometry by Frans van Schooten in 1649. From what Torricelli is writing one has real doubts that he had seen loops for elongated cycloids, that is for cycloids described by a fixed point outside the generating circle.

[162] There is some confusion in the various texts with the terminology about elongated or shortened cycloid, even if one compares the radius or the perimeter of the proper generating circle to the base of the classical cycloid.

[163] I have difficulties in making sense to this paragraph, due to the use of the adjective "double". In the commentary, I have used integral calculus to validate some of the results on the generalized cycloids. These results had already been clearly mentioned by Mersenne in his intervention in the *Harmonie universelle*, in the case of a shortened cycloid only.

[164] The original edition reversed the italicization of the two sentences in this paragraph, and the correct order is restored.

[165] From this presentation we see that the construction is not the one Descartes had given with the instantaneous center of rotation. In fact, a whole tradition will follow, with Huygens, and later with Boscovich.

[166] Perhaps Torricelli wishes to imply that he is only concerned with geometry, and not with mechanics. It is interesting to know that later works on the cycloid, by Pascal for example, will avoid speaking about mechanical description of the cycloid. Implicitly there is here a question about purity of method.

Annex 3. An English Translation of the Original Two Pages from Mersenne About the *Roulette* in *Harmonie Universelle*, 2nd Vol.

This text appeared in the second volume published in Paris in 1637 by Pierre Ballard, in a specific part devoted to *Physical and Mathematical Observations* (beginning in his *Harmonie universelle* with its own pagination, pp. 24–25).

XI. Obſeruation.

De la ligne décrite par la reuolution d'vn cercle ſur vn plan droit.

DANS la 9 Prop. du 2 liure des Mouuemens, j'auois dit que le cercle ou le globe roulant ſur vn plan horizontal, & faiſant ſon tour entier, décriuoit la moitié d'vne Ellipſe, parce qu'en effet la ligne eſtant décrite mechaniquement, comme j'ay dit, approche ſi pres de la moitié d'vne Ellipſe, qu'il eſt malaiſé d'en apperceuoir la difference à l'œil. Mais lors qu'on examine cette ligne geometriquement, elle n'eſt ny Ellipſe, ny Helice, ny Quadratrice, car elle eſt entredeux, & ſe peut décrire en pluſieurs façons, premierement comme j'ay dit dans ladite 9 Propoſition : en ſecond lieu, par les ſinus verſes & droits du cercle qui décrit la ligne ; car ſi on éleue les ſinus verſes ſur la ligne décrite ſur le plan horizontal par la circonference du cercle, à laquelle elle eſt égale, & que l'on applique tellement aux ſommets des ſinus verſes les ſinus droits, qu'ils ſoient tirez à coſté gauche, & paralleles au plan, la ligne courbe décrite par le cercle, paſſera par toutes les extremitez des ſinus droits, comme par autant de points, qui monſtrent la maniere de la décrire. La 3 maniere de la décrire a eſté trouuée par vn excellent Geometre, par le moyen d'vn cercle diuiſé en tant de parties que l'on voudra, par l'entremiſe de pluſieurs diametres, dont l'vn eſtant vis à vis de la ligne du plan, les autres ſe trouuent tellement diſpoſez à l'égard de ladite ligne droite du plan, qu'autant de lignes droites, comme il y a de ſemidiametres décrits dans le cercle, eſtans tellement appliquées deſſus ou deſſous la ligne du plan, qu'elles ſoient paralleles, & égales auſdits ſemidiametres, les extremitez de ces lignes marquent les endroits, ou les points, par leſquels cette ligne courbe, laquelle on peut nommer *roulette*, doit paſſer.

XI. Observation

About the line described by the revolution of a circle on a right plane

In the ninth proposition for the second book on movements, I had said that the circle under the globe rolling on a horizontal plane, when turning a whole round, described half of an ellipse.[167] Because in fact the line is, as I said, mechanically described so close to half an ellipse that it is difficult to visually perceive the difference. But when we geometrically[168] examine this line, it is neither an Ellipse, a Helix, nor a Quadratrix, for it is in between, and can be described in several ways. First of all, as I said in the ninth proposition.[169] Secondly, by the right and verse sines of the circle which describe the line. For if the verse sines are elevated on the line described on the horizontal plane by the circumference of the circle to which it is equal, and the right sines are so much applied left and parallel to the plane to the tops of the verse sines, the curved line described by the circle will pass through all the ends of the right sines, as by so many points, which show the manner of constructing it.[170] The third manner of describing it has been found by an excellent geometer,[171] by means of a circle divided into as many parts as one may wish with several diameters, one of which is opposite to the straight line of the plane and the others are so disposed with respect to the said straight line of the plane, and so many straight lines as there are semi-diameters described in the circle, being applied in such a way above or below the straight line of the plane that they are parallel and equal to the said diameters, the extremities of these lines mark the places, or points, by which this curved line, which may be called *roulette*[172] must go through[173]

[*Source* Bibliothèque Nationale de France—Gallica. Public domain]

(continued)

[167] This is to be found in Marin Mersenne, *Harmonie universelle*, 1637.

[168] The adverb "geometrically" is a direct translation of the Latin "*geometrice*", meaning the kind of rigor required in Greek geometry, but not implying the use of figurative representations; in fact Mersenne is using it for two characterization of the curve, but not necessarily for all constructions. By naming a globe and a plane, Mersenne is using the three dimensions of mechanics, but he reduces the appearance of the curve to a plane: the construction is rigorous, but not in the sense of Euclid.

[169] This mode of description is experimental: if one revolves, a circle on a straight line while maintaining a pen on some hole made in the area of the circle.

[170] The verb used is in fact "describe", that Mersenne frequently employed but in many senses. Therefore, I preferred to use the construction meaning. As it is an analytic one, so that every point of the cycloid can be constructed, with some approximation, by using a numerical table of sines and verse sines.

[171] It is clear from what follows that this geometer is Gilles Personne de Roberval.

[172] It is the first mention of such a name, roulette, and with good reasons we may think that Mersenne is responsible for it. We do not have direct information about what Galileo had chosen as a name for the curve he claimed to have been studying before 1600. Dati, in his History of the cycloid in 1663, affirmed that Galileo spoke of "*linea arcuate*" in a letter dated 24 February 1640 to Cavalieri. Perhaps this explains why Torricelli has used the expression *arcuato* for a special curvilined parallelogram.

[173] The third construction provided by Mersenne corresponds remarkably to a drawing found in the papers by Roberval as published in 1693 (Roberval, 1693d).

Physiques & Mathematiques. 25

Or entre plusieurs proprietez de cette ligne, celle que le sieur Roberual a rencontré, est fort remarquable, à sçauoir que l'espace compris par la ligne droite égale à la circonference du cercle, & par la ligne courbe, est triple dudit cercle qui décrit la ligne courbe: de sorte que celuy qui pourra quarrer cét espace, c'est à dire, qui luy trouuera vn quarré égal, aura quarré le cercle. Or les autres points pris en la circonference du cercle, & qui ne touchent pas le plan, décriuent vne portion de cette mesme ligne: mais chaque point plus proche du centre décrit vne ligne differente, quoy que sa base soit égale, n'y ayant que les seuls points qui sont dans l'axe d'vne sphere, qui décriuent des lignes droites, lors que l'vn de ses grands cercles roule sur vne ligne droite.

Quant aux cercles de differente grandeur, ils décriuent tousiours vne ligne courbe semblable, lors que les circonferences des plus grands ont mesme raison à leurs lignes droites, que les circonferences des moindres à leurs lignes droites, & non autrement, comme il arriue aux cercles, dont parle Aristote dans la 25 Question Mechanique, dont le plus grand porte le moindre, ou est porté par le moindre: car pour lors la ligne courbe faite par l'vn, n'est pas semblable à la ligne courbe faite par l'autre, quoy que la ligne droite faite par leur roulement, soit égale, & que la maniere de les décrire toutes par les sinus susdits trouuez par le sieur Roberual, serue également pour tous. Mais lors que la circonference est sous-double de la ligne droite, l'espace compris par la ligne droite & la courbe est quintuple du cercle qui décrit lesdites lignes: si la ligne droite est deux fois & demie plus grande que la circonference, l'espace est sextuple du cercle, & ainsi des autres. En diminuant, la ligne droite estant moindre de moitié que la circonference, l'espace est seulement double du cercle, sans qu'il puisse iamais arriuer que l'espace soit égal au cercle, lequel il surpassera tousiours, encore que l'on diminuë tousiours la ligne droite iusques à l'infiny.

Physics & Mathematics

Now, among the several properties of this line, that which the Sieur Roberval has encountered is very remarkable, namely that the space comprised between the straight line that is equal to the circumference of the circle, and the curved line, is three times the circle which describes the curved line. So that anyone who can square this space, that is to say who will find an equal square, will have squared the circle.[174] Now the other points taken on the circumference of the circle, which do not touch the plane, describe a portion of the same line.[175] But each point closer to the center describes a different line, although its base is equal, having only the points that are on the axis of a sphere, which describe straight lines, when one of its great circle rolls on a straight line[176]

As for the circles of different magnitudes, they describe a similar curved line when the circumferences of the greatest have the same ratio to their straight lines, and no other, as the circumferences from less to their straight lines, and not otherwise.[177] As it happens to the circles of which Aristotle speaks in the 25th Mechanical Question,[178] the greatest of which carries the least, or is borne by the least. For then the curved line made by one is not similar to the curved line made by the other, although the straight line made by their rolling[179] is equal, and the manner of describing them all by the aforesaid sines found by the Sieur Roberval, also serves for all.[180] But when the circumference is sub-double of the straight line, the space comprised between the straight line and the curve is fivefold the circle which describes the said lines. If the straight line is two and a half times greater than the circumference, space is sixfold the circle, and similarly for all others. When diminishing, the straight line being less than half the circumference, the space is only double of the circle.[181] It can never happen that the space will be equal to the circle, which it will always surpass, even if one always diminishes the straight line to infinity

[*Source* Bibliothèque Nationale de France—Gallica. Public domain]

[174] Recall that to formally square a space requires to construct the side of the square by ruler and compass from the data.

[175] I am not too sure what Mersenne wished to determine here, just perhaps that the cycloid is the same curve by translation, even if it is not an identical curve for the various fixed points on the circumference of the circle.

[176] It is important that Mersenne felt concerned by such other curves of the cycloid kind, as it provides a link to mechanics. A year later, those curves will be drawn in a very long letter from Descartes to Mersenne (23 August 1638, see Ms BNF, fonds fr 5160, fol. 13–20 and *Oeuvres* de Descartes, 1996, II, pp. 307–343). Perhaps Mersenne is trying to say that the portion of the straight line on which those curves are observed (the definition set if we speak in function graph terms) remains the same, having the length of the perimeter of the rotating circle, or generating circle as Torricelli will later call it (see Annex 2).

[177] Such remarks on similar curves became frequent for conics at the end of the sixteenth century. They seem new for a curve like the cycloid, and express the dependance upon the radius of the generating circle.

[178] Sometimes numbered 24, this Question in *Problemata mechanica* attributed to Aristotle (that I prefer to call Problems on mechanics, as in the Latin usual version, *Quaestiones mechanicae*), expresses the paradoxical question of how shorter or larger circles than the one rolling without sliding on the straight line behave on the horizontal line to which they abstractly maintain a punctual contact. This is the so-called Aristotle's wheel paradox, on which Galileo was to discuss in the first pages of his *Discorsi* in 1638. In a strange way, Mersenne who had earlier in the mechanical part of his book explained the difference between rolling without sliding, did not try to use here again this vocabulary. But one may be sure that he very well grasped the way he had solved the old riddle.

[179] Mersenne only used here the verb to roll, even if he earlier made a difference between rolling without slipping, and slipping.

[180] It is clear that Mersenne claimed the Roberval's analytical method using verse sines and sines for finding the area of the different shortened or elongated roulette is the same as the one used for the regular curve. But it is also clear that for elongated roulette, that is a roulette with a base larger than the perimeter of the rotating circle, the curve comes backwards and it presents a loop. So in this case the meaning is not clear of what area is being measured, and from which base. When the roulette is shortened, that is with a smaller perimeter, then the notion of space under the curve is meaningful, as soon as one makes precise the base. We then may easily reconstruct the choices being made so that the values of areas make sense.

[181] See Sect. 3.3 for an explanation (and a verification) of the results here described on shortened cycloids.

Annex 4. Partial Translation of the *Tractatus duo* of John Wallis, Published in 1659 and Concerning the Torricelli-Roberval Quarrel

This text provides Wallis' answer to Pascal on the subject of the cycloid. It is entitled *De cycloide*, was published in 1659, and also occurred in volume I of Wallis' *Opera mathematica* (Wallis 1695, pp. 542–545; Scott 1981). In the preface to *De Cycloide*, Wallis had already explained what he called the Torricelli-Roberval quarrel, then paralleling it to the treatment that Pascal had reserved to him in 1658 for the challenge for the roulette. Here he made precise that if he knew the result on the surface of the arch of the cycloid, from the published work of Torricelli in 1644, he still was unaware of any other Roberval's work on this subject. In the preface written in the form of a letter addressed to Christiaan Huygens, Wallis explains things in more detail. It is clear that Jacob Bernoulli in his speech, which is translated after this one, copied a lot from this text, and the first extract is here reproduced just to signal this.

In this respect, I would certainly have preferred the author of the short Histoire de la roulette, at least in that he says against Torricelli (against Torricelli, for I care less about Lalouvère, since he is still alive and is able to make his own apology), refrains from insulting a man of the greatest merit, who has been dead for many years. Torricelli, in any case, has made himself known to us by his writings as a scientist and a mathematician, and among those who have best served mathematics: moreover, I believe, an honest man. I do not see what might be found in him, which might excite the bile of the most illustrious man, or of Roberval, of whom he takes the side.

Wallis goes on and Jacob Bernoulli will repeat it in his decanate speech, in particular, about Beaugrand's "robbery" of a text by Roberval that Beaugrand had brought to Galileo, and which Torricelli could have seen. It is therefore on Torricelli and Roberval that Wallis continues, and he now speaks about French mathematicians.

On the contrary, they say (this is the argument to which they give the palm), they have a letter written from his [Torricelli] hand,[182] which they still retain as a sort of treasure (as if it were of such a price), where he concedes priority to Roberval in solving this problem. It was then that this honest man, having at last learned what he did not know at the time of the publication of his book,[183] that Roberval had also (without his knowledge[184]) proved the same thing, although he had not disclosed it through print (which he has not yet done, I believe), had recognized it without any bitterness. But, that he had known about it at the time when he published his book, they do not say that he avowed it, and they do not affirm it themselves. They blame him, unless they want to consider it criminal that someone else finds what Roberval, perhaps to reserve the secret of this knowledge, squeezes at home, or likes better to make known to his friends[185] only. As to what concerns Mersenne, to whom, according to them, people deprived of what is given to Galileo, this is based, it seems to me, on a non-existent or very thin foundation.[186]

Wallis goes on about Mersenne and Galileo, and this will be reproduced verbatim by Jacob Bernoulli, up to the passage on the subject of laughing in France. Then Wallis comes to a fairly general reflection.

But perhaps it will be judged that I am stretched to excess on a matter which does not concern me. However may I be forgiven by these illustrious men (whom I respect, as well as their Mersenne) if I might have preferred that they could have used with more parsimony words by which they attack without any necessity the reputation of a man, very learned and who once served great services. I would rather have wished, if they make so great fuss of what they do not wish to allow being ravished by others, that they link it more surely to them, and for the greater advantage of the public utility, they would have published it themselves, so that they would not have to reclaim it for them at the end by invoking a right of return. At least if, while they neglect it, others learn it by other ways, they do not pretend that something wrong was done to them, or they should understand that they have done the wrong to themselves. But let us leave this subject and return to our Dettonville.

Wallis speaks about Pascal, i.e. Dettonville (as Pascal chose to disguise his name), and about the roulette challenge, defending himself from speaking out of spite. He then comes to a general matter.

But only to show by the facts how little it is surprising, on such subjects, that several persons, living in very distant places, and without having informed each other, when they treat the same geometrical matters, also arrive at the same views. And, consequently, they must not accuse each other of plagiarism (especially in the present century so fertile in genius), nor, because one was the first to find, may insult without foundation the other one. For invention depends on intelligence, but the first invention depends on Fortune. And it is not uncommon that with such subtlety or intelligence, one finds afterwards what another (without having it known) had been the first to find.

[182] No doubt that the "secret" document is the letter written by Torricelli to Roberval, on October 1st 1643. It has been published in 1693, after this Wallis' text appeared for the first time, and is translated here in English as the first text of the annex.

[183] This information is false. Wallis does not seem to know that Torricelli heard about properties of the cycloid proved by Roberval before the publication of *Opera Geometrica.*

[184] Wallis has maintained this assertion even for the publication of his Works in 1695, two years then after Torricelli's and Roberval's letters were published in Paris by the Academy of sciences.

[185] There certainly is a critical appraisal of Mersenne's group by using the words "friends", and no conviviality meaning involved.

[186] Wallis ignores, or wants to ignore, Mersenne's remarks in *Harmonie universelle* in the second volume published in 1637, or possibly 1638.

Wallis revisits his method by "inscriptions and circumscriptions" to compare rectification and integration of two curves, the parabola and the spiral. They were an object of a proof by Pascal as well. He concludes on a remark about concision, thus giving the reputation of vain brilliance to the way according to the Ancients. It is a turning point in the use of indivisibles.

> I could have composed the whole work in a much more extensive way, and by seeking more brilliance, which perhaps would have been more agreeable to beginners and less experienced people in these matters. Had I have wished to progress according to the ceremony of definitions, lemmas, problems, theorems, scholia, and to decorate long appendices of constantly linear demonstrations. But you do not less appreciate, I believe, a harmonious brevity, with few words and many things. You who, simply guided by the finger stretched out towards principles, no less grasp the meaning of a succinctly reported proof than if it were extended at length with the greatest solemnity.

Annex 5. Speech by Jacob Bernoulli on the History of the Cycloid

The text in Latin is from Ms UB Basel LI a 749 A3 (6 folios). It has been reproduced in *Die Werke von Jakob Bernoulli*, Bd 4, *Reihentheorie*, with a presentation by André Weil, Birkhäuser Verlag, 1993, p. 254–266. It had already been published in 1927 by Friedrich Weiss, *Archiv for Geschichte der Mathematik, der Naturwissenschaften und der Technik*, Bd. 10, 1927/28, p. 345–355, in the historical movement to recover an image of the production of the Bernoulli family in Basel. It is the first English translation.

> Presentation made on June 26th, 1701 when I had inaugurated my third decanate in Philosophy[187]
>
> When two centuries ago and during the last one, Divine Providence had given mortals the free gift that is the discovery of Typographic Art, the crass barbarity and ignorance of Fine Arts, which infested the preceding centuries, began to retreat little by little and to dissipate, those who dealt with literature worked at first to save from death the writings and works of the Ancients. In which they believed that the treasury of all the sciences (*omnis sapientiae thesaurus*) was to be found. They reconstituted those that were lost, and rectified those who were corrupted. So that scholars employed all their time to perfect editions, versions, interpretations and commentaries, and they had no other occupation throughout the whole sixteenth century. Then as erudition during this period of time was restricted to the bounds of the study of languages and philology alone, positive sciences (*scientiae reales*) were not increased in any way. This increase was reserved for the following century only, the seventeenth. Then, indeed, the secrets of the Ancients had already been sufficiently excavated and toiled so that the work, which was to be consecrated to this task, might now be regarded as useless. And yet nothing was found there that marked a very great knowledge of Nature, or which could satisfy a mind eager for knowledge. Under these conditions, here and there more sagacious masters came, who set to themselves the task of knowing for themselves, and, leaving aside opinions of the Ancients, carefully scrutinized Nature by reading it in its very book.[188] It is, therefore, thanks to them that the sciences were not only recovered and sanitized in their first foundations,[189] but still prolonged in an immense way beyond the antique limits already reached, and daily enriched with new increments.
>
> It would take a long time, gentlemen listeners, to go through the details and make an inventory of everything that can be considered as recently acquired as a discovery in the natural sciences, medical arts, mathematics and other disciplines. By neglecting the dreams of those who consider that there is no price for the knowledge of which there is no trace in the writings of the Ancients. Of course, even suspecting something of that kind in the mathematical disciplines is highly ridiculous, while obviously they made more progress in this century alone than before in a total of two thousand years.
>
> That is why now, since it is now necessary for me to speak from this chair again, I thought that I should not spend more pleasantly this little hour with you, and would not be able to render more useful filial duties to a very learned century who had not too long ago been dying in our arms if, among the innumerable inventions with which he honors himself, I selected one aside, and if I recounted in a brief exposition the history of the very famous cycloid curve. Born at the beginning of the century, grown in its midst, it reached its apogee at the end of this century. But it was also the occasion of the burning brawls between the first among Geometers. While I am going to relate all this, gentlemen listeners, there is no reason why I should fear that you would lend a less attentive and patient ear to that celebrated battle of prominent masters, as you once read the ridiculous Batrachomyomachia[190]!
>
> To-day all philosophers agree that nothing is known of things apart from the various relations which they have between themselves and of which (especially for material things) the principal is that by which they can be compared according to the more and to the less.[191] It follows that, since there is nothing more capable of representing any curve than the so-called coordinates, we can easily guess the importance of the consideration of curves, and we shall not be surprised at seeing that their study mobilizes the whole of Geometry, and that the greatest Geometers have been opposed one another with always a great rivalry of ingenuity.

---

[187] The original text is not divided into paragraphs, and they are added here to help reading.

[188] The expression, *naturae libro*, the book of Nature, seems to be here a direct allusion to Galileo's book *Il Saggiatore*, published in 1623.

[189] *Fundamentis instauratae et repurgatae*.

[190] Βατραχομυομαχία, or the Battle of Frogs and Mice, is a parody of the *Illiad*, that has been attributed by Plutarch to Pigres of Halicarnassus. It is used as a synonym of a silly dispute.

[191] A definition of mathematics, that seemed usual in the late seventeenth century, was that it had to do with what is linked with "the more and the less".

The Cyclois (or Cycloid), which others call Trochoid, the French Roulette[192] (or rather small wheel), is a Line Curve of the number of those called Mechanical, or Transcendant.[193] It is that traced in the air by a nail fixed to the circumference of a wheel that evolves by turning on a plane. It is commonly believed that the one who first discovered it and was the father of this curve is Galileo. Although Wallis thinks that it was well before Galileo, and affirms that it was already traced in a work of cardinal de Cuse,[194] in the collection of the Savilian Library[195] which exists among his manuscripts proper, transferred about 1454 by donation of this cardinal. But it was described wrongly in the printed copies and subsequently, it is said to be mentioned again in about 1520 by Charles de Bovelles,[196] a no less famous French geometer.

In fact, it may be thought that these authors rather pointed to the curve than they actually found and studied it. Galileo was without contest the first of the mathematicians of his time, of a very penetrating and quite divine genius, and he was remarkable both to solve problems and to raise some of a new genre in all disciplines, as he excelled particularly in Optics and Astronomy. The army of the literati was in admiration of Galileo, when with the new mechanism of his tubular machine, he brought the sky closer to their eyes and showed new satellites around Jupiter and new constellations, which he later called "Mediceans".[197] They gave to this master, in every place, such renown and such glory that he was reputed to be the Archimedes of his age. This with other first-rate discoveries is a proof that he could find the Cycloid line anyway by himself (and without the help of Charles de Bovelles whose fame is inferior to his own[198]). Indeed, as for us, we have no intention of discussing the glory of this invention in a case that has not yet been clarified. But precisely one hundred years ago today or one hundred and two years (as Torricelli testifies), Galileo had first invented this curve in Pisa on the occasion of the construction of an arch bridge in this city.[199] He gave it the name Cycloïs, following the example of Cissoid[200] (and yet in conformity with the Greek language, it would be better to say Cycloid than Cyclois). He soon began to study the nature and the properties of this line, and more particularly to look for the measure of the area under the curve.[201] It is quite strange for this inquiry that this penetrating Master was quite tortured, and incessantly embarrassed. In this childhood of Geometry,[202] we will rightly forgive a Master. By instituting a mechanical experiment, he explored the question through the areas of figures that were suspended to a lever. So he had suspected that this area was three times that of the generating circle, but he will never reach the truth by a geometrical reasoning.[203] Finally, when he had sweated blood and water over this problem forty whole years without result, and entirely despaired of any success, he transmitted the problem to Bonaventura Cavalieri[204] by a letter which he wrote to him in the year thirty-nine of the last century, and invited him to solve the knot which he had not been able to untie.[205]

From the skill and genius of Cavalieri, it was hoped that he would satisfy Galileo's expectation and show the desired measure of the cycloidal area, especially since Cavalieri had discovered the first threads for the Analysis of the infinitely small,[206] from which it did not appear that difficult to penetrate such Mysteries. But as well, the measure of the Cycloid was not within Cavalieri's ability. Because of the difficulty, he left aside the measure of this curve and never returned to it thereafter. What Cavalieri himself did not hide in the letter he wrote to Torricelli in 1643. It is thus that, happier in that than Cavalieri himself, the first of all who measured the area of the cycloid, or, if you

---

[192] The word "roulette" is attested in 1637 by Mersenne, in the second volume of *Harmonie universelle* (see Annex 3). The word cycloid is attested in 1644 in Torricelli's *Opera geometrica*. In 1638, in letters, Fermat and Descartes used no specific word for the curve for which they provided results on areas and tangents. Torricelli assumed that Roberval was to know this curve under the name of "trochoid".

[193] In 1637, Descartes explained that he was dealing with some plane curves only, not the mechanical ones, and used a definition with two independent variables polynomial equations. We now call such curves algebraic, and Leibniz used then the word "transcendant" for all other curves.

[194] The famous mathematician, philosopher and theologian, cardinal Nicolas de Cues, had left a large collection of manuscripts, some of them went to Saville's collection. Such explanations were provided in the previous text from Wallis, and he even reproduced figures of the cycloid from this time in Wallis, 1697.

[195] The Saville collection was established by Sir Henry Saville for the private use of persons occupying two chairs of mathematics and astronomy according to his will. Manuscripts in this collection went to the Bodleian Library in 1884.

[196] Charles de Bovelles (1479–1566) was a professor to the college Cardinal Lemoine in Paris, famous in particular for his "Géométrie" in French published in 1511, and also for a "Géometrie practique" in 1547. One has not found a mention relevant to the roulette in such books.

[197] By the use of the telescope, Galileo had seen four of Jupiter's satellites, and named such objects "Mediceans" in 1610, in honor of the ducal family of Florence. His *Sidereus nuncius* was published in 1610. He became mathematician and philosopher of the prince.

[198] Grading reputation seems to be a habit in Bernoulli's family.

[199] It is only through Torricelli's testimony that we know Galileo had the project of building a bridge with an arch having the form of a Cycloid. Moreover, it is difficult to understand how this curve may have had mechanical properties similar to that of the catenary, which at the time was thought to be a parabola.

[200] The word "cissoids" coming from the Greek name for ivy (if one is following the indication by Proclos), seemed to be linked to cycloid because the two curves have cusps. It does not seem that Jacob Bernoulli was aware a cissoid might be seen as a roulette in the case of a parabola rolling without slipping on another one. It seems difficult that this might have been the reason for Galileo to have invented the name for the curve.

[201] No explanation is provided for the reason why Galileo was interested to find the area of an arch of cycloid. For Bernoulli it did not seem more urgent to find the general form of the curve, at least with the determination of tangents, and particularly vertical tangents.

[202] This is certainly an expression borrowed to Carlo Dati's book, except it was restricted to the curve, and is extended by Bernoulli to the whole of geometry.

[203] *Geometrico ratiocinio*: it certainly means using geometry, perhaps the method of exhaustion, but not in general the kind of rigor the adverb *geometrice* would have been implying.

[204] Bonaventura Cavalieri (1598–1647) is famous for his book published in 1635 on the "Geometry of the continuous by the Indivisibles".

[205] If we do not have proofs of Galileo's old work on the cycloid, at least we have good testimonies to his relations with Cavalieri. The quoted letter is dated 1639, after Descartes, Roberval, and Fermat at least had written rather precise letters on the subject to Mersenne.

[206] "prima Analyseos infinite parvum". By first avoiding the word "indivisible", Jacob Bernoulli considers the discovery of the Calculus to be a result of the Moderns along the seventeenth century.

will, unveiled its measure to the world, is Evangelista Torricelli.[207] This mathematician of Florence used these few rudiments of Geometry of the Indivisibles, which Bonaventura Cavalieri had already exposed. Thanks to those, he found that the area of the Cycloid is exactly three times that of its circle. What Torricelli himself, after discovering it, took care to publish by printing the following year in 1644.

However, while all this was going on in Italy, in France Mersenne had proposed the same problem to his French compatriots and to Italians.[208] And on this problem, once thrown as a bone of contention[209] to the crowd, the most prestigious Geometers of this time: Lalouvère, Roberval, Descartes, Fermat,[210] and others, applied themselves with envy and with the greatest rivalry of ingenuity. So that it was not only the area of the Cycloid, but also its tangents, its center of gravity, so even the cycloidic volumes and their centers of gravity, which were provided. Hence, as usual, disputes became much worse, mainly between Roberval and Torricelli. For these scholars were above all ambitious of the title of first inventor of the first properties of the Cycloid, and hardly accepted that one of them should share with the other a part of the honor. They fought miserably, and, what a horror, falsely accused one another of plagiarism.[211] This can be read in the pamphlets published on both sides, and the long letter of protest from Roberval to Torricelli can be found in the *Recueil de mathématiques et de physique* printed in Paris, eight years ago.[212] Into which Roberval's discoveries are contained - some are talented enough, but in some others the author made unpardonable logical mistakes, so that I am surprised the editor had not noticed and suppressed them.[213] Although it does not seem to me worth arguing to find out who first found these Theorems, from Torricelli or Roberval, while what was discovered about the Cycloid at the time of his childhood does not seriously matter (Cfr. Nardi, 1994). But since we know that the quarrel between these Diumvirs was not pushed with less ardor than if it had implied the salvation of Greece, the glory of Italy or France,[214] it will not be out of place to provide some details on this.

A few geometrical problems had been proposed in France by Father Mersenne of the Minims order. He was a lucky man, and never weary to propose problems, less fortunate to solve them.[215] Among other things, the problem of the area of the Cycloid seemed to him very difficult, on which he had perhaps heard that Galileo had formerly been fighting in vain. Being himself little able to penetrate the area measure, he invited famous mathematicians from all over to solve it, having adroitly omitted, as usual, the name of Galileo.[216] But he invited Roberval first, a Royal professor of Mathematics,[217] whose skill he hoped to get a solution among all others. After a certain period of time, Roberval told him that he had solved the problem. But they both agreed to retain this discovery for themselves as they wanted to suggest the same question to other mathematicians, so that if those could not solve this difficult problem, the glory both for the one who had posed it and for the one who had resolved it, could have been increased.[218] Mersenne, happy with this success and made more audacious, challenged the whole universe and arrogantly proposed to the first geometers of Italy and France to solve the problem of the Trochoid (indeed this is how he called it[219]), as if he had been the first person to have thought of this line and no one had been able to overcome it. He wrote among other things to Galileo and Torricelli, and asked them the dimension of the curve.[220] Galileo quickly recognized his cycloid in the Trochoid proposed to him. It was the curve he had studied many years ago, but it had been set aside for a long time, and abandoned because of the difficulty it presented. Still recollecting this fact, and from lack of self-confidence, he refused to look again at the problem, but confided the business to the famous geometricians, Cavalieri and Torricelli. The first of the two was not happier than Galileo, but the second, having had better success, found the solution and published it. He did not know at all what the French were doing in the meanwhile,

---

207 The letter from Evangelista Torricelli (1608–1647) to Roberval, that was ignored by Wallis, had been published in 1693.

208 Jacob Bernoulli forgot to explain that Mersenne's propaganda in letters towards the roulette were in 1638, while Galilo's letter to Cavalieri was in 1640.

209 Bernoulli wrote "pomo Eridos", and it may have given to Montucla the idea of calling the curve the Helena's of geometers.

210 Antoine de La Louvère (1600–1664), a friend of Fermat (1601–1665) in Toulouse had, in 1660, published a rather thick book about the cycloid. From Roberval (1602–1675), a Treatise on Indivisibles will be published as late as 1693 in an academic compendium, and Jacob Bernoulli will soon quote it. Bernoulli knew that he was mixing two periods of time, the one in the 1640s with Mersenne, Roberval, Descartes, Fermat, and the one in the 1660s with Fermat as well, the others being dead, but also new ones like Lalouvère, Pascal and Wallis.

211 This phraseology of Jacob Bernoulli is not correct: there were discussions, but no such accusations.

212 This very long letter from Roberval to Torricelli, published in 1693 without any indication of a date, and probably sent in July 1646, can be found in *Recueil*, pp. 284–302.

213 One is surprised to see corrective notes appearing on the text by Roberval in 1693. They are from Roberval himself, as is explained in a note placed at the beginning of the collection of texts (see Fig. 9 in the commentary).

214 Even if it was in the form of a joke, Bernoulli is highlighting national issues, where previously only cities, schools, or perhaps regions were mentioned.

215 Such a judgment on Mersenne is certainly inspired by the reading of Pascal's *Traité de la roulette* in 1658, and may certainly have been inspired by Roberval. Mersenne never pretended to be a true geometer, and Descartes was even mocking him by saying that Roberval was Mersenne's preferred geometer (Cfr. Letter 1638).

216 It is not so easy to interpret what Jacob Bernoulli wished to emphasize by mentioning the absence of Galileo's name in Mersenne's letters to various correspondants. Has it a link with Galileo's condemnation? Surprisingly, Bernoulli seems to say letters were sent by Mersenne on the area of an arch before Mersenne asked Roberval to work on the curve. See an interpretation of this kind in Mersenne's correspondance in July 1630.

217 Roberval became royal professor in 1634, so the date of an intervention of Roberval on the cycloid cannot be earlier in this narrative.

218 It seems interesting that Bernoulli put on the same level, Roberval, the discoverer of the area, and Mersenne, the man who asked a question that is not precisely known about the curve. I have tried to explain in the commentary that the discovery of rolling without slipping was due to Mersenne, even if Galileo mentioned it in his Discourse in 1638. But precisely Bernoulli in his narrative is not interested by mathematical details of this kind.

219 We have seen that in 1637 Mersenne publicly used the word "roulette and Torricelli in 1643 and in a letter to Roberval, claimed that Roberval was used to call it trochoid.

220 This means the area of the arch of the curve.

and what they concealed in their privacy, since nothing printed so far had been published by them.[221] In order to satisfy Mersenne's wish, Torricelli, in all good faith, for he was an honest man, let him know, with the rest of his discoveries, the solution also of the problem of the cycloid. Frightened by this news, and indignant that another might have pierced the mystery, Roberval accused the Italian of plagiarism, and alleged that his secrets had secretly been taken from him by Jean Beaugrand[222] and sent to Italy.[223] Learning this, Torricelli rejected such an accusation and the two scholars were to plead their cause by mutual letters and each was to attribute to himself what the other considered as his own (or believed as such). Mersenne took care of the letters, playing the role of intermediary between the geometers of each of the two nations. Sometimes for two years, without saying a word about Mersenne's discoveries or Torricelli's discoveries, Roberval astutely concealed his own discoveries, and during this time he was able to ruminate and to put some order in his own head. In the meantime, while all those exchanged "words" were printed by rivalry, a book by Blaise Pascal[224] was published in France under the title – Histoire de la Roulette, in which the author supported the cause of the French, and attributed the benefit of the discovery of the curve to Mersenne, and to Roberval that of the proof. The summary of Pascal's book tends to the following.

Mersenne first found the Line around 1615 and, after a delay of about twenty years, in 1634 precisely, proposed it to Roberval. The latter having realized this wish invited the other Geometers to participate in the research, not without having indicated to them the triple ratio of the area of the cycloid to its generating circle. Very quickly, two of them found proofs, namely Fermat and Descartes, but Beaugrand, a plagiarist,[225] sent these proofs to Galileo in 1638 along with other discoveries of third parties,[226] suppressing authors' names, although no doubt without attributing results to himself. However, he used such terms that, for readers who were not attentive, it might seem that only modesty, in any case, had prevented him from admitting he was the author. Therefore, a little to disguise the whole thing, he changed the word Trochoid into Cycloid. A little while later, Galileo and Beaugrand being both dead, Torricelli, who had found Galileo's manuscripts, and thought himself justified to take possession of them all in total impunity, annexed the discovery for himself. He published a book in 1644, which was not received in France without some laughter, since it attributed discoveries to Torricelli for which Roberval had been known to be the father, among those still living and for eight years already. And other affirmations of the same kind.

But since the tales may be suspected, of a master who may have been carried away by too great a love of the Fatherland, a Frenchman especially to the cause of a countryman and a great friend, it is worth listening too the version of foreigners which we presume to be always less subject to partiality. Among all these testimonies and before all others, what the Dutchman Huygens related about Pascal and his History of the Trochoid, deserves to be cited. Huygens proclaims at length in favor of Torricelli and restores to him what belongs to him. And certainly, I would prefer, he says, if only out of respect, that the author of the History of Roulette would have abstained from speaking of Torricelli rather than sully a man of great merit, who had already died for several years.[227] In any case, we know from his works that Torricelli was a learned mathematician of value, and one who deserves the highest degree for his mathematics; I also believe him a honest man. And I do not see what he could be guilty of, to the point of heating Roberval's bile. Torricelli published in 1644 his proofs on the area of the cycloid being the triple of the area of the circle. I do not see why it would have been forbidden! One does not deny that these proofs were his and it is not alleged that he claimed for his own anything from Roberval.

And somewhat further, Huygens goes on[228]: we owe more to Torricelli, who publicly disclosed the proofs than to Roberval, who has hitherto kept his own. And we do not judge at all in a correct way, it is certain, by asserting that if Roberval did not want to publish his proofs, then Torricelli would have no right to publish his own. But, it is said, Torricelli attributed to Galileo what is due to Mersenne and attributed to himself what is due to Roberval! Please, let us speak short but well; I do not notice either of the two assertions. In any case, Roberval was the author of these solutions and Torricelli was not less the author of his own. If, on the other hand, Roberval thought it was important for him that the universe knew that his proofs were the first to be, whatever Torricelli might not have ignored, he would have been good mannered to provide an indication, but by no means was it necessary to sully Torricelli for the reason he ignored it, or to charge him with unwarranted suspicions. But it may be that among the Galileo's manuscripts he had seen a writing from Beaugrand, in which Beaugrand had shown to Galileo Roberval's proofs, without naming Roberval. From which he could quickly seize the opportunity in view of his own solutions. Here are their suspicions.[229] But do they consider this fact to be established, I do not know. And unless Torricelli himself had acknowledged (which they do not affirm), they do not specify how this fact seem established to them. But, in any case, they do not really allege that Torricelli claimed for himself proofs that were surreptitiously concealed, or they do not refuse proofs produced by Torricelli as not being his own. What I do not understand at all, therefore, is where is the crime. They bring as proof nothing firmly established on this subject, and just awkward suspicions. And again, with few lines, Huygens goes on like this. That is why I do not know what their accusation is, unless by chance they want to impute to crime that someone else may discover or publish what Roberval keeps jealously for himself, known secretly of himself alone or prefers to reserve to his only friends. And finally he concludes. Although I do not wish to resume against him his: "It was a subject of laughter in France," we are some who are admittedly less likely to laugh than his French compatriots, at least we ask what is so important to motivate such moans, when we compare the Torricelli's writings with this little story.

---

[221] Jacob Bernoulli wished to ignore the publication by Mersenne in 1637, in the second volume of *Harmonie universelle*. Or he thought that the parametrization of the curve, that is provided there, was not a sufficient proof.

[222] Jean de Beaugrand (1590–1640), a mathematician patronized by Gaston d'Orléans, the king's turbulent brother, and holding an official function of royal secretary after 1634, is certainly the one that had organized the link between Mersenne and Fermat. He went to Italy in 1636, met Cavalieri and Galileo, and liked to organize polemics. He was the one to have written on Vieta. In a rather unanimous way he has been considered as the traitor in all stories about the cycloid.

[223] This is not exactly what can be read in the strange letter Roberval sent to Torricelli in 1646.

[224] Blaise Pascal (1623–1662) is the younger one in this story about the cycloid.

[225] Jacob Bernoulli never wrote that either Torricelli or Roberval were plagiarists: he is without any doubt sure of Beaugrand's theft.

[226] It is not clear who these other persons might be, as no other names were quoted about the properties of the cycloid.

[227] These sentences are from Wallis (see Annex 4). It is certain a pen mistake from Bernoulli to claim it came form Huygens.

[228] Bernoulli is once more confusing a reference to Wallis with one to Huygens; he seems here more to express his own thoughts.

[229] "Their" is here used to relate to the French geometers.

Let us leave Huygens. The English Wallis is entirely of his opinion. He who in the preface to his Treatise on the Cycloid, published in 1659, in which he pleads the cause of Torricelli against Pascal, says in particular: I had learned from Torricelli (from Torricelli, I say, for that Roberval had done some study on this subject, see my misfortune, I did not even think of it in a dream). In short, I had learned from this master, on the one hand, that the area of the Cycloid was three times that of the circle, and, on the other hand, what was the method for constructing tangents to it.[230] That someone had done more extensive research on the cycloid, I was totally unaware of it, and as well as myself (as far as I know) all my compatriots knew nothing about it. And the more we are not supposed to know it in any way. Indeed, whatever he claims to have found, Roberval has concealed it in his private coffers, or at least communicated it only to a few close friends, as I have said. If it were necessary to fight by quoting authorities, one could add many testimonies of prestigious masters who attribute to Torricelli the glory of having explored the curve demonstratively: Schooten in Holland, Tacquet in Flanders, Leibniz in Germany.[231] And there is no shortage of scholars among the French themselves who subscribe to the cause of Torricelli: their testimony is not to be disdained since they were on the spot. From all others they are supposed to have known at best what actually happened in France. For example, the very learned Lalouvère who is not afraid to put the discovery of the measure of the cycloid on Torricelli's account as the first to disclose it. As for us, having weighed minutely what is said from the two parties, here is what we think and say. It is certain that Galileo thought of the cycloid before Mersenne; it is possible that Torricelli and Roberval have found the measure of an arch, each in his own way. Who first solved the problem[232] is uncertain but certain that Torricelli first published his results and likely that Roberval was helped by Torricelli's discoveries in order to exploit and perfect his own.

But I resume the thread of my story by leaving Torricelli and Roberval. Their skirmishes, however, paved the way, and, as it were, prepared a much more violent struggle, which resumed shortly afterwards on the occasion of certain problems on the cycloid proposed by Dettonville (that is, Pascal) with reward at stake. He who is ignorant of the universally known account of it, should read the Preface of Wallis to his Treatise on the Cycloid,[233] and the long letters which the same wrote to Huygens on this subject. We thus will be brief. Here is the occasion of such grabbings. It was at this time that Pascal, at the instigation of his friends, reflected on the proofs of the truth of the Christian religion. So, those having heard that he had mastered such difficult problems as it seemed then, thought that the fame of this solved problem could also contribute to the authority of things over religion, especially if they were sure that nobody else could be as distinguished on this ground. That is why they urged him to publicly propose his problems, convinced that no other was capable of solving them. Pascal proposed the following problems on a special sheet printed in Paris in June 1658. Given a line parallel to the base of a cycloid, find the measure of the area, Its center of gravity, the volumes generated by its revolution, and the centers of gravity of these volumes and half volumes. The anonymous author demanded the solution of these problems to the most remarkable mathematicians throughout the world, proposing at the same time a reward in these terms. The one who will have solved and proved the problems before October 1st, 1658, we name him a great Apollo. The former will receive the value of forty Spanish doubles of gold, which the Spaniards themselves call doublons, and the French call pistoles, and the second one twenty gold double of the same kind.[234] If there is only one who finds the solution, he alone will have sixty Spanish doubles of gold. Other conditions were added later, according to which it was asked to make known the thing within strict limits of time, by an official document to the famous M. de Carcavi.[235] The sum of money in question was deposited at the Duke of Liancourt[236] place, so great was the solemnity of the framework of the affair thus conducted. That is why Wallis, having suddenly learned this by a letter from Digby, set about this search and found the solution. Not so much motivated by the proposed price, which seemed to exist only for display, but only by the desire of the famous Chevalier. And because he saw that a good part of the time had already passed and that a good part of the remaining time had to be amputated (which was enough at least to write down and to send the letter). Pressed by the shortness of time, he summed up everything in a few lines, and sent to Carcavi an abridgment of his method, provided, however, with the hand of a notary public. Carcavi (or rather in his absence Pascal to whom he had entrusted this office) received the letter in Paris on the 20th of September, that is, ten whole days before the deadline. Thereupon the jealous Frenchmen, who did not wish to attribute the prize to a foreigner, being unable to dispute it against Wallis under another pretext, sought an escape in the person of the notary. They alleged that it would have been more consistent with the intention of the anonymous proponent that the attestation would have been made by notaries in Paris than by notaries in Oxford, because it was the Parisians who had to authenticate the reception of M. de Carcavi, and the Oxonians were quite incapable of doing so. As if, in truth, Carcavi could have ignored his good reception without the testimony of the Parisian notaries or also what was the day! But let us listen to Wallis himself speak and complain so violently of this frivolous subterfuge. As for the price, he says, about which the French seem to care very much, that is what matters least to me. As for the thing itself, I leave it to the judgment of mathematicians. For what is said of the official document, I would think that either it is necessary to give credit to Mr. Carcavi from the moment when any solution has been found elsewhere, or there is no need of it. Indeed, to attest that he received this solution or what day he received it, why did it need public notaries? But one attaches the meaning one wants to the words of the Anonymous, and that one decides what one wants of the price, I do not oppose it and even that before all the others fellows one grants only French citizens. I refer myself to the fact and do not care to dispute the rule. During this time Wallis wrote a second and a third letter to Pascal, but he received no reply. Instead of this, he wrote—as I have already mentioned above—the Treatise on the History of Roulette, in

---

[230] It is strange that Torricelli was not more precise on the subject of tangents: we cannot deduce from what he wrote what was the method he learned from Viviani, and how Viviani found it.

[231] The demonstration by van Schooten in 1649 for tangents is the Descartes' method with the instantaneous center of rotation, so it uses a mobile circle, and the proof by Leibniz in 1690, while looking like the baroque frame of reference, used a functional representation.

[232] Bernoulli is certain that the first problem was that of the area of the arch.

[233] See a partial translation in Annex 4.

[234] Such precisions are certainly important in a family of merchants like the Bernoullis.

[235] Pierre de Carcavi became in 1667 the secretary for the King's library in Paris, having been earlier counsellor to the Parliament of Toulouse where he met Pierre de Fermat as a colleague. He entered the Académie des sciences as a mathematician. Carcavi was on the side of Roberval against Descartes. He came to Paris at the service of Duc de Liancourt, and this is the reason why he intervened for the Pascal's challenge.

[236] Roger du Plessis, Duc de Liancourt, who died in 1676, had been obliged to stay in his domains for his political choices during the Fronde. He then became an austere Jansenist, having a place at Port Royal des champs, and so was close to Pascal.

which everywhere a profound silence reigns over the solutions of Wallis. So that Wallis, finding himself estranged, was obliged to publish himself what he had found. Solutions to the Pascalian problems were too given by Lalouvère,[237] S.J, and a French himself, and others: but in view of the limited bounds of time, no one obtained the prize.

This, however, had the effect that there had been no one since that time, who may have seem to know it more than the average Mathematician, who did not attempt to sharpen his forces to this stone of genius and to produce, consequently, something new. Even better, the first-rate geometers were no less assiduous in studying thoroughly what the others had left intact and impracticable. Among them, because of his talent, it is necessary to celebrate Christopher Wren,[238] geometer emeritus and architect of the king of England, who first performed the rectification of the cycloid curve and moreover observed that elongated or contracted cycloids, were equivalent to lines of Ellipses.[239] Let us add the incomparable Dutch Mathematician Christiaan Huygens, who realized the absolute quadrature of a cycloidal segment, a rather important discovery, that Wallis, the diligent defender of English glory, asserted to his disciple Wren (and not once) to be absolutely original. And let us not forget the honor of our Germany,[240] Leibniz, who, not long afterwards, showed another oblique segment capable of absolute quadrature.[241] But above all, the inventions later made by the very ingenious Huygens, are worthy to pass to posterity, because of their much greater utility in ordinary life. He had first adapted the use of pendulums to the mechanisms of clock-making, but had remarked that, under all circumstances, the equality of oscillations could not be obtained, since it was observed that long elongations were slower than the short ones. He began to find out on what curve the pendulums should oscillate so that their vibrations could be made in exactly the same time, and he found that our cycloid enjoyed this property. But as there was still another difficulty, that the pendulum may oscillate along this curve (while the pendulum wire, when it is stretched straight, can give the pendulum no other movement than circular), he continued to seek what curvature of the wire, moreover, enabled this to be obtained. And he again found that this curvature was cycloidal. To this end he taught the artifice of suspending the pendulum between two plates of cycloidal form, in order to obtain, by their alternate winding and unwinding, a precisely equal movement of the mechanism. He thus discovered two remarkable properties of the cycloid. The first, Mechanical, is that a grave descending along it from any place more or less elevated, arrives at the bottom in the same time.[242] The other, Geometric, that its evolute has the same description. By applying these properties to the manufacture of clocks, how much this highly renowned master has deservedly deserved from the human race, those well know who studied thoroughly what impulse this has given to the perfection of all Geography and Nautical Science. It is enough for this invention to arrive at what usually happens to the most remarkable of them (and which Huygens himself had predicted in advance): there were many people who wanted to be the authors or, if not themselves, somebody from their nation rather than accept to honor him. However, in his prolegomena at the *Horologium Oscillatorium*, the author refutes them abundantly and proves not only that he has found the construction of this mechanism, that he has busied himself in perfecting it by an original reflection, and that he published his drawing and description by printing it. He had sent copies on all sides at a time when no mention of such clocks had been made in any person's writing, or no rumor ran about such clock.

From that time on, because mathematicians were hypnotized by the contemplation of this very beautiful invention, hardly anything important happened to the cycloid until the last decade of the last century, when this very noble curve finally reached the pinnacle of its perfections, thanks to the labors of some, who revealed in it a great number of remaining properties, not in any way yielding to those that had hitherto been recorded. Thanks to them, finally, it was known that the cycloid, as it is its own Evolute, is also its own Caustic. That the same is remarkable besides by its tautochronism discovered by Huygens, also by its brachistochronism or the fastest descent. That it conceals within it not only two, but a multitude of segments, sectors and zones of which one can make the quadrature. And many other peculiarities, of which I refrain from making the list lest, by recounting personal facts, I might become unpleasant to my Audience.

You have thus heard, gentlemen listeners, the History of the Cycloid. No other curve at this time is more famous or more known among Mathematicians. You have seen it been taken care of by non vulgar men, but by remarkable masters, for the splendor of their birth, the renown of their knowledge, the repercussion of their virtues and their authority; and by far more unrivaled great masters. You have seen them bitterly fighting and not being dispossessed of anything without enduring it very badly. What, then, do you conclude, is hidden in such works but divine, capable of filling the minds of those who know it with incredible voluptuousness? Do not count on me, at least, either to appease your taste, or to go where these great men have already arrived: it is only possible to an extremely small number. As for me, I think I would have not wasted my time if I obtain for some of you, who with strong laughings are ridiculing a science that you do not know, and such great men, that do not apply what popular ones say: "The ignorant despises what he cannot reach."

So I said.

---

[237] Lalouvère was a jesuit doing mathematics, close to Pierre de Fermat.

[238] The famous architect Christopher Wren, responsible for the rebuilding of baroque St Paul's cathedral in London, and a founding member of the Royal Society, provided the rectification of an arch of the cycloid in 1658, communicated his result to Pascal, who used it immediately.

[239] Using a parametric representation of the cycloid, and integral calculus, it is easy to show the link with the ellipse for the rectification of the cycloid, and to obtain 4R for its length for one arch. Wren had to work with the exhaustion method with inscribed figures. Pascal was astonished when he received direct notification of Wren's results, and so went to modify the conditions of his challenge, by precisely using Wren's method.

[240] Even if he is from Basle, Bernoulli considered himself as a German, for linguistic reasons for sure.

[241] With such enterprises on segments of the cycloid that can be expressed in rational terms, depending on the radius, and not from pi, that Leibniz began, and then the two Bernoullis, Johann and Jacob, a completely new phase of Calculus is beginning, that is the one to classify integrals.

[242] A certain number of mechanical problems, implying what we name calculus of variations, all implied the cycloid (looking for the tautochrone, the brachistochrone, etc.).

References

Alexander A (2014) Infinitesimal. How a dangerous mathematical theory shaped the modern world. Scientific American/Farrar, Straus and Giroux, New York.

Aristotle ([1936]1955) Minor works W. S. Hett, Trans. Translated by Hett WS. The Harvard University Press–Heinemann W LTD, Cambridge MA–London.

Baron ME ([1969]1987) The origins of the infinitesimal calculus. Dover, New York.

Bascelli T (2015) Torricelli's indivisibles. In: Jullien V (ed) Seventeenth century indivisibles revisited. Birkhäuser, Basel, pp. 105–136.

Beaugrand J (1636) Geostatice, seu de vario pondere gravium secundam varia a terra (centro) intervalle dissertation mathematica. Toussaint du Bray, Paris.

Bernoulli J ([1701]1993) Oratio de Historia Cycloidis, Ms UB Basel, L1a 749 A. 3 (8 fol.) [published in 1927 by Friedrich Weiss, and reproduced in: Die Werke von Jakob Bernoulli, Bd 4, Birkhäuser, 1993, pp. 257–266].

Bernoulli J (1742a) Propositiones variae mechanico–dynamicae. In: Bernoulli 1742b, vol IV, pp. 253–386.

Bernoulli J (1742b) Opera omnia, tam antea sparsim edita, quam hactenus inedita [...] quibus continentur ea, quae ab A. 1690 usque ad A. 1727 prodierunt. Accedunt Lectiones Mathematicae de Calculo Integralium atque Anedokta. I–IV vols, Lausannae & Genevae, Bousquet [Facsimile 1968. Reprint Curavit Hofman JE. Georg Olms Verlag Hildeshe].

Bos HJM (2001) Redefining geometrical exactness. Descartes' transformation of the early modern concept of construction. Springer, Berlin.

Bourdieu P (2001) Science de la science et réflexivité. Raisons d'agir Editions, Paris, p. 91 and seq.

Braudel F ([1989]1994) Le modèle italien. Champs, Paris.

Cantor M (1900) Vorlesungen über Geschichte der Mathematik, Zweiter Band von 1200–1668. Zweite Edition, Teubner, Leipzig.

Cavalieri B (1635) Geometria indivisibilibus continuorum nova quadam ratione promota […]. Clemente Ferroni, Bologna.

Comte A (1830–1842) Cours de philosophie positive. Bachelier, Paris.

Comte A (1843) Traité élémentaire de géométrie analytique. Carillian-Goeury et Dalmont, Paris.

Comte A (2007) Premiers cours de philosophie positive. Clément-Colas Y (ed). PUF, Paris.

Costabel P (1962) Le De cycloïde de R. Boscovich. Revue d'histoire des Sciences et de ses applications 15(1):31–42.

Costabel P (1964) Essais sur les secrets des Traités de la Roulette. Taton E (ed) L'Œuvre scientifique de Pascal. PUF, Paris, pp. 169–206.

Dati C (1919) Lettera a' Filateti. Della vera storia della cicloide […]. Firenze, 1663 [see also: Opere di Evangelista Torricelli, I, Parte II, pp. 441–482].

De Duillier NF (1699) Lineæ brevissimæ descensus investigatio geometrica duplex, cui addita est investigatio geometrica solidi rotundi in quo minima fiat resistentia, London.

De Gandt F (1984–1985) Naissance et métamorphose d'une théorie mathématique : la géométrie des indivisibles en Italie (Galilée, Cavalieri, Torricelli). Sciences et Techniques en Perspective 9:179–229.

De l'Hôpital GFA (1696) Analyse des infiniment petits, pour l'intelligence des lignes courbes. Imp. royale, Paris.

Descartes R (1637) Florimond de Beaune, Johannes Hudde, Henricus van Heuraet, Johan de Witt, Erasmus Bartholin, Frans van Schooten, Geometria à Renato Des Cartes anno 1637 Gallicè edita; postea autem unà cum notis Florimondi de Beaune […] Gallicè conscriptis in Latinam linguam versa, & commentariis illustrata, operâ atque studio Francisci à Schooten, Florimond de Beaune (trad.), Frans van Schooten (éd. sc.), Leyde, Jean Maire, 1649, in-4° [Id., 1659, 2nd edn, pp. 265–271].

Descartes (1996) Œuvres de Descartes. New Edition. Vrin, Paris.

Dhombres (1988) Une histoire de l'objectivité scientifique et le concept de postérité. In: Guesnerie R, Hartog F (eds) Des Sciences et des techniques: un débat. Cahier des Annales 45:127–148.

Dhombres J (1996) Entries: Sciences et savants [and] Académie des sciences. In: Dictionnaire de l'Ancien regime. PUF, Paris.

Dhombres J (1999) Entry: Académies. Dictionnaire d'histoire et de philosophie des sciences. PUF, Paris.

Dhombres J (2006) Exactitude et esthétique. La mise en scène du savoir au temps de Peiresc. Sciences et Techniques en Perspective 10(1):15–74.

Dhombres (2010) La trajectoire d'une parabole. Métamorphoses de la philosophie naturelle sous l'effet des mathématiques, XIIIe Entretiens de la Garenne Lemot. In: Pigeaud J (ed) Métamorphose(s), PUR, Rennes, pp. 213–241.

Dhombres (2011a) La réception en France de Galilée. In: Pepe L (ed) Les rôles différents de Peiresc et de Mersenne auprès de Gassendi, de Descartes et de Jacques le Tenneur, Galileo e la scuola galileiana nelle Università del Seicento. CLUEB, Bologna, pp. 47–60.

Dhombres (2011b) Le jet d'eau et l'arc-en-ciel à l'âge baroque : réalisation des mathématiques, mathématisation de la philosophie naturelle et représentation des phénomènes. In: Cousinie NFC (ed) L'artiste et le philosophe. L'histoire de l'art à l'épreuve de la philosophie au XVIIe siècle. PUR, Rennes, pp. 151–196.

Dhombres J (2015) An epistemological path through the historiography on Indivisibles. In: Jullien V (ed) Seventeenth century indivisibles revisited. Birkhäuser, Basel, pp. 391–450.

Dhombres J (2023a) Les savoirs mathématiques et leurs pratiques culturelles. Hermann, Paris, forthcoming.

Dhombres J (2023b) Confrontation de deux manières d'apprentissage du calcul différentiel et intégral: Johann Bernoulli et Guillaume de l'Hôpital (1691–1696). Sciences et Techniques en Perspective, forthcoming.

Dhombres J, Radelet de Grave P (2008) Une mécanique donnée à voir. Les thèses présidées par Grégoire de Saint-Vincent en 1624. Brepols, Turnhout.

Dhombres J, Sakarovitch J (eds) (1994) Desargues en son temps. Blanchard, Paris.

Ellis M (2004) The coffee-house: a cultural history. Weidenfeld, London.

Fermat P (1891–1912) Œuvres de Fermat, 4 vols. Tannery P, Henry C (eds). Gauthier-Villars, Paris.

Galilei G (1638) Discorsi e dimostrazioni matematiche. In: Favaro A (ed) Galilei 1890–1909, VIII, pp. 39–362. Le Opere di Galileo Galilei: Edizione nazionale sotto gli auspici di sua maestà il re d'Italia, Barbera, Firenze.

Gröning J (1701) Historia cycloidis. Qua Genesis et proprietates Lineae Cyloidalis praecipuae, secundum Ejus Infantium, Adolescentium & Juventutum, Ordine chronologico recesentur […]. Hamburg, Gottfried Liebezeit.

Hall R (1980) The Quarrel between Newton and Leibniz. Cambridge University Press, Cambridge.

Hazard P (1935) La crise de la conscience européenne (1680–1715). Librairie générale, Paris.
Huygens C (1678) Christiani Hugenii Horologium Oscillatorium. Apud F. Muguet, Paris.
Huygens C (1888–1950) Œuvres complètes de Christiaan. Huygens publiées par la Société hollandaise des sciences. Nijhoff, La Haye.
Huygens C (1901) Œuvres completes de Christiaan Huygens. Hollandsche Maatschappij der Wetenschappen, vol. IX. Nijhoff, La Haye.
Huygens C (1920) Œuvres completes de Christiaan Huygens. Hollandsche Maatschappij der Wetenschappen, vol XIV. Nijhoff, La Haye.
Huygens C (1932) Œuvres completes de Christiaan Huygens. Hollandsche Maatschappij der Wetenschappen, vol XVII. Nijhoff, La Haye.
Huygens C (1934) Œuvres completes de Christiaan Huygens. Hollandsche Maatschappij der Wetenschappen, vol XVIII. Nijhoff, La Haye.
Itard J (1975) La lettre de Torricelli à Roberval d'octobre 1643. Revue d'Histoire des sciences 28(2):114–124.
Jacob R (1977) Robert Boyle and the English revolution. Franklin, New York.
Jullien V (2015) Roberval's indivisibles. In: Jullien V (ed) Seventeenth century indivisibles revisited. Birkhäuser, Berlin, pp. 177–210.
Kuhn T (1977) The essential tension. Selected studies in scientific tradition and change. The University of Chicago Press, Chicago.
Latour B (1988) Ramsès II est-il mort de la tuberculose? La Recherche 309:84–85.
Leibniz GW (2012) Sämtliche Schriften und Briefe, Siebente Reihe, Mathematische Schriften, Sechster Band, 1673–1676. Band. Arithmetische Kreisquadratur. Akademie der Wissenschaften zu Göttingen. Brandenburgische Akademie der Wissenschaften, Berlin.
Lenoble R ([1943]1971) Mersenne ou la naissance du mécanisme. Vrin, Paris.
Mancosu P, Vailati E (1991) Torricelli's infinitely long solid and its philosophical reception in the seventeenth-century. Isis 82(1):50–70.
Mersenne M (1624) L'impiété des déistes, athées et libertins de ce temps, combattue et renversée de point en point par raisons tirées de la philosophie et de la théologie, ensemble la réfutation du Poème des déistes. P. Bilaine, Paris.
Mersenne M (1634) Les Mécaniques de Galilée, mathématicien et ingénieur du duc de Florence, traduites de l'italien en françois par L.P.M.M., Paris.
Mersenne M (1636) Harmonie universelle, contenant la théorie et la pratique de la musique, où il est traité de la Nature des Sons, & des Mouvemens, des Consonances, des Dissonances, des Genres, des Modes, de la Composition, de la voix, des Chants, & de toutes sortes d'Instruments Harmoniques. Sébastien Cramoisy, Paris.
Mersenne M (1637) Harmonie universelle, second volume, Nouvelles observations mathématiques et physiques. Pierre Ballard [The new pagination begins after book 6 that stops at page 408. It is quite possible that the addition (XXth book as described by Mersenne himself) only appeared in 1638].
Mersenne M (1639) Nouvelles pensées de Galilée, A paris. Guenon.
Mersenne M (1644) F. Marini Mersenni [...] Cogitata physico-mathematica, in quibus tam naturae quam artis effectus admirandi certissimis demonstrationibus explicantur. De Mensuris, ponderibus et numeris. De hydraulico-pneumaticis phaenomenis. De arte nautica, seu Histiodromia et hydrostatica (cum tractatu de magnetis proprietatibus. De musica theorica et practica. De mechanicis phaenomenis. De ballisticis, seu acontismologicis phaenomenis, Paris, aux frais de A. Bertier.
Merton RK (1968) Sociological theory. American Journal of Sociology 50:462–479 [reprinted: Id., 1968, Social theory and Social structure. Free Press].
Merton RK (1973) The sociology of science. The University of Chicago Press, Chicago.
Meskens AD (2005) Joannes della Faille, s.j. Mathematics, modesty and missed opportunities. Belgisch historisch instituut te Rome, Bruxelles.
Montucla E (1758) Histoire des mathématiques. Charles-Antoine Jombert, Paris.
Montucla E (1799–1802) Histoire des mathématiques. Agasse, Paris.
Nardi A (1994) Théorème de Torricelli ou théorème de Mersenne. Etudes Philosophiques 1(2):87–118.
Newton I (1687) The mathematical principles of natural philosophy, by Sir Isaac Newton. Translated into English by Motte A, Symonds, London.
Osler MJ (ed) (2000) Rethinking the scientific revolution. Cambridge University Press, Cambridge.
Pascal B (1963) Œuvres complètes de Pascal. Lafuma L (ed). Seuil, Paris.
Pascal B (1992) Œuvres complètes de Pascal. Mesnard J (ed). Brouwer, Paris.
Roberval GP de (1636) Traité de mechanique des poids soustenus par des puissances sur les plans inclinez. Charlemagne, Paris.
Roberval GP de (1693a) De trochoïde ejusque spatio. In: Divers ouvrages de mathématiques et de physique par MMrs les savants de l'Académie Royale des Sciences. Imprimerie royale, Paris, pp. 246–278.
Roberval GP de (1693b) Observation sur la composition des mouvemens, et sur le moyen de trouver les touchantes des ligne courbes. In: Divers ouvrages de mathématiques et de physique par MMrs les savants de l'Académie Royale des Sciences. Imprimerie royale, Paris, pp. 69–111.
Roberval GP de (1693c) Traité des indivisibles. In: Divers ouvrages de mathématiques et de physique par MMrs les savants de l'Académie Royale des Sciences, pp 190–245. Imprimerie royale [English free translation: Walker E (1932) A study of the Traité des indivisibles of Gilles Personne de Roberval. Teachers College, New York].
Roberval GP de (1693d) Traité de la roulette. In: Divers ouvrages de mathématiques et de physique par MMrs les savants de l'Académie Royale des Sciences. Imprimerie royale, Paris.
Roberval GP de (1693e) Divers ouvrages de mathématiques et de physique par MMrs les savants de l'Académie Royale des Sciences. Imprimerie royale, Paris.
Rochot B (1965) Sur Mersenne et le mot cycloïde. Appendice IV, Correspondance du P. Marin Mersenne, vol IX. Cornelius de Waard/CNRS, Paris, pp. 584–586.
Scott JF (1981) The mathematical work of John Wallis [first ed. 1938] 2nd edn. Chelsea Publishing, New York.
Shapin SS (1985) Leviathan and the air pump: Hobbes, Boyle, and the experimental life. Princeton University Press, Princeton.
Souffrin P (1993) Galilée, Torricelli et la loi fondamentale de la dynamique scolastique: La proportionnalité velocitas-momentum revisitée. Sciences et Techniques en Perspective 25:122–134.
Taton R (1951) L'œuvre mathématique de G. Desargues. PUF, Paris.
Taton R (1955) Causalité et accidents de la découverte scientifique. Masson, Paris. [English translation: Id., 1962. Reason and Chance in Scientific Discovery].
Teixeira FG (1909) Traité des courbes spéciales remarquables planes et gauches, Coïmbre, Imp. de l'Université, II, Chapitre X.

Wallis J (1659) Tractatus duo, Prior, de Cycloide et corporibux inde genitis, Posterior, Epistolaries in qua agitur de Cissoide [...]. Lichtfeld [It has been reproduced in Wallis 1695, pp. 499–547; I have not checked if some lines have not been modified. The proof by Christopher Wren on the rectification doc the cycloid is to be found in Wallis, 1659].

Wallis J (1693) Algebra Tractatus. Historicus & Practicus. Anno 1685 Angelice editus. Opera Mathematicorum Volumen alterum. Oxoniae. Sheldonian Theatro Sheldoniano.

Wallis J (1695) Opera mathematica. Sheldonian Theatre, Oxford.

Weber M (1904–1905) Die Protestantische Ethik und der Geist des Kapitalismus. Archiv für Sozialwissenschaften und Sozialpolitik, Bd. XX und XXI.

Weil A (1993) Introduction to Oratio de Historia Cycloidis. Die Werke von Jakob Bernoulli, Bd 4. Birkhäuser, Berlin, pp. 254–255.

Check for updates

5 Historical and Methodological Details on the *De Motu Gravium Naturaliter Descendentium* in Torricelli's *Opera Gometrica* (1644)

Raffaele Pisano and Paolo Bussotti

Abstract

In this paper, we deal with Torricelli's principle in mechanics according to which two heavy bodies linked together cannot move by themselves unless their common centre of gravity descends. At the beginning of the *De motu gravium naturaliter descendentium*, in his *Opera geometrica* (1644), Evangelista Torricelli (1608–1647) expressed this principle in a very brief section entitled *Praemittamus*. Our aim is to sustain the idea that Torricelli was constructing a general view of a kind of physics based on general principles, of which the previous one is the most important and explicit statement. In our discussion, the following themes are addressed: (1) a comparison between the Torricelli's principle and the virtual work principle; (2) the role of Torricelli principle within the history of the virtual work principle; (3) the logic behind the demonstrations offered in physics relying on general principles such as that of Torricelli.

Keywords

Torricelli • Physics-mathematics relationship • *Praemittamus* • Equilibrium • *De motu gravium* • Principle of virtual work

1 An Outline

In *De motu gravium naturaliter descendentium* of the *Opera geometrica* (1644), Torricelli begins his explanations starting from the following statement:

Premise: Two heavy bodies joined together cannot move by themselves unless their common centre of gravity descends.[1]

[1] "*Praemittamus*: Duo gravia simul coniuncta ex se moveri non posse, nisi centrum commune gravitatis ipsorum discendat". (Torricelli 1644, p. 99; *Id*., 1919–1944, p. 105; authors' italic; our translation). This paper includes advanced–structured reorganization of a short self–sufficient interlude and spots–ideas of the Torricelli's Principle presented in some previous publications (i.e., Pisano 2017; Pisano and Capecchi 2013, 2016; Gillispie and Pisano 2014; Pisano et al. 2021; Capecchi and Pisano 2007, 2008, 2009a, b, c, 2010a, b); we warmly thank the publishers. Briefly, for the historical-mathematical contributions on Torricelli see Castelnuovo (1962), Festa (1992), De Gandt (1992), Baroncelli (1993), Krarup and Vajda (1997), Saccheri (1708), pp. 74–75. On Torricelli's correspondence and life see: Gliozzi 1970–1990; Segre (1983), Medolla (1993), Tenca (1956, 1958, 1960), Nonnoi (1982), Bertoni (1985, 1987, 2003), Valilati (1987), Caverni (1891–1900), Vols. I–IV. For studies concerning Torricelli in the spehere of the history of mathematics including inedited works, see Loria (1897, 1903, 1909, 1915, 1917, 1919, 1922a, b, 1923, 1932, 1938, 1944, 1945, 1947), Podetti (1914), Agostini (1930a, b, c, 1931, 1949, 1950, 1951a, b, c, d, e, Bortolotti (1921, 1922, 1923, 1924, 1925a, b, 1928a, b, c, 1930, 1931, 1937, 1939, 1947), Loria and Vassura (Torricelli 1919–1944), Vassura (1945), Visani (1952), Procissi (1952, 1953), Taton (1963), Belloni (Torricelli 1975), Fabbri (1947–1997). See also: Toscano (2008, 2014), Favino (2009, 2019), Rossini (1959) and their suggested biographical references, as well.

R. Pisano (✉)
HOPAST at IEMN, Lille University, CNRS, Lille, France
e-mail: raffaele.pisano@univ-lille.fr

P. Bussotti
DIUM, Udine University, Udine, Italy
e-mail: paolo.bussotti@uniud.it

R. Pisano et al., *Homage to Evangelista Torricelli's Opera Geometrica 1644–2024*,
Logic, Epistemology, and the Unity of Science 55,
https://doi.org/10.1007/978-3-031-06963-5_5

The *Praemittamus* (premise)—which is nowadays called *Torricelli's Principle* in mechanics—is nevertheless rarely used in the *De motu gravium naturaliter descendentium* (hereafter *De motu*). To be precise Torricelli resorts to it only in the proofs of Proposition I and Proposition II (Torricelli 1644, pp. 99–100). In these proofs, *Torricelli's Principle* is used by means of a *reductio ad absurdum*. In our opinion, this means that the epistemic character of the principle is that of a Greek axiom (even if it is not declared) holding for Torricelli's entire theory. Furthermore, Torricelli uses a principle, with which we will deal, to prove the intriguing Proposition XVI of the *De motu ac mometis varia* (This proposition is missing in 1644–edition; see Torricelli, 1919–1944, pp. 231–249, Proposition XVI, pp. 243–246). The proof of this proposition is more complex than those of the Propositions I and Proposition II in *De motu*.

In the literature, important authors, such as Ernst Mach (1838–1916), tried to reduce Torricelli's principle to the ones expounded or, anyway, used by Galileo Galilei (1564–1642) in *Le mecaniche*[2] (Galilei 1890–1909, II, pp. 147–191) and other works. Furthermore, the *Praemittamus* is considered to be a specific–basic–early expression of the *Principle of virtual work*.[3] Besides Galileo's theory on dynamics, Torricelli presents problems, which, in his opinion, were still unsolved. His main concern is to prove a supposition by Galileo, which states: the degrees of speed for a body moving on an inclined plane are directly proportional to the inclination of the plane (Galilei 1890–1909, VIII, p. 205; see also below). In his *Opera geometrica*, Torricelli suggests that this supposition might be *proved* beginning with a "theorem" according to which "the momentum of equal bodies on planes unequally inclined are to each other as the perpendicular lines of equal parts of the same planes" (Torricelli, 1644, p. 99). Moreover, Torricelli also assumes that this theorem has not yet been demonstrated.

In fact, Torricelli's principle can be used to prove Galileo's theorem. Let us consider that, in the first edition of Galileo's *Discorsi e dimostrazioni intorno a due nuove scienze* (1638) there is no proof of this "theorem". Actually, Torricelli knew the demonstration that Vincenzo Viviani (1622–1703) added in the second edition of his work (1656, *Opere di Galileo Galilei linceo*). However, the "theorem" clearly appears as an important object of discussion in some letters by Torricelli to Galileo (Torricelli, , III, pp. 48, 51, 55, 58, 61).

Let us summarize the situation concerning this problem: Galileo moves from a static concept of virtual motions toward a dynamic one. This step was favoured by the Galilean concept of "momento" [momentum] (Cfr. Galluzzi, 1970, pp. 153–408). Nonetheless, the implications of this passage and the connection with Torricelli's principle are anything but granted. For, Nardi (1598, 1648?) expressed his perplexity about the possibility of demonstrating Galileo's theorem as an early reasoning concerning the principle of virtual work (Fig. 1).[4]

---

[2] *Le Mecaniche* have two versions: a short version (4 manuscripts) and longer version (14 manuscript). See: Gatto (2002), Pisano (2023–2024), Pisano and Bussotti (2012, 2017a). *Le Mecaniche* are an interesting series of Galileo's lectures around first years of 1590. Mersenne published this work in French language. It presents slight differences in respect to the original Italian version. An extensive study on the two *Le Mecaniche* including the complete translation and transcription side–by–side texts into English is pre-print (Pisano 2023–2024). For the first time short—and—long versions (manuscripts 1594/fl. 17th; 1623, 1634) of *Le Mecaniche* are discussed and published in English (Pisano 2023–2024, 2009a, b, c, 2009d; Pisano and Bussotti 2017a, c, 2015a, 2012; see also Pisano et al. 2017a, b).

[3] We have no room to deal with this correlated important principle in Lagrange and into history of science. For, we refer to a selected literature: Lagrange (1762, 1764, 1788, 1870–1873, 1892), Fossombroni (1794), Fraser (1983), Bussotti (2003). More recently: Pisano (2017).

[4] Cfr. Nardi fl. 1642s. On Nardi see Milighetti (2004, 2006). See also: Torricelli (1975), p. 157, Fts. 2–3, Capecchi and Pisano (2010a), Pisano and Capecchi (2016).

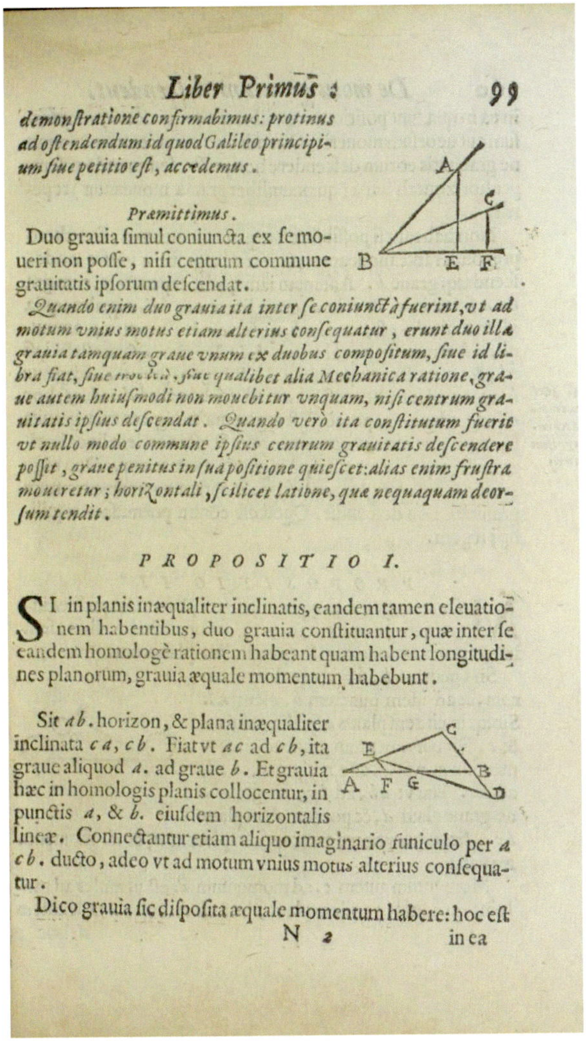

Liber Primus : 99

demonſtratione confirmabimus: protinus ad oſtendendum id quod Galileo principium ſiue petitio eſt, accedemus.

Præmittimus.

Duo grauia ſimul coniuncta ex ſe moueri non poſſe, niſi centrum commune grauitatis ipſorum deſcendat.

Quando enim duo grauia ita inter ſe coniunctà fuerint, vt ad motum vnius motus etiam alterius conſequatur, erunt duo illa grauia tamquam graue vnum ex duobus compoſitum, ſiue id libra fiat, ſiue troclea, ſiue qualibet alia Mechanica ratione, graue autem huiuſmodi non mouebitur vnquam, niſi centrum grauitatis ipſius deſcendat. Quando verò ita conſtitutum fuerit vt nullo modo commune ipſius centrum grauitatis deſcendere poſſit, graue penitus in ſua poſitione quieſcet: alias enim fruſtra moueretur; horizontali, ſcilicet latione, quæ nequaquam deorſum tendit.

PROPOSITIO I.

SI in planis inæqualiter inclinatis, eandem tamen eleuationem habentibus, duo grauia conſtituantur, quæ inter ſe eandem homologè rationem habeant quam habent longitudines planorum, grauia æquale momentum habebunt.

Sit *ab*. horizon, & plana inæqualiter inclinata *ca*, *cb*. Fiat vt *ac* ad *cb*, ita graue aliquod *a*. ad graue *b*. Et grauia hæc in homologis planis collocentur, in punctis *a*, & *b*. eiuſdem horizontalis lineæ. Connectantur etiam aliquo imaginario funiculo per *acb*. ducto, adeo vt ad motum vnius motus alterius conſequatur.

Dico grauia ſic diſpoſita æquale momentum habere: hoc eſt

N 2 in ea

Premise.

Two heavy bodies linked together [simul coniuncta = jointed each other] cannot move by themselves unless their common centre of gravity descends.

Proposition I.

If in [two] planes unequally inclined but with the same elevation, two heavy bodies are considered which are in the same ratio as the length of the planes, these bodies will have the same momentum.

Fig. 1 Torricelli's Premise and Proposition I (Torricelli 1644, p. 99). *Source* Courtesy University of Oklahoma, History of Science Collections[5]

Therefore, Torricelli's Proposition II (Fig. 3) is, as a matter of fact, Galileo's theorem[6] (Fig. 2 and as cited in Fig. 3).

---

[5] "Propositio I. Si in [two] planis inaequaliter inclinatis, eandem tamen elevationem habentibus, duo gravia constituantur, quae inter se eandem homologè rationem habeant quam habent longitudines planorum, gravia aequale momentum habebunt." (Torricelli, 1644, p. 99; authors' translation).

[6] Galilei 1890–1909, VIII, Day III, p. 205. Though this is not the main subject of our paper, we also point out that Torricelli's principle might be associated—as a case study—to the *Proposition I* on two bodies already argued by Guido Ubaldo del Monte: "Propositione I. Se il peso sará sostenuto nel centro della sua gravezza da linea diritta non si fermerà giamai, se quella istessa linea sarà á piombo del l'orizonte." (Del Monte 1581, p. 3; *Id.*, 1615, p. 6).

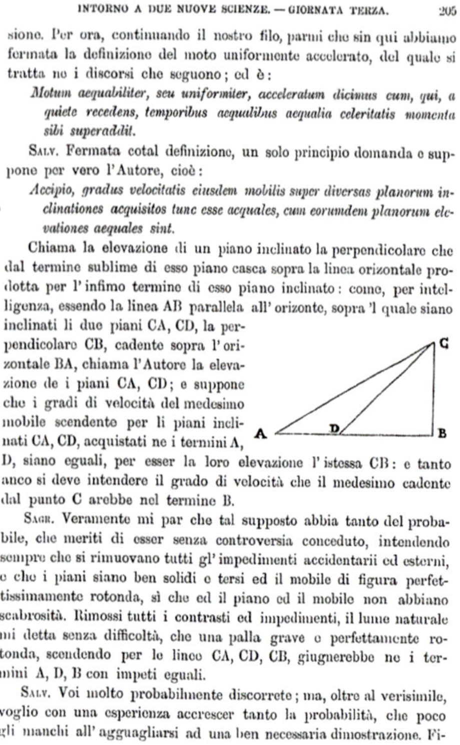

INTORNO A DUE NUOVE SCIENZE. — GIORNATA TERZA. 205

sione. Per ora, continuando il nostro filo, parmi che sin qui abbiamo fermata la definizione del moto uniformente accelerato, del quale si tratta ne i discorsi che seguono; ed è:

Motum aequabiliter, seu uniformiter, acceleratum dicimus eum, qui, a quiete recedens, temporibus aequalibus aequalia celeritatis momenta sibi superaddit.

Salv. Fermata cotal definizione, un solo principio domanda e suppone per vero l'Autore, cioè:

Accipio, gradus velocitatis eiusdem mobilis super diversas planorum inclinationes acquisitos tunc esse aequales, cum eorumdem planorum elevationes aequales sint.

Chiama la elevazione di un piano inclinato la perpendicolare che dal termine sublime di esso piano casca sopra la linea orizontale prodotta per l'infimo termine di esso piano inclinato: come, per intelligenza, essendo la linea AB parallela all'orizonte, sopra 'l quale siano inclinati li due piani CA, CD, la perpendicolare CB, cadente sopra l'orizontale BA, chiama l'Autore la elevazione de i piani CA, CD; e suppone che i gradi di velocità del medesimo mobile scendente per li piani inclinati CA, CD, acquistati ne i termini A, D, siano eguali, per esser la loro elevazione l'istessa CB: e tanto anco si deve intendere il grado di velocità che il medesimo cadente dal punto C arebbe nel termine B.

Sagr. Veramente mi par che tal supposto abbia tanto del probabile, che meriti di esser senza controversia conceduto, intendendo sempre che si rimuovano tutti gl'impedimenti accidentarii ed esterni, e che i piani siano ben solidi e tersi ed il mobile di figura perfettissimamente rotonda, sì che ed il piano ed il mobile non abbiano scabrosità. Rimossi tutti i contrasti ed impedimenti, il lume naturale mi detta senza difficoltà, che una palla grave e perfettamente rotonda, scendendo per le linee CA, CD, CB, giugnerebbe ne i termini A, D, B con impeti eguali.

Salv. Voi molto probabilmente discorrete; ma, oltre al verisimile, voglio con una esperienza accrescer tanto la probabilità, che poco gli manchi all'agguagliarsi ad una ben necessaria dimostrazione. Fi-

"[Sagr] […] seem that up to the present we have established the definition of uniformly accelerated motion which is expressed as follows: A motion is said to be equally or uniformly accelerated when, starting from rest, its momentum [*celeritatis momenta*] receives equal increments in equal times. Salv. This definition established, the Author makes a single assumption, namely. The speeds acquired by one and the same body moving down planes of different inclinations are equal when the height of these planes are equal."

Fig. 2 Galileo's Theorem (Galilei 1890–1909, VIII, Day III, p. 205). *Source* Google Books Public Domain[7]

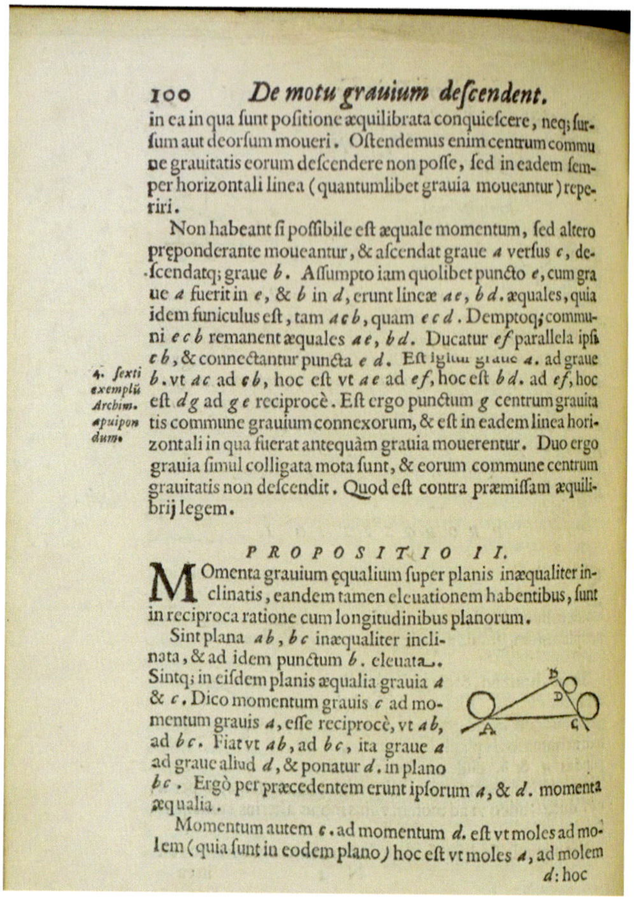

100 *De motu grauium deſcendent.*

in ea in qua ſunt poſitione æquilibrata conquieſcere, neq; ſurſum aut deorſum moueri. Oſtendemus enim centrum commune grauitatis eorum deſcendere non poſſe, ſed in eadem ſemper horizontali linea (quantumlibet grauia moueantur) reperiri.

Non habeant ſi poſſibile eſt æquale momentum, ſed altero preponderante moueantur, & aſcendat graue *a* verſus *c*, deſcendatq; graue *b*. Aſſumpto iam quolibet puncto *e*, cum graue *a* fuerit in *e*, & *b* in *d*, erunt lineæ *ae*, *bd*. æquales, quia idem funiculus eſt, tam *acb*, quam *ecd*. Demptoq; communi *ecb* remanent æquales *ae*, *bd*. Ducatur *ef* parallela ipſi *cb*, & connectantur puncta *e d*. Eſt igitur graue *a*. ad graue *b*. vt *ac* ad *cb*, hoc eſt vt *ae* ad *ef*, hoc eſt *bd*. ad *ef*, hoc eſt *dg* ad *ge* reciprocè. Eſt ergo punctum *g* centrum grauitatis commune grauium connexorum, & eſt in eadem linea horizontali in qua fuerat antequàm grauia mouerentur. Duo ergo grauia ſimul colligata mota ſunt, & eorum commune centrum grauitatis non deſcendit. Quod eſt contra præmiſſam æquilibrij legem.

4. ſexti exemplũ Archim. æquipondum.

P R O P O S I T I O I I.

Momenta grauium ęqualium ſuper planis inæqualiter inclinatis, eandem tamen eleuationem habentibus, ſunt in reciproca ratione cum longitudinibus planorum.

Sint plana *ab*, *bc* inæqualiter inclinata, & ad idem punctum *b*. eleuata. Sintq; in eiſdem planis æqualia grauia *a* & *c*. Dico momentum grauis *c* ad momentum grauis *a*, eſſe reciprocè, vt *ab*, ad *bc*. Fiat vt *ab*, ad *bc*, ita graue *a* ad graue aliud *d*, & ponatur *d*. in plano *bc*. Ergò per præcedentem erunt ipſorum *a*, & *d*. momenta æqualia.

Momentum autem *c*. ad momentum *d*. eſt vt moles ad molem (quia ſunt in eodem plano) hoc eſt vt moles *a*, ad molem *d*: hoc

[…]

Proposition II.

Momenta of equal bodies on planes which are unequally inclined, but which have the same elevation, are related to each other in the same ratio as the lengths of the [same] planes.

Fig. 3 Torricelli's Proposition II (Torricelli, 1644, p. 100). *Source* Courtesy University of Oklahoma, History of Science Collections[8]

[7] "Salv. Fermata cotal definizione [*motus aequaliter, seu uniformiter acceleratum dicimus eum, qui, a quiete recedens, temporibus aequalibus, aequalis celeritatis momenta sibi superaddit*] un solo principio domanda e suppone per vero l'Autore, cioè: *Accipio, Gradus velocitatis eiusdem mobilis super diversas planorum inclinationes acquisitos tunc esse aequales, cum eorumdem planorum elevation aequales sint.* Ex hac petitione dependet quasi universa illius doctrina de motu tùm accelerato, tum proiectorum." (Galilei 1809–1909, VIII, Day III, p. 205; author's italic. Translation retrieved from Galilei, 1954, p. 169).

[8] "*Propositio II.* Momenta gravium aequalium super planis inaequaliter inclinatis, eandem tamen elevationem habentibus [with respect to the ground], sunt in reciproca ratione cum longitudinibus planorum." (Torricelli 1644, p. 100; authors' translation).

Particularly, Torricelli named Galileo's theorem "*Supposit. Galilei* […] *Galileus principium* […]" (Fig. 4, see right side) and claimed:

> To afford the naturally accelerated Motion, Galileo assumes hypothetically a principle which he himself does not consider completely evident [$\neq$ he considers it experimentally evident], because he tries to prove it with the not very exact experiment of pendulum, that is: […].[9]
>
> Because we did not previously encounter this Theorem, at first we will confirm it with a proof, then we will arrive to show what in Galileo is a principle or a petition.[10]

Later on, he added an important lemma:

> The total *momentum* of a heavy body is to the *momentum* the body assumes on an inclined plane as the length of the inclined plane [its hypotenuse] is to the perpendicular.[11]

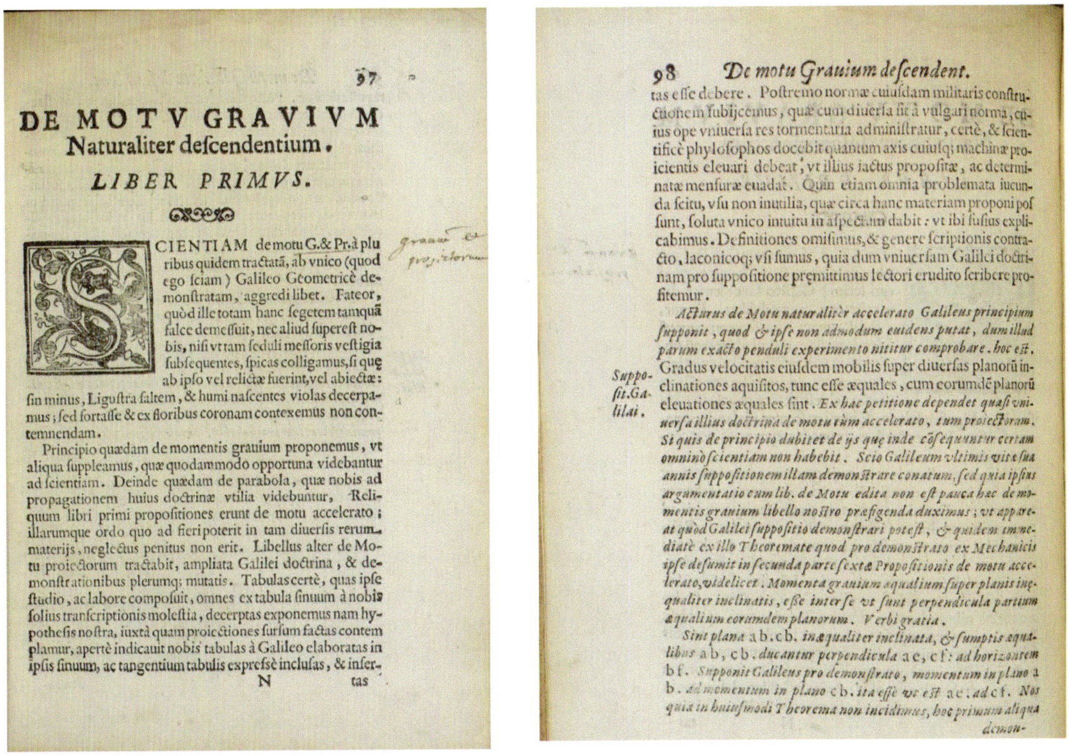

97

DE MOTV GRAVIVM
Naturaliter descendentium.
LIBER PRIMVS.

SCIENTIAM de motu G.& Pr. à pluribus quidem tractatã, ab vnico (quod ego sciam) Galileo Geometricè demonstratam, aggredi libet. Fateor, quòd ille totam hanc segetem tamquã falce demessuit, nec aliud superest nobis, nisi vt tam seduli messoris vestigia subsequentes, spicas colligamus, si quę ab ipso vel relictæ fuerint, vel abiectæ: sin minus, Ligustra saltem, & humi nascentes violas decerpamus; sed fortasse & ex floribus coronam contexemus non contemnendam.

Principio quædam de momentis grauium proponemus, vt aliqua suppleamus, quæ quodammodo opportuna videbantur ad scientiam. Deinde quædam de parabola, quæ nobis ad propagationem huius doctrinæ vtilia videbuntur. Reliquum libri primi propositiones erunt de motu accelerato; illarumque ordo quo ad fieri poterit in tam diuersis rerum materijs, neglectus penitus non erit. Libellus alter de Motu proiectorum tractabit, ampliata Galilei doctrina, & demonstrationibus plerumq; mutatis. Tabulas certè, quas ipse studio, ac labore composuit, omnes ex tabula sinuum à nobis solius transcriptionis molestia, decerptas exponemus nam hypothesis nostra, iuxtà quam proiectiones sursum factas contemplamur, apertè indicauit nobis tabulas à Galileo elaboratas in ipsis sinuum, ac tangentium tabulis expressè inclusas, & inser-

N tas

98 *De motu Grauium descendent.*

tas esse debere. Postremo normæ cuiusdam militaris constructionem subijcemus, quæ cum diuersa sit à vulgari norma, cuius ope vniuersa res tormentaria administratur, certè, & scientificè phylosophos docebit quantum axis cuiusq; machinæ proicientis eleuari debeat, vt illius iactus propositæ, ac determinatæ mensuræ euadat. Quin etiam omnia problemata iucunda scitu, vsu non inutilia, quæ circa hanc materiam proponi possunt, soluta vnico intuitu in aspectum dabit: vt ibi fusius explicabimus. Definitiones omisimus, & genere scriptionis contracto, laconicoq; vsi sumus, quia dum vniuersam Galilei doctrinam pro suppositione pręmittimus lectori erudito scribere profitemur.

Acturus de Motu naturalitèr accelerato Galileus principium supponit, quod & ipse non admodum euidens putat, dum illud parum exacto penduli experimento nititur comprobare. hoc est. Gradus velocitatis eiusdem mobilis super diuersas planorũ inclinationes aquisitos, tunc esse æquales, cum eorumdẽ planorũ eleuationes æquales sint. *Ex hac petitione dependet quasi vniuersa illius doctrina de motu tùm accelerato, tum proiectorum. Si quis de principio dubitet de ijs quę inde cõsequuntur certam omninò scientiam non habebit. Scio Galileum vltimis vitæ suæ annis suppositionem illam demonstrare conatum, sed quia ipsius argumentatio cum lib. de Motu edita non est pauca hæc de momentis grauium libello nostro præfigenda duximus; vt appareat quòd Galilei suppositio demonstrari potest, & quidem immediatè ex illo Theoremate quod pro demonstrato ex Mechanicis ipse desumit in secunda parte sextæ Propositionis de motu accelerato, videlicet. Momenta grauium æqualium super planis inęqualiter inclinatis, esse inter se vt sunt perpendicula partium æqualium eorumdem planorum. Verbi gratia.*

Suppo-sit. Ga-lilai.

Sint plana ab. cb. *inæqualiter inclinata, & sumptis æqualibus* ab, cb. *ducantur perpendicula* ae, cf: *ad horizontem* bf. *Supponit Galileus pro demonstrato, momentum in plano* ab. *ad momentum in plano* cb. *ita esse vt est* ae. *ad* cf. *Nos quia in huiusmodi Theorema non incidimus, hoc primum aliqua*

demon-

Fig. 4 Torricelli's reasoning upon Galileo's principle and a commented transcription (Torricelli 1644, pp. 97–98). *Source* Courtesy University of Oklahoma, History of Science Collections

[9] "*Acturus de Motu naturalitèr accelerato Galileus principium supponit, quod & ipse non admodum euidens putat, dum illud parum exacto penduli experimento nititur comprobare. hoc est.* (Torricelli 1644, p. 98; authors' translation).

[10] "*Nos quia in huiusmodi Theorema non incidimus, hoc primum aliqua demonstratione confirmabimus: protinus ad ostedendum [we are going to demonstrate by means of] id quod Galileo principium sive petitio est, accedemus*" (Torricelli 1644, pp. 98–99; authors' translation).

[11] "*Lemma.* Momentum totale gravius ad momentum quod habet in plano inclinato, est ut longitudo ispius plani inclinati ad perpendiculum" (Torricelli, 1644, p. 101; authors' translation). […]. Hic [without ad absurdum proof] Propositio secunda iterum. Momenta gravium aequalium super planis inaequaliter inclinatis sunt in reciproca ratione cum longitudinibus planorum […]." (Torricelli 1644, p. 102; authors' translation). Recent studies highlighted the significant connections between Torricelli's principle and Galileo's theorem (Capecchi and Pisano, 2007a, 2009, 2010a).The authors developed a logical–historical research on Torricelli's *Opera geometrica*.

> Scientiam de motu G. et Pr. a pluribus quidem tractatam, ab unico (quod egosciam) Galileo. Geometricè demonstratam, aggredi libet. [...] (*Ivi*, p. 97).
> [...].
> Definitiones omisimus, & genere scriptionis contracto, laconicoque; usi sumus, quia dum universam Galilei doctrinam pro suppositione praemittimus lectori erudito scribere profitemur.
> *Supposit. Galilai.*
>
> > *Acturus de Motu naturaliter accelerato Galileus principiu supponit, quod & ipse non admodum evidens putat, dum illud parum exacto penduli experimento nititur comprobare, hoc est*
> > Gradus velocitatis eiufdem mobilis fuper diuerfas planorũ inclinationes aquifitos, tunc effe æquales, cum eorumdẽ planorũ eleuationes æquales fint. *Ex hac petitione dependet quafi vniuerfa illius doctrina de motu tùm accelerato, tum proiectorum. Si quis de principio* [Galileo's] *dubitet de ijs quae inde consequuntur certam omnino scientiam non habebit.* [Torricelli quotes Galileo's theorem from The mechanics]. [...] *Momenta gravium aequalium super planis inequaliter inclinatis, esse inter se ut sunt perpendicula partium aequalium eorumdem planorum. Verbi gratia.* [From which, according to Torricelli, one can try to demonstrate, that is]. [...]. *Nos quia in huiusmodi Theorema non incidimus, hoc primum aliqua demonstratione confirmabimus: protinus ad ostendendum id quod Galileo principium sive petitio est accedemus.* [However Torricelli is going to demonstrate this theorem. To this end he puts first the] *Praemittimus* [of equilibrium]. Duo gravia sumu coniuncta ex se moveri non posse, nisi centrum commune gravitatis ipsorum descendat [with which he demonstrates the two propositions].
> > *Propositio I.* [Equality of the momentum of two bodies placed on differently inclined planes.
> > [...]
> > *Propositio II.* [Reciprocal proportion of bodies placed on differently incline planes]
> > [...]
> > *Aliter.* [Another proof for Proposition II]
>
> (*Ivi*, pp. 98–101)

1.1 The Structure of This Paper

As clarified in our Outline, in this paper we deal with a specific proposition stated by Torricelli at the beginning of his *De motu gravium naturaliter descendentium* (Torricelli 1644, pp. 95–243; *Id.*, 1919–1944, II), the already mentioned *Praemittamus*. This assertion with the explanation added by Torricelli is remarkable for at least four reasons that open four perspectives with regard to the logic and history of science. In the course of this research, the four following items will be addressed.

(1) The way in which the *Praemittamus* was used by Torricelli within his work to deal with questions connected with the movement of the bodies—and, in particular, of their barycentres[12]—under the action of gravity.
(2) The history of the *virtual work* principle. For, Torricelli's principle might also be interpreted as an expression of the virtual work principle. We will try to clarify what its role within the history of the virtual work principle was.
(3) The Torricelli's principle is an attempt to prove Galileo's theorem, which seems to Torricelli an *unjustified thesis*, a sort of *petitio principii.*
(4) The *Praemittamus* is used to prove physical propositions by the *reductio ad absurdum*. Thus, it can be seen as one of the possible keys to grasp the logical and methodological nature of the *ad absurdum* demonstrations within physics (mechanics at that time), especially when general principles, such as that of the *virtual work*, are used.[13]

---

[12] Taking into account the different historical approaches to deal with centre of gravity from Antiquity (Aristotelian) to Renaissance (Aristotelian school, Archimedean approaches, de Nemore and Tartaglia's speeches, Commandino, etc.), we take a few lines to introduce it to the reader who has no familiarity with these issues (Pisano 2020; Pisano and Capecchi, 2016 and recent Pisano's works). In modern terms (and in general) the centre of gravity (or centre of mass) of a rigid body is the point of application of the force–weight of a body. Of course, for those times, particular cases (such as the centre of mass of a hollow body) and the applications that are solved with advanced mathematical techniques (for example, the momentum of the centre of mass) should be excluded. On centre of gravity see also (very important): Commandino (1565, 1575).

[13] It can be interpreted as an early expression of the modern principle of virtual work (also works). Galileo—before Torricelli—already dealt with it. See sections below.

2 An Introduction to Torricelli's Principle and Its Physical-Epistemological *Status*

Probably, Torricelli's principle takes its origin from a common knowledge principle, which holds that a body is in equilibrium if its centre of gravity cannot fall in any of its possible motions. Since the time of Greek science, the centre of gravity of a body has been defined in two distinct ways which we could call dynamic and static respectively. In the dynamic definition, the centre of gravity of a body is a point, which moves naturally towards the centre of the Universe. For the static definition the centre of gravity is a point of a body that provides a state of indifferent equilibrium, e.g., if the body is hung from its centre of gravity (for example, by a spherical hinge) it will be in a state of equilibrium whatever its configuration is. It is necessary to note that in both definitions, it is understood that only one gravity centre exists.

At a first glance, the static definition may emerge as a logical consequence of the dynamic one. Indeed, one can see that in the case of a single body, the statement that it can be in equilibrium only if its centre of gravity cannot fall is implicit in the dynamic definition of the centre of gravity.[14] However, even though the dynamic definition is richer than the static one, the latter is simpler and offers better *rational criteria* to calculate the centre of gravity for a system of bodies.[15]

When the scientists of the sixteenth century began to formalise the conditions for the equilibrium of single and compound bodies, they were aware of the two above–mentioned criteria:

(1) The Archimedean or statics criterion, according to which a system of bodies is in equilibrium and does not fall if its centre of gravity is constrained.
(2) The Aristotelian, or dynamical perspective (Aristotle 1955b, see also 1949, 1955a, c, 1984, 1999; Winter 2007) according to which a physical system of bodies is in equilibrium if the weights and the *tendencies to motion* of the bodies are mutually balanced.

As frequently happens in scientific thought, two different criteria of a theory will conflict with each other even though they are mathematically equivalent. This fact can be easily seen in the case-study of the equilibrium of a lever when it has equal arms and equal weights. This case has been a subject of discussion since the sixteenth century to argue whether the system is in a state of stable, instable or indifferent equilibrium. Direct empirical experience made it difficult to find a doubt-free solution due to the unavoidable presence of friction, which influences the measurement of the relevant parameters. In order to understand the problem adequately, we cite the reasoning of both Guido Ubaldo del Monte (1547–1607; Renn and Damerow 2010), a supporter of the Archimedean approach, and Simon Stevin for the Aristotelian approach. In his *Mechanicorum Liber,* del Monte (Pisano and Capecchi, 2016) sets up two premises: (a) the centre of gravity of a body is unique; (b) the centre of gravity is determined by the rules explained by Archimedes. Following these premises, the two equal weights applied at the extremities of the lever can be replaced by a single weight (sum of the two) applied to the fulcrum. From this, it follows that the horizontal lever is in an indifferent state of equilibrium.

Simon Stevin, however, does not begin from the existence of a centre of gravity. On the contrary, he begins by reformulating this concept as a problem. Through a suggestive analysis and using the concept of gravity of position (*gravitas secundum situm*),[16] he calls into question if the centre of gravity, defined according to the Archimedean static criteria, can effectively be established. He concludes it cannot be defined and the *complex* horizontal lever-weights are in an unstable equilibrium state. According to Stevin, the centre of gravity as a point of suspension of indifferent equilibrium would only exist if the falling directions of natural bodies were all parallel. Because he thought the difference between the real situation (converging direction) and the ideal (parallel directions) to be empirically negligible, (Cfr. Dijksterhuis, 1970, pp. 48–59) the indifferent equilibrium can be admitted from a practical point of view.

[14] For example, see the case of a point of a body which moves towards the centre of the universe; bonded by a certain degrees of freedom (i.e., rotation, pivots side to side).

[15] Of course, once the centre of gravity for each body is known.

[16] The *gravitas secundum situm* concept was introduced by Jordanus de Nemore (fl. end XIII century). According to it, the gravity of a body depends not only on its weight but also on the way the body is constrained. That is, the closer the body to vertical (fall down) motion, the more the *gravitas secudum situm.* For a more in depth examination on *gravitas secundum situm,* and about Jordanus de Nemore see Pisano and Capecchi (2016), Dijksterhuis (1955–1958), I, Duhem (1905, II), Clagett ([1959]1961), Thomson (1976), Brown (1967–1968), Hoyrup (1988).

2.1 The Structure of *De Motu Gravium Naturaliter Descendentium*

The complete title of this work by Torricelli as referred to in his *Opera geometrica* is *De motu gravium naturaliter descendentium, et de proiectorum libri duo, in quibus ingenium naturae circa parabolicam lineam ludentis per motum ostenditur, et universa proiectorum doctrina unius descriptione semicirculi, absolvitur.* The text, published in 1644 is divided into two books: in the first book, Torricelli proposes a series of theoretical propositions on the motion of the bodies under the action of gravity, while in the second book he deals with several specific cases concerning the motions of the projectiles. The first book consists of 58 propositions, the second of 37 propositions to which a sort of appendix entitled *De motu aquarum* follows. It is to be noted that a series of tables related to the degree of elevation with which a projectile is shot is added to the text. The second book is closed by a brief treatise entitled *Della squadra.* The language is Latin, but the explanations concerning the use of the tables and the treatise *Della squadra* are written in Italian.

In the introduction to his book, Torricelli explicitly claims that his work has to be interpreted as an extension of Galileo's results and an improvement of some slight inaccuracies existing in Galileo's work. Thence, this book, at least in the explicit intentions of Torricelli, enters within the Galilean tradition, as pointed out in our Outline. Torricelli introduces his book reminding the reader that Galileo was the only one to treat the science of motion by means of geometrical demonstrations, that is, in a perspicuous manner. In particular, Torricelli has the intention to improve and to extend the theory *De motu locali* expounded by Galileo in the third day of *Discorsi e dimostrazioni intorno a due nuove scienze* (1638) and the *De motu projectorum,* the fourth day of that masterpiece. This second subject is mainly dealt with in the second book of *De motu*. The method is geometrical, where both synthetic and analytic procedures in the classical sense of these terms are used. There is no use of algebra.

Our thesis is that Torricelli goes beyond Galileo from several points of view. Of course, at that time, the relationship (calculated by proportions) between mass and velocity, as well as that among force, mass and velocity was an immature matter, for Galileo, too (Cfr. Pisano and Bussotti 2022). But, in *Le Mecaniche* (short version 1594 and longer one 17th; 1623; 1634) Galileo deals with *momentum* as *virtual motion*. It is the case of a motion caused by an unstable condition of equilibrium and by a mass:

> Momentum is the propension to go downwards, caused not so much on by gravity of a heavy [moveable] body, but by the mutual disposition of the different heavy bodies. By this momentum you will see many times a less heavy body counterbalances another heavier one […].[17]

In *Discorsi e dimostrazioni matematiche intorno a due nuove scienze* Galileo performed his idea of *Momentum* which seems quite similar to Torricelli's sentence on the *principle of virtual work* as reported above (section one):

> Salv. […] Then I say that all the momentum of the falling body is along the vertical and is a maximum when it falls in that direction; [see below Fig. 6] […]. Finally on the horizontal plane the momentum vanishes altogether; the body finds itself in a condition of indifference as to motion or rest; has no inherent tendency to move in any direction, and offers no resistance to being set in motion. For just as a heavy body

---

[17] "Momento è la propensione di andare al basso, cagionata non tanto dalla gravità del mobile, quanto dalla disposizione che abbino tra di loro diversi corpi gravi; mediante il qual momento si vedrà molte volte un corpo men grave contrapesare un altro di maggior gravità […]." (*Le Mecaniche* in Galilei 1890–1909, II, p. 159; see also Galluzzi 1970, Galluzzi and Torrini 1975–198). We specify that, of course, we refer to *Galilean momentum* as a physical property. We do not refer to modern moment (or rotation), even if Galileo calls it as "Momento" (*Ibidem*). The moment, generally speaking, corresponds to a measure of an effect caused by a physical quantity around a certain axis. Nowadays, in mechanics, it is a vector as the product of the mass and vector velocity of a particle's mass: $p = mv$. Other explicit comments regarding the *principle of virtual work* are referred to by Galileo (for *virtual velocities*) in *Le Mecaniche* (Galilei 1890–1909, II, pp. 155–191) and in *Discorsi intorno alle cose che stanno in sul'acqua* (Galilei 1890–1909, IV, pp. 3–141). Particularly, in this latter manuscript, Galileo clearly attributed his Momentum (in other modern words the *virtual velocity*) to Aristotle ([1936]1955, 847a 10–15, 847b 10, pp. 329–332; see also *Id.*, 1853, 1949, 1955a, 1984, 1999) also adding that the idea of what nowadays we call the *principle of virtual work* was born thanks to the observation of the motion of points, which rotate along a circumference. Galileo also dealt with the law of *virtual displacement* in more than one situation (Galilei 1890–1909 II, pp. 240–242, IV, pp. 68–69; VIII, pp. 310–331, pp. 329–330). Concerning Torricelli's principle, one can also consider John Wallis's assumptions (Wallis 1693), and Pierre Varignon's (1654–1722; see Varignon, 1725) who aimed at founding the entire statics upon an evident geometric principle: the composition of forces. In this sense, it is also an alternative to the *principle of virtual work.* Let us remark that in his letter to Johann Bernoulli (1667–1748), Varignon also dealt with concept of *virtual velocities*, as components of (vectorial) *virtual infinitesimal displacements* towards the direction of the forces (Bernoulli 1742a, b, II). After Bernoulli, the most significant contribution to the development of the *principle of virtual work* is probably due to Vincenzo Riccati (1707–1775) who tried to establish it upon simple principles easily accepted by his contemporaries, introducing *Principles of actions* in *Dialogo di Vincenzo Riccati della compagnia di Gesù* (Riccati 1749) and in *De' principi della meccanica* (Riccati, 1772). See also Ampère's demonstrations (Ampère 1806). Recently, Pisano (2015, 2017), Pisano and Capecchi (2010, 2013, 2015b). We also suggest to check the literature on/around the two Carnots: Carnot (1780, 1786, 1803a, b, 1813, 1990), Clapeyron (1834), Clausius (1850), Pisano (2010), Gillispie (1970–1979, 1971, 1976), Gillispie and Youschkevitch (1979), Gillispie and Pisano (2014), Nabonnand (2010, 2011).

> or system of bodies cannot of itself move upwards, recede from the common center [*comun centro*] toward which all heavy things tend, so it is impossible for any body of its own accord to assume any motion other than one which carries it nearer to the aforesaid common center.[18]

Coming back to Torricelli, the content of the first book of the *Opera geometrica*[19] can be summarised by dividing its propositions into groups concerning similar subjects:

(1) Propositions I–V: basic statements on the movement on an inclined plane.
(2) Propositions VI–IX: geometrical properties of parabola useful in mechanics, in particular, the determination of some features of the *latus rectum* are expounded.
(3) Propositions X–XVI: fall-time of the bodies in different circumstances especially along chords of circles and chords of parabolas. In these propositions, some of the classical Galilean results are referred to and specified. For example, the propositions X and XI concern the property that in a parabolic motion under the action of gravity the velocities are as the squares of times. Here instead of velocity Torricelli uses the polysemic and not always well explained term *impetus*,[20] but in this context *impetus* can be reasonably translated by velocity.
(4) Propositions XVII–XX: *Impetus* and parabolic motion. Here the decomposition of the parabolic motion is dealt with. In particular, in proposition XVIII Torricelli claims that in any point of a parabola there are two *impetus*. The one is horizontal. It is constant. The other one is perpendicular and it continuously increases.
(5) Propositions XXI–LVIII: all these propositions concern the time of composed movements, part of which is a fall movement and part of which is a uniform movement carried out with the velocity the body held at the end of the fall. Many interesting cases are dealt with, in which the properties of the conic sections are exploited and many significant theoretical problems are proposed and solved. For, example, proposition XLII is remarkable. Torricelli considers a segment AB and a longer segment F, both of the segments are horizontal (as a matter of fact, the position of F is not relevant, only its length is important). Torricelli poses the problem to find a point E of space so that A, B and E are collinear, such that $AE + EB = F$ and such that the time for a body, under the only action of gravity, to descend along AE is the same as the time necessity for its descent along EB. An interesting analytical method, in the Greek sense of the term, is used. After the rigorous proof, Torricelli also offers a practical solution to solve the problem which is quicker, but less perspicuous.

2.2 The *Praemittamus* Within Torricelli's Frameworks

This section is composed of two subsections: in the first one, we will briefly draw the scientific picture in which the *Praemittamus* is inserted. In the second one, we will enter into Torricelli's argumentation.

---

[18] "Salv. [...] dico, l'impeto massimo e totale [*momentum*] del grave per descendere essere per la perpendicolare [see figure p. 215] [see Fig. 6] [...] poichè, sì come è impossibile che un grave o un composto di essi si muova naturalmente all'in su, discostandosi dal comun centro verso dove cospirano tutte le cose gravi, così è impossibile che egli spontaneamente si muova, se con tal moto il suo proprio centro di gravità non acquista avvicinamento al suddetto centro comune [...]." (*Discorsi e dimostrazioni matematiche sopra due nuove scienze* in Galilei 1890–1909, VIII, Day III, p. 215; retrieved from Galilei, 1954, p. 181).

[19] Although the second book is in itself quite interesting, our work concerns themes developed by Torricelli in the first book. Thence, for our aims, it has been enough to summarize the content of this book.

[20] Particularly, we remark Jean Buridan's medieval theory (Buridan 1509; see also 1513, 1942; Del Monte 1581, 1588, 1615): (1) The impetus varies with the speed of the projectile and with its (in modern terms) mass. Paraphrasing Buridan, we can say, the more the speed is impressed by the motor on the mobile, the stronger is the transmitted impetus. (2) The impetus is a permanent quality different from both motion and *heavy body* (and so different from the weight) of the projectile which it was linked with. The *impetus* might be weakened by the movement, i.e., by the air resistance and the degree of inclination of the launch. This involves a case-study on the trajectory of projectiles which scholars (e.g., Tartaglia) had not yet sufficiently and theoretically developed (Pisano and Capecchi 2016). Perhaps, it was also not yet sufficiently—and—theoretically developed by other scholars in the sixteenth century who more or less based their reasoning on the medieval theory of impetus (Buridan 1509). Nicolas d'Oresme's (ca. 1320/1325–1382) version should also be considered. An English study is in *The Science of Mechanics in the Middle Ages* (Clagett 1959). Buridanus' work is reproduced by Maier (1968) who included—with some modifications—the Parisian edition dated 1509. For some good introductory comments of *Subtilissimae Quaestiones* one can see Clagett (1959 and secondary literatures cited). Clagett dated Buridan's manuscript around 1357. It is archived at the *Vatican Library in Roma* (Vat. Lat. 2136, 1r.).

The *Praemittamus* and Its Context

In order to introduce the *Praemittamus*, we highlight that, from a conceptual point of view, Torricelli, in his *Opera geometrica* seems to bring Galileo's position to an extreme by assuming that the bodies could be placed not on the Earth, but at "infinite distance from it" or that the cords, by which heavy bodies are suspended, are actually parallel. In fact, for Galileo the physical results obtained by assuming an infinite distance were to be considered empirically approximate. For Torricelli they are complete and only justified by means of mathematical calculation[21]; "[...] but not for this the proportion of the figure already demonstrated fails." This is what Torricelli claims:

> The following objection is very common among quite serious scholars, too: "Archimedes has supposed something false when he considered parallel the threads to which the magnitudes hanging from a lever are suspended". Actually, I (for the peace of those learned men) have thought that the foundations of mechanics should be considered in a completely different manner. I admit that, if physical magnitudes are freely suspended to the lever, the material threads will converge when they are directed to the centre of the earth. However, if that lever or those bodies are not considered on the earth's surface, but in very high regions, beyond the sun's orbit, then those threads (though directed to the centre of the earth) will not tend to converge, but will be almost mutually equidistant. Let us conceive that this mechanical lever is brought beyond the stellar lever of the firmament, to an infinite distance. In this case, who will not understand that the threads are not convergent anymore, but they are exactly parallel? When I consider the lever, which weights geometrical figures, I do not conceive that it is posed among the sheets of the books, where I see it depicted, neither I suppose that the point, to which the magnitudes tend, is the centre of the earth.[22]

Since the cords by which bodies are "simul coniuncta" [jointed] to the lever are infinitely long, i.e., not finite, it appears that the author chooses a point at the infinite in his mathematical theory. Specifically:

> It is particularly fruitful for a geometer to develop all the operations by means of the intellect, resorting to the abstraction. For, who will deny that, if one is allowed to consider the figures as hanging from a lever, it will be also legitimate to consider the lever as posed beyond the limit of the world, posed at an infinite distance? [...] All the advantages are lost if one considers those figures as tending to the sole Earth's centre.[23]

After all Torricelli, differently from Galileo, seems to be more interested in the mathematical aspect of the theory and shows all his capability of abstraction, as well as his ability to relate it to reality. He wastes no time reporting theoretical offences:

> Thus, the following fundamental proposition of mechanics, namely that the threads of the lever are parallel, can be considered false when the magnitude hanging from the lever are real, physical objects and tend to the centre of the Earth. In contrast to this, such a proposition will not be false when the magnitudes (both in the cases they are abstract or concrete) do not tend to the centre of the Earth either to any other point close to the lever, but to a point posed at an infinite distance.[24]

---

[21] The references concern the results obtained by Archimedes with regard to the areas' calculation by means of his mechanical method (Capecchi and Pisano 2009).

[22] "*Principiò, vulgatiſſima eft etiam apud grauiſſimos viros obiectio illa, videlicet. Archimedem ſuppoſuiſſe aliquod falſum, dum fila magnitudinum ex librá pendentium conſiderauit tanquam inter ſe parallela, cum tamen re vera in ipſo tetræ centro concurrere debeant. Ego verò (quod pace clariſſimorum virorum dictum ſit) crediderim fundamentum Mecanicum longè alia ratione eſſe conſiderandum. Concedo ſi Fiſicæ magnitudines ad libram liberè ſuſpendantur, quòd fila materialia ſuſpenſionum conuergentia erunt; quandoquidem ſingula ad centrum terræ reſpiciunt. Verumtamen ſi eadem libra, licet corporea, conſideretur non in ſuperficie terræ, ſed in altiſſimís regionibus vltrâ orbem Solis; tum fila (dummodo adhuc ad terræ centrum reſpiciant) multò minùs conuergentia inter ſe erunt; ſed quaſi æquidistantia. Concipiamus iam ipſam libram Mecanicam vltrà stellatam libram firmamenti in infinitam diftantiam eſſe prouectam; quis non intelligit fila ſuſpenſionum iam non ampliùs conuergentia, ſed exactè parallela fore? Quando ego conſidero libram, figuras Geometricas ponderantem, non cõcipio illam eſſe inter cartas librorum in quibus depicta conſpicitur; ne que ſuppono punctum, ad quod magnitudines ipſius tendunt, eſſe centrum terræ;* [...]" (Torricelli 1644, *De Dimensione Parabolæ*, p. 9. *Id.*, 1919–1944, I, p. 96; authors' translation).

[23] "*Peculiare quoddam beneficium habet Geometra, cum ipſe abstractionis ope, omnes operationes ſuas mediante intellectu exequatur. Quis igitur mihi hoc negauerit, ſi libeat conſiderare figuras appenſas ad libram, quæ quidem libra vltra mundi confinium in infinitam distantiã remota ſupponatur?* [...]. *Destruit ergò beneficium ſuum quiſquis figuras illas, tamquam ad vnicum terræ centrum tendentes, contemplatur.*" (Torricelli 1644, *De Dimensione Parabolæ*, p. 10. *Id.*, 1919–1944, I, p. 97).

[24] "*Tunc itaque falſum dici poterit fundamentum Mecanicum, nempe fila libræ parallela eſſe, quando magnitudines ad libram appenſæ fiſicæ ſint, realeſque, & ad terræ centrum conſpirantes. Non autem falſum erit, quando magnitudines (ſiue abftractæ, ſiue concretæ ſint) non ad centrum terræ, neque ad aliud punctum propinquum libræ reſpiciant; ſed ad aliquod punctum infinitè distans* [from the Earth] commitantur." (Torricelli 1644, *De Dimensione Parabolæ*, p. 11. *Id.*, 1919–1944, I, pp. 97–98).

We can also note, understandably, a certain caution expressing his ideas too explicitly:

> For the rest, to facilitate the reading of this work, we will not depart from the usual terms. Thence, we will call "Centre of the Earth" that point to which the magnitudes hanging from the lever are thought to tend. We will call "Horizon" the plane erected perpendicularly to the line connecting the centre of the Earth with the centre of the lever.[25]

As a reflection and not a conclusion, it is reasonable to assume that—with respect to the common experimental approach of that time, which represented the framework for his physical–mathematical studies—Torricelli seems to have developed a good logical-mathematical skill. He had guessed that a new and idealised kind of mathematics was necessary to deal with mechanics (Cfr. Grattan-Guinness, 1990). Later on, Isaac Newton grasped all the consequences of this idealization while constructing the foundations for his *Principia* (1687; see Pisano and Bussotti listed works on the subject).

Analysis of the *Praemittamus*

The complete formulation of the *Praemittamus* is followed by a commentary, which is an explanation of this *premise* and sounds like this:

> Two heavy bodies linked together cannot move by themselves unless their common centre of gravity descends. [Explanation]. For, when two heavy bodies are linked so that if one moves, the other one moves too, these bodies can be considered as composing a sole body. In any case, when the two bodies are constrained to form a sole body by means of a lever or a pulley or any other machine, that body will never move unless its gravity centre descends. When the situation is such that the common centre of gravity can descend in no way, the body remains at rest in its position. Otherwise, it will move in vain, that is, its movement will be a horizontal one, which never tends upwards.[26]

Therefore, Torricelli is describing a criterion for the stability of a system of two (or, in fact, several) bodies composing a sole body.

The statement, according to which, if no external action modifies the constraints of the system, no transversal movement is possible unless the centre of gravity can descend, describes the condition of vertical equilibrium of the system itself. For, Torricelli does not exclude that, for some reasons, a horizontal movement could subsist. However, it is clear that, according to Torricelli, under the action of gravity, the movement perpendicular to the gravity direction is not connected to gravity, therefore it is not connected with the centre of gravity of the body or system of constrained bodies.

At least two considerations seem appropriate:

(1) From a historical point of view, we avoid using the expression *work of a force*, because the subject with which we are dealing precedes the publication of the Newton's *Principia*.[27] Nonetheless, Torricelli gives the impression, in this circumstance, to grasp, at least intuitively, such a concept. For, in his mind, it is clear that—to use a modern language—gravity exerts no work along a horizontal direction.
(2) Furthermore, though the concepts of mass and weight were not clearly distinguished in Torricelli's age (this will be another merit of Newton), he seems intuitively to understand an important truth: the gravity centre of a physical body is a fundamental element when a gravity attraction exists. Since the action of gravity is downwards and not upwards, if the constraints of the system are so constituted to compensate the action of gravity (so that the centre of gravity cannot descend), the body can, in no way, move along the vertical.

A Historical and Terminological Annotation

While speaking of the horizontal movement, Torricelli uses the word *frustra* (*in vain*). This terminology, induces one to think that he was interested in analysing movements (or the components of the movements, as in the inclined planes) along the vertical, which is rather obvious. Probably these studies were also connected to the improvement of the funicular engines, which is not surprising given the epoch when Torricelli wrote his works.[28]

---

[25] "*Cæterum, breuitatis, & facilitatis gratià à vocabulis confuetis non difcedemus; punctumque illud ad quod magnitudines libræ contendere fupponuntur, Centrum terræ nominabimus; Planum verò illud, quod erectum eft ad lineam cõnectentem prædictum punctum cum centro libræ, Horizontem de more appellabimus*" (Torricelli 1644, *De Dimensione Parabolæ*, p. 11. *Id.*, 1919–1944, I, p. 98).

[26] See above for the original text. (Torricelli 1644, p. 99; Torricelli, 1919–1944, p. 105).

[27] On the *Principia*'s subject see recently: Bussotti and Pisano (2014a b, c, 2017), Pisano and Bussotti (2016, 2017b). It is well known that the quantity *Work* was not a Newtonian concept, but of course, before speaking of the Work of a force, it is necessary to define what a force is.

[28] With regard to the relations between the development of the machines and the development of mechanics, as well as other parts of physics, such as heat theory, the literature is abundant. We limit references to: Gillispie and Pisano (2014), Pisano and Bussotti (2014a, b, 2015b), Pisano (2015), Reuleaux (1876), Singer (1954–1958). Particularly, in our works it is also possible to find more extensive references to the literature.

Finally, the *Praemittamus* represents one of the first examples of a *general principle* posed *explicitly* at the basis of physics (and of part of structural mechanics). The term *general* is appropriate because the principle concerns all the situations in which a constrained motion of a body exists. Therefore, it regards very general situations. This generality is clearly detectable in his audacious idea to extend the concept of centre of gravity from a single body to an aggregate of two (or potentially more than two) bodies. This extension might appear not well founded, but it is the unquestionable mark of the generality of Torricelli's view with regard to the movements which take place under the action of gravity.

The principle poses a clear distinction between propositions, which are assumed to be true without any demonstration and propositions which have to be proved. This distinction is obvious if one refers to the structure of Euclid's *Elements* or to the works by Archimedes, where axioms or principles are clearly distinguished from proved theorems and/or solved problems. However, such a distinction was far less clearly stated in the period in which Torricelli worked. For, the principles and the premises of the reasoning were often implicitly assumed, rather than explicitly formulated in advance. Taking into account that Archimedes was, jointly with Galileo, the main source of Torricelli's inspiration (Pisano and Capecchi, 2010b) it is not unlikely that he drew inspiration from this great scientist as to the form with which to present a treatise (on this point, see Sect. 2.3).

Let us now see the two propositions of *De Motu Gravium* where the *Praemittamus* is used. In Proposition I, Torricelli is going to prove that (Fig. 5).

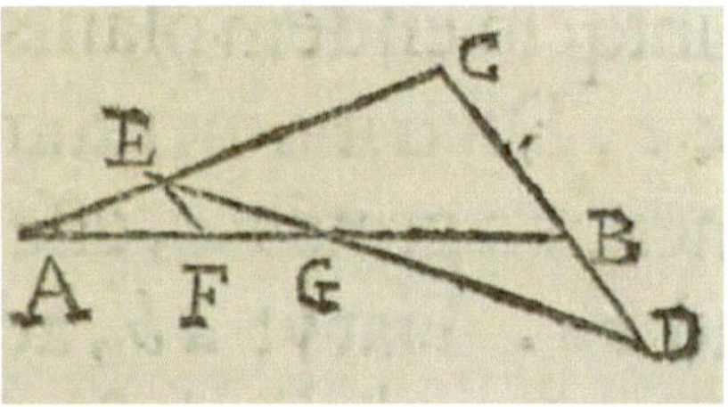

If two heavy bodies [*A* and *B*] are posed on two planes having a different inclination but the same altitude, so that the weights of the two bodies have the same ratio as the longitudes of the planes, the two bodies have the same *momentum*.

Fig. 5 Particular of Proposition I used by Torricelli to prove the statement (Torricelli 1644, p. 99). *Source* Courtesy University of Oklahoma, History of Science Collections[29]

Referring to Torricelli's picture, it is necessary to prove that, given two inclined planes *AC* and *BC* with the same altitude, but different inclinations and indicating the weights of the two heavy bodies by *A* and *B*, if

$$AC : A = BC : B$$

holds, then there is equilibrium (the bodies have the same *momentum*). The proof conceived by Torricelli runs like this: let us suppose that there is no equilibrium and that, for example, *A* goes up in *E* towards *C*. Hence *B* will descend in *D* along the direction *CB*, being $AE = BD$. Thus, $AC + CB = EC + CD$. Let *EF* be drawn parallel to *CD*. It holds:

$A{:}B = AC{:}CB$ (Hypothesis)
$AC{:}CB = AE{:}EF$ (Similarity of the triangles *ACB* and *AEF*)
$AC{:}CB = BD{:}EF$ $(BD = AE)$
$BD{:}EF = DG{:}EG$ (Similarity of the triangles *BDG* and *EFG*)
$A{:}B = DG{:}EG$.

Hence, *G* is the new centre of gravity. This means that the centre of gravity moves in a line parallel to the horizon. But the premise claims that a movement of a system of two joined bodies on which no external action apart from gravity is exerted can move only if its gravity centre goes downwards. However, in this circumstance the centre of gravity would move horizontally. Therefore, because of the *Praemittamus*, the supposed movement is impossible, and the two bodies remain at rest. Thus, an equilibrium is obtained, both bodies have the same *momentum*.

[29] "*PROPOSITIO I.* Si in planis inæqualiter inclinatis, eandem tamen eleuationem habentibus, duo grauia conftituantur, quæ inter fe eandem homologè rationem habeant quam habent longitudines planorum, grauia æquale momentum habebunt". (Torricelli 1644, Propositio I, p. 99).

This proposition is not new. In fact, it is one of those proved by Galileo (see Fig. 10), but what is significant is the method of demonstration and the novelty of Torricelli's *Praemittamus*.

The second circumstance in which Torricelli uses his premise is Proposition II. Here Torricelli proves another classical proposition by Galileo: The *momenta* of two equally heavy bodies posed on two planes with different inclinations, but having the same altitude, are inversely as the plane-lengths. The premise is not used directly, but only in the sense that the first proposition, proved by the premise, is exploited.

2.3 Archimedes's Influence on Torricelli

Archimedes's influence was relevant for most scholars of the seventeenth century, and particularly for Torricelli (Pisano and Bussotti 2014b). Indeed, it is detectable Torricelli's attitude to confront geometric matter both with the methods of the ancients, in particular the exhaustion method, and with the indivisibles, so attempting to compare the two, as clearly stated by his correspondence with Cavalieri (Torricelli 1919, 1944; see mainly vol. 3). Torricelli, in particular, proposed twenty-one different ways of squaring a parabola (Heath, 2002; *Quadrature of the parabola*, Propositions 17 and 24, pp. 246, 251), eleven times with exhaustion, ten with indivisibles. The *reductio ad absurdum* proof is always present while dealing with exhaustion. It is known that Archimedes solved this problem in two different manners: (1) by means of his mechanical method; (2) by the use of exhaustion. The influence of Archimedes on Torricelli is detectable also considering the two general approaches of both scientists, because Torricelli considered axioms in a similar way to Archimedes's and different from Euclid's. That is, axioms could have their origin not only in evidence, but also in experience. Additionally, they could be local axioms, namely unproved propositions valid in a given theory, whose validity could be non-universal, differently from the axioms used by Euclid in his geometry. For, Euclid's axioms were, at that time, believed universally valid. This poses an interesting difference between geometry, as a universal doctrine of space, and mechanics, a more specific doctrine in which other particular axioms are added to the Euclidean structure.

Torricelli exploited all Archimedes's techniques in an original manner. Thence, the knowledge of Archimedes's contribution is also fundamental in an historical study of Torricelli's mechanics. Archimedes was the first scientist to set rational criteria for determining centres of gravity of bodies and his work contains physical concepts formalised on a mathematical basis. In the Book I of the *On the Equilibrium of Planes* (Heath, 2002) Archimedes, besides studying the rule governing the law of the lever, also finds the centres of gravity of various geometrical plane figures (Heath, 2002; Clagett, 1964–1984; Heiberg, 1881). Archimedes's typical method of arguing in mechanics was by the use of the *reductio ad absurdum*. Torricelli in his study on the centres of gravity resumes the same approach.

Archimedes in *On the Equilibrium of Planes* presents a system of seven axioms and the structure of the text is similar—but not identical—to that of Euclid's *Elements*, despite the different content.[30]

However, Torricelli's *Praemittamus* is a single principle and not a system of principles explained in a complete axiomatic structure. On the other hand, the first book of another fundamental work by Archimedes concerning physics, *On Floating Bodies*, is opened by a single premise, which has precisely the style of Torricelli's. For, Archimedes writes:

[30] In *On the Equilibrium of planes* Archimedes does not follow, in some respects, a Euclidean structure. For, we do not find something like this: definition of centre of gravity, concepts of balance and determination- calculation of centres of gravity, mathematical procedure to calculate centre of gravity for different geometric configurations, replacing the section of the balance with a body, observation/relations between numbers/interpreted by a ratio, geometrical figures to which numbers values were designed, geometrical figures presented also as masses of balance, the ideal parallelism of the direction of forces-weight, etc. Archimedes's approach, in this context, mainly does not present axioms and/or general theory, but he focuses on a centre of gravity by one, two or more compounded magnitudes. Particularly, he focuses on ratios and rational criteria to be used to prove theorems. On that, he uses many *reductio ad absurdum* proofs. In addition, Archimedes does not intend to develop a complete theory of mechanical science, or to write a detailed treatise. Instead, he concentrates on one single, important problem: the determination of centres of gravity of bodies of any shape. Archimedes based his theory on Euclid's theories as expounded in the *Elements*, which were widely accepted at the time. Nonetheless, the order of the propositions follows a basic logic, similar, but not identical, to that of Euclid, which proceeds systematically from simpler cases to more complex ones. In this work by Archimedes, the axiomatic structure and the concatenation of the propositions are less evident. The first four propositions for example are demonstrated directly from the suppositions and are not connected in a serial order; the second after the first, the third after the second, and so on. The main difference between the organization of Euclidean and Archimedean theories is in the nature of the principles. Indeed, not all Archimedean principles are *necessary* with respect to *necessary* attributes by Euclid; some of them are indeed *logically true*, or *empirically evident*. Archimedes introduces significant statements based upon experience: the most interesting empirical assertion is that if equal weights are positioned on a lever, at different distances from fulcrum, there is no equilibrium but instead a tendency for the more distant weight to move downward. It has a clear empirical nature: "[…] equal weights suspended at unequal distances [from fulcrum] are not in equilibrium but [they] incline towards the weight suspended at the greater distance [from fulcrum]" (Archimedes's *On the Equilibrium of planes* in Heath 2002, Book I, p. 189). On the subject, see Pisano and Capecchi (2013, 2010b).

Let it be supposed that a fluid is of such character that, its parts lying evenly and being continuous, that part which is thrust the less is driven along by that which is thrust the more; and that each of its parts is thrust by the fluid which is above it in a perpendicular direction if the fluid be sunk in anything and compressed by anything else.[31]

Torricelli *Praemittamus* resembles the style of Archimedes's statement: it concerns general physical situations; it is a single proposition; it is a general principle on the behaviour of bodies. Thence, it is reasonable to think that *On the Floating Bodies* inspired Torricelli. More explicitly, Torricelli declares and explains his Archimedean background.

[...] Among all works, pertaining to the mathematical disciplines, the inventions and discoveries by Archimedes seem to claim for themselves the role of a Prince. These works, with their miraculous sharpness scare the minds.[32]

However, the Archimedean influence on Torricelli goes further. The well-known books *De sphaera et solidis sphaeralibus* (Torricelli 1644, *Liber primus*, pp. 11–46) present an extension of the Archimedean proofs of books I–II of *On the sphere and cylinder* (Heath, 2002, pp. 1–90). For, we read:

Two books on sphere and spherical solids. In which Archimedes's doctrine of the sphere and cylinder is once again composed and more widely developed and more universally extended to any kind of solids, which can be generated either around or within a sphere from the rotation of a regular polygon.[33]

In other parts of his work, Torricelli confronted a set of problems not yet solved by Archimedes, or by the other ancient mathematicians. With the same style as Archimedes, he does not try to arrive at the first principles of the theory and does not restrict himself to a single way of demonstrating (Torricelli 1644, *De solido hyperbolico acuto problema secundum,* p. 116).

2.4 On Galilean *Aggregatum* and Torricellian *Simul Coniuncta*

The two main sources of inspiration for Torricelli were Archimedes and Galilei. Thence, after having detected Archimedes's influence on Torricelli, it seems appropriate to highlight Galileo's.

In the Renaissance[34] the idea of centre of gravity of a single body was intuitively extended to Galilean "aggregatum" and to Torricellian "simul coniuncta" bodies. In both cases, the two scientists deal with compounded bodies.

In the Third Day-*Discorsi*, Galileo proposes (Pisano and Bussotti 2012; Bussotti and Pisano 2023) an early approach to the centre of gravity of *aggregate* bodies that, when subjected to their heaviness, only move toward the centre of the Earth. The arguments appear in a context, different from the centre of gravity theory, in which Galileo faces the dynamic behaviour of bodies moving on differently inclined planes.

In *Le Mecaniche*, when he deals with the equilibrium of bodies moving down an inclined plane, Galileo compares the weights and the movements of two bodies connected by a rope. One moves on the inclined plane and the other moves on the vertical one. He concludes that for an *equilibrium status*, the drawdown (on the vertical ones) is inversely proportional to the bodies' weights. In other words, Galileo once again presents an early thesis, according to which, the centre of mass of an aggregate of bodies, subjected to weight only, tends to move toward the centre of the Earth:

[31] Archimedes (1897), p. 253; see also Archimedes (2002).

[32] "*Proemium.* Inter omnia opera ad Mathematicas difciplinas pertinentia, iure optimo Principem fibi locum vindicare videntur Archimedis inuenta; quæ quidem ipfo fubtilitatis miraculo terrent animos." (Torricelli 1644, *Proemium*, p. 5).

[33] De Sphaera et Solidis Sphaeralibus Libri Duo. In quibus Archimedis Doctrina de Sphaera and cylindro denuo componitur, latius promovetur, *Et in omni specie solidorum, quae vel circa, vel intra Sphaeram, ex conversione polygonorum regularium gigni possint, universalius Propagatur* (Torricelli, 1644, *De Sphaera et Solidis Sphaeralibus*, p. 2).

[34] On the role played by physical laws in the history of science (from antiquity to the Renaissance) see Poinsot (1806, 1838), Bussotti and Pisano (2014a; 2013), Pisano (2010, 2012, 2013a, b), Pisano and Robarts (2022, 2024), Gaukroger et al., (2000), Schuster (2000, 2013a, b), Fourier (1878, 1888–1890), Jardine (1988), Knobloch et al. (2001), Kragh (1987), Lindberg (1976).

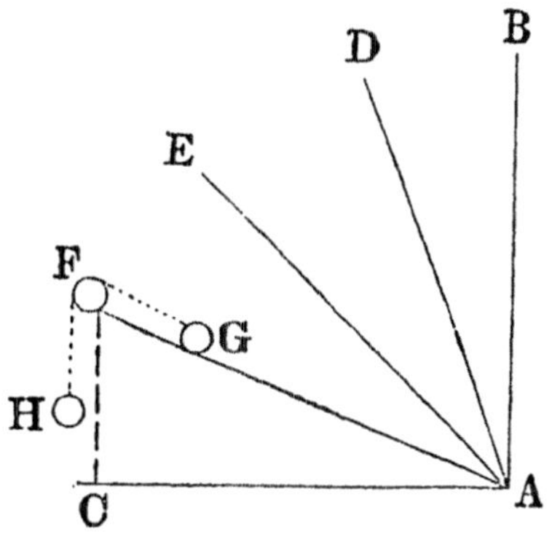

Salv. Since then I have your permission, let us first consider this notable fact, that the momenta or speeds ["i momenti o le velocita"] of one and the same moving body vary with the inclination of the plane. The speed reaches a maximum along a vertical direction, and for other directions diminishes as the plane diverges from the vertical. Therefore the impetus, ability, energy, ["l'impeto, il talento, l'energia"] or, one might say, the momentum ["il momento"] of descent of the moving body is diminished by the plane upon which it is supported and along which it rolls. For the sake of greater clearness erect the line *AB* perpendicular to the horizontal *AC*; next draw *AD*, *AE*, *AF*, etc., at different inclinations to the horizontal; I say that that all the maximum and total impetus [momentum] of the falling body is along the vertical and is a maximum when it falls in that direction; the momentum is less along *DA* and still less along *EA*, and even less yet along the more inclined plane *FA*. Finally on the horizontal plane the momentum vanishes altogether; the body finds itself in a condition of indifference as to motion or rest; has no inherent tendency to move in any direction, and offers no resistance to being set in motion. For just as a heavy body or system of bodies cannot of itself move upwards, or recede from the common centre ["comun centro"] toward which all heavy things tend, so it is impossible for any body of its own accord to assume any motion other than one which carries it nearer to the aforesaid common centre. Hence, along the horizontal, by which we understand a surface, every point of which is equidistant from this same common centre, the body will have no momentum whatever.

Galilei 1890-1909, VIII, III. p. 215 (retrieved from Galilei 1954, p. 181)

Fig. 6 Particular on centre of gravity related to an inclined plane (Galilei 1890–1909, VIII, III, p. 215). *Source* Google Books Public Domain

This argumentation appears in a passage in which Galileo discusses the dynamic behaviour of bodies that move on planes having different inclinations. Therefore, the situation is observed from an external position in respect to the centre of gravity.

In *Le Mecaniche*, when discussing the equilibrium of a body on an inclined plane, Galileo compares the weights and movements of two bodies linked by a cord; as already pointed out, one moves on the inclined plane and the other moves vertically. This, to conclude that an equilibrium configuration–and–the lowering of both bodies (if they are measured-calculated vertically) are inversely proportional to their respective weights (Fig. 6):

> [...] it should be noticed that although the movable body *E* will have passed over all the line *AC* in the same time that the other heavy body *F* will have fallen through an equal interval, nevertheless the heavy *E* will not have been removed from the common centre of heavy things more than the distance along the perpendicular *CB*; while the heavy body *F*, descending perpendicularly will have dropped by a space equal to the whole line *AC*. In addition, since heavy bodies do not have any resistance to transverse motions except in proportion to their removal from the centre of the earth, then the movable body *E* not being raised more than the distance *CB*, in the whole motion *AC*, while *F* has dropped perpendicularly as much as the whole length of *AC*, we may rightly say that the travel of the force *F* has the same ratio to the travel of the force *E* as the line *AC* has to the line *CB*, or as the weight *E* has to the weight *F*. Therefore it is very important to consider along what lines the motions are made, and especially of inanimate heavy bodies, whose momenta have their whole power and their entire resistance in the line perpendicular to the horizon; and in the others lines, transversely rising or falling, they only have that power, impetus, or resistance which is greater or lesser according as the inclination approach more or less to the perpendicular elevation [see Fig. 10 below]. (Galilei 1890–1909, II, pp. 185–186; authors' translation/adapted from Seeger (1966, 2013, pp. 149–150).

In what follows, two more Galilean reasonings, drawn from *Discorsi e dimostrazioni*, such as examples are presented (Fig. 7).[35]

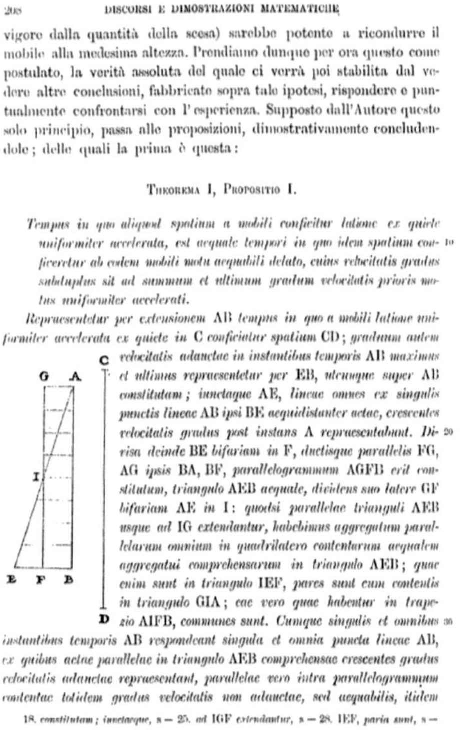

208 DISCORSI E DIMOSTRAZIONI MATEMATICHE

vigore dalla quantità della scesa) sarebbe potente a ricondurre il mobile alla medesima altezza. Prendiamo dunque per ora questo come postulato, la verità assoluta del quale ci verrà poi stabilita dal vedere altre conclusioni, fabbricate sopra tale ipotesi, rispondere e puntualmente confrontarsi con l'esperienza. Supposto dall'Autore questo solo principio, passa alle proposizioni, dimostrativamente concludendole; delle quali la prima è questa:

Theorema I, Propositio I.

Tempus in quo aliquod spatium a mobili conficitur latione ex quiete uniformiter accelerata, est aequale tempori in quo idem spatium conficeretur ab eodem mobili motu aequabili delato, cuius velocitatis gradus subduplus sit ad summum et ultimum gradum velocitatis prioris motus uniformiter accelerati.

Repraesentetur per extensionem AB *tempus in quo a mobili latione uniformiter accelerata ex quiete in* C *conficiatur spatium* CD; *graduum autem velocitatis adauctae in instantibus temporis* AB *maximus et ultimus repraesentetur per* EB, *utcunque super* AB *constitutam; iunctaque* AE, *lineae omnes ex singulis punctis lineae* AB *ipsi* BE *aequidistanter actae, crescentes velocitatis gradus post instans* A *repraesentabunt. Divisa deinde* BE *bifariam in* F, *ductisque parallelis* FG, AG *ipsis* BA, BF, *parallelogrammum* AGFB *erit constitutum, triangulo* AEB *aequale, dividens suo latere* GF *bifariam* AE *in* I: *quodsi parallelae trianguli* AEB *usque ad* IG *extendantur, habebimus aggregatum parallelarum omnium in quadrilatero contentarum aequalem aggregatui comprehensarum in triangulo* AEB; *quae enim sunt in triangulo* IEF, *pares sunt cum contentis in triangulo* GIA; *eae vero quae habentur in trapezio* AIFB, *communes sunt. Cumque singulis et omnibus instantibus temporis* AB *respondeant singula et omnia puncta lineae* AB, *ex quibus actae parallelae in triangulo* AEB *comprehensae crescentes gradus velocitatis adauctae repraesentant, parallelae vero intra parallelogrammum contentae totidem gradus velocitatis non adauctae, sed aequabilis, itidem*

18. *constitutam; iunctaque*, s — 25. *ad* IGF *extendantur*, s — 28. IEF, *paria sunt*, s —

[...] *for if the parallel lines in the triangle* AEB *are extended to* GI, *then the sum* [aggregatum] *of all the parallels contained in the quadrilateral is equal to the sum* [aggregatui] *of those contained in the triangle* AEB.[35]

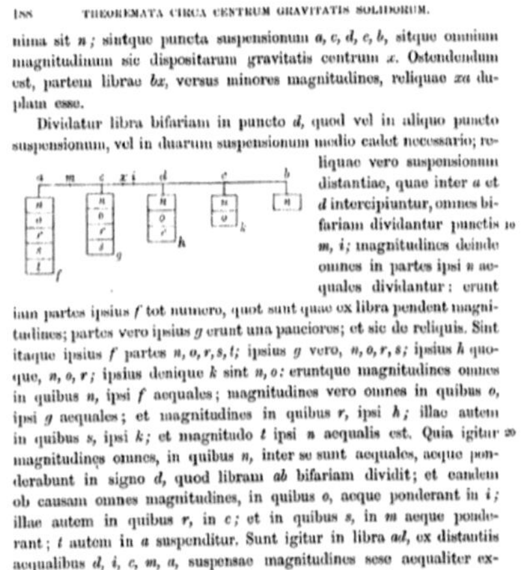

188 THEOREMATA CIRCA CENTRUM GRAVITATIS SOLIDORUM.

nima sit *n*; sintque puncta suspensionum *a*, *c*, *d*, *e*, *b*, sitque omnium magnitudinum sic dispositarum gravitatis centrum *x*. Ostendendum est, partem librae *bx*, versus minores magnitudines, reliquae *xa* duplam esse.

Dividatur libra bifariam in puncto *d*, quod vel in aliquo puncto suspensionum, vel in duarum suspensionum medio cadet necessario; reliquae vero suspensionum distantiae, quae inter *a* et *d* intercipiuntur, omnes bifariam dividantur punctis *m*, *i*; magnitudines deinde omnes in partes ipsi *n* aequales dividantur: erunt iam partes ipsius *f* tot numero, quot sunt quae ex libra pendent magnitudines; partes vero ipsius *g* erunt una pauciores; et sic de reliquis. Sint itaque ipsius *f* partes *n*, *o*, *r*, *s*, *t*; ipsius *g* vero, *n*, *o*, *r*, *s*; ipsius *h* quoque, *n*, *o*, *r*; ipsius denique *k* sint *n*, *o*: eruntque magnitudines omnes in quibus *n*, ipsi *f* aequales; magnitudines vero omnes in quibus *o*, ipsi *g* aequales; et magnitudines in quibus *r*, ipsi *h*; illae autem in quibus *s*, ipsi *k*; et magnitudo *t* ipsi *n* aequalis est. Quia igitur magnitudines omnes, in quibus *n*, inter se sunt aequales, aeque ponderabunt in signo *d*, quod libram *ab* bifariam dividit; et eandem ob causam omnes magnitudines, in quibus *o*, aeque ponderant in *i*; illae autem in quibus *r*, in *c*; et in quibus *s*, in *m* aeque ponderant; *t* autem in *a* suspenditur. Sunt igitur in libra *ad*, ex distantiis aequalibus *d*, *i*, *c*, *m*, *a*, suspensae magnitudines sese aequaliter excedentes, et quarum excessus minimae aequatur: maxima autem, quae est composita ex omnibus *n*, pendet ex *d*; minima, quae est *t*, pendet ex *a*; et reliquae ordinate dispositae sunt. Estque rursus alia libra *ab*; in qua magnitudines aliae, praedictis numero et magnitudine aequales, eodem ordine dispositae sunt: quare librae *ab*, *ad* a centris omnium magnitudinum secundum eandem rationem dividentur. Est autem centrum gravitatis dictarum magnitudinum *x*: quare *x* dividit libras *ba*, *ad* sub eadem ratione, ita ut sicut *bx* ad *xa*, ita *xa* ad *xd*; quare *bx* dupla est ipsius *xa*, ex lemmate supra posito. Quod erat probandum.

18. *f aequatur* — 19. *g aequatur* — 27. *minime* —

Nevertheless the centre of gravity of up above cited magnitudes is *x*; thus *x* divides the balances *ba* and *ad* in the same proportions, so that, *bx* is to *xa*, as *xa* is to *xd*; thus, for the up cited lemma it is twice the size of *xa*, which was to be demonstrated.

[...] the centre of gravity of the compounded by all of them [The magnitudes circumscribed] shall divide the line *xm* so that the part toward *x* is twice the size of other one [...].[36]

Fig. 7 Examples of Galilean reasoning upon *aggregatum* bodies. *Source* Google Books Public Domain[36]

[35] "[...] *quodsi parallelae trianguli* AEB *usque ad* IG *extendatur, habebimus aggregatum parallelarum omnium in quadrilatero contentarum aequalem aggregatui comprehensarum in triangulo* AEB;" (Galilei 1890–1909, VIII, Day III, p. 208; Translation retrieved from Galilei, 1954, p. 173).

[36] "Est autem centrum gravitatis dictarum magnitudinum *x*; quare *x* dividit libras *ba*, *ad* sub eadem ratione, ita ut sicut *bx* and *ax*, ita *xa* ad *xd*; quare *bx* dupla est ipsius *xa*, ex lemmate supra posito. Quod erat probandum. (Galilei 1890–1909, I, pp. 188–189). We remark that Galileo used the term "magnitudinum" ("magnitudo") as heavy body. In the *Theoremata*, Galileo proposed his ideas on scaling linked by geometrical and mathematical interpretation essentially based on *ad absurdum* proofs and the theory of proportions. From his standpoint, these ideas are a finite and demonstrated argument, as shown by his famous final acronymum Q.E.D. for *quod erat demonstrandum* (Galilei 1890–1909, I, p. 188). Galileo works on the centre of gravity of a compound balance associated in an ideal geometric composition. On the Archimedean structure of Galileo's *Theoremata*, see Pisano and Bussotti (2012).

Based on the two previous quoted passages, it is understandable that, according to Galileo, the gravity of a heavy body remains concentrated in its centre of gravity, which exists and is unique. Then one might wonder if his reasoning adopted for a single body, could also be applied to a system of *n* bodies. In effect, based on our research (Pisano and Bussotti 2012 and their listed work below), a generalization to *n* bodies does not emerge. In this regard, Ernst Mach (1838–1916) wrote, while dealing with the compound pendulum (Huygens 1673) by Huygens[37] (1629, 1695):

> 1. The next in succession of the great mechanical inquires is Huygens, who in every respect must be ranked as Galileo's peer. [...] Huygens not only continued the research which Galileo had begun, but he also solved the first problems in the *dynamics of several masses*, whereas Galileo had throughout restricted himself to dynamic of *single* bodies (Mach 1883, p. 155).
>
> 11. Questions respecting the motions of bodies that exert a reciprocal influence on each other, cannot be solved by Galileo's principles alone. A new principle is necessary that will define this mutual action. Such as was that restored by Huygens in his investigation to the centre of oscillation (*Ivi*, p. 198).

The Galilean concept of "aggregatum" recalls the idea of contiguity: a set of bodies forms an "aggregatum", when they are also adjoined; it is not necessary that the components of the "aggregatum" are non–detachable; it is only necessary for them to be adjacent.

The crucial novelty of Torricelli's principle compared to Galilean "aggregatum" principle consists in affirming that the criteria of equilibrium are also valid for the centre of gravity of an "aggregatum" of bodies which are not adjacent. This idea by Torricelli might be considered as derived from a reasonable hypothesis, based on an analogy with what happens in the case of adjacent bodies. Finally, Torricelli, instead of Galilean "aggregatum", uses "simul coniuncta" that is "linked together" (in some not specified manner) and describes a movement in a particular physical-geometrical situation (Torricelli 1644, p. 99). Taking into account (the modern idea of *constructed system of bodies* and) a particular mobile mechanism, he also describes a motion of the entire compounded system:

> [...] those two heavy bodies will be such as a sole heavy body composed of two bodies, which might be a balance, or a trickle, or any other mechanical tool.[38]

Torricelli then associates the idea of Galilean "aggregatum" to that particular mechanism by which the same bodies are "simul coniuncta" = jointed, also linked among themselves. On that in another section, he stresses this concept also detailing the operative terms. Torricelli specifies the manner to link two bodies, by a little rope [*funiculo*; in Italian as *funicella*]. On that, he writes:

> [...] by means of an imagined little rope [funiculo] so that [given two heavy bodies] the motion of one follows the motion of the other.[39]

We duly remark that he suggests, not only that a mobile mechanism is needed for the movement of one body to follow the movement of the other, but he also explicitly declares the instrument, the "rope", with which the connection and relative movement occur.[40] We underline that when Torricelli studies the stability of the lever, he shows some uncertainties regarding the concept of centre of gravity of "simul coniuncta" (linked) bodies. He correctly considers the centre of gravity as a geometrical point, which is posed under the fulcrum, and in the course of the demonstration, he uses the centre of gravity of a "simul coniuncta" that do not rise when the lever rotates. Later on, however, when he comments on his results, he attributes a physical nature to the centre of gravity of two "simul coniuncta" bodies. Despite the fundamental progress towards the correct use of the notion of gravity centre in a physical context, the nature of the equilibrium of the lever of equal arms still presented problems for Torricelli. For, while dealing with this question, he wrote (Fig. 8).

---

[37] See also la *Démonstration de l'équilibre de la balance* (Huygens 1669). It deals with the equality between the height of fall of heavy bodies and the height of ascension of the centre of gravity (a sort of living force theorem).

[38] "[*Praemittamus*] *erunt duo illa gravia tamquam grave unum ex duobus compositum, siue id libra siat, siue trocleà. ſiue qualibet alia Mechanica ratione* [...]." (Torricelli 1644, D*e motu Grauium deſcendent, Liber Primus*, pp. 99; authors' translation).

[39] "Connectantur etiam aliquo imaginario funiculo [little rope] [...] adeo ut ad motum unius motus alterius consequatur". (Torricelli 1644, p. 99; authors' translation).

[40] In this case, the centre of gravity of the "aggregatum" or of "simul coniuncta" bodies is no longer the centre of gravity of a body obtained from the union of single bodies, but we could suggest the hypothesis of a modern system of bodies. The idea of a centre of gravity of a system of bodies, as understood today, will only arise within collision theory, in particular with Huygens (1629–1695, 1673), in which it will become a very powerful theoretical instrument (Cfr. Mach 1883, p. 178).

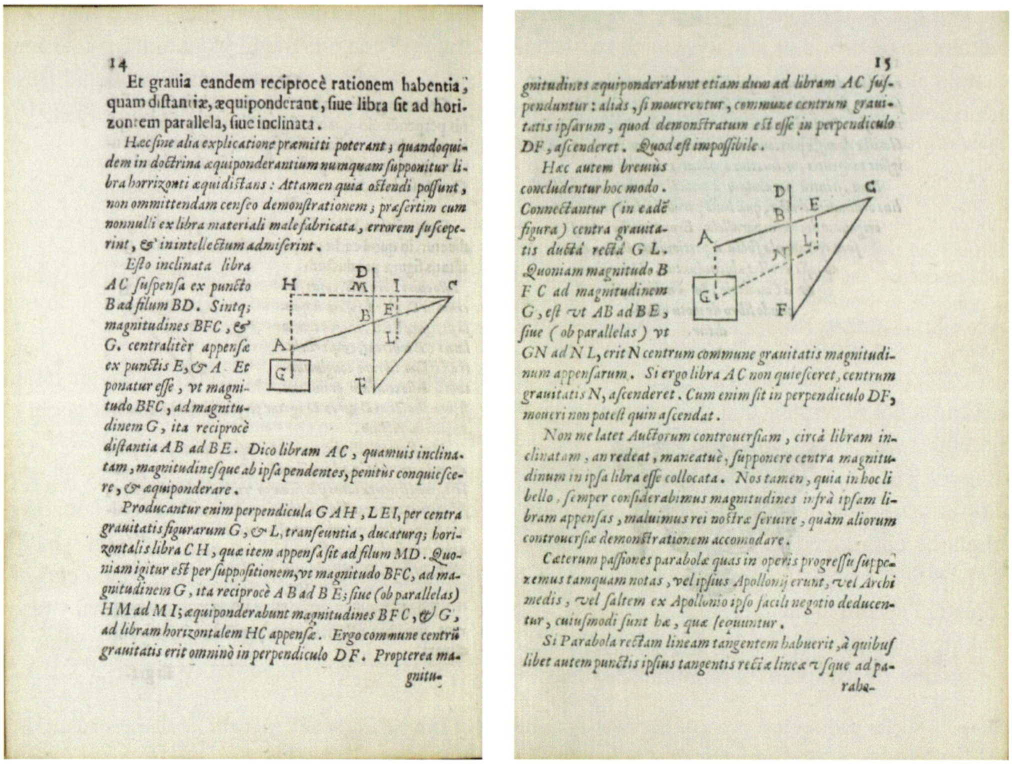

14

Et grauia eandem reciprocè rationem habentia, quam distantiæ, æquiponderant, siue libra sit ad horizontem parallela, siue inclinata.

Hæc sine alia explicatione præmitti poterant; quandoquidem in doctrina æquiponderantium numquam supponitur libra horrizonti æquidistans: Attamen quia ostendi possunt, non ommittendam censeo demonstrationem; præsertim cum nonnulli ex libra materiali malefabricata, errorem susceperint, & inintellectum admiserint.

Esto inclinata libra AC suspensa ex puncto B ad filum BD. Sintq; magnitudines BFC, & G. centralitèr appensæ ex punctis E, & A. Et ponatur esse, vt magnitudo BFC, ad magnitudinem G, ita reciprocè distantia AB ad BE. Dico libram AC, quamuis inclinatam, magnitudinesque ab ipsa pendentes, penitùs conquiescere, & æquiponderare.

Producantur enim perpendicula GAH, LEI, per centra grauitatis figurarum G, & L, transeuntia, ducaturq; horizontalis libra CH, quæ item appensa sit ad filum MD. Quoniam igitur est per suppositionem, vt magnitudo BFC, ad magnitudinem G, ita reciprocè AB ad BE; siue (ob parallelas) HM ad MI; æquiponderabunt magnitudines BFC, & G, ad libram horizontalem HC appensæ. Ergo commune centrũ grauitatis erit omninò in perpendiculo DF. Propterea magnitu-

15

gnitudines æquiponderabunt etiam dum ad libram AC suspenduntur: aliàs, si mouerentur, commune centrum grauitatis ipsarum, quod demonstratum est esse in perpendiculo DF, ascenderet. Quod est impossibile.

Hæc autem breuiùs concludentur hoc modo. Connectantur (in eadẽ figura) centra grauitatis ductâ rectâ GL. Quoniam magnitudo BFC ad magnitudinem G, est vt AB ad BE, siue (ob parallelas) vt GN ad NL, erit N centrum commune grauitatis magnitudinum appensarum. Si ergo libra AC non quiesceret, centrum grauitatis N, ascenderet. Cum enim sit in perpendiculo DF, moueri non potest quin ascendat.

Non me latet Auctorum controuersiam, circà libram inclinatam, an redeat, maneatuè, supponere centra magnitudinum in ipsa libra esse collocata. Nos tamen, quia in hoc libello, semper considerabimus magnitudines infrà ipsam libram appensas, maluimus rei nostræ seruire, quàm aliorum controuersiæ demonstrationem accomodare.

Cæterum passiones parabolæ quas in operis progressu supponemus tamquam notas, vel ipsius Apollonij erunt, vel Archimedis, vel saltem ex Apollonio ipso facili negotio deducentur, cuiusmodi sunt hæ, quæ sequuntur.

Si Parabola rectam lineam tangentem habuerit, à quibus libet autem punctis ipsius tangentis rectæ lineæ vsque ad pa- rabo-

I do not conceal that, concerning the controversy among the authors about the inclined balance on the question if it is able to go back or stay in its position, the centres of [gravity] of heavy bodies are supposed to be placed in the same balance. Therefore, since in this booklet we always consider the heavy bodies suspended under [by]a balance, [then] we prefer to take care ofour aims only, rather thantoadapt our demonstrations to adispute which concernsothers.

Fig. 8 Torricelli's equilibrium of the lever in the *De dimensione parabola* (Torricelli 1644, pp. 14–15). *Source* Courtesy University of Oklahoma, History of Science Collections[41]

Torricelli's reasoning would be correct only if two bodies were "simul coniuncta" in a rigid way to the lever axis.

Torricelli seems to be in difficulty when he tries to apply his principle to a new physical situation, which he had not theoretically dealt with in detail. In Torricelli's reasoning, the lever-weight system is in stable equilibrium since its centre of gravity is under the lever and therefore rises for a possible movement of the lever itself. Therefore, we can summarize the following Torricellian items:

- Torricelli's principle is valuable when it is applied to two linked heavy bodies.
- The heavy bodies can also be non-contiguous (unbounded).
- The bodies can be detachable from each other.
- A running mechanism (machineries) is necessary to produce a consequential movement (the motion of one follows the motion of the other).

[41] Torricelli (1644), *De dimensione parabolae, problemata duo*, pp. 14–15. The reader should pay attention to the numeration of pages. Sometimes, like in this case, the numbering in the 1644—edition starts over. Cfr. Torricelli (1975), *Sulla misura della parabola, Libri due*, pp. 336–337.

Finally, taking into account "aggregatum" and "simul coniuncta", we propose the following parallel view of shared features of Galileo's and Torricelli's concepts[42]:

Galileo—*Aggregatum*	Torricelli—*Simul coniuncta*
Geometrical-static concept for the application of the definition of the centre of gravity	Geometrical-static concept for building the equilibrium among two or more bodies connected to a certain movement
Two or more heavy bodies are mutually linked without describing how to do that	Two or more heavy bodies can *also* be mutually linked by a running mechanisms (machineries)

2.5 Another Method of Application of Torricelli's Principle

Torricelli, after the second proposition, adds an interesting brief commentary. He writes:

> *In a different manner.* This same proposition we have proved relying upon our first proposition, deduced by assuming the principle [the *Praemittamus*] and resorting to an *ad absurdum* reasoning, can be proved in an absolute and direct manner by the principles themselves of mechanics.[43]

Therefore, Torricelli, in these circumstances, focuses on the use of his principle and on the *ad absurdum* method of demonstration implied in its use rather than on the results of the two propositions in themselves.[44]

It might appear surprising that Torricelli resorts to his principle in no other circumstance within his *De motu gravium*, which is a clear indication that, though he had understood the novelty of his approach, he did not ascribe to it the fundamental importance which it deserves.

We try to offer a more complete picture of the fact that Torricelli was probably conceiving the idea that some parts of mechanics might be addressed by taking into account principles connected to the possible movements of the gravity centre of systems of bodies. What follows does not concern directly the *Praemittamus*, but a similar and interesting theoretical situation.

It is appropriate to refer to a series of manuscripts (Torricelli, 1919, 1944, II, p. 235) which Gino Benedetto Loria (1862–1954) and Giuseppe Vassura (1866–1949) co-editors of Torricelli's works, joined under the title *De Motu ac Momentis Varia* (Torricelli 1919, II, pp. 233–249). These manuscripts are made up of sixteen propositions. In fact, Vincenzo Viviani had prepared their copy ready to be printed (the title itself is due to Viviani), but he excluded exactly the proof of Proposition XVI, which is the one interesting in our context. Loria and Vassura first collected this important proof.

Torricelli in proposition XVI (Torricelli 1919–1944, II, pp. 242–246) considers a beam *BF* posed as in Fig. 9. In modern terms, he poses the problem to find the proportion of the beam's weight to the force applied in *F* and directed towards *E* which can equilibrate the *momentum* of the beam to slide along the direction *EF*.

In practice, Torricelli is trying to find a force applied in *F* and directed towards *E*, which is able to maintain the beam in equilibrium. The force can be thought of as produced by a weight *Q* posed in the position *Z* (Fig. 9):

[42] See also Huygens' principle concerning the laws of motion on the collision of bodies (1669) and Démonstration de l'équilibre (1686).

[43] "*Aliter. Idem quod hic demonftrauimus ex Prima Propofitione, quæ fumpto principio deducebat ad impoffibile, ostendetur etiam abfolutè, & affermatiuè ex ipfis Mechanicæ principijs*". (Torricelli 1644, *De motu Grauium defcendent, Liber Primus*, p. 101; Torricelli 1919–1944, II, p. 106; authors' translation and italics).

[44] We will consider the problem of the *ad absurdum* demonstration within physics in the last section of our paper.

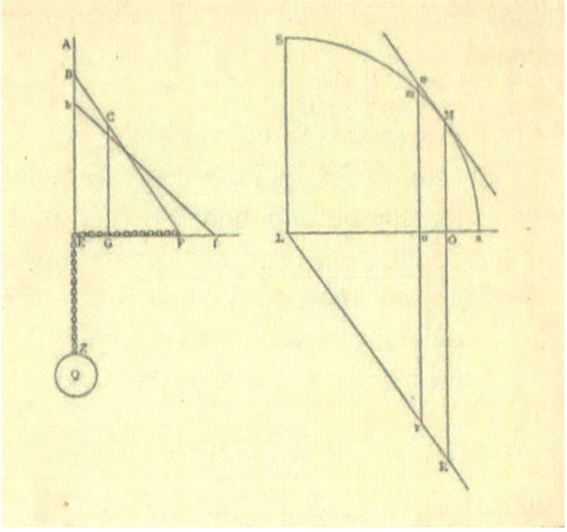

Fig. 9 The image used to prove Torricelli's Proposition XVI. The figure is not original by Torricelli. It is due to Torricelli's editors, Loria and Vassura (Torricelli 1919, 1944, II, p. 244)

After a long series of deductions, Torricelli claims:

> If the distance of the gravity centre of those two bodies was equal to *nr* rather than to *mr*, the common centre of gravity would have moved along the horizontal line *Mn*. In the analysed case the gravity centre describes a circular arc *Mm*, which is tangent to that horizontal line, hence due to a *preceding lemma (lemma precedente)*, the gravity centre will remain at rest. (Torricelli 1909–1944, IV, pp. 244–245).

The problem is to grasp to what lemma Torricelli was referring. Loria and Vassura write that in the whole of Torricelli's works there is no trace of such a lemma (*Ivi*, p. 234). This is right because, although the lemma deals with the movement of the centre of gravity of bodies under the action of gravity, it cannot be assimilated to the *Praemittamus* of the *De motu gravium*; it is not the same lemma. We propose the idea that this lemma is the famous proposition of classical physics: "the gravity centre of a system of bodies is at rest or move in uniform rectilinear motion, unless external forces act on the system".

In the specific case: at the end of his proof, Torricelli created a situation in which the effect of gravity force had been annihilated. Therefore, the gravity centre of the two bodies' system is either at rest or moves in rectilinear uniform motion (and furthermore horizontally, in the specific case). However, as a result of the reasoning, if the gravity centre moved, such a movement would take place on an arc of a circle, but this is impossible because of the previous principle. Therefore, it seems to us plausible that the "mysterious lemma" by Torricelli is the famous principle on the movement of the gravity centre. If this is the case, Torricelli would have discovered a treasure, whose value he had not completely grasped and appreciated. On the other hand, this is not surprising because when a unifying concept or a series of unifying concepts is lacking (in this case the unifying concepts are Newton's concept of force and his gravitational theory), it can happen that the value of genial and innovative discoveries is not fully appreciated by the discoverers themselves.

The idea that Torricelli was thinking of a general series of principles seems confirmed by what he wrote at the beginning of his work *De dimensione parabola solidique hyperbolici problemata duo*. This text begins with *Suppositiones et definitiones*, the first of which says:

> Pose that the nature of the gravity centre is such that a body of a [given] magnitude freely suspended from anyone of its points is never at rest unless its gravity centre reaches the lowest point of its sphere.[45]

This statement is something in the middle between a definition of gravity centre and the enunciation of a physical principle. For sure, the context is a physical one, because while referring to the centre of gravity in the mere geometrical meaning, it makes no sense to speak of its descent in the terms used by Torricelli. It makes sense within a gravitational field. From a logical point of view this definition-principle expresses the property that the necessary condition for a body to be at rest is that its gravity centre reaches the lowest point of its sphere. We might interpret the "of its sphere" as "compatible with its constraints". This proposition is, thence, the complement of the *Praemittamus*, which can be expressed as: "the necessary condition for a system of bodies to move is that their gravity centre descends". The logical conjunction of these two

[45] "Ponatur eam eſſe centri grauitatis naturam, vt magnitudo liberè ſuſpenſa ex quolibet ſui puncto nunquam quieſcat niſi cum centrum grauitatis ad infimum ſuæ ſphærę punctum peruenerit." (Torricelli 1644, *De dimensione parabolae*, p. 11; authors' translation).

propositions is that the necessary and sufficient condition for a body or a system of bodies to be at rest is that its gravity centre has reached the lowest position compatible with the constraints.

Thus, Torricelli was constructing a system of principles applicable to several physical situations for the movements of bodies within a gravitational field.

We are going to close this section with a comparison between Torricelli's *Praemittamus* and some results by Galileo, which might induce one to think that Torricelli's principle was already existing in Galileo's works, whereas we think this is not the case. It seems to us more appropriate to pose here this comparison rather than in Sect. 2.4, where some general conceptions by Galileo and Torricelli have been compared.

For, Galileo, in his *Le Mecaniche* (as lessons given in Padua around 1595; Galilei 1595, II, pp. 147–191; Pisano and Bussotti 2012, 2015a, 2017a) expounds a reasoning, which, in some way, is similar to Torricelli's. Given the inclined plane *ABC* and two bodies *F* and *E* (Fig. 10) Galileo considers a dynamical situation based on a clear, but not explicitly formulated, static presupposition: if $F{:}CB = E{:}AC$, and the height of *E* on the horizon is the same as *F*'s, then there is equilibrium. This is *Propositio* I of Torricelli, which Galileo had proved. If we move downwards the body *F*, and consequently the body *E* moves upwards, the body *F* descends of a certain height *h*. Hence *E* moves on the inclined plane by *h* and its ascent on the horizon is less than *h*. Galileo calculates it and shows that the descent of *F* is to the vertical ascent of *E* as *AC* to *CB*, namely as *E* to *F*.

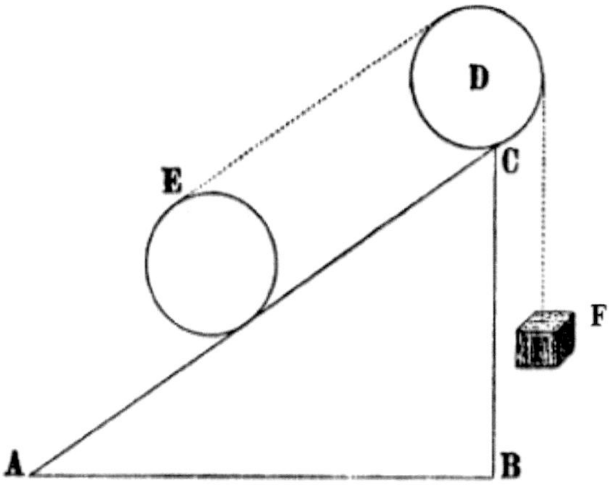

Fig. 10 Galileo on his theorem (1595[1634] *Le Mechaniche*, in Galilei 1890–1909, II, p. 185). *Source* Google Books Public Domain

Almost at the end of his proof, Galileo claims:

> It is here necessary to point out that, though the mobile *E* will have traversed the entire line *AC* in the same time in which the other heavy body *F* will have gone downwards, the body *E* will not go far from the common centre of the heavy bodies more than the perpendicular *CB*. While, the body *F*, descending perpendicularly, will have gone downwards for a space equivalent to the whole line *AC*. In addition, since *the heavy bodies do not resist the transversal motions, unless during this motion they withdraw from the centre of the Earth* [...].[46]

The proposition "*The heavy bodies do not resist the transversal motions, unless during this motion, they withdraw from the centre of the Earth*" is quite interesting because it is equivalent to claiming that a body (or a system of bodies) can produce a resistance to a transversal motion only if it withdraws from the centre of the earth. While if the body (of the system of bodies) descends towards the centre of the Earth it does not resist a transversal motion. But:

[46] "Ma qui però si deve avvertire che, se bene il mobile *E* averà passata tutta la linea *AC* nel tempo medesimo che l'altro grave *F* si sarà per eguale intervallo abbassato, niente di meno il grave *E* non si sarà discostato dal centro comune delle cose gravi più di quello che sia la perpendicolare *CB*; ma però il grave *F*, discendendo a perpendicolo, si sarà abbassato per spazio eguale a tutta la linea *AC*. *E perchè i corpi gravi non fanno resistenza a i moti transversali, se non in quanto in essi vengono a discostarsi dal centro della terra* [...]". (Galilei 1595 in Galilei 1890–1909, II, pp. 185–186; authors' translation and italic).

(a) This is the same as to claim that the system Earth-body can move—if no external force is applied—only if the centre of gravity of the system descends. In this formulation, Galileo's principle seems physically equivalent to Torricelli's.
(b) The differences are, nonetheless, relevant: first of all, Galileo speaks, while formulating this principle, of a single body; we have added in brackets "system of bodies" because this is physically true and because it seems implicit in Galileo's reasoning, but, in fact, Galileo speaks of a sole body. Furthermore, the assertion by Galileo is not posed at the beginning of the treatise as a principle from which to draw consequences, but almost as an incidental observation. Moreover, the difference between the intuition of a principle and its complete and independent formulation is fundamental.

With this, we have no intention to deny that Galileo was on the road to grasp the principle, but he was on the way, while Torricelli arrived at the end of this way and formulated this principle explicitly.

3 The *Praemittamus* and the Principle of Virtual Work

In modern terms, to define the *principle of virtual work,* one can specify that a *displacement* is possible if it is compatible with the fixed constraints. In addition, it is *virtual* if it is compatible with the constraints even if they are moving. Specifically, *virtual displacement* δs of a point is any arbitrary infinitesimal change in the position of the point consistent with the constraints imposed on the motion of the point. This displacement can be only imagined.

If one only considers the case of time–independent constraints, one can also derive a *possible displacement* (i.e., a rotation as well); and if the *virtual displacements* are mathematically independent variables, they are also arbitrary.

The *principle of virtual work*—which arises in the application of the *Principle of least action* to the study of forces and movement of a mechanical system—is a law of mechanics for which there is not any generally accepted epistemological and ontological account: it can be seen both as a principle and as a theorem to be proved. The *principle of virtual work* is used in statics for a solution of a special class of problems involved in a system in equilibrium.

Generally speaking, the necessary and sufficient condition for equilibrium of a mechanical system without friction is that the virtual work done by the externally applied forces *f* is zero, like that[47]:

$$\sum f_i \cdot dr_i = 0$$

Expressing the concept in words:

(a) The necessary and sufficient condition for the equilibrium of a body(particle) is that the virtual work done by all working forces acting on the body during any *virtual displacement* (δs) consistent with the constraints imposed on the particle is zero;
(b) The necessary and sufficient condition for the equilibrium of a rigid body is also that the virtual work done by all external forces acting on the body during any δs consistent with the constraints imposed on the body is equal to zero.

Particularly, if the *principle of virtual work* is applied to a system of rigid bodies (i.e., mechanism) no *virtual work* is done by internal forces, by reactions in smooth constraints, or by forces normal to the direction of motion. The virtual work is done by reactions when friction is present. In statics, it is possible to calculate a force *F* implied in an equilibrium state concerning, e.g., a crank-slider mechanism in the position given by a certain known angle. For example, knowing a) the position of the action-points (a physical action described/applied by/to an infinitesimal point) of the applied forces implied in a given mechanical system, b) the coordinates-position (α, β, γ) of the crank (and distances and other data of the problem) then, according to the *Principle of virtual work*, an immediate equilibrium equation can be written using $\delta\alpha$, $\delta\beta$ and $\delta\gamma$.

Let us briefly remind the reader the concepts behind the principle of virtual work (Pisano 2017) starting from the notion of *virtual displacement*: given any physical constrained system, the *virtual displacements* are those displacements representing all the possible states of the system, which are compatible with the constraints. Such displacements are usually considered infinitesimal. This means that they are interpreted as little perturbations in respect to the initial state of the system. The *virtual work* is the work done by the forces associated to the virtual displacements. Expressed in an intuitive manner: *the principle deals with the relation between the active forces and the constraint–reactions*. Until the action of the active forces

[47] Or, that is the same: $\delta W = \sum_{j=1}^{m} F_j \cdot r_j \leq 0$ δW is the variation of the virtual work *W*, while *m* is the number of the generalized coordinates of the system, F_j is the resultant of the *m* forces acting in the direction r_j.

is compensated by the constraint–reaction the work is virtual, potential and, if necessary to solve physical problems, it can be considered infinitesimal. Therefore, there is equilibrium and the displacements are virtual. As soon as the action of the active forces overcomes the constraint–reactions, the forces do a real work, the displacements become finite and there is not equilibrium. The principle has a vast application in several branches of physics where problems of equilibrium under certain constraints are treated (Cfr. Oliveira and Pisano, 2022).

Let us transcribe into formal terms what Torricelli claims in his *Praemittamus*: be m the mass of the system of bodies he is considering; be z_0 the vertical coordinate of the system's centre of gravity, g the gravity acceleration. Now: let us assume that the centre of gravity of the body descends and, hence, $\delta z > 0$. In this case, when the body descends the direction of the gravity force is the same as the direction of the body and, hence, the quantity $mg\delta z$ is greater than zero. Torricelli claims that the case in which $mg\delta z > 0$ is the only one in which the bodies can move. In any other case, that is, when $mg\delta z \leq 0$, no movement is possible. But, this is exactly the expression assumed by the principle of virtual work when the situation is that described by Torricelli. Thence, *the two principles are equivalent as far as the physical situations to which Torricelli's principle is applicable are concerned.*

Modern physics has enormously extended the field of application of the principle of virtual work beyond mechanics, to include other field of physics. Within mechanics, the treatment has become extremely refined considering several kinds of constraints, differentiated according to their dependence on time, on the relation between the variables describing the equation of the constraints and on the presence or absence of friction. This has brought the number of possible applications of this principle far beyond what Torricelli might have imagined. However, by restricting the cases to those for which Torricelli's principle is applicable, the two statements are equivalent.

With regard to the principle of virtual work, at least two considerations are necessary.

(1) First of all, the principle of virtual work also assumes different forms and names. The three best known alternative forms are those of *principle of virtual velocities*, *principle of virtual displacements* and *principle of D'Alembert*, which is, in general, considered as an extension of the principle of virtual work.[48] The four "virtual" principles (works, velocities, displacements, D'Alembert[49]) have some slight differences. However, they belong to a common root. It consists in the idea of considering equilibrium not as a completely passive state, but as a state in which some quantities exist (work, displacement, velocity) which are incompatible with equilibrium when they become finite (and hence, actual, not virtual), but are compatible if they are virtual and, given this condition, produce infinitesimal works, velocities or displacements.

(2) The fiction of "virtual" is useful to approach a vast series of physical problems.[50] It seems established that Aristotle, while dealing with a series of problems connected to the lever claimed a proposition resembling the principle of virtual velocities (Cfr. Baldi 1621; Cartelon 1975). In the course of history several other authors, the most important of whom are probably Jordanus de Nemore (fl. XIII—ca. 1260) and Niccolò Tartaglia (1499?–1557), based some of their achievements on principles similar to those of virtual works, velocities or displacements.[51] However: (a) no one of these formulations was completely precise, explicit and correct; (b) these formulations were directly referred to movements of bodies constrained to a lever or laying on a scale, whereas Torricelli's formulation is correct and more general, more abstract. For, he does not deal with the centre of gravity of single bodies, he treats a general conjunction between these bodies, not restricted, in principle, to a conjunction by a lever or of a scale. In fact, his principle works even in the case in which the two bodies are not joined by a material constraint. His proposition has the form of a general principle. This means that Torricelli's principle is a fundamental step in the history of virtual works principles. It represents a break with respect to the conceptions of the other authors we have mentioned. Therefore, also referring to considerations on the use made by Torricelli of his principle and the features of Galileo's demonstrative procedures in mechanics, we do not agree with Mach's opinion. For, according to Mach:

> The Torricellian form of expression, however, contains in no respect more than the Galilean.[52]

---

[48] Mach (1883, 1919), p. 53. Mach deals with the principle of virtual work, which he calls of virtual velocities or displacements, at pp. 49–77 of his fundamental work, where he also faces D'Alembert principle (*Ivi*, pp. 331–343). See also Pisano (2017), Pisano and Capecchi (2016, 2013), Pisano and Bussotti (2015c, 2015d), p. 72, 2015c; Euler ([1730–1]1738, [1773]1774, 1736, 1749, [1750]1752, 1757, 1775); see also Maupertuis (1746).

[49] D'Alembert ([1743]1758, 1751–1780, 1767), Hankins (1970); see also Ampère (1806).

[50] As to the role of "fiction" in some aspects of the history of science, see Pisano and Bussotti (2017b).

[51] With regard to this history, see Pisano (2017), Pisano and Capecchi (2016) and recently a didactic standpoint by Sparavigna (2015).

[52] Mach (1883), p. 53.

We have already tried to explain why Torricelli's principle represents a conspicuous advancement in respect to the system of principles used by Galileo.

Though it goes beyond the aim of our paper, it may be interesting to refer to the letter addressed, in 1717, by Johann Bernoulli (1677–1748) to Pierre Varignon (1654–1722), where Bernoulli explicitly stated the principle of virtual work. As referred to by Varignon, Johann Bernoulli wrote:

> General Proposition. Theorem XL. In every equilibrium of any given force, independently of how these forces are applied and under what angles they mutually act, directly or indirectly, the sum of the positive energies is equal to the sum of the negative energies, positively taken.[53]

Energy is explicitly defined by Bernoulli as the product of the force times the distance in its direction. With this formulation, the modern history of the principle of virtual work begins. Torricelli represents one of the highest points in the ancient history, if not proto–history, of this principle.

4 The Methodological Status of Torricelli's Principle

This section is composed of two sub-sections: in the former, we will consider some of the physical and logical features of the way in which Torricelli presents his results on constrained motions under the action of gravity. In the latter, the logic of the *ad absurdum* proofs based upon general principles, such as those of Torricelli, within physics will be dealt with.

4.1 The Physics of the Principles and Forces. Conceptual Developments

Torricelli's principle is a paradigmatic example of a way of approaching physics, which is different from that connoting the work by Newton. In Newton's *Principia* the principle of conservation of momentum is stated. However, his physics is the physics of the forces, not of the principles, in the precise sense that, for any given problem or theorem, Newton decomposes all the acting forces along certain directions and solves that problem or proves that theorem by means of such decompositions (Pisano 2010, 2011). The idea behind this method is to arrive at general results starting from a, so to say, microscopic analysis of any single force acting in the considered system. The concept is to decompose a general situation into a series of specific situations, to determine the physics of these specific situations and to recompose the picture as a sum of single pieces of which, after an appropriate analysis, we know the behaviour. What is general is obtained as a sum of what is specific.

The physics of the principles has a completely different logical way of argumentation: some general principles are given. A problem regarding a system is decomposed into sub-problems regarding some subsystems making up the original one, only until a general principle is applicable. Once arrived in this situation, the solution is achieved working on the general principle (or principles), without taking into account the behaviour of any single force. It is a completely different way to see physics: the physical behaviour of a system is reconstructed by principles, which refer to the universe in its entirety, therefore to a level of reality, which is absolutely general. The behaviour of a whole system is grasped by referring to even more general situations and principles, not by a "microscopic" decomposition of the forces. An important note seems to us necessary: the difference between the physics of the forces and the physics of the principles deals with a procedural-epistemological aspect, that is on the way in which the physicians make physics; it is not necessarily connected to the way in which they consider the physical world in itself. For, there is no doubt that Newton[54] believed in general laws (it is not by *chance* that he is the discoverer of universal gravitation). It is probably true that the physicists who, between the end of the seventeenth century and through the first half of the eighteenth century, applied either the physics of the principles or the physics of the forces also had different conceptions as to the nature of the physical world and of its laws. This is surely true for Newton and Leibniz (Bussotti 2015; Bussotti and Pisano 2020, 2017; Leibniz 2009), who were respectively the

[53] "Proposition générale. Théorème XL. En toute équilibre de forces quelconques, en quelque manière qu'elles soient appliquées, et suivant quelques directions qu'elles agissent les uns sur les autres, ou médiatement, ou immédiatement, la somme se énergies affirmatives sera égale à la somme des énergies négatives, prises affirmativement". (Varignon, 1725, p. 176; see also see also Bernoulli ([1748]1850, 1733, 1703, 1727, 1742a, 1742b), Radelet de Grave (1998). Authors' translation. It is very interesting to note that this Bernoulli's statement also appears such as "Lemma" in a military trait on the *Seconde Hypotese—Pour la recherche de la poussées des voûtes* (Frezier, 1739, p. 357).

[54] Newton ([1671]1736, [1686–1687]1803, [1713]1729, [1726][1739–1742]1822, 1714, 1730, 1746), Pisano and Bussotti (2016, 2017b), Bussotti and Pisano (2014a, b, c), Cohen and Smith (2002).

"champions" of the physics of the forces and the physics of the principles (Pisano and Capecchi 2013; Capecchi and Pisano 2007a, b, c, 2008, 2009a, b, c, d, 2010a, b).

Nonetheless, starting from the 10s–20s of the eighteenth century, most physicists adopted both approaches. At the beginning, the Newtonian one was far preponderant, but progressively, when new and general principles were discovered and specified, the physics of the principles became a continuously more extended methodology.

With regard to Torricelli, we may ascribe to him the principle expressed by the *Praemittamus* and, according to our interpretation, the principle that the gravity centre of a bodies-system on which no external action is applied is at rest or moves with a uniform rectilinear motion. Here the situation is rather interesting because Torricelli worked in a period in which the concept of force had not yet been defined and the physics of the principles was, at best, in a gestational phase.

Well, Torricelli acts in a very clever manner: when the action of a force is required, he replaces the abstract notion of force and of its direction with the use of a concrete weight, engine or device, which offers the necessary conditions to solve the problem he poses. In the example of the beam-equilibrium we have analysed (Fig. 2), where we would use the concept of force, without caring about the origin of such a force, Torricelli poses the weight Q, which replaces such an abstract force, at least with regard to the problems faced by Torricelli. Our abstract concept, the force, is replaced by a specific force, the weight, whose action is equivalent to the required force. What in our eyes is a concretization of an abstract notion, was, in fact, an almost unavoidable means in a period in which the abstract notion of force was not yet defined. The studies on the inclined plane by Galileo and by Torricelli himself are more elementary examples than the beam-equilibrium, which show how a concretization of force can lead to right results with regard to the decomposition of the physical actions existing in a certain system. Therefore, several interesting results regarding the forces can be achieved also without this abstract notion. Obviously, when the problems become more complicated, when actions different from those existing on an inclined plane or on a beam are considered, the abstract concept of force assumes progressively a more prominent role. However, a physics of forces was, in part, possible also before the definition of this concept given by Newton.

With regard to the physics of the principles, the situation is even more interesting. We repeat that Torricelli probably did not completely grasp the importance of his principle/principles, because he used them on very few occasions. Nonetheless, their use is very modern and perspicuous: he utilizes these principles as a limitation to the possible motions: if a motion would imply their violation, such a motion is not possible. The principles mark a limitation to the possible motions and, more generally, to the possible physical situations in nature. This is their aim. In the case of Torricelli, this is particularly evident because, when adopting his principle/s, he resorts to *ad absurdum* demonstrations, but this is always true, independently of the method of proof used. When the principle of energy conservation or of the least action, or any other, is applied, among the possible situations in nature, only those compatible with that principle are selected. In this sense, the principles have a huge economic value (in Mach's meaning) within physics; Torricelli guessed this feature.

4.2 An Interplay Between Galileo and Torricelli upon Physics and Mathematics Relationship

It is well known that Galileo tried to introduce, by means of a particular formalism, mathematics into physical science: this issue involved him deeply, as emerges from his reflections and wide debate in the *Dialogues concerning the two chief world system* (Galilei 1890–1909, VII, pp. 1–520). However, the theoretical level of mathematics in the epoch of Galileo was not adequate to encourage the author towards this choice. Some examples may shed light with regard to the difficulty of dealing with physics when the mathematical means are rather poor; e.g., for a heavy body, it was necessary to establish the instant just after the body started to move (that is velocity $\neq 0$), rather than identifying when its instant velocity was zero (state of rest).[55]

As an example to describe this situation, let us think of the classical problem concerning the composed balance. This will shed light on the difficulty of dealing with the concept of infinity in correct mathematical terms. Taking into account the physical–mathematical aspects at that time, we can describe the behaviour of a composed balance and of its centre of gravity. In *Le Mecaniche*, on this subject, Galileo wrote:

[55] For example, for a heavy body at rest it was necessary to establish the instant t when it started to move. In other words, it was necessary to determine at which instant t its configuration passed from a *velocity* $= 0$ to a *velocity* $\neq 0$. Nowadays, this problem is solved by the passage to the limit. This is a matter of debate within the history of science and teaching, typically NoS (Nature of Science).

[...] And this [the centre of gravity of the heavy body] is that point that would join with the centre of the universe of heavy things, that is with that of the earth if, by means of a free tool it might descend. From this we will draw this supposition: Every heavy body moves downwards so that its gravity centre never leaves the straight line which, from that centre, placed at beginning of the motion, is produced until the centre of the universe [in Aristotelian school, it is coincident with the centre of the earth] [...].[56]

In *Le Mecaniche,* a motion of a not-constrained heavy body is related to the movement of its centre of gravity along the straight line extending to the centre of the Earth (Aristotelian approach). This reasoning can also be applied to a composed system of bodies. For example, looking for the equilibrium of a balance from which several bodies are suspended. We should remark that in the description of the compounded balance[57] (Pisano and Bussotti 2012) Galileo assumed a parallelism of the threads that converge towards the centre of the earth. While in the *Discorsi*, he wrote:

Salv. [...]. I may add that, in all their discussions, Archimedes and the others considered themselves as located at an infinite distance from the centre of the earth, in which case their assumptions were not false, and therefore their conclusions were absolutely correct. When we wish to apply our proven conclusions to distances which, though finite, are very large, it is necessary for us to infer, on the basis of demonstrated truth, what correction is to be made for the fact that our distance from the centre of the earth is not really infinite, but merely very great in comparison with the small dimensions of our apparatus [...].[58]

It looks as if Galileo considers these cords as directions, parallel to each other, and on which the centre of gravity of each heavy body moves. This Galilean reasoning is very important because it appears as a theoretical premise (a necessary physical condition) to a reasoning to determine (from a mathematical point of view) the total centre of gravity of a composed balance. However, we remark that the threads are finite lines and at finite distance from each other. Thus, it seems that Galileo considered a finite number of threads, which have an infinite length (even if related to the Earth). This attitude, at a first analysis, suggests an approach (at least in this case) in which both actually and potentially infinite quantities are considered. With regard to this question, part of the secondary literature argued that *Galileo made an operational use of mathematics*, which appears more compatible with a potential conception of infinity. On the other side, i.e., Settle, presents a Galilean *infinite indefinite quanta as constitutive elemets of a segment*, that is a choice for an actual infinite. The argument is, still nowadays, a matter of reflection, so we should admit the difficulties of deciding the Galilean choice, also taking into account the historical scientific context (Cfr. Galluzzi 1979). In particular, Attilio Frajese (1902–1986), as to the uniform accelerated motion wrote:

Galileo compares two infinities (the number of points [geometry] of a segment *AB* and the number of instants [time] of the duration of the motion) and states that they are equivalent based on what we already called "biunivocal correspondence" ("To all single instants of time along the motion from *A* to *B* all points of the segment *AB* correspond").[59]

[...] here [*Discorsi e dimostrazioni matematiche*, VIII, pp. 208-209] there is another comparison of infinites, which is described by an equality relationship: Substantially, Galileo claims that if – concerning accelerated motion –all the segments of the parallel lines (which are infinite in number) are added, and if the same calculus is done for the segments (whose number is also infinite) – relative to that uniform motion – than the two(infinite) sums are equal.[60]

---

[56] "[...] E questo [il centro di gravità o baricentro del corpo] è quel punto il quale andrebbe a unirsi col centro dell'universale delle cose gravi, ciò è con quello della terra, quando in qualche mezzo libero potesse discendervi. Dal che caveremo noi questa supposizione: Qualunque grave muoversi al basso così che il centro della sua gravità non esca mai fuori di quella linea retta, che da esso centro, posto nel primo termine del moto, si produce insino al centro dell'universo [...]". (Galilei 1890–1909, II, p. 160; authors' translation).

[57] The authors specify that the bodies suspended from a balance by means of threads are subjected only to the strength-weight and that the strength-weight term is a modern interpretation of the concept of "peso" written in Galileo's theory. Then it would be also necessary to specify when Galileo by "peso" intended something like our mass and, when he would have intended what today we define *strength-weight*; then suitable only with the word hung. Nevertheless, even though interesting, for brevity of exposure, we devoted space to such matter in others works (i.e., see recently Pisano 2011, 2017; Capecchi and Pisano 2007a, 2010a).

[58] "Salv. [...] Aggiungo qui, che noi possiamo dire che Archimede e gli altri supposero nelle loro contemplazioni, esser costituiti per infinita lontananza remoti dal centro, nel qual caso i loro assunti non erano falsi, e che però concludevano con assoluta dimostrazione. Quando poi noi vogliamo praticar in distanza terminata le conclusioni dimostrate col suppor lontananza immensa, doviamo diffalcar dal vero dimostrato quello che importa il non esser la nostra lontananza dal centro realmente infinita, ma ben tale che domandar si può immensa in comparazione della piccolezza de gli artificii praticati da noi [...]". (Galilei 1890–1909, VIII, day IV, p. 275. Translation in Galilei 1954, pp. 251–252).

[59] "Galileo paragona tra loro due infiniti (il numero dei punti del segmento *AB* e il numero degli istanti della durata del moto) ed afferma che essi sono uguali; in base a quella che chiamammo a suo tempo "corrispondenza biunivoca" ("A tutti e singoli gli istanti del tempo *AB* corrispondono tutti e singoli i punti del segmento *AB*")." (Frajese 1969, p. 114. Authors' translation).

[60] "[...] qui [*Discorsi e dimostrazioni matematiche*, VIII, pp. 208–209] avviene un altro confronto tra infiniti che pure si risolve in una relazione di uguaglianza: Galileo viene in sostanza ad affermare che si addizionano tutti i tratti di linee parallele (che sono in numero infinito) relativi al moto accelerato, e si compie la stessa operazione per i tratti (pure in numero infinito) relativi a quel moto uniforme, le due somme (infinite) sono uguali." (*Ivi*, p. 115. Authors' translation).

Though with some differences, Torricelli seems to take this position of Galileo to the extreme. He also considered that the heavy bodies could be posed in the external space, not on the Earth, that is they could be also considered to be at an infinite distance from the Earth. In effect, while for Galileo the results obtained by admitting an infinite distance were to be considered empirically approximate, for Torricelli these results seem to be justified by a mathematical calculus.[61] Torricelli, on the subject wrote (Fig. 11).

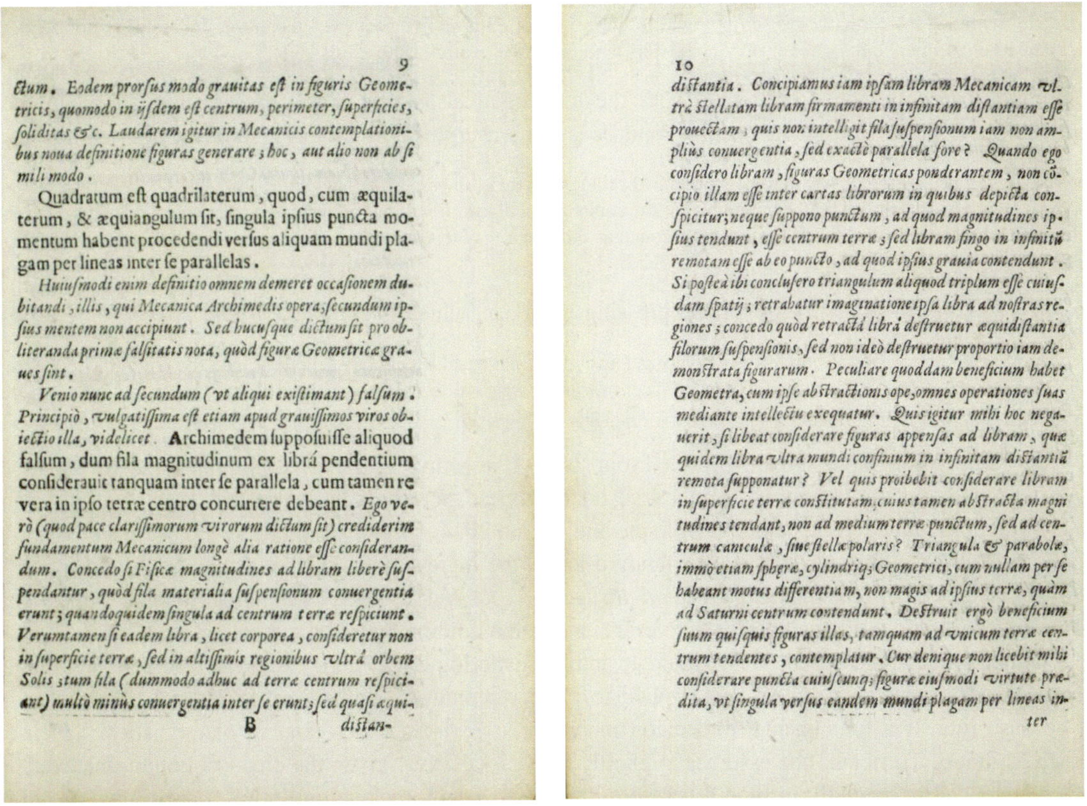

9

ctum. Eodem prorsus modo grauitas est in figuris Geometricis, quomodo in iisdem est centrum, perimeter, superficies, soliditas &c. Laudarem igitur in Mecanicis contemplationibus noua definitione figuras generare; hoc, aut alio non absimili modo.

Quadratum est quadrilaterum, quod, cum æquilaterum, & æquiangulum sit, singula ipsius puncta momentum habent procedendi versus aliquam mundi plagam per lineas inter se parallelas.

Huiusmodi enim definitio omnem demeret occasionem dubitandi, illis, qui Mecanica Archimedis opera, secundum ipsius mentem non accipiunt. Sed hucusque dictum sit pro obliteranda primæ falsitatis nota, quòd figuræ Geometricæ graues sint.

Venio nunc ad secundum (vt aliqui existimant) falsum. Principiò, vulgatissima est etiam apud grauissimos viros obiectio illa, videlicet. Archimedem supposuisse aliquod falsum, dum fila magnitudinum ex librá pendentium considerauit tanquam inter se parallela, cum tamen re vera in ipso terræ centro concurrere debeant. *Ego verò (quod pace clarissimorum virorum dictum sit) crediderim fundamentum Mecanicum longè alia ratione esse considerandum. Concedo si Fisicæ magnitudines ad libram liberè suspendantur, quòd fila materialia suspensionum conuergentia erunt; quandoquidem singula ad centrum terræ respiciunt. Verumtamen si eadem libra, licet corporea, consideretur non in superficie terræ, sed in altissimis regionibus vltrá orbem Solis; tum fila (dummodo adhuc ad terræ centrum respiciant) multò minùs conuergentia inter se erunt; sed quasi æqui-*

B distan-

10

distantia. Concipiamus iam ipsam libram Mecanicam vltrà stellatam libram firmamenti in infinitam distantiam esse prouectam; quis non intelligit fila suspensionum iam non ampliùs conuergentia, sed exactè parallela fore? Quando ego considero libram, figuras Geometricas ponderantem, non cõcipio illam esse inter cartas librorum in quibus depicta conspicitur; neque suppono punctum, ad quod magnitudines ipsius tendunt, esse centrum terræ; sed libram fingo in infinitũ remotam esse ab eo puncto, ad quod ipsius grauia contendunt. Si posteà ibi conclusero triangulum aliquod triplum esse cuiusdam spatij; retrahatur imaginatione ipsa libra ad nostras regiones; concedo quòd retractá librá destruetur æquidistantia filorum suspensionis, sed non ideò destruetur proportio iam demonstrata figurarum. Peculiare quoddam beneficium habet Geometra, cum ipse abstractionis ope, omnes operationes suas mediante intellectu exequatur. Quis igitur mihi hoc negauerit, si libeat considerare figuras appensas ad libram, quæ quidem libra vltra mundi confinium in infinitam distantiã remota supponatur? Vel quis proibebit considerare libram in superficie terræ constitutam, cuius tamen abstracta magnitudines tendant, non ad medium terræ punctum, sed ad centrum canculæ, siue stellæ polaris? Triangula & parabolæ, immò etiam sphęræ, cylindriq; Geometrici, cum nullam per se habeant motus differentiam, non magis ad ipsius terræ, quàm ad Saturni centrum contendunt. Destruit ergò beneficium suum quisquis figuras illas, tamquam ad vnicum terræ centrum tendentes, contemplatur. Cur denique non licebit mihi considerare puncta cuiuscunq; figuræ eiusmodi virtute prædita, vt singula versus eandem mundi plagam per lineas in-

ter

I will immediately admit that even the most obedient scholars accepted the objection that Archimedes made a false assumption when he considered the threads of magnitudes hanging from a balance, as if they were parallel, when in fact they converge toward the centre of the Earth. But I believe (in agreement with what some notable scholars claim) that the [Galilean] principles of Mechanics have to be considered in a completely different way. I admit that, if some physical magnitudes are freely hung [without the action of an external force, but by force-weight only] to a balance the material [fili] wires [that is by not of the ideal wires] will be convergent toward the centre of the earth. If the same balance, even if real, is considered not on the surface of the earth, but in the very far regions beyond the orbit of the sun, then the wires, even though they tend toward the centre of the earth, will be much less convergent, and will be almost parallel. Let us now imagine moving our mechanical balance beyond the firmament, that is to an infinite distance. Who does not understand that the suspension of wires will be no longer converge, but will be exactly parallel to each other? [...].

Fig. 11 Torricelli on the relationship physics-mathematics. *Source* Courtesy University of Oklahoma, History of Science Collections[62]

[61] In effect, we remark that exactitude concerns the mathematical result on the areas obtained by Archimedes's mechanical method (Pisano and Capecchi 2010b).
[62] Torricelli 1644, *De dimensione parabolae, problemata duo*, pp. 9–10; see also *Id.*, 1975, pp. 330–331.

Therefore, the assertion that the wires which suspend the magnitudes (heavy bodies) are infinitely long, seems—at this stage of the theory—to be a choice for using another kind of infinite in mathematics. In particular:

> The geometer [mathematician] has only a peculiar privilege, which is to perform, by virtue of abstraction, every operation by his intellect. Therefore, who will deny me the possibility, if I like so, to consider the bodies [heavy bodies] hanging to a balance, and to suppose the latter [infinitely] far, beyond the boundaries of the world [that is] to an infinite distance? Or, who will prevent me to consider the balance, placed on the surface of the Earth, but, abstractly to think the earth itself, so that the magnitudes [heavy bodies=masses] do not tend to the centre [of the Earth], but to the centre of the constellation of the Dog, or of the Polar Star? In fact, the triangles and the parabolas, as well as the spheres and the cylinders geometrically considered, since they are completely indifferent [and abstracted with respect] to motion, do not tend toward the centre of the earth, as they do not tend to the centre of Saturn. The geometer, therefore, destroys his privilege [to think abstractly] when he considers the figures [heavy bodies] as tending to the centre of the Earth, only.[63]

Unlike Galileo, Torricelli—most concerned with the mathematical aspects of his theory—reveals all his capacity of abstraction, which is typical for a mathematician, and does not waste it on criticizing theoretical proscriptions:

> [...] Therefore, it will be possible to say that the mechanical principle is false, in case of parallel wires and when the magnitudes suspended [heavy bodies] are physical and real, and tending toward the centre of the Earth. It will not be false [the mechanical principle] in case of the magnitudes [heavy bodies=masses] whether abstract or concrete and when they tend – not to the center of the Earth or to another point near the balance – but to some point infinitely far from the Earth.[64]

We can see a certain prudence in explicitly expressing his ideas, which is comprehensible:

> [...] Moreover, for sake of brevity and comfort, we do not get away from the usual vocabulary [allowed], and the point to which is supposed tend to the heavy bodies of the balance, will be called *Centre of the Earth* and the plane perpendicular to the line which connects to the above point with the centre of the balance, as usual, we will call, *Horizon*.[65]

Finally, Torricelli, as we have shown, proves Proposition I resorting to the *Praemittamus* and to an *ad absurdum* reasoning. In the course of the history of science and mathematics, there have been long discussions with regard to the logical status of *ad absurdum* argumentations (Gillispie and Pisano 2014, Chap 7; Rahman 2005). In this context, we do not enter these discussions. We only recall that, for standard logic, the logical status of the *ad absurdum* reasoning is exactly equivalent to that of any other demonstration. The *ad absurdum reasoning* is based on a logical scheme, which is the *modus tollens* and on the connected third excluded principle. It is a scheme holding the same logical value as the *modus ponens*, on which direct demonstrations are based. With regard to intuitionistic logic, the third excluded principle and, hence, the *ad absurdum* reasoning is acceptable only if it is applied to finitary premises. With regard to physics, the situation is rather interesting because time is a fundamental dimension in physics which does not exist in mathematics. Specifically, in a principle, such as that of Torricelli, one assumes implicitly that its validity—given the physical conditions expressed by its hypothesis—is extended unlimitedly in time. Therefore, this creates such a situation: let us suppose that an *ad absurdum* proof is conceived by using a principle such as that of Torricelli. All the elements are finite: there are few bodies, few *momenta*, few potential movements and so on. Therefore, given the explicit elements, such an *ad absurdum* proof should also be accepted within the schemes of intuitionistic logic. For, if all elements are finite, the *third excluded principle* is applicable within intuitionistic logic. However, an infinitary assumption is implicit in such a proof because it is based on a principle, which foresees its value along an unlimited interval of time, to be more precise: for the whole of eternity. Therefore, when a principle is applied in an *ad absurdum* proof within the physics of the principles, the proof, rigorously speaking, is not intuitionistic because the third excluded principle is applied to an infinitary assertion. This is fully coherent with what we claimed as regards the nature of the principles. They say something with regard to the entire structure of space and time and since, at least in classical physics, time in its entirety has to be assumed as an infinite magnitude, this implies the non–intuitionistic character of the *ad absurdum* proofs within this context.

Finally, it is reasonable to hypothesize that—with respect to the scientific and experimental Galilean context—Torricelli already had enough mathematical logic-mathematical attitudes towards that kind of idealized mathematics (differential calculus, etc.) that, later on, will be developed by other scholars like Leibniz, Newton etc. (Newton 1714, 1730 [1686–1687] 1803; Bussotti and Pisano 2014b, c, 2017; Bussotti 2015; Pisano and Bussotti 2016).

---

[63] Torricelli (1644), *De dimensione parabolae, problemata duo*, pp. 9–10; see also *Id.*, 1975, pp. 330–331.

[64] Torricelli (1644), *De dimensione parabolae, problemata duo*, pp. 10–11; see also *Id.*, 1975, p. 332.

[65] *Ibidem.*

5 Concluding Remarks

In this paper, we have discussed some aspects of Torricelli's gravity centre theory. In particular, we have pointed out:

(a) The role of the *Praemittamus* within his theorization;
(b) More in general: the role of the principles in Torricelli's approach to physics;
(c) The way in which Archimedes and Galileo, Torricelli's two main sources, influenced his work;
(d) The role of the *Praemittamus* within the history of the principle of virtual work;
(e) The complex situation concerning the use of mathematics within physics between the 20s and the 50s of the seventeenth century, in particular in reference to Galileo and Torricelli;
(f) The way in which the *reductio ad absurdum* is used in a physics based on principles.

The picture is rather interesting: Torricelli was an important crossroad for the development of physics, in particular, for the physics of the principles, and for the improvement of the use of mathematics within physics. Furthermore, his work is also significant for the logic of physics, as it is shown by the way in which he resorted to the *ad absurdum* reasoning.

Rather than summarizing the content of our paper, we would like to conclude with a curiosity, a sort of historiographical thesis claimed by Torricelli: the ancients knew some mathematical-physical methods they kept secret. Once again, Torricelli's reference point is Archimedes. We read:

> Among all the works pertinent to Mathematical disciplines that we have, it seems to us that Archimedes rightly vindicates the position of a Prince. The subtlety of his works almost terrifies the soul.[66]

As we remarked above, Archimedes, in the *Quadratura parabolae*, first obtains results using the mechanical approach and then reconsiders the discourse with the classical methods of geometry to confirm the correctness of his results in a rigorous way. Similarly Torricelli, with the compelling idea of duplicating the procedure, devotes many pages to prove certain theorems on the "parabolic segment", by following the geometry used in antiquity (Knobloch 2005; Torricelli 1644), *Quadratura parabolae pluris modis per duplicem positionem more antiquorum absoluta*, pp. 17–54) and then proving the validity of the thesis also with the "indivisibiles" (Archimedes 2002, *Quadratura parabolae,* pp. 253–252, 55–84; Torricelli 1644, *De solido acuto hyperbolico problema alterum*, pp. 93–135).

In this respect, it is interesting to note that he underlines the "concordantia" (Torricelli 1644, *De solido acuto hyperbolico problema alterum*, p. 103) of methods and reasonings (Torricelli 1644, *Quadratura parabolae per novam indivisibilium Geometriam pluribus modis absoluta*, p. 55). The main idea was not only the desire to give the reader results and methods, but also to say that the *indivisibles technique was not completely unknown to the ancient Greek scholars*. Besides, Torricelli seems to hold onto the idea that the methods of demonstration of the ancients, such as Archimedes's method, *was intentionally kept secret*. He states that the ancient geometers worked according to a method "in invenzione" suitable "ad occultandum artis arcanum" (Torricelli 1644, *Quadratura parabolae per novam indivisibilium Geometriam pluribus modis absoluta*, p. 55).

In part, this thesis was due to the fascination exerted on Torricelli and on several scientists living in the first part of the seventeenth century by the results the ancients, and in particular Archimedes, obtained. However, Torricelli was right, as Heiberg's discovery of Archimedes's method in 1906 has shown. Perhaps, Hellenistic science will be able to surprise us and to provoke our further admiration by means of methods and procedures which are kept hidden in some manuscripts. We cannot exclude that Torricelli was right!

Acknowledgements We thank the anonymous reviewers for their careful reading of our manuscript and their many insightful comments and suggestions. We also express gratitude to Dr. Julie Robarts (University of Melbourne, Australia) for her observations on English language; eventual remaining English mistakes are our own.

---

[66] "*Proemium*. Inter omnia opera ad Mathematicas diſciplinas pertinentia, iure optimo Principem ſibi locum vindicare videntur Archimedis inuenta; quæ quidem ipſo ſubtilitatis miraculo terrent animos." (Torricelli 1644, p. 5; authors' translation).

References

Agostini A (1930a) Dimostrazione di una proposizione di Torricelli sulla spirale logaritmica. Periodico di Matematica 168/4–10:143–151.

Agostini A (1930b) Un brano inedito di Torricelli sulla rettificazione della spirale logaritmica. Il Bollettino di Matematica 170/2–9:XXV–XXXVIII.

Agostini A (1930c) Il problema inverso delle tangenti nell'opera di Torricelli. Archeion 12/1:33–37.

Agostini A (1931) Il problema inverso delle tangenti e l'integrazione indefinita nelle opere del Torricelli. Archeion 13:55–59.

Agostini A (1949) La memoria di Torricelli sopra la spirale logaritmica riordinata e completata. Accademia Navale, Livorno.

Agostini A (1950) Il metodo delle tangenti fondato sopra la dottrina dei moti nelle opere di Torricelli. Periodico di Matematiche 28/4:141–158.

Agostini A (1951a) I baricentri trovati da Torricelli. Bollettino Unione Matematici Italiani:149–159.

Agostini A (1951b) Massimi e minimi nella corrispondenza di Torricelli con Michelangelo Ricci. Atti del IV Congresso dell'U.M.I.

Agostini A (1951c) Il "De tactionibus" di Evangelista Torricelli. Bollettino Unione Matematici Italiani:319–321.

Agostini A (1951d) Problemi di massimo e minimo nella corrispondenza di E. Torricelli. Rivista Matematica dell'Università di Parma:265–275.

Agostini A (1951e) Gli studi di Torricelli sulla cicloide. Accademia Navale, Livorno.

Ampère AM (1806) Démonstration du principe des vitesses virtuelles, dégagée de toute considération des infiniment petits. Journal de l'Ècole Polytechnique 6/13:247–269.

Archimedes (1897) On the floating bodies. Heath TL (ed). The works of Archimedes. The Cambridge University Press, Cambridge.

Archimedes (2002) On the equilibrium of planes, Books I–II. In: Heath TL (ed). The works of Archimedes. Dover Publications Inc., New York, pp. 189–220.

Aristotle (1853) On the Definition and Division of Principles. In: Octavius Freire Porphyry Owen (ed). The Organon, Or Logical Treatises, of Aristotle. Bohn H G, London, Vol 1, pp. 263–266.

Aristotle ([1936]1955) Minor Works. Translated by Hett WS. The Harvard University Press–Heinemann W LTD, Cambridge MA–London.

Aristotle (1949) Aristotle's Prior and Posterior Analytics. A Revised Text with Introduction and Commentary Ross WD (ed). The Oxford University Press, Oxford.

Aristotle (1955a) De Caelo. Translation into English by Stocks JL. The Tech Classics Archive of M.I.T.

Aristotle (1955b) Mechanical problems. In: Hett WS (ed) Aristotle. Minor works. William Heinemann, Cambridge, pp. 328–414.

Aristotle (1955c) Minor Works. Transaltion by Hett WS. The Harvard University Press. Heinemann W LTD, Cambridge MA–London.

Aristotle (1984) The Complete Works of Aristotle. Barnes (ed) The Princeton University Press, Princeton, NJ.

Aristotle (1999) Physics. Translation by Waterfield R. The Oxford University Press, Oxford.

Baldi B (1621) In Mechanica Aristotelis Problemata Exercitationes: adiecta succinta narratione de autoris vita et scriptis. Ioannis Albini, Moguntiae.

Baroncelli G (1993) On the Invention of the Geometric Spiral: an Unpublished Letter of Torricelli to Michelangelo Ricci. Nuncius 8/2:601–606.

Bernoulli D ([1733]1738) Theoremata de oscillationibus corporum filo flexili connexorum et catenae verticaliter suspense. Novi commentarii Academiae Scientiarum Imperialis Petropolitanae, pp. 108–123.

Bernoulli D ([1748]1750) Remarques sur le principe de la conservation des forces vives pris dans un sens général. Mémoire de l'Académie Royale des Sciences et des Lettres de Berlin, vol IV, pp. 356–364.

Bernoulli Jakob (1703) Démonstration générale du centre de balancement et d'oscillation, tirée de la nature du levier. Mémoire de l'Académie Royale des Sciences Paris, pp. 78–84.

Bernoulli Johann (1727) Theoremata selecta pro conservatione virium vivarum demonstranda et esperimenta confirmanda. Novi commentarii Academiae Scientiarum Imperialis Petropolitanae 2, pp. 200–207.

Bernoulli Johann (1742a) Propositiones variae mechanico–dynamicae. In: Bernoulli 1742b, Vol IV, pp. 253–386.

Bernoulli Johann (1742b) Opera omnia, tam antea sparsim edita, quam hactenus inedita ... quibus continentur ea, quae ab A. 1690 usque ad A. 1727 prodierunt. Accedunt Lectiones Mathematicae de Calculo Integralium atque Anedokta. I–IV vols, Lausannae & Genevae, Bousquet [Facsimile 1968. Reprint Curavit Hofman JE. Georg Olms Verlag Hildeshe].

Bertoni G (1985) Breve storia della 'vexata quaestio' del luogo natale di Evangelista Torricelli. Torricelliana. Bollettino della Società Torricelliana di Scienze e Lettere Faenza 36:330–345.

Bertoni G (1987) La faentinità di Evangelista Torricelli e il suo vero luogo di nascita. Torricelliana. Bollettino della Società Torricelliana di Scienze e Lettere Faenza 38:85–94.

Bertoni G (2003) La faentinit di Evangelista Torricelli e il suo vero luogo di nascita. Studi e ricerche del Liceo Torricelli. Ragazzini, Faenza, pp. 5–12.

Bortolotti E (1992) Applicazioni del calcolo integrale alla determinazione del centro di gravità di figure geometriche. Accademia di Bologna, Bologna.

Brown EJ (1967–1968). The Scientia de Ponderibus in the later Middle Ages, Ph.D. dissertation. Tutor Clagett M. Madison: University of Winscosin [available via: UMI – Proquest Company.

Buridan J (1509) Subtilissimae Quaestiones super octo Physicorum libros Aristotelis, Paris. [Reprint: Id. (1964) Kommentar zur Aristotelischen Physik. Minerva, Frankfurt].

Buridan J (1513) Quaestiones super decem libros Ethicorum Aristotelis ad Nicomachum, Paris. [Reprint: Id. (1968) Super decem libros Ethicorum. Minerva, Frankfurt].

Buridan J (1942) Iohannis Buridani Quaestiones super libris quattuor De caelo et mundo. In: Moody EA (ed). Medieval Academy of America. Studies and documents 6/40.

Bussotti P (2003) On the Genesis of the Lagrange Multipliers. Journal of optimization theory and applications 117/3:453–459.

Bussotti P (2015) The Complex Itinerary of Leibniz's Planetary Theory. Springer–Birkhäuser, Basel.

Bussotti P, Pisano R (2013) On the Conceptual Frames in René Descartes' *Physical Works*. Advances in Historical Studies 2/3:106–125.

Bussotti P, Pisano R (2014a) Newton's Philosophiae Naturalis Principia Mathematica "Jesuit" Edition: The Tenor of a Huge Work. Accademia Nazionale Lincei–Matematica e Applicazioni 25/4:413–444.

Bussotti P, Pisano R (2014b) On the Jesuit Edition of Newton's Principia. Science and Advanced Researches in the Western Civilization. Newton Special Issue (Pisano ed): History and Historical Epistemology of Science. *Advances in Historical Studies* 3/1:33–55.

Bussotti P, Pisano R (2014c) On the Jesuit edition of Newton's Principia. Science and advanced researches in the western civilization. In: Pisano R (ed) Isaac Newton and his scientific heritage: new studies in the history and historical epistemology of science. Special Issue. Advances in Historical Studies 3/1:33–55.

Bussotti P, Pisano R (2017) Historical and Philosophical Details on Leibniz's Planetary Movements as Physical–Structural Model. In: Pisano, Fichant, Oliveira and Bussotti 2017 (eds), pp. 49–92.

Bussotti P, Pisano R (2020) Historical and Foundational Details on the Method of Infinite Descent: Every Prime Number of the Form 4n+1 is the Sum of Two Squares. Foundations of Science (Springer) 25:671–702.

Bussotti P, Pisano R (2023) Details in Galileo's Mechanism of Inertia as New Intellectual Historical Insights. In: Pisano R (ed). A History of Physics: Phenomena, Ideas & Mechanisms. Essays in Honor of Salvo D'Agostino Dordrecht, Springer, *forthcoming*.

Capecchi D, Pisano R (2007a) La teoria dei baricentri di Torricelli come fondamento della statica. Physis XLIV:1–29.

Capecchi D, Pisano R (2007b) Il principio di Torricelli prima di Torricelli. In Preziosi, B. (ed.). Proceedings of XXIV SISFA Congress. Bibliopolis, Napoli, pp. 107–112.

Capecchi D, Pisano R (2008) La meccanica in Italia nei primi anni del Cinquecento. Il contributo di Niccolò Tartaglia. In Proceedings of XXV SISFA Congress. University of Milano, Milano, C17.1–C17.6.

Capecchi D, Pisano R (2010a) Reflections on Torricelli's Principle in Mechanics. Organon 42:81–98.

Capecchi D, Pisano R (2010b) Scienza e tecnica nell'architettura del Rinascimento. CISU, Roma.

Carnot L (1780) Mémoire sur la théorie des machines pour concourir au prix que l'Académie Royale des Sciences de Paris doit adjuger en 1781. The manuscript is dated from Béthune 15 July 1780. It is conserved in the Archives de l'Académie des sciences, Institut de France, and consists of 191 sections in 106 folios. Sections 101–160 are reproduced in: Gillispie (1971), Appendix C, pp. 299–343.

Carnot L (1786) Essai sur les machines en général. Defay, Dijon.

Carnot L (1803a) Principes fondamentaux de l'équilibre et du mouvement. Deterville, Paris.

Carnot L (1803b) Géométrie de position. Duprat, Paris.

Carnot L (1808a) Essai sur les machines en général (I part) In: Philosophical Magazine: comprehending the various branches of science, the liberal and fine arts, agriculture, manufactures, and commerce. Tilloch A (ed). Vol. XXX. Murray J, London, pp. 8–14; pp. 154–158; pp. 208–221; pp. 310–320. [very interesting is the avant–titre– page of the book where a portrait of Lazare Carnot is published].

Carnot L (1808b) Essai sur les machines en général (II Part). In: Philosophical Magazine: comprehending the various branches of science, the liberal and fine arts, agriculture, manufactures, and commerce. Tilloch A (ed). Vol. XXXI. Murray J, London, pp. 136–145; pp. 220–228; 295–305.

Carnot L (1813) Réflexions sur la métaphysique du calcul infinitésimal. Courcier, Paris.

Carnot L (1990) Rapport sur une nouvelle machine inventée par MM. Niepce, et nommée par eux Pyrélophore; par MM. Berthollet et Carnot. Lu le 15 décembre 1806, imprimé en 1807, 1er semestre. In: Charnay JP (ed) Lazare Carnot ou le savant–citoyen, La Herne, Paris, Presses de L'Université de Paris–Sorbonne, 1990 – Centre d'études et de recherches sur les stratégies et les conflits, Colloque en Sorbonne, 1988).

Cartelon H (1975) Does Aristotle Have A Mechanics? in Barnes et al.: Articles on Aristotle. Vol. I: Science, Duckworth (ed.), London.

Castelnuovo G (1962) Le origini del calcolo infinitesimale nell'era moderna, con scritti di Newton, Leibniz, Torricelli. Feltrinelli, Milano.

Caverni R (1891–1900) Storia del metodo sperimentale in Italia. 6 Vols. Civelli, Firenze.

Clagett M (1959) The science of mechanics in the Middle Ages. The University of Wisconsin Press, Madison.

Clagett M, Moody E ([1952]1960) The medieval science of weights. Scentia de ponderibus. The University of Wisconsin Press, Madison.

Clapeyron ÉBP (1834) Mémoire sur la puissance motrice du feu. Journal de l'École Royal Polytechnique XXIIIème/XIV:153–190.

Clausius RJE (1850) Über die bewegende Kraft der Wärme und die Gesetze, welche sich daraus für die Wärmelehre selbst ableiten lassen. Annalen der Physik und Chemie 79:368–397, 500–524. (English trans: *Id.*, On the Motive Power of Heat and on the Laws which can be deduced from it for the theory of heat itself. In: Mendoza 1960, pp. 73–74, pp. 109–152.

Cohen IB, Smith GE (2002) (eds) The Cambridge Companion to Newton. The Cambridge University Press, Cambridge.

Commandino F (1565) Federici commandini vrbinatis liber de centro gravitatis solidorum Bononiae: Ex Officina Alexandri Benacii.

Commandino F (1575) De gli elementi di Euclide libri quindici. Con gli scholii antichi. Tradotti prima in lingua latina da m. Federico Commandino da Vrbino, & con commentarij illustrati, et hora di ordine delli istesso trasportati nella nostra vulgare, et da lui riueduti in Vrbino : appresso Domenico Frisolino In Vrbino : in casa di Federico Commandino.

D'Alembert JLR ([1743]1758) Traité de dynamique. Nouvelle édition, revue et fort augmentée par l'Auteur Chez David Libraire, Paris.

D'Alembert JLR (1751–1780) Encyclopédie ou dictionnaire raisonné des sciences, des arts et des métiers, Diderot & D'Alembert. Briasson–David–Le Breton–Durand, Paris.

D'Alembert JLR (1767) Mélanges de littérature, d'Histoire et de philosophie. Tome V. Chatelain & Fils. Amsterdam, pp. 3–274.

De Gandt F (1992) L'évolution de la théorie des indivisibles et l'apport de Torricelli. Biblioteca Nuncius Studi Testi X:103–118.

Del Monte G ([1577]1581). Le Meccaniche dell'Illustrissimo Sig. Guido Ubaldo dè Marchesi del Monte, tradotto in volgare dal Sig. Filippo Pigafetta. Evangelista Deuchino, Venezia [1577: Mechanicorum Liber].

Del Monte G (1588) In duos Archimedis aequeponderantium libros paraphrasis. Hieronymum Concordiam, Pisauri.

Del Monte G (1615) Le mechaniche dell'illustriss. Sig. Guido Ubaldo de' marchesi del Monte tr. Tradotto in volgare da Filippo Pigafetta (1581). Evangelista Deuchino, Venezia.

Dijksterhuis EJ (1955–1958) (ed.) The Principal Works of Simon Stevin, N.V. Swets & Zeitlinger, Amsterdam, Vol. I.

Duhem PM (1905–1906). Les origines de la statique. 2 vols, Hermann, Paris.

Euler L ([1730–1731]1738) De communicatione motus in collision corporum. Commentarii academiae scientiarum Petropolitanae 5:159–168.

Euler L ([1773]1774) De pressione ponderis in planum in cui incumbit. Novi Commentarii academiae scientiarum Petropolitanae, vol XVIII, pp. 289–329.

Euler L (1736) Mechanica sive motus scientia analytice exposita. Auctore Leonhardo Eulero academiae imper. scientiarum membro et matheseos sublimioris professore. Tomus I. Instar supplementi ad commentar. acad. scient. imper. Petropoli. Ex typographia academiae scientarum.

Euler L (1749) Recherches sur les mouvements des corps célestes en général (1747). Mémoire de l'Académie Royale des Sciences et des Lettres de Berlin, Vol. I, pp. 93–143.
Euler L ([1750]1752) Découverte d'un nouveau principe de Mécanique. Mémoires de l'académie des sciences de Berlin. Vol. VI, 1752, pp. 185–217.
Euler L (1757) Principes généraux du mouvement des fluides. Mémoires de l'Academie des Sciences de Berlin 11:274–315.
Euler L (1775) Nova methodus motum corporum rigidorum determinandi. Novi Commentarii academiae scientiarum Petropolitanae, vol XX, pp. 208–238.
Fabbri S (1947–1997) La Società Torricelliana di Scienze e Lettere di Faenza nel Cinquantenario della Fondazione. Bollettino della Società Torricelliana di Scienze e Lettere s.n.
Favino F (2009) In Urbe mathematicus: Torricelli a Roma. Galilaeana VI:1000–1032.
Favino F (2019) Entry: Torricelli. Dizionario biografico degli italiani. Vol. 96. Via: https://www.treccani.it/enciclopedia/evangelista-torricelli_%28Dizionario-Biografico%29/.
Fermat P (1891–1912) Œuvres de Fermat publiées par P. Tannery et Ch. Henry. Gauthier–Villars, Paris.
Festa E (1992) La notion d'agrégat d'indivisibles' dans la constitution de la cinématique galiléenne: Cavalieri, Galilée, Torricelli, Revue d'Histoire des Sciences 45/2–3:307–336.
Fossombroni V (1794) Memoria sul princpio delle velocità virtuali. Gaetano Cambiagi, Firenze.
Fourier JBF (1878) Analytical theory of heat. Translated by Freeman A. The Cambridge University Press, London.
Fourier JBJ (1888–1890) Œuvres de Fourier par les soins de M. Gaston Darboux 2 vols, Gauthier–Villars, Paris.
Fox R (1974) The rise and fall of Laplacian physics. Historical Studies in the Physical Sciences 4:89–136.
Fox R (1995) Science, Industry and the Social Order in Post–Revolutionary France. Variorum Collected Studies Series 489, Ashgate Variorum.
Frajese A (1969) Attraverso la storia della matematica. Le Monnier, Firenze.
Fraser CG (1983) J.L. Lagrange's Early Contribution to the Principles and Methods of Mechanics. Archive for History of Exact Sciences 28/3:197–241.
Frezier M (1739) La théorie et la pratique coupe des pierres et des bois pour la construction des voutes et autres parties des bâtimen[t]s Civils & Militaires. Traité de Stéréotomie à l'usage de l'Architecture. Par M. Frezier, Chevalier de l'Ordre Militaire de Saint Louis. Ingénieur ordinaire du Roy en Chef à Landau. Strasbourg, Doulsseker & Fils & Paris, Jombert, Tome III.
Galilei G Ms. B – Trattato delle fortificazioni, con disegni diversi. Biblioteca Ambrosiana. D: 296 Inf.– 38499.
Galilei G ([1595]1634) Le mecaniche. In: Galilei 1890–1909, II, pp. 147–191.
Galilei G (1890–1909) Breve instruzione all'architettura militare. In: 1890–1909, II, pp. 15–75.
Galilei G (1890–1909) Le opere di Galileo Galilei: Edizione nazionale sotto gli auspici di sua maestà il re d'Italia. Favaro A (ed.), Barbera, Firenze.
Galilei G (1890–1909) Trattato di fortificazione. In: Galilei 1890–1909, II, pp. 77–146.
Galilei G (1954) Dialogues Concerning Two New Sciences, translated by Henry Crew & Alfonso de Salvio, with an introduction by Antonio Favaro, Dover Publications, Inc., New York [Originally published in 1904 by the MacMillan company].
Galilei G Ms. A – Trattato di fortificazioni e modo d'espugnare la città con disegni e piante di fortificazioni. Biblioteca Ambrosiana. D: 328 Inf.– 38640.
Galilei G Ms. m – Trattato di fortificazioni, con disegni e figure (acephalous). Biblioteca Ambrosiana. N: 281 Sup.– 2– 80277.
Galluzzi P (1979) Momento: Studi galileiani. Ateneo & Bizzarri, Roma.
Galluzzi P, Torrini M (1975–1984) (eds) The works of Galileo Galilei's disciples. Correspondence (in Italian). Giunti–Barbera, Firenze, 2 Vols.
Gatto R (2002) Galileo Galilei. Le mecaniche. Edizione critica e saggio introduttivo. Olschki, Firenze.
Gaukroger S, Schuster J, Sutton J (2000) (eds) Descartes' natural philosophy. Routledge London and New York.
Gillispie CC (1970–1979) Carnot Lazare–Nicolas–Marguerite. III. In: Gillispie (ed) 1970–1980, pp. 70–79.
Gillispie CC (1971) Lazare Carnot Savant. A monograph treating Carnot's scientific work, with facsimile reproduction of his unpublished writings on mechanics and on the calculus, and an essay concerning the latter by Youschkevitch AP, Princeton University Press, Princeton, NJ.
Gillispie CC (1976) The scientific work of Lazare Carnot, and its influence on that of his son. In: Taton (ed) 1976, pp. 23–33.
Gillispie CC (ed) (1970–1980) Dictionary of Scientific Biography. Charles Scribner's Sons, New York.
Gillispie CC, R Pisano (2014) Lazare and Sadi Carnot. A Scientific and Filial Relationship, 2nd edition. Springer, Dordrecht.
Gillispie CC, Youschkevitch AP (1979) Lazare Carnot savant et sa contribution à la théorie de l'infini mathématique. Vrin, Paris [Collection des Travaux de l'Académie Internationale d'Histoire des Sciences, 20].
Gliozzi M (1970–1990) Biography in Dictionary of Scientific Biography. Charles Scribner's Sons, New York. http://www.encyclopedia.com/topic/Evangelista_Torricelli.aspx [the reader should pay attention to the place of birth cited. Nowadays, it is known that he was born in Roma, Italy).
Grattan–Guinness I (1990) Convolutions in French Mathematics, Birkhäuser, Basel.
Hankins TL (1970) Jean D'Alembert. Science and the Enlightenment. The Clarendon Press, Oxford.
Henry J (2011) Mathematics made no contribution to public weal: why jean Frenel (1497–1558) became a physician. Centaurus 53/3:193–220.
Hoyrup J (1988) Jordanus de Nemore, 13th Century Mathematical Innovator. Archive for History of Exact Science 38:307–363.
Huygens C (1669) Démonstration de l'équilibre de la balance. [Leu dans l'Académie le 2 décembre 1673]. In Huygens 1629–1695, XVII, pp. 42–47 [see also Huygens 1937, XIX, pp. 40–47].
Huygens C (1673) Christiani Hugenii Zulichemii, Const. F. Horologium Oscillatorium sive De motu pendulorum ad horologia aptato demonstrationes geometricae. Parisiis, apud F. Muguet, regis et illustrissimi archiepiscopi typographum, via Citharae, ad insigne trium Regum.
Huygens C (1629–1695) Œuvres complètes de Christiaan Huygens. Société hollandaise des sciences. Martinus Nijhoff, La Haye.
Huygens C (1937) Christiaan Huygens, Oeuvres complètes. Tome XIX. Mécanique théorique et physique 1666–1695. Edited by Vollgraff JA). Martinus Nijhoff, Den Haag. [via: http://www.dbnl.org/tekst/huyg003oeuv19_01/colofon.php].
Jammer M (1957) Concepts of force: A study in the foundations of dynamics, Harvard University Press, Cambridge MA.
Jaouiche K (1976) Le livre du Qarastun de Thabit Ibn Qurra. Brill, Leiden.
Jardine N (1988) Epistemologies of the sciences. In: Schmitt CB and Skinner O (eds). Cambridge History of the Renaissance philosophy. The Cambridge University Press, Cambridge.

Knobloch E (2005) Géométrie pratique, Géométrie savante. Albertiana 8:27–56.
Knobloch E, Vasoli C, Siraisi N (2001) Il Rinascimento. Medioevo, Rinascimento. Storia della scienza. Vol. IV. Istituto della Enciclopedia Italiana, Roma, pp. 605–1044.
Kragh H (1987) An Introduction to the Historiography of Science. The Cambridge University Press, Cambridge.
Krarup J, Vajda S (1997) On Torricelli's Geometrical Solution to a Problem of Fermat, Duality in Practice. Journal of Mathematics Applied in Business and Industry 8/3:215–224.
Lagrange JL (1762) Application de la méthode exposée dans le mémoire précédente à la solution des différents problèmes de dynamique. In: Lagrange 1870–1873, I, pp. 363–468.
Lagrange JL (1764) Recherches sur la libration de la Lune dans lesquelles on tache de résoudre la question proposes par l'Académie Royale des Sciences, pour le Prix de l'année 1764. Prix de l'académie Royale des Sciences de Paris, IX (1777), pp. 1–50. See also Lagrange 1870–1873, Tome VI, pp. 5–61.
Lagrange JL (1788) Mécanique Analytique. Desaint, Paris.
Lagrange JL (1870–1873) Œuvres de Lagrange. Seconde édition, Courcier (ed), I–XIV vols. (in X) Paris Gauthier–Villars, Paris.
Lagrange JL (1892) [Lagrange's letter to Euler on November 24th 1759 in:] Œuvres de Lagrange. Tome XIV. Darboux G (ed). Gauthier–Villars, Paris, pp. 170–174.
Laplace PS (1816) Sur la vitesse du son dans l'air et dans l'eau. Annales de chimie et de physique III:238–241.
Leibniz GW (2009) Gottfried Wilhelm Leibniz Naturwissenschaftlich–medizinisch–technische Schriften Reihe VIII der Akademieausgabe. Berlin–Brandenburgische Akademie der Wissenschaften, Российская Академия Наук, in Zusammenarbeit mit dem Centre National de la Recherche Scientifique, der Gottfried Wilhelm Leibniz Bibliothek Hannover und der Herzog August Bibliothek Wolffenbüttel. Knobloch E (project leader editor). There is a printed version of the volume(s) and there is an online–version available via: http://leibnizviii.bbaw.de/.
Lindberg DC (1976) Theories of Vision from Al–Kindi to Kepler. The Chicago University Press, Chicago.
Loria G (1897) Evangelista Torricelli e la prima rettificazione di una curva. Rendiconti della Regia Accademia dei Lincei 5/6:318–323.
Loria G (1903) Un'impresa nazionale di universale interesse. Pubblicazione delle opere di Evangelista Torricelli. Atti II Congresso Internazioanle delle Scienze Storiche, pp. 23–27.
Loria G (1909) In memoria di Evangelista Torricelli. Discorso. Rivista Ligure di Scienze. Lettere ed arti, Genova.
Loria G (1915) L'infinito e l'infinitesimo secondo i matematici dell'antichità e secondo i matematici moderni anteriori al secolo XVIII. Scientia 12 [see also *Id.*, 1916, Scientisa I].
Loria G (1917) Una generazione delle coniche a centro ideata da E. Torricelli. Periodico di Matematiche 32:189–192.
Loria G (1919) E. Torricelli nella storia della geometria. Rendiconti della Regia Accademia dei Lincei 5/28:409–415.
Loria G (1922a) Una lettera inedita di E. Torricelli al P. Mersenne. Archivi di storia della scienza 3:272–276.
Loria G (1922b) L'opera geometrica di Evangelista Torricelli. Bollettino di Matematica 2/1:I–VII.
Loria G (1923) Evangelista Torricelli. Gli scienziati italiani. Vol. I, Parte II, pp. 299–311.
Loria G (1932) Una lettera inedita di Evangelista Torricelli. Archeion XIV:12–14.
Loria G (1938) Le prétendu larcin de Torricelli. Archeion XIX.
Loria G (1944) Il III centenario del barometro. Bollettino della Società Torricelliana di Scienze e Lettere s.n.
Loria G (1945) Prime linee di una bibliografia torricelliana. Bollettino della Società Torricelliana di Scienze e Lettere s.n.
Loria G (1947) L'avverso destino di Evangelista Torricelli. Rendiconti dell'Accademia Nazionale dei Lincei 8/3:665–669.
Lorini B (1596) Delle Fortificationi di Bvonaivto Lorini. Nobile Fiorentino. Libri Cinque. Ne' quali si mostra con le più facili regole la Scienza con la ratica, di fortificare le città, & altri luoghi sopra diversi siti [...] Nuovamente dati in luce. Con Priuilegio. In Vinegia, appresso Gio. Antonio Rampanzetto, MDXCVI. BNCF - Palatino 11. 3 .7. 37.
Lorini B (1609) Le Fortificationi di Bvonaivto Lorini. Nobile Fiorentino. Nnuovamente ristampate, Corrette & Ampliate [...] Con l'Aggiunta del Sesto Libro. Con Privilegi. In Venetia, presso Francesco Rampazetto. fortificare le città, & altri luoghi sopra diversi siti [...]. Nuovamente dati in luce. In Vinegia, MDCIX. Presso Francesco Rampanzetto.
Mach E ([1883]1919) The Science of Mechanics. A Critical and Historical Account of its Development, Chicago-London, The Open Court Publishing Co.
Maupertuis PL (1746) Les loix du mouvement et du repos déduits d'un principe métaphysique. Histoire de l'Académie Royale des Sciences et des Belles Lettres, pp. 267–294.
Medolla G (1993) Alcuni documenti inediti relativi alla vita di Evangelista Torricelli. Bollettino Unione Matematici Italiani XIII/2:287–295.
Mendoza E (ed) (1960) Reflections on the motive power of fire and other papers on the second law of thermodynamics by E Clapeyron and R Clausius. Dover, New York.
Milighetti MC (2004) Da Antonio Nardi (1598–1649 ca.) a Francesco Redi (1626–1697): l'eredità culturale di un Galileiano. Annali Aretini XII:221–238.
Milighetti MC (2006) Sophia e Mathesis negli scritti di Antonio Nardi. Bollettino di storia delle matematiche 26/1:9–31.
Nabonnand P (2010) L'argument de la généralité chez Carnot, Poncelet et Chasles. Via: http://hal.archives-ouvertes.fr/hal-00637385/fr/.
Nabonnand P (2011) Une géométrie sans figure? La Figure et la Lettre, pp. 99–120.
Nardi A (fl 1642s) Le scene. Codice Galileano 130. Biblioteca Nazionale di Firenze.
Newton I ([1671]1736) The method of fluxions and infinite series. Henry Woodfall, London.
Newton I ([1686–1687]1803) The Mathematical Principles Of Natural Philosophy, by Sir Isaac Newton. Translated into English by Motte A, Symonds, London, Vol. I.
Newton I ([1713]1729) The mathematical principles of natural philosophy. Translated by Motte Andrew. Motte B, London.
Newton I (1714) Philosophiae Naturalis Principia Mathematica. Reprints of 2nd English edition [1713, Cotes R (ed), Cantabrigiae], Rogerus Cotes Editoris Praefatio, Sumptibus Societatis, Amstaelodami.
Newton I ([1726] [1739–1742]1822) Philosophiae Naturalis Principia Mathematica, auctore Isaaco Newtono, Eq. Aurato. Perpetuis commentariis illustrate, communi studio pp. Thomae le Seur et Francisci Jacquier ex Gallicana Minimorum Familia, matheseos professorum. Editio nova, summa cura recensita. A. et J. Duncan, Glasgow.

Newton I (1730) Queries. In: Opticks, 4th English Edition corrected, Innys W (ed), London.

Newton I (1746) Methodus fluxionum et serierum infinitarum, in Opuscula mathematica. Marcum Michaelem Bousque, Lausanne et Genevae, I, 29–200.

Nonnoi G (1982) Evangelista Torricelli, la pressione dell'aria e gli Autori de' crepuscoli. Torricelliana. Bollettino della Societa` Torricelliana di Scienze e Lettere Faenza:17–46.

Oliveira ARE, Pisano R (2022) Historical Details of Coulomb's Torsional Forces in Metal Wires In: Ceccarelli M, López-García R (eds). Explorations in the History and Heritage of Machines and Mechanisms. Vol. 40. Dordrecht, Springer, pp. 242–254.

Pisano R (2007) Brief history of centre of gravity theory. Epistemological notes. In: Kokowski M (ed). Proceedings of the 2nd ESHS Congress, Krakow, Polish Academy of Science, Krakow, pp. 934–941.

Pisano R (2009a) Il ruolo della scienza archimedea nei lavori di meccanica di Galilei e di Torricelli. In: Giannetto E, Giannini G, Capecchi D, Pisano R, (eds). Da Archimede a Majorana: La fisica nel suo divenire. Guaraldi Editore, Rimini, pp. 65–74.

Pisano R (2009b) On method in Galileo Galilei' mechanics. In: Hunger H (ed). Proceedings of ESHS 3rd Conférence. The Austrian Academy of Science, Vienna, pp. 147–186.

Pisano R (2009c) Continuity and discontinuity. On method in Leonardo da Vinci' mechanics. Organon 41:165–182.

Pisano R (2009d) Galileo Galilei. Riflessioni epistemologiche sulla resistenza dei corpi. In Giannetto E, Giannini G, Toscano M (eds). Relatività, Quanti Chaos e altre Rivoluzioni della Fisica. Guaraldi Editore Rimini, pp. 61–72.

Pisano R (2010) On Principles in Sadi Carnot's Thermodynamics (1824). Epistemological Reflections. Almagest 2/2010:128–179.

Pisano R (2011) Physics-Mathematics Relationship. Historical and Epistemological notes. In Barbin E, Kronfellner M, Tzanakis C, (eds), European Summer University History And Epistemology In Mathematics. Vienna: Verlag Holzhausen GmbH-Holzhausen Publishing Ltd., pp. 457–472.

Pisano R (2012) On Lazare and Sadi Carnot. A synthetic view of a historical epistemological research program. In: Proceedings of XXX SISFA Congress, Urbino, Argalia Editore, Urbino, pp. 147–153.

Pisano R (2013a) On Lagrangian in Maxwell's electromagnetic theory. Scientiatum VI. História das Ciências e das Técnicas e Epistemologia. The University of Federate University of Rio de Janeiro Press, Brazil, pp. 44–59.

Pisano R (2013b) Historical Reflections on Physics Mathematics Relationship In Electromagnetic Theory In: Barbin E., Pisano R., (eds). The Dialectic Relation between Physics And Mathematics In The XIXth Century. Springer, Dordrecht, pp. 31–57.

Pisano R (2014) Note sulle Fortificazioni nei Quesiti di Tartaglia. Libro Sesto e sua Gionta. In: D'Agostino S, Fabricatore G (eds). Proceedings of the International Conference in History of Engineering. Vol. II. Cuzzolin, Napoli, pp. 815–825.

Pisano R (2015) (ed) A Bridge between Conceptual Frameworks, History of Mechanism and Machine Science. Dordrecht, Springer.

Pisano R (2016) Details on the mathematical interplay between Leonardo da Vinci and Luca Pacioli? Journal of the British Society for the History of Mathematics 31/2:104–111.

Pisano R (2017) A Development of the Principle of Virtual Laws and its Framework in Lazare Carnot's Mechanics as Manifest Relationship between Physics and Mathematics. Transversal–International Journal for Historiography of science 2:166–203.

Pisano R (2020) A Tale of Tartaglia's Libro Sesto & Gionta in Quesiti et Inventioni diverse (1546–1554). Exploring the Historical and Cultural Foundations. Foundations of Science 25/2:477–50.

Pisano R (2023–2024) Galileo Galilei: The Two Manuscripts *On Mechanics*. Translations, Text and Commentarie. Dordrecht, Springer, *forthcoming*.

Pisano R, Bussotti P (2012) Galileo and Kepler. On Theoremata Circa Centrum Gravitatis Solidorum and Mysterium Cosmographicum. History Research 2/2:110–145.

Pisano R, Bussotti P (2014a) Historical and Epistemological Reflections on the Culture of Machines around the Renaissance: How Science and Technique Work? Acta Baltica Historiae et Philosophiae Scientiarum 2/2:20–42.

Pisano R, Bussotti P (2014b) Notes on mechanics and mathematics in Torricelli as physics mathematics relationships in the history of science. Problems of Education in the 21st Century International Journal. Special Issue 61/61:88–97.

Pisano R, Bussotti P (2015a) Historical and Epistemological Reflections on the Culture of Machines around the Renaissance: Machines, Machineries and Perpetual Motion. Acta Baltica Historiae et Philosophiae Scientiarum 3/1:69–87.

Pisano R, Bussotti P (2015b) Galileo in Padua: architecture, fortifications, mathematics and « practical » science, Lettera Matematica Pristem International (Springer):209–221.

Pisano R, Bussotti P (2015c) The Emergencies of Mechanics and Thermodynamics in the Western Society during 18th–19th Century. In: Pisano R (ed) 2015, pp. 399–436.

Pisano R, Bussotti P (2015d) Fibonacci and the bacus Schools in Italy. Mathematical conceptual streams, education and its changing relationship with society. Almagest 6/2:126–165.

Pisano R, Bussotti P (2016) A Newtonian Tale Details on Notes and Proofs in Geneva Edition of Newton's Principia. Bulletin BSHM-Journal of the British Society for the History of Mathematics 31/3:160–178.

Pisano R, Bussotti P (2017a) (eds) Homage to Galileo Galilei 1564–2014. Reading Iuvenilia Galilean Works within History and Historical Epistemology of Science. Philosophiae Scientiae 21/1.

Pisano R, Bussotti P (2017b) The Fiction of Infinitesimals in Newton's Works. On the Metaphoric Use of the Infinitesimals in Newton. In Marcacci F, Rossi MG (eds). Isonomia-Epistemologica. Reasoning, Metaphor and Science 9:141–160.

Pisano R, Bussotti P (2017c) On the Conceptualization of Force in Johannes Kepler's Corpus: an Interplay between Physics/Mathematics and Metaphysics. In: Pisano R, Agassi J and Drozdova D (eds). Hypothesis and Perspective in HPS. Homage to Alexandre Koyré 1892–1964. Springer, pp. 295–346.

Pisano R, Bussotti P (2022) Conceptual Frameworks on the Relationship between Physics–Mathematics in the Newton *Principia Geneva Edition* (1822). Foundations of Sciences (Springer) 27/3:1127–1182.

Pisano R, Capecchi, D (2008) La meccanica in Italia nei primi anni del Cinquecento. Il contributo di Niccolò Tartaglia. In: Tucci P (ed). Proceedings of XXV SISFA Congress. Milano, pp. C17.1–C17.6 Via: http://www.brera.unimi.it/sisfa/atti/index.html.

Pisano R, Capecchi D (2009) Il ruolo della meccanica ne Le fortificationi di Buonaiuto Lorini In: D'Agostino, S. (ed). III Convegno di Storia dell'ingegneria, II. Napoli: Cuzzolin editore, pp. 797–808.
Pisano R, Capecchi D (2010a) On Archimedean Roots in Torricelli Mechanics. In: Ceccarelli M, Paipetis S (eds). The Genius of Archimedes. Springer, Dordrecht, pp. 17–28.
Pisano R, Capecchi D (2010b) Galileo Galilei: Notes on Trattato di Fortificazione. In: Altamore A, Antonini G (eds), Galileo and the renaissance scientific discourse. Edizioni nuova cultura Roma, pp. 28–41.
Pisano R, Capecchi D (2012) Historical Reflections on Scale Ratio in Galilean Tratattto di Fortificazione. In: Ceccarelli M (ed) International Symposium on the History of Machines and Mechanisms, Springer, pp. 463–473.
Pisano R, Capecchi D (2013) Conceptual and mathematical structures of mechanical science between 18th and 19th centuries. Almagest 2/4:86–121.
Pisano R, Capecchi D (2016) Tartaglia's science weights. Mechanics in Sixteen-Century. Selections from Books VII–VIII. Dordrecht, Springer.
Pisano R, Gaudiello I (2009) Continuity and discontinuity. An epistemological inquiry based on the use of categories in history of science. Organon 41:245–265.
Pisano R, Robarts J (2022) A Short Comparative Historiography of Science & Technology of Mechanics into Engineering—and—Architecture Literatures During the Renaissance. In: Ceccarelli M, López-García R (eds). Explorations in the History and Heritage of Machines and Mechanisms. Vol. 40. Dordrecht, Springer, pp. 227–240.
Pisano R, Robarts J (2024) Briefs on Buonaiuto Lorini's Fortificationi (1609): Geometry, Machines & Mechanics into Engineering during the Renaissance. In: Ceccarelli M. et al. (eds). 8th HMM. Springer, Cham, *forthcoming*.
Pisano R, Agassi J, Drozdova D (eds) (2017d) Hypothesis and perspective in history and philosophy of science. Homage to Alexandre Koyré 1892–1964. Springer, Dordrecht.
Pisano R, Cooppersmith J, Peake M (2021) Essay on Machines in General (1786). Text, Translations and Commentaries. Lazare Carnot's Mechanics – Vol. 1. Springer.
Pisano R, Fichant M, Bussotti P, Oliveira ARE (2017). (eds) The Dialogue between Sciences, Philosophy and Engineering. New Historical and Epistemological Insights. Homage to Gottfried W. Leibniz 1646–1716. The College Publications, London.
Podetti F (1914) La teoria delle proporzioni in un manoscritto inedito di Evangelista Torricelli. Bollettino di bibliografia e storia delle scienze matematiche A/16–3:65–71.
Poinsot L (1806) La Théorie générale de l'équilibre et du mouvement des systèmes. Journal de l'Ècole Polytechnique 6/13:206–241.
Poinsot L (1838) Note sur une certaine démonstration du principe des vitesses virtuelles, qu'on trouve au chapitre III de livre 1er de la "Mécanique céleste". Journal de mathématiques pures et appliquées 1/3:244–248.
Procissi A (1952) Nuove ricerche bibliografiche torricelliane. Bollettino della Società Torricelliana di Scienze e Lettere 3:13–15.
Procissi A (1953) Su l'inviluppo delle parabole in Torricelli e sulla nozione di inviluppo di una famiglia di curve piane. Periodico di matematiche 4/31–1:113–116.
Radelet de Grave P (1996) Entries: Stevin, Kepler, Leibniz, Huygens. In: Dictionnaire du patrimoine littéraire européen, Patrimoine Littéraire Européen. vol. 8, Avènement de l'Equilibre européen 1616–1720. De Boeck Université, p. 18, pp. 745–755, pp. 1020–1027.
Radelet de Grave P (1998) La moindre action comme lien entre la philosophie naturelle et la mécanique analytique: continuités d'un questionnement. LLULL 21:439–484.
Radelet de Grave P (2007) Kepler (1571–1630) et Cavalieri (1598–1647) astrologues, ou le logarithme au secours de l'astrologie. In Mélanges offerts à Hossam Elkadem par ses amis et ses élèves, Bruxelles. Archives et Bibliothèques de Belgique:297–315.
Radelet de Grave P (2009) The Problem of the Elastica Treated by Jacob Bernoulli and the Further Development of this Study by Leonhard Euler. In: Werner L, Kurrer KE (eds). Proceedings of the Third International Congress on Construction History (Cottbus, May 2009). Brandenburg University of Technology Cottbus, Verlag Service GmbH. Germany, pp. 1209–1219.
Rahman S (2005) The Dialogic of just being different. Hintikka's new approach to the notion of episteme and its impact on "second generatio" dialogics. In: Kolak D, Symons J (eds). Quantifiers, Questions and Quantum Physics. Essays in Honour of Jaakko Hintikka. Springer Verlag, pp. 57–187.
Renn J (1995) Historical epistemology and interdisciplinary. In: Gavroglu K J.J. Stachel JJ, Wartofsky MW (eds). Physics, Philosophy, and the Scientific Community: Essays in the Philosophy and History of the Natural Sciences and Mathematics In Honor of Robert S. Cohen Kluwer, Doertrcth, pp. 241–251.
Renn J, Damerow P (2010) The equilibrium controversy. Guidobaldo del Monte's critical notes on the mechanics of Jordanus and Benedetti and their historical and conceptual background. Edition open access, Berlin.
Renn J, Damerow P, McLaughlin P (2003) Aristotle, Archimedes, Euclid, and the Origin of Mechanics: The Perspective of Historical Epistemology. The Max Planck Institute for the history of science of Berlin, print n. 239, Berlin, pp. 43–59.
Reuleaux F (1876) Kinematics of Machinery: Outlines of Theory of Machines, London: Macmillan & Co.
Riccati V (1749) Dialogo di Vincenzo Riccati della Compagnia di Gesù dove ne' congressi di più giornate delle forze vive e dell'azioni delle forze morte si tien discorso. Stamperia di Lelio dalla Volpe, Bologna.
Riccati V (1772) De' Principi della Meccanica. Stamperia Coleti, Venezia.
Rossini G (1959) La famiglia di Evangelista Torricelli. In: AA.VV. Convegno di studi torricelliani in occasione del 350 anniversario della nascita di Evangelista Torricelli, 19–20 ottobre 1958, Lega, Faenza. pp. 133–149.
Saccheri G (1708) Neo-statica. Mediolani : Ex Typographia Josephi Pandulfi Malatestae, MDCCVIII.
Schuster JA (2000) Descartes opticien: the construction of the law of refraction and the manufacture of its physical rationales, 1618–1629. In: Gaukroger, Schuster and Sutton 2000, pp. 258–312.
Schuster JA (2013a) Descartes–Agonistes. Physico–mathematics, Method and Corpuscular–Mechanism 1618–1633. Springer, Dordrecht.
Schuster JA (2013b) Cartesian Physics. In: Fox R *et al.* (eds). The Oxford Handbook of the History of Physics, The Oxford University Press, Oxford, pp. 56–95.
Seeger RJ ([1966]2013) Men of Physics: Galileo Galilei, His Life and His Works: The Commonwealth and International Library: Selected Readings in Physics. [Pergamon Press] Elsevier, Amsterdam.

Segre M (1983) Torricelli's Correspondence on Ballistics. Annals of Science 40/5:489–499.
Singer C (1954–1958) A History of technology, vols. 2–4, Oxford: The Clarendon Press.
Sparavigna CA (2015) Teaching Physics During the 17th Century: Some Examples From the Works of Evangelista Torricelli. International Journal of Sciences 4/4:50–58.
Tartaglia N ([1554]1959) La nuova edizione dell'opera Quesiti et inventioni diverse de Nicolo Tartaglia brisciano Riproduzione in facsimile dell'edizione del 1554. Masotti A (ed), Commentari dell'Ateneo di Brescia, Tipografia La Nuova cartografica, Brescia.
Tartaglia N (1537) La noua scientia de Nicolo Tartaglia: con una gionta al terzo libro, per Stephano Da Sabio, Venetia.
Tartaglia N (1565a) Archimedes De insidentibus aqueae, Lib. I et II, in Iordani Opvscvlvm de Ponderositate, apud Curtium Troianum Navò, Venetia.
Tartaglia N (1565b) Iordani Opvscvlvm de Ponderositate, Nicolai Tartaleae Stvdio Correctvm Novisqve Figvrisavctvm. Cvm Privilegio Traiano Cvrtio, Venetiis, Apvd Curtivm Troianvm. MDLxv.
Taton R (1963) L'annonce de l'expérience barométrique en France. Revue d'histoire des sciences et leurs applications 16:77–83.
Taton R (ed) (1976) Sadi Carnot et l'essor de la thermodynamique, Table Ronde du Centr National de la Recherche Scientifique. École Polytechnique, 11–13 Juin 1974. Éditions du Centre National de la Recherche Scientifique, Paris.
Tenca L (1956) Su una svista di stampa in 'de Dimensione Parabolae' di Evangelista Torricelli notata da Stefano Angeli. Bollettino Unione Matematici Italiani 3/11:258–259.
Tenca L (1958) L'attivita` matematica di Evangelista Torricelli. Periodico di Matematiche 36/4:251–263.
Tenca L (1960) I presunti contrasti fra Evangelista Torricelli e Vincenzio Viviani. Periodico di Matematiche 4/38:87–94.
Thomson RB (1976) Jordanus de Nemore: Opera. Mediaeval Studies 38:97–144.
Torricelli E (1644) Opera geometrica, Masse & de Landis, Florentia.
Torricelli E (1919–1944) Opere di Evangelista Torricelli. Loria G, Vassura G (eds). 4 Vols., Montanari, Faenza.
Torricelli E (1975) Opere scelte. Belloni L (ed). UTET, Padova.
Toscano F (2008) L'erede di Galileo. Vita breve e mirabile di Evangelista Torricelli. Sironi, Milano.
Toscano F (2014) Evangelista Torricelli: grandi scoperte e autoconsure. Nuova Civiltà delle Macchine 3/1:31–46.
Truesdell C (1968a) Essay in the history of mechanics. Springer, New York.
Truesdell CA (1968b) Whence the law of the moment of momentum? [Lecture 202, Whence the law of moment of momentum? Colloquium in Engineering Science. Mechanical Engineering Department, Columbia University, 1963, January 8th] In: Essays in the history of mechanics. Springer, New York, pp. 239–271.
Truesdell CA (1970) The tragicomedy of classical thermodynamics. Springer, Berlin.
Truesdell CA (1976) Termodinamica razionale. Accademia Nazionale dei Lincei, Contributi del Centro Linceo Interdisciplinare di Scienze Matematiche e loro Applicazioni. Anno CCCLXXXIII, N. 20, Roma, pp. 208–235.
Truesdell CA (1980) The tragicomical history of thermodynamics. 1822–1854. Springer, Berlin.
Truesdell CA, Bharatha S (1977) The concepts and logic of classical thermodynamics as a theory of heat engines: rigorously constructed upon the foundation laid by S Carnot and F Reech (Theoretical and mathematical physics). Springer, Berlin/Heidelberg/New York.
Vailati G (1987) Scritti di Vailati. Quaranta M (ed). Sala Bolognese, Forni editore.
Varanini SL (2001) Antonio Nardi: note in margine ai manoscritti. Bruniana & Campanelliana 7/2:469–498.
Varignon P (1725) La nouvelle Mécanique ou statique. Jobert C, Paris.
Vassura G (1945) Umanità in Evangelista Torricelli. Pubblicato dalla commissione per le onoranze a Evangelista Torricelli III per l'anno 1944. Centenario della scoperta del barometro. Bollettino della Società Torricelliana di Scienze e Lettere:15–18.
Visani A (1952) Notizie sulla salute e sull'ultima malattia di Evangelista Torricelli. Bollettino della Società Torricelliana di Scienze e Lettere 3:5–12.
Wallis J (1693) Geometriae professoris Saviliani, in celeberrima Academia Oxoniensi Opera mathematica by Wallis, John, 1616–1703. Oxoniae. E Theatro Sheldoniano.
Westman RS (1980) The Astronomer's Role in the Sixteenth Century: A Preliminary Study. History of Science 18:105–147.
Winter TN (2007) The Mechanical Problems in the corpus of Aristotle. Digital Commons. The University of Nebraska–Lincoln, pp. 1–43 [Retrieved March 16, 2008, available via http://digitalcommons.unl.edu/classicsfacpub/68].

6 Torricelli's Proof of Galileo's Assumption on the Velocity Acquired at the End of a Free Fall

Patricia Radelet de Grave

Abstract

Galileo's *Discorsi e dimostrazioni matematiche intorno a due nuove scienze attenenti alla meccanica e i movimenti locali* (hereafter *Discorsi*) are dated 1638 (Galilei 1890–1909, VIII, pp. 39–362). In 1639, Vincenzo Viviani added significant studies (Galilei 1890–1909, VIII, pp. 215–219, pp. 442–445). Since Galileo's *Discorsi* (1638) and Viviani's additions (1639) are both published in *Discorsi* (1638), volume VIII of the *Le opere di Galileo Galilei* (with also some ad hoc editorial specifications in the footnotes), henceforward for cross–matching references, I will refer to 1638 only, adding their corresponding pages in *Le opere di Galileo Galilei*. By means of my comments, this will make clear the difference between Galileo's work (1638) and *Viviani's additions* (1639). In 1979, Paolo Galluzzi, published "*Momento*". In this article, he identified a confusion made by Galileo between Archimedean Momentum and his own new *Momentum*. This confusion, he said, occurs in his *Proofs Concerning the Inclined Plane*. The aim of this paper is to show that there was no confusion in Galileo's mind. It will be shown studying the free fall as well as the fall on an inclined plane in various of his own works such as *De motu antiquiora* (1589–1592), *Le mecaniche* (1593), the various *de motu* and the *Discorsi e i movimenti locali* (1638) as well as Viviani's addition to the *Discorsi*. But our comparision will be extended to other authors such as Jordanus de Nemore in his *De ratione ponderis* (fl. 13th), Gregorius van Sto Vincentio, *Theoremata mathematica scientiae staticae de ductu ponderum per planitiem recta and oblique horizontem decussantem* (Van Sto Vincentio Gregorius (1624) Theoremata mathematica scientiae staticae de ductu ponderum per planitiem recta et oblique horizontem decussantem, Louvain.), Torricelli with his *de motu gravium* (1644), and a quick look at a letter of Johann Bernoulli, all concerning the same arguments. The work of Torricelli plays an important role in this story because he exchanged letters with Galileo and was his assistant for the three last monthes of Galileo's life.

Keywords

Torricelli • Galilei • Viviani • *De motu gravium* • Opera Geometrica • *Discorsi e dimostrazioni matematiche* • Gregorius a St. Vincentio • *Praemittimus* • Free fall • Archimedean momentum • *Le mecaniche*

1 Introduction

Before studying the proof of Evangelista Torricelli (1608–1647), let us recall some biographical facts. In Rome, Torricelli was a pupil of a friend of Galileo Galilei (1564–1642), the Benedictine monk Benedetto Castelli (1578–1643). In 1641, on March the 2nd, three years after the publication of the *Discorsi e dimostrazioni matematiche, intorno due nuove scienze attenenti alla mecanica & i movimenti locali*, Castelli wrote to Galileo, who was confined to Arcetri:

P. Radelet de Grave (✉)
Mathematics and Physics Research Institute, Catholic University of Louvain–la–Neuve, Ottignies-Louvain-La-Neuve, Belgium
e-mail: patricia.radelet@uclouvain.be

R. Pisano et al., *Homage to Evangelista Torricelli's Opera Geometrica 1644–2024*,
Logic, Epistemology, and the Unity of Science 55,
https://doi.org/10.1007/978-3-031-06963-5_6

> I hope [...] to bring you a book, and maybe also a second one, written by one of my disciples, who received the first principles of geometry ten years ago in my school, and who then made such advances, that he proved many propositions, already proved by you, from that *De motu*, but in another way.[1]

Castelli talks here about Torricelli's manuscript *De motu gravium*, that will be published three years later. After that, letters were exchanged between Galileo and Torricelli and the latter explained that the second book of his *De motu* was not yet finished. Nevertheless, in October of the same year, Torricelli arrived at Arcetri where he became Galileo's assistant until Galileo's death, only three months later.

Torricelli's work was published in 1644 in his *Opera Geometrica*, but his main ideas and certainly the first five propositions that we will study here, must have been written before Castelli's letter was sent to Galileo. Torricelli's first idea must have been to give a proof to Galileo's fundamental assumption on which rests the whole third day of his *Discorsi*. Galileo had also been trying to prove that assumption and Torricelli's book grew out of this research and of his critical study of the *Discorsi*.

The complete title of the text written between 1639 and 1641, is: "About the motion of weights falling down naturally, and of projectiles, in two books.[2] The particular titles of the two books are:

- First book, the motion of heavy bodies.
- Second book, the motion of projectiles.[3]

Two titles that are almost identical to those of the third and fourth days in Galileo's *Discorsi*.

- Third day: Naturally accelerated motion.
- Fourth day: Motion of projectiles.[4]

However, the content is rather different, and Torricelli presumes that his reader perfectly knows Galileo's results and proofs perfectly and he only refers, in *marginalia*, to places in Galileo's text. Just as one did it with Euclid.

2 Galileo's Assumption and Torricelli's Reaction to It

At the beginning of the third day of the *Discorsi*, Galileo gives both a definition and a principle. He defines uniformly accelerated motion by using an unexpected term; namely *momentum*:

> *We say that a motion is equally or uniformly accelerated, if starting from rest, it acquires in equal times equal momentum of velocity.*
> Salv.: Such a definition being fixed, the Author asks and supposes but one principle, namely: I admit that the degrees of velocity the same moving body acquires on planes differently inclined are equal when the elevations of those planes are equal.[5]

The last sentence gives the assumption that Torricelli intends to prove. It is well known and I will simply call it "Galileo's assumption". Only in 1639, after publication of the *Discorsi*, will Galileo find a proof of his assumption. Only a copy by Vincenzo Viviani (1622–1703) remains (Galilei 1638, VIII, pp. 442–445). I shall call it "Galileo's manuscript". Therefore, in 1639 this proof was elaborated by Viviani using this Galilean manuscript. I shall call it *Viviani's addition* (Galilei 1638, VIII, pp. 215–219). This text is reproduced in a footnote, together with the *Discorsi*, in the VIIIth volume of the *Edizione Nazionale*, precisely at the place where Galileo wanted it to be printed in a new edition of the *Discorsi*. At the beginning of "Galileo's manuscript", Galileo writes:

---

[1] "Spero [...] di portargli un libro, e forsi ancora il secondo libro, fatto da un mio discepolo, il quale, havendo hauti i primi principii di geometria dieci anni sono dalla mia scola, ha poi fatto tale progresso, che ha dimostrate molte proposizioni di quelle de motu dimostrate già da V.S., ma diversamente." (Castelli 1641, p. 303). This letter tells us that in March 1641, Torricelli's *De motu* was almost finished. This was just one part of his *Opera geometrica* (Torricelli 1644).

[2] *De motu gravium naturaliter descendentium, et projectorum libri duo* (Torricelli 1644, pp. 95–153).

[3] *Liber primus, De motu gravium naturaliter descendentium. Liber secundus, De motu projectorum.*

[4] *Giornata terza: De motu naturaliter accelerata. Giornata quarta: De motu projectorum.*

[5] *Motum aequabiliter seu uniformiter acceleratum dicimus eum, qui a quiete recedens temporibus aequalibus aequalia celeritatis momenta sibi superaddit.* Salv. Fermata cotal definizione un solo principio domanda, e suppone per vero l'Autore, cioè; Accipio, gradus velocitatis ejusdem mobilis super diversarum planorum inclinationes acquisitos tunc esse aequales, cum eorumdem planorum elevations aequales sint. (Galilei 1638, VIII, p. 205; Author's translation; Galileo's italics).

> That weights falling down from the same height along diversely inclined planes have acquired, when having reached the horizon, same degrees of velocity (a proposition until now only supposed by me, or only confirmed experimentally or with only probable discourse), we will be able to prove it demonstratively in what follows.[6]

It will be interesting to compare Galileo's and Torricelli's proofs. But to understand those proofs correctly we shall have to go back to earlier texts of Galileo, in order to get acquainted with his vocabulary, also used by Torricelli. We shall have to understand what *momentum* exactly means.

The same property of the velocity acquired by a moving body after free fall, had been discovered (but not proved) by Gregorius a St. Vincentio (hereafter Gregorius) and published in his *Theoremata mathematica scientiae staticae de ductu ponderum per planitiem recta and oblique horizontem decussantem* that appeared in Louvain in 1624 (Fig. 1).[7]

Transported on a slope, straight or obliquely [see Fig. 1.], weights of equal[a] *momentum* accumulate equal forces when traversing an equal measure of the vertical distance, whether they implement them in a slow fall or a very speedy one, whether they cross the horizon in a run or more oblique or more straight, whether they are near the centre of the universe or if they are far away from it.
(Author's translation).

In the footnote "a", Gregorius adds:

> a. The *Virtus* that *A* acquires falling towards *B*, is equal to the *Virtus* that the same *A* acquires falling along an inclined plane from *C* to *D*.

Fig. 1 Gregorius's illustration of *Theorem 13* (van Sto Vincentio 1624, p. 14). *Source* Courtesy Royal Library of Belgium, Belgium

Gregorius uses *Momentum* in an Archimedean meaning. The weights have equal *momenta* means that they analogously equilibrate a balance. Thus *momentum* has there the dimension of a weight multiplied by a length. As most of Gregorius' manuscripts were destroyed during the sac of Prague, we don't know if he proved that theorem. But its Aristotelian origin is clear in his work as well as in Galileo's. Gregorius claims:

[6] Che i gravi descendenti dalla medesima sublimità sopra l'orizonte per linea rette in qualsivoglia modo inclinato, si trovino, giunti che sieno all'orizonte, aver acquistato eguali gradi di velocità (proposizione sin qua da me supposta, o solo con esperienze e probabili discorsi confermata), potremo nel seguente modo dimostramente provare. (Galilei 1638, VIII, p. 442; Author's translation).

[7] Parium[A] momentorum pondera, sive recta, seu in obliquum declive ferantur, pares coaceruant vires, per parem perpendiculi mensuram, sive id lapsu lento, sive pernicissimo expediant, sive obliqiore, sive rectiore cursu finitorem decussent, sive centro universi propinqua sint, sive longo intervallo dissita. (a.) Virtus quam acquirit *A* lanbendo, in *B*, aequalis est virtuti, ejusdem *A* labentis per inclinatum ex *C* in *D*. […] (Gregorius, Theorema 13 in: Dhombres and Radelet de Grave 2008b, p. 314; see also *Id.*, 2008a).

> As long as one avoids moving away from the centre of the universe the centre of the mass that one pushes, it allows oneself to be directed at will, even if it is enormous.[8]

As we shall see, Galileo often expresses the same idea, as for example in Viviani's addition:

> Because by such a motion, the body neither gains nor loses with respect to its distance from the common centre of heavy things, that with respect to the horizon remains always the same.[9]

A modern expression would be that one only has to expend work for elevating a weight. And once this is admitted, it is clear that the only important measure is the distance along the vertical. Jordanus de Nemore (fl. XIII—ca. 1260) already gave a particular importance to the height of fall in order to find the apparent weight on an inclined plane (De Nemore 13th, pp. 190–191). After having recalled Galileo's assumption, Torricelli explains[10]:

> On this axiom depends almost all his [Galileo's] doctrine of motion, the natural fall as well as the motion of a projectile. If one doubts about this principle, the whole science that then follows will not be secured.[11]

Torricelli's task will thus be to prove "Galileo's assumption". But he knows that it is a consequence of a more easily proved theorem.

2.1 Torricelli's *Praemittimus*

I know that during the last years of his life, Galileo tried hard to prove that supposition, …; as it seems possible to prove Galileo's supposition, and even that this supposition can be immediately derived from that theorem he considered as mechanically proved in the second part of the sixth proposition of *de motu accelerato*, that is to say: *Momenta* of equal weights moving on unequally inclined planes are to each other in the same ratio as the elevations of equal parts of those planes. *Verbi gratia* (Fig. 2).[12]

---

[8] Quam diu enim caves, centrum quam agitas molis, ab universi medio se iungere, sequeretur quo dirigis nutum, immanis etiam moles. 1624, Gregorius à St. Vincentio, Theorème 2. (Dhombres and Radelet de Grave 2008b, pp. 292–293; Author's translation).

[9] Non facendo con tal moto perdita alcuna, ne meno acquisto, in riguardo della propria distanza dal comun centro delle cose gravi, che nell'orizonte si conserva sempre l'istessa (Galilei 1638, VIII, p. 217; Author's translation).

[10] The notations of Galileo and Torricelli are not uniform at all. That is the reason why I try to introduce some uniformity. For the same reason, I give formalized versions of their propositions, and at the end, put together the principal relations they found, with uniform notations. But my aim is to keep those formalized expressions consistent with the mathematics they know, namely proportion theory. That is the reason why I avoid modern expressions of their properties, because they are never completely equivalent.

[11] "Ex hac petitione dependet quasi universa illius doctrina de motu, tum accelerato, tum projectorum. Si quis de principio dubitet de ijs quae inde consequuntur certam omnino scientiam non habebit" (Torricelli 1644, p. 98; Author's translation).

[12] "Scio Galileum ultimis vitae sua annis suppositionem illam demonstrare conatum, […]; ut appareat quod Galilei suppositio demonstrari potest, et quidem immediate ex illo Theoremate quod pro demonstrato ex Mechanicis ipse desumit in secunda parte sextae Propositionis de motu accelerato, videlicet. Momenta gravium aequalium super planis inaequaliter inclinatis, esse inter se ut sunt perpendicula partium aequalium eorumdem planorum. Verbi gratia. Sint plana *AB*, *CB* inaequaliter inclinata, et sumptis aequalibus *AB*; *CD* ducantur perpendicula *AE*; *CF*: ad horizontem *BF*. Supponit Galileus pro demonstrato, momentum in plano *AB* ad momentum in plano *CB* ita esse ut est *AE* ad *CF*. Nos quia in huius modi Theorema non incidimus, hoc primum aliqua demonstratione confirmabimus: protinus ad ostendendum id quo Galileo principium sive petitio est, accedemus" (Torricelli 1644, pp. 98–99; Author's translation).

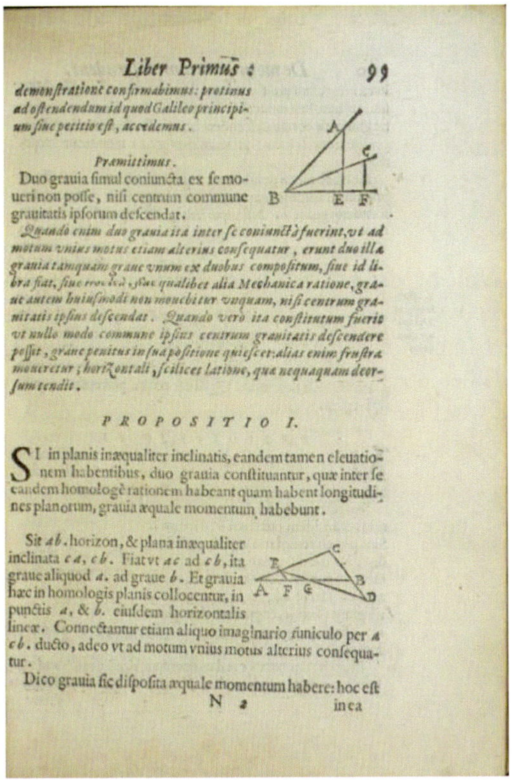

Liber Primus. 99

demonstratione confirmabimus: protinus ad ostendendum id quod Galileo principium sine petitio est, accedemus.

Præmittimus.

Duo grauia simul coniuncta ex se moueri non posse, nisi centrum commune grauitatis ipsorum descendat.

Quando enim duo grauia ita inter se coniuncta fuerint, vt ad motum vnius motus etiam alterius consequatur, erunt duo illa grauia tamquam graue vnum ex duobus compositum, siue id libra fiat, siue troclea, siue qualibet alia Mechanica ratione, graue autem huiusmodi non mouebitur vnquam, nisi centrum grauitatis ipsius descendat. Quando vero ita constitutum fuerit vt nullo modo commune ipsius centrum grauitatis descendere possit, graue penitus in sua positione quiescet: alias enim frustra moueretur; horizontali, scilicet latione, quæ nequaquam deorsum tendit.

PROPOSITIO I.

SI in planis inæqualiter inclinatis, eandem tamen eleuationem habentibus, duo grauia constituantur, quæ inter se eandem homologè rationem habeant quam habent longitudines planorum, grauia æquale momentum habebunt.

Sit *ab.* horizon, & plana inæqualiter inclinata *ca*, *cb*. Fiat vt *ac* ad *cb*, ita graue aliquod *a*. ad graue *b*. Et grauia hæc in homologis planis collocentur, in punctis *a*, & *b*. eiusdem horizontalis lineæ. Connectantur etiam aliquo imaginario funiculo per *acb*. ducto, adeo vt ad motum vnius motus alterius consequatur.

Dico grauia sic disposita æquale momentum habere: hoc est

N 2 in ea

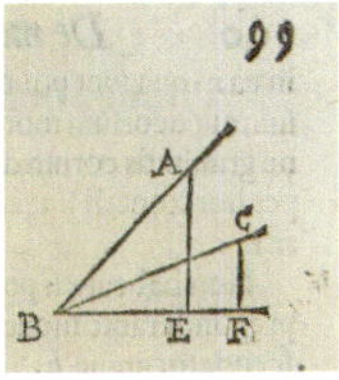

Let the planes *AB*, *CB* [Fig. 2.] be unequally inclined, and supposing *AB* and *BC* to be equal, draw the perpendiculars *AE*, *CF* on the horizon *BF*. Galileo supposes to be proved that the *moment* on the plane *AB* is to the *moment* on the plane *CB* as *AE* is to *CF*. As we don't accept the theorem in this way, we shall first confirm it with a proof: Let us immediately proceed to show what Galileo's principle or assumption is.

$$\frac{M_{AB}}{M_{BC}} = \frac{AE}{CF} \tag{1}$$

(Author's translation).

Fig. 2 The *Praemittimus* (Torricelli 1644, p. 99). *Source* Courtesy University of Oklahoma, History of Science Collections

Galileo and Torricelli consider the various *momenta* of equal weights running down diversely inclined planes. As those weights are not in equilibrium, those *momenta* are different from the "Archimedean momentum" we just read in Gregorius. Torricelli replaces Galileo's assumption by a new one that is called today "Torricelli's principle".

3 Galileo's So Called "Confusion" Between *Archimedean Momentum* and His New *Momentum* in Proofs Concerning the Inclined Plane

In his book about *Momento*, Paolo Galluzzi also underlines that "confusion"[13] in 1979:

> The term *momento*, that appears for the first time in *Le mecaniche*, is used to indicate that concept [the efficiency of the weight] of which we underlined the duplicate signification: the one, Archimedean, rigorously "statical" and the other one substantially "dynamical" corresponding, the first, to the geometrical scheme of the balance, the second, to the model of the inclined plane.[14]

In the text of his *Theoremata*, quoted above, Gregorius a Sto Vincentio avoids Galileo's "confusion" when he says:

> [...] weights of equal *momentum* accumulate equal forces when traversing an equal measure of the vertical distance [...].[15]

[13] The aim of my paper is to show that it is not a confusion.

[14] "Il termine momento, che fa la sua prima comparsa ne Le mecaniche, viene utilizzato per indicare tale concetto [l'efficacia del peso], del quale abbiamo sottolineato una duplice valenza: una, archimedea, rigorosamente "statica", e l'altra sostanzialmente "dinamica" corrispondenti, la prima, allo schema geometrico della bilancia, la seconda, al modello del piano inclinato". (Galluzzi 1979, p. 223; Author's translation).

[15] "Parium momentorum pondera, sive recta, seu in obliquum declive ferantur, pares concervant vires, per parem perpendiculi mensuram [...]." (Gregorius, Theorema 13 in: Dhombres and Radelet de Grave 2008b, pp. 314–315; Author's translation).

He avoids using *momentum* and he uses *vis* or *virtus* for what I'll call "Galileo's momentum" in order to emphasize the difference. For Gregorius *virtus* means something like our kinetic energy, something that enables the moving body to return to the height it started from. If Galileo's idea is similar, why does he use the word *momentum* recalling Archimedes law for the lever? Galileo makes this "confusion" in at least three texts concerning the inclined plane: *de motu antiquiora* (1589–1592), *Le mecaniche* (1593) and in the manuscript copied by Viviani, (with the proof of his assumption). The vocabulary of Viviani's addition (1639) is the same.

3.1 *De motu antiquiora* (1589–1592). The Apparent Weight on an Inclined Plane

To learn more about that "confusion", we have to read the first text Galileo wrote on the inclined plane. It is in his first *De motu*, generally called *De motu antiquiora* (1589–1592). There Galileo does not use the misleading word *momentum* and one understands better the relation he establishes, using the same word for two different things, different from our actual point of view, as well as the relation between his *momentum* and the word *impeto* he will use later. Galileo starts saying:

> [...] it is manifest that what is heavy is carried downward with as much force, as would be necessary for pulling it upward by force; that is, it is carried downward with as much force as that with which it resists going up.[16]

These two lines, similar to Gregorius' remark on the *virtus*, express a fact that we explain today by saying that the kinetic energy acquired during the fall is enough to pull the moving body back to its initial height. This particular case of conservation of energy is explained by the conversion of potential energy into kinetic energy and vice–versa. But Galileo's intention, in this text, is to compare the different weights a body seems to have moving on differently inclined planes. Today, we call them, "apparent weights",[17] using a vocabulary invented by Stevin (Fig. 3).

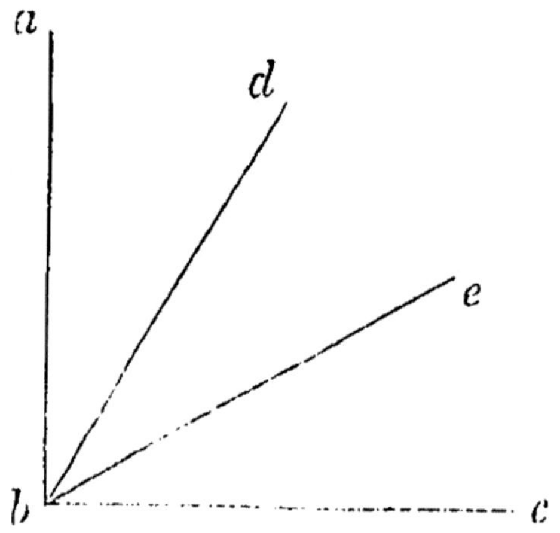

If then we find with how much less force the heavy thing is pulled up by force on line *bd* than on line *ba* [Fig. 3.], it will then have been found with how much more force the same heavy thing goes down on line *ab* than on line *bd*; and, similarly, if we find how much more force is required in order to impel the mobile upwards on line *bd* than on *be*, it will then have been discovered with how much more force it will go down on *bd* than on *be*. But we shall know how much less force is required to pull the mobile up by force on *bd* than on *be*, when we shall have found out how much greater will be the heaviness of this same mobile on the plane along line *bd*, than on the plane along line *be*. Let us proceed then to a careful investigation of this heaviness. (Author's translation).

Fig. 3 The trajectories Galilei uses to ask about the vertical free–fall motion compared to trajectories with different slopes (Galilei 1589–1592, p. 296). *Source* Gallica, Bibliothèque nationale de France. Public Domain

[16] "[…] manifestum est, grave deorsum ferri tanta vi, quanta esset necessaria ad illud sursum trahendum; hoc est, fertur deorsum tanta vi, quanta resistit ne ascendat." (Galilei 1589–1592, pp. 296–298; Author's translation). In 1986, the text was improved with the manuscript by Pierre Souffrin.

[17] "Si itaque inveniamus quanto minori vi trahitur sursum grave per lineam *bd* quam per lineam *ba*, erit iam inventum quanto maiori vi descendat idem grave per lineam *ab* quam per lineam *bd*; et, similiter, si inveniamus quanto maior vis requiritur ad sursum impellendum mobile per lineam *bd* quam per *be*, erit iam compertum quanto maiori vi descendet per *bd* quam per *be*. Sed tunc sciemus quanto minor vis requiratur ad sursum trahendum mobile per *bd* quam per *be*, quando cognoverimus quanto eiusdem mobilis maior erit gravitas in plano secundum lineam *bd*, quam in plano secundum lineam *be*. Procedamus itaque ad inquisitionem talis gravitatis." (Galilei 1589–1592, p. 297; Author's translation).

Galileo explains that:

(1) To pull up a body, along an inclined plane, one needs a force equal to the force the body gets falling down along the same plane;
(2) This force is less along *be* than along *bd* and less along *bd* than along *ba*.

This idea will be central in all of his three proofs but other words will be used instead of force (*vis*). He tries then to weigh this force with help of a balance. There, precisely, we find the origin of his use of the word "momentum". Galileo considers a fictitious balance *cad* (Fig. 4) in equilibrium with two equal weights hanging symmetrically at *c* and *d*. Then he hangs a weight in *s*, and considers a bent lever *cas* having one arm *as* orthogonal to the inclined plane *gh* and says that what we call its apparent weight on the inclined plane *gh* and what he calls its weight on the plane *gh*, is equal to the product: $W \times ap$ in the balance, *W* being its weight and *ap* its distance to the fulcrum *a*, that is what I call the "Archimedean momentum".

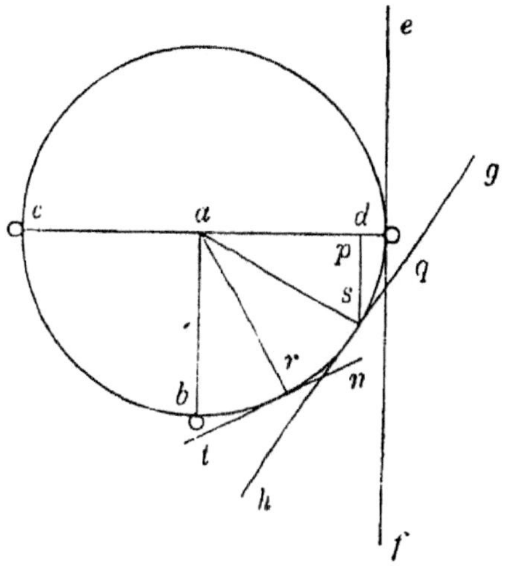

And let us consider a balance *cd* [Fig. 4.], which centre is *a*, and at point *c* a weight equal to another weight that is at point *d*. If we also imagine that the line *ad*, maintained in *a*, moved towards *b*, in the first point *d* of the descent of the body will be along the line *ef*; that is the reason why along the line *ef*, the descent of the moving body will follow the weight of the moving body at the point *d*. Inversely when the moving body is at point *s*, at the first point *s* its descent is as along the line *gh*; that is the reason why along the line *gh* the motion follows the weight that the moving body has at point *s*. And thus inversely when the moving body is at point *r*, then its descent at the first point *r* is as along the line *tn*; that is the reason why along the line *tn* the motion follows the weight that the moving body has at point *r*.

Fig. 4 The construction Galilei uses to prove the tangent motion to a circumference of a weight (Galilei 1589–1592, p. 297). *Source* Gallica, Bibliothèque nationale de France. Public Domain[18]

Galileo makes here a crucial restriction. It is only at the point *s* or *r*, when the motion starts, that his comparison holds. Once left alone, it will be accelerated as in any fall vertical or along an inclined plane.

> But then again, when the mobile will be at point *s*, its descent at the initial point *s* will be as if on line *gh*; that is why the motion of the mobile on line *gh* will be according to the heaviness the mobile has at point *s*. And again, when the mobile will be at point *r*, then its descent at the initial point *r* will be as if on line *tn*; that is why the mobile will be moved on line *tn* according to the heaviness it has at point *r*. If, then, we show that the mobile at point *s* is less heavy than at point *d*, it will then be manifest that its motion will be slower on line *gh* than on *ef*: and if, again, we show that at *r* the mobile is even less heavy than at point *s*, it will then be manifest that the motion will be slower on line *nt* than on *gh*. Now it is already manifest, that the mobile at point *r* exerts less weight than at point *s*; and less at *s*, than at *d*. For the weight at point *d* weighs equally with the weight at point *c*, since the distances *ca* and *ad* are equal: but the weight at point *s* does not weigh equally with the weight at *c*.[19]

[18] "Et intelligatur libra *cd*, cuius centrum *a*, et in puncto *c* pondus aequale ponderi alii quod sit in puncto *d*. Si itaque intelligamus, lineam *ad*, manente puncto *a*, moveri versus *b*, in primo puncto *d* descensus mobilis erit veluti per lineam *ef*; quare per lineam *ef* descensus mobilis erit secundum gravitatem mobilis in puncto *d*. Rursus, quando mobile erit in puncto *s*, in primo puncto *s* suus descensus erit veluti per lineam *gh*; quare mobilis per lineam *gh*, motus erit secundum gravitatem quam habet mobile in puncto *s*. Et rursus, quando mobile erit in puncto *r*, tunc illius descensus in primo puncto *r* erit veluti per lineam *tn*; quare mobile per lineam *tn* movebitur secundum gravitatem quam habet in puncto *r*." (Galilei 1589–1592, p. 297; Author's translation).

[19] "Rursus, quando mobile erit in puncto *s*, in primo puncto *s* suus descensus erit veluti per lineam *gh*; quare mobilis per lineam *gh* motus erit secundum gravitatem quam habet mobile in puncto *s*. Et rursus, quando mobile erit in puncto *r*, tunc illius descensus in primo puncto *r* erit veluti per lineam *tn*; quare mobile per lineam *tn* movebitur secundum gravitatem quam habet in puncto *r*. Si itaque ostendamus, mobile in puncto *s* minus esse grave quam in puncto *d*, erit iam manifestum quod illius motus erit tardior per lineam *gh* quam per *ef*: quod si, rursus, ostendamus, in *r* mobile adhuc minus esse grave quam in puncto *s*, erit iam manifestum quod tardior erit motus per lineam *nt* quam per *gh*. Atque iam manifestum est, mobile in puncto *r* minus gravare quam in puncto *s*; et in *s*, quam in *d*. Pondus enim in puncto *d* aequeponderat ponderi in puncto *c*, cum distantiae *ca*, ad sint aequales: sed pondus in puncto *s* non aequiponderat ponderi *c*." (Galilei 1589–1592, p. 297; Author's translation).

This paragraph is clear. The more the plane is inclined the less is the weight of the body sitting on it. Important also is the other fact that the lighter the body is, the slower it will behave. When the weight was hanging at *d*, it was equilibrated or had the same *momentum* as *c* hanging on the other side at the same distance of the fulcrum. Galileo then considers that when the same weight is put in *s* its weight is the same as the "apparent weight" it would have on an inclined plane coinciding with the tangent *gh*. Galileo has now a crooked lever *cas* and he compares the weight of the body, its *momentum* at *s* with that of the same body placed at the extremity *c*. What that weight is, he tells us in the next part of the text.

> For if a line is drawn from point *s* perpendicular to *cd*, the weight at *s*, as compared with the weight at *c*, is as if it were suspended from *p*; but the weight at *p* exerts less weight than the weight at *c*, since the distance *pa* is smaller than the distance *ac*. And, similarly, a weight at *r* exerts less weight than a weight at *s*: which will likewise be evident if a perpendicular is drawn from *r* on *ad*, which will intersect this same *ad* between points *a* and *p*. Consequently it is manifest that the mobile will go down with a greater force on line *ef* than on line *gh* and on *gh* than on *nt*.[20]

Galileo draws *sp* perpendicular to *cd* and says that the weight at *s* is the one that in *d* equilibrated *c*, but now hanging at *p* nearer to the fulcrum *a*, his "Archimedean momentum" is smaller than it was when suspended at *d*. And indeed it is lighter at *s* and lighter at *r* because the lever arm becomes shorter and shorter. The product: $W \times l$, *W* being what we call its weight, *l* the distance to the fulcrum *a*, its lever arm, it is clearly the "Archimedean momentum", Galileo calls it *weight* and deduces that $W_s/W_c = pa/ca$. Today one sees that $pa/ca = \sin \alpha$ where α is the inclination of the plane *gh* or the angle in *s* of the triangle *aps*. Note also that angle *asp* is equal to the inclination of the plane *gh*.

> It is thus evident that the moving body descends with greater force along the line *ef* than along the line *gh*, and along the line *gh* than along the line *nt*.[21]

The moving body goes down with less force or velocity along *gh* than along *ef*, and lesser on *nt*.

> But how much greater is the force along *ef* than along *gh*, we will know the following way: extend, thus the line *ad* outside the circle, that cuts the line *gh* at the point *q*. And as the moving body goes down more easily along the line *ef* than along the line *gh*, the heavier it is at *d* than at point *s*; it is also that heavier at point *d* than at point *s*, as longer is the line *da* than the line *ap*; thus the moving body goes as easier along the line *ef* than along the line *gh*, as the line *da* is longer than that line *pa*. We thus have the same proportion between the velocity along *ef* and the velocity along *gh*, as between the line *da* and the line *pa*. And *da* is also to *ap* as *qs* is to *sp*, it is to say as the oblique descent to the vertical one.[22]

To know how much heavier it is in *d* than in *s*, he considers the triangle *psq*, that is not complete on the figure, where *q* is the intersection of the lever arm *ad* (extended outside the circle) with the inclined plane *gh*. Than he finds that as *da* is to *ap* so is *qs* to *sp* or $da/ap = qs/sp$. Where, as we know today, $sp = qs \sin \alpha$ if α is the inclination of the plane or the angle in *q* in the triangle *sqp*.

We could formalise the last sentence in the following way: $v_{ef}/v_{gh} = da/pa = qs/sp$. The velocities are inversely proportional to the lengths of the planes. Galileo speaks here of the "velocity along a segment", it is thus different from our velocity. It is not the value of a function taken at one point but the velocity in the sense of Aristotles' velocity. Namely: having a higher velocity means either to go through the same distance in less time, or to go through a longer distance in the same time. In our terms, it is a mean velocity on a certain distance.

> [...] One observes thus that the force that pulls up the weight along an inclined plane is smaller than the one pulling it along the vertical as the strait vertical is less than the oblique one; and consequently that the force to go down of the same body is that greater along the vertical than along the inclined plane as the length of the way down along inclined plane is longer than the way along the vertical.[23]

---

[20] "Ducta enim linea ex puncto *s* perpendiculari super *cd*, pondus in *s*, respectu ponderis in *c*, est *ac* si penderet ex *p*; sed pondus in *p* minus gravat quam pondus in *ck*, cum distantia *pa* sit minor distantia *ac*. Et, similiter, pondus in *r* minus gravat quam pondus in *s*: quod ibidem patebit ducta perpendiculari ex *r* super *ad*, quae secabit ipsam ad inter puncta *a*, *p*." (Galilei 1589–1592, p. 298; Author's translation).

[21] "Manifestum est igitur quod mobile maiori vi descendet per lineam *ef* quam per lineam *gh*, et per *gh* quam per *nt*." (Galilei 1589–1592, p. 298; Author's translation).

[22] "Sed quanto maiori vi moveatur per *ef* quam per *gh* ita innotescet: extensa, scilicet, linea *ad* extra circulum, quae secet lineam *gh* in puncto *q*. Et quia tanto facilius descendit mobile per lineam *ef* quam per *gh*, quanto gravius est in puncto *d* quam in puncto *s*; est autem tanto gravius in puncto *d* quam in *s*, quanto longior est linea *da* quam linea *ap*; ergo mobile eo facilius descendet per lineam *ef* quam per *gh*, quo linea *da* longior est ipsa *pa*. Eandem ergo proportionem habebit celeritas in *cf* ad celeritatem in *gh*, quam linea, *da* ad lineam *pa*. Est autem sicut *da* ad *ap* ita *qs* ad *sp*, hoc est obliquus descensus ad rectum descensum" (Galilei 1589–1592, p. 298; Author's translation).

[23] "[...] constat igitur, tanto minori vi trahi sursum idem pondus per inclinatum ascensum quam per rectum, quanto rectus ascensus minor est obliquo; et, consequenter, tanto maiori vi descendere idem grave per rectum descensuiu quam per inclinatum, quanto maior est inclinatus descensus quam rectus" (Galilei 1589–1592, p. 298; Author's translation).

Galileo says that the descent of the moving body will follow the weight of the moving body at [each] point. Whatever the descent means, it has to do with the velocity and the direction of the motion. The velocity will "follow the weight", the direction the weight would take, if left to itself. Just as important is the repetition of the expression: *in the first point*. These points are important because they will be hidden by the new vocabulary Galileo will introduce in his next texts on the same argument.

3.2 *Le mecaniche* (1593), the *Momentum to Go Down*

A bit later, in *Le mecaniche*, Galileo gives a similar reasoning but he changes his vocabulary, saying now *Momento o impeto ad andare a basso* or "Momentum or impetus to go down" instead of force to go down.

This is not a simple question of vocabulary. First, he almost repeats the explanation he had given in *De motu* with the bent lever. But then he interprets the result differently. He thought that motion of the weight to occur along the quarter of a circle instead of being at the end of a bent lever. Having in mind that in every point of the circle it lies on a plane that is differently inclined. Its inclination being the one of the tangent in that particular point (Fig. 5).

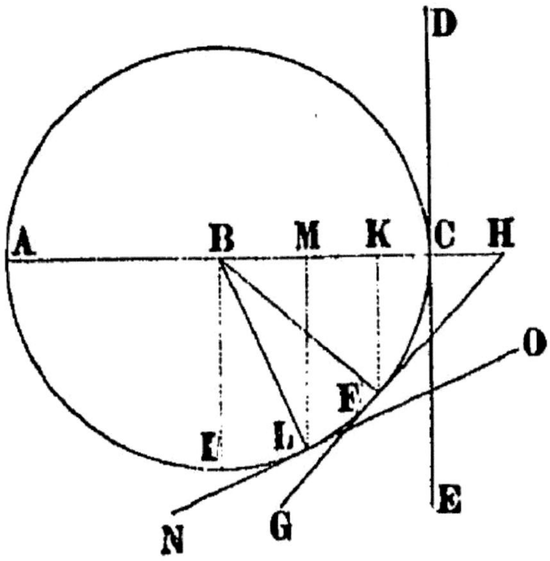

But to consider the weight that goes down and which is more or less sustained by the semi–diameters *BF*, *BL* [Fig. 5.] and constrained to move on the circumference *CFLI*, is not different from imagining the same circumference *CFLI* to be an incurved surface sustaining the same mobile, so that it is obliged to go down leaning and pressing on it; [...] So that undoubtedly, we may assert that when it goes down starting in the first point *C* along the circumference *CFLI*, the weight will first, in *C* have its total moment to go down; because it is in no way sustained by the circumference, and in this first point *C* and it doesn't undergo any other solicitation than one it would have if it were free, along the perpendicular and tangent *DCE*. But if the mobile is driven to point *F*, then the circular way against which it is leaning, sustains part of its gravity and its momentum to go down is diminished in the same proportion as the line *BK* surpasses the line *BC*; but when the mobile is in *F*, in the first point of its new motion, things happen as if it were on an inclined plane along the tangent *GFH* because the inclination of the circumference in point *F* is not different from the inclination of the tangent *FG* [...].

Fig. 5 Galilei's libra as described in *Le mecaniche* (Galilei 1593, p. 181). *Source* Gallica, Bibliothèque nationale de France. Public Domain[24]

[24] "Ma il considerare questo grave discendente, e sostenuto dalli semidiametri *BF*, *BL* ora meno e ora più, e constretto a caminare per la circonferenza *CFLI*, non è diverso da quello che saria imaginarsi la medesima circonferenza *CFLI* esser una superficie così piegata, e sottoposta al medesimo mobile, si che, appoggiandovisi egli sopra, fosse constretto a descendere in essa; perchè se nell'uno e nell'altro modo disegna il mobile il medesimo viaggio, niente importerà s'egli sia sospeso dal centro *B* e sostenuto dal semidiametro del cerchio, o pure se, levato tale sostegno, s'appoggi e camini su la circonferenza *CFLI*. [...] Onde indubitatamente potremo affermare, che, venendo al basso il grave dal punto c per la circonferenza *CFLI*, nel primo punto c il suo momento di discendere sia totale ed integro; perchè non viene in parte alcuna sostenuto dalla circonferenza, e non è, in esso primo punto c in disposizione a moto diverso di quello, che libero farebbe nella perpendicolare e contingente *DCE*. Ma se il mobile sarà costituito nel punto *F*, allora dalla circolare via, che gli è sottoposta, viene in parte la gravità sua sostenuta, ed il suo momento d'andare al basso diminuito con quella proporzione, con la quale la linea *BK* è superata dalla *BC*: ma quando il mobile è in *F*, nel primo punto di tale suo moto è come se fosse nel piano elevato secondo la contingente linea *GFH*, perciò che l'inclinazione della circonferenza nel punto F non differisce dall'inclinazione della contingente *FG* [...]" (Galilei 1593, p. 181; Author's translation).

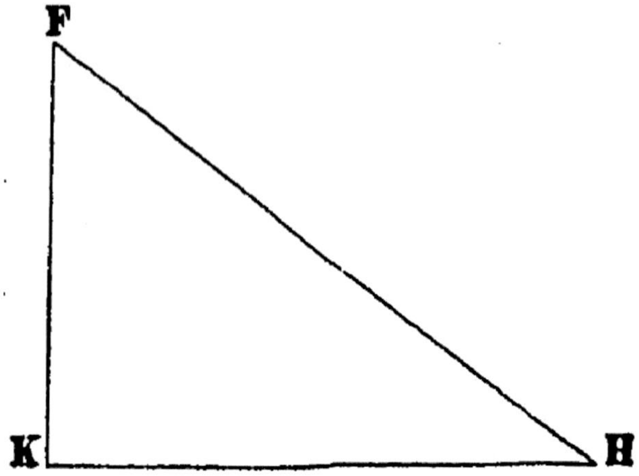

So that, taking a better figure, that is here [Fig. 6.], the "momentum to come down" that the moving body has along the inclined plane *FH*, has to its total momentum, with which it gravitates along the perpendicular to the horizon *FK*, the same proportion than that line *KF* to *FH*. A clear link between those two "momentum" is established by the fact that the total moment of a weight or "momentum along the vertical" is equal to the weight itself.

Fig. 6 The inclined plane Galilei applies to prove proportionality of momenta to the ratio between the vertical line and the *FH* line (Galilei 1593, p. 183). *Source* Gallica, Bibliothèque nationale de France. Public Domain[25]

Here Galileo seems to transfer the word *Momentum* meaning $P \times l$, thus a statical notion, into a dynamical one, linked to the velocity, as was his force in *de motu*. In fact he describes the function giving the velocity in every point of the quarter of a circle. In any of those points, the weight has its velocity at the first instant of the motion on an inclined plane coinciding with the tangent to the quarter of a circle. He does not mention $sin\ \alpha$ where α would be the angle of inclination of the plane. But that is what he describes. This trigonometrical aspect will be much clearer by Torricelli. Using Fig. 6, we call $M(W)_{fh}$, the moment of the weight, along the plane *fh*, we could formalise Galileo's sentence, as:

$$\frac{M(W)_{fh}}{M(W)_{fk}} = \frac{fk}{fh} \tag{2}$$

where $M(W)_{fk}$ is the total weight W, or:

$$M(W)_{fh} = W \sin \alpha \tag{3}$$

with α being the inclination of the plane *KHF*. Galileo goes further saying:

> And if it is so, it is manifest that, as the force that sustains the weight on the perpendicular *fk* must be equal to it [the weight], and to sustain it along the inclined plane *fh*, it will be enough that it be as minor as the perpendicular *fk* is minor than the line *fh* ... And we shall conclude with that universal proposition: on the elevated plane the force is to the weight in the same proportion as the perpendicular from the extremity of the plane to the horizon, is to the length of the plane.[26]

He seems to turn back to his previous vocabulary, using force instead of *momentum*. But remember that, as he said in *de Motu* in the first instant of the motion, the force to go down equals the *Archimedean momentum*. But during its motion on the quarter of the circle, the weight changes inclined plane so that its *Archimedean momentum* changes in every point together with its *momentum to go down*. All that would be much clearer, had Galileo mentioned the function giving the change of *momentum*, namely the sine function.

---

[25] "Sì che, passando a più distinta figura, quale è la presente [Fig. 6], il momento di venire al basso che ha il mobile sopra il piano inclinato *fh*, al suo totale momento, con lo qual gravita nella perpendicolare all'orizonte *fk*, ha la medesima proporzione che essa linea *kf* alla *fh*. E se così è, resta manifesto che, sì come la forza che sostiene il peso nella perpendicolare *fk* deve essere ad esso eguale, così per sostenerlo nel piano inclinato *fh* basterà che sia tanto minore, quanto essa perpendicolare *fk* manca dalla linea *fh* ... però concluderemo questa universale proposizione: sopra il piano elevato la forza al peso avere la medesima proporzione, che la perpendicolare dal termine del piano tirata all'orizonte, alla lunghezza d'esso piano [...]" (Galilei 1593, p. 183; Author's translation).

[26] "E se così è, resta manifesto che, sì come la forza che sostiene il peso nella perpendicolare *fk* deve essere ad esso eguale, così per sostenerlo nel piano inclinato *fh* basterà che sia tanto minore, quanto essa perpendicolare *fk* manca dalla linea *fh*. E perchè, come altre volte s'è avvertito, la forza per muover il peso basta che insensibilmente superi quella che lo sostiene, però concluderemo questa universale proposizione: sopra il piano elevato la forza al peso avere la medesima proporzione, che la perpendicolare dal termine del piano tirata all'orizonte, alla lunghezza d'esso piano [...]" (Galilei 1593, pp. 180–183; Author's translation).

De l'Hôpital (1661–1704) and Johann Bernoulli (1667–1748) will do something similar when talking of the fall in their correspondence around 1694. They talk about "the velocity in the first instant", meaning acceleration at the beginning of the motion. Johann adds an important characteristic to that notion, he writes:

> [...] the velocity in the first instant will nevertheless be infinitely small [...].[27]

Galileo knew he missed a point and did not publish *Le mecaniche*. Torricelli will give the solution as we will see.

3.3 Viviani's Addition[28] (1639) to the *Discorsi*: *De motu naturaliter accelerato*

Speaking of the various *momenta* the same weight has on variously inclined planes, Viviani or Galileo says referring to his *de motu antiquioria* (Fig. 7).

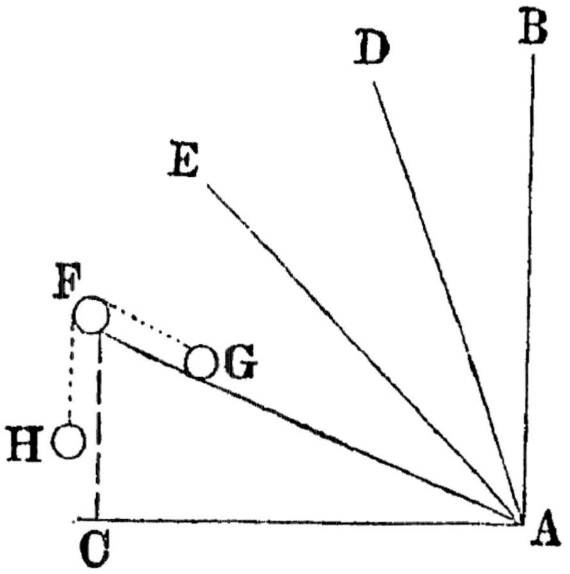

This change of *momentum* [on planes variously inclined, tangent to the quarter of a circle the weight is going down] being clear, it is here my duty to explain what was proved at length and conclusively, in a very old treatise on mechanics, our Academician [Galileo himself] wrote when in Padua, solely for the use of his students, when considering the origin and nature of that marvellous machine, the screw. What he proved is in what proportion the *impetus* changes on various planes having different inclinations, as for instance for the plane *FA* [Fig. 7.], tracing its elevation upon the horizon, that is *FC*. [...] This direction *FC* is that along which the *impetus* of a heavy body and its *momentum* of descent is a maximum; let us discover what proportion this *momentum* bears to the *momentum* of the same body moving along the inclined plane *FA*. This proportion, I say, is the inverse of that of the aforesaid lengths. This is the lemma preceding the theorem which I hope to prove a little later.

Fig. 7 Picture Galilei sketched to prove that momenta of the same weight along different inclined planes are inversely proportional to the slopes (Galilei 1638, VIII, p. 215). *Source* Gallica, Bibliothèque nationale de France. Public Domain[29]

To this *Lemma* proved as in *Le mecaniche*, Galileo gives a new proof, based on the fact that, in order to get an equilibrium, both bodies are linked with an inextensible rope. In that situation, the length the one body travels along the vertical *FC* is always equal to the length the other body travels along the inclined plane *FA*. But that new proof cancels all traces of the Archimedean origin of the momentum together with the balance that was used in the previous proofs. Only the word *momentum* remains. But: *What he proved is in what proportion the impetus changes on various planes having different inclinations* and a new word will be introduced, the *impetus*.

---

[27] [...] la vitesse au premier instant n'en sera pourtant qu'infiniment petite. In Der Briefwechsel von Johann Bernoulli [...]. (Bernoulli 1955, p. 243; Author's translation).

[28] Reproduced in a long footnote in Galilei 1638, VIII, pp. 214–219.

[29] "Appresa questa mutazione d'impeto, mi fa qui mestier esplicare quello che in un antico trattato di mecaniche, scritto già, in Padova dal nostro Academico sol per uso de' suoi discepoli, fu diffusamente o concludentemente dimostrato, in occasione di considerare l'origine e natura del maraviglioso strumento della vite; ed è con qual, proporzione si faccia tal mutazione d'impeto per diverse inclinazioni di piani: come, per esempio, del piano inclinato *AF* tirando la sua elevazione sopra l'orizonte, cioè la linea *FC*, per la quale l'impeto d'un grave ed il momento del descendere è il massimo, cercasi qual proporzione abbia questo momento al momento dell'istesso mobile por l'inclinata *FA*; qual proporzione dico esser reciproca delle dette lunghezze: e questo sia il lemma da premettersi al teorema, che dopo io spero di poter dimostrare." (Galilei 1638, VIII, p. 216; Author's translation).

It is clear that the *impetus* to descent of a body is equal to the resistance or least force sufficient to prevent it or to stop it: for such a force or resistance, and in order to measure it, I propose to use the weight of another body. Let us suppose, placed upon the plane *FA*, a body *G* linked with a cord running over the point *F* to the weight *H*; and let us consider that the space that this one will ascend or descend, along its perpendicular, is always equal to the complete ascend or descend of the other moving body *G* along the inclined plane *AF*.[30]

The problem of the length of the travel is clear, but from the point of view of the resistance the problem is different. When *G* falls along the inclined plane *FA*, it does not resist as it would when falling the same length *FA* along the vertical. It resists only as if it were falling along the vertical component, as we would say today, of *FA*, this is to say *FC*. Things are different for *H* that falls vertically from the distance *FC* and resists along the whole distance *FC*.

[…] but not to the rise or fall along the vertical in which direction alone *G* (as all other bodies) exerts its resistance. This is clear. I observe considering the motion of the body *G*, for example in the direction of the top, from *A* to *F*, in the triangle *AFC* to be composed of a horizontal transversal *AC* and of a perpendicular *CF*, and as along the horizontal there is no resistance of the body to be moved (because by such a motion the body neither gains nor loses with respect to its distance from the common centre of heavy things, that in the horizon remains always the same) it follows that resistance is met only with respect of the body rising through the perpendicular distance *CF*.[31]

After a few words of explanation on the variation of the *momentum* "function" of the length of the planes or as we would say today, of the angle of inclination of those planes, Galileo explains the results he had given in *Le mecaniche*. In a *Lemma*, he intends to prove that $M(H)/M(G) = FA/FC$, or as we would write it today $M(H) = M(G) \sin \alpha$, where α is the inclination of the plane. But instead of considering the same weight on two differently inclined planes, he considers now two different weights, *G* and *H*. The first lying on an inclined plane and the other on the vertical, as shown in Fig. 7.

Since then the body *G* in moving from *A* to *F* only offers resistance as it rises through the perpendicular distance *CF*, while the other body *H* must necessary fall vertically through the entire distance *FA*, and that this ratio of rise and fall is maintained the same whether the motion of the two bodies being large or small (because their connection is inextensible); we are to assert positively that, in case of equilibrium, that is when the bodies are at rest, the momenta, the velocities, or their tendency to motion, i.e., the spaces which would be traversed by them in equal times, must be in the inverse ratio to their weights. Following what one proves in every case of mechanical motion.[32]

Knowing thus that the vertical displacement of *G* is *FC* and that of *H* is *FA*, Galileo may apply the law of the lever as given by Aristotle and already applied to the inclined plane by Jordanus (XIIIth c) in a similar situation.

So that it will suffice to prevent the descent of *G*, that *H* be as less heavy than *G*, in the same proportion as space *CF* is less than space *FA*. Let thus be that *FA* be to *FC*, as weight *G* is to weight *H*; then the equilibrium will follow, so that the weights *G* and *H* will have the same *momentum* and their motion will stop.[33]

We could formalise this criteria of equilibrium, for which $M(W(G)) = M(W(H))$, in the following way: $W(G)/W(H) = FA / FC$ or $W(H) = W(G) \sin\alpha$, where α is the inclination of the plane. Thus the *momenta*, linked to the velocities or "Galilean momenta" are in the same proportion as the weights, as the "Archimedean momenta". The preceding passage shows us clearly Galileo's "confusion". At the equilibrium: the *momenta*, the velocities, or the tendency to motion, i.e., the spaces which would be traversed by them in equal times, must be in the inverse ratio to their weights. Those weights being measured by the balance. In that same passage some authors, such as Duhem, see in the use of a conditional in the expression "would be traversed" (passerebbero) one of the origins of the principle of virtual velocities. But I don't agree with them. In

---

30 "Qui è manifesto, tanto esser l'impeto del descendere d'un grave, quanta è la resistenza o forza minima che basta per proibirlo e fermarlo: per tal forza o resistenza, e sua misura, mi voglio servire della gravità d'un altro mobile. Intendasi ora, sopra il piano *FA* posare il mobile *G*, legato con un filo che, cavalcando sopra l'*f*, abbia attaccato un peso *H*; e consideriamo che lo spazio della scesa o salita a perpendicolo di esso è ben sempre eguale a tutta la salita o scesa dell'altro mobile *G* per l'inclinata *AF*." (Galilei 1638, VIII, p. 216; Author's translation).

31 "[…] ma non già alla salita o scesa a perpendicolo, nella qual sola esso mobile *G* (si come ogn'altro mobile) esercita la sua resistenza. Il che è manifesto. Imperochè considerando, nel triangolo *afc* il moto del mobile *G*, per esempio all'in su da *A* in *F*, esser composto del trasversale orizontale *AC* o del perpendicolare *CF*; ed essendo che quanto all'orizontale, nessuna, come s'è detto, è la resistenza del medesimo ad l'esser mosso (non facondo con tal moto perdita alcuna, nè meno acquisto, in riguardo della propria distanza dal comun centro delle cose gravi, che nell'orizonte si conserva sempre l'istessa); resta, la resistenza esser solamente rispetto al dover salire la perpendicolare *CF*." (Galilei 1638, VIII, pp. 216–217; Author's translation).

32 "Mentre che dunque il grave *G*, movendosi da *A* in *F*, resiste solo, nel salire, lo spazio perpendicolare *CF*, ma che l'altro grave *H* scende a perpendicolo necessariamente quanto tutto lo spazio *FA*, e che tal proporzione di salita e scesa si mantien sempre l'istessa, poco o molto che sia il moto de i detti mobili (per esser collegati insieme); possiamo assertivamente affermare, che quando debba seguire l'equilibrio, cioè la quiete tra essi mobili, i momenti, le velocità, o le lor propensioni al moto, cioè gli spazii che da loro si passerebbero nel medesimo tempo, devon rispondere reciprocamente alle loro gravità, secondo quello che in tutti i casi de' movimenti mecanici si dimonstra: […]" (Galilei 1638, VIII, p. 217; Author's translation).

33 "[…] si che basterà, per impedire la scesa del *G*, che lo *H* sia tanto men grave di quello, quanto a proporzione lo spazio *CF* è minore dello spazio *FA*. Sia fatto, dunque, come *FA* ad FC, così il grave *G* al grave *H*; chè allora seguirà l'equilibrio, cioè i gravi *H*, *G* averanno momenti eguali, e cesserà il moto de i detti mobili." (Galilei 1638, VIII, p. 217; Author's translation).

my opinion, Galileo's *would* replaces the expression "in the first point" of the motion, as we saw in *de motu*. As Bernoulli and de l'Hôpital's "in the first instant" of its motion, the conditional "would" makes the transition between rest and motion, between statics and dynamics. This part has been added by Viviani to Galileo's manuscript. That is the reason why I also reproduce Galileo's original Italian text here:

> Thus while the moving body *G*, moves, for example from *A* to *F*, only the perpendicular space *CF* resists to its going up, but the other moving body *H* goes up along the all space *FA*, we may thus reasonably state that velocity as well as spaces gone through during the same times by those moving bodies must correspond to their weights; and it will be enough to prevent the descent of the weight *H*, that the weight *G* be as more heavy than that of *H*, as the inclined *AF* is greater than the perpendicular *FC*.[34]

Galileo does not argue here what happens after the first instant, or how that velocity increases when the weight rolls along the inclined plane. He only states this in the *Discorsi: De motu naturaliter accelerato*. There he shows that the velocity, during the fall, increases adding in every instant the same increment of velocity, something we would call *dv* (Galilei 1638, VIII, p. 171). As I announced it at the beginning, in its dynamical sense, *momentum* has to do with *dv*.

4 Torricelli's Proof of Galileo's Assumption

Now that we are acquainted with Galileo's and Torricelli's *momentum*, let us try to understand the five (initial) propositions of *De motu gravium naturaliter descendensium et proiectorum* published by Torricelli in 1644. There he proves Galileo's assumption that the degrees of velocity the same moving body acquires falling down on differently inclined planes having the same elevation, are the same when it has descended the same vertical height. Torricelli first proposes to replace this assumption by the following one:

4.1 Torricelli's Principle

Torricelli claimed:

> We set out as a principle that two weights linked together cannot move by themselves, unless their common centre of gravity goes down. Indeed, when two weights are linked together in such a way that the motion of one of them causes the motion of the other, be the binding produced by means of a balance or pulley or any other mechanism, those two weights will behave as a single one composed of two parts; but such a weight will never move, unless its centre of gravity descends. But then, when it will be made in such a manner that its common centre of gravity can not descend in any way, it remains at rest in the position it occupies. Otherwise, in fact, it would move in vain, because it would take a horizontal motion that tends no way down.[35]

We could also say that a body remains in equilibrium if its centre of gravity cannot descend. Torricelli will use the principle in that sense. In fact, the essence of the principle, that we already found in the works of Jordanus de Nemore (1225–1260) and of Gregorius, and whose Aristotelian origin is obvious, was also evocated by Galileo in Viviani's addition that follows theorem II of the *de motu naturaliter accelerato*. Theorem II gives the relation between spaces and squares of times and in the addition, Galileo (or Viviani) writes.

[34] "Mentre dunque il mobile *G*, muovendosi, per esempio, da *A* in *F*, resiste solo, nel salire, lo spazio perpendicolare *cf*, ma che l'altro grave *H* scende a perpendicolo quanto è tutto lo spazio *FA*, possiamo ragionevolmente affermare, le velocità e gli spazi passati nel medesimo tempo da tali mobili dover risponder reciprocamente alle loro gravità; e basterà, per impedir la scesa del peso *H*, che il gravo *G* sia tanto più grave di quello *H*, quanto l'inclinata *AF* è maggiore della perpendicolare *FC*." (Galilei 1638, VIII, p. 444; Author's translation). As we can see, in Galileo's original text, there is no trace of any principle of virtual velocities.

[35] "Praemittimus, Duo gravia simul conjuncta ex se moveri non posse, nisi centrum commune gravitates ipsorum descendat. Quando enim duo gravia inter se coniuncta fuerint, ut ad motum unius motus etiam alterius consequatur, erunt duo illa gravia tamquam grave unum ex duobus compositum, sive id libra fiat, sive trochlea, sive qualibet alia Mechanica ratione, grave autem hujiusmodi non movebitur unquam, nisi centrum gravitates ipsius descendat. Quando vero ita constitutum fuerit ut nullo modo commune ipsius centrum gravitates descendere possit, grave penitus in sua positione quiescet: alias enim frustra moveretur; horizontali, scilicet latione, quae nequaquam deorsum tendit." (Torricelli 1644, p. 99; Author's translation).

> [...] because, just as a weight or a set of weights will not move naturally upwards, moving away from the common centre to which tend all weights, it is impossible as well that a weight moves spontaneously if by that motion its centre of gravity does not approach the common centre in question: thus, on an horizontal plane, that is on a surface that remains equally distant from the same centre, and therefore without inclination, the *impeto* or *momentum* of that moving body will be zero.[36]

This is an application of Aristotle's idea that all weights tend to a unique centre. Let us underline here that *impetus* and *momentum* are synonymous as we often saw before with Galileo.

4.2 Torricelli's *Proposition I*

> If two weights are placed on two planes, unequally inclined but having the same elevation and if those weights are between them as the lengths of the planes, those weights will have the same *momentum*.[37]

We could formulate this proposition, using Fig. 8 of his proposition *II*.

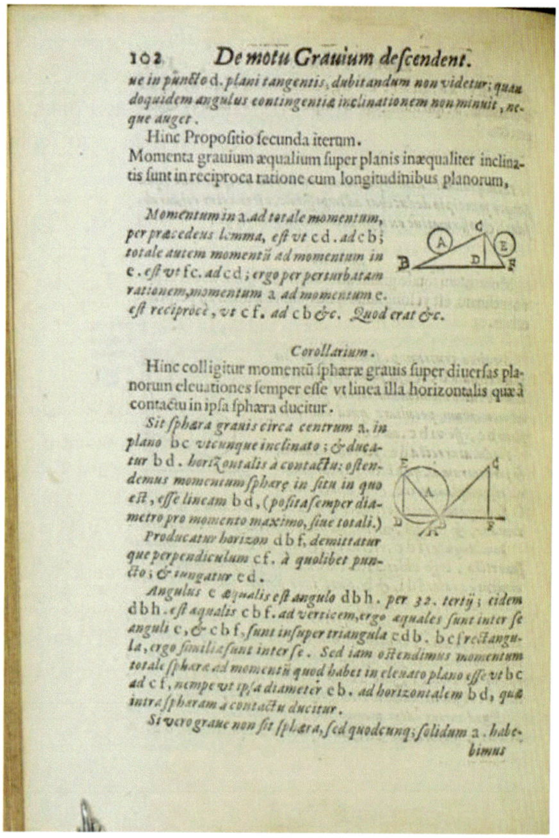
102 De motu Grauium deſcendent.
ne in puncto d. plani tangentis, dubitandum non videtur; quandoquidem angulus contingentiæ inclinationem non minuit, neque auget.
Hinc Propoſitio ſecunda iterum.
Momenta grauium æqualium ſuper planis inæqualiter inclinatis ſunt in reciproca ratione cum longitudinibus planorum,
Momentum in a. ad totale momentum, per præcedens lemma, eſt vt cd. ad cb; totale autem momentũ ad momentum in e. eſt vt fc. ad cd; ergo per perturbatam rationem, momentum a. ad momentum e. eſt reciprocè, vt cf. ad cb &c. Quod erat &c.
Corollarium.
Hinc colligitur momentũ ſphæræ grauis ſuper diuerſas planorum eleuationes ſemper eſſe vt linea illa horizontalis quæ à contactu in ipſa ſphæra ducitur.
Sit ſphæra grauis circa centrum a. in plano bc vtcumque inclinato; & ducatur bd. horizontalis à contactu: oſtendemus momentum ſphæræ in ſitu in quo eſt, eſſe lineam bd, (poſita ſemper diametro pro momento maximo, ſiue totali.)
Producatur horizon dbf, demittatur que perpendiculum cf. à quolibet puncto; & iungatur cd.
Angulus c æqualis eſt angulo dbh. per 32. tertij; eidem dbh. eſt æqualis cbf. ad verticem, ergo æquales ſunt inter ſe anguli c, & cbf, ſunt inſuper triangula cdb. bcf rectangula, ergo ſimilia ſunt inter ſe. Sed iam oſtendimus momentum totale ſphæræ ad momentũ quod habet in eleuato plano eſſe vt bc ad cf, nempe vt ipſa diameter cb. ad horizontalem bd, quæ intra ſphæram à contactu ducitur.
Si vero graue non ſit ſphæra, ſed quodcunq; ſolidum a. habebimus

Let two bodies *A* and *E* lay on two planes *CB* and *CF* differently inclined but having the same elevation *CD*. We call W_A, W_E the weights of *A* and *E* and M_A, M_E their *momenta*. Than Torricelli's proposition I can be written in the following way:

$$\text{If } W_A / W_E = CB / CF, \text{ then } M_A = M_E$$

I say that if the weights are in such a position, their *momentum* are equal: that means that the position in which they are is equilibrated and none of them can move neither up nor down.

Fig. 8 Torricelli's proof of $M_A/M_E = CF/CB$ (Torricelli 1644, p. 102). *Source* Courtesy University of Oklahoma, History of Science Collections[38,39]

---

[36] "[...] poichè, si come è impossibile che un grave o un composto di essi si muova naturalmente all'in su, discostandosi dal comun centro verso dove conspirano tutte le cose gravi, così è impossibile che egli spontaneamente si muova, se non con tal moto il suo proprio centro di gravità non acquista avvicinamento al sudetto centro comune: onde sopra l'orizontale, che qui s'intende per una superficie equalmente lontana dal medesimo centro, e perciò affato priva d'inclinazione, nullo sarà l'impeto o momento di detto mobile." (Galilei 1638, VIII, p. 215; Author's translation).

[37] "Si in planis inaequaliter inclinatis, eandem tamen elevationem habentibus, duo gravia constituantur, quae inter se eandem homologe rationem habeant quam longitudinem planorum, gravia aequale momentum habebunt." (Torricelli 1644, p. 99; Author's translation).

[38] In order to introduce some uniformity in the notations, I take everywhere the notations of Fig. 8 even if the weights A and E are sometimes equal sometimes not.

[39] "Dico gravia sic disposta aequale momentum habere: hoc est in ea in qua sunt positione aequilibrata conquiescere, neque sursum aut deorsum moveri." (Torricelli 1644, p. 99; Author's translation).

The last part of the proposition, "those weights will have the same *momentum*" clearly means here that they are in equilibrium. There is no motion in that proposition and the *momentum* is derived from Archimedes law for the equilibrium of a lever; it is the product of a weight by a distance. The *fulchrum* being at their common centre of gravity. This proposition corresponds to the last one we showed in Galileo's *Discorsi* but Torricelli proves it by contradiction. He shows that their centre of gravity cannot go down and consequently using his principle given in the *Praemittimus*, that they are in equilibrium. As Galileo in *De motu antiquioria* and in *Le mecaniche*, Torricelli starts considering an equilibrium and using the *Archimedean momentum*. Torricelli doesn't seem to know Galileo's proof, that was only given by Viviani in the addition to the *Discorsi* and known to us via Viviani's manuscript reproduced in the *Edizione Nazionale*.

4.3 Torricelli's *Proposition II*

Momenta of equal heavy bodies lying on unequally inclined planes having the same elevation, are in reciprocal ratio with the length of the planes.[40]

Still using Fig. 8 of the *Proposition II,* we could formulate this proposition:

Let two weights *A* and *E* lay on two planes *CB* and *CF* differently inclined but having the same elevation *CD*. We call P_A, P_E the weights of *A* and *E* and M_A, M_E their *momenta.*

$$\text{If } W_A = W_E$$

then we have

$$M_A / M_E = CF / CB$$

After having considered the case of equilibrium, where the momenta must be equal, he considers equal weights. In such a situation, we don't have any equilibrium. We know that if weights are equal they cannot be in equilibrium unless the situation is completely symmetrical and *CF* = *CB*. What does *momentum* mean here? Does Torricelli make the same "confusion" as Galileo? The question is not easy to answer as we showed in the previous section with Galileo's texts. The difficulty is also clear in what Salviati says in Viviani's addition to the beginning of third day:

Salv. Since then I have your permission, let us first of all consider as a notable fact, that the momenta or velocities of one and the same moving body vary with the inclination of the plane and that the velocity reaches a maximum along a line erected perpendicularly to the horizon, and that for another direction that velocity diminishes the more the plane diverges from the vertical, that is the more obliquely it inclines. Therefore the *impetus*, ability, energy, or, one might say, the *momentum* of descent of the moving body is diminished by the plane upon which it is supported and along which it rolls.[41]

Contrarily to *Proposition I* where Torricelli had an equilibrium and could deduce a relation between *momenta.* Here he finds the *momenta* (Fig. 8) that restore equilibrium. Those are no *Archimedean momenta.* And in Torricelli's *Proposition II* the weights *A* and *E* are not in equilibrium either. But one could find a certain weight, that Torricelli calls *D* which would have the same momentum as *A* and so would be in equilibrium with *A*. But says Torricelli:

So the *momentum* of *E* is to the *momentum* of *D* as the *moles* to the *moles* (because they lay on the same plane) that is as *moles* of *A* [because weights *A* and *E* are equal] is to *moles* of *D*, that is as *CB* is to *CF*. So the *momentum* of *E* is to that of *D*, or to the *momentum* of *A*, that is equal to the *momentum* of *D*, as *CB* to *BF*. QED[42]

$$\frac{M_E}{M_D} = \frac{W_E}{W_D} = \frac{W_A}{W_D} = \frac{CB}{CF} \tag{4}$$

[40] "Momenta gravium aequalium super planis inaequaliter inclinatis, eandem tamen elevationem habentibus, sunt in reciproca ratione cum longitudinibus planorum." (Torricelli 1644, p. 100; Author's translation).

[41] SALVIATI: Già che dunque me ne date licenza, considerisi in primo luogo, come effetto notissimo, che i momenti o le velocità d'un istesso mobile son diverse sopra diverse inclinazioni di piani, e che la massima è per la linea perpendicolarmente sopra l'orizonte elevata, e che per l'altre inclinate si diminuisce tal velocità, secondo che quelle più dal perpendiculo si discostano, cioè più obliquamente s'inclinano; onde l'impeto, il talento, l'energia, o vogliamo dire il momento, del descendere vien diminuito nel mobile dal piano soggetto, sopra il quale esso mobile s'appoggia e descende. (Galilei 1638, VIII, p. 215; Author's translation).

[42] "Momentum autem *E* ad momentum *D* est ut moles ad moles (quia sunt in eodem plano) hoc est ut moles *A*, ad molem *D*, hoc est ut *CB* ad *CF*. Est ergo momentum *E* ad *D*, vel ad momentum *A* ipsi momento *D* aequale, ut *CB* ad *BF*. Quod erat &c." (Torricelli 1644, pp. 100–101; Author's translation).

and as $M_E = M_A$, because A and D are in equilibrium:

$$\frac{M_E}{M_D} = \frac{M_E}{M_A} = \frac{CB}{CF} \tag{5}$$

We find here a transition by Torricelli, similar to the one we saw by Galileo. The one between an "Archimedean momentum" as expressed by the equilibrium of a balance to a dynamical one. But Torricelli does not seem satisfied with his proof and in his *Aliter*, he announces that he will prove the same as in *Proposition II,* without using proposition I because that one was only proved by contradiction. To do this, he introduces a *Lemma.*

4.4 Torricelli's *Lemma* to *Proposition II*

The total *momentum* of a weight is to the *momenta* it has on an inclined plane as the length of that plane is to the vertical.[43]

To prove this, Torricelli, uses once more both notions of *momentum*, the Archimedean and the Galilean one, giving us more information about the way he considers *momenta* (Fig. 9).

Let a sphere with centre A lay on the inclined plane BC and CE be vertical, I say that the total momentum of weight A is to its particular *momentum* on plane BC as BC is to CE.

Torricelli takes (Fig. 9) a point F on DA perpendicular to the plane, and draws FG horizontal and thus parallel to BE. FG being equal to FD and FH equal to FA. LA and LI are parallel and orthogonal to BE. He shows than that both rectangular triangles FDI and BCE are equal. Than he considers that point F is fixed and he imagines the weight, with centre in A, in two situations just as Galileo did in *De motu Antiquior a* and in *Le mecanich e*.

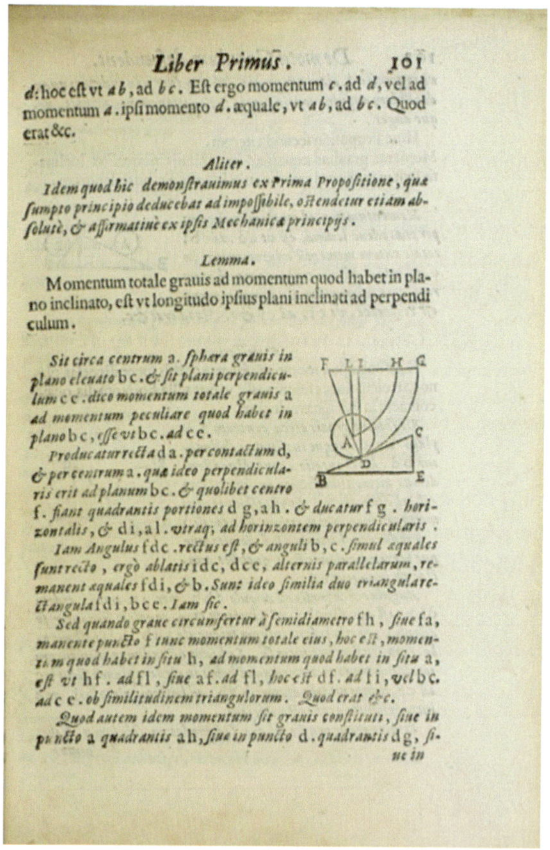

Liber Primus. 101

d: hoc eſt vt ab, ad bc. Eſt ergo momentum c. ad d, vel ad momentum a. ipſi momento d. æquale, vt ab, ad bc. Quod erat &c.

Aliter.

Idem quod hic demonſtrauimus ex Prima Propoſitione, quæ ſumpto principio deducebat ad impoſſibile, oſtendetur etiam abſolutè, & affirmatiuè ex ipſis Mechanicæ principijs.

Lemma.

Momentum totale grauis ad momentum quod habet in plano inclinato, eſt vt longitudo ipſius plani inclinati ad perpendiculum.

Sit circa centrum a. *ſphæra grauis in plano eleuato* bc. *& ſit plani perpendiculum* ce. *dico momentum totale grauis* a *ad momentum peculiare quod habet in plano* bc, *eſſe vt* bc. *ad* ce.

Producatur recta da. *per contactum* d, *& per centrum* a. *quæ ideo perpendicularis erit ad planum* bc. *& quolibet centro* f. *fiant quadrantis portiones* dg, ah. *& ducatur* fg. *horizontalis, &* di, al. *vtraq; ad horizontem perpendicularis.*

Iam Angulus fdc. *rectus eſt, & anguli* b, c. *ſimul æquales ſunt recto, ergò ablatis* idc, dce, *alternis parallelarum, remanent æquales* fdi, *&* b. *Sunt ideo ſimilia duo triangula rectangula* fdi, bce. *Iam ſic.*

Sed quando graue circumfertur à ſemidiametro fh, *ſiue* fa, *manente puncto* f *tunc momentum totale eius, hoc eſt, momentum quod habet in ſitu* h, *ad momentum quod habet in ſitu* a, *eſt vt* hf. *ad* fl, *ſiue* af. *ad* fl, *hoc eſt* df. *ad* fi, *vel* bc. *ad* ce. *ob ſimilitudinem triangulorum. Quod erat &c.*

Quod autem idem momentum ſit grauis conſtituti, ſiue in puncto a *quadrantis* ah, *ſiue in puncto* d. *quadrantis* dg, *ſiue in*

Fig. 9 Torricelli relates momenta on horizontal and inclined planes to the length of the inclined plane and its height (Torricelli 1644, p. 101). *Source* Courtesy University of Oklahoma, History of Science Collections[44]

[43] "Momentum totale gravis ad momenta quod habet in plano inclinato, est ut longitude ipsius plani inclinati ad perpendiculum." (Torricelli 1644, p. 101; Author's translation).

[44] "Sit circa centrum A sphaera gravis in plano elevato BC and sit plani perpendiculum CE, dico momentum totale gravis A ad momentum peculiare quod habet in plano BC, esse ut BC ad CE." (Torricelli 1644, p. 101; Author's translation).

He just forgets to mention the weight that should equilibrate the weight *A* when hanging at *H*, considering Galileo's text as known. In the first situation the weight *A* has its full *Momentum* as if it were hanging to the lever at point *H* and in the second it lies on the inclined plane *BC*. Considering the first situation, the total *momentum* of the weight, the "Archimedean one", when hanging at *H* is $W_A \times FH$. Considering the second situation, he says that the *momentum* of the same weight is the one it would have hanging at *L*. Therefore, both *momenta* are to each other as *HF* to *LF*. Using the similarity of the triangles *FDI*, *CBE* and *FAL*, he finds that *HF* is to *LF* as *BC* to *CE*, thus as the length of the plane is to the length of the vertical. That is $\sin\alpha$ where α is the inclination of the plane.

Torricelli proves that the ratio of the momentum of a weight laying on the inclined plane is to his total momentum as $M_{CB}/M_{Total} = HF/LF = CB/CE$. We could also write $M_{CB} = M_{Total} \sin\alpha$ where α is the inclination of the plane. Here *momentum* has the Archimedean meaning of weight times distance. But what is the meaning of that momentum with respect to the inclined plane and to the *momentum* of the definition of the uniformly accelerated motion that we recalled at the beginning of this article?

We have here a first aspect of the relation between the *Archimedean momentum* and Galileo's one. The similarity of the triangles *FDI*, *CBE* and *FAL* shows that lever arms are linked to the lengths of the planes. But let us return to Torricelli's proof. Having proved the relation between the *moment of the body moving vertically* to that of the same body moving along an inclined plane, Torricelli proves his *Proposition II*, generalizing this result to any two planes having the same elevation. This condition is essential and we will say more about it in the conclusion (Fig. 10).

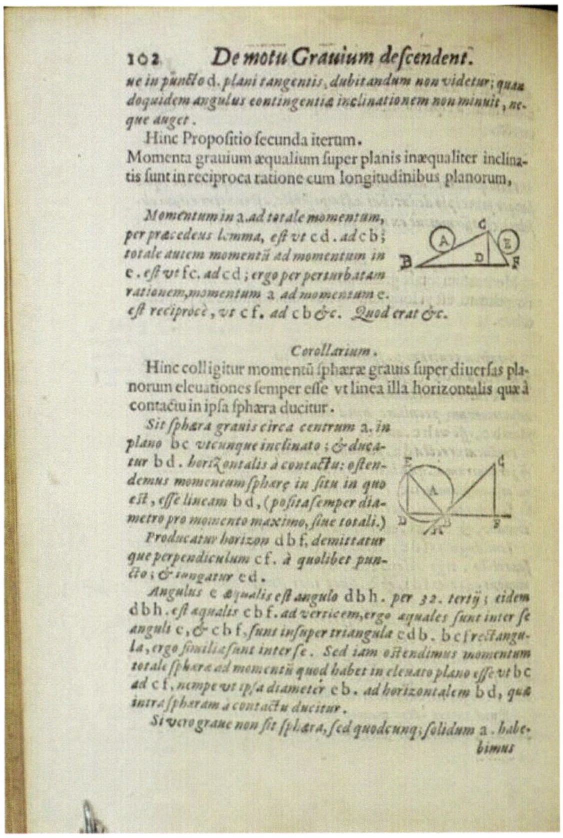

102 *De motu Grauium deſcendent.*

ne in puncto d. *plani tangentis, dubitandum non videtur; quandoquidem angulus contingentiæ inclinationem non minuit, neque auget.*

Hinc Propoſitio ſecunda iterum.

Momenta grauium æqualium ſuper planis inæqualiter inclinatis ſunt in reciproca ratione cum longitudinibus planorum,

Momentum in a. *ad totale momentum, per præcedens lemma, eſt vt* cd. *ad* cb; *totale autem momentū ad momentum in* e. *eſt vt* fc. *ad* cd; *ergo per perturbatam rationem, momentum* a *ad momentum* e. *eſt reciprocè, vt* cf. *ad* cb &c. *Quod erat* &c.

Corollarium.

Hinc colligitur momentū ſphæræ grauis ſuper diuerſas planorum eleuationes ſemper eſſe vt linea illa horizontalis quæ à contactu in ipſa ſphæra ducitur.

Sit ſphæra grauis circa centrum a. *in plano* bc *vtcunque inclinato; & ducatur* bd. *horizontalis à contactu: oſtendemus momentum ſphæræ in ſitu in quo eſt, eſſe lineam* bd, *(poſita ſemper diametro pro momento maximo, ſiue totali.)*

Producatur horizon dbf, *demittatur que perpendiculum* cf. *à quolibet puncto; & iungatur* cd.

Angulus c *æqualis eſt angulo* dbh. *per 32. tertij; eidem* dbh. *eſt æqualis* cbf. *ad verticem, ergo æquales ſunt inter ſe anguli* c, & cbf, *ſunt inſuper triangula* cdb. bcf *rectangula, ergo ſimilia ſunt inter ſe. Sed iam oſtendimus momentum totale ſphæræ ad momentū quod habet in eleuato plano eſſe vt* bc *ad* cf, *nempe vt ipſa diameter* cb. *ad horizontalem* bd, *quæ intra ſphæram à contactu ducitur.*

Si vero graue non ſit ſphæra, ſed quodcunq; ſolidum a. *habebimus*

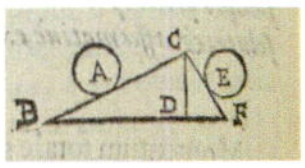

Torricelli takes any triangle like *BFC* (We still use Fig. 10 taken from *Proposition II*), draws the height *CD* and applies twice his Lemma. He finds:

$$\frac{M_A}{M_E} = \frac{CB}{CF} \tag{6}$$

Fig. 10 Picture used by Torricelli for Proposition *II* (Torricelli 1644, p. 102). *Source* Courtesy University of Oklahoma, History of Science Collections

4.5 Torricelli's *Corollarium* to *Proposition II*

In his *Corollarium*, Torricelli gives a rather unusual theorem. We will try to explain his reasons later (Fig. 11).

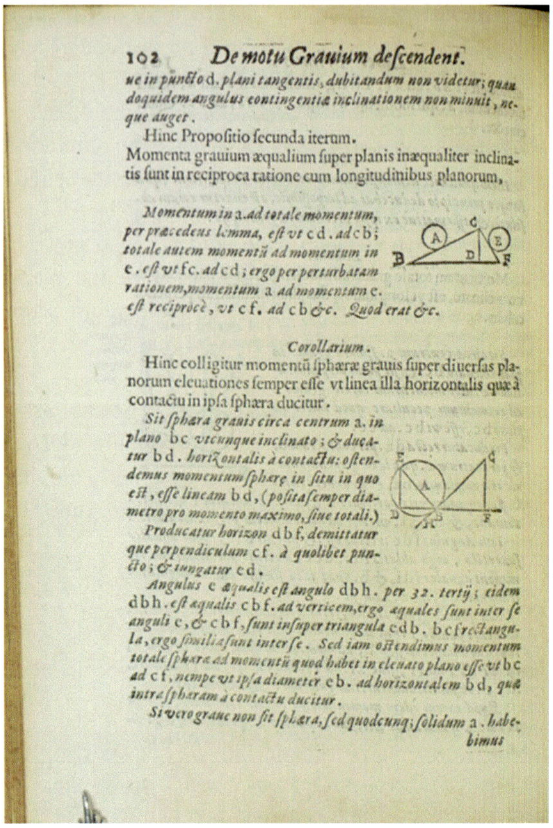

102 *De motu Grauium deſcendent.*

ne in pũncto d. *plani tangentis, dubitandum non videtur; quandoquidem angulus contingentiæ inclinationem non minuit, neque auget.*

Hinc Propoſitio ſecunda iterum.

Momenta grauium æqualium ſuper planis inæqualiter inclinatis ſunt in reciproca ratione cum longitudinibus planorum,

Momentum in a. *ad totale momentum, per præcedens lemma, eſt vt* c d. *ad* c b; *totale autem momentũ ad momentum in* e. *eſt vt* f c. *ad* c d; *ergo per perturbatam rationem, momentum* a *ad momentum* e. *eſt reciprocè, vt* c f. *ad* c b *&c. Quod erat &c.*

Corollarium.

Hinc colligitur momentũ ſphæræ grauis ſuper diuerſas planorum eleuationes ſemper eſſe vt linea illa horizontalis quæ à contactu in ipſa ſphæra ducitur.

Sit ſphæra grauis circa centrum a. *in plano* b c *vtcunque inclinato; & ducatur* b d. *horizontalis à contactu: oſtendemus momentum ſphæræ in ſitu in quo eſt, eſſe lineam* b d, *(poſita ſemper diametro pro momento maximo, ſiue totali.)*

Producatur horizon d b f, *demittaturque perpendiculum* c f. *à quolibet puncto; & iungatur* c d.

Angulus e *æqualis eſt angulo* d b h. *per 32. tertij; eidem* d b h. *eſt æqualis* c b f. *ad verticem, ergo æquales ſunt inter ſe anguli* e, *&* c b f, *ſunt inſuper triangula* e d b. b c f *rectangula, ergo ſimilia ſunt inter ſe. Sed iam oſtendimus momentum totale ſphæræ ad momentũ quod habet in eleuato plano eſſe vt* b c *ad* c f, *nempe vt ipſa diameter* e b. *ad horizontalem* b d, *quæ intra ſphæram à contactu ducitur.*

Si vero graue non ſit ſphæra, ſed quodcunq; ſolidum a. *habebimus*

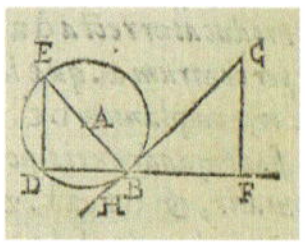

So one deduces that the momentum of a spherical weight lying on various planes having different elevations is always as the horizontal line [*BD*] drawn through the contact point [*B*] [having taken the total *momentum* as equal to the diameter].

In the picture the corollarium (particular of the picture 11) the weight is represented by the circle *EDB* centred in *A*. *B* is the contact point and *BD* is horizontal. Torricelli shows that the triangles *EDB* and *CFB* are similar, so that the total *momentum* is to the *momentum* plane on the *BC* as *BC* is to *CF* or as the horizontal *BD* is to the diameter *EB* (see below Eq. 7).

Fig. 11 How to deduce the momentum of a spherical mass (Torricelli 1644, p. 102). *Source* Courtesy University of Oklahoma, History of Science Collections[45]

$$\frac{M_{Total}}{M_{BC}} = \frac{BC}{CF} = \frac{EB}{DB} \tag{7}$$

The advantage for introducing the circle and the new triangle is to show the sine in the circle. We have:

$$M_{BC} = M_{Total}\sin a$$

where the angle of inclination of the plane α is angle *B* in the triangle *CBF* or angle *E* in the similar triangle *BED*. With the help of the next figure (Fig. 12), Torricelli generalizes to a non–spherical weight. He supposes that the total *momentum* or force able to sustain the moving body on a vertical plane, its weight, to be the diameter *BC* and claims that its *momentum* in the plane *AB* is *BD*.

[45] "Hinc colligitur momentum sphaerae gravis super diversas planorum elevationes semper esse ut linea illa horizontalis quae a contactu in ipsa sphaera ducitur." (Torricelli 1644, p. 102; Author's translation).

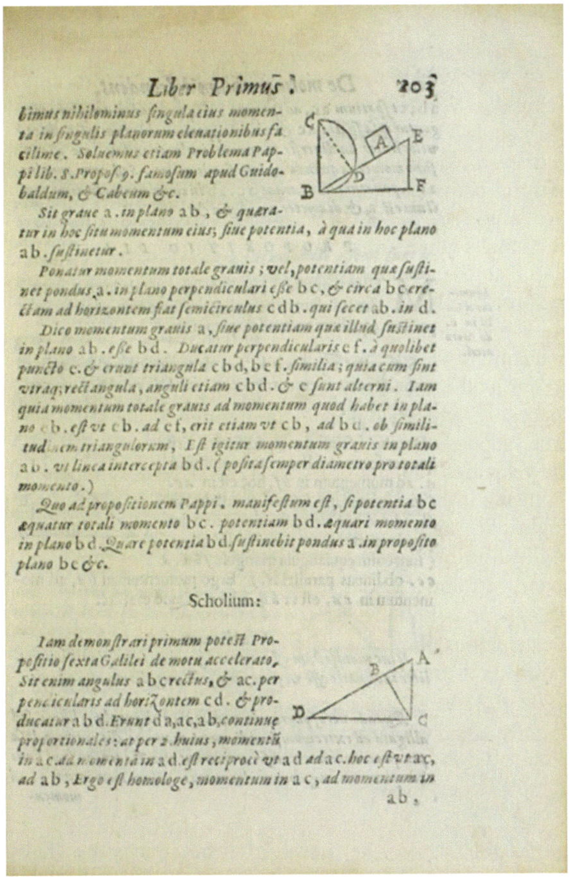

Liber Primus. 103

bimus nihilominus ſingula eius momenta in ſingulis planorum eleuationibus facillime. Soluemus etiam Problema Pappi lib. 8. Propoſ. 9. famoſum apud Guidobaldum, & Cabeum &c.

Sit graue a. in plano ab, & quæratur in hoc ſitu momentum eius; ſiue potentia, à qua in hoc plano ab. ſuſtinetur.

Ponatur momentum totale grauis; vel, potentiam quæ ſuſtinet pondus a. in plano perpendiculari eſſe bc, & circa bc erectam ad horizontem fiat ſemicirculus cdb. qui ſecet ab. in d.

Dico momentum grauis a, ſiue potentiam quæ illud ſuſtinet in plano ab. eſſe bd. Ducatur perpendicularis ef. à quolibet puncto e. & erunt triangula cbd, bef. ſimilia; quia cum ſint vtraq; rectangula, anguli etiam cbd. & e ſunt alterni. Iam quia momentum totale grauis ad momentum quod habet in plano eb. eſt vt eb. ad ef, erit etiam vt cb, ad bd. ob ſimilitudinem triangulorum, Eſt igitur momentum grauis in plano ab. vt linea intercepta bd. (poſita ſemper diametro pro totali momento.)

Quo ad propoſitionem Pappi. manifeſtum eſt, ſi potentia bc æquatur totali momento bc. potentiam bd. æquari momento in plano bd. Quare potentia bd. ſuſtinebit pondus a. in propoſito plano bc &c.

Scholium:

Iam demonſtrari primum poteſt Propoſitio ſexta Galilei de motu accelerato. Sit enim angulus abc rectus, & ac. perpendicularis ad horizontem cd. & producatur abd. Erunt da, ac, ab, continuę proportionales: at per 2. huius, momentũ in ac ad momenta in ad. eſt reciprocè vt ad ad ac. hoc eſt vt ac, ad ab, Ergo eſt homologe, momentum in ac, ad momentum in ab,

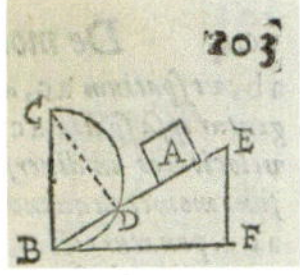

Let a weight be in the plane *AB* [we would call it *BE*], and one asks its *momentum* in that situation or the force able to sustain it in that plane *AB*. […] I say that the *momentum* of the weight *A*, or force that supports it in the plane *AB* is *BD*.

Fig. 12 Torricelli's construction to get the momentum of the body A (Torricelli 1644, p. 103). *Source* Courtesy University of Oklahoma, History of Science Collections

The triangles *CBD* and *BEF* are similar[46] so that *EB* is to *EF* as *BD* is to *BC* and the *momentum* in the plane *BE* is as the line *BD*.

$$\frac{M_{Total}}{M_{BC}} = \frac{EB}{EF} = \frac{BC}{BD} \tag{8}$$

But why does Torricelli do all this?

4.6 Torricelli's *Scholium* to *Proposition II*

Torricelli claims to be able to prove the sixth proposition of Galileo's *de motu naturaliter accelerato* without using Galileo's supposition on the equality of the speeds at the bottom of equally elevated but differently inclined planes (see above) (Fig. 13).

[46] "Sit grave in plano *AB*, & quaeritur in hoc situ momentum eius; sive potentia, a qua in hoc plano *AB* sustenitur ... Dico momentum gravis *A*, sive potentiam quae illud sustinet in plano *AB* esse *BD*." (Torricelli 1644, p. 103; Author's translation).

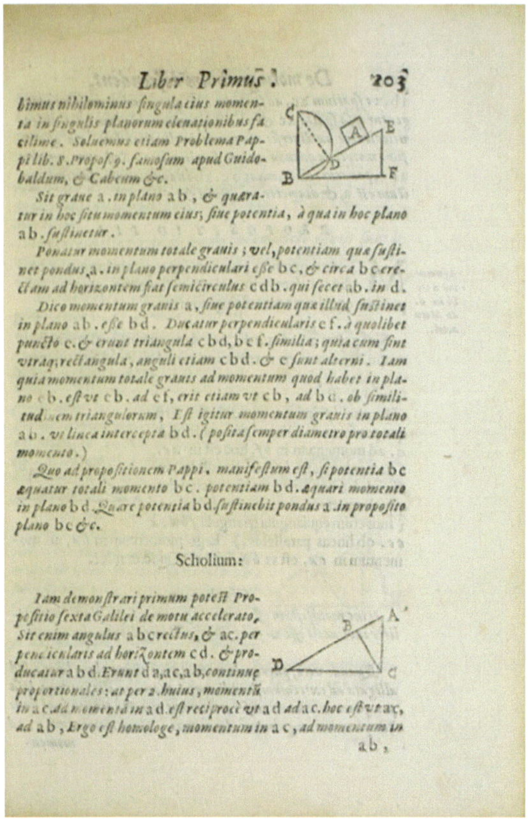

Liber Primus. 103

bimus nihilominus singula eius momenta in singulis planorum elevationibus facilime. Soluemus etiam Problema Pappi lib. 8. Propos. 9. famosum apud Guidobaldum, & Cabeum &c.

Sit grave a. in plano ab, & quaeratur in hoc situ momentum eius; sive potentia, à qua in hoc plano ab. sustinetur.

Ponatur momentum totale gravis; vel, potentiam quae sustinet pondus a. in plano perpendiculari esse bc, & circa bc erectam ad horizontem fiat semicirculus cdb. qui secet ab. in d.

Dico momentum gravis a, sive potentiam quae illud sustinet in plano ab. esse bd. Ducatur perpendicularis cf. à quolibet puncto c. & erunt triangula cbd, bcf. similia; quia cum sint utraq; rectangula, anguli etiam cbd. & c sunt alterni. Iam quia momentum totale gravis ad momentum quod habet in plano cb. est ut cb. ad cf, erit etiam ut cb, ad bd. ob similitudinem triangulorum, Est igitur momentum gravis in plano ab. ut linea intercepta bd. (posita semper diametro pro totali momento.)

Quo ad propositionem Pappi, manifestum est, si potentia bc aequatur totali momento bc. potentiam bd. aequari momento in plano bd. Quare potentia bd. sustinebit pondus a. in proposito plano bc &c.

Scholium:

Iam demonstrari primum potest Propositio sexta Galilei de motu accelerato, Sit enim angulus abc rectus, & ac. perpendicularis ad horizontem cd. & producatur abd. Erunt da, ac, ab, continue proportionales: at per 2. huius, momentum in ac ad momentum in ad. est reciproce ut ad ad ac. hoc est ut ac, ad ab, Ergo est homologe, momentum in ac, ad momentum in ab,

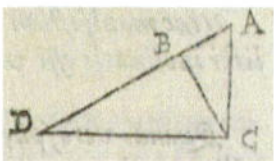

Scholium. We are now able to give a first proof of the sixth Proposition of Galileo's *de motu accelerato*. Let thus *ABC* (Fig. 13) be a right angle, and *AC* perpendicular to the horizontal *CD*, and let us draw *ABD*. *DA*, *AC* and *AB* will be in a continuous geometrical proportion; but proposition II says that, the *moment* on *AC* is to the *moment* on *AD* reciprocally as *AD* is to *AC*, or [as *DA*, *AC* and *AB* are in a continuous proportion] as *AC* is to *AB*. Thus, in a similar way, the *moment* on *AC* is to the *moment* on *AB* as the space *AC* to the space *AB*. Thus those spaces *AC* and *AB* will be gone through in the same time, supposing here, as does Galileo himself, that the velocities on planes differently inclined are as the momenta if the weights are equal. But *ABC* is a right angle, *BC* and *AB* are chords of a circle which highest point is *A* and diameter is *AC*. Thus those spaces *AC* and *AB* will be gone through in the same time, supposing here, as does Galileo himself, that the velocities on planes differently inclined are as the momenta if the weights are equal. But *ABC* is a right angle, *BC* and *AB* are chords of a circle which highest point is A and diameter is *AC*. CQFD.[47]

Fig. 13 The picture Torricelli draws to prove Galilei's sixth Proposition in *De Motu* (Torricelli 1644, p. 103). *Source* Courtesy University of Oklahoma, History of Science Collections

As *DA*, *AC* and *AB* are in a continuous proportion, we have:

$$DA\,/\,AC = AC\,/\,AB.$$

This is a rather common theorem concerning the height of a rectangular triangle. We will meet it several times below. The proposition certainly concerns the Galilean *momentum* exclusively. It says that velocities on various planes, differently inclined, are as the "Galilean momenta" if the *moles* or weights are equal. Velocities having the Aristotelian meaning of mean velocity. Galileo uses "degree of velocity" and later in his text, Torricelli will also use the same phrase for the velocity acquired at the bottom of an inclined plane, making a difference between the mean velocity along a certain distance of Aristotle and the velocity in one point, but a particular one, the end of the fall.

Now we could try to formulate the previous sentences as follows.

The *Proposition II* says that, for a same weight on differently inclined planes but having the same height, $M_{AC}/M_{AD} = AD/AC$, but, geometry tells us that $DA/AC = AC/AB$ thus $M_{AC}/M_{AD} = AC/AB$. But $M_{AB} = M_{AD}$ because *AB* and *AD* are on the same plane, thus $M_{AC}/M_{AB} = AC/AB$. Torricelli deduces from the last analogy that *AC* and *AB* are gone through in the same time. This must be explained. He does it "[…] supposing here, as does Galileo himself, that the velocities on differently inclined planes are as the momenta if their weights are equal […]".

[47] "Scholium. Iam demonstrari primum potest Propositio sexta Galilei de motu accelerato. Sit enim angulus *ABC* rectus & ac perpendicularis ad horizontem *CD* & producatur *ABD*. Erunt *DA*, *AC*, *AB* continue proportionales: at per II huius, momentum in *AC* ad momentum in *AD* est reciproce ut *AD* ad *AC*, hoc est ut *AC* ad *AB*. Ergo est homologe, momentum in *AC*, ad momentum in *AB*, ut spatium *AC* ad spatium *AB*; quare eodem tempore peragentur ipsa spatia *AC* et *AB*. Supponimus hic cum ipso Galileo, velocitates in diversis planorum inclinationibus, ita esse ut sunt momenta quando eadem fuerit moles. Sed cum angulus *ABC* ponatur rectus, erunt *BC*, *AB* in circulo cuius sublime punctum est *A*, & diameter *AC*. Quod &c." (Torricelli 1644, p. 103; Author's translation).

Then if their weights are equal, their *momenta* are as the velocities and if their *momentum* also are as spaces, times must be equal, because $v = Space/Time$.

$$\frac{M_{AC}}{M_{BA}} = \frac{Sp_{AC}}{Sp_{BA}} \times \frac{T_{BA}}{T_{AC}} \tag{9}$$

and as we already know that the *momenta* are as the lengths, the times must be equal. Conclusion, if the weights are equal: $T_{AC} = T_{AB}$. The last sentence

> But as angle *ABC* is a right angle, *BC* and *AB* are on a circle which highest point is *A*, and diameter[48] *AC*.

makes the link with Galileo's expression of the same theorem. Galileo's theorem is usually called the "theorem of chords" because he considers the cords of a circle as Torricelli shows in that last sentence. Torricelli will come back to Galileo's circle later.

4.7 Torricelli's Proposition *III*: Proof of His *Praemittimus* Key to the Proof of Galileo's Assumption

> Momentum of equal weights on unequally inclined planes are in homologous ratio with the perpendiculars corresponding to equal parts of their planes.[49]

Let *a* and *e* be two equal weights laying on two differently inclined planes [Fig. 14.] *CA* and *BA*, and let *CE* and *BD* be the perpendiculars corresponding to equal parts *AC* and *AB* of those planes.

If

$$W_{AC} = W_{AB}$$

then we have

$$M_{AC} / M_{AB} = CE / BD.$$

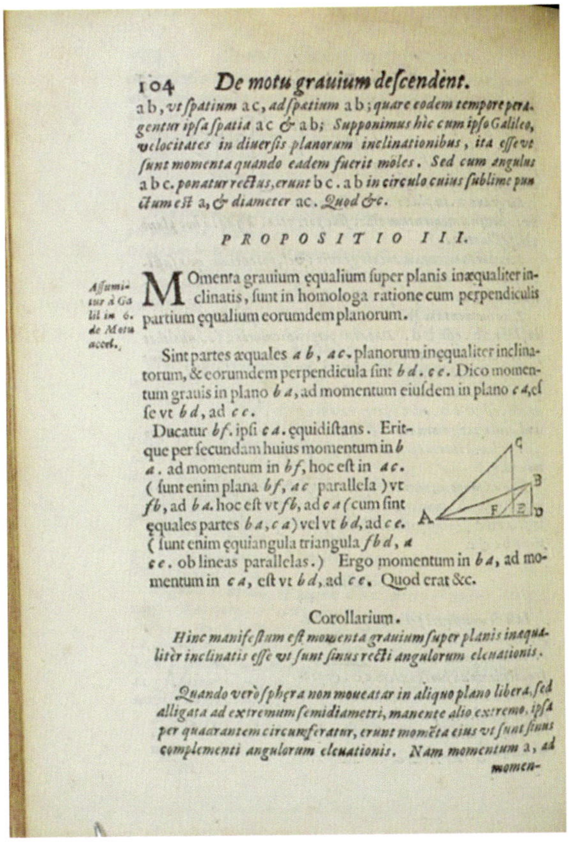

104 De motu grauium descendent.

ab, vt spatium ac, ad spatium ab; quare eodem tempore peragentur ipsa spatia ac & ab; Supponimus hic cum ipso Galileo, velocitates in diuersis planorum inclinationibus, ita esse vt sunt momenta quando eadem fuerit moles. Sed cum angulus abc. ponatur rectus, erunt bc. ab in circulo cuius sublime punctum est a, & diameter ac. Quod &c.

PROPOSITIO III.

Assumitur à Galil in 6. de Motu accel.

Momenta grauium ęqualium super planis inæqualiter inclinatis, sunt in homologa ratione cum perpendiculis partium ęqualium eorumdem planorum.

Sint partes æquales ab, ac. planorum inęqualiter inclinatorum, & eorumdem perpendicula sint bd. ce. Dico momentum grauis in plano ba, ad momentum eiusdem in plano ca, esse vt bd, ad ce.

Ducatur bf. ipsi ca. ęquidistans. Eritque per secundam huius momentum in ba. ad momentum in bf, hoc est in ac. (sunt enim plana bf, ac parallela) vt fb, ad ba. hoc est vt fb, ad ca (cum sint ęquales partes ba, ca) vel vt bd, ad ce. (sunt enim ęquiangula triangula fbd, ace. ob lineas parallelas.) Ergo momentum in ba, ad momentum in ca, est vt bd, ad ce. Quod erat &c.

Corollarium.

Hinc manifestum est momenta grauium super planis inequaliter inclinatis esse vt sunt sinus recti angulorum elevationis.

Quando verò sphęra non moueatur in aliquo plano libera, sed alligata ad extremum semidiametri, manente alio extremo, ipsa per quadrantem circumferatur, erunt momẽta eius vt sunt sinus complementi angulorum elevationis. Nam momentum a, ad momen-

Fig. 14 Torricelli's construction to *Proposition III* (Torricelli 1644, p. 104). *Source* Courtesy University of Oklahoma, History of Science Collections

Torricelli adds in the margin:

> Admitted by Galileo in the 6th prop. of the de motu accelerato.[50]

[48] "Sed cum angulus *abc* ponatur rectus, erunt *BC*, ab in circulo cuius sublime punctum est *A*, & diameter *AC*." (Torricelli 1644, p. 104; Author's translation).

[49] "Momenta gravium aequalium super planis inaequaliter inclinatis, sunt homologa ratione cum perpendiculis partium aequalium eorumdem planorum." (Torricelli 1644, p. 104; Author's translation).

[50] "Assumitur a Galil. In 6. De motu accel." (Torricelli 1644, p. 104; Author's translation).

But comparing Torricelli's formulation with Galileo's we observe that it is a bit different. Galileo says:

I admit that the degrees of velocity a mobile acquires on unequally inclined planes are equal when the elevations of those planes are equal.[51]

In fact, there are two differences namely, first Torricelli calls *Momentum* what Galileo calls "acquired degrees of velocity" and this is consistent with the words *celeritatis momenta* Galileo uses in his definition of uniformly accelerated motion given supra. Thus, Torricelli uses here exactly the same word *momentum* for both *Archimedian momentum* and Galilean *celeritatis momentum*. The reason has to be found in the proof of Torricelli's *Lemma* to *Proposition II* as we showed before.

The second difference is that Torricelli's formulation is more general. He says that the *momenta* are in the same ratio as the elevations of the planes when Galileo only says that they are equal when their elevation is the same. To prove his Proposition *III*, Torricelli uses Fig. 14 with its two unequally inclined planes having the same length $AC = AB$. He knows, by proposition II, that $M_{BA}/M_{BF} = M_{BA}/M_{AC}$, because BF is parallel to AC. So that $M_{AC}/M_{AB} = BA/FB = CA\,/CA = CE/BD$.

4.8 Torricelli's *corollarium* to Proposition *III*

This is a crucial point! We saw that Galileo was in fact looking for a function that would have helped him to express his *momentum* in "the first point" on any inclined plane in a more elegant way. Here Torricelli shows the way to do it using the sine function. Torricelli proceeds and explains his expression using the *Archimedean notion of momentum* (Fig. 15).

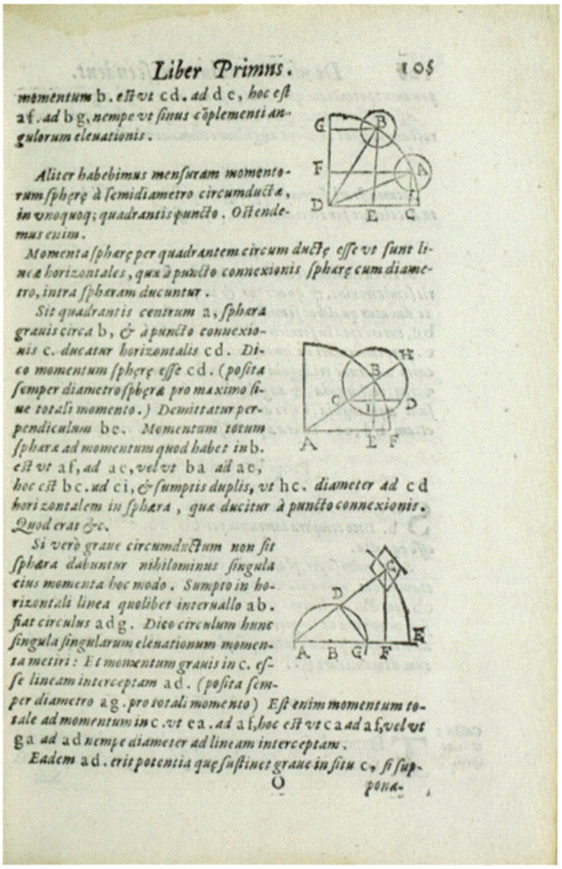

Liber Primus. 105

momentum b. est ut cd. ad dc, hoc est af. ad bg, nempe ut sinus cõplementi angulorum eleuationis.

Aliter habebimus mensuram momentorum sphærę à semidiametro circumductæ, in unoquoq; quadrantis puncto. Ostendemus enim.

Momenta sphærę per quadrantem circum ductę esse ut sunt lineæ horizontales, quæ à puncto connexionis sphærę cum diametro, intra sphæram ducuntur.

Sit quadrantis centrum a, sphæra grauis circa b, & à puncto connexionis c. ducatur horizontalis cd. Dico momentum sphærę esse cd. (posita semper diametro sphęræ pro maximo siue totali momento.) Demittatur perpendiculum bc. Momentum totum sphæræ ad momentum quod habet in b. est ut af, ad ac, vel ut ba ad ac, hoc est bc. ad ci, & sumptis duplis, ut hc. diameter ad cd horizontalem in sphæra, quæ ducitur à puncto connexionis. Quod erat &c.

Si verò graue circumductum non sit sphæra dabuntur nihilominus singula eius momenta hoc modo. Sumpto in horizontali linea quolibet internallo ab. fiat circulus adg. Dico circulum hunc singula singularum eleuationum momenta metiri: Et momentum grauis in c. esse lineam interceptam ad. (posita semper diametro ag. pro totali momento) Est enim momentum totale ad momentum in c. ut ea. ad af, hoc est ut ca ad af, vel ut ga ad ad nempe diameter ad lineam interceptam.

Eadem ad. erit potentia quę sustinet graue in situ c, si suppona-

O

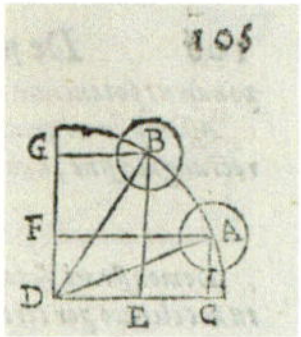

Corollary: It is now evident that the *momentum* of weights lying on unequally inclined planes are as the *sinus recti* of the angle of elevation.

We could write (Fig. 15)

$$M_B\,/\,M_A = \sin BDE\,/\sin ADC.$$

Fig. 15 How to calculate the momentum of two spheres, according to Torricelli (Torricelli 1644, p. 105). *Source* Courtesy University of Oklahoma, History of Science Collections[52]

---

[51] "Accipio, gradus velocitatis ejusdem mobilis super diversarum planorum inclinationes acquisitos tunc esse aequales, cum eorumdem planorum elevations aequales sint." (Galilei 1638, VIII, p. 205; Author's translation).

[52] "Corollarium: Hinc manifestum est momenta gravium super planis inaequaliter inclinatis esse ut sinus recti angulorum elevationis." (Torricelli 1644, p. 104; Author's translation).

When the sphere is not moved freely in some plane, but is linked at the extremity of a semi–diameter, fixed at the other extremity, and moved around by it on a quarter of a circle, its momentum are as the *sines* of the complement of the angles of elevation. But momentum *A* [Fig. 15] is to momentum *B* is as *CD* to *DE*, that is as *AF* to *BG*, or as the sines of the complements of the elevations.[53]

It becomes clearer and clearer that for him both notions, *Archimedean* and *Galilean momenta* are equivalent, if point after point you consider the motion as a motion in a first instant.

In what follows note the words in *unoquoque quadrantis puncto*. For him, the *sine* is a function that gets different values in every point of the quarter of the circle, thus defined on the quarter of circle. The importance this fact has for him, justifies the rather strange corollaries he gave just before. They give the possibility of using the sine "function". Of course Torricelli doesn't have a full notion of function, but he sees that in this problem for any two points of the quarter of the circle, the *momenta* are proportional to the sines of the angles defined by those point and the horizontal *DC*. From the point of view of the notion of function, this is an important step (Figs. 16 and 17).

We will have a different measure of the *momentum* of the sphere moved around by the semi–diameter, in every point of the quadrant. We thus show that: The momentum of the sphere moved around the quadrant are as the horizontal line, that is drawn from the point of intersection [*C*] of the sphere with the diameter.[54]

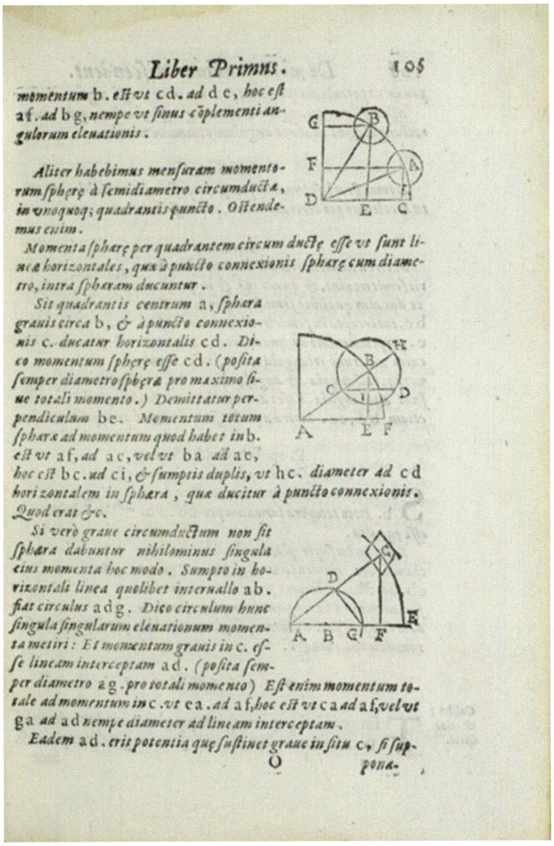

Liber Primus. 105

momentum b. est ut cd. ad dc, hoc est af. ad bg, nempe ut sinus cõplementi angulorum elevationis.

Aliter habebimus mensuram momentorum sphęrę à semidiametro circumductæ, in unoquoq; quadrantis puncto. Ostendemus enim.

Momenta sphærę per quadrantem circum ductę esse ut sunt lineæ horizontales, quæ à puncto connexionis sphærę cum diametro, intra sphæram ducuntur.

Sit quadrantis centrum a, sphæra gravis circa b, & à puncto connexionis c. ducatur horizontalis cd. Dico momentum sphęrę esse cd. (posita semper diametro sphęræ pro maximo sive totali momento.) Demittatur perpendiculum be. Momentum totum sphæræ ad momentum quod habet in b. est ut af, ad ac, vel ut ba ad ae, hoc est bc. ad ci, & sumptis duplis, ut hc. diameter ad cd horizontalem in sphæra, quæ ducitur à puncto connexionis. Quod erat &c.

Si verò grave circumductum non sit sphæra dabuntur nihilominus singula eius momenta hoc modo. Sumpto in horizontali linea quolibet intervallo ab. fiat circulus adg. Dico circulum hunc singula singularum elevationum momenta metiri: Et momentum gravis in c. esse lineam interceptam ad. (posita semper diametro ag. pro totali momento) Est enim momentum totale ad momentum in c. ut ca. ad af, hoc est ut ca ad af, vel ut ga ad ad nempe diameter ad lineam interceptam.

Eadem ad. erit potentia quę sustinet grave in situ c, si suppona-

O

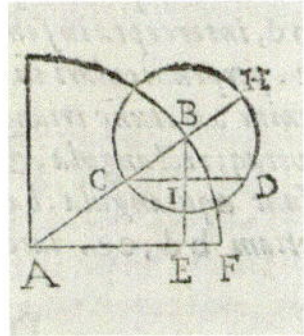

In the Fig. 16 the *momentum* of the sphere, whose centre is *B*, is *CD*, assuming the diameter *CH* to be the "maximal or total momentum". This means that the *momentum on a vertical plane* for Torricelli, as well as for Galileo, and the *momentum* of a weight hanging to a lever whose arms are horizontal, Archimedean *momentum* as I called it, are the same. So the total *momentum* of the sphere is to the *momentum* it has on *AH* as *AF* to *AE*, or as *BC* to *CI* or *CH* to *CD*.

Fig. 16 The way Torricelli defines *CD* as the momentum of a sphere (Torricelli 1644, p. 105). *Source* Courtesy University of Oklahoma, History of Science Collections

$$\frac{M_{Total}}{M_{AH}} = \frac{AF}{AE} = \frac{BC}{CI} = \frac{CH}{CD} \tag{10}$$

[53] "Quando vero sphaera non moveatur in aliquo plano libera, sed alligata ad extremum semidiametri, manente alio extremo, ipsa per quadrantem circumferatur, erunt momenta eius ut sunt sinus complementi angulorum elevationis. Nam momentum *A*, ad momentum *B* est ut *CD* ad *DE*, hoc est [ut] *AF* ad *BG*, nempe ut sinus complementi angulorum elevationis." (Torricelli 1644, pp. 104–105; Author's translation).

[54] "Aliter habebimus mensuram momentorum sphaerae a semidiametro circumductae, in unoquoque quadrantis puncto. Ostendemus enim. Momenta sphaerae quadrantem circum ductae esse ut linea horizontales, quae a puncto connexionis sphaerae cum diametro, intra sphaeram ducuntur." (Torricelli 1644, p. 105; Author's translation).

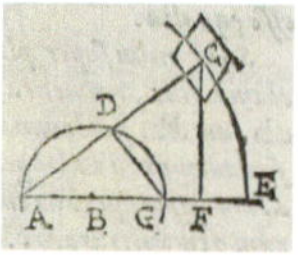

Here also as in the previous proposition, Torricelli gives another expression valid also for weights that are not spherical (Fig. 17). He introduces another circle, whose diameter corresponds to the *total momentum*, in order to read the *sinus rectus AD.* In his Scholium, Torricelli is now able to give another proof of the sixth proposition of Galileo's *De motu accelerato* but before, he adds a Lemma.

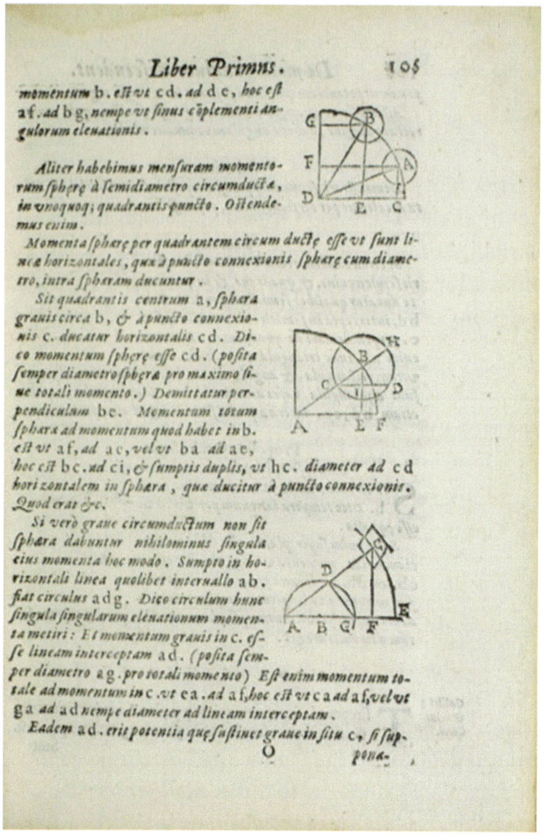
Liber Primus. 105

momentum b. est vt cd. ad dc, hoc est af. ad bg, nempe vt sinus cõplementi angulorum elevationis.

Aliter habebimus mensuram momentorum sphęrę à semidiametro circumducta, in vnoquoq; quadrantis puncto. Ostendemus enim.

Momenta sphęrę per quadrantem circum ductę esse vt sunt lineæ horizontales, quæ à puncto connexionis sphęrę cum diametro, intra spheram ducuntur.

Sit quadrantis centrum a, sphæra grauis circa b, & à puncto connexionis c. ducatur horizontalis cd. Dico momentum sphęrę esse cd. (posita semper diametro sphęræ pro maximo siue totali momento.) Demittatur perpendiculum bc. Momentum totum sphæræ ad momentum quod habet in b. est vt af, ad ac, vel vt ba ad ae, hoc est bc. ad ci, & sumptis duplis, vt hc. diameter ad cd horizontalem in sphæra, quæ ducitur à puncto connexionis. Quod erat &c.

Si verò graue circumductum non sit sphæra dabuntur nihilominus singula eius momenta hoc modo. Sumpto in horizontali linea quolibet internallo ab. fiat circulus adg. Dico circulum hunc singula singularum eleuationum momenta metiri: Et momentum grauis in c. esse lineam interceptam ad. (posita semper diametro ag. pro totali momento) Est enim momentum totale ad momentum in c. vt ca. ad af, hoc est vt ca ad af, vel vt ga ad ad nempe diameter ad lineam interceptam.

Eadem ad. erit potentia quę sustinet graue in situ c, si suppona-

O

Fig. 17 The way Torricelli evaluates the momentum of a non-spherical weight (Torricelli 1644, p. 105). *Source* Courtesy University of Oklahoma, History of Science Collections

4.9 Torricelli's *Lemma* to *Proposition III*

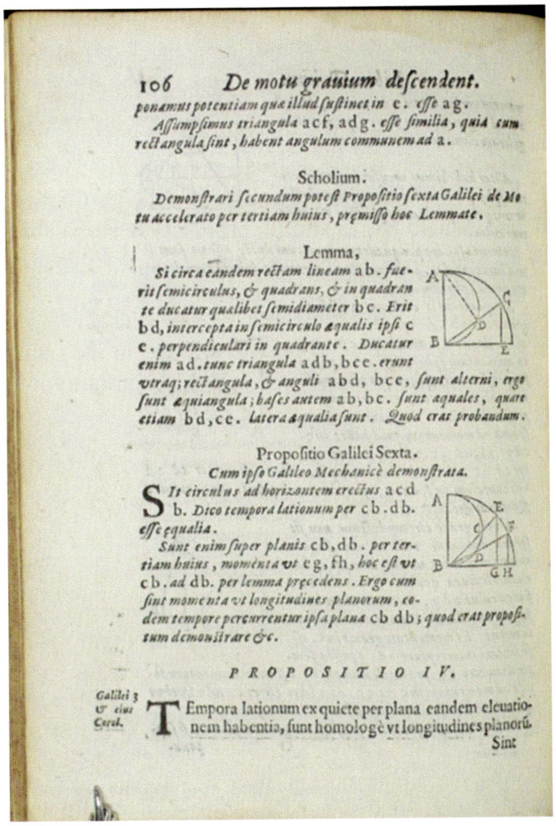

106 *De motu grauium descendent.*

ponamus potentiam quæ illud sustinet in c. *esse* ag.

Assumpsimus triangula acf, adg. *esse similia, quia cum rectangula sint, habent angulum communem ad* a.

Scholium.

Demonstrari secundum potest Propositio sexta Galilei de Motu accelerato per tertiam huius, præmisso hoc Lemmate.

Lemma,

Si circa eandem rectam lineam ab. *fuerit semicirculus, & quadrans, & in quadrante ducatur qualibet semidiameter* bc. *Erit* bd, *intercepta in semicirculo æqualis ipsi* ce. *perpendiculari in quadrante. Ducatur enim* ad. *tunc triangula* adb, bce. *erunt vtraq; rectangula, & anguli* abd, bce, *sunt alterni, ergo sunt æquiangula; bases autem* ab, bc. *sunt æquales, quare etiam* bd, ce. *latera æqualia sunt. Quod erat probandum.*

Propositio Galilei Sexta.

Cum ipso Galileo Mechanicè demonstrata.

S*It circulus ad horizontem erectus* acd b. *Dico tempora lationum per* cb. db. *esse æqualia.*

Sunt enim super planis cb, db. *per tertiam huius, momenta vt* eg, fh, *hoc est vt* cb. *ad* db. *per lemma præcedens. Ergo cum sint momenta vt longitudines planorum, eodem tempore percurrentur ipsa plana* cb db; *quod erat propositum demonstrare &c.*

PROPOSITIO IV.

Galilei 3 & eius Corol. TEmpora lationum ex quiete per plana eandem eleuationem habentia, sunt homologè vt longitudines planorũ.

Sint

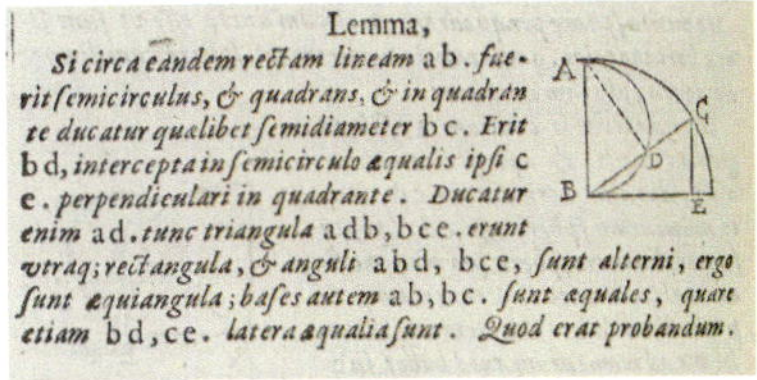

Lemma,

Si circa eandem rectam lineam ab. *fuerit semicirculus, & quadrans, & in quadrante ducatur qualibet semidiameter* bc. *Erit* bd, *intercepta in semicirculo æqualis ipsi* ce. *perpendiculari in quadrante. Ducatur enim* ad. *tunc triangula* adb, bce. *erunt vtraq; rectangula, & anguli* abd, bce, *sunt alterni, ergo sunt æquiangula; bases autem* ab, bc. *sunt æquales, quare etiam* bd, ce. *latera æqualia sunt. Quod erat probandum.*

Lemma. Let *BD*, intercepted by the half circle, equal to *CE* perpendicular in the quadrant. Tracing *AD*, the triangles *ADB* and *BCE* are both rectangular, and the angles *ABD*, *BCE* are equal; as the bases *AB*, *BC* are equal, the sides *BD*, *CE* are also equal.

Thus

$$BC / CE = AB / BD.$$

In addition, in the Fig. 18, both triangles *ABD* and *BEC* are equal and so are *CE* and *BD*.

Fig. 18 Geometrical description for Lemma in *Proposition III* of Torricelli's *Opera Geometrica* (Torricelli 1644, p. 106). *Source* Courtesy University of Oklahoma, History of Science Collections[55]

4.10 Torricelli's Second Part of the Proof of Galileo's Theorem of Chords

Torricelli gives then the second part of his proof of the sixth proposition of Galileo's *de motu accelerato* which says:

> If from the highest or lowest point in a vertical circle there be drawn any inclined planes meeting the circumference, the times of descent along these chords are equal to the other.[56]

> Torricelli's proof is the following (Fig. 19):

Galileo's sixth proposition, that Galileo himself proves mechanically.[57]

[55] Lemma. "Erit *BD*, intercepta in semicirculo aequalis ipsi *CE* perpendiculari in quadrante. Ducatur enim *AD* tunc triangula *ADB*, *BCE* erunt utraque rectangula, et anguli *ADB*, *BCE*, sunt alterni, ergo sunt aequiangula; bases autem *AB*, *BC* sunt aequales, quare etiam *BD*, *CE*." (Torricelli 1644, p. 106; Author's translation).

[56] "Si a puncto sublimi, vel imo circuli ad horizontem electi ducantur qualibet plana usque ad circumferentiam inclinata, tempora descensuum per ipsa erunt aequalia." (Galilei 1638, VIII, p. 221; Author's translation).

[57] "Propositio Galilei sexta. Cum ipso Galileo Mechanice proves. Sit circulus ad horizontem erectus *ACDB*. Dico tempora lationum per *CB*, *DB* esse aequalia. Sunt enim super planis *CB*, *DB* per tertiam huius, momenta ut *EG*, *FH* hoc est ut *CB* ad *DB* per lemma praecedens. Ergo cum sint momenta ut longitudines planorum, eodem tempore percurrentur ipsa plana *CB*, *DB*; quod erat propositum demonstrare &c." (Torricelli 1644, p. 106; Author's translation).

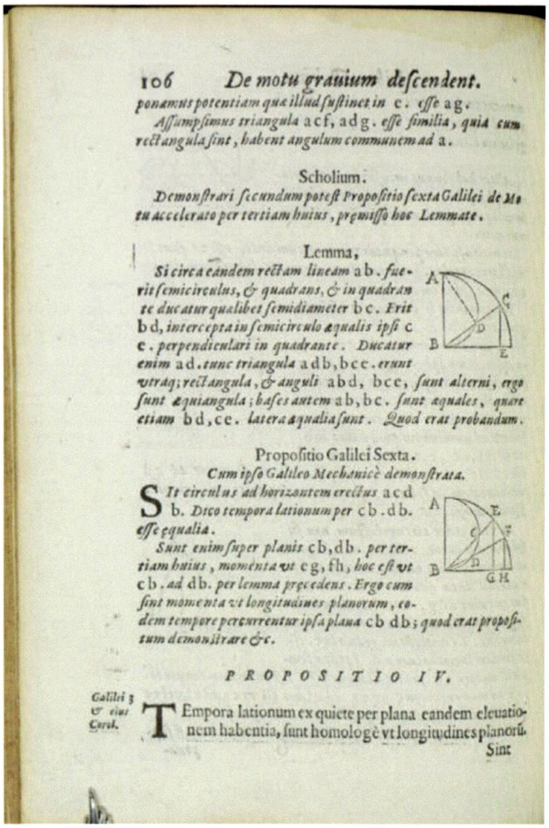

106 *De motu grauium descendent.*

ponamus potentiam quæ illud sustinet in c. *esse* ag.

Assumpsimus triangula acf, adg. *esse similia, quia cum rectangula sint, habent angulum communem ad* a.

Scholium.

Demonstrari secundum potest Propositio sexta Galilei de Motu accelerato per tertiam huius, præmisso hoc Lemmate.

Lemma,

Si circa eandem rectam lineam ab. *fuerit semicirculus, & quadrans, & in quadrante ducatur qualibet semidiameter* bc. *Erit* bd, *intercepta in semicirculo æqualis ipsi* ce. *perpendiculari in quadrante. Ducatur enim* ad. *tunc triangula* adb, bce. *erunt vtraq; rectangula, & anguli* abd, bce, *sunt alterni, ergo sunt æquiangula; bases autem* ab, bc. *sunt æquales, quare etiam* bd, ce. *latera æqualia sunt. Quod erat probandum.*

Propositio Galilei Sexta.

Cum ipso Galileo Mechanicè demonstrata.

S*It circulus ad horizontem erectus* acd b. *Dico tempora lationum per* cb. db. *esse æqualia.*

Sunt enim super planis cb, db. *per tertiam huius, momenta vt* eg, fh, *hoc est vt* cb. *ad* db. *per lemma præcedens. Ergo cum sint momenta vt longitudines planorum, eodem tempore percurrentur ipsa plana* cb db; *quod erat propositum demonstrare &c.*

PROPOSITIO IV.

Galilei 3 & eius Coral.

TEmpora lationum ex quiete per plana eandem eleuationem habentia, sunt homologè vt longitudines planorũ.

Sint

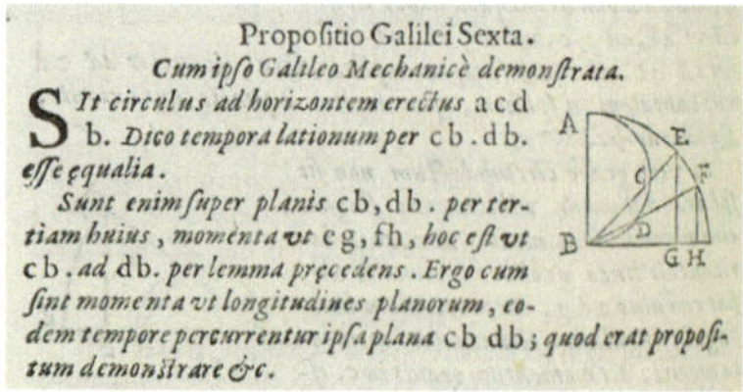

Propositio Galilei Sexta.

Cum ipso Galileo Mechanicè demonstrata.

S*It circulus ad horizontem erectus* acd b. *Dico tempora lationum per* cb. db. *esse æqualia.*

Sunt enim super planis cb, db. *per tertiam huius, momenta vt* eg, fh, *hoc est vt* cb. *ad* db. *per lemma præcedens. Ergo cum sint momenta vt longitudines planorum, eodem tempore percurrentur ipsa plana* cb db; *quod erat propositum demonstrare &c.*

Let *ACDB* (Fig. 19) be a circle perpendicular to the horizon. I say that the times of fall through *CB* and *DB* are equal. By the third proposition supra, the momentum on planes *CB*, *DB* are as *EG*, *FH*, that is as *CB* to *DB* by the preceding lemma. So as the momentum are as the length of the planes, those planes *CB* and *DB* will be gone through in the same time. (Torricelli 1644, p. 106; Author's translation).

Therefore we have:

$$\frac{M_{CB}}{M_{DB}} = \frac{EG}{FH} = \frac{CB}{DB} \tag{11}$$

Fig. 19 The construction Torricelli draws to prove the sixth Proposition of Galilei (Torricelli 1644, p. 106). *Source* Courtesy University of Oklahoma, History of Science Collections

Once more:

$$M_{CB}/M_{DB} = (e_{CB}/e_{DB}) \times (t_{DB}/t_{CB})$$

and as we just showed that the *momenta* are as the spaces, the times must be equal.

4.11 Torricelli's *Proposition IV*

The times of fall, starting from rest along planes having the same elevation are homologue to the lengths of the planes.[58]

Given two planes *AB* and *AC* (Fig. 20) having different inclinations but the same elevation, let the times of fall be T_{AB}, T_{AC}, then

[58] "Tempora lationum ex quiete per plana eandem elevationem habentia, sunt homologe ut longitudines planorum." (Torricelli 1644, pp. 106–107; Author's translation).

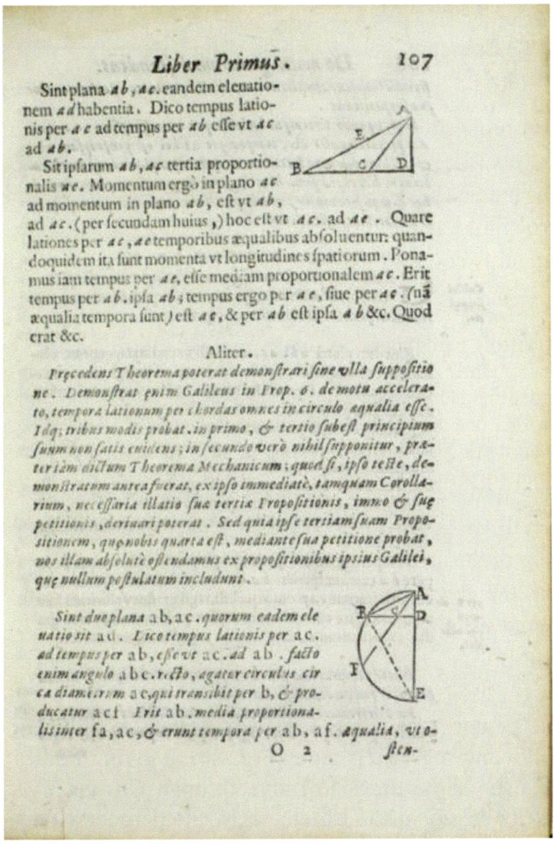

Liber Primus. 107

Sint plana *ab*, *ac*. eandem eleuationem *ad* habentia. Dico tempus lationis per *ac* ad tempus per *ab* eſſe vt *ac* ad *ab*.

Sit ipſarum *ab*, *ac* tertia proportionalis *ae*. Momentum ergò in plano *ac* ad momentum in plano *ab*, eſt vt *ab*, ad *ac*. (per ſecundam huius,) hoc eſt vt *ac*. ad *ae*. Quare lationes per *ac*, *ae* temporibus æqualibus abſoluentur: quandoquidem ita ſunt momenta vt longitudines ſpatiorum. Ponamus iam tempus per *ae*, eſſe mediam proportionalem *ac*. Erit tempus per *ab*. ipſa *ab*; tempus ergo per *ac*, ſiue per *ae*. (nã æqualia tempora ſunt) eſt *ac*, & per *ab* eſt ipſa *ab* &c. Quod erat &c.

Aliter.

Præcedens Theorema poterat demonſtrari ſine vlla ſuppoſitione. Demonſtrat enim Galileus in Prop. 6. de motu accelerato, tempora lationum per chordas omnes in circulo æqualia eſſe. Idq; tribus modis probat. in primo, & tertio ſubeſt principium ſuum non ſatis euidens; in ſecundo verò nihil ſupponitur, præter iam dictum Theorema Mechanicum; quod ſi, ipſo teſte, demonſtratum antea fuerat, ex ipſo immediatè, tamquam Corollarium, neceſſaria illatio ſuæ tertiæ Propoſitionis, immo & ſuæ petitionis, deriuari poterat. Sed quia ipſe tertiam ſuam Propoſitionem, quæ nobis quarta eſt, mediante ſua petitione probat, nos illam abſolutè oſtendamus ex propoſitionibus ipſius Galilei, quæ nullum poſtulatum includunt.

Sint due plana ab, ac. *quorum eadem eleuatio ſit* ad. *Dico tempus lationis per* ac. *ad tempus per* ab, *eſſe vt* ac. *ad* ab. *facto enim angulo* abc. *recto, agatur circulus circa diametrum* ac, *qui tranſibit per* b, *& producatur* acf. *Erit* ab. *media proportionalis inter* fa, ac, *& erunt tempora per* ab, af. *æqualia, vt oſten-*

O 2

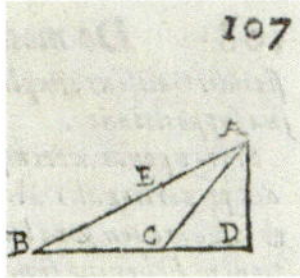

$$\frac{T_{AC}}{T_{AB}} = \frac{AC}{AB} \tag{12}$$

Fig. 20 The picture Torricelli uses to prove the Eq. (11) (Torricelli 1644, p. 107). *Source* Courtesy University of Oklahoma, History of Science Collections

Which is Galileo *Proposition III* as Torricelli adds in a *marginalia*: *Galilei III & eius Corol.* Galileo's *Proposition III* is (Fig. 21)[59]:

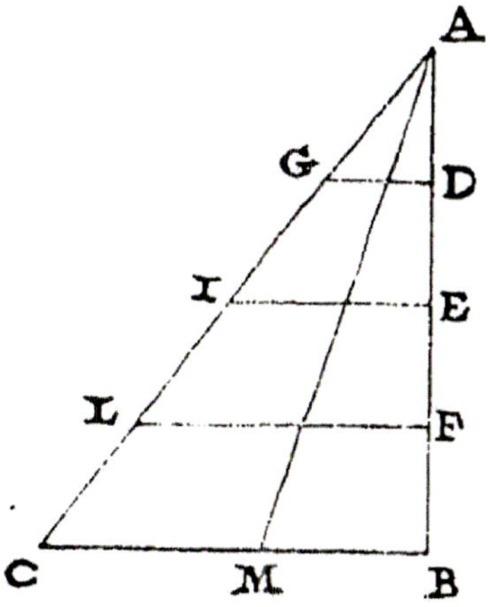

Proposition *III*: If one and the same body (Fig. 21) starting from rest, falls along an inclined plane and also along a vertical, each having the same height, the times of fall will be to each other as the lengths of the inclined plane and the vertical.
Corollary: Hence we may infer that the times of descent along planes having different inclinations, but the same vertical height stand to one another in the same ratio as the lengths of the planes.

$$\frac{T_{ac}}{T_{ab}} = \frac{ac}{ab} \quad \text{and} \quad \frac{T_{am}}{T_{ab}} = \frac{am}{ab} \tag{13}$$

Fig. 21 Galilei's construction for *Theorema III* (Galilei 1638, VIII, p. 216). *Source* Gallica, Bibliothèque nationale de France. Public Domain

[59] "Propositio *III*: Si super plano inclinato, atque in perpendiculo, quorum eadem sit altitudo, feratur ex quiete idem mobile; tempora lationum erunt inter se, ut plani ipsius, & perpendiculi longitudines. Corollarium: Hinc colligitur, tempora descensuum super planis diversimode inclinatis, dum tamen eorum eadem sit elevatio, esse inter se, ut eorum longitudines." (Galilei 1638, VIII, p. 215; Author's translation).

The Corollary is:

$$\frac{T_{ac}}{T_{am}} = \frac{ac}{am} \tag{14}$$

Torricelli takes the point *E* (Fig. 20) such that *AE* is the third proportional to *AB* and *AC*, thus AB/AC = AC/AE and using the same theorem as in his scholium to *Proposition II,* he gets that *EA* = *AD* the height of the triangle *ABD*. But we know by his *Proposition II* that $M_{AB}/M_{AC} = AB/AC = AC/AE$ thus the body goes down through *AB* and *AC* in the same time, when the *momenta* are as the length of the planes. If the time through *AE* is the proportional mean *AE*. Let the time through *AB* be *AB*, then the time through *AE* or *AC*, because they are equal, is equal to *AC*. Torricelli adds a different proof that is based on Galileo's theorem of cords. In another way:

> It is possible to prove the precedent theorem without any supposition. Galileo's sixth proposition of the *de motu accelerato* proves indeed that times of fall along all chords of a circle are equal. And he proves it from three different manners. In the first and in the third ones, he uses his principle that is not sufficiently evident; in the second one he does not suppose anything despite his mechanical theorem as we already qualified it; but if, by his own witness, it was already proven before, one may immediately conclude out of it, as a corollary, the necessity of his third proposition as well as of its petition. But since the third of his propositions, that is our fourth, implies its petition, we show it [the petition] perfectly, out of propositions of Galileo that contain no postulate.[60]

Galileo's *Theorema mechanicum*, mentioned here by Torricelli, is given in his Prop 6 s part and it corresponds to Torricelli's *Proposition III,* given supra. It says that the *momentum* of the same body placed on differently inclined planes as *AC* and *AD*, in Galileo's figure (Fig. 22) are in the same ratio as the elevations of those planes.

that we may formalise in the following way, if we put the same weight on both planes,

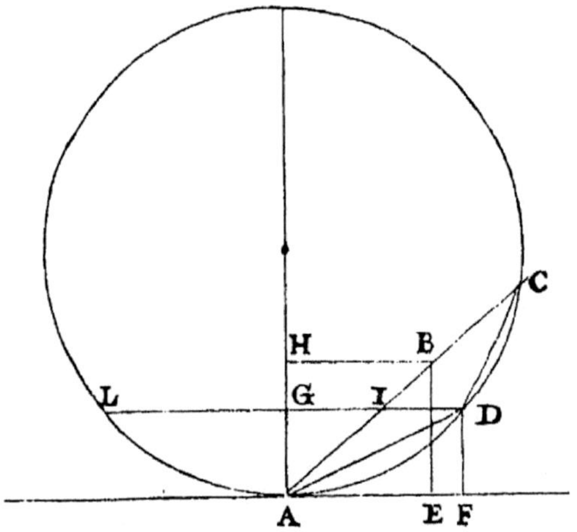

> The same may be proved using mechanics. Following the next figure [Fig. 20]: Moving bodies need the same times to move through *CA* and *DA*. Let *BA* be equal to *DA*, and draw the perpendiculars *BE* and *DF*; it follows from the principle of mechanics that the *momentum* of the weight on the plane along the line *ABC* is to its total *momentum* as *BE* is to *BA*; in the same manner the *momentum* along the plane *AD* is to its total *momentum* as *DF* is to *DA*, or to *BA*. Therefore the *momentum* of this same weight along the plane *DA* is to that along the plane *ABC* as the length *DF* is to the length *BE* (Galilei 1638, VIII, pp. 221–222; Author's translation).

Fig. 22 Galilei's construction for Theorema *IV, Propositio IV* (Galilei 1638, VIII, p. 222). *Source* Gallica, Bibliothèque nationale de France. Public Domain

$$W_A B = W_A D \text{ then } \frac{M_{AB}}{M_{total}} = \frac{BE}{BA} \text{ and } \frac{M_{AD}}{M_{total}} = \frac{DF}{DA} = \frac{DF}{BA} \text{ thus } \frac{M_{DA}}{M_{BA}} = \frac{DF}{BE}$$

[60] "Aliter: Praecedens Theorema poterat demonstrari sine ulla suppositione. Demonstrat aenim Galileus in Prop. 6. de motu accelerato, tempora lationum per chordas omnes in circulo aequalia esse. Idque tribus modis probat. In primo, & tertio subest principium suum non satis evidens; in secundo vero nihil supponitur, praeter iam dictum Theorema Mechanicum; quod si, ipso teste, demonstratum antea fuerat, ex ipso immediate, tamquam Corollarium, necessaria illatio suae tertiae Propositionis, immo & suae petitionis, derivari poterat. Sed quia ipse tertiam suam Propositionem, quae nobis quarta est, mediante sua petitione probat, nos illam absolute ostendamus ex propositionibus ipsius Galilei, quae nullum postulatum includunt." (Torricelli 1644, p. 107; Author's translation).

> […] for this reason, this same weight will in equal times, according to the second proposition of the first book, traverse spaces along the planes *CA* and *DA* which are to each other as the lengths *BE* and *DF*. But it can be shown that *CA* is to *DA* as *BE* is to *DF*. Hence the falling body will traverse the two paths *CA* and *DA* in equal times.[61]

It follows from what precedes that if we take the same time $T_{AC} = T_{AD} \equiv T$, then $S_p(T)_{AC}/S_p(T)_{DA} = BE/DF$. But it can be shown, and Galileo does it, that $CA/DA = BE/DF$, thus $S_p(T)_{CA} = S_p(T)_{DA}$.

4.12 Torricelli's Second Proof of *Proposition IV*

In Fig. 23, *AB* and *AC* are planes having the same elevation *AD* and Torricelli must show that

$$\frac{T_{AC}}{T_{AB}} = \frac{AC}{AB} \tag{15}$$

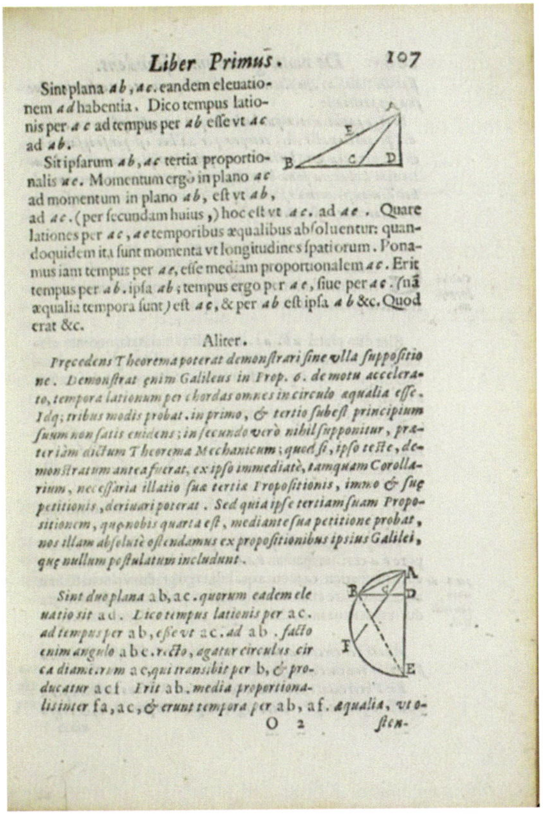
Liber Primus. 107

Sint plana *ab*, *ac*. eandem eleuationem *ad* habentia. Dico tempus lationis per *ac* ad tempus per *ab* eſſe vt *ac* ad *ab*.

Sit ipſarum *ab*, *ac* tertia proportionalis *ae*. Momentum ergò in plano *ac* ad momentum in plano *ab*, eſt vt *ab*, ad *ac*. (per ſecundam huius,) hoc eſt vt *ac*. ad *ae*. Quare lationes per *ac*, *ae* temporibus æqualibus abſoluentur: quandoquidem ita ſunt momenta vt longitudines ſpatiorum. Ponamus iam tempus per *ae*, eſſe mediam proportionalem *ac*. Erit tempus per *ab*. ipſa *ab*; tempus ergo per *ae*, ſiue per *ac*. (nã æqualia tempora ſunt) eſt *ac*, & per *ab* eſt ipſa *ab* &c. Quod erat &c.

Aliter.

Præcedens Theorema poterat demonſtrari ſine vlla ſuppoſitione. Demonſtrat enim Galileus in Prop. 6. de motu accelerato, tempora lationum per chordas omnes in circulo æqualia eſſe. Idq; tribus modis probat. in primo, & tertio ſubeſt principium ſuum non ſatis euidens; in ſecundo verò nihil ſupponitur, præter iàm dictum Theorema Mechanicum; quod ſi, ipſo teſte, demonſtratum antea fuerat, ex ipſo immediatè, tamquam Corollarium, neceſſaria illatio ſuæ tertiæ Propoſitionis, immo & ſuę petitionis, deriuari poterat. Sed quia ipſe tertiam ſuam Propoſitionem, quę nobis quarta eſt, mediante ſua petitione probat, nos illam abſolutè oſtendamus ex propoſitionibus ipſius Galilei, quę nullum poſtulatum includunt.

Sint duo plana ab, ac. *quorum eadem eleuatio ſit* ad. *Dico tempus lationis per* ac. *ad tempus per* ab, *eſſe vt* ac. *ad* ab. *facto enim angulo* abe. *recto, agatur circulus circa diametrum* ac, *qui tranſibit per* b, *& producatur* acf. *Erit* ab. *media proportionalis inter* fa, ac, *& erunt tempora per* ab, af. *æqualia, vt oſten-*

O 2

Fig. 23 Torricelli's construction to recall Galilei's demonstration of *Propositio IV* (Torricelli 1644, p. 107). *Source* Courtesy University of Oklahoma, History of Science Collections

If thus we put that the time of fall through *AC* is equal to *AC* itself, then the mean proportional *AB*, is the time through *AF* that is the time through *AB* itself. etc. That is the reason why the time of fall, starting from rest, through planes having the same elevation are homolog to the lengths of the planes and we proved it without that petition, the truth of witch we will prove in the next theorem.

(Torricelli 1644, p. 107; Author's translation).

---

[61] "Idem aliter demonstratur ex mechanicis: nempe sequenti figura [Fig. 20]: Mobile temporibus aequalibus pertransire *CA*, *DA*. Sit enim *BA* aequalis ipsi *DA*, & ducantur perpendiculares *BE*, *DF*, constat ex elementis mechanicis, momentum ponderis super plano secundum lineam *ABC* elevato ad momentum suum totale esse ut *BE* ad *BA*, ejusdemque ponderis momentum super elevatione ad ad totale suum momentum esse, ut *DF* ad *DA* vel *BA*; ergo ejusdem ponderis momentum super plano secundum *DA* inclinato ad momentum super inclinatione secundum *ABC* est, ut linea *DF* ad linea *BE*; quare spatia, quae pertransibit idem pondus temporibus aequalibus super inclinationibus *CA*, *DA*, erunt inter se ut lineae *BE*, *DF*, ex propositione secunda primi libri. Verum ut *BE* ad *DF*, ita demonstratur se habere *AC* ad *DA*; ergo idem mobile temporibus aequalibus pertransibit lineas *CA*, *DA*." (Galilei 1938, pp. 221–222; Author's translation).

Torricelli takes the time along *AC* as unit of time.[62] He says that $T_{AC} \equiv AC$, then as *AB* is the mean proportional between *FA* and *AC*, we have *FA*/*AB* = *AB*/*AC*, then $AB = T_{AF} = T_{AB}$ because $T_{AF} = T_{AB}$ by Galileo's theorem of chords or by Torricelli's Proposition *III*.

4.13 Torricelli's *Proposition V*

Torricelli comes now to the proof of Galileo's assumption concerning the equality of the degrees of velocity in *D*, *B*, *C* at the end of the fall (Fig. 24).

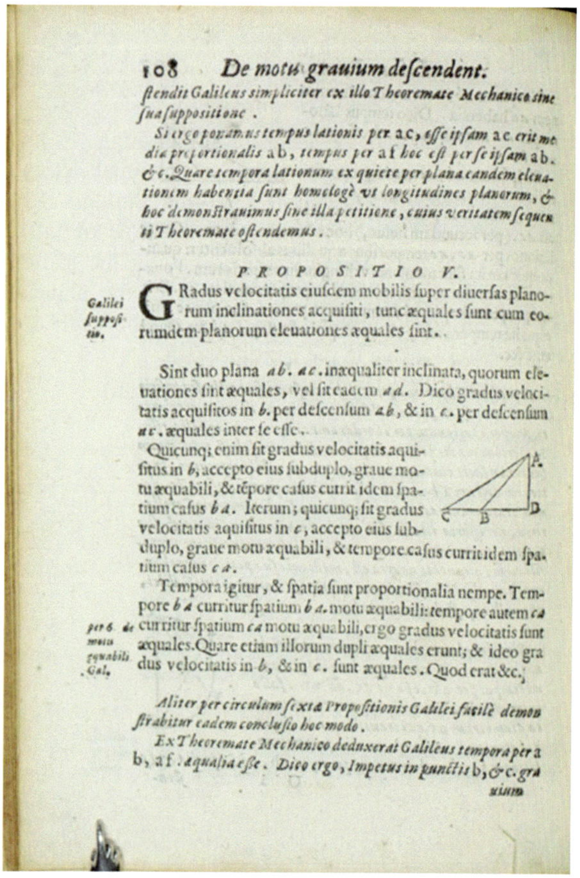

108 De motu grauium deſcendent.

ſtendit Galileus ſimpliciter ex illo Theoremate Mechanico ſine ſua ſuppoſitione.

Si ergo ponamus tempus lationis per ac, eſſe ipſam ac erit media proportionalis ab, tempus per af hoc eſt per ſe ipſam ab. &c. Quare tempora lationum ex quiete per plana eandem eleuationem habentia ſunt homologè vt longitudines planorum, & hoc demonſtrauimus ſine illa petitione, cuius veritatem ſequenti Theoremate oſtendemus.

PROPOSITIO V.

Galilei ſuppoſitio.

GRadus velocitatis eiuſdem mobilis ſuper diuerſas planorum inclinationes acquiſiti, tunc æquales ſunt cum eorumdem planorum eleuationes æquales ſint.

Sint duo plana ab. ac. inæqualiter inclinata, quorum eleuationes ſint æquales, vel ſit eadem ad. Dico gradus velocitatis acquiſitos in b. per deſcenſum ab, & in c. per deſcenſum ac. æquales inter ſe eſſe.

Quicunq; enim ſit gradus velocitatis aquiſitus in b, accepto eius ſubduplo, graue motu æquabili, & tēpore caſus currit idem ſpatium caſus ba. Iterum; quicunq; ſit gradus velocitatis aquiſitus in c, accepto eius ſubduplo, graue motu æquabili, & tempore caſus currit idem ſpatium caſus ca.

Tempora igitur, & ſpatia ſunt proportionalia nempe. Tempore ba curritur ſpatium ba. motu æquabili: tempore autem ca curritur ſpatium ca motu æquabili, ergo gradus velocitatis ſunt æquales. Quare etiam illorum dupli æquales erunt; & ideo gradus velocitatis in b, & in c. ſunt æquales. Quod erat &c.

per 6. de motu æquabili Gal.

Aliter per circulum ſextæ Propoſitionis Galilei facilè demonſtrabitur eadem concluſio hoc modo.

Ex Theoremate Mechanico deduxerat Galileus tempora per ab, af. æqualia eſſe. Dico ergo, Impetus in punctis b, &c. grauium

The degrees of velocity acquired by the same mobile on differently inclined planes are equal when the elevations of those planes are equal.

Let the weight *E* (Fig. 24) run successively through two planes *AC* et *AB* differently inclined but having the same elevation *AD* then $v_{AC} = v_{AB}$ and $v(C) = v(B)$. Whatever is the degree of velocity acquired by the mobile in *B*, it will be twice the velocity the mobile would need to go through *AB* in the same time with a uniform motion. In addition, it is the same for the degree of velocity in *C*, whatever it is.

Fig. 24 The triangle of proportions Torricelli draws to *Propositio V* in his *Opera Geometrica* (Torricelli 1644, p. 108). *Source* Courtesy University of Oklahoma, History of Science Collections[63]

This is Merton's theorem as it was also given by Galileo in his *Theorem I*, *Proposition I* of *de motu accelerato*.

> The time during which some space is gone through by a moving body starting from rest and falling with a uniformly accelerated motion is equal to the time during which the same space would be gone through by the same mobile with a uniform motion, which degree of velocity would be half of the greatest and last degree of velocity reached during the preceding uniformly accelerated motion.[64]

[62] "Si ergo ponamus tempus lationis per *AC*, esse ipsam *AC* erit media proportionalis *AB*, tempus per *AF* hoc est per se ipsam *AB*. etc. Quare tempora lationum ex quiete per plana eandem elevationem habentia sunt homologe ut longitudines planorum, & hoc demonstravimus sine illa petitione, cuius veritatem sequenti theoremate ostendemus." (Torricelli 1644, p. 108; Author's translation).

[63] "Gradus velocitatis eiusdem mobilis super diversas planorum inclinationes acquisiti, tunc aequales sunt cum eorumdem planorum elevationes aequales sint." (Torricelli 1644, p. 108; Author's translation).

[64] "Tempus, in quo aliquod spatium a Mobili conficitur latione ex quiete uniformiter accelerata, est aequale tempori in quo idem spatium conficeretur ab eodem mobili motu aequabili delato, cujus velocitatis gradus subduplus sit ad summum & ultimum gradum velocitatis prioris motus uniformiter accelerati." (Galilei 1638, VIII, p. 208; Author's translation).

Then Torricelli uses Galileo's *Theorem VI, Proposition VI* of *de motu aequabilis*, as he says in a *marginalia*. This proposition says:

> If two moving bodies are moved with a uniform motion, the ratio of their velocities will be compounded of the ratio of the spaces run through and of the inverse ratio of the times.[65]

$$\frac{T_{AC}}{T_{AB}} = \frac{AC}{AB} \quad \frac{v_{AC}}{v_{AB}} = \frac{AC}{AB} \times \frac{AB}{AC} = 1 \tag{16}$$

So $v_{AC} = v_{AB}$ and the square of those velocities, that is the degrees of velocity at B and C are also equal.

Aliter: Using the Galileo's circle of prop. *VI* of *de motu accelerato*, Torricelli proves the same with help of a simplified version of (Fig. 23). Galileo has proved that the times taken to go through AB and AC or through any other chord, are equal.

5 Galileo's Proof of His Supposition

As we already said, we have two versions of Galileo's proof of his assumption. Galileo's manuscript copied by Viviani that we called Galileo's manuscript[66] and a more elaborated version that was composed by Viviani for a second edition of the *Discorsi*. This last version has been published in the *Edizione Nazionale* and we already used it. Now we shall analyse the direct copy of Galileo's manuscript that is more concise. I shall also add some remarks concerning Viviani's addition to the *Discorsi*.

5.1 In Galileo's Manuscript in Possession of Viviani, Written After the Publication of the *Discorsi*

After having presented the situation of Fig. 25 and the criteria of equilibrium that I already formalized in the first part: if we place the same body on two differently inclined planes, so that
$M(W_G) = M(W_H)$, Galileo's criteria is: $W_G/W_H = FA/FC$. He goes then on as follows[67]:

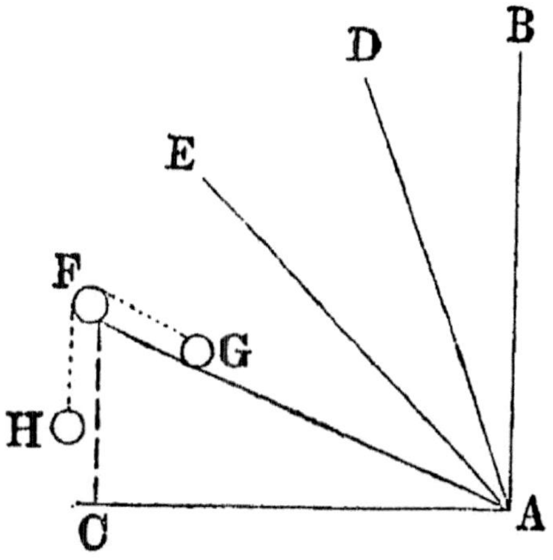

> And since we agreed that the *impetus*, energy, or *momentum* [or propensity] to move of the moving body is as great as the minimal force or sufficiently resistant to stop it, we concluded, that the *impetus* along the inclined plane is to the maximum *impetus* along the perpendicular, as that perpendicular, that is the elevation of that same inclined plane, to the length of that inclined plane.

Fig. 25 The construction for *impetus* in the discussion between Salviati and Sagredo (Galilei 1638, VIII, p. 215). *Source* Gallica, Bibliothèque nationale de France. Public Domain

[65] "Si duo mobilia aequabili motu ferantur, ratio velocitatum ipsorum composite erit ex ratione spatiorum peractorum, & ex ratione temporum contrarie sumptorum." (Galilei 1638, VIII, p. 196; Author's translation).

[66] I think that Viviani when he writes that the copy he makes comes from an original of Galileo's hand.

[67] Some words are missing in the manuscript, I inserted them with help of Viviani's addition and put them between square brackets.

Galileo gives a relation concerning *impetus* but he recalls it, one could call it *momentum*.[68] Thus he shows that:

$$M_{FA}/M_{total} = FA/FC.$$

This point established, and supposed that the moving body, starting from rest and going down naturally, increases its velocity adding equal increases in equal times as I defined the accelerated motion in my book; thus, as I intend to prove here, the spaces gone through are in duplicated proportion of times, and consequently for the degrees of velocity, which as was said, increases in the same proportion as time; We shall prove our conclusion, that is: the degrees of velocity in the horizon are equal, if acquired by a moving body starting from rest at whatever height, along whatever inclined plane and reaching the horizon.[69]

First, we have to notice the introduction of the terms degrees of velocity, where the velocity is no more the "mean velocity" but the velocity at a point. Galileo recalls here, his definition of the uniformly accelerated motion, namely a motion that increases its velocity adding equal increases of velocity in equal times[70] and his conclusions about that motion. First that spaces are in duplicated proportion of times: $e \propto t^2$, proved in his *Proposition II* and secondly that the degrees of velocity are in the same proportion as times: $v(e) \propto t$, that he deduces thanks to the relation $e = v(e)\, t$. With these premises, he will prove his assumption about velocities down a certain height. The formulation of that assumption given by Viviani in the addition is clearer, we reproduce it here:

[...] the degrees of velocity of a moving body falling down from the same height along planes having any inclination, are all equal when reaching the horizon, all impediments having been removed.[71]

Galileo recalls that the velocity, in the motion he considers, increases "regularly" with time.

We must here draw attention on the fact that on whatever inclination, the body from the beginning when it was at rest, goes increasing its velocity in proportion to time and consequently the quantity of impetus [increases its velocity in proportion of time]; as the *impetus* were during the first motion, such will be the degrees of velocity gained during the same time, as they grow both in the same proportion during the same time.[72]

We have here a description of the acceleration as proportional to the *impetus*, gained during the first motion, one should say during the first instant of the motion, that is proportional to the degree of velocity that will be added in equal times to the velocity. That is exactly what he called *momentum* and what Bernoulli–de l'Hôpital will call "velocity in the first instant". What Galileo means here is clearly similar to the *dv* that is regularly added to the velocity. Having recalled all this, Galileo gives his proof, starting with the description of his notations on a triangle representing any inclined plane (Fig. 26).

---

[68] I kept the *M* of momentum in the formula, in order to facilitate comparisons.

[69] "E perchè siamo convenuti, che tanto sia l'impeto, l'energia o il momento, [o la propensione] al moto del mobile, quanto è la forza o resistenza minima che basta a fermarlo, concludiamo, l'impeto per l'inclinata all'impeto massimo per la perpendicolare stare com'essa perpendicolare, cioè come l'elevazione della medesima inclinata, alla lunghezza dell'inclinata. Stabilito ciò, e posto che il mobile grave, partendosi dalla quiete e naturalmente scendendo, vada con eguali aggiunte in tempi uguali accrescendo la sua velocità, come vien da me definito il moto accelerato nel mio Libro; onde, come quivi io dimostro, gli spazi passati sono in duplicata proporzione de' tempi, ed in conseguenza de i gradi di velocità, la quale, come si è detto, cresce con la proporzione del tempo; dimostreremo la nostra conclusione, cioè i gradi di velocità nell'orizonte esser eguali, quelli però acquistati dal mobile che dalla quiete si parta da qualunque altezza, e per quali si sieno inclinazioni pervenga all'orizzonte." (Galilei 1638, VIII, p. 444; Author's translation).

[70] Galilei (1638, VIII, p. 205) and supra at the beginning of our text on Galileo's assumptions.

[71] "I gradi di velocità d'un mobile descendente con moto naturale dalla medesima sublimità per piani in qualsivoglia modo inclinati, all'arrivo all'orizonte son sempre eguali, rimossi gl'impedimenti." (Galilei 1638, VIII, p. 218; Author's translation).

[72] "E qui devesi avvertire, che posto che in quali si voglino inclinazioni il mobile dalla partita dalla quiete vada crescendo la velocità, con la proporzione del tempo, e in conseguenza la quantità dell'impeto; quali furono gl'impeti nella prima mossa, tali saranno i gradi delle velocità guadagnati nello stesso tempo, poi che e questi e quelli crescono con la medesima proporzione nel medesimo tempo." (Galilei 1638, VIII, p. 444; Author's translation).

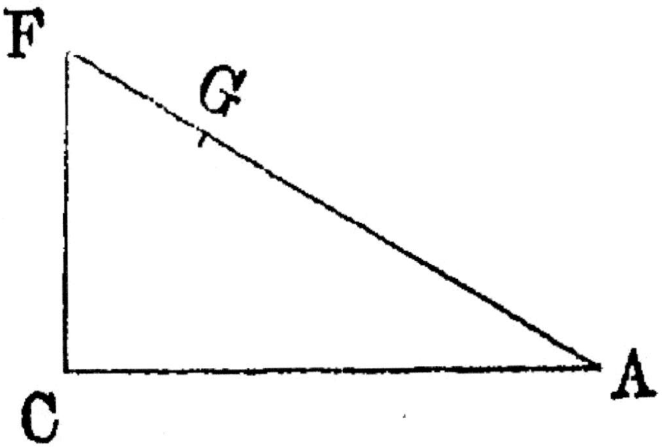

Let now FA (Fig. 26) be the inclined plane, its elevation above the horizon the perpendicular FC, and the horizontal CA; and on the inclined plane FA take FG, third proportional of AF, FC: and because the impetus along the perpendicular FC will be to the impetus along the inclined plane FA as FA to FC; Therefore, the moving body will spend the same time on a [perpendicular] space [FC as on a space] equal to FG along the inclined plane FA, and the degree of velocity in C to the degree of velocity in G would have the same proportion as FC to FG. But the degree of velocity in A to the degree in G has the same proportion as AF to FC, that is as FC to FG, and thus the degrees in A and in C have the same proportion to the degree in G, and therefore they are equal to each other: this is what was to prove.

Fig. 26 Galilei's model of inclined plane (Galilei 1638, VIII, p. 445). *Source* Gallica, Bibliothèque nationale de France. Public Domain

Galileo uses here[73] the same geometrical theorem as Torricelli in the *scholium* to his *Proposition II*. He takes FG, along FA, third proportional to AF, FC, thus CG is the height of the rectangle triangle FCA. He recalls also that $M_{total}/M_{FA} = FA/FC$, so that, he says, the moving body spends the same time on the perpendicular FC as on the distance FG along the inclined plane: $T_{FC} = T_{FG}$. This last relation is deduced, without mentioning it, from his theorem of chords. He deduces then that: $v(C)/v(G) = FC/FG$, where $v(C)$ and $v(G)$ are the velocities in the points C and G and no more the Aristotelean mean velocities on particular distances. Progressively, he approaches the idea of a "function velocity".

Galileo goes on saying that $v(C)/v(G) = AF/FC = FC/FG$, thus $v(A)/v(G) = v(C)/v(G)$ and $v(A) = v(C)$. Let us note that in this proof, Galileo prefers to use the word *impetus* instead of *momentum* making a difference between the *Archimedean momentum* and his own, leaving behind the word *momentum* and its static flavour.

5.2 In Viviani's Addition to the *Discorsi*

And since we agreed that the *impetus, energy, momentum or propensity to move* of a body is as great as the minimal force or resistance sufficient to stop it, and we concluded that the weight H is sufficient to prevent the motion of weight G, it follows that the minor weight H [Fig. 27], that exerts its maximum momentum along the perpendicular FC, will be the exact measure of the partial momentum exerted by the major weight G along the inclined plane fa; but the measure of the *total momentum* of the same weight G is its own weight (because to prevent the vertical fall of a weight one needs the opposition of a same weight, which must be free to move perpendicularly); thus the impetus or partial momentum of G along the inclined plane FA to the maximum and total impetus of the same G along the perpendicular FC, will be as the weight H to the weight G, that is through the construction, as that perpendicular FC, elevation of the inclined plane, to FA itself: That is the lemma that we intended to prove and which, as we will see, has been assumed by our Author in the second part of the sixth proposition of the present treatise.[74]

[73] "Ora sia il piano inclinato FA, la sua elevazione sopra l'orizonte la perpendicolare FC, e l'orizontale CA; e prendasi nell'inclinata FA la FG, terza proporzionale dopo AF, FC: e perchè l'impeto per la perpendicolare FC all'impeto per l'inclinata FA sta come FA ad FC, i medesimi impeti staranno come le FC, FG. Il mobile, dunque, nell'istesso tempo che passasse uno spazio [perpendicolare FC passerà ancora uno spazio] eguale all'FG nell'inclinata FA, ed il grado di velocità in C al grado di velocità in G averebbe la medesima proporzione della FC alla FG. Ma il grado di velocità in A al grado in G ha la medesima proporzione che la AF alla FC, cioè che la FC alla FG; adunque i gradi in A ed in C al grado in G anno la medesima proporzione, e però son fra loro eguali; che è quel che doveva dimostrarsi" (Galilei 1638, VIII, p. 444; Author's translation).

[74] "E perchè siamo convenuti, che di un mobile tanto sia l'impeto, l'energia, il momento, o la propensione al moto, quanta è la forza o resistenza minima che basta a fermarlo, e si è concluso che il grave H è bastante a proibire il moto al grave G, adunque il minor peso H che nella perpendicolare *fc* esercita il suo momento totale, sarà la precisa misura del momento parziale che il maggior peso G esercita per il piano inclinato *fa*; ma la misura del total momento del medesimo grave G è egli stesso (poichè per impedire la scesa perpendicolare d'un grave si chiede il contrasto d'altrettanto grave, che pur sia in libertà di moversi perpendicolarmente); adunque l'impeto o momento parziale del G per l'inclinata FA, all'impeto massimo e totale dell'istesso G per la perpendicolare FC, starà come il peso H al peso G, cioè, per la costruzione, come essa perpendicolare FC, elevazione dell'inclinata, alla medesima inclinata FA: che è quello che per lemma si propose di dimostrare, e che dal nostro Autore, come vedranno, vien supposto per noto nella seconda parte della sesta proposizione del presente trattato." (Galilei 1638, VIII, p. 217; Author's translation).

$$\frac{M(G)_{FA}}{M(G)_{total}} = \frac{W(H)}{W(G)} = \frac{FC}{FA} \tag{17}$$

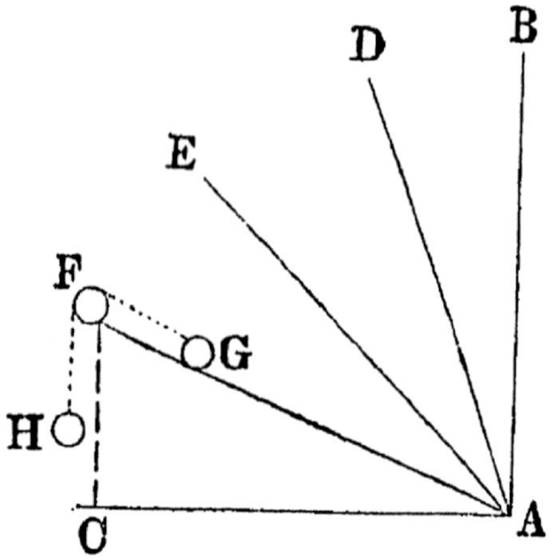

Sagr. From what you have shown thus far, it appears to me that one might infer ex aequalis and by exchanged proportion, that the momentum of descent of the same body along planes differently inclined, as *FA* and *F[C]*, but having the same elevation [*FC*], are in inversed ratio of the lengths of the same planes.

Fig. 27 Picture from the discussion from Salviati confuting Simplicio (Galilei 1638, VIII, p. 215). *Source* Gallica, Bibliothèque nationale de France. Public Domain[75]

With the notations in Fig. 27, Sagredo finds the result of Torricelli's *Proposition II.*

$$\frac{M(G)_{FA}}{M(G)_{FC}} = \frac{FC}{FA} \tag{18}$$

Salv. Perfectly right. This point established, I pass to the proof of the following theorem, that is that: the degrees of velocity of a moving body falling down from the same height along planes having any inclination, are all equal when reaching the horizon, all impediments being removed.[76]

This is Galileo's assumption and it says that (Fig. 28): $v(E) = v(B)$, with the conditions that the initial points are at the same height and the final ones on a same horizontal. To prove it Galileo reminds the properties of the naturally accelerated motion.

First, we must here draw attention on the fact, that in whatever inclination the body starting from rest, gains from the beginning, velocity or quantity of *impeto* (sic in Galileus) in proportion to time (in agreement with the definition given by the Author for the naturally accelerated motion), hence, as he has proved in the preceding proposition [that is *Proposition II* of the *Discorsi*], the spaces traversed are proportional to squares of times and consequently to the square of the degrees of velocity; the degrees of velocity gained during the same time will be proportional to the *impetus* of the first motion, as they both grow with the same proportion in the same time.[77]

[75] "Sagr. Da questo che V.S. ha concluso fin qui, parmi che facilmente si possa dedurre, argumentando ex aequali con la proporzione perturbate, che i momenti dell'istesso mobile per piano diversamente inclinati, come FA, FI, che abbiamo l'istessa elevazione, son fra di loro in reciproca propozione de' medesimi piani." (Galilei 1638, VIII, p. 217; Author's translation).

[76] "Salv. Verissima conclusione. Fermato questo, passerà adesso a dimostrare il teorema, cioè che: I gradi di velocità d'un mobile descendente con moto naturale dalla medesima sublimità per piani in qualsivoglia modo inclinati, all'arrivo all'orizonte son sempre eguali, rimossi gl'impedimenti." (Galilei 1638, VIII, p. 218; Author's translation).

[77] "Qui devesi prima avvertire, che stabilito che in qualsivoglino inclinazioni il mobile dalla partita dalla quiete vada crescendo la velocità, o quantità dell'impeto, con la proporzione del tempo (secondo la definizione data dall'Autore al moto naturalmente accelerato), onde, com'egli ha per l'antecedente proposizione dimostrato, gli spazii passati sono in duplicata proporzione de' tempi, e conseguentemente de' gradi di velocità; quali furono gl'impeti nella prima mossa, tali proporzionalmente saranno i gradi delle velocità guadagnati nell'istesso tempo, poichè e questi e quelli crescono con la medesima proporzione nel medesimo tempo." (Galilei 1638, VIII, p. 218; Author's translation).

$$e \propto t^2 \propto v(E)^2 \tag{19}$$

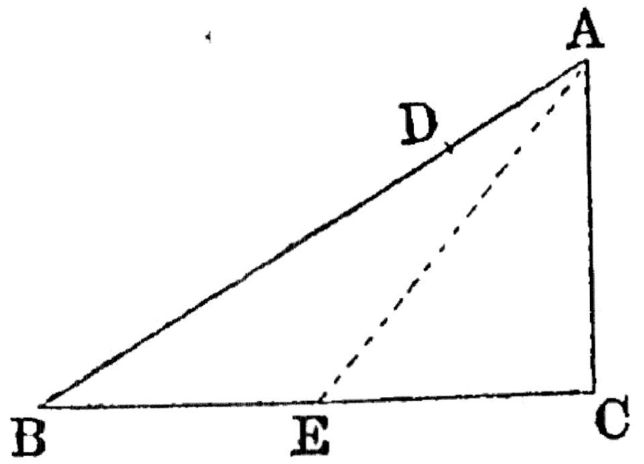

Let now *AB* be the inclined plane, its elevation above horizon the perpendicular *AC*, and the horizontal *BC*; and as the *impetus* of a moving body, as we just saw, along the perpendicular *AC* is to the *impetus* of the same body along the inclined plane *AB* as *AB* is to *AC*. On the inclined plane *AB* take the third proportional of *AB*, *AC*: the *impetus* along *AC* will be to the *impetus* along *AB*, or along *AD* as *AC* to *AD*; and therefore, the moving body will spend the same time on the perpendicular space *AC* as on the space *ad* along the inclined plane *AB* (as *momenta* are as spaces), and the degree of velocity in *C* will have the same proportion to the degree of velocity in *D* as *AC* to *AD*.
But the degree of velocity in *B* to the same degree in *D* is as the time along *AB* to the time along *AD*, according to the definition of accelerated motion, and the time along *AB* to the time along *AD* are as the same *AC*, mean proportional between *BA* and *AD*, to *AD*, according to the last corollary of the second proposition; and thus the degrees of velocity in *B* and in *C* have the same proportion to the degree of velocity in *D* than *AC* to *AD* and therefore they are equal: this is the theorem that one intended to prove.

Fig. 28 The triangle Salviati uses to prove Sagredo Galilei's thesis (Galilei 1638, VIII, p. 218).[78] *Source* Gallica, Bibliothèque nationale de France. Public Domain

After having described the figure and given names to the various segments, Galileo reminds the following result: $M(G)_{AC}/M(G)_{AB} = AB/AC$. He takes *AD* to be the third proportional to *AB*, *AC* so that *DC* is the height of the triangle *ABC* and we have $AB/AC = AC/AD$. Then we have $M(G)_{AC}/M(G)_{AB} = M(G)_{AC}/M(G)_{AD} = AC/AD$. So that, he says, $t_{AC} = t_{AD}$. A deduction that was not clear in Galileo's manuscript, but Viviani explains it, adding between brackets "[...] as *momenta* are as spaces [...]". Thus as I already supposed, Galileo's *momentum* looks like something proportional to $v = et$ so that, as we already showed, if *momentum* are as spaces, times must be equal. So that one has $v(C)/v(D) = AC/AD$, but $v(B)/v(D) = t_{AB}/t_{AD} = AC/AD$, and thus $v(C) = v(B)$. But to do that proof, Galileo had to transform his Archimedean vision of a statical *momentum*, that he replaces by the impetus (Fig. 28).

[78] "Ora sia il piano inclinato *AB*, la sua elevazione sopra l'orizonte la perpendicolare *AC*, e l'orizontale *CB*; e perchè, come poco fa si è concluso, l'impeto d'un mobile per la perpendicolare *AC*, all'impeto del medesimo per l'inclinata *AB*, sta come *AB* ad *AC*, prendasi nell'inclinata *AB* la *AD*, terza proporzionale delle *AB*, *AC*: l'impeto dunque per *AC* all'impeto per la *AB* cioè par la *AD*, sta come la *AC* all'*AD*; e perciò il mobile nell'istesso tempo che passerebbe lo spazio perpendicolare *AC*, passerà ancora lo spazio ad nell'inclinata *AB* (essendo i momenti come gli spazii), ed il grado di velocità in *C* al grado di velocità in *D* averà la medesima proporzione della *AC* alla *AD*. Ma il grado di velocità in *B* al medesimo grado in *D* sta come il tempo per *AB* al tempo per *AD*, per la definizione del moto accelerato, ed il tempo per *AB* al tempo per *AD* sta come la medesima *AC*, media tra le *BA*, *AD*, alla *AD*, per l'ultimo corollario della seconda proposizione; adunque i gradi in *B* ed in *C* al grado in *D* hanno la medesima proporzione della *AC* alla *AD*, e però sono eguali: che è il teorema che intesi di dimostrare." (Galilei 1638, VIII, p. 218; Author's translation).

6 Conclusion

We shall never know, I presume, what Torricelli and Galileo discussed during the three months they spent together in Arcetri. But they surely talked about motion. Those discussions, as well as Galileo's *Discorsi*, certainly deeply influenced Torricelli's *De motu* (as I showed in this article). But in this conclusion, I would like to underline two particular points. The first concerns the *Galilean momentum* that both Galileo and Torricelli used; the second concerns our notion of function as anticipated by both Torricelli and Galileo and to show how different they are.

6.1 On *Galilean Momentum*

In his first text, *De motu antiquiora*, Galileo finds the apparent weight of a weight in equilibrium on an inclined plane with help of a balance or a lever. This a result of pure statics he obtains using the "Archimedian momentum".

In the next text, *Le mecaniche*, Galileo tries to transform his problem of statics into a dynamical one. He speaks now of *the momentum to go down* that I called the "Galilean momentum". He keeps the link with statics saying that the *Archimedean momentum* is the *momentum in the first instant*, without knowing exactly what that new *moment* means.

In the first part of Viviani's addition to the *Discorsi*, Galileo modifies his experiment and links together two weights with a rope, one hanging vertically along *FC* and the other one lying on an inclined plane *FA* (Fig. 7). Now, the space travelled by both weights must be equal. But in fact what he calculates is an equilibrium so that he finds the same result as in *Le mechaniche*. Thus the *momenta*, linked to the velocities or "Galilean momenta" are in the same proportion as the weights, as the "Archimedean momenta". But this is not true because he neglects the acceleration the falling body acquires. In fact it is only true as he often says, in the first point or in the first moment.

In these three proofs, Galileo is unable to introduce motion into his proofs. He only obtains results of statics. But his initial Definition and Principle, the one Torricelli wants also to prove, concerns motion. It is only after the publication of the *Discorsi* containing his *De motu accelerato*, that in the manuscript he left to Viviani, Galileo will face an accelerated motion whose increment is given by the *Archimedian momentum* in the first instant, in order to prove his "assumption". As we can read in the sentence of Viviani's addition to the *Discorsi* that we already quoted: "And since we agreed that the impetus, energy, momentum or propensity to move of a body is as great as the minimal force or resistance sufficient to stop it".

Torricelli's principle is formulated as a negation: *We pose as a principle that two weights linked together cannot move by themselves, unless their common centre of gravity goes down.*

His first two propositions are results of statics, there he generalizes Galileo's proposition of the *Discorsi* to two weights in equilibrium on two differently inclined planes. His *Lemma* and its proof are similar to those of Galileo. Only the two propositions of his *Corollarium* are different, but look like an exercise.

6.2 On the Notion of Function

In Galileo's texts, we witness his struggle to describe the motion of a weight along an inclined plane. And he certainly knows Aristotle's text on motion where he says that a motion is always executed in time, and today we would add that to describe a motion in time, we need a function.

In *Le mecaniche* is the remarkable fact that he considers explicitly the motion at one point, because the inclination of the tangent changes in every point. More, when he gives the *momentum* to go down, he always adds that he gives it *in the first point of its motion*. But we must keep in mind that the body is on a circle and that it passes just an instant on each particular inclined plane. Galileo is not speaking of a motion in successive points of one and the same inclined plane. However, he aims to do it. His problem is that *Archimedean momentum* only gives him the *impetus in the first point*. Only after having published the *Discorsi* where he explains that during the accelerated motion of a weight on an inclined plane, the weight receives in every instant one and the same quantity of velocity, thus the same as the one he can calculate: the *impetus* in the first instant.

Torricelli takes a completely different way to reach the same result. He immediately sees in that problem the role of one the only "functions" together with the logarithm, known in his time: the sine. Torricelli shows this in a corollary to his central *Praemittimus*, the one he uses to prove Galileo's assumption. In that way he bypasses Galileo's difficulty and on all his figures, he represents the circle giving the value of the sine. And even if his proof, in *Proposition V*, of Galileo's assumption seems a bit old fashion using Merton's theorem, his proof of his *Praemittimus* intended to replace Galileo's assumption is more modern since he uses the sine function. The Corollary shows it plainly: "It is now evident that the *momentum* of weights lying on unequally inclined planes are as the *sinus recti* of the angle of elevation".

References

Bernoulli J (1955) Der Briefwechsel von Johann Bernoulli, Vol. I. Birkhäuser, Basel.

Castelli B (1641) Benedetto Castelli a Galileo Galilei in Firenze. Roma 2 marzo 1641. In: Antonio Favaro, Le Opere di Galileo Galilei, Prima Edizione, Vol. XVIII, Edizione Nazionale, Barbera Editore, Firenze (1890).

Clagett M (1964–1984) Archimedes in the Middle Ages, Madison–Philadelphia, Memoirs of the American Philosophical Society, 5 Vols. 10 Tomes. The Clarendon University Press, Oxford.

De Nemore J (fl. 13th) De ratione ponderis. In: Clagett 1964–1984, pp. 190–191.

Dhombres J, Radelet de Grave P (2008a) Liber amicorum, Réminesciences, Brepols, Turnhout.

Dhombres J, Radelet de Grave P (2008b) Une mécanique donnée à voir, Les thèses illustrées défendues à Louvain en juillet 1624 par Grégoire de Saint-Vincent S.J., in De diversis artibus, Collection de travaux de l'Académie internationale d'Histoire des sciences. Brepols, Turnhout.

Galilei G (1589–1592) De motu antiquiora In: Galilei 1890–1909, I, pp. 251–419.

Galilei G (1593) Le mecaniche. In: Galilei 1890–1909, II, pp. 147–191.

Galilei G (1638) Discorsi e dimostrazioni matematiche intorno a due nuove scienze attenenti alla meccanica e i movimenti locali, del signor Galileo Galilei linceo. Elsevirii, Leida. In: Galilei 1890–1909, VIII, pp. 39–362. [1639 Viviani's additions: Ivi, VIII, pp. 215–219, pp. 442–445; see also: *Ivi, Frammenti attinenti ai Discorsi e dimostrazioni matematiche intorno a due nuove scienze* VIII, pp. 363–448].

Galilei G (1890–1909) Le opere di Galileo Galilei: Edizione nazionale sotto gli auspici di sua maestà il re d'Italia. Favaro A (ed), Barbera, Firenze.

Galluzzi P (1979) Momento. Studi Galileiani, ILIESI, Roma.

Torricelli E (1644) Opera geometrica. Masse & de Landis, Florentiae.

Van Sto Vincentio Gregorius (1624) Theoremata mathematica scientiae staticae de ductu ponderum per planitiem recta et oblique horizontem decussantem, Louvain.

Why Publish a Book of Geometry in Seventeenth-Century Europe? Science, Status, and Print Culture in Evangelista Torricelli's *Opera Geometrica* (1644)

7

J. B. Shank

Abstract

What led Evangelista Torricelli to publish his one and only printed book in 1644? This article answers that question by situating Torricelli and his work within the social milieu that joined mathematicians and geometry with elite status as intellectuals in seventeenth-century Europe, showing how book publication and authorship did and did not advance those connections. Moving in four parts, it shows first how Galileo was not, as he was in so many other aspects of Torricelli's work, the model for his idea of book authorship, and then turns instead to several other seventeenth-century examples–Roberval, Huygens, and Newton in particular—as better comparative examples for understanding Torricelli. The article then turns to the precise logics informing Torricelli's career as a published author, setting up his book as a culminating event in his ascent as the esteemed mathematician in service to the court of the Medici Grand Duke of Tuscany. The article concludes by using this frame to read the contents of Torricelli's opus, namely his puzzling inclusion of 21 different proofs of the same proposition already demonstrated by Archimedes.

Keywords

Geometry • Books • Authorship • Print publication • Manuscript publication • Courtly mathematics • Ancient and modern geometry

1 Introduction

If asked to offer an account of how a particular book of geometry published last year made its way into print, a very specific cultural process would need to be made visible, one that in this case ties mathematical work to writing and texts; mathematical texts to would be published authors; and would be authors to publishers, printers, and the marketplace for mathematical books. Were we to narrow our inquiry further by focusing on the origins of a recently published book of advanced geometrical science—there are many other kinds of geometry books published each year—a further set of cultural conditions would need to be revealed. We might start by noting the anomalous character of such a book since advanced geometrical science is rarely the subject of extended format print publications. In other sciences, books are often the primary vehicle for circulating the latest research, but in mathematics cutting-edge geometric research is generally published via articles in specially crafted journals addressed to highly specialized audiences, texts that must survive the evaluative scrutiny of specially trained editors and other expert mathematicians before they are published. "Why publish your research in a book," our would be geometer-author might be asked by a scientific colleague or publisher? And continuing the analogy by assuming that she is, like most published mathematicians today, a professionally credentialed mathematician employed by a college, university, or professional scientific institution, our would be geometer-author might indeed be led to reconsider whether publishing a book of geometry is in fact the best way for her to disseminate her science and advance research in her field.

J. B. Shank (✉)
University of Minnesota, Minneapolis, USA
e-mail: jbshank@umn.edu

R. Pisano et al., *Homage to Evangelista Torricelli's Opera Geometrica 1644–2024*,
Logic, Epistemology, and the Unity of Science 55,
https://doi.org/10.1007/978-3-031-06963-5_7

Whatever the wisdom of her choice, let us imagine that our imagined contemporary geometer has in fact chosen to publish her geometrical science in a book, and hers is the published book whose life course we are tracing. As a scientific book, the text would have first won the approval of the properly credentialed guarantors of academic knowledge that define modern professional scientific fields like mathematics. In this respect, mathematical articles published in journals are not that different from mathematical books since whatever the precise necessities of each genre, all published mathematical science carries with it the traces, often manifest through footnote citations or intertextual references, to the other books and articles that are the professional foundation of this and any other modern scientific field. Before publishing her work, our geometer might have previously presented her science at a professional meeting where similarly credentialed expert mathematicians exchanged research and debated its merits in an environment governed by highly internalized protocols of collegiality and professional decorum. She might even have personally known some of those who were asked to evaluate the quality of her book manuscript from encounters at one of these meetings. Whatever her actual relation to the mathematicians who evaluated her book, she would trust that the same standards and norms operative in these professional gatherings would also be operative when conceiving how her book should be written, read and judged. As an established professional herself, and a credentialed member the community that would ultimately serve as the intended audience for and judge of her work, our geometer-author would also not have worried at all about the cost of publishing her book since she would count on the existence of an academic press who would bear these costs so long as the expert professionals in her field determined her book to be worthy of publication.

If we instead ask why Evangelista Torricelli decided in 1643 to compile some of his extant mathematical work into his *Opera Geometrica* published at Florence in 1644, no analogous explanation of his rationale is possible. To cite just one major difference between his position and the one just described, authors in seventeenth-century Europe were responsible for the costs of book production, not the printers or publishers, and among Torricelli's calculations would have been the financial burden of bringing his book into print. None of the professional institutions, protocols, norms, and standards driving mathematical publishing today were present in 1644 either, and if we are to understand the rationale motivating Torricelli's publication, we need to reconstruct in precise historical detail the very different understanding of the place of books in the life and profession of the "mathematician" in seventeenth-century Europe—or more precisely, the vocation of the mathematician since no professionalized scientific discipline of mathematics akin to the one just described existed at this time.

Such a perspective is especially important with respect to early modern mathematicians like Torricelli since too many historians of mathematics still view early modern mathematical book publication through the anachronistic, and thus highly distorting, lens of our modes of professionalized, scientific publication today. It is in fact still the norm to see Torricelli's motivations for publishing his work described according to a set of institutionalized professional logics for publication that we know to be completely alien to actual scientific life in seventeenth-century Europe. To treat the *Opera Geometrica* as if it were merely the result of Torricelli deciding to circulate his mathematical ideas among his professional colleagues, and to contribute to the advancement of debates in his field, is to completely misconstrue the actual nature of early modern mathematical work, and the purpose of book publishing within it. To take seriously the real historical agendas motivating Torricelli's publication is therefore to go a long way toward fully historicizing Torricelli's mathematical work within its early modern context. For that reason, this article will reconstruct the historical publishing strategies that led to the *Opera Geometrica* in all of their historical difference as a way to fully historicize its author vocation as a mathematician and his celebrated book's actual mathematical contributions.

2 Mathematics and Book Publication in the 17th Century: The Anomalous Case of Galileo

Given Torricelli's self-declared identity as a member of the *Galileisti*,[1] one approach for discerning his mathematical publishing strategies would be to compare them with those used by his model and early mentor, Galileo Galilei. Unfortunately, as is the case with so many other aspects of Torricelli's life and work, an assumption of direct Galileian influence distorts rather than clarifies the agendas one finds in Torricelli's life and work. Galileo's relationship to books and print publication was also, like so many other aspects of his career, highly exceptional and idiosyncratic. Torricelli certainly found many things to follow in the life of his master, but he did not find in Galileo a model of how to connect mathematical work with book publication.

The exceptional nature of Galileo's relationship to book publication stems from his particular career trajectory, and the way in which it shaped his continuously changing understanding of the reason for and purpose of printed mathematical

[1] *Opere di Evangelista Torricelli* (hereafter *OET*, 3, p. 35) by Giuseppe Vassura and Gino Loria eds., 5 vols in 4 (Torricelli 1919–1944).

books. To understand these shifts, it helps to divide Galileo's life and work into two periods: the first, before 1610, when Galileo was a university professor of mathematics at Pisa and then at Padua, and was at the same time a practitioner of the mechanical mathematical arts; and the second phase of his life, after 1610, when he became the Principle Mathematician and Philosopher to the Grand Duke of Tuscany.

During the first period, Galileo's relationship to book publishing was fairly straightforward. Published books either served for him as gifts demonstrating his talents to patrons, who might then be moved to reward him with pensions; or they served commercial ends as products useful for advancing his material livelihood (Biagioli 2016). Examples of the first are his mathematical writings, prepared under the guidance of Ostillio Ricci and Guidabaldo del Monte, which earned him his faculty appointments first at Pisa in 1589 and then in Padua in 1592. In these works, Galileo demonstrated his mathematical acumen, especially as it related to the mechanics of Archimedes, and it was through the circulation of these textual demonstrations that he secured his youthful status as a titled mathematician (Heilbron 2010, pp. 34–56; Valleriani 2010, pp. 12–20; Bertoloni Meli 2006, Chaps. 2 and 3).

Yet importantly for the arguments of this article, these were not printed mathematical books, even if they were, contrary to the designation offered by Jochen Büttner, Peter Damerow, and Jurgen Renn, "published" rather than "unpublished mathematical treatises" (Büttner et al. 2004, pp. 99–117, 2001, pp. 183–202).

The distinction involves the difference between works that are circulated publicly whether in print or in manuscript, and works which are never intended to leave the study, such as the private notebooks of a mathematician. We know that Galileo presented his work *La bilancetta* to Guidobaldo al Monte in the late 1580s, along with the work contained in the *Theoremata circa centrum gravitatis solidorum,* which was later published in print as an appendix to his *Discorsi* of 1638 (Valleriani 2010, pp. 15–19). Yet while these works were not printed at first, they were, I would argue, published since they were circulated to a broader audience, and served as supporting materials for Galileo's job application dossier when he was appointed as a university mathematics professor. In this sense, these were published mathematical works even if they were not printed books.

Galileo's only printed mathematical book from this period, *Le operazioni del compasso geometrico et militare* from 1606, illustrates the other side of his publishing strategies before 1610.[2] Produced out of his own workshop, which was also his home in Padua, and designed as a manual that would accompany the actual instrument described in the text, this book was designed as a commercial product that would enhance the notoriety of his invention and thus increase its sales. It was very often sold together with the instrument itself, and the book served a commercial purpose that shaped the nature of its contents. In particular, since Galileo was eager to protect the secrets that made his instrument uniquely powerful, his printed manual did not reveal the full extent of his knowledge, only that portion of it that made his instrument desirable and useful to his customers. The success of this mathematical book, therefore, depended as much on what Galileo held back from his readers, forcing them to use of his instrument instead, than any full revelation of his complete mathematical knowledge (Biagioli 2016, pp. 1–20).

Such were the calculations that mathematical artisans made when deciding what to publish in their printed mathematical books. Galileo's initial work with his telescope at first followed a similar pattern. As Albert Van Helden and Owen Gingerich show, Galileo at first began composing his telescopic observations as an empirical chronicle in Italian. But at a certain moment, apparently as he realized that he had material of much more dramatic importance, he shifted into Latin and began to compose the work that would later appear in print in 1610 as *Sidereus Nuncius*. The scholarship on this epochal work is too great to summarize here, but for our purposes two issues are most important (Van Helden and Gingerich 2003, pp. 251–267). First, Galileo used this printed book to secure a new international status for himself as a Copernican astronomer, an identity not before cultivated by Galileo. He also used it to secure a newly lucrative patronage situation, making himself the "messenger of the stars" who revealed the Jovian planets to their rightful owner, the Grand Duke of Tuscany. Galileo had used earlier print works to secure the patronage he needed, and in this case he used a spectacular printed book, prepared with great care and cunning, to garner the favor of a powerful and wealthy patron accomplishing an elevation of his esteem and status as a savant.[3]

Once he began working as the mathematician and philosopher to the Grand Duke of Tuscany, Galileo's publishing strategies changed in step with his new social position. No longer in need of patronage, his books lost their character as ploys to attract either institutional sponsors or material rewards. They also began to serve as vehicles for his new status as a

[2] Galileo was behind several books related to the debates surrounding the so-called new star of 1604, and even though Galileo's name did not appear on the title page these could certainly be studied as examples of his book publishing strategies. But since the focus here is on Torricelli, not Galileo, and Torricelli produced no publications of this sort, I will say no more about them (Drake 1976).

[3] The *locus classicus* for this history is Biagioli 1992, Chaps. 2 and 3. See also Wilding 2014, pp. 89–116, 2011, pp. 221–233.

courtier. As Mario Biagioli has shown most fully, works like his *Discorso intorno alle cose che stanno in su l'acqua o che in quella si muovono* were the result of scientific debates staged at the Ducal court. The transfer of the in-person debate into textual publication was primarily designed to reveal Galileo as the brilliant courtier that his patron believed him to be. In short, these were books that staged for a wider public Galileo's identity as a brilliant court savant, and to be successful they needed sparkling wit and clever argumentation at least as much as they needed sober mathematical reasoning. Their point was less to advance mathematical truth or knowledge (although being right was essential to them) than to win the scientific argument at hand. The rhetoric of such books starkly separated scientific friends from foes, and right thinking from error, and in their overwhelmingly contentiousness, they stand less as works that secure knowledge than as exercizes in spirited intellectual dispute (Biagioli 1992).

Ironically, the *Accademia dei Lyncei,* which is often described as the first modern scientific academy, and which Galileo certainly joined because of its zeal for scientific investigation, encouraged rather than checked these polemical tendencies in Galileo's print publications. In his *Istoria e dimostrazioni intorno alle macchie solari* of 1613 and in his *Il Saggiatore* of 1624, both works published through the direct support and intervention of Frederico Cesi and the Lyncei, Galileo further cultivated his polemical manner of using printed books to sustain and initiate combat with scientific rivals. Cesi, the leader of the Lyncei, encouraged this tendency, not least by footing the publication costs of these and other Galileian works (Freedberg 2003, Chaps. 4 and 5). Galileo's life changing *Dialogo sopra i due massimi sistemi del mondo* of 1632 was similarly conceived and supported by the Lyncei in this way. Taken as a whole, what is revealed in these books is Galileo's tendency to use books as vehicles for staging perceived battles between him and his mathematical rivals, and the encouragement in this direction, both materially and intellectually, that he received from his friends and patrons among the Lyncei.

The change in his status brought on by his prosecution as a heretic in 1633 changed his publication habits in the final years of his life, and his *Discorsi e dimostrazioni matematiche intorno a due nuove scienze* marks a new departure for Galileo, the last before his death in 1642 (Heilbron 2010, pp. 329–58). As a printed book that neither seeks to offer gifts to patrons nor to stage polemics with rivals, and as a work published in Leiden far away from the Lyncei and the Italian intellectual politics that they encouraged, it is a different kind of work than the others in the Galileian corpus. For Torricelli, this may have been the mathematical book that most appealed to him, but even here, Galileo used the dialogue form that he had deployed in his *Dialogo,* along with the characters featured in that text, in developing his scientific presentation. And even if Galileo used this final text to reveal his life work in geometry and mechanics in a manner less clouded by intellectual politics, he still couched his material within a rhetorical idiom, however subdued, that evoked Galileo's characteristic identity in his published books as a controversialist and a wit.

Torricelli was a very different kind of mathematician than Galileo, and he lived a very different life. He accordingly adopted a very different genre and idiom for his mathematical publications. Even if he shared with Galileo a position as a Tuscan mathematician courtier, he also lived and worked after the events of the Galileo Affair, an episode that no doubt shaped his idea of the appropriate conduct for a mathematician and a courtier. Yet it is too simple to attribute Torricelli's avoidance of the polemical style of Galileo, and his reticence about natural philosophical contestation, to his fear of running afoul of church authority and crossing the line of heterodoxy in the manner of Galileo. Torricelli was simply a different kind of mathematician with a very different agenda and career trajectory. If his publishing strategies look very different from Galileo's, this may be because he fashioned himself as a different kind of mathematician, and experienced accordingly a very different set of career pressures, and a very different set of questions regarding book publication than Galileo did.

Some of the mathematics published in Torricelli's *Opera* was the material that he circulated via Benedetto Castelli in pursuit of a position alongside Galileo at Arcetri in 1640–1641. In this case, Torricelli's publishing strategies mirror Galileo's at the very beginning of his career at least in this one narrow respect. For in this instance, he used manuscript publication to elevate himself within a respected mathematical circle as part of a project of professional elevation within Italian scientific institutions. In every other case, however, the agendas that led to the publication of the *Opera Geometrica* are not to be found in the publication strategies that motivated Galileo. Torricelli's agendas were informed like Galileo's by his position and obligations as a Tuscan mathematician courtier, but the publication differences that these produced reveal the error of treating the role of the courtly mathematician in seventeenth-century Italy as an undifferentiated monolith.

3 Mathematics and Book Publication in the 17th Century—Other Models

The publishing strategies of Galileo, courtier were not those of Torricelli, courtier. These differences should also not be explained, I argue, by any imagined over-determining influence of the Galileo Affair, the Church, or its authority on Torricelli's youthful life and work. It is much simpler and more plausible to see Torricelli as a different kind of intellectual with a different set of mathematical and scientific agendas, and to view his printed book publications (or rather publication, since he only published one print book) in terms of his own understanding of the courtly role and his own strategy for realizing it. For Galileo, staging polemics served his preferred identity, but Torricelli found different ways to be a mathematical courtier, and his publication practices reveal this different, if no less successful, approach.

Besides those offered by Galileo, what other models of mathematical publication might Torricelli have drawn upon? The first thing to stress about the exceptional nature of the Galileian model is the abundance of printed book publications he authored under his name. To this he added other works published under pseudonyms and by others who put their name on his work.[4] Viewed comparatively, Galileo was simply an exceptional mathematical author in this period in being extremely prolific and highly diverse in his publishing practices. Far more common in the seventeenth century were mathematicians who acquired esteemed reputations despite rather than because of active programs of book publication.

To start, consider the case of Gilles Personne de Roberval (1602–1672), a contemporary of roughly the same age as Torricelli, and one of the most important of his mathematical interlocutors.[5] Roberval rose to become one of the most well-known and influential European mathematicians of the middle of the seventeenth century, but he published almost nothing while achieving that status. He came from very humble origins, and his career was launched largely by accident. A local parish priest started his ascent when he discerned Roberval's talents and taught him mathematics and Latin before sending him off, laden with recommendation letters, on a set of travels. These led him among other places to Fermat's door and then to Paris, where he met Mersenne and made connections with the French crown. In 1632, he was appointed mathematics professor at the Collège Gervais in Paris, and he used the modest stipend allotted him to rent two small second floor rooms at the college. He resided in this simple flat until the end of his life. In 1634, he competed for and won the Ramus Chair at the Collège Royale, arguably the most prestigious and lucrative mathematical professorship in Europe, and certainly a plum position in Paris. The written charter regulating this professorship stipulated that its holder must submit himself to a public mathematical competition every three years to retain the post. After he won the competition in 1634, Roberval met this challenge every three years until his death. After Gassendi's death in 1655, Roberval was also awarded his chair in mathematics at the Collège Royale as well, yet these appointments did not lead Roberval to change his living arrangements. As David Sturdy writes, he lived his entire life in his two-room apartment:

> [...] which had only beds, tables and chairs [...] of poor quality. No pictures hung on the walls, and there were no ornaments. However, he did have some books in his rooms by authors such as Euclid, Archimedes, Viète, Torricelli, Gassendi, Descartes, Mersenne, Kepler, Vitruvius, Herodotus, Cicero and Quintillian. He also had several dictionaries, books on grammar, and (despite not being a religious man) a Latin Bible. When he died not even a bottle of wine was found in his rooms, but a large amount of cash was found, something of the order of eight times his annual salary. (Sturdy 1995, p. 106)

The presence of Torricelli's *Opera Geometrica* in Roberval's modest personal library speaks to the importance of this work in Roberval's life, along with his status as one of Torricelli's most important mathematical readers. But more to the point is the relative dearth of mathematical books possessed by Roberval, and the eclectic nature of his library overall. Roberval also expressed his highly representative relationship to books in his decisions regarding mathematical authorship. His various appointments, esteemed though they were, carried with them no obligation to publish, and Roberval only published two books in his lifetime: his *Traité de Mécanique des Poids Soutenus par des Puissances sur des Plans Inclinés à l'Horizontale* published in 1636, soon after he assumed the Ramus Chair, and a work on Aristarchus and heliocentric cosmology published in 1644 (Roberval 1636, 1644). It is beyond the scope of this paper to explore Roberval's precise publishing strategies in these two cases, but the point to emphasize is the absence of any diligent program of book production despite his esteemed mathematical status, and the failure of Roberval's invisibility in the print public sphere to produce any problems for him when securing his status as a leading mathematician.

---

[4] Two examples are Cecco di Ronchitti, *Dialogo in perpuosito de la stella nova*, which was a work by Galileo published under this pseudonym (Galileo 1605) and Mario Guiducci, *Discorso delle comete* (Guiducci 1619), which was largely written by Galileo although published under the name of his student.

[5] The most comprehensive account of Roberval's life and work remains Auger 1962. See Entry "Gilles Personne de Roberval," by Hara in *Dictionary of Scientific Biography* (hereafter *DSB*; Gillispie, 1970–1990).

Remembering Roberval's modest financial situation and the burden upon authors to front the costs of their own book publication, his decision not to publish printed books makes perfect sense. His possession of a prestigious professorship that conveyed upon him institutional authority further supports this thinking, especially since there was no demand in seventeenth-century Europe that university professors publish books to prove their worth. From this perspective, Isaac Newton also illustrates well the very unmodern relationship between life as a university professor and print publication in seventeenth-century Europe. Like Roberval, Newton published very little mathematical work in his lifetime, and much of what he did publish he released reluctantly while evincing no real desire to become a published author at all.[6] After his appointment as the Lucasian Professor in Mathematics at Cambridge University in 1669, Newton, like Roberval, also found his burden to publish further minimized rather than increased. The professorship gave him a secure position with a steady stipend and no need to author printed books to sustain his status. His failure to publish, therefore, like Roberval's, was less a personal idiosyncrasy than a common outcome for seventeenth-century mathematics professors given the economic and professional realities of the day.

The connection of both Roberval and Torricelli to Marin Mersenne also helps to explain their seemingly counter-intuitive, at least by contemporary standards, relationship to mathematical book publishing. If one thinks of mathematical publication as a means for expert mathematicians to communicate their specialized research with one another, then printed texts were not in the seventeenth century the best way to accomplish this task. Unlike today, where highly regulated professional print publications channel expert knowledge to the expert readers intended as the audience for this scientific research, print culture in the seventeenth century was a kind of Wild West with lots of mathematical gunslingers around and no sheriffs present to maintain order. Printing and circulating mathematical knowledge through books was also expensive, and while it certainly held the promise of a wide dissemination of one's work, it also brought with it the risk of responses from a wide array of individuals, including those would-be mathematicians ill-equipped (or so an author might claim) to criticize the work in question (Johns 1998).

If the goal of publication centered on securing priority for an important discovery or an original mathematical innovation, then print was even less likely to serve ones' needs. As Adrian Johns has shown most fully, the re-packaging and reprinting of material published elsewhere under new names, and with new imprimaturs, was widespread in early modern Europe, and while laws protecting copyright were beginning to be written and enforced, there was absolutely no guarantee that a printed book would secure for its author stable and secure possession of the contents found therein (Johns 2010). For this reason, mathematicians seeking to establish ownership of particular mathematical ideas tended to avoid printing their work in books. Rarely were publishers viewed as esteemed authorities capable of securing intellectual credibility to its authors, and as the unrivaled esteem of the cloistered friar Merin Mersenne made clear, it was most often those far removed from the book trade who were most able to accomplish this credit-building ratification. By contrast, using print publication to secure priority and esteem was more likely to provoke frustrating and distracting debates with the kind of readers who were neither the intended audience for this expert work nor the sort of esteemed peers who warranted critical response.

Far easier and more secure, and far more common as a result given the realities of seventeenth-century print culture, was for mathematical work to be published narrowly and with strategic care through circulation in manuscript to select readers within more restricted networks of mathematical exchange. Here the crucial role played by Marin Mersenne as a "specialized publisher," working almost entirely with unprinted manuscripts to secure exactly this kind of restricted and elite mathematical community cannot be overstated. Connected to all of the expert and respected mathematicians that mattered, and an esteemed mathematician and man of unimpeachable *honnêteté* himself, the manuscript network that Mersenne managed and facilitated was the closest thing that seventeenth-century Europe had to a modern professional mathematical community (Grosslight 2013; Crosbie, Entry "Marin Mersenne," by Crosbie in *DSB*).

Torricelli's first mathematical publication, in fact, appeared through the Mersenne network in May or June 1643. Torricelli started the process by sending a one page set of mathematical claims, without accompanying proofs, to Jean-Francois Niçeron, a friend and fellow Minim Friar close to Mersenne. This short mathematical publication contained Torricelli's famous claim about the finite volume of an infinite hyperbolic solid, and as he intended, the work first reached Mersenne and then through him Roberval and the other expert mathematicians in his circle. This was Torricelli's precise intended audience. As Paolo Mancosu and Ezio Vailati have shown, this work, published in manuscript through the channels managed by Mersenne, marks the real beginning of Torricelli's public career as a mathematician over a year before the print publication of the *Opera Geometrica* (Mancosu and Vailati 1991). It also created the context for the arrival the next year of Torrielli's magnum opus, a work that served many agendas for Torricelli, but one that was neither an effort to introduce his

[6] See Entry "Issac Newton" by Cohen (*DSB*). He summarizes Newton's print publications.

mathematical ideas for the first time into the public sphere, nor an effort to secure priority for his original mathematical claims. In seventeenth-century Europe, such specialized and expert mathematical agendas were poorly realized within the unregulated and boisterous world of early modern print, and to understand why Torricelli published the book he did in 1644 —his only print book published in his lifetime—we need to look beyond questions of expert mathematical communication for answers.

4 Torricelli and the Publication of the *Opera Geometrica*

To understand the motives that led Torricelli toward the publication of his one and only print publication, we need to let go of the idea that Torricelli needed to print this book in order to secure his place as an expert mathematician. He also did not need this book to reveal his mathematical findings to others, or to secure his research claims among expert colleagues. Torricelli felt neither pressure in 1643, and to find the reasons why he did feel the urge to become a published author of a printed book, we need to look elsewhere.

To understand the book's genesis, it is helpful to break Torricelli's brief yet prolific life into a set of periods that each presented him with different challenges leading to different rationales for publishing.

The beginning of the first period can be dated to Torricelli's second and third extant letters of January and June 1640. Torricelli's first letter from September 1632 is famous since it is written to Galileo and expresses the young Torricelli's proclamation to be a *Galileisto* at the moment when Galileo's trial for heresy was just beginning. The twenty-four year old Torricelli could realize this epistolary act of devotion because he was in Rome at the time working for Galileo's disciple Benedetto Castelli. Eight years later, in 1640, Torricelli sent his second and third extant letters containing examples of his mathematical work developed under Castelli over the previous decade (*OET*, 3, pp. 37–42). His third letter in June 1640 was addressed to Castelli explicitly, and it contained mathematics that was inspired by the example found in Galileo's *Discorsi* published in 1638. Castelli passed the letter on to Galileo, and it in turn opened up exchanges about Torricelli between Castelli and the imprisoned Galileo at Arcetri, a dialogue which led to Galileo's invitation to Torricelli to join him at his Florentine villa cum prison in June 1641. The would-be acolyte's move to Florence in October of that year, starting what would become an illustrious Tuscan career, began this first period which ended with Galileo's death in January 1642. Galileo's death left both the thirty-four year old Torricelli and the much younger Vincenzo Viviani, who also resided with Galileo at Arcetri during these years—he celebrated his twentieth birthday there in April 1642—as the likely heirs to Galileo as the Principal Mathematician and perhaps Philosopher to the Grand Duke of Tuscany.

At the center of the exchanges in 1640–1641 was Torricelli's mathematical work on the motion of heavy bodies and projectiles that would later be published in print as Books III and IV of Torricelli's *Opera Geometrica*. Since there is no evidence that this work was circulated beyond the circle of Castelli and Galileo (and perhaps Viviani, who was already at Arcetri at the time), it is probably inaccurate to call this an example of manuscript publication. But it was certainly a use of written texts to present and circulate Torricelli's work to an expert audience, albeit in pursuit of patronage as much as the public advancement of mathematical knowledge. Nevertheless, since these two motivations were inseparably entangled in seventeenth-century Europe in the same way that professional career advancement within a university is entangled today with mathematical publication today, one should mark this moment as Torricelli's debut as an up and coming young mathematician seeking a place within the expert mathematical community of his day.

The second period in Torricelli's career begins sometime after January 1642 when he was recognized as Galileo's successor as the Principle Mathematician of the Grand Duke of Tuscany. Since Galileo was in Padua when he won the same appointment in 1610, he left a correspondence trail indicating the precise negotiation over the terms of his appointment. Residing in Florence at the time of Galileo's death, Torricelli left no such paper trail, so it is impossible to pinpoint exactly when he was made the Medici court mathematician. Whenever it happened, it likely occurred before he used Jean-François Niçeron to communicate a sample of his mathematical work to Mersenne and his network in June 1643 eighteen months after Galileo's death. This was the mathematics that made Torricelli an international mathematical celebrity, and it was also the work that provoked his debates with Roberval and others, arguments that would dominate the remainder of his short life (he died four years later in October 1647).

What is intriguing about the events of the summer of 1643 with respect to the arguments of this article is the coincidence of a number of changes in Torricelli's life that all conspired to generate a very complex and fluid set of calculations about publishing. Framing everything was Torricelli's elevation as the Medici court mathematician, a promotion that brought with it new expectations regarding his public intellectual work. Galileo presented one model of how to proceed, but even if his scandals with the church are left aside—an interpretive move that is too rarely made, leading to the common overstatement of the influence of the Galileo Affair on Torricelli's life and science—Galileo's publishing strategies as courtier were risky at best, and difficult to emulate in any case given the centrality of the Galileian personality to them. Torricelli also made clear his interest in mathematics alone as opposed to the physico-mathematics and natural philosophy that Galileo preferred, and this too was the source of all of Galileo's problems.[7] To become the public mathematician that Torricelli needed to become as courtier, he needed to publish, but he did not need to follow Galileo's model, nor even less become the brash and sparkling controversialist that was Galileo's hallmark.

If he eschewed the model of Galileo, what other models presented themselves to Torricelli in 1642–1643? Christian Huygens offers one example (Andriesse 2005; Bos 1986). The son of a well-bred Dutch statesman and poet, Huygens began his scientific career already secure in his social and intellectual position. Like Torricelli in 1642–1643, he did not, therefore, need to publish gifts that would secure him social elevation or status, nor did he need to establish himself as an authoritative human being of respectable status. What Huygens (and by comparison Torricelli) lacked was recognition among mathematicians for his research, and ownership of mathematical work that would secure his name among these precise experts. Accordingly, between 1649 and 1659 he published both in manuscript and in print a series of short works in Latin on everything from geometry and mechanics to planetary astronomy. These included a pamphlet size work on his theory of the cycloid pendulum and the clock based on it.[8]

It was based on this work, but equally so on his esteemed family pedigree, that Huygens was invited to join the newly founded *Académie Royale des Sciences* in 1666, a new institution created by the ambitious young monarch Louis XIV and his equally energetic minister Jean-Baptiste Colbert. The project of the new academy was effectively to centralize the ordinary royal patronage of scientific savants by collecting them all in a single company joined to the court as one unit. Huygens was the most esteemed of this new coterie of French court *mathématiciens et physiciens* appointed in 1666, and he was given a gigantic six thousand *livres* pension as a special mark of his favor. The point to note in this context is that after Huygens entered the academy, he stopped the prolific publishing of the preceding decade, producing only two books between 1666 and 1690, and only one individually authored mathematical book (Huygens 1673, 1684).

As a member of the *Académie Royale des Sciences,* Huygens was beholden to a collective ethos that, at least at first, treated academicians as members of a collective company, and channeled publishing agendas away from individual research and publishing programs toward collective work in the service of the crown (all this changed in the 1690s and was codified in an academic reform of 1699 that started to demand publication as a marker of service; Hahn 1971, Chaps. 1 and 2; Stroup 1990; Shank 2018, Chap. 3). Huygens served as the de facto head of this company at first, and no evidence suggests that he was anything other than happy with this arrangement, especially given the financial benefits that went with it. He also agreed fully, and was even the inspiration for the move of the company in a particular publishing direction when it decided to focus on natural history for the academy's initial collective publications (Stroup 1990, p. 70).

Denis Dodart was named as the editor on the title page when the French royal press published the French academy's first book, a lavish, supersized volume with exceptionally large and beautifully rendered plates printed in 1676 (Dodart 1676). But it was the members of the royal academy as a collective, also listed prominently next to Dodart's name, who were the real authors of this book. The academy presented Louis XIV with a copy of this ornate volume when he made his one and only visit to the royal academy in 1682, and the book itself, with its focus on French plants and those exotic specimens present in the royal collections at Versailles, along with its extravagance as a material object, exemplified the court orientation of French academic science during Huygens's time there.

Huygens's own more fully developed account of his pendulum clock published in 1673 should also be viewed as his individual contribution to this same collectivist academic esprit. His *Horlogium Oscillatarium* was also published by a royally financed printer, like the academy's work on plants, and given the ever present cost burdens associated with book publication, it is important to note that a specially designated royal printer assigned to the French royal academy was one of the new perks added to the new academy structure that Colbert and the French crown innovated in 1666. Printing, however, remained costly, even in the court of the Sun King, and the royal academy did not use its printer at first as a vehicle for the wide circulation of individual scientific work by royal academicians (this too would change in the 1690s). The royal press was used, rather, as a mechanism for publishing a few select books that demonstrated the academy's worthiness in the eyes of its royal patron and his publics. Huygens's publishing record while a member of the *Académie des Sciences* illustrates well the results that such publication structures produced for mathematicians beholden to such courtly institutions.

---

[7] See Torricelli's letter to Michelangelo Ricci, 10 February 1646 (*OET*, 3, pp. 357–358).

[8] A complete list of all of Huygens' print publications is found in *Oeuvres Complètes de Christiaan Huygens* (Huygens 1888–1950, XXII, pp. 375–38).

Although no work on courtly scientific publishing in Tuscany has been done that would allow for explicit comparisons with its counterpart in France, it is appropriate to see the French case as an example of what any early modern sovereign might have done, and expected in return, had he had the resources which the Sun King possessed. The case of Vincenzo Viviani can illustrate the point. When Torricelli died in 1647, only five years after Galileo, the Medici again found themselves in need of a court mathematician. Viviani appeared as a logical successor, yet he was still only twenty-five years old in 1647, and even as he matured he does not appear to have been recognized with this or a comparable title even as he came to play something like this role in practice during the rest of his long life (he died in 1703). There was no obligation on the part of the Grand Duke to appoint a court mathematician, and as the changes in France show, the nature of European courtly science began to change dramatically after 1650. Similar changes occurred in Tuscany as well, not least the creation of the Tuscan *Accademia del Cimento* in 1657, which served as a model for new royal scientific institutions in London, Paris and elsewhere around Europe and a company that also instituted its own print publication.

The title pages of Viviani's first four printed books published between 1659 and 1677 (he only published four other print books during the remaining twenty-five years of his life) illustrate well the shifting sands and his shifting identity within them and Tuscan court science. His first book, *De maximis et minimis qeometrica divination in quintum Conicorum Apollonii Pergaei, adhuc desideratum* published at Florence in 1659 was a scholarly restoration of the fifth book of Apollonius' work on conic sections (Viviani 1659). It was dedicated to Grand Duke Ferdinando II de Medici, but Viviani was only listed as "Auctore" on the tite page. Fifteen years later in his Italian *Quinto libro degli elementi d' Euclide, ovvero scienza universale delle proporzioni, spiegata colla dottrina del Galileo,* Viviani brought out what would become his trademark identity, declaring himself on the title page to be "[...] the last disciple of Galileo." (Viviani 1674). This title may in fact have captured best Viviani's ultimate role with respect to mathematics at the Tuscan court since among his major contributions was to serve as Galileo's first historian, as well as the executor of his memory and legacy. Vivaini's effort to produce a published collection of Galileo's works in these years was certainly an important part of his work, and in this way Viviani served the Tuscan court as a mathematician by securing the memory and ongoing legacy of their most important and famous offspring, Galileo and his mathematical science (Galluzzi 1998; Segre 1989).

He was also involved in the work of the *Accademia del Cimento* after 1657, including its pioneering publications *I Saggi*, books that did much to create and sustain the idea of a Florentine "Galileian science" (Boschiero 2007). In his fourth publication, *Enodatio problematum universis gemoetricis praepositorum a D. Claudio Comiers* printed at Florence in 1677, he showed the development (or not) of his identity by listing himself merely as one of the mathematicians to his Highness the Grand Duke of Tuscany (Viviani 1677). These are not markers of Viviani's elevation to a titled position at the Tuscan court, but it should be noted that however slim and scattered his publishing record might appear, it was enough to attract the attention of Louis XIV and Colbert, who attempted to recruit Viviani to join their new *Académie Royale des Sciences* in 1666. The French crown offered him a lucrative pension, but he ultimately chose to stay home in Tuscany, and this example illustrates well the error of trying to read mathematical reputation in seventeenth-century Europe from the number and nature of the authored mathematical books that a mathematician published.

Viviani's relation to print publishing was more complex than Torricelli's since he lived much longer than his predecessor and worked at a time when the court structures supporting mathematicians were being transformed, especially by the rise of the new academies and the new understandings of state science that supported them (Biagioli 1995, 1996). Returning to Torricelli, several lessons can be drawn from these cases that explain the rationale behind the publication of the *Opera Geometrica*.

The changes that Torricelli experienced in 1642 set the context for his publication decisions over the next two years, and his elevation to a titled position at the Tuscan court produced his most important reason to publish since it gave him both the reason to produce a printed book and the means to overcome the financial burdens involved. Bonaventura Cavalieri recognized these motivations in his letter of congratulations to Torricelli sent in May 1643. He praised him for working to bring the book to completion, but he also praised through Torricelli the Grand Duke, singling out his love of letters and his generosity in sustaining the costs necessary to bring Torricelli's work to the public. Torricelli also suggested that his book was primarily conceived as a gift-like work of service to his sovereign patrons (*OET*, 3, pp. 123–124). He dedicated one part of it to Grand Duke Ferdinando, and another to the Grand Duke's brother Leopoldo, the man who did so much a decade later to create and support the *Accademia del Cimento*. Much of the book's contents also suggest a treatise designed to serve as a compendium of Torricelli's previous mathematical achievements, one that would demonstrate his established merit and package his mathematical labors as a single circulatable bundle conveying the glory of the Medici and Tuscany.

Books I and II, for example, entitled *De sphaera et solidis sphaeralibus* reconnected with the venerable work of Archimedes and extended it in light of recent thinking by Galileo especially. This part of Torricelli's work was resonant with Viviani's work on Apollonius and Euclid published a decade later, and it reflects the way that mathematics in the seventeenth century was still strongly attached to the Humanist project of ancient mathematical reconstruction and interpretation. Books III and IV continued in this historical vein, publishing in print for the first time *De motu gravium naturaliter descendentium* and *De motu proeicturum,* which Torricelli had developed under Castelli's supervision in 1640–1641. This also established a connection with Galileo's *Discorsi,* which had served as his calling card when he won a place alongside Viviani and Galileo at Arcetri in 1641. Castelli died in April 1643, after Torricelli's *Opera* had been accepted for publication by the Medici, but before it went into press, and Torricelli connected himself with the Castelli thread of Galileian history as well by including an appendix to his treatise entitled *De motu acquarum.* Here he paid homage to his early teacher and mentor by adding his own extensions to Castelli's work in hydraulics published in 1628 as *Della misura dell'acque correnti.*[9]

In all these ways, Torricelli's *Opera Geometrica* is best seen as a retrospective compendium summarizing Torricelli's previous mathematical achievements in a manner suitable for establishing his reputation as a mathematician worthy of his title at the Medici court. Paid for by the Medici as a gift to the public, the printed book is also best described as a work of service that showed to the public at large Torricelli's estimable worthiness and honor to his patrons, while also justifying the privileges acquired from them. Such were the linkages that tied mathematics to court patronage in seventeenth-century Europe, and the strong influence of these ties in the case of Torricelli's *Opera* links it to many other mathematical books published for similar reasons in the same period.

Not every aspect of the *Opera Geometrica* can be fully explained by these early modern court-driven publication logics, however. In particular, Books V and VI of the *Opera* entitled *De Dimensione parabole solidique hyperbolici* present us with a very different set of issues. In these books, Torricelli deployed Cavalieri's geometry of indivisibles, entering with this material into the debates about infinitesimalist approaches to geometry that were heated at the time. He also included in his book his demonstration of the finite volume equal to the infinite volume of a hyperbolic solid, which used the method of indivisibles. This was the proposition that he had sent via Niçeron to Mersenne in June 1643, and the work that had marked his public mathematical debut.

Torricelli had already received word from the Medici regarding their willingness to support the publication of a mathematical book under his name before he sent his one page list of mathematical propositions to Paris in 1643. His decision, therefore, to first publish his infintesimalist mathematics in manuscript via the Mersenne circle was not made out of fear of not getting his work published. He likely calculated at this time that manuscript publication would offer a more immediate way to intervene in the mathematical debates of the moment, and to make an immediate splash within them. Book production in early modern Europe was a lengthy and tedious process (even more so than today!), a fact illustrated by the passage of over a year between the commitment of the Medici to publish Torricelli's book in April 1643, and its appearance as a printed treatise in September 1644.[10] During those sixteen months, Torricelli saw his finite volume infinite solid debated by some of the leading mathematicians in Europe, and he also saw his work deployed in the ongoing debate about the validity of using indivisibles in geometry, a debate that had recently become focused on exchanges between his Galileian colleague Cavalieri and the Jesuit Paul Guldin. By the time Torricelli's book was published in late 1644, this debate had already taken new turns, and the rationale for including this work in his *Opera* cannot be reduced simply to Torricelli's desire to get his work "out there" and to participate in this debate. His manuscript publication accomplished that goal, so other reasons need to be found regarding why Torricelli decided to include this material in what was otherwise a historical, retrospective compendium of Torricelli's previous mathematical accomplishments.

Torricelli's correspondence with Cavalieri after the death of Galileo in 1642 offers a revealing window into these deliberations. Cavalieri held a professorship in mathematics at the University of Bologna, a position that did not require book publication, but one that made publication both rewarding and cost effective since the university possessed a press that professors could use to publish their work. Cavalieri used these institutional supports in 1635 to publish at Bologna *Geometria indivisibilibus continuorum nova quadam ratione promota*, the book that introduced his geometry of indivisibles to the world.[11] Among its readers was Paul Guldin, a child of Jewish parents who had converted to Protestantism, and who was in 1635 a mathematics professor at Graz who had undergone his own conversion to Catholicism that culminated in his entrance into the Jesuit order and his mathematical study under Christopher Clavius at the Jesuit *Collegio Romano*[12] (*DBS*). Guldin was one of the fierce Jesuit critics of infinitesimal mathematics that Amir Alexander has recently studied, and he responded directly and critically to Cavalieri's work in his *Centrobaryca seu de centro gravitatis trium specierum quantitatis continuae*

[9] Castelli 1628. See Entry "Benedetto Castelli" by Drake (*DSB*; Maffioli 1994, Chap. 1).
[10] The step by step details of the path into publication of Torricelli's *Opera Geometrica* are summarized in Toscano 2008, pp. 111–120.
[11] On Cavalieri and his method of indivisibles, see Andersen 1985.
[12] On the Guldin-Cavalieri dispute, see Mancosu 1996, Chap. 2.

published in four volumes at the Viennese printing house supported by the Society of Jesus between 1635 and 1641 (Guldin 1635–1641; Alexander 2014, Chaps. 4 and 5). Volume II published in 1640 criticized Cavalieri's work directly, and by the time that Torricelli and Cavalieri were exchanging letters, the latter had read the work and was planning his response (*OET*, 3, p. 65).

Cavalieri received a pre-publication copy of Torricelli's demonstration of the finite volume infinite solid in this context, and given its use of the method of indivisibles to achieve its results, he immediately saw in the work a resource for challenging Guldin and "liberating him from his mistakes."[13] Galileo's death had occurred eleven months earlier, and by the start of 1643 Torricelli had certainly experienced his elevation as Medici court mathematician, and was therefore confronted by his need to make a public demonstration of his mathematical merit. His calculations in this regard got entangled with his interest in Cavalieri's mathematical disputes with Guldin, and with his participation and interest in the reception of the geometry of indivisibles more generally.

In January 1643, Cavalieri wrote to Torricelli explaining his decision to add a response to Guldin in his *Triganometria plana, et sphaerica, linearis et logarithmica*, which he was preparing to publish at Bologna in the coming year. Wanting to add impact to the response, he asked Torricelli if he could append his demonstration of the finite volume infinite solid to this work under his own name, a request that he framed as a favor. This was an indicator of how mathematical exchange occurred in this period as a kind of gift exchange rather than as a proto-professional sharing of mathematical research. He also mentioned his hope of recruiting other mathematicians who might give him similar favors, namely Antonio Nardi and Antonio Rocca. He further raised the possibility that these mathematicians, and perhaps Torricelli as well, might publish their own works in the geometry of indivisibles independently as a further statement of support for his position against Guldin's. He lastly raised the prospect of authoring a published dialogue that might stage the debate with Guldin in print and resolve it in his favor.[14]

All of this talk is resonant with Galileo's strategies when pursuing one of his mathematical battles, especially the idea of a dialogue where Cavlaieri could stage the dispute in a way friendly to his position and hostile to the thinking of his enemies. Yet importantly from the perspective of Torricelli's own self-fashioning as a mathematician and a courtier, he did not respond favorably to Cavalieri's overtures. In February, Torricelli responded that he had still not succeeded in finding a copy of Book II of Guldin's *Centrobaryca,* so he was still not sure exactly what Guldin was criticizing and what the repudiation amounted to.[15] He also reported soon after about his meeting with the Medici, and their decision to support the publication of his book. By June, the decision had been made to publish Torricelli's finite volume infinite solid via manuscript through the Mersenne network, and not in the way that Cavalieri had originally requested. And while Cavalieri continued to talk about the idea of a dialogue that would address the dispute directly, his requested favors from Nardi and Rocca did not materialize and he ultimately reconceived the idea. In November 1643, Guldin died, and by early 1644 Torricelli was sending Cavalieri the text of his forthcoming treatise. Meanwhile, Cavalieri began working on a Latin work, eventually published as part of his 1647 *Exercitationes geometricae sex*, where he disputed directly with the deceased Guldin and defended his geometry against the criticisms made of it in the *Centrobaryca.*

In May 1644, Cavalieri wrote to Torricelli praising what he had read from Torricelli's forthcoming treatise and celebrating the forthcoming *Opera Geometrica* as the more important book with respect to their joint project of refuting Guldin. In June, Cavalieri expressed some concern about the relative absence of the geometry of indivisibles in the text he was reading, and he also lamented with Torricelli the exceedingly slow publication process since he wanted to have Torricelli's treatise in print so he could use it as a base for correcting his own dialogue with Guldin, which he was trying to bring into print.[16] When the *Opera Geometrica* was finally published in September 1644, complete with the full presentation of Torricelli's work on the finite volume infinite solid and other work deploying the method of indivisibles, Cavalieri was thrilled. Roberval among others indicated that Torricelli's treatise was now the work to reckon with regarding the geometry of indivisibles, and while the treatise contained much more than this, and its reception was not restricted to its use of the geometry of indivisibles alone, the presence of this treatise with its strong demonstration of the value of indivisibles satisfied Cavalieri. His own dialogue with Guldin was accordingly shelved, and it only appeared in print in a compendium published right before his death in November 1647, a month after Torricelli's own death in Florence (Cavalieri 1647, *III. In Paulum Gulidinum è Societate Iesu dicta Indivisibilia oppugnatem*).

---

[13] Cavalieri to Torricelli, 29 December 1642 (*OET*, 3, p. 90).

[14] Cavalieri to Torricelli, 13 January 1643 (*OET*, 3, pp. 92–96).

[15] Torricelli to Cavalieri, 21 February 1643 (*OET*, 3, pp. 104–105).

[16] Cavalieri to Torricelli, 7 May 1644 (*OET*, 3, pp. 179–180).

The publication delays of Torricelli's better-supported and "more important" (in Cavalieri's words) treatise had accidentally created the timing of his precise intervention in the debate about indivisible geometry. Yet, ironically, the later, print publication of this material, after it had already created a stir in manuscript, made this more timely section of the mathematical treatise compatible with the more historical and retrospective contents of the rest of the book. By appearing when it did, after the death of Guldin, and a year and a half after the manuscript publication of Torricelli's finite volume infinite solid, and at a moment when the debates that it had triggered had begun to solidify, the *Opera Geometrica* suggested a framing of this material in terms of a battle won, and a mathematical achievement accomplished.

This effect was especially reinforced by the other mathematical material published alongside it, which was all authoritative work free of any hint of controversy, work that testified to Torricelli's established accomplishment as a mathematician. The appearance of this precise geometry in this precise way also allowed Torricelli to accomplish the self-fashioning goals desired through his treatise in a doubly beneficial way. He was able to claim the sparkling *sprezzatura* needed in a court mathematician while also steering clear of the controversies that had often accompanied Galileo's own claim upon the same, controversies that became the source for his ultimate downfall.

In sum, Torricelli accomplished, however unintentionally and accidentally, a kind of institutional squaring of the circle through his *Opera Geometrica.* He was able to use his treatise, supported by his sovereign patrons and dedicated to them, to both possess the cake of established respectability and reputation while also consuming its sugary sweetness by having spectacularly innovative and sparklingly controversial mathematical work circulate under his name in it, and all this without the taint of polemical combat. All of this together raised his esteem as court mathematician.

5 Conclusion. Reading the *Opera Geometrica* in Light of Its Print Agendas

How does this understanding of the origins of the *Opera Geometrica* change our reading of the contents found therein? And how is the place of the book in the history of seventeenth-century mathematics changed when viewed from this perspective? Most important is the perspective it offers on one of the most puzzling aspects of the treatise: Toricelli's decision to include twenty-one different demonstrations of the same truth already demonstrated indubitably by Archimedes almost two thousand years earlier in his account of *De Dimensione parabolae*. Why perform such a seemingly unnecessary and apparently redundant mathematical exercise? Two explanations for this content follow from the rationale for the book as outlined above.

The first involves Torricelli's relation to Archimedes and the ancient tradition of geometry that he exemplified. For a modern, the point of mathematical research is to discover new truths and to expand the body of extant, authoritative mathematical knowledge. From this perspective, demonstrating in a new way what is already known indubitably is only valuable if the exercise generates new insights beyond what is contained in the original claim of the proof itself. But Torricelli did not approach Archimedes in this way. He approached the venerable Syracusan as a mathematician with a seventeenth-century understanding of the relationship between ancient and modern mathematics. One difference was to see value in the simple reconstruction of ancient mathematical thought. We saw this urge in Viviani's devotion to the mathematical reconstruction of Apollonius and Euclid, and while one motivation involved simply recovering lost mathematics for the present, the re-production and consequent re-interpretation of ancient mathematics was also valued because of the insights it could offer into mathematical knowledge generally.

The Ancient author Pappus had suggested in his mathematical commentaries that the Ancients possessed powerful mathematical methods since lost to posterity (Bos 2001). Recovering lost books was one way to recover these methods, but another was to reproduce the work of the ancients in new and imaginative ways. Early modern mathematicians from Viète to Newton produced what they claimed was new and modern mathematics through attempts at reconstructing ancient thought, and Torricelli's relationship to Archimedes is closely aligned to this tradition.[17] From this perspective, his twenty-one re-workings of Archimedes' understanding of the area under the parabola can be understood as an early modern mathematical research project in pursuit of the deeper insights contained in the limited textual record of the ancient past.

Such a relationship to ancient mathematics and to Archimedes in general also served Torricelli's ambitions as a courtier. Caught between the early modern esteem for the ancients as possessors of universal knowledge and wisdom, and the new urges in the period pushing savants to claim identities as original innovators and *virtuosi* with exceptional individual talent, a complex mix of reverence and rebellion characterized the attitude of many ambitious early modern intellectuals when facing

[17] On Newton in this respect, see Guicciardini 2009.

the legacy of antiquity. Torricelli exemplified this dynamic attitude in his own work. On the one hand, he welcomed the appellation "the new Archimedes," which was often given to him, and he worked hard to cultivate a reputation as a reverent follower of this ancient mathematician and sage. At the same time, he also showed his contemporary virtuosity by trying to exceed the limits of the past, and by building new and better mathematical achievements that superseded those of the ancients. In this way, he cultivated a highly potent and successful double relationship to Archimedes as both his heir and supercessor, a stance that made him both an antique hero and a new and modern talent who was capable of exceeding the merit of even the greatest geniuses of the past.

Torricelli's twenty-one different demonstrations of the Archimedean truth that the area under the parabola equals 4/3 the area of the inscribed triangle composed upon its base was a perfect illustration of this double relation. In a treatise designed to demonstrate his worthiness as the mathematician of the Grand Duke of Tuscany, the logic behind this performance also makes perfect sense. Also in step with it was his use in this part of his treatise of Cavalieri's method of indivisibles to perform some of his most spectacular work.

At least three agendas were served by using the geometry of indivisibles to show in different ways the Archimedean truth about the area under the parabola. One was strictly epistemological and directed at Guldin's challenges regarding the validity of the method. If the method of indivisibles produced in its own way the same truths guaranteed by rigorous Euclidean methods, then epistemological arguments challenging the validity of infinitesimal methods were undermined. Later in the century, during the debates about the validity of the Leibnizian differential calculus, practitioners of the new analysis would often demonstrate a proposition twice, once using the new analysis and a second time using traditional Euclidean methods. The point was to show the validity of the first through its agreement with solutions produced by the rigorous methods of the second.[18] Torricelli's juxtaposition of rigorous Euclidean proofs with similar results obtained using the innovative method of indivisibles was similarly motivated.

The later practices of Johann Bernoulli and others regarding infinitesimalist mathematics also illustrates another aspect of Torricelli's agenda. When arguing why the differential calculus was better than traditional geometry even though it was no less rigorous than the Euclidean methods, Bernoulli and Leibniz often praised the elegance and economy of the solutions achieved in this new analytic way. The unnecessary tedium of letting traditional fetishes about exceptional rigor stand in the way of sparkling innovation was also ridiculed as a backward looking avoidance of brilliant innovation.[19] Torricelli was not as articulate as Bernoulli in expressing these values, yet I would suggest that a similar set of aesthetic concerns were also at play in Torricelli's work. If the point had only been to demonstrate the epistemological equality of the Archimedean solution with the one achieved through Cavalieri's indivisibles, then a few demonstrations making the point would have been sufficient. So why then did he offer twenty one? Perhaps because this Baroque display of mathematical *sprezzatura* allowed Torricelli to make manifest the sparkle of his inventiveness and the abundance of his *ingegno,* and to make clear his mathematical superiority when compared with Archimedes.

Performances of scientific prowess and intellectual *virtù* were essential features of any successful self-fashioning of any early modern mathematical courtier, and this was especially true in mid-seventeenth-century Tuscany given the way that Galileo set the standard by which all others measured themselves. Is it not therefore plausible to imagine Torricelli conceiving of such mathematical work as a virtuoso aesthetic performance, one designed to show the sublime genius of his creative mind? Such a reading is especially suggested by the presence of this prodigious mathematical ingenuity in his magnum opus, the treatise which he gave as a gift to his esteemed patrons, and which he authored to show his own worthiness to receive the special marks of distinction he had been granted. Such were the cultural logics driving the professional career management of ambitious seventeenth-century mathematicians like Torricelli, and I would argue, by way of conclusion, that the mathematical results present in his court-supported *Opera Geometrica* were profoundly shaped, both in general intent and precise substance, by the cultural dynamics of early modern courtly science and especially the patterns for these present in Tuscany around 1650.

---

[18] Bernoulli did this in his presentation of the solutions to his "Brachistochrone" challenge problem of 1693 (Costabel and Peiffer 1988).

[19] Pierre Varignon was an especially vigorous adavocate for athe aesthetic beauty and economy of infinitesimal analysis (Mancosu 1996, Chap. 6.3; Blay 1992).

References

Alexander A (2014) Infinitesimal: How a Dangerous Mathematical Theory Shaped the Modern World. Scientific American/Farrar, Straus and Giroux, New York.

Andersen K (1985) Cavalieri's Method of Indivisibles. Archive for History of Exact Sciences 31/4:291–367.

Andriesse CD (2005) Huygens: The Man Behind the Principle. Cambridge University Press, Cambridge.

Auger L (1962) Un savant méconnu: Gilles Personne de Roberval (1602–1675): son activité intellectuelle dans les domaines mathématique, physique, mécanique et philosophique. Blanchard, Paris.

Bertoloni Meli D (2006) Thinking with Objects: The Transformation of Mechanics in the Seventeenth Century. Johns Hopkins University Press, Baltimore.

Biagioli M (1992) Galileo, Courtier: The Practice of Science in the Culture of Absolutism. University of Chicago Press, Chicago.

Biagioli M (1995) Le prince et les savants: la civilité scientifique au XVIIe siècle. Annales Histoire Sciences Sociales 50/6:1417–1453.

Biagioli M (1996) Etiquette, Interdependence, and Sociability in Seventeenth-Century Science. Critical Inquiry 22/2:193–238.

Biagioli M (2016) Galileo's Instruments of Credit: Telescopes, Images, Secrecy. University of Chicago Press, Chicago.

Blay M (1992) La Naissance de la mécanique analytique. Presses Universitaires de la France, Paris.

Bos HJM (1986) Introduction. In: Richard J, Blackwell RJ (trans.) Christian Huygens' The Pendulum Clock, or Geometrical Demonstrations Concerning the Motion of Pendula as Applied to Clocks. Ames. Iowa State University Press, pp. xi–xxx.

Bos HJM (2001) Redefining Geometric Exactness: Descartes' Transformation of the Early Modern Concept of Construction. Springer, Dordrecht.

Boschiero L (2007) Experiment and Natural Philosophy in Seventeenth-Century Tuscany: The History of the Accademia del Cimento. Springer, Dordrecht.

Busard HLL (1970–1990) Entry: Paul Guldin. In: DBS.

Büttner J, Damerow P, Renn J (2001) Traces of an Invisible Giant: Shared Knowledge in Galileo's Unpublished Treatises. In: Montesinos J, Solis C (eds). Largo campo di filosofare. Eurosymposium Galileo 2001. La Orotava: Fondaciòn Canaria Orotava de Historia de la Ciencia, pp. 183–202.

Büttner J, Damerow P, Renn J (2004) Galileo's Unpublished Treatises. In: Palmerino CR, Thijssen JMMH (eds). The Reception of the Galileian Science of Motion in Seventeenth-Century Europe. Springer, Dordrecht, pp. 99–117.

Castelli B (1628) Della misura dell'acque correnti. Stamparia Camerale, Roma.

Cavalieri B (1647) Exercitationes geometricae sex. Jacobi Montii, Bologna.

Costabel P, Peiffer J (eds) (1998) Der Briefwechsel von Johann I Bernoulli. Band 2: Der Briefwechsel mit Pierre Varignon: Erster Teil: 1692–1702. Basel, Birkhauser.

Crosbie AC (1970–1990) Entry: Marin Mersenne. In: DBS.

DBS see: Gillispie 1970–1980.

Di Ronchitti C (1605) Dialogo in perpuosito de la stella nova. Tiozzi, Padova.

Dodart M (1676) Mémoires pour servir à l'histoire des plantes. Dressez par M. Dodart de l'Académie Royale des Sciences, Paris de l'Imprimerie Royale, Paris.

Drake S (1970–1990) Entry: Benedetto Castelli. In: DBS.

Drake S (1976) Galileo against the philosphers in his dialogue of Cecco di Roncitti (1605) and Considerations of Alimberto Mauri (1606). Zeitlin & Ver Brugge, Los Angeles.

Freedberg D (2003) The Eye of The Lynx: Galileo, His Friends, and the Beginnings of Modern Natural History. University of Chicago Press, Chicago.

Galilei G (as Cecco di Ronchitti) 1605 Dialogo in perpuosito de la stella nova. Tiozzi, Padua.

Galluzzi P (1998) The Sepulchres of Galileo: The 'Living' Remains of a Hero of Science. In: Machamer P (ed). The Cambridge Companion to Galileo. Cambridge University Press, Cambridge, pp. 417–47.

Gillispie CC (ed) (1970–1980) Dictionary of scientific biography. Charles Scribner's Sons, New York.

Grosslight J (2013) Small Skills, Big Networks: Marin Mersenne as Mathematical Intelligencer. History of Science 51:337–74.

Guicciardini N (2009) Isaac Newton on Mathematical Method and Certainty. MIT Press, Cambridge–MA.

Guiducci M (1619) Discorso delle comete. Pietro Cecconcelli, Firenze.

Guldin P (1634–1641) Centrobaryca seu de centro gravitatis trium specierum quantitatis continuae, 4 vols. G. Gelbhaar & M. Cosmerovii, Vienna.

Hahn R (1971) Anatomy of a Scientific Institution: The Paris Academy of Sciences, 1666–1803. University of California Press, Berkeley.

Hara K (1970–1990) Entry: Gilles Personne de Roberval. In: DBS.

Heilbron J (2010) Galileo. Oxford University Press, Oxford.

Huygens C (1673) Christiani Hugenii Zulichemii, Const. F. Horologium Oscillatorium sive De motu pendulorum ad horologia aptato demonstrationes geometricae. Parisiis, apud F. Muguet, regis et illustrissimi archiepiscopi typographum, via Citharae, ad insigne trium Regum.

Huygens C (1684) Astroscopia Compendiaria tubi optici molimine liberata. Arnold Leers, The Hague.

Huygens C (1888–1950) Œuvres Complètes de Christiaan Huygens, 22 vols. Martinus Nijhoff, The Hague.

Johns A (1998) The Nature of the Book: Print and Knowledge in the Making. University of Chicago Press, Chicago.

Johns A (2010) Piracy: The Intellectual Property Wars from Guttenberg to Gates. University of Chicago Press, Chicago.

Maffioli C (1994) Out of Galileo: The Science of Waters, 1628–1718. Erasmus Publishing, Rotterdam.

Mancosu P (1996) Philosophy of Mathematics and Mathematical Practice in the Seventeenth Century. Oxford University Press, Oxford.

Mancosu P, Vailati E (1991) Torricelli's Infinitely Long Solid and its Philosophical Reception in the Seventeenth Century. Isis 82/1:50–70.

Meli DB (2006) Thinking with Objects: The Transformation of Mechanics in the Seventeenth Century. Johns Hopkins University Press, Baltimore.

OET see: Torricelli 1919–1944.

Roberval GP (1636) Traité de mechanique. Des poids soustenus par des puissances sur les plans inclinez à l'horizon. Des puissances qui soustiennent un poids suspendu à deux chordes. Richard Charlemagne, Paris.

Roberval GP (1644) Aristarchi Samii de mundi systemate, partibus et motibus ejusdem libellus. Adjectae sunt AE. de Roberval notae in eundem libellum. A. Bertier, Paris.
Segre M (1989) Viviani's Life of Galileo. Isis 80/2:206–231.
Shank JB (2018) Before Voltaire: the French origins of "Newtonian" mechanics (1680–1715). University of Chicago Press, Chicago.
Stroup A (1990) A Company of Scientists: Botany, Patronage, and Community at the Seventeenth-Century Parisian Royal Academy of Sciences. University of California Press, Berkeley.
Sturdy DJ (1995) Science and Social Status: The Members of the Académie des Sciences 1666–1750. Boydell & Brewer, Woodbridge.
Torricelli E (1919–1944) Opere di Evangelista Torricelli edite in occasione del III centenario della nascita col concorso del Comune di Faenza da Gino Loria e Giuseppe Vassura. Montanari. Faenza.
Toscano F (2008) L'erede di Galileo. Vita breve e mirabile di Evangelista Torricelli. Sironi, Milano.
Valleriani M (2010) Galileo Engineer. Springer, Dordrecht.
Van Helden A, Gingerich O (2003) From "Occhiale" to Printed Page: The Making of Galileo's *Sidereus Nuncius*. Journal for the History of Astronomy 34:251–267.
Viviani V (1659) De maximis et minimis geometrica divination in quintum Conicorum Apollonii Pergaei, adhuc desideratum. Giuseppe Coccini, Firenze.
Viviani V (1674) Quinto libro degli elementi d'Euclide, ovvero scienza universale delle proporzioni, spiegata colla dottrina del Galileo. Condotta, Firenze.
Viviani V (1677) Enondatio problematum universis gemoetricis praepositorum a D. Claudio Comiers. Gugliantini, Firenze.
Wilding N (2011) Manuscripts in Motion: The Diffusion of Galilean Copernicanism. Italian Studies 66/2:221–33.
Wilding N (2014) Transalpine Messengers. Galileo's Idol: Gianfrancesco Sagredo and the Politics of Knowledge. University of Chicago Press, Chicago, pp. 89–116.

Part 3

Critical Transcription of the *Opera Geometrica*, 1644

Opera Geometrica (1644) Transcription: General Considerations

8

Raffaele Pisano

This section provides information and background on the selection of *Opera geometrica* (Fig. 1). We have worked on transcription from Latin and Italian vernacular language, which is the original language of the Torricelli's text. Further facsimile texts are added for readers and our critical comments can be found as footnotes to this chapter.

Source In situ & Credit
History of Science Collections and *University of Oklahoma Libraries* at University of Oklahoma, U.S.A.

Fig. 1 Torricelli 1644, *Opera Geometrica*, Florentiae. Masse and de Landis, Florentiae (Torricelli 1644, 3rd of 401 *scanned* images–pages). *Source* Courtesy University of Oklahoma, History of Science Collections

R. Pisano (✉)
HOPAST at IEMN, Lille University–CNRS, Lille, France
e-mail: raffaele.pisano@univ-lille.fr

R. Pisano et al., *Homage to Evangelista Torricelli's Opera Geometrica 1644–2024*,
Logic, Epistemology, and the Unity of Science 55,
https://doi.org/10.1007/978-3-031-06963-5_8

The transcription is entirely original and accurate. We respected Torricelli's wording, including original Latin and vernacular Italian characters. We also corrected some misprinting, which mostly derived from the difficulty of reproduction of the 1644 edition; we regularized the spelling of words, by adopting the most used form, where required.

Below I present the main content of the edition that we used for this transcription:

Preliminary matters	Part 0	*Serenissimo Magno Duci Etruria* (pp. 3–4)
		Proemium (pp. 5–10)
Chapters on motion and mechanics	Part 1	*De sphaera et solidis sphaearalibus, Liber primus* (pp. 11–46)
		De solidis sphaeralibus, Liber secundus (pp. 47–94)
	Part 2	*De motu graviun naturaliter descendetium, Liber primus* (pp. 97–153)
		De motu proiectorum, Liber secundus (pp. 154–243)[1]
Chapters on geometry and infinitesimal	Part 3	*De dimensione parabolae solidique hyperboloci, problemata duo* (pp. 1–84)[2]
		Ad Serenissima Principem Leopoldum ab Etruria (pp. 3–4)
		Ad Lectorem Proemium (pp. 5–16)
		Quadratura parabolae, Problema primum (pp. 17–84)[3]
		Appendix De dimensione cycloidis (pp. 85–90)
		Appendix Scholium De cycloidibus aliarum specierum (pp. 91–92)
	Part 4	*De solido acuto hyperbolico, Problema alterum* (pp. 93–112)
		De solido acuto hyperbolico, Problema Secudnum (pp. 113–135)[4]
		Appendix De dimensione Cochlea (pp. [136]145–150)[5]

In detail, we did not simplify, modernize, or in any other way correct the wording of the *Opera geometrica*. We respected Torricelli's wording, both contents and editing. The reader will find that we preserved the character of the original manuscripts adding commentary in the footnotes to make them more accessible and comprehensible for a modern reader, where required. Thus we made a few changes from the original text; most of them are simply typographical adjustments, such as the bold and italic style, the resolution of "ſ", "v", "u", join the letters of the geometrical figures (e.g., "A BC" as "ABC", "A B" as "AB", etc.) when required—appropriate and following the contemporary standard rule of transcription, as well. Therefore, we attempted to reflect the beauty and clarity of Torricelli's prose.

The page-numbers and original running header are displayed. The figures themselves are not displayed; the original ones are conserved on the left side pages. Missing and/or unclear characters are rendered as "?" in square brackets–boxes. Handwritten text (non-belonged to Torricelli) is noted and, when possible it is transcribed.

Additional necessary explanations of language, and obvious printers' errors are indicated–corrected and scientifically annotated in the footnotes.

The critical transcription is an entirely original production within a unique research—and—source volume devoted to Torricelli's *Opera geometrica* and other of his works.

Acknowledgements I thank the anonymous reviewers for their careful reading of our manuscript and their many insightful comments and suggestions. I also express gratitude to Dr. Julie Robarts (University of Melbourne, Australia) for her observations on English language; eventual remaining English mistakes are my own.

[1] N.B.: The final part is in Italian vernacular language (pp. 216–243).

[2] N.B.: The numbering of pages restarts from 1.

[3] N.B.: The pages number 55 and 56 are repeated. The numbering of pages is broken: 79 and 80 lack. Nevertheless, the content is saved.

[4] N.B.: The numbering of pages restarts from 129. Thus, it would be pp. 129–135.

[5] N.B.: The numbering of pages starts with p. 136 and just after with p. 145, so the numeration continues consequently. The last page is 150, but, in fact, the appendix is only seven pages long.

Opera Geometrica (1644) Transcription: Image–Text Side by Side

9

Raffaele Pisano, Paolo Bussotti and Patricia Radelet de Grave

Abstract

This section provides the first full edited transcription—from Latin and Italian vernacular language—of the *Opera geometrica*. Further, as above said, facsimile texts are added, including our critical comments as footnotes to this chapter.

R. Pisano (✉)
HOPAST at IEMN, Lille University, Lille, France
e-mail: raffaele.pisano@univ-lille.fr

P. Bussotti
DIUM, Udine University, Udine, Italy
e-mail: paolo.bussotti@uniud.it

P. Radelet de Grave
Mathematics and Physics Research Institute, Catholic University of Louvain–la–Neuve, Ottignies-Louvain-la-Neuve, Belgium
e-mail: patricia.radelet@uclouvain.be

R. Pisano et al., *Homage to Evangelista Torricelli's Opera Geometrica 1644–2024*,
Logic, Epistemology, and the Unity of Science 55,
https://doi.org/10.1007/978-3-031-06963-5_9

[Front Cover]

[Credit]

THE · LIBRARY · OF · THE · UNIVERSITY · OF · OKLAHOMA
The
E. DeGOLYER
COLLECTION
in the History of
SCIENCE AND
TECHNOLOGY

[A dedicatory, 1876]

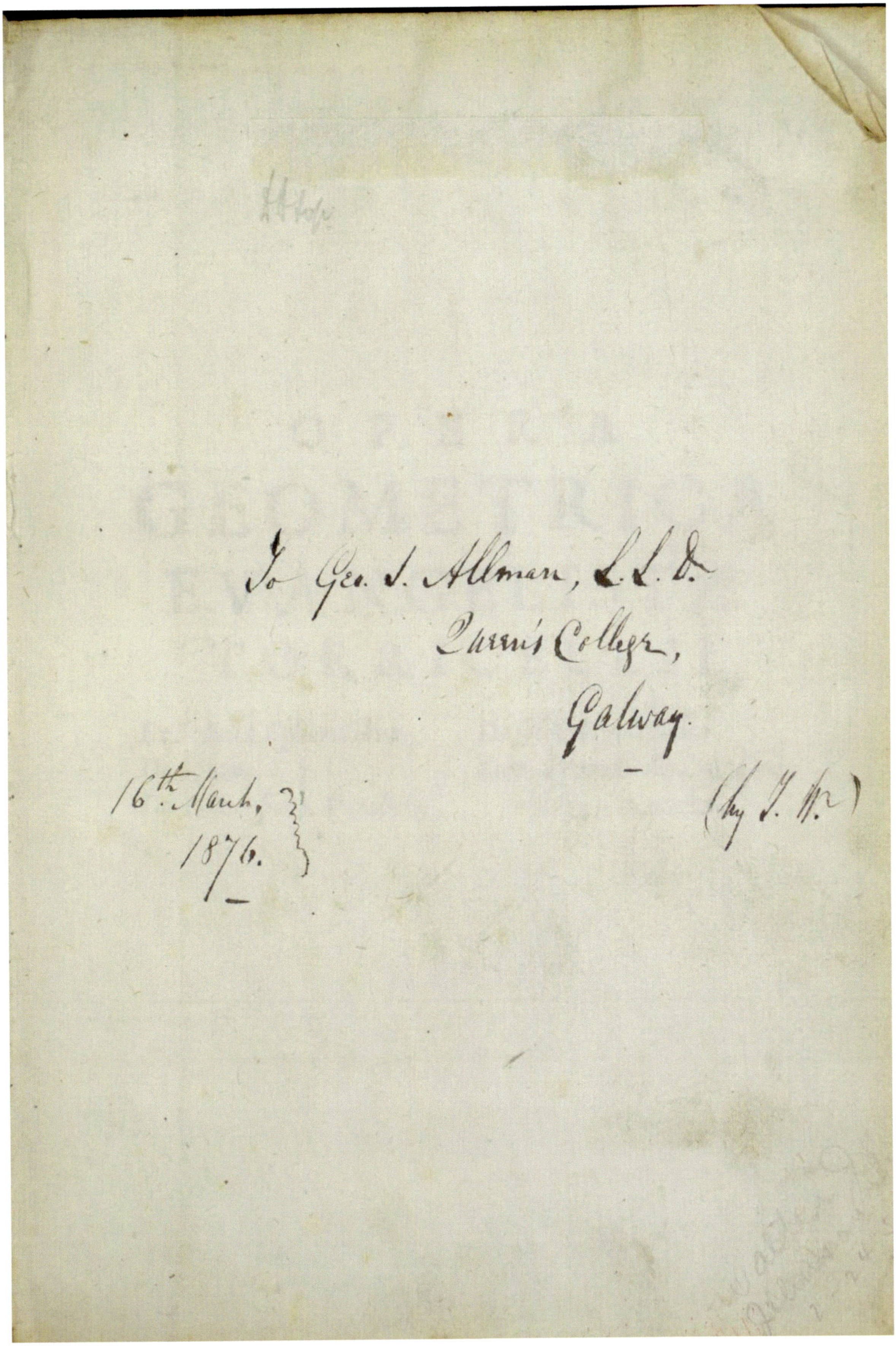
To Geo. J. Allman, LL.D.
Queen's College,
Galway.

16th March,
1876.

(by J. W.)

[verso]

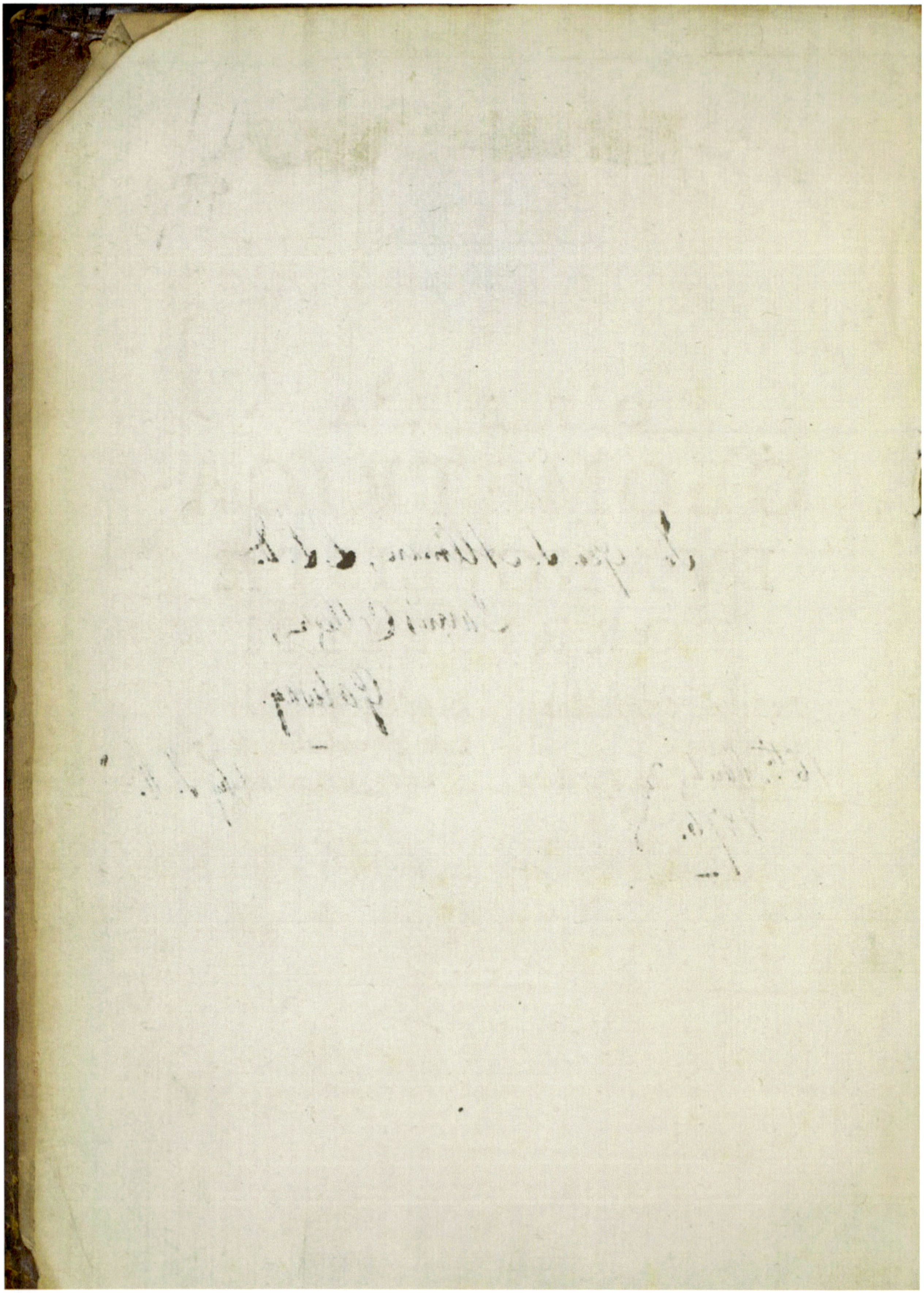

[Frontispiece *Opera geometrica*]

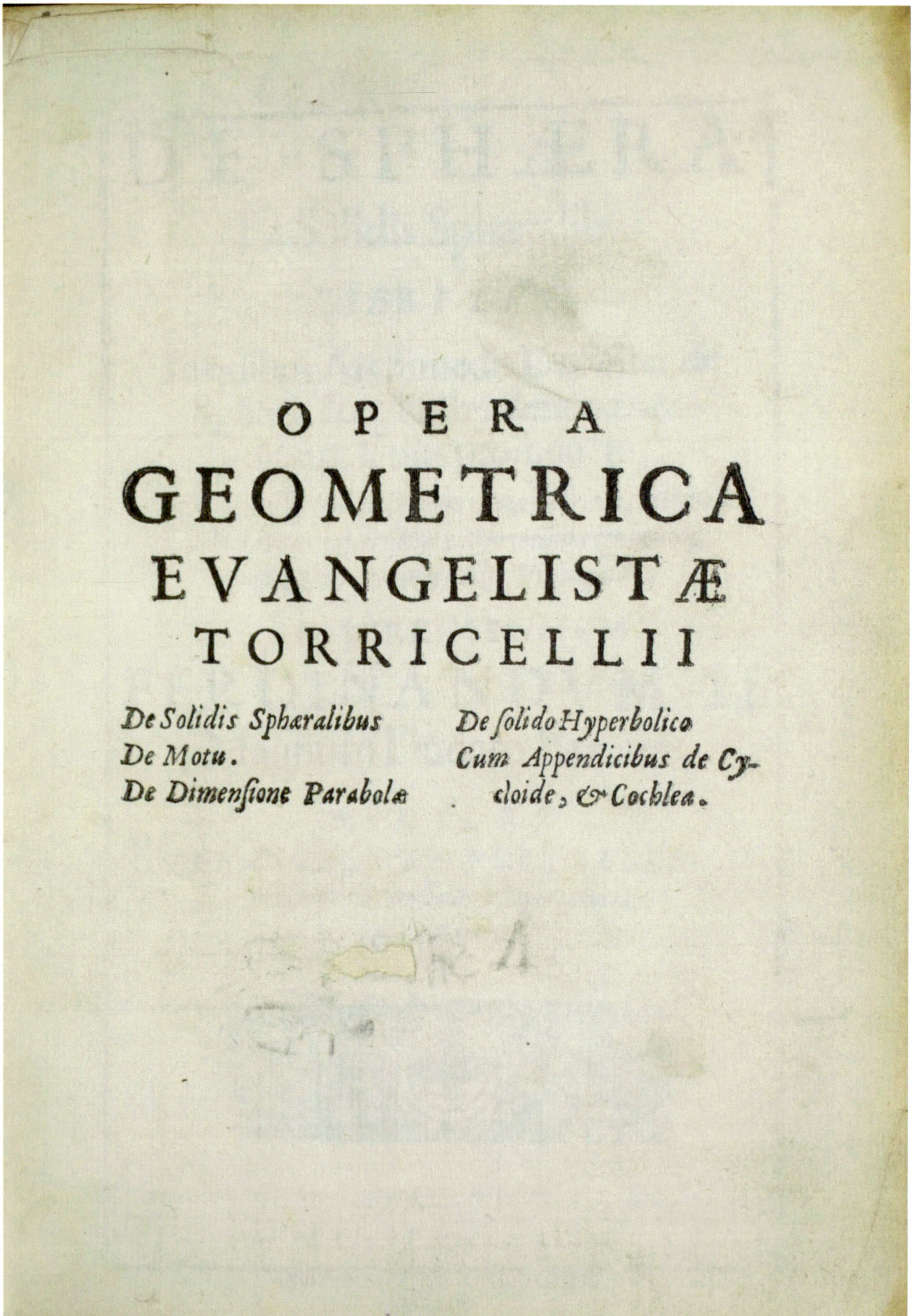

OPERA
GEOMETRICA
EVANGELISTÆ
TORRICELLII

De Solidis Sphæralibus
De Motu.
De Dimensione Parabolæ
De solido Hyperbolico
Cum Appendicibus de Cycloide, & Cochlea.

OPERA

GEOMETRICA
EVANGELISTÆ
TORRICELLII

De Solidis Sphæralibus
De Motu
De Dimensione Parabolæ
De Solido Hyperbolico
Cum Appendicibus de Cy–
cloide, & Cochlea

[Frontispiece De Sphaera et Soldiis Sphaeralibus]

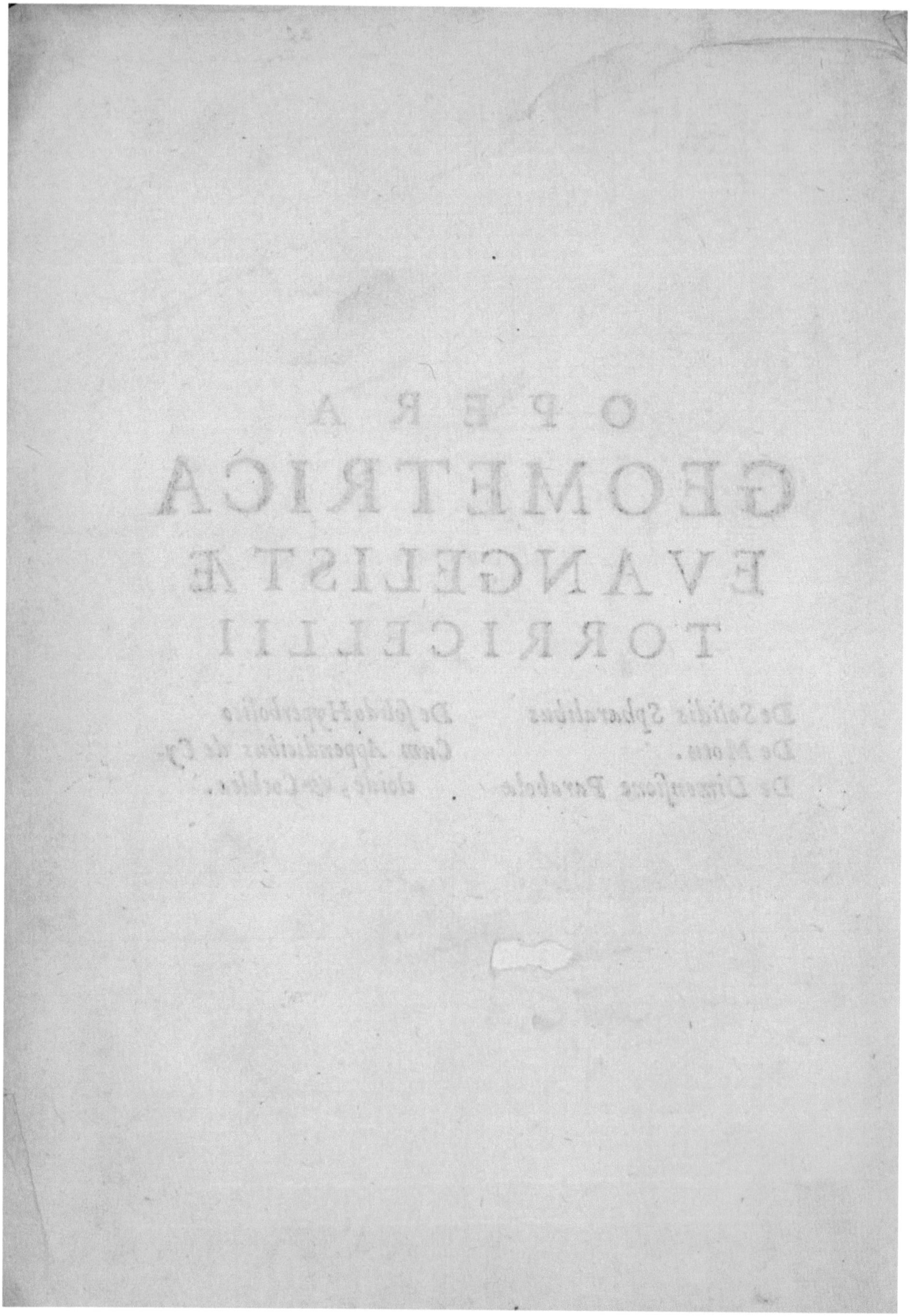
OPERA
GEOMETRICA
EVANGELISTÆ
TORRICELLII

De Solidis Sphæralibus
De Motu.
De Dimensione Parabolæ.
De solido Hyperbolico
Cum Appendicibus de Cycloide, & Cochlea.

[Frontispiece *De* Sphaera *et Soldiis Sphaeralibus*]

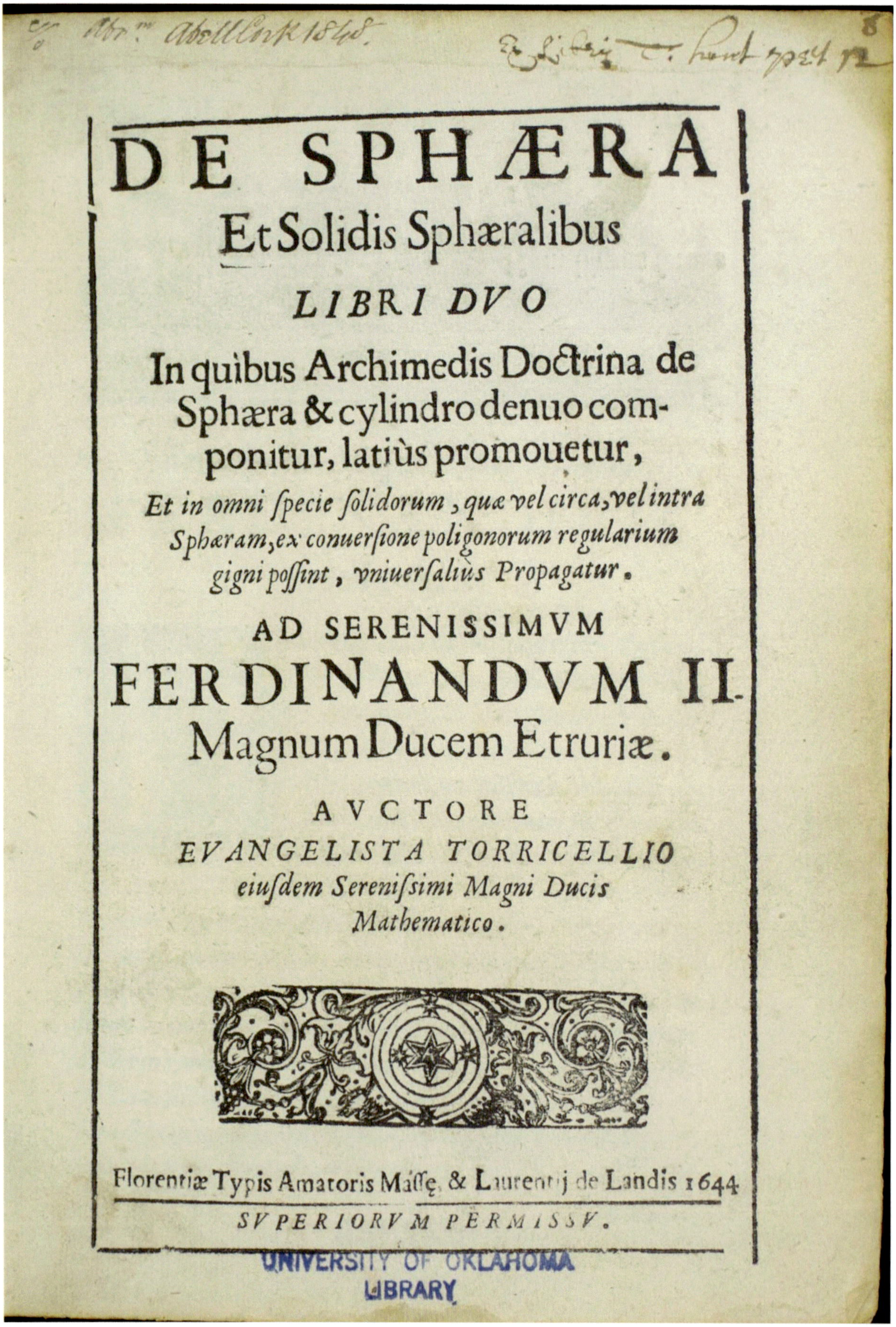

DE SPHÆRA
Et Solidis Sphæralibus
LIBRI DVO
In quibus Archimedis Doctrina de Sphæra & cylindro denuo componitur, latiùs promouetur,
Et in omni ſpecie ſolidorum, quæ vel circa, vel intra Sphæram, ex conuerſione poligonorum regularium gigni poſſint, vniuerſaliùs Propagatur.
AD SERENISSIMVM
FERDINANDVM II.
Magnum Ducem Etruriæ.
AVCTORE
EVANGELISTA TORRICELLIO
eiuſdem Sereniſsimi Magni Ducis Mathematico.

Florentiæ Typis Amatoris Maſſę, & Laurentij de Landis 1644.
SVPERIORVM PERMISSV.

D E S P H Æ R A
Et Solidis Sphæralibus
LIBRI DVO

In quibus Archimedis[1] Doctrina de
Sphæra & cylindro denuo com–
ponitur, latiùs promouetur,
Et in omni specie solidorum, quæ vel circa, vel intra
Sphæram, ex conuersione poligonorum regolarium
gigni possint, vniuersalius Propagatur.

AD SERENISSIMVM
F E R D I N A N D V M I I
Magnum Ducem Etruriæ

AVCTORE
EVANGELISTA TORRICELLIO
eiusdem Serenissimi Magni Ducis
Mathematico

Florentiæ Typis Amatoris Masse & Laurentij de Landis 1644
SVPERIORVM PERMISSV.

[1] Archimedean influence in Torricelli's *Opera geometrica* is evident. This section, *De sphaera et solidis sphaeralibus* (Torricelli 1644, I, 1, *Liber primus*, pp. 11–46) presents an extension of the Archimedean proofs in *On the sphere and cylinder* (Archimedes 1897, II, pp. 1–90. Cfr. Pisano and Bussotti's chapters above).

[verso]

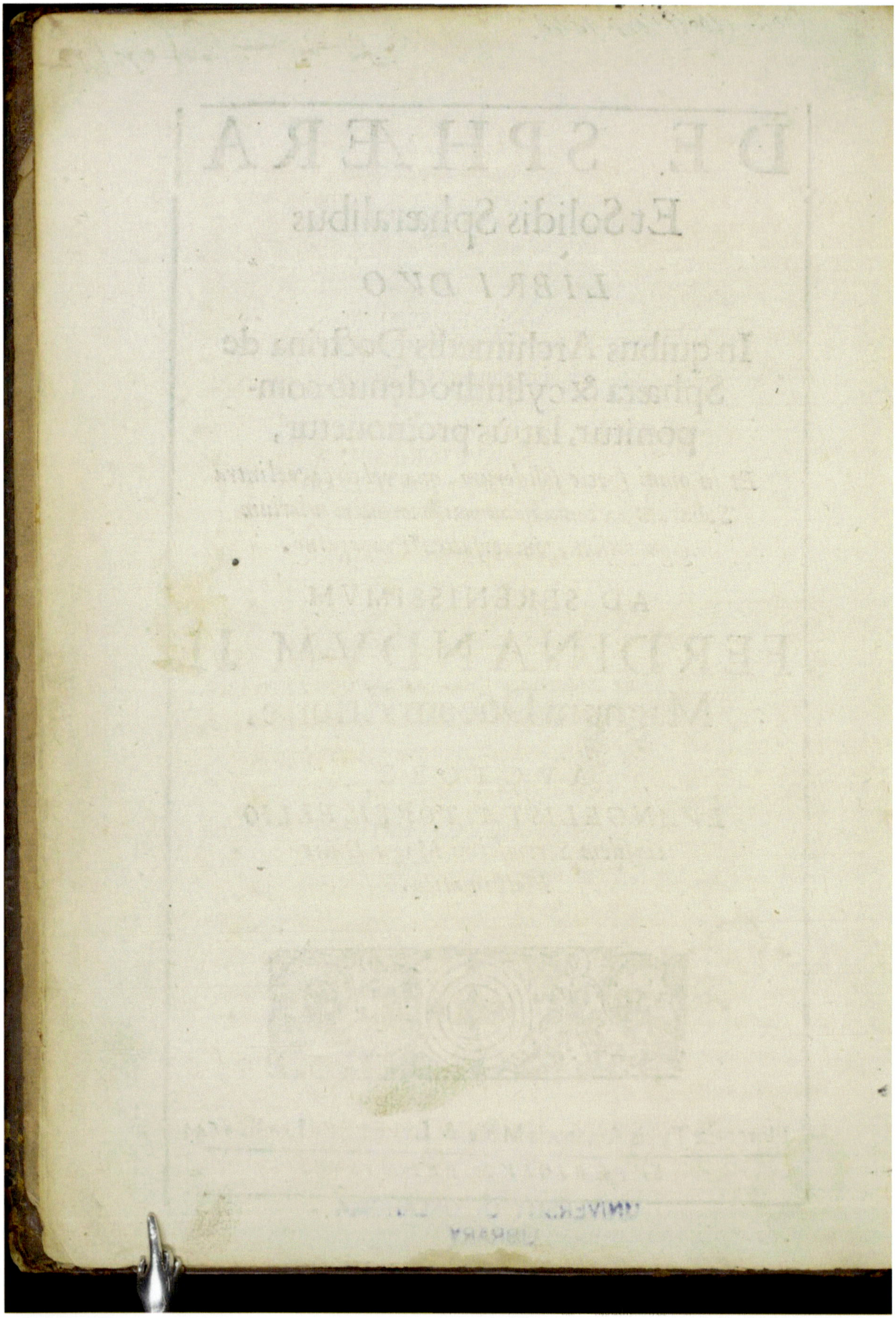

3

Sereniſsimo Magno Duci Etruriæ

FERDINANDO II.

ERVBESCEREM *profectò, Sereniſſime Magne Dux, oblaturus libellum hunc Sereniſſimæ Celſitudini Tuæ, niſi haberem maxima Archimedis, et Galilei nomina, quæ prætendere poſſim audaciæ meæ: Exigua enim ſunt opuſcula hæc, & de rebus ætate noſtra neglectis, nempe Geometricis. Attamen, niſi fallor, duo maxima Geometriæ opera promouebunt, cum veterem De Sphæra, et Cylindro, nouamque De Motu ſcientiam exequantur. Sed ego fruſtra Geometriam excuſo apud eum Principem, cui non ſolum hæreditaria, ſed etiam ingenita eſt Mathematicarum diſciplinarum protectio. Sereniſſimus enim Coſmus II. Pater Tuus ſtipendijs celeberrimo Galileo oblatis; deinde Ser. C. Tua, beneficijs maximis in huiuſmodi ſcientiæ cultores collocatis, optime demonſtrauit intelligere, quanti momenti ſint*

A 2 *Mathe-*

[513 T630] 3

Serenissimo Magno Duci Etruriae

FERDINANDO II

ERUBESCEREM *profectò, Serenissime Magne Dux, oblaturus libellum hunc Sereniisimæ Celſitudini Tuæ, niſi haberem maxima Archimedis, et Galilei nomina, quæ prætendere poſſim audaciæ meæ: Exigua enim ſunt opuſcula hæc, & de rebus ætate nostra neglectis, nempe Geometricis. Attamen, niſi fallor, duo maxima Geometriæ opera promouebunt, cum veterem De Sphæra, et Cylindro, nouamque De Motu ſcientiam exequantur. Sed ego fruſtra Geometriam excuſo apud eum Principem, cui non ſolum hæreditaria, ſed etiam ingenita eſt Mathematicarum diſciplinarum protectio. Sereniſſimus enim Coſmus II. Pater Tuus ſtipendijs celeberrimo Galileo oblatis; deinde Ser. C. Tua, beneficijs maximis in huiuſmodi ſcientiæ cultores collocatis, optime demonſtrauit intelligere, quanti momenti sint*

A 2

Mathe-

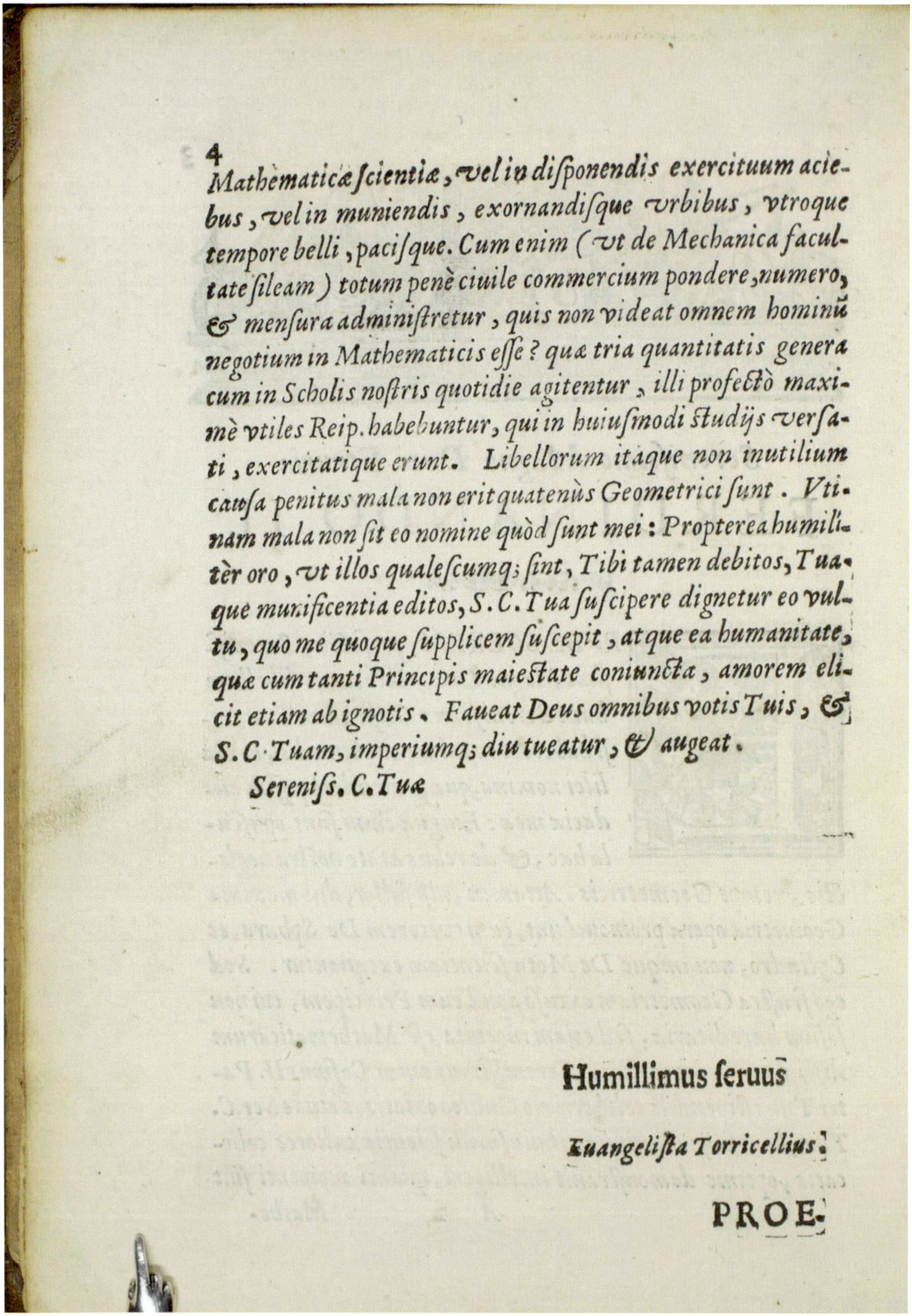

4

Mathematicæ scientiæ, vel in disponendis exercituum acie-
bus, vel in muniendis, exornandisque vrbibus, vtroque
tempore belli, pacisque. Cum enim (vt de Mechanica facul-
tate sileam) totum penè ciuile commercium pondere, numero,
& mensura administretur, quis non videat omnem hominũ
negotium in Mathematicis esse? quæ tria quantitatis genera
cum in Scholis nostris quotidie agitentur, illi profectò maxi-
mè vtiles Reip. habebuntur, qui in huiusmodi studijs versa-
ti, exercitatique erunt. Libellorum itaque non inutilium
causa penitus mala non erit quatenùs Geometrici sunt. Vti-
nam mala non sit eo nomine quòd sunt mei: Propterea humili-
tèr oro, vt illos qualescumq; sint, Tibi tamen debitos, Tua-
que munificentia editos, S. C. Tua suscipere dignetur eo vul-
tu, quo me quoque supplicem suscepit, atque ea humanitate,
quæ cum tanti Principis maiestate coniuncta, amorem eli-
cit etiam ab ignotis. Faueat Deus omnibus votis Tuis, &
S. C. Tuam, imperiumq; diu tueatur, & augeat.

Serenifs. C. Tuæ

Humillimus seruus

Euangelista Torricellius.

PROE-

4

Mathematicæ ſcientiæ, vel in diſponendis exercituum aciebus, vel in muniendis, exornandiſque vrbibus, vtroque tempore belli, paciſque. Cum enim (vt de Mechanica facultate ſileam) totum penè ciuile commercium pondere, numero, & menſura adminiſtretur, quis non videat omnem hominũ negotium in Mathematicis eſſe? quæ tria quantitatis genera cum in Scholis noſtris quotidie agitentur, illi profectò maximè vtiles Reip. habebuntur, qui in huiuſmodi studijs verſati, exercitatique erunt. Libellorum itaque non inutilium cauſa penitus mala non erit quatenùs Geometrici ſunt. Vtinam mala non ſit eo nomine quòd ſunt mei: Propterea humilitèr oro, vt illos qualeſcumq; ſint, Tibi tamen debitos, Tuaque munificentia editos, S. C. Tua ſuſcipere dignetur eo vultu, quo me quoque ſupplicem ſuſcepit, atque ea humanitate, quæ cum tanti Principis maiestate coniuncta, amorem elicit etiam ab ignotis. Faueat Deus omnibus votis Tuis, & S. C. Tuam, imperiumq; diu tueatur, et augeat.
Sereniſs. C. Tuæ.

Humillimus ſeruus
Euangeliſta Torricellius.

PROE-

3

PROEMIVM.

INTER omnia opera ad Mathematicas diſciplinas pertinentia, iure optimo Principem ſibi locum vindicare videntur Archimedis inuenta; quæ quidem ipſo ſubtilitatis miraculo terrent animos. Verùm inter omnes libros egregij Authoris longè eminet ille, qui De Sphæra, & cylindro inſcribitur: neque enim poſteritatis tantùm conſenſu, ſed etiam ipſius Scriptoris iudicio primas tenet. Certè hunc ipſe in titulum ſepulcri elegit, dignumq; prę cęteris iudicauit, qui tanti viri tumulum exornaret, oſtenderetque. Hunc tamen ſi quis attentiùs conſiderare, & perpendere velit, magnum quidem inuentum fateatur neceſſe eſt, ſed fortaſſe non abſolutum. Loquor equidem de primo tantùm libro, in quo partem operis Theorematicam, & omnem doctrinæ inuentionem exequitur: quo veluti iacto fundamento, in ſecunda parte poſtea, quaſi coronidis loco, problemata quędam tamquam corollaria ad eam rem ſpectantia ipſe ſubnectit. Titulus libri eſt De Sphæra, & Cylindro; quę quidem verba apud nos idem ſonant, ac ſi dixiſſet De Sphæra, atque vnico ſolido ſphęrali; ſed ſphæralia ſolida, quorum vnum eſt cylindrus, multitudine ſunt infinita, vt mox patebit. Ergo abſolutior fortaſſe contemplatio videri potuiſſet, ſi eximius Author proportionem, non tantùm eam, quæ eſt inter ſphæram, vnicumque ex ſphęralibus ſolidis perquiſiſſet, verumetiam omnem aliam rationem, quæ inter ſphęram ipſam, & vnumquodq; ex infinitis ſphæralibus ſolidis intercedit, oſtendendam ſibi aſſumpſiſſet. Hoc itaque propoſitum

erit,

5

PROEMIUM[2]

INTER omnia opera ad Mathematicas diſciplinas pertinentia, iure optimo Principem ſibi locum vindicare videntur Archimedis inuenta; quæ quidem ipſo ſubtilitatis miraculo terrent animos. Verùm inter omnes libros egregij Authoris longè eminet ille, qui De Sphæra, & cylindro inſcribitur: neque enim poſteritatis tantùm conſenſu, ſed etiam ipſius Scriptoris iudicio primas tenet. Certè hunc ipſe in titulum ſepulcri elegit, dignumq; prę cęteris iudicauit, qui tanti viri tumulum exornaret, oſtenderetque. Hunc tamen ſi quis attentiùs conſiderare, & perpendere velit, magnum quidem inuentum fateatur neceſſe eſt, ſed fortaſſe non abſolutum. Loquor equidem de primo tantùm libro, in quo partem operis Theorematicam, & omnem doctrinæ inuentionem exequitur: quo veluti iacto fundamento, in ſecunda parte poſtea, quaſi coronidis loco, problemata quędam tamquam corollaria ad eam rem ſpectantia ipſe ſubnectit. Titulus libri eſt De Sphæra, & Cylindro; quę quidem verba apud nos idem ſonant, ac ſi dixiſſet De Sphæra, atque vnico ſolido ſphęrali; ſed ſphæralia ſolida, quorum vnum eſt cylindrus, multitudine ſunt infinita, vt mox patebit. Ergo abſolutior fortaſſe contemplatio videri potuiſſet, ſi eximius Author proportionem, non tantùm eam, quæ eſt inter ſphæram, vnicumque ex ſphęralibus ſolidis perquiſiſſet, verumetiam[3] omnem aliam rationem, quæ inter ſphęram ipſam, & vnumquodq; ex infinitis ſphæralibus ſolidis inter cedit, oſtendendam ſibi aſſumpſiſſet. Hoc itaque propoſitum

erit,

---

[2] We let note the role played by Archimedes's mechanics in Torricelli's *Opera geometrica*. It was truly a scientific inspiration for Torricelli as he himself declared from the beginning (Cfr. Pisano and Bussotti's chapters above).

[3] Verumetiam] verum etiam.

6

& inſtitutum meum in præſenti libello. Doctrinam non ſolum de Sphæra, & cylindro, ſed de ſphæra, & ſphæralibus ſolidis omnibus proſequemur: Mutatiſq; plerumque Archimedæis fundamentis, vniuerſaliori demonſtratione illam complecti conabimur, atque in omni ſpecie ſolidorum, vel intrà, vel circà ſphęram deſcriptorum, propagabimus.

Ex libro Archimedis De Sphæra & Cylindro duo hæc colliguntur ſpectantia ad illa ſolida, quæ nos ſphæralia appellamus: Primum, quòd ſphæra dupla eſt inſcripti ſibi rombi ſolidi æquilateri; quod quidem vnum eſt ex ſolidis ſphæralibus, genitum ex reuolutione quadrati inſcripti, & circa diagonalem conuerſi. Alterum; quòd cylindrus ad inſcriptam ſibi ſphæram eſt ſeſquialter. quod quidem & vnum ex ſolidis ſphæralibus eſt, genitum ex conuerſione quadrati circumſcripti, & circa ipſius catetum reuoluti. Stantibus his, contemplatione dignum mihi videbatur vniuerſalius aliquod problema huiuſmodi.

21. Sph et cyl

Manf ex 31

Dato poligono quocunque regulari ſiuè intrà, ſiuè circà circulum deſcripto, & ſiue circa diagonalem, ſiue circa catetum reuoluto; proportionem dicere, quam factum ex polygono ſolidum habeat, ad factam ex circulo ſphæram.

Penitus autem ex voto ſucceſſit inſtituta contemplatio. Nam inuenta proportione, ſex iſta inferiùs adnotata Theoremata ita ſe habere comperi, quemadmodum hìc ſubijciuntur.

Prima ſolidorum ſphæralium ſpecies.

Si intrà circulum deſcriptum fuerit poligonum regulare habens latera numero parià, & conuertatur figura circa catetum B. Quæritur ratio ſphæræ ad factum ſolidum.

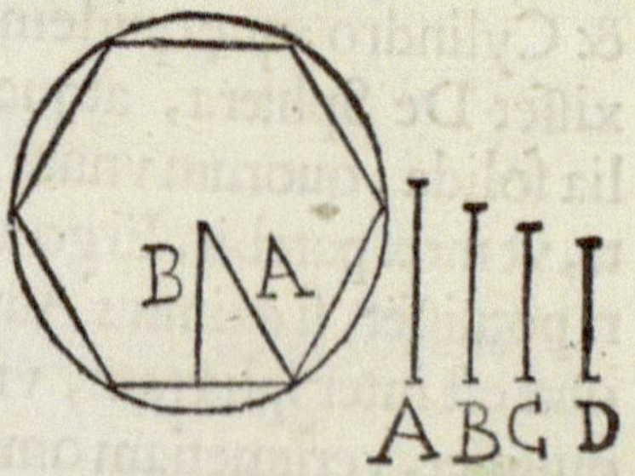

Continuetur ratio radij poligoni ad catetum eiuſdem, nempe A ad B in quatuor terminis A, B, C, D. Erit que ſphæra ad ſolidum inſcriptum, vt diameter ſphæræ, hoc eſt vt dupla ipſius A, ad vtramq; ſimul B, & D.

Theor. 14. lib ij

Secun

6 [Proemium]

& inftitutum meum in præfenti libello. Doctrinam non folum de Sphęra, & cylindro, fed de fphæra, & fphæralibus folidis omnibus profequemur: Mutatifq; plerumque Archimedæis fundamentis, vniuerfaliori demonftratione illam complecti conabimur, atque in omni fpecie folidorum, vel intrà, vel circà fphęram defcriptorum, propagabimus.

Ex libro Archimedis De Sphæra & Cylindro duo hæc colliguntur fpectantia ad illa folida, quæ nos fphæralia appellamus: [31 Sph et Cyl] Primum, quòd fphæra dupla eft infcripti fibi rombi folidi æquilateri; quod quidem vnum eft ex folidis fphæralibus, genitum ex reuolutione quadrati infcripti, & circa diagonalem conuerfi. Alterum; quòd cylindrus ad infcriptam fibi fphæram eft [Manus ex 31] fefquialter. quod quidem & vnum ex folidis fphæralibus eft, genitum ex conuerfione quadrati circumfcripti, & circa ipfius catetum reuoluti. Stantibus his, contemplatione dignum mihi videbatur vniuerfalius aliquod problema huiufmodi.

Dato poligono quocunque regulari fiuè intrà, fiuè circà circulum defcripto, & fiue circa diagonalem, fiue circa catetum reuoluto; proportionem dicere, quam factum ex polygono folidum habeat, ad factam ex circulo fphæram.

Penitus autem ex voto fucceffit inftituta contemplatio. Nam inuenta proportione, fex ifta inferiùs adnotata Theoremata ita fe habere comperi, quemadmodum hìc fubijciuntur.

Prima folidorum fphæralium fpecies

Si intrà circulum defcriptum fuerit poligonum[4] regulare habens latera numero parià, & conuertatur figura circa catetum B. Quæritur ratio fphæræ ad factum folidum.

Theor 14. *lib ij* Continuetur ratio radij poligoni ad catetum eiufdem, nempe A ad B in quatuor terminis A, B, C, D. Erit que fphæræ[5] ad folidum infcriptum, vt diame-
ter[6] fphæra,[7] hoc eft vt dupla ipfius A, ad vtramq; fimul B, & D.

Secun

[4] Poligonum] Polygonum. As announced we decided to respect original Torricelli's wording so, hereafter, we let Poligonum or poligonum.
[5] fphæræ] fphæra
[6] diame-ter] diameter. We remind that we respect the original wording-pagination, punctuation etc. of the original document displayed (e.g., see above). Of course, we insert needed corrections.
[7] fphæræ] fphæra

Secunda species.

Si intra circulum descriptum fuerit poligonum regulare habens latera numero paria, & conuertatur figura circà diagonalem A B. Quæritur ratio sphęrę ad factum sphærale solidum.

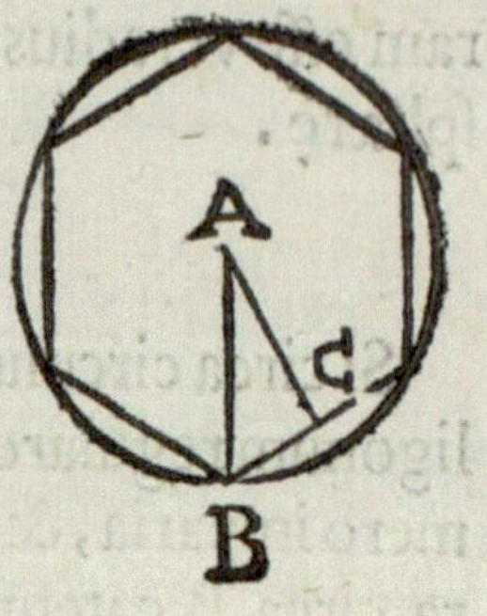

Ostenditur. Sphæram esse ad solidum, vt quadratum A B. ad quadratum cateti A C. *Theor. 7. lib. ij.*

Tertia species.

Si intrà circulum describatur poligonum regulare habens latera numero imparia,& conuertatur figura circa catetum B. Quæritur ratio sphęrę ad factum sphęrale solidum.

Continuetur ratio radij A. ad catetum B. in quatuor terminis A, B, C, D. Eritq; sphæra ad solidum, vt quadrupla ipsius A. ad B. semel, C. bis, & D. semel simulq; sumptas. *Theor. xjx. lib. ij.*

Quarta species.

Si circà circulum describatur poligonum regulare, habens latera numero paria, & conuertatur figura circa catetum C. Quæritur ratio solidi ad sphæram.

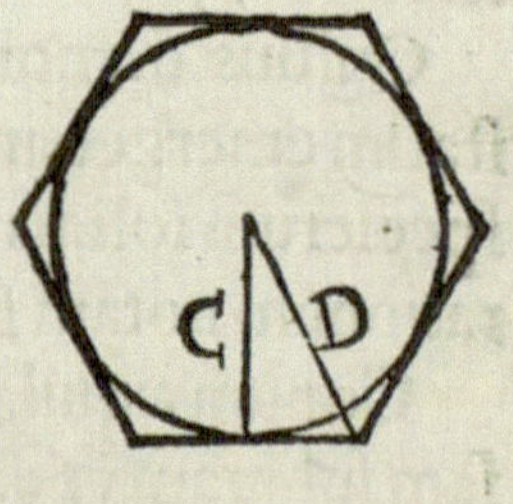

Ostenditur solidum esse ad inscriptam sibi sphæram, vt duo simul quadrata, quorũ vnum sit ex radio D. alterum ex cateto C, ad duplum quadrati C. *Theor. xiij lib. ij.*

Quinta Species.

Si circà circulum describatur poligonum regulare habens latera numero paria; & conuertatur figura circa diagonalem A. Quæritur ratio solidi ad sphæram.

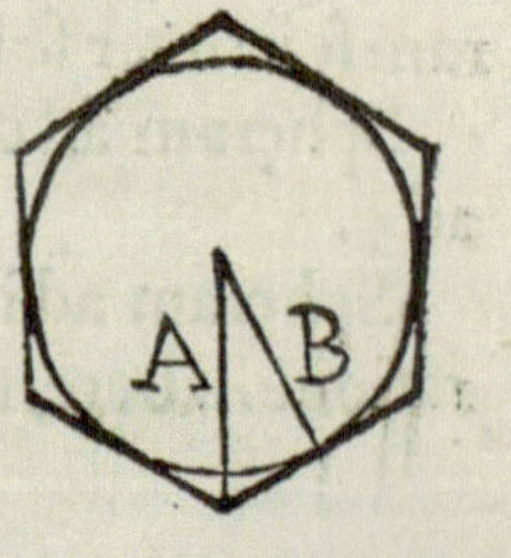

Ostenditur solidum ad inscriptã sibi sphæram

ram

[Proemium] 7

Secunda species.

Si intra circulum deſcriptum fuerit poligonum regulare habens latera numero paria, & conuertatur figura circà diagonalem AB. Quæritur ratio ſphęrę ad factum ſphærale solidum. *Theor.*

Oſtenditur. Sphæram eſſe ad ſolidum, vt quadratum AB. ad quadratum cateti AC. *7. lib. ij.*

Tertia species.

Si intrà circulum deſcribatur poligonum regulare habens latera numero imparia, & conuertatur figura circa catetum B. Quæritur ratio ſphęrę ad factum ſphęrale ſolidum.

Continuetur ratio radij A. ad catetum B. in quatuor terminis A, B, C, D. Eritq; ſphæra ad ſolidum, vt quadrupla ipſius A. ad B. ſemel, C. bis, & D. ſemel ſimulq; ſumptas.

Theor. xix. lib. ij.

Quarta species.

Si circà circulum deſcribatur poligonum regulare, habens latera numero paria, & conuertatur figura circa catetum C. Quæritur ratio ſolidi ad ſphæram.

Oſtenditur ſolidum eſſe ad inſcriptam ſibi ſphæram, vt duo ſimul quadrata, quorũ vnum ſit ex radio D. alterum ex cateto C, ad duplum quadrati C.

Quinta Species.

Theor. xiij. lib. ij.

Si circà circulum deſcribatur poligonum regulare habens latera numero paria; & conuertatur figura circa diagonalem A. Quæritur ratio ſolidi ad ſphæram.

Oſtenditur ſolidum ad inſcriptã ſibi ſphæ-

ram.

ram esse vt radius A ad catetum B.hoc est vt axis solidi ad axem sphærę.

Theor.6. lib.2.

Sexta, & vltima Species.

Si circa circulum describatur poligonum regulare habens latera numero imparia, & conuertatur figura circa B catetum. Quæritur ratio solidi ad Sphæram.

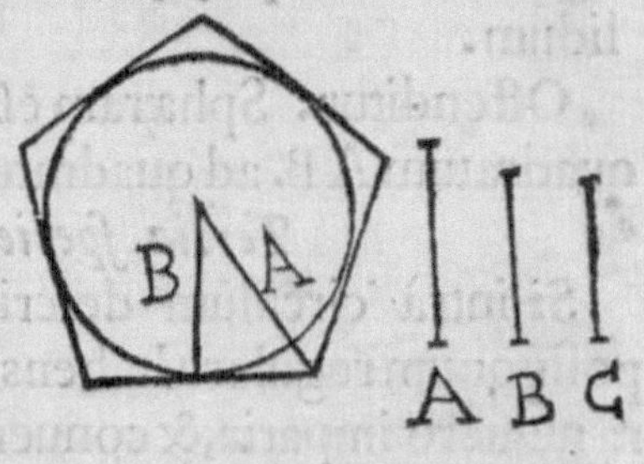

Continuetur ratio radij A ad catetum poligoni B, in tribus terminis A, B, C. Eritque solidum ad sphæram, vt A semel, B bis, & C semel simulque sumptæ, ad quadruplam ipsius C.

Theor. xviij. lib ij.

Solidorum itaq: sphęralium species omninò sex emergunt, & vniuscuiusq; speciei ratio ad suam sphæram innotescit. Possent fortasse videri tres tantum solidorum species, si solida absolutè, ac sine suis sphæris considerentur. Verùm si illa ad sphæram referantur, statim relatio variatur, & proportio alia consurgit, prout cognata solidis ipsis sphæra inscripta fuerit, vel circumscripta.

Quibus demonstratis, varia pro Corollarijs Theoremata statim emergebant; cuiusmodi sunt. Datis ex prędictarum sex specierum solidis duobus quibuscunque, alterius ad alterum rationem notam facere.

Conum æquilaterum circa sphæram descriptum, esse ad ipsam sphęram vt 9. ad 4. Nempe duplum sesqui quartum. Propterea si circa eandem sphęram conus, cylindrusq; æquilateri descripti sint, tria solida, nempe conum, cylindrum, & sphęram fore inter se in continua proportione sesquialtera.

Sphęram ad conum ęquilaterum sibi inscriptum esse vt 32. ad 9.

Sphęram ad inscriptum cylindrum ęquilaterum inefabilem rationē habere, nēpe vt diameter quadrati alicuius ad $\frac{3}{4}$ late-

ris

8 [Proemium]

Theor. 6. *lib.* 2. ram eſſe vt radius A ad catetum B. hoc eſt vt axis ſolidi ad axem ſphærę.

Sexta, & ultima Species.

Si circa circulum deſcribatur poligonum regulare habens latera numero imparia, & conuertatur figura circa B catetum. Quæritur ratio ſolidi ad Sphæram.

Continuetur ratio radij A ad catetum poligoni B, in tribus terminis A, B, C. Eritque ſolidum ad ſphæram, vt A ſemel, B bis, & C ſemel ſimulque ſumptæ, ad quadruplam ipſius C.

Theor. xviij. lib. ij.

Solidorum itaq: ſphęralium ſpecies omninò ſex emergunt, & vniuſcuiuſq; ſpeciei ratio ad ſuam ſphæram innoteſcit. Poſſent fortaſſe videri tres tantum ſolidorum ſpecies, ſi ſolida abſolutè, ac ſine ſuis ſphæris conſiderentur. Verum ſi illa ad ſphæram referantur, ſtatim relatio variatur, & proportio alia conſurgit, prout cognata ſolidis ipſis ſphæra inſcripta fuerit, vel circumſcripta.

Quibus demonſtratis, varia pro Corollarijs Theoremata ſtatim emergebant; cuiuſmodi ſunt. Datis ex prędictarum ſex ſpecierum ſolidis duobus quibuſcunque, alterius ad alterum rationem notam facere.

Conum æquilaterum circa ſphæram deſcriptum, eſſe ad ipſam ſphęram vt 9. ad 4. Nempe duplum ſeſquiquartum. Propterea ſi circa eandem ſphęram conus, cylindruſq; æquilateri deſcripti ſint, tria ſolida, nempe conum, cylindrum, & ſphęram fore inter ſe in continua proportione ſeſquialtera.

Sphęram ad conum ęquilaterum ſibi inſcriptum eſſe vt 32. ad 9.

Sphęram ad inſcriptum cylindrum ęquilaterum inefabilem rationẽ habere, nẽpe vt diameter quadrati alicuius ad 3 late-

4

ris

9

ris eiusdem.

Rombum solidum æquilaterum sphærę circumscriptum ad eandem sphæram incomensuràbilem esse, nempe vt diameter quadrati alicuius ad latus eiusdem.

Sphærale solidum exagonale circa catetum reuolutum esse ad inscriptam sibi sphæram sesquisextum,

Sphęram autem ad exagonale solidum sibi inscriptum, & circà diagonalem reuolutum, esse sesquitertiam.

Et alia huiusmodi, quę quidem altiùs perscrutanti innumera patebunt. Interim satis superque mihi erit aliqua apposuisse, quæ propria claritate vltrò se se offerunt etiam aspernanti. Horum maxima pars Corollaria esse poterant præcedentium sex Theorematum; attamen illa demonstrabimus ex sola etiam Euclidis doctrina, sine ope illorum quę de sphæralibus præmiseramus; Vt videre est ad Propositiones 30. & 9. seqq. in secundo libro. Cæterum huiùs contemplationis occasionem, mox etiã & scriptionis incitamentum præbuit mihi acutissimus librorum Archimedis scrutator Antonius Nardus Aretinus: huic enim refero, atque ipsius eruditis colloquijs, si quid verè Geometricũ in hac scriptura exciderit mihi.

S verò pleraque mala erunt, & fortasse omnia, hoc vnum culpandus erit ornatissimus vir, & genere, doctrinà, moribusq; conspicuus Andreas Arrighettus Florentinus, qui post magna in me collata beneficia, editionem mali libri non suasit, sed iussit.

B DEFI-

ris eiusdem.

Rombum ſolidum æquilaterum ſphærę circumſcriptum ad eandem ſphæram incomenſuràbilem eſſe, nempe vt diameter quadrati alicuius ad latus eiuſdem.

Sphærale ſolidum exagonale circa catetum reuolutum eſſe ad inſcriptam ſibi ſphæram ſeſquiſextum,

Sphęram autem ad exagonale ſolidum ſibi inſcriptum, & circà diagonalem reuolutum, eſſe ſeſquitertiam.

Et alia huiuſmodi, quę quidem altiùs perſcrutanti innumera patebunt. Interim ſatis ſuperque mihi erit aliqua appoſuiſſe, quæ propria claritate vltrò ſe ſe offerunt etiam aſpernanti. Horum maxima pars Corollaria eſſe poterant præcedentium ſex Theorematum; attamen illa demonſtrabimus ex ſola etiam Euclidis doctrina, ſine ope illorum quę de ſphæralibus præmiſeramus; Vt videre eſt ad Propoſitiones 30. & 9. ſeqq. in ſecundo libro. Cæterum huiùs contemplationis occaſionem, mox etiã & ſcriptionis incitamentum præbuit mihi acutiſſimus librorum Archimedis ſcrutator Antonius Nardus Aretinus: huic enim refero, atque ipſius eruditis colloquijs, ſi quid verè Geometricũ in hac ſcriptura exciderit mihi.

S[8] verò pleraque mala erunt, & fortaſſe omnia, hoc vnum culpàndus erit ornatiſſimus vir, & genere, doctrinà, moribuſq; conſpicuus Andreas Arrighettus Florentinus, qui poſt magna in me collata beneficia, editionem mali libri non ſuaſit, ſed iuſſit.

B DEFI-

[8] S] Si.

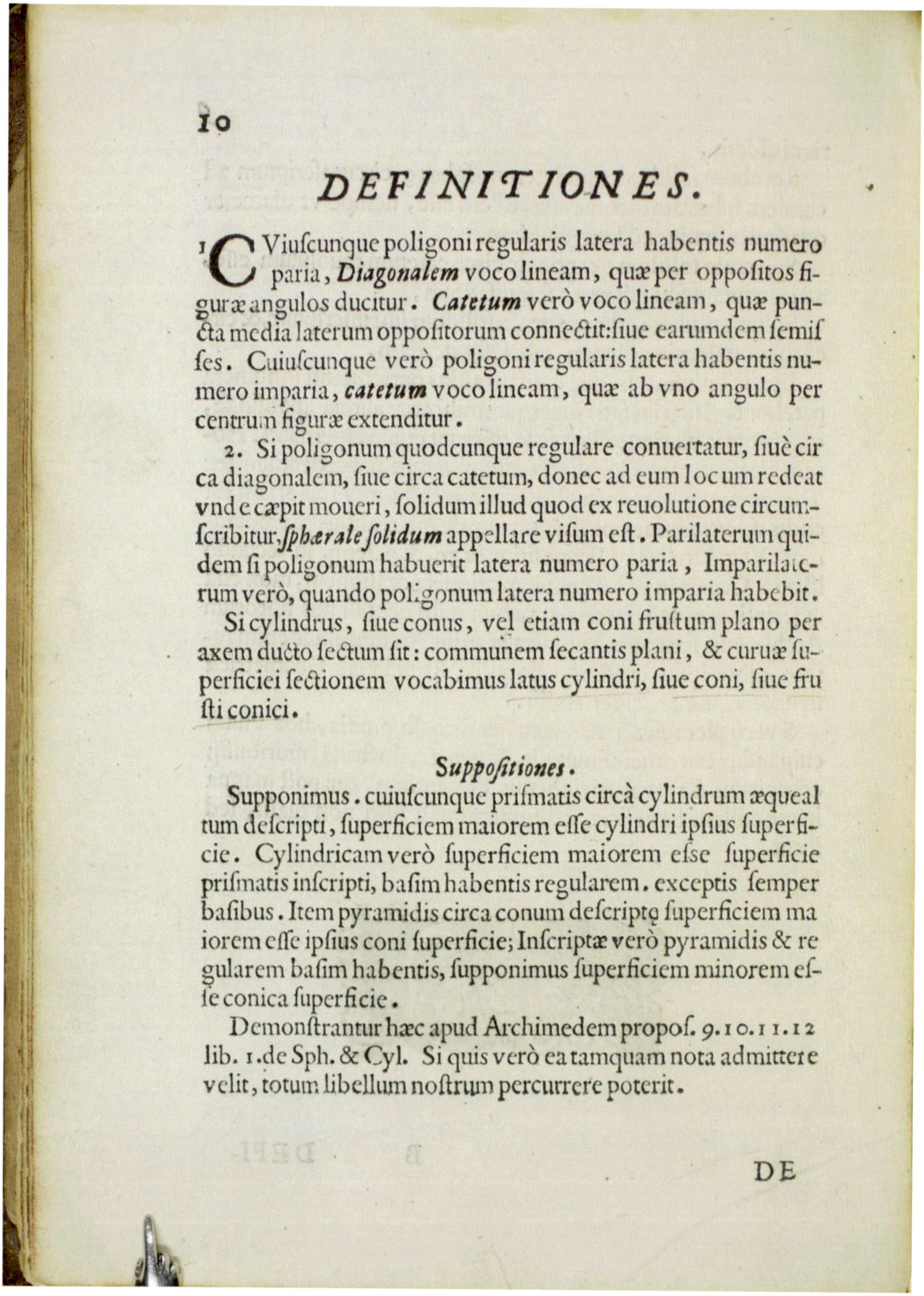

10

DEFINITIONES.

1 CViuſcunque poligoni regularis latera habentis numero paria, ***Diagonalem*** voco lineam, quæ per oppoſitos figuræ angulos ducitur. ***Catetum*** verò voco lineam, quæ puncta media laterum oppoſitorum connectit: ſiue earumdem ſemiſſes. Cuiuſcunque verò poligoni regularis latera habentis numero imparia, ***catetum*** voco lineam, quæ ab vno angulo per centrum figuræ extenditur.

2. Si poligonum quodcunque regulare conuertatur, ſiuè circa diagonalem, ſiue circa catetum, donec ad eum locum redeat vnde cæpit moueri, ſolidum illud quod ex reuolutione circumſcribitur, ***ſphærale ſolidum*** appellare viſum eſt. Parilaterum quidem ſi poligonum habuerit latera numero paria, Imparilaterum verò, quando poligonum latera numero imparia habebit.

Si cylindrus, ſiue conus, vel etiam coni fruſtum plano per axem ducto ſectum ſit: communem ſecantis plani, & curuæ ſuperficiei ſectionem vocabimus latus cylindri, ſiue coni, ſiue fruſti conici.

Suppoſitiones.

Supponimus. cuiuſcunque priſmatis circà cylindrum æquealtum deſcripti, ſuperficiem maiorem eſſe cylindri ipſius ſuperficie. Cylindricam verò ſuperficiem maiorem eſse ſuperficie priſmatis inſcripti, baſim habentis regularem. exceptis ſemper baſibus. Item pyramidis circa conum deſcripto ſuperficiem maiorem eſſe ipſius coni ſuperficie; Inſcriptæ verò pyramidis & regularem baſim habentis, ſupponimus ſuperficiem minorem eſſe conica ſuperficie.

Demonſtrantur hæc apud Archimedem propoſ. 9.10.11.12 lib. 1. de Sph. & Cyl. Si quis verò ea tamquam nota admittere velit, totum libellum noſtrum percurrere poterit.

DE

10 [Proemium]

DEFINITIONES.

1. CViuſcunque poligoni regularis latera habentis numero paria, ***Diagonalem*** voco lineam, quæ per oppoſitos figuræ angulos ducitur. ***Catetum*** verò voco lineam, quæ puncta media laterum oppoſitorum connectit: ſiue earumdem ſemiſſes. Cuiuſcunque verò poligoni regularis latera habentis numero imparia, *catetum* voco lineam, quæ ab vno angulo per centrum figuræ extenditur.

2. Si poligonum quodcunque regulare conuertatur, ſiuè circa diagonalem, ſiue circa catetum, donec ad eum locum redeat vnde cæpit moueri, ſolidum illud quod ex reuolutione circumſcribitur, ***ſphærale ſolidum*** appellare viſum eſt. Parilaterum quidem ſi poligonum habuerit latera numero paria, Imparilaterum verò, quando poligonum latera numero imparia habebit.

Si cylindrus, ſiue conus, vel etiam coni fruſtum plano per axem ducto ſectum ſit: communem ſecantis plani, & curuæ ſuperficiei ſectionem vocabimus latus cylindri, ſiue coni, ſiue fruſti conici.

Suppositiones.

Supponimus. cuiuſcunque priſmatis circà cylindrum æquealtum deſcripti, ſuperficiem maiorem eſſe cylindri ipſius ſuperficie. Cylindricam verò ſuperficiem maiorem eſse ſuperficie priſmatis inſcripti, baſim habentis regularem. exceptis ſemper baſibus. Item pyramidis circa conum deſcriptọ ſuperficiem maiorem eſſe ipſius coni ſuperficie; Inſcriptæ verò pyramidis & regularem baſim habentis, ſupponimus ſuperficiem minorem eſſe conica ſuperficie.

Demonſtrantur hæc apud Archimedem propoſ. 9. 10. 11. 12 lib. 1. de Sph. & Cyl. Si quis verò ea tamquam nota admittere velit, totum libellum noſtrum percurrere poterit.

DE

11

DE SOLIDIS SPHAERALIBVS LIBER PRIMVS.

PROPOSITIO PRIMA.

SI Cylindri recti ſuperficies ſecetur plano oppoſitis baſibus parallelo; erunt ſegmenta ſuperficiei cylindricæ inter ſe, vt ſegmenta axis, ſiue lateris cylindri, homologe ſumpta.

Eſto cylindrus rectus ABCD, ſeceturq; plano E F oppoſitis baſibus parallelo; Dico cylindricam ſuperficiem AEFD, ad cylindricam EBCF, eſse vt axis ad axem, ſiue vt latus AE, ad latus EB.

Producatur vtrimque in infinitum cylindrus, & accipiatur recta E G multiplex ipſius EA, iuxtà quamlibet multiplicitatem, ſectaque EG in partes ipſi E A æquales, agantur per puncta diuiſionum H, I, G; plana oppoſitis baſibus parallela. Eritque tam multiplex recta GE ipſius EA: quàm multiplex eſt cylindrica ſuperficies E L, ſuperficiei E D.

Sumatur etiam recta E M multiplex ipſius EB, iux-

11

DE SOLIDIS SPHAERALIBUS LIBER PRIMVS

PROPOSITIO PRIMA.

SI[9] Cylindri recti ſuperficies ſecetur plano oppoſitis baſibus parallelo; erunt ſegmenta ſuperficiei cylindricæ inter ſe, vt ſegmenta axis, ſiue lateris cylindri, homologe ſumpta.

Eſto cylindrus rectus ABCD, ſeceturq; plano EF oppoſitis baſibus parallelo; Dico cylindricam ſuperficiem AEFD, ad cylindricam EBCF, eſse vt axis ad axem, ſiue vt latus AE, ad latus EB.

Producatur vtrimque in infinitum cylindrus, & accipiatur recta EG multiplex ipſius EA, iuxtà quamlibet multiplicitatem, ſectaque EG in partes ipſi EA æquales, agantur per puncta diuiſionum H, I, G; plana oppoſitis baſibus parallela. Eritque tam multiplex recta GE ipſius EA: quàm multiplex eſt cylindrica ſuperficies EL, ſuperficiei ED.

Sumatur etiam recta EM multiplex ipſius EB, iux-

B 2 ta

---

[9] SI] Si.

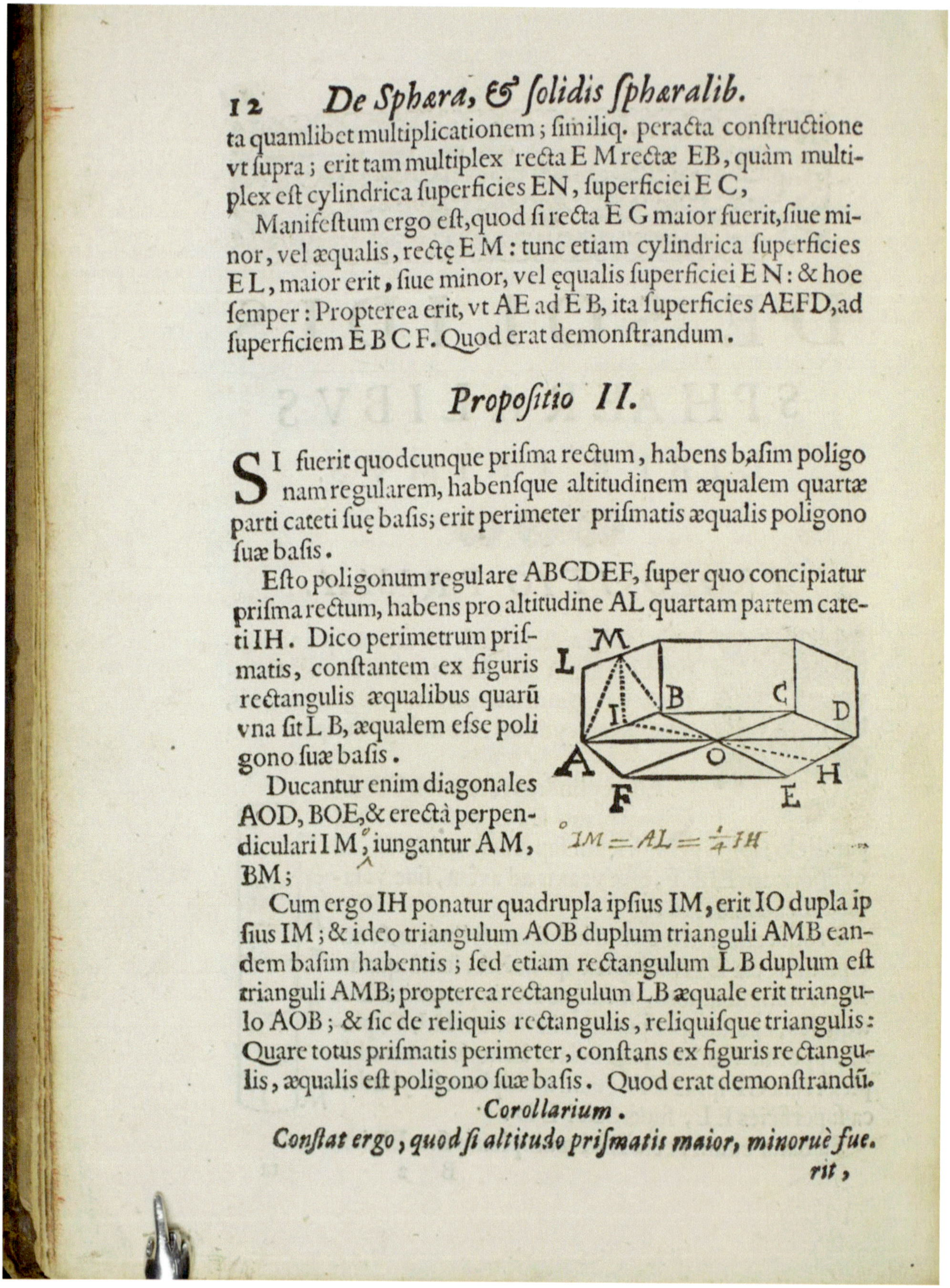

12 *De Sphæra, & ſolidis ſphæralib.*

ta quamlibet multiplicationem; ſimiliq. peracta conſtructione vt ſupra; erit tam multiplex recta E M rectæ EB, quàm multiplex eſt cylindrica ſuperficies EN, ſuperficiei E C,

Manifeſtum ergo eſt, quod ſi recta E G maior fuerit, ſiue minor, vel æqualis, rectę E M: tunc etiam cylindrica ſuperficies E L, maior erit, ſiue minor, vel ęqualis ſuperficiei E N: & hoc ſemper: Propterea erit, vt AE ad E B, ita ſuperficies AEFD, ad ſuperficiem E B C F. Quod erat demonſtrandum.

Propoſitio II.

SI fuerit quodcunque priſma rectum, habens baſim poligonam regularem, habenſque altitudinem æqualem quartæ parti cateti ſuę baſis; erit perimeter priſmatis æqualis poligono ſuæ baſis.

Eſto poligonum regulare ABCDEF, ſuper quo concipiatur priſma rectum, habens pro altitudine AL quartam partem cateti IH. Dico perimetrum priſmatis, conſtantem ex figuris rectangulis æqualibus quarũ vna ſit L B, æqualem eſse poligono ſuæ baſis.

Ducantur enim diagonales AOD, BOE, & erectâ perpendiculari I M, iungantur A M, BM;

Cum ergo IH ponatur quadrupla ipſius IM, erit IO dupla ipſius IM; & ideo triangulum AOB duplum trianguli AMB eandem baſim habentis; ſed etiam rectangulum L B duplum eſt trianguli AMB; propterea rectangulum LB æquale erit triangulo AOB; & ſic de reliquis rectangulis, reliquiſque triangulis: Quare totus priſmatis perimeter, conſtans ex figuris rectangulis, æqualis eſt poligono ſuæ baſis. Quod erat demonſtrandũ.

Corollarium.

Conſtat ergo, quod ſi altitudo priſmatis maior, minoruè fuerit,

ta quamlibet multiplicationem; ſimiliq. peracta conſtructione vt ſupra; erit tam multiplex recta EM rectæ EB, quàm multiplex eſt cylindrica ſuperficies EN, ſuperficiei EC,[10].

Manifeſtum ergo eſt, quod ſi recta EG maior fuerit, ſiue minor, vel æqualis, rectę EM: tunc etiam cylindrica ſuperficies EL, maior erit, ſiue minor, vel ęqualis ſuperficiei EN: & hoc ſemper: Propterea erit, vt AE ad EB, ita ſuperficies AEFD, ad ſuperficiem EBCF. Quod erat demonſtrandum.

Propoſitio II.

SI[11] fuerit quodcunque priſma rectum, habens baſim poligonam regularem, habenſque altitudinem æqualem quartæ parti cateti ſuę baſis; erit perimeter priſmatis æqualis poligono ſuæ baſis.

Eſto poligonum regulare ABCDEF, ſuper quo concipiatur priſma rectum, habens pro altitudine AL quartam partem cateti IH. Dico perimetrum priſmatis, conſtantem ex figuris rectangulis æqualibus quarũ vna ſit LB, æqualem eſse poligono ſuæ baſis. [*IM* = *AL* = ¼ *IH*].

Ducantur enim diagonales AOD, BOE, & erectà perpendiculari IM, iungantur AM, BM;

Cum ergo IH ponatur quadrupla ipſius IM, erit IO dupla ipſius IM; & ideo triangulum AOB duplum trianguli AMB eandem baſim habentis; ſed etiam rectangulum LB duplum eſt trianguli AMB; propterea rectangulum LB æquale erit triangulo AOB; & ſic de reliquis rectangulis, reliquiſque triangulis: Quare totus priſmatis perimeter, conſtans ex figuris rectangulis, æqualis eſt poligono ſuæ baſis. Quod erat demonſtrandũ.

Corollarium

Conſtat ergo, quod ſi altitudo priſmatis maior, minoruè fue.[12]

rit,

[10] ,] .
[11] SI] Si.
[12] Fue.] fue-].

rit, quàm quarta pars cateti suæ basis, erit perimeter prismatis maior, minorùe quàm poligonum suæ basis.

Propositio III.

SI fuerit cylindrus rectus, cuius altitudo ęqualis sit quartæ parti diametri suę basis; erit cylindrica superficies ęqualis circulo suæ basis.

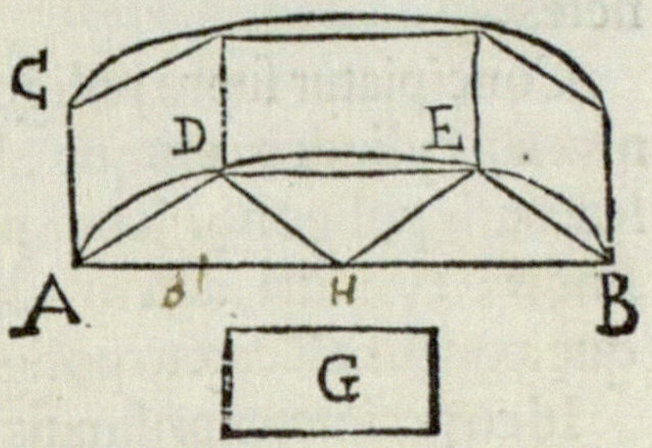

Esto cylindrus rectus, cuius basis circulus circa diametrum A B descriptus; altitudo verò A C, ęqualis sit quartę parti diametri A B.

Dico cylindricam superficiem ęqualem esse circulo suæ basis A B.

Si enim ęqualis non est; erit circulus vel maior, uel minor cylindricà superficie.

Sit primùm circulus maior quàm cylindri superficies; & supposità differentia G, describatur intrà circulum aliquod poligonum ADEB, quod quidem deficiat à circulo minori defectu, quàm sit spatium G; & ideo erit poligonum inscriptum adhuc maius quàm cylindrica superficies (quomodo fiat hoc constat ex Commentarijs in Archimedem, & ex XII. Euclidis:) Tum supra poligonum ADEB concipiatur prisma rectum eiusdem cum cylindro altitudinis.

Cùm ergò altitudo prismatis eadem sit ac cylindri, nempe quarta pars rectæ A B, erit altitudo prismatis maior quàm quarta pars cateti suę basis poligonæ, & ideo perimeter prismatis maior erit quàm poligonum suę basis, & multo maior, quàm cylindrica superficies (factum enim est poligonum maius cylindrica superficie.) Quod est absurdum: est enim contra pręmissas suppositiones. *ex Corol lar. præced.*

Ponatur deinde circulus minor quàm cylinrdica superficies: & supposità differentia G, describatur circa circulum aliquod

C poli-

Liber Primus. 13

rit, quàm quarta pars cateti ſuæ baſis, erit perimeter priſmatis maior, minoruè quàm poligonum ſuæ baſis.

Propositio III.

SI[13] fuerit cylindrus rectus, cuius altitudo ęqualis ſit quartæ parti diametri ſuę baſis; erit cylindrica ſuperficies ęqualis circulo ſuæ baſis.
Eſto cylindrus rectus, cuius baſis circulus circa diametrum AB deſcriptus; altitudo verò AC, ęqualis ſit quartę parti diametri AB.

Dico cylindricam ſuperficiem ęqualem eſse circulo ſuæ baſis AB.

Si enim ęuqualis[14] non eſt; erit circulus vel maior, uel minor cylindricà ſuperficie.

Sit primùm circulus maior quàm cylindri ſuperficies; & ſuppoſità differentia G, deſcribatur intrà circulum aliquod poligonum ADEB, quod quidem deficiat à circulo minori defectu, quàm ſit ſpatium G; & ideo erit poligonum inſcriptum adhuc maius quàm cylindrica ſuperficies (quomodo fiat hoc conſtat ex Commentarijs in Archimedem, & ex XII. Euclidis:) Tum ſupra poligonum ADEB concipiatur priſma rectum eiuſdem cum cylindro altitudinis.

Cùm ergò altitudo priſmatis eadem ſit ac cylindri, nempe quarta pars rectæ AB, erit altitudo priſmatis maior quàm quarta pars cateti ſuę baſis poligonæ, & ideo perimeter priſmatis maior erit quàm poligonum ſuę baſis, & multo maior, quàm cylindrica ſuperficies (factum enim eſt poligonum maius cylindrica superficie.) Quod eſt abſurdum: eſt enim contra prę miſsas ſuppoſitiones. *ex Corol lar prae- ced*

Ponatur deinde circulus minor quàm cylindrica ſuperficies: & ſuppoſità differentia G, deſcribatur circa circulum aliquod

C poli-

[13] SI] Si.

[14] ęuqualis] equalis.

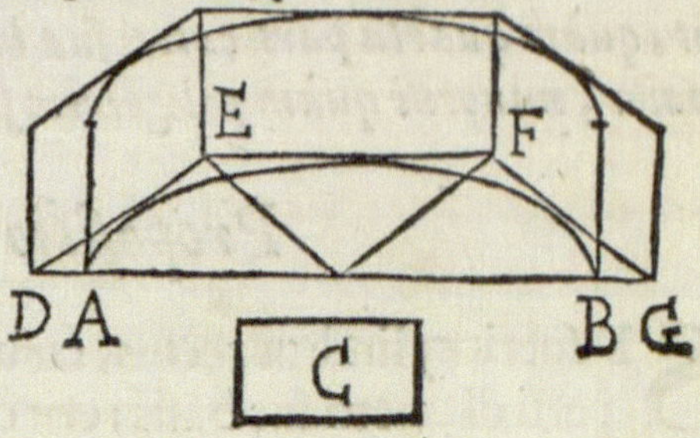

poligonum regulare D E F G, quod excedat circulum ſpatio minori quàm ſit C. (quomodo hoc fiat conſtat apud Commentarios in Archim. & in XII. Euclidis.) eritq; etiam poligonum minus quàm cylindrica ſuperficies.

Concipiatur ſuprà poligonum erigi priſma eiuſdem altitudinis cum cylindro; eritque altitudo priſmatis quarta pars cateti ſuæ baſis poligonę. (cum priſmatis altitudo eadem ſit atq: cylindri; cylindri autem altitudo eſt quarta pars rectæ A B, quę æqualis eſt cateto poligoni, quod eſt baſis priſmatis.)

per ij. huius. Ideo perimeter priſmatis æqualis erit poligono ſuę baſis; & propterea minor quàm cylindrica ſuperficies. Quod eſt contra præmiſſas ſuppoſitiones.

Erit ergò ſuperficies cylindrica æqualis circulo ſuę baſis, Quod erat demonſtrandum.

Propoſitio IV.

CYlindri recti ſuperficies ad circulum ſuę baſis eſt vt latus cylindri ad quartam partem diametri eiuſdem baſis.

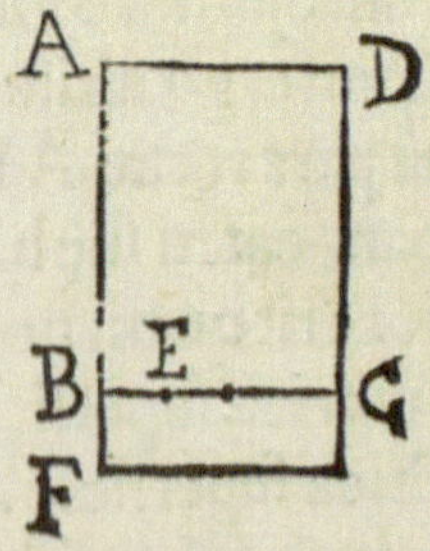

Eſto cylindrus rectus, cuius rectangulum per axem ſit ABCD; ſumptaq; BE, quæ quarta pars ſit ipſius BC; Dico cylindricam ſuperficiem ABCD ad circulum ſuæ baſis eſſe, vt AB ad BE.

Producatur cylindrus verſus F, ſectàque B F æquali ipſi B E, erit per præcedentem, cylindrica ſuperficies F C æqualis circulo ſuæ baſis BC.

per j. huius. Iam: cylindrica ſuperficies BD, ad cylindricam ſuperficiem F C eſt

poligonum regulare DEFG, quod excedat circulum ſpatio minori quàm ſit C. (quomodo hoc fiat conſtat apud Commentarios in Archim. & in XII. Euclidis.) eritq; etiam poligonum minus quàm cylindrica ſuperficies.

Concipiatur ſuprà poligonum erigi priſma eiuſdem altitudinis cum cylindro; eritque altitudo priſmatis quarta pars cateti ſuæ baſis poligonę. (cum priſmatis altitudo eadem ſit atq: cylindri; cylindri autem altitudo eſt quarta pars rectæ A B, quę æqualis eſt cateto poligoni, quod eſt baſis priſmatis.)

per ij. Hu[-] ius. Ideo perimeter priſmatis æqualis erit poligono ſuę baſis; & propterea minor quàm cylindrica ſuperficies. Quod eſt contra præmiſſas ſuppoſitiones.

Erit ergò ſuperficies cylindrica æqualis circulo ſuę baſis, Quod erat demonſtrandum.

Propositio IV.

CYlindri recti ſuperficies ad circulum ſuę baſis eſt vt latus cylindri ad quartam partem diametri eiuſdem baſis

Eſto cylindrus rectus, cuius rectangulum per axem ſit ABCD; ſumptaq; BE, quæ quarta pars ſit ipſius BC; Dico cylindricam ſuperficiem ABCD ad circulum ſuæ baſis eſſe, vt AB ad BE.

Producatur cylindrus verſus F, ſectàque

per j. hu- ius.. BF æquali ipſi BE, erit per præcedentem, cylindrica ſuperficies FC æqualis circulo ſuæ basis BC.

Iam: cylindrica ſuperficies BD, ad cylindricam ſuperficiem

FC est

F C eſt vt AB ad B F; ſuperficies verò FC ad circulum BC (ob ęqualitatem) eſt vt F B ad B E; Ergo ex ęquo erit cylindrica ſuperficies BD ad circulum B C, vt AB ad BE, nempe vt latus cylindri ad ¼ diametri baſis eiuſdem. Quod erat oſtendendum.

$\frac{1}{4}$

Propoſitio V.

CYlindri recti ſuperficies ad circulum quemlibet, eſt vt rectangulum per axem cylindri ad quadratum ſemidiametri ipſius circuli.

Eſto cylindrus rectus cuius rectangulum per axem ſit AB, & centrum baſis H. Ponatur autem circulus quilibet cuius ſemidiameter CD. Dico cylindricam ſuperficiem ad circulum ex CD, eſſe vt rectangulum AB ad quadratum CD.

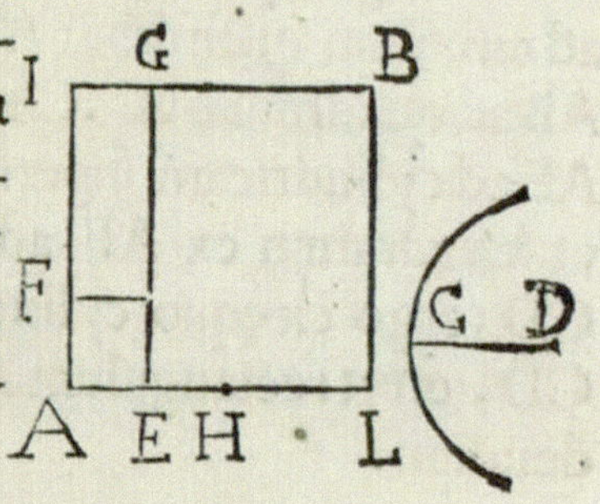

Fiat ex AE (quæ quidem 4. pars ſit rectę AL) quadratum FE, producaturque EG.

Erit ergò cylindrica ſuperficies A B ad circulum ſuę baſis, vt IA ad AE, hoc eſt vt IA ad AF, hoc eſt vt rectangulum I E ad quadratum FE; ſiue, ſumptis quadruplis, vt rectangulum A B ad quadratum ex AH. Circulus verò baſis A L ad circulum ex CD, eſt vt quadratum ex AH ad quadratum ex CD; ergò ex æquo erit cylindrica ſuperficies ad circulum ex CD, vt rectangulum per axem ad quadratum CD. Quod erat demonſtrandum. *per præced.* *Prim. 6.* *2. duodecimi.*

Corollarium.

Pro Corollario erit Propoſitio XIII. lib. 1. Archim. de Sphæra & Cylindro. Conſtat enim quòd ſi CD. media fuerit proportionalis inter IA, AL; quadratum ex CD æquale erit rectangulo AB. & propterea, ex demonſtratis, cylindricam ſuperficiem AIBL æqualem eſſe circulo ex CD neceſſe eſt.

Propo-

FC est AB ad BF; ſuperficies verò FC ad circulum BC (ob ęqualitatem) eſt vt FB ad BE; Ergo ex ęquo erit cylindrica ſuperficies BD ad circulum BC, vt AB ad BE, nempe vt latus cylindri ad ¼ diametri baſis eiuſdem. Quod erat oſtendendum.

Propositio V.

CYlindri recti ſuperficies ad circulum quemlibet, eſt vt rectangulum per axem cylindri ad quadratum ſemidiametri ipſius circuli.

Eſto cylindrus rectus cuius rectangulum per axem ſit AB, & centrum baſis H. Ponatur autem circulus quilibet cuius ſemidiameter CD. Dico cylindricam ſuperficiem ad circulum ex CD, eſſe vt rectangulum AB ad quadratum CD.

Fiat ex AE (quæ quidem 4. pars ſit rectę AL) quadratum FE, producaturque EG.
Erit ergò cylindrica ſuperficies AB ad circulum ſuę baſis, vt IA ad AE, hoc eſt vt IA ad AF, hoc eſt vt rectangulum IE ad quadratum FE; ſiue, ſumptis quadraplis, vt rectangulum AB ad quadratum ex AH. Circulus verò baſis AL ad circulum ex CD, eſt vt quadratum ex AH ad quadratum ex CD; ergò ex æquo erit cylindrica ſuperficies ad circulum ex CD, vt rectangulum per axem ad quadratum CD. Quod erat demonſtrandum.

per prae-ced Prim. 6 2.duode-cimi.

Corollarium.

Pro Corollario erit Propoſitio XIII. lib. 1. *Archim. de Sphæra & Cylindro. Conſtat enim quòd ſi CD. media fuerit proportionalis inter IA, AL; quadratum ex CD æquale erit rectangulo AB. & propterea, ex demonſtratis, cylindricam ſuperficiem AIBL æqualem eſſe circulo ex CD neceſſe eſt.*

Propo-

Prodoſitio VI.

CYlindrorum ſuperficies inter ſe ſunt vt eorumdem rectangula per axem homologè ſumpta.

Sint cylindri recti quorum rectangula per axem ſint AB,CD. Dico cylindricam ſuperficiem AB, ad cylindrieam CD eſſe, vt rectangulum AB ad rectangulum CD.

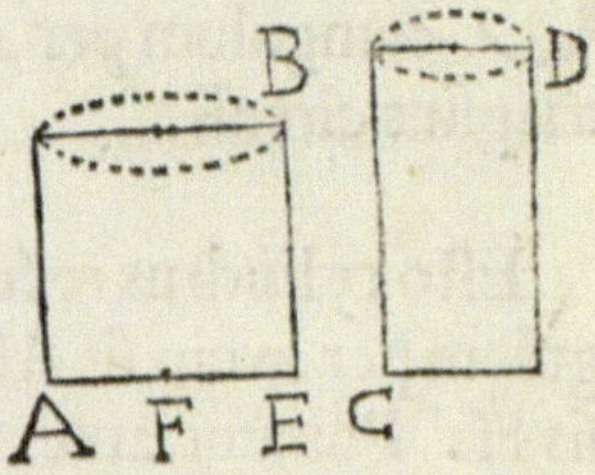

Accipiatur pro circulo quolibet, circulus circa diametrum A E.

per præced. Erit ergò cylindrica ſuperficies AB ad circulum quemlibet AE, ut rectang. AB ad quadratum AF. Circulus verò ex AF ad cylindricam ſuperficiem CD eſt *per præced.* vt quadratum ex AF ad rectangulum CD; ergo ex ęquo cylindrica ſuperficies AB ad cylindricam CD, eſtvt rectangulum A B ad rectang. CD. Quod erat oſtendendum.

Propoſitio VII.

SI recta pyramis baſim habuerit poligonam regularemque erit baſis pyramidis ad reliquam ipſius ſuperficiem, vt ſemi catetus baſis ad catetum ſuperficiei.

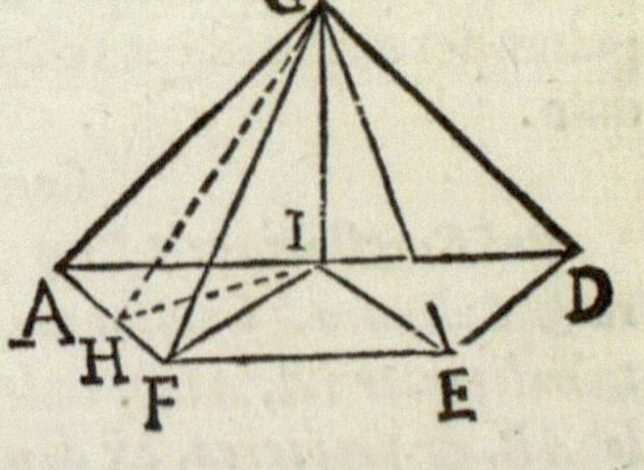

Eſto pyramis recta, cuius baſis poligonum regulare AFED. vertex verò G, & centrum baſis ſit I. Secto deinde vno latere bifariam in H, iunctiſq; GH, IH, erit GH catetus ſuperficiei pyramidis; IH vero ſemicatetus baſis; quandoquidem omnia triangula in ſuperficie ſunt æquicruria, & æqualia inter ſe; quod etiam verum eſt & in baſi.

Dico

Prodositio[15] VI.

CYlindrorum ſuperficies inter ſe ſunt vt eorumdem rectangula per axem homologè ſumpta.

Sint cylindri recti quorum rectangula per axem ſint AB, CD. Dico cylindricam ſuperficiem AB, ad cylindricam CD eſſe, vt rectangulum AB ad rectangulum CD.

per praeced Accipiatur pro circulo quolibet, circulus circa diametrum AE.

E it[16] ergò cylindrica ſuperficies AB ad circulum quemlibet AE, ut rectang. AB ad quadratum AF. Circulus verò ex AF ad cylindricam ſuperficiem CD eſt vt quadratum ex AF ad rectangulum CD; ergo ex ęquo cylindrica ſuperficies AB ad cylindricam CD, *per praeced* eſt vt rectangulum AB ad rectang. CD. Quod erat oſtendendum.

Propositio VII.

SI recta pyramis baſim habuerit poligonam regularemque erit baſis pyramidis ad reliquam ipſius ſuperficiem, vt ſemi catetus baſis ad catetum ſuperficiei.

Eſto pyramis recta, cuius baſis poligonum regulare AFED. vertex verò G, & centrum baſis ſit I. Secto deinde vno latere bifariam in H, iunctiſq; GH, IH, erit GH catetus ſuperficiei pyramidis; IH vero ſemicatetus baſis; quandoquidem omnia triangula in ſuperficie ſunt æquicruria, & æqualia inter ſe; quod etiam verum eſt & in baſi.

Dico

---

[15] *Prodositio*] *Propositio.*

[16] it] sit.

Dico basim ad superficiem esse vt I H ad HG.

Triangulum enim A I F, ad triangulum A G F (cum sint in ea dem basi) est vt IH, ad H G, ergo etiam ipsorum ęquemultiplicia, nempe basis, & superficies pyramidis, in eadem ratione erunt, nempe vt I H ad HG. Quod erat ostendendum. 15. *quinti.*

Propositio VIII.

CONI recti basis ad reliquam conicam superficiem, est vt semidiameter basis ad latus coni.

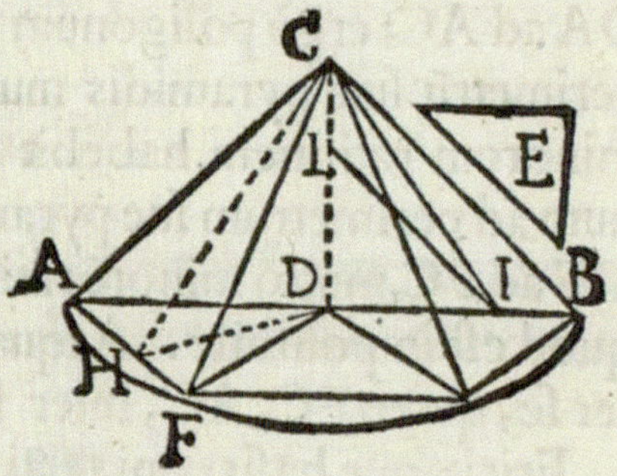

Esto conus rectus, cuius basis AB, vertex verò C, axis CD.

Dico circulum basis, ad reliquam conicam superficiem, esse vt DA, ad AC.

Si enim ita non est; erit circulus AB vel maior, vel min. quã oportet esse, vt ad conicam superficiem sit quemadmodum DA ad AC.

Sit primùm maior; & ponatur tantò maior quantum est spatium E. Inscribatur in circulo poligonum deficiens â circulo, minori defectu quàm spatium E; habebitq; huiusmodi poligonum ad conicam superficiem adhuc maiorem rationem, quàm DA ad AC. Secto deinde vno poligoni latere AF bifariam in H, iungantur DH, CH; & super poligono concipiatur pyramis quæ verticem habeat in C; seceturque DI ęqualis ipsi DH, & ducatur I L paralella ad BC, iungaturq. IC.

Cum itaq. poligonum ad conicam superficiem maiorem habeat rationem quàm DA ad AC; multò maiorem rationem habebit ad superficiem suæ pyramidis, quàm DA ad AC, vel DB ad BC. Sed poligonum ad superficiem pyramidis, per præcedentem, est vt DH ad HC; habebit ergo DH ad HC, siue DI ad IC, multò maiorem rationem quàm DB ad BC, vel quàm DI ad IL. Et propterea IC minor esset quàm IL. absurdum.

C Nam

Dico bafim ad fuperficiem effe vt IH ad HG. 15. *quinti*

Triangulum enim AIF, ad triangulum AGF (cum fint in eadem bafi) eft vt IH, ad HG, ergo etiam ipforum ęquemultiplicia, nempe bafis, & fuperficies pyramidis, in eadem ratione erunt, nempe vt IH ad HG. Quod erat oftendendum.

Propositio VIII.

COni recti bafis ad reliquam conicam fuperficiem, eft vt femidiameter bafis ad latus coni.

Efto conus rectus, cuius bafis AB, vertex verò C, axis CD.

Dico circulum bafis, ad reliquam conicam fuperficiem, effe vt DA, ad AC.

Si enim ita non eft; erit circulus AB vel maior, vel min. quã oportet effe, vt ad conicam fuperficiem fit quemadmodum DA ad AC.

Sit primùm maior; & ponatur tantò maior quantum eft fpatium E. Infcribatur in circulo poligonum deficiens à circulo, minori defectu quàm fpatium E; habebitq; huiufmodi poligonum ad conicam fuperficiem adhuc maiorem rationem, quàm DA ad AC. Secto deinde vno poligoni latere AF bifariam in H, iungantur DH, CH; & fuper poligono concipiatur pyramis quæ verticem habeat in C; feceturque DI ęqualis ipfi DH, & ducatur IL paralella[17] ad BC, iungaturq. IC.

Cum itaq. poligonum ad conicam fuperficiem maiorem habeat rationem quàm DA ad AC; multò maiorem rationem habebit ad fuperficiem fuæ pyramidis, quàm DA ad AC, vel DB ad BC. Sed poligonum ad fuperficiem pyramidis, per præcedentem, eft vt DH ad HC; habebit ergo DH ad HC, fiue DI ad IC, multò maiorem rationem quàm DB ad BC, vel quàm DI ad IL. Et propterea IC minor effet quàm IL. abfurdum.

C Nam

[17] paralela] parallela.

Nam quadratum I C æquale est duobus quadratis I D, D C; cum quadratum IL æquale sit tantùm duobus ID, DL. Ponatur deinde circulus basis AB minor quàm oportet esse vt ad conicam superficiem sit quemadmodum recta DA ad AC, sitque tantò minor quantum est spatium E. Circumscribatur circulo AB poligonum aliquod excedens circulum minori excessu quàm sit spatium E. Habebitq. poligonum ad conicam superficiẽ, adhuc minorem rationem quàm DA ad AC; ergò poligonum ad perimetrũ suæ pyramidis multò minorem rationem habebit quàm D A ad AC. Sed poligonum ad perimetrum suę pyramidis est vt DF ad FC; propterea DF ad FC, multò minorẽ rationem habebit quàm DA ad AC; quod est impossibile. Aequales etenim sunt tam DF, DA, inter se, quàm FC, AC, inter se.

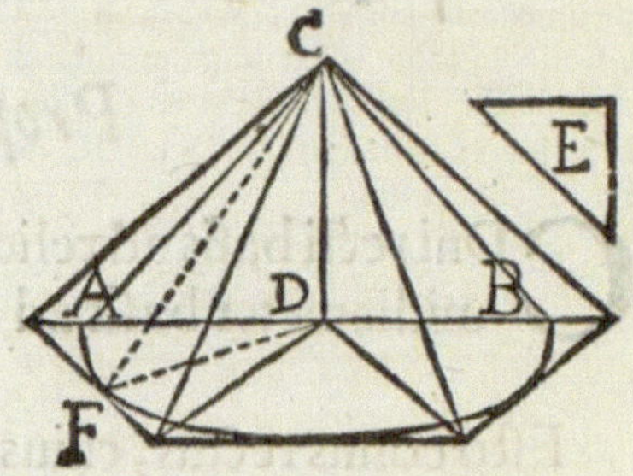

per 7. huius.

Erit itaque basis coni recti àd reliquam superficiem, vt D A ad AC. Quod erat demonstrandum.

Corollarium.

Hinc patet quòd curua superficies coni, ęqualis est circulo cuidam, cuius semidiameter med. prop. sit inter C A, AD. nempe, inter latus, & semidiametrum basis coni. Nam sumpta media inter CA, AD. erit circulus qui fit ex media, ad circulum qui fit ex AD. vt CA ad AD. Sed etiam curua coni superficies, ad circulum ex AD. est vt C A ad AD. Ergo æqualis est curua coni superficies, circulo, cuius semidiameter media proportionalis sit inter CA, AD.

per præced.

Propositio IX.

CViuslibet coni recti superficies, ad superficiem cuiuscunq; cylindri recti demptis basibus, est vt rectangulum sub latere, & semidiametro basis coni, ad rectangulum per axem cylindri,

Esto

Nam quadratum IC æquale eſt duobus quadratis ID, DC; cum quadratum IL æquale ſit tantùm duobus ID, DL. Ponatur deinde circulus baſis AB minor quàm oportet eſſe vt ad conicam ſuperficiem ſit quemadmodum recta DA ad AC, ſitque tantò minor quantum eſt ſpatium E. Circumſcribatur circulo AB poligonum aliquod excedens circulum minori exceſsu quàm ſit ſpatium E. Habebitq. poligonum ad conicam ſuperficiẽ, adhuc minorem rationem quàm DA ad AC; ergò poligonum ad perimetrũ ſuæ pyramidis multò minorem rationem habebit quàm DA ad AC. Sed poligonum ad perimetrum ſuę pyramidis eſt vt DF ad FC; propterea DF ad FC, multò minorẽ rationem habebit quàm DA ad AC; quod eſt impoſſibile. Aequales etenim ſunt tam DF, DA, inter ſe, quàm FC, AC, inter ſe.

per 7. huius.

Erit itaque baſis coni recti àd reliquam ſuperficiem, vt DA ad AC. Quod erat demonſtrandum.

Corollarium.

Hinc patet quòd curua ſuperficies coni, ęqualis eſt circulo cuidam, cuius ſemidiameter med. prop. ſit inter CA, AD. nempe, inter latus, & ſemidiametrum baſis coni. Nam ſumpta media inter CA, AD. erit circulus qui ſit ex media, ad circulum qui ſit ex AD. vt CA ad AD. Sed etiam curva coni superficies, ad circulum ex AD. eſt vt CA ad AD. Ergo æqualis eſt curua coni ſuperficies, circulo, cuius ſemidiameter media proportionalis ſit inter CA, AD.

per praeced.

Propoſitio IX.

CViuslibet coni recti ſuperficies, ad ſuperficiem cuiuſcunq; cylindri recti demptis baſibus, eſt vt rectangulum ſub latere, & ſemidiametro baſis coni, ad rectangulum per axem cylindri,

Esto

Esto conus ABC,cuius basis AC,axis vero B H; & cylindrus cuius rectangulum per axem sit DE. Dico conicam superficiem ad cylindricam esse,vt rectangulum BAH,ad rectangulum DE.

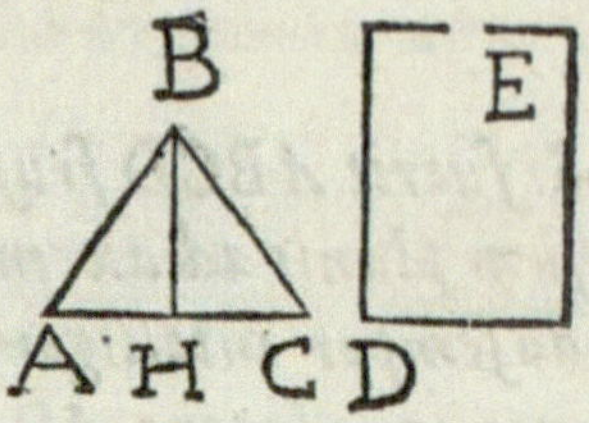

Nàm conica superficies ad circulum suę basis est vt AB, àd AH, siue vt rectā gulum BAH ad quadratum AH. circulus autem ex AH, ad cylindricam superficiem DE, est vt quadratum AH,ad rectangulum DE. Propterea,ex æquo, erit conica superficies ABC ad cylindricam DE, vt rectangulum BAH ad rectangulum DE. Quod erat ostendendum.

per 8. huius.
per 5. huius.

Propositio X.

Conicæ superficies,demptis basibus, inter se sunt vt rectangula sub lateribus conorum, & sub semidiametris basiū comprænensa.

Sint duo coni recti ABC,DEF. quorum axes BG, EH. Dico curuam coni ABC superficiem, ad curuam superficiem coni DEF, esse vt rectangulum B AG, ad rectangulum EDH. quę nimirum sub lateribus conorum, & semidiametris basium comprænenduntur.

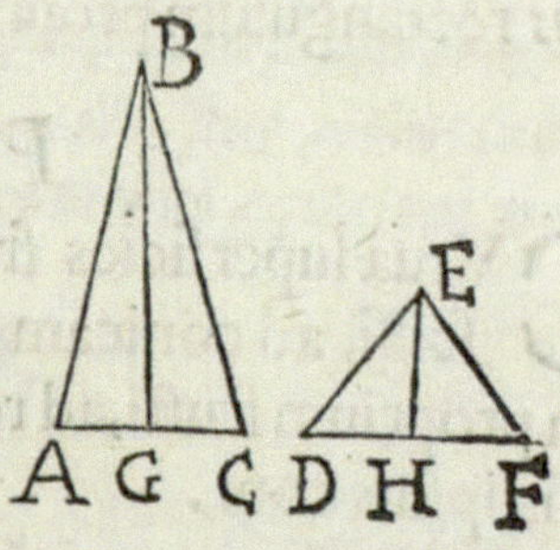

Conica enim superficies ABC, ad circulum AC,est vt recta BA ad AG,siue vt rectangulum BAG; ad quadratum AG. Circulus verò AC ad DF circulum,est vt quadratum AG, ad DH; denique circulus DF ad conicam superficiem DEF, est vt quadratum DH,ad rectangulum EDH. ergò ex æquo curua coni superficies ABC ad curuam DEF , erit vt rectangulum BAG, ad rectangulum E D H. Quod erat ostendendum.

per 8. huius.
per 8. huius.

Efto conus ABC, cuius bafis AC, axis vero BH; & cylindrus cuius rectangulum per axem fit DE. Dico conicam fuperficiem ad cylindricam effe, vt rectangulum BAH, ad rectangulum DE. *per* 8. *hu ius*

Nàm conica fuperficies ad circulum fuę bafis eft vt AB, àd AH, fiue vt rectãgulum BAH ad quadratum AH. circulus autem ex AH, ad cylindricam fuperficiem DE, eft vt quadratum AH, ad rectangulum DE. Propterea, ex æquo, erit conica fuperficies ABC ad cylindricam DE, vt rectangulum BAH ad rectangulum DE. Quod erat oftendendum. *per* 5. *hu ius.*

Propofitio X.

COnicæ fuperficies, demptis bafibus, inter fe funt vt rectangula fub lateribus conorum, & fub femidiametris bafiũ compræhenfa.

Sint duo coni recti ABC, DEF. quorum axes BG, EH. Dico curuam coni ABC fuperficiem, ad curuam fuperficiem coni DEF, effe vt rectangulum BAG, ad rectangulum EDH. quę nimirum fub lateribus conorum, & femidiametris bafium compræhenduntur. *per* 8. *hu ius*

Conica enim fuperficies ABC, ad circulum AC, eft vt recta BA ad AG, fiue vt rectangulum BAG; ad quadratum AG. Circulus verò AC ad DF circulum, eft vt quadratum AG, ad DH; denique circulus DF ad conicam fuperficiem DEF, eft vt quadratum DH, ad rectangulum EDH. ergò ex æquo curua coni fuperficies ABC ad curuam DEF, erit vt rectangulum BAG, ad rectangulum EDH. Quod erat oftendendum. *per* 8. *hu ius.*

C 2 Lem-

20 *De Sphæra, & solidis sphæralib.*

Lemma.

Si fuerit ABCD frustum coni recti, abscissum planis ad axem erectis (hoc enim modo semper intelligemus frusta conica) secenturque latera AB, DC bifariam in punctis E, & H. iungaturq; EH. Dico rectam EH componi ex vtràque BL, AI, nempe ex semidiametris basium oppositarum frusti conici.

Iungantur BD, EI, LH; Et quoniam AI, ID æquales sunt; item AE, EB, æquales: erunt parallelæ EI, BD. & ideo in parallelogrammo æqualia erunt latera ID, EM. Ob eandem causam æqualia sunt BL, MH. Ergo tota EH æqualis erit ipsis ID, BL simul sumptis. Quod erat, &c. per 2. sexti.

Definitiones.

Med: arith: *Vocabimus imposterum breuitatis causa lineam EH mediã Aritmeticam frusti conici.*

Rect: proprium *Rectangulum verò sub EH & AB latere frusti conici, dicemus* rectangulum proprium frusti conici.

Propositio XI.

Curua superficies frusti conici, planis ad axem erectis abscissi, ad conicam quamlibet superficiem, est vt rectangulum proprium frusti, ad rectangulum sub latere, & semidiametro basis ipsius coni.

Esto frustum conicũ ABCD abscissum planis ad axem erectis, sitque conus quilibet EFG, cuius axis FH. Dico curuam frusti AC superficiem, ad curuã coni EFG superficiem, esse, vt rectãgulum sub AB, & sub utraque AL, BI contentum, ad rectangulũ FEH, Proprium

Com-

Lemma.

Si fuerit ABCD fruſtum coni recti, abſciſſum planis ad axem erectis (hoc enim modo ſemper intelligemus fruſta conica) ſecenturque latera AB, DC bifariam in punctis E, & H. iungaturq; EH. Dico rectam EH componi ex vtràque BL, AI, nempe ex ſemidiametris baſium oppoſitarum fruſti conici.

per 2. ſexti.

Iungantur BD, EI, LH; Et quoniam AI, ID. æquales ſunt; item AE, EB, æquales: erunt parallelæ EI, BD. & ideo in parallelogrammo æqualia erunt latera ID, EM. Ob eandem cauſam æqualia ſunt BL, MH. Ergo tota EH æqualis erit ipſis ID, BL ſimul ſumptis. Quod erat &c.

[Mid:arith: Rect: proprium]

Definitiones.

Vocabimus impoſterum breuitatis cauſa lineam EH media Aritmeticam fruſti conici.

Rectangulum verò ſub EH & AB latere fruſti conici, dicemus rectangulum proprium fruſti conici.

Propoſitio XI.

[Proprium]

CVrua ſuperficies fruſti conici, planis ad axem erectis abſciſſi, ad conicam quamlibet ſuperficiem, eſt vt rectangulum proprium fruſti, ad rectangulum ſub latere, & ſemidiametro baſis ipſius coni.

Eſto fruſtum conicũ ABCD abſciſſum planis ad axem erectis, ſitque conus quilibet EFG, cuius axis FH. Dico curuam fruſti AC ſuperficiem, ad curuã coni EFG ſuperficiem, eſſe, vt rectãgulum ſub AB, & ſub utraque AL, BI contentum, ad rectangulũ FEH,

Com-

Compleatur conus AMD. cuius datum erat frustum, factoque angulo MAN recto, & secta AN equali ipsi AL. compleatur rectangulum AP. Ducto deinde diametro MN, & facta BO parallela ad AN. erit BO æqualis ipsi BI. compleatur etiam figura BQ.

Iam superficies curua coni AMD ad superficiem curuam coni BMC est vt rectangnlum LAM ad rectangulum IBM; nempe vt rectangulum AP ad BQ; & diuidendo, erit curua frusti conici ABCD superficies, ad superficiem coni BMC, vt gnomon AOP, ad rectangulum BQ. hoc est vt rectangulum sub AB; & vtraque AN, BO, siue AL, BI, ad rectangulum IBM. Curua verò superficies coni BMC ad curuam coni EFG, est vt rectangul. IBM ad rect. FEH. ergò ex æquo curua frusti conici ABCD superficies ad curuam coni EFG superficiem est vt rectan. contentum sub AB, & vtraque AL, BI ad rectangulum FEH. *per 10. huius.*

Corollarium.

Patet ergò quod frusti conici ABCD superficies sine basibus ad superficiem coni EFG. est vt rectangulum proprium frusti ad rectangulum FEH. Rectangulum autem proprium frusti comprehenditur sub recta AB, & sub vtraq; AL, BI, siue potiùs sub AB, & media Aritmetica, quam demonstrauimus æqualem vtrisque AL, BI. lemmate

Propositio XII.

CViuscunque frusti conici superficies ad superficiem cylindri recti, est vt rectangulum proprium frusti ad rectangulum per axem cylindri.

Esto frustum conicum ABCD, & cylindrus cuius rectangulum per axem sit EF. Secetur AB bifariā in H, & agatur media Aritmetica HI equidistanter ad BC. Dico conicam frusti superficiem, ad cylindricam EF, esse vt rectangulum sub HI, & AB, ad rectangulum EF.

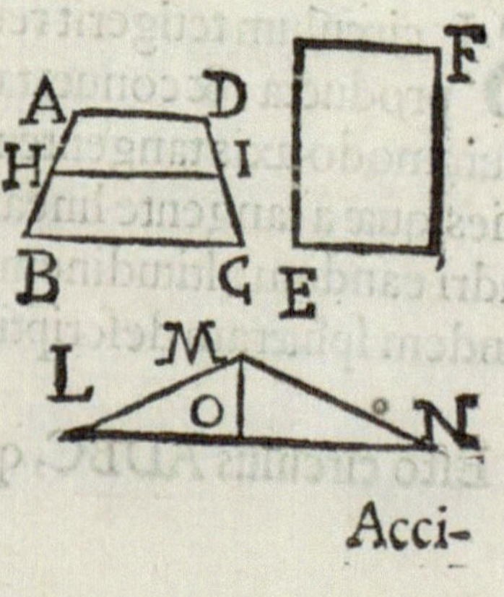

Acci-

Compleatur conus AMD. cuius datum erat fruftum, factoque angulo MAN recto, & fecta AN ęquali ipfi AL. compleatur rectangulum AP. Ducto deinde diametro MN, & facta BO parallela ad AN. erit BO æqualis ipfi BI. compleatur etiam figura BQ.

Iam fuperficies curua coni AMD ad fuperficiem curuam coni BMC eft vt rectangulum LAM ad rectangulum IBM; nempe vt rectangulum AP ad BQ; & diuidendo, erit curua frufti conici ABCD fuperficies, ad fuperficiem coni BMC, vt gnomon AOP, ad rectangulum B Q. hoc eft vt rectangulum fub A B; & vtraque AN, BO, fiue AL, BI, ad rectangulum IBM. Curua verò fuperficies coni BMC ad curuam coni EFG, eft vt rectangul. IBM ad rect. FEH. ergò ex æquo curua frufti conici ABCD fuperficies ad curuam coni EFG fuperficiem eft vt rectan. contentum fub AB, & vtraque AL, BI ad rectangulum FEH. *per* 10 *huius*

Corollarium..

Patet ergò quod frufti conici ABCD fuperficies fine bafibus ad fuperficiem coni EFG eft vt rectangulum proprium frufti ad rectangulum FEH. Rectangulum autem proprium frufti comprehenditur fub recta AB, & fub vtraq; AL, BI, fiue potiùs fub AB, & media Aritmetica, quam demonftrauimus æqualem vtrifque AL, BI [*Lemmate*]

Propofitio XII..

CViufcunque frufti conici fuperficies ad fuperficiem cylindri recti, eft vt rectangulum proprium frufti ad rectangulum per axem cylindri.

Efto fruftum conicum ABCD, & cylindrus cuius rectangulum per axem fit EF. Secetur AB bifariã in H, & agatur media Aritmetica HI ęquidiftanter ad BC. Dico conicam frufti fuperficiem, ad cylindricam EF, effe vt rectangulum fub HI, & AB, ad rectangulum EF.

Acci-

Accipiatur conus quilibet LMN, cuius axis MO. Eritq; curua frusti superficies ad conicam curuam LMN, vt rectangulum sub AB, HI, ad rectangulum MLO; sed curua coni LMN ad curuam cylindri EF superficiem, est vt rectangulum MLO, ad rectangulum EF; ergo ex æquo curua frusti conici superficies, ad curuam superficiem cylindri, est vt rectangulum sub AB, & HI, nempe vt rectangulum proprium frusti, ad rectangulum EF per axem cylindri. Quod erat ostendendum.

per præced.

9. huius

Corollarium.

Curua superficies cuiuscunq; frusti conici ABCD æqualis demonstratur circulo cuidam, cuius quidem circuli semidiameter E media proportionalis sit inter latus AB frusti conici, & inter FH mediã Aritmeticam eiusdem frusti,

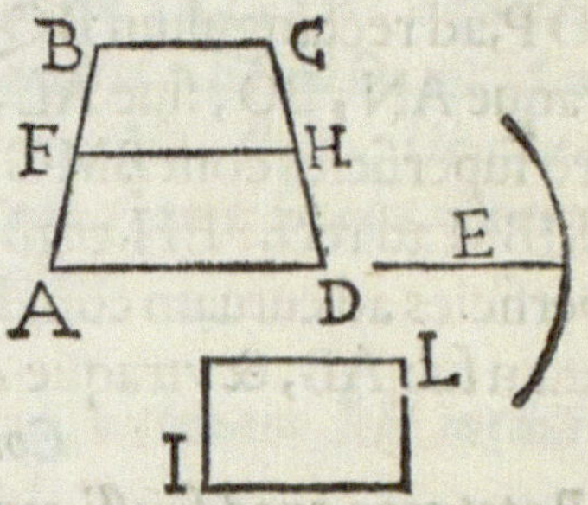

Esto quadratum E æquale rectangulo sub BA, FH. sumaturque cylindrus quilibet IL; & erit curua frusti conici superficies ad curuam cylindricam IL, vt rectangulum sub BA, FH ad rectangulum IL; siue vt quadratum E ad rectangulum IL; hoc est vt circulus ex radio E, ad curuam cylindricam IL. Aequales ergò sunt inter se curua superficies frusti conici AC, & circulus ex radio E factus. Quæ quidem Archimedis Propositio est 16. libri primi de Sph. & cyl.

per præced.

5. huius.

Propositio XIII.

SI circulum tetigerit recta quępiam linea æqualitèr vtrimq; producta; & conuertatur circulus circa quemlibet sui axẽ (dummodo axis tangentem non secet) erit conici frusti superficies, quæ à tangente linea describitur, æqualis superficiei cylindri eandem altitudinem cum frusto conico habentis, & circa eandem sphæram describtibilis.

Esto circulus ADBC, quem duæ diametri AB, CD secent ad angu-

per prae- ed. Accipiatur conus quilibet LMN, cuius axis MO. Eritq; curua fruſti ſuperficies ad conicam curuam LMN, vt rectangulum ſub AB, HI, ad rectangulum MLO; ſed curua coni LMN ad curuam cylindri EF ſuperficiem, eſt vt rectangulum MLO, ad rectangulum EF; ergo ex æquo curua fruſti conici ſuperficies, ad curuam ſuperficiem cylindri, eſt vt rectangulum ſub AB, & HI, nempe vt [9 huius] rectangulum proprium fruſti, ad rectangulum EF per axem cylindri. Quod erat oſtendendum.

Corollarium.

Curua ſuperficies cuiuſcunq; fruſti conici ABCD æqualis demonſtratur circulo cuidam, cuius quidem circuli ſemidiameter E media proportionalis ſit inter latus AB fruſti conici, & inter FH mediã Aritmeticam eiuſdem fruſti,

per prae– ced *5. huius.* Eſto quadratum E æquale rectangulo ſub BA, FH. ſumaturque cylindrus quilibet IL; & erit curua fruſti conici ſuperficies ad curuam cylindricam IL, vt rectangulum ſub BA, FH ad rectangulum IL; ſiue vt quadratum E ad rectangulum IL; hoc eſt vt circulus ex radio E, ad curuam cylindricam IL. Aequales ergò ſunt inter ſe curua ſuperficies fruſti conici AC, & circulus ex radio E factus. Quæ quidem Archimedis Propoſitio eſt 16. libri primi de Sph. & cyl.

Propoſitio XIII.

SI circulum tetigerit recta quępiam linea æqualitèr vtrimq; producta; & conuertatur circulus circa quemlibet ſui axẽ (dummodo axis tangentem non ſecet) erit conici fruſti ſuperficies, quæ à tangente linea deſcribitur, æqualis ſuperficiei cylindri eandem altitudinem cum fruſto conico habentis, & circa eandem ſphæram deſcriptibilis.

Eſto circulus ADBC, quem duæ diametri AB, CD ſecent ad.

angu-

angulos rectos. Duas insuper tangentes habeat alteram DG in extremitate diametri CD, alteram verò vbicunque in I, & ęqualitèr producătur hinc inde IL IM; dumodo axem AB productum non secent. Agantur deinde per L, & per M parallelæ ad CD, rectæ LE, MF. tum figura conuertatur circa axem AB. Tangens GH describet cylindricam quandam superficiem cuius rectă gulum per axem erit EFHG: Tangens verò LM designabit frustũ conicę superficiei; deniq; circulus ipse sphæram circumscribet.

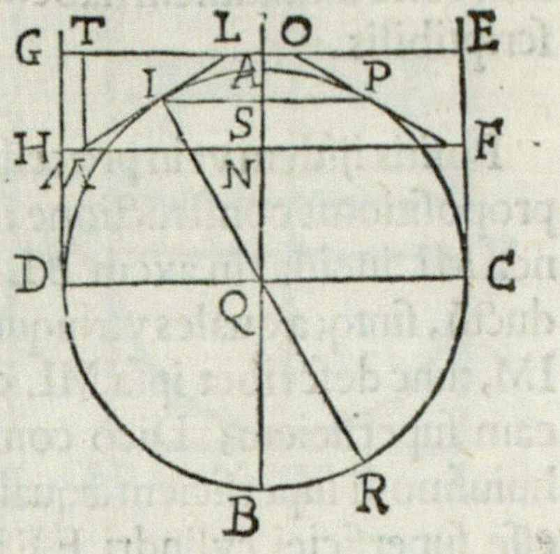

Dico cylindricam superficiem à linea GH descriptam, & conicam superficiem à linea LM factam æquales esse inter se.

Ducatur IP media Aritmetica conici frusti; & agatur IR per centrum Q: eritq; IR perpendicularis ad LM: Ducatur etiam MT perpendicularis ad EG.

Quoniam duo anguli TMI, T LM vni recto sunt æquales, nempe ipsi L IQ, demptis alternis TLM, LIS, erunt æquales reliqui TML, SIQ. ideoque triangula TML, SIQ. cum rectangulasint, similia erunt; Ergò vt TM ad ML ita SI ad IQ. hoc est (sumptis duplis) PI ad IR: & ideo rectangulum sub TM, IR (quod quidem est rectangulum EFHG.) ęquale erit rectangulo sub ML, IP. quod proprium vocamus frusti conici. Proptereà per præcedentem æqualis erit superficies conici frusti, quę à linea ML describitur, superficiei cylindri EFHG, eădem altitudinem cum ipso frusto habentis, & circà eandẽ sphęram ADBC. describtibilis. Quod &c.

Propositio XIV.

SI circulum tetigerit recta linea æqualitèr vtrinq; producta, & conuertatur circulus circa axem, qui cum tangente conueniat

angulos rectos. Duas inſuper tangentes habeat alteram DG in extremitate diametri CD, alteram verò vbicunque in I, & ęqualitèr producãtur hinc inde IL IM; dumodo axem AB productum non ſecent. Agantur deinde per L, & per M parallelæ ad CD, rectæ LE, MF. tum figura conuertatur circa axem AB. Tangens GH deſcribet cylindricam quandam ſuperficiem cuius rectãgulum per axem erit EFHG: Tangens verò LM deſignabit fruſtũ conicę ſuperficiei; deniq; circulus ipſe ſphæram circumſcribet. Dico cylindricam ſuperficiem à linea GH deſcriptam, & conicam ſuperficiem à linea LM factam æquales eſſe inter ſe.

Ducatur IP media Aritmetica conici fruſti; & agatur IR per centrum Q: eritq; IR perpendicularis ad LM: Ducatur etiam MT perpendicularis ad EG.

Quoniam duo anguli TMI, TLM vni recto ſunt æquales, nempe ipſi LIQ, demptis alternis TLM, LIS, erunt æquales reliqui TML, SIQ. ideoque triangula TML, SIQ, cum rectangula ſint, ſimilia erunt; Ergò vt TM ad ML ita SI ad IQ. hoc eſt (ſumptis duplis) PI ad IR: & ideo rectangulum ſub TM, IR (quod quidem eſt rectangulum EFHG.) ęquale erit rectangulo ſub ML, IP. quod proprium vocamus fruſti conici. Propterea per præcedentem æqualis erit ſuperficies conici fruſti, quę à linea ML deſcribitur, ſuperficiei cylindri EFHG, eãdem altitudinem cum ipſo fruſto habentis, & circà eandẽ ſphęram ADB*C*. deſcriptibilis. Quod &c.

Propoſitio XIV.

SI circulum tetigerit recta linea æqualitèr vtrinq; producta, & conuertatur circulus circa axem, qui cum tangente con-

ueniat

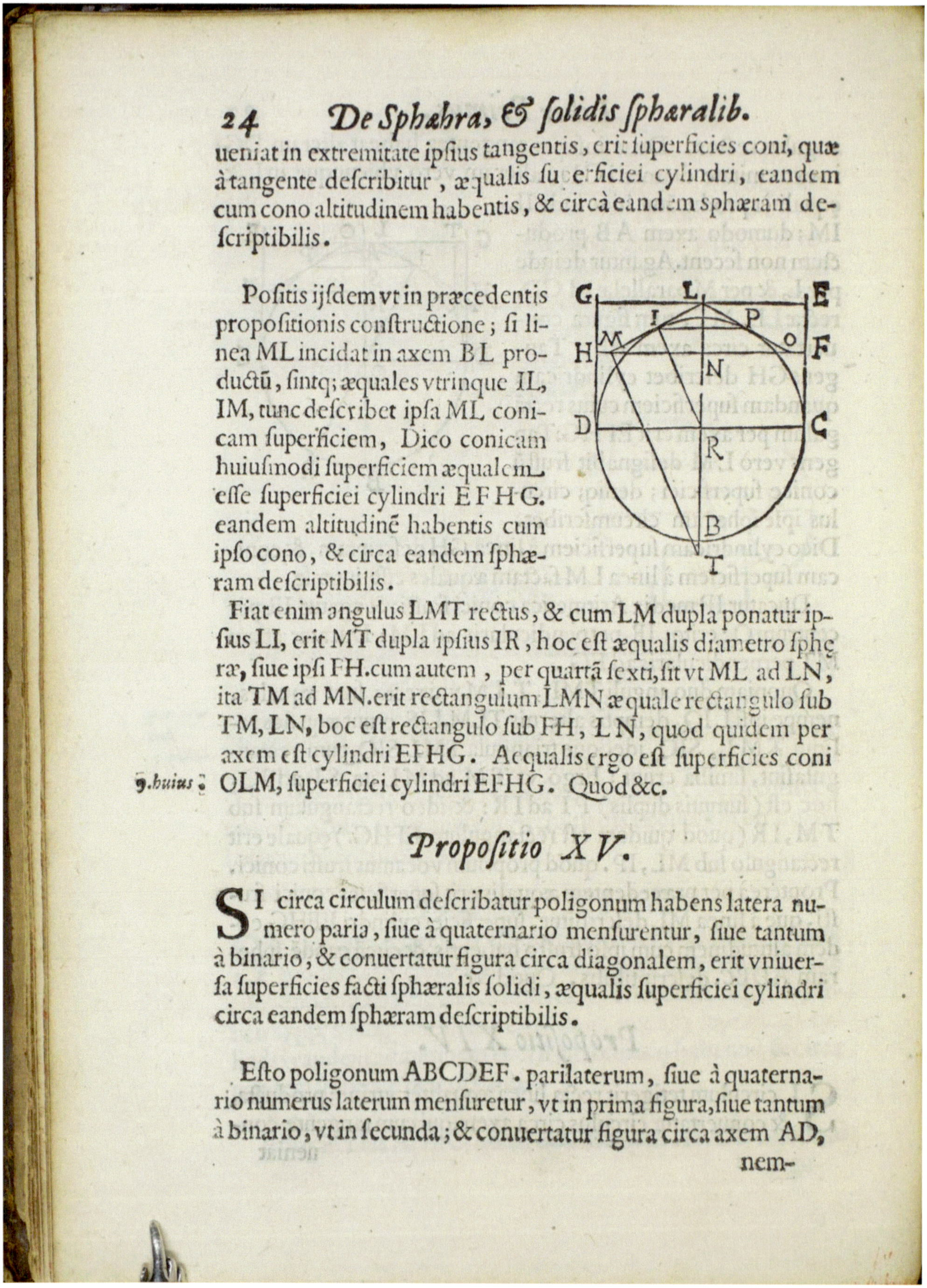

24 De Sphæra, & solidis sphæralib.
ueniat in extremitate ipsius tangentis, erit superficies coni, quæ à tangente describitur, æqualis superficiei cylindri, eandem cum cono altitudinem habentis, & circà eandem sphæram describtibilis.
Positis ijsdem vt in præcedentis propositionis constructione; si linea ML incidat in axem BL productũ, sintq; æquales vtrinque IL, IM, tunc describet ipsa ML conicam superficiem, Dico conicam huiusmodi superficiem æqualem esse superficiei cylindri EFHG. eandem altitudinẽ habentis cum ipso cono, & circa eandem sphæram describtibilis.
G L E
I P
H M O F
N
D C
R
B
T
Fiat enim angulus LMT rectus, & cum LM dupla ponatur ipsius LI, erit MT dupla ipsius IR, hoc est æqualis diametro sphæræ, siue ipsi FH. cum autem, per quartã sexti, sit vt ML ad LN, ita TM ad MN. erit rectangulum LMN æquale rectangulo sub TM, LN, hoc est rectangulo sub FH, LN, quod quidem per axem est cylindri EFHG. Aequalis ergo est superficies coni
9. huius OLM, superficiei cylindri EFHG. Quod &c.
Propositio XV.
SI circa circulum describatur poligonum habens latera numero paria, siue à quaternario mensurentur, siue tantum à binario, & conuertatur figura circa diagonalem, erit vniuersa superficies facti sphæralis solidi, æqualis superficiei cylindri circa eandem sphæram descriptibilis.
Esto poligonum ABCDEF. parilaterum, siue à quaternario numerus laterum mensuretur, vt in prima figura, siue tantum à binario, vt in secunda; & conuertatur figura circa axem AD,
nem-

ueniat in extremitate ipſius tangentis, erit ſuperficies coni, quæ à tangente deſcribitur, æqualis ſuperficiei cylindri, eandem cum cono altitudinem habentis, & circà eandem sphæram deſcriptibilis.

Poſitis ijſdem vt in præcedentis propoſitionis conſtructione; ſi linea ML incidat in axem BL productũ, ſintq; æquales vtrinque IL, IM, tunc deſcribet ipſa ML conicam ſuperficiem, Dico conicam huiuſmodi ſuperficiem æqualem eſſe ſuperficiei cylindri EFHG. eandem altitudinẽ habentis cum ipſo cono, & circa eandem ſphæram deſcriptibilis.

Fiat enim angulus LMT rectus, & cum LM dupla ponatur ipſius LI, erit MT dupla ipſius IR, hoc eſt æqualis diametro ſphęræ, ſiue ipſi FH. cum autem, per quartã ſexti, ſit vt ML ad LN, ita TM ad MN. erit rectangulum LMN æquale rectangulo ſub TM, LN, hoc eſt
9. huius. rectangulo ſub FH, LN, quod quidem per axem eſt cylindri EFHG. Aequalis ergo eſt ſuperficies coni OLM, ſuperficiei cylindri EFHG. Quod &c.

Propoſitio XV.

SI circa circulum deſcribatur poligonum habens latera numero paria, ſiue à quaternario menſurentur, ſiue tantum à binario, & conuertatur figura circa diagonalem, erit vniuerſa ſuperficies facti ſphæralis ſolidi, æqualis ſuperficiei cylindri circa eandem ſphæram deſcriptibilis.

Eſto poligonum ABCDEF. parilaterum, ſiue à quaternario numerus laterum menſuretur, vt in prima figura, ſiue tantum à binario, vt in ſecunda; & conuertatur figura circa axem AD,

nem–

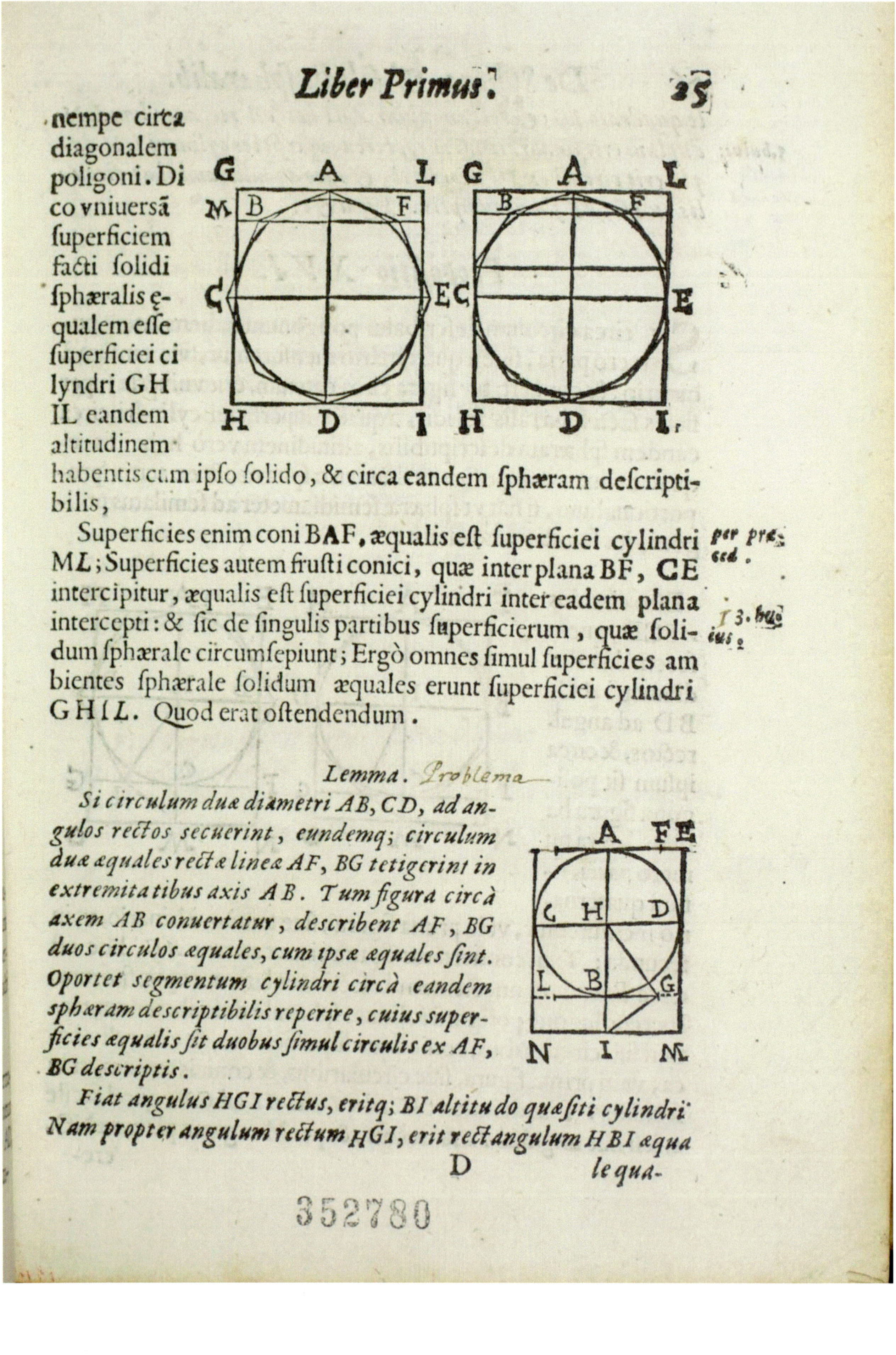

nempe circa diagonalem poligoni. Dico vniuerſã ſuperficiem facti ſolidi ſphæralis ęqualem eſſe ſuperficiei cilyndri GHIL eandem altitudinem habentis cum ipſo ſolido, & circa eandem ſphæram deſcriptibilis,

Superficies enim coni BAF, æqualis eſt ſuperficiei cylindri ML; Superficies autem fruſti conici, quæ inter plana BF, CE intercipitur, æqualis eſt ſuperficiei cylindri inter eadem plana intercepti: & ſic de ſingulis partibus ſuperficierum, quæ ſolidum ſphærale circumſepiunt; Ergò omnes ſimul ſuperficies ambientes ſphærale ſolidum æquales erunt ſuperficiei cylindri GHIL. Quod erat oſtendendum. *per præced.* *13. huius.*

Lemma. Problema

Si circulum duæ diametri AB, CD, ad angulos rectos secuerint, eundemq; circulum duæ æquales rectæ lineæ AF, BG tetigerint in extremitatibus axis AB. Tum figura circà axem AB conuertatur, describent AF, BG duos circulos æquales, cum ipsæ æquales ſint. Oportet segmentum cylindri circà eandem sphæram descriptibilis reperire, cuius superficies æqualis ſit duobus ſimul circulis ex AF, BG descriptis.

Fiat angulus HGI rectus, eritq; BI altitudo quæſiti cylindri. Nam propter angulum rectum HGI, erit rectangulum HBI æquale qua-

D

nempe circa diagonalem poligoni. Dico vniuersã ſuperficiem facti ſolidi ſphæralis ęqualem eſſe ſuperficiei cilyndri GHIL eandem altitudinem habentis cum ipſo ſolido, & circa eandem ſphæram deſcriptibilis, *per prae– ced.*

Superficies enim coni BAF, æqualis eſt ſuperficiei cylindri ML; Superficies autem fruſti conici, quæ inter plana BF, CE intercipitur, æqualis eſt ſuperficiei cylindri inter eadem plana intercepti: & ſic de ſingulis partibus ſuperficierum, quæ ſolidum ſphærale circumſepiunt; Ergò omnes ſimul ſuperficies ambientes ſphærale ſolidum æquales erunt ſuperficiei cylindri GHIL. Quod erat oſtendendum. 3. *hu- ius.*

[Problema]

Lemma.

Si circulum duæ diametri AB, CD, ad angulos rectos secuerint, eundemq; circulum duæ æquales rectæ lineæ AF, BG tetigerint in extremitatibus axis AB. Tum figura circà axem AB conuertatur, describent AF, BG duos circulos æquales, cum ipsæ æquales ſint. Oportet segmentum cylindri circà eandem sphæram descriptibilis reperire, cuius superficies æqualis ſit duobus ſimul circulis ex AF, BG descriptis.

Fiat angulus HGI rectus, eritq; BI altitudo quæſiti cylindri[18] *Nam propter angulum rectum HGI, erit rectangulum HBI æqua*
D *le qua-*

[18] *cylindri*] *cylindri.*

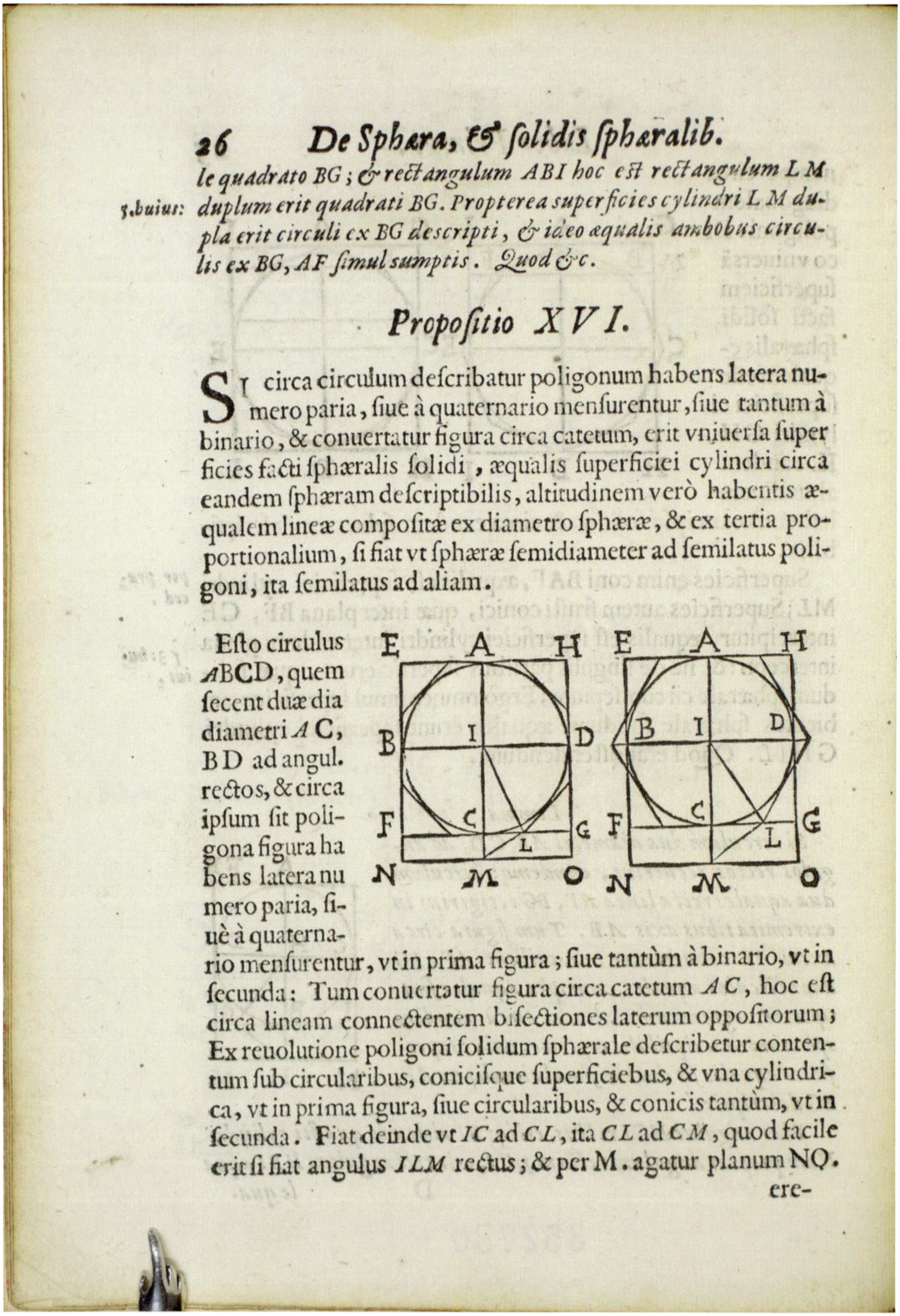

26 *De Sphæra, & ſolidis ſphæralib.*

le quadrato BG; & rectangulum ABI hoc eſt rectangulum LM [5. huius:] *duplum erit quadrati BG. Propterea ſuperficies cylindri LM dupla erit circuli ex BG deſcripti, & ideo æqualis ambobus circulis ex BG, AF ſimul ſumptis. Quod &c.*

Propoſitio XVI.

SI circa circulum deſcribatur poligonum habens latera numero paria, ſiue à quaternario menſurentur, ſiue tantum à binario, & conuertatur figura circa catetum, erit vniuerſa ſuperficies facti ſphæralis ſolidi, æqualis ſuperficiei cylindri circa eandem ſphæram deſcriptibilis, altitudinem verò habentis æqualem lineæ compoſitæ ex diametro ſphæræ, & ex tertia proportionalium, ſi fiat vt ſphæræ ſemidiameter ad ſemilatus poligoni, ita ſemilatus ad aliam.

Eſto circulus ABCD, quem ſecent duæ diametri AC, BD ad angul. rectos, & circa ipſum ſit poligona figura habens latera numero paria, ſiuè à quaternario menſurentur, vt in prima figura; ſiue tantùm à binario, vt in ſecunda: Tum conuertatur figura circa catetum AC, hoc eſt circa lineam connectentem biſectiones laterum oppoſitorum; Ex reuolutione poligoni ſolidum ſphærale deſcribetur contentum ſub circularibus, coniciſque ſuperficiebus, & vna cylindrica, vt in prima figura, ſiue circularibus, & conicis tantùm, vt in ſecunda. Fiat deinde vt IC ad CL, ita CL ad CM, quod facile erit ſi fiat angulus ILM rectus; & per M. agatur planum NO. ere-

le quadrato BG; & rectangulum ABI hoc est rectangulum LM duplum erit quadrati BG. Propterea superficies cylindri LM dupla erit
5.huius: *circuli ex BG descripti, & ideo æqualis ambobus circulis ex BG, AF ſimul sumptis. Quod &c.*

Propoſitio XVI.

SI circa circulum deſcribatur poligonum habens latera numero paria, ſiue à quaternario menſurentur, ſiue tantum à binario, & conuertatur figura circa catetum, erit vniuerſa ſuperficies facti ſphæralis ſolidi, æqualis ſuperficiei cylindri circa eandem ſphæram deſcriptibilis, altitudinem verò habentis æqualem lineæ compoſitæ ex diametro ſphæræ, & ex tertia proportionalium, ſi fiat vt ſphæræ ſemidiameter ad ſemilatus poligoni, ita ſemilatus ad aliam.

Eſto circulus *A*BCD, quem ſecent duæ dia[19] diametri[20] *A*C, BD ad angul. rectos, & circa ipſum ſit poligona figura habens latera numero paria, ſiuè à quaternario menſurentur, vt in prima figura; ſiue tantùm à binario, vt in ſecunda: Tum conuertatur figura circa catetum *AC,* hoc eſt circa lineam connectentem biſectiones laterum oppoſitorum; Ex reuolutione poligoni ſolidum ſphærale deſcribetur contentum ſub circularibus, coniciſque ſuperficiebus, & vna cylindrica, vt in prima figura, ſiue circularibus, & conicis tantùm, vt in ſecunda. Fiat deinde vt *IC* ad *CL,* ita *CL* ad *CM,* quod facile erit ſi fiat angulus *ILM* rectus; & per M. agatur planum NO.

ere-

[19] dia]
[20] dia diametri] diametri.

erectum ad axem. Dico vniuersam superficiem solidi sphæralis æqualem esse superficiei cylindri ENOH.

Hòc autem patet ex præmissis; Nam tota sphęralis solidi superficies, demptis circulis oppositis, æqualis est superficiei cylindricæ inter plana EH, FG compræhensæ. Duo verò circuli oppositi quorum centra A, & C æquales sunt (per præcedens lemma) superficiei cylindricæ inter duo plana FG. NO. contentę. Propterea vniuersa simul sphæralis solidi superficies æqualis erit superficiei cylindri ENOH. circa eandem sphæram descripti, & altitudinem habentis AM, quæ componitur ex diametro sphæræ *AC*, & ex recta C M, quæ quidem tertia proportionalis est ad semidiametrum I*C*, & semilatus, *CL*. Quod &c. *deducitur ex 13. huius.*

Lemma.

Si circulum ABCD duæ diametri AC, BD secent ad angulos rectos; recta autem linea CE eũdem contingat in extremitate axis AC & conuertatur figura circa AC; ipsa CE circulum describet. Oportet segmentum cylindri circa eandem sphæram descripti reperire, cuius superficies æqualis sit circulo ex CE descripto.

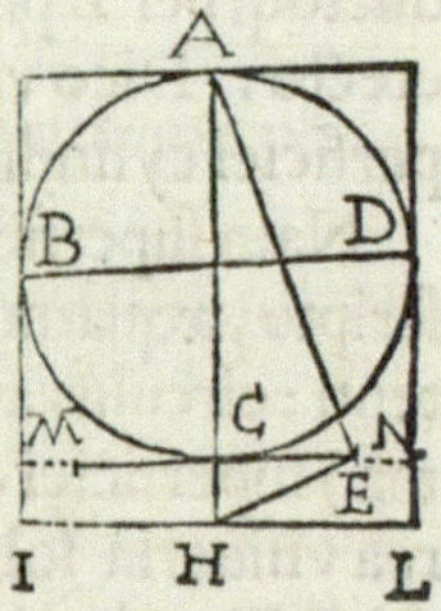

Fiat angulus AEH rectus, ductoque plano per H ad axem erecto. Dico cylindricam superficiem MILN. æquari circulo ex CE. Est enim ob angulum rectum AEH, rectangulum ACH, hoc est rectangulum ML, æquale quadrato CE. Propterea superficies cylindri MILN æqualis erit circulo ex CE. Quod &c. *per 5. huius.*

Propositio XVII.

SI circa circulum describatur poligonum habens latera numero imparia, & conuertatur figura circà catetum poligoni: erit vniuersa superficies facti sphæralis solidi æqualis superficiei cylindri circà eandem sphęram descriptibilis, altitudinẽ

erectum ad axem. Dico vniuerſam ſuperficiem ſolidi ſphæralis æqualem eſſe ſuperficiei cylindri ENOH.

Hoc autem patet ex præmiſſis; Nam tota ſphęralis ſolidi ſuperficies, demptis circulis oppoſitis, æqualis eſt ſuperficiei cylindricæ inter plana EH, FG compræhenſæ. Duo verò circuli oppoſiti quorum centra A, & C æquales ſunt (perpræcedens lemma) ſuperficiei cylindricæ inter duo plana FG. NO. contentę. Propterea vniuerſa ſimul ſphæralis ſolidi ſuperficies æqualis erit ſuperficiei cylindri ENOH. circa eandem ſphæram deſcripti, & altitudinem habentis AM, quæ componitur ex diametro ſphæræ *AC*, & ex recta *C*M, quæ quidem tertia proportionalis eſt ad ſemidiametrum I*C*, & ſemilatus, *CL*. Quod &c. *Deduci-tur ex- I 3. hu-ius.*

Lemma.

Si circulum ABCD duæ diametri AC, BD secent ad angulos rectos; recta autem linea CE eũdem contingat in extremitate axis AC & conuertatur figura circa AC; ipsa CE circulum describet. Oportet segmentum cylindri circa eandem sphæram descripti reperire, cuius superficies æqualis ſit circulo ex CE descripto. *per 5. hu*

Fiat angulus AEH rectus, ductoque plano per H ad axem erecto. Dico cylindricam superficiem MILN. æquari circulo ex CE. Eſt enim ob angulum rectum AEH, rectangulum ACH, hoc eſt rectangulum ML, æquale quadrato C E. Propterea ſuperficies cylindri MILN æqualis erit circulo ex CE. Quod &c. [-] *ius.*

Propoſitio XVII.

SI circa circulum deſcribatur poligonum habens latera numero imparia, & conuertatur figura circà catetum poligoni: erit vniuerſa ſuperficies facti ſphæralis ſolidi æqualis ſuperficiei cylindri circà eandem ſphęram deſcriptibilis, altitudinẽ.

D 2 verò

verò habentis æqualem lineæ compositæ ex cateto poligoni, & ex tertia proportionalium, si fiat vt diameter circuli ad semilatus poligoni, ità semilatus ad aliam.

Esto circulus *ABCD*, circa quem sit poligonum E*F*GHI. habens latera numero imparia; & conuertatur figura circa catetum EC, nempe circa lineã, quæ ab vno angulo E perducitur ad bisectionem lateris oppositi; oriaturq; solidum sphærale contentum sub conicis superficiebus, vnicoque circulo.

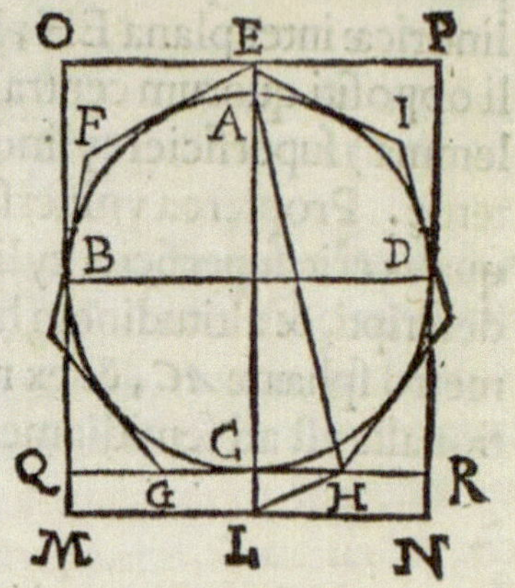

*F*acto deinde angulo recto *AHL*, ductoq; per *L* plano MN ad axem erecto. Dico vniuersam solidi superficiem æqualem esse superficiei cylindri OMNP.

13.huius. Nam superficies solidi sphæralis, dempto circulo ex *C*H. descripto, æquatur superficiei cylindri iuter plana OP, QR contenti: circulus autem ex *C* H factus ęqualis est (præcedens lemma) superficiei cylindri inter plana QR, MN contenti: Propterea vniuersa solidi superficies ęqualis erit superficiei cylindri OMNP. qui quidem circa eandem sphęram cum ipso solido describitur, altitudinem verò habet lineam *E L*, quę componitur ex cateto *EC*, & ex linea *CL*, quæ tertia proportionalis est, si fiat vt *AC* diameter sphæræ, ad *CH* semilatus poligoni, ita *CH* ad aliam. Quod erat &c.

Propositio XVIII.

HEmisphærij superficies ęqualis est superficiei curuę cylindri eandem ipsi basim, & eandem altitudinem habentis.

*E*sto hemisphærium *ABC*, & circa ipsum cylindrus eiusdem altitudinis, *ADEC*.

Dico

verò habentis æqualem lineæ compoſitæ ex cateto poligoni, & ex tertia proportionalium, ſi fiat vt diameter circuli ad ſemilatus poligoni, ità ſemilatus ad aliam.

Eſto circulus *ABCD,* circa quem ſit poligonum EFGHI. habens latera numero imparia; & conuertatur figura circa catetum EC, nempe circa lineã, quæ ab vno angulo E perducitur ad biſectionem lateris oppoſiti; orieturq; ſolidum ſphærale contentum ſub conicis ſuperficiebus, vnicoque circulo.

*F*acto deinde angulo recto *AHL,* ductoq; per *L* plano MN ad axem erecto. Dico vniuerſam ſolidi ſuperficiem æqualem eſſe ſuperficiei cylindri OMNP.

13. huius. Nam ſuperficies ſolidi ſphæralis, dempto circulo ex *C*H. deſcripto, æquatur ſuperficiei cylindri iuter[21] plana OP, QR contenti: circulus autcm ex *C*H factus ęqualis eſt (præcedens lemma) ſuperficiei cylindri inter plana QR, MN contenti: Propterea vniuerſa ſolidi ſuperficies ęqualis erit ſuperficiei cylindri OMNP. qui quidem circa eandem ſphęram cum ipſo ſolido deſcribitur, altitudinem verò habet lineam *EL,* quę componitur ex cateto *EC,* & ex linea *CL,* quæ tertia proportionalis eſt, ſi fiat vt *AC* diameter ſphæræ, ad *CH* ſemilatus poligoni, ita *CH* ad aliam. Quod erat &c.

Propoſitio XVIII.[22]

HÆmisphærij ſuperficies ęqualis eſt ſuperficiei curuę cylindri eandem ipſi baſim, & eandem altitudinem habentis.

*E*ſto hemisphærium *ABC,* & circa ipſum cylindrus eiuſdem altitudinis, *ADEC.*

Dico

---

[21] Iuter] inter

[22] *Ad Absurdum* Proof.

Dico ſuperficiem hemiſphęrij ęqualem eſſe ſuperficiei cylindri *ADEC*.

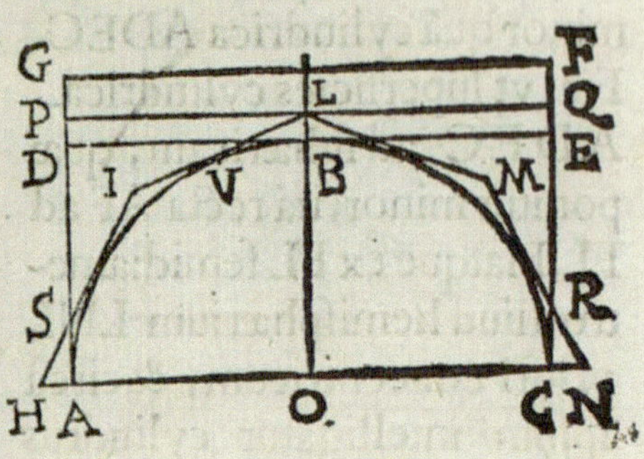

• Si enim non eſt æqualis, vel maior erit, vel minor. Ponatur primùm ſphęrica ſuperficies maior: fiatque vt cylindri ſuperficies ad ſuperficiem hemiſpherij, quæ maior ponitur, ita recta AD ad AG: intelligaturq; cylindrus productus vſque ad GF. Secetur deinde arcus AB bifariam, iterumq; portiones eius bifariam, & hoc ſemper, donec poligoni circà ſemicirculum A B C deſcripti ſemilatus VL minus ſit quam recta DG. (quod fieri poſſe conſtat ex prima Decimi; ſemilatera enim poligonorum circulo circumſcriptorum ex continua arcuum biſectione ſemper minuuntur pluſquam pro medietate, vt ab alijs oſtenſum eſt.) Factum ergò ſit; & eſto poligonum HILMN, conuerſàque figura circa axem LO, fiat ex poligono, ſemiſolidum ſphærale ſub conicis ſuperficiebus compræhenſum. Cum itaque recta DG maior ſit quã ſemilatus LV, multo maior eadem erit quàm LB, & propterea planum PQ productum per L intra puncta D & G cadet.

Iam quia ſuperficies cylindri AE ad ſuperficiem hemiſphærij eſt vt A D ad A G, hoc eſt vt cylindrica ſuperficies A E ad cylindricam A F, erit cylindrica ſuperficies A F æqualis ſphæricæ. Propterea, ſi ſphærica ſuperficies æqualis ſit cylindricæ AF. maior erit quam cylindrica A Q, hoc eſt quam conicæ omnes HI*L*MN, multoq; maior quam omnes A S I L M R C, quod eſt abſurdum. Eſt enim contrà principium ab Archimede præmiſsum. *6. huius* *ex XV. huius.*

Aſſumpſimus conicam quæ deſcribitur à linea HS maiorem eſſe quàm illa ſuperficies, quæ deſcribitur à linea AS. quod patet ex 12. huius. Rectangulum enim proprium conicæ ſuperficiei multò maius eſt quam rectangulum per axem cylindricę, quando quidem ſub maioribus lateribus continetur.

Pona-

Dico ſuperficiem hemiſphęrij ęqualem eſſe ſuperficiei cylindri *ADEC.*

Si enim non eſt æqualis, vel maior erit, vel minor. Ponatur primùm ſphęrica ſuperficies maior: fiatque vt cylindri ſuperficies ad ſuperficiem hemiſpherij, quæ maior ponitur, ita recta AD ad AG: intelligaturq; cylindrus productus vſque ad GF. Secetur deinde arcus AB bifariam, iterumq; portiones eius bifariam, & hoc ſemper, donec poligoni circà ſemicirculum ABC deſcripti ſemilatus VL minus ſit quam recta DG. (quod fieri poſſe conſtat ex prima Decimi; ſemilatera enim poligonorum circulo circumſcriptorum ex continua arcuum biſectione ſemper minuuntur pluſquam pro medietate, vt ab alijs oſtenſum eſt.) Factum ergò ſit; & eſto poligonum HILMN, conuersàque figura circa axem LO, fiat ex poligono, ſemiſolidum ſphærale ſub conicis ſuperficiebus compræhenſum. Cum itaque recta DG maior ſit quã ſemilatus LV, multo maior eadem erit quàm LB, & propterea planum PQ productum per L intra puncta D & G cadet. *6. huius ex xv huius.*

Iam quia ſuperficies cylindri AE ad ſuperficiem hemiſphærij eſt vt AD ad AG, hoc eſt vt cylindrica ſuperficies AE ad cylindricam AF, erit cylindrica ſuperficies AF æqualis ſphæricæ. Propterea, ſi ſphærica ſuperficies æqualis ſit cylindricæ AF. maior erit quam cylindrica AQ, hoc eſt quam conicæ omnes HILMN, multoq; maior quam omnes ASILMRC, quod eſt abſurdum. Eſt enim contrà principium ab Archimede præmiſsum.

Aſſumpſimus conicam quæ deſcribitur à linea HS maiorem eſſe quàm illa ſuperficies, quæ deſcribitur à linea AS. quod patet ex 12. *huius. Rectangulum enim proprium conicæ ſuperficiei multò maius est quam rectangulum per axem cylindricę, quando quidem ſub maioribus lateribus continetur.*

Pona-.

30 *De Sphæra, & ſolidis ſphæralib.*

Ponatur iam ſphærica ABC minor quã cylindrica ADEC. Fiat vt ſuperficies cylindrica ADEC ad ſphæricam, quæ ponitur minor; ita recta AF ad FL. Fiatque ex FL ſemidiametro aliud hemiſphærium LNI, priori concentricum, & circà ipſum intelligatur cylindrus LHMI: Inſcribatur etiam intrà ſemicirculum ABC. figura laterum æqualium, ita vt latera ipſius non tangant ſemicirculum LNI. (quod fieri poſſe conſtat ex Euclide.) Deſcribaturq; alius ſemicirculus ſemidiametro FO, qui contingat ſingula latera factæ figuræ, & conuertatur vniuerſa figura circa FB. ita vt fiat ſemiſolidum ſphærale AVBTC conicis ſuperficiebus circumſeptum; ex ſemicirculo autem FO fiat aliud hemiſphærium, circà quod concipiatur cylindrus RQSP.

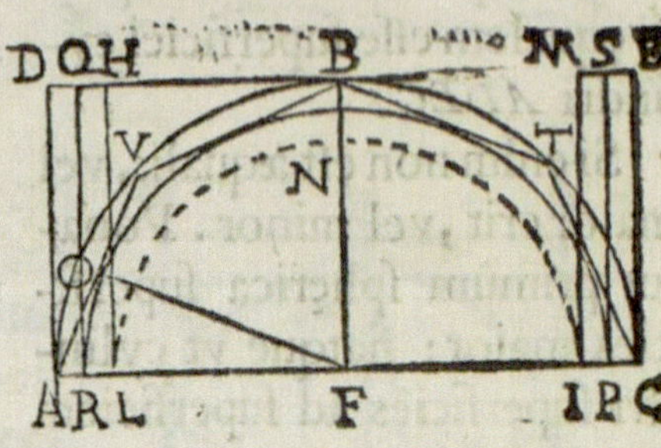

Iam ſic; ſuperficies cylindri ADEC ad ſuperficiem hemiſphęrij eſt, per conſtructionem, vt AF ad FL, hoc eſt vt AC ad LI, hoc eſt vt rectangulum AE ad rectangulum LM, hoc eſt vt cylindrica AE ad cylindricam LM. Quare ſphęrica ſuperficies ęqualis erit cylindricæ LM, & propterea minor quàm cylindrica RS, hoc eſt quàm omnes conicę AVBTC, abſurdum. ſphęrica enim ſuperficies ABC maior eſt quàm omnes conicę AVBTC.

6.huius. *ob 15.huius.*

Hemiſphęrij ergò ſuperficies ęqualis erit ſuperficiei cylindri eandem ipſi baſim, eandemq; altitudinem habentis. Cum demonſtratum ſit neque maiorem eſſe, neque minorem. Quod erat &c.

Propoſitio XIX.

CViuſcunque minoris portionis Sphæræ ſuperficies ęqualis eſt curuæ ſuperficiei cylindri circà integram ſphęram deſcripti,

Ponatur iam ſphærica ABC minor quã cylindrica ADEC. Fiat vt ſuperficies cylindrica ADEC ad ſphæricam, quæ ponitur minor; ita recta AF ad FL. Fiatque ex FL ſemidiametro aliud hemiſphærium LNI, priori concentricum, & circà ipſum intelligatur cylindrus LHMI: Inſcribatur etiam intrà ſemicirculum ABC. figura laterum æqualium, ita vt latera ipſius non tangant ſemicirculum LNI. (quod fieri poſſe conſtat ex Euclide.) Deſcribaturq; alius ſemicirculus ſemidiametro FO, qui contingat ſingula latera factæ figurę, & conuertatur vniuerſa figura circa FB. ita vt fiat ſemiſolidum ſphærale AVBTC conicis ſuperficiebus circumſeptum; ex ſemicirculo autem FO fiat aliud hemiſphærium, circà quod concipiatur cylindrus RQSP.

6. *huius.* Iam ſic; ſuperficies cylindri ADEC ad ſuperficiem hemiſphęrij eſt, per conſtructionem, vt AF ad FL, hoc eſt vt AC ad LI, hoc eſt vt rectangulum AE ad rectangulum LM, hoc eſt vt cylindrica AE ad cylindricam LM. Quare ſphęrica ſuperficies ęqualis erit cylindricæ LM, & propterea minor quàm cylindrica R S, hoc eſt quàm omnes conicę AVBTC, abſurdum. ſphęrica enim ſuperficies ABC maior eſt quàm omnes conicę AVBTC. *ob* 15. *hu*[-]*ius.*

Hemiſphęrij ergò ſuperficies ęqualis erit ſuperficiei cylindri eandem ipſi baſim, eandemq; altitudinem habentis. Cum demonſtratum ſit neque maiorem eſſe, neque minorem. Quod erat &c.

Propoſitio XIX.[23]

CViuſcunque minoris portionis Sphæræ ſuperficies ęqualis eſt curuæ ſuperficiei cylindri circà integram ſphęram de-

ſcripti

[23] *Ad Aburdum* Proof.

ſcripti, & eandem altitudinem cum ipſa portione habentis.

Eſto minor ſphærę portio A B C, & portio cylindri FDEG; circa integram ſphęram deſcripti, eandẽ tamen altitudinem HB cũ ipſa portione ſphęrica habentis. Dico ſphęricam ſuperficiem A B C æqualem eſſe ſuperficiei cylindri FDEG.

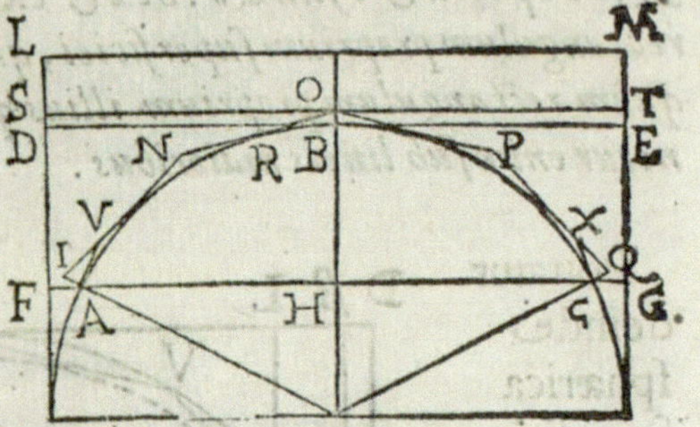

Si enim non eſt ęqualis, vel maior erit vel minor.

Ponatur primum maior; & ipſi ſphęricę ſuperficiei A B C. conſtruatur ęqualis (vt in pręcedenti) cylindrica FLMG: ſecto deinde arcu AB bifariam, & portiones eius iterum bifariam, & ſic ſemper, circumſcribatur arcui A B C figura multorum laterum I N O P Q, terminata ad diametros, quę ducuntur per puncta A & C. Sitque per prędictam biſectionem arcuum, ſemilatus R O minus quàm recta D L, vt propterea planum S T, ductum per O, cadat intra puncta D, & L. Quemadmodum in pręcedenti &c. Conuertatur deinde figura vniuerſa circà OH, & ex conuerſione figurę I N O P Q naſcetur portio ſolidi ſphęralis ſub conicis ſuperficiebus contenta.

Iam ſic. Quia ſphęrica ſuperficies A B C. ęqualis eſt per cõſtructionem cylindricae F L M G, maior eadem erit quàm cylindrica FSTG, & multò maior quàm omnes conicae INOPQ, multoq; etiam maior quam omnes conicae AVNOPXC. Quod eſt abſurdum, & contrà principia Archimedis.

Aſſumpſimus cylindricam ſuperficiem F S T G maiorem eſſe omnibus conicis INOP Q. Hoc enim patet. Nam ex 13. 14. & 15. huius colligi poteſt, conicas I N O P Q æquales eſſe ſuperficiei cylindricæ contentæ inter planum ST, & planum quod duceretur per puncta I Q.

Aſſumpſimus etiam, ductà tangente A V. conicam ſuperficiem,

ſcripti, & eandem altitudinem cum ipſa portione habentis. 1 et 7

Eſto minor ſphærę portio ABC, & portio cylindri FDEG; circa integram ſphęram deſcripti, eandē tamen altitudinem HB cū ipſa portione ſphęrica habentis. Dico ſphęricam ſuperficiem ABC æqualem eſſe ſuperficiei cylindri FDEG. [...] ti.

Si enim non eſt ęqualis, vel maior erit vel minor.

Ponatur primum maior; & ipſi ſphęricę ſuperficiei ABC, conſtruatur ęqualis (vt in pręcedenti) cylindrica FLMG: ſecto deinde arcu AB bifariam, & portiones eius iterum bifariam, & ſic ſemper, circumſcribatur arcui ABC figura multorum laterum INOPQ, terminata ad diametros, quę ducuntur per puncta A & C. Sitque per prędictam biſectionem arcuum, ſemilatus RO minus quàm recta DL, vt propterea planum ST, ductum per O, cadat intra puncta D, & L. Quemadmodum in pręcedenti &c. Conuertatur deinde figura vniuerſa circà OH, & ex conuerſione figurę INOPQ naſcetur portio ſolidi ſphęralis ſub conicis ſuperficiebus contenta.

Iam ſic. Quia ſphęrica ſuperficies ABC. ęqualis eſt per cõſtructionem cylindricae FLMG, maior eadem erit quàm cylindrica FSTG, & multò maior quàm omnes conicae INOPQ, multoq; etiam maior quam omnes conicae AVNOPXC. Quod eſt abſurdum, & contrà principia Archimedis.

Aßumpſimus cylindricam ſuperficiem FSTG maiorem eſſe omnibus conicis INOPQ Hoc enim patet. Nam ex 13. 14. & 15. huius colligi poteſt, conicas INOPQ æquales eſſe ſuperficiei cylindricæ contentæ inter planum ST, & planum quod duceretur per puncta IQ Aſſumpſimus etiam, ductà tangente AV. conicam ſuperfi-

ciem,

32 *De Sphæra, & solidis sphæralib.*

ciem, quę fit à linea IV, maiorem eſſe quàm illa quæ fit linea AV. Quod quidem demonſtratur apud Archimedem ad Propoſitionem 37. de Sphęra & cylindro. Sed & ex noſtris deduci poteſt. Nam rectangulum proprium ſuperficiei, quæ fit à linea IV, maius eſt quàm rectangulum proprium illius quæ fit à linea AV. Continetur enim ſub lineis maioribus.

Ponatur deinde ſphærica ſuperficies portionis ABC min. quã cylindrica FDEG.

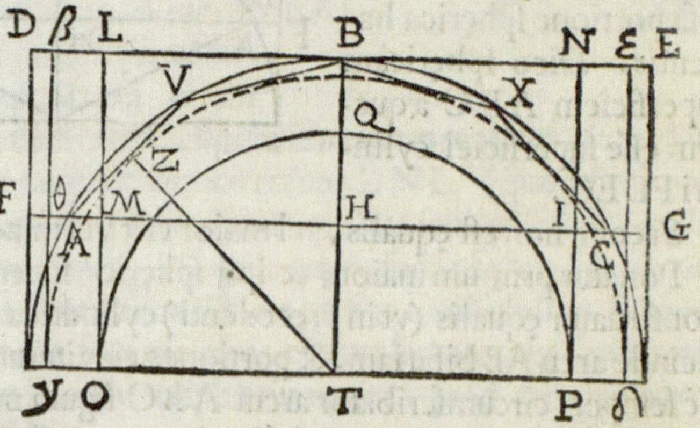

Fiat vt cylindrica F D E G ad ſphæricam ſuperficiem ABC, quæ minor ponitur, ita FH ad HM. & centro T ſemidiametro autem HM fiat hemiſphærium OQP, circa quod intelligatur cylindrus OLNP Intra arcum autem ABC figura inſcribatur multorum laterum A V B X C per continuam biſectionem arcuum ita vt latera ipſius non tangant ſemicirculum OQP, & conuertatur vniuerſa figura circa axem BT. Intelligatur autem radio TZ (quæ recta perpendicularis ſit ad vnum latus figuræ inſcriptæ) deſcribi ſphæram, quæ tangat ſingula figuræ A V B X C latera, & circa huiuſmodi ſphæram deſcriptus concipiatur ſuus cylindrus γβεδ.

6. huius. Iam ſic. cylindrica ſuperficies FDEG per conſtructionem eſt ad ſphæricam ABC, vt FH ad HM, hoc eſt vt FG ad MI. hoc *ex 6 huius.* eſt vt rectangulum FE ad rectangulum MN, hoc eſt vt eadem cylindrica FE, ad cylindricam MN. Erit ideò ſphærica ſuper- *explicatur infra* ficies ABC æqualis cylindricæ M N nempè minor cylindrica θε. hoc eſt minor omnib. conicis AVBXC; quod eſt abſurdũ.

Aſſum-

ciem, quę ſit à linea IV, maiorem eße quàm illa quæ ſit linea AV.
Quod quidem demonſtratur apud Archimedem ad Propoſitionem 37. de Sphęra & cylindro. Sed & ex noſtris deduci poteſt. Nam rectangulum proprium ſuperficiei, quæ ſit à linea IV, maius eſt quàm rectangulum proprium illius quæ ſit à linea AV. Continetur enim ſub lineis maioribus.

Ponatur deinde ſphærica ſuperficies portionis ABC min. quã cylindrica FDEG.

Fiat vt cylindrica FDEG ad ſphæricam ſuperficiem ABC, quæ minor ponitur, ita FH ad HM. & centro T ſemidiametro autem HM fiat hemiſphærium OQP, circa quod intelligatur cylindrus OLNP Intra arcum autem ABC figura inſcribatur multorum laterum AVBXC

6. huius per continuam biſectionem arcuum ita vt latera ipſius non tangant ſemicirculum OQP, & conuertatur vniuerſa figura circa axem BT.

ex 6 huius. Intelligatur autem radio TZ (quæ recta perpendicularis ſit ad vnum latus figuræ inſcriptæ) deſcribi ſphæram, quæ tangat ſingula figuræ AVBXC latera, & circa huiuſmodi ſphæram deſcriptus concipiatur ſuus cylindrus γβεδ[24]

Iam ſic. cylindrica ſuperficies FDEG per conſtructionem eſt ad ſphæricam ABC, vt FH ad HM, hoc eſt vt FG ad MI. hoc eſt vt

explicatur infra rectangulum FE ad rectangulum MN, hoc eſt vt eadem cylindrica FE, ad cylindricam MN. Erit ideò ſphærica ſuperficies ABC æqualis cylindricæ MN nempè minor cylindrica θε. hoc eſt minor omnib. conicis AVBXC; quod eſt abſurdũ.

Assum-

[24] γβεδ] γβεδ.

Assumpsimus cylindricam superficiem θι æqualem esse omnib. conicis AVBXC. Quod patet ex demonstratis. Sunt enim tam cylindrus θι, quàm omnes illæ conicæ eiusdem altitudinis HB; & circà eandem sphæram γ δ describuntur. 11. huius.

Constat ergò superficiem ABC ęqualem esse cylindricæ DFGE. cum demonstratum sit neque maiorem esse, neque minorem. Quod &c.

Corollarium I.

Ex prima duarum præmissarum Propositionum patet superficiem integram sphærę, æqualem esse superficiei cylindri sibi circumscripti, & eiusdem cum ipsa sphæra altitudinis.

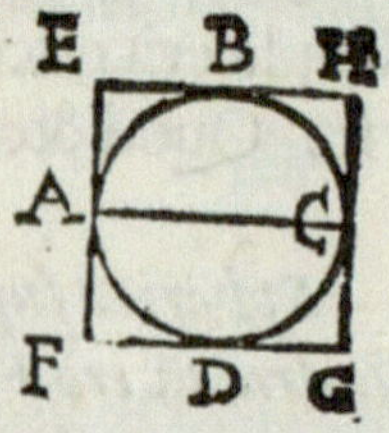

Cum enim hæmisphærium ABC superficiem habeat æqualem superficiei cylindri AEHC, & item hemisphærium alterum ADC, superficiem habeat æqualem superficiei cylindri AFGC, erit coniunctim tota sphæræ superficies ęqualis superficiei cylindri FEHG; exceptis semper basibus.

Corollarium II.

Manifestum etiam est ex vltima propositione, superficiem maioris sphæræ portionis, æqualem esse superficiei cylindri eandem cum portione altitudinem habentis, & circà eandem sphæram descriptibilis.

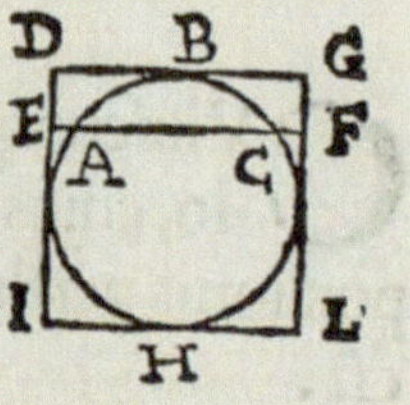

Cum enim integra sphęrę superficies æqualis sit superficiei cylindri IDGL, & demonstratum sit superficiem segmenti minoris ABC ęqualem esse superficiei cylindri EDGF, erit reliqua superficies sphæræ AHC, æqualis reliquæ superficiei EILF. Quod oportebat &c. Corolar. praced.

Propositio XX.

SVperficies sphærę quadrupla est maximi circuli in eadem sphæra descriptibilis.

E Sit

Aſſumpſimus cylindricam ſuperficiem θε æqualem eſſe omnib. conicis AVBXC. Quod patet ex demonstratis. Sunt enim tam cylindrus θε, quàm omnes illæ conicæ eiuſdem altitudinis HB; & circà eandem ſphæram γδ deſcribuntur. 11. *huius*
Conſtat ergò ſuperficiem ABC ęqualem eſſe cylindricæ DFGE. cum demonſtratum ſit neque maiorem eſſe, neque minorem. Quod &c

Corollarium I.

Ex prima duarum præmiſſarum Propoſitionum patet ſuperficiem integram ſphærę, æqualem eſſe ſuperficiei cylindri ſibi circumſcripti, & eiuſdem cum ipſa ſphæra altitudinis.

Cum enim hæmiſphærium ABC ſuperficiem habeat æqualem ſuperficiei cylindri AEHC, & item hemiſphærium alterum ADC, ſuperficiem habeat æqualem ſuperficiei cylindri AFGC, erit coniunctim tota ſphæræ ſuperficies ęqualis ſuperficiei cylindri FEHG; exceptis ſemper baſibus.

Corollarium II.

Manifestum etiam eſt ex vltima propoſitione, ſuperficiem maioris ſphæræ portionis, æqualem eſſe ſuperficiei cylindri eandem cum portione altitudinem habentis, & circà eandem ſphæram deſcriptibilis.

Cum enim integra ſphęrę ſuperficies æqualis ſit ſuperficiei cylindri IDGL, & demonſtratum ſit ſuperficiem ſegmenti minoris ABC ęqualem eſſe ſuperficiei cylindri EDGF, erit reliqua ſuperficies ſphæræ AHC, æqualis reliquæ ſuperficiei EILF. Quod oportebat &c. *Corolar-praeced.*

Propoſitio XX.

SVperficies ſphærę quadrupla eſt maximi circuli in eadem ſphæra deſcriptibilis.

E Sit

Sit sphæra ABCD cuius diameter AC; & circà ipsam intelligatur cylindrus eiusdem altitudinis FEHG.

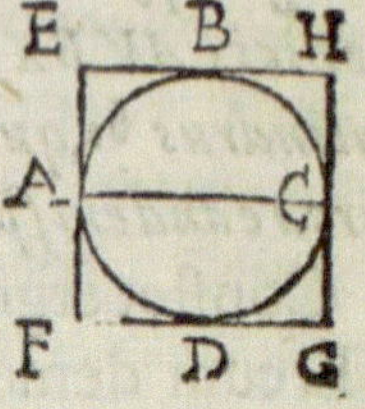

Dico superficiem sphęræ quadruplam esse maximi circuli in ea describtibilis.

Superficies enim cylindri FEHG sine basibus, est ad circulum suę basis circa FG, siue circa A C descriptum, vt EF ad quar. partem ipsius FG, hoc est vt FG ad quar. partem ipsius FG; hoc est quadrupla. Propterea etiam superficies sphaerae, quae cylindricae est aequalis, quadrupla erit circuli circa AC descripti, qui in sphaera maximus est. Quod &c.

4. huius.

ex prim. Corolar. praeced.

Aliter.

Sphęrica superficies ABCD ęqualis est cylindricę FEHG; cylindrica verò FEHG ad circulum, cuius semidiameter sit AC, est vt rectangulum per axem EG, ad quadratum ex semidiametro AC, nempe ad quadratum EG; & ideò ęqualis: propterea etiam sphęrica superficies ęqualis erit circulo cuius semidiameter sit AC; ergò quadrupla erit circuli cuius diameter sit AC. Quod &c.

5. huius.

Propositio XXI.

CViuscunq; portionis sphaerę superficies aequalis est circulo, cuius semidiameter aequalis sit lineae quae ex polo portionis perducitur ad circulum, qui in eiusdem portionis basi est.

Esto sphaerae portio siue minor siue maior ABC. cuius ex polo ducta sit recta AB. Dico superficiem portionis aequalem esse circulo qui fit ex AB tamquam semidiametro.

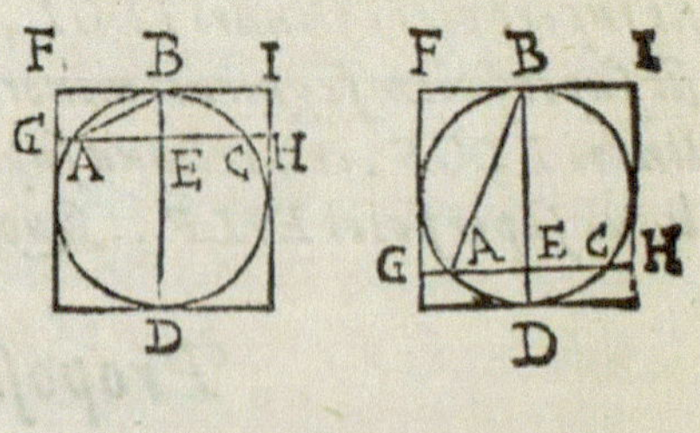

Cum enim quadratum AB aequale sit rectangulo DBE ob circulum, aequale erit & rectangulo GFIH, quod idem est ac rectangulum DBE. Propterea circu-

Sit ſphæra ABCD cuius diameter AC; & circà ipſam intelligatur cylindrus eiuſdem altitudinis FEHG.

Dico ſuperficiem ſphęræ quadruplam eſſe maximi circuli in ea deſcriptibilis.

4. huius. Superficies enim cylindri FEHG ſine baſibus, eſt ad circulum ſuę baſis circa FG, ſiue circa AC deſcriptum, vt EF ad quar. partem
ex prim. ipſius FG, hoc eſt vt FG ad quar. partem ipſius FG; hoc eſt quadrupla. Propterea etiam ſuperficies ſphaerae, quae cylindricae eſt
Corolar. aequalis, quadrupla erit circuli circa AC deſcripti, qui in ſphaera maximus eſt. Quod &c.
praeced.

Aliter.

Sphęrica ſuperficies ABCD ęqualis est cylindricę FEHG; cylindrica verò FEHG ad circulum, cuius ſemidiameter ſit AC, eſt vt rectangulum per axem EG, ad quadratum ex ſemidiametro AC, nempe ad quadratnm EG; & ideò ęqualis: propterea etiam ſphęrica
5. huius. *ſuperficies ęqualis erit circulo cuius ſemidiameter ſit AC; ergò quadrupla erit circuli cuius diameter ſit AC. Quod &c*

Propoſitio XXI.

CViuſcunq; portionis ſphaerę ſuperficies aequalis eſt circulo, cuius ſemidiameter aequalis ſit lineae quae ex polo portionis perducitur ad circulum, qui in eiuſdem portionis baſi eſt.

Eſto ſphaerae portio ſiue minor ſiue maior ABC. cuius ex polo ducta ſit recta AB. Dico ſuperficiem portionis aequalem eſſe circulo qui fit ex AB tamquam ſemidiametro.

Cum enim quadratum AB aequale ſit rectangulo DBE ob circulum, aequale erit & rectangulo GFIH, quod idem eſt ac rectangulum DBE. Propterea

circu-

circulus ex AB aequalis erit superficiei cylindri, cui per axem sit rectang. GFIH, & ideo aequalis etiam superficiei sphaericae portionis ABC. Quod &c. *5 huius*

Tria hæc Theoremata, quæ sequuntur, ex Archimede desumpta sunt; quod quidem fecimus ne lector Archimedem adire cogeretur, sed vniuersam hanc doctrinam in hoc libello haberet.

Propositio XXII.

SInt duo coni recti ABC, DEF. Sitq; curuæ coni ABC superficiei ęqualis circulus DF; nempe basis alterius coni DEF; rectæ verò IH, quę ex centro I ducitur perpendiculariter ad latus AB, æqualis sit altitudo EL: Dico conos ABC, DEF, esse ęquales. *per Cor. praeced.*

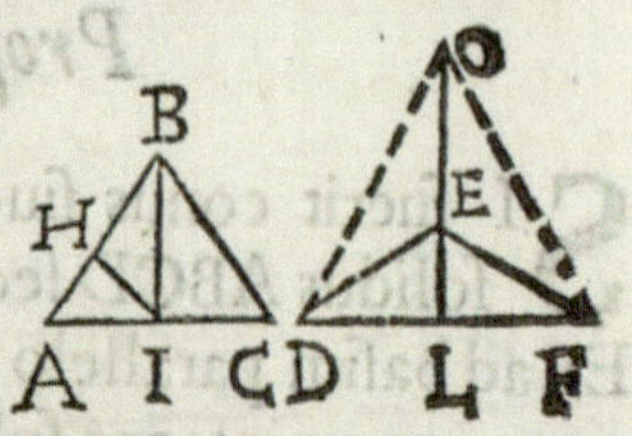

Nam altitudo BI ad altitudinem EL est vt BI ad IH (ob ęqualitatem) siue vt BA, ad AI, nempe vt curua superficies ABC ad basim AC; siue vt basis DF ad basim AC. reciprocè. Quare æquales erunt coni ABC, DEF. Quod erat &c. *per 4. se xii. 8. huius* *15. 12. Elem.*

Corollarium.

Hinc patet quòd si conus aliquis, puta DOF. basim quidem habeat DF æqualem curuæ superficiei ABC, altitudinem verò OL non ęqualem perpendiculari IH; Ita fore conum ABC ad conũ DOF, vt est IH ad OL. Nam conus DEF ad conum DOF. est vt EL. ad LO. Ergo (sumptis antecedentium æqualibus) conus ABC ad conum DOF, erit vt IH ad OL.

Propositio XXIII.

SI fuerit rombus solidus ABCD, ex duobus conis rectis compositus; Sitq; conus EFG. habens basim EG æqualem superficiei curuæ alterius conorum rombi, puta, BAD; altitudinem verò FH æqualem rectæ CL, quę quidem ex vertice reliqui coni BCD ducitur perpendiculariter in latus AB productũ

circulus ex AB aequalis erit ſuperficiei cylindri, cui per axem ſit rectang. GFIH, & ideo aequalis etiam ſuperficiei ſphaericae portionis ABC. Quod &c. 5 *huius:*

Tria hæc Theoremata, quæ ſequuntur, ex Archimede deſumpta ſunt; quod quidem fecimus ne lector Archimedem adire cogeretur, ſed vniuerſam hanc doctrinam in hoc libello haberet.

Propoſitio XXII.[25]

SInt duo coni recti ABC, DEF. Sitq; curuæ coni ABC ſuperficiei ęqualis circulus DF; nempe baſis alterius coni DEF; rectæ verò IH, quę ex centro I ducitur perpendiculariter ad latus AB, æqualis ſit altitudo EL: Dico conos ABC, DEF, eſſe ęquales.

Nam altitudo BI ad altitudinem EL eſt vt BI ad IH (ob ęqualitatem) ſiue vt BA, ad AI, nempe vt curua ſuperficies ABC ad baſim AC; ſiue vt baſis DF ad baſim AC. reciprocè. Quare æquales erunt coni ABC, DEF. Quod erat &c. *per* 4. *Se xti.* *8. huius*

Corollarium.

Hinc patet quòd ſi conus aliquis, puta DOF. baſim quidem habeat DF æqualem curuæ ſuperficiei ABC, altitudinem verò OL non ęqualem perpendiculari IH; Ita fore conum ABC ad conũ DOF, vt eſt IH ad OL. Nam conus DEF ad conum DOF. eſt vt EL. ad LO. Ergo (ſumptis antecedentium æqualibus) conus ABC ad conum DOF, erit vt IH àd OL. [15. 12 El..]

Propoſitio XXIII.[26]

SI fuerit rombus ſolidus ABCD, ex duobus conis rectis compoſitus; Sitq; conus EFG. habens baſim EG æqualem ſuperficiei curuæ alterius conorum rombi, puta, BAD; altitudinem verò FH æqualem rectæ CL, quę quidem ex vertice reliqui coni BCD ducitur perpendiculariter in latus AB productũ

E 2 alte-

---

[25] *Ex æquo* proof. (Cfr. Pisano and Bussotti's chapters above).

[26] *Ibidem.*

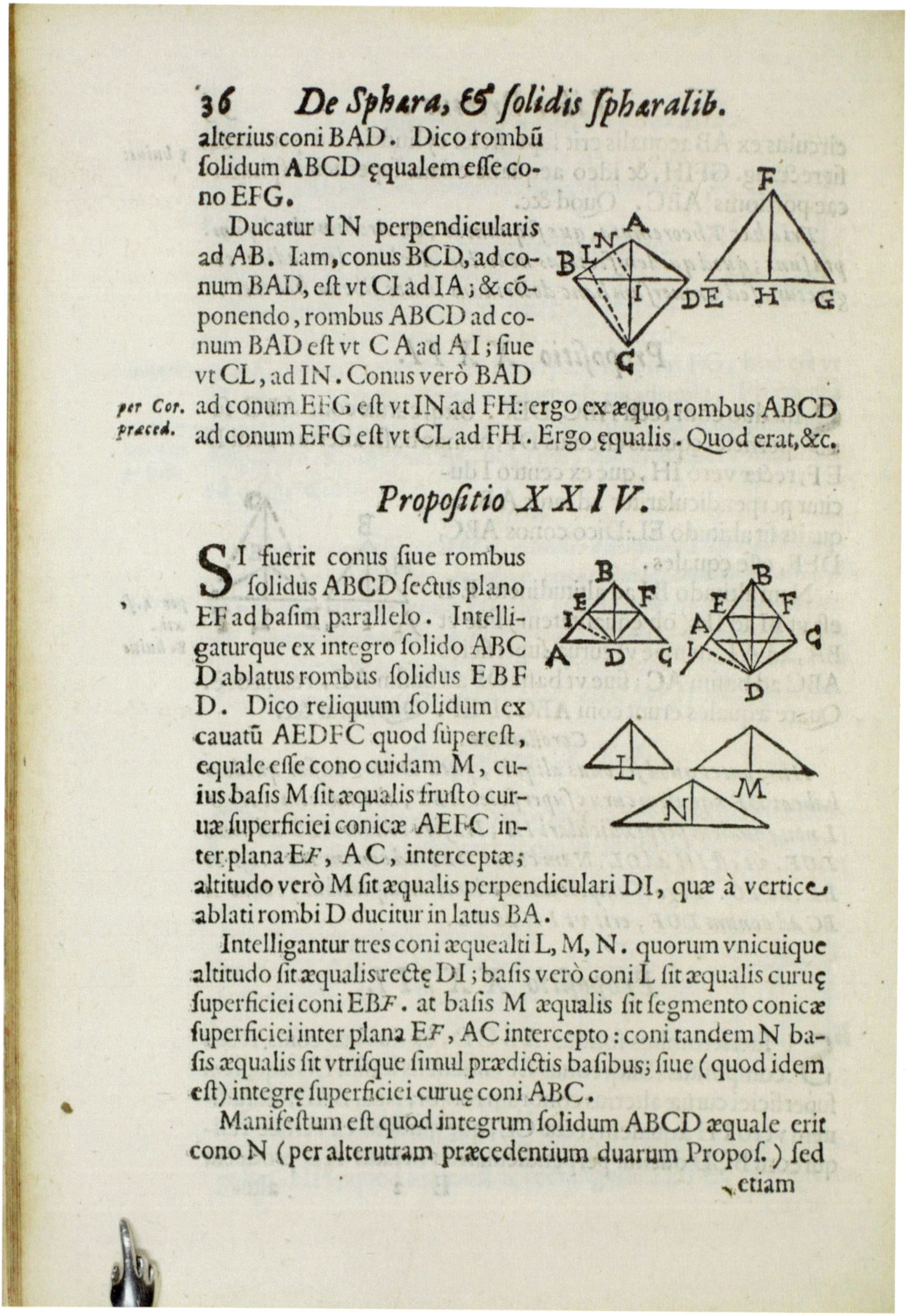

36 *De Sphæra, & ſolidis ſphæralib.*

alterius coni BAD. Dico rombũ ſolidum ABCD ęqualem eſſe cono EFG.

Ducatur IN perpendicularis ad AB. Iam, conus BCD, ad conum BAD, eſt vt CI ad IA; & cõponendo, rombus ABCD ad conum BAD eſt vt CA ad AI; ſiue vt CL, ad IN. Conus verò BAD ad conum EFG eſt vt IN ad FH: ergo ex æquo rombus ABCD ad conum EFG eſt vt CL ad FH. Ergo ęqualis. Quod erat, &c.

per Cor. præced.

Propoſitio XXIV.

SI fuerit conus ſiue rombus ſolidus ABCD ſectus plano EF ad baſim parallelo. Intelligaturque ex integro ſolido ABCD ablatus rombus ſolidus EBFD. Dico reliquum ſolidum excauatũ AEDFC quod ſupereſt, equale eſſe cono cuidam M, cuius baſis M ſit æqualis fruſto curuæ ſuperficiei conicæ AEFC inter plana EF, AC, interceptæ; altitudo verò M ſit æqualis perpendiculari DI, quæ à vertice ablati rombi D ducitur in latus BA.

Intelligantur tres coni æquealti L, M, N. quorum vnicuique altitudo ſit æqualis rectę DI; baſis verò coni L ſit æqualis curuę ſuperficiei coni EBF. at baſis M æqualis ſit ſegmento conicæ ſuperficiei inter plana EF, AC intercepto: coni tandem N baſis æqualis ſit vtriſque ſimul prædictis baſibus; ſiue (quod idem eſt) integrę ſuperficiei curuę coni ABC.

Manifeſtum eſt quod integrum ſolidum ABCD æquale erit cono N (per alterutram præcedentium duarum Propoſ.) ſed

etiam

alterius coni BAD. Dico rombũ ſolidum ABCD ęqualem eſſe cono EFG.

Ducatur IN perpendicularis ad AB. Iam, conus BCD, ad conum BAD, eſt vt CI ad IA; & cõponendo, rombus ABCD ad conum BAD eſt vt CA ad AI; ſiue vt CL, ad IN. Conus verò BAD ad conum EFG eſt vt IN ad FH: ergo ex æquo rombus ABCD ad conum EFG eſt vt CL ad FH. Ergo ęqualis. Quod erat, &c.

per *Cor praeced*

Propoſitio XXIV.

SI fuerit conus ſiue rombus ſolidus ABCD ſectus plano EF ad baſim parallelo. Intelligaturque ex integro ſolido ABCD ablatus rombus ſolidus EBFD. Dico reliquum ſolidum excauatũ AEDFC quod ſupereſt, equale eſſe cono cuidam M, cuius baſis M ſit æqualis fruſto curuæ ſuperficiei conicæ AEFC inter plana EF, AC, interceptæ; altitudo verò M ſit æqualis perpendiculari DI, quæ à vertice ablati rombi D ducitur in latus BA.

Intelligantur tres coni æquealti L, M, N. quorum vnicuique altitudo ſit æqualis rectę DI; baſis verò coni L ſit æqualis curuę ſuperficiei coni EBF. at baſis M æqualis ſit ſegmento conicæ ſuperficiei inter plana EF, AC intercepto: coni tandem N baſis æqualis ſit vtriſque ſimul prædictis baſibus; ſiue (quod idem eſt) integrę ſuperficiei curuę coni ABC.

Manifeſtum eſt quod integrum ſolidum ABCD æquale erit cono N (per alterutram præcedentium duarum Propoſ.) ſed

etiam

etiam duo coni *L* & M ſimul ſumpti æquales ſunt eidem cono N. ergo integrum ſolidum ABCD æquale erit duobus conis L & M ſimul ſumptis. Demptis itaque, rombo EB*F*D, & cono L, qui per præcedentem ſunt æquales, reliquum ſolidum excauatum AED*F*C ęquale erit reliquo cono M. Quod erat &c. *ex 24. Quinti.*

Propoſitio XXV

SI ex cylindro auferatur conus eandem ipſi baſim, & eandem altitudinem habens, erit reliquum excauatum ſolidum, quod ex cylindro ſupereſt, æquale cono cuidam, cuius baſis æqualis ſit ſuperficiei curuæ cylindri, altitudo verò æqualis ſemidiametro baſis ipſius cylindri.

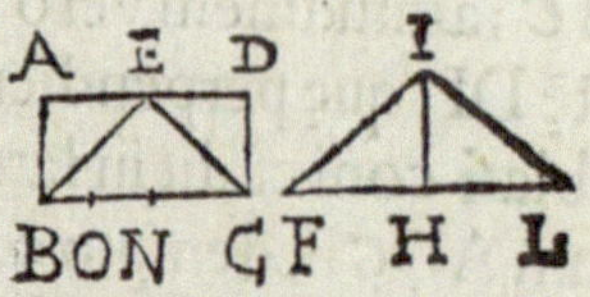

Eſto cylindrus, cuius rectangulum per axem ſit ABCD. & ex ipſo auferatur conus BEC, vt dictum eſt. Sumatur autem alius conus *F*IL, cuius baſis *F*L æqualis ſit ſuperficiei curuę cylindri, altitudo æqualis rectę NB. hoc eſt ſemidiametro baſis cylindri. Dico reliquum ex cylindro ſolidum, dempto cono BEC, ęquale eſſe cono *F*IL.

Secetur BN bifariam in O. Conus ergo *F*IL ad conum BEC, rationem habet compoſitam ex ratione altitudinum HI ad BA, hoc eſt NB ad BA, & ex ratione baſium, hoc eſt baſis quæ circa *F*L ad baſim quæ circa BC, ſiue quod idem eſt, ſuperficiei cylindricę ad baſim propriam quæ circa BC, hoc eſt, lineę AB ad BO. Erit ergò conus *F*IL ad conum BEC, vt NB ad BO, nempe duplus: ſolidum etiam cylindricum excauatum, dempto cono BEC, duplum eſt eiuſdem coni BEC. Propterea ſolidum cylindricum excauatum æquale erit cono *F*IL, cuius baſis æquatur ſuperficiei cylindri, altitudo verò æqualis eſt ſemidiametro baſis cylindri. Quod &c. *4. huius*

Pro-

etiam duo coni *L* & M ſimul ſumpti æquales ſunt eidem cono N. ergo integrum ſolidum ABCD æquale erit duobus conis L & M ſimul ſumptis. Demptis itaque, rombo EB*F*D, & cono L, qui per præcedentem ſunt æquales, reliquum ſolidum excauatum AEDFC ęquale erit reliquo cono M. Quod erat &c. *ex 24 Quinti.*

Propoſitio XXV.

SI ex cylindro auferatur conus eandem ipſi baſim, & eandem altitudinem habens, erit reliquum excauatum ſolidum, quod ex cylindro ſupereſt, æquale cono cuidam, cuius baſis æqualis ſit ſuperficiei curuæ cylindri, altitudo verò æqualis ſemidiametro baſis ipſius cylindri. *4. huius.*

Eſto cylindrus, cuius rectangulum per axem ſit ABCD. & ex ipſo auferatur conus BEC, vt dictum eſt. Sumatur autem alius conus FIL, cuius baſis FL æqualis ſit ſuperficiei curuę cylindri, altitudo æqualis rectę NB. hoc eſt ſemidiametro baſis cylindri. Dico reliquum ex cylindro ſolidum, dempto cono BEC, ęquale eſſe cono FIL.

Secetur BN bifariam in O. Conus ergo FIL ad conum BEC, rationem habet compoſitam ex ratione altitudinum HI ad BA, hoc eſt NB ad BA, & ex ratione baſium, hoc eſt baſis quæ circa FL ad baſim quæ circa BC, ſiue quod idem eſt, ſuperficiei cylindricę ad baſim propriam quæ circa BC, hoc eſt, lineę AB ad BO. Erit ergò conus FIL ad conum BEC, vt NB ad BO, nempe duplus: ſolidum etiam cylindricum excauatum, dempto cono BEC, duplum eſt eiuſdem coni BEC. Propterea ſolidum cylindricum excauatum æquale erit cono FIL, cuius baſis æquatur ſuperficiei cylindri, altitudo verò æqualis eſt ſemidiametro baſis cylindri. Quod &c.

Pro-

Propositio XXVI.

SI ex cono conus auferatur eandem habens basim altitudinem verò minorem, erit excauatum solidum conicũ, quod relinquitur, æquale cono cuidam, cuius quidem basis æqualis sit curuæ superficiei totius prioris coni, altitudo verò ęqualis perpendiculari, quæ ex vertice ablati coni demittitur in latus maioris coni.

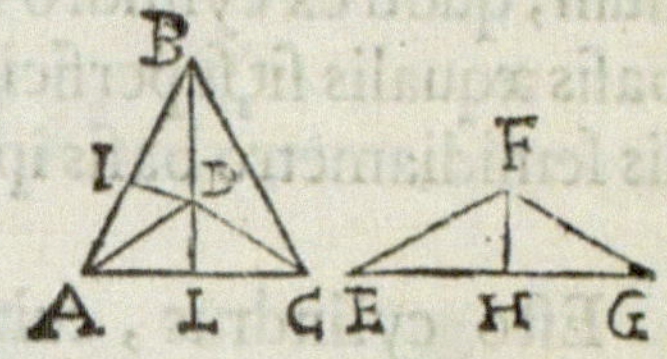

Esto conus rectus ABC ex quo auferatur conus ADC, vti dictum est. Ponatur autem conus EFG, habens basin EG, ęqualem curuę superficiei coni ABC; altitudinem verò HF ęqualem rectę DI, quę perpendiculariter à vertice ablati coni cadit in latus AB. Dico solidum conicum excauatum ADCB, dempto cono ADC, ęquale esse cono EFG.

Nam cum triangula BLA, BID, rectangula sint, habeantque angulum communem ABL, similia erunt. Sed conus EFG ad conum ADC rationem habet compositam ex ratione basium, nempe circuli circa EG, siue superficiei curuę coni ABC, ad circulum circa AC, hoc est rectę BA ad AL; siue BD ad DI, & ex ratione altitudinum, nempe HF ad DL, siue DI ad DL. Conus ergo EFG, ad conum ADC erit vt linea BD ad LD. Sed conus ABC ad conum ADC est vt BL ad LD, & diuidendo, etiam solidum excauatum ADCB ad conum ADC est vt linea BD ad DL. Propterea constat solidum excauatum ADCB ęquale esse cono EFG. Quod &c.

8. huius. 4. sexti.

Lemma.

Si ab eadem magnitudine AB duæ magnitudines inæquales auferantur AC, maior, & AD minor fueritq; DC, nempe differētia inter ablatas, æqualis differentię siue excessui, quo maius residuum BD superat quandam magnitudinem E. Dico ipsam E mino-

Propoſitio XXVI.

SI ex cono conus auferatur eandem habens baſim altitudinem verò minorem, erit excauatum ſolidum conicũ, quod relinquitur, æquale cono cuidam, cuius quidem baſis æqualis ſit curuæ ſuperficiei totius prioris coni, altitudo verò ęqualis per pendiculari, quæ ex vertice ablati coni demittitur in latus maioris coni.

Eſto conus rectus AB*C* ex quo auferatur conus AD*C*, vti dictum eſt. Ponatur autem conus E*F*G, habens baſin EG, ęqualem curuę ſuperficiei coni ABC; altitudinem verò H*F* ęqualem rectę DI, quę perpendiculariter à vertice ablati coni cadit in latus AB. Dico ſolidum conicum excauatum AD*C*B, dempto cono AD*C*, ęquale eſſe cono EFG.

8. *huius.* Nam cum triangula BLA, BID, rectangula ſint, habeantque angulum communem ABL, ſimilia erunt. Sed conus EFG ad conum
4. *Sexti.* ADC rationem habet compoſitam ex ratione baſium, nempe circuli circa EG, ſiue ſuperficiei curuę coni ABC, ad circulum circa A*C*, hoc eſt rectę BA ad AL; ſiue BD ad DI, & ex ratione altitudinum, nempe HF ad DL, ſiue DI ad DL. *C*onus ergo EFG, ad conum ADC erit vt linea BD ad LD. Sed conus ABC ad conum ADC eſt vt BL ad LD, & diuidendo, etiam ſolidum excauatum ADCB ad conum ADC eſt vt linea BD ad DL. Propterea conſtat ſolidum excauatum ADCB ęquale eſſe cono EFG. Quod &c.

Lemma.

Si ab eadem magnitudine AB duæ magnitudines inæquales auferantur AC, maior, & AD minor fueritq; DC, nempe differẽtia inter ablatas, æqualis differentię ſiue exceſſui, quo maius reſiduum BD ſuperat quandam magnitudinem E. Dico ipſam E

mino-

minori residuo CB equalem esse.

Patet hoc. Cum enim maius residuum DB superet magnitudinem E excessu DC; si excessus abijciatur, erit reliqua CB equalis magnitudini E. Propterea magnitudo E equalis est minori residuo. Quod &c.

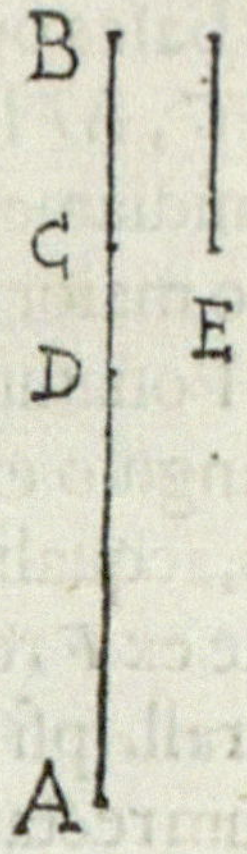

Propositio XXVII.

SI ex conico frusto conus auferatur, qui pro basi habeat maiorem frusti basim, altitudinem verò eandem cum frusto; Erit reliquum excauatum solidum æquale cono cuidam, qui basim habeat æqualem superficiei curuę frusti, altitudinem verò ęqualem perpendiculari quę ducitur ex vertice ablati coni in latus alterum conici frusti.

Esto conicum frustum ABCD, cuius maior basis sit circulus circa BC. Et ex ipso auferatur conus BEC, cuius basis sit idem circulus circa BC; altitudo verò FE eadem cum frusto. Dico reliquū solidum excauatū dempto cono BEC, ęquale esse cono cuidam, cuius basis aequalis sit curuae superficiei conici frusti ABCD. altitudo vero sit linea EH. quae nimirùm ex E vertice ablati coni cadit perpendicularitèr in AB latus conici frusti.

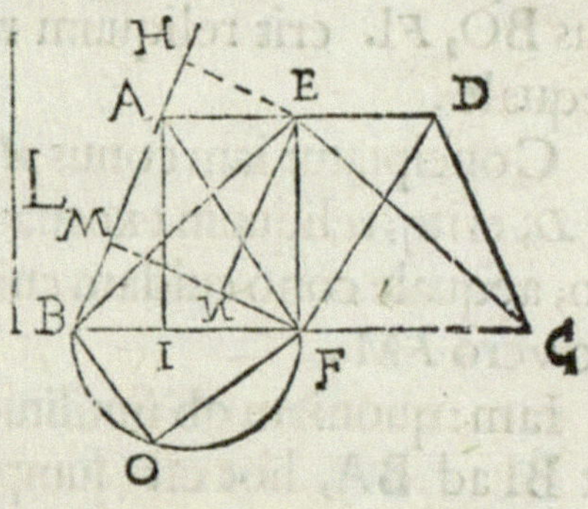

Inscribatur alius conus AFD habens basim circà AD, & verticem in F. Ducaturque AI parallela ad EF, eritque tota IC aequalis vtrique simul semidiametro basium, nempe ipsi EA, ipsiq; FB. Fiat deinde circa FB semicirculus FOB, in quo applicetur BO aequalis ipsi FI, siue ipsi EA; critq; circulus ex semidiametro FO differentia inter duos circulos, quorum semidiametri sint, FB, BO, siue FB, & EA; nempe differentia in-

ter

minori reſiduo CB ęqualem eße.

Patet hoc. Cum enim maius reſiduum DB ſuperet magnitudinem E exceſſu DC; ſi exceſſus abijciatur, erit reliqua CB ęqualis magnitudini E. Propterea magnitudo E ęqualis eſt minori reſiduo. Quod &c.

Propoſitio XXVII.

SI ex conico fruſto conus auferatur, qui pro baſi habeat maiorem fruſti baſim, altitudinem verò eandem cum fruſto; Erit reliquum excauatum ſolidum æquale cono cuidam, qui baſim habeat æqualem ſuperficiei curuę fruſti, altitudinem verò ęqualem perpendiculari quę ducitur ex vertice ablati coni in latus alterum conici fruſti.

Eſto conicum fruſtum ABCD, cuius maior baſis ſit circulus circa *BC*. Et ex ipſo auferatur conus *BEC,* cuius baſis ſit idem circulus circa *BC*; altitudo verò *F*E eadem cum fruſto. Dico reliquũ ſolidum excauatũ dempto cono *BEC,* ęquale eſſe cono cuidam, cuius baſis aequalis ſit curuae ſuperficiei conici fruſti *ABCD.* altitudo vero ſit linea EH. quae nimirùm ex E vertice ablati coni cadit perpendicularitèr in A*B* latus conici fruſti

Inſcribatur alius conus AFD habens baſim circà AD, & verticem in *F.* Ducaturque AI parallela ad *E*F, eritque tota I*C* aequalis vtrique ſimul ſemidiametro baſium, nempe ipſi *E*A, ipſiq; *FB. F*iat deinde circa *FB* ſemicirculus *FOB,* in quo applicetur *B*O aequalis ipſi *F*I, ſiue ipſi *EA*; eritq; circulus ex ſemidiametro *F*O differentia inter duos circulos, quorum ſemidiametri ſint, *FB, B*Q, ſiue *FB,* & *E*A; nempe differentia in-
ter

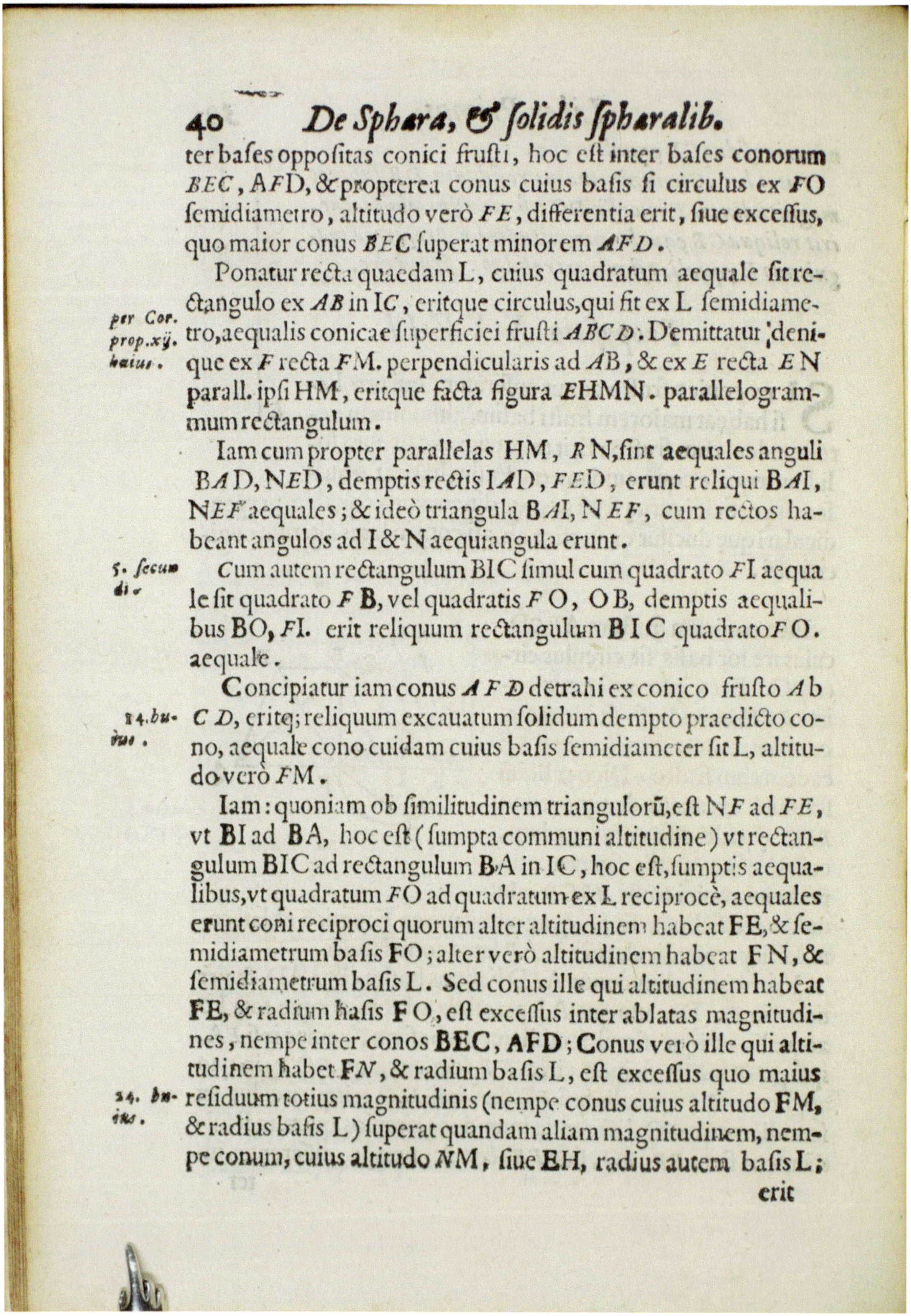

40 *De Sphæra, & ſolidis ſphæralib.*

ter baſes oppoſitas conici fruſti, hoc eſt inter baſes conorum *BEC*, *AFD*, & propterea conus cuius baſis ſi circulus ex *FO* ſemidiametro, altitudo verò *FE*, differentia erit, ſiue exceſſus, quo maior conus *BEC* ſuperat minorem *AFD*.

Ponatur recta quaedam L, cuius quadratum aequale ſit rectangulo ex *AB* in I*C*, eritque circulus, qui fit ex L ſemidiametro, aequalis conicae ſuperficiei fruſti *ABCD*. Demittatur deni- *per Cor. prop. xij. huius.* que ex *F* recta *F*M. perpendicularis ad *A*B, & ex *E* recta *E* N parall. ipſi HM, eritque facta figura *E*HMN. parallelogrammum rectangulum.

Iam cum propter parallelas HM, *R* N, ſint aequales anguli B*A*D, N*E*D, demptis rectis I*A*D, *FE*D, erunt reliqui B*A*I, N*EF* aequales; & ideò triangula B*A*I, N*EF*, cum rectos habeant angulos ad I & N aequiangula erunt.

5. ſecundi. Cum autem rectangulum BIC ſimul cum quadrato *F*I aequale ſit quadrato *F* B, vel quadratis *F* O, O B, demptis aequalibus BO, *F*I. erit reliquum rectangulum B I C quadrato *F* O. aequale.

Concipiatur iam conus *A F D* detrahi ex conico fruſto *A* b *24. huius.* *C D*, eritq; reliquum excauatum ſolidum dempto praedicto cono, aequale cono cuidam cuius baſis ſemidiameter ſit L, altitudo verò *F*M.

Iam: quoniam ob ſimilitudinem triangulorū, eſt N*F* ad *FE*, vt BI ad BA, hoc eſt (ſumpta communi altitudine) vt rectangulum BIC ad rectangulum BA in IC, hoc eſt, ſumptis aequalibus, vt quadratum *F*O ad quadratum ex L reciprocè, aequales erunt coni reciproci quorum alter altitudinem habeat FE, & ſemidiametrum baſis FO; alter verò altitudinem habeat F N, & ſemidiametrum baſis L. Sed conus ille qui altitudinem habeat FE, & radium baſis F O, eſt exceſſus inter ablatas magnitudines, nempe inter conos BEC, AFD; Conus verò ille qui altitudinem habet F*N*, & radium baſis L, eſt exceſſus quo maius *24. huius.* reſiduum totius magnitudinis (nempe conus cuius altitudo FM, & radius baſis L) ſuperat quandam aliam magnitudinem, nempe conum, cuius altitudo *N*M, ſiue EH, radius autem baſis L;

erit

ter baſes oppoſitas conici fruſti, hoc eſt inter baſes conorum *BEC,* A*F*D, & propterea conus cuius baſis ſi circulus ex *FO* ſemidiametro, altitudo verò *FE,* differentia erit, ſiue exceſſus, quo maior conus *BEC* ſuperat minorem *AFD.*

Ponatur recta quaedam L, cuius quadratum aequale ſit rectangulo ex *AB* in I*C,* eritque circulus, qui fit ex L ſemidiametro, aequalis conicae ſuperficiei fruſti *ABCD.* Demittatur denique ex *F* recta *F*M. perpendicularis ad *A*B, & ex *E* recta *E*N parall. ipſi HM, eritque facta figura *E*HMN. parallelogrammum rectangulum.

per Cor prop. xij huius

Iam cum propter parallelas HM, *R*N, ſint aequales anguli BD, N*E*D, demptis rectis IAD, FED, erunt reliqui B*A*I, NEF aequales; & ideò triangula B*A*I, N*EF,* cum rectos habeant angulos ad I & N aequiangula erunt

5. secum di

*C*um autem rectangulum BIC ſimul cum quadrato FI aequale ſit quadrato FB, vel quadratis FO, OB, demptis aequalibus BO, FI. erit reliquum rectangulum BIC quadrato FO. aequale.

Concipiatur iam conus *AFD* detrahi ex conico fruſto *A*b*CD,*[27] eritq; reliquum excauatum ſolidum dempto praedicto cono, aequale cono cuidam cuius baſis ſemidiameter ſit L, altitudo verò *F*M.

24. hu- ius

Iam: quoniam ob ſimilitudinem triangulorũ, eſt N*F* ad *FE,* vt BI ad BA, hoc eſt (ſumpta communi altitudine) vt rectangulum BIC ad rectangulum BA in IC, hoc eſt, ſumptis aequalibus, vt quadratum *F*O ad quadratum ex L reciprocè, aequales erunt coni reciproci quorum alter altitudinem habeat FE, & ſemidiametrum baſis FO; alter verò altitudinem habeat FN, & ſemidiametrum baſis L. Sed conus ille qui altitudinem habeat FE, & radium haſis[28] FO, eſt exceſſus inter ablatas magnitudines, nempe inter conos BEC, AFD; Conus verò ille qui altitudinem habet F*N,* & radium baſis L, eſt exceſſus quo maius reſiduum totius magnitudinis (nempe conus cuius altitudo FM, & radius baſis L) ſuperat quandam aliam magnitudinem, nempe conum, cuius altitudo *N*M, ſiue EH, radius autem baſis L;

24. hu- ius

erit

[27] AbCD] ABCD
[28] haſis] baſis

erit itaque hæc magnitudo, per Lemma præmissum, æqualis minori residuo; ergò conus prædictus, cuius altitudo EH, & basis circulus ex L æqualis superficiei conici frusti, æqualis erit minori residuo, hoc est reliquo conici frusti ABCD. dempto cono BEC. Quod erat &c.

Aliter.

Sed conemur idem ostendere minus laboriosa demonstratione; si possibile erit ex difficultate materiæ, & veriùs ex tenuitate ingenij.

Sit conicum frustum ABCD cuius maior basis BC, & ex ipso auferatur conus BEC, altitudinem habens eandem cum frusto, & pro basi, maiorem ipsius frusti basim. Compleatur conus BGC, cuius datum erat frustum, ductaque EH ad angulos rectos ipsi BG,

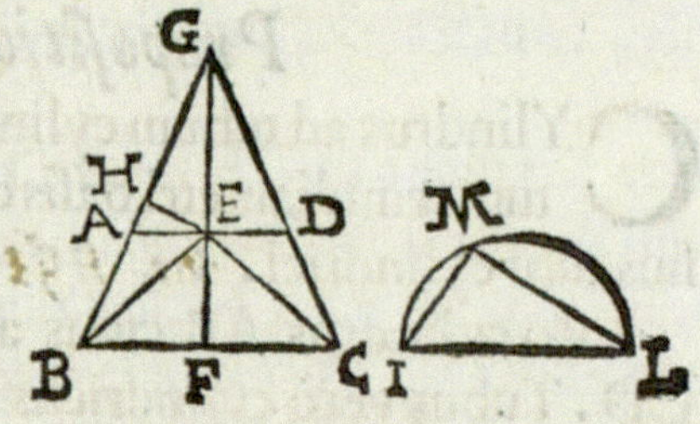

ponatur IL media proportionalis inter GB, BF, eritq; circulus ex IL semidiametro descriptus, æqualis superficiei coni BGC. fiat circa IL semicirculus IML, in quo aptetur IM media proportionalis inter GA, AE, eritq; circulus ex semidiametro IM factus æqualis superficiei coni AGD; Reliquus circulus ex semidiametro ML factus, æqualis erit superficiei conicæ frusti ABCD. (si enim ab æqualibus æqualia demas reliqua sunt æqualia.) per Cor. 8. huius. per Cor. 8. huius.

Dico reliquum solidum frusti conici ABCD, ablato cono BEC, æquale esse cono cuidam, cuius altitudo sit EH; basis verò æqualis superficiei conicæ ipsius frusti; hoc est circulus ex semidiametro ML descriptus.

Cum .n. duo circuli ex radijs IM, LM facti æquales sint circulo ex IL descripto, si altitudo vnicuique eadem assumatur EH, erunt duo coni simul (quorum altitudo communis EH, bases verò circuli ex radijs IM, LM) æquales cono, cuius altitudo eadem EH, basis verò circulus ex IL; iste vero conus æqualis est solido conico BECG, dempto cono BEC, ergo duo illi coni æquales erunt solido BECG. Proptereà ablatis vtrinque æqualibus conis, nempè co- 26 huius

F

erit itaque hæc magnitudo, per Lemma præmiſſum, æqualis minori reſiduo; ergò conus prædictus, cuius altitudo EH, & baſis circulus ex L æqualis ſuperficiei conici fruſti, æqualis erit minori reſiduo, hoc eſt reliquo conici fruſti ABCD. dempto cono BEC. Quod erat &c.

Aliter.

Sed conemur idem ostendere minus laborioſa demonstratione; ſi poſſibile erit ex difficultate materiæ, & veriùs ex tenuitate ingenij.

Sit conicum fruſtum ABCD cuius maior baſis BC, & ex ipſo auferatur conus BEC, altitudinem habens eandem cum fruſto, & pro baſi, maiorem ipſius fruſti baſim. Compleatur conus BGC, cuius datum erat fruſtum, ductaque EH ad angulos rectos ipſi BG, ponatur IL media proportionalis inter GB, BF, eritq; circulus ex IL ſemidiametro deſcriptus, æqualis ſuperficiei coni BGC. fiat circa IL ſemicirculus IML, in quo aptetur I M media proportionalis inter GA, AE, eritq; circulus ex ſemidiametro IM factus æqualis ſuperficiei coni AGD; Reliquus circulus ex ſemidiametro ML factus, æqualis erit ſuperficiei conicæ fruſti ABCD. (ſi enim ab æqualibus æqualia demas reliqua ſunt æqualia.) per Cor 8. huius. per Cor 8. huius.

Dico reliquum ſolidum fruſti conici ABCD, ablato cono BEC, æquale eſſe cono cuidam, cuius altitudo ſit EH; baſis verò æqualis ſuperficiei conicæ ipſius frusti; hoe eſt circulus ex ſemidiametro ML deſcriptus. 26 *huius*

Cum. n. duo circuli ex radijs IM, LM facti xquales[29] *ſint circulo ex IL deſcripto, ſi altitudo vnicuique eadem aſſumatur EH, erunt duo coni ſimul (quorum altitudo communis EH, baſes vero circuli ex radijs IM, LM) æquales cono, cuius altitudo eadem EH, baſis verò circulus ex IL; iste vero conus æqualis est ſolido conico BECG, dempto cono BEC, ergo duo illi coni ęquales erunt ſolido BECG. Propterea ablatis vtrinque æqualibus conis, nem*

F *pè co-*

[29] *xquales*] *equales.*

pe cono, cuius baſis ex IM eſt, altitudo EH, & cono AGD (ſunt enim æquales per 22. huius) remanebunt ęqualia, ſolidum nempe excauatum fruſti ABCD, detracto cono BEC, & conus cuius altitudo EH, baſis circulus ex LM radio factus, qui quidem ęqualis eſt ſuperficiei conicę fruſti ABCD. Quod &c.

Definitio.

Si ex cylindro cylindrus auferatur æquealtus, & circa eundẽ axem deſcriptus, ſolidum excauatum quod relinquitur, Tubum cylindricum appellabimus.

Propoſitio XXVIII.

CYlindrus ad tubum cylindricum æquealtum, eſt vt quadratum ſemidiametri baſis cylindri ad rectangulum baſis ipſius tubi cylindrici. viz EGI

Eſto cylindrus AB cuius axis CD. Tubus verò cylindricus EF (dempto nimirum cylindro GH) ęquealtus ſit cum cylindro AB. Dico cylindrum AB ad tubum EF eſſe vt quadratum AC ſemidiametri baſis cylindri, ad rectangulum EGI, nempe ad rectangulum baſis tubi, hoc eſt quod fit à differentia EG. & ab aggregato GI ſemidiametrorum baſis ipſius tubi.

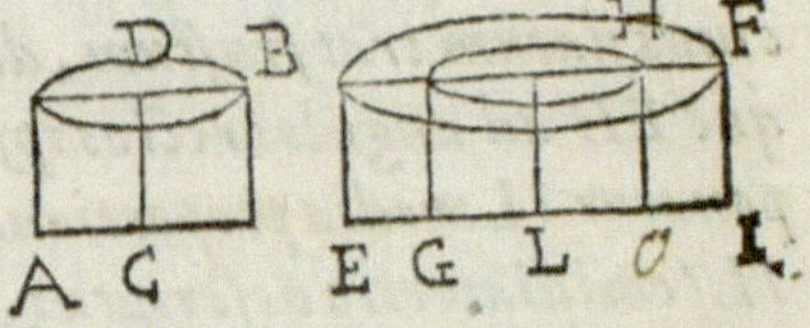

Nam cylindrus integer EF ad cylindrum GH, eſt vt quadratum EL ad LG. quadratum. Et diuidendo, Tubus cylindricus EF ad cylindrum GH eſt vt rectangulum EGI ad quadratum GL. Sed cylindrus GH ad AB cylindrum eſt vt quadratum GL ad quadratum BC. Ergo ex æquo erit tubus cylindricus EF ad cylindrum AB vt rectangulum EGI ad quadratum AC. Conuertendo igitur patet quod propoſitum erat.

Propoſitio XXIX.

DAtę figuræ ſolidæ rotundæ figuram inſcribere, alteramque circumſcribere ex cylindris æquealtis, ita vt deſcriptarum differentia minor ſit quolibet dato ſolido.

Eſto

pe cono, cuius baſis ex IM est, altitudo EH, & cono AGD (ſunt enim æquales per 22. huius) remanebunt ęqualia, ſolidum nempe excauatum fruſti ABCD, detracto cono BEC, & conus cuius altitudo EH, baſis circulus ex LM radio factus, qui quidem ęqualis eſt ſuperficiei conicę frusti ABCD. Quod &c.

Definitio.

Si ex cylindro cylindrus auferatur æquealtus, & circa eundẽ axem deſcriptus, ſolidum excauatum quod relinquitur, Tubum cylindricum appellabimus.

Propoſitio XXVIII.

CYlindrus ad tubum cylindricum æquealtum, eſt vt quadratum ſemidiametri baſis cylindri ad rectangulum baſis ipſius tubi cylindrici.

Eſto cylindrus AB cuius axis CD. Tubus verò cylindricus EF (dempto nimirum cylindro GH) ęquealtus ſit cum cylindro AB. Dico cylindrum AB ad tubum EF eſſe vt quadratum AC ſemidiametri baſis cylindri, ad rectangulum EGI, nempe ad rectangulum baſis tubi, hoc eſt quod fit à differentia EG. & ab aggregato GI ſemidiametrorum baſis ipſius tubi.

Nam cylindrus integer EF ad cylindrum GH, eſt vt quadratum EL ad LG. quadratum. Et diuidendo, Tubus cylindricus EF ad cylindrum GH eſt vt rectangulum EGI ad quadratum GL. Sed cylindrus GH ad AB cylindrum eſt vt quadratum GL ad quadratum BC. Ergo ex æquo erit tubus cylindricus EF ad cylindrum AB vt rectangulum EGI ad quadratum A C. Conuertendo igitur patet quod propoſitum erat.

Propoſitio XXIX.

DAtę figuræ ſolidæ rotundæ figuram inſcribere, alteramque circumſcribere ex cylindris æquealtis, ita vt deſcriptarum differentia minor ſit quolibet dato ſolido.

Esto

Esto cylindrus ABC D, cuius axis EF: datoq; intra cylindrum solido AED circa eundem axē EF reuoluto, siue hemisphęriū, siue conus, vel conoides sit, oportet ipsi solido AED duas figuras ex cylindris æquealtis cō positas, alteram quidem

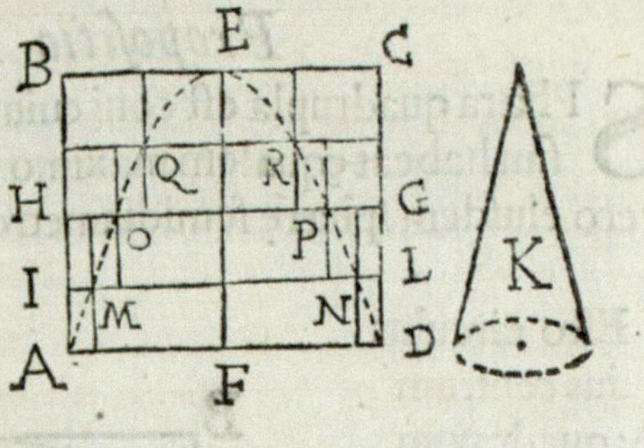

inscribere, alteram verò circumscribere ita vt circumscripta superet inscriptam minori excessu quā sit quodlibet datum solidum K.

Secetur bifariam cylindrus AC plano HG ad axem EF erecto; iterumq; cylindrus HD bifariam secetur plano IL; & hoc fiat semper donec cylindrus aliquis puta AL minor remaneat quàm solidum K. Tunc diuiso toto cylindro AC in cylindros æquealtos ac ipse AL, oriantur in solido AED sectiones MN, OP. QR. Concipiamus super vnoquoq; circulorum MN, OP, QR, duos cylindros, alterum quidem versus E, alterum autem versus partes F conuersum. Eruntq; omnes simul cylindri quī verticem habent versus F, æquales omnibus simul cylindris verticem versus E habentibus (cum singuli singulis ęquales sint) Ergo si omnibus cylindris qui verticem habent versus E, addas cylindrum AL, superabit iam figura circa solidum AED descripta, figuram eidem inscriptam, differentià AL; Nempe minori excessu quàm sit solidum K. Quod erat &c.

Corollarium.

Hinc patet quòd data figura solida, siue hemisphęrium sit, siue conus, siue conoides &c. ipsi duę figurę solidę ex cylindris æquealtis compositæ altera inscribi potest, altera vero circumscribi; ità vt differentia inter datam solidam figuram, & descriptarū alterutram, minor sit quolibet dato solido K,

Differentia enim inter figuram datam & alteram descriptarum minor vtiq; erit quam differentia inter descriptas (est enim

F 2 pars

Efto cylindrus ABCD, cuius axis EF: datoq; intra cylindrum ſolido AED circa eundem axẽ EF reuoluto, ſiue hemiſphęriũ, ſiue conus, vel conoides ſit, oportet ipſi ſolido AED duas figuras ex cylindris æquealtis cõpoſitas, alteram quidem inſcribere, alteram verò circumſcribere ita vt circumſcripta ſuperet inſcriptam minori exceſſu quã ſit quodlibet datum ſolidum K.

Secetur bifariam cylindrus AC plano HG ad axem EF erecto; iterumq; cylindrus HD bifariam ſecetur plano IL; & hoc fiat ſemper donec cylindrus aliquis puta AL minor remaneat quàm ſolidum K. Tunc diuiſo toto cylindro AC in cylindros æquealtos ac ipſe AL, oriantur in ſolido AED ſectiones MN, OP. QR. Concipiamus ſuper vnoquoq; circulorum MN, OP, QR, duos cylindros, alterum quidem verſus E, alterum autem verſus partes F conuerſum. Eruntq; omnes ſimul cylindri qui verticem habent verſus F, æquales omnibus ſimul cylindris verticem verſus E habentibus (cum ſinguli ſingulis ęquales ſint) Ergo ſi omnibus cylindris qui verticem habent verſus E, addas cylindrum AL, ſuperabit iam figura circa ſolidum AED deſcripta, figuram eidem inſcriptam, differentià AL; Nempe minori exceſſu quàm ſit ſolidum K. Quod erat &c.

Corollarium.

Hinc patet quòd data figura ſolida, ſiue hemiſpherium ſit, ſiue conus, ſiue conoides &c. ipſi duę figurę ſolidę ex cylindris æquealtis compoſitæ altera inſcribi poteſt, altera vero circumſcribi; ità vt differentia inter datam ſolidam figuram, & deſcriptarũ alterutram, minor ſit quolibet dato ſolido K,

Differentia enim inter figuram datam & alteram deſcriptarum minor vtiq; erit quam differentia inter deſcriptas (eſt enim

F 2 pars

pars eiusdem) ergo multò minor quàm solidum K.

Propositio XXX.

Sphæra quadrupla est coni cuiusdam, qui quidem conus basim habeat æqualem maximo sphæræ circulo, altitudinem vero eiusdem sphæræ semidiametro æqualem.

Esto circulus cuius centrum A; quadratum ipsi circumscriptum sit BCDE; iunctisque EA, AD. conuertatur figura circa axem FG ita vt à quadrato fiat cylindrus, a circulo sphæra ~~circulus~~; à triangulo EAD, conus EAD.

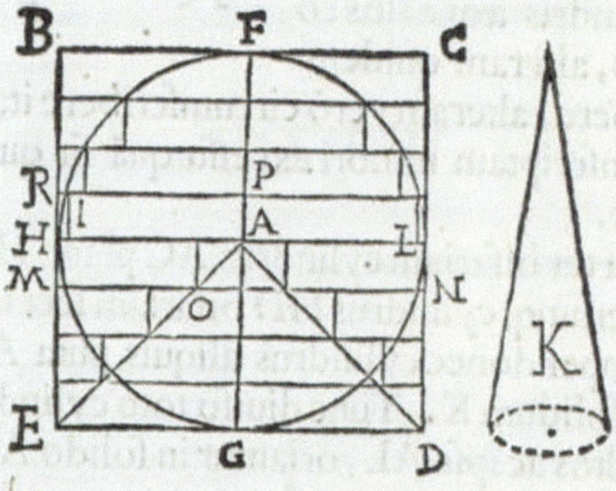

Dico sphæram quadruplam esse coni EAD. Nisi enim quadrupla sit, non erit hæmisphærium æquale solido, quod describitur à triangulo EHA. circa axem FG. conuerso (cum hoc solidum duplum sit coni EAD.) Erit itaq; hemisphærium vel maius, vel minus solido trianguli EHA.

Esto primùm maius, si potest esse; sitque excessus æqualis solido k. Inscribatur in hemisphærio figura ex cylindris æquealtis constans ita vt ab hemisphærio deficiat minori defectu quã sit solidum K. Et erit figura inscripta adhuc maior quàm solidum trianguli EHA. Secetur etiam axis AG in tot partes æquales in quot sectus erit AF. Ductisq; per puncta sectionum planis ad axem erectis, intelligatur in solido trianguli EHA. inscripta figura ex tubis cylindricis æquealtis constans, quorum vnus sit, cuius sectio est rectangulum HO. *(per 29. huius)*

Iam cylindrus IL ad tubum cylindricum HO, est vt quadratum

pars eiuſdem) ergo multò minor quàm ſolidum K

Propositio XXX.

SPhæra quadrupla eſt coni cuiuſdam, qui quidem conus baſim habeat ęqualem maximo ſphæræ circulo, altitudinem vero eiuſdem ſphærę ſemidiametro æqualem.

Eſto circulus cuius centrum A; quadratum ipſi circumſcriptum ſit BCDE; iunctiſque EA, AD. conuertatur figura circa axem FG ita vt à quadrato fiat cylindrus, ad ſphæra circulus[30]; à triangulo EAD, conus EAD.

Dico ſphæram quadruplam eſſe coni EAD. Niſi enim quadrupla ſit, non erit hæmiſphærium ęquale ſolido, quod deſcribitur à triangulo EHA. circa axem FG. conuerſo (cum hoc ſolidum duplum ſit coni EAD.) Erit itaq; hemiſphærium vel maius, vel minus ſolido trianguli EHA.

per 29 huius Eſto primùm maius, ſi poteſt eſſe; ſitque exceſſus æqualiſ ſolido k. Inſcribatur in hemiſphærio figura ex cylindris æquealtis conſtans ita vt ab hemiſphærio deficiat minori defectu quã ſit ſolidum K. Et erit figura inſcripta adhuc maior quàm ſolidum trianguli EHA. Secetur etiam axis AG in tot partes æquales in quot ſectus erit AF. Ductiſq; per puncta ſectionum planis ad axem erectis, intelligatur in ſolido trianguli EHA. inſcripta figura ex tubis cylindricis æquealtis conſtans, quorum vnus ſit, cuius ſectio eſt rectangulum HO.

Iam cylindrus IL ad tubum cylindricum HO, eſt vt quadra-

tum

[30] ad ſphaera circulus] à ſphaera circulo

tum IP ad rectangulum MON. Sed quadratum IP æquale est rectangulo FPG, nempe ipsi MON (nam F P æqualis est rectæ BR, siuè ME, siue MO, & reliqua PG reliquę ON) ergo cylindrus IL æqualis est tubo cylindrico HO. Hoc modo procedendo ostenduntur omnes cylindri in hæmispherio æquales omnibus tubis in solido trianguli EHA. Quare figura in hemisphęrio inscripta ex cylindris constans, æqualis erit figuræ in solido trianguli EHA descriptæ ex tubis cylindricis compositę. Sed figura in hemisphærio descripta maior erat integro solido trianguli EHA. Ergò necesse est quod figura inscripta in solido E HA eodem solido maior sit. pars suo toto. Quod esse non potest. *per 28 huius.*

Esto deinde, si fieri potest, hemisphæriū minus solido trianguli E H A; sitq; defectus æqualis solido k Circumscribatur ipsi hemisphærio figura solida ex cylindris æquealtis constans, ita vt excessus figuræ super hemisphærium minus sit solido K. Tunc enim circumscripta figura adhuc minor erit solido trianguli EHA. Concipiamus deinde solido trianguli EHA aliquam figuram esse circumscriptam cōstantem ex tubis cylindricis æquealtis ac cylindri ex quibus cōstat figura hæmisphęrio circumscripta. *per 29. huius.*

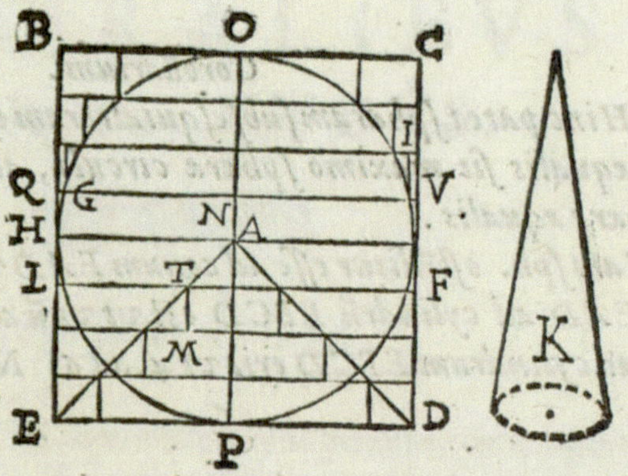

Iam primus cylindrus HV figuræ circa hemisphęrium descriptæ, æqualis est primo tubo cylindrico figuræ circumscriptæ solido trianguli EHA; nam & iste tubus, cylindrus est HF.

Secundus cylindrus G I ad secundum tubum LM, est vt quadratum GN ad rectangulum LTF, nempe æqualis (quadratum enim GN, æquale est rectangulo ONP, siue LTF, nam recta O *per 28. huius.*

N re-

tum IP ad rectangulum MON. Sed quadratum IP æquale eſt rectangulo FPG, nempe ipſi MON (nam FP æqualis eſt rectæ BR, ſiuè ME, ſiue MO, & reliqua PG reliquę ON) ergo cylindrus IL æqualis eſt tubo cylindrico HO. Hoc modo procedendo oſtenduntur omnes cylindri in hæmiſpherio æquales omnibus tubis in ſolido trianguli EHA. Quare figura in hemiſphęrio inſcripta ex cylindris conſtans, æqualis erit figuræ in ſolido trianguli EHA deſcriptæ ex tubis cylindricis compoſitę. Sed figura in hemiſphærio deſcripta maior erat integro ſolido trianguli EHA. Ergò neceſſe eſt quod figura inſcripta in ſolido EHA eodem ſolido maior ſit. pars ſuo toto. Quod eſſe non poteſt. *per* 28 *huius.*

Eſto deinde, ſi fieri poteſt, hemiſphæriũ minus ſolido trianguli EHA; ſitq; defectus æqualis ſolido K[31] *per* 29 *huius.*

Circumſcribatur ipſi hemiſphærio figura ſolida ex cylindris æquealtis conſtans, ita vt exceſſus figuræ ſuper hemiſphærium minus ſit ſolido K. Tunc enim circumſcripta figura adhuc minor erit solido trianguli EHA. Concipiamus deinde ſolido trianguli EHA aliquam figuram eſſe circumſcriptam cõſtantem ex tubis cylindricis æquealtis ac cylindri ex quibus cõſtat figura hæmiſphęrio circumſcripta.

Iam primus cylindrus HV figuræ circa hemiſphęrium deſcriptæ, æqualis eſt primo tubo cylindrico figuræ circumſcriptæ ſolido trianguli EHA; nam & iſte tubus, cylindrus eſt HF. *per* 8. *huius.*

Secundus cylindrus GI ad ſecundum tubum LM, eſt vt quadratum GN ad rectangulum LTF, nempe æqualis (quadratum enim GN, æquale eſt rectangulo ONP, ſiue LTF, nam recta O

N re-

[31] K] K.

N rectæ BQ, siue L E, siue LT; æqualis est, & reliqua NP reliquæ TF.)

Ergo omnes simul cylindri figuræ circa hemisphęrium descriptæ, hoc est eadem figura, æqualis erit omnibus simul tubis cylindricis circa solidum trianguli EHA descriptis, cum singuli singulis æquales sint. Sed figura circa hemisphærium descripta minor erat solido trianguli EHA. Necesse igitur est quòd solidum trianguli EHA maius sit, quam figura sibi circumscripta. pars suo toto. Quod esse non potest.

Hemisphærium igitur neque maius, neque minus erit solido trianguli EHA, sed ipsi æquale, solidum verò trianguli EHA duplum est coni EAD, ergò hemisphæriũ duplum erit coni EAD, Sphæra verò eiusdem quadrupla erit, Quod erat propositum.

Corollarium.

Hinc patet sphæram subsesquialteram esse cylindri, cuius basis æqualis sit maximo sphæræ circulo, altitudo verò diametro sphæræ æqualis.

Nam sph. ostẽditur esse ad conum EAD vt 4, ád vnũ, conus verò EAD ad cylindrũ EBCD est vt vnũ ad 6. ergo ex æquo sphæra ad cylindrum EBCD erit vt 4. ad 6. Nempe subsesquialtera.

DE

N rectæ BQ, ſiue LE, ſiue LT; æqualis eſt, & reliqua NP reliquæ TF.)

Ergo omnes ſimul cylindri figuræ circa hemiſphęrium deſcriptæ, hoc eſt eadem figura, æqualis erit omnibus ſimul tubis cylindricis circa ſolidum trianguli EHA deſcriptis, cum ſinguli ſingulis æquales ſint. Sed figura circa hemiſphærium deſcripta minor erat ſolido trianguli EHA. Neceſſe igitur eſt quòd ſolidum trianguli EHA maius ſit, quam figura ſibi circumſcripta. pars ſuo toto. Quod eſſe non poteſt. Hemiſphærium igitur neque maius, neque minus erit ſolido trianguli EHA, ſed ipſi æquale, ſolidum verò trianguli EHA duplum eſt coni EAD, ergò hemiſphæriũ duplum erit coni EAD, Sphæra verò eiuſdem quadrupla erit, Quod erat propoſitum.

Corollarium.

Hinc patet ſphæram ſubſeſquialteram eſſe cylindri, cuius baſis æqualis ſit maxìmo ſphæræ circulo, altitudo verò diametro ſphæræ æqualis.

Nam ſph. oſtẽditur eſſe ad conum EAD vt 4, ad vnũ, conus vero EAD ad cylindrũ EBCD est vt vnũ ad 6. ergo ex æquo ſphæra ad cylindrum EBCD erit vt 4. ad 6. Nempe ſubſeſquialtera.

DE

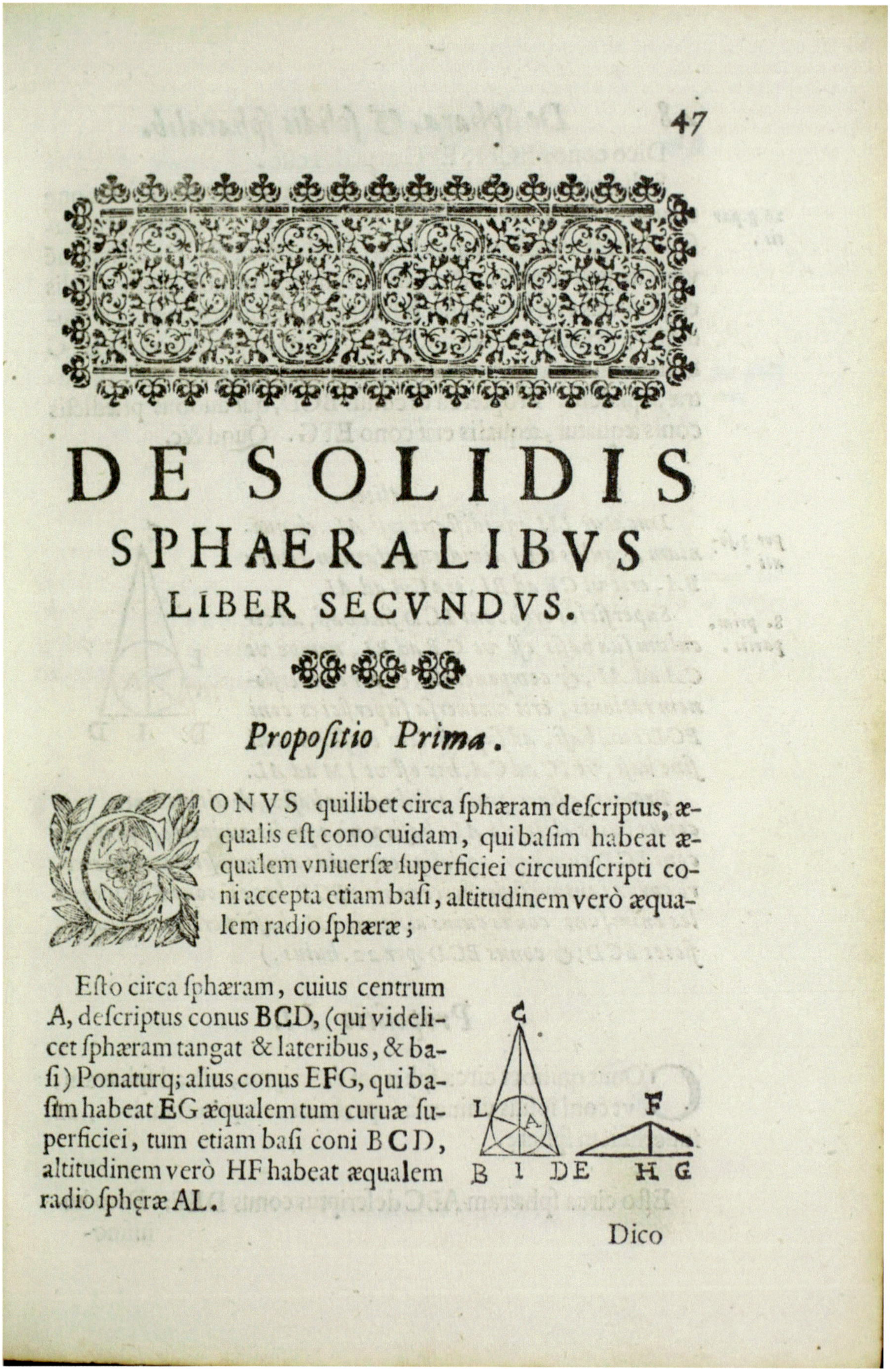

47

DE SOLIDIS SPHAERALIBVS LIBER SECVNDVS.

Propoſitio Prima.

CONVS quilibet circa ſphæram deſcriptus, æqualis eſt cono cuidam, qui baſim habeat æqualem vniuerſæ ſuperficiei circumſcripti coni accepta etiam baſi, altitudinem verò æqualem radio ſphæræ;

Eſto circa ſphæram, cuius centrum A, deſcriptus conus BCD, (qui videlicet ſphæram tangat & lateribus, & baſi) Ponaturq; alius conus EFG, qui baſim habeat EG æqualem tum curuæ ſuperficiei, tum etiam baſi coni BCD, altitudinem verò HF habeat æqualem radio ſphęræ AL.

Dico

47
(continued)

DE SOLIDIS SPHAERALIBVS LIBER SECVNDVS.

Propoſitio Prima.

CONVS quilibet circa ſphæram deſcriptus, æqualis eſt cono cuidam, qui baſim habeat æqualem vniuerſæ ſuperficiei circumſcripti coni accepta etiam baſi, altitudinem verò æqualem radio ſphæræ;

Eſto circa ſphæram, cuius centrum A, deſcriptus conus BCD, (qui videlicet ſphæram tangat & lateribus, & baſi) Ponaturq; alius conus EFG, qui baſim habeat EG æqualem tum curuæ ſuperficiei, tum etiam baſi coni BCD, altitudinem verò HF habeat æqualem radio ſphęræ AL.

Dico

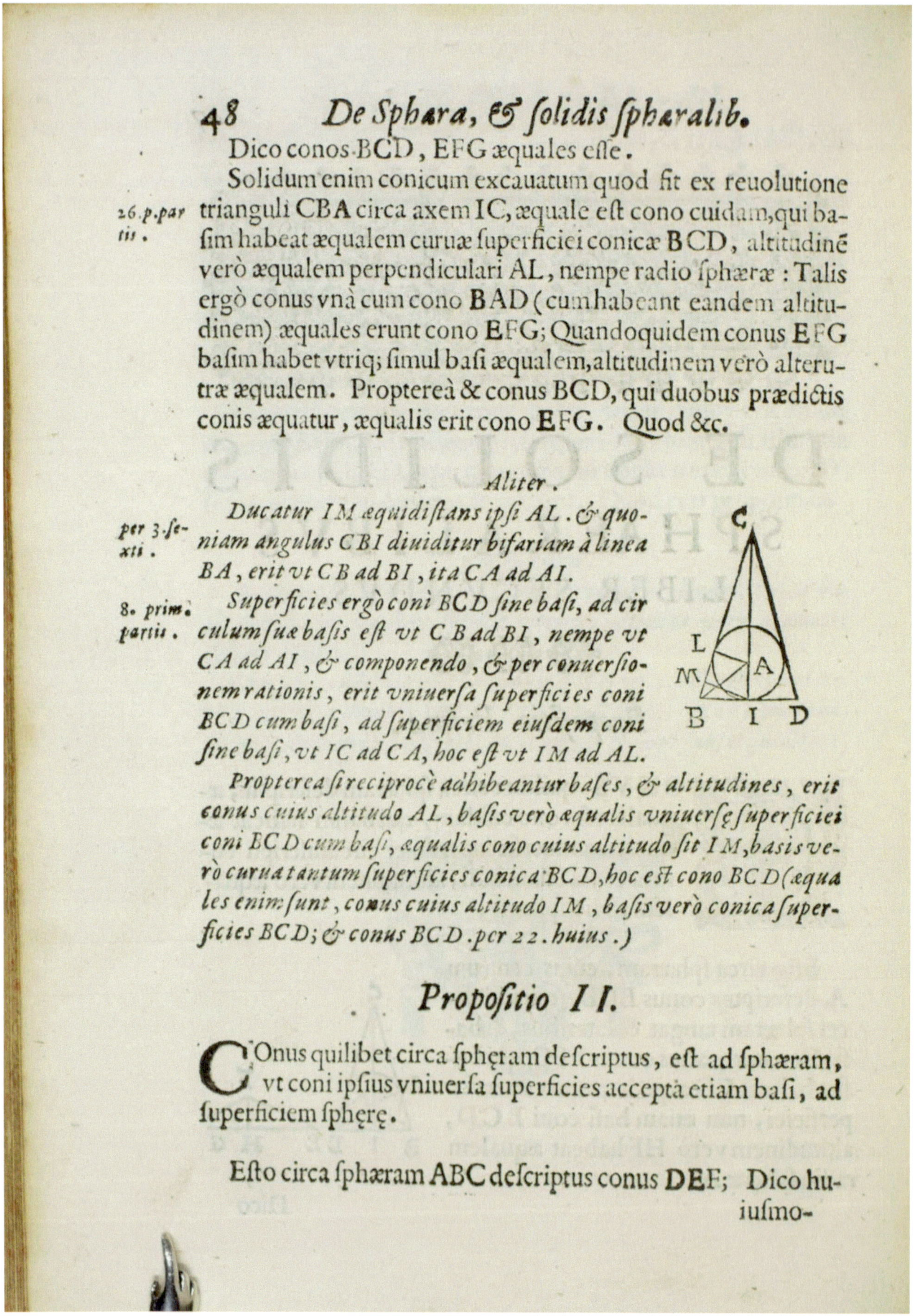

48 *De Sphæra, & ſolidis ſphæralib.*

Dico conos BCD, EFG æquales eſſe.

Solidum enim conicum excauatum quod fit ex reuolutione trianguli CBA circa axem IC, æquale eſt cono cuidam, qui baſim habeat æqualem curuæ ſuperficiei conicæ BCD, altitudinẽ verò æqualem perpendiculari AL, nempe radio ſphæræ: Talis ergò conus vnà cum cono BAD (cum habeant eandem altitudinem) æquales erunt cono EFG; Quandoquidem conus EFG baſim habet vtriq; ſimul baſi æqualem, altitudinem verò alterutræ æqualem. Propterea & conus BCD, qui duobus prædictis conis æquatur, æqualis erit cono EFG. Quod &c. *26.p.partis.*

Aliter.

Ducatur IM æquidiſtans ipſi AL. & quoniam angulus CBI diuiditur bifariam à linea BA, erit vt CB ad BI, ita CA ad AI. *per 3.ſexti.*

Superficies ergò coni BCD ſine baſi, ad circulum ſuæ baſis eſt vt CB ad BI, nempe vt CA ad AI, & componendo, & per conuerſionem rationis, erit vniuerſa ſuperficies coni BCD cum baſi, ad ſuperficiem eiuſdem coni ſine baſi, vt IC ad CA, hoc eſt vt IM ad AL. *8. primæ partis.*

Propterea ſi reciprocè adhibeantur baſes, & altitudines, erit conus cuius altitudo AL, baſis verò æqualis vniuerſę ſuperficiei coni BCD cum baſi, æqualis cono cuius altitudo ſit IM, baſis verò curua tantum ſuperficies conica BCD, hoc eſt cono BCD (æquales enim ſunt, conus cuius altitudo IM, baſis verò conica ſuperficies BCD; & conus BCD. per 22. huius.)

Propoſitio II.

Conus quilibet circa ſphęram deſcriptus, eſt ad ſphæram, vt coni ipſius vniuerſa ſuperficies acceptà etiam baſi, ad ſuperficiem ſphęrę.

Eſto circa ſphæram ABC deſcriptus conus DEF; Dico huiuſmo-

26.*p.par Tis* Dico conos BCD, EFG æquales eſſe.

Solidum enim conicum excauatum quod fit ex reuolutione trianguli CBA circa axem IC, æquale eſt cono cuidam, qui baſim habeat æqualem curuæ ſuperficiei conicæ BCD, altitudinẽ verò æqualem perpendiculari AL, nempe radio ſphæræ: Talis ergò conus vnà cum cono BAD (cum habeant eandem altitudinem) æquales erunt cono EFG; Quandoquidem conus EFG baſim habet vtriq; ſimul baſi æqualem, altitudinem verò alterutræ æqualem. Propterea & conus BCD, qui duobus prædictis conis æquatur, æqualis erit cono EFG. Quod &c.

Aliter

per 3.*ſe-xti* *Ducatur IM æquidiſtans ipſi AL. & quoniam angulus CBI diuiditur bifariam à linea BA, erit vt CB ad BI, ita CA ad AI.*

8.*prim partis* *Superficies ergò coni*[32] *BCD ſine baſi, ad circulum ſuæ baſis eſt vt CB ad BI, nempe vt CA ad AI, & componendo, & per conuerſionem rationis, erit vniuerſa ſuperficies coni BCD cum baſi, ad ſuperficiem eiuſdem coni ſine baſi, vt IC ad CA, hoc eſt vt IM ad A L.*

Propterea ſi reciprocè adhibeantur baſes, & altitudines, erit conus cuius altitudo AL, baſis verò æqualis vniuerſę ſuperficiei coni BCD cum baſi, æqualis cono cuius altitudo ſit IM, basis verò curua tantum ſuperficies conica BCD, hoc est cono BCD (æquales enim ſunt, conus cuius altitudo IM, baſis verò conica ſuperficies BCD; & conus BCD. per 22. huius.)

Propoſitio II

COnus quilibet circa ſphęram deſcriptus, eſt ad ſphæram, vt coni ipſius vniuerſa ſuperficies accepta etiam baſi, ad ſuperficiem ſphęrę.

Eſto circa ſphæram ABC deſcriptus conus DEF; Dico hu-

iusmo-

[32] *coni*] *coni.*

iuſmodi conum eſſe ad ſphæram, vt coni ſuperficies vna cum baſi, ad ſuperficiem ſphęrę.

Ponatur conus HIL vt in præcedēti, cuius baſis ęqualis ſit integro perimetro coni D E F vna cum baſi, altitudo verò P I ęqualis radio ſphęrę O C, eritq; conus HIL ęqualis cono DEF.

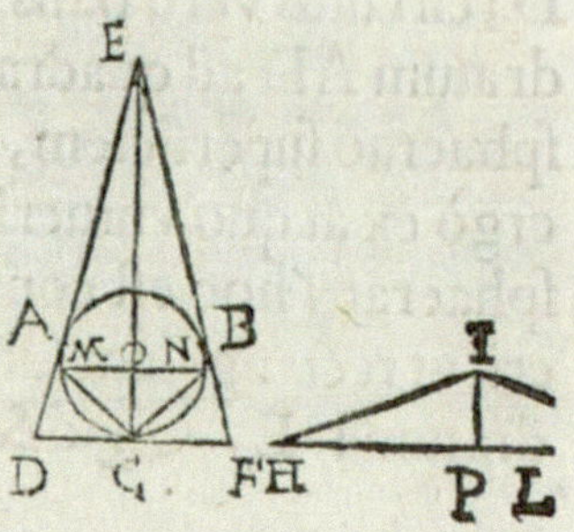

per præced.

Agatur per centrum O planum M N ad axem erectum, & in hemiſphęrio M CN concipiatur conus MCN.

Iam conus DEF ad conum HIL (ob ęqualitatem) eſt vt totus perimeter coni DEFD ad baſim HL, conus autem H I L ad conum MCN, (cum eandem habeant altitudinē) eſt vt baſis HL ad baſim MN, conus denique MCN ad ſphęrã, eſt vt baſis MN ad ſuperficiem ſphęrę (nempe in ratione ſub quadrupla) quare ex ęquo erit conus DEF ad ſphęram, vt vniuerſus perimeter coni DEF ad ſuperficiem ſphęrę. Quod &c.

20. et 30 p. partis

Propoſitio III.

Conus quilibet circa ſphęram deſcriptus, eſt ad ſphęram, vt rectangulum contentum ſub latere & ſemibaſi coni tamquàm vna linea, & ſub ſemibaſi, ad quadratũ diametri ſphęrę.

Eſto circa ſphęram, cuius diameter DE, deſcriptus conus quilibet A B C. Dico conum ad ſphęram eſſe vt rectangulum ſub BAD tamquàm vnà linea, & ſub AD compraehenſum, ad quadra-DE.

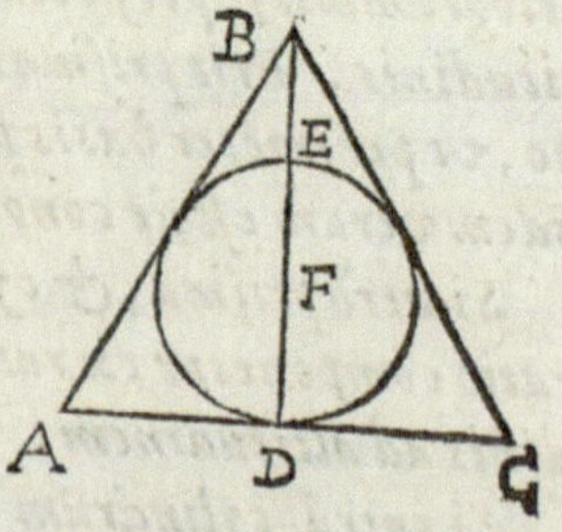

Curua enim ſuperficies coni A B C ad circulum ſuae baſis eſt vt BA ad A D, & componendo erit totus coni perimeter ad eundem circulum baſis vt BA, AD ſimul ad AD; hoc

8. p. pars

G eſt vt

iufmodi conum eſſe ad ſphæram, vt coni ſuperficies vna cum baſi, ad ſuperficiem ſphęrę.

Ponatur conus HIL vt in præcedẽti, cuius baſis ęqualis ſit integro perimetro coni DEF vna cum baſi, altitudo verò PI ęqualis radio ſphęrę OC, eritq; conus HIL ęqualis cono DEF.. *per prae ced.*

Agatur per centrum O planum M N ad axem erectum, & in hemiſphęrio MCN concipiatur conus MCN.

Iam conus DEF ad conum HIL (ob ęqualitatem) eft vt totus perimeter coni DEFD ad baſim HL, conus autem HIL ad conum MCN, (cum eandem habeant altitudinẽ) eft vt baſis HL ad baſim MN, conus denique MCN ad ſphęrã, eft vt baſis MN ad ſuperficiem ſphęrę (nempe in ratione ſub quadrupla) quare ex ęquo erit conus DEF ad ſphęram, vt vniuerſus perimeter coni DEF ad ſuperficiem ſphęrę. Quod &c. 20.*et* 30 *p. partis*

Propoſitio III.

COnus quilibet circa ſphęram deſcriptus, eft ad ſphęram, vt rectangulum contentum ſub latere & ſemibaſi coni tamquàm vna linea, & ſub ſemibaſi, ad quadratũ diametri ſphęrę.

Efto circa ſphęram, cuius diameter DE, deſcriptus conus quilibet ABC.

Dico conum ad ſphęram eſſe vt rectangulum ſub BAD tamquàm vnà linea, & ſub AD compraehenſum, ad quadra DE. 8.*p.pars*

Curua enim ſuperficies coni ABC ad circulum ſuae baſis eft vt BA ad AD, & componendo erit totus coni perimeter ad eundem circulum baſis vt BA, AD ſimul ad AD; hoc

G eſt vt

eſt vt rectangulum ſub linea BAD, & ſub AD ad quadratum A D; circulus verò baſis coni, ad circulum circa D E, eſt vt quadratum AD ad quadratum DF, circulus denique circa DE ad ſphaerae ſuperficiem, eſt vt quadratum DF ad quadratum DE, ergò ex aequo vniuerſus coni ABCA perimeter ad ſuperficiem ſphaerae (hoc eſt conus ipſe ad ſphaeram per praecedentem) erit vt rectangulum ſub recta linea BAD, & ſub AD, ad quadratum DE. Quod &c.

Corollarium.

Pro Corollario poteſt oſtendi conum æquilaterum ad inſcriptã ſphæram, eſſe vt 9. ad 4. Poſito enim latere A C. *6. erit rectangulum ſub latere cum ſemibaſi, & ſemibaſi 27. quadratum verò BD 27. & quadratum DE 12. ergo conus ad ſphæram erit vt 27 ad 12. ſiue vt 9: ad 4.*

Scholium.

Poſſent hic Theoremata non pauca proponi circa ſolidorum circumſcriptionem, & inſcriptionem: qualia ſunt.

Si circa ſphęram priſma concipiatur, quod ſingulis ſuis parallelogrammis ſphæram contingat; ſitque eiuſdem altitudinis, Erit priſma ad ſphæram, vt perimeter baſis priſmatis ad duas tertias periphæriç maximi circuli ſphæræ.

Si verò non eiuſdem ſit altitudinis; ratio priſmatis ad sphęram componetur ex prædicta, & ex ratione altitudinum; altitudo autem ſphęræ diameter eſt.

Si cylindro circumſcribatur priſma, quod ſingulis ſuis parallelogrammis ſuperficiem cylindri contingat; ſintq; eiuſdem altitudinis. Erit priſma ad cylindrum, vt baſis ad baſim: nempe, vt perimeter baſis priſmatis, ad periphæriam baſis cylindri: idem verum eſt de cono, & pyramidibus circumſcriptis.

Si verò priſma, & cylindrus non eiuſdem altitudinis fuerint; ratio componetur ex ratione perimetrì ad periphęriam, & altitudinis ad altitudinem.

Si intra cylindrum inſcribatur priſma eiuſdem altitudinis, habens baſim poligonam, regularem, & parilateram; Erit cylindrus ad priſma, vt periphęria baſis cylindri ad perimetrum poligoni

goni

eſt vt rectangulum ſub linea BAD, & ſub AD ad quadratum AD; circulus verò baſis coni, ad circulum circa DE, eſt vt quadratum AD ad quadratum DF, circulus denique circa DE ad ſphaerae ſuperficiem, eſt vt quadratum DF ad quadratum DE, ergò ex aequo vniuerſus coni ABCA perimeter ad ſuperficiem ſphaerae (hoc eſt conus ipſe ad ſphaeram per praecedentem) erit vt rectangulum ſub recta linea BAD, & ſub AD, ad quadratum DE. Quod &c.

Corollarium.

Pro Corollario poteſt oſtendi conum æquilaterum ad inſcriptã ſphæram, eſſe vt 9. ad 4. Poſito enim latere AC. 6. erit rectangulum ſub latere cum ſemibaſi, & ſemibaſi 27. quadratum verò BD 27. & quadratum DE 12. ergo conus ad ſphæram erit vt 27 ad 12. ſiue vt 9: ad 4.

Scholium.

Poſſent hic Theoremata non pauca proponi circa ſolidorum circumſcriptionem, & inſcriptionem: qualia ſunt.

Si circa ſphęram priſma concipiatur, quod ſingulis ſuis parallelogrammis ſphæram contingat; ſitque eiuſdem altitudinis, Erit priſma ad ſphæram, vt perimeter baſis priſmatis ad duas tertias periphærię maximi circuli ſphæræ.

Si verò non eiuſdem sit altitudinis; ratio priſmatis ad sphęram componetur ex prædicta, & ex ratione altitudinum; altitudo autem ſphęræ diameter eſt.

Si cylindro circumſcribatur priſma, quod singulis ſuis parallelogrammis ſuperficiem cylindri contingat; sintq; eiuſdem altitudinis. Erit priſma ad cylindrum, vt basis ad basim: nempe, vt perimeter basis priſmatis, ad periphæriam basis cylindri: idem verum eſt de cono, & pyramidibus circumſcriptis.

Si verò priſma, & cylindrus non eiuſdem altitudinis fuerint; ratio componetur ex ratione perimetrì ad periphęriam, & altitudinis ad altitudinem.

Si intra cylindrum inſcribatur priſma eiuſdem altitudinis, habens basim poligonam, regularem, & parilateram; Erit cylindrus ad priſma, vt periphęria basis cylindri ad perimetrum poli-

goni

goni regularis in eodem circulo descripti, quod habeat latera multitudine subdupla poligoni basis prismatis. Quæ uera sunt etiam de cono, & pyramidibus inscriptis.

Quando uerò basis prismatis parilatera fueris, siue regularis, siue irregularis: Erit cylindrus ad inscriptum prisma, ut peripheria basis cylindri ad omnes sinus arcuum à lateribus basis prismatis subtensorum. Dummodo nullus arcus semicirculo maior sit. Quando uerò arcus aliquis semicirculo maior sit; & quando figurarum altitudo non sit eadem, & alia huiusmodi, omnia demonstrari possunt facili quidem negotio; sed institutum nostrum est non omnem solidorum inscriptionem, & circumscriptionem prosequi; sed illam, tantum, quę circa sphęram est, uel intra ipsam; Pròpterea ad inceptum reuertamur.

Propositio IV.

SI circà circulum describatur poligonum habens latera numero paria, siue à quaternario, siue à binario mensurata, & reuoluatur figura circa diagonalem, erit factum sphaerale solidum aequale cono cuidam qui basim habeat aequalem superficiei solidi, altitudinem verò semidiametro sphaerae aequalē.

Hoc autem quandò numerus laterum mensuratur à quaternario demonstratum fuit ab Archimede Prop. 29. siue mauis 25. lib. p. de sph. & cylin. Quando verò laterū numerus etiam à binario tantum mensuratur, ostendemus sic, eritque demonstratio (exceptis quæ de vltimo solido cylindrico dicentur) eadem cum ea quam affert Archimedes,

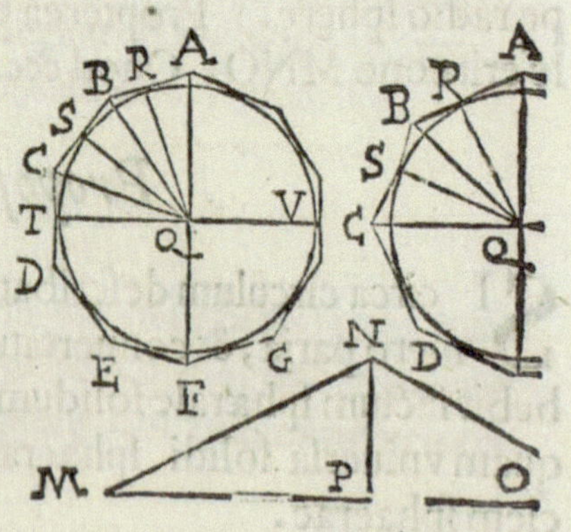

Esto poligonum ABCDEFG habens latera à binario tantum mensurata, vt in prima figura. Ergò semipoligonum AB

goni regularis in eodem circulo deſcripti, quod habe at latera multitudine ſubdupla poligoni basis priſmatis. Quæ ueræ ſunt etiam de cono, & pyramidibus inſcriptis.

Quando uerò basis priſmatis imparilatera fueris, siue regularis, siue irregularis: Erit cylindrus ad inſcriptum priſma, ut peripheria basis cylindri ad omnes sinus arcuum à lateribus basis priſmatis ſubtenſorum. Dummodo nullus arcus ſemicirculo maior sit. Quando uerò arcus aliquis ſemicirculo maior sit; & quando figurarum altitudo non sit eadem, & alia huiuſmodi, omnia demonstrari poſſunt facili quidem negotio; ſed inſtitutum noſtrum eſt non omnem ſolidorum inſcriptionem, & circumſcriptionem proſequi; ſed illam, tantum, quę circa ſphęram est, uel intra ipſam; Pròpterea ad inceptum reuertamur.

Propoſitio IV.[33]

SI circà circulum deſcribatur poligonum habens latera numero paria, ſiue à quaternario, ſiue à binario menſurata, & reuoluatur figura circa diagonalem, erit factum ſphaerale ſolidum aequale cono cuidam qui baſim habeat aequalem ſuperficiei ſolidi, altitudinem verò ſemidiametro ſphaerae aequalẽ.

Hoc autem quandò numerus laterum menſuratur à quaternario demonſtratum fuit ab Archimede Prop. 29. ſiue mauis 25. lib. p. de ſph. & cylin. Quando verò laterũ numerus etiam à binario tantum menſuratur, oſtendemus ſic, eritque demonſtratio (exceptis quæ de vltimo ſolido cylindrico dicentur) eadem cum ea quam affert Archimedes,

Eſto poligonum ABCDEFG habens latera à binario tantum menſurata, vt in prima figura. Ergò ſemipoligonum AB

[33] The same proof is developed by Archimedes (Archimedes 1897, *On the sphere and cylinder*, II, Proposition 25, p. 32). A difference can be remarked. Archimedes proved the case-study, 4–multipolygons, while Torricelli proved the case-study 2–multipolygons. (Cfr. Pisano and Bussotti's chapters above).

CDEF habebit latera numero imparia, latusque vnum tanget circulum in puncto T, atq; ideo cylindricam superficiem in cõuersione describet. Intelligatur conus M N O, cuius basis sit circulus MO æqualis vniuersæ superficiei solidi sphæralis, altitudo verò PN, æqualis sit radio sphæræ. Dico sphærale solidum æquale esse cono MNO.

Rombus enim solidus factus in conuersione figuræ à triangu
23.p.par tis. lo ABQ, æqualis est cono cuidam cuius basis æqualis sit conicæ superficiei descriptæ à linea AB, altitudo verò sit radius QR. Solidum autem excauatum factum in conuersione à triangulo
24.p.par tis. BCQ, æquatur cono cuidam cuius basis æqualis sit conicæ superficiei descriptæ à linea BC. altitudo verò æqualis radio sphęræ QS. & sic semper procedatur. Vltimum denique solidum
25.p.par tis. cylindricum excauatum descriptum à triangulo C T Q, æquale est cono cuidam, cuius basis æqualis sit superficiei cylindricæ à linea CT factæ, altitudo verò æqualis sit semidiametro cylindri, QT; Et sic de solidis circa alterum hemisphærium TFV descriptis. Ergo vniuersum sphærale solidum, æquale erit omnibus prædictis conis simul sumptis: ijsdem autem omnibus prædictis conis æqualis est conus MNO (cum basim habeat omnibus simul illorum basibus æqualem, nempe superficiei solidi sphęralis, altitudinem verò vnicuique illorum ęqualem, nempe radio sphęrę.) Propterea prędictum solidum sphærale ęquale erit cono MNO. Quod &c.

Propositio V.

SI circa circulum describatur poligonum habens latera numero paria, & conuertatur figura circa diagonalem: habebit factum sphærale solidum ad sphæram suam eam rationẽ, quam vniuersa solidi sphaeralis superficies habet ad superficiem sphaerae.

Manente praecedentis Propositionis constructione; Esto sphaerale solidum cuius diagonalis, atque axis sit AB, centrum autem sphaerae sit C. Dico sphaerale solidum ad inscriptam

sibi

CDEF habebit latera numero imparia, latuſque vnum tanget circulum in puncto T, atq; ideo cylindricam ſuperficiem in cõuerſione deſcribet. Intelligatur conus MNO, cuius baſis ſit circulus MO æqualis vniuerſæ ſuperficiei ſolidi ſphæralis, altitudo verò PN, æqualis ſit radio ſphæræ. Dico ſphærale ſolidum æquale eſſe cono MNO.

23.p.partis. Rombus enim ſolidus factus in conuerſione figuræ à triangulo ABQ, æqualis eſt cono cuidam cuius baſis æqualis ſit conicæ ſuperficiei deſcriptæ à linea AB, altitudo verò ſit radius QR.

Solidum autem excauatum factum in conuerſione à triangulo BCQ, æquatur cono cuidam cuius baſis æqualis ſit conicæ ſuperficiei *24.p.partis.* deſcriptæ à linea BC. altitudo verò æqualis radio ſphęræ QS. & ſic ſemper procedatur. Vltimum denique ſolidum cylindricum excauatum deſcriptum à triangulo CTQ, æquale eſt cono cuidam, cuius baſis æqualis ſit ſuperficiei cylindricæ à linea CT factæ, altitudo verò æqualis ſit ſemidiametro cylindri, QT; Et ſic de ſolidis circa alterum hemiſphærium TFV deſcriptis. Ergo vniuerſum *25.p.partis.* ſphærale ſolidum, æquale erit omnibus prædictis conis ſimul ſumptis: ijſdem autem omnibus prædictis conis æqualis eſt conus MNO (cum baſim habeat omnibus ſimul illorum baſibus æqualem, nempe ſuperficiei ſolidi ſphęralis, altitudinem verò vnicuique illorum ęqualem, nempe radio ſphęrę.) Propterea prędictum ſolidum ſphærale ęquale erit cono MNO. Quod &c.

Propoſitio V.

SI circa circulum deſcribatur poligonum habens latera numero paria, & conuertatur figura circa diagonalem: habebit factum ſphærale ſolidum ad ſphæram ſuam eam rationẽ, quam vniuerſa ſolidi ſphaeralis ſuperficies habet ad ſuperficiem ſphaerae.

Manente praecedentis Propoſitionis conſtructione; Eſto ſphaerale ſolidum cuius diagonalis, atque axis ſit AB, centrum autem ſphaerae ſit C. Dico ſphaerale ſolidum ad inſcriptam

sibi

ſibi ſphaeram eſ-ſe, vt ſuperficies ſolidi ad ſuperficiem ſphaerae.

Inſcribatur .n. in hemiſphaerio conus DEF,& ponatur conus GIH cuius baſis G H aequalis ſit vniuerſae ſuperficiei ſolidi ſphaeralis vt in praecedenti altitudo verò L I aequalis radio ſphaerae,& erit per praecedentem ſphaerale ſolidum aequale cono G I H.

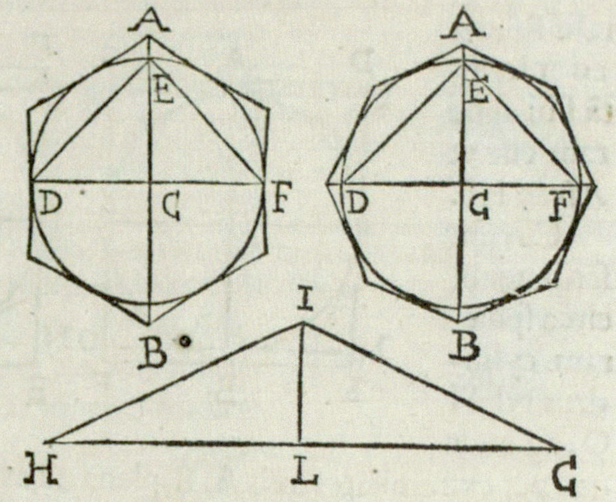

Propter aequalitatem ergò, erit ſphaerale ſolidum ad conũ GIH vt ſuperficies vniuerſa ſphaeralis ſolidi ad baſim coni G I H; conus autem GIH ad conum DEF (ob aequalem altitudinẽ) eſt vt baſis circa GH ad baſim circa DF; conus denique D E F ad ſphaeram, eſt vt baſis circa DF ad ſuperficiem ſphaerae (nẽpe in ratione ſubquadrupla.) Propterea erit ex aequo ſphaerale ſolidum ad inſcriptam ſibi ſphaeram vt vniuerſa ſphaeralis ſolidi ſuperficies ad ſuperficiem ſphaerae. Quod &c.

Propoſitio VI.

SI circa circulum deſcribatur poligonum habens latera numero paria, & conuertatur figura circa diagonalem, erit factum ſphaerale ſolidum ad inſcriptam ſibi ſphaeram vt axis ſolidi ad axem ſphaerae.

Manente praecedentium conſtructione; eſto ſphaerale ſolidum, cuius diagonalis, atque axis ſit AB, centrum verò ſphaeræ ſit C, & diameter HI.

Dico

ſibi ſphaeram eſſe, vt ſuperficies ſolidi ad ſuperficiem ſphaerae.

Inſcribatur.n. in hemiſphaerio conus DEF, & ponatur conus GIH cuius baſis GH aequalis ſit vniuerſae ſuperficiei ſolidi ſphaeralis vt in praecedenti altitudo verò LI aequalis radio ſphaerae, & erit per praecedentem ſphaerale ſolidum aequale cono GIH.

Propter aequalitatem ergò, erit ſphaerale ſolidum ad conũ GIH vt ſuperficies vniuerſa ſphaeralis ſolidi ad baſim coni GIH; conus autem GIH ad conum DEF (ob aequalem altitudinẽ) eſt vt baſis circa GH ad baſim circa DF; conus denique DEF ad ſphaeram, eſt vt baſis circa DF ad ſuperficiem ſphaerae (nẽpe in ratione ſubquadrupla.) Propterea erit ex aequo ſphaerale ſolidum ad inſcriptam ſibi ſphaeram vt vniuerſa ſphaeralis ſolidi ſuperficies ad ſuperficiem ſphaerae. Quod &c.

Propoſitio VI.

SI circa circulum deſcribatur poligonum habens latera numero paria, & conuertatur figura circa diagonalem, erit factum ſphaeràle ſolidum ad inſcriptam ſibi ſphaeram vt axis ſolidi ad axem ſphaerae.

Manente praecedentium conſtructione; eſto ſphaerale ſolidum, cuius diagonalis, atque axis ſit AB, centrum verò ſphaeræ ſit C, & diameter HI.

Dico

Dico sphę rale solidũ ad inscriptã sibi sphę ram esse vt AB ad HI. Circum scribatur n: circa sphæram cylindrus NLMO. agantur

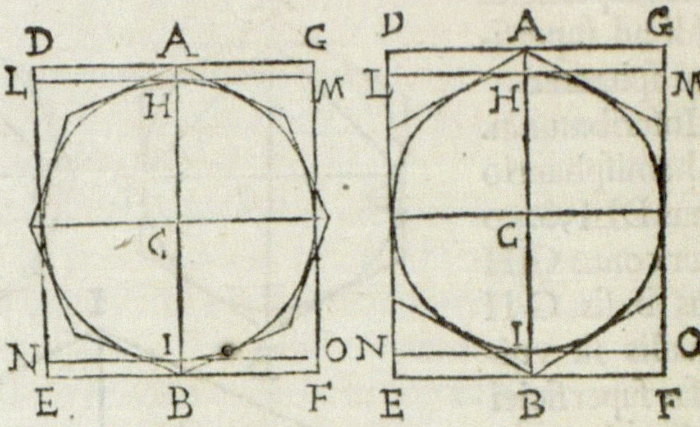

que per extremitates axis A, B, plana ad axem erecta DG, EF. per extremitates verò diametri HI. plana LM, NO.

Erit, per præcedentem, sphærale solidum ad sphæram vt superficies sphæralis solidi ad superficiem sphæræ; hoc est, (sumptis æqualibus) vt superficies cylindri DEFG, ad superficiem cylindri LNOM, hoc est vt AB ad HI. Quare sphærale solidum ad sphæram est vt axis solidi ad diametrum sphæræ. Quod &c.

per 15. p.partis. 18.p.partis. prima p. partis.

Propositio VII.

SI intra circulum describatur poligonum habens latera numero paria, & conuertatur figura circa diagonalem, erit sphæra ad inscriptum sibi sphærale solidum, vt quadratum diametri sphæræ, ad quadratum cateti poligoni.

Sit.n .circ. cuius cent. A, & diamet. BC poligonum regulare, cuius diagonalis sit linea BC, & conuertatur figura circa BC. Dico sphæram circumscriptam ad inclusum sphærale solidum, esse vt quadratum AC, ad quadratum cateti poligoni AD. Ducatur enim DE ex D. perpendicularis ad BC, & EF perpendicularis ad AD, eruntq; in continua proportione quatuor rectę AC, AD, AE, AF. Concipiatur etiam radio AD aliam sphæram describi,

4. sexti.

Dico ſphęrale ſolidũ ad inſcriptã ſibi ſphęram eſſe vt AB ad HI.

Circumſcribatur n: circa ſphæram cylindrus NLMO. agantur que per extremitates axis A, B, plana ad axem erecta DG, EF. per extremitates verò diametri HI. plana LM, NO.

per 15. *p:partis.* 18.*p.partis* *prima p. partis.*

Erit, per præcedentem, ſphærale ſolidum ad ſphæram vt ſuperficies ſphæralis ſolidi ad ſuperficiem ſphæræ; hoc eſt, (ſumptis æqualibus) vt ſuperficies cylindri DEFG, ad ſuperficiem cylindri LNOM, hoc eſt vt AB ad HI. Quare ſphærale ſolidum ad ſphæram eſt vt axis ſolidi ad diametrum ſphæræ. Quod &c.

Propoſitio VII.

SI intra circulum deſcribatur poligonum habens latera numero paria, & conuertatur figura circa diagonalem, erit ſphæra ad inſcriptum ſibi ſphærale ſolidum, vt quadratum diametri ſphæræ, ad quadratum cateti poligoni.

4.*fexsi.*[34]

Sit.n. circ. cuius cent. A, & diamet. BC poligonum regulare, cuius diagonalis ſit linea BC, & conuertatur figura circa BC. Dico ſphæram circumſcriptam ad incluſum ſphærale ſolidum, eſſe vt quadratum AC, ad quadratum cateti poligoni AD. Ducatur enim DE ex D. perpendicularis ad BC, & EF perpendicularis ad AD, eruntq; in continua proportione quatuor rectę AC, AD, AE, AF. Concipiatur etiam radio AD aliam ſphæram deſcri–
bi,

[34] *fexsi*] *fexti*.

bi, quæ ſingulas conicas ſuperficies ſolidi ſphæralis continget; necnon cylindricam, ſi quam huiuſmodi ſphærale ſolidum habebit.

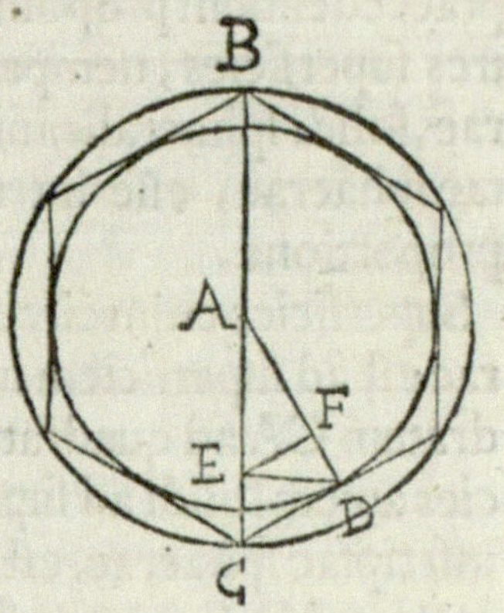

Erit itaque ſphæra maior ad ſphærã minorem, vt C A ad AF; minor verò ſphæra ad ſphęrale ſolidum, quod ſibi circumſcribitur (per præcedentem) eſt vt DA ad AC, hoc eſt, vt A F ad A E; Propterea ex æquo erit circumſcripta ſphæra maior, ad inſcriptum ſolidum ſphærale, vt CA ad AE; nempe vt quadratum CA ad quadratum AD. Quod erat &c. *vltimæ duodecimi.*

Propoſitio VIII.

SI circa ſphærale ſolidum, natum ex reuolutione poligoni circà diagonalem reuoluti, ſphæra circumſcribatur, & altera inſcribatur. Habebit circumſcripta ſphæra ad ſolidum, duplicatam rationem illius, quam habet ſolidum ad inſcriptam ſphæram.

Repetita figura Propoſitionis præcedentis; cum ſit circumſcripta ſphęra ad ſolidum vt quadratum CA ad quadratum AD; ſolidum verò ad inſcriptam ſibi minorem ſphæram, vt CA ad A D; patet rationem circumſcriptæ ſphæræ ad ſolidum ſphærale duplicatam eſſe illius quam ſolidum habet ad inſcriptam ſphæram. Quod &c. *6. huius.*

Propoſitio IX.

SI circa ſphærale ſolidum, natum ex reuolutione poligoni circà diagonalem reuoluti, ſphęra circumſcribatur, & altera inſcribatur: Erit ſuperficies ſolidi ſphęralis media proportionalis inter ſuperficies duarum ſphærarum.

Manan-

bi, quæ ſingulas conicas ſuperficies ſolidi ſphæralis continget; necnon cylindricam, ſi quam huiuſmodi ſphærale ſolidum habebit.

vltimae. duodeci-

Erit itaque ſphæra maior ad ſphærã minorem, vt CA ad AF; minor verò ſphæra ad ſphęrale ſolidum, quod ſibi circumſcribitur (per præcedentem) eſt vt DA ad AC, hoc eſt, vt AF ad AE;

mi.

Propterea ex æquo erit circumſcripta ſphæra maior, ad inſcriptum ſolidum ſphærale, vt CA ad AE; nempe vt quadratum CA ad quadratum AD. Quod erat &c.

Propoſitio VIII.

SI circa ſphærale ſolidum, natum ex reuolutione poligoni circà diagonalem reuoluti, ſphæra circumſcribatur, & altera inſcribatur. Habebit circumſcripta ſphæra ad ſolidum, duplicatam rationem illius, quam habet ſolidum ad inſcriptam ſphæram.

Repetita figura Propoſitionis præcedentis; cum ſit circumſcripta ſphęra ad ſolidum vt quadratum CA ad quadratum AD; ſolidum verò ad inſcriptam ſibi minorem ſphæram, vt CA ad AD; patet rationem circumſcriptæ ſphæræ ad ſolidum ſphærale duplicatam eſſe illius quam ſolidum habet ad inſcriptam ſphæram. Quod &c.

6.huius.

Propoſitio IX.

SI circa ſphærale ſolidum, natum ex reuolutione poligoni circà diàgonalem reuoluti, ſphęra circumſcribatur, & altera inſcribatur: Erit ſuperficies ſolidi ſphęralis media proportionalis inter ſuperficies duarum ſphaerarum.

Manan-

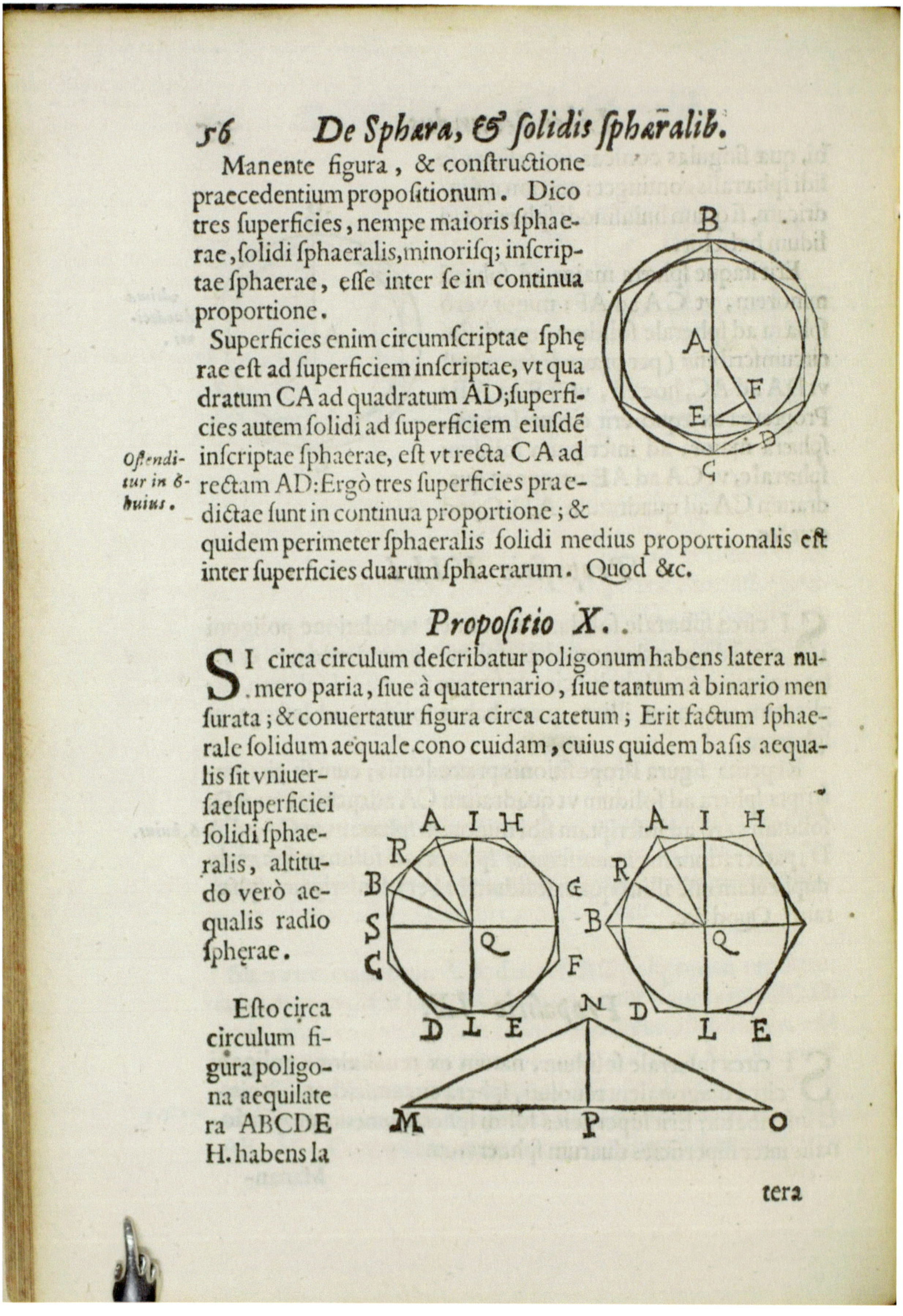

56 *De Sphæra, & ſolidis ſphæralib.*

Manente figura, & conſtructione praecedentium propoſitionum. Dico tres ſuperficies, nempe maioris ſphaerae, ſolidi ſphaeralis, minoriſq; inſcriptae ſphaerae, eſſe inter ſe in continua proportione.

Superficies enim circumſcriptae ſphærae eſt ad ſuperficiem inſcriptae, vt quadratum CA ad quadratum AD; ſuperficies autem ſolidi ad ſuperficiem eiuſdẽ inſcriptae ſphaerae, eſt vt recta CA ad rectam AD: Ergò tres ſuperficies praedictae ſunt in continua proportione; & quidem perimeter ſphaeralis ſolidi medius proportionalis eſt inter ſuperficies duarum ſphaerarum. Quod &c.

Oſtenditur in 8. huius.

Propoſitio X.

SI circa circulum deſcribatur poligonum habens latera numero paria, ſiue à quaternario, ſiue tantum à binario menſurata; & conuertatur figura circa catetum; Erit factum ſphaerale ſolidum aequale cono cuidam, cuius quidem baſis aequalis ſit vniuerſae ſuperficiei ſolidi ſphaeralis, altitudo verò aequalis radio ſphærae.

Eſto circa circulum figura poligona aequilatera ABCDEH. habens la

tera

Manente figura, & conſtructione praecedentium propoſitionum. Dico tres ſuperficies, nempe maioris ſphaerae, ſolidi ſphaeralis, minoriſq; inſcriptae ſphaerae, eſſe inter ſe in continua proportione.

Ostenditur in 6-huius.

Superficies enim circumſcriptae ſphęrae eſt ad ſuperficiem inſcriptae, vt quadratum A ad quadratum AD; ſuperficies autem ſolidi ad ſuperficiem eiuſdẽ inſcriptae ſphaerae, eſt vt recta CA ad rectam AD: Ergò tres ſuperficies praedictae ſunt in continua proportione; & quidem perimeter ſphaeralis ſolidi medius proportionalis eſt inter ſuperficies duarum ſphaerarum. Quod &c.

Propoſitio X.

SI circa circulum deſcribatur poligonum habens latera numero paria, ſiue à quaternario, ſiue tantum à binario menſurata; & conuertatur figura circa catetum; Erit factum ſphaerale ſolidum aequale cono cuidam, cuius quidem baſis aequalis ſit vniuerſae ſuperficiei ſolidi ſphaeralis, altitudo verò aequalis radio ſphęrae.

Eſto circa circulum figura poligona aequilatera ABCDEH. habens la

tera

tera numero paria, & conuertatur figura circa catetum IL, oriet urq; solidum contentum sub conicis, circularibus, & vna cylindrica superficie, quando numerus laterum à quaternario mẽsuratur; quandò verò à binario tantum, tunc erit solidum sphærale sub conicis, & circularibus tantum superficiebus compræhensum. Dico vtrumq; sphærale solidum æquale esse cono cuidam MNO, qui basim habeat æqualem vniuersæ solidi sphæralis superficiei, altitudinem verò PN æqualem radio sphæræ.

Hoc ostendetur similiter vt propositione 4. factum est. Nam conus qui fit à triangulo IAQ in conuersione circa axem IL, æquatur cono qui basim habeat ęqualem circulo qui fit ex radio IA, altitudinem verò aequalem radio sphæræ QI, quia idem prorsus est. Solidum autem excauatum, quod fit à triangulo ABQ, æquale probatur cono cuidam, cuius basis æqualis sit conicæ superficiei factæ à linea AB, altitudo verò sit QR. radius sphæræ. Vltimum denique cylindricum solidum excauatum, factum à triangulo BQS (quando poligoni latera à quaternario mensurantur, aliàs cylindricum solidum nullum est) æquatur cono cuius basis ęqualis sit cylindricę superficiei factæ à linea BS. altitudo verò sit QS; & sic de altero hemisphęrio. Propterea vniuersum sphærale solidum ęquale erit omnibus praedictis conis simul sumptis; & ideo aequale erit etiam cono MNO, qui omnibus illis simul sumptis aequiualet; (quandoquidem basim habet omnibus simul illorum basibus aequalem ex suppositione; altitudinem verò vnicuique illorum aequalem, nempe radium sphaerae.) Quod &c.

[Margin notes: 34. p. pātis. — 35. p. pātis.]

Propositio XI.

SI circa circulum describatur poligonum habens latera numero paria, & conuertatur figura circa catetum, habebit factum sphaerale solidum ad inscriptam sibi sphaeram eam rationem, quam vniuersa solidi sphaeralis superficies habet ad superficiem sphaerae.

Manente praecedentis propositionis cõstructione, esto sphae-

H rale

tera numero paria, & conuertatur figura circa catetum IL, orieturq; ſolidum contentum ſub conicis, circularibus, & vna cylindrica ſuperficie, quando numerus laterum à quaternario mẽſuratur; quandò verò à binario tantum, tunc erit ſolidum ſphærale ſub conicis, & circularibus tantum ſuperficiebus compræhenſum. Dico vtrumq; ſphærale ſolidum æquale eſſe cono cuidam MNO, qui baſim habeat æqualem vniuerſæ ſolidi ſphæralis ſuperficiei, altitudinem verò PN æqualem radio ſphæræ.

Hoc oſtendetur ſimiliter vt propoſitione 4. factum eſt. Nam conus qui fit à triangulo IAQ in conuerſione circa axem IL, æquatur cono qui baſim habeat ęqualem circulo qui fit ex radio IA, altitudinem verò aequalem radio ſphæræ QI, quia idem prorſus eſt. Solidum 24.
autem excauatum, quod fit à triangulo ABQ, æquale probatur cono cuidam, cuius baſis æqualis ſit conicæ ſuperficiei factæ à linea AB, *p.par.*
altitudo verò ſit QR. radius ſphæræ. Vltimum denique cylindricum ſolidum excauatum, factum à triangulo BQS (quando poligoni latera *tis.*
à quaternario menſurantur, aliàs cylindricum ſolidum nullum eſt) æquatur cono cuius baſis ęqualis ſit cylindricę ſuperficiei factæ à linea BS. altitudo verò ſit QS; & ſic de altero hemiſphęrio. 25.
Propterea vniuerſum ſphærale ſolidum ęquale erit omnibus praedictis conis ſimul ſumptis; & ideo aequale erit etiam cono MNO, qui *p.par.*
omnibus illis ſimul ſumptis aequiualet; (quandoquidem baſim habet omnibus ſimul illorum baſibus aequalem ex ſuppoſitione; *tis.*
altitudinem verò vnicuique illorum aequalem, nempe radium ſphaerae.) Quod &c.

Propoſitio XI.

SI circa circulum deſcribatur poligonum habens latera numero paria, & conuertatur figura circa catetum, habebit factum ſphaerale ſolidum ad inſcriptam ſibi ſphaeram eam rationem, quam vniuerſa ſolidi ſphaeralis ſuperficies habet ad ſuperficiem ſphaerae.

Manente praecedentis propoſitionis cõſtructione, eſto ſphae-

H rale

rale ſolidũ cuius catetus, & axis ſit AB; centrum autem ſphaerae ſit C. Dico ſphaerale ſolidum ad inſcriptam ſibi ſphaeram eſſe vt vniuerſa ipſius ſolidi ſuperficies ad ſuperficiem ſphaerae.

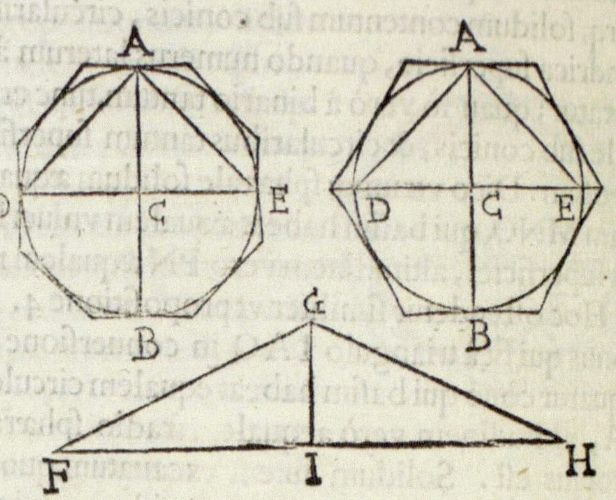

Concipiatur enim in hemiſphaerio conus DAE,& intelligatur alius conus FGH, cuius baſis FH aequalis ſit vniuerſae ſuperficiei ſolidi ſphaeralis,altitudo verò IG ęqualis radio ſphaerae; & erit per praecedentem ſphaerale ſolidum aequale cono FGH.

Propter aequalitatem ergo, erit ſphaerale ſolidum ad conũ FGH, vt ſuperficies vniuerſa ſphæralis ſolidi, ad baſim coni FGH, conus autem FGH, ad conum DAE (ob æqualem altitudinem) eſt vt baſis circa FH, ad baſim circa DE; denique conus DAE, ad ſphæram, eſt vt baſis circa DE ad ſuperficiem ſphærae (nẽpe in ratione ſubquadrupla:)Propterea erit ex ęquo ſphærale ſolidum ad inſcriptam ſibi ſphæram,vt vniuerſa ſphæralis ſolidi ſuperficies ad ſuperficiem ſphæræ. Quod &c.

Propoſitio XII.

SI circa circulum deſcribatur poligonum habens latera numero paria, & conuertatur figura circa catetum; Habebit factum ſphærale ſolidum ad inſcriptam ſibi ſphæram, eam rationem, quam habet compoſita recta linea ex diametro ſphæræ, & ex tertia proportionali (ſi fiat vt ſemidiameter ſphæræ ad ſemi-

rale ſolidũ cuius catetus, & axis ſit AB; centrum autem ſphaerae ſit C. Dico ſphaerale ſolidum ad inſcriptam ſibi ſphaeram eſſe vt vniuerſa ipſius ſolidi ſuperficies ad ſuperficiem ſphaerae.

Concipiatur enim in hemiſphaerio conus DAE, & intelligatur alius conus FGH, cuius baſis FH aequalis ſit vniuerſae ſuperficiei ſolidi ſphaeralis, altitudo verò IG ęqualis radio ſphaerae; & erit per praecedentem ſphaerale ſolidum aequale cono FG H.

Propter aequalitatem ergo, erit ſphaerale ſolidum ad conũ FGH, vt ſuperficies vniuerſa ſphæralis ſolidi, ad baſim coni FGH, conus autem FGH, ad conum DAE (ob æqualem altitudinem) eſt vt baſis circa FH, ad baſim circa DE; denique conus DAE, ad ſphæram, eſt vt baſis circa DE ad ſuperficiem ſphærae (nẽpe in ratione ſubquadrupla:) Propterea erit ex ęquo ſphærale ſolidum ad inſcriptam ſibi ſphæram, vt vniuerſa ſphæralis ſolidi ſuperficies ad ſuperficiem ſphæræ. Quod &c.

Propoſitio XII.

SI circa circulum deſcribatur poligonum habens latera numero paria, & conuertatur figura circa catetum;

Habebit factum ſphærale ſolidum ad inſcriptam ſibi ſphæram, eam rationem, quam habet compoſita recta linea ex diametro ſphæræ, & ex tertia proportionali (ſi fiat vt ſemidiameter ſphæræ ad

ſemi-

ſemilatus poligoni, ita ſemilatus ad aliam,) ad diametrum ſphę-ræ.

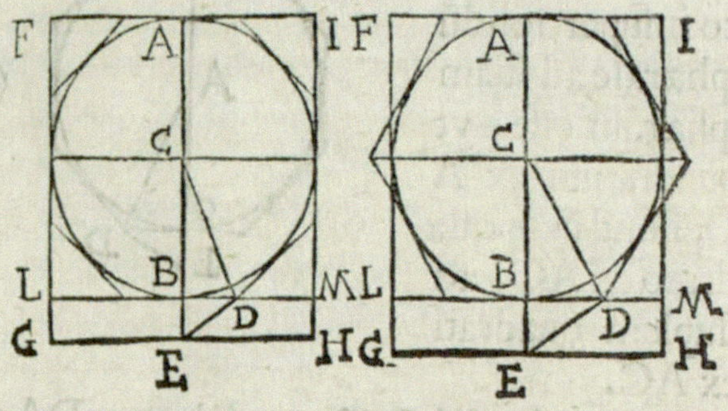

Manente præ dentium propoſitionum cōſtructione, eſto ſphęrale ſolidū cuius catetus, & axis ſit AB; centrum autem ſphærę ſit C. Fiat angulus CDE rectus, eritq; BE tertia proportionalis ad ſemidiametrum CB, & ſemilatus poligoni B D. Dico ſphærale ſolidum ad inſcriptã ſibi ſphærã eſſe vt EA ad AB; nempe vt compoſita ex diametro ſphæræ AB, & tertia proportionali BE, ad diametrum ſphæræ AB. Concipiatur circa ſphæram deſcriptus cylindrus FLMI, & per puncta A; B; E. producantur plana FI, LM, GH, ad axem erecta.

Erit ergo, per præcedentem, ſphærale ſolidum ad inſcriptã ſibi ſphæram, vt ſuperficies ſolidi ad ſuperficiem ſphæræ; hoc eſt, ſumptis æqualibus, vt ſuperficies cylindri FGHI. ad ſuperficiem cylindri FLMI; hoc eſt vt linea AE ad AB. Quod &c. 16. pri. partis. 18. p. partis.

Propoſitio XIII.

SI circà circulum deſcribatur poligonum habens latera numero paria, & conuertatur figura circa catetum; erit factũ ſphęrale ſolidum ad ſuam ſphæram, vt duo quadrata, nempe vt quadratum diagonalis, & quadratum cateti ſimul, ad duplum quadrati eiuſdem cateti.

Eſto circa circulum, cuius centrum A, deſcriptum poligonum habens latera numero paria, & conuertatur figura circa catetum BC: factoq; angulo recto ADE, erit (per præcedentem)

ſemilatus poligoni, ita ſemilatus ad aliam,) ad diametrum ſphęræ.

Manente prædentium propoſitionum cōſtructione, eſto ſphęrale ſolidũ cuius catetus, & axis ſit AB; centrum autem ſphærę ſit C. Fiat angulus CDE rectus, eritq; BE tertia proportionalis ad ſemidiametrum CB, & ſemilatus poligoni BD. Dico ſphærale ſolidum ad inſcriptã ſibi ſphærã eſſe vt EA ad AB; nempe vt compoſita ex diametro ſphæræ AB, & tertia proportionali BE, ad diametrum ſphæræ AB. Concipiatur circa ſphæram deſcriptus cylindrus FLMI, & per puncta A; B; E. producantur plana FI, LM, GH, ad axem erecta. *16.pre. partis.*

Erit ergo, per præcedentem, ſphærale ſolidum ad inſcriptã ſibi ſphæram, vt ſuperficies ſolidi ad ſuperficiem ſphæræ; hoc eſt, ſumptis æqualibus, vt ſuperficies cylindri FGHI. ad ſuperficiem cylindri FLMI; hoc eſt vt linea AE ad AB. Quod &c. *18.p. p.par. tis.*

Propoſitio XIII.

SI circà circulum deſcribatur poligonum habens latera numero paria, & conuertatur figura circa catetum; erit factũ ſphęrale ſolidum ad ſuam ſphæram, vt duo quadrata, nempe vt quadratum diagonalis, & quadratum cateti ſimul, ad duplum quadrati eiuſdem cateti.

Eſto circa circulum, cuius centrum A, deſcriptum poligonum habens latera numero paria, & conuertatur figura circa catetum BC: factoq; angulo recto ADE, erit *(*per præcedentem*)*.

H 2 foli-

60 *De Sphæra, & ſolidis ſphæralib.*

ſolidum ſphærale ad ſuam ſphæram vt EB ad BC. Dico inſuper ſolidũ ſphærale ad ſuam ſphæram eſſe, vt quadratum ex AD, ſimul cum quadrato ex AC, ad duplum quadrati ex AC.

Nam EA ad AC eſt vt quadratum DA ad quadratum AC; & componendo, erunt EA, & AC ſimul, hoc eſt tota EB, ad AC, vt duo quadrata DA, AC ſimul ad quadratum AC; ſumptiſque conſequentium duplis, erit EB ad BC (hoc eſt ſolidum ſphærale ad ſphæram) vt duo quadrata DA, AC ſimul, ad duplum quadrati ex AC. Quod &c.

per præced.

Propoſitio XIV.

SI intrà circulum deſcribatur poligonum habens latera numero paria, & connertatur figura circà catetum; erit ſphæra ad inſcriptum ſibi ſolidum, vt integra diameter ſphęrę, ad ſecundam ſimul, & quartam proportionalium, in ratione ſemidiametri ſphaerae ad ſemicatetum poligoni.

Sit in circulo cuius diameter AB poligonum habens latera numero paria, & conuertatur figura circa catetũ CD: Ducanturque perpendiculares DF ad rectã HE, & FI ad HD; & erunt quatuor lineae EH, HD, HF, HI, in continua ratione ſemidiametri HE ad ſemicatetum HD. Dico ſphaerã ad inſcriptum ſolidum eſſe, vt dupla HE ad v-

tramq;

60 *De Sphæhra, & ſolidis ſphæralib.*

ſolidum ſphærale ad ſuam ſphæram vt EB ad BC. Dico inſuper ſolidũ ſphærale ad ſuam ſphæram eſſe, vt quadratum ex AD, ſimul cum quadrato ex AC, ad duplum quadrati ex AC.

Nam EA ad AC eſt vt quadratum DA ad quadratum AC; & componendo, erunt EA, & AC ſimul, hoc eſt tota EB, ad AC, vt duo quadrata DA, AC ſimul ad quadratum AC; ſumptiſque conſequentium duplis, erit EB ad BC (hoc eſt ſolidum ſphærale ad ſphæram) vt duo quadrata DA, AC ſimul, ad duplum quadrati ex AC. Quod &c.

per prę-ced.

Propoſitio XIV.

SI intrà circulum deſcribatur poligonum habens latera numero paria, & connertatur[35] figura circà catetum; erit ſphæra ad inſcriptum ſibi ſolidum, vt integra diameter ſphęrę, ad ſecundam ſimul, & quartam proportionalium, in ratione ſemidiametri ſphaerae ad ſemicatetum poligoni.

Sit in circulo cuius diameter AB poligonum habens latera numero paria, & conuertatur figura circa catetũ CD: Ducanturque perpendiculares DF ad rectã HE, & FI ad HD; & erunt quatuor lineae EH, HD, HF, HI, in continua ratione ſemidiametri HE ad ſemicatetum HD. Dico ſphaerã ad inſcriptum ſolidum eſſe, vt dupla HE ad v-

tramq;

[35] connertatur] convertatur] conuertatur.

tramq; ſimul DH, HI. Vel vt integra diameter ſphæræ ad CI.

Intelligatur alia ſphæra intra ſolidum inſcripta. Erit ergo exterior ſphæra ad interiorem, vt EH ad HI, ſiue vt dupla EH ad duplam HI; interior verò ſphæra ad ſolidum ſphærale eſt, vt duo quadrata ex HD, ad duo quadrata HD, HE, hoc eſt vt duo quadrata ex HI, ad duo quadrata ex HI, HF, hoc eſt (vt infrà oſtendemus) vt dupla HI ad HI, HD; Propterea erit ex æquo ſphæra exterior ad inſcriptum ſibi ſphærale ſolidum, vt dupla HE, hoc eſt integra diameter ſphæræ, ad HI, & HD ſimul; quæ quidem ſunt ſecunda, & quarta in ratione ſemidiam. ſphęræ ad ſemicatetum poligoni. Quod &c. *Vltima duodecimi.* *per præc.*

Quod autem aſſumptum eſt oſtendemus. Dico vt duo quadrata ex HI ad duo quadrata ſimul HI, HF. ita eſſe duplam HI ad HI, HD.

Nam ob angulum rectum HFD, erit vt quadratum FH ad quadratum HI, ita recta DH ad HI, & componendo, ſumptiſque conſequentium duplis, erit vt quadrata FH, HI, ad duo quadrata ex HI, ita duæ rectæ DH, HI, ad duplam HI. Conuertendo ergò, erunt duo quadrata ex HI, ad duo quadrata HI, HF. vt dupla HI, ad HI, HD ſimul. Quod erat &c.

Propoſitio XV.

SI circà circulum deſcribatur poligonum habens latera numero imparia, & conuertatur figura circa catetum poligoni, erit factum ſphærale ſolidum æquale cono cuidam, cuius baſis æqualis ſit vniuerſæ ſuperficiei ſolidi, altitudo verò radio ſphæræ ſit æqualis.

Eſto circuli centrum A, polig. verò perimeter BCDEFGH. Et ſint latera eius numero imparia; conuertaturque figura circa catetum BI, vt oriatur ſolidum ſphærale contentum ſub conicis ſuperficiebus vnicoque circulo circa diametrum EF deſcripto. Ponatur iam conus LMN, qui baſim habeat æqualem vniuerſæ ſuperficiei ſphæralis ſolidi, altitudinem verò OM æqualem radio

tramq; ſimul DH, HI. Vel vt integra diameter ſphæræ ad CI

Intelligatur alia ſphæra intra ſolidum inſcripta. Erit ergo exterior ſphæra ad interiorem, vt EH ad HI, ſiue vt dupla EH ad duplam HI; interior verò ſphæra ad ſolidum ſphærale eſt, vt duo quadrata ex HD, ad duo quadrata HD, HE, hoc eſt vt duo quadrata ex HI, ad duo quadrata ex HI, HF, hoc eſt (vt infrà oſtendemus) vt dupla HI ad HI, HD; Propterea erit ex æquo ſphæra exterior ad inſcriptum ſibi ſphærale ſolidum, vt dupla HE, hoc eſt integra diameter ſphæræ, ad HI, & HD ſimul; quæ quidem ſunt ſecunda, & quarta in ratione ſemidiam. ſphęræ ad ſemicatetum poligoni. Quod &c. *Vltima. duodecimi.*

Quod autem aſſumptum est oſtendemus. Dico vt duo quadrata ex HI ad duo quadrata ſimul HI, HF. ita eſſe duplam HI ad HI, HD. *per pręc.*

Nam ob angulum rectum HFD, erit vt quadratum FH ad quadratum HI, ita recta DH ad HI, & componendo, ſumptiſque conſequentium duplis, erit vt quadrata FH, HI, ad duo quadrata ex HI, ita duæ rectæ DH, HI, ad duplam HI. Conuertendo ergò, erunt duo quadrata ex HI, ad duo quadrata HI, HF. vt dupla HI, ad HI, HD ſimul. Quod erat &c.

Propoſitio XV.

SI circà circulum deſcribatur poligonum habens latera numero imparia, & conuertatur figura circa catetum poligoni, erit factum ſphærale ſolidum æquale cono cuidam, cuius baſis æqualis ſit vniuerſæ ſuperficiei ſolidi, altitudo verò radio ſphæræ ſit æqualis.

Eſto circuli centrum A, polig. verò perimeter BCDEFGH.

Et ſint latera eius numero imparia; conuertaturque figura circa catetum BI, vt oriatur ſolidum ſphærale contentum ſub conicis ſuperficiebus vnicoque circulo circa diametrum EF deſcripto.

Ponatur iam conus LMN, qui baſim habeat æqualem vniuerſæ ſuperficiei ſphæralis ſolidi, altitudinem verò OM æqualem

radio

radio sphæræ AI. Dico solidum sphærale æquale esse cono LMN.

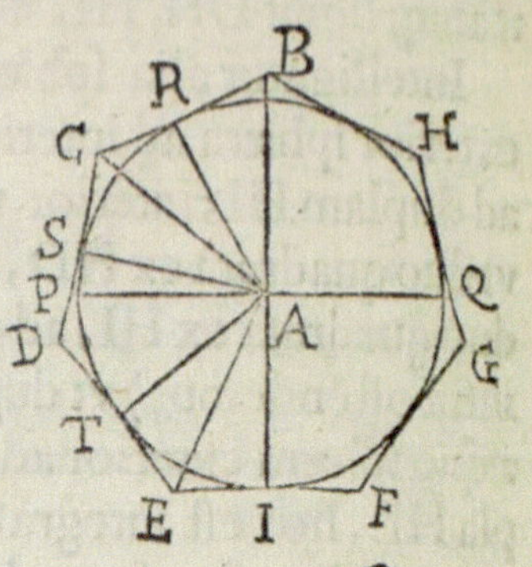

Agatur per centrum sphærę planum PQ ad axem erectum, quod transuersè, secabit aliquod latus poligoni, puta CD.

Erit iam rombus solidus, factus à conuersione triang. BCA, æqualis cono, qui basim habeat æqualem conicæ superficiei factæ à linea BC; altitudinem autem ęqualem radio sphęrę AR. Solidum verò conicū excauatum quod fit ex gyro trianguli CPA, æquale erit cono, qui basim habeat ęqualem superficiei, quę fit à linea CP altitudinem verò ęqualem radio sphęrę AS. Solidum quoque excauatum, factum ex reuolutione trianguli PDA, ęquatur cono, qui basim habeat ęqualem superficiei conicę quę fit à motu lineę PD, altitudinem autem ęqualem radio sphęrę AS. Eadem prorsum eodem modo dicuntur de solido conico excauato, facto à triangulo DAE; & de vltimo cono facto à reuolutione trianguli EIA. Propterea totum sphęrale solidum ęquale erit omnibus prędictis conis simul sumptis, vel cono L M N, qui omnibus illis prędictis ęquiualet: (habet enim basim omnibus simul illorum basibus ęqualem, altitudinem verò ęqualem vnicuiq; illorum.) Quod &c.

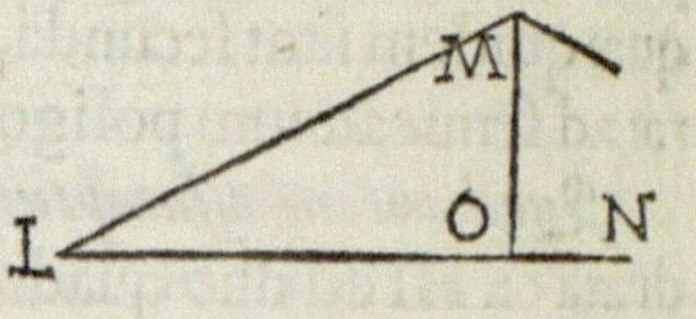

23.p.partis. *24.p.partis.* *27.p.partis.*

Scholium.

Attulimus in hac Propositione Theor. 23. 24. & 27. p. partis; Nam ex gyro trianguli BCA oritur rombus solidus vt in 23. p. partis. Ex gyro trianguli CPA oritur solidum quoddam excauatū, quale relinquitur si ex cono auferatur rombus solidus: vt in 24. p. partis. Denique ex conuersione trianguli DPA oritur solidum quoddam excauatum habens basim circularem PQ: quale relinquitur si ex frusto conico conus auferatur habens basim eandem

cum

radio ſphæræ AI. Dico ſolidum ſphærale æquale eſſe cono LMN.

Agatur per centrum ſphærę planum PQ ad axem erectum, quod tranſuersè, ſecabit aliquod latus poligoni, puta CD.

23.*p.par.* *tis.* Erit iam rombus ſolidus, factus à conuerſione triang. BCA, æqualis cono, qui baſim habeat æqualem conicæ ſuperficiei factæ à linea BC; altitudinem autem ęqualem radio ſphęrę AR. Solidum verò conicũ excauatum quod fit ex gyro trianguli CPA, æquale erit cono, qui baſim habeat ęqualem ſuperficiei, quę fit à linea CP altitudinem verò ęqualem radio ſphęrę AS. Solidum quoque excauatum, 24.*p.par.* *tis.* factum ex reuolutione trianguli PDA, ęquatur cono, qui baſim habeat ęqualem ſuperficiei conicę quę fit à motu lineę PD, altitudinem autem ęqualem radio ſphęrę AS. Eadem prorſum eodem modo dicuntur de ſolido conico excauato, facto à triangulo DAE; & de vltimo cono facto à reuolutione trianguli EIA. Propterea totum ſphęrale ſolidum ęquale erit omnibus prędictis conis ſimul ſumptis, vel 27.*p.par.* *tis.* cono LMN, qui omnibus illis prędictis ęquiualet: (habet enim baſim omnibus ſimul illorum baſibus ęqualem, altitudinem verò ęqualem vnicuiq; illorum.) Quod &c.

Scholium.

Attulimus in hac Propoſitione Theor. 23. 24. & 27. p. partis;

Nam ex gyro trianguli BCA oritur rombus ſolidus vt in 23. p. partis. Ex gyro trianguli CPA oritur ſolidum quoddam excauatũ, quale relinquitur ſi ex cono auferatur rombus ſolidus: vt in 24. p. partis. Denique ex conuerſione trianguli DPA oritur ſolidum quoddam excauatum habens baſim circularem PQ: quale relinquitur ſi ex fruſto conico conus auferatur habens baſim eandem

cum

cum maiore basi frusti conici, altitudinem quoque eandem vt in Prop. 27. p. partis.

Propositio XVII.

SI circa circulum describatur poligonum habens latera numero imparia, & conuertatur figura circa catetum; habebit factum sphęrale solidum ad inscriptam sibi sphęram, eam rationem quam vniuersa sphęralis soli di superficies habet, ad superficiem sphęrę.

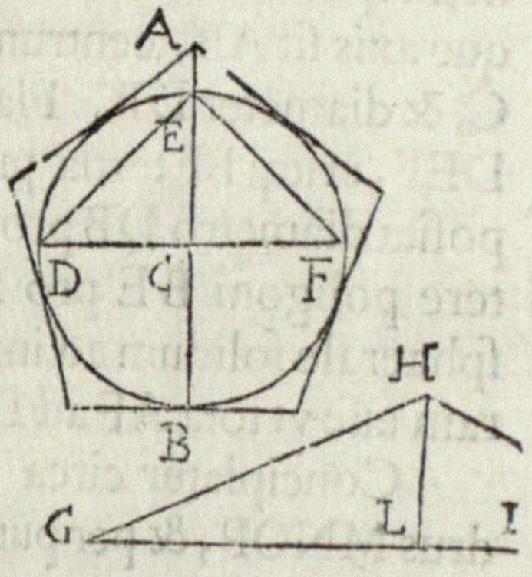

Manente pręcedentis propositionis constructione, sit sphęrale solidum cuius catetus, siue axis sit AB, centrum verò sphaerae sit C. Dico sphaerale solidum ad inscriptã sibi sphaeram esse, vt ipsius solidi integra superficies ad superficiem sphaerae.

Concipiatur in hemisphaerio conus DEF; & intelligatur conus GHI cuius basis GI ęqualis sit vniuersae superficiei solidi sphaeralis, altitudo verò LH aequalis sit radio sphaerae, & erit per praecedentem, sphaerale solidũ aequale cono GHI.

Propter aequalitatem ergò, erit sphaerale solidum ad conũ GHI, vt superficies vniuersa sphaeralis solidi, ad basim coni GHI; conus autem GHI ad conum DEF (ob aequalem altitudinem) est vt basis circa GI, ad basim circa DF. conus denique DEF, ad sphaeram est, vt basis circa DF ad superficiem sphaerae (nempe in ratione subquadrupla.) Propterea erit ex aequo, sphaerale solidum ad inscriptam sibi sphaerã, vt vniuersa sphaeralis solidi superficies ad superficiem sphaerae. Quod &c.

Pro-

cum maiore baſi fruſti conici, altitudinem quoque eandem vt in Prop. 27. p. partis.

Propoſitio XVII.

SI circa circulum deſcribatur poligonum habens latera numero imparia, & conuertatur figura circa catetum; habebit factum ſphęrale ſolidum ad inſcriptam ſibi ſphęram, cam rationem quam vniuerſa ſphęralis ſolidi ſuperficies habet, ad ſuperficiem ſphęrę.

Manente pręcedentis propoſitionis conſtructione, ſit ſphęrale ſolidum cuius catetus, ſiue axis ſit AB, centrum verò ſphaerae ſit C. Dico ſphaerale ſolidum ad inſcriptã ſibi ſphaeram eſſe, vt ipſius ſolidi integra ſuperficies ad ſuperficiem ſphaerae.

Concipiatur in hemiſphaerio conus DEF; & intelligatur conus GHI cuius baſis GI ęqualis ſit vniuerſae ſuperficiei ſolidi ſphaeralis, altitudo verò LH aequalis ſit radio ſphaerae, & erit per praccedentem, ſphaerale ſolidũ aequale cono GHI.

Propter aequalitatem ergò, erit ſphaerale ſolidum ad conũ GHI, vt ſuperficies vniuerſa ſphaeralis ſolidi, ad baſim coni GHI; conus autem GHI ad conum DEF (ob aequalem altitudinem) eſt vt baſis circa GI, ad baſim circa DF. conus denique DEF, ad ſphaeram eſt, vt baſis circa DF ad ſuperficiem ſphaerae (nempe in ratione ſubquadrupla.) Propterea erit ex aequo, ſphaerale ſolidum ad inſcriptam ſibi ſphaerã, vt vniuerſa ſphaeralis ſolidi ſuperficies ad ſuperficiem ſphaerae. Quod &c.

Pro-

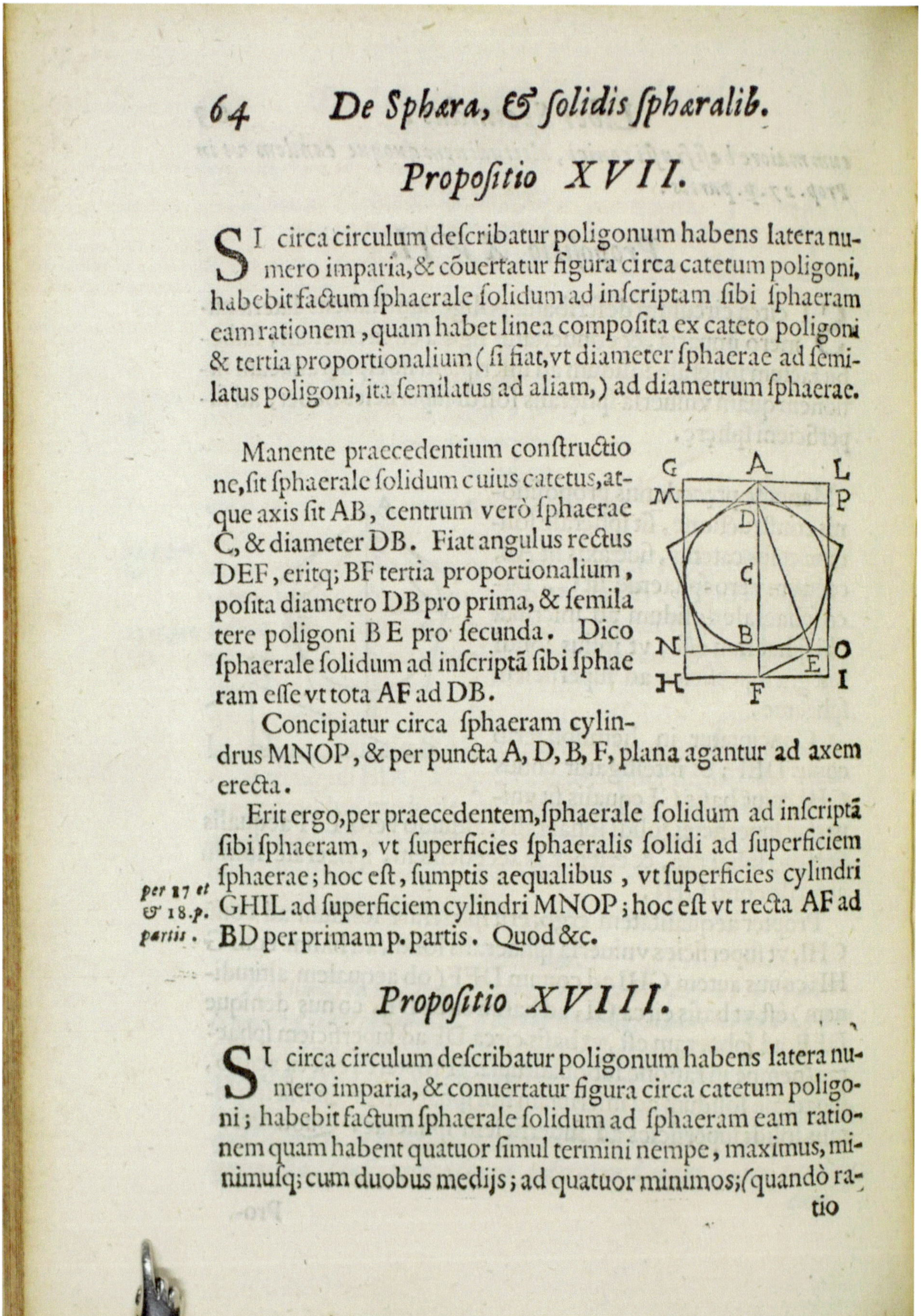

64 De Sphæra, & ſolidis ſphæralib.

Propoſitio XVII.

SI circa circulum deſcribatur poligonum habens latera numero imparia, & cõuertatur figura circa catetum poligoni, habebit factum ſphaerale ſolidum ad inſcriptam ſibi ſphaeram eam rationem, quam habet linea compoſita ex cateto poligoni & tertia proportionalium (ſi fiat, vt diameter ſphaerae ad ſemilatus poligoni, ita ſemilatus ad aliam,) ad diametrum ſphaerae.

Manente praecedentium conſtructione, ſit ſphaerale ſolidum cuius catetus, atque axis ſit AB, centrum verò ſphaerae C, & diameter DB. Fiat angulus rectus DEF, eritq; BF tertia proportionalium, poſita diametro DB pro prima, & ſemilatere poligoni B E pro ſecunda. Dico ſphaerale ſolidum ad inſcriptã ſibi ſphaeram eſſe vt tota AF ad DB.

Concipiatur circa ſphaeram cylindrus MNOP, & per puncta A, D, B, F, plana agantur ad axem erecta.

Erit ergo, per praecedentem, ſphaerale ſolidum ad inſcriptã ſibi ſphaeram, vt ſuperficies ſphaeralis ſolidi ad ſuperficiem ſphaerae; hoc eſt, ſumptis aequalibus, vt ſuperficies cylindri GHIL ad ſuperficiem cylindri MNOP; hoc eſt vt recta AF ad BD per primam p. partis. Quod &c.

per 17 et & 18.p. partis.

Propoſitio XVIII.

SI circa circulum deſcribatur poligonum habens latera numero imparia, & conuertatur figura circa catetum poligoni; habebit factum ſphaerale ſolidum ad ſphaeram eam rationem quam habent quatuor ſimul termini nempe, maximus, minimuſq; cum duobus medijs; ad quatuor minimos; (quandò ratio

Propoſitio XVII.

SI circa circulum deſcribatur poligonum habens latera numero imparia, & cõuertatur figura circa catetum poligoni, habebit factum ſphaerale ſolidum ad inſcriptam ſibi ſphaeram eam rationem, quam habet linea compoſita ex cateto poligoni & tertia proportionalium (ſi fiat, vt diameter ſphaerae ad ſemilatus poligoni, ita ſemilatus ad aliam,) ad diametrum ſphaerae.

Manente praecedentium conſtructione, ſit ſphaerale ſolidum cuius catetus, atque axis ſit AB, centrum verò ſphaerae C, & diameter DB. Fiat angulus rectus DEF, eritq; BF tertia proportionalium, poſita diametro DB pro prima, & ſemilatere poligoni B E pro ſecunda. Dico ſphaerale ſolidum ad inſcriptã ſibi ſphaeram eſſe vt tota AF ad DB.

per 17 et. & 18. p. partis.

Concipiatur circa ſphaeram cylindrus MNOP, & per puncta A, D, B, F, plana agantur ad axem erecta.

Erit ergo, per praecedentem, ſphaerale ſolidum ad inſcriptã ſibi ſphaeram, vt ſuperficies ſphaeralis ſolidi ad ſuperficiem ſphaerae; hoc eſt, ſumptis aequalibus, vt ſuperficies cylindri GHIL ad ſuperficiem cylindri MNOP; hoc eſt vt recta AF ad BD per primam p. partis. Quod &c.

Propoſitio XVIII.

SI circa circulum deſcribatur poligonum habens latera numero imparia, & conuertatur figura circa catetum poligoni; habebit factum ſphaerale ſolidum ad ſphaeram eam rationem quam habent quatuor ſimul termini nempe, maximus, minimuſq; cum duobus medijs; ad quatuor minimos; (quandò ra-

tio

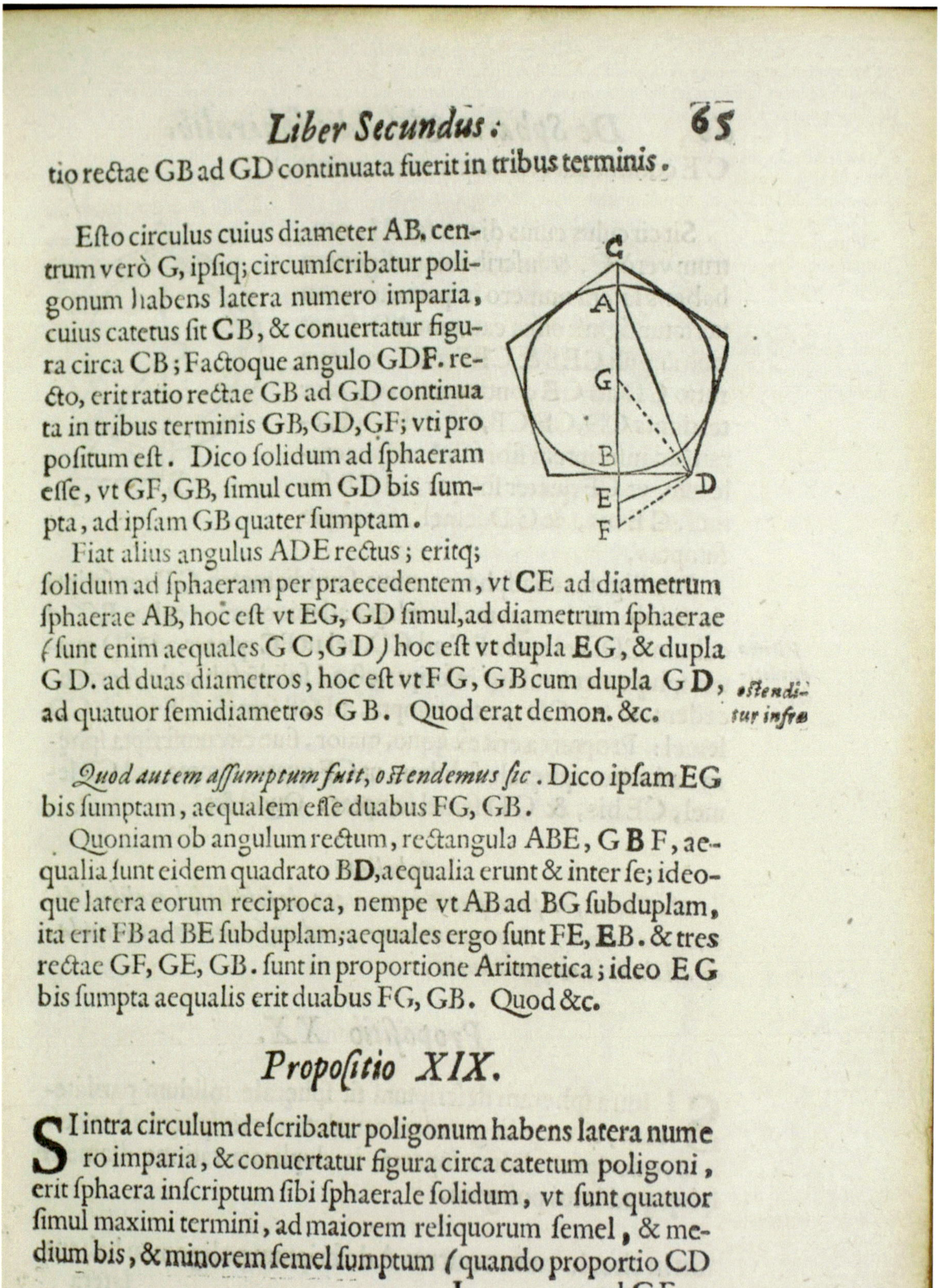

Liber Secundus. 65

tio rectae GB ad GD continuata fuerit in tribus terminis.

Esto circulus cuius diameter AB, centrum verò G, ipsiq; circumscribatur poligonum habens latera numero imparia, cuius catetus sit CB, & conuertatur figura circa CB; Factoque angulo GDF. recto, erit ratio rectae GB ad GD continua ta in tribus terminis GB, GD, GF; vti pro positum est. Dico solidum ad sphaeram esse, vt GF, GB, simul cum GD bis sumpta, ad ipsam GB quater sumptam.

Fiat alius angulus ADE rectus; eritq; solidum ad sphaeram per praecedentem, vt CE ad diametrum sphaerae AB, hoc est vt EG, GD simul, ad diametrum sphaerae (sunt enim aequales GC, GD) hoc est vt dupla EG, & dupla GD. ad duas diametros, hoc est vt FG, GB cum dupla GD, ad quatuor semidiametros GB. Quod erat demon. &c. *ostenditur infra*

Quod autem assumptum fuit, ostendemus sic. Dico ipsam EG bis sumptam, aequalem esse duabus FG, GB.

Quoniam ob angulum rectum, rectangula ABE, GBF, aequalia sunt eidem quadrato BD, aequalia erunt & inter se; ideoque latera eorum reciproca, nempe vt AB ad BG subduplam, ita erit FB ad BE subduplam; aequales ergo sunt FE, EB. & tres rectae GF, GE, GB. sunt in proportione Aritmetica; ideo EG bis sumpta aequalis erit duabus FG, GB. Quod &c.

Propositio XIX.

SI intra circulum describatur poligonum habens latera numero imparia, & conuertatur figura circa catetum poligoni, erit sphaera inscriptum sibi sphaerale solidum, vt sunt quatuor simul maximi termini, ad maiorem reliquorum semel, & medium bis, & minorem semel sumptum (quando proportio CD

I ad CE

tio rectae GB ad GD continuata fuerit in tribus terminis.

Efto circulus cuius diameter AB, centrum verò G, ipſiq; circumſcribatur poligonum habens latera numero imparia, cuius catetus ſit CB, & conuertatur figura circa CB; Factoque angulo GDF. recto, erit ratio rectae GB ad GD continuata in tribus terminis GB, GD, GF; vti propoſitum eft. Dico ſolidum ad ſphaeram eſſe, vt GF, GB, ſimul cum GD bis ſumpta, ad ipſam GB quater ſumptam.

Fiat alius angulus ADE rectus; eritq; ſolidum ad ſphaeram per praecedentem, vt CE ad diametrum ſphaerae AB, hoc eft vt EG, GD ſimul, ad diametrum ſphaerae (ſunt enim aequales GC, GD) hoc eft vt dupla EG, & dupla GD. ad duas diametros, hoc eft vt FG, GB cum dupla GD ad quatuor ſemidiametros GB. Quod erat demon. &c. *ostendi- . tur infra.*

Quod autem aſſumptum fuit, ostendemus ſic. Dico ipſam EG bis ſumptam, aequalem eſſe duabus FG, GB.

Quoniam ob angulum rectum, rectangula ABE, GBF, aequalia ſunt eidem quadrato BD, aequalia erunt & inter ſe; ideoque latera eorum reciproca, nempe vt AB ad BG ſubduplam, ita erit FB ad BE ſubduplam; aequales ergo ſunt FE, EB. & tres rectae GF, GE, GB. ſunt in proportione Aritmetica; ideo EG bis ſumpta aequalis erit duabus FG, GB. Quod &c.

Propoſitio XIX.

SI intra circulum deſcribatur poligonum habens latera numero imparia, & conuertatur figura circa catetum poligoni, erit ſphaera inſcriptum ſibi ſphaerale ſolidum, vt ſunt quatuor ſimul maximi termini, ad maiorem reliquorum ſemel, & medium bis, & minorem ſemel ſumptum (quando proportio CD

I ad CE

C E continuata erit in quatuor terminis.)

Sit circulus cuius diameter AI, centrum verò C, & inscribatur poligonum habens latera numero imparia; tum conuertatur figura circa catetum AD. Fiant que anguli CEB & CBF recti, eritque ratio CD ad CE continuata in quatuor terminis CD, CE, CB, CF. Dico sphaeram ad inscriptum sibi solidum sphaerale esse, vt CF quater sumpta, ad CB semel, CE bis, & CD semel, simulque sumptas,

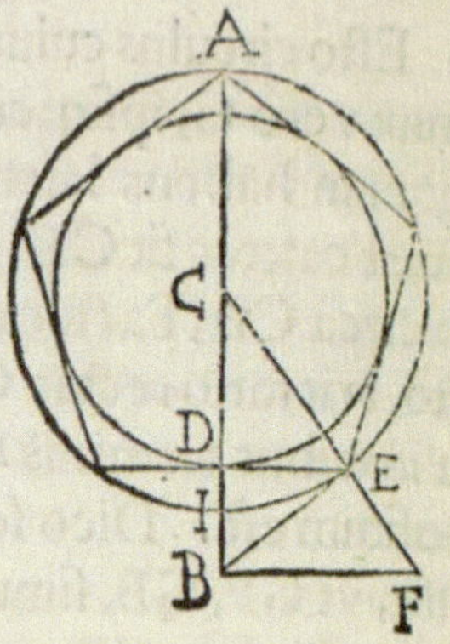

Intelligatur alia sphaera cuius semidiameter CD: inscripta in solido. Erit ergo maior sphęra ad minorem vt cubus EC ad

Vltima duodecimi. cubum CD. vel recta FC ad CD, uel ut FC quater, ad CD quater; sphęra uerò minor inscripta, est ad solidũ sphęrale, (perpręcedentem) ut CD quater sumpta, ad CB semel, CE bis, & CD semel: Propterea erit ex ęquo, maior, siue circumscripta sphęra, ad suum sphęrale solidum, ut CF quater sumpta, ad CB semel, CE bis, & CD semel sumptas. Quod &c.

Scholium.

Hactenus sex pracipua Theoremata de solidis sphæralibus demonstrata sunt. Sequuntur nunc quædam scitu non iniucunda, & ad doctrinam spectantia.

Propositio XX.

SI intra sphęram descriptum sit sphęrale solidum parilaterum, circaq; diagonalem reuolutum: erit sphęra ad excessum, quo ipsa solidum superat, in duplicata ratione diametri sphęrę ad latus poligoni.

Sit in circulo cuius centrum A descriptum poligonum habens

latera

CE continuata erit in quatuor terminis.)

Sit circulus cuius diameter AI, centrum verò C, & inſcribatur poligonum habens latera numero imparia; tum conuertatur figura circa catetum AD. Fiant que anguli CEB & CBF recti, eritque ratio CD ad CE continuata in quatuor terminis CD, CE, CB, CF. Dico ſphaeram ad inſcriptum ſibi ſolidum ſphaerale eſſe, vt CF quater ſumpta, ad CB ſemel, CE bis, & CD ſemel, ſimulque ſumptas,

Intelligatur alia ſphaera cuius ſemidiameter C D: inſcripta in ſolido. Erit ergo maior ſphęra ad minorem vt cubus EC ad cubum CD. vel recta FC ad CD, uel ut FC quater, ad CD quater; ſphęra uerò minor inſcripta, eſt ad ſolidũ ſphęrale, (per præcedentem) ut CD quater ſumpta, ad CB ſemel, CE bis, & CD ſemel: Propterea erit ex ęquo, maior, ſiue circumſcripta ſphęra, ad ſuum ſphęrale ſolidum, ut CF quater ſumpta, ad CB ſemel, CE bis, & C D ſemel ſumptas. Quod &c.

Vltima. duodecimi.

Scholium.

Hactenus ſex præcipua Theoremata de ſolidis ſphæralibus demonſtrata ſunt. Sequuntur nunc quædam ſcitu non iniucunda, & ad doctrinæm ſpectantia.

Propoſitio XX.

SI intra ſphęram deſcriptum ſit ſphęrale ſolidum parilaterum, circaq; diagonalem reuolutum: erit ſphęra ad exceſſum, quo ipſa ſolidum ſuperat, in duplicata ratione diametri ſphęrę ad latus poligoni.

Sit in circulo cuius centrum A deſcriptum poligonum habens

latera

latera numero paria, & conuertatur circa diagonalem B C. Dico sphęram ad excessum, quo ipsa solidũ superat, esse ut quadratum BC ad quadratum CD.

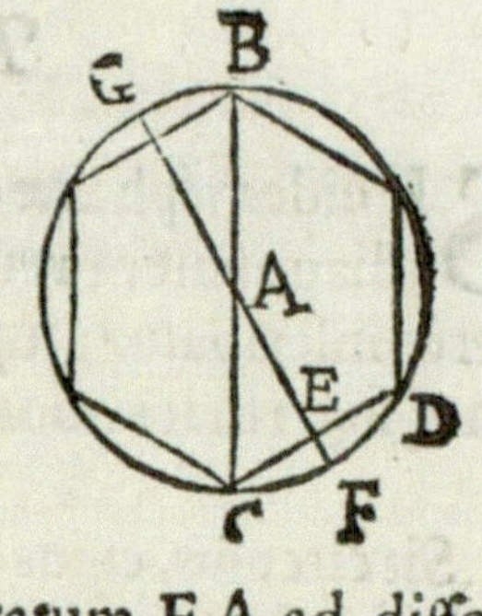

Ducatur AE ex centro perpendicularis ad latus CD, & producatur.

Quoniam per demonstrata, est ut sphęra ad solidum sphęrale ita quadratum F A ad quadratum AE, erit per conuersionem rationis sphæra ad excessum, vt quadratum F A ad differentiam quadratorum FA, AE, hoc est ad rectangulum F E G, siue ad quadratum EC. Constat ergò sphæram ad excessum quo ipsa superat inscriptum sphæıale solidum esse vt quadratum FA ad quadratum EC, siue vt quadratum B C ad quadratum C D. Quod &c. *q. huius.*

Propositiò XXI.

SI in eadem sphæra duo solida parilatera, & circa diagonalem reuoluta, concipiantur, erit differentia vnius à sphęra, ad differentiam alterius à sphæra, homologè in duplicata ratione laterum.

Sint in circulo cuius diameter AB duo semipoligona ACB, ADB; & conuertatur figura circà diagonalem AB. Dico differentiam inter sphęram, & sphærale solidũ ACB, ad differentiam inter sphæram & sphærale solidum ADB, esse vt quadratum CB ad quadratum BD.

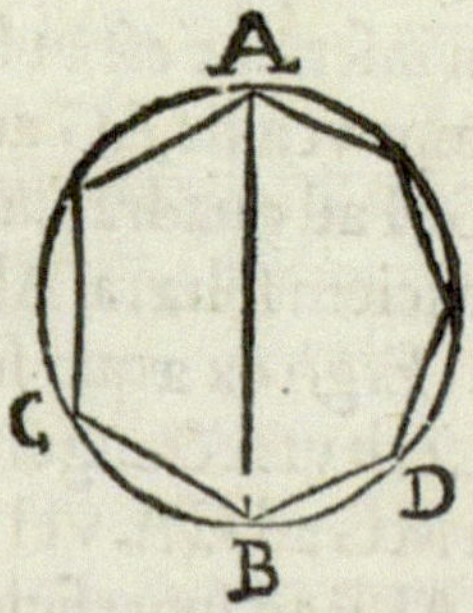

Demonstratum enim est differentiam ACB, esse ad sphæram, vt quadratum B C, ad quadratum AB, sed sphæra ad differentiam ADB, est vt quadratum AB ad quadratum B D, ergo ex æquo erit differentia A C B ad differentiam A D B vt quadratum BC ad quadratum B D. Quod &c. *per præced.* *per eandem.*

I 2 Propo-

latera numero paria, & conuertatur circa diagonalem BC. Dico ſphęram ad exceſſum, quo ipſa ſolidũ ſuperat, eſſe ut quadratum BC ad quadratum CD

Ducatur AE ex centro perpendicularis ad latus CD, & producatur.

Quoniam per demonſtrata, eſt ut ſphęra ad ſolidum ſphęrale ita quadratum FA ad quadratum AE, erit per conuerſionem rationis ſphæra ad exceſſum, vt quadratum FA ad differentiam quadratorum FA, AE, hoc eſt ad rectangulum FEG, ſiue ad quadratum EC. Conſtat ergò ſphæram ad exceſſum quo ipſa ſuperat inſcriptum ſphærale ſolidum eſſe vt quadratum FA ad quadratum EC, ſiue vt quadratum BC ad quadratum CD. Quod &c. *7.huius.*

Propoſitio XXI.

SI in eadem ſphæra duo ſolida parilatera, & circa diagonalem reuoluta, concipiantur, erit differentia vnius à ſphęra, ad differentiam alterius à ſphæra, homologè in duplicata ratione laterum,

Sint in circulo cuius diameter AB duo ſemipoligona ACB, ADB; & conuertatur figura circà diagonalem AB. Dico differentiam inter ſphęram, & ſphærale ſolidũ ACB, ad differentiam inter ſphæram & ſphærale ſolidum ADB, eſſe vt quadratum CB ad quadratum BD. *per praeced.*

Demonſtratum enim eſt differentiam ACB, eſſe ad ſphæram, vt quadratum BC, ad quadratum AB, ſed ſphæra ad differentiam ADB, eſt vt quadratum AB ad quadratum BD, ergo ex æquo erit differentia ACB ad differentiam ADB vt quadratum BC ad quadratum BD. Quod &c. *per ean-. dem.*

Propoſitio XXII.

SI eidem ſphæræ duo ſolida parilatera, & ſimilia, circaque diagonalem reuoluta, alterum circumſcribatur, alterum verò inſcribatur; ſuperficies ſphæræ media proportionalis erit inter ſuperficies duorum ſolidorum.

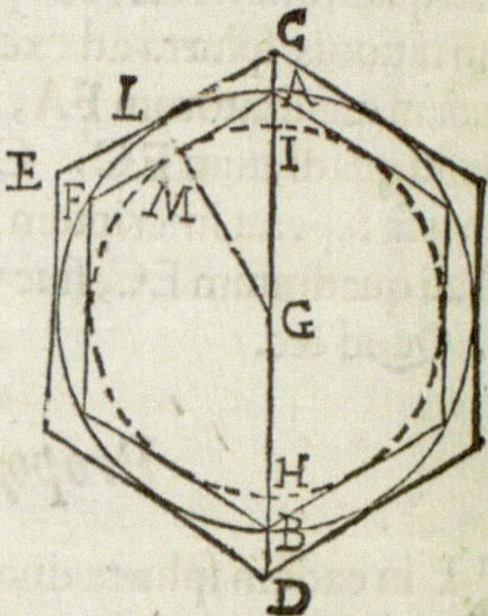

Sit circulus, cuius diameter AB atque ipſi duo poligona, alterum circumſcribatur, alterum verò inſcribatur, habeatq; vtrumq; latera numero paria, & ſit numerus laterum vnius æqualis numero laterũ alterius, vt ſphęralia ſolida ſimilia euadant. Tum conuertatur figura circa diagonalem CD.

Dico ſuperficiem factæ ſphæræ mediam proportionalem eſſe inter ſuperficies factorum ſolidorũ. Ducatur ex centro G recta GL ad contactus M & L, & radio GM fiat ſphæra IMH.

Iam ſuperficies ſolidi AF ad ſuperficiem ſphæræ IM intra ipſum inſcriptæ eſt vt ſolidum AF ad ſphæram IM, per 5. huius, nempe vt axis AG ad GM, per 6. huius, hoc eſt vt rectangulum AGM ad quadratum GM; Superficies verò ſphæræ IM ad ſuperficiem ſphæræ ALF eſt vt quadratum GM ad quadratum GA. Ergò ex æquo ſuperficies ſolidi AF ad ſuperficiem ſphæræ AL erit vt rectangulum AGM ad quadratum GA, nempe vt recta MG ad GA. vel vt recta LG ad GC. Sed ſuperficies ſphæræ ALF ad ſuperficiem ſolidi CE eſt vt LG ad GC. (quod probatur eodem mo do vt factum fuit ſupra) ergo in continua proportione ſunt ſuperficies vniuerſa ſolidi AMF, ſuperficies ſphæræ ALF. & ſuperficies ſolidi CE. Quod erat &c.

Corol-

Propoſitio XXII.

SI eidem ſphæræ duo ſolida parilatera, & ſimilia, circaque diagonalem reuoluta, alterum circumſcribatur, alterum verò inſcribatur; ſuperficies ſphæræ media proportionalis erit inter ſuperficies duorum ſolidorum.

Sit circulus, cuius diameter AB atque ipſi duo poligona, alterum circumſcribatur, alterum verò inſcribatur, habeatq; vtrumq; latera numero paria, & ſit numerus laterum vnius æqualis numero laterũ alterius, vt ſphęralia ſolida ſimilia euadant. Tum conuertatur figura circa diagonalem CD.

Dico ſuperficiem factæ ſphæræ mediam proportionalem eſſe inter ſuperficies factorum ſolidorũ. Ducatur ex centro G recta GL ad contactus M & L, & radio GM fiat ſphæra IMH.

Iam ſuperficies ſolidi AF ad ſuperficiem ſphæræ IM intra ipſum inſcriptæ eſt vt ſolidum AF ad ſphæram IM, per 5. huius, nempe vt axis AG ad GM, per 6. huius, hoc eſt vt rectangulum AGM ad quadratum GM; Superficies verò ſphæræ IM ad ſuperficiem ſphæræ ALF eſt vt quadratum GM ad quadratum GA. Ergò ex æquo ſuperficies ſolidi AF ad ſuperficiem ſphæræ AL erit vt rectangulum AGM ad quadratum GA, nempe vt recta MG ad GA. vel vt recta LG ad GC. Sed ſuperficies ſphæræ ALF ad ſuperficiem ſolidi CE eſt vt LG ad GC. (quod probatur eodem modo vt factum fuit ſupra) ergo in continua proportione ſunt ſuperficies vniuerſa ſolidi AMF, ſuperficies ſphæræ ALF. & ſuperficies ſolidi CE. Quod erat &c.

Corol-

Corollarium.

Hinc patet etiam quod si eidem solido sphærali parilatero circa diagonalem reuoluto duæ sphæræ, altera circumscribatur, altera verò inscribatur, tres superficies in continua proportione erunt inter se.

Propositio XXIII.

SPhæralia solida parilatera circa diagonalem reuoluta, & eidem sphæræ, vel æqualibus sphæris circumscripta, inter se sunt vt axes.

Sint circa circulum cuius centrum A duo poligona dissimlia, quorum latera numero paria sint, & conuertantur circa diagonalem. Sitq; alterius factorum solidorum B FC, axis BC; alterius verò nempe DGE, esto axis DE, Dico solidum BFC, ad solidum DGE esse vt BC ad DE.

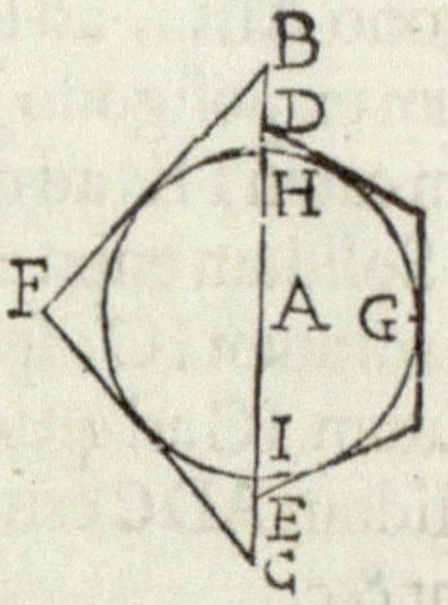

Hoc autem patet. Quoniam solidum B FC ad sphæram est vt BC ad diametrum HI; sphæra vero ad alterum solidum D G E est vt diameter HI ad axem DE, erit ex æquo, solidum BFC. ad solidum DGE, vt BC ad DE. Quod erat &c. *6. huius.* *ex eod.*

Hinc facile ostendi potest excessum, quo solidum BFC superat sphæram, ad excessum quo solidum DGE superat eandem sphæram, esse vt BH, ad HD.

Cum enim solidum BFC ad sphæram sit vt BA ad AH, erit diuidendo excessus BFC ad sphæram, vt BH ad HA. Eadem ratione sphæra ad excessum DGE erit, vt HA ad HD; ergo ex æquo, excessus BFC ad excessum DGE, supra sphæram erit vt BH, ad HD. Quod &c. *6. huius.*

Pro-

Corollarium.

Hinc patet etiam quod ſi eidem ſolido ſphærali parilatero circa diagonalem reuoluto duæ ſphæræ, altera circumſcribatur, altera verò inſcribatur, tres ſuperficies in continua proportione erunt inter ſe.

Propoſitio XXIII.

SPhæralia ſolida parilatera circa diagonalem reuoluta, & eidem ſphæræ, vel æqualibus ſphæris circumſcripta, inter ſe ſunt vt axes.

Sint circa circulum cuius centrum A duo poligona diſſimilia, quorum latera numero paria ſint, & conuertantur circa diagonalem. Sitq; alterius factorum ſolidorum BFC, axis BC; alterius verò nempe DGE, eſto axis DE, Dico ſolidum BFC, ad ſolidum DGE eſſe vt BC ad DE.

Hoc autem patet. Quoniam ſolidum BFC ad ſphæram eſt vt BC ad diametrum HI; ſphæra vero ad alterum ſolidum DGE eſt vt diameter HI ad axem DE, erit ex æquo, ſolidum BFC. ad ſolidum DGE, vt BC ad DE. Quod erat &c. 6.*huius.* *ex.cod.*

Hinc facile oſtendi poteſt exceſſum, quo ſolidum BFC ſuperat ſphæram, ad exceßum quo ſolidum DGE ſuperat eandem ſphæram, eſſe vt BH, ad HD.

Cum enim ſolidum BFC ad ſphæram ſit vt BA ad AH, erit diuidendo exceſſus BFC ad ſphæram, vt BH ad HA. Eadem ratione ſphæra ad exceſſum DGE erit, vt HA ad HD; ergo ex æquo, exceſſus BFC ad exceſſum DCE, ſupra ſphæram erit vt BH, ad HD. Quod &c. 6.*huius.*

Pro-

Propoſitio XXIV.

SOlida ſphęralia parilatera, eidem, vel æqualibus ſphæris inſcripta, & circa diagonalem reuoluta, ſunt inter ſe in duplicata ratione catetorum.

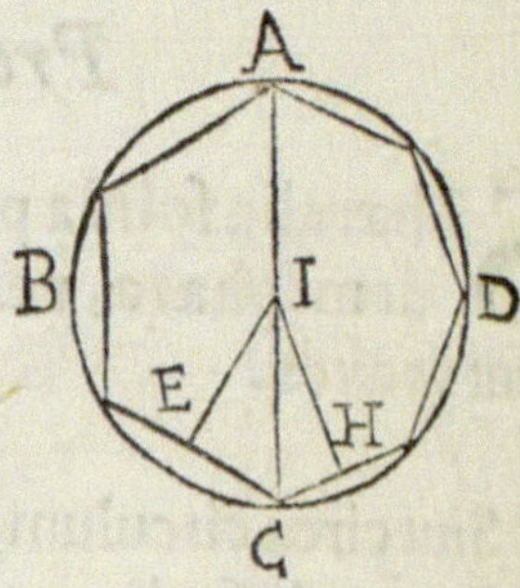

Inſcribantur in circulo cuius diameter AC duo ſemipoligona ABC, ADC, & conuertarur figura circa diagonalem AC, vt deſcribantur duo ſolida ſphæralia vt imperatum eſt.

Dico ſolidum ſphærale factum ex poligono ABC, ad ſolidum ſphærale factum ex poligono ADC, eſſe vt quadratum cateti I E, ad quadratum cateti IH.

7. huius. per eandem. Solidum enim ex ABC ad ſphæram, eſt vt quadratum I E ad quadratum IC; ſphæra autem ad ſolidum ADC, eſt vt quadratum IC ad quadratum IH; ergo ex æquo ſolidum ABC ad ſolidum ADC erit, vt quadratum IE ad quadratum IH. Qqod erat &c.

Propoſitio XXV.

SI intra æquales, vel eandem ſphæram, cuius diameter AB, deſcripta fuerint duo ſolida ſphæralia parilatera, quorum duo latera ſint B C, BD; demittanturque ex punctis C, D, perpendiculares CE, DF ad diametrum; erit ſolidum cuius latus BC, ad ſolidum cuius latus BD, vt AE ad AF.

Ducantur enim ex centro I ad latera BC, BD perpendiculares IG, IH.

Recta EA ad rectam AB, eſt vt quadratum A C ad quadratum AB (ob angulum in ſemicirculo rectum ACB) recta autem BA ad AF, eſt vt quadratum AB, ad quadratum AD, ergo ex

æquo

Propoſitio XXIV.

SOlida ſphęralia parilatera, eidem, vel æqualibus ſphæris inſcripta, & circa diagonalem reuoluta, ſunt inter ſe in duplicata ratione catetorum.

Inſcribantur in circulo cuius diameter AC duo ſemipoligona ABC, ADC, & conuertatur figura circa diagonalem AC, vt deſcribantur duo ſolida ſphæralia vt imperatum eſt.

Dico ſolidum ſphærale factum ex poligono ABC, ad ſolidum ſphærale factum ex poligono ADC, eſſe vt quadratum cateti IE, ad quadratum cateti IH.

7.huius. per eandem. Solidum enim ex ABC ad ſphæram, eſt vt quadratum I E ad quadratum IC; ſphæra autem ad ſolidum ADC, eſt vt quadratum IC ad quadratum IH; ergo ex æquo ſolidum ABC ad ſolidum ADC erit, vt quadratum IE ad quadratum IH. Qqod[36] erat &c.

Propoſitio XXV.

SI intra æquales, vel eandem ſphæram, cuius diameter AB, deſcripta fuerint duo ſolida ſphæralia parilatera, quorum duo latera ſint BC, BD; demittanturque ex punctis C, D, perpendiculares CE, DF ad diametrum; erit ſolidum cuius latus BC, ad ſolidum cuius latus BD, vt AE ad AF.

Ducantur enim ex centro I ad latera BC, BD perpendiculares IG, IH.

Recta EA ad rectam AB, eſt vt quadratum A C ad quadratum AB (ob angulum in ſemicirculo rectum ACB*)* recta autem BA ad AF, eſt vt quadratum AB, ad quadratum AD, ergo ex

æquo

[36] Qqod] Quod.

æquo recta EA ad rectam AF, est vt quadratum AC ad quadratum AD, hoc est vt quadratum IG ad quadratum IH, hoc est vt solidum cuius latus est BC ad solidum cuius latus est BD. Quod erat &c. *per præced.*

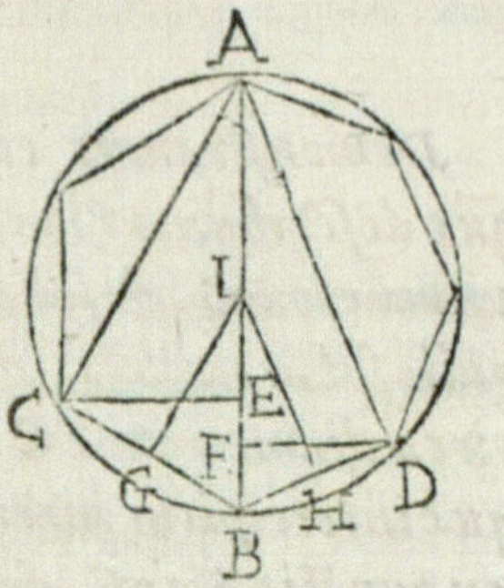

Propositio XXVI.

SI intra sphæram cuius diameter AB descriptum sit solidum sphærale parilaterum, & circa diagonalem reuolutum; demittaturque ab extremitate lateris BC quod diàmetrum contingit, recta CD perpendicularis ad diametrum circuli AB, erit conus cuius basis circulus AFCBE altitudo verò sit AD, subduplus solidi sphęralis; conus verò, cuius eadem sit basis, & altitudo DB, erit subduplus differentiæ, quæ inter sphæram, & solidum sphærale est.

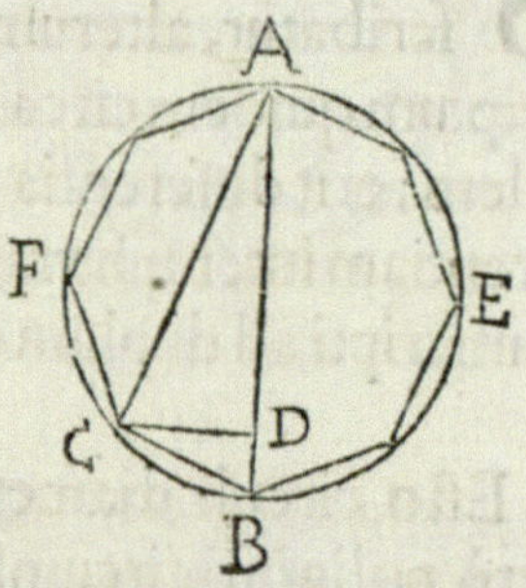

Sphæra enim ad inscriptum solidum est vt quadratum diametri ad quadratum cateti AC (est enim AC ob angulum rectum ACB, æqualis catęto poligoni,) hoc est vt BA recta ad rectam AD. *7. huius.*

Iam quia conus, cuius basis AFCBE altitudo verò sit AB; æqualis est hæmisphærio in eadem basi constituto; erit dictus conus, hoc est hemisphærium, ad conum cuius basis eadem AFCBE, altitudo verò AD, vt AB ad AD. Sed hemisphęrium etiam ad semisolidum est vt AB ad AD; vt ostendimus supra. Propterea conus cuius basis circulus AFCBE, altitudo autem AD, erit æqualis semisolido sphærali, siue subduplus solidi sphęralis. Quod &c. *ex 30. p. partis.*

Similiter inferetur, conum cuius basis eadem AFCBE, altitudo verò DB, subduplum esse excessus illius, quo sphæra solidum superat.

Scho-

æquo recta EA ad rectam AF, eft vt quadratum AC ad quadratum AD, hoc eft vt quadratum IG ad quadratum IH, hoc eft vt folidum cuius latus eft BC ad folidum cuius latus eft BD. Quod erat &c. *per prę-. ced.*

Propofitio XXVI.

SI intra fphæram cuius diameter AB defcriptum fit folidum fphærale parilaterum, & circa diagonalem reuolutum; demittaturque ab extremitate lateris BC quod diàmetrum contingit, recta CD perpendicularis ad diametrum circuli AB, erit conus cuius bafis circulus AFCBE altitudo verò fit AD, fubduplus folidi fphęralis; conus verò, cuius eadem fit bafis, & altitudo DB, erit fubduplus differentiæ, quæ inter fphæram, & folidum fphærale eft.

Sphæra enim ad infcriptum folidum eft vt quadratum diametri ad quadratum cateti AC (eft enim AC ob angulum rectum ACB, æqualis cateto poligoni,) hoc eft vt BA recta ad rectam AD. *7.huius.*

Iam quia conus, cuius bafis AFCBE altitudo verò fit AB; æqualis eft hæmifphærio in eadem bafi conftituto; erit dictus conus, hoc eft hemifphærium, ad conum cuius bafis eadem AFCBE, altitudo verò AD, vt AB ad AD. Sed hemifphęrium etiam ad femifolidum eft vt AB ad AD; vt oftendimus fupra. *ex.30.p. partis.*

Propterea conus cuius bafis circulus AFCBE, altitudo autem AD, erit æqualis femifolido fphærali, fiue fubduplus folidi fphęralis. Quod &c.

Similiter inferetur, conum cuius bafis eadem AFCBE, altitudo verò DB, fubduplum effe exceffus illius, quo fphæra folidum fuperat.

Scho-

Scholium.

Demonstramus etiam singula illa solida rotunda annularia, quæ describuntur in reuolutione figuræ à bilineis mixtis, quale vnum est FC, & solidum sphærale circundant; æqualia esse singulis sphæroidibus, quarum vniuscuiusq; maximus circulus sit circa diametrum FC. Axis verò æqualis sit portioni rectæ ex AB. quæ intercipitur inter duas perpendiculares ad ipsam AB ductas ex punctis F & C. & sic de reliquis. Sed hoc alibi.

Propositio XXVII.

SI eidem circulo duo poligona parilatera alterum circumscribatur, alterum verò inscribatur; & conuertatur circum scriptum quidem circa catetum, inscriptum verò circa diagonalem; erit differentia inter circumscriptum & sphæram, ad differentiam inter sphæram & inscriptum, vt quadratum lateris circumscripti ad duplum quadrati lateris inscripti.

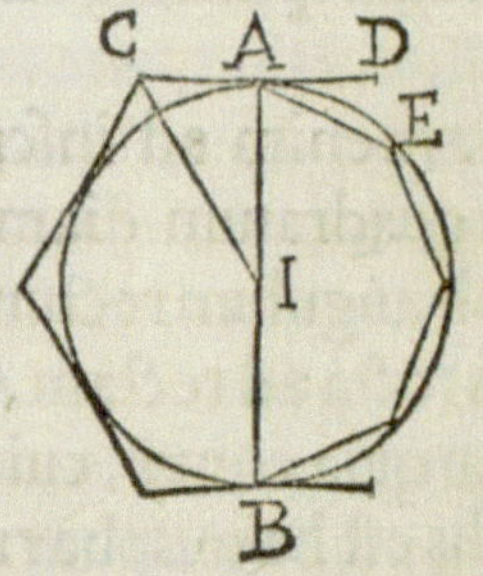

Esto circuli diameter AB, latus verò poligoni circumscripti CD. & inscripti AE. Dico excessum, quo maius solidum sphæram superat, ad excessum, quo sphæra superat minus esse vt quadratum CD ad duo quadrata ex AE.

13. huius. Solidum enim circumscriptum est ad sphæram vt duo quadrata CI, IA ad duplum quadrati ex IA; ergo diuidendo, erit excessus solidi

20. huius supra sphæram, ad ipsam sphęram, vt quadratum CA ad duplum quadrati ex IA, siue vt quadr. CD, ad duplũ quadr. ex AB Sphæra autem ad excessum, quo ipsa superat minus solidum, est vt quadratũ AB ad quadratum AE, vel vt duo quadrata ex AB ad duo quadrata ex AE. Propterea ex æquo excessus solidi maioris supra sphęram, ad excessum sphęræ supra minus solidum,

erit

Scholium.

Demonſtramus etiam ſingula illa ſolida rotunda annularia, quæ deſcribuntur in reuolutione figuræ à bilineis mixtis, quale vnum est FC, & ſolidum ſphærale circundant; æqualia eſſe ſingulis ſphæroidibus, quarum vniuſcuiuſq; maximus circulus ſit circa diametrum FC. Axis verò æqualis ſit portioni rectæ ex AB. quæ intercipitur inter duas perpendiculares ad ipſam AB ductas ex punctis F & C. & ſic de reliquis. Sed hoc alibi.

Propoſitio XXVII.

SI eidem circulo duo poligona parilatera alterum circumſcribatur, alterum verò inſcribatur; & conuertatur circumſcriptum quidem circa catetum, inſcriptum verò circa diagonalem; erit differentia inter circumſcriptum & ſphæram, ad differentiam inter ſphæram & inſcriptum, vt quadratum lateris circumſcripti ad duplum quadrati lateris inſcripti.

Eſto circuli diameter AB, latus verò poligoni circumſcripti CD. & inſcripti AE. Dico exceſſum, quo maius ſolidum ſphæram ſuperat, ad exceſſum, quo ſphæra ſuperat minus eſſe vt quadratum CD ad duo quadrata ex AE.

13.*hu-. ius.* Solidum enim circumſcriptum eſt ad ſphæram vt duo quadrata CI, IA ad duplum quadrati ex IA; ergo diuidendo, erit exceſſus ſolidi ſupra ſphæram, ad ipſam ſphęram, vt quadratum CA ad duplum quadrati ex IA, ſiue vt quadr. CD, ad duplũ quadr. ex AB.

20.*hu- ius.* Sphæra autem ad exceſſum, quo ipſa ſuperat minus ſolidum, eſt vt quadratũ AB ad quadratum AE, vel vt duo quadrata ex AB ad duo quadrata ex AE. Propterea ex æquo exceſſus ſolidi maioris ſupra ſphęram, ad exceſſum ſphęræ ſupra minus ſolidum,

erit

erit vt quadratum ex CD ad duo quadrata ex AB. Quod &c.

Propositio XXVIII.

QVodlibet sphærale solidum circa diagonalẽ reuolutum (cuius latera numero quidem paria sint, sed nullo modo à quaternario mensurentur, vt sunt 6. 10. 14. 18. 22. &c.) inscripti sibi rombi solidi duplum est.

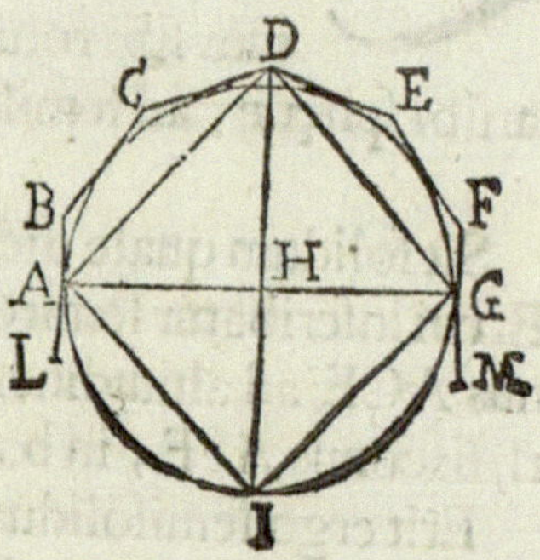

Sit solidum quale dictum est AB CDEFG. circa axẽ siue diagonalem DI reuolutum. Manifestũ est quod duo latera opposita BL. FM. contingent sphæram in extremitatibus A, G, diametri AG, quæ quidem perpendicularis sit ad DI.; quãdoquidẽ laterum numerus à binario tantum mensuratur, non autem à quaternario.

Inscribantur iam duo coni; nempe ADG in semisolido, habens altitudinem HD.; conus verò AIG in hemisphærio. Erit igitur semisolidum ABCDEFG ad hemisphærium vt axis ad axem, nempe vt DH ad HI, hoc est vt conus ADG ad conum AIG (cum sint in eadem basi;) & permutando semisolidum ad suum conum ADG, erit vt hemisphærium ad suum conum AIG; quare duplum erit. Propterea omne solidum, quale dictum est duplum erit inscripti sibi rombi solidi, Quod &c. 6. huius.

Lemma.

Si hemisphærium ABC, & conus quicumque rectus DBE eandem altitudinem habuerint FB; erit hemisphærium ad prædictum conum vt duplum basis hemisphęrij ad basim eiusdem coni.

Sit vt ponitur: Et inscribatur in hemisphærio conus ABC. Erit ergo conus ABC ad conum DBE vt basis AC ad basim DE;

K *sum-*

erit vt quadratum ex CD ad duo quadrata ex A E. Quod &c.

Propoſitio XXVIII.

QVodlibet ſphærale ſolidum circa diagonalẽ reuolutum (cuius latera numero quidem paria ſint, ſed nullo modo à quaternario menſurentur, vt ſunt 6. 10. 14. 18. 22. &c.) inſcripti ſibi rombi ſolidi duplum eſt.

Sit ſolidum quale dictum eſt ABCDEFG. circa axẽ ſiue diagonalem DI reuolutum. Manifeſtũ eſt quod duo latera oppoſita BL. FM. contingent ſphæram in extremitatibus A, G, diametri AG, quæ quidem perpendicularis ſit ad DI.; quãdoquidẽ laterum numerus à binario tantum menſuratur, non autem à quaternario.

Inſcribantur iam duo coni; nempe ADG in ſemiſolido, habens altitudinem HD.; conus verò AIG in hemiſphærio. Erit igitur ſemiſolidum ABCDEFG ad hemiſphærium vt axis ad axem, nempe vt DH ad HI, hoc eſt vt conus ADG ad conum AIG (cum ſint in eadem baſi;) & permutando ſemiſolidum ad ſuum conum ADG, erit vt hemiſphærium ad ſuum conum AIG; quare duplum erit. Propterea omne ſolidum, quale dictum eſt duplum erit inſcripti ſibi rombi ſolidi, Quod &c. 6.*huius.*

Lemma.

Si hemiſphærium ABC, & conus quicumque rectus DBE eandem altitudinem habuerint FB; erit hemiſphærium ad prædictum conum vt duplum baſis hemiſphęrij ad baſim eiuſdem coni.

Sit vt ponitur: Et inſcribatur in hemiſphærio conus ABC. Erit ergo conus ABC ad conum DBE vt baſis AC ad baſim DE;

K *ſum-*

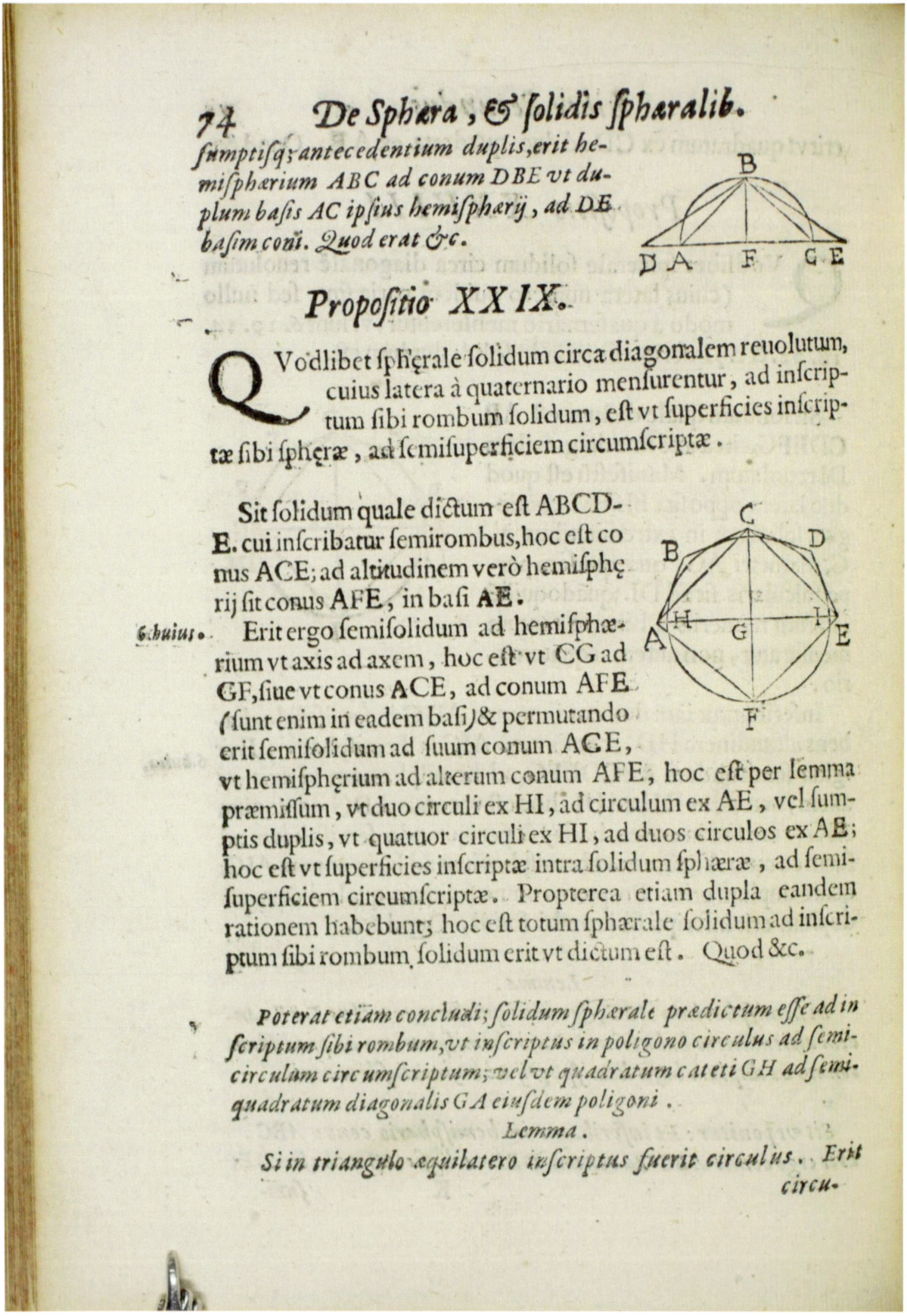

74 *De Sphæra, & ſolidis ſphæralib.*

ſumptiſq; antecedentium duplis, erit hemiſphærium ABC ad conum DBE vt duplum baſis AC ipſius hemiſphærij, ad DE baſim coni. Quod erat &c.

Propoſitio XXIX.

QVodlibet ſphærale ſolidum circa diagonalem reuolutum, cuius latera à quaternario menſurentur, ad inſcriptum ſibi rombum ſolidum, eſt vt ſuperficies inſcriptæ ſibi ſphæræ, ad ſemiſuperficiem circumſcriptæ.

Sit ſolidum quale dictum eſt ABCDE. cui inſcribatur ſemirombus, hoc eſt conus ACE; ad altitudinem verò hemiſphærij ſit conus AFE, in baſi AE.

6. huius. Erit ergo ſemiſolidum ad hemiſphærium vt axis ad axem, hoc eſt vt CG ad GF, ſiue vt conus ACE, ad conum AFE (ſunt enim in eadem baſi) & permutando erit ſemiſolidum ad ſuum conum ACE, vt hemiſphærium ad alterum conum AFE, hoc eſt per lemma præmiſſum, vt duo circuli ex HI, ad circulum ex AE, vel ſumptis duplis, vt quatuor circuli ex HI, ad duos circulos ex AE; hoc eſt vt ſuperficies inſcriptæ intra ſolidum ſphæræ, ad ſemiſuperficiem circumſcriptæ. Propterea etiam dupla eandem rationem habebunt; hoc eſt totum ſphærale ſolidum ad inſcriptum ſibi rombum ſolidum erit vt dictum eſt. Quod &c.

Poterat etiam concludi; ſolidum ſphærale prædictum eſſe ad inſcriptum ſibi rombum, vt inſcriptus in poligono circulus ad ſemicirculum circumſcriptum; vel vt quadratum cateti GH ad ſemiquadratum diagonalis GA eiuſdem poligoni.

Lemma.

Si in triangulo æquilatero inſcriptus fuerit circulus. Erit circu-

ſumptiſq; antecedentium duplis, erit hemiſphærium ABC ad conum DBE vt duplum baſis AC ipſius hemiſphærij, ad DE baſim coni. Quod erat &c.

Propoſitio XXIX.

QVodlibet ſphęrale ſolidum circa diagonalem reuolutum, cuius latera à quaternario menſurentur, ad inſcriptum ſibi rombum ſolidum, eſt vt ſuperficies inſcriptæ ſibi ſphęræ, ad ſemiſuperficiem circumſcriptæ.

Sit ſolidum quale dictum eſt ABCDE. cui inſcribatur ſemirombus, hoc eſt conus ACE; ad altitudinem verò hemiſphęrij ſit conus AFE, in baſi AE.

6.*huius.* Erit ergo ſemiſolidum ad hemiſphærium vt axis ad axem, hoc eſt vt CG ad GF, ſiue vt conus ACE, ad conum AFE *(*ſunt enim in eadem baſi*)* & permutando erit ſemiſolidum ad ſuum conum ACE, vt hemiſphęrium ad alterum conum AFE, hoc eſt per lemma præmiſſum, vt duo circuli ex HI, ad circulum ex AE, vel ſumptis duplis, vt quatuor circuli ex HI, ad duos circulos ex AE; hoc eſt vt ſuperficies inſcriptæ intra ſolidum ſphæræ, ad ſemiſuperficiem circumſcriptæ. Propterea etiam dupla eandem rationem habebunt; hoc eſt totum ſphærale ſolidum ad inſcriptum ſibi rombum ſolidum erit vt dictum eſt. Quod &c.

Poterat etiam concludi; ſolidum ſphærale prædictum eſſe ad inſcriptum ſibi rombum, vt inſcriptus in poligono circulus ad ſemicirculum circumſcriptum; vel vt quadratum cateti GH ad ſemiquadratum diagonalis GA eiuſdem poligoni.

Lemma.

Si in triangulo æquilatero inſcriptus fuerit circulus. Erit.

circu-

circulus alter cuius diameter sit latus trianguli, triplus inscripti circuli.

Inscribatur circulus ABC in triang. æquilatero DEF. Sitque G punctum, centrum & circuli, & trianguli; propterea DG dupla ipsius GC, hoc est ipsius GA. Ergo quadr. DG quadruplũ est quadrati ex GA. & quadratum DA triplum erit eiusdem GA; Quare etiam circulus cuius semidiameter sit DA triplus erit circuli cuius semidiameter sit CA. Quod erat &c.

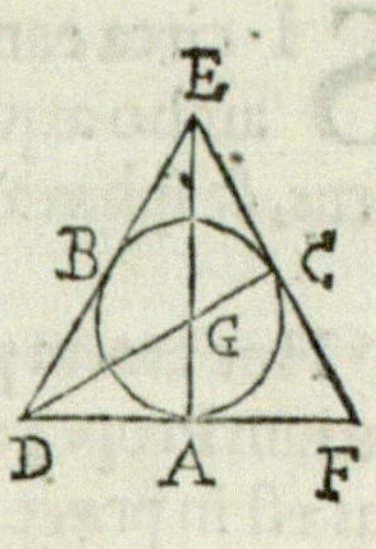

Propositio XXX.

SI circa circulum descriptum fuerit triangulum æquilaterum & reuoluatur figura, erit factus conus æquilaterus ad inscriptam sibi sphæram vt 9. ad 4.

Esto circa circulum ABC triangulum æquilaterum DEF, & conuertatur figura. Dico factum conum çquilaterum esse ad inscriptam sphæram in proportione dupla sesquiquarta, nempe vt 9. ad 4.

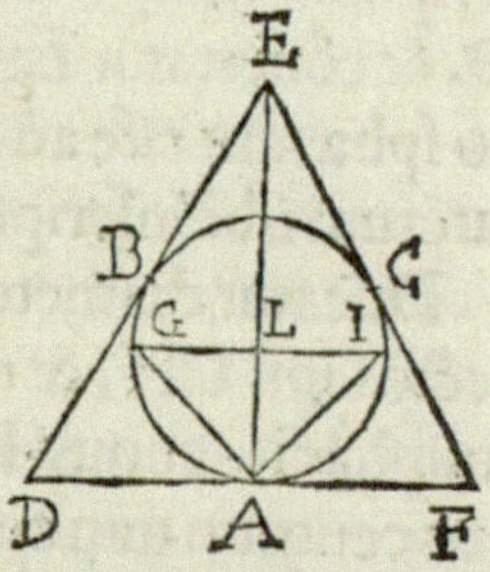

Concipiatur in hemisphęrio GAI conus GAI. Erit iam per lemma præcedens circulus cuius diameter DF triplus circuli cuius diameter GI; sed conus DEF ad conum GAI rationem habet compositam ex ratione altitudinum EA ad AL; quæ tripla est: Et ex ratione basium, nempe circuli DF ad circulum GI quæ similiter tripla est: quare conus DEF ad conum GAI erit vt 9. ad vnum, sumptisq; consequentium quadruplis, erit conus DEF ad sphæram sibi inscriptam, vt 9. ad 4. Quod erat &c.

circulus alter cuius diameter ſit latus trianguli, triplus inſcripti circuli.

Inſcribatur circulus ABC in triang. æquilatero DEF. Sitque G punctum, centrum & circuli, & trianguli; propterea DG dupla ipsius GC, hoc eſt ipsius GA. Ergo quadr. DG quadruplũ est quadrati ex GA. & quadratum DA triplum erit eiuſdem GA; Quare etiam circulus cuius ſemidiameter sit DA triplus erit circuli cuius ſemidiameter sit CA. Quod erat &c.

Propoſitio XXX.

SI circa circulum deſcriptum fuerit triangulum æquilaterum & reuoluatur figura, erit factus conus æquilaterus ad inſcriptam ſibi ſphæram vt 9. ad 4.

Eſto circa circulum ABC triangulum æquilaterum DEF, & conuertatur figura.

Dico factum conum ęquilaterum eſſe ad inſcriptam ſphæram in proportione dupla ſeſquiquarta, nempe vt 9. ad 4.

Concipiatur in hemiſphęrio GAI conus GAI. Erit iam per lemma præcedens circulus cuius diameter DF triplus circuli cuius diameter GI; ſed conus DEF ad conum GAI rationem habet compoſitam ex ratione altitudinum EA ad AL; quæ tripla eſt: Et ex ratione baſium, nempe circuli DF ad circulum GI quæ ſimiliter tripla eſt: quare conus DEF ad conum GAI erit vt 9. ad vnum, ſumptiſq; conſequentium quadruplis, erit conus DEF ad ſphæram ſibi inſcriptam, vt 9. ad 4. Quod erat &c.

Propositio XXXI.

SI circa eandem sphæram descripti sint conus, & cylindrus, ambo æquilateri; erunt tria solida, nempe conus, cylindrus, & sphæra in continua proportione sesquialtera.

Hoc autem patet. Posita enim sphæra vt 4. erit(per Corollarium Prop. 30. p. partis) cylindrus vt 6; conus autem ostensus est in præcedenti esse vt 9. Quare tria solida erunt inter se in continua proportione sesquialtera. Quod &c.

Propositio XXXII.

SPHæra adinscriptum sibi conum æquilaterum est in ratione numeri 32. ad 9.

Sit in circulo cuius centrum A inscriptum triangulum ęquilaterum C B D. & cōuertatur figura circa CH. Dico sphæram esse ad factum conū ęquilaterum sibi inscriptum vt 32. ad 9.

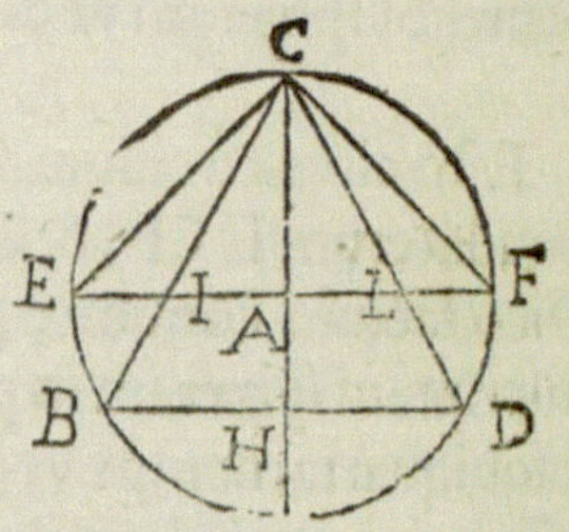

Ducatur diameter EF ad angulos rectos ipsi CH, & concipiatur in hemisphærio conus ECF: Punctum A erit centrum tum circuli, tum etiam trianguli æquilateri BCD, propterea C H sesquialtera erit ipsius CA.

Sed cum etiam ICL sit triangulum æquilaterum, erit CA potentià tripla ipsius AI, ergò & circulus ex CA, siue ex AE triplus erit circuli ex AI; ideoq; conus ECF, triplus coni ICL. videlicet vt 24. ad 8. Conus autem I C L ad conum BCD ob similitudinem, est vt cubus A C ad cubum CH, nimirum vt 8. ad 27. Quare ex æquo erit conus ECF ad conum B C D vt 24. ad 27. Reductaque ratione ad minimos terminos, erit conus ECF ad conum BCD vt 8. ad 9. Sumptis igitur antecedentium

quadru-

Propoſitio XXXI.

SI circa eandem ſphæram deſcripti ſint conus, & cylindrus, ambo æquilateri; erunt tria ſolida, nempe conus, cylindrus, & ſphæra in continua proportione ſeſquialtera.

Hoc autem patet. Poſita enim ſphæra vt 4. erit(per[37] Corollarium Prop. 30. p. partis) cylindrus vt 6; conus autem oftenſus eft in præcedenti eſſe vt 9. Quare tria ſolida erunt inter ſe in continua proportione ſeſquialtera. Quod &c.

Propoſitio XXXII.

SPHæra adinſcriptum ſibi conum æquilaterum eft in ratione numeri 32. ad 9.

Sit in circulo cuius centrum A inſcriptum triangulum ęquilaterum CBD. & cõuertatur figura circa CH. Dico ſphæram eſſe ad factum conũ ęquilaterum ſibi inſcriptum vt 32. ad 9.

Ducatur diameter EF ad angulos rectos ipſi CH, & concipiatur in hemiſphærio conus ECF: Punctum A erit centrum tum circuli, tum etiam trianguli æquilateri BCD, propterea C H ſeſquialtera erit ipſius CA.

Sed cum etiam ICL ſit triangulum æquilaterum, erit CA potentià tripla ipſius AI, ergò & circulus ex CA, ſiue ex AE triplus erit circuli ex AI; ideoq; conus ECF, triplus coni ICL. videlicet vt 24. ad 8. Conus autem ICL ad conum BCD ob ſimilitudinem, eft vt cubus AC ad cubum CH, nimirum vt 8. ad 27. Quare ex æquo erit conus ECF ad conum BCD vt 24. ad 27. Reductaque ratione ad minimos terminos, erit conus ECF ad conum BCD vt 8. ad 9. Sumptis igitur antecedentium

quadru-

[37] erit(per] erit (per.

quadruplis ſphęra ad inſcriptum ſibi conum æquilaterum erit vt 32. ad 9. Quod erat &c.

Propoſitio XXXIII.

ROmbus ſolidus æquilaterus circa ſphæram deſcriptus eſt ad ipſam ſphæram vt diameter quadrati ad latus eiuſdem.

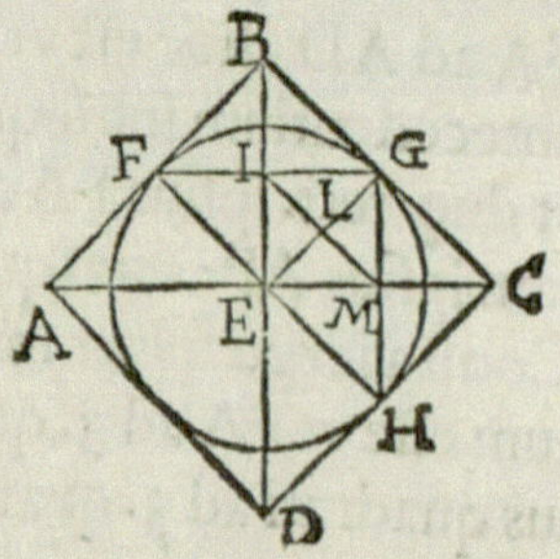

Eſto quadratum ABCD circa circulum cuius centrum E; & voluatur figura circa diagonalem BD; Dico rombũ ſolidum æquilaterum factum ex reuolutione, eſſe ad ſphæram vt diameter quadrati ad latus eiuſdem.

Intelligatur in hemiſphęrio conus FGH, cuius baſis FH, altitudo EG, & ducatur IM.

Erit iam conus ABC cuius baſis AC, ſimilis cono FGH, vterque enim rectus & rectangulus eſt. Ergo conus ABC ad conũ FGH erit vt cubus BE. ad cubum EG, nempe vt recta BE ad EL. (ſunt enim EB, EG, EI, EL in continua ratione) ſumptis autem conſequentium duplis, erit conus ABC ad hemiſphæriũ, vt BE ad EG, & propterea totus rombus ſolidus ad totam ſphęram ſibi inſcriptam erit vt BE ad EG, hoc eſt vt diameter alicuius quadrati ad latus eiuſdem. Quod &c.

Propoſitio XXXIV.

SPhæra ad inſcriptum ſibi cylindrum æquilaterum eſt vt diameter quadrati ad 3. quart. lateris eiuſdem.

Deſcribatur intra circulum cuius centrum A quadratum BCDE, & voluatur figura circa catetum AG. Dico ſphęram ad cylindrum BCDE, eſſe vt diameter alicuius quadrati ad 3. quart. lateris eiuſdem.

Intel-

quadruplis ſphęra ad inſcriptum ſibi conum æquilaterum erit vt 32. ad 9. Quod erat &c.

Propoſitio XXXIII.

ROmbus ſolidus æquilaterus circa ſphæram deſcriptus eſt ad ipſam ſphæram vt diameter quadrati ad latus eiuſdem.

Eſto quadratum ABCD circa circulum cuius centrum E; & voluatur figura circa diagonalem BD; Dico rombũ ſolidum æquilaterum factum ex reuolutione, eſſe ad ſphæram vt diameter quadrati ad latus eiuſdem.

Intelligatur in hemiſphęrio conus FGH, cuius baſis FH, altitudo EG, & ducatur IM.

Erit iam conus ABC cuius baſis AC, ſimilis cono FGH, vterque enim rectus & rectangulus eſt. Ergo conus ABC ad conũ FGH erit vt cubus BE. ad cubum EG, nempe vt recta BE ad EL. (ſunt enim EB, EG, EI, EL in continua ratione) ſumptis autem conſequentium duplis, erit conus ABC ad hemiſphæriũ, vt BE ad EG, & propterea totus rombus ſolidus ad totam ſphęram ſibi inſcriptam erit vt BE ad EG, hoc eſt vt diameter alicuius quadrati ad latus eiuſdem. Quod &c.

Propoſitio XXXIV.

SPhæra ad inſcriptum ſibi cylindrum æquilaterum eſt vt diameter quadrati ad 3. quart. lateris eiuſdem.

Deſcribatur intra circulum cuius centrum A quadratum BCDE, & voluatur figura circa catetum AG. Dico ſphęram ad cylindrum BCDE, eſſe vt diameter alicuius quadrati ad 3. quart. lateris eiuſdem.

Intel-

78 *De Sphæra, & solidis sphæralib.*

Intelligatur circa sphæram alter cylindrus æquilaterus FILM. & producta AM iungantur AD, GO. Erunt ob similitudinem triangulorum, in continua ratione FA, AD, AG, AP; Et quia cylindri sunt similes, nempe æquilateri, erit cylindrus IFML ad cylindrum BCDE vt cubus FM ad

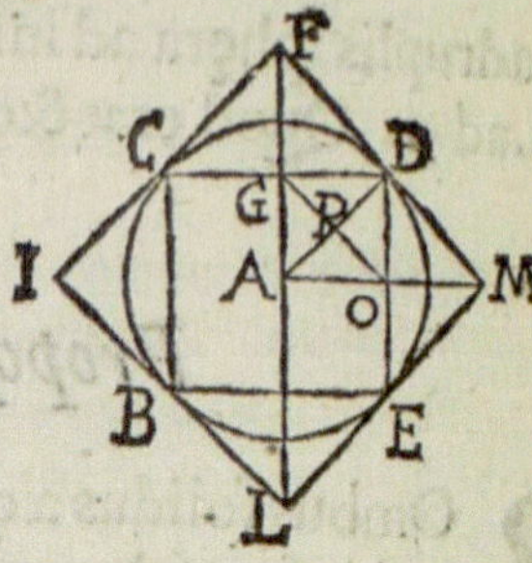

cubum CD, hoc est vt cubus FD ad cubum DG, siue vt cubus FA ad AD, hoc est vt recta FA ad quartam AP. Sumptisque antecedentium subsequialteris, erit sphæra ad cylindrũ BCDE vt duę tert. ipsius FA ad AP; hoc est vt tota FA ad sesquialteram ipsius AP; siue (quod idem est) vt FA ad 3. quar. rectæ AD. Constat ergo sphæram ad inscriptum sibi cylindrum æquilaterum esse vt FA ad 3. quar. ipsius AD; hoc est vt diameter alicuius quadrati ad 3. quar. lateris eiusdem. Quod &c.

Propositio XXXV.

Solidum exagonale; hoc est sphærale solidum genitum ab exagono circa catetum reuoluto, septuplum est coni eandem sibi basim, & altitudinem habentis.

Esto exagonum æquilaterum, & equiangulum ACDEFB & conuertatur circa catetum HI; inscribaturq; conus AIB. Dico exagonale solidum factum ex reuolutione, septuplum esse coni AIB.

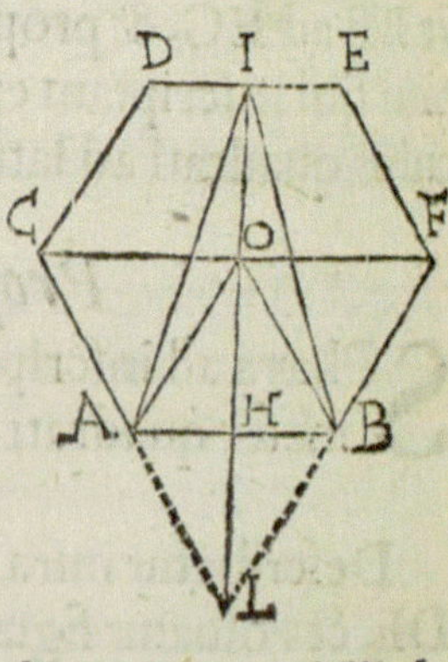

Producantur CA, FB donec concurrant in aliquo puncto L, eruntque ob exagonum, quatuor triangula æquilatera OCA, OAB, OBF, ABL, æqualia inter se. Concipiatur ergo conus CLF perfectus; eritque conus AIB duplus coni ALB, quandoquidem eandẽ habet basim AB, sed altitu-

Intelligatur circa ſphæram alter cylindrus æquilaterus FILM. & producta AM iungantur AD, GO. Erunt ob ſimilitudinem triangulorum, in continua ratione FA, AD, AG, AP;

Et quia cylindri ſunt ſimiles, nempe æquilateri, erit cylindrus IFML ad cylindrum BCDE vt cubus FM ad cubum CD, hoc eſt vt cubus FD ad cubum DG, ſiue vt cubus FA ad AD, hoc eſt vt recta FA ad quartam AP. Sumptiſque antecedentium ſubſequialteris, erit ſphæra ad cylindrũ BCDE vt duę tert. ipſius FA ad AP; hoc eſt vt tota FA ad ſeſquialteram ipſius AP; ſiue (quod idem eſt) vt FA ad 3. quar. rectæ AD

Conſtat ergo ſphæram ad inſcriptum ſibi cylindrum æquilaterum eſſe vt FA ad 3. quar. ipſius AD; hoc eſt vt diameter alicuius quadrati ad 3. quar. lateris eiuſdem. Quod &c.

Propoſitio XXXV.

SOlidum exagonale; hoc eſt ſphærale ſolidum genitum ab exagono circa catetum reuoluto, ſeptuplum eſt coni eandem ſibi baſim, & altitudinem habentis.

Eſto exagonum æquilaterum, & ęquiangulum ACDEFB & conuertatur circa catetum HI; inſcribaturq; conus AIB. Dico exagonale ſolidum factum ex reuolutione, ſeptuplum eſſe conì AIB.

Producantur CA, FB donec concurrant in aliquo puncto L, eruntque ob exagonum, quatuor triangula æquilatera OCA, OAB, OBF, ABL, æqualia inter ſe.

Concipiatur ergo conus CLF perfectus; eritque conus AIB duplus coni ALB, quandoquidem eandẽ habet baſim AB, ſed al-

titu-

titudinem habet HI duplam ipſius HL.

Iam conus CLF. ad conum ALB, erit ob ſimilitudinem, vt cubus CL ad cubum LA, nempe vt 8. ad 1; & diuidendo ſemiſolidum CABF erit ad conum ALB, vt ſeptem ad vnum. Propterea etiam dupla eandem rationem habebunt, hoc eſt ſolidũ exagonale integrum ſeptuplum erit coni AIB. Quod erat &c.

Propoſitio XXXVI.

SI circa circulum deſcribatur exagonum, & reuoluatur figura circa catetum; erit ſphæra ſextupla coni, qui eandem baſim, & eandem altitudinem cum ſolido habeat.

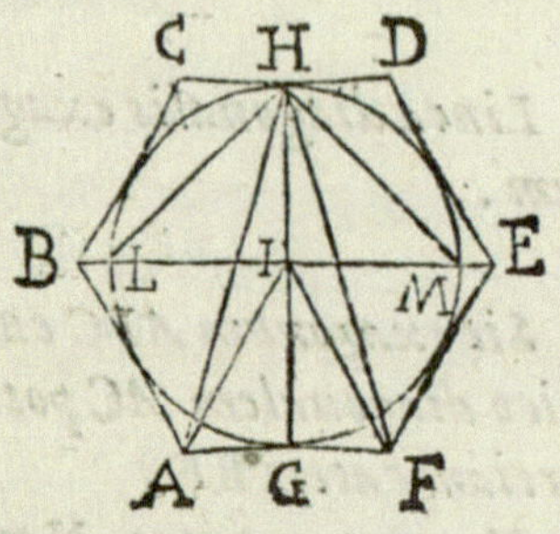

Eſto circa circulum cuius centrũ I exagonum ABCDEF. & conuertatur circa catetum GH; inſcribatur que in facto ſolido exagonali conus AHF, qui baſim habeat circulum circa AF, altitudinem verò GH eandem cum ſolido. Dico ſphæram ſextuplam eſſe coni AHF.

Concipiantur duo alij coni; nempe LHM in hemiſphærio, & AIF ſuper baſi AF conſtitutus ad centrum I.

Erit ergò propter exagonum, triangulum AIF æquilaterum, & ideo ipſa IG tripla erit potentià ipſius GA. Conſtat igitur quod circulus cuius diameter LM (dupla ſcilicet ipſius IG) triplus erit circuli cuius diameter AF, & propterea conus L H M triplus erit coni AIF. Sphæra autem duodecupla erit coni AIF, & ideo ſextupla coni AHF. Quod erat &c.

Propoſitio XXXVII.

SI circa circulum deſcribatur exagonum, & voluatur figura circa catetum; erit factum ſolidum ad factam ſphæram ſeſquiſextum.

Eſto

titudinem habet HI duplam ipſius HL.

Iam conus CLF. ad conum ALB, erit ob ſimilitudinem, vt cubus CL ad cubum LA, nempe vt 8. ad 1; & diuidendo ſemiſolidum CABF erit ad conum ALB, vt ſeptem ad vnum. Propterea etiam dupla eandem rationem habebunt, hoc eſt ſolidũ exagonale integrum ſeptuplum erit coni AIB. Quod erat &c.

Propoſitio XXXVI.

SI circa circulum deſcribatur exagonum, & reuoluatur figura circa catetum; erit ſphæra ſextupla coni, qui eandem baſim, & eandem altitudinem cum ſolido habeat.

Eſto circa circulum cuius centrũ I exagonum ABCDEF. & conuertatur circa catetum GH; inſcribaturque in facto ſolido exagonali conus AHF, qui baſim habeat circulum circa AF, altitudinem verò GH eandem cum ſolido. Dico ſphæram ſextuplam eſſe coni AHF.

Concipiantur duo alij coni; nempe LHM in hemiſphærio, & AIF ſuper baſi AF conſtitutus ad centrum I.

Erit ergò propter exagonum, triangulum AIF æquilaterum, & ideo ipſa IG tripla erit potentià ipſius GA. Conſtat igitur quod circulus cuius diameter LM (dupla ſcilicet ipſius IG) triplus erit circuli cuius diameter AF, & propterea conus LHM triplus erit coni AIF. Sphæra autem duodecupla erit coni AIF, & ideo ſextupla coni AHF. Quod erat &c.

Propoſitio XXXVII.

SI circa circulum deſcribatur exagonum, & voluatur figura circa catetum; erit factum ſolidum ad factam ſphæram ſeſquiſextum

Esto

Eſto circa circulum cuius centrum I exagonum ABCDEF. & conuertatur figura circa catetum GH. Dico ſolidum ſphærale factum, eſſe ad ſphęram vt 7. ad *6.*

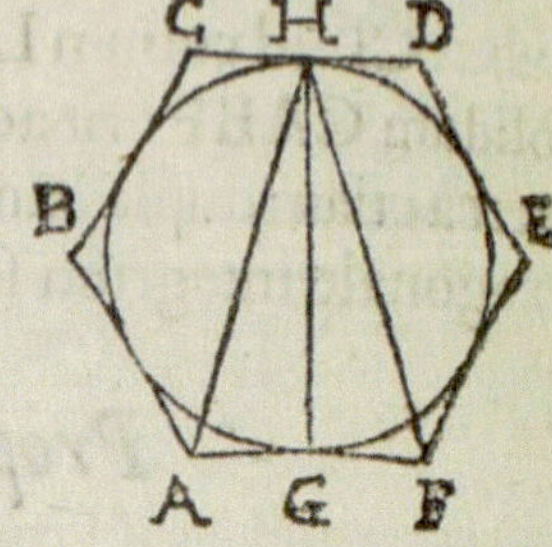

Concipiatur enim in ſolido conus AHF, vt in duabus pręcedentibus propoſitionibus.

Erit ergo (per 35.huius) ſolidum exagonale ad conum AHF vt 7.ad vnum, conus autem AHF ad ſphærã eſt vt 1. ad *6.*; quare ex æquo erit ſolidum ad ſphæram vt 7. ad *6.* Quod &c.

per præced.

Lemma.

Linea diagonalis exagoni potentià ſeſquitertia eſt cateti eiuſdem.

Sit exagonum ABC cuius centrum D. Dico diagonalem AC potentià eſſe ſeſquitertiam cateti EF.

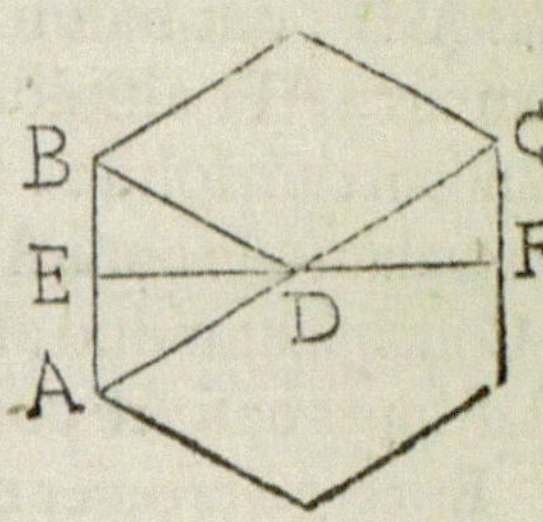

Hoc autem patet. Nam ducta DB. erit ABD triangulum æquilaterum, ob exagonum; & AD latus erit potentià ſeſquitertium perpendicularis DE; ergò ſumptis lineis duplis, etiam AC ſeſquitertia erit potentià ipſius EF. Quod &c.

Propoſitio XXXVIII.

SPhæra inſcripti ſibi ſolidi exagonalis circa diagonalem reuoluti, ſeſquitertia eſt.

Sit in circulo cuius centrum A deſcriptum exagonum BCDEFG; iunctiſq; DH, DL, DM, DI, conuertatur figura circa diagonalem DG. Dico ſphæram inſcripti ſolidi exagonalis ſeſquiter-

Eſto circa circulum cuius centrum I exagonum ABCDEF. & conuertatur figura circa catetum GH. Dico ſolidum ſphærale factum, eſſe ad ſphęram vt 7. ad 6.

Concipiatur enim in ſolido conus AHF, vt in duabus pręcedentibus propoſitionibus.

per præced. Erit ergo (per 35. huius) ſolidum exagonale ad conum AHF vt 7. ad vnum, conus autem AHF ad ſphærã eſt vt 1. ad 6.; quare ex æquo erit ſolidum ad ſphæram vt 7. ad 6. Quod &c.

Lemma.

Linea diagonalis exagoni potentià ſeſquitertia est cateti eiuſdem.

Sit exagonum ABC cuius centrum D.

Dico diagonalem AC potentià eſſe ſeſquitertiam cateti EF.

Hoc autem patet. Nam ducta DB. erit ABD triangulum æquilaterum, ob exagonum; & AD latus erit potenti à ſeſquitertium perpendicularis DE; ergò ſumptis lineis duplis, etiam AC ſeſquitertia erit potentià ipſius EF. Quod &c.

Propoſitio XXXVIII.

SPhæra inſcripti ſibi ſolidi exagonalis circa diagonalem reuoluti, ſeſquitertia eſt.

Sit in circulo cuius centrum A deſcriptum exagonum BCDEFG; iunctiſq; DH, DL, DM, DI, conuertatur figura circa diagonalem DG. Dico ſphæram inſcripti ſolidi exagonalis ſeſqui-

ter-

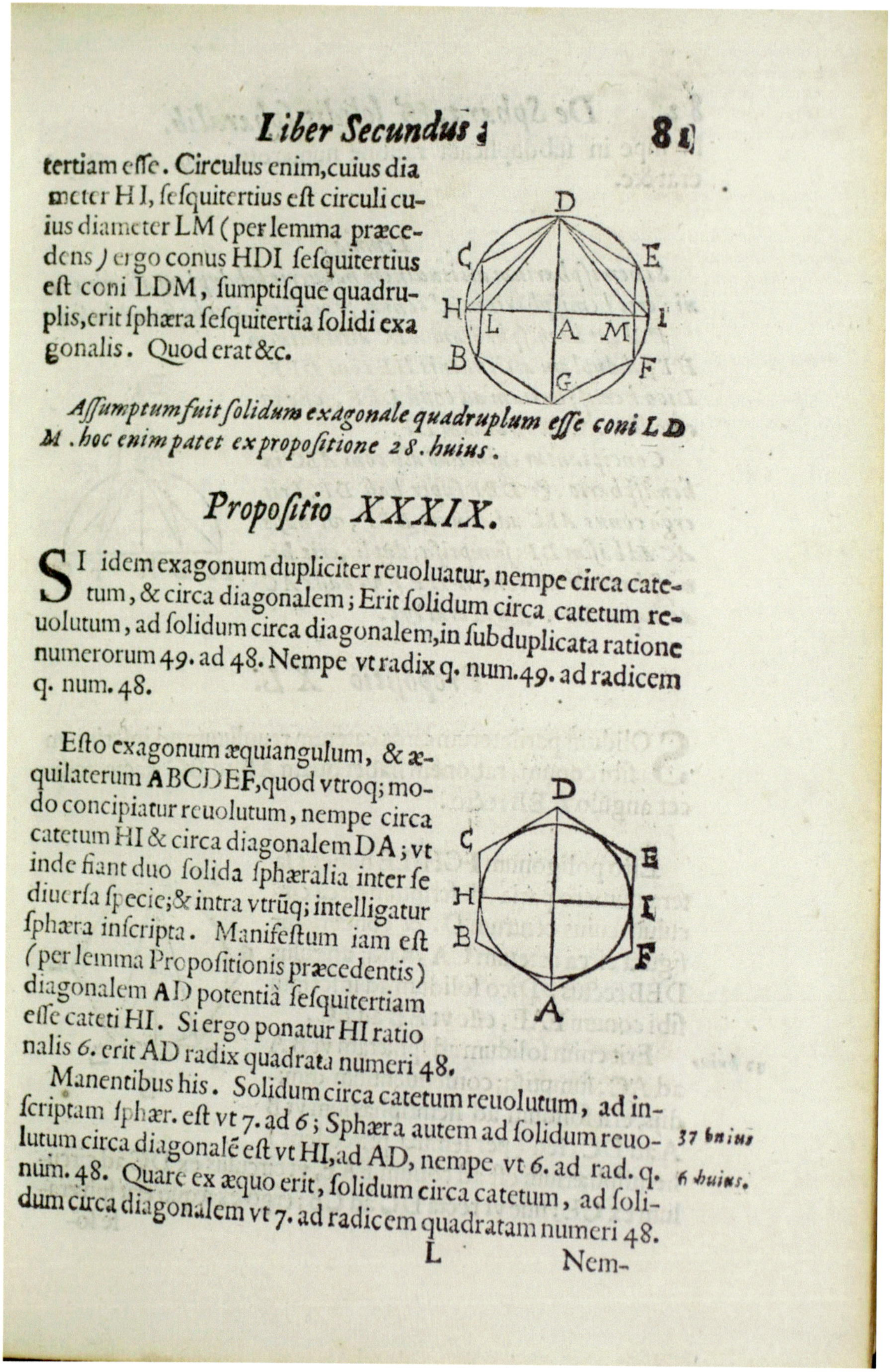

Liber Secundus. 81

tertiam eſſe. Circulus enim, cuius diameter HI, ſeſquitertius eſt circuli cuius diameter LM (per lemma præcedens) ergo conus HDI ſeſquitertius eſt coni LDM, ſumptiſque quadruplis, erit ſphæra ſeſquitertia ſolidi exagonalis. Quod erat &c.

Aſſumptum fuit ſolidum exagonale quadruplum eſſe coni LDM. hoc enim patet ex propoſitione 28. huius.

Propoſitio XXXIX.

SI idem exagonum dupliciter reuoluatur, nempe circa catetum, & circa diagonalem; Erit ſolidum circa catetum reuolutum, ad ſolidum circa diagonalem, in ſubduplicata ratione numerorum 49. ad 48. Nempe vt radix q. num. 49. ad radicem q. num. 48.

Eſto exagonum æquiangulum, & æquilaterum ABCDEF, quod vtroq; modo concipiatur reuolutum, nempe circa catetum HI & circa diagonalem DA; vt inde fiant duo ſolida ſphæralia inter ſe diuerſa ſpecie; & intra vtrũq; intelligatur ſphæra inſcripta. Manifeſtum iam eſt (per lemma Propoſitionis præcedentis) diagonalem AD potentiâ ſeſquitertiam eſſe cateti HI. Si ergo ponatur HI rationalis 6. erit AD radix quadrata numeri 48.

Manentibus his. Solidum circa catetum reuolutum, ad inſcriptam ſphær. eſt vt 7. ad 6; Sphæra autem ad ſolidum reuolutum circa diagonalẽ eſt vt HI, ad AD, nempe vt 6. ad rad. q. num. 48. Quare ex æquo erit, ſolidum circa catetum, ad ſolidum circa diagonalem vt 7. ad radicem quadratam numeri 48. 37 huius 6 huius.

L Nem-

tertiam eſſe. Circulus enim, cuius diameter HI, ſeſquitertius eſt circuli cuius diameter LM (per lemma præcedens) ergo conus HDI ſeſquitertius eſt coni LDM, ſumptiſque quadruplis, erit ſphæra ſeſquitertia ſolidi exagonalis. Quod erat &c.

Aſſumptum fuit ſolidum exagonale quadruplum eſſe coni LDM. hoc enim patet ex propoſitione 28. huius.

Propoſitio XXXIX.

SI idem exagonum dupliciter reuoluatur, nempe circa catetum, & circa diagonalem; Erit ſolidum circa catetum reuolutum, ad ſolidum circa diagonalem, in ſubduplicata ratione numerorum 49. ad 48. Nempe vt radix q. num. 49. ad radicem q. num. 48.

Eſto exagonum æquiangulum, & æquilaterum ABCDEF, quod vtroq; modo concipiatur reuolutum, nempe circa catetum HI & circa diagonalem DA; vt inde fiant duo ſolida ſphæralia inter ſe diuerſa ſpecie; & intra vtrũq; intelligatur ſphæra inſcripta. Manifeſtum iam eſt (per lemma Propoſitionis præcedentis) diagonalem AD potentià ſeſquitertiam eſſe cateti HI. Si ergo ponatur HI rationalis 6. erit AD radix quadrata numeri 48. 37 *huius*

Manentibus his. Solidum circa catetum reuolutum, ad inſcriptam ſphær. eſt vt 7. ad 6; Sphæra autem ad ſolidum reuolutum circa diagonalẽ eſt vt HI, ad AD, nempe vt 6. ad rad. q.num. 48. Quare ex æquo erit, ſolidum circa catetum, ad ſolidum circa diagonalem vt 7. ad radicem quadratam numeri 48. 6 *huius.*

L Nem-

Nempe in subduplicata ratione numerorum 49. 48. Quod erat &c.

Lemma.

Si hemisphærium altitudinem habuerit subduplam alicuius coni: erit hemisphærium ad conum prædictum, vt basis ad basim.

Habeat hæmisphærium ABC altitudinem HB subduplam altitudinis HE coni DEF. Dico hemisphærium ad conũ DEF, esse vt circulus AC ad circulum DF.

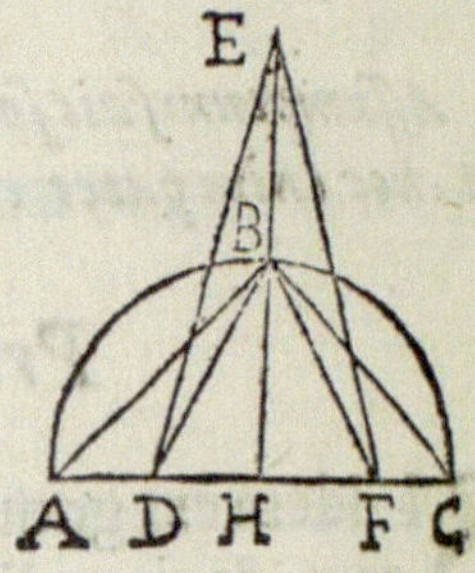

Concipiantur enim duo alij coni ABC in hemisphærio, & DBF super basi DF. Erit ergò conus ABC ad conum DBF, vt basis AC ad basim DF; sumptisq; duplis, erit hemisphærium ad conum DEF vt basis AC ad basim DF. Quod erat &c.

Propositio XL.

Solidum parilaterum circa catetum reuolutum ad inscriptum sibi conum, rationem habet quam AB ad BC; facto scilicet angulo DEB recto.

Esto poligonum FGHILE habens latera numero paria, descriptum circa circulum cuius centrum D. & conuertatur figura circa catetum CA, fiatq; angulus DEB rectus. Dico solidum ad inscriptũ sibi conum FAE, esse vt AB ad BC.

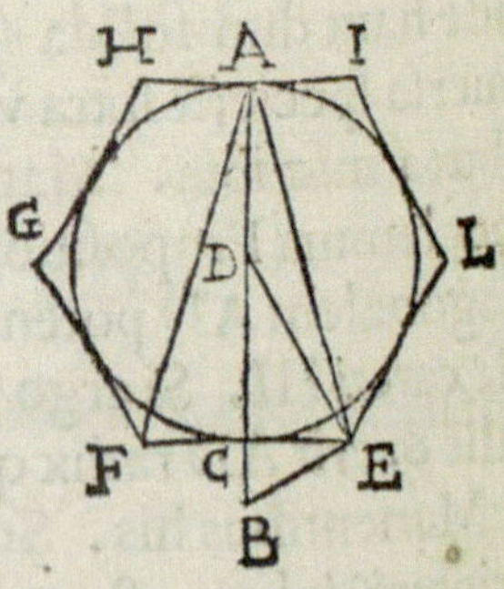

32 huius Erit enim solidum ad sphæram vt BA ad AC, sumptisq; consequentium dimidijs, erit solidum ad hemisphærium vt BA ad DC, sed (per lemma præcedens) hemisphærium est ad conum FAE, vt circulus ex DC ad circulum ex CE; siue vt recta DC ad CB; ergò ex æquo erit sphærale so-

Nempe in ſubduplicata ratione numerorum 49. 48. Quod erat &c.

Lemma.

Si hemiſphærium altitudinem habuerit ſubduplam alicuius coni: erit hemiſphærium ad conum prædictum, vt baſis ad baſim.

Habeat hæmiſphærium ABC altitudinem HB ſubduplam altitudinis HE coni DEF.

Dico hemiſphærium ad conũ DEF, eßе vt circulus AC ad circulum DF.

Concipiantur enim duo alij coni ABC in hemiſphærio, & DBF ſuper baſi DF. Erit ergò conus ABC ad conum DBF, vt baſis AC ad baſim DF; ſumptiſq; duplis, erit hemiſphærium ad conum DEF vt baſis AC ad baſim DF. Quod erat &c.

Propoſitio XL.

SOlidum parilaterum circa catetum reuolutum ad inſcriptum ſibi conum, rationem habet quam AB ad BC; facto ſcilicet angulo DEB recto.

Eſto poligonum FGHILE habens latera numero paria, deſcriptum circa circulum cuius centrum D. & conuertatur figura circa catetum CA, fiatq; angulus DEB rectus. Dico ſolidum ad inſcriptũ ſibi conum FAE, eſſe vt AB ad BC.

32 *huius.* Erit enim ſolidum ad ſphæram vt BA ad AC, ſumptiſq; conſequentium dimidijs, erit ſolidum ad hemiſphærium vt BA ad DC, ſed (per lemma præcedens) hemiſphærium eſt ad conum FAE, vt circulus ex DC ad circulum ex CE; ſiue vt recta DC ad CB; ergò ex æquo erit ſphæra-

le ſo-

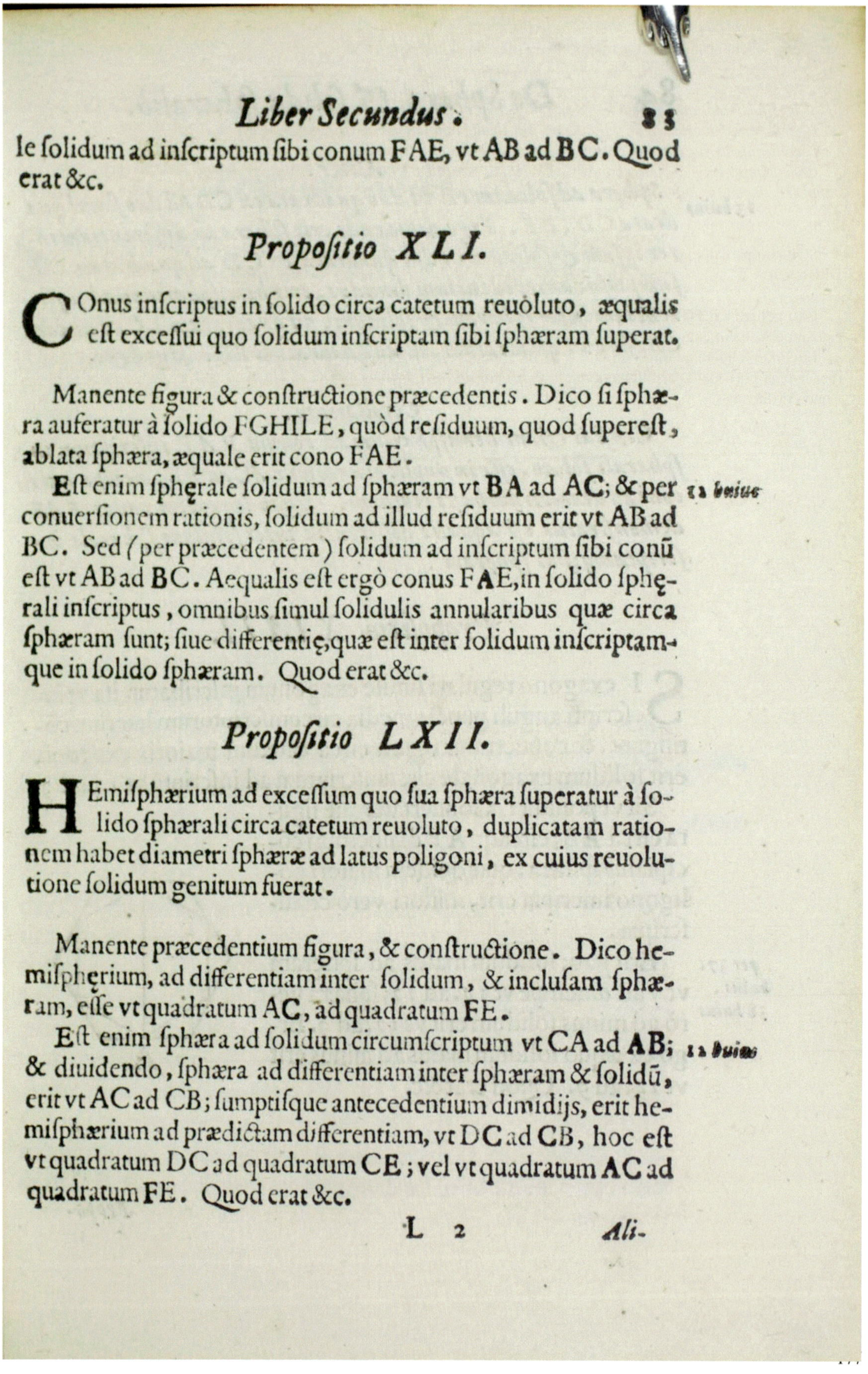

Liber Secundus. 83

le ſolidum ad inſcriptum ſibi conum FAE, vt AB ad BC. Quod erat &c.

Propoſitio XLI.

Conus inſcriptus in ſolido circa catetum reuoluto, æqualis eſt exceſſui quo ſolidum inſcriptam ſibi ſphæram ſuperat.

Manente figura & conſtructione præcedentis. Dico ſi ſphæra auferatur à ſolido FGHILE, quòd reſiduum, quod ſupereſt, ablata ſphæra, æquale erit cono FAE.

Eſt enim ſphęrale ſolidum ad ſphæram vt BA ad AC; & per conuerſionem rationis, ſolidum ad illud reſiduum erit vt AB ad BC. Sed (per præcedentem) ſolidum ad inſcriptum ſibi conũ eſt vt AB ad BC. Aequalis eſt ergò conus FAE, in ſolido ſphęrali inſcriptus, omnibus ſimul ſolidulis annularibus quæ circa ſphæram ſunt; ſiue differentię, quæ eſt inter ſolidum inſcriptamque in ſolido ſphæram. Quod erat &c. 11 huius

Propoſitio LXII.

Hemiſphærium ad exceſſum quo ſua ſphæra ſuperatur à ſolido ſphærali circa catetum reuoluto, duplicatam rationem habet diametri ſphæræ ad latus poligoni, ex cuius reuolutione ſolidum genitum fuerat.

Manente præcedentium figura, & conſtructione. Dico hemiſphęrium, ad differentiam inter ſolidum, & incluſam ſphæram, eſſe vt quadratum AC, ad quadratum FE.

Eſt enim ſphæra ad ſolidum circumſcriptum vt CA ad AB; & diuidendo, ſphæra ad differentiam inter ſphæram & ſolidũ, erit vt AC ad CB; ſumptiſque antecedentium dimidijs, erit hemiſphærium ad prædictam differentiam, vt DC ad CB, hoc eſt vt quadratum DC ad quadratum CE; vel vt quadratum AC ad quadratum FE. Quod erat &c. 11 huius

L 2 *Ali-*

le ſolidum ad inſcriptum ſibi conum FAE, vt AB ad BC. Quod erat &c.

Propoſitio XLI.

COnus inſcriptus in ſolido circa catetum reuoluto, æqualis eſt exceſſui quo ſolidum inſcriptam ſibi ſphæram ſuperat.

Manente figura & conſtructione præcedentis. Dico ſi ſphæra auferatur à ſolido FGHILE, quòd reſiduum, quod ſupereſt, ablata ſphæra, æquale erit cono FAE.

Eſt enim ſphęrale ſolidum ad ſphæram vt BA ad AC; & per conuerſionem rationis, ſolidum ad illud reſiduum erit vt AB ad BC. 22
Sed (per præcedentem) ſolidum ad inſcriptum ſibi conũ eſt vt AB ad BC. Aequalis eſt ergò conus FAE, in ſolido ſphęrali inſcriptus, [23] *huius*
omnibus ſimul ſolidulis annularibus quæ circa ſphæram ſunt; ſiue differentię, quæ eſt inter ſolidum inſcriptamque in ſolido ſphæram. Quod erat &c.

Propoſitio LXII.

HEmiſphærium ad exceſſum quo ſua ſphæra ſuperatur à ſolido ſphærali circa catetum reuoluto, duplicatam rationem habet diametri ſphæræ ad latus poligoni, ex cuius reuolutione ſolidum genitum fuerat.

Manente præcedentium figura, & conſtructione. Dico hemiſphęrium, ad differentiam inter ſolidum, & incluſam ſphæram, eſſe vt quadratum AC, ad quadratum FE.

Eſt enim ſphæra ad ſolidum circumſcriptum vt CA ad AB; & diuidendo, ſphæra ad differentiam inter ſphæram & ſolidũ, erit vt
AC ad CB; ſumptiſque antecedentium dimidijs, erit hemiſphærium ad prædictam differentiam, vt DC ad CB, hoc eſt vt quadratum 12 *huius.*
DC ad quadratum CE; vel vt quadratum AC ad quadratum FE. Quod erat &c

L 2 *Ali-*

Aliter.

13 huius *Sphęra ad solidum est vt duo quadrata ex CD ad duo simul quadrata CD, DE. Ergo diuidendo erit sphæra ad differentiam inter ipsam & solidum vt duo quadrata ex CD ad quadratum CE sumptisq; antecedentium dimidijs, erit hemisphærium ad differentiam inter sphæram & solidum, vt quadratum DC ad quadr. CE, siue vt quadratum AC ad quadratum FE. Quod &c.*

Corollarium.

Constat etiam hemisphærium ad conum FAE inscriptum in sphærali solido, esse in duplicata ratione AC ad FE, nempe axis coni ad diametrum basis eiusdem. Quandoquidem conus FAE demonstratus est æqualis differentiæ inter solidum sphærale inscriptamq; sibi sphęram.

Propositio XLIII.

SI exagono regulari simile exagonum inscribatur, ita vt inscripti anguli puncta media circumscriptorum laterum cõtingant, & conuertatur figura circà catetum maioris exagoni, erit solidum exagonale circumscriptum ad inscriptũ vt 14. ad 9.

Sit vt ponitur: Conuertaturque figura circà AB; circaq; AB diametrum cõcipiatur sphæra, quæ quidem maiori poligono inscripta erit, minori verò circũscripta.

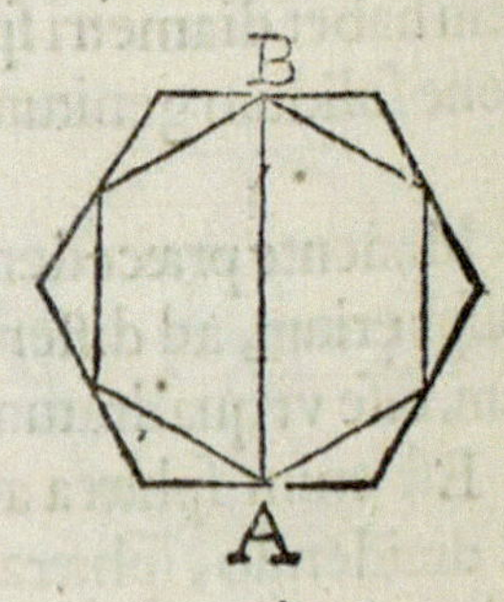

per 37. huius. 38 huius Erit itaq; solidum maius ad sphęram vt 7 ad 6. nempe vt 14. ad 12; sphæra verò ad minus solidum erit vt 12. ad 9. Ergò ex æquo solidum maius ad minus erit vt 14. ad 9. Quod erat &c.

Propo-

Aliter.

Sphęra ad ſolidum est vt duo quadrata ex CD ad duo ſimul quadrata CD, DE. Ergo diuidendo erit ſphæra ad differentiam inter
13 huius. *ipſam & ſolidum vt duo quadrata ex CD ad quadratum CE ſumptiſq; antecedentium dimidijs, erit hemiſphærium ad differentiam inter ſphæram & ſolidum, vt quadratum DC ad quadr. CE, ſiue vt quadratum AC ad quadratum FE. Quod &c..*

Corollarium.

Conſtat etiam hemiſphærium ad conum FAE inſcriptum in ſphærali ſolido, eſſe in duplicata ratione AC ad FE, nempe axis coni ad diametrum baſis eiuſdem. Quandoquidem conus FAE demonſtratus eſt æqualis differentiæ inter ſolidum ſphærale inſcriptamq; ſibi ſphęram.

Propoſitio XLIII.

SI exagono regulari ſimile exagonum inſcribatur, ita vt inſcripti anguli puncta media circumſcriptorum laterum cõtingant, & conuertatur figura circà catetum maioris exagoni, erit ſolidum exagonale circumſcriptum ad inſcriptũ vt 14. ad 9.

per 37 *huius* Sit vt ponitur: Conuertaturque figura circà AB; circaq; AB diametrum cõcipiatur ſphæra, quæ quidem maiori poligono inſcripta erit, minori verò circũſcripta.

38 huius. Erit itaq; ſolidum maius ad ſphęram vt 7 ad 6. nempe vt 14. ad 12; ſphæra verò ad minus ſolidum erit vt 12. ad 9. Ergò ex æquo ſolidum maius ad minus erit vt 14. ad 9. Quod erat &c.

Propo-

Propoſitio XLIV.

Solidum ſphærale factum ex reuolutione alicuius poligoni circa diagonalem, ad ſolidum ex reuolutione eiuſdem poligoni circà catetum; eſt vt rectangulum ſub diagonali, & cateto, bis ſumptum, ad duo ſimul quadrata, quorum alterum ex diagonali fit, alterum autem ex cateto.

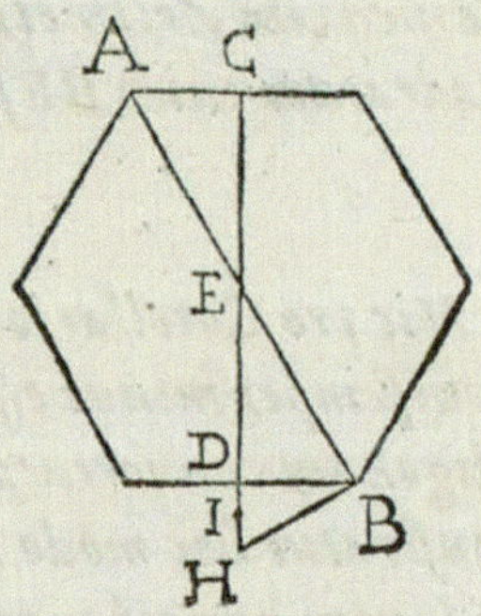

Eſto poligonum regulare quodcumque, habens latera nnmero paria, cuius diagonalis ſit AB, catetus verò CD. Et concipiatur poligonum conuerti duplici axe; nempe primùm circà diagonalem AB; & iterum circa catetum CD. Dico ſolidum ex diagonali ad ſolidum ex cateto eſſe, vt rectangulum BED bis ſumptum, ad quadrata ex BE, & ex ED: ſiue vt eorum quadrupla.

Fiat angulus EBH rectus, ſeceturq; bifariam DH in I; eritq. EI media Aritmetica inter ED, EH: Iam ſolidum ex diagonali ad inſcriptam ſibi ſphæram eſt, vt AB, ad CD; ſphæra verò ad ſolidum ex cateto, eſt vt CD, ad CH; ergò ex æquo ſolidum ex diagon. ad ſolidum ex cateto, erit ut AB ad CH, ſiue ut EB ad EI, (ſunt enim ſemiſſes rectarum AB, CH.) Cum autem BE media Geometrica ſit inter HE, ED; ipſa uerò EI media Aritmetica ſit inter eaſd. erit ſolidum ex diagonali ad ſolidum ex cateto ut media Geomet. ad mediam Aritmet. inter rectas HE, ED Sed ratio rectæ HE ad ED, ead. eſt ac quadr. BE ad quadr. ED: propterea erit ſolidum ex diagonali ad ſolidum ex cateto, ut ſpatium medium proportionale Geometricum ad ſpatium medium Aritmeticum inter quadrata BE, ED. Spatium autem mediũ Geometricum inter quadrata BE, ED. eſt rectangulum BED; medium verò Aritmeticum eſt quadratũ ED, cum ſemiſſe quadrati DB. Ergo ſolidum ex diagonali ad ſolidum ex cateto erit

6. huius
11 huius

vt re-

Propoſitio XLIV.

SOlidum ſphærale factum ex reuolutione alicuius poligoni circa diagonalem, ad ſolidum ex reuolutione eiuſdem poligoni circà catetum; eſt vt rectangulum ſub diagonali, & cateto, bis ſumptum, ad duo ſimul quadrata, quorum alterum ex diagonali fit, alterum autem ex cateto.

Eſto poligonum regulare quodcumque, habens latera nnmero[38] paria, cuius diagonalis ſit AB, catetus verò CD. Et concipiatur poligonum conuerti duplici axe; nempe primùm circà diagonalem AB; & iterum circa catetum CD. Dico ſolidum ex diagonali ad ſolidum ex cateto eſſe, vt rectangulum BED bis ſumptum, ad quadrata ex BE, & ex ED: ſiue vt eorum quadrupla.

Fiat angulus EBH rectus, ſeceturq; bifariam DH in I; eritq.

EI media Aritmetica inter ED, EH: Iam ſolidum ex diagonali ad inſcriptam ſibi ſphæram eſt, vt AB, ad CD; ſphæra verò ad ſolidum ex cateto, eſt vt CD, ad CH; ergò ex æquo ſolidum ex diagon. ad ſolidum ex cateto, erit ut AB ad CH, ſiue ut EB ad EI, (ſunt enim ſemiſſes rectarum AB, CH.) Cum autem BE media Geometrica ſit inter HE, ED; ipſa uerò EI media Aritmetica sit inter eaſd. erit ſolidum ex diagonali ad ſolidum ex cateto ut media Geomet. ad mediam Aritmet. inter rectas HE, ED[39] *6. huius.* *12 huius.*

Sed ratio rectæ HE ad ED, ead. eſt ac quadr. BE ad quadr. ED: propterea erit ſolidum ex diagonali ad ſolidum ex cateto, ut ſpatium medium proportionale Geometricum ad ſpatium medium Aritmeticum inter quadrata BE, ED. Spatium autem mediũ Geometricum inter quadrata BE, ED. eſt rectangulum BED; medium verò Aritmeticum eſt quadratũ ED, cum ſemiſſe quadrati DB. Ergo ſolidum ex diagonali ad ſolidum ex cateto erit

vt re–

[38] nnmero] numero.
[39] ED] ED.

vt rectangulum BED; ad quadratum ED cum semisse quadrati DB; Vel (sumptis duplis) vt rectangulum BED, bis sumptum, ad quadratum ED bis, cum integro quadrato DB. Siue vt rectangulum BED bis sumptum, ad quadrata BE, ED. Quod erat &c.

Assumpsimus rectangulum BED, medium proportionale esse inter quadrata BE, ED. Hoc enim patet in propositis quibuscunque rectis duabus lineis.

Assumpsimus etiam quadratum ED cum semisse quadrati DB, esse medium Aritmeticum inter quadrata BE, ED. Quod patet quadratum enim BE superat quadratum ED quadrato BD.

Corollarium.

Hic pro Corollario demonstrari potest, solidum ex diagonali factum semper minus esse solido, quod fit ex cateto; quando idem poligonum conuertatur circa diagonalem, & circa catetum. Demonstratur hoc modo.

Quoniam rectangulum BED bis sumptum, minus est duobus quadratis BE, ED (sunt enim in continua ratione quadratum EB, rectangulum DEB, & quadratum E D, ideoq; dupla mediæ, minor est duabus extremis magnitudinibus.) Et est vt rectangulum BED bis sumptum ad quadr. BE, E D simul, ità solidum ex diagonali ad solidum ex cateto; Erit solidum ex diagonali minus quam solidum ex cateto. Quod erat &c.

Si quis autem quærat, quo excessu solidum ex cateto superet solidum ex diagonali. Hoc modo illum proportione notum habebit.

Faciat vt duo quadrata BE, ED simul, ad quadratum quod fit ex differenti a rectarum BE, ED, ità maius solidum ad aliud: Et habebit excessum quo maius solidum superat minus.

Propo-

vt rectangulum BED; ad quadratum ED cum ſemiſſe quadrati DB; Vel (ſumptis duplis) vt rectangulum BED, bis ſumptum, ad quadratum ED bis, cum integro quadrato DB. Siue vt rectangulum BED bis ſumptum, ad quadrata BE, ED. Quod erat &c.

Aſſumpſimus rectangulum BED, medium proportionale eße inter quadrata BE, ED. Hoc enim patet in propoſitis quibuſcunque rectis duabus lineis.

Aſſumpſimus etiam quadratum ED cum ſemiſſe quadrati DB, eße medium Aritmeticum inter quadrata BE, ED. Quod patet quadratum enim BE ſuper at quadratum ED quadrato BD.

Corollarium.

Hic pro Corollario demonſtrari poteſt, ſolidum ex diagonali factum ſemper minus eſſe ſolido, quod fit ex cateto; quando idem poligonum conuertatur circa diagonalem, & circa catetum. Demonſtratur hoc modo.

Quoniam rectangulum BED bis ſumptum, minus eſt duobus quadratis BE, ED (ſunt enim in continua ratione quadratum EB, rectangulum DEB, & quadratum ED, ideoq; dupla mediæ, minor eſt duabus extremis magnitudinibus.) Et eſt vt rectangulum BED bis ſumptum ad quadr. BE, ED ſimul, ità ſolidum ex diagonali ad ſolidum ex cateto; Erit ſolidum ex diagonali minus quam ſolidum ex cateto. Quod erat &c.

Si quis autem quærat, quo exceſſu ſolidum ex cæteto ſuperet ſolidum ex diagonali. Hoc modo illum proportione notum habebit.

Faciat vt duo quadrata BE, ED ſimul, ad quadratum quod fit ex differentia rcctarum[40] *BE, ED, ità maius ſolidum ad aliud: Et habebit exceßum quo maius ſolidum ſuperat minus.*

Propo-

[40] *rcctarum*] *rectarum.*

Propositio XLV.

SI intra poligonum regulare parilaterum inſcribatur ſimile poligonum, ità vt anguli inſcripti biſectiones laterum circumſcripti contingant;conuertaturq; figura circa catetum maioris poligoni: Erit maius ſolidum ſphęrale ad minus, vt ſunt duo ſimul quadrata duarum diagonalium, ad duo quadrata minoris cateti.

Eſto poligonum parilaterum ABC &c. intra quod inſcribatur ſimile poligonum AIC &c. vti dictum eſt. Conuertaturq; figura circa AC catetum maioris poligoni. Dico ſolidum ſphęrale ABC,ad ſolidum AIC eſſe vt duo quadrata ſimul duarum diagonalium, nempe BD, DC. ad duo quadrata minoris cateti DI. Circumſcribatur ſolido AIC ſua ſphęra,quę alteri ſolido inſcripta erit.

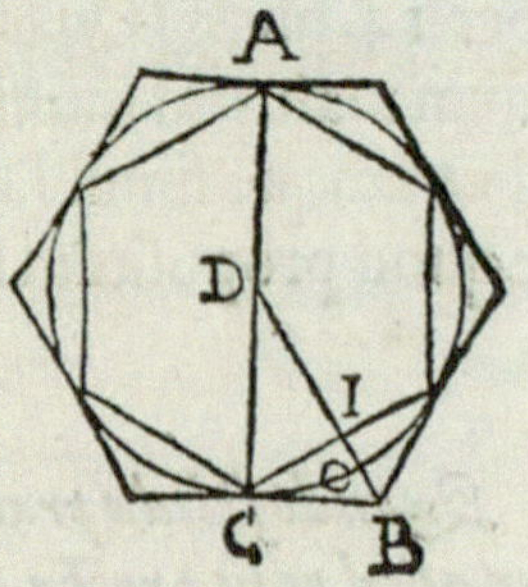

Iam ſolidum ABC ad inſcriptam ſphæram, eſt vt duo quadrata ſimul BD,DC ad duplum quadrati DC (per 13. huius.) Sphæra verò ad inſcriptum ſolidum eſt,vt duplum quadrati DC ad duplum quadrati DI (per 7. huius) Ergo ex æquo maius ſolidum ſphærale ad minus erit vt duo ſimul quadrata BD,DC ad duplum quadrati DI. Quod erat &c.

Propositio XLVI.

IIſdem poſitis: ſi conuertatur figura circa diagonalem maioris poligoni GC. Erit maius ſolidum ad minus, vt integer axis AC maioris ſolidi, ad vtramque ſimul, nempe ſemicatetum DG minoris, & quartam proportionalium GF; ſi fiat vt ſemidiagonalis minoris ad ſemicatetum; ita ſemicatetus ad tertiam, & tertia ad quartam.

Eſto ſolidum quale poſitum eſt ABCH. cui inſcriptum ſit ſolidum

Propoſitio XLV.

SI intra poligonum regulare parilaterum inſcribatur ſimile poligonum, ità vt anguli inſcripti biſectiones laterum circumſcripti contingant; conuertaturq; figura circa catetum maioris poligoni: Erit maius ſolidum ſphęrale ad minus, vt ſunt duo ſimul quadrata duarum diagonalium, ad duo quadrata minoris cateti.

Eſto poligonum parilaterum ABC &c. intra quod inſcribatur ſimile poligonum AIC &c. vti dictum eſt. Conuertaturq; figura circa AC catetum maioris poligoni. Dico ſolidum ſphęrale ABC, ad ſolidum AIC eſſe vt duo quadrata ſimul duarum diagonalium, nempe BD, DC. ad duo quadrata minoris cateti DI. Circumſcribatur ſolido AIC ſua ſphęra, quę alteri ſolido inſcripta erit.

Iam ſolidum ABC ad inſcriptam ſphæram, eſt vt duo quadrata ſimul BD, DC ad duplum quadrati DC (per 13. huius.) Sphæra verò ad inſcriptum ſolidum eſt, vt duplum quadrati DC ad duplum quadrati DI (per 7. huius) Ergo ex æquo maius ſolidum ſphærale ad minus erit vt duo ſimul quadrata BD, DC ad duplum quadrati DI. Quod erat &c.

Propoſitio XLVI.

IIſdem poſitis: ſi conuertatur figura circa diagonalem maioris poligoni GC. Erit maius ſolidum ad minus, vt integer axis AC maioris ſolidi, ad vtramque ſimul, nempe ſemicatetum DG minoris, & quartam proportionalium GF; ſi fiat vt ſemidiagonalis minoris ad ſemicatetum; ita ſemicatetus ad tertiam, & tertia ad quartam.

Eſto ſolidum quale poſitum eſt ABCH. cui inſcriptum ſit ſo-

lidum

lidum IBD. vti dictum eſt. Ducatur, DE perpendicularis ad GB,& EF ad GC;eruntq; in continua proportione CG, GB,GD, GE GF.ob angulos rectos.

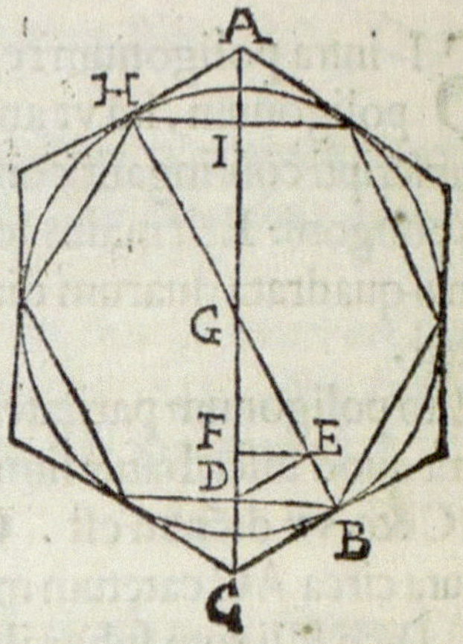

Iam ſolidum maius ad ſphæram eſt vt AC ad HB (per 6.huius) ſphęra autem ad ſolidum minus eſt vt HB ad vtramque ſimul DG. GF (per 14. huius) Quare ex æquo ſolidum maius ad minus erit vt AC ad vtramque ſimul DG. GF. nempe quod propoſitum fuerat.

Corollarium.

Quando ſolida prædicta ab exagono genita fuerint: demonſtratur quod poſita recta AC 32. DG. & GF notæ ſunt. nempe DG. 12. & GF. 9. Ergo in hoc caſu ſolidum maius ad minus eſſet vt 32. ad 21.

Supereſt nunc vt ſolida ſphæralia abſolutè conſiderata interſe conferamus, & hoc quot modis fieri poterit: quemadmodum in proœmio operis nos eſſe facturos promiſeramus.

Propoſitio XLVII.

SOlida ſphęralia parilatera circa diagonalem reuoluta, inter ſe ſunt vt parallelepipeda baſi quadr. cateti, altitudine vero diagonali eorumdem.

Sint duo ſolida ſphæralia parilatera circa diagonales AC, DF reuoluta. Sintq; HI,LV perpēdiculares ad latera CB, FE, Dico ſolidum ſphærale ABC ad ſolidum DEF, eſſe vt parallelepipedum baſi quadrato HI altitudine uerò HC, ad paralle-

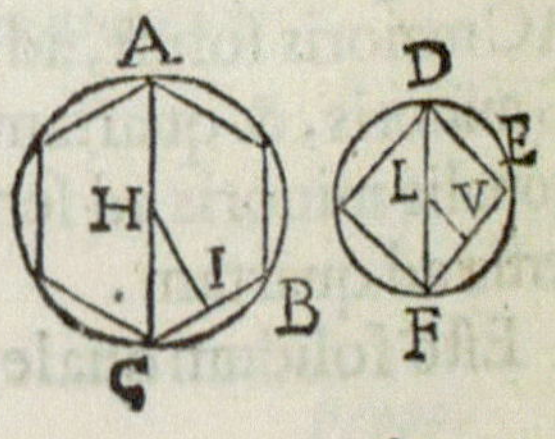

lep:

lidum IBD. vti dictum eſt. Ducatur, DE perpendicularis ad GB, & EF ad GC; eruntq; in continua proportione CG, GB, GD, GE GF. ob angulos rectos.

Iam ſolidum maius ad ſphæram eſt vt AC ad HB (per 6. huius) ſphęra autem ad ſolidum minus eſt vt HB ad vtramque ſimul DG. GF (per 14. huius) Quare ex æquo ſolidum maius ad minus erit vt AC ad vtramque ſimul DG. GF. nempe quod propoſitum fuerat.

Corollarium.

Quando ſolida prædicta ab exagono genita fuerint: demonſtratur quod poſita recta AC 32. DG. & G F notæ ſunt. nempe DG. 12. & GF. 9. Ergo in hoc caſu ſolidum maius ad minus eßet vt 32. ad 21.

Supereſt nunc vt ſolida ſphæralia abſolutè conſiderata inter ſe conferamus, & hoc quot modis fieri poterit: quemadmodum in proẽmio operis nos eſſe facturos promiſ eramus.

Propoſitio XLVII.

SOlida ſphęralia parilatera circa diagonalem reuoluta, inter ſe ſunt vt parallelepipeda baſi quadr. cateti, altitudine vero diagonali eorumdem.

Sint duo ſolida ſphæralia parilatera circa diagonales AC, DF reuoluta. Sintq; HI, LV perpẽdiculares ad latera CB, FE, Dico ſolidum ſphærale ABC ad ſolidum DEF. eſſe vt parallelepipedum baſi quadrato HI altitudine uerò HC, ad paralle-

lep:

ep. baſi quadrato LV, altitudine LF.

Intelligatur vtrique circumſcripta ſphæra ſua. Tunc enim ſolidum ABC ad ſphæram ſuam erit vt quadratum IH ad quadratum HC, ſiue (ſumpta communi altitudine CH) vt parallelepipedum baſi quadrato IH, altitudine HC, ad cubum HC. Sphęra autem ABC ad ſphæram DEF, eſt vt cubus HC ad cubum LF. At ſphæra DEF, (vt nuper in altera oſtendebamus) ad ſolidum ſuum DEF. eſt vt cubus LF, ad parallelepipedum baſi quadrato LV, altitudine LF: ergo ex æquo erit ſolidum ABC, ad ſolidum ſphærale DEF, vt parallelepipedum baſi quãdrato HI, altitudine HC; ad parallelepipedum baſi quadrato LV, altitudine LF. Quod erat &c. *7. huius.*

Scholium.

Idem concludetur etiamſi concipiantur ſphæræ iuxta 6. huius intra data ſolida inſcriptæ; ſiue altera tantum inſcripta, altera verò circumſcripta iuxta 6. & 7. huius ſicut experienti patebit.

Propoſitio XLVIII.

Solida ſphæralia parilatera circa catetum reuoluta inter ſe ſunt, vt parallelepipeda baſi quadrato diagonalis, altitudine verò quæ ſit æqualis cateto, & quartæ proportionalium, ſi fiat vt diagonalis ad catetum, ita catetus ad tertiam, & ita tertia ad quartam.

Sint duo ſolida ſphæralia circa catetos B, & D reuoluta. Continueturque ratio A ad B in quatuor terminis A, B, E, F. Itẽ ratio diagonalis C ad catetum D continuetur in quatuor terminis C, D, H, I. Dico, primum ſolidum ad ſecundum eſſe vt parallelepipedum baſi quadrato A, altitudine verò B & F; ad parallelepipedum baſi quadrato C. altitudine verò D & I.

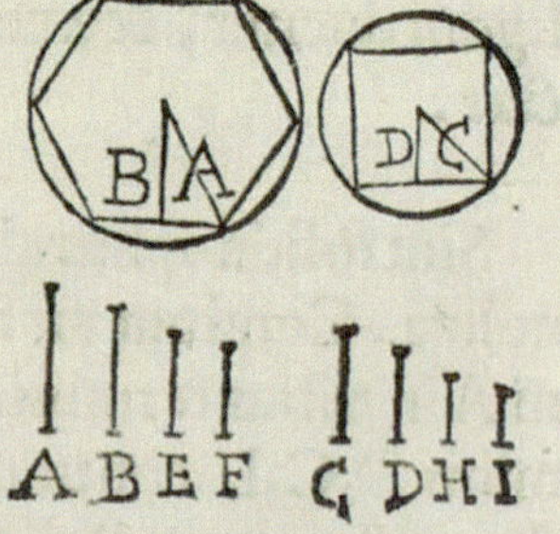

Nam primum ſolidum ad ſphęram ſuam eſt vt B & F ſimul ad M Abis *14 huius*

ep. baſi quadrato LV, altitudine LF.

Intelligatur vtrique circumſcripta ſphæra ſua. Tunc enim ſolidum ABC ad ſphæram ſuam erit vt quadratum IH ad quadratum HC, ſiue (ſumpta communi altitudine CH) vt parallelepipedum baſi quadrato IH, altitudine HC, ad cubum HC. Sphęra autem ABC ad ſphæram DEF, eſt vt cubus HC ad cubum LF. At ſphæra DEF, (vt nuper in altera oſtendebamus) ad ſolidum ſuum DEF. eſt vt cubus LF, ad parallelepipedum baſi quadrato LV, altitudine LF: ergo ex æquo erit ſolidum ABC, ad ſolidum ſphærale DEF, vt parallelepipedum baſi quadrato HI, altitudine HC; ad parallelepipedum baſi quadrato LV, altitudine LF. Quod erat &c. 7 *huius.*

Scholium.

Idem concludetur etiam ſi concipiantur ſphæræ iuxta 6. huius intra data ſolida inſcriptæ; ſiue altera tantum inſcripta, altera verò circumſcripta iuxta 6. & 7. huius ſicut experienti patebit.

Propoſitio XLVIII.

SOlìda ſphæralia parilatera circa catetum reuoluta inter ſe ſunt, vt parallelepipeda baſi quadrato diagonalis, altitudine verò quæ ſit æqualis cateto, & quartæ proportionalium, ſi fiat vt diagonalis ad catetum, ita catetus ad tertiam, & ita tertia ad quartam.

Sint duo ſolida ſphæralia circa catetos B, & D reuoluta. Continueturque ratio A ad B in quatuor terminis A, B, E, F. Itẽ ratio diagonalis C ad catetum D continuetur in quatuor terminis, C, D, H, I. Dico, primum ſolidum ad ſecundum eſſe vt parallelepipedum baſi quadrato A, altitudine verò B & F; ad parallelepipedum baſi quadrato C. altitudine verò D & I.

Nam primum ſolidum ad ſphęram ſuam eſt vt B& F ſimul ad 14 *huius.*

M A bis

90 *De Sphæra, & ſolidis ſphæralib.*

A bis ſumptam: acceptaq; communi baſi quadrato A; erit ſolidum primum ad ſphæram ſuam, vt parallelepipedum baſi quadrato A, altitudine verò B & F ſimul, ad duos cubos A. Sphæra autem prima ad ſecundam ſphærã eſt vt duo cubi A ad duos cubos C. Sphæra tandem ſecunda ad ſolidum ſuum, eſt vt duo cubi C, ad parallelepipedum baſi quadrato C altitudine verò D, & I ſimul (quod oſtenditur vt nuper factum eſt in prima ſphæ ra) ergo ex æquo primum ſolidum ſphærale ad ſecundum, erit vt parallelepipedum baſi quadrato A, altitudine B & F ſimul, ad parallelepipedum baſi quadrato C altitudine verò D & I ſimul. Quod erat &c.

Scholium.

Idem concludi poteſt ſi ſphæræ concipiantur intra ipſa ſolida inſcriptæ iuxta Propoſitionem 13. huius; ſiue altera inſcripta, altera verò circumſcripta iuxta 13. & 14. huius. Quando verò termini proportionis alij euadant à propoſitis, vt in hac, & in ſequentibus, ſcias proportionem ſemper eandem eſſe, in quibuſcun que tandem terminis eueniat.

Propoſitio IL.

SOlida ſphæralia imparilatera ſunt inter ſe vt parallelepipeda, baſi quadrato perpendicularis, quæ ex centro poligoni ducitur in latus eiuſdem, altitudine verò æquali prædictæ perpendiculari, vna cum dupla eius, quæ ex centro ad angulum po ligoni ducitur, & cum tertia proportionalium ad duas prædictas.

Sint ſolida ſphæralia imparilatera, circa catetos B, & D. reuoluta. Continuetur ratio perpendicularis B ad radium poligoni A in tribus terminis B, A, E. Item ratio D. ad C in tribus terminis D, C, I, continuata ſit. Dico ſolidum primum ad ſecundum eſſe vt parallelepipedum baſi quadrato B, altitudine verò æquali B ſemel, A bis, & E. ſemel, ſimulq; ſumptis, ad parallelepipe-

A bis ſumptam: acceptaq; communi baſi quadrato A; erit ſolidum primum ad ſphæram ſuam, vt parallelepipedum baſi quadrato A, altitudine verò B & F ſimul, ad duos cubos A. Sphæra autem prima ad ſecundam ſphærã eſt vt duo cubi A ad duos cubos C. Sphæra tandem ſecunda ad ſolidum ſuum, eſt vt duo cubi C, ad parallelepipedum baſi quadrato C altitudine verò D, & I ſimul (quod oſtenditur vt nuper factum eſt in prima ſphęra) ergo ex æquo primum ſolidum ſphærale ad ſecundum, erit vt parallelepipedum baſi quadrato A, altitudine B & F ſimul, ad parallelepipedum baſi quadrato C altitudine verò D & I ſimul.

Quod erat &c.

Scholium.

Idem concludi poteſt ſi ſphæræ concipiantur intra ipſa ſolida inſcriptæ iuxta Propoſitionem 13. huius; ſiue altera inſcripta, altera verò circumſcripta iuxta 13. & 14. huius. Quando verò termini proportionis alij euadant à propoſitis, vt in hac, & in ſequentibus, ſcias proportionem ſemper eandem eſſe, in quibuſcunque tandem terminis eueniat.

Propoſitio IL.

SOlida ſphæralia imparilatera ſunt inter ſe vt parallelepipeda, baſi quadrato perpendicularis, quæ ex centro poligoni ducitur in latus eiuſdem, altitudine verò æquali prædictæ perpendiculari, vna cum dupla eius, quæ ex centro ad angulum poligoni ducitur, & cum tertia proportionalium ad duas prædictas.

Sint ſolida ſphæralia imparilatera, circa catetos B, & D. reuoluta. Continuetur ratio perpendicularis B ad radium poligoni A in tribus terminis B, A, E. Item ratio D. ad C in tribus terminis D, C, I, continuata ſit. Dico ſolidum primum ad ſecundum eſſe vt parallelepipedum baſi quadrato B, altitudine verò æquali B ſemel, A bis, & E. ſemel, ſimulq; ſumptis, ad paralle-

lepipe-

lepipedum baſi quadr. D. altitudine verò æquali D. ſemel, C. bis, & I ſemel ſimulq; ſumptis.

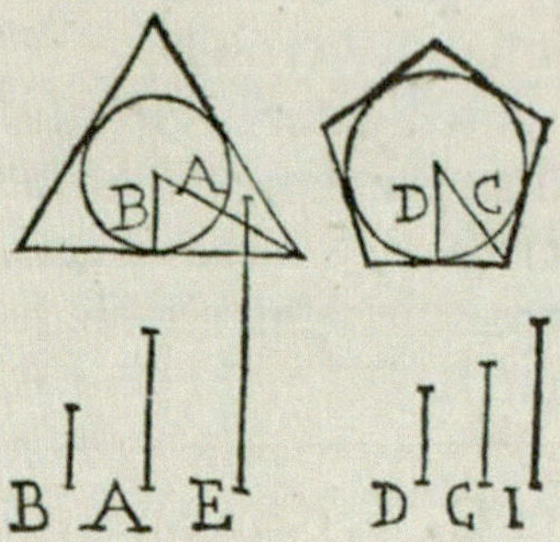

Concipiatur in vtroq; ſolido ſphęrali ſua ſphęra inſcripta, eritq; ſolidum primum ad ſphæram ſuam vt B & E ſimul cum dupla ipſius A ad quadruplam B. ſumptaque communi baſi quadrato B. erit ſolidum primum ad ſphæram ſuam vt parallelepipedũ baſi quadrato B. altitudine verò B & E cum dupla A. ad quatuor cubos B. Sphæra autem prima ad ſecundam eſt, vt quatuor cubi B ad quatuor cubos D; Sphæra tandem ſecunda ad ſolidum ſuum eſt, vt quatuor cubi D. ad parallelepipedum baſi quadrato D. altitudine D & I cum dupla ipſius C (quod oſtenditur vt nuper factum eſt) ergo ex æquo patet quod propoſitum fuerat. &c. *18 huius*

Propoſitio L.

SOlidum ſphærale parilaterum circa diagonalem reuolutũ, ad ſolidum ſphęrale parilaterum circa catetum reuolutum, eſt vt parallelepipedum baſi quadrato cateti, altitudine diagonali bis ſumptum, ad parallelepipedum baſi quadrato cateti ſimul diagonaliſque, altitudine verò cateti.

Sint duo ſolida ſphæralia, quorũ alterum circa diagonalem A ſit reuolutum, alterum verò circa catetum C. Dico ſolidum primũ circa diagonalem, ad ſolidũ ſecundum circa catetum, eſſe vt parallelepipedũ baſi quadr. B altitudine A

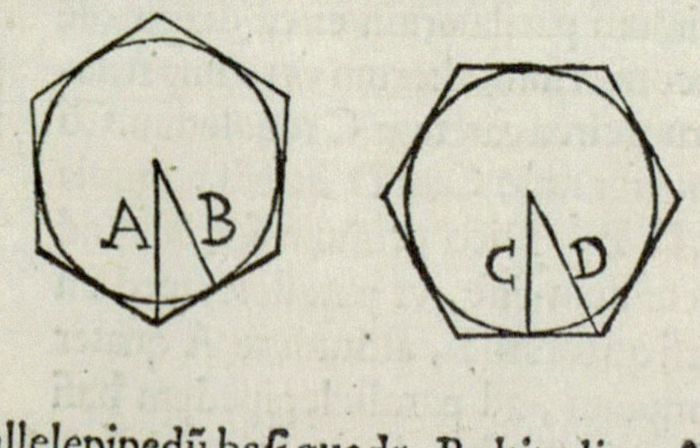

M 2 bis

lepipedum baſi quadr. D. altitudine verò æquali D. ſemel, C. bis, & I ſemel ſimulq; ſumptis. 18
Concipiatur in vtroq; ſolido ſphęrali ſua ſphæra inſcripta, eritq; ſolidum primum ad ſphæram ſuam vt B & E ſimul cum dupla ipſius A ad quadruplam B. ſumptaque communi baſi quadrato B. erit ſolidum primum ad ſphæram ſuam vt parallelepipedũ baſi quadrato B. altitudine verò B & E cum dupla A. ad quatuor cubos B. Sphæra autem prima ad ſecundam eſt, vt quatuor cubi B ad quatuor cubos D; Sphæra tandem ſecunda ad ſolidum ſuum eſt, vt quatuor cubi D. ad parallelepipedum baſi quadrato D. altitudine D & I cum dupla ipſius C (quod oſtenditur vt nuper factum eſt) ergo ex æquo patet quod propoſitum fuerat. &c. *huius*

Propoſitio L.

SOlidum ſphærale parilaterum circa diagonalem reuolutũ, ad ſolidum ſphęrale parilaterum circa catetum reuolutum, eſt vt parallelepipedum baſi quadrato cateti, altitudine diagonali bis ſumptum, ad parallelepipedum baſi quadrato cateti ſimul diagonaliſque, altitudine verò cateti.

Sint duo ſolida ſphæralia, quorũ alterum circa diagonalem A ſit reuolutum, alterum verò circa catetum C. Dico ſolidum primũ circa diagonalem, ad ſolidũ ſecundum circa catetum, esse vt parallelepipedũ basi quadr. B altitudine A bis ſumptum, ad parallelepipedum baſi æquali quadratis C,D, altitudine verò C.

6. huius. Intelligatur in vtroque ſolido inſcripta ſua ſphæra. Et erit ſolidum primum ad ſphæram ſuam, vt recta A ad B; ſumptaq; eadem baſi quadrato B; erit ſolidum primum ad ſphæram ſuam, vt parallelepipedum baſi quadrato B altitudine verò A, ad cubum B. ſiue vt duplum dicti parallelepipedi ad duos cubos B. Sphæra verò prima ad ſecundam eſt, vt duo cubi B, ad duos cu-
13 huius bos C. Sphæra tandem ſecunda ad ſolidum ſuum eſt, vt duo quadrata ex C, ad duo quadrata C, & D; ſumptaque communi altitudine C, eſt, vt duo cubi C, ad parallelepipedum baſi æquali quadratis C & D. altitudine verò C. Propterea ex æquo patet quod propoſitum erat.

Propoſitio LI.

Solidum ſphęrale parilaterum circa diagonalem reuolutum, ad ſolidum ſphærale imparilaterum eſt, vt parallelepipedum baſi quadrato cateti, altitudine diagonali quater ſumptum; ad parallelepipedum baſi quadrato rectæ illius quæ ex cenrro poligoni imparilateri perpendiculariter ducitur in latus eiuſdẽ; altitudine verò æquali prædictæ perpendiculari, vna cum dupla illius quę ex centro ad angulum ducitur, & cum tertia proportionalium ad duas prædictas.

Sint duo ſolida ſphæralia, nempe primum parilaterum circa diagonalẽ A conuerſum, alterum verò imparilaterum circa catetum C reuolutum. Cõtinuetur ratio C ad D in trib. terminis C,D,E. Dico primum ſolidum ad ſecundum eſſe, vt parallelepipedum baſi quadrato B, altitudine A quater ſumptum, ad parallelepipedum baſi

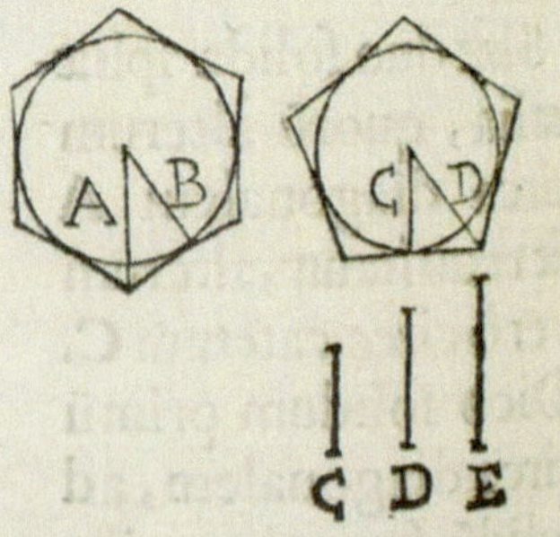

quadrato C, altitudine verò æquali rectis C, & E cum dupla D.

ſimul

bis ſumptum, ad parallelepipedum baſi æquali quadratis C, D, altitudine verò C.

6. huius. Intelligatur in vtroque ſolido inſcripta ſua ſphæra. Et erit ſolidum primum ad ſphæram ſuam, vt recta A ad B; ſumptaq; eadem baſi quadrato B; erit ſolidum primum ad ſphæram ſuam, vt parallelepipedum baſi quadrato B altitudine verò A, ad cubum B. ſiue vt duplum dicti parallelepipedi ad duos cubos B.

Sphæra verò prima ad ſecundam eſt, vt duo cubi B, ad duos cubòs C. Sphæra tandem ſecunda ad ſolidum ſuum eſt, vt duo quadrata ex C, ad duo quadrata C, & D; ſumptaque communi altitudine C, eſt, vt duo cubi C, ad parallelepipedum baſi æquali quadratis C & D. altitudine verò C. Propterea ex æquo patet quod propoſitum erat.

Propoſitio LI.

13 huius SOlidum ſphęrale parilaterum circa diagonalem reuolutum, ad ſolidum ſphærale imparilaterum eſt, vt parallelepipedum baſi quadrato cateti, altitudine diagonali quater ſumptum; ad parallelepipedum baſi quadrato rectæ illius quæ ex centro poligoni imparilateri perpendiculariter ducitur in latus eiuſdẽ; altitudine verò æquali prædictæ perpendiculari, vna cum dupla illius quę ex centro ad angulum ducitur, & cum tertia proportionalium ad duas prædictas.

Sint duo ſolida ſphæralia, nempe primum parilaterum circa diagonalẽ A conuerſum, alterum verò imparilaterum circa catetum C reuolutum. Cõtinuetur ratio C ad D in trib. terminis C, D, E. Dico primum ſolidum ad ſecundum eſſe, vt parallelepipedum baſi quadrato B, altitudine A quater ſumptum, ad parallelepipedum baſi quadrato C, altitudine verò æquali rectis C, & E cum dupla D.

simul

ſimul ſumptis.

Nam ſolidum primum ad ſphæram ſuam eſt, vt recta A ad B; ſiue ſumpta còmmuni baſi quadrato B; vt parallelepipedum baſi quadrato B altitudine A, ad cubum B; Vel ut parallelepipedum prædictum quater ſumptum, ad cubum B quater ſumptum ſphæra uerò prima ad ſecundam eſt ut quatuor cubi B ad quatuor cubos C, Sphæra denique ſecunda ad ſolidum ſuum (ut oſtenſum eſt in 49. huius) eſt ut quatuor cubi C, ad parallelepipedum baſi quadrato C, altitudine uerò æquali rectis C & E cum dupla D. ſimul ſumptis. Propterea ex æquo patet quod propoſitum erat.

Propoſitio LII.

SOlidum ſphærale parilaterum circa catetum reuolutum, ad ſolidum ſph. imparilaterum, eſt ut parallelepipedum baſi æquali quadratis diagonalis & cateti altitudine cateti bis ſumptum, ad parallelepipedum baſi quadrato lineæ quę ex centro ducitur perpendiculariter in latus poligoni imparilateri, altitudine uerò æquali prędictæ lineæ, una cum illa quę ex centro ad unum angulum perducitur, & cum tertia proportionalium ad duas prædictas.

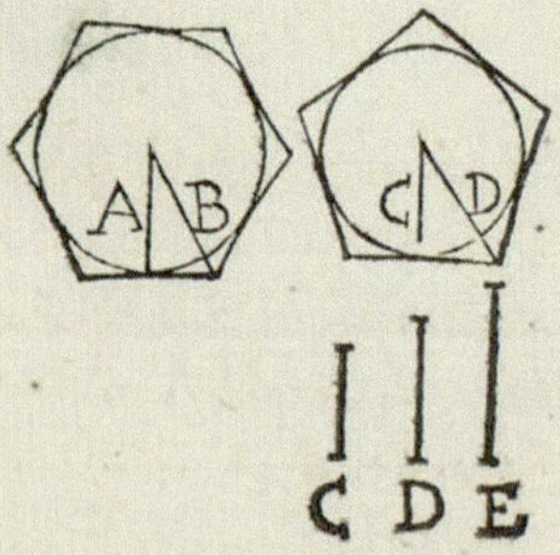

Sint duo ſolida ſphæralia; alterum parilaterum circa catetum A reuolutum; alterum imparilaterum circa C conuerſum. Et ratio C ad D, continuetur in tribus terminis C, D, E. Dico primum ſolidum ad ſecundum eſſe; vt parallelepipedum baſi æquali quadratis B & A, altitudine verò A, bis ſumptum; ad parallelepipedum baſi quadrato C, altitudine verò æquali C, & E, cum dupla ipſius D.

Nam ſolidum primum ad ſphæram ſuam eſt, vt duo quadrata B 13 *huius*

ta B

ſimul ſumptis.

Nam ſolidum primum ad ſphæram ſuam eſt, vt recta A ad B; ſiue ſumpta còmmuni baſi quadrato B; vt parallelepipedum baſi quadrato B altitudine A, ad cubum B; Vel ut parallelepipedum prædictum quater ſumptum, ad cubum B quater ſumptum ſphæra uerò prima ad ſecundam eſt ut quatuor cubi B ad quatuor cubos C, Sphæra denique ſecunda ad ſolidum ſuum (ut oſtenſum eſt in 49. huius) eſt ut quatuor cubi C, ad parallelepipedum baſi quadrato C, altitudine uerò æquali rectis C & E cum dupla D. ſimul ſumptis. Propterea ex æquo patet quod propoſitum erat.

Propoſitio LII.

SOlidum ſphærale parilaterum circa catetum reuolutum, ad ſolidum ſph. imparilaterum, eſt ut parallelepipedum baſi æquali quadratis diagonalis & cateti altitudine cateti bis ſumptum, ad parallelepipedum baſi quadrato lineæ quę ex centro ducitur perpendiculariter in latus poligoni imparilateri, altitudine uerò æquali prędictæ lineæ, una cum illa quę ex centro ad unum angulum perducitur, & cum tertia proportionalium ad duas prædictas.

Sint duo ſolida ſphæralia; alterum parilaterum circa catetum A reuolutum; alterum imparilaterum circa C conuerſum. Et ratio C ad D, continuetur in tribus terminis C, D, E. Dico primum ſolidum ad ſecundum eſſe, vt parallelepipedum baſi æquali quadratis B & A, altitudine verò A, bis ſumptum; ad parallelepipedum baſi quadrato C, altitudine verò æquali C, & E, cum dupla ipſius D.

Nam ſolidum primum ad ſphæram ſuam eſt, vt duo quadra- *13 huius*

ta B

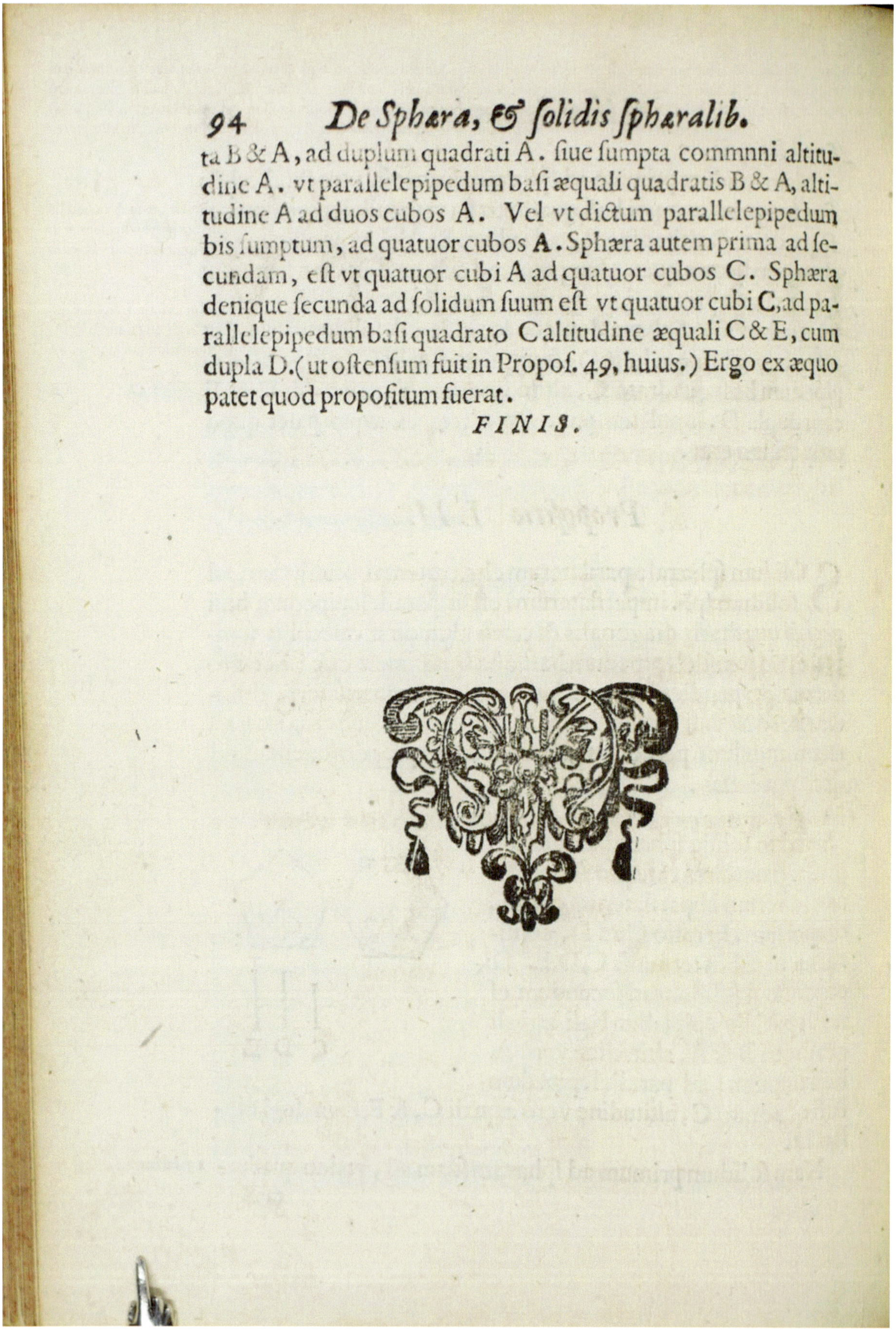

94 *De Sphæra, & solidis sphæralib.*

ta B & A, ad duplum quadrati A. siue sumpta commnni altitudine A. vt parallelepipedum basi æquali quadratis B & A, altitudine A ad duos cubos A. Vel vt dictum parallelepipedum bis sumptum, ad quatuor cubos A. Sphæra autem prima ad secundam, est vt quatuor cubi A ad quatuor cubos C. Sphæra denique secunda ad solidum suum est vt quatuor cubi C, ad parallelepipedum basi quadrato C altitudine æquali C & E, cum dupla D. (ut ostensum fuit in Propos. 49. huius.) Ergo ex æquo patet quod propositum fuerat.

FINIS.

ta B & A, ad duplum quadrati A. ſiue ſumpta communi altitudine A. vt parallelepipedum baſi æquali quadratis B & A, altitudine A ad duos cubos A. Vel vt dictum parallelepipedum bis ſumptum, ad quatuor cubos A. Sphæra autem prima ad ſecundam, eſt vt quatuor cubi A ad quatuor cubos C. Sphæra denique ſecunda ad ſolidum ſuum eſt vt quatuor cubi C, ad parallelepipedum baſi quadrato C altitudine æquali C & E, cum dupla D. (ut oſtenſum fuit in Propoſ. 49. huius.) Ergo ex æquo patet quod propoſitum fuerat.

FINIS.

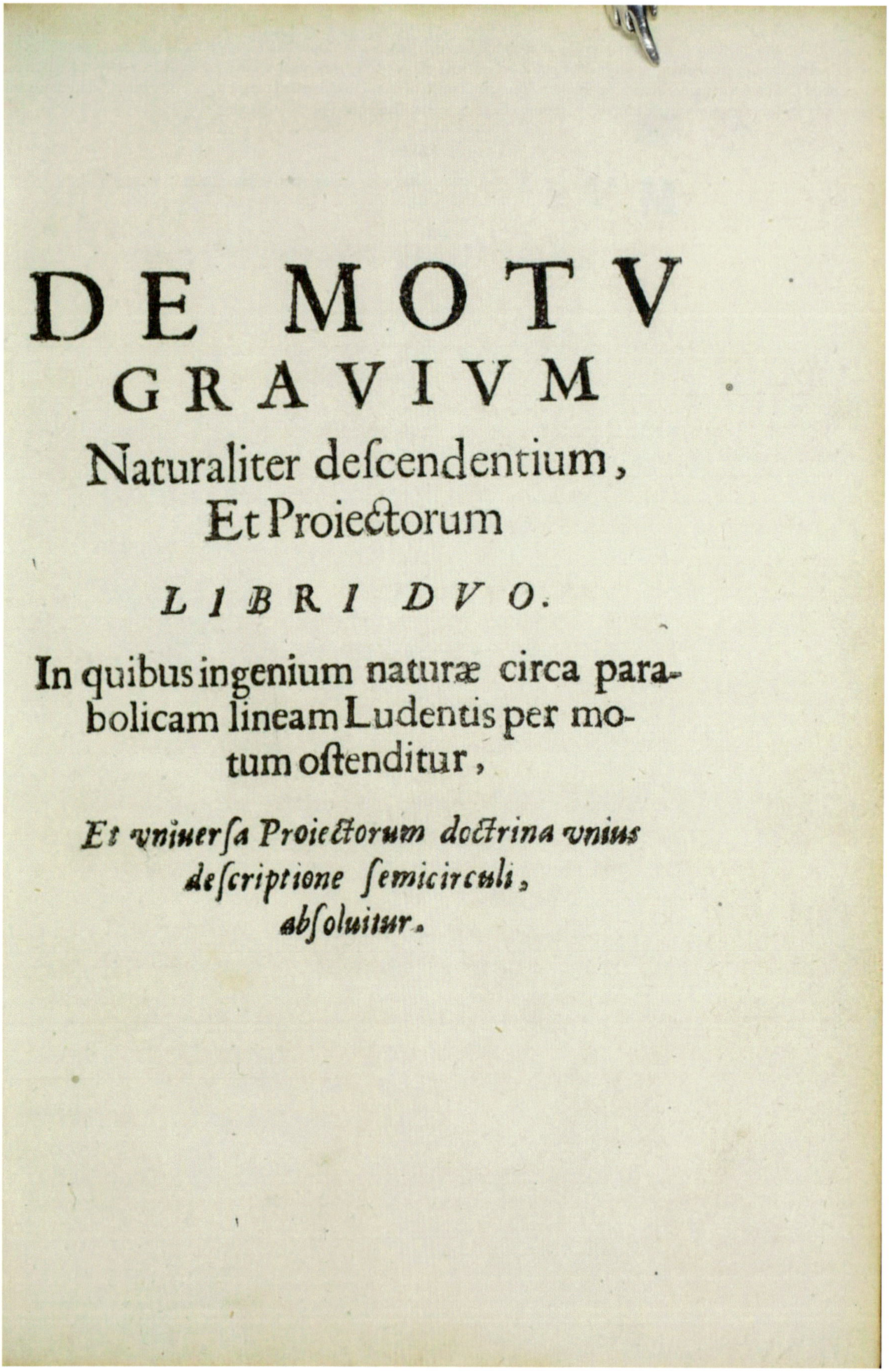

DE MOTV
GRAVIVM
Naturaliter deſcendentium,
Et Proiectorum

LIBRI DVO.

In quibus ingenium naturæ circa parabolicam lineam Ludentis per motum oſtenditur,

Et vniuerſa Proiectorum doctrina vnius deſcriptione ſemicirculi, abſoluitur.

DE MOTV GRAVIVM

Naturaliter deſcendentium,
Et Proiectorum

LIBRI DVO.

In quibus ingenium naturæ circa parabolicam lineam Ludentis per motum oſtenditur,

Et vniuerſa Proiectorum doctrina vnius deſcriptione ſemicirculi, abſoluitur.

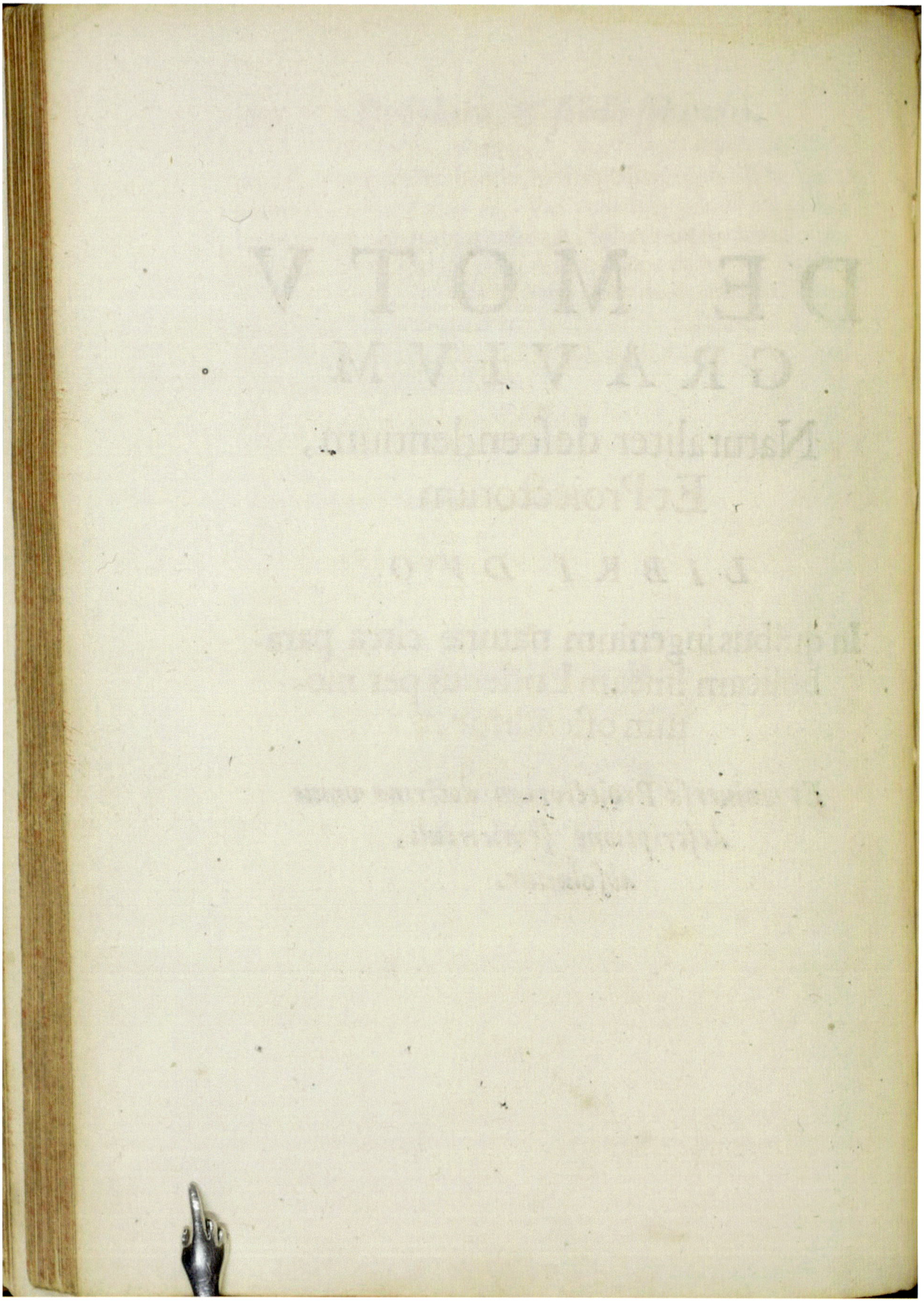

DE MOTV GRAVIVM Naturaliter deſcendentium.

LIBER PRIMVS.

SCIENTIAM de motu G.& Pr. à pluribus quidem tractatã, ab vnico (quod ego ſciam) Galileo Geometricè demonſtratam, aggredi libet. Fateor, quòd ille totam hanc ſegetem tamquã falce demeſſuit, nec aliud ſupereſt nobis, niſi vt tam ſeduli meſſoris veſtigia ſubſequentes, ſpicas colligamus, ſi quę ab ipſo vel relictæ fuerint, vel abiectæ: ſin minus, Liguſtra ſaltem, & humi naſcentes violas decerpamus; ſed fortaſſe & ex floribus coronam contexemus non contemnendam.

Principio quædam de momentis grauium proponemus, vt aliqua ſuppleamus, quæ quodammodo opportuna videbantur ad ſcientiam. Deinde quædam de parabola, quæ nobis ad propagationem huius doctrinæ vtilia videbuntur, Reliquum libri primi propoſitiones erunt de motu accelerato; illarumque ordo quo ad fieri poterit in tam diuerſis rerum materijs, neglectus penitus non erit. Libellus alter de Motu proiectorum tractabit, ampliata Galilei doctrina, & demonſtrationibus plerumq; mutatis. Tabulas certè, quas ipſe ſtudio, ac labore compoſuit, omnes ex tabula ſinuum à nobis ſolius tranſcriptionis moleſtia, decerptas exponemus nam hypotheſis noſtra, iuxtà quam proiectiones ſurſum factas contemplamur, apertè indicauit nobis tabulas à Galileo elaboratas in ipſis ſinuum, ac tangentium tabulis expreſsè incluſas, & inſer-

N tas

97

DE MOTV GRAVIVM[41]

Naturaliter defcendentium.

LIBER PRIMVS.

SCIENTIAM de motu G. & Pr. à pluribus quidem tractatã, ab vnico (quod ego fciam) Galileo Geometricè demonftratam, aggredi libet. Fateor, quòd ille totam hanc fegetem tamquã falce demeffuit, nec aliud fupereft nobis, nifi vt tam feduli mefforis veftigia fubfequentes, fpicas colligamus, fi quę ab ipfo vel relictæ fuerint, vel abiectæ: fin minus, Liguftra faltem, & humi nafcentes violas decerpamus; fed fortaffe & ex floribus coronam contexemus non contemnendam. [gravium et projectorum]

Principio quædam de momentis grauium proponemus, vt aliqua fuppleamus, quæ quodammodo opportuna videbantur ad fcientiam. Deinde quædam de parabola, quæ nobis ad propagationem huius doctrinæ vtilia videbuntur, Reliquum libri primi propofitiones erunt de motu accelerato; illarumque ordo quo ad fieri poterit in tam diuerfis rerum materijs, neglectus penitus non erit. Libellus alter de Motu proiectorum tractabit, ampliata Galilei doctrina, & demonftrationibus plerumq; mutatis. Tabulas certè, quas ipfe ftudio, ac labore compofuit, omnes ex tabula finuum à nobis folius tranfcriptionis moleftia, decerptas exponemus nam hypothefis noftra, iuxtà quam proiectiones furfum factas contemplamur, apertè indicauit nobis tabulas à Galileo elaboratas in ipfis finuum, ac tangentium tabulis exprefsè inclufas, & infer-

N tas

[41] *De motu gravium naturaliter descendentium, et de proiectorum libri duo, in quibus ingenium naturae circa parabolicam lineam ludentis per motum ostenditur, et universa proiectorum doctrina unius descriptione semicirculi, absolvitur.* The text is divided into two books: in the first book, Torricelli proposes a series of theoretical propositions on the motion of the bodies under the action of gravity, while in the second book he deals with several specific cases concerning the motions of the projectiles. The first book consists of 58 propositions, the second of 37 propositions to which a sort of appendix entitled *De motu aquarum* follows. It is to remark that a series of tables related to the degree of elevation with which a projectile is shot are added to the text. The second book is closed by a brief treatise entitled *Della squadra.* The language is Latin, but the explanations concerning the use of the tables and the treatise *Della squadra* are written in Italian (Cfr. Pisano and Bussotti's chapters above).

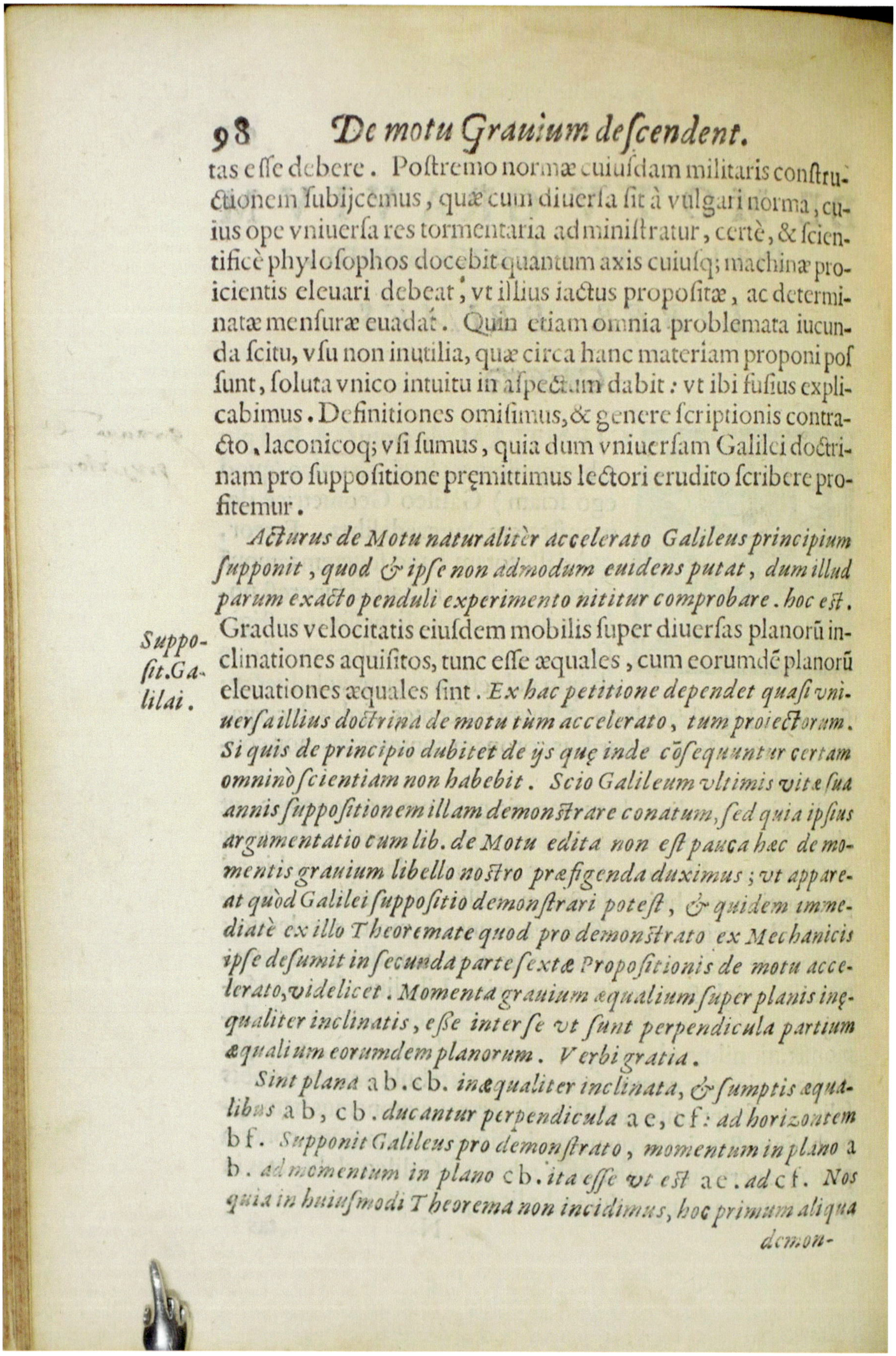

98 De motu Grauium deſcendent.

tas eſſe debere. Poſtremo normæ cuiuſdam militaris conſtructionem ſubijcemus, quæ cum diuerſa ſit à vulgari norma, cuius ope vniuerſa res tormentaria adminiſtratur, certè, & ſcientificè phyloſophos docebit quantum axis cuiuſq; machinæ proicientis eleuari debeat, vt illius iactus propoſitæ, ac determinatæ menſuræ euadat. Quin etiam omnia problemata iucunda ſcitu, vſu non inutilia, quæ circa hanc materiam proponi poſſunt, ſoluta vnico intuitu in aſpectum dabit: vt ibi fuſius explicabimus. Definitiones omiſimus, & genere ſcriptionis contracto, laconicoq; vſi ſumus, quia dum vniuerſam Galilei doctrinam pro ſuppoſitione præmittimus lectori erudito ſcribere profitemur.

Acturus de Motu naturalitèr accelerato Galileus principium ſupponit, quod & ipſe non admodum euidens putat, dum illud parum exacto penduli experimento nititur comprobare. hoc eſt.

Suppoſit. Galilai.

Gradus velocitatis eiuſdem mobilis ſuper diuerſas planorū inclinationes aquiſitos, tunc eſſe æquales, cum eorumdē planorū eleuationes æquales ſint. *Ex hac petitione dependet quaſi vniuerſa illius doctrina de motu tùm accelerato, tum proiectorum. Si quis de principio dubitet de ijs quę inde cōſequuntur certam omninò ſcientiam non habebit. Scio Galileum vltimis vitæ ſuæ annis ſuppoſitionem illam demonſtrare conatum, ſed quia ipſius argumentatio cum lib. de Motu edita non eſt pauca hæc de momentis grauium libello noſtro præfigenda duximus; vt appareat quòd Galilei ſuppoſitio demonſtrari poteſt, & quidem immediatè ex illo Theoremate quod pro demonſtrato ex Mechanicis ipſe deſumit in ſecunda parte ſextæ Propoſitionis de motu accelerato, videlicet. Momenta grauium æqualium ſuper planis inæqualiter inclinatis, eſſe inter ſe vt ſunt perpendicula partium æqualium eorumdem planorum. Verbi gratia.*

Sint plana a b. c b. *inæqualiter inclinata, & ſumptis æqualibus* a b, c b. *ducantur perpendicula* a e, c f: *ad horizontem* b f. *Supponit Galileus pro demonſtrato, momentum in plano* a b. *ad momentum in plano* c b. *ita eſſe vt eſt* a e. *ad* c f. *Nos quia in huiuſmodi Theorema non incidimus, hoc primum aliqua demon-*

tas eſſe debere. Poſtremo normæ cuiuſdam militaris conſtructionem ſubijcemus, quæ cum diuerſa ſit à vulgari norma, cuius ope vniuerſa res tormentaria adminiſtratur, certè, & ſcientificè phyloſophos docebit quantum axis cuiuſq; machinæ proficientis eleuari debeat, vt illius iactus propoſitæ, ac determinatæ menſuræ euadat. Quin etiam omnia problemata iucunda ſcitu, vſu non inutilia, quæ circa hanc materiam proponi poſſunt, ſoluta vnico intuitu in aſpectum dabit: vt ibi fuſius explicabimus. Definitiones omiſimus, & genere ſcriptionis contracto, laconicoq; vſi ſumus, quia dum vniuerſam Galilei doctrinam pro ſuppoſitione pręmittimus lectori erudito ſcribere profitemur.

Acturus de Motu naturalitèr accelerato Galileus principium ſupponit, quod & ipſe non admodum euidens putat, dum illud parum exacto penduli experimento nititur comprobare. hoc est.[43]

Suppo-. sit.Galilai.[42] Gradus velocitatis eiuſdem mobilis ſuper diuerſas planorũ inclinationes aquiſitos, tunc eſſe æquales, cum eorumdẽ planorũ eleuationes æquales ſint. *Ex hac petitione dependet quaſi vniuerſa illius doctrina de motu tùm accelerato, tum proiectorum. Si quis de principio dubitet de ijs quę inde cõſequuntur certam omninò ſcientiam non habebit. Scio Galileum vltimis vitæ ſua annis ſuppoſitionem illam demonstrare conatum, ſed quia ipſius argumentatio cum lib. de Motu edita non eſt pauca hæc de momentis grauium libello nostro præfigenda duximus; vt appareat quòd Galilei ſuppoſitio demonſtrari poteſt, & quidem immediatè ex illo Theoremate quod pro demonstrato ex Mechanicis ipſe deſumit in ſecunda parte ſextæ Propoſitionis de motu accelerato, videlicet. Momenta grauium æqualium ſuper planis inęqualiter inclinatis, eßе inter ſe vt ſunt perpendicula partium æqualium eorumdem planorum. Verbi gratia.*

Sint plana ab. cb. *inæqualiter inclinata, & ſumptis æqualibus* ab, cb. *ducantur perpendicula* ae, cf*: ad horizontem* bf. *Supponit Galileus pro demonſtrato, momentum in plano* ab. *ad momentum in plano* cb. *ita eſſe vt est* ae. *ad* cf. *Nos quia in huiuſmodi Theorema non incidimus, hoc primum aliqua*

demon-

[42] Torricelli named Galileo's theorem "*Supposit. Galilei* […] *Galileus principium* […]" (see also below, p. 100. Cfr. Pisano and Bussotti's chapters above).

[43] Torricelli's reasoning upon Galileo's principle: "To afford the naturally accelerated Motion, Galileo assumes hypothetically a principle which he himself does not consider completely evident [≠ he considers it experimentally evident], because he tries to prove it with the not very exact experiment of pendulum, that is: […]. Because we did not previously encounter this Theorem, as first we will confirm it with some proof, then we will arrive to show what in Galileo is a principle or a petition […]." (Torricelli 1644, pp. 98-99; authors' translation; Cfr. *Ibidem*).

demonstratione confirmabimus: protinus ad ostendendum id quod Galileo principium siue petitio est, accedemus.

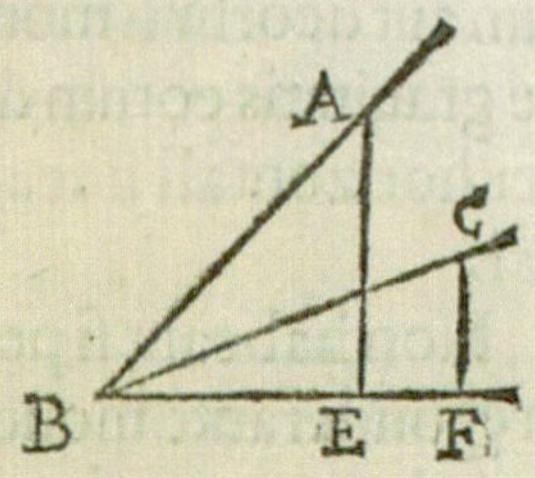

Præmittimus.

Duo grauia simul coniuncta ex se moueri non posse, nisi centrum commune grauitatis ipsorum descendat.

Quando enim duo grauia ita inter se coniuncta fuerint, vt ad motum vnius motus etiam alterius consequatur, erunt duo illa grauia tamquam graue vnum ex duobus compositum, siue id libra fiat, siue trochlea, siue qualibet alia Mechanica ratione, graue autem huiusmodi non mouebitur vnquam, nisi centrum grauitatis ipsius descendat. Quando verò ita constitutum fuerit vt nullo modo commune ipsius centrum grauitatis descendere possit, graue penitus in sua positione quiescet: alias enim frustra moueretur; horizontali, scilicet latione, quæ nequaquam deorsum tendit.

PROPOSITIO I.

SI in planis inæqualiter inclinatis, eandem tamen eleuationem habentibus, duo grauia constituantur, quæ inter se eandem homologè rationem habeant quam habent longitudines planorum, grauia æquale momentum habebunt.

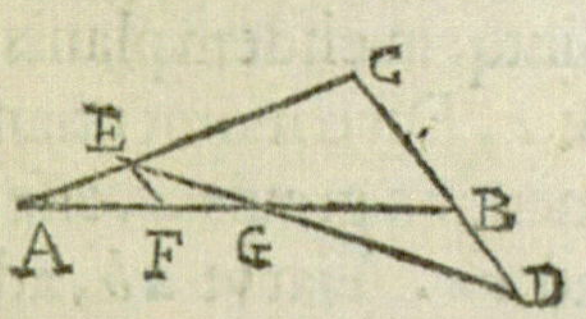

Sit *ab*. horizon, & plana inæqualiter inclinata *ca*, *cb*. Fiat vt *ac* ad *cb*, ita graue aliquod *a*. ad graue *b*. Et grauia hæc in homologis planis collocentur, in punctis *a*, & *b*. eiusdem horizontalis lineæ. Connectantur etiam aliquo imaginario funiculo per *acb*. ducto, adeo vt ad motum vnius motus alterius consequatur.

Dico grauia sic disposita æquale momentum habere: hoc est

demonſtratione confirmabimus: protinus ad oſtendendum id quod Galileo principium ſiue petitio eſt, accedemus.

Praemittimus.

Duo grauia ſimul coniucta ex ſe moueueri non poſſe, niſi centrum commune grauitatis ipſorum deſcendat.[44]

Quando enim duo grauia ita inter ſe coniunctà fuerint, vt ad motum vnius motus etiam alterius conſequatur, erunt duo illæ grauia tamquam graue vnum ex duobus compoſitum, ſiue id libra fiat, ſiue qualibet[45] *alia Mechanica ratione, graue autem huiuſmodi non mouebitur vnquam, niſi centrum grauitatis ipſius deſcendat. Quando verò ita conſtitutum fuerit vt nullo modo commune ipſius centrum grauitatis deſcendere poſſit, graue penitus in ſua poſitione quieſcet: alias enim fruſtra moueretur; horizontali, ſcilicet latione, quæ nequaquam deorſum tendit.*

PROPOSITIO I.[46]

SI in planis inæqualiter inclinatis, eandem tamen eleuationem habentibus, duo grauia conſtituantur, quæ inter ſe eandem homologè rationem habeant quam habent longitudines planorum, grauia æquale momentum habebunt.

Sit *ab.* horizon, & plana inæqualiter inclinata *ca, cb.* Fiat vt *ac* ad *cb,* ita graue aliquod *a.* ad graue *b.* Et grauia hæc in homologis planis collocentur, in punctis *a, & b.* eiuſdem horizontalis lineæ. Connectantur etiam aliquo imaginario funiculo per *acb.* ducto, adeo vt ad motum vnius motus alterius conſequatur.

Dico grauia ſic diſpoſita æquale momentum habere: hoc eſt

N 2 in ea

[44] "*Premise*: Two heavy bodies joined together cannot move by themselves unless their common centre of gravity descends". (Authors' translation. Cfr. Pisano and Bussotti's chapters above). The *Praemittamus* (*premise*) – which is nowadays called *Torricelli's Principle* – in mechanics – is nevertheless rarely used in the *De motu gravium naturaliter descendentium.* To be precise Torricelli resorts to it only in the proofs of *Proposition I* and *Proposition II* (Torricelli 1644, pp. 99-100). In these proofs, Torricelli's Principle is used by means of a *reductio ad absurdum.* Torricelli brings Galileo's position to the extreme by assuming that the bodies could be placed not on the Earth, but at "[in reality] infinite distance from it" or that the cord, by which heavy bodies are suspended, are actually parallel. In fact, for Galileo the physical results obtained by assuming an infinite distance were to be considered empirically approximate. For Torricelli they are complete and only justified by means of mathematical calculation. The *Praemittamus* represents one of the first examples of a *general principle* posed *explicitly* at the basis of physics (and of part of structural mechanics). The term *general* is appropriate because the principle concerns all the situations in which a constrained motion of a body exists. Therefore, it regards very general situations. (Cfr. Pisano and Bussotti's chapters above).

[45] *ſiue qualibet*] *ſiue trocleà.*

[46] "If in [two] planes unequally inclined but with the same elevation, two heavy bodies are considered which are in the same ratio as the length of the planes, these bodies will have the same momentum." (Authors' translation. Cfr. *Ibidem*).

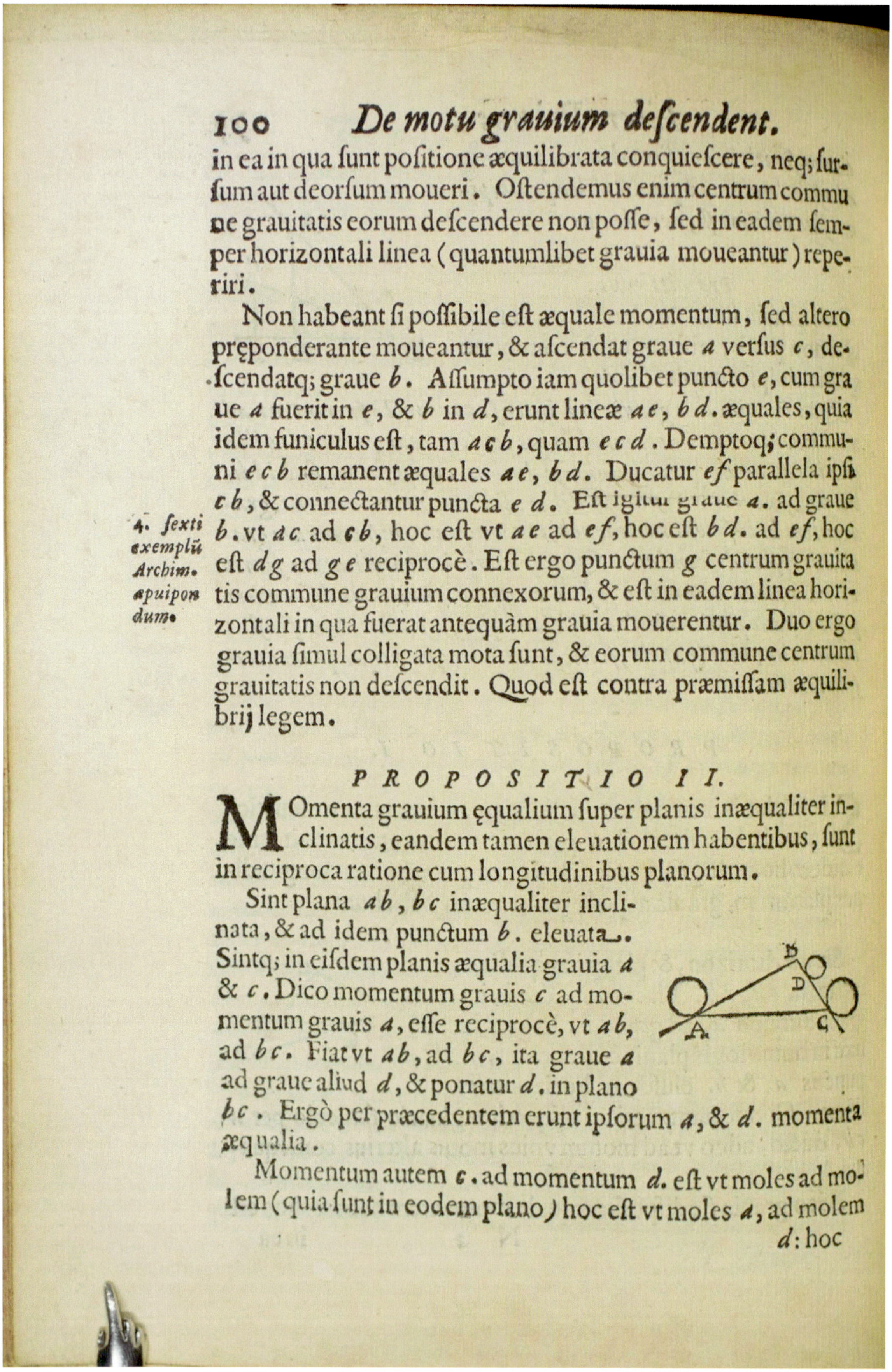

100 *De motu grauium descendent.*

in ea in qua sunt positione æquilibrata conquiescere, neq; sursum aut deorsum moueri. Ostendemus enim centrum commune grauitatis eorum descendere non posse, sed in eadem semper horizontali linea (quantumlibet grauia moueantur) reperiri.

Non habeant si possibile est æquale momentum, sed altero pręponderante moueantur, & ascendat graue *a* versus *c*, descendatq; graue *b*. Assumpto iam quolibet puncto *e*, cum graue *a* fuerit in *e*, & *b* in *d*, erunt lineæ *ae*, *bd*. æquales, quia idem funiculus est, tam *acb*, quam *ecd*. Demptoq; communi *ecb* remanent æquales *ae*, *bd*. Ducatur *ef* parallela ipsi *cb*, & connectantur puncta *e d*. Est igitur graue *a*. ad graue *b*. vt *ac* ad *cb*, hoc est vt *ae* ad *ef*, hoc est *bd*. ad *ef*, hoc est *dg* ad *ge* reciprocè. Est ergo punctum *g* centrum grauitatis commune grauium connexorum, & est in eadem linea horizontali in qua fuerat antequàm grauia mouerentur. Duo ergo grauia simul colligata mota sunt, & eorum commune centrum grauitatis non descendit. Quod est contra præmissam æquilibrij legem.

4. sexti exemplū Archim. æquipondum.

PROPOSITIO II.

Momenta grauium ęqualium super planis inæqualiter inclinatis, eandem tamen eleuationem habentibus, sunt in reciproca ratione cum longitudinibus planorum.

Sint plana *ab*, *bc* inæqualiter inclinata, & ad idem punctum *b*. eleuata. Sintq; in eisdem planis æqualia grauia *a* & *c*. Dico momentum grauis *c* ad momentum grauis *a*, esse reciprocè, vt *ab*, ad *bc*. Fiat vt *ab*, ad *bc*, ita graue *a* ad graue aliud *d*, & ponatur *d*. in plano *bc*. Ergò per præcedentem erunt ipsorum *a*, & *d*. momenta æqualia.

Momentum autem *c*. ad momentum *d*. est vt moles ad molem (quia sunt in eodem plano) hoc est vt moles *a*, ad molem *d*: hoc

100 *De motu grauium*[47] *defcendent.*

in ea in qua funt pofitione æquilibrata conquiefcere, neq; furfum aut deorfum moueri. Oftendemus enim centrum commune grauitatis eorum defcendere non poffe, fed in eadem femper horizontali linea (quantumlibet grauia moueantur) reperiri.

Non habeant fi poffibile eft æquale momentum, fed altero prȩponderante moueantur, & afcendat graue *a* verfus *c,* defcendatq; graue *b.* Affumpto iam quolibet puncto *e,* cum graue *a* fuerit in *e,* & *b* in *d,* erunt lineæ *ae, bd.* æquales, quia idem funiculus eft, tam *acb,* quam *ecd.* Demptoq; communi *ecb* remanent æquales *ae, bd.* Ducatur *ef* parallela ipfi *cb,* & connectantur puncta *ed.* Eft igitur graue *a.* ad graue *b.* vt *ac* ad *cb,* hoc eft vt *ae* ad *ef,* hoc eft *bd.* ad *ef,* hoc eft *dg* ad *ge* reciprocè. Eft ergo punctum *g* centrum grauitatis commune grauium connexorum, & eft in eadem linea horizontali in qua fuerat antequàm grauia mouerentur. Duo ergo grauia fimul colligata mota funt, & eorum commune centrum grauitatis non defcendit. Quod eft contra præmiffam æquilibrij legem.

4. *fexti. exemplũ Archim. æpuipon dum.*

PROPOSITIO II.[48]

MOmenta grauium ȩqualium fuper planis inæqualiter inclinatis, eandem tamen eleuationem habentibus, funt in reciproca ratione cum longitudinibus planorum. Sint plana *a b, b c* inæqualiter inclinata, & ad idem punctum *b.* eleuata.

Sintq; in eifdem planis æqualia grauia *a* & *c.* Dico momentum grauis *c* ad momentum grauis *a,* effe reciprocè, vt *ab,* ad *bc.* Fiat vt *ab,* ad *bc,* ita graue *a* ad graue aliud *d,* & ponatur *d.* in plano *bc.* Ergò per præcedentem erunt ipforum *a,* & *d.* momenta æqualia.

Momentum autem *c.* ad momentum *d.* eft vt moles ad molem (quia funt in eodem plano*)* hoc eft vt moles *a,* ad molem

d: hoc

[47] *motu grauium*] *motu Grauium.*

[48] Torricelli's *Proposition II* is accurately Galileo's theorem: "[Sagr.] [...] seem that up to the present we have established the definition of uniformly accelerated motion which is expressed as follows: A motion is said to be equally or uniformly accelerated when, starting from rest, its momentum [*celeritatis momenta*] receives equal increments in equal times. Salv. This definition established, the Author makes a single assumption, namely. The speeds acquired by one and the same body moving down planes of different inclinations are equal when the height of these planes are equal heights of these planes are equal." (Galilei 1890-1909, VIII, Day III, p. 205; author's italic. Translation retrieved from Galilei, 1954, p. 169). Torricelli's "*Proposition II.* Momenta of equal bodies on planes which are unequally inclined, but which have the same elevation, are related to each other in the same ratio as the lengths of the planes themselves." (Authors' translation. Cfr. Pisano and Bussotti's chapters above). Torricelli named Galileo's theorem "*Supposit. Galilei* [...] *Galileus principium* [...]" (Torricelli 1644, pp. 97–98. author's italic. Cfr. *Ibidem*).

d: hoc est vt *ab*, ad *bc*. Est ergo momentum *c*. ad *d*, vel ad momentum *a*. ipsi momento *d*. æquale, vt *ab*, ad *bc*. Quod erat &c.

Aliter.

Idem quod hic demonstrauimus ex Prima Propositione, quæ sumpto principio deducebat ad impossibile, ostendetur etiam absolutè, & affirmatiuè ex ipsis Mechanicæ principijs.

Lemma.

Momentum totale grauis ad momentum quod habet in plano inclinato, est vt longitudo ipsius plani inclinati ad perpendiculum.

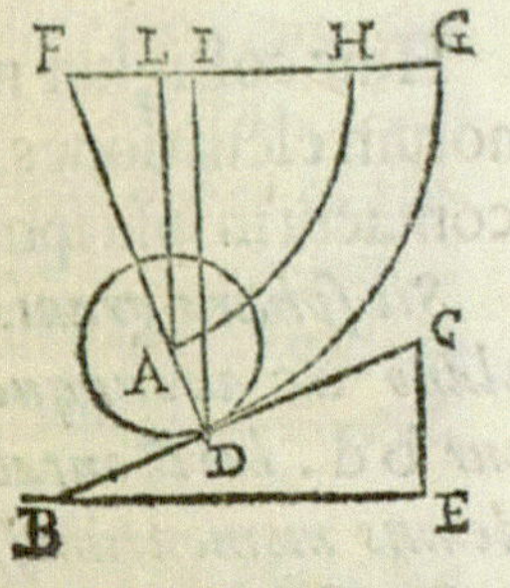

Sit circa centrum a. *sphæra grauis in plano eleuato* b c. *& sit plani perpendiculum* c e. *dico momentum totale grauis* a *ad momentum peculiare quod habet in plano* b c, *esse vt* b c. *ad* c e.

Producatur recta d a. *per contactum* d, *& per centrum* a. *quæ ideo perpendicularis erit ad planum* b c. *& quolibet centro* f. *fiant quadrantis portiones* d g, a h. *& ducatur* f g. *horizontalis, &* d i, a l. *vtraq; ad horinzontem perpendicularis.*

Iam Angulus f d c. *rectus est, & anguli* b, c. *simul æquales sunt recto, ergò ablatis* i d c, d c e, *alternis parallelarum, remanent æquales* f d i, *&* b. *Sunt ideo similia duo triangula rectangula* f d i, b c e. *Iam sic.*

Sed quando graue circumfertur à semidiametro f h, *siue* f a, *manente puncto* f *tunc momentum totale eius, hoc est, momentum quod habet in situ* h, *ad momentum quod habet in situ* a, *est vt* h f. *ad* f l, *siue* a f. *ad* f l, *hoc est* d f. *ad* f i, *vel* b c. *ad* c e. *ob similitudinem triangulorum. Quod erat &c.*

Quod autem idem momentum sit grauis constituti, siue in puncto a *quadrantis* a h, *siue in puncto* d. *quadrantis* d g, *siue in*

d: hoc eft vt *ab*, ad *bc*. Eft ergo momentum *c*. ad *d*, vel ad momentum *a*. ipſi momento *d*. æquale, vt *ab*, ad *bc*. Quod erat &c.

Aliter.

Idem quod hic demonſtrauimus ex Prima Propoſitione, quæ ſumpto principio deducebat ad impoſſibile, ostendetur etiam abſolutè, & affermatiuè ex ipſis Mechanicæ principijs.[49]

Lemma.

Momentum totale grauis ad momentum quod habet in plano inclinato, eft vt longitudo ipſius plani inclinati ad perpendiculum.

Sit circa centrum a. *ſphæra grauis in plano eleuato* bc. *& ſit plani perpendiculum* c e. *dico momentum totale grauis* a *ad momentum peculiare quod habet in plano* bc, *eſſe vt* bc. *ad* ce.

Producatur recta da. *per contactum* d, *& per centrum* a. *quæ ideo perpendicularis erit ad planum* bc. *& quolibet centro* f. *fiant quadrantis portiones* dg, ah. *& ducatur* fg. *horizontalis, &* di, al. *vtraq; ad horinzontem perpendicularis.*

Iam Angulus fdc. *rectus eſt, & anguli* b, c. *ſimul æquales ſunt recto, ergò ablatis idc, dce, alternis parallelarum, remanent æquales fdi, & b. Sunt ideo ſimilia duo triangula rectangula fdi, bce. Iam ſic.*

Sed quando graue circumfertur à ſemidiametro fh, *ſiue* fa, *manente puncto* f *tunc momentum totale eius, hoc est, momentum quod habet in ſitu* h, *ad momentum quod habet in ſitu* a, *eſt vt* hf. *ad* fl, *ſiue* af. *ad* fl, *hoc est* df. *ad* fi, *vel* bc. *ad* ce. *ob ſimilitudinem triangulorum. Quod erat &c.*

Quod autem idem momentum ſit grauis conſtituti, ſiue in puncto a *quadrantis* ah, *ſiue in puncto* d. *quadrantis* dg, *ſi-*

ue in

[49] "*Lemma*. The total *momentum* of a heavy body is to the *momentum* the body assumes on an inclined plane as the length of the inclined plane [its hypotenuse] is to the perpendicular." (See also next page. Authors' translation. Cfr. Pisano and Bussotti's chapters above).

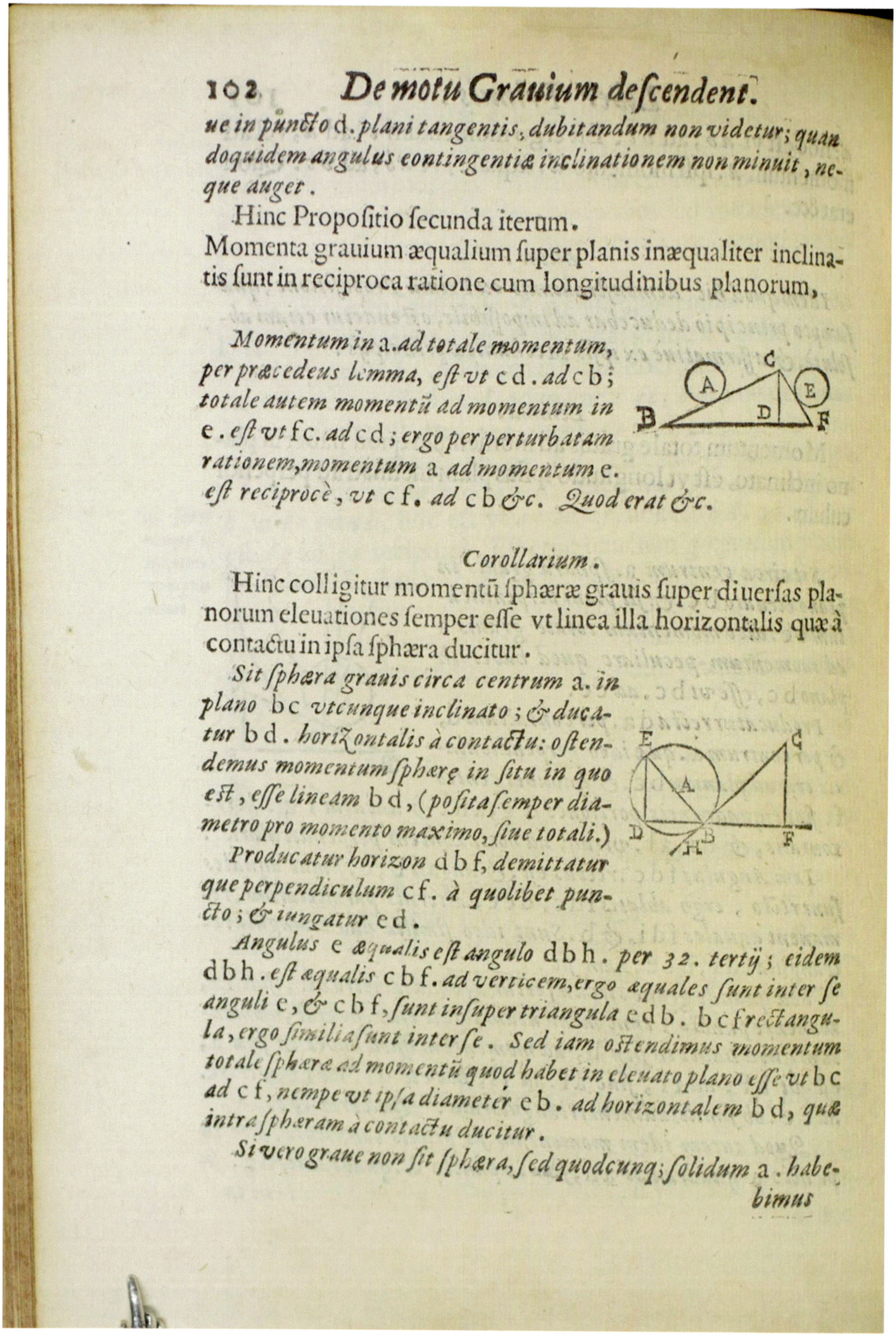

102 *De motu Grauium descendent.*

ue in puncto d. *plani tangentis, dubitandum non videtur; quandoquidem angulus contingentiæ inclinationem non minuit, neque auget.*

Hinc Propositio secunda iterum.

Momenta grauium æqualium super planis inæqualiter inclinatis sunt in reciproca ratione cum longitudinibus planorum,

Momentum in a. *ad totale momentum, per præcedens lemma, est vt* c d. *ad* c b; *totale autem momentũ ad momentum in* e. *est vt* f c. *ad* c d; *ergo per perturbatam rationem, momentum* a *ad momentum* e. *est reciprocè, vt* c f. *ad* c b *&c. Quod erat &c.*

Corollarium.

Hinc colligitur momentũ sphæræ grauis super diuersas planorum eleuationes semper esse vt linea illa horizontalis quæ à contactu in ipsa sphæra ducitur.

Sit sphæra grauis circa centrum a. *in plano* b c *vtcunque inclinato; & ducatur* b d. *horizontalis à contactu: ostendemus momentum sphęrę in situ in quo est, esse lineam* b d, *(posita semper diametro pro momento maximo, siue totali.)*

Producatur horizon d b f, *demittaturque perpendiculum* c f. *à quolibet puncto; & iungatur* e d.

Angulus e *æqualis est angulo* d b h. *per 32. tertij; eidem* d b h. *est æqualis* c b f. *ad verticem, ergo æquales sunt inter se anguli* e, *&* c b f, *sunt insuper triangula* e d b. b c f *rectangula, ergo similia sunt inter se. Sed iam ostendimus momentum totale sphæræ ad momentũ quod habet in eleuato plano esse vt* b c *ad* c f, *nempe vt ipsa diameter* e b. *ad horizontalem* b d, *quæ intra sphæram à contactu ducitur.*

Si vero graue non sit sphæra, sed quodcunq; solidum a. *habebimus*

ue in puncto d. *plani tangentis, dubitandum non videtur; quandoquidem angulus contingentiæ inclinationem non minuit, neque auget.*

Hinc Propoſitio ſecunda iterum.

Momenta grauium æqualium ſuper planis inæqualiter inclinatis ſunt in reciproca ratione cum longitudinibus planorum,

Momentum in a. *ad totale momentum, per præcedeus lemma, eſt vt* cd. *ad* cb; *totale autem momentũ ad momentum in* e. *eſt vt* fc. *ad* cd*; ergo per perturbatam rationem, momentum* a *ad momentum* e. *eſt reciprocè, vt* cf. *ad* cb *&c. Quod erat &c.*[50]

Corollarium.

Hinc colligitur momentũ ſphæræ grauis ſuper diuerſas planorum eleuationes ſemper eſſe vt linea illa horizontalis quæ à contactu in ipſa ſphæra ducitur.

Sit ſphæra grauis circa centrum a. *in plano* bc *vtcunque inclinato; & ducatur* bd. *horizontalis à contactu: oſtendemus momentum ſphærę in ſitu in quo est, eſſe lineam* bd, *(poſita ſemper diametro pro momento maximo, ſiue totali.).*

Producatur horizon dbf, *demittaturque perpendiculum* cf. *à quolibet puncto; & iungatur* ed.

Angulus e *æqualis eſt angulo* dbh. *per 32. tertij; eidem* dbh. *eſt æqualis* cbf. *ad verticem, ergo æquales ſunt inter ſe anguli* e, & cbf, *ſunt inſuper triangula* edb. bcf *rectangula, ergo ſimilia ſunt inter ſe. Sed iam ostendimus momentum totale ſphæræ ad momentũ quod habet in eleuato plano eſſe vt* bc *ad* cf, *nempe vt ipſa diameter* eb. *ad horizontalem* bd, *quæ intra ſphæram à contactu ducitur.*

Si vero graue non ſit ſphæra, ſed quodcunq; ſolidum a. *habe-*

bimus

[50] Cfr. Pisano and Bussotti's chapters above.

bimus nihilominus singula eius momenta in singulis planorum eleuationibus facilime. Soluemus etiam Problema Pappi lib. 8. Propos. 9. famosum apud Guidobaldum, & Cabeum &c.

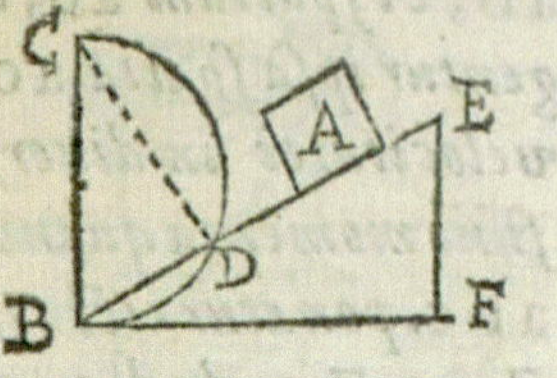

Sit graue a. *in plano* a b, *& quæratur in hoc situ momentum eius; siue potentia, à qua in hoc plano* a b. *sustinetur.*

Ponatur momentum totale grauis; vel, potentiam quæ sustinet pondus a. *in plano perpendiculari esse* b c, *& circa* b c *erectam ad horizontem fiat semicirculus* c d b. *qui secet* a b. *in* d.

Dico momentum grauis a, *siue potentiam quæ illud sustinet in plano* a b. *esse* b d. *Ducatur perpendicularis* e f. *à quolibet puncto* e. *& erunt triangula* c b d, b e f. *similia; quia cum sint vtraq; rectangula, anguli etiam* c b d. *&* e *sunt alterni. Iam quia momentum totale grauis ad momentum quod habet in plano* e b. *est vt* e b. *ad* e f, *erit etiam vt* c b, *ad* b d. *ob similitudinem triangulorum, Est igitur momentum grauis in plano* a b. *vt linea intercepta* b d. *(posita semper diametro pro totali momento.)*

Quo ad propositionem Pappi. manifestum est, si potentia b c *æquatur totali momento* b c. *potentiam* b d. *æquari momento in plano* b d. *Quare potentia* b d. *sustinebit pondus* a. *in proposito plano* b e *&c.*

Scholium:

Iam demonstrari primum potest Propositio sexta Galilei de motu accelerato. Sit enim angulus a b c *rectus, &* ac. *perpendicularis ad horizontem* c d. *& producatur* a b d. *Erunt* d a, a c, a b, *continuę proportionales: at per 2. huius, momentū in* a c. *ad momentū in* a d. *est reciprocè vt* a d *ad* a c. *hoc est vt* a c, *ad* a b, *Ergo est homologe, momentum in* a c, *ad momentum in*

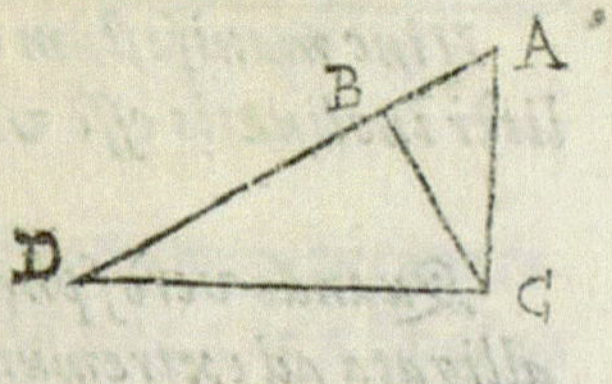

a b,

bimus nihilominus ſingula eius momenta in ſingulis planorum eleuationibus facilime. Soluemus etiam Problema Pappi lib. 8. Propoſ. 9. famoſum apud Guidobaldum, & Cabeum &c.

Sit graue a. *in plano* ab, *& quæratur in hoc ſitu momentum eius; ſiue potentia, à qua in hoc plano* ab. *ſuſtinetur.*

Ponatur momentum totale grauis; vel, potentiam quæ ſuſtinet pondus. a. *in plano perpendiculari eßе* bc, *& circa* bc *erectam ad horizontem fiat ſemicirculus* cdb. *qui ſecet* ab. *in* d.

Dico momentum grauis a, *ſiue potentiam quæ illud ſustinet in plano* ab. *eßе* bd. *Ducatur perpendicularis* ef. *à quolibet puncto* e. *& erunt triangula* cbd, bef. *ſimilia; quia cum ſint vtraq; rectangula, anguli etiam* cbd. *&* e *ſunt alterni. Iam quia momentum totale grauis ad momentum quod habet in plano* eb. *eſt vt* eb. *ad* ef, *erit etiam vt* cb, *ad* bd. *ob ſimilitudinem triangulorum, Eſt igitur momentum grauis in plano* ab. *vt linea intercepta* bd. *(poſita ſemper diametro pro totali momento.)*

Quo ad propoſitionem Pappi. manifeſtum eſt, ſi potentia bc *æquatur totali momento* bc. *potentiam* bd. *æquari momento in plano* bd. *Quare potentia* bd. *ſuſtinebit pondus* a. *in propoſito plano* be *&c.*

Scholium:

Iam demonſtrari primum potest Propoſitio ſexta Galilei de motu accelerato. Sit enim angulus abc *rectus, &* ac. *per pendicularis ad horizontem* cd. *& producatur* abd. *Erunt* da, ac, ab, *continuę proportionales: at per 2. huius, momentũ in* ac. *ad momentã in* ad. *eſt reciproeè vt* ad *ad* ac. *hoc eſt vt* ac, *ad* ab, *Ergo eſt homologe, momentum in* ac, *ad momentum in* a b,[51]

[51] a b,] ab,

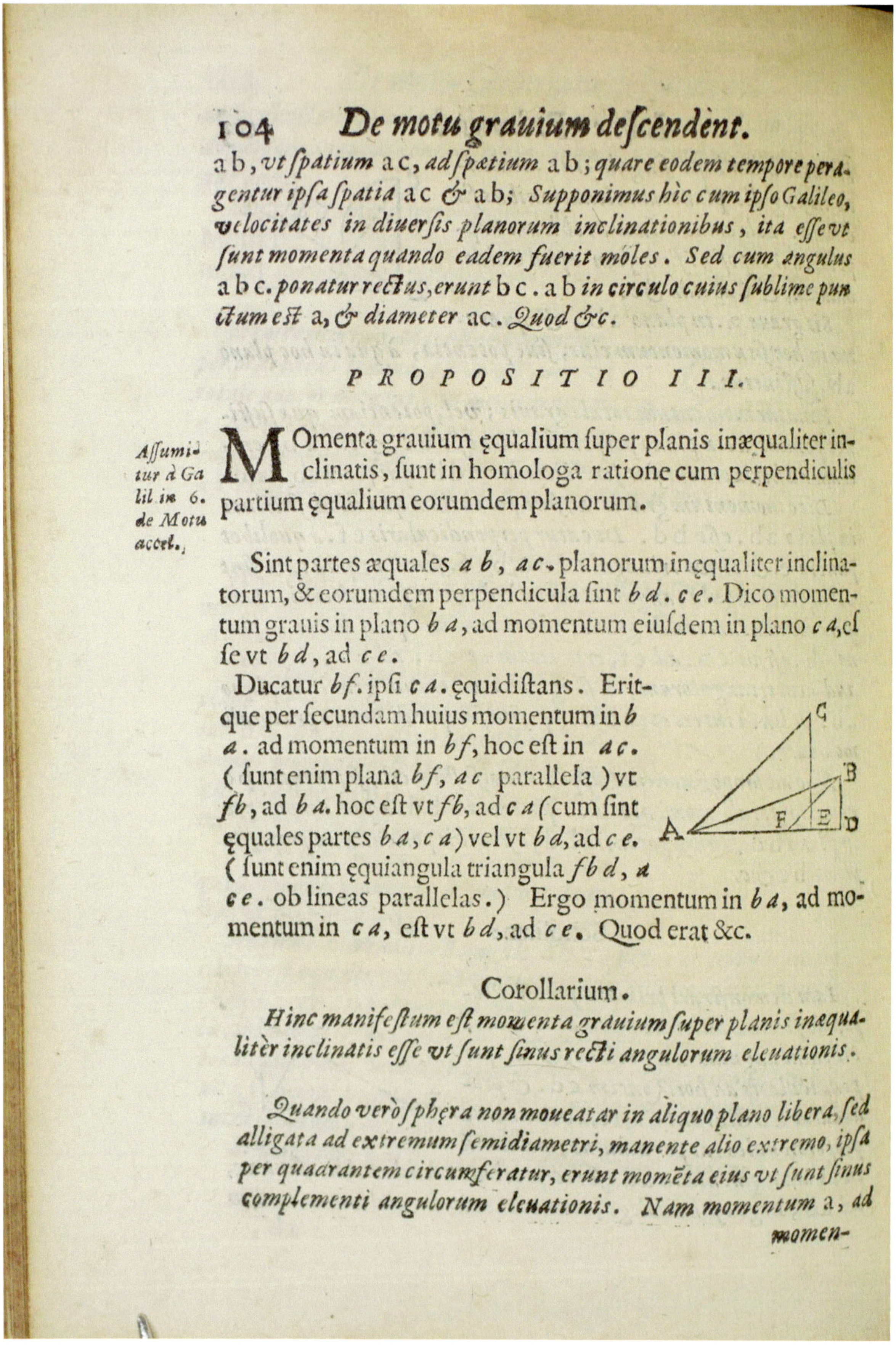

104 *De motu grauium deſcendent.*

a b, *vt ſpatium* a c, *ad ſpatium* a b; *quare eodem tempore peragentur ipſa ſpatia* a c *&* a b; *Supponimus hìc cum ipſo Galileo, velocitates in diuerſis planorum inclinationibus, ita eſſe vt ſunt momenta quando eadem fuerit moles. Sed cum angulus* a b c. *ponatur rectus, erunt* b c. a b *in circulo cuius ſublime punctum eſt* a, *& diameter* a c. *Quod &c.*

PROPOSITIO III.

Aſſumitur à Galil. in 6. de Motu accel.

Momenta grauium ęqualium ſuper planis inæqualiter inclinatis, ſunt in homologa ratione cum perpendiculis partium ęqualium eorumdem planorum.

Sint partes æquales *a b*, *a c*. planorum inęqualiter inclinatorum, & eorumdem perpendicula ſint *b d*. *c e*. Dico momentum grauis in plano *b a*, ad momentum eiuſdem in plano *c a*, eſſe vt *b d*, ad *c e*.

Ducatur *bf*. ipſi *c a*. ęquidiſtans. Eritque per ſecundam huius momentum in *b a*. ad momentum in *bf*, hoc eſt in *a c*. (ſunt enim plana *bf*, *a c* parallela) vt *fb*, ad *b a*. hoc eſt vt *fb*, ad *c a* (cum ſint ęquales partes *b a*, *c a*) vel vt *b d*, ad *c e*. (ſunt enim ęquiangula triangula *f b d*, *a c e*. ob lineas parallelas.) Ergo momentum in *b a*, ad momentum in *c a*, eſt vt *b d*, ad *c e*. Quod erat &c.

Corollarium.

Hinc manifeſtum eſt momenta grauium ſuper planis inæqualiter inclinatis eſſe vt ſunt ſinus recti angulorum eleuationis.

Quando verò ſphęra non moueatur in aliquo plano libera, ſed alligata ad extremum ſemidiametri, manente alio extremo, ipſa per quadrantem circumferatur, erunt momēta eius vt ſunt ſinus complementi angulorum eleuationis. Nam momentum a, *ad momen-*

ab, *vt ſpatium* ac, *ad ſpatium* ab; *quare eodem tempore peragentur ipſa ſpatia* ac & ab; *Supponimus hìc cum ipſo Galileo, velocitates in diuerſis planorum inclinationibus, ita eſſe vt ſunt momenta quando eadem fuerit moles. Sed cum angulus* abc. *ponatur rectus, erunt* bc. ab *in circulo cuius ſublime punctum est* a, & *diameter* ac. *Quod &c.*

PROPOSITIO III.

MOmenta grauium ęqualium ſuper planis inæqualiter inclinatis, ſunt in homologa ratione cum perpendiculis partium ęqualium eorumdem planorum.[53]

Aſſumitur à Galil in 6. *de Motu accel.*

Sint partes æquales *ab, ac.* planorum inęqualiter inclinatorum, & eorumdem perpendicula ſint *bd. ce.* Dico momentum grauis in plano *ba,* ad momentum eiuſdem in plano *ca,* eſſe vt *bd,* ad *ce.*

Ducatur *bf.* ipſi *ca.* ęquidiſtans. Eritque per ſecundam huius momentum in *ba.* ad momentum in *bf,* hoc eſt in *ac.* (ſunt enim plana *bf, ac* parallela) vt *fb,* ad *ba.* hoc eſt vt *fb,* ad *ca* (cum ſint ęquales partes *ba, ca*) vel vt *bd,* ad *ce.* (ſunt enim ęquiangula triangula *fbd, ace.* ob lineas parallelas.) Ergo momentum in *ba,* ad momentum in *ca,* eſt vt *bd,* ad *ce.* Quod erat &c.

Corollarium.

Hinc manifeſtum eſt momenta grauium ſuper planis inæqualitèr inclinatis eſſe vt ſunt ſinus recti angulorum eleuationis.

Quando verò ſphęra non moueatur[54] *in aliquo plano libera, ſed alligata ad extremum ſemidiametri, manente alio extremo, ipſa per quaarantem circumferatur, erunt momēta eius vt ſunt ſinus complementi angulorum eleuationis. Nam momentum* a, *ad*

momen-

---

[52] *motu grauium*] *motu Grauium.*

[53] Cfr. Pisano and Bussotti's chapters above.

[54] *moueatar*] *moueatur..*

momentum b. *est vt* cd. *ad* de, *hoc est* af. *ad* bg, *nempe vt sinus cōplementi angulorum eleuationis.*

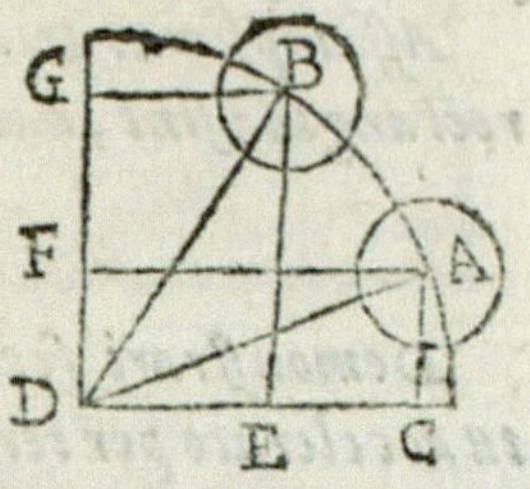

Aliter habebimus mensuram momentorum sphęrę à semidiametro circumducta, in vnoquoq; quadrantis puncto. Ostendemus enim.

Momenta sphæræ per quadrantem circum ductę esse vt sunt lineæ horizontales, quæ à puncto connexionis sphærę cum diametro, intra sphæram ducuntur.

Sit quadrantis centrum a, *sphæra grauis circa* b, *& à puncto connexionis* c. *ducatur horizontalis* cd. *Dico momentum sphęrę esse* cd. *(posita semper diametro sphęræ pro maximo siue totali momento.) Demittatur perpendiculum* be. *Momentum totum sphæræ ad momentum quod habet in* b. *est vt* af, *ad* ac, *vel vt* ba *ad* ae, *hoc est* bc. *ad* ci, *& sumptis duplis, vt* hc. *diameter ad* cd *horizontalem in sphæra, quæ ducitur à puncto connexionis. Quod erat &c.*

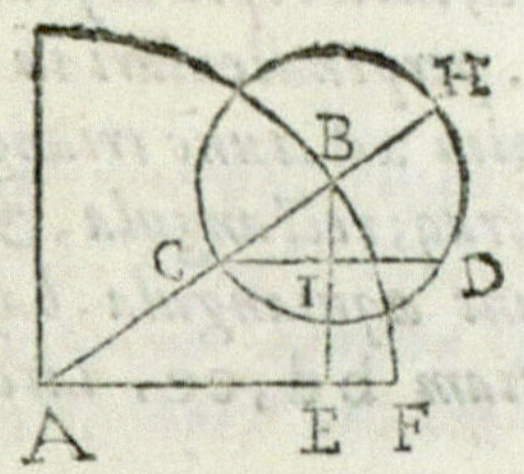

Si verò graue circumductum non sit sphæra dabuntur nihilominus singula eius momenta hoc modo. Sumpto in horizontali linea quolibet interuallo ab. *fiat circulus* adg. *Dico circulum hunc singula singularum eleuationum momenta metiri: Et momentum grauis in* c. *esse lineam interceptam* ad. *(posita semper diametro* ag. *pro totali momento) Est enim momentum totale ad momentum in* c. *vt* ea. *ad* af, *hoc est vt* ca *ad* af, *vel vt* ga *ad* ad *nempe diameter ad lineam interceptam.*

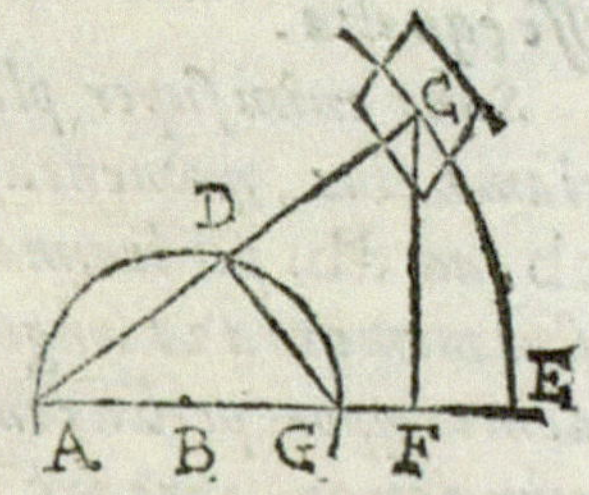

Eadem ad. *erit potentia quę sustinet graue in situ* c, *si suppona-*

O *pona-*

momentum b. *est vt* cd. *ad* de, *hoc eſt* af. *ad* bg, *nempe vt ſinus cõplementi angulorum eleuationis.*

Aliter habebimus menſuram momentorum ſphęrę à ſemidiametro circumductæ, in vnoquoq; quadrantis puncto. Ostendemus enim.

Momenta ſphærę per quadrantem circum ductę eſſe vt ſunt lineæ horizontales, quæ à puncto connexionis ſphærę cum diametro, intra ſphæram ducuntur.

Sit quadrantis centrum a, *ſphæra grauis circa* b, *& à puncto connexionis* c. *ducatur horizontalis* cd. *Dico momentum ſphęrę eſſe* cd. *(poſita ſemper diametro ſpbęræ*[55] *pro maximo ſiue totali momento.) Demittatur perpendiculum* be. *Momentum totum ſphæræ ad momentum quod habet in* b. *est vt* af, *ad* ae, *vel vt* ba *ad* ae, *hoc est* bc. *ad* ci, *& ſumptis duplis, vt* hc. *diameter ad* cd *horizontalem in ſphæra, quæ ducitur à puncto connexionis. Quod erat &c.*

Si verò graue circumductum non ſit ſphæra dabuntur nihilominus ſingula eius momenta hoc modo. Sumpto in horizontali linea quolibet interuallo ab. *fiat circulus* adg. *Dico circulum hunc ſingula ſingularum eleuationum momenta metiri: Et momentum grauis in* c. *eſſe lineam interceptam* ad. *(poſita ſemper diametro* ag. *pro totali momento)*[56] *Eſt enim momentum totale ad momentum in* c. *vt* ea. *ad* af, *hoc est vt* ca *ad* af, *vel vt* ga *ad* ad *nempe diameter ad lineam interceptam.*

Eadem ad. *erit potentia quę ſuſtinet graue in ſitu* c, *ſi ſup-*

O *pona-*

[55] *ſpbęræ*] *ſphęræ.*
[56] *momento)*] *momento).*

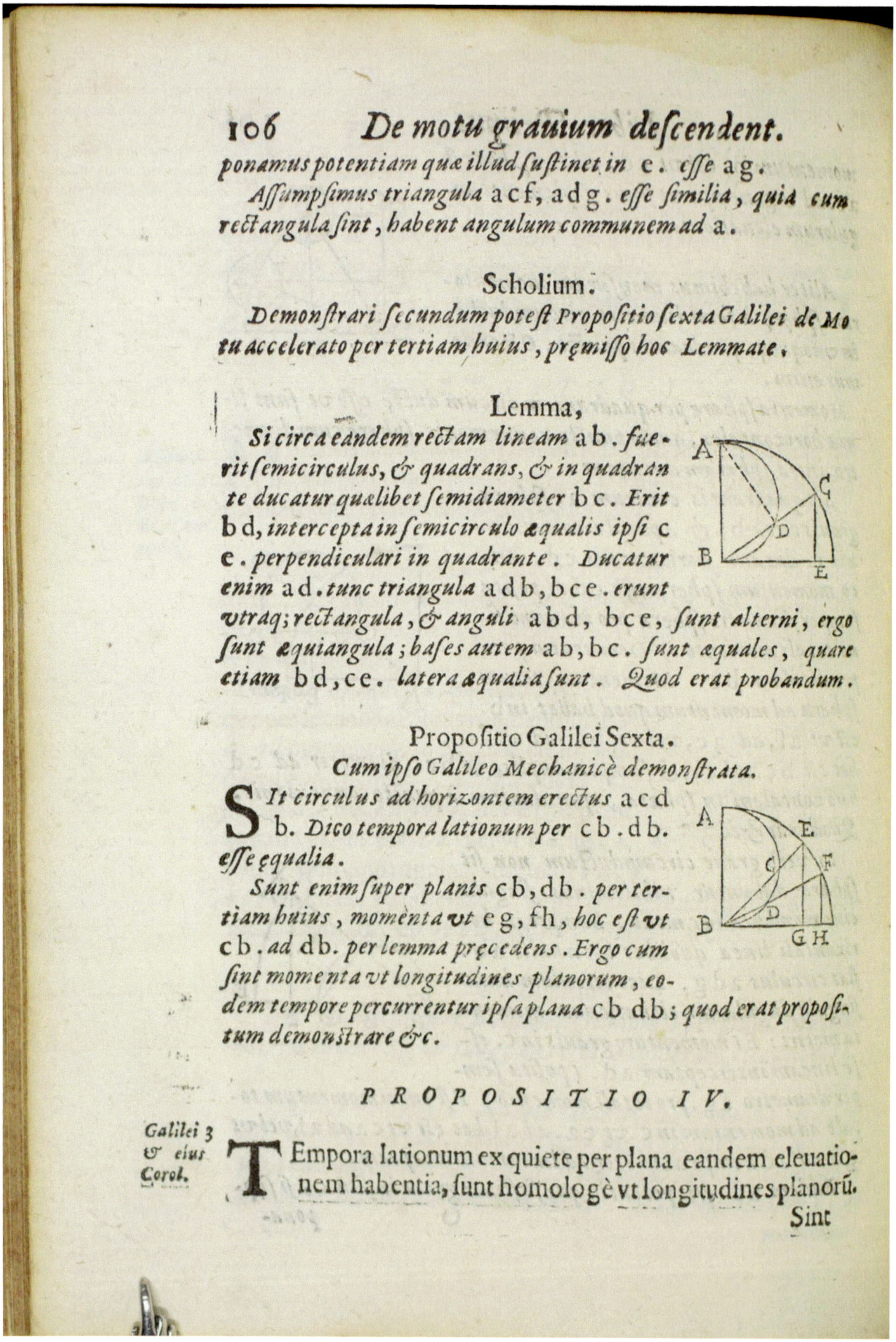

106 *De motu grauium descendent.*

ponamus potentiam quæ illud sustinet in c. *esse* ag.

Assumpsimus triangula acf, adg. *esse similia, quia cum rectangula sint, habent angulum communem ad* a.

Scholium.

Demonstrari secundum potest Propositio sexta Galilei de Motu accelerato per tertiam huius, premisso hoc Lemmate.

Lemma,

Si circa eandem rectam lineam ab. *fuerit semicirculus, & quadrans, & in quadrante ducatur qualibet semidiameter* bc. *Erit* bd, *intercepta in semicirculo æqualis ipsi* ce. *perpendiculari in quadrante. Ducatur enim* ad. *tunc triangula* adb, bce. *erunt vtraq; rectangula, & anguli* abd, bce, *sunt alterni, ergo sunt æquiangula; bases autem* ab, bc. *sunt æquales, quare etiam* bd, ce. *latera æqualia sunt. Quod erat probandum.*

Propositio Galilei Sexta.

Cum ipso Galileo Mechanicè demonstrata.

S*It circulus ad horizontem erectus* acdb. *Dico tempora lationum per* cb. db. *esse equalia.*

Sunt enim super planis cb, db. *per tertiam huius, momenta vt* eg, fh, *hoc est vt* cb. *ad* db. *per lemma precedens. Ergo cum sint momenta vt longitudines planorum, eodem tempore percurrentur ipsa plana* cb db; *quod erat propositum demonstrare &c.*

PROPOSITIO IV.

Galilei 3 & eius Corol.

TEmpora lationum ex quiete per plana eandem eleuationem habentia, sunt homologè vt longitudines planorũ.

Sint

106 *De motu grauium*[57] *defcendent.*

ponamus potentiam quæ illud fuftinet in e. *effe* ag.

Affumpfimus triangula acf, adg. *effe fimilia, quia cum rectangula fint, habent angulum communem ad* a.

Scholium.

Demonftrari fecundum poteft Propofitio fexta Galilei de Motu accelerato per tertiam huius, prȩmiffo hoc Lemmate.

Lemma,

Si circa eandem rectam lineam ab. *fuerit femicirculus, & quadrans, & in quadrante ducatur quælibet femidiameter* bc. *Erit* bd, *intercepta in femicirculo æqualis ipfi* ce. *perpendieulari*[58] *in quadrante. Ducatur enim* ad. *tunc triangula* adb, bce. *erunt vtraq; rectangula, & anguli* abd, bce, *funt alterni, ergo funt æquiangula; bafes autem* ab, bc. *funt æquales, quare etiam* bd, ce. *latera æqualia funt. Quod erat probandum.*

Propofitio Galilei Sexta.

Cum ipfo Galileo Mechanicè demonftrata.

S*It circulus ad horizontem erectus* acdb. *Dico tempora lationum per* cb. db. *effe ȩqualia.*

Sunt enim fuper planis cb, db. *per tertiam huius, momènta vt* eg, fh, *hoc eft vt* cb. *ad* db. *per lemma prȩcedens. Ergo cum fint momenta vt longitudines planorum, eodem tempore percurrentur ipfa plana* cbdb; *quod erat propofitum demonstrare &c.*

Galilei 3 *& eius Corol*

PROPOSITIO IV.

TEmpora lationum ex quiete per plana eandem eleuationem habentia, funt homologè vt longitudines planorũ.

Sint

[57] *motu grauium*] *motu Grauium.*
[58] *perpendieulari*] *perpendiculari.*

Sint plana *ab*, *ac*. eandem eleuationem *ad* habentia. Dico tempus lationis per *a c* ad tempus per *ab* esse vt *ac* ad *ab*.

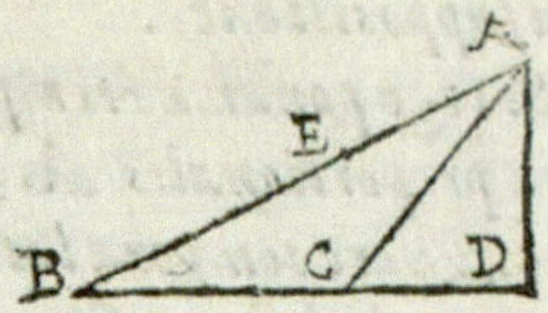

Sit ipsarum *ab*, *ac* tertia proportionalis *ae*. Momentum ergò in plano *ac* ad momentum in plano *ab*, est vt *ab*, ad *ac*. (per secundam huius,) hoc est vt *ac*. ad *ae*. Quare lationes per *ac*, *ae* temporibus æqualibus absoluentur: quandoquidem ita sunt momenta vt longitudines spatiorum. Ponamus iam tempus per *ae*, esse mediam proportionalem *ac*. Erit tempus per *ab*. ipsa *ab*; tempus ergo per *ae*, siue per *ac*. (nã æqualia tempora sunt) est *ac*, & per *ab* est ipsa *ab* &c. Quod erat &c.

Aliter.

Pręcedens Theorema poterat demonstrari sine vlla suppositione. Demonstrat enim Galileus in Prop. 6. de motu accelerato, tempora lationum per chordas omnes in circulo æqualia esse. Idq; tribus modis probat. in primo, & tertio subest principium suum non satis euidens; in secundo verò nihil supponitur, prætar iàm dictum Theorema Mechanicum; quod si, ipso teste, demonstratum antea fuerat, ex ipso immediatè, tamquam Corollarium, necessaria illatio suæ tertiæ Propositionis, immo & suę petitionis, deriuari poterat. Sed quia ipse tertiam suam Propositionem, quę nobis quarta est, mediante sua petitione probat, nos illam absolutè ostendamus ex propositionibus ipsius Galilei, quę nullum postulatum includunt.

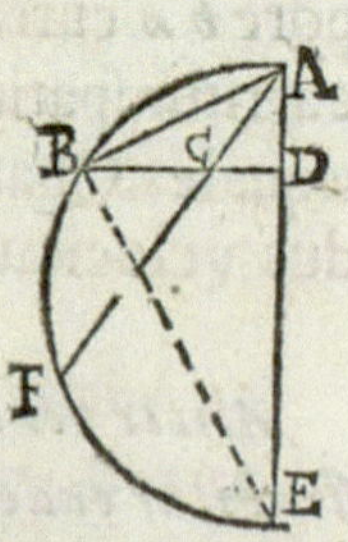

Sint duo plana a b, a c. *quorum eadem eleuatio sit* a d. *Dico tempus lationis per* a c. *ad tempus per* a b, *esse vt* a c. *ad* a b. *facto enim angulo* a b e. *recto, agatur circulus circa diametrum* a e, *qui transibit per* b, *& producatur* a c f *Erit* a b. *media proportionalis inter* f a, a c, *& erunt tempora per* a b, a f. *æqualia, vt o-*

 sten-

Sint plana *ab, ac.* eandem eleuationem *ad* habentia. Dico tempus lationis per *ac* ad tempus per *ab* eſſe vt *ac* ad *ab.*

Sit ipſarum *ab, ac* tertia proportionalis *ae.* Momentum ergò in plano *ac* ad momentum in plano *ab,* eſt vt *ab,* ad *ac.* (per ſecundam huius,) hoc eſt vt *ac.* ad *ae.* Quare lationes per *ac, ae* temporibus æqualibus abſoluentur: quandoquidem ita ſunt momenta vt longitudines ſpatiorum. Ponamus iam tempus per *ae,* eſſe mediam proportionalem *ac.* Erit tempus per *ab.* ipſa *ab*; tempus ergo per *ae,* ſiue per *ac. (*nã æqualia tempora ſunt*)* eſt *ac,* & per *ab* eſt ipſa *ab* &c. Quod erat &c.

Aliter.

Pręcedens Theorema poterat demonſtrari ſine vlla ſuppoſitione. Demonſtrat ęnim Galileus in Prop. 6. de motu accelerato, tempora lationum per chordas omnes in circulo æqualia eſſe. Idq; tribus modis probat. in primo, & tertio ſubeſt principium ſuum non ſatis euidens; in ſecundo verò nihil ſupponitur, præter iàm dictum Theorema Mechanicum; quod ſi, ipſo teste, demonstratum antea fuerat, ex ipſo immediatè, tamquam Corollarium, neceſſaria illatio ſuæ tertiæ Propoſitionis, immo & ſuę petitionis, deriuari poterat. Sed quia ipſe tertiam ſuam Propositionem, quę nobis quarta eſt, mediante ſua petitione probat, nos illam abſolutè oſtendamus ex propoſitionibus ipsius Galilei, quę nullum poſtulatum includunt.

Sint duo plana ab, ac. *quorum eadem eleuatio sit* ad. *Dico tempus lationis per* ac. *ad tempus per* ab, *eβe vt* ac. *ad* ab. *facto enim angulo* abe. *recto, agatur circulus circa diametrum* ac, *qui transibit per* b, *& producatur* acf[59] *Erit* ab. *media proportionalis inter* fa, ac, *& erunt tempora per* ab, af. *æqualia, vt o-*

O 2 *ſten-*

[59] acf] acf.

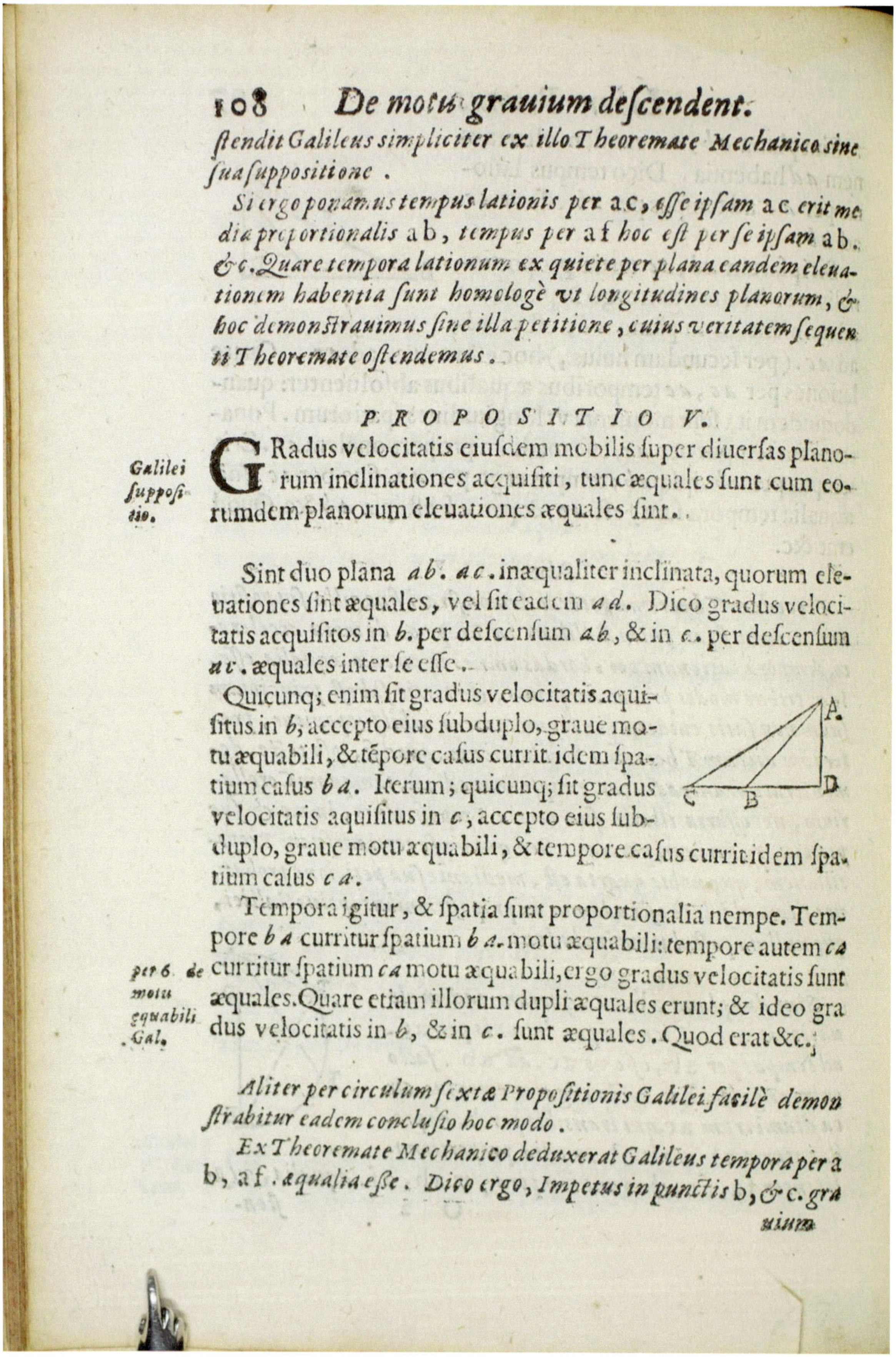

108 *De motu grauium descendent.*

stendit Galileus simpliciter ex illo Theoremate Mechanico sine sua suppositione.

Si ergo ponamus tempus lationis per a c, *esse ipsam* a c *erit media proportionalis* a b, *tempus per* a f *hoc est per se ipsam* a b. *&c. Quare tempora lationum ex quiete per plana eandem eleuationem habentia sunt homologè vt longitudines planorum, & hoc demonstrauimus sine illa petitione, cuius veritatem sequenti Theoremate ostendemus.*

PROPOSITIO V.

Galilei suppositio. GRadus velocitatis eiusdem mobilis super diuersas planorum inclinationes acquisiti, tunc æquales sunt cum eorumdem planorum eleuationes æquales sint.

Sint duo plana *ab. ac.* inæqualiter inclinata, quorum eleuationes sint æquales, vel sit eadem *ad.* Dico gradus velocitatis acquisitos in *b.* per descensum *ab*, & in *c.* per descensum *ac.* æquales inter se esse.

Quicunq; enim sit gradus velocitatis aquisitus in *b*, accepto eius subduplo, graue motu æquabili, & tẽpore casus currit idem spatium casus *ba.* Iterum; quicunq; sit gradus velocitatis aquisitus in *c*, accepto eius subduplo, graue motu æquabili, & tempore casus currit idem spatium casus *ca.*

Tempora igitur, & spatia sunt proportionalia nempe. Tempore *ba* curritur spatium *ba.* motu æquabili: tempore autem *ca* curritur spatium *ca* motu æquabili, ergo gradus velocitatis sunt æquales. Quare etiam illorum dupli æquales erunt; & ideo gradus velocitatis in *b*, & in *c.* sunt æquales. Quod erat &c. *per 6. de motu æquabili Gal.*

Aliter per circulum sextæ Propositionis Galilei facilè demonstrabitur eadem conclusio hoc modo.

Ex Theoremate Mechanico deduxerat Galileus tempora per a b, a f. *æqualia esse. Dico ergo, Impetus in punctis* b, *& c. grauium*

ftendit Galileus simpliciter ex illo Theoremate Mechanico sine fua fuppositione.

Si ergo ponamus tempus lationis per ac, *effe ipfam* ac *erit media proportionalis* ab, *tempus per* af *hoc eft per fe ipfam* ab. *&c. Quare tempora lationum ex quiete per plana eandem eleuationem habentia funt homologè vt longitudines planorum, & hoc demonstrauimus fine illa petitione, cuius veritatem fequenti Theoremate oftendemus.*

PROPOSITIO V.

Galilei fuppofitio.

GRadus velocitatis eiufdem mobilis fuper diuerfas planorum inclinationes acquifiti, tunc æquales funt cum eorumdem planorum eleuationes æquales fint.[61]

Sint duo plana *ab. ac.* inæqualiter inclinata, quorum eleuationes fint æquales, vel fit eadem *ad.* Dico gradus velocitatis acquifitos in *b.* per defcenfum *ab,* & in *c.* per defcenfum *ac.* æquales inter fe effe.

Quicunq; enim fit gradus velocitatis aquifitus in *b,* accepto eius fubduplo, graue motu æquabili, & tēpore cafus currit idem fpatium cafus *ba.* Iterum; quicunq; fit gradus velocitatis aquifitus in *c,* accepto eius fubduplo, graue motu æquabili, & tempore cafus currit idem fpatium cafus *ca.*

per 6. de motu ęquabili Gal.

Tempora igitur, & fpatia funt proportionalia nempe. Tempore *ba* curritur fpatium *ba.* motu æquabili: tempore autem *ca* curritur fpatium *ca* motu æquabili, ergo gradus velocitatis funt æquales. Quare etiam illorum dupli æquales erunt; & ideo gradus velocitatis in *b,* & in *c.* funt æquales. Quod erat &c.

Aliter per circulum fextæ Propofitionis Galilei facilè demonftrabitur eadem conclufio hoc modo.

Ex Theoremate Mechanico deduxerat Galileus tempora per ab, af. æqualia eβe. Dico ergo, Impetus in punctis b, & c. *gra*

uium

[60] *motu grauium*] *motu Grauium.*

[61] Cfr. Pisano and Bussotti's chapters above.

uium ab eadem altitudine, & ex quiete in a *descendentium, æquales esse.*

Quia enim a b, *&* a f. *æquali tempore peraguntur, erunt impetus in* b, *&* f. *punctis, vt sunt spatia peracta* a b, a f. *(Acceptis enim eorum subduplis æquali tempore, & motu æquabili curruntur spatia* b a . f a, *quare subdupli illi impetus sunt vt spatia, & propterea etiam illorum dupli vt eadem spatia erunt.) Impetus ergò* b. *ad impetum* f. *est vt* a b, *ad* a f, *impetus verò in* f. *ad impetum in* c. *est vt* f a. *ad* a b. *(nempe vt tempora, quia* a b, *media proportionalis est inter* f a. a c.) *Ergo ex æquali, impetus in* b. *ad impetum in* c. *est vt* a b, *ad ipsammet* a b. *Quare impetus in* b, *&* c. *sunt æquales. &c. Vel sic.*

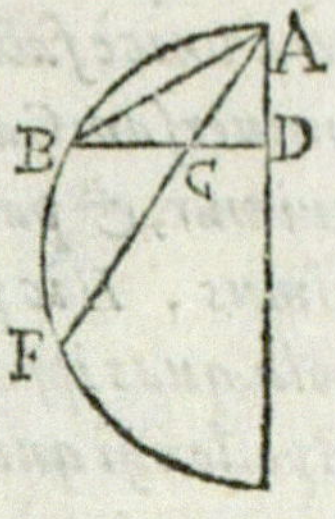

per prim. Gal. de m. acc.

Gradus impetus in c. *ad gradum in* f, *est vt* c a. *ad* a b, *vel* b a, *ad* a f, *cum tres* c a, a b, a f. *sint in continua proportione. Sed gradus etiam impetus in* b. *ad gradum in* f, *est vt* b a *ad* a f *(vt supra demonstrauimus.) Quare vterque gradus* c, *&* b. *ad eundem* f. *eandem rationẽ habet; & ideo æquales sunt. Quod &c.*

Corollarium.

Hinc pro Corollario extrhaemus id quod in ipso progressu demonstrationis ostensum est, nempe. Impetus grauium in fine chordarum circuli, quæ ex puncto sublimi descẽdant, ita esse vt sunt ipsæmet chordæ: hoc est impetus in punctis b, c, d, *ita esse vt sunt* a b, a c. a d. *&c.*

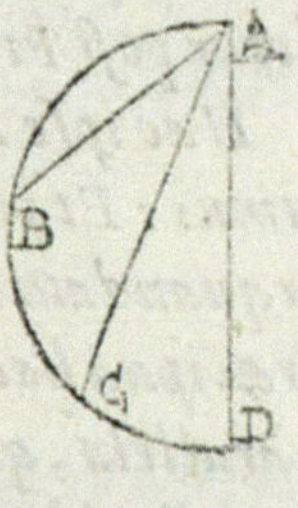

Scholium:

Cum deinceps futurus sit sermo de lineis quas parabolas vocant, non erit inconueniens, antequam illarum passiones in ordine ad motum consideremus, pauca quædam necessaria nobis præmonstrare. Sic enim fiet vt paratiores accedere possimus ad

con-

uium ab eadem altitudine, & ex quiete in a *deſcendentium, æquales eße.*

Quia enim ab, & af. *æquali tempore peraguntur, erunt impetus in* b, & f. *punctis, vt ſunt ſpatia peracta* ab, af. *(Acceptis enim eorum ſubduplis æquali tempore, & motu æquabili curruntur ſpatia* ba. fa, *quare ſubdupli illi impetus ſunt vt ſpatia, & propterea etiam illorum dupli vt eadem ſpatia erunt.) Impetus ergò* b. *ad impetum* f. *eſt vt* ab, *ad* af, *impetus verò in* f. *ad impetum in* c. *est vt* fa. *ad* ab. *(nempe vt tempora, quia* ab, *media proportionalis eſt inter* fa. ac.) *Ergo ex æquali, impetus in* b. *ad impetum in* c. *eſt vt* ab, *ad ipſammet* ab. *Quare impetus in* b, & c. *ſunt æquales. &c. Vel ſic.* *per prim. Gal. de m. aec.*

Gradus impetus in c. *ad gradum in* f, *eſt vt* ca. *ad* ab, *vel* ba, *ad* af, *cum tres* ca, ab, af. *ſint in continua proportione. Sed gradus etiam impetus in* b. *ad gradum in* f, *eſt vt* ba *ad* af (*vt ſupra demonſtrauimus.) Quare vterque gradus* c, & b. *ad eundem* f. *eandem rationẽ habet; & idea æquales ſunt. Quod &c.*

Corollarium.

Hinc pro Corollario extrhaemus id quod in ipſo progreſſu demonſtrationis ostenſum eſt, nempe. Impetus grauium in ſine chordarum circuli, quæ ex puncto ſublimi deſcẽdant, ita eſſe vt ſunt ipſæmet chordæ: hoc eſt impetus in punctis b, c, d, *ita eſſe vt ſunt* ab, ac. ad. *&c.*

Scholium:

Cum deinceps futurus ſit ſermo de lineis quas parabolas vocant, non erit inconueniens, antequam illarum paſſiones in ordine ad motum conſideremus, pauca quædam neceſſaria nobis præmonſtrare. Sic enim fiet vt paratiores accedere poſſimus ad

con-

contemplandam lineam pro motibus non solum proiectorum, sed etiam (quod non scripsit Galileus) naturaliter cadentium, à natura vnicè factam. Premittimus semper tamquam suppositum vniuersam Galilei doctrinam de motu: illius enim vestigia sequimur, & paucula quædam Theoremata ab ipso neglecta colligimus. Hic præcipuè supponuntur duæ Propositiones de Parabola, quas ipse operi suo de Motu Proiectorum, prefigit, alteram Apolloniji quidem sed Marte proprio demonstratam à Galileo; alteram verò penitus ex Apollonij lib. 1. prop. 33. desumptam, & demonstratam.

Prima est huiusmodi.

Lineæ, quæ intra parabolam basi parallelæ ducuntur, sunt in subdupla ratione portionum diametri ad verticem parabolę interceptarum.

Secunda verò est hec.

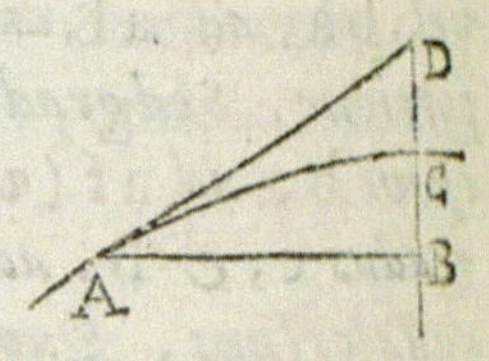

Si in parabola aliquod punctum *a* sumatur ex quo linea ducatur basi parallela *a b*, & portioni diametri *b c*. ad verticem interceptę, æqualis recta linea *c* d. ponatur in directum. Recta linea *d a*, quæ ab extremo positæ lineæ termino *d*: ad punctum *a* in parabola sumptum, ducitur, parabolam continget, *quod & à nobis aliquo modo ostēdetur post Prop. 10.*

Hæc ipse. His premissis reliqua nobis oportuna demonstrabimus: Et primò animaduertendum est, quod vnaquæq; parabola quandam rectam lineam peculiarem habet, cuius proprietas præcipua hæc est. Ducta intra parabolam quacunq; linea basi parallela, quadratum ductæ æquale est rectangulo, quod sub illa peculiari linea, & portione diametri ad verticem parabolę abscissa, continetur. Exempl. g. Quadratum rectæ a b *equale est rectangulo sub* b c *& illa peculiari linea contento; & hoc semper vbicumq; fuerit ducta* a b

Vocatur autem peculiaris illa linea Latus Rectum.

Quæ verò ducuntur æquidistantes basi, Ordinatim applicatæ dicuntur.

PRO-

contemplandam lineam pro motibus non ſolum proiectorum, ſed etiam (quod non ſcripſit Galileus) naturaliter cadentium, à natura vnicè factam. Prĕmittimus ſemper tamquam ſuppoſitum vniuerſam Galilei doctrinam de motu: illius enim veſtigia ſequimur, & paucula quædam Theoremata ab ipſo neglecta colligimus. Hic præcipuè ſupponuntur auæ[62] *Propoſitiones de Parabola, quas ipſe operi ſuo de Motu Proiectorum, prĕfigit, alteram Apolloniji quidem ſed Marte proprio demonſtratam à Galileo; alteram verò penitus ex Apollonij lib. 1. prop. 33. deſumptam, & demonſtratam*

Prima est huiuſmodi.

Lineæ, quæ intra parabolam baſi parallelæ ducuntur, ſunt in ſubdupla ratione portionum diametri ad verticem parabolĕ interceptarum.

Secunda verò eſt hec.[63]

Si in parabola aliquod punctum *a* ſumatur ex quo linea ducatur baſi parallela *ab*, & portioni diametri *bc*. ad verticem interceptĕ, æqualis recta linea *cd*. ponatur in directum. Recta linea *da*, quæ ab extremo poſitæ lineæ termino *d*: ad punctum *a* in parabola ſumptum, ducitur, parabolam continget, *quod & à nobis aliquo modo ostẽdetur poſt Prop. 10.*

Hæc ipſe. His prĕmiſſis reliqua nobis oportuna demonſtrabimus: Et primò animaduertendum est, quod vnaquæq; parabola quandam rectam lineam peculiarem habet, cuius proprietas præcipua hæc est. Ducta intra parabolam quacunq; linea baſi parallela, quadratum ductæ æquale eſt rectangulo, quod ſub illa peculiari linea, & portione diametri ad verticem parabolĕ abſciſſa, continetur. Exempl. g. Quadratum rectæ ab *ĕquale est rectangulo ſub* bc *& illa peculiari linea contento; & hoc ſem per vbicumq; fuerit ducta* ab

Vocatur autem peculiaris illa linea Latus Rectum.

Quæ verò ducuntur æquidiſtantes baſi, Ordinatim applicatæ dicuntur.

P R O–

---

[62] *auæ*] *quæ*.
[63] *hec.*] *haec*.

PROPOSITIO VI.

Parabolæ latus rectum demonstrare.

MAnente figura, & constructione eadem quam ponit Galileus in prima iam dictarum propositionum de Parabola.

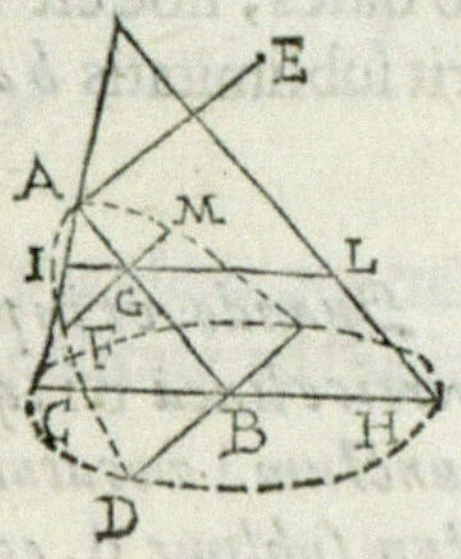

Fiat vt *ab*, ad *bc*. ita *bb*. ad *ae*. Dico *ac*. esse latus rectum.

Sumatur enim quodlibet punctum in parabola quod sit *f*, & ducatur *fg*. parallela ipsi *db*, item per *g*. agatur *igl*. parallela ad *ch*. eritq; *bglh* parallelogramum. Et quia factum est vt *ab*, ad *bc* ita *bh* ad *ae* erit *ag* ad *gi*, vt *ab* ad *bc*. hoc est vt *bh* ad *ae* siue vt *gl*, ad *ae*. Rectangulum ergo *gae*. æquale erit rectangulo *igl*, hoc est quadrato *fg*, Est ergo *ea*. latus rectum.

Corollarium.

Hinc manifestum est, si linea fg *. ordinatim applicata producatur vsq; ad vlteriorem semiparabolam in* m; *ipsam* gm *æqualem fore ipsi* fg. *eodem enim modo ostenditur quadratum* gm, *atq; quadratum* gf. *æquari rectangulo* igl *etc.*

PROPOSITIO VII.

SVblimitas parabolæ apud Galileum, quarta pars est lateris recti eiusdem parabolæ.

Maneat constructio Propositionis V. de Motu proiectorũ Galilei, qua ipse reperit sublimitatem parabolæ.

Factum ibi fuit vt *ab*. altitudo, ad *bc* (æqualem nempe dimidio basis *ae*,) ita *bc*. ad aliam, quæ sit *bd*, & hæc erat sublimitas apud Galileum, probabimusq; ipsam esse quartam lateris recti partem.

Qua-

PROPOSITIO VI.

Parabolæ latus rectum demonſtrare.

MAnente figura, & conſtructione eadem quam ponit Galileus in prima iam dictarum propoſitionum de Parabola.

Fiat vt *ab,* ad *bc.* ita *bb.* ad *ae.* Dico *ae.* eſſe latus rectum.

Sumatur enim quodlibet punctum in parabola quod ſit *f,* & ducatur *fg.* parallela ipſi *db,* item per *g.* agatur *igl.* parallela ad *ch.* eritq; *bglh* parallelogramum. Et quia factum eſt vt *ab,* ad *bc* ita *bh* ad *ae* erit *ag* ad *gi,* vt *ab* ad *bc.* hoc eſt vt *bh* ad *ae* ſiue vt *gl,* ad *ae.* Rectangulum ergo *gae.* æquale erit rectangulo *igl,* hoc eſt quadrato *fg,* Eſt ergo *ea.* latus rectum.

Corollarium.

Hinc manifeſtum eſt, ſi linea fg. *ordinatim applicata producatur vſq; ad vlteriorem ſemiparabolam in* m; *ipſam* gm *ęqualem fore ipſi* fg. *eodem enim modo oſtenditur quadratum* gm, *atq; quadratum* gf. *ęquari rectangulo* igl etc.

PROPOSITIO VII.

SVblimitas parabolæ apud Galileum, quarta pars eſt lateris recti eiuſdem parabolæ.

Maneat conſtructio Propoſitionis V. de Motu proiectorũ Galilei, qua ipſe reperit ſublimitatem parabolæ.

Factum ibi fuit vt *ab.* altitudo, ad *bc (*æqualem nempe dimidio baſis *ae,)* ita *bc.* ad aliam, quæ ſit *bd,* & hæc erat ſublimitas apud Galileum, probabimuſq; ipſam eſſe quartam lateris recti partem

Qua-

Quadratum *ea*. quadruplum est quadrati *bc*. hoc est rectanguli *abd*, (per constructionem) ergo etiam rectangulum sub *ab*, & latere recto quadruplum erit eiusdem *abd*; sed communis est *ab* altitudo rectangulorum, ergo bases, hoc est latus rectum, quadruplum erit sublimitatis *bd*. Quod &c.

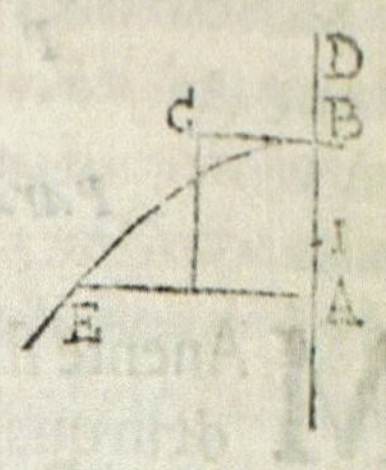

Definitio.

Quando, vt in praecedenti figura, sumitur in axe parabolę ex vertice linea bi. *quę aequalis sit quartae parti lateris recti, tunc punctum* i *vocatur focus parabolę. Manifestum ergo est; punctum sublime* d, *et focum* i, *aequaliter distare à vertice parabolae; nempe tantum vtrinq; quanta est quarta pars lateris recti.*

PROPOSITIO VIII.

Recta linea quae ex foco parabolae ordinatim applicatur, dupla est portionis axis ad verticem interceptae, Vel. aequalis est semissi lateris Recti.

Sit latus rectum *ab*, & focus *d*. Rectangulum *bad*. quadruplum est quadrati *ad*. (quia cum habeant communem altitudinem *ad*, basis *ba*. quadrupla est basis *ad*.) ergo etiã quadratum *cd*. quadruplum est eiusdem quadrati *ad*. Est igitur *cd* dupla ipsius *da*; siue aequalis semissi lateris recti. Quod erat &c.

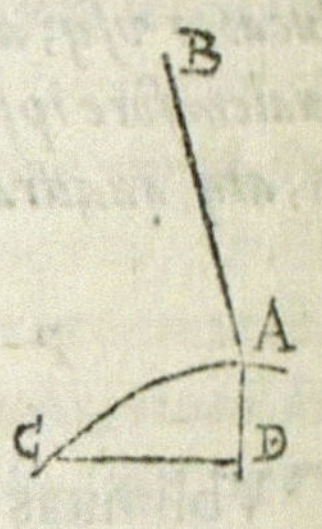

PROPOSITIO IX.

Si Parabolae quotcunq; circa eandem diametrum sint, lineę quae in illis ordinatim ducuntur, proportionales erunt.

Sit diameter communis *ab*. ordinatim autem ductae sint *cd ce*, & *bf*. *bg*. Dico esse vt *ec* ad *gb*, ita *de*. ad *fb*.

Sunt

Quadratum *ea.* quadruplum eſt quadrati *bc.* hoc eſt rectanguli *abd,* (per conſtructionem) ergo etiam rectangulum ſub *ab,* & latere recto quadruplum erit eiuſdem *abd*; ſed communis eſt *ab* altitudo rectangulorum, ergo baſes, hoc eſt latus rectum, quadruplum erit ſublimitatis *bd.* Quod &c

Definitio.

Quando, vt in præcedenti figura, ſumitur in axe parabolę ex vertice linea bi. *quę æqualis ſit quartæ parti lateris recti, tunc punctum* i *vocatur focus parabolę. Manifestum ergo est; punctum ſublime* d, *et focum* i, *æqualiter distare à vertice parabolæ; nempe tantum vtrinq; quanta eſt quarta pars lateris recti.*

PROPOSITIO VIII.

REcta linea quæ ex foco parabolæ ordinatim applicatur, dupla eſt portionis axis ad verticem interceptæ, Vel. æqualis eſt ſemiſſi lateris Recti.

Sit latus rectum *ab.* & focus *d.* Rectangulum *bad.* quadruplum eſt quadrati *ad.* (quia cum habeant communem altitudinem *ad,* baſis *ba.* quadrupla eſt baſis *ad.*) ergo etiã quadratum *cd.* quadruplum eſt eiuſdem quadrati *ad.* Eſt igitur *cd* dupla ipſius *da*; ſiue æqualis ſemiſſi lateris recti. Quod erat &c.

PROPOSITIO IX.

SI Parabolæ quotcunq; circa eandem diametrum ſint, lineę quæ in illis ordinatim ducuntur, proportionales erunt.

Sit diameter communis *ab.* ordinatim autem ductæ ſint *cdce,* & *bf. bg.* Dico eſſe vt *ec* ad *b,* ita *dc.* ad *fb.*

Sunt

Sunt enim quadrata, *ec*, ad *gb*, vt recta *ac*. ad *ab*. Quadrata etiam *dc* ad *fb* sunt vt *ac*, ad *ab*. Ergo in eadem ratione sunt quadrata inter se; quare vt recta *ec* ad *gb*, ita est *dc* ad *fb*. Quod erat &c.

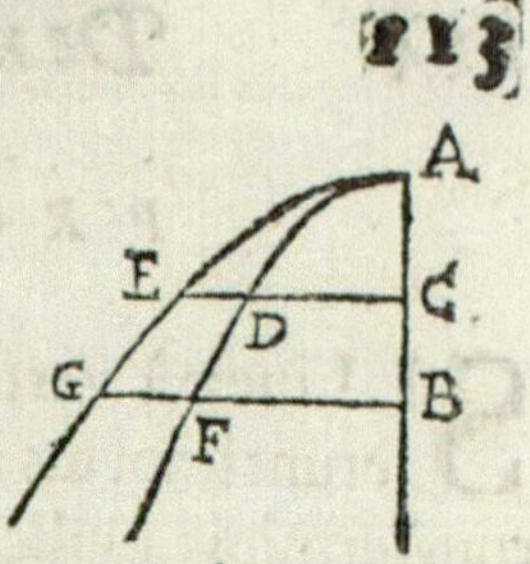

PROPOSITIO X.

TEmpora lationum, quæ ex quiete fiunt per plana quęcunque sunt inter se vt lineæ in parabola applicatę ad spatia per quæ grauia descenderunt.

Sint spatia quælibet *ab*, *ac*, siue perpendicularia siue inclinata, & circa diametrum *ac* fiat parabola quælibet *ade*. atq; ordinatim ducantur *bd*. *ce*. Dico tempus per *ab* ad tempus per *ac*. esse vt *bd*, ad *ce*. Sunt enim tempora in subdupla ratione spatiorum ex Galileo, sed lineæ *db*, *ec*. sunt in subdupla ratione spatiorum (quia quadrata earũ sunt vt *ab*, ad *ac*.) ergo eadem ratio est & temporũ, & linearũ ordinatim ad spatia applicatarũ. Quod erat. &c.

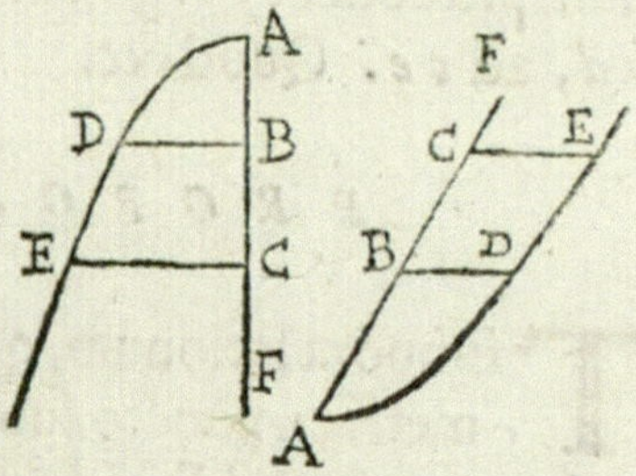

Aliter.

Si ponamus tempus lationis per latus rectum af *esse ipsummet latus rectum, erit tempus per* ab *. media proportionalis inter* fa, ab, *nempe ipsa* bd. *& tempus per* ac *media proportionalis* ce *&c. & sic de singulis. Quare &c.*

Corollarium.

Hinc manifestum est impetus grauium in fine portionum diametri parabolæ, esse inter se vt lineæ, quæ ordinatim applicantur ad extrema ipsarum portionum puncta. Sunt enim ex Galileo impetus vt ipsa tempora, sed ordinatim ductæ sunt vt ipsa tempora, ergo impetus sunt vt ordinatim ductæ &c.

P PRO-

Sunt enim quadrata, *ec,* ad *gb,* vt recta *ac.* ad *ab.* Quadrata etiam *dc* ad *fb* ſunt vt *ac,* ad *ab.* Ergo in eadem ratione ſunt quadrata inter ſe; quare vt recta *ec* ad *b,* ita eſt *dc* ad *fb.* Quod erat &c.

PROPOSITIO X.

TEmpora lationum, quæex[64] quiete fiunt per plana quęcunque ſunt inter ſe vt lineæ in parabola applicatę ad ſpatia per quæ grauia deſcenderunt.

Sint ſpatia quælibet *ab, ac,* ſiue perpendicularia ſiue inclinata, & circa diametrum *ac* fiat parabola quælibet *ad e.* atq; ordinatim ducantur *bd. ce.* Dico tempus per *ab* ad tempus per *ac.* eſſe vt *bd,* ad *ce.* Sunt enim tempora in ſubdupla ratione ſpatiorum ex Galileo, ſed lineæ *db, ec.* ſunt in ſubdupla ratione ſpatiorum (quia quadrata earũ ſunt vt *ab,* ad *ac.*) ergo eadem ratio eſt & temporũ, & linearũ ordinatim ad ſpatia applicatarũ. Quod erat. &c.

Aliter.

Si ponamus tempus lationis per latus rectum af *eſſe ipſummet latus rectum, erit tempus per* ab. *media proportionalis inter* fa, ab, *nempe ipſa* bd. *& tempus per* a c *media proportionalis* ce *&c. & ſic de ſingulis. Quare &c.*

Corollarium.

Hinc manifeſtum est impetus grauium in fine portionum diametri parabolæ, eſſe inter ſe vt lineæ, quæ ordinatim applicantur ad extrema ipſarum portionum puncta. Sunt enim ex Galileo impetus vt ipſa tempora, ſed ordinatim ductæ ſunt vt ipſa tempora, ergo impetus ſunt vt ordinatim ductæ &c.

P *P R O-*

[64] quæex] quæ ex.

PROPOSITIO XI.

SI Lineæ à vertice parabolæ vſq; ad ſectionem ducantur, erunt impetus in fine linearum, vt ſunt ordinatim ex ipſarum terminis applicatæ.

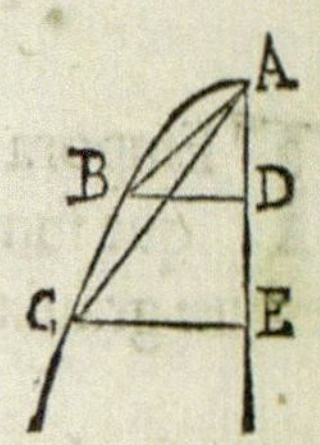

Sit parabola cuius vertex *a*. & ducantur ex vertice *ab*, *ac*. Dico impetus in *b*, & *c*. eſſe vt *bd* ad *ce*. Manifeſtum eſt; Quia impetus in *b*, & *c*. ſunt ijdem ac in *d*. & *e*. In *d*. autem & in *e*. ſunt vt *bd*. ad *ce*. (per Corollarium præcedens) ergo etiam in *b*, & *c*. ſunt vt *bd*, ad *ce*. Quod &c.

PROPOSITIO XII.

TEmpora lationum, quæ in circulo fiunt per portiones diametri ex puncto ſublimi, ſunt inter ſe vt chordæ quæ ex eodem puncto ſublimi ducuntur, ad puncta peripheriæ in quæ incidunt ordinatim ductæ ex terminis dictarum portionum.

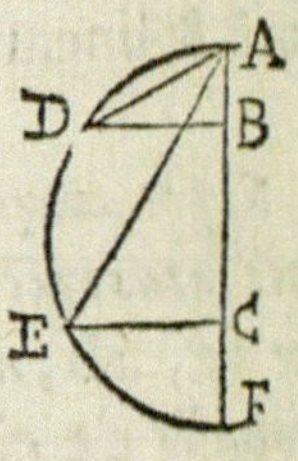

Sumantur vtcunq; *ab*, *ac*. & ducantur ordinatim *bd*, *ce*. iunganturq; *ad*, *ae*. Dico tempora lationum per *ab*, *ac* portiones diametri ita eſſe inter ſe vt ſunt *ad*, *ae*.

Si enim ponatur tempus per *af* eſſe *af* erit tempus per *ab* ipſa *ad*. cum ſit media proportionalis. Et cum tempus per *af*. ſit *af*, erit tẽpus per *ac*. ipſa *ae* media proportionalis. Quare &c.

Lemma.

Tempora lationum ex quiete per vnumquodq; latus trianguli rectanguli, cuius baſis ad horizontem erecta ſit, æqualia ſunt inter ſe.

Hoc

PROPOSITIO XI.

SI Lineæ à vertice parabolæ vſq; ad ſectionem ducantur, erunt impetus in fine linearum, vt ſunt ordinatim ex ipſarum terminis applicatæ.

Sit parabola cuius vertex *a.* & ducantur ex vertice *ab, ac.* Dico impetus in *b,* & *c.* eſſe vt *bd* ad *ce.* Manifeſtum eſt; Quia impetus in *b,* & *c.* ſunt ijdem ac in *d.* & *e.* In *d.* autem & in *e.* ſunt vt *bd.* ad *ce.* (per Corollarium præcedens) ergo etiam in *b,* & *c.* ſunt vt *bd,* ad *ce.* Quod &c.

PROPOSITIO XII.

TEmpora lationum, quæ in circulo fiunt per portiones diametri ex puncto ſublimi, ſunt inter ſe vt chordæ quæ ex eodem puncto ſublimi ducuntur, ad puncta peripherię in quæ incidunt ordinatim ductæ ex terminis dictarum portionum.

Sumantur vtcunq; *ab, ac.* & ducantur ordinatim *bd, ce.* iunganturq; *ad, ae.* Dico tempora lationum per *ab, ac* portiones diametri ita eſſe inter ſe vt ſunt *ad, ae.*

Si enim ponatur tempus per *af* eſſe *af* erit tempus per *ab* ipſa *ad.* cum ſit media proportionalis. Et cum tempus per *af.* ſit *af,* erit tẽpus per *ac.* ipſa *ae* media proportionalis. Quare &c.

Lemma.

Tempora lationum ex quiete per vnumquodq; latus trianguli rectanguli, cuius baſis ad horizontem erecta ſit, æqualia ſunt inter ſe.

Hoc

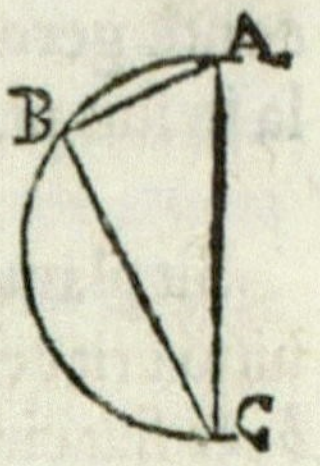

Hoc est tamquam Corollarium sextæ propositionis Galilei de motu accelerato. Sit enim semicirculus abc, *cuius punctum sublime sit* a. *& chordæ* ab, bc. *in semicirculo ad idem punctum* b. *coaptatę. Dico tempora per* ab, *rectam ex quiete in* a, *& per* bc. *ex quiete in* b. *esse æqualia. Vtrumq; enim tempus per* ab, *& per* bc. *æquale est tempori per diametrum* ac *ex Galileo ergo sunt æqualia inter se tempora per tres lineas* ac, ab, bc. *Quod &c.*

PROPOSITIO XIII.

Tempora lationum per semidiametros quadrantis erecti æqualia sunt temporibus tum secantium, tum etiam tangentium angulorum complementi eleuationis, qua fuerint eleuatæ dictæ semidiametri.

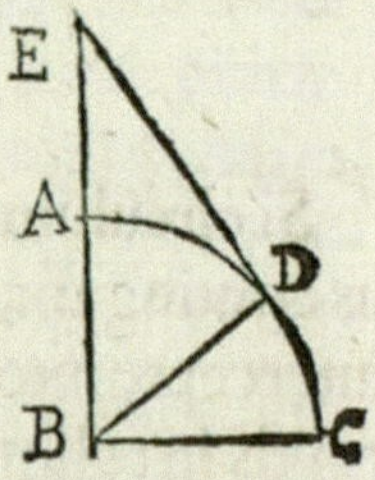

Sit quadrans erectus *abc.* semidiameter, quæcunq; *bd.* & tangens *de.* quæ angulum *bde.* rectum efficiet. Dico tempus per *db,* æquale esse tempori per secantem *eb,* siue per tangentem *ed.* quarum vtraq;, nempe *eb* est secans, *ed* tangens anguli *abd,* nempe complementi eleuationis semidiametri *db.*

Propositum autem manifestum est per lemma præcedens cũ triangulum. *bde.* sit rectangulum, & ideo tempora per vnum quodq; latus ęqualia sint.

PROPOSITIO XIV.

S*I ab aliquo puncto plana diuersimodè inclinata digrediantur, & ab illo eodem puncto dimittantur grauia eodem tempore simul; ostendit Galileus ex Corollario Propos. 6. de motu accelerato, grauia huiusmodi semper in peripheria alicuius circuli in immensum crescentis reperiri. Stante hoc dicimus.*

P 2 Si

PROPOSITIO XV.

SI Plana diuerſimodè inclinata ad vnum punctum concurrant, & grauia dimittantur eodem ſimul tempore ex aliqua circuli peripheria, cuius infimum punctum ſit concurſus planorum, ipſa grauia ſemper in aliquo circulo ſimul diſpoſita commeabunt.

Sint plana *ab* perpendiculum, & *cb.* vtcunque inclinatum, quæ concurrant in *b,* & per cõcurſum *b* tranſeat quælibet peripheria *bca.* ita vt *b.* ſit infimum punctum ipſius. Dico grauia ex *a,* & *c.* eodem tempore demiſſa ſemper in aliqua circuli peripheria commeare, quæ tranſeat per *b.*

Si enim ponamus graue *a* deſcendiſſe vſq; in *e.* producatur *bd.* æqualis ipſi *ae,* & per *e.* & per *d* agantur duo circuli qui priorem ciculum[67] contingant in puncto *b.* Erit per Lem. præcedens vt *ae,* ad *cf.* ita *bd.* ad *bg.* ergo *cf,* & *bg.* ſunt æquales. Sed *bd, bg.* ex quiete in *b.* eodem tempore peraguntur, ergo etiam *ae, cf.* eodem tempore peragentur (ſunt enim æquales longitudine, & inclinatione ipſis *bd, bg.*) Quare grauia *a, c,* & infinita alia demiſſa ſimul ex peripheria *bca,* ſemper in aliqua peripheria ſimul diſpoſita reperientur. Quod &c.

PROPOSITIO XVI.

DAtis quotcunq; ſpatijs. deinceps in directum continuatis, vnicuiq; ſuæ lationis tempus adſcribere.

Sint ſpatia quotcunq; deinceps, ſiue æqualia ſiue inæqualia *ab, bc, cd.* vt apparet in prima figura & circa diametrum *ad* fiat parabola *ag.* ducanturq; ordinatim *be, cf, dg.* & ſint parallelæ diametro *eh. fi.*
Dico *db, hi, ig,* tempora eſſe

ſpa-

[67] ciculum] circulum.

spatiorum respectiuè *ab*, *bc*, *cd*. Hoc enim patet. Nam *be*, siue *dh*. tempus est ipsius *ab*. & *cf*. vel *di*. tempus est ipsius *ac*. quare *hi* tempus est spatij *bc*. &c. & sic de reliquis.

Potest etiam sine parabola idem perfici hoc modo. vt in secunda figura. Ponatur ae *æqualis ipsi* ab. *primo spatiorum*; & *ducatur* ao *perpendicularis ad* ad. *fiat deinde semicirculi circa diametros* eb, ec, ed. &c. *qui secent rectam* ao, *in punctis* f, i. o. *Dico* af, fi, io. *esse respectiuè tempora quæsita spatiorum* ab, bc, cd. &c. *Nam si ponamus tempus per* ab, *esse* ab, *hoc est* af, *erit tempus per* ac. *media proportionalis, quæ est* ai, *ergo tempus differentię* bc. *erit* fi; *eodem modo ostendetur tempus per* cd, *esse* io. *Quare patet* &c.

Propositum sit nobis; punctum sublime parabolarum (quod optimè, & ingeniosè reperit Galileus) diuerso modo considerare. Fiet enim vt nobis plus lucis afferat, ad impetus in singulis parabolæ punctis determinandos, & clarius mente concipiendos. Placuit has propositiones sub titulo De motu accelerato ponere, licet sapiant aliquid de proiectione, quia sunt circa parabola de genere earum, quæ initium habent ex vertice, & ab ipso motu naturaliter accelerato deriuare concipiuntur, sine vlla instrumentorum impellentium ope.

Lemma,

Si mobile aliquod ex a *pūcto demissum, eodem tempore peragat duo spatia* ab, bc. *Dico ipsum omnino transire per punctum* c; *quamcunq; lineā descripserit præcedenti latione. Transeat enim si possibile est per* d. *ergo quia discessit ab* a. *eodem tempore confecit spatia* ab, & bd, *quod est contra suppositum. Quo enim tempore conficit* ab, *non absoluit* bd, *sed ipsam* bc. *Quare constat* &c.

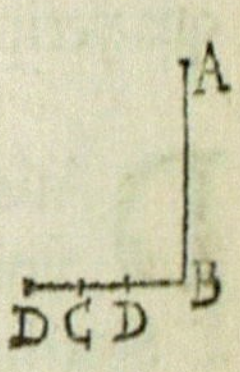

PROPOSITIO XVII.

Proposita qualibet parabola cuius vertex sit *b*. oportet punctum aliquod sublime reperire, ex quo si graue cadat vsq; in *b*,

ſpatiorum reſpectiuè *ab, bc, cd.* Hoc enim patet. Nam *be,* ſiue *dh.* tempus eſt ipſius *ab.* & *cf.* vel *di.* tempus eſt ipſius *ac.* quare *hi* tempus eſt ſpatij *bc.* &c. & ſic de reliquis.

Poteſt etiam ſine parabola idem perfici hoc modo. vt in ſecunda figura. Ponatur a e *æqualis ipſi* ab. *primo ſpatiorum; & ducatur* ao *perpendicularis ad* ad. *fiãt deinde ſemicirculi circa diametros* eb, ec, ed. *&c. qui ſecent rectam* ao, *in punctis* f, i.o. *Dico* af, fi, io. *eſſe reſpectiuè tempora quæſita ſpatiorum* ab, bc, cd. *& c. Nam ſi ponamus tempus per* ab, *eſſe* ab, *hoc eſt* af, *erit tempus per* ac. *media proportionalis, quæ eſt* ai, *ergo tempus differentię* bc. *erit* fi*; eodem modo oſtendetur tempus per* cd, *eſſe* io. *Quare patet &c.*

Propoſitum ſit nobis; punctum ſublime parabolarum (quod optimè, & ingenioſè reperit Galileus) diuerſo modo conſiderare. Fiet enim vt nobis plus lucis afferat, ad impetus in ſingulis parabolæ punctis determinandos, & clarius mente concipiendos. Placuit has propoſitiones ſub titulo De motu accelerato ponere, licet ſapiant aliquid de proiectione, quia ſunt circa parabola de genere earum, quæ initium habent ex vertice, & ab ipſo motu naturaliter accelerato deriuare concipiuntur, ſine vlla instrumentorum impellentium ope.

Lemma,

Si mobile aliquod ex a *pũcto demiſſum, eodem tempore peragat duo ſpatia* ab, bc. *Dico ipſum omnino tranſire per punctum* c; *quamcunq; lineã deſcripſerit præcedenti latione. Tranſeat enim ſi poſſibile est per* d. *ergo quia diſceſſit ab* a. *eodem tempore confecit ſpatia* ab, & bd, *quod eſt contra ſuppoſitum. Quo enim tempore conficit* ab, *non abſoluit* bd, *ſed ipſam* bc. *Quare constat &c.*

PROPOSITIO XVII.

PRopoſita qualibet parabola cuius vertex ſit *b.* oportet punctum aliquod ſublime reperire, ex quo ſi graue cadat vſq

in *b,*

in *b*, & ex puncto *b*. cum impetu iam concepto, horrizontaliter conuertatur, ipsam propositam parobolam describat.

Sit quælibet parabola *bhe*. Ponantur *ab*, & *bc*. vtraq. æqualis quartæ parti lateris recti propositæ parabolę. Sumatur iam in parabola quantumuis producta aliquod punctum *e*. Dico graue post casum per *ab*, horizontaliter conuersum in *b*. cum impetu iam concepto, per ipsum *e* punctum transire. Debet autem post horizontalem conuersionem in puncto *b* factam, grauitas suam descensus operationem inchoare. Ducantur ordinatim *ch*, *ef*. Ponamusq; tempus casus per *bc*. esse *ch*. Ergo graue horizontaliter conuersum in *c*, decurret motu æquabili tépore casus duplum ipsius casus spatium. hoc est tempore *ch*. ipsam *cb*, vel tempore *fe* (eadem velocitate) ipsam *fc*. Graue igitur impetu per *bc*, siue per *ab* acquisito conficit horizontalem *fe*. tempore *fe*. *8 huius.* *x. huius*

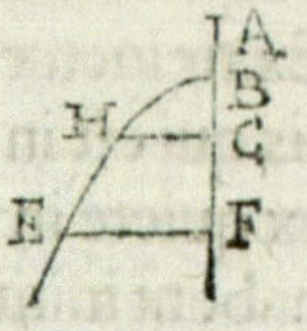

Sed eodem tempore *fe*. decurrit etiam perpendicularem *bf* (quando grauitas incipit operari in *b* vt in casu nostro) ergo eo dé tépore conficit *bf*, & *fe*. Quare graue omnino transibit per *e*. *per lem. præced.*

Transit ergo graue post casum *ab* per singula propositæ parabolæ puncta.

PROPOSITIO XVIII.

IN *Quolibet puncto parabolæ duo simul impetus insunt. Alter horizontalis, qui semper idem est, & ęqualis sibi ipsi; Alter perpendicularis, qui nunquam idem est, sed semper augetur.*

Dico horizontalem semper esse tamquam linea quę ex foco ordinatim ducitur. perpendicularem vero esse vt linea quæ ordinatim ducitur ex eo puncto, quod examinatur.

In figura præcedentis Propositionis.

Sit parabola cuius vertex *b*. & ponantur *ba*, *bc*. æquales vtraq; quartæ parti lateris recti. Quia impetus sunt vt tépora, erit impetus cadétis ex *a* in *b*, siue ex *b* in *c*, vt ipsa *ch*. per Propositionem *Ex defini. Galilei.*

in *b,* & ex puncto *b.* cum impetu iam concepto, horrizontaliter[68] conuertatur, ipſam propoſitam parobolam deſcribat.

Sit quælibet parabola *bhe.* Ponantur *ab,* & *bc.* vtraq. æqualis quartæ parti lateris recti propoſitæ parabolę. Sumatur iam in parabola quantumuis producta aliquod punctum *e.* Dico graue poſt caſum per *ab,* horizontaliter conuerſum in *b.* cum impetu iam concepto, per ipſum *e* punctum tranſire. Debet autem poſt horizontalem conuerſionem in puncto *b* factam, grauitas ſuam deſcenſus operationem inchoare. Ducantur ordinatim *ch, ef.* Ponamuſq; tempus caſus per *bc.* eſſe *ch.* Ergo graue horizontaliter conuerſum in *c,* decurret motu æquabili tēpore caſus duplum ipſius caſus ſpatium. hoc eſt tempore *ch.* ipſam *ch,* vel tempore *fe (*eadem velocitate*)* ipſam *fe.* Graue igitur impetu per *bc,* ſiue per *ab* acquiſito conficit horizontalem *fe.* tempore *fe.*

8 huius.

x. huius.

Sed eodem tempore *fe.* decurrit etiam perpendicularem *bf* (quando grauitas incipit operari in *b* vt in caſu noſtro) ergo eo dē tēpore conficit *bf,* & *fe.* Quare graue omnino tranſibit per *e.*

per lem. præced.

Tranſit ergo graue poſt caſum *ab* per ſingula propoſitæ parabolæ puncta.

PROPOSITIO XVIII.

IN *Quolibet puncto parabolæ duo ſimul impetus inſunt. Alter horizontalis, qui ſemper idem eſt, & ęqualis ſibi ipſi; Alter perpendicularis, qui nunquam idem eſt, ſed ſemper augetur.*

Dico horizontalem ſemper eſſe tamquam linea quę ex foco ordinatim ducitur. perpendicularem vero eſſe vt linea quæ ordinatim ducitur ex eo puncto, quod examinatur.

Ex defi- [nitio]*ne Gali-. lei*

In figura præcedentis Propoſitionis.

Sit parabola cuius vertex *b.* & ponantur *ba, c.* æquales vtraq; quartæ parti lateris recti. Quia impetus ſunt vt tēpora, erit impetus cadētis ex *a* in *b,* ſiue ex *b* in *c,* vt ipſa *ch.* per Propoſitio

nem

[68] horrizontaliter] horizontaliter.

nem X. huius,& eius Corollarium. Vbicunq; ergo ſumatur punctum in parabola, impetus horizontalis in eo erit vt *c h*. Quandoquidem impetus horizontalis eſt indelebiliter æquabilis. Examinetur iam punctum quodlibet *e*. Impetus perpendicularis qui eſt in *e*, eſt idem ac impetus naturaliter cadentis per *bf*. ex quiete in *b*. Vterq; enim deſcenſus venit ab altitudine *b*, vbi habent initium accelerationis. Impetus autem cadentis ex *b*. in *f*. eſt *fe*. ergo impetus perpendicularis in puncto *e* parabolæ erit *fe*. Quare in eodem puncto parabolæ ſunt duo impetus, alter vt *c h*. quę eſt ex foco, alter vt *fe* quæ ex puncto examinato applicatur. Quod &c.

Scholium.

Hinc poſſet oſtendi demonſtratione directa proprietas tangentis parabolę, ſiue Theorema mauis, ſiue Problema poſito prius hoc principio.

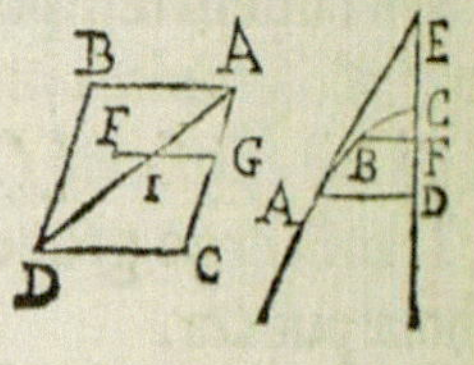

Si mobile aliquod *a* in prima figura ex angulo parallelogrãmi alicuius, vel ex quolibet puncto diametri feratur æquabilitèr duplici ſimul latione, nempe progreſſiua ſecundum lineam *a c*, & laterali ſecundum *ab* vtcunq; inclinatã. ſitq; proportio duarum velocitatum eadem ac proportio laterum *ac* ad *ab* homologè. Dico mobile iturum eſſe ſecundũ diametrum *ad* hoc eſt per ipſam diametrum.

Si enim poſſibile eſt feratur mobile extra diametrum per aliquod punctum *e*, ducaturq; *eg* parallela ad *ab*. Ergo quam proportionem habent ſpatia peracta à mobili, eam habebunt & impetus: nempe vt ſpatium progreſſiuum peractum *ag* ad laterale peractum *ge*, ita impetus progreſſiuus ad impetum lateralem, ideoq; vt *ag* ad *ge* ita *ac* ad *ab* ob ſuppoſitionem, ſiue *ac* ad *cd*, ſiue *ag* ad *gi*, eſſent ergo æquales *ge* & *gi*. totum & pars.

Eſto

nem X. huius, & eius Corollarium. Vbicunq; ergo ſumatur punctum in parabola, impetus horizontalis in eo erit vt *ch.* Quandoquidem impetus horizontalis eſt indelebiliter æquabilis. Examinetur iam punctum quodlibet *e.* Impetus perpendicularis qui eſt in *e,* eſt idem ac impetus naturaliter cadentis per *bf:* ex quiete in *b.* Vterq; enim deſcenſus venit ab altitudine *b,* vbi habent initium accelerationis. Impetus autem cadentis ex *b.* in *f.* eſt *fe.* ergo impetus perpendicularis in puncto *e* parabolæ erit *fe.* Quare in eodem puncto parabolæ ſunt duo impetus, alter vt *ch.* quę eſt ex foco, alter vt *fe* quæ ex puncto examinato applicatur. Quod &c

Scholium.

Hinc poſſet oſtendi demonſtratione directa proprietas tangentis parabolę, ſiue Theorema mauis, ſiue Problema poſito prius hoc principio.

Si mobile aliquod *a* in prima figura ex angulo parallelogrãmi alicuius, vel ex quolibet puncto diametri feratur æquabilitèr duplici ſimul latione, nempe progreſſiua ſecundum lineam *ac,* & laterali ſecundum *ab* vtcunq; inclinatã. ſitq; proportio duarum velocitatum eadem ac proportio laterum *ac* ad *ab* homologè. Dico mobile iturum eſſe ſecundũ diametrum *ad* hoc eſt per ipſam diametrum.

Si enim poſſibile eſt feratur mobile extra diametrum per aliquod punctum *e,* ducaturq; *eg* parallela ad *ab.* Ergo quam proportionem habent ſpatia peracta à mobili, eam habebunt & impetus: nempe vt ſpatium progreſſiuum per actum *ag* ad laterale peractum *ge,* ita impetus progreſſiuus ad impetum lateralem, ideoq; vt *ag* ad *ge* ita *ac* ad *ab* ob ſuppoſitionem, ſiue *ac* ad *cd,* ſiue *ag* ad *gi,* eſſent ergo æquales *ge* & *gi.* totum & pars.

Eſto

Esto iam in secunda figura quodlibet punctum *a* in curua parabolica *abc*, & applicata *ad*, factisq; æqualibus *dc*, *ce*, ducatur *ae*, quam dico tangentem esse. Esto focus *f*, & applicata ex foco recta *fb*. Erunt iam in *a* duo impetus alter progressiuus deorsum secundum directionem lineæ *cd*, alter lateralis secundum *da*, estq; progressiui impetus ad lateralem ratio vt *ad* ad *bf*, per præcedentem Proposit. siue vt *ed* ad *da* (cum ęquale sit rectangulum sub *ed* & semisse lateris recti *fb*, quadrato *da*) Ergo mobile dum est in puncto *a* feretur secundum diametralem *ae*; sed fertur etiam iuxtà parabolicam lineam quã percurrens describit, ergo recta *ae*, & parabolica nõ se intersecant in puncto *a*, sed tangunt: &c. Hæc demonstratio peculiaris est pro parabola; sed & vniuersalem habemus pro qualibet sectione Conica, consideratis æqualibus velocitatibus vnius puncti, quod ęqualiter mouetur in vtraq; linea quę ex focis procedit.

Eadem ratione demonstratur Propositio 18. de lineis spiralibus Archimedis vnica breuiq; demonstratione, non solum quando tangens consideratur ad extremum primæ reuolutionis punctum; sed vbicunq; punctum sit in curua spirali semper ostenditur periphæria, quæ per punctũ contactus ducitur ęqualis cuidam rectæ lineæ &c. Quæ Propositiuncula cum olim inter amicos à me vulgata fuisset, Clar. Virum Galileum meruit habere laudatorem. vt extant ipsius epistolæ apud me.

Immò & hac ratione ostenduntur etiam vnico Theoremate tangentes quarumdam curuarum, inter quas, omnium linearũ Cycloidalium, vt breuiter attingemus ad finem libri de Quadratura Parabolæ, omittentes demonstrationem tam tangentium, quàm etiam solidorum, & centrorum grauitatis ipsius Cycloidis ad euitandam molem. Satis sit interea lectorem hic admonuisse quòd si Cycloidis spatium circa basim conuertatur, erit solidum ad cylindrum circumscriptum vt 5. ad 8. si uerò circa tangentem basi parallelam ut 7. ad 8. Centrum Cycloidis axem secat ita ut partes sint ut 7. ad 5. Demonstratur etiam ratio solidi circa axem ad cylindrum circumscriptum; item in qua

Q linea

Efto iam in fecunda figura quodlibet punctum *a* in curua parabolica *abc,* & applicata *ad,* factifq; æqualibus *dc, ce,* ducatur *ae,* quam dico tangentem effe. Efto focus *f,* & applicata ex foco recta *fb.* Erunt iam in *a* duo impetus alter progreffiuus deorfum fecundum directionem lineæ *cd,* alter lateralis fecundum *da,* eftq; progreffiui impetus ad lateralem ratio vt *ad* ad *bf,* per præcedentem Propofit. fiue vt *ed* ad *da* (cum ęquale fit rectangulum fub *ed* & femiffe lateris recti *fb,* quadrato *da*) Ergo mobile dum eft in puncto *a* feretur fecundum diametralem *ae*; fed fertur etiam iuxtà parabolicam lineam quã percurrens defcribit, ergo recta *ae,* & parabolica nõ fe interfecant in puncto *a,* fed tangunt: &c. Hæc demonftratio peculiaris eft pro parabola; fed & vniuerfalem habemus pro qualibet fectione Conica, confideratis æqualibus velocitatibus vnius puncti, quod ęqualiter mouetur in vtraq; linea quę ex focis procedit

Eadem ratione demonftratur Propofitio 18. de lineis fpiralibus Archimedis vnica breuiq; demonftratione, non folum quando tangens confideratur ad extremum primæ reuolutionis punctum; fed vbicunq; punctum fit in curua fpirali femper oftenditur periphæria, quæ per punctũ contactus ducitur ęqualis cuidam rectæ lineæ &c. Quæ Propofitiuncula cum olim inter amicos à me vulgata fuiffet, Clar. Virum Galileum meruit habere laudatorem. vt extant ipfius epiftolæ apud me

Immò & hac ratione oftenduntur etiam vnico Theoremate tangentes quarumdam curuarum, inter quas, omnium linearũ Cycloidalium, vt breuiter attingemus ad finem libri de Quadratura Parabolæ, omittentes demonftrationem tam tangentium, quàm etiam folidorum, & centrorum grauitatis ipfius Cycloidis ad euitandam molem. Satis fit interea lectorem hic admonuiffe quòd fi Cycloidis fpatium circa bafim conuertatur, erit folidum ad cylindrum circumfcriptum vt 5. ad 8. fi uèro circa tangentem bafi parallelam ut 7. ad 8. Centrum Cycloidis axem fecat ita ut partes fint ut 7. ad 5. Demonftratur etiam ratio folidi circa axem ad cylindrum circumfcriptum; item in qua linea axi parallela sit centrum semicycloidis. Clar. Vir Antonius Nardius ostendit quòd si Cyclois circa tangentem axi parallelam conuertatur solidum ad suum cylindrum erit subsesquitertium; quæ omnia fortasse aliquando edentur, interea ad opus reuertamur.

Lemma.

Si in diametro parabolæ æquales sint *ac* ex vertice, & *db.* non ex vertice, Erit quadratum *be.* æquale quadratis *df. cg.*

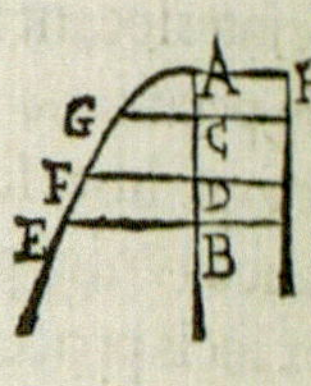

Sit ah. *latus rectum, & compleantur rectangula* cb, dh. *Quia* ac. db. *ponuntur equales, erit rectangulum* bh. *æquale duobus rectangulis* dh, ch, *seu (quod idem est) quadratum* eb. *duobus quadratis* df. cg. *equale erit. Quod &c.*

PROPOSITIO XIX.

IMpetus in punctis parabolæ sunt inter se vt lineæ ordinatim applicatæ non ad ipsamet puncta, sed tanto longius à vertice quanta est quarta pars lateris recti.

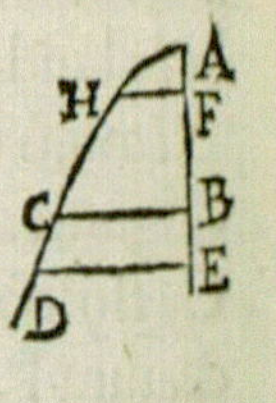

Sit parabola cuius vertex *a.* focus *f.* Sumptoque in ea quolibet puncto *c*, Dico impetum in *c* esse vt *de*, quæ applicata sit tanto longius à vertice, quàm ipsa *cb*, quanta est *af.* nempe quarta pars lateris recti.

Impetus enim qui simul sunt in *c.* sunt *cb, hf,* ergo momentum impetus ex ipsis compositum debet esse potentia ipsis equale. Sed & recta *de.* equatur potentia ipsis *cb, hf.* per lemma præcedens, ergo momentum *de.* est momentum siue impetus cōpositus ex duobus illis qui sunt in puncto *c.* Quod &c.

per 2: Galil: de motu accel:

PRO-

linea axi parallela ſit centrum ſemicycloidis. Clar. Vir Antonius Nardius oſtendit quòd ſi Cyclois circa tangentem axi parallelam conuertatur ſolidum ad ſuum cylindrum erit ſubſeſqui tertium; quæ omnia fortaſſe aliquando edentur, interea ad opus reuertamur.

Lemma.

Si in diametro parabolæ æquales ſint *ac* ex vertice, & *db.* non ex vertice, Erit quadratum *be.* æquale quadratis *df. cg.*

Sit ah. *latus rectum, & compleantur rectangula* cb, dh.

Quia ac. db. *ponuntur ęquales, erit rectangulum* bh. *æquale duobus rectangulis* dh, ch, *ſeu* (*quod idem eſt*) *quadratum* eb. *duobus quadratis* df. cg. *ęquale erit. Quod &c.*

PROPOSITIO XIX.

IMpetus in punctis parabolæ ſunt inter ſe vt lineæ ordinatim applicatæ non ad ipſamet puncta, ſed tanto longius à vertice quanta eſt quarta pars lateris recti.

Sit parabola cuius vertex *a.* focus *f.* Sumptoque in ea quolibet puncto *c,* Dico impetum in *c* eſſe vt *de,* quæ applicata ſit tanto longius à vertice, quàm ipſa *cb,* quanta eſt *af.* nempe quarta pars lateris recti.

per 2: Galil: de motu accel.

Impetus enim qui ſimul ſunt in *c.* ſunt *cb, hf,* ergo momentum impetus ex ipſis compoſitum debet eſſe potentia ipſis ęquale. Sed & recta *de.* ęquatur potentia ipſis *cb, hf.* per lemma præcedens, ergo momentum *de.* eſt momentum ſiue impetus cõpoſitus ex duobus illis qui ſunt in puncto *c.* Quod &c.

P R O-

PROPOSITIO XX.

IMpetus in quolibet parabolæ pũcto idem eſt ac impetus gra uis naturalitèr cadentis ex ſublimitate ſimul, & altitudine ciuſdem parabolę. *Gal. de motu ac.*

Sit parabola cuius altitudo *c b*, ſublimitas *a b*. Dico impetum in puncto *h* eundem eſſe ac naturalitèr cadentis ex *a*. in *c*.

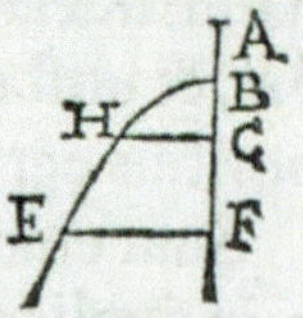

Sumatur *c f*. æqualis quartę parti lateris recti, hoc eſt ipſi *a b*. Erit impetus in puncto *h*: ut linea *e f*. per Præcedentem. At impetus etiam cadentis naturalitèr per *b f*, ſiue per *a c*, eſt eadem linea *e f*, ergo idem impetus eſt in puncto *h* parabolæ, ac in puncto *c*. grauis delapſi ex ſublimitate ſimul & altitudine *a b*, *b c*. Quod erat.

PROPOSITIO XXI.

TEmpora lationum per datas lineas horizontales poſt caſus è perpendiculo, æqualia erunt, quando altitudines perpendiculorum duplicatam rationem habuerint illius, quam horizontales lineæ habent.

Sint horizontales lineæ datæ *a b*, *c d*. & altitudines perpendiculares ſint *e a*, *f c*. Sitq; *e a*. ad *f c*. in duplicata ratione illius quam *a b* habet ad *c d*. Dico poſt caſus *e a*, *f c*, eodem tempore peragi *a b*, & *c d*.

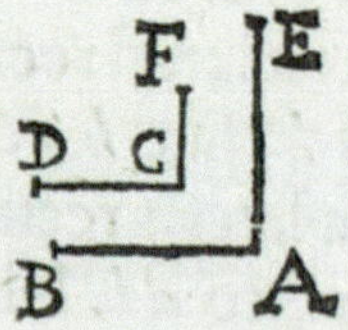

Hoc autem patet. quia cum ſint *a b*, *c d*. in ſubdupla ratione ipſarum *e a*, *f c*, erunt etiam vt tempora caſuum, & ideo vt velocitates, ſiue vt impetus qui ſunt in *a*, & *c*. Propterea cum ſint velocitates in *c*, & *a*, vt ipſa ſpatia *c d*, *a b* homologè, eodem tempore ipſa ſpatia peragentur. Quod &c,

PROPOSITIO XX.

IMpetus in quolibet parabolæ pũcto idem eſt ac impetus grauis naturalitèr cadentis ex ſublimitate ſimul, & altitudine eiuſdem parabolę.

Gal. dc. motu ac.

Sit parabola cuius altitudo *cb,* ſublimitas *ab.* Dico impetum in puncto *h* eundem eſſe ac naturalitèr cadentis ex *a.* in *c.*

Sumatur *cf.* æqualis quartę parti lateris recti, hoc eſt ipſi *ab.* Erit impetus in puncto *h*: ut linea *ef.* per Præcedentem. At impetus etiam cadentis naturalitèr per *bf,* ſiue per *ac,* eſt eadem linea *ef,* ergo idem impetus eſt in puncto *h* parabolæ, ac in puncto *c.* grauis delapſi ex ſublimitate ſimul & altitudine *ab, bc.* Quod erat.

PROPOSITIO XXI.

TEmpora lationum per datas lineas horizontales poſt caſus è perpendiculo, æqualia erunt, quando altitudines perpendiculorum duplicatam rationem habuerint illius, quam horizontales lineæ habent.

Sint horizontales lineæ datæ *ab, cd.* & altitudines perpendiculares ſint *ea, fc.* Sitq; *ea.* ad *fc.* in duplicata ratione illius quam *ab* habet ad *cd.* Dico poſt caſus *ea, fc,* eodem tempore peragi *ab,* & *cd.*

Hoc autem patet. quia cum ſint *ab, cd.* in ſubdupla ratione ipſarum *ea, fc,* erunt etiam vt tempora caſuum, & ideo vt velocitates, ſiue vt impetus qui ſunt in *a,* & *c.* Propterea cum ſint velocitates in *c,* & *a,* vt ipſa ſpatia *cd, ab* homologè, eodem tempore ipſa ſpatia peragentur. Quod &c,

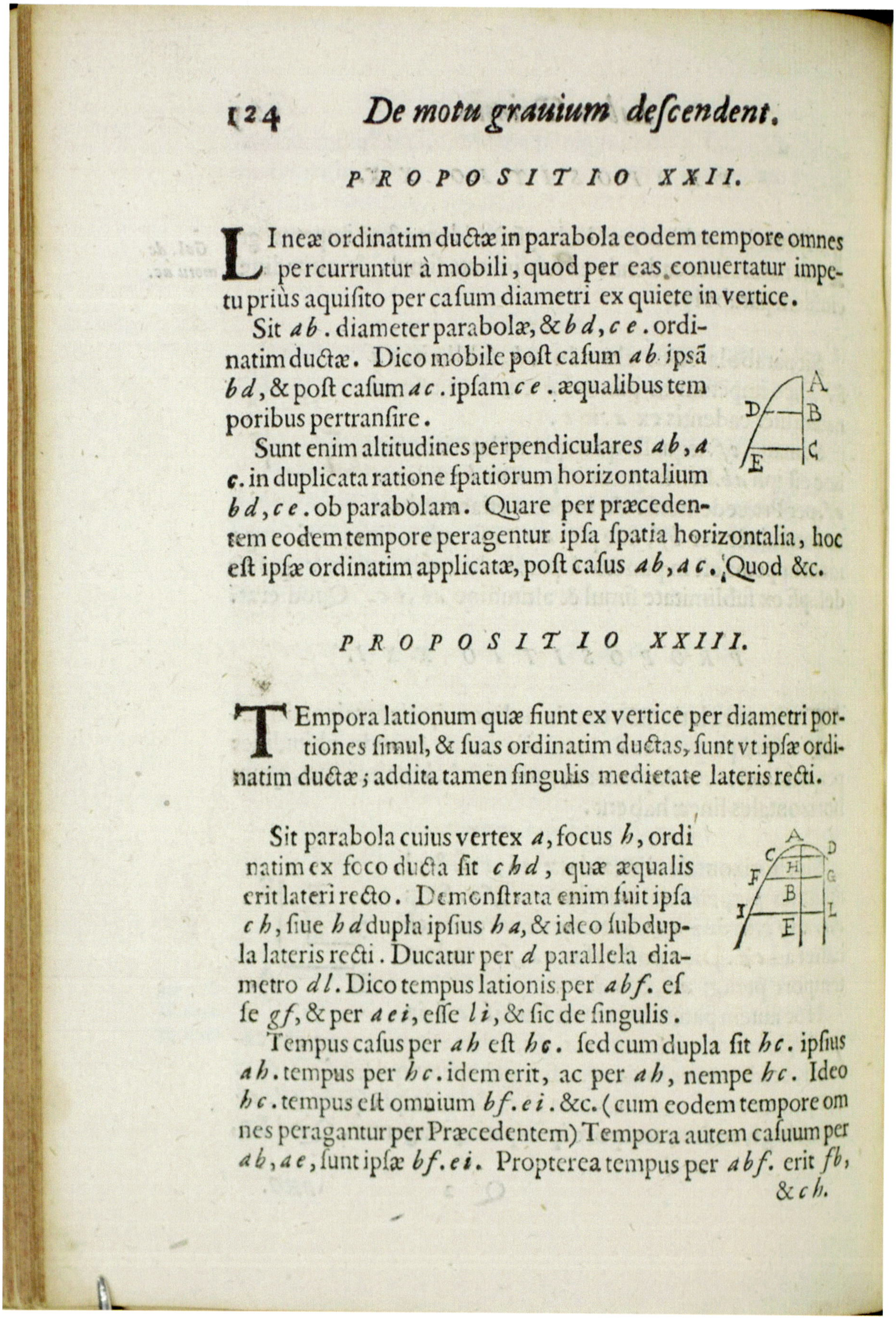

124 *De motu grauium descendent.*

PROPOSITIO XXII.

LIneæ ordinatim ductæ in parabola eodem tempore omnes percurruntur à mobili, quod per eas conuertatur impetu priùs aquisito per casum diametri ex quiete in vertice.

Sit *ab*. diameter parabolæ, & *bd*, *ce*. ordinatim ductæ. Dico mobile post casum *ab* ipsã *bd*, & post casum *ac*. ipsam *ce*. æqualibus temporibus pertransire.

Sunt enim altitudines perpendiculares *ab*, *ac*. in duplicata ratione spatiorum horizontalium *bd*, *ce*. ob parabolam. Quare per præcedentem eodem tempore peragentur ipsa spatia horizontalia, hoc est ipsæ ordinatim applicatæ, post casus *ab*, *ac*. Quod &c.

PROPOSITIO XXIII.

TEmpora lationum quæ fiunt ex vertice per diametri portiones simul, & suas ordinatim ductas, sunt vt ipsæ ordinatim ductæ; addita tamen singulis medietate lateris recti.

Sit parabola cuius vertex *a*, focus *h*, ordinatim ex foco ducta sit *chd*, quæ æqualis erit lateri recto. Demonstrata enim fuit ipsa *ch*, siue *hd* dupla ipsius *ha*, & ideo subdupla lateris recti. Ducatur per *d* parallela diametro *dl*. Dico tempus lationis per *abf*. esse *gf*, & per *aei*, esse *li*, & sic de singulis.

Tempus casus per *ah* est *hc*. sed cum dupla sit *hc*. ipsius *ah*. tempus per *hc*. idem erit, ac per *ah*, nempe *hc*. Ideo *hc*. tempus est omnium *bf*. *ei*. &c. (cum eodem tempore omnes peragantur per Præcedentem) Tempora autem casuum per *ab*, *ae*, sunt ipsæ *bf*. *ei*. Propterea tempus per *abf*. erit *fb*, & *ch*.

PROPOSITIO XXII.

LIneæ ordinatim ductæ in parabola eodem tempore omnes percurruntur à mobili, quod per eas conuertatur impetu priùs aquifito per ca fum diametri ex quiete in vertice.

Sit *ab.* diameter parabolæ, & *bd, ce.* ordinatim ductæ. Dico mobile poft cafum *ab* ipfã *bd,* & poft cafum *ac.* ipfam *ce.* æqualibus temporibus pertranfire.

Sunt enim altitudines perpendiculares *ab, ac.* in duplicata ratione fpatiorum horizontalium *bd, ce.* ob parabolam. Quare per præcedentem eodem tempore peragentur ipfa fpatia horizontalia, hoc eft ipfæ ordinatim applicatæ, poft cafus *ab, ac.* Quod &c.

PROPOSITIO XXIII.

TEmpora lationum quæ fiunt ex vertice per diametri portiones fimul, & fuas ordinatim ductas, funt vt ipfæ ordinatim ductæ; addita tamen fingulis medietate lateris recti.

Sit parabola cuius vertex *a,* focus *h,* ordinatim ex foco ducta fit *chd,* quæ æqualis erit lateri recto. Demonftrata enim fuit ipfa *ch,* fiue *hd* dupla ipfius *ha,* & ideo fubdupla lateris recti. Ducatur per *d* parallela diametro *dl.* Dico tempus lationis per *abf.* effe *gf,* & per *aei,* effe *li,* & fic de fingulis.

Tempus cafus per *ah* eft *hc.* fed cum dupla fit *hc.* ipfius.
ah. tempus per *hc.* idem erit, ac per *ah,* nempe *hc.* Ideo *hc.* tempus eft omnium *bf. ei.* &c. (cum eodem tempore omnes peragantur per Præcedentem) Tempora autem cafuum per *ab, ae,* funt ipfæ *bf. ei.* Propterea tempus per *abf.* erit *fb,*

& c h

[69] *motu grauium*] *motu Grauium.*

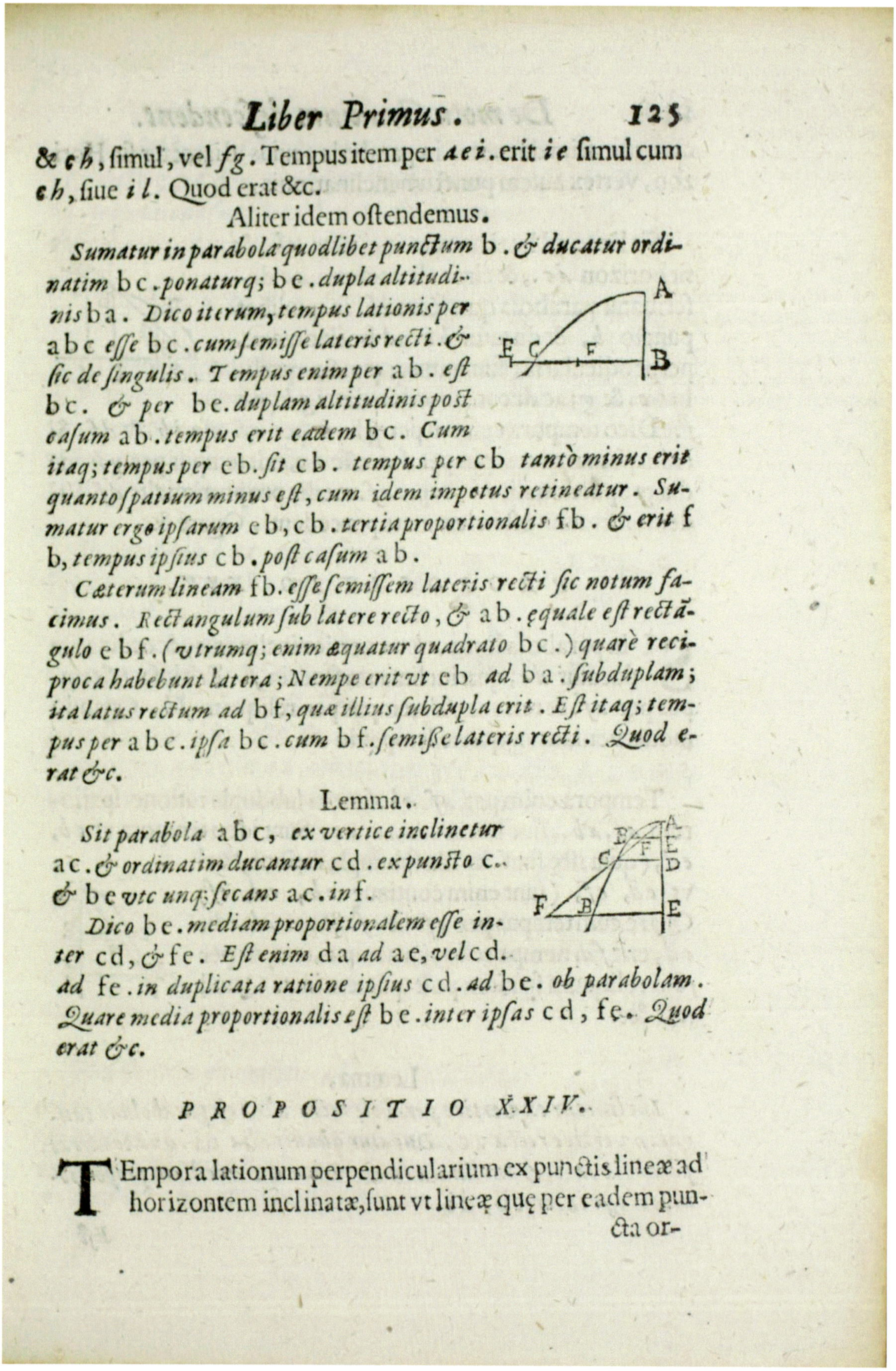

Liber Primus. 125

& *eh*, simul, vel *fg*. Tempus item per *aei*. erit *ie* simul cum *eh*, siue *il*. Quod erat &c.

Aliter idem ostendemus.

Sumatur in parabola quodlibet punctum b. *& ducatur ordinatim* bc. *ponaturq;* be. *dupla altitudinis* ba. *Dico iterum, tempus lationis per* abc *esse* bc. *cum semisse lateris recti. & sic de singulis. Tempus enim per* ab. *est* bc. *& per* be. *duplam altitudinis post casum* ab. *tempus erit eadem* bc. *Cum itaq; tempus per* eb. *sit* cb. *tempus per* cb *tantò minus erit quanto spatium minus est, cum idem impetus retineatur. Sumatur ergo ipsarum* eb, cb. *tertia proportionalis* fb. *& erit* fb, *tempus ipsius* cb. *post casum* ab.

Cæterum lineam fb. *esse semissem lateris recti sic notum facimus. Rectangulum sub latere recto, &* ab. *ęquale est rectangulo* ebf. *(vtrumq; enim æquatur quadrato* bc.*) quarè reciproca habebunt latera; Nempe erit vt* eb *ad* ba. *subduplam; ita latus rectum ad* bf, *quæ illius subdupla erit. Est itaq; tempus per* abc. *ipsa* bc. *cum* bf. *semiße lateris recti. Quod erat &c.*

Lemma.

Sit parabola abc, *ex vertice inclinetur* ac. *& ordinatim ducantur* cd. *ex puncto* c. *&* be *vtcunq; secans* ac. *in* f.

Dico be. *mediam proportionalem esse inter* cd, *&* fe. *Est enim* da *ad* ae, *vel* cd. *ad* fe. *in duplicata ratione ipsius* cd. *ad* be. *ob parabolam. Quare media proportionalis est* be. *inter ipsas* cd, fe. *Quod erat &c.*

PROPOSITIO XXIV.

TEmpora lationum perpendicularium ex punctis lineæ ad horizontem inclinatæ, sunt vt lineæ quę per eadem puncta or-

& *ch,* ſimul, vel *fg.* Tempus item per *aei.* erit *ie* ſimul cum *ch,* ſiue *il.* Quod erat &c.

Aliter idem oſtendemus.

Sumatur in parabola quodlibet punctum b. *& ducatur ordinatim* bc. *ponaturq*; be. *dupla altitudinis* ba. *Dico iterum, tempus lationis per* abc *eſſe* bc. *cum ſemiſſe lateris recti. & ſic de ſingulis. Tempus enim per* ab. *eſt* bc. *& per* be. *duplam altitudinis post caſum* ab. *tempus erit eadem* bc. *Cum itaq; tempus per* eb. *ſit* cb. *tempus per* cb *tantò minus erit quanto ſpatium minus eſt, cum idem impetus retineatur. Sumatur ergo ipſarum* eb, cb. *tertia proportionalis* fb. *& erit* fb, *tempus ipſius* cb. *poſt caſum* ab.

Cæterum lineam fb. *eſſe ſemiſſem lateris recti ſic notum facimus. Rectangulum ſub latere recto, &* ab. *ęquale eſt rectãgulo* ebf. *(vtrumq; enim æquatur quadrato* bc.*) quarè reciproca habebunt latera; Nempe erit vt* eb *ad* ba. *ſubduplam; ita latus rectum ad* bf, *quæ illius ſubdupla erit. Eſt itaq; tempus per* abc. *ipſa* bc. *cum* bf. *ſemiße lateris recti. Quod erat &c.*

Lemma.

Sit parabola abc, *ex vertice inclinetur* ac. *& ordinatim ducantur* cd. *ex punsto*[70] c. *&* b e *vtcunq; ſecans* ac. *in* f.

Dico be. *mediam proportionalem eſſe inter* cd, *&* fe. *Eſt enim* da *ad* ae, *vel* cd. *ad* fe. *in duplicata ratione ipſius* cd. *ad* be. *ob parabolam.*

Quare media proportionalis eſt be. *inter ipſas* cd, fe. *Quod erat &c.*

PROPOSITIO XXIV.

TEmpora lationum perpendicularium ex punctis lineæ ad horizontem inclinatæ, ſunt vt lineæ quę per eadem pun-

cta or-

[70] *punsto*] *puncto.*

cta ordinatim ducuntur in parabola, cuius diameter ſit Horizon, vertex autem punctum inclinationis.

Sit linea ad horizontem inclinata *ab*, & ſit horizon *ac*., & circa diametrum *ac* deſcribatur parabola quæ ſecet *ab*. in quolibet puncto *b*, & ducatur *bc*. ad horizontem perpendicularis; ſumanturq; puncta quælibet *e*. & *g*; ac ducantur ordinatim *igh*, *de f*. Dico tempora caſuum per *g*h, & per *ef*. eſſe *ih*, & *df*. Si enim ponamus tempus per *bc*. eſſe *bc*, erit tempus per *gh*. media proportionalis *ih*; & per *ef*. media proportionalts *df*. Vti demonſtratum eſt in lemmate præcedenti. Quare &c.

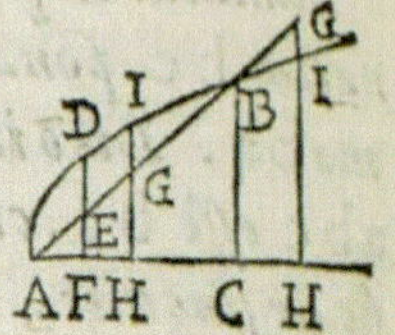

PROPOSITIO XXV.

SI in parabola linea *ab* ex vertice inclinetur, & ordinatim ducatur *de*, quæ inclinatam ſecet in *f*. Erit ipſa *ef*. tempus per *af*. & reliqua *fd*. tempus per reliquam *fb*. Quando motus veniant ex quiete ſemper in *a*.

Tempora enim per *af*. *ab*. ſunt in ſubdupla ratione ſpatiorum *af*, *ab*., ſiue linearum *ae*, *ac*. Sunt ideo tempora vt *cb*, *ed*, (quia iſtę ſunt in illa ſubdupla ratione) vel vt *ed*, *ef*, (ſunt enim continuę *cb*, *cd*, *ef*) Quare cum tempus per *af*. ſit *ef*. & per *ab*. ſit *ed*, erit *fd*. nempe reliquum tempus, tempus per *fb*. reliquum ſpatium poſt quietem in *a*. Quod erat &c.

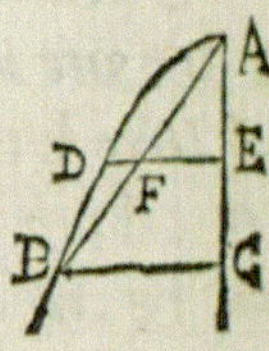

Lemma,

Inclinetur ex vertice parabolę recta ab, *& parabolam tangat in vertice recta* ac. *Ducatur quæuis alia* di. *quæ occurrat parabolæ in* i, *& inclinatæ in* e. *Dico* de. *mediam proportionalem eſſe inter* cb, di.

Eſt

cta ordinatim ducuntur in parabola, cuius diameter ſit Horizon, vertex autem punctum inclinationis.

Sit linea ad horizontem inclinata *ab,* & ſit horizon *ac.,* & circa diametrum *ac* deſcribatur parabola quæ ſecet *ab.* in quolibet puncto *b,* & ducatur *bc.* ad horizontem perpendicularis; ſumanturq; puncta quælibet *e.* & *g*; ac ducantur ordinatim *igh, def.* Dico tempora caſuum per gh, & per *f.* eſſe *ih,* & *df.* Si enim ponamus tempus per *bc.* eſſe *bc,* erit tempus per *gh.* media proportionalis *ih*; & per *ef.* media proportionalts *df.* Vti demonſtratum eſt in lemmate præcedenti. Quare &c.

PROPOSITIO XXV.

SI in parabola linea *ab* ex vertice inclinetur, & ordinatim ducatur *de,* quæ inclinatam ſecet in *f.* Erit ipſa *ef.* tempus per *af.* & reliqua *fd.* tempus per reliquam *fb.* Quando motus veniant ex quiete ſemper in *a.*

Tempora enim per *af. ab.* ſunt in ſubdupla ratione ſpatiorum *af, ab.,* ſiue linearum *ae, ac.* Sunt ideo tempora vt *cb, ed,* (quia iſtę ſunt in illa ſubdupla ratione) vel vt *ed, ef,* (ſunt enim continuę *cb, cd, ef*)
Quare cum tempus per *af.* ſit *ef.* & per *ab.* ſit *ed,* erit *fd.* nempe reliquum tempus, tempus per *fb.* reliquum ſpatium poſt quietem in *a.* Quod erat &c.

Lemma,

Inclinetur ex vertice parabolę recta ab, *& parabolam tangat in vertice recta* ac. *Ducatur quæuis alia* di. *quæ occurrat parabolæ in* i, *& inclinatæ in* e. *Dico* de. *mediam proportionalem eſſe inter* cb, di.

Eſt

Est enim bc. *ad* di *longitudine ut* ca. *ad* ad, *vel* bc *ad* de. *potentiâ. Quare in continua proportione sunt* bc. de. di *&c. &* de. *media est, Quod oportebat ostendere*,

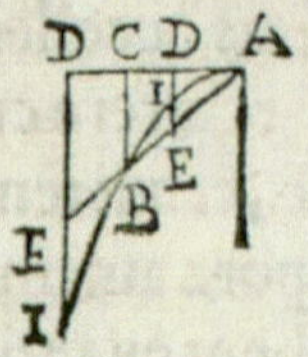

PROPOSITIO XXVI.

SI Horizontalis linea parabolam contingat, tempora casuam ex punctis parabolę vsq; ad horizontē; siue ex punctis horizontis, vt in secunda figura, vsq; ad parabolam; erunt ut lineæ parallelę inter horizontem, & quamlibet aliam ex contactu inclinatam interceptę.

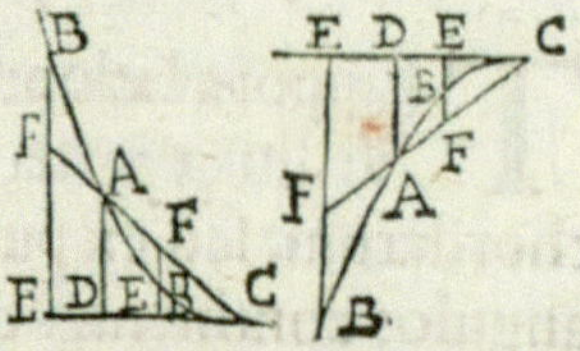

Sit parabola *bac*. cuius uertex *c* & eam in *c* tangat horizon *cd*. & ex contactu *c*. inclinetur utcunq; *ca* ducaturq; ex puncto *a*. *ad*. horizonti perpendicularis. Ducta iam qualibet perpendiculari *ef*. Dico tempus casus per *eb*. esse *fe*. &c. & sic de singulis. Ponamus tempus per *ad*. esse *ad*, erit tempus per *be*. ipsa *fe*. media proportionalis uti demonstratum est. Quare &c.

PROPOSITIO XXVII.

TEmpora lationum per chordas ex uertice parabolę inclinatas, compositam rationem habent, ex ratione longitudinum chordarum, & ex ratione ordinatim applicatarum, cōtrariè tamen sumptarum.

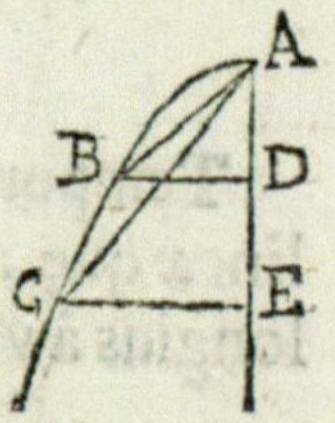

Sint chordæ ex uertice *ab*, *ac*. & ordinatim ducantur *bd*, *ce*. Dico tempus per *ab*. ad tempus per *ac*. habere rationem compositam ex ratione *ab* ad *ac*, & ex ratione *ce* ad *bd*. Si enim concipiamus lationes il-

las

Eſt enim bc. *ad* di *longitudine vt* ca. *ad* ad, *vel* bc *ad* de. *potentià. Quare in continua proportione ſunt* bc. de. di &c. & de. *media est, Quod oportebat oſtendere,*

PROPOSITIO XXVI.

SI Horizontalis linea parabolam contingat, tempora caſuam ex punctis parabolę vſq; ad horizontẽ; ſiue ex punctis horizontis, vt in ſecunda figura, vſq; ad parabolam; erunt ut lineæ parallelę inter horizontem, & quamlibet aliam ex contactu inclinatam interceptę.

Sit parabola *bac.* cuius uertex *c* & eam in *c* tangat horizon *cd.* & ex contactu *c.* inclinetur utcunq; *ca* ducaturq; ex puncto *a. ad.* horizonti perpendicularis. Ducta iam qualibet perpendiculari *ef.* Dico tempus caſus per *eb.* eſſe *fe.* &c. & ſic de ſingulis. Ponamus tempus per *ad.* eſſe *ad,* erit tempus per *be.* ipſa *fe.* media proportionalis uti demonſtratum eſt. Quare &c.

PROPOSITIO XXVII.

TEmpora lationum per chordas ex uertice parabolę inclinatas, compoſitam rationem habent, ex ratione longitudinum chordarum, & ex ratione ordinatim applicatarum, cõtrariè tamen ſumptarum.

Sint chordæ ex uertice *ab, ac.* & ordinatim ducantur *bd, ce.* Dico tempus per *ab.* ad tempus per *ac.* habere rationem compoſitam ex ratione *ab* ad *ac,* & ex ratione *ce* ad *bd.* Si enim concipiamus lationes il-

las

as accelleratas æquabiles fieri, & grauia per ſpatia *ba*, *ca* ecurrere cum gradu ſubduplo impetus quem habebant in *b*, & *c*, erunt tempora recurſuum eadem ac tempora caſuum. Tempora autem lationum æquabilium compoſitam rationem habent ex ratione longitudinum ſpatiorum *ab*, ad *ac*; & ex ratione velocitatum contrariè ſumptarum *ce*, ad *bd*. (ſunt enim velocitates in *b*, & *c* eædem ac in *d*, & *e*, & velocitates in *d*, & *e* ſunt vt tempora *bd*. *ce*.) Ergo etiam tempora caſuum naturaliter acceleratorum per *ab*, *ac*. compoſitam rationem habebunt ex iiſdem rationibus *ab*. ad *ac*. & *ce*. ad *bd*. Quod erat &c.

1. Gal. de motu a. c. per Galil: de motu æqua.

PROPOSITIO XXVIII.

TEmpora lationum per chordas ex uertice parabolę, ſunt vt lineæ quæ ordinatim applicantur non ex terminis chordarum, ſed ex punctis diametri in quæ cadunt lineæ rectos angulos continentes cum ipſis chordis,

Sit parabolæ diameter *ag*. & chordæ ex vertice ſint *ab*. *ad*. fiantq: anguli *abf*, *adg*. recti, & ordinatim ad puncta *f*. *g*. applicentur *fh*, *gi*. Dico tempora lationum per *ab*, *ad*. eſſe ipſas ordinatim applicatas *fh*, *gi*,

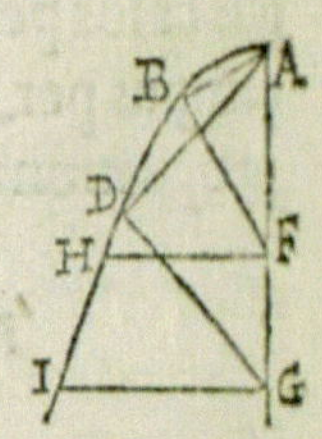

Tempus enim per *ab*. æquale eſt tempori per *af*. exiſtente angulo *abf*. recto. Item tempus per *ad*. ob eandem cauſam æquatur tempori per *ag*. Tempora autem per *af*. *ag*. ſunt ipſæ *hf*. *ig*. Ergo tempora lationum per chordas *ab*. *ad*. ſunt *hf*, & *gi*. Quod erat &c.

Proponetur etiam hoc modo.

Tempora lationum per chordas ex vertice parabolę ſunt vt lineæ quę applicantur non ex terminis chordarum, ſed tanto longius à vertice quanta eſt lateris recti longitudo.

Oſten-

as accelleratas æquabiles fieri, & grauia per ſpatia *ba, ca* ecurrere[71] cum gradu ſubduplo impetus quem habebant in *b*, & *c*, erunt tempora recurſuum eadem ac tempora caſuum. Tempora autem lationum æquabilium compoſitam rationem habent ex ratione longitudinum ſpatiorum *ab*, ad *ac*; & ex ratione velocitatum contrariè ſumptarum *ce*, ad *bd*. (ſunt enim velocitates in *b*, & *c* eædem ac in *d*, & *e*, & velocitates in *d*, & *e* ſunt vt tempora *bd. ce.*) Ergo etiam tempora caſuum naturaliter acceleratorum per *ab, ac.* compoſitam rationem habebunt ex iiſdem rationibus *ab.* ad *ac.* & *ce.* ad *bd.* Quod erat & c.

1. *Gal. dc motu ac. per Galil: de mo tu æqua*

PROPOSITIO XXVIII.

TEmpora lationum per chordas ex uertice parabolę, ſunt vt lineæ quæ ordinatim applicantur non ex terminis chordarum, ſed ex punctis diametri in quæ cadunt lineæ rectos angulos continentes cum ipſis chordis,

Sit parabolæ diameter *ag.* & chordæ ex vertice ſint *ab. ad.* fiantq: anguli *abf, adg.* recti, & ordinatim ad puncta *f. g.* applicentur *fh, gi.* Dico tempora lationum per *ab, ad.* eſſe ipſas ordinatim applicatas *fh, gi,*[72]

Tempus enim per *ab.* æquale eſt tempori per *af.* exiſtente angulo *abf.* recto. Item tempus per *ad.* ob eandem cauſam æquatur tempori per *ag.* Tempora autem per *af. ag.* ſunt ipſæ *hf. ig.* Ergo tempora lationum per chordas *ab. ad.* ſunt *hf,* & *gi.* Quod erat &c.

Proponetur etiam hoc modo.

Tempora lationum per chordas ex vertice parabolę ſunt vt lineæ quę applicantur non ex terminis chordarum, ſed tanto longius à vertice quanta eſt lateris recti longitudo.

Oſten-

---

[71] ecurrere] recurrere.
[72] *gi,*] *gi.*

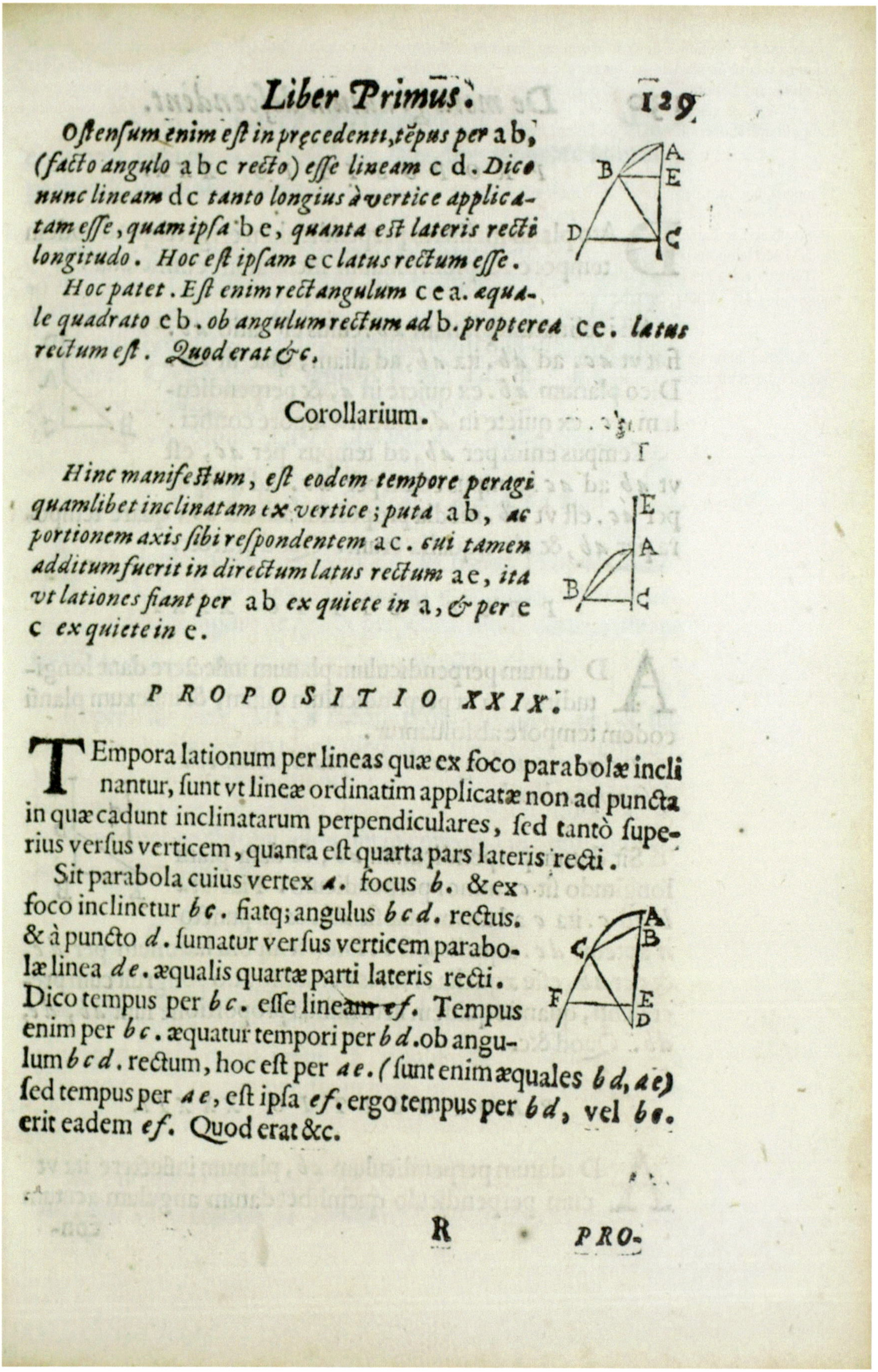

Liber Primus. 129

Ostensum enim est in precedenti, tēpus per a b, *(facto angulo* a b c *recto) esse lineam* c d. *Dico nunc lineam* d c *tanto longius à vertice applicatam esse, quam ipsa* b e, *quanta est lateris recti longitudo. Hoc est ipsam* e c *latus rectum esse.*

Hoc patet. Est enim rectangulum c e a. *æquale quadrato* e b. *ob angulum rectum ad* b. *propterea* c e. *latus rectum est. Quod erat &c.*

Corollarium.

Hinc manifestum, est eodem tempore peragi quamlibet inclinatam ex vertice; puta a b, *ac portionem axis sibi respondentem* a c. *cui tamen additum fuerit in directum latus rectum* a e, *ita vt lationes fiant per* a b *ex quiete in* a, *& per* e c *ex quiete in* e.

PROPOSITIO XXIX.

TEmpora lationum per lineas quæ ex foco parabolæ inclinantur, sunt vt lineæ ordinatim applicatæ non ad puncta in quæ cadunt inclinatarum perpendiculares, sed tantò superius versus verticem, quanta est quarta pars lateris recti.

Sit parabola cuius vertex *a.* focus *b.* & ex foco inclinetur *b c.* fiatq; angulus *b c d.* rectus. & à puncto *d.* sumatur versus verticem parabolæ linea *d e.* æqualis quartæ parti lateris recti. Dico tempus per *b c.* esse lineam *e f.* Tempus enim per *b c.* æquatur tempori per *b d.* ob angulum *b c d.* rectum, hoc est per *a e.* (sunt enim æquales *b d, a e)* sed tempus per *a e,* est ipsa *e f.* ergo tempus per *b d,* vel *b c.* erit eadem *e f.* Quod erat &c.

R PRO-

Oftenſum enim eſt in pręcedenti, tẽpus per ab, *(facto angulo* abc *recto) eſſe lineam* cd. *Dico nunc lineam* dc *tanto longius à vertice applicatam eſſe, quam ipſa* be, *quanta est lateris recti longitudo. Hoc eſt ipſam* e c *latus rectum eſſe.*

Hoc patet. Eſt enim rectangulum c e a. *æquale quadrato* e b. *ob angulum rectum ad* b. *propterea* ce. *latus rectum eſt. Quod erat &c*

Corollarium.

Hinc manifestum, eſt eodem tempore peragi quamlibet inclinatam ex vertice; puta ab, *ac portionem axis ſibi reſpondentem* ac. *cui tamen additum fuerit in directum latus rectum* ae, *ita vt lationes fiant per* ab *ex quiete in* a, *& per* ec *ex quiete in* e.

PROPOSITIO XXIX.

TEmpora lationum per lineas quæ ex foco parabolæ inclinantur, ſunt vt lineæ ordinatim applicatæ non ad puncta in quæ cadunt inclinatarum perpendiculares, ſed tantò ſuperius verſus verticem, quanta eſt quarta pars lateris recti.

Sit parabola cuius vertex *a.* focus *b.* & ex foco inclinetur *bc.* fiatq; angulus *bcd.* rectus. & à puncto *d.* ſumatur verſus verticem parabolæ linea *de.* æqualis quartæ parti lateris recti.

Dico tempus per *bc.* eſſe lineam *ef.* Tempus enim per *bc.* æquatur tempori per *bd.* ob angulum *bcd.* rectum, hoc eſt per *ae.* (ſunt enim æquales *bd, ae)* ſed tempus per *ae,* eſt ipſa *ef.* ergo tempus per *bd,* vel *bc erit eadem ef. Quod erat &c.*

R *P R O-*

PROPOSITIO XXX.

DAto plano inclinato perpendiculum erigere, quod eodem tempore ac ipsum planum inclinatum conficiatur.

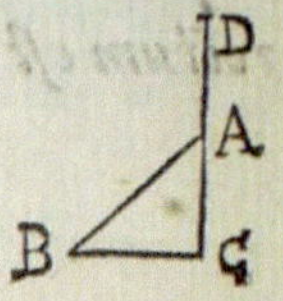

Sit inclinatum planum *ab*. cuius eleuatio *ac*. fiat vt *ac*. ad *ab*. ita *ab*, ad aliam, quæ sit *cd*. Dico planum *ab*. ex quiete in *a*, & perpendiculum *dc*. ex quiete in *d*. eodem tempore confici.

Tempus enim per *ab*. ad tempus per *ac*, est vt *ab* ad *ac*: tempus etiam per *dc*. ad tempus per *ac*. est vt *ab*. media proportionalis ad *ac*; quare tempora per *ab*, & *dc* æqualia erunt. Quod erat &c.

PROPOSITIO XXXI.

AD datum perpendiculum planum inflectere datæ longitudinis, ita vt perpendiculum ipsum, & inflexum planũ eodem tempore absoluantur.

Debet autem longitudo dati plani minor esse ipso perpendiculo.

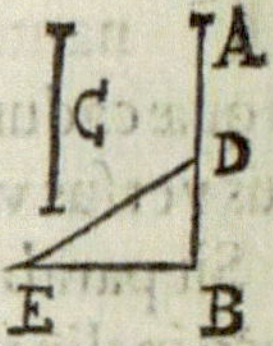

Sit datum perpendiculum *ab*, & data plani longitudo sit *c*. minor perpendiculo. Fiat vt *a b*. ad *c*. ita *c* ad aliam quæ sit *db*. & ex puncto *d* aptetur *de*. æqualis ipsi *c*. Dico tempora lationum per *de*, & per *ab*. esse æqualia. Huius demonstratio congruit cum præcedenti, quandoquidem in continua proportione sunt *ab*, *de*, *db*. Quod &c.

PROPOSITIO XXXII.

AD datum perpendiculum *ab*, planum inflectere ita vt cum perpendiculo quemlibet datum angulum acutum con-

PROPOSITIO XXX.

DAto plano inclinato perpendiculum erigere, quod eodem tempore ac ipfum planum inclinatum conficiatur.

Sit inclinatum planum *ab.* cuius eleuatio *ac.* fit vt *ac.* ad *ab.* ita *ab,* ad aliam, quæ fit *cd.* Dico planum *ab.* ex quiete in *a,* & perpendiculum *dc.* ex quiete in *d.* eodem tempore confici.

Tempus enim per *ab.* ad tempus per *ac,* eft vt *ab* ad *ac:* tempus etiam per *dc.* ad tempus per *ac.* eft vt *ab.* media proportionalis ad *ac*; quare tempora per *ab,* & *dc* æqualia erunt. Quod erat &c

PROPOSITIO XXXI.

AD datum perpendiculum planum inflectere datæ longitudinis, ita vt perpendiculum ipfum, & inflexum planũ eodem tempore abfoluantur.

Debet autem longitudo dati plani minor eße ipfo perpendiculo.

Sit datum perpendiculum *ab,* & data plani longitudo fit *c.* minor perpendiculo. Fiat vt *ab.* ad *c.* ita *c* ad aliam quæ fit *db.* & ex puncto *d* aptetur *de.* æqualis ipfi *c.* Dico tempora lationum per *de,* & per *ab.* effe æqualia. Huius demonftratio congruit cum prȩcedenti, quandoquidem in continua proportione funt *ab, de, db.* Quod &c.

PROPOSITIO XXXII.

AD datum perpendiculum *ab,* planum inflectere ita vt cum perpendiculo quemlibet datum angulum acutum

con-

[73] *motu grauium*] *motu Grauium.*

contineat, puta æqualem ipsi *bac*, & eodem tempore ac ipsum perpendiculum absoluatur.

Fiat circa diametrum *ab*. circulus qui secet *ac* in *d*. Demissàq; perpendiculari *de*. compleatur parallelogrammum *adef*.

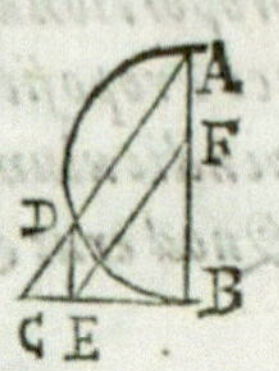

Manifestum est planum *fe*. quæsito nostro satisfacere. Cum enim *fe*. equalis sit ipsi *ad*. & æqualiter inclinata ob parallelogrammum, eodem tempore absoluentur *fe*, *ad* vel *ab*. Quod erat &c.

PROPOSITIO XXXIII.

AD datum perpendiculum *dc*, in figura Propositionis XXX. & ex dato in eo puncto *a*. planum inflectere, quod eodem tempore ac ipsum perpendiculum conficiatur ex quiete.

Reperiatur inter *dc*, *ca* media proportionalis *ab*, & habebimus longitudinem plani alicuius. Applicetur hęc longitudo ex *a*, sitq; illa iam applicata *ab*: Manifestum est ex præcedentibus Propositionibus ipsam *ab*. & perpendiculum *dc* eodem tempore absolui. cum sint in continua proportione *dc ab ac*. Quod erat &c.

PROPOSITIO XXXIV.

SI ad perpendiculum aliquod *ab*, planum *cd* inflexum sit ad angulũ semirectum. Erit tempus per *cd*. æquale tempori perpendiculi, quod ipsius *cb*. duplum sit. Proponetur etiã hoc modo. Tempus per diametrum quadrati erecti, æquale est tempori per duplum lateris erecti.

Sit c d. *planum vt supponitur. vel sit diameter quadrati cuius latus* c b. *erectum sit. ponaturq;* a b. *dupla ipsius* c b. *Di-*

continineat, puta æqualem ipſi *bac,* & eodem tempore ac ipſum perpendiculum abſoluatur.

Fiat circa diametrum *ab.* circulus qui ſecet *ac* in *d.* Demiſsàq; perpendiculari *de.* compleatur parallelogrammum *adef.*
Manifeſtum eſt planum *fe.* quæſito noſtro ſatisfacere. Cum enim *fe.* ęqualis ſit ipſi *ad.* & æqualiter inclinata ob parallelogrammum, eodem tempore abſoluentur *fc, ad* vel *ab.* Quod erat &c.

PROPOSITIO XXXIII.

AD datum perpendiculum *dc,* in figura Propoſitionis *XXX.* & ex dato in eo puncto *a.* planum inflectere, quod eodem tempore ac ipſum perpendiculum conficiatur ex quiete.

Reperiatur inter *dc, ca* media proportionalis *ab,* & habebimus longitudinem plani alicuius. Applicetur hęc longitudo ex *a,* ſitq; illa iam applicata *ab:* Manifeſtum eſt ex præcedentibus Propoſitionibus ipſam *ab.* & perpendiculum *dc* eodem tempore abſolui. cum ſint in continua proportione *dc ab ac.* Quod erat &c.

PROPOSITIO XXXIV.

SI ad perpendiculum aliquod *ab,* planum *cd* inflexum ſit ad angulũ ſemirectum. Erit tempus per *cd.* æquale tempori perpendiculi, quod ipſius *cb.* duplum ſit. Proponetur etiã hoc modo. Tempus per diametrum quadrati erecti, æquale eſt tempori per duplum lateris erecti.

Sit cd. *planum vt ſupponitur. vel ſit diameter quadrati cuius latus* cb. *erectum ſit. ponaturq;* ab. *aupla*[74] *ipſius* cb. *Di-*

[74] *aupla*] *dupla.*

132 *De motu Grauium descendent.*

co tempora per c d, a b, *esse ęqualia. Quia* d c. *ad* c b *potentia est vt* a b *ad eandem* c b. *longitudine, nempe in ratione dupla, erunt continuæ proportionales* a b, c d, c b. *Quare per præcedentes Propositiones eodem tempore absoluentur perpendiculum* a b, *& planum inclinatum* c d. *Quod erat &c.*

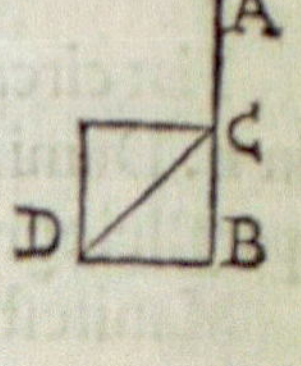

PROPOSITIO XXXV.

AD datum perpendiculum *ab*, planum, vel plana inclinare ad datum in horizonte punctum *c*, ita vt inclinata plana & perpendiculum ipsum eodem tempore absoluantur.

Debet autem punctum *c*. Distare à puncto *b*. nō amplius quàm sit semissis ipsius *ab*.

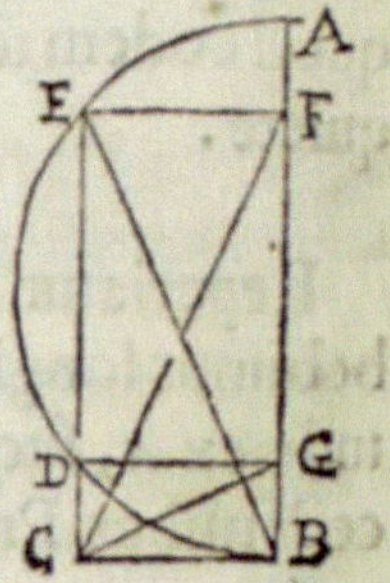

Fiat circa *ab*. circulus *aedb*, & erigatur *ce*. quæ omninò incidet in circulum. (aliàs problema insolubile esset) incidat in *d*. & *e*. Ductisq; *ef*. *dg*. parallelis horizonti *bc*. Dico plana *fc*, *gc*. ad punctum *c* inclinata, eodem tempore absolui.

Cum enim *eb*. *fc*. sint diametri figuræ rectangulę erectę, & ideo æquales, & ęqualiter inclinatę, eodem tempore peragentur. Ergo tempus per *ab*, per *eb*, vel per *fc*. vnum atq; idem est.

Eodem modo infertur tempus per *gc* ęquale esse tempori per *ab*. Quod erat &c.

PROPOSITIO XXXVI.

IN dato circulo cuius centrum est *a*. Diametrum aptare ita vt tempus per aptatam diametrum ęquale sit cuilibet dato tempori.

Debet autem datum tempus maius esse tempore casus per dia-

co tempora per cd, ab, *eſſe ęqualia. Quia* dc. *ad* cb *potentia est vt* ab *ad eandem* cb. *longitudine, nempe in ratione dupla, erunt continuæ proportionales* ab, cd, cb. *Quare per præcedentes Propoſitiones eodem tempore abſoluentur perpendiculum* ab, *& planum inclinatum* cd. *Quod erat &c.*

PROPOSITIO XXXV.

AD datum perpendiculum *ab,* planum, vel plana inclinare ad datum in horizonte punctum *c,* ita vt inclinata plana & perpendiculum ipſum eodem tempore abſoluantur.

Debet autem punctum *c.* Diſtare à puncto *b.* nõ amplius quàm ſit ſemiſſis ipſius *ab.*

Fiat circa *ab.* circulus *aedb,* & erigatur *ce.* quæ omninò incidet in circulum. (aliàs problema inſolubile eſſet) incidat in *d.* & *e.* Ductiſq; *ef. dg.* parallelis horizonti *bc.* Dico plana *fc, gc.* ad punctum *c* inclinata, eodem tempore abſolui.

Cum enim *eb. fc.* ſint diametri figuræ rectangulę erectę, & ideo æquales, & ęqualiter inclinatę, eodem tempore peragentur. Ergo tempus per *ab,* per *eb,* vel per *fc.* vnum atq; idem eſt.

Eodem modo infertur tempus per *gc* ęquale eſſe tempori per *ab.* Quod erat &c.

PROPOSITIO XXXVI.

IN dato circulo cuius centrum eſt *a.* Diametrum aptare ita vt tempus per aptatam diametrum ęquale ſit cuilibet dato tempori.

Debet autem datum tempus maius eſſe tempore caſus per

dia-

diametrum perpendicularem.

Ponamus tempus per diametrum perpendicularem *c d*. eſſe *c d*. & tempus datum ſit *e d*. Reperiatur ipſarum *c d*, *e d* tertia proportionalis quę ſit *f d*. & circa *f d* fiat circulus *f h d*. in quo ex puncto *d*. aptetur *d h*. ęqualis ipſi *c d*. Poſtremo ipſi *h d*. agatur per *a*. parallela *i l*. Dico diametrum *i l*. dato tempore *e d*. abſolui.

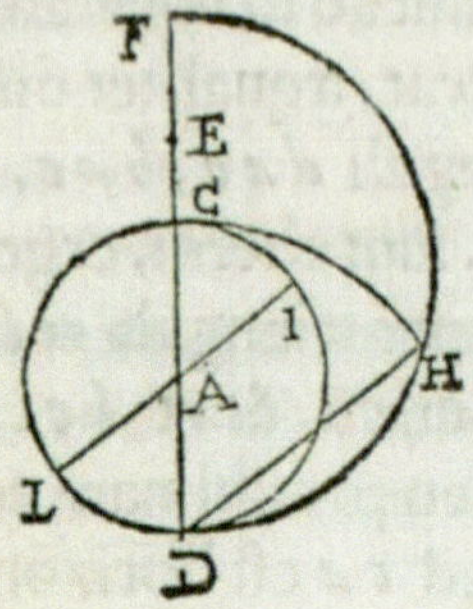

Cum enim tempus per *c d*, ſit *c d*, erit *e d*, (quia media proportionalis eſt) tempus per *f d*, hoc eſt per *h d*. (per ſextam Galilei de motu accelerato) hoc eſt per *i l*. quę ęqualis & parallela eſt ipſi *h d*. Tempus igitur per diametrum *i l*. eſt *e d*. Quod erat &c.

PROPOSITIO XXXVII.

SI fuerint *a b*. *d c*. ad horizontem perpendiculares, & ſumatur vbicunq; punctum *e*, ſiue intra, ſiue extra parallelas, fiatque ad *e*. angulus *d e c* rectus. Dico per interceptas *a d*, *b c*. ſemper eſſe tempora lationum ex quiete ęqualia. Ducatur enim per *a af*. parallela ipſi *b c*. erit angulus *f a d*. ęqualis angulo *e*; & ideo rectus. Quare tempora per *f a*. *a d*. latera trianguli rectanguli, cuius baſis erecta eſt, æqualia erunt inter ſe. & ideo etiam per *b c*. *a d*. (ſunt enim *af*. *b c*. latera oppoſita parallelogrammi, quæ ſemper eodem tempore peraguntur.) Quod erat &c.

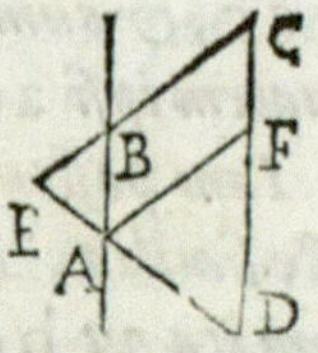

per lem. 13 huius

PROPOSITIO XXXVIII.

SI ab eodem horizontis puncto *a*. ad idem planum perpendiculare *b c*. duo plana inclinentur *a b*. *a d* ęqualiter ab inclinatione ſemirecta diſtantia, tempora lationum per ipſa plana inclinata, ęqualia erunt inter ſe.

Erigatur ex *a* perpendiculum *a e*. Fiatq; circulus circa triangulum *a b d*.

Quia

diametrum perpendicularem.

Ponamus tempus per diametrum perpendicularem *cd.* eſſe *cd.* & tempus datum ſit *ed.* Reperiatur ipſarum *cd, ed* tertia proportionalis quę ſit *fd.* & circa *fd* fiat circulus *fhd.* in quo ex puncto *d.* aptetur *dh.* ęqualis ipſi *cd.* Poſtremo ipſi *d.* agatur per *a.* parallela *il.* Dico diametrum *il.* dato tempore *e d.* abſolui.

Cum enim tempus per *cd,* ſit *cd,* erit *ed,* (quia media proportionalis eſt) tempus per *fd,* hoc eſt per *hd.* (per ſextam Galilei de motu accelerato) hoc eſt per *il.* quę ęqualis & parallela eſt ipſi *hd.* Tempus igitur per diametrum *il.* eſt *ed.* Quod erat &c.

PROPOSITIO XXXVII.

SI fuerint *ab. dc.* ad horizontem perpendiculares, & ſumatur vbicunq; punctum *e,* ſiue intra, ſiue extra parallelas, fiatque ad *e.* angulus *dec* rectus. Dico per interceptas *ad, bc.* ſemper eſſe tempora lationum ex quiete ęqualia. Ducatur enim per *aaf.* parallela ipſi *bc.* erit angulus *fad.* ęqualis angulo *e*; & ideo rectus. Quare tempora per *fa. ad.* latera trianguli rectanguli, cuius baſis erecta eſt, æqualia erunt inter ſe. & ideo etiam per *bc. ad.* (ſunt enim *af. bc.* latera oppoſita parallelogrammi, quæ ſemper eodem tempore peraguntur.) Quod erat &c

PROPOSITIO XXXVIII.

SI ab eodem horizontis puncto *a.* ad idem planum perpendiculare *bc.* duo plana inclinentur *ab. ad* ęqualiter ab inclinatione ſemirecta diſtantia, tempora lationum per ipſa plana inclinata, ęqualia erunt inter ſe.

Erigatur ex *a* perpendiculum *ae.* Fiatq; circulus circa triangulum *abd.*

Quia

Quia lineæ *ab.ad.* per hypot. æqualiter distant ab illa quæ angulum rectum *eac.* bifariam secat, æqualiter distabunt etiam ab ipsis *ea*, *ac*,& anguli *dac*, *bae*. æquales erunt, sed *bae*, & *abd*. sunt alterni, ergo æquales erunt *dac*, & *abd*; quare triangula rectangula *acb*, *adc*. æquiangula erunt, & vt *bc*. ad *ca*, ita erit *ca*. ad *cd*; & ideo per vltimam terti Eucl. recta *ca*. circulum continget. Sed *ca* est horizontalis, ergo punctum *a*. est punctum infimũ circuli, & ideo tempora lationum per *ba*, *da* æqualia erunt. Quod erat &c.

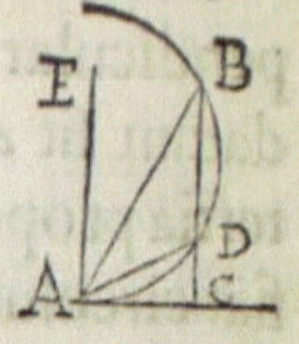

Alitèr.

Hoc idem ostendemus sine circulo, curiosa quadam inuersione. Sint eadem plana ab, ad *quamuis* bc *non sit perpendiculum, dummodo inclinata plana faciant angulos cum horizonte* ac, *& cum plano* bc *permutatim æquales, hoc est* cab *æqualem ipsi* adc, *&* cad *ipsi* abc.

Iam positum est triangula abc, adc. *esse similia. Imaginemur iam conuerti figuram ita vt* bc *sit horizon, &* ac *facta sit perpendiculum. Habebunt in illo situ plana* ab, ad. *easdem inclinationes quas ante inuersionem habebant, permutatim tamen. nam* ab *minus decliue erit, &* ad. *magis; habebuntque plana in eo situ eandem communem eleuationem. Ergo per 2. huius, erit in eo situ inuerso Momentum inclinationis maioris* ad, *ad momentum inclinationis minoris* ab, *vt* ab. *ad* ad. *Restituamus nunc figuram in pristinum, & habebimus (permutatis planis) easdem inclinationes. Dicamus igitur iterum. Momentum inclinationis maioris* ab. *ad momentũ inclinationis minoris* ad. *est vt* ab. *ad* ad. *Quare cum sint momenta vt spatia, eodem tempore absoluentur* ab. ad. *Quod erat &c.*

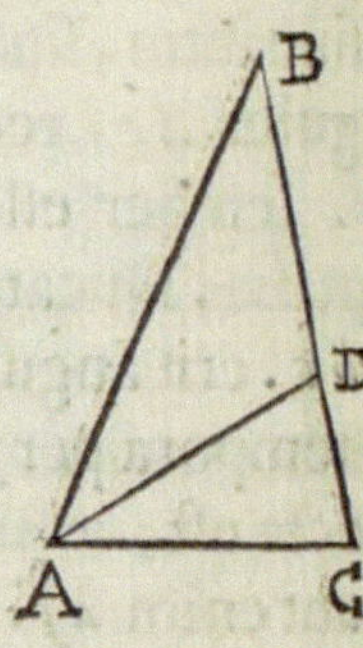

Poterat etiam proponi sic. Si ab eodem horizontis puncto *a* duo

Quia lineæ *ab. ad.* per hypot. æqualiter diftant ab illa quæ angulum rectum *eac.* bifariam fecat, æqualiter diftabunt etiam ab ipfis *ea, ac,* & anguli *dac, bae.* æquales erunt, fed *bae, & abd.* funt alterni, ergo æquales erunt *dac, & abd*; quare triangula rectangula *acb, adc.* æquiangula erunt, & vt *bc.* ad *ca,* ita erit *ca.* ad *cd*; & ideo per vltimam terti Eucl. recta *ca.* circulum continget. Sed *ca* eft horizontalis, ergo punctum *a.* eft punctum infimũ circuli, & ideo tempora lationum per *ba, da* æqualia erunt. Quod erat &c.

Alitèr.

Hoc idem oftendemus fine circulo, curiofa quadam inuerfione. Sint eadem plana ab, ad *quamuis* bc *non fit perpendiculum, dummodo inclinata plana faciant angulos cum horizonte* ac, *& cum plano* bc *per mutatim æquales, hoc est* cab *æqualem ipfi* adc, & cad *ipfi* abc.

Iam pofitum eft triangula abc, adc. *eße fimilia. Imaginemur iam conuerti figuram ita vt* bc *fit horizon, &* ac *facta fit perpendiculum. Habebunt in illo fitu plana* ab, ad. *eafdem inclinationes quas ante inuerfionem habebant, permutatim tamen. nam* ab *minus decliue erit, &* ad. *magis; habebuntque plana in eo fitu eandem commune eleuationem. Ergo per 2. huius, erit in eo fitu inuerfo Momentum inclinationis maioris* ad, *ad momentum inclinationis minoris* ab, *vt* ab. *ad* ad. *Reftituamus nunc figuram in priftinum, & habebimus (permutatis planis) eafdem inclinationes. Dicamus igitur iterum. Momentum inclinationis maioris* ab. *ad momentũ inclinationis minoris* ad. *est vt* ab. *ad* ad. *Quare cum fint momenta vt fpatia, eodem tempore abfoluentur ab. ad. Quod erat &c.*

Poterat etiam proponi fic. Si ab eodem horizontis puncto

a duo

[75] *motu grauium*] *motu Grauium.*

a duo plana ad aliquod planum *bc* inflectantur, ita vt *ab* ad *ad* sit vt *bc* ad *ca*. erunt tempora lationum per vtrumq; inclinatum planum æqualia.

PROPOSITIO XXXIX.

SI fuerit quodcunq; planum eleuatum *ab*, & quodcunque horizontale spatium *ac*. sectum bifariam in *d*. Dico si ponatur tempus per *ab*, esse *ab*. tempus per *ac* post casum *ba* esse semissem ipsius *ac*, nempe *ad*.

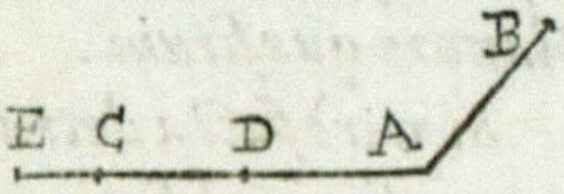

ex Gal.

Ponatur enim *ac*. dupla ipsius *ab*. Iam si supponamus tempus per *ab*. esse *ab*. erit tempus per *ae* eadem *ab*. Sed si spatij *ae* est tempus *ab*, erit spatij *ac* tempus *ad*. (est enim vt spatium *ae*, ad spatiũ *ac*, ita tempus *ab*, ad *ad*.) Quare cum tempus per planũ eleuatum *ab*. sit ipsa *ab*. erit post casum *ba*, tempus per *ac*. dimidia *ac*. Quod oportebat &c.

Hæc Propositio re ipsa congruit cum propos. 25. Galilei de Motu accellerato. Nos illam diuerso modo proposuimus consulentes opportunitati eorum quæ hinc sequuntur, vt infra apparebit.

PROPOSITIO XL.

SI ex terminis *a* & *b*. alicuius lineæ horizontalis duo plana inæqualia ad idem punctum *c*. composita fuerint, *ca* maius, *cb* minus. & differentia longitudinis planorum ęqualis sit semissi horizontalis *ab*. Erit tempus lationis directæ ex *c* vsq; in *a*. æquale tempori lationis inflexæ ex *c* per *b*. vsq; in eundem terminum *a* horizontalis spatij.

Diuidatur *ab*. bifariam in *d*. erunt ergo *cb*, *bd*. ęquales ipsi *ca*. Ponamus tempus per *cb* esse *cb*. hoc supposito erit tem-

a duo plana ad aliquod planum *bc* inflectantur, ita vt *ab* ad *ad* ſit vt *bc* ad *ca.* erunt tempora lationum per vtrumq; inclinatum planum æqualia.

PROPOSITIO XXXIX.

SI fuerit quodcunq; planum eleuatum *ab,* & quodcunque horizontale ſpatium *ac.* ſectum bifariam in *d.* Dico ſi ponatur tempus per *ab,* eſſe *ab.* tempus per *ac* poſt caſum *ba* eſſe ſemiſſem ipſius *ac,* nempe *ad.*

Ponatur enim *ae.* dupla ipſius *ab.*
Iam ſi ſupponamus tempus per *ab.* eſſe *ab.* erit tempus per *ae* eadem *ab.* Sed ſi ſpatij *ae* eſt tempus *ab,* erit ſpatij *ac* tempus *ad.* (eſt enim vt ſpatium *ae,* ad ſpatiũ *ac,* ita tempus *ab,* ad *ad.*) Quare cum tempus per planũ eleuatum *ab.* ſit ipſa *ab.* erit poſt caſum *ba,* tempus per *ac.* dimidia *ac.* Quod oportebat &c. *ex Gal*

Hæc Propoſitio re ipſa congruit cum propoſ. 25. Galilei de Motu accellerato. Nos illam diuerſo modo propoſuimus conſulentes opportunitati eorum quæ hinc ſequuntur, vt infra apparebit.

PROPOSITIO XL.

SI ex terminis *a* & *b.* alicuius lineæ horizontalis duo plana inæqualia ad idem punctum *c.* compoſita fuerint, *ca* maius, *cb* minus. & differentia longitudinis planorum ęqualis ſit ſemiſſi horizontalis *ab.* Erit tempus lationis directæ ex *c* vſq; in *a.* æquale tempori lationis inflexæ ex *c* per *b.* vſq; in eundem terminum *a* horizontalis ſpatij.

Diuidatur *ab.* bifariam in *d.* erunt ergo *cb, bd.* ęquales ipſi *ca.* Ponamus tempus per *cb* eſſe *cb.* hoc ſuppoſito erit

tem-

tempus *ca* ipsa *ca*; & per duas *cb*, *ba*, tempus, per Præcedentem, erit *cbd*; nempe æquale ipsi tempori *ca*. Quod erat &c.

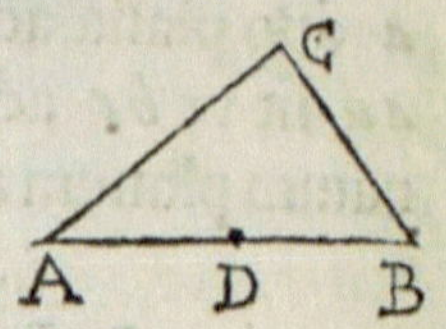

Problema.

Hinc manifesta est solutio problematis; dato plano inclinato ac, *& horizonte* ab. *oportet minus aliquod planum, inuenire* cd. *ita vt si ex terminis* a, *&* b. *ad vnum punctum duo illa plana componantur, tempus lationis per maius æquale sit tempori lationis per minus, & horizontem simul, Dematur ex plano* ac. *pars* cd *æqualis ipsi* ae *semissi horizontalis spatij. reliquum* ad. *erit planum quæsitum.*

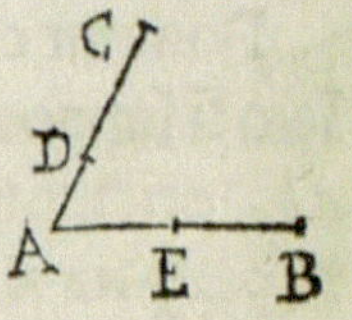

Si verò facta detractione ex ca *nihil reliquum sit, vel fieri nullo modo possit, problema insolubile erit. Demonstratio patet ex Præcedenti.*

Dato vero minori plano ad, *& spatio horizontali* ab *in eadem figura, si ipsi* ad. *addatur* dc. *quæ æqualis sit* ae, *semissi horizontis, maius planum quæsitum erit* ac. *Debent autem vtraq; plana* ad; ac. *simul, maiora esse spatio* ab, *alias insolubile esset problema; nam duo latera trianguli reliquo debent esse maiora.*

Quando data fuerint ipsa duo plana inæqualia ad, ac, *& quæratur quantum sit spatium horizontale, ex cuius extremis punctis erigi data plana possint, & ad vnum punctum componi, ita vt tempus lationis per maius planum, æquale sit tempori lationis per minus & per horizontem simul; Accipietur differentia planorum* dc, *quę duplicata spatium horizontale quęsitum* ab *exibebit. Debent autem tres lineæ* ab, ac, ad *tales esse vt triang. possint cōponere; alias probl. esset insolubile. Horum omnium demonstratio cum illa præcedentis Propositionis congruit; ideo rem indicasse satis duximus. Libet hic obiter recensere quasdam propositiunculas, quamquam ex 3. Conicorum dependeat ipsarum demonstratio: apparebit enim ex ijs naturam etiam*

circa

tempus *ca* ipſa *ca*; & per duas *cb, ba,* tempus, per Præcedentem, erit *cbd*; nempe ęquale ipſi tempori *ca.* Quod erat &c.

Problema.

Hinc manifeſta eſt ſolutio problematis; dato plano inclinato ac, *& horizonte* ab. *oportet minus aliquod planum, inuenire* cd. *ita vt ſi ex terminis* a, *&* b. *ad vnum punctum duo illa plana componantur, tempus lationis per maius æquale ſit tempori lationis per minus, & horizontem ſimul, Dematur ex plano* ac. *pars* cd *æqualis ipſi* ae *ſemiſſi horizontalis ſpatij. reliquum* ad. *erit planum quæſitum.*

Si verò facta detractione ex ca *nihil reliquum ſit, vel fieri nullo modo poſſit, problema inſolubile erit. Demonſtratio patet ex Præcedenti.*

Dato vero minori plano ad, *& ſpatio horizontali* ab *in eadem figura, ſi ipſi* ad. *addatur* dc. *quæ æqualis ſit* ae, *ſemiſſe horizontis, maius planum quæſitum erit* ac. *Debent autem vtraq; plana* ad; ac. *ſimul, maiora eſſe ſpatio* ab, *alias inſolubile eſſet problema; nam duo latera trianguli reliquo debent eſſe maiora.*

Quando data fuerint ipſa duo plana inęqualia ad, ac, *& quæratur quantum ſit ſpatium horizontale, ex cuius extremis punctis erigi data plana poſſint, & ad vnum punctum componi, ita vt tempus lationis per maius planum, æquale ſit tempori lationis per minus & per horizontem ſimul; Accipietur differentia planorum* dc, *quę duplicata ſpatium horizontale quęſitum* ab *exibebit. Debent autem tres lineæ* ab, ac, ad *tales eſſe vt triang. poſſint cõponere; alias probl. eſſet inſolubile. Horum omnium demonſtratio cum illa præcedentis Propoſitionis congruit; ideo rem indicaſſe ſatis duximus. Libet hic obiter recenſere quaſdam propoſitiunculas, quamquam ex 3. Conicorum dependeat ipſarum demonſtratio: apparebit enim ex ijs naturam etiam*

circa

circa hyperbolen quasdam nugas meditatam fuisse ad motum spectantes. Si cui conica non placent, digressione hac prætermissa, pauca hæc euitare poterit, & ad Propositionem 44. se conferre. Materia precedentium hanc continuationem nimis expostulabat.

PROPOSITIO XLI.

SI recta linea *ab*. in quatuor æqales partes *ac*, *cd*, *de*, *eb*. diuisa fuerit, & ex punctis *c*, *e*, excitentur duæ hyperbolæ, quæ sectiones oppositæ dicuntur, quarum foci sint *a*, & *b*. Sumpto in altera earum quolibet puncto *f*. erit tempus per *fa*, æquale tempori per *fb*, *ba*.

Hoc enim patet ex præcedentibus. Nam propter hyperbolam linea *fa*, æqualis est ipsis *fb*, *ce*. per 51. tertij Conicorum. Sed *ce*. semissis est spatij horizontalis *ab*. per hypotesim, ergo æqualia sunt tempora lationum tam per *fa*, quam per *fb*, *ba*. Quod erat &c.

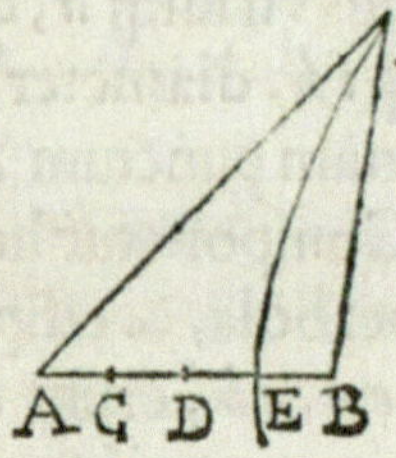

PROPOSITIO XLII.

SI datum sit horizontale spatium *ab* terminatum, & longitudo alicuius plani *f*. data sit maior quam *ab*. Secare oportet planum *f*. in duas partes inæquales ea lege, vt si ex terminis *a*, *b*, facta plana ad idem punctum componantur, tempus lationis per maius planum, æquale sit tempori lationis per minus planum & per horizontalem simul.

Hoc duplici modo absoluemus. Primum contemplatiuè, siue per resolutionem, deinde practicè.

Resolutiuè hoc modo. factum iam sit quod faciendum est. & sint duo plana *ea*, *eb*, vt imperatum est, nempe æqualia si-

S mul

circa hyperbolen quaſdam nugas meditatam fuiſſe ad motum ſpectantes. Si cui conica non placent, digreſſione hac prætermiſſa, pauca hæc euitare poterit, & ad Propoſitionem 44. ſe conferre. Materia pręcedentium hanc continuationem nimis expoſtala. bat.[76]

PROPOSITIO XLI.

SI recta linea *ab.* in quatuor æqales[77] partes *ac, cd, de, eb.* diuiſa fuerit, & ex punctis *c, e,* excitentur duæ hyperbolæ, quæ ſectiones oppoſitæ dicuntur, quarum foci ſint *a,* & *b.* Sumpto in altera earum quolibet puncto *f.* erit tempus per *fa,* æquale tempori per *fb, ba.*

Hoc enim patet ex præcedentibus. Nam propter hyperbolam linea *fa,* æqualis eſt ipſis *fb, ce.* per 51. tertij Conicorum. Sed *ce.* ſemiſſis eſt ſpatij horizontalis *ab.* per hypoteſim, ergo æqualia ſunt tempora lationum tam per *fa,* quam per *fb, ba.* Quod erat &c.

PROPOSITIO XLII.

SI datum ſit horizontale ſpatium *ab* terminatum, & longitudo alicuius plani *f.* data ſit maior quam *ab.* Secare oportet planum *f.* in duas partes inæquales ea lege, vt ſi ex terminis *a, b,* facta plana ad idem punctum componantur, tempus lationis per maius planum, æquale ſit tempori lationis per minus planum & per horizontalem ſimul.

Hoc duplici modo abſoluemus. Primum contemplatiuè, ſiue per reſolutionem, deinde practicè.

Reſolutiuè hoc modo. factum iam ſit quod faciendum eſt. & ſint duo plana *ea, eb,* vt imperatum eſt, nempe æqualia ſi-

S mul

---

[76] *expoſtala. bat.*] *expostulabat.* In the text, *expostulabat* is correctly written, even if it is divided into parts. The "u" seems to be an "a" because the letter is not clear. By the way, we have marked it for the reader.

[77] æqales] aequales.

138 *De motu Grauium deſcendent.*

mul ipſi *f*, & eiuſmodi vt tempus per *e a*, æquale ſit tempori per *e b*, *b a*. Producatur *a b*, vtrimq; in *c*, & *d*. ita vt tota *c d*. æqualis ſit ipſi *f*, & *c a*, *b d*. æquales ſint inter ſe. Certum eſt quia *a e*, *e b*, ſimul æquales ſunt ipſi *c d*, punctum *e* eſſe in ellipſi, cuius axis maior eſt *c d*, & foci ſunt *a*,*b*, puncta. Certum etiam eſt, quia tempus per ea æquale eſt tempori per *e b*, *b a*, idem punctum *e* eſſe in hyperbola cuius foci ſint *a*, *b*, & vertex *h*. (diuisà nempe *a b*, in quatuor partes æquales, quarum vna ſit *h b*.) hoc autem demonſtratum eſt in præcedenti. Erit ergo punctum *e*. in communi concurſu duarum ſectionū, ſed duæ ſectiones datæ ſunt; quandoquidem dantur foci communes vtriuſq; *a*, & *b*, & *c d*. data eſt axis maior ellipſis; daturq; *i h*. diameter hyperbolæ, nempe ſemiſſis ipſius *a b*; quare etiam punctum *e*. datum erit.

ex 3. l. Conicor.

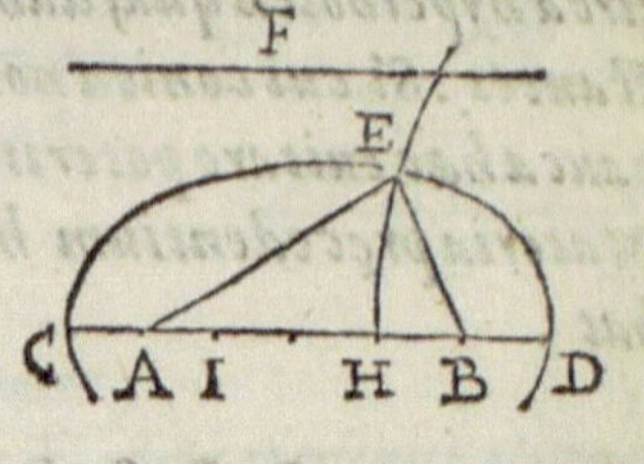

Componetur hoc modo. Factis igitur duabus ſectionibus hyperbola, & ellipſi, quæ concurrant in *e*, ſi à puncto *e* ducantur *e a*, *e b*, erunt *e a*, *e b*, ſimul æquales ipſi *f*, & erit tempus per *e a*, æquale tempori per *e b*, *b a*, ſimul, ob hyperbolam. Quod oportebat &c.

Facilius tamen hoc modo practicè in figura ſequenti.

Secetur ab ipſa linea F. pars *l m*, quæ æqualis ſit ſemiſſi ſpatij horizontalis *a b*. & reliqua *l n*. diuidatur bifariam in *o*. Dico *n o*, *o m* eſſe plana quæſita, quæ ſi à punctis *a b* inclinentur ad vnū punctum, æqualia facient tempora lationum, tam per maius planum *o m*, quàm per minus *n o* ſimul cum horizonte *b a*. Hoc autem perſpicuum eſt ex Propoſitione 41. cum differentia longitudinis planorum *l m*, ſit per conſtrutionem æqualis ſemiſſi ſpatij horizontalis *a b*. Quod erat &c.

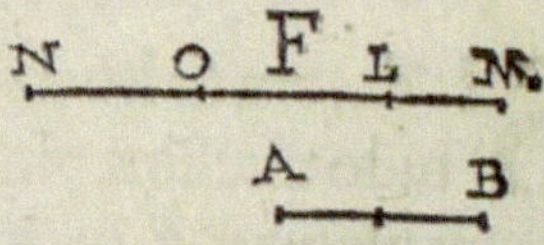

PRO-

mul ipſi *f,* & eiuſmodi vt tempus per *e a,* æquale ſit tempori per *eb, ba.* Producatur *ab,* vtrimq; in *c,* & *d.* ita vt tota *cd.* æqualis ſit ipſi *f,* & *ca, bd.* æquales ſint inter ſe.

ex 3. l. Conicor.

Certum eſt quia *ae, eb,* ſimul æquales ſunt ipſi *cd,* punctum *e* eſſe in ellipſi, cuius axis maior eſt *cd,* & foci ſunt *a, b,* puncta. Certum etiam eſt, quia tempus per ea æquale eſt tempori per *eb, ba,* idem punctum *e* eſſe in hyperbola cuius foci ſint *a, b,* & vertex *h.* (diuisà nempe *ab,* in quatuor partes æquales, quarum vna ſit *hb.*) hoc autem demonſtratum eſt in præcedenti. Erit ergo punctum *e.* in communi concurſu duarum ſectionũ, ſed duæ ſectiones datæ ſunt; quandoquidem dantur foci communes vtriuſq; *a,* & *b,* & *cd.* data eſt axis maior ellipſis; daturq; *ih.* diameter hyperbolæ, nempe ſemiſſis ipſius *ab*; quare etiam punctum *e.* datum erit.

Componetur hoc modo. Factis igitur duabus ſectionibus hyperbola, & ellipſi, quæ concurrant in *e,* ſi à puncto *e* ducantur *ea, eb,* erunt *ea, eb,* ſimul æquales ipſi *f,* & erit tempus per *ea,* æquale tempori per *eb, ba,* ſimul, ob hyperbolam. Quod oportebat &c.

Facilius tamen hoc modo practicè in figura ſequenti.

Secetur ab ipſa linea F. pars *lm,* quæ æqualis ſit ſemiſſi ſpatij horizontalis *ab.* & reliqua *ln.* diuidatur bifariam in *o.* Dico *no, om* eſſe plana quæſita, quæ ſi à punctis *ab* inclinentur ad vnũ punctum, æqualia facient tempora lationum, tam per maius planum *om,* quàm per minus *no* ſimul cum horizonte *ba.* Hoc autem perſpicuum eſt ex Propoſitione 41. cum differentia longitudinis planorum *lm,* ſit per conſtrutionem æqualis ſemiſſi ſpatij horizontalis *ab.* Quod erat &c.

P R O-

PROPOSITIO XLIII.

DAto horizontali ſpatio *ab*, datoq; angulo *bac*, qui minor ſit angulo trianguli æquilateri, (alias enim problema inſolubile eſſet) oportet triangulum conſtituere quod habeat baſim *ab*, & angulum *bac*. & tempus per latus *ca*, æquale ſit tempori per *cb*, *ba*.

Diuidatur *ab*. in quatuor æquales partes quarum vna ſit *db* & ex vertice *d*, focis *a*, *b*, fiat *dc* hyperbola quæ ſecet rectam *ac*, in *c*. (ſecabit enim omnino, vt infra demonſtrabimus.) Dico *c* eſſe tertium quæſiti trianguli punctum. Ducta enim *cb*. patet differentiam inter *ac*, *cb*, eſſe ſemiſſem ipſius *ab*, propter hyperbolam, & propter diuiſionem lineæ *ab* in quatuor æquales partes. Quare tempus per *ca*, æquale erit tempori per *cb*, *ba*: Quod erat &c.

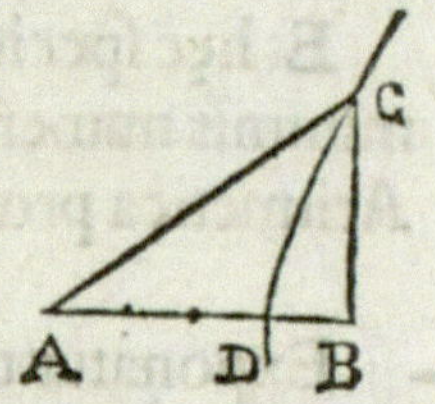

Quod autem in præcedenti figura (ſuppoſito angulo bac. *minore quam ſit angulus trianguli æquilateri) linea* ac *cum hyperbola conueniat, ſic demonſtrabimus in ſequenti figura.*

Sit aſymptotos bc. *Ergo rectangulum* eaf *æquale eſt quartę parti figuræ per 45. tertij Conicorum. Quadratum etiam* fc. *æquale eſt eidem quartæ parti figurę, per primam ſecundi Conicorum: erunt ergo æqualia inter ſe rectangulum* eaf, *& quadratum* fc. *Quare vt* af. *ad* fc. *ita* fc. *ad* ac, *hoc eſt ad* fd. *Sunt igitur puncta* dca *in ſemicirculo cuius diameter eſt* da, *& centrū* b. *Sed cū ſint ęquales* bf. af. *per conſtructionē in propoſitione pręcedenti; & anguli ad* f. *recti ſint, eſt.n* ba *axis, &* fc *ad axem applicata. erunt æquales* cb, ca. *& triangulum* abc, *æquilaterum erit. Qualibet ergo linea quæ*

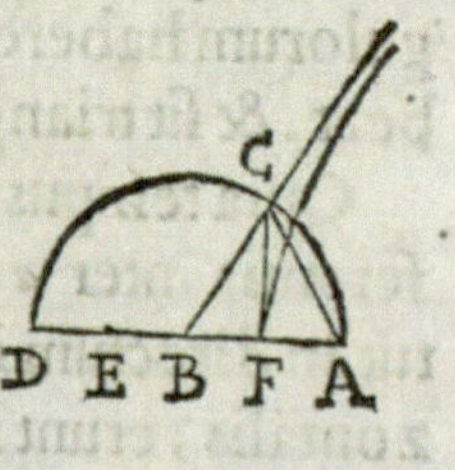

PROPOSITIO XLIII.

DAto horizontali ſpatio *ab,* datoq; angulo *bac,* qui minor ſit angulo trianguli æquilateri, (alias enim problema inſolubile eſſet) oportet triangulum conſtituere quod habeat baſim *ab,* & angulum *bac.* & tempus per latus *ca,* ęquale ſit tempori per *cb, ba.*

Diuidatur *ab.* in quatuor æquales partes quarum vna ſit *db* & ex vertice *d,* focis *a, b,* fiat *dc* hyperbola quæ ſecet rectam *ac,* in *c.* (ſecabit enim omnino, vt infra demonſtrabimus.) Dico *c* eſſe tertium quæſiti trianguli punctum. Ducta enim *cb.* patet differentiam inter *ac, cb.* eſſe ſemiſſem ipſius *ab,* propter hyperbolam, & propter diuiſionem lineæ *ab* in quatuor æquales partes. Quare tempus per *ca,* æquale erit tempori per *cb, ba*: Quod erat &c.

Quod autem in præcedenti figura (ſuppoſito angulo bac. *minore quam ſit angulus trianguli æquilateri) linea* ac *cum hyperbola conueniat, ſic demonstrabimus in ſequenti figura.*

Sit aſymptotos bc. *Ergo rectangulum* eaf *æquale eſt quartę parti figuræ per 45. tertij Conicorum. Quadratum etiam* fc. *æquale est eidem quartæ parti figurę, per primam ſecundi Conicorum: erunt ergo æqualia inter ſe rectangulum* eaf, *& quadratum* fc. *Quare vt* af. *ad* fc. *ita* fc. *ad* ae, *hoc est ad* fd. *Sunt igitur puncta* dca *in ſemicirculo cuius diameter est* da, *& centrũ* b. *Sed cũ ſint ęquales* bf. af. *per conſtructionẽ in propoſitione prędecenti; & anguli ad* f. *recti ſint, est.* nba *axis, &* fc *ad axem applicata. erunt æquales* cb, ca. *& triangulum* abc, *æquilaterum erit. Quælibet ergo linea quæ*

 ad pun-

ad punctum d. *angulum contineat cum* da. *minorem angulo* abc, *trianguli æquilateri, conueniet omnino cum* bc; *Quare in figura præcedentis propoſitionis linea* ac *conueniet cum aſymptoto, & ideo etiam cum hyperbola.*

PROPOSITIO XLIV.

EX infinitis ſpeciebus triangulorum rectangulorum, vna tantum eſt quę habeat hanc prerogatiuam, quod ſcilicet tempus per hypotenuſam ęquale ſit tempori per reliqua duo latera.

Et hęc ſpecies illa eſt quæ prima omnium, hoc eſt, quæ in minimis numeris habet tria latera comenſurabilia; Nempe in Aritmetica proportione numerorum 3. 4. & 5.

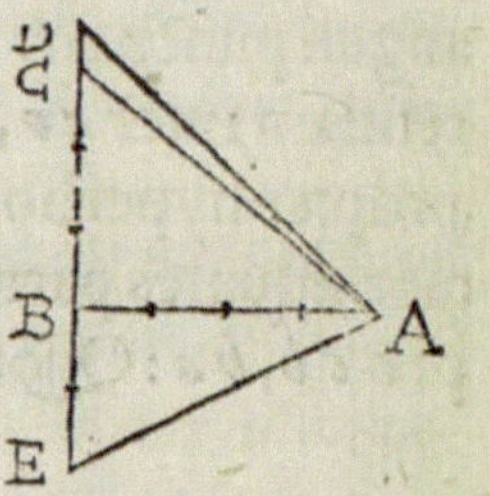

Exponatur triangulum *abc*; cuius latus *ab* horizontale ſit 4. & *bc*. erectum ſit 3. hypotenuſa autem *ac*. ſit 5. Perſpicuum eſt angulum *cba*. rectum eſſe; cum quadratum *ac*. 25. æquale ſit duobus quadratis *cb*, *ba*. 9. & 16. Manifeſtum etiam eſt tempus per *ca*. ęquari tempori per *cb*, *ba*. cum differentia inter *ac*. *cb*. ſit 2. ſemiſſis ſpatij horizontalis *ab*. quod eſt 4.

Dico præterea nullam aliam ſpeciem triangulorum rectangulorum habere illam proprietatem. Nam ſi poſſibile eſt, habeat. & ſit triangulum illius ſpeciei ipſum *adb*.

Quia tempus per *da*. æquale eſt tempori per *dba*. erit differentia inter *ad*. *db*. æqualis ſemiſſi horizontalis *ab*. Ponatur in directum ipſi *db*. linea *be*, quę ęqualis ſit ſemiſſi horizontalis; erunt iam *ad*, *de*. ęquales inter ſe; & *ac*, *ce*, ob eãdem cauſã æquales inter ſe; quod impoſſibile eſt. Iunctâ enim *ae*. eſſet vterq; angulus *dae*, *cae*, ęqualis angulo *e*. quod eſt abſurdum. Nulla ergo ſpecies triangulorũ rectangulorum repe ritur, præter iam dictam quę habeat ſuperiùs enarratam proprietatem.

Poſ-

ad punctum d. *angulum contineat cum* da. *minorem angulo* abc, *trianguli æquilateri, conueniet omnino cum* bc; *Quare in figura præcedentis propoſitionis linea* ac *conueniet cum aſymptoto, & ideo etiam cum hyperbola.*

PROPOSITIO XLIV.

EX infinitis ſpeciebus triangulorum rectangulorum, vna tantum eſt quę habeat hanc prerogatiuam, quod ſcilicet tempus per hypotenuſam ęquale ſit tempori per reliqua duo latera.

Et hęc ſpecies illa eſt quæ prima omnium, hoc eſt, quæ in minimis numeris habet tria latera comenſurabilia; Nempe in Aritmetica proportione numerorum 3. & 5.[79]

Exponatur triangulum *abc*; cuius latus *ab* horizontale ſit 4. & *bc.* erectum ſit 3. hypotenuſa autem *ac.* ſit 5. Perſpicuum eſt angulum *cba.* rectum eſſe; cum quadratum *ac.* 25. æquale ſit duobus quadratis *cb, ba.* 9. & 16. Manifeſtum etiam eſt tempus per *ca.* ęquari tempori per *cb, ba.* cum differentia inter *ac. cb.* ſit 2. ſemiſſis ſpatij horizontalis *ab.* quod eſt 4.

Dico præterea nullam aliam ſpeciem triangulorum rectangulorum habere illam proprietatem. Nam ſi poſſibile eſt, habeat. & ſit triangulum illius ſpeciei ipſum *adb.*

Quia tempus per *da.* æquale eſt tempori per *dba.* erit differentia inter *ad. db.* æqualis ſemiſſi horizontalis *ab.* Ponatur in directum ipſi *db.* linea *be,* quę ęqualis ſit ſemiſſi horizontalis; erunt iam *ad, de.* ęquales inter ſe; & *ac, ce,* ob eãdem cauſã æquales inter ſe; quod impoſſibile eſt. Iunctà enim *ae.* eſſet vterq; angulus *dae, cae,* ęqualis angulo *e.* quod eſt abſurdum. Nulla ergo ſpecies triangulorũ rectangulorum reperitur, præter iam dictam quę habeat ſuperiùs enarratam proprietatem.

Poſ-

[78] *motu grauium*] *motu Grauium.*
[79] 3. & 5.] 3. 4 . & 5.

Possemus etiam demonstrare ex infinitis speciebus triangulorum obliquangulorum, quæ vnum angulum datum habeant, puta 40. graduum, vnam tantum speciem esse quæ prædictam proprietatem habeat. Quinetiam ostenderetur ex infinitis hyperbolarum speciebus, vnam tantum speciem esse quæ illam habeat prærogatiuam. Sed non est tanti omnia hæc minuta enucleatim percensere, vt lectoris patientia, beneuolentiaq; vlterius abutamur.

PROPOSITIO XLV.

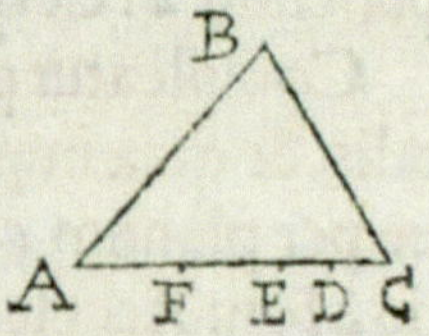

SI Fuerit quodcunq; triangulum *abc*, habens latera *ab*, *bc* inæqualia, puta *ab* maius, *bc* minus & basim horizontalem. Dico eodem tempore fieri lationem per *ba*. solam, & per *bc*, simul cum tanto horizontali spatio quanta est bis differentia inter ipsa latera. Sit enim differentia inter latera *cd*, cuius dupla ponatur *ce*. Perspicuum est *bc*, *cd*, simul æquari ipsi *ba*. Iam si supponamus tempus per *ba*, esse *ba*, erit tempus per *bc*. ipsa *bc*, & post casum *bc*, tempus per *ce*. erit dimidia ipsius *ce*, hoc est *cd*. Aequale est igitur tempus per *ba* tempori per *bc*, *ce* simul. Quod erat &c.

Ijsdem positis: quando *ce* (in eadem figura) minor fuerit quam basis trianguli. Dico duo grauia eodem temporis momento demissa ex *b* per latera *ba*, *bc* post conuersionem horizontalem factam in *a*, & *c*. conuenire in pũcto basis *f*. quod quidem bifariam secet ipsam *ae*. Ostensum enim est eodem tempore peruenire duo grauia ad puncta *a*. & *e*. ergo etiam reliqua spatia *af*, *ef*, æquali tempore peragentur, cum sint æqualia per hypothesim, & gradus velocitatis æquales sint. per V. huius.

PRO-

Poſſemus etiam demonstrare ex infinitis ſpeciebus triangulorum obliquangulorum, quæ vnum angulum datum habeant, puta 40. graduum, vnam tantum ſpeciem eße quæ prædictam proprietatem habeat. Quinetiam oſtenderetur ex infinitis hyperbolarum ſpeciebus, vnam tantum ſpeciem eße quæ illam habeat prærogatiuam. Sed non eſt tanti omnia hæc minuta enucleatim percenſere, vt lectoris patientia, beneuolentiaq; vlterius abutamur.

PROPOSITIO XLV.

SI Fuerit quodcunq; triangulum *abc,* habens latera *ab, bc* inæqualia, puta *ab* maius, *bc* minus & baſim horizontalem. Dico eodem tempore fieri lationem per *ba.* ſolam, & per *bc,* ſimul cum tanto horizontali ſpatio quanta eſt bis differentia inter ipſa latera. Sit enim differentia inter latera *cd,* cuius dupla ponatur *ce.* Perſpicuum eſt *bc, cd,* ſimul æquari ipſi *ba.* Iam ſi ſupponamus tempus per *ba,* eſſe *ba,* erit tempus per *bc.* ipſa *bc,* & poſt caſum *bc,* tempus per *ce.* erit dimidia ipſius *ce,* hoc eſt *cd.* Aequale eſt igitur tempus per *ba* tempori per *bc, ce* ſimul. Quod erat &c.

Ijſdem poſitis*:* quando *ce (*in eadem figura) minor fuerit quam baſis trianguli. Dico duo grauia eodem temporis momento demiſſa ex *b* per latera *ba, bc* poſt conuerſionem horizontalem factam in *a,* & *c.* conuenire in pũcto baſis *f.* quod quidem bifariam ſecet ipſam *ae.* Oſtenſum enim eſt eodem tempore peruenire duo grauia ad puncta *a.* & *e.* ergo etiam reliqua ſpatia *af, ef,* æquali tempore peragentur, cum ſint æqualia per hypotheſim, & gradus velocitatis æquales ſint. per V. huius.

P R O-

PROPOSITIO XLVI.

POſito quolibet triangulo *abc*, cuius baſis *cb* horizonti parallela ſit. Si graue ex quiete in vertice *a* per alterum latus *ac* cadat, & inde per baſim *cb* cum impetu concepto conuertatur, baſiq; per acta cum eodem impetu per alterum latus *ba* aſcendat, impetus ille perducet graue per aſcenſum *ba*. vſque ad idem punctum *a*. ex quo diſceſſerat.

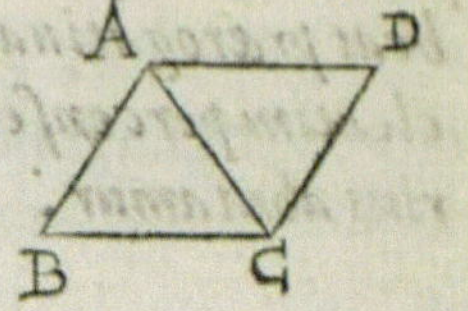

Compleatur parallelogrãmum *abcd*. eritq; *ad*, horizontalis, & quia impetus aquiſitus per deſcenſum *ac*. perducit graue per planum *cd*. vſq; ad *d*. per Scholium Prop. 23. Galilei de Motu Accelerato, idem impetus (poſt tranſmiſſam baſim motu æquabili) perducet mobile ex *b* vſq; in *a*, ſunt enim *ba*, *cd* æquales, & æqualiter inclinatæ. Quare &c.

Lemma.

Si inter parallelas horizontales ab, cd. *duæ lineæ fuerint* bd, da. *erit tempus caſus per vnam* bd, *ad tempus aſcenſus per alteram* da, *vt eſt ipſa* bd *ad* da. *Eſt enim tempus caſus per quamcumq; lineam ex Galileo, æquale tempori aſcenſus per eandem, quando ſiat aſcenſus cum impetu per deſcenſum aquiſito. Sed tempora caſuum per* bd, *&* ad *ſunt vt* bd *ad* ad, *ergo etiam tempus caſus per* bd. *ad tempus aſcenſus per* da, *erit vt* bd: *ad* da. *Quod &c.*

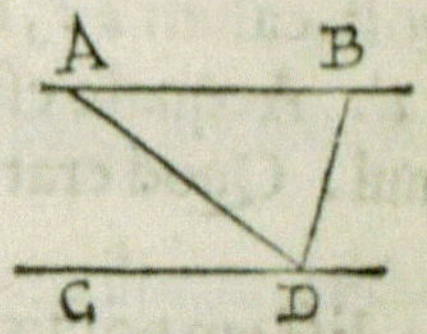

PROPOSITIO XLVII.

POſito quolibet triangulo *abc*. cuius baſis *bc* horizontalis ſit, ſi ſiant lationes ex quiete in vertice *a*. vtrinq; per tria latera,

PROPOSITIO XLVI.

POſito quolibet triangulo *abc,* cuius baſis *cb* horizonti parallela ſit. Si graue ex quiete in vertice *a* per alterum latus *ac* cadat, & inde per baſim *cb* cum impetu concepto conuertatur, baſiq; per acta cum eodem impetu per alterum latus *ba* aſcendat, impetus ille perducet graue per aſcenſum *ba.* vſque ad idem punctum *a.* ex quo diſceſſerat.

Compleatur parallelogrãmum *abcd.* eritq; *ad,* horizontalis, & quia impetus aquiſitus per deſcenſum *ac.* perducit graue per planum *cd.* vſq; ad *d.* per Scholium Prop. 23. Galilei de Motu Accelerato, idem impetus (poſt tranſmiſſam baſim motu æquabili) perducet mobile ex *b* vſq; in *a,* ſunt enim *ba, cd* æquales, & æqualiter inclinatæ. Quare &c.

Lemma.

Si inter parallelas horizontales ab, cd. *duæ lineæ fuerint* bd, da. *erit tempus caſus per vnam* bd, *ad tempus aſcenſus per alteram* da, *vt eſt ipſa* bd *ad* da. *Est enim tempus caſus per quamcumq; lineam ex Gælileo, æquale tempori aſcenſus per eandem, quando fiat aſcenſus cum impetu per deſcenſum aquiſito. Sed tempora caſuum per* bd, & ad *ſunt vt* bd *ad* ad, *ergo etiam tempus caſus per* bd. *ad tempus aſcenſus per* da, *erit vt* bd*: ad* da. *Qnod*[80] *&c.*

PROPOSITIO XLVII.

POſito quolibet triangulo *abc.* cuius baſis *bc* horizontalis ſit, ſi fiant lationes ex quiete in vertice *a.* vtrinq; per tria

latera,

[80] *Qnod*] *Quod.*

latera, erit tempus lationum per *a c*, *c b*, *b a*, æquale tempori lationum per *a b*, *b c*, *c a*.

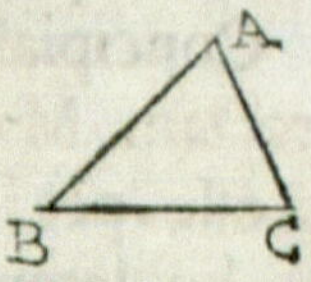

Ponamus enim tempus casus per *a b*. esse ipsã *a b*, erit tempus per *b c*. semissis ipsius *b c*. per Proposit. 40. huius, cum *b c* sit horizontalis: transmissa vero basi motu æquabili, tempus ascensus per *c a* erit ipsa *c a* per lem. præced.. Eodem modo: cum sit tẽpus casus per *a b* ipsa *a b*, erit tẽpus casus per *a c* ipsa *a c*, & per horizontalem *c b* erit semissis ipsius *c b*, inde per ascensum *b a* erit *b a*. per lemma præcedens. Est ergò tẽpus per vtramq; viã tamquam duo latera trianguli simul cum dimidia basi. Quare tempora per vtramq; viam, siue *a b c a*, siue *a c b a*. æqualia inter se erunt. Quod &c.

Possent similia demonstrari de figuris poligonis, & irregularibus: Sed cum hęc omnia solita breuitate exequi non possint, existimaui eorum demonstrationem apud eruditos plus molestiæ allaturam, quam doctrinæ.

PROPOSITIO XLVIII.

AD aliquod perpendiculum data duo plana diuersæ longitudinis ab eodem horizontis puncto inflectere, ita vt tempora per infexa plana ęqualia sint. Vel.

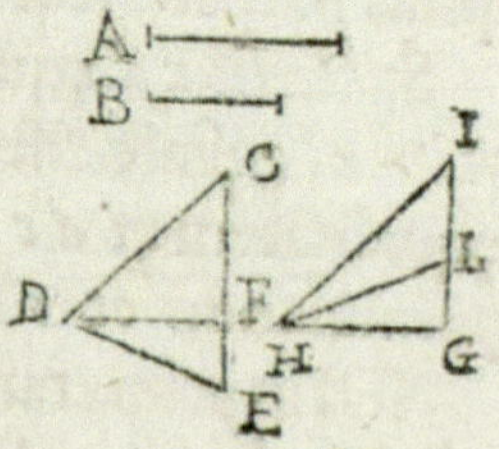

Proponit aliquis geminos asseres *a*, & *b*. diuersę longitudinis, ea lege vt ab vno eodemq; puncto in pauimento inflecti debeant ad parietem, & grauia ex fastigijs eorũ eodem tempore demissa, simul eodemq; tempore ferantur in terram.

Componantur *a*, & *b*. ad angulũ rectum; sintq; *c d*, *d e*, & productà *c e*, ipsi perpendicularis sit *d f*. Accipiatur iam in pauimento distantia *g h* à pariete *g i*, quæ æqualis sit ipsi *f d*; Tum à puncto *h*. inflectantur ad parietem *h i*, *h l*. ęquales ipsis *a*, *b*, vel ipsis *c d*, *d e*.

Dico

latera, erit tempus lationum per *ac, cb, ba,* æquale tempori lationum per *ab, bc, ca.*

Ponamus enim tempus caſus per *ab.* eſſe ipsã *ab,* erit tempus per *bc.* ſemiſſis. ipſius *bc.* per Propoſit. 40. huius, cum *bc* ſit horizontalis: tranſmiſſa vero baſi motu æquabili, tempus aſcenſus per *ca* erit ipſa *ca* per lem. præced. Eodem modo: cum ſit tẽpus caſus per *ab* ipſa *ab,* erit tẽpus caſus per *ac* ipſa *ac,* & per horizontalem *cb* erit ſemiſſis ipſius *cb,* inde per aſcenſum *ba* erit *ba.* per lemma præcedens. Eſt ergò tẽpus per vtramq; viã tamquam duo latera trianguli ſimul cum dimidia baſi. Quare tempora per vtramq; viam, ſiue *abca,* ſiue *acba.* æqualia inter ſe erunt. Quod &c.

Poſſent ſimilia demonſtrari de figuris poligonis, & irregularibus: Sed cum hęc omnia ſolita breuitate exequi non poſſint, exiſtimaui eorum demonſtrationem apud eruditos plus molestiæ allaturam, quam octrinæ.[81]

PROPOSITIO XLVIII.

AD aliquod perpendiculum data duo plana diuerſæ longitudinis ab eodem horizontis puncto inflectere, ita vt tempora per infexa plana ęqualia ſint. Vel.

Proponit aliquis geminos aſſeres *a,* & *b.* diuerſę longitudinis, ea lege vt ab vno eodemq; puncto in pauimento inflecti debeant ad parietem, & grauia ex faſtigijs eorũ eodem tempore demiſſa, ſimul eodemq; tempore ferantur in terram.

Componantur *a,* & *b.* ad angulũ rectum; ſintq; *cd, de,* & productà *ce,* ipſi perpendicularis ſit *df.* Accipiatur iam in pauimento diſtantia *gh* à pariete *gi,* quæ æqualis ſit ipſi *fd*; Tum à puncto *h.* inflectantur ad parietem *hi, hl.* ęquales ipſis *a,b,* vel ipſis *cd, de.*

Dico

[81] *octrinæ*] *doctrinæ.*

144 *De motu Grauium descendent.*

Dico tempora per *ih*, & per *lh*. æqualia esse.

Concipiamus basim *ce*. trianguli *cde*, esse ad horizontem erectam. Manifestum est tempora lationum per *cd*, *de*. ęqualia esse. per lemma Propos. 13. Sed cum duo latera *cd*, *df*. duobus lateribus *ih*. *hg*. æqualia sint vtrumq; vtrique, & anguli *cfd*, *igh*. recti, si ex quadratis ęqualibus *dc*. *hi*. demantur quadrata æqualia *df*. *hg*. remanebunt æqualia quadrata *fc*, *gi*, & ideo *fc*. *gi*. lineæ ęquales erunt; & propterea integra triangula *cdf*. *ihg*. æqualia, & similia erunt. & tempus per *ih*. æquale tempori per *cd*.

Eodem modo ostendetur tempus per *lh*. æquale tempori per *de*. Quare cum æqualia sint tempora per *cd*. *de*. æqualia erunt etiam per *ih*. *hl*. Quod erat &c.

PROPOSITIO IL.

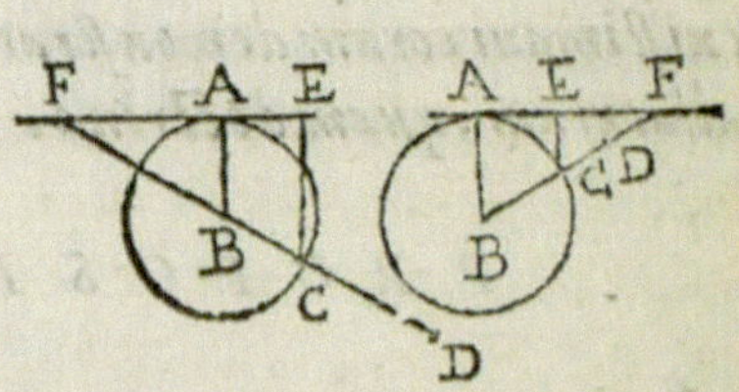

SI ex *a* puncto sublimiori circuli ad horizontem erecti graue cadat vsque in centrum *b*, & inde per quodcumq; planum siue eleuatum, siue decliue, conuertatur cum impetu iam concepto; graue huiusmodi tempore casus *ab* absoluet spatium *bd*; quod nempe æquale sit vtrisq. tum semidiametro *bc*, tum etiã ipsius perpendiculo *ce*.

Secta sit *cd* ęqualis ipsi *ce*. Dico tempus per *ab*, ex quiete in *a*, ęquale esse tempori per *bd*. post casum *ab*. Est .n. ob ęqualitatem vt *dc* ad *cb*, ita *ce* ad *ba*, hoc est *cf* ad *fb*, & permutando vt *cd* ad *cf*. ita *cb* ad *bf*. in vtraq. figura.

Sed in prima tantùm erit componendo vt *df* ad *cf*, ita *cf* ad *bf*. In secunda verò erit. Conuertendo, per conuersionẽ rationis, & iterum conuertendo, vt *df* ad *cf*, ita *cf* ad *bf*. Quare in vtroq. casu tres lineę *df*, *cf*, *bf*. sunt in continuà proportione.

Iam si tempus per *ab* ponatur esse *ab*, erit tempus per *fb* ipsa *fb*,

Dico tempora per *ih,* & per *lh.* æqualia eſſe.

Concipiamus baſim *c e.* trianguli *cdc,* eſſe ad horizontem erectam. Manifeſtum eſt tempora lationum per *cd, de.* ęqualia eſſe. per lemma Propoſ. 13. Sed cum duo latera *cd, df.* duobus lateribus *ih. hg.* æqualia ſint vtrumq; vtrique, & anguli *cfd, igh.* recti, ſi ex quadratis ęqualibus *dc. hi.* demantur quadrata æqualia *df. hg.* remanebunt æqualia quadrata *fc, gi,* & ideo *fc. gi.* lineæ ęquales erunt; & propterea integra triangula *cdf. ihg.* æqualia, & ſimilia erunt. & tempus per *ih.* æquale tempori per *cd.*

Eodem modo oſtendetur tempus per *lh.* æquale tempori per *de.* Quare cum æqualia ſint tempora per *cd. de.* æqualia erunt etiam per *ih. hl.* Quod erat &c.

PROPOSITIO IL

SI ex *a* puncto ſublimiori circuli ad horizontem erecti graue cadat vſque in centrum *b,* & inde per quodcumq; planum ſiue eleuatum, ſiue decliue, conuertatur cum impetu iam concepto; graue huiuſmodi tempore caſus *ab* abſoluet ſpatium *bd;* quod nempe æquale ſit vtriſq. tum ſemidiametro *bc,* tum etiã ipſius perpendiculo *ce.*

Secta ſit *cd* ęqualis ipſi *ce.* Dico tempus per *ab,* ex quiete in *a,* ęquale eſſe tempori per *bd.* poſt caſum *ab.* Eſt. n. ob ęqualitatem vt *dc* ad *cb,* ita *ce* ad *ba,* hoc eſt *cf* ad *fb,* & permutando vt *cd* ad *cf.* ita *cb* ad *bf.* in vtraq. figura.

Sed in prima tantùm erit componendo vt *df* ad *cf,* ita *cf* ad *bf.* In ſecunda verò erit. Conuertendo, per conuerſionẽ rationis, & iterum conuertendo, vt *df* ad *cf,* ita *cf* ad *bf.* Quare in vtroq. caſu tres lineę *df, cf, bf.* ſunt in continuà proportione.

Iam ſi tempus per *ab* ponatur eſſe *ab,* erit tempus per *fb* ip

ſa *fb,*

ſa *fb*, & per *fd* tempus erit media proportionalis *fc*. Quare tempus per reliquum lineæ, nempe per *b d*, erit reliquum temporis, nempe *b c*. Idem ergo tempus eſt lationis per *a b* ex quiete in *a*, & per *b d* poſt caſum *a b*.

In hac propoſitione re ipſa demonſtrantur duo Theoremata Galilei, De Motu accelerato: ſed quia valde ad rem noſtram faciunt, eadem diuerſa iterum ratione contemplabimur, vt lucem ſequenti Corollario præferant.

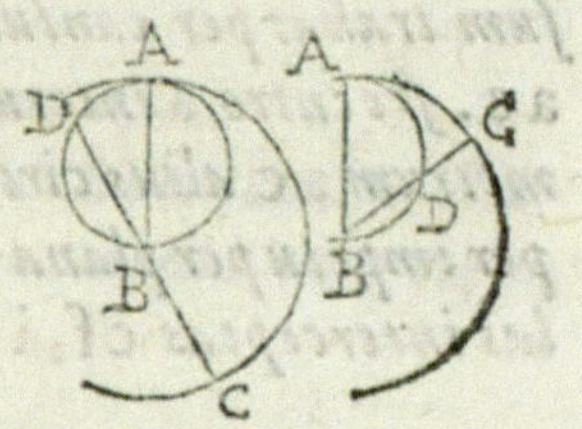

Si graue naturaliter cadat ex a *in* b. *& ex* b *cum impetu concepto, per quodlibet planum* b c *conuertatur. Quæritur quantum ſpatij per planum* b c *abſoluat mobile tempore caſus* a b.

Fiat circa diametrum a b. *circulus* a d b, *centroq;* b, *& interuallo* b a *circulus* a c. *Dico graue delapſum per* a b, *ſi ex puncto* b *cum impetu concepto conuertatur per planum inclinatum* b c, *tempore æquali tempori caſus, percurrere ſpatium æquale vtriſq; ſimul* b c, c d.

Si enim poſt caſum a b *graue conuerteretur per planum quodcunq;* b c, *motuq; æquabili procederet, graue huiuſmodi per planum* b c *tempore æquali tempori caſus ſpatium perageret duplũ ipſius* b a, *ergo percurreret ſpatium duplum ipſius* b c. *tempore caſus, ſi poſt caſum æquabili motu procederet. Sed ſuperueniẽte operatione grauitatis, mobile non procedet motu æquabili ſuper plano* b c; *Quin immò tempore caſus* a b, *grauitas promouebit mobile ſuper plano* b c *tantum ſpatium quanta eſt inclus a in circulo linea* d b *(quo enim tempore grauitas trahit mobile ex* a *in* b. *eodem tempore trahit etiam ex* d *in* b *per* 6. *Galilei de Motu Accelerato.) Ergo duplæ* b c *in prima figura addenda erit* d b, *conſpirant enim deorſum tam motus æquabilis, quam motus grauitatis; In ſecunda vero figura à dupla* b c *detrahenda erit* d b *(quia motus grauitatis contrarius eſt motui æquabili) & ſic* ex Gal.

T *con-*

ſa *fb*, & per *fd* tempus erit media proportionalis *fc*. Quare tempus per reliquum lineæ, nempe per *bd*, erit reliquum temporis, nempe *bc*. Idem ergo tempus eſt lationis per *ab* ex quiete in *a*, & per *bd* poſt caſum *ab*.

In hac propoſitione re ipſa demonſtrantur duo Theorematæ Galilei,[82] *De Motu accelerato: ſed quia valde ad rem nostram faciunt, eadem diuerſa iterum ratione contemplabimur, vt lucem ſequenti Corollario præferant.*

Si graue naturaliter cadat ex a *in* b. *& ex* b *cum impetu concepto, per quodlibet planum* bc *conuertatur. Quæritur quantum ſpatij per planum* bc *abſoluat mobile tempore caſus* ab.

Fiat circa diametrum ab. *circulus* adb, *centroq;* b, *& interuallo* ba *circulus* ac. *Dico graue delapſum per* ab, *ſi ex puncto* b *cum impetu concepto conuertatur per planum inclinatum* bc, *tempore æquali tempori caſus, percurrere ſpatium æquale vtriſq; ſimul* bc, cd.

Si enim poſt caſum ab *graue conuerteretur per planum quodcunq;* bc, *motuq; æquabili procederet, graue huiuſmodi per planum* bc *tempore æquali tempori caſus ſpatium perageret duplũ ipſius* ba, *ergo percurreret ſpatium duplum ipſius bc. tempore caſus, ſi post caſum æquabili motu procederet. Sed ſuperueniẽte operatione grauitatis, mobile non procedet motu æquabili ſuper plano* bc; *Quin immò tempore caſus* ab, *grauitas promouebit mobile ſuper plano* bc *tantùm ſpatium quanta est inclusa in circulo linea* db *(quo enim tempore grauitas trahit mobile ex* a *in* b. *eodem tempore trahit etiam ex* d *in* b *per 6. Galilei de Motu Accelerato.) Ergo duplæ* bc *in prima figura addenda erit* db, *conſpirant enim deorſum tam motus æquabilis, quam motus grauitatis; In ſecunda vero figura à dupla* bc *detrahenda erit* db *(quia motus grauitatis contrarius eſt motui æquabili) & ſic* *ex Gal.*

T *con-*

[82] Cfr. Pisano and Bussotti's chapters above.

concludemus mobile post conuersionem, tempore æquali tempori casus a b, *percurrere spatia* b c, c d. *in vtraq; figura*.

Corollarium.

Pro Corollario animaduertimus quod si graue aliquod ex puncto a *impellatur motu velocissimo per horizontalem* a b, *peragatq; certo aliquo tempore spatium* a b; *atque eodem tempore quo fit latio* a b *grauitas motu naturali deorsum trahat per tantum spatij quãta est* a c. *si centro* a *interuallo* a b *fiat circulus* b d e f, *& circa diametrum* a c *alius circulus* a h i c, *mobile impulsum ab eodem semper impetu per plana* a d, a e, a f, *eodem tempore peraget singulas interceptas* c f, i e, h d, a b. *&c. etiam infra horizontem*.

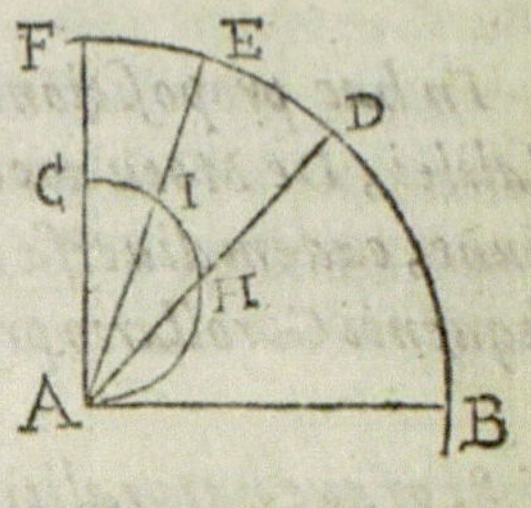

PROPOSITIO L.

SI duo grauia demittantur eodem temporis momento ex diuersis plani eleuati punctis, & post casum per eandem horizontalem lineam conuertantur; grauia in quodam puncto simul conuenient, quod in horizontali tantum distat à plano eleuato quanta est dupla mediæ proportionalis inter altitudines casuum.

Sit planum eleuatum *ab*. in quo sumantur duo quælibet puncta *a, c*. ex quibus duo grauia demittantur eodem simul tempore. Sit autem *b e* media inter *ab*, *bc*. & ipsius *be*. dupla sit horizontalis *bd*. Dico grauia eod. tẽporis puncto demissa ex *a*. & *c*. in puncto *d*. conuenire.

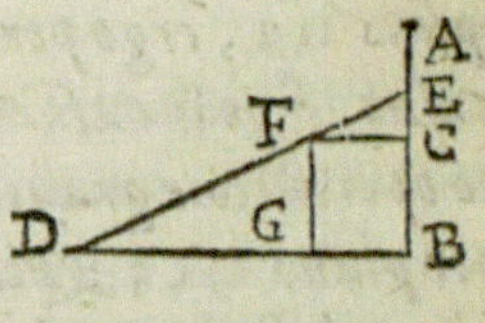

Iungatur enim *e d*. & compleatur parallelogrammum *b c f g*. Cumq; sit *b d*. dupla ipsius *b e*, erit *f c*. hoc est *g b*. dupla *c e*. Iam sic. Mobile post casum *c b*. suo impetu currit horizontaliter tempore casus *c b* duplam *c b*. ergo tempore *e c*. curret eodem impetu duplam *e c*. hoc est ipsam *b g*. Tempore igitur inte-

concludemus mobile poft conuerfionem, tempore æquali tempori cafus ab, *percurrere fpatia* bc, cd. *in vtraq; figura.*

Corollarium.

Pro Corollario animaduertimus quod fi graue aliquod ex puncto a *impellatur motu velociffimo per horizontalem* ab, *peragatq; certo aliquo tempore fpatium* ab; *atque eodem tempore quo fit latio* ab *grauitas motu naturali deorfum trahat per tantum fpatij quãta eft* ac. *fi centro* a *interuallo* ab *fiat circulus* bdef, *& circa diametrum* ac *alius circulus* ahic, *mobile impulfum ab eodem femper impetu per plana* ad, ae, af, *eodem tempore peraget fingulas interceptas* cf, ie, hd, ab. &c. *etiam infra horizontem.*

PROPOSITIO L.

SI duo grauia demittantur eodem temporis momento ex diuerfis plani eleuati punctis, & poft cafum per eandem horizontalem lineam conuertantur; grauia in quodam puncto fimul conuenient, quod in horizontali tantum diftat à plano eleuato quanta eft dupla medię proportionalis inter altitudines cafuum.

Sit planum eleuatum *ab.* in quo fumantur duo quælibet puncta *a, c.* ex quibus duo grauia demittantur eodem fimul tempore. Sit autem *be* media inter *ab, bc.* & ipfius *be.* dupla fit horizontalis *bd.* Dico grauia eod. tẽporis puncto demiffa ex *a.* & *c.* in puncto *d.* conuenire.

Iungatur enim *ed.* & compleatur parallelogrammum *bcfg.* Cumq; fit *bd.* dupla ipfius *be,* erit *fc.* hoc eft *gb.* dupla *ce.* Iam fic. Mobile poft cafum *cb.* fuo impetu currit horizontaliter tempore cafus *cb* duplam *cb.* ergo tempore *ec.* curret eodem impetu duplam *ec.* hoc eft ipfam *bg.* Tempore igitur

inte-

integro *e b* fiunt lationes per *c b*, & *b g*, & eodem tempore fit casus per *a b*, quare eodẽ tẽporis momento erunt grauia alterũ quidem in *g*; alterum autem in *b*. Sed reliqua etiam spatia *b d*, *g d*. ęqualibus temporibus peraguntur. (Velocitates enim sunt vt tempora casuum, hoc est vt *e b*. ad *f g*. sed spatia *b d*, *g d* ob similitudinem triangulorum sunt vt velocitates, quare vti dictum est ęqualibus temporibus peragentur.)

Sunt ergo coniunctim tempora per *a b*, *b d*. ęqualia temporibus coniunctim per *c b*, *b d*. Quare duo grauia &c. conuenient in puncto *d*. Quod erat &c.

Idem aliter demonstrabimus.

Sumptis vtcunq; altitudinibus a c b c. *Demittantur duo grauia eodem tempore ex* a, *&* b. *Sitq;* d c. *media inter casus, cuius dupla ponatur horizontalis* c e. *Dico tempora lationum* a c e. b c e. *æqualia esse*.

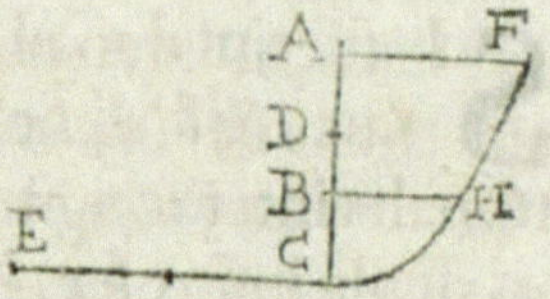

Fiat circa diametrum c a *parabola quęcunque, quæ verticem habeat in* c. *ducanturq; ordinatim* a f. b h. *Notum est in parabola ita esse* a f. *ad* b h. *vt est* a c *ad* c d, *vel vt* c d, *ad* c b.

Iam. Tempus per a c. *est* a f, *& per* c e. *post casum* a c. *est* b h. (*si enim tempore* f a. *graue currit duplam* a c, *tempore* b h, *curret duplam* d c *hoc est ipsam* c e *cum sint proportionales* f a *ad* a c, *vt* h b, *ad* d c.) *Eodem modo. Tempus per* b c. *est* b h, *& per* c e. *post casum* b c. *est* a f. (*si .n. tempore* b h. *currit duplàm* b c, *tempore* a f. *curret duplam* d c. *hoc est ipsam* c e. *quia sunt proportionales vt* b h. *ad* b c. *ita* a f *ad* d c.) *Ergo tempora lationum* a c e. *sunt lineæ* a f. b h. *Tempora autem lationum* b c e. *sunt lineæ* b h, a f. *Quare coniunctim tempora per* a c, c e. *æqualia sunt temporibus per* b c. c e. *coniunctim. Quod erat &c.*

Corollarium Primum.

Hinc manifestum est dato quolibet spatio horizontali *a c.*

integro *eb* fiunt lationes per *cb,* & *bg,* & eodem tempore fit cafus per *ab,* quare eodẽ tẽporis momento erunt grauia alterũ quidem in *g*; alterum autem in *b.* Sed reliqua etiam ſpatia *bd, gd.* ęqualibus temporibus peraguntur. (Velocitates enim ſunt vt tempora caſuum, hoc eft vt *eb.* ad *fg.* ſed ſpatia *bd, gd* ob ſimilitudinem triangulorum ſunt vt velocitates, quare vti dictum eft ęqualibus temporibus peragentur.)

Sunt ergo coniunctim tempora per *ab, bd.* ęqualia temporibus coniunctim per *cb, bd.* Quare duo grauia &c. conuenient in puncto *d.* Quod erat &c.

Idem aliter demonſtrabimus.

Sumptis vtcunq; altitudinibus ac bc. *Demittantur duo grauia eodem tempore ex* a, & b. *Sitq;* dc. *media inter caſus, cuius dupla ponatur horizontalis* ce. *Dico tempora lationum* ace. bce. *æqualia eſſe.*

Fiat circa diametrum c a *parabola quęcunque, quæ verticem habeat in* c. *ducanturq; ordinatim* af. bh. *Notum eſt in parabola ita eſſe* af. *ad* bh. *vt eſt ac* ad *cd, vel vt cd,* ad *cb.*

Iam. Tempus per ac. *eſt af, & per* ce. *post caſum* ac. *eſt* bh. *(ſi enim tempore* fa. *graue currit duplam* ac, *tempore* bh, *curret duplam* dc *hoc eſt ipſam* ce *cum ſint proportionales* fa *ad* ac, *vt* hb, *ad* dc.*) Eodem modo. Tempus per* bc. *eſt* bh, *& per* ce. *poſt caſum* bc. *eſt* af. *(ſi. n. tempore* bh. *currit duplàm* bc, *tempore* af. *curret duplam* dc. *hoc eſt ipſam* ce. *quia ſunt proportionales vt* bh. *ad* bc. *ita* af *ad* dc.*) Ergo tempora lationum* ace. *ſunt lineæ* af. bh. *Tempora autem lationum* bce. *ſunt lineæ* bh, af. *Quare coniunctim tempora per* ac, ce. *æqualia ſunt temporibus per* bc. ce. *coniunctim. Quod erat &c.*

Corollarium Primum.

Hinc manifeftum eft dato quolibet ſpatio horizontali *ac.*

T 2 cuius

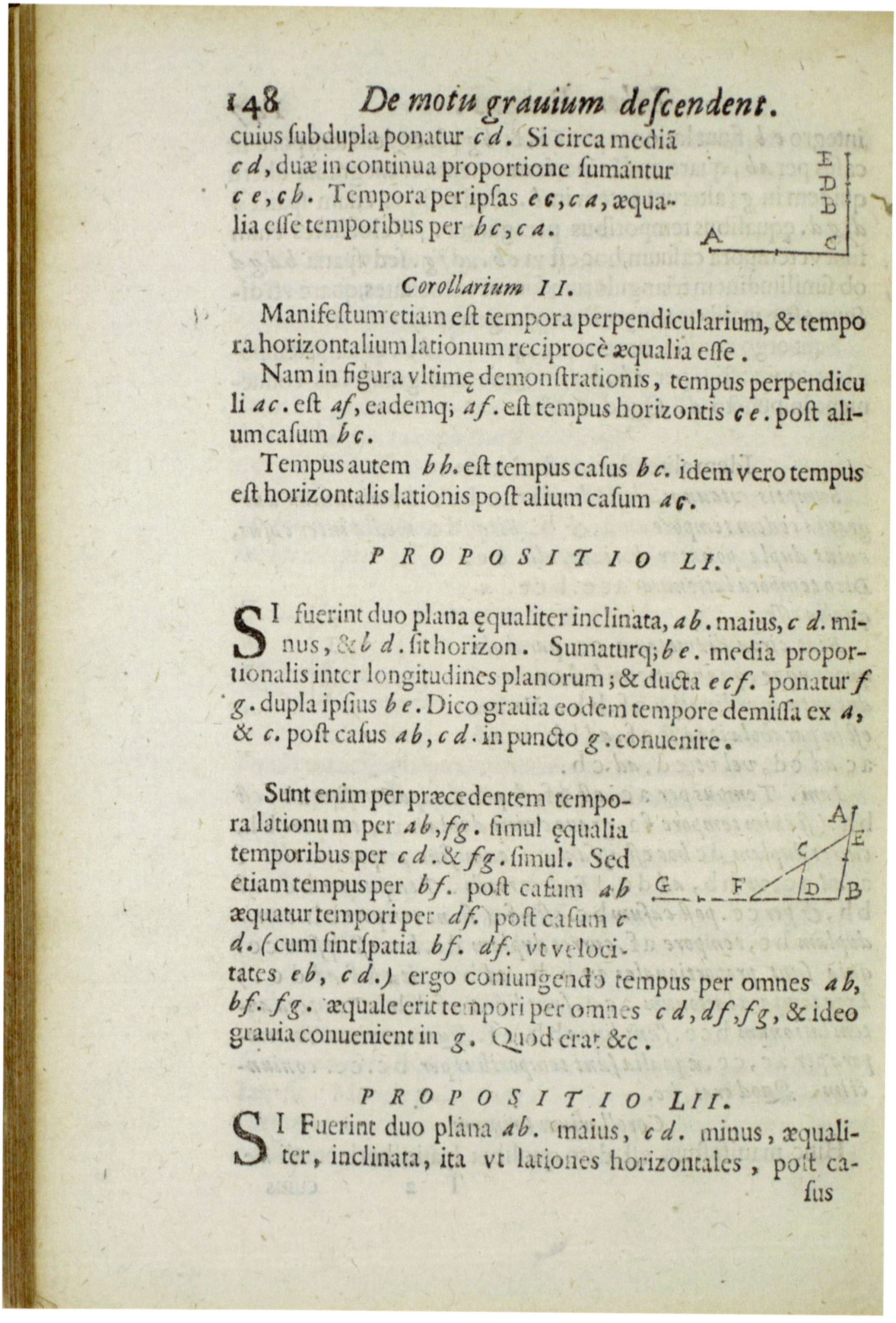

148 De motu grauium descendent.
cuius subdupla ponatur *c d*. Si circa mediã *c d*, duæ in continua proportione sumantur *c e*, *c b*. Tempora per ipsas *e c*, *c a*, æqualia esse temporibus per *b c*, *c a*.

Corollarium II.
Manifestum etiam est tempora perpendicularium, & tempora horizontalium lationum reciprocè æqualia esse.
Nam in figura vltimę demonstrationis, tempus perpendiculi *ac*. est *af*, eademq; *af*. est tempus horizontis *c e*. post alium casum *b c*.
Tempus autem *b h*. est tempus casus *b c*. idem vero tempus est horizontalis lationis post alium casum *a c*.

PROPOSITIO LI.
SI fuerint duo plana ęqualiter inclinata, *a b*. maius, *c d*. minus, & *b d*. sit horizon. Sumaturq; *b e*. media proportionalis inter longitudines planorum; & ducta *e c f*. ponatur *f g*. dupla ipsius *b e*. Dico grauia eodem tempore demissa ex *a*, & *c*. post casus *a b*, *c d*. in puncto *g*. conuenire.
Sunt enim per præcedentem tempora lationum per *a b*, *f g*. simul ęqualia temporibus per *c d*. & *f g*. simul. Sed etiam tempus per *b f*. post casum *a b* æquatur tempori per *d f*. post casum *c d*. (cum sint spatia *b f*. *d f*. vt velocitates *e b*, *c d*.) ergo coniungendo tempus per omnes *a b*, *b f*. *f g*. æquale erit tempori per omnes *c d*, *d f*, *f g*, & ideo grauia conuenient in *g*. Quod erat &c.

PROPOSITIO LII.
SI Fuerint duo plana *ab*. maius, *c d*. minus, æqualiter, inclinata, ita vt lationes horizontales, post casus

cuius fubdupla ponatur *cd.* Si circa mediã *cd,* duæ in continua proportione fumantur *ce, cb.* Tempora per ipfas *ec, ca,* æqualia effe temporibus per *bc, ca.*

Corollarium II.

Manifeftum etiam eft tempora perpendicularium, & tempora horizontalium lationum reciprocè æqualia effe.

Nam in figura vltimę demonftrationis, tempus perpendiculi *ac.* eft *af,* eademq; *af.* eft tempus horizontis *ce.* poft alium cafum *bc.* Tempus autem *bh.* eft tempus cafus *bc.* idem vero tempus eft horizontalis lationis poft alium cafum *ac.*

PROPOSITIO LI.

SI fuerint duo plana ęqualiter inclinata, *ab.* maius, *cd.* minus, & *bd.* fit horizon. Sumaturq; *be.* media proportionalis inter longitudines planorum; & ducta *ecf.* ponatur *fg.* dupla ipfius *be.* Dico grauia eodem tempore demiffa ex *a,* & *c.* poft cafus *ab, cd.* in puncto *g.* conuenire.

Sunt enim per præcedentem tempora lationum per *ab, fg.* fimul ęqualia temporibus per *cd.* & *fg.* fimul. Sed etiam tempus per *bf.* poft cafum *ab* æquatur tempori per *df.* poft cafum *cd.* (cum fint fpatia *bf. df.* vt velocitates *eb, cd.*) ergo coniungendo tempus per omnes *ab, bf. fg.* æquale erit tempori per omnes *cd, df, fg,* & ideo grauia conuenient in *g.* Quod erat &c.

PROPOSITIO LII.

SI Fuerint duo plana *ab.* maius, *cd.* minus, æqualiter, inclinata, ita vt lationes horizontales, poft ca-

fus

[83] *motu grauium*] *motu Grauium.*

ſus in prima figura contrariȩ inuicem ſint;in ſecunda verò ad eaſdem partes. Sumaturq; media proportionalis inter longitudines planorum *be*, & ſit *df* dupla *ge*, hoc eſt differentiȩ inter mediam *be*, & minus planum;ducta deinde *fh*, æquali ipſi *cd*, & parallela ad *ab*, iungatur *eh*. quæ ſecet horizontem in *i*

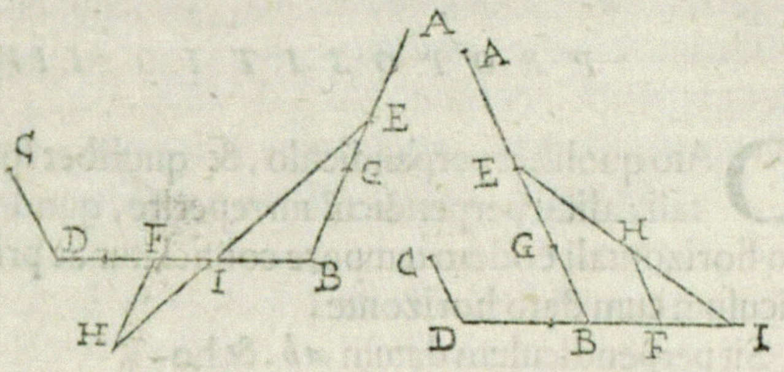

Dico grauia ex *a*.& *c*.eodem tẽpore demiſſa, ſi verſus *i* conuertantur in pũcto *i* conuenire.

Ponamus tempus per *cd*. eſſe *cd*. vel *bg*. ſibi æqualem. Ergo in horizonte graue *c*. tempore *bg*. curret duplam *bg*, & tempore *ge*. curret duplam *ge*, nempe *df*. Eſt itaq; tota *eb*. tempus per *cdf*. Eadem quoq; *eb*. tempus eſt per *ab*. Quare eodem tempore peragentur *cdf*. & *ab*. Reliquæ autem *fi*.*bi*. eodem tempore peraguntur (cum propter ſimilitudinem triangulorum ſpatia *fi*, *bi*, ſint vt velocitates *hf*, *eb*.) ergo coniunctim idem tempus erit tam per *cdi*, quam per *abi*. Quare grauia conuenient in *i*. Quod erat &c.

In ſecunda figura non debent lationes horizontales eſſe contrariæ, nam grauia nunquam conuenirent: ſed ambȩ verſus partes i.

PROPOSITIO LIII.

DAtis duobus perpendiculis *ab*, *cb*, inuenire ſpatium horizontale quod cum alterutro datorum perpendiculorũ eodem tempore conficiatur.

Ponatur *bd*. æqualis ipſi *bc*, & circa *ad*. fiat ſemicirculus: ponaturq; horizontalis *be*. dupla ipſius *bf*. Dico latio nes *a*

be,

ſus in prima figura contrarię inuicem ſint; in ſecunda verò ad eaſdem partes. Sumaturq; media proportionalis inter longitudines planorum *be,* & ſit *df* dupla *ge,* hoc eſt differentię inter mediam *be,* & minus planum; ducta deinde *fh,* æquali ipſi *cd,* & parallela ad *ab,* iungatur *eh.* quæ ſecet horizontem in *i*[84]

Dico grauia ex *a.* & *c.* eodem tẽpore demiſſa, ſi verſus *i* conuertantur in pũcto *i* conuenire.

Ponamus tempus per *cd.* eſſe *cd.* vel *bg.* ſibi æqualem. Ergo in horizonte graue *c.* tempore *bg.* curret duplam *bg,* & tempore *ge.* curret duplam *ge,* nempe *df.* Eſt itaq; tota *eb.* tempus per *cdf.* Eadem quoq; *eb.* tempus eſt per *ab.* Quare eodem tempore peragentur *cdf.* & *ab.* Reliquæ autem *fi. bi.* eodem tempore peraguntur (cum propter ſimilitudinem triangulorum ſpatia *fi, bi,* ſint vt velocitates *hf, eb.*) ergo coniunctim idem tempus erit tam per *cdi,* quam per *abi.* Quare grauia conuenient in *i.* Quod erat &c.

In ſecunda figura non debent lationes horizontales eſſe contrariæ, nam grauia nunquam conuenirent: ſed ambę verſus partes i.

PROPOSITIO LIII.

DAtis duobus perpendiculis *ab, cb,* inuenire ſpatium horizontale quod cum alterutro datorum perpendiculorũ eodem tempore conficiatur.

Ponatur *bd.* æqualis ipſi *bc,* & circa *ad.* fiat ſemicirculus: ponaturq; horizontalis *be.* dupla ipſius *bf.* Dico lationes *a*

be,

[84] *i*] *i.*

be, *cbe*. eodem tempore absolui.

Hoc enim patet per Corollarium primũ Propositionis 51. Nam altitudines perpendiculares *ab*, *bc*. sunt continuæ proportionales circa *bf*. semissem spatij horizontalis. Quare factum est quod &c.

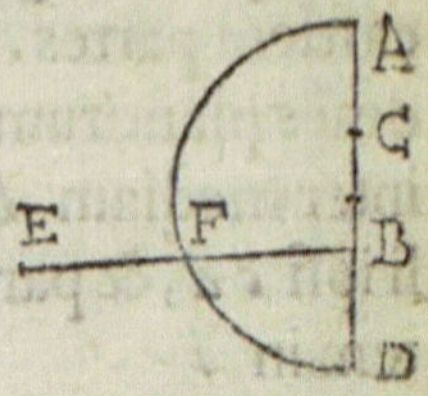

PROPOSITIO LIV.

DAto quolibet perpendiculo, & quolibet spatio horizontali; aliud perpendiculum reperire, quod cum dato spatio horizontali eodem tempore conficiatur ac primum perpendiculum cum dato horizonte.

Sit perpendiculum datum *ab*. & horizon *bc*. cuius semissis sit *bd*. Iungatur *ad*. fiatq; angulus *bdf*. ad horizontẽ æqualis angulo *bad*, qui est ad perpendiculum. Dico tempora lationum per *fb*. *bc* simul, & per *ab*, *bc*. simul, æqualia esse. Triangula enim rectangula *fbd*, *dab*. facta sunt æquiangula. Quare vt *fb*. ad *bd*. ita *bd*. ad *ba*. Et cum *bd*. semissis horizontalis spatij media sit proportionalis inter perpendicula *fb*. *ab*. erunt tempora per *fbc*, & per *abc*. æqualia. Quod erat &c.

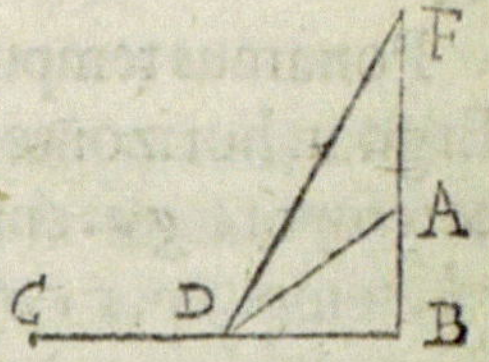

PROPOSITIO LV.

Gal. 19. de motu Accel

SI fuerit horizontalis *ab*. dupla perpendiculi *ae*, Dico ipsas *eab*. post casum *ea* breuiori tempore percurri, quã aliud quodcunq; perpendiculum cum eodem spatio horizontali *ab*.

Erigatur *bd*. perpendicularis ad *ab*, & per *e*, *b*. puncta, circa diametrum *ea* agatur parabola, *ecb*. cuius focus erit *a*. (po-

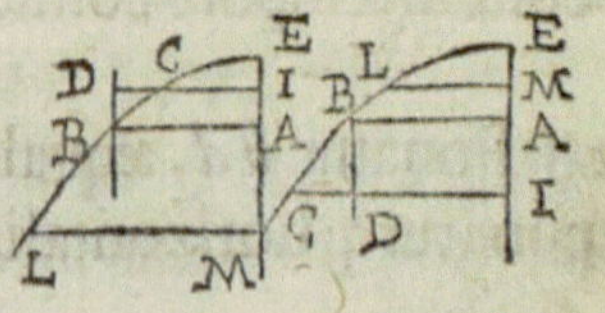

be, cbe. eodem tempore abſolui.

Hoc enim patet per Corollarium primũ Propoſitionis 51. Nam altitudines perpendiculares *ab, bc.* ſunt continuæ proportionales circa *bf.* ſemiſſem ſpatij horizontalis. Quare factum eſt quod &c.

PROPOSITIO LIV.

DAto quolibet perpendiculo, & quolibet ſpatio horizontali; aliud perpendiculum reperire, quod cum dato ſpatio horizontali eodem tempore conficiatur ac primum perpendiculum cum dato horizonte.

Sit perpendiculum datum *ab.* & horizon *bc.* cuius ſemiſſis ſit *bd.* Iungatur *ad.* fiatq; angulus *bdf.* ad horizontẽ æqualis angulo *bad,* qui eſt ad perpendiculum. Dico tempora lationum per *fb. bc* ſimul, & per *ab, bc.* ſimul, æqualia eſſe. Triangula enim rectangula *fbd, dab.* facta ſunt æquiangula. Quare vt *fb.* ad *bd.* ita *bd.* ad *ba.* Et cum *bd.* ſemiſſis horizontalis ſpatij media ſit proportionalis inter perpendicula *fb. ab.* erunt tempora per *fbc,* & per *abc.* æqualia. Quod erat &c.

PROPOSITIO LV.

Gal. 29. *de motu Accel*

SI fuerit horizontalis *ab.* dupla perpendiculi *ae,* Dico ipſas *eab.* poſt caſum *ea* breuiori tempore percurri, quã aliud quodcunq; perpendiculum cum eodem ſpatio horizontali *ab.*

Erigatur *bd.* perpendicularis ad *ab,* & per *e, b.* puncta, circa diametrum *ea* agatur parabola, *ecb.* cuius focus erit

a. (po-

a. (posita enim est *ab*. dupla ipsius *ae*.) Sumatur iam quodlibet aliud perpendiculum *ei*, & ducatur horizontalis *id*.

Tempus per *ea*. est *ab*. tempus autem per *ab*. est idem ac tempus casus, ergo tempus per *eab*. est ipsa *ab* bis sumpta. Sed tempus per *ei*. est *ic*. tempus autem per *id*. quantum sit, sic venabimur. Velocitate *ab*. tempore *ab*. curritur *ab*. Sed velocitate *ci*. tempore *ab*. non curretur eadem *ab*. Fiat igitur vt velocitas *ci*. ad velocitatem *ab*, ita tempus *ab*. ad aliud *ml*. Et erit *ml*. tempus ipsius *di*. post casum *ei*. Patet ergo *ml*. *ci*. primam & tertiam proportionalium, maiores esse quam dupla mediæ, hoc est quam *ab*. bis sumpta. Quare &c.

PROPOSITIO LVI.

SI *ab*. horizontalis dupla fuerit eleuatę *ae*. Dico, quò longius à puncto *e*. demittatur graue, eo tardius lationem suam vsq; in *b* absoluere.

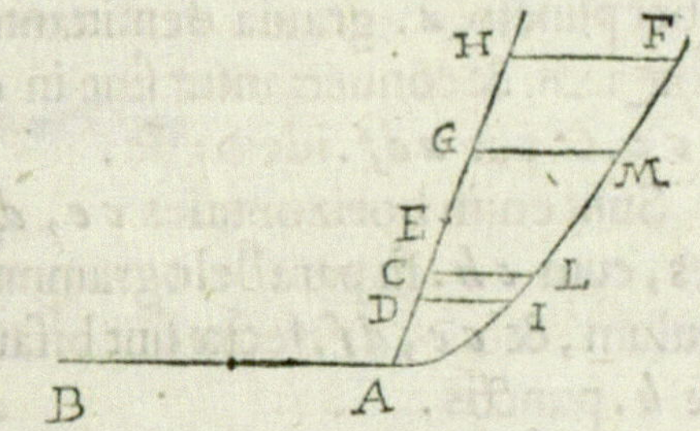

Demittātur ex punctis *c*, & *d*. Duo grauia; ostendendum est maiori tempore fieri lationem per *dab*. quam per *cab*. Fiat circa diametrum *ae*, parabola *amf*. & ipsis *ca*. *ae*. sit tertia proportionalis *ag*. Ipsis autem *da*, *ae*. tertia sit *ah*, & ducantur ordinatim lineæ ex punctis *d*, *c*, *g*, *h*.

Quia quadrato eidem *ae*, æquale est vtrumq; rectangulum *dah*, *cag*. erunt hęc eadem rectangula æqualia inter se; propterea latera reciprocè proportionalia habebunt. nempe vt *ha* ad *ag*. ita *ca*, ad *ad*. Sed in hac eadem proportione ob parabolam, sunt quadrata, *hf*. ad *gm*, & *cl* ad *di*, ergo proportionalia sunt etiam latera, nempe vt *hf*. ad *gm*. ita *cl* ad *di*. extremæ autem *hf*, *di* maiores sunt quam mediæ *gm*, *cl*; & extremæ simul sunt tempus lationis *dab*, at mediæ sunt tempus lationis *cab*. Quare tardius absoluetur latio per *dab*, quam per *cab*. Quod erat &c. *vide 2. demōstrationem Prop. 50*

Idem

a. (pofita enim eft *ab.* dupla ipfius *ae.*) Sumatur iam quodlibet aliud perpendiculum *ei,* & ducatur horizontalis *id.*

Tempus per *ea.* eft *ab.* tempus autem per *ab.* eft idem ac tempus cafus, ergo tempus per *eab.* eft ipfa *ab* bis fumpta. Sed tempus per *ei.* eft *ic.* tempus autem per *id.* quantum fit, fic venabimur. Velocitate *ab.* tempore *ab.* curritur *ab.* Sed velocitate *ci.* tempore *ab.* non curretur eadem *ab.* Fiat igitur vt velocitas *ci.* ad velocitatem *ab,* ita tempus *ab.* ad aliud *ml.* Et erit *ml.* tempus ipfius *di.* poft cafum *ei.* Patet ergo *ml. ci.* primam & tertiam proportionalium, maiores effe quam dupla mediæ, hoc eft quam *ab.* bis fumpta. Quare &c.

PROPOSITIO LVI.

SI *ab.* horizontalis dupla fuerit eleuatę *ae.* Dico, quò longius à puncto *e.* demittatur graue, eo tardius lationem fuam vfq; in *b* abfoluere.

Demittãtur ex punctis *c,* & *d.* Duo grauia; oftendendum eft maiori tempore fieri lationem per *dab.* quam per *cab.* Fiat circa diametrum *ae,* parabola *amf.* & ipfis *ca. ae.* fit tertia proportionalis *ag.* Ipfis autem *da, ae.* tertia fit *ah,* & ducantur ordinatim lineæ ex punctis *d, c, g, h.*

Quia quadrato eidem *ae,* æquale eft vtrumq; rectangulum *dah, cag.* erunt hęc eadem rectangula æqualia inter fe; propterea latera reciprocè proportionalia habebunt. nempe vt *ha* ad *ag.* ita *ca,* ad *ad.* Sed in hac eadem proportione ob parabolam, funt quadrata, *hf.* ad *gm,* & *cl* ad *di,* ergo proportionalia funt etiam latera, nempe vt *hf.* ad *gm.* ita *cl* ad *di.* extremæ autem *hf, di* maiores funt quam mediæ *gm, cl*; & extremæ fimul funt tempus lationis *dab,* at mediæ funt tempus lationis *cab.* Quare tardius abfoluetur latio per *dab,* quam per *cab.* Quod erat &c.

vide 2 demoftra[-]*tionem Prop.* 50

Idem

Idem infertur etiam de punctis *g*, & *h*, ſupra ipſum *e*. ſumptis. Sunt enim tempora eorum, æqualia temporibus punctorum *c*. & *d*. vtrumque vtriq; &c.

vide eandem.

PROPOSITIO LVII.

SI ab aliquo puncto lineæ circulum tangentis in puncto ſublimi, grauia cadant in periphæriam & inde per chordas horizontales conuertantur. erunt tempora lationum per vtramque chordam & eius perpendiculum, æqualia.

Tangat linea *ab*, circulum erectum, in puncto ſublimi *b*. Et ſic tangens horizontalis omnino erit. Sumpto deinde quolibet puncto *a*. grauia demittantur perpendiculariter in periphæriam, & conuertantur ſiue in *c*, ſiue in *d*. Dico tempus per *ace*. & per *adf*. idem eſſe.

Sunt enim horizontales *ce*, *df*. æquales, cum *ch*. ſit parallelogrammum rectãgulum, & *ce*, *df*. ſectæ ſint bifariam in *g* & *h*. punctis.

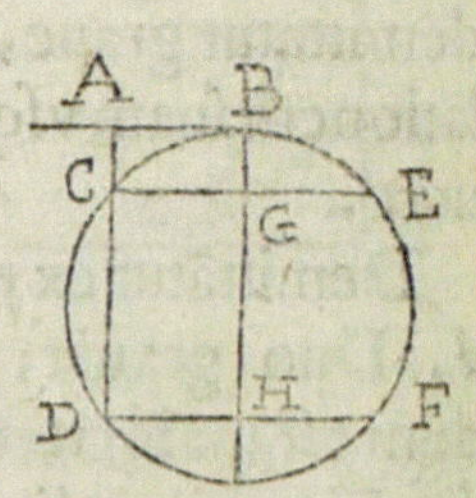

Quia ergo *ab*. ſemiſſis ſpatij horizontalis media proportionalis eſt inter altitudines perpendiculares *ac*, *ad*. (linea enim *ab*. tangit, & *ad*. circulum ſecat) erunt per Corollarium primum Propoſitionis 50. huius, tempora lationum *ace*, *adf*. æqualia. Quod erat &c.

Et per ſecundum eiuſdem Propoſitionis Corollarium eadem tempora reciprocè ęqualia ſunt &c.

PROPOSITIO LVIII.

TEmpus per axem parabolæ, & eius ordinatim applicatam ſimul, æquale eſt tempori per quartam lateris recti partem, & eandem ordinatim applicatam.

Sit

vide ean dem. Idem infertur etiam de punctis *g,* & *h,* ſupra ipſum *e.* ſumptis.
Sunt enim tempora eorum, æqualia temporibus punctorum *c.* & *d.* vtrumque vtriq; &c.

PROPOSITIO LVII.

SI ab aliquo puncto lineæ circulum tangentis in puncto ſublimi, grauia cadant in periphæriam & inde per chordas horizontales conuertantur. erunt tempora lationum per vtramque chordam & eius perpendiculum, æqualia.

Tangat linea *ab,* circulum erectum, in puncto ſublimi *b.* Et ſic tangens horizontalis omnino erit. Sumpto deinde quolibet puncto *a.* grauia demittantur perpendiculariter in periphæriam, & conuertantur ſiue in *c,* ſiue in *d.* Dico tempus per *ace.* & per *adf.* idem eſſe.

Sunt enim horizontales *ce, df.* æquales, cum *ch.* ſit parallelogrammum rectãgulum, & *ce, df.* ſectæ ſint bifariam in *g* & *h.* punctis.

Quia ergo *ab.* ſemiſſis ſpatij horizontalis media proportionalis eſt inter altitudines perpendiculares *ac, ad.* (linea enim *ab.* tangit, & *ad.* circulum ſecat) erunt per Corollarium primum Propoſitionis 50. huius, tempora lationum *ace, adf.* æqualia. Quod erat &c.

Et per ſecundum eiuſdem Propoſitionis Corollarium eadem tempora reciprocè ęqualia ſunt &c.

PROPOSITIO LVIII.

TEmpus per axem parabolæ, & eius ordinatim applicatam ſimul, æquale eſt tempori per quartam lateris recti partem, & eandem ordinatim applicatam.

Sit

[85] *motu grauium*] *motu Grauium.*

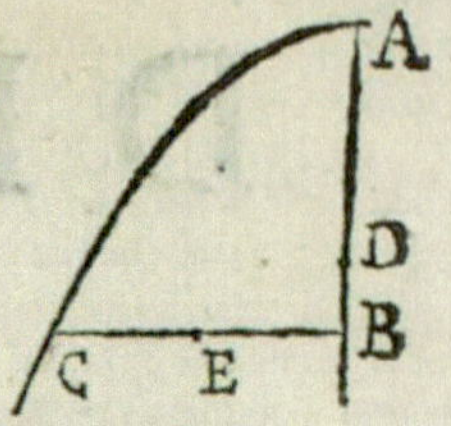

Sit axis parabolæ, *ab*, eius ordinatim applicata *bc*. Et ſecetur *bd*. æqualis quartæ parti lateris recti. Dico tempora per *abc*. & per *dbc*. æqualia eſſe inter ſe. Diuidatur *bc*. bifariam in *e*.

Erunt tum quadratum *eb*, tum rectangulum *abd*, ſubquadrupla quadrati *cb*, Sunt ideo æqualia inter ſe, & ipſa *eb*. media proportionalis eſt inter *ab*, *bd*. Quare per Corollarium primum Propoſitionis 50. huius, tempora per *abc*. & per *dbc* æqualia ſunt &c.

Sunt etiam per ſecundum eiuſdem Propoſitionis Corollarium, reciprocè æqualia. Quod ſatis ſit oſtendiſſe circa motum grauium naturaliter deſcendentium.

Finis Primi Libri.

V DE MO-

Sit axis parabolæ, *ab,* eius ordinatim applicata *bc*. Et ſecetur *bd.* æqualis quartæ parti lateris recti. Dico tempora per *abc.* & per *dbc.* æqualia eſſe inter ſe. Diuidatur *bc.* bifariam in *e.*

Erunt tum quadratum *eb,* tum rectangulum *abd,* ſubquadrupla quadrati *cb,* Sunt ideo æqualia inter ſe, & ipſa *eb.* media proportionalis eſt inter *ab, bd.* Quare per Corollarium primum Propoſitionis 50. huius, tempora per *abc.* & per *dbc* æqualia ſunt &c.

Sunt etiam per ſecundum eiuſdem Propoſitionis Corollarium, reciprocè æqualia. Quod ſatis ſit oſtendiſſe circa motum grauium naturaliter deſcendentium.

Finis Primi Libri.

V DE MO

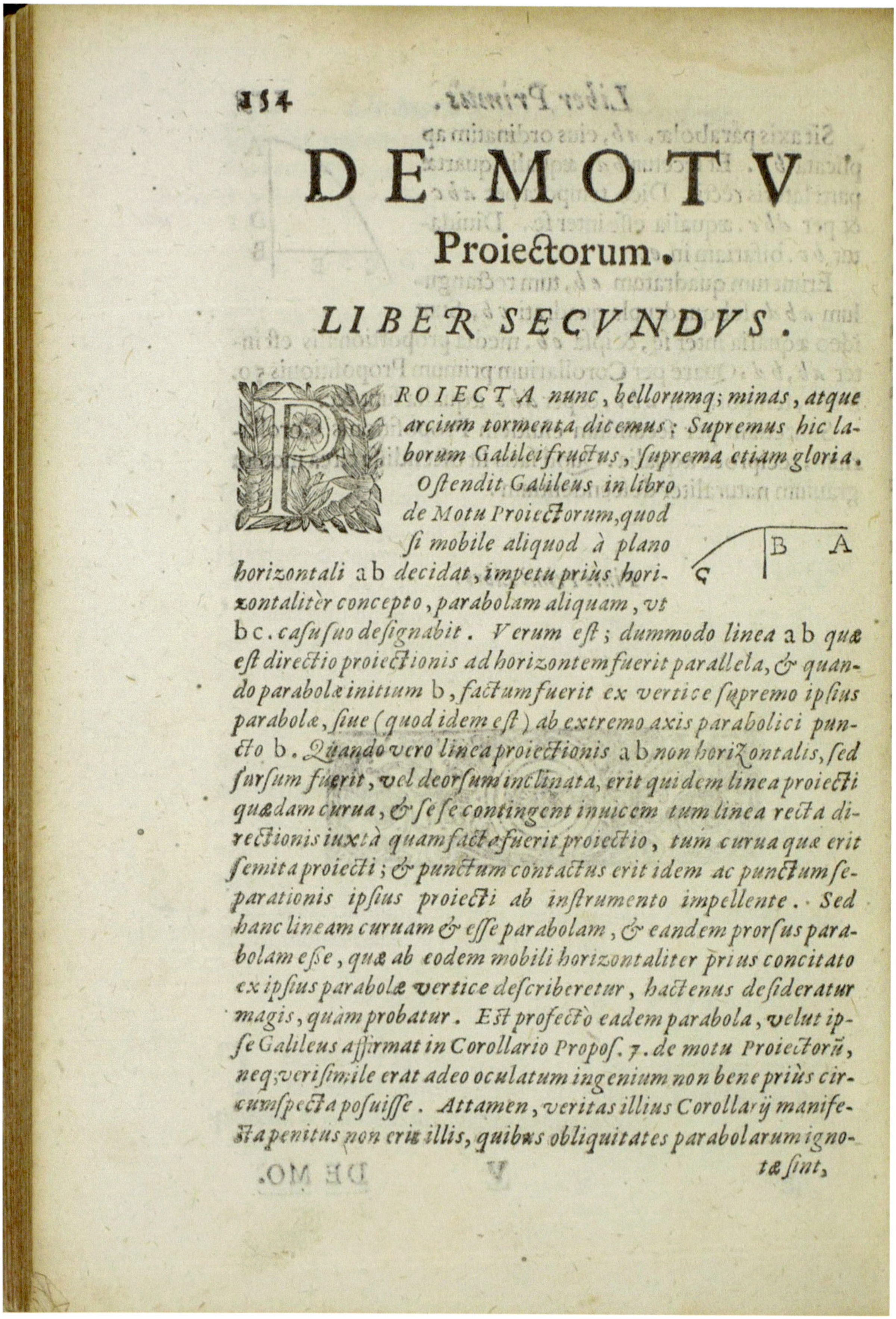

154

DE MOTV
Proiectorum.
LIBER SECVNDVS.

PROIECTA nunc, bellorumq; minas, atque arcium tormenta dicemus; Supremus hic laborum Galilei fructus, suprema etiam gloria.

Ostendit Galileus in libro de Motu Proiectorum, quod si mobile aliquod à plano horizontali ab *decidat, impetu priùs horizontalitèr concepto, parabolam aliquam, vt* bc. *casu suo designabit. Verum est; dummodo linea* ab *quæ est directio proiectionis ad horizontem fuerit parallela, & quando parabolæ initium* b, *factum fuerit ex vertice supremo ipsius parabolæ, siue (quod idem est) ab extremo axis parabolici puncto* b. *Quando vero linea proiectionis* ab *non horizontalis, sed sursum fuerit, vel deorsum inclinata, erit quidem linea proiecti quædam curua, & sese contingent inuicem tum linea recta directionis iuxtà quam facta fuerit proiectio, tum curua quæ erit semita proiecti; & punctum contactus erit idem ac punctum separationis ipsius proiecti ab instrumento impellente. Sed hanc lineam curuam & esse parabolam, & eandem prorsus parabolam esse, quæ ab eodem mobili horizontaliter prius concitato ex ipsius parabolæ vertice describeretur, hactenus desideratur magis, quàm probatur. Est profectò eadem parabola, velut ipse Galileus affirmat in Corollario Propos. 7. de motu Proiectorũ, neq; verisimile erat adeo oculatum ingenium non bene priùs circumspecta posuisse. Attamen, veritas illius Corollarij manifesta penitus non erit illis, quibus obliquitates parabolarum ignotæ sint,*

DE MOTV

Proiectorum

LIBER SECVNDVS.

PROJECTA nunc, bellorumq; minas, atque arcium tormenta dicemus: Supremus hic laborum Galilei fructus, ſuprema etiam gloria. Oſtendit Galileus in libro de Motu Proiectorum, quod ſi mobile aliquod à plano horizontali ab *decidat, impetu priùs horizontaliter concepto, parabolam aliquam, vt* bc. *caſu ſuo deſignabit. Verum eſt; dummodo linea* ab *quæ eſt directio proiectionis ad horizontem fuerit parallela, & quando parabolæ initium* b, *factum fuerit ex vertice ſupremo ipſius parabolæ, ſiue (quod idem eſt) ab extremo axis parabolici puncto* b. *Quando vero linea proiectionis* ab *non horizontalis, ſed ſurſum fuerit, vel deorſum inclinata, erit quidem linea proiecti quædam curua, & ſeſe contingent inuicem tum linea recta directionis iuxtà quam facta fuerit proiectio, tum curua quæ erit ſemita proiecti; & punctum contactus erit idem ac punctum ſeparationis ipſius proiecti ab inſtrumento impellente. Sed hanc lineam curuam & eſſe parabolam, & eandem prorſus parabolam eßе, quæ ab eodem mobili horizontaliter prius concitato ex ipſius parabolæ vertice deſcriberetur, hactenus deſideratur magis, quàm probatur. Est profectò eadem parabola, velut ipſe Galileus affirmat in Corollario Propoſ. 7. de motu Proiectorũ, neq; veri ſimile erat adeo oculatum ingenium non bene priùs circumſpecta poſuiſſe. Attamen, veritas illius Corollarij manifesta penitus non erit illis, quibus obliquitates parabolarum igno-*

tæ sint,

tę sint, & vltra duas à Galileo præmissas propositiones Apollonius familiaris non fuerit. Cum itaq; proiectiones vt plurimum fiant per lineas ad horizontem inclinatas, ex quibus oriuntur parabolæ obliquæ, non habentes initium ex vertice, quales frequētissime occurrunt in omnibus ferè iactibus machinarum, immò etiam neq; verticem, neq; axem habentes, quales sunt proiectiones inclinatę deorsum, lucem Corollario Galilei afferre conabimur, & cuiusmodi sit linea curua proiectorum vniuersalius determinabimus.

Definitio.

Directio proiectionis dicitur linea recta quę tangit lineam curuam proiecti in primo puncto eiusdem lineæ curuæ. quæ quidem directio in tormentis bellicis est eadem ac ipsius machinæ axis.

PROPOSITIO PRIMA.

SI graue sursum proiectum ex *a.* ascendat motu naturalitèr Deficiente vsq; ad sublimius punctum suæ lationis *b.* idem verò mobile æquali tempore, & eadem velocitate quam in pūcto *a* habebat, sed motu æquabili ascendat vsq; in *c.* Dico *a c.* duplam esse ipsius *ab.*

Si enim non est dupla, ponamus aliquam *ad* duplam esse ipsius *ab.*

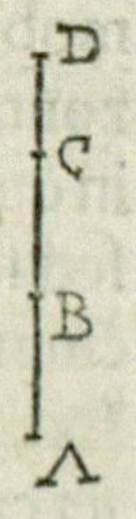

Concipiamus iam cadere naturalitèr mobile ex *b* in *a.* Gradus ille impetus acquisiti post casum ex *b* in *a.* ęst ille prorsus qui vehit mobile ad eandem altitudinem *b.* eodem tempore, & motu naturaliter deficiēte. Idem verò gradus impetus eodem tempore, sed motu æquabili perducit mobile ad altitudinem *ad* duplam casus *ba.* Sed ille idem impetus qui per suppositionem perducit mobile ex *a* in *b* motu naturaliter deficiente, illud perducebat etiam motu ęquabili eodemq; tempore ex *a* in *c.* Vnus ergò

Galil. in scholio prop. 23.

Galil. ib.

tę ſint, & vltra duas à Galileo præmiſſas propoſitiones Apollonius familiaris non fuerit. Cum itaq; proiectiones vt plurimum fiant per lineas ad horizontem inclinatas, ex quibus oriuntur parabolæ obliquæ, non habentes initium ex vertice, quales frequẽtiſſime occurrunt in omnibus ferè iactibus machinarum, immò etiam neq; verticem, neq; axem habentes, quales ſunt proiectiones inclinatę deorſum, lucem Corollario Galilei afferre conabimur, & cuiuſmodi ſit linea curua proiectorum vniuerſalius determinabimus.

Definitio.

Directio proiectionis dicitur linea recta quę tangit lineam curuam proiecti in primo puncto eiuſdem lineæ curuæ. quæ quidem directio in tormentis bellicis eſt eadem ac ipſius machinæ axis.

PROPOSITIO PRIMA.

SI graue ſurſum proiectum ex *a.* aſcendat motu naturalitèr

Deficiente vſq; ad ſublimius punctum ſuæ lationis *b.* idem verò mobile æquali tempore, & eadem velocitate quam in pũcto *a* habebat, ſed motu æquabili aſcendat vſq; in *c.* Dico *ac.* duplam eſſe ipſius *ab.*

Si enim non eſt dupla, ponamus aliquam *ad* duplam eſſe ipſius *ab.*

Concipiamus iam cadere naturalitèr mobile ex *b* in *a.* Gradus ille impetus acquiſiti poſt caſum ex *b* in *a.* eſt ille prorſus qui vehit mobile ad eandem altitudinem *b.* eodem tempore, & motu naturaliter deficiẽte. Idem verò gradus impetus eodem tempore, ſed motu æquabili perducit mobile ad altitudinem *ad* duplam caſus *ba.* Sed ille idem impetus qui per ſuppoſitionem perducit mobile ex *a* in *b* motu naturaliter deficiente, illud perducebat etiam motu ęquabili eodemq; tempore ex *a* in *c.* Vnus ergò

Galil. in scholio prop.23.

Galil.ib

V 2 idemq

idemq;gradus impetus eodem tempore; motuq; æquabili perducit mobile per duo ſpatia inæqualia *ac*, *ad*. Quod eſt abſurdum.

PROPOSITIO II.

SEmita proiectorum, quęcunq; illa ſit, ſublimiori ſui puncto bifariam ſecat perpendiculum quod inter horizontem, & lineam directionis intercipitur.

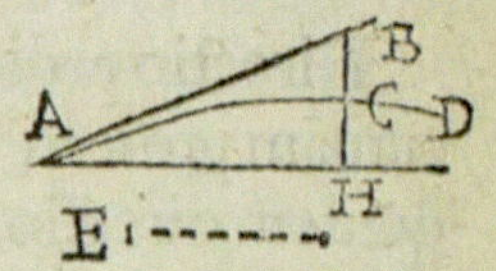

Proiciatur mobile ex *a* iuxtà directionē vtcumq; eleuatam *ab*. Patet quod ſine tractione grauitatis procederet mobile motu recto, & æquabili per lineam directionis *a b*. Sed grauitate intus operante ab ipſa directione ſtatim declinare incipiet, creſcente ſemper deuiationis menſura; & deſcribet aliquam lineam curuam *a c d*. quæcunq; ſit. Hæc linea punctum aliquod ſublimius cæteris habet; illud nempe quod eſt aſcenſionis extremum, & primum deſcenſionis. Sit huiuſmodi punctum *c*, & per *c* ducatur perpendiculum *bch*. Dico *hb*. duplam eſſe ipſius *hc*.

Abſtrahamus motum horizontalem; hic enim motus, quo ad lationem perpendicularem de qua agemus eſt tamquam non eſſet; cum illam neq; iuuet, neq; impediat. Concipiamus etiam mobile habere ſemper ſecum ſuum perpendiculum *hb*. horizontali quadam latione vna cum ipſo translatum ex *a* verſus *b*; in quo perpendiculo aſcendit graue, motu quodam continuo, ſed ſemper magis ac magis deficiente, à puncto *h* vſq; ad punctum *c*. Conficit ergo mobile in ſuo perpendiculo tempore Exempli gratia *e*. ſpatium *hc*. ſed ſi motu æquabili aſcendiſſet cum impetu & tempore eodem, reperiretur in *b* (deberet enim ob motum æquabilem eſſe ſemper in communi ſectione linearum *ab*, *hb*.) Quare per præcedentem, ſpatium *hb*. ipſius *hc* duplum eſt. Quod erat &c.

PRO-

idemq; gradus impetus eodem tempore; motuq; æquabili perducit mobile per duo ſpatia inæqualia *ac, ad.* Quod eſt abſurdum.

PROPOSITIO II.

SEmita proiectorum, quęcunq; illa ſit, ſublimiori ſui puncto bifariam ſecat perpendiculum quod inter horizontem, & lineam directionis intercipitur.

Proiciatur mobile ex *a* iuxtà directionẽ vtcumq; eleuatam *ab.* Patet quod ſine tractione grauitatis procederet mobile motu recto, & æquabili per lineam directionis *ab.* Sed grauitate intus operante ab ipſa directione ſtatim declinare incipiet, creſcente ſemper deuiationis menſura; & deſcribet aliquam lineam curuam *acd.* quæcunq; ſit. Hæc linea punctum aliquod ſublimius cæteris habet; illud nempe quod eſt aſcenſionis extremum, & primum deſcenſionis. Sit huiuſmodi punctum *c,* & per *c* ducatur perpendiculum *bch.* Dico *hb.* duplam eſſe ipſius *hc.*

Abſtrahamus motum horizontalem; hic enim motus, quo ad lationem perpendicularem de qua agemus eſt tamquam non eſſet; cum illam neq; iuuet, neq; impediat. Concipiamus etiam mobile habere ſemper ſecum ſuum perpendiculum *hb.* horizontali quadam latione vna cum ipſo translatum ex *a* verſus *h*; in quo perpendiculo aſcendit graue, motu quodam continuo, ſed ſemper magis ac magis deficiente, à puncto *h* vſq; ad punctum *c.* Conficit ergo mobile in ſuo perpendiculo tempore Exempli gratia *e.* ſpatium *hc.* ſed ſi motu æquabili aſcendiſſet cum impetu & tempore eodem, reperiretur in *b (*deberet enim ob motum æquabilem eſſe ſemper in communi ſectione linearum *ab, hb.)* Quare per præcedentem, ſpatium *hb.* ipſius *h c* duplum eſt. Quod erat &c.

PRO-

PROPOSITIO III.

LInea curua, quæ describitur à mobili secundum quamlibet eleuationem proiecto, parabola est, & prorsus eadem, quam describeret mobile si cum horizontali impetu proiceretur à vertice eiusdem lineæ curuæ.

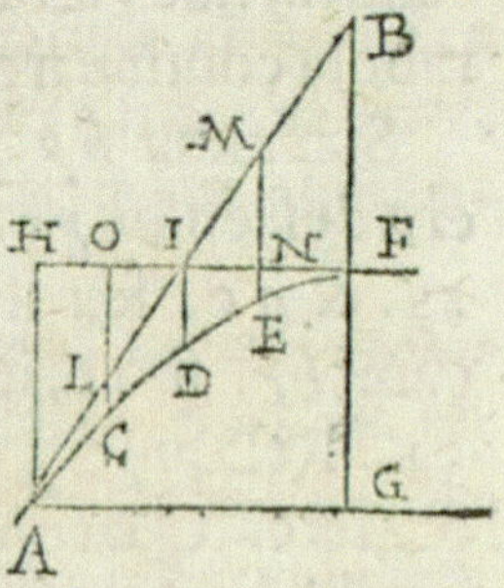

Sit linea proiectionis directiua *ab.* vtcumq; eleuata, & linea curua *acde*, cuius sublimius punctum sit *f.* Ducatur perpendiculum *bfg.* & erunt per præcedentia æquales *bf. fg.* Ducatur horizontalis *fh*, & perpendicularis *ab.* erunt iterũ æquales *fi. ih*, & *ai, ib.* Diuidatur *ab.* in quotcunq; partes ęquales *al, li, im mb.* & agantur perpendiculares per puncta *l, i, m.* Manifestum est quod spatia æqualia *al, li, im, mb*, percurrerentur à mobili temporibus æqualibus, si motu æquabili, & sine accessu noui motus deorsum, ab interna grauitate procedentis, moueretur. Sed quia ei statim atq; à proiiciente dimittitur in *a.* superaduenit attractio grauitatis, incipiet cõtinuo à linea directionis deorsum deuiare, & deuiationes tales erunt vt linea *lc.* descensus vnius temporis sit vt vnum: linea verò *id.* descensus duorum temporum sit vt quatuor, & *me.* trium temporum vt nouem, *bf.* quatuor temporum vt 16., & *ex Gal.*
sic deinceps ea lege vt semper descensuum spatia sint vt tempo rum quadrata. Quia verò *al, li, im, mb*, sunt æquales, erunt *ho. oi, in, nf.* (quòd inter easdem parallelas sint) æquales; & cum sit *bf*, 16, erit *mn.* 8. ergo reliqua *ne* est vnum. (quandoquidem tota *me* erat 9.) Ipsa verò *id.* posita fuerat vt 4. nec immutatur. *ol* autem ęqualis ipsi *mn.* est 8. & addita *lc*, quæ posita fuerat vnum, erit tota *oc* vt 9. At *ha.* ęqualis ipsi *bf* erit 16. Ergò cum spatia *fn, ni, io, ob.* sint æqualia; & *ne, id, oc, ha.* sint vt 1, 4, 9. 16. & sic deinceps vt reliqui semper numeri quadrati, erit linea procedens ex *f.* per puncta *e. d. c. a.* parabola

PROPOSITIO III.

LInea curua, quæ deſcribitur à mobili ſecundum quamlibet eleuationem proiecto, parabola eſt, & prorſus eadem, quam deſcriberet mobile ſi cum horizontali impetu proiceretur à vertice eiuſdem lineæ curuæ.

Sit linea proiectionis directiua *ab.* vtcumq; eleuata, & linea curua *acde,* cuius ſublimius punctum ſit *f.* Ducatur perpendiculum *bfg.* & erunt per præcedentia æquales *bf. fg.* Ducatur horizontalis *fh,* & perpendicularis *ah.* erunt iterũ æquales *fi. ih,* & *ai, ib.* Diuidatur *ab.* in quotcunq; partes ęquales *al, li, im*[86] *mb.* & agantur perpendiculares per puncta *l, i, m.* Manifeſtum eſt quod ſpatia æqualia *al, li, im, mb,* percurrerentur à mobili temporibus æqualibus, ſi motu æquabili, & ſine acceſſu noui motus deorſum, ab interna grauitate procedentis, moueretur. Sed quia ei ſtatim atq; à proiciente dimittitur in *a.* ſuperaduenit attractio grauitatis, incipiet cõtinuo à linea directionis deorſum deuiare, & deuiationes tales erunt vt linea *lc.* deſcenſus vnius temporis ſit vt vnum: linea verò *id.* deſcenſus duorum temporum ſit vt quatuor, & *me.* trium temporum vt nouem, *bf.* quatuor temporum vt 16., & ſic deinceps ea lege vt ſemper deſcenſuum ſpatia ſint vt temporum quadrata. Quia verò *al, li, im, mb,* ſunt æquales, erunt *ho. oi, in, nf.* (quòd inter eaſdem parallelas ſint) æquales*:* & cum ſit *bf,* 16, erit *mn.* 8. ergo reliqua *ne* eſt vnum. (quandoquidem tota *m e* erat 9.) Ipſa verò *id.* poſita fuerat vt 4. nec immutatur. *ol* autem ęqualis ipſi *mn.* eſt *8.* & addita *lc,* quæ poſita fuerat vnum, erit tota *oc* vt *9.* At *ha.* ęqualis ipſi *bf* erit *16.* Ergò cum ſpatia *fn, ni, io, ob.* ſint æqualia; & *ne, id, oc, ha.* ſint vt 1, 4, 9. 16. & ſic deinceps vt reliqui ſemper numeri quadrati, erit linea procedens ex *f.* per puncta *e. d. c. a.* pa- *ex Gal.*

rabola

[86] *im*] *im,*

158 *De motu Proiectorum*

rabola recta cuius vertex *f*. & de qua agit Galileus. Sed hęc eadem linea eſt tractus proiectionis obliquę ex *a*. factę per ſuppoſitionem noſtram. Ergò linea curua, quę deſcribitur à mobili ſecundum quamlibet eleuationem proiecto, eadem parabola eſt quam deſignaret ſi cum impetu horizontali opportuno ex vertice ipſius proiectum fuiſſet.

Manente eadem conſtructione, & figura, dico etiam poſt culmen, ſiue verticem *f*. mobile ex *a* proiectum, in eadem parabola continuare motum ſuum.

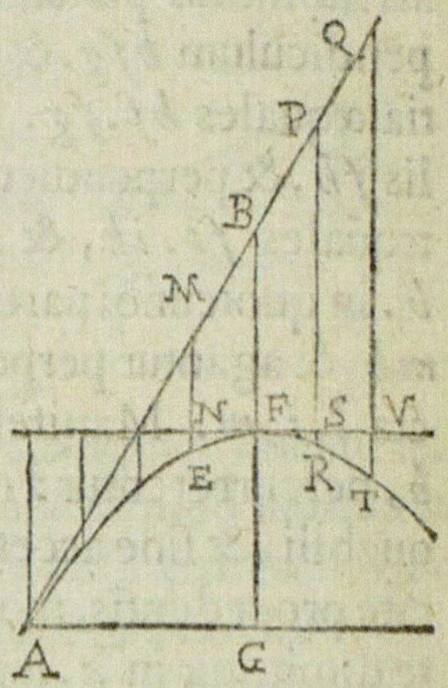

Sumantur *bp*. *pq*. ipſi *bm* ęquales; erit deſcenſus *pr*. quinq; temporum vt 25. & *qt*. ſex temporum vt 36. Sed cum *bf*. ſit 16, ipſa *pf*. eſt 24. & *qu* 32. Reliquę ergo *fr*, *ut*, ſunt vt unum & 4. &c. Quare puncta *f*, *r*, *t*. per quę incedit mobile ſunt in eadem continuata parabola in qua ſunt *e*, & *f*.

Linea etiam curua, quę deſcribitur à mobili ſecundum quamlibet directionem deorſum proiecto, parabola eſt, & eadem prorſus quam deſcriberet mobile ſi horizontaliter concitatum à uertice ipſius proiceretur.

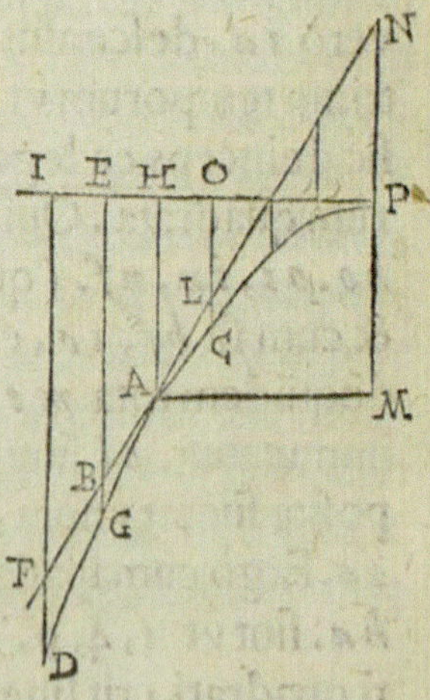

Manente eadem figura propoſitionis tertię. Sit linea proiectionis deorſum factæ *af*., & ſit impetus idem qui fuerat in proiectione *al*. ſurſum. Manifeſtum eſt quod mobile ſi motu ęquabili moueretur percurreret lineam rectam *af*. Sumantur iam *ab*, *bf*. ęquales tum inter ſe, tum etiam ipſi *al*. patet etiam quod ipſę *ab*. *bf*. motu ęquabili, temporibus ęqualibus abſoluerentur cum ęquales ſint. Sed quia grauitas ſtatim incipit de-

orſum

rabola recta cuius vertex *f.* & de qua agit Galileus. Sed hęc eadem linea eſt tractus proiectionis obliquę ex *a.* factę per ſuppoſitionem noſtram. Ergò linea curua, quę deſcribitur à mobili ſecundum quamlibet eleuationem proiecto, eadem parabola eſt quam deſignaret ſi cum impetu horizontali opportuno ex vertice ipſius proiectum fuiſſet.

Manente eadem conſtructione, & figura, dico etiam poſt culmen, ſiue verticem *f.* mobile ex *a* proiectum, in eadem parabola continuare motum ſuum.

Sumantur *bp. pq.* ipſi *bm* ęquales; erit deſcenſus *pr.* quinq; temporum vt 25. & *qt.* ſex temporum vt 36. Sed cum *bf.* ſit 16, ipſa *pſ.* eſt 24. & *qu* 32. Reliquę ergo *ſr, ut,* ſunt vt unum & 4. &c. Quare puncta *f, r, t.* per quę incedit mobile ſunt in eadem continuata parabola in qua ſunt *e, &f.*

Linea etiam curua, quę deſcribitur à mobili ſecundum quamlibet directionem deorſum proiecto, parabola eſt, & eadem prorſus quam deſcriberet mobile ſi horizontaliter concitatum à uertice ipſius proiceretur.

Manente eadem figura propoſitionis tertię. Sit linea proiectionis deorſum factæ *af.,* & ſit impetus idem qui fuerat in proiectione *al.* furſum. Manifeſtum eſt quod mobile ſi motu ęquabili moueretur percurreret lineam rectam *af.* Sumantur iam *ab, bf.* ęquales tum inter ſe, tum etiam ipſi *al.* patet etiam quod ipſę *ab. bf.* motu ęquabili, temporibus ęqualibus abſoluerentur cum ęquales ſint. Sed quia grauitas ſtatim incipit de-

orsum

orsum trahere, mobile à linea recta *af* deuiabit, & erit descensus *bg*. unius temporis ut unum, eritq; ęqualis ipsi *lc* qui fuerat descensus unius eorumdem temporum. Descensus autem *fd*. duorum temporum erit ut 4. & sic semper deinceps. Quia uero *ha*. est 16., erit *eb*. 24. & addita *bg*, tota *eg* erit 25. Eodem modo; *if* est 32, & *fd*. 4. ergo tota *id*. est 36.

Cum itaq. sint ęquales *ob*, *he*, *ei*. & *oc*. *ha*, *eg*, *id* sint (cõtinuato ordine numerorum quadratorum,) ut 9. 16. 25. 36. erit linea *cagd*. portio eiusdem continuatæ parabolę: ergò linea curua quę describitur à mobili deorsum proiecto parabola est, & prorsus eadem quam descripsisset si à uertice ipsius, cum horizontali impetu oportuno proiectum fuisset.

Diximus cum horizontali impetu opportuno, quia si mobile cum eodem impetu proiceretur ex p *horizontaliter deorsum, & ex* a *secundum* a l *sursum, nequaquam eandem parabolam describeret vtraq; latione. Requiritur enim maior impetus in proiectione ex* a *sursum facta ad hoc vt eandem parabolam describat quam designauisset si ex* p *horizontaliter proiectum fuisset. Ratio verò vnius impetus ad alium vt eadem parab. euadat, erit hæc.*

Si mobile horizontaliter proiectum ex p. *quolibet impetu descripsit parabolam* p c a, *ad hoc vt ex* a *proiectum describat eandem, debet impetus ex* a *ad impetum ex* p. *esse vt* a n *ad* a m.

Tunc enim si mobile iuxtà tangentem a n *proiciatur cum impetu dicto eandem parabolam* a c p. *percurret.*

Si quis autem, propter numerorum applicationem, ea quæ attulimus non demonstrationem putet, sed computum, vel exemplum, habeat hic demonstrationem puram, præmisso hoc lemmate.

Lemma.

Si fuerit vt a c *ad* a b. *potentia; ita* c e *ad* b d. *longitudine, & sint parallelæ* c e, b d. *Dico (coniunctà* a e*) ipsam* b f. *mediam proportionalem esse inter duas* c e, b d.

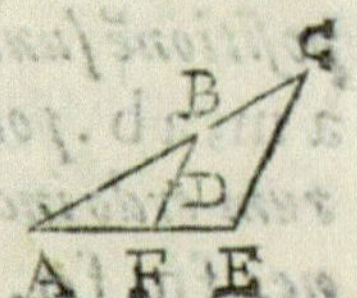

Est enim c e *ad* b d. *longitudine vt* c a *ad* a b, *hoc*

orſum trahere, mobile à linea recta *af* deuiabit, & erit deſcenſus *bg*. unius temporis ut unum, eritq; ęqualis ipſi *lc* qui fuerat deſcenſus unius eorumdem temporum. Deſcenſus autem *fd.* duorum temporum erit ut 4. & ſic ſemper deinceps. Quia uero *ha.* eſt 16., erit *eb.* 24. & addita *bg,* tota *e g* erit 25. Eodem modo; *if* eſt 32, & *fd.* 4. ergo tota *id.* eſt 36.

Cum itaq. ſint ęquales *oh, he, ei.* & *oc. ha, eg., id* ſint (cõtinuato ordine numerorum quadratorum,) ut 9. 16. 25. 36. erit linea *cagd.* portio eiuſdem continuatæ parabolę: ergò linea curua quę deſcribitur à mobili deorſum proiecto parabola eſt, & prorſus eadem quam deſcripſiſſet ſi à uertice ipſius, cum horizontali impetu oportuno proiectum fuiſſet.

Diximus cum horizontali impetu opportuno, quia ſi mobile cum eodem impetu proiceretur ex p *horizontaliter deorſum, & ex* a *ſecundum* al *ſurſum, nequaquam eandem parabolam deſcriberet vtraq; latione. Requiritur enim maior impetus in proiectione ex* a *ſurſum facta* ad hoc *vt eandem parabolam à eſcribat*[87] *quam deſignauiſſet ſi ex* p *horizontaliter proiectum fuiſſet. Ratio verò vnius impetus ad alium vt eadem parab. euœdat, erit hœc.*

Si mobile horizontaliter proiectum ex p. *quolibet impetu deſcripſit parabolam* pca, ad hoc *vt ex* a *proiectum deſcribat eandem, debet impetus ex* a *ad impetum ex* p. *eſſe vt* an *ad* am.

Tunc enim ſi mobile iuxtà tangentem an *proiciatur cum impetu dicto eandem parabolam* acp. *percurret.*

Si quis autem, propter numerorum applicationem, ea quœ attulimus non demonstrationem putet, ſed computum, vel exemplum, habeat hic demonſtrationem puram, prœmiſſo hoc lemmate.

Lemma.

Si fuerit vt ac *ad* ab. *potentia; ita* ce *ad* bd. *longitudine, & ſint parallelœ* ce, bd. *Dico (coniunctà* ae*) ipſam* bf. *mediam proportionalem eſſe inter duas* ce, bd.

Eſt enim ce *ad* bd. *longitudine vt* ca *ad* ab,

hoc

---

[87] *eſcribat*] *deſcribat.*

hoc est ce; *ad* bf. *potentià. Quare* bf. *media est inter* ce, bd, *&c.*

Si verò, vt sunt ab, ac, ad. *potentià, ità fuerint parallelæ* be, cf, dg. *longitudine, sintq;* ah, ig. *æquales, Dico etiam* he, if, *æquales esse. Est enim* ga *ad* ai *vt* gd *ad* ic, *vel per præcedens, vt* ic *ad* cf. *ergo diuidendo erit vt* gi *ad* ia, *ita* if *ad* fc. *eodem modo ostendemus esse vt* gh *ad* ha, *ita* he *ad* eb. *His demonstratis. Est* fc *ad* be. *longitudine, vt* ca *ad* ab, *vel* ia *ad* ah, *vel* gh *ad* ha, *vel* he *ad* eb. *potentià. Ergò* he *media est inter* fc. be. *Iterum. Est* fc *ad* be *longitudine vt* ca *ad* ab, *vel* ia *ad* ah, *vel* ai *ad* ig, *vel* cf *ad* fi *potentià. ergo* fi *media est inter* cf, be. *Inter easdem verò media erat etiam* eh, *ergo* fi, eh. *æquales sunt. Quod erat præmittendum.*

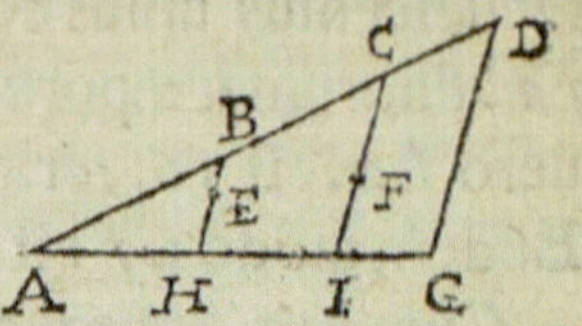

Resumpta iam figura propositionis 3. huius, fiat vt antea expositum est proiectio per lineam acdef. *ducaturq; recta* af. *& accipiantur æquales* ap. fq. *Erunt ob descensionem naturaliter acceleratam spatia* lc. me, bf. *in duplicata ratione temporum* al, am, ab. *& ideo per præcedens lemma æquales erunt* cp, eq, *& propterea remanent æquales* lc, ne *(nam tota* lp, *toti* nq. *æqualis est, producta enim* idr. *erunt æquales* ar, rf; *& ipsa* ir *eandem rationem habebit ad* lp, *& ad* nq, *nempe quam habet* ra *ad* ap. *vel* rf *ad* fq.) *Ergo omnes lineæ quæ ex* f *versus* id *successiuè descendunt à linea* fh. *sunt æquales omnibus & singulis illis respectiuè, quæ ex* a *versus lineam* id. *successiuè descendunt à linea* ab, *singulæ singulis (quod enim ostensum est de sola* ne. *ostendi potest de omnibus.) Sed omnes istæ, quarum series ex* a *incipit per suppositionẽ sunt inter se longitudine vt sunt* al, am, ab. *potentià, ergo etiam omnes illæ quarum ordo incipit ex* f. *erunt longitudine vt omnes* fn, fo, fh. *potentià. Quare linea curua*

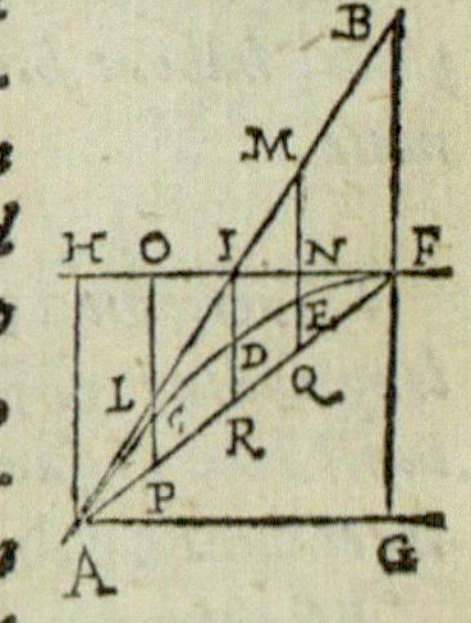

acef.

hoc eſt ce; *ad* bf. *potentià. Quare* bf. *media est inter* ce, bd, *&c.*

Si verò, vt ſunt ab, ac, ad. *potentià, ità fuerint parallelæ* be, cf, dg. *longitudine, ſintq;* ah, ig. *æquales, Dico etiam* he, if, *æquales eſſe. Eſt enim* ga *ad* ai *vt* gd *ad* ic, *vel per præcedens, vt* ic *ad* cf. *ergo diuidendo erit vt* gi *ad* ia, *ita* if *ad* fc. *eodem modo oſtendemus eſſe vt* gh *ad* ha, *ita* he *ad* eb. *His demonstratis. Eſt* fc *ad* be. *longitudine, vt* ca *ad* ab, *vel* ia *ad* ah, *vel* gh *ad* ha, *vel* he *ad* eb. *potentià. Ergò* he *media eſt inter* fc. e. *Iterum. Est* fc *ad* be *longitudine vt* ca *ad* ab, *vel* ia *ad* ah, *vel* ai *ad* ig, *vel* cf *ad* fi *potentià. ergo* fi *media eſt inter* cf, be. *Inter eaſdem verò media erat etiam* eh, *ergo* fi, eh. *æquales ſunt. Quod erat præmittendum.*

Reſumpta iam figura propoſitionis 3. huius, fiat vt antea expoſitum est proiectio per lineam acdef. *ducaturq; recta* af. *& accipiantur æquales* ap. fq. *Erunt ob deſcenſionem naturaliter acceleratam ſpatia* lc. me, bf. *in duplicata ratioue temporum* al, am, ab. *& ideo per præcedens lemma æquales erunt* cp, eq, *& propterea remanent æquales* lc, ne *(nam tota* p, *toti* nq. *æqualis eſt, producta enim* idr. *erunt æquales* ar, rf; *& ipſa* ir *eandem rationem habebit ad* lp, *& ad* nq, *nempe quam habet* ra *ad* ap. *vel* rf *ad* fq.) *Ergo omnes lineæ quæ ex* f *verſus* id *ſucceſſiuè deſcendunt à linea* fh. *ſunt æquales omnibus & ſingulis illis reſpectiuè, quæ ex* a *verſus lineam* id. *ſucceſſiuè deſcendunt à linea* ab, *ſingulæ ſingulis (quod enim ostenſum eſt de ſola* ne. *oſtendi potest de omnibus.) Sed omnes istæ, quarum ſeries ex* a *incipit per ſuppoſitionẽ ſunt inter ſe longitudine vt ſunt* al, am, ab. *potentià, ergo etiam omnes illæ quarum ordo incipit ex* f. *erunt longitudine vt omnes* fn, fo, fh. *potentià. Quare linea curuæ*

a c e f

a c e f. &c. quæ describitur à mobili secundum quamlibet e leuationem proiecto, eadem parabola est, quam designaret si ex vertice f cum impetu horizontali oportuno proiectum fuisset.

Lemma.

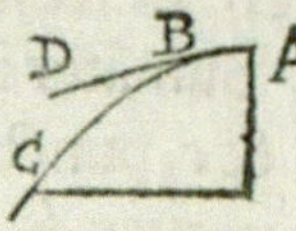

Si mobile proiectum, dum parabolam a b c *percurrit, in aliquo ipsius puncto* b. *omni grauitate spoliaretur, tunc proculdubio per lineam rectam* b d. *tangentem parabolæ lationem suam cōtinuaret motu semper æquabili. Quandoquidem dempta ei esset omnis causa quæ motum aut inflectere posset, aut accelerare, vel retardare. Manifestum etiam est, impetum ipsius mobilis in qualibet portione tangentis* b d, *eundem semper futurum fore qui fuerat in puncto* b.

PROPOSITIO IV.

IMpetus in punctis parabolæ vt sunt portiones tangentiū, inter duas parallelas diametro interceptæ.

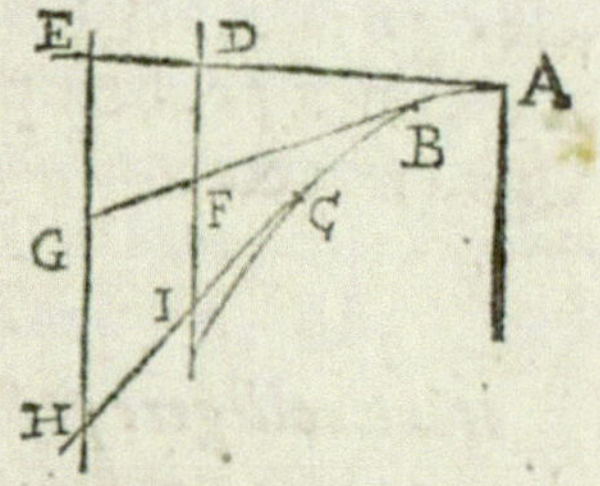

Proposita parabola *abc.* ducantur tangentes *ae*, *bg*, *ch.* à quibuscunque punctis *a.b.c.* tum duæ lineæ parallelę diametro vbicunque sint *di*, *eh.* Dico lineas interceptas *de.fg*, *ih.* ipsos impetus qui sunt in punctis *a.b.c.* proportione repræsentare.

Vnaquæq. enim ipsarum *de*, *fg*, *ih*. eodem tempore absolueretur à mobili; quandoquidē progressio horizontalis quæ inter duas parallelas *di*, *eh*. est, eodem semper tempore debeat absolui, vbicunq. reperiatur mobile, & per quamcunq; inclinationem procedat. Sed motus in ipsis lineis interceptis sunt æquabiles, ergo impetus erunt vt spatia. Quare impetus ipsius *de*, vel puncti *a*. erit vt linea *de*. Ipsius autem *fg*, vel puncti *b*. erit vt *fg*. & sic deinceps. Quod erat &c. *Lem. præced.*

X PRO-

ace f. *&c. quæ deſcribitur à mobili ſecundum quam libet eleuationem proiecto, eadem parabola eſt, quam deſignaret ſi ex vertice* f *cum impetu horizontali oportuno proiectum fuiſſet.*

Lemma.

Si mobile proiectum, dum parabolam abc *percurrit, in aliquo ipſius puncto* b. *omni grauitate ſpoliaretur, tunc proculdubio*[88] *per lineam rectam* bd. *tangentem parabolæ lationem ſuam cõtinuaret motu ſemper æquabili. Quandoquidem dempta ei eſſet omnis cauſa quæ motum aut inflectere poſſet, aut accelerare, vel retardare. Manifeſtum etiam est, impetum ipſius mobilis in qualibet portione tangentis* bd, *eundem ſemper futurum fore qui fuerat in puncto* b.

PROPOSITIO IV.

IMpetus in punctis parabolæ vt ſunt portiones tangentiũ, inter duas parallelas diametro interceptæ.

Propoſita parabola *abc.* ducantur tangentes *ae, bg, ch.* à quibuſcunque punctis *a. b. c.* tum duæ lineæ parallelę diametro vbicunque ſint *di, eh.* Dico lineas interceptas *de. fg, ih.* ipſos impetus qui ſunt in punctis *a. b. c.* proportione repreſentare.

Vnaquæq. enim ipſarum *de, fg, ih.* eodem tempore abſolueretur à mobili; quandoquidẽ progreſſio horizontalis quæ inter duas parallelas *di, eh.* eſt, eodem ſemper tempore debeat abſolui, vbicunq. reperiatur mobile, & per quamcunq; inclinationem procedat. *Lem. pre.* Sed motus in ipſis lineis interceptis ſunt æquabiles, ergo impetus erunt vt ſpatia. Quare impetus ipſius *de,* vel puncti *a.* erit vt linea *de.* *ced.* Ipſius autem *fg,* vel puncti *b.* erit vt *fg.* & ſic deinceps. Quod erat &c.

X P R O-

[88] *proculdubio*] *procul dubio.*

PROPOSITIO V.

IMpetus in punctis parabolæ æqualitèr vtrimq; à vertice distantibus, æquales sunt inter se, licet alter ascendat, alter vero descendat.

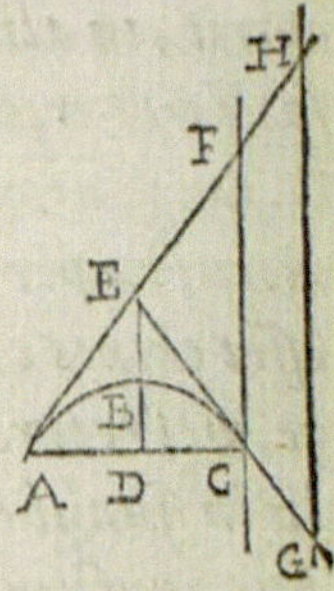

Sumãtur in parabola proiectionis factæ versus *b* & *c*, puncta quæuis *a*, *c*, quæ æqualitèr distent à vertice *b*, hoc est, quæ sint in eadem horizontali linea *a c*. Dico impetus in *a* & *c* æquales esse. Accipiatur *b e* æqualis ipsi *b d*, & ducantur *a e*, *c e*, quarum vtraq; tangens erit. Ducantur etiã lineæ *c f*, *g h*. diametro parallelę ubicunq; uisum fuerit; & producantur tangentes *a h*, *e g*. Erit ergò per præcedentem impetus in *a* ut *f h*. & in *c* ut *c g*, quæ si ęquales fuerint, æquales erunt impetus in punctis *a* & *c*.

Latera *a d*, *d c*. sunt æqualia, & *d e* commune; anguli autẽ ad *d* recti, ergo anguli *a e d*, *d e c* sunt ęquales. Añgulo autem *a e d* ęqualis est *e h g*. ob parallelas, & ipsi *d e g*. ęqualis est *c g h*, item ob parallelas; est ergo triangulum *e h g* ęquicrure, & linea *f c*. basi parallela, quare *f h*, *c g* ęquales sunt. Quod erat &c.

Corollarium.

Hinc colligere possumus facta proiectione ab a, *quod si mobile ex puncto* c *reflectatur retrorsum per eandem suam viam cum impetu eodem, eademq; directione quam habet in puncto* c, *per eandem parabolam recurrere debere; Habet enim in* c, *vt ostendimus eundem impetum, & eandem directionem quam habebat in* a, *quare eandem parabolam designare debet quam designauerat ex* a.

PROPOSITIO VI.

SIT parabola *a b* cuius altitudo *a c*, & sublimitas *a d*; ostendendum est alitèr ac in primo libro, eundem esse impetum

PROPOSITIO V.

IMpetus in punctis parabolæ æqualitèr vtrimq; à vertice diftantibus, æquales ſunt inter ſe, licet alter aſcendat, alter vero deſcendat.

Sumãtur in parabola proiectionis factæ verſus *b* & *c,* puncta quæuis *a, c,* quæ æqualitèr diftent à vertice *b,* hoc eft, quæ ſint in eadem horizontali linea *ac.* Dico impetus in *a* & *c* æquales eſſe. Accipiatur *be* æqualis ipſi *bd,* & ducantur *ae, ce,* quarum vtraq; tangens erit. Ducantur etiã lineæ *cf, gh.* diametro parallelę ubicunq; uiſum fuerit; & producantur tangentes *ah, eg.* Erit ergò per præcedentem impetus in *a* ut *fh.* & in *c* ut *cg,* quæ ſi ęquales fuerint, æquales erunt impetus in punctis *a* & *c.*

Latera *ad, dc.* ſunt æqualia, & *de* commune; anguli autẽ ad *d* recti, ergo anguli *aed, dec* ſunt ęquales. Angulo autem *aed* ęqualis eft *ehg.* ob parallelas, & ipſi *deg.* ęqualis eft *cgh,* item ob parallelas; eft ergo triangulum *chg* ęquicrure, & linea *fc.* baſi parallela, quare *fh, cg* ęquales ſunt.

Quod erat &c.

Corollarium.

Hinc colligere poßumus facta proiectione ab a, *quod ſi mobile ex puncto* c *reflectatur retrorſum per eandem ſuam viam cum impetu eodem, eademq; directione quam habet in puncto* c, *per eandem parabolam recurrere debere; Habet enim in* c, *vt oſtendimus eundem impetum, & eandem directionem quam habebat in* a, *quare eandem parabolam deſignare debet quam deſignauerat ex* a.

PROPOSITIO VI.

SIT parabola *ab* cuius altitudo *ac,* & ſublimitas *ad*; oftendendum eft alitèr ac in primo libro, eundem eſſe im-

petum

petum parabolæ in *b*, ac grauis cadentis naturaliter ex puncto ſublimitatis *d*, uſq. in *c*.

Ducantur tangentes *ae*, *bf*, & ſit *be* parallela ipſi *ac*. Notum ergo eſt per 4. huius impetum parabolę in *b*. ad impetum parabolę in *a*, eſſe ut *bf* ad *ae*, cum ſint *bf*, & *ae* inter eaſdem ad diametrum parallelas interceptę. Agatur per *d* & *e* alia parabola *deh*. & impetus caſus per *da*, ad impetum caſus per *dc*. erit ut applicata *ae* ad applicatam *ch*. Si ergo fuerint ęquales tangens *bf* & appplicata *ch*, erit ex ęquo, impetus in *b* ad impetum in *c* ut *bf* ad *ch*. nempe æqualis. Oſtendo *bf* *ch*. ęquales eſſe, ſic. Secetur *cb* biſariam in *i*, & erit per demonſtrata in pręcedenti libro, *ci* media proportionalis inter *ca*, & *ad*. Iam ſic. Quadratum *ca* ad rectangulum *cad* ſub eadem altitudine, eſt ut *ca* ad *ad*, ergo quadratum *ca*, ad quadratum *ci* eſt ut *ca* ad *ad*. Sumptis autem quadratorum quadruplis erit quadratum *fc* ad quadratum *cb* ut *ca* ad *ad*. & componendo quadrata *fc*, *cb*, uel quadratum *fb* ad quadratum *cb* erit ut *cd* ad *da*, hoc eſt ut quadratum *ch* ad quadratum *ae*: ſed quadrata *cb* & *ae*. ęqualia ſunt, ergo etiam quadrata *fb*, *ch*. Quare æquales ſunt lineę *fb*, *ch*. Quod erat &c.

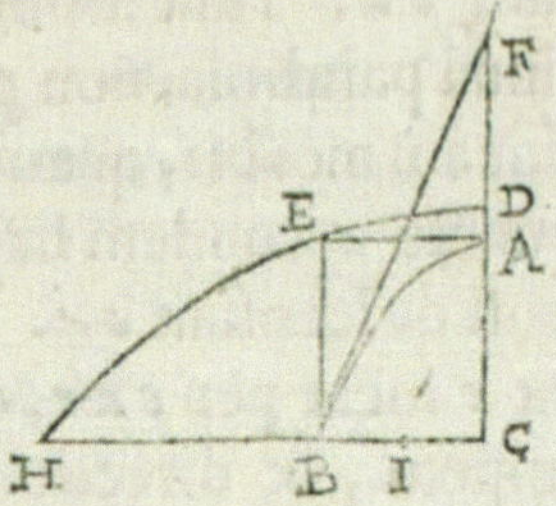

PROPOSITIO VII.

SI ab eodem puncto, cum eodem impetu, & eadem directione fiant proiectiones utrinque, ſurſum nempe, & deorſum: mobile utrinque per portiones unius eiuſdemq. continuatæ parabolę percurret.

Fiat ex puncto *a* cum directione *ab*. proiectio ſurſum *ac*. & ab eodem puncto directione *ba* fiat proiectio deorſum *ad*.

petum parabolæ in *b,* ac grauis cadentis naturaliter ex puncto ſublimitatis *d,* uſq. in *c.*

Ducantur tangentes *ae, bf,* & ſit *be* parallela ipſi *ac.* Notum ergo eſt per 4. huius impetum parabolę in *b.* ad impetum parabolę in *a,* eſſe ut *bf* ad *ae,* cum ſint *bf,* & *ae* inter eaſdem ad diametrum parallelas interceptę. Agatur per *d* & *e* alia parabola *deh.* & impetus caſus per *da,* ad impetum caſus per *dc.* erit ut applicata *ae* ad applicatam *ch.* Si ergo fuerint ęquales tangens *bf* & appplicata[89] *ch,* erit ex ęquo, impetus in *b* ad impetum in *c* ut *bf* ad *ch.* nempe æqualis. Oſtendo *bfch.* ęquales eſſe, ſic. Secetur *cb* bifariam in *i,* & erit per demonſtrata in prędedenti libro, *ci* media proportionalis inter *ca,* & *ad.* Iam ſic. Quadratum *ca* ad rectangulum *cad* ſub eadem altitudine, eſt ut *ca* ad *ad,* ergo quadratum *ca,* ad quadratum *ci* eſt ut *ca* ad *ad.* Sumptis autem quadratorum quadruplis erit quadratum *fc* ad quadratum *cb* ut *ca* ad *ad.* & componendo quadrata *fc, cb,* uel quadratum *fb* ad quadratum *cb* erit ut *cd* ad *da,* hoc eſt ut quadratum *ch* ad quadratum *ae:* ſed quadrata *cb* & *ae.* ęqualia ſunt, ergo etiam quadrata *fb, ch.* Quare æquales ſunt lineę *fb, ch.* Quod erat &c.

PROPOSITIO VII.

SI ab eodem puncto, cum eodem impetu, & eadem directione fiant proiectiones utrinque, ſurſum nempe, & deorſum: mobile utrinque per portiones unius eiuſdemq. continuatæ parabolę percurret.

Fiat ex puncto *a* cum directione *ab.* proiectio ſurſum *ac.* & ab eodem puncto directione *ba* fiat proiectio deorſum *ad.*

 Dico

[89] appplicata] applicata.

Dico *c a d* vnam, & eandem continuatam parabolam eſſe. Si enim continua non eſt, demittatur mobile ex vertice *c* verſus *a*. per parabolam *c a*. Tunc mobile cum *c a d*. non ſit continua parabola, non per ipſam *a d*, ſed per aliã lineam meabit, quæ ſit *a e*. Verùm mobile in puncto *a* eundem habet impetum ſiue ante aſcenſum *a c*, ſiue poſt deſcenſum *c a*. Mobile ergo ex punctò *a*. quando venit ex *c* meat per *c a e*, quando verò proicitur ex *a* cum eodem impetu, & directione currit per *a d*. Quod eſt abſurdum. Cum enim in vtroque caſu diſcedat ab *a* cum eodem impetu, eademq; directione, debet etiam in vtroq; caſu per eandem lineam *a d* ambulare. Quod &c.

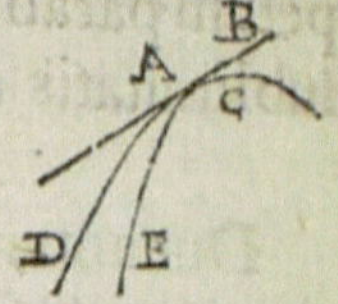

PROPOSITIO VIII.

DAta qualibet parabola à mobili ſurſum proiecto deſcripta, proiectio perpendicularis ſurſum eiuſdem mobilis facta cum eodem impetu, tantum aſcendet, quantum eſt aggregatum altitudinis, & ſublimitatis ſimul datæ parabolę.

Sit parabola *a b*, cuius altitudo *c b*, & ſublimitas *b d*. ponaturq; *a e* æqualis & parallela ipſi *c d*. Facta autem ſit parabola à proiectione ex *a* verſus *b*. Dico ſi fiat proiectio cum eodẽ impetu per lineam *a e* ſurſum, mobile vſq; ad punctum *e* peruenturum eſſe. Impetus enim parabolę in *a*, ſiue fiat proiectio ex *a*, in *b*, ſiue ex *b* in *a*. idem eſt (vt oſtendimus.) at ex Galileo idem eſt ac naturalitèr cadentis ex *d* in *c*, ſed impetus naturaliter cadentis ex *d* in *c* ille eſt qui reuehit mobile ex *c* in *d*, ergò etiam ex *a* in *e*. Quod &c.

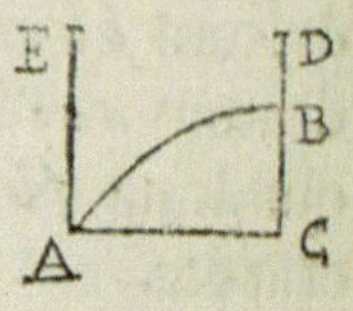

Pro- iect.

Definitio.

Impoſterum quando datum impetum nominabimus, illum in ſpatijs determinabimus, vt Galileo viſum eſt; alia enim ratione ſub

Dico *cad* vnam, & eandem continuatam parabolam eſſe. Si enim continua non eſt, demittatur mobile ex vertice *c* verſus *a.* per parabolam *ca.* Tunc mobile cum *cad.* non ſit continua parabola, non per ipſam *ad,* ſed per aliã lineam meabit, quæ ſit *ae.* Verùm mobile in puncto *a* eundem habet impetum ſiue ante aſcenſum *ac,* ſiue poſt deſcenſum *ca.* Mobile ergo ex punctò *a.* quando venit ex *c* meat per *cae,* quando verò proicitur ex *a* cum eodem impetu, & directione currit per *ad.* Quod eſt abſurdum. Cum enim in vtroque caſu diſcedat ab *a* cum eodem impetu, eademq; directione, debet etiam in vtroq; caſu per eandem lineam *ad* ambulare. Quod &c.

PROPOSITIO VIII.

DAta qualibet parabola à mobili ſurſum proiecto deſcripta, proiectio perpendicularis ſurſum eiuſdem mobilis facta cum eodem impetu, tantum aſcendet, quantum eſt aggregatum altitudinis, & ſublimitatis ſimul datæ parabolę.

Sit parabola *ab,* cuius altitudo *cb,* & ſublimitas *bd.* ponaturq; *ae* æqualis & parallela ipſi *ed.* Facta autem ſit parabola à proiectione ex *a* verſus *b.* Dico ſi fiat proiectio cum eodẽ impetu per lineam *ae* ſurſum, mobile vſq; ad punctum *e* peruenturum eſſe. Impetus enim
v. *Pro-.* parabolę in *a,* ſiue fiat proiectio ex *a,* in *b,* ſiue ex *b* in *a.* idem eſt *(*vt oſtendimus.*)* at ex Galileo idem eſt ac naturalitèr cadentis ex *d* in
iect. *c,* ſed impetus naturaliter cadentis ex *d* in *c* ille eſt qui reuehit mobile ex *c* in *d,* ergò etiam ex *a* in *e.* Quod &c.

Definitio.

Impoſterum quando datum impetum nominabimus, illum in ſpatijs determinabimus, vt Galileo viſum eſt; alia enim ratione

sub

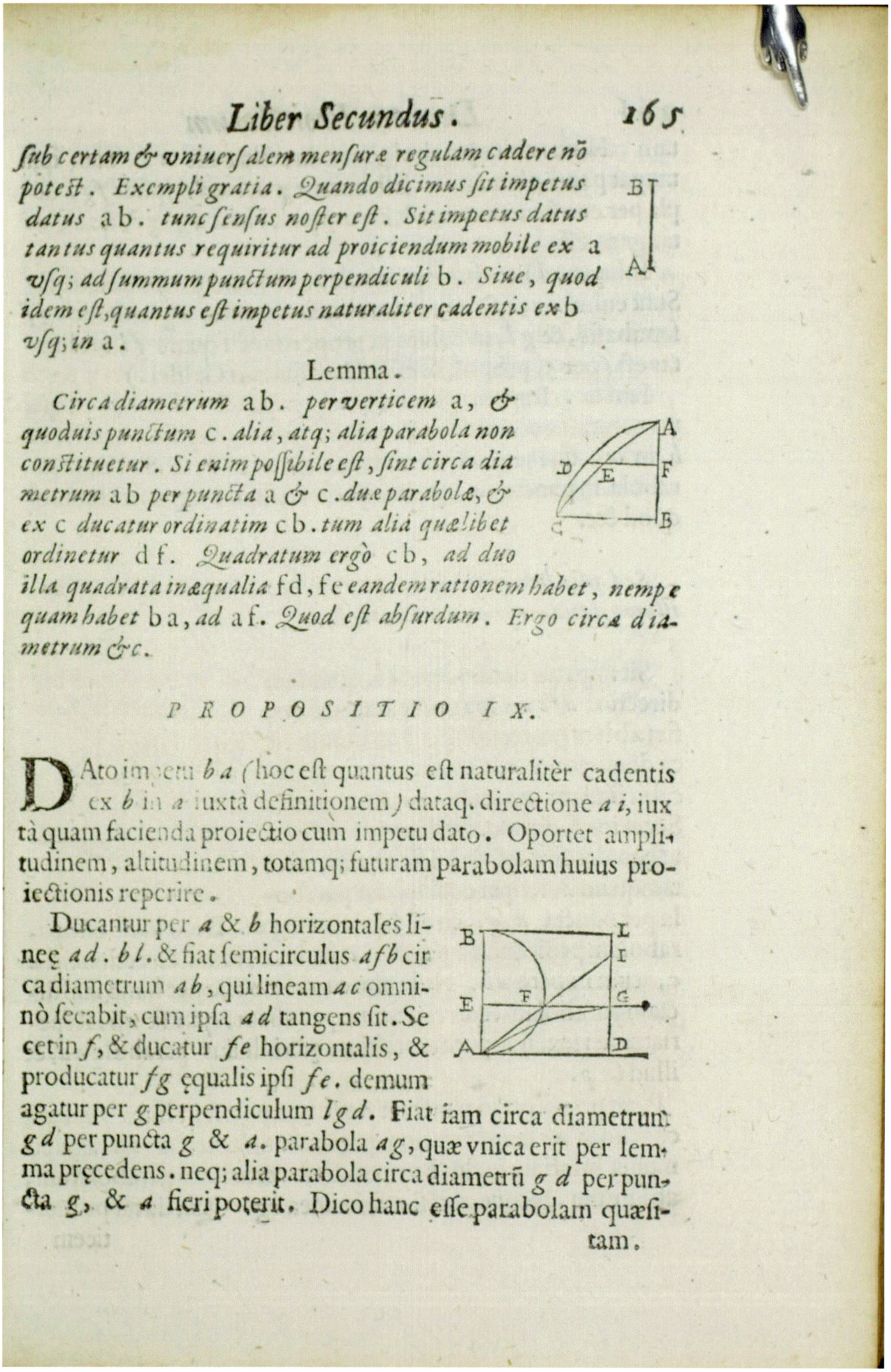

Liber Secundus. 165

sub certam & uniuersalem mensuræ regulam cadere nõ potest. Exempli gratia. Quando dicimus sit impetus datus a b. *tunc sensus noster est. Sit impetus datus tantus quantus requiritur ad proiciendum mobile ex* a *vsq; ad summum punctum perpendiculi* b. *Siue, quod idem est, quantus est impetus naturaliter cadentis ex* b *vsq; in* a.

Lemma.

Circa diametrum a b. *per verticem* a, *& quoduis punctum* c. *alia, atq; alia parabola non constituetur. Si enim possibile est, sint circa dia metrum* a b *per puncta* a *&* c. *duæ parabolæ, & ex* c *ducatur ordinatim* c b. *tum alia quælibet ordinetur* d f. *Quadratum ergò* c b, *ad duo illa quadrata inæqualia* f d, f e *eandem rationem habet, nempe quam habet* b a, *ad* a f. *Quod est absurdum. Ergo circa diametrum &c.*

PROPOSITIO IX.

Dato impetu *b a* (hoc est quantus est naturalitèr cadentis ex *b* in *a* iuxtà definitionem) dataq. directione *a i*, iuxtà quam facienda proiectio cum impetu dato. Oportet amplitudinem, altitudinem, totamq; futuram parabolam huius proiectionis reperire.

Ducantur per *a* & *b* horizontales lineę *a d*. *b l*. & fiat semicirculus *a f b* circa diametrum *a b*, qui lineam *a c* omninò secabit, cum ipsa *a d* tangens sit. Secet in *f*, & ducatur *f e* horizontalis, & producatur *f g* ęqualis ipsi *f e*. demum agatur per *g* perpendiculum *l g d*. Fiat iam circa diametrum *g d* per puncta *g* & *a*. parabola *a g*, quæ vnica erit per lemma pręcedens. neq; alia parabola circa diametrũ *g d* per puncta *g*, & *a* fieri poterit. Dico hanc esse parabolam quæsitam.

ſub certam & vniuerſalem menſuræ regulam cadere nõ potest. Exempli gratia. Quando dicimus ſit impetus datus ab. *tunc ſenſus noſter eſt. Sit impetus datus tantus quantus requiritur ad proiciendum mobile ex* a *vſq; ad ſummum punctum perpendiculi* b. *Siue, quod idem eſt, quantus eſt impetus naturaliter cadentis ex* b *vſq; in* a.

Lemma.

Circa diametrum ab. *per verticem* a, *& quoduis punctum* c. *alia, atq; alia parabola non constituetur. Si enim poſſibile eſt, ſint circa diametrum* ab *per puncta* a *&* c. *duæ parabolæ, & ex* c *ducatur ordinatim* cb. *tum alia quælibet ordinetur* df. *Quadratum ergò* cb, *ad duo illa quadrata inæqualia* fd, fe *eandem rationem habet, nempe quam habet* ba, *ad* af. *Quod eſt abſurdum. Ergo circæ diametrum &c.*

PROPOSITIO IX.

DAto impetu *ba (*hoc eſt quantus eſt naturalitèr cadentis ex *b* in *a* iuxtà definitionem*)* dataq. directione *a i,* iuxtà quam facienda proiectio cum impetu dato. Oportet amplitudinem, altitudinem, totamq; futuram parabolam huius proiectionis reperire.

Ducantur per *a* & *b* horizontales lineę *ad. bl.* & fiat ſemicirculus *afb* circa diametrum *ab,* qui lineam *ac* omninò ſecabit, cum ipſa *ad* tangens ſit. Secet in *f,* & ducatur *fe* horizontalis, & producatur *fg* ęqualis ipſi *fe.* demum agatur per *g* perpendiculum *lgd.* Fiat iam circa diametrum *gd* per puncta *g* & *a.* parabola *ag,* quæ vnica erit per lemma pręcedens. neq; alia parabola circa diametrũ *gd* per puncta *g,* & *a* fieri poterit. Dico hanc eſſe parabolam quæſi-

tam

tam. Huius enim parabolę linea directiua est *a i.* cum ipsa tangat parabolam in *a*. Est enim *e g* uel *a d.* ipsius *f g*. dupla per constructionem, & ideo ęquales sunt *d g*, *g i*, quare *a i* tangens est.

Insuper. Dico hanc parabolam ab impetu dato describi. Sunt enim *a e*, *e f*, *e b*, siue tres ipsis æquales *d g* altitudo, *g f* semibasis, & *g l*, in continua proportione: quare *g l* sublimitas est (per 5. propos. & eius Corollarium Galilei.)

Iam sic. Impetus parabolę *a g* in puncto *a* tantus est quantus naturaliter cadentis ex *l* in *d*. per 10. Galilei. hoc est ex *b* in *a*. siue proiecti ascendentis ex *a* in *b*. Habet ergo parabola in puncto *a* etiam impetum datum. Quare factum est quod &c.

Sed quia hæc propositio magni erit momenti pro sequentibus, illam ostendamus etiam alio modo,

Sit impetus datus idem *a b*. & eadem directio *a f c*. Quæritur parabola quę fiet a b hac proiectione. Fiat ut ante circa diametrum *a b* semicirculus, qui secabit lineam *a c*, cum *a d*, sit tangens. Secet in *f*, ductaq. horizontali *e f g* ita ut ęquales sint *e f*, *f g*, describatur, siue tamquam descripta concipiatur parabola per puncta *a*, *g*, circa diametrum *g d*. Dico hanc esse parabolam proiecti, si à puncto *a* iaciatur, iuxta directionem *a c*, cum impetu *a b*. Nisi enim currat mobile per hanc iam dictam parabolam, curret omnino per aliam, quæ sit *a p*. Reperiatur uertex, siue punctum altius cęteris huius parabolę *a d*, & illud sit *p*.

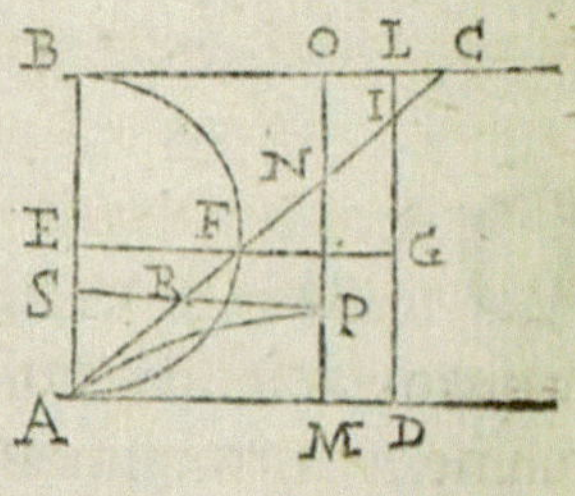

Patet primò quòd punctum *p* esse non potest in linea *l d*, quia cum linea *a c* tangat utramq. parabolam, secaretur *i d* axis cõmunis bifariam in duobus punctis à uerticibus parabolarum, absurdum. Neque potest esse in linea *e g*. Quia ducta per uer-

ticem

tam. Huius enim parabolę linea directiua eſt *ai.* cum ipſa tangat parabolam in *a.* Eſt enim *eg* uel *ad.* ipſius *fg.* dupla per conſtructionem, & ideo ęquales ſunt *dg, gi,* quare *ai* tangens eſt.

Inſuper. Dico hanc parabolam ab impetu dato deſcribi.

Sunt enim *ae, ef, eb,* ſiue tres ipſis æquales *dg* altitudo, *gf* ſemibaſis, & *gl,* in continua proportione: quare *gl* ſublimitas eſt (per 5. propoſ. & eius Corollarium Galilei.)

Iam ſic. Impetus parabolę *a g* in puncto *a* tantus eſt quãtus naturaliter cadentis ex *l* in *d.* per 10. Galilei. hoc eſt ex *b* in *a.* ſiue proiecti aſcendentis ex *a* in *b.* Habet ergo parabola in puncto *a* etiam impetum datum. Quare factum eſt quod &c.

Sed quia hæc propoſitio magni erit momenti pro ſequentibus, illam ostendamus etiam alio modo,

Sit impetus datus idem *ab.* & eadem directio *afc.* Quæritur parabola quę fiet ab hac proiectione. Fiat ut ante circa diametrum *a b* ſemicirculus, qui ſecabit lineam *ac,* cum *ad,* ſit tangens. Secet in *f,* ductàq. horizontali *efg* ita ut ęquales ſint *ef, fg,* deſcribatur, ſiue tamquam deſcripta concipiatur parabola per puncta *a,g,* circa diametrum *gd.* Dico hanc eſſe parabolam proiecti, ſi à puncto *a* iaciatur, iuxta directionem *ac,* cum impetu *ab.* Niſi enim currat mobile per hanc iam dictam parabolam, curret omnino per aliam, quæ ſit *ap.* Reperiatur uertex, ſiue punctum altius cęteris huius parabolę *ad,* & illud ſit *p.*

Patet primò quòd punctum *p* eſſe non poteſt in linea *ld,* quia cum linea *ac* tangat utramq. parabolam, ſecaretur *id* axis cõmunis bifariam in duobus punctis à uerticibus parabolarum, abſurdum. Neque poteſt eſſe in linea *eg.* Quia ducta per uer-

ticem

cem diametro, putà, *mn*. bifariam ſecaretur *mn* à linea *eg*. abſurdum: ſola enim *id*. ſecatur bifariam ex omnibus ſibi parallelis in angulo *cad*.

Sit iam punctum *p*. vbicunq; ducaturq; *prſ*. horizontalis. Quia *pn*, *pm*. ſunt æquales per ſecundam huius, erunt *nr*, *ra*. & *pr*, *rſ*. æquales; & quia parabola *ap* impetum habet *ba*. hoc eſt *om*, erit *o* punctum ſuæ ſublimitatis, & ideo lineæ *op*, *pr*, *pm*. in continua proportione erunt; & rectangulum *opm*. quadrato *pr* æquale, commutatiſq; lineis cum ſibi æqualibus, rectangulum *bſa* æquale erit quadrato *ſr*. Punctum ergo *r* eſt in ſemicirculi peripheria. Quod eſt abſurdum, recta enim linea *aſ* in duobus tantum punctis periphæriæ occurrit. Quare &c.

Corollaria.

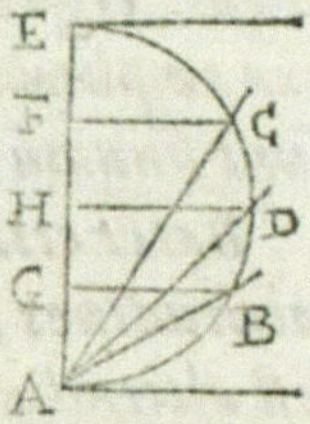

Hinc manifeſtum eſt, dato impetu alicuius machinæ qui ſit verbi gratia ea. *Si deſcribatur circa* ea *ſemicirculus* ade, *dari altitudines, & amplitudines omnium proiectionum, quæ ab eadem machina fieri poſſunt. Exempli gratia. Manente ſemper eodem impetu* ea. *fiant proiectiones per lineas diuerſimodè eleuatas* ac, ad, ab. *Proiectio facta ſecundum directionem* ac *aſcendet vſq; ad horizontalem* fc *productam; iactus autem factus iuxta directionem* ad *apicem habebit in linea* hd *producta. Proiectionis autem ſecundum lineam* ab *factæ, maxima altitudo erit in horizontali* gb *producta.*

In libro Galilei de motu naturaliter accelerato oſtēditur, proiecta ab eodem impetu ex a. *ſi à planis diuerſimodè inclinatis fulciantur, ſemper ad vnum idemq; planum horizontale peruenire. Hic verò apparet ſingulas proiectorum aſcenſiones variari, quando per aerem purum ſine vllo ſubiecto fulcro proiciuntur iuxtà diuerſas eleuationes: Minus enim aſcendit mobile quod per lineam* ab. *minus eleuatam emittitur, quàm illud quod per lineam* ad. *magis eleuatam proiectum fuerit.*

Patet etiam nullam altitudinem adeò aſcendere poſſe, vt ad

ipſam

ticem diametro, putà, *mn.* bifariam ſecaretur *mn* à linea *eg.* abſurdum: ſola enim *id.* ſecatur bifariam ex omnibus ſibi parallelis in angulo *cad.*

Sit iam punctum *p.* vbicunq; ducaturq; *prſ.* horizontalis.

Quia *pn, pm.* ſunt æquales per ſecundam huius, erunt *nr, ra.* & *pr, rſ.* æquales; & quia parabola *ap* impetum habet *ba.* hoc eſt *om,* erit *o* punctum ſuæ ſublimitatis, & ideo lineæ *op, pr, pm.* in continua proportione erunt; & rectangulum *opm.* quadrato *pr* æquale, commutatiſq; lineis cum ſibi æqualibus, rectangulum *bſa* æquale erit quadrato *ſr.* Punctum ergo *r* eſt in ſemicirculi peripheria. Quod eſt abſurdum, recta enim linea *af* in duobus tantum punctis periphæriæ occurrit. Quare &c.

Corollaria.

Hinc manifestum est, dato impetu alicuius machinæ qui ſit verbi gratia ea. *Si deſcribatur circa* ea *ſemicirculus* ade, *dari altitudines, & amplitudines omnium proiectionum, quæ ab eadem machina fieri poſſunt. Exempli gratia. Manente ſemper eodem impetu* ea. *fiant proiectiones per lineas diuerſimodè eleuatas* ac, ad, ab. *Proiectio facta ſecundum directionem* ac *aſcendet vſq; ad horizontalem* fc *productam; iactus autem factus iuxta directionem* ad *apicem habebit in linea* hd *producta. Proiectionis autem ſecundum lineam* ab *factæ, maxima altitudo erit in horizontali* gb *producta.*

In libro Galilei de motu naturaliter accelerato oſtẽditur, proiecta ab eodem impetu ex a. *ſi à planis diuerſimodè inclinatis fulciantur, ſemper ad vnum idemq; planum horizontale peruenire. Hic verò apparet ſingulas proiectorum aſcenſiones variari, quando per aerem purum ſine vllo ſubiecto fulcro proiciuntur iuxtà diuerſas eleuationes: Minus enim aſcendit mobile quod per lineam* ab. *minus eleuatam emittitur, quàm illud quod per lineam* ad. *magis eleuatam proiectum fuerit.*

Patet etiam nullam altitudinem adeò aſcendere poſſe, vt ad

ipsam

168 *De motu Proiectorum*

ipſam parallelam horizontalem, quæ ducitur per ſummum proiectionis perpendicularis punctum e, *peruenire poſſit.*

Manifeſtum etiam eſt amplitudines omnes ſemper augeri ab illa proiectione quæ dicitur tiro di punto in bianco *vſq; ad proiectionem factam ad angulum ſemirectum.*

A ſemirecta vero vſque ad perpendicularem, ſemper minui, donec penitus euaneſcant; quod accidit in proiectione perpendiculari, quæ nullam habet amplitudinem.

Denique obſeruare licet amplitudines parabolarum ab eodem impetu factarum, quarum eleuationes æqualiter ab angulo ſemirecto diſtent, inter ſe æquales eſſe.

Cum enim lineæ a b, a c. *æqualiter diſtent ab eleuatione ſemirecta, erunt arcus* d b, d c *æquales. quibus inſiſtunt æquales anguli; & ideo arcus* b a, c e *reliqui ex quadrantibus æquales erunt, ergo etiam ſinus eorum* b g, c f *æquales erunt, & propterea amplitudines integræ parabolarum, quæ quidem quadruplæ ſunt ſinuum* b g, c f. *erunt æquales.*

Patet etiam proiectionum æqualiter à ſemirecta diſtantium altitudines, & ſublimitates reciprocè inter ſe æquales eſſe, hoc eſt altitudinem vnius, ſublimitati alterius æquari.

Corollarium ergò erit quod Galileo Theorema ſatis arduum fuerat, nempe proiectionem ſemirectam omium maximam eſſe ab eodem impetu factarum. Si enim ponatur angulus c a d *ſemirectus erit* c d *ſemidiameter, hoc eſt maximus omnium ſinuum qui in ſemicirculo dari poſſint.*

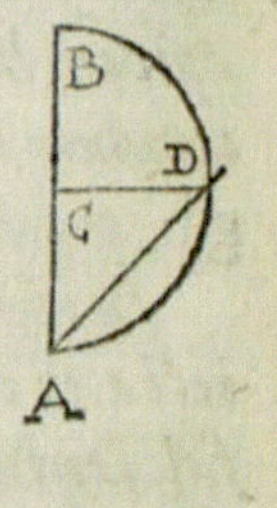

Patet etiam integram amplitudinem parabolæ ſemirectæ duplam eſſe lineæ ſublimitatis, ſiue impetus a b *quia demonſtrata eſt quadrupla rectæ* c d, *hoc eſt dupla ipſius* a b.

PROPOSITIO X.

DAto impetu & altitudine inuenienda ſit directio iuxta quã facta fuit proiectio: inuenienda etiam ſit amplitudo proiectionis. Sit in præcedenti figura, impetus datus *a b*, & data altitu-

ipſam parallelam horizontalem, quæ ducitur per ſummum proiectionis perpendicularis punctum e, *peruenire poſſit.*

Manifeſtum etiam eſt amplitudines omnes ſemper augeri ab illa proiectione quæ dicitur tiro di punto in bianco *vſq; ad proiectionem factam ad angulum ſemirectum.*

A ſemirecta vero vſque ad perpendicularem, ſemper minui, donec penitus euaneſcant; quod accidit in proiectione perpendiculari, quæ nullam habet amplitudinem.

Denique obſeruare licet amplitudines parabolarum ab eodem impetu factarum, quarum eleuationes æqualiter ab angulo ſemirecto diſtent, inter ſe æquales eſſe.

Cum enim lineæ ab, ac. *æqualiter distent ab eleuatione ſemirecta, erunt arcus* db, dc *æquales. quibus inſiſtunt æquales anguli; & ideo arcus* ba, ce *reliqui ex quadrantibus ęquales erunt, ergo etiam ſinus eorum* bg, cf *æquales erunt, & propterea amplitudines integræ parabolarum, quæ quidem quadruplæ ſunt ſinuum* bg, cf. *erunt æquales.*

Patet etiam proiectionum æqualiter à ſemirecta distantium altitudines, & ſublimitates reciprocè inter ſe æquales eſſe, hoc eſt altitudinem vnius, ſublimitati alterius æquari.

Corollarium ergò erit quod Galileo Theorema ſatis arduum fuerat, nempe proiectionem ſemirectam omium[90] *maximam eſſe ab eodem impetu factarum. Si enim ponatur angulus* cad *ſemirectus erit* cd *ſemidiameter, hoc eſt maximus omnium ſinuum qui in ſemicirculo dari poſſint.*

Patet etiam integram amplitudinem parabolæ ſemirectæ duplam eße lineæ ſublimitatis, ſiue impetus a b quia demonstrata est quadrupla rectæ cd, hoc eſt dupla ipſius ab.

PROPOSITIO X.

DAto impetu & altitudine inuenienda ſit directio iuxta quã facta fuit proiectio: inuenienda etiam ſit amplitudo proiectionis. Sit in præcedenti figura, impetus datus *ab,* & data

Altitu-

[90] *omium*] *omnium.*

altitudo sit *ac*. fiat circa *ab* semicirculus *adb*, & ducantur *cd*. horizontalis, *ad* autem ad punctum *d*. Manifestum est ex prę cedentibus, directionem quæsitam esse *ad*, amplitudinem verò integram esse *cd*. quater sumptam. Nulla enim parabola prę- ter illam quæ fit iuxtà directionem *ad*. cum habeat impetum *ab*, habebit altitudinem *ac*.

PROPOSITIO XI.

Dato impetu, & amplitudine inuenienda fit directio iuxta quam facta fuit parabola; inuenienda etiam sit altitudo.

Sit datus impetus *ab*, & sit *ad* quarta pars datæ amplitudinis. Fiat circa *ab* semicirculus *acb*, & erigatur *dce* (quæ si in semicirculum non incidit problema impossibile est,) secetq; semicirculum in punctis *c*. & *e*. Dico vtramq; directionem siue *ac*, siue *ae*, si datus impetus *ab*, adhibeatur, parabolam designare, cuius amplitudo quadrupla erit lineæ *ad*. Hoc enim ex prę- cedentibus liquet. Nam proiectiones factæ cum impetu *ab* iuxta directiones *ac* vel *ae* amplitudinem habent quadruplam ipsius *ge*, vel *fc*. vel *ad*. quæ inter se æquales sunt. Altitudo verò esse potest tùm linea *af*, tum etiam *ag*. Vt apparet &c.

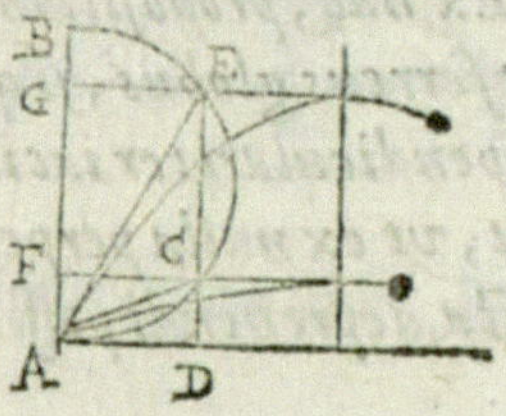

PROPOSITIO XII.

Data amplitndine *ab*, & directione *ac*. inueniendus sit impetus, & altitudo parabolę.

Datis ijsdem, inuenienda sit mensura lineę perpendicularis ad cuius apicem ascenderet mobile si cum eodem impetu sursum perpendicularitèr proiceretur.

Sumatur *ad*, quarta pars ipsius *ab*, & erigantur perpendicula-

altitudo ſit *ac.* fiat circa *ab* ſemicirculus *adb,* & ducantur *cd.* horizontalis, *ad* autem ad punctum *d.* Manifeſtum eſt ex pręcedentibus, directionem quæſitam eſſe *ad,* amplitudinem verò integram eſſe *cd.* quater ſumptam. Nulla enim parabola pręter illam quæ fit iuxtà directionem *ad.* cum habeat impetum *ab,* habebit altitudinem *ac.*

PROPOSITIO XI.

DAto impetu, & amplitudine inuenienda ſit directio iuxta quam facta fuit parabola; inuenienda etiam ſit altitudo.

Sit datus impetus *ab,* & ſit *ad* quarta pars datæ amplitudinis. Fiat circa *ab* ſemicirculus *acb,* & erigatur *dce* (quæ ſi in ſemicirculum non incidit problema impoſſibile eſt,) ſecetq; ſemicirculum in punctis *c.* & *e.* Dico vtramq; directionem ſiue *ac,* ſiue *ae,* ſi datus impetus *ab,* adhibeatur, parabolam deſignare, cuius amplitudo quadrupla erit lineæ *ad.* Hoc enim ex pręcedentibus liquet. Nam proiectiones factæ cum impetu *ab* iuxta directiones *ac* vel *ae* amplitudinem habent quadruplam ipſius *ge,* vel *fc.* vel *ad.* quæ inter ſe æquales ſunt. Altitudo verò eſſe poteſt tùm linea *af,* tum etiam *ag.* Vt apparet &c.

PROPOSITIO XII.

DAta amplitndine *ab,* & directione *ac.* inueniendus ſit impetus, & altitudo parabolę.

Datis ijſdem, inuenienda ſit menſura lineę perpendicularis ad cuius apicem aſcenderet mobile ſi cum eodem impetu ſurſum perpendicularitèr proiceretur.

Sumatur *ad,* quarta pars ipſius *ab,* & erigantur perpendi-

Y cula-

culares *dc. ae.* fiatq; angulus *ace* rectus. Dico *ae.* esse impetum proiectionis, & *dc.* altitudinem. Semicirculus enim circa diametrum *ae.* transit per angulum rectum *ace.* Ergo parabolæ, cuius amplitudo sit *ab*, & directio *ac*, impetus est *ea*.

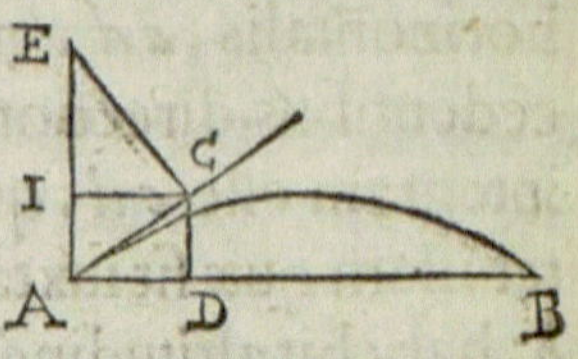

Patet etiam altitudinem parabolę esse lineam *dc.* vel *ai.* Cum autem impetus sit *ae* manifestum est proiectionem perpendicularem sursum ex *a* factam ascensuram esse vsque ad *e* punctum, si proiciatur mobile cum eodem impetu à quo facta fuit parabola *ab*.

Ex hac, propositione colligere possumus quantũ spatij ascendat ferreus globus, si quando ab æneo quolibet tormento sursum perpendiculariter iaciatur: cuius quidem spatij mensura tanta erit, vt ex nulla perpendiculari altitudine siue arte, siue natura facta, deprehendi possit; aut aliter experimento subiacere.

PROPOSITIO XIII.

Data altitudine *ab*, & directione *ac*. reliqua reperire.

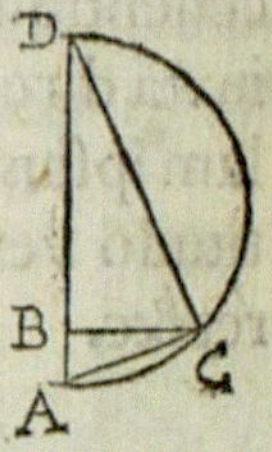

Ducatur per punctum *b*. horizontalis *bc* quę incidat in ipsam *ac* in puncto *c*. Fiatq; angulus *acd* rectus: & circa triangulum rectangulum *acd*. transibit semicirculus propositionis 9. huius. Amplitudo ergo quadrupla erit ipsius *bc*. & impetus erit *ad*. Quod &c.

PROPOSITIO XIV.

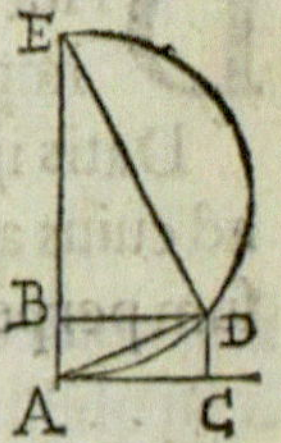

Data altitudine *ab*, & basi (cuius tamen quarta pars sit *ac*) Reliqua reperire.

Compleatur rectangulum *bacd*. & diameter *ad* directionem indicabit. Facto deinde angulo *ade*

recto

culares *dc. ae.* fiatq; angulus *ace* rectus. Dico *ae.* eſſe impetum proiectionis, & *dc.* altitudinem. Semicirculus enim circa diametrum *ae.* tranſit per angulum rectum *ace.* Ergo parabolæ, cuius amplitude ſit *ab,* & directio *ac,* impetus eſt *ea.*

Patet etiam altitudinem parabolę eſſe lineam *dc.* vel *ai.*

Cum autem impetus ſit *ae* manifeſtum eſt proiectionem perpendicularem ſurſum ex *a* factam aſcenſuram eſſe vſque ad *e* punctum, ſi proiciatur mobile cum eodem impetu à quo facta fuit parabola *ab.*

Ex hac, propoſitione colligere poſſumus quantũ ſpatij aſcendat ferreus globus, ſi quando ab æneo quolibet tormento ſurſum perpendiculariter iaciatur: cuius quidem ſpatij menſura tanta erit, vt ex nulla perpendiculari altitudine ſiue arte, ſiue natura facta, deprehendi poſſit; aut aliter experimento ſubiacere.

PROPOSITIO XIII.

DAta altitudine *ab,* & directione *ac.* reliqua reperire.

Ducatur per punctum *b.* horizontalis *bc* quę incidat in ipſam *ac* in puncto *c.* Fiatq; angulus *acd* rectus: & circa triangulum rectangulum *acd.* tranſibit ſemicirculus propoſitionis 9. huius. Amplitutudo ergo quadrupla erit ipſius *bc.* & impetus erit *ad.* Quod &c.

PROPOSITIO XIV.

DAta altitudine *ab,* & baſi (cuius tamen quarta pars ſit *ac*) Reliqua reperire.

Compleatur rectangulum *bacd.* & diameter *ad* directionem indicabit. Facto deinde angulo *ade.*

recto

recto, erit *ae* impetus, vt facile ex præcedentibus colligitur &c.

PROPOSITIO XV.

PRoiectio perpendicularis ſurſum æqualis eſt dimidiæ baſi proiectionis ſemirectæ, ſi fuerit ab eodem impetu facta vtraq; tam perpendicularis, quam ſemirecta proiectio.

Sit parabola ſemirecta *abc*, & ſuper *ad* media amplitudine fiat quadratum *adef*, erit *ae* diameter ipſa directio ſemirecta. Factoq; circa *af* ſemicirculo, tranſibit ſemicirculus per *i* centrum quadrati, & erit *af* impetus; quare proiectio perpendicularis ſurſum vſq; ad *f* punctum aſcendet. Patet ego propoſitum.

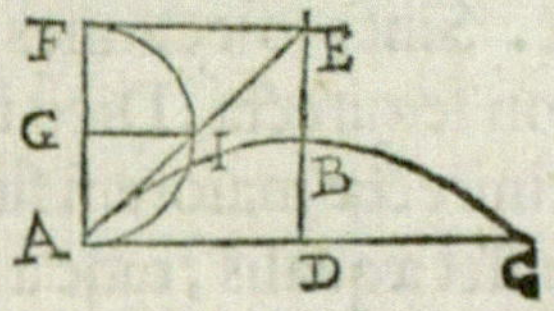

PROPOSITIO XVI.

SI facta fuerit proiectio ad eleuationem anguli ſemirecti, amplitudo integræ proiectionis erit latus rectum deſcriptæ parabolæ,

Sit eleuatio ſemirecta iuxtà lineam *ab*, factaque ſit à proiecto quęlibet parabola *acd*. Dico *ad*. eſſe latus rectum huius parabolæ. Cum enim angulus *eab* ſemirectus ſit, *aeb* rectus, erunt ęqualia latera *ae*, *eb*. ergo *ae*. dupla erit ipſius *ec*. Sed cum *ae* media proportionalis ſit inter latus rectum & *ec*. erit latus rectum duplum ipſius *ae*. nempe æquale ipſi *ad*. Quod erat &c.

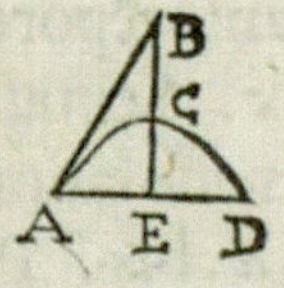

recto, erit *ae* impetus, vt facile ex præcedentibus colligitur &c.

PROPOSITIO XV.

PRoiectio perpendicularis ſurſum æqualis eſt dimidiæ baſi proiectionis ſemirectæ, ſi fuerit ab eodem impetu facta vtraq; tam perpendicularis, quam ſemirecta proiectio.

Sit parabola ſemirecta *abc,* & ſuper *ad* media amplitudine fiat quadratum *adef,* erit *ae* diameter ipſa directio ſemirecta. Factoq; circa *af* ſemicirculo, tranſibit ſemicirculus per *i* centrum quadrati, & erit *af* impetus; quare proiectio perpendicularis ſurſum vſq; ad *f* punctum aſcendet. Patet ego propoſitum.

PROPOSITIO XVI.

SI facta fuerit proiectio ad eleuationem anguli ſemirecti, amplitudo integræ proiectionis erit latus rectum deſcriptæ parabolæ,

Sit eleuatio ſemirecta iuxtà lineam *ab,* factaque ſit à proiecto quę libet parabola *acd.* Dico *ad.* eſſe latus rectum huius parabolæ. Cum enim angulus *eab* ſemirectus ſit, *aeb* rectus, erunt ęqualia latera *ae, eb.* ergo *ae.* dupla erit ipſius *ec.* Sed cum *ae* media proportionalis ſit inter latus rectum & *ec.* erit latus rectum duplum ipſius *ae.* nempe æquale ipſi *ad.*

Quod erat &c.

PROPOSITIO XVII.

AD proiectiones æquales faciendas, minor impetus requiritur in ea, quæ ad eleuationem ſemirectam fieri debeat,

Demonſtratum iam eſt, ſi ab eodem impetu factæ ſint proiectiones, longius procedere eam quæ ad angulum ſemirectum fuerit exploſa. Sint proiectiones *ab*. ſemirecta, & *c.c.* non ſemirecta. Dico impetum ipſius *c.c*. non ſemirectæ maiorem fuiſſe quam ipſius *b*. ſemirectæ. Si enim fuiſſet æqualis, tunc amplitudo iactus *ac* ex demonſtratis fuiſſet minor quam *ae*. vt verbi gratia *ad*; ſed cum æqualis ponatur amplitudo, maior omnino impetus fuit per *ac*. quam per *ab*; vel minor impetus requiritur in ſemirecta quam in alia. Quod erat &c.

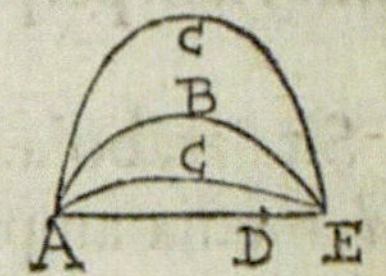

PROPOSITIO XVIII.

TEmpus, ſiue durationem vniuſcuiuſque proiectionis definire.

Conſtructà ſolita præcedentium figura, ſit proiectio *abc*. oportet tempus, ſiue durationem eius reperire; hoc eſt quanto tẽpore fiat latio per parabolam *abc*. Scimus iam ex Galileo idem tempus eſſe lationis *abc*, & cadentis ex *b* in *d*. bis. Ponamus ergo tempus cadentis naturaliter ex *f* in *a*, eſſe *fa*; eritq; tempus per *ea* media proportionalis *ag*. & hoc ſemper. ergo linea *ag* metitur tempus lationis per *ea*, ſiue per *bd*, ſiue per ſemiparabolam *bc*, vel per integram etiã parabolam *abc*, Eandem enim rationem inter ſe habebunt durationes parabolarum, quam habent ſemiparabolarũ: & nos loquimur de proportionibus, non de menſuris.

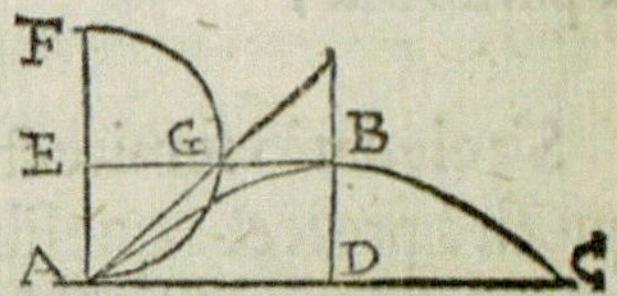

PRO-

PROPOSITIO XVII.

AD proiectiones æquales faciendas, minor impetus requiritur in ea, quæ ad eleuationem ſemirectam fieri debeat,

Demonſtratum iam eſt, ſi ab eodem impetu factæ ſint proiectiones, longius procedure eam quæ ad angulum ſemirectum fuerit exploſa. Sint proiectiones *ab.* ſemirecta, & *c.c.* non ſemirecta. Dico impetum ipſius *c.c.* non ſemirectæ maiorem fuiſſe quam ipſius *b.* ſemirectæ. Si enim fuiſſet æqualis, tunc amplitudo iactus *ac* ex demonſtratis fuiſſet minor quam *ae.* vt verbi gratia *ad*; ſed cum æqualis ponatur amplitudo, maior omnino impetus fuit per *ac.* quam per *ab*; vel minor impetus requiritur in ſemirecta quam in alia. Quod erat &c.

PROPOSITIO XVIII.

TEmpus, ſiue durationem vniuſcuiuſque proiectionis definire.

Conſtructà ſolita præcedentium figura, ſit proiectio *abc.* oportet tempus, ſiue durationem eius reperire; hoc eſt quanto tēpore fiat latio per parabolam *abc.* Scimus iam ex Galileo idem tempus eſſe lationis *abc,* & cadentis ex *b* in *d.* bis. Ponamus ergo tempus cadentis naturaliter ex *f* in *a,* eſſe *fa*; eritq; tempus per *ea* media proportionalis *ag.* & hoc ſemper. ergo linea *ag* metitur tempus lationis per *ea,* ſiue per *bd,* ſiue per ſemiparabolam *bc,* vel per integram ęriã parabolam *abc,* Eandem enim rationem inter ſe habebunt durationes parabolarum, quam habent ſemiparabolarũ: & nos loquimur de proportionibus, non de menſuris.

P R O-

PROPOSITIO XIX.

DVrationes proiectionum sunt vt lineę ordinatim applicatę in aliqua parabola ad suam vniuscuiusq; altitudinem.

Sint altitudines duarum parabolarum *ab*, *ac*. (à quocunq; impetu siue eodem, siue nõ, factæ sint, & quascunq; bases habeant, siue ęquales, siue inęquales.) Fiat circa *ca*. parabola inuersa *afd*, & semicirculus. *age*. Eruntque tempora parabolarum vt sunt tempora cadentium per *ca*, *ba*; hoc est vt *cd*, *bf* per *X* pręcedentis libri. Quod erat &c.

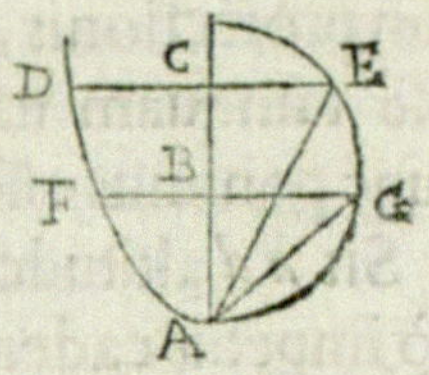

Sunt insuper *ea*, *ga*. in eadem ratione ac *cd*, *bf*, quare etiam *ea*, *ga*. chordæ in semicirculo erunt vt tempora parabolarum &c.

PROPOSITIO XX.

PArabolarum æqualem basim habentium impetus in puncto sublimiori sunt in contraria ratione temporum, siue durationum earumdem.

Sint duæ parabolę *ac*. *af*. quæ eandem habeant basim, & eundem axem *bc*. fiatq; circa *bc*. parabola inuersa *bde*. Dico impetum in *c* ad impetum in *f* esse ut *fd*, ad *ce*. Impetus enim in punctis *c*; & *f* sunt puri illi impetus horizontales, secundum quos conficitur latio horizontalis *ab*.

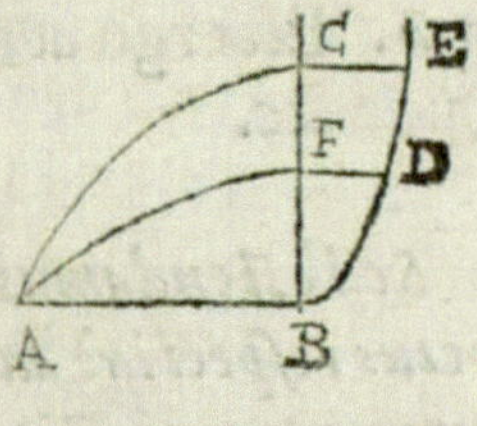

Cum itaq; eadem latio horizontalis *ab* absoluatur temporibus *ce*, *fd*, erunt impetus horizontales reciprocè ut *fd*, ad *ce*. per 3. Propos. Gal. De Motu naturaliter accelerato. *per prac.*

PRO-

PROPOSITIO XIX.

DVrationes proiectionum ſunt vt lineę ordinatim applicatę in aliqua parabola ad ſuam vniuſcuiuſq; altitudinem.

Sint altitudines duarum parabolarum *ab, ac.* (à quocunq; impetu ſiue eodem, ſiue nõ, factæ ſint, & quaſcunq; baſes habeant, ſiue ęquales, ſiue inęquales.) Fiat circa *ca.* parabola inuerſa *afd,* & ſemicirculus. *age.* Eruntque tempora parabolarum vt ſunt tempora cadentium per *ca, ba*; hoc eſt vt *cd, bf* per *X* pręcedentis libri. Quod erat &c.

Sunt inſuper *ea, ga.* in eadem ratione ac *cd, bf,* quare etiam *ea, ga.* chordæ in ſemicirculo erunt vt tempora parabolarum &c.

PROPOSITIO XX.

PArabolarum æqualem baſim habentium impetus in puncto ſublimiori ſunt in contraria ratione temporum, ſiue durationum earumdem.

Sint duæ parabolę *ac. af.* quæ eandem habeant baſim. & eundem axem *bc.* fiatq; circa *bc.* parabola inuerſa *bde.* Dico impetum in *c* ad impetum in *f* eſſe ut *fd,* ad *ce.* Impetus enim in punctis *c;* & *f* ſunt puri illi impetus horizontales, ſecundum quos conficitur latio horizontalis *ab.* Cum itaq; eadem latio horizontalis *ab* abſoluatur temporibus *ce, fd,* erunt impetus horizontales reciprocè ut *fd,* ad *ce.* *per prae.*
per 3. Propoſ. Gal. De Motu naturaliter accelerato.

P R O-

PROPOSITIO XXI.

IMpetum purum horizontalem, qui inuariabilis eſt ſemper idem in unoquoq; parabolę puncto definire.

Item etiam & perpendicularem uariabilem. Repetita præcedentium propoſitionum figura, ſupponimus, ut ſemper, impetum totalem, ſiue compoſitum proiectionis, quem habet mobile in puncto *a*, eſſe tamquam naturaliter cadentis ex *b* in *a*. & hunc ponimus eſſe ut linea *ba*.

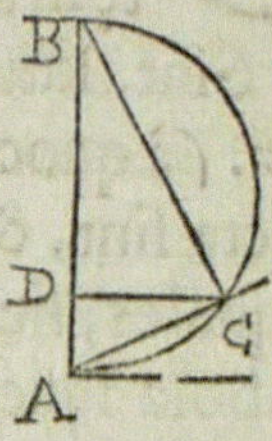

Sit *ad* altitudo, & *bd* ſublimitas parabolę Ergò impetus cadentis per *bd* ſublimitatem parabolę, erit ut linea *bc*, media proportionalis inter *ab*, *bd*.

At iſte impetus cadentis ex *b* in *d* eſt ille purus horizontalis qui lationi ineſt in quolibet puncto parabolę, & eſt inuariabilis. Quare in vnoquoq; puncto parabolę impetus horizontalis erit ut linea *bc*.

Perpendicularis uerò impetus qui eſt in primo lationis puncto ſic determinabitur. Manente ſemper unica ſuppoſitione, impetum ſilicet caſus per *ea*. eſſe ipſam *ea*. Impetus perpendicularis in fine parabolæ *h*, eſt tamquam naturaliter cadentis ex *b* in *c*, uel ex *d* in *a*. Eſt ergo ut media proportionalis *af*. Quod &c.

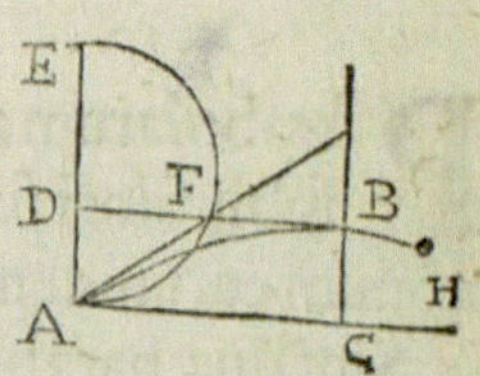

Sed oſtendamus etiam pro Corollario, quomodo varietur impetus reſpectiuè ad horizontem, eiuſdem globi ferrei ab eodem tormento proiecti; creſcit enim impetus perpendicularis non quē admodum creſcunt eleuationes tormenti, aut altitudines parabolæ, ſed ea ratione qua creſcit in ſemicirculo chorda af. *Hinc animaduertere licet futurum fore vt idem globus ferreus eodem tormento exploſus dum ad horizontem redit aliquãdo tecta fornicesq; domorum traiciat, quandoq; verò neque glaciem alicuius lacunæ lędere poſſit.*

Nota-

PROPOSITIO XXI.

IMpetum purum horizontalem, qui inuariabilis eſt ſemper idem in unoquoq; parabolę puncto definire.

Item etiam & perpendicularem uariabilem. Repetita præcedentium propoſitionum figura, ſupponimus, ut ſemper, impetum totalem, ſiue compoſitum proiectionis, quem habet mobile in puncto *a,* eſſe tamquam naturaliter cadentis ex *b* in *a.* & hunc ponimus eſſe ut linea *ba.*

Sit *ad* altitudo, & *bd* ſublimitas parabolę Ergò impetus cadentis per *bd* ſublimitatem parabolę, erit ut linea *b c,* media proportionalis inter *ab, bd.*

At iſte impetus cadentis ex *b* in *d* eſt ille purus horizontalis qui lationi ineſt in quolibet puncto parabolę, & eſt inuariabilis. Quare in vnoquoq; puncto parabolę impetus horizontalis erit ut linea *bc.*

Perpendicularis uerò impetus qui eſt in primo lationis puncto ſic determinabitur. Manente ſemper unica ſuppoſitione, impetum ſilicet caſus per *ea.* eſſe ipſam *ea.* Impetus perpendicularis in fine parabolæ *h,* eſt tamquam naturaliter cadentis ex *b* in *c,* uel ex *d* in *a.* Eſt ergo ut media proportionalis *af.* Quod &c.

Sed ostendamus etiam pro Corollario, quomodo varietur impetus reſpectiuè ad horizontem, eiuſdem globiferrei[91] *ab eodem tormento proiecti; creſcit enim impetus perpendicularis non quĕadmodum creſcunt eleuationes tormenti, aut altitudines parabolæ, ſed ea ratione qua creſcit in ſemicirculo chorda* af. *Hinc animaduertere licet futurum fore vt idem globus ferreus eodem tormento exploſus dum ad horizontem redit aliquãdo tecta forniceſq; domorum traiciat, quandoq; verò neque glaciem alicuius lacunæ lędere poſſit*

Nota-

[91] *globiferrei*] *globi ferrei.*

Notabimus etiam vim in percutiendo, grauium per se naturaliter cadentium per spatia b a, c a. *esse* (*facto semicirculo quolibet* a e d *per punctum* a) *vt sunt* a e, a f. *Sunt enim* a e, a f. *tempora casuũ cum sint in subdupla ratione spatiorum; ergò sunt etiam mensuræ, siue indices momentorum velocitatis, aggregatorum per spatia* b a, c a.

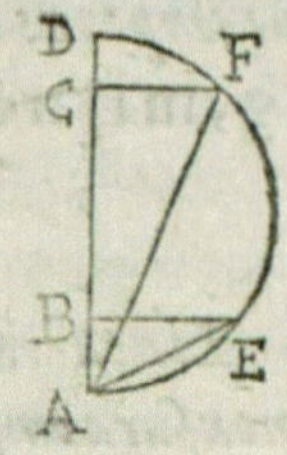

PROPOSITIO XXII.

IMpetum compositum, siue absolutum, quantus sit in quolibet puncto parabolę demonstrare.

In solita præcedentium propositionũ figura, sumatur quodlibet punctum *a* in parabola *g h a*. & ducatur horizontalis *a b i d*. Ducaturq; *c b*. Dico (facta semper eadem suppositione impetum silicet per *c g* esse *c g*) quod impetus compositus in *a*, siue in *i*, est recta *c b*. Cum enim impetus in puncto parabolę *a* sit vt naturaliter cadentis ex puncto sublimitatis *e* vsq; in *f*, vel ex *c* vsq. in *d*. erit ille impetus vt *c b* media proportionalis inter *g c*, *cd*. Quod erat &c.

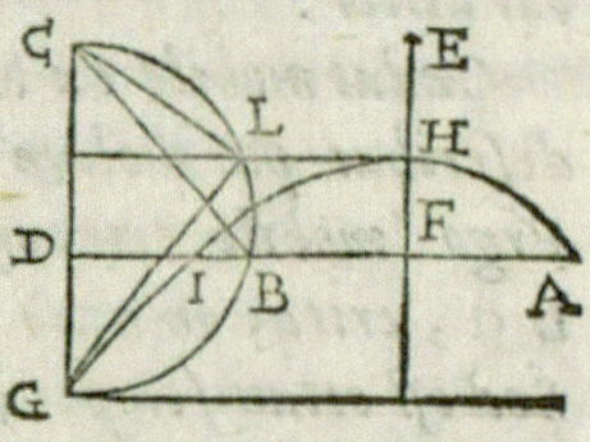

Idem etiam hoc alio modo considerabimus.

Si ex puncto sublimi alicuius datæ parabolę alia parabola circa eandem diametrum describatur, lineę ordinatim ductæ in descripta, determinabunt impetus absolutos in singulis punctis parabolę datę.

Sit data parabola a b, *cuius punctum sublime sit* c. *Circa communem diametrum fiat per* c *parabola quælibet* c d, *Ductisque ordinatim quotcunq; lineis* a e, f g. h i, b d. *Dico impetus in punctis* a, f, h, b, *esse vt sunt lineę* a e, l g. m i, & n d. *Hoc.n.patet. Quia impetus in* a. f. h. b. *pũctis sunt*

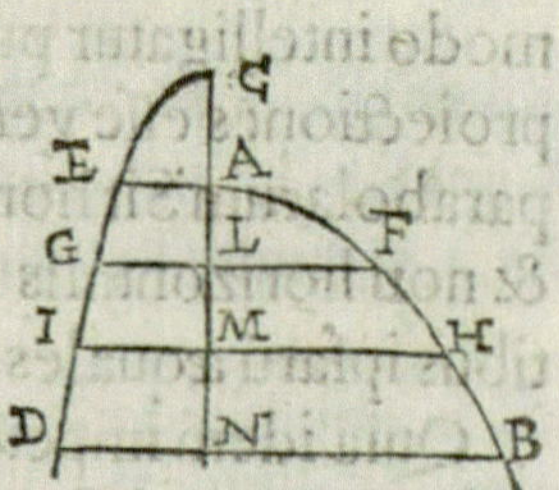

vt im-

Notabimus etiam vim in percutiendo, grauium per ſe naturaliter cadentium per ſpatia ba, ca. *eſſe (facto ſemicirculo quolibet* aed *per punctum* a) *vt ſunt* ae, af. *Sunt enim* ae, af. *tempora caſuū cum ſint in ſubdupla ratione ſpatiorum; ergò ſunt etiam menſuræ, ſiue indices momentorum velocitatis, aggregatorum per ſpatia* ba, ca.

PROPOSITIO XXII.

IMpetum compoſitum, ſiue abſolutum, quantus ſit in quolibet puncto parabolę demonſtrare.

In ſolita præcedentium propoſitionū figura, ſumatur quodlibet punctum *a* in parabola *gha.* & ducatur horizontalis *abid.* Ducaturq; *cb.* Dico (facta ſemper eadem ſuppoſitione impetum ſilicet per *cg* eſſe *cg*) quod impetus compoſitus in *a,* ſiue in *i,* eſt recta *cb.* Cum enim impetus in puncto parabolę *a* ſit vt naturaliter cadentis ex puncto ſublimitatis *e* vſq; in *f,* vel ex *c* vſq. in *d.* erit ille impetus vt *cb* media proportionalis inter *gc, cd.* Quod erat &c.

Idem etiam hoc alio modo conſiderabimus.

Si ex puncto ſublimi alicuius datæ parabolę alia parabola circa eandem diametrum deſcribatur, lineę ordinatim ductæ in deſcripta, determinabunt impetus abſolutos in ſingulis punctis parabolę datę.

Sit data parabola ab, *cuius punctum ſublime ſit* c. *Circa communem diametrum fiat per* c *parabola quælibet* cd, *Ductiſque ordinatim quotcunq; lineis* ae, fg. hi, bd. *Dico impetus in punctis* a, f, h, b, *eſſe vt ſunt lineę* ae, lg. mi, & nd. *Hoc. n. patet. Quia impetus in* a. f. h. b. *pūctis, ſunt*

ut im-

vt impetus cadentium per c a, c l, c m, c n, *nempe vt lineę* a e, l g, m i, n d. *Quod erat &c.*

Lemma.

Linea a b. *quam in praecedentibus pro mensura impetus ponebamus, & circa quam semicirculum describebamus, quarta pars est lateris recti parabolæ* b c, *factæ ab horizontali proiectione. Patet hoc ex primo libro, cum* a b *sit impetus, hoc est sublimitas parabolæ* b c *secundum Galileum. attamen demonstretur alitèr.*

Cadat mobile ex a *in* b, *& inde horizontalitèr conuersum describat parabolam* b c. *Sumatur* b d. *dupla ipsius* b a. *Ergo mobile tempore casus percurret horizontale spatium* b d, *eritq; omninò post tempus casus in perpendiculo* d c. *Sed est etiam semper in parabola* b c, *ergo in* c *communi concursu erit. Cum ergo facta sit descensio* d c *tempore casus, erit* d c *æqualis* b a. *Patet autem quadratum* b d *æquari rectangulo sub* c d, *et quadrupla* b a (*cum vtraq;* c d, b a *semissis sit ipsius* b d.) *Est ergo* b a *quarta pars lateris recti parabolę horizontalis* b c. *Quod &c.*

PROPOSITIO XXIII.

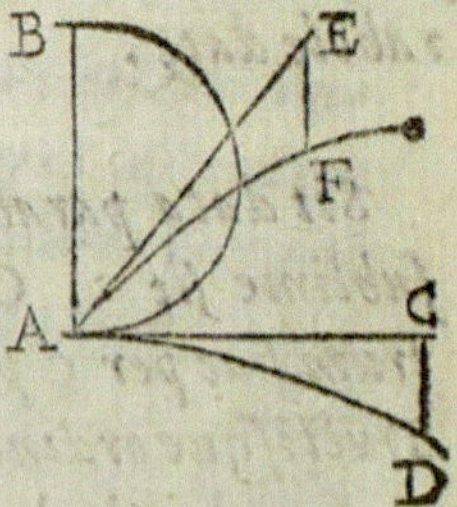

Omnes parabolæ ab eodem impetu *ab* factę idem habent latus rectum. (dummodo intelligatur punctum *a*, ex quo fiunt proiectiones esse vertex omniũ obliquarũ. parabolarum Sit horizontalis parabola *ad*, & non horizontalis *af*: sumantur in tangentibus ipsarũ æquales *ac*, *ae*.

Quia idem impetus est per *ac*, & per *ae*, ipsæ *ac*, *ae*, abstracta grauitatis operatione, eodem tempore absoluerentur; essentq. grauia eodem tẽpore

vt impetus cadentium per ca, cl, cm, cn, *nempe vt lineę* ae, lg, mi, nd. *Quod erat &c.*

Lemma.

Linea ab. *quam in præcedentibus pro menſura impetus ponebamus, & circa quam ſemicirculum deſcribebamus, quarta pars est lateris recti parabolæ* bc, *factæ ab horizontali proiectione. Patet hoc ex primo libro, cum* ab *ſit impetus, hoc est ſublimitas parabolæ* bc *ſecundum Galileum. attamen demonstretur alitèr.*

Cadat mobile ex a *in* b, *& inde horizontalitèr conuerſum deſcribat parabolam* bc. *Sumatur* bd. *dupla ipſius* ba. *Ergo mobile tempore caſus percurret horizontale ſpatium* bd, *eritq; omninò post tempus caſus in perpendiculo* dc. *Sed eſt etiam ſemper in parabola* bc, *ergo in* c *communi concurſu erit. Cum ergo facta ſit deſcenſio* dc *tempore caſus, erit* dc *æqualis* ba. *Patet autem quadratum* bd *æquari rectangulo ſub* cd, *et quadrupla* ba (*cum vtraq;* cd, ba *ſemiſſis ſit ipſius* bd.) *Eſt ergo* ba *quarta pars lateris recti parabolę horizontalis* bc. *Quod &c.*

PROPOSITIO XXIII.

OMnes parabolæ ab eodem impetu *ab* factę idem habent latus rectum. (dum modo intelligatur punctum *a,* ex quo fiunt proiectiones eſſe vertex omniũ obliquarũ. parabolarum Sit horizontalis parabola *ad,* & non horizontalis *af:* ſumantur in tangentibus ipſarũ æquales *ac, ae.*

Quia idem impetus eſt per *ac,* & per *ae,* ipſæ *ac, ae,* abſtracta grauitatis operatione, eodem tempore abſoluerentur; eſſentq. grauia eodem tẽ-pore

pore in *e*. & *e*: sed cum grauitas operetur, & idem sit tempus, æquales erunt descensus *ef*, *cd*. Quadratum autem *ac*. æquatur rectangulo sub *cd* & quadrupla *ab*, cum demonstratum sit in Lem. præced. rectam *ab* esse quartam partem lateris recti parabolę *ad*, quare etiam quadratum *ae* æquale erit rectangulo sub *ef*, & quadrupla *ab*. Est igitur eadem *ab* quarta pars lateris recti omnium parabolarum ab eodem impetu factarum. Quod erat &c.

Corollarium.

Hinc manifestum est semper sublimitatem parabolarum ascendentium, siue lineam impetus, quartam partem esse lateris recti illius portionis parabolæ obliquæ, quæ verticem habeat in puncto separationis proiecti ab instrumento impellente,

Verbi gratia. Si mobile post casum a b *ex quiete in* a, *conuertatur non horizontalitèr, sed per quamlibet inclinatam* b c, *parabolamq; describat* b d. *Patet lineam impetus, siue sublimitatem* a b *esse quartam partem lateris recti parabolæ* b d. *considerata tamen parabola obliqua* b d *ita vt eius vertex sit punctum* b, *& applicatarum regula sit tangens* b c.

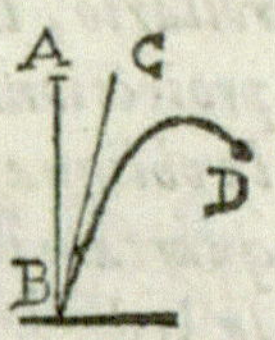

Quod autem hęc conueniant cum doctrina Conicorum, sic demonstrabimus. Si parabola a b *duas tangentes habuerit* a c *per verticem, &* b c *non per verticem: sumptaq; fuerit* a d *quarta pars lateris recti, Dico iunctam* d c *angulos rectos facere cum* b c.

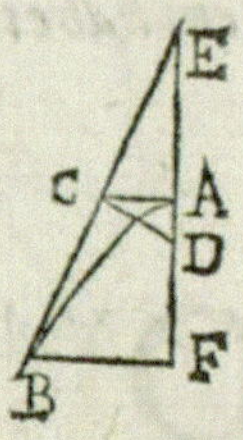

Agatur b f *ordinatim. Cum sint æquales* f a, a e. *erit quadratum* b f. *quadruplum quadrati* c a. *sed idem quadratum* b f *quadruplum est rectanguli* f a d, *hoc est rectanguli* e a d. *æqualia ergo sunt quadratum* c a, *& rectangulum* e a d; *angulusq;* e c d. *rectus.*

Hic, nisi penitus abs re nostra esset, facillimè eliceremus demonstrationem foci. Si enim produceretur b d. *essent per quartam primi elementorum æquales anguli* d e c, d b c. *sed ad rem nostram.*

Z His

pore in *e.* & *e:* ſed cum grauitas operetur, & idem ſit tempus, æquales erunt deſcenſus *ef, cd.* Quadratum autem *ac.* æquatur rectangulo ſub *cd* & quadrupla *ab,* cum demonſtratum ſit in Lem. præced. rectam *ab* eſſe quartam partem lateris recti parabolę *ad,* quare etiam quadratum *a e* æquale erit rectangulo ſub *ef,* & quadrupla *ab.* Eſt igitur eadem *ab* quarta pars lateris recti omnium parabolarum ab eodem impetu factarum.

Quod erat &c.

Corollarium.

Hinc manifeſtum eſt ſemper ſublimitatem parabolarum aſcendentium, ſiue lineam impetus, quartam partem eſſe lateris recti illius portionis paræbolæ obliquæ, quæ verticem habeat in puncto ſeparationis proiecti ab instrumento impellente,

Verbi gratia. Si mobile poſt caſum ab *ex quiete in* a, *conuertatur non horizontalitèr, ſed per quamlibet inclinatam* bc, *parabolamq; deſcribat* bd. *Patet lineam impetus, ſiue ſublimitatem* ab *eſſe quartam partem lateris recti parabolæ* bd. *conſiderata tamen parabola obliqua* bd *ita vt eius vertex ſit punctum* b, *& applicatarum regula ſit tangens* bc.

Quod autem hęc conueniant cum doctrina Conicorum, ſic demonstrabimus. Si parabola ab *duas tangentes habuerit* ac *per verticem, &* bc *non per verticem: ſumptaq; fuerit* ad *quarta pars lateris recti, Dico iunctam* dc *angulos rectos facere cum* bc.

Agatur bf *ordinatim. Cum ſint æquales* fa, ae. *erit quadratum* bf. *quadruplum quadrati* ca. *ſed idem quadratum* bf *quadruplum eſt rectanguli* fad, *hoc eſt rectanguli* ead. *æqualia ergo ſunt quadratum* ca, *& rectangulum* ead; *anguluſq;* ecd. *rectus.*

Hic, niſi penitus abs re noſtra eſſet, facillimè eliceremus demonstrationem foci. Si enim produceretur bd. *eſſent per quartam primi elementorum æquales anguli* dec, dbc. *ſed ad rem noſtram.*

Z *His*

His demonstratis, data parabola ab *cuius axis* bd, *& focus sit* c. *si sumatur quodlibet punctum* a *in sectione, & ordinatim ducatur* ad. *Dico latus rectum parabolæ obliquæ verticem habentis in* a, *esse quadruplum linearum simul* db, bc; *siue lineæ* ce, *siue lineæ* ca *si producatur. Ducatur tangens* ae, *& ex* b, *agatur* bf *parallela tangenti* ae; *&* af *ducatur parallela axi* bd. *Erit per iam demonstrata quadratum* he. *æquale rectangulo* bec; *quadrupla etiam æqualia erunt; hoc est quadratum* ae, *vel* fb, *æquale erit rectangulo sub* be, *& quadrupla* ec, *vel sub* fa, *& quadrupla eiusdem* ec. *Quære ipsa* ec, *vel* db, bc *simul sunt quarta pars lateris recti parabolæ obliquæ cuius vertex sit* a. *& diameter* af.

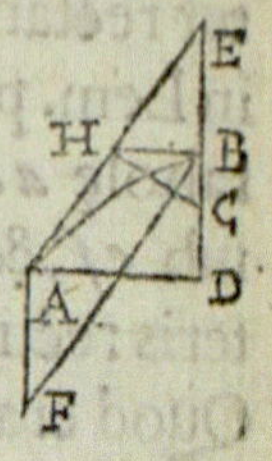

Nos autem dicebamus in præcedenti Corollario, lineam ab, *quæ metitur impetum proiectionum, siue quæ sublimitas est parabolæ obliquæ* bd *verticem habentis in* b, *esse quartam partem lateris recti eiusdem parabolæ* bd. *Quod esse verum confirmauimus etiã ex doctrina Apollonij, cum linea* ab *constet ex* de, *& ex sublimitate, vel quarta parte lateris recti parabolæ rectæ quæ verticem habet* d.

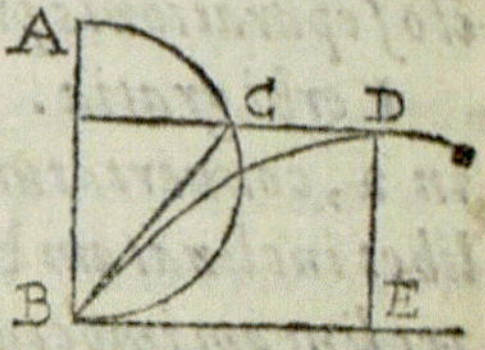

PROPOSITIO XXIV.

QVælibet parabola infinitas habet sublimitates.

Si enim per punctum sublime *d*. quod reperit Galileus, linea horizontalis *de* producatur; quælibet linea perpendicularis quæ ex hac demittatur in parabolam sublimitas eiusdem parabolæ erit, dummodo impetus à mobili per descensum aquisitus conuertatur non per lineam horizontalem, sed tangentem.

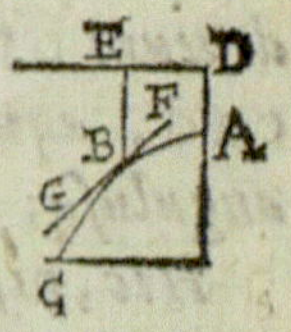

Pro-

His demonſtratis, data parabola ab *cuius axis* bd, *& focus ſit* c. *ſi ſumatur quodlibet punctum* a *in ſectione, & ordinatim ducatur* ad. *Dico latus rectum parabolæ obliquæ verticem habentis in* a, *eßе quadruplum linearum ſimul* db, bc; *ſiue lineæ* ce, *ſiue lineæ* ca *ſi producatur. Ducatur tangens* ae, *& ex* b, *agatur* bf *parallela tangenti* ae; *&* af *ducatur parallela axi* bd. *Erit per iam demonſtrata quadratum* he. *æquale rectangulo* bec; *quadrupla etiam æqualia erunt; hoc eſt quadratum* ae, *vel* fb, *æquale erit rectangulo ſub* be, *& quadrupla* ec, *vel ſub* fa, *& quadrupla eiuſdem* ec. *Quære ipſa* ec, *vel* db, bc *ſimul ſunt quarta pars lateris recti parabolæ obliquæ cuius vertex ſit* a. *& diameter* af.

Nos autem dicebamus in præcedenti Corollario, lineam ab, *quæ metitur impetum proiectionum, ſiue quæ ſublimitas est parabolæ obliquæ* bd *verticem habentis in* b, *eſſe quartam partem lateris recti eiuſdem parabolæ* bd. *Quod eſſe verum confirmauimus etiã ex doctrina Apollonij, cum linea* ab *conſtet ex* de, *& ex ſublimitate, vel quarta parte lateris recti parabolæ rectæ quæ verticem habet* d.

PROPOSITIO XXIV.

QVælibet parabola infinitas habet ſublimitates.

Si enim per punctum ſublime *d.* quod reperit Galileus, linea horizontalis *de* producatur; quælibet linea perpendicularis quæ ex hac demittatur in parabolam ſublimitas eiuſdem parabolæ erit, dummodo impetus à mobili per deſcenſum aquiſitus conuertatur non per lineam horizontalem, ſed tangentem.

Pro-

Propoſita ſit parabola *a b c*. cuius ſublimitas *a d*. & per *d* agatur horizonti ęquidiſtans *d e*. Demittatur iam quælibet *e b*. parallela ipſi *d a*. Dico *e b* ſublimitatem eſſe parabolæ *a b c*, dummodo mobile in puncto *b*. conuenienti modo conuertatur, hoc eſt per tangentem in puncto *b*. Vel. Dico graue poſt caſum *e b* per tangentem *bf*. ſiue *b g*. conuerſum, propoſitam parabolam percurrere. Eſt enim idem impetus cadentis ex *e* in *b*, ac venientis ex *d* per *a* in *b*. Cum ergo in vtroque caſu reperiatur in *b* idem impetus, eademq; directio, ſiue venerit mobile ex *e* in *b*, ſiue ex *d* per *a*. in *b*. continuabit mobile per eandem lineam *b c* curſum ſuum. Eadem dicemus de conuerſione per *b a* poſt eundem caſum *e b*. Quare *e b* ſublimitas eſt parabolæ *a b c*. Quod &c.

PROPOSITIO XXV.

DAtis binis quibuſcunq;, ſiue impetu & directione, ſiue impetu & amplitudine, ſiue amplitudine, & directione; focum parabolæ reperire.

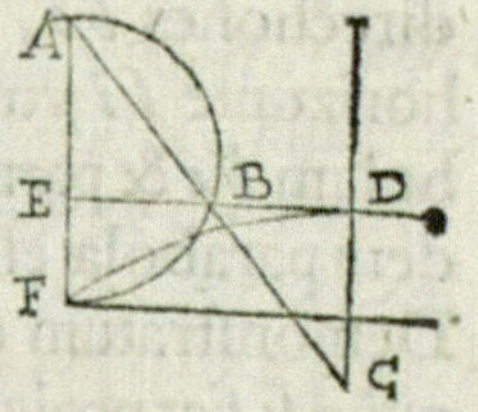

Iuxtà duo data conſtruatur figura propoſitionum præcedentium, & producatur *a b c* donec concurrat cum axe paraboię *d c*. Dico *c* focum eſſe parabolæ. Cum enim per conſtructionem æquales ſint *e b*, *b d*, æquales erunt etiam *a b*, *b c* inter eaſdem paralle las, & *e a*. *d c*. Sed *e a* ſublimitas eſt parabolæ *f d*, ergo *d c* quarta pars eſt lateris recti, & propterea *c*. focus erit. Quod &c.

Corollarium.

Hinc patet parabolas quæ ſemirectæ ſint habere focum in horizontali linea; Minores vero portiones quam ſemirecta focum habere ſub horizonte, & maiores ſupra horizontem.

Propoſita ſit parabola *abc.* cuius ſublimitas *ad.* & per *d* agatur horizonti ęquidiſtans *de.* Demittatur iam quælibet *eb.* parallela ipſi *da.* Dico *eb* ſublimitatem eſſe parabolæ *abc,* dummodo mobile in puncto *b.* conuenienti modo conuertatur, hoc eſt per tangentem in puncto *b.* Vel. Dico graue poſt caſum *eb* per tangentem *bf.* ſiue *bg.* conuerſum, propoſitam parabolam percurrere. Eſt enim idem impetus cadentis ex *e* in *b,* ac venientis ex *d* per *a* in *b.* Cum ergo in vtroque caſu reperiatur in *b* idem impetus, eademq; directio, ſiue venerit mobile ex *e* in *b,* ſiue ex *d* per *a.* in *b.* continuabit mobile per eandem lineam *bc* curſum ſuum. Eadem dicemus de conuerſione per *ba* poſt eundem caſum *eb.* Quare *eb* ſublimitas eſt parabolæ *abc.* Quod &c.

PROPOSITIO XXV.

DAtis binis quibuſcunq;,[92] ſiue impetu & directione, ſiue impetu & amplitudine, ſiue amplitudine, & directione; focum parabolæ reperire.

Iuxtà duo data conſtruatur figura propoſitionum præcedentium, & producatur *abc* donec concurrat cum axe parabole *dc.* Dico *c* focum eſſe parabolæ. Cum enim per conſtructionem æquales ſint *eb, bd,* æquales erunt etiam *ab, bc* inter eaſdem parallelas, & *ea. dc.* Sed *ea* ſublimitas eſt parabolæ *fd,* ergo *dc* quarta pars eſt lateris recti, & propterea *c.* focus erit. Quod &c.

Corollarium.

Hinc patet parabolas quæ ſemirectæ ſint habere focum in horizontali linea; Minores vero portiones quam ſemirecta focum habere ſub horizonte, & maiores ſupra horizontem.

[92] quibuſcunq;,] quibuſcunq,

PROPOSITIO XXVI.

PArabola proiectionis horizontalis maxima omnium est, quæ fieri possint ab eodem impetu.

Sit impetus *ab*. fiatq; circulus *adb*. Sit etiam parabola iactus horizontalis *bc*, & alia parabola *be*. Dico maiorem esse parabolam *bc*. quàm *be*. Est enim *ba* recta sub limitas parabolę *bc*, & *ha* recta sublimitas parabolę *be*. Quare *bc* omnium maxima erit; cum maiorem habeat sublimitat em, ideoq. maius latus re ctum. Quod erat &c.

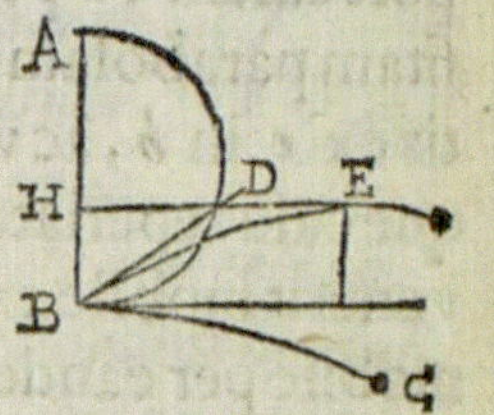

PROPOSITIO XXVII.

PArabolę ab eodem impetu factæ, quarũ directiones æqualiter ab horizonte vtrimq; distant sursum, & deorsum eiusdem parabolę portiones sunt,

Sit impetus *ab* & fiant proiectiones iuxtà directiones *bc*, *bd*, ęqualibus angulis ab horizonte *bi* vtrimq; distantes, Dico parabolam *be* & parabolam *bf*. portiones eiusdem parabolæ esse. Producatur enim *cbg*. Demonstratum est Propositione 7. huius, quòd si fiat proiectio cum eodẽ impetu iuxtà directionem *bc*, siue *bg*, parabolæ harum proiectionum vnam eandemq; continuatam parabolam efficient. Erit ergo *bh* eadem parabola ac *be*; quare etiam *bf* eadem parabola erit ac *be*; quandoquidem æqualitèr inclinantur directiones *bg*, *bd*, idemq; est impetus.

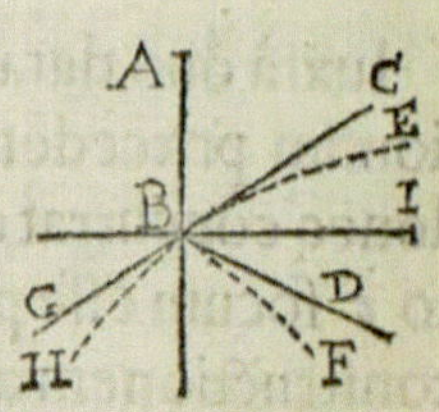

PROPOSITIO XXVIII.

SI ab eodem puncto *b*, cum eodem impetu *ab*, eodem tẽporis momento simul proiciantur grauia per diuersas

incli-

PROPOSITIO XXVI.

PArabola proiectionis horizontalis maxima omnium eft, quæ fieri poffint ab eodem impetu.

Sit impetus *ab.* fiatq; circulus *adb.* Sit etiam parabola iactus horizontalis *bc,* & alia parabola *be.* Dico maiorem effe parabolam *bc.* quàm *be.* Eft enim *ba* recta fublimitas parabolę *bc,* & *ha* recta fublimitas parabolę *be.* Quare *bc* omnium maxima erit; cum maiorem habeat fublimitatem, ideoq. maius latus rectum. Quod erat &c.

PROPOSITIO XXVII.

PArabolę ab eodem impetu factæ, quarũ directiones æqualiter ab horizonte vtrimq; diftant furfum, & deorfum eiufdem parabolę portiones funt,

Sit impetus *ab* & fiant proiectiones iuxtà directiones *bc, bd,* ęqualibus angulis ab horizonte *bi* vtrimq; diftantes, Dico parabolam *be* & parabolam *bf.* portiones eiufdem parabolæ effe. Producatur enim *cbg.* Demonftratum eft Propofitione 7. huius, quòd fi fiat proiectio cum eodẽ impetu iuxtà directionem *bc,* fiue *bg,* parabolæ harum proiectionum vnam eandemq; continuatam parabolam efficient. Erit ergo *bh* eadem parabola ac *be*; quare eriam *bf* eadem parabola erit ac *be*; quandoquidem æqualitèr inclinantur directiones *bg, bd,* idemq; eft impetus.

PROPOSITIO XXVIII.

SI ab eodem puncto *b,* cum eodem impetu *ab,* eodem tẽporis momento fimul proiciantur grauia per diuerfas

Incli-

inclinationes sursum vel deorsum, erunt omnia grauia semper in periphęria alicuius circuli cuius centrum erit in perpendiculo *ab*.

Sit facta proiectio horizontalis *be*, & non horizontalis quælibet alia *bh* iuxta directionem *bd*. Sumptoq; in horizontali parabola quouis puncto *e*, ducatur perpendiculum *ec*, & horizontalis recta *ef*. Seceturq; *bd* ęqualis *bc*, & demittatur perpendiculum *dh* æquale ipsi *ce*, vel *bf*. Dico grauia eodem simul tempore esse in *e* & in *h*. Cum enim æquales sint *bc*. *bd*. eodem simul tempore grauia essent in *c* & in *d*. si ęquabili motu procederent. Sed cum grauitas operetur, erunt grauium descensus eiusdem temporis, ęquales. at per suppositionē descensus vnius est *ce*, ergo descensus alterius erit *dh*. Quare grauia simul erunt in *e* & in *h*. & propterea in periphęria circuli, cuius centrum est *f*, nam *fe*, *fh* ęquales sunt, cum sint latera parallelogrammorum opposita lateribus *bc*, *bd*. ęqualibus.

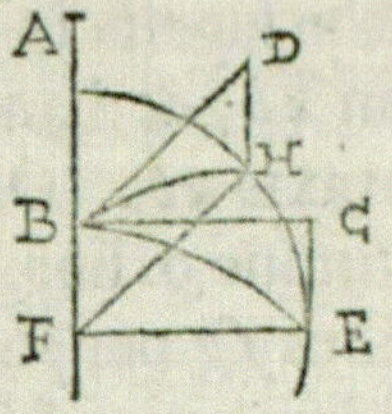

Verum ergo est, non solum grauia cadentia ab eodem puncto per diuersas planorum inclinationes, sed etiam proiecta semper esse in eiusdem circuli periphæria. Exempli gratia; siquis ex aliquo puncto grauia proiceret cum eodem impetu per diuersas inclinationes, aliudq; graue emitteret eodem temporis instanti ex quiete, & ab eodem puncto; videret grauia proiecta semper in aliquo circulo disposita commeare, & huiusmodi circulus semper haberet centrum in graui illo, quod naturalitèr descendit emissum ex quiete. vti dictum est.

PROPOSITIO XXIX.

SI ab eodem puncto, & cum eodem semper impetu fiant proiectiones; vertices parabolarum, siue quod idem est, extrema ascensionum puncta erunt in sphæroidis superficie cu-

ius

inclinationes ſurſum vel deorſum, erunt omnia grauia ſemper in periphęria alicuius circuli cuius centrum erit in perpendiculo *ab.*

Sit facta proiectio horizontalis *be,* & non horizontalis quælibet alia *bh* iuxta directionem *bd.* Sumptoq; in horizontali parabola quouis puncto *e,* ducatur perpendiculum *ec,* & horizontalis recta *ef.* Seceturq; *bd* ęqualis *bc,* & demittatur perpendiculum *dh* æquale ipſi *ce,* vel *bf.* Dico grauia eodem ſimul tempore eſſe in *e* & in *h.* Cum enim æquales ſint *bc. bd.* eodem ſimul tempore grauia eſſent in *c* & in *d.* ſi ęquabili motu procederent. Sed cum grauitas operetur, erunt grauium deſcenſus eiuſdem temporis, ęquales. at per ſuppoſitionẽ deſcenſus vnius eſt *ce,* ergo deſcenſus alterius erit *dh.* Quare grauia ſimul erunt in *e* & in *h.* & propterea in periphęria circuli, cuius centrum eſt *f,* nam *fe, fh* ęquales ſunt, cum ſint latera parallelogrammorum oppoſita lateribus *bc, bd.* ęqualibus.

Verum ergo eſt, non ſolum grauia cadentia ab eodem puncto per diuerſas planorum inclinationes, ſed etiam proiecta ſemper eſſe in eiuſdem circuli periphæria. Exempli gratia; ſiquis[93] *ex aliquo puncto grauia proiceret cum eodem impetu per diuerſas inclinationes, aliudq; graue emitteret eodem temporis inſtanti ex quiete, & ab eodem puncto; videret grauia proiecta ſemper in aliquo circulo diſpoſita commeare, & huiuſmodi circulus ſemper haberet centrum in graui illo, quod naturalitèr deſcendit emiſſum ex quiete. vti dictum est.*

PROPOSITIO XXIX.

SI ab eodem puncto, & cum eodem ſemper impetu fiant proiectiones; vertices parabolarum, ſiue quod idem eſt, extrema aſcenſionum puncta erunt in ſphæroidis ſuperficie cu-

ius

[93] *ſiquis*] *ſi quis.*

ius quidem maior diameter horizontalis ſit, & dupla minoris.

Sit impetus *ab*, & circa *ab*. Fiat ſemicirculus *adb*. tum fiant proiectiones iuxtà tangentes *ad*, *ae*, Dico vertices parabolarũ eſſe in ſupeificie ſphæroidis; quæ habeat axem *ab*, & diametrum horizontalem duplam axis *ab*. Demonſtratum enim eſt Propoſitione 9. huius, quod productis horizontalibus *fg* per punctum *d*, & *hi* per punctum *e*, quæ duplæ ſint linearum *fd*, *he*, demonſtratum inquam eſt puncta *g* & *i* eſſe vertices parabolarum. Sed puncta *g* & *i* ſunt in ſphæroidis ſuperficie, de qua diximus (eſt enim vt *gf* ad *fd*, ita *ih* ad *he*; ergo patet propoſitum.

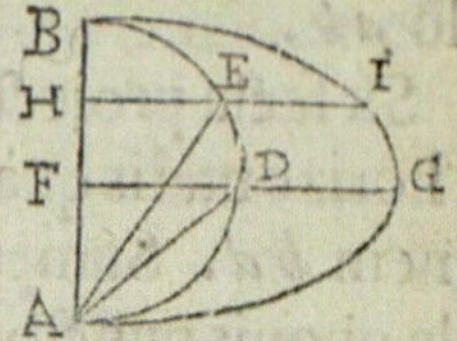

Sphera ergo actiuitatis aſcendentis proiectorum eſt in ſuperficie ſpheroidis illius ſpeciei que diametrum habeat duplam axis.

Lemma I.

Si recta linea duas parabolas contingat in eodem puncto, ſintque parabolarum diametri parallele, ipſe pat abolæſe mutuo contingent in illo eodem puncto.

Sit recta linea ab *quæ in puncto* b *duas parabolas* cbd, fbh. *contingat, & habeant parabolæ parallelas diametros, Dico huiuſmodi parabolas ſe mutuo contingere. Si enim non contingunt, ſecent; & intelligatur alteram parabolarum eſſe* cbh. *alteram verò* fbd. *Agatur* bi *parallela diametris, &* cd. *parallela tangenti. Erunt ergo æquales* ci, ih; *item equales* fi, id. *Quod eſt impoſſibile, Quare &c.*

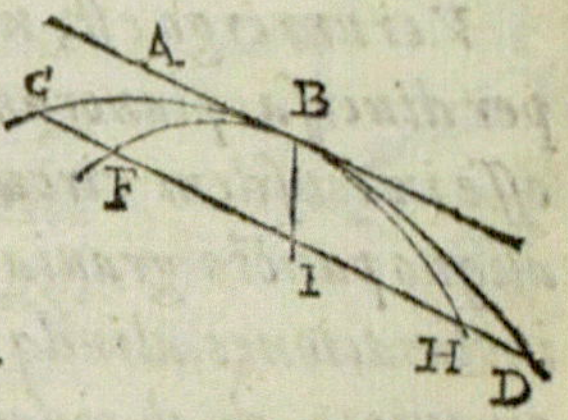

Lemma *II.*

Si due parabolæ abc, dbe. *ſe mutuò contingant in* b, *& habeant diametros parallelas; Dico has parabolas numquam amplius conuenire.*

Si enim poſſibile eſt conueniant in f. *& ducatur* bh *parallela dia-*

ius quidem maior diameter horizontalis ſit, & dupla minoris.

Sit impetus *ab,* & circa *ab.* Fiat ſemicirculus *adb.* tum fiant proiectiones iuxtà tangentes *ad, ae,* Dico vertices parabolarũ eſſe in ſuperficie ſphæroidis; quæ habeat axem *ab,* & diametrum horizontalem duplam axis *ab.* Demonſtratum enim eſt Propoſitione 9. huius, quod productis horizontalibus *fg* per punctum *d,* & *hi* per punctum *e,* quæ duplæ ſint linearum *fd, he,* demonſtratum inquam eſt puncta *g* & *i* eſſe vertices parabolarum. Sed puncta *g* & *i* ſunt in ſphæroidis ſupęrficie, de qua diximus (eſt enim vt *gf* ad *fd,* ita *ih* ad *he*; ergo pater propoſitum.

Sphęra ergo actiuitatis aſcendentis proiectorum est in ſuperſicie ſphęroidis illius ſpeciei quę diametrum habe at duplam axis.

Lemma I.

Si recta linea duas parabolas contingat in eodem puncto, ſintque parabolarum diametri parallelę, ipſę parabolæ ſe mutuo contingent in illo eodem puncto.

Sit recta linea ab *quæ in puncto* b *duas parabolas* cbd, fbh. *contingat, & habeant parabolæ parallelas diametros, Dico huiuſmodi parabolas ſe mutuo contingere. Si enim non contingunt, ſecent; & intelligatur alteram parabolarum eſſe* cbh. *alteram verò* fbd. *Agatur* bi *parallela diametris, &* cd. *parallela tangenti. Erunt ergo æquales* ci, ih*; item ęquales* fi, id. *Quod eſt impoſſibile, Quare &c.*

Lemma *II.*

Si duę parabolæ abc, dbe. *ſe mutuò contingant in* b, *& habeant diametros parallelas; Dico has parabolas numquam amplius conuenire.*

Si enim poſſibile eſt conueniant in f. *& ducatur* bh *paralle-*

la di a-

la diametris, & fh *ordinatim; tum alia quauis* ci *ordinatim producatur. Habebit ergo quadratum* fh *eandem rationem ad duo quadrata* ci, ei, *nempe quã habet* hb, *ad* bi. *Quod est impossibile ergo &c.*

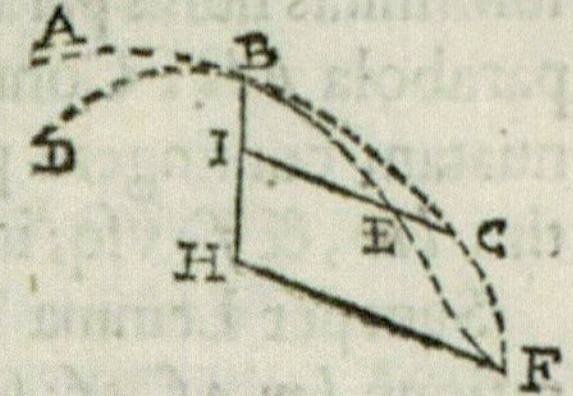

PROPOSITIO XXX.

SI ab eodem puncto cum eodem semper impetu proiectiones fiant, parabolę omnes contingent superficiem conoidis parabolici, cuius latus rectum quadruplum sit proiectionis sursum perpendiculariter factæ.

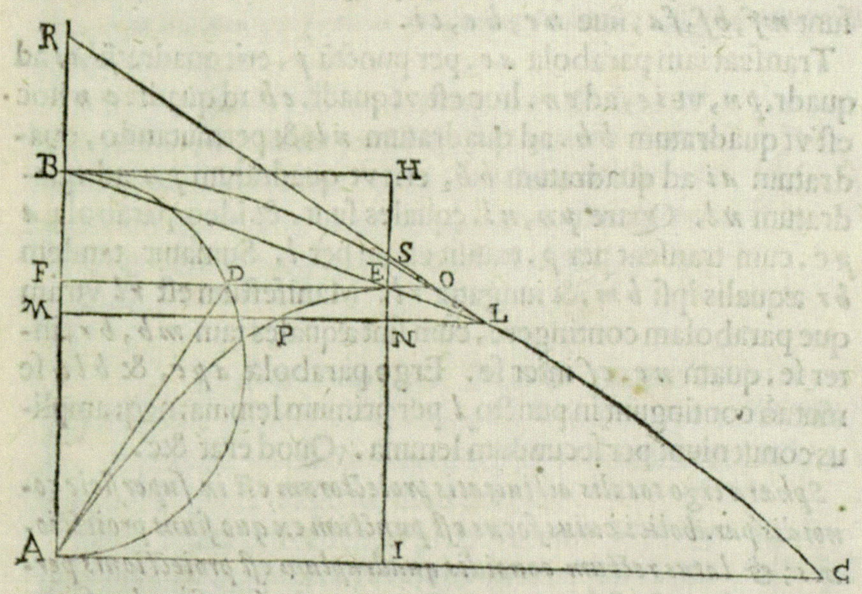

Sit impetus *ab*. & circa *ab* fiat circulus *adb*. circaq; axem *ab* fiat ex vertice *b* parabola *blc*. cuius focus sit *a*. Fiat iam proiectio iuxtà quamlibet eleuationem *ad*, sumaturq; *de* æqualis ipsi *df*. & demisso perpendiculo *hei*. fiat sua parabola circa diametrum *ei* ex puncto *a*, eritq; hęc parabola semita proiecti ab impetu *ba* iuxtà lineam directiuam *ad*, & erit *he* subli-

la diametris, & fh *ordinatim; tum alia quæuis* ci *ordinatim producatur. Habebit ergo quadratum* fh *eandem rationem ad duo quadrata* ci, ei, *nempe quã habet* hb, *ad* bi. *Quod est impoſſibile ergo &c.*

PROPOSITIO XXX.

SI ab eodem puncto cum eodem ſemper impetu proiectiones fiant, parabolę omnes contingent ſuperficiem conoidis parabolici, cuius latus rectum quadruplum ſit proiectionis ſurſum perpendicularitèr factæ.

Sit impetus *ab.* & circa *ab* fiat circulus *adb.* circaq; axem *ab* fiat ex vertice *b* parabola *blc.* cuius focus ſit *a.* Fiat iam proiectio iuxtà quamlibet eleuationem *ad,* ſumaturq; *de* æqualis ipſi *df.* & demiſſo perpendiculo *hei.* fiat ſua parabola circa diametrum *ei* ex puncto *a,* eritq; hęc parabola ſemita proiecti ab impetu *ba* iuxtà lineam directiuam *ad,* & erit *he*

Subli-

p.Conic. ſublimitas huius parabolę. Ducatur *be*, quæ conueniet cum parabola *blc*. Conueniat in *l*. Dico parabolam *ape* continuatam contingere parabolam *blc* in *l*. Ducantur ordinatim *lm*, & *fe* vſq; in *o*.

Sunt per Lemma Propoſ. 24. præcedentis libri in continua ratione *lm*, *of*, *ef*; Quadratum *of*. quadruplum eſt rectanguli *abf* ob parabolam cuius focus eſt *a*. & quadratum *ef* cum ſit per conſtructionem quadruplum quadrati *df*, quadruplum etiam erit rectanguli *afb*. His demonſtratis ſic procedemus. Recta *mb* ad rectam *bf* per 4. ſexti eſt vt *ml* ad *fe*, ſiue vt quadratum *fo* ad quadratum *fe* (ſumptiſq; eorum ſubquadruplis) vt rectangulum *abf* ad rectangulum *afb*, hoc eſt (omiſſa communi altitudine) vt recta *ba* ad *af*; Quare diuidendo erit vt *mf* ad *fb*, ita *bf* ad *fa*, & propterea in continua ratione ſunt *mf*, *bf*, *fa*; ſiue *ne*, *he*, *ei*.

Tranſeat iam parabola *ae*. per punctũ *p*. erit quadratũ *ai* ad quadr. *pn*, vt *ie*, ad *en*. hoc eſt vt quadr. *eh* ad quadr. *en* hoc eſt vt quadratum *bh*. ad quadratum *nl*; & permutando, quadratum *ai* ad quadratum *bh*, erit vt quadratum *pn* ad quadratum *nl*. Quare *pn*, *nl*. ęquales ſunt. & ideo parabola *ape*. cum tranſeat per *p*. tranſit etiam per *l*. Sumatur tandem *br* æqualis ipſi *bm*. & iungatur *rl*. Manifeſtum eſt *rl* vtramque parabolam contingere, cum ſint æquales tam *mb*, *br*, inter ſe, quam *ne*, *ef* inter ſe. Ergo parabolæ *ape*, & *blc* ſe mutuò contingunt in puncto *l* per primum lemma; neq; amplius conueniunt per ſecundum lemma. Quod erat &c.

Sphæra ergo totalis actiuitatis proiectorum eſt in ſuperficie conoidis parabolici, cuius focus eſt punctum ex quo fiunt proiectiones; & latus rectum conoidis quadruplum eſt proiectionis perpendicularis ſurſum. Demonſtratum enim fuit ſingulas ſingularum proiectionum parabolas huiuſmodi conoidis ſuperficiem attingere, numquam excedere. Proiecta igitur, eodem tempore ſunt in ſphæræ ſuperficie, in fine aſcenſionis ſunt in ſphæroidis ſuperficie; ſuprema illorum actiuitas eſt in conoidis parabolici ſuperficie.

Lem-

p. Conic. ſublimitas huius parabolę. Ducatur *be,* quæ conueniet cum parabola *blc.* Conueniat in *l.* Dico parabolam *ape* continuatam contingere parabolam *blc* in *l.* Ducantur ordinatim *lm,* & *fe* vſq; in *o.*

Sunt per Lemma Propoſ. 24. præcedentis libri in continua ratione *lm, of, ef*; Quadratum *of.* quadruplum eſt rectanguli *abf* ob parabolam cuius focus eſt *a.* & quadratum *ef* cum ſit per conſtructionem quadruplum quadrati *df,* quadruplum etiam erit rectanguli *afb.* His demonſtratis ſic procedemus. Recta *mb* ad rectam *bf* per 4. ſexti eſt vt *ml* ad *fe,* ſiue vt quadratum *fo* ad quadratum *fe* (ſumptiſq; eorum ſubquadruplis) vt rectangulum *abf* ad rectangulum *afb,* hoc eſt (omiſſa communi altitudine) vt recta *ba* ad *af*; Quare diuidendo erit vt *mf* ad *fb,* ita *bf* ad *fa,* & proptera in continua ratione ſunt *mf, bf, fa*; ſiue *ne, he, ei.*

Tranſeat iam parabola *ae.* per punctũ *p.* erit quadratũ *ai* ad quadr. *pn,* vt *ie,* ad *en.* hoc eſt vt quadr. *eh* ad quadr. *en* hoc eſt vt quadratum *bh.* ad quadratum *nl*; & permutando, quadratum *ai* ad quadratum *bh,* erit vt quadratum *pn* ad quadratum *nl.* Quare *pn, nl.* ęquales ſunt. & ideo parabola *ape.* cum tranſeat per *p.* tranſit etiam per *l.* Sumatur tandem *br* æqualis ipſi *bm.* & iungatur *rl.* Manifeſtum eſt *rl* vtramque parabolam contingere, cum ſint æquales tam *mb, br,* inter ſe, quam *ne, ef* inter ſe. Ergo parabolæ *ape,* & *blc* ſe mutuò contingunt in puncto *l* per primum lemma; neq; amplius conueniunt per ſecundum lemma. Quod erat &c.

Sphæra ergo totalis actiuitatis proiectorum est in ſuperficie conoidis parabolici, cuius focus eſt punctum ex quo fiunt proiectiones; & latus rectum conoidis quadruplum eſt proiectionis perpendicularis ſurſum. Demonstratum enim fuit ſingulas ſingularum proiectionum parabolas huiuſmodi conoidis ſuperficiem attingere, numquam excedere. Proiecta igitur, eodem tempore ſunt in ſphęræ ſuperficie, in fine aſcenſionis ſunt in ſphęroidis ſuperficie; ſuprema illorum actiuitas est in conoidis parabolici ſuperficie.

Lem-

Lemma.

Parabolę æquales abc, def, *& circa eandem diametrum* ah *descriptæ asymptoti sunt; hoc est cum semper magis accedant inuicem, nunquam tamen conueniunt.*

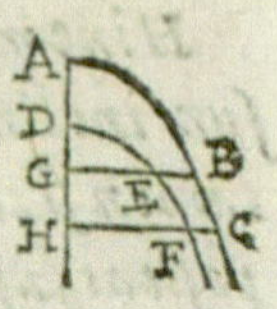

Rectangulum sub ad *& latere recto differentia est inter quadrata* bg, gc, *item etiam inter quadrata* ch, hf. *Ergo rectangula etiam sub* eb, *&* bg e *tamquam vna linea, & sub* fc, chf *tamquam vna linea, æqualia erunt inter se, cum sint differentiæ quadratorum; reciproca ergo habebunt latera, nempe vt* eb *ad* fc *ita erit* chf *linea ad* bge *lineam. est autem* chf *linea maior quam* bge, *& ideo* eb *maior erit quam* fc. *Parabolæ ergo semper magis accedunt. Quod nunquam conueniant patet.*

Nam si possibile est, conueniant in a. *& ducatur ordinatim* ab. *Cum parabolę sint ęquales habebunt idem latus rectum, eritq; quadratum* ab. *æquale vtriq; rectangulo quod continetur sub latere recto, & alterutra ipsarum* cb, db. *Quod est impossibile.*

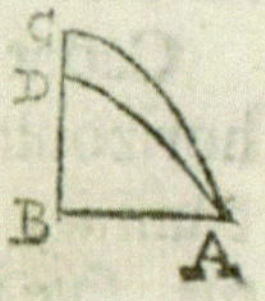

PROPOSITIO XXXI.

PArabola proiectionis horizontalis nunquam conuenit cũ superficie conoidis præcedentis propositionis, etiam si semper magis ac magis ad illud accedat.

Sit in figura præcedentis propositionis impetus *ab*; parabola genitrix conoidis sit *bc*, parabola autem horizontalis proiectionis sit *ad*. Dico has parabolas semper quidem accedere, nunquam tamen conuenire. Sunt enim circa eandem diametrum *ab*; & sunt æquales; quandoquidem recta *ab* est quarta pars lateris recti parabolæ *bc*, per constructionem, & parabolæ *ad*, quia est ipsius sublimitas. Ergo per lemma præcedens asymptoti erunt. Quod erat &c.

Aa Coro-

Lemma.

Pœrabolę æquales abc, def, *& circa eandem diametrum* ah *deſcriptæ aſymptoti ſunt; hoc eſt cum ſemper magis accedant inuicem, nunquam tamen conueniunt.*

Rectangulum ſub ad *& latere recto differentia eſt inter quadrata* bg, ge, *item etiam inter quadrata* ch, hf. *Ergo rectangula etiam ſub* eb, *&* bge *tamquam vna linea, & ſub* fc, chf *tamquam vna linea, æqualia erunt inter ſe, cum ſint differentiæ quadratorum*; *reciproca ergo habebunt latera, nempe vt* eb *ad* fc *ita erit* chf *linèa ad* bge *lineam. eſt autem* chf *linea maior quam* bge, *& ideo* eb *maior erit quam* fc. *Parabolæ ergo ſemper magis accedunt. Quod nunquam conueniant patet.*

Nam ſi poſſibile est, conueniant in a. *& ducatur ordinatim* ab. *Cum parabolę ſint ęquales habebunt idem latus rectum, eritq; quadratum* ab. *æquale vtriq; rectangulo quod continetur ſub latere recto, & alterutra ipſarum* cb, db. *Quod eſt impoſſibile.*

PROPOSITIO XXXI.

PArabola proiectionis horizontalis nunquam conuenit cũ ſuperficie conoidis præcedentis propoſitionis, etiam ſi ſemper magis ac magis ad illud accedat.

Sit in figura præcedentis propoſitionis impetus *ab*; parabola genitrix conoidis ſit *bc,* parabola autem horizonralis[94] proiectionis ſit *ad.* Dico has parabolas ſemper quidem accedere, nunquam tamen conuenire. Sunt enim circa eandem diametrum *ab*; & ſunt æquales*;* quandoquidem recta *ab* eſt quarta pars lateris recti parabolæ *bc,* per conſtructionem, & parabolæ *ad,* quia eſt ipſius ſublimitas. Ergo per lemma præcedens aſymptoti erunt. Quod erat &c.

Aa Coro-

[94] horizonralis] horizontalis.

186 *De motu Proiectorum*

Corollarium.

Hinc manifestum est parabolas factas iuxtà directiones deorsum inclinatas numquam contingere superficiem conoidis; Attamen si continuatę intelligantur illud contingent ad partes oppositas superiores. Demonstrauimus enim Propos. 7. & 27. Parabolas directionum deorsum vergentium easdem esse ac directionum sursum vergentium, dummodo lineę directionum ęqualitèr ab horizonte distent vtrimq;

PROPOSITIO XXXII.

DAto impetu siue sublimitate *ac*, cuius proiectio semirecta sit parabola *aeb*, Dico, si proiectio fiat cum eodem impetu horizontalitèr ex puncto sublimitatis *c*, iactum, siue parabolam cadere in *b*.

Cadat enim, si possibile est, iactus horizontaliter factus ex puncto *c* in punctum *d*. Et quia parabolæ *cd* impetus, siue sublimitas ponitur recta *ac*, erit *ac*. quarta pars lateris recti parabolæ *cd*; ergo *ad* applicata ex foco dupla erit ipsius *ac*. Sed etiam *ab* dupla erat ipsius *ac*; cum supponatur *ab* amplitudo facta à proiectione semirecta, ergo æquales essent *ad*, *ab*, impossibile. Patet ergo propositum.

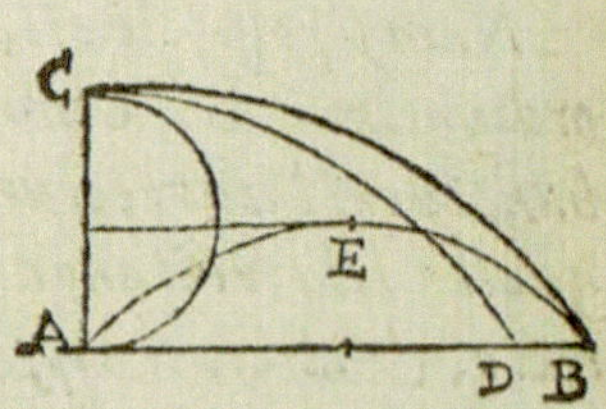

Patet etiam quod iactus *cb* describit parabolam genitricem illius Conoidis, cuius superficiem tangunt omnes proiectiones factæ ex puncto *a* cum eodem impetu.

PROPOSITIO XXXIII.

DAto impetu, & quocunque plano siue erecto, siue ad horizontem inclinato, reperire in dato plano remotissimũ, siue altissimum punctum ad quod cum dato impetu fieri possit iactus.

Item reperire directionem quæ altissimum, siue remotissimũ illum iactum faciat.

Sit

Corollarium

Hinc manifeſtum eſt parabolas factas iuxtà directiones deorſum inclinatas numquam contingere ſuperficiem conoidis; Attamen ſi continuatę intelligantur illud contingent ad partes oppoſitas ſuperiores. Demonstrauimus enim Propoſ. 7. & 27. Parabolas directionum deorſum vergentium easdem eſſe ac directionum ſurſum vergentium, dummodo lineę directionum ęqualitèr ab horizonte distent vtrimq;.

PROPOSITIO XXXII.

DAto impetu ſiue ſublimitate *ac,* cuius proiectio ſemirecta ſit parabola *aeb,* Dico, ſi proiectio fiat cum eodem impetu horizontalitèr ex puncto ſublimitatis *c,* iactum, ſiue parabolam cadere in *b.*

Cadat enim, ſi poſſibile eſt, iactus horizontaliter factus ex puncto *c* in pũctum *d.* Et quia parabolæ *cd* impetus, ſiue ſublimitas ponitur recta *ac,* erit *ac.* quarta pars lateris recti parabolæ *cd*; ergo *ad* applicata ex foco dupla erit ipſius *ac.* Sed etiam *ab* dupla erat ipſius *ac*; cum ſupponatur *ab* amplitudo facta à proiectione ſemirecta, ergo æquales eſſent *ad, ab,* impoſſibile. Patet ergo propoſitum.

Patet etiam quod iactus *cb* deſcribit parabolam genitricem illius Conoidis, cuius ſuperficiem tangunt omnes proiectiones factæ ex puncto *a* cum eodem impetu.

PROPOSITIO XXXIII.[95]

DAto impetu, & quocunque plano ſiue erecto, ſiue ad horizontem inclinato, reperire in dato plano remotiſſimũ, ſiue altiſſimum punctum ad quod cum dato impetu fieri poſſit iactus.

Item reperire directionem quæ altiſſimum, ſiue remotiſſimũ illum iactum faciat

Sit

[95] Proof of Archimedes's proposition on *Conoids and spheroids* (Archimedes 1897, pp. 99-150; Cfr. Pisano and Bussotti's chapters above).

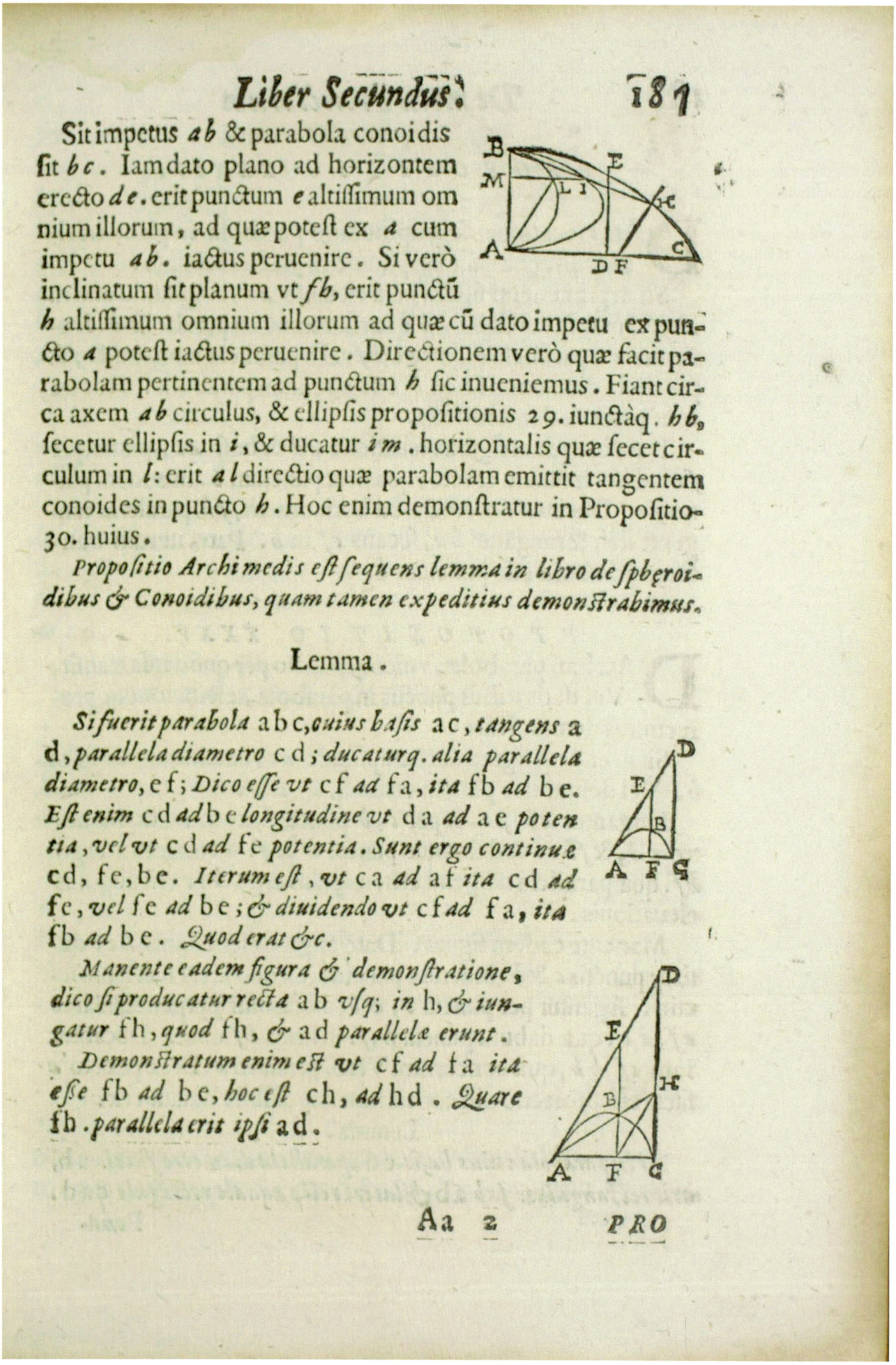

Liber Secundus. 181

Sit impetus *ab* & parabola conoidis ſit *bc.* Iam dato plano ad horizontem erecto *de.* erit punctum *e* altiſſimum omnium illorum, ad quæ poteſt ex *a* cum impetu *ab.* iactus peruenire. Si verò inclinatum ſit planum vt *fh,* erit punctũ *h* altiſſimum omnium illorum ad quæ cũ dato impetu ex puncto *a* poteſt iactus peruenire. Directionem verò quæ facit parabolam pertinentem ad punctum *h* ſic inueniemus. Fiant circa axem *ab* circulus, & ellipſis propoſitionis 29. iunctàq. *hb,* ſecetur ellipſis in *i,* & ducatur *im.* horizontalis quæ ſecet circulum in *l*: erit *al* directio quæ parabolam emittit tangentem conoides in puncto *h.* Hoc enim demonſtratur in Propoſitio 30. huius.

Propoſitio Archimedis eſt ſequens lemma in libro de ſpbęroidibus & Conoidibus, quam tamen expeditius demonstrabimus.

Lemma.

Si fuerit parabola abc, *cuius baſis* ac, *tangens* ad, *parallela diametro* cd; *ducaturq. alia parallelæ diametro,* ef; *Dico eſſe vt* cf *ad* fa, *ita* fb *ad* be. *Eſt enim* cd *ad* be *longitudine vt* da *ad* ae *potentia, vel vt* cd *ad* fe *potentia. Sunt ergo continuæ* cd, fe, be. *Iterum eſt, vt* ca *ad* af *ita* cd *ad* fe, *vel* fe *ad* be; *& diuidendo vt* cf *ad* fa, *ita* fb *ad* be. *Quod erat &c.*

Manente eadem figura & demonſtratione, dico ſi producatur recta ab *vſq; in* h, *& iungatur* fh, *quod* fh, *&* ad *parallelæ erunt.*

Demonstratum enim est vt cf *ad* fa *ita eße* fb *ad* be, *hoc eſt* ch, *ad* hd. *Quare* fh. *parallela erit ipſi* ad.

PROPOSITIO XXXIV.

DAta eleuatione & amplitudine parabolæ in plano horizontali, quæritur amplitudo in plano inclinato.

Sit in præcedenti figura data eleuatio *a d*, amplitudo autem *a c*, planumq; datum sit *a b* quæritur transitus parabolæ *b*. Ducantur *c d* parallela diametro; *h f* verò parallela tangenti,& *f b*. parallela diametro; Dico *b* esse transitum parabolæ. Hoc autem patet ex demonstratis.

Data in eadem figura eleuatione *a d* & basi *a c*, planoque *f e* ad horizontem erecto, quæritur punctum *b*. in eodem plano *f e*. Ducatur *c d* erecta ad horizontem, *f h* parallela tangenti *a d*, & iungatur *h a*, secans *e f* in *b*. Patet iterum transitum parabolæ esse punctum *b*.

PROPOSITIO XXXV.

DAta basi parabolæ, vnicoq; puncto per quod ipsa transit; Vel datis tribus punctis in parabola, eleuationem proiectionis demonstrare.

Sit in eadem figura data amplitudo *a c*, datumque punctum *b*. Vel dentur tria puncta vtcumq. *a*, *b*. *c*.

Iungantur *a b*, *a c*. & per *c*, *b* puncta sint parallelæ diametro *c d*, & *f b e*. & dabuntur puncta *h* & *f*. Producatur ergo *h f*. quæ parallela tangenti erit. angulus ergo *h e c* erit angulus eleuationis.

Manente eadem figura. Dato angulo eleuationis *h e c* datisq; punctis *c* & *b*. inuenire punctum ex quo facta fuerit proiectio. Agantur per puncta *c*, & *b*, horizonti perpendiculares *e f*, *d c*, quæ dabunt puncta *h*, *f*. in linea *h f* data. Ducantur iam *c f*, *h b*, quæ concurrant Verbi gratia in *a*. Et ex puncto *a* facta erit proiectio nisi concurrant impossibile datam erit.

Lemma.

Si in parabola cuius basis c d, *parallela diametro fuerit* a b, *erit rectangulum sub* a b *& latere recto æquale rectãgulo* c a d.

Pona-

PROPOSITIO XXXIV.

DAta eleuatione & amplitudine parabolæ in plano horizontali, quæritur amplitudo in plano inclinato.

Sit in præcedenti figura data eleuatio *ad,* amplitudo autem *ac,* planumq; datum ſit *ah* quæritur tranſitus parabolæ *b.* Ducantur *cd* parallela diametro; *hf* verò parallela tangenti, & *fb.* parallela diametro; Dico *b* eſſe tranſitum parabolæ. Hoc autem patet ex demonſtratis.

Data in eadem figura eleuatione *ad* & baſi *ac,* planoque *fe* ad horizontem erecto, quæritur punctum *b.* in eodem plano *fe.* Ducatur *cd* erecta ad horizontem, *fh* parallela tangenti *ad,* & iungatur *ha,* ſecans *ef* in *b.* Patet iterum tranſitum parabolæ eſſe punctum *b.*

PROPOSITIO XXXV.

DAta baſi parabolæ, vnicoq; puncto per quod ipſa tranſit;

Vel datis tribus punctis in parabola, eleuationem proiectionis demonſtrare.

Sit in eadem figura data amplitudo *ac,* datumque punctum *b.* Vel dentur tria puncta vtcumq. *a, b. c.* Iungantur *ab, ac.* & per *c, b* puncta ſint parallelæ diametro *cd,* & *fbe.* & dabuntur puncta *h* & *f.* Producatur ergo *hf.* quæ parallela tangenti erit. angulus ergo *hec* erit angulus eleuationis.

Manente eadem figura. Dato angulo eleuationis *hec* datisq; punctis *c* & *b.* inuenire punctum ex quo facta fuerit proiectio. Agantur per puncta *c,* & *b,* horizonti perpendiculares *ef, dc,* quæ dabunt puncta *h, f.* in linea *hf* data. Ducantur iam *ef, hb,* quæ concurrant Verbi gratia in *a.* Et ex puncto *a* facta erit proiectio niſi concurrant impoſſibile datam erit.

Lemma.

Si in parabola cuius baſis cd, *parallela diametro fuerit* ab, *erit rectangulum ſub* ab *& latere recto æquale rectãgulo* cad.

Pona-

Ponatur enim e f *latus rectũ, & complea tur rectangulum* e i. *Quia* c d *secta est bifariam & non bifariam, erit quadratũ* l d, *hoc est rectangulum* e i *æquale rectã gulo* c a d *& quadrato* l a. *Demptis æqualibus (nempe hinc quadrato* l a, *siue* m b, *& inde rectangu lo* c h) *reliqua æqualia erunt, hoc est rectangula* m i, *&* c a d. *Quod &c.*

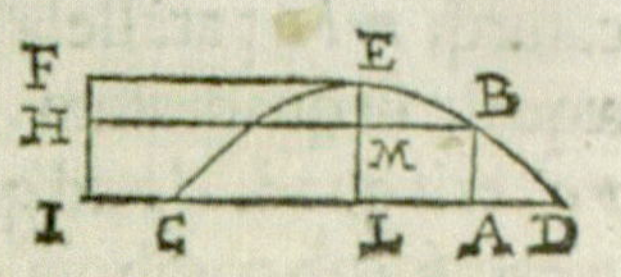

Iam si rectæ a b, c d. *fuerint parallelæ diametro, erit rectangulum* e a f *ad rectangulum* e c f *vt* a b, *ad* c d.

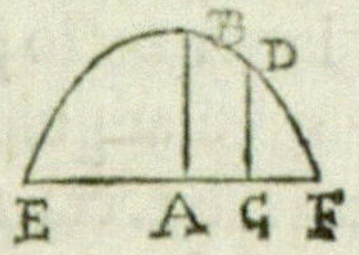

Sunt enim illa rectangula æqualia rectangulis sub a b *& latere recto, & sub* c d *ac latere recto respectiue, ista vero cum habeant æqualem altitudinem, erunt vt bases* a b, c d. *Quare etiam rectangula* e a f, e c f. *erunt vt* a b, *ad* c d.

PROPOSITIO XXXVI.

DAta directione *a b*, & basi *a d*. data est altitudo parabolæ supra quoduis punctum *c*. Diuidatur bifariam *a d* in *f*, & erigatur *f h*. Quoniam datur angulus *b a d* directionis, & basis *a d*, dabitur in triangulo rectangulo latus *b d*, & ideo *f h*, quæ quidem est quarta pars ipsius *b d* ob parabolam. Fiat ergo vt rectangulum *a f d*, siue quadratum semibasis *a f* ad rectangulum *a c d*, ita altitudo *f h*. ad aliam. & quarta reperta erit altitudo quæsita *c e*. Quod erat &c.

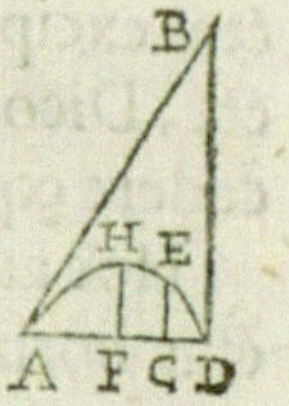

Lemma.

Si conoides parabolicum *a b c* secetur plano *d e f* æquidistantèr axi, sectio parabola erit, & æqualis semper ei quæ conoides generauit, hoc est æquale latus rectum habẽs. Sumpto enim quolibet puncto *i* in sectione *d i e*, applicetur *i l*, ducaturq.

Ponatur enim ef *latus rectũ, & compleatur rectangulum* ei. *Quia* cd *ſecta eſt bifariam & non bifariam, erit quadratũ* ld, *hoc est rectangulum* ei *æquale rectãgulo* cad *& quadrato* la. *Demptis ęqualibus (nempe hinc quadrato* la, *ſiue* mb, *& inde rectangulo* eh) *reliqua æqualia erunt, hoc est rectangula* mi, *&* cad. *Quod &c.*

Iam ſi rectæ ab, cd. *fuerint parallelæ diametro, erit rectangulum* eaf *ad rectangulum* ecf *vt* ab, *ad* cd.

Sunt enim illa rectangula æqualia rectangulis ſub ab *& latere recto, & ſub* cd *ac latere recto reſpectiue, iſta vero cum habeant æqualem altitudinem, erunt vt baſes* ab, cd. *Quare etiam rectangula* eaf, ecf. *erunt vt* ab, *ad* cd.

PROPOSITIO XXXVI.

DAta directione *ab,* & baſi *ad.* data eſt altitudo parabolæ ſupra quoduis punctum *c.* Diuidatur bifariam *ad* in *f,* & erigatur *fh.* Quoniam datur angulus *bad* directionis, & baſis *ad,* dabitur in triangulo rectangulo latus *bd,* & ideo *fh,* quæ quidem eſt quarta pars ipſius *bd* ob parabolam. Fiat ergo vt rectangulum *afd,* ſiue quadratum ſemibaſis *af* ad rectangulum *acd,* ita altitudo *fh.* ad aliam. & quarta reperta erit altitudo quæſita *ce.* Quod erat &c.

Lemma.

Si conoides parabolicum *abc* ſecetur plano *def* æquidiſtantèr axi, ſectio parabola erit, & æqualis ſemper ei quæ conoides generauit, hoc eſt æquale latus rectum habẽs. Sumpto enim quolibet puncto *i* in ſectione *die,* applicetur *il,* du-

caturq

caturq; *mln* parallela ad *ac*. Iam: cum

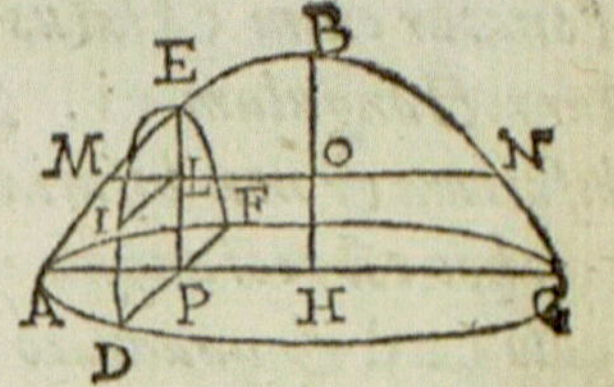

æquale ſit quadratum. *dp* rectangulo *a pc*, erit quad. *dp* ad quad. *ah*, vt recta *pe* ad *hb* ob parabolã *abc*, ſed quad. *ah* ad quad. *mo* eſt vt recta *hb* ad *bo*, & quad. *mo* ad rectangulum *mln*, ſiue ad quad. *il* eſt vt recta *ob* ad *le*; ergo ex æquo, quad. *dp*. ad *il*, eſt vt recta *pe* ad *el*. Propterea ſectio *die* parabola erit.

Amplius. Quoniam verò rectangulum ſub diametro *pe*, & latere recto parabolę *def* æquale eſt quadrato applicatæ *dp* ſiue rectangulo *apc*, cui rectangulo *apc* æquale eſt rectangulum ſub recta *pe*, & latere recto parabolæ *abc*. (per lemma Propoſ. præced.) æqualia erunt inter ſe illa prędicta rectangula; ſed altitudo *pe* eadem eſt vtriq;, ergo baſes æquales erunt, nempe latus rectum parabolæ *die*, æquale erit lateri recto parabolæ *abc*. Quod &c.

PROPOSITIO XXXVII.

SI ab eodem puncto, cum eodem ſemper impetu, proiectiones fiant per omnes lineas horizontales, omneſq; iactus excipiantur in aliqua ſuperficie plana ad horizontem erecta, Dico omnes illos iactus in quandam lineam parabolicam cadere ęqualem ſemper parabolæ proiectionis.

Hoc autem patet ex lemmate præmiſſo. Nam omnes illi iactus horizontales ſuperficiem quandam deſcribunt conoidis parabolici, quam ſuperficiem ſecat planũ illud erectum in quo feriunt iactus, ergo ſectio in quam cadunt iactus, erit parabola æqualis parabolæ genitrici conoidis; Propterea patet propoſitum.

Si verò iactus omnes terminentur in horizonte, ſectio circulus erit; quando verò in planis inclinatis, ſectiones erunt ellipſes, quod facile colligi poteſt ex demonſtrationibus antiquorum, qui Demonſtrauerunt obliquam ſectionem conoidis ellipſim eſſe.

DE

caturq; *mln* parallela ad *ac.* Iam: cum æquale ſit quadratum. *dp* rectangulo *apc,* erit quad. *dp* ad quad. *ah,* vt recta *pe* ad *hb* ob parabolã *abc,* ſed quad. *ah* ad quad. *mo* eſt vt recta *hb* ad *bo,* & quad. *mo* ad rectangulum *m ln,* ſiue ad quad. *il* eſt vt recta *ob* ad *le;* ergo ex æquo, quad. *dp.* ad *il,* eſt vt recta *pe* ad *el.* Propterea ſectio *die* parabola erit.

Amplius. Quoniam verò rectangulum ſub diametro *pe,* & latere recto parabolę *def* æquale eſt quadrato applicatæ *dp* ſiue rectangulo *apc,* cui rectangulo *apc* æquale eſt rectangulum ſub recta *pe,* & latere recto parabolæ *abc.* (per lemma Propoſ. præced.) æqualia erunt inter ſe illa prędicta rectangula; ſed altitudo *pe* eadem eſt vtriq;, ergo baſes æquales erunt, nempe latus rectum parabolæ *die,* æquale erit lateri recto parabolæ *abc.* Quod &c.

PROPOSITIO XXXVII.

SI ab eodem puncto, cum eodem ſemper impetu, proiectiones fiant per omnes lineas horizontales, omneſq; iactus excipiantur in aliqua ſuperficie plana ad horizontem erecta, Dico omnes illos iactus in quandam lineam parabolicam cadere ęqualem ſemper parabolæ proiectionis.

Hoc autem patet ex lemmate præmiſſo. Nam omnes illi iactus horizontales ſuperficiem quandam deſcribunt conoidis parabolici, quam ſuperficiem ſecat planũ illud erectum in quo feriunt iactus, ergo ſectio in quam cadunt iactus, erit parabola æqualis parabolæ genitrici conoidis; Propterea patet propoſitum.

Si verò iactus omnes terminentur in horizonte, ſectio circulus erit; quando verò in planis inclinatis, ſectiones erunt ellipſes, quod facile colligi poteſt ex demonſtrationibus antiquorum, qui Demonſtrauerunt obliquam ſectionem conoidis ellipſim eſſe.

D E

DE MOTV AQVARVM.

Iam verò & de aquis aliquam huic libello contemplationem inserere non erit inconueniens: aquis enim pręceteris corporibus sublunaribus adeò peculiaris, & cognatus videtur motus, vt ferè nunquam quiescant. Omitto magnum illum nutantis maris motum; Prętereo etiam omnem fluminum, aquarumq; currentiũ tum mensuram, tum vsum, quarum omnis doctrina reperta primum fuit ab Abbate Benedicto Castellio præceptore meo. Scripsit ille scientiam suam, & illam non solum demonstratione, verũ etiam opere confirmauit, maxima cum Principum & populorum vtilitate, maiore cum admiratione phylosophorum. Extat illius liber, vere aureus. Nos minuta quædam, & plerumq; inutilia, non tamen penitus incuriosa circa hanc materiam prosequemur.

Supponimus.

Aquas violentèr erumpentes in ipso eruptionis puncto eundẽ impetum habere, quem haberet graue aliquod, siue ipsius aquæ gutta vna, si ex suprema eiusdem aquæ superficie vsq; ad orificium eruptionis naturalitèr cecidisset.

Exempli gratia. Si tubus ab conuenientis capacitatis, hoc est magnæ laxitatis, intelligatur semper aqua plenus vsq; ad libellam a, et perforetur angusto orificio in b. Supponimus aquam ex b erumpentem eundem impetum habere, quem haberet graue aliquod si naturalitèr ex a in b cecidisset.

Hoc ratione quodammodo confirmari posse videtur nam si ad osculum b alius tubus inseratur, et exquisitè coaptetur, aqua ex b influens in tubum b c, tantam vim habet vt se ipsam euehat vsq; ad eãdem libellam horizontalem c a ductam per orificium a.

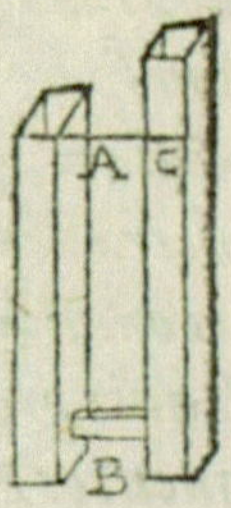

Quare verisimile videtur etiam quando ipsa ex b libera erumpit, habere vim redeñdi vsq; ad horizõtalem lineam quę per a ducitur; vel quod idem est habere tantum impetum quantus est grauis alicuius,

siue

DE MOTV AQVARVM.[96]

IAm verò & de aquis aliquam huic libello contemplationem inſerere non erit inconueniens: aquis enim pręceteris corporibus ſublunaribus adeò peculiaris, & cognatus videtur motus, vt ferè nunquam quieſcant. Omitto magnum illum nutantis maris motum; Pręterea etiam omnem fluminum, aquarumq; currentiũ tum menſuram, tum vſum, quarum omnis doctrina reperta primum fuit ab Abbate Benedicto Castellio præceptore meo. Scripſit ille ſcientiam ſuam, & illam non ſolum demonſtratione, verũ etiam opere confirmauit, maxima cum Principum & populorum vtilitate, maiore cum admiratione phyloſophorum. Extat illius liber, vere aureus. Nos minuta quædam, & plerumq; inutilia, non tamen penitus incurioſa circa hanc materiam proſequemur.

Supponimus.

Aquas violentèr erumpentes in ipſo eruptionis puncto eundẽ impetum habere, quem haberet graue aliquoà,[97] *ſiue ipſius aquæ gutta vna, ſi ex ſuprema eiuſdem aquæ ſuperficie vſq; ad orificium eruptionis naturalitèr cecidiſſet.*

Exempli gratia. Si tubus ab *conuenientis capacitatis, hoc eſt magnæ laxitatis, intelligatur ſemper aqua plenus vſq; ad libellam* a, *et perforetur anguſto orificio in* b. *Supponimus aquam ex* b *erumpentem eundem impetum habere, quem haberet graue aliquod ſi naturalitèr ex* a *in* b *cecidiſſet.*

Hoc ratione quodammodo confirmari poſſe videtur nam ſi ad oſculum b *alius tubus inſeratur, et exquiſtè coaptetur, aqua ex* b *influens in tubum* bc, *tantam vim habet vt ſe ipſam euehat vſq; ad eãdem libellam horizontalem* c a *ductam per orificium* a.

Quare veriſimile videtur etiam quando ipſa ex b *libera erumpit, habere vim redeãdi*[98] *vſq; ad horizõtalem lineam quę per* a *ducitur; vel quod idem eſt habere tantum impetum quantus eſt grauis alicuius,*

ſiue

[96] The results obtained in *De Motu Aquarum* can be interpreted as an appendix of Torricelli's studies on the fall of the heavy bodies because they depend on the achievements obtained in the two books *De motu gravium naturaliter descendentium*. No new principle is added (Cfr. Pisano and Bussotti's chapters above).

[97] *aliquoà*] *aliquoàd*. In the text, the distinction between "à" and "d" is not immediately clear. By the way, we have marked it for the reader.

[98] *redeãdi*] *redendi*.

ſiue vnius guttæ liberè cadentis ex a in b.

Experimentum etiam aliquo modo principium noſtrum probat, quamquam aliqua ex parte reprobare videatur. Nam ſi oſculum b ſurſum dirigatur, & ſit aptè rotundum, & leuigatum, ſitq; reliqua totius tubi latitudo multo capacior quàm orificiũ b, videbimus aquã ſalientẽ per lineã b c, quaſi ad libellam ſuam a d aſcẽdere. Defectionis autẽ c d cauſam adſcribere poſſumus partim impedimento aeris qui cõtra quodcunq; corpus mobile luctatur; partim etiã ipſimet aquæ, quæ dum ex faſtigio c reditum affectat deorſum, ſe ipſam venientem impedit, & retardat, neque ſinit ſubeuntes guttas ad illud ipſum ſignum ad quod ſuo impetu peruenirent, aſcendere poſſe. Hoc manifeſtè patebit, quando oppoſita manu foramen b penitus occludatur; deinde retracta quam citiſſimè manu repente aperiatur: videbuntur enim primæ, & præeuntes guttæ altius peruenire, quàm ſit deinde culmen aquæ c. poſtquam aqua deorſum fluere cœperit. illæ enim priores guttæ præcedentem aquam non habent, quæ contra ipſas refluens motum ipſarum in fine aſcenſionis impediat ſuppono enim ductum b c perpendicularem.

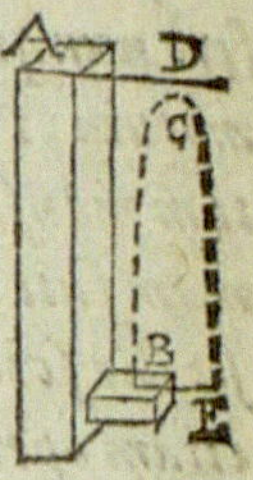

Adde etiam quod ſi quis obſeruet aerem ipſi aquæ b c circũfuſum, reperiet ipſum agitari, & ſurſum moueri, quæ quidem latio non fit ſine vi, & propterea cum impedimento motus aquæ aſcendentis. Vnde eſt, quod ſi quis velit de hoc principio experimentum facere, ſumendum eſſet argentum viuum, quod ob intimam grauitatem magis aptum eſt, & ad conſeruandum diu cõceptum impetum, & ad ſuperandam aeris reſiſtentiam. Aqua autem ob leuitatem multum aberrare videbitur, & præcipuè ſi tubus magnæ fuerit altitudinis: tunc enim ob maximum impetum ſpargitur in guttulas minutiſſimas tamquam roris, neq; dimidiam, & fortaſſe tertiam, quartamue partem aſcendit illius interualli quod re ipſa, theoricè loquendo, & remotis impedimẽtis omnibus concepto impetu totum exæquare deberet. Cęterum ſi quis prędictis rationibus non acquieſcat, videat an interſequentes Propoſitiones vllam probet; quod ſi ita erit, facile per reſolu-

ſiue vnius guttæ liberè cadentis ex a *in* b.

Experimentum etiam aliquo modo principium noſtrum probat, quamquam aliqua ex parte reprobare videatur. Nam ſi oſculum b *ſurſum dirigatur, & ſit aptè rotundum, & leuigatum, ſitq; reliqua totius tubi latitudo multo capacior quàm orificiũ* b, *videbimus aquã ſalientẽ per lineã* bc, *quaſi ad libellam ſuam* ad *aſcẽdere. Defectionis autẽ* cd *cauſam adſcribere poſſumus partim impedimento aeris qui cõtra quodcunq; corpus mobile luctatur; partim etiã ipſimet aquæ, quæ dum ex faſtigio* c *reditum affectat deorſum, ſe ipſam venientem impedit, & retardat, neque ſinit ſubeuntes guttas ad illud ipſum ſignum ad quod ſuo impetu peruenirent, aſcendere poſſe. Hoc manifeſtè patebit, quando oppoſita manu foramen* b *penitus occludatur; deinde retracta quam citiſſimè manu repente aperiatur: videbuntur enim primæ, & præeuntes guttæ altius peruenire, quàm ſit deinde culmen aquæ* c. *poſtquam aqua deorſum fluere cæperit. illæ enim priores guttæ præcedentem aquam non habent, quæ contra ipſas refluens motum ipſarum in fine aſcenſionis impediat ſuppono enim ductum* bc *perpendicularem.*

Adde etiam quod ſi quis obſeruet aerem ipſi aquæ bc *circũfuſum, reperiet ipſum agitari, & ſurſum moueri, quæ quidem latio non ſit ſine vi, & propterea cum impedimento motus aquæ aſcendentis. Vnde eſt, quod ſi quis velit de hoc principio experimentum facere, ſumendum eſſet argentum viuum, quod ob intimam grauitatem magis aptum eſt, & ad conſeruandum diu cõceptum impetum, & ad ſuperandam aeris reſiſtentiam. Aqua autem ob leuitatem multum aberrare videbitur, & præcipuè ſi tubus magnæ fuerit altitudinis: tunc enim ob maximum impetum ſpargitur in guttulas minutiſſimas tamquam roris, neq; dimidiam, & fortaße tertiam, quartamue partem aſcendit illius interualli quod re ipſa, theoricè loquendo, & remotis impedimẽtis omnibus concepto impetu totum exæquare deberet. Cęterum ſi quis prędictis rationibus non acquieſcat, videat an inter ſequentes Propoſitiones vllam probet; quod ſi ita erit, facile per re-*

ſolu-

lutionem ex approbata propositione primam suppositionem demõstrabimus; sin minus totam hanc appendicem de motu aquarum vel saltu prætermittat, vel funditus è libello euellat, quod equidem libentissimè concedo, etsi factum experimentum omni diligentia magnam partem sequentium propositionum exactissimè confirmauit.

His expositis consideremus aquam recidiuam in e, nempe in plano horizontali ducto per libellam orificij b. Ex Galileo habemus impetum aquę cadentis ex c in e tantum esse quantus vehere potest eandem aquam ex e in c. Ergo impetus in e idem est ac in b, sed in e impetus est tamquam grauis cadentis ex c in e. vel ex a in b (diximus enim quod punctum c reipsa deberet esse in libella a d abstractis impedimentis quæ aquam retardant) ergo impetus in b est tamquam grauis cadentis naturalitèr ex a in b.

His suppositis quædam demonstrabimus de aquis erumpentibus, quæ mirè cum doctrina proiectorum conuenire videntur

PRimùm manifestum est omnes aquas erumpentes ex foraminibus tubi alicuius perforati, parabolas describere. Primę enim guttæ scaturientes e tubo sunt de natura proiectorum quandoquidem ipsæ, quamquam liquidæ attamen sunt sphęrulæ graues & coherentes, & ideo parabolam certè designabunt. Omnes autem subsequentes, quæ cum eodem impetu emittuntur (supponimus enim tubos semper aqua plenos) semitam pręcædentium percurrent; quare continuus ille aquæ fluentis tractus parabola erit.

Obijcet fortasse aliquis hoc non videri, presertim quando tubi orificium valde angustum erit, & impetus vehemens. Tunc enim (vt videre est in linea illa aquea, quę ex fontium fistulis violentiùs erumpit) prior pars orbitæ illius ascendentis magis tensa apparet, et ad parabolam veriùs conformata; posterior verò, hoc est ea quam aqua descendens percurrit, magis prona, et vt ita dicam, languida atq; curua conspicitur. Obiectioni respondetur; non solùm precedentem propositiunculam, sed etiam

solutionem ex approbata propoſitione primam ſuppoſitionem demõſtrabimus; ſin minus totam hanc appendicem de motu aquarum vel ſaltu prætermittat, vel funditus è libello euellat, quod equidem libentiſſimè concedo, et ſi factum experimentum omni diligentia magnam partem ſequentium propoſitionum ex actiſſimè confirmauit.

His expoſitis conſideremus aquam recidiuam in e, *nempe in plano horizontali ducto per libellam orificij* b. *Ex Gælileo habemus impetum aquę cadentis ex* c *in* e *tantum eſſe quantus vehere poteſt eandem aquam ex* e *in* c. *Ergo impetus in* e *idem est ac in* b, *ſed in* e *impetus est tamquam grauis cadentis ex* c *in* e. *vel ex* a *in* b *(diximus enim quod punctum* c *reipſa deberet eſſe in libella* ad *abstractis impedimentis quæ aquam retardant) ergo impetus in* b *eſt tamquam grauis cadentis naturalitèr ex* a *in* b.

His ſuppoſitis quædam demonstrabimus de aquis erumpentibus, quæ mirè cum doctrina proiectorum conuenire videntur

P*Rimùm manifestum eſt omnes aquas erumpentes ex foraminibus tubi alicuius perforati, parabolas deſcribere. Primę enim guttæ ſcaturientes e tubo ſunt de naturæ proiectorum quandoquidem ipſæ, quamquam liquidæ attamen ſunt ſphęrulæ graues & coherentes, & ideo parabolam certè deſignabunt. Omnes autem ſubſequentes, quæ cum eodem impetu emittuntur (ſupponimus enim tubos ſemper aqua plenos) ſemitam pręcædentium percurrent; quare continuus ille aquæ fluentis tractus parabola erit.*

Obijcet fortaſſe aliquis hoc non videri, pręſertim quando tubi orificium valde anguſtum erit, & impetus vehemens. Tunc enim (vt videre eſt in linea illa aquea, quę ex fontium fiſtulis violentiùs erumpit) prior pars orbitæ illius aſcendentis magis tenſa apparet, et ad parabolam veriùs conformata; poſterior verò, hoc est ea quam aqua deſcendens percurrit, magis prona, et vt ita dicam, languida atq; curua conſpicitur. Obiectioni reſpondetur; non ſolùm pręcedentem propoſitiunculam, ſed etiam

Bb *maio-*

maiorem partem sequentium huic instantię subiacere. Causa est impedimentum medij, quod ad momentum corporis mobilis valdè sensibilem habet rationem, multoq; maiorem quàm in proiectionibus quę fiunt à machinis bellicis. Siquidem illic materia proiectionis sunt globi plumbei, ferrei, vel saltem marmorei; hic verò linea est, & quidem aquea. Nulli igitur mirum sit, quòd cum fundamentum huius doctrinę ęquè verũ sit Thoricè loquẽdo, ac in proiectis Galilei, practicè tamen multum ab ipsis contemplationibus aberrent experimenta, quæ ad hoc vt exactiora euaderent, vel fieri deberent in medio non obstante, vel saltem grauissima materia esset adhibenda. Quamquam si quis modica altitudine, solertiq; diligentia experiri hęc omnia velit, minimum quoddam, & plerumq; insensibile deesse comperiet. Experimentum, quod nobis confirmauit has penè omnes speculatiunculas, factũ fuit tubo quodam, imò capsula parallelepipeda, cuius altitudo passum Geometricum excedebat, cuius basis vno palmo quadrato non erat minor. Foramina verò erant rotunda, circuloque humanæ pupillę maiora, non perperam facta, sed solertissimè excauata in lamellis cupreis, tenuibus, & ad horizontem erectis. Aqua enim violenter erumpens semper directione exit perpendiculari ad illud planum ex quo erumpit, ideoq; fiebat vt emissiones nostri tubi horizontales essent.

DAto tubo *a b* semper pleno, & aptè perforato foraminibus *c d e*. hoc est quæ sint figuræ circularis, sitq; illorũ ductus horizontalis, hoc est in tenui lamella plana pendiculari. Datoq; horizonte quolibet *b g*, inuenire amplitudinẽ vniuscuiusq; parabolæ. Fiat circa *a b* diametrũ semicirculus *a h b*. Eritq; parabolæ fluẽtis, ex *e* amplitudo dupla lineæ *e i* quæ horizontaliter ducitur in semicirculo. Et amplitudo parabolæ erumpentis ex *d* erit dupla lineæ *d h*. Et hoc probatur, quia cum aqua sit velut proiectum quoddam, sitq; (per suppositum) ipsius punctũ sublime *a*, erunt per Propositionem 5. Galilei, semisses amplitudinum medio loco proportiona-

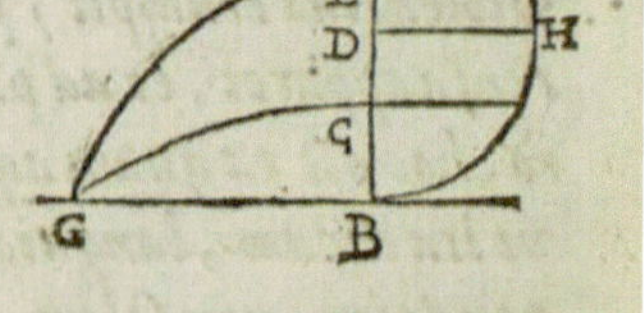

les

maiorem partem ſequentium huic instantię ſubiacere. Cauſa eſt impedimentum medij, quod ad momentum corporis mobilis valdè ſenſibilem habet rationem, multoq; maiorem quàm in proiectionibus quę fiunt à machinis bellicis. Siquidem illic materia proiectionis ſunt globi plumbei, ferrei, vel ſaltem marmorei; hic verò linea est, & quidem aquea. Nulli igitur mirum ſit, quòd cum fundamentum huius doctrinę ęquè verũ ſit Thoricè[99] *loquẽdo, ac in proiectis Galilei, practicè tamen multum ab ipſis contemplationibus aberrent experimenta, quæ* ad hoc *vt exactiora euaderent, vel fieri deberent in medio non obſtante, vel ſaltem grauiſſima materia eſſet adhibenda. Quamquam ſi quis modica altitudine, ſolertiq; diligentia experiri hęc omnia velit, minimum quoddam, & plerumq; inſenſibile deeſſe comperiet. Experimentum, quod nobis confirmauit has penè omnes ſpeculatiunculas, factũ fuit tubo quodam, imò capſula parallelepipeda, cuius altitudo paſſum Geometricum excedebat, cuius baſis vno palmo quadrato non erat minor. Foramina verò erant rotunda, circuloque humanæ pupillę maiora, non perperam facta, ſed ſolertiſſimè excauata in lamellis cupreis, tenuibus, & ad horizontem erectis. Aqua enim violenter erumpens ſemper directione exit perpendiculari ad illud planum ex quo erumpit, ideoq; fiebat vt emiſſiones nostri tubi horizontales eſſent.*

DAto tubo *ab* ſemper pleno, & aptè perforato foraminibus *cde.* hoc eſt quæ ſint figuræ circularis, ſitq; illorũ ductus horizontalis, hoc eſt in tenui lamella plana pendiculari.[100] Datoq; horizonte quolibet *bg,* inuenire amplitudinẽ vniuſcuiuſq; parabolæ. Fiat circa *ab* diametrũ ſemicirculus *ahb.* Eritq; parabolæ fluẽtis, ex *e* amplitudo dupla lineæ *ei* quæ horizontaliter ducitur in ſemicirculo. Et amplitudo parabolæ erumpentis ex *d* erit dupla lineæ *dh.* Et hoc probatur, quia cum aqua ſit velut proiectum quoddam, ſitq; (per ſuppoſitum) ipſius punctũ ſublime *a,* erunt per Propoſitionem 5. Galilei, ſemiſſes amplitudinum medio loco proportiona-

.les

[99] *Thoricè*] *Theoricè.*
[100] pendiculari] perpendiculari.

les inter sublimitatem, & altitudinem; quare semisses amplitudinum æquales erunt lineis *e i*, *d h*.

Corollaria.

Hinc manifestum est quod si tubus a b *perforetur in* d *puncto medio altitudinis, tunc emissionem ex* d *factam longiùs quã quælibet alia cadere.*

Foramina verò quæ æqualitèr à puncto medio d. *distant, patet æquales amplitudines facere.*

Manifestum etiam est inferiores parabolas semper superioribus maiores esse, cum habeant maiorem sublimitatem, hoc est maius latus rectum, est enim sublimitas quarta pars lateris recti, vt ostensum est.

DAto dolio, siue tubo *a b* quod aptè perforatum sit in *c*, & emissionem faciat *c d*. Inuenienda sit aquę in tubo latenris libella horizontalis, siue superficies suprema.

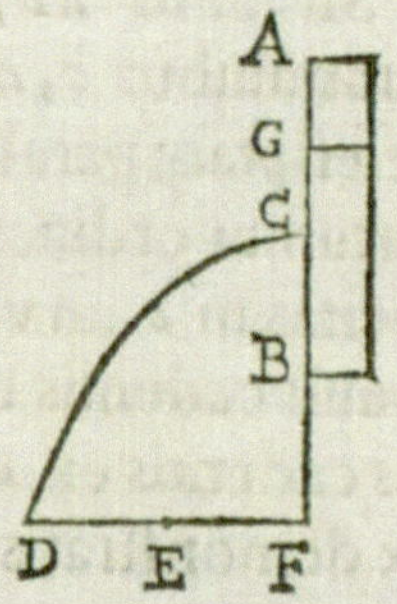

Sit horizon *d f*, & producatur *c b* in *f*, & secetur bifariam *f d* in *e*, fiatq; vt *c f* altitudo, ad *f e*. semibasim, ita *f e*, ad aliam, quæ erit sublimitas *c g*. Patebit ergo libel lam aquæ in tubo latentis esse per punctum *g*.

SI tubus *a b* aptè perforetur vbicunq; in *c*, emissio fluentis aquæ coni rectanguli superficiem continget, cuius axis sit ipse tubus, vertex verò sit in aquæ libella.

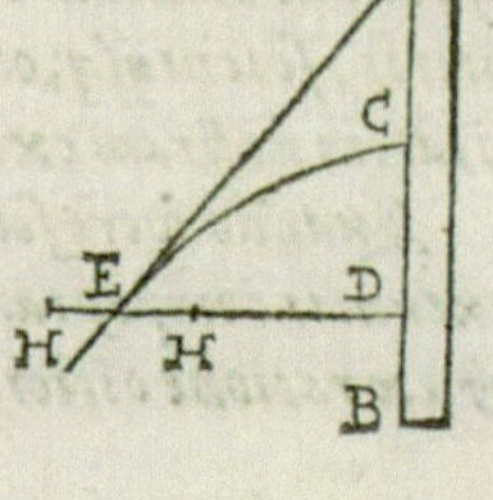

Sit angulus coni *b a e*. semirectus, & *a b* tubus, hoc est linea ea in quà sunt foramina, ponatur axis coni. Sumatur æqualis *c d* ipsi *c a*. Ducaturq; horizon talis *d e*. Dico parabolam transire per *e*. Si enim potest, transeat per *h*. & cum aquæ sublimitas sit *c a*. erit semissis lineę

les inter ſublimitatem, & altitudinem; quare ſemiſſes amplitudinum æquales erunt lineis *ei, dh.*

Corollaria.

Hinc manifestum eſt quod ſi tubus ab *perforetur in* d *puncto medio altitudinis, tunc emiſſionem ex* d *factam longiùs quã quælibet alia cadere.*

Foramina verò quæ æqualitèr à puncto medio d. *diſtant, patet æquales amplitudines facere.*

Manifeſtum etiam est inferiores parabolas ſemper ſuperioribus maiores eſſe, cum habeant maiorem ſublimitatem, hoc eſt maius latus rectum, est enim ſublimitas quarta pars lateris recti, vt oſtenſum eſt.

DAto dolio, ſiue tubo *ab* quod aptè perforatum ſit in *c,* & emiſſionem faciat *cd.* Inuenienda ſit aquę in tubo latentis libella horizontalis, ſiue ſuperficies ſuprema.

Sit horizon *df,* & producatur *cb* in *f,* & ſecetur bifariam *fd* in *e,* fiatq; vt *cf* altitudo, ad *fe.* ſemibaſim, ita *fe,* ad aliam, quæ erit ſublimitas *cg.* Patebit ergo libellam aquæ in tubo latentis eſſe per punctum *g.*

SI tubus *ab* aptè perforetur vbicunq; in *c,* emiſſio fluentis aquæ coni rectanguli ſuperficiem continget, cuius axis ſit ipſe tubus, vertex verò ſit in aquæ libella.

Sit angulus coni *bae.* ſemirectus, & *ab* tubus, hoc eſt linea ea in quà ſunt foramina, ponatur axis coni. Sumatur æqualis *cd* ipſi *ca.* Ducaturq; horizontalis *de.* Dico parabolam tranſire per *e.* Si enim poteſt, tranſeat per *h.* & cum aquæ ſublimitas ſit *ca.* erit ſemiſſis lineę

 hd me-

hd media proportionalis inter duas æquales *dc*, *ca*. & propterea tota *dh*. æqualis erit ipsi *da*, vel *dc*. quod est absurdum. Si ergo parabola transit per punctum *e*. ipsa *ea* tangens est, cum æquales sint *ac*, *cd*.

Hinc manifestum est, quòd si tubus in omnibus suis punctis aptè perforatus fuerit, omnes emissiones quodammodo conspirare videbuntur ad formandam coni rectanguli speciem. Si verò non tubus, sed sphærula in vertice ipsius posita, aptè perforata sit in omnibus suis punctis, emissiones omnes cuiusdam conoidis parabolici imaginem conformabunt, ex Propos. 30. huius.

AQuarum ex tubo *ab* perforato erumpentium velocitates sunt vt lineæ in parabola applicatę ad suam vniuscuiusq; sublimitatem.

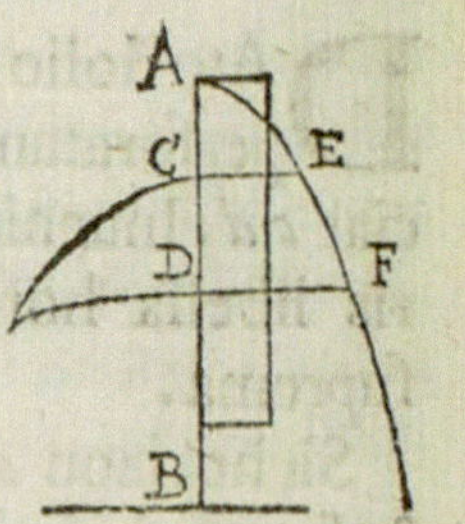

Sit tubus *ab* semper aqua plenus, & ex foraminibus *c*, *d* erumpant fluentes lineæ: descriptaq; parabola *aef* circa axem *ab* ducantur ordinatim *ce*, *df*. Erit ergo velocitas in *c* ad velocitatem in *d*, vt impetus grauis cadentis ex *a* in *c*. ad impetum grauis cadentis ex *a* in *d*, nempe vt *ce*, ad *df*. ex demonstratis in primo libro de motu.

Corollarium.

Hinc sequitur ex doctrina Abbatis Castellij quantitatē aquæ exeuntis per ostium c. *ad quantitatē aquę exeuntis per* d *(quādo foramina fuerint ęqualia) esse vt* c e, *ad* d f. *Hoc est aquas erūpentes ex foraminibus æqualibus esse in subdupla ratione sublimitatum, siue altitudinum suarum. Veritatem huius Corollarij primus omnium experimento indagauit eruditiss. vir, æquè literis, scientijsq; omnibus ornatus,* Raphael Magiottus, *& veritatem nostram exitus felicitate confirmauit.*

Quando verò foramina inæqualia erunt, quantitates aquæ exeuntis compositam rationem habebunt ex ratione velonitatū, & ex ratione ostiorum.

Si tu-

hd media proportionalis inter duas æquales *dc, ca.* & propterea tota *d h.* æqualis erit ipſi *da,* vel *de.* quod eſt abſurdum. Si ergo parabola tranſit per punctum *e.* ipſa *e a* tangens eſt, cum æquales ſint *ac, cd.*

Hinc manifestum eſt quòd ſi tubus in omnibus ſuis punctis aptè perforatus fuerit, omnes emiſſiones quodammodo conſpirare videbuntur ad formandam coni rectanguli ſpeciem. Si verò non tubus, ſed ſphærula in vertice ipſius poſita, aptè perforata ſit in omnibus ſuis punctis, emiſſiones omnes cuiuſdam conoidis parabolici imaginem conformabunt, ex Propoſ. 30. huius.

AQuarum ex tubo *ab* perforato erumpentium velocitates ſunt vt lineæ in parabola applicatę ad ſuam vniuſcuiuſq; ſublimitatem.

Sit tubus *ab* ſemper aqua plenus, & ex foraminibus *c, d* erumpant fluentes lineæ*:* deſcriptaq; parabola *aef* circa axem *ab* ducantur ordinatim *ce, df.* Erit ergo velocitas in *c* ad velocitatem in *d,* vt impetus grauis cadentis ex *a* in *c.* ad impetum grauis cadentis ex *a* in *d,* nempe vt *ce,* ad *df.* ex demonſtratis in primo libro de motu.

Corollarium.

Hinc ſequitur ex doctrina Abbatis Caſtellij quantitatẽ aquæ exeuntis per ostium c. *ad quantitatẽ aquę exeuntis per* d *(quã do foramina fuerint ęqualia) eſſe vt* ce, *ad* df. *Hoc eſt aquas erũpentes ex foraminibus æqualibus eße in ſubdupla ratione ſublimitatum, ſiue altitudinum ſuarum. Veritatem huius Corollarij primus omnium experimento indagauit eruditiſs. vir, æquè literis, ſcientijſq; omnibus ornatus,* Raphael Magiottus, *& veritatem nostram exitus felicitate confirmauit.*

Quando verò foramina inæqualia erunt, quantitates aquæ exeuntis compoſitam rationem habebunt ex ratione velocitatũ, & ex ratione ostiorum.

Si tu-

SI tubus *ab* cylindricus, ſiue priſmaticus perforatus in fundo *b* fluat, neque alius humor ſuperinfundatur, velocitates ſupremæ ſuperficiei humoris latentis decreſcent cum eadem ratione, qua decreſcunt etiam lineæ ordinatim applicatæ in parabola *bd*, quæ axem habeat *ba* verticem verò *b*. Hoc manifeſtum eſt. Nam quando aquæ ſumma ſuperficies erit *c*, velocitas erit *cd* & quando ſumma ſuperficies erit *e*, velocitas erit *ef*. ex iam demonſtratis; & hoc modo ſemper.

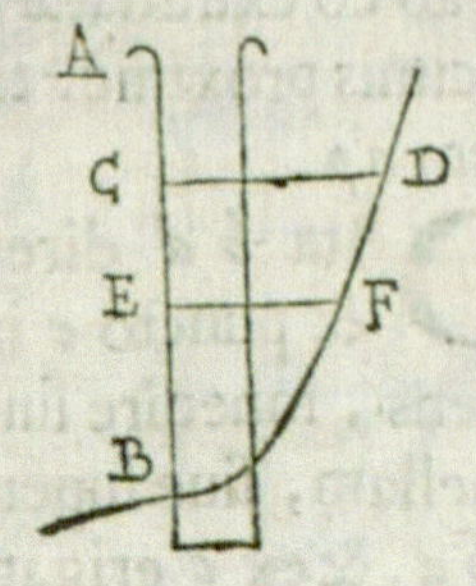

CViuſmodi ſit ſolidum ab aquis cadentibus conformatum inueſtigare.

Sit vas aqua ſemper plenum *ab* ampliſſimum, cuius foramen in fundo circulare ſit *cd*, ſolidum autem aquæ ex eo fluentis ſit *copd*, & ſolidi axis ſit *fb*. Dico lineam *dnp* ſolidi huius genitricem talem eſſe, vt numerus biquadratus diametri *cd*, ad biquadratum diametri *op*. ſit reciprocè vt altitudo *fb* ad altitudinem *fi*.

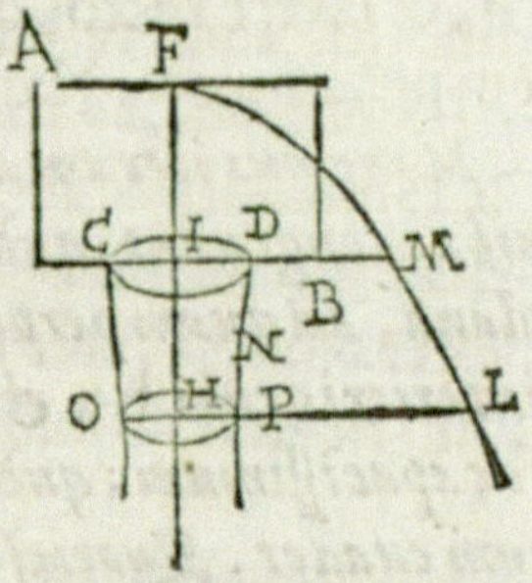

Oſtendit Abbas Caſtellius ſectionem *cd* ad ſectionem *op*, eſſe reciprocè vt velocitas in *op*, ad velocitatem in *cd*. nempe vt *hl* ad *im* in parabola *fml*. His præmiſſis. Quadratus numerus diametri *cd* ad quadratum *op* eſt vt circulus *cd*, ad circulum *op*. nempe vt *hl* ad *im*. Numerus autem quadratus ex *hl* ad quadratum ex *im* eſt vt *hf* ad *fi*. ergo biquadratus numerus diametri *cd*. ad biquadratum *op* eſt reciprocè vt altitudo *fh*, ad altitudinem *fi*. Quod &c.

Data ſit eadem figura altitudo *fi* 100. *fh* 160. dataq; ſit diameter foraminis *cd* 50. Quęritur quãta futura ſit ſolidi diameter *op*. Fiat vt *fh* ad *fi*, nempe vt 160. ad 100. ita numerus biquadratus diametri *cd*, nempe 6250000. ad alium, qui erit 3906250; idemq. erit numerus biquadratus diametri *op*: ſi ergo ab

SI tubus *ab* cylindricus, ſiue priſmaticus perforatus in fundo *b* fluat, neque alius humor ſuper infundatur, velocitates ſupremæ ſuperficiei humoris latentis decreſcent cum eadem ratione, qua decreſcunt etiam lineæ ordinatim applicatæ in parabola *bd,* quæ axem habeat *ba* verticem verò *b.* Hoc manifeſtum eſt. Nam quando aquæ ſumma ſuperficies erit *c,* velocitas erit *cd* & quando ſumma ſuperficies erit *e,* velocitas erit *ef.* ex iam demonſtratis; & hoc modo ſemper.

CViuſmodi ſit ſolidum ab aquis cadentibus conformatum inueſtigare.

Sit vas aqua ſemper plenum *ab* ampliſſimum, cuius foramen in fundo circulare ſit *cd,* ſolidum autem aquæ ex eo fluentis ſit *copd,* & ſolidi axis ſit *fh.* Dico lineam *dnp* ſolidi huius genitricem talem eſſe, vt numerus biquadratus diametri *cd,* ad biquadratum diametri *op.* ſit reciprocè vt altitudo *fh* ad altitudinem *fi.*

Oſtendit Abbas Caſtellius ſectionem *cd* ad ſectionem *op,* eſſe reciprocè vt velocitas in *op,* ad velocitatem in *cd.* nempe vt *hl* ad *im* in parabola *fml.* His præmiſſis. Quadratus numerus diametri *cd* ad quadratum *op* eſt vt circulus *cd,* ad circulum *o p.* nempe vt *hl* ad *im.* Numerus autem quadratus ex *hl* ad quadratum ex *im* eſt vt *hf* ad *fi.* ergo biquadratus numerus diametri *cd.* ad biquadratum *op* eſt reciprocè vt altitudo *fh,* ad altitudinem *fi.* Quod &c.

Data ſit eadem figura altitudo *fi* 100. *fh* 160. dataq; ſit diameter foraminis *cd* 50. Quęritur quãta futura ſit ſolidi diameter *op.* Fiat vt *fh* ad *fi,* nempe vt 160. ad 100. ita numerus biquadratus diametri *cd,* nempe 6,250,000. ad alium, qui erit 3,906,250; idemq. erit numerus biquadratus diametri *op*: ſi er-

go ab

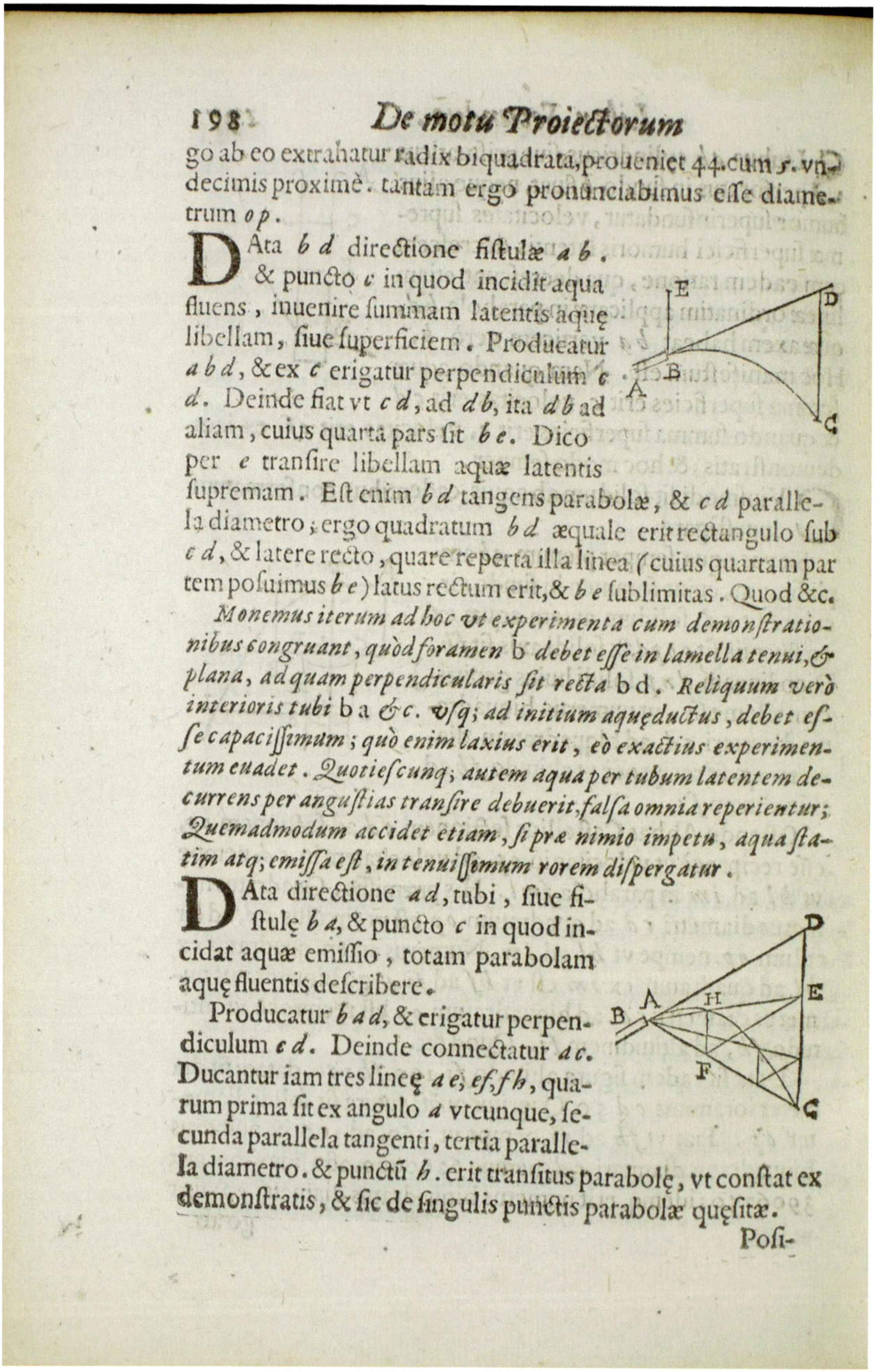

198 De motu Proiectorum

go ab eo extrahatur radix biquadrata, proueniet 44. cum 5. vndecimis proximè. tantam ergo pronunciabimus esse diametrum *o p*.

DAta *b d* directione fistulæ *a b*. & puncto *c* in quod incidit aqua fluens, inuenire summam latentis aquę libellam, siue superficiem. Producatur *a b d*, & ex *c* erigatur perpendiculum *c d*. Deinde fiat vt *c d*, ad *d b*, ita *d b* ad aliam, cuius quarta pars sit *b e*. Dico per *e* transire libellam aquæ latentis supremam. Est enim *b d* tangens parabolæ, & *c d* parallela diametro; ergo quadratum *b d* æquale erit rectangulo sub *c d*, & latere recto, quare reperta illa linea (cuius quartam partem posuimus *b e*) latus rectum erit, & *b e* sublimitas. Quod &c.

Monemus iterum ad hoc vt experimenta cum demonstrationibus congruant, quòd foramen b *debet esse in lamella tenui, & plana, ad quam perpendicularis sit recta* b d. *Reliquum verò interioris tubi* b a *&c. vsq; ad initium aquęductus, debet esse capacissimum; quò enim laxius erit, eò exactius experimentum euadet. Quotiescunq; autem aqua per tubum latentem decurrens per angustias transire debuerit, falsa omnia reperientur; Quemadmodum accidet etiam, si præ nimio impetu, aqua statim atq; emissa est, in tenuissimum rorem dispergatur.*

DAta directione *a d*, tubi, siue fistulę *b a*, & puncto *c* in quod incidat aquæ emissio, totam parabolam aquę fluentis describere.

Producatur *b a d*, & erigatur perpendiculum *c d*. Deinde connectatur *a c*. Ducantur iam tres lineę *a e*, *e f*, *f h*, quarum prima sit ex angulo *a* vtcunque, secunda parallela tangenti, tertia parallela diametro. & punctũ *h*. erit transitus parabolę, vt constat ex demonstratis, & sic de singulis punctis parabolæ quęsitæ.

Posi-

go ab eo extrahatur radix biquadrata, proueniet 44. cum *5.* vndecimis proximè. tantam ergo pronunciabimus eſſe diametrum *op.*

DAta *bd* directione fiſtulæ *ab.* & puncto *c* in quod incidit aqua fluens, inuenire ſummam latentis aquę libellam, ſiue ſuperficiem. Producatur *abd,* & ex *c* erigatur perpendiculum *cd.* Deinde fiat vt *cd,* ad *db,* ita *db* ad aliam, cuius quarta pars ſit *be.* Dico per *e* tranſire libellam aquæ latentis ſupremam. Eſt enim *bd* tangens parabolæ, & *cd* parallela diametro; ergo quadratum *bd* æquale erit rectangulo ſub *cd,* & latere recto, quare reperta illa linea *(*cuius quartam partem poſuimus *be)* latus rectum erit, & *be* ſublimitas. Quod &c.

Monemus iterum ad hoc *vt experimenta cum demonſtrationibus congruant, quòd foramen* b *debet eſſe in lamella tenui, & plana, ad quam perpendicularis ſit recta* bd. *Reliquum verò interioris tubi* ba *&c. vſq; ad initium aquęductus, debet eſſe capaciſſimum; quò enim laxius erit, eò exactius experimentum euadet. Quotieſcunq; autem aqua per tubum latentem decurrens per anguſtias tranſire debuerit, falſa omnia reperientur; Quemadmodum accidet etiam, ſi præ nimio impetu, aqua ſtatim atq; emiſſa eſt, in tenuiſſimum rorem diſpergatur.*

DAta directione *ad,* tubi, ſiue fiſtulę *ba,* & puncto *c* in quod incidat aquæ emiſſio, totam parabolam aquę fluentis deſcribere.

Producatur *bad,* & erigatur perpendiculum *cd.* Deinde connectatur *ac.* Ducantur iam tres lineę *ae, ef, fh,* quarum prima ſit ex angulo *a* vtcunque, ſecunda parallela tangenti, tertia parallela diametro. & punctũ *h.* erit tranſitus parabolę, vt conſtat ex demonſtratis, & ſic de ſingulis punctis parabolæ quęſitæ.

Posi-

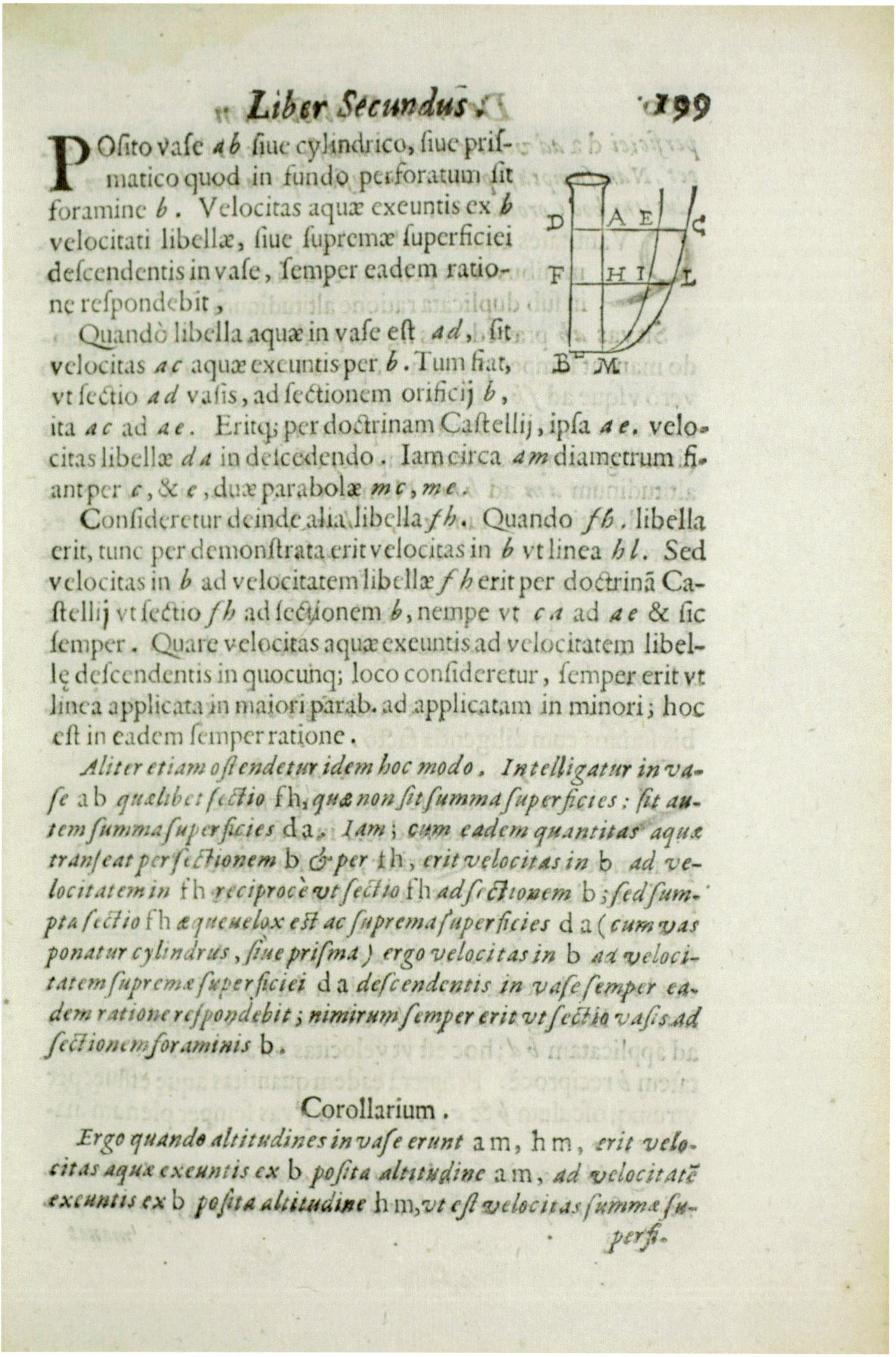

Liber Secundus. 199

Posito vase *ab* siue cylindrico, siue prismatico quod in fundo perforatum sit foramine *b*. Velocitas aquæ exeuntis ex *b* velocitati libellæ, siue supremæ superficiei descendentis in vase, semper eadem ratione respondebit,

Quandò libella aquæ in vase est *ad*, sit velocitas *ac* aquæ exeuntis per *b*. Tum fiat, vt sectio *ad* vasis, ad sectionem orificij *b*, ita *ac* ad *ae*. Eritq; per doctrinam Castellij, ipsa *ae*. velocitas libellæ *da* in descedendo. Iam circa *am* diametrum fiant per *c*, & *e*, duæ parabolæ *mc*, *me*.

Consideretur deinde alia libella *fh*. Quando *fh*. libella erit, tunc per demonstrata erit velocitas in *b* vt linea *hl*. Sed velocitas in *b* ad velocitatem libellæ *fh* erit per doctrinã Castellij vt sectio *fh* ad sectionem *b*, nempe vt *ca* ad *ae* & sic semper. Quare velocitas aquæ exeuntis ad velocitatem libellę descendentis in quocunq; loco consideretur, semper erit vt linea applicata in maiori parab. ad applicatam in minori; hoc est in eadem semper ratione.

Aliter etiam ostendetur idem hoc modo. Intelligatur in vase ab *qualibet sectio* fh, *quæ non sit summa superficies: sit autem summa superficies* da. *Iam; cum eadem quantitas aquæ transeat per sectionem* b *& per* fh, *erit velocitas in* b *ad velocitatem in* fh *reciprocè vt sectio* fh *ad sectionem* b; *sed sumpta sectio* fh *æquevelox est ac suprema superficies* da *(cum vas ponatur cylindrus, siue prisma) ergo velocitas in* b *ad velocitatem supremæ superficiei* da *descendentis in vase semper eadem ratione respondebit; nimirum semper erit vt sectio vasis ad sectionem foraminis* b.

Corollarium.

Ergo quando altitudines in vase erunt am, hm, *erit velocitas aquæ exeuntis ex* b *posita altitudine* am, *ad velocitatẽ exeuntis ex* b *posita altitudine* hm, *vt est velocitas summæ superfi-*

perfi-

POſito vaſe *ab* ſiue cylindrico, ſiue priſmatico quod in fundo perforatum ſit foramine *b.* Velocitas aquæ exeuntis ex *b* velocitati libellæ, ſiue ſupremæ ſuperficiei deſcendentis in vaſe, ſemper eadem ratione reſpondebit,

Quandò libella aquæ in vaſe eſt *ad,* ſit velocitas *ac* aquæ exeuntis per *b.* Tum fiat, vt ſectio *ad* vaſis, ad ſectionem orificij *b,* ita *ac* ad *ae.* Eritq; per doctrinam Caſtellij, ipſa *ae.* velocitas libellæ *da* in deſcedendo. Iam circa *a m* diametrum fiant per *c,* & *e,* duæ parabolæ *mc, me.*

Conſideretur deinde alia libella *fh.* Quando *fh.* libella erit, tunc per demonſtrata erit velocitas in *b* vt linea *hl.* Sed velocitas in *b* ad velocitatem libellæ *fh* erit per doctrinã Caſtellij vt ſectio *fh* ad ſectionem *b,* nempe vt *ca* ad *ae* & ſic ſemper. Quare velocitas aquæ exeuntis ad velocitatem libellę deſcendentis in quocunq; loco conſideretur, ſemper erit vt linea applicata in maiori parab. ad applicatam in minori; hoc eſt in eadem ſemper ratione.

Aliter etiam oſtendetur idem hoc modo. Intelligatur in vaſe ab *quælibet ſectio* fh, *quæ non ſit ſumma ſuperficies: ſit autem ſumma ſuperficies* da. *Iam; cum eadem quantitas aquæ tranſeat per ſectionem* b *& per* fh, *erit velocitas in* b *ad velocitatem in* fh *reciprocè vt ſectio* fh *ad ſectionem* b*; ſed ſumpta ſectio* fh *æqueuelox est ac ſuprema ſuperficies* da *(cum vas ponatur cylindrus, ſiue priſma) ergo velocitas in* b *ad velocitatem ſupremæ ſuperficiei* da *deſcendentis in vaſe ſemper eadem ratione reſpondebit; nimirum ſemper erit vt ſectio vaſis adſectionem foraminis* b.

Corollarium.

Ergo quando altitudines in vaſe erunt am, hm, *erit velocitas aquæ exeuntis ex* b *poſita altitudine* am, *ad velocitatẽ exeuntis ex* b *poſita altitudine* hm, *vt eſt velocitas ſummæ ſu-*

perſi

perficiei d a *ad velocitatem ſummæ ſuperficiei* f h. *hoc enim patet. Nam ſumpta ſuperiori concluſione permutando tantum deducitur hoc Corollarium.*

QVantitates aquarum ab eodem, ſiue ab æqualibus foraminibus erumpentium eodem tempore, ſunt inter ſe in ſub duplicata ratione altitudinum.

Sit vas *ab* præcedentis figuræ perforatum in *b*. & aliquando maneat ſemper plenum vſque ad ſignum *da*; aliquando verò vſque ad *fh*. Dico quantitatem aquæ exeuntis quando altit. eſt *am*, ad quantitatem aquę exeuntis quando altit. eſt *hm* (intellige ſemper eodem tempore) eſſe in ſubduplicata ratione altitudinum *am* ad *hm*. Nempe vt recta *ac* ad *hl*. Nam quando altitudines ſunt *am*, & *hm*, Velocitates in *b* ſunt ex Coroll. præced. vt velocitas ſummę ſuperficiei *da*, ad velocitatem ſummæ ſuperficiei *fh*; ſiue vt applicata *ae* ad *hi*. Ergo quantitates aquarum erumpentiũ ex eodem foramine *b* erunt vt *ae* ad *hi*, nempe in ſubduplicata ratione altitudinum *am*. ad *hm*.

Hæc ſpeculatio conuenit exactiſſimè cum experimẽto à nobis ſumma cum diligentia facto.

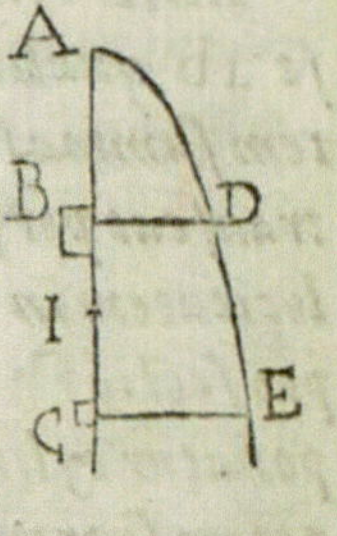

QVoddam vas cuius ſummitas *a* perforatum eſt foramine *b* ita vt ſuperinfluẽte quodam aquę ductu in *a* ſemper plenum permaneat. Quæritur, quo foramine perforari debeat in *c* vt eadem ſuperinfluẽte aqua plenum prçciſè ſicut antea permaneat. Sumatur inter *ab*, *ac* media *ai*. Fiatq; vt altitudo *ca* ad mediam *ai* ita oſculum *b* ad oſculum *c*. Erit ergo oſculum *b*. ad oſculum *c*, vt applicata *ce*, ad applicatam *bd*; hoc eſt vt velocitas foraminis *c* ad velocitatem *b* reciprocè. Propterà eadem quantitas aquę effluet per vtrumq; oſculum *b* & *c*; propoſitumq; vas ſemper plenum manebit.

Quoddam verò vas a b *cum perforatum ſit in fundo foramine* b, *ſuperinfluente quodam dato aquæ ductu* d, *plenum permanet*

perficiei da *ad velocitatem ſummæ ſuperficiei* fh. *hoc enim patet. Nam ſumpta ſuperiori concluſione permutando tantum deducitur hoc Corollarium.*

QVantitates aquarum ab eodem, ſiue ab æqualibus foraminibus erumpentium eodem tempore, ſunt inter ſe in ſub duplicata[101] ratione altitudinum.

Sit vas *ab* præcedentis figuræ perforatum in *b*. & aliquando maneat ſemper plenum vſque ad ſignum *da*; aliquando verò vſque ad *fh*. Dico quantitatem aquæ exeuntis quando altit. eſt *m*, ad quantitatem aquę exeuntis quando altit. eſt *hm* (intellige ſemper eodem tempore) eſſe in ſubduplicata ratione altitudinum *am* ad *hm*. Nempe vt recta *ac* ad *hl*. Nam quando altitudines ſunt *am*, & *hm*, Velocitates in *b* ſunt ex Coroll. præced. vt velocitas ſummę ſuperficiei *da*, ad velocitatem ſummæ ſuperficiei *fh*; ſiue vt applicata *a* ad *hi*. Ergo quantitates aquarum erumpentiũ ex eodem foramine *b* erunt vt *ae* ad *hi*, nempe in ſubduplicata ratione altitudinum *am*. ad *hm*.

Hæc ſpeculatio conuenit exactiſſimè cum experimẽto à nobis ſumma cum diligentia facto.

QVoddam vas cuius ſummitas *a* perforatum eſt foramine *b* ita vt ſuperinfluẽte quodam aquę ductu in *a* ſemper plenum permaneat. Quæritur, quo foramine perforari debeat in *c* vt eadem ſuperinfluẽte aqua plenum pręcisè ſicut antea permaneat.

Sumatur inter *ab*, *ac* media *ai*. Fiatq; vt altitudo *ca* ad mediam *ai* ita oſculum *b* ad oſculum *c*, Erit ergo oſculum *b*. ad oſculum *c*, vt applicata *ce*, ad applicatam *bd*; hoc eſt vt velocitas foraminis *c* ad velocitatem *b* reciprocè. Propterà eadem quantitas aquę effluet per vtrumq; oſculum *b* & *c*; propoſitumq; vas ſemper plenum manebit.

Quoddam verò vas ab *cum perforatum ſit in fundo foramine* b, *ſuperinfluente quodam dato aquæ ductu* d, *plenum per-*

manet

[101] ſub duplicata] ſubduplicata.

manet vſq; ad ſignum c. *Quæritur quantitas aquæ in idem vas ingerendæ ad hoc vt repleatur vſq; ad ſignum* a. *Sumatur inter* a b, b c *media* b e: *fiatq; vt* e b *ad* b a, *ita quantitas aquæ datæ* d *ad aliam quantitatem; quæ ingeſta omninò vas replebit vſq; ad ſignum* a, *neq; illud excedet. Quod cum multis alijs huius generis facilè demonſtratur ex præcedentibus.*

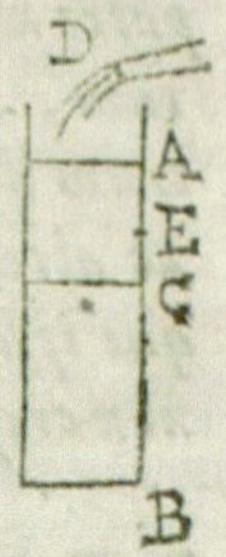

AQuarum fluentium (quę tamen aliquo vaſe excipi poſſint) proportionem dicere, ſine vlla temporis, velocitatis, ſectioniſq; menſura.

Sumatur vt in præcedenti figura, quodcunque uas *ab*, cuiuſcunq; figuræ ſit, ita tamen perforatum in fundo, ut minor ex datis aquis fluentibus ingeſta non effluat ſtatim tota, ſed increſcat, & aliquam altitudinem faciat in uaſe, puta altitudinem *bc* & deinde non creſcat amplius; ſed tantum aquæ prorſus uas emittat, quantum recipit. Maior uerò aquæ quantitas altitudinem faciat *ab*. Patet ex præcedentibus aquam maiorem ad minorem eſſe in ſubduplicata ratione rectæ *ab* ad *bc*. Nam cum utraque aqua tranſeat per eandem ſectionem *b*, & altera earum altitudinem habeat *ab* altera uero *cb*, erunt uelocitates aquarum per dictam ſectionem exeuntium in ſubduplicata ratione *ab* ad *cb*. Ergo & quantitates aquarum fluentium erunt in ſubduplicata ratione factarum altitudinum *ab*, *bc*.

Lemma.

Sit diameter alicuius parabolæ a b, *& mobile aliquod moueatur per* a b *ea lege, vt in quocunq; puncto lineę* a b *conſideretur ſemper impetus eius ſit ut linea ordinatim ex illo puncto intra aliquam parabolam applicata Dico hunc motum eundem eſſe ac grauium naturaliter cadentium. Intelligatur enim aliquod graue moueri ex* a *in* b *motu naturaliter accelerato, & concipiatur eius momentum eiuſmodi ut tam graue quam etiam mobile ſimul dimiſſa ex* a. *eodem tempore perueniant ad punctum* b

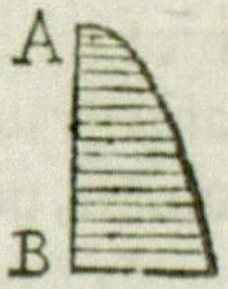

manet vſq; ad ſignum c. *Quæritur quantitas aquæ in idem vas ingerendæ* ad hoc *vt repleatur vſq; ad ſignum* a. *Sumatur inter* ab, bc. *media* be*: fiatq; vt* eb *ad* ba, *ita quantitas aquæ datæ* d *ad aliam quantitatem; quæ ingeſta omninò vas replebit vſq; ad ſignum* a, *neq; illud excedet. Quod cum multis alijs huius generis facilè demonstratur ex præcedentibus.*

AQuarum fluentium *(*quę tamen aliquo vaſe excipi poſſint) proportionem dicere, ſine vlla temporis, velocitatis, ſectioniſq; menſura.

*S*umatur vt in præcedenti figura, quodcunque uas *ab,* cuiuſcunq; figuræ ſit, ita tamen perforatum in fundo, ut minor ex datis aquis fluentibus ingeſta non effluat ſtatim tota, ſed increſcat, & aliquam altitudinem faciat in uaſe, puta altitudinem *bc* & deinde non creſcat amplius; ſed tantum aquæ prorſus uas emittat, quantum recipit. Maior uerò aquæ quantitas altitudinem faciat *ab.* Patet ex præcedentibus aquam maiorem ad minorem eſſe in ſubduplicata ratione rectæ *ab* ad *bc.* Nam cum utraque aqua tranſeat per eandem ſectionem *b,* & altera earum altitudinem habeat *ab* altera uero *cb,* erunt uelocitates aquarum per dictam ſectionem exeuntium in ſubduplicata ratione *ab* ad *cb.* Ergo & quantitates aquarum fluentium erunt in ſubduplicata ratione factarum altitudinum *ab, bc.*

Lemma.

Sit diameter alicuius parabolæ ab, *& mobile aliquod moueatur per* ab *ea lege, vt in quocunq; puncto lineę* ab *conſideretur ſemper impetus eius ſit ut linea ordinatim ex illo puncto intra aliquam parabolam applicata. Dico hunc motum eundem eſſe ac grauium naturaliter cadentium. Intelligatur enim aliquod graue moueri ex* a *in* b *motu naturaliter accelerato, & concipiatur eius momentum eiuſmodi ut tam graue quam etiam mobile ſimul dimiſſa ex* a. *eodem tempore perueniant ad punctum* b

Cc *patet*

202 *De motu Proiectorum*

patet amborum mobilium vnum atq; eundem futurum eſſe motum, Nam in quocunq; puncto lineæ a b *conſideretur alterutrum dictorum ſiue mobile, ſiue graue, eundem impetum habebit ac alterum, quare pariter etiam tranſibunt ſpatium* a b, *parteſque ipſius. & hoc uerum etiam erit ſi mobile moueatur ex* b *in* a, *non creſcente, ſed decreſcente impetu.*

VAſa cylindrica ſiue Priſmatica in fundo perforata ea lege exhauriuntur, ut diuiſo toto tempore in partes æquales, emiſſio ultimi temporis ſit ut unum, emiſſio autem penultimi tēporis ſit ut 3. antepenultimi temporis ut 5. & ſic deinceps ut numeri impares ab unitate.

Sit uas ut poſitum eſt; perforatum in fundo, ipſiq; adſcribatur parabola *c d*. Iam demonſtrauimus fluente ex fundo aqua, libellam *a e* ita deſcendere ut ſemper uelocitas ipſius ſit ut linea ſibi reſpondens in parabola, nempe impetus in *e a*, ſit ut *a d*, in *f h* ſit ut *h i*, & ſic ſemper; erit ergo motus libellæ *e a* tamquam motus deficiens grauium ſurſum reflexorum, ſiue proiectorum; & diuiſo toto tempore emiſſionis in partes æquales, erit ſpatium *i c* decurſum à libella ultimo tempore, ut unum; ſpatium autem *h i* ut tres, & *h a* ut quinque. Nam, ex lemmate pręmiſſo, motus libellæ *a e* eſt tamquam motus grauium non cadentium, ſed ſurſum perpendiculariter proiectorum (quod idem eſt) ergo motus libellæ *a e* eadem ſpatia tranſibit temporibus ęqualibus, atq; graue aliquod ſurſum proiectum, nempe ultimo tempore unum, penultimo tria; & ſic deinceps.

ex lem. præced.

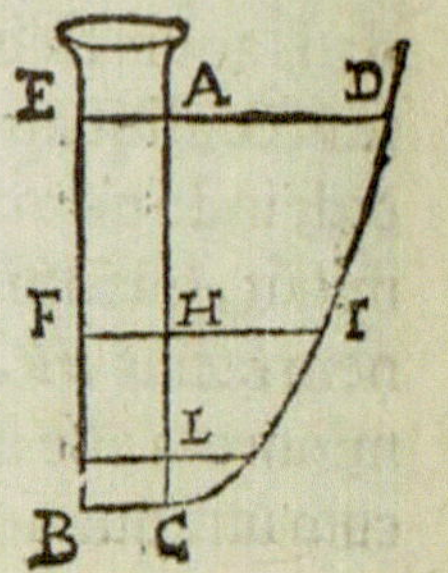

Si fiat uas conoidale parabolicum cuius axis ſit a b, *& perforatum ſit in fundo* b *uideri poterit emiſſio eius eiuſmodi ut motus ſupremæ ſuperficiei deſcēdentis, æquabilis ſit: hoc eſt ut æqualibus temporibus æquales altitudine moles exhauriantur; quod tamen falſum eſt. Sunt enim conoidea parabolica*

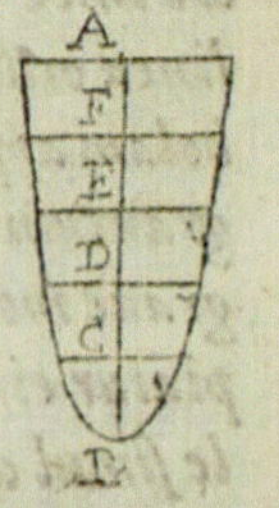

tamen

patet amborum mobilium vnum atq; eundem futurum esse motum, Nam in quocunq; puncto lineæ ab *consideretur alterutrum dictorum siue mobile, siue graue, eundem impetum habebit ac alterum, quare pariter etiam transibunt spatium* ab, *partesque ipsius. & hoc uerum etiam erit si mobile moueatur ex* b *in* a, *non crescente, sed decrescente impetu.*

VAsa cylindrica siue Prismatica in fundo perforata ea lege exhauriuntur, ut diuiso toto tempore in partes æquales, emissio ultimi temporis sit ut unum, emissio autem penultimi tẽporis sit ut 3. antepenultimi temporis ut 5. & sic deinceps ut numeri impares ab unitate.

Sit uas ut positum est; perforatum in fundo, ipsiq; adscribatur parabola *cd.* Iam demonstrauimus fluente ex fundo aqua, libellam *a e* ita descendere ut semper uelocitas ipsius sit ut linea sibi respondens in parabola, nempe impetus in *ea,* sit ut *ad,* in *fh* sit ut *hi,* & sic semper; erit ergo motus libellæ *ea* tamquam motus deficiens grauium sursum reflexorum, siue proiectorum; & diuiso toto tempore emissionis in partes æquales, erit spatium *ic* decursum à libella ultimo tempore, ut unum; spatium autem *hi* ut. tres, & *ha* ut quinque. Nam, ex lemmate prẹmisso, motus libellæ *ae* est tamquam motus grauium non cadentium, sed sursum perpendiculariter proiectorum (quod idem est) ergo motus libellæ *ae* eadem spatia transibit temporibus ẹqualibus, atq; graue aliquod sursum proiectum, nempe ultimo tempore unum, penultimo tria; & sic deinceps.

ex lem. praeced.

Si fiat uas conoidale parabolicum cuius axis sit ab, *& perforatum sit in fundo* b *uideri poterit emissio eius eiusmodi ut motus supremæ superficiei descẽdentis, æquabilis sit: hoc est ut æqualibus temporibus æquales altitudine moles exhauriantur;quod tamen falsum est. Sunt enim conoidea parabolica*

inter

inter se ut quadrata axium, siue altitudinum. Si ergo diuidamus totam ab *in partes æquales, erit conoides* cb *ut unum, &* db *ut quatuor; ipsumq;* eb *ut 9. & sic deinceps semper ut numeri quadrati. Erunt ergo conoides* cb *ut unum, differentia autem* cd *ut tria,* de *ut 5.* ef*, ut 7. & sic de inceps differentiæ erunt ut numeri ab unitate impares. Quare uidebitur alicui quod singulæ huiusmodi differentię æqualibus temporibus exauriri debeant per iā demonstrata in præcedenti; sed quoniam in huiusmodi uacuatione plurimi refert cuius figuræ sit ipsum uas, absolutè falsum hoc esse pronuntiamus; demonstrationemq. unusquisq. colligere poterit ex his quæ sequuntur.*

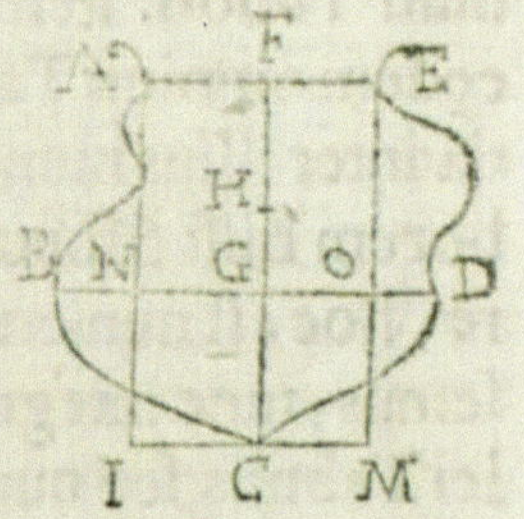

Esto uas irregulare *a b c d e* perforatum in fundo foramine *c*; & considerentur duæ ipsius sectiones *a e*, *b d*. Dico uelocitatem summæ superficiei aquæ descendentis, quando erit *a e*, ad velocitatem superficiei, quando erit *b d*, rationem habere compositam ex ratione subduplicata altitudinū *f c* ad *c g*, & reciproca sectionum, nempe sectionis *b d* ad *a e*. Concipiatur enim super basi sectionis *a e* quęcunq. illa sit, uas prismaticum *a i m e* cuius altitudo sit *f c*. Iam velocitas sectionis prismaticæ *a e* ad *n o* erit ut recta *f c* ad *c h* mediam inter altitudines. Velocitas uerò sectionis *n o* ad uelocitatem sectionis *b d*, cum eandem altitudinem habeant, est reciprocè ut sectio *b d* ad *n o*. Ergo patet quòd ratio uelocitatis sectionis *a e* ad uelocitatem sectionis *b d* componitur ex ratione rectæ *f c* ad *c h*, & ex ratione sectionis *b d* ad *n o*, siue *b d* ad *a e*.

Hinc manifestum est quod nuper de Conoide parabolico dicebamus, nempe motum supremæ superficiei descendentis non esse æquabilem; sed subinde acceleratum. Qua uerò ratione acceleretur; & qua ratione uarientur uelocitates supremæ superficiei aquę descendentis in sphęra perforata, sphæroide, atq. alijs uasibus regularibus, facilè ex contēplatione pręcedenti patebit.

inter ſe ut quadrata axium, ſiue altitudinum. Si ergo diuidamus totam ab *in partes æquales, erit conoides* cb *ut unum, &* db *ut quatuor; ipſumq;* eb *ut 9. & ſic deinceps ſemper ut numeri quadrati. Erunt ergo conoides* cb *ut unum, differentia autem* cd *ut tria,* de *ut 5.* e f, *ut 7. & ſic deinceps differentiæ erunt ut numeri ab unitate impares. Quare uidebitur alicui quod ſingulæ huiuſmodi differentię æqualibus temporibus exauriri debeant per iã demonſtrata in præcedenti; ſed quoniam in huiuſmodi uacuatione plurimi refert cuius figuræ ſit ipſum uas, abſolutè falſum hoc eße pronuntiamus; demonstrationemq. unuſquiſq. colligere poterit ex his quæ ſequuntur.*

ESto uas irregulare *abcde* perforatum in fundo foramine *c;* & conſiderentur duæ ipſius ſectiones *ae, bd.* Dico uelocitatem ſummæ ſuperficiei aquæ deſcendentis, quando erit *ae,* ad velocitatem ſuperficiei, quando erit *bd,* rationem habe re compoſitam ex ratione ſubduplicata altitudinũ *fc* ad *cg,* & reciproca ſectionum, nempe ſectionis *bd* ad *ae.* Concipiatur enim ſuper baſi ſectionis *ae* quęcunq. illa ſit, uas priſmaticum *aime* cuius altitudo ſit *fc.* Iam velocitas ſectionis priſmaticæ *ae* ad *no* erit ut recta *fc* ad *ch* mediam inter altitudines. Velocitas uerò ſectionis *no* ad uelocitatem ſectionis *bd,* cum eandem altitudinem habeant, eſt reciprocè ut ſectio *bd* ad *no.* Ergo patet quòd ratio uelocitatis ſectionis *ae* ad uelocitatem ſectionis *b d* componitur ex ratione rectæ *fc* ad *ch,* & ex ratione ſectionis *bd* ad *no,* ſiue *bd* ad *ae.*

Hinc manifeſtum eſt quod nuper de Conoide parabolico dicebamus, nempe motum ſupremæ ſuperficiei deſcendentis non eſſe æquabilem; ſed ſubinde acceleratum. Qua uerò ratione acceleretur; & qua ratione uarientur uelocitates ſupremæ ſuperficiei aquę deſcendentis in ſphęra perforata, ſphæroide, atq. alijs uaſibus regularibus, facilè ex contẽplatione pręcedenti patebit.

DE TABVLIS.

SEquuntur Tabulæ non quidem doctis calculi vigilijs elaboratæ, vt à Galileo factum eſt, ſed ex ipſa Tabula ſinuũ ac Tangentium facili breuiq; negotio tranſcriptæ. Quocunq. tamen modo collectæ fuerint, non minus augent Galilei gloriam, quàm laborem noſtrum comminuerint. Cuius enim induſtriæ tanta ſolertia eſt, vt per innumeras multiplicationum, diuiſionum, & radicum ambages, adeoſdem pene numeros appellere potuerit, quos ex Tabula deſumere nobis conceſſum fuit? Prædictum hoc volo, nos ſupponere voluiſſe eandem maximam amplitudinem ſemiparabolarum cum Galileo partium 10000. item maximam altitudinem partium 10000. vt eędem omnino Tabulæ euaderent, & aliqua interdum differentia inter illius numeros & noſtros appareret. Ideo in ſolum laborem biſſectionum incidimus. Si verò ſuppoſitionem variare, hoc eſt numerum hunc duplum 20000. ſupponere voluiſſemus, tunc integræ Tabulæ diuerſę quidem euaſiſſent à Galilei Tabulis, ſed numeri poterant ſine vlla biſſectione ex ſinibus, & Tangentibus prout ibi leguntur mutuari.

Tabu-

DE TABVLIS.

SEquuntur Tabulæ non quidem doctis calculi vigilijs elaboratæ, vt à Galileo factum eſt, ſed ex ipſa Tabula ſinuũ ac Tangentium facili breuiq; negotio tranſcriptæ. Quocunq. tamen modo collectæ fuerint, non minus augent Galilei gloriam, quàm laborem noſtrum comminuerint. Cuius enim induſtriæ tanta ſolertia eſt, vt per innumeras multiplicationum, diuiſionum, & radicum ambages, ad eoſdem pene numeros appellere potuerit, quos ex Tabula deſumere nobis conceſſum fuit? Prædictum hoc volo, nos ſupponere voluiſſe eandem maximam amplitudinem ſemiparabolarum cum Galileo partium 10,000. item maximam altitudinem partium 10,000. vt eędem[102] omnino Tabulæ euaderent, & aliqua interdum differentia inter illius numeros & noſtros appareret. Ideo in ſolum laborem biſſectionum incidimus. Si verò ſuppoſitionem variare, hoc eſt numerum hunc duplum 20,000. ſupponere voluiſſemus, tunc integræ Tabulæ diuerſę quidem euaſiſſent à Galilei Tabulis, ſed numeri poterant ſine vlla biſſectione ex ſinibus, & Tangentibus prout ibi leguntur mutuari.

Tabu-

[102] eędem] eadem.

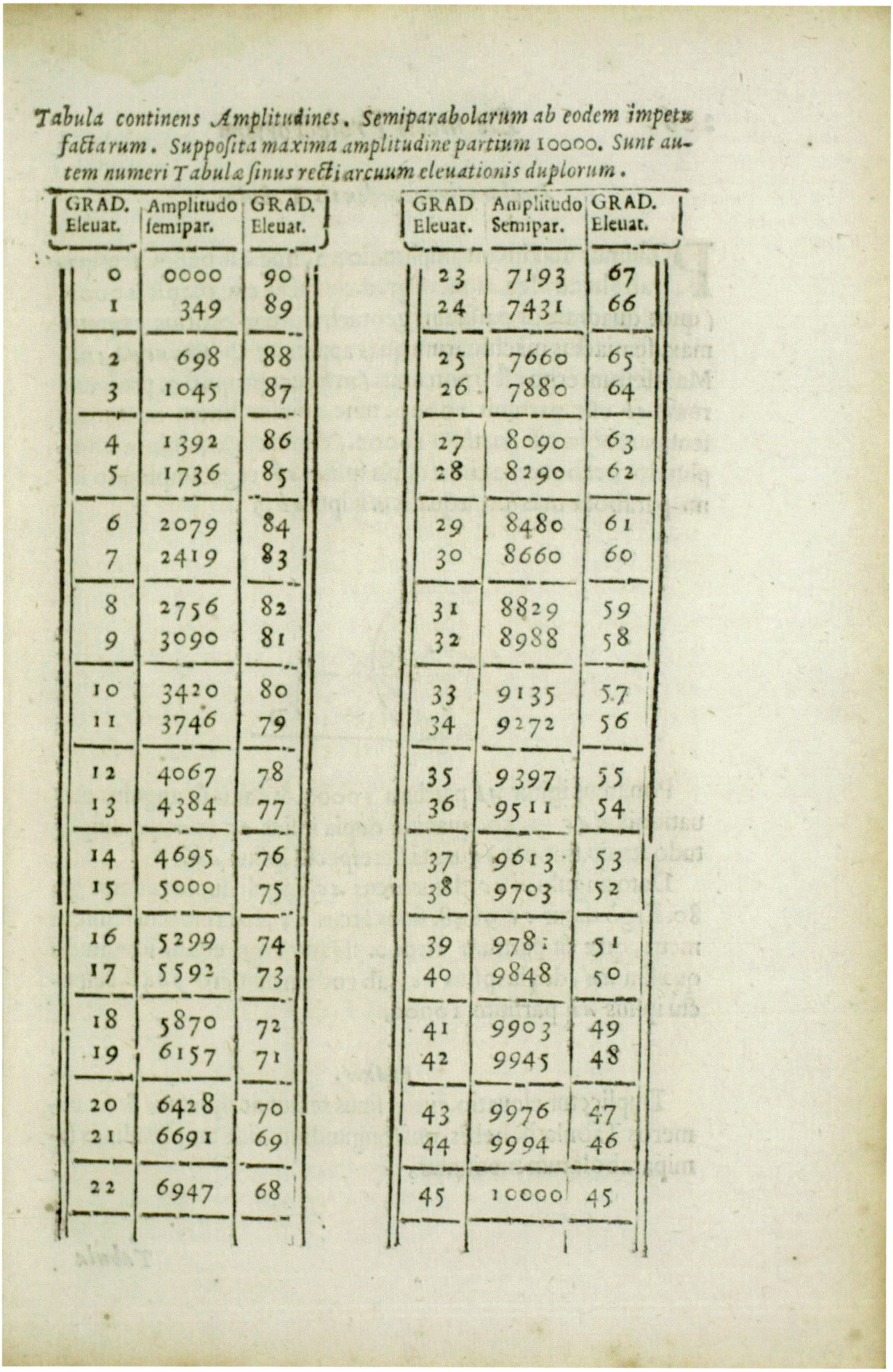

Tabula continens Amplitudines. Semiparabolarum ab eodem impetu factarum. Supposita maxima amplitudine partium 10000. Sunt autem numeri Tabulæ sinus recti arcuum eleuationis duplorum.

GRAD. Eleuat.	Amplitudo semipar.	GRAD. Eleuat.	GRAD Eleuat.	Amplitudo Semipar.	GRAD. Eleuat.
0	0000	90	23	7193	67
1	349	89	24	7431	66
2	698	88	25	7660	65
3	1045	87	26	7880	64
4	1392	86	27	8090	63
5	1736	85	28	8290	62
6	2079	84	29	8480	61
7	2419	83	30	8660	60
8	2756	82	31	8829	59
9	3090	81	32	8988	58
10	3420	80	33	9135	57
11	3746	79	34	9272	56
12	4067	78	35	9397	55
13	4384	77	36	9511	54
14	4695	76	37	9613	53
15	5000	75	38	9703	52
16	5299	74	39	9781	51
17	5592	73	40	9848	50
18	5870	72	41	9903	49
19	6157	71	42	9945	48
20	6428	70	43	9976	47
21	6691	69	44	9994	46
22	6947	68	45	10000	45

[*Liber Secundus.*] [205]

Tabula continens Amplitudines. Semiparabolarum ab eodem impetu factarum. Suppoſita maxima amplitudine partium 10000. *Sunt autem numeri Tabulæ ſinus recti arcuum eleuationis duplorum.*

GRAD. Eleuat.	Amplitudo ſemipar.	GRAD. Eleuat.
0	0000	90
1	349	89
2	698	88
3	1045	87
4	1392	86
5	1736	85
6	2079	84
7	2419	83
8	2756	82
9	3090	81
10	3420	80
11	3746	79
12	4067	78
13	4384	77
14	4695	76
15	5000	75
16	5299	74
17	5592	73
18	5870	72
19	6157	71
20	6428	70
21	6691	69
22	6947	68

GRAD Eleuat.	Amplitudo Semipar.	GRAD. Eleuat.
23	7193	67
24	7431	66
25	7660	65
26	7880	64
27	8090	63
28	8290	62
29	8480	61
30	8660	60
31	8829	59
32	8988	58
33	9135	57
34	9272	56
35	9397	55
36	9511	54
37	9613	53
38	9703	52
39	9781	51
40	9848	50
41	9903	49
42	9945	48
43	9976	47
44	9994	46
45	10000	45

[103] A series of tables (Torricelli 1644, pp. 204–226) appears. Torricelli connects different elements of the heavy bodies' motion. (Cfr. Pisano and Bussotti's chapters above).

Declaratio precedentis Tabulæ.

POsuimus maximam amplitudinem, hoc est basim semiparabolæ ad eleuationem grad.45. factæ esse partiũ 10000: (quot quadrantes passuum geometricorum prorsus exequat max. semiactus machinarum quas appellant *Colubrine da 30.*) Manifestum ergo est si ponamus (in figura propos. 9. proiectorum) *ab* esse partium 10000. tunc amplitudinem maximam semiparabolæ esse partium 10000. (demonstrata enim est amplitudo parabolæ maximæ dupla ipsius *ab*, ergo amplitudo semi-parabolæ maximę æqualis erit ipsi *ab*.)

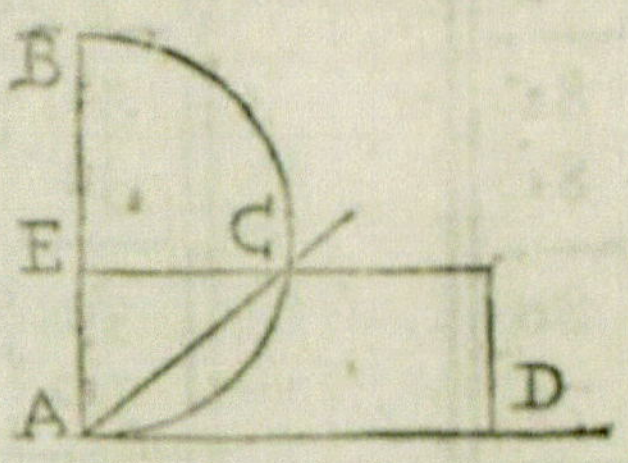

Ponamus igitur *ab* partium 10000. & datus sit angulus eleuationis *dac* gr. 40. quæritur dupla ipsius *ec* (nempe amplitudo semiparabolæ) quanta sit respectu ipsius *ab*.

Dato angulo *dac* datur arcus *ac* anguli duplus nempe gr. 80. Ergò datur *ce* 9848. sinus arcus *ac*, sed respectu semidia metri, quæ sit partium 10000. Datur ergo etiam longitudo quæsita *ad*, dupla ipsius *ec*, ab eodem numero 9848. respectu ipsius *ab* partium 10000,

Praxis.

Duplicetur eleuatio, eiusq. sinus rectus accipiatur; & sic numeros Tabulæ habebis, qui longitudines, siue amplitudines semiparabolarum metiuntur,

Tabula

Declaratio prȩcedentis Tabulæ.

POſuimus maximam amplitudinem, hoc eſt baſim ſemiparabolæ ad eleuationem grad. 45. factæ eſſe partiũ 10,000. (quot quadrantes paſſuum geometricorum prorſus exȩquat max. ſemiactus machinarum quas appellant *Colubrine da 30.*) Manifeſtum ergo eſt ſi ponamus (in figura propoſ. 9. proiectorum) *ab* eſſe partium 10,000. tunc amplitudinem maximam ſemiparabolæ eſſe partium 10,000. (demonſtrata enim eſt amplitudo parabolæ maximæ dupla ipſius *ab*, ergo amplitudo ſemiparabolæ maximȩ æqualis erit ipſi *ab*.)

Ponamus igitur *ab* partium 10,000. & datus ſit angulus eleuationis *dac* gr. 40. quæritur dupla ipſius *ec* (nempe amplitudo ſemiparabolæ) quanta ſit reſpectu ipſius *ab*.

Dato angulo *dac* datur arcus *ac* anguli duplus nempe gr.80. Ergò datur *ce* 9848. ſinus arcus *ac*, ſed reſpectu ſemidiametri, quæ ſit partium 10,000. Datur ergo etiam longitudo quæſita *ad*, dupla ipſius *ec*, ab eodem numero 9848. reſpectu ipſius *ab* partium 10,000,

Praxis.

Duplicetur eleuatio, eiuſq. ſinus rectus accipiatur; & ſic numeros Tabulæ habebis, qui longitudines, ſiue amplitudines ſemiparabolarum metiuntur,

Tabula

Tabula continens Altitudines semiparabolarũ quarũ impetus idẽ sit cũ impetu præcedentis Tabulæ. Supponitur maxima altitudo part. 10000. Sunt autem numeri Tabulæ semisses sinuum versorum arcuum eleuationis duplorum.

GRAD Eleuat.	ALTIT. semipar.	GRAD Eleuat.	ALTIT. Semipar.	GRAD Eleuat.	ALTIT. Semipar.
1	3	31	2653	61	7650
2	12	32	2808	62	7796
3	27	33	2966	63	7939
4	49	34	3127	64	8078
5	76	35	3290	65	8214
6	109	36	3455	66	8346
7	149	37	3622	67	8473
8	194	38	3790	68	8597
9	245	39	3960	69	8716
10	302	40	4132	70	8830
11	364	41	4304	71	8940
12	432	42	4477	72	9045
13	506	43	4651	73	9145
14	585	44	4826	74	9240
15	670	45	5000	75	9330
16	760	46	5174	76	9415
17	855	47	5349	77	9494
18	955	48	5523	78	9568
19	1060	49	5696	79	9636
20	1170	50	5868	80	9698
21	1284	51	6040	81	9755
22	1403	52	6210	82	9806
23	1527	53	6378	83	9851
24	1654	54	6545	84	9891
25	1786	55	6710	85	9924
26	1922	56	6873	86	9951
27	2061	57	7034	87	9973
28	2204	58	7192	88	9988
29	2350	59	7347	89	9997
30	2500	60	7500	90	10000

[*Liber Secundus.*] [207]

Tabula continens Altitudines ſemiparabolarũ quarũ impetus idẽ ſit cũ impetu pracedentis Tabulæ. Supponitur maxima altitudo part. 10000. *Sunt autem numeri Tabulæ ſemiſſes ſinuum verſorum arcuum eleuationis duplorum.*

GRAD Eleuat.	ALTIT. ſemipar.
1	3
2	1 2
3	27
4	49
5	76
6	109
7	149
8	194
9	245
10	302
11	364
12	432
13	506
14	585
15	670
16	760
17	855
18	955
19	1060
20	1170
21	1284
22	1403
23	1527
24	1654
25	1786
26	1922
27	2061
28	2204
29	2350
30	2500

GRAD Eleuat.	ALTIT. Semipar.
31	2653
32	2808
33	2966
34	3127
35	3290
36	3455
37	3622
38	3790
39	3960
40	4132
41	4304
42	4477
43	4651
44	4826
45	5000
46	5174
47	5349
48	5523
49	5696
50	5868
51	6940
52	6210
53	6378
54	6545
55	6710
56	6873
57	7034
58	7192
59	7347
60	7500

GRAD Eleuat.	ALTIT. Semipar.
61	7650
62	7796
63	7939
64	8078
65	8214
66	8346
67	8473
68	8597
69	8716
70	8830
71	8940
72	9045
73	9145
74	9240
75	9330
76	9415
77	9494
78	9568
79	9636
80	9698
81	9755
82	9806
83	9851
84	9891
85	9924
86	9951
87	9973
88	9988
89	9997
90	10000

[104] Torricelli offers the relation between the initial inclination angle of a projectile's trajectory and the amplitude of the half-parabola which is the trajectory itself. The condition is that the projectile always starts with the same speed for every inclination angle and the maximal length of the half-parabola is posed equal to 10,000 (obtained when the inclination is 45 degrees). (Cfr. Pisano and Bussotti's chapters above).

Explanatio præcedentis Tabulæ.

POnimus maximam altitudinem omnium proiectionum ab eodem impetu factarum esse partium 10000. Ponimus ergo in subiecta figura lineam *ab* esse 10000. partium. Dato deinde angulo *dac* eleuationis gr. 40. quæritur quanta sit altitudo *ae* respectu ipsius *ab* quæ est 10000.

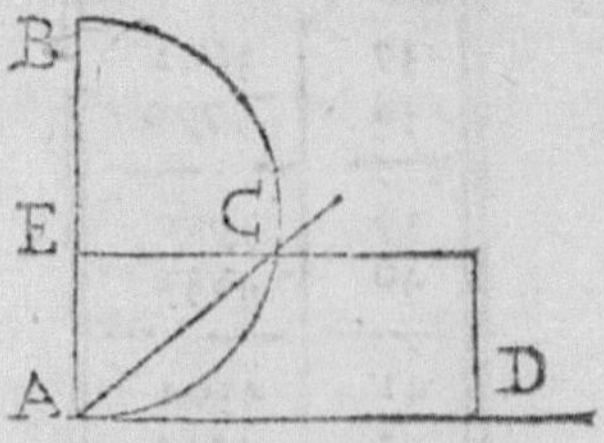

Datur quidem *ae* 8264. ex Tabula sinuum, cum sit sinus versus arcus *ac* gr. 80. qui arcus duplus est anguli eleuationis *dac*. Sed datus numerus rectæ *ae* 8264. erit respectu semidiametri, quę sit partium 10000. Cum verò nos ponamus totam diametrum *ab* esse partium 10000. tunc *ae* erit 4132. hoc est tantummodo semissis illius numeri ex Tabula sinuum versorum excerpti.

Praxis.

Duplicetur data eleuatio; illiusq. sinus versi semissis accipiatur; & sic habebis numeros Tabulæ præcedentis, qui altitudines parabolarum, siue proiectionum metiuntur.

Decla-

Explanatio præcedentis Tabulæ.

POnimus maximam altitudinem omnium proiectionum ab eodem impetu factarum eſſe partium 10,000. Ponimus ergo in ſubiecta figura lineam *ab* eſſe 10,000. partium.
Dato deinde angulo *dac* eleuationis gr. 40. quęritur quanta ſit altitudo *a e* reſpectu ipſius *ab* quæ eſt 10,000.

Datur quidem *ae* 8264. ex Tabula ſinuum, cum ſit ſinus verſus arcus *ac* gr. 80. qui arcus duplus eſt anguli eleuationis *dac.* Sed datus numerus rectæ *ae* 8264. erit reſpectu ſemidiametri, quę ſit partium 10,000. Cum verò nos ponamus totam diametrum *ab* eſſe partium 10,000. tunc *ae* erit 4132.[105] hoc eſt tantummodo ſemiſſis illius numeri ex Tabula ſinuum verſorum excerpti.

Praxis.

Duplicetur data eleuatio; illiuſq. ſinus verſi ſemiſſis accipiatur; & ſic habebis numeros Tabulæ præcedentis, qui altitudines parabolarum, ſiue proiectionum metiuntur.

Decla-

105 4132] 4132.

Declaratio Sequentis Tabulæ.

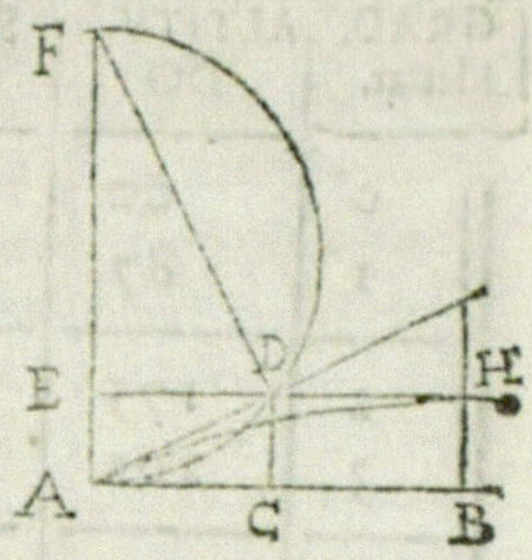

PОnimus *ab* amplitudinem omniũ semiparabolarũ esse partiũ 10000 Data iam eleuatione *cad* gr. 30. iuxtà quam dirigendum est tormentum. Quæritur parabolæ *ab* Altitudo, & sublimitas.

Secetur *ab* bifariam in *c*, erigaturque ad angulos rectos *cd*. Fiatq. angulus *adf* rectus. Manifestum est circa diametrum *af* describi semicirculum qui transeat per *d* cum rectus sit angulus ad *d* donec concurrat cum directione *ad*. Cum ergo fiat proiectio cum impetu *fa*, & directione *ad*, erit amplitudo semiparabolæ linea dupla ipsius *ed*, nempe ipsa *ab*; altitudo verò *bh*, vel *ae*, sublimitas *ef*. Quæritur ergo quantitas linearum *ae*, *ef*.

Cum *ab* sit partium 10000. erit *ed* nempe semissis ipsius, partium 2500. semper, quæcunque sit eleuatio. Si ergo *ed* sit sinus totus, erit altitudo *ea* 5774. tangens anguli eleuationis *cad*, hoc est *eda*. Sublimitas verò *ef* erit 17320. tangens complementi eiusdem anguli. Hæc autem vera sunt quãdo *ed* sit 10000. sed quia in casu nostro *ed* est tantummodo partium 2500, nempe semissis sinus totius, erunt *ea*, *ef* semisses dictarum tangentium; hoc est altitudo *ae* 2887. ipsa verò sublimitas *ef* 8660. propterea tormentum illud, quod eleuabitur gr. 30. ad hoc vt faciat amplitudinem semiparabolę partium 10000. debebit habere sublimitatem, siue impetũ 8660. Quod quidem idem est, ac si diceremus quòd impetus proiectionis tantus esse debet, quantus est grauis alicuius naturalitèr cadentis ab altitudine 8660. earumdem partium. Altitudo verò talis parabolę erit 2887.

Praxis.

Pro altitudine, sume semissem tangentis anguli eleuationis.
Pro sublimitate, sume semissem tangentis complementi anguli eleuationis.

Dd Tabu-

Declaratio Sequentis Tabulæ.

POnimus *ab* amplitudinem omniũ ſemiparabolarũ eſſe partiũ 10,000
Data iam eleuatione *cad* gr. 30. iuxtà quam dirigendum eſt tormentum. Quæritur parabolæ *ah* Altitudo, & ſublimitas.

Secetur *ab* bifariam in *c,* erigaturque ad angulos rectos *cd.* Fiatq. angulus *adf* rectus. Manifeſtum eſt circa diametrum *af* deſcribi ſemicirculum qui tranſeat per *d* cum rectus ſit angulus ad *d* donec concurrat cum directione *ad.*

Cum ergo fiat proiectio cum impetu *fa,* & directione *ad,* erit amplitudo ſemiparabolæ linea dupla ipſius *ed,* nempe ipſa *ab;* altitudo verò *bh,* vel *ae,* ſublimitas *ef.* Quæritur ergo quantitas linearum *ae, ef.*

Cum *ab* ſit partium 10,000. erit *ed* nempe ſemiſſis ipſius, partium 2500. ſemper, quæcunque ſit eleuatio. Si ergo *ed* ſit ſinus totus, erit altitudo *ea* 5774. tangens anguli eleuationis *cad,* hoc eſt *eda.* Sublimitas verò *ef* erit 17,320. tangens complementi eiuſdem anguli. Hæc autem vera ſunt quãdo *ed* ſit 10,000. ſed quia in caſu noſtro *ed* eſt tantummodo partium 2500, nempe ſemiſſis ſinus totius, erunt *ea, ef* ſemiſſes dictarum tangentium; hoc eſt altitudo *ae* 2887. ipſa verò ſublimitas *f* 8660. propterea tormentum illud, quod eleuabitur gr. *30.*[106] ad hoc vt faciat amplitudinem ſemiparabolę partium 10,000. debebit habere ſublimitatem, ſiue impetũ 8660.

Quod quidem idem eſt, ac ſi diceremus quòd impetus proiectionis tantus eſse debet, quantus eſt grauis alicuius naturalitèr cadentis ab altitudine 8660. earumdem partium. Altitudo verò talis parabolę erit 2887.

Praxis.

Pro altitudine, ſume ſemiſſem tangentis anguli eleuationis.

Pro ſublimitate, ſume ſemiſſem tangentis complementi anguli eleuationis.

Dd Tabu-

[106] *30*] 30.

210

Tabula continens Altitudines, & sublimitates Semiparabolarum quarum amplitudines æquales sint. Partium ſilicet ſemper 10000. Sunt

GRAD. Eleuat.	ALTITVDO	SVBLIMITAS	GRAD. Eleuat.	ALTITVDO	SVBLIMITAS
0	00	infinita	23	2122	11779
1	87	286450	24	2226	11230
2	175	143186	25	2332	10723
3	262	95406	26	2439	10252
4	350	71503	27	2548	9813
5	437	57150	28	2659	9404
6	525	47572	29	2772	9020
7	614	40722	30	2887	8660
8	703	35577	31	3004	8321
9	792	31569	32	3124	8002
10	882	28356	33	3247	7699
11	972	25723	34	3373	7413
12	1063	23523	35	3501	7141
13	1154	21657	36	3633	6882
14	1247	20054	37	3768	6635
15	1340	18660	38	3906	6400
16	1434	17437	39	4049	6175
17	1529	16354	40	4196	5959
18	1625	15388	41	4346	5752
19	1722	14521	42	4502	5553
20	1820	13737	43	4663	5362
21	1919	13025	44	4828	5178
22	2020	12375	45	5000	5000

Tabula continens Altitudines, & ſublimitates Semiparabolarum quarum amplitudines æquales ſint. Partium ſilicet ſemper 10000. Sunt

GRAD. Eleuat.	ALTITV-DO	SVBLIMI-TAS
0	00	infinita
1	87	286450
2	175	143186
3	262	95406
4	350	71503
5	437	57150
6	525	47572
7	614	40722
8	703	35577
9	792	31569
10	882	28356
11	972	25723
12	1063	23523
13	1154	21657
14	1247	20054
15	1340	18660
16	1434	17437
17	1529	16354
18	1625	15388
19	1722	14521
20	1820	13737
21	1919	13025
22	2020	12375

GRAD. Eleuat.	ALTITV-DO	SVBLIMI-TAS
23	2122	11779
24	2226	11230
25	2332	10723
26	2439	10252
27	2548	9813
28	2659	9404
29	2772	9020
30	2887	8660
31	3004	8321
32	3124	8002
33	3247	7699
34	3373	7413
35	3501	7141
36	3633	6882
37	3768	6635
38	3906	6400
39	4049	6175
40	4196	5959
41	4346	5752
42	4502	5553
43	4663	5362
44	4828	5178
45	5000	5000

autem, Altitudines semisses tangentium angulorum eleuationis At Sublimitates sunt semisses Tangentium complementorum eleuationis.

GRAD. Eleuat.	ALTITVDO.	SVBLIMIT.
46	5178	4828
47	5362	4663
48	5553	4502
49	5752	4346
50	5959	4196
51	6175	4049
52	6400	3906
53	6635	3768
54	6882	3633
55	7141	3501
56	7413	3373
57	7699	3247
58	8002	3124
59	8321	3004
60	8660	2887
61	9021	2772
62	9404	2659
63	9813	2548
64	10252	2439
65	10723	2332
66	11230	2226
67	11779	2122
68	12375	2020

GRAD. Eleuat.	ALTITVDO.	SVBLIMIT.
69	13025	1919
70	13737	1820
71	14521	1722
72	15388	1625
73	16354	1529
74	17437	1434
75	18660	1340
76	20054	1247
77	21657	1154
78	23523	1063
79	25723	972
80	28356	882
81	31569	792
82	35577	703
83	40722	614
84	47572	525
85	57150	437
86	71503	350
87	95406	262
88	143186	175
89	286450	87
90	*infinita.*	00

intera, Altitudines ſemiſſes tangentium angulorum eleuationis At Sublimitates ſunt ſemiſſes Tangentium complementorum eleuationis.

GRAD. Eleuat.	ALTI-TVDO.	SVBLI MIT.
46	5178	4828
47	5362	4663
48	5553	4502
49	5752	4346
50	5959	4196
51	6175	4049
52	6400	3906
53	6635	3768
54	6882	3633
55	7141	3501
56	7413	3373
57	7699	3247
58	8002	3124
59	8321	3004
60	8660	2887
61	9021	2772
62	9404	2659
63	9813	2548
64	10252	2439
65	10723	2332
66	11230	2226
67	11779	2122
68	12375	2020

GRAD. Eleuat.	ALTI-TVDO.	SVBLI MIT.
69	13025	1919
70	13737	1820
71	14521	1722
72	15388	1625
73	16354	1529
74	17437	1434
75	18660	1340
76	20054	1247
77	21657	1154
78	23523	1063
79	25723	972
80	28356	882
81	31569	792
82	35577	703
83	40722	614
84	47572	525
85	57150	437
86	71503	350
87	95406	262
88	143186	175
89	286450	87
90	*infinita.*	00

212

Tabula continens durationes, siue impetus ad horizontem comparatos proiectionum ab eodem impetu factarum. Supponitur maxima duratio, siue impetus maximus esse 10000. *Sunt autem numeri Tabulæ sinus recti eleuationum.*

GRAD Eleuat.	Duratio siue Imp.	GRAD Eleuat.	Duratio siue Imp.	GRAD Eleuat.	Duratio vel Imp.
1	75	31	5150	61	8746
2	349	32	5299	62	8892
3	523	33	5446	63	8910
4	698	34	5592	64	8988
5	872	35	5736	65	9063
6	1045	36	5878	66	9135
7	1219	37	6018	67	9205
8	1392	38	6157	68	9272
9	1564	39	6293	69	9336
10	1736	40	6428	70	9397
11	1908	41	6561	71	9455
12	2079	42	6691	72	9510
13	2250	43	6820	73	9563
14	2419	44	6947	74	9613
15	2588	45	7071	75	9659
16	2756	46	7193	76	9703
17	2924	47	7314	77	9744
18	2090	48	7431	78	9781
19	3256	49	7547	79	9816
20	3420	50	7660	80	9848
21	3584	51	7771	81	9878
22	3746	52	7880	82	9903
23	3907	53	7986	83	9925
24	4067	54	8090	84	9945
25	4226	55	8192	85	9962
26	4384	56	8290	86	9976
27	4540	57	8387	87	9986
28	4695	58	8480	88	9994
29	4848	59	8572	89	9998
30	5000	60	8660	90	10000

Tabula continens durationes, ſiue impetus ad horizontem comparatos proiectio num ab eodem impetu factarum. Supponitur maxima duratio, ſiue impetus maximus eſſe 10000. *Sunt autem numeri Tabulæ ſinus recti eleuationum.*

GRAD Eleuat.	Duratio ſiue Imp.
1	75
2	349
3	523
4	698
5	872
6	1045
7	1219
8	1392
9	1564
10	1736
11	1908
12	2079
13	2250
14	2419
15	2588
16	2756
17	2924
18	2090
19	3256
20	3420
21	3584
22	3746
23	3907
24	4067
25	4226
26	4384
27	4540
28	4695
29	4848
30	5000

GRAD Eleuat.	Duratio ſiue Imp.
31	5150
32	5299
33	5446
34	5592
35	5736
36	5878
37	6018
38	6157
39	6293
40	6428
41	6561
42	6691
43	6820
44	6947
45	7071
46	7193
47	7314
48	7431
49	7547
50	7660
51	7771
52	7880
53	7986
54	8090
55	8192
56	8290
57	8387
58	8480
59	8572
60	8660

GRAD Eleuat.	Duratio vel Imp.
61	8746
62	8892
63	8910
64	8988
65	9063
66	9135
67	9205
68	9272
69	9336
70	9397
71	9455
72	9510
73	9563
74	9613
75	9659
76	9703
77	9744
78	9781
79	9816
80	9848
81	9878
82	9903
83	9925
84	9945
85	9962
86	9976
87	9986
88	9994
89	9998
90	10000

Expositio præcedentis Tabulę.

IVxtà ea, quæ demonstrata sunt in propositionibus 18. & 21. Proiectorum, quando *ab* fuerit maxima duratio, siue maximus impetus ad horizontem comparatus, erit *ac* intercepta in semicirculo, duratio, siue impetus eleuationis *ac.* ad horizontem comparatus.

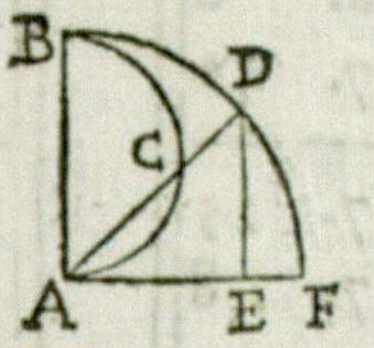

Supponimus *ab* esse partium 10000. nempe sinum totum. & dato angulo eleuationis *eac* gr. 30. quærimus *ac.*

Manifestum est, facto quadrante *bdf*, quando *ab* fuerit sinus totus, tunc *ca* esse sinum rectum eleuationis, hoc est anguli *eac*, quandoquidem *ca* ęqualis est ipsi *de*. Vt ostendimus in præcedenti libello. Ergo cum *df* arcus sit gr. 30. erit recta *de*, siue *ac*, 5000. respectu ipsius *ab* quæ est 10000. Hoc autem significat, quòd impetus grauis naturaliter cadentis ex *b* in *a* ad impetum parabolæ in extremo puncto (dummodo ad horizontem tantùm comparetur) erit vt 10000. ad 5000. Duratio verò, siue tempus lationis naturalis per perpendiculũ *ba* ad tempus, siue durationem parabolæ erit vt eadem *ba* ad *ac*, nempe vt 10000. ad 5000.

Praxis.

Sume ipsos sinus rectos eleuationũ, & habebis numeros Tabulę, exhibentes durationes, & impetus parabolarum ad horizontem comparatos.

Tabu-

Expoſitio præcedentis Tabulę.

IVxtà ea, quæ demonſtrata ſunt in propoſitionibus *18.*[107] & 21.
Proiectorum, quando *ab* fuerit maxima duratio, ſiue maximus impetus ad horizontem comparatus, erit *ac* intercepta in ſemicirculo, duratio, ſiue impetus eleuationis *ac.* ad horizontem comparatus.

Supponimus *ab* eſſe partium 10,000. nempe ſinum totum. & dato angulo eleuationis *eac* gr. 30. quærimus *ac.*

Manifeſtum eſt, facto quadrante *bdf,* quando *ab* fuerit ſinus totus, tunc *ca* eſſe ſinum rectum eleuationis, hoc eſt anguli *eac,* quandoquidem *ca* ęqualis eſt ipſi *de.* Vt oſtendimus in præcedenti libello. Ergo cum *df* arcus ſit gr. 30. erit recta *de,* ſiue *ac,* 5000. reſpeſtu ipſius *ab* quæ eſt 10,000. Hoc autem ſignificat, quòd impetus grauis naturaliter cadentis ex *b* in *a* ad impetum parabolæ in extremo puncto (dummodo ad horizontem tantùm comparetur) erit vt 10,000. ad 5000. Duratio verò, ſiue tempus lationis naturalis per perpendiculũ *ba* ad tempus, ſiue durationem parabolæ erit vt eadem *ba* ad *ac,* nempe vt 10,000. ad 5000.

Praxis.

Sume ipſos ſinus rectos eleuationũ, & habe bis numeros Tabulę, exhibentes durationes, & impetus parabolarum ad horizontem comparatos.

Tabu-

[107] *8*] 8.

Tabula continens Gradus eleuatinis aquæ debet adhiberi, vt Amplitudo proiectionis fiat datæ mensuræ.

Spatia.	GRAD. Eleuat.	Complemen.
10	17	89. 43
20	34	89. 26
30	52	89. 8
40	1. 9	88. 51
50	1. 26	88. 34
60	1. 43	88. 17
70	2. 0	88. 0
80	2. 18	87. 42
90	2. 35	87. 25
100	2. 52	87. 8
110	3. 9	86. 51
120	3. 27	86. 33
130	3. 44	86. 16
140	4. 1	85. 59
150	4. 19	85. 41
160	4. 36	85. 24
170	4. 54	85. 6
180	5. 11	84. 49
190	5. 29	84. 31
200	5. 46	84. 14
210	6. 4	83. 56
220	6. 21	83. 39
230	6. 39	83. 21
240	6. 57	83. 3
250	7. 14	82. 46
260	7. 32	82. 28
270	7. 50	82. 10
280	8. 8	81. 52
290	8. 26	81. 34
300	8. 44	81. 16
310	9 2	80. 58
320	9. 20	80. 40
330	9. 38	80. 22
340	9. 56	80. 4
350	10. 14	79 46
360	10. 33	79. 27
370	10. 51	79. 9
380	11. 10	78. 50
390	11. 29	78. 31
400	11. 47	78. 13
410	12. 6	77. 54
420	12. 25	77. 35
430	12. 44	77. 16
440	13. 3	76. 57
450	13. 22	76. 38
460	13. 42	76. 18
470	14. 1	75. 59
480	14. 21	75. 39
490	14 40	75. 20
500	15. 0	75. 00

Spatia siue incrementa proiectionum æqualia.

[214] [*De motu Proiectorum*]

Tabula continens Gradus eleuationis aquæ debet adhiberi, vt Amplitudo proiectionis fiat datæ menſuræ.

Spatia ſiue incrementa proiectionum æqualia.

Spatia.	GRAD. Eleuat.		Comple-men.
10		17	89. 43
20		34	89. 26
30		52	89. 8
40	1.	9	88. 51
50	1.	26	88. 34
60	1.	43	88. 17
70	2.	0	88. 0
80	2.	18	87. 42
90	2.	35	87. 25
100	2.	52	87. 8
110	3.	9	86. 51
120	3.	27	86. 33
130	3.	44	86. 16
140	4.	1	85. 59
150	4.	19	85. 41
160	4.	36	85. 24
170	4.	54	85. 6
180	5.	11	84. 49
190	5.	29	84. 31
200	5.	46	84. 14
210	6.	4	83. 56
220	6.	21	83. 39
230	6.	39	83. 21
240	6.	57	83. 3
250	7.	14	82. 46
260	7.	32	82. 28
270	7.	50	82. 10
280	8.	8	81. 52
290	8.	26	81. 34
300	8.	44	81. 16
310	9.	2	80. 58
320	9.	20	80. 40
330	9.	38	80. 22
340	9.	56	80. 4
350	10.	14	79 46
360	10.	33	79. 27
370	10.	51	79. 9
380	11.	10	78. 50
390	11.	29	78. 31
400	11.	47	78. 13
410	12.	6	77. 54
420	12.	25	77. 35
430	12.	44	77. 16
440	13.	3	76. 57
450	13.	22	76. 38
460	13.	42	76. 18
470	14.	1	75. 59
480	14.	21	75. 39
490	14	40	75. 20
500	15.	0	75. 00

Supponimus proiectiones omnes eundem impetum habere, hoc est, esse eiusdem machinæ. & maximam proiectionem ponimus part. 4000.

Spatia siue incrementa proiectionum æqualium.

Spatia.	GRAD Eleuat.	Complemen.	Spatia.	GRAD. Eleuat.	Complemen.
510	15. 20	74. 40	760	24. 44	65. 16
520	15. 40	74. 20	770	25. 11	64. 49
530	16. 0	74. 0	780	25. 38	64. 22
540	16. 21	73. 39	790	26. 6	63. 54
550	16. 41	73. 19	800	26. 34	63. 26
560	17. 2	72. 58	810	27. 3	62. 57
570	17. 23	72. 37	820	27. 33	62. 27
580	17. 44	72. 16	830	28. 3	61. 57
590	18. 5	71. 55	840	28. 34	61. 26
600	18. 26	71. 34	850	29. 6	60 54
610	18. 48	71. 12	860	29. 39	60. 21
620	19. 10	70. 50	870	30. 14	59 46
630	19. 32	70. 26	880	30. 50	59. 10
640	19. 54	69. 6	890	31. 27	58. 33
650	20. 16	69. 44	900	32. 5	57. 55
660	20. 39	69. 21	910	32. 45	57. 15
670	21. 2	68. 58	920	33. 28	56 32
680	21. 25	68. 35	930	34. 13	55 47
690	21. 49	68. 21	940	35. 2	54. 58
700	22. 13	67. 47	950	35. 54	54. 6
710	22. 37	67. 23	960	36. 52	53 8
720	23. 2.	66. 58	970	37. 58	52. 2
730	23. 27	66. 33	980	39. 16	50. 44
740	23. 52	66. 8	990	40. 57	49. 3
750	24. 18	65. 42	1000	45. 00	45. 00

[*Liber Secundus.*] [215]

Supponimus proiectiones omnes eundem impetum habere, hoc eſt, eſſe eiuſdem machinæ. & maximam proiectionem ponimus part. 4000.

Spatia ſiue incrementa proiectionum æqualium.

Spatia.	GRAD Eleuat.		Comple-men.	
510	15.	20	74.	40
520	15.	40	74.	20
530	16.	0	74.	0
540	16.	21	73.	39
550	16.	41	73.	19
560	17.	2	72.	58
570	17.	23	72.	37
580	17.	44	72.	16
590	18.	5	71.	55
600	18.	26	71.	34
610	18.	48	71.	12
620	19.	10	70.	50
630	19.	32	70.	26
640	19.	54	69.	6
650	20.	16	69.	44
660	20.	39	69.	21
670	21.	2	68.	58
680	21.	25	68.	35
690	21.	49	68.	21
700	22.	13	67.	47
710	22.	37	67.	23
720	23.	2.	66.	58
730	23.	27	66.	33
740	23.	52	66.	8
750	24.	18	65.	42

Spatia.	GRAD. Eleuat.		Comple-men.	
760	24.	44	65.	16
770	25.	11	64.	49
780	25.	38	64.	22
790	26.	6	63.	54
800	26.	34	63.	26
810	27.	3	62.	57
820	27.	33	62.	27
830	28.	3	61.	57
840	28.	34	61.	26
850	29.	6	60[.]	54
860	29.	39	60.	21
870	30.	14	59[.]	46
880	30.	50	59.	10
890	31.	27	58.	33
900	32.	5	57.	55
910	32.	45	57.	15
920	33.	28	56[.]	32
930	34.	13	55[.]	47
940	35.	2	54.	58
950	35.	54	54.	6
960	36.	52	53[.]	8
970	37.	58	52.	2
980	39.	16	50.	44
990	40.	57	49.	3
1000	45.	00	45.	00

216 *De motu Proiectorum*

Explicatio praecedentis Tabulæ.

DVM supponitur maxima proiectio esse partium 4000. tunc supponitur quarta ipsius pars, hoc est semidiameter circuli Propositionis 9. proiectorum esse 1000. qui numerus supponitur etiam pro sinu toto in Tabula sinuum.

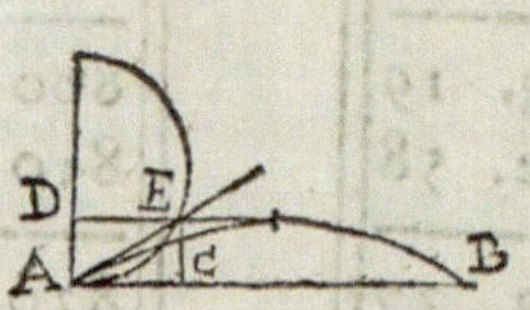

Quando ergo data erit futura amplitudo *ab* partiū 2120.(quę maior non sit numero 4000.) dabitur etiam *ac* 530.ipsius *ab* quarta pars,hoc est *de* erit 530. Ergo ex Tabulis sinuum habebitur quantitas arcus *ae* gr. 32. cuius semissis gr. 16. erit mensura anguli *cae*. Nempe eleuatio quæsita, iuxtà quam fiet proposita amplitudo *ab* partium 2120. talium qualium maxima proiectio integra sit 4000.

Praxis pro construenda Tabula; & ad hoc vt quiuis per solas sinuum tabulas probl. soluere possit.

Datę futuræ Amplitudinis quartam partem sume. hanc in tabula sinuum quære, arcumq. ipsi respondentem bifariam seca. Sic habebis eleuationem, quæ amplitudinem quæsitam facit. Sed fusius hæc in sequentibus explicabimus.

Vso della

Explicatio prȩcedentis Tabulæ.

DVM ſupponitur maxima proiectio eſſe partium 4000. tunc ſupponitur quarta ipſius pars, hoc eſt ſemidiameter circuli Propoſitionis 9. proiectorum eſſe 1000. qui numerus ſupponitur etiam pro ſinu toto in Tabula ſinuum.

Quando ergo data erit futura amplitudo *ab* partiũ 2120. *(*quȩ maior non ſit numero 4000.*)* dabitur etiam *ac* 530. ipſius *ab* quarta pars, hoc eſt *de* erit 530. Ergo ex Tabulis ſinuum habebitur quantitas arcus *ae* gr. 32. cuius ſemiſſis gr. 16. erit menſura anguli *cae.* Nempe eleuatio quæſita, iuxtà quam fiet propoſita amplitudo *ab* partium 2*1*20.[108] talium qualium maxima proiectio integra ſit 4000.

Praxis pro conſtruenda Tabula;* & ad hoc *vt quiuis per ſolas ſinuum tabulas probl. ſoluere poſſit.

Datȩ futuræ Amplitudinis quartam partem ſume. hanc in tabula ſinuum quære, arcumq. ipſi reſpondentem bifariam ſeca. Sic habebis eleuationem, quæ amplitudinem quæſitam facit. Sed fuſius hæc in ſequentibus explicabimus.

Vſo della

108 2*1*20] 2120.

Vſo della præcedente Tauola.

SVpponiamo che il maſſimo tiro, cioè il tiro fatto all'eleuazione del ſeſto punto della ſquadra da vna colubrina ſia per eſempio 4000. paſſi geometrici. Voglio fare con la medeſima vn tiro di maniera tale, che rieſca per appunto longo paſſi 2360. Piglio la quarta parte di 2360. la quale è 590. e guardo ſu la tauola, e vedo dirimpetto ad eſſo numero la eleuazione da darſi à detto pezzo eſſer gradi 18. e minuti 5. ouero gradi 71. e minuti 55. ſuo complemento. E dico per le coſe dimoſtrate, che il ſudetto pezzo con vna di queſte due eleuazioni tirerà la palla lontano paſſi 2360. ſopra l'orizonte. Se bene quelle eleuazioni, le quali paſſano il ſeſto punto della ſquadra, non ſi pongono per l'artiglierie, ma ſolamente per l'vſo de' mortari, o trabocchi, o ſaltamartini. Deue però auuertirſi che con quella prima eleuazione la palla farà vna ſtrada baſſa, ma veloce, come la linea ſegnata a, *e con impeto grande orizontale opportuno per sfondare muraglie, o dare altro impulſo laterale. Ma con l'altra eleuazione farà la ſtrada* b, *la quale ſarà pigra di moto orizontale, ma con aſſai impeto perpendicolare nel fine, oportuno per sfondare volte, tetti, e far altre paſſate perpendicolari all'orizonte; ouero per gettar robbe in vn certo determinato ſegno, come ſarebbero ſacchetti imballati con corde, pieni di zolfo, o ſalnitro, o farina; ouero palle con lettere, & altro dentro. In ſomma l'vna, e l'altra eleuazione, che egualmẽte ſia diſtante dal ſeſto punto porterà la palla nello ſteſſo luogo, però con la prima, e minor' eleuazione caderà in terra (come eſſi dicono)* Di ſtriſcio; *e con la ſeconda e maggior' eleuazione batterà quaſi perpendicolare.*

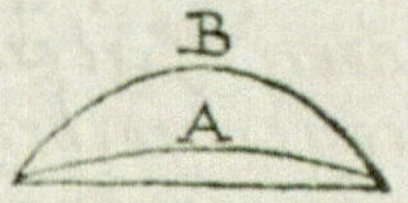

Sò che rariſſime volte, e forſi anco mai s'incontrerà che il maſſimo tiro d'vn'artiglieria ſia per appunto que' paſſi 4000. come par che ſi ſupponga nel calcolo della Tauola noſtra, & anco in quelle del Sig. Galileo. però la detta Tauola potrebbe parere inu-

E e *tile.*

Vſo della præcedente Tauola[109].

SVpponiamo che il maſſimo tiro, cioè il tiro fatto all'eleuazione del ſeſto punto della ſquadra da vna colubrina ſia per eſempio 4000. paſſi geometrici. Voglio fare con la medeſima vn tiro di maniera tale, che rieſca per appunto longo paſſi 2360. Piglio la quarta parte di 2360. la quale è 590. e guardo ſu la tauola, e vedo dirimpetto ad eſſo numero la eleuazione da darſi à detto pezzo eſſer gradi 18. e minuti 5. ouero gradi 71. e minuti 55. ſuo complemento. E dico per le coſe dimostrate, che il ſudetto pezzo con vna di queſte due eleuazioni tirerà la palla lontano paſſi 2360. ſopra l'orizonte. Se bene quelle eleuazioni, le quali paſſano il ſesto punto della ſquadra, non ſi pongono per l'artiglierie, ma ſolamente per l'vſo de' mortari, o trabocchi, o ſaltamartini. Deue però auuertirſi che con quella prima eleuazione la palla farà vna ſtrada baſſa, ma veloce, come la linea ſegnata a, *e con impeto grande orizontale opportuno per sfondare muraglie, o dare altro impulſo laterale. Ma con l'altra eleuazione farà la ſtrada* b, *la quale ſarà pigra di moto orizontale, ma con aſſai impeto perpendicolare nel fine, oportuno per sfondare volte, tetti, e far altre paſſate perpendicolari all'orizonte; ouero per gettar robbe in vn certo determinato ſegno, come ſarebbero ſacchetti imballati con corde, pieni di zolfo, o ſalnitro, o farina; ouero palle con lettere, & altro dentro. In ſomma l'vna, e l' altra eleuazione, che egualmẽte ſia distante dal ſeſto punto porterà la palla nello steſſo luogo, però con la prima, e minor' eleuazione caderà in terra (come eſſi dicono)* Di ſtriſcio; *e con la ſeconda e maggior' eleuazione batterà quaſi perpendicolare.*

Sò che rariſſime volte, e forſi anco mai s'incontrerà che il maſſimo tiro a'vn'artiglieria ſia per appunto que'paſſi 4000. come par che ſi ſupponga nel calcolo della Tauola noſtra, & anco in quelle del Sig. Galileo. però la detta Tauola potrebbe parere inu-

E c *tile*

[109] The following part is written in Italian vernacular language.

tile. Ma noi mostreremo che il numero supposto di 4000. per ciò non serue ad alcuna machina particolare, acciò possa seruir' a tutte vniuersalmente. Bisogna dunque auuertire che quel numero supposto di 4000. non è di passi, ne di canne, o braccia, ne di altre determinate misure, ma si bene di parti astratte, tali quali esse si siano, che però potendo conuertirsi in tutte le sorti di misure possibili, fanno la Tauola generale tanto per le colubrine, quanto per i mortari, o balleStre. E per dare vno esempio come ella si possa addattare a tutte le spezie dell'artiglierie, e ridurre le parti aStratte in passi geometrici, faremo così.

Il massimo tiro di vn Cannone per esperienza fatta trouo che è per esempio passi 2300. e voglio con lo stesso fare vn tiro ilquale sia di passi 860. faccio così. Se il massimo tiro 2300. mi dà 860. tiro quesito, il numero 1000. massimo della Tauola che mi darà, faccio l'operazione, e trouo 374. il qual numero cercato nella Tauola si ritroua fra 370. e 380. Però adoprando a giudizio la parte proporzionale trouerassi l'arco della sua eleuazione douer' essere gradi 11. in circa, ouero 79. suo complemento. E così è certo che quella tale artiglieria, la quale eleuata a' 6. punti tiraua passi 2300. eleuata gradi 11. ouero 79. del quadrante tirerà passi 860. come desiderauamo.

Perche poi ad alcuno potrebbe parer difficile il trouar per esperienza il tiro massimo dell'artiglieria, mostreremo.

Come da qualunque tiro fatto anco casualmente, si possa trouar' il tiro massimo di vn pezzo d'artiglieria.

Sia vn pezzo dirizzato conforme la linea *ac*, della quale sia l'eleuazione l'angolo *bac*, qualunque si sia.

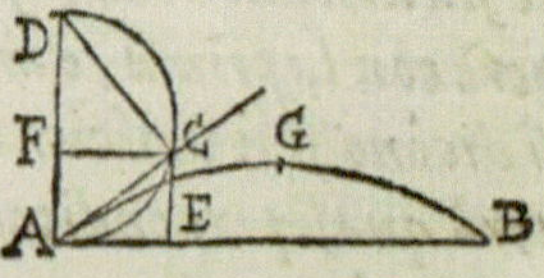

Si misuri detto angolo con la squadra, e trouisi per esempio gr. 30. poi si spari l'artiglieria, e vada la palla fino al punto *b*, e si misuri diligentemente la linea *ab*, che sia per esempio 2400. passi Geometrici. Dico che date queste due cose, cioè la eleuazione, e la lun-

tile. Ma noi moftreremo che il numero fupposto di 4000. per ciò non ferue ad alcuna machina particolare, acciò poffa feruir' a tutte vniuerfalmente. Bifogna dunque auuertire che quel numero fuppofto di 4000. non è di paffi, ne di canne, o braccia, ne di altre determinate mifure, ma fi bene di parti aftratte, tali quali effe fi fiano, che però potendo conuertirfi in tutte le forti di mifure poffibili, fanno la Tauola generale tanto per le colubrine, quanto per i mortari, o ballestre. E per dare vno efempio come ella fi poffa addattare a tutte le fpezie dell'artiglierie, e ridurre le parti astratte in paffi geometrici, faremo così.

Il maffimo tiro di vn Cannone per efperienza fatta trouo che è per efempio paffi 2300. e voglio con lo fteßo fare vn tiro il quale fia di paffi 860. faccio così. Se il maffimo tiro 2300. mi dà 860. tiro quefito, il numero 1000. maffimo della Tauola che mi darà, faccio l'operazione, e trouo 374. il qual numero cercato nella Tauola fi ritroua fra 370. e 380. Però adoprando a giudizio la parte proporzionale trouerafi l'arco della fua eleuazione douer' eßere gradi 11. in circa, ouero 79. fuo complemento. E cosi è certo che quella tale artiglieria; la quale eleuata a' 6. punti tiraua paffi 2300. eleuata gradi 1 1.[110] *ouero 79. del quadrante tirerà paffi 860. come defiderauamo.*

Perche poi ad alcuno potrebbe parer difficile il trouar per efperienza il tiro maffimo dell'artiglieria, moftreremo.

Come da qualunque tiro fatto anco cafualmente, fi poffa trouar' il tiro maffimo di vn pezzo d'artiglieria.

SIa vn pezzo dirizzato conforme la linea *ac,* della quale fia l'eleuazione l'angolo *bac,* qualunque fi fia.

Si mifuri detto angolo con la fquadra, e trouifi per efempio gr. 30. poi fi fpari l'artiglieria, e vada la palla fino al punto *b,* e fi mifuri diligentemente la linea *ab,* che fia per efempio 2400. paffi Geometrici. Dico che date quefte due cofe, cioè la eleuazione, e

la lun-

[110] *1* 1.] 11.

219

la lunghezza del tiro casuale *agb*. viene ad esser data anco la linea *ad*, la quale è la metà del massimo tiro, conforme si è dimostrato nell'vltimo Corollario della Proposizione 9. de proietti.

Essendo dato l'angolo della eleuazione *cab*, gr. 30. sarà il triangolo rettangolo *eac* dato in spezie: e perche è data la *ab* in passi, sarà data la *ae*. quarta parte di essa, cioè 600. passi. Opereremo dunque così per trouar la quantità di *ad*. per via di calcolo, e de i seni.

Facciasi. Come il seno retto 86602. dell'angolo *ace* gr. 60. cioè del supplemento della eleuazione, al lato *ae*, che è 600. Così il seno totale 100000. ad vn quarto numero 693. E così la hipotenusa *ac* sarà passi 693. Ma per che anco il triãgolo rettangolo *acd* è dato in spezie, facciasi di nuouo. Come il seno retto 50000. dell'angolo *adc*, il quale è eguale all'angolo dato della eleuazione *eac*, alla retta *ac* che si trouò 693. passi, così il seno totale ad vn quarto numero 1386. & così la retta cercata *ad* sarà passi 1386. Ma perche *ad* essendo la linea dell'impeto, o sublimità è eguale alla metà del massimo tiro, se noi raddoppieremo 1386. verrà a farsi il numero di 2772. passi, che tanta sarà la lunghezza del massimo tiro, che si cercaua di quella machina, la quale eleuata gr. 30. si trouò tirar passi 2400.

Ma con molto maggior breuità, e con vn calcolo solo potremo operare così. Pongasi che il seno totale sia *cf*, saranno *fa*, & *fd* le tangenti vna dell'angolo della eleuazione, l'altra del suo supplemento. Facciasi dunque. Come il seno totale, alla *cf* che è 600. così 230940. (che è la somma d'ambidue le sudette Tangenti) ad vn quarto numero 1386. e così la retta *ad* si trouerà come prima 1386. pasfi. la quale raddoppiata darà come sopra la misura del tiro semiretto, o massimo, come vogliam chiamarlo.

Per Corollario si può auuertire che questo è il modo di argomẽtare, da qualsiuoglia tiro d'vn'artiglieria, quanto la medesima sia per tirare all'insu per linea perpendiculare; che sarà quanto

Ee 2 *la li-*

la lunghezza del tiro caſuale *agb.* viene ad eſſer data anco la linea *ad,* la quale è la metà del maſſimo tiro, conforme ſi è dimoſtrato nell'vltimo Corollario della Propoſizione 9. de proietti.

Eſſendo dato l'angolo della eleuazione *cab,* gr. 30. ſarà il triangolo rettangolo *eac* dato in ſpezie: e perche è data la *ab* in paſſi, ſarà data la *ae.* quarta parte di eſſa, cioè 600. paſſi. Opereremo dunque così per trouar la quantità di *ad.* per via di calcolo, e de i ſeni.

Facciaſi. Come il ſeno retto 86,602. dell'angolo *ace* gr.60. cioè del ſupplemento della eleuazione, al lato *ae,* che è 600. Così il ſeno totale 100,000. ad vn quarto numero 693. E così la hipotenuſa *ac* ſarà paſſi 693. Ma per che anco il triãgolo rettangolo *acd* è dato in ſpezie, facciaſi di nuouo. Come il ſeno retto *50,000.* dell'angolo *adc,* il quale è eguale all'angolo dato della eleuazione *eac,* alla retta *ac* che ſi trouò 693. paſſi, così il ſeno totale ad vn quarto numero 1386. & così la retta cercata *ad* ſarà paſſi 1386. Ma perche *ad* eſſendo la linea dell'impeto, o ſublimità è eguale alla metà del maſſimo tiro, ſe noi raddoppieremo 1386. verrà a farſi il numero di 2772. paſſi, che tanta ſarà la lunghezza del maſſimo tiro, che ſi cercaua di quella machina, la quale eleuata gr. 30. ſi trouò tirar paſſi 2400.

Ma con molto maggior breuità, e con vn calcolo ſolo potremo operare così. Pongaſi che il ſeno totale ſia *cf,* ſaranno *fa,* & *fd* le tangenti vna dell'angolo della eleuazione, l'altra del ſuo ſupplemento. Facciaſi dunque. Come il ſeno totale, alla *cf* che è 600. così 230,940. (che è la ſomma d'ambidue le ſudette Tangenti) ad vn quarto numero 1386. e così la retta *ad* ſi trouerà come prima 1386.[111] pasſi. la quale raddoppiata darà come ſopra la miſura del tiro ſemiretto, o maſſimo, come vogliam chiamarlo.

Per Corollario ſi può auuertire che questo è il modo di argomẽtare, da qualſiuoglia tiro d'vn'artiglieria, quanto la medeſima ſia per tirare all'insu per linea perpendiculare; che ſarà quanto

E c 2 *la li-*

[111] 1386.] 1386.

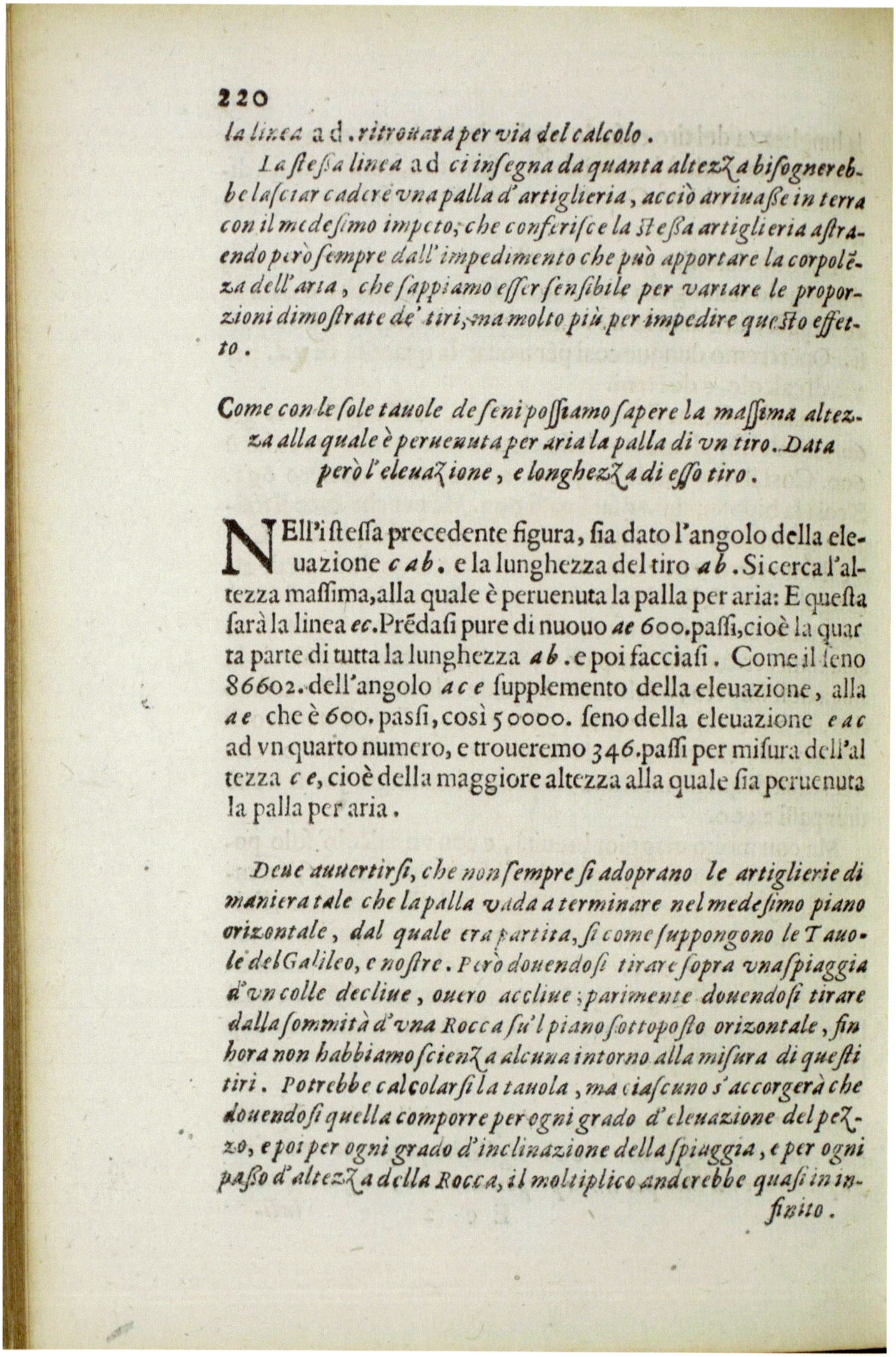

220

la linea a d. *ritrouata per via del calcolo.*

La stessa linea a d *ci insegna da quanta altezza bisognerebbe lasciar cadere vna palla d'artiglieria, acciò arriuasse in terra con il medesimo impeto; che conferisce la stessa artiglieria astraendo però sempre dall'impedimento che può apportare la corpolenza dell'aria, che sappiamo esser sensibile per variare le proporzioni dimostrate de' tiri, ma molto più per impedire questo effetto.*

Come con le sole tauole de seni possiamo sapere la massima altezza alla quale è peruenuta per aria la palla di vn tiro. Data però l'eleuazione, e longhezza di esso tiro.

NEll'istessa precedente figura, sia dato l'angolo della eleuazione *c a b.* e la lunghezza del tiro *a b*. Si cerca l'altezza massima, alla quale è peruenuta la palla per aria: E questa sarà la linea *ec*. Prēdasi pure di nuouo *ae* 600. passi, cioè la quarta parte di tutta la lunghezza *a b*. e poi facciasi. Come il seno 86602. dell'angolo *a c e* supplemento della eleuazione, alla *a e* che è 600. pasi, così 50000. seno della eleuazione *e a c* ad vn quarto numero, e troueremo 346. passi per misura dell'altezza *c e*, cioè della maggiore altezza alla quale sia peruenuta la palla per aria.

Deue auuertirsi, che non sempre si adoprano le artiglierie di maniera tale che la palla vada a terminare nel medesimo piano orizontale, dal quale era partita, si come suppongono le Tauole del Galileo, e nostre. Però douendosi tirare sopra vna spiaggia d'vn colle decliue, ouero accliue; parimente douendosi tirare dalla sommità d'vna Rocca su'l piano sottoposto orizontale, fin hora non habbiamo scienza alcuna intorno alla misura di questi tiri. Potrebbe calcolarsi la tauola, ma ciascuno s'accorgerà che douendosi quella comporre per ogni grado d'eleuazione del pezzo, e poi per ogni grado d'inclinazione della spiaggia, e per ogni passo d'altezza della Rocca, il moltiplico anderebbe quasi in infinito.

la linea ad. *ritrouata per* via *del calcolo.*

La fteßa linea ad *ci infegna da quanta altezza bifognerebbe lafciar cadere vna palla d'artiglieria, acciò arriuaße in terra con il medefimo impeto, che conferifce la steßa artiglieria aftraendo però fempre dall'impedimento che può apportare la corpolēza dell'aria, che fappiamo effer fenfibile per variare le proporzioni dimoftrate de' tiri, ma molto più per impedire questo effetto.*

Come con le fole tauole de feni poffiamo fapere la maffima altezza alla quale è peruenuta per aria la palla di vn tiro. Data però l'eleuazione, e longhezza di effo tiro.

NEll'ifteffa precedente figura, fia dato l'angolo della eleuazione *cab.* e la lunghezza del tiro *ab.* Si cerca l'altezza maffima, alla quale è peruenuta la palla per aria: E quefta farà la linea *ec.* Prēdafi pure di nuouo *ae* 600. paffi, cioè la quarta parte di tutta la lunghezza *ab.* e poi facciafi. Come il feno 86602. dell'angolo *ace* fupplemento della eleuazione, alla *ae* che è 600. pasfi, così 50,000. feno della eleuazione *eac* ad vn quarto numero, e troueremo 346. paffi per mifura dell'altezza *ce,* cioè della maggiore altezza alla quale fia peruenuta la palla per aria.

Deue auuertirfi, che non fempre fi adoprano le artiglierie di maniera tale che la palla vada a terminare nel medefimo piano orizontale, dal quale era partita, fi come fuppongono le Tauole del Galileo, e noftre. Però douendofi tirare fopra vna fpiaggia d'vn colle decliue, ouero accliue; parimente douendofi tirare dalla fommità d'vna Rocca fu'l piano fottopofto orizontale, fin hora non habbiamo fcienza alcuna intorno alla mifura di quefti tiri. Potrebbe calcolarfi la tauola, ma ciafcuno s'accorgerà che douendofi quella comporre per ogni grado d'eleuazione del pezzo, e poi per ogni grado d'inclinazione della fpiaggia, e per ogni paßo d'altezza della Rocca, il moltiplico anderebbe quafi in in-

finito

finito. Mostreremo però per regola generale il modo di calcolare i sudetti tiri quando occorrerà.

Vna tale artiglieria con la direzzione *ab*. fà il tiro *acd*. sopra il piano horizontale *ad*. Ma io voglio tirare sopra il piano *ac* inclinato, e cerco quãto sarà la lunghezza *ac* del detto tiro sopra questo piano. Si tirino *bd* per il punto *d*, & *hf* per il punto *c*. perpendicolari all'orizonte; e si congiunga *fe*. quale per le cose mostrate è parallela alla *ab*.

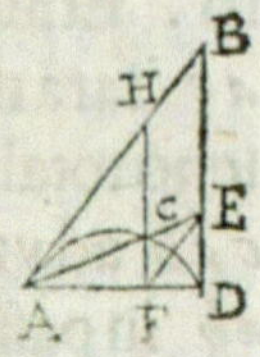

Misurisi con qualche strumento l'angolo *dae*, cioè la eleuazione della spiaggia, e dalla Tauola delle amplitudini si troui la lunghezza del tiro orizontale *ad*. Dopo questo facciasi. Come *db* tangente dell'angolo dell'eleuazione dell'Artiglieria, alla *be*, che è la differenza delle tangenti de duoi angoli *dab*, *dae* noti, essendo vno la eleuazione dell'artiglieria; l'altro la eleuazione della spiaggia sopra l'orizonte; così *da* nota in passi, ad vn quarto numero; e si trouerà la retta *fa* in passi. Facciasi poi di nuouo. Come il seno totale, a quel trouato quarto numero che è la misura di *af* in passi, così *ac* secante dell'angolo *fac*, ad vn quarto, e così aueremo noto il numero de i passi, i quali misureranno la linea *ac*, cioè la lunghezza del tiro che farà quella tale artiglieria sopra il piano *ae* quando si tira all'insù.

Ma quando dal punto *a* bisognasse tirare giù per vna spiaggia descendente come *ab*. Così troueremo la quantità di *ab*, cioè doue vada a ferire la palla. Sia data la direzzione *ae*, cioè, sia dato l'angolo della eleuazione dei pezzo, *cae*, sia dato ancora l'angolo della inclinazione della spiaggia *cab*. Immaginiamoci l'orizonte *ac*. e tiriamo perpendicolari ad esso le *be*, & *bdf*, e congiunghiamo *ch*. la quale sarà parallela alla *ae*. Hora su la tauola delle amplitudini troueremo quanti passi sia *ad*, ma noi cerchiamo quanto sia *ab*. Però facciasi il calcolo così. Come *fd* tangente dell'angolo

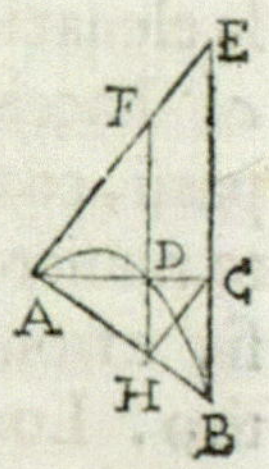

della

finito. Moſtreremo però per regola generale il modo di calcolare i ſudetti tiri quando occorrerà.

Vna tale artiglieria con la direzzione *ab.* fà il tiro *acd.* ſopra il piano horizontale *ad.* Ma io voglio tirare ſopra il piano *ac* inclinato, e cerco quãto ſarà la lunghezza *ac* del detto tiro ſopra queſto piano. Si tirino *bd* per il punto *d,* & *hf* per il punto *c.* perpendicolari all'orizonte; e ſi congiunga *fe.* quale per le coſe moſtrate è parallela alla *ab.*

Miſuriſi con qualche ſtrumento l'angolo *dae,* cioè la eleuazione della ſpiaggia, e dalla Tauola delle amplitudini ſi troui la lunghezza del tiro orizontale *ad.* Dopo queſto facciaſi. Come *db* tangente dell'angolo dell'eleuazione dell'Artiglieria, alla *be,* che è la differenza delle tangenti de duoi angoli *dab, dae* noti, eſſendo vno la eleuazione dell'artiglieria; l'altro la eleuazione della ſpiaggia ſopra l'orizonte; così *da* nota in paſſi, ad vn quarto numero; e ſi trouerà la retta *fa* in paſſi. Facciaſi poi di nuouo. Come il ſeno totale, a quel trouato quarto numero che è la miſura di *af* in paſſi, così *ac* ſecante dell'angolo *fac,* ad vn quarto, e così aueremo noto il numero de i paſſi, i quali miſureranno la linea *ac,* cioè la lunghezza del tiro che farà quella tale artiglieria ſopra il piano *ae* quando ſi tira all'insù.

Ma quando dal punto *a* biſognaſſe tirare giù per vna ſpiaggia deſcendente come *ab.* Così troueremo la quantità di *ab,* cioè doue vada a ferire la palla. Sia data la direzzione *ae,* cioè, ſia dato l'angolo della eleuazione del pezzo, *cae,* ſia dato ancora l'angolo della inclinazione della ſpiaggia *cab.* Immaginiamoci l'orizonte *ac.* e tiriamo perpendicolari ad eſſo le *be,* & *hdf,* e congiunghiamo *ch.* la quale ſarà parallela alla *ae.* Hora ſu la tauola delle amplitudini troueremo quanti paſſi ſia *ad,* ma noi cerchiamo quanto ſia *ab.* Però facciaſi il calcolo così. Come *fd* tangente dell'angolo

della

della eleuazione dell'artiglieria, alla *db* tangente della inclinazione del piano, così *ad* nota in passi, ad vn quarto numero. Et aueremo la misura di *dc* in passi; e però anco tutta la *ac* sarà nota in passi. Facciasi dunque di nuouo. Come *ac* seno totale, alla *ac* nota in passi, così *ab* secante dell'angolo *cab* ad vn quarto numero; e sarà la misura cercata della retta *ab* in passi; cioè la lunghezza del tiro su la spiaggia descendente *ab*.

Occorre ancora spesse volte di tirare in piani perpendicolari all'orizonte, come in muraglie di Città, o di Torri, o d'altro. Però anco in questo caso soggiungeremo il calcolo per trouar l'altezza di quel punto doue nel sudetto muro ferirà la palla.

Sia la direzzione del pezzo la linea *ab*, e l'orizonte *ac*, & il muro della torre *de* perpendicolare all'orizonte; e sia la distanza *ad* nota in passi. Immaginiamoci che la palla passi libera senza battere nel muro, e vada a colpire nell'orizonte in *c*. la Tauola delle amplitudini dà la quantità della *ac*. Ma noi cerchiamo l'altezza *de*. Tirisi *cb* perpendicolare all'orizonte, & *df* parallela ad *ab*, e poi congiungasi *fa*, la quale passerà per la comun sezzione della parabola, e del muro, come si può raccorre dalle cose già mostrate. Facciasi hora, come *ca* lunghezza del tiro orizontale, alla *cd* differenza tra le linee *ac*, *ad*, già note, Così la *bc* tangente dell'angolo della eleuazione dell'artiglieria; alla *cf* tangente dell'angolo *fac*. Facciasi di nuouo. Come il seno totale, alla *ad* nota in pasi, così la già trouata tãgente dell'angolo *fac*, ad vn quarto numero; il quale sarà la cercata misura della retta *de* in passi, e troueremo il punto *e*, nel quale anderebbe à ferire quel tiro. Lo stesso calcolo si può anco ridurre quando il muro *de* non sia perpendicolare, ma a scarpa, come quelli delle moderne fortezze; ma dubitando di apportar più tedio, che vtile lascierò la cura di ciò a quel Geometra che se ne curerà.

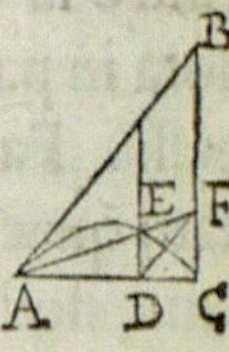

Le amplitudini delle parabole, delle quali tratta il Galileo, & anco noi, suppongono che il tiro non termini su'l piano della campagna,

della eleuazione dell'artiglieria, alla *dh* tangente della inclinazione del piano, così *ad* nota in paſſi, ad vn quarto numero. Et aueremo la miſura di *dc* in paſſi; e però anco tutta la *ac* ſarà nota in paſſi. Facciaſi dunque di nuouo. Come *ac* ſeno totale, alla *ac* nota in paſſi, così *ab* ſecante dell'angolo *cab* ad vn quarto numero; e ſarà la miſura cercata della retta *ab* in paſſi; cioè la lunghezza del tiro ſu la ſpiaggia deſcendente *ab*.

Occorre ancora ſpeſſe volte di tirare in piani perpendicolari all'orizonte, come in muraglie di Città, o di Torri, o d'altro. Però anco in queſto caſo ſoggiungeremo il calcolo per trouar l'altezza di quel punto doue nel ſudetto muro ferirà la palla.

Sia la direzzione del pezzo la linea *ab*, e l'orizonte *ac*, & il muro della torre *de* perpendicolare all'orizonte; e ſia la diſtanza *ad* nota in paſſi. Immaginiamoci che la palla paſſi libera ſenza battere nel muro, e vada a colpire nell'orizonte in *c*. la Tauola delle amplitudini dà la quantità della *ac*. Ma noi cerchiamo l'altezza *de*. Tiriſi *cb* perpendicolare all'orizonte, & *df* parallela ad *ab*, e poi congiungaſi *fa*, la quale paſſerà per la comun ſezzione della parabola, e del muro, come ſi può raccorre dalle coſe già moſtrate. Facciaſi hora, come *ca* lunghezza del tiro orizontale, alla *cd* differenza tra le linee *ac*, *ad*, già note, Così la *bc* tangente dell'angolo della eleuazione dell'artiglieria; alla *cf* tangente dell'angolo *fac*. Facciaſi di nuouo. Come il ſeno totale, alla *ad* nota in pasſi, così la già trouata tāgente dell'angolo *fac*, ad vn quarto numero; il quale ſarà la cercata miſura della retta *de* in paſſi, e troueremo il punto *e*, nel quale anderebbe à ferire quel tiro. Lo fteſso calcolo ſi può anco ridurre quando il muro *de* non ſia perpendicolare, ma a ſcarpa, come quelli delle moderne fortezze; ma dubitando di apportar più tedio, che vtile laſcierò la cura di ciò a quel Geometra che ſe ne curerà.

Le amplitudini delle parabole, delle quali tratta il Galileo, & anco noi, ſuppongono che il tiro non termini ſu'l piano della campagna,

223

pagna, ma in quel piano horizontale, che passa per la bocca del pezzo. Questo non sarà vero, se non quando l'artiglieria si mettesse con le ruote in vna fossa, si che la bocca venisse per appunto nel liuello della campagna: Ma per che ciò non si costuma, e per che i tiri vanno a terminare nell'orizonte, che tocca l'infimo punto delle ruote, cercheremo Geometricamente quanto possa prolungarsi vn tiro liuellato, o vogliam dire orizontale, per cagione dell'altezza della bocca del pezzo sopra il piano della campagna. Pare che il semidiametro delle ruote, e la grossezza del metallo cagionino che la bocca dell'artiglierie ordinarie venga alta sopra il sito orizontale intorno a due braccia. Suppongasi dunque in a la bocca d'vna colubrina, e sia l'horizonte bc. altezza della bocca sia la retta ab supposta 2. braccia, & la parabola acd sia il tiro liuellato; si cerca la retta bc. Sia il tiro semiretto, o massimo della medesima artiglieria, la parabola aef, e pongasi che af sia 5000 passi Geometrici, cioè 15000. braccia Fiorentine. Facciasi il semicircolo solito della Proposizione 9. ahi. & prodotta al eguale alla ai, si applichi ld. Per le cose mostrate sarà la retta ai impeto della parabola aef, ouero della acd (perche sono della medesima machina) però al sarà la quarta parte del lato retto della parabola acd. adunque ld sarà doppia di la; ma anche af si mostrò già doppia di ai, però sono eguali ld, & af; & vengono ad esser date tre linee, cioè la 7500 ld 15000, & ab 2. Dunque se noi faremo per la regola del tre. Come la retta la ad ab note, così il quadrato della retta ld ad vn'altro nu. troueremo 60000 il quale sarà il quad. della retta bc; e cauatane la radice quadra, troueremo che la retta bc sarà 245. braccia. Concludasi dunque che quella machina, la quale fa il massimo tiro di 15000. braccia, se hauerà la bocca solleuata due braccia sopra l'orizonte, farà il tiro liuellato, cioè con niente d'eleuazione, lungo in ogni modo braccia 245.

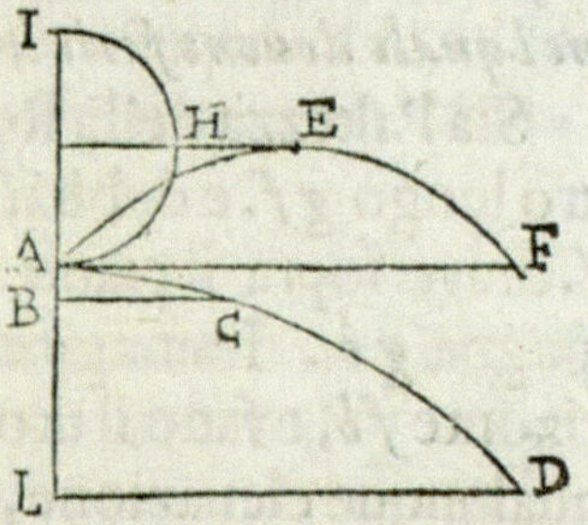

Quan-

pagna, ma in quel piano horizontale, che paßa per la bocca del pezzo. Questo non ſarà vero, ſe non quando l'artiglieria ſi metteſſe con le ruote in vna foſsa, ſi che la bocca veniſse per appunto nel liuello della campagna: Ma per che ciò non ſi coſtuma, e per che i tiri vanno a terminare nell'orizonte, che tocca l'infimo punto delle ruote, cercheremo Geometricamente quanto poſſa prolungarſi vn tiro liuellato, o vogliam dire orizontale, per cagione dell'altezza della bocca del pezzo ſopra il piano della campagna. Pare che il ſemidiametro delle ruote, e la groſſezza del metallo cagionino che la bocca dell'artiglierie ordinarie venga alta ſopra il ſito orizontale intorno a due braccia. Suppongaſi dunque in a *la bocca d'vna colubrina, e ſia l'horizonte* bc. *altezza della bocca ſia la retta* ab *ſuppoſta 2. braccia, & la parabola* acd *ſia il tiro liuellato; ſi cerca la retta* bc. *Sia il tiro ſemiretto, o maſſimo della medeſima artiglieria, la parabola* aef, *e pongaſi che* af *ſia 5000 paſſi Geometrici, cioè 15,000. braccia Fiorentine. Facciaſi il ſemicircolo ſolito della Propoſizione 9.* ahi. *& prodotta* al *eguale alla* ai, *ſi applichi* ld. *Per le coſe moſtrate ſarà la retta* ai *impeto della parabola* aef, *ouero della* acd *(perche ſono della medesima machina) però* al *ſarà la quarta parte del lato retto della parabola* acd. *adunque* ld *ſarà doppia di* la; *ma anche* af *ſi moſtrò già doppia di* ai, *però ſono eguali* ld, & af; *& vengono ad eſſer date tre linee, cioè* la *7500* ld *15,000, &* ab 2. *Dunque ſe noi faremo per la regola del tre. Come la retta* la *ad* ab *note, così il quadrato della retta* ld *ad vn'altro nu. troueremo 60,000 il quale ſarà il quad della retta* bc; *e cauatane la radice quadra, troueremo che la retta* bc *ſarà 245. braccia. Concludaſi dunque che quella machina, la quale fa il maſſimo tiro di 15,000. braccia, ſe hauerà la bocca ſolleuata due braccia ſopra l'orizonte, farà il tiro liuellato, cioè con niente d'eleuazione, lungo in ogni modo braccia 245.*

Quan-

224

Quanto poi poſſa prolungarſi qualunque altro tiro, non orizontale, ma inclinato in ſu, ouero in giù, per cagione dell'altezza delle ruote, o d'vn baſtione, o d'vna Rocca, o di qualunque altro ſito, che la ſollieui ſopra il piano horizontale, ſi cercherà in queſto modo.

E certo che douendoſi tirare dalla cima d'vna Rocca, ouero di vn caſtello poſto in cima d'vn ſaſſo, o da qualunque luogo alto ſu'l piano orizontale della campagna ſottopoſta, i tiri riuſciranno aſſai più lunghi che i notati ſopra la tauola delle amplitudini; e queſta differenza ſarà maggiore tanto più quanto più alta ſarà la ſituazione dell'artiglieria ſopra quel piano orizontale nel quale deuono ferir le palle, e terminare i tiri.

Sia l'altezza della Rocca, o d'altro luogo *gf.* e debbaſi dal punto *f.* tirare ſopra il piano della campagna *ge.* Immaginiamoci l'orizonte *fb,* e fatto il tiro *fcbe* con qualunque eleuazione, ſi cerca la miſura di *ge.*

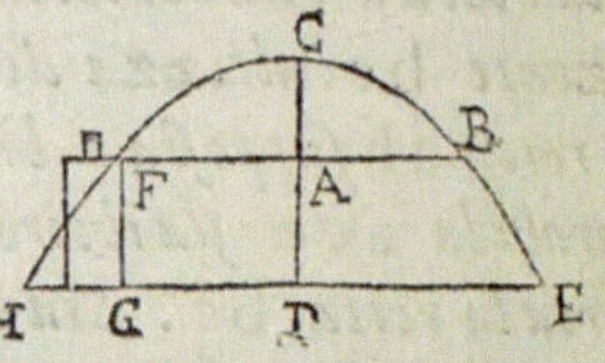

Dalla Tauola delle amplitudini ſi trouerà la quantità di *ab* e dalla Tauola delle altezze ſi trouerà la *ac.* altezza della parabola. La pratica poi del calcolo ſi potrà fare in più modi.

Quadretur numerus ab; *quadratuſq; diuidatur per* ac, *& quotus erit latus rectum parabolæ* fcb. *Ducatur deinde quotus iam dictus, in* cd, *& producti radix quadrata dabit* de.

Ouero potremo operare così.

Ducantur ſimul numeri dc, ca, *& producti radix quadrata erit medio loco proportionalis inter* dc, ca. *Fiat vt* ca. *ad prædictam radicem, ita* ab *ad alium numerum. & quartus numerus erit iterum* de.

Ouero finalmente a queſto modo,

Fiat vt numerus ca. *altitudo parabolę ex tabula. Ad numerum* cd *altitud. parabolæ & arcis ſimul, Ita* ab. *ſemiamplitudo parabolæ ex Tabula, Ad alium quartum numerum.*

Sumatur deinde numerus medio loco proportionalis inter dictum

Quanto poi poſſa prolungarſi qualunque altro tiro, non orizontale, ma inclinato in ſu, ouero in giù, per cagione dell'altezza delle ruote, o d'vn bastione, o d'vna Rocca, o di qualunque altro ſito, che la ſollieui ſopra il piano horizontale, ſi cercherà in queſto modo.

E certo che douendoſi tirare dalla cima d'vna Rocca, ouero di vn castello poſto in cima d'vn ſaßo, o da qualunque luogo alto ſu'l piano orizontale della campagna ſottopoſta, i tiri riuſciranno aſſai più lunghi che i notati ſopra la tauola delle amplitudini; e queſta differenza ſarà maggiore tanto più quanto più alta ſarà la ſituazione dell'artiglieria ſopra quel piano orizontale nel quale deuono ferir le palle, e terminare i tiri.

Sia l'altezza della Rocca, o d'altro luogo *gf.* e debbaſi dal punto *f.* tirare ſopra il piano della campagna *ge.* Immaginiamoci l'orizonte *fb,* e fatto il tiro *fcbe* con qualunque eleuazione, ſi cerca la miſura di *ge.*

Dalla Tauola delle amplitudini ſi trouerà la quantità di *abe* dalla Tauola delle altezze ſi trouerà la *ac.* altezza della parabola. La pratica poi del calcolo ſi potrà fare in più modi.

Quadretur numerus ab; *quadratuſq; diuidatur per* ac, *& quotus erit latus rectum parabolæ* fcb. *Ducatur deinde quotus iam dictus, in* cd, *& producti radix quadrata dabit* de.

Ouero potremo operare coſi.

Ducantur ſimul numeri dc, ca, *& producti radix quadrata erit medio loco proportionalis inter* dc, ca. *Fiat vt* ca. *ad prædictam radicem, ita* ab *ad alium numerum. & quartus numerus erit iterum* de.

Ouero finalmente a queſto modo,

Fiat vt numerus ca. *altitudo parabolę ex tabula. Ad numerum* cd *altitud. parabolæ & arcis ſimul, Ita* ab. *ſemiamplitudo parabolæ ex Tabula, Ad alium quartum numerum.*

Sumatur deinde numerus medio loco proportionalis inter di-

ctum

quartum numerum, *& inter* a b, *quia medius ille proportionalis exhibebit ipſum* d c. *Et cum* d g *equalis ipſi* a b *nota ſit, erit tota* g c *nota*.

Ma potrebbe ſoggiungere alcuno che dalla ſommità *gf* più ſpeſſo forſi occorrerà tirare con l'artiglieria inchinata all'ingiù che all'insù, però ſarebbe neceſſario ſapere per regola Geometrica la lunghezza de' tiri, il che ſi auerà in queſto modo, Sia il tiro da farſi all'ingiù il ſegnato *fh* con qualunque angolo d'inclinazione ſotto l'orizonte, ſi cerca *gh*. fingaſi con l'immaginazione che il tiro abia da farſi ſopra l'orizonte con la medeſima inclinazione per l'appunto, & per le regole precedenti ſi troui la quantità di *de* come ſopra, ouero di *dh*, dalla quale ſe leueremo la già nota *fa*, ouero *dg* rimarrà nota la quantità cercata g *h*.

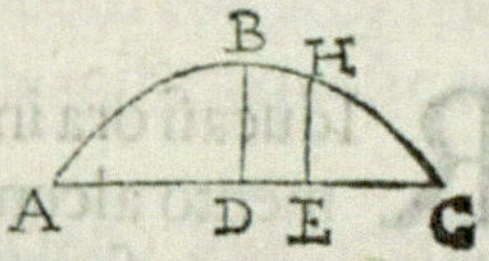

Ma ſe data la eleuazione gr. 40. del tiro *abc*, e la baſe *ac* 1600. paſſi noi voleſſimo ſapere tutte le diuerſe altezze del tranſito della palla ſopra qualunque punto della linea *ac*. Faremo così. Diuiſa per mezzo la *ac*, & alzata *db*. queſta ſarà l'altezza ſuprema, e ſi trouerà ſu la Tauola delle altezze, e delle amplitudini, operando in queſto modo. Nella tauola delle amplitudini dirimpetto alli gradi 40. di eleuazione trouo la linea *ad* eſſere parti 9848. ma nella Tauola delle altezze trouo la linea *bd* eſſere parti 4132. Poi per la regola del tre, dico. Se *ad* 9848. mi dà paſſi 800. conforme alla ſuppoſizione; *db* che è parti 4132. quanti paſſi darà? e ritrouo che la retta *bd* è paſſi 336. Sia ora propoſto qualunque punto *e* ſopra di cui ſi vuole ſaper l'altezza del tranſito della palla, cioè la linea *eh*. Suppongaſi che la retta *ae* ſia 1000, & la *ec* 600. e facciaſi di nuouo la regola del tre in queſto modo. Se il numero quadrato di *ad*, che è 640000. Mi dà il numero rettangolo delle rette *ae*, *ec*. che è 600000, Il numero *bd*, che fu trouato 336. che mi darà? E ritrouo 315 paſſi adunque l'altezza della parabola ſopra il punto *e* fu paſſi 315. Che è quello che ſi cercaua.

ctum quartum numerum, & inter ab, *quia medius ille proportionalis exhibebit ipſum* de. *Et cum* dg *ęqualis ipſi* ab *nota ſit, erit tota* ge *nota.*

Ma potrebbe ſoggiungere alcuno che dalla ſommità *gf* più ſpeſſo forſi occorrerà tirare con l'artiglieria inchinata all'ingiù che all'insù, però ſarebbe neceſſario ſapere per regola Geometrica la lunghezza de'tiri, il che ſi auerà in queſto modo,

Sia il tiro da farſi all'ingiù il ſegnato *fh* con qualunque angolo d'inclinazione ſotto l'orizonte, ſi cerca *gh.* fingaſi con l'immaginazione che il tiro abia da farſi ſopra l'orizonte con la medeſima inclinazione per l'appunto, & per le regole precedenti ſi troui la quantità di *de* come ſopra, ouero di *dh,* dalla quale ſe leueremo la già nota *fa,* ouero *dg* rimarrà nota la quantità cercata *gh.*

Ma ſe data la eleuazione gr. 40. del tiro *abc,* e la baſe *ac* 1600. paſſi noi voleſſimo ſapere tutte le diuerſe altezze del tranſito della palla ſopra qualunque punto della linea *ac.* Faremo coſi. Diuiſa per mezzo la *ac,* & alzata *db.* queſta ſarà l'altezza ſuprema, e ſi trouerà ſula Tauola delle altezze, e delle amplitudini, operando in queſto modo. Nella tauola delle amplitudini dirimpetto alli gradi 40. di eleuazione trouo la linea *ad* eſſere parti 9848. ma nella Tauola delle altezze trouo la linea *bd* eſſere parti 4132. Poi per la regola del tre, dico. Se *ad* 9848. mi dà paſſi 800. conforme alla ſuppoſizione; *db* che è parti 4132. quanti paſſi darà? e ritrouo che la retta *bd* è paſſi 336. Sia ora propoſto qualunque punto *e* ſopra di cui ſi vuole ſaper l'altezza del tranſito della palla, cioè la linea *eh.* Suppongaſi che la retta *ae* ſia 1000, & la *ec* 600. e facciaſi di nuouo la regola del tre in queſto modo. Se il numero quadrato di *a d,* che è 640,000. Mi dà il numero rettangolo delle rette *ae, ec.* che è 600,000, Il numero *bd,* che fu trouato 336. che mi darà? E ritrouo 315 paſſi adunque l'altezza della parabola ſopra il punto *e* fu paſſi 315. Che è quello che ſi cercaua.

Ff *Ba-*

226

Basterà l'auer' accennato questo poco per il calcolo di alcune varietà le quali possono occorrere intorno a questi tiri. Poteuano porsi altri casi simili a questi, e particolarmente i conuersi loro, ma dalla intelligenza di questi si possono facilmente dedurre quelli, e l'ingegno di qualunque Geometra applicandoui trouerà minor difficoltà nello sciorre molti di questi problemi da se medesimo, che nel passare le lunghezze, & le oscurità delle nostre esplicazioni. Però passeremo alla fabbrica della squadra, la quale pare veramente appropriata, anzi fatta dalla natura a posta per misurar scientificamente, & Geometricamente i tiri de proietti.

DELLA SQVADRA.

Riducasi ora in pratica, e sciolgasi per mezzo di vno strumento alcune delle già dimostrate proposizioni. Fabricheremo vna squadra militare, la quale con certezza inuariabile insegni (almeno alli Filosofi Geometri, se non a' Bombardieri pratici) quanta eleuazione debba darsi a qualsiuoglia machina, acciò la lunghezza del tiro riesca della proposta misura. Sciorremo anco per mezzo d'essa tutti i Problemi, che sopra il tirar delle artiglierie si possino formare; quali già furono promessi dal Tartaglia, e poi ridotti in Tauole dal Galileo, cõ alcun'altro di più. Si accorse l'industria militare, che l'vso di vna machina tanto nobile, e di tanta conseguenza, quanto è l'artiglieria, sarebbe stato troppo ristretto, & di poco benefi-cio, se quella nõ si fusse potuta adoperare se non dentro a quella poca distanza, ch'ella tira di punto in bianco, o vogliam dire di mira senza dargli con la squadra aiuto vantaggioso di alcuna eleuazione. Fù però pensato come si potesse fare, acciò con quel medesimo pezzo, il quale per se stesso non tiraua più che 200. ouero 250. passi Geometrici si potesse tirarne e 400. & anco 600. e più, e più, fino alla lunghezza del massimo tiro, che possa farsi da quel tale pezzo. L'inuenzione fu questa; Comin-

Bafterà l'auer' accennato quefto poco per il calcolo di alcune varietà le quali poſſono occorrere intorno a questi tiri. Poteuano porſi altri caſi ſimili a quefti, e particolarmente i conuerſi loro, ma dalla intelligenza di questi ſi poſſono facilmente dedurre quelli, e l'ingegno di qualunque Geometra applicandoui trouerà minor difficoltà nello ſciorre molti di quefti problemi da ſe medeſimo, che nel paſſare le lunghezze, & le oſcurità delle nostre eſplicazioni. Però paſſeremo alla fabbrica della ſquadra, la quale pare veramente appropriata, anzi fatta dalla natura a posta per miſurar ſcientificamente, & Geometricamente i tiri de proietti.

DELLA SQVADRA.

RIducaſi ora in pratica, e ſciolgaſi per mezzo di vno ftrumento alcune delle già dimoftrate propoſizioni. Fabricheremo vna ſquadra militare, la quale con certezza inuariabile inſegni (almeno alli Filoſofi Geometri, ſe non a' Bombardieri pratici) quanta eleuazione debba darſi a qualſiuoglia machina, acciò la lunghezza del tiro rieſca della propofta miſura. Sciorremo anco per mezzo d'eſſa tutti i Problemi, che ſopra il tirar delle artiglierie ſi poſſino formare; quali già furono promeſſi dal Tartaglia, e poi ridotti in Tauole dal Galileo, cõ alcun'altro di più. Si accorſe l'induftria militare, che l'vſo di vna machina tanto nobile, e di tanta conſeguenza, quanto è l'artiglieria, ſarebbe ftato troppo riftretto, & di poco beneficio, ſe quella nõ ſi fuſſe potuta adoperare ſe non dentro a quella poca diftanza, ch'ella tira di punto in bianco, o vogliam dire di mira ſenza dargli con la ſquadra aiuto vantaggioſo di alcuna eleuazione. Fù però penſato come ſi poteſſe fare, acciò.

con quel medeſimo pezzo, il quale per ſe fteſſo non tiraua più che 200. ouero 250. paſſi Geometrici ſi poteſſe tirarne e 400.& anco 600. e più, e più, ſino alla lunghezza del maſſimo tiro, che poſſa farſi da quel tale pezzo. L'inuenzione fu quefta; Co-

min-

minciarono ad aiutare il pezzo con l'eleuazione; cioè non lo dirizzauano a dirittura verſo l'oggetto in che doueua colpire, ma tenendolo nello ſteſſo verticale dell'oggetto, lo eleuauano ſopra quella linea retta, la quale và dal pezzo all'oggetto: e ciò faceuano ora più, & ora meno, conforme che la sforzatura del tiro doueua eſſere maggiore, o minore. Artifizio che fino dal principio del mondo è ſtato noto anco a i putti ineſperti. Vediamo che douendo eſſi con vna palla di neue, o d'altro, colpire in vn ſegno viciniſſimo, la ſcagliano a dirittura verſo eſſo ſegno; ma douendo poi giuocare a chi tira più lontano, ouero fare a ſaſſi tra di loro, non tirano già orizontalmente, ne a dirittura verſo i loro contrarij, ma volgendo i colpi a mezz'aria, ſenza auer fatto altra ſpeculazione, tirano tutti all'eleuazione del quinto, & anco del ſeſto punto della ſquadra militare a loro ignota. I Bombardieri poi ebbero col progreſſo del tempo vno ſtrumento, ilquale facilmente miſura queſte eleuazioni.

Fu inuentata da Niccolò Tartaglia Breſciano Matematico inſigne vna ſquadra con le gambe diſuguali congiunta con il quadrante, la quale già più di cento anni è ſempre ſtata in vſo, & è ancora l'vnica regolatrice de' Bombardieri, non ſolo per adoprar l'artiglieria, & alzarla in quei tiri, che eſſi chiamano di volata, ma anco per liuellarla negli orizontali. Diuiſe il Tartaglia quel quadrante in 12. parti eguali, cominciando la numerazione di eſſe dalla gamba minore; ſuddiuiſe anco ciaſcuna di eſſe in altre 12. parti eguali, nominando quelle prime Punti, e queſte ſeconde Minuti della ſquadra. Ponghiamo la figura della ſquadra, e moſtriamo come eſsa miſuri l'eleuazione del pezzo.

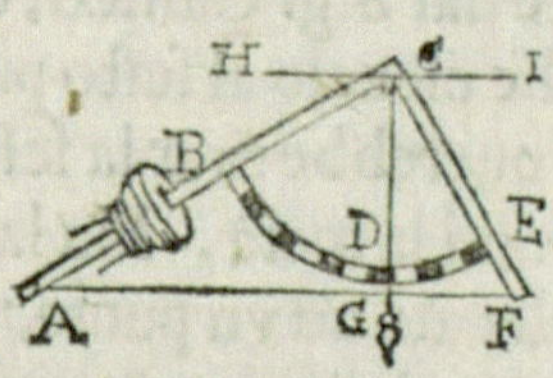

Sia l'anima del Cannone ab *fermo in qualche poſitura; Mettaſi in bocca d'eſſo la maggior gamba della ſquadra* ca *. ſi che ſi addatti ſu'l fondamento di detta anima, e caſchi il piombo in* d *. Io dico che l'angolo* ccd*, cioè l'arco* cd*, è la miſura della eleuazione*

minciarono ad aiutare il pezzo con l'eleuazione; cioè non lo dirizzauano a dirittura verſo l'oggetto in che doueua colpire, ma tenendolo nello fteſſo verticale dell'oggetto, lo eleuauano ſopra quella linea retta, la quale và dal pezzo all'oggetto: e cio faceuano ora più, & ora meno, conforme che la ſforzatura del tiro doueua eſſere maggiore, o minore. Artifizio che fino dal principio del mondo è ftato noto anco a i putti ineſperti. Vediamo che douendo eſſi con vna palla di neue, o d'altro, colpire in vn ſegno viciniſſimo, la ſcagliano a dirittura verſo eſſo ſegno; ma douendo poi giuocare a chi tira più lontano, ouero fare a ſaſſi tra di loro, non tirano già orizontalmente, ne a dirittura verſo i loro contrarij, ma volgendo i colpi a mezz'aria, ſenza auer fatto altra ſpeculazione. tirano tutti all'eleuazione del quinto, & anco del ſefto punto della ſquadra militare a loro ignota. I Bombardieri poi ebbero col progreſſo del tempo vno ftrumento, il quale facilmente miſura quefte eleuazioni.

Fu inuentata da Niccolò Tartaglia[112] Breſciano Matematico inſigne vna ſquadra con le gambe diſuguali congiunta con il quadrante, la quale già più di cento anni è ſempre ftata in vſo, & è ancora l'vnica regolatrice de' Bombardieri, non ſolo per adoprar l'artiglieria, & alzarla in quei tiri, che eſſi chiamano di volata, ma anco per liuellarla negli orizontali. Diuiſe il Tartaglia quel quadrante in 12. parti eguali, cominciando la numerazione di eſſe dalla gamba minore; ſuddiuiſe anco ciaſcuna di eſſe in altre 12. parti eguali, nominando quelle prime.

Punti, e quefte ſeconde Minuti della ſquadra. Ponghiamo la figura della ſquadra, e moftriamo come eſsa miſuri l'eleuazioe del pezzo.

Sia l'anima del Cannone ab *fermo in qualche poſitura; Mettaſi in bocca d'eſſo la maggior gamba della ſquadra* ca. *ſi che ſi addatti ſu'l fondamento di detta anima, e caſchi il piombo in* d. *Io dico che l'angolo* ecd, *cioè l'arco* ed, *è la miſura della eleua-*

Ff 2 *zione*

[112] Cfr. Pisano 2020; Pisano and Capecchi 2016 (historical analysis, transcription and English translation of *Quesiti*, 1546; 1544).

zione del pezzo. Tirisi vna orizontale af. *saranno gl'angoli intorno al punto* g, *retti, ma anco l'angolo* acf. *è retto, adunque gl'angoli* caf, *&* fcg *sono eguali per l'8. del sesto. Ouero così. Tirisi per* c *l'orizontale* hi. *Se da gli angoli retti* hcd, ace, *si leuerà il comune* acd, *resterà l'angolo* ecd *della squadra eguale all'angolo dell'artiglieria sotto l'orizonte* hi, *o sopra l'orizonte* af, *che è lo stesso per essere alterni.*

Col mezzo poi di questa squadra si è fatta dalli Bombardieri con lunghe osseruazioni vna pratica tale, ch'essi sanno quanti punti debba eleuarsi verbigrazia vna Colubrina da 40. per colpire in vn segno lontano per esempio passi 700. geometrici, o in qualunque altra distanza.

Ma vaglia il vero, le osseruazioni sono tanto fallaci; sono così pochi i Bombardieri che le abbiano fatte, e le abbiano fatte esquisitamente, che l'vso dell'artiglieria, leuatone il tiro di punto in bianco, non può auere se non pochissimo di certezza. Volendosi acquistare qualche scienza sicura intorno alla squadra ordinaria, sarebbe necessario di fare l'esperienze non solo con tutte le sorti di palle, e con tutte le differenze di poluere, ma in tutte le spezie de i pezzi, & anco in tutti quelli, che essendo della medesima spezie, sono differenti di grandezza, e poi a tutti i gradi delle eleuazioni possibili. Moltiplico, che quasi anderebbe in infinito, E notiamo, che conuerrebbe fare queste esperienze tutte ad vna ad vna; poiche nõ è vero che per via di proporzioni si possa da tre, o quattro tiri di vn Cannone, fatti a diuersa eleuazione, argomentare alcun'altro, ne pur dello stesso Cannone caricato con la stessa poluere, e palla. Che questo sia così, si dimostra per mezzo delle Tauole poste dal Sig. Galileo, e da noi. Per esempio. Quel Cannone che eleuato al sesto punto tira passi 4000. eleuato ad vn punto douerebbe tirar la sesta parte, & a due punti la terza, & a tre punti la metà. Ma la cosa passa molto diuersamente. Perche eleuato ad vn punto, tira 1032. in cambio di *666*. che è la sesta parte del sudetto massimo tiro 4000. Al secondo punto poi (& osseruisi che con questa eleuazione l'artiglierie tirano sem-

no sem-

zione del pezzo. Tirifi vna orizontale af. *faranno gl'angoli intorno al punto* g, *retti, ma anco l'angolo* acf. *è retto, adunque gl'angoli* caf, *&* fcg *fono eguali per l' 8. del fefto. Ouero cosi. Tirifi per* c *l'orizontale* hi. *Se da gli angoli retti* hcd, ace, *fi leuerà il comune* acd, *refterà l'angolo* ecd *della fquadra eguale all'angolo dell'artiglieria fotto l'orizonte* hi, *o fopra l'orizonte* af, *che è lo fteffo per effere alterni.*

Col mezzo poi di quefta fquadra fi è fatta dalli Bombardieri con lunghe offeruazioni vna pratica tale, ch'effi fanno quanti punti debba eleuarfi verbigrazia vna Colubrina da 40.per colpire in vn fegno lontano per efempio paffi 700. geometrici, o in qualunque altra diftanza.

Ma vaglia il vero, le offeruazioni fono tanto fallaci; fono cofi pochi i Bombardieri che le abbiano fatte, e le abbiano fatte efquifitamente, che l'vfo dell'artiglieria, leuatone il tiro di punto in bianco, non può auere fe non pochiffimo di certezza. Volendofi acquiftare qualche fcienza ficura intorno alla fquadra ordinaria, farebbe neceffario di fare l'efperienze non folo con tutte le forti di palle, e con tutte le differenze di poluere, ma in tutte le fpezie de i pezzi, & anco in tutti quelli, che effendo della medefima fpezie, fono differenti di grandezza, e poi a tutti i gradi delle eleuazioni poffibili. Moltiplico, che quafi anderebbe in infinito, E notiamo, che conuerrebbe fare quefte efperienze tutte ad vna ad vna; poiche nõ è vero che per via di proporzioni fi poffa da tre, o quattro tiri di vn Cannone, fatti a diuerfa eleuazione, argomentare alcun'altro, ne pur[113] dello fteffo Cannone caricato con la ftefsa poluere, e palla. Che quefto fia cofi, fi dimoftra per mezzo delle Tauole pofte dal Sig. Galileo, e da noi. Per efempio. Quel Cannone che eleuato al fefto punto tira paffi 4000. eleuato ad vn punto douerebbe tirar la fefta parte, & a due punti la terza, & a tre punti la metà. Ma la cofa paffa molto diuerfamente. Perche eleuato ad vn punto, tira 1032. in cambio di 666. che è la fefta parte del fudetto maffimo tiro 4000. Al fecondo punto poi (& offeruifi che con quefta eleuazione l'artiglierie tira-
no fem-

[113] ne pur] nepur] neppure.

no ſempre la metà del maſſimo tiro)nel caſo noſtro tirerà 2000 in cambio di 1333. che è la terza parte. Al terzo punto tirerà 2824. in cambio di 2000. che è la metà del maſſimo tiro Al quarto punto tirerà 3464. in cambio di 2666. Al quinto tirerà 3860. in cambio di 2333, che ſono cinque ſeſti di quel maſſimo. Vedeſi dunque come accreſcendo egualmente le eleuazioni del pezzo, cioè tirando prima ad vn punto ſolo, poi a due, & a tre, e quattro &c. fino al ſeſto, gl'accreſcimenti della lunghezza de i tiri non creſcono egualmente, cioè con la medeſima proporzione con la quale creſcono le eleuazioni: Ma mentre il primo punto tira 1032. il ſecondo accreſce ſopra eſſo, 968. Il terzo accreſce 824; il quarto 640. Il quinto 396. Il ſeſto 140. Per cauar dunque qualche regola dalle eſperienze, era neceſſario il farle eſattamente, a tutti i gradi della eleuazione, in tutte le ſorti de i pezzi, con tutte le varietà delle polueri, e le diuerſe materie delle palle, e forſi anco direi che era neceſſario che le faceſſe ogni Bombardiere da ſe ſteſſo. Coſe quaſi impoſſibili a ridurſi ſotto regole, e cauarne certezza alcuna, ſe la Teorica, e la Geometria non ce ne daua manifeſta ſcienza mediante quell'vnica propoſizione del Galileo, nella quale primo di tutti egli hà auuertito, & inſegnato a noi, che i proietti camminano tutti per vna linea parabolica. Su queſta ſuppoſizione fonderemo lo ſtrumento promeſſo: ſe poi per l'impedimento del mezzo le parabole venghino troppo deformate, o per molti altri accidenti i tiri rieſcono incoſtantisſimi, ci baſterà auer ſodisfatto indubitatamente alla ſcuola de Matematici, ſe non a quella de Bombardieri.

Noi auanti di porre la fabbrica della noſtra ſquadra, quale non cōſiſte in altro che nel deſcriuere vn ſolo ſemicircolo, diuideremo la ſquadra ordinaria in punti diſuguali, di maniera tale che miſurino non le eleuazioni del pezzo, ma le lunghezze de i tiri, che è quello di che l'vſo noſtro hà biſogno. Così aueremo certezza che l'artiglieria, ſe ſarà alzata ad vn punto di eſſa ſquadra, tirerà alla lunghezza d'vn' tale ſpazio, qualunque ſi ſia; alzata poi a due punti raddoppierà preciſamente

quel

no ſempre la metà del maſſimo tiro) nel caſo noſtro tirerà 2000 in cambio di 1333. che è la terza parte. Al terzo punto tirerà 2824. in cambio di 2000. che è la metà del maſſimo tiro. Al quarto punto tirerà 3464. in cambio di 2666. Al quinto tirerà 3860. in cambio di 2333, che ſono cinque ſeſti di quel maſſimo. Vedeſi dunque come accreſcendo egualmente le eleuazioni del pezzo, cioè tirando prima ad vn punto ſolo, poi a due, & a tre, e quattro &c. fino al ſeſto, gl'accreſcimenti della lunghezza de i tiri non creſcono egualmente, cioè con la medeſima proporzione con la quale creſcono le eleuazioni: Ma mentre il primo punto tira 1032. il ſecondo accreſce ſopra eſſo, 968. Il terzo accreſce 824; il quarto 640. Il quinto 396. Il ſeſto 140. Per cauar dunque qualche regola dalle eſperienze, era neceſſario il farle eſattamente, a tutti i gradi della eleuazione, in tutte le ſorti de i pezzi, con tutte le varietà delle polueri, e le diuerſe materie delle palle, e forſi anco direi che era neceſſario che le faceſſe ogni Bombardiere da ſe ſteſſo. Coſe quaſi impoſſibili a ridurſi ſotto regole, e cauarne certezza alcuna, ſe la Teorica, e la Geometria non ce ne daua manifeſta ſcienza mediante quell'vnica propoſizione del Galileo, nella quale primo di tutti egli hà auuertito, & inſegnato a noi, che i proietti camminano tutti per vna linea parabolica. Su queſta ſuppoſizione fonderemo lo ſtrumento promeſſo: ſe poi per l'impedimento del mezzo le parabole venghino troppo deformate, o per molti altri accidenti i tiri rieſcono incoſtantisſimi, ci baſterà auer ſodisfatto indubitatamente alla ſcuola de Matematici, ſe non a quella de Bombardieri.

Noi auanti di porre la fabbrica della noſtra ſquadra, quale non cõſiſte in altro che nel deſcriuere vn ſolo ſemicircolo, diuideremo la ſquadra ordinaria in punti diſuguali, di maniera tale che miſurino non le eleuazioni del pezzo, ma le lunghezze de i tiri, che è quello di che l'vſo noſtro hà biſogno. Coſi aueremo certezza che l'artiglieria, ſe ſarà alzata ad vn punto di eſſa ſquadra, tirerà alla lunghezza d'vn'tale ſpazio, qualunque ſi ſia; alzata poi a due punti raddoppierà preciſamente

quel

230

tiro; ſe a tre punti, tirerà tre di quei ſpazij, ſe a quattro e mezzo, tirerà quattro e mezzo: ſe a cinque & vn quarto, tirerà cinque & vn quarto; e così fino al ſesto punto creſceranno ſempre nello ſteſso modo, e con la ſteſsa proporzione i punti della ſquadra nello ſtrumento, e gli ſpazij de i tiri nella campagna, e dal ſeſto fino al duodecimo punto anderanno nella ſteſſa maniera decreſcendo. La coſtruzione, e dimoſtrazione e Geometrica, cauandoſi dalla propoſizione da noi poſta al numero XI. de proietti, la quale dalla data amplitudine inſegna trouar l'eleuazione. E ſerue comunemente per qualſiuoglia ſorte d'artiglieria, e di mortari, per qualunque ſpezie di palla, o di poluere.

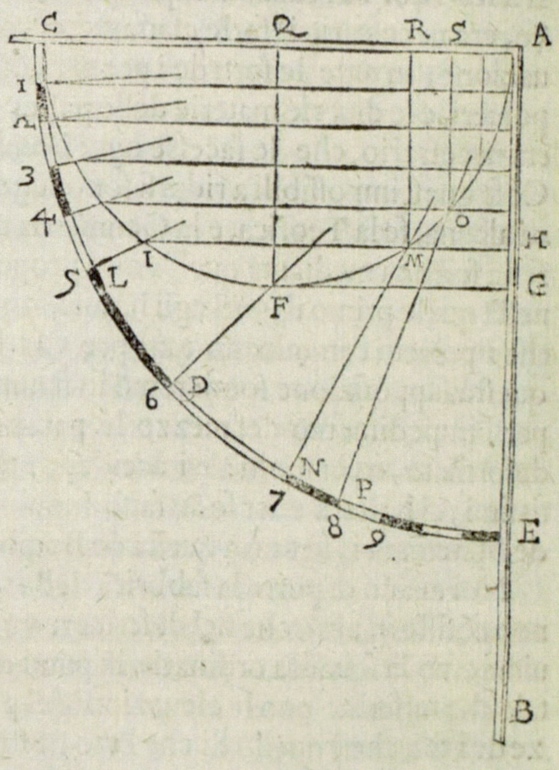

Siano le gambe della ſquadra a b *la maggiore, &* a c *la minore: poi fatto cẽtro in* a *facciaſi il quadrante* c d e. *ſopra il quale ſi hanno a notare i punti diſuguali; & intorno al diametro* a c. *facciaſi il ſemicircolo* a f c *e tirata la* f g *perpẽdicolare ad* a b, *e tãgente al ſemicircolo diuidaſi* a g *in ſei parti vguali per auere i ſei punti della ſquadra, e poi ciaſcuna parte in 12. per auere i minuti (quãdo però la grandezza dello ſtromento ſarà capace di queſta ſeconda diuiſione.) Hora ſia vna delle ſei parti la* g h. *Alziſi la* h m i. *parallela a* g f. *la qua-*

tiro; ſe a tre punti, tirerà tre di quei ſpazij, ſe a quattro e mezzo, tirerà quattro e mezzo: ſe a cinque & vn quarto, tirerà cinque & vn quarto; e coſi fino al ſeſto punto creſceranno ſempre nello ſteſso modo, e con la ſteſsa proporzione i punti della ſquadra nello ſtrumento, e gli ſpazij de i tiri nella campagna, e dal ſeſto fino al duodecimo punto anderanno nella ſteſſa maniera decreſcendo. La coſtruzione, e dimoſtrazione e Geometrica, cauandoſi dalla propoſizione da noi poſta al numero XI. de proietti, la quale dalla data amplitudine inſegna trouar l'eleuazione. E ſerue comunemente per qualſiuoglia ſorte d'artiglieria, e di mortari, per qualunque ſpezie di palla, o di poluere.

Siano le gambe della ſquadra ab *la maggiore, &* ac *la minore: poi fatto cẽtro in* a *facciaſi il quadrante* cde. *ſopra il quale ſi hanno a notare i punti diſuguali; & intorno al diametro* ac. *faccia ſi il ſemicircolo* afc *e tirata la* fg *perpẽdicolare ad* ab, *e tãgente al ſemicircolo diuidaſi* ag *in ſei parti vguali per auere i ſei punti della ſquadra, e poi ciaſcuna parte in 12. per auere i minuti (quãdo però la grandezza dello ſtromento ſarà capace di queſta ſeconda diuiſione.) Hora ſia vna delle ſei parti la* gh. *Alziſi la* hmi. *parallela a* gf.

la qua

la quale seghi il semicircolo ne i punti m, i. *Tirisi poi dal centro* a *la retta* a f d, *& il punto* d *sarà il sesto della squadra. Tirisi la* a i l, *& il punto* l *sarà il quinto della squadra; tirisi la* a m n, *& il punto* n. *sarà il settimo, e così di tutti gli altri. Auertasi però che l'operazione sarà più giusta, se dopo auer trouato i punti 1. 2. 3. &c. formeremo con la trasportazione d'essi il nono, decimo, & vndecimo. I mezzi punti, i quarti, & i minuti si trouerranno nello stesso modo, col diuidere in mezzo, o in quattro parti, ouero in dodici ciascuna delle porzioni della linea* a g. *con alzar le perpendicolari dalli punti delle diuisioni; segheranno dette perpendicolari il semicircolo, & per i punti delle sezioni si tireranno i semidiametri nel quadrante, che questi segheranno il quadrante ne i punti desiderati, de' mezzi punti, de' quarti, o minuti.*

Hora è manifesto per la Proposizione IX. nostra, che se la linea della direzione, o vogliam dire della eleuazione del pezzo sarà *ao*, ouero *ap*; la amplitudine e lunghezza del tiro sarà come la quadrupla di *so*; e se la direzione sarà *am*, ouero *an*. il tiro sarà come la quadrupla di *rm*: e quando la eleuazione fusse secondo la linea *afd*, il tiro sarà come la quadrupla di *qf*. Ma le linee *so*, *rm*, *qf*. per la costruzione nostra egualmente si eccedono, e però anco le loro quadruple, ouero i tiri sopradetti egualmente si eccederanno l'vn l'altro.

Vso della predetta diuisione, fatta nella squadra ordinaria.

Siaci proposta qualunque artiglieria, o mortaro, e con essa facciasi vna sola esperienza; cioè sia eleuata a qualunque punto, come per esempio al quinto. Sparisi, e si misuri la lunghezza del tiro, e trouisi, verbigrazia, essere 2000. parti. fatto questo possiamo sapere quanto tirerà la medesima artiglieria caricata nello stesso modo, & eleuata a qualsiuoglia punto, o minuto, che sarà facile per la regola del tre, essendo in questo strumento tanto i punti quanto la lunghezza de i tiri proporzionali. La pratica è questa. Voglio sapere quanto tira il sesto

punto.

la quale feghi il femicircolo ne i punti m, i. *Tirifi poi dal centro* a *la retta* afd, *& il punto* d *farà il festo della fquadra. Tirifi la* ail, *& il punto* l *farà il quinto della fquadra; tirifi la* amn, *& il punto* n. *farà il fettimo, e cosi di tutti gli altri. Auuertafi però che l'operazione farà più giufta, fe dopo auer trouato i punti 1. 2. 3. &c. formeremo con la trafportazione d'effi il nono, decimo, & vndecimo. I mezzi punti, i quarti, & i minuti fi troueranno nello fteffo modo, col diuidere in mezzo, o in quattro parti, ouero in dodici ciafcuna delle porzioni della linea* a g. *con alzar le perpendicolari dalli punti delle diuifioni; fegheranno dette perpendicolari il femicircolo, & per i punti delle fezioni fi tireranno i femidiametri nel quadrante, che quefti fegheranno il quadrante ne i punti defiderati, de' mezzi punti, de' quarti, o minuti.*

Hora è manifefto per la Propofizione *IX.* noftra, che fe la linea della direzione, o vogliam dire della eleuazione del pezzo farà *ao,* ouero *ap*; la amplitudine e lunghezza del tiro farà come la quadrupla di *fo*; e fe la direzione farà *am,* ouero *an.* il tiro farà come la quadrupla di *rm*: e quando la eleuazione fuffe fecondo la linea *afd,* il tiro farà come la quadrupla di *qf.* Ma le linee *fo, rm, qf.* per la coftruzione noftra egualmente fi eccedono, e però anco le loro quadruple, ouero i tiri fopradetti egualmente fi eccederanno l'vn l'altro.

Vfo della predetta diuifione, fatta nella fquadra ordinaria.

SIaci propofta qualunque artiglieria, o mortaro, e con effa facciafi vna fola efperienza; cioè fia eleuata a qualunque punto, come per efempio al quinto. Sparifi, e fi mifuri la lunghezza del tiro, e trouifi, verbigrazia, effere 2000. parti. fatto quefto poffiamo fapere quanto tirerà la medefima artiglieria caricata nello fteffo modo, & eleuata a qualfiuoglia punto, o minuto, che farà facile per la regola del tre, effendo in quefto ftrumento tanto i punti quanto la lunghezza de i tiri proporzionali. La pratica è quefta. Voglio fapere quanto tira il fefto

punto

232

punto. Fò così; se 5. punti diedero 2000 passi quanto daranno 6. punti? e trouo 2400. passi. Dico dunque che quella artiglieria al sesto punto, cioè col massimo tiro, tirerà due mila e quattrocento di quelle parti delle quali al 5. punto ne tiraua 2000.

Auuertasi però che in cambio di fare questa operazione cō i punti 7. 8. 9. 10. 11. & 12. si fa con i loro complementi, i quali sono 5. 4. 3. 2. 1. & 0.

Ma se ci fusse comandato (& importa molto più) che noi eleuassimo il sudetto pezzo in tal modo, che la lunghezza del tiro douesse riuscire per esempio passi 1300. opereremo così. Se 2000. passi furono fatti da 5. punti, o per dir meglio da 60. minuti di squadra, 1300. passi da quanti minuti si faranno? ecco l'operazione 2000. 60. 1300.? 39. E troueremo che per fare il tiro di lunghezza di passi 1300. bisognerebbe dare all'artiglieria l'eleuazione di minuti trentanoue di squadra, ouero di punti tre & vn quarto.

Modo per fabbricare la squadra nostra.

MA se noi volessimo formare vno strumento, il quale nō solo misurasse la lunghezza de i tiri fatti a diuerse eleuazioni, ma anco l'altezza della parabola, la durazione, o tempo del viaggio, la sublimità, e l'altre cose dimostrate nel precedente libro de proietti; tutto si farà col solo, e semplice semicircolo della proposizione 9. Ma venghiamo alla costruzione.

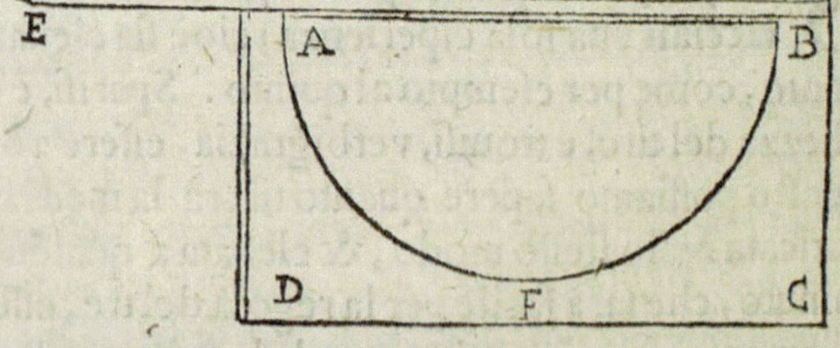

Prendasi la lamina rettangola *a b c d*. di ottone, o d'altra soda materia laquale habbia la gamba *a e* lunga da poter mettersi in bocca del pezzo. Facciasi so

pra il

punto. Fò coſi; ſe 5. punti diedero 2000 paſſi quanto daranno 6. punti? e trouo 2400. paſſi. Dico dunque che quella artiglieria al ſefto punto, cioè col maſſimo tiro, tirerà due mila e quattrocento di quelle parti delle quali al 5. punto ne tiraua 2000.

Auuertaſi però che in cambio di fare queſta operazione cõ i punti 7. 8. 9. 10. *11.*[114] & 12. ſi fa con i loro complementi, i quali ſono 5. 4. 3. 2. 1. & 0.

Ma ſe ci fuſſe comandato (& importa molto più) che noi eleuaſſimo il ſudetto pezzo in tal modo, che la lunghezza del tiro doueſse riuſcire per eſempio paſſi 1300. opereremo coſi. Se 2000. paſſi furono fatti da 5. punti, o per dir meglio da 60. minuti di ſquadra, 1300. paſſi da quanti minuti ſi faranno*?* ecco l'operazione 2000. 60. 1300.*? 39*. E troueremo che per fare il tiro di lunghezza di paſſi 1300. biſognerebbe dare all'artiglieria l'eleuazione di minuti trentanoue di ſquadra, ouero di punti tre & vn quarto.

Modo per fabbricare la ſquadra noſtra.

MA ſe noi voleſſimo formare vno ſtrumento, il quale nõ ſolo miſuraſse la lunghezza de i tiri fatti a diuerſe eleuazioni, ma anco l'altezza della parabola, la durazione, o tempo del viaggio, la ſublimità, e l'altre coſe dimoſtrate nel precedente libro de proietti; tutto ſi farà col ſolo, e ſemplice ſemicircolo della propoſizione 9. Ma venghiamo alla coſtruzione.

Prendaſi la lamina rettangola *abcd.* di ottone, o d'altra ſoda materia la quale habbia la gamba *ae* lunga da poter metterſi in bocca del pezzo. Facciaſi ſo-

pra il

[114] *11.*] 11.

pra il diametro *ab* vno semicircolo *afb*, che sarà il semicircolo della proposizione 9. de proietti, & in *b* pongasi il filo con il piombo, e diuidasi il semicircolo *afb* in 90. parti eguali, che saranno li 90. gradi del quadrante; ouero in 144. parti eguali, che saranno i punti, e minuti eguali della squadra ordinaria. Mostriamo ora Geometricamēte come questa squadra sia atta a misurare con somma simplicità le lunghezze, l'altezze de i tiri, il tempo delle durazioni, le sublimità delle parabole, e le eleuazioni de' pezzi. E poi porremo la diuisione delle linee in essa, senza auer bisogno di Tauola alcuna per operar detta squadra.

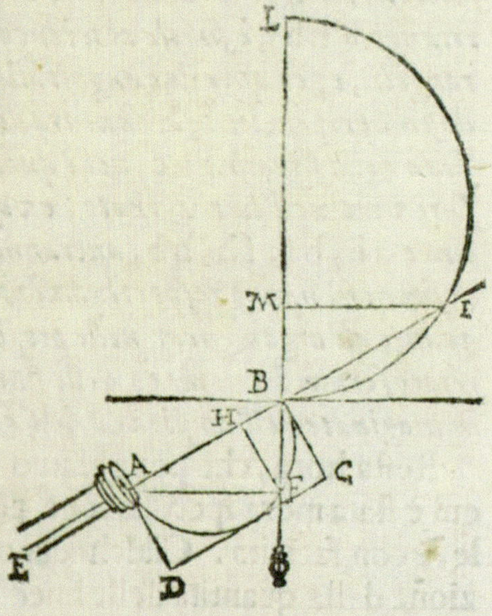

Pongasi, come nella seguēte figura, la squadra sopradetta cabcd. *in bocca di vn pezzo* ea. *qualunque si sia, e caschi il piombo su'l pūto* f *del semicircolo* afb *diuiso in 90. parti eguali. E certo primieramēte, che l'arco* bf *misura l'eleuazione del pezzo* ea *sopra l'orizonte. imperò che hauendo noi con la diuisione del semicircolo in 90. parti solamente valutato ogni due gradi per vn solo, abbiamo fatto che l'arco* bf *sia misura dell'angolo* baf, *cioè della eleuazione del pezzo sopra l'orizonte; quale orizonte sarà sempre la linea* af. *Dico di più, che se noi fingeremo che la linea* ab. *diametro del semicircolo sia l'impeto della proposta artiglieria, ouero la metà del massimo tiro; la linea* fh, *perpendicolare al diametro, sarà la quarta parte della amplitudine, o lunghezza del tiro; la* bh. *sarà l'altezza suprema della parabola; la* ah, *sarà la sublimità; la* bf. *sarà il tempo o durazione del tiro.*

pra il diametro *ab* vno ſemicircolo *afb,* che ſarà il ſemicircolo della propoſizione 9. de proietti, & in *b* pongaſi il filo con il piombo, e diuidaſi il ſemicircolo *afb* in 90. parti eguali, che ſaranno li 90. gradi del quadrante; ouero in 144. parti eguali, che ſaranno i punti, e minuti eguali della ſquadra ordinaria. Moſtriamo ora Geometricamẽte come queſta ſquadra ſia atta a miſurare con ſomma ſimplicità le lunghezze, l'altezze de i tiri, il tempo delle durazioni, le ſublimità delle parabole, e le eleuazioni de' pezzi. E poi porremo la diuiſione delle linee in eſsa, ſenza auer biſogno di Tauola alcuna per operar detta ſquadra.

Pongaſi, come nella ſeguẽte ſigura, la ſquadra ſopradetta eabcd. *in bocca di vn pezzo* ea. *qualunque ſi ſia, e caſchi il piombo ſu'l pũto* f *del ſemicircolo* afb *diuiſo in 90. parti eguali. E certo primieramẽte, che l'arco* bf *miſura l'eleuazione del pezzo* ea *ſopra l'orizonte. imperò che hauendo noi con la diuiſione del ſemicircolo in 90. parti ſolamente valutato ogni due gradi per vn ſolo, abbiamo fatto che l'arco* bf *ſia miſura dell'angolo* baf, *cioè della eleuazione del pezzo ſopra l'orizonte; quale orizonte ſarà ſempre la linea* af. *Dico di più, che ſe noi ſingeremo che la linea* ab. *diametro del ſemicircolo ſia l'impeto della propoſta artiglieria, ouero la metà del maſſimo tiro; la linea* fh, *perpendicolare al diametro, ſarà la quarta parte della amplitudine, o lunghezza del tiro; la* bh. *ſarà l'altezza ſuprema della parabola; la* ah, *ſarà la ſublimità; la* bf. *ſarà il tempo o durazione del tiro.*

234

Che tutto questo sia vero si dimostrerà così (accennando la Proposizione 9. de proietti, e le sue annesse.) Prolunghisi la linea della direzione a b i, *e parimente il perpendicolo* f b l *indefinitamente. Prendasi poi con l'immaginazione la* b l *di tanta lunghezza che sia eguale realmente alla metà del massimo tiro della nostra proposta artiglieria; & intorno al diametro* b l, *facciasi col pensiero il gran semicircolo* b i l, *che seghi la* b i *in qualunque punto* i; *e tirisi la orizontale* i m. *E manifesto, per la citata Proposizione 9. de proietti, che la linea* m i *sarà la quarta parte reale della lunghezza del tiro: parimente che* d m *sarà l'altezza non finta, ma reale d'esso tiro; e così l'altre misure nel semicircolo* b i l *saranno tutte vere, e reali. Hora notisi che il triangolo* h b f, *è simile al triangolo* b m i, *per essere ambidue rettangoli, e per auer due angoli alla cima* b. *Adunque nelle medesime proporzioni saranno tra di loro tutte le misure piccole, e finte della squadra* a c, *nelle quali proporzioni sono tutte le misure vere nell'immaginato, e vasto semicircolo* b i l. *Cioè, le linee* a b, b f, f h, h b, *aueranno fra di loro le proporzioni medesime che hanno respettiuamente, le* l b, b i, i m, m b. *Però quanto all'argomentar nelle proporzioni, potremo senz' alcuno errore seruirci non meno delle finte su la squadra, che delle vere immaginate nell'ampiezza dell'aria.*

Resta hora, che ponghiamo come questa dottrina, che fin qui è stata mera speculazione, possa ridursi in pratica manouale, e con facilità. Ciascheduno vede che per auer noi cognizione della quantità delle linee *ab*, *bf*, *fh*, *hb*, e loro proporzioni nella precedente figura, sarebbe necessario, che tutte le predette linee fussero diuise in parti minutissime con qualche comune misura. A questo effetto però diuideremo in parti eguali, e minutissime il diametro *ab*, & il semidiametro *ed*, come appare nella seguente figura, doue dipinghiamo la squadra finita; e poi daremo a ciascuna diuisione della circonferenza le sue guide parallele a detti diametri, acciò in essi si possa leggere il numero, e la quantità delle linee, che saranno indici della lunghezza, & altezza de tiri; e nel punto dell'angolo

del

Che tutto questo ſia vero ſi dimoſtrerà così (accennando la Propoſizione 9. de proietti, e le ſue anneſſe.) Prolunghiſi la linea della direzione abi, *e parimente il perpendicolo* fbl *indefinitamente. Prendaſi poi con l'immaginazione la* bl *di tanta lunghezza che ſia eguale realmente alla metà del maſſimo tiro della noſtra propoſta artiglieria; & intorno al diametro* bl, *facciaſi col penſiero il gran ſemicircolo* bil, *che ſeghi la* bi *in qualunque punto* i; *e tiriſi la orizontale* im. *E manifeſto, per la citata Propoſizione 9. de proietti, che la linea* mi *ſarà la quarta parte reale della lunghezza del tiro: parimente che* dm *ſarà l'altezza non finta, ma reale d'eſſo tiro; e coſì l'altre miſure nel ſemicircolo* bil *ſaranno tutte vere, e reali. Hora notiſi che il triangolo* hbf, *è ſimile al triangolo* bmi, *per eßere ambidue rettangoli, e per auer due angoli alla cima* b. *Adunque nelle medeſime proporzioni ſaranno tra di loro tutte le miſure piccole, e finte della ſquadra* ac, *nelle quali proporzioni ſono tutte le miſure vere nell'immaginato, e vaſto ſemicircolo* bil. *Cioè, lé linee* ab, bf, fh, hb, *aueranno fra di loro le proporzioni medeſime che hanno reſpettiuamente, le* lb, bi, im, mb. *Però quanto all'argomentar nelle proporzioni, potremo ſenz'alcuno errore ſeruirci non meno delle finte ſu la ſquadra, che delle vere immaginate nell'ampiezza dell'aria.*

Reſta hora, che ponghiamo come queſta dottrina, che fin quì è ſtata mera ſpeculazione, poſſa ridurſi in pratica manouale, e con facilità. Ciaſcheduno vede che per auer noi cognizione della quantità delle linee *ab, bf, fh, hb,* e loro proporzioni nella precedente figura, ſarebbe neceſſario, che tutte le predette linee fuſſero diuiſe in parti minutiſſime con qualche comune miſura. A queſto effetto però diuideremo in parti eguali, e minutiſſime il diametro *ab,* & il ſemidiametro *ed,* come appare nella ſeguente figura, doue dipinghiamo la ſquadra finita; e poi daremo a ciaſcuna diuiſione della circonferenza le ſue guide parallele a detti diametri, acciò in eſſi ſi poſſa leggere il numero, e la quantità delle linee, che ſaranno indici della lunghezza, & altezza de tiri; e nel punto dell'angolo

del

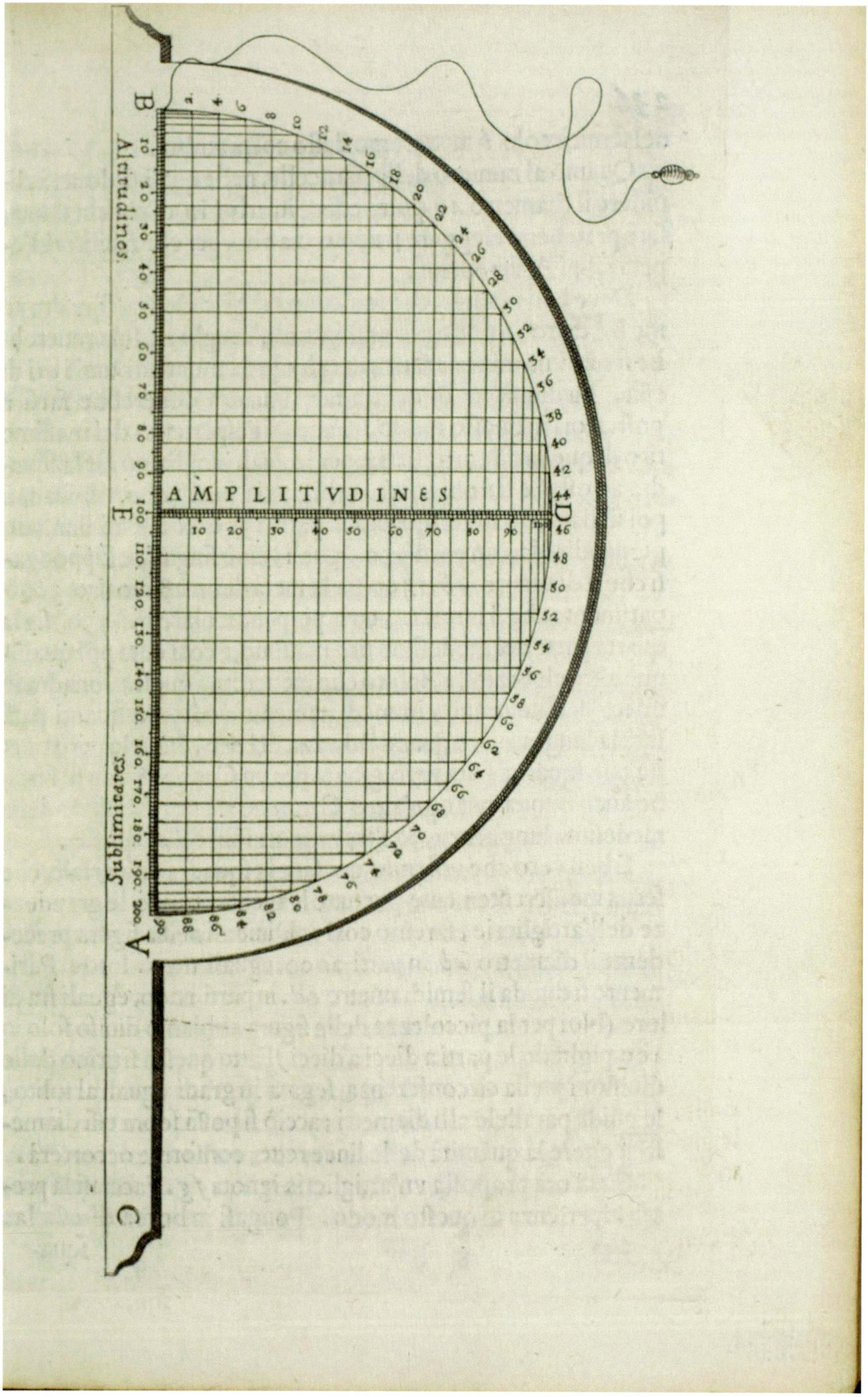
B
Altitudines.
AMPLITVDINES
E
D
Sublimitates.
A
C

236

del semicircolo *b* metteremo il filo col piombo.

Quanto al numero delle particelle, nelle quali si douerà diuidere il diametro *ab* potrà essere in arbitrio di ciascheduno; sarà però bene eleggere il numero 2000. per che faciliterà l'operazioni Aritmetiche.

Deue ben notarsi, che se alcuno fabbricasse vna squadra come si è detto, a posta per vna spezie d'artiglierie sola, auerebbe senza vna minima fatica di calcolo la misura di tutti i tiri di essa. La diuisione di questa tale squadra douerebbe farsi a posteriori in questo modo. Facciasi l'esperienza del massimo tiro di quella tale artiglieria, per la quale vogliamo far la squadra a posta; e si troui essere verbigrazia passi 3000. Diuadasi poi il diametro della squadra in parti 1500. & il semidiametro perpendicolare in parti 750. eguali; cioè fingasi, e suppongasi che il diametro *ab* 1500. sia la metà del massimo tiro 3000. parimente che il semidiametro perpendicolare *cd* 750. sia la quarta parte del medesimo tiro massimo, e così data poi qualũque altra eleuazione, subito che metteremo questa squadra in bocca del suo pezzo, immediatamente vedremo quanti passi sarà la lunghezza, e quanti l'altezza del tiro, &c. Ma però questa tale squadra fatta verbigrazia per vn Cannone da 60. sarebbe anco buona per ogn'altro Cannone da 60. che fusse della medesima lunghezza, & altre proporzioni come quello.

E ben vero che volendo noi fare la squadra vniuersale, che serua indifferentemente per tutte le spezie, e tutte le grandezze dell'artiglierie, faremo così. Diuidasi nella figura precedente il diametro *ab* in parti 2000. eguali tra di loro. Parimente si diuida il semidiametro *cd*. in parti 1000, eguali fra di loro. (Noi per la piccolezza della figura abbiamo diuiso solo in 100. pigliãdo le parti a dieci a dieci.) Fatto questo si tirino dalle diuisioni della circonferenza segata in gradi eguali al solito, le guide parallele alli diametri; acciò si possa sopra essi diametri leggere la quantità delle linee rette, conforme occorrerà.

Siaci ora proposta vn'artiglieria ignota *fg*. Facciasi la premia esperienza in questo modo. Pongasi in bocca di essa la squa-

del ſemicircolo *b* metteremo il filo col piombo.

Quanto al numero delle particelle, nelle quali ſi douerà diuidere il diametro *ab* potrà eſſere in arbitrio di ciaſcheduno; ſarà però bene eleggere il numero 2000. per che faciliterà l'operazioni Aritmetiche.

Deue ben notarſi, che ſe alcuno fabbricaſſe vna ſquadra come ſi è detto, a poſta per vna ſpezie d'artiglierie ſola, auerebbe ſenza vna minima fatica di calcolo la miſura di tutti i tiri di eſſa. La diuiſione di queſta tale ſquadra douerebbe farſi a poſteriori in queſto modo. Facciaſi l'eſperienza del maſſimo tiro di quella tale artiglieria, per la quale vogliamo far la ſquadra a poſta; e ſi troui eſſere verbigrazia paſſi 3000. Diuadaſi poi il diametro della ſquadra in parti 1500. & il ſemidiametro perpendicolare in parti 750. eguali; cioè fingaſi, e ſuppongaſi che il diametro *ab* 1500. ſia la metà del maſſimo tiro 3000. parimente che il ſemidiametro perpendicolare *cd* 750. ſia la quarta parte del medeſimo tiro maſſimo, e coſì data poi qualũque altra eleuazione, ſubito che metteremo queſta ſquadra in bocca del ſuo pezzo, immediatamente vedremo quanti paſſi ſarà la lunghezza, e quanti l'altezza del tiro, &c. Ma però queſta tale ſquadra fatta verbigrazia per vn Cannone da *60*. ſarebbe anco buona per ogn'altro Cannone da 60. che fusſe della medeſima lunghezza, & altre proporzioni come quello.

E ben vero che volendo noi fare la ſquadra vniuerſale, che ſerua indifferentemente per tutte le ſpezie, e tutte le grandezze dell'artiglierie, faremo coſi. Diu[i]daſi nella figura precedente il diametro *ab* in parti 2000. eguali tra di loro. Parimente ſi diuida il ſemidiametro *ed.* in parti 1000, eguali fra di loro. (Noi per la piccolezza della ſigura abbiamo diuiſo ſolo in 100. pigliãdo le parti a dieci a dieci.*)* Fatto queſto ſi tirino dalle diuiſioni della circonferenza ſegata in gradi eguali al ſolito,le guide parallele alli diametri; acciò ſi poſſa ſopra eſſi diametri leggere la quantità delle linee rette, conforme occorrerà.

Siaci ora propoſta vn'artiglieria ignota *fg*. Facciaſi la previa eſperienza in queſto modo. Pongaſi in bocca di eſſa la

iqua-

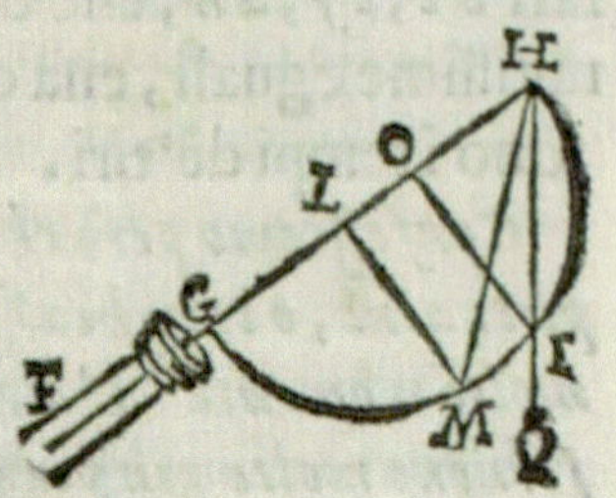

ſquadra, e caſchi il filo in qualunque luogo in *i*. Leggaſi per via della ſua guida la quantità di *io* ſu'l ſemidiametro diuiſo, e ſi tenga a memoria, e poi ſpariſi l'artiglieria, e ſi miſuri il tiro, che ſia paſſi (per eſempio) 1250. Carichiſi di nuouo l'artiglieria nello ſteſſo modo, e s'inalzi diuerſamente tanto che il filo caſchi altroue in *m*. ſi cerca la lunghezza di queſto tiro.

Facciaſi così. Se il numero di *io* dà la lunghezza di 1250 paſſi, il numero di *ml*, che ſi legge ſu'l ſemidiametro diuiſo, quanti paſſi darà? e trouerai parimente la lunghezza del tiro numerata in paſſi.

Chi voleſſe l'altezze, e non le lunghezze de' tiri, faccia la ſteſſa operazione come s'è detto, ma non con le linee *io*, *ml* che danno le lunghezze, ma ſi bene con le *ho*, *hl*, le quali danno le altezze. Se poi voleſſimo le ſublimità, biſognerebbe operare con le *go*, *gl*. Ma quello che importa più, ſe alcuno dopo fatta la preuia eſperienza, voleſſe che quella medeſima artiglieria faceſſe vn tiro propoſtoci, longo per eſempio 2200. paſſi. Si cerca quanta eleuazione debba darſi al pezzo. Operiſi così. Se li paſſi 1250. della preuia eſperienza mi danno *io* numerata. li paſſi 2200. che mi daranno; e trouerai vn numero il quale ſia per eſempio nella ſquadra aſcritto alla linea *ml*. Si douerà dunque alzar l'artiglieria tanto che il filo paſſi per il punto *m*. & allora il tiro riuſcirà di paſſi 2200.

I tempi ouero durazioni de i tiri ſi danno da le linee *hi*, *hm*. e per auer la quantità di queſte ſi può far in due modi. Primo per via di calcolo, perche il quadrato del tẽpo *hi* (nella paſſata figura) è ſempre eguale alli due quadrati, dell'altezza *ho*, e della quarta parte della lunghezza *io*. Secondo, con fare a tutte le diuiſioni della periferia *b*, *c*, *d*, (nella preſente figura) dal centro *a*. le guide circo-

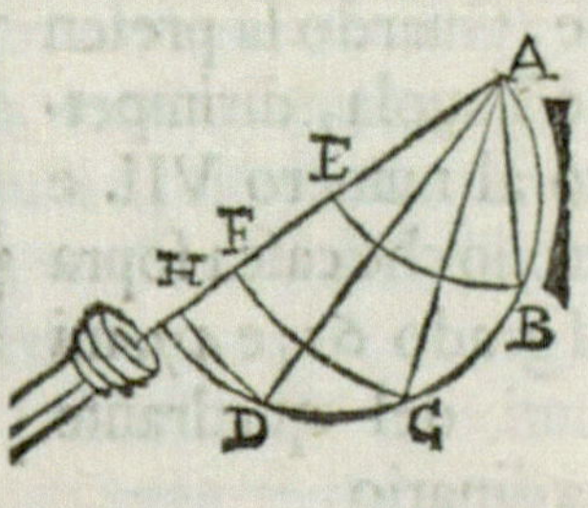

ſquadra, e caſchi il filo in qualunque luogo in *i.* Leggaſi per via della ſua guida la quantità di *io* ſu'l ſemidiametro diuiſo, e ſi tenga a memoria, e poi ſpariſi l'artiglieria, e ſi miſuri il tiro, che ſia paſſi (per eſempio) 1250. Carichiſi di nuouo l'artiglieria nello fteſſo modo, e ſ'inalzi diuerſamente tanto che il filo caſchi altroue in *m.* ſi cerca la lunghezza di queſto tiro.

Facciaſi coſi. Se il numero di *io* dà la lunghezza di 1250 paſſi, il numero di *ml,* che ſi legge ſu'l ſemidiametro diuiſo, quanti paſſi darà? e trouerai parimente la lunghezza del tiro numerata in paſſi.

Chi voleſſe l'altezze, e non le lunghezze de'tiri, faccia la fteſſa operazione come ſ'è detto, ma non con le linee *io, ml* che danno le lunghezze, ma ſi bene con le *ho, hl,* le quali danno le altezze. Se poi voleſſimo le ſublimità, biſognerebbe operare con le *go, gl.* Ma quello che importa più, ſe alcuno dopo fatta la preuia eſperienza, voleſſe che quella medeſima artiglieria faceſſe vn tiro propoſtoci, longo per eſempio 2200. paſſi. Si cerca quanta eleuazione debba darſi al pezzo. Operiſi coſi. Se li paſſi 1250. della preuia eſperienza mi danno *io* numerata. li paſſi 2200. che mi daranno; e trouerai vn numero il quale ſia per eſempio nella ſquadra aſcritto alla linea *ml.* Si douerà dunque alzar l'artiglieria tanto che il filo paſſi per il punto *m.* & allora il tiro riuſcirà di paſſi 2200.

I tempi ouero durazioni de i tiri ſi danno da le linee *hi, hm.* e per auer la quantità di queſte ſi può far in due modi. Primo per via di calcolo, perche il quadrato del tẽpo *hi* (nella paſſata figura) è ſempre eguale alli due quadrati, dell'altezza *ho,* e della quarta parte della lunghezza *io.* Secondo, con fare a tutte le diuiſioni della periferia *b, c, d,* (nella preſente figura) dal centro *a.* le guide circolari *be*, *cf*, *dh*, che così essendo la *ab* diuisa già in parti minutissime eguali, essa ci misurerà tutte le rette *ab*, *ac*, *ad*. che sono i tempi de' tiri.

Confessiamo però che quanto all'vso militare, solamente l'amplitudini, o lunghezze de i tiri pare che importino, e siano di molto momento L'altre sono tutte curiosità accessorie, le quali seruono molto più per gusto di Geometria, che per vtile di guerra. Però chi volesse la squadra solo per questo rispetto delle lunghezze; io prenderei il semicircolo abc *di ottone (come nella presente figura) il quale hauesse la gamba* ad, *e col semidiametro* eb *diuiso in parti minutissime & eguali facendo il principio della numeratione dal punto* e. *Di più darei à tutti i punti della periferia* f, g, h, i, *le loro guide* gh, fi. *parallele alla* ac, *e così s'hauerebbero sopra la* eb. *diuise, e numerate tutte le* fn, go, *le quali seruono per le amplitudini, o lunghezze de tiri.*

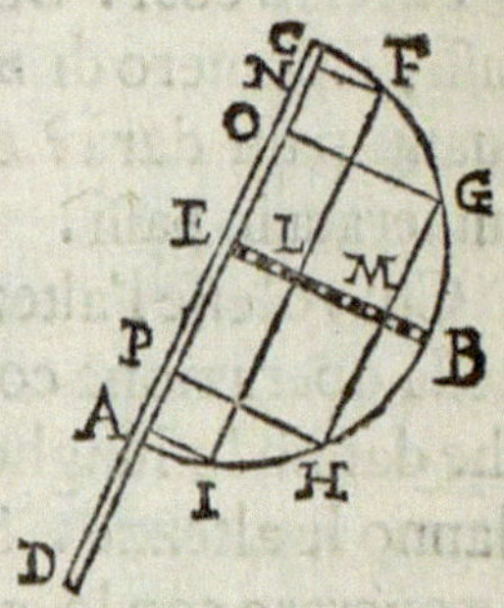

Tauola La quale mostra quanti gradi e minuti del quadrante ordinario contenga ciaschedun punto della nostra squadra, che hà i punti disuguali, posta a car. 230.

Per esempio si cerca doue caschi la diuisione del settimo punto nostro disuguale. Guardo la presente Tauola, dirimpetto al numero VII. e trouo che casca sopra il grado 61. e 47. minuti del quadrante ordinario.

Punti della squadra disuguali	Gradi del Quadrante ordinario	Punti della squadra disuguali	Gradi del Quadrante ordinario
mezzo I	2. 23 4. 48	mezzo VII	56. 46 61. 47
mezzo II	7. 15 9. 44	mezzo VIII	65. 42 69. 6
mezzo III	12. 19 15. 0	mezzo IX	72. 10 75.
mezzo IV	17. 50 20. 54	mezzo X	77. 41 80. 16
mezzo V	24. 18 28. 13	mezzo XI	82. 45 85. 12
mezzo VI	33. 14 45. 0	mezze XII	87. 37 90.

Mà

[*De motu Proiectorum*]

lari *be, cf, dh,* che cosi eſſendo la *ah* diuiſa già in parti minutiſſime eguali, eſſa ci miſurerà tutte le rette *ab, ac, ad.* che ſono i tempi de'tiri.

Confeſſiamo però che quanto all'vſo militare, ſolamente l'amplitudini, o lunghezze de i tiri pare che importino, e ſiano di molto momento. L'altre ſono tutte curioſità acceſſorie, le quali ſeruono molto più per guſto di Geometria, che per vtile di guerra. Però chi voleſſe la ſquadra ſolo per questo riſpetto delle lunghezze; io prenderei il ſemicircolo abc *di ottone (come nella preſente figura) il quale haueſſe la gamba* ad, *e col ſemidiametro* eb *diuiſo in parti minutiſſime & eguali facendo il principio della numerazione dal punto* e. *Di più darei à tutti i punti della periferia* f, g, h, i, *le loro guide* gh, fi. *parallele alla* ac, *e coſi ſ'hauerebbero ſopra la* eb. *diuiſe, e numerate tutte le* fn, go, *le quali ſeruono per le amplitudini, o lunghezze de tiri.*

Tauola La quale moſtra quanti gradi e minuti del quadrante ordinario contenga ciaſchedun punto della noſtra ſquadra, che hà i punti diſuguali, posta a car. 230.

Per eſempio ſi cerca doue caſchi la diuiſione del ſettimo punto noſtro diſuguale. Guardo la preſente Tauola, dirimpetto al numero VII. e trouo che caſca ſopra il grado 61. e 47. minuti del quadrante ordinario.

Punti della ſquadra diſuguali [left side]

Gradi del Quadrante ordinario. [right side]

mezzo I	2. 23 4. 48		mezzo VII	56. 46 61. 47
mezzo II	7. 15 9. 44		mezzo VIII	65. 42 69. 6
mezzo III	12. 19 15. 0		mezzo IX	72. 10 75.
mezzo IV	17. 50 20. 54		mezzo X	77. 41 80. 16
mezzo V	24. 18 28. 13		mezzo XI	82. 45 85. 12
mezzo VI	33. 14 45. 0		mezze XII	87. 37 90.

Mà

MA Già che ſiamo entrati a conſiderare il moto, e l'impeto de' proietti, non ſi può sfuggir l'occaſione di ſoggiungere qualche coſa circa la varietà delle forze loro, nel battere ſopra le ſuperficie reſiſtenti, ora con maggiore, ora con minor angolo d'inclinazione. Il Galileo contempla l'impeto d'eſſi proietti in ogni punto della loro parabola, e lo miſura ſolamente quanto è in ſe medeſimo, cioè riſpetto a quel piano, in cui perpendicolarmente egli percuoteſſe.

Noi ſupponendo che vn'impeto mentre arriua a percuotere, quanto a ſe ſia ſempre l'iſteſſo, lo conſidereremo, e miſureremo quanto egli ſia riſpetto al piano reſiſtente, variato, ſolamente dalla diuerſità de gl'angoli dell'incidenza. Non è Bōbardiero tanto ineſperto, il quale non ſappia, che le palle dell'artiglieria, mentre percuotono in vn muro, hanno ſempre minore, e minor forza, (data ogn'altra parità) quanto l'angolo dell'incidenza ſarà più e più acuto. Si che, ſe quel cannone con ſeſſanta libbre di ferro, e quaranta di poluere, non ſolo sfonda ma ancora ſconquaſſa con il tiro perpendicolare vna cortina; appena poi la offenderà, benche abbia la medeſima carica, e la medeſima diſtanza, con proiezzione del tiro, ch'eſſi chiamano di ſtriſcio. *Il Problema, per quanto io ſappia, è intatto. Però ſe ſi produrrà qualche coſa meno ſuſſiſtente, e non pura geometrica, o ſi compatiſca fin che altri tratti meglio la dottrina, o ſi rifiuti affatto, che poco importa.*

Suppoſizioni.

1. Parleremo ſolamente per i tiri dell'artiglierie: però Supponiamo, che quella porzione della linea, che fà la palla poco prima e poco dopo al colpire, ſia come linea retta. Sò che ſi tratta di linea veramente curua; ma auendo queſta (ſe fuſſe intera) la ſua lunghezza di più di tre mila paſſi geometrici, ſi potrà bene conſiderarne vn braccio, ouero vn palmo ſolo, ouero vn dito ſenza errore ſenſibile, per linea retta.

2. Supponiamo ſecondariamente che le forze, ouero impeti de proietti ſiano come ſono i ſpazij, che nello ſteſſo tempo ſi

po ſi

[*Liber Secundus.*] 239

MA *Già che ſiamo entrati a conſiderare il moto, e l'impeto de'proietti, non ſi può sfuggir l'occaſione di ſoggiungere qualche coſa circa la varietà delle forze loro, nel battere ſopra le ſuperficie reſiſtenti, ora con maggiore, ora con minor angolo d'inclinazione. Il Galileo contempla l'impeto d'eſſi proietti in ogni punto della loro parabola, e lo miſura ſolamente quanto è in ſe medeſimo, cioè riſpetto a quel piano, in cui perpendicolarmente egli percuoteſſe.*

Noi ſupponendo che vn'impeto mentre arriua a percuotere, quanto aſe ſia ſempre l'iſteſſo, lo conſidereremo, e miſureremo quanto egli ſia riſpetto al piano reſiſtente, variato, ſolamente dalla diuerſità de gl'angoli dell'incidenza. Non è Bõbardiero tanto ineſperto, il quale non ſappia, che le palle dell'artiglieria, mentre percuotono in vn muro, hanno ſempre minore, e minor forza, (data ogn'altra parità) quanto l'angolo dell'incidenza ſarà più e più acuto. Si che, ſe quel cannone con ſeſſanta libbre di ferro, e quaranta di poluere, non ſolo sfonda ma ancora ſconquaſſa con il tiro perpendicolare vna cortina; appena poi la offenderà, benche abbia la medeſima carica, e la medeſima diſtanza, con proiezzione del tiro, ch'eſſi chiamano di ſtriſcio. *Il Problema, per quanto io ſappia, e intatto. Però ſe ſi produrrà qualche coſa meno ſuſſiſtente, e non pura geometrica, o ſi compatiſca fin che altri tratti meglio la dottrina, o ſi rifiuti affatto, che poco importa.*

Suppoſizioni.

1. Parleremo ſolamente per i tiri dell'artiglierie: però Supponiamo, che quella porzione della linea, che fà la palla poco prima e poco dopo al colpire, ſia come linea retta. Sò che ſi tratta di linea veramente curua; ma auendo queſta (ſe fuſſe intera) la ſua lunghezza di più di tre mila paſſi geometrici, ſi potrà bene conſiderarne vn braccio, ouero vn palmo ſolo, ouero vn dito ſenza errore ſenſibile, per linea retta.

2. Supponiamo ſecondariamente che le forze, ouero impeti de proietti ſiano come ſono i ſpazij, che nello ſteſſo tem-

po si

Gal. 2. de motu æquabili — po ſi ſcorrono. cioè Se gli ſpazij *a. b. c.* ſaranno ſcorſi dal mobile nel medeſimo tempo, gl'impeti, o forze nel colpire ſaranno come gli ſpazij *a. b. c.* reſpettiuamente.

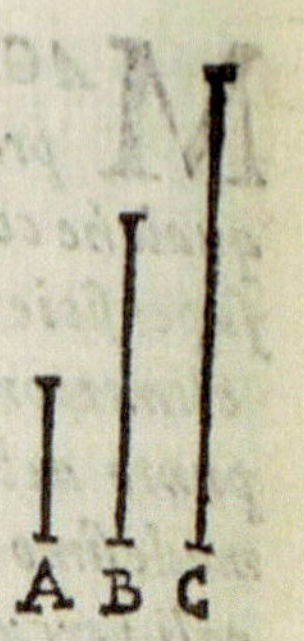

3. Ma ſe il medeſimo ſpazio verrà ſcorſo dal mobile in diuerſi tempi, gl'impeti o forze del mobile in percuotere aueranno la proporzione reciproca de i tempi. cioè Se il medeſimo ſpazio *d.* ſarà ſcorſo vna volta nel tempo *e*, & vn'altra nel tempo *f*. la forza della prima ſarà come *f*, e della ſeconda come *e*. — Gal. 3. de motu æquabili

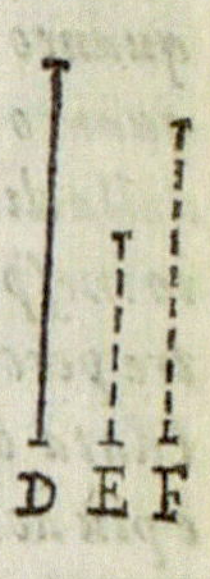

4. Supponiamo poi che tutti i tiri abbino, quanto a ſe ſteſſi, ſempre il medeſimo impeto, ilche ſeguirebbe quando ſtãdo ferma l'artiglieria ſempre nel medeſimo luogo, con la medeſima carica, medeſima eleuazione, e diſtanza &c. ſolo ſi variaſſe l'obliquità del muro.

Suppoſto queſto: mentre vna palla di cannone ſi auuicina al muro oppoſto, la linea, e dirittura del tiro, o è perpendicolare, al muro, o nò. Se è perpendicolare, la percoſſa opera con vna tal foza (che prouer emo eſſer la maſſima che poſſa auer quel tiro.) Se ſarà ad angoli obliqui, come la linea *ab.* alla parete *bc.* Io noto che riſpetto alla parete *bc.* ſono nella linea *ab* del proietto due moti inſieme compoſti: vno cioè, di auuicinamento perpendicolare alla parete, l'altro di paſſaggio laterale, o parallelo alla ſteſſa. Il perpendicolare ci viene, e moſtrato, e miſurato dalla linea *ac*. il parallelo dalla linea *cb*; poi che nel medeſimo tempo vengono paſſati dalla palla ambi gli ſpazij *ac*, *cb*.

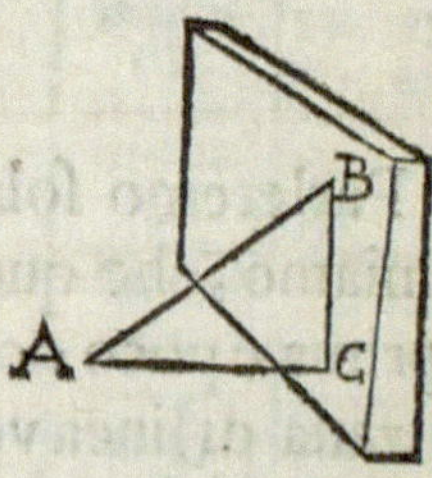

Hora oſſeruiamo che di queſte due ſorti d'impeto, vna ſola è a propoſito, per accreſcer le forze riſpetto a rompere il muro, & internar la palla in eſſo, cioè l'impeto della lazione perpen-

Gal. 2 *de motu æqua ili*[115].

Gal. 3. *de motu. æquabili.*

po ſi ſcorrono. cioè Se gli ſpazij *a. b. c.* ſaranno ſcorſi dal mobile nel medeſimo tempo, gl'impeti, o forze nel colpire ſaranno come gli ſpazij *a. b. c.* reſpettiuamente.

3. Ma ſe il medeſimo ſpazio verra ſcorſo dal mobile in diuerſi tempi, gl'impeti o forze del mobile in percuotere aueranno la proporzione reciproca de i tempi. cioè Se il medeſimo ſpazio *d.* ſarà ſcorſo vna volta nel tempo *e,* & vn'altra nel tempo *f.* la forza della prima ſarà come *f,* e della ſeconda come *e.*

4. Supponiamo poi che tutti i tiri abbino, quanto a ſe ſteſſi, ſempre il medeſimo impeto, il che ſeguirebbe quando ſtãdo ferma l'artiglieria ſempre nel medeſimo luogo, con la medeſima carica, medeſima eleuazione, e diſtanza &c. ſolo ſi variaſſe l'obliquità del muro.

Suppoſto queſto: mentre vna palla di cannone ſi auuicina al muro oppoſto, la linea, e dirittura del tiro, o è perpendicolare, al muro, o nò. Se è perpendicolare, la percoſſa opera con vna tal foza[116] *(*che proueremo eſſer la maſſima che poſſa auer quel tiro.*)* Se ſarà ad angoli obliqui, come la linea *ab.* alla parete *bc.* Io noto che riſpetto alla parete *bc.* ſono nella linea *ab* del proietto due moti inſieme compoſti: vno cioè, di auuicinamento perpendicolare alla parete, l'altro di paſſaggio laterale, o parallelo alla ſteſſa. Il perpendicolare ci viene, e moſtrato, e miſurato dalla linea *ac.* il parallello dalla linea *cb;* poi che nel medeſimo tempo vengono paſſati dalla palla ambi gli ſpazij *ac, cb.*

Hora oſſeruiamo che di queſte due ſorti d'impeto, vna ſola è a propoſito, per accreſcer le forze riſpetto a rompere il muro, & internar la palla in eſso, cioè l'impeto della lazione per-

pen-

[115] *æqua ili*] *æquabili.*

[116] foza] forza.

pendicolare *ac*. L'altro, ancor che fusse infinito, non accrescerà mai la forza del proietto contro alla resistenza del muro, se però non gli accellerasse anco la lazione perpēdicolare. Anzi se fusse l'orizontale semplice, e solo, senza mistura alcuna del perpendicolare, che altro farebbe la palla, se non correre equidistante dal muro, senza mai toccarlo, non che romperlo, se bene fusse vn sottilissimo vetro? Quando dunque data la dirittura di qualsiuoglia proiezione, noi sapremo quanto di questo impeto perpendicolare entra nella composizione del moto, sapremo anco l'attiuità, o momento del proietto verso la resistenza della muraglia contraposta.

Sia la linea di qualsiuoglia incidenza, *ab*. sopra il piano *bf*, presa con qualunque inclinanzione, ma però sia la porzione *ab* tanto piccola che possa considerarsi per retta. Tirisi perpendicolare al piano la *ac*., e si congiunga *cb*. Tanto dunque di moto parallelo sarà nella linea *ab*. rispetto alla parete *bf*., quanta è la linea *cb*. Ma di questo non facciamo stima; perche moltiplicato non aiuta, e diminuito nō debilita il momento, mentre l'altro impeto non alterato resti il medesimo. Di perpendicolare poi nella stessa, sarà quanto la linea *ac*. e la forza del colpo sarà magiore, e minore, secondo che nello stesso tempo sarà scorsa la *ac*. maggiore o minore.

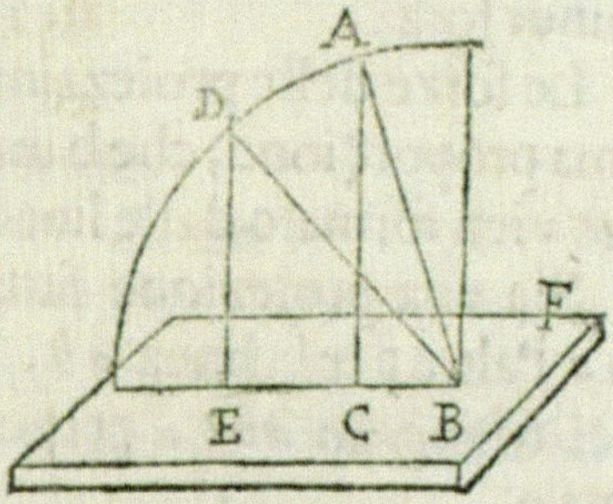

Supponiamo hora che la forza del incidēza *ab*. sia come *ac*. Per saper la forza di qualunq; altra incidenza *db*; prendasi *db*. eguale a *ba*, e tirata la *de*. perpendicolare al piano sarà la forza di questa incidenza come l'istessa linea *de*. Poiche se *ab*. *db*. sono eguali, e sono i tiri dell'istesso cannone, saranno scorse nel medesimo tempo. Adunque anco le *ac*. *de*. sono fatte nell'istesso tempo: però gli impeti rispetto al muro sono come *ac*. *de*. Per la 2. supposizione.

Inse-

pendicolare *ac.* L'altro, ancor che fuſſe infinito, non accreſcerà mai la forza del proietto contro alla reſiſtenza del muro, ſe però non gli accelleraſſe anco la lazione perpẽdicolare. Anzi ſe fuſse l'orizontale ſemplice, e ſolo, ſenza miſtura alcuna del perpendicolare, che altro farebbe la palla, ſe non correre equidiſtante dal muro, ſenza mai toccarlo, non che romperlo, ſe bene fuſse vn ſottilisſimo vetro*?* Quando dunque data la dirittura di qualſiuoglia proiezione, noi ſapremo quanto di queſto impeto perpendicolare entra nella compoſizione del moto, ſapremo anco l'attiuità, o momento del proietto verſo la reſiſtenza della muraglia contrapoſta.

Sia la linea di qualſiuoglia incidenza, *ab.* ſopra il piano *bf,* preſa con qualunque inclinanzione, ma però ſia la porzione *ab* tanto piccola che poſsa conſiderarſi per retta. Tiriſi perpendicolare al piano la *ac.,* e ſi congiunga *cb.* Tanto dunque di moto parallelo ſarà nella linea *a b.* riſpetto alla parete *bf.,* quanta è la linea *cb.* Ma di queſto non facciamo ſtima; perche moltiplicato non aiuta, e diminuito nõ debilita il momento, mentre l'altro impeto non alterato reſti il medeſimo. Di perpendicolare poi nella ſteſsa, ſarà quanto la linea *ac.* e la forza del colpo ſarà magiore, e minore, ſecondo che nello ſteſso tempo ſarà ſcorſa la *ac.* maggiore o minore.

Supponiamo hora che la forza del incidẽza *ab.* ſia come *ac.*

Per ſaper la forza di qualunq; altra incidenza *db*; prendaſi *db.* eguale a*ba,* e tirata la *de.* perpendicolare al piano ſarà la forza di queſta incidenza come l'iſteſsa linea *de.* Poiche ſe *ab. db.* ſono eguali, e ſono i tiri dell'iſteſso cannone, ſaranno ſcorſe nel medeſimo tempo. Adunque anco le *ac. de.* ſono fatte nell'iſteſso tempo*:* però gli impeti riſpetto al muro ſono come *ac. de.* *Per la 2 ſupposi-. zione.*

Inse-

Inferiremo dunque che, le attiuità, o momenti de i tiri diuersamente inclinati sono come i seni retti degli angoli d elle incidenze.

Si caua di quì per Corollario che la inciden za perpendicolare *ab*. hà maggior forza di tutte le altre, esse ndo la forza di essa come il seno totale. E la proiezione parallela non ne hà niente, essendo la forza sua come seno nullo. L'incidenza *db*. ad angolo di 30. gradi hà la metà della forza totale, essendo il seno suo la metà del semidiametro. Le altre poi, conforme aueranno maggiore o minor seno retto, aueranno maggiore, o minor forza.

Le forze delle proiezzioni hanno reciprocamente la medesima proporzione, che hanno i lati del triangolo, che sul piano vien formato dalle linee dell'incidenze.

Sia vna proiezione fatta per la linea *ac* l'altra per la linea *ab*. E sia il piano del triangolo *abc*. perpendicolare al muro.

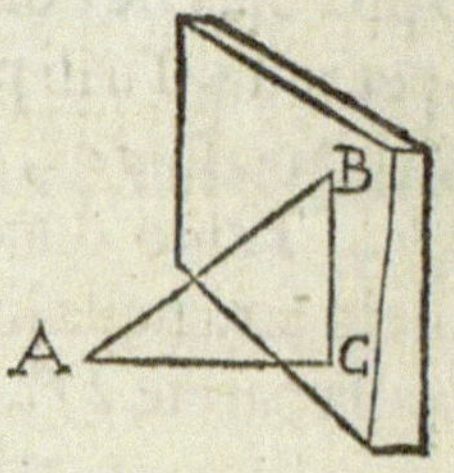

Perche dunque lo spazio *ac*. si corre dalla palla nel tempo *ac*; e lo spazio *ab* cioè (leuato il moto parallelo) lo stesso spazio *ac* si corre nel tempo *ab*. saranno le forze reciproche de i tempi. Cioè la forza per *ac*, sarà come *ab*. e per *ab* sarà come *ac*.

Per la 3. suppositione.

Allora i proietti aueranno la stessa forza nel percuotere, quando gl'impeti saranno come le secanti de gli angoli del complemento delle incidenze.

Sia l'impeto per la perpendicolare *ab* come *ab*. & abbia vna tal forza; Acciò l'impeto per l'inclinata *ac*. abbia la medesima forza, dico che l'impeto per *ac* all'impeto per *ab* deue essere come *ac* alla *ab*; la quale *ac* è secante dell'angolo *bac* complemento dell'inclinazione.

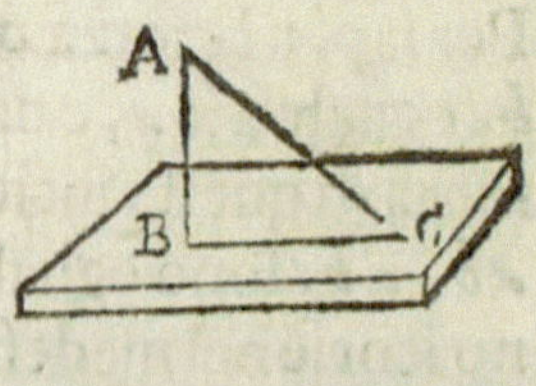

Poiche

Inferiremo dunque che, le attiuità, o momenti de i tiri diverſamente inclinati ſono come i ſeni retti degli angoli delle incidenze.

Si caua di quì per Corollario che la incidenza perpendicolare *ab.* hà maggior forza di tutte le altre, eſsendo la forza di eſsa come il ſeno totale. E la proiezione parallela non ne hà niente, eſsendo la forza ſua come ſeno nullo. L'incidenza *db.* ad angolo di 30. gradi hà la metà della forza totale, eſsendo il ſeno ſuo la metà del ſemidiametro. Le altre poi, conforme aueranno maggiore o minor ſeno retto, aueranno maggiore, o minor forza.

Le forze delle proiezzioni hanno reciprocamente la medeſima proporzione, che hanno i lati del triangolo, che ſul piano vien formato dalle linee dell'incidenze.

Sia vna proiezione fatta per la linea *ac* l'altra per la linea *ab.* E ſia il piano del triangolo *abc.* perpendicolare al muro.

Perche dunque lo ſpazio *ac.* ſi corre dalla palla nel tempo *ac;* e lo ſpazio *ab* cioè (leuato il moto parallelo) lo fteſso ſpazio *ac* ſi
Per la 3. corre nel tempo *ab.* ſaranno le forze reciproche de i tempi. Cioè la forza per *ac,* ſarà come *ab.* e per *ab* ſarà come *ac.*
ſuppoſi-. Allora i proietti aueranno la fteſsa forza nel percuotere, quando gl'impeti ſaranno come le ſecanti de gli angoli del complemento
zione. delle incidenze.

Sia l'impeto per la perpendicolare *ab* come *ab.* & abbia vna tal forza; Acciò l'impeto per l'inclinata *ac.* abbia la medeſima forza, dico che l'impeto per *ac* all'impeto per *ab* deue eſsere come *ac* alla *ab*; la quale *ac* è ſecante dell'angolo *bac* complemento dell'inclinazione.

Poiche

243

Poiche se saranno gl'impeti per *ab*, & *ac*. come sono gli spazij *ab*, & *ac*. i mobili scorreranno nel medesimo tempo le due linee *ab*. *ac*. cioè lo stesso auuicinamento perpendicolare *ab*. Dunque aueranno la medesima forza contro al muro. *Per la 2 supposizione.*

Di più se col tale cannone, e per la linea *cb*. la palla s'internasse tutta per l'appunto nel muro; adunque per tutte le linee eleuate non solo s'immergerà tutta nella solidità, ma farà sempre maggior passata, perche hà maggior forza. Ma delle meno eleuate, per che ciscuna auerà minor forza, niuna entrerà totalmente nella parete, ma alcune anco risalteranno, e sfuggiranno dall'altra parte.

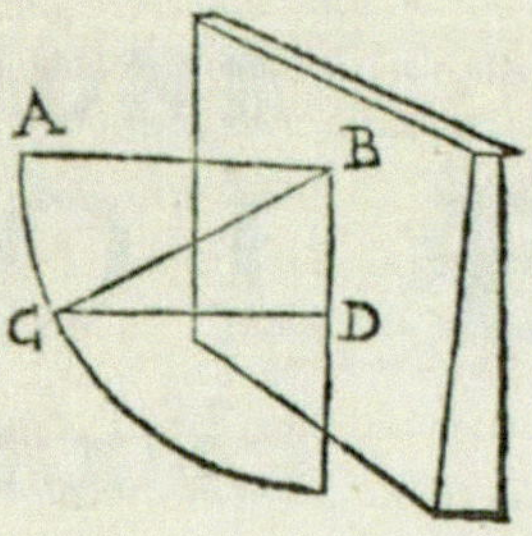

Sia però detto tutto questo astraendo da vn certo effetto di piegamento, o refrazzione che fanno i proietti nel passar con inclinazione dal mezzo raro al mezzo più denso, incuruandosi la linea al contrario della refrazzione della luce, e spezie visibili.

FINE DE' LIBRI DEL MOTO.

DE DIMEN-

Poiche ſe ſaranno gl'impeti per *ab*, & *ac*. come ſono gli ſpazij *ab*, & *ac*. i mobili ſcorreranno nel medeſimo tempo le due linee *ab*. *ac*. cioè lo fteſso auuicinamento perpendicolare *ab*. Dunque aueranno la medeſima forza contro al muro.

Per la 2 ſuppoſi-. zione.

Di più ſe col tale cannone, e per la linea *cb*. la palla ſ'internaſse tutta per l'appunto nel muro; adunque per tutte le linee eleuate non ſolo ſ'immergerà tutta nella ſolidità, ma farà ſempre maggior paſsata, perche hà maggior forza. Ma delle meno eleuate, per che ciſcuna[117] auerà minor forza, niuna entrerà totalmente nella parete, ma alcune anco riſalteranno, e sfuggiranno dall'altra parte.

Sia però detto tutto queſto astraendo da vn certo effetto di piegamento, o refrazzione che fanno i proietti nel paſſar con inclinazione dal mezzo raro al mezzo più denſo, incuruandoſi la linea al contrario della refrazzione della luce, e ſpezie viſibili.

FINE DE' LIBRI DEL MOTO.

DE DIMEN-

[117] ciſcuna] ciaſcuna.

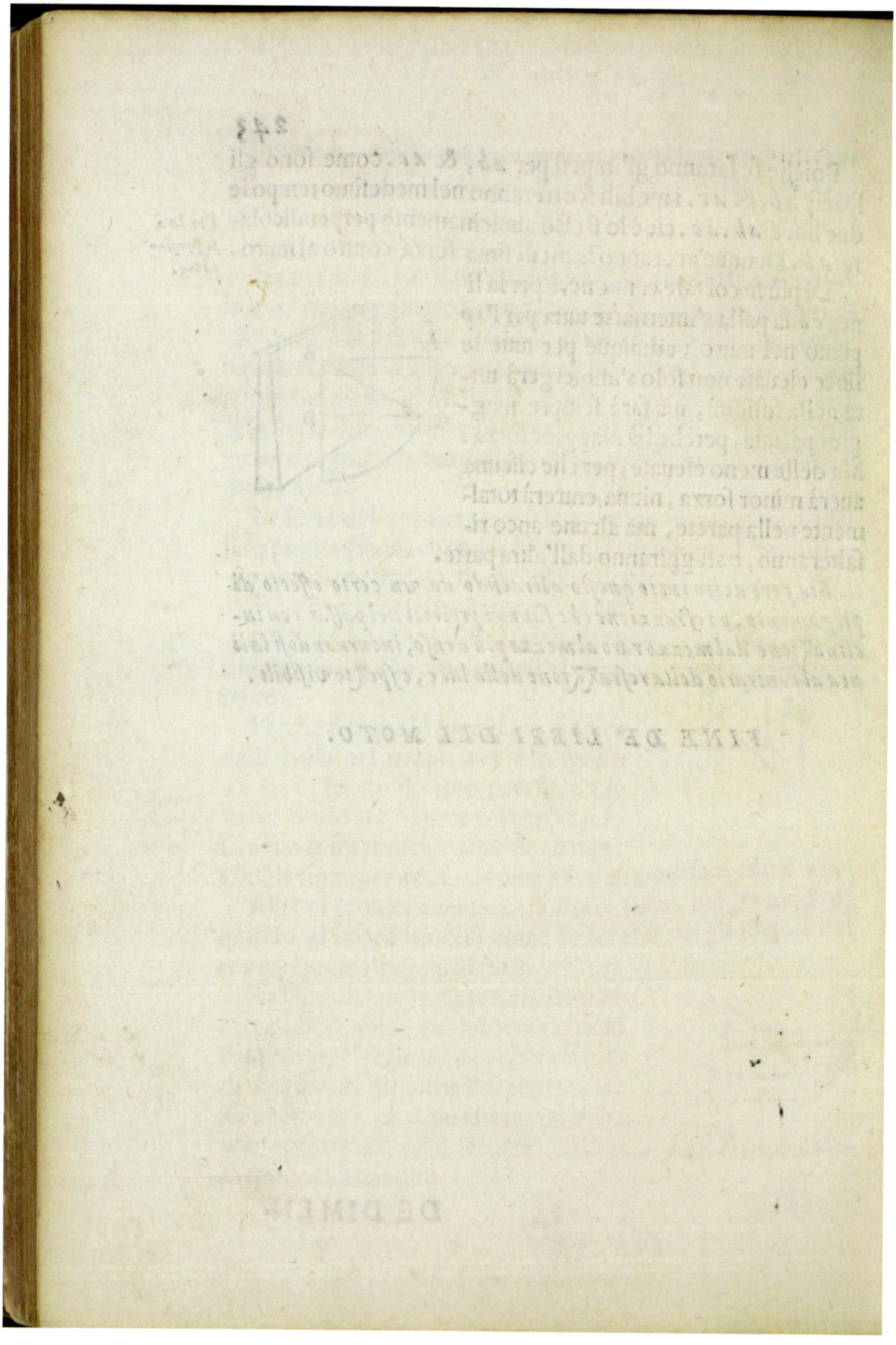

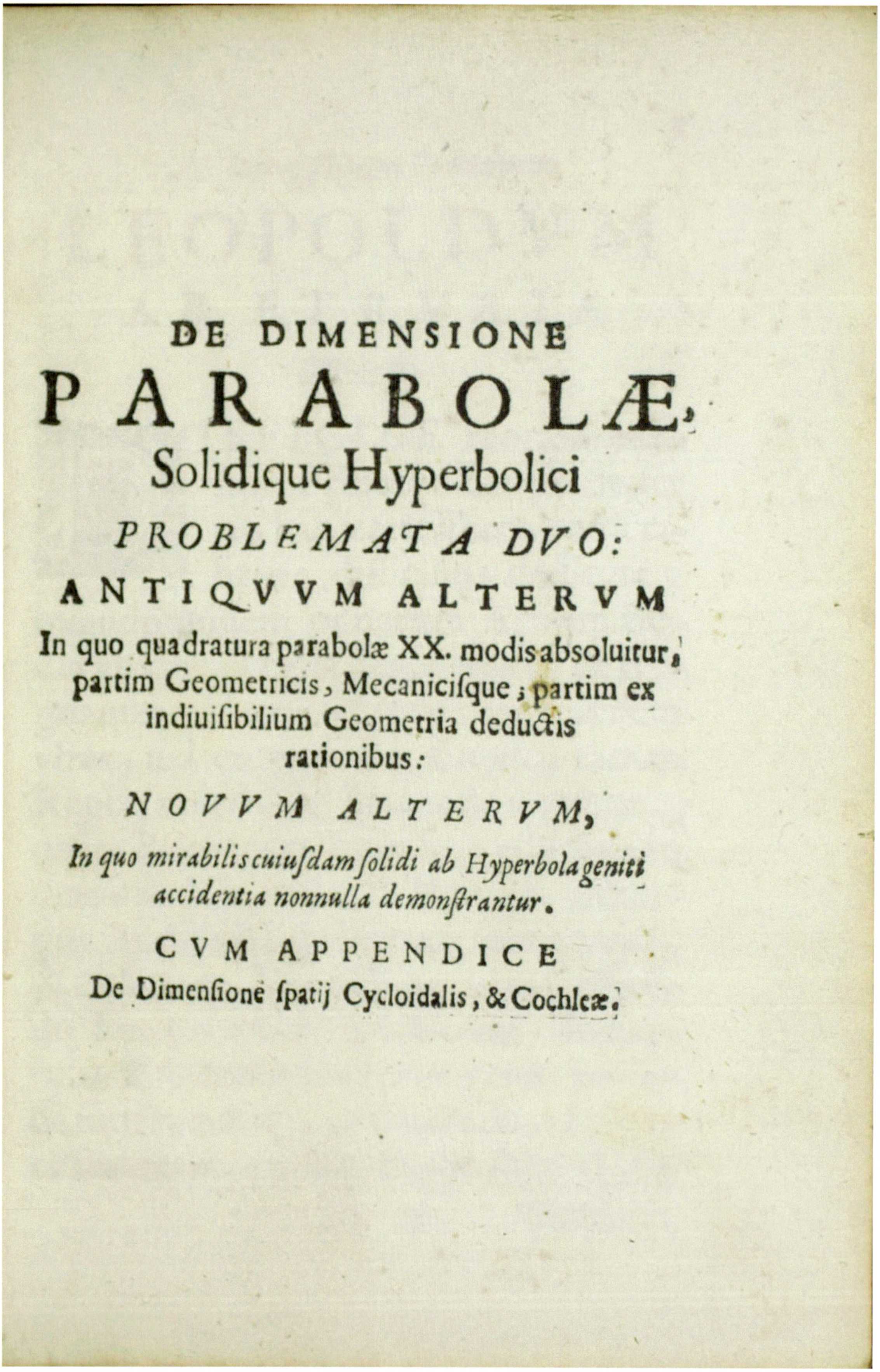

DE DIMENSIONE

PARABOLÆ,

Solidique Hyperbolici

PROBLEMATA DVO:

ANTIQVVM ALTERVM

In quo quadratura parabolæ XX. modis absoluitur, partim Geometricis, Mecanicisque; partim ex indiuiſibilium Geometria deductis rationibus:

NOVVM ALTERVM,

In quo mirabilis cuiuſdam ſolidi ab Hyperbola geniti accidentia nonnulla demonſtrantur.

CVM APPENDICE

De Dimenſione ſpatij Cycloidalis, & Cochleæ.

[1]

DE DIMENSIONE PARABOLÆ, [118].

Solidique Hyperbolici.

PROBLEMATA DVO:
ANTIQVVM ALTERVM.

In quo quadratura parabolæ XX. modis absoluitur,
partim Geometricis, Mecanicifque; partim ex
indiuifibilium Geometria deductis
rationibus:.
NOVVM ALTERVM,
In quo mirabilis cuiufdam folidi ab Hyperbola geniti
accidentia nonnulla demonftrantur..

CVM APPENDICE
De Dimenfione fpatij Cycloidalis, & Cochleæ.

[118] The numbering of pages restarts from p. 1 to p. 84.

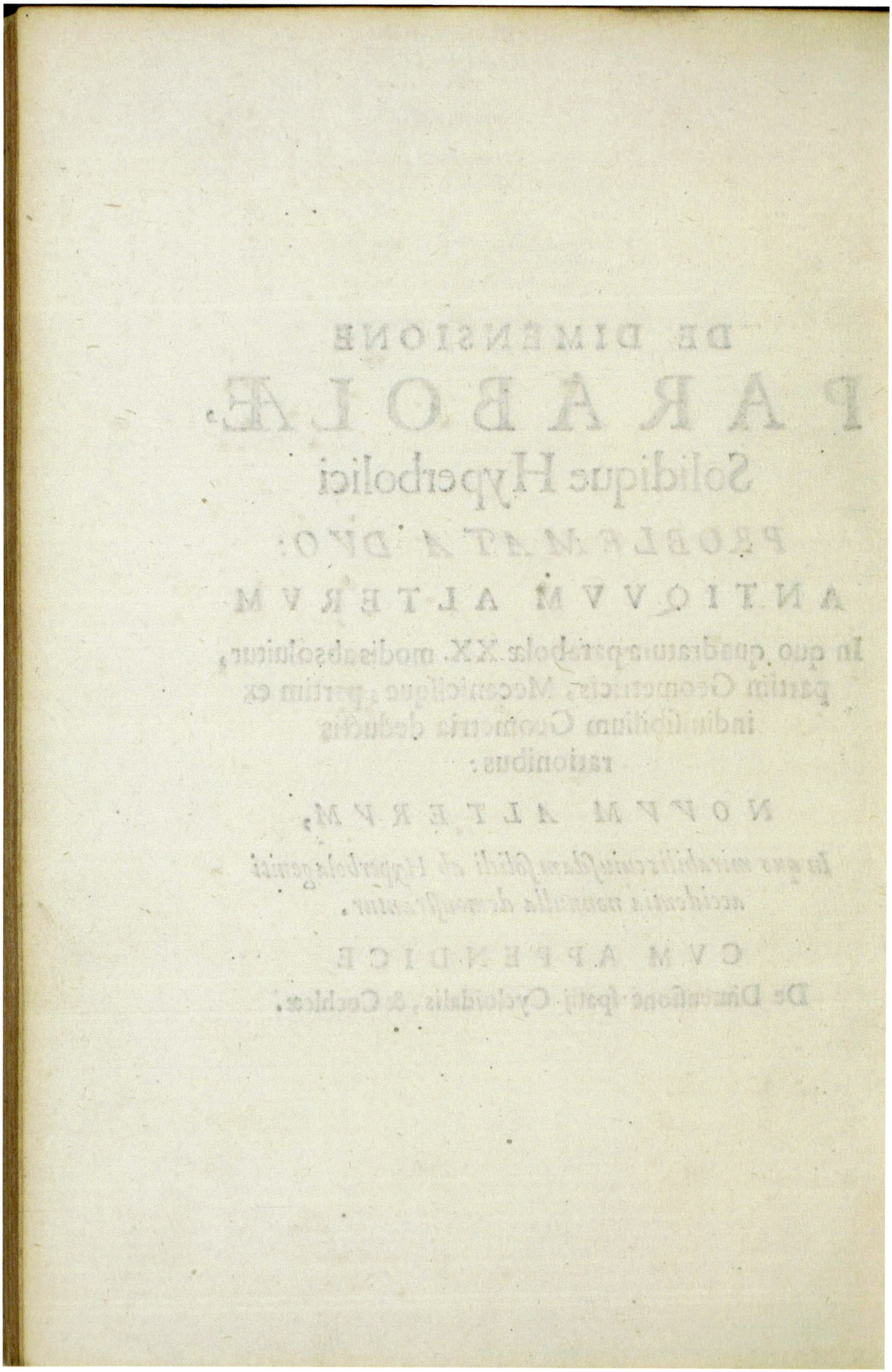

[2] [*De Dimensione Parabolæ*]

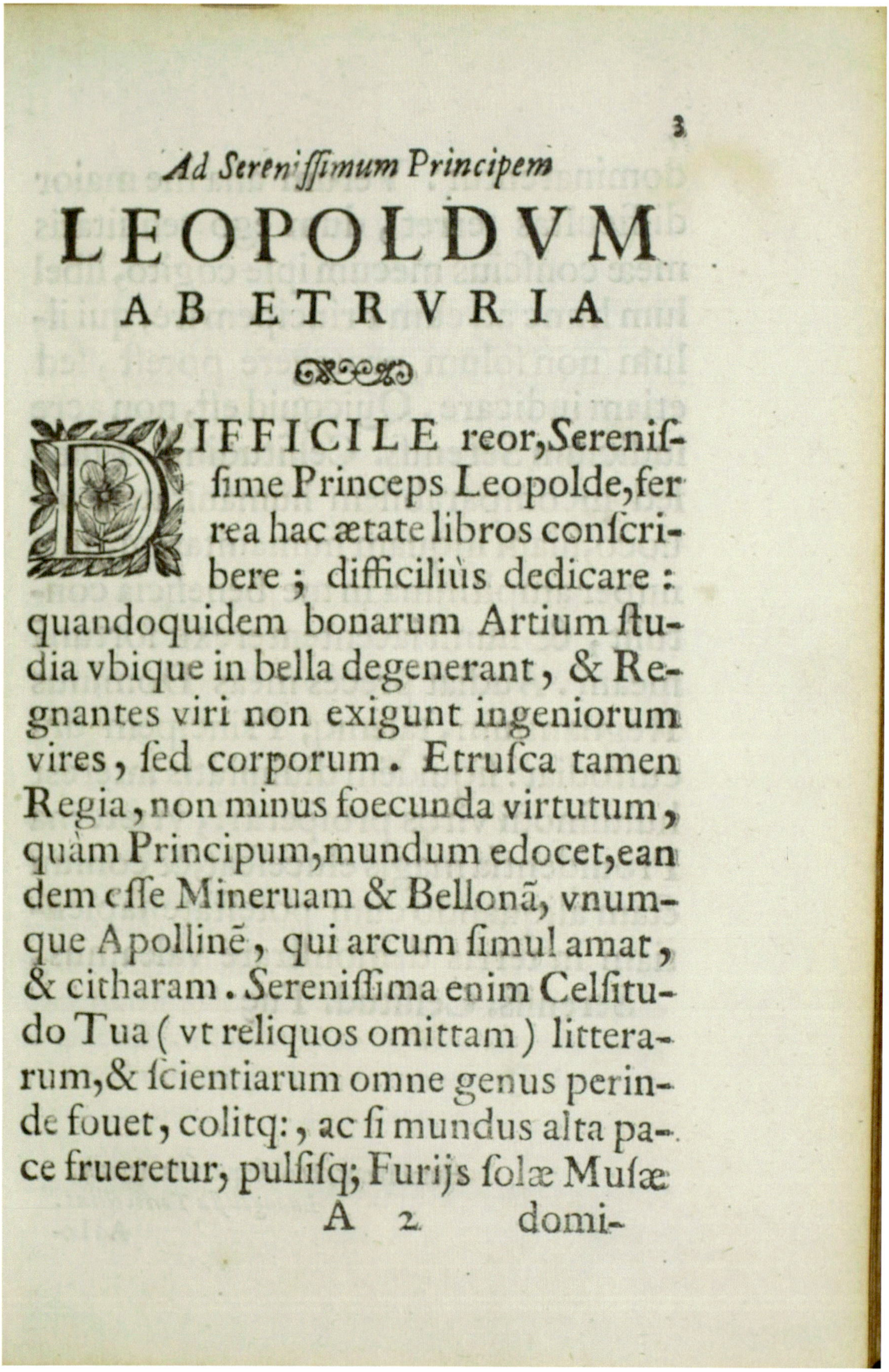
3

Ad Sereniſſimum Principem

LEOPOLDVM
AB ETRVRIA

DIFFICILE reor, Sereniſſime Princeps Leopolde, ferrea hac ætate libros conſcribere; difficiliùs dedicare: quandoquidem bonarum Artium ſtudia vbique in bella degenerant, & Regnantes viri non exigunt ingeniorum vires, ſed corporum. Etruſca tamen Regia, non minus foecunda virtutum, quàm Principum, mundum edocet, eandem eſſe Mineruam & Bellonã, vnumque Apollinẽ, qui arcum ſimul amat, & citharam. Sereniſſima enim Celſitudo Tua (vt reliquos omittam) litterarum, & ſcientiarum omne genus perinde fouet, colitq:, ac ſi mundus alta pace frueretur, pulſiſq; Furijs ſolæ Muſæ

A 2 domi-

[*De Dimensione Parabolæ*] 3

Ad Sereniſſimum Principem

LEOPOLDVM
AB ETRVRIA

DIFFICILE reor, Sereniſſime Princeps Leopolde, ferrea hac ætate libros conſcribere; difficiliùs dedicare: quandoquidem bonarum Artium ſtudia vbique in bella degenerant, & Regnantes viri non exigunt ingeniorum vires, ſed corporum. Etruſca tamen Regia, non minus foecunda virtutum, quàm Principum, mundum edocet, eandem eſſe Mineruam & Bellonã, vnumque Apollinẽ, qui arcum ſimul amat, & citharam. Sereniſſima enim Celſitudo Tua (vt reliquos omittam) litterarum, & ſcientiarum omne genus perinde fouet, colitq:, ac ſi mundus alta pace frueretur, pulſiſq; Furijs ſolæ Muſæ

A 2 domi-

4
dominarentur. Verùm alia me maior difficultas terret, dum ego tenuitatis meæ conſcius mecum ipſe cogito, libellum hunc ad eum Principem ire, qui illum non ſolum protegere poteſt, ſed etiam iudicare. Quicquid eſt, non acre iudicium Sereniſs. Celſitudinis Tuæ, ſed incomparabilem humanitatem inuoco, illam inquam humanitatem, quæ nuper ampliſſima in me beneficia contulit, & humi iacentem erexit fortunã meam. Audiat preces meas Dominus Regnantium, talemq; Principem diu cuſtodiat: ſiquidem diuinitatis interest huiuſmodi viros proſperari, vt æterna Prouidentia magis eluceſcat, & coniunctam aliquando cum poteſtate ſapientiam in terris demonſtrare valeamus.

Sereniſs. Celſitud. Tuę

Humillimus, & obſequentiſs. ſeruus
Euangeliſta Torricellius.

Ad Le-

dominarentur. Verùm alia me maior difficultas terret, dum ego tenuitatis meæ conſcius mecum ipſe cogito, libellum hunc ad eum Principem ire, qui illum non ſolum protegere poteſt, ſed etiam iudicare. Quicquid eſt, non acre iudicium Sereniſs. Celſitudinis Tuæ, ſed incomparabilem humanitatem inuoco, illam inquam humanitatem, quæ nuper ampliſſima in me beneficia contulit, & humi iacentem erexit fortunã meam. Audiat preces meas Dominus Regnantium, talemq; Principem diu cuſtodiat: ſiquidem diuinitatis intereſt huiuſmodi viros proſperari, vt æterna Prouidentia magis eluceſcat, & coniunctam aliquando cum poteſtate ſapientiam in terris demonſtrare valeamus.

Sereniſs. Celſitud. Tuę

Humillimus, & obſequentiſs. ſeruus
Euangeliſta Torricellius.

Ad Le-

AD LECTOREM
Proemium.

NVllus in vniuerſo Mathematicarum diſciplinarum Theatro fortaſſe tritior puluis reperitur, quàm parabolæ quadratura. Quarè ergò (inquis amice Lector) circà tritum argumentum tàm diù deſudaſti? libentér equidem excipio obiectiones tuas; ſed vtinam vltimus deſudauerim. Quam tamen veniam mihi negas, ſcias eandem plurimis, & egregiè laudatis Scriptoribus te denegare. Obiectum enim de parabolæ quadratura, quod noſtra hac ætate confiteor mihi nimis iam inueteraſſe, crediderim neq; nouum fuiſſe Caualerio, Galileo, Lucæ Valerio, & alijs. Quin immò ipſum Archimedem accuſat, quicumque improbat lucubrationes circà ſubiectum vetus inſtitutas. Audiamus ipſum in Proëmio Quadraturæ parabolæ, vbi ſcribens Doſitheo inquit. Eorum enim, qui antehac Geometriæ operam dederunt, nonnulli id inueſtigare, & memoriæ mandare ſtuduerunt, circulo dato, vel circuli portione quacunque, ſpatium rectilineum æquale illi poſſe inueniri. Item ſpatium à coni totius rectanguli ſectione compræhenſum & lineâ rectâ, ad quadrati formam & menſuram reducere conati ſunt; ſumentes non facilé conceſſibilia fundamenta. *Quibus verbis diſertiſſimè fatetur Geometrarũ Princeps, argumentum librorum De dimenſione circuli, & de quadratura parabolæ, neque ſuum fuiſſe, neque nouum. Sed ſi quis attentè conſideret Proëmialem epiſtolam, libro de lineis ſpiralibus præfixam, intel-*

liget

[*De Dimensione Parabolæ*] 5

AD LECTOREM
Proemium.

NVllus *in vniuerſo Mathematicarum diſciplinarum Theatro fortaſſe tritior puluis reperitur, quàm parabolæ quadratura. Quarè ergò (inquis amice Lector) circà tritum argumentum tàm diù deſudaſti? libentér equidem excipio obiectiones tuas; ſed vtinam vltimus deſudauerim. Quam tamen veniam mihi negas, ſcias eandem plurimis, & egregiè laudatis Scriptoribus te denegare. Obiectum enim de parabolæ quadratura, quod nostra hac ætate confiteor mihi nimis iam inueteraſſe, crediderim neq; nouum fuiſſe Caualerio, Galileo, Lucæ Valerio, & alijs. Quin immò ipſum Archimedem accuſat, quicumque improbat lucubrationes circà ſubiectum vetus institutas. Audiamus ipſum in Proëmio Qnadraturæ parabolæ, vbi ſcribens Doſitheo inquit.* Eorum enim, qui ante hac Geometriæ operam dederunt, nonnulli id inueſtigare, & memoriæ mandare ſtuduerunt, circulo dato, vel circuli portione quacunque, ſpatium rectilineum æquale illi poſſe inueniri. Item ſpatium à coni totius rectanguli ſectione compræhenſum & lineâ rectá, ad quadrati formam & menſuram reducere conati ſunt; ſumentes non facilè conceſſibilia fundamenta. *Quibus verbis diſertiſſimè fatetur Geometrarũ Princeps, argumentum librorum De dimenſione circuli, & de quadratura parabolæ, neque ſuum fuiße, neque nouum. Sed ſi quis attentè conſideret Proëmialem epiſtolam, libro de lineis ſpiralibus præfixam, intel-*

liget

6

liget præcipua Archimedis Theoremata, aliorum inuenta fuisse, & magna ex parte Cononis. Maximæ enim Propositiones librorum De Sphæra & Cylindro; De conoidibus, & sphæroidibus, & De lineis spiralibus *(qui libri inter opera Archimedis Principem locum tenent) Cononis sunt:* Qui *(vt inquit Auctor)* non satis temporis ad hæc excogitanda sortitus, vitam permutauit, & ipsa reliquit inexplicata; cum illa inuenisset, & alia quamplurima perquisisset, ac multum adeò Geometricas facultates ampliasset. *Si ergò licuit admirabili, ac propè diuino Auctori, circa aliorum inuenta laborare; quis negabit ignoscendum ingeniolo meo mutuata theoremata contemplanti? Sed esto quòd conclusio antiqua sit; argumenta certè, quibus illa confirmabitur, vt plurimum noua erunt, & inaudita: Immò cum ad alteram partem libelli accedemus, in quà de solido acuto hyperbolico dicendum est, non solum ipsum Theorema inexcogitatum, & vt ita dicam paradoxicum erit, sed etiam demonstrandi ratio inusitata, & penitus noua. Verùm (inquis) reliqui Scriptores, qui huiusmodi quadraturam aggressi sunt, vel singulas, vel ad summum binas prodiderunt; neque tamen mediocrem laudem consequuti sunt. Fateor; sed nec ego libellum hunc ex professo institui, composuique: immò quod & alijs, mihi quoq; accidit; singulas hasce quadraturas diuersis temporibus inueni, quas in vnum collectas nunc demum volentibus simul exhibeo. Tu tamen exclamas; heu nimis est: quotus enim quisq; reperietur tam famelicus Geometra, qui legat penè vicies repetitam propositionem, cum numero lemmatum ferè duplo? Huic sanè obiectioni libet contradicere. Cum enim libellus in Propositiones, vt plurimum*

non

6 [*De Dimensione Parabolæ*]

liget præcipua Archimedis Theoremata, aliorum inuenta fuiſſe, & magna ex parte Cononis. Maximæ enim Propoſitiones librorum De Sphæra & Cylindro; De conoidibus, & ſphæroidibus, & De lineis ſpiralibus *(qui libri inter opera Archimedis Principem locum tenent) Cononis ſunt:* Qui *(vt inquit Auctor)* non ſatis temporis ad hæc excogitanda ſortitus, vitam permutauit, & ipſa reliquit inexplicata; cum illa inueniſſet, & alia quamplurima perquiſiſſet, ac multum adeò Geometricas facultates ampliaſſet. *Si ergò licuit admirabili, ac propè diuino Auctori, circa aliorum inuenta laborare; quis negabit ignoſcendum ingeniolo meo mutuata theoremata contemplanti? Sed eſto quòd concluſio antiqua ſit; argumenta certè, quibus illa confirmabitur, vt plurimum noua erunt, & inaudita: Immò cum ad alteram partem libelli accedemus, in quà de ſolido acuto hyperbolico dicendum eſt, non ſolum ipſum Theorema inexcogitatum, & vt ita dicam paradoxicum erit, ſed etiam demonſtrandi ratio inuſitata, & penitus noua. Verùm (inquis) reliqui Scriptores, qui huiuſmodi quadraturam aggreſſi ſunt, vel ſingulas, vel ad ſummum binas prodiderunt; neque tamen mediocrem laudem conſequuti ſunt. Fateor; ſed nec ego libellum hunc ex profeſſo inſtitui, compoſuique: immò quod & alijs, mihi quoq; accidit; ſingulas haſce quadraturas diuerſis temporibus inueni, quas in vnum collectas nunc demum volentibus ſimul exhibeo. Tu tamen exclamas; heu nimis eſt: quotus enim quiſq, reperietur tam famelicus Geometra, qui legat penè vicies repetitam propoſitionem, cum numero lemmatum ferè duplo? Huic ſanè obiectioni libet contradicere. Cum enim libellus in Propoſitiones, vt plurimum*

non

7

non coherentes digestus sit; sed ita dispositas ut ubicunque libuerit initium facere possis, & finem, dicam cum Martiale

..... tibi carta plicetur

Altera; diuisum sic breue fiet opus,

Si verò mauis probare consilium eorum qui vnam, aut alteram tantum quadraturam edidere; quis prohibet? & in hoc legere potes vnam, aut alteram; si tamen hoc quoq; nimis videbitur, nullam. Vtilitatem exigis? concedo; & in hanc partem libellum excusare non ausim. attamen non deerit fortasse aliquis qui penitus inutilem non existimet, cum Geometricus, sit. Sola enim Geometria inter liberales disciplinas acritér exacuit ingenium, idoneumque reddit ad ciuitates exornandas in pace, & in bello defendendas: cæteris enim paribus, ingenium quod exercitatum sit in Geometrica palestra, peculiare quoddam, & virile robur habere solet: præstabitq; semper, & antecellet, circà studia Architecturæ, rei bellicæ, nauticæq; siue mauis circà Aritmeticam, artemq; metiendi, vnde totum ciuile commercium dependet, regiturque. Quinetiam circà ministeria fluuiorum, & aquarum stagnantium, vnde non nisi magna percipiuntur siue damna, siue beneficia; pro vt bene, vel male intellecta fuerit huiusmodi rerum natura. Sed esto quòd inutilis penitus habeatur libellus; siue quia Reipub. nihil interest parabolæ quadraturæ; siue quia multis ab hinc sæculis excogitata fuerat, & demonstrata. Huic verò obiectioni respondeat Reuerendiss. D. Benedictus Castellius Magister meus. Ipse enim dicet quòd si Principes terræ, solam illam vulgarem, & apparentem vtilitatem in Artibus magni facerent, exigerentque, damnandi penitus essent Sculptores, Celatoresque egregij; eijcien-

di ci-

non coherentes digeftus ſit; ſed ita diſpoſitas vt vbicunque libuerit initium facere poſſis, & finem, dicam cum Martiale

..... tibi carta plicetur

Altera; diuiſum ſic breue fiet opus,

Si verò mauis probare conſilium eorum qui vnam, aut alteram tantum quadraturam edidere; quis prohibet? & in hoc legere potes vnam, aut alteram; ſi tamen hoc quoq; nimis videbitur, nullam. Vtilitatem exigis? concedo; & in hanc partem libellum excuſare non auſim. attamen non deerit fortaſſe aliquis qui penitus inutilem non exiſtimet, cum Geometricus, ſit. Sola enim Geometria inter liberales diſciplinas acritér exacuit ingenium, idoneumque reddit ad ciuitates exornandas in pace, & in bello defendendas: cæteris enim paribus, ingenium quod exercitatum ſit in Geometrica palestra, peculiare quoddam, & virile robur habere ſolet: præstabitq; ſemper, & antecellet, circà ſtudia Architecturæ, rei bellicæ, nauticæq; ſiue mauis circà Aritmeticam, artemq; metiendi, vnde totum ciuile commercium dependet, regiturque. Quinetiam circà miniſteria fluuiorum, & aquarum ſtagnantium, vnde non niſi magna percipiuntur ſiue damna, ſiue beneficia; pro vt bene, vel male intellecta fuerit huiuſmodi rerum natura. Sed eſto quòd inutilis penitus habeatur libellus; ſiue quia Reipub. nihil intereſt parabolæ quadratura; ſiue quia multis ab hinc ſæculis excogitata fuerat, & demonſtrata. Huic verò obiectioni reſpondeat Reuerendiſs. D. Benedictus Caſtellius Magister meus. Ipſe enim dicet quòd ſi Principes terræ, ſolam illam vulgarem, & apparentem vtilitatem in Artibus magni facerent, exigerentque, damnandi penitus eſſent Sculptores, Celatoreſque egregij; eijcien-

dici-

8

di ciuitatibus Pictores, Musici, Citharædi, Poëtæ, atque id genus alij. Contrà verò ditandi essent, atq; opibus, officijsq; omnibus demerendi pistores, quorum vtilitati nulla alia par est; caupones, sutores, & quicunque artem colunt vitæ hominum summoperè vtilem. Quinetiam si vtilitas sola attendatur, damnandus erit vini vsus, & detestanda cultura vinearum. At in summo prætio habenda aqua, cuius vtilitates tàm facilè est numerare, quàm difficilè sit ijs non indigere.

Vtcumque ea res sese habeat, veniamus ad obiectiones quæ circà artis fundamenta versantur. Indignor equidem Lucam Valerium, verè nostri sæculi Archimedem, cum optimam causam suscepisset, pessimá defensione vsum fuisse. Solent ab eruditis culpari figurarum Geometricarum dimensiones, quæ Mecanicis fundamentis innixæ stabiliuntur, tamquàm duplex falsum supponant: alterum, quòd superficies grauitatem non habentes, habere tamen concipiuntur: *alterum verò,* quòd fila quæ magnitudines ad libram suspendunt æquidistantia supponuntur, cum tamen in centro terræ concurrere debeant. *Ego verò in ea sum sententia, vel nullam ex his suppositionib. esse falsam, vel reliqua omnia principia Geometriæ falsa existere eodem modo. Falsum enim est, quòd circulus habeat centrum, sphæra superficiem, conus soliditatem. Loquor de figuris abstractis quales Geometria considerare solet; non autem de fisicis, & concretis. Necesse igitur erit fateri quòd circuli centrum, superficies sphæræ, soliditas coni, & reliqua huiusmodi non controuersa, nullam aliam habeant existentiam, præter illam quam accipiunt per definitionem, & per intellectum.*

di ciuitatibus Pictores, Muſici, Citharœdi, Poëtæ, atque id genus alij. Contrà verò ditandi eſſent, atq; opibus, officijſq; omnibus demerendi pistores, quorum vtilitati nulla alia pareſt; caupones, ſutores, & quicunque artem colunt vitæ hominum ſummoperè vtilem. Quinetiam ſi vtilitas ſola attendatur, damnandus erit vini vſus, & detestanda cultura vinearum. At in ſummo prætio habenda aqua, cuius vtilitates tàm facilè eſt numerare, quàm difficilè ſit ijs non indigere.

Vtcumque ea res ſeſe habeat, veniamus ad obiectiones quæ circà artis fundamenta verſantur. Indignor equidem Lucam Valerium, verè noſtri ſæculi Archimedem, cum optimam cauſam ſuſcepiſſet, peſſimá defenſione vſum fuiſſe. Solent ab eruditis culpari figurarum Geometricarum dimenſiones, quæ Mecanicis fundamentis innixæ stabiliuntur, tam quàm duplex falſum ſupponant: alterum, quòd ſuperficies grauitatem non habentes, habere tamen concipiuntur: *alterum verò,* quòd fila quæ magnitudines ad libram ſuſpendunt æquidiſtantia ſupponuntur, cum tamen in centro terræ concurrere debeant. *Ego verò in ea ſum ſententia, vel nullam ex his ſuppoſitionib. eſſe falſam, vel reliqua omnia principia Geometriæ falſa existere eodem modo. Falſum enim est, quòd circulus habeat centrum, ſphæra ſuperficiem, conus ſoliditatem. Loquor de figuris abſtractis quales Geometria conſiderare ſolet; non autem de fiſicis, & concretis. Neceſſe igitur erit fateri quòd circuli centrum, ſuperficies sphæræ, ſoliditas coni, & reliqua huiuſmodi non controuerſa, nullam aliam habeant existentiam, præter illam quam accipiunt per definitionem, & per intelle-*

ctum

ctum. *Eodem prorſus modo grauitas eſt in figuris Geometricis, quomodo in iiſdem eſt centrum, perimeter, ſuperficies, ſoliditas &c. Laudarem igitur in Mecanicis contemplationibus noua definitione figuras generare; hoc, aut alio non abſimili modo.*

Quadratum eſt quadrilaterum, quod, cum æquilaterum, & æquiangulum ſit, ſingula ipſius puncta momentum habent procedendi verſus aliquam mundi plagam per lineas inter ſe parallelas.

Huiuſmodi enim definitio omnem demeret occaſionem dubitandi, illis, qui Mecanica Archimedis opera, ſecundum ipſius mentem non accipiunt. Sed hucuſque dictum ſit pro obliteranda primæ falſitatis nota, quòd figuræ Geometricæ graues ſint.

Venio nunc ad ſecundum (vt aliqui exiſtimant) falſum. Principiò, vulgatiſſima eſt etiam apud grauiſſimos viros obiectio illa, videlicet. Archimedem ſuppoſuiſſe aliquod falſum, dum fila magnitudinum ex librá pendentium conſiderauit tanquam inter ſe parallela, cum tamen re vera in ipſo terræ centro concurrere debeant. *Ego verò (quod pace clariſſimorum virorum dictum ſit) crediderim fundamentum Mecanicum longè alia ratione eſſe conſiderandum. Concedo ſi Fiſicæ magnitudines ad libram liberè ſuſpendantur, quòd fila materialia ſuſpenſionum conuergentia erunt; quandoquidem ſingula ad centrum terræ reſpiciunt. Verumtamen ſi eadem libra, licet corporea, conſideretur non in ſuperficie terræ, ſed in altiſſimis regionibus vltrà orbem Solis; tum fila (dummodo adhuc ad terræ centrum reſpiciant) multò minùs conuergentia inter ſe erunt; ſed quaſi æqui-*

 diſtan-

ctum. Eodem prorſus modo grauitas eſt in figuris Geometricis, quomodo in ijſdem eſt centrum, perimeter, ſuperficies, ſoliditas &c. Laudarem igitur in Mecanicis contemplationibus noua definitione figuras generare; hoc, aut alio non ab ſimili modo.

Quadratum eſt quadrilaterum, quod, cum æquilaterum, & æquiangulum ſit, ſingula ipſius puncta momentum habent procedendi verſus aliquam mundi plagam per lineas inter ſe parallelas.

Huiuſmodi enim definitio omnem demeret occaſionem dubitandi, illis, qui Mecanica Archimedis opera, ſecundum ipſius mentem non accipiunt. Sed hucuſque dictum ſit pro obliteranda primæ falſitatis nota, quòd figuræ Geometricæ graues ſint.

Venio nunc ad ſecundum (vt aliqui exiſtimant) falſum. Principiò, vulgatiſſima eſt etiam apud grauiſſimos viros obiectio illa, videlicet. Archimedem ſuppoſuiſſe aliquod falſum, dum fila magnitudinum ex librá pendentium conſiderauit tanquam inter ſe parallela, cum tamen revera in ipſo terræ centro concurrere debeant. *Ego verò (quod pace clariſſimorum virorum dictum ſit) crediderim fundamentum Mecanicum longè alia ratione eſſe conſiderandum. Concedo ſi Fiſicæ magnitudines ad libram liberè ſuſpendantur, quòd fila materialia ſuſpenſionum conuergentia erunt; quandoquidem ſingula ad centrum terræ reſpiciunt. Verumtamen ſi eadem libra, licet corporea, conſideretur non in ſuperficie terræ, ſed in altiſſimís regionibus vltrâ orbem Solis; tum fila (dummodo adhuc ad terræ centrum reſpiciant) multò minùs conuergentia inter ſe erunt; ſed quaſi æqui-*

B *distan-*

10

distantia. Concipiamus iam ipsam libram Mecanicam vltrà stellatam libram firmamenti in infinitam distantiam esse prouectam; quis non intelligit fila suspensionum iam non ampliùs conuergentia, sed exactè parallela fore? Quando ego considero libram, figuras Geometricas ponderantem, non cõcipio illam esse inter cartas librorum in quibus depicta conspicitur; neque suppono punctum, ad quod magnitudines ipsius tendunt, esse centrum terræ; sed libram fingo in infinitũ remotam esse ab eo puncto, ad quod ipsius grauia contendunt. Si posteà ibi conclusero triangulum aliquod triplum esse cuiusdam spatij; retrahatur imaginatione ipsa libra ad nostras regiones; concedo quòd retractâ librâ destruetur æquidistantia filorum suspensionis, sed non ideò destruetur proportio iam demonstrata figurarum. Peculiare quoddam beneficium habet Geometra, cum ipse abstractionis ope, omnes operationes suas mediante intellectu exequatur. Quis igitur mihi hoc negauerit, si libeat considerare figuras appensas ad libram, quæ quidem libra vltra mundi confinium in infinitam distantiã remota supponatur? Vel quis proibebit considerare libram in superficie terræ constitutam, cuius tamen abstracta magnitudines tendant, non ad medium terræ punctum, sed ad centrum caniculæ, siue stellæ polaris? Triangula & parabolæ, immò etiam sphęræ, cylindriq; Geometrici, cum nullam per se habeant motus differentiam, non magis ad ipsius terræ, quàm ad Saturni centrum contendunt. Destruit ergò beneficium suum quisquis figuras illas, tamquam ad vnicum terræ centrum tendentes, contemplatur. Cur denique non licebit mihi considerare puncta cuiuscunq; figuræ eiusmodi virtute prædita, vt singula versus eandem mundi plagam per lineas in-

ter

diſtantia. Concipiamus iam ipſam libram Mecanicam vltrà ſtellatam libram firmamenti in infinitam diſtantiam eſſe prouectam; quis non intelligit fila ſuſpenſionum iam non ampliùs conuergentia, ſed exactè parallela fore? Quando ego conſidero libram, figuras Geometricas ponderantem, non cõcipio illam eſſe inter cartas librorum in quibus depicta conſpicitur; neque ſuppono punctum, ad quod magnitudines ipſius tendunt, eſſe centrum terræ; ſed libram fingo in infinitũ remotam eſſe ab eo puncto, ad quod ipſius grauia contendunt. Si poſteà ibi concluſero triangulum aliquod triplum eſſe cuiuſdam ſpatij; retrahatur imaginatione ipſa libra ad noſtras regiones; concedo quòd retractâ librâ deſtruetur æquidiſtantia filorum ſuſpenſionis, ſed non ideò deſtruetur proportio iam demonſtrata figurarum. Peculiare quoddam beneficium habet Geometra, cum ipſe abſtractionis ope, omnes operationes ſuas mediante intellectu exequatur. Quis igitur mihi hoc negauerit, ſi libeat conſiderare figuras appenſas ad libram, quæ quidem libra vltra mundi confinium in infinitam diſtantiã remota ſupponatur? Vel quis proibebit conſiderare libram in ſuperficie terræ conſtitutam, cuius tamen abſtracta magnitudines tendant, non ad medium terræ punctum, ſed ad centrum caniculæ, ſiue ſtellæ polaris? Triangula & parabolæ, immò etiam ſphęræ, cylindriq; Geometrici, cum nullam per ſe habeant motus differentiam, non magis ad ipſius terræ, quàm ad Saturni centrum contendunt. Deſtruit ergò beneficium ſuum quiſquis figuras illas, tamquam ad vnicum terræ centrum tendentes, contemplatur. Cur denique non licebit mihi conſiderare puncta cuiuſcunq; figuræ eiuſmodi virtute prædita, vt ſingula verſus eandem mundi plagam per lineas in-

ter

ter se parallelas æquali momento contendant? His ita suppositis, quæ vera sunt, quemadmodum sunt veræ passiones figurarum, quæ in definitionibus adhibentur, vera etiam erunt quæcunq; Theoremata per Mecanicas rationes ab ipsis abstrahentibus fuerint considerata, neque per falsas positiones demonstrabuntur. Tunc itaque falsum dici poterit fundamentum Mecanicum, nempe fila libræ parallela esse, quando magnitudines ad libram appensæ fisicæ sint, realesque, & ad terræ centrum conspirantes. Non autem falsum erit, quando magnitudines (siue abstractæ, siue concretæ sint) non ad centrum terræ, neque ad aliud punctum propinquum libræ respiciant; sed ad aliquod punctum infinitè distans connitantur.

Cæterum, breuitatis, & facilitatis gratià à vocabulis consuetis non discedemus; punctumque illud ad quod magnitudines libræ contendere supponuntur, Centrum terræ nominabimus; Planum verò illud, quod erectum est ad lineam cõnectentem prædictum punctum cum centro libræ, Horizontem de more appellabimus.

Suppositiones, & definitiones.

I.

Ponatur eam esse centri grauitatis naturam, vt magnitudo liberè suspensa ex quolibet sui puncto nunquam quiescat nisi cum centrum grauitatis ad infimum suæ sphærę punctum peruenerit.

Concipiamus figuram ABC, suspensam ex sui puncto D, median-

ter ſe parallelas æquali momento contendant? His ita ſuppoſitis, quæ vera ſunt, quemadmodum ſunt veræ paſſiones figurarum, quæ in definitionibus adhibentur, vera etiam erunt quæcunq; Theoremata per Mecanicas rationes ab ipſis abstrahentibus fuerint conſiderata, neque per falſas poſitiones demonſtrabuntur. Tunc itaque falſum dici poterit fundamentum Mecanicum, nempe fila libræ parallela eſſe, quando magnitudines ad libram appenſæ ſiſicæ ſint, realeſque, & ad terræ centrum conſpirantes. Non autem falſum erit, quando magnitudines (ſiue abſtractæ, ſiue concretæ ſint) non ad centrum terræ, neque ad aliud punctum propinquum libræ reſpiciant; ſed ad aliquod punctum infinitè distans connitantur.

Cæterum, breuitatis, & facilitatis gratià à vocabulis conſuetis non diſcedemus; punctumque illud ad quod magnitudines libræ contendere ſupponuntur, Centrum terræ nominabimus; Planum verò illud, quod erectum eſt ad lineam cõnectentem prædictum punctum cum centro libræ, Horizontem de more appellabimus.

Suppoſitiones, & definitiones.

I.

POnatur eam eſſe centri grauitatis naturam, vt magnitudo liberè ſuſpenſa ex quolibet ſui puncto nunquam quieſcat niſi cum centrum grauitatis ad infimum ſuæ ſphærę punctum peruenerit.

Concipiamus figuram ABC, ſuſpenſam ex ſui puncto D,

B 2 *median-*

12

mediante filo ED; liberè; hoc est, ita ut in quamcumque partem conuerti possit. Sit centrum grauitatis F. ponamusque rectam EDG. perpendicularem esse ad horizontem.

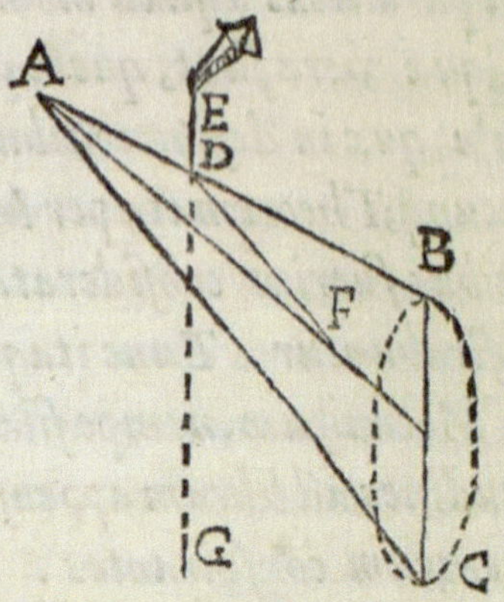

Certum est, donec centrum F fuerit extrà perpendiculum EG, figuram ipsam numquam mansuram esse. Quando verò punctum F. fuerit in perpendiculo suspensionis EG, tunc figura omninò quiescet: Centrum enim grauitatis ipsius nusquàm poterit ampliùs inferiùs descendere: Quin immò si figura moueretur, centrum ipsum ascenderet, quod esse non potest. Si quis enim centro E, interuallo EDF. tamquam unà recta linea, sphæram concipiat esse descriptam; ipsa erit sphæra, in cuius superficie feretur punctum F, quando EDF. extensa fuerit, & ad rectitudinem redacta. certumque est infimũ punctum huiusmodi sphæræ esse in perpendiculo EG.

II.

Æquiponderare sibi ipsi figura dicetur, quæ ab aliquo sui puncto liberè suspensa maneat, & ad nullam sui partem inclinetur.

III.

Æquiponderat sibi ipsi figura, quando (cum liberè suspensa sit) in ipso suspensionis perpendiculo centrum grauitatis reperitur. si enim adhuc moueretur, centrum grauitatis ascenderet. Quod est impossibile.

Cen-

mediante filo ED; liberè; hoc eſt, ita vt in quamcumque partem conuerti poſſit. Sit centrum grauitatis F. ponamuſque rectam EDG. perpendicularem eſſe ad horizontem.

Certum eſt, donec centrum F fuerit extrà perpendiculum EG, figuram ipſam numquam manſuram eſſe. Quando verò punctum F. fuerit in perpendiculo ſuſpenſionis EG, tunc figura omninò quieſcet: Centrum enim grauitatis ipſius nuſquàm poterit ampliùs inferiùs deſcendere: Quin immò ſi figura moueretur, centrum ipſum aſcenderet, quod eſſe non poteſt. Si quis enim centro E, interuallo EDF. tamquam vnà recta linea, ſphæram concipiat eſſe deſcriptam; ipſa erit ſphæra, in cuius ſuperficie feretur punctum F, quando EDF. extenſa fuerit, & ad rectitudinem redacta. certumque eſt infimū punctum huiuſmodi ſphæræ eſſe in perpendiculo EG.

II.

Æquiponderare ſibi ipſi figura dicetur, quæ ab aliquo ſui puncto liberè ſuſpenſa maneat, & ad nullam ſui partem inclinetur.

III.

Æquiponderat ſibi ipſi figura, quando (cum liberè ſuſpenſa ſit) in ipſo ſuſpenſionis perpendiculo centrum grauitatis reperitur. ſi enim adhuc moueretur, centrum grauitatis aſcenderet. Quod eſt impoſſibile.

Cen-

IV.

Centrum grauitatis tunc reperitur in ipso suspensionis perpendiculo, quando figura liberè suspensa sibi ipsi æquiponderat. Alias enim figura quiesceret, & centrum grauitatis ipsius posset adhuc inferiùs descendere. Quod est absurdum.

V.

Centralitér ad illud libræ punctum appendi figura dicetur, in quod cadit perpendiculum, ex centro grauitatis figuræ productum.

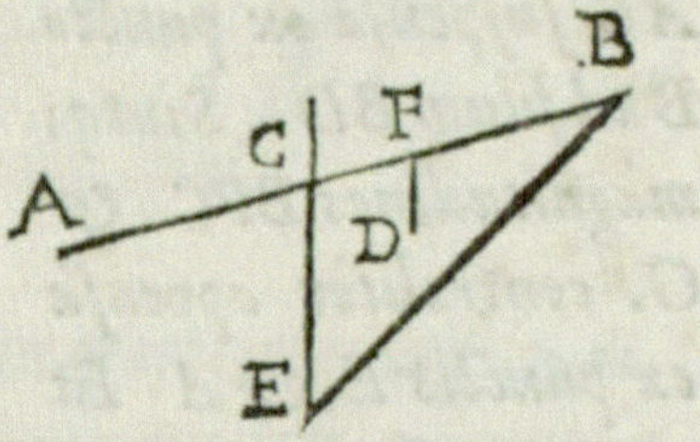

Esto enim libra AB, cuius fulcrum sit C, & ad ipsam appensa sit figura CEB, ita vt totum latus CB cohereat, & sit veluti ad ipsam libram conglutinatum. Esto centrum grauitatis figuræ punctum D, & ex D agatur perpendiculum DF ad horizontem erectum.

Iam figura CEB dicetur, & considerabitur, tamquam appensa centralitér ad punctum F. Constat enim ex prædictis, quòd si figuræ latus CB soluatur vndiq; à brachio libræ, solumque remaneat filum connexionis DF, nihilo tamen minus figura adhuc manebit vt prius manebat, eandemq; seruabit versus libram positionem, quam antea habebat. Vide Arch. Prop. 6. De Quadratura parabolæ.

VI.

Æqualia grauia ex æqualibus distantijs æquiponderant, siue libra ad horizontem parallela fuerit, siue inclinata.

Et gra-

[*De Dimensione Parabolæ*] 13

IV.

Centrum grauitatis tunc reperitur in ipſo ſuſpenſionis perpendiculo, quando figura liberè ſuſpenſa ſibi ipſi æquiponderat. Alias enim figura quieſceret, & centrum grauitatis ipſius poſſet adhuc inferiús deſcendere. Quod eſt abſurdum.

V.

Centralitér ad illud libræ punctum appendi figura dicetur, in quod cadit perpendiculum, ex centro grauitatis figuræ productum.

Eſto enim libra AB, cuius fulcrum ſit C, & ad ipſam appenſa ſit figura CEB, ita vt totum latus CB cohereat, & ſit veluti ad ipſam libram conglutinatum, Eſto centrum grauitatis figuræ punctum D & ex D agatur perpendiculum DF ad horizontem erectum.

Iam figura CEB dicetur, & conſiderabitur, tamquam appenſa centralitèr ad punctum F. Constat enim ex prædictis, quòd ſi figuræ latus CB ſoluatur vndiq; à brachio libræ, ſolumque remaneat filum connexionis DF, nihilo tamen minus figura adhuc manebit vt prius manebat, eandemq; ſeruabit verſus libram poſitionem, quam antea habebat. Vide Arch. Prop. 6. De Quadratura parabolæ.

VI.

Æqualia grauia ex æqualibus diſtantijs æquiponderant, ſiue libra ad horizontem parallela fuerit, ſiue inclinata.

Et gra-

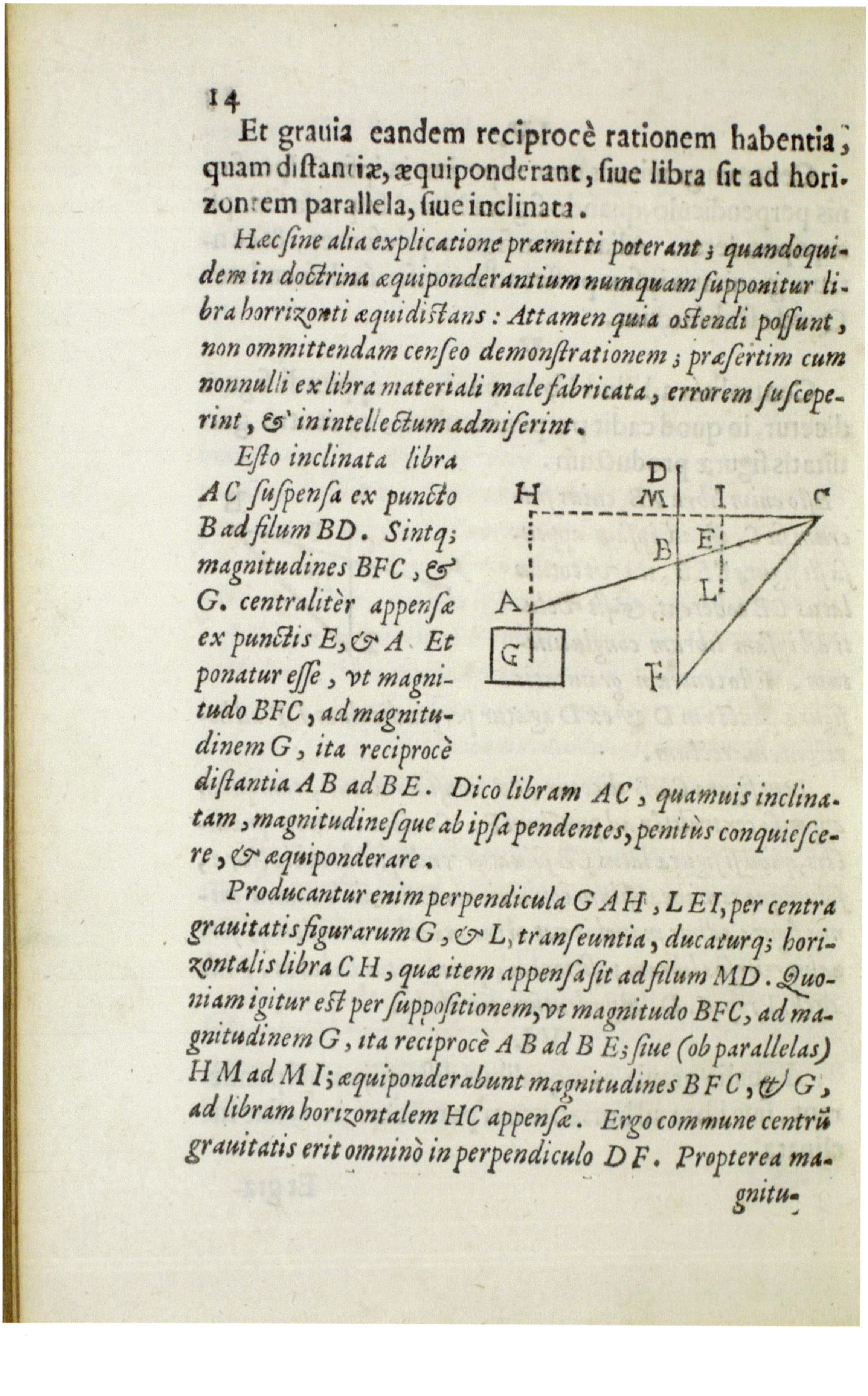

14

Et grauia eandem reciprocè rationem habentia, quam distantiæ, æquiponderant, siue libra sit ad horizontem parallela, siue inclinata.

Hæc sine alia explicatione præmitti poterant; quandoquidem in doctrina æquiponderantium numquam supponitur libra horrizonti æquidistans: Attamen quia ostendi possunt, non ommittendam censeo demonstrationem; præsertim cum nonnulli ex libra materiali male fabricata, errorem susceperint, & in intellectum admiserint.

Esto inclinata libra AC suspensa ex puncto B ad filum BD. Sintq; magnitudines BFC, & G. centralitèr appensæ ex punctis E, & A. Et ponatur esse, vt magnitudo BFC, ad magnitudinem G, ita reciprocè distantia AB ad BE. Dico libram AC, quamuis inclinatam, magnitudinesque ab ipsa pendentes, penitùs conquiescere, & æquiponderare.

Producantur enim perpendicula GAH, LEI, per centra grauitatis figurarum G, & L, transeuntia, ducaturq; horizontalis libra CH, quæ item appensa sit ad filum MD. Quoniam igitur est per suppositionem, vt magnitudo BFC, ad magnitudinem G, ita reciprocè AB ad BE; siue (ob parallelas) HM ad MI; æquiponderabunt magnitudines BFC, & G, ad libram horizontalem HC appensæ. Ergo commune centrū grauitatis erit omninò in perpendiculo DF. Propterea magnitu-

14 [*De Dimensione Parabolæ*]

Et grauia eandem reciprocè rationem habentia, quam diftantiæ, æquiponderant, fiue libra fit ad horizontem parallela, fiue inclinata.

Hæc fine alia explicatione præmitti poterant; quandoquidem in doctrina æquiponderantium numquam fupponitur libra horrizonti[119] *æquidistans: Attamen quia ostendi poffunt, non ommittendam*[120] *cenfeo demonftrationem; præfertim cum nonnulli ex libra materiali male fabricata, errorem fufceperint, & in intellectum admiferint.*

Efto inclinata libra AC fufpenfa ex puncto B ad filum BD. Sintq; magnitudines BFC, & G. centralitèr appenfæ ex punctis E, & A. Et ponatur effe, vt magnitudo BFC, ad magnitudinem G, ita reciprocè diftantia AB ad BE. Dico libram AC, quamuis inclinatam, magnitudinefque ab ipfa pendentes, penitùs conquiefcere, & æquiponderare.

Producantur enim perpendicula GAH, LEI, per centra grauitatis figurarum G, & L, tranfeuntia, ducaturq; horizontalis libra CH, quæ item appenfa fit ad filum MD. Quoniam igitur est per fuppofitionem, vt magnitudo BFC, ad magnitudinem G, ita reciprocè AB ad BE; fiue (ob parallelas) HM ad MI; æquiponderabunt magnitudines BFC, & G, ad libram horizontalem HC appenfæ. Ergo commune centrũ.

grauitatis erit omninò in perpendiculo DF. Propterea ma-

gnitu-

[119] *horrizonti*] *horizonti.*
[120] *ommittendam*] *omittendam.*

15

gnitudines æquiponderabunt etiam dum ad libram AC suspenduntur: aliàs, si mouerentur, commune centrum grauitatis ipsarum, quod demonstratum est esse in perpendiculo DF, ascenderet. Quod est impossibile.

Hæc autem breuiùs concludentur hoc modo. Connectantur (in eadẽ figura) centra grauitatis ductâ rectâ GL. Quoniam magnitudo B F C ad magnitudinem G, est vt AB ad BE, siue (ob parallelas) vt GN ad NL, erit N centrum commune grauitatis magnitudinum appensarum. Si ergo libra AC non quiesceret, centrum grauitatis N, ascenderet. Cum enim sit in perpendiculo DF, moueri non potest quin ascendat.

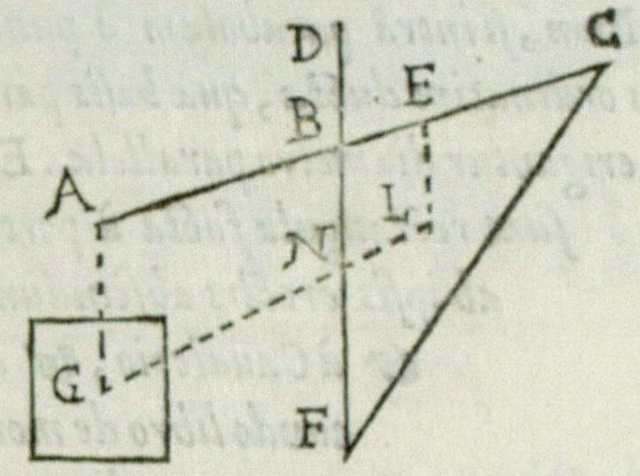

Non me latet Auctorum controuersiam, circà libram inclinatam, an redeat, maneatuè, supponere centra magnitudinum in ipsa libra esse collocata. Nos tamen, quia in hoc libello, semper considerabimus magnitudines infrà ipsam libram appensas, maluimus rei nostræ seruire, quàm aliorum controuersiæ demonstrationem accomodare.

Cæterum passiones parabolæ quas in operis progressu supponemus tamquam notas, vel ipsius Apollonij erunt, vel Archimedis, vel saltem ex Apollonio ipso facili negotio deducentur, cuiusmodi sunt hæ, quæ sequuntur.

Si Parabola rectam lineam tangentem habuerit, à quibuslibet autem punctis ipsius tangentis rectæ lineæ vsque ad para-

rabo-

gnitudines æquiponderabunt etiam dum ad libram AC ſuſpenduntur: aliàs, ſi mouerentur, commune centrum grauitatis ipſarum, quod demonstratum est eſſe in perpendiculo DF, aſcenderet. Quod eſt impoſſibile.

Hæc autem breviùs concludentur hoc modo. Connectantur (in eadē figura) centra grauitatis ductâ rectâ GL. Quoniam magnitudo BFC ad magnitudinem G, eſt vt AB ad BE, ſiue (ob parallelas) vt GN ad NL, erit N centrum commune grauitatis magnitudinum appenſarum. Si ergo libra AC non quieſceret, centrum grauitatis N, aſcenderet. Cum enim ſit in perpendiculo DF, moueri non poteſt quin aſcendat.

Non me latet Auctorum controuerſiam, circá libram inclinatam, an redeat, maneatuè, ſupponere centra magnitudinum in ipſa libra eſſe collocata. Nos tamen, quia in hoc libello, ſemper conſiderabimus magnitudines infrà ipſam libram appenſas, maluimus rei nostræ ſeruire, quàm aliorum controuerſiæ demonstrationem accomodare.

Cæterum paſſiones parabolæ quas in operis progreſſu ſuppenemus tamquam notas, vel ipſius Apollonij erunt, vel Archimedis, vel ſaltem ex Apollonio ipſo facili negotio deducentur, cuiuſmodi ſunt hæ, quæ ſequuntur.

Si Parabola rectam lineam tangentem habuerit, à quibuſlibet autem punctis ipſius tangentis rectæ lineæ vſque ad pa-

rabo-

16

rabolam demittantur æquidistantes diametro, erunt demissæ inter se longitudine vt sunt portiones tangentis potentià inter se. Deducitur enim hoc ex 20. prim. Conic. Nàm rectæ illæ demissæ portionibus diametri respondent; at partes ipsius tangentis, ordinatim applicatis æquales sunt.

Item, si intrà parabolam à punctis quibuslibet rectæ illius ordinatim ductæ, quæ basis parabolæ dicitur, rectæ lineæ erigantur diametro parallelæ. Erunt erectæ inter se vt sunt rectangula facta à portionibus basis, quæ ab ipsis erectis abscinduntur. Hoc enim & à Caualerio, & à nobis in secundo libro de motu ostenditur.* & 29 prop. lib. 4. Conic. Mydorgij

16 [*De Dimensione Parabolæ*]

rabolam demittantur æquidistantes diametro, erunt demiſſæ inter ſe longitudine vt ſunt portiones tangentis potentià inter ſe. Deducitur enim hoc ex 20. *prim. Conic. Nàm rectæ illæ demiſſæ portionibus diametri reſpondent; at partes ipſius tangentis, ordinatim applicatis æquales ſunt.*

Item, ſi intrà parabolam à punctis quibuslibet rectæ illius ordinatim ductæ, quæ baſis parabolæ dicitur, rectæ lineæ erigantur diametro parallelæ. Erunt erectæ inter ſe vt ſunt rectangula facta à portionibus baſis, quæ ab ipſis erectis abſcinduntur. Hoc enim & à Caualerio, & à nobis in

ſecundo libro de motu
ostenditur.

[Cf. 29 prop
lib.A.conic.
Mydorgij]

QVADRATVRA PARABOLAE.

Pluribus modis per duplicem positionem, more antiquorum, absoluta.

Lemma Primum.

SI PARABOLA duas tangentes habuerit, alteram ex termino basis, alteram verò ex vertice: tangens, quę ad basim est, bifariàm secabitur ab illa, quæ per verticem ducitur.

Esto parabola a b c, *cuius diameter* b i, *ordinatìm verò applicata (siue basis) sit* a c; *tangens ex termino basis sit* c d; *per verticem verò tangens* b e. *Dico ipsam* c d *bifariàm secari in puncto* e.

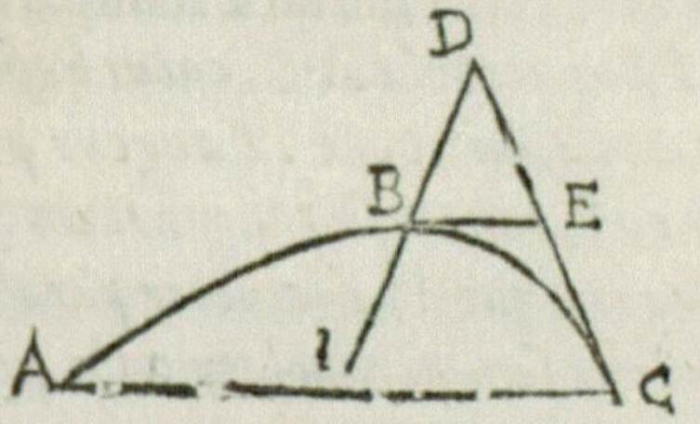

Cum.n. c d *sit tangens,* d i *diameter, erunt æquales inter se* d b. b i. *Et quia* a c *ordinatìm applicata* 35 primi Conic.

C

QVADRATVRA PARABOLAE

Pluribus modis per duplicem poſitionem, more antiquorum, abſoluta.

Lemma Primum.

SI PARABOLA duas tangentes habuerit, alteram ex termino baſis, alteram verò ex vertice: tangens, quę ad baſim eſt, bifariàm ſecabitur ab illa, quæ per verticem ducitur. 35 *primi Conic.*

Eſto parabola abc, *cuius diameter* bi, *ordinatìm verò applicata (ſiue baſis) ſit* ac; *tangens ex termino baſis ſit* cd; *per verticem verò tangens* be. *Dico ipſam* cd *bifariàm ſecari in puncto* e.

Cum. n. cd *ſit tangens,* di *diameter, erunt æquales inter ſe* db. bi. *Et quia* ac *ordinatìm appli-*

C *cata*

per 32. eiusd. per 2 sexti. *cata est ad diametrum* b i, *ipsa verò* b e *tangit in puncto* b, *erunt parallelę* a c, b e. *Et ideo erit vt* d b *ad* b i, *ita* d e *ad* e c. *Quare æquales erunt etiam* d e, e c. *Quod erat ostendendendum &c.*

Lemma II.

Sì parabola duas tangentes habuerit ex basis terminis; recta linea quæ ab occursu duarum tangentium ducitur diametro parallela, propositæ parabolæ diameter erit.

Esto parabola a b c, *cuius ex punctis* a *&* c, *duæ tangentes sint* a d, c d, *concurrentes in* d. *Ex puncto autem* d, *recta ducatur* d e *diametro parallela. Dico ipsam* d e *propositæ parabolæ diametrum esse.*

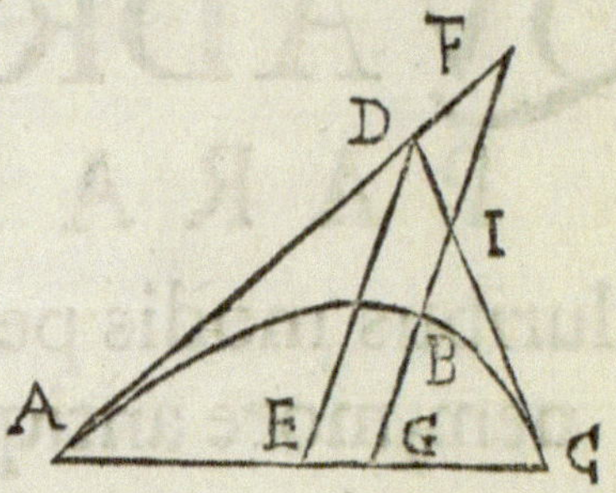

Sit enim, si possibile est, diameter f g. *Erunt ergò ob tangentem* a f *æquales inter se diametri portiones* f b, b g. *Iterum ob tangentem* c i, *æquales erunt* i b, b g. *Et ideò æquales erunt inter se ipsæ* f b, b i: *totum, & pars. quod fieri non potest. Non est ergò alia diameter præter ipsam* d e. *Quod erat ostendendum &c.*

per 35. 1. Conicorum.

Lemma III.

Si parabola tres tangentes habuerit; duas ad basim, & tertiam per verticem; erit triangulum sub tangentibus compræhensum octuplum trianguli, quod oritur ex ductu quartæ tangentis per verticem alterutræ semiparabolæ.

Esto parabola a b c, *cuius basis* a c, *diameter* b d; *duæ tangentes ad basim* a e, c e. *Tangens per verticem sit* f b g. *Demittatur ex* f, *concursu tangentium* a f, g f, *recta* f i, *diametro parallela: eritque* f i *diameter parabolę* a i b. *Ducatur denique* l m, *tangens semiparabolam* a i b *per verticem* i. *Dico triangulum* f e g, *sub tangentibus compræhensum, octuplum esse trianguli* l f m,

per Lem. præced.

18 De Dimenſione Parabolæ

per 32 eiuſd. per 2 ſexti. *cata eſt ad diametrum* bi, *ipſa verò* be *tangit in puncto* b, *erunt pœrallelę* ac, be. *Et ideo erit vt* db *ad* bi, *ita* de *ad* ec. *Quare æquales erunt etiam* de, ec. *Quod erat oſtendendendum &c.*

Lemma II[121]

Sì parabola duas tangentes habuerit ex baſis terminis; recta linea quæ ab occurſu duarum tangentium ducitur diametro parallela, propoſitæ parabolæ diameter erit.

Eſto parabola abc, *cuius ex punctis* a *&* c, *duæ tangentes ſint* ad, cd, *concurrentes in* d. *Ex puncto autem* d, *recta ducatur* de *diametro parallela. Dico ipſam* de *propoſitæ parabolæ diametrum eſſe.*

per 35. 1. Conicorum. *Sit enim, ſi poſſibile eſt, diameter* fg. *Erunt ergò ob tangentem* af *æquales inter ſe diametri portiones* fb, bg. *Iterum ob tangentem* ci, *æquales erunt* ib, bg. *Et ideò æquales erunt inter ſe ipſæ* fb, bi*: totum, & pars. quod fieri non potest. Non eſt ergò alia diameter præter ipſam* de. *Quod erat ostendendum &c.*

Lemma III.

Si parabola tres tangentes habuerit; duas ad baſim, & tertiam per verticem; erit triangulum ſub tangentibus compræhenſum octuplum trianguli, quod oritur ex ductu quartæ tangentis per verticem alterutræ ſemiparabolæ.

per Lem. præced. *Eſto parabola* abc, *cuius baſis* ac, *diameter* bd; *duæ tangentes ad baſim* ae, ce. *Tangens per verticem ſit* fbg. *Demittatur ex* f, *concurſu tangentium* af, gf, *recta* fi, *diametro parallela: eritque* fi *diameter parabolę* aib. *Ducatur denique* lm, *tangens ſemiparabolam* aib *per verticem* i. *Dico triangulum* feg, *ſub tangentibus compræhenſum, octuplum eſſe trianguli*

lfm,

[121] *Ad absurdum* proof. (Cfr. Pisano and Bussotti's chapters above).

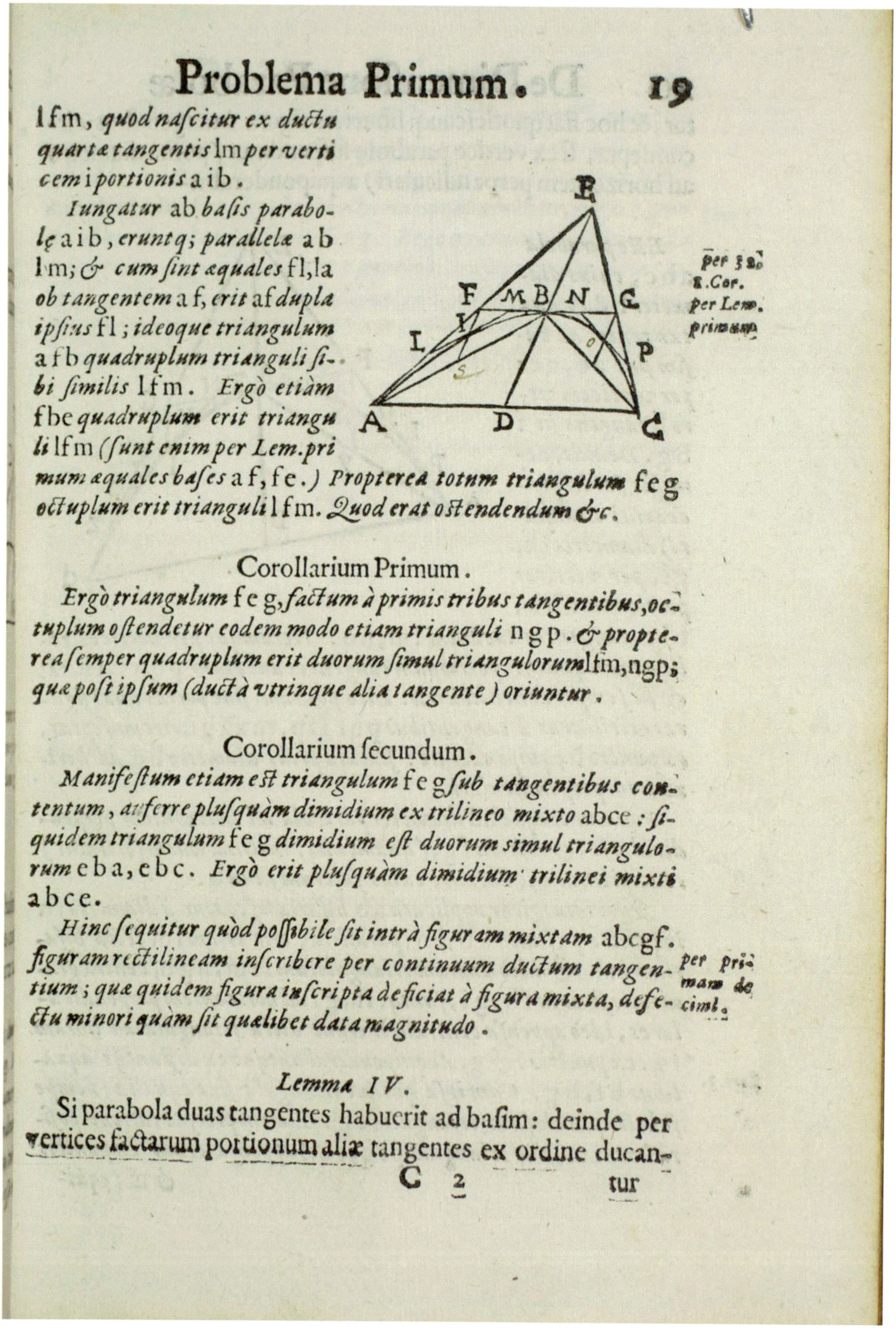

Problema Primum. 19
lfm, *quod nascitur ex ductu quartæ tangentis* lm *per verticem portionis* aib.
Iungatur ab *basis parabolę* aib, *eruntq; parallelæ* ab lm; *& cum sint æquales* fl, la *ob tangentem* af, *erit* af *dupla ipsius* fl; *ideoque triangulum* afb *quadruplum trianguli sibi similis* lfm. *Ergò etiàm* fbe *quadruplum erit trianguli* lfm (*sunt enim per Lem. primum æquales bases* af, fe.) *Propterea totum triangulum* feg *octuplum erit trianguli* lfm. *Quod erat ostendendum &c.*
per 32. 1. Cor. per Lem. primum
E
F M B N G
I
L
O
P
S
A D C
Corollarium Primum.
Ergò triangulum feg, *factum à primis tribus tangentibus, octuplum ostendetur eodem modo etiam trianguli* ngp. *& proptereà semper quadruplum erit duorum simul triangulorum* lfm, ngp; *quæ post ipsum (ductà vtrinque alia tangente) oriuntur.*
Corollarium secundum.
Manifestum etiam est triangulum feg *sub tangentibus contentum, auferre plusquàm dimidium ex trilineo mixto* abce: *siquidem triangulum* feg *dimidium est duorum simul triangulorum* eba, ebc. *Ergò erit plusquàm dimidium trilinei mixti* abce.
Hinc sequitur quòd possibile sit intrà figuram mixtam abcgf. *figuram rectilineam inscribere per continuum ductum tangentium; quæ quidem figura inscripta deficiat à figura mixta, defectu minori quàm sit qualibet data magnitudo.* per primam decimi.
Lemma IV.
Si parabola duas tangentes habuerit ad basim: deinde per vertices factarum portionum aliæ tangentes ex ordine ducantur
C 2

Problema Primum 19

lfm, *quod nafcitur ex ductu quartæ tangentis* lm *per verticem* i *portionis* aib. *per* 32.

Iungatur ab *bafis parabolę* aib, *eruntq; parallelæ* ablm; *& cum fint æquales* fl, la *ob tangentem* af, *erit* af *dupla ipfius* fl; *ideoque triangulum* afb *quadruplum trianguli fibi fimilis* lfm. *Ergò etiàm* fbe *quadruplum erit trianguli* lfm *(funt enim per Lem. primum æquales bafes* af, fe.*) Propterea totum triangulum* feg *octuplum erit trianguli* lfm. *Quod erat ostendendum &c.* 8[...] *Cor. per Lem. primum*

Corollarium Primum.

Ergò triangulum feg, *factum à primis tribus tangentibus, octuplum oftendetur eodem modo etiam trianguli* ngp. *& propterea femper quadruplum erit duorum fimul triangulorum* lfm, ngp; *quæ poft ipfum (ductà vtrinque alia tangente) oriuntur.*

Corollarium fecundum.

Manifeftum etiam est triangulum feg *fub tangentibus contentum, auferre plufquàm dimidium ex trilineo mixto* abce*: fiquidem triangulum* feg *dimidium eft duorum fimul triangulorum* eba, ebc. *Ergò erit plufquàm dimidium trilinei mixti* abce.

Hinc fequitur quòd poffibile fit intrà figuram mixtam abcgf. *figuram rectilineam infcribere per continuum ductum tangentium; quæ quidem figura infcripta deficiat à figura mixta, defectu minori quàm fit quælibet data magnitudo.*

Lemma IV.

Si parabola duas tangentes habuerit ad bafim: deinde per vertices factarum portionum aliæ tangentes ex ordine ducan-

C 2 tur *per primam de ciml.*

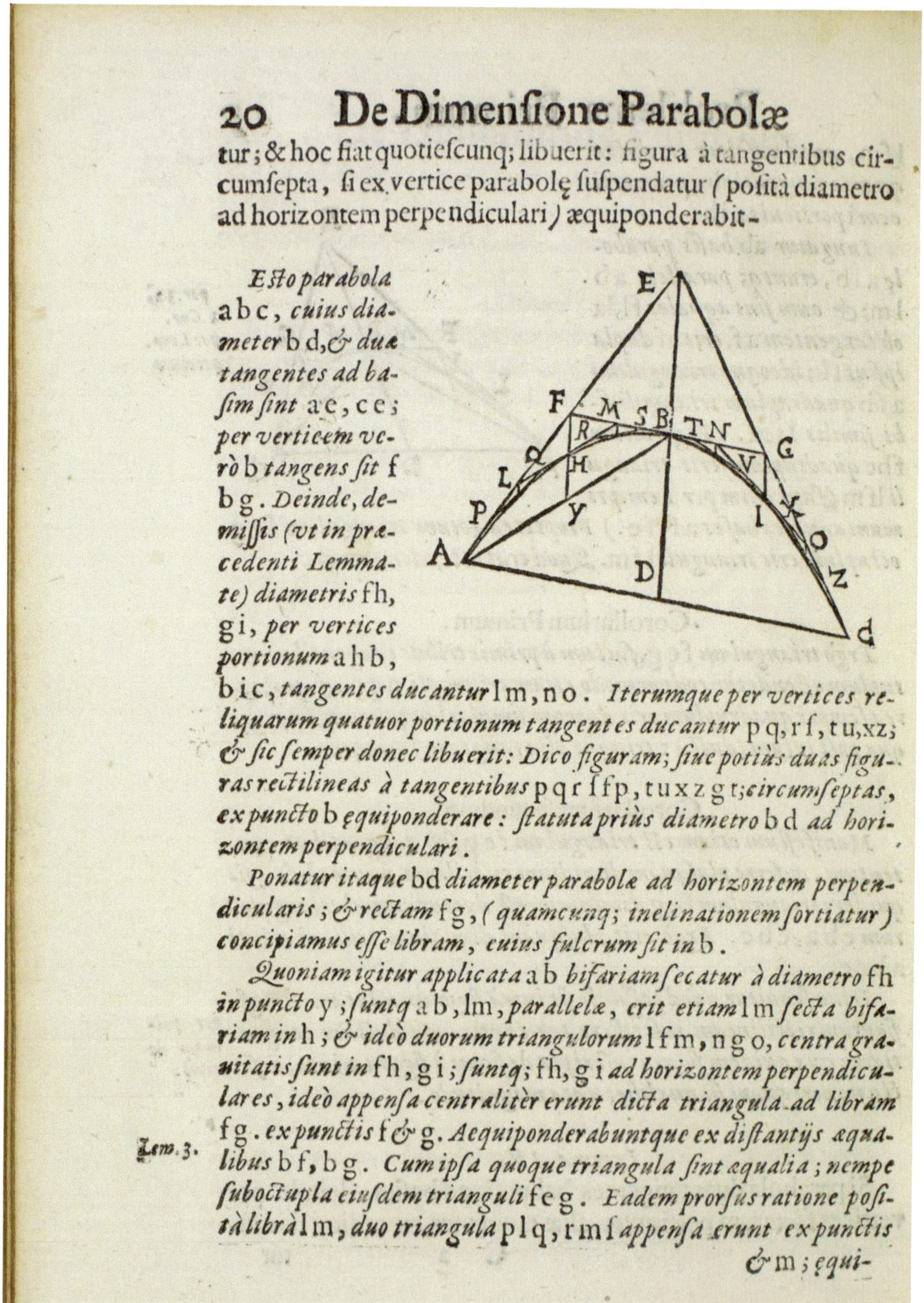

20 De Dimensione Parabolæ

tur; & hoc fiat quotiescunq; libuerit: figura à tangentibus circumsepta, si ex vertice parabolę suspendatur (positâ diametro ad horizontem perpendiculari) æquiponderabit.

Esto parabola a b c, *cuius diameter* b d, *& duæ tangentes ad basim sint* a e, c e; *per verticem verò* b *tangens sit* f b g. *Deinde, demissis (vt in præcedenti Lemmate) diametris* f h, g i, *per vertices portionum* a h b, b i c, *tangentes ducantur* l m, n o. *Iterumque per vertices reliquarum quatuor portionum tangentes ducantur* p q, r s, t u, x z; *& sic semper donec libuerit: Dico figuram; siue potiùs duas figuras rectilineas à tangentibus* p q r s f p, t u x z g t; *circumseptas, ex puncto* b *ęquiponderare: statuta priùs diametro* b d *ad horizontem perpendiculari.*

Ponatur itaque bd *diameter parabolæ ad horizontem perpendicularis; & rectam* f g, *(quamcunq; inclinationem sortiatur) concipiamus esse libram, cuius fulcrum sit in* b.

Quoniam igitur applicata a b *bifariam secatur à diametro* f h *in puncto* y; *suntq;* a b, l m, *parallelæ, erit etiam* l m *secta bifariam in* h; *& ideò duorum triangulorum* l f m, n g o, *centra grauitatis sunt in* f h, g i; *suntq;* f h, g i *ad horizontem perpendiculares, ideò appensa centraliter erunt dicta triangula ad libram* f g. *ex punctis* f *&* g. *Aequiponderabuntque ex distantijs æqualibus* b f, b g. *Cum ipsa quoque triangula sint æqualia; nempe suboctupla eiusdem trianguli* f e g. *Eadem prorsus ratione positâ librâ* l m, *duo triangula* p l q, r m s *appensa erunt ex punctis* Lem. 3.

& m; *ęqui-*

tur; & hoc fiat quotieſcunq; libuerit: figura à tangentibus circumſepta, ſi ex vertice parabolę ſuſpendatur (poſità diametro ad horizontem perpendiculari) æquiponderabit.

Esto parabola abc, *cuius diameter* bd, *& duæ tangentes ad baſim ſint* ae, ce; *per verticem verò* b *tangens ſit* fbg. *Deinde, demiſſis (vt in præcedenti Lemmate) diametris* fh, gi, *per vertices portionum* ahb, bic, *tangentes ducantur* lm, no. *Iterumque per vertices reliquarum quatuor portionum tangentes ducantur* pq, rſ, tu, xz; *& ſic ſemper donec libuerit: Dico figuram; ſiue potiùs duas figuras rectilineas à tangentibus* pqrſſp, tuxzgt; *circumſeptas, ex puncto* b *ęquiponderare: ſtatuta priùs diametro* bd *ad horizontem perpendiculari.*

Ponatur itaque bd *diameter parabolæ ad horizontem perpendicularis; & rectam* fg, *(quamcunq; inclinationem ſortiatur) concipiamus eſſe libram, cuius fulcrum ſit in* b.

Quoniam igitur applicata ab *bifariam ſecatur à diametro* fh *in puncto* y; *ſuntq* ab, lm, *parallelæ, erit etiam* lm *ſecta bifariam in* h; *& ideò duorum triangulorum* lfm, ngo, *centra grauitatis ſunt in* fh, gi; *ſuntq;* fh, gi *ad horizontem perpendiculares, ideò appenſa centralitèr erunt dicta triangula ad libram* fg. *ex punctis* f & g. *Aequiponderabuntque ex diſtantijs æqualibus* bf, bg. *Cum ipſa quoque triangula ſint æqualia; nempe ſuboctupla eiuſdem trianguli* feg. *Eadem prorſus ratione poſità librà* lm, *duo triangula* plq, rmſ *appenſa*

Lem.3. *erunt ex punctis*

& m; *equi-*

& m; æquiponderabuntq; ex puncto h, *& ideo appensa erunt ex puncto* f. *(quandoquidem filum suspensionis* fh *perpendiculare est ad horizontem.) Duo verò triangula* tnu, xoz, *prædictis æqualia (cum sint singula suboctupla æqualium* lfm, ngo.) *pon derabunt simul ambo ex puncto* g. *Ergo quatuor simul prædicta triangula æquiponderabunt ex puncto* b, *nempe medio totius libræ* fg: *Eodem modo concludemus reliqua triangula, quotcunque sint, ex puncto* b *æquiponderare. Vniuersa ergo figura à tangentibus circumscipta ex puncto* b *æquiponderabit. Quod &c.* Lem 3.

Corollarium I.

Hinc pro Corollario animaduertemus centrum grauitatis prędictę figurę, à tangentibus compræhensę, esse in diametro parabolæ. Cum enim figura prędicta ęquiponderet ex puncto b, *erit centrum grauitatis illius in linea quę ex puncto* b *ducitur perpendicularis ad horizontem; quapropter erit in* bd *diametro parabolę.*

Corollarium II.

Colligemus etiam centrum grauitatis omnium trilineorū mixtorum, quę sub linea parabolica abc, *& sub omnibus tangentibus* apqrstuxzc, *comprębenduntur, semper in diametro parabolę existere. Patebit autem hoc modo. Centrum trapetij* afgc, 15 primi *est in diametro; centrum etiam parabolę est in diametro; ergò centrum reliquæ figurę mixtę erit in diametro. Si ergo centrum huiusmodi figuræ est in diametro; centrum etiam figurę à tangentibus circumscptę demonstratum est esse in diametro; propterea cētrum omnium simul trilineorum, quę continētur sub tangentibus & linea parabolica, erit in diametro per 8. prim. Aequipond.* æquipon der. 4. secundi eiusd. 8. primi eiusd.

Lemma V.

Si parabola duas tangentes habuerit alteram per verticem, alteram verò ad basim, & ex altera parte basis habeat parallelam diametro; figura sub tribus prædictis rectis lineis, & curuà parabolicà compræhensa, æquiponderabit ex puncto tangentis ver-

Problema Primum. 21

& m; æquiponderabuntq; ex puncto h, *& ideo appenſa erunt ex puncto* f. *(quandoquidem filum ſuſpenſionis* fh *perpendiculare eſt ad horizontem.) Duo verò triangula* tnu, xoz, *prædictis æqualia (cum ſint ſingula ſuboctupla ęqualium* lfm, ngo.) *ponderabunt ſimul ambo ex puncto* g. *Ergo quatuor ſimul prædicta triangula æquiponderabunt ex puncto* b, *nempe medio totius libræ* fg*:* *Lem* 3. *Eodem modo concludemus reliqua triangula, quotcunque ſint, ex puncto* b *æquiponderare. Vniuersa ergo figura à tangentibus circumſepta*[122] *ex puncto* b *æquiponderabit. Quod &c.*

Corollarium I.

Hinc pro Corollario animaduertemus centrum grauitatis prędictę figurę, à tangentibus compræhenſę, eſſe in diametro parabolæ. Cum enìm figura prędicta ęquiponderet ex puncto b, *erit centrum grauitatis illius in linea quę ex puncto* b *ducitur perpendicularis ad horizontem; quapropter erit in* bd *diametro parabolę.*

Corollarium II.

Colligemus etiam centrum grauitatis omnium trilineorũ mixtorum, quę ſub linea parabolica abc, *& ſub omnibus tangentibus* apqrſtuxzc, *compręhenduntur, ſemper in diametro parabolę exiſtere. Patebit autem hoc modo. Centrum trapetij* afgc, *eſt in diametro; centrum etiam parabolę eſt in diametro; ergò centrum reliquæ figurę mixtę erit in diametro. Si ergo centrum huiuſmodi figuræ est in diametro; centrum etiam figurę à tangentibus circumſeptę demonſtratum eſt eſſe in diametro; propterea cẽtrum omnium ſimul trilineorum, quę continẽtur ſub tangentibus & linea parabolica, erit in diametro per 8. prim. Aequipond.* 15 *primi æquipon der.* 4. *ſecundi eiuſd.* 8. *primi. eiuſd.*

Lemma V[123] .

Si parabola duas tangentes habuerit alteram per verticem, alteram verò ad baſim, & ex altera parte baſis habeat parallelam diametro; figura ſub tribus prædictis rectis lineis, & curuà parabolicà compræhenſa, æquiponderabit ex puncto tangen-

tis ver-

---

[122] *circumſepta*] *circumſcripta.*

[123] *Ad absurdum* proof. (Cfr. Pisano and Bussotti's chapters above).

tis verticalis, in quo ea ſic diuiditur, vt pars ad reliquam tangentem terminata, dupla ſit illius quæ ad parallelam diametro terminatur.

Eſto parabola a b c, *cuius tangens ad baſim ſit* c d; *per verticem verò* f b g; *&* a g *ſit parallela diametro. Secetur deinde* f g *in* e, *ita vt* f e *dupla ſit reliquæ* e g. *Dico figurã* a b c f g (*ſtatutà diametro ad horiz. perpendiculari*) *æquiponderare ex puncto* e.

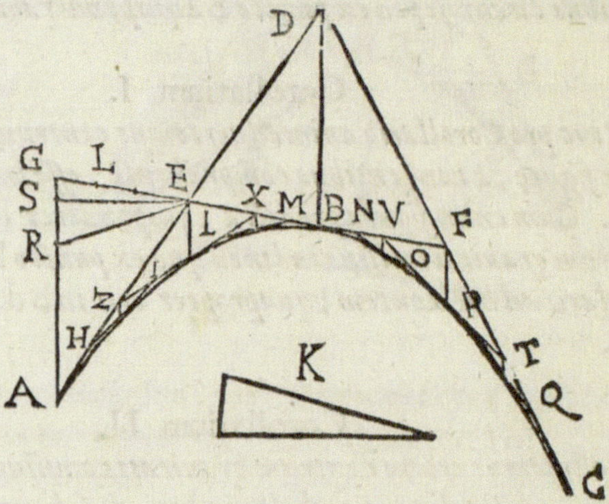

Concipiamus enim diametr. parabolæ eſſe horizonti perpendicularem (*hoc autem modo sẽper intelligendum eſt*) *quamcunq; tãdem inclinationem ſortiatur libra* g f. *Et ducta tangente* a d (*quæ omnino tranſibit per* e, *vt infrà oſtendemus*) *intelligatur* g f. *librã eſſe, cuius fulcrum eſt* e; *ex qua pendent ab vna parte triangulum* a g e; *ab alterà verò, figura mixta* a b c f e. *Quæ quidem figuræ ſi inter ſe non aequiponderant, ponamus alteram ipſarũ praeponderare. Eſto igitur; & praeponderet primò* a b c f e, *tanto exceſſu quantum eſt ſpatium* k.

Inſcribatur intrà ipſam alia figura à tangentibus h i l m n o p q f e h, *terminata, ita vt reliquae portiunculae ſub dictis tangentibus & curua parabolica contentae, ſimul minores ſint ſpatio* k (*quod fieri poſſe conſtat ex Corollario Secundo Lemmatis Tertij.*) *Praeponderabit igitur adhuc figura ſub tangentibus compraehenſa; quandoquidem pars ablata minor eſt exceſſu* K, *& in eodem puncto* b *ponderat ſimul cum tota magnitudine; tam enim ablatae, quàm totius, centrum grauitatis eſt in diametro,*

vt oſten-

tis verticalis, in quo ea ſic diuiditur, vt pars ad reliquam tangentem terminata, dupla ſit illius quæ ad parallelam diametro terminatur.

Eſto parabola abc, *cuius tangens ad baſim ſit* cd; *per verticem verò* fbg; *&* ag *ſit parallela diametro. Secetur deinde* fg *in* e, *ita vt* fe *dupla ſit reliquæ* eg. *Dico figurã* abcfg *(statutà diametro ad horiz. perpendiculari) æquiponderare ex puncto* e.

Concipiamus enim diametr. parabolæ eſſe horizonti perpendicularem (hoc autem modo sẽper intelligendum est) quamcunq; tãdem inclinationem ſortiatur libra gf. *Et ducta tangente* ad *(quæ omnino tranſibit per* e, *vt infrà ostendemus) intelligatur* gf. *Librã eſſe, cuius fulcrum eſt* e; *ex qua pendent ab vna parte triangulum* age; *ab alterà verò, figura mixta* abcfe. *Quæ quidem figuræ ſi inter ſe non aequiponderant, ponamus alteram ipſarũ praeponderare. Eſto igitur; & praeponderet primò* abcfe, *tanto exceſſu quantum est ſpatium* k.

Inſcribatur intrà ipſam alia figura à tangentibus hilmnopqfeh, *terminata, ita vt reliquae portiunculae ſub dictis tangentibus & curua parabolica contentae, ſimul minores ſint ſpatio* k *(quod fieri poſſe conſtat ex Corollario Secundo Lemmatis Tertij.) Praeponderabit igitur adhuc figura ſub tangentibus compraehenſa; quandoquidem pars ablata minor eſt exceſſu* K, *& in eodem puncto* b *ponderat ſimul cum tota magnitudine; tam enim ablatae, quàm totius, centrum grauitatis eſt in diametro,*

vt osten-

Problema Primum. 23

vt ostendimus ad Corollarium Secundum Lemmatis Quarti.

Accipiatur iàm g r *quarta pars totius* g a ; *ducaturq;* e r . *Sumatur etiam* e l *dupla reliquae* l g ; *& ex puncto* l *centralitèr suspensum erit quodlibet triangulum habens verticem in* e *puncto, & basim in recta* g a, *quae ad horizontem recta ponitur.* — *Vide Archim. de Quadratura parab. Propos. 6. 8. 10. et 12.*

Iam sic : duo triangula z e x, u f t, *ad triangulum* e d f . *sunt vt duo ad 8. & ad triangul.* e b d *vt 2. ad 4. & ad aequale* a g e. *vt 2. ad 4. ergò ad triangulum* a r e . *erunt vt 2. ad 3., nempe vt* l e *ad* e g, *hoc est vt* l e *ad* e b *reciprocè. Quare in libra* l b *duo praedicta triangula* z e x, u f t *aequiponderant triangulo* a r e *ex puncto* e. — *per Corol prim Lem. 3.*

Sumatur iterum g ſ *quarta pars totius* g r, *& iungatur* e ſ. *Cũ ergo* g r e *sit quarta pars ipsius* g a e, *vel ipsius* e d b, *erit* g r e *octaua pars totius* e d f . *Quapropter aequale erit* g r e, *alterutro ipsorum* z e x, u f t. *Sed quoniam* h z i *est octaua pars ipsius* z e x, *erunt quatuor simul triangula* h z i, l x m, n u o, p t q. *(quoniam aequalia sunt inter se) ad triangulum* z e x *vt 4. ad 8; siue vt 2. ad 4: & propterea etiam ad triangulum* g r e, *erunt vt 2. ad 4; ad ipsum verò* ſ r e *erunt vt 2. ad 3, nempe vt* l e *ad* e b. *Equiponderant igitur ex puncto* e *hinc triangulum* ſ r e, *& inde quatuor praedicta triangula* h z i, l x m, n u o, p t q. *Eodem prorsus modo, si sub quatuor his triang. alia fuerint in residuis portiũculis triang. ex ordine descripta, ostendentur aequiponderare ex eodem puncto* e, *cum quodam triang. cuius vertex sit* e, *basis vero contineat 3. quar. ipsius* g ſ *&c. Sed in nostro casu, cum demonstratum sit prima duo triangula* z e x, u f t, *aequiponderare triangulo* a r e. *Reliqua item quatuor triangula, quorum vnũ est* h z i. *aequiponderare triangulo* ſ r e ; *Aequiponderabit tota simul figura ex praedictis triangulis composita, triangulo* a e ſ, *ex puncto* e. *Sed demonstratum fuit, eandem figuram praeponderare triangulo* a e g ; *necesse igitur est vt triangulum* a e g *minus sit triangulo* a e ſ ; *totum sua parte ; Quod est impossibile.* — *Ex Lemmate 3.*

Si vero ponamus praeponderare triangulum a e g *figurae* a b c f e. *Esto ; & sit excessus quo praeponderat, spatium* k . *Accipiatur* g r e *quarta pars trianguli* a g e : *& iterum accipiatur* g ſ e,

quar-

vt ostendimus ad Corollarium Secundum Lemmatis Quarti.

Accipiatur iàm gr *quarta pars totius* ga; *ducaturq;* er. *Sumatur etiam* el *dupla reliquae* lg; *& ex puncto* l *centralitèr ſuſpenſum erit quodlibet triangulum habens verticem in* e *puncto, & baſim in recta* ga, *quae ad horizontem recta ponitur.*

Iam ſic: duo triangula zex, uft, *ad triangulum* edf. *ſunt vt duo ad 8. & ad triangul.* ebd *vt 2. ad 4. & ad aequale* age. *vt 2. ad 4. ergò ad triangulum* are. *erunt vt 2. ad 3., nempe vt* le *ad* eg, *hoc eſt vt* le *ad* eb *reciprocè. Quare in libra* lb *duo praedicta triangula* zex, uft *aequiponderant triangulo* are *ex puncto* e.

Sumatur iterum gſ *quarta pars totius* gr, *& iungatur* eſ. *Cũ ergo* gre *ſit quarta pars ipsius* gae, *vel ipſius* edb, *erit* gre *octaua pars totius* edf. *Quapropter aequale erit* gre, *alterutro ipſorum* zex, uft. *Sed quoniam* hzi *eſt octaua pars ipſius* zex, *erunt quatuor ſimul triangula* hzi, lxm, nuo, ptq. *(quoniam aequalia ſunt inter ſe) ad triangulum* zex *vt 4. æd 8; ſiue vt 2. ad 4: & propterea etiam ad triangulum* gre, *erunt vt 2. ad 4; ad ipſum verò* ſre *erunt vt 2. ad 3, nempe vt* le *ad* eb. *Equiponderant igitur ex puncto* e *hinc triangulum* ſre, *& inde quatuor praedicta triangula* hzi, lxm, nuo, ptq. *Eodem prorſus modo, si ſub quatuor his triang. alia fuerint in reſiduis portiũculis triang. ex ordine deſcripta, oſtendentur aequiponderare ex eodem puncto* e, *cum quodam triang. cuius vertex ſit* e, *baſis vero contine at 3. quar. ipſius* gſ *&c. Sed in noſtro caſu, cum demonſtratum ſit prima duo triangula* zex, uft, *aequiponderare triangulo* are. *Reliqua item quatuor triangula, quorum vnũ eſt* hzi. *aequiponderare triangulo* ſre; *Aequiponderabit tota ſimul figura ex praedictis triangulis compoſita, triangulo* aeſ, *ex puncto* e. *Sed demonstratum fuit, eandem figuram praeponderare triangulo* aeg; *neceſſe igitur est vt triangulum* aeg *minus ſit triangulo* aeſ; *totum ſua parte; Quod est impoſſibile.*

Si vero ponamus praeponderare triangulum aeg *figurae* abcfe. *Esto; & ſit exceſſus quo praeponderat, ſpatium* k. *Accipiatur* gre *quarta pars trianguli* age*: & iterum accipiatur* gſe,

quar-

Vide Archim. de Quadratura parab. Propoſ. 6. 8 10. es 12

per Corol prim Lem. 3.

Ex Lemmate 3.

quarta pars trianguli g r e; *& hoc semper fiat, donec veniatur ad aliquod triangulum* g ſ e, *quod minus sit spatio* k. *Tunc enim triangulum* a ſ e *adhuc praeponderabit figurae* a b c f e. *Sed eodem modo, vt supra demonstrabimus triangulum ipsum* a ſ e *aequiponderare alicui figurae rectilineae inscriptae intra figuram mixtam* a b c f e. *Necesse ergo iterum erit vt inscripta figura rectilinea maior sit quàm figura mixta* abcfe, *cui ipsa inscribitur; pars suo toto. Quod est absurdum &c.*

possibile est per 1. Decimi.

Aequiponderat ergo ex puncto e *vniuersa figura* a b c f g, *quae sub curua parabolica, duabusq. tangentibus, & linea ipsi diametro parallela continetur. Quod &c.*

Quod assumptum est ita ostendemus: Nempe tangentem a d *transire per punctum* e: *hoc est, ita secare rectam* f g, *vt pars* f e, *dupla sit reliquae* e g. *Secet enim* a d *tangens rectam* f g *vtcunq; in* e. *Iam; cum parallelæ sint* a g, b d, *& æquales* a e, e d; *erunt æquales etiam* g e, e b, *Sed æquales sunt* f b, b e, *ergo* f e *dupla est ipsius* e g. *Ideo* a d *transit per illud* e *punctum, quod ab initio dixeramus.*

Propositio Prima.

QVAELIBET parabola ſesquitertia eſt trianguli eandem ipſi baſim, & eandem altitudinem habentis.

Eſto parabola ABC, cuius diameter BD, iunganturq. AB, BC, Dico parabolam ſesquitertiam eſſe trianguli ABC, eandem cum ipſa baſim, & eandem altitudinem habentis,

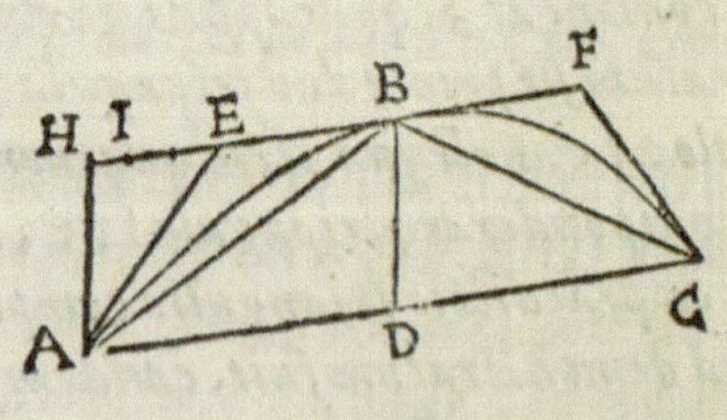

Ducantur tangentes AE, CF ad baſim: FH verò per verticem B; & AH ſit ipſi diametro parallela. concipiamuſq. parabolæ diametrum erectam eſſe ad horizontem. Iam ſecta HE in I, ita vt EI dupla ſit ipſius IH, erit triangulum HAE centralitèr appenſum ad punctum I (habet enim

quarta pars trianguli gre; *& hoc ſemper fiat, donec veniatur ad aliquod triangulum* gſe, *quod minus ſit ſpatio* k. *Tunc enim triangulum* aſe *adhuc praeponderabit figurae* abcfe. *Sed eodem modo, vt ſupra demonſtrabimus triangulum ipſum* aſe *aequiponderare alicui figurae rectilineae inſcriptae intra figuram mixtam* abcfe. *Neceſſe ergo iterum erit vt inſcripta figura rectilinea maior ſit quàm figura mixta* abcfe, *cui ipſa inſcribitur; pars ſuo toto. Quod eſt abſurdum &c.*

poſſihile eſt per 1. *Decimi.*

Aequiponderat ergo ex puncto e *vniuerſa figura* abcfg, *quae ſub curua parabolica, duabuſq. tangentibus, & linea ipſi diametro parallela continetur. Quod &c.*

Quod aßumptum eſt ita oſtendemus: Nempe tangentem ad *tranſire per punctum* e*: hoc est, ita ſecare rectam* fg, *vt pars* fe, *dupla ſit reliquae* eg. *Secet enim* ad *tangens rectam* fg *vtcunq; in* e. *Iam; cum parallelæ ſint* ag, bd, *& æquales* ae, ed; *erunt æquales etiam* ge, eb, *Sed æquales ſunt* fb, be, *ergo* fe *dupla est ipſius* eg. *Ideo* ad *transit per illud* e *punctum, quod* ab initio *dixeramus.*

Propoſitio Prima.

QVAELIBET parabola ſeſquitertia eſt trianguli eandem ipſi baſim, & eandem altitudinem habentis.

Eſto parabola ABC, cuius diameter BD, iunganturq. AB, BC, Dico parabolam ſeſquitertiam eſſe trianguli ABC, eandem cum ipſa baſim, & eandem altitudinem habentis,

Ducantur tangentes AE, CF ad baſim: FH verò per verticem B; & AH ſit ipſi diametro parallela. concipiamuſq. parabolæ diametrum erectam eſſe ad horizontem. Iam ſecta HE in I, ita vt EI dupla ſit ipſius IH, erit triangulum HAE centralitèr appenſum ad punctum I (habet e-

nim

nim centrum grauitatis in recta quæ ex I ducitur parallela ad H A, & propterea ad horizontem perpendiculari.) Erit insuper figura mixta A B C F E centraliter appensa ad punctum B. (quandoquidem habet centrum grauitatis in diametro B D ad horizontem perpendiculari.) Sed vniuersa magnitudo, composita ex dicto triangulo H A E, dictaq; figura mixta, æquiponderat ex puncto E; erit ergo triangulum H A E ad figuram mixtam ABCFE, vt reciprocè B E ad E I, nempe vt 3. ad 2. Propterea trapezium AEFC sextuplum dicti trianguli, erit ad figuram mixtam ABCFE, vt 18. ad 2. & per conuersionem rationis, ad parabolam erit vt 18, ad 16. Qualium ergo partium parabola est 16, earum trapezium AEFC est 18. & triangulum A BC 12. Quare parabola ad triangulum ABC erit vt 16, ad 12 nempe sesquitertia. Quod erat ostendendum.

Vide Corol 1. Lemmat. 4.

Lemm. præced.

ostendetur inf.

ostenditur inf.

Quod trapezium a e f c *sextuplum sit trianguli* h a e, *patet. Nam parallelogrammũ* hd, *duplum est trianguli* h a b *& propterea quadruplum trianguli* h a c. *ergò trapezium* a e b d *triplum erit trianguli* h a e. *totum verò trapezium* a e f c. *sextuplum dicti trianguli* h a e. *Quod &c.*

Cum autem trapezium a e f c *sextuplum sit trianguli* h a e, *erit sextuplum etiam trianguli* e a b ; *& ideo triplum duorum* e a b, b c f, *Nempe vt 18. ad 6. Per conuersionem verò rationis, erit ad triangulum* a b c *vt 18. ad 12. Quod &c.*

Lemma VI.

Si duæ parabolæ vtraque duas tangentes ad basim habuerit; erunt inter se trilinea mixta sub tangentibus, & curuis parabolicis contenta, vt sunt ipsa triangula sub tangentibus compræhensa,

Sint duæ parabolæ a b c, d e f. *quarum vtraque duas tangentes ad basim habeat* a g, g c *prioris, &* d h, f h. *posterioris parabolæ. Dico trilineum mixtum* a b c g, *ad trilineum mixtum* d e f h, *esse vt triangulum* a g c, *ad triangulum* d h f.

D *Si enim*

nim centrum grauitatis in recta quæ ex I ducitur parallela ad HA, & propterea ad horizontem perpendiculari.) Erit inſuper figura mixta ABCFE centraliter appenſa ad punctum B. (quandoquidem habet centrum grauitatis in diametro BD ad horizontem perpendiculari.*)* Sed vniuerſa magnitudo, compoſita ex dicto triangulo HAE, dictaq; figura mixta, æquiponderat ex puncto E; erit ergo triangulum HAE ad figuram mixtam ABCFE, vt reciprocè BE ad EI, nempe vt 3. ad 2. Propterea trapezium AEFC ſextuplum dicti trianguli, erit ad figuram mixtam ABCFE, vt 18. ad 2. & per conuerſionem rationis, ad parabolam erit vt 18, ad 16. Qualium ergo partium parabola eſt 16, earum trapezium AEFC eſt 18. & triangulum ABC 12. Quare parabola ad triangulum ABC erit vt 16, ad 12 nempe ſeſquitertia. Quod erat oſtendendum.

Vide Co. rol 1. *Le* [m]*mat.* 4.

Lemm. præced.

ostende- tur inf.

oſtendi-. tur inf.

Quod trapezium aefc *ſextuplum ſit trianguli* hae, *patet. Nam parallelogrammũ* hd, *duplum est trianguli* hab *& propterea quadruplum trianguli* hae. *ergò trapezium* aebd *triplum erit trianguli* hae. *totum verò trapezium* aefc. *ſextuplum dicti trianguli* hae. *Quod &c.*

Cum autem trapezium aefc *ſextuplum ſit trianguli* hae, *erit ſextuplum etiam trianguli* eab; *& ideo triplum duorum* eab, bcf, *Nempe vt 18. ad 6. Per conuerſionem verò rationis, erit ad triangulum* abc *vt 18. ad 12. Quod &c.*

Lemma VI.[124]

Si duæ parabolæ vtraque duas tangentes ad baſim habuerit; erunt inter ſe trilinea mixta ſub tangentibus, & curuis parabolicis contenta, vt ſunt ipſa triangula ſub tangentibus compræhenſa,

Sint duæ parabolæ abc, def. *quarum vtraque duas tangentes ad baſim habeat* ag, gc *prioris, &* dh, fh, *posterioris parabolæ. Dico trilineum mixtum* abcg, *ad trilineum mixtum.* defh, *eſſe vt triangulum* agc, *ad triangulum* dhf.

D Si enim

[124] *Ad absurdum* proof. (Cfr. Pisano and Bussotti's chapters above).

26 **De Dimensione Parabolæ**

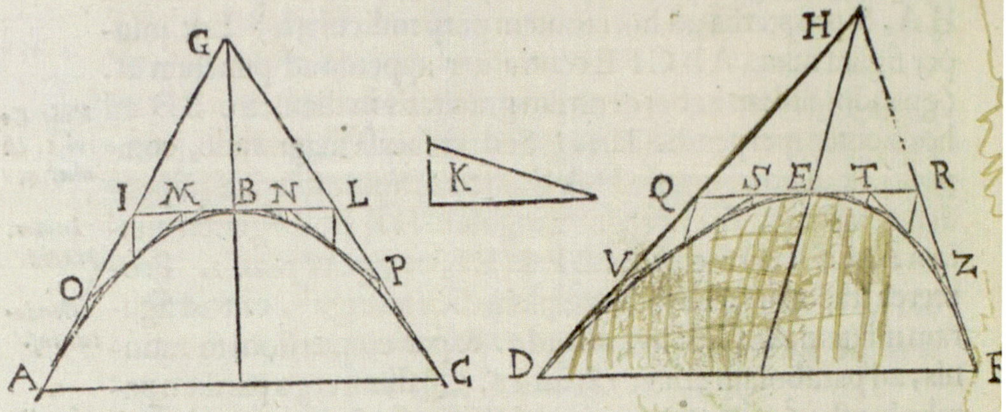

Si enim non est ita : habebit alterum ex trilineis, puta a b c g,
ad reliquum , maiorem rationem quàm triangulum a g c, *ad* dhf.
Esto spatium k *excessus, quo trilineum* a b c g, *maius est quàm*
vt sit in ratione triangulorum.
Ducatur per verticem b, *tangens* i l ; *demissisq; ex punctis* i,
Lem. 2. *&* l, *lineis diametro parallelis (quæ diametri semiparabolarum*
erunt) ducantur tangentes o m, n p : *& ex punctis* o ; m ; n, p,
demittantur aliæ diametri vt supra ; ducanturq; aliæ tangentes:
Et hoc semper fiat, quousq; reliquę simul omnes portiunculæ, quæ
sub tangentibus , & curua parabolica continentur ; minores sint
Corol. 1. *spatio* k. *Tunc.n.vniuersa figura tangentibus circumsepta, & in*
Lem. 3. *trilineo mixto* a b c g *inscripta , habebit adhuc ad trilineum* d e
f h , *rationem maiorem, quā triang.* a g c, *ad triangulum* d h f.
Inscribatur iam etiam in altero trilineo mixto d e f h. *figura*
totidem laterum ; ductis nimirùm tangentibus toties , quoties
ductæ fuerint in priori trilineo.
Quoniam verò est , vt triangulum i g l *ad triangulum* q r h,
ita duo simul triangula o i m, n l p, *ad duo simul* u q s, t r z.
15. quinti. *(sunt enim partes cum pariter multiplicibus in eadem ratione .)*
Et vt duo simul triangula o i m, n l p, *ad duo* u q s, t r z, *ita*
quatuor triangula quæ sunt infra puncta o, m, n, p, *ad quatuor*
illa, quæ sunt sob punctis u, s, t, z ; *ob eandem causam &c. Erunt*
12. quinti. *etiam omnia antecedentia simul (nempe figura inscripta in prio-*
ri tri-

Si enim non eſt ita: habebit alterum ex trilineis, puta abcg, *ad reliquum, maiorem rationem quàm triangulum* agc, *ad* dhf. *Esto ſpatium* k *exceſſus, quo trilineum* abcg, *maius eſt quàm. vt ſit in ratione triangulorum.*

Lem. 2. *Ducatur per verticem* b, *tangens* il; *demiſſiſq; ex punctis* i, & l, *lineis diametro parallelis (quæ diametri ſemiparabolarum erunt) ducantur tangentes* om, np*: & ex punctis* o; m; n, p, *demittantur aliæ diametri vt ſupra; ducanturq; aliæ tangentes:*

Corol. 1. *Et hoc ſemper fiat, quouſq; reliquę ſimul omnes portiunculæ, quæ ſub tangentibus, & curua parabolica continentur; minores*
*Lem.*3. *ſint ſpatio* k. *Tunc.n. vniuerſa figura tangentibus circumſepta, & in trilineo mixto* abcg *inſcripta, habebit adhuc ad trilineum* defh, *rationem maiorem, quã triang.* agc, *ad triangulum* dhf.

Inſcribatur iam etiam in altero trilineo mixto defh. *figura totidem laterum; ductis nimirùm tangentibus toties, quoties ductæ fuerint in priori trilineo.*

25. *quin.* *Quoniam verò eſt, vt triangulum* igl *ad triangulum* qrh, *ita duo ſimul triangula* oim, nlp, *ad duo ſimul* uqſ, trz. *(ſunt enim*
[ti]. *partes cum pariter multiplicibus in eadem ratione.)*

12. *quin.* *Et vt duo ſimul triangula* oim, nlp, *ad duo* uqſ, trz, *ita quatuor triangula quæ ſunt infra puncta* o, m, n, p, *ad quatuor illa, quæ*
[ti]. *ſunt ſub punctis* u, ſ, t, z; *ob eandem cauſam &c. Erunt etiam omnia antecedentia ſimul (nempe figura inſcripta in prio-*

ri tri-

ri trilineo mixto,) ad omnia consequentia simul (nempe ad figuram inscriptam in posteriori trilineo mixto) vt vnum ad vnum; nempe vt igl, *ad* qhr. *Siue sumptis eorum quadruplis, vt* agc *ad* dhf. *Sed eadem inscripta figura habebat ad trilineum* defh *maiorem rationem quàm sit trianguli* agc *ad* dhf. *Minus ergo erit trilineum mixtum* defh, *quàm figura sibi inscripta: totum sua parte. Quod est impossibile. Trilinea ergò sub tangentibus, & curuà parabolicà comprehensa, sunt inter se vt triangula sub iisdem tangentibus & basibus contenta.* x. quinti

Propositio II.

PArabola sesquitertia est trianguli eandem ipsi basim, & eandem altitudinem habentis.

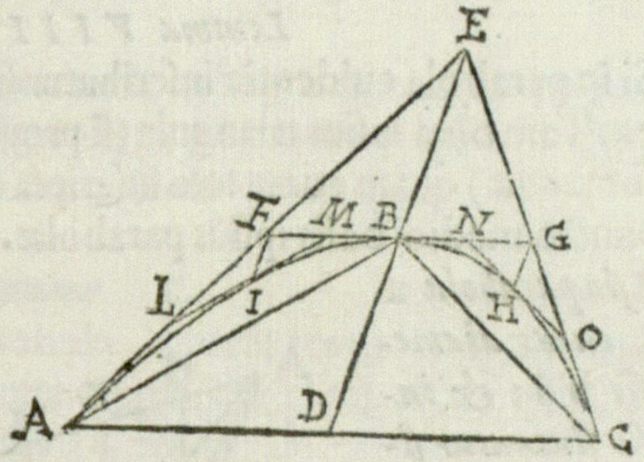

Sit parabola ABC, cuius diameter BD; & sit inscriptum triangulum ABC. Dico parabolam sesquitertiam esse trianguli ABC,

Ducantur duæ tangentes ad basim, quæ sint AE, CE. & FG. tãgat per verticem B Demissis deinde FI, GH diametro parallelis, vt sint diametri portionum AIB, BHC; ducantur per I, & H tangentes LM, NO. 2. Lem.

Erit ergo per Lemma præcedens, trilineum ABCE, ad trilineum AIBF, vt est triangulum AEC, ad ABF. siue ad FBE. Idem verò trilineum ABCE ad aliud trilineum BHCG, erit vt idem triangulum AEC, ad triangulum BGC; hoc est ad BGE, Coniunctim ergo, erit trilineum ABCE ad duo trilinea AIBF, BHCG, vt triangulum AEC ad triangulum FEG, nempe vt 4, 24. quinti.

D 2 ad vnum

ri trilineo mixto,) ad omnia confequentia fimul (nempe ad figuram infcriptam in posteriori trilineo mixto) vt vnum ad vnum; nempe vt igl, *ad* qhr. *Siue fumptis eorum quadruplis, vt* agc *ad* dhf. *Sed eadem infcripta figura habebat ad trilineum* defh *maiorem rationem quàm fit trianguli* agc *ad* dhf. *Minus ergo erit trilineum mixtum* defh, *quàm figura fibi infcripta: totum fua parte. Quod eft impoffibile. Trilinea ergò fub tangentibus, & curuà parabolicà comprœhenfa, funt inter fe vt triangula fub ijfdem tangentibus & bafibus contenta.* *x. quin* [ti]

Propofitio II.

PArabola fefquitertia eft trianguli eandem ipfi bafim, & eandem altitudinem habentis.

Sit parabola ABC, cuius diameter BD; & fit infcriptum triangulum ABC. Dico parabolam fefquitertiam effe trianguli ABC,

Ducantur duæ tangentes ad bafim, quæ fint AE, CE. & FG. tãgat per verticem B Demiffis deinde FI, GH diametro parallelis, vt fint diametri portionum AIB, BHC; ducantur per I, & H tangentes LM, NO. 3.*Le*[m]

Erit ergo per Lemma præcedens, trilineum ABCE, ad trilineum AIBF, vt eft triangulum AEC, ad ABF. fiue ad FBE. Idem verò trilineum ABCE ad aliud trilineum BHCG, erit vt idem triangulum AEC, ad triangulum BGC; hoc eft ad BGE, Coniunctim ergo, erit trilineum ABCE ad duo trilinea AIBF, BHCG, vt triangulum AEC ad triangulum FEG, nempe vt 4, 24. *quin. ti.*

ad vnum, & diuidendo, erit triangulum F E G ad duo trilinea AIBF, BHCG, vt 3, ad vnum. Trapezium autem AFCG, ad eadem trilinea erit vt 9. ad vnum; & per conuersionem rationis, ad parabolã erit vt 9. ad 8. & ad triangulum A B C, vt 9. ad 6, Qualium ergò partium parabola est 8, talium triangulum ABC est 6. Constat ergo parabolam inscripti sibi trianguli sesquitertiam esse. Quod erat &c.

Lemma VII.

Si in parabola inscribatur triangulum: eandem habens cum parabola basim, eandemq; altitudinem. Inscribantur etiam pariter & in reliquis portionibus duo alia triangula: Erit triangulum primò inscriptum, octuplum alterutri posteriùs inscripti trianguli.

Demonstratur hoc Lemma ab Archimede Prop. 21. De Quadratura parabolæ.

Lemma VIII.

Si in parabola euidentèr inscribatur figura ex triangulis constans. Tam bina ipsius triangula (si prout sibi mutuò respondent ita sumantur) quàm etiam tota inscripta figura, æquiponderabit ex puncto medio basis ipsius parabolæ.

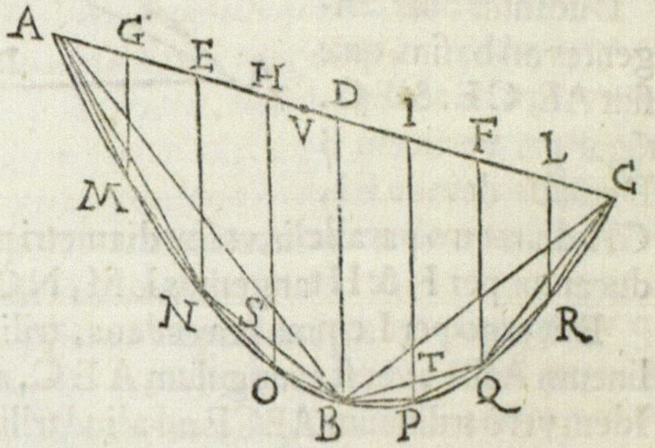

Esto parabola a b c, *cuius diameter sit* b d; *& inuersa statuatur figura, ita vt diameter ad horizontem sit perpendicularis. Sectà deinde vtraq;* ad, dc *bifariam in* e f; *iterumque sectis partibus bifariã in* g, h, i, l. *&c. Ducantur* gm, en, ho, ip, fq, lr. *&c. Parallelæ ad diametrum.*

Inscri-

ad vnum, & diuidendo, erit triangulum FEG ad duo trilinea AIBF, BHCG, vt 3, ad vnum. Trapezium autem AFCG, ad eadem trilinea erit vt 9. ad vnum; & per conuerſionem rationis, ad parabolã erit vt 9. ad 8. & ad triangulum ABC, vt 9. ad 6, Qualium ergò partium parabola eſt 8, talium triangulum ABC eſt 6. Conſtat ergo parabolam inſcripti ſibi trianguli ſeſquitertiam eſſe. Quod erat &c.

Lemma VII.

Si in parabola inſcribatur triangulum: eandem habens cum parabola baſim, eandemq; altitudinem. Inſcribantur etiam pariter & in reliquis portionibus duo alia triangula: Erit triangulum primò inſcriptum, octuplum alterutri poſteriùs inſcripti trianguli.

Demonſtratur hoc Lemma ab Archimede Prop. 21. De Quadratura parabolæ.

Lemma VIII.

Si in parabola euidentèr inſcribatur figura ex triangulis conſtans. Tam bina ipſius triangula (ſi prout ſibi mutuò reſpondent ita ſumantur) quàm etiam tota inſcripta figura, æquiponderabit ex puncto medio baſis ipſius parabolæ.

Eſto parabola abc, *cuius diameter ſit* bd; *& inuerſa ſtatuatur figura, ita vt diameter ad horizontem ſit perpendicularis. Sectà deinde vtraq;* ad, dc *bifariam in* ef; *iterumque ſectis partibus bifariã in* g, h, i, l. *&c. Ducantur* gm, en, ho, ip, fq, lr. *&c. Parallelæ ad diametrum.*

Inſcri-

Inſcribaturque in parabolà figura amnobpqrc. (*quæ dicitur euidentèr inſcribi.*) *Dico triangula quæ figuram inſcriptam componunt, ſi bina, & prout ſibi mutuò reſpondent ita ſumãtur, æquiponderare ex puncto* d. *Preterea vniuerſam figuram inſcriptam, ex ipſis triangulis compoſitam, ab eodem puncto* d, *ęquiponderare.*

Sumantur enim duo triangula ſibi mutuò reſpondentia, puta, nob, bpq, *quæ interſe æqualia erunt; cum triangula* anb, bqc *ſuboctupla. ſint eiuſdem trianguli* abc: *ipſa vero* nob, bpq, *ſuboctupla ſint æqualium triangulorum* anb, bqc. *Habebunt inſuper centra grauitatis in rectis* of, pt, *quæ quidem ab angulis* o, p, *ducuntur ad puncta media baſium,* nb, bq, *Cum verò* osh, pti *rectę ad horizontem poſitę ſint perpendiculares, erunt prædicta triangula* nob, bpq *centralitèr appenſa ex punctis* h, & i. *Quamobrem ab ęqualibus diſtantijs* hd, di, *ęquiponderabunt. Et ſic de reliquis figurę triangulis. Quod erat primò propoſitum.*

Lem. 7. *13 primi æquipon der.* *4. ſecundi eiuſd. 8. primi eiuſd.*

Figura autem vniuerſa euidentèr inſcripta componitur ex partibus æquiponderantibus à puncto d; *quarè etiam ipſa ex* d *puncto æquiponderabit. Quod erat oſtendendum, &c.*

Lemma IX.

Poſitis ijſdem. SI à parabola dematut vuiuerſa figura euidentèr inſcripta, etiàm omnia ſegmenta parabolica, quæ circumrelinquuntur, ex puncto D. æquiponderabunt.

Repetita enim eadem figura demonſtratum eſt figuram inſcriptam æquiponderare ex puncto d. *Ergo figura inſcripta centrũ grauitatis habet in perpendiculo horizontali* db. (*per 4. ſuppoſitionem*) *ſed etiam parabola centrum grauitatis habet in diametro* db, (*per 4. ſecundi ęquiponderantium*) *ergo centrum omnium reliquorum ſegmentorum erit in diametro* db. *Quare ex puncto* d *ęquiponderabunt. per 3. ſuppoſitionem. Quod &c.*

Cotollarium;

Conſtat etiam eodem prorſus argumenro, reliquum figuræ euiden-

Problema Primum. 29

Inſcribaturque in parabolà figura amnobpqrc. (*quæ dicìtur euidentèr inſcrìbi.) Dico triangula quæ figuram inſcriptam componunt, ſi bina, & prout ſibi mutuò reſpondent ita ſumãtur, æquiponderare ex puncto* d. *Prçterea vniuerſam figuram inſcriptam, ex ipſis triangulis compoſitam, ab eodem puncto* d, *çquiponderare.* Lem.7.

Sumantur enìm duo triangula ſibi mutuò reſpondentia, puta, nob, bpq, *quæ interſe*[125] *æqualia erunt; cum triangula* anb, bqc *ſuboctupla. ſint eiuſdem trianguli* abc*: ipſa vero* nob, bpq, *ſuboctupla ſint æqualium triangulorum* anb, bqc. *Habebunt inſuper contra grauitatis in rectis* oſ, pt, *quæ quidem ab angulis* o, p, *ducuntur ad puncta media baſium,* nb, bq, *Cum verò* osh, pti *rectç ad horizontem poſitç ſint perpendiculares, erunt prædicta triangula* nob, bpq *centralitèr appenſa* ex punctis h, & i. *Quamobrem ab çqualibus diſtantijs* hd, di, *çquiponderabunt. Et ſic de reliquis figurç triangulis. Quod erat primò propoſitum.* 13 *primi. æquipon. der.* 4. *ſecun. di eiuſd.*

Figura autem vniuerſa euidentèr inſcripta componitur ex partibus æquiponderantibus à puncto d; *quarè etiam ipſa ex* d *puncto æquiponderabit. Quod erat oſtendendum, &c.* 8. *primi. eiuſd.*

Lemma IX.

Poſitis ijſdem. SI à parabola dematut vuiuerſa figura euidentèr inſcripta, etiàm omnia ſegmenta parabolica, quæ circum relinquuntur, ex puncto D. æquiponderabunt.

Repetita enim eadem figura demonſtratum eſt figuram inſcriptam æquiponderare ex puncto d. *Ergo figura inſcripta centrũ grauitatis habet in perpendiculo horizontali* db. *(per 4. ſuppoſitionem) ſed etiam parabola centrum grauitatis habet in diametro* db, *(per 4. ſecundi çquiponderantium) ergo centrum omnium reliquorum ſegmentorum erit in diametro* db. *Quare ex puncto* d *çquiponderabunt. per 3. ſuppoſitionem. Quod &e.*

Corollarium;

Conſtat etiam eodem prorſus argumenro,[126] *reliquum figuræ eui-*

den-

[125] *interſe*] *inter ſe.*
[126] *argumenro*] *argumento.*

denter inscriptæ, detracto priùs triangulo a b c, *æquiponderare ex puncto* d. *Item reliquum parabolæ, dempto triangulo* a b c, *æquiponderare ex* d.

Lemma X.

Si ex parabola auferatur dimidium trianguli inscripti, tota reliqua figura mixta æquiponderabit ex puncto basis reliqui trianguli, in quo sic ea diuiditur, vt pars ad curuam terminata quadrupla sit illius, quæ terminatur ad diametrum;

Esto parabola abc *inuersa; eiusq; diameter* bd *ita statuatur vt ad horizõtem sit perpendicularis; Detractoq; semitriangulo inscripto* dbc; *seecetur ad basis reliqui semitrianguli, in quinq; partes æquales; quarum vna sit* de. *Dico huiusmodi figuram ex puncto* e *suspensam, æquiponderare.*

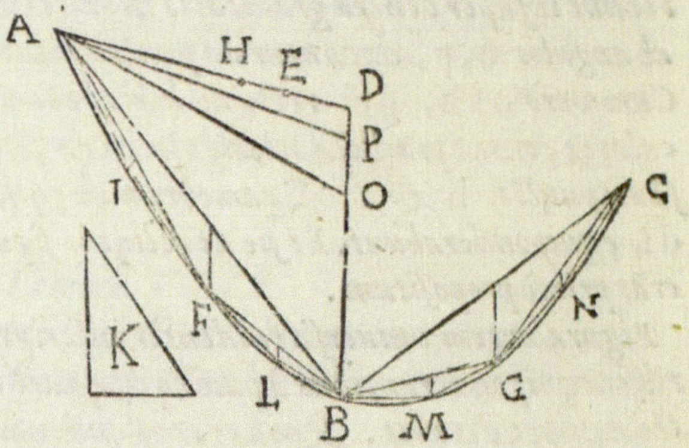

Nisi enim æquiponderet; Cum recta ad *sit libra, cuius fulcrũ est in* e, *& magnitudo* a fb gc, *constans ex duabus portionibus* Coroll. Lem. 9. *parabolicis, appensa sit ad punctum* d *secundum centrum grauitatis ipsius: Reliquum autem triangulum* abd *altera magnitudo appensa sit ad punctum* h *(sumptà* dh *tertia parte totius* da;) *Altera ex his duabus magnitudinib. præponderare necesse erit.*

Ponamus primò præponderare duas portiones afb, bgc; *& sit excessus quo præponderant, æqualis spatio* K.

Inscribatur euidentèr intrà duas portiones parabolicas figura multilatera, ita vt omnia simul segmenta parabolica circumrelicta minora sint spatio K. *Tunc enim præponderabit adhuc figura inscripta multilatera* aiflbmgncba.

Acci-

dentèr inſcriptæ, detracto priùs triangulo abc, *æquiponderare ex puncto* d. *Item reliquum parabolæ, dempto triangulo* abc, *ęquiponderare ex* d.

Lemma X.[127]

Si ex parabola auferatur dimidium trianguli inſcripti, tota reliqua figura mixta æquiponderabit ex puncto baſis reliqui trianguli, in quo ſic ea diuiditur, vt pars ad curuam terminata quadrupla ſit illius, quæ terminatur ad diametrum;

Eſto parabola abc *inuerſa; eiuſq; diameter* bd *ita ſtatuatur vt ad horizõtem ſit pęrpendicularis; Detractoq; ſemitriangulo inſcripto* dbc; *ſecetur* ad *baſis reliqui ſemitrianguli, in quinq; partes ęquales; quarum vna ſit* de. *Dico huiuſmodi figuram ex puncto* e *ſuſpenſam, æquiponderare.*

Niſi enim æquiponderet; Cum recta ad *ſit libra, cuius fulcrũ eſt in* e, *&* *magnitudo* afbgc, *conſtans ex duabus portionibus parabolicis, appenſa ſit ad punctum* d *ſecundum centrum grauitatis ipſius: Reliquum autem triangulum* abd *altera magnitudo appenſa*
Coroll *ſit ad punctum* h *(ſumptà* dh *tertia parte totius* da;*) Altera ex his duabus magnitudinib. præponderare neceſſe erit.*
Lem.9. *Ponamus primò præponderare duas portiones* afb, bgc; *& ſit exceſſus quo præponderant, æqualis ſpatio* K.

Inſcribatur euidentèr intrà duas portiones parabolicas figura multilatera, ita vt omnia ſimul ſegmenta parabolica circumrelicta minora ſint ſpatio K. *Tunc enim præponderabit adhuc figura inſcripta multilatera* a iflbmgncba.

Acci-

[127] *Ad absurdum* proof. (Cfr. Pisano and Bussotti's chapters above).

Accipiatur do *quarta pars totius* db; *& ductâ* ao, *non solum triangulum* abo, *æquiponderabit sibi ipsi ex puncto* h; *sed etiam quodcumq; aliud triangulum habens verticem in* a, *& basim in recta* db, *sibi ipsi æquiponderabit ex puncto eodem* h.

Iam sic: Qualium partium ad *est 15*, dh *est 5*, *&* dc *est 3. Ergo* de *ad* eh, *erit vt 3. ad 2. Cum autem demonstratum sit duo triangula* afb, bgc, *æquiponderare ex puncto* d; *triangulum verò* boa, *ex puncto* h; *& cum duo prædicta triangula sint ad totum triangulum* abd *vt duo ad 4.; erunt eadem ad triangulum* abo, *vt 2. ad 3.; nempe vt* he *ad* ed *reciprocè. Quamobrem duo illa triangula* afb, bgc, *cum triangulo* abo, *æquiponderabunt suspensa ex puncto* e. Lem. 8. ex Lem. 7. 6. Aequipond.

Sumatur deinde dp *quarta pars ipsius* do; *ducaturque* ap. *Iam; quia duo triangula* flb, bmg *æquiponderant ex* d *itemque duo* aif, gnc, *aquiponderant ab eodem puncto* d; *omnia simul prædicta quatuor triangula æquiponderabunt ex puncto* d; *Quatuor autem prędicta triangula ad triangulũ* afb *sunt vt duo ad 4. Sunt autem* afb, aod, *subquadrupla eiusdem trianguli* abd *& propterea ad triangulum* aop, *erunt vt 2. ad 3. nempe vt* he *ad* ed, *reciproè. Aequiponderant ergo quatuor illa triangula cum triangulo* aop, *ex puncto* e. *Ergo vniuersa simul figura euidenter inscripta* aiflbmgncba *æquiponderat triangulo* abp. *Sed eadem prçponderabat triangulo* abd, *Minus ergò est triangulum* abd *quàm triangulum* abp. *totum sua parte: quod est impossibile.* Lem, 8. ex Lem. 7. 6. Aequipond.

Ponamus deinde prçponderare triangulum abd; *& sit excessus quo prçponderat æqualis spatio* K.

Accipiatur aod *quarta pars totius trianguli* abd; *iterumque sumatur* apd, *quarta pars trianguli* aod; *et hoc semper fiat, donec veniatur ad aliquod triangulum, puta* apd, *quod minus sit quàm spatium* K. *Tunc enim reliquum* abp *adhuc prçponderabit duabus portionibus parabolicis* afb, bgc. *Sed idem triangulum ostendetur (eodem prorsus modo vt supra) çquiponderare alicui figurç intrà parabolicas portiones inscriptç: necesse igitur erit quòd portiones parabolicç minores sint quàm figura* illa

Accipiatur do *quarta pars totius* db; *& ductà* ao, *non ſolum triangulum* abo, *æquiponderabit ſibi ipſi ex puncto* h; *ſed etiam quodcumq; aliud triangulum habens verticem in* a, *& baſim in recta* db, *ſibi ipſi æquiponderabit ex puncto eodem* h. Lem.8.

Iam ſic: Qualium partium ad *est 15,* dh *eſt 5, &* de *est 3. Ergo* de *ad* eh, *erit vt 3. ad 2. Cum autem demonſtratum ſit duo triangula* afb, bgc, *æquiponderare ex puncto* d; *triangulum verò* boa, *ex puncto* h; *& cum duo pradicta*[128] *triangula ſint ad totum triangulum* abd *vt duo ad 4.; erunt eadem ad triangulum* abo, *vt 2. ad 3.; nempe vt* he *ad* ed *reciprocè. Quamobrem duo illa triangula* afb, bgc, *cum triangulo* abo, *æquiponderabunt ſuſpenſa ex puncto* e. ex Lem 7. 6. *Ae–quipond.*

Sumatur deinde dp *quarta pars ipſius* do; *ducaturque* ap.

Iam; quia duo triangula flb, bmg *ęquiponderant ex* d *itemque duo* aif, gnc, *aquiponderant ab eodem puncto* d; *omnia ſimul prædicta quatuor triangula ęquiponderabunt ex puncto* d; *Quatuor autem prędicta triangula ad triangulũ* a fb *ſunt vt duo ad 4. Sunt autem* afb, aod, *ſubquadrupla eiuſdem trianguli* abd *& propterea ad triangulum* aop, *erunt vt 2. ad 3. nempe vt* he *ad* ed, *reciprocè. Aequiponderant ergo quatuor illa triangula cum triangulo* aop, *ex puncto* e. *Ergo univerſa ſimul figura euidenter inſcripta* aiflbmgncba *æquiponderat triangulo* abp. *Sed eadem pręponderabat triangulo* abd, *Minus ergò est triangulum* abd *quàm triangulum* abp. *totum ſua parte: quod eſt impoſſibile.* Lem.8. ex Lem 7.

Ponamus deinde pręponderare triangulum abd; *& ſit exceſſus quo pręponderat æqualis ſpatio* K.

Accipiatur aod *quarta pars totius trianguli* abd; *iterumque ſumatur* apd, *quarta pars trianguli* aod; *et hoc ſemper fiat, donec veniatur ad aliquod triangulum, puta* apd, *quod minus ſit quàm ſpatium* K. *Tunc enim reliquum* abp *adhuc pręponderabit duabus portionibus parabolicis* afb, bgc. *Sed idem triangulum oſtendetur (eodem prorſus modo vt ſupra) ęquiponderare alicui figurę intrà parabolicas portiones inſcriptę: neceſſe igitur erit quòd portiones parabolicę minores ſint quàm figura* 6.*Ae–quipond.*

illa

[128] *pradicta*] *prędicta*.

illa sibi inscripta; totum sua parte. Quod est impossibile. Aequiponderant ergo parabola inuersa (dempto semitriangulo inscripto) ex puncto quod dictum est, Quod erat ostendend. &c.

Corollarium

Hinc inferre possumus, quòd si ex puncto c, *recta ducetur diametro aquidistans, centrum praedictae figurae erit in producta. Siquidem figura ex puncto* c *aequiponderat, & linea ex* c *ducta aequidistans diametro, est ad horizontem perpendicularis. Posset etiam demonstrari, nisi extrà rem esset, centrum praedictae figurae dictam parallelā ita secare, vt pars quae terminatur ad curuam sit ad reliquam vt* 11. *ad* 12.

Propositio III.

PArabola sesquitertia est trianguli eandem sibi basim, & eandem altitudinem habentis.

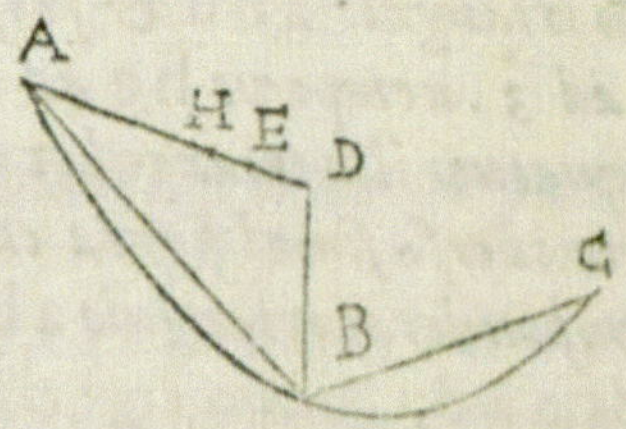

Esto parabola ABC, ex qua demptum sit dimidium trianguli inscripti: Sumptaq; DH, quæ sit tertia pars totius DA & DE quinta pars eiusdem; si parabola huiusmodi statuatur inuersa, ita vt diameter sit horizonti perpendicularis, æquiponderabit figura ex puncto E, Sed triangulum ABD appensum est secundum centrum grauitatis ad punctum H libræ HD. Duæ autem parabolicæ portiones residuæ appensæ sunt secundum centrum grauitatis ad punctum D; Ergo triangulum ABD, ad duas reliquas portiones erit vt DE ad EH, reciprocè, nempe vt 3. ad 2: Sumptis autem antecedentiũ duplis erit totum inscriptum triangulum ad reliquas portiones vt 6. ad 2. Conuertendo igitur, & componendo, erit ipsa parabola ad inscriptum sibi triangulum vt 8, ad 6. Nempe sesquitertia. Quod &c.

praced. Lem.

Libet

illa ſibi inſcripta; totum ſua parte. Quod eſt impoſſibile. Aequiponderant ergo parabola inuerſa (dempto ſemitriangulo inſcripto) ex puncto quod dictum eſt, Quod erat ostendend. &c.

Corollarium.

Hinc inferre poßumus, quòd ſi ex puncto e, *recta ducetur diametro aequidistans, centrum praedictae figurae erit in producta. Siquidem figura ex puncto* e *aequiponderat, & linea ex* e *ducta aequidiſtans diametro, est ad horizontem perpendicularis. Poſſet etiam demonſtrari, niſi extrà rem eſſet, centrum praedictae figurae dictam parallelã ita ſecare, vt pars quae terminatur ad curuam ſit ad reliquam vt 11. ad 12.*

Propoſitio III.[129]

PArabola ſeſquitertia eſt trianguli eandem ſibi baſim, & eandem altitudinem habentis.

Eſto parabola ABC, ex qua demptum ſit dimidium trianguli inſcripti: Sumptaq; DH, quæ ſit tertia pars totius DA & DE quinta
præced. Lem. pars eiuſdem; ſi parabola huiuſmodi ſtatuatur inuerſa, ita vt diameter ſit horizonti perpendicularis, æquiponderabit figura ex puncto E, Sed triangulum ABD appenſum eſt ſecundum centrum grauitatis ad punctum H libræ HD. Duæ autem parabolicæ portiones reſiduæ appenſæ ſunt ſecundum centrum grauitatis ad punctum D; Ergo triangulum ABD, ad duas reliquas portiones erit vt DE ad EH, reciprocè, nempe vt 3. ad 2: Sumptis autem antecedentiũ duplis erit totum inſcriptum triangulum ad reliquas portiones vt 6. ad 2. Conuertendo igitur, & componendo, erit ipſa parabola ad inſcriptum ſibi triangulum vt 8, ad 6. Nempe ſeſquitertia. Quod &c.
Libet

[129] This is Torricelli's *Proposition III* on *Quadrature of the Parabola*. As to Archimedes, see the *Proposition XXIV* on *Quadrature of the Parabola* (Archimedes 1897, Proposition 24, p. 251. Cfr. Pisano and Bussotti's chapters above). The kind of proof is *Componendo*. Cfr. *Ibidem.*

Problema Primum. 33

Libet hic demonstrare Lemma Lucæ Valerij, nostro tamen modo, diuersisque penitùs Mechanicæ principijs. Ipse enim vtitur Propositione illa, quà ante demonstrauerat centrum grauitatis hemisphærij. Nos autem simili ratione, ac in præcedentibus, demonstrabimus & Lemma, & ipsam Valerij conclusionem.

Lemma XI.

Omnis semiparabola æquiponderat ex puncto basis, in quo sic ea diuiditur vt pars ad curuam terminata sit ad reliquam vt quinque ad tria:

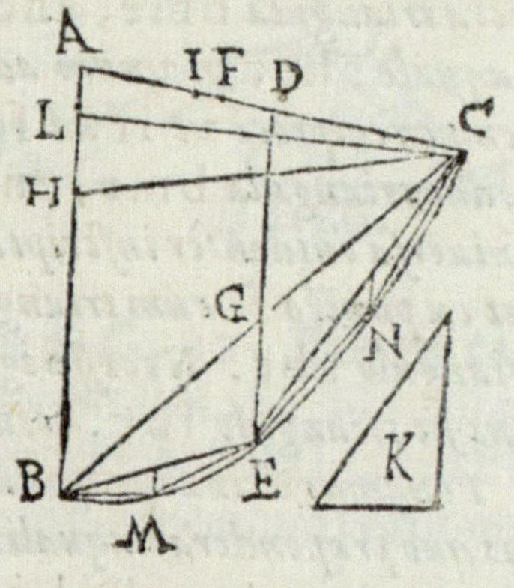

Esto semiparabola abc, *cuius diameter* ab *statuatur ad horizontem perpendicularis: Sectâ deinde* ac *in* f, *ita vt* cf *ad* fa, *sit vt 5. ad 3. vel vt 15. ad 9. Dico figuram ex puncto* f *suspensam æquiponderare.*

Secetur iterùm ac *bifariàm in* d, *& demissâ* de *parallela diametro, erit ipsa* de, *diameter parabolę* bec. *Sumatur iam* ai *tertia pars totius* ac. *Qualium igitur partium* ac *est 24. taliũ* ad *est 12.* af *9. &* ai *8, Ergo* df *tres, &* fi *vna. Iam si figura non æquiponderat ex puncto* f; *Cum* id *sit libra quædam cuius fulcrum est* f, *& ad punctum* i *appensum sit triangulum* abc; *ex puncto verò* d *appensa sit parabola* bec; *altera ex his figuris præponderabit. Ponamus primò præponderare parabolam* bec, *sitq; excessus quo præponderat æqualis spatio* K. *ex supp. quinta.*

Inscribatur euidentèr intra parabolam bec *figura rectilinea, ita vt omnes simul residuæ portiunculæ quibus parabola excedit inscriptam sibi figuram, minores sint spatio* K. *Manifestũ est, quòd inscripta euidentèr figura adhuc præponderabit triangulo* abc.

Accipiatur ahc *quarta pars totius trianguli* abc. *Cum autem* dc *sit ad horizontem perpendicularis, & triangulum* bec

E *habeat*

Libet hic demonſtrare Lemma Lucæ Valerij, nostro tamen modo, diuerſiſque penitùs Mechanicæ principijs. Ipſe enim vtitur Propoſitione illa, quà ante demonſtrauerat centrum grauitatis hemiſphærij. Nos autem ſimili ratione, ac in præcedentibus, demonstrabimus & Lemma, & ipſam Valerij concluſionem[130].

Lemma XI.[131] .

Omnis ſemiparabola æquiponderat ex puncto baſis, in quo ſic ea diuiditur vt pars ad curuam terminata ſit ad reliquam vt quinque ad tria:

Eſto ſemiparabola abc, *cuius diameter* ab *ſtatuatur ad horizontem perpendicularis: Sectà deinde* ac *in* f, *ita vt* cf *ad* fa, *ſit vt 5. ad 3. vel vt 15. ad 9. Dico figuram ex puncto* f *ſuſpenſam ęquiponderare.*

Secetur iterùm ac *bifariàm in* d, *& demiſsà* de *parallela diametro, erit ipſa* de, *diameter parabolę* bec. *Sumatur iam* ai *tertia pars totius* ac. *Qualium igitur partium* ac *eſt 24. taliũ* ad *eſt 12.* af *9. &* ai *8, Ergo* df *tres, &* fi *vna. Iam ſi figura non æquiponderat ex puncto* f; *Cum* id *ſit libra quædam cuius fulcrum eſt f, & ad punctum* i *appenſum ſit triangulum* abc; *ex puncto verò* d *appenſa ſit parabola* bec; *altera ex his figuris præponderabit. Ponamus primò præponderare parabolam* bec, *ſitq; exceſſus quo præponderat æqualis ſpatio* K. *ex ſuppo quinta.*

Inſcribatur euidentèr intra parabolam bec *figura rectilinea, ita vt omnes ſimul reſiduæ portiunculæ quibus parabola excedit inſcriptam ſibi figuram, minores ſint ſpatio* K. *Manifeſtũ eſt, quòd inſcripta euidentèr figura adhuc præponderabit triangulo* abc.

Accipiatur ahc *quarta pars totius trianguli* abc. *Cum autem* de *ſit ad horizontem perpendicularis, & triangulum* bec

E *habeat*

[130] In *Proposition III*, Torricelli quotes Luca Valerio's *De centris gravitatis solidorum libri tres* (Valerio 1604, II, p. 12).

[131] *Ad absurdum* proof. (Cfr. Pisano and Bussotti's chapters above).

k 3 primi æquipon. *habeat centrum grauitatis in recta* g e; *erit dictam triangulum appensum ad* d. *Triangulum verò* b h c *appensum ad punctum* i; *quandoquidem* a i *tertia pars est totius* a c, *ipsa vero* a b *perpendicularis ad horizontem constituta est*, *Quoniam autem* b e c *ad* a b c *est vt vnum ad 4., erit idem* b e c *ad* h b c *vt vnũ*
Lem. 7. *ad 3. nempe reciprocè vt* i f *ad* f d, *Aequiponderant ergo ex puncto* f, *triangula* b e c, *&* h b c.

Sumatur iterum a l c *quarta pars trianguli* a h c; *Et quoniã duo triangula* b m e, e n c *æquiponderant ex* g (*vti demonstra*
Lem. 8. *tum est*) *æquiponderabunt etiam suspensa ex* d. *Cum autem duo*
Lem. 7. *dicta triangula* b m e, e n c, *sint ad triangulum* b e c, *siue ad ipsi æquale* a h c, *vt vnum ad 4.; erunt ad* l h c, *vt vnum ad 3; nempe reciprocè vt* i f *ad* f d. *Aequiponderant ergo ex puncto* f. *duo triangula* b m e, e n c, *cum triangulo* l h c. *Figura ergò vniuersa euidentèr inscripta intrà parabolam* b e c. *æquiponderat ex puncto* f. *cum triangulo* l b c. *Sed eadem præponderabat triangulo* a b c. *Necesse igitur est quòd triangulum* a b c *minus sit triangulo* l b c. *totum sua parte*, *Quod est absurdum*.

Ponamus deinde præponderare triangulum a b c, *& sit excessus quo præponderat æqualis spatio* K. *Sumatur* a h c *quarta pars totius trianguli* a b c. *Iterùm sumatur* a l c *quarta pars trianguli* a h c, *Et hoc semper fiat, donec ventum fuerit ad aliquod triangulum, puta* a l c, *minus spatio* K. *Tunc enim triangulum* l b c *adhuc præponderabit parabolę* b e c. *Sed eodem modo, quo supra, demonstrabimus dictum triangulum* l b c *æquiponderare cuidam figuræ euidentèr inscriptæ intrà parabolam* b e c. *Vnde sequeretur ipsam parabolam* b e c *minorem esse aliqua figurà sibi inscriptà; totum videlicet sua parte*. *Quod est absurdum*. *Aequiponderat ergo semiparabola, vti dictum est constituta, & ex puncto* f *suspensa*. *Quod &c.*

Corollarium.

Hinc patet, quòd (cum semiparabola æquiponderet ex puncto f.) *si ab* f *demittatur recta horizonti perpendicularis, in hac de-*
Suppositio 4. *missà erit centrum grauitatis semiparabolæ; aliàs enim non ęqui*
pon-

3 *primi.* *habeat centrum grauitatis in recta* ge; *erit dictum triangulum appenſum ad* d. *Triangulum verò* bhc *appenſum ad punctum* i;
ęquipon. *quandoquidem* ai *tertia pars eſt totius* ac, *ipſa vero* ab *perpendicularis ad horizontem conſtituta eſt, Quoniàm autem* bec *ad* abc *eſt vt*
Lem 7. *vnum ad 4., erit idem* bec *ad* hbc *vt vnũ ad 3. nempe reciprocè vt* if *ad* fd, *Aequiponderant ergo ex puncto* f, *triangula* bec, *&* hbc.

Sumatur iterum alc *quarta pars trianguli* ahc; *Et quoniã duo triangula* bme, enc *æquiponderant ex* g (*vti demonstratum est*) *æquiponderabunt etiam ſuſpenſa ex* d. *Cum autèm duo dicta triangula* bme, enc, *ſint ad triangulum* bec, *ſiue ad ipſi aquale* ahc, *vt vnum ad 4.; erunt ad* lhc, *vt vnum ad 3; nempe reciprocè vt* if *ad* fd. *Aequiponderant ergo ex puncto* f. *duo triangula* bme, enc, *cum*
*Lem.*8 *triangulo* lhc. *Figura ergò vniuerſa euidentèr inſcripta intrà parabolam* bec. *æquiponderat ex puncto* f. *cum triangulo* lbc. *Sed eadem*
*Lem.*7. *præponderabat triangulo* abc. *Neceſſe igitur est quòd triangulum* abc *minus ſit triangulo* lbc. *totum ſua parte, Quod est abſurdum.*

Ponamus deinde præponderare triangulum abc, *& ſit exceſſus quo præponderat æqualis ſpatio* K. *Sumatur* ahc *quarta pars totius trianguli* abc. *Iterùm ſumatur* alc *quarta pars trianguli* ahc, *Et hoc ſemper fiat, donec ventum fuerit ad aliquod triangulum, puta* alc, *minus ſpatio* K. *T unc enim triangulum* lbc *adhuc præponderabit parabolę* bec. *Sed eodem modo, quo ſupra, demonstrabimus dictum triangulum* lbc *æquiponderare cuidam figuræ euidentèr inſcriptæ intrà parabolam* bec. *Vnde ſequeretur ipſam parabolam* bec *minorem eſſe aliqua figurà ſibi inſcriptà; totum videlicet ſua parte. Quod eſt abſurdum.*

Aequiponderat ergo ſemiparabola, vti dictum est conſtituta, & ex puncto f *ſuſpenſa. Quod &c.*

Corollarium.

Hinc patet, quòd (cum ſemiparabola æquiponderet ex puncto f.) *ſi ab* f *demittatur recta horizonti perpendicularis, in hac demiſſà*
Suppoſi- *erit centrum grauitatis ſemiparabolæ; aliàs enim non ęqui*
tio 4. *pon-*

ponderaret ex f. Verùm quoniam etiam diameter parabolę ad horizontem perpendicularis constituta est, concludemus; quòd à recta quæ ex puncto f ducitur diametro ęquidistans, transit per centrum semiparabolæ.

Propositio IV.

Parabola sesquitertia est trianguli eandem ipsi basim, & eandem altitudinem habentis.

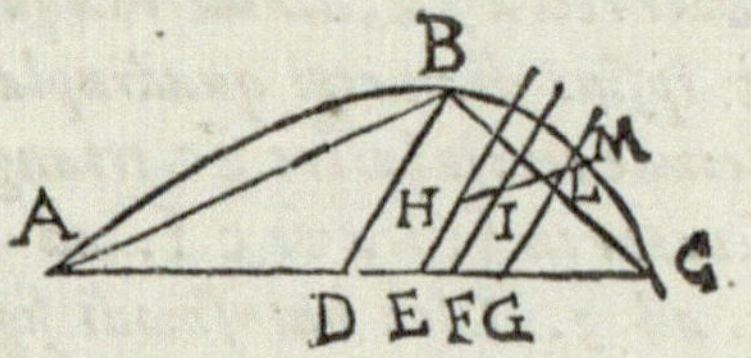

Esto parabola ABC, cuius diameter BD, triangulum verò inscriptum sit ABC, Dico parabolam dicti trianguli esse sesquitertiam.

Sumatur, qualium partium tota DC est 24. talium DE 8.; DF 9; & DG, 12. Eritque earundem EF vna, & FG tres, Ductis verò EH, FI, GL, diametro parallelis, erit in EH centrum trianguli BDC; in FI centrum semiparabolæ DBMC, & in GL centrum portionis BMC, *13. primò æquip.* *Lem præced.*

Ponatur centrum trianguli esse punctum quodcumque H. Item centrum semiparabolæ esse punctum quodcunque I (quamquàm huiusmodi puncta extrà ipsas figuras vbicunq; libuerit sumantur, tamen verum semper eodē modo inferemus.) iunctà deinde HI, & productà, in ipsa HI erit centrum portionis parabolicę BMC; quod cum sit etiam in recta GM producta, necessariò erit in communi concursu L. Parabola ergò BMC ad triangulum DBC erit reciprocè vt HI ad IL, hoc est, vt EF ad FG, nempe vt vnum ad 3. Componendo ergo, sumptisq; duplis, erit tota parabola ad totum triangulum vt 4. ad 3. Nempe sesquitertia. Quod erat propositum, &c. *4. sec. æquip.* *8. primi æquipond.*

Poterat hęc demonstratio produci etiam hoc modo, pręmisso hoc Lemmate.

ponderaret ex f. *Verùm quoniam etiam diameter parabolę ad horizontem perpendicularis constituta eſt, concludemus; quòd recta quæ ex puncto* f *ducitur diametro ęquidistans, tranſit per centrum ſemiparabolæ.*

Propoſitio IV[132]

PArabola ſeſquitertia eſt trianguli eandem ipſi baſim, & eandem altitudinem habentis.

Eſto parabola AB*C*, cuius diameter BD, triangulum verò inſcriptum ſit ABC, Dico parabolam dicti trianguli eſſe ſeſquitertiam.

Sumatur, qualium partium tota DC eſt 24. talium DE 8.; DF 9; & DG, 12. Eritque earundem EF vna, & FG tres, Ductis verò EH, FI, GL, diametro parallelis, erit in EH centrum trianguli BDC; in FI centrum ſemiparabolæ DBMC, & in GL centrum portionis BMC, i 3. *prim. æquìp. Lem præ. ced.*

Ponatur centrum trianguli eſſe punctum quodcumque H.

Item centrum ſemiparabolæ eſſe punctum quodcunque I (quamquàm huiuſmodi puncta extrà ipſas figuras vbicunq; libuerit ſumantur, tamen verum ſemper eodẽ modo inferemus.) iunctà deinde HI, & productà, in ipſa HI erit centrum portionis parabolicę BMC; quod cum ſit etiam in recta GM producta, neceſſariò erit in communi concurſu L. Parabola ergò BMC ad triangulum DBC erit reciprocè vt HI ad IL, hoc eſt, vt EF ad FG, nempe vt vnum ad 3. Componendo ergo, ſumptiſq; duplis, erit tota parabola ad totum triangulum vt 4. ad 3. Nempe ſeſquitertia. Quod erat propoſitum, &c. 4. *ſee. æquip.* 8. *prim. Aequipon* [d]*i*

Poterat hęc demonstratio produci etiam hoc modo, pręmiße hoc Lemmate.

E 2 *Si pa-*

---

[132] *Ad absurdum* proof. (Cfr. Pisano and Bussotti's chapters above).

36 De Dimensione Parabolæ

Si parabola ad extremum basis lineam habuerit diametro parallelam, & diametri quadruplam, ductoq; tertio latere, compleatur triangulum. Vniuersa hæc figura æquiponderabit ex pũcto tertij lateris, in quo sic diuiditur vt pars ad curuam terminata sit ad reliquam vt 5. ad 3.

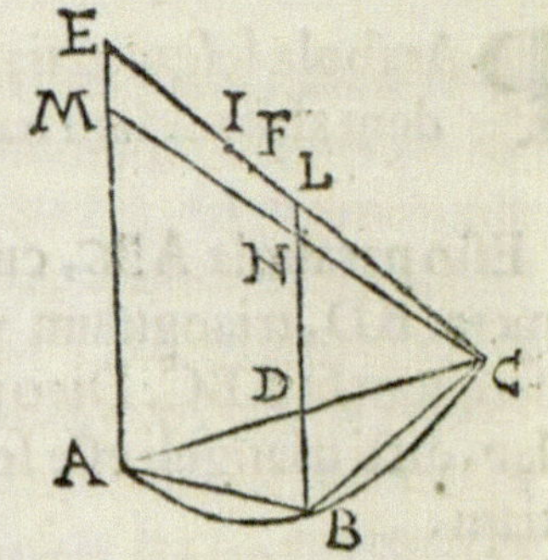

Esto parabola abc, *cuius diameter* db *statuatur ad horizontem perpendicularis; consideretur que ipsa parabola inuersa: Tum ad alterutrum basis* ac *extremum, puta ad punctum* a, *adiungatur recta* ae, *diametro æquidistans, & ipsius diametri quadrupla. Ducto deinde tertio latere* ec *trianguli* eac, *seceturin* f, *ita vt* cf. *ad* fe *sit vt 5. ad 3. Dico huiusmodi figuram ex puncto* f *æquiponderare. Quoniam enim* ce *ordinatim applicatur ad diametrum* ae; *erit tota figura* eabc *semiparabola. Ergo ijsdem rationibus, eodemq; progressu, quo vsi sumus in lemmate 11, ostendemus totam figuram æquiponderare ex* f. *Sumatur iam* ei *octo earum partium, qualium tota* ec *est 24. &* el *12. &* ef *9, Eritq;* if *earumdem vna, &* fl *3. Ergo cum parabola* abc *pendeat ex puncto* l, *appensa ad ipsum secundum centrum grauitatis; triangulum verò* aec *ex puncto* i; *erit parabola* abc *ad triangulum* aec *vt reciprocè* if *ad* fl, *nempe vt vnum ad 3.; siue ut 4. ad 12. & propterea ad triangulum* abc *ut 4. ad 3. &c. Est enim* abc *quarta pars ipsius* ace *&c. Constat ergo parabolam sesquitertiã esse inscripti sibi trianguli.*

Ostendetur infra

Quod assumptum est, nempe rectam ce *ordinatim applicari ad diametrum* ae, *ostendemus hoc modo.*

ob commune basim AC altitudinem verò in ratione quadrupla.

Si enim non est ordinatim applicata ce, *applicetur ordinatim* cm; *eritq;* mabc *semiparabola; & quia sunt æquales* ad, dc, *erit* mc *secta bifariam in* n. *Ergo* ma *sesquitertia est ipsius* nb; *sed etiam* ea *ob constructionem sesquitertia est ipsius* lb;

ob semiparabolã

ergo

Si parabola ad extremum bafis lineam habuerit diametro parallelam, & diametri quadruplam, ductoq; tertio latere, compleatur triangulum. Vniuerfa hæec figura ęquiponderabit ex pũcto tertij lateris, in quo fic diuiditur vt pars ad curuam terminata fit ad reliquam vt 5. ad 3.

Ostende tur infra

ob commune ba[-] fim A C altitudinem verò in ratione quad rupla.

Efto parabola abc, *cuius diameter* db *ftatuatur ad horizontem perpendicularis; confidereturque ipfa parabola inuerfa: Tum ad alterutrum bafis* a c *extremum, puta ad punctum* a, *adiungatur recta* ae, *diametro æquidiftans, & ipfius diametri quadrupla. Ducto deinde tertio latere* ec *trianguli* eac, *fecetur in* f, *ita vt* cf. *ad* fe *fit vt 5. ad 3. Dico huiufmodi figuram ex puncto* f *ęquiponderare. Quoniam enim* ce *ordinatim applicatur ad diametrum* ae; *erit tota figura* eabc *femiparabola. Ergo ijfdem rationibus, eodemq; progreffu, quo vfi fumus in lemmate 11, ostendemus totam figuram æquiponderare ex* f. *Sumatur iam* ei *octo earum partium, qualium tota* ec *est 24. &* el *12. &* e f *9, Eritq;* if *earumdem vna, &* fl *3. Ergo cum parabola* abc *pendeat ex puncto* l, *appenfa ad ipfum fecundum centrum grauitatis; triangulum verò* aec *ex puncto* i; *erit parabola* abc *ad triangulum* aec *vt reciprocè* if *ad* fl, *nempe vt vnum ad 3.; fiue ut 4. ad 12. & propterea ad triangulum* abc *ut 4. ad 3. & c. Eft enim* abc *quarta pars ipfius* ace *&c. Conftat ergo parabolam fefquitertiã effe infcripti fibi trianguli.*

ob femiparabolã

Quod affumptum est, nempe rectam ce *ordinatìm applicari ad diametrum* ae, *oftendemus hoc modo.*

Si enim non eft ordinatìm applicata ce, *applicetur ordinatìm* cm; *eritq;* mabc *femiparabola; & quia funt ęquales* ad, dc, *erit* mc *fecta bifariam in* n. *Ergo* ma *fefquitertia eft ipfius* nb; *fed etiam* ea *ob constructionem fefquitertia est ipfius* lb;

ergo

Problema Primum. 37

ergo & reliqua em *ſeſquitertia eſt reliquę* ln; & ec *ſeſqui-* 19 quinti.
tertia ipſius cl. *quod eſt impoſſibile. Eſt enim dupla, non autem ſeſquitertia. Quare nulla alia pręter* ce *ex puncto* c *ordinatim applicari poteſt ad diametrum* ae. *Quod &c.*

Propoſitio V.

PArabola ſeſquitertia eſt trianguli eamdem ipſi baſim, eandemq; altitudinem habentis.

Eſto parabola ABC, cuius diameter, BD, triangulum inſcriptum ABC; Dico parabolam eſſe ſeſquitertiam trianguli ABC. ſibi inſcripti.

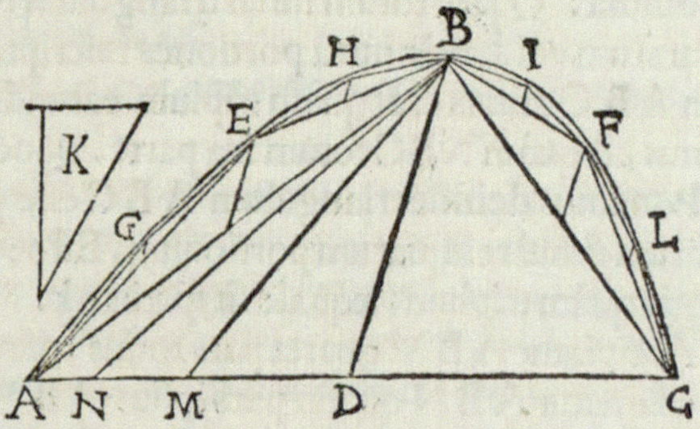

Si enim ita non eſt, neq; triangulum ABC erit triplum duarum ſimul reliquarum portionum AEB, BFC Sed erit vel magis quàm triplum, ſiue minus quam triplum.

Sit primò minus quam triplum, eruntq; duæ reliquæ portiones magis quàm tertia pars trianguli ABC. Eſto exceſſus æqualis ſpatio K, & inſcribantur intrà portiones primùm triangula AEB, BFC; iterumque in reliquis portiunculis quatuor triangula AGE, EHB, BIF, FLC; deinde octo &c. & hoc ſemper donec exceſſus portionum ſupra inſcriptas euidentèr figuras ſit minor ſpatio k. Tunc .n. erunt inſcriptæ figuræ adhuc maiores quàm tertia pars trianguli ABC.

Sumatur iam triangulum ABM quarta pars totius trianguli ABC. Et quoniam ABC quadruplum eſt tam trianguli ABM, ex Lem. 7.
quam triangulorum AEB, BFC ſimul ſumptorum, æquale erit

trian-

ergo & reliqua em *ſeſquitertia est reliquę* ln; *&* ec *ſeſquitertia ipſius* cl. *quod eſt impoſſibile. Est enim dupla, non autem ſeſquitertia. Quare nulla alia pręter* ce *ex puncto* c *ordinatim applicari poteſt ad diametrum* ae. *Quod &c.* 19 *quinti.*

Propoſitio V.[133]

PArabola ſeſquitertia eſt trianguli eamdem ipſi baſim, eandemq; altitudinem habentis.

Eſto parabola ABC, cuius diameter, BD, triangulum inſcriptum ABC; Dico parabolam eſſe ſeſquitertiam trianguli ABC. ſibi inſcripti.

Si enim ita non eſt, neq; triangulum ABC erit triplum duarum ſimul reliquarum portionum AEB, BFC Sed erit vel magis quàm triplum, ſiue minus quam triplum.

Sit primò minus quam triplum, eruntq; duæ reliquæ portiones magis quàm tertia pars trianguli ABC. Eſto exceſſus æqualis ſpatio K, & inſcribantur intrà portiones primùm triangula AEB, BFC; iterumque in reliquis portiunculis quatuor triangula AGE, EHB, BIF, FLC; deinde octo &c. & hoc ſemper donec exceſſus portionum ſupra inſcriptas euidentèr figuras ſit minor ſpatio k. Tunc.n. erunt inſcriptæ figuræ adhuc maiores quàm tertia pars trianguli ABC.

Sumatur iam triangulum ABM quarta pars totius trianguli ABC. Et quoniam ABC quadruplum eſt tam trianguli ABM quam triangulorum AEB, BFC ſimul ſumptorum, æquale erit *ex Lem.* 7.

trian-

[133] *Ad absurdum proof* and *Componendo*. (Cfr. Pisano and Bussotti's chapters above).

triangulum A B M duobus simul triangulis A E B, B F C; & propterea triangulum M B C triplum erit duorum simul triangulorum A E B, B F C.

Accipiatur iterùm triangulum A B N quarta pars totius trianguli A B M. Cum ergò A B M quadruplum sit ipsius A B N; &

Lem. 7. duo A E B, B F C, quadrupla sint quatuor simul subsequentiũ triangulorum A G E, E H B, B I F, F L C; cumque antecedentia sint æqualia, æqualia erunt etiam consequentia; & propterea cum triangulum N B M triplum sit trianguli A B N, triplum etiã erit idem triangulum N B M. quatuor simul triangulorum A G

12 Quinti. E, E H B, B I F, F L C. Et vt vnum ad vnum ita omnia simul ad omnia. Quare totum simul triangulum N B C, triplum erit figurarum euidentèr intrà portiones inscriptarum. Sed triangulum A B C minus erat quàm triplum earundem: Ergo A B C minus est quàm N B C totum sua parte. Quod est absurdum &c.

Ponamus deinde triangulum A B C esse plus quam triplum duarum simul reliquarum portionum. Esto: & excessui, quo est magis quàm triplum, æquale sit spatium k.

Accipiatur A B M quarta pars totius trianguli A B C. Iterum sumatur A B N quarta pars ipsius A B M. Et hoc semper fiat donec veniatur ad aliquod triangulum, puta A B N, quod minus sit spatio K. Eritq: adhuc triangulum N B C magis quàm triplum duarum portionum. Sed eàdem prorsus ratione, & ordine quo supra, ostendemus triangulum N B C triplum esse cuiusdam figuræ intrà portiones euidentèr inscriptæ; Necesse igitur erit quòd portiones ipsæ minores sint quàm figuræ intrà ipsas descriptę: Totum sua parte. quod est impossibile.

Triangulum ergo A B C duarum reliquarum portionum triplum est; & componendo; & per conuersionem rationis parabola ad suum triangulum erit vt 4. ad 3. Nempe sesquitertia. Quod erat propositum &c.

Lemma XII.

Si parabola tres tangentes habuerit, duas ad basim, tertiam verò per verticem: Erit triangulum sub tangentibus compræhensum, reliquæ figurę (demptà parabolà) triplum.

Esto

triangulum ABM duobus ſimul triangulis AEB, BFC; & propterea triangulum MBC triplum erit duorum ſimul triangulorum AEB, BFC.

Accipiatur iterùm triangulum ABN quarta pars totius trianguli ABM. Cum ergò ABM quadruplum ſit ipſius ABN; & duo AEB, *Lem* 7. BFC, quadrupla ſint quatuor ſimul ſubſequentiũ triangulorum AGE, EHB, BIF, FLC; cumque antecedentia ſint æqualia, æqualia erunt etiam conſequentia; & propterea cum triangulum NBM triplum ſit trianguli ABN, triplum etiã erit idem triangulum NBM. quatuor ſimuI triangulorum AGE, EHB, BIF, FLC. Et vt vnum ad vnum ita omnia ſimul ad omnia. Quare totum ſimul triangulum NBC, 12 *Quinti.* triplum erit figurarum euidentèr intrà portiones inſcriptarum. Sed triangulum ABC minus erat quàm triplum earundem: Ergo ABC minus eſt quàm NBC totum ſua parte. Quod eſt abſurdum &c.

Ponamus deinde triangulum ABC eſſe plus quam triplum duarum ſimul reliquarum portionum. Eſto: & exceſſui, quo eſt magis quàm triplum, æquale ſit ſpatium k.

Accipiatur ABM quarta pars totius trianguli ABC. Iterum ſumatur ABN quarta pars ipſius ABM. Et hoc ſemper fiat donec veniatur ad aliquod triangulum, puta ABN, quod minus ſit ſpatio K. Eritq: adhuc triangulum NBC magis quàm triplum duarum portionum. Sed eàdem prorſus ratione, & ordine quo ſupra, oſtendemus triangulum NBC triplum eſſe cuiuſdam figuræ intrà portiones euidentèr inſcriptæ; Neceſſe igitur erit quòd portiones ipſæ minores ſint quàm figuræ intrà ipſas deſcriptę: Totum ſua parte. quod eſt impoſſibile.

Triangulum ergo ABC duarum reliquarum portionum triplum eſt; & componendo; & per conuerſionem rationis parabola ad ſuum triangulum erit vt 4. ad 3. Nempe ſeſquitertia.

Quod erat propoſitum &c.

Lemma XII.[134]

Si parabola tres tangentes habuerit, duas ad baſim, tertiam verò per verticem: Erit triangulum ſub tangentibus compræhenſum, reliquæ figurę (demptà parabolà) triplum.

Esto

[134] *Ad absurdum* proof. (Cfr. Pisano and Bussotti's chapters above).

Problema Primum. 39

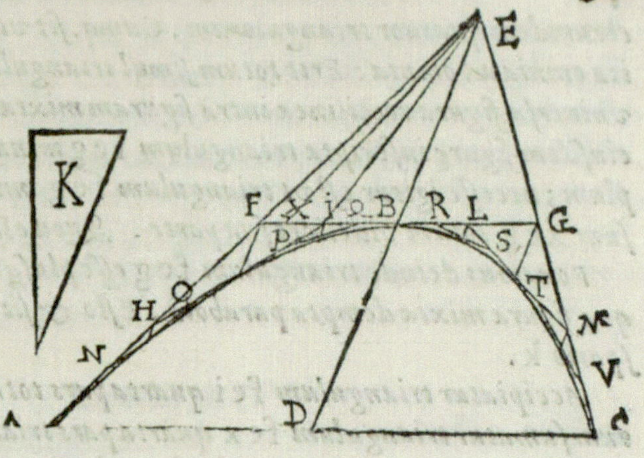

Esto parabola abc, *cuius diameter* bd; *tangentes ad basim* ae,ce; *per uerticem verò* fbg.

Dico triangulum feg, *sub tangẽtibus comprẹhensum reliquæ figurẹ mixtae* abcgf *(dẽpta scilicet parabola) triplum esse.*

Si enim non est triplum, erit certè uel magis, uel minùs quàm triplum.

Sit primò minùs quàm triplum; eritq; reliqua figura mixta abcgf, *magis quàn tertia pars trianguli* feg. *Sit excessus* K. *Ducanturq; per uertices abscissarum portionum tangentes* hi, lm; *Iterumque per uertices subsequentium portionum, tãgentes agantur* no, pq, rs, tu. *& hoc semper; donec excessus figurae mixtae* abcgf, *supra figuram ex triangulis constantem* nopqrstugf, *minus aliquando relinquatur quàm spatium* K. *Tunc enim erit adhuc figura ex triangulis inscripta maior quàm tertia pars trianguli* feg.

Accipiatur triangulum fei *quarta pars trianguli* feg; *eritque triangulum* fei *aequale duobus simul triangulis* hfi, lgm: *(cum tam ista duo, quàm illud solum, subquadrupla sint eiusdẽ* *Lem. 3.* *trianguli* feg) *Ergo triangul.* ieg *triplùm erit duorum simul triangulorum* hfi, lgm.

Sumatur iterùm triangulum fex *quarta pars ipsius* fei. *Cumque sit* fei *quadruplum trianguli* fex, *duo verò triangula* hfi, lgm *quadrupla sint quatuor simul triangulorum* nho, piq *Lem 3.* rlf, tmu, *& antecedentia æqualia; etiam consequentia æqualia erunt; eritq; triangulum* fex, *æquale quatuor prædictis triangulis* nho, piq, rls, tmu: *& propterea* xei *triplum erit*

eorum-

Esto parabola abc, *cuius diameter* bd; *tangentes ad basim* ae, ce*; per uerticem verò* fbg.

Dico triangulum feg, *sub tangẽtibus comprehensum reliquæ figurę mixtae* abcgf *(dẽpta scilicet parabola) triplum esse.*

Si enim non est triplum, erit certè uel magis, uel minùs quàm triplum.

Si primò minùs quàm triplum; eritq; reliqua figura mixta abcgf, *magis quàm tertia pars trianguli* feg *Sit excessus* K. *Ducanturq; per uertices abscissarum portionum tangentes* hi, lm*; Iterumque per uertices subsequentium portionum, tãgentes agantur* no, pq, rs, tu. *& hoc semper; donec excessus figurae mixtae* abcgf, *supra figuram ex triangulis constantem* nopqrstugf, *minus aliquando relinquatur quàm spatium* K. *Tunc enim erit adbuc figura ex triangulis inscripta maior quàm tertia pars trianguli* feg.

Accipiatur triangulum fei *quarta pars trianguli* feg*; eritque triangulum* fei *aequale duobus simul triangulis* hfi, lgm*: (cum tam ista duo, quàm illud solum, subquadrupla sint eiusdẽ trianguli* feg*) Ergo triangul.* ieg *triplum erit duorum simul triangulorum* hfi, lgm. *Lem.*3.

Sumatur iterùm triangulum fex *quarta pars ipsius* fei. *Cumque sit* fei *quadruplum trianguli* fex, *duo verò triangula* hfi, lgm *quadrupla sint quatuor simul triangulorum* nho, piq, rls, tmu, *& antecedentia æqualia; etiam consequentia æqualia erunt; eritq;* *Lem.*3. *triangulum* fex, *æquale quatuor prædictis triangulis* nho, piq, rls, tmu*: & propterea* xei *triplum erit*

eorum-

12 Quinti. *corumdem quatuor triangulorum. Cumq. sit vt vnum ad vnum, ita omnia ad omnia: Erit totum simul triangulum* x e g *triplum vniuersæ figuræ rectilineæ intrà figuram mixtam inscriptæ. Sed eiusdem figurę inscriptæ triangulum* f e g *minus erat quàm triplum; necesse igitur est vt triangulum* f e g *minus sit quàm ipsum* x e g *totum videlicet sua parte. Quod est impossibile.*

Ponamus deinde triangulum f e g *esse plusquàm triplum reliquę figuræ mixtæ dempta parabolà. Esto & sit excessus ęqualis spatio* k.

Accipiatur triangulum f e i *quarta pars totius* f e g: *& iterum sumatur triangulum* f e x *quarta pars trianguli* f e i: *& hoc fiat semper donec veniatur ad aliquod triangulum, puta* f e x, *quod minus sit spatio* K. *Eritq. triangulum* x e g *adhuc maius quàm triplum reliquæ figuræ mixtæ* a b c g f. *Sed eadem penitùs ratione, atque ordine vt supra, ostendemus triangulum* xeg *esse triplum cuiusdam figurę intrà figuram mixtam* a b c g f. *descriptę. Necesse ergò erit, vt figura mixta* a b c g f. *minor sit quàm aliqua figura sibi inscripta; totum sua parte. Quod est absurdum.*

Si ergò parabola tres tangentes habuerit, vt positum est, erit triangulum sub tangentibus contentum, reliquæ figurae, dempta parabola triplum. Quod erat propositum &c.

Propositio VI.

Parabola sesquitertia est trianguli eandem ipsi basim, & eandem altitudinem habentis.

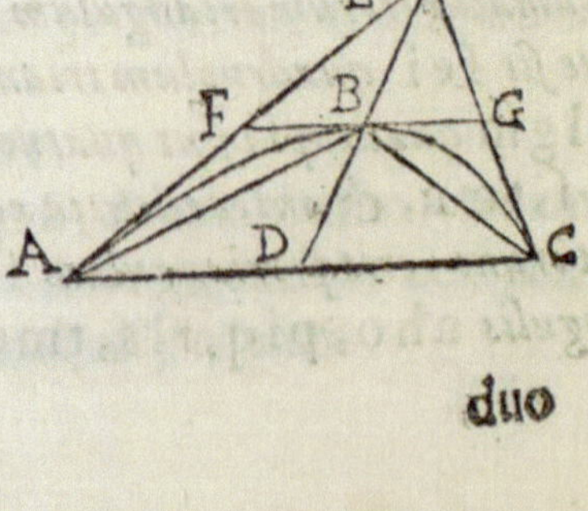

Esto parabola ABC, cuius diameter BD; duæ tangentes AE, CE ad basim, & tertia FBG per verticem. Dico parabolam sesquitertiam esse inscripti sibi trianguli ABC. Triangulum enim FEG ad

duo

12 *Quin ti.* *eorumdem quatuor triangulorum. Cumq. ſit vt vnum ad vnum, ita omnia ad omnia: Erit totum ſimul triangulum* xeg *triplum vniuerſæ figuræ rectilineæ intrà figuram mixtam inſcriptæ. Sed eiuſdem figurę inſcriptæ triangulum* feg *minus erat quàm triplum; neceſſe igitur eſt vt triangulum* feg *minus ſit quàm ipſum* xeg *totum videlicet ſua parte. Quod est impoſſibile.*

Ponamus deinde triangulum feg *eſſe pluſquàm triplum reliquę figuræ mixtæ dempta parabolà. Eſto & ſit exceſſus ęqualis ſpatio* k.

Accipiatur triangulum fei *quarta pars totius* feg*: & iterum ſumatur triangulum* fex *quarta pars trianguli* fei*: & hoc fiat ſemper donec veniatur ad aliquod triangulum, puta* fex, *quod minus ſit ſpatio* K. *Eritq. triangulum* xeg *adhuc maius quàm triplum reliquæ figuræ mixtæ* abcgf. *Sed eadem penitùs ratione, atque ordine vt ſupra, oſtendemus triangulum* xeg *eſſe triplum cuiuſdam figurę intrà figuram mixtam* abcgf. *deſcriptę. Neceſſe ergò erit, vt figura mixta* abcgf. *minor ſit quàm aliqua figura ſibi inſcripta; totum ſua parte. Quod eſt abſurdum.*

Si ergò parabola tres tangentes habuerit, vt poſitum eſt, erit triangulum ſub tangentibus contentum, reliquæ figurae, dempta parabola triplum. Quod erat propoſitum &c.

Propoſito VI.

PArabola ſeſquitertia eſt trianguli eandem ipſi baſim, & eandem altitudinem habentis.

Eſto parabola ABC, cuius diameter BD; duæ tangentes AE, CE ad baſim, & tertia FBG per verticem. Dico parabolam ſeſquitertiam eſſe inſcripti ſibi trianguli ABC.

Triangulum enim FEG ad

duo

duo trilinea mixta AFB, BGC per præcedens lemma, est vt 3. ad vnum. Ergò trapezium. AFGC (cum triplum sit trianguli FEG) ad duo eadem trilinea mixta erit vt 9. ad vnū. Et ad parabolam erit (per conuersionem rationis) vt 9. ad 8. & ad triangulum ABC, erit vt 9. ad 6. Qualium ergo partium parabola est octo, talium triangulum ABC est 6; Quare parabola ad inscriptum sibi triangulum est vt 8. ad 6. nempe sesquitertia. Quod erat &c.

Lemma XIII.

Si parabola tres tangentes habuerit; duas ad basim, tertiam verò per verticem, & ex vniuersa figura dempta sit parabola, dimidiumq; trianguli sub tangentibus contenti. Reliqua figura æquiponderabit ex quodam puncto, quod ita integram tangentem lateralem diuidit, vt pars quæ ad contactum curuæ terminatur sit ad reliquam vt 9. ad vnum.

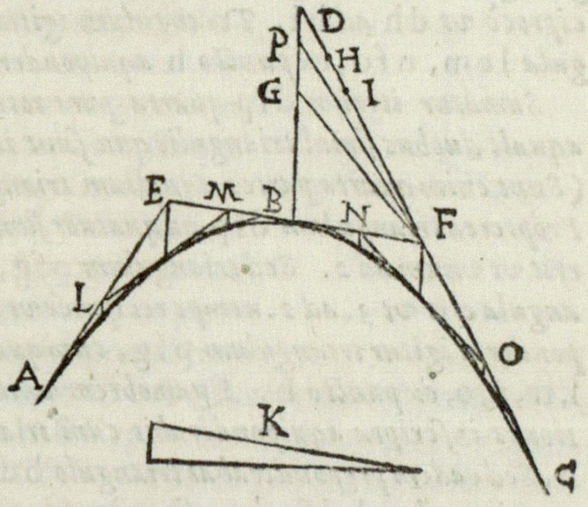

Esto parabola abc, *cuius diameter* bd *concipiatur ad horizontem perpendicularis; Sintque duæ tangētes ad basim* ae, cd, *verticalis verò tangens* ebf. *Sectâ deinde laterali* cd *in* h, *ita vt* ch *ad* hd *sit vt 9. ad vnum; Dico figuram huiusmodi (demptâ parabolâ, & semitriangulo verticali* ebd*) æquiponderare ex puncto* h.

Sumatur di *quinque partium earum, quarum* df *est 15. siue quarum* dh *est 3. Eritq;* dh *ad* hi *vt 3. ad 2. Cum autem* bd *sit ad horizontem perpendicularis, portiones mixtę* abe,

F bcf,

duo trilinea mixta AFB, BGC per præcedens lemma, eft vt 3. ad vnum. Ergò trapezium. AFGC (cum triplum ſit trianguli FEG) ad duo eadem trilinea mixta erit vt 9. ad vnũ. Et ad parabolam erit (per conuerſionem rationis) vt 9. ad 8. & ad triangulum ABC, erit vt 9. ad 6. Qualium ergo partium parabola eft octo, talium triangulum ABC eft 6; Quare parabola ad inſcriptum ſibi triangulum eft vt 8. ad 6. nempe ſeſquitertia. Quod erat &c.

Lemma XIII.[135]

Si parabola tres tangentes habuerit; duas ad baſim, tertiam verò per verticem, & ex vniuerſa figura dempta ſit parabola, dimidiumq; trianguli ſub tangentibus contenti. Reliqua figura æquiponderabit ex quodam puncto, quod ita integram tangentem lateralem diuidit, vt pars quæ ad contactum curuæ terminatur ſit ad reliquam vt 9. ad vnum.

Esto parabola abc, *cuius diameter* bd *concipiatur ad horizontem perpendicularis; Sintque duœ tangẽtes ad baſim* ae, cd, *verticalis verò tangens* ebf. *Sectà deinde laterali* cd *in* h, *ita vt* ch *ad* hd *ſit vt 9. ad vnum; Dico figuram huiuſmodi (demptà parabolà, & ſemitriangulo verticali* ebd*) æquiponderare ex puncto* h.

Sumatur di *quinque partium earum, quarum* df *eſt 15. ſiue quarum* dh *est 3. Eritq;* dh *ad* hi *vt 3. ad 2. Cum autem* bd *ſit ad horizontem perpendicularis, portiones mixtę* abe,

F b c f,

[135] *Ad absurdum* proof. (Cfr. Pisano and Bussotti's chapters above).

bcf, *appensę erunt secundum centrum grauitatis ad punctum* b, *siue ad punctum* d. *Triangulum verò* bdf *ob eandem causam, & eodem modo pendebit centralitèr ex puncto* i. *(quandoquidem* fi *dupla est ipsius* id; *& ipsa* db *ad horizontem perpendicularis.) Iam si istæ magnitudines non æquiponderant ex* h *puncto libræ* di, *altera ipsarum præponderabit. Esto; & præponderent primò duæ portiones mixtę* abe, bcf. *Sitq; excessus quo præponderant, æqualis spatio* K.

Inscribatur intrà mixtas portiones figura ex tangentibus, vt iàm sæpè factum est. Donec excessus portionum suprà figuram rectilineam inscriptam minor sit spatio K. *Tunc enim figura inscripta adhuc præponderabit triangulo* bdf.

Accipiatur triangulum dfg *quarta pars totius trianguli* df

ex Lem. 3. b; *eritq; triangulum* dfg *æquale triangulo* nfo *(cum ambo sint subquadrupla eiusdem trianguli* dfb*) & proptere a triangulum* dfg *ad duo triangula* lem, nfo, *erit vt vnum ad 2. ergò* bgf *ad duo triangula* lem, nfo, *erit vt 3. ad 2. nempe reciprocè vt* dh *ad* hi. *Triangulum igitur* bgf, *& duo triangula* lem, nfo, *ex puncto* h *æquiponderant inuicem.*

Sumatur iterùm dfp *quarta pars totius* dfg, *eritq;* dfp

ex Lem. 3. *æquale duobus simul triangulis quæ sunt infrà puncta* n, *&* o. *(Sunt enim quartæ partes æqualium triangulorum* dfg, nfo, *Propterea triangulum* dfp *ad quatuor simul triangula* l, m, no, *erit vt vnum ad 2. Sed triangulum* pfg, *ad eadem quatuor triangula erit vt 3. ad 2. nempe reciprocè vt* dh, *ad* hi. *Aequiponderat igitur triangulum* pfg, *cum quatuor dictis triangulis* l, m, n, o, *ex puncto* h. *Quamobrem vniuersa figura intrà portiones inscripta æquiponderabit cum triangulo* bpf *ex puncto* h. *Sed eadem pręponderabat triangulo* bdf. *Necesse igitur est vt triangulum* bdf *minus sit quàm triangulum* bpf, *totum sua parte. Quod esse non potest,*

Ponamus deinde pręponderare triangulum bdf *duabus simul portionibus mixtis* abe, bcf; *& ponatur excessus quo præponderat, ęqualis spatio* K.

Accipiatur triangulum dfg *quarta pars ipsius* dfb. *& iterum*

bcf, *appenſę erunt ſecundum centrum grauitatis ad punctum* b, *ſiue ad punctum* d. *Triangulum verò* bdf *ob eandem cauſam, & eodem modo pendebit centralitèr ex puncto* i. *(quadoquidem* fi *dupla est ipſius* id; *& ipſa* db *ad horizontem perpendicularis.) Iam ſi istæ magnitudines non æquiponderant ex* h *puncto libræ* di, *altera ipſarum præponderabit. Eſto; & præponderent primò duæ portiones mixtę* abe, bcf. *Sitq; exceſſus quo præponderant, æqualis ſpatio* K.

Inſcribatur intrà mixtas portiones figura ex tangentibus, vt iàm ſæpè factum eſt. Donec exceſſus portionum ſuprà figuram rectilineam inſcriptam minor sit ſpatio K. *Tunc enim figura inſcripta adhuc præponderabit triangulo* bdf.

ex Lem. 3. *Accipiatur triangulum* dfg *quarta pars totius trianguli* dfb; *eritq; triangulnm* dfg *æquale triangulo* nfo *(cum ambo ſint ſubquadrupla eiuſdem trianguli* dfb) *& propterea triangulum* dfg *ad duo triangula* lem, nfo, *erit vt vnum ad 2. ergò* bgf *ad duo triangula* lem, nfo, *erit vt 3. ad 2. nempe reciprocè vt* dh *ad* hi. *Triangulum igitur* bgf, *& duo triangula* lem, nfo, *ex puncto* h *æquiponderant inuicem.*

Sumatur iterùm dfp *quarta pars totius* dfg, *eritq;* dfp *æquale duobus ſimul triangulis quæ ſunt infrà puncta* n, *&* o. (*Sunt enim quartæ partes æqualium triangulorum* dfg, nfo,[136] *Propterea triangulum* dfp *ad quatuor ſimul triangula* l, m, n, o, *erit vt vnum ad 2. Sed triangulum* pfg, *ad eadem quatuor triangula erit vt 3. ad 2. nempe reciprocè vt* dh, *ad* hi. *Aequiponderat igitur triangulum* pfg,

ex Lem. 3. *cum quatuor dictis triangulis* l, m, n, o, *ex puncto* h. *Quamobrem vniuerſa figura intrà portiones inſcripta æquiponderabit cum triangulo* bpf *ex puncto* h. *Sed eadem pręponderabat triangulo* bdf. *Neceſſe igitur est vt triangulum* bdf *minus ſit quàm triangulnm* bpf, *totum ſua parte. Quod eſſe non potest,*

Ponamus deinde pręponderare triangulum bdf *duabus ſimul portionibus mixtis* abe, bcf; *& ponatur exceſſus quo præponderat, ęqualis ſpatio* K.

Accipiatur triangulum dfg *quarta pars ipſius* dfb. *& ite-*

rum

[136] nfo,] nfo).

rum sumatur d t p *quarta pars ipsius* d f g, *& sic semper, donec veniatur ad aliquod triangulū, puta* d f p *minus spatio* K. *Tunc enim reliquum triangulū adhuc preponderabit portionibus mixtis* a b e, b c f. *Sed ostendemus eodem penitus argumento, atq; ordine ut supra, idem triangulum* p f b *equiponderare alicui figure intrà portiones* a b e, b c f, *descripte. Necesse ergo erit quòd ipsæ duae portiones mixte minores sint quàm aliqua sibi inscripta figura; totum sua parte; quod est absurdum. Constat ergò quod propositum fuerat.*

Propositio VII.

PArabola sesquitertia est trianguli eandem ipsi basim, & eādem altitudinem habentis.

Esto, ut in præcedenti lemmate, parabola A B C. cum duabus tangentibus lateralibus, siue ad basim, AE, CD; atque EBF per uerticem. Concipiaturq; diameter ad horizontem perpendicularis; & ablatà parabola, detractoque dimidio uerticalis trianguli; accipiatur D I tertia pars totius DF, & sit DH sesquialtera ipsius HI. Aequiponderant ergo (per lemma præcedens) ex puncto H libræ D I, due magnitudines. Nempe hinc duæ portiones A B C F E appensæ ad punctum D; inde uerò triangulum D B F appensum ad punctum I. Quamobrem D B F ad A B C F E, erit ut reciprocè DH ad HI, nempe ut 3. ad 2. Sumptisq; antecedentium duplis, erit totum uerticale triangulum E D F ad reliquam figuram mixtam triplum. Propterea (ut in Propositione sexta demonstratum est) parabola inscripti sibi trianguli sesquitertia erit. Quod erat propositum demonstrare.

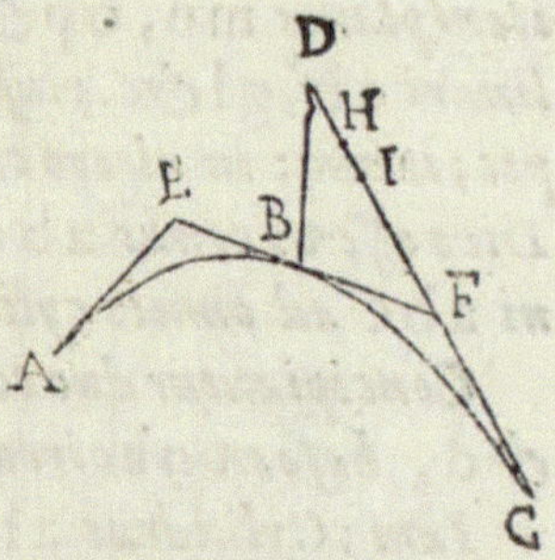

rum ſumatur dfp *quarta pars ipſius* dfg, *& ſic ſemper, donec veniatur ad aliquod triangulũ, puta* dfp *minus ſpatio* K. *Tunc enìm reliquum triangulũ adhuc prępondersbit portionibus mixtis* abe, bcf. *Sed oſtendemus eodem penitus argumento, atq; ordine vt ſupra, idem triangulum* pfb *ęquiponderare alicui figurę intrà portiones* abe, bcf, *deſcriptę. Neceße ergo erit quòd ipſæ duae portiones mixtę minores sint quàm aliqua ſibi inſcripta figura; totum ſua parte; quod eſt abſurdum. Constat ergò quod propoſitum fuerat.*

Propoſitio VII.

PArabola ſeſquitertia eſt trianguli eandem ipſi baſim, & eãdem altitudinem habentis.

Eſto, ut in præcedenti lemmate, parabola ABC. cum duabus tangentibus lateralibus, ſiue ad baſim, AE, CD; atque EBF per uerticem. Concipiaturq; diameter ad horizontem perpendicularis; & ablatà parabola, detractoque dimidio uerticalis trianguli; accipiatur DI tertia pars totius DF, & ſit DH ſeſquialtera ipſius HI. Aequiponderant ergo (per lemma præcedens) ex puncto H libræ DI, duę magnitudines. Nempe hinc duæ portiones ABCFE appenſæ ad punctum D; inde uerò triangulum DBF appenſum ad punctum I. Quamobrem DBF ad ABCFE, erit ut reciprocè DH ad HI, nempe ut 3. ad 2. Sumptiſq; antecedentium duplis, erit totum uerticale triangulum EDF ad reliquam figuram mixtam triplum, Propterea (ut in Propoſitione ſexta demonſtratum eſt) parabola inſcripti ſibi trianguli ſeſquitertia erit. Quod erat propoſitum demonſtrare.

 Lem-

Lemma XIV.

Si duorum conorum latera trianguli per axem secta fuerint in partes æquales numero, & magnitudine, ductisque per puncta sectionum planis basi parallelis, super sectionum circulis intelligantur cylindri æquealti intrà conos descripti: Erit ut primus conus ad secundum, ita omnes cylindri primi coni, ad omnes cylindros secundi coni,

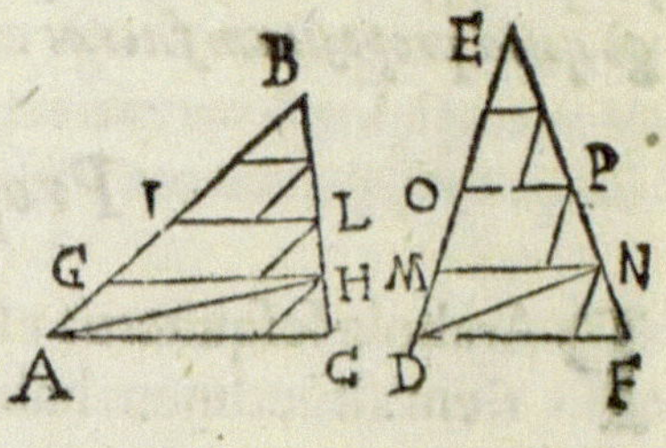

Sint duorum conorum triangula per axem abc, def, *& duo eorum latera, puta* ab, de, *secentur in partes numero æquales; nempe in totidem partes diuidatur tàm* ab, *quàm* de; *sintq; partes lateris* ab *equales inter se, & partes* de *item æquales inter se. Ductis deinde per singula sectionum puncta planis* gh, il *&c. basi* ac *parallelis; item planis* mn, op *&c. basi* df *parallelis; Concipiantur cylindri* ah, gl *&c. eiusdem altitudinis intrà conum* abc *descripti; itemq; in altero cono alij cylindri æquealti intelligantur; Dico esse vt conus* abc *ad conum* def, *ita omnes cylindros coni* abc *ad omnes cylindros coni* def,

Concipiantur duo coni gah, mdn; *quorum vertices sint* a, *&* d, *bases verò circuli* gh, mn.

Iàm; Cylindrus ah *ad conum* gah, *est vt cylindrus* dn *ad conum* mdn. *(nempe in ratione tripla) conus verò* gah *ad conum* gbh *in eadem basi, est vt* ag *ad* gb; *siue (propter diuisionem in constructione adhibitam) vt* dm *ad* me, *hoc est vt conus* mdn *ad conum* men. *Conus denique* gbh *ad conum similem* abc, *est vt cubus* gb *ad cubum* ba; *siuè (propter constructionem) vt cubus* me *ad cubum* ed, *nēpe vt conus* men *ad conum similem* def. *Quarè ex æquo cylindrus* ah *ad conū* abc, *erit ut cylindrus* dn *ad conum* def. *Et permutando cylindrus* ah *ad cylindrum* dn *erit vt conus* abc *ad conum* def.

Vlterius. Cylindrus etiam gl *ad cylindrum* mp. *eodem penitus*

Lemma XIV.[137]

Si duorum conorum latera trianguli per axem ſecta fuerint in partes æquales numero, & magnitudine, ductiſque per puncta ſectionum planis baſi parallelis, ſuper ſectionum circulis intelligantur cylindri æquealti intrà conos deſcripti: Erit ut primus conus ad ſecundum, ita omnes cylindri primi coni, ad omnes cylindros ſecundi coni,

Sint duorum conorum triangula per axem abc, def, *& duo eorum latera, puta* ab, de, *ſecentur in partes numero æquales; nempe in totidem partes diuidatur tàm* ab, *quàm* de; *ſintq; partes lateris* ab *ęquales inter ſe, & partes* de *item æquales inter ſe. Ductis deinde per ſingula ſectionum puncta planis* gh, il *&c. baſi* ac *parallelis; item planis* mn, op *&c. baſi* df *parallelis; Concipiantur cylindri* ah, gl *&c. eiuſdem altitudinis intrà conum* abc *deſcripti; itemq; in altero cono alij cylindri æquealti intelligantur; Dico eſſe vt conus* abc *ad conum* def, *ita omnes cylindros coni* abc *ad omnes cylindros coni* def,

Concipiantur duo coni gah, mdn; *quorum vertices ſint* a *&* d, *baſes verò circuli* gh, mn.

Iàm; Cylindrus ah *ad conum* gah, *est vt cylindrus* dn *ad conum* mdn. *(nempe in ratione tripla) conus verò* gah *ad conum* gbh *in eadem baſi, eſt vt* ag *ad* gb; *ſiue (propter diuiſionem in constructione adhibitam) vt* dm *ad* me, *hoc est vt conus* mdn *ad conum* men. *Conus denique* gbh *ad conum ſimilem* abc, *est vt cubus* gb *ad cubum* ba; *ſiuè (propter conſtructionem) vt cubus* me *ad cubum* ed, *nẽpe vt conus* men *ad conum ſimilem* def. *Quarè ex æquo cylindrus* ah *ad conũ* abc, *erit ut cylindrus* dn *ad conum* def. *Et permutando cylindrus* ah *ad cylindrum* dn *erit vt conus* abc *ad conum* def.

Vlterius. Cylindrus etiam gl *ad cylindrum* mp. *eodem pe-.*

nitus

[137] *Ex æquo, dividendo* and *permutando* proofs. (Cfr. Pisano and Bussotti's chapters above).

nitus modo demonstratur esse vt conus g b h *ad conum* m e n, *siue vt conus* a b c *ad conum* d e f; *& hoc modo semper. Propterea vt vnus cylindrus* a h *ad vnum* d n, *ita quilibet antecedentium ad quemlibet consequentium, ergo vt vnus ad vnum, nempe vt conus* a b c *ad conum* d e f, *ita omnes simul cylindri coni* a b c, *ad omnes simul cylindros coni* d e f. *Quod &c.*

11 Quinti.
12 Quinti.

Lemma XV.

Dato trilineo mixto, sub lineà parabolica, eiusq; tangente, & alià rectà diametro parallela compræhenso; possibile est in dato trilineo figuram inscribere constantem ex parallelogrammis æquealtis, quæ figura deficiat à trilineo mixto minori differentià quàm sit quæcumq; data magnitudo.

Esto linea parabolica a b c, *cuius tangens* c d, *& diametro æquidistans sit* a d. *Dico intrà trilineū mixtum* a b c d. *describi posse figuram constantem ex parallelogrammis æquealtis, quae figura deficiat à trilineo mixto, minori defectu quàm sit spatium quodcunque datum* K.

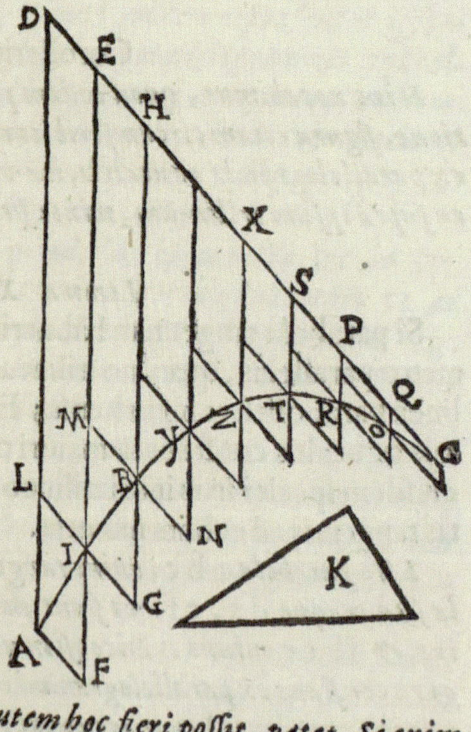

Secetur enim d c *bifariam in* x; *iterumq; partes bifariàm diuidantur in* h *& in* p; *semperq; hoc fiat, donec veniatur ad sectionem aliquam puta* d e, *eiusmodi, vt parallelogram.* a d e, *minus sit spatio* K. (*Quod autem hoc fieri possit, patet. Si enim compleatur parallelogrammum* a d c, *ex ipso per continuam bisectionem semper detrahitur dimidium; ergò tandem remanebit*

nitus modo demonſtratur eſſe vt conus gbh *ad conum* men, *ſiue vt conus* abc *ad conum* def; *& hoc modo ſemper. Propterea vt vnus cylindrus* ah *ad vnum* dn, *ita quilibet antecedentium ad quemlibet conſequentium, ergò vt vnus ad vnum, nempe vt conus* abc *ad conum* def, *ità omnes ſimul cylindri coni* abc, *ad omnes ſimul cylindros coni* def. *Quod &c.* *11 Quinti.* 12 *Quinti.*

Lemma XV.

Dato trilineo mixto, ſub lineà parabolica, eiuſq; tangente, & alià rectà diametro parallela compræhenſo; poſſibile eſt in dato trilineo figuram inſcribere conſtantem ex parallelogrammis æquealtis, quæ figura deficiat à trilineo mixto minori differentià quàm ſit quæcumq; data magnitudo.

Eſto linea parabolica abc, *cuius tangens* cd, *& diametro æquidiſtans ſit* ad. *Dico intrà trilineũ mixtum* abcd. *deſcribi poſſe figuram conſtantem ex parallelogrammis ęquealtis, quae figura deficiat à trilineo mixto, minori defectu quàm ſit ſpatium quodcunque datum* K.

Secetur enim dc *biſariam in* x; *iterumq; partes biſariàm diuidantur in* h *& in* p; *ſemperq; hoc fiat, donec veniatur ad ſectionem aliquam puta* de, *eiuſmodi, vt parallelogram.* ade, *minus ſit ſpatio* K. (*Quod autem hoc fieri poſſit, patet. Si enim compleatur parallelogrammum* adc, *ex ipſo per continuam biſectionem ſemper detrahitur dimidium; ergò tandem remane-*

bit

bit a e *minus quolibet dato spatio.) Ducantur deinde ex punctis sectionum rectę* e f, h g *&c. æquidistantes ipsi* d a; *per puncta autem* i, b *&c. vbi parallelę secant parabolam, ducantur* l g, m n *&c. æquidistantes tangenti* c d. *Et factum erit quod oportebat.*

36. Primi. *Parallelogrammum enim* c o, *æquale est ipsi* o p, *& addito communi* o r, *erunt duo* c o, o r, *æqualia ipsi* r q, *siue ipsi* r s: *additoq; cōmuni* r t, *erunt tria* c o, o r, r t, *ęqualia ipsi* t p, *hoc est ipsi* t x, *additoq; cōmuni* t z. *& sic semper procedendo, erunt deniq. omnia simul parallelogrāma* c o r t z y b i a *æqualia ipsi parallelogrāmo* a e. *nempe minora spatio* k. *Multò igitur minor erit defectus figuræ inscriptę ex parallelogrammis æquealtis compositę, à trilineo mixto* a b c d, *quàm sit propositum spatium* K. *Quod erat &c.*

Corollarium.

Hinc notabimus, quod eodem prorsus modo, eàdemq; operatione, figura etiam circumscribitur datò trilineo mixto, constans ex parallelogrāmis æquealtis, ita vt excessus figuræ circumscriptæ suprà ipsum trilineum, minor sit quocunque spatio dato K.

Lemma XVI.

Si parabola tangentem habuerit: & insuper duas rectas diametro parallelas, quæ duo trilinea abscindant sub tangente, & lineà parabolicà compræhensa; Erit figura ex parallelogrammis æquealtis constans in maiori trilineo descripta, ad figuram eiusdem speciei in minori trilineo descriptam, vt cubus maioris tangentis ad cubum minoris.

Esto parabola a b c, *cuius tangens* c d; *& diametro parallela sit vtraque* d a, e f; *vt fiant duo trilinea mixta* a b c d *maius, &* f b c e *minus. Dico, si in utroque trilineo inscribatur figura constans ex parallelogrammis æqualibus utrimque numero, (ut in præcedenti Lemmate expositum est) figuram trilinei* a b c d, *ad figuram trilinei* f b c e, *esse ut cubus* d c *ad cubum* c e.

Concipiamus, (ad euitandam linearum multitudinem, & confusio-

bit ae *minus quolibet dato ſpatio.) Ducantur deinde ex punctis ſectionum rectę* ef, hg *& c. æquidistantes ipſi* da; *per puncta autem* i, b *&c. vbi parallelę ſecant parabolam, ducantur* lg, mn *&c. æquidiſtantes tangenti* cd. *Et factum erit quod oportebat.*

36 *Primi.* *Parallelogrammum enim* co, *æquale eſt ipſi* op, *& addito communi* or, *erunt duo* co, or, *æqualia ipſi* rq, *ſiue ipſi* rf*: additoq; cõmuni* rt, *erunt tria* co, or, rt, *ęqualia ipſi* tp, *hoc eſt ipſi* tx, *additoq; cõmuni* tz. *& ſic ſemper procedendo, erunt deniq. omnia ſimul parallelogrãma* cortzybia *æqualia ipſi parallelogrãmo* ae. *nempe minora ſpatio* k. *Multò igitur minor erit defectus figuræ inſcriptę ex parallelogrammis æquealtis compoſitę, à trilineo mixto* abcd, *quàm ſit propoſitum ſpatium* K.

Quod erat &c.

Corollarium.

Hinc notabimus, quod eodem prorſus modo, eàdemq; operatione, figura etiam circumſcribitur dato trilineo mixto, constans ex parallelogrãmis æquealtis, ita vt exceſſus figuræ circumſcriptæ ſuprà ipſum trilineum, minor ſit quocunque ſpatio dato K.

Lemma XVI.[138] .

Si parabola tangentem habuerit: & inſuper duas rectas diametro parallelas, quæ duo trilinea abſcindant ſub tangente, & lineà parabolicà compræhenſa; Erit figura ex parallelogrammis æquealtis conſtans in maiori trilineo deſcripta, ad figuram eiuſdem ſpeciei in minori trilineo deſcriptam, vt cubus maioris tangentis ad cubum minoris.

Esto parabola abc, *cuius tangens* cd; *& diametro parallela ſit vtraque* da, ef; *vt fiant duo trilinea mixta* abcd *maius, &* fbce *minus. Dico, ſi in utroque trilineo inſcribatur figura conſtans ex parallelogrammis æqualibus utrimque numero, (ut in præcedenti Lemmate expoſitum eſt) figuram trilinei* abcd, *ad figuram trilinei* fbce, *eſſe ut cubus* dc *ad cubum* ce.

Concipiamus, (ad euitandam linearum multitudinem, & con-

fusio-

[138] *Ex æquo* proof. (Cfr. Pisano and Bussotti's chapters above).

Problema Primum. 47

fusionem) triangulum gec *cum sua portione parabolę intercepta* fbc, *transferri, et esse idem quod positum est sub signis* hil. *trilineumq.* fbce *esse idem cũ trilineo* mnli.

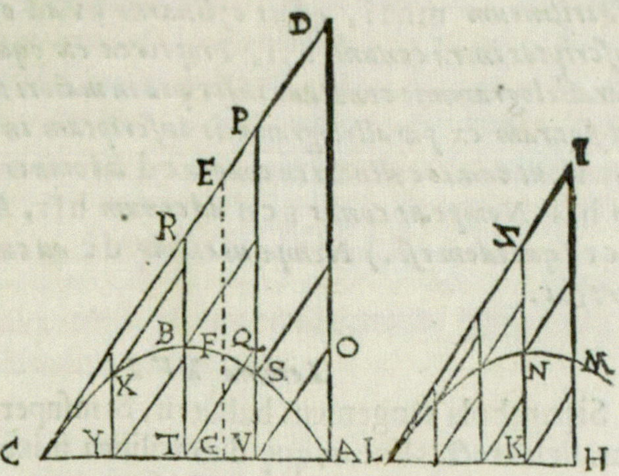

Inscribatur iam in utroque trilineo abcd, *&* mnli, (*quod quidem repræsentat ipsum* fbce *trāslatum) figura constans ex parallelogrammis æquealtis; & sit idem numerus parallelogrammorum in utroq. trilineo. Intelligatur etiam conus, cuius uertex* c, *siue* l; *& diameter basis sit, hinc quidem* ad, *inde uerò* hi. *Sintque in singulis coni segmentis cylindri ęquealti* op, qr*&c.*

Iam parallelogrammum bp *ad* ſd, *est ut recta* br *ad* ſp, 1. senti.
hoc est ut quadratum rc *ad* cp: *hoc est ut quadratum* rt *ad* ob parabolam.
quadratum pu: *hoc est ut cylindrus* qr *ad cylindrum* op. *Eodem modo, erit parallelogrammum* xr *ad* ſd, *ut cylindrus* yr *ad* ud. *Ergo erunt duo simul parallelogramma* bp, xr, 14 Quinti.
ad ſd; *ut duo simul cylindri* tp, yr, *ad cylindrum* ud. *Procedendo itaque semper hoc modo, & denique componendo, erit tota inscripta figura ex parallelogrammis constans in trilineo* abcd, *ad parallelogrammum* ſd, *ut omnes simul cylindri, qui in cono* acd, *ad cylindrum* ud.

Amplius; parallelogrammum ſd *ad* ni *compositam habet rationem, ex ratione rectę* ſp *ad* nz, *siue quadrati* pc *ad* zl (*sunt enim duæ figuræ, sed circa eandem parabolam translatam) siue quadrati* pu *ad* zK; *Et ex ratione rectæ* dp *ad* iz. *Est ergò parallelogrammum* ſd *ad* ni *ut cylindrus* ud *ad* ki,

Deni-

fuſionem) triangulum gec *cum ſua portione parabolę intercepta* fbc, *transferri, et eſſe idem quod poſitum eſt ſub ſignis* hil. *trilineumq.* fbce *eſſe idem cũ trilineo* mnli. 1.*ſexti.*

Inſcribatur iam in utroque trilineo abcd, & mnli, (*quod quidem repræſentat ipſum* fbce *trãſlatum) figura conſtans ex parallelogrammis æquealtis; & ſit idem numerus parallelogrammorum in utroq. trilineo. Intelligatur etiam conus, cuius uertex* c, *ſiue* l; *& diameter baſis ſit, hinc quidem* ad, *inde uerò* hi. *Sìntque in ſingulis coni ſegmentis cylindri ęquealti* op, qr &c. *ob parabolam.*

Iam parallelogrammum bp *ad* ſd, *eſt ut recta* br *ad* ſp, *hoc eſt ut quadratum* rc *ad* cp*: hoc est ut quadratum* rt *ad quadratum* pu*: hoc eſt ut cylindrus* qr *ad cylindrum* op. *Eodem modo, erit parallelogrammum* xr *ad* ſd, *ut cylindrus* yr *ad* ud. *Ergo erunt duo ſimul parallelogramma* bp, xr, *ad* ſd; *ut duo ſimul cylindri* tp, yr, *ad cylindrum* ud. *Procedendo itaque ſemper hoc modo, & denique componendo, erit tota inſcripta figura ex parallelogrammis conſtans in trilineo* abcd, *ad parallelogrammum* ſd, *ut omnes ſimul cylindri, qui in cono* acd, *ad cylìndrum* ud. 24 *Quiti.*

Amplius; parallelogrammum ſd *ad* ni *compositam habet rationem, ex ratione rectę* ſp *ad* nz, *siue quadrati* pc *ad* zl (*ſunt enim duæ figuræ, ſed circa eandem parabolàm translatam) siue quadrati* pu *ad* zK; *Et ex ratione rectæ* dp *ad* iz. *Est ergò parallelogrammum* ſd *ad* ni *ut cylindrus* ud *ad* ki,

Deni-

Denique parallelogrammum n i *ad totam figuram inscriptam intrà trilineum* m n l i, *est ut cylindrus* k i *ad omnes cylindros inscriptos intrà conum* h l i, *Propterea ex ęquo erit figura ex parallelogrammis constans inscripta in maiori trilineo* a b c d, *ad figuram ex parallelogrammis inscriptam in minori trilineo* m n l i, *ut omnes cylindri in cono* a c d *ad omnes cylindros in cono* h l i. *Nempe ut conus* a c d *ad conum* h l i, *hoc est ad conum* g c e *(qui idem est.) Nempe ut cubus* d c *ad cubum* c e. *Quod erat &c.*

ostendetur eodē modo ut in altera figura.

Lem. 14.

Lemma XVII.

Si parabola tangentem habuerit, & insuper duas diametro parallelas rectas lineas, quę duo trilinea mixta abscindant; Erunt inter se abscissa trilinea ut cubi suarum tangentium.

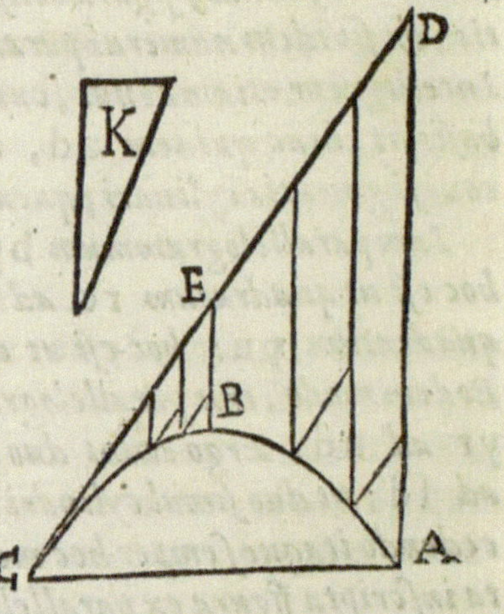

Esto parabola a b c, *cuius tangens* c d : *& diametro parallela sit utraque* d a, e b. *Dico trilineum mixtum* a b c d *ad trilineum mixtum* b c e, *esse ut cubus tangentis* d c, *ad cubum tagentis* c e.

Si enim ità non est, sit alterum illorum, si possibile est, maius quā ut habeat dictam proportionem ad reliquum ; & ponamus illud esse a b c d, *maius quàm quod esse deberet excessu* K.

Inscribatur intrà trilineum a b c d *figura ex parallelogrammis ęquealtis constans ; ita ut à trilineo deficiat minori defectu quàm sit spatium* k *(hęc autem fieri posse ostendimus) Habebitque adhuc figura inscripta ad reliquum trilineum* b c e *maiorē rationem quàm cubus* d c *ad cubum* c e.

Lem. 15

Inscribatur intrà alterum trilineum b c e *figura eiusdem speciei, & eiusdem numeri parallelogrammorum cùm descripta intra*

48 De Dimenſione Parabolæ

oftendetur eodê modo vt in altera figura.

Denique parallelogrammum ni *ad totam figuram inſcriptam intrà trilineum* mnli, *eſt ut cylindrus* ki *ad omnes cylindros inſcriptos intrà conum* hli, *Propterea ex ęquo erit figura ex parallelogrammis constans inſcripta in maiori trilineo* abcd, *ad figuram ex parallelogrammis inſcriptam in minori trilineo* mnli, *ut omnes cylindri in cono* acd *ad omnes cylindros in cono* hli. *Nempe ut conus* acd *ad conum* hli, *hoc eſt ad conum* gce *(qui idem eſt.) Nempe ut cubus* dc *ad cubum* ce. *Quod erat &c.*

Lemma XVII.[139]

Si parabola tangentem habuerit, & inſuper duas diametro parallelas rectas lineas, quę duo trilinea mixta abſcindant; Erunt inter ſe abſciſſa trilinea vt cubi ſuarum tangentium.

Lem. 14. *Eſto parabola* abc, *cuius tangens* cd*: & diametro parallela ſit utraque* da, eb. *Dico trilineum mixtum* abcd *ad trilineum mixtum* bce, *eſſe ut cubus tangentis* dc, *ad cubum tangentis* ce.

Si enim ità non est, sit alterum illorum, si poſſibile eſt, maius quã ut habeat dictam proportionem ad reliquum; & ponamus illud eſſe abcd, *maius quàm quod eße deberet exceſſu* K.

Inſcribatur intrà trilineum abcd *figura ex parallelogrammis ęquealtis constans; ita ut à trilineo deficiat minori defectu quàm sit ſpatium* k *(hęc autem fieri poſſe oſtendimus) Habebitque adhuc figura inſcriptæ ad reliquum trilineum* bc e *maiorẽ rationem quàm cubus* dc *ad cubum* ce.

Lem. 15 *Inſcribatur intrà alterum trilineum* bce *figura eiuſdem ſpeciei, & eiuſdem numeri parallelogrammorum cùm deſcripta in-*

tra

[139] *Ad absurdum* proof. (Cfr. Pisano and Bussotti's chapters above).

tra trilineum a b c d. *Erit ergò figura inſcripta trilineo* a b c d *ad figuram inſcriptam trilineo* b c e *vt cubus* d c *ad cubum* c e. *Sed eadem figura inſcripta trilineo* a b c d *ad trilineum* b c e *habet maiorem rationem quàm cubus* d c *ad* c e. *Minus ergo eſt trilineum* b c e *quàm inſcripta ſibi figura. totum ſua parte. Quod eſt impoſſibile. Conſtat ergo propoſitum.* *Lem præcedens.*

Propoſitio VIII.

PArabola ſeſquitertia eſt trianguli eandem ipſi baſim, & eandem altitudinem habentis.

Eſto parabola A B C, cuius diameter B E, tangentes verò A F, C F, productæ eouſque, donec occurrant ipſis A D, C H, diametro parallelis. Iunganturq; rectę lineę A B, B C. (licet in figura omiſſę ſint.) Dico parabolã trianguli A B C eſse ſeſquitertiã.

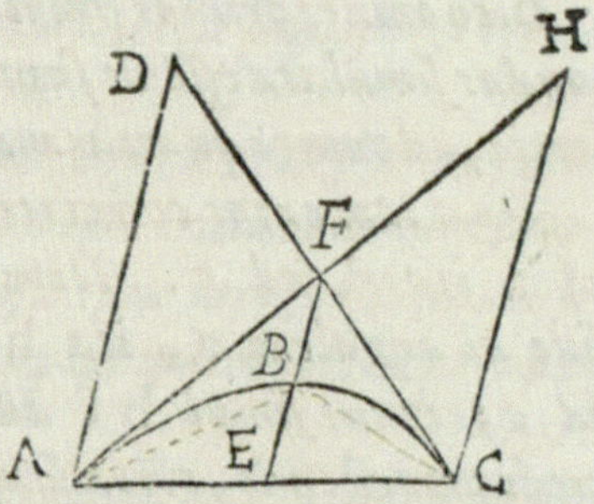

Erit enim A B C D ad trilineum B C F, ut cubus D C ad cubum C F, nempe ut octo ad unum. (cum enim ſit ut A E ad E C, ita D F ad F C, erit D F æqualis ipſi F C; cubuſq; D C octuplus cubi C F.) Item trilineum C B A H ad trilineum B A F, eſt ut octo ad unum. Coniunctim ergò erunt duo trilinea A B C D, C B A H, ad ſpatium A B C F. ut octo ad unum. Et diuidendo bis, erunt duo triangula A F D, C F H, ad ſpatium A B C F, ut 6. ad unum. Quamobrem triangulum A F D, ſiue A F C ad ſpatium A B C F erit ut 3, ad unum; & ad parabolam erit ut 3. ad 2. uel ut 6. ad 4. Propterea parabola erit ad triangulum A B C ut 4. ad 3. Nempe ſeſquitertia. Quod erat propoſitum demonſtrare, &c. *Lem. præcedens. 2. Sexti.*

G *Lem-*

tra trilineum abcd. *Erit ergò figura inſcripta trilineo* abcd *ad figuram inſcriptam trilineo* bce *vt cubus* dc *ad cubum* ce. *Sed eadem figura inſcripta trilineo* abcd *ad trilineum* bce *habet maiorem rationem quàm cubus* dc *ad* ce. *Minus ergo eſt trilineum* bce *quàm inſcripta ſibi figura. totum ſua parte. Quod eſt impoſſibile. Constat ergo propoſitum.* *Lem. præ cedens*

Propoſitio VIII.

PArabola ſeſquitertia eſt trianguli eandem ipſi baſim, & eandem altitudinem habentis.

Eſto parabola ABC, cuius diameter BE, tangentes verò AF, CF, productæ eouſque, donec occurrant ipſis AD, CH, diametro parallelis. Iunganturq; rectę lineę AB, BC. (licet in figura omiſſę ſint.) Dico parabolã trianguli ABC eſse ſeſquitertiã.

Erit enim ABCD ad trilineum BCF, ut cubus DC ad cubum CF, nempe ut octo ad unum. (cum enim ſit ut AE ad EC, ita DF ad FC, erit DF æqualis ipſi FC; cubuſq; DC octuplus cubi CF.) Item trilineum CBAH ad trilineum BAF, eſt ut octo ad unum. *Lem.* Coniunctim ergò erunt duo trilinea ABCD, CBAH, ad ſpatium ABCF. ut octo ad unum. Et diuidendo bis, erunt duo triangula *præ* AFD, CFH, ad ſpatium ABCF, ut 6. ad unum. Quamobrem triangulum AFD, ſiue AFC ad spatium ABCF erit ut 3, ad unum; & ad *cedens* parabolam erit ut 3. ad 2. uel ut 6. ad 4. Propterea parabola erit ad triangulum ABC ut 4. ad 3. Nempe ſeſquitertia. Quod erat 2.*Sexti* propoſitum demonſtrare, &c.

G *Lem-*

Lemma XVIII.

Si fuerit vt prima magnitudo ad ſecundam, ita tertia ad quartam; Et hoc quotieſcunq; libuerit. Fuerintq; omnes primæ inter ſe, item omnes tertiæ magnitudines inter ſe æquales. Erunt omnes primæ ſimul ad omnes ſecundas, vt ſunt omnes tertiæ ſimul, ad omnes quartas magnitudines.

Eſto vt a *prima ad* b *ſecundam, ita* c *tertia ad* d *quartam. Et iterum vt* e *prima ad* f *ſecundam, ita* g *tertia ad* h *quartam; Et ſic quotieſcunq; libuerit. Sintq; omnes primæ* a, e, i, *&c. item omnes tertiæ* c, g, m, *&c. inter ſe æquales.*

Dico omnes primas ſimul ad omnes ſecundas ſimul, ita eſſe vt ſunt omnes ſimul tertiæ, ad omnes quartas magnitudines.

Quoniam enim conuertendo eſt vt b *ad* a *ita* d *ad* c. *Item vt* f *ad* e; *ſiue ad æqualem* a, *ita* h *ad* g, *ſiue ad* c; *erunt ſimul* b f *ad* a, *vt ſunt* d h *ſimul ad* c. *Hoc*

24. quinti. *modo procedendo, oſtendemus omnes ſecundas ſimul eſſe ad* a, *vt ſunt omnes quartæ ſimul ad ipſam* c. *Ipſa verò* a *ad omnes*

15 quinti. *primas eſt vt* c *ad omnes tertias (ſunt enim æqueſubmultiplices) Ergo ex æquo omnes ſecundæ ad omnes primas, ſunt vt omnes quartæ ſimul ad omnes tertias. Conuertendo igitur cōſtat quod erat propoſitum demonſtrare.*

Lemma XIX.

Si parabola tangentem habuerit ad baſim; ex alia verò parte rectam diametro parallelam. Erit triangulum ſub tangente, & parallela diametro, ipſâq; baſi compræhenſum, ipſius parabolæ triplum.

Eſto

Lemma XVIII[140] .

Si fuerit vt prima magnitudo ad ſecundam, ita tertia ad quartam; Et hoc quotieſcunq; libuerit. Fuerintq; omnes primæ inter ſe, item omnes tertiæ magnitudines inter ſe æquales. Erunt omnes primæ ſimul ad omnes ſecundas, vt ſunt omnes tertiæ ſimul, ad omnes quartas magnitudines.

Eſto vt a *prima ad* b *ſecundam, ita* c *tertia ad* d *quartam. Et iterum vt* e *prima ad* f *ſecundam, ita* g *tertia ad* h *quartam; Et ſic quotieſcunq; libuerit. Sintq; omnes primæ* a, e, i, *&c. item omnes tertiæ* c, g, m, *&c. inter ſe æquales.*

Dico omnes primas ſimul ad omnes ſecundas ſimul, ita eſſe vt ſunt omnes ſimul tertiæ, ad omnes quartas magnitudines.

24.*quin ti.* *Quoniam enim conuertendo eſt vt* b *ad* a *ita* d *ad* c. *Item vt* f *ad* e; *ſiue ad æqualem* a, *ita* h *ad* g, *ſiue ad* c; *erunt ſimul* bf *ad* a, *vt ſunt* dh *ſimul ad* c. *Hoc modo procedendo, oſtendemus omnes ſecundas ſimul eſſe ad* a, *vt ſunt omnes quartæ ſimul ad ipſam* c. *Ipſa verò* a *ad omnes primas eſt vt* c *ad omnes tertias (ſunt enim æqueſubmultiplices)*

15.*quint- ti.* *Ergo ex ęquo omnes ſecundæ ad omnes primas, ſunt vt omnes quartæ ſimul ad omnes tertias. Conuertendo igitur cõstat quod erat propoſitum demonſtrare.*

Lemma XIX.[141] .

Si parabola tangentem habuerit ad baſim; ex alia verò parte rectam diametro parallelam. Erit triangulum ſub tangente, & parallela diametro, ipſâq; baſi compræhenſum, ipſius parabolæ triplum.

Esto

[140] *Ex æquo* proof. (Cfr. Pisano and Bussotti's chapters above).

[141] *Ex æquo* and *Ad absurdum* proofs. (Cfr. *Ibidem*).

Esto parabola abc, *cuius tangens* cd, *parallela diametro sit* ad; *Dico triangulum* adc *esse parabolæ ipsius* abc, *triplum*.

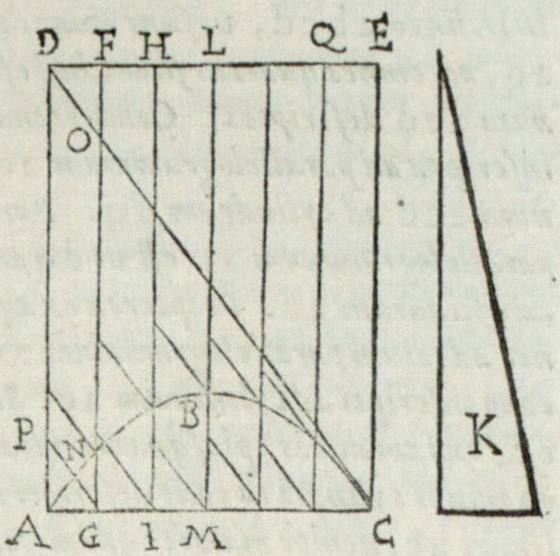

Si enim non est triplum parabolæ, per conuersionẽ rationis, non erit sesquialterum trilinei abcd; *& propterea (duplicato antecedente) totum parallelogrammum* ae *non erit triplum trilinei* abcd.

Trilineum ergo abcd *erit vel plus, vel minus quàm tertia pars parallelogrammi* ae. *Ponatur primùm esse plus quàm tertia pars, & sit excessui æquale spatium* K.

Inscribatur intrà trilineum abcd, *figura constans ex parallelogrammis æquealtis, deficiensque ab ipso trilineo minori defectu quàm sit ipsum spatium* K. *Et inscripta iam sit eiusmodi figura. Erit ergò adhuc figura inscripta plus quàm tertia pars parallelogrammi* ae.

Concipiatur circa rectam ad, *circulus, qui sit basis cuiusdam coni verticẽ habentis in puncto* c. *& super eàdẽ basi intelligatur cylindrus* ae *eiusdem altitudinis cum ipso cono; sectusque sit tam conus quàm cylindrus planis basi parallelis per singulas rectas* fg, hi, lm *&c. ductis. Concipiantur etiam intrà conum* acd *cylindri æquealti* po, oi *&c.* 1. *sexti.*

Iam sic: Parallelogrammum af *ad* nd, *est vt recta* da *ad* on; *nempe vt quadratum* dc *ad quadratum* co; *siue, vt quadratum* da *ad quadratum* og, *nempe, vt cylindrus* af *ad cylindrum* po, *Et sic, semper. Suntq; omnes primæ magnitudines æquales parallelogrammo* af, *& ideo æquales inter se; omnes autem tertiæ magnitudines æquales cylindro* af, *atque ideo inter se. Erunt ergo omnes primæ simul, hoc est parallelogram-* b *parabolam.* o *similitud. triãgul.*

G 2 mum

Esto parabola abc, *cuius tangens* cd, *parallela diametro ſit* ad; *Dico triangulum* adc *eſſe parabolæ ipſius* abc, *triplum.*

Si enim non eſt triplum parabolæ, per conuerſione rationis, non erit ſeſquialterum trilinei abcd; *& propterea (duplicato antecedente) totum parallelogrammum* a e *non erit triplum trilinei* abcd.

Trilineum ergo abcd *erit vel plus, vel minus quàm tertia pars parallelogrammi* ae. *Ponatur primùm eßе plus quàm tertia pars, & ſit exceſſui ęquale ſpatium* K.

Inſcribatur intrà trilineum abcd, *figura conſtans ex parallelogrammis æquealtis, deficienſque ab ipſo trilineo minori defectu quàm ſit ipſum ſpatium* K. *Et inſcripta iam ſit eiuſmodi figura. Erit ergò adhuc figura inſcripta plus quàm tertia pars parallelogrammi* ae.

Concipiatur circa rectam ad, *circulus, qui ſit baſis cuiuſdam coni verticẽ habentis in puncto* c. *& ſuper eàdẽ baſi intelligatur cylindrus* a e *eiuſdem altitudinis cum ipſo cono; ſectuſque ſit tam conus quàm cylindrus planis baſi parallelis per ſingulas rectas* fg, hi, lm *& c. ductis. Concipiantur etiam intrà conum* acd *cylindri æquealti* po, oi *&c.* 1. *ſexti* h *parabolam*

Iam ſic: Parallelogrammum af *ad* nd, *eſt vt recta* da *ad* on*: nempe vt quadratum* dc *ad quadratum* co; *ſiue, vt quadratum* da *ad quadratum* og, *nempe, vt cylindrus* af *ad cylindrum* po, *Et ſic, ſemper. Suntq; omnes primæ magnitudines ęquales parallelogrammo* af, *& ideo æquales inter ſe; omnes autem tertiæ magnitudines æquales cylindro* af, *atque ideo inter ſe. Erunt ergo omnes primæ ſimul, hoc eſt parallelogram-* *ob ſimilitud. triãgul.*

mum a q, *ad omnes secundas simul, nempe ad figuram inscriptã in trilineo* a b c d, *vt sunt omnes tertiæ simul, nempe cylindrus* a q, *ad omnes quartas simul, hoc est ad omnes cylindros intrà conum* a c d *descriptos. Conuertendo igitur; erit figura trilineo inscripta ad parallelogrammum* a q *vt omnes cylindri intra conum* a c d *ad cylindrum* a q. *Parallelogrammum verò* a q *ad parallelogrammum* a e *est vt* d q *ad* d e, *hoc est vt cylindrus* a q *ad cylindrum* a e. *Propterea ex æquo, figura inscripta in trilineo ad totum parallelogrammum* a e, *erit vt omnes cylindri in cono inscripti ad cylindrum* a e. *Sed figura inscripta in trilineo est (ex iam dictis) plusquam tertia pars parallelogrammi* a c, *ergò omnes cylindri in cono descripti erunt plusquã tertia pars cylindri* a e, *nempe maiores quàm conus* a c d. *pars videlicet suo toto. Quod est impossibile.*

Sed ponamus nunc trilineum a b c d *esse minus quàm tertia pars parallelogrammi* a e; *sitq; defectus ęqualis spatio* K. *Circumscribatur trilineo* a b c d *figura constans ex parallelogrammis æquealtis excedensq; minori excessu quàm sit spatium* K; *& erit figura circumscripta adhuc minor quàm tertia pars parallelogrammi* ae.

Corol. Lẽmat. 15.

Concipiatur iterum circa rectam a d *circulus pro basi coni, qui verticẽ habeat* c; *itemq; pro basi cylindri* a c e d *eiusdẽ altitudinis cum ipso cono* a c d.

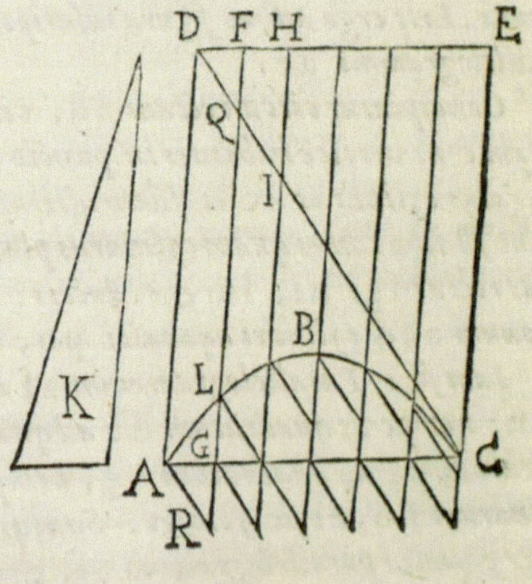

Intelligatur insuper circa conum descripta figura solida constans ex cylindris æquealtis a q, g i *&c.*

Iam parallelogrammum a f *ad parallelogrammum* a q *(ob æqualitatem) est vt cylindrus* a d f g *ad cylindrum* a d q r. *Amplius. Parallelogram*

mum

mum aq, *ad omnes ſecundas ſimul, nempe ad figuram inſcriptã in trilineo* abcd, *vt ſunt omnes tertiæ ſimul, nempe cylindrus* aq, *ad omnes quartas ſimul, hoc eſt ad omnes cylindros intrà conum* acd *deſcriptos. Conuertendo igitur; erit figura trilineo inſcripta ad parallelogrammum* aq *vt omnes cylindri intra conum* acd *ad cylindrum* aq. *Parallelogrammum verò* aq *ad parallelogrammum* ae *eſt vt* dq *ad* de, *hoc eſt vt cylindrus* aq *ad cylindrum* ae. *Propterea ex æquo, figura inſtripta in trilineo ad totum parallelogrammum* ae, *erit vt omnes cylindri in cono inſcripti ad cylindrum* ae. *Sed figura inſcripta in trilineo eſt (ex iam dictis) pluſquam tertia pars parallelogrammi* ae, *ergò omnes cylindri in cono deſcripti erunt pluſquã tertia pars cylindri* ae, *nempe maiores quàm conus* acd. *pars videlicet ſuo toto. Quod eſt impoſſibile.*

Sed ponamns[142] *nunc trilineum* abcd *eße minus quàm tertia pars parallelogrammi* ae; *ſitq; defectus ęqualis ſpatio* K.

Corol. Lẽ mat. 15 *Circumſcribatur trilineo* abcd *figura constans ex parallelogrammis æquealtis excedenſq; minori exceſſu quàm ſit ſpatium* K; *& erit figura circumſcripta adhuc minor quàm tertia pars parallelogrammi* ae.

Concipiatur iterum circa rectam ad *circulus pro baſi coni, qui verticẽ habeat* c; *itemq; pro baſi cylindri* aced *eiuſdẽ altitudinis cum ipſo cono* acd.

Intelligatur inſuper circa conum deſcripta figura ſolida conſtans ex cylindris æquealtis aq, gi *&c.*

Iam parallelogrammum af *ad parallelogrammum* aq *(ob æqualitatem) est vt cylindrus* adfg *ad cylindrum* adqr. *Amplius. Parallelogram-*

mum

[142] *ponamns*] *ponamus*.

mum gh *ad parallelogrammum* li *eſt vt* gf, *ſiue* ad *ad* lq; *nempe vt quadratum* dc *ad quadratum* cq, *ſiue vt quadratum* da, *vel* fg, *ad quadratum* gq; *nempe vt cylindrus* gh *ad cylindrum* gi. *&c. & hoc modo ſemper. Suntq; omnes ſingillatim primæ magnitudines æquales parallelogrammo* af, *& ideò interſe: item omnes tertię æquales cylindro* af, *& ob id interſe; ergo erunt omnes primæ ſimul, hoc eſt parallelogrammum* ae, *ad omnes ſecundas ſimul, hoc eſt ad figuram trilineo circumſcriptã, vt omnes tertiæ ſimul, nempe cylindrus* ae, *ad omnes quartas ſimul, nempe ad cylindros conum* acd *circumſcribentes. Conuertendo igitur, erit figura circumſcripta trilineo, ad parallelogrammum* ae, *vt omnes cylindri circumſcribentes conũ ad cylindrum* ae. *Sed figura trilineo circumſcripta minor eſt quàm tertia pars parallogrammi* ae; *ergo etiam omnes cylindri circũſcribentes conum minores erunt quàm tertia pars cylindri* ae; *Nempe minores cono* acd. *Totum ſua parte: quod eſſe non poteſt. Triangulum ergo* adc *ipſius parabolæ omninò triplum erit. Quod propoſitum fuerat.*

1. ſexti. ob parabolam. ob ſimilitud. triãgul.

Propoſitio IX.

PArabola ſeſquitertia eſt trianguli eandem ipſi baſim, & eandem altitudinem habentis.

Eſto parabola ABC, cuius diameter EB, triangulum inſcriptum ſit ABC, Dico parabolam trianguli ABC eſſe ſeſquitertiam; Ducatur enim tangens CD, & ſit recta AD diametro æquidiſtans: Erit ergò per præcedens lemma, triangulum ACD parabolæ triplũ; & propterea erit parabola partes quatuor earum, quarum triangulum ADC eſt duodecim, nempe qualiũ

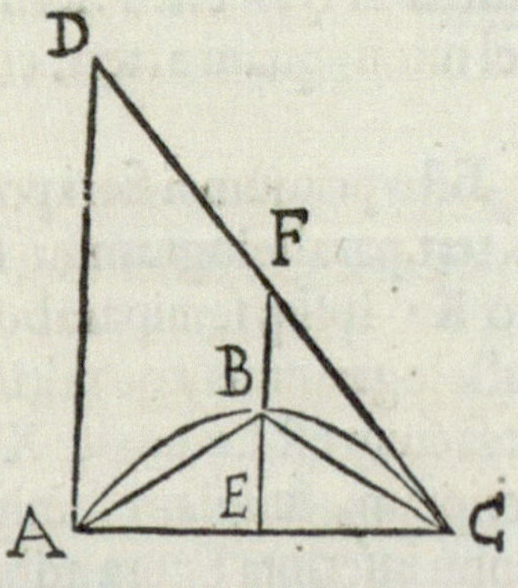

trian-

mum gh *ad parallelogrammum* li *eft vt* gf, *fiue* ad *ad* lq; *nempe vt quadratum* dc *ad quadratum* cq, *fiue vt quadratum* da, *vel* fg, *ad quadratum* gq; *nempe vt cylindrus* gh *ad cylindrum* gi. *&c. & hoc modo femper. Suntq; omnes fingillatim primæ magnitudines æquales parallelogrammo* af, *& ideò inter fe: item omnes tertię æquales cylindro* af, *& ob id inter fe; ergo erunt omnes primæ fimul, hoc est parallelogrammum* ae, *ad omnes fecundas fimul, hoc eft ad figuram trilineo circumfcriptã, vt omnes tertiæ fimul, nempe cylindrus* ae, *ad omnes quartas fimul, nempe ad cylindros conum* acd *circumscribentes. Conuertendo igitur, erit figura circumfcripta trilineo, ad parallelogrammum* ae, *vt omnes cylindri circumfcribentes conũ ad cylindrum* ae. *Sed figura trilineo circumfcripta minor eft quàm tertia pars parallogrammi* ae; *ergo etiam omnes cylindri circũfcribentes conum minores erunt quàm tertia pars cylindri* ae; *Nempe minores cono* acd. *Totum fua parte: quod effe non poteft. Triangulum ergo* adc *ipfius parabolæ omninò triplum erit. Quod propofitum fuerat.*

1. *fexti ob paræbolam ob simili tud. triãngal.*

Propofitio IX.[143]

PArabola fefquitertia eft trianguli eandem ipfi bafim, & eandem altitudinem habentis.

Efto parabola ABC, cuius diameter EB, triangulum infcriptum fit ABC, Dico parabolam trianguli ABC effe fefquitertiam;

Ducatur enim tangens CD, & fit recta AD diametro æquidiftans:

Erit ergò per præcedens lemma, triangulum ACD parabolæ triplũ; & propterea erit parabola partes quatuor earum, quarum triangulum ADC eft duodecim, nempe qualiũ

trian-

[143] *Ex æquo* and *Ad absurdum* proofs. (Cfr. Pisano and Bussotti's chapters above).

triangulum A B C eſt tres. (triangulum enim A B C æquale eſt triangulo E F C, cum vtrumq; duplum ſit trianguli E B C, ergo triangulum A B C quarta pars erit totius A D C.) Conſtat ergo parabolam ad inſcriptum ſibi triangulum eſſe vt 4. ad 3. Nempe ſeſquitertiam. Quod &c.

Propoſitio X.

Parabola ſeſquitertia eſt trianguli eandem ſibi baſim, eandemq; altitudinem habentis.

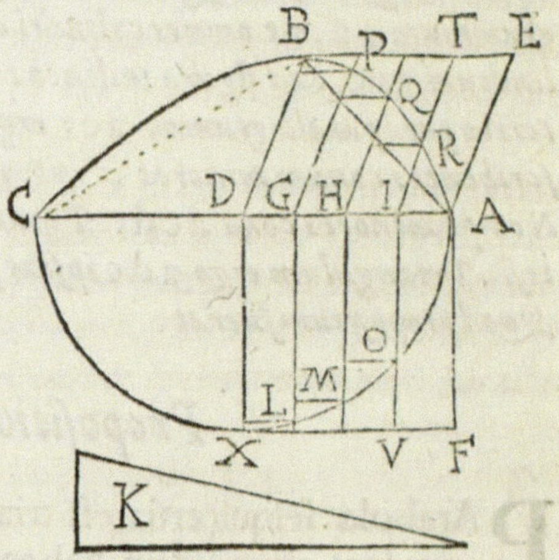

Eſto parabola A B C cuius diameter B D. Dico parabolam A B C inſcripti ſibi trianguli eſſe ſeſquitertiam.

Compleatur parallelogrammum A D B E, & niſi parabola ſeſquitertia ſit trianguli ſibi inſcripti, neque (ſumptis dimidijs) ſemiparabola A B D ſeſquitertia erit trianguli A B D; neq; eadem ſemiparabola A B D erit 2. tert. parallelogrammi E D, ſed vel plus, vel minùs quàm 2. tert. eiuſdem.

Eſto primùm ſi fieri poteſt ſemiparabola A B D magìs quã 2. tert. parallelogrammi E D; & ponatur exceſſus æqualis ſpatio K. Ipſiq; ſemiparabolæ figura inſcribatur conſtans ex parallelogrammis æquealtis (more apud Geometras vſitato, prout factum eſt Lemmate XV.) ita vt differentia inter figuram inſcriptam, & ipſam ſemiparabolam minor ſit ſpatio K. Tunc enim inſcripta figura adhuc maior erit quàm 2. ter. parallelogrã mi A D B E.

Duca-

triangulum ABC eſt tres. (triangulum enim ABC æquale eſt triangulo EFC, cum vtrumq; duplum ſit trianguli EBC, ergo triangulum ABC quarta pars erit totius ADC.) Conſtat ergo parabolam ad inſcriptum ſibi triangulum eſſe vt 4.ad 3. Nempe ſeſquitertiam. Quod &c.

Propoſitio X.

PArabola ſeſquitertia eſt trianguli eandem ſibi baſim, eandemq; altitudinem habentis.

Eſto parabola ABC cuius diameter BD. Dico parabolam ABC inſcripti ſibi trianguli eſſe ſeſquitertiam.

Compleatur parallelogrammum ADBE, & niſi parabola ſeſquitertia ſit trianguli ſibi inſcripti, neque (ſumptis dimidijs) ſemiparabola ABD ſeſquitertia erit trianguli ABD; neq; eadem ſemiparabola ABD erit 2. tert. parallelogrammi ED, ſed vel plus, vel minùs quàm 2. tert. eiuſdem.

Eſto primùm ſi fieri poteſt ſemiparabola ABD magìs quã 2. tert. parallelogrammi ED; & ponatur exceſſus æqualis ſpatio K. Ipſiq; ſemiparabolæ figura inſcribatur conſtans ex parallelogrammis æquealtis (more apud Geometras vſitato, prout factum eſt Lemmate XV.) ita vt differentia inter figuram inſcriptam, & ipſam ſemiparabolam minor ſit ſpatio K. Tunc enim inſcripta figura adhuc maior erit quàm 2. ter. Parallelogrãmi ADBE.

Duca-

Ducatur circa diametrum A C semicirculus A X C, completoq; rectangulo, siue quadrato A F X D. ducantur G L, H M, I O perpendiculares ad A C, & compleantur rectangula D L, G M, H O; Tum intelligatur figura A F X D circumuerti circa axem A D; ita vt quadrans A D X hemisphærium describat, quadratum verò A F X D, cylindrum; & rectangula in quadrante inscripta totidem cylindros faciant in ipso hemisphærio compræhensos.

Iam parallelogrammum B G ad P D, est vt B D ad G P, siue vt rectangulum C D A ad rectangulū C G A; siue vt quadratum X D ad quadratū L G; siue vt cylindrus X G ad L D. Et hoc modo semper. Suntque omnes primæ magnitudines æquales parallelogrammo B G, & omnes tertiæ æquales cylindro X G. Ergo erunt omnes primæ simul, hoc est parallelogrammum T D, ad omnes secundas simul, nempe ad figuram inscriptam in semiparabola, vt sunt omnes tertię simul, nempe cylindrus V D ad omnes quartas simul, hoc est ad omnes cylindros in hemisphærio inscriptos. Parallelo grammum verò T D ad E D est vt cylindrus V D ad F D, ergo ex æquo, erit parallelogrammum E D ad figuram in semiparabola inscriptam vt cylindrus F D ad omnes cylindros in ipso hemisphærio cōpræhensos. Sed parallelogrammum E D minus est quam sesquialterum figuræ intra semiparabolam inscriptæ; Ergò cylindrus F D minor erit quàm sesquialter omnium cylindrorum in hemisphærio descriptorum. Quod est absurdum. Scimus enim dictum cylindrum hemisphærij esse sesquialterum. *Lem. 11.*

Esto deinde (si fieri potest) semiparabola minor quam 2. tert. ipsius parallelogrammi E D. Ponaturq; defectus æqualis spatio K.

Tum ipsi semiparabolæ figura quædam circumscribatur, constans ex parallelogrammis æquealtis (more solito, vt factum est in Lemmate X V. [eiusque Corollario) ita vt differentia inter circumscriptam figuram ipsamq; semiparabolam minor sit spatio K. Tunc enim manifestum est, quòd figura circumscripta

adhuc

Ducatur circa diametrum AC ſemicirculus AXC, completoq; rectangulo, ſiue quadrato AFXD. ducantur GL, HM, IO perpendiculares ad AC, & compleantur rectangula DL, GM, HO; Tum intelligatur figura AFXD circumuerti circa axem AD; ita vt quadrans ADX hemiſphærium deſcribat, quadratum verò AFXD, cylindrum; & rectangula in quadrante inſcripta totidem cylindros faciant in ipſo hemiſphærio compræhenſos. *Lem.* 18

Iam parallelogrammum BG ad PD, eſt vt BD ad GP, ſiue vt rectangulum CDA ad rectangulũ CGA; ſiue vt quadratum XD ad quadratũ LG; ſiue vt cylindrus XG ad LD. Et hoc modo ſemper. Suntque omnes primæ magnitudines æquales parallelogrammo BG, & omnes tertiæ æquales cylindro XG. Ergo erunt omnes primæ ſimul, hoc eſt parallelogrammum TD, ad omnes ſecundas ſimul, nempe ad figuram inſcriptam in ſemiparabola, vt ſunt omnes tertię ſimul, nempe cylindrus VD ad omnes quartas ſimul, hoc eſt ad omnes cylindros in hemiſphærio inſcriptos. Parallelogrammum verò TD ad ED eſt vt cylindrus VD ad FD, ergo ex æquo, erit parallelogrammum ED ad figuram in ſemiparabola inſcriptam vt cylindrus FD ad omnes cylindros in ìpſo hemiſphærio cõpræhenſos. Sed parallelogrammum ED minus eſt quam ſeſquialterum figuræ intra ſemiparabolam inſcriptæ; Ergò cylindrus FD minor erit quàm ſeſquialter omnium cylindrorum in hemiſphærio deſcriptorum. Quod eſt abſurdum. Scimus enim dictum cylindrum hemiſphærij eſſe ſeſquialterum.

Eſto deinde (ſi fieri poteſt) ſemiparabola minor quam 2. tert. ipſius parallelogrammi ED. Ponaturq; defectus æqualis ſpatio K.

Tum ipſi ſemiparabolæ figura quædam circumſcribatur, conſtans ex parallelogrammis æquealtis (more ſolito, vt factum eſt in Lemmate XV. (eiuſque Corollario) ita vt differentia inter circumſcriptam figuram ipſamq; ſemiparabolam minor ſit ſpatio K. Tunc enim manifeſtum eſt, quòd figura circumſcripta

adhuc

adhuc minor erit quam 2. ter. parallelogrammi.ED.

Fiat circa diametrum A C semicirculus, vt in descriptione præcedentis constructionis, completoque quadrato A O F D, perficiantur reliqua rectangula F L, G M, H N, I A. circa quadrantem descripta. Tum reuoluatur figura AF circa axem A D, ita vt solida generentur iam dicta: nempe hemisphærium ex quadrante, cylindrus ex quadrato A F; totidemque cylindri quot rectangula erunt ipsi quadranti circumscripta.

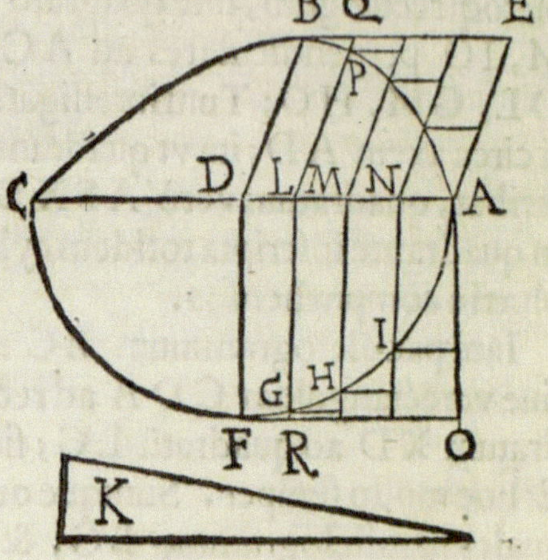

Iàm parallelogrammum B L ad se ipsum est vt cylindrus factus ex F L ad se ipsum. Amplius. Parallelogrammum Q M ad P M; est vt Q L ad L P; siue B D ad L P, siue vt rectang. C D A ad C L A, siue vt quadratum F D ad L G, siue vt quadratum R L ad L G; nempe vt cylindrus factus ex R M ad cylindrum ex G M: & hoc modo semper. Suntq; omnes primæ magnitudines æquales parallelogrammo B L, omnesq; tertiæ

Lem.18. æquales cylindro facto ex F L. Ergo erunt omnes primæ simul nempe parallelogrammum A B ad omnes simul secundas, nēpe ad figuram semiparabolæ circumscriptam, vt sunt omnes tertiæ simul, nempe cylindrus ex O D factus, ad omnes quartas, nempe ad cylindros hemisphærio circumscriptos. Sed parallelogrammum E D magìs est quàm sesquialterum figuræ circumscriptæ ad semiparabolam, ergo cylindrus ex OD magis quàm sesquialter erit ad omnes cylindros hemisphærio circumscriptos. Quod est absurdum: Scimus enim cylindrum hemisphærio circumscriptum ipsius hemisphærij esse sesquialterum.

Patet itaq; parallelogr. ED sesquialterū esse ad semiparabolam ABD; & ideo semiparab. sesquitertia trianguli ABD,

QVA-

adhuc minor erit quam 2. ter. parallelogrammi. ED.

Fiat circa diametrum AC ſemicirculus, vt in deſcriptione prędcedentis conſtructionis, completoque quadrato AOFD, perficiantur reliqua rectangula FL, GM, HN, IA. circa quadrantem deſcripta. Tum reuoluatur figura AF circa axem AD, ita vt ſolida generentur iam dicta: nempe hemiſphærium ex quadrante, cylindrus ex quadrato AF; totidemque cylindri quot rectangula erunt ipſi quadranti circumſcripta.

Iàm parallelogrammum BL ad ſe ipſum eſt vt cylindrus factus ex FL ad ſe ipſum. Amplius. Parallelogrammum QM ad PM; eſt vt QL ad LP; ſiue BD ad LP, ſiue vt rectang. CDA ad CLA, ſiue vt quadratum FD ad LG, ſiue vt quadratum RL ad LG; nempe vt cylindrus factus ex RM ad cylindrum ex GM: & hoc modo ſemper. Suntq; omnes primæ magnitudines æquales parallelogrammo BL, omneſq; *Lem.* 18 tertiæ ęquales cylindro facto ex FL. Ergo erunt omnes primæ ſimul nempe parallelogrammum AB ad omnes ſimul ſecundas, nẽpe ad figuram ſemiparabolæ circumſcriptam, vt ſunt omnes tertiæ ſimul, nempe cylindrus ex OD factus, ad omnes quartas, nempe ad cylindros hemiſphærio circumſcriptos. Sed parallelogrammum ED magìs eſt quàm ſeſquialterum figurę circumſcriptę ad ſemiparabolam, ergo cylindrus ex OD magis quàm ſeſquialter erit ad omnes cylindros hemisphærio circumſcriptos. Quod eſt abſurdum: Scimus enim cylindrum hemiſphærio circumſcriptum ipſius hemiſphærij eſſe ſeſquialterum.

Patet itaq; parallelogr. ED ſeſquialterũ eſſe ad ſemiparabolam ABD; & ideo ſemiparab. ſeſquitertia trianguli ABD,

Q V A-

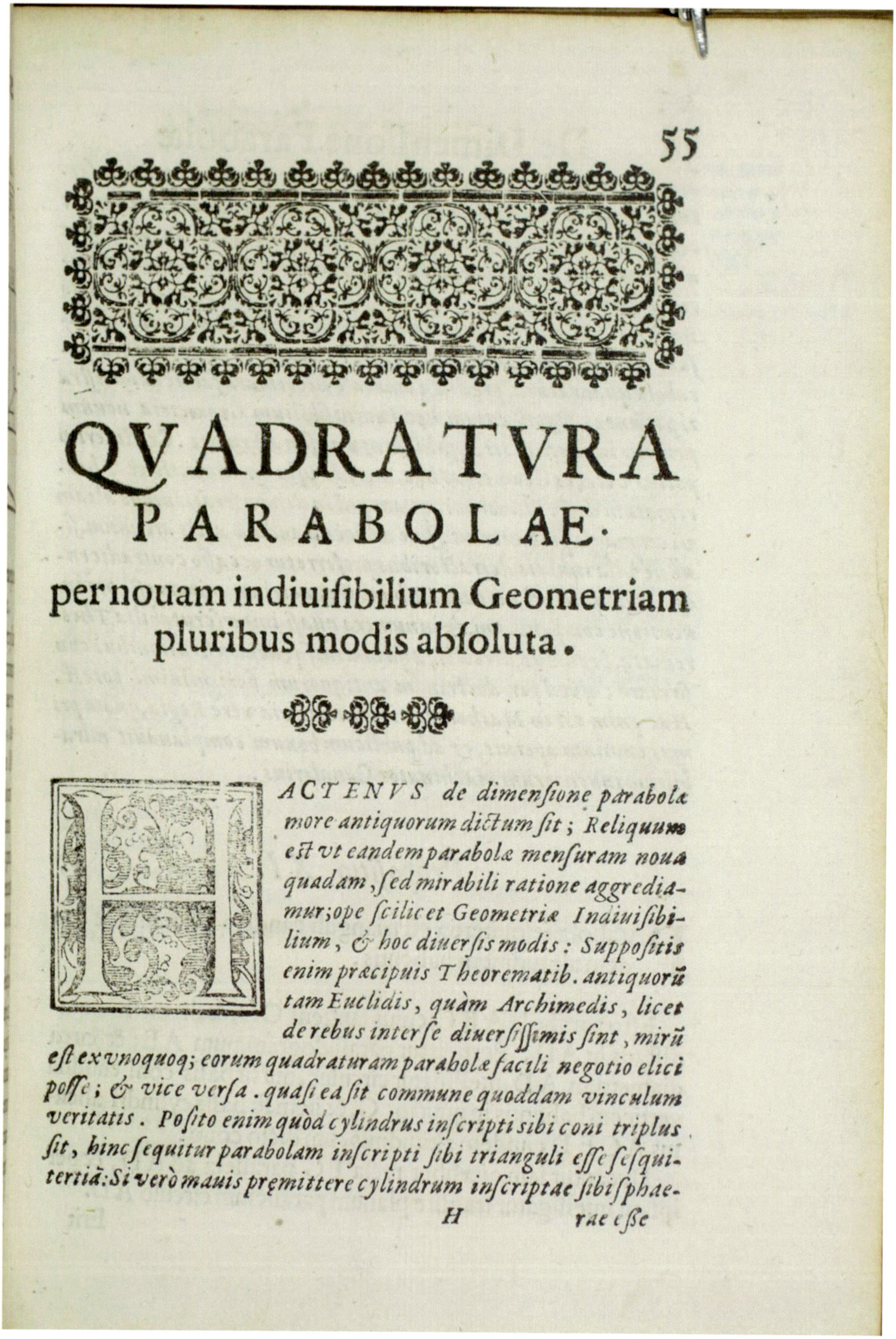

55

QVADRATVRA PARABOLAE.

per nouam indiuisibilium Geometriam pluribus modis absoluta.

HACTENVS de dimensione parabolæ more antiquorum dictum sit; Reliquum est vt eandem parabolæ mensuram noua quadam, sed mirabili ratione aggrediamur; ope scilicet Geometriæ Indiuisibilium, & hoc diuersis modis: Suppositis enim præcipuis Theorematib. antiquorũ tam Euclidis, quàm Archimedis, licet de rebus inter se diuersissimis sint, mirũ est ex vnoquoq; eorum quadraturam parabolæ facili negotio elici posse; & vice versa. quasi ea sit commune quoddam vinculum veritatis. Posito enim quòd cylindrus inscripti sibi coni triplus sit, hinc sequitur parabolam inscripti sibi trianguli esse sesquitertiã: Si verò mauis præmittere cylindrum inscriptae sibi sphae-

H rae esse

55

QVADRATVRA. PARABOLAE.[144].

per nouam indiuifibilium Geometriam pluribus modis abfoluta.

HACTENUS de dimenfione parabolæ more antiquorum dictum fit; Reliquum est vt eandem parabolæ menfuram noua quadam, fed mirabili ratione aggrediamur; ope fcilicet Geometriæ Indiuifibilium, & hoc diuerfis modis: Suppofitis enim præcipuis Theorematib. antiquorũ tam Euclidis, quàm Archimedis, licet de rebus inter fe diuerfiffimis fint, mirũ eft ex vnoquoq; eorum quadraturam parabolæ facili negotio elici poffe; & vice verfa. quafi ea fit commune quoddam vinculum veritatis. Pofito enim quòd cylindrus infcripti sibi coni triplus fit, hinc fequitur parabolam infcripti fibi trianguli effe fefquitertiã: Si verò mauis pręmittere cylindrum infcriptae fibi fphae-

H *rae esse*

[144] With respect to previous section, the page number 55 is repeated. We also remark that Torricelli, with the compelling idea of duplicating the procedure, devoted many pages to prove certain theorems on the *parabolic segment*, by following the geometry used in antiquity (Torricelli 1644, pp. 17–54) and then also proving the validity of his proofs with the *indivisibilium* method (Torricelli 1644, IV, pp. 93–135. Cfr. Pisano and Bussotti's chapters above).

æræ esse sesquialterum, continuò parabolæ quadratura infertur. Eadem concluditur supposita demonstratione, quæ probat centrum grauitatis coni positum esse in axe, ita vt pars quæ ad verticem est, reliquę sit tripla. Parabola non minus quadratur etiam supponendo spatium à linea spirali in prima reuolutione descripta, & à recta quæ initium est reuolutionis, comprehensum, subtriplum esse primi circuli. Contrà verò; supposità parabolæ quadratura, prędicta omnia Theoremata facilè demonstrari possunt. Quod autem hęc Indiuisibilium Geometria nouum penitus inuentum sit, equidem non ausim affirmare. Crediderim potius veteres Geometras hac metodo vsos in inuentione Theorematum difficillimorum, quamquā in demonstrationibus aliam viam magis probauerint, siue ad occultandum artis arcanum, siue ne vlla inuidis detractoribus proferretur occasio contradicendi. Quicquid est, certum est hanc Geometriam mirum esse pro inuentione compendium, & innumera quasi imperscrutabilia Theoremata, breuibus, directis, affirmatiuisq; demonstrationibus confirmare; quod per doctrinam antiquorum fieri minimè potest. Hæc enim est in Mathematicis spinetis via verè Regia, quàm primus omnium aperuit, & ad publicum bonum complanauit mirabilium inuentorum machinator Caualerius.

Propositio XI.

PArabola sesquitertia est trianguli eandem ipsi basim, & eandem altitudinem habentis.

Esto parabola A B C, cuius tangens C D, & diametro æquidistans sit A D. Perficiatur parallelogrammum A E; & circa diametrum A D intelligatur circulus, qui sit basis coni cuiusdā verticem habentis in puncto C, & item sit basis cylindri alicuius A C E D eiusdem altitudinis cum dicto cono.

Ducantur iam quælibet recta F G parallela ad A D, & per ipsam intelligatur transire planum parallelum circulo A D.

Erit

rae eſſe ſeſquialterum, continuò parabolæ quadratura infertur. Eadem concluditur ſuppoſita demonſtratione, quæ probat centrum grauitatis coni poſitum eße in axe, ita vt pars quæ ad verticem eſt, reliquę ſit tripla. Parabola non minus quadratur etiam ſupponendo ſpatium à linea ſpirali in prima reuolutione deſcripta, & à recta quæ initium eſt reuolutionis, compręhenſum, ſubtriplum eſſe primi circuli. Contrà verò; ſuppoſità parabolæ quadratura, prędicta omnia Theoremata facilè demonſtrari poſſunt. Quod autem hęc Indiuiſibilium Geometria nouum penitus inuentum sit, equidem non auſim affirmare. Crediderim potius veteres Geometras hac metodo vſos in inuentione Theorematum difficillimorum, quamquã in demonstrationibus aliam viam magis probauerint, ſiue ad occultandum artis arcanum, ſiue ne vlla inuidis detractoribus proferretur occaſio contradicendi. Quicquid eſt, certum est hanc Geometriam mirum eſſe pro inuentione compendium, & innumera quaſi imperſcrutabilia Theoremata, breuibus, directis, affirmatiuiſq; demonſtrationibus confirmare; quod per doctrinam antiquorum fieri minimè potest. Hæc enim est in Mathematicis ſpinetis via *verè Regia, quàm primus omnium aperuit, & ad publicum bonum complanauit mirabilium inuentorum machinator Caualerius.*

Propoſitio XI.

PArabola ſeſquitertia eſt trianguli eandem ipſi baſim, & eandem altitudinem habentis.

Eſto parabola ABC, cuius tangens CD, & diametro æquidiſtans ſit AD. Perficiatur parallelogrammum AE; & circa diametrum AD intelligatur circulus, qui ſit baſis coni cuiuſdã verticem habentis in puncto C, & item ſit baſis cylindri alicuius ACED eiuſdem altitudinis cum dicto cono.

Ducantur iam quælibet recta FG parallela ad AD, & per ipſam intelligatur tranſire planum parallelum circulo AD.

Erit

[145] With respect to previoous section, the page number 56 is repeated.

Erit ergò F G ad I B vt recta D A ad I B hoc est vt quadratum D C ad quadratum C I, siue vt quadratum D A ad I G, hoc est vt circulus D A ad circulum I G nempe vt circulus F G ad eundem I G. Et hoc semper; suntque omnes primæ magnitudines æquales rectæ D A. & ideò inter se; omnes etiam tertiæ æquales circulo D A, & ob id inter se; ergo per Lemma 18, erunt omnes primæ simul, nempe parallelogrammum A E, ad omnes secundas simul, nempe ad trilineum A B C D, vt sunt omnes tertiæ simul, nempe cylindrus A E, ad omnes quartas simul hoc est ad conum A C D. Est igitur parallelogrammum A E triplum trilinei A B C D. Sumptoque dimidio, erit triangulum A C D sesquialterum trilinei A B C D; & per conuersionem rationis, erit triangulum A C D triplum ipsius parabolæ. Propterea, ex demonstratione propositionis 9. erit parabola inscripti sibi trianguli sesquitertia. Quod erat &c.

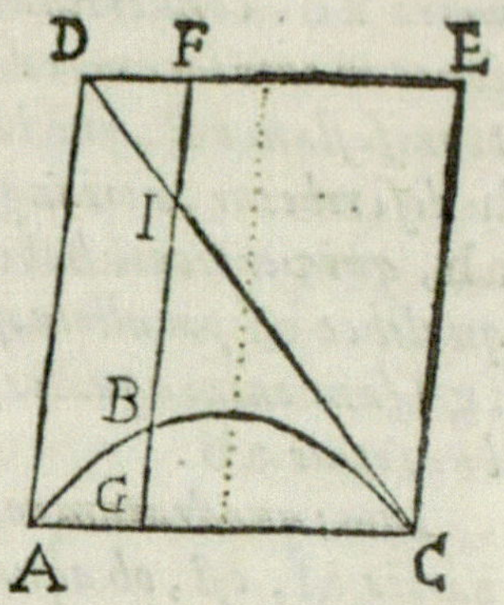

Alia quoque ratione parabolam quadrabimus, demonstratis priùs, quà fieri poterit breuitate, indiuisibilium principijs. Declinabimus autē ab immenso Caualerianę Geometriæ oceano, minori audacia radentes terram. Qui volet, hæc omnia videre poterit (in fonte dicam, an in pelago?) circa medium secundi libri Geometriæ Indiuisibilium Caualerij.

Lemma XX.

Quadrata omnium partium cuiuscunq; rectæ lineæ subtripla sunt totidem quadratorum totius.

Esto qualibet recta linea a b. *Dico omnia simul quadrata omnium partium recta* a b *esse subtripla totidem quadratorum eiusdem rectæ lineæ* a b.

Erit ergò FG ad IB vt recta DA ad IB hoc eſt vt quadratum DC ad quadratum CI, ſiue vt quadratum DA ad IG, hoc eſt vt circulus DA ad circulum IG nempe vt circulus FG ad eundem IG. Et hoc ſemper; ſuntque omnes primæ magnitudines æquales rectæ DA. & ideò inter ſe; omnes etiam tertiæ æquales circulo DA, & ob id inter ſe; ergo per Lemma 18, erunt omnes primæ ſimul, nempe parallelogrammum AE, ad omnes ſecundas ſimul, nempe ad trilineum ABCD, vt ſunt omnes tertiæ ſimul, nempe cylindrus AE, ad omnes quartas ſimul hoc eſt ad conum ACD. Eſt igitur parallelogrammum AE triplum trilinei ABCD. Sumpoque dimidio, erit triangulum ACD ſeſquialterum trilinei ABCD; & per conuerſionem rationis, erit triangulum ACD triplum ipſius parabolæ. Propterea, ex demonſtratione propoſitionis 9. erit parabola inſcripti ſibi trianguli ſeſquitertia.
Quod erat &c.

ob para-bolam
*ob ſimi-litud.*tri[-]ang.
[*non ſtatim Ex parabola appar. sid ex coroll ad 1 prop.lib.*4 *Geo Cavallerij*][146]

Alia quoque ratione parabolam quadrabimus, demonſtratis priùs, quà fieri poterit breuitate, indiuiſibilium principijs. Declinabimus autẽ ab immenſo Caualerianę Geometriæ oceano, minori audacia radentes terram. Qui volet, hæc omnia videre poterit (in fonte dicam, an in pelago?) circa medium ſecundi libri Geometriæ Indiuiſibilium Caualerij.

Lemma XX.

Quadrata omnium partium cuiuſcunq; rectæ lineæ ſubtripla ſunt totidem quadratorum totius.

Esto quælibet recta linea ab. *Dico omnia ſimul quadrata omnium partium rectæ* ab *eſſe ſubtripla totidem quadratorum eiuſdem rectæ lineæ* ab

H 2 *Fiat*

[146] It seems to be a reference to Cavalieri's fourth Book of Geometry.

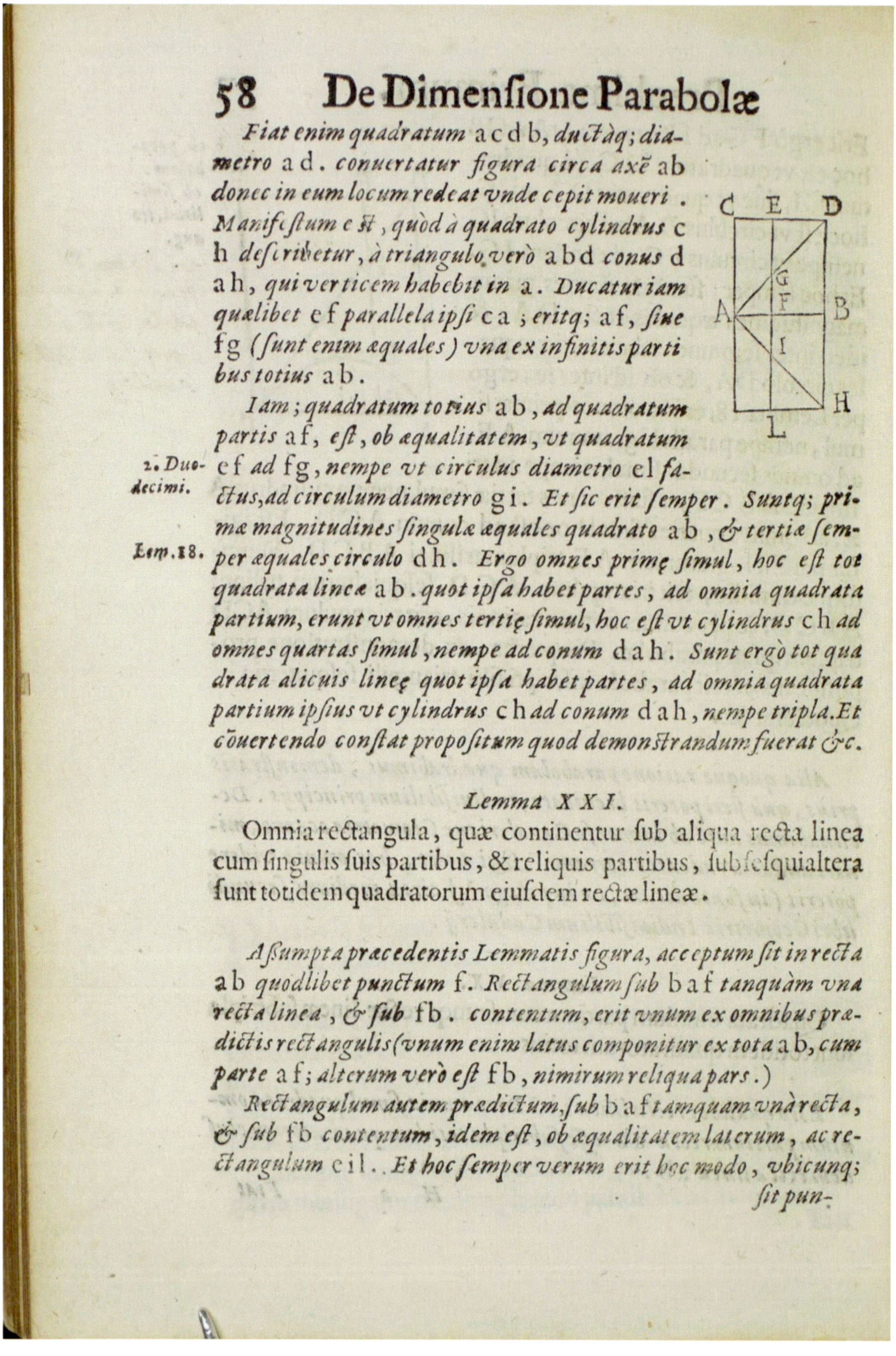
58 De Dimenſione Parabolæ

Fiat enim quadratum a c d b, ductáq; diametro a d. conuertatur figura circa axẽ ab donec in eum locum redeat vnde cepit moueri. Manifeſtum eſt, quòd à quadrato cylindrus c h deſcribetur, à triangulo verò a b d conus d a h, qui verticem habebit in a. Ducatur iam quælibet e f parallela ipſi c a; eritq; a f, ſiue f g (ſunt enim æquales) vna ex infinitis partibus totius a b.

C E D
G
F
A B
I
H
L

Iam; quadratum totius a b, ad quadratum partis a f, eſt, ob æqualitatem, vt quadratum e f ad f g, nempe vt circulus diametro e l factus, ad circulum diametro g i. Et ſic erit ſemper. Suntq; primæ magnitudines ſingulæ æquales quadrato a b, & tertiæ ſemper æquales circulo d h. Ergo omnes primę ſimul, hoc eſt tot quadrata lineæ a b. quot ipſa habet partes, ad omnia quadrata partium, erunt vt omnes tertię ſimul, hoc eſt vt cylindrus c h ad omnes quartas ſimul, nempe ad conum d a h. Sunt ergò tot quadrata alicuis linee quot ipſa habet partes, ad omnia quadrata partium ipſius vt cylindrus c h ad conum d a h, nempe tripla. Et cõuertendo conſtat propoſitum quod demonſtrandum fuerat &c.

2. Duodecimi.
Lem. 18.

Lemma XXI.

Omnia rectangula, quæ continentur ſub aliqua recta linea cum ſingulis ſuis partibus, & reliquis partibus, ſubſeſquialtera ſunt totidem quadratorum eiuſdem rectæ lineæ.

Aſſumpta præcedentis Lemmatis figura, acceptum ſit in recta a b quodlibet punctum f. Rectangulum ſub b a f tanquàm vna recta linea, & ſub f b. contentum, erit vnum ex omnibus prædictis rectangulis (vnum enim latus componitur ex tota a b, cum parte a f; alterum verò eſt f b, nimirum reliqua pars.)

Rectangulum autem prædictum, ſub b a f tamquam vnà recta, & ſub f b contentum, idem eſt, ob æqualitatem laterum, ac rectangulum e i l. Et hoc ſemper verum erit hoc modo, vbicunq; ſit pun-

Fiat enim quadratum acdb, *ductàq; diametro* ad. *conuertatur figura circa axẽ* ab *donec in eum locum redeat vnde cepit moueri. Manifeſtum est, quòd à quadrato cylindrus* ch *deſcribetur, à triangulo verò* abd *conus* dah, *qui verticem habebit in* a. *Ducatur iam quælibet* ef *parallela ipſi* ca; *eritq;* af, *ſiue* fg *(ſunt enim æquales) vna ex infinitis partibus totius* ab.

Iam; quadratum totius ab, *ad quadratum partis* af, *eſt, ob æqualitatem, vt quadratum* ef *ad* fg, *nempe vt circulus diametro* el *factus, ad circulum diametro* gi. *Et ſic erit ſemper. Suntq; primæ magnitudines ſingulæ æquales quadrato* ab, *& tertiæ ſemper*
2. *Duo-* *æquales circulo* dh. *Ergo omnes primę ſimul, hoc eſt tot quadrata lineæ* ab. *quot ipſa habet partes, ad omnia quadrata partium, erunt*
decimi. *vt omnes tertię ſimul, hoc eſt vt cylindrus* ch *ad omnes quartas ſimul, nempe ad conum* dah. *Sunt ergò tot quadrata alicuis linec quot ipſa habet partes, ad omnia quadrata partium ipſius vt cylindrus* ch *ad conum* dah, *nempe tripla. Et cõuertendo conſtat propoſitum*
Lem. 18. *quod demonstrandum fuerat &c.*

Lemma XXI.

Omnia rectangula, quæ continentur ſub aliqua recta linea cum ſingulis ſuis partibus, & reliquis partibus, ſubſeſquialtera ſunt totidem quadratorum eiuſdem rectæ lineæ.

Aßumpta præcedentis Lemmatis figura, acceptum ſit in recta ab *quodlibet punctum* f. *Rectangulum ſub* baf *tanquàm vna recta linea, & ſub* fb. *contentum, erit vnum ex omnibus prædictis rectangulis (vnum enim latus componitur ex tota* ab, *cum parte* af; *alterum verò eſt* fb, *nimirum reliqua pars.)*

Rectangulum autem prædictum, ſub baf *tamquam vnà recta, & ſub* fb *contentum, idem eſt, ob æqualitatem laterum, ac rectangulum* eil. *Et hoc ſemper verum erit hoc modo, vbicunq;*

sit pun-

ſit punctum f. *Sed omnia rectangula ſub rectis interceptis in trapezio* cahd *(qualium vna eſt* ei*) & ſub reliquis, qualium vna eſt* il*; vna cum omnibus quadratis intermediarum ſectionũ (qualium vna eſt* fi*) æquantur (propter* v *ſecundi elementorũ) omnibus quadratis dimidiarum, qualium vna eſt* fl. *Omnia verò quadrata intermediarum ſectionum (qualiam vna eſt* fi*) ad omnia quadrata dimidiarum (qualium vna eſt* fl,*) ſunt vt vnũ ad 3. Si ergo demantur omnia quadrata intermediarum, remanebunt omnia rectangula, quorum vnum eſt* eil, *ſiue omnia rectangula contenta ſub* ab *cum ſingulis ſuis partibus, & reliquis partibus, ſubſeſquialtera omnium quadratorum, quæ ſiunt à dimidijs, ſiue totidem quadratorum totius* ab. *Quod fuerat oſtendendum &c.* *præced. Lem.*

Propoſitio XII.

PArabola ſeſquitertia eſt trianguli eandem ipſi baſim, & eãdem altitudinem habentis.

Eſto parabola ABC cuius diameter BE, & circa parabolã ſit parallelogrammum DC. Ducatur quælibet FG diametro parallela; eritq; FG. ad GI, vt BE ad GI, ſiue vt rectangulum CEA, ad CGA, hoc eſt vt quadratum CE ad rectangulũ CGA. Et hoc modo ſemper; Suntq; primæ magnitudines ſemper ęquales rectæ BE; tertiæ autem ſemper æquales quadrato CE. Ergo omnes primæ ſimul, hoc eſt parallelogrammum AB, ad omnes ſecundas ſimul, nempe ad ſemiparabolã AIBE; erunt vt omnes ſimul tertiæ, videlicet tot quadrata lineæ CE quot ipſa habet partes, ad omnes quartas ſimul, nempe ad omnia rectangula ſub CE cum ſingulis ſuis partibus, & ſub reliquis partibus. Ergo (ex præcedenti lemmate) parallelogrammum AB erit ipſius ſemiparabolę ſeſquialterum: Totumq; *Lem. 18.*

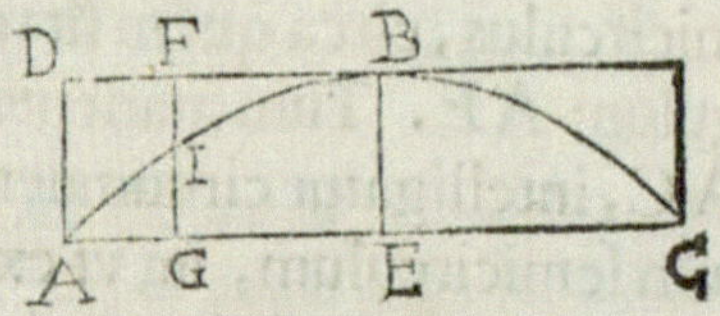

ſit punctum f. *Sed omnia rectangula ſub rectis interceptis in trapezio* cahd *(qualium vna eſt* ei*) & ſub reliquis, qualium vna eſt* il; *vna cum omnibus quadratis intermediarum ſectionũ (qualium vna eſt* fi*) æquantur (propter* v *ſecundi elementorũ) omnibus quadratis dimidiarum, qualium vna eſt* fl. *Omnia verò quadrata intermediarum ſectionum (qualium vna eſt* fi*) ad omnia quadrata dimidiarum (qualium vna est* fl,*) ſunt vt vnã ad 3. Si ergo demantur omnia quadrata intermediarum, remanebunt omnia rectangula, quorum vnum* *præced.* *eſt* eil, *ſiue omnia rectangula contenta ſub* ab *cum ſingulis ſuis partibus, & reliquis partibus, ſubſeſquialtera omnium quadratorum,* *Lem.* *quæ ſiunt à dimìdijs, ſiue totidem quadratorum totius* ab. *Quod fuerat oſtendendum &c.*

Propoſitio XII.

PArabola ſeſquitertia eſt trianguli eandem ipſi baſim, & eãdem altitudinem habentis.

Eſto parabola ABC cuius diameter BE, & circa parabolã ſit parallelogrammum DC. Ducatur quælibet FG diametro parallela; eritq; FG. ad GI, vt BE ad GI, ſiue vt rectangulum CEA, ad CGA, hoc eſt vt quadratum CE ad rectangulũ CGA. Et hoc modo ſemper; Suntq; primæ magnitudines ſemper ęquales rectæ BE; tertiæ autem ſemper æquales quadrato CE. Ergo omnes primæ ſimul, hoc eſt parallelogrammum AB, ad omnes ſecundas ſimul, nempe ad ſemiparabolã AIBE; erunt vt omnes ſimul tertiæ, videlicet tot quadrata *Lem.*18. lineæ CE quot ipſa habet partes, ad omnes quartas ſimul, nempe ad omnia rectangula ſub CE cum ſingulis ſuis partibus, & ſub reliquis partibus. Ergo (ex præcedenti lemmate) parallelogrammum AB erit ipſius ſemiparabolę ſeſquialterum: To-

tumq;

tumque parallelogrammum D C erit totius parabolæ sesquialterum, nempe vt 6. ad 4. Propterea parabola ad inscriptum sibi triangulum (quod quidem parallelogrammi D C sub duplum est) erit vt 4. ad 3. Nempe sesquitertia. Quod erat &c.

Possumus sine molestia illorum lemmatum, parabolam quadrare eodem argumento, diuersis tamen principijs, nempe per suppositionem proportionis, quam cylindrus habet ad sphęram sibi inscriptam; quæ quidem proportio sesquialtera est, vt ostenditur ex Archimede; libro Primo de Sphęra & Cylindro.

Propositio XIII.

Parabola sesquitertia est trianguli eandem ipsi basim, & eādem altitudinem habentis.

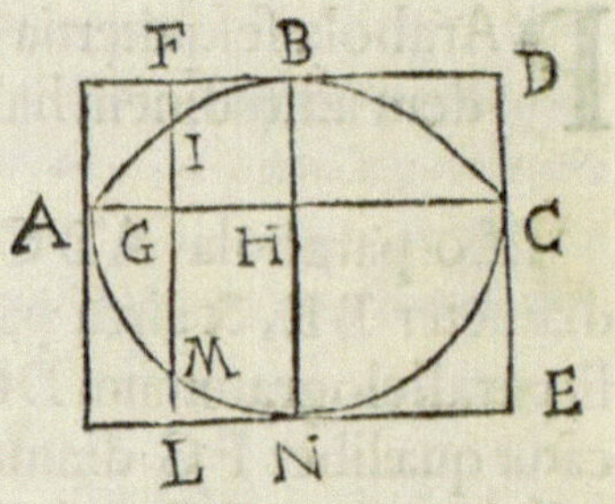

Esto parabola ABC, circa quam sit parallelogrammū AD; & circa diametrum A C fiat semicirculus, circa quem sit rectangulum A E. Tum manente axe A C, intelligatur circumuerti ipsum semicirculum, ita vt ex ipsius reuolutione Sphæra circumscribatur: ex cōuersione verò rectang. A E cylindrus nascatur.

Sumpto iam quolibet puncto G. ducatur recta G F parallela diametro H B; & per idem punctum G agatur planum G L erectum ad axem A C.

3. 4 lib Caualieri Ind

Erit recta F G ad G I, vt B H ad G I (ob æqualitatem) hoc est vt rectangulum C H A, ad rectangulum C G A, siue vt quadratum H N ad quadratum G M (ob circulum) siue vt quadratum G L ad quadratum G M; nempe vt circulus ex semidiametro G L in cylindro, ad circulum ex semidiametro G M in sphæra. Et hoc semper, vbicunque sumatur punctum G. Sunt autem æquales inter se tàm omnes primæ, quàm omnes tertiæ

magni

tumque parallelogrammum DC erit totius parabolæ ſeſquialterum, nempe vt 6. ad 4. Propterea parabola ad inſcriptum ſibi triangulum (quod quidem parallelogrammi DC ſub duplum eſt) erit vt 4. ad 3. Nempe ſeſquitertia. Quod erat &c.

Poſſumus ſine molestia illorum lemmatum, parabolam quadrare eodem argumento, diuerſis tamen principijs, nempe per ſuppoſitionem proportionis, quam cylindrus habet ad ſphęram ſibi inſcriptam; quæ quidem proportio ſeſquialtera eſt, vt ostenditur ex Archimede; libro Primo de Sphęra & Cylindro.

Propoſitio XIII.

PArabola ſeſquitertia eſt trianguli eandem ipſi baſim, & eădem altitudinem habentis.

Eſto parabola ABC, circa quam ſit parallelogrammũ AD; & circa diametrum AC fiat ſemicirculus, circa quem ſit rectangulum AE. Tum manente axe AC, intelligatur circumuerti ipſum ſemicirculum, ita vt ex ipſius reuolutione Sphæra circumſcribatur: ex cõuerſione verò rectang. AE cylindrus naſcatur.

Sumpto iam quolibet puncto G. ducatur recta GF parallela diametro HB; & per idem punctum G agatur planum GL erectum ad axem AC.

Erit recta FG ad GI, vt BH ad GI (ob æqualitatem) hoc eſt vt rectangulum CHA, ad rectangulum CGA, ſiue vt quadratum HN ad
[3. 4 ?
?] quadratum GM (ob circulum) ſiue vt quadratum GL ad quadratum GM; nempe vt circulus ex ſemidiametro GL in cylindro, ad circulum ex ſemidiametro GM in ſphæra. Et hoc ſemper, vbicunque ſumatur punctum G. Sunt autem æquales inter ſe tàm omnes primæ, quàm omnes tertiæ

magni

magnitudines. Ergò omnes primæ, nempe parallelogrammũ AD ad omnes ſecundas, nempe ad parabolam ABC, erunt vt omnes tertiæ, hoc eſt cylindrus, ad omnes ſimul quartas, videlicet ad ſphæram. Sed cylindrus ad ſphæram eſt ſeſquialter; ergò parallelogrammum etiam AD parabolæ ſeſquialterum erit: & ipſa parabola inſcripti ſibi trianguli ſeſquitertia; vt in præcedenti concluſum eſt. Quod &c. *Lem.18.*

Lemma XXII.

Si magnitudines quotcunque ad libram appenſæ fuerint ex quibuſcunq; punctis: totidemq; magnitudines alterius ordinis ex ijſdem punctis pendeant, pariter cum prædictis magnitudinibus proportionales. Erit vnum idemq; libræ punctum centrum æquilibrij vtriusque ordinis magnitudinum.

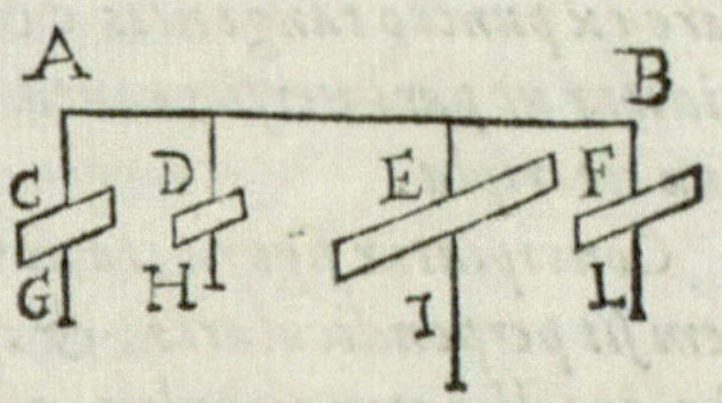

Sint ad libram ab *magnitudines primi ordinis quotcunque* c, d, e, f, *ex quibuſcunque punctis appenſę. Totidemque magnitudines* g, h, i, l, *ſecundi ordinis pendeant ex ijſdem punctis; & ſint proportionales: nempe: Vt* c *ad* d, *ita ſit* g *ad* h. *Iterum ut* c *ad* e, *ita ſit* g *ad* i. *&c. Dico idem punctum libræ eſſe centrum commune æquilibrij utriuſque ordinis magnitudinum ſuſpenſarum.*

Cum enim ſit ut c *ad* d, *ita* g *ad* h, *ex eodem puncto ęquiponderabunt, tam duę magnitudines* c *&* d, *quàm duę* g *&* h.

Ampliùs. Cum ſit ut c *ad* d *ita* g *ad* h, *erit conuertendo & componendo* dc *ad* c, *ut* hg *ad* g. c *autem ad* e *eſt ut* g *ad* i; *ergò ex ęquo* cd *ſimul ad* e, *erit ut* gh *ſimul ad* i. *Quare magnitudines* cd, *&* e, *ex eodem puncto ęquiponderabunt, ex quo ęquiponderant duę* gh, *&* i.

Vlteriùs. Cum autem per iam dicta, ſit ut cd *ad* e, *ita* gh *ad* i, *erit componendo* cde *ad* e, *ut* ghi *ad* i. *Sed* e *ad* c *eſt ut* i *ad* g; *&* c *ad* f, *ut* g *ad* l. *Quare ex ęquo* cde *ſimul ad* f,

erit

magnitudines. Ergò omnes primæ, nempe parallelogrammũ AD ad omnes ſecundas, nempe ad parabolam AB C, erunt vt omnes tertiæ, hoc eſt cylindrus, ad omnes ſimul quartas, videlicet ad ſphæram. Sed cylindrus ad ſphæram eſt ſeſquialter; ergò parallelogrammum etiam AD parabolæ ſeſquialterum erit: & ipſa parabola inſcripti ſibi trianguli ſeſquitertia; vt in præcedenti concluſum eſt. Quod &c. *Lem.*18.

Lemma XXII.[147]

Si magnitudines quotcunque ad libram appenſæ fuerint ex quibuſcunq; punctis: totidemq; magnitudines alterius ordinis ex ijſdem punctis pendeant, pariter cum prædictis magnitudinibus proportionales. Erit vnum idemq; libræ punctum centrum æquilibrij vtriusque ordinis magnitudinum.

Sint ad libram ab *magnitudines primi ordinis quotcunque* c, d, e, f, *ex quibuſcunque punctis appenſę. Totidemque magnitudines* g, h, i, l, *ſecundi ordinis pendeant ex ijſdem punctis; & ſint proportionales: nempe: Vt* c *ad* d, *ita ſit* g *ad* h. *Iterum ut* c *ad* e, *ita ſit* g *ad* i. *&c. Dico idem punctum libræ eſſe centrum commune æquilibrij utriuſque ordinis magnitudinum ſuſpenſarum.*

Cum enim ſit ut c *ad* d, *ita* g *ad* h, *ex eodem puncto ęquiponderabunt, tam duę magnitudines* c *&* d, *quàm duę* g *&* h.

Ampliùs. Cum ſit ut c *ad* d *ita* g *ad* h, *erit conuertendo & componendo* dc *ad* c, *ut* hg *ad* g. c *autem ad* e *est ut* g *ad* i; *ergò ex ęquo* cd *ſimul ad* e, *erit ut* gh *ſimul ad* i. *Quare magnitudines* cd, & e, *ex eodem puncto ęquiponderabunt, ex quo ęquiponderant duę* gh, &i.

Vlteriùs. Cum autem per iam dicta, ſit ut c d *ad* e, *ita* gh *ad* i, *erit componendo* cde *ad* e, *ut* ghi *ad* i. *Sed* e *ad* c *eſt ut* i *ad* g; *&* c *ad* f, *ut* g *ad* l. *Quare ex ęquo* cde *ſimul ad* f,

erit

[147] *Ex æquo* proof. (Cfr. Pisano and Bussotti's chapters above).

erit ut g h i *simul ad* l. *Ergo duę magnitudines* c d e, *&* f. *habebunt idem punctum æquilibrij, quod habent duę magnitudines* g h i *&* l. *Et sic etiam si sint plures magnitudines, usque in infinitum, quod erat propositum &c.*

Lemma XXIII.

Si parabola tangentem habuerit ad basim, ex altera verò parte lineam diametro parallelam. Trilineum compræhensum sub curuà parabolicà, sub tangente, & sub parallela prædictà, æquiponderabit ex puncto tangentis vbi ea sic diuiditur, vt pars ad contactum terminata reliquæ sit tripla.

Esto parabola a b c, *euius tangens ad basim sit* c d; *ęquidistans diametro sit* a d *Dico trilineum mixtum* a b c d *ęquiponderare ex puncto tangentis* c d, *ubi ea sic diuiditur ut pars uersus contactum* c, *reliqua sit tripla.*

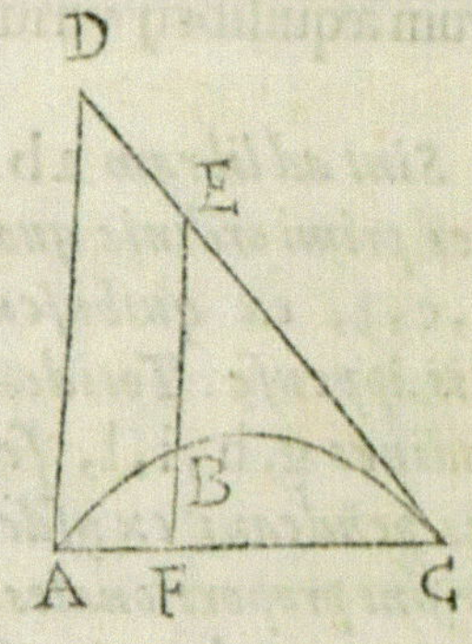

Concipiatur figura ita ut d a *ad horizontem sit perpendicularis; & circa diametrū* d a *intelligatur circulus, qui sit basis coni uerticem habentis in puncto* c.

Sumpto iam quolibet puncto e *ducatur* e f *ęquidistans ipsi* d a; *& per ipsam transeat planum parallelum basi coni.*

ob parabolam. *Erit ergò recta* d a *ad* e b, *ut quadratum* d c *ad* c e; *siue ut quadratum* d a *ad* e f, *hoc est ut circulus* d a *ad* e f. *Et hoc semper, ubicunq; sit punctum* e. *Ergò cum ad libram* d c *pendeant ab ijsdem punctis magnitudines duorum ordinum proportionales ut in praecedenti lemmate imperatum est, habebunt omnes magnitudines simul primi ordinis (hoc est omnes lineae trilinei* a b c d, *siue ipsum trilineum) idem punctum aequilibrij, quod habent omnes magnitudines simul secundi ordinis (hoc est omnes circuli coni* a c d, *siue idem conus.) Conus autem aequiponderat ex puncto quod secat* c d *ita ut pars ad* c *reliquae sit tripla*

erit ut ghi *ſimul ad* l. *Ergo duę magnitudines* cde, *&* f. *habebunt idem punctum æquilibrij, quod habent duę magnitudines* ghi *&* l. *Et sic etiam si sint plures magnitudines, uſque in infinitum, quod erat propoſitum &c.*

Lemma XXIII.

Si parabola tangentem habuerit ad baſim, ex altera verò parte lineam diametro parallelam. Trilineum compræhenſum ſub curuà parabolicà, ſub tangente, & ſub parallela prædictà, æquiponderabit ex puncto tangentis vbi ea ſic diuiditur, vt pars ad contactum terminata reliquæ ſit tripla.

Esto parabola abc, *euius*[148] *tangens ad baſim ſit* cd; *ęquidiſtans diametro ſit* ad

Dico trilineum mixtum abcd *ęquiponderare ex puncto tangentis* cd, *ubi ea ſic diuiditur ut pars uerſus contactum* c, *reliquæ sit tripla.*

Concipiatur figura ita ut da *ad horizontem ſit perpendicularis; & circa diametrũ* da *intelligatur circulus, qui ſit baſis coni uerticem habentis in puncto* c.

Sumpto iam quolibet puncto e *ducatur* ef *ęquidiſtans ipſi* da; *& per ipſam tranſeat planum parallelum baſi coni.*

ob parabolam. *Erit ergò recta* da *ad* eb, *ut quadratum* dc *ad* ce; *siue ut quadratum* da *ad* ef, *hoc eſt ut circulus* da *ad* ef. *Et hoc ſemper, ubicunq; ſit punctum* e. *Ergò cum ad libram* dc *pendeant ab ijſdem punctis magnitudines duorum ordinum proportionales ut in praecedenti lemmate imperatum eſt, habebunt omnes magnitudines ſimul primi ordinis (hoc est omnes lineae trilinei* abcd, *ſiue ipſum trilineum) idem punctum aequilibrij, quod habent omnes magnitudines simul ſecundi ordinis (hoc eſt omnes circuli coni* acd, *siue idem conus.) Conus autem aequiponderat ex puncto quod ſecat* cd *ita ut pars ad* c *reliquae sit*

tripla

[148] *euius*] *cuius.*

tripla, quandoquidem recta da *est ad horizontem perpendicularis; ergo etiam trilineum* abcd *æquiponderabit ex eodem puncto. Quod erat propositum &c.*

Propositio XIV.

PArabola sesquitertia est trianguli eandem ipsi basim, & eãdem altitudinem habentis.

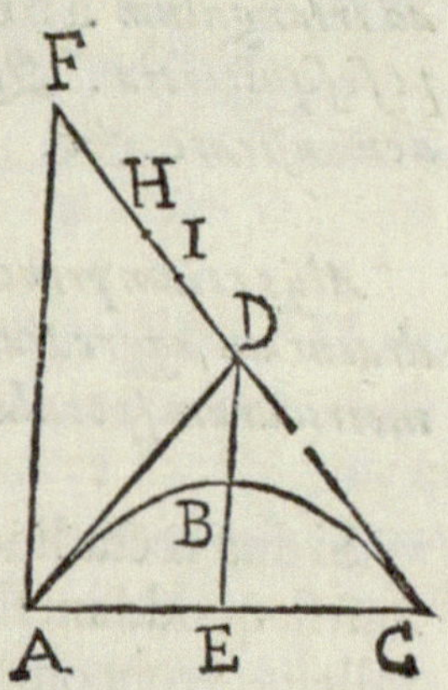

Esto parabola ABC, cuius diameter DE intelligatur ad horizontem perpendicularis; sintque CF, & AD tangentes; ipsa vero AF diametro æquidistans.

Sumatur deinde FH quarta pars totius FC; & ex puncto H (per Lemma præcedens) æquiponderabit trilineum mixtum ABCF. Accipiatur etiam FI tertia pars totius FC, & ex I æquiponderabit totum triangulum AFC. Parabola vero, cum habeat centrum in diametro, æquiponderat ex D. Ergo trilineum ABCF ad ipsam parabolam erit reciprocè vt DI ad IH, nempe duplum (qualium enim partium FC ast 12. talium ipsa FD est 6. FI vero 4. & FH 3. & ideò DI 2, & IH vna.) Propterea componendo erit totum triangulum AFC, parabolæ triplum. Reliquum quadraturę absoluitur vt in Propositione IX. factum est. Quod erat &c.

Aliter.

Positis iisdem, vt suprà, sumatur fh, *quarta pars totius* fc, *æquiponderabitq; ex puncto* h *trilineum mixtum* abcf. *Sumatur etiam* fi, *tertià pars ipsius* fd; *tunc enim æquiponderabit ex puncto* i *triangulum* fda. *Trilineum uerò mixtum* abcd, *æquiponderat ex puncto* d. (*nam triangulum totum* adc *æquiponderat ex puncto* d; *parabola etiam ablata ex eodem puncto* d

I

tripla, quandoquidem recta da *eſt ad horizontem perpendicularis; ergo etiam trilineum* abcd *æquiponderabit ex eodem puncto. Quod erat propositum &c.*

Propoſitio XIV.

PArabola ſeſquitertia eſt trianguli eandem ipſi baſim, & eãdem altitudinem habentis.

Eſto parabola ABC, cuius diameter DE intelligatur ad horizontem perpendicularis; ſintque CF, & AD tangentes; ipſa vero AF diametro æquidiſtans.

Sumatur deinde FH quarta pars totius FC; & ex puncto H (per Lemma præcedens*)* æquiponderabit trilineum mixtum ABCF. Accipiatur etiam FI tertia pars totius FC, & ex I æquiponderabit totum triangulùm AFC. Parabola vero, cum habeat centrum in diametro, æquiponderat ex D. Ergo trilineum ABC F ad ipſam parabolam erit reciprocè vt DI ad IH, nempe duplum *(*qualium enim partium FC aſt[149] 12. talium ipſa FD eſt 6. FI vero 4. & FH 3. & ideò DI 2, & IH vna.) Propterea componendo erit totum triangulum AFC, parabolæ triplum. Reliquum quadraturę abſoluitur vt in Propoſitione IX. factum eſt. Quod erat &c.

Aliter.

Positis iiſdem, vt ſuprà, ſumatur fh, *quarta pars totius* fc, *æquiponderabitq; ex puncto* h *trilineum mixtum* abcf. *Sumatur etiam* fi, *tertià pars ipsius* fd; *tunc enim æquiponderabit ex puncto* i *triangulum* fda. *Trilineum uerò mixtum* abcd, *æquiponderat ex puncto* d. *(nam triangulum totum* adc *æquiponderat ex puncto* d; *parabola etiam ablata ex eodem pun-*

I *cto* d

[149] aſt] eſt.

cto d *ęquiponderat, ergò etiam reliquū trilineum* a b c d *ex puncto* d *æquiponderare necesse est.*) *Erit itaque triangulum* f d a *ad trilineum* a b c d *ut reciprocè* d h *ad* h i *nempe vt 3. ad unum; & per conuersionem rationis triangulum* a d c *ad parabolam erit ut 3. ad 2. siue ut 6. ad 4. Quarè parabola ad triangulum* a b c *erit ut 4. ad 3. Nēpe sesquitertia. Quod erat propositum demonstrare. &c.*

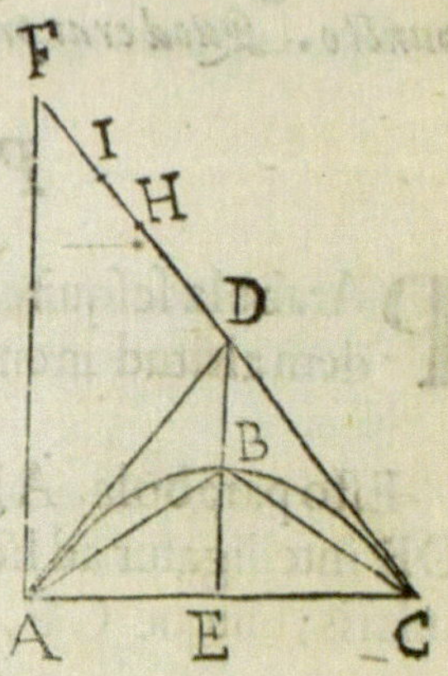

Alijs etiam principijs parabolæ quadraturam aggrediamur, præmissa sequenti progressionum Geometricarum speculatione.

Lemma XXIV.

Si duæ rectæ lineæ inuicem concurrant, & inter ipsas descriptum sit quoddam flexilineum constans ex lineis alternatim parallelis; erunt omnes lineæ, quæ inter se parallelæ sunt, in continua proportione.

Concurrant inuicem duae rectæ lineę, a b, c b *in puncto* b; *& inter ipsas descriptum sit flexi lineum* c a d e f g. *&c. itaut* c a, d e, f g *&c. sint inter se parallelę; item* a d, e f, *& reliquae uicisim sumptae inter se parallelae sint. Dico* a c, e d, g f, *esse in continua proportione.*

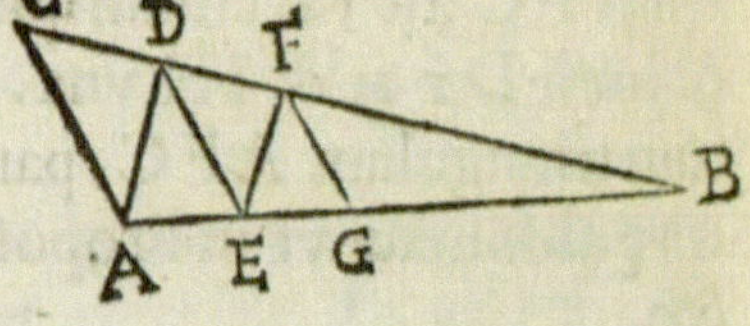

Est enim, ob parallelas, ut a c *ad* e d, *ita* a b *ad* b e, *siue*
2. & 4. Sexti. d b *ad* b f, *hoc est* e d *ad* g f. *Constat ergò quod propositum fuerat.*

Lemma XXV.

Positis duabus rectis lineis inuicem concurrentibus, vt suprà; Si inter

cto d *ęquiponderat, ergò etiam reliquũ trilineum* abcd *ex puncto* d *æquiponderare neceſſe eſt.*) *Erit itaque triangulum* fda *ad trilineum* abcd *ut reciprocè* dh *ad* hi *nempe vt 3. ad unum; & per conuersionem rationis triangulum* adc *ad parabolam erit ut 3. ad* 2. *siue ut 6. ad 4. Quarè parabola ad triangulum* abc *erit ut 4. ad 3. Nẽpe ſeſquitertia. Quod erat propositum demonſtrare. &c.*

Alijs etiam principijs parabolæ quadraturam aggrediamur, præmiſsa ſequenti progreſſionum Geometricarum ſpeculatione.

Lemma XXIV.

Si duæ rectæ lineæ inuicem concurrant, & inter ipſas deſcriptum ſit quoddam flexilineum conſtans ex lineis alternatìm parallelis; erunt omnes lineæ, quæ inter ſe parallelæ ſunt, in continua proportione.

Concurrant inuicem duae rectæ lineę, ab, cb *in puncto* b; *& inter ipſas deſcriptum sit flexilineum* ca def g. *&c. ita ut* ca, de, fg *&c. sint inter ſe parallelę; item* ad, ef, *& reliquae uicisim ſumptae inter ſe parallelae sint. Dico* ac, ed, gf, *eſſe in continua proportione.*

2. & 4. *Sextì.* *Eſt enim, ob parallelas, ut* ac *ad* ed, *ita* ab *ad* be, *siue* db *ad* bf, *hoc est* ed *ad* gf. *Constat ergò quod propositum fuerat.*

Lemma XXV.

Poſitis duabus rectis lineis inuicem concurrentibus, vt ſuprà;

Si inter

si inter ipsas fuerint duæ parallelæ AC, DE, & iunctà CD, continuatum intelligatur flexilineum ACDE in infinitum vsque ad punctum concursus B. Dico in huiusmodi flexilineo esse omnes, & singulos ad vnguem terminos qui sunt in progressione proportionis AC ad DE. in infinitum continuatæ.

Ponatur f *equalis ipsi* a c, *&* g *æqualis ipsi* d e: *Et concipiatur proportio* f *ad* g *continuata in infinitis suis terminis* f H.

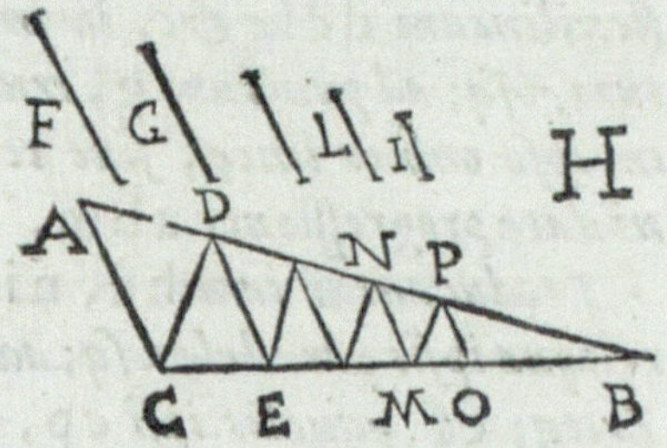

Iam, si possibile est, aliquem, siue aliquos terminos esse in progressione f H, *qui nō reperiantur in flexilineo. Esto: & sit maximus terminus* i, *illorum, qui cùm sint in progressione* f H, *non sunt in flexilineo. Erit ergo terminus* l *ipsi precedens, in flexilineo. Sit ille* m n. *Et quoniam* l *ad* i *est vt* f *ad* g, *siue vt* a c *ad* d e, *siue vt* n m *ad* p o *proximè sequentem, suntq; equales* l, *&* n m; *erunt æquales etiam* i *&* p o. *Terminus ergo* i *qui ponebatur non esse in flexilineo, in eodem repertus est.*

Eodem penitus modo demonstrabimus nullum terminum esse in flexilineo, qui non sit etiam in progressione f H. *&c. Concludemus igitur esse in flexilineo omnes precisè terminos proportionis* a c *ad* d e *in infinitum continuate, cum demonstratum sit nullum in flexilineo terminum desiderari qui sit in progressione* f H; *neq; vllum superabundare, qui non reperiatur etiam in progressione* f H. *&c.*

Lemma XXVI.

Suppositis infinitis rectis lineis in continua proportione maioris inæqualitatis, rectam lineam, quæ prædictis omnibus sit æqualis reperire.

Ponantur primæ duæ linee date progressionis esse a, b: *quib. ponantur equales;* c d *maiori* a, *&* e f *minori* b. *Sintq;* c d, e f

ſi inter ipſas fuerint duæ parallelæ AC, DE, & iunctà CD, continuatum intelligatur flexilineum ACDE in infinitum vſque ad punctum concurſus B. Dico in huiuſmodi flexilineo eſſe omnes, & ſingulos ad vnguem terminos qui ſunt in progreſſione proportionis AC ad DE. in infinitum continuatæ.

Ponatur f *ęqualis ipſi* ac, *&* g *æqualis ipſi* de: *Et concipiatur proportio* f *ad* g *continuata in infinitis ſuis terminis* fH.

Iam, ſi poſſibile eſt, aliquem, ſiue aliquos terminos eſſe in progreſſione fH, *qui nõ reperiantur in flexilineo. Esto: & ſit maximus terminus* i, *illorum, qui cùm ſint in progreſſione* fH, *non ſunt in flexilineo. Erit ergo terminus* l *ipſi pręcedens, in flexilineo. Sit ille* mn. *Et quoniam* l *ad* i *est vt* f *ad* g, *ſiue vt* ac *ad* de, *ſiue vt* nm *ad* po *proximè ſequentem, ſuntq; ęquales* l, *&* n m; *erunt æquales etiam* i *&* po. *Terminus ergo* i *qui ponebatur non eſſe in flexilineo, in eodem repertus eſt.*

Eodem penitus modo demonstrabimus nullum terminum eſſe in flexilineo, qui non ſit etiam in progreſſione fH. *&c. Concludemus igitur eſſe in flexilineo omnes pręcisè terminos proportionis* ac *ad* de *in infinitum continuatę, cum demonſtratum ſit nullum in flexilineo terminum deſiderari qui ſit in progreſſione* fH; *neq; vllum ſuperabundare, qui non reperiatur etiam in progreſſione* fH. *&c.*

Lemma XXVI.

Suppoſitis infinitis rectis lineis in continua proportione maioris inæqualitatis, rectam lineam, quæ prædictis omnibus ſit æqualis reperire.

Ponantur primæ duæ lineę datę progreſſionis eſſe a, b*: quib. ponantur ęquales;* cd *maiori* a, *&* ef *minori* b. *Sintq;* cd, ef

I 2 *paral-*

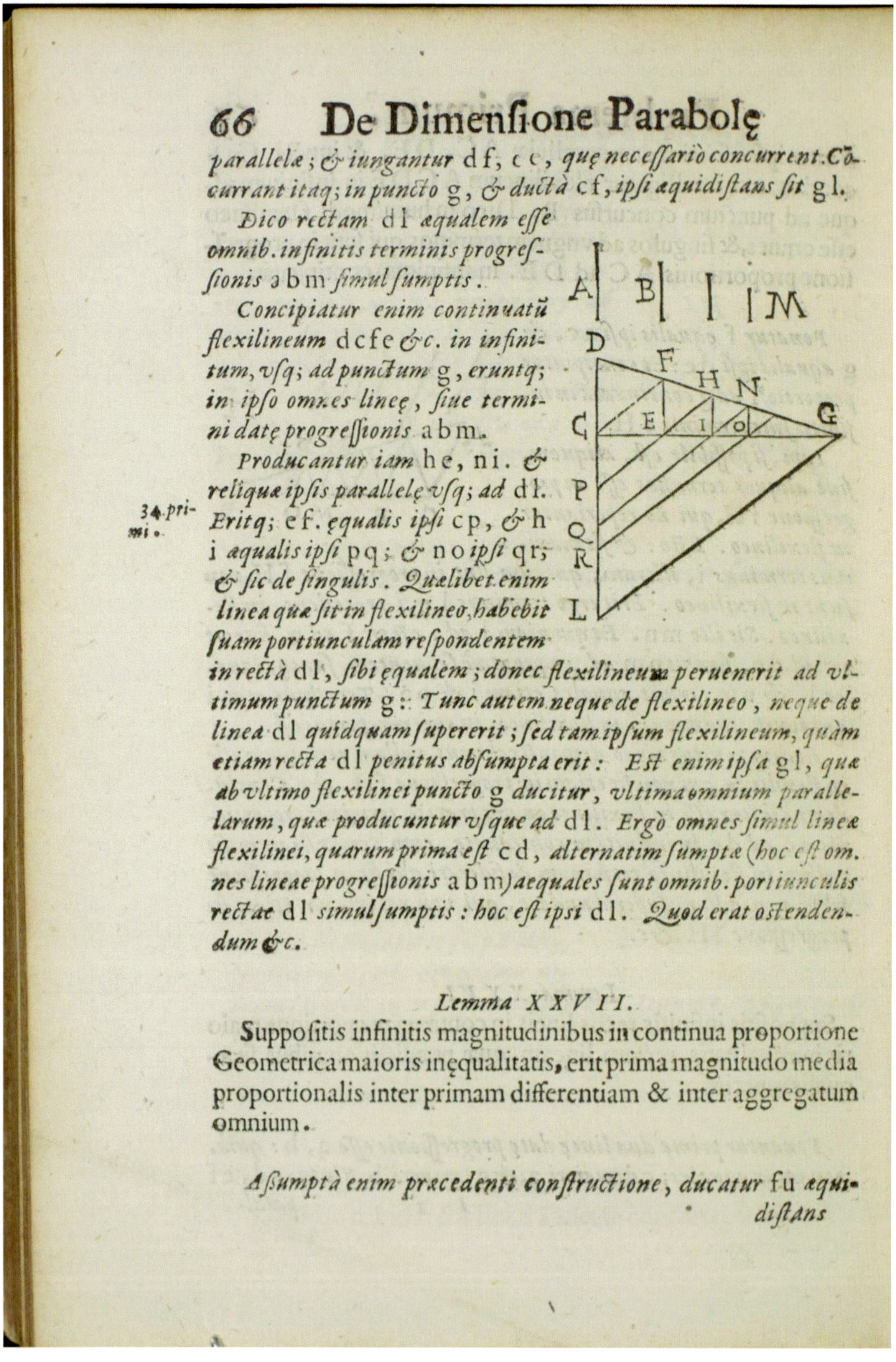

66 De Dimensione Parabolę

parallelæ; & iungantur d f, c e, *quę necessariò concurrent. Concurrant itaq; in puncto* g, *& ductà* c f, *ipsi æquidistans sit* g l.

Dico rectam d l *æqualem esse omnib. infinitis terminis progressionis* a b m *simul sumptis.*

Concipiatur enim continuatũ flexilineum d c f e *&c. in infinitum, vsq; ad punctum* g, *eruntq; in ipso omnes lineę, siue termini datę progressionis* a b m.

Producantur iam h e, n i. *& reliquæ ipsis parallelę vsq; ad* d l.

34. primi. *Eritq;* e f. *ęqualis ipsi* c p, *&* h i *æqualis ipsi* p q; *&* n o *ipsi* q r; *& sic de singulis. Quælibet enim linea quæ sit in flexilineo, habebit suam portiunculam respondentem in rectà* d l, *sibi ęqualem; donec flexilineum peruenerit ad vltimum punctum* g: *Tunc autem neque de flexilineo, neque de linea* d l *quidquam supererit; sed tam ipsum flexilineum, quàm etiam recta* d l *penitus absumpta erit: Est enim ipsa* g l, *quæ ab vltimo flexilinei puncto* g *ducitur, vltima omnium parallelarum, quæ producuntur vsque ad* d l. *Ergò omnes simul lineæ flexilinei, quarum prima est* c d, *alternatim sumptæ (hoc est omnes lineae progressionis* a b m) *aequales sunt omnib. portiunculis rectae* d l *simul sumptis: hoc est ipsi* d l. *Quod erat ostendendum &c.*

Lemma XXVII.

Suppositis infinitis magnitudinibus in continua proportione Geometrica maioris inęqualitatis, erit prima magnitudo media proportionalis inter primam differentiam & inter aggregatum omnium.

Assumptà enim præcedenti constructione, ducatur f u *æquidistans parallelæ; & iungantur* df, ce, *quę neceſſariò concurrent. Cõcurrant itaq; in puncto* g, *& ductà* cf, *ipſi æquidiſtans ſit* gl.

Dico rectam dl *æqualem eſſe omnib. infinitis terminis progreſſionis* abm *ſimul ſumptis.*

Concipiatur enim continuatũ flexilineum dcfe *&c. in infinitum, vſq; ad punctum* g, *eruntq; in ipſo omnes lineę, ſiue termini datę progreſſionis* abm.

Producantur iam he, ni. *& reliquæ ipſis parallelę vſq; ad* dl. *Eritq;* ef. *ęqualis ipſi* cp, & hi *æqualis ipſi* pq; & no *ipſi* qr; *& ſic de ſingulis. Quælibet enim linea quæ ſit in flexilineo, habebit ſuam portiunculam reſpondentem in rectà* dl, *ſibi ęqualem; donec flexilineum peruenerit ad vltimum punctum* g: *Tunc autem neque de flexilineo, neque de lineæ* dl *quídquam ſupererit; ſed tam ipſum flexilineum,*
34.*pri-. quàm etiam recta* dl *penitus abſumpta erit: Est enim ipſa* gl, *quæ ab vltimo flexilinei puncto* g *ducitur, vltima omnium parallelarum,*
mi. quæ producuntur vſque ad dl. *Ergò omnes ſimul lineæ flexilinei, quarum prima eſt* cd, *alternatim ſumptæ* (*hoc eſt om. nes*[150] *lineae progreſſionis* abm*) aequales ſunt omnib. portiunculis rectae* dl *simul ſumptis: hoc eſt ipsi* dl. *Quod erat ostendendum &c.*

Lemma XXVII.

Suppoſitis infinitis magnitudinibus in continua proportione Geometrica maioris inęqualitatis, erit prima magnitudo media proportionalis inter primam differentiam & inter aggregatum omnium.

Aßumptà enim præcedenti conſtructione, ducatur fu *æqui-*

distans

[150] *om. nes*] *omnes.*

distans ipsi g c : *& erit* d u *prima differentia. Sed* d u *ad primam magnitudinem* d c *est ut* f d *ad* d g, *hoc est ut* d c *ad* d l *aggregatum omnium. Quod erat demonstrandum &c.*

4. Sexti.

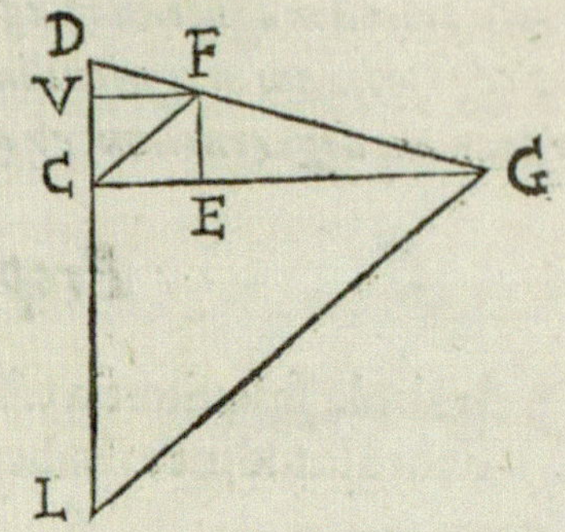

SCHOLIVM.

Hoc esse uerum etiam in numeris, & cuiuscunq. generis magnitudinibus non dubitabimus affirmare. Afferremus etiam uniuersaliorem demonstrationem, praecipuè cum admodum breuis sit. Huius ueritatis conclusio cum à nobis obiter celeberrimo Caualerio collata fuisset, ipse etiā idem Theorema sequenti demonstratione, quę à nobis iam in prima inuentione adhibita fuerat, confirmauit.

Pręmittitur hoc. Quòd si fuerint quotcunq. magnitudines siue finitę numero, siue infinitę, quarum antecedens semper sequente maior sit, erit prima omnium magnitudo ęqualis omnib. differentijs simul cum ipsa minima magnitudine sumptis.

Notum est hoc apud Geometras, demonstraturq. ut à nobis factum est in lemmate 15. Vbi ostendimus parallelogrammum a e *ęquale esse omnibus differentis inter sequentia parallelogramma, & minimo parallelogrammo* o c.

Supponantur iàm infinitae numero magnitudines in continua proportione Geometrica maioris inaequalitatis; manifestum est quòd minima omnium magnitudo vel non erit, uel punctum erit. Ergo in hoc casu erit prima magnitudo aequalis omnibus tantum differentijs.

Cum autem ponantur magnitudines in continua proportione Geometrica, erunt etiam differentiæ in eadem ratione proportionales; & ideo (factà conuersione) erit vt prima differentia ad primam magnitudinem, ita secunda differentia ad secundam magnitudinem, & sic semper. Propterea vt vna ad vnam, ita collectim erunt omnes ad omnes. Nempe vt prima differentia ad

pri-

distans ipsi gc*: & erit* du *prima differentia. Sed* du *ad primam magnitudinem* dc *est ut* fd *ad* dg, *hoc eſt ut* dc *ad* dl *aggregatum omnium. Quod erat demonstrandum &c.* *4.Sexti.*

SCHOLIVM.

Hoc eßе uerum etiam in numeris, & cuiuſcunq. generis magnitudinibus non dubitabimus affirmare. Afferremus etiam uniuerſaliorem demonſtrationem, praecipuè cum admodum breuis sit. Huius ueritatis conclusio cum à nobis obiter celeberrimo Caualerio collata fuiſſet, ipſe etiã idem Theorema ſequenti demonſtratione, quę à nobis iam in prima inuentione adhibita fuerat, confirmauit.

Pręmittitur hoc. Quòd si fuerint quotcunq. magnitudines siue finitę numero, siue infinitę, quarum antecedens ſemper ſequente maior sit, erit prima omnium magnitudo ęqualis omnib. differentijs simul cum ipſa minima magnitudine ſumptis.

Notum eſt hoc apud Geometras, demonſtraturq. ut à nobis factum eſt in lemmate 15. Vbi ostendimus parallelogrammum ae *ęquale eſſe omnibus differentis inter ſequentia parallelogramma, & minimo parallelogrammo* OC.

Supponantur iàm infinitae numero magnitudines in continua proportione Geometrica maioris inaequalitatis; manifeſtum est quòd minima omnium magnitudo vel non erit, uel punctum erit. Ergo in hoc caſu erit prima magnitudo aequalis omnibus tantum differentijs.

Cum autem ponantur magnitudines in continua proportione Geometrica, erunt etiam differentiæ in eadem ratione proportionales; & ideo (factà conuerſione) erit vt prima differentia ad primam magnitudinem, ita ſecunda differentia ad ſecundam magnitudinem, & ſic ſemper. Propterea vt vna ad vnam, ita collectim erunt omnes ad omnes. Nempe vt prima differentia ad

pri-

primam magnitudinem, ita erunt omnes simul differentiæ (hoc est ipsa prima magnitudo) ad omnes magnitudines simul. Constat ergò primam magnitudinem mediam proportionalem esse inter primam differentiam, & aggregatum omnium,

Propositio XV.

Parabola sesquitertia est trianguli eandem ipsi basim, & eã dem altitudinem habentis.

Esto parabola ABC in quâ inscriptum sit triangulum ABC. Dico parabolam trianguli ABC esse sesquitertiam.

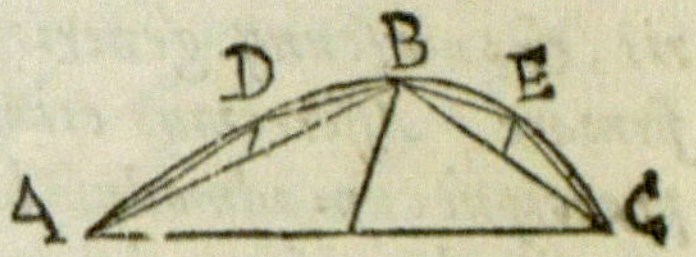

Inscribantur enim etiam in reliquis portionibus ADB, BEC, duo triangula ADB, BEC. Eritq; triangulum ABC quadruplum duorum simul triangulorum ADB, BEC. *Lem.7.* Concipiantur etiam in reliquis quatuor portiunculis AD, DB, BE, EC. inscripta quatuor triangula; eruntq; duo simul triangula ADB, BEC quadrupla prædictorum simul quatuor subsequentium triangulorum; *Lem.7.* & hoc modo semper. Parabola igitur nihil aliud est quàm aggregatum quoddam infinitarum numero magnitudinum in proportione quadrupla, quarum prima est triangulum ABC, secunda verò constat ex duobus triangulis ADB, BEC. Propterea prima magnitudo ABC media proportionalis erit inter primã differentiam, & aggregatum omnium, nempe parabolam.

Ponatur itaq; triangulum ABC esse vt 4. & ideo duo simul triangula ADB, BEC erunt vt vnum: eritq; prima differentia (nimirum inter 4. & vnum) vt 3. Ergo aggregatum omnium infinitarum magnitudinum, nempe ipsa parabola, erit (per lemma 27.) ad primam magnitudinem, hoc est ad inscriptũ triangulum ABC, vt prima ipsa magnitudo ad primam differentiam; videlicet vt 4. ad 3. nempe sesquitertia. Quod erat propositum demonstrare &c.

Aliter.

primam magnitudinem, ita erunt omnes ſimul differentiæ (hoc eſt ipſa prima magnitudo) ad omnes magnitudines ſimul. Constat ergò primam magnitudinem mediam proportionalem eße inter primam differentiam, & aggregatum omnium,

Propoſitio XV.

PArabola ſeſquitertia eſt trianguli eandem ipſi baſim, & eãdem altitudinem habentis.

Esto parabola ABC in qua inſcriptum ſit triangulum ABC. Dico parabolam trianguli ABC eſſe ſeſquitertiam.

Lem.7. Inſcribantur enim etiam in reliquis portionibus ADB, BEC, duo triangula ADB, BEC. Eritq; triangulum ABC quadruplum duorum ſimul triangulorum ADB, BEC. Concipiantur etiam in reliquis quatuor portiunculis AD, *Lem.7* DB, BE, EC. inſcripta quatuor triangula; eruntq; duo ſimul triangula ADB, BEC quadrupla prædictorum ſimul quatuor ſubſequentium triangulorum; & hoc modo ſemper. Parabola igitur nihil aliud eſt quàm aggregatum quoddam infinitarum numero magnitudinum in proportione quadrupla, quarum prima eſt triangulum ABC, ſecunda verò conſtat ex duobus triangulis ADB, BEC. Propterea prima magnitudo ABC media proportionalis erit inter primã differentiam, & aggregatum omnium, nempe parabolam.

Ponatur itaq; triangulum ABC eſſe vt 4. & ideo duo ſimul triangula ADB, BEC erunt vt vnum: eritq; prima differentia (nimirum inter 4. & vnum) vt 3. Ergo aggregatum omnium infinitarum magnitudinum, nempe ipſa parabola, erit (per lemma 27.) ad primam magnitudinem, hoc eſt ad inſcriptũ triangulum ABC, vt prima ipſa magnitudo ad primam differentiam; videlicet vt 4. ad 3. nempe ſeſquitertia. Quod erat propoſitum demonſtrare &c.

Aliter

Aliter.

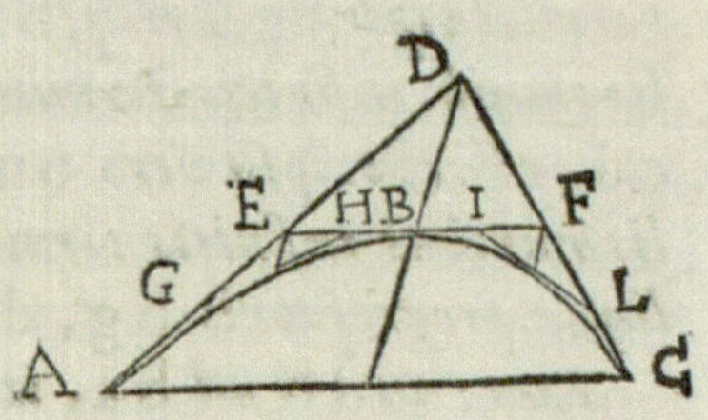

Esto parabola abc. *cuius diameter* db, *tangentes ad basim* ad, cd, *per verticem verò* ef. *Inscribantur autem in reliquis trilineis* abe, bcf, *duo triangula* geh, ifl, *(vt imperatum fuit pro constructione lemmatum Tertij, & Quarti.) Item in reliquis quatuor trilineis mixtis, quatuor triangula concipiantur; & hoc modo semper. Eritq; vniuersum trilineum* abcd *nihil aliud quàm aggregatum quoddã infinitarum multitudine magnitudinum in proportione quadrupla, quarum prima est triangulum* edf, *secunda verò constat ex duobus triangulis* geh, ifl; *tertia verò ex quatuor sequentibus &c. Propterea aggregatum omnium, nempe trilineum mixtum* abcd, *ad primam magnitudinem, nempe ad triangulum* edf, *erit vt ipsa prima magnitudo ad primam differentiam, videlicet vt 4. ad 3.* Coroi. 1. Lem. 3. Lem. 27.

Cum itaque trilineum abcd *ad triangulum* edf, *sit vt 4. ad tria, erit idem trilineum ad triangulum* adc *vt 4. ad 12. & ideo parabola ad triangulum* adc *erit vt 8. ad 12. & ad inscriptum sibi triangulum vt 8. ad 6. Nempe sesquitertia. Quod erat demonstrandum &c.*

Lemma XXVIII.

Si fuerint infinitæ numero rectæ lineę AB, CD, EF. &c. in continua proportione Geometrica maioris inæqualitatis: altera autem ponatur progressio BG, DH, FI &c. ita vt sit quemadmodum AB prima ad BG primam, itã CD secunda ad DH secundam: & ita tertia EF ad tertiam FI & sic semper. Dico vniuersum aggregatum progressionis AB, CD, EF, &c. ad aggregatum progressionis BG, DH, FI, esse vt AB ad BG.

Intel-

Aliter.

Eſto parabola abc. *cuius diameter* db, *tangentes ad baſim* ad, cd, *per verticem verò* ef. *Inſcribantur autem in reliquis trilineis* abe, bcf, *duo triangula* geh, ifl, *(vt imperatum fuit pro conſtructione lemmatum Tertij, & Quarti.) Item in reliquis quatuor trilineis mixtis, quatuor triangula concipiantur; & hoc modo ſemper. Eritq; vniuerſum trilineum* abcd *nihil aliud quàm aggregatum quoddã infinitarum multitudine magnitudinum in proportione quadrupla, quarum prima eſt triangulum* edf, *ſecunda verò conſtat ex duobus* Coroi.1. *triangulis* geh, ifl; *tertia verò ex quatuor ſequentibus &c. Propterea aggregatum omnium, nempe trilineum mixtum* abcd, *ad primam* Lem.3. *magnitudinem, nempe ad triangulum* edf, *erit vt ipſa prima magnitudo ad primam differentiam, videlicet vt 4. ad 3.*

Cum itaque trilineum abcd *ad triangulum* edf, *ſit vt 4. ad tria, erit idem trilineum ad triangulum* a d c *vt 4. ad 12. & ideo* Lem.27. *parabola ad triangulum* adc *erit vt 8. ad 12 & ad inſcriptum ſibi triangulum vt 8. ad 6. Nempe ſeſquiteria. Quod erat demonſtrandum &c.*

Lemma XXVIII.

Si fuerint infinitæ numero rectæ lineę AB, CD, EF. &c. in continua proportione Geometrica maioris inæqualitatis: altera autem ponatur progreſſio BG, DH, FI &c. ita vt ſit quemadmodum AB prima ad BG primam, ità CD ſecunda ad DH ſecundam: & ita tertia EF ad tertiam FI & ſic ſemper. Dico vniuerſum aggregatum progreſſionis AB, CD, EF, &c. ad aggregatum progreſſionis BG, DH, FI, eſſe vt AB ad BG.

Intel-

70 De Dimensione Parabolæ

Intelligantur omnes termini duarum progressionum esse in flexilineis &c. iuxta Lēma 25. *iunctisq;* ad, gd, *ducatur* ol *parallela ipsi* ad, *&* om *parallela ipsi* dg, *Eritq;* bl *æqualis omnib. infinitis terminis* ab, (Lém.26.) cd, ef. *&c. ipsa vero* om *æqualis omnibus infinitis terminis reliquæ progressionis* bg, dh, fi. (Lem.26.)

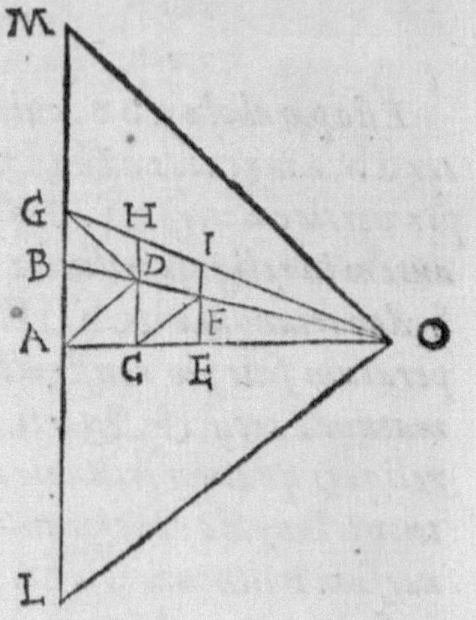

Iam: vt lb *ad* ba, *ita est* ob *ad* bd, *hoc est* mb. *ad* bg. (4. sexti.) *Permutando igitur, Aggregatum* lb, *ad aggregatum* bm, *est vt* ab *ad* bg; *nempe vt vna magnitudo ad vnam. Quod erat &c.*

Hoc Theorema poterat supponi tamquam demonstratum in Propositione 12. libri V. Euclidis: vnum enim atq; idem est cum Theoremate dictæ Propositionis: Verùm, quoniam ferè omnes opinantur Euclidem ibi supponere multitudinem magnitudinum finitam, voluimus auxilio flexilineorum vti,

Propositio XVI.

Parabola sesquitertia est trianguli eandem ipsi basim, & eandem altitudinem habentis.

Sit parabola ABC, cuius diameter DE, tangentes ad basim AD, CD: per verticem vero FBG. triangulum inscriptum ABC. Dico parabolam trianguli ABC. esse sesquitertiam.

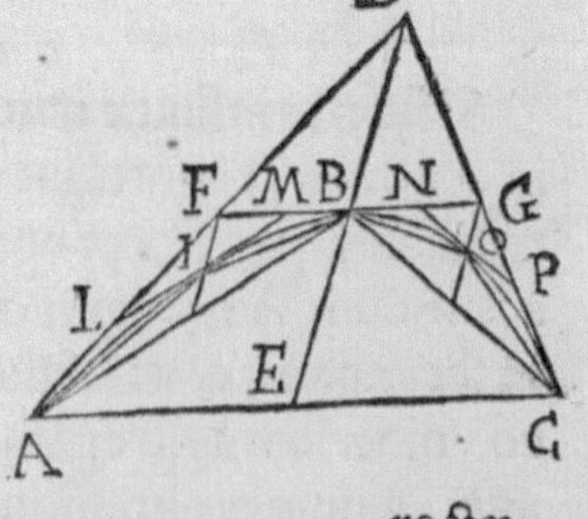

ob parabolam. Cum enim ipsa EB æqualis sit ipsi BD, recta verò AC dupla

rectæ

iuxsa[151] *Lẽ ma* 25. *Intelligantur omnes termini duarum progreſſionum eſſe in flexilineis &c. iunctiſq;* ad, gd, *ducatur* ol *parallela ipſi* ad, & om *parallela ipſi* dg, *Eritq;* bl *æqualis omnib. infinitis terminis* ab, cd, ef. *&c. ipſa vero* om *æqualis omnibus infinitis terminis reliquæ progreſſionis* bg, dh, fi.

Lém. 26. Lem. 26. *Iam: vt* lb *ad* ba, *ita eſt* ob *ad* bd, *hoc eſt* mb. *ad* bg. *Permutando igitur, Aggregatum* lb, *ad aggregatum* bm, *eſt vt* ab *ad* bg; *nempe vt vna magnitudo ad vnam. Quod erat &c.*

4.*ſexti.* *Hoc Theorema poterat ſupponi tamquam demonſtratum in Propoſitione 12. libri V. Euclidis: vnum enim atq; idem est cum Theoremate dictæ Propoſitionis: Verùm, quoniam ferè omnes opinantur Euclidem ibi ſupponere multitudinem magnitudinum finitam, voluimus auxilio flexilineorum vti,*

Propoſitio XVI.

PArabola ſeſquitertia eſt trianguli eandem ipſi baſim, & eandem altitudinem habentis.

Sit parabola ABC, cuius diameter DE, tangentes ad baſim AD, CD: per verticem vero FBG. triangulum inſcriptum ABC. Dico parabolam trianguli ABC. eſſe ſeſquitertiam.

ob parabolam Cum enim ipſa EB æqualis ſit ipſi BD, recta verò AC dupla

rectae

[151] *iuxsa*] *iuxsta.*

rectæ F G; erit inſcriptum triangulum A B C duplum trianguli F D G ſub tangentibus compræhenſi. Et hoc ſemper verum eſt etiam circa reliquas portiones parabolicas AIB, BOC; (eſt enim AIB parabola, cuius tangentes ad baſim ſunt AF, BF, ideoq; triangulum inſcriptum AIB duplum erit trianguli tangentium LFM. Idemq; verum etiam eſt ex alterà parte: Ergo duo ſimul triangula AIB, BOC, dupla ſunt duorũ ſimul LFM, NGP.) ergò cum ſint duæ progreſſiones vtraq; in proportione continuata magnitudinum infinitarum multitudine, (altera nempe intrà parabolam, cuius primus terminus eſt triangulum A B C, ſecundus verò, duo triangula ſimul AI B, BOC &c. altera verò progreſſio extrà parabolam, cuius nempe primus terminus eſt triangulum F D G; ſecundus autem duo ſimul triangula L F M, N G P. &c,) ſuntq; ſinguli termini progreſſionis, quæ intrà parabolam eſt, dupli ſingulorum terminorum progreſſionis, quæ extrà eſt: Erit ergo aggregatum vniuerſum primę progreſſionis duplum totius aggregati ſecundæ progreſſionis; Nempe ipſa parabola dupla erit trilinei mixti ABCD. Componendo igitur, & per conuerſionem rationis, erit triangulum A D C ipſius parabolę ſeſquialterum, nempe vt 6. ad 4. ideoq; parabola ad triangulum A B C erit vt 4: ad 3. videlicet ſeſquitertia. Quod erat oſtendendum &c. *Lem. 28.*

Parabolę quadratura haberi poteſt ſumptis alijs principijs, opē tamen indiuiſibilium. Supponimus quę Archimedes demonſtrauit in libro de lineis Spiralibus ad Propoſitiones 14. & 25. Præmiſſo Lemmate huiuſmodi.

Lemma XXIX.

Si fuerit vt prima magnitudo ad ſecundam, ita tertia ad quartam, & hoc quotieſcunq; libuerit: fuerintq; omnes primę, item & omnes tertię eodem ordine proportionales: Erunt omnes primæ ſimul ad omnes ſecundas, vt ſunt omnes tertię ſimul ad omnes quartas.

K *Sit*

rectæ FG; erit inſcriptum triangulum ABC duplum trianguli FDG ſub tangentibus compræhenſi. Et hoc ſemper verum eſt etiam circa reliquas portiones parabolicas AIB, BOC; (eſt enim AIB parabola, cuius tangentes ad baſim ſunt AF, BF, ideoq; triangulum inſcriptum AIB duplum erit trianguli tangentium LFM. Idemq; verum etiam eſt ex alterà parte*:* Ergo duo ſimul triangula AIB, BOC, dupla ſunt duorũ ſimul LFM, NGP.*)* ergò cum ſint duæ progreſſiones vtraq; in proportione continuata magnitudinum infinitarum multitudine, (altera nempe intrà parabolam, cuius primus terminus eſt triangulum ABC, ſecundus verò, duo triangula ſimul AIB, BOC &c. altera verò progreſſio extrà parabolam, cuius nempe primus terminus eſt triangulum FDG; ſecundus autem duo ſimul triangula LFM, NGP. &c,) ſuntq; ſinguli termini progreſſionis, quæ intrà parabolam eſt, dupli ſingulorum terminorum progreſſionis, quæ extrà eſt: Erit *Lem.*28
ergo aggregatum vniuerſum primę progreſſionis duplum totius aggregati ſecundæ progreſſionis; Nempe ipſa parabola dupla erit trilinei mixti ABCD. Componendo igitur, & per conuerſionem rationis, erit triangulum ADC ipſius parabolę ſeſquialterum, nempe vt *6.* ad 4. ideoq; parabola ad triangulum ABC erit vt 4: ad 3. videlicet ſeſquitertia. Quod erat oſtendendum &c.

Parabolę quadratura haberi poteſt ſumptis alijs principijs, ope tamen indiuiſibilium. Supponimus quę Archimedes demonſtrauit in libro de lineis Spiralibus ad Propoſitiones 14. & 25. Præmiſſo Lemmate huiuſmodi.

Lemmæ XXIX[152]

Si fuerit vt prima magnitudo ad ſecundam, ita tertia ad quartam, & hoc quotieſcunq; libuerit: fuerintq; omnes primę, item & omnes tertię eodem ordine proportionales: Erunt omnes primæ ſimul ad omnes ſecundas, vt ſunt omnes tertię ſimul ad omnes quartas.

K *Sit*

[152] *Ex æquo* proof. (Cfr. Pisano and Bussotti's chapters above).

72 **De Dimensione Parabolæ**

Sit a *prima ad* b *secundam, vt* c *tertia ad* d *quartã; &* e *ad* f, *vt* g *ad* h; *& hoc quotiescunq; libuerit. Sintque omnes primæ* a, e, i, *&c. & omnes tertię* c, g, m, *&c. proportionales ex ordine, Nempe vt* a *ad* e *ita sit* c *ad* g. *Amplius: vt* a *ad* i, *ita sit* c *ad* m *&c. et sic semper. Dico omnes primas simul* a, e, i, *etc. ad omnes secundas simul* b, f, l *etc. esse vt sunt omnes tertiæ simul* c, g, m, *etc. ad omnes quartas simul* d, h, n *etc.*

Accipiantur o, p, q; *singulæ æquales primæ primarum, hoc est ipsi* a; *et sint totidem quot sunt omnes primæ* a, e, i; *etc. Item sumantur* r, s, t; *totidem quot sunt, omnes tertię; et sint singulę* r, s, t, *æquales primę tertiarum nempe ipsi* c.

Iam ob ęqualitatem erit vt o *ad* a, *ita* r *ad* c. *Amplius: Cum* p *sit æqulis ipsi* a, *et* s *ipsi* c, *erit (propter suppositionem) vt* p *ad* e, *ita* s *ad* g. *et hoc semper. sunt*

Lem. 18. *que omnes* o, p, q *æquales, itemq; omnes* r, s, t, *ęquales, ergo erunt omnes simul* o, p, q, *etc. ad omnes* a, e, i, *etc. ut omnes* r, s, t, *simul, ad omnes* c, g, m. *Denique conuertendo, omnes* a, e, i, *ad omnes* o, p, q, *erunt ut omnes* c, g, m, *ad omnes* r, s, t. *Quod memento.*

Quoniam uerò ut o *ad* a, *ita* r *ad* c: *et ut* a *ad* b, *ita* c *ad* d: *erit ex æquo* o *ad* b, *ut* r *ad* d: *Eadem penitus ratione concludemus ex æquo esse ut* p *ad* f, *ita* s *ad* h: *et sic de cęteris.*

Lem. 18. *Erunt ergo omnes simul* o, p, q, *etc. ad omnes* b, f, l, *etc. ut sunt omnes simul* r, s, t, *etc. ad omnes* d, h, n; *etc. Quare ex ęquo erunt omnes* a, e, i, *etc. ad omnes* b, f, l, *etc. ut omnes* c, g, m, *etc. ad omnes* d, h, n. *etc. Quod erat ostendendum*

Propo-

Sit a *prima ad* b *ſecundam, vt* c *tertia ad* d *quartã; &* e *ad* f, *vt* g *ad* h*; & hoc quotieſcunq; libuerit. Sintque omnes primæ* a, e, i, *&c. & omnes tertię* c, g, m, *&c. proportionales ex ordine, Nempe vt* a *ad ita ſit* c *ad* g. *Amplius: vt* a *ad* i, *ita ſit* c *ad* m *&c. et ſic ſemper. Dico omnes primas ſimul* a, e, i, etc. *ad omnes ſecundas ſimul* b, f, l etc. *eſſe vt ſunt omnes tertiæ ſimul* c, g, m, etc. *ad omnes quartas ſimul* d, h, n etc.

Accipiantur o, p, q*; ſingulæ æquales primæ primarum, hoc est ipſi* a*; et ſint totidem quot ſunt omnes primæ* a, e, i*;* etc. *Item ſumantur* r, ſ, t*; totidem quot ſunt, omnes tertię; et ſint ſingulę* r, ſ, t, *æquales primę tertiarum nempe ipſi* c.

Iam ob ęqualitatem erit vt o *ad* a, *ita* r *ad* c. *Amplius: Cum* p *ſit æqualis ipſi* a, *et* ſ *ipſi* c, *erit (propter ſuppoſitionem) vt* p *ad* e, *ita* ſ *ad* g. *et hoc ſemper. ſuntque omnes* o, p, q *æquales, itemq; omnes* r, ſ, t, *ęquales, ergo erunt omnes ſimul* o, p, q, etc. *ad omnes* a, e, i,
*Lem.*18. etc. *ut omnes* r, ſ, t, *ſimul, ad omnes* c, g, m. *Denique conuertendo, omnes* a, e, i, *ad omnes* o, p, q, *erunt ut omnes* c, g, m, *ad omnes* r, ſ, t. *Quod memento.*

Quoniam uerò ut o *ad* a, *ita* r *ad* c*: et ut* a *ad* b, *ita* c *ad* d*: erit ex æquo* o *ad* b, *ut* r *ad* d*: Eadem penitus ratione concludemus ex æquo eſse ut* p *ad* f, *ita* ſ *ad* h*: et ſic de cęteris.*

*Lem.*18. *Erunt ergo omnes ſimul* o, p, q, etc. *ad omnes* b, f, l, etc. *ut ſunt omnes ſimul* r, s, t, etc. *ad omnes* d, h, n*;* etc. *Quare ex ęquo erunt omnes* a, e, i, etc. *ad omnes* b, f, l, etc. *ut omnes* c, g, m, etc. *ad omnes* d, h, n. etc. *Quod erat ostendendum.*

Propo-

Propoſitio XVII.

PArabola ſeſquitertia eſt trianguli eandem ipſi baſim, & eã dem altitudinem habentis.

Sit parabola A B C, cuius tangens ſit A E; diametro vero æquidiſtans ſit C E; & ducatur quælibet F D, parallela ipſi C E, Eritq; E C ad F B. longitudine, vt E A ad A F, ſiue E C ad F D potentia. Propterea erunt in cõtinua proportione. E C, F D, F B.

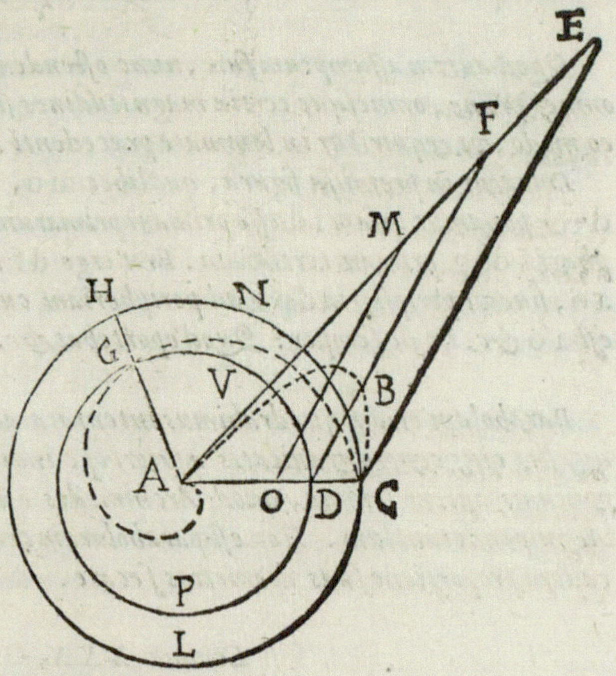

eb para. bolem.

Fiant deinde centro A, interuallis A C, A D, duo circuli; & ponatur Elicis initium ex ſemidiametro A C. Sitq; ipſa elix A G C.

Erit itaq; D F ad F B, vt C E ad D F; ſiue vt C A, ad A D, hoc eſt vt C A ad A G, ſiue vt peripheria tota C L H C, ad arcum C L H: hoc eſt vt peripheria tota D P G D, ad arcum D P G. Atque hoc erit ſemper, vbicunque ſumatur punctum D. Suntq; omnes primæ, item omnes tertiæ magnitudines, eo modo quo debent proportionales (vt infrà oſtendemus.) Quare omnes primæ ſimul, nempe triangulum A E C, ad omnes ſecundas ſimul nempe ad trilineum mixtum A B C E, erit vt omnes tertiæ ſimul, nempe vt circulus C L H, ad omnes quar-

14. Spiralium.

Lemma praced.

Propoſitio XVII.

PArabola ſeſquitertia eſt trianguli eandem ipſi baſim, & eãdem altitudinem habentis. *ob para. bolam*

Sit parabola ABC, cuius tangens ſit AE; diametro vero æquidiſtans ſit CE; & ducatur quælibet FD, parallela ipſi CE, Eritq; EC ad FB. longitudine, vt EA ad AF, ſiue EC ad FD potentia. Propterea erunt in cõtinua proportione. EC, FD, FB.

Fiant deinde centro A, interuallis AC, AD, duo circuli; & ponatur Elicis initium ex ſemidiametro AC. Sitq; ipſa elix AGC. 14. *ſpira lium..*

Erit itaq; DF ad FB, vt CE ad DF; ſiue vt CA, ad AD, hoc eſt vt CA ad AG, ſiue vt periphęria tota CLHC, ad arcum CLH: hoc eſt vt periphęria tota DPGD, ad arcum DPG. Atque hoc erit ſemper, vbicunque ſumatur punctum D. Suntq; omnes primæ, item omnes tertiæ magnitudines, eo modo quo debent proportionales (vt infrà oſtendemus.) Quare omnes primæ ſimul, nempe triangulum AEC, ad omnes ſecundas ſimul nempe ad trilineum mixtum ABCE, erit vt omnes tertiæ ſimul, nempe vt circulus CLH, ad omnes quar- *Lemma præced.*

tas ſimul, hoc eſt ad reliquum ipſius circuli, dempto helicis ſpatio CAGC. Circulus autem CLH, dicti ſpatij, dempto helicis ſpatio, ſeſquialter eſt; Ergò etiam triangulum ACE ſeſquialterum erit trilinei mixti ABCE: Et per conuerſionem rationis, triangulum ACE, triplum erit parabolę ABC. Reliquum quadraturæ abſoluetur vt in 9. Propoſitione factum eſt.

25. de lineis ſpiralibus.

Quod autem aſſumptum fuit, nunc oſtendemus; ſcilicet quòd omnes primę, omneſque tertiæ magnitudines ſint proportionales eo modo, vt requiritur in lemmate præcedenti.

Ducatur in premiſſa figura, qualibet m o, *æquidiſtans ipſi* f d; *& ponamus ipſam* f d *eſſe primam primarum; ipſam verò peripheriā* d p g *primam tertiarum. Erit ergò* d f *ad* o m, *vt* d a *ad* a o, *ſiue vt peripheria* d p g *ad peripheriam cuius ſemidiameter eſt* a o *&c. Et ſic ſemper. Quod oportebat &c.*

Parabolam etiam quadrabimus intentata adhuc via; nimirū quæſito eius centro grauitatis à priori ope indiuiſibilium. Supponimus autem lemma, quod Archimedes oſtendit in ſecundo Aequiponderantium. Hoc eſt parabolarum centra grauitatis, in eadem proportione ſuas diametros ſecare.

Lemma XXX.

Centrum grauitatis parabolæ diametrum ita diuidit, vt pars ad verticem terminata, reliquæ ſit ſeſquialtera.

Eſto conus quilibet a b c, *cuius baſis* a m c, *axis* b d, *triāgulum verò per axem ſit* a b c; *& ſectus ſit conus plano* e f g, *vt iubetur in XI. Propoſitione lib. primi Conicorum. Eritque ſectio quæ vocatur parabola, illiuſque diameter erit* f h, *Eſto iam centrum grauitatis parabolæ* e f g, *quoduis punctum, pùta* i. *Oſtendendum eſt rectam* f i *ſeſquialteram eſſe ipſius* i h.

Agatur per punctum i *recta* a i l; *ſeceturque conus alio plano* m n o, *ipſi* e f g *parall: eritq. ſectio* m n o *parabola, & eius centrum grauitatis erit* p *(eſt enim ob parallelas, vt* f i *ad* i h,

ita

25. *de lineis ſpi-rxlibus*[153]

tas ſimul, hoc eſt ad reliquum ipſius circuli, dempto helicis ſpatio CAGC. Circulus autem CLH, dicti ſpatij, dempto helicis ſpatio, ſeſquialter eſt; Ergò etiam triangulum ACE ſeſquialterum erit trilinei mixti ABCE: Et per conuerſionem rationis, triangulum ACE, triplum erit parabolę ABC. Reliquum quadraturæ abſoluetur vt in *9.* Propoſitione factum eſt.

Quod autem aſſumptum fuit, nunc oſtendemus; ſcilicet quòd omnes primę, omneſque tertiæ magnitudines ſint proportionales eo modo, vt requiritur in lemmate præcedenti.

Ducatur in prępmiſſa figura, quælibet mo, *æquidiſtans ipſi* fd*; & ponamus ipſam* fd *eſſe primam primarum; ipſam verò periphæriã* dpg *primam tertiarum. Erit ergò* df *ad* om, *vt* da *ad* ao, *ſiue vt periphęria* dpg *ad peripbęriam*[154] *cuius ſemidiameter eſt* ao *&c. Et ſic ſemper. Quod oportebat &c.*

Parabolam etiam quadrabimus intentata adhuc via*; nimirũ quæſito eius ceutro grauitatis à priori ope indiuiſibilium. Supponimus autem lemma, quod Archimedes oſtendit in ſecundo Aequiponderantium. Hoc eſt parabolarum centra grauitatis, in eadem proportione ſuas diametros ſecare.*

Lemma XXX.[155]

Centrum grauitatis parabolæ diametrum ita diuidit, vt pars ad verticem terminata, reliquæ ſit ſeſquialtera.

Eſto conus quilibet abc, *cuius baſis* amc, *axis* bd, *triãgulum verò per axem ſit* abc; *& ſectus ſit conus plano* efg, *vt iubetur in* XI. *Propoſitione lib. primi Conicorum. Eritque ſectio quæ vocatur parabola, illiuſque diameter erit* fh, *Eſto iam centrum grauitatis parabolæ* efg, *quoduis punctum, pùta* i. *Oſtendendum eſt rectam* fi *ſeſquialteram eße ipſius* ih.

Agatur per punctum i *recta* ail*; ſeceturque conus alio plano* mno, *ipſi* efg *parall: eritq. ſectio* mno *parabola, & eius centrum grauitatis erit* p *(eſt enim ob parallelas, vt* fi *ad* ih,

ita

[153] *ſpi-rxlibus*] *ſpi-ralibus.*
[154] *peripbęriam*] *periphęriam.*
[155] This proof is based on those done by Archimedes, *Equiponderants.* (Cfr. Pisano and Bussoti chapters above).

ita n p *ad* p r ; *sed* i *ponitur cẽtrum grauitatis parabolę* e f g ; *ergo per Proposit. 7. lib. secundi æquiponderantium* p *centrum grauitatis erit parabolę* m n o.) *Et sic semper, vbicunq. sit planum* m n o. *Omnium ergò singillatim parabolarum quę sunt in cono* a b c, *centra grauitatis reperiuntur in recta* a l : *Quare etiam commune centrum grauitatis omnium earumdem simul prędictarum parabolarum erit in recta* a l. *Omnes autem parabolę, atq. ipse conus idem sunt ; ergò centrum coni est in recta* a l ; *quod cum sit etiam in axe* b d : *erit centrum coni in communi concursu* l, *ideòq.* b l *erit ipsius* l d *tripla*.

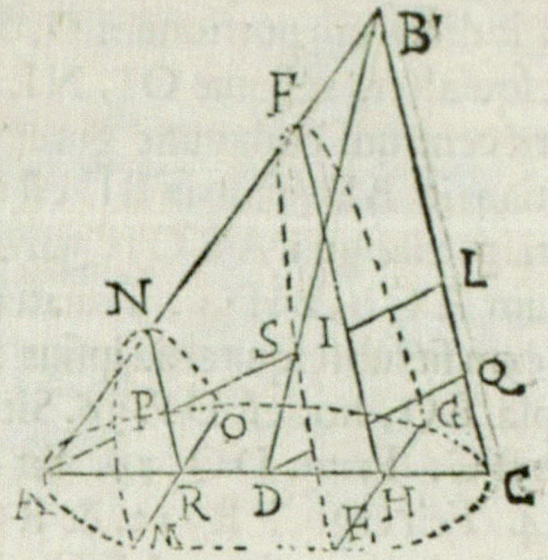

Ducatur ex centro basis recta d q, *ęquidistans ipsi* a l; *erunt que ęquales* c q, q l. *Cum autem ob centrum coni ipsa* b s *tripla sit ipsius* l d, *erit etiam* b l *tripla ipsius* l q : *& ideo* b l *sesquialtera ipsius* l c: *Quare etiam* f i *sesquialtera erit ipsius* i h. *Quod erat propositum &c*, *ob 2. sexti.* *2. sexli.*

Propositio XVIII.

PArabola sesquitertia est trianguli eandem ipsi basim, & eãdem altitudinem habentis.

Esto parabola ABC, cuius diameter BD: inscriptum verò triangulum ABC. Dico parabolam sesquitertiam esse trianguli ABC.

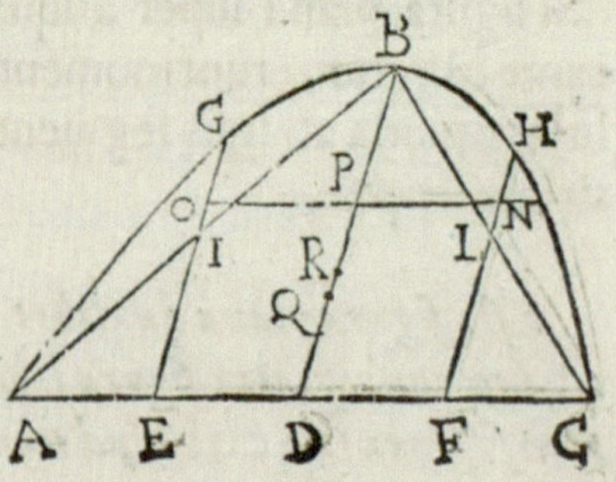

Secentur bifariam AD, DC in punctis E, & F: ductæq. EG, FH, diametro æquidistantes, ipsę dic-

ita np *ad* pr*; ſed* i *ponitur cẽtrum grauitatis parabolę* efg*; ergo per Propoſit. 7. lib. ſecundi æquiponderantium* p *centrum grauitatis erit parabolę* mno.) *Et ſic ſemper, vbicunq. ſit planum* mno. *Omnium ergò ſingillatim parabolarum quę ſunt in cono* abc, *centra grauitatis reperiuntur in recta* al*: Quare etiam commune centrum grauitatis omnium earumdem ſimul prędictarum parabolarum erit in recta* al. *Omnes autem parabolę, atq. ipſe conus idem ſunt; ergò centrum coni est in recta* al*; quod cum ſit etiam in axe* bd*: erit centrum coni in communi concurſu* ſ*, id eòq.* bſ *erit ipſius* ſd *tripla.*

Ducatur ex centro baſis recta dq, *ęquidiſtans ipſi* al*; eruntque ęquales* cq, ql. *Cum autem ob centrum coni ipſa* bs *tripla ſit ipſius* ſd, *erit etiam* bl *tripla ipſius* lq: *& ideo* bl *ſeſquialtera ipſius* lc*: Quare etiam* ſi *ſeſquialtera erit ipsius* ih. *Quod erat propositum &c,* *ob 2. ſex. ti.* *2. ſexti.*

Propoſitio XVIII.

PArabola ſeſquitertia eſt trianguli eandem ipſi baſim, & eãdem altitudinem habentis.

Eſto parabola ABC, cuius diameter BD: inſcriptum verò triangulum ABC. Dico parabolam ſeſquitertiam eſſe trianguli ABC. Secentur bifariam AD, DC in punctis E, & F: ductæq. EG, FH, diametro æquidiſtantes, ip-

se diclæ diametri erunt portionũ AGB, BHC. Sint centra grauitatis dictarum portionum O, & N; eruntq; vtraq. GO, HN, sesquialtera reliquæ OI, NL. Iungatur ON, & in ipsa ON erit centrum commune grauitatis duarum portionum: sed est etiam in BD (nàm in BD est tam centrum totius parabolæ, quã etiam trianguli ABC) Quare punctum P. centrum erit portionum AGB, BHC. Ponatur BD partium 60. eritque GE, (cum sit subsesquitertia ipsius BD) partium 45. ipsa IE 30. & ipsa EO, hoc est DP. 36. Sit Q. centrum grauitatis trianguli ABC. Eritq. DQ. 20. Sit R centrum parabolę, eritq. RD 24. Erit ergo PR, 12. & RQ, 4. Sed vt PR ad RQ ita reciprocè triangulum ABC ad duas portiones AGB, BHC. Quarè triangulum ABC ad duas portiones AGB, BHC erit vt 12. ad 4. nempe vt 3. ad vnum; Componendòque, & per conuersionem rationis, erit parabola ABC ad inscriptum sibi triangulum vt 4. ad 3. Nempe sesquitertia. Quod erat propositum &c.

Lem præcedens. *8. primi æquip.* *Lem præced.*

Noua adhuc ratione quadraturam parabolæ inuademus sumpto sequenti lemmate, quod quidem è Schola Caualeriana prodijsse relatum est. Inseruiebat enim mensurę cuiusdam solidi ab ipsa parabola circà ordinatim applicatam reuolutæ, geniti. Est autem lemma huiusmodi, Authore Io. Antonio Roccha præstanti Geometra.

Lemma XXXI.

Si figura plana super aliquà sui rectà lineà figuram ipsam secante libretur, erunt momenta segmentorum figuræ, vt sunt solida rotunda ab ipsis segmentìs, circa secantem lineam reuolutis, descripta.

Esto figura plana qualibet acdbfc, *quam secet recta linea* ab: *& concipiatur figura librari super rectà* ab. *Dico momentum segmenti* acdb, *ad momentum segmenti* aefb. *esse vt solidum rotundum genitum ex reuolutione segmenti* acdb *circa*

axem

Lem prę cedens. ſæ diametri erunt portionũ AGB, BHC. Sint centra grauitatis dictarum portionum O, & N; eruntq; vtraq. GO, HN, ſeſquialtera reliquæ OI, NL. Iungatur ON, & in ipſa ON erit centrum commune grauitatis duarum portionum: ſed eſt etiam in BD (nàm in BD eſt tam *8. primi æquip.* centrum totius parabolæ, quã etiam trianguli ABC) Quare punctum P. centrum erit portionum AGB, BHC. Ponatur BD partium 60. eritque GE, (cum ſit ſubſeſquitertia ipſius BD) partium 45. ipſa I E 30. & ipſa EO, hoc eſt DP. 36. Sit Q. centrum grauitatis trianguli ABC. Eritq. DQ. 20. Sit R centrum parabolę, eritq. RD 24. Erit ergo PR, 12. & RQ, 4. Sed vt PR ad RQ ita reciprocè triangulum ABC ad duas portiones AGB, BHC. Quarè triangulum ABC ad duas portiones AGB, BHC erit vt 12. ad 4. nempe vt 3. ad vnum; *Lem*[.] *præ ced.* Componendòque, & per conuerſionem rationis, erit parabola ABC ad inſcriptum ſibi triangulum vt 4. ad 3. Nempe ſeſquitertia. Quod erat propoſitum &c.

Noua adhuc ratione quadraturam parabolæ inuademus ſumpto ſequenti lemmate, quod quidem è Schola Caualeriana prodijſſe relatum est. Inſeruiebat enim menſurę cuiuſdam ſolidi ab ipſa parabola circà ordinatim applicatam reuolutæ, geniti. Est autem lemma huiuſmodi, Authore Io. Antonio Roccha præſtanti Geometra.

Lemma XXXI[156] .

Si figura plana ſuper aliquà ſui rectà lineà figuram ipſam ſecante libretur, erunt momenta ſegmentorum figuræ, vt ſunt ſolida rotunda ab ipſis ſegmentìs, circa ſecantem lineam reuolutis, deſcripta.

Eſto figura plana quælibet acdbfe, *quam ſecet recta linea* ab: *& concipiatur figura librari ſuper rectà* ab. *Dico momentum ſegmenti* acdb, *ad momentum ſegmenti* aefb. *eſſe vt ſolidum rotundum genitum ex reuolutione ſegmenti* acdb *circa*

Axem

[156] This proof is based on those done by Archimedes, *Equiponderants*. (Cfr. Pisano and Bussoti chapters above).

axem ab, *ad solidum rotundum genitum ex conuersione reliqui segmenti circa eũdem axem reuoluti.*

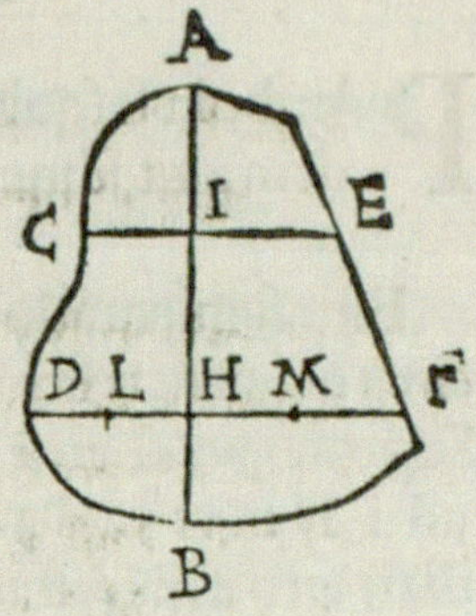

Sumptis enim duobus quibuscunq. punctis h, *&* i. *in recta* ab: *ducantur per* h *& per* i, *rectæ* ce, df. *perpendiculares ad ipsam* ab: *secenturq. bifariam segmenta* dh, hf, *in punctis* l, *&* m.

Habebit ergo momentum rectæ dh *ad momentum rectae* hf, *rationem compositam ex ratione magnitudinum* dh *ad* hf, *& ex ratione distantiarum* lh *ad* hm; *siue* dh *ad* hf. *Propterea momentum recta* dh *ad momentum* hf *erit vt quadratum* dh *ad quadratum* hf.

Eodem modo ostendetur momentum rectae ci, *ad momentum rectae* ie, *esse vt quadratum* ci *ad quadratum* ie, *& sic semper.*

Amplius momentum dh *ad momentum* ci, *est (ob eandem rationem vt supra) vt quadratum* dh *ad quadratum* ci: *& hoc semper, Erunt ergo omnes primae simul magnitudines, nempe omnia momenta figurae* acdb, *ad omnes secundas simul, nempe ad omnia momenta reliquę figurę* aefb: *vt sunt omnes tertię simul, nempe omnia quadrata figurae* acdb, *ad omnia quadrata reliquae figurae. Siue vt sunt omnes circuli figurae* acdb *(nempe solidum rotundum ex ipsius conuersione circa axem* ab *descriptum) ad omnes circulos reliquae figurae* aefb *(nempe ad solidum rotundum ex ipsius reuolutione circa eundem axem* ab, *genitum.) Quod erat ostendendum &c.* *Lem.29.*

Hoc pręmisso (quod quidẽ vti suprà ediximus penitus ab aliȷs desumptum est, & hic insertum tamquam alienum, neque quod ego sciam adhuc vulgatum) parabolam quadrabimus, supposita demonstratione, quã multis modis probatur Cylindrum inscripti sibi conoidis parabolici esse duplum.

Propo-

axem ab, *ad ſolidum rotundum genitum ex conuersione reliqui ſegmenti circa eũdem axem reuoluti.*

Sumptis enim duobus quibuſcunq. punctis h, & i. *in recta* ab*: ducantur per* h *& per* i, *rectæ* ce, df. *perpendiculares ad ipſam* ab: *ſecenturq, bifariam ſegmenta* dh, hf, *in punctis* l, & m.

Habebit ergo momentum rectæ dh *ad momentum rectae* hf, *rationem compositam ex ratione magnitudinum* dh *ad* hf, *& ex ratione distantiarum* lh *ad* hm*; siue* dh *ad* hf. *Propterea momentum rectæ* dh *ad momentum* hf *erit vt quadratum* dh *ad quadratum* hf.

Eodem modo ostendetur momentum rectae ci, *ad momentum rectae* ie, *eſſe vt quadratum* ci *ad quadratum* ie, *& sic ſemper.*

Amplius momentum dh *ad momentum* ci, *eſt (ob eandem rationem vt ſupræ) vt quadratum* dh *ad quadratum* ci: *& hoc ſemper, Erunt ergo omnes primae simul magnitudines, nempe omnia momenta figurae* acdb, *ad omnes ſecundas simul, nempe ad omniæ momenta reliquę figurę* aefb*: vt ſunt omnes tertię ſimul, nempe omnia quadrata figurae* acdb, *ad omnia quadrata reliquae figurae.* *L em.29.* *Siue vt ſunt omnes circuli figurae* acdb *(nempe ſolidum rotundum ex ipsius conuersione circa axem* ab *deſcriptum) ad omnes circulos reliquae figurae* aefb *(nempe ad ſolidum rotundum ex ipsius reuolutione circa eundem axem* ab, *genitum.) Quod erat ostendendum &c.*

Hoc prẹmiſſo (quod quidẽ vti ſuprà ediximus penitus ab alijs deſumptum est, & hic inſertum tamquam alienum, neque quod ego ſciam adhuc vulgatum) parabolam quadrabimus, ſuppoſitæ demonſtratione, quà multis modis probatur Cylindrum inſcripti ſibi conoidis parabolici eſſe duplum.

Propo-

Propoſitio XIX.

PArabola ſeſquitertia eſt trianguli eandem ipſi baſim, & eandem altitudinem habentis.

Eſto ſemiparabola ABCD, circa quam ſit rectangulum DE. Sumatur punctum F, ita vt AF. ad FD ſit vt 5.ad 3. ductaq. FG

Lem.11. diametro æquidiſtans, erit in ipſa FG centrum grauitatis ſemiparabolæ. Eſto illud punctum quodlibet, puta I, & per I ducatur LIM parallela ad AD, accipiaturq; IN. æqualis ipſi IM.

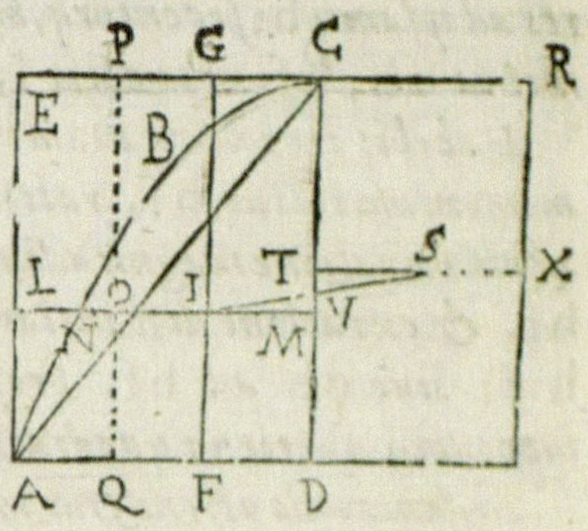

Intelligatur etiam producta PQ parallela diametro CD (vbicunq; cadat) ita vt parallelogrammum rectangulum DP, æquale ſit ipſi ſemiparabolæ. Tum concipiatur applicatum ad rectã CD, rectangulum DR, ita vt æquiponderet ſemiparabolæ facta libratione ſuper recta CD. Sitq; centrum dicti rectanguli punctum S; & ductà TSX parallela ipſi AD iungatur recta IS.

Iam; manifeſtum eſt ex lemmate præmiſſo quòd cylindrus factus à rectangulo DR circa axem DC reuoluto, æqualis erit conoidali parabolico facto à conuerſione ſemiparabolæ ACD, circa eundem axem CD reuolutę; cum æqualia ſupponantur figurarum planarum momenta. Erit ergo cylindrus à rectangulo DR factus, ſubduplus cylindri à rectangulo DE facti, & ideo quadratum TX ſubduplum erit quadrati ML (cylindri enim æquealti ſunt inter ſe vt baſium quadrata) quod memento.

4. ſexti. Verùm MN ad TX, eſt vt IM ad TS (ſunt enim ſubduplæ earumdem), ſiue vt IV ad VS; nempe (quia æquiponderant figuræ planæ ſuper linea CD, ſiue ex puncto V) vt rectan gulum

Propoſitio XIX.

PArabola ſeſquitertia eſt trianguli eandem ipſi baſim, & eandem altitudinem habentis.

Eſto ſemiparabola ABCD, circa quam ſit rectangulum DE.

*Lem.*11. Sumatur punctum F, ita vt AF. ad FD ſit vt 5. ad 3. ductaq. FG diametro æquidiſtans, erit in ipſa FG centrum grauitatis ſemiparabolæ. Eſto illud punctum quodlibet, puta I, & per I ducatur LIM parallela ad AD, accipiaturq; IN. æqualis ipſi IM. Intelligatur etiam producta PQ parallela diametro CD (vbicunq; cadat) ita vt parallelogrammum rectangulum DP, æquale ſit ipſi ſemiparabolæ. Tum concipiatur applicatum ad rectã CD, rectangulum DR, ita vt æquiponderet ſemiparabolæ factà libratione ſuper recta CD. Sitq; centrum dicti rectanguli punctum S; & ductà TSX parallela ipſi AD iungatur recta IS.

Iam; manifeſtum eſt ex lemmate præmiſſo quòd cylindrus factus à rectangulo DR circa axem DC reuoluto, æqualis erit conoidali parabolico facto à conuerſione ſemiparabolæ ACD, circa eundem axem CD reuolutę; cum æqualia ſupponantur figurarum planarum momenta. Erit ergo cylindrus à rectangulo DR factus, ſubduplus cylindri à rectangulo DE facti, & ideo quadratum TX ſubduplum erit quadrati ML *(*cylindri enim æquealti ſunt inter ſe vt baſium quadrata*)* quod memento.

Verùm MN ad TX, eſt vt IM ad TS *(*ſunt enim ſubduplæ earumdem*)* ſiue vt IV ad VS; nempe (quia æquiponderant figuræ planæ

4. *ſexti.* ſuper linea D, ſiue ex puncto V) vt rectan

gulum

gulum D R ad ſemiparabolam reciprocè, ſiue ad rectangulum D P. ipſi ſemiparabolæ æquale: ſiue vt eorum baſes T X ad M O. Ergo T X media proportionalis eſt inter M N, MO: Quare rectangulum N M O, cum æquale ſit quadrato T X, ſubduplum erit quadrati L M.

Ratio verò quadrati L M, ad rectangulum N M O, componitur ex ratione L M ad M N (quæ ſeſquitertia eſt per conſtructionem; ſumpſimus enim punctum F; ita vt A F ad F D, eſſet vt 5. ad 3.) & ex ratione L M ad M O; quæ quidem ignot. erat, ſed neceſſariò ſeſquialtera nunc apparet. Ratio enim dup. a cõponitur ex ſeſquitertia, & ſeſquialtera, vt ipſis etiam Cantoribus vulgatum eſt; vt videre eſt in his tribus numeris 4. 3. 2.

Rectangulum ergo D E ad ipſum D P, ſiue ad ſemiparabolam, ſeſquialterum erit; & ipſa ſemiparabola ad triangulum A C D. ſeſquitertia erit. Quod erat oſtendendum &c.

Lemma XXXII.

Sit parabola A B C, cuius baſis A C, tangens C D; diametro æquidiſtans ſit A D. Sumpto quolibet puncto E. ducatur E F diametro æquidiſtans. Dico eſſe vt F E ad E B, ita C A ad A E.

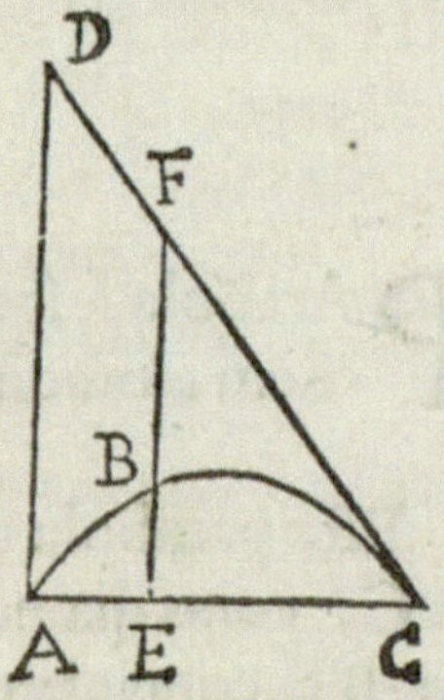

ob parabolam.

Eſt enim D A ad F B longitudine, vt D C ad C F potentiã, ſiue vt D A ad F E potentiã; Sunt ergo in continuã ratione D A, F E, F B. Quod memento.

Iam vt A C ad C E ita eſt A D ad E F, ſiue E F ad F B; & per conuerſionem rationis, vt C A ad A E, ita eſt F E ad E B. Quod erat oſtendend. &c.

Lemma XXXIII.

Quælibet parabola æqualis eſt duabus parabolis ſimul ſumptis, quæ quidem æqualem ipſi baſim habeant, diametrum verò ſub-

gulum DR ad ſemiparabolam reciprocè, ſiue ad rectangulum DP. ipſi ſemiparabolæ æquale: ſiue vt eorum baſes TX ad MO. Ergo TX media proportionalis eſt inter MN, MO: Quare rectangulum NMO, cum æquale ſit quadrato TX, ſubduplum erit quadrati LM.

Ratio verò quadrati LM, ad rectangulum NMO, componitur ex ratione LM ad MN (quæ ſeſquitertia eſt per conſtructionem; ſumpſimus enim punctum F; ita vt AF ad FD, eſſet vt 5. ad 3.) & ex ratione LM ad MO; quæ quidem ignot. erat, ſed neceſſariò ſeſquialtera nunc apparet. Ratio enim dupla cõponitur ex ſeſquitertia, & ſeſquialtera, vt ipſis etiam Cantoribus vulgatum eſt; vt videre eſt in his tribus numeris 4. 3. 2.

Rectangulum ergo DE ad ipſum DP, ſiue ad ſemiparabolam, ſeſquialterum erit; & ipſa ſemiparabola ad triangulum ACD. ſeſquitertia erit. Quod erat oſtendendum &c.

Lemma XXXII.

Sit parabola ABC, cuius baſis AC, tangens CD; diametro æquidiſtans ſit AD. Sumpto quolibet puncto E. ducatur EF diametro æquidiſtans. Dico eſſe vt FE ad EB, ita CA ad AE. *ob para.*[-]

Eſt enim DA ad FB longitudine, vt DC ad CF potentià, ſiue vt DA ad FE potentià i[158] Sunt ergo in continuà ratione DA, FE, FB. Quod memento. *bolam.*

Iam vt AC ad CE ita eſt AD ad EF, ſiue EF ad FB; & per conuerſionem rationis, vt CA ad AE, ita eſt FE ad EB. Quod erat oſtendend. &c.

Lemma XXXIII.

Quælibet parabola æqualis eſt duabus parabolis ſimul ſumptis, quæ quidem ęqualem ipſi baſim habeant, diametrum ve-

L rò sub-

[157] The numbering of pages is broken: 79 and 80 are missing. Neverthless the content is saved.

[158] potentià i] potentià.

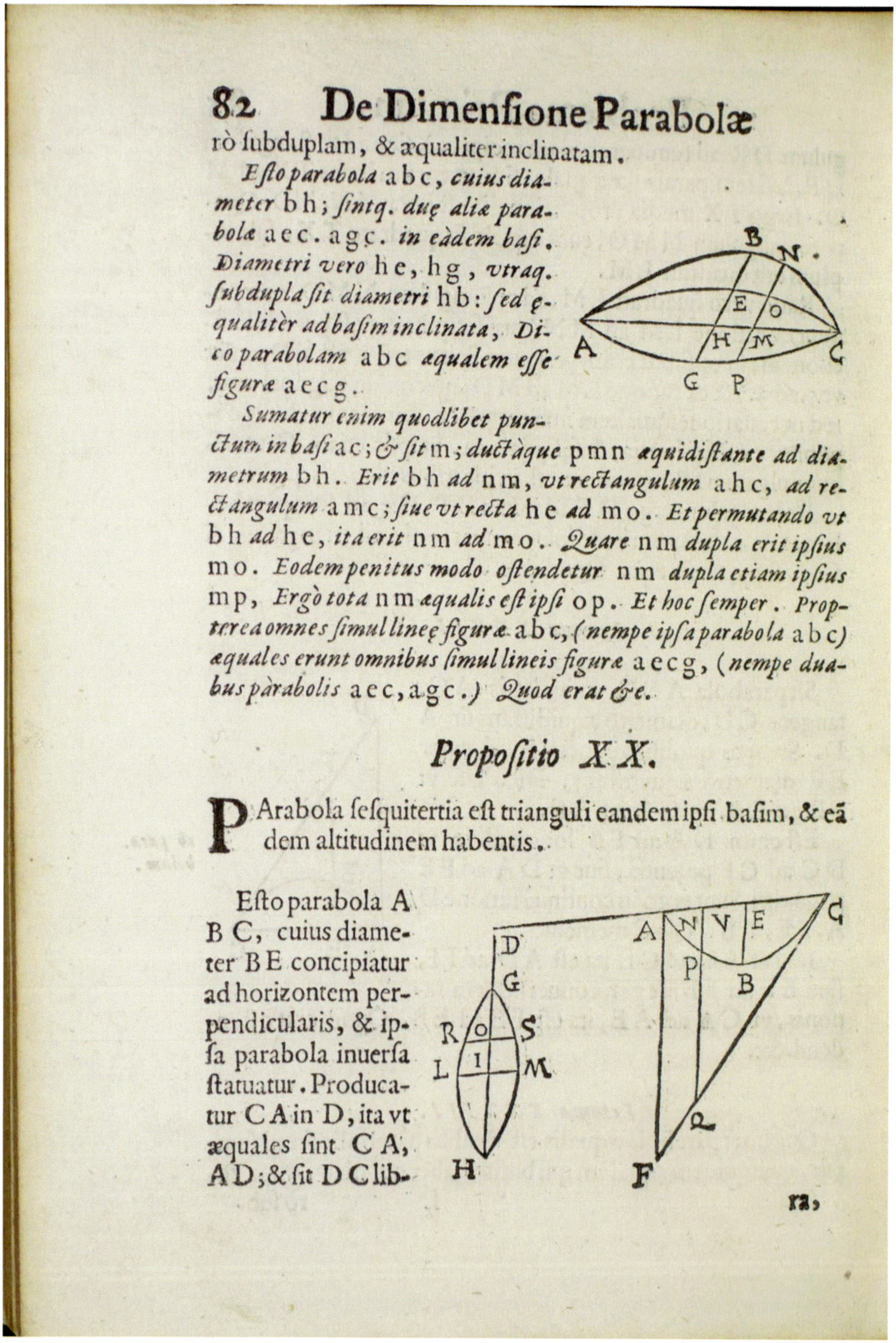

82 De Dimensione Parabolæ

rò subduplam, & æqualiter inclinatam.

Esto parabola abc, *cuius diameter* bh; *sintq. duę aliæ parabolæ* aec. agc. *in eàdem basi. Diametri vero* he, hg, *vtraq. subdupla sit diametri* hb: *sed ęqualitèr ad basim inclinata, Dico parabolam* abc *æqualem esse figuræ* aecg.

Sumatur enim quodlibet punctum in basi ac; *& sit* m; *ductàque* pmn *æquidistante ad diametrum* bh. *Erit* bh *ad* nm, *vt rectangulum* ahc, *ad rectangulum* amc; *siue vt recta* he *ad* mo. *Et permutando vt* bh *ad* he, *ita erit* nm *ad* mo. *Quare* nm *dupla erit ipsius* mo. *Eodem penitus modo ostendetur* nm *dupla etiam ipsius* mp, *Ergò tota* nm *æqualis est ipsi* op. *Et hoc semper. Propterea omnes simul lineę figuræ* abc, (*nempe ipsa parabola* abc) *æquales erunt omnibus simul lineis figuræ* aecg, (*nempe duabus parabolis* aec, agc.) *Quod erat &c.*

Propositio XX.

Parabola sesquitertia est trianguli eandem ipsi basim, & eandem altitudinem habentis.

Esto parabola ABC, cuius diameter BE concipiatur ad horizontem perpendicularis, & ipsa parabola inuersa statuatur. Producatur CA in D, ita vt æquales sint CA, AD; & sit DC libra,

rò ſubduplam, & æqualiter inclinatam.

Eſto parabola abc, *cuius diameter* bh; *ſintq; duę aliæ parabolæ* aec. agc. *in eàdem baſi. Diametri vero* he, hg, *vtraq. ſubdupla ſit diametri* hb: *ſed ęqualitèr ad baſim inclinata, Dico parabolam* abc *æqualem eſſe figuræ* aecg.

Sumatur enim quodlibet punctum in baſi ac; *& ſit* m*; ductàque* pmn *æquidiſtante ad diametrum* bh. *Erit* bh *ad* nm, *vt rectangulum* ahc, *ad rectangulum* amc*; ſiue vt recta* he *ad* mo. *Et permutando vt* bh *ad* he, *ita erit* nm *ad* mo. *Quare* nm *dupla erit ipſius* mo. *Eodem penitus modo oſtendetur* nm *dupla etiam ipſius* mp, *Ergò tota* nm *æqualis eſt ipſi* op. *Et hoc ſemper. Propterea omnes ſimul lineę figuræ* abc, *(nempe ipſa parabola* abc*) æquales erunt omnibus ſimul lineis figuræ* aecg, (*nempe duabus parabolis* aec, agc.*) Quod erat &c.*

Propoſitio XX.

PArabola ſeſquitertia eſt trianguli eandem ipſi baſim, & eãdem altitudinem habentis.

Eſto parabola ABC, cuius diameter BE concipiatur ad horizontem perpendicularis, & ipſa parabola inuerſa ſtatuatur. Producatur CA in D, ita vt æquales ſint CA, AD; & ſit DC lib-

ra,

bra, cuius fulcrum est A. Ducatur CF tangens parabolam; & AF diametro EB æquidistans. Ponatur etiam GH æqualis ipsi AC, & diuisa bifariam GH in I, sit vtraq. IL, IM, sub dupla rectę EB. & æqualitèr ad basim inclinata vt est ipsa EB ad AC. Fiantq; duæ parabolæ GLH, GMH, quæ (per lemma præced.) simul æquales erunt parabolæ ABC; Et suspendatur figura GLHM ex puncto D.

Accipiãtur iam puncta O, & N æqualiter distantia à punctis I, & E respectiuè. Ductisq. NQ æquidistantèr ad EB, & ROS ad LM; Erit vt in præcedenti lemmate NP ęqualis ipsi RS.

Iam QN ad RS est (ob æqualitatem) vt QN ad NP, siue vt DA ad AN reciprocè. Aequiponderant ergo rectæ QN, & RS. & sic semper. Ergo omnes simul lineæ trianguli AFC (nempe ipsum triangulum) æquiponderant omnibus simul lineis figuræ GLHM, (nempe ipsi figurę GLHM.) *Lem. 3.*

Accipiatur AV tertiã pars totius AC. Manifestum est, quod si ex V demittatur recta ęquidistans ipsi AF. in ipsa erit centrum grauitatis trianguli AFC; eritq. ipsa ad horizontem perpendicularis. Propterea erit triangulum AFC. appensum centralitèr ex puncto V. Eritq, triangulum AFC ad spatium GLHM. reciprocè vt DA ad AV, nempe triplum.

Cum autem spatium GLHM æquale sit parabolæ ABC; erit triangulum AFC triplum etiam parabolæ ABC.

Reliquum quadraturę absoluitur vt in Propositione IX. Factum est, Quod &c.

Propositio XXI.

Parabola sesquitertia est trianguli eandem ipsi basim, & eandem altitudinem habentis.

Esto semiparabola ABC, cuius diameter CE, ordinata AE, tangens verò CD, & compleatur parallelogrammum AECD. Manifestum est quod omnes lineæ trilinei mixti D

bra, cuius fulcrum eft A. Ducatur CF tangens parabolam; & AF diametro EB æquidiftans. Ponatur etiam GH æqualis ipſi AC, & diuiſa bifariam GH in I, ſit vtraq. IL, IM, ſub dupla rectę EB. & æqualitèr ad baſim inclinata vt eft ipſa EB ad AC. Fiantq; duæ parabolæ GLH, GMH, quæ (per lemma præced.) ſimul æquales erunt parabolæ ABC; Et ſuſpendatur figura GLHM ex puncto D.

Accipiãtur iam puncta O, & N æqualiter diftantia à punctis I, & E reſpectiuè. Ductiſq; NQ æquidiftantèr ad EB, & ROS ad LM; Erit vt in præcedenti lemmate NP ęqualis ipſi RS.

Iam QN ad RS eft (ob æqualitatem) vt QN ad NP, ſiue vt DA ad AN reciprocè. Aequiponderant ergo rectæ QN, & RS. & ſic ſemper. Ergo omnes ſimul lineæ trianguli AFC (nempe ipſum triangulum) æquiponderant omnibus ſimul lineis figuræ GLHM, (nempe *Lem.* 32.
ipſi figurę GLHM.)

Accipiatur AV tertià pars totius AC. Manifeftum eft, quod ſi ex V demittatur recta ęquidiftans ipſi AF. in ipſa erit centrum grauitatis trianguli AFC; eritq; ipſa ad horizontem perpendicularis. Propterea erit triangulum AFC. appenſum centralitèr ex puncto V. Eritq; triangulum AFC ad ſpatium GLHM. reciprocè vt DA ad AV, nempe triplum.

Cum autem ſpatium GLHM æquale ſit parabolæ ABC; erit triangulum AFC triplum etiam parabolæ ABC.

Reliquum quadraturę abſoluitur vt in Propoſitione IX. *Factum est, Quod &c.*

Propoſitio XXI.

PArabola ſeſquitertia eft trianguli eandem ipſi baſim, & eandem altitudinem habentis.

Efto ſemiparabola ABC, cuius diameter CE, ordinata AE, tangens verò CD, & compleatur parallelogrammum AECD. Manifeftum eft quod omnes lineæ trilinei mixti D

L 2 ABC,

A B C, quæ quidem diametro parallelæ ſint, inter ſe ſunt in eadem ratione in qua ſunt omnes circuli coni alicuius, qui axem habeat D C, & verticem C. *Lem. 22.* Ergo centrum grauitatis omnium linearum trilinei D A B C, erit in illa, quæ diuidit libram D C; quemadmodum diuidit eandem centrum grauitatis coni; nempe vt pars ad C terminata, reliquæ ſit tripla. Fiat ergo C F tripla ipſius F D. & ducta F M parallela ad C E, erit centrum grauitatis trilinei D A B C in recta F M. vbicunq. ſit.

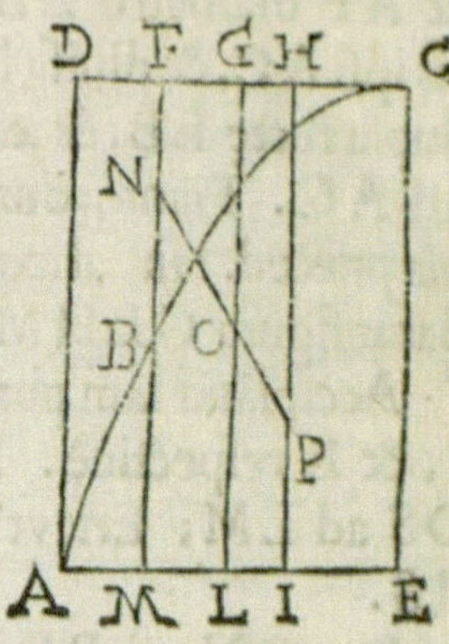

Item, omnes lineæ, quæ in ſemiparabola A B C E ducuntur ad diametrum parallelæ, inter ſe ſunt in eadem ratione, in qua ſunt omnes circuli alicuius hemiſphærij, cuius axis ſit A E, *Lem. 22.* vertex verò A. Ergò centrum grauitatis omnium linearum ad libram A E appenſarum, ſiue ipſius ſemiparabolæ, erit in illa, quæ libram A E ſic diuidit vt diuidit eandem centrum grauitatis hemiſphærij; Nempe vt pars ad A terminata, ſit ad reliquam vt 5. ad 3. Fiat ergo A I ad I E vt 5. ad 3.; & ducta I H parallela ad C E, erit centrum ſemiparabolæ in recta I H, vbicunque ſit. Ducatur tandem G L, quæ bifariam ſecet latera A E, D C. & in G L erit centrum grauitatis parallelogrammi D E. quod ſit O. Ponatur centrum grauitatis ſemiparabolę eſſe punctũ quoduis P. ductaq. P O, producatur in N; & erit N. centrũ grauitatis trilinei D A B C. Iam, ſemiparabola ad trilineum eſt vt N O ad O P, ſiue vt M L ad L I; nempe vt 2. ad vnum; (qualium enim partium tota A E eſt 8, talium A M eſt 2, M L eſt 2, L I eſt vna, & reliqua I E 3. per conſtructionem) Ergo ſemiparabola ad parallelogrammum erit vt 2. ad 3, ſiue vt 4. ad 6; & ſemiparabola ad triangulum inſcriptum vt 4. ad 3. Nempe ſeſquitertia. Quod &c.

F I N I S.

APPEN-

ABC, quæ quidem diametro parallelæ ſint, inter ſe ſunt in eadem ratione in qua ſunt omnes circuli coni alicuius, qui axem habeat
*Lem.*22. DC, & verticem C. Ergo centrum grauitatis omnium linearum trilinei DABC, erit in illa, quæ diuidit libram DC; quemadmodum diuidit eandem centrum grauitatis coni; nempe vt pars ad C terminata, reliquæ ſit tripla. Fiat ergo CF tripla ipſius FD. & ducta FM parallela ad CE, erit centrum grauitatis trilinei DABC in recta FM. vbicunq. ſit.

Item, omnes lineæ, quæ in ſemiparabola ABCE ducuntur ad diametrum parallelæ, inter ſe ſunt in eadem ratione, in qua ſunt
*Lem.*22. omnes circuli alicuius hemiſphærij, cuius axis ſit AE, vertex verò A. Ergò centrum grauitatis omnium linearum ad libram AE appenſarum, ſiue ipſius ſemiparabolæ, erit in illa, quæ libram AE ſic diuidit vt diuidit eandem centrum grauitatis hemiſphærij; Nempe vt pars ad A terminata, ſit ad reliquam vt 5. ad 3. Fiat ergo AI ad IE vt 5. ad 3.; & ducta IH parallela ad CE, erit centrum ſemiparabolæ in recta IH, vbicunque ſit. Ducatur tandem GL, quæ bifariam ſecet latera AE, DC. & in GL erit centrum grauitatis parallelogrammi DE. quod ſit O. Ponatur centrum grauitatis ſemiparabolę eſſe punctũ quoduis P. ductaq. PO, producatur in N, & erit N. centrũ grauitatis trilinei DABC. Iam, ſemiparabola ad trilineum eſt vt NO ad OP, ſiue vt ML ad LI; nempe vt 2. ad vnum; (qualium enim partium tota AE eſt 8, talium AM eſt 2, ML eſt 2, LI eſt vna, & reliqua IE 3. per conſtructionem) Ergo ſemiparabola ad parallelogrammum erit vt 2. ad 3, ſiue vt 4. ad 6; & ſemiparabola ad triangulum inſcriptum vt 4. ad 3. Nempe ſeſquitertia. Quod &c.

FINIS.

APPEN-

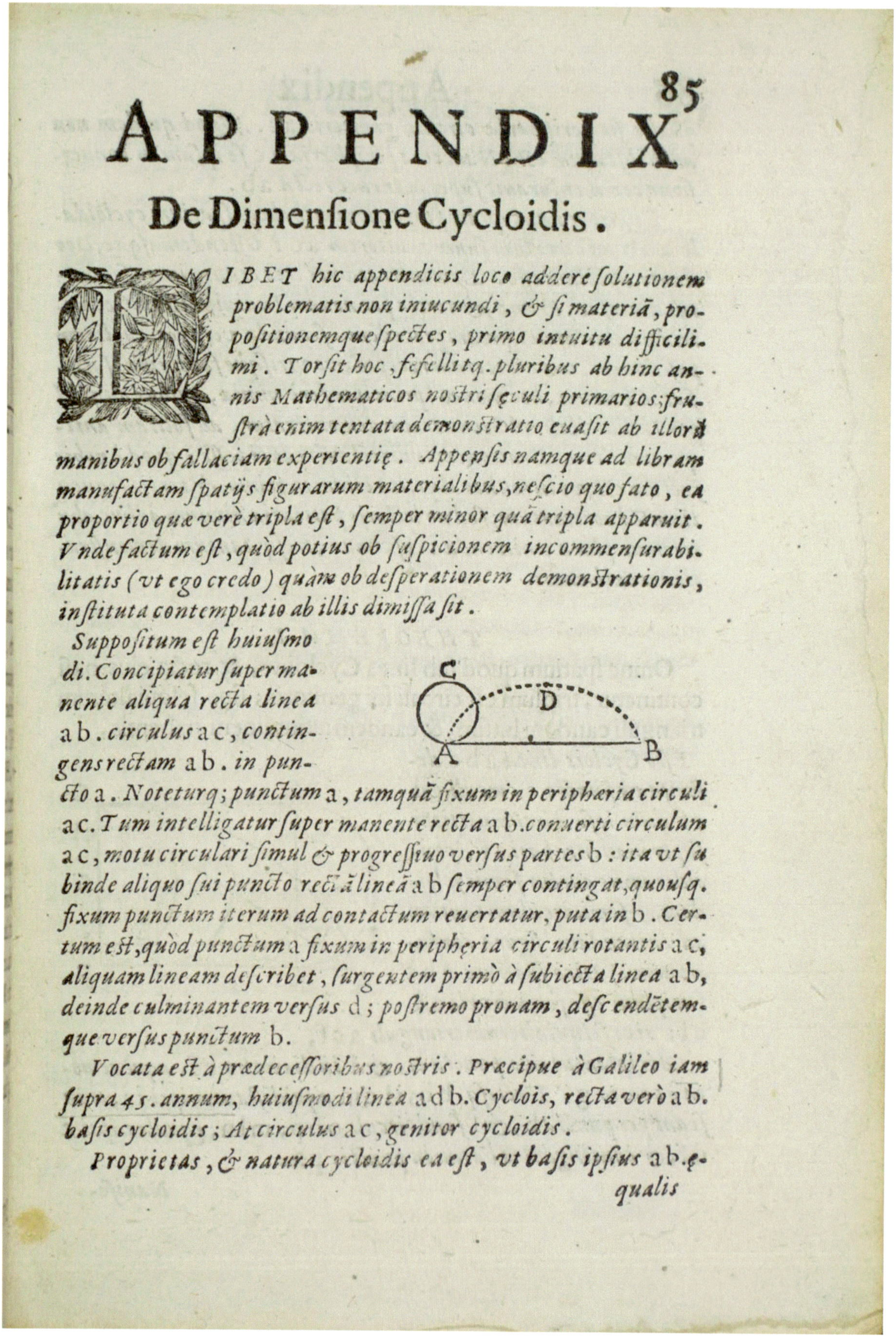

85

APPENDIX
De Dimenſione Cycloidis.

LIBET hic appendicis loco addere ſolutionem problematis non iniucundi, & ſi materiã, propoſitionemque ſpectes, primo intuitu difficilimi. Torſit hoc, feſellitq. pluribus ab hinc annis Mathematicos noſtri ſeculi primarios; fruſtrà enim tentata demonſtratio euaſit ab illorũ manibus ob fallaciam experientię. Appenſis namque ad libram manufactam ſpatijs figurarum materialibus, neſcio quo fato, ea proportio quæ verè tripla eſt, ſemper minor quã tripla apparuit. Vnde factum eſt, quòd potius ob ſuſpicionem incommenſurabilitatis (vt ego credo) quàm ob deſperationem demonſtrationis, inſtituta contemplatio ab illis dimiſſa ſit.

Suppoſitum eſt huiuſmodi. Concipiatur ſuper manente aliqua recta linea a b. circulus a c, contingens rectam a b. in puncto a. Noteturq; punctum a, tamquã fixum in peripharia circuli a c. Tum intelligatur ſuper manente recta a b. conuerti circulum a c, motu circulari ſimul & progreſſiuo verſus partes b: ita vt ſubinde aliquo ſui puncto rectã lineã a b ſemper contingat, quouſq. fixum punctum iterum ad contactum reuertatur, puta in b. Certum eſt, quòd punctum a fixum in peripheria circuli rotantis a c, aliquam lineam deſcribet, ſurgentem primò à ſubiecta linea a b, deinde culminantem verſus d; poſtremo pronam, deſcendẽtemque verſus punctum b.

Vocata eſt à prædeceſſoribus noſtris. Præcipue à Galileo iam ſupra 45. annum, huiuſmodi linea a d b. Cyclois, recta verò a b. baſis cycloidis; At circulus a c, genitor cycloidis.

Proprietas, & natura cycloidis ea eſt, vt baſis ipſius a b. ęqualis

APPENDIX
De Dimenfione Cycloidis.

LIBET *hic appendicis loco addere folutionem problematis non iniucundi, & fi materiã, propofitionemque fpectes, primo intuitu difficilimi. Torfit hoc, fefellitq. pluribus ab hinc annis Mathematicos nostri fęculi primarios; fruftrà enim tentata demonstratio euafit ab illorũ manibus ob fallaciam experientię. Appenfis namque ad libram manufactam fpatijs figurarum materialibus, nefcio quo fato, ea proportio quæ verè tripla eft, femper minor quã tripla apparuit. Vnde factum eft, quòd potius ob fufpicionem incommenfurabilitatis (vt ego credo) quàm ob defperationem demonstrationis, inftituta contemplatio ab illis dimiffa fit.*

Suppofitum eft huiufmodi. Concipiatur fuper manente aliqua recta line a ab. *circulus* ac, *contingens rectam* ab. *in puncto* a. *Noteturq; punctum* a, *tamquã fixum in periphæria circuli* ac. *Tum intelligatur fuper manente recta* ab. *conuerti circulum* ac, *motu circulari fimul & progreffiuo verfus partes* b*: ita vt fubinde aliquo fui puncto rectã lineã* ab *femper contingat, quoufq. fixum punctum iterum ad contactum reuertatur, puta in* b. *Certum est, quòd punctum* a *fixum in periphęria circuli rotantis* ac, *aliquam lineam defcribet, furgentem primò à fubiecta linea* ab, *deinde culminantem verfus* d*; poftremo pronam, defcendẽtemque verfus punctum* b.

Vocata est à prædeceffforibus nostris. Præcipue à Galileo iam fupra 45. annum, huiufmodi linea adb. *Cyclois, recta verò* ab. *bafis cycloidis; At circulus* ac, *genitor cycloidis.*

Proprietas, & natura cycloidis ea eft, vt bafis ipfius ab. ę-

qualis

86 Appendix

qualis ſit peripheriæ circuli genitoris a c. *Quod quidem non adeò obſcurum eſt: Nam tota periphæria* a c *ſe ipſam in conuerſione commenſurauit ſuper manente recta* a b.

Quæritur nunc quam proportionem habeat ſpatium cycloidale a d b *ad circulum ſuum genitorem* a c *? Oſtendemuſque, Deo dante, triplum eſſe. Demonſtrationes tres erunt, inter ſe penitus diuerſae. Prima, & tertia per nouam Indiuiſibilium Geometriam nobis amiciſſimam procedent: ſecunda verò per duplicem poſitionem, more veterum recepto; vt vtriſque fautoribus ſatiſfiat. Cęterum, hoc moneo; principia ferè omnia, quibus aliquid per Indiuisibilium Geometriam demonſtratur, ad ſolitam antiquorum demonſtrationem indirectam reduci poſſe: quod à nobis factum eſt, vt in multis alijs, ita etiam in primo, & tertio ſequentium Theorematum; ſed ne lectoris patientia nimium adhuc abuteremur plura omittenda cenſuimus, treſq; tantum demonſtrationes exib emus*

THEOREMA I.

Omne ſpatium quod ſub linea Cycloide, & recta eius baſi continetur, triplum eſt circuli ſui genitoris; ſiue ſeſquialterum trianguli eandem baſim, & eandem altitudinem habentis.

Eſto Cyclois linea a b c *deſcripta à puncto* c *circuli* c d e f *dum ipſe circumuertitur ſuper manente baſi* a f. (*conſideramus autem ſemicycloidem, & ſemicirculum tantum ad euitandam figuræ confuſionem.*) *Dico ſpatium* a b c f *triplum eſſe ſemicirculi* c d e f; *siue ſeſquialterum trianguli* a c f,

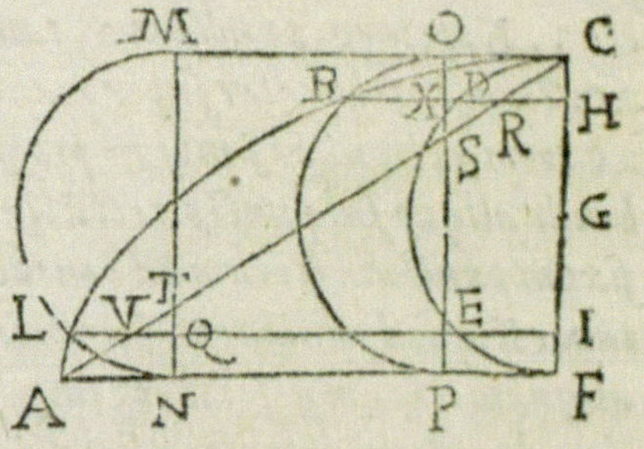

Accipiantur duo puncta h, *&* i *in diametro* c f. *æquè remota à centro* g. *Ductiſq.* h b, i l c m *æquidiſtanter ipsi* f a, *tranſcant per puncta* b, *&* l *ſemicirculi* o b p, m l n, *ęquales ipsi* c d f, *& contingentes basim in punctis* p n.

Naniſe-

qualis ſit periphœriæ circuli genitoris ac. *Quod quidem non adeò obſcurum est: Nam tota periphœria* ac *ſe ipſam in conuerſione commenſurauit ſuper manente recta* ab.

Quæritur nunc quam proportionem habeat ſpatium cycloidale adb *ad circulum ſuum genitorem* ac*? Oſtendemuſque, Deo dante, triplum eſſe. Demonstrationes tres erunt, inter ſe penitus diuerſae. Prima, & tertia per nouam Indiuiſibilium Geometriam nobis amiciſſimam procedent: ſecunda verò per duplicem poſitionem, more veterum recepto; vt vtriſque fautoribus ſatiſfiat. Cęterum, hoc moneo; principia ferè omnia, quibus aliquid per Indiuisibilium Geometriam demonstratur, ad ſolitam antiquorum demonstrationem indirectam reduci poſſe: quod à nobis factum est, vt in multis alijs, ita etiam in primo, & tertio ſequentium Theorematum; ſed ne lectoris patientia nimium adhuc abuteremur plura omittenda cenſuimus, treſq; tantum demonſtrationes exibemus*[159].

THEOREMA I.

Omne ſpatium quod ſub linea Cycloide, & recta eius baſi continetur, triplum eſt circuli ſui genitoris; ſiue ſeſquialterum trianguli eandem baſim, & eandem altitudinem habentis.

Eſto Cyclois linea abc *deſcripta à puncto* c *circuli* cdef *dum ipſe circumuertitur ſuper manente basi* af. (*consideramus autem ſemicycloidem, & ſemicirculum tantum ad euitandam figuræ confusionem.*) *Dico ſpatium* abcf *triplum eſſe ſemicirculi* cdef*; siue ſeſquialterum trianguli* acf,

Accipiantur duo puncta h, & i *in diametro* cf. *æquè remota à centro* g. *Ductiſq.* hb, ilcm *æquidiſtanter ipsi* fa, *tranſeant per puncta* b, & l *ſemicirculi* obp, mln, *ęquales ipsi* cdf, *& contingentes basim in punctis* pn.

Nanife-[160]

[159] *exibemus*] *exibemus*.
[160] *Nanife-*] *Manife-*.

Manifestum est rectas hd, ie, xb, ql *aequales esse, per 14 Tertij, aequalesq. erunt arcus* ob, ln. *Item cum ęquales sint* ch if, *ęquales erunt* cr, ua *ob parallelas*.

Tota periphęria mln, *ob cycloidem, aequalis est rectę* af. *itēque arcus* ln *rectae* an *ob eandem causam, cum arcus* ln *seipsum super recta* an *commensurauerit; ergo reliquus arcus* lm, *reliquae rectę* nf *ęqualis erit. Eadem ratione arcus* bp. *rectę* ap, *& arcus* bo *rectę* pf, *aequalis erit*,

Iam recta an *aequalis est arcui* ln, *siue arcui* bo, *siue rectę* pf. *Ergo ob parallelas, ęquales erunt* at, sc. *Verum quia aequales erant etiam* cr, au. *reliquę* ut, sr *ęquales erunt. Propterea in triangulis aequiangulis* utq, rsx, *aequalia erunt latera homologa* uq, xr. *Patet itaque quod duae rectae* lu, br *simul sumptae aequales erunt duabus rectis* lq, bx, *nempe ipsis* ei, dh, *& hoc semper verum erit vbicunq. sumantur duo puncta* h, *&* i, *dummodo aequaliter a centro sint remota. Ergo omnes lineae figurae* albca *aequales sunt omnibus lineis semicirculi* cdef; *& ideò figura bilinearis* albca *aequalis erit semicirculo* cdef.

Sed triangulum acf *duplum est semicirculi* cdef. *(nam triangulum* acf *reciprocum est triangulo Propos. pr. Arch. de dimens. circ. cum latus* af *semiperiphaeriae, latus verò* fc *diametro sit aequale, vnde sequitur triangulum* acf *aequale esse integro circulo cuius diameter sit* cf.*) Ergo componendo, totum cycloidale spatium sesquialterum erit trianguli inscripti* acb; *Triplum verò semicirculi* cdef. *Quod erat*.

Lemma I.

Si super lateribus oppositis alicuius rectanguli AF, duo semicirculi descripti sint, EIF, AGD erit figura sub periphærijs, & sub reliquis lateribus compręhensa ęqualis prędicto rectangulo.

Vocetur autem talis figura Arcuatum; tam si fuerit integra, quàm etiam ipsius partes, quando secta fuerit à linea ipsi fd *parallela*.

Demon-

Manifestum eſt rectas hd, ie, xb, ql *æquales eſſe, per 14 Tertij, æqualeſq. erunt arcus* ob, ln. *Item cum ęquales sint* chif, *ęquales erunt* cr, ua *ob parallelas.*

Tota periphęria mln, *ob cycloidem, æqualis eſt rectę* af. *itẽque arcus* ln *rectæ* an *ob eandem cauſam, cum arcus* ln. *ſeipſum ſuper recta* an *commenſurauerit; ergo reliquus arcus* lm, *reliquæ rectę* nf *ęqualis erit. Eadem ratione arcus* bp. *rectę* ap, *& arcus* bo *rectę* pf, *æqualis erit,*

Iam recta an *æqualis est arcui* ln, *ſiue arcui* bo, *ſiue rectę* pf. *Ergo ob parallelas, ęquales erunt* at, ſc. *Verum quia æquales erant etiam* cr, au. *reliquę* ut, ſr *ęquales erunt. Propterea in triangulis aequiangulis* utq, rſx, *aequalia erunt latera homologa* uq, xr. *Patet itaque quod duae rectæ* lu, br *ſimul ſumptae aequales erunt duabus rectis* lq, bx, *nempe ipſis* ei, dh, *& hoc ſemper verum erit vbicunq. ſumantur duo puncta* h, *&* i, *dumodo aequaliter a centro ſint remota. Ergo omnes lineae figurae* albca *aequales ſunt omnibus lineis ſemicirculi* cdef; *& ideò figura bilinearis* albca *aequalis erit ſemicirculo* cdef.

Sed triangulum acf *duplum eſt ſemicirculi* cdef. *(nam triangulum* acf *reciprocum eſt triangulo Propoſ. pr. Arch. de dimenſ. circ. cum latus* af *ſemiperiphaeriae, latus verò* fc *diametro ſit æquale, vnde ſequitur triangulum* a cf *aequale eſſe integro circulo cuius diameter ſit* cf.*)* *Ergo componendo, totum cycloidale ſpatium ſeſquialterum erit trianguli inſcripti* acb;.

Triplum verò ſemicirculi cdef. *Quod erat*

Lemma I.

Si ſuper lateribus oppoſitis alicuius rectanguli AF, duo ſemicirculi deſcripti ſint, EIF, AGD erit figura ſub periphærijs, & ſub reliquis lateribus compręhenſa ęqualis prędicto rectangulo.

Vocetur autem talis figura Arcuatum; tam ſi fuerit integræ, quàm etiam ipſius partes, quando ſecta fuerit à linea ipsi fd *parallela.*

Demon-

Demonstratur; quoniam cum sint aequales semicirc. dempto communi segmento b g c, *additisque communibus trilineis* e b a, c f d. *clarum erit propositum.*

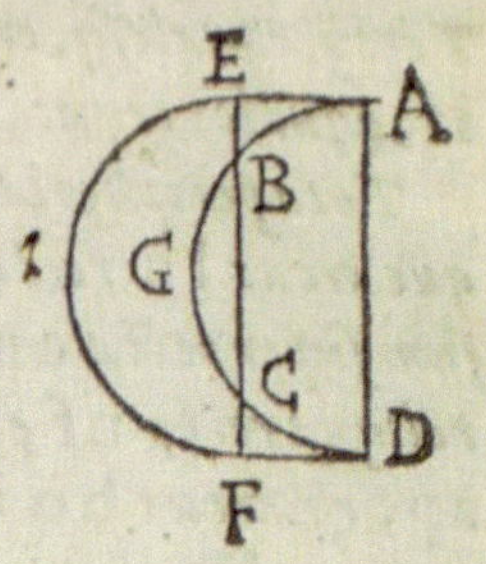

Quando verò detur casus quod segmentum nullum sit, tunc breuior, faciliorq. demonstratio erit. Facilè etiam per eandem prostapheresim ostenditur arcuatum sectum à linea ipsi fd *parallela equale esse rectangulo aequealto, & super eadem basi constituto.*

Lemma II.

Esto linea cycloidalis a b c *descripta à puncto* c *semicirculi* c d e *dum conuertitur super manente* a e. *Compleatur rectangulũ* a f c e, *fiatq. circa diametrum* a f *semicirculus* a g f. *Dico cycloidem* a b c *secare bifariam arcuatum* a g f c d e.

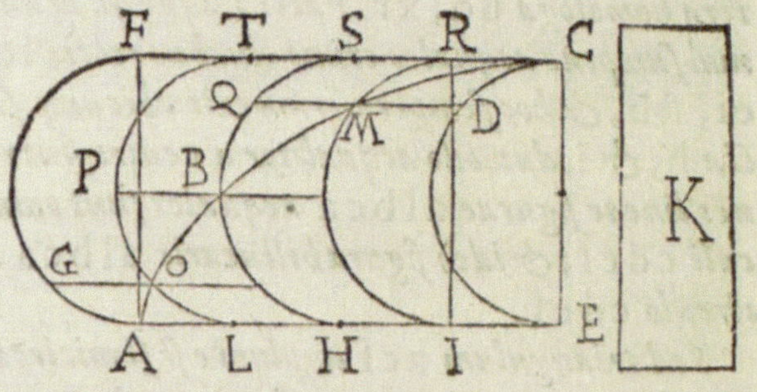

Si enim ita non est, erit vtiq. alterum ex duobus trilineis f g a b c, a b c d e, *magis quam dimidium eiusdem arcuati. Esto & ponatur alterum ex ipsis (quodcunq. sit) puta* a b c d e *maius quàm dimidium arcuati. Sitq. excessus, quo trilineum superat semissem arcuati, aequalis spatio cuidam* K.

Secetur bifariam a e *in* h; *& iterùm* h e *in* i: *& sic fiat semper donec rectangulum aliquod* i e c *minus reperiatur spatio* K. *Tunc diuidatur integra* a e *in particulas equales ipsi* i e, *& per puncta diuisionum* l, h, i, *transeant semicirculi aequales ipsi* c d e *semicirculo, tangentes basim in punctis* l, h, i. *secantesq. cycloidem in* o, b, m, *per quae puncta agantur recta* g o, p b, q m d *aequidistantes basi* a e.

Eri

Demonstratur; quoniam cum sint aequales ſemicirc. dempto communi ſegmento bgc, *additisque communibus trilineis* eba, cfd. *clarum erit propositum.*

Quando verò detur caſus quod ſegmentum nullum sit, tunc breuior, faciliorq. demonſtratio erit. Facilè etiam per eandem proſtapthereſim oſtenditur arcuatum ſectum à linea ipſi fd *parallela ęquale eße rectangulo æquealto, & ſuper eadem baſi constituto.*

Lemma II.

Eſto linea cycloidalis abc *deſcripta à puncto* c *ſemicirculi* cde *dum conuertitur ſuper manente* ae. *Compleatur rectangulũ* afce, *fiatq. circa diametrum* af *ſemicirculus* agf. *Dico cycloidem* abc *ſecare bifariam arcuatum* agfcde.

Si enim ita non eſt, erit vtiq. alterum ex duobus trilineis fgabc, abcde, *magis quam dimidium eiuſdem arcuati. Eſto & ponatur alterum ex ipsis (quodcunq. sit) puta* abcd e *maius quàm dimidium arcuati. Sitq. exceſſus, quo trilineum ſuperat ſemiſſem arcuati, aequalis ſpatio cuidam* K.

Secetur bifariam ae *in* h*; & iterùm* he *in* i*: & sic ſiat ſemper donec rectangulum aliquod* iec *minus reperiatur ſpatio* K. *Tunc diuidatur integra* ae *in particulas æquales ipsi* ie, *& per puncta diuisionum* l, h, i, *tranſeant ſemicirculi aequales ipsi* cde *ſemicirculo, tangentes basim in punctis* l, h, i. *ſecanteſq. cycloidem in* o, b, m, *per quae puncta agantur rectæ* go, pb, qmd *æequidiſtantes basi* ae.

Eri

Erit itaque arcuatum oh ęquale ipsi gl: arcuatum verò bi ęquale arcuato ph: & arcuatum me æquale arcuato qi. Propterea vniuersa figura inscripta in trilineo abcde constans ex arcuatis, ęqualis erit figuræ eidem trilineo circumscriptę, excepto tamen arcuato imrcde. Quod si figuræ circumscriptæ addas suum arcuatum imrcde, superabit circumscripta figura ipsam inscriptam excessu prædicti arcuati, siue rectangulo re, nempe minori excessu quam sit spatium K. Propterea inscripta in trilineo figura adhuc erit plusquàm dimidium arcuati agfcde. & ideo maior quam trilineum fgabc. Sed eadem æqualis est alteri figurę ex arcuatis compositę & in trilineo fgabc descriptæ: ergò hęc inscripta figura maior esset suo trilineo fgabc. pars suo toto. quod esse non potest. -stenditur infra

Quod inscriptæ figuræ sint ęquales patet. Nam arcus ol ęqualis est rectæ la, hoc est rectae ie, hoc est arcui rm (ob cycloidem.) Ergo arcuatum oh aequale erit arcuato mf. & sic de singulis.

Si verò supponeremus trilineum fgabc maius quàm dimidium arcuati agfcde, constructio figurae, & demonstratio penitus eadem erit. Ergò concludemus cycloidem lineam abc bifariam secare arcuatum agfcde. Quod erat propositum.

THEOREMA II.

Spatium cycloidale triplum est circuli sui genitoris.

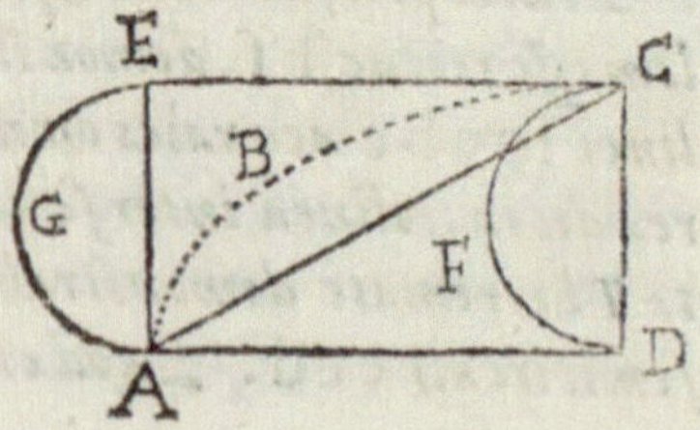

Esto cyclois abc descripta à puncto c circuli cfd. dico spatium abcd triplũ esse semicirculi cfd.

Compleatur rectangulum adce; factoq. super ae semicirculo age, ducatur ac.

Triangulum adc duplum est semicirculi cfd (nam basis ad aequalis est periphaeriae cfd ob cycloidem, altitudo verò dc aqualis diametro) ideò rectangulum ed quadruplum erit eius-

M dem

Erit itaque arcuatum oh *ęquale ipſi* gl*: arcuatum verò* bi *ęquale arcuato* ph: *& arcuatum* me *æquale arcuato* qi. *Propterea vniuerſa figura inſcriptæ in trilineo* abcde *conſtans ex arcuatis, ęqualis erit figuræ eidem trilineo circumſcriptę, excepto tamen arcuato* imrcde. *Quod ſi figuræ circumſcriptæ addas ſuum arcuatum* imrcde, *ſuperabit circumſcripta figura ipſam inſcriptam exceſſu prædicti arcuati, ſiue rectangulo* re, *nempe minori exceſſu quam ſit ſpatium* K. *Propterea inſcripta in trilineo figura adhuc erit pluſquàm dimidium arcuati* agfcde. *& ideo maior quam trilineum* fgabc. *Sed eadem æqualis est alteri figurę ex arcuatis compoſitę & in trilineo* fgabc *deſcriptæ: ergò hęc inſcripta figura maior eſſet ſuo trilineo* fgabc. *pars ſuo toto. quod eſſe non poteſt.*

Quod inſcriptæ figuræ ſint ęquales patet. Nam arcus ol *ęqualis est rectæ* la, *hoc eſt rectae* ie, *hoc eſt arcui* rm (*ob cycloidem.) Ergo arcuatum* oh *aequale erit arcuato* mſ. *& ſic de ſingulis.* *oſtendi-*

Si verò ſupponeremus trilineum fgabc *maius quàm dimidium arcuati* agfcde, *conſtructio figurae, & demonstratio penitus eadem erit. Ergò concludemus cycloidem lineam* abc *bifariam ſecare arcuatum* agfcde. *Quod erat propoſitum.* *tur infra*

THEOREMA II.

Spatium cycloidale triplum eſt circuli ſui genitoris.

Eſto cyclois abc *deſcripta à pũcto* c *circuli* cfd. *dico ſpatium* abcd *triplũ eßе ſemicirculi* cfd.

Compleatur rectangulum adce*; factoq. ſuper* ae *ſemicirculo* age, *ducatur* ac.

Triangulum adc *duplum eſt ſemicirculi* cfd (*nam baſis* ad *aequalis eſt periphaeriae* cfd *ob cycloidem, altitudo verò* dc *aqualis diametro) ideò rectangulum* ed *quadruplum erit eiuſ-*

M *dem*

dem ſemicirculi cfd. *Ergo arcuatum* agecfd *quadruplum erit eiuſdem ſemicirculi: propterea trilineum* abcfd *(per lemma praecedens) duplum erit ſemicirculi, & componendo ſpatium* abcd *triplum erit eiuſdem ſemicirculi* cfd.

THEOREMA III.

Omne ſpatium cycloidale triplum eſt circuli ſui genitoris.

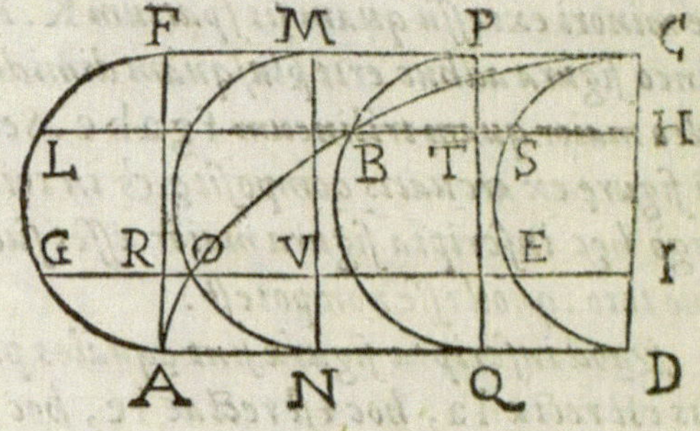

Eſto cycloidalis linea abc *deſcripta à puncto* c *ſemicirculi* ced. *Dico ſpatium* abcd *triplum eſſe ſemicirc.* ced.

Compleatur rectangulum afcd; *factoq. ſemicirculo* agf, *accipiātur duo puncta* h, & i *in diametro* cd *eque remota à centro, & ducantur* hl, ig *aequidiſtantes ad* ad. *quae cycloidem ſecent in quibuſuis punctis* b, & o. *Agantur denique per* b, & o *duo ſemicirculi* pbq, mon, *vt in precedentibus factum eſt.*

Iam recta go, *equalis eſt recte* ru *(cum àquales ſint* gr, ou, *& communis* ro*) ſiue aequalis eſt recte* an, *nempe arcui* on *(ob cycloidem) vel arcui* pb, *ſiue recte* pc, *vel* th, *vel* bſ.

Eodem prorſus modo, quo demonſtrauimus rectam go *aequalem eſſe rectae* bſ, *demonſtrantur omnes & ſingulae lineae trilinei* fgabc *aequales omnibus lineis trilinei* abced. *Propterea dicta trilinea inter ſe aequalia erunt. Ergo vt in praecedenti Theoremate demonſtrabitur cycloidale ſpatium triplum eſſe ſemicirculi* ced. *Quod erat &c.*

FINIS.

SCHO-

dem ſemicirculi cfd. *Ergo arcuatum* agecfd *quadruplum erit eiuſdem ſemicirculi: propterea trilineum* abcfd (*per lemma præcedens) duplum erit ſemicirculi, & componendo ſpatium* abcd *triplum erit eiuſdem ſemicirculi* cfd.

THEOREMA III.

Omne ſpatium cycloidale triplum eſt circuli ſui genitoris.

Esto cycloidalis linea abc *deſcripta à puncto* c *ſemicirculi* ced. *Dico ſpatium* abcd *triplum eſſe ſemicirc.* ced.

Compleatur rectangulum afcd*; factoq. ſemicirculo* agf, *accipiãtur duo puncta* h, & i *in diametro* cd *ęque remota à centro, & ducantur* hl, ig *æquidistantes ad* ad. *quæ cycloidem ſecent in quibuſuis punctis* b, & o. *Agantur denique per* b, & o *duo ſemicirculi* pbq, mon, *vt in prędentibus factum est.*

Iam recta go, *equalis eſt rectę* ru (*cum æquales ſint* gr, ou, *& communis* ro*) ſiue æqualis eſt rectę* an, *nempe arcui* on *(ob cycloidem) vel arcui* pb, *ſiue rectę* pc, *vel* th, *vel* bſ.

Eodem prorſus modo, quo demonstrauimus rectam go *æqualem eſſe rectae* bſ, *demonstrantur omnes & ſingulae lineae trilinei* fgabc *aequales omnibus lineis trilinei* abced. *Propterea dicta trilinea inter ſe aequalia erunt. Ergo vt in praecedenti Theoremate demonstrabitur cycloidale ſpatium triplum eſſe ſemicirculi* ced. *Quod erat &c.*

FINIS.

SCHO-

91

SCHOLIVM

De Cycloidibus aliarum specierum.

Hactenus de Cycloide dictum sit: vlterius enim contemplationem hanc demonstrando protrahere odiosum esset, & ex appendice liber fieret. Proponi tamen poterant adhuc non pauca circa hanc figuram planam, quam Cycloidem Primariam appellare non esset inconueniens; quandoquidem infinitae aliae species huiusmodi figurarum ab ipsa iam considerata primaria Cycloide oriuntur. Concipiamus enim (in figura paginae 85.) non solum periphaeriam circuli a c *aequabili conuersione rotari, sed etiam vniuersum planum tam internum, quàm externum ipsius circuli* a c *in infinitum extensi. Manifestum est quòd circuli centrum rectam lineam describet ipsi* a b *aequidistantem. Puncta verò, quae intra periphaeriam* a c *sunt constituta, Cycloides describent humiliores quàm ipsa primaria* a d b. *quasdam etiam (quod incredibile quasi videtur) flexuosas, & non ad easdem partes concauum habentes: tales autem fient à punctis prope cẽtrum circuli rotantis* a c *existentibus. Puncta verò, quae extra periphaeriam* a c *erunt, Cycloides describent, ipsa primaria altiores, & vsque in infinitum crescentes.*

Circulum cuiuscunq. cycloidis proprium genitorem dicere possumus eum, cuius periphaeria concentrica sit periphaeriae a c, *transeatq. per punctum cycloidem ipsam describens.*

In hoc conueniunt omnes, quod aequalibus basibus insistunt; humiliores tamen cycloides basim habent genitrici periphaeria maiorem: altiores verò minorem genitrici periphaeria basim habent.

Ratio, quam vnaquaeq. cycloidalis figura habet ad suum triangulum, vel ad circulum suum proprium genitorem, semper est maioris inęqualitatis, & variatur in infinitum. Si tamen vtrum

M 2 que

SCHOLIVM
De Cycloidibus aliarum ſpecierum.

HA*ctenus de Cycloide dictum ſit: vlterius enim contemplationem hanc demonſtrando protrahere odioſum eßet, & ex appendice liber fieret. Proponi tamen poterant adhuc non pauca circa hanc figuram planam, quam Cycloidem Primariam appellare non eſset inconueniens; quandoquidem infinitae aliae ſpecies huiuſmodi figurarum ab ipſa iam conſiderata primaria Cycloide oriuntur. Concipiamus enim(in figura paginae 85.) non ſolum periphaeriam circuli* ac *aequabili conuerſione rotari, ſed etiam vniuerſum planum tam internum, quàm ex ternum ipſius circuli* ac *in infinitum extenſi. Manifeſtum eſt quòd circuli centrum rectam lineam deſcribet ipſi* ab *aequidistantem. Puncta verò, quae intra periphaeriam* ac *ſunt conſtituta, Cycloides deſcribent humiliores quàm ipſa primaria* adb. *quaſdam etiam (quod incredibile quaſi videtur) flexuoſas, & non ad eaſdem partes concauum habentes: tales autem fient à punctis prope cẽtrum circuli rotantis* ac *existentibus. Puncta verò, quae extra periphaeriam* ac *erunt, Cycloides deſcribent, ipſa primaria altiores, & vſque in infinitum creſcentes.*

Circulum cuiuſcunq. cycloidis proprium genitorem dicere poſſumus eum, cuius periphaeria concentrica ſit pèriphaeriae ac, *tranſeatq. per punctum cycloidem ipſam deſcribens.*

In hoc conueniunt omnes, quod aequalibus baſibus inſiſtunt; humiliores tamen cycloides baſim habent genitrici periphaeria maiorem: altiores verò minorem genitrici periphaeria baſim habent. *M* 2

Ratio, quam vnaquaeq. cycloidalis figura habet ad ſuum triangulum, vel ad circulum ſuum proprium genitorem, ſemper eſt maioris inęqualitatis, & variatur in infinitum. Si tamen vtrum *que quę simul consideres & triangulum, & circulum, ęqualitatis ratio erit.*

Omne spatium sub qualibet cycloide linea, & rectà eius basi contentum, ad triangulum super eadem basi, & sub eàdem altitudine constitutum, est vt periphæria circuli proprij genitoris vna cum duplo basis cycloidis, ad duplum basis cycloidis. Ad circulum verò proprium genitorem vnumquodque cycloidale spatium est vt duplum basis cycloidis vna cum periphæria genitoris circuli ad eiusdem circuli periphæriam.

Hinc problemati locus pararetur, datà quacunq; ratione maioris inęqualitatis, cycloidale spatium inuenire, quod ad triangulum, siue circulum suum sit in data ratione, & in data basi.

Cuiuscunq. cycloidalis spatij ad quodlibet spatium cycloidale (*etiam si non ab eadem primaria cycloide ortum ducant*) ratio componitur ex ratione altitudinis ad altitud., & ex ratione dupli basis cum periphærià genitrice, ad duplum basis cum peripheria genitrice.

Tangentem ad quodlibet imperatum punctum dari posse certũ est; peculiari primùm ratione pro Cycloide primaria; deinde vniuersali etiam pro omnibus alijs. Tangens ad datum quodlibet punctum primariæ cycloidis ducitur ex puncto sublimiori genitoris circuli per ipsum datum punctum transeuntis.

Tangens ad datum punctum cuiuscunque cycloidis ducitur hoc modo. Transeat per datum punctum cycloidis circulus ipsius genitor, quem in dato eodem puncto contingat recta conueniens vel cum basi cycloidis, vel cum alia ipsi æquidistante. Fiatq; vt radius circuli poprij ad radium circuli primarij, ita tangens prædicta inter datum punctum, & basim, vel æquidistantẽ intercepta, ad aliam quandam lineam aptè sumendam à termino tangentis in ipsa vel basi, vel æquidistante. Tum ab extremitate huius assumptæ tangens ad imperatum punctum cycloidis emittatur.

Nonnulla etiam Theoremata pro Mecanicis contemplatoribus ex hac figura deriuari possent, nisi consulendum iam tandem esset ne simul cum molestia tedium fiat,

FINIS.

que ſimul conſideres & triangulum, & circulum, ę qualitatis ratio erit.

Omne ſpatium ſub qualibet cycloide linea, & rectà eius baſi contentum, ad triangulum ſuper eadem baſi, & ſub eàdem altitudine conſtitutum, eſt vt periphæria circuli proprij genitoris vna cum duplo baſis cycloidis, ad duplum baſis cycloidis. Ad circulum verò proprium genitorem vnumquodque cycloidale ſpatium eſt vt duplum baſis cycloidis vna cum periphæria genitoris circuli ad eiuſdem circuli periphæriam.

Hinc problemati locus pararetur, datà quacunq; ratione maioris inęqualitatis, cycloidale ſpatium inuenire, quod ad triangulum, ſiue circulum ſuum ſit in data ratione, & in data baſi.

Cuiuſcunq. cycloidalis ſpatij ad quodlibet ſpatium cycloidale (*etiam ſi non ab eadem primaria cycloide ortum ducant*) ratio componitur ex ratione altitudinis ad altitud., & ex ratione dupli baſis cum periphærià genitrice, ad duplum baſis cum periphęria genitrice.

Tangentem ad quodlibet imperatum punctum dari poſſe certũ eſt; peculiari primùm ratione pro Cycloide primaria; deinde vniuerſali etiam pro omnibus alijs. Tangens ad datum quodlibet punctum primariæ cycloidis ducitur ex puncto ſublimiori genitoris circuli per ipſum datum punctum tranſeuntis.

Tangens ad datum punctum cuiuſcunque cycloidis ducitur hoc modo. Tranſeat per datum punctum cycloidis circulus ipſius genitor, quem in dato eodem puncto contingat recta conueniens vel cum baſi cycloidis, vel cum alia ipſi æquidiſtante.

Fiatq; vt radius circuli poprij ad radium circuli primarij, ita tangens prædicta inter datum punctum, & baſim, vel æquidiſtantẽ intercepta, ad aliam quandam lineam aptè ſumendam à termino tangentis in ipſa vel baſi, vel æquidiſtante. Tum ab extremitate huius aſſumptæ tangens ad imperatum punctum cycloidis emittatur.

Nonnulla etiam Theoremata pro Mecanicis contemplatoribus ex hac figura deriuari poſſent, niſi conſulendum iam tandem eſſet ne ſimul cum moleſtia tedium fiat,

FINIS.

DE SOLIDO ACVTO HYPERBOLICO

Problema alterum.

Proemium ad Lectorem.

AGGREDIOR iam opus quod ipsis Geometriæ candidatis non solum difficile videatur verùm etiam impossibile. Hactenus enim in Mathematicis Scholis repertæ sunt dimensiones figurarum ab omni parte finem habentium: quandóquidem inter omnia solida, quæ ab antiquis, & modernis Auctoribus multiplici conatu ad mensuram redacta sunt, nullum adhuc, quod ego sciam, vllam dimensionem habuit extensione infinitam. Imò statim atq. proponatur siue solidum aliquod, siue figura plana, cuius aliqua extensio in infinitam distantiam procedat, vnusquisque cogitabit huiusmodi figuram infinitę magnitudinis esse debere. Attamen solidum habet Geometria, longitudine quidem infinitum, sed tanta præditum subtilitate, vt licet in infinitum producatur, exigui tamen cylindri molem non excedat. Tale erit solidum illud ab hyperbola genitum, quod huius libelli contemplatione prosequemur; intactum hucusque ab alijs, & multiplici, curiosaque Theorematum varietate fæcundissimum; eò vsq. vt, nisi me fal-

lat

DE SOLIDO ACVTO HYPERBOLICO
Problema alterum.

Proemium ad Lectorem.

AGGREDIOR iam opus quod ipſis Geometriæ candidatis non ſolum difficile videatur verùm etiam impoſſibile. Hactenus enim in Mathematicis Scholis repertæ ſunt dimenſiones figurarum ab omni parte finem habentium: quandoquidem inter omnia ſolida, quæ ab antiquis, & modernis Auctoribus multiplici conatu ad menſuram redacta ſunt, nullum adhuc, quod ego ſciam, vllam dimenſionem habuit extenſione infinitam. Imò ſtatim atq. proponatur ſiue ſolidum aliquod, ſiue figura plana, cuius aliqua extenſio in infinitam diſtantiam procedat, vnuſquiſque cogitabit huiuſ modi figuram infinitę magnitudinis eſſe debere. Attamen ſolidum habet Geometria, longitudine quidem infinitum, ſed tanta præditum ſubtilitate, vt licet in infinitum producatur, exigui tamen cylindri molem non excedat. Tale erit ſolidum illud ab hyperbola genitum, quod huius libelli contemplatione proſequemur; intactum hucuſque ab alijs, & multiplici, curioſaque Theorematum varietate fæcundiſſimum; eò vſq. vt, niſi me fal-

lat

lat affectus, vniuersa Geometria inter hactenus consideratas figuras nullam habeat curiositatis abundantiorem.

Quo ad methodum demonstrandi, vnicum quidem, & precipuum Theorema duplici conatu ostendemus, & per indiuisibilia, & more Veterum. Quamquam (vt vera fateamur) primò inuentum sit per Indiuisibilium Geometriam; qui sanè verus est demonstrandi modus scientificus, semper directus, & ipsi naturæ germanus. Miseret me veteris Geometriæ, quæ cũ Indiuisibilium doctrinam, siue non nouerit, siue non admiserit, circà dimensionem solidorum adeò paucas veritates inuenit, vt ipsà penurià infelix ad ætatem nostram peruenerit. Antiquorum enim Theoremata circà doctrinam solidorum, quota pars sunt contemplationum, quas mirabilis nostro æuo Caualerius (omissis alijs) instituit, circà tot classes solidorum, specie differẽtium, multitudine abundantium? Methodus nostra, quam vsurpaturi sumus in præfato Theoremate, procedet per Indiuisibilia curua, sine aliorum exemplo, non tamen sine præmissa Geometriæ approbatione, Considerabimus enim omnes cylindricas superficies circà communem axem in nostro solido descriptibiles. Cuius rei cum nullum Caualerius ipse tradiderit in sua Geometria elementum, existimauimus nostram arguendi rationem exemplis aliquot esse corroborandam. Quamquam hoc apud me superfluum sit; cum iam totum huius libelli progressum ratum habeam, eò quòd ipsum admiserit, probaueritque doctissimus, & eruditissimus vir *Raphael Magiottus*; cui, vt in plurimis alijs scientijs, artibusq;, ita & in Mathematicis disciplinis neminem quis iure anteposuerit. Præmittemus itaque ante ipsum opus, sub Exemplorum nomine, quasdam Geometriç propositiones iam pridem notas, sed à nobis per Indiuisibilia curua demonstratas: Sic enim magis manifestum fiet hunc modum demonstrandi non esse negligendum, præsertim cum in rebus difficillimis maximum ipsius momentum reperiatur. Indiuisibilia verò curua quæ ad huiusmodi demonstrationes idonea sunt, in planis quidem figuris solæ circulorum periphæriç se se offerunt; in solidis autem, superficies sphæricæ, cylin-

lat affectus, vniuerſa Geometria inter hactenus conſideratas figuras nullam habeat curioſitatis abundantiorem.

Quo ad methodum demonſtrandi, vnicum quidem, & pręcipuum Theorema duplici conatu oſtendemus, & per indiuiſibilia, & more Veterum. Quamquam (vt vera fateamur) primò inuentum ſit per Indiuiſibilium Geometriam; qui ſanè verus eſt demonſtrandi modus ſcientificus, ſemper directus, & ipſi naturæ germanus. Miſeret me veteris Geometriæ, quæ cũ Indiuiſibilium doctrinam, ſiue non nouerit, ſiue non admiſerit, circà dimenſionem ſolidorum adeò paucas veritates inuenit, vt ipsà penurià infelix ad ætatem noſtram peruenerit. Antiquorum enim Theoremata circà doctrinam ſolidorum, quota pars ſunt contemplationum, quas mirabilis noſtro æuo Caualerius (omſſis alijs) inſtituit, circà tot claſſes ſolidorum, ſpecie differẽtium, multitudine abundantium? Methodus noſtra, quam vſurpaturi ſumus in præfato Theoremate, procedet per Indiuiſibilia curua, ſine aliorum exemplo, non tamen ſine præmiſſa Geometriæ approbatione, Conſiderabimus enim omnes cylindricas ſuperficies circà communem axem in noſtro ſolido deſcriptibiles. Cuius rei cum nullum Caualerius ipſe tradiderit in ſua Geometria elementum, exiſtimauimus noſtram arguendi rationem exemplis aliquot eſſe corroborandam. Quamquam hoc apud me ſuperfluum ſit; cum iam totum huius libelli progreſſum ratum habeam, eò quòd ipſum admiſerit, probaueritque doctiſſimus, & eruditiſſimus vir *Raphael Magiottus*; cui, vt in plurimis alijs ſcientijs, artibuſq;, ita & in Mathematicis diſciplinis neminem quis iure antepoſuerit. Præmittemus itaque ante ipſum opus, ſub Exemplorum nomine, quaſdam Geometrię propoſitiones iam pridem notas, ſed à nobis per Indiuiſibilia curua demonſtratas: Sic enim magis manifeſtum fiet hunc modum demonſtrandi non eſſe negligendum, præſertim cum in rebus difficillimis maximum ipſius momentum reperiatur. Indiuiſibilia verò curua quæ ad huiuſmodi demonſtrationes idonea ſunt, in planis quidem figuris ſolæ circulorum periphærię ſe ſe offerunt; in ſolidis autem, ſuperficies ſphæricæ, cy-

lin-

lindricæ, conicæque. Quandoquidem istæ tantum considerabiles sunt, tamquam ipsas figuras perfectè adæquantes, & vndique æqualis, vniformisque (vt ita dicam) spissitudinis. Pręmittimus igitur ante operis aggressionem, promissa aliquot Theorematum Geometricorum Exempla.

EXEMPLVM PRIMVM.

Esto circulus, cuius centrũ a, *semidiameter* a b, *tangens verò sit* b c, *quæ supponatur æqualis peripheriæ* b d. *Tũ coniũgatur* a c. *Dico circulum* b d. *triangulo* a b c *esse ęqualem*.

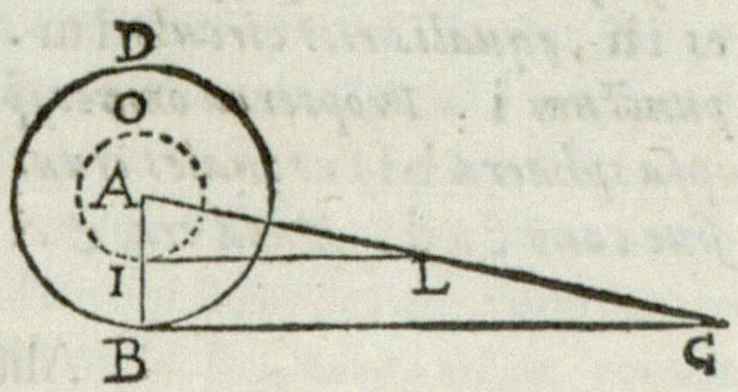

Sumatur in semidiametro a b, *quodlibet punctum* i; *& per* i *agantur, peripheria* i o *circa idem centrum* a, *& recta* i l *parallela ad* b c. *Erit itaque peripharia* b d, *ad peripheriam* i o, *vt semidiameter* b a, *ad* a i. *(demonstratur enim hoc à priori, non supposità circuli dimensione) siue vt* b c *ad* i l; *& permutando; erit peripharia* b d, *ad rectam* b c, *vt peripheria* i o, *ad rectam* i l. *Ergò peripheria* i o, *rectæ* i l *erit æqualis: & hoc semper, vbicunque sit punctum* i. *Quarè & omnes peripheriæ simul sumptę, omnibus rectis simul sumptis æquales erunt: nempe circulus ipse* b d, *æqualis erit triangulo* a b c. *Quod erat &c.* Ex 11. 5 Pappi

Concordat cum hoc Theoremate Propositio Prima Archim. De Dimensione circuli.

Exemplu m II.

Esto circulus, cuius radius a b, *tangensque* b c *sit æqualis diametro; & coniunctà* a c *conuertatur figura circà* a b, *ita vt fiat sphæra* b f, *& conus rectus* c a d. *Dico sphęram* b f, *cono* c a d *esse æqualem. Sumatur enim in* a b *quoduis punctũ* i, *& per ipsum* i *transeat superficies sphęrica* i h, *circà centrum* a;

circu-

lindricæ, conicæque. Quandoquidem iftæ tantum confiderabiles funt, tamquam ipfas figuras perfectè adæquantes, & vndique æqualis, vniformifque (vt ita dicam) fpiffitudinis. Pręmittimus igitur ante operis aggreffionem, promiffa aliquot Theorematum Geometricorum Exempla.

EXEMPLVM PRIMVM.[161]

E*Sto circulus, cuius centrũ* a, *femidiameter* ab, *tangens verò fit* bc, *quæ fupponatur æqualis periphæriæ* bd. *Tũ coniũgatur* ac. *Dico circulum* bd. *triangulo* abc *effe ęqualem.*

Sumatur in femidiametro ab, *quodlibet punctum* i*; & per* i *agantur, periphęria* io *circa idem centrum* a, *& recta* il *parallela ad* bc. *Erit itaque periphæria* bd, *ad periphæriam* io, *vt femidiameter* ba, *ad* ai. *(demonftratur enim hoc à priori, non fuppofità circuli dimenfione) fiue vt* bc *ad* il; *& permutando; erit periphæria* bd, *ad rectam* bc, *vt periphæria* io, *ad rectam* il. *Ergò periphęria* io, *rectæ* il *erit æqualis: & hoc femper, vbicunque fit punctum* i. *Quarè & omnes periphæriæ fimul fumptę, omnibus rectis fimul fumptis æquales erunt: nempe circulus ipfe* bd, *æqualis erit triangulo* abc. *Quod erat &c.* [Ex Pappi] II.5

Concordat cum hoc Theoremate Propofitio Prima Archim. De Dimenfione circuli.

Exemplum II.[162]

E*Sto circulus, cuius radius* ab, *tangenfque* bc *fit æqualis diametro; & coniunctà* ac *conuertatur figura circà* ab, *ita vt fiat fphæra* bf, *& conus rectus* cad. *Dico fphęram* bf, *cono* cad *effe æqualem. Sumatur enim in* ab *quoduis punctũ* i, *& per ipfum* i *tranfeat fuperficies fphęrica* ih, *circà centrum* a;

circu-

[161] This proof is in agreement *cum doctrina Archimedis* (Cfr. Pisano and Bussotti's chapters above).

[162] *Ibidem.*

96

20. 1. sphær. circulusq; lim in cono cad. *Iam: superficies sphęrica* bf *aequalis erit circulo* cd. *sphęrica verò* bf, *ad sphaericam* ih, *est vt quadratum* ba, *ad quadratũ* ai; *siue vt quadratum* bc *ad quad.* il; *nempe vt circulus* cd, *ad circulum* lm. *Sed antecedentes ęquales sunt; ergò etiã consequentes: nempe sphęrica superficies* ih, *ęqualis erit circulo* lm. *Et hoc semper, vbicumque sit punctum* i. *Propterea omnes sphaericae superficies simul (siue ipsa sphaera* bf*) aequales erunt omnibus circulis simul sumptis, siue cono* cad. *Quod erat &c.*

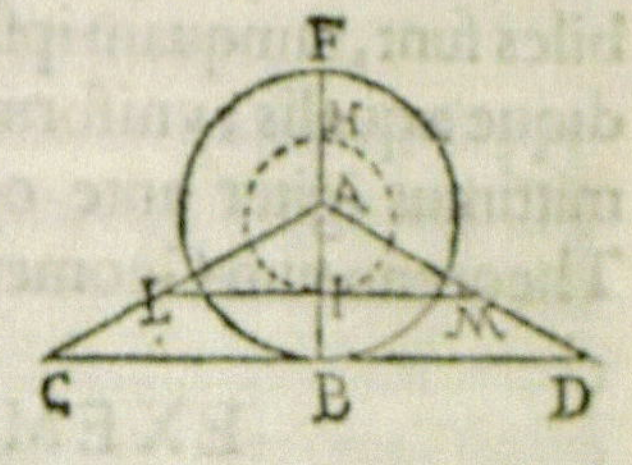

Aliter.

E*sto sphaera, cuius diameter* ab, *tangensq.* bd *sit aequalis semidiametro sphaerae: Et coniunctà* ad, *conuertatur triangul.* adb *circà axem* bd. *ita vt fiat conus rectus* adc.

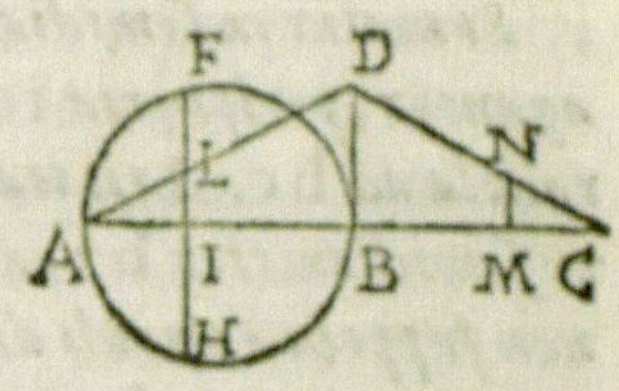

Dico sphaeram ab *aequalem esse cono* adc. *Sumatur enim in diametro* ab *quoduis punctum* i, *per quod transeat circulus* fh, *ad axem erectus in sphaera; & superficies cylindrica* limn, *circà axem* db *in cono*.

4. sexti. *Iam: cum* ab *dupla sit ipsius* bd, *erit* ai *dupla* il, *ergò quadratum* fi, *quod aequale est rectangulo* aib, *duplum erit*

5. p. de solidis sphær. *rectanguli* lib, *& aequale rectangulo* lim. *Propterea erit circulus* fh *aequalis superficiei cylindricae* limn. *Et hoc semper, vbicunque sit punctum* i. *Ergo omnes circuli simul, siue ipsa sphęra, ęquales erunt omnibus superficiebus cylindricis simul sumptis, nempe ipsi cono* adc, *Quod concordat cum 32. lib. 1. De Sphaera & Cylindro Archimedis.*

EXEM-

96 [*Proemium.*]

[20.i. Sphaer] *circulusq;* lim *in cono* cad. *Iam: superficies sphęrica* bf *æqualis erit circulo* cd. *sphęrica verò* bf, *ad sphæricam* ih, *est vt quadratum* ba, *ad quadratũ* ai; *siue vt quadratum* bc *ad quad.* il; *nempe vt circulus* cd, *ad circulum* lm. *Sed antecedentes ęquales sunt; ergò etiã consequentes: nempe sphęrica superficies* ih, *ęqualis erit circulo* lm. *Et hoc semper, vbicumque sit punctum* i. *Propterea omnes sphaericae superficies simul (siue ipsa sphaera* bf*) aequales erunt omnibus circulis simul sumptis. siue cono* cad. *Quod erat &c.*

Aliter.

E*Sto sphaera, cuius diameter* ab, *tangensq.* bd *sit aequalis semidiametro sphaeræ: Et coniunctà* ad, *conuertatur triangul.* adb *circà axem* bd. *it a vt fiat conus rectus* adc.

Dico sphaeram ab *aequalem esse cono* adc. *Sumatur enim in diametro* ab *quoduis punctum* i, *per quod transeat circulus* fh, *ad*
4. *sexti.* *axem erectus in sphaera; & superficies cylindrica* limn, *circà axem* db *in cono.*
5.*p.de.* *Iam: cum* ab *dupla sit ipsius* bd, *erit* ai *dupla* il, *ergò quadratum* fi, *quod aequale est rectangulo* aib, *duplum erit rectanguli* lib,
solidis. *& aequale rectangulo* lim. *Propterea erit circulus* fh *aequalis superficiei cylindricae* limn. *Et hoc semper, vbicunque sit punctum* i.
sphaer. *Ergo omnes circuli simul, siue ipsa sphęra, ęquales erunt omnibus superficiebus cylindricis simul sumptis, nempe ipsi cono* adc, *Quod concordat cum 32. lib. 1.*

De Sphaera & Cylindro Archimedis.

E X *E* M-

97

Exemplum III.

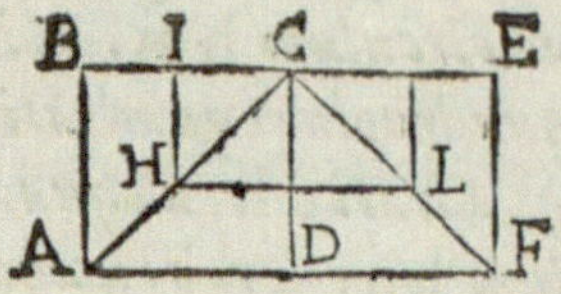

ESto quadratum abcd (nisi enim quadratum supponatur, ratiocinatio falsa euaderet, ob inæqualem superficierum spissitudinem; siue ob diuersitatem transitus) cuius quadrati esto diameter ac; & conuertatur figura circà axem cd, ita vt fiat cylindrus bf, & conus acf. Sumatur deinde in rectà ac, quoduis punctum h; per quod intelligatur actus circulus hl, intrà conum comprehensus; & insuper superficies cylindrica, cuius sectio sit hi, axis verò cd. Erit ergò superficies cylindrica hi ad circulum hl suam basim, vt recta hi, ad quartam partem diametri hl Et hoc verum erit semper, vbicunq; sit punctum h. Ergò omnes simul superficies cylindricæ (nempe solidum, quod ex cylindro relinquitur, dempto cono acf) ad omnes simul circulos (hoc est, ad conum acf) erunt, vt sunt omnes simul rectę trianguli abc, ad quartam partem omnium rectarum trianguli acf; nempe in ratione dupla. Quod concordat cum Theoremate X. lib. XII. Euclidis. 4. primò de Solid. Sphær.

Exemplum IV.

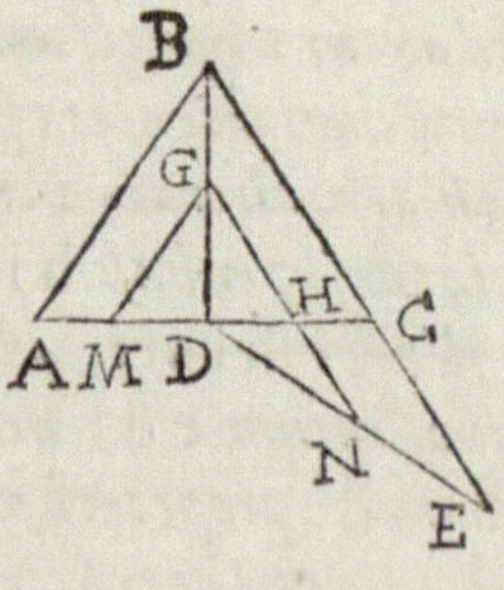

ESto conus rectus abc, cuius axis bd; & productà bc in e, ita vt circulus, cuius diameter ce, sit ęqualis curuæ superficiei coni abc, concipiatur circà diametrum ce circulus erectus ad planum abc; & super circulo ce, intelligatur alter conus cde, habens verticem in d. Dico conum abc, cono cde, esse ęqualem.

Sumatur in rectà dc quoduis punctum h, per quod ductà gh n parallela ad be, intelligatur per hg superficies conica mgh; cir-

N

Exemplum III.[163]

ESto *quadratum* abcd *(nisi enim quadratum ſupponatur, ratiocinatio falſa euaderet, ob inæqualem ſuperficierum ſpiſſitudinem; ſiue ob diuerſitatem tranſitus) cuius quadrati eſto diameter* ac; *& conuertatur figura circà axem* cd, *ita vt fiat cylindrus* bf, *& conus* acf.

Sumatur deinde in rectà ac, *quoduis punctum* h; *per quod intelligatur actus circulus* hl, *intrà conum compræhenſus; & inſuper ſuperficies cylindrica, cuius ſectio ſit* hi, *axis verò* cd. *Erit ergò ſuperficies cylindrica* hi *ad circulum* hl *ſuam baſim, vt recta* hi, *ad quartam partem diametri* hl[164] *Et hoc verum erit ſemper, vbicunq; ſit punctum* h. *Ergò omnes ſimul ſuperficies cylindricæ (nempe ſolidum, quod ex cylindro relinquitur, dempto cono* acf*) ad omnes ſimul circulos (hoc est, ad conum* acf*) erunt, vt ſunt omnes ſimul rectę trianguli* abc, *ad quartam partem omnium rectarum trianguli* acf*; nempe in ratione dupla. Quod concordat cum Theoremate X. lib.* XII. *Euclidis.*

4. *primi de Solid. Sphae*

Exemplum IV.[165]

ESto *conus rectus* abc, *cuius axis* bd*; & productà* bc *in* e, *ita vt circulus, cuius diameter.* ce, *ſit ęqualis curuæ ſuperficiei coni* abc, *concipiatur circà diametrum* ce *circulus erectus ad planum* abc*; & ſuper circulo* ce, *intelligatur alter conus* cde, *habens verticem in* d. *Dico conum* abc, *cono* cde, *eſſe ęqualem.*

Sumatur in rectà dc *quoduis punctum* h, *per quod ductà* ghn *parallela ad* be, *intelligatur per* hg *ſuperficies conica* m-

N g h; *cir-*

[163] This proof is in agreement *cum Theoremate Euclidis* (Cfr. Pisano and Bussotti's chapters above).

[164] hl] hl.

[165] This proof is in agreement *cum Doctrina Archimedis* (Cfr. *Ibidem*).

98

gh; *circàq; ipsam* hn *circulus parallelus circulo* ce. *Iam: conica superficies* abc, *ad circulum suæ basis* ac, *est vt recta* bc *ad* cd, *siue vt* gh *ad* hd; *nempe, vt conica superficies* mgh, *ad circulum* hm. *Circulus autem* ac, *ad circulum* ce, *est vt quadruplum quadrati* dc, *ad quadr.* ce; *siue vt quadruplum quadrati* dh, *ad quadrat* hn; *hoc est vt circulus* mh, *ad circulum* hn. *Ergò ex æquo, erit conica superficies* abc, *ad circulum* ce, *vt conica* mgh, *ad circulum* hn. *Et hoc semper verum erit, vbicunq; fuerit punctum* h. *Ergò omnes simul conicę superficies (nempe conus* abc*) æquales erunt omnibus simul circulis, nempe cono* cde. *Quod erat &c.*

Concordat cum hoc Theoremate Propos. xvij lib. primi De Sphæra, & cylindro Archim.

Exemplum V.

E*Sto circulus, cuius diameter* ab, *ponaturque tangens* bc *diamet. ęqualis, & iunctà* ac, *conuertatur figura circa axē* ab, *ita vt fiat sphæra* aebf, *& conus rectus* can.

Dico conum can, *ipsius sphæræ duplum esse.*

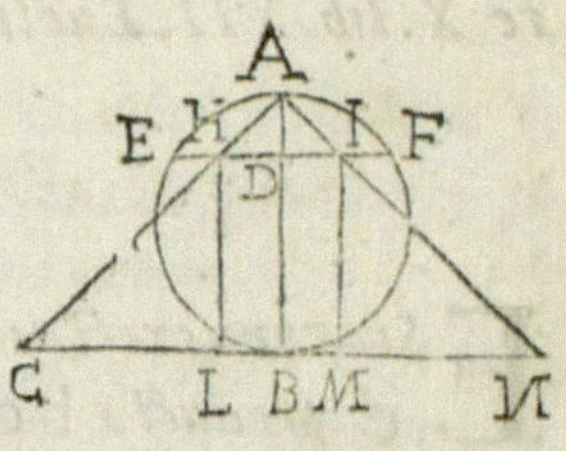

Accipiatur in diametro ab *quodlibet punctum* d, *per quod agatur planum* ef *ad axem* ab *erectum; quod quidem planum duos circulos effi iet, alterum* ef *in sph ęra, alterum verò* hi *in cono; Concipiatur super basi* hi *cylindrus rectus* hlmi. *Iam: superficies cylindri* hlmi, *ad circulum* ef, *est vt rectangulum* li, *ad quadratum* ed; *nempe dupla. Et hoc semper; vbicunq. sit punctum* d: *propterea, vt vna ad vnā, ita omnes ad omnes. Erunt ergò omnes superficies cylindricæ, nempe conus* can, *ad omnes circulos, nempe ad spheram* aebf, *in ratione dupla. Quod &c.*

5. p: de soli: sph.

Concordat cum hoc Theoremate Proposit. xxxij. De Sphera, & cylindro.

Exem-

98 [*Proemium.*]

gh; *circàq; ipſam* hn *circulus parallelus circulo* ce. *Iam*: *conica ſuperficies* abc, *ad circulum ſuæ baſis* c, *est vt recta* bc *ad* cd, *ſiue vt* gh *ad* hd; *nempe, vt conica ſuperficies* mgh, *ad circulum* hm. *Circulus autem* ac, *ad circulum* ce, *est vt quadruplum quadrati* dc, *ad quadr.* ce; *ſiue vt quadruplum quadrati* dh, *ad quadrat* hn; *hoc est vt circulus* mh, *ad circulum* hn. *Ergò ex æquo, erit conica ſuperficies* abc, *ad circulum* ce, *vt conica* mgh, *ad circulum* hn. E*t hoc ſemper verum erit, vbicunq; fuerit punctum* h. *Ergò omnes ſimul conicę ſuperficies* (*nempe conus* abc) *æquales erunt omnibus ſimul circulis, nempe cono* cde. *Quod erat &c.*

Concordat cum hoc Theoremate Propoſ. xvij lib. primi De Sphæra, & cylindro Archim.

Exemplum V.[166]

E*Sto circulus, cuius diameter* ab, *ponaturque tangens* bc *diamet. ęqualis, & iunct à* ac, *conuertatur figura circà axẽ* ab, *ita vt fiat ſphæra* aebf, *& conus rectus* can.

Dico conum can, *ipſius ſphæræ duplum eſſe.*

5. *p: de ſoli: sph:* *Accipiatur in diametro* ab *quodlibet punctum* d, *per quod agatur planum* ef *ad axem* ab *erectum; quod quidem planum duos circulos effi iet,*[167] *alterum* ef *in ſphæra, alterum verò* hi *in cono; Concipiatur ſuper baſi* hi *cylindrus rectus* hlmi. *Iam: ſuperficies cylindri* hlmi, *ad circulum* ef, *est vt rectangulum* li, *ad quadratum* ed; *nempe dupla. Et hoc ſemper; vbicunq. ſit punctum* d*: propterea, vt vna ad vnã, ita omnes ad omnes. Erunt ergò omnes ſuperficies cylindricæ, nempe conus* can, *ad omnes circulos, nempe ad ſphęram* aebf, *in ratione dupla. Quod &c.*

Concordat cum hoc Theoremate Propoſit. xxxij. De Sphera, & cylindro.

Exem-

[166] This proof is in agreement *cum doctrina Archimedis* (Cfr. Pisano and Bussotti's chapters above).
[167] *effi iet*] *efficiet.*

Exemplum VI.

Esto circulus, cuius diameter ab, *tangenſq;* ac *ponatur diametro æqualis coniunctaq;* ec, *conuertatur figura circa axem* ef *æquidiſtantem tangenti* ac, *ita vt à circulo deſcribatur ſphæra, à triangulo verò* ace *ſolidum quoddam cylindricum excauatum dempto cono* ced. *Dico ſphęram prædicto ſolido excauato eſſe æqualem.*

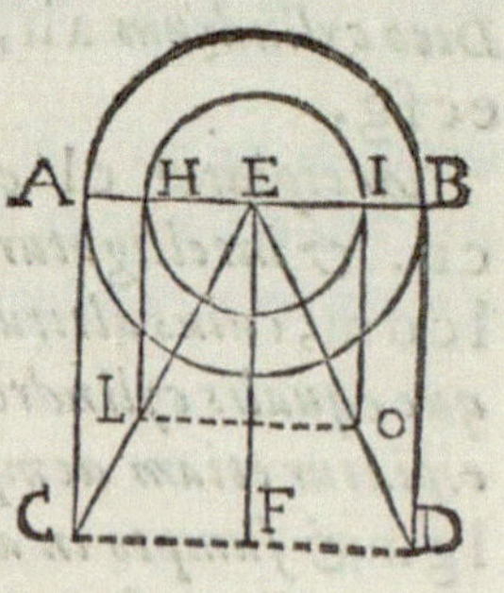

Sumatur in diametro ab *quoduis punctum* h, *per quod intelligatur ſuperficies ſphærica* hi, *priori ſuperficiei ſphęricæ concentrica; & inſuper ſuperficies cylindrica, quę deſcribitur à recta* hl *tangenti* ac *parallela, circa axem* ef *reuoluta. Iam:* ca *ad* ae *eſt vt* lh *ad* he, *& ſumptis conſequentium duplis,* ca *ad* ab, *eſt vt* lh *ad* hi. *Propterea,* lh *ad* hi *æqualis erit; & ideò ſuperficies curua cylindri* lhio *æqualis erit ſuperficiei ſphæricæ* hi. *Et hoc ſemper, vbi unq; fuerit punctum* h. *Propterea omnes omnibus, nempe omnes ſuperficies ſphęricę ſimul, ſiue ſphęra* ab, *ęquales erunt omnibus ſuperficiebus cylindricis ſimul, hoc eſt ſolido excauato* cabd. *dempto cono* ced. *Quod &c.*

Concordat cum hoc Theor. Propoſitio 32. de Sphæra & cylindro.

Lemma.

Superficies cuiuſcunque cylindri recti ab *(intelligo ſemper ſine baſibus) ad ſuperficiem curuam cuiuſcunque ſegmenti ſphærici* cde, *eſt vt rectangulum per axem cylindri, ad rectangulum* fdi, *ſub catето ſegmenti, & diametro ſphęræ.*

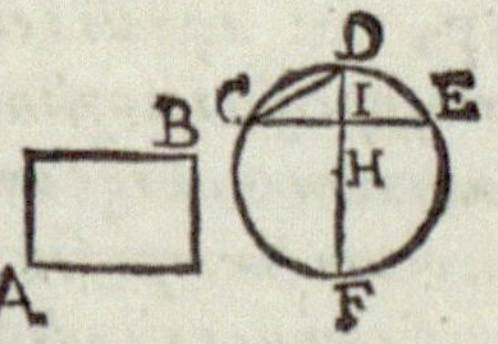

Nam, ſuperficies cylindrica ab, *ad circulum cuius ſemidiameter ſit linea ex polo* dc, *eſt vt rectangulum* ab, *ad quadratum* 5. p. de ſ. li. ſph.

Exemplum VI.

E*Sto circulus, cuius diameter* ab, *tangenſq;* ac *ponatur diametro æqualis coniunctaq;* ec, *conuertatur figura circa axem* ef *æquidiſtantem tangenti* ac, *ita vt à circulo deſcribatur ſphæra, à triangulo verò* ace *ſolidum quoddam cylindricum excauatum dempto cono* ced. *Dico ſphęram prædicto ſolido excauato eſſe æqualem.*

Sumatur in diametro ab *quoduis punctum* h, *per quod intelligatur ſuperficies ſphærica* hi, *priori ſuperficiei ſphęricæ concentrica; & inſuper ſuperficies cylindrica, quę deſcribitur à recta* hl *tangenti* ac *parallela, circa axem* ef *reuoluta. Iam:* ca *ad* ae *eſt vt* lh *ad* he, *& ſumptis conſequentium duplis,* ca *ad* ab, *eſt vt* lh *ad* hi. *Propterea,* lh *ad* hi *æqualis erit; & ideò ſuperficies curua cylindri* lhio *æqualis erit ſuperficiei ſphæricæ* hi. *Et hoc ſemper, vbi unq; fuerit punctum* h. *Propterea omnes omnibus, nempe omnes ſuperficies ſphęricę ſimul, ſiue ſphęra* ab, *ęquales erunt omnibus ſuperficiebus cylindricis ſimul, hoc eſt ſolido excauato* cabd. *dempto cono* ced. *Quod &c.*

Concordat cum hoc Theor. Propoſitio* 32. *de Sphæra & cylindro.

Lemma.

S*Vperficies cuiuſcunque cylindri recti* ab *(intelligo ſemper ſine baſibus) ad ſuperficiem curuam cuiuſcunque ſegmenti ſphærici* cde, *eſt vt rectangulum per axem cylindri, ad rectangulum* fdi, *ſub cateto ſegmenti, & diametro ſphęræ.* 5. *p. de. ſli. ſph.*

Nam, ſuperficies cylindrica ab, *ad circulum cuius ſemidiameter ſit linea ex polo* dc, *eſt vt rectangulum* ab, *ad quadra-*.

N 2 *tum*

tum dc: *Ergò, sumptis consequentium ęqualibus, erit cylindrica superficies, ad curuam sphærici segmenti* cde *superficiē, vt rectangulum* ab *ad rectangulum* fdi. *Quod erat &c.*

Exemplum VII.

E*Sto sphæra, vnà cum cylindro sibi circumscripto, quorum axis sit recta* cd. *Secenturq; plano* ab *ad axem erecto. Dico cylindrum* ah, *sesquialterum esse solidi sectoris sphęrici* ecfg.

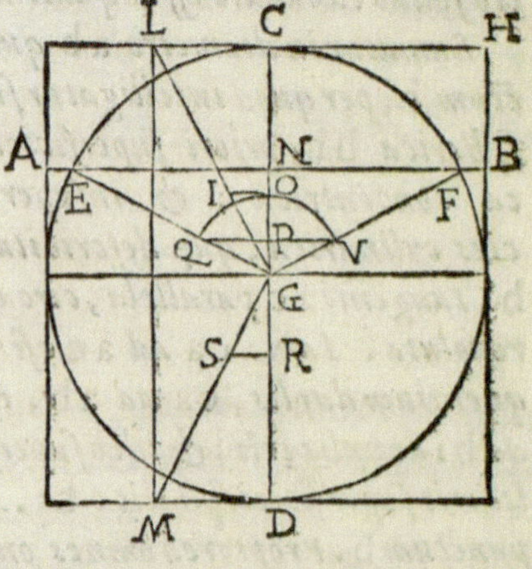

Accipiatur cl *ęqualis ipsi* cn. *& intelligatur cylindrus* lcdm, *cuius altitudo* lc. *eritque ęqualis cylindro* ah. *Concipiatur etiam demptus conus* lgm; *& sumpto in axe* cg *quouis puncto* o *fiant æquales* go, gr, *& transeat per ipsum punctū* o *sphęrica superficies* qou *in sectore; & cylindrica* iorf, *in solido cylindrico* lcdm *excauato ablatione coni* lgm; *sitque tam sphęricę superficiei quam cylindricę transeuntis per* o, *diameter ipsa* or.

Iam: tota cg, *ad totam rectam* go, *est vt* eg *ad* gq, *nempe, vt ablata* ng, *ad* gp; *ideò reliqua* cn, *ad* op, *erit vt tota* cg, *ad* go; *siue vt* cl, *ad* oi. *Sed antecedentes sunt æquales, ergò rectę* op, oi *ęquales erunt. Propterea rectangula* rop, roi *æqualia erunt; & superficies sphęrica* qou *æqualis erit superficiei cylindricę* iorf. *Et hoc semper, vbicunque sit punctum* o. *Ergò omnes superficies omnium segmentorum sphæricorum (nempe solidus sector* ecfg) *æquales erunt omnibus superficiebus cylindricis simul sumptis, hoc est solido excauato* lcdm. *Cui si addatur conus iam ablatus* lgm, *patebit propositum: nempe cylindrum* lcdm, *siue* ah, *sesquialterum esse sectoris sphęrici* ecfg. *Quod &c.*

Lem. prę ced.

Lem-

tum dc: *Ergò, ſumptis conſequentium ęqualibus, erit cylindrica ſuperficies, ad curuam ſphæric ſegmenti* cde *ſuperficiẽ, vt rectangulum* ab *ad rectangulum* fdi. *Quod erat &c.*

Exemplum VII.

E*Sto ſphæra, vnà cum cylindro ſibi circumſcripto, quorum axis ſit recta* cd. *Secenturq; plano* ab *ad axem erecto. Dico cylindrum* ah, *ſeſquialterum eſſe ſolidi ſectoris ſphęrici* ecfg.

Accipiatur cl *ęqualis ipſi* cn. *& intelligatur cylindrus* lcdm, *cuius altitudo* lc. *eritque ęqualis cylindro* ah. *Concipiatur etiam demptus conus* lgm; *& ſumpto in axe* cg *quouis puncto* o *fiant æquales* go, gr, *& tranſeat per ipſum punctũ* o *ſphęrica ſuperficies* qou *in ſectore; & cylindrica* iorſ, *in ſolido cylindrico* lcdm *excauato ablatione coni* lgm*; ſit que tam ſphęricę ſuperficiei quam cylindricę tranſeuntis per* o, *diameter ipſa* or.

Iam: tota cg, *ad totam rectam* go, *est vt* eg *ad* gq, *nempe, vt ablata* ng, *ad* gp; *ideò reliqua* cn, *ad* op, *erit vt tota* cg, *ad* go*; ſiue vt* cl, *ad* oi. *Sed antecedentes ſunt æquales, ergò rectę* op, oi *ęquales erunt. Propterea rectangula* rop, roi *æqualia erunt; & ſuperficies ſphęrica.* qou *æqualis erit ſuperficiei cylindricę* iorſ. *Et hoc ſemper, vbicunque ſit punctum* o. *Ergò omnes ſuperficies omnium ſegmentorum ſphæricorum (nempe ſolidus ſector* ecfg) *æquales erunt omnibus ſuperficiebus cylindricis ſimul ſumptis, hoc eſt ſolido excauato* lcdm. *Cui ſi addatur conus iam ablatus* lgm, *patebit propoſitum: nempe cylindrum* lcdm, *ſiue* ah, *ſeſquialterum eſſe ſectoris ſphęrici* ecfg. *Quod &c.*

Lem pr e. *ced:*

Lem-

Lemma.

Arcus circulorum ab, cd. *inter se rationem habent compositam ex ratione semidiametrorum* ag *ad* cf, *& ex ratione angulorum* agb *ad* cfd. *Nam, fiat angulus* cfh *aequalis angulo* agb. *Erit igitur arcus* ab *ad* ch, *ut semidiameter* ag *ad* cf; *sed arcus* ch *ad* cd. *est ut angulus* cfh, *vel* agb, *ad angulum* cfd. *Ergò patet arcum* ab, *ad* cd, *rationem habere compositam ex rationibus semidiametrorum* ag *ad* cf, *& angulorum* agb *ad* cfd.

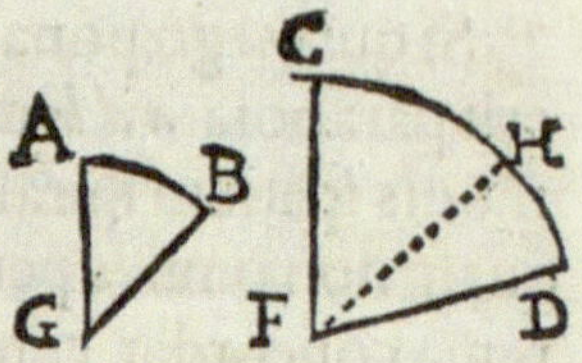

Exemplum VIII.

Esto circulus, cuius semidiameter ab *sit initium lineae spiralis* aeib. *Secetur bifariam* ab *in* c; *& erectâ perpendiculari* cd, *quantacunque, fiat per puncta* adb *parabola, cuius diameter* cd. *& centro* a, *interuallo* ac *fiat arcus* ce. *Dico spatium sub ipsâ spirali, & rectâ* ab *comprahensum, ad factā parabolam* adb, *esse ut arcus* ce, *ad rectam* cd.

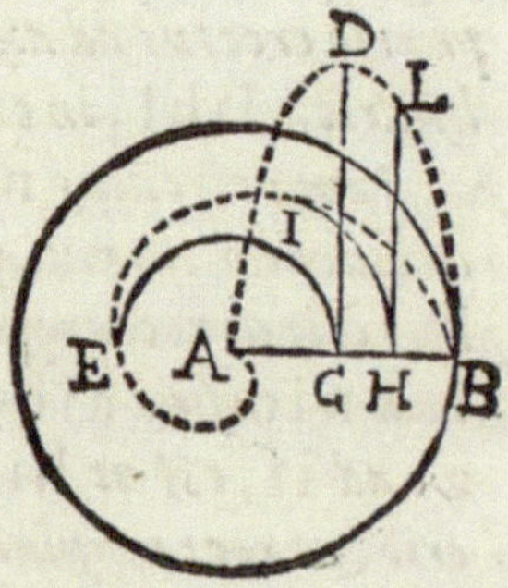

Sumatur in ab *quoduis punctum aliud à puncto* c, *puta* h; *& per* h *fiat arcus* hi *in spatio spiralis*, *& recta* hl *in parabola, ipsius diametro equidistans*.

Iam: arcus ce *ad* hi *rationem habet compositam ex rationem semidiametrorum* ca *ad* ah; *& ex ratione angulorum, siue (quod idem est ob lineam spiralem) ex ratione temporum, nempe rectae* cb *ad* bh. *Ergo arcus* ce, *ad* hi, *est ut rectangulum* acb, *ad rectangulum* ahb, *siue ut recta* cd, *ad* hl, *ob parabolam. Permutando igitur, arcus* ce, *ad rectam* cd, *est ut arcus* hi *ad rectam* hl; *& hoc modo semper;* ubicun-

℞ lema

Lemma.

A*Rcus circulorum* ab, cd. *inter ſe rationem habent compoſitam ex ratione ſemidiametrorum* a g *ad* c f, *& ex ratione angulorum* agb *ad* cfd. *Nam, fiat angulus* cfh *æqualis angulo* agb. *Erit igitur arcus* ab *ad* ch, *ut ſemidiameter* ag *ad* cf; *ſed arcus* ch *ad* cd. *eſt vt angulus* cfh, *vel* agb, *ad angulum* cfd. *Ergò patet arcum* ab, *ad* cd, *rationem habere compoſitam ex rationibus ſemidiametrorum* ag *ad* cf, *& angulorum* agb *ad* cfd.

Exemplum VIII.

E*Sto circulus, cuius ſemidiameter* ab *ſit initium lineæ ſpiralis* aeib. *Secetur bifariam* ab *in* c*; & erectà perpendiculari* cd, *quantacunque, fiat per puncta* adb *parabola, cuius diameter* cd. *& centro* a, *interuallo* ac *fiat arcus* ce. *Dico ſpatium ſub ipsà ſpirali, & rectà* ab *compræhenſum, ad factã parabolam* adb, *eſſe vt arcus* ce, *ad rectam* cd.

Sumatur in ab *quoduis punctum aliud à pũcto* c, *puta* h; *& per* h *fiat arcus* hi *in ſpatio ſpiralis, & recta* hl *in parabola, ipſius diametro equidiſtans.*

Iam: arcus ce *ad* hi *rationem habet compoſitam ex rationem ſemidiametrorum* ca *ad* ah*; & ex ratione angulorum, ſiue* (*quod idem eſt ob lineam ſpiralem*) *ex ratione temporum, nempe rectæ* cb *ad* bh. *Ergo arcus* ce, *ad* hi, *eſt vt rectangulum* acb, *ad rectangulum* ahb, *ſiue vt recta* cd, *ad* hl, *ob parabolam. Permutando igitur, arcus* ce, *ad rectam* cd, *eſt vt arcus* hi *ad rectam* hl; *& hoc modo ſemper*;. [R Lem ?]

ubicun-

102

ubicunque sit punctum h. *Ergò omnes simul arcus, siuè spatiũ spiralis, ad omnes simul rectas, nempe ad parabolam, erunt ut unus arcus* c e, *ad unam rectam* c d. *Quod erat &c.*

Si quis ergò ponat rectam *c d* ęqualem arcui semicirculi *c e*, erit parabola *a d b* æqualis spatio spirali. Quinque adhuc aliis modis spatium spiralis lineę, in parabolam transformatur, quã-quam non omnes per curua Indiuisibilia procedant. Et Theorema concordat cum 25. de lineis spiralibus Archimedis.

Exemplum IX.

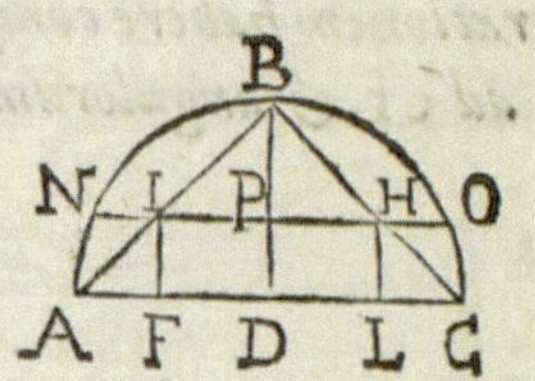

E*sto hæmisphærium* a b c, *cuius axis* b d. *conus uerò inscriptus* a b c. *Dico, hemisphærium ipsius coni esse duplum. Sumatur in rectà* a b *punctum quoduis* i; *per quod trãseat circulus* n o *in hemisphærio erectus ad axem; & superficies cylindrica* f i h l, *in cono circa axem* p d.

2. duodecimi: *Iam: circulus* n o, *ad* i h, *est ut quadratum* n p *ad* p i: & *diuidendo, armilla circularis, cuius latitudo* n i, *ad circulum* i h, *erit ut rectangulum* n i o, *ad quadratum* i p: *sed rectangu-*

16 Sexti. *lum* n i o *siue* a i b, *ęquale est rectangulo* f h (*nam. per 4. sexti,* a i *ad* i f, *est ut* h i *ad* i b.) *Ergò, armilla* n i, *ad circulum* i h,

5: l: pi de soli: sph. *erit, ut rectangulum* f h, *ad quadratum* i p; *siue ut cylindrica superficies* f i h l, *ad eundem circulum* i h. *Aequales ergò sunt armilla circularis, cuius latitudo* n i, *& superficies cylindrica* f i h l: *& hoc semper, ubicunq. sit punctum* i. *Ergò omnes simul armillę, nempe solidum hæmisphæricum excauatum dempto cono* a b c, *ęquales erunt omnibus simul superficiebus cylindricis, nẽpe ipsi cono* a b c. *Propterea coniungendo, patet hemisphærium inscripti coni duplum esse. Quod &c.*

Exemplum X,

Q*Vodlibet minus segmentum sphaericum* a b c, *aequale est conoidi cuidam hyperbolico* e d f: *eandem altitudinẽ* b d *ha-*

ubicunque ſit punctum h. *Ergò omnes ſimul arcus, ſiuè ſpatiũ ſpiralis, ad omnes ſimul rectas, nempe ad parabolam, erunt ut unus arcus* ce, *ad unam rectam* cd. *Quod erat &c.*

Si quis ergò ponat rectam *cd* ęqualem arcui ſemicirculi *ce,* erit parabola *adb* æqualis ſpatio ſpirali. Quinque adhuc alijs modis ſpatium ſpiralis lineę, in parabolam transformatur, quãquam non omnes per curua Indiuiſibilia procedant. Et Theorema concordat cum 25. de lineis ſpiralibus Archimedis.

Exemplum IX.

E*Sto hæmiſphærium* abc, *cuius axis* bd. *conus uerò inſcriptus* abc. *Dico, hemiſphærium ipſius coni eſſe duplum.*

3.*duodecimi:* *Sumatur in rectà* ab *punctum quoduis* i*; per quod trãſeat circulus* no *in hemiſphærio erectus ad axem; & ſuperficies cylindrica* fihl, *in cono circa axem* pd.

16 *Sex-. ti.* *Iam: circulus* no, *ad* ih, *est ut quadratum* np *ad* pi*: & diuidendo, armilla circularis, cuius latitudo* ni, *ad circulum* ih, *erit ut rectangulum* nio, *ad quadratum* ip*: ſed rectangulum* nio *ſiue* aib, *ęquale eſt rectangulo* fh *(nam, per 4. ſexti,* ai *ad* if, *est ut* hi *ad* ib.) *Ergò, armilla* ni, *ad circulum* ih, *erit, ut rectangulum* fh, *ad quadratum* ip*; ſiue ut cylindrica ſuperficies* fihl, *ad eundem circulum* ih.

5:1:*pi de ſolid: ſph.* *Aequales ergò ſunt armilla circularis, cuius latitudo* ni, *& ſuperficies cylindrica* fihl*: & hoc ſemper, ubicunq. ſit punctum* i. *Ergò omnes ſimul armillę, nempe ſolidum hæmiſphæricum excauatum dempto cono* abc, *ęquales erunt omnibus ſimul ſuperficiebus cylindricis, nẽpe ipſi cono* abc. *Propterеà coniungendo, patet hemiſphærium inſcripti coni duplum eſſe. Quod &c.*

Exemplum X,[168]

Q*Vodlibet minus ſegmentum ſphaericum* abc, *aequale eſt conoidi cuidam hyperbolico* edf*: eandem altitudinẽ*

bd *ha-*

[168] *Ex æquo* Proof. (Cfr. Pisano and Bussotti's chapters above).

b d *habenti, super basim uerò* e f, *equalem curuae superficiei segmenti, constituto: cuius latus uersum sit* d g, *scilicet differentia inter catetum segmenti, & radium sphaerae.*

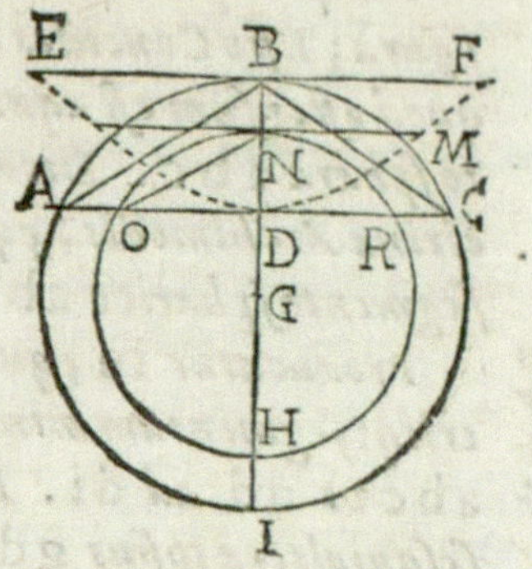

Nam, sumatur in sagitta b d *quoduis punctum* n, *per quod transeat sphaerica superficies* o n r, *priori concentrica in segmento, & circulus cuius radius* n m, *basi parallelus in conoide,*

Eritq. curua superficies a b c, *ad curuam* o n r, *ut cir ulus ex radio* a b, *ad circulum ex radio* o n, *ob aequalitatem: siue ut quadratum* a b *ad* o n, *uel ut rectangulum* i b d, *ad rectangulum* h n d. *siue, in subduplis, ut rectangulum* g b d, *ad* g n d; *siue (ob hyperbolam) ut quadratū* b f, *ad quadratum* n m; *siue ut circulus radio* b f, *ad circulum ex radio* n m. *Sed antecedentia sunt aequalia per suppositionē, ergo aequalis erit superficies curua* o n r, *circulo cuius radius* n m. *Et hoc semper; ergo omnes omnibus; nempe sphaerae segmentum minus aequale erit conoidi hyperbolico. Quod erat &c.*

Quandò verò segmentum sphæræ fuerit hemisphærium, demonstratur æquale cono, qui basim habeat, equalem curuæ superficiei hæmisphęrij, & altitudinem eandem.

Quando verò fuerit segmentum sphæræ maius, tunc ostendetur æquale duobus solidis nempe frusto cuidam recto conoidis hyperbolici, cuius maior basis sit æqualis curuæ superficiei segmenti sphærici, latus versum sit excessus sagittæ segmenti suprà radium sphæræ, altitudo verò excessus diametri suprà sagittam. Et cono cuidam, super minori basi prædicti frusti cōstituto, cum altitudine, quæ sit æqualis lateri verso ipsius frusti. Facilis demonstratio est, quamquam propositio difficilis videatur.

Concordantia praecedentis demonstrationis cum doctrina Archimedis.

Sum-

b d *habenti, ſuper baſim uerò* ef, *ęqualem curuae ſuperficiei ſegmenti, conſtituto*: *cuius latus uerſum ſit* dg, *ſcilicet differentia inter catetum ſegmenti, & radium ſphaerae.*

Nam, ſumatur in ſagitta bd *quoduis punctum* n, *per quod tranſeat ſphaerica ſuperfieies* onr, *priori concentrica in ſegmento, & circulus cuius radius* nm, *baſi parallelus in conoide,*

Eritq. *curua ſuperficies* abc, *ad curuam* onr, *ut circulus ex radio* ab, *ad circulum ex radio* on, *ob aequalitatem: ſiue ut quadratum* ab *ad* on, *uel ut rectangulum* ibd, *ad rectangulum* hnd. *ſiue, in ſubduplis, ut rectangulum* gbd, *ad* gnd*; ſiue (ob hyperbolam) ut quadratũ* bf, *ad quadratum* nm*; ſiue ut circulus radio* bf, *ad circulum ex radio* nm. *Sed antecedentia ſunt aequalia per ſuppoſitionẽ, ergo aequalis erit ſuperficies curua* onr, *circulo cuius radius* nm. *Et hoc ſemper; ergo omnes omnibus; nempe ſphaerae ſegmentum minus aequale erit conoidi hyperbolico. Quod erat &c.*

Quandò verò ſegmentum ſphæræ fuerit hemiſphærium, demonſtratur æquale cono, qui baſim habeat, ęqualem curuæ ſuperficiei hæmiſphęrij, & altitudinem eandem.

Quando verò fuerit ſegmentum ſphæræ maius, tunc oſtendetur æquale duobus ſolidis nempe fruſto cuidam recto conoidis hyperbolici, cuius maior baſis ſit æqualis curuæ ſuperficiei ſegmenti ſphærici, latus verſum ſit exceſſus ſagittæ ſegmenti ſuprà radium ſphæræ, altitudo verò exceſſus diametri ſuprà ſagittam. Et cono cuidam, ſuper minori baſi prædicti fruſti cõſtituto, cum altitudine, quæ ſit æqualis lateri verſo ipſius fruſti. Facilis demonſtratio eſt, quamquam propoſitio difficilis videatur

Concordantia praecedentis demonstrationis cum doctrina Archimedis.[169]

Sum-

[169] The Archimedean influence in Torricelli's *Opera geometrica* – as announced at the beginning – continues to be extremely evident. (Cfr. Pisano and Bussotti's chapters above).

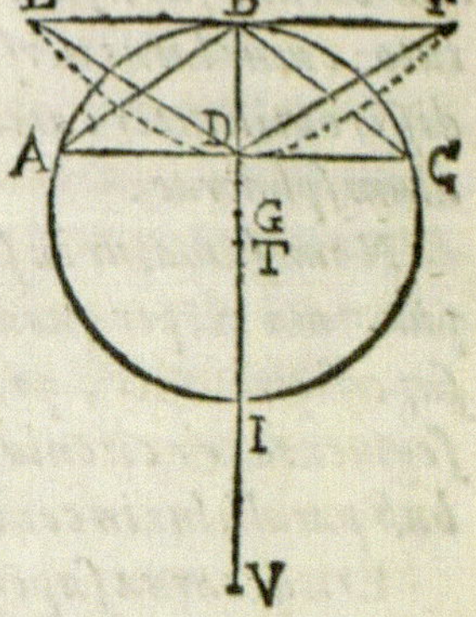

Sumpta praecedenti constructione & figura; Esto Conoides hyperbolicum e d f *quod ostensum est aequale minori segmento sphaerae* a b c. *Dico illud, etiam ex doctrina Archimedis, equale esse praedicto segmento spharico* a b c.

31: de Conoid: & Sphæroid: 21 eiusdem;

Producatur i u *equalis radio sphaerae; eritq; segmentum minus* a b c, *ad conum* a b c *vt* u d *ad* d i. *Ponatur etiam* t d. *sesquialtera ipsius* g d. *Eritq; conoides* e d f, *ad conum* e d f, *vt* t b *ad* b g.

Iam: segmentum sphaericum, ad conum suum a b c, *est vt* u d *ad* d i: *conus autem* a b c, *ad conũ aeque altum* e d f, *est vt quadratum* a d, *ad quadratum* e b; *siue ad quadr.* a b; *nempe vt rectangulum* i d b, *ad rectang.* i b d, *sicut eorumdẽ altitudines,* d i *ad* i b; *Ergò ex equo, segmentum sphaerae* a b c, *ad conum* e d f, *est ut* u d, *ad* i b: *siue (sumptis earumdem rectarum subduplis) ut* t b *ad* b g. *Nempe ut Conoides, ad eundem conum.* e d f. *Aequantur ergo segmentũ sphaerae, & ipsum Conoid: etiam ex doctrina Archimedis.*

Assumpsimus rectas *t b, b g*. esse semisses rectarum *u d, i b*, respectiuè. & hoc patet. Nàm, *u d* constat ex duabus semidiametris, & ex ipsa *g d*; sed *t b* constat ex vnica semidiametro, & semisse *d g*, ob constructionem. Reliquum manifestum est.

Latus Rectum praedicti Conoidis non est necessarium, quandoquidem datur latus Versum, & semidiameter basis, sed si quis illud requirat inueniet duplum esse lateris versi.

Exemplum XI.

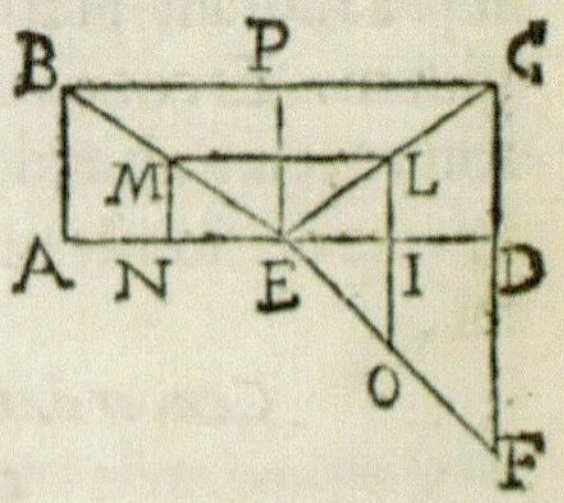

Esto cylindrus rectus a b c d, *cuius basis* a d, *axis verò* p e, *intelligaturque ablatus ab ipso conus* b e c, *ita ut relinquatur cylindrus excauatus. Producatur deinde* c d *in* f, *ita ut* d f *possit dup-*

lum

104 [*Proemium.*]

Sumpta præcedenti conſtructione & figura; *Eſto Conoides hyperbolicum* edf *quod oſtenſum eſt æquale minori ſegmento ſphæræ* abc. *Dico illud, etiam ex doctrina Archimedis, ęquale eſſe prædicto ſegmento ſphærico* abc.

Producatur iu *ęqualis radio ſphæræ; eritq; ſegmentum minus* abc, *ad conum* abc *vt* ud *ad* di. *Ponatur etiam* td. *ſeſquialtera ipſius* gd. *Eritq*; *conoides* edf, *ad conum* edf, *vt* tb *ad* bg.

31*: de Co noíd: & ſphęroid:* 27 *eiuſ-dem*;

Iam: ſegmentum ſphæricum, ad conum ſuum abc, *est vt* ud *ad* di*: conus autem* abc, *ad conũ æque altum* edf, *eſt vt quadratum* ad, *ad quadratum* eb; *ſiue ad quadr.* ab*; nempe vt rectangulum* idb, *ad rectang.* ibd, *ſiue vt eorumdẽ altitudines,* di *ad* ib*; Ergò ex ęquo, ſegmentum ſphæræ* abc, *ad conum* edf, *est ut* ud, *ad* ib*: ſiue (ſumptis earumdem rectarum ſubduplis) ut* tb *ad* bg. *Nempe ut Conoides, ad eundem conum.* edf. *Aequantur ergo ſegmentũ ſphæræ, & ipſum Conoid: etiam ex doctrinà Archimedis.*

Aſſumpſimus rectas *tb, bg.* eſſe ſemiſſes rectarum *ud, ib,* reſpectiuè. & hoc patet. Nàm, *ud* conſtat ex duabus ſemidiametris, & ex ipſa *gd*; ſed *tb* conſtat ex vnica ſemidiametro, & ſemiſſe *dg,* ob conſtructionem. Reliquum manifeſtum eſt.

Latus Rectum prædicti Conoidis non eſt neceſſarium, quandoquidem datur latus Verſum, & ſemidiameter baſis, ſed ſi quis illud requirat inueniet duplum eſſe lateris verſi.

Exemplum XI.[170]

Esto cylindrus rectus abcd, *cuius baſis* ac, *axis verò* pe, *intelligaturque ablatus ab ipſo conus* bec, *ita ut relinquatur cylindrus excauatus. Producatur deinde* cd *in* f, *ita ut* df *poſſit dup-*

lum

[170] This proof is in agreement *cum Theoremate Eiclidis* (Cfr. Pisano and Bussotti's chapters above).

lum rectanguli cde, *& iunctà* ef *conuertatur triang.* cdf *(ſaltem imaginatione, nam figura non eſt perfecta) ita vt oriatur conus, cuius baſis ſemidiameter ſit* df, *axis verò* ed. *Dico talem conum æqualem eſſe prędicto cylindro excauato. Sumatur enim in axe* ed, *quoduis punctum* i, *& per ipſum tranſeat ſuperficies cylindrica* ilmn, *circà axem* ep *in ſolido excauato cylindrico; & circulus cuius radius* io *in eo cono, qui axem habet* ed.

Iam circulus ex radio df, *ad circulum ex radio* io *eſt vt quadratum* df *ad* io, *ſiue vt quadratum* de *ad* ei, *ſiue vt rectangulum* cde *ad rectang.* lie; *Sed quadratum* df *ponitur duplum rectanguli* cde; *ergò quadratum* io *duplum erit rectanguli* lie: *& ideo æquale rectangulo* linm. *Propterea circulus ex radio* io *æqualis erit ſuperficiei cylindricæ* linm; *& hoc ſemper vbicunq; ſit punctum* i. *Ergo omnes circuli ſimul, ſiue conus cuius axis eſt* ed, *æquales erunt omnibus ſuperficiebus cylindricis ſimul, ſiue ſolido cylindrico excauato* abecd. *Quod erat &c.*

2. Duodecimi.

5. p. de ſoli: ſph.

Quod autem concordet cum Euclide l. 12. oſtenditur. Nam conus *bec*, ad conum eum qui habet axem *ed*, rationem habet compoſitam ex ratione altitudinum, nempe rectæ *pe* ad *ed*, ſiue rectanguli *ped*, ad quadratum *ed*, & ex ratione baſium, nempe quadrati *ed*, ad quadratum *df*. Ergo conus *bec*, ad conum cuius axis eſt *ed*, eſt vt rectangulum *ped*, ad quadratum *df*, nempe ſubduplus, ob conſtructionem; ſed idem conus *bec* ſubduplus eſt ſolidi excauati *abecd*, ergo etiam ex doctrina Euclidis patet ſolidum cylindricum excauatum *abecd*, æquale eſſe cono, cuius axis eſt *ed*, radius verò baſis *df*. &c.

Exemplum XII.

QVilibet cylindrus rectus ab, *cuius axis ſit* cf, *æqualis eſt conoidi parabolico, cuius altitudo ſit* cd; *ſemidiameter verò baſis ſit* de, *quæ quidem potentià ſit æqualis rectangulo* ab; *& erit circulus ex radio* de *æqualis ſuperficiei cylindricę* ab.

5. p. de ſolid. ſph.

O *Intel-*

lum rectanguli cde, *& iunctà* ef *conuertatur triang.* cdf (*ſaltem imaginatione, nam figura non est perfecta*) *ita vt oriatur conus, cuius baſis ſemidiameter ſit* df, *axis verò* ed. *Dico talem conum æqualem eſſe prędicto cylindro excauato. Sumatur enim in axe* ed, *quoduis punctum* i, *& per ipſum tranſeat ſuperficies cylindrica* ilmn, *circà axem* ep *in ſolido excauato cylindrico; & circulus cuius radius* io *in eo cono, qui axem habet* ed. *2. Duodecimi.*

Iam circulus ex radio df, *ad circulum ex radio* io *eſt vt quadratum* df *ad* io, *ſiue vt quadratum* de *ad* ei, *ſiue vt rectangulum* cde *ad rectang.* lie*; Sed quadratum* df *ponitur duplum rectanguli* cde; *ergò quadratum* io *duplum erit rectanguli* lie: *& ideo æquale rectangulo* linm. *Propterea circulus ex radio* io *æqualis erit ſuperficiei cylindricæ* linm*; & hoc ſemper vbicunq; ſit punctum* i. *Ergo omnes circuli ſimul, ſiue conus cuius axis eſt* ed, *æquales erunt omnibus ſuperficiebus cylindricis ſimul, ſiue ſolido cylindrico excauato* abecd. *Quod erat &c.* *5.p.de. ſoli: ſph.*

Quod autem concordet cum Euclide l. 12. oſtenditur. Nam conus *bec,* ad conum cum qui habet axem *ed,* rationem habet compoſitam ex ratione altitudinum, nempe rectæ *pe* ad *ed,* ſiue rectanguli *ped,* ad quadratum *ed,* & ex ratione baſium, nempe quadrati *ed,* ad quadratum *df.* Ergo conus *bec,* ad conum cuius axis eſt *ed,* eſt vt rectangulum *ped,* ad quadratum *df,* nempe ſubduplus, ob conſtructionem; ſed idem conus *bec* ſubduplus eſt ſolidi excauati *abecd,* ergo etiam ex doctrina Euclidis patet ſolidum cylindricum excauatum *abecd,* æquale eſſe cono, cuius axis eſt *ed,* radius verò baſis *d f.* &c.

Exemplum XII.

QVilibet *cylindrus rectus* ab, *cuius axis ſit* cf, *æqualis est conoidi parabolico, cuius altitudo ſit* cd*; ſemidiameter verò baſis ſit* de, *quæ quidem potentià ſit æqualis rectangulo* ab*; & erit circulus ex radio* de *æqualis ſuperficiei cylindricę* ab *5. p. de ſo lid. ſph.*

O *Intel-*

106

Intelligatur conuerti ſęmiparabola ec d *circà axem* c d, *itàvt prædictum conoides oriatur. ſumpto deinde in axe* c d, *quolibet puncto* i, *per ipſum tranſeat in cylindro ſuperficies cylindrica* il, *circà axem* cf; *at in conoide, circulus, cuius ſemidiameter ſit* ih, *baſi parallelus.*

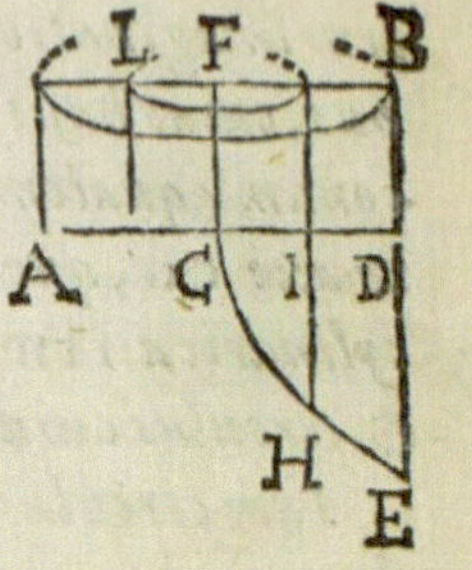

Iam: ſuperficies cylindrica ab: *ad cylindricam* il, *eſt vt rectangulum* ab, *ad rectangulum* il, *ſiue vt eorundem ſemibaſes*, dc *ad* ci, *ſiue vt quadratum* de *ad* ih; *nempe vt circulus, ex radio* de *ad circulum ex radio* ih. *Sed antecedentes ponuntur æquales, ergo etiam conſequentes; nempe ſuperficies cylindrica* il, *æqualis erit circulo ex radio* ih: *& hoc ſemper, vbicunque ſit punctum* i. *Propterea omnes cylindricæ ſimul ſuperficies omnibus circulis ęquales erunt. videlicet cylindrus conoidi. Quod &c.*

6.p.de ſo lid ſph ob para bolam.

Demonſtratur concordare cum Archimede hoc modo. cylindrus *ab* ad conum in conoide inſcriptum, rationem habet compoſitam ex ratione altitudinum, nempe ex ratione *fc* ad tertiam partem *cd*, (pro cono inſcripto, accipio cylindrum in eàdem quidem baſi, ſed cum altitudine ſubtripla) & ex ratione baſium, nempe quadrati *cd* ad *de*, ſiue quadrati *cd*, ad rectangulum *ab*, ſiue rectæ *cd* ad duplam *cf*, ſiue in ſubtriplis vt tertia pars *cd*, ad duas tertias ipſius *cf*. Propterea cylindrus *ab*, ad conum in conoide inſcriptum, erit vt *fc*, ad duas tertias ipſius *fc*, nempe ſeſquialter. Concordat itaq; cum 23. de Conoid. & ſphæroid.

Exemplum XIII.

Q*Vilibet conus rectus* abc, *cuius axis ſit* bd, *ęqualis eſt ſphæroidi, quæ axem habeat* dc, *nempe ſemidiametrum baſis coni; & ſecta* dc *bifariam in* f, *ſemidiameter ſphęroidis* fc *potentià ſit ſubdupla trianguli* abc.

Compleatur rectangulum fhlu: *erit q., ob ſuppoſitionem, recta* fc

106 [*Proemium.*]

Intelligatur conuerti ſęmiparabola ecd *circà axem* cd, *ità vt prædictum conoides oriatur. ſumpto deinde in axe* cd, *quolibet puncto* i, *per ipſum tranſeat in cylindro ſuperficies cylindrica* il, *circà axem* cf; *at in conoide, circulus, cuius ſemidiameter ſit* ih, *baſi parallelus.*

Iam: ſuperficies cylindrica ab*: ad cylindricam* il, *est vt rectangulum* ab, *ad rectangulum* il, *ſiue vt eorundem ſemibaſes,* dc *ad* ci, *ſiue vt quadratum* de *ad* ih*; nempe vt circulus, ex radio* de *ad circulum ex radio* ih. *Sed antecedentes ponuntur æquales, ergo etiam conſequentes; nempe ſuperficies cylindrica* il, *æqualis erit circulo ex radio* ih*: & hoc ſemper, vbicunque ſit punctum* i. *Propterea omnes cylindricæ ſimul ſuperficies omnibus circulis ęquales erunt. videlicet cylindrus conoidi. Quod &c.*

6. p. de ſolid. ſph ob parabolam.

Demonſtratur concordare cum Archimede hoc modo. cylindrus *ab* ad conum in conoide inſcriptum, rationem habet compoſitam ex ratione altitudinum, nempe ex ratione *fc* ad tertiam partem *cd,* (pro cono inſcripto, accipio cylindrum in eàdem quidem baſi, ſed cum altitudine ſubtripla*)* & ex ratione baſium, nempe quadrati *cd* ad *de,* ſiue quadrati *cd,* ad rectangulum *ab,* ſiue rectæ *cd* ad duplam *cf,* ſiue in ſubtriplis vt tertia pars *cd,* ad duas tertias ipſius *cf.* Propterea cylindrus *ab,* ad conum in conoide inſcriptum, erit vt *fc,* ad duas tertias ipſius *fc,* nempe ſeſquialter. Concordat itaq; cum 23. de Conoid. & ſphæroid.

Exemplum XIII.

Q*Vilibet conus rectus* abc, *cuius axis ſit* bd, *ęqualis eſt ſphæroidi, quæ axem habeat* dc, *nempe ſemidiametrum baſis coni; & ſectà* dc *bifariam in* f, *ſemidiameter ſphęroidis* fe *potentià ſit ſubdupla trianguli* abc.

Compleatur rectangulum fhlu*: eritq., ob ſuppoſitionem, re-*

cta f e

cta fc *equalis potentià rectangulo* fl *: ideoq. circulus, cuius radius* fe, *æqualis erit ſuperficiei cylindricæ, quę tranſit per* fh *circà axem* bd. 5 p. de ſolid: ſph:

Sumatur iam qdodlibet punctũ i *in axe* dc. *& per* i *tranſeat ſuperficies cylindrica* imno*: & circulus in ſphæroide, cuius radius ſit* ip. *Superficies itaque cylindrica* fl, *ad cylindricam* in, *eſt vt rectangulum* fl *ad* in. *Nempe, rationem habet cõpoſitam ex ratione* fh *ad* im, *ſiue* fc *ad* ci; *& ex ratione* fu *ad* io, *vel* fd *ad* di. 6 p: de ſolid; ſdb:

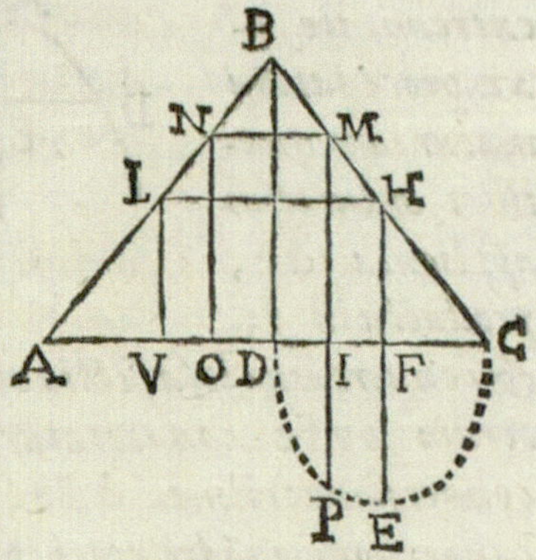

erit itaq. cylindrica fl, *ad cylindricam* in, *vt rectang.* dfc *ad rectang.* dic; *ſiue vt quadratum* fe *ad* ip, *nempe vt circulus ex radio* fe, *ad circulum ex radio* ip. *Sed antecedentes æquales ſunt, ergo etiam conſequentes: nimirum, ſuperficies cylindrica* imno, *æqualis erit circulo ex radio* ip. *& hoc ſemper, vbicunq. punctum* i. *Ergò omnes omnibus: hoc eſt conus* abc, *æqualis erit ſphæroidi prædictæ. Quod erat &c.* ob ellipſim.

Concordare cum Archimede oſtendemus. Nàm; conus *abc*, ad conum in hemiſphæroide inſcriptum, rationem habet compoſitam ex ratione baſium, nempe quadrati *dc*, ad quadratum *fe*; vel quadrati *dc* ad rectangulum *fl*; ſiue (cum rectangula habeant æqualem baſim) rectæ *dc*, ad *fh*; ſiue *ac* ad *db*. Et ex ratione altitudinum, nempe *bd* ad *df*. Erit ergò conus *abc*, ad conum in hemiſphæroide inſcriptum, vt recta *ac* ad rectam *df*, nempe quadruplus. Concordat ergò cum Prop. 29 de Conoid. & ſphæroid.

Exemplum XIV.

E*ſto parabola, vel hyperbola, vel ellipſis, vel circuli circumferentia, cuius axis* ab; *ſemilatus rectum* ac *ſit ad angulos*

cta fe *ęqualis potentià rectangulo* fl*: ideoq. circulus, cuis*[171] *radius* fe, *æqualis erit ſuperficiei cylindricæ, quę tranſit per* fh *circà axem* bd. *5. p. de ſolid: ſph:*

Sumatur iam quodlibet punctũ i *in axe* dc. *& per* i *tranſeat ſuperficies cylindrica* imno*: & circulus in ſphæroide, cuius radius ſit* ip. *Superficies itaque cylindrica* fl, *ad cylindricam* in, *est vt rectangulum* fl *ad* in. *Nempe, rationem habet cõpoſitam ex ratione* fh *ad* im, *ſiue* fc *ad* ci*; & ex ratione* fu *ad* io, *vel* fd *ad* di. *erit itaq. cylindrica* fl, *ad cylindricam* in, *vt rectang.* dfc *ad rectang.* dic*; ſiue vt quadratum* fe *ad* ip, *nempe vt circulus ex radio* fe, *ad circulum* ex *radio* ip. *Sed antecedentes æquales ſunt, ergo etiam conſequentes: nimirum, ſuperficies cylindrica* imno, *æqualis erit circulo* ex *radio* ip. *& hoc ſemper, vbicunq. punctum* i. *Ergò omnes omnibus: hoc eſt conus* abc, *æqualis erit ſphæroidi prædictæ. Quod erat &c.* *6 p: de ſolid; ſph: ob ellipſim*

Concordare cum Archimede oſtendemus. Nàm; conus *abc,* ad conum in hemiſphæroide inſcriptum, rationem habet compoſitam ex ratione baſium, nempe quadrati *dc,* ad quadratum *fe*; vel quadrati *dc* ad rectangulum *fl*; ſiue (cum rectangula habeant æqualem baſim) rectæ *dc,* ad *fh*; ſiue *ac* ad *db.* Et ex ratione altitudinum, nempe *bd* ad *df.* Erit ergò conus *abc,* ad conum in hemiſphæroide inſcriptum, vt recta *ac* ad rectam *df,* nempe quadruplus. Concordat ergò cum Prop. 29 de Conoid. & ſphæroid.

Exemplum XIV.

E*Sto parabola, vel hyperbola, vel ellipſis, vel circuli circumferentia, cuius axis* ab*; ſemilatus rectum* ac *ſit ad an-*

O 2 *gulos*

[171] *cuis*] *cuius.*

108

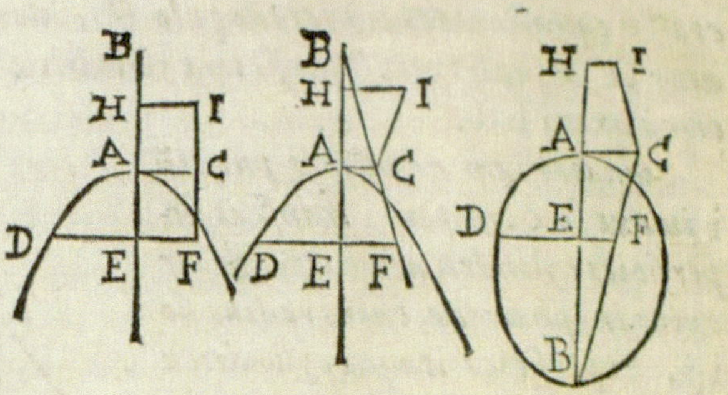

gulos rectos cum axe ab: *& coniuncta* bc *ab extremitate axis procedat. Sumatur iam quælibet ordinatim applicata* de, *producta in* f; *& conuertatur ipsa sectio conica circà axem* ac; *sed quadrilaterum* aefc *conuertatur circà* ac. *Dico solidum factum à conuersione trilinei* dae, *ęquale esse solido* efcih, *facto à conuersione quadrilateri* aefc, *circà axem* ac *reuoluti*.

Nam; cum ac *sit semilatus rectum, erit quadratum applicatæ* de, *duplum rectanguli* aef, *& ideò æquale rectangulo* hef. *Propterea, circulus, cuius radius sit* de, *æqualis erit superficiei cylindricę, quæ describitur à rectà* ef, *circà axem* ac *conuersà. Et hoc semper, vbicunq; sit punctum* e. *Ergo omnes omnibus. Nempe omnes circuli simul, siue solidum conoidale, æquale erit omnibus superficiebus cylindricis simul sumptis, nempe solido descripto à quadrilatero* aefc, *circà axem* ac *conuerso. Quod &c.*

Ɉ: p; de solid: sph:

Scholium.

SI quis verò dubitet, an præcedens Theorema concordet cum Propositionibus Archimedis, omnem dubitandi occasionem delebunt tres sequentes demonstrationes.

Concordantia pro Conoide parabolico.

Esto conoides parabolicum *abc*. Ostendit Archim. Prop 23. de Conoid. & Sphæroid. Conoides *abc*, esse sesquialterum coni *abc*.

Esto

gulos rectos cum axe ab*: & coniuncta* bc *ab extremitate axis procedat. Sumatur iam quælibet ordinatim applicata* de, *producta in* f; *& conuertatur ipſa ſectio conica circà axem* ae; *ſed quadrilaterum* aefc *conuertatur circà* ac. *Dico ſolidum factum à conuerſione trilinei* dae, *ęquale eſſe ſolido* efcih, *facto à conuerſione quadrilateri* aefc, *circà axem* a c *reuoluti.*

Nam; cum ac *ſit ſemilatus rectum, erit quadratum applicatæ* de, *duplum rectanguli* aef, *& ideò æquale rectangulo* hef. *Propterea, circulus, cuius radius ſit* de, *æqualis erit ſuperficiei cylindricę, quæ deſcribitur à rectà* ef, *circà axem* ac *conuersà. Et*
5: p; de ſo *hoc ſemper, vbicunq; ſit punctum* e. *Ergo omnes omnibus. Nempe omnes circuli ſimul, ſiue ſolidum conoidale, æquale erit omnibus*
Lid: ſph: *ſuperficiebus cylindricis ſimul ſumptis, nempe ſolido deſcripto à quadrilatero* aefc, *circà axem* ac *conuerſo. Quod &c.*

Scholium.

SI quis verò dubitet, an præcedens Theorema concordet cum Propoſitionibus Archimedis, omnem dubitandi occaſionem delebunt tres ſequentes demonſtrationes.

Concordantia pro Conoide parabolico.

Eſto conoides parabolicum *abc.* Oſtendit Archim. Prop 23. de Conoid. & Sphæroid. Conoides *abc,* eſſe ſeſquialterum coni *abc*
Esto

Eſto ſolidum, quale diſcriptum eſt à quadrilatero *d b h g*, in pręcedenti conſtructione; quod quidem ſolidum in parabola, erit cylindrus. Secetur in tres partes ęquales tàm *b h*, quàm etiam *b d*. Eritq; conus *a b c* ęqualis cylindro ſuper eàdem baſi *a c* conſtituto, ſub altitudine verò *d l*; conſiderabimusq; cylindrum hunc, pro dicto cono *a b c*.

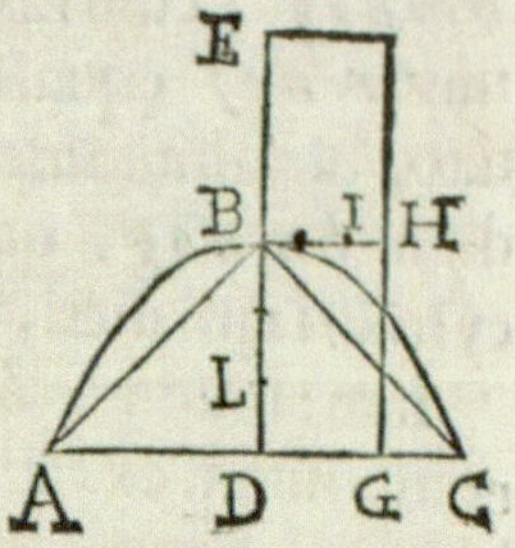

Iam: cylindrus *g e*, ad conum *a b c*, ſiue ad cylindrum eius vicarium, rationem habet compoſitam, ex ratione altitudinum *h b* ad *l d*. & ex ratione baſium, nempe circuli *e d*, ad circulum *a c*, ſiue quadrati *b d*, ad *d a*; ſiue rectę *b d*, ad duplam *b h*. cum enim *b h* ſit ſemilatus rectum, erit quadratum *a d* ęquale rectangulo ſub *b d*, & dupla *b h*) ſiue in ſubtriplis, rectę *l d* ad *b i* duas tert. ipſius *b h*. Eſt ergò cylindrus *g e*, ad conum *a b c*, vt *h b* ad *b i*; nempe ſeſquialter. Quod concludit etiam Archimedes de Conoide parabolico.

Pro Conoide Hyperbolico.

Eſto deinde conoides hyperbolicum *a b c*, cuius latus verſum *b e*; ſitq; *f b* ſeſquialtera ipſius *b e*. Oſtendit Archimedes Prop: 27. de Conoid. & ſphęroid. quòd conoides *a b c*, ad conum *a b c*, eſt vt *f g* ad *g e*. Dico etiam ſolidum *h m n o g* genitum in Exemplo 14. ad conum *a b c* eſſe, vt *f g*, ad *g e*.

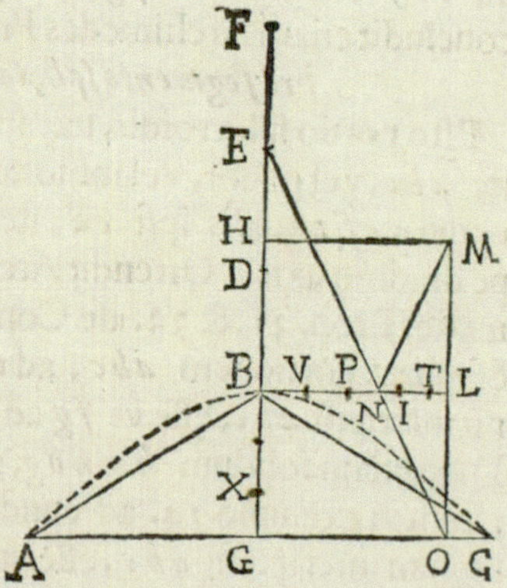

Secentur in tres partes æquales rectæ *b g*, *b n*, *n l*, eritq; conus *a b c* ęqualis cylindro cuidam, cuius baſis ſit eadem *a c*, altitudo verò ſubtripla, nempe *g x*. At ſolidum

h m n o g

[*Proemium.*] 109

Efto ſolidum, quale diſcriptum eſt à quadrilatero *dbhg,* in pręcedenti conſtructione; quod quidem ſolidum in parabola, erit cylindrus. Secetur in tres partes ęquales tàm *bh,* quàm etiam *bd.* Eritq; conus *abc* ęqualis cylindro ſuper eàdem baſi *ac* conſtituto, ſub altitudine verò *dl*; conſiderabimusq; cylindrum hunc, pro dicto cono *abc.*

Iam: cylindrus *ge,* ad conum *abc,* ſiue ad cylindrum eius vicarium, rationem habet compoſitam, ex ratione altitudinum *hb* ad *ld.* & ex ratione baſium, nempe circuli *ed,* ad circulum *ac,* ſiue quadrati *bd,* ad *da*; ſiue rectę *bd,* ad duplam *bh.* cum enim *bh* ſit ſemilatus rectum, erit quadratum *ad* ęquale rectangulo ſub *bd,* & dupla *bh*) ſiue in ſubtriplis, rectę *ld* ad *bi* duas tert. ipſius *bh.* Eſt ergò cylindrus *ge,* ad conum *abc,* vt *hb* ad *bi*; nempe ſeſquialter. Quod concludit etiam Archimedes de Conoide parabolico.

Pro Conoide Hyperbolico.

Eſto deinde conoides hyperbolicum *abc,* cuius latus verſum *be*; ſitq; *fb* ſeſquialtera ipſius *be.* Oſtendit Archimedes Prop: 27. de Conoid. & ſphęroid. quòd conoides *abc,* ad conum *abc,* eſt vt *fg* ad *ge.* Dico etiam ſolidum *hmnog* genitum in Exemplo 14. ad conum *abc* eſſe, vt *fg,* ad *ge.*

Secentur in tres partes æquales rectæ *bg, bn, nl,* eritq; conus *abc* ęqualis cylindro cuidam, cuius baſis ſit eadem *ac,* altitudo verò ſubtripla, nempe *gx.* At ſolidum

hmnog

110

hmnog (cum nil aliud sit, nisi cylindrus quidam cui deest conus *mno*) ęquale erit cylindro super eadem basi *hg* constituto, cum altitudine verò *bt*. Considerabimus igitur tam solidum *hmnog*, quàm etiam conum *abc*, tanquam si essent cylindri iam dicti, eorumdem solidorum vicarij.

Iam: solidum *hmnog*, ad conum *abc*, rationem habet compositam ex ratione altitudinum *bt* ad *gx*, & ex ratione basium, nempe circuli *hg*, ad circulum *ac*; siue quadrati *bg*, ad quadratum *ag*; siue rectæ *bg*, ad duplam ipsius *go* (cum enim *bn*. sit semilatus rectum, erit quadratum *ag* æquale rectangulo sub *bg*, & dupla ipsius *go*.) siue, sumptis subtriplis, vt *gx*, ad duas tertias ipsius *go*, vel ad duas tertias *bl*. Erit ergò solidum *hmnog*, ad conum *abc*, vt *bt* ad *tu*. Quod memento.

Recta *be* ad *eg*, est vt *bn* ad *go*, siue vt *bn* ad *bl*, siue (in subsesquialteris) vt *nu* ad *ut*. Sumptis ergo antecedentium dimidijs, erit *fe* ad *eg*, vt *bu*, ad *ut*. & componendo, *fg*, ad *ge*, vt *bt*, ad *tu*. Propterea solidum *hmnog*, ad conum *abc*, (quod iàm ostendimus esse vt *bt*, ad *tu*) erit etiam vt *fg*, ad *ge*. Quod prorsus de conoide concludit etiam Archimedes Prop. 27. de Conoid. & sphęroid.

Pro segmento sphęroidali, vel sphærico.

Esto portio sphæroidis, siue sphęræ *abc*, vel maior, vel minor; ponaturq; *ef* æqualis ipsi *ed*; nempe dimidio axis. Ostendit Archimedes Prop. 31. & 33. de Conoi. & sphær. portionem *abc*, ad conũ inscriptũ *abc*, esse vt *fg* ad *ge*. Dico etiam solidum *hmnog*, genitum in exemplo 14. ad eundem conum inscriptum *abc*, esse vt *fg* ad *ge*. Secentur in tres partes æquales, rectæ *bg*, *bl*, *ln*. eritque conus *abc* æqualis cylindro, cuius basis

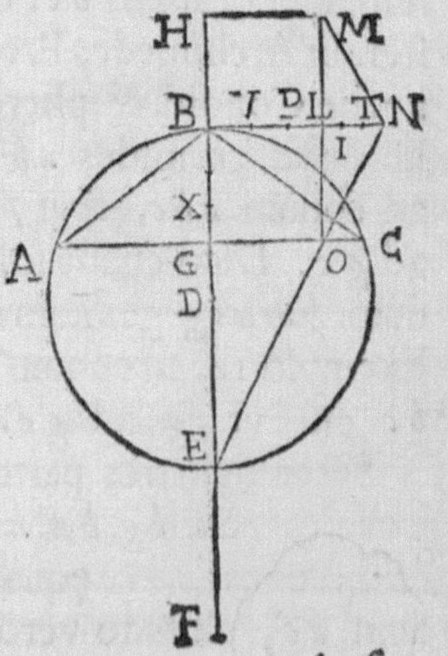

hmnog (cum nil aliud ſit, niſi cylindrus quidam cui deeſt conus *mno*) ęquale erit cylindro ſuper eadem baſi *hg* conſtituto, cum altitudine verò *bt*. Conſiderabimus igitur tam ſolidum *hmnog*, quàm etiam conum *abc*, tanquam ſi eſſent cylindri iam dicti, eorumdem ſolidorum vicarij.

Iam: ſolidum *hmnog*, ad conum *abc*, rationem habet compoſitam ex ratione altitudinum *bt* ad *gx*, & ex ratione baſium, nempe circuli *hg*, ad circulum *ac*; ſiue quadrati *bg*, ad quadratum *ag*; ſiue rectæ *bg*, ad duplam ipſius *go* (cum enim *bn*. ſit ſemilatus rectum, erit quadratum *ag* æquale rectangulo ſub *bg*, & dupla ipſius *go*.) ſiue, ſumptis ſubtriplis, vt *gx*, ad duas tertias ipſius *go*, vel ad duas tertias *bl*. Erit ergò ſolidum *hmnog*, ad conum *abc*, vt *bt* ad *tu*. Quod memento.

Recta *be* ad *eg*, eſt vt *bn* ad *go*, ſiue vt *bn* ad *bl*, ſiue (in ſubſeſquialteris) vt *nu* ad *ut*. Sumptis ergo antecedentium dimidijs, erit *fe* ad *eg*, vt *bu*, ad *ut*. & componendo, *fg*, ad *ge*, vt *bt*, ad *tu*. Propterea ſolidum *hmnog*, ad conum *abc*, (quod iàm oſtendimus eſſe vt *bt*, ad *tu*) erit etiam vt *fg*, ad *ge*. Quod prorſus de conoide concludit etiam Archimedes Prop. 27. de Conoid. & ſphęroid.

Pro ſegmento ſphęroidali, vel ſphærico.

Eſto portio ſphæroidis, ſiue ſphęræ *abc*, vel maior, vel minor; ponaturq; *ef* æqualis ipſi *ed*; nempe dimidio axis. Oſtendit Archimedes Prop. 31. & 33. de Conoi. & ſphær. portionem *abc*, ad conũ inſcriptũ *abc*, eſſe vt *fg* ad *ge*. Dico etiam ſolidum *hmnog*, genitum in exemplo 14. ad eundem conum inſcriptum *abc*, eſſe vt *fg* ad *ge*. Secentur in tres partes æquales, rectæ *bg*, *bl*, *ln*. eritque conus *a b c* æqualis cylindro, cuius

basis

basis eadem sit cum cono, nempe *ac*; altitudo autem subtripla, nempe *gx*. Solidum verò *hmnog*, quia componitur ex cylindro *hmog*, & ex cono *mno*, æquale erit cylindro super eadem basi *hg* constituto, cum altitudine *bi*. Considerabimus igitur tam solidum *hmnog*, quàm etiam conum *abc*, tanquã si essent cylindri iam dicti eorundem solidorum vicarij.

Iam: solidum *hmnog*, ad conum *abc*, rationem habet compositam ex ratione altitudinum *bi* ad *gx*; & ex ratione basium, nempe circuli *hg* ad *ac*, siue quadrati *bg* ad quadratum *ga*, siue rectæ *bg* ad duplam ipsius *go*. (cum enim *bn*, sit semilatus rectum, erit quadratum *ag* æquale rectangulo sub *bg*, & dupla *go*) siue sumptis subtriplis, vt *gx* ad duas tertias ipsius *go*, vel ad duas tertias *bl*. Ergo solidum *hmnog* ad conum *abc*, erit vt *ib* ad *bp*. quod memento.

Recta *bg* ad *ge*, est vt *no* ad *oe*, siue vt *nl* ad *lb*, siue (in subsesquialteris) vt *tl* ad *lu*; Componendo autem *be* ad *eg*, erit vt *tu* ad *ul*; sumptisq; antecedentium dimidijs, erit *fe* ad *eg*, vt *ip* ad *ul*, siue ad *pb*: Et componendo, *fg* ad *ge*, erit vt *ib* ad *bp*. Propterea, solidum *hmnog*, ad conum *abc* (quod iam ostendimus esse: vt *ib* ad *bp*) erit etiam vt *fg* ad *ge*. Quod prorsus de portione sphæroidis concludit etiam Archimedes Prop. 31 & 33. de Conoid. & sphæroidibus.

Plura adhuc exhibere poteram exempla demonstrationum per Indiuisibilia curua procedentium, nisi superflua, immò etiã & molesta existimassem. Hoc vnum admoneo lectorem, in magna parte præcedentium Theorematum me facilitatis gratia fecisse casum Propositionis particularem, cum tamen facere potuissem vniuersalissimum. Exempli causa. Poteram (in figura primi exempli) supponere tangentem *bc* cuiuscunq; longitudinis, & deinde ostendere ita esse circulum ad triangulum, vt peripharia ad tangentem: sed faciliorem conclusionem iudicaui æqualitatem inferre, quàm proportionalitatem; præsertim cum in solido Hyperbolico de æqualitate tantùm ratio habeatur. Si itaq; Corollaria limitata plerumq; demonstraui, vice Theorematum vniuersalium, scias datà operà factum esse.

Defi-

bafis eadem fit cum cono, nempe *a c*; altitudo autem fubtripla, nempe *gx*. Solidum verò *hmnog,* quia componitur ex cylindro *hmog,* & ex cono *mno,* æquale erit cylindro fuper eadem bafi *hg* conftituto, cum altitudine *bi.* Confiderabimus igitur tam folidum *hmnog,* quàm etiam conum *abc,* tanquã fi effent cylindri iam dicti eorumdem folidorum vicarij.

Iam: folidum *hmnog,* ad conum *abc,* rationem habet compofitam ex ratione altitudinum *bi* ad *gx*; & ex ratione bafium, nempe circuli *hg* ad *ac,* fiue quadrati *bg* ad quadratum *ga,* fiue rectæ *bg* ad duplam ipfius *go.* (cum enim *bn,* fit femilatus rectum, erit quadratum *ag* æquale rectangulo fub *bg,* & dupla *go)* fiue fumptis fubtriplis, vt *gx* ad duas tertias ipfius *go,* vel ad duas tertias *bl.* Ergo folidum *hmnog* ad conum *abc,* erit vt *ib* ad *bp.* quod memento.

Recta *bg* ad *ge,* eft vt *no* ad *oe,* fiue vt *nl* ad *lb,* fiue *(*in fubfefquialteris*)* vt *tl* ad *lu*; Componendo autem *be* ad *eg,* erit vt *tu* ad *ul*; fumptifq; antecedentium dimidijs, erit *fe* ad *eg,* vt *ip* ad *ul,* fiue ad *pb*: Et componendo, *fg* ad *ge,* erit vt *ib* ad *bp.* Propterea, folidum *hmnog,* ad conum *abc (*quod iam oftendimus effe: vt *ib* ad *bp*) erit etiam vt *fg* ad *ge.* Quod prorfus de portione fphæroidis concludit etiam Archimedes Prop. 31 & 33. de Conoid. & fphæroidibus.

Plura adhuc exhibere poteram exempla demonftrationum per Indiuifibilia curua procedentium, nifi fuperflua, immò etiã & molefta exiftimaffem. Hoc vnum admoneo lectorem, in magna parte præcedentium Theorematum me facilitatis gratia feciffe cafum Propofitionis particularem, cum tamen facere potuiffem vniuerfaliffimum. Exempli caufa. Poteram *(*in figura primi exempli*)* fupponere tangentem *bc* cuiufcunq; longitudinis, & deinde oftendere ita effe circulum ad triangulum, vt periphæria ad tangentem: fed faciliorem conclufionem iudicaui æqualitatem inferre, quàm proportionalitatem; præfertim cum in folido Hyperbolico de æqualitate tantùm ratio habeatur.
Si itaq; Corollaria limitata plerumq; demonftraui, vice Theorematum universalium, fcias datà operà factum effe.

Defi-

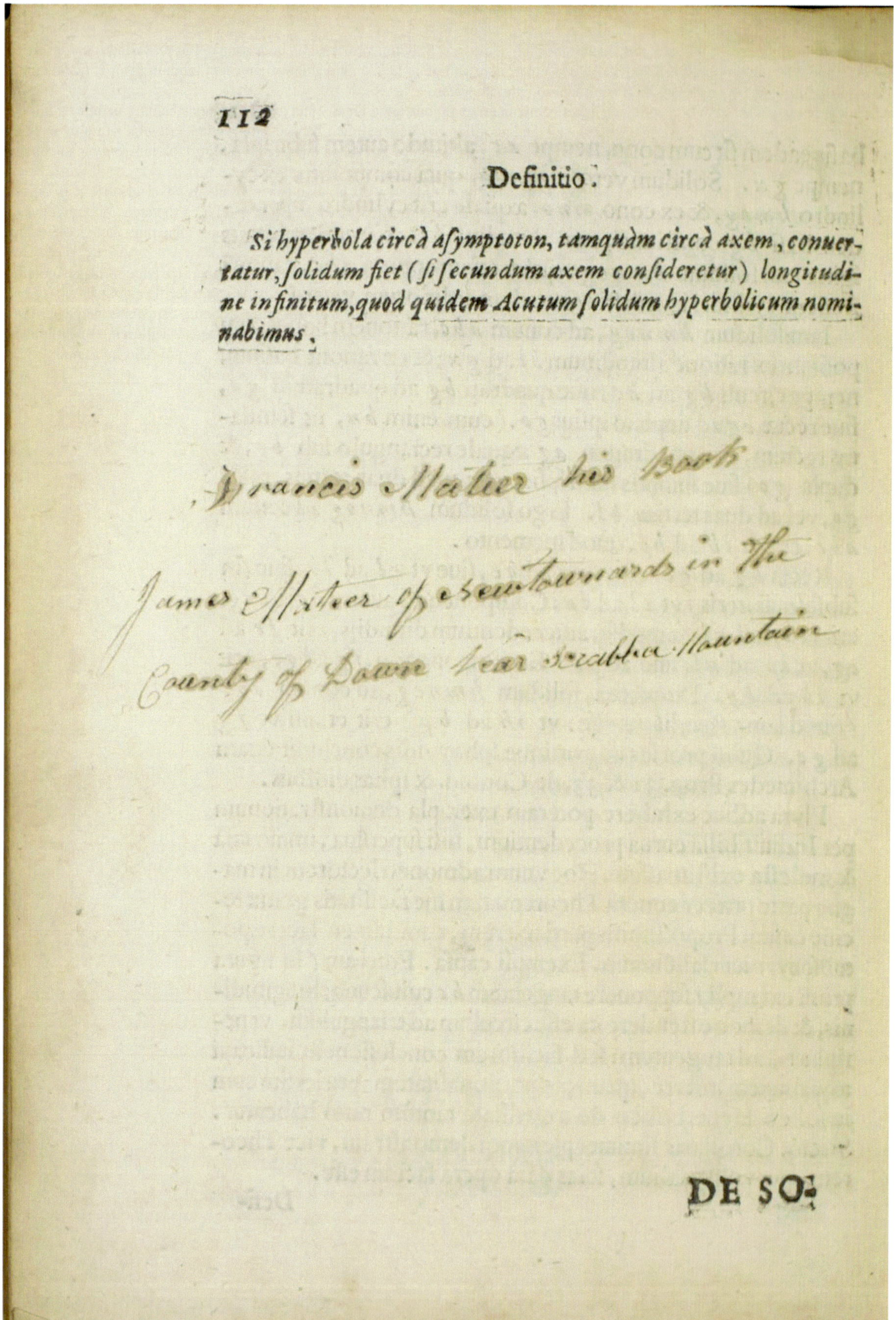

112

Definitio.

Si hyperbola circà asymptoton, tamquàm circà axem, conuertatur, solidum fiet (si secundum axem consideretur) longitudine infinitum, quod quidem Acutum solidum hyperbolicum nominabimus.

Francis Mater his Book

James Mater of Newtownards in the County of Down Near Scrabba Mountain

DE SO-

112 [*Proemium.*]

Definitio.

Si hyperbola circà aſymptoton, tamquàm circà axem, conuertatur, ſolidum fiet (ſi ſecundum axem conſideretur) longitudine infinitum, quod quidem Acutum ſolidum hyperbolicum nominabimus

DE SO-

DE SOLIDO

Hiperbolico Acuto.

Lemma Primum.

ESTO hyperbola, cuius asymptoti sint ab, ac, *angulum rectum continentes; & reuolutà figurà circà axem* ab, *factum supponatur solidum acutum hyperbolicum infinitè longũ versus* b; *quemadmodum definitum est. Intelligatur iàm intrà ipsum acutum solidum, rectangulum aliquod per axem* ab *ductum, puta* defg. *Dico hoc rectangulum æquale esse quadrato semiaxis ipsius hyperbolæ.*

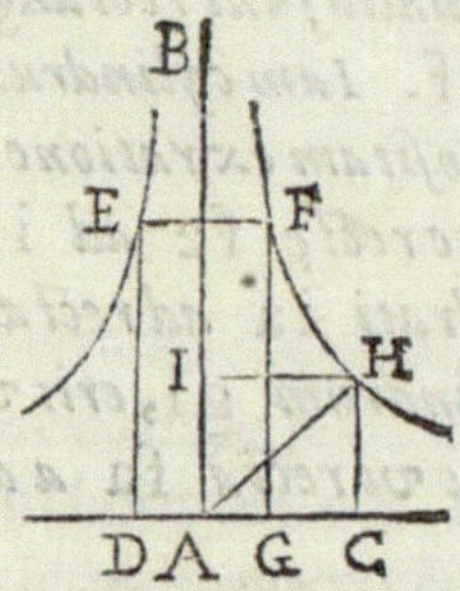

Ducatur ex a *centro hyperbolæ, semiaxis* ah, *qui angulum* bac *bifariam secabit; fiatq; rectang.* aihc; *quod omninò quadratum erit (nam cum rectangula figura sit, angulus* a *bifariam ab axe* ah *diuiditur). Ergò quadratum rectæ* ah *duplum erit quadrati* aihc, *siue duplum rectanguli* af. *& ideò æquale rectangulo* defg, *Quod erat propositum &c.* ob 12. secundi Conic.

Lemma II.

OMnes cylindri, circà cõmunem axem intrà solidum acutum hyperbolicum descripti, isoperimetri sunt. intellige semper sine basibus. Esto acutum solidum, cuius axis ab, *& intra ipsum intelligantur descripti circà cõmunem axem* ab *quotlibet cylindri* cdef, ghli. *Eruntq; æqualia rectã-*

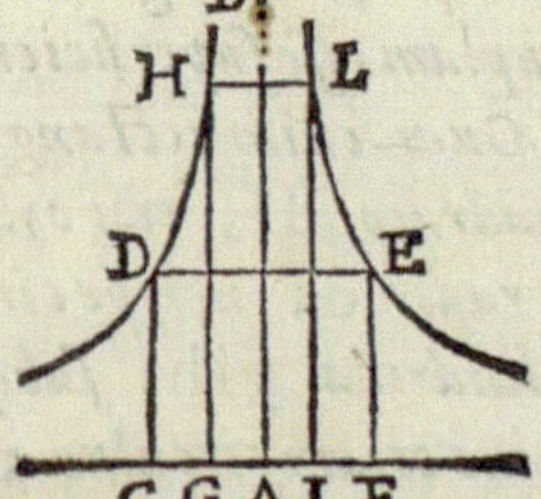

P gula

DE SOLIDO.
Hiperbolico Acuto.

Lemma Primum.

E*STO hyperbola, cuius aſymptoti ſint* ab, ac, *angulum rectum continentes; & reuolutà figurà circà axem* ab, *factum ſupponatur ſolidum acutum hyperbolicum infinitè longũ verſus* b; *quemadmodum definitum est. Intelligatur iàm intrà ipſum acutum ſolidum, rectangulum aliquod per axem* ab *ductum, puta* defg. *Dico hoc rectangulum æquale eſſe quadrato ſemiaxis ipſius hyperbolæ.* *ob* 12. *ſe-*.

Ducatur ex a *centro hyperbolæ, ſemiaxis* ah, *qui angulum* bac *bifariam ſecabit; fiatq; rectang.* aihc; *quod omninò quadratum erit (nam cum rectangula figura ſit, angulus* a *bifariam ab axe* ah *diuiditur). Ergò quadratum rectæ* ah *duplum erit quadrati* aihc, *ſiue duplum rectanguli* af. *& ideò ęquale rectangulo* defg, *Quod erat propoſitum &c.* *cunds Con* [-] *ic.*

Lemma II.

O*Mnes cylindri, circà cõmunem axem intrà ſolidum acutum hyperbolicum deſcripti, iſoperimetri ſunt. intellige ſemper ſine baſibus. Esto acutum ſolidum, cuius axis* ab, *& intra ipſum intelligantur deſcripti circà cõmunem axem* ab *quotlibet cylindri* cdef, ghli. *Eruntq; æqualiarectã-*

P *gula*

12 secundi Conic. 6 p. de solid. sph. gula per axem ce, gl. ergò æquales erunt etiam curuæ cylindrorum superficies. Quod erat.

Lemma III.

OMnes isoperimetri cylindri (cuiusmodi sunt illi, qui in acuto solido hyperbolico describuntur) inter se sunt vt diametri suarum basium. Quoniam enim, in præcedenti figurà, æqualia sunt rectangula ac, al; erit vt fe ad il, ita ia ad af. Iam cylindrus ce ad cylindrum gl, rationem habet compositam ex ratione quadrati fa ad quadratum ai; & ex ratione rectę fe ad il; siue ex ratione rectæ ia ad af, vel quadrati ia ad rectangulum iaf. Propterea cylindrus ce ad cylindrum gl, erit vt quadratum fa ad rectangulum iaf; nempe vt rectà fa ad ai. Quod &c.

Lemma IV.

ESto solidum Acutum abc, cuius axis db, & centrum hyperbolæ sit punctum d. in quo scilicet asymptoti conueniunt. axis autem hyperbolę sit df. Intelligatur ex centro d, ad interuallum df descripta sphæra aefc, quæ maxima erit omnium intra acutũ solidum describtibilium ex centro d. Sumptoq; cylindro quocunque intrà acutum solidum descripto, puta gihl. Dico cylindri gh superficiem subquadruplam esse superficiei sphæræ aefc.

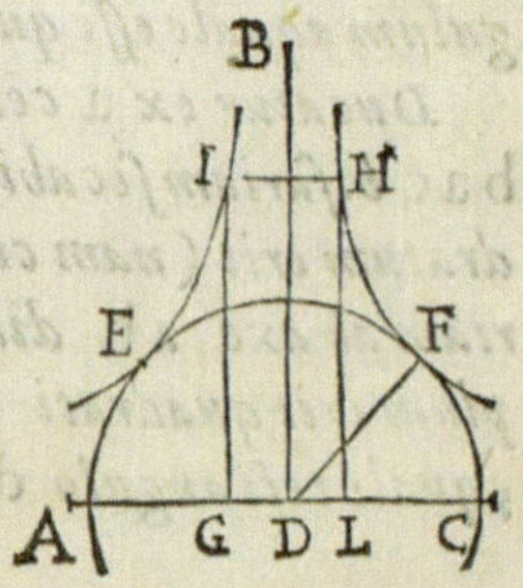

Cum enim rectangulum gh per axem cylindri, ęquale sit quadrato df, erit cylindrica superficies æqualis circulo qui fit ex radio df nempe circulo aefc: Propterea eadem superficies cylindrica gihl subquadrupla erit superficiei sphæræ aefc, cuius etiam circulus aefc subquadruplus est. Quod &c.

4. de sol. sphær.

Lem-

12 *ſecun di Conic.*
6 *p. de ſo lid. ſph.*

gula per axem ce, gl. *ergò æquales erunt etiam curuæ cylindrorum ſuperficies. Quod erat.*

Lemma III.

O*Mnes iſoperimetri cylindri (cuiuſmodi ſunt illi, qui in acuto ſolido hyperbolico deſcribuntur) inter ſe ſunt vt diametri ſuarum baſium. Quoniam enim, in præcedenti figurà, æqualia ſunt rectangula* ae, al; *erit vt* fe *ad* il, *ita* ia *ad* af. *Iam cylindrus* ce *ad cylindrum* gl, *rationem habet compoſitam ex ratione quadrati* fa *ad quadratum* ai; *& ex ratione rectę* fe *ad* il; *ſiue ex ratione rectæ* ia *ad* af, *vel quadrati* ia *ad rectangulum* iaf. *Propterea cylindrus* ce *ad cylindrum* gl, *erit vt quadratum* fa *ad rectangulum* iaf; *nempe vt rectà* fa *ad* ai. *Quod &c.*

Lemma IV.

E*Sto ſolidum Acutum* abc, *cuius axis* db, *& centrum hyperbolæ ſit punctum* d. *in quo ſcilicet aſymptoti conueniunt. axis autem hyperbolę ſit* df. *Intelligatur ex centro* d, *ad interuallum* df *deſcripta ſphæra* aefc, *quæ maxima erit omnium intra acutũ ſolidum deſcriptibilium ex centro* d. *Sumptoq; cylindro quocunque intrà acutum ſolidum deſcripto, puta* gihl. *Dico cylindri* gh *ſuperficiem ſubquadruplam eſſe ſuperficiei ſphæræ* aefc.

4*: de ſol. ſphær.*

Cum enim rectangulum gh *per axem cylindri, ęquale ſit quadrato* df, *erit cylindrica ſuperficies æqualis circulo qui ſit ex radio* df *nempe circulo* aefc*: Propterea eadem ſuperficies cylindrica* gihl *ſubquadrupla erit ſuperficiei ſphæræ* aefc, *cuius etiam circulus* aefc *ſubquadruplus eſt. Quod &c.*

Lem-

Lemma V.

Cuiuscunque cylindri ghil *intra solidum acutum descripti (vt in præcedenti figura) superficies sine basibus æqualis est circulo cuius semidiameter sit linea* df. *nempe semiaxis, siue semilatus versum ipsius hyperbolæ. Hoc enim in ipso progressu præcedentis lemmatis demonstratum est.*

Theorema.

Solidum acutum hyperbolicum infinitè longum, sectum plano ad axem erecto, vnà cum cylindro suæ basis, æquale est cylindro cuidam recto, cuius basis diameter sit latus versum, siue axis hyperbolæ, altitudo verò sit æqualis semidiametro basis ipsius acuti solidi.

Esto hyperbola cuius asymptoti *ab*, *ac* angulum rectum contineant; sumptoq; in hyperbola quolibet puncto *d*, ducatur *dc* æquidistans ipsi *ab*, & *dp* æquidistans *ac*. Tũ conuertatur vniuersa figura circa axẽ *ab*. ità vt fiat solidum acutum hyperbolicum *ebd*, vnà cum cylindro suæ basis *fedc*. Poducatur *ba* in *h*. ita vt *ah*. æqualis sit integro axi, siue lateri verso hyperbolæ. Et circa diametrum *ah* intelligatur circulus erectus ad asymptoton *ac*: & super basi *ah* concipiatur cylindrus rectus *acgh*, cuius altitudo sit *ac*, nempe semidiameter basis acuti solidi. Dico solidum vniuersum *febdc*, quanquam sine fine longum, æquale tamen esse cylindro *acgh*.

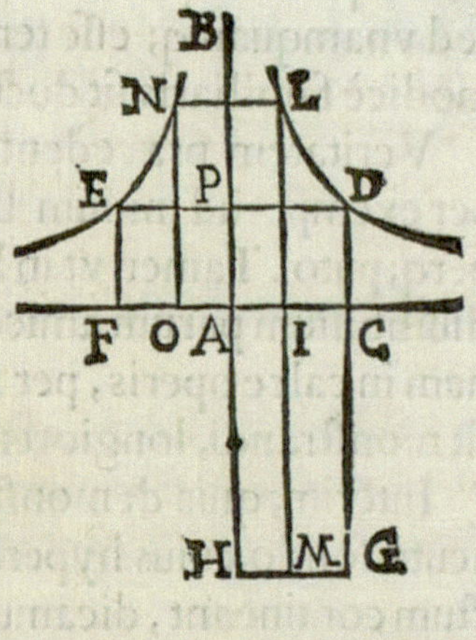

Accipiatur in recta *ac* quodlibet punctum *i*, & per *i* intelligatur ducta superficies cylindrica *onli* in solido acuto

Lemma V.

C*Viuſcunque cylindri* ghil *intra ſolidum acutum deſcripti (vt in prędecenti figura) ſuperficies ſine baſibus æqualis eſt circulo cuius ſemidiameter ſit linea* df. *nempe ſemiaxis, ſiue ſemilatus verſum ipſius hyperbolæ. Hoc enim in ipſo progreſſu præcedentis lemmatis demonſtratum eſt.*

Theorema.

SOlidum acutum hyperbolicum infinitè longum, ſectum plano ad axem erecto, vnà cum cylindro ſuæ baſis, æquale eſt cylindro cuidam recto, cuius baſis diameter ſit latus verſum, ſiue axis hyperbolæ, altitudo verò ſit æqualis ſemidiametro baſis ipſius acuti ſolidi.

Eſto hyperbola cuius aſymptoti *ab, ac* angulum rectum contineant; ſumptoq; in hyperbola quolibet puncto *d,* ducatur *dc* æquidiſtans ipſi *ab,* & *dp* æquidiſtans *ac.* Tũ conuertatur vniuerſa figura circa axẽ *ab.* ità vt fiat ſolidum acutum hyperbolicum *ebd,* vna cum cylindro ſuæ baſis *fedc.* Poducatur[172] *ba* in *h.* ita vt *ah.* æqualis ſit integro axi, ſiue lateri verſo hyperbolæ. Et circa diametrum *ah* intelligatur circulus erectus ad aſymptoton *ac*: & ſuper baſi *ah* concipiatur cylindrus rectus *acgh,* cuius altitudo ſit *ac,* nempe ſemidiameter baſis acuti ſolidi. Dico ſolidum vniuerſum *febdc,* quanquam ſine fine longum, æquale tamen eſſe cylindro *acgh.*

Accipiatur in recta *ac* quodlibet punctum *i,* & per *i* intelligatur ducta ſuperficies cylindrica *onli* in ſolido acuto

[172] Poducatur] Producatur.

compræhensa circa axem *ab*: item circulus *im* in cylindro *acgh* æquidistans basi *ah*.

Erit ergo prædicta superficies cylindrica *onli* ad circulum *im*, vt rectangulum per axem *ol*, ad quadratum radij circuli *im*; nempe vt rectangulum *ol*, ad quadratum semiaxis hyperbolæ; & ideo æqualis ex lemmate. Et hoc semper verum erit, vbicunq; sumatur punctum *i*. Propterea omnes simul superficies cylindricæ, hoc est ipsum solidum acutum *ebd*, vna cum cylindro basis *fedc*, æquale erit omnibus circulis simul, hoc est cylindro *acgh*. Quod erat &c.

Scholium.

Incredibile videri potest, cum solidum hoc infinitam longitudinem habeat, nullam tamen ex illis superficiebus cylindricis quas nos consideramus, infinitam longitudinem habere; sed vnamquamq; esse terminatam; vt vnicuiq; patebit, cui vel modicè familiaris sit doctrina Conicorum.

Veritatem præcedentis Theorematis satis per se claram, & per exempla ad initium libelli proposita confirmatam satis superq; puto. Tamen vt in hac parte satisfaciam lectori etiam Indiuisibilium parùm amico, iterabo hanc ipsam demonstrationem in calce operis, per solitam veterum Geometrarum viam demonstrandi, longiorem quidem, sed non ideo mihi certiorē.

Interim, quia demonstrationes exhibebimus de illo tantùm acuto solido, cuius hyperbolæ genitricis asymptoti angulum rectum contineant, dicamus hic obitèr, omissà demonstratione, quibus figuris æqualia sint acuta solida; quando asymptoton angulus obtusus fuerit, vel acutus.

Demonstrationes, quas ad euitandam molem prętcrimus, sibi lector industrius facili negotio comparabit.

Esto hyperbola cuius asymptoti *ab*, *ac* angulum obtusū contineant; & reuoluta figura circa axem *ab* fiat solidum acutum

compræhenſa circa axem *ab*: item circulus *im* in cylindro *acgh* æquidiftans baſi *ah*.

Erit ergo prædicta ſuperficies cylindrica *onli* ad circulum *im*, vt rectangulum per axem *ol*, ad quadratum radij circuli *im*; nempe vt rectangulum *ol*, ad quadratum ſemiaxis hyperbolæ; & ideo æqualis ex lemmate. Et hoc ſemper verum erit, vbicunq; ſumatur punctum *i*. Propterea omnes ſimul ſuperficies cylindricæ, hoc eft ipſum ſolidum acutum *ebd*, vna cum cylindro baſis *fedc*, æquale erit omnibus circulis ſimul, hoc eft cylindro *acgh*. Quod erat &c.

Scholium.

Incredibile videri poteft, cum ſolidum hoc infinitam longitudinem habeat, nullam tamen ex illis ſuperficiebus cylindricis quas nos conſideramus, infinitam longitudinem habere; ſed vnamquamq; eſſe terminatam; vt vnicuiq; patebit, cui vel modicè familiaris ſit doctrina Conicorum.

Veritatem præcedentis Theorematis ſatis per ſe claram, & per exempla ad initium libelli propoſita confirmatam ſatis ſuperq; puto. Tamen vt in hac parte ſatisfaciam lectori etiam Indiuiſibilium parùm amico, iterabo hanc ipſam demonftrationem in calce operis, per ſolitam veterum Geometrarum viam demonftrandi, longiorem quidem, ſed non ideo mihi certiorẽ.

Interim, quia demonftrationes exhibebimus de illo tantùm acuto ſolido, cuius hyperbolæ genitricis aſymptoti angulum rectum contineant, dicamus hic obitèr, omiſsà demonftratione, quibus figuris æqualia ſint acuta ſolida; quando aſymptoton angulus obtuſus fuerit, vel acutus.

Demonſtrationes, quas ad euitandam molem pręterimus, ſibi lector industrius fàcili negotio comparabit.

Efto hyperbola cuius aſymptoti *ab, ac* angulum obtuſũ contineant; & reuoluta figura circa axem *ab* fiat ſolidum acu-

tum

tũ infinitè lon-gum verſus *b.* ſeceturq;(vt in prima fig.)pla no *de* ad axẽ erecto. Erit ſolidum acutum *dbe* æquale cylindro *dile*,

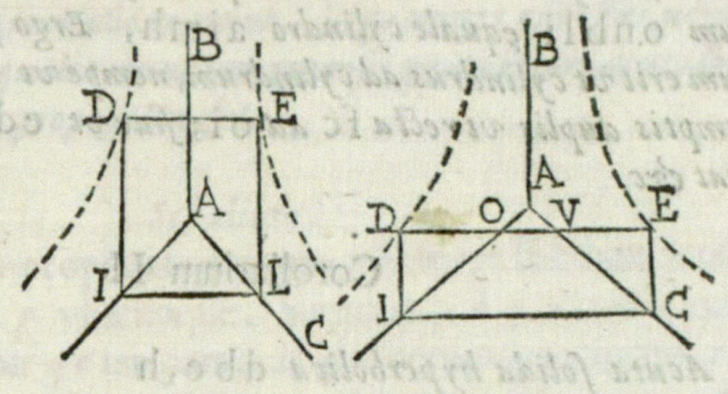

& cono *ial*. In ſecunda verò figura ſit planum ſecans *de*. erit ſolidum acutum vniuerſum quod imponitur ſuper circulo *de* ſumpto etiam cono *oau*, æquale cylindro *ie*, & cono *iac* ſimul ſumptis.

Quando verò angulus aſymptoton acutus ponatur, & ſit planum ſecans *cd* in prima figura. Erit ſolidum acutum *chd* vna cum cono *eai* æquale cylindro *ceid*. At in ſecunda figura erit vniuerſum ſolidum acutum fa-

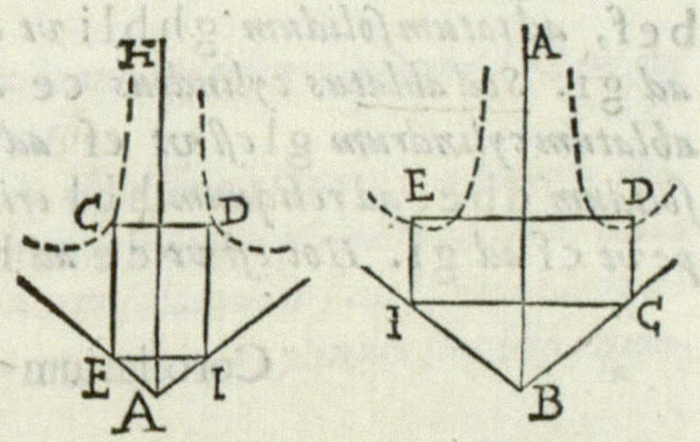

ctum ex conuerſione quadrilinei mixti *abcda* ſine fine longi, duplum cylindri *iedc*.

Sequuntur iàm ſub nomine Corollariorum Propoſitiones quædam ex præcedenti Theoremate promanãtes; quæ quidem aliquot prærogatiuas huius acuti ſolidi hyperbolici fortaſſe nõ contemnendas demonſtrabunt.

Corollarium Primum.

Acuta ſolida hyperbolica e b d, n b l, *quæ in figura pag.* 115 *ex ſectionib.* e d, n l *ad axẽ erectis fiunt, vna cum cylindris ſuarum baſium, inter ſe ſunt vt diametri earundem baſium, nempe vt recta* e d *ad* n l.

Nam reſumpta præcedentis Theorematis figura, & constru-ctione

tũ infinitè longum verſus *b.* ſeceturq; (vt in prima fig.) plano *de* ad axẽ erecto. Erit ſolidum acutum *dbe* æquale cylindro *dile,* & cono *ial.* In ſecunda verò figura ſit planum ſecans *de.* erit ſolidum acutum vniuerſum quod imponitur ſuper circulo *de* ſumpto etiam cono *oau,* æquale cylindro *ie,* & cono *iac* ſimul ſumptis.

Quando verò angulus aſymptoton acutus ponatur, & ſit planum ſecans *cd* in prima figura. Erit ſolidum acutum *chd* vna cum cono *eai* æquale cylindro *ceid.* At in ſecunda figura erit vniuerſum ſolidum acutum factum ex conuerſione quadrilinei mixti *abcda* ſine fine longi, duplum cylindri *iedc.*

Sequuntur iàm ſub nomine Corollariorum Propoſitiones quædam ex præcedenti Theoremate promanãtes; quæ quidem aliquot prærogatiuas huius acuti ſolidi hyperbolici fortaſſe nõ contemnendas demonſtrabunt.

Corollarium Primum.

Acuta ſolida hyperbolica ebd, nbl, *quæ in figura pag. 115 ex ſectionib.* ed, nl *ad axẽ erectis ſiunt, vna cum cylindris ſuarum baſium, inter ſe ſunt vt diametri earundem baſium, nempe vt recta* ed *ad* nl.

Nam reſumpta pręcedentis Theorematis figura, & conſtru-

ctione

118 De Solido Hyperbolico

ctione, erit solidum feb dc, *æquale cylindro* ac̄ gh. *& solidum* onbli *ęquale cylindro* aimh, *Ergo solidum ad solidum erit vt cylindrus ad cylindrum, nempe vt* ca *ad* ai, *siue sumptis duplis vtrecta* fc *ad* oi; *siue vt* ed *ad* nl. *Quod erat &c.*

Corollarium II.

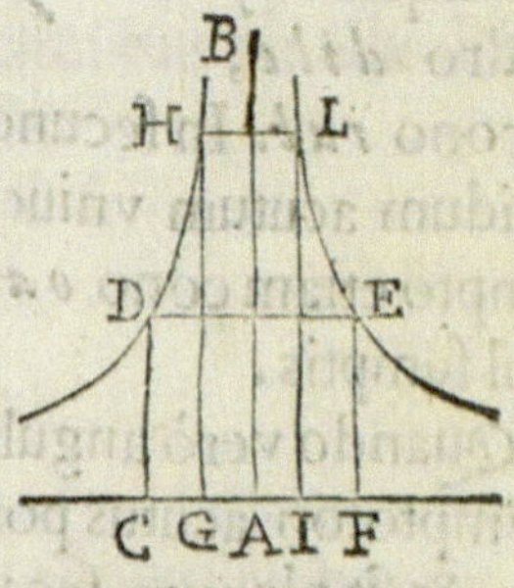

Acuta solida hyperbolica dbe, hbl, *etiam sine cylindris suarum basium sumpta, inter se sunt vt diametri earumdem basium, nempe vt* de *ad* hl. *Descriptis enim basium cylindris* cdef, ghli, *erit totum solidum* cdbef, *ad totum solidum* ghbli *vt* cf *ad* gi. *Sed ablatus cylindrus* ce *ad ablatum cylindrum* gl *est vt* cf *ad* gi. *Ergo reliquum etiam solidum* dbe, *ad reliquum* hbl *erit vt totum ad totum; nempe vt* cf *ad* gi. *Hoc est vt* de *ad* hl. *Quod &c.*

3. l.ᵐᵃ huius

Corollarium III.

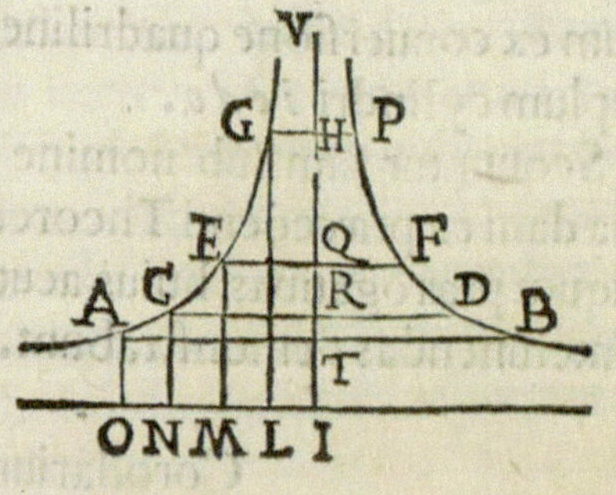

Esto solidum acutum sectum planis ab, cd, ef, gp *ità vt sectionum semidiametri sint vt numeri naturalitèr ab vnitate progredientes (quod facile fiet, si accepta ad libitum* il, *ęquales ipsi* il *secentur* lm, mn, no *&c. ductisq; lg, me,* nc *&c. ad axem parallelis, per puncta* g, *&* e, *&* c *&c. agantur secantia plana.) Dico omnia frusta intercepta æqualia esse tum inter se, tum etiam acuto solido* gup.

Patet hoc. Nam cum acuta solida sint vt diametri basium; & in hoc casu diametri basium ponantur vt numeri naturalitèr ab

vni-

ctione, erit ſolidum febdc, *æquale cylindro* acgh. *& ſolidum* onbli *ęquale cylindro* aimh, *Ergo ſolidum ad ſolidum erit vt cylindrus ad cylindrum, nempe vt* ca *ad* ai, *ſiue ſumptis duplis vt recta* fc *ad* oi; *ſiue vt* ed *ad* nl. *Quod erat &c.*

Corollarium II.

Acuta ſolida hyperbolica db e, hbl, *etiam ſine cylindris ſuarum baſium ſumpta, inter ſe ſunt vt diametri earumdem* [3. Lemma huius] *baſium, nempe vt* de *ad* hl. *Deſcriptis enim baſium cylindris* cdef, ghli, *erit totum ſolidum* cdbef, *ad totum ſolidum* ghbli *vt* cf *ad* gi. *Sed ablatus cylindrus* ce *ad ablatum cylindrum* gl *eſt vt* cf *ad* gi. *Ergo reliquum etiam ſolidum* dbe, *ad reliquum* hbl *erit vt totum ad totum; nempe vt* cf *ad* gi. *Hoc eſt vt* de *ad* hl. *Quod &c.*

Corollarium III.

Eſto ſolidum acutum ſectum planis ab, cd, ef, gp *ità vt ſectionum ſemidiametri ſint vt numeri naturalitèr ab vnitate progrediẽtes (quod facile fiet, ſi accepta* ad libitum il, *ęquales ipſi* il *ſecentur* lm, mn, no *&c. ductisq;* lg, me, nc *&c. ad axem parallelis, per puncta* g, *&* e, *&* c *&c. agantur ſecantia plana.) Dico omnia fruſta intercepta æqualia eſſe tum inter ſe, tum etiam acuto ſolido* gup.

Patet hoc. Nam cum acuta ſolida ſint vt diametri baſium; & in hoc caſu diametri baſium ponantur vt numeri naturalitèr ab

uni-

vnitate progredientes, etiam acuta solida gup, euf, cud *&c. in eadem Aritmetica ratione erunt. Ergo omnes excessus, nempe omnia frusta æqualia erunt tam inter se, quàm etiam acuto solido* gup. *vt eràt propositum &c.*

Scholium.

Poterat etiam proponi hoc modo. Si fuerit solidum acutũ sectum plano *g p* vbicunque. Sumaturq. *h q* semissis axis *h i*. Deinde sumatur *q r* ter. pars axis *q i*; iterumq; accipiatur *r t* quar. pars axis *r i*: postea accipiatur quinta pars reliqui axis. & hoc semper; & per puncta sectionum plana agantur; erunt ead. ut supra &c.

Corollarium IV.

Acutum solidum hyperbolicum abscissum plano ad axem erecto, æquale est cylindro suæ basis.

Esto solidum acutum a b c *abscissum plano* a c *ad axem erecto (hoc enim modo intelligemus semper plana secantia, quod oportet meminisse) & supponatur solidum infinitè productum ad partes* b. *Dico solidum* a b c, *æquale esse cylindro suæ basis nempe* d a c e.

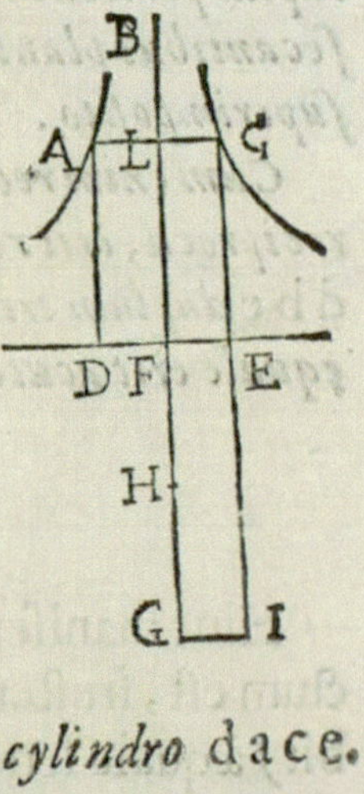

Fiat enim cylindrus f e i g *vt in Theoremate pag. 115. Eritq; totum solidum* d a bc e, *ex demonstratis æquale cylindro* f i. *Iam cylindrus* f i, *ad cylindrum* d c, *rationem habet compositam ex ratione quadrati* h f *ad* fe *& ex ratione rectæ* fe *ad* e c; *siue quadrati* f e *ad rectangulum* f e c. *Cylindrus itaq;* f i *ad cylindrum* d c, *est vt quadratum* f h *ad rectangulum* f c, *nempe duplus. Propterea solidum vniuersum* d a b c e *(cum æquale sit cylindro* f i*) duplum erit cylindri* d c. *Et diuisim, erit solidum acutũ* a b c *æquale suæ basis cylindro* d a c e. *Quod &c.*

Corol.

vnitate progredientes, etiam acuta ſolida gup, euf, cud *&c. in eadem Aritmetica ratione erunt. Ergo omnes exceſſus, nempe omnia fruſta æqualia erunt tam inter ſe, quàm etiam acuto ſolido* gup. *vt eràt propoſitum &c.*

Scholium.

Poterat etiam proponi hoc modo. Si fuerit ſolidum acutũ ſectum plano *gp* vbicunque. Sumaturq. *hq* ſemiſſis axis *hi.* Deinde ſumatur *qr* ter. pars axis *qi*; iterumq; accipiatur *rt* quar. pars axis *ri*: poſtea accipiatur quinta pars reliqui axis. & hoc ſemper; & per puncta ſectionum plana agantur; erunt ead. ut ſupra &c.

Corollarium IV.

Acutum ſolidum hyperbolicum abſciſsum plano ad axem erecto, æquale est cylindro ſuæ baſis.

Esto ſolidum acutum abc *abſciſſum plano* ac *ad axem erecto (hoc enim modo intelligemus ſemper plana ſecantia, quod oportet meminiſſe) & ſupponatur ſolidum infinitè productum ad partes* b. *Dico ſolidum* abc, *æquale eſſe cylindro ſuæ baſis nempe* dace.

Fiat enim cylindrus feig *vt in Theoremate pag. 115. Eritq; totum ſolidum* dabce, *ex demonſtratis ęquale cylindro* fi. *Iam cylindrus* fi, *ad cylindrum* dc, *rationem habet compoſitam ex ratione quadrati* hf *ad* fe *& ex ratione rectę* fe *ad* ec; *ſiue quadrati* fe *ad rectangulum* fec. *Cylindrus itaq*; fi *ad cylindrum* dc, *est vt quadratum* fh *ad rectangulum* fc, *nempe duplus. Propterea ſolidum vniuerſum* dabce *(cum æquale ſit cylindro* fi*) duplum erit cylindri* dc. *Et diuiſim, erit ſolidum acutũ* abc *æquale ſuæ baſis cylindro* dace. *Quod &c.*

Corol

Corollarium V.

Maximum hemisphærium fbe, *intra solidum acutum inscriptibile ex* d *centro hyperbolæ, subsesquialterum est vniuersi solidi* fhace *ipsum hemisphærium ambientis. Solidum autem* fhace *constat ex acuto solido infinitè longo* ha c, *& ex cylindro basis hemisphærium tangente* fhce.

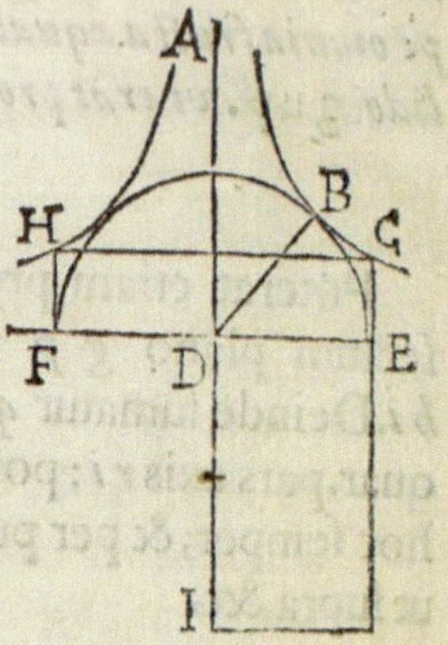

Facto enim cylindro ie *vt in Theoremate pag. 115. erit hemisphærium* fbe *subsesquialterum cylindri* ie; *Cum eandem altitudinem habeat, & basim eandem, nempe circulum cuius radius est semiaxis* db. *Subsesquialterum ergo erit ipsum hæmisphærium etiam solidi* fhace, *quod æquale demonstratum est cylindro* ie. *&c.*

Corollarium VI.

Esto solidum acutum cuius axis ab, *(in figura pag. 118.) sectum vbicunq; plano* dc. *Secetur verò & altero plano* hl, *quod capiat portionem axis duplam. Dico frustum solidum* dhle, *à secantibus planis interceptum æquale esse solido acuto* hbl *sibi superimposito.*

Cum enim rectangula ce, gl *sint æqualia, & latera eorum reciproca, erit recta* de *dupla ipsius* hl, *& ideo solidum acutũ* dbe *duplum erit acuti solidi* hbl, *& diuidendo, frustum* dhle *æquale erit acuto solido* hbl. *Quod &c.*

Scholium.

Hinc manifestum est, quòd si acutum solidum secetur vti dictum est, frustum interceptum *dhle* (quod duas bases habebit) æquale semper erit cylindro minoris basis *ghli*. Subduplum uerò erit cylindri maioris basis *cdef*.

Corol-

Corollarium V.

Maximum hemiſphærium fbe, *intra ſolidum acutum inſcriptibile ex* d *centro hyperbolæ, ſubſeſquialterum eſt vniuerſi ſolidi* fhace *ipſum hemiſphærium ambientis. Solidum autem* fhace *constat ex acuto ſolido infinitè longo* hac, *& ex cylindro baſis hemiſphærium tangente* fhce.

Facto enim cylindro ie *vt in Theoremate pag. 115. erit hemiſphęrium* fbe *ſubſeſquialterum cylindri* ie; *Cum eandem altitudinem habeat, & baſim eandem, nempe circulum cuius radius est ſemiaxis* db. *Subſeſquialterum ergo erit ipſum hæmiſphærium etiam ſolidi* fhace, *quod ęquale demonstratum eſt cylindro* ie. *&c.*

Corollarium VI.

Esto ſolidum acutum cuius axis ab, *(in figura pag. 118.) ſectum vbicunq; plano* de. *Secetur verò & altero plano* hl, *quod capiat portionem axis duplam. Dico fruſtum ſolidum* dhle, *à ſecantibus planis interceptum æquale eſſe ſolido acuto* hbl *ſibi ſuperimpoſito.*

Cum enim rectangula ce, gl *ſint æqualia, & latera eorum reciproca, erit recta* de *dupla ipſius* hl, *& ideo ſolidum acutũ* dbe *duplum erit acuti ſolidi* hbl, *& diuidendo, fruſtum* dhle *ęquale erit acuto ſolido* hbl. *Quod &c.*

Scholium.

Hinc manifeſtum eſt, quòd ſi acutum ſolidum ſecetur vti dictum eſt, fruſtum interceptum *dhle (*quod duas baſes habebit*)* æquale ſemper erit cylindro minoris baſis *ghli.* Subduplum uerò erit cylindri maioris baſis *cdef.*

Corol-

Corollarium VII.

Esto solidum acutum sectum à tribus planis ab; cd, ef, *secantibus axem solidi proportionaliter; hoc est, sit vt* gh *ad* gi, *ita* gi *ad* gl. *Dico frustum* acdb *ad frustũ* cefd *esse vt* li *ad* ih. *nempe in reciproca ratione altitudinum.*

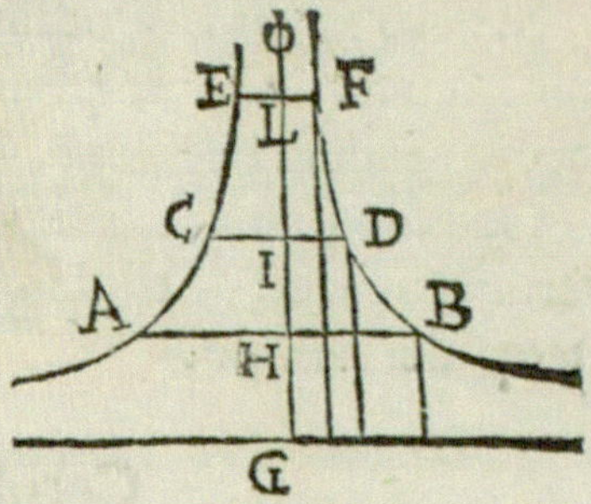

Cum enim rectangula gf, gd, gb. *sint equalia, & latera eorum reciproca, erunt tres recta* hb, id, lf, *in eadem continua proportione in quà sunt* gl, gi, gh. *Sed solida acuta* aob, cod, eof *sunt vt basium semidiametri* hb, id, lf, *siue vt* gl, gi, gh, *ergo excessus solidorum inter se erunt vt excessus linearum. Nempe frustum solidum* acdb, *ad frustum* cefd *erit vt* li *ad* ih. *Quod &c.*

Scholium.

Ex demonstratis patet primò, quomodo datum frustum *a ef b* secari possit plano *c d*, ita vt factæ portiones inter se sint vt altitudines, reciprocè tamen sumptæ. Quod quidem fit sumendo *g i* mediam proportionalem inter *g l*, *g h*. *in fig precéd.*

Manifestum etiam est, quòd si sumatur quodlibet segmẽtum axis, puta *a b*. & secetur bifariam in *c*. deinde *a c* secetur bifariam in *d*; reliquum autem *a d* bisecetur in *e*; & sic semper. Erunt frusta solida intercepta à planis per *b*, *c*, *d*, *e*, ductis, in continua proportione in qua axes, siue axium differentię. Eritq; primum, & subtilissimũ frustũ *fi* æquale acuto solido sibi superimposito. At secundum frustũ duplum erit primi; tertiũ quadruplum primi, 4. verò octuplum, quintum

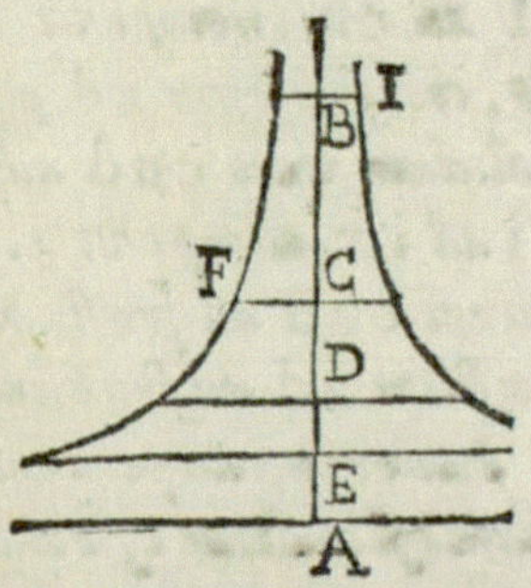

Q

Corollarium VII.

Efto folidum acutum fectum à tribus planis ab; cd, ef, *fecantibus axem folidi proportionaliter; hoc est, fit vt* gh *ad* gi, *ita* gi *ad* gl. *Dico fruftum* acdb *ad fruftũ* cefd *effe vt* li *ad* ih. *nempe in reciproca ratione altitudinum.*

Cum enim rectangula gf, gd, gb. *fint ęqualia, & latera eorum reciproca, erunt tres rectæ* hb, id, lf, *in eadem continua proportione in quà funt* gl, gi, gh. *Sed folida acuta* aob, cod, eof *funt vt bafium femidiametri* hb, id, lf, *fiue vt* gl, gi, gh, *ergo exceffus folidorum inter fe erunt vt exceffus linearum. Nempe frustum folidum* acdb, *ad frustum* cefd *erit vt* li *ad* ih. *Quod &c.*

Scholium.

Ex demonftratis patet primò, quomodo datum fruftum *aefb* fecari poffit plano *cd,* ita vt factæ portiones inter fe fint vt altitudines, reciprocè tamen fumptæ. Quod quidem fit fumendo *gi* mediam proportionalem inter *gl, gh.* *in fig pre* [-] *ced.*

Manifeftum etiam eft, quòd fi fumatur quodlibet fegmẽtum axis, puta *ab.* & fecetur bifariam in *c.* deinde *ac* fecetur bifariam in *d*; reliquum autem *ad* bifecetur in *e*; & fic femper. Erunt frufta folida intercepta à planis per *b, c, d, e,* ductis, in continua proportione in qua axes, fiue axium differentię. Eritq; primum, & fubtiliffimũ fruftũ *fi* æquale acuto folido fibi fuperimpofito. At fecundum fruftũ duplum erit primi; tertiũ quadruplum primi, 4. verò octuplum, quin-

Q tum

tum ſedecuplum: & ſic ſemper; quò magis ad centrum *a* accedemus, maiora præcedentibus erunt fruſta, & multiplicia ſecundum numeros in proportione dupla progredientes ab vnitate.

Si verò ſumatur quodlibet ſegmentum axis *a e*, cuius duplum ponatur *a d*; & ipſius *a d* duplum ſecetur *a c*, & ſic deinceps; eadem euenient, vt ſupra dictum eſt.

Quæcunq; autem diximus exemplo allato de ratione dupla, verum etiam eſt de tripla, quadrupla; ſeſquialtera, & de quacunq; alia ratione.

Corollarium VIII.

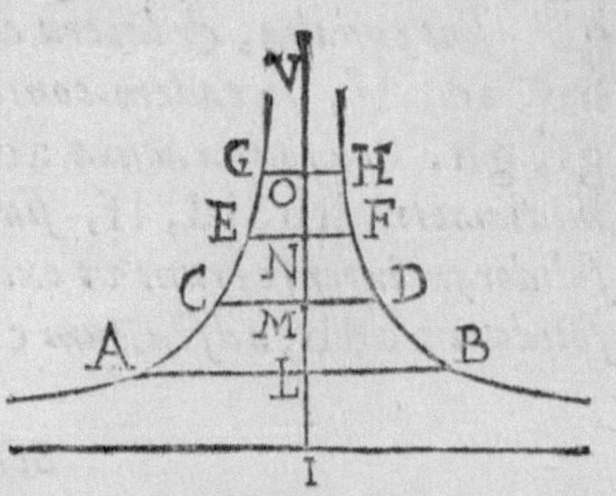

Si ſolidum acutum ſectum fuerit planis a b; c d, e f, g h, *&c. ita vt axis portiones à cẽtro* i *incipientes, nempe* i l, l m, m n, n o *&c. æquales ſint; erit primũ fruſtum* a d, *ad ſecundum* c f *vt 3. ad vnum: ſecundum verò fruſtum ad tertiũ erit vt 4. ad 2; Tertium ad quartum erit vt 5. ad 3. quartum ad quintum vt 6. ad 4.; Et ſic ſemper vt numeri binario differentes; addita ſcilicet ſemper vnitate vtrique termino rationis.*

Nam ſolidum acutum a u b *ad ſolidum acutum* c u d, *eſt vt* a l *ad* c m, *nempe vt* m i, *ad* i l, *hoc eſt duplum. Et diuidendo, erit fruſtum* a d *ęquale ſolido acuto* c u d. *ſiue vt 3. ad 3. Solidum vero* c u d *ad ſolidum* e u f *eſt vt* c m *ad* e n, *ſiue vt* n i *ad* i m, *nempe vt 3. ad 2. Et per conuerſionẽ rationis erit ſolidum* c u d *ad fruſtum* c f *vt 3. ad vnum. Ergo ex ęquo erit fruſtum* a d *ad fruſtum* c f *vt 3. ad unum. Quod &c.*

Eodem modo penitus ratio reliquorum fruſtorum conſequentium oſtenditur eſſe talis qualis propoſita eſt.

Scholium.

Patet in progreſſu demonſtrationis primum fruſtum *a d* æ-

quale

tum ſedecuplum: & ſic ſemper; quò magis ad centrum *a* accedemus, maiora præcedentibus erunt fruſta, & multiplicia ſecundum numeros in proportione dupla progredientes ab vnitate.

Si verò ſumatur quodlibet ſegmentum axis *ae,* cuius duplum ponatur *ad*; & ipſius *ad* duplum ſecetur *ac,* & ſic deinceps; eadem euenient, vt ſupra dictum eſt.

Quæcunq; autem diximus exemplo allato de ratione dupla, verum etiam eſt de tripla, quadrupla; ſeſquialtera, & de quacunq; alia ratione.

Corollarium VIII.

Si ſolidum acutum ſectum fuerit planis ab; cd, ef, gh, *&c. ita vt axis portiones à cẽtro* i *incipientes, nempe* il, lm, mn, no *&c. æquales ſint; erit primũ fruſtum* ad, *ad ſecundum* cf *vt 3. ad vnum: ſecundum verò fruſtum ad tertiũ erit vt 4. ad 2; Tertium ad quartum erit vt 5. ad 3. quartum ad quintum vt 6. ad 4.; Et sic ſemper vt numeri binario differentes; addita ſcilicet ſemper vnitate vtrique termino rationis.*

Nam ſolidum acutum aub *ad ſolidum acutum* cud, *est vt* a l *ad* cm, *nempe vt* mi, *ad* il, *hoc eſt duplum. Et diuidendo, erit frustum* ad *ęquale ſolido acuto* cud. *ſiue vt* 3. *ad 3. Solidum vero* cud *ad ſolidum* euf *est vt* cm *ad* en, *ſiue vt* ni *ad* m, *nempe vt 3. ad 2. Et per conuerſionẽ rationis erit ſolidum* cud *ad fruſtum* cf *vt 3. ad vnum. Ergo ex ęquo erit fruſtum* ad *ad fruſtum* cf *vt* 3. *ad unum. Quod &c.*

Eodem modo penitus ratio reliquorum fruſtorum conſequentium oſtenditur eße talis qualis propoſita est.

Scholium.

Patet in progreſſu demonſtrationis primum fruſtum *ad* æ-

quale

quale esse solido acuto sibi imposito *cud*. At secundum frustũ *cf* duplum est solidi *euf* sibi impositi; Tertium verò triplum; quartum quadruplum. & sic in infinitum.

Corollarium IX.

Si solidum acutum à cylindricis superficiebus diuisum fuerit, erunt solida annularia inter cylindricas superficies intercepta, inter se, vt sunt portiones asymptoti ab ipsis cylindricis superficiebus abscissę.

Sit hyperbola abc, *& lineæ quotcunq;* ad, be, cf. *parallelæ asymptoto* hi; *& conuertatur figura circa asymptotõ* hi. *Dico solidum descriptum à quadrilineo* ebcf, *ad solidum descriptum à quadrilineo* dabe, *esse vt recta* fe *ad* ed.

Fiat enim cylindrus lf, *vt in Theoremate pag. 115. eritq; solidum* nmicf *æquale cylindro* lf. *Et solidum* poibe, *æquale cylindro* le. *ablatis ergo æquali bus, remanebit cylindrus* lf *ęqualis solido sibi respondenti facto à quadrilineo* ebcf. *Pari ratione cylindrus* te *ęqualis ostendetur solido sibi respondenti facto a quadrilineo* dabe; *erit igitur ob æqualitatem, solidum quadrilinei* ebcf *ad solidum quadrilinei* dabe *vt cylindrus* lf *ad cylindrum* te, *nempe vt recta* fe *ad* ed. *Quod &c.*

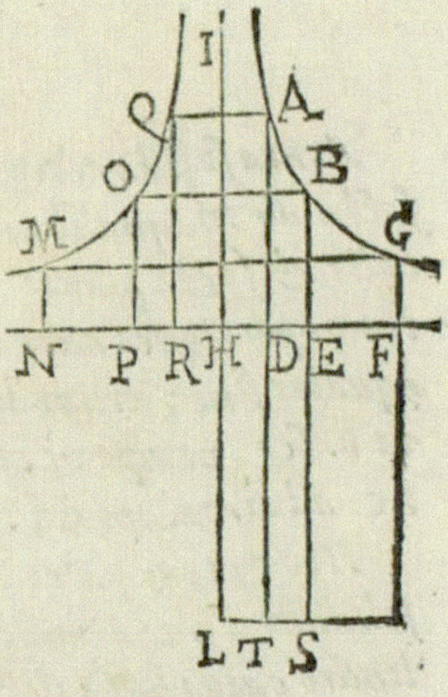

Corollarium X.

Acuta solida abc, def. *super basibus æqualibus* ac, df *constituta, & à conuersione inæqualium hyperbolarum descripta, sunt inter se in duplicata ratione axium suarum hyperbolarum.*

Intelligantur enim sub basibus solidorum cylindri hc, lf, *erit*

quale eſſe ſolido acuto ſibi impoſito *cud.* At ſecundum fruſtũ *cf* duplum eſt ſolidi *euf* ſibi impoſiti; Tertium verò triplum; quartum quadruplum. & ſic in infinitum.

Corollarium IX.

Si ſolidum acutum à cylindricis ſuperficiebus diuiſum fuerit, erunt ſolida annularia inter cylindricas ſuperficies intercepta, inter ſe, vt ſunt portiones aſymptoti ab ipſis cylindricis ſuperficiebus abſciſſę.

Sit hyperbola abc, *& lineæ quotcunq;* ad, be, cf. *parallelæ aſymptoto* hi; *& conuertatur figura circa aſymptotõ* hi. *Dico ſolidum deſcriptum à quadrilineo* ebcf, *ad ſolidum deſcriptum à quadrilineo* dabe, *eſſe vt recta* fe *ad* ed.

Fiat enim cylindrus lf, *vt in Theoremate pag. 115. eritq; ſolidum* nmicf *æquale cylindro* lf. *Et ſolidum* poibe, *æquale cylindro* le. *ablatis ergo æqualibus, remanebit cylindrus* ff *ęqualis ſolido ſibi reſpondenti facto à quadrilineo* ebcf. *Pari ratione cylindrus* t e *ęqualis oſtendetur ſolido ſibi reſpondenti facto a quadrilinco* dabe; *erit igitur ob æqualitatem, ſolidum quadrilinei* ebcf *ad ſolidum quadrilinei* dabe *vt cylindrus* ff *ad cylindrum* te, *nempe vt recta* fe *ad* ed. *Quod &c.*

Corollarium X.

Acuta ſolida abc, def. *ſuper baſibus æqualibus* ac, df *conſtituta, & à conuerſione inæqualium hyperbolarum deſcripta, ſunt inter ſe in duplicata ratione axium ſuarum hyperbolarum.*

Intelligantur enim ſub baſibus ſolidorum cylindri hc, lf,

L 2 *erit*

124 De Solido Hyperbolico

eritq; solidum abc æquale cylindro hc; & solidum def. æquale cylind. lf. Propterea solidū abc ad solidum def, erit vt cylind. hc ad cylind. lf. siue (cum æquales bases habeant) vt altitudo ha ad altitudinem ld. siue vt rectangulum hc ad rectangulum lf. hoc est, sumptis æqualibus, vt quadratum axis in ad quadratum axis mo. Quod &c.

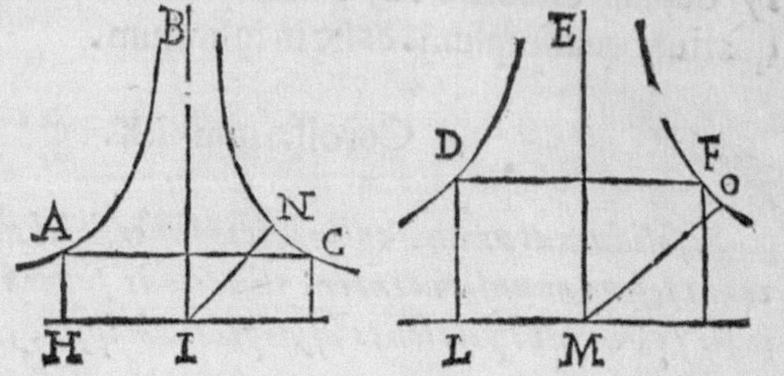

Corollarium XI.

Acuta solida abc, def facta ab inæqualibus hyperbolis, & secta planis ac, df ita vt portiones axis lh, oi æquales sint; erunt inter se vt bases, nempe vt circulus ac ad circulum df.

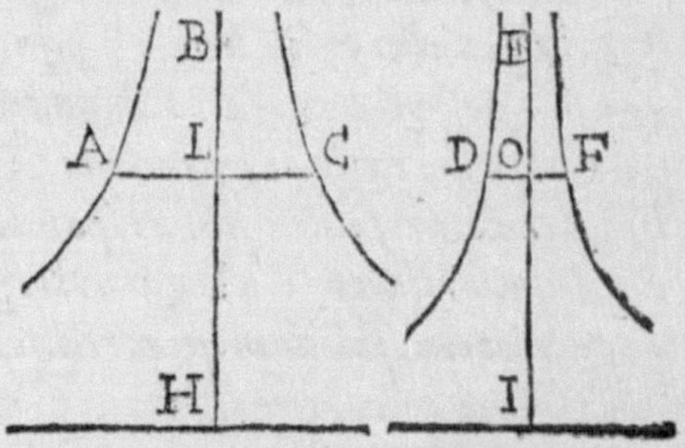

Hoc autem patet. Nam solidum abc æquale est cylindro cuius basis sit ac altitudo verò lh. & solidum def æquale est cylindro cuius basis sit df altitudo vero oi. Ergo solidum abc ad solidum def erit vt prædictus cylindrus ad dictum cylindrum, nempe (cum æquales altitudines habeant) vt basis ac ad basim df. Quod erat &c.

Corollarium XII.

Acuta solida quæcunq; sint abc, def; in p. fig. huius pag. inter se sunt vt solida rectang. basi quadrato axium hyperbolarum; alti-

eritq; ſolidum abc *æquale cylindro* hc; *& ſolidum* def. *æquale cylind.* lf. *Propterea ſolidũ* abc *ad ſolidum* def, *erit vt cylind.* hc *ad cylind.* lf. *ſiue (cum æquales baſes habeant) vt altitudo* ha *ad altitudinem* ld. *ſiue vt rectangulum* hc *ad rectangulum* lf. *hoc eſt, ſumptis æqualibus, vt quadratum axis* in *ad quadratum axis* mo. *Quod &c.*

Corollarium XI.

Acuta ſolida abc, def *facta ab inæqualibus hyperbolis, & ſecta planis* ac, df *ita vt portiones axis* lh, oi *ęquales ſint; erunt inter ſe vt baſes, nempe vt circulus* ac *ad circulum* df.

Hoc autem patet. Nam ſolidum abc *æquale est cylindro cuius baſis ſit* ac *altitudo verò* lh. *& ſolidum* def *æquale eſt cylindro cuius baſis ſit* df *altitudo vero* oi. *Ergo ſolidum* abc *ad ſolidum* def *erit vt prædictus cylindrus ad dictum cylindrum, nempe (cum æquales altitudines habeant) vt baſis* ac *ad baſim* df. *Quod erat &c.*

Corollarium XII.

Acuta ſolida quæcunq; ſint abc, def; *in p. fig. huius pag. inter ſe ſunt vt ſolida rectang baſi quadrato axium hyperbolarum;*

alti-

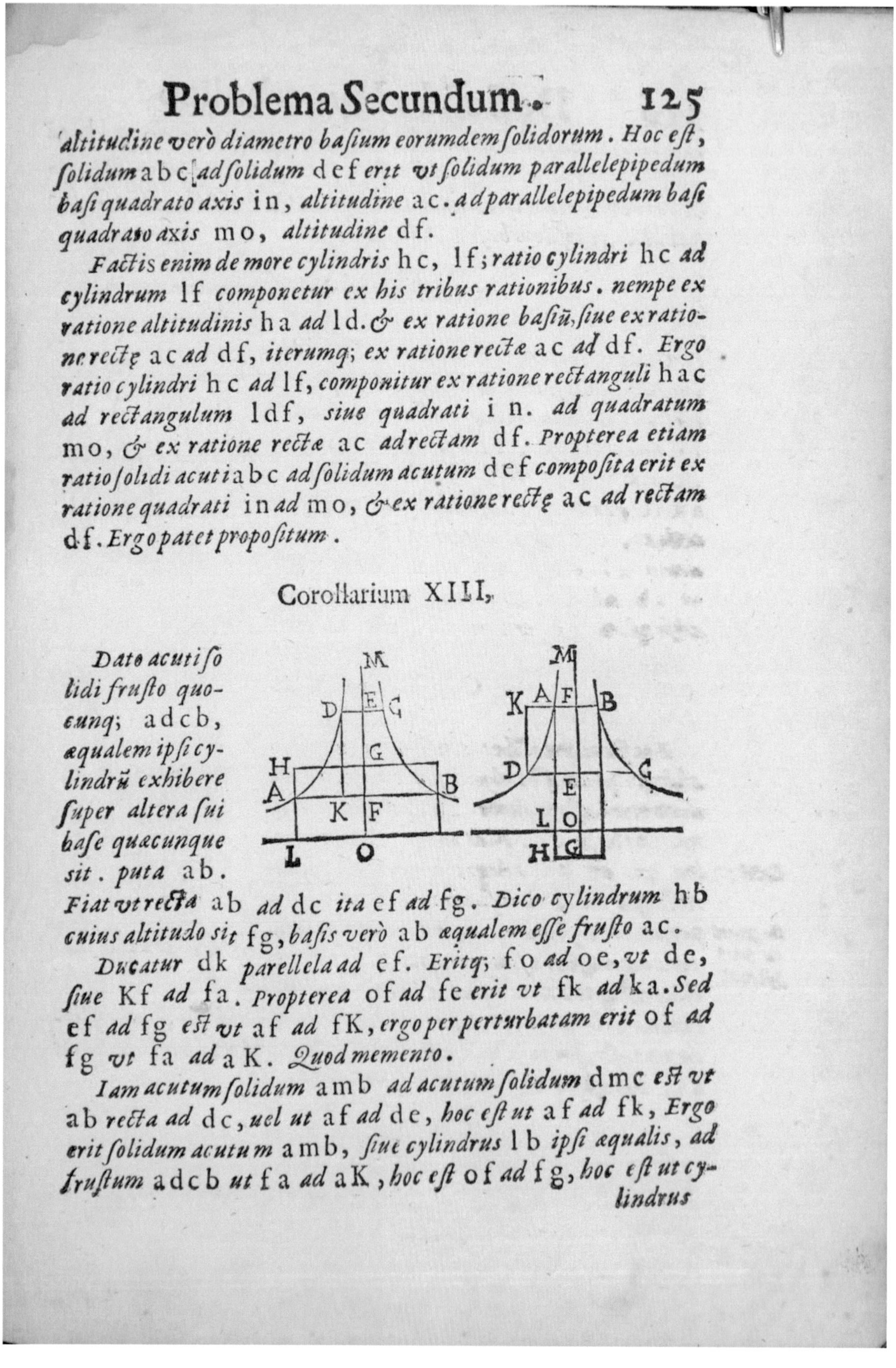

Problema Secundum. 125

altitudine verò diametro basium eorumdem solidorum. Hoc est, solidum abc ad solidum def erit vt solidum parallelepipedum basi quadrato axis in, altitudine ac. ad parallelepipedum basi quadrato axis mo, altitudine df.

Factis enim de more cylindris hc, lf; ratio cylindri hc ad cylindrum lf componetur ex his tribus rationibus. nempe ex ratione altitudinis ha ad ld. & ex ratione basiũ, siue ex ratione rectę ac ad df, iterumq; ex ratione rectæ ac ad df. Ergo ratio cylindri hc ad lf, componitur ex ratione rectanguli hac ad rectangulum ldf, siue quadrati in. ad quadratum mo, & ex ratione rectæ ac ad rectam df. Propterea etiam ratio solidi acuti abc ad solidum acutum def composita erit ex ratione quadrati in ad mo, & ex ratione rectę ac ad rectam df. Ergo patet propositum.

Corollarium XIII.

Dato acuti solidi frusto quocunq; adcb, æqualem ipsi cylindrũ exhibere super altera sui base quæcunque sit. puta ab. Fiat vt recta ab ad dc ita ef ad fg. Dico cylindrum hb cuius altitudo sit fg, basis verò ab æqualem esse frusto ac.

Ducatur dk parellela ad ef. Eritq; fo ad oe, vt de, siue Kf ad fa. Propterea of ad fe erit vt fk ad ka. Sed ef ad fg est vt af ad fK, ergo per perturbatam erit of ad fg vt fa ad aK. Quod memento.

Iam acutum solidum amb ad acutum solidum dmc est vt ab recta ad dc, uel ut af ad de, hoc est ut af ad fk, Ergo erit solidum acutum amb, siue cylindrus lb ipsi æqualis, ad frustum adcb ut fa ad aK, hoc est of ad fg, hoc est ut cylindrus

altitudine verò diametro baſium eorumdem ſolidorum. Hoc eſt, ſolidum abc *ad ſolidum* def *erit vt ſolidum parallelepipedum baſi quadrato axis* in, *altitudine* ac. *ad parallelepipedum baſi quadrato axis* mo, *altitudine* df.

Factis enim de more cylindris hc, lf; *ratio cylindri* hc *ad cylindrum* lf *componetur ex his tribus rationibus. nempe ex ratione altitudinis* ha *ad* ld. *& ex ratione baſiũ, ſiue ex ratione rectę* ac *ad* df, *iterumq; ex ratione rectæ* ac *ad* df. *Ergo ratio cylindri* hc *ad* lf, *componitur ex ratione rectanguli* hac *ad rectangulum* ldf, *siue quadrati* in. *ad quadratum* mo, *& ex ratione rectæ* ac *ad rectam* df. *Propterea etiam ratio ſolidi acuti* abc *ad ſolidum acutum* def *compoſita erit ex ratione quadrati* in *ad* mo, *& ex ratione rectę* ac *ad rectam* df. *Ergo patet propoſitum.*

Corollarium XIII,

Dato acuti ſolidi fruſto quocunq; adcb, *æqualem ipſi cylindrũ exhibere ſuper altera ſui baſe quæcunque sit. puta* ab. *Fiat vt recta* ab *ad* dc *ita* ef *ad* fg. *Dico cylindrum* hb *cuius altitudo sit* fg, *baſis verò* ab *æqualem eſſe fruſto* ac.

Ducatur dk *parellela ad* ef. *Eritq*; fo *ad* oe, *vt* de, *ſiue* K f *ad* fa. *Propterea* of *ad* fe *erit vt* fk *ad* ka. *Sed* ef *ad* fg *est vt* af *ad* fK, *ergo per perturbatam erit* of *ad* fg *vt* fa *ad* aK. *Quod memento.*

Iam acutum ſolidum amb *ad acutum ſolidum* dmc *est vt* ab *recta ad* dc, *uel ut* af *ad* de, *hoc eſt ut* af *ad* fk, *Ergo erit ſolidum acutum* amb, *ſiue cylindrus* lb *ipſi æqualis, ad fruſtum* adcb *ut* fa *ad* aK, *hoc eſt* of *ad* fg, *hoc eſt ut cy-*

lindrus

lindrus lb *ad cylindrum* bh. *Constat igitur cylindrum* lb *eādem habere rationem & ad frustum* adcb, *& ad cylindrum* bh. *Quare cylindrus* bh *æqualis erit dato frusto acuti solidi, & super altera eiusdem basi. Quod &c.*

Corollarium XIV,

Circumscriptus cylindrus aefb *ad frustum acuti solidi* adcb. *est ut diameter* ab *maioris basis ad diametrum* dc *minoris basis. Fiat enim ut* ab *ad* dc *ita* gh *ad* hi, *& erit cylindrus* amlb *æqualis frusto solido. per Cor. precedens. Cylindrus autem* af *ad cylindrum* al *est ut* gh *ad* hi; *hoc est ut* ab *ad* dc. *Quare cylindrus circumscriptus* aefb *etiā ad frustum* ac *erit ut recta* ab *ad* dc. *Quod &c.*

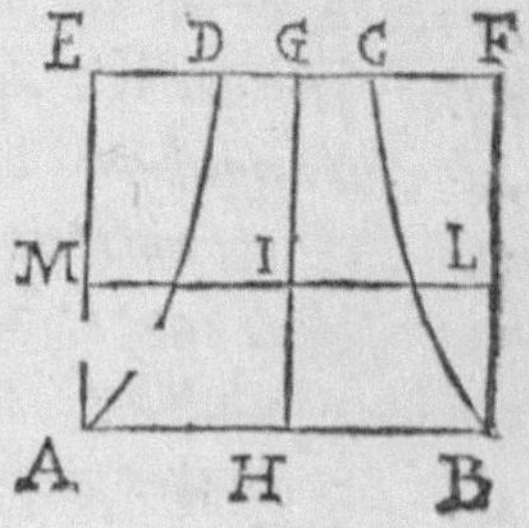

Corollarium XV.

Frustum quodlibet acuti solidi adcb *ad inscriptum sibi clyndrum* edcf, *est ut diameter basis maioris* ab, *ad diametrum minoris basis* dc. *Fiat enim vt* ab *ad* dc. *ita* gh *ad* hi, *eritq; cylindrus* al *æqualis frusto* ac. *Erit insuper cylindrus* al *isoperimeter cylindro* ec, *quandoquidem latera eorum facta sunt reciproca, & ideo rectangula per axem æqualia. Erit ergo (per lemma 3. huius) cylindrus* al *siue frustum* adcb, *ad cylindrum inscriptum* ec *vt diametri basium, siue vt recta* ab *ad* ef, *hoc est vt* ab *ad* dc. *Quod &c.*

Corol. 23

6. primi de solid. sphæral.

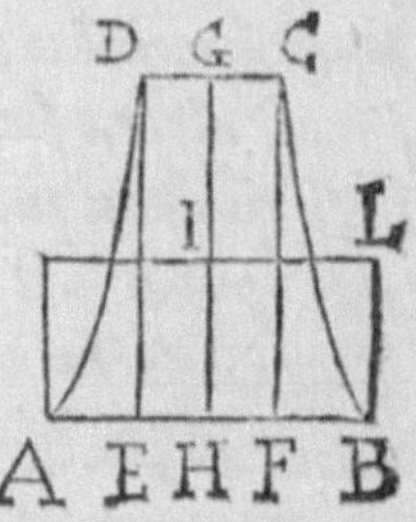

Corol-

lindrus lb *ad cylindrum* bh. *Constat igitur cylindrum* lb. *eãdem habere rationem & ad frustum* adcb, *& ad cylindrum* bh. *Quare cylindrus* bh *æqualis erit dato frufto acuti ſolidi, & ſuper altera eiuſdem baſi. Quod &c.*

Corollarium XIV,

Circumſcriptus cylindrus aefb *ad fruftum acuti ſolidi* adcb. *est ut diameter* ab *maioris basis ad diametrum* dc *minoris baſis. Fiat enim ut* ab *ad* dc *ita* gh *ad* hi, *& erit cylindrus* amlb *ęqualis frufto ſolido. per Cor. pręcedens. Cylindrus autem* af *ad cylindrum* al *est ut* gh *ad* hi; *hoc eſt ut* ab *ad* dc. *Quare cylindrus circumſcriptus* aefb *etiã ad fruftum* ac *erit ut recta* ab *ad* dc. *Quod &c.*

Corol.23 **Corollarium XV.**

Fruftum quodlibet acuti ſolidi adcb *ad inſcriptum ſibi cylindrum* edcf, *eſt ut diameter baſis maioris* ab, *ad diametrum minoris baſis*
6. primi dc. *Fiat enim vt* ab *ad* dc. *ita* gh *ad* hi, *eritq; cylindrus* al *æqualis frufto* ac. *Erit inſuper cylindrus* al *iſoperimeter cylindro* ec,
de ſolid. *quandoquidem latera eorum facta ſunt reciproca, & ideo rectangula per axem æqualia. Erit ergo (per lemma 3. huius) cylindrus* al
ſphæral. *ſiue frustum* adcb, *ad cylindrum inſcriptum* ec *vt diametri baſium, ſiue vt recta* ab *ad* ef, *hoc eſt vt* ab *ad* dc. *Quod &c.*

Corol-

Corollarium XVI,

Frustum quodlibet acuti solidi adcb. *medium proportionale est inter inscriptum, & circumscriptum sibi* cylindrum.

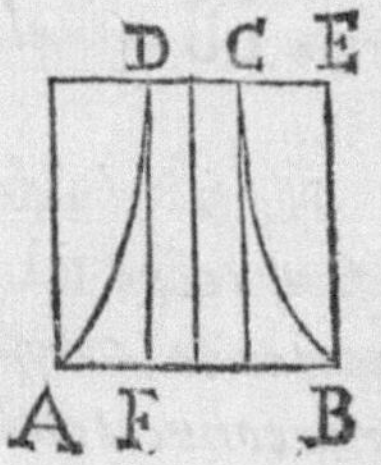

Demonstratum enim est in duobus præceden tibus Coroll. quòd circumscriptus cylindrus ae *ad frustum* adcb *est vt recta* ab *ad* dc. *Frustum verò* adcb *ad inscriptum cylindrũ est vt* ab *ad* dc. *Ergo constat quòd frustum est mediũ proportionale inter duos cylindros. Quod erat &c.*

Corollarium XVII.

Datum acutum solidum aeb *in data ratione secare vt* f *ad* g. *Fiat vt* g *ad* f *ita data* hi *ad* il. *& per* l *agatur planum* cd. *Eritq; conuertendo, & cõponendo* f *&* g, *simul ad* g *vt* lh *ad* hi, *siue vt* ab *ad* cd, *vel ut solidum* aeb *ad solidum* ced; *& diuidendo patet propositum.*

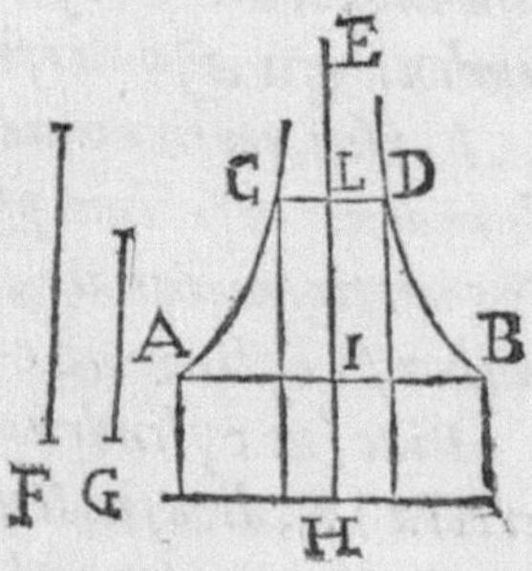

Si verò basis acuti solidi sit cd, *& oporteat illud secare iterum inferius versus hyperbolę centrum plano* ab, *ita vt frustũ* acdb *ad reliquum solidum* ced *quamlibet datam rationem habeat ut* f *ad* g. *Ita imperata exequemur. Fiat ut* f *&* g *simul ad* g, *ita data* lh *ad* hi; *& per* i *ducatur planum* ab. *erit que ut* f, *&* g *simul ad* g, *ita* ab *ad* cd; *siue solidum* aeb *ad* ced; *& diuidendo patet propositum. Quod erat &c.*

Corollarium XVIII.

Dato solido acuto secto plano ab. *frustum accipere* cabd *uersus* n *centrum hyperbolæ, quod sit æquale cuicumq; dato cylindro* gh *molis etiam immensæ.*

Fiat

Corollarium XVI,

Fruftum quodlibet acuti folidi adcb. *medium proportionale eft inter infcriptum, & circumfcriptum fibi cylindrum.*

Demonftratum enim eft in duobus præcedentibus Coroll. quòd circumfcriptus cylindrus a e *ad frustum* adcb *eft vt recta* ab *ad* dc. *Fruftum verò* adcb *ad infcriptum cylindrũ est vt* ab *ad* dc. *Ergo constat quòd fruftum est mediũ proportionale inter duos cylindros. Quod erat &c.*

Corollarium XVII.

Datum acutum folidum aeb *in data ratione fecare vt* f *ad* g. *Fiat vt* g *ad* f *ita data* hi *ad* il. *& per* l *agatur planum* cd. *Eritq; conuertendo, & cõponendo* f *&* g, *simul ad* g *vt* lh *ad* hi, *fiue vt* ab *ad* cd, *vel ut folidum* aeb *ad folidum* ced; *& diuidendo patet propofitum.*

Si verò bafis acuti folidi fit cd, *& oporteat illud fecare iterum inferius verfus hyperbolę centrum plano* ab, *ita vt fruftũ* acdb *ad reliquum folidum* ced *quamlibet datam rationem habeat ut* f *ad* g. *Ita imperata exequemur. Fiat ut* f *&* g *fimul ad* g, *ita data* lh *ad* hi; *& per* i *ducatur planum* ab. *eritque ut* f, *&* g *fimul ad* g, *ita* ab *ad* cd; *fiue folidum* aeb *ad* ced; *& diuidendo patet propofitum. Quod erat &c.*

Corollarium XVIII.

Dato folido acuto fecto plano ab. *frustum accipere* cabd *uerfus* n *centrum hyperbolæ, quod fit æquale cuicumq; dato cylindro* gh *molis etiam immenfæ.*

Fiat

Fiat ut cylindrus al *ad cylindrum* gh *ita recta* nl *data ad rectam* lf. *& erecta* fd *ductoq; plano* dc. *Dico frustum* cb *aequale esse cylindro* gh.

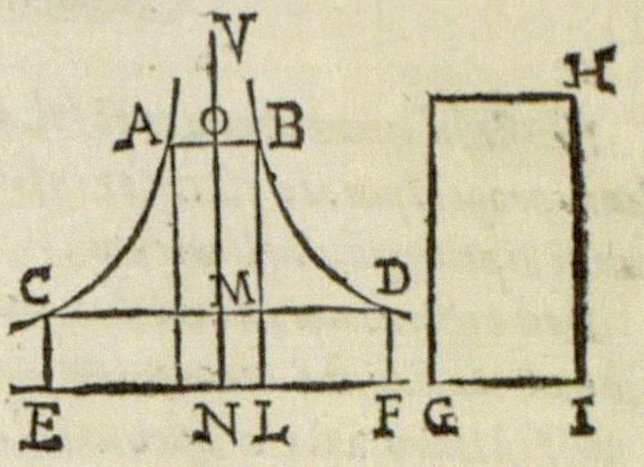

Nam cylindrus al *ad* gh, *est ut recta* nl *ad* lf, *& conuertendo, componendo, iterumque conuertendo, erit cylindrus* al *ad cylindros* al, gh, *ut* ln *ad* nf, *siue ut* ob *ad* md; *siue ut solidum acutum* aub *ad solidum acutum* cud; *siue ut cylindrus* al *ad solidum acutum* cud. *Aequales ergo sunt duo simul cylindri* al *&* gh, *acuto solido* cud. *Demptisq. aequalibus, nempe cylindro* al *& solido acuto* aub, *remanet cylindrus* gh *aequalis frusto* cabd. *Quod &c.*

Versus uerticem uerò limitatione opus est. Esto Datum solidum acutum sectum plano cd, *debeatq; sumi frustum* cabd *uersus uerticem, aequale cylindro dato* gh *(dummodo cylindrus* gh *minor sit cylindro* ecdf.)

Fiat, ut cylindrus ed *ad* gh, *ita recta* nf *data, ad* fl, *& erecta* lb, *dico frustum* cabd *aequale esse cylindro dato* gh.

Nam recta fn *ad* nl, *est ut* dm *ad* bo, *siue ut acutum solidum* cud *ad acutum* aub; *& per conuersionem rationis* nf *ad* fl, *erit ut acutum solidum* cud, *siue ut cylindrus* ed *ad frustum* cabd. *Sed ut* nf *ad* fl, *ita est etiam cylindrus idem* ed *ad* gh, *aequantur ergo frustum* cabd, *et cylindrus* gh. *Quod &c.*

Scholium.

Ex priori parte huius demonstrationis patet solidum hyperbolicum versus infinitam planitiem *ef* magnitudine infinitum esse. potest enim ex ipso sumi pars ipsius quae aequalis sit cuicunque magnitudini datae:

Corol

Fiat ut cylindrus al *ad cylindrum* gh *ita recta* nl *data ad rectam* lf. *& erecta* fd *ductoq; plano* dc. *Dico fruſtum* cb *æquale eſſe cylindro* gh.

Nam cylindrus al *ad* gh, *eſt ut recta* nl *ad* lf, *& conuertendo, componendo, iterũque conuertendo, erit cylindrus* al *ad cylindros* al, gh, *ut* ln *ad* nf, *ſiue ut* ob *ad* md; *ſiue ut ſolidum acutum* aub *ad ſolidum acutum* cud; *ſiue ut cylindrus* al *ad ſolidum acutum* cud. *Aequales ergo ſunt duo ſimul cylindri* al *&* gh, *acuto ſolido* cud. *Demptiſq. æqualibus, nempe cylindro* al *& ſolido acuto* aub, *remanet cylindrus* gh *æqualis fruſto* cabd. *Quod &c.*

Verſus uerticem uerò limitatione opus eſt. Esto Datum ſolidum acutum ſectum plano cd, *debeatq; ſumi frustum* cabd *uerſus uerticem, æquale cylindro dato* gh *(dummodo cylindrus* gh *minor ſit cylindro* ecdf.)

Fiat, ut cylindrus ed *ad* gh, *ita recta* nf *data, ad* fl, *& erecta* lb, *dico frustum* cabd *æquale eſſe cylindro dato* gh.

Nam recta fn *ad* nl, *eſt ut* dm *ad* bo, *ſiue ut acutum ſolidum* cud *ad acutum* aub; *& per conuerſionem rationis* nf *ad* fl, *erit ut acutum ſolidum* cud, *siue ut cylindrus* ed *ad frustum* cabd. *Sed ut* nf *ad* fl, *ita eſt etiam cylindrus idem* ed *ad* gh, *æquantur ergo fruſtum* cabd, *et cylindrus* gh.

Quod &c.

Scholium.

Ex priori parte huius demonſtrationis patet ſolidum hyperbolicum verſus infinitam planitiem *ef* magnitudine infinitum eſſe. poteſt enim ex ipſo ſumi pars ipſius quæ æqualis ſit cuicũque magnitudini datæ:

Corol

Problema Secundum. 129

Corollarium XIX.

Esto solidum acutum sectum plano ab. *Oportet illud secare iterũ alio plano* pi, *ita vt frustum* api b *ad cylindrum sibi circumscriptũ, sit vt* c *ad* d; *dummodo ratio* c *ad* d, *sit minoris inæqualitatis.*

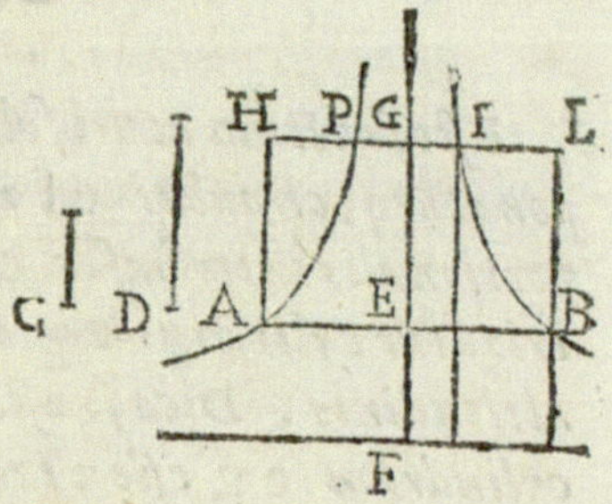

Fiat, vt c *ad* d, *ita data* ef *ad* fg; *& per* g *ducatur planum* hl. *Eritq;* c *ad* d, *vt* ef *ad* fg, *nempe(ob æqualia rectangula) vt* ig *ad* be, *hoc est vt frustum* ai *ad cylindrum* al. *Quod &c.*

Si verò datum planum secans sit pi, *& solidum secandum sit inferius versus* f *iterum eadem lege, ita procedemus. Fiat vt* c *ad* d, *ita* ef *ad datam* fg. *& per* e *ducatur planum* ab. *Eritq; frustum* ai *ad cylindrum* al, *vt* gi *ad* eb, *siue vt* ef *ad* fg, *hoc est vt* c *ad* d. *Quod erat &c.*

Corollarium XX.

Esto solidum acutum sectum plano ab. *oportet illud iterum secare versus* f. *ita vt frustum intersectiones comprahensum, ad inscriptum sibi cylindrum quamlibet datam rationem maioris inæqualitatis habeat, vt* c *ad* d.

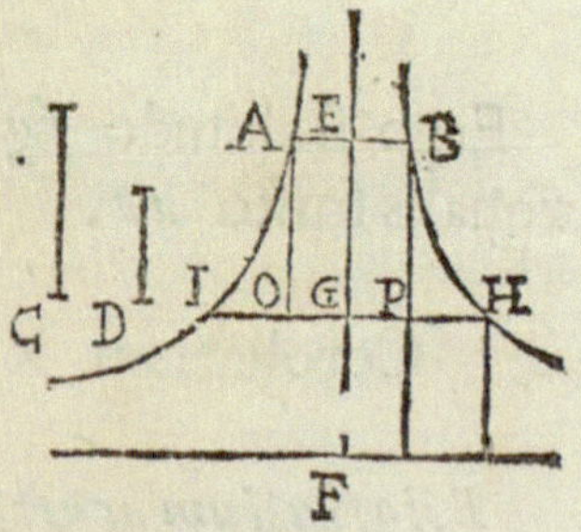

Fiat, vt c *ad* d, *ita data* ef *ad* fg; *ductoq; per* g *plano* ih. *Erit frustum* ib *ad cylindrum inscriptum* ob, *vt* gh *ad* eb, *siue vt* ef *ad* fg, *siue vt* c *ad* d. *Quod &c.*

Si vero planum secans datum sit ih, *& secandum sit solidũ iterum eadem lege versus infinitam longitudinem. fiat vt* c *ad*

R d *ita*

Corollarium XIX.

Efto ſolidum acutum ſectum plano ab. *Oportet illud ſecare iterũ alio plano* pi, *ita vt fruſtum* apib *ad cylindrum ſibi circumſcriptũ, ſit vt* c *ad* d; *dummodo ratio* c *ad* d, *ſit minoris inæqualitatis.*

Fiat, vt c *ad* d, *ita data* ef *ad* fg; *& per* g *ducatur planum* hl. *Eritq*; c *ad* d, *vt* ef *ad* fg, *nempe (ob ęqualia rectangula) vt* ig *ad* be, *hoc est vt frustum* ai *ad cylindrum* al. *Quod &c.*

Si verò datum planum ſecans ſit pi, *& ſolidum ſecandum ſit inferius verſus* f *iterum eadem lege, ita procedemus. Fiat vt* c *ad* d, *ita* ef *ad datam* fg. *& per* e *ducatur planum* ab. *Eritq; frustum* ai *ad cylindrum* al, *vt* gi *ad* eb, *ſiue vt* ef *ad* fg, *hoc eſt vt* c *ad* d. *Quod erat &c.*

Corollarium XX.

Efto ſolidum acutum ſectum plano ab. *oportet illud iterum ſecare verſus* f. *ita vt frustum inter ſectiones compræhenſum, ad inſcriptum ſibi cylindrum quamlibet datam rationem maioris inæqualitatis habeat, vt* c *ad* d.

Fiat, vt c *ad* d, *ita data* ef *ad* fg; *ductoq; per* g *plano* ih. *Erit frustum* ib *ad cylindrum inſcriptum* ob, *vt* gh *ad* eb, *ſiue vt* ef *ad* fg, *ſiue vt* c *ad* d. *Quod &c.*

Si vero planum ſecans datum ſit ih, *& ſecandum ſit ſolidũ iterum eadem lege verſus infinitam longitudinem. fiat vt* c *ad*

R d *ita*

d *ita* ef *ad datam* fg. *Eritq; frustum* ib *ad cylindrum* ob *vt* ef *ad* fg, *nempe vt* c *ad* d. *Quod &c.*

Corollarium XXI.

Esto frustum acuti solidi abcd *ponaturq; circulus* ef *medius proportionalis intrr bases* ad, bc, *& erigatur cylindrus* eg *cuiuscunq; altitudinis. Dico frustum* ac *ad cylindrum* eg *esse vt recta* il *ad* gf.

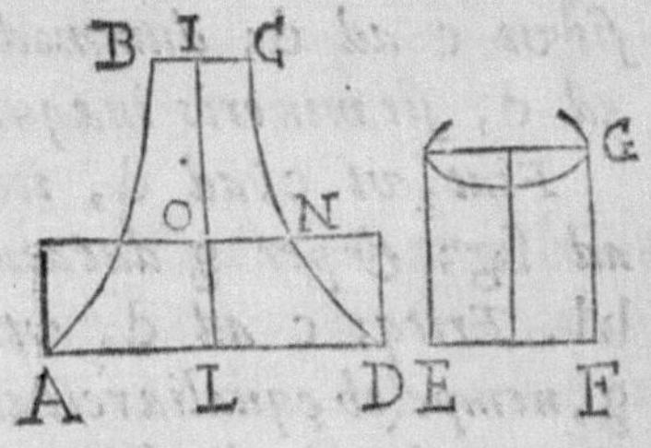

Fiat enim vt recta ad *ad* bc *ita* il *ad* lo, *& ad altitudinẽ* lo *erigatur cylindrus* an, *qui ęqualis erit frusto* ac (*per coroll. 13.*) *Iam cylindrus* an *ad cylindrum* eg, *rationem habet compositam ex ratione basium, nempe quadrati* ad *ad* ef; *hoc est ex ratione rectæ* ad *ad* bc; *siue potius rectę* il *ad* lo *& ex ratione altitudinum, nempe* lo *ad* fg. *Ergo cylindrus.* an *ad* eg, *erit vt recta* il *ad* fg; *Propterea etiam frustũ* ac *ad cylindrum* eg *erit vt* il *ad* fg. *Quod &c.*

Scholium.

Ergo si altitudo *fg* fiat æqualis ipsi *il* erit cylindrus *eg* æqualis frusto *ac*.

Corollarium XXII.

Esto frustum acuti solidi abcd, *quod habeat alteram ex suis basibus; quæcunq; illa sit) puta* ad, *ęqualem basi* em *cylindri* eg. *Dico frustum* ac *ad cylindrum* eg *esse vt rectãgulum sub diametro inę*

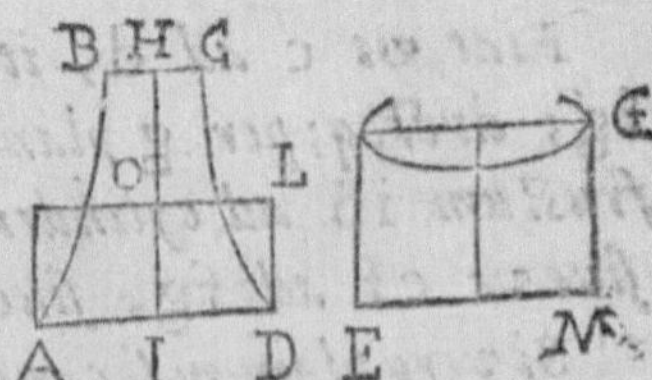

qualis

d *ita* ef *ad datam* fg. *Eritq; fruſtum* ib *ad cylindrum* ob *vt* ef *ad* fg, *nempe vt* c *ad* d. *Quod &c.*

Corollarium XXI.

Eſto frustum acuti ſolidi abcd *ponaturq; circulus* ef *medius proportionalis inter baſes* ad, bc, *& erigatur cylindrus* eg *cuiuſcunq; altitudinis. Dico fruſtum* ac *ad cylindrum* eg *eſse vt recta* il *ad* gf.

Fiat enim vt recta ad *ad* bc *ita* il *ad* lo, *& ad altitudinẽ* lo *erigatur cylindrus* an, *qui ęqualis erit fruſto* ac (*per coroll. 13.*) *Iam cylindrus* an *ad cylindrum* eg, *rationem habet compoſitam ex ratione baſium, nempe quadrati* ad *ad* ef; *hoc est ex ratione rectæ* ad *ad* bc; *ſiue potius rectę* il *ad* lo *& ex ratione altitudinum, nempe* lo *ad* fg. *Ergo cylindrus.* an *ad* eg, *erit vt recta* il *ad* fg; *Propterea etiam fruſtũ* ac *ad cylindrum* eg *erit vt* il *ad* fg. *Quod &c.*

Scholium.

Ergo ſi altitudo *fg* fiat æqualis ipſi *il* erit cylindrus *eg* æqualis frufto *ac.*

Corollarium XXII.

Esto frustum acuti ſolidi abcd, *quod habeat alteram ex ſuis baſibus quæcunq*[173]*; illa ſit) puta* ad, *ęqualem baſi* em *cylindri* eg. *Dico fruſtum* ac *ad cylindrum* eg *eſſe vt rectãgulum ſub diametro inę*

qualis

[173] *quæcunq*] *(quæcunq.*

qualis basis, & sub altitudine frusti, ad rectangulum per axem cylindri. Nempe vt rectangulum bc, hi *ad rectangulum* eg.

Fiat vt ad *ad* bc *ita* hi *ad* io; *erectoq; cylindro* al *cum altitudine* io, *erit frustum* ac *æquale cylindro* al. *Iam cylindrus* al *ad* eg, *ob æquales bases, est vt* oi *ad* gm, *Sed ratio rectæ* oi *ad* gm, *componitur ex ratione rectæ* oi *ad* ih, *siue* bc *ad* ad, *hoc est* bc *ad* em; *& ex ratione* hi *ad* gm. *Ergo ratio* oi *ad* gm *erit eadem quæ est rectang.* bc, hi *ad rectangulum sub* em, mg. *Propterea etiam cylindrus* al, *siue frustum* ac *ad cylindrum* eg *erit vt rectangulum* bc, hi, *ad rectangulum* emg. *Quod &c.*

Corollarium XXIII,

Si frustum acuti solidi abcd *& cylindrus* ef *æquales altitudines habuerint. Erit frustum* ac *ad cylindrum* ef *vt rectangulum sub* bc, ad, *ad quadratum* eg.

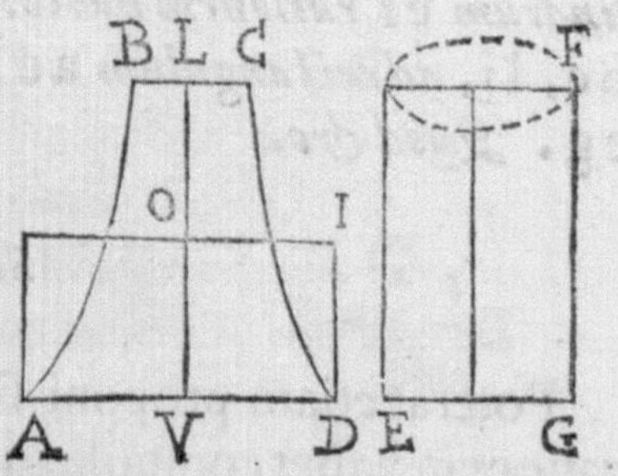

Fiat vt ad *ad* bc, *ita* lu *ad* uo. *eritq; frustum* ac *æquale cylindro* ai *cuius altitudo sit* uo. *Iam cylindrus* ai *ad cylindrum* ef, *rationem habet cōpositam ex ratione altitudinum* uo *ad* gf; *siue* uo *ad* ul *siue* bc *ad* ad; *nempe ex ratione rectang.* bc, ad, *ad quadratum* ad. *Et ex ratione basium; nempe quadrati* ad *ad* eg. *Ergo cylindrus* ai, *siue frustum* ac, *ad cylindrum* ef, *erit ut rectang. sub* bc, ad, *ad quadratum* eg. *Quod &c.*

Corollarium XXIV.

Frustum acuti solidi abcd, *ad cylindrum quemlibet* ef, *rationem habet compositam ex ratione rectanguli* bc, li *ad rectangulum* ad, gf; *& ex ratione quadr.* ad *ad quadratū* eg.

qualis bafis, & fub altitudine frusti, ad rectangulum per axem cylindri. Nempe vt rectangulum bc, hi *ad rectangulum* eg.

Fiat vt ad *ad* bc *ita* hi *ad* io; *erectoq; cylindro* al *cum altitudine* io, *erit fruftum* ac *æquale cylindro* al. *Iam cylindrus* al *ad* eg, *ob æquales bafes, eft vt* oi *ad* gm, *Sed ratio rectæ* oi *ad* gm, *componitur ex ratione rectæ* oi *ad* ih, *fiue* bc *ad* ad, *hoc eft* bc *ad* em; *& ex ratione* hi *ad* gm. *Ergo ratio* oi *ad* gm *erit eadem quæ eft rectang.* bc, hi *ad rectangulum fub* em, mg. *Propterea etiam cylindrus* al, *fiue frustum* ac *ad cylindrum* eg *erit vt rectangulum* bc, hi, *ad rectangulum* emg. *Quod &c.*

Corollarium XXIII,

Si frustum acuti folidi abcd *& cylindrus* ef *æquales altitudines habuerint. Erit frustum* ac *ad cylindrum* ef *vt rectangulum fub* bc, ad, *ad quadratum* eg.

Fiat vt ad *ad* bc, *ita* lu *ad* uo. *eritq; fruftum* ac *æquale cylindro* ai *cuius altitudo fit* uo. *Iam cylindrus* ai *ad cylindrum* ef, *rationem habet cōpofitam ex ratione altitudinum* uo *ad* gf; *fiue* uo *ad* ul *fiue* bc *ad* ad; *nempe ex ratione rectang.* bc, ad, *ad quadratum* ad. *Et ex ratione bafium; nempe quadrati* ad *ad* eg. *Ergo cylindrus* ai, *fiue fruftum* ac, *ad cylindrum* ef, *erit ut rectang. fub* bc, ad, *ad quadratum* eg. *Quod &c.*

Corollarium XXIV.

Fruftum acuti folidi abcd, *ad cylindrum quemlibet* ef, *rationem habet compofitam ex ratione rectanguli* bc, li *ad rectangulum* ad, gf; *& ex ratione quadr.* ad *ad quadratũ* eg.

R 2 *Fiat*

Fiat ut ad *ad* bc, *ita* li *ad* io; *eritque cylindrus* au *aequalis frusto* ac. *Iam recta* io *ad rectam* gf, *est ut rectangulum sub* bc, li *ad rectangulum sub* ad, gf. (*nam ratio rectae* io *ad* gf, *componitur ex ratione* io *ad* il, *sive* bc *ad* ad; *& ex ratione* il *ad* gf. *Ergo recta* io *ad* gf, *est ut rectangulum* bc, il *ad rectangulum* ad, gf.) *Sed cylindrus* au *ad cylindrum* ef *rationem habet compositam ex ratione* io *ad* gf, *nempe ex ratione rectanguli* bc, li, *ad rectangulum* ad, gf; *& ex ratione quadrati* ad *ad* eg. *Propterea etiam frustum* ac *ad cylindrum* ef *rationem habebit compositam ex ratione rectanguli* bc, li, *ad rectangulum* ad, gf; *& ex ratione quadrati* ad *ad* eg. *Quod &c.*

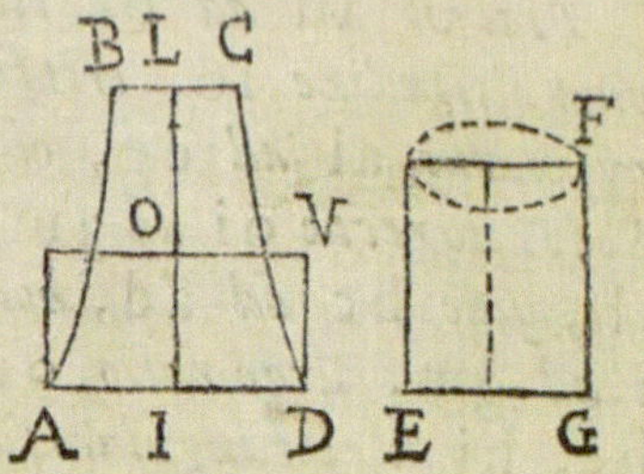

Scholium.

Poterat etiam proponi sic. Frustum *ac* ad cylindrum *ef*, rationem habet compositam ex ratione rectanguli *ad*, *il*, ad rectangulum *bc*, *fg*; & ex ratione quadrati *bc*, ad *eg*.

Corollarium XXV.

Sint duo frusta acutorum solidorum qualiacunque. Dico frustum hbce *ad frustum* dfga, *habere rationem compositam ex ratione rectangulorum basium, & ex ratione altitudinum; nempe ex ratione rectanguli* bc, he *ad rectangulum* fg, da; *& ex ratione rectae* in *ad* ml.

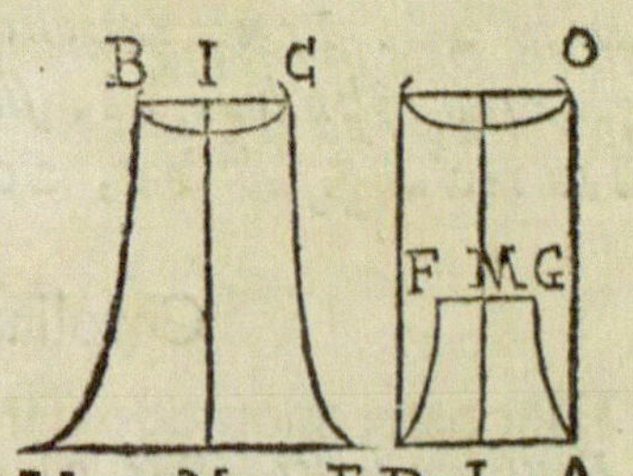

Fiat

Fiat vt ad *ad* bc, *ita* li *ad* io; *eritque cylindrus* au *æqualis frusto* ac. *Iam recta* io *ad rectam* gf, *eſt vt rectangulum ſub* bc, li *ad rectangulum ſub* ad, gf. *(nam ratio rectæ* io *ad* gf, *componitur ex ratione* io *ad* il, *ſiue* bc *ad* ad; *& ex ratione* il *ad* gf. *Ergo recta* io *ad* gf, *est vt rectangulum* bc, il *ad rectangulum* ad, gf.*) Sed cylindrus* au *ad cylindrum* ef *rationem habet compoſitam ex ratione* io *ad* gf, *nempe ex ratione rectanguli* bc, li, *ad rectangulum* ad, gf; *& ex ratione quadrati* ad *ad* eg. *Propterea etiam fruſtum* ac *ad cylindrum* ef *rationem habebit compoſitam ex ratione rectanguli* bc, li, *ad rectangulum* ad, gf; *& ex ratione quadrati* ad *ad* eg. *Quod &c.*

Scholium.

Poterat etiam proponi ſic. Fruftum *ac* ad cylindrum *ef,* rationem habet compoſitam ex ratione rectanguli *ad, il,* ad rectangulum *bc, fg*; & ex ratione quadrati *bc,* ad *eg.*

Corollarium XXV.

Sint duo fruſta acutorum ſolidorum qualiacunque. Dico frustum hbce *ad fruſtum* dfga, *habere rationem compoſitam ex ratione rectangulorum baſium, & ex ratione altitudinum; nempe ex ratione rectanguli* bc, he *ad rectangulum* fg, da; *& ex ratione rectæ* in *ad* ml.
Fiat

Problema Secundum. 133

Fiat enim ſuper baſi da *cylindrus* do *cum altitudine* ao, *quæ ſit æqualis ipſi* ni. *Eritq; (per Coroll. 23.) fruſtum* hc *ad cylindrum* do, *vt rectangulum* bc, he *ad quadratum* da. *Cylindrus autem* do *ad fruſtum* dg *eſt (per Coroll. 22.) vt rectangulum* dao *ad rectangulum* fg, ml. *Nempe ad illud, rationem habet compoſitam ex ratione rectæ* da *ad* fg, *ſiue ex ratione quadrati* da *ad rectangulum* da, fg. *Et ex ratione rectæ* oa *ad* ml, *ſiue* in *ad* ml. *Ratio itaq; fruſti* hc *ad fruſtum* dg *componitur ex rationibus, rectanguli* bc, he, *ad quadratum* da; *& ex ratione quadrati* da *ad rectangulum* da, fg: *& ex ratione rectæ* in *ad* ml. *Demptoq. medio illo termino ſuperfluo nempe quadrato* da. *Erit ratio fruſti* hc *ad fruſtum* dg *compoſita ex ratione rectanguli* bc, he, *ad rectangulum* da, fg; *& ex ratione rectæ* in *ad* ml. *Quod erat &c.*

Corollarium XXVI.

Eſto fruſtum ſolidi acuti abcd *ſectũ plano* hl; *ducaturq;* bn *parallela ad axẽ*. *Dico, totum fruſtum* abcd *ad partem* hbcl, *eſſe vt* an *ad* hi.

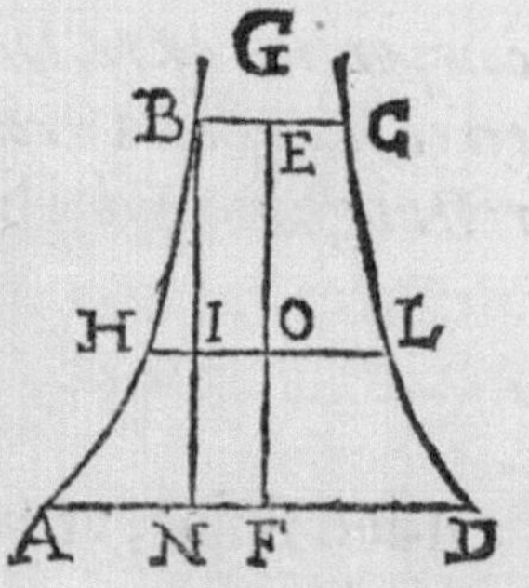

Nam ſolidum acutum agd, *ad ſolidũ* bgc, *eſt vt* af *ad* be, *ſiue vt* af *ad* fn; *& diuidendo fruſtum* abcd *ad ſolidum acutum* bgc *erit vt* an *ad* nf, *ſiue vt* an *ad* io. *Solidum verò* bgc *ad fruſtum* hc *(ſimili argumento) eſt vt* oi *ad* ih. *Ergo ex æquo fruſtum* ac, *ad* hc *erit vt* an *ad* hi. *Quod &c.*

Scholium.

Hinc patet quomodo datum fruſtum acuti ſolidi in data ratione ſecari poſſit, quod tamen ad finem Corollariorum elegantiori problemate exequemur.

Eſto

Fiat enim ſuper baſi da *cylindrus* do *cum altitudine* ao, *quæ ſit æqualis ipſi* ni. *Eritq; (per Coroll. 23.) fruſtum* hc *ad cylindrum* do, *vt rectangulum* bc, he *ad quadratum* da. *Cylindrus autem* do *ad frustum* dg *eſt (per Coroll. 22.) vt rectangulum* d a o *ad rectangulum* fg, ml. *Nempe ad illud, rationem habet compoſitam ex ratione rectæ* da *ad* fg, *ſiue ex ratione quadrati* da *ad rectangulum* da, fg. *Et ex ratione rectæ* oa *ad* ml, *ſiue* in *ad* ml. *Ratio itaq; fruſti* he *ad fruſtum* dg *componitur ex rationibus, rectanguli* bc, he, *ad quadratum* da; *& ex ratione quadrati* da *ad rectangulum* da, fg: *& ex ratione rectæ* in *ad* ml. *Demptoq. medio illo termino ſuperfluo nempe quadrato* da. *Erit ratio frusti* hc *ad frustum* dg *compoſita ex ratione rectanguli* bc, he, *ad rectangulum* da, fg; *& ex ratione rectæ* in *ad* ml. *Quod erat &c.*

Corollarium XXVI.

Eſto fruſtum ſolidi acuti abcd *ſectũ plano* hl; *ducaturq;* bn *parallela ad axẽ. Dico, totum fruſtum* abcd *ad partem* hbcl, *eſſe vt* an *ad* hi.

Nam ſolidum acutum agd, *ad ſolidũ* bgc, *eſt vt* af *ad* be, *ſiue vt* af *ad* fn; *& diuidendo fruſtum* abcd *ad ſolidum acutum* bgc *erit vt* an *ad* nf, *ſiue vt* an *ad* io. *Solidum verò* bgc *ad fruſtum* hc *(ſimili argumento) eſt vt* oi *ad* ih. *Ergo ex æquo. fruſtum* ac, *ad* hc *erit vt* an *ad* hi. *Quod &c.*

Scholium.

Hinc patet quomodo datum fruſtum acuti ſolidi in data ratione ſecari poſſit, quod tamen ad finem Corollariorum elegantiori problemate exequemur.

Esto

134 De Solido Hyperbolico

Corollarium XXVII,

Esto frustum solidi acuti abcd. *cuius axis* mi. *sitq; centrum hyperbolæ punctū* h. *Secetur deinde frustum* ac *plano quocunq;* ef *ad axem erecto. Dico frustum* af, *ad frustum* ec. *esse vt rectangulum sub* il, hm, *ad rectangulum sub* hi, lm.

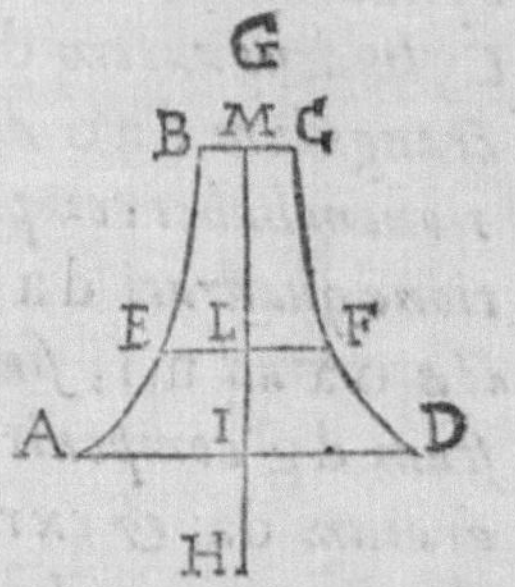

Nam frustum af *ad frustum* ec, *rationem habet compositam ex ratione frusti* af *ad acutum solidum* egf; *& ex ratione solidi longi* egf, *ad frustum* ec. *Sed quia solidum acutum* agd *ad acutum solidum* egf *est vt recta* ai *ad* el; *siue vt recta* lh *ad* hi, *erit diuidendo frustum* af *ad solidum* egf *vt* li *ad* ih. *Amplius: Solidum* egf *ad solidum* bgc *est vt* el *ad* bm, *siue vt* mh *ad* hl; *& per conuersionem rationis, erit solidum* egf *ad frustum* ec, *vt* hm *ad* ml. *Patet ergò quòd ratio frusti* af *ad frustum* ec, *componitur ex ratione* li *ad* ih, *& ex ratione* hm *ad* ml *Proptereà frustum* af *ad* ec, *erit vt rectangulum sub* li, hm, *ad rectangulum sub* ih, lm. *Quod &c.*

Scholium.

Ideò si fiat, vt *mh* ad *hi*, ita *ml*, ad *li*. Bifariam secabitur frustum *ac* à plano per punctum *l* ducto. Aequalia enim erunt ipsa rectangula.

Corollarium XXVIII.

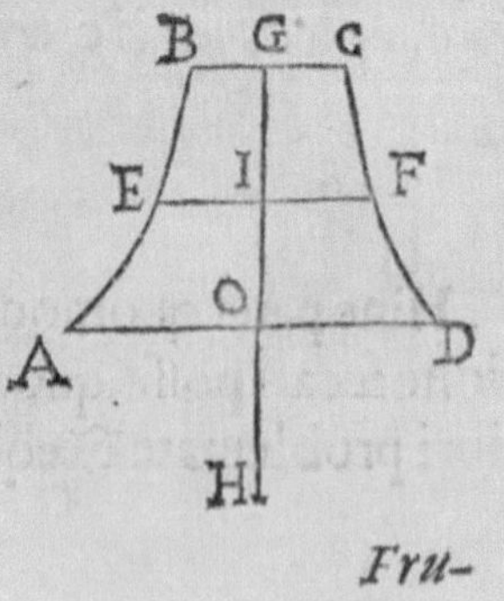

Si axis frusti abcd *bifariam secetur à plano* ef. *Erunt portiones inter se, nempe* af. *ad* ec *vt recta* ad *ad* bc. *scilicet vt diametri basium remotarum.*

Fru-

Corollarium XXVII,

Efto fruftum folidi acuti abcd. *cuius axis* mi. *fitq; centrum hyperbolæ punctũ* h. *Secetur deinde fruftum* ac *plano quocunq;* ef *ad axem erecto. Dico frustum* af, *ad fruftum* ec. *effe vt rectangulum fub* il, hm, *ad rectangulum fub* hi, lm.

Nam frustum af *ad fruftum* ec, *rationem habet compofitam ex ratione frufti* af *ad acutum folidum* egf; *& ex ratione folidi longi* egf, *ad fruftum* ec. *Sed quia folidum acutum* agd *ad acutum folidum* egf *eft vt recta* ai *ad* el; *fiue vt recta* lh *ad* hi, *erit diuidendo fruftum* af *ad folidum* egf *vt* li *ad* ih. *Amplius: Solidum* egf *ad folidum* bgc *est vt* el *ad* bm, *fiue vt* mh *ad* hl; *& per conuerfionem rationis, erit folidum* egf *ad frustum* ec, *vt* hm *ad* ml. *Patet ergò quòd ratio frufti* af *ad frustum* ec, *componitur ex ratione* li *ad* ih, *& ex ratione* hm *ad* ml[174] *Propterea frustum* af *ad* ec, *erit vt rectangulum fub* li, hm, *ad rectangulum fub* ih, lm. *Quod &c.*

Scholium.

Ideò fi fiat, vt *mh* ad *hi,* ita *ml,* ad *li.* Bifariam fecabitur fruftum *ac* à plano per punctum *l* ducto. Aequalia enim erunt ipfa rectangula.

Corollarium XXVIII

Si axis frusti abcd *bifariam fecetur à plano* ef. *Erunt portiones inter fe, nempe* af. adec *vt recta* ad *ad* bc. *fcilicet vt diametri bafium remotarum.*

Fru-

[174] ml] ml.

Fruſtum enim a f *ad* e c, *eſt vt rectangulum ſub* h g, o i *ad rectangulum ſub* h o, i g *per præced: Sed* o i, *&* i g, *altitudines rectangulorum ſunt æquales, Ergò fruſtum* a f *ad* e c, *erit vt* g h *ad* h o, *ſiue vt* a d *ad* b c. *Quod &c.*

Scholium.

Hinc patet; quò d ſi in ſolido longo hyperbolico quotcunq; ſumantur axis portiones deinceps æquales *a*, *b*, *c*, *d*, *e*. vbicunq; fiat initium. Erit fruſtum *fg* ad *gh* vt recta *fa* ad *hc*. Fruſtum verò *gh* ad *hi* erit ut *mb* ad *nd*. & fruſtum *hi* ad *il* ut *hc*, ad *le*. & ſic in infinitum.

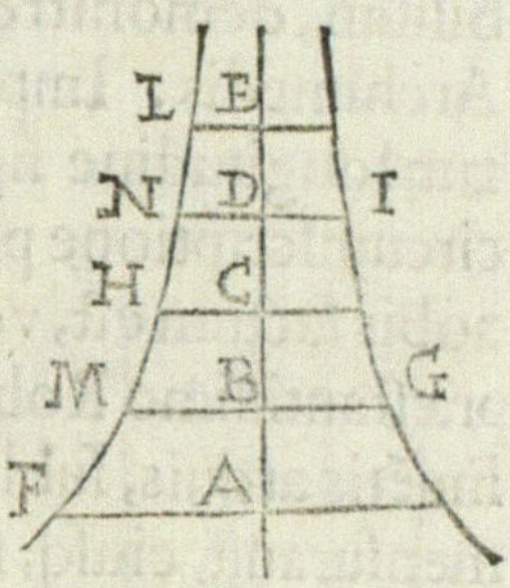

Corollarium XXIX.

Datum acuti ſolidi fruſtum a b c d *in data ratione ſecare; puta vt* e *ad* f.

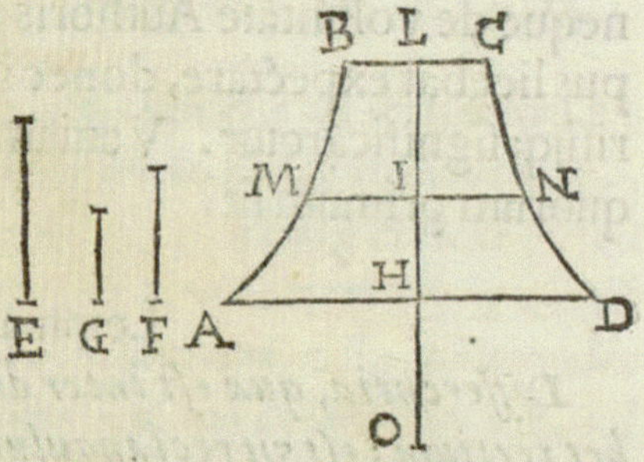

Fiat, vt recta a d *ad* b c, *ita* e *ad aliam quæ ſit* g. *Deinde fiat, vt* g *ad* f, *ita* h i *ad* i l, *& per* i *ducatur planũ* m n

Iam fruſtum a n *ad* m c *eſt vt rectangulum* l o, i h, *ad rectangulum* l i, o h. *Ergò ratio fruſti* a n *ad* m c *componitur ex ratione laterum* l o *ad* o h, *ſiue* a d *ad* b c, *ſiue* e *ad* g. *Et ex ratione laterum* h i *ad* i l, *ſiue* g *ad* f. *Ergò ratio fruſti* a n *ad* m c. *componitur ex ratione* e *ad* g, *&* g *ad* f. *Proptereà erit* a n *fruſtum ad* m c *vt* e *ad* f. *Quod &c.*

Iam iſta ſufficiat demonſtrauiſſe, ex plurimis Theorematibus, quæ ex fæcundiſſimo hoc ſolido deriuari poterant. Interim ad promiſſam demonſtrationem accedamus, quam tamen præterire poterit quicunq; iam allatà contentus fuerit.

DE

Fruftum enim af *ad* ec, *est vt rectangulum fub* hg, oi *ad rectangulum fub* ho, ig *per præced: Sed* oi, & ig, *altitudines rectangulorum funt æquales, Ergò frustum* af *ad* ec, *erit vt* gh *ad* ho, *fiue vt* ad *ad* bc. *Quod &c.*

Scholium.

Hinc patet; quòd fi in folido longo hyperbolico quotcunq; fumantur axis portiones deinceps æquales *a, b, c, d, e.* vbicunq; fiat initium. Erit fruftum *fg* ad *gh* vt recta *fa* ad *hc*. Fruftum verò *gh* ad *hi* erit ut *mb* ad *nd*. & fruftum *hi* ad *il* ut *hc,* ad *le.* & fic in infinitum.

Corollarium XXIX.

Datum acuti folidi fruftum abcd *in data ratione fecare; puta vt* e *ad* f.

Fiat, vt recta ad *ad* bc, *ita* e *ad aliam quæ fit* g. *Deinde fiat, vt* g *ad* f, *ità* hi *ad* il, *& per* i *ducatur planũ* mn[175].

Iam fruftum an *ad* mc *eft vt rectangulum* lo, ih, *ad rectangulum* li, oh. *Ergò ratio frufti* an *ad* mc *componitur ex ratione laterum* lo *ad* oh, *fiue* ad *ad* bc, *fiue* e *ad* g. *Et ex ratione laterum* hi *ad* il, *fiue* g *ad* f. *Ergò ratio frufti* an *ad* mc. *componitur ex ratione* e *ad* g, *&* g *ad* f. *Propterea erit* an *fruftum ad* mc *vt* e *ad* f. *Quod &c.*

Iam ifta fufficiat demonftrauiffe, ex plurimis Theorematibus, quæ ex fæcundiffimo hoc folido deriuari poterant. Interim ad promiffam demonftrationem accedamus, quam tamen præterire poterit quicunq; iam allatà contentus fuerit.

D *E*

[175] mn] mn.

136

De Dimenſione Acuti ſolidi Hyperbolici iuxtà methodum Antiquorum.

SVpereſt nunc vt Theorema illud, quod poſt lemma Quintum oſtendimus per methodum, & doctrinam Indiuiſibilium, demonſtremus iterùm more Antiquorum, & præcipuè Archimedis. Impoſſibile enim quodammodo videtur, infinitam longitudine figuram ſub ſolita figurarum inſcriptione, & circumſcriptione poſſe comprehendi. Tamen id non ſolum à nobis factum eſt, verum etiam à Clariſſimo viro, & Geometra præſtantiſſimo Roberuallio, qui noſtrum ſolidum hyperbolicū inuētis arduis, ſublimibus, acutiſſimis, & vt breuitèr dicam ſuis, menſurauit, eiuſq; fruſtum in data ratione diſſecuit. Abſtineo ab illius demōſtrationis editione inuitus. comparuit enim eius epiſtola eo prorſus tempore, quo iàm hęc pręlis ſubijcerentur, neque de voluntate Authoris ſatis conſtabat, neque iam per tēpus licebat expectare, donec illius beneplacitum ex Gallia Pariſijq; ſignificaretur. Veniamus itaq; ad lemmata opportuna, quorum primum ſit.

Lemma Primum.

Differentia, quæ eſt inter duos circulos, ad circulum quemlibet tertium; eſt vt rectangulum comprahenſum ſub differentia, & aggregato ſemidiametrorum eorundem circulorum, ad quadratum ſemidiametri tertij illius circuli.

Vocetur autem talis differentia duorum circulorum, quando concentrici fuerint, Armilla.

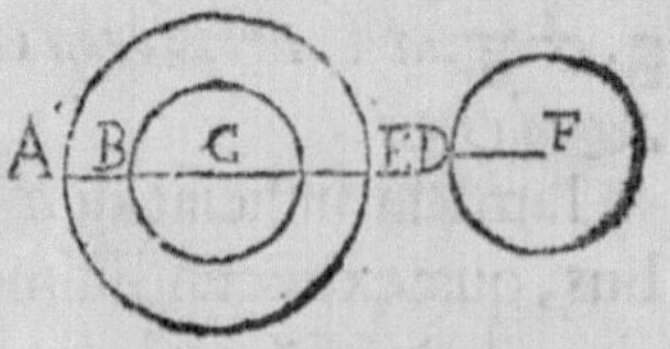

Eſto Armilla, ſiue differentia duorum circulorum concentricorum, illa cuius latitudo ab, *centrum verò* c. *Dico Armillam* ab, *ad circulū quemlibet* df; *eſſe vt rectangulum*

abc,

136

De Dimenſione Acuti ſolidi Hyperbolici iuxtà methodum Antiquorum.

SVpereſt nunc vt Theorema illud, quod poſt lemma Quintum oſtendimus per methodum, & doctrinam Indiuiſibilium, demonſtremus iterùm more Antiquorum, & præcipuè Archimedis. Impoſſibile enim quodammodo videtur, infinitam longitudine figuram ſub ſolita figurarum inſcriptione, & circumſcriptione poſſe compræhendi. Tamen id non ſolum à nobis factum eſt, verum etiam à Clariſſimo viro, & Geometra præſtantiſſimo Roberuallio, qui noſtrum ſolidum hyperbolicũ inuẽtis arduis, ſublimibus, acutiſſimis, & vt breuitèr dicam ſuis, menſurauit, eiuſq; fruſtum in data ratione diſſecuit. Abſtineo ab illius demõſtrationis editione inuitus. comparuit enim eius epiſtola eo prorſus tempore, quo iàm hęc pręlis ſubijcerentur, neque de voluntate Authoris ſatis conſtabat, neque iam per tẽpus licebat expectare, donec illius beneplacitum ex Gallia Pariſijq; ſignificaretur. Veniamus itaq; ad lemmata opportuna, quorum primum ſit.

Lemma Primum.

Differentia, quæ eſt inter duos circulos, ad circulum quemlibet tertium; eſt vt rectangulum compræhenſum ſub differentia, & aggregato ſemidiametrorum eorundem circulorum, ad quadratum ſemidiametri tertij illius circuli.

Vocetur autem talis differentia duorum circulorum, quando concentrici fuerint, Armilla.

Eſto Armilla, ſiue differentia duorum circulorum concentricorum, illa cuius latitudo ab, *centrum verò c. Dico Armillam* ab, *ad circulũ quemlibet* df; *eſſe vt rectangulum*

a b c,

abc, *ad quadratum semidiametri* df.

Nam circulus ex radio ac, *ad circulum ex radio* cb, *est* per 1. duodecimi.
vt quadratum ac, *ad quadratum* cb; *& diuidendo Armilla*
ab, *ad circulum ex radio* cb; *erit vt rectangulum* abc, *ad* ex 5. secundi.
quadratum cb. *Circulus verò ex radio* cb, *ad circulum ex radio* df, *est vt quadratum* cb, *ad quadratum* df. *Ergò ex æquò, erit Armilla* ab, *ad circulum* df, *vt rectangulum* abc, *ad quadratum* df. *Quod erat &c.*

Lemma II.

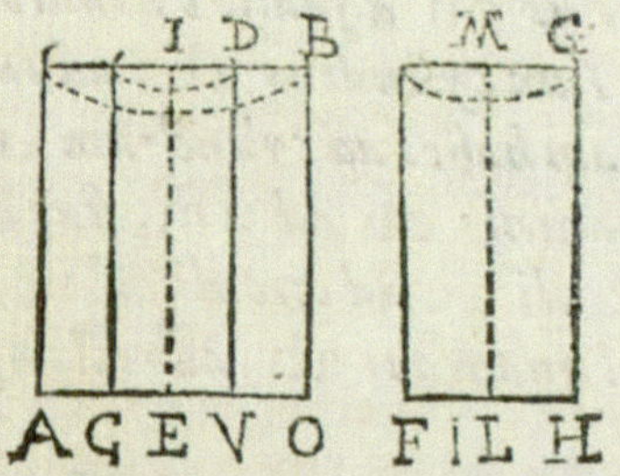

Si ex cylindro recto ab, *ablatus fuerit cylindrus* cd, *circa communem axem* ie *constitutus; reliquum solidum excauatum quod remanet, æquale erit cylindro cuidam recto* fg, *cuius quidem basis* fh *æqualis sit Armillæ, quę circa centrum* e *latitudinem habet* ac; *altitudo verò* lm *æqualis sit altitudini* ei.

Vocetur autem tale solidum excauatum, tubus cylindricus.

Quoniam tres cylindri ab, cd, fg, *æquealti sunt; Erit cylindrus* ab *ad* cd, *vt circulus* ao *ad circulum* cu. *& diuidendo erit tubus cylindricus ad cylindrum* cd, *vt armilla* ac *ad circulum* cu; *sed cylindrus* cd *ad cylindrum* fg, *est vt circulus* cu *ad circulum* fh. *Ergo ex æquo erit tubus cylindricus* ab, *ad cylindrum* fg, *vt armilla* ac *ad circulum* fh. *Sed armilla* ac *circulo* fh *supponitur æqualis; ergò & tubus cylindricus* ab, *ęqualis erit cylindro* fg. *Quod erat &c.*

Lemma III.

Quilibet cylindrus rectus ab, *ad quemlibet tubum cylindricum rectum* cd, *rationem habet compositam ex ratione al-*

S *titudi-*

abe, *ad quadratum ſemidiametri* df. *per L. duedecimi.*

Nam circulus ex radio ac, *ad circulum ex radio* cb, *eſt vt quadratum* ac, *ad quadratum* cb; *& diuidendo Armilla* ab, *ad circulum ex radio* cb; *erit vt rectangulum* abe, *ad quadratum* cb. *Circulus verò ex radio* cb, *ad circulum ex radio* df, *eſt vt quadratum* cb, *ad quadratum* df. *Ergò ex æquò, erit Armilla* ab, *ad circulum* df, *vt rectangulum* abe, *ad quadratum* df. *Quod erat &c.* *ex 5. ſecundi.*

Lemma II.

Si ex cylindro recto ab, *ablatus fuerit cylindrus* cd, *circa communem axem* ie *constitutus; reliquum ſolidum excauatum quod remanet, æquale erit cylindro cuidam recto* fg, *cuius quidem baſis* fh *æqualis ſit Armillæ, quę circa centrum* e *latitudinem habet* ac; *altitudo verò* lm *æqualis ſit altitudini* ei.

Vocetur autem tale ſolidum excauatum, tubus cylindricus.

Quoniam tres cylindri ab, cd, fg, *æquealti ſunt; Erit cylindrus* ab *ad* cd, *vt circulus* ao *ad circulum* cu. *& diuidendo erit tubus cylindricus ad cylindrum* cd, *vt armilla* ac *ad circulum* cu; *ſed cylindrus* cd *ad cylindrum* fg, *est vt circulus* cu *ad circulum* fh. *Ergo ex æquo erit tubus cylindricus* ab, *ad cylindrum* fg, *vt armilla* ac *ad circulum* fh. *Sed armilla* ac *circulo* fh *ſupponitur æqualis; ergò & tubus cylindricus* ab, *ęqualis erit cylindro* fg. *Quod erat &c.*

Lemma III.

Quilibet cylindrus rectus ab, *ad quemlibet tubum cylindricum rectum* cd, *rationem habet compoſitam ex ratione al-*

S *titudi-*

[176] The numbering of pages restarts from 129. Thus this section would be pp. 129–135.

130 De Solido Hyperbolico

titudinum, nempe e b *ad* f d, *& ex ratione baſium, nempe ex ratione quadrati* a h *ad re-*
Lem 1. *ctangulum* c i f. (*demonſtratum enim eſt ita eſſe circulum* a e *ad armillam* c i, *vt quadratum* a h *ad rectangulum* c i f.)

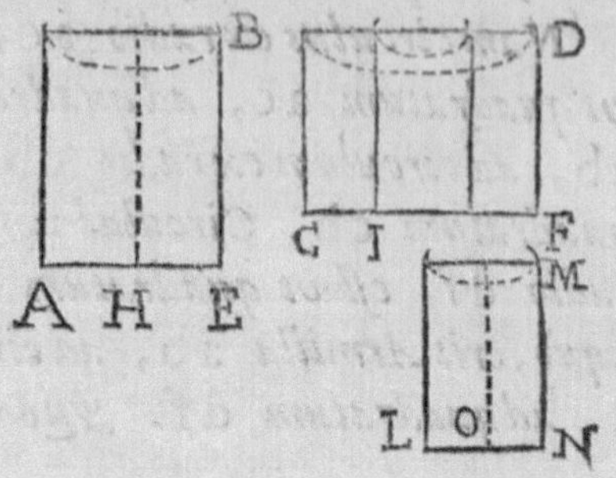

Ponatur cylindrus l m, *cuius altitudo* n m *ſit æqualis altitudini* f d; *baſis verò* l n, *æqualis ſit armillæ* c i; *Et erit, per præcedens lemma, tubus cylindricus* c d *æqualis cylindro* l m.

Iam cylindrus a b, *ad tubum* c d *eandem habebit rationem quam habet ad cylindrum* l m; *nempe compoſitam ex ratione altitudinis* e b *ad* n m, *ſiue ad* f d; *& ex ratione baſium, hoc eſt circuli* a e *ad circulum* l n; *ſiue quadrati* a h *ad quadratum* l o, *vel quadrati* a h, *ad rectangulum* c i f. *Quod erat &c.*

Lemma IV.

Eſto hyperbola cuius aſymptoti ſint a b, b c, *angulum rectum comprehendentes; ſitque hyperbolæ ſemiaxis* b d. (*ſemiaxem appello, quia* b *punctum in quo aſymptoti concurrunt, centrum hyperbolæ eſt.*) *Dico quadratum rectæ* b d, *duplum eſſe cuiuſcunque rectanguli* a c, *inter aſymptotos, & hyperbolam ipſam comprehenſi.*

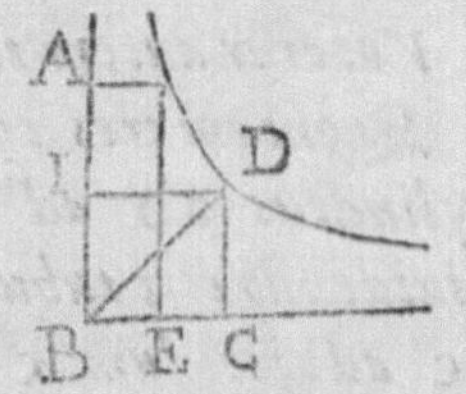

Ducantur d c, d i *aſymptoti æquidiſtantes; eritq; figura* b i d c, *quadratum: cum anguli ad* b *ſemirecti ſint; ſed ad* c, *& ad* i *recti. Ideo quadratum lineæ* b d, *duplum erit quadrati*
ex 12. sec. Con. b i d c; *ſiue rectanguli* a e, *inter aſymptotos, & hyperbolam ipſam comprehenſi. Quod erat &c.*

Lemma V.

Eſto hyperbola a b, *cuius aſymptoti angulum rectum continen-*

titudinum, nempe eb *ad* fd, *& ex ratione baſium, nempe ex ratione quadrati* ah *ad rectangulum* cif. *(demonstratum enim eſt ita eſſe*
Lem 1: *circulum* ae *ad armillam* ci, *vt quadratum* ah *ad rectangulum* cif.)

Ponatur cylindrus lm, *cuius altitudo* nm *ſit æqualis altitudini* fd; *baſis verò* ln, *æqualis ſit armillæ* ci; *Et erit, per præcedens lemma, tubus cylindricus* cd *æqualis cylindro* lm.

Iam cylindrus ab, *ad tubum* cd *eandem habebit rationem quam habet ad cylindrum* lm; *nempe compoſitam ex ratione altitudinis* eb *ad* nm, *ſiue ad* fd; *& ex ratione baſium, hoc eſt circuli* ae *ad circulum* ln; *ſiue quadrati* ah *ad quadratum* lo *vel quadrati* ah, *ad rectangulum* cif. *Quod erat &c.*

Lemma IV.

Esto hyperbola cuius aſymptoti ſint ab, bc, *angulum rectum comprẹhendentes; ſitque hyperbolæ ſemiaxis* bd. (*ſemiaxem appello, quia* b *punctum in quo aſymptoti concurrunt, centrum hyperbolæ est.*) *Dico quadratum rectæ* bd, *duplum eßе cuiuſcunque rectanguli* ae, *inter aſymptotos, & hyperbolam ipſam compræhenſi.*

Ducantur dc, di *aſymptoti aquidistantes; eritq; figura* bidc, *quadratum: cum anguli ad* b *ſemirecti ſint; ſed ad* c, *& ad* i *recti. Ideo*
ex 12. *quadratum lineæ* bd, *duplum erit quadrati* bidc; *ſiue rectanguli* ae, *inter aſymptotos, & hyperbolam ipſam compræhenſi. Quod erat*
ſec. Con. *&c.*

Lemma V.

Esto hyperbola ab, *cuius aſymptoti angulum rectum conti-*

nen-.

uentes sint cd, de; *Sumptisq; duobus punctis* a, b, *vtcumq; in hyperbola, ducantur duæ rectæ* be, ai, *asymptoto* cd. *æquidistantes, &* an, bm *alteri asymptoto* de *parallelę, quæ concurrant in* l. *Tum conuertatur vniuersa figura circa axẽ* cd.

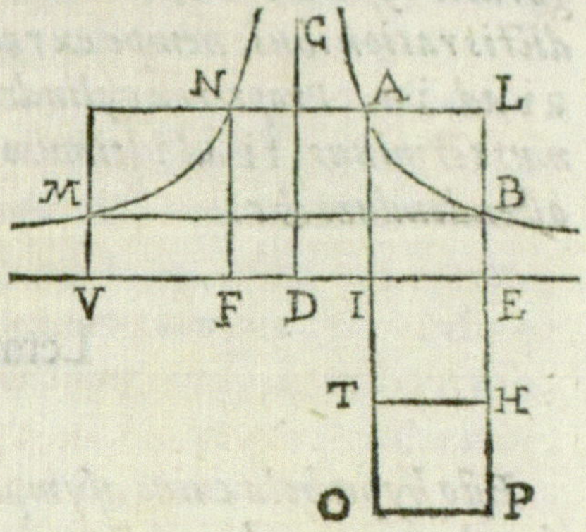

Dico cylindrum quemdam i epo *(cuius quidem basis* io *habeat semidiametrum* it *æqualem semiaxi hyperbolæ; altitudo verò sit intercepta* ie.*) maiorem esse tubo illo cylindrico, qui fit ex conuersione rectanguli* ib *circa axem* cd; *Minorem verò tubo illo qui fit ex conuersione rectanguli* il, *circa eundem axem reuoluti.*

In primis; quia it *est æqualis semiaxi hyperbolæ, erit quadratum* it *duplum rectanguli* db, *siue æquale rectangulo* ub. *Iam: cylindrus* oe, *ad tubum qui fit ex rectangulo* ib *(intellige semper circa axem* cd*) rationem habet compositam ex ratione basium; nempe ex ratione quadrati* it, *siue rectanguli* ub, *ad rectangulum* uie. *Hoc est (abiectis rectangulis) ex ratione lateris* ue *ad* ei: *& ex ratione lateris* eb *ad* iu. *Et insuper ex ratione altitudinum; nempe rectę* ei *ad* eb. *Ergo ratio cylindri* oe *ad tubum* ib, *componitur ex prædictis tribus rationibus; scilicet. ex ratione rectæ* ue *ad* ei: *& ex ratione* ei *ad* eb; *& ex ratione* eb *ad* iu. *propterea cylindrus* oe, *ad tubum* ib, *erit vt primus terminus ad vltimum; nempe vt recta* ue *ad* iu; *hoc est minor. Quod erat ostendendum primò.* *Lem. præcedens.* *Lem 3.*

Ratio verò cylindri oe, *ad tubum, qui fit ex rectangulo* il, *cõponitur ex ratione basium, scilicet ex ratione quadrati* it, *vel rectanguli* in, *ad rectangulum* uie; *hoc est (abiectis rectangulis) ex ratione lateris* fi, *ad* ie; *& ex ratione reliqui lateris* ai, *ad* iu. *Et insuper ex ratione altitudinum, nempe* ie *ad* ai. *Er-*

nentes ſint cd, de; *Sumptiſq; duobus punctis* a, b, *vtcumq; in hyperbola, ducantur duæ rectæ* be, ai, *aſymptoto* cd. *æquidiſtantes, &* an, bm *alteri aſymptoto*[177] de *parallelę, quæ concurrant in* l. *Tum conuertatur vniuerſa figura circa axẽ* cd.

Dico cylindrum quemdam iepo *(cuius quidem baſis* io *habeat ſemidiametrum* it *æqualem ſemiaxi hyperbolæ; altitudo verò ſit intercepta* ie.*) maiorem eßе tubo illo cylindrico, qui fit ex conuerſione rectanguli* ib *circa axem* cd; *Minorem verò tubo illo qui fit ex conuerſione rectanguli* il, *circa eundem axem reuoluti.*

In primis; quia it *est æqualis ſemiaxi hyperbolæ, erit quadratum* it *duplum rectanguli* db, *ſiue æquale rectangulo* ub. *Iam: cylindrus* oe, *ad tubum qui fit ex rectangulo* ib (*intellige ſemper circa axem* cd) *rationem habet compoſitam ex ratione baſium; nempe ex ratione quadrati* it, *ſiue rectanguli* ub, *ad rectangulum* uie. *Hoc eſt (abiectis rectangulis) ex ratione lateris* ue *ad* ei: *& ex ratione lateris* eb *ad* iu. *Et inſuper ex ratione altitudinum; nempe rectę* ei *ad* eb. *Ergo ratio cylindri* oe *ad tubum* ib, *componitur ex prædictis tribus rationibus; ſcilicet. ex ratione rectæ* ue *ad* ei*: & ex ratione* ei *ad* eb; *& ex ratione* eb *ad* iu. *propterea cylindrus* oe, *ad tubum* ib, *erit vt primus terminus ad vltimum; nempe vt recta* ue *ad* iu; *hoc eſt minor. Quod erat oſtendendum primò.*

Lem. pr e [-] *cedens*

*Lem.*3

Ratio verò cylindri oe, *ad tubum, qui fit ex rectangulo* il, *cõponitur ex ratione baſium, ſcilicet ex ratione quadrati* it, *vel rectanguli* in, *ad rectangulum* uie; *hoc eſt (abiectis rectangulis) ex ratione lateris* fi, *ad* ie; *& ex ratione reliqui lateris* ai, *ad* iu. *Et inſuper ex ratione altitudinum, nempe* ie *ad* ai. *Er-*

S 2 *go ra-*

[177] *aſymptoto*] *aſymptote.*

gò ratio cylindri o c, *ad tubum* i l, *componitur ex his tribus prædictis rationibus; nempe ex ratione* f i *ad* i e; *&* i e *ad* a i; *&* a i *ad* i u. *Propterea cylindrus* o e, *ad tubum* i l, *erit vt primus terminus* f i *ad vltimum* i u. *& ideo minor. Quod erat ostendendum &c.*

Lemma VI.

Esto hyperbola cuius asymptoti c d, d e *angulum rectum comprehendant, sumptisq. in hyperbola vtcumque duobus punctis* a *&* b; *ducantur* a i, b e *asymptoto* c d *parallelæ.*

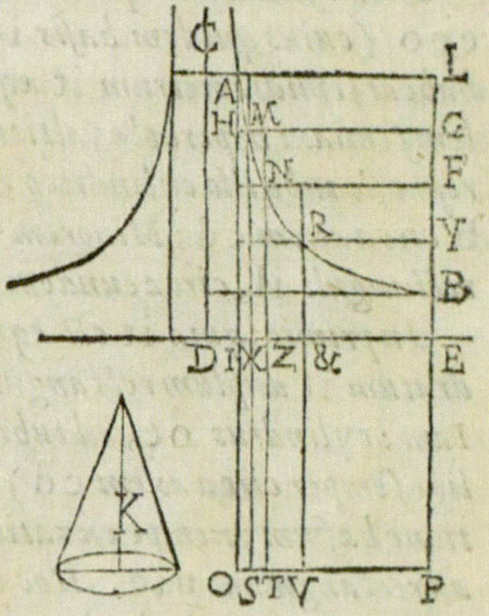

Dico solidum illud annulare quod describitur ex conuersione quadrilinei mixti i a b e, *circa axem* c d *reuoluti, æquale esse cuidam cylindro recto* i e p o. *Debet autem huius cylindri altitudo esse* i e; *diameter verò basis* i o, *debet esse æqualis integro axi ipsius hyperbolæ.*

Sit enim (si possibile est) solidum illud annulare factum ex quadrilineo i a b e, *circa axem* c d *reuoluto, minus cylindro* o e: *& ponatur defectus æqualis cuidam solido* K.

Secetur b l *bifariam in* f. *deinde reliqua* f l *secetur bifariã in* g; *Et hoc fiat semper donec tubus aliquis cylindricus, qui describitur ex reuolutione rectanguli* a l g, *minor sit solido* K. *Tunc enim sectà totà* b l, *in partes æquales vltimę* g l, *ducantur à singulis punctis diuisionum, rectæ* g h, f n, y r, *æquidistantes ipsi* d e. *Ex punctis verò* m, n r. *in quibus prædictę parallelę hyperbolam secant, demittantur rectæ, siue potius plana* m s, n t, r u. *ad asymptoton* d e *erecta. Denique ex conuersio-*

gò ratio cylindri oe, *ad tubum* il, *componitur ex his tribus prædictis rationibus; nempe ex ratione* fi *ad* ie; *&* ie *ad* ai; *&* ai *ad* iu. *Propterea cylindrus* oe, *ad tubum* il, *erit vt primus terminus* fi *ad vltimum* iu. *& ideo minor. Quod erat oſtendendum &c.*

Lemma VI.

Eſto hyperbola cuius aſymptoti cd, de *angulum rectum compræhendant, ſumptiſq. in hyperbola vtcumque duobus punctis* a *&* b; *ducantur* ai, be *aſymptoto* cd *parallelæ.*

Dico ſolidum illud annulare quod deſcribitur ex conuerſione quadrilinei mixti iabe, *circa axem* cd *reuoluti, æquale eſſe cuidam cylindro recto* iepo. *Debet autem huius cylindri altitudo eſſe* ie; *diameter verò baſis* io, *debet eſſe æqualis integro axi ipſius hyperbolæ.*

Sit enim (ſi poſſibile eſt) ſolidum illud annulare factum ex quadrilineo iabe, *circa axem* cd *reuoluto, minus cylindro* oe*: & ponatur defectus æqualis cuidam ſolido* K.

Secetur bl *bifariam in* f. *deinde reliqua* fl *ſecetur bifariã in* g; *Et hoc fiat ſemper donec tubus aliquis cylindricus, qui deſcribitur ex reuolutione rectanguli* alg, *minor ſit ſolido* K.

Tunc enim ſectà totà bl, *in partes æquales vltimę* gl, *ducantur à ſingulis punctis diuiſionum, rectæ* gh, fn, yr, *æquidiſtantes ipſi* de. *Ex punctis verò* m, nr. *in quibus prædictę parallelę hyperbolam ſecant, demittantur rectæ, ſiue potius plana* mſ, nt, ru. *ad aſymptoton* de *erecta. Denique ex con-*

uersio-

uersione singulorum rectangulorum æqualium, quorum vnum est a g, *totidem tubi cylindrici describantur circa axem* c d.

Iam: tubus qui fit à rectangulo r b *(intellige semper circa axem* c d *) ob æqualem altitudinem, eandemque basim, æqualis erit tubo* r f. *additoq; communi tubo* r n. *erunt duo tubi* b r, r n *simulsumpti æquales tubo* n y, *siue tubo* n g. *Additoq. communi* n m. *erunt tres tubi* b r n m, *æquales tubo* m f, *siue* m l; *& addito communi vltimo* m a, *erunt omnes tubi simul* b r n m a, *æquales tubo* a g, *nempe minores solido* K. *ob constructionem. Propterea vniuersa figura solida constans ex tubis* e r, & n, z m x a, *circumscripta solido annulari facto à quadrilineo* i a b e, *minus addit supra ipsum solidum annulare, quàm sit solidum* K. *Ergo ipsa figura circumscripta adhuc minor erit cylindro* o e. *Quod est absurdum. Nam tubus* a x. *superat cylindrum* x o; *Tubus* Lem. 5. *item* m z *superat cylindrum* z s. *& sic de reliquis per lemma 5.*

Ponatur deinde (si possibile est) solidum annulare genitum ex quadrilineo i a b e, *maius esse cylindro* o e. *ponaturq. excessus æqualis solido cuidam* K.

Peragatur similis constructio, vt supra; ita vt omnes tubi cylindrici b r n m a, *minores iterum ostendantur solido* k. *Tunc enim figura inscripta in solido annulari prædicto, constans ex tubis* & b, z r, x n, i m, *minus deficiet ab ipso solido annulari, quà sit solidum* k. *Propterea eadem inscripta figura adhuc maior erit cylind.* o e. *Quod est absurdum. Nam tubus* x h *minor est cylin-* Lem. 5. *dro* x o; *& tubus* x n *minor est cylindro* x t. *Et sic de reliquis.*

Patet ergo, quòd solidum annulare genitum ex conuersione quadrilinei i a b e, *circa axem* c d, *æquale est cylindro* o e. *Siquidem ostensum est, neq; minus, neq. maius esse posse.*

Lemma VII.

Esto hyperbola, cuius asymptoti angulum rectum continentes sint a b, b c; *& conuertatur figura circa axem* a b, *ita ut fiat solidum hyperbolicum, cuius infinita sit longitudo versus partes* a. *Secto deinde huiusmodi solido, plano* d e *ad axem erecto,*

uerſione ſingulorum rectangulorum ęqualium, quorum vnum eſt ag, *totidem tubi cylindrici deſcribantur circa axem* cd.

Iam: *tubus qui ſit à rectangulo* rb *(intellige ſemper circa axem* cd*) ob æqualem altitudinem, eandemque baſim, æqualis erit tubo* rf. *additoq; communi tubo* rn. *erunt duo tubi* br, rn *ſimul ſumpti æquales tubo* ny, *ſiue tubo* ng. *Additoq. communi* nm. *erunt tres tubi* brnm, *æquales tubo* mf, *ſiue* ml; *& addito communi vltimo* ma, *erunt omnes tubi ſimul* brnma, *ęquales tubo* ag, *nempe minores ſolido* K. *ob constructionem. Propterea vniuerſa figura ſolida constans ex tubis* er, & n, zmxa, *circumſcripta ſolido annulari facto à quadrilineo* iabe, *minus addit ſupra ipſum ſolidum annulare, quàm ſit ſolidum* K. *Ergo ipſa figura circumſcripta adhuc minor erit* *Lem.*5. *cylindro* oe. *Quod eſt abſurdum. Nam tubus* ax. *ſuperat cylindrum* xo; *Tubus item* mz *ſuperat cylindrum* zſ. *& ſic de reliquis per lemma 5.*

Ponatur deinde (ſi poſſibile eſt) ſolidum annulare genitum ex quadrilineo iabe, *maius eſſe cylindro* oe. *ponaturq. exceßus æqualis ſolido cuidam* K.

Peragatur ſimilis constructio, vt ſupra; ita vt omnes tubi cylindrici brnma, *minores iterum ostendantur ſolido* k. *Tunc enim figura inſcripta in ſolido annulari prædicto, conſtans ex tubis* & b, zr, xn, im, *minus deficiet ab ipſo ſolido annulari, qua ſit ſolidum* k. *Propterea eadem inſcripta figura adhuc maior erit cylind.* oe. *Quod est abſurdum. Nam tubus* xh *minor eſt cylindro* xo; *& tubus* xn *minor eſt cylindro* xt. *Et ſic de reliquis.* *Lem.* 5.

Patet ergo, quòd ſolidum annulare genitum ex conuerſione quadrilinei iabe, *circa axem* cd, *æquale eſt cylindro* oe. *Siquidem ostenſum eſt, neq; minus, neq. maius eſſe poſſe.*

Lemma VII.

Eſto hyperbola, cuius aſymptoti angulum rectum continentes ſint ab, bc; *& conuertatur figura circa axem* ab, *ita ut fiat ſolidum hyperbolicum, cuius infinita ſit longitudo verſus partes* a. *Secto deinde huiuſmodi ſolido, plano* de *ad axem ere-*

cto,

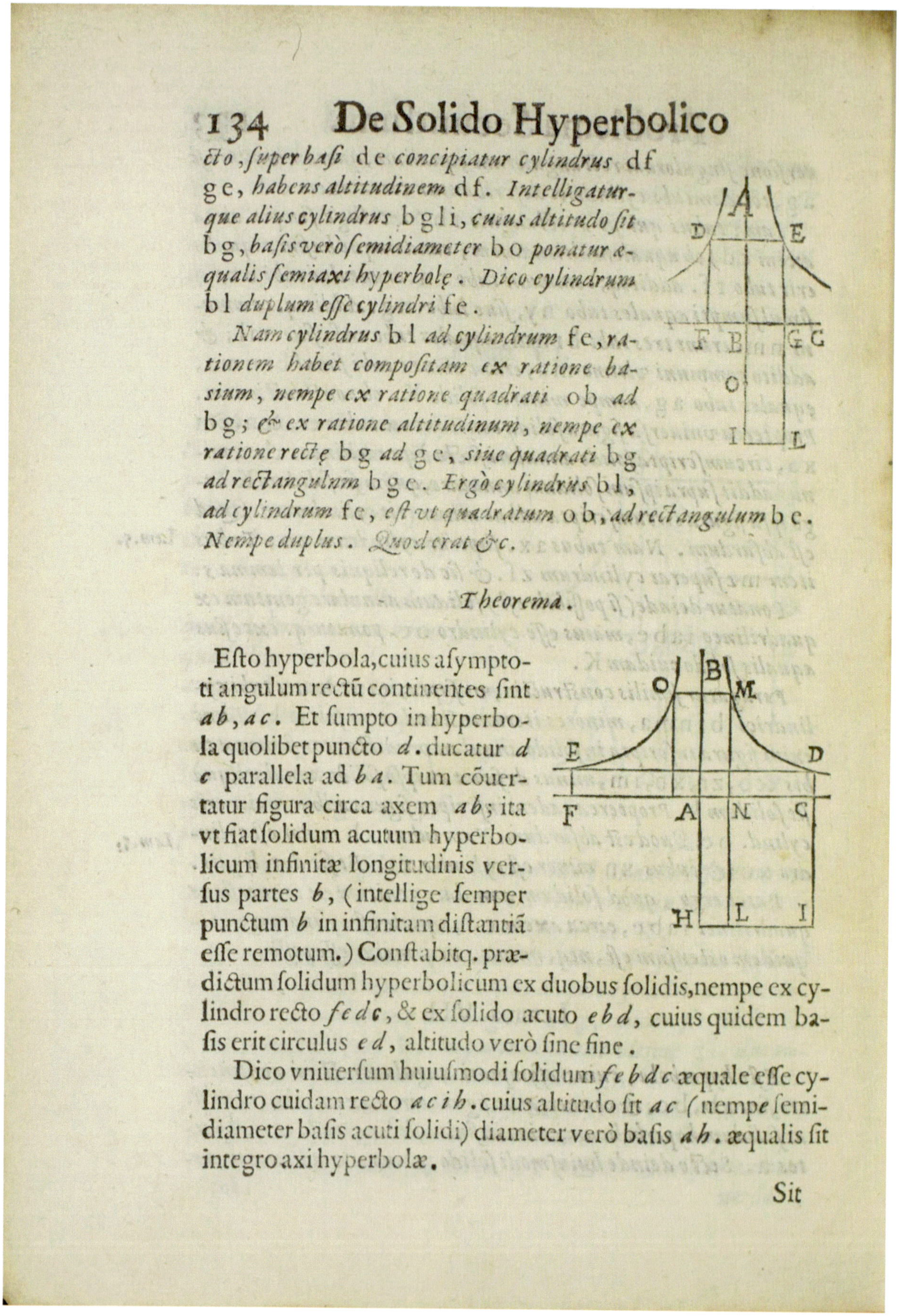

134 De Solido Hyperbolico

cto, super basi de *concipiatur cylindrus* df ge, *habens altitudinem* df. *Intelligaturque alius cylindrus* bgli, *cuius altitudo sit* bg, *basis verò semidiameter* bo *ponatur æqualis semiaxi hyperbolę. Dico cylindrum* bl *duplum esse cylindri* fe.

Nam cylindrus bl *ad cylindrum* fe, *rationem habet compositam ex ratione basium, nempe ex ratione quadrati* ob *ad* bg; *& ex ratione altitudinum, nempe ex ratione rectę* bg *ad* ge, *siue quadrati* bg *ad rectangulum* bge. *Ergò cylindrus* bl, *ad cylindrum* fe, *est vt quadratum* ob, *ad rectangulum* bc. *Nempe duplus. Quod erat &c.*

Theorema.

Esto hyperbola, cuius asymptoti angulum rectũ continentes sint *ab*, *ac*. Et sumpto in hyperbola quolibet puncto *d*. ducatur *dc* parallela ad *ba*. Tum cõuertatur figura circa axem *ab*; ita vt fiat solidum acutum hyperbolicum infinitæ longitudinis versus partes *b*, (intellige semper punctum *b* in infinitam distantiã esse remotum.) Constabitq. prædictum solidum hyperbolicum ex duobus solidis, nempe ex cylindro recto *fedc*, & ex solido acuto *ebd*, cuius quidem basis erit circulus *ed*, altitudo verò sine fine.

Dico vniuersum huiusmodi solidum *febdc* æquale esse cylindro cuidam recto *acih*. cuius altitudo sit *ac* (nempe semidiameter basis acuti solidi) diameter verò basis *ah*. æqualis sit integro axi hyperbolæ.

Sit

cto, ſuper baſi de *concipiatur cylindrus* dfge, *habens altitudinem* df. *Intelligaturque alius cylindrus* bgli, *cuius altitudo ſit* bg, *baſis verò ſemidiameter* bo *ponatur æqualis ſemiaxi hyperbolę. Dico cylindrum* bl *duplum eſſe cylindri* fe.

Nam cylindrus bl *ad cylindrum* fe, *rationem habet compoſitam ex ratione basium, nempe ex ratione quadrati* ob *ad* bg; *& ex ratione altitudinum, nempe ex ratione rectę* bg *ad* ge, *siue quadrati* bg *ad rectangulum* bge. *Ergò cylindrus* bl, *ad cylindrum* fe, *eſt vt quadratum* ob, *ad rectangulum* be.

Nempe duplus. Quod erat &c.

Theorema.

Eſto hyperbola, cuius aſymptoti angulum rectũ continentes ſint *ab, ac.* Et ſumpto in hyperbola quolibet puncto *d.* ducatur *dc* parallela ad *ba.* Tum cõuertatur figura circa axem *ab*; ita vt fiat ſolidum acutum hyperbolicum infinitæ longitudinis verſus partes *b,* (intellige ſemper punctum *b* in infinitam diſtantiã eſſe remotum.) Conſtabitq. prædictum ſolidum hyperbolicum ex duobus ſolidis, nempe ex cylindro recto *fedc,* & ex ſolido acuto *ebd,* cuius quidem baſis erit circulus *ed,* altitudo verò ſine fine.

Dico vniuerſum huiuſmodi ſolidum *febdc* æquale eſſe cylindro cuidam recto *acih.* cuius altitudo ſit *ac (*nempe ſemidiameter baſis acuti ſolidi) diameter verò baſis *ah.* æqualis ſit integro axi hyperbolæ

Sit

Problema Secundum. 135

Sit enim (ſi poſſibile eſt) ſolidum hyperbolicum *febdc* minus cylindro *ai*. Ponaturq; ex cylindro *ai* cylindrus aliquis *ncil*, qui æqualis ſit ſolido hyperboblico: & producatur *lnm* donec hyperbolæ occurrat in *m*. (occurret enim, cum aſymptoto *ab* ſupponatur parallela.)

Iam cylindrus *ni*, æqualis erit ſolido annulari, quod deſcribitur à reuolutione quadrilinei mixti *nmdc*; & propterea minus omninò erit ſolido integro hyperbolico *febdc*. Non ergo eidem eſt æqualis. Quod eſt contra ſuppoſitum. *Lem.6.*

Ponatur deinde (ſi poſſibile eſt) ſolidum hyperbolicum *febdc* maius cylindro *ai*. Quoniam igitur ſolidum hyperbolicum *febdc*. (ſiue finitæ magnitudinis ſit, ſiue infinitę) maius ſupponitur quàm cylindrus *ai*. Erit aliquod ipſius ſegmentũ, puta *feomdc*, æquale cylindro *ai*. Quod eſt abſurdum. Nã ſolidum annulare factum à reuolutione quadrilinei *nmdc*, ęquale eſt cylindro *ni*; Cylindrus autem *on* ſubduplus eſt cylindri *nh*. Ergò tota portio ſolidi hyperbolici *feomdc*, minor erit cylindro *ai*. *Lem:6:* *Lem:7:*

Patet ergo, quòd vniuerſum ſolidum acutum hyperbolicum *febdc*, quamquam infinitæ longitudinis ſit, æquale tamen eſt prædicto cylindro *ai*. Quandoquidem neque minus, neq; maius eſſe poteſt. Quod erat oſtendendum &c.

APPEN-

Sit enim *(*ſi poſſibile eſt*)* ſolidum hyperbolicum *febdc* minus cylindro *ai.* Ponaturq; ex cylindro *ai* cylindrus aliquis *ncil,* qui æqualis ſit ſolido hyperboblico*:* & producatur *lnm* donec hyperbolæ occurrat in *m.* (occurret enim, cum aſymptoto *ab* ſupponatur parallela.).

Iam cylindrus *ni,* æqualis erit ſolido annulari, quod deſcribitur à reuolutione quadrilinei mixti *nmdc*; & propterea minus omninò erit ſolido integro hyperbolico *febdc.* Non ergo eidem eſt æqualis. Quod eſt contra ſuppoſitum. *Lem.*6.

Ponatur deinde (ſi poſſibile eſt*)* ſolidum hyperbolicum *febdc* maius cylindro *ai.* Quoniam igitur ſolidum hyperbolicum *febdc.* (ſiue finitæ magnitudinis ſit, ſiue infinitę*)* maius ſupponitur quàm cylindrus *ai.* Erit aliquod ipſius ſegmentũ, puta *feomdc,* æquale cylindro *ai.* Quod eſt abſurdum. Nã ſolidum annulare factum à reuolutione quadrilinei *nmdc,* ęquale eſt cylindro *ni*; Cylindrus autem *on* ſubduplus eſt cylindri *nh.* Ergò tota portio ſolidi hyperbolici *feomdc,* minor erit cylindro *ai.* *Lem.*6*:* *Lem.*7*:*

Patet ergo, quòd vniuerſum ſolidum acutum hyperbolicum *febdc,* quamquam infinitæ longitudinis ſit, æquale tamen eſt prædicto cylindro *ai.* Quandoquidem neque minus, neq; maius eſſe poteſt. Quod erat oſtendendum &c.

A P P E N-

APPENDIX
De Dimensione Cochleæ.

CVM adhuc à nemine, quod ego sciam, Geometrica consideratione examinatum sit solidum vulgatum, & antiquissimum, meoq; iudicio aliqua animaduersione non indignũ (Cochleam intelligo,) non abs re fore iudicaui illud breui cõtemplatione prosequi. Non enim aliena erit à præcedenti libello præsens speculatio, quæ per Indiuisibilia curua, superficiesq; cylindricas procedit. Neq; ingratum Geometris opus futurum existimo, si demonstrauero cui figuræ notæ iam dimensionis, æquale sit solidum quiddam neque rectum, neque rotundum, sed spirali reuolutione contortum, quale nullum adhuc inter mensuratas figuras possidet Geometria. Præmissa itaq; definitione veniamus ad lemmata, qua fieri poterit breuitate, expedienda.

Definitio.

SI eodem tempore moueantur duæ planæ figuræ, quæ semper in eodẽ plano consistant, nempe rectãgulum *a b c d*. circa axem *a b* motu circulari æquabili, & figura quæcunq; *d e* motu progressiuo super latere *d c*. Solidum quod à figura genitrice *d e* describitur, Cochleam appello.

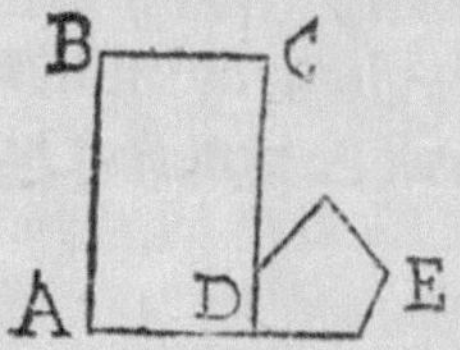

Lemma Primum.

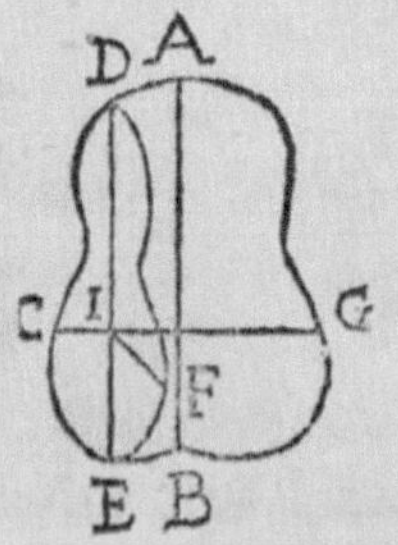

Esto solidum quodlibet rotundum a c b g; *cuius axis sit* a b, *figura genitrix* a b c; *sectũq. sit plano* d f e *æquidistantèr axi, & ad figuram genitricem erecto, quod quidem faciat in superficie solidi rotundi semisectionem lineam* d f e. *Dico*

soli-

136

APPENDIX

De Dimenſione Cochleæ.[178]

CVM adhuc à nemine, quod ego ſciam, Geometrica conſideratione examinatum ſit ſolidum vulgatum, & antiquiſſimum, meoq; iudicio aliqua animaduerſione non indignũ (Cochleam intelligo,) non abs re fore iudicaui illud breui cõtemplatione proſequi. Non enim aliena erit à præcedenti libello præſens ſpeculatio, quæ per Indiuiſibilia curua, ſuperficieſq; cylindricas procedit. Neq; ingratum Geometris opus futurum exiſtimo, ſi demonſtrauero cui figuræ notæ iam dimenſionis, æquale ſit ſolidum quiddam neque rectum, neque rotundum, ſed ſpirali reuolutione contortum, quale nullum adhuc inter menſuratas figuras poſſidet Geometria. Prę̨miſſa itaq; definitione veniamus ad lemmata, qua fieri poterit breuitate, expedienda.

Definitio.

SI eodem tempore moueantur duæ planæ figuræ, quæ ſemper in eodẽ plano conſiſtant, nempe rectãgulum *ab cd.* circa axem *ab* motu circulari æquabili, & figura quæcunq; *de* motu progreſſiuo ſuper latere *dc.* Solidum quod à figura genitrice *de* deſcribitur, Cochleam appello.

Lemma Primum.

Eſto ſolidum quodlibet rotundum acbg; *cuius axis ſit* ab, *figura genitri*x abc; *ſectũq ſit plano* dfe *æquidiſtantèr axi, & ad figuram genitricem erecto, quod quidem faciat in ſuperficie ſolidi rotundi ſemiſectionem lineam* dfe. *Dico.*

ſoli-

[178] The numbering of pages starts with p. 136 and just after with p. 145, so the numeration continues consequently.

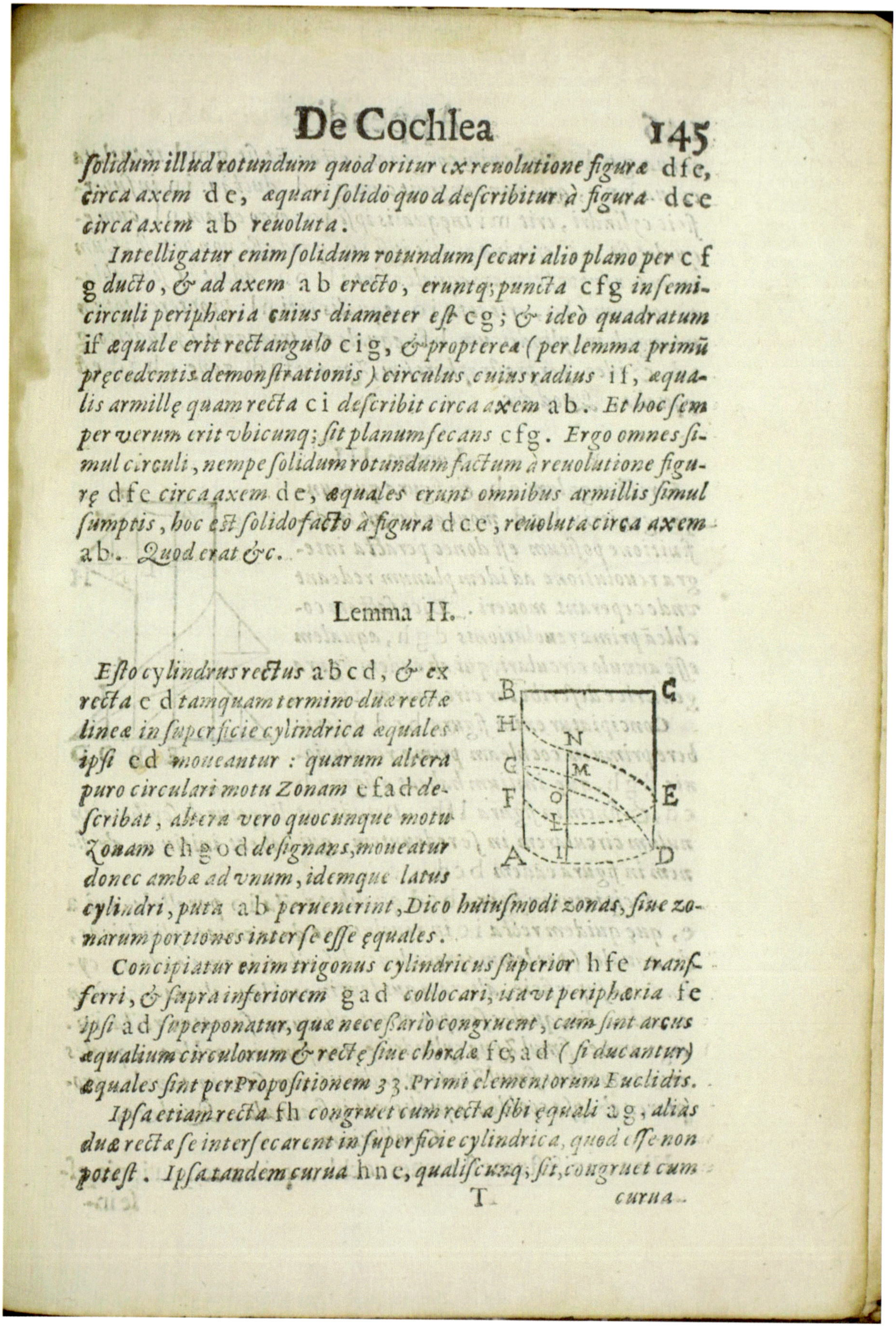

De Cochlea 145

solidum illud rotundum quod oritur ex reuolutione figuræ dfe, circa axem de, æquari solido quod describitur à figura dce circa axem ab reuoluta.

Intelligatur enim solidum rotundum secari alio plano per cfg ducto, & ad axem ab erecto, eruntq; puncta cfg in semicirculi periphæria cuius diameter est cg; & ideò quadratum if æquale erit rectangulo cig, & propterea (per lemma primũ pręcedentis demonstrationis) circulus cuius radius if, æqualis armillę quam recta ci describit circa axem ab. Et hoc semper verum erit vbicunq; sit planum secans cfg. Ergo omnes simul circuli, nempe solidum rotundum factum à reuolutione figurę dfe circa axem de, æquales erunt omnibus armillis simul sumptis, hoc est solido facto à figura dce, reuoluta circa axem ab. Quod erat &c.

Lemma II.

Esto cylindrus rectus abcd, & ex recta cd tamquam termino duæ rectæ lineæ in superficie cylindrica æquales ipsi cd moueantur: quarum altera puro circulari motu Zonam efad describat, altera vero quocunque motu Zonam ehgod designans, moueatur donec ambæ ad vnum, idemque latus cylindri, puta ab peruenerint, Dico huiusmodi zonas, siue zonarum portiones inter se esse ęquales.

Concipiatur enim trigonus cylindricus superior hfe transferri, & supra inferiorem gad collocari, ita vt periphæria fe ipsi ad superponatur, quæ necessariò congruent, cum sint arcus æqualium circulorum & rectę siue chordæ fe, ad (si ducantur) æquales sint per Propositionem 33. Primi elementorum Euclidis.

Ipsa etiam recta fh congruet cum recta sibi ęquali ag, aliàs duæ rectæ se intersecarent in superficie cylindrica, quod esse non potest. Ipsa tandem curua hne, qualiscunq; sit, congruet cum

T curua

ſolidum illud rotundum quod oritur ex reuolutione figuræ dfe, *circa axem* de, *æquari ſolido quod deſcribitur à figura* dce *circa axem* ab *reuoluta.*

Intelligatur enim ſolidum rotundum ſecari alio plano per cfg *ducto, & ad axem* ab *erecto, eruntq; puncta* cfg *in ſemicirculi periphæria cuius diameter eſt* cg; *& ideò quadratum* if *æquale erit rectangulo* cig, *& propterea (per lemma primũ pręcedentis demonſtrationis) circulus cuius radius* if, *æqualis armillę quam recta* ci *deſcribit circa axem* ab. *Et hoc ſemper verum erit vbicunq; ſit planum ſecans* cfg. *Ergo omnes ſimul circuli, nempe ſolidum rotundum factum à reuolutione figurę* dfe *circa axem* de, *æquales erunt omnibus armillis ſimul ſumptis, hoc est ſolido facto à figura* dce, *reuoluta circa axem*a b. *Quod erat &c.*

Lemma II.

Eſto cylindrus rectus abcd, *& ex recta* ed *tamquam termino duæ rectæ lineæ in ſuperficie cylindrica æquales ipſi* ed *moueantur: quarum altera puro circulari motu Zonam* efad *deſcribat, altera vero quocunque motu zonam* ehgod *deſignans, moueatur donec ambæ ad vnum, idemque latus cylindri, puta* ab *peruenerint, Dico huiuſmodi zonas, ſiue zonarum portiones inter ſe eſſe ęquales.*

Concipiatur enim trigonus cylindricus ſuperior hfe *tranſferri, & ſupra inferiorem* gad *collocari, ita vt periphæria* fe *ipſi* ad *ſuperponatur, quæ neceßariò congruent, cum ſint arcus æqualium circulorum & rectę ſiue chordæ* fe, ad *(ſi ducantur) æquales ſint per Propoſitionem 33. Primi elementorum Euclidis.*

Ipſa etiam recta fh *congruet cum recta ſibi ęquali* ag, *aliàs duæ rectæ ſe interſecarent in ſuperficie cylindrica, quod eſſe non poteſt. Ipſa tandem curua* hne, *qualiſcunq; ſit, congruet cum*

T *curua-*

[179] The numbering of pages starts with p. 145; see above. The last page is 150, but, in fact, the appendix is only seven pages long.

curua g o d. *Nisi enim congruat; esto: Et sit* g m d *translata curua* h n e, *quæ non congruit cum* g o d. *ductaq. in insuperficie cylindri, erit* m i *inęqualis ipsi* i o; *ergo etiam* n l, *cum ęqualis sit* m i, *erit inæqualis ipsi* i o, *quod esse non potest; Cum enim per suppositionem ęquales sint* i l, o n, *additáq; siue ablatà communi* l o, *erit tota* i o, *æqualis toti* n l. *Propterea totum triangulum cylindricum* h f e *ęquale est triangulo cylindrico* g a d. & *ideo, per prosthaphęresim, Zona* e f a d, *zonę* e h g d *est ęqualis. Quod &c.*

Lemma III.

Si rectangulum a b, & *figura quæcunque genitrix* b c d *moueantur, vt in definitione positum est donec peracta integra reuolutione ad idem planum redeant vnde ceperant moueri. Dico factam cochleã primæ reuolutionis* d g h, *æqualem esse annulo circulari, qui ab eadem figura genitrice describetur circa axem* a e.

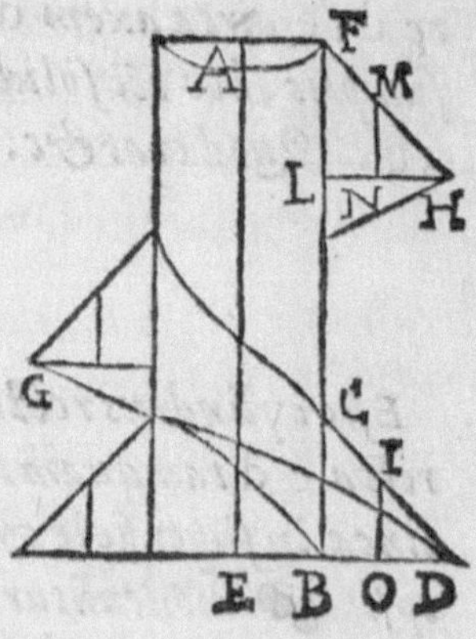

Concipiatur enim figura b c d *describere primum cochleam primæ reuolutionis* d g h, *quę initium habeat à figura* b c d, & *finem in figura* l f h. *Deinde intelligatur describere annulum circularem in se redeuntem, qui habeat initium, & finem in figura eadem* b c d.

Accipiatur in figura b c d *quælibet recta* i o *parallela axi* a e, *quę quidem recta* i o *in reuolutione duas zonas cylindricas, & æquales (per lemma pręcedens) describet, in vna eademq; cylindrica superficie, alteram quidem in cochlea, alteram verò in annulo. Et ęquales semper erunt, vbicunq; sumatur recta* i o. *ergo omnes simul Zonę cylindricę quę sunt in cochlea, ęquales erunt omnibus simul Zonis cylindricis quę sunt in annulo, propterea & ipsa cochlea ęqualis erit ipsi annulo. Quod &c,*

Corollarium.

Hinc manifestum est omnes cochleas primæ reuolutionis esse in-

curua god. *Nifi enim congruat; esto: Et fit* gmd *translata curua* hne, *quæ non congruit cum* god. *ductàq.* in *in fuperficie cylindri, erit* mi *inęqualis ipfi* io; *ergo etiam* nl, *cum ęqualis fit* mi, *erit inæqualis ipfi* io, *quod eβe non poteft; Cum enim per fuppofitionem ęquales fint* il, on, *additàq; fiue ablatà communi* lo, *erit tota* io, *æqualis toti* nl. *Propterea totum triangulum cylindricum* hfe *ęquale est triangulo cylindrico* gad. *& ideo, per proftraphęrefim, zona* efad, *zonę* ehgd *eft ęqualis. Quod &c.*

Lemma III.

Si rectangulum ab, *& figura quæcunque genitrix* bcd *moueantur, vt in definitione pofitum eft donec peracta integra reuolutione ad idem planum redeant vnde ceperant moueri. Dico factam cochleã primæ reuolutionis* dgh, *æqualem effe annulo circulari, qui ab eadem figura genitrice defcribetur circa axem* ae.

Concipiatur enim figura bcd *defcribere primum cochleam primæ reuolutionis* dgh, *quę initium habeat à figura* bcd, *& finem in figura* lfh. *Deinde intelligatur defcribere annulum circularem in fe redeuntem, qui habeat initium, & finem in figura eadem* bcd.

Accipiatur in figura bcd *quælibet recta* io *parallela axi* ae, *quę quidem recta* io *in reuolutione duas zonas cylindricas, & æquales (per lemma prędecens) defcribet, in vna eademq; cylindrica fuperficie, alteram quidem in cochlea, alteram verò in annulo. Et ęquales femper erunt, vbicunq; fumatur recta* io. *ergo omnes fimul zonę cylindricę quę funt in cochlea, ęquales erunt omnibus fimul zonis cylindricis quę funt in annulo, propterea & ipfa cochlea ęqualis erit ipfi annulo. Quod &c,*

Corollarium.

Hinc manifeftum eft omnes cochleas primæ reuolutionis ef-

ſe inſe inter ſe æquales, quandoquidem ſingulę eidem annulo circulari æquales ſunt.

Lemma IV.

Manentibus ijs quæ Apollonius ſupponit in XI, XII, & XIII, primi Conicorum. Eſto conus abc, *ſectus plano non verticali per* fnr, *faciente in ſuperficie coni ſectionem* fnr, *quaecunq; illa ſit; cuius diameter eſto* fe. *Ducaturq;* fi *aequidiſtans ipſi* ac. *Tum fiat, vt* fe *ad* ea *(partem baſis trianguli per axē à vertice coni auuerſam) ita* if *ad* fl. *Dico* fl *eſſe latus rectum ſectionis.*

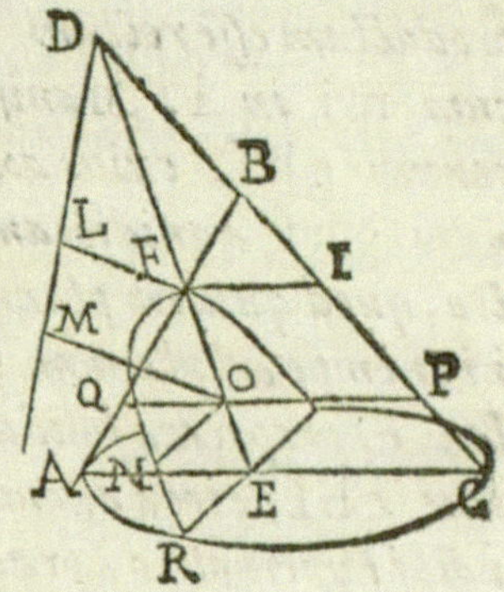

Ponatur fl *ad punctum* f *vtcumque, & ducatur* dl *ab extremitate axis: Accepto deinde quolibet puncto* n *in ſectione, applicetur* no, *& per* o *agatur* qp *aequidiſtans ipſi* ac; *at* om *ducatur parallela ad* fl. *Erit iam* fo *ad* oq, *vt* fe *ad* ea, *ſiue vt* if *ad* fl, *nempe vt* po *ad* om, *ob parallelas; Ergo rectangula* fom, poq *ſunt æqualia; quamobrem rectangulum* fom *æquale erit quadrato* on, *& propterea* fl *rectum figuræ latus. Quod &c.*

Licet hoc verum ſit in omni ſectione coni, ſolam hyperbolam depiximus, quoniam ſola hyperbola facit ad rem noſtram.

Lemma V.

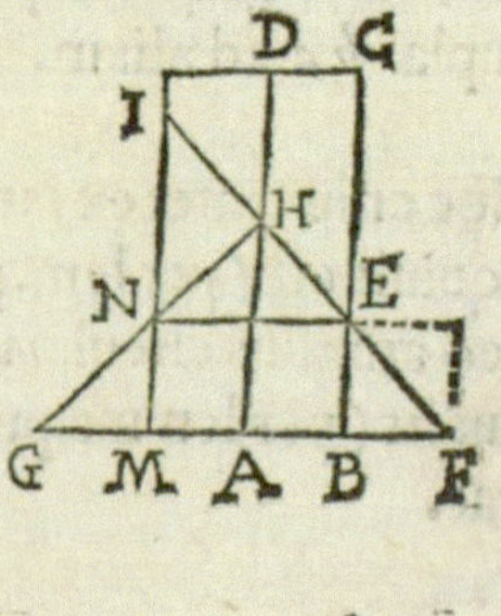

Si rectangulum ac, *in eodem exiſtens plano cum triangulo orthogonio* ebf. *conuertatur circa manens latus* ad *donec ad locum redeat vnde cœpit moueri. Dico annulum circularem deſcriptum à triangulo* ebf *æqualem eſ-*

ſe inter ſe æquales, quandoquidem ſingulę eidem annulo circulari æquales ſunt.

Lemma IV.

Manentibus ijs quæ Apollonius ſupponit in XI, XII, & XIII, *primi Conicorum. Esto conus* abc, *ſectus plano non verticali per* fnr, *faciente in ſuperficie coni ſectionem* fnr, *quaecunq; illæ ſit; cuius diameter esto* fe. *Ducaturq;* fi *aequidiſtans ipſi* ac. *Tum fiat, vt* fe *ad* ea *(partem baſis trianguli per axẽ à vertice coni auuerſam) ita* if *ad* fl. *Dico* fl *eſſe latus rectum ſectionis.*

Ponatur fl *ad punctum* f *vtcumque, & ducatur* dl *ab extremitate axis: Accepto deinde quolibet puncto* n *in ſectione, applicetur* no, *& per* o *agatur* qp *aequidiſtans ipſi* ac; *at* om *ducatur parallela ad* fl. *Erit iam* fo *ad* oq, *vt* fe *ad* ea, *ſiue vt* if *ad* fl, *nempe vt* po *ad* om, *ob parallelas; Ergo rectangula* fom, poq *ſunt æqualia; quamobrem rectangulum* fom *æquale erit quadrato* on, *& propterea* fl *rectum figuræ latus. Quod &c.*

Licet hoc verum ſit in omni ſectione coni, ſolam hyperbolam depiximus, quoniam ſola hyperbola facit ad rem nostram.

Lemma V.

Si rectangulum ac, *in eodem existens plano cum triangulo orthogonio* ebf. *conuertatur circa manens latus* ad *donec ad locum redeat vnde cæpit moueri. Dico annulum circularem deſcriptum à triangulo* eb f *æqualem eſ-.*

T a se co-

ſe conoidi cuidam hyperbolico, cuius altitudo ſit be; *cuius latus rectum ſit quarta proportionalium ſi fiat vt* eb *ad* bf *ita dupla* ba *ad aliam. Verſum verò latus quarta ſit proportionalium, ſi fiat vt* fb *ad* be, *ita dupla* ba *ad aliam,*

Conuertatur figura vti dictum eſt, & rectangulum ac *deſcribat cylindrum cuius ſectio per axem* em; *intelligaturque productam eſſe rectam* fe, *donec cum axe conueniat in* h, *& cum* mi *in* i. *Manifeſtum eſt triangulum* haf *deſcribere conum* ghf, *cuius axis eſt* ah. *Concipiatur iàm ſecari conum* ghf *æquidiſtantèr axi plano per* eb, *siue per* inm *ducto, quod quidem planum erectum ſit ad figuram genitricem coni, nempe ad planum* ghf. *Eritq. ſectio in cono* ghf *hyperbola; Et propterea ſolidum quod deſcribitur à triangulo* mng, *siue* ebf, *circa axem* ad, *æquale erit (per lemma primum) conoidi hyperbolico à prædicta hyperbola deſcripto. Huius autem conoidis, siue huius hyperbolæ latus rectum habetur (per lemm. præced.) ſi fiat vt* nm, *ad* mg, *ita* en, *siue dupla* ba *ad aliã.* per 4. ſexti. *Verſum verò, quod eſt* ni, *habebitur, ſi fiat vt* gm *ad* mn, *ita* en, *siue dupla* ba *ad aliam quæ erit* ni. *Quod erat &c.*

Theorema.

Cochlea primæ reuolutionis, quæ deſcribitur à triangulo *ebf* in præcedenti figura, æqualis eſt conoidi cuidam hyperbolico, cuius altitudo ſit *eb*; latus rectum ſit quarta proportionalium, ſi fiat vt *eb* ad *bf*, ita dupla *ba* ad aliam. Verſum verò latus ſit quarta proportionalium, ſi fiat vt *fb* ad *be*, ità dupla *ba* ad aliam.

Hoc enim patet ex iam demonſtratis. Prædicta enim cochlea æqualis eſt (per lem. primum) annullo facto à triangulo *ebf*. Sed annulus circularis trianguli *ebf* prædicto conoidi eſt æqualis (per lemma præcedens.) Ergo patet quod propoſitum erat.

Scho-

148 Appendix

ſe conoidi cuidam hyperbolico, cuius altitudo ſit be; *cuius latus rectum ſit quarta proportionalium ſi fiat vt* eb *ad* bf *ita dupla* ba *ad aliam. Verſum verò latus quarta sit proportionalium, si fiat vt* fb *ad* be, *ita dupla* ba *ad aliam,*

Conuertatur figura vti dictum eſt, & rectangulum ac *deſcribat cylindrum cuius ſectio per axem* cm; *intelligaturque productam eſſe rectam* fe, *donec cum axe conueniat in* h, *& cum* mi *in* i. *Manifeſtum eſt triangulum* haf *deſcribere conum* ghf, *cuius axis est* ah. *Concipiatur iàm ſecari conum* ghf *æquidistantèr axi plano per* eb, *siue per* inm *ducto, quod quidem planum erectum sit ad figuram genitricem coni, nempe ad planum* ghf. *Eritq. ſectio in cono* ghf *hyperbola; Et propterea ſolidum quod deſcribitur à triangulo* mng, *ſiue* ebf, *circa axem* ad, *æquale erit (per lemma primum) conoidi hyperbolico à prædicta hyperbola deſcripto. Huius autem conoidis, siue huius hyperbolæ latus rectum habetur (per lemm. præced.) si fiat vt* nm, *ad* mg, *ita* en, *siue dupla* ba *ad aliã. Verſum verò, quod eſt* ni, *habebitur, si fiat vt* gm *ad* mn, *ita* en, *siue dupla* ba *ad aliam quæ crit*[180] ni. *Quod erat &c.*

per 4. ſex[-] ti

Theorema.

Cochlea primæ reuolutionis, quæ deſcribitur à triangulo *ebf* in præcedenti figura, æqualis eſt conoidi cuidam hyperbolico, cuius altitudo ſit *eb*; latus rectum ſit quarta proportionalium, ſi fiat vt *eb* ad *bf,* ita dupla *ba* ad aliam. Verſum verò latus ſit quarta proportionalium, ſi fiat vt *fb* ad *be,* ità dupla *ba* ad aliam.

Hoc enim patet ex iam demonſtratis. Prædicta enim cochlea æqualis eſt (per lem. primum) annullo facto à triangulo e*bf.* Sed annulus circularis trianguli *ebf* prædicto conoidi eſt æqualis (per lemma pręcedens.) Ergo patet quod propoſitum erat

Scho-

[180] *crit*] *erit.*

Scholium.

Cochlea verò cuius figura genitrix parallelogrammum rectangulum sit, æqualis est cylindro cuius altitudo sit e b, *eadem cum altitudine figuræ genitricis, semidiameter verò basis media proportionalis sit inter* f b, *& rectam compositam ex ex* f a, a b.

Si verò figura genitrix circulus fuerit, erit facta cochlea primæ reuolutionis ad sphæram circuli genitoris, vt periphæria quæ describitur à radio, qui sit æqualis vtrique, nempe rectę a b *in præcedenti figura, semidiametroq; circuli genitoris, ad duas tertias diametri eiusdem circuli genitoris.*

R*Eliquum esset vt Mechanica etiam Theoremata horum solidorum exequeremur, præsertim quando Cochlea gignitur à triangulo: Centrum enim grauitatis in axe est, diuiditq; portiunculam quandam ipsius axis (æqualem abscindendam lateri* e b, *& circa punctum medium ipsius axis collocandam) veluti conoidis cuiusdam hyperbolici centrum secat propriam diametrum; siue prædictæ portiunculę semissem ita diuidit, vti eandem secaret centrum grauitatis cuiusdam segmenti sphærici duplam habentis altitudinē, basimq; dato cuidam circulo æqualem. Sed tanti non est singulas istas nugas longiùs protrahere, ut te beneuolum Lectorem vlteriùs adhuc torqueamus. Fortasse etiam fiet, nisi vniuersa hæc, quæ in istis libellis continentur, tibi displicuisse comperiam, vt ea quæ hic desiderantur, & multò plura circa grauitatem, ipsiusq; centrum, peculiari libello Geometricè comprehendam. Interim scio me patrocinium debere longissimæ tot mensium desidiæ: cum iam supra annum, ex quo opuscula hæc maximis Geometris promissa sunt, producatur lentissima eorum impressio, quod quidem pluribus de causis factum est; neq; hoc tam negligentiæ meæ imputandum est, quàm fortuitis quibusdam casibus, insperatisque. Accidit enim intermedio hoc tempore, vt plurium mensium studio atq; labore inciderim in solutionem optici illius Problematis tamdiù perqui-*

Scholium.

Cochlea verò cuius figura genitrix parallelogrammum rectangulum ſit, ęqualis est cylindro cuius altitudo ſit eb, *eadem cum altitudine figuræ genitricis, ſemidiameter verò baſis media proportionalis ſit inter* fb, *& rectam compoſitam ex ex*[181] fa, ab.

Si verò figura genitrix circulus fuerit, erit facta cochlea primæ reuolutionis ad ſphæram circuli genitoris, vt periphæria quæ deſcribitur à radio, qui ſit æqualis vtrique, nempe rectę ab *in præcedenti figura, ſemidiametroq; circuli genitoris, ad duas tertias diametri eiuſdem circuli genitoris.*

R*Eliquum eſſet vt Mechanica etiam Theoremata horum ſolidorum exequeremur, præſertim quando Cochlea gignitur à triangulo: Centrum enim grauitatis in axe eſt, diuiditq; portiunculam quandam ipſius axis (æqualem abſcindendam lateri* eb, *& circa punctum medium ipſius axis collocandam) veluti conoidis cuiuſdam hyperbolici centrum ſecat propriam diametrum; ſiue prædictæ portiunculę ſemiſſem ita diuidit, vti eandem ſecaret centrum grauitatis cuiuſdam ſegmenti ſphærici duplam habentis altitudinẽ, baſimq; dato cuidam circulo æqualem. Sed tanti non est ſingulas iſtas nugas longiùs protrahere, ut te beneuolum Lectorem vlteriùs adhuc torqueamus. Fortaße etiam fiet, niſi vniuerſa hæc, quæ in istis libellis continentur, tibi diſplicuiſſe comperiam, vt eaquæ hic deſiderantur, & multò plura circa grauitatem, ipſiuſq; centrum, peculiari libello Geometricè comprehendam. Interim ſcio me patrocinium debere longiſſimæ tot menſium deſidiæ: cum iam ſupra annum, ex quo opuſcula hæc maximis Geometris promiſsa ſunt, producatur lentiſſima eorum impreſſio. quod quidem pluribus de cauſis factum eſt; neq; hoc tam negligentiæ meæ imputandum eſt, quàm fortuitis quibuſdam caſibus, inſperatiſque. Accidit enim intermedio hoc tempore, vt plurium menſium studio atq; labore inciderim in ſolutionem optici illius Problematis tamdiù per-*

qui-

[181] *ex ex*] *ex*.

quisiti, cuius videlicet figuræ esse debeant superficies vitrorum, quæ ad vsum Telescopij elaborantur. Exitus demonstrationem confirmauit. quamquam enim neque optatam figuram (vt credibile est) perfectè haberent, neque vndequaque absoluta, & perpolita à Tirone adhuc inexperto, & inexercitato viderentur, ope tamen, & vi figuræ illius ad quam proximè tantùm accedebant, ad eum vsq; perfectionis gradum peruenerunt, vt Telescopia optimi cuiusq; artificis, cuius ad hunc diem fama in hac Vrbe innotuerit, superauerint. Neque iudicium hoc perperàm prolatum est; sed repetitis sæpius, summaq; cum diligentia varijs experimentis, nocte, dieque, & adhibitis eruditissimis testibus, quorum iudicium nemo iure damnauerit. Certè, qualecunq; fuerit inuentum, nescio plusne gaudij, laudisq; mihi attulerit, an præmij: quandoquidem Serenissimi Magni Ducis effusa, & verè Regia liberalitas magno auri pondere donatum me non semel voluit. Mirum itaq; videri non debet quòd omissà per integrum semestre libellorum curà, totam operam nouo inuẽto, mihiq; in primis exoptatissimo, ne dicam vtilissimo, impenderim. Factum etiam est vt hac de causa libelli minus castigati euaserint; authore nimirùm distracto, & ad alia, eaq; diuersissima, conuerso. Quapropter orandus etiam atq; etiam es beneuole lector, vt hæc qualiacunq; æqui, boniq; facias, & errata vel toleres, vel corrigas. presertim cum tam manifesta plerunq. sint, vt neminem fugere valeant, sed vltrò sese ipsa offerant; vt videre est in prima statim epistola nuncupatoria, & subinde satis frequenter in ijs quæ sequuntur. Correctiones non addemus in fine operis, vt pleriq. solent: quia neque satis vacauit temporis ad mendosa omnia adnotanda, neq. volumus mutilà breuiq. recensione aliquot erratorum, omnem deinde excusationi meę locum erripere; dum tacita prętermissio eorum, quę censum effugissent, tamquam approbationis quoadam genus mihi potuisset imputari.

FINIS.

quiſiti, cuius videlicet figuræ eſſe debeant ſuperficies vitrorum, quæ ad vſum Teleſcopij elaborantur. Exitus demonstrationem confirmauit. quamquam enim neque optatam figuram (vt credibile eſt) perfectè haberent, neque vndequaque abſoluta, & perpolita a Tirone adhuc inexperto, & inexercitato viderentur, ope tamen, & vi figuræ illius ad quam proximè tantùm accedebant, ad eum vſq; perfectionis gradum peruenerunt, vt Teleſcopia optimi cuiuſq; artificis, cuius ad hunc diem fama in hac Vrbe innotuerit, ſuperauerint. Neque iudicium hoc perperàm prolatum eſt; ſed repetitis ſæpius, ſummaq; cum diligentia varijs experimentis, nocte, dieque, & adhibitis eruditiſſimis testibus, quorum iudicium nemo iure damnauerit. Certè, qualecunq; fuerit inuentum, neſcio pluſne gaudij, laudiſq; mihi attulerit, an præmij: quandoquidem Sereniſſimi Magni Ducis effuſa, & verè Regia liberalitas magno auri pondere donatum me non ſemel voluit. Mirum itaq; videri non debet quòd omiſsà per integrum ſemeſtre libellorum curà, totam operam nouo inuēto, mihiq; in primis ex optatiſſimo, ne dicam vtiliſſimo, impenderim. Factum etiam eſt vt hac de cauſa libelli minus caſtigati euaſerint; aut hore nimirùm diſtracto, & ad alia, eaq; diuerſiſſima, conuerſo. Quapropter orandus etiam atq; etiam es beneuole lector, vt hæc qualiacunq; æqui, boniq; facias, & errata vel toleres, vel corrigas. pręſertim cum tam manifeſta plerunq. ſint, vt neminem fugere valeant, ſed vltrò ſeſe ipſa offerant; vt videre eſt in prima ſtatim epiſtola nuncupatoria, & ſubinde ſatis frequenter in ijs quæ ſequuntur. Correctiones non addemus in fine operis, vt pleriq. ſolent: quia neque ſatis vacauit temporis ad mendoſa omnia adnotanda, neq. voluimus mutilà breuiq. recenſione aliquot erratorum, omnem deinde excuſationi meę locum erripere; dum tacita prętermiſſio eorum, quę cenſum effugiſſent, tamquam approbationis quoddam genus mihi potuiſſet imputari.

FINIS.

499

[499]

Back Cover

513
T63o

Back Cover – Spine

Acknowledgements We thank the anonymous reviewers for their careful reading of our manuscript and their many insightful comments and suggestions. We also express gratitude to Dr. Julie Robarts (University of Melbourne, Australia) for her observations on English language; eventual remaining English mistakes are our own.

Bibliography: Evangelista Torricelli, Main Sources

Raffaele Pisano

Abstract

Taking into account this book and its aim, below I provided a selection of the main literature of Evangelista Torricelli.

Torricelli E ([1644]1897) Esperienza dell'argento vivo, Accademia del Cimento. In: Gustav Hellmann G (ed) Instrumenti per conoscer l'alterazioni dell'aria, Neudrucke von schriften und karten über meteorologieund erdmagnetismus. Asher & Co., Berlin.

Torricelli E (1644) Opera geometrica. Masse & de Landis, Florentiae.

Torricelli E (1715) Lezioni Accademiche d'Evangelista Torricelli Mattematico, e Filosofo del Sereniss. In: di Toscana FGD, Guiducci J, Franchi S (eds). S.A.R., Firenze.

Torricelli E (1823) Lezioni accademiche. In: Silvestri G (ed) Biblioteca scelta di opere italiane antiche e moderne 139 Milano.

Torricelli E (1864) Lettere fin qui inedite di Evangelista Torricelli precedute dalla vita di lui. In: Ghinassi G (ed). Faenza.

Torricelli E (1919–1944) Opere di Evangelista Torricelli edite in occasione del III centenario della nascita col concorso del Comune di Faenza da Gino Loria e Giuseppe Vassura. Montanari. Faenza.

Torricelli E (1955) De infinitis spiralibus. Edited by Carruccio E. Domus Galilaeana, Pisa.

Torricelli E (1975) Opere scelte. Belloni L (ed). UTET, Torino.

R. Pisano (✉)
HOPAST at IEMN, Department of Physics, Lille University–CNRS, Villeneuve d'Ascq, France
e-mail: raffaele.pisano@univ-lille.fr

R. Pisano et al., *Homage to Evangelista Torricelli's Opera Geometrica 1644–2024*,
Logic, Epistemology, and the Unity of Science 55,
https://doi.org/10.1007/978-3-031-06963-5

Bibliography

AA.VV. (1768) Raccolta d'Áutori che Trattatano del Moto delle Acque. Seconda Edizione. Tomo Quarto. In Firenze MDCCLXVIII. Nella Stamperia di Sua Altezza Relae Con Licenza de' Superiori.

AA.VV. (2009) Galileiana. Journal of Galilean Studies VI:3–70.

Agostini A (1930a) Dimostrazione di una proposizione di Torricelli sulla spirale logaritmica. Periodico di Matematiche 168/4–10:143–151.

Agostini A (1930b) Un brano inedito di Torricelli sulla rettificazione della spirale logaritmica. Il Bollettino di Matematica 170/2–9:XXV–XXXVIII.

Agostini A (1930c) Il problema inverso delle tangenti nell'opera di Torricelli. Archeion 12/1:33–37.

Agostini A (1931) Il problema inverso delle tangenti e l'integrazione indefinita nelle opere del Torricelli. Archeion 13:55–59.

Agostini A (1949) La memoria di Torricelli sopra la spirale logaritmica riordinata e completata. Accademia Navale, Livorno.

Agostini A (1950) Il metodo delle tangenti fondato sopra la dottrina dei moti nelle opere di Torricelli. Periodico di Matematiche 28/4:141–158.

Agostini A (1951a) I baricentri trovati da Torricelli. Bollettino Unione Matematici Italiani:149–159.

Agostini A (1951b) Massimi e minimi nella corrispondenza di Torricelli con Michelangelo Ricci. Atti del IV Congresso dell'U.M.I.

Agostini A (1951c) Il "De tactionibus" di Evangelista Torricelli. Bollettino Unione Matematici Italiani:319–321.

Agostini A (1951d) Problemi di massimo e minimo nella corrispondenza di E. Torricelli. Rivista Matematica dell'Università di Parma:265–275.

Agostini A (1951e) Gli studi di Torricelli sulla cicloide. Accademia Navale, Livorno.

Alexander A (2014) Infinitesimal. How a Dangerous Mathematical Theory Shaped the Modern World. Scientific American/Fakkar. Straus ad Giroux, New York.

Ampère AM (1806) Démonstration du principe des vitesses virtuelles, dégagée de toute considération des infiniment petits. Journal de l'Ècole Polytechnique 6/13:247–269.

Andersen K (1985) Cavalieri's Method of Indivisibles. Archive for the History of Exact Sciences 31:291–367.

Andersen K, Giusti E, Jullien V (2015) Cavalieri's Indivisibles. In: Jullien V (ed). Seventeenth-Century Indivisibles Revisited. Springer, Dordrecht, pp. 31–55.

Andriesse CD (2005) Huygens: the man behind the principle. Cambridge University Press, Cambridge.

Archimedes (1897) The Works of Archimedes. Edited by T.L. Heath. Cambridge. The Cambridge University Press, Cambridge.

Archimedes (1912) The Method of Archimedes. Recently Discovered by Heiberg. A Supplement to The Works of Archimedes 1897. Edited by T. L. Heath. Cambridge. The Cambridge University Press, Cambridge.

Archimedes (2002) On the equilibrium of planes, Books I-II. In: Heath TL (ed). The works of Archimedes. Dover Publications Inc., New York, pp. 189–220.

Archimedes (2018) Metodo. Nel laboratorio del genio. Boringhieri, Torino.

Aristotle (1853) On the Definition and Division of Principles. In: Octavius Freire Porphyry Owen (ed). The Organon, Or Logical Treatises, of Aristotle. Bohn H G, London, Vol 1, pp. 263–266.

Aristotle ([1936]1955) Minor Works Translated by Hett WS. The Harvard University Press–Heinemann W LTD, Cambridge MA–London.

Aristotle (1949) Aristotle's Prior and Posterior Analytics. A Revised Text with Introduction and Commentary Ross WD (ed). The Oxford University Press, Oxford.

Aristotle (1955a) De Caelo. Translation into English by Stocks JL. The Tech Classics Archive of M.I.T.

Aristotle (1955b) Mechanical problems. In: Hett WS (ed) Aristotle. Minor works. William Heinemann, Cambridge, pp. 328–414.

Aristotle (1955c) Minor Works. Transaltion by Hett WS. The Harvard University Press. Heinemann W LTD, Cambridge MA–London.

Aristotle (1984) The Complete Works of Aristotle. Barnes (ed) The Princeton University Press, Princeton, NJ.

Aristotle (1999) Physics. Translation by Waterfield R. The Oxford University Press, Oxford.

Auger L (1962) Un savant méconnu: Gilles Personne de Roberval (1602–1675): son activité intellectuelle dans les domaines mathématique, physique, mécanique et philosophique. Blanchard, Paris.

Bachmakova IG (1963) Les Méthodes différentielles d'Archimède. Archive for History of Exact Sciences 2:87–107.

Baldi B (1621) In Mechanica Aristotelis Problemata Exercitationes: adiecta succinta narratione de autoris vita et scriptis. Ioannis Albini, Moguntiae.

Baron ME ([1969]1987) The Origins of the Infinitesimal Calculus. Dover, NY.

Baroncelli G (1993) On the Invention of the Geometric Spiral: an Unpublished Letter of Torricelli to Michelangelo Ricci. Nuncius 8/2:601–606.

Barrow I ([1683]1976) Lectiones Mathematicae. Whewell W (ed). The Mathematical Works. Verlag–Hildesheim, New York.

Bascelli T (2015) Torricelli's Indivisibles. In Jullien 2015, pp. 105–136.

R. Pisano et al., *Homage to Evangelista Torricelli's Opera Geometrica 1644–2024*,
Logic, Epistemology, and the Unity of Science 55,
https://doi.org/10.1007/978-3-031-06963-5

Beaugrand J (1636) Geostatice, seu de vario pondere gravium secundam varia a terra (centro) intervalle dissertation mathematica. Toussaint du Bray, Paris.

Berlingozzi R (1902) Raffaello Maggiotti e la sua opera scientifica nel secolo XVII. Tipografia Varchi, Montevarchi.

Bernoulli D ([1733]1738) Theoremata de oscillationibus corporum filo flexili connexorum et catenae verticaliter suspense. Novi commentarii Academiae Scientiarum Imperialis Petropolitanae, pp. 108–123.

Bernoulli D ([1748]1750) Remarques sur le principe de la conservation des forces vives pris dans un sens général. Mémoire de l'Académie Royale des Sciences et des Lettres de Berlin, vol IV, pp. 356–364.

Bernoulli J (1955) Der Briefwechsel von Johann Bernoulli, Vol. I. Birkhäuser, Basel.

Bernoulli Jacob ([1701]1993) Oratio de Historia Cycloidis, Ms UB Basel, L1a 749 A. 3 (8 fol.) [Die Werke von Jakob Bernoulli, Bd 4, Birkhäuser, 1993, pp. 257–266].

Bernoulli Jacob (1703) Démonstration générale du centre de balancement et d'oscillation, tirée de la nature du levier. Mémoire de l'Académie Royale des Sciences Paris, pp. 78–84.

Bernoulli Johann (1727) Theoremata selecta pro conservatione virium vivarum demonstranda et esperimenta confirmanda. Novi commentarii Academiae Scientiarum Imperialis Petropolitanae 2, pp. 200–207.

Bernoulli Johann (1742a) Propositiones variae mechanico–dynamicae. In: Bernoulli 1742b, Vol IV, pp. 253–386.

Bernoulli Johann (1742b) Opera omnia, tam antea sparsim edita, quam hactenus inedita ... quibus continentur ea, quae ab A. 1690 usque ad A. 1727 prodierunt. Accedunt Lectiones Mathematicae de Calculo Integralium atque Anedokta. I–IV vols, Lausannae & Genevae, Bousquet [Facsimile 1968. Reprint Curavit Hofman JE. Georg Olms Verlag Hildeshe].

Bertoloni Meli D (2006) Thinking with Objects: The Transformation of Mechanics in the Seventeenth Century. Johns Hopkins University Press, Baltimore.

Bertoni G (1985) Breve storia della 'vexata quaestio' del luogo natale di Evangelista Torricelli. Bollettino della Società Torricelliana di Scienze e Lettere 36:330–345.

Bertoni G (1987) La faentinità di Evangelista Torricelli e il suo vero luogo di nascita. Bollettino della Società Torricelliana di Scienze e Lettere 38:85–94.

Bertoni G (2003) La faentinit di Evangelista Torricelli e il suo vero luogo di nascita. Studi e ricerche del Liceo Torricelli. Ragazzini, Faenza, pp. 5–12.

Biagioli M (1992) Galileo, Courtier: The Practice of Science in the Culture of Absolutism. University of Chicago Press, Chicago.

Biagioli M (1995) Le prince et les savants: la civilité scientifique au XVIIe siècle. Annales Histoire Sciences Sociales 50/6:1417–1453.

Biagioli M (1996) Etiquette, Interdependence, and Sociability in Seventeenth-Century Science. Critical Inquiry 22/2:193–238.

Biagioli M (2016) Galileo's Instruments of Credit: Telescopes, Images, Secrecy. University of Chicago Press, Chicago.

Bolzano B (1851) Paradoxien des Unendliches. Reclam, Leipzig.

Boncompagni B (1875) Intorno ad alcune lettere di Evangelista Torricelli, del P. Marino Mersenne e di Francesco de Verdus. Bullettino di bibliografia e di storia delle scienze matematiche e fisiche VIII:353–456.

Bortolotti E (1921) Gli inviluppi di linee curve e i primordi del metodo inverso delle tangenti. Periodico di Matematiche 4/1:263–276.

Bortolotti E (1922) Le prime applicazioni del calcolo integrale alla determinazione del centro di gravità delle figure geometriche. Rendiconti della R. Accademia di Bologna, Bologna.

Bortolotti E (1923) Notizia storica su la "Regola di Torricelli" per la misura dei volumi. Periodico di Matematiche 3/4.

Bortolotti E (1924) La scoperta e le successive generalizzazioni di un teorema fondamentale di calcolo integrale. Archeion 5:205–227.

Bortolotti E (1925a) La memoria "De infinitis hyperbolis" di Torricelli. Archeion 6/1:49–58; see also 6/2:139–152.

Bortolotti E (1925b) La memoria "De infinitis hyperbolis" di Torricelli. Leonardo da Vinci, Roma.

Bortolotti E (1928a) I progressi del metodo infinitesimale nell'Opera geometrica di Evangelista Torricelli. Periodico di Matematiche 8/1:18–59.

Bortolotti E (1928b) Le prime rettificazioni di un arco di curva nella memoria "De infinitis spiralibus" di Torricelli. Rendiconti della Reale Accademia di Bologna.

Bortolotti E (1928c) Studi e ricerche sulla storia della matematica in Italia nei secoli XVI e XVII. Zanichelli, Bologna.

Bortolotti E (1930) Le coniche di Apollonia e il problema inverso delle tangenti di Torricelli. Archeion XII/3–4:267–271.

Bortolotti E (1931) L'integrazione indefinita e il teorema di inversione nell'opera di Torricelli. Archeion 1/13:60–64.

Bortolotti E (1937) Il metodo infinitesimale nell'Opera geometrica di Evangelista Torricelli. Atti del primo Congresso dell'Unione matematica italiana. Zanichelli, Bologna, pp. 459–477.

Bortolotti E (1939a) L'infinito e il limite nel Rinascimento italiano. Bollettino della Unione Matematica Italiana Serie II, Anno I.

Bortolotti E (1939b) L'Opera geometrica di Evangelista Torricelli. Monatsschrift für Mathematik und Physik 48:457–486.

Bortolotti E (1947) La storia della matematica nella università di Bologna. Zanichelli, Bologna.

Bortolotti E (1992) Applicazioni del calcolo integrale alla determinazione del centro di gravità di figure geometriche. Accademia di Bologna, Bologna.

Bos HJM (1986) Introduction. In: Richard J, Blackwell RJ (trans.) Christian Huygens' The Pendulum Clock, or Geometrical Demonstrations Concerning the Motion of Pendula as Applied to Clocks. Ames. Iowa State University Press, pp. xi–xxx.

Bos HJM (2001) Redefining Geometrical exactness. Descartes' Transformation of the Early Modern Concept of Construction, Springer Verlag, Berlin.

Boschiero L (2007) Experiment and Natural Philosophy in Seventeenth-Century Tuscany: The History of the Accademia del Cimento. Springer, Dordrecht.

Bourdieu P (2001) Science de la science et réflexivité. Raisons d'agir Editions, Paris, p. 91 and seq.

Braudel F ([1989]1994) Le modèle italien. Champs, Paris.

Brown EJ (1967–1968). The Scientia de Ponderibus in the later Middle Ages, Ph.D. dissertation. Tutor Clagett M. Madison: University of Winscosin [available via: UMI – Proquest Company.

Buridan J (1509) Subtilissimae Quaestiones super octo Physicorum libros Aristotelis, Paris. [Reprint: Id. (1964) Kommentar zur Aristotelischen Physik. Minerva, Frankfurt].

Buridan J (1513) Quaestiones super decem libros Ethicorum Aristotelis ad Nicomachum, Paris. [Reprint: Id. (1968) Super decem libros Ethicorum. Minerva, Frankfurt].
Buridan J (1942) Iohannis Buridani Quaestiones super libris quattuor De caelo et mundo. In: Moody EA (ed). Medieval Academy of America. Studies and documents 6/40.
Busard HLL (1970–1990) Entry: Paul Guldin. In: DBS.
Buscelli T (2010) I fondamenti della nuova scienza del moto: la cinematica di Galileo e la geometria di Torricelli. PhD Dissertation. Scuola di Dottorato di Ricerca in Filosofia. Indirizzo: Filosofia Teoretica. University of Padua, Padova.
Bussotti P (2003) On the Genesis of the Lagrange Multipliers. Journal of optimization theory and applications 117/3:453–459.
Bussotti P (2015) The Complex Itinerary of Leibniz's Planetary Theory. Springer–Birkhäuser, Basel.
Bussotti P (2019) The Calculations of Areas and Volumes Using the Method of Archimedes. Some Didactic Considerations. Journal of Baltic Science Education 18/6:812–815.
Bussotti P, Pisano R (2013) On the Conceptual Frames in René Descartes' *Physical Works*. Advances in Historical Studies 2/3:106–125.
Bussotti P, Pisano R (2014a) Newton's Philosophiae Naturalis Principia Mathematica "Jesuit" Edition: The Tenor of a Huge Work. Accademia Nazionale Lincei–Matematica e Applicazioni–History of Mathematics 25/4:413–444.
Bussotti P, Pisano R (2014b) On the Jesuit Edition of Newton's Principia. Science and Advanced Researches in the Western Civilization. Newton Special Issue (Pisano ed): History and Historical Epistemology of Science. *Advances in Historical Studies* 3/1:33–55.
Bussotti P, Pisano R (2017) Historical and Philosophical Details on Leibniz's Planetary Movements as Physical–Structural Model. In: Pisano, Fichant, Oliveira and Bussotti 2017 (eds), pp. 49–92.
Bussotti P, Pisano R (2020) Historical and Foundational Details on the Method of Infinite Descent: Every Prime Number of the Form 4n+1 is the Sum of Two Squares. Foundations of Science (Springer) 25:671–702.
Bussotti P, Pisano R (2023) Details in Galileo's Mechanism of Inertia as New Intellectual Historical Insights. In: Pisano R (ed). A History of Physics: Phenomena, Ideas & Mechanisms. Essays in Honor of Salvo D'Agostino Dordrecht, Springer, *forthcoming*.
Büttner J, Damerow P, Renn J (2001) Traces of an Invisible Giant: Shared Knowledge in Galileo's Unpublished Treatises. In: Montesinos J, Solis C (eds). Largo campo di filosofare. Eurosymposium Galileo 2001. La Orotava: Fondacìon Canaria Orotava de Historia de la Ciencia, pp. 183–202.
Büttner J, Damerow P, Renn J (2004) Galileo's Unpublished Treatises. In: Palmerino CR, Thijssen JMMH (eds). The Reception of the Galileian Science of Motion in Seventeenth-Century Europe. Springer, Dordrecht, pp. 99–117.
Cantor M (1900) Vorlesungen über Geschichte der Mathematik, Zweiter Band von 1200–1668. Zweite Edition, Teubner, Leipzig.
Cantor G (1932) Gesamelte Abhandlungen. Springer, Berlin.
Cantor G, Dedekind R (1937) Briefwechsel. Edited by E. Noether and J. Cavaillès. Hermann, Paris.
Capecchi D, Pisano R (2007a) La teoria dei baricentri di Torricelli come fondamento della statica. Physis XLIV:1–29.
Capecchi D, Pisano R (2007b) Il principio di Torricelli prima di Torricelli. In Preziosi, B. (ed.). Proceedings of XXIV SISFA Congress. Bibliopolis, Napoli, pp. 107–112.
Capecchi D, Pisano R (2008) La meccanica in Italia nei primi anni del Cinquecento. Il contributo di Niccolò Tartaglia. In Proceedings of XXV SISFA Congress. University of Milano, Milano, C17.1–C17.6.
Capecch D, Pisano R (2010a) Reflections on Torricelli's Principle in Mechanics. Organon 42:81–98.
Capecchi D, Pisano R (2010b) Scienza e tecnica nell'architettura del Rinascimento. CISU, Roma.
Carnot L (1780) Mémoire sur la théorie des machines pour concourir au prix que l'Académie Royale des Sciences de Paris doit adjuger en 1781. The manuscript is dated from Béthune 15 July 1780. It is conserved in the Archives de l'Académie des sciences, Institut de France, and consists of 191 sections in 106 folios. Sections 101–160 are reproduced in: Gillispie (1971), Appendix C, pp. 299–343.
Carnot L (1786) Essai sur les machines en général. Defay, Dijon.
Carnot L (1803a) Principes fondamentaux de l'équilibre et du mouvement. Deterville, Paris.
Carnot L (1803b) Géométrie de position. Duprat.
Carnot L (1808a) Essai sur les machines en général (I part) In: Philosophical Magazine: comprehending the various branches of science, the liberal and fine arts, agriculture, manufactures, and commerce. Tilloch A (ed). Vol. XXX. Murray J, London, pp. 8–14; pp. 154–158; pp. 208–221; pp. 310–320. [very interesting is the avant–titre–page of the book where a portrait of Lazare Carnot is published].
Carnot L (1808b) Essai sur les machines en général (II Part). In: Philosophical Magazine: comprehending the various branches of science, the liberal and fine arts, agriculture, manufactures, and commerce. Tilloch A (ed). Vol. XXXI. Murray J, London, pp. 136–145; pp. 220–228; 295–305.
Carnot L (1813) Réflexions sur la métaphysique du calcul infinitésimal. Courcier, Paris.
Carnot L (1990) Rapport sur une nouvelle machine inventée par MM. Niepce, et nommée par eux Pyrélophore; par MM. Berthollet et Carnot. Lu le 15 décembre 1806, imprimé en 1807, 1er semestre. In: Charnay JP (ed) Lazare Carnot ou le savant–citoyen, La Herne, Paris, Presses de L'Université de Paris–Sorbonne, 1990 – Centre d'études et de recherches sur les stratégies et les conflits, Colloque en Sorbonne, 1988).
Carroll T, Dougherty ST, Perkins D (2013) Infinitesimal and a Tale of Seventeenth-Century Mathematics. Mathematics Magazine 86/4:239–254.
Cartelon H (1975) Does Aristotle Have A Mechanics? in Barnes et al.: Articles on Aristotle. Vol. I: Science, Duckworth (ed.), London.
Castelli B (1628) Della misura dell'acque correnti. Stamperia Camerale, Roma.
Castelli B (1641) Benedetto Castelli a Galileo Galilei in Firenze. Roma 2 marzo 1641. In: Antonio Favaro, Le Opere di Galileo Galilei, Prima Edizione, Vol. XVIII, Edizione Nazionale, Barbera Editore, Firenze (1890).
Castelnuovo G (1962) Le origini del calcolo infinitesimale nell'era moderna con scritti di Newton, Leibniz, Torricelli, Feltrinelli, Milano [First edition *Id.*, 1938, Zanichelli, Bologna].
Cavalieri B (1635) Geometria indivisibilibus continuorum nova quadam ratione promota [...] Bologne, Clemente Ferroni.
Cavalieri B (1647) Exercitationes geometricae sex. Jacobi Montii, Bologna.
Cavalieri B (1966) Geometria degli indivisibili. UTET, Torino.
Clagett M (1959) The science of mechanics in the middle ages. The University of Wisconsin Press, Madison.
Clagett M (1964–1984) Archimedes in the Middle Ages, Madison–Philadelphia, Memoirs of the American Philosophical Society, 5 Vols. 10 Tomes. The Clarendon University Press, Oxford.

Clagett M, Moody E ([1952]1960) The medieval science of weights. Scentia de ponderibus. The University of Wisconsin Press, Madison.

Clapeyron ÉBP (1834) Mémoire sur la puissance motrice du feu. Journal de l'École Royal Polytechnique XXIIIème/XIV:153–190.

Clausius RJE (1850) Über die bewegende Kraft der Wärme und die Gesetze, welche sich daraus für die Wärmelehre selbst ableiten lassen. Annalen der Physik und Chemie 79:368–397, 500–524. English trans: *Id.*, On the Motive Power of Heat and on the Laws which can be deduced from it for the theory of heat itself. In: Mendoza 1960, pp. 73–74, pp. 109–152.

Cohen IB, Smith GE (2002) The Cambridge companion to Newton. The Cambridge University Press, Cambridge.

Commandino F (1565) Federici commandini vrbinatis liber de centro gravitatis solidorum Bononiae: Ex Officina Alexandri Benacii.

Commandino F (1575) De gli elementi di Euclide libri quindici. Con gli scholii antichi. Tradotti prima in lingua latina da m. Federico Commandino da Vrbino, & con commentarij illustrati, et hora di ordine delli istesso trasportati nella nostra vulgare, et da lui riueduti in Vrbino : appresso Domenico Frisolino In Vrbino : in casa di Federico Commandino.

Comte A (1830–1842) Cours de philosophie positive. Bachelier, Paris.

Comte A (1843) Traité élémentaire de géométrie analytique. Carillian-Goeury et Dalmont, Paris.

Comte A (2007) Premiers cours de philosophie positive. Clément-Colas Y (ed). PUF, Paris.

Costabel P (1962) Le De cycloïde de R. Boscovich. Revue d'Histoire des Sciences et de leurs Applications 15/1:31–42.

Costabel P (1964) Essais sur les secrets des *Traités de la Roulette*. In: Taton E (ed) L'Œuvre scientifique de Pascal, PUF, Paris, pp. 169–206.

Costabel P, Peiffer J (eds) (1998) Der Briefwechsel von Johann I Bernoulli. Band 2: Der Briefwechsel mit Pierre Varignon: Erster Teil: 1692–1702. Basel, Birkhauser.

Crosbie AC (1970–1990) Entry: Marin Mersenne. In: DBS.

D'Alembert JLR ([1743]1758) Traité de dynamique. Nouvelle édition, revue et fort augmentée par l'Auteur Chez David Libraire, Paris.

D'Alembert JLR (1751–1780) Encyclopédie ou dictionnaire raisonné des sciences, des arts et des métiers, Diderot & D'Alembert. Briasson–David–Le Breton–Durand, Paris.

D'Alembert JLR (1767) Mélanges de littérature, d'Histoire et de philosophie. Tome V. Chatelain & Fils. Amsterdam, pp. 3–274.

Dati C (1919) Lettera a' Filateti. Della vera storia della cicloide [...]. Firenze, 1663 [see also: Opere di Evangelista Torricelli, I, parte II, pp. 441–482.

Dauben J (1979) Georg Cantor: His mathematics and Philosophy of the Infinite. Princeton, NJ.

Davis M (2005) Applied Nonstandard Analysis. Dover, NY [Reprinted with corrections from *Id.*, 1977, Interscience-Wiley, NY].

Davis M (2012) The universal computer: the road from Leibniz to Turing. CRC Press, Boca Raton.

DBS see: Gillispie 1970–1980.

De Duillier NF (1699) Lineæ brevissimæ descensus investigatio geometrica duplex, cui addita est investiga-tio geometrica solidi rotundi in quo minima fiat resistentia, London.

De Gandt F (1984–1985) Naissance et métamorphose d'une théorie mathématique : la géométrie des indivisibles en Italie (Galilée, Cavalieri, Torricelli). Sciences et Techniques en Perspective, 9:179–229.

De Gandt F (1987) L'oeuvre de Torricelli: science galiléenne et nouvelle géométrie. Les Belles Lettres, Paris.

De Gandt F (1991) Cavalieri's Indivisible and Euclid's Canons. Barker P, Ariew R (eds). Revolution and Continuity. Essays in the History and Philosophy of Early Modern Science. The Catholic University of America, Washington, pp. 157–182.

De Gandt F (1992) L'évolution de la théorie des indivisibles et l'apport de Torricelli. Biblioteca Nuncius Studi Testi X:103–118.

De l'Hôpital GFA (1696) Analyse des infiniment petits, pour l'intelligence des lignes courbes, Paris, Imp. royale.

De Nemore J (fl. 13th) De ratione ponderis. In: Clagett 1964–1984, pp. 190–191.

De Waard C (1936) L'expérience barométrique ses antécédents, et ses explications. Etude historique. Imprimerie nuovelle, Thouars.

Del Monte G ([1577]1581) Le Meccaniche dell'Illustrissimo Sig. Guido Ubaldo dè Marchesi del Monte, tradotto in volgare dal Sig. Filippo Pigafetta. Evangelista Deuchino, Venezia [1577: Mechanicorum Liber].

Del Monte G (1588) In duos Archimedis aequeponderantium libros paraphrasis. Hieronymum Concordiam, Pisauri.

Del Monte G (1615) Le mechaniche dell'illustriss. Sig. Guido Ubaldo de' marchesi del Monte tr. Tradotto in volgare da Filippo Pigafetta (1581). Evangelista Deuchino, Venezia.

Descartes (1996) Œuvres de Descartes. New Edition. Vrin, Paris.

Descartes R (1637) Florimond de Beaune, Johannes Hudde, Henricus van Heuraet, Johan de Witt, Erasmus Bartholin, Frans van Schooten, Geometria à Renato Des Cartes anno 1637 Gallicè edita; postea autem unà cum notis Florimondi de Beaune [...] Gallicè conscriptis in Latinam linguam versa, & commentariis illustrata, operâ atque studio Francisci à Schooten, Florimond de Beaune (trad.), Frans van Schooten (éd. sc.), Leyde, Jean Maire, 1649, in-4° [*Id.*, 1659, 2nd ed., pp. 265–271].

Descartes R (1897–1913) Oeuvres de Descartes. 12 vols. Adams C, Tannery P (eds). Paris; Discours de la méthode et Essais, Specimina philosophiae. vol VI Principia philosophiae, Latin version vol VIII, Principia philosophiae French translation vol IX; Physico-mathematica vol X, Le Monde ou Traité de la lumière, vol XI (*Id*, 1964- 1974 par Rochot B, Costabel P, Beaude J et Gabbery A, Paris).

Dhombres (1988) Une histoire de l'objectivité scientifique et le concept de postérité. In: Guesnerie R, Hartog F (eds). Des Sciences et des techniques : un débat. Cahier des Annales 45:127–148.

Dhombres (2010) La trajectoire d'une parabole. Métamorphoses de la philosophie naturelle sous l'effet des mathématiques, XIIIe Entretiens de la Garenne Lemot. In: Pigeaud J (ed). Métamorphose(s), PUR, Rennes, pp. 213–241.

Dhombres (2011a) La réception en France de Galilée. Les rôles différents de Peiresc et de Mersenne auprès de Gassendi, de Descartes et de Jacques le Tenneur, Galileo e la scuola galileiana nelle Università del Seicento. Pepe L (ed). CLUEB, Bologna, pp. 47–60.

Dhombres (2011b) Le jet d'eau et l'arc-en-ciel à l'âge baroque : réalisation des mathématiques, mathématisation de la philosophie naturelle et représentation des phénomènes, In: Cousinie, Nau FC (eds). L'artiste et le philosophe. L'histoire de l'art à l'épreuve de la philosophie au XVIIe siècle. PUR, Rennes, 2011, pp. 151–196.

Dhombres J (1996) Entries: Sciences et savants [and] Académie des sciences. In: Dictionnaire de l'Ancien régime, PUF, Paris.

Dhombres J (1999) Entry: Académies. Dictionnaire d'histoire et de philosophie des sciences, PUF, Paris.

Dhombres J (2006) Exactitude et esthétique. La mise en scène du savoir au temps de Peiresc, Sciences et Techniques en Perspective 10/1:15–74.

Dhombres J (2015) An epistemological path through the historiography on Indivisibles. In: Jullien V (ed), Seventeenth Century Indivisibles Revisited. Birkhäuser, Basel, pp. 391–450.

Dhombres J (2023a) Les savoirs mathématiques et leurs pratiques culturelles. Hermann, forthcoming.

Dhombres J (2023b) Confrontation de deux manières d'apprentissage du calcul différentiel et intégral: Johann Bernoulli et Guillaume de l'Hôpital (1691–1696). Sciences et Techniques en Perspective, forthcoming.

Dhombres J, Radelet de Grave P (2008a) Liber amicorum, Réminesciences, Brepols, Turnhout.

Dhombres J, Radelet de Grave P (2008b) Une mécanique donnée à voir, Les thèses illustrées défendues à Louvain en juillet 1624 par Grégoire de Saint-Vincent S.J., in De diversis artibus, Collection de travaux de l'Académie internationale d'Histoire des sciences. Brepols, Turnhout.

Dhombres J, Sakarovitch J (1994) (ed) Desargues en son temps. Blanchard, Paris.

Di Ronchitti C (1605) Dialogo in perpuosito de la stella nova. Tiozzi, Padova.

Dijksterhuis EJ (1955–1958) (ed.) The Principal Works of Simon Stevin, N.V. Swets & Zeitlinger, Amsterdam, Vol. I.

Dodart M (1676) Mémoires pour servir à l'histoire des plantes. Dressez par M. Dodart de l'Académie Royale des Sciences, Paris de l'Imprimerie Royale, Paris.

Drake S (1970–1990) Entry: Benedetto Castelli. In: DBS.

Drake S (1976) Galileo Against the Philosphers in his Dialogue of Cecco di Roncitti (1605) and Considerations of Alimberto Mauri (1606). Zeitlin & Ver Brugge, Los Angles.

Duhem PM (1905–1906). Les origines de la statique. 2 vols, Hermann, Paris.

Ellis M (2004) The Coffee-House: a Cultural History. Weidenfeld, London, 2004.

Enderton H (1977) Elements of Set Theory, Academic Press, NY.

Euclid (2007) Euclid's Elements of geometry. The Greek text of J.L. Heiberg (1883–1885) from Euclidis Elementa, edidit et Latine interpretatus est I. L. Heiberg, in aedibus B.G. Teubneri, 1883–1885. Edited and provided with a modern English translation by Richard Fitzpatrick. Via: http://farside.ph.utexas.edu/Books/Euclid/Elements.pdf.

Euler L ([1730–1731]1738) De communicatione motus in collision corporum. Commentarii academiae scientiarum Petropolitanae 5:159–168.

Euler L ([1773]1774) De pressione ponderis in planum in cui incumbit. Novi Commentarii academiae scientiarum Petropolitanae, vol XVIII, pp. 289–329.

Euler L (1736) Mechanica sive motus scientia analytice exposita. Auctore Leonhardo Eulero academiae imper. scientiarum membro et matheseos sublimioris professore. Tomus I. Instar supplementi ad commentar. acad. scient. imper. Petropoli. Ex typographia academiae scientarum.

Euler L (1749) Recherches sur les mouvements des corps célestes en général (1747). Mémoire de l'Académie Royale des Sciences et des Lettres de Berlin, Vol. I, pp. 93–143.

Euler L (1750]1752) Découverte d'un nouveau principe de Mécanique. Mémoires de l'académie des sciences de Berlin. Vol. VI, 1752, pp. 185–217.

Euler L (1757) Principes généraux du mouvement des fluides. Mémoires de l'Academie des Sciences de Berlin 11:274–315.

Euler L (1775) Nova methodus motum corporum rigidorum determinandi. Novi Commentarii academiae scientiarum Petropolitanae, vol XX, pp. 208–238.

Fabbri S (1947–1997) La Società Torricelliana di Scienze e Lettere di Faenza nel Cinquantenario della Fondazione. Bollettino della Società Torricelliana di Scienze e Lettere s.n.

Favino F (2009) In Urbe mathematicus: Torricelli a Roma. Galilaeana VI:1000–1032.

Favino F (2019) Entry: Torricelli. Dizionario biografico degli italiani. Vol. 96. Via: https://www.treccani.it/enciclopedia/evangelista-torricelli_%28Dizionario-Biografico%29/.

Fermat P (1891–1912) Œuvres de Fermat, 4 vols. Tannery P, Henry C (eds). Gauthier-Villars, Paris.

Festa E (1992) La notion d'«agrégat d'indivisibles» dans la constitution de la cinématique galiléenne : Cavalieri, Galilée, Torricelli. Revue d'histoire des sciences 45/2–3:307–336.

Fossombroni V (1794) Memoria sul princpio delle velocità virtuali. Gaetano Cambiagi, Firenze.

Fourier JBF (1878) Analytical theory of heat. Translated by Freeman A. The Cambridge University Press, London.

Fourier JBJ (1888–1890) Œuvres de Fourier par les soins de M. Gaston Darboux 2 vols, Gauthier–Villars, Paris.

Fox R (1974) The rise and fall of Laplacian physics. Historical Studies in the Physical Sciences 4:89–136.

Fox R (1995) Science, Industry and the Social Order in Post–Revolutionary France. Variorum Collected Studies Series 489, Ashgate Variorum.

Frajese A (1969) Attraverso la storia della matematica. Le Monnier, Firenze.

Fraser CG (1983) J.L. Lagrange's Early Contribution to the Principles and Methods of Mechanics. Archive for History of Exact Sciences 28/3:197–241.

Freedberg D (2003) The Eye of The Lynx: Galileo, His Friends, and the Beginnings of Modern Natural History. University of Chicago Press, Chicago.

Frezier M (1739) La théorie et la pratique coupe des pierres et des bois pour la construction des voutes et autres parties des bâtimen[t]s Civils & Militaires. Traité de Stéréotomie à l'usage de l'Architecture. Par M. Frezier, Chevalier de l'Ordre Militaire de Saint Louis. Ingénieur ordinaire du Roy en Chef à Landau. Strasbourg, Doulsseker & Fils & Paris, Jombert, Tome III.

Gàbici S, Toscano F (2006) Scienziati di Romagna. Sironi, Milano.

Galilei G Ms. B—Trattato delle fortificazioni, con disegni diversi. Biblioteca Ambrosiana. D: 296 Inf.—38499.

Galilei G ([1595]1634) Le mecaniche. In: Galilei 1890–1909, II, pp. 147–191.

Galilei G (1589–1592) De motu antiquiora In: Galilei 1890–1909, I, pp. 251–419.

Galilei G (1593) Le mecaniche. In: Galilei 1890–1909, II, pp. 147–191.

Galilei G (1632) Dialogo sopra i due massimi sistemi del mondo. In: Galilei 1890–1909, VII, 21–520.

Galilei G (1638) Discorsi e dimostrazioni matematiche intorno a due nuove scienze attenenti alla meccanica e i movimenti locali, del signor Galileo Galilei linceo. Elsevirii, Leida. In: Galilei 1890–1909, VIII, pp. 39–362. [1639 Viviani's additions: *Ivi,* VIII, pp. 215–219, pp. 442–445; see also: *Ivi, Frammenti attinenti ai Discorsi e dimostrazioni matematiche intorno a due nuove scienze* VIII, pp. 363–448].

Galilei G (1890–1909) Breve instruzione all'architettura militare. In: 1890–1909, II, pp. 15–75.

Galilei G (1890–1909) Le opere di Galileo Galilei: Edizione nazionale sotto gli auspici di sua maestà il re d'Italia. Favaro A (ed.), Barbèra, Firenze.

Galilei G (1890–1909) Trattato di fortificazione. In: Galilei 1890–1909, II, pp. 77–146.

Galilei G (1954) Dialogues Concerning Two New Sciences, translated by Henry Crew & Alfonso de Salvio, with an introduction by Antonio Favaro, Dover Publications, Inc., New York [Originally published in 1904 by the MacMillan company].
Galilei G (as Cecco di Ronchitti) 1605 Dialogo in perpuosito de la stella nova. Tiozzi, Padua.
Galilei G Ms. A – Trattato di fortificazioni e modo d'espugnare la città con disegni e piante di fortificazioni. Biblioteca Ambrosiana. D: 328 Inf.– 38640.
Galilei G Ms. m – Trattato di fortificazioni, con disegni e figure (acephalous). Biblioteca Ambrosiana. N: 281 Sup.– 2– 80277.
Galluzzi P (1976) Evangelista Torricelli: concezione della matematica e segreto degli occhiali. Annali dell'Istituto e Museo di storia della scienza di Firenze I:71–95.
Galluzzi P (1979) Momento: Studi Galileiani. Ateneo e Bizzarri, Roma.
Galluzzi P (1998) The Sepulchres of Galileo: The 'Living' Remains of a Hero of Science. In: Machamer P (ed). The Cambridge Companion to Galileo. Cambridge University Press, Cambridge, pp. 417–47.
Galluzzi P, Torrini M (1975–1984) (eds) Le opere dei discepoli di Galileo Galilei. Carteggio. Giunti–Barbera, Firenze, 2 Vols.
Gassendi P ([1649]1964) Syntagma. In: Petrus Gassendi, Opera Omnia, I, Verlag, Stuttgart–Bas Cannstatt.
Gatto R (2002) Galileo Galilei. Le mecaniche. Edizione critica e saggio introduttivo. Olschki, Firenze.
Gaukroger S, Schuster J, Sutton J (2000) (eds) Descartes' natural philosophy. Routledge London and New York.
Gerhardt C (1978) (ed) Die Philosophischen Schriften von G. W. Leibniz. 7 Vols. Georg Olms Verlag, Hildesheim–NY.
Ghinassi G (1864) Lettere fin qui inedite di Evangelista Torricelli precedute dalla vita di lui scritta da Giovanni Ghinassi. Pietro Conti, Faenza.
Gillispie CC (1970–1979) Carnot Lazare–Nicolas–Marguerite. III. In: Gillispie (ed) 1970–1980, pp. 70–79.
Gillispie CC (1971) Lazare Carnot Savant. A monograph treating Carnot's scientific work, with facsimile reproduction of his unpublished writings on mechanics and on the calculus, and an essay concerning the latter by Youschkevitch AP, Princeton University Press, Princeton, NJ.
Gillispie CC (1976) The scientific work of Lazare Carnot, and its influence on that of his son. In: Taton (ed) 1976, pp. 23–33.
Gillispie CC (ed) (1970–1980) Dictionary of scientific biography. Charles Scribner's Sons, New York.
Gillispie CC, Pisano R (2014) Lazare and Sadi Carnot. A Scientific and Filial Relationship, 2nd edition. Springer, Dordrecht.
Gillispie CC, Youschkevitch AP (1979) Lazare Carnot savant et sa contribution à la théorie de l'infini mathématique. Vrin, Paris [Collection des Travaux de l'Académie Internationale d'Histoire des Sciences, 20].
Giusti E (1980) Bonaventura Cavalieri and the Theory of Indivisibles. Cremonese, Bologna.
Gliozzi M (1931) Origini e sviluppi dell'esperienza Torricelliana. Giappichelli, Torino.
Gliozzi M (1970–1990) Biography in Dictionary of Scientific Biography. Charles Scribner's Sons, New York. http://www.encyclopedia.com/topic/Evangelista_Torricelli.aspx [the reader should pay attention to the place of birth cited. Nowadays, it is known that he was born in Roma, Italy).
Gödel K ([1933]1995) The Present Situation in the Foundations of Mathematics. In: Feferman S *et al.* (eds). Kurt Gödel Collected Works. Vol. III. Oxford University Press, NY, pp. 45–53.
Gödel K ([1947]1990) What is Cantor's Continuum Problem? In: Feferman S *et al.* (eds). Kurt Gödel Collected Works. Vol. II. Oxford University Press, NY, pp. 176–187 [See also revised and expanded version, *Id.*, pp. 254–270].
Gödel K ([1951]1995) Some Basic Theorems on the Foundations of Mathematics and Their Implications. In: Feferman S *et al.* (eds). Kurt Gödel Collected Works. Vol. III. Oxford University Press, NY, pp. 304–323.
Grattan–Guinness I (1990) Convolutions in French Mathematics, Birkhäuser, Basel.
Gröning J (1701) Historia cycloidis. Qua Genesis et proprietates Lineae Cyloidalis praecipuae, secundum Ejus Infantium, Adolescentium & Juventutum, Ordine chronologico recesentur [...]. Hamburg, Gottfried Liebezeit.
Grosslight J (2013) Small Skills, Big Networks: Marin Mersenne as Mathematical Intelligencer. History of Science 51:337–74.
Guiducci M (1619) Discorso delle comete. Pietro Cecconcelli, Firenze.
Guldin P (1634–1641) Centrobaryca seu de centro gravitatis trium specierum quantitatis continuae, 4 vols. G. Gelbhaar & M. Cosmerovii, Vienna.
Hahn R (1971) Anatomy of a Scientific Institution: The Paris Academy of Sciences, 1666–1803. University of California Press, Berkeley.
Hall R (1980) The Quarrel between Newton and Leibniz. Cambridge University Press, New York.
Hankins TL (1970) Jean D'Alembert. Science and the Enlightenment. The Clarendon Press, Oxford.
Hara K (1970–1990) Entry: Gilles Personne de Roberval. In: DBS.
Hazard P (1935) La crise de la conscience européenne (1680–1715). Librairie générale, Pars.
Heilbron J (2010) Galileo. Oxford University Press, Oxford.
Henry J (2011) Mathematics made no contribution to public weal: why jean Frenel (1497–1558) became a physician. Centaurus 53/3:193–220.
Hoyrup J (1988) Jordanus de Nemore, 13th Century Mathematical Innovator. Archive for History of Exact Science 38:307–363.
Huygens C (1629–1695) Œuvres complètes de Christiaan Huygens. Société hollandaise des sciences. Martinus Nijhoff, La Haye.
Huygens C (1669) Démonstration de l'équilibre de la balance. [Leu dans l'Académie le 2 décembre 1673].In Huygens 1629–1695, XVII, pp. 42–47 [see also Huygens 1937, XIX, pp. 40–47].
Huygens C (1673) Christiani Hugenii Zulichemii, Const. F. Horologium Oscillatorium sive De motu pendulorum ad horologia aptato demonstrationes geometricae. Parisiis, apud F. Muguet, regis et illustrissimi archiepiscopi typographum, via Citharae, ad insigne trium Regum.
Huygens C (1678) Christiani Hugenii Horologium Oscillatorium. Apud F. Muguet, Paris.
Huygens C (1684) Astroscopia Compendiaria tubi optici molimine liberata. Arnold Leers, The Hague.
Huygens C (1888–1950) Œuvres complètes de Christiaan Huygens publiées par la Société hollandaise des sciences. Nijhoff, La Haye.
Huygens C (1901) Œuvres completes de Christiaan Huygens. Hollandsche Maatschappij der Wetenschappen, IX, Nijhoff, La Haye.
Huygens C (1920) Œuvres completes de Christiaan Huygens. Hollandsche Maatschappij der Wetenschappen, XIV, Nijhoff, La Haye.
Huygens C (1932) Œuvres completes de Christiaan Huygens. Hollandsche Maatschappij der Wetenschappen, XVII. Nijhoff, La Haye.
Huygens C (1934) Œuvres completes de Christiaan Huygens. Hollandsche Maatschappij der Wetenschappen, XVIII. Nijhoff, La Haye.
Huygens C (1937) Christiaan Huygens, Oeuvres complètes. Tome XIX. Mécanique théorique et physique 1666–1695. Edited by Vollgraff JA). Martinus Nijhoff, Den Haag. [via: http://www.dbnl.org/tekst/huyg003oeuv19_01/colofon.php].
Itard J (1975) La lettre de Torricelli à Roberval d'octobre 1643. Revue d'Histoire des Sciences 28(2):114–124.
Jacob R (1977) Robert Boyle and the English revolution. Franklin, New York.

Jammer M (1957) Concepts of force: a study in the foundations of dynamics. Harvard University Press, Cambridge MA.
Jaouiche K (1976) Le livre du Qarastun de Thabit Ibn Qurra. Brill, Leiden.
Jardine N (1988) Epistemologies of the sciences. In: Schmitt CB and Skinner O (eds). Cambridge History of the Renaissance philosophy. The Cambridge University Press, Cambridge.
Jech T (1991) Set Theory. Academic Press, NY [Second edition with corrections from: Id., 1978, Springer, Berlin–Heidelberg].
Johns A (1998) The Nature of the Book: Print and Knowledge in the Making. University of Chicago Press, Chicago.
Johns A (2010) Piracy: The Intellectual Property Wars from Guttenberg to Gates. University of Chicago Press, Chicago.
Jullien V (2015) (ed) Seventeenth-Century Indivisibles Revisited. Springer, Dordrecht.
Jullien V (2015) Roberval's indivisibles. In: Jullien 2015, pp. 177–210.
Knobloch E (2005) Géométrie pratique, Géométrie savante. Albertiana 8:27–56.
Knobloch E, Vasoli C, Siraisi N (2001) Il Rinascimento. Medioevo, Rinascimento. Storia della scienza. Vol. IV. Istituto della Enciclopedia Italiana, Roma, pp. 605–1044.
Kragh H (1987) An Introduction to the Historiography of Science. The Cambridge University Press, Cambridge.
Krarup J, Vajda S (1997) On Torricelli's Geometrical Solution to a Problem of Fermat, Duality in Practice. Journal of Applied and Industrial Mathematics 8/3:215–224.
Kuhn T (1977) The Essential Tension. Selected Studies in Scientific Tradition and Change. The University of Chicago Press, Chicago and London.
Lagrange JL (1762) Application de la méthode exposée dans le mémoire précédente à la solution des différents problèmes de dynamique. In: Lagrange 1870–1873, I, pp. 363–468.
Lagrange JL (1764) Recherches sur la libration de la Lune dans lesquelles on tache de résoudre la question proposes par l'Académie Royale des Sciences, pour le Prix de l'année 1764. Prix de l'académie Royale des Sciences de Paris, IX (1777), pp 1–50. See also Lagrange 1870–1873, Tome VI, pp. 5–61.
Lagrange JL (1788) Mécanique Analytique. Desaint, Paris.
Lagrange JL (1870–1873) Œuvres de Lagrange. Seconde édition, Courcier (ed), I–XIV vols. (in X) Paris Gauthier–Villars, Paris.
Lagrange JL (1892) [Lagrange's letter to Euler on November 24th 1759 in:] Œuvres de Lagrange. Tome XIV. Darboux G (ed). Gauthier–Villars, Paris, pp. 170–174.
Laplace PS (1816) Sur la vitesse du son dans l'air et dans l'eau. Annales de chimie et de physique III:238–241.
Latour B (1988) Ramsès II est-il mort de la tuberculose ? La Recherche 309:84–85.
Leibniz GW (2009) Gottfried Wilhelm Leibniz Naturwissenschaftlich–medizinisch–technische Schriften Reihe VIII der Akademieausgabe. Berlin–Brandenburgische Akademie der Wissenschaften, Российская Академия Наук, in Zusammenarbeit mit dem Centre National de la Recherche Scientifique, der Gottfried Wilhelm Leibniz Bibliothek Hannover und der Herzog August Bibliothek Wolffenbüttel. Knobloch E (project leader editor). There is a printed version of the volume(s) and there is an online–version available via: http://leibnizviii.bbaw.de/.
Leibniz GW (2012) Sämtliche Schriften und Briefe, Siebente Reihe, Mathematische Schriften, Sechster Band, 1673–1676. Band. Arithmetische Kreisquadratur. Akademie der Wissenschaften zu Göttingen. Brandenburgische Akademie der Wissenschaften, Berlin.
Lenoble R ([1943]1971) Mersenne ou la naissance du mécanisme. Vrin, Paris.
Lindberg DC (1976) Theories of vision from Al–Kindi to Kepler. The Chicago University Press, Chicago.
Loria G (1897) Evangelista Torricelli e la prima rettificazione di una curva. Rendiconti della Regia Accademia dei Lincei 5/6:318–323.
Loria G (1900) Le ricerche inedite di Evangelista Torricelli sopra la curva logaritmica. Biblioteca matematica III/1:75–89.
Loria G (1903) Un'impresa nazionale di universale interesse. Pubblicazione delle opere di Evangelista Torricelli. Atti II Congresso Internazioanle delle Scienze Storiche, pp. 23–27.
Loria G (1909) In memoria di Evangelista Torricelli. Discorso. Rivista Ligure di Scienze. Lettere ed arti, Genova.
Loria G (1915) L'infinito e l'infinitesimo secondo i matematici dell'antichità e secondo i matematici moderni anteriori al secolo XVIII. Scientia 12 [see also *Id.*, 1916, Scientisa I].
Loria G (1917) Una generazione delle coniche a centro ideata da E. Torricelli. Periodico di Matematiche 32:189–192.
Loria G (1919) E. Torricelli nella storia della geometria. Rendiconti della Regia Accademia dei Lincei 5/28:409–415.
Loria G (1922a) Una lettera inedita di E. Torricelli al P. Mersenne. Archivi di storia della scienza 3:272–276.
Loria G (1922b) L'opera geometrica di Evangelista Torricelli. Bollettino di Matematica 2/1:I–VII.
Loria G (1923) Evangelista Torricelli. Gli scienziati italiani. Vol. I, Parte II, pp. 299–311.
Loria G (1932) Una lettera inedita di Evangelista Torricelli. Archeion XIV:12–14.
Loria G (1938) Le prétendu larcin de Torricelli. Archeion XIX.
Loria G (1944) Il III centenario del barometro. Bollettino della Società Torricelliana di Scienze e Lettere s.n.
Loria G (1945) Prime linee di una bibliografia torricelliana. Bollettino della Società Torricelliana di Scienze e Lettere s.n.
Loria G (1947) L'avverso destino di Evangelista Torricelli. Rendiconti dell'Accademia Nazionale dei Lincei 8/3:665–669.
Lorini B (1596) Delle Fortificationi di Bvonaivto Lorini. Nobile Fiorentino. Libri Cinque. Ne' quali si mostra con le più facili regole la Scienza con la ratica, di fortificare le città, & altri luoghi sopra diversi siti [...] Nuovamente dati in luce. Con Priuilegio. In Vinegia, appresso Gio. Antonio Rampanzetto, MDXCVI.
Lorini B (1609) Le Fortificationi di Bvonaivto Lorini. Nobile Fiorentino. Nnuovamente ristampate, Corrette & Ampliate [...] Con l'Aggiunta del Sesto Libro. Con Privilegi. In Venetia, presso Francesco Rampazetto. fortificare le città, & altri luoghi sopra diversi siti [...]. Nuovamente dati in luce. In Vinegia, MDCIX. Presso Francesco Rampanzetto.
Mach E ([1883]1919) The Science of Mechanics. A Critical and Historical Account of its Development, Chicago-London, The Open Court Publishing Co.
Maffioli C (1994) Out of Galileo: The Science of Waters, 1628–1718. Erasmus Publishing, Rotterdam.
Mancosu P (1996) Philosophy of Mathematics & Mathematical Practice in the Seventeenth Century. Oxford University Press, NY–Oxford.
Mancosu P, Vailati E (1991) Torricelli's Infinitely Long Solid and Its Philosophical Reception in the Seventeenth-Century. Isis 82/1:50–70.
Maupertuis PL (1746) Les loix du mouvement et du repos déduits d'un principe métaphysique. Histoire de l'Académie Royale des Sciences et des Belles Lettres, pp. 267– 294.

Medolla G (1993) Alcuni documenti inediti relativi alla vita di Evangelista Torricelli. Bollettino Unione Matematici Italiani XIII/2:287–295.

Meli DB (2006) Thinking with Objects: The Transformation of Mechanics in the Seventeenth Century. Johns Hopkins University Press, Baltimore.

Mendoza E (ed) (1960) Reflections on the motive power of fire and other papers on the second law of thermodynamics by E Clapeyron and R Clausius. Dover, New York.

Mersenne M (1624) L'impiété des déistes, athées et libertins de ce temps, combattue et renversée de point en point par raisons tirées de la philosophie et de la théologie, ensemble la réfutation du Poème des déistes, P. Bilaine, Paris.

Mersenne M (1634) Les Mécaniques de Galilée, mathématicien et ingénieur du duc de Florence, traduites de l'italien en françois par L.P.M.M., Paris.

Mersenne M (1636) Harmonie universelle, contenant la théorie et la pratique de la musique, où il est traité de la Nature des Sons , & des Mouvemens, des Consonances, des Dissonances, des Genres, des Modes, de la Composition, de la voix, des Chants, & de toutes sortes d'Instruments Harmoniques, 1636, Sébastien Cramoisy, Paris.

Mersenne M (1637) Harmonie universelle, second volume, Nouvelles observations mathématiques et physiques, Paris, Pierre Ballard. [The new pagination begins after book 6 that stops at page 408. It is quite possible that the addition (XXth book as described by Mersenne himself) only appeared in 1638].

Mersenne M (1639) Nouvelles pensées de Galilée, A paris, Guenon.

Mersenne M (1644) F. Marini Mersenni [...] Cogitata physico-mathematica, in quibus tam naturae quam artis effectus admirandi certissimis demonstrationibus explicantur. De Mensuris, ponderibus et numeris. De hydraulico-pneumaticis phaenomenis. De arte nautica, seu Histiodromia et hydrostatica (cum tractatu de magnetis proprietatibus. De musica theorica et practica. De mechanicis phaenomenis. De ballisticis, seu acontismologicis phaenomenis, Paris, aux frais de A. Bertier.

Merton RK (1968) Sociological Theory, American Journal of Sociology 50:462–479 [reprinted: *Id*., 1968, Social theory and Social structure. Free Press, NY].

Merton RK (1973) The Sociology of Science. Theoretical and Empirical Investigation, The University of Chicago Press, Chicago.

Merzbach UC, Boyer CB (2011) A History of Mathematics. 3rd edition. edition. Wiley, Hoboken.

Meskens AD (2005) Joannes della Faille, s.j., Mathematics, Modesty and Missed Opportunities. Belgisch historisch instituut te Rome, Bruxelles.

Middleton WEK (1963) The Place of Torricelli in the History of Barometer. Isis 54/1:11–28.

Middleton WEK (1964) The History of the Barometer. Johns Hopkins. Baltimore.

Milighetti MC (2004) Da Antonio Nardi (1598–1649 ca.) a Francesco Redi (1626–1697): l'eredità culturale di un Galileiano. Annali Aretini XII:221–238.

Milighetti MC (2006) Sophia e Mathesis negli scritti di Antonio Nardi. Bollettino di Storia delle Matematiche 26/1:9–31.

Miniati M, Van Helden A, Greco V, Molesini G (2002) Seventeenth-century telescope optics of Torricelli, Divini, and Campani. Applied Optics 41/4:644–647.

Molesini G (2010) Testing telescope optics of seventeenth-century Italy. In: Van Helde A, Dupré S, van Gent-R, Zuidervaart H (eds). The origins of the telescope. KNAW Press, Amsterdam, pp. 271–280.

Montucla E (1758) Histoire des mathématiques. Charles-Antoine Jombert, Paris.

Montucla E (1799–1802) Histoire des mathématiques. Agasse, Paris.

Moore AW (2001) The Infinite. Second Edition, Routledge, London.

Nabonnand P (2010) L'argument de la généralité chez Carnot, Poncelet et Chasles. Via: http://hal.archives-ouvertes.fr/hal-00637385/fr/.

Nabonnand P (2011) Une géométrie sans figure? La Figure et la Lettre, pp. 99–120.

Nardi A (1994) Théorème de Torricelli ou théorème de Mersenne. Etudes philosophiques 1/2:87–118.

Nardi A (fl 1642s) Le scene. Codice Galileano 130. Biblioteca Nazionale di Firenze.

Nelli GBC (1793) Vita e commercio letterario di Galileo Galilei Nobile e Patrizio Fiorentino Matematico e Filosofo Sopraordinario de' Grand Duchi di Toscana Cosimo e Ferdinando II, 2 Vols. Gia' Ghetti Sinibaldi da Montecuccoli, Losanna.

Newton I ([1671]1736) The method of fluxions and infinite series. Henry Woodfall, London.

Newton I ([1686–1687]1803) The Mathematical Principles Of Natural Philosophy, by Sir Isaac Newton. Translated into English by Motte A, Symonds, London, Vol. I

Newton I ([1713]1729) The Mathematical principles of natural philosophy. Translated by Motte Andrew. Motte B, London.

Newton I ([1726][1739–1742]1822) Philosophiae Naturalis Principia Mathematica, auctore Isaaco Newtono, Eq. Aurato. Perpetuis commentariis illustrate, communi studio pp. Thomae le Seur et Francisci Jacquier ex Gallicana Minimorum Familia, matheseos professorum. Editio nova, summa cura recensita. A. et J. Duncan, Glasgow.

Newton I ([1726]1729) The mathematical principles of natural philosophy. Transl. by Motte Andrew. London, Motte.

Newton I (1687) Philosophiae naturalis principia mathematica. Imprimatur S. Pepys. Reg. Soc. Preses. Julii 5. 1686. Londini, Jussi Societatus Regiae ac Typis Josephi Streater. Prostat apud plures Bibliopolas. Anno MDCLXXXVII.

Newton I (1714) Philosophiae Naturalis Principia Mathematica. Reprints of 2nd English edition [1713, Cotes R (ed), Cantabrigiae], Rogerus Cotes Editoris Praefatio, Sumptibus Societatis, Amstaelodami.

Newton I (1730) Queries. In: Opticks, 4th English Edition corrected, Innys W (ed), London.

Newton I (1746) Methodus fluxionum et serierum infinitarum, in Opuscula mathematica. Marcum Michaelem Bousque, Lausanne et Genevae, I, 29–200.

Nonnoi G (1982) Evangelista Torricelli, la pressione dell'aria e gli Autori de' crepuscoli. Bollettino della Società Torricelliana di Scienze e Lettere Faenza:17–46.

Nonnoi G (1997) Evangelista Torricelli, la pressione dell'aria e gli 'autori dei crepuscoli'. Bollettino della Società torricelliana di Scienze e Lettere:17–45.

OET see: Torricelli 1919–1944.

Oliveira ARE, Pisano R, (2022) Historical Details of Coulomb's Torsional Forces in Metal Wires In: Ceccarelli M, López-García R (eds). Explorations in the History and Heritage of Machines and Mechanisms. Vol. 40. Dordrecht, Springer, pp. 242–254.

Oresme N (1486) Tractatus de latitudinibus formarum. Mathaeus Cerdonis, Padua.

Osler MJ (ed) (2000) (ed) Rethinking the Scientific Revolution. Cambridge University Press, Cambridge.

Palmieri P (2009) Superimposition: on Cavalieri's practice of Mathematics. Archive for History of Exact Sciences 63:471–495.

Pascal B (1963) Œuvres complètes de Pascal. Lafuma L (ed). Seuil, Paris.

Pascal B (1992) Œuvres complètes de Pascal. Mesnard J (ed). Brouwer, Paris.

Pascal B (1998–2000) Traités de l'équilibre des liqueurs et de la pesanteur de la mas de l'air, contenant l'explication des causes de divers effets de la Nature qui n'avaient point été bien connus jusques ici, et particulièrement ceux que l'on avait attribués à l'horreur du vide. In: Œuvres complètes de Blaise Pascal. Vol 1. Gallimard, Paris, pp. 459–543.

Paternoster G, Rinzivillo R, Schettino E (1996). Studio di una lente per cannocchiale di grandi dimensioni lavorata da Evangelista Torricelli. Nuncius XI:123–134.

Pisano R (2007) Brief history of centre of gravity theory. Epistemological notes. In: Kokowski M (Ed). Proceedings of the 2nd ESHS Congress, Polish Academy of Science, Krakow, pp. 934–941.

Pisano R (2009a) Il ruolo della scienza archimedea nei lavori di meccanica di Galilei e di Torricelli. In: Giannetto E, Giannini G, Capecchi D, Pisano R, (eds). Da Archimede a Majorana: La fisica nel suo divenire. Guaraldi Editore, Rimini, pp. 65–74.

Pisano R (2009b) On method in Galileo Galilei' mechanics. In: Hunger H (ed). Proceedings of ESHS 3rd Conférence. The Austrian Academy of Science, Wien, pp. 147–186.

Pisano R (2009c) Continuity and discontinuity. On method in Leonardo da Vinci' mechanics. Organon 41:165–182.

Pisano R (2009d) Galileo Galilei. Riflessioni epistemologiche sulla resistenza dei corpi. In Giannetto E, Giannini G, Toscano M (eds). Relatività, Quanti Chaos e altre Rivoluzioni della Fisica. Guaraldi Editore Rimini, pp. 61–72.

Pisano R (2010) On Principles in Sadi Carnot's Thermodynamics (1824). Epistemological Reflections. Almagest 2/2010:128–179.

Pisano R (2011) Physics–Mathematics Relationship. Historical and Epistemological notes. In Barbin E, Kronfellner M, Tzanakis C, (eds), European Summer University History And Epistemology In Mathematics. Vienna: Verlag Holzhausen GmbH-Holzhausen Publishing Ltd., pp. 457–472.

Pisano R (2012) On Lazare and Sadi Carnot. A synthetic view of a historical epistemological research program. In: Proceedings of XXX SISFA Congress, Urbino, Argalia Editore, Urbino, pp. 147–153.

Pisano R (2013a) On Lagrangian in Maxwell's electromagnetic theory. Scientiatum VI. História das Ciências e das Técnicas e Epistemologia. The University of Federate University of Rio de Janeiro Press, Brazil, pp. 44–59.

Pisano R (2013b) Historical Reflections on Physics Mathematics Relationship In Electromagnetic Theory In: Barbin E., Pisano R., (eds). The Dialectic Relation between Physics And Mathematics In The XIXth Century. Springer, Dordrecht, pp. 31–57.

Pisano R (2014) Note sulle Fortificazioni nei Quesiti di Tartaglia. Libro Sesto e sua Gionta. In: D'Agostino S, Fabricatore G (eds). Proceedings of the International Conference in History of Engineering. Vol. II. Cuzzolin, Napoli, pp. 815–825.

Pisano R (2015) (ed) A Bridge between Conceptual Frameworks, History of Mechanism and Machine Science. Dordrecht, Springer.

Pisano R (2016) Details on the mathematical interplay between Leonardo da Vinci and Luca Pacioli. Journal of the British Society for the History of Mathematics 31/2:104–111.

Pisano R (2017) A Development of the Principle of Virtual Laws and its Framework in Lazare Carnot's Mechanics as Manifest Relationship between Physics and Mathematics. Transversal 2:166–203.

Pisano R (2020) A Tale of Tartaglia's Libro Sesto & Gionta in Quesiti et Inventioni diverse (1546–1554). Exploring the Historical and Cultural Foundations. Foundations of Science 25/2:477–50.

Pisano R (2023–2024) Galileo Galilei: The Two Manuscripts *On Mechanics*. Translations, Text and Commentaries. Dordrecht, Springer, *forthcoming*.

Pisano R, Agassi J, Drozdova D (2017) (eds) Hypothesis and Perspective in History and Philosophy of Science. Homage to Alexandre Koyré 1892–1964. Dordrecht, Springer.

Pisano R, Bussotti P (2012) Galileo and Kepler. On Theoremata Circa Centrum Gravitatis Solidorum and Mysterium Cosmographicum. History Research 2/2:110–145.

Pisano R, Bussotti P (2014a) Historical and Epistemological Reflections on the Culture of Machines around the Renaissance: How Science and Technique Work? Acta Baltica Historiae et Philosophiae Scientiarum 2/2:20–42.

Pisano R, Bussotti P (2014b) Notes on mechanics and mathematics in Torricelli as physics mathematics relationships in the history of science. Problems of Education in the 21st Century International Journal. Special Issue 61/61:88–97.

Pisano R, Bussotti P (2015a) Galileo in Padua: architecture, fortifications, mathematics and « practical » science, Lettera Matematica Pristem International (Springer):209–221.

Pisano R, Bussotti P (2015c) The Emergencies of Mechanics and Thermodynamics in the Western Society during 18th–19th Century. In: Pisano R (ed) 2015, pp. 399–436.

Pisano R, Bussotti P (2015c) Historical and epistemological reflections on the culture of machines around the renaissance: machines. Machineries and Perpetual Motion. Acta Baltica Historiae et Philosophiae Scientiarum 3/1:69–87.

Pisano R, Bussotti P (2015c) Fibonacci and the Abacus Schools in Italy. Mathematical Conceptual Streams, Education and its Changing Relationship with Society. Almagest 6/2:126–165.

Pisano R, Bussotti P (2016) A Newtonian Tale Details on Notes and Proofs in Geneva Edition of Newton's Principia. Bulletin Journal of the British Society for the History of Mathematics 31/3:160–178.

Pisano R, Bussotti P (2017a) (eds) Homage to Galileo Galilei 1564–2014. Reading Iuvenilia Galilean Works within History and Historical Epistemology of Science. Philosophiae Scientiae 21/1.

Pisano R, Bussotti P (2017b) The Fiction of Infinitesimals in Newton's Works. On the Metaphoric Use of the Infinitesimals in Newton. In Marcacci F, Rossi MG (eds). Isonomia-Epistemologica. Reasoning, Metaphor and Science 9:141–160.

Pisano R, Bussotti P (2017c) On the Conceptualization of Force in Johannes Kepler's Corpus: an Interplay between Physics/Mathematics and Metaphysics. In: Pisano R, Agassi J and Drozdova D (eds). Hypothesis and Perspective in History and Philosophy of Science. Homage to Alexandre Koyré 1892–1964. Springer, pp. 295–346.

Pisano R, Bussotti P (2022) Conceptual Frameworks on the Relationship between Physics–Mathematics in the Newton *Principia Geneva Edition* (1822). Foundations of Sciences (Springer) 27/3:1127–1182.

Pisano R, Bussotti P (2023a) Historical and Methodological Details on the *De motu gravium naturaliter descendentium* in Torricelli's *Opera Gometrica* (1644) In: Pisano R, Dhombres J, Radelet de Grave P, Bussotti P (eds). Homage to Evangelista Torricelli's Opera Geometrica 1644–2024. Text, Transcription, Commentaries and Selected Essays as New Historical Insights, Springer, Cham, pp. 189–224.

Pisano R, Bussotti P (2023b) The Physics–Mathematics Interplay in Newton Principia Geneva Edition (1822): A New Case Study of the Three-body Problem, Proposition LXVI–Theorem XXVI. Submitted to Philosophical Transactions A.

Pisano R, Capecchi D (2009) Il ruolo della meccanica ne Le fortificationi di Buonaiuto Lorini In: D'Agostino, S. (ed). III Convegno di Storia dell'ingegneria, II. Napoli: Cuzzolin editore, pp. 797–808.

Pisano R, Capecchi D (2010a) On Archimedean Roots in Torricelli Mechanics. In: Ceccarelli M, Paipetis S (eds). The Genius of Archimedes. Springer, Dordrecht, pp. 17–28.

Pisano R, Capecchi D (2010b) Galileo Galilei: Notes on Trattato di Fortificazione. In: Altamore A, Antonini G (eds), Galileo and the renaissance scientific discourse. Edizioni nuova cultura Roma, pp. 28–41.

Pisano R, Capecchi D (2012) Historical reflections on scale ratio in Galilean Tratattto di Fortificazione. In: Ceccarelli M (ed) International symposium on the history of machines and mechanisms. Springer, Berlin, pp. 463–473

Pisano R, Capecchi D (2013) Conceptual and mathematical structures of mechanical science between 18th and 19th centuries. Almagest 2/4:86–121.

Pisano R, Capecchi D (2016) Tartaglia's science weights. Mechanics in Sixteen-Century. Selections from Books VII-VIII. Dordrecht, Springer.

Pisano R, Capecchi, D (2008) La meccanica in Italia nei primi anni del Cinquecento. Il contributo di Niccolò Tartaglia. In: Tucci P (ed). Proceedings of XXV SISFA Congress. Milano, pp. C17.1- C17.6 via: http://www.brera.unimi.it/sisfa/atti/index.html.

Pisano R, Cooppersmith J, Peake M (2021) Essay on Machines in General (1786). Text, Translations and Commentaries. Lazare Carnot's Mechanics – Vol. 1. Springer, Cham.

Pisano R, Fichant M, Bussotti P, Oliveira ARE (2017) (eds) The Dialogue between Sciences, Philosophy and Engineering. New Historical and Epistemological Insights. Homage to Gottfried W. Leibniz 1646–1716. The College Publications, London.

Pisano R, Gaudiello I (2009) Continuity and discontinuity. An epistemological inquiry based on the use of categories in history of science. Organon 41:245–265.

Pisano R, Robarts J (2022) A Short Comparative Historiography of Science & Technology of Mechanics into Engineering—and—Architecture Literatures During the Renaissance. In: Ceccarelli M, López-García R (eds). Explorations in the History and Heritage of Machines and Mechanisms. Vol. 40. Dordrecht, Springer, pp. 227–240.

Pisano R, Robarts J (2024) Briefs on Buonaiuto Lorini's Fortificationi (1609): Geometry, Machines & Mechanics into Engineering during the Renaissance. In: Ceccarelli M et al. (eds). 8th HMM Ankara, Springer, Cham, *forthcoming*.

Podetti F (1914) La teoria delle proporzioni in un manoscritto inedito di Evangelista Torricelli. Bollettino di Bibliografia e Storia delle Scienze Matematiche A/16–3:65–71.

Poinsot L (1806) La Théorie générale de l'équilibre et du mouvement des systèmes. Journal de l'Ècole Polytechnique 6/13:206–241.

Poinsot L (1838) Note sur une certaine démonstration du principe des vitesses virtuelles, qu'on trouve au chapitre III de livre 1er de la "Mécanique céleste". Journal de mathématiques pures et appliquées 1/3:244–248.

Prager FD (1981) Berti's devices and Torricelli's barometer from 1641 to 1643. Annali dell'Istituto e Museo di Storia della Scienza di Firenze 2:35–53.

Procissi A (1952) Nuove ricerche bibliografiche torricelliane. Bollettino della Società Torricelliana di Scienze e Lettere 3:13–15.

Procissi A (1953) Su l'inviluppo delle parabole in Torricelli e sulla nozione di inviluppo di una famiglia di curve piane. Periodico di matematiche 4/31–1:113–116.

Radelet de Grave P (1996) Entries: Stevin, Kepler, Leibniz, Huygens. In: Dictionnaire du patrimoine littéraire européen, Patrimoine Littéraire Européen. vol. 8, Avènement de l'Equilibre européen 1616–1720. De Boeck Université, p. 18, pp. 745–755, pp. 1020–1027.

Radelet de Grave P (1998) La moindre action comme lien entre la philosophie naturelle et la mécanique analytique: continuités d'un questionnement. LLULL 21:439–484.

Radelet de Grave P (2007) Kepler (1571–1630) et Cavalieri (1598–1647) astrologues, ou le logarithme au secours de l'astrologie. In Mélanges offerts à Hossam Elkadem par ses amis et ses élèves, Bruxelles. Archives et Bibliothèques de Belgique, pp. 297–315.

Radelet de Grave P (2009) The Problem of the Elastica Treated by Jacob Bernoulli and the Further Development of this Study by Leonhard Euler. In: Werner L, Kurrer KE (eds). Proceedings of the Third International Congress on Construction History (Cottbus, May 2009). Brandenburg University of Technology Cottbus, Verlag Service GmbH. Germany, pp. 1209–1219.

Rahman S (2005) The Dialogic of just being different. Hintikka's new approach to the notion of episteme and its impact on "second generatio" dialogics. In: Kolak D, Symons J (eds.). Quantifiers, Questions and Quantum Physics. Essays in Honour of Jaakko Hintikka. Springer Verlag, pp. 57–187.

Renn J (1995) Historical epistemology and interdisciplinary. In: Gavroglu K J.J. Stachel JJ, Wartofsky MW (eds). Physics, Philosophy, and the Scientific Community: Essays in the Philosophy and History of the Natural Sciences and Mathematics In Honor of Robert S. Cohen Kluwer, Doertrcth, pp. 241–251.

Renn J, Damerow P (2010) The equilibrium controversy. Guidobaldo del Monte's critical notes on the mechanics of Jordanus and Benedetti and their historical and conceptual background. Edition open access, Berlin.

Renn J, Damerow P, McLaughlin P (2003) Aristotle, Archimedes, Euclid, and the Origin of Mechanics: The Perspective of Historical Epistemology. The Max Planck Institute for the history of science of Berlin, print n. 239, Berlin, pp. 43–59.

Reuleaux F (1876) Kinematics of Machinery: Outlines of Theory of Machines, London: Macmillan & Co.

Riccati V (1749) Dialogo di Vincenzo Riccati della Compagnia di Gesù dove ne' congressi di più giornate delle forze vive e dell'azioni delle forze morte si tien discorso. Stamperia di Lelio dalla Volpe, Bologna.

Riccati V (1772) De' Principi della Meccanica. Stamperia Coleti, Venezia.

Roberval GP (1636) Traité de mechanique. Des poids soustenus par des puissances sur les plans inclinez à l'horizon. Des puissances qui soustiennent un poids suspendu à deux chordes. Richard Charlemagne, Paris.

Roberval GP de (1644) Aristarchi Samii de mundi systemate, partibus et motibus ejusdem libellus. Adjectae sunt AE. de Roberval notae in eundem libellum. A. Bertier, Paris.

Roberval GP de (1636) Traité de mechanique des poids soustenus par des puissances sur les plans inclinez, Charlemagne, Paris.

Roberval GP de (1693a) De trochoïde ejusque spatio. In: Divers ouvrages de mathématiques et de physique par MMrs les savants de l'Académie Royale des Sciences, Imprimerie royale, Paris, pp. 246–278.

Roberval GP de (1693b) Observation sur la composition des mouvemens, et sur le moyen de trouver les touchantes des ligne courbes. In: Divers ouvrages de mathématiques et de physique par MMrs les savants de l'Académie Royale des Sciences, Imprimerie royale, Paris, 1693, pp. 69–111.

Roberval GP de (1693c) Traité des indivisibles. In: Divers ouvrages de mathématiques et de physique par MMrs les savants de l'Académie Royale des Sciences, Imprimerie royale, Paris, pp. 190–245.

Roberval GP de (1693d) Traité de la roulette In: Divers ouvrages de mathématiques et de physique par MMrs les savants de l'Académie Royale des Sciences, Imprimerie royale, Paris.

Roberval GP de (1693e) Divers ouvrages de mathématiques et de physique par MMrs les savants de l'Académie Royale des Sciences, Imprimerie royale, Paris.

Robinson A (1970) Non–Standard Analysis. North–Holland, Amsterdam.

Robinson P (1994) Evangelista Torricelli. The Mathematical Gazette, 78, 481: 37–47.

Rochot B (1965) Sur Mersenne et le mot cycloïde. Appendice IV, Correspondance du P. Marin Mersenne, ed. Cornelius de Waard, Paris, CNRS, Paris, IX, pp. 584–586.

Roda M (2017) Entry: Rocca, Giovanni Antonio. Dizionario Biografico degli Italiani. Via https://www.treccani.it/enciclopedia/giovanni-antonio-rocca_(Dizionario-Biografico)/.

Rolle M (1703) Du nouveau système de l'infini par M. Rolle. In: Histoire de l'Académie Royale des Sciences Avec les Memoires de Mathematique & de Physique pour la même Anne. A Paris, Chez Jean Boudot, Imprimerie Ordinaire du Roy, & de de l'Académie Royale des Sciences, pp. 312–336.

Ronchi V (1924) Sopra una lente di Evangelista Torricelli. Universo 5/2:1–3.

Ronchi V (ed) (1968) Scritti di ottica. Scritti di Tito Lucrezio Caro, Leonardo da Vinci, G. Rucellai, G. Fracastoro, G. Cardano, D. Barbaro, F. Maurolico, G. B. Della Porta, G. Galilei, F. Sizi, E. Torricelli, F. M. Grimaldi, G. B. Amici. Il Polifilo, Milano.

Rossini G (1959) La famiglia di Evangelista Torricelli. In: AA.VV. Convegno di studi torricelliani in occasione del 350 anniversario della nascita di Evangelista Torricelli, 19–20 ottobre 1958, Lega, Faenza. pp. 133–149.

Rougier L (2010) De Torricelli à Pascal. Philosophia Scientiae 14/2.

Rufini E (1926) Il "Metodo" di Archimede e le origini del calcolo infinitesimale nell'antichità. Zanichelli, Bologna.

Saccheri G (1708) Neo-statica. Mediolani : Ex Typographia Josephi Pandulfi Malatestae, MDCCVIII.

Saito K (2013) Archimede e la doppia riduzione all'assurdo. Lettera Matematica Pristem 86:21–28.

Schuster JA (2000) Descartes opticien: the construction of the law of refraction and the manufacture of its physical rationales, 1618–1629. In: Gaukroger, Schuster and Sutton 2000, pp. 258–312.

Schuster JA (2013a) Descartes–Agonistes. Physico–mathematics, Method and Corpuscular–Mechanism 1618–1633. Springer, Dordrecht.

Schuster JA (2013b) Cartesian Physics. In: Fox R *et al.* (eds). The Oxford Handbook of the History of Physics, The Oxford University Press, Oxford, pp. 56–95.

Scott JF (1981) The Mathematical work of John Wallis. Chelsea Pub. [first ed. 1938] 2nd Ed. New York.

Seeger RJ ([1966]2013) Men of Physics: Galileo Galilei, His Life and His Works: The Commonwealth and International Library: Selected Readings in Physics. [Pergamon Press] Elsevier, Amsterdam.

Segre M (1983) Torricelli's Correspondence on Ballistics. Annals of Science 40/5:489–499.

Segre M (1989) Viviani's Life of Galileo. Isis 80/2:206–231.

Shakespeare W ([1600]1980) Hamlet. In: Bevington D (ed). The Complete Works of Shakespeare. Third Edition. Collins NY.

Shank JB (2018) Before Voltaire: the French origins of "Newtonian" mechanics (1680–1715). University of Chicago Press, Chicago.

Shapin, Schaffer S (1985) Leviathan and the Air pump: Hobbes, Boyle, and the Experimental Life. Princeton University Press, Princeton.

Singer C (1954–1958) A History of technology, vols. 2–4, Oxford: The Clarendon Press.

Souffrin P (1993) Galilée, Torricelli et la loi fondamentale de la dynamique scolastique: la proportionnalité velocitas-momentum revisitée. Sciences et Techniques en Perspective 25:122–134.

Sparavigna CA (2015) Teaching Physics During the 17th Century: Some Examples From the Works of Evangelista Torricelli. International Journal of Sciences 4/4:50–58.

Stroup A (1990) A Company of Scientists: Botany, Patronage, and Community at the Seventeenth-Century Parisian Royal Academy of Sciences. University of California Press, Berkeley.

Sturdy DJ (1995) Science and Social Status: The Members of the Académie des Sciences 1666–1750. Boydell & Brewer, Woodbridge.

Tartaglia N (1537) La noua scientia de Nicolo Tartaglia: con una gionta al terzo libro, per Stephano Da Sabio, Venetia.

Tartaglia N ([1554]1959) La nuova edizione dell'opera Quesiti et inventioni diverse de Nicolo Tartaglia brisciano Riproduzione in facsimile dell'edizione del 1554. Masotti A (ed.). Commentari dell'Ateneo di Brescia, Tipografia La Nuova cartografica, Brescia.

Tartaglia N (1565) Iordani Opvscvlvm de Ponderositate, Nicolai Tartaleae Stvdio Correctvm Novisqve Figvrisavctvm. Cvm Privilegio Traiano Cvrtio, Venetiis, Apvd Curtivm Troianvm. M D Lxv.

Tartaglia N (1565a) Archimedes De insidentibus aqueae, Lib. I et II, in Iordani Opvscvlvm de Ponderositate, apud Curtium Troianum Navò, Venetia.

Tartaglia N (1565b) Iordani Opvscvlvm de Ponderositate, Nicolai Tartaleae Stvdio Correctvm Novisqve Figvrisavctvm. Cvm Privilegio Traiano Cvrtio, Venetiis, Apvd Curtivm Troianvm. MDLxv.

Taton R (1951) L'œuvre mathématique de G. Desargues. PUF, Paris.

Taton R (1955) Causalité et accidents de la découverte scientifique, Masson, Paris [English translation: Id., 1962. Reason and Chance in Scientific Discovery].

Taton R (1963) L'annonce de l'expérience barométrique en France. Revue d'histoire des sciences et leurs applications 16:77–83.

Taton R (ed) (1976) Sadi Carnot et l'essor de la thermodynamique, Table Ronde du Centr National de la Recherche Scientifique. École Polytechnique, 11–13 Juin 1974. Éditions du Centre National de la Recherche Scientifique, Paris.

Teixeira FG (1909) Traité des courbes spéciales remarquables planes et gauches, Coïmbre, Imp. de l'Université, II, Chapitre X.

Tenca L (1956) Su una svista di stampa in 'de Dimensione Parabolae' di Evangelista Torricelli notata da Stefano Angeli. Bollettino Unione Matematici Italiani 3/11:258–259.

Tenca L (1958) L'attivita` matematica di Evangelista Torricelli. Periodico di Matematiche 36/4:251–263.

Tenca L (1960) I presunti contrasti fra Evangelista Torricelli e Vincenzio Viviani. Periodico di Matematiche 4/38:87–94.

Thomson RB (1976) Jordanus de Nemore: Opera. Mediaeval Stud 38:97–144.

Torricelli E (1644) Opera geometrica. Florentiae. Masse & de Landis, Florentiae.

Torricelli E (1644) De Solido Hyperbolico Acuto Problema Secundum. In: *Id.*, (1919–1944) Opera geometrica [Vol. I, pp. 113–135]. Opere di Evangelista Torricelli. Loria G, Vassura G (eds). 4 Vols., Montanari, Faenza, pp. 191–221.

Torricelli E ([1644]1897) Esperienza dell'argento vivo, Accademia del Cimento. In Gustav Hellmann G (ed): Instrumenti per conoscer l'alterazioni dell'aria, Neudrucke von schriften und karten über meteorologieund erdmagnetismus, Asher & Co, Berlin.

Torricelli E (1715) Lezioni Accademiche d'Evangelista Torricelli Mattematico, e Filosofo del Sereniss. Ferdinando II Gran Duca di Toscana, Jacopo Guiducci, Santi Franchi (eds), S.A.R., Firenze.

Torricelli E (1823) Lezioni accademiche. In: Silvestri G (ed). Biblioteca scelta di opere italiane antiche e moderne 139 Milano.

Torricelli E (1864) Lettere fin qui inedite di Evangelista Torricelli precedute dalla vita di lui. G. Ghinassi (ed.), Faenza.

Torricelli E (1919–1944) Opere di Evangelista Torricelli edite in occasione del III centenario della nascita col concorso del Comune di Faenza da Gino Loria e Giuseppe Vassura. Montanari, Faenza.

Torricelli E (1955) De infinitis spiralibus. Edited by Carruccio E. Domus Galilaeana, Pisa.

Torricelli E (1975) Opere scelte. Belloni L (ed). UTET, Torino.

Toscano F (2008) L'erede di Galileo. Vita breve e mirabile di Evangelista Torricelli. Sironi Editore, Milano.

Toscano F (2014) Evangelista Torricelli: grandi scoperte e autoconsure. Nuova Civiltà delle Macchine 3/1:31–46.

Truesdell C (1968a) Essay in the history of mechanics. Springer, New York.

Truesdell CA (1968b) Whence the law of the moment of momentum? [Lecture 202, Whence the law of moment of momentum? Colloquium in Engineering Science. Mechanical Engineering Department, Columbia University, 1963, January 8th] In: Essays in the history of mechanics. Springer, New York, pp. 239–271.

Truesdell CA (1970) The tragicomedy of classical thermodynamics. Springer, Berlin.

Truesdell CA (1976) Termodinamica razionale. Accademia Nazionale dei Lincei, Contributi del Centro Linceo Interdisciplinare di Scienze Matematiche e loro Applicazioni. Anno CCCLXXXIII, N. 20, Roma, pp. 208–235.

Truesdell CA (1980) The tragicomical history of thermodynamics. 1822–1854. Springer, Berlin.

Truesdell CA, Bharatha S (1977) The concepts and logic of classical thermodynamics as a theory of heat engines: rigorously constructed upon the foundation laid by S Carnot and F Reech (Theoretical and mathematical physics). Springer, Berlin/Heidelberg/New York.

Vailati G (1987) Scritti di Vailati. Quaranta M (ed). Sala Bolognese, Forni editore.

Valerio L (1606) Quadratura parabolae per simplex falsum: et altera quam secunda Archimedis expiditior ad martium columnam. Apud Lepidum Facium, Romae.

Van Helden A, Gingerich O (2003) From "Occhiale" to Printed Page: The Making of Galileo's *Sidereus Nuncius*. Journal for the History of Astronomy 34:251–267.

Van Sto Vincentio Gregorius (1624) Theoremata mathematica scientiae staticae de ductu ponderum per planitiem recta et oblique horizontem decussantem, Louvain.

Varanini SL (2001) Antonio Nardi: note in margine ai manoscritti. Bruniana & Campanelliana 7/2:469–498.

Varignon P (1725) La nouvelle Mécanique ou statique. Jobert C, Paris.

Vassura G (1945) Umanità in Evangelista Torricelli. Pubblicato dalla commissione per le onoranze a Evangelista Torricelli III per l'anno 1944. Centenario della scoperta del barometro. Bollettino della Società Torricelliana di Scienze e Lettere:15–18.

Visani A (1952) Notizie sulla salute e sull'ultima malattia di Evangelista Torricelli. Bollettino della Società Torricelliana di Scienze e Lettere 3:5–12.

Vita V (1972) Gli indivisibili curvi di Bonaventura Cavalieri. Archimede XXIV/1–2:16–24.

Viviani V (1659) De maximis et minimis geometrica divination in quintum Conicorum Apollonii Pergaei, adhuc desideratum. Giuseppe Coccini, Firenze.

Viviani V (1674) Quinto libro degli elementi d'Euclide, ovvero scienza universale delle proporzioni, spiegata colla dottrina del Galileo. Condotta, Firenze.

Viviani V (1677) Enondatio problematum universis gemoetricis praepositorum a D. Claudio Comiers. Gugliantini, Firenze.

Viviani V (2001) Vita di Galileo. Salerno Editrice, Roma.

Wallis J (1659) Tractatus duo, Prior, de Cycloide et corporibux inde genitis, Posterior, Epistolaries in qua agitur de Cissoide [...] Oxford, Lichtfeld [Reproduced in Wallis 1695, pp. 499–547].

Wallis J (1693) Algebra Tractatus. Historicus & Practicus. Anno 1685 Angelice editus. Opera Mathematicorum Volumen alterum. Oxoniae. Sheldonian Theatro Sheldoniano.

Wallis J (1693) Geometriae professoris Saviliani, in celeberrima Academia Oxoniensi Opera mathematica by Wallis, John, 1616–1703. Oxoniae. E Theatro Sheldoniano.

Wallis J (1695) *Opera mathematica*, Sheldonian Theatre, Oxford.

Weber M (1904–1905) Die Protestantische Ethik und der Geist des Kapitalismus. Archiv fur *Archiv für Sozialwissenschaften und Sozialpolitik*, Bd. XX und XXI.

Weil A (1993) Introduction to Oratio de Historia Cycloidis, Die Werke von Jakob Bernoulli, Bd 4. Birkhäuser, Berlin, pp. 254–255.

West JB (2013) Torricelli and the Ocean of Air: The First Measurement of Barometric Pressure. Physiology 28:66–73.

Westman RS (1980) The Astronomer's Role in the Sixteenth Century: A Preliminary Study. History of Science 18:105–147.
Wilding N (2011) Manuscripts in Motion: The Diffusion of Galilean Copernicanism. Italian Studies 66/2:221–33.
Wilding N (2014) Transalpine Messengers. Galileo's Idol: Gianfrancesco Sagredo and the Politics of Knowledge. University of Chicago Press, Chicago, pp. 89–116.
Winter TN (2007) The Mechanical Problems in the corpus of Aristotle. Digital Commons. The University of Nebraska–Lincoln, pp. 1–43 [Retrieved March 16, 2008, available via http://digitalcommons.unl.edu/classicsfacpub/68].
Zeuthen HG (1896) Geschichte der Mathematik im Altertum und Mittelalter. Höst, Kopenhagen.

Author Index

R. Pisano et al., *Homage to Evangelista Torricelli's Opera Geometrica 1644–2024*, Logic, Epistemology, and the Unity of Science 55,
https://doi.org/10.1007/978-3-031-06963-5

Subject Index

R. Pisano et al., *Homage to Evangelista Torricelli's Opera Geometrica 1644–2024*,
Logic, Epistemology, and the Unity of Science 55,
https://doi.org/10.1007/978-3-031-06963-5

GPSR Compliance

The European Union's (EU) General Product Safety Regulation (GPSR) is a set of rules that requires consumer products to be safe and our obligations to ensure this.

If you have any concerns about our products, you can contact us on ProductSafety@springernature.com

In case Publisher is established outside the EU, the EU authorized representative is:

Springer Nature Customer Service Center GmbH
Europaplatz 3
69115 Heidelberg, Germany

Batch number: 10756588

Printed by Printforce, the Netherlands